10	11	12	13 III	14 IV	15 V	16 VI		
								2 **He** 4.00
			5 **B** 10.81	6 **C** 12.01	7 **N** 14.01	8 **O** 16.00	9 **F** 19.00	10 **Ne** 20.18
			13 **Al** 26.98	14 **Si** 28.09	15 **P** 30.97	16 **S** 32.06	17 **Cl** 35.45	18 **Ar** 39.95
28 **Ni** 58.71	29 **Cu** 63.54	30 **Zn** 65.37	31 **Ga** 69.72	32 **Ge** 72.59	33 **As** 74.92	34 **Se** 78.96	35 **Br** 79.91	36 **Kr** 83.80
46 **Pd** 106.4	47 **Ag** 107.87	48 **Cd** 112.40	49 **In** 114.82	50 **Sn** 118.69	51 **Sb** 121.75	52 **Te** 127.60	53 **I** 126.90	54 **Xe** 131.30
78 **Pt** 195.09	79 **Au** 196.97	80 **Hg** 200.59	81 **Tl** 204.37	82 **Pb** 207.19	83 **Bi** 208.98	84 **Po** 210	85 **At** 210	86 **Rn** 222

← Metals | Nonmetals →

Metalloids

62 **Sm** 150.35	63 **Eu** 151.96	64 **Gd** 157.25	65 **Tb** 158.92	66 **Dy** 162.50	67 **Ho** 164.93	68 **Er** 167.26	69 **Tm** 168.93	70 **Yb** 173.04	Lanthanides
94 **Pu** 239.05	95 **Am** 241.06	96 **Cm** 247.07	97 **Bk** 249.08	98 **Cf** 251.08	99 **Es** 254.09	100 **Fm** 257.10	101 **Md** 258.10	102 **No** 255	Actinides

GENERAL CHEMISTRY

▼ ▼ ▼ ▼

Most elemental sulfur is obtained as a by-product of the petroleum industry. This huge mound of sulfur was produced by the reaction of hydrogen sulfide and sulfur dioxide. The golden staircase rising through the mound symbolizes the progress that you will make in chemistry as you work through this text.

▲ ▲ ▲ ▲

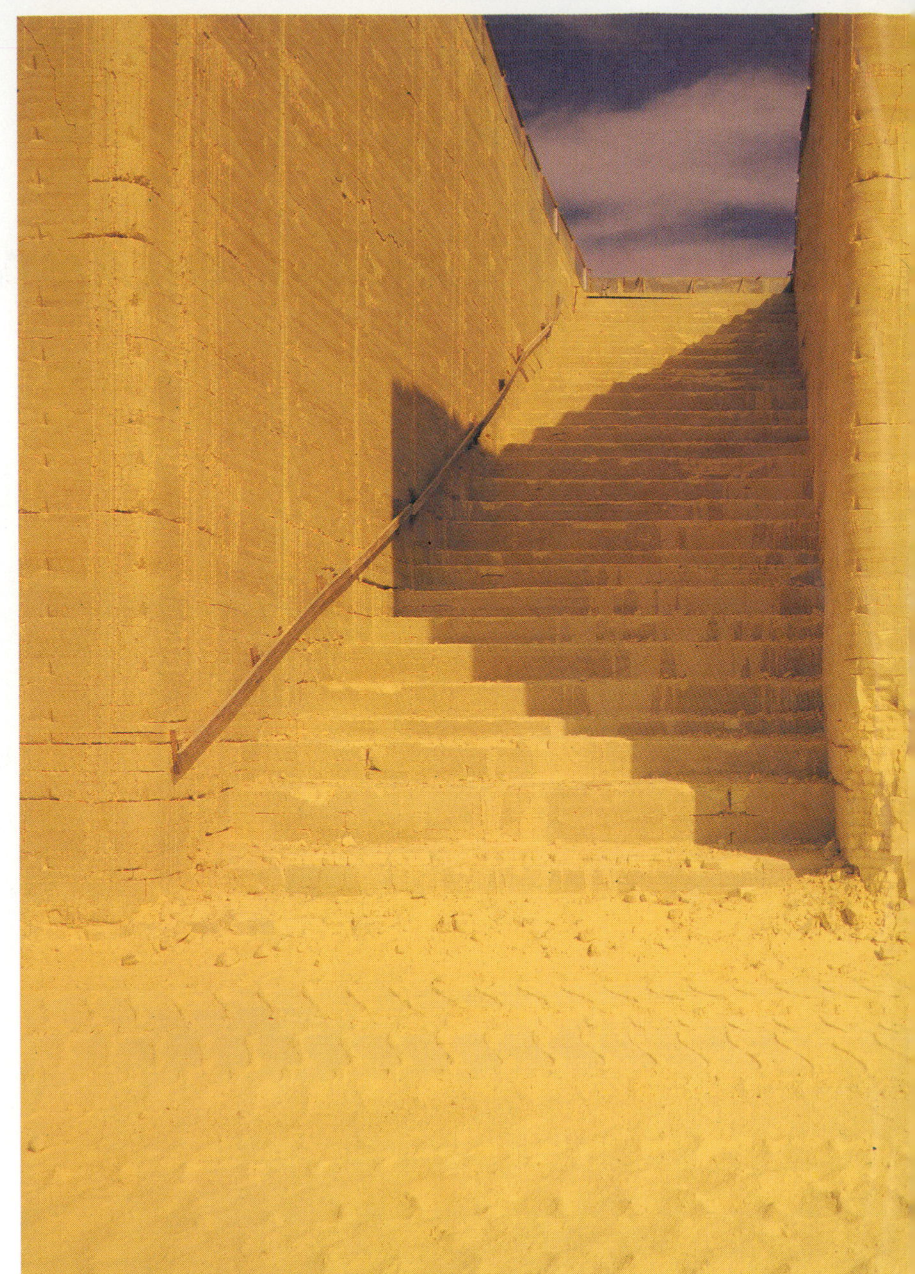

GENERAL CHEMISTRY

SECOND EDITION

ANNOTATED INSTRUCTOR'S VERSION

P. W. ATKINS

Oxford University

J. A. BERAN

Texas A&I University

SCIENTIFIC AMERICAN BOOKS Distributed by W. H. Freeman and Company

Front cover image:

C. Bryan Jones © 1990

Back cover images:

(Top) At high temperatures, a mixture of powdered iron and sulfur will react to produce a number of iron sulfides. As the reaction proceeds, there is a vigorous evolution of energy as heat. (Ken Karp)

(Middle) The head of a "strike-anywhere" match is composed of potassium chlorate, $KClO_3$, which acts as an oxidizing agent, and tetraphosphorus trisulfide, P_4S_3, which acts as a reducing agent. It also contains a binder and ground glass (to enhance the frictional heating when the match is struck). (Chip Clark)

(Bottom) When cooled very slowly, rhombic sulfur forms large crystals. Here the crystals are shown embedded in cooled magma, the molten rock that lies beneath the crust of the Earth, that has risen to the surface and solidified. (Chip Clark © 1989)

Library of Congress Cataloging-in-Publication Data

Atkins, P. W. (Peter William), 1940–
General chemistry / P. W. Atkins and J. A. Beran.—2nd ed.
p. cm.
Includes index.
ISBN 0-7167-2234-8
ISBN 0-7167-2265-8 (ISE ed.)
ISBN 0-7167-2284-4 (Annotated Instructor's Version)
1. Chemistry. I. Beran, J. A. (Jo Allen). II. Title.
QD31.2.A75 1992 91-559
540—dc20 CIP

Printed in the United States of America.

Scientific American Library
A Division of HPHLP
New York

Scientific American Books is a subsidiary of Scientific American, Inc.
Distributed by W.H. Freeman and Company.
41 Madison Avenue, New York, New York 10010

1 2 3 4 5 6 7 8 9 0 KP 9 9 8 7 6 5 4 3 2

CONTENTS IN BRIEF

CONTENTS

PART III RATES AND EQUILIBRIUM

ABOUT THE AUTHORS

A leading chemist, science writer, and educator, **P. W. Atkins** lectures in physical chemistry at the Oxford University and is a fellow of Lincoln College, Oxford. Among his many publications are *Physical Chemistry, Fourth Edition,* the leading textbook in that field, *Quanta* (Oxford University Press), *Inorganic Chemistry* (with D.F. Shriver and C.H. Langford), the leading textbook in its field, and three volumes for the Scientific American Library *(Molecules, The Second Law,* and *Atoms, Electrons, and Change).*

Dr. Atkins has held visiting professorships in Israel, France, Japan, and China and frequently travels throughout the world lecturing on chemical education.

J. A. Beran, recognized for his outstanding teaching credentials and contributions to chemical education, is Professor of Chemistry at Texas A&I University in Kingsville where, since 1968, he has coordinated the general chemistry program. He has helped develop programs in environmental chemistry and has worked with similar programs at the University of Texas, the University of Colorado, and the University of California, San Diego. He is also involved with the Advanced Placement and International Baccalaureate Programs in Chemistry.

Dr. Beran is the author of the bestselling *Laboratory Manual for General Chemistry,* and *Laboratory Manual for Fundamentals of Chemistry.*

PREFACE

The guiding principle in revising GENERAL CHEMISTRY remains our desire to develop in students a scientific attitude—what we have come to think of as "a sense of the reasonable." The result, within the chapters, the examples, and the exercises, is a consistent and thoughtful focus on learning chemistry by reasoning one's way to an answer. We aim to show students how to develop problem-solving skills by asking the right questions rather than merely applying the right formulas.

We believe that our collaboration has produced a text that combines thoughtful, tested pedagogy with a sense of the integrity of the scientific enterprise, and does so in a visually pleasing and highly readable way.

CHANGES IN THE SECOND EDITION

Of the many changes in this second edition of GENERAL CHEMISTRY, perhaps the most significant is our coauthorship. We have worked closely together during the preparation of this edition, continually drawing upon each other's skills and experience, and we believe instructors and students alike will benefit from our collaboration.

Our approach was not merely to graft on new material. Rather, we dismantled the book and, on the basis of three years of firsthand classroom use and feedback from many users of the first edition, decided what succeeded. Then we brought in new approaches that we knew from experience worked in the classroom, and rewrote and reassembled the parts—continually integrating feedback from our many readers—into a text that we believe is highly teachable and responsive to both instructors and students. With the help of a corps of reviewers, many of whom used the first edition, we have reconsidered literally every word, as well as the overall structure of the book. The result is a newly invigorated text that clearly communicates chemistry to students.

Recognizing that many students have had little or no exposure to the scientific method, we have made sure that the early chapters proceed gradually. Throughout the revision process we carefully evaluated and agreed on what is teachable when and how a topic is best introduced, with the result that the more advanced material now appears later in the book than it did in the first edition.

We emphasize throughout that frequently a uniform mode of thought can be deployed to solve a wide variety of problems. For instance, we have adopted in this edition a consistent way of setting up equilibrium calculations and a more consistent (and more natural) way of dealing with conversion factors.

We also have paid considerable attention to the uniformity of level. Chapter 3, which introduces the different classes of chemical reactions, is now pitched at a more introductory level than it was in the first edition. Chapters 8 and 9, on molecular structure, have been entirely reorganized so that one chapter (8) takes the reader from octet completion through to molecular shapes in terms of VSEPR theory; then Chapter 9 develops the details of electron distributions, treating valence bond theory in much greater detail than in the first edition and concluding with a section on molecular orbitals.

A similar major reorganization of material has taken place in Chapters 14 and 15 on acids and bases. We have grouped all the elementary material in Chapter 14 and have moved the more advanced material into Chapter 15. This arrangement allows instructors to select more easily the depth at which they prefer to cover acids, bases, and equilibrium. In particular, throughout these chapters we have adopted a uniform style of setting up and solving equilibrium problems, which the student should find easy to apply in a consistent way.

Other important changes within the text include a reorganized discussion of the role of electron pairs by way of Lewis acids and bases in Chapter 8, which allows instructors to go on to the theory of bonding in the context of real chemical reactions. Thermochemistry and enthalpy change are now the exclusive focus of Chapter 6; the first and second laws of thermodynamics are discussed together in the redrafted Chapter 16. Recent developments in a variety of subjects are incorporated throughout the text and the problems, including the latest on superconductivity, ozone depletion, smog control, DNA research, and new species such as the buckminsterfullerene.

NEW SUPPORT FOR STUDENTS

In addition to smoothing the level of exposition, particularly in the important early chapters, we have added human interest with two innovations in this edition. The first is the series of **Guest Boxes,** contributions from leaders in a variety of important fields that place their work in the context of the given chapter and indicate some of the unsolved problems working chemists and researchers face. We intentionally avoided biographical or philosophical essays, for we want to show students how the ideas they are meeting in the book are developed and applied in the real world. The second innovation is the **Cases** that conclude each chapter. These develop an interesting theme of applied or descriptive chemistry with a brief set of questions, often drawing upon previous chapters' skills, to test the student's ability to go beyond the basics. As the last item in each chapter, the Cases, which are often illustrated, function as a magnet of interest.

All the material at the end of each chapter has been reconsidered for this edition. For instance, the exercises, many of which are new, are now divided into two groups. **Classified exercises** cover the material systematically; they are paired, with the answers to the first (i.e., odd-numbered) of each pair given at the end of the text. Classified exercises are followed by the unpaired **Unclassified exercises,** which span the chapter without reference to the topic headings.

All defined terms are now in bold face type, and the fundamental ones are set off from the text. As in the first edition, the **Glossary** at the end of the book gives not only definitions but also examples of the use of many important terms. We have also included a number of in-text **Boxes** that describe interesting and motivating material that would otherwise interrupt the text. The expanded appendices include a review of basic mathematical skills and tables of data for help in problem solving.

ENDURING STRENGTHS OF THE TEXT

In making the many changes described here we were also mindful of the strengths that won praise for the first edition, and we planned carefully to retain them. The clarity of exposition, the striking artwork, and the sensitivity to students' needs continue to be hallmarks of the text.

Users of the first edition uniformly praised its readable style. We believe the chapters are even more readable now. Many readers praised the illustrations for being informative as well as attractive: we have retained and even expanded upon them. Many line drawings have been redone for even greater clarity, and many new photographs have been added. The first edition was also admired for the strategy sections within the worked examples, and we have retained them. These sections set the scene, suggest to students how to organize their thoughts for solving the problem, and indicate how to estimate the order of magnitude of the answer. In the second edition, the worked examples are more closely concentrated in the principles chapters than before, better to reinforce important skills.

The extensive work we have done in this revision

has built upon the strengths of the first edition. Our ideas have been shaped by our own experience and by feedback from previous users, students, and reviewers. Underlying our efforts has been our commitment to providing students with a text that will convey not only the content of chemistry but the excitement of the subject as well.

SUPPLEMENTS TO THE TEXT

We are fortunate to have been joined in this revision by a group of experienced and dedicated teachers who have worked hard and imaginatively to produce an excellent set of supplementary materials. These supplements are:

- **Study Guide** by David Becker, Oakland Community College and Oakland University. Identifies key words and concepts, important chemical and mathematical equations, and common pitfalls in problem-solving. Includes a generous selection of self-test questions and special sections that highlight descriptive chemistry the students should remember.
- **Student's Solutions Manual** by Joseph Topping, Towson State University, and Charles Trapp, University of Louisville. Contains detailed solutions for the exercises and case questions not answered in the text.
- **Instructor's Manual** by Joseph Topping and Charles Trapp. Contains detailed solutions to all exercises and case questions in the text, with chapter outlines, objectives, review questions and solutions, and teaching hints.
- **Test Bank** by Robert Balahura, University of Guelph. Available in hard copy, IBM, and Macintosh formats. Provides approximately 1000 test items in multiple choice format.
- **Overhead Transparencies**. Includes over 200 sharply rendered figures and tables from the text, most in full color.
- **Videotaped Lecture Demonstrations** by Ted Baldwin, Industrial Words and Images (Dewey Carpenter, Louisiana State University, consultant). A four-hour set of vivid presentations of chemical phenomena. The convenient way to demonstrate a wide variety of experiments that are elaborate, potentially hazardous, or expensive to create.
- **Videodisc of Lecture Demonstrations and Chemical Illustrations** by Ted Baldwin, Industrial Words and Images, and Windsor Digital. A two-sided laserdisc containing highlights of the 56 demonstrations in our videotape series and many new demonstrations, plus approximately 1000 figures and photographs.
- **Laboratory Separates** by Julian L. Roberts, Jr., J. Leland Hollenberg, and James M. Postma. A complete package of individual laboratory experiments that allows flexibility in designing a laboratory sequence.

THE ANNOTATED INSTRUCTOR'S VERSION

This annotated version of the text has been prepared principally for instructors in collaboration with David Becker, of Oakland Community College and Oakland University (and author of the accompanying Study Guide). We have aimed to produce a version that does more than merely indicate where additional pedagogical information may be found: many of the annotations are enhancements of the text, points that show what lies behind the principles, insights into the authors' organization and presentation of the material, and ways to enrich the student understanding. We have broadly classified the annotations as follows:

E Enrichment: an extension of the text, a point of interest, or an explanation.

N Note: a suggestion to the instructor about how the material may be used or additional material obtained.

D Demonstration: a reference to a tested demonstration in the literature, or an easily organized demonstration. We consider that even very simple demonstrations can enrich a lecture, and have provided some suggestions that can be implemented with very little preparation.

? Question: a suggestion for a question that could be asked to start a class thinking about the material.

V Video: a reference to a demonstration on the videotape or laserdisc provided as supplementary material.

T Overhead Transparency: a suggestion of an appropriate transparency to use at this point in an exposition.

U Unclassified Exercise: an indication of which unclassified exercises it would be appropriate to assign.

Standard journal references are used; the other abbreviations are as follows:

CD1 and CD2: *Chemical Demonstrations, a Sourcebook for Teachers,* Vols. 1 and 2, edited by L. R. Summerlin, C. L. Borgford, and J. B. Ealy. American Chemical Society, Washington, D.C. (1988).

BZS1, BZS2, and BZS3: *Chemical Demonstrations, A handbook for Teachers of Chemistry* by B. Z. Shakhashiri, University of Wisconsin Press, Madison, Wisconsin (1989).

TDC: *Tested Demonstrations in Chemistry,* 6 ed., edited by H. N. Alyea and F. B. Dutton. The Journal of Chemical Education, Easton, Pennsylvania (1969).

ACKNOWLEDGMENTS

As so often in an enterprise of this magnitude, there are an enormous number of people to thank; space allows us to mention only those whom we encountered directly, but the others are far from forgotten. We have been very well served by our publisher, and are very grateful to Elisa Adams, our development editor, who guided us with so much good advice throughout the preparation of the revision. We are also particularly grateful to Georgia Lee Hadler, our project editor, whose skill and care controlled the immensely complex project once it had gone into production. Jodi Simpson, our copy editor, kept our literacy under firm control and made a number of helpful suggestions, and we are very grateful to her for her care and attention to detail. Behind the scenes, but not unnoticed, were Sheila Anderson and Mara Kasler who coordinated the production and the illustrations, respectively. The design and art are a major feature of the book, and we are particularly grateful to Alison Lew and Megan Higgins, the designer and the art director, to Travis Amos, the photo researcher, to John Hatzakis, the production artist, to Tomo Narashima, the illustrator, and to Chip Clark, who provided the new photographs. All of them contributed so much in helping to turn a daunting task into a pleasure.

No book of this scope could have been written without the advice of many wise, experienced, and well-informed reviewers. Many contributed to the first edition; their influence is still apparent and our gratitude to them remains undiminished:

David L. Adams, Bradford College
John E. Adams, University of Missouri, Columbia
Martin Allen, College of St. Thomas (retired)
Norman C. Baenziger, University of Iowa
John E. Bauman, University of Missoui, Columbia
David Becker, Oakland Community College
James P. Birk, Arizona State University
Luther K. Brice, American University
J. Arthur Campbell, Harvey Mudd College
Dewey K. Carpenter, Louisiana State University
Geoffrey Davies, Northeastern University
Walter J. Deal, University of California, Riverside
John DeKorte, Northern Arizona University
Fred M. Dewey, Metropolitan State University
John H. Forsberg, St. Louis University
Marjorie H. Gardner, Lawrence Hall of Science
Gregory D. Gillespie, North Dakota State University
L. Peter Gold, Pennsylvania State University
Michael Golde, University of Pittsburgh
Thomas J. Greenbowe, Southeastern Massachusetts University
Robert N. Hammer, Michigan State University (consultant)
Joe S. Hayes, Mississippi Valley State University (retired)
Henry Heikkinen, University of Northern Colorado

Forrest C. Hentz, Jr., North Carolina State University
Jeffrey A. Hurlbut, Metropolitan State University
Earl S. Huyser, University of Kansas
Murray Johnston, University of Colorado, Boulder
Philip C. Keller, University of Arizona
Robert Loeschen, California State University, Long Beach
David G. Lovering, Royal Military College of Science, Shrivenham
James G. Malik, San Diego State University
Saundra McGuire, Cornell University
Amy E. Stevens Miller, University of Oklahoma
E. A. Ogryzlo, University of British Columbia
M. Larry Peck, Texas A&M University
Lee G. Pedersen, University of North Carolina, Chapel Hill
W. D. Perry, Auburn University
Everett L. Reed, University of Massachusetts
Don Roach, Miami-Dade Community College
E. A. Secco, St. Francis Xavier University
R. L. Stern, Oakland University
R. Carl Stoufer, University of Florida
Billy L. Stump, Virginia Commonwealth University
James E. Sturm, Lehigh University
James C. Thompson, University of Toronto
Donald D. Titus, Temple University
Joseph J. Topping, Towson State University
Patrick A. Wegner, California State University, Fullerton

For help in preparing the second edition we wish to thank:

M. C. Banta, Sam Houston State University
David Becker, Oakland University/Oakland Community College
J. M. Bellama, University of Maryland—College Park
Richard Bivens, Allegheny College
P. M. Boorman, University of Calgary
Larry Bray, Miami-Dade Community College
Mark Chamberlain, Glassboro State College
Corinna Czekaj, Oklahoma State University
Walter Dean, Lawrence Technological University
James A. Dix, State University of New York—Binghamton
Craig Donahue, University of Michigan, Dearborn
Wendy Elcesser, Indiana University of Pennsylvania
Grover Everett, University of Kansas
Gordon Ewing, New Mexico State University
Bruce Garatz, Brooklyn Polytechnic University
L. Peter Gold, Pennsylvania State University
Michael Golde, University of Pittsburgh
Stanley Grenda, University of Nevada
Kevin Grundy, Dalhousie University
Kenneth I. Hardcastle, California State University at Northridge
Larry W. Houk, Memphis State University
Robert Hubbs, De Anza College
Brian Humphrey, Montclair State College
Robert Jacobson, Iowa State University
Charles Johnson, University of North Carolina
Delwin Johnson, St. Louis Community College at Forest Park
Stanley Johnson, Orange Coast College
Murray Johnston, University of Delaware
Andrew Jorgensen, University of Toledo
Herbert Kaesz, University of California at Los Angeles
Manickam Krishnamurthy, Howard University
M. Steven McDowell, South Dakota School of Mines and Technology
Gloria J. Meadors, University of Tulsa
Patricia Metz, Texas Tech University
Ronald E. Noftle, Wake Forest University
Jane Joseph Ott, Furman University
Dick Potts, University of Michigan, Dearborn
B. Ken Robertson, University of Missouri
Larry Rosenhein, Indiana State University
Patricia Samuel, Boston University
Henry Shanfield, University of Houston
Charles Trapp, University of Louisville
Arlen Viste, Augustana College
Tom Weaver, Northwestern University
Rick White, Sam Houston State University

We welcome comments and suggestions from students and instructors alike; please write to us in care of Scientific American Books.

P. W. A.
J. A. B.

GENERAL CHEMISTRY

In this chapter we meet the basic language of chemistry, see a little of chemistry's methods and aims, and learn the ground rules for reporting and using measurements. The illustration shows the nebula observed in 1987, the most recent one that has been detected and photographed. Such explosions scatter newly made elements through space and make them available on planets.

1 PROPERTIES, MEASUREMENTS, AND UNITS

Chemistry begins in the stars. The stars are the source of the chemical elements, which are the building blocks of matter and the core of our subject. Within the stars, intense heat causes atoms of hydrogen, the smallest particles of the simplest element, to smash together, merge, and become atoms of other elements—carbon, oxygen, iron, and the rest. This merging releases even more heat, which generates starlight; so starlight, including sunlight, is a sign that the elements are still being formed.

The planet Earth is made up of elements forged long ago inside ancient stars. Many million years after a star is formed and begins to cool, its outer layers may collapse, like a falling roof, into its exhausted core. This mighty quake produces such large shock waves that the star shrugs off its outer layers and sends them into space in a huge explosion called a supernova. Six such explosions have been detected in our galaxy in the past 1000 years, the most recent in 1987. The shock of explosion raises the temperature in the star even higher than before: the Crab nebula (produced by the supernova of 1054) was visible in broad daylight for three weeks. At such high temperatures, even the bigger atoms collide violently enough to merge and become still heavier atoms. The very heavy elements now found on Earth, including uranium and gold, were made in this way.

Changes (we shall come to know them as chemical changes) have taken place among the elements and have converted them into the raw materials we find on Earth; thus the elements formed in stars have become rocks, ocean, air, vegetation, and flesh. We humans have discovered how to change these raw materials into substances that are well suited to our particular tasks.

In the early days of civilization, stone was replaced by metal as the material from which tools and weapons were made. At first, metals were blended together by accident. Then, it was found that certain mixtures were easier to cast and mold, easier to use, or more durable, and blending was done to achieve specific ends. The economic impact

D Introducing chemistry. CD2, 1–4.

of the new materials, together with humanity's spirit of inquiry, eventually gave rise to **science**, the systematically collected and organized body of knowledge based on experimentation, observation, and careful reasoning. In particular, that spirit of inquiry led to the branch of science known as chemistry. Greed admittedly drove that spirit originally, for chemistry sprang from the alchemists' vain struggle to convert lead into gold. Although the alchemists failed to copy the stars, in the course of their failure they discovered that they could achieve many other conversions of matter, for example, rocks into acids and fats into soaps. Thus they opened the road to modern chemistry.

Chemists continue to discover through **experiments**—tests done under carefully controlled conditions—how to convert one kind of material into another. Their experiments, the way chemists collect, organize, and use their observations and measurements, are where we begin our journey. In the course of this journey, we shall meet the elements and see the combinations they form. We shall end it where chemistry merges with biology, with combinations of elements so elaborate that we regard them as being alive.

THE PROPERTIES OF SUBSTANCES

E The name chemistry is derived from *alchemy*, which in turn descends from *al* (the) and *khemia*, a Greek form of a native name for Egypt, where alchemy originated.

The subject of this text is chemistry:

> **Chemistry** is the branch of science concerned with the study of matter and the changes matter can undergo.

By **matter** we mean anything that has mass and takes up space. It includes the bricks and wood used for houses, the metal of airplanes, and the flesh and bone of human bodies. Matter includes water, air, earth, drugs, fertilizers, microchips, plastics, explosives, and food. Matter does not include light or abstract concepts like beauty (although chemistry, and chemicals, can be beautiful) because these do not take up space and have no mass.

Chemists collect information about matter by making careful observations of a **sample**, a representative part of a whole. A sample of an ocean might be some seawater in a test tube (Fig. 1.1); a sample of a rock might be a tiny fragment on a microscope slide. As in everyday life, in chemistry we distinguish between different **substances**, or different kinds of matter. Pure water is one substance, iron another. We recognize different substances by their distinct **properties**, or characteristics. Some properties are unique to that substance, for example, its color, its smell, and its behavior when it is heated or when it is mixed with other substances. Other properties are not unique but vary with time and place, for example, mass, volume, and temperature.

FIGURE 1.1 Chemists study the properties of matter by making observations on a sample.

1.1 PHYSICAL AND CHEMICAL PROPERTIES

In this section we start to sharpen concepts that are already familiar from everyday experience. That is a common feature of science: scientists accept much of everyday experience, express it precisely, and then explore the implications of what they have found. Sharp and precise definitions provide an excellent basis for organizing observations and then making discoveries by noticing patterns or misfits in these patterns. As we shall see, the Russian chemist Dmitri Mendeleev made what is perhaps the greatest of all discoveries in chemistry, that the elements can be organized in a "periodic table" (which we describe in Chapter 2). Once he had recognized the pattern of the table, he was able to predict new elements that would fill the gaps in it.

We classify properties as either chemical or physical according to whether or not the formation of other substances is involved.

A **physical property** of a substance is a characteristic that we observe or measure without changing the identity of the substance.

A **chemical property** of a substance is a characteristic that we observe or measure only by changing the identity of the substance.

That gold is yellow, conducts electricity, and melts at 1063°C are three of its physical properties, because no new substances are formed. That natural gas—which is mainly methane—burns to produce carbon dioxide and water is one of methane's chemical properties because new substances are formed when methane burns.

FIGURE 1.2 The three physical states of matter are easily distinguished by noting whether the substance retains its shape (solid), fills the bottom of the container (liquid), or fills the container completely however much is present (gas).

Physical state. Often the first characteristic of a substance we notice is its **physical state**, that is, whether it is a solid, a liquid, or a gas at a particular temperature. We do this—almost unconsciously because it is such a familiar distinction—by noting which of the following physical characteristics the substance has (Fig. 1.2):

A **solid** is a rigid form of matter that maintains the same shape, whatever the shape of its container.

Common solids (at ordinary temperatures) include brass, granite, quartz, and the elements copper, titanium, vanadium, and silicon.

E A solid can also be defined as a substance in which there is internal long-range order (as in the regular array of atoms in a crystal); a liquid can be defined as a condensed phase in which there is only short-range order. In a gas, there is no spatial order. A *mesophase* (a liquid crystal) is a fluid in which there is short-range order in one direction and long-range order in another direction.

A **liquid** is a fluid form of matter (one that flows) that takes the shape of the part of the container it occupies.

Common substances that are liquids at room temperature include benzene (which is obtained from petroleum) and water. The only elements that are liquid at room temperature are mercury and bromine, but the elements gallium and cesium melt to liquids at body temperature.

D Heat sulfur in a test tube, and note the color changes and changes of physical state; finally, pour the molten sulfur into water, making plastic sulfur.

D The three states: Gas, liquid, and solid. TDC, 5.1.

A **gas** is a fluid form of matter that *fills* any container it occupies and can easily be compressed into a much smaller volume.

The term **vapor** is also widely used to mean a gas, but it usually implies that both the gas and the liquid state of the substance are present. For instance, we refer to the water vapor that fills the space above the liquid in a kettle. Eleven elements are gases at room temperature. The gases nitrogen and oxygen are the two principal components of the atmosphere. Helium, neon, and argon are three of the so-called noble gases. Helium is the gas used to fill the Goodyear blimp. The gaseous elements also include the pale yellow-green gas chlorine, which is used as a disinfectant in public swimming pools.

TABLE 1.1 Changes of state

Initial state	Change	Final state
solid	melting or fusion ⇌ freezing	liquid
liquid	vaporization ⇌ condensation	vapor
solid	sublimation ⇌ deposition	vapor

Physical change. The change of a substance from one of its physical states to another state in which it can exist is called a **change of state**. For example, a change of state occurs when liquid water freezes and becomes ice. Various changes of state are listed in Table 1.1; they include the familiar processes of melting and freezing. One striking observation is that a change of state of a substance always occurs at a specific temperature. The temperature at which a physical change occurs is a physical property of the substance. When we heat a solid such as ice, it remains solid right up to its **melting point**, the temperature at which it melts; then some of it changes to a liquid. The temperature of the melting ice remains the same as we continue heating it, but more of it melts. Only when all the solid has melted does the temperature of the substance rise. Similarly, when a liquid is cooled, it remains liquid until the temperature has fallen to the **freezing point**, the temperature at which it freezes; then it begins to solidify. Once again, the temperature of the substance remains constant until all the liquid portion of the substance has frozen. For all common materials, the melting point is exactly the same as the freezing point; so ice melts to water at 0°C and water freezes to ice at 0°C.

N Which terms in this section should the students commit to memory?

Another change in the physical state of a substance occurs when the liquid is heated to its **boiling point**, the temperature at which it boils. Under typical everyday conditions, water boils when the temperature has reached 100°C; then suddenly bubbles of vapor start to form throughout its bulk. As in freezing and melting, the temperature of the boiling water remains constant until all the liquid has vaporized: the boiling water in a kettle stays at 100°C until all the water is gone. Substances that boil at temperatures lower than 100°C are said to be **volatile**. Alcohol (ethanol) boils at 78°C, so it is more volatile than water. Ether (diethyl ether) is even more volatile because it boils at 34.5°C. The substances we call gases are so volatile that their boiling points lie well below room temperature.

E Later, we distinguish between a *melting point* and *melting temperature*. The former is the melting temperature at a given pressure (typically 1 atm), so it corresponds to a single point on the phase diagram. Similar remarks apply to the boiling temperature and the boiling point.

That different substances melt and boil at characteristic temperatures means that we can use these physical properties as guides to the identities of substances. For example, one white powder may look very much like any other, but we may suspect that one is aspirin if it melts at 135°C (the melting point of aspirin). Several white substances may melt close to that temperature, but the identification can be confirmed by carrying out other tests.

One physical change listed in Table 1.1 that may be less familiar than the others is **sublimation**, the conversion of a solid directly into a vapor. Solid carbon dioxide is one substance that sublimes, or vaporizes without first melting, and this property gives it its name "dry ice." Another is frost on a cold, dry morning, which disappears as it changes to water vapor without first melting. The reverse of sublimation—a vapor condensing directly to a solid—is called **deposition**.

V Video 3: Sublimation.

E More precisely, a vapor is a gaseous phase of a substance at a temperature lower than its critical temperature; a gas is a substance above its critical temperature.

U Exercise 1.63.

Chemical change. The formation of one substance from another substance is called a **chemical change**. Chemical changes include the very complicated reactions that occur when food is cooked and the substances that contribute to its flavor and aroma are formed. The extraction of metals from their ores makes use of chemical changes, as does the production of synthetic fibers (for example, nylon, acrylics such as Orlon, and polyesters such as Dacron) from air, coal, and petroleum. Chemical changes occur during **chemical reactions**, in which one substance responds ("reacts") to the presence of another, to a change of temperature, or to some other influence. For example, the conversion of silica to silicon, the material used in the manufacture of microchips, is a chemical change brought about by a reaction between silica and carbon. Chemists respond daily to the challenge to discover the chemi-

D Differentiating physical and chemical change. *J. Chem. Ed.*, **1970**, *47*, 689.

? What are some naturally occurring chemical changes? Examples are the turning of leaves in the fall, the rotting of wood, the blooming of a flower, the action of rainwater on limestone.

FIGURE 1.3 Interior view of an aluminum production plant, where the electrolysis operation is conducted on a large scale. The pot room, the vast chamber shown here, contains the electrolytic cells that produce the aluminum.

cal reactions that produce a desired substance from other, more readily available starting materials.

A chemical change can be brought about by passing an electric current through liquids (including molten substances). This process, called **electrolysis**, has led to a number of important discoveries. For example, in 1807 the English chemist Humphry Davy, who made a systematic study of electrolysis, discovered two new metallic elements—sodium and potassium—within the period of a few days. Electrolysis is the foundation of several great chemical industries. It is used on a huge scale to produce aluminum metal from aluminum oxide (Fig. 1.3). It is also used to produce sodium hydroxide and chlorine from brine. Electrolysis (which is discussed further in Chapter 17) is the only method we have for producing the gas fluorine, some of which is used for refining the uranium used in the nuclear power industry.

1.2 SUBSTANCES AND MIXTURES

One very important task that chemists have to carry out is **chemical analysis**, the determination of the chemical composition of samples of substances. Chemical analyses make use of physical techniques, which are based on physical properties, and chemical techniques, which are based on chemical properties. Common physical techniques are listed in Table 1.2, and we shall describe them in more detail shortly. Chemical techniques include electrolysis, mixing and heating the unknown substance with known ones, burning it in air, and testing its reaction to hydrochloric acid and sulfuric acid. We shall encounter numerous chemical techniques as we progress through the text.

E Fluorine can be produced by a purely chemical procedure, but it is not economically viable. Fluorine is normally produced by the electrolysis of a cooled solution of potassium fluoride in anhydrous liquid hydrogen fluoride.

T OHT of Chart in the Summary.

In everyday language the term *substance* refers to any sample of matter; in scientific usage, however, it has a special meaning:

A **substance** is a single type of matter.

Water, sodium chloride (common salt), and all the elements are each a single substance. The first step in a chemical analysis is determining

T OHT of Table 1.2.

TABLE 1.2 Methods for separating the components of mixtures

Separation method	Physical property used	Procedure
filtration	solubility	pouring of solid + liquid mixture onto a filter; solid trapped by filter; liquid passes through filter
crystallization	solubility	slow crystallization of solid from solution
distillation	volatility	boiling off the more volatile component of a liquid mixture
chromatography	ability to adsorb to surface	passing liquid or gaseous mixture over paper or through a column coated or packed with adsorbent
centrifugation	density	rotation of liquid-solid mixture at high speed in a centrifuge; solid collects at bottom of sample tube

V Video 1: Mixtures

D Separating metallic iron from cereal, CD2, 34.

N At this stage we have not defined the term *atom*, and so must give the classical, empirical definition of the term *element*. Later, once atoms have been introduced and the concept of atomic number established, it will be appropriate to adopt the direct definition of an element as a substance that consists of only one type of atom (that is, of atoms of the same atomic number).

E There are about 109 known elements; two are liquids, eleven are gases.

V Video 4: Elements.

E What is an element? *J. Chem. Ed.*, **1989**, *66*, 66.

N Similarly, later we shall be able to define a compound as a specific combination of atoms of different elements.

E Molten sodium chloride can be electrolyzed. Carbon dioxide can be reduced to carbon by passing it over hot zinc.

D Mixtures and compounds. TDC, 1.5.

U Exercise 1.62.

whether a sample consists of a single substance or a mixture of substances:

> A **mixture** consists of more than one substance; and these components may be separated by making use of their different physical properties.

Examples of mixtures are air (which consists of nitrogen, oxygen, and other gases), seawater (water containing many dissolved substances, particularly sodium chloride), and brass (which consists of copper and zinc).

The second step in an analysis is determining whether each (pure) substance is an element or a compound:

> An **element** is a substance that cannot be broken down into simpler components by using chemical techniques.

The elements are the building blocks of compounds. Examples of elements are hydrogen (a gas), bromine (a liquid), and iron and phosphorus (both solids). None of these substances can be separated into simpler components by using the chemical techniques of electrolysis, burning, exposure to acids, and so on. Compounds, however, *can* be broken down into elements by using chemical techniques:

> A **compound** is a specific combination of elements that can be broken down by using chemical techniques.

Examples of compounds are the solid sodium chloride (table salt, a compound of the elements sodium and chlorine) and the gas carbon dioxide (one of the gases in the atmosphere, a compound of carbon and oxygen). Neither of these substances can be broken down into its elements by using physical techniques such as freezing or vaporization, but they can be broken down into elements by using chemical techniques.

▼ EXAMPLE 1.1 Distinguishing a compound from a mixture

When seawater is boiled, the water vaporizes and a solid residue is left behind. When the water is condensed and an electric current is passed through it (the process of electrolysis), the gases hydrogen and oxygen are formed. Identify the mixture and the compound.

STRATEGY We identify a *mixture* by the observation that its components can be separated on the basis of their different physical properties. We identify a *compound* as a substance that can be broken down into its elements by chemical techniques.

SOLUTION Seawater is separated into components, the water and the residue (which is itself a mixture) by the physical technique of boiling (which corresponds to distillation; see Table 1.2); so seawater is a mixture. Water cannot be broken down into its elements by using physical techniques, so it is a single substance. The observation that water decomposes into its elements when subjected to the chemical technique of electrolysis shows that it is a compound consisting of hydrogen and oxygen.

EXERCISE E1.1 Is silica, from which silicon is obtained by heating with carbon, a mixture or a compound?

▲ [*Answer*: a compound (of silicon and oxygen)]

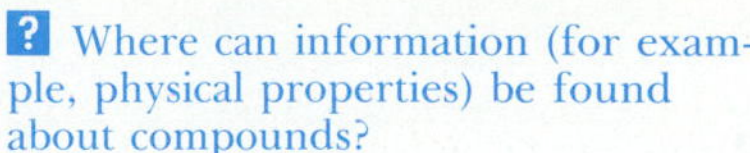

? Where can information (for example, physical properties) be found about compounds?

D Electrolyze water, collect the gases, and demonstrate properties of hydrogen and oxygen.

Differences between mixtures and compounds are summarized in Table 1.3. The most important distinction at this stage is that *whereas the physical properties of the individual components can usually be recognized in a mixture, the properties of a compound are usually strikingly different from the properties of its constituent elements.* Water, a compound, is totally unlike the gases hydrogen and oxygen from which it is formed. In contrast, a mixture of sugar and sand is both sweet (from the sugar) and gritty (from the sand). When shaken with water, the sugar dissolves but the sand does not.

The classification of mixtures. In some cases we can identify the different components of a mixture visually, by their colors or their shapes. A mixture like this, a patchwork of aggregates of different substances, is called a **heterogeneous mixture**. Many of the rocks that form the landscape are heterogeneous mixtures (Fig. 1.4). A mixture of sugar and sand, no matter how thoroughly it is mixed, is a heterogeneous mixture at the microscopic level. Milk is a heterogeneous mixture, because

TABLE 1.3 Differences between mixtures and compounds

Mixture	Compound
components can be separated by using physical techniques	components cannot be separated by using physical techniques
composition is variable	composition is fixed
properties are related to those of its components	properties are unlike those of its components
little heat is produced during formation*	considerable heat is produced during formation†

*There are exceptions: mixing sulfuric acid with water produces a lot of heat.
†In some cases a chemical change is accompanied by considerable *absorption* of heat.

FIGURE 1.4 This piece of rock is a heterogeneous mixture of many substances.

under the microscope we can see the individual globules of butterfat, that is, we see aggregates of fat molecules floating in the liquid.

? What kind of mixtures are tap water, river water, soil, and human flesh?

In other mixtures we cannot make out distinct regions, even with a very powerful microscope: the components are so well mixed that the composition of the mixture is the same throughout the sample. Such a mixture is called a **homogeneous mixture**. We cannot tell visually whether a sample is a homogeneous mixture or a pure substance until attempts have been made to separate its components. This is the case with a sample of air—a homogeneous mixture of several colorless gases—for we cannot see that air is a mixture, even under a microscope. However, if we liquefy a sample of air, then we can separate its components by collecting them in a series of flasks as they boil off at their respective (and different) boiling points.

Homogeneous mixtures are also called **solutions**, although this name generally refers to mixtures in which one component (a liquid) is much more abundant than the others are. Beer is a solution of alcohol in water, together with the substances responsible for its flavor. Seawater is a solution of salt (sodium chloride) and many other substances in water. In each case the most abundant component of a solution—water, in these examples—is called the **solvent** and the dissolved substances are the **solutes**.

? Why is the ice in partially frozen ice cubes clear, but then cloudy when the entire cube is frozen? (Assume that tap water is used.)

D The silicate garden. CD1, 23.

We normally dissolve a solid (that is, prepare a solution in which it is the solute) by shaking it up with the solvent or by stirring, as we do when we dissolve sugar in a cup of coffee. The process that is the opposite of dissolving is called either crystallization or precipitation. **Crystallization**, the process in which the solute comes out of solution slowly as crystals, often occurs as a solvent slowly evaporates. **Precipitation**, the process in which the solute comes out of solution rapidly as a finely divided powder, a "precipitate," can be almost instantaneous (Fig. 1.5). The distinction between crystallization and precipitation is not very profound, because the precipitating powder is usually a mass of very small crystals.

FIGURE 1.5 Precipitation occurs when an insoluble substance is formed. Here yellow lead iodide, which is insoluble, precipitates when we mix solutions of lead nitrate and potassium iodide.

Beer and seawater are examples of **aqueous solutions**, that is, solutions in which the solvent is water (the Latin word *aqua* means "water"). Aqueous solutions are very common in everyday life and in chemical laboratories; consequently, most of the solutions we discuss in this text will be aqueous. There are also many examples of **nonaqueous solutions**, that is, solutions in which the solvent is not water. In "dry cleaning," the grease and dirt on fabrics are dissolved in the nonaqueous solvent tetrachloroethylene, which is a compound of carbon and chlorine and dissolves grease far better than water does.

Techniques for separating the components of mixtures. In **filtration**, we shake the sample with a liquid and then pour the mixture through a fine mesh, the filter. The soluble material in the sample dissolves in the liquid and passes through the filter. The insoluble material in the sample does not dissolve and is captured by the filter. We could use filtration to separate common salt from powdered glass because salt is soluble in water but glass is not. Filtration is one step in the treatment of domestic water supplies because it removes the particles of insoluble matter from the water.

Crystallization is another technique used for separating and purifying the components of mixtures, because impurities (the other dissolved components in the mixture) stay in the solution as the crystals

T OHT: Figs. 1.6 and 1.7.

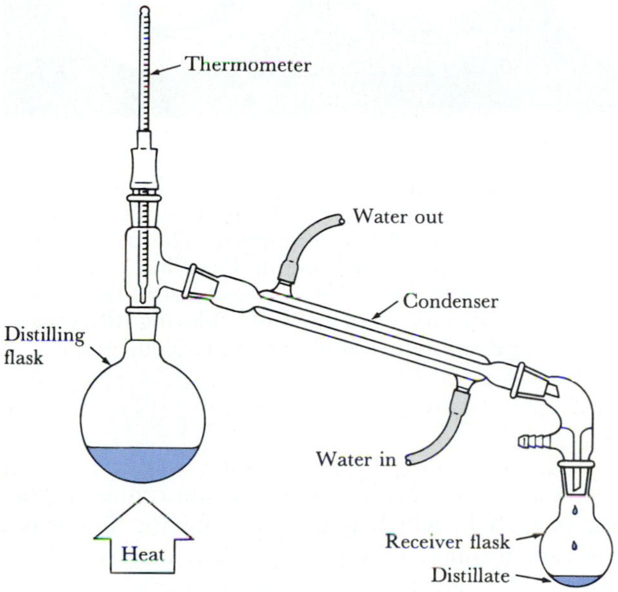

Thermometer
Water out
Condenser
Distilling flask
Water in
Heat
Receiver flask
Distillate

FIGURE 1.6 A simple distillation apparatus. As the liquid mixture is heated, the more volatile component boils off, condenses in the water-jacketed tube (the condenser), and is collected. The product in the receiver flask is called the distillate.

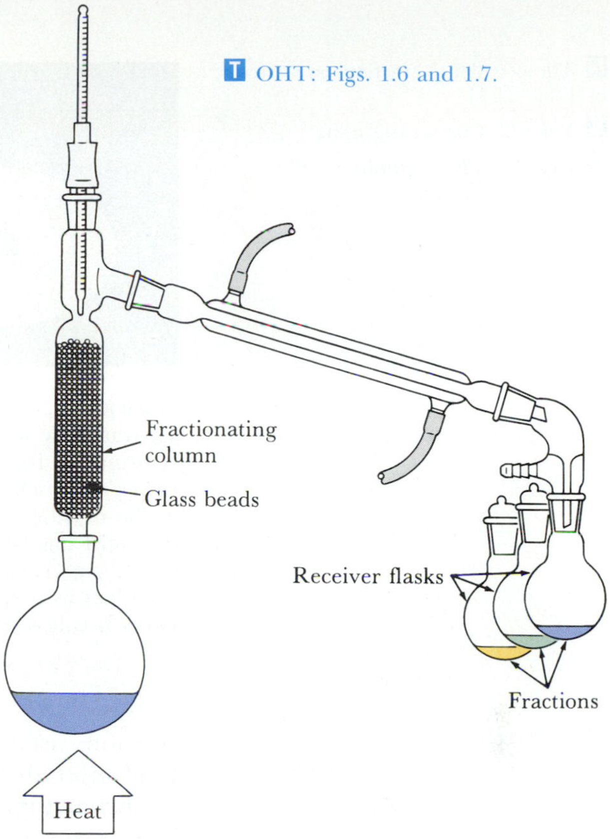

Fractionating column
Glass beads
Receiver flasks
Fractions
Heat

FIGURE 1.7 The apparatus used for fractional distillation includes a fractionating column packed with glass beads. A small portion of the distillate is collected in a series of flasks, and the content of each flask is called a fraction.

form. This purification process occurs on a grand scale when ice forms from seawater. Sea ice is largely fresh water because the salt remains behind in the ocean.

The technique of **distillation** separates the components of a mixture by vaporizing one or more of the components (Fig. 1.6). It can be used to remove water (which boils at 100°C) from salt, which boils at a much higher temperature and is therefore left behind when the water evaporates. **Fractional distillation** (Fig. 1.7) is used for separating the components of mixtures of liquids that have different boiling points: the mixture is heated and the rising vapor passes up through a fractionating column packed with glass beads. The vaporized component with the highest boiling point condenses on the beads near the bottom of the column and drips back into the boiling mixture. The vaporized component with the lowest boiling point continues to rise through the column without condensing and passes out into the condenser. Therefore, the first "fractions" to be collected consist of the most volatile of the mixed liquids; when that component has all boiled off the sample, the next most volatile component comes through, and so on. Giant fractionating columns (Fig. 1.8) are used to separate the various components of petroleum, ranging from the volatile fractions used as gasoline (boiling in the range 50 to 200°C) and kerosene (175 to 325°C) to the less volatile

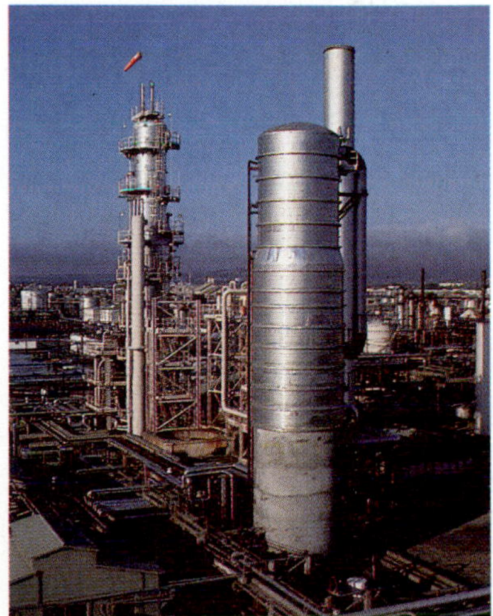

FIGURE 1.8 The fractionating columns in an oil refinery work on a principle similar to that used in the laboratory, but on a much larger scale.

N The petroleum fractions are listed in Table 23.3.

V Video 2: Chromatography.

D Paper chromatography. TDC, 24–54s, p. 173 and p. 217.

FIGURE 1.9 In paper chromatography the components of a mixture are separated by washing them along a paper—the "support"—with a solvent. A primitive form of the technique is shown here. On the left is a dry filter paper to which a drop of food coloring has been applied. The filter paper in the middle was spotted and then placed in solvent. The filter paper on the right has been allowed to dry after the solvent spread out to the edges of the paper, carrying the two components of the food coloring different distances as it spread. The dried support showing the separated components is called a chromatogram.

? What physical separation techniques could be used to produce potable water from river water? (See Case 3.)

E Modern chromatography uses a variety of supports and moving phases. As well as those detailed in the text, they include *paper chromatography* (where the stationary phase is held in the pores of the paper), *thin-layer chromatography* (in which the stationary phase is a finely divided solid, perhaps coated with liquid, held on a glass plate), *gel chromatography* (where the stationary phase is held in the interstices of a polymeric solid), and *ion-exchange chromatography* (in which the solid stationary phase is a finely divided ion-exchange resin in a column).

T OHT: Fig. 1.10.

fractions used as diesel fuel (above 275°C). The residue remaining behind after distillation is asphalt, which is used for paving highways.

Chromatography is a separation technique that relies on the abilities of surfaces to **adsorb**, or bind, substances. There are various versions of the technique. In the simplest, the sample is washed out from the center of a disk of filter paper (Fig. 1.9). Substances that are adsorbed most weakly to the paper move further than others and, if they are colored, give rise to separate patches of color on the paper. (This appearance is the origin of the name chromatography, which comes from the Greek word meaning "color writing.")

In **gas-liquid chromatography** the sample is vaporized and carried in a stream of helium, the carrier gas, through a long narrow tube. The tube is coated or packed with aluminum oxide (or a similar adsorbent material) soaked in the stationary phase, a liquid of low volatility. The component that is adsorbed *least* strongly by the stationary phase emerges first from the other end (Fig. 1.10). As time passes, the other components of the original mixture emerge, and each is detected electronically and signaled by a peak on a chart called a **chromatogram**. In

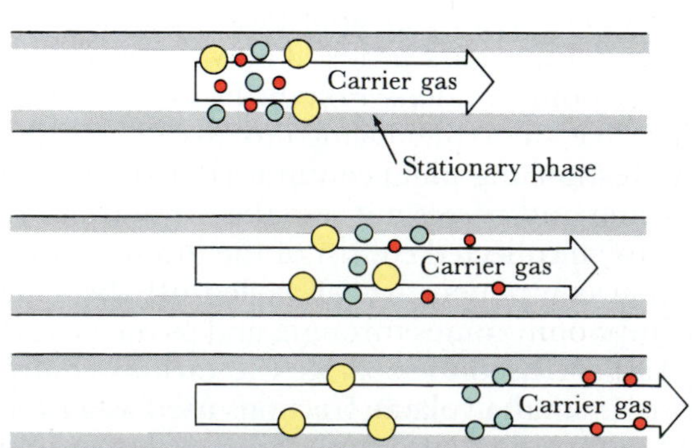

FIGURE 1.10 In gas-liquid chromatography (GLC), the gaseous mixture is separated as it passes through a long coated tube. Some molecules adsorb to the surface more strongly than others and therefore emerge later, but eventually all pass through. The less strongly adsorbing component (yellow) emerges first, and the more strongly adsorbing component (red) emerges last.

practice, a chromatogram is any kind of record of the composition of a mixture (Fig. 1.11). A chromatogram has a unique pattern and can be likened to a fingerprint, for a mixture can be identified by the pattern of its chromatogram. Gas-liquid chromatography is used to detect narcotics and explosives in airline baggage; the equipment "sniffs" the baggage and is highly sensitive to very small amounts of substance.

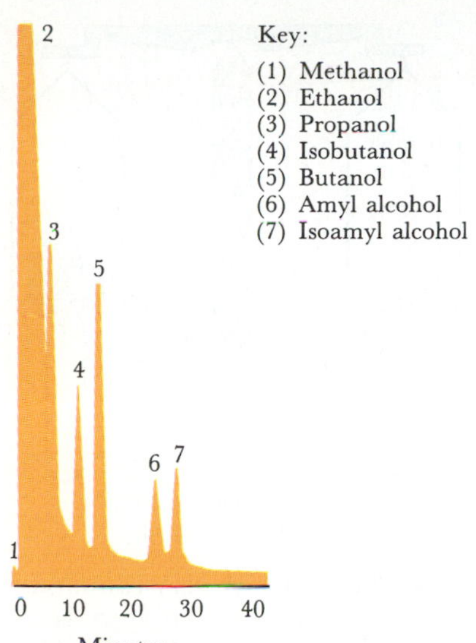

FIGURE 1.11 A gas chromatogram of bourbon whiskey, showing the components that contribute to bourbon's flavor. Mixing the chemicals does not, it is said, recreate the flavor.

MEASUREMENTS AND UNITS

Many of our observations of physical properties of substances are qualitative, or purely descriptive. However, advances in science often depend on quantitative observations, which rely on measurements and the reporting of numerical information (Fig. 1.12). Much of chemistry relies on the measurement of mass m, volume V, and temperature T:

The **mass** of a sample is the quantity of matter it contains.

The **volume** of a sample is the amount of space it occupies.

The **temperature** of a sample indicates how hot or cold it is.

Chemists also measure time t when they need to know how long it takes for the properties or composition of a sample to change.

More precisely, the mass of an object is a measure of its resistance to a change in its state of motion: a ball with a large mass takes more effort to throw than one with a small mass. It is also important to distinguish mass from weight, the gravitational pull on a sample. An astronaut has the same mass (is built from the same amount of matter) on Earth, in space, and on the Moon. The astronaut's weight is different in each case because the gravitational pull on the astronaut is different in the three locations. Weight is proportional to mass, however; so we can measure the mass of an object by comparing the pull exerted on that object by the Earth with the pull the Earth exerts on an object of known mass. If the two pulls are the same, the masses of the unknown and known objects are the same. This principle lies behind the design of the balance and the term *weighing* (Fig. 1.13).

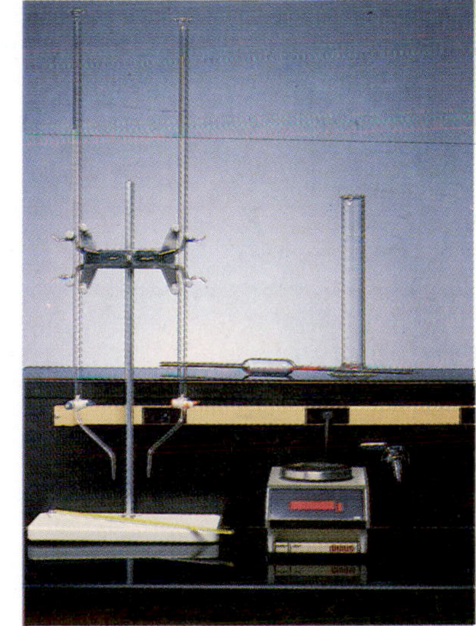

FIGURE 1.12 A typical laboratory bench and some of the measuring instruments often used by chemists. Clockwise from lower right: balance for determining mass, thermometer for measuring temperature; two burets and a pipet for transferring specific volumes of liquids; graduated cylinder for measuring volumes.

1.3 THE INTERNATIONAL SYSTEM OF UNITS

We measure a quantitative property of a sample by comparing the measured property with a standard unit of that property. An example of a unit of length is "1 cm," and the length of, say, a uranium rod may be reported as 3.5 cm only if everyone knows what is meant by "1 cm" (Fig. 1.14). To make sure that everyone does, the different units are carefully defined at international meetings and then publicized around the world.

Scientists have found the metric system of units to be a convenient one. This system was introduced immediately after the French Revolution when, with revolutionary fervor, the French did away with their old units as well as their former leaders. The original metric system defined what we now call the **base units**, from which all the other units are constructed (Table 1.4). The complete collection of rules, symbols,

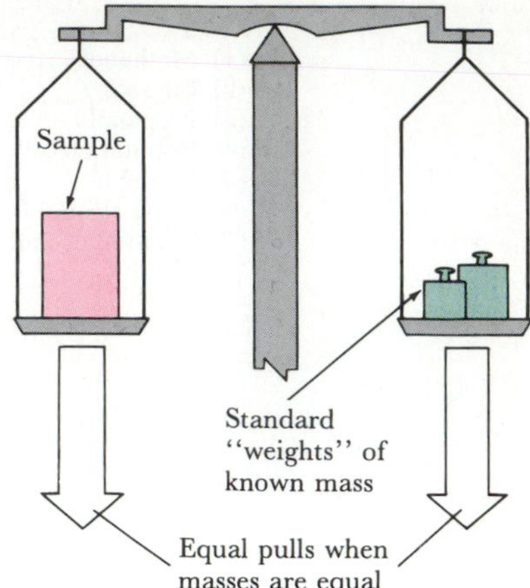

FIGURE 1.13 A very common measurement in a laboratory is that of the mass of a sample. The principle of the balance is that the gravitational pull of the Earth on a sample (its "weight") is proportional to the mass of the sample. When the pull on the known masses in the pan on the right is equal to that on the sample on the left, the beam from which they hang is horizontal.

T OHT Tables 1.4 and 1.5.

D Measure the mass of common objects using a top-loader balance.

E Standards of measurement. *Sci. Amer.*, **1968**, June, 50.

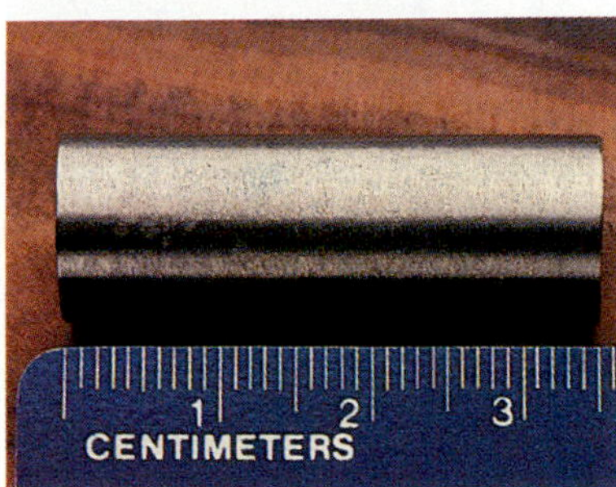

FIGURE 1.14 The process of measurement consists of comparing an unknown with a standard and reporting the number of standard units of the property. Here we are measuring the length of a uranium rod by comparison with units (centimeters) marked on a ruler. The length of the rod is 3.5 cm.

TABLE 1.4 SI base units

Property	Symbol for property	Base unit	Symbol for unit	Further information (page)
mass	m	kilogram	kg	12
length	l	meter	m	13
time	t	second	s	13
temperature	T	kelvin	K	13
amount of substance	n	mole	mol	56
electric current	I	ampere	A	686
luminous intensity	I_v	candela	cd	not used in this text

and definitions is called the **International System of Units**. This title is normally shortened to SI, from the French *Système International.* The great advantage of the system is that it enables scientists to communicate easily with one another wherever they work and whatever their field of interest (Box 1.1). Chemists, for example, can exchange information readily with biologists, engineers, physicists, and anyone else who needs their knowledge. The relationships between SI units and some older units (which are sometimes called English units and are still widely used in the United States) are summarized in Table 1.5 and in more detail inside the back cover.

Mass. The SI base unit of mass is the kilogram (abbreviated kg). One kilogram (1 kg) is defined as the mass of a certain block of platinum-iridium alloy kept at the International Bureau of Weights and Measures, at Sèvres, just outside Paris.

Kilograms are fairly convenient for everyday use. This book, for example, has a mass of about 2 kg. Laboratory samples are typically much smaller, however, with masses of thousandths or ten-thousandths of a kilogram. Because it is simpler to use a unit that is similar in size to the thing being measured, smaller units of mass are widely used. One of these smaller units is the gram (g), which is defined so that one thousand grams make one kilogram:

$$1 \text{ kg} = 10^3 \text{ g}$$

A penny has a mass of 3 g, which is more convenient to write (and remember) than 0.003 kg. Note that in using 10^3, which stands for $10 \times 10 \times 10 = 1000$, we are using "scientific notation." This notation is reviewed in Appendix 1A.

A still smaller unit of mass is the milligram (mg):

$$1 \text{ mg} = 10^{-3} \text{ g}$$

In this expression, 10^{-3} stands for $\frac{1}{(10 \times 10 \times 10)} = \frac{1}{1000}$, as explained in Appendix 1A. A piece of human hair about one-half inch long has a mass of about 1 mg. The first samples of penicillin that were isolated in the early 1940s had masses of only a few milligrams.

As we shall see, the International System makes use of a series of prefixes like kilo- (k) and milli- (m) that may be applied to any unit.

Each prefix multiplies its unit by some power of 10. The most important prefixes are listed in Table 1.6.

Length. The SI unit of length is the meter (m). Originally, the meter was defined as $\frac{1}{10,000,000}$ of the distance from the North Pole to the Equator. However, that distance has been found to change slightly under the influence of the Moon and the tides. Today the meter is defined as the distance light travels in just over $\frac{1}{300,000,000}$ s. (The precise definition is the distance it travels in exactly $\frac{1}{299,792,458}$ s.) This fraction was chosen so that the distance would be very close to the meter already in use.

A meter is a convenient measure of length for everyday use. Most people are between 1.5 and 2 m tall. However, most laboratory samples have dimensions much smaller than 1 m, typically around 0.01 m. Therefore, a more convenient unit for laboratory work is the centimeter (cm), with

$$1 \text{ cm} = 10^{-2} \text{ m}$$

A penny has a diameter of about 2 cm.

Time. The SI unit of time is the second (s). The second was originally defined as $\frac{1}{86,400}$ of the length of the day. However, the Earth does not rotate at a constant rate; as a result, by that definition seconds are slightly longer on some days and shorter on others. The second is now defined as the duration of just under 10 billion (specifically, 9,192,631,770) oscillations of the light waves emitted by cesium atoms. That peculiar number of oscillations was chosen so that the duration would be very close to the second already in use. The name of the unit comes from the division of the hour into small, minute parts called minutes, which were then divided a second time into seconds.

Temperature. Temperature is important in chemistry because it affects the properties of substances, including the physical state of a sample and the ability to undergo chemical change. Temperatures in chemistry are often reported on the **Celsius scale**, named for Anders Celsius, the eighteenth-century Swedish astronomer who devised it. On the Celsius scale, water freezes at 0°C (zero degrees Celsius) and boils at

TABLE 1.5 Relations between units*

Common unit	SI unit
1 pound (lb)	453.6 grams (g)
1 inch (in)	**2.54** centimeters (cm)
1 quart (qt)	0.946 liter (L)

*Additional conversions are given inside the back cover. A number in bold type is exact.

N Are students to memorize Table 1.5?

U Exercise 1.65.

D Show some common laboratory equipment: balance, meter stick, thermometer, beakers, flasks, pipets, and burets.

E The specific transition in cesium is one in which the spin of an electron reverses in the magnetic field of the nucleus. The division of the hour into 60 minutes and of minutes into 60 seconds is a relic of the Babylonian number system.

? If time can be measured so accurately, why on occasion do we need leap seconds and leap years?

T OHT: Table 1.6.

TABLE 1.6 SI prefixes

	Factor	Prefix*	Symbol
Multiples	1,000,000,000 (10^9, billion)	giga-	G
	1,000,000 (10^6, million)	mega-	M
	1,000 (10^3)	kilo-	k
Subdivisions	$\frac{1}{10}$ (10^{-1})	deci-	d
	$\frac{1}{100}$ (10^{-2})	centi-	c
	$\frac{1}{1,000}$ (10^{-3})	milli-	m
	$\frac{1}{1,000,000}$ (10^{-6})	micro-	μ (Greek mu)
	$\frac{1}{1,000,000,000}$ (10^{-9})	nano-	n
	$\frac{1}{1,000,000,000,000}$ (10^{-12})	pico-	p

*Other prefixes are also available but are less commonly used.

N The information in Table 1.6 should be memorized in the form that reflects the meaning of the prefix. For example, 1 cm = 10^{-2} m; not 1 m = 100 cm. Imprecise presentation of SI conversions only adds to the difficulties that students encounter.

U Exercises 1.74, 1.81, 1.90.

D Show liquid nitrogen and its effect on various objects (such as rubber hose, a flower, a rubber ball, a balloon). Pour liquid nitrogen in a tilted beaker and collect liquid oxygen.

D How hot is hot water from a faucet (in °C, °F, and K)?

N Description of Fahrenheit's thermometer. *J. Chem. Ed.*, **1969**, *46*, 192.

E Fahrenheit originally proposed 0° for the lowest temperature that he could achieve using a mixture of salt and ice, and 96° for another readily transportable and reproducible object, the human body. In 1724 he suggested replacing these fixed points by the freezing and boiling points of water, 32°F and 212°F, respectively.

T OHT: Figs. 1.15 and 1.16

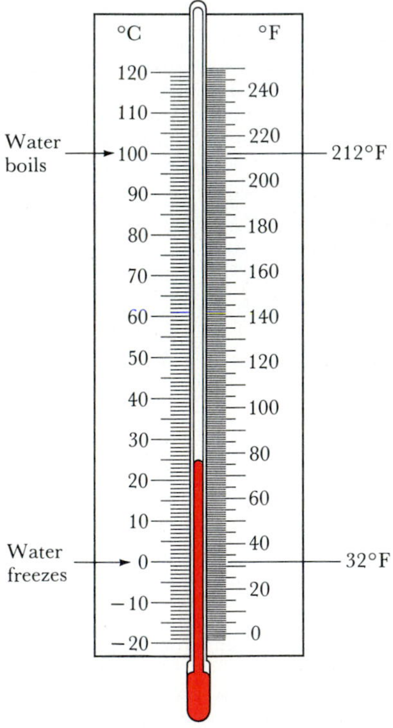

FIGURE 1.15 The Celsius and Fahrenheit temperature scales. Note that the freezing point of water is 0°C (32°F) and its boiling point is 100°C (212°F).

100°C. The Celsius scale is different from the Fahrenheit scale, which is named for Gabriel Daniel Fahrenheit (the German who proposed it in about 1714). On the Fahrenheit scale, water freezes at 32°F and boils at 212°F (Fig. 1.15). We convert a temperature on one scale into the equivalent temperature on the other by using the relation

$$°F = 1.8 \times °C + 32 \qquad (1)$$

EXAMPLE 1.2 Converting between Celsius and Fahrenheit temperatures

Express body temperature, about 99°F, on the Celsius scale.

STRATEGY Estimate the required temperature by referring to Fig. 1.15. Because Fahrenheit degrees are smaller than Celsius degrees, we expect a temperature on the Fahrenheit scale to be numerically greater than the equivalent temperature on the Celsius scale (so long as the temperature is greater than −40°C, where the scales coincide). For the precise value, we rearrange Eq. 1 to

$$°C = \frac{°F - 32}{1.8}$$

and insert the Fahrenheit temperature.

SOLUTION A reading of 99 on the Fahrenheit scale is opposite 37 on the Celsius scale, so body temperature is 37°C. From the equation,

$$°C = \frac{99 - 32}{1.8} = 37$$

That is, 99°F corresponds to 37°C.

EXERCISE E1.2 Convert (a) 25°C to the Fahrenheit scale and (b) −10°F to the Celsius scale.

[*Answer*: (a) 77°F; (b) −23°C]

In the International System, temperatures are reported on the **Kelvin scale**, which is named for Lord Kelvin, the British physicist who invented it in 1848. The SI unit of temperature is called the kelvin (K); the word *degree* is not used in this unit name. The kelvin is the same size as the degree Celsius, so there are 100 kelvins between the temperatures at which water freezes and boils. However, the zero on the Kelvin scale is 273.15°C below the freezing point of water:

$$K = 273.15 + °C \qquad (2)$$

For example, 100.00°C corresponds to 373.15 K. The relation between the scales is illustrated in Fig. 1.16, where the Kelvin scale begins opposite −273.15°C and has no negative values.

Kelvin chose −273.15°C for the starting point of his scale partly because (as we shall see in Chapter 5) the volume of a gas decreases as it is cooled and experiments suggested that the volume of any sample of gas reaches zero at −273.15°C. Because nothing can have a negative volume, −273.15°C is the **absolute zero** of temperature, the lowest possible value. For scientific calculations, using a scale with 0 for that point makes sense, even though it leads to some unfamiliar numbers for common temperatures—such as 273 K for the freezing point of water and 373 K for its boiling point. The Kelvin scale has many advantages;

whenever we use the symbol T in any equation in this text, we shall mean a temperature on the Kelvin scale.

E The degree sign, °, was dropped from the symbol for the kelvin in 1968.

Derived units. The volume, V, of a rectangular solid, the amount of space it occupies, can be found by combining three measurements:

$$V = \text{length} \times \text{width} \times \text{height}$$

Recall that we have defined the unit of length as the meter. This formula suggests that we can define the unit of volume as the volume of a cube with each side equal to 1 m:

$$V = 1\ \text{m} \times 1\ \text{m} \times 1\ \text{m} = 1\ \text{m}^3$$

The unit of volume—the cubic meter (m^3)—is an example of a **derived unit**, a combination of the base units.

The cubic meter is a very large unit of volume compared to the volumes of the samples we typically meet in the laboratory. Often a more convenient unit is the volume of a cube with sides 1 cm in length:

$$V = 1\ \text{cm} \times 1\ \text{cm} \times 1\ \text{cm} = 1\ \text{cm}^3$$

This unit of volume—the cubic centimeter—is sometimes abbreviated 1 cc (for cubic centimeter). Notice that cm^3 stands for $(cm)^3$.

Although 1 cm^3 is a convenient unit for small samples, it can be too small for describing larger volumes of liquids and gases used in laboratories and in everyday life. The liter (L), defined as exactly 1000 cm^3, is sometimes used instead:

$$1\ \text{L} = 10^3\ \text{cm}^3$$

The SI prefixes can be applied to the liter, and chemists commonly use the milliliter (mL) to express the sizes of volumes used in laboratory experiments:

$$1\ \text{mL} = 10^{-3}\ \text{L}$$

Therefore,

$$1\ \text{mL} = 1\ \text{cm}^3$$

exactly. Because 1 mL is exactly the same as 1 cm^3, either unit may be used in calculations. In this book we use liters for the volumes of gases and large samples of liquids (including solutions), milliliters for small liquid or gaseous samples, and cubic centimeters for solids.

In deriving the unit of volume, we combined three measurements of the same property (length). However, many properties are measured using combinations of different units. An example is **density** d, or mass per volume:

$$d = \frac{\text{mass}}{\text{volume}} \qquad (3)$$

Substances with a high density (for example, lead) have a much larger amount of matter in a given volume than do substances with a low density (for example, aluminum).

We can derive the units of density by thinking of a sample of 1 unit of mass (for convenience, 1 g) occupying 1 unit of volume (1 cm^3), so

$$d = \frac{1\ \text{g}}{1\ \text{cm}^3} = 1\ \text{g/cm}^3$$

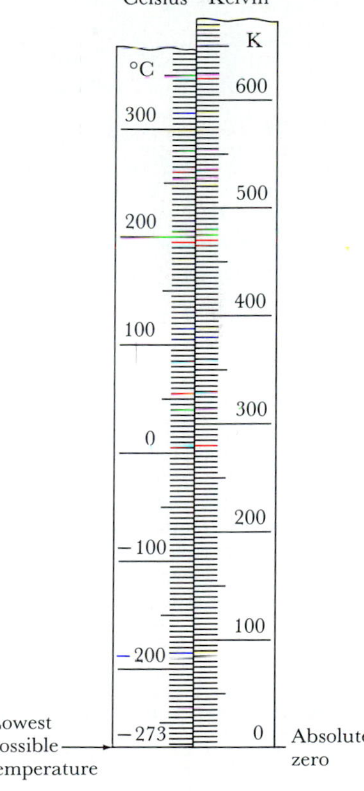

FIGURE 1.16 The Kelvin and Celsius temperature scales. The lowest possible temperature—absolute zero—is 0 on the Kelvin scale and about −273 on the Celsius scale.

BOX 1.1 ▸ UNITS OF MEASUREMENT

JOHN W. LYONS, *Director, National Institute of Standards and Technology*

Scientists agree in many cases, that we cannot really interpret an observation until we measure it or put a number on it. For most chemistry experiments and chemical applications, we express much of our data numerically. To do so requires measurements and measuring devices that determine the number of units or fractional units needed. If the measurement is to mean anything to other scientists, then we have to agree on definitions and values of the units. In other words, we have to standardize our units.

That the standardization of masses, lengths, and volumes began thousands of years ago is shown by the discovery all over the world of the artifacts used as such standards, some dating from several millennia B.C. Moreover, kings and emperors of long ago established units of length based on the lengths of their own forearms, thumbs, and feet. In our work, however, we are much more precise than that.

Of the many unit systems devised over the centuries, the one based on powers of ten has become the international standard. This modern standardization process began in 1791, when France adopted the meter as the standard unit of length. The term *meter* (in French, *mètre*) gave rise to the common name of the entire system of units—the metric system. Part of the impetus for the adoption of the metric system came from the work of the French chemist Antoine Lavoisier, who was at that time engaged in careful determinations of the mass of a given volume of water. In due course his work led to the first definition of the kilogram as the unit of mass. In 1875 seventeen nations signed the *Convention du mètre* at Paris and established the *Bureau International des Poids et Mesures* to develop, maintain, and enhance the international system of measures. Now forty-seven nations have signed the treaty, and all but three nations use the system in daily commerce.

Although scientists have long since adopted the international system, engineers have been slower to give up the earlier systems—presumably because of the long lifetimes of engineered works. Thus, engineers in the United States still use the inch-pound system in much of their work. Gradual change, however, is occurring. Members of

(An alternative way of writing the same result is

$$d = \frac{1\ \text{g}}{1\ \text{cm}^3} = 1\ \text{g cm}^{-3}$$

This exponential notation for units is becoming much more widely used, and we shall use it occasionally in this text.) The derived unit of density is grams per cubic centimeter (g/cm^3 or g cm^{-3}).

N Exponential notation is particularly useful when calculations are done in which compound units appear in the numerator and denominator of expressions: then denominators in denominators are avoided and cancellations are much easier to perform.

Suppose we find that the mass of a 2-cm^3 block of iron is 16 g. Its density is

$$d = \frac{16\ \text{g}}{2\ \text{cm}^3} = 8\ \text{g/cm}^3$$

D The mysterious sunken ice cube. CD2, 7.

A similar calculation gives grams per milliliter (g/mL or g mL^{-1}) as the unit of density when the volume is expressed in milliliters (as it is for liquids). The density of water at 4°C, for example, is 1.0 g/mL. The volumes of gases are normally reported in liters, so a convenient unit of density for them is grams per liter (g/L). The density of air at sea level

the electrical community, who grew up with the evolving international system, use the units of that system exclusively.

The definitions of the standard units have largely been established by physicists. When the current system, which is known as the *Système International d'Unités* or the International System of Units, was first adopted in 1954, there were six base units: meter, kilogram, second, ampere, kelvin, and candela. There were also two supplemental units describing, respectively, the plane angle (the radian) and the solid angle (the steradian). Not until 1971 were chemists able to convince the physicists to add the mole to the set of base physical units. At that time, the mole was defined very carefully in terms of numbers of atoms in exactly 12 grams of carbon-12 (see the text); it should be used exactly as the international community has specified.

It is important to remember, however, that the practical implementation of the International System of Units (for example, determining how many atoms there are in 12 grams of carbon-12) is not fixed for all time but changes in response to advances in science and technology. For example, in January 1990, the values of electrical units were changed slightly to reflect better experimental measurements. This most recent change affects only manufacturers and users of high-precision instruments and was a change not in definition but in the experimental determination of the units.

In the past two decades, the definitions of most of the SI units have been changed from artifacts such as the meter bar to measurements based on fundamental natural phenomena such as the distance that light travels through space in $\frac{1}{299,792,458}$ second for the meter and the oscillations of cesium-133 atoms for the second.

The only base unit that remains dependent on an arbitrary artifact is the kilogram. Efforts are now under way to replace the kilogram artifact, a platinum-iridium cylinder, by a definition connected to more fundamental phenomena such as the density and the spacing between atoms of a sample of ultrapure silicon or the measurement of the force generated by the flow of electric current in two wires. So far, however, the artifact and the balances used with it have proved to be more precise.

and 25°C is 1.2 g/L, about a thousand times less than that of water: air is a much less "concentrated" form of matter than water is. The density of lead is 11.3 g/cm^3, about four times that of aluminum (2.70 g/cm^3); this observation indicates that lead is a more concentrated form of matter than aluminum is.

U Exercise 1.80

E Other derived units include those for measuring pressure, force, work, energy, velocity, acceleration, viscosity, and surface tension.

In some calculations we shall meet derived units like kg·m^2/s^2 (read "kilogram meter-squared per second-squared"), where the raised dot signifies that the units kg and m^2 are multiplied together. In other words, the raised dot means that 1 kg is multiplied by (1 m/s)^2:

$$1\ \text{kg} \times (1\ \text{m/s})^2 = 1\ \text{kg·m}^2\text{/s}^2$$

If a product of units appears in the denominator, then it is enclosed in parentheses; for example,

E Density of antifreeze-water mixture. *J. Chem. Ed.*, **1990**, *67*, 1068.

$$\frac{1\ \text{kg}}{1\ \text{L} \times 1\ \text{s}} = 1\ \text{kg/(L·s)}$$

The expression on the right is read as 1 kilogram per liter-second.

E Extensive properties include mass, volume, amount, internal energy, enthalpy, entropy, heat capacity, and free energy. Intensive properties include temperature, pressure, density, and all the molar quantities, such as molar heat capacity.

N Intensive and extensive properties. *J. Chem. Ed.*, **1970**, *47*, 154.

U Exercises 1.64 and 1.83.

1.4 EXTENSIVE AND INTENSIVE PROPERTIES

Chemists keep their collection of knowledge manageable by organizing it in several ways. Different branches of science often emerge from such organization (biology and the classification of living things into species, is a notable example), for it enables patterns to be detected. The organization of knowledge is a first step in finding explanations and making predictions. We have already talked about the distinction between physical properties and chemical properties (see Section 1.1). We can also divide the properties of substances into the categories intensive and extensive, where

> An **extensive property** is a physical property that depends on the size (the "extent") of the sample.
>
> An **intensive property** is a physical property that is independent of the size of the sample.

The mass of a sample of sugar is an extensive property, because the mass is larger when the sample is larger. The temperature of a sample of water taken from a thoroughly mixed heating tank is an intensive property, because the same value is obtained whatever the size of the sample (Fig. 1.17). The density of a sample is intensive, because, although the mass and the volume increase as the size of the sample is increased, their *ratio* remains the same. The mass of a 2-cm^3 sample of lead (22.6 g) is twice the mass of a 1-cm^3 sample (11.3 g), but their densities are equal (11.3 g/cm^3).

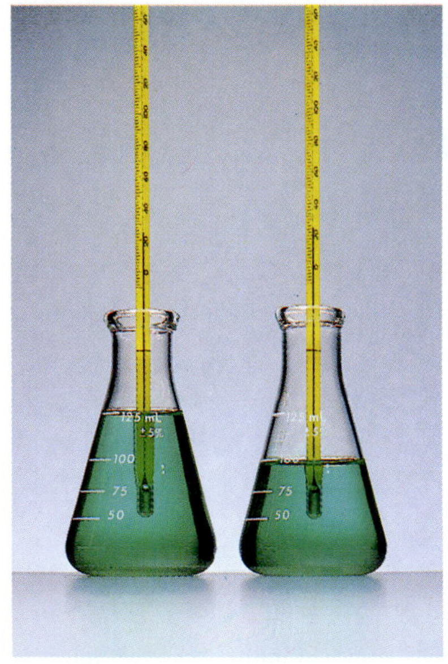

FIGURE 1.17 Mass is an extensive property, but temperature is intensive. These two samples of iron(II) sulfate solution were both taken from the same well-mixed supply: they have different masses but the same temperature.

EXAMPLE 1.3 Distinguishing extensive and intensive properties

Which properties in the following statement are extensive and which intensive? "The density of water, which is a colorless liquid at room temperature, was determined by measuring the mass and volume of a sample."

STRATEGY Decide which properties depend on the size of the sample (extensive properties) and which do not (intensive properties). In some cases a property may be independent of size because an increase in one component property cancels an increase in another, as we illustrated with density. In other cases, by its nature the property, like temperature, may be independent of the sample size.

SOLUTION Density, color, and temperature are all intensive. Mass and volume are extensive.

EXERCISE E1.3 Classify the properties in the following statement: "Lead is a soft, dense metal with a low melting point."

[*Answer*: softness, density, and melting point are all intensive]

Intensive properties can be used to identify a substance, no matter what the size of the sample (Fig. 1.18). If a chemist finds that a metal object has a density of 11.3 g/cm^3, then it is certainly not aluminum (which has a density of 2.7 g/cm^3), but it is possible that the metal is lead. An extensive property, however, depends on the sample size and thus is not a reliable guide to the sample's identity: we cannot conclude

FIGURE 1.18 Some substances can be differentiated on sight, but others have similar appearances and can be identified only by carrying out tests. Most chemists will recognize the blue crystals as copper(II) sulfate. However, careful tests are needed to determine the identity of the two white powders. One is aspirin and the other is benzoic acid.

that a lump of metal is aluminum simply by measuring its mass. It is strictly meaningless to say, for example, that "lead is heavier than aluminum," because mass is extensive and 1 m^3 of aluminum is much heavier than 1 cm^3 of lead. It *is* meaningful to say that "lead is denser than aluminum," because density is intensive and any sample of lead is denser than any sample of aluminum. A feature of chemistry (and of science in general) is that it provides a language that helps to turn everyday expressions into precise, unambiguous statements.

USING MEASUREMENTS

One feature of chemistry, and of science in general, is its ability to use the result of the measurement of one property to calculate the value of another property. For instance, we might want to calculate the distance that a car could travel on a tank of gasoline, given the measurement of its gasoline consumption. In this text we shall do all such calculations by multiplying and dividing units as though they were numbers. A simple example will make this practice clear.

N The techniques we describe here are often called *factor analysis*, *unit analysis*, and (least appropriately, for it has another meaning) *dimensional analysis*.

N A flowchart for dimensional analysis. *J. Chem. Ed.*, **1986**, *63*, 527.

▼ EXAMPLE 1.4 Changing the units of a measurement

An automobile is known to have a gasoline consumption of 35 mi/gal. How far can it travel on 15 gal of fuel?

STRATEGY The distance is the product of the number of gallons available and the consumption. However, the point of this example is to show how the units (in this case, gal and mi/gal) are multiplied and divided in the same way we multiply and divide numbers.

SOLUTION

$$\text{Distance (mi)} = 15\ \cancel{\text{gal}} \times 35 \frac{\text{mi}}{\cancel{\text{gal}}} = 525\ \text{mi}$$

Note that the gallons units cancel and that the distance is obtained in the correct units (miles) automatically.

U Exercises 1.68 and 1.78.

EXERCISE E1.4 The density of ice is 0.96 g/cm^3. What is the mass of a 50-cm^3 block of ice?

[*Answer*: 48 g]

Another type of calculation that frequently occurs in science is the conversion of one set of units into another. For example, suppose the speed of a molecule is reported as 200 meters per second (200 m/s) and that of another molecule as 1000 miles per hour (1000 mi/h). Which is going faster? To answer this kind of question, we have to compare like with like; so we must express the two speeds in the same units. We also need to make conversions between units when a property is reported in one set of units but the information is needed in another. We might know the mass and volume of a sample of gasoline in pounds and gallons but need to find its density in grams per milliliter.

1.5 CONVERSION FACTORS

? What is the purpose of writing down "Information required"?

Suppose a catalog gives the diameter of a piece of glass tubing as 1.50 inches (1.50 in). How can we convert that measurement to centimeters? We begin by writing

$$\text{Diameter (cm)} = \text{diameter (in)} \times \text{number of centimeters per inch}$$

The number of centimeters per inch is an example of a **conversion factor**, a ratio that is used to convert values from one unit to another. We see in Table 1.5 that

$$1 \text{ in} = 2.54 \text{ cm}$$

N To convert a measurement within the International System (for example, from μL to mL), it is safest to use only the conversions listed in Table 1.6. For example, it is safer to use

$$\text{Volume (mL)} = 20\ \mu\text{L} \times \frac{10^{-6}\text{ L}}{1\ \mu\text{L}} \times \frac{1\text{ mL}}{10^{-3}\text{ L}}$$

than to use

$$\text{Volume (mL)} = 20\ \mu\text{L} \times \frac{1\text{ mL}}{10^{3}\ \mu\text{L}}$$

or an equivalent expression.

so we know that there are 2.54 cm per inch. Hence,

$$\underset{\substack{\uparrow \\ \text{Information} \\ \text{required}}}{\text{Diameter (cm)}} = \underset{\substack{\uparrow \\ \text{Information} \\ \text{given}}}{1.50\ \cancel{\text{in}}} \times \underset{\substack{\uparrow \\ \text{Conversion} \\ \text{factor}}}{\frac{2.54\text{ cm}}{1\ \cancel{\text{in}}}} = 3.81\text{ cm}$$

The information given is always the measurement that is unique to a particular problem. In this case, the diameter in inches—1.50 in—is unique to the problem; the conversion factor is the same whatever the value of the diameter.

If we were converting from centimeters to inches (for example, converting a length quoted as 15 cm to inches), we would need the number of inches per centimeter:

$$\text{Length (in)} = \text{length (cm)} \times \text{number of inches per centimeter}$$

We would then use the notation 1 in = 2.54 cm to form a conversion factor:

$$\underset{\substack{\uparrow \\ \text{Information} \\ \text{required}}}{\text{Length (in)}} = \underset{\substack{\uparrow \\ \text{Information} \\ \text{given}}}{15.0\ \cancel{\text{cm}}} \times \underset{\substack{\uparrow \\ \text{Conversion} \\ \text{factor}}}{\frac{1\text{ in}}{2.54\ \cancel{\text{cm}}}} = 5.91\text{ in}$$

We use the same pattern whenever we make a conversion. In each case (and in everything that follows), we proceed by writing

$$\text{Information required} = \text{information given} \times \text{conversion factor}$$

and obtain the conversion factor by rearranging the expression

$$\text{Units given} = \text{units required}$$

into the conversion factor that gives the number of units required per units given:

$$\text{Conversion factor} = \frac{\text{units required}}{\text{units given}}$$

The relations between units, given in Tables 1.5 and 1.6 and inside the back cover, can be used to set up the conversion factors required for all the exercises in this text. Once we have set up the conversion factors, we carry out the calculation by applying the rule mentioned previously: units are multiplied and divided like numbers.

N Whenever we write a relation of this form, the conversion is from the units on the left to the units on the right (in the direction we read).

▼ EXAMPLE 1.5 Converting units

A supplier packages hydrochloric acid in quarts (qt), but you were instructed to buy a certain number of liters. What is the volume in liters of 1.85 qt of hydrochloric acid?

STRATEGY A liter is slightly larger than a quart (Fig 1.19), so we expect the number of liters in our answer to be slightly *smaller* than the equivalent number of quarts. For the precise answer, we need to set up the expression

$$\text{Information required} = \text{information given} \times \text{conversion factor}$$

The information given (the information unique to the problem) is the measured volume. We find the conversion factor (the number of liters per quart) from the information in Table 1.5.

SOLUTION From Table 1.5, we find that 1 qt = 0.946 L; so the conversion factor is

$$\text{Conversion factor} = \frac{\text{units required}}{\text{units given}} = \frac{0.946\ \text{L}}{1\ \text{qt}}$$

Therefore,

$$\underset{\substack{\uparrow \\ \text{Information required}}}{\text{Volume (L)}} = \underset{\substack{\uparrow \\ \text{Information given}}}{1.85\ \cancel{\text{qt}}} \times \underset{\substack{\uparrow \\ \text{Conversion factor}}}{\frac{0.946\ \text{L}}{1\ \cancel{\text{qt}}}} = 1.75\ \text{L}$$

Note that, for the same volume, the number of liters is indeed smaller than the number of quarts.

EXERCISE E1.5 Calculate the mass in ounces of a 250-g package of breakfast cereal. Use 1 oz = 28.35 g.

[*Answer*: 8.82 oz] ▲

N It is more logical (and helpful) to begin the analysis of a problem with "Information given" than with "Conversion factor."

U Exercises 1.61, 1.67, 1.69, 1.70–73, 1.77, 1.84, 1.89, and 1.93.

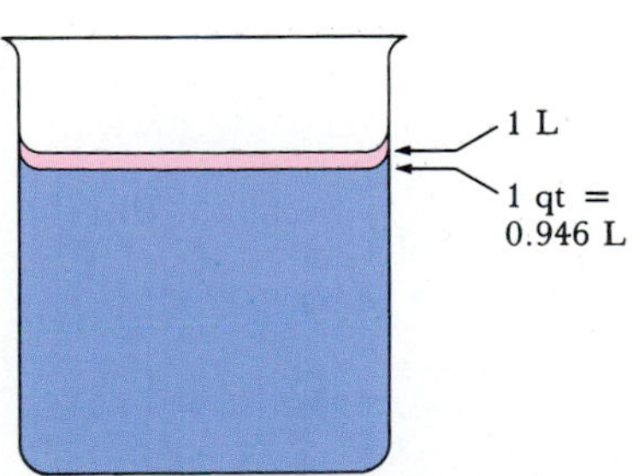

FIGURE 1.19 The relative sizes of 1 L and 1 qt; a quart is about 5% smaller than a liter.

It is often necessary to convert a unit raised to a power. For example, we might wish to convert a surface area expressed as 256 cm^2 into square meters (m^2). We adopt the same procedure as before but use the

N In the conversion from centimeters to meters, use 1 cm = 10^{-2} m rather than 100 cm = 1 m, in the spirit of Table 1.6.

appropriate *power* of the conversion factor. Therefore, to convert (centimeters)^2 to (meters)^2, we use the *square* of the conversion factor from centimeters to meters. The conversion factor is derived from the relation 1 cm = 10^{-2} m:

$$\underset{\substack{\uparrow\\ \text{Information}\\ \text{required}}}{\text{Area (m}^2)} = \underset{\substack{\uparrow\\ \text{Information}\\ \text{given}}}{256\ \text{cm}^2} \times \underset{\substack{\uparrow\\ \text{Conversion}\\ \text{factor}}}{\left(\frac{10^{-2}\ \text{m}}{1\ \text{cm}}\right)^2} = 256 \times 10^{-4}\ \text{m}^2$$

That is, the area is 2.56×10^{-2} m^2. Note how the cm^2 of the information given cancels the cm^2 that comes from squaring the conversion factor. The original units always cancel like this when the conversion has been set up correctly.

U Exercise 1.85.

▼ **EXAMPLE 1.6 Converting derived units: I**

A meter monitoring the supply of oxygen gas recorded 0.254 in^3. What is the volume of oxygen in cm^3?

STRATEGY Because 1 cm^3 is smaller than 1 in^3, we expect the number of cubic centimeters to be larger than the equivalent number of cubic inches. We must form the conversion factor to convert from inches to centimeters and then use the cube of that ratio.

SOLUTION The relation 1 in = 2.54 cm gives us a conversion factor of 2.54 cm/(1 in), so we write

$$\text{Volume (cm}^3) = 0.254\ \text{in}^3 \times \left(\frac{2.54\ \text{cm}}{1\ \text{in}}\right)^3 = 4.16\ \text{cm}^3$$

If desired, we could go on to convert this volume to milliliters by using the relation 1 mL = 1 cm^3. The volume of oxygen would then be 4.16 mL.

EXERCISE E1.6 The surface area of a sample of copper was 1.22 in^2. Express the area in square centimeters.

▲ [*Answer*: 7.87 cm^2]

In many cases we need to convert the denominator of a derived unit. For example, the conversion of 1.5 kilometers per second (1.5 km/s) to kilometers per hour (km/h) requires the conversion of seconds to hours in the denominator. We can carry out the conversion exactly as we did before, but we must make certain that the units of seconds cancel. We can use the relation 3600 s = 1 h to form two conversion factors, 3600 s/(1 h) and 1 h/(3600 s), but only the former cancels the seconds in the denominator of the information given. Therefore, we write

$$\underset{\substack{\uparrow\\ \text{Information}\\ \text{required}}}{\text{Speed (km/h)}} = \underset{\substack{\uparrow\\ \text{Information}\\ \text{given}}}{1.5\,\frac{\text{km}}{\cancel{\text{s}}}} \times \underset{\substack{\uparrow\\ \text{Conversion}\\ \text{factor}}}{\frac{3600\ \cancel{\text{s}}}{1\ \text{h}}} = 5400\,\frac{\text{km}}{\text{h}}$$

N An alternative way of presenting this material is to use exponential notation for the units and to treat the conversion of a unit in a denominator as the conversion of a unit raised to a power (in this instance, the power −1). For this example,

$$\begin{aligned}\text{Speed (km h}^{-1}) &= 1.5\ \text{km s}^{-1} \times 3600\ \text{s h}^{-1}\\ &= 5400\ \text{km h}^{-1}\end{aligned}$$

That is, 1.5 km/s corresponds to 5400 km/h. The simple way to verify that the correct conversion factor has been used is to make sure (as we have in this example) that the units cancel to give the desired result.

▼ EXAMPLE 1.7 Converting derived units: II

The number of aspirin tablets per bottle is 125. Given that there are 36 bottles per case, what is the total number of aspirin tablets per case?

STRATEGY The answer, $36 \times 125 = 4500$ tablets per case, is probably obvious, but it is often useful to see how an obvious conclusion can be reached systematically, for that simple exercise is good practice for making conversions that are less obvious. In this example we must convert tablets per bottle to tablets per case. The conversion factor from bottles to cases is obtained from

$$36 \text{ bottles} = 1 \text{ case}$$

and should be arranged so that the unit *bottle* in 125 tablets per bottle is canceled and replaced by the unit *case*.

SOLUTION We use

$$\text{Conversion factor} = \frac{36 \text{ bottles}}{1 \text{ case}}$$

Then

$$\underset{\substack{\uparrow \\ \text{Information required}}}{\text{Number (tablets per case)}} = \underset{\substack{\uparrow \\ \text{Information given}}}{125\,\frac{\text{tablets}}{\cancel{\text{bottle}}}} \times \underset{\substack{\uparrow \\ \text{Conversion factor}}}{\frac{36\ \cancel{\text{bottles}}}{1 \text{ case}}}$$

$$= 4500 \text{ tablets per case}$$

(Do not worry about the final *s* in bottles not canceling: abbreviations for units, such as cm and kg, are never written as plurals, so the problem does not arise in practice.)

EXERCISE E1.7 The number of atoms per gram of carbon is 5.0×10^{22} atoms/g. How many atoms are there per microgram (μg) of carbon?

[*Answer*: 5.0×10^{16} atoms/μg] ▲

U Exercises 1.76, 1.86, and 1.87.

Converting more complex derived units. The same principles apply to the conversion of more complex combinations of units. Suppose we want to compare the speeds mentioned earlier, the 200 m/s of one molecule and the 1000 mi/h of another molecule, and we decide to convert 200 m/s to miles per hour. The relation we need is

$$\begin{aligned} \underset{\substack{\uparrow \\ \text{Information required}}}{\text{Speed (mi/h)}} = \underset{\substack{\uparrow \\ \text{Information given}}}{\text{speed (m/s)}} &\times \text{number of kilometers per meter} \\ &\times \text{number of miles per kilometer} \\ &\times \underset{\substack{\uparrow \\ \text{Conversion factors}}}{\text{number of seconds per hour}} \end{aligned}$$

? What is the minimum number of SI to English conversion factors for mass, length, and volume that must be memorized? (See Table 1.5: one in each case.)

Note that more complex conversions like this one require two or more conversion factors. Although they can be set up in a line so that all the multiplications and cancellations are done in one step, there is nothing wrong in going step by step. Indeed, that method sometimes gives greater insight into the significance of each conversion. However, in this case, we will demonstrate the one-line approach. We derive the three conversion factors from the relations

$$10^3 \text{ m} = 1 \text{ km} \qquad 1.609 \text{ km} = 1 \text{ mi} \qquad 3600 \text{ s} = 1 \text{ h}$$

Then

$$\text{Speed (mi/h)} = 200\,\frac{\text{m}}{\text{s}} \times \frac{1\ \text{km}}{10^3\ \text{m}} \times \frac{1\ \text{mi}}{1.609\ \text{km}} \times \frac{3600\ \text{s}}{1\ \text{h}} = 447\ \text{mi/h}$$

The first molecule is traveling at 447 mi/h, so it is moving more slowly than the second molecule.

▼ **EXAMPLE 1.8 Converting more complex derived units**

The density of mercury is 13.6 g/cm^3. Express this density in kilograms per cubic meter (kg/m^3).

STRATEGY Treat the multiple unit as a product and use the information in Table 1.6 to set up a string of conversion factors that will produce the units required. Because we need one conversion factor for each power of the unit (cm^3, for instance, requires three), we convert a unit raised to some power by the conversion factor raised to the same power.

N As before, for consistency, we continue to use the relations 10^{-2} m = 1 cm rather than 1 m = 100 cm.

U Exercises 1.88 and 1.92.

SOLUTION From Table 1.6, we find that

$$10^3\ \text{g} = 1\ \text{kg} \qquad 1\ \text{cm} = 10^{-2}\ \text{m}$$

Then

$$\underset{\substack{\uparrow\\ \text{Information}\\ \text{required}}}{\text{Density (kg/m}^3)} = \underset{\substack{\uparrow\\ \text{Information}\\ \text{given}}}{13.6\,\frac{\text{g}}{\text{cm}^3}} \times \underset{\substack{\uparrow\\ \text{Conversion}\\ \text{factor from}\\ \text{g to kg}}}{\frac{1\ \text{kg}}{10^3\ \text{g}}} \times \underset{\substack{\uparrow\\ \text{Conversion}\\ \text{factor from}\\ 1/\text{cm}^3 \text{ to } 1/\text{m}^3}}{\left(\frac{1\ \text{cm}}{10^{-2}\ \text{m}}\right)^3}$$

$$= 13.6 \times 10^3\,\frac{\text{g}\cdot\text{kg}\cdot\text{cm}^3}{\text{cm}^3\cdot\text{g}\cdot\text{m}^3} = 1.36 \times 10^4\ \text{kg/m}^3$$

EXERCISE E1.8 Express 6.5 g/mm^3 in micrograms per cubic nanometer (μg/nm^3).

▲ [*Answer*: $6.5 \times 10^{-12}\ \mu$g/nm^3]

1.6 THE RELIABILITY OF MEASUREMENTS AND CALCULATIONS

An important aspect of the public, international character of science is the honesty with which scientists report and use measurements. A simple example of how information should *not* be used is the following calculation of the density of sodium chloride (table salt) given that its mass is 2.5 g and its volume is 1.14 cm^3:

$$d = \frac{2.5\ \text{g}}{1.14\ \text{cm}^3} = 2.19298\ \text{g/cm}^3$$

But why is this incorrect? As you will see, the answer hinges on the significance of the data, 2.5 g and 1.14 cm^3.

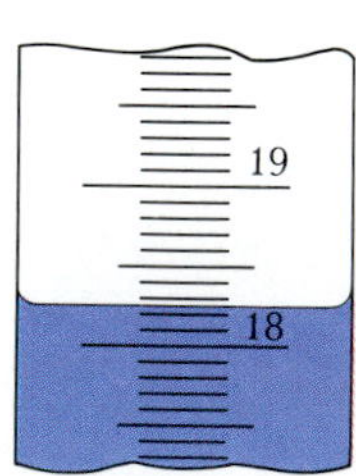

FIGURE 1.20 The volume of liquid could be reported as 18.26 mL, but the last digit is uncertain.

Significant figures. Suppose we are measuring the volume of a liquid in a buret, and the level of the liquid is as shown in Fig. 1.20. We could report it as 18.26 mL; but the last digit, the 6, is little more than a guess. Three people looking at the same setup might report the volume as

18.25 mL, 18.27 mL, or 18.24 mL. All three people would agree that it was 18.2*something* mL, but might disagree about the *something*. Even if we averaged the readings to, say, 18.26 mL, the last digit would be uncertain.

Scientists report the range of uncertainty in this measurement by writing 18.26 ± 0.02 mL. This notation indicates that the true value is estimated to lie between (18.26 − 0.02) mL = 18.24 mL and (18.26 + 0.02) mL = 18.28 mL. This method is cumbersome, however; so, by convention, the last digit in any reported measurement is assumed to be uncertain by ±1. We interpret 18.26 mL, for example, as meaning that the true volume lies between 18.25 and 18.27 mL. As we can see, this convention may underestimate the range within which the true value lies, so it must be applied cautiously.

The digits in the measurement up to and including the first uncertain digit (the 6 in 18.26 mL) are the **significant figures** (written sf) in the measurement. There are four significant figures (written 4 sf) in 18.26 mL. In a volume reported as 18 mL (signifying that it lies between 17 and 19 mL), there would be 2 sf.

The significance of zeros in a measurement, such as 22.0 mL, 80.1 g, or 0.0025 kg, is sometimes the source of difficulty. The trailing zero in 22.0 mL is significant because it signifies that the volume lies between 21.9 and 22.1 mL. Hence, 22.0 mL has 3 sf. The zero in 80.1 g counts as an ordinary digit, so 80.1 g also has 3 sf. However, the leading zeros in 0.0025 kg are not significant. We can see this by changing the units from kilograms to grams and noting that 0.0025 kg is exactly the same as 2.5 g, a measurement with 2 sf.

To find the number of significant figures in a measurement, we rewrite the measurement in scientific notation with one nonzero digit on the left of the decimal point. Then all the digits in the number multiplying the power of 10 are significant. For example, by writing 0.0025 kg as 2.5×10^{-3} kg, we see that it has 2 sf—as we noted earlier when we wrote it as 2.5 g. This procedure also works for other measurements: the volume 22.0 mL would be rewritten as 2.20×10^{1} mL, and, as noted earlier, has 3 sf.

Special care must be taken when using a measurement reported as a round number of tens (such as 30 mL or 200 g). Are the zeros significant, implying that the mass lies between 199 and 201 g? Or are they space fillers, implying that the mass lies between 190 and 210 g, or even between 100 and 300 g? Reporting the measurement in scientific notation avoids ambiguity. Thus, 200 g would be written as 2.00×10^{2} g, 2.0×10^{2} g, or 2×10^{2} g, depending on the uncertainty of the measurement. Another way of showing that the zeros are significant, but one that is rarely used, is to write the decimal point with no zeros after it. Then 200. g means the same as 2.00×10^{2} g, a value with 3 sf.

E Meter sticks in the demonstration of error measurements. *J. Chem. Ed.*, **1989**, *66*, 437.

? What determines the number of significant figures in the measurement of the mass of a sample and the volume of a liquid sample? (The type of balance and volumetric glassware used, as could be demonstrated.)

E Atlantic-Pacific sig. figs. *J. Chem. Ed.*, **1989**, *66*, 12.

D The length of a pestle: A class exercise in measurement and statistics. *J. Chem. Ed.*, **1986**, *63*, 894.

? What do you need to know to assess the significance of the report that "the attendance at the event was 21,000"? (Is the number an estimate from the broadcast booth, from gate receipts, a turnstile count?)

U Exercise 1.82.

N Throughout this text, unless stated otherwise in particular instances, we shall suppose that all zeros are significant, so 200 g is a measurement with 3 sf.

▼ EXAMPLE 1.9 Counting significant figures

Determine the number of significant figures in (a) 50.00 g; (b) 0.00501 m; (c) 0.0100 mm.

STRATEGY Write each measurement in scientific notation, keeping all trailing zeros but no leading zeros. Count the number of digits in each.

SOLUTION In the following, the significant figures are in bold type.

(a) **50.00** g = **5.000** × 10^1 g; 4 sf

(b) 0.00**501** m = **5.01** × 10^{-3} m; 3 sf

(c) 0.0**100** mm = **1.00** × 10^{-2} mm; 3 sf

EXERCISE E1.9 Determine the number of significant figures in (a) 2.1010 kg; (b) 100.000°C; (c) 0.000 000 1 K.

[*Answer*: (a) 5 sf; (b) 6 sf; (c) 1 sf]

T OHT: Figs. 1.21 and 1.22.

We can now see why the density of sodium chloride that we calculated at the beginning of this section—2.19298 g/cm³—is misleading. The mass of the sample, which we are told is 2.5 g, actually lies somewhere in the range 2.4 to 2.6 g. The volume—1.14 cm³—lies somewhere in the range 1.13 to 1.15 cm³. The density therefore lies somewhere in the range calculated from the smallest mass and largest volume,

$$d = \frac{2.4\text{ g}}{1.15\text{ cm}^3} = 2.1\text{ g/cm}^3$$

and that calculated from the largest mass and smallest volume,

$$d = \frac{2.6\text{ g}}{1.13\text{ cm}^3} = 2.3\text{ g/cm}^3$$

Quoting the density as 2.2 g/cm³ (signifying that it lies in the range 2.1 to 2.3 g/cm³) is justified; reporting it as 2.19298 g/cm³ is not.

Integers and definitions. The results of counting are *exact.* There is no uncertainty in the report "12 eggs": it means exactly 12, not 12 ± 1. The *integer* (whole number) 12 could in fact be taken to mean 12.000 . . . , with the zeros continuing forever, and we could think of it as having an infinite number of significant figures. The value printed in bold type in Table 1.5 (for the conversion between inches and centimeters) is also exact, even though it has digits after a decimal point. The relation is a *definition,* and 1 in = 2.54 cm defines an inch exactly. In this case 2.54 is shorthand for 2.5400 . . . , with the zeros continuing forever. In contrast, 1 lb = 453.6 g is only an approximation to the exact (defined) value, 1 lb = 453.59237 g. The factor of 1.8 in Eq. 1 (the relation between the Celsius and Fahrenheit scales) and the 273.15 in Eq. 2 (the relation between the Kelvin and Celsius scales) are also exact. *All definitions are exact and have an infinite number of significant figures.*

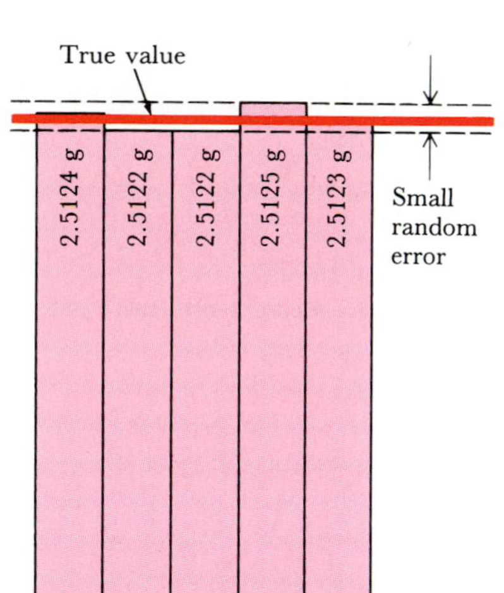

FIGURE 1.21 The variation between the individual determinations of mass of a sample is small and random, and so the measurements are precise. In this case, the measurements are also accurate because they are all close to the true value of the mass of the sample.

Errors in measurements. All measurements are uncertain, but some are more uncertain than others. Suppose we weigh a sample of magnesium several times and get the following results (Fig. 1.21):

2.5124 g   2.5122 g   2.5122 g   2.5125 g   2.5123 g

These measurements are all very close to the average value (2.5123 g), so we say there is small **random error**—the variation from measurement to measurement, which sometimes gives a high value and sometimes a low one. When the random error is small, we say that the measurements are precise. The five measurements above are indeed precise, and we could feel reasonably safe in reporting the mass as 2.5123 g, with 5 sf.

Now suppose that there is a speck of dust of mass 0.0100 g on the balance pan. Our measurements now might be

2.5224 g 2.5222 g 2.5222 g 2.5225 g 2.5223 g

with an average value 2.5223 g (Fig. 1.22). The measurements are precise, but there is a **systematic error**—an error that appears in every measurement and does not average out. Measurements without a systematic error are said to be accurate, those with a systematic error are inaccurate. That is, accurate measurements are close to the accepted value, inaccurate measurements are not. The accuracy of a measurement depends on both the quality of the apparatus and the skill of the person using it. We can summarize these remarks as follows:

Precise measurements have small random error and are reproducible in repeated trials.

Accurate measurements have small systematic error and give a result close to the accepted value of the property.

U Exercise 1.94.

N Another example of a source of systematic error could be the use of a pipet at a temperature other than that at which it had been calibrated.

? How can random errors and systematic errors be minimized?

1.7 SIGNIFICANT FIGURES IN CALCULATIONS

E Significant figure rules for general arithmetic functions. *J. Chem. Ed.*, **1989**, *66*, 573.

E Propagation of significant figures. *J. Chem. Ed.*, **1985**, *62*, 693.

E Significant figures and error propagation. *J. Chem. Ed.*, **1989**, *66*, 272.

We shall often need to calculate the value of a property from other information, the **data**. A simple example is our earlier calculation of the density, where we used the mass and volume of the sample as data. In such cases we need to decide how many significant figures in the calculated value of a property are justified by the number of significant figures in the data. The rules we shall now discuss have been designed to keep us from writing 2.19298 g/cm^3 for the density of sodium chloride with data as imprecise as a mass measured as 2.5 g and a volume measured as 1.14 cm^3.

Rule for addition and subtraction. Whenever we add or subtract, *the number of decimal places in the result should be the same as the smallest number of decimal places in the data.* Suppose we need the total volume of three blocks of copper with measured volumes 1.12 cm^3, 1.2 cm^3, and 2.028 cm^3; then according to this rule, only one digit should follow the decimal point:

$$\begin{array}{l} 1.12 \text{ cm}^3 \\ 1.2 \\ \underline{2.028} \\ 4.348 \text{ cm}^3\text{: round to } 4.3 \text{ cm}^3 \end{array}$$

In calculations of this kind, we use the following rules for rounding:

. . . 100000 . . .	through	. . . 149999 . . .	rounds to	. . . 1
. . . 150000 . . .	through	. . . 199999 . . .	rounds to	. . . 2
. . . 200000 . . .	through	. . . 249999 . . .	rounds to	. . . 2
. . . 250000 . . .	through	. . . 299999 . . .	rounds to	. . . 3

and so on for other digits. Thus, 4.348 cm^3 rounds to 4.3 cm^3 when the answer should have 2 sf, but 4.348 cm^3 would be rounded to 4.35 cm^3 when 3 sf are appropriate. The rounding rules imply that 4.348 cm^3 should be rounded in one step to 4.3 cm^3, not first to 4.35 cm^3 and then to 4.4 cm^3.

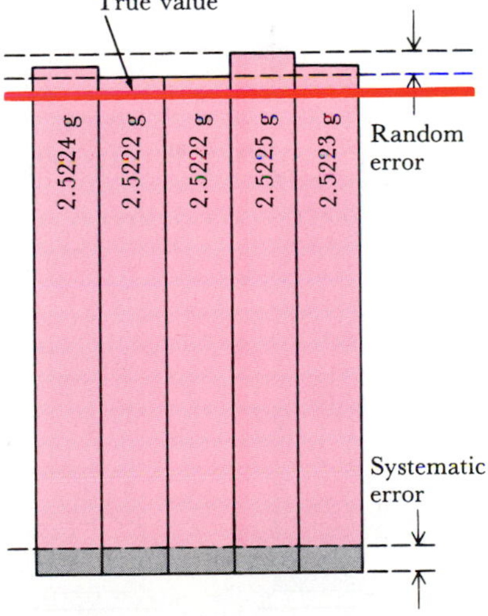

FIGURE 1.22 The variation is also small in this series of measurements, but there is a systematic error common to them all, and although the measurements are precise, they are not accurate.

? Comment on the report that the attendance at the beginning of a game was 21,000 but the number had dwindled to 16,214 by the end of the game.

E When figures signify nothing. *J. Chem. Ed.*, **1991**, *68*, 469.

▼ **EXAMPLE 1.10 Adding and subtracting with the correct number of significant figures**

Report (a) the total volume of a sample of water prepared by adding 25.6 mL to 50 mL; (b) the temperature in kelvins corresponding to the boiling point of sulfur, 444.674°C.

STRATEGY Add or subtract as required, identify the data value with the smallest number of digits following the decimal point, and round the result to that number.

SOLUTION (a) The smallest number of decimal places in the data is zero (for the volume 50 mL), so the total volume is

$$25.6\text{ mL} + 50\text{ mL} = 75.6\text{ mL: round to }76\text{ mL}$$

(b) The conversion formula (Eq. 2) is

$$K = {}^\circ C + 273.15$$

where 273.15 is exact; in other words, it has an infinite number of digits after the decimal point. Therefore, we round to the number of digits in the given Celsius temperature. Because

$$444.674 + 273.15000\ldots = 717.824$$

we record the boiling point of sulfur as 717.824 K.

EXERCISE E1.10 Report (a) the total mass of the sample prepared from 1.001 g sugar, 2.05 g salt, 5.0 g water; (b) the Celsius temperature corresponding to the melting point of iron, 1813 K.

▲ [*Answer*: (a) 8.1 g; (b) 1540°C]

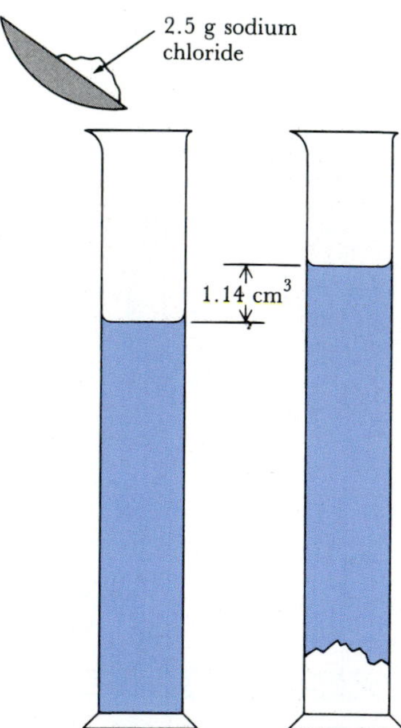

FIGURE 1.23 One way to measure the volume of a solid sample (such as salt grains) is to add it to a liquid in which it is insoluble (cyclohexane, for instance), and to observe the change in volume of the liquid.

The addition rule is based on the fact that the range of uncertainty of the least precise measurement is so great that it determines the precision of the sum. The rule is only a rough guide, and it may in some cases give too optimistic a range of uncertainty. The only reliable procedure, the one used by professional scientists, is to calculate the least and greatest values from the data, and to report the full range of uncertainty.

Rule for multiplication and division. Whenever we multiply or divide, *the number of significant figures in the result should be the same as the smallest number of significant figures in the data.* This rule leads to the proper precision in the sodium chloride density example: for a mass of 2.5 g (2 sf) and a volume of 1.14 cm^3 (3 sf; Fig. 1.23), the least number of significant figures is two and

$$d = \frac{2.5\text{ g}}{1.14\text{ cm}^3} = 2.19298\text{ g/cm}^3\text{: round to }2.2\text{ g/cm}^3$$

The basis of this rule is the same as that for the previous rule. The widest range of uncertainty in the data dominates and leads to a precision that is *approximately* that given by the rule. Again, the rule can be optimistic, and a more reliable procedure is to calculate and report the full range of uncertainty.

▼ **EXAMPLE 1.11 Multiplying and dividing with the correct number of significant figures**

Calculate the mass of a 250 mL (3 sf) sample of carbon dioxide that has a density of 2.095 g/L (4 sf). What would the density of the gas be if the same sample were in a container of volume 155 mL (3 sf)?

STRATEGY The first part of the calculation is a conversion (of volume to mass), using density as the conversion factor, so we use the rule for significant figures set out above. (In practice, the intermediate answer is stored in the memory of a calculator and we do not round the answer until the end of the calculation.) Our answer cannot have more than three significant figures. The density in the second part is the mass calculated in the first part divided by the new volume. Because the same mass is confined to a smaller volume, we expect a greater density. In other words, our answer should be a number larger than 2.095.

SOLUTION The volume of the sample is converted to mass using the density and the relation 1 mL = 10^{-3} L:

$$\text{Mass (g)} = \text{volume} \times \text{conversion factor} \times \text{density}$$

$$= 250\ \cancel{\text{mL}} \times \frac{10^{-3}\ \cancel{\text{L}}}{1\ \cancel{\text{mL}}} \times 2.095\ \frac{\text{g}}{\cancel{\text{L}}} = 0.524\ \text{g}$$

Hence, the mass of the sample is 0.524 g. For the density of the gas in the smaller container of volume 155 mL, we calculate

$$\text{Density (g/L)} = \frac{\text{mass (g)}}{\text{volume (L)}}$$

$$= \frac{0.524\ \text{g}}{155\ \cancel{\text{mL}}} \times \frac{1\ \cancel{\text{mL}}}{10^{-3}\ \text{L}} = 3.38\ \text{g/L}$$

Note that our answer is rounded to 3 sf.

U Exercises 1.66, 1.79, and 1.91.

EXERCISE E1.11 Calculate the volume occupied by 1.04 mg of oxygen gas, given that its density is 1.5 g/L. What would its density be if the same sample occupied 0.755 mL?

[*Answer*: 0.69 mL; 1.38 g/L]

Multiplication by an integer is a disguised form of addition. The total length of two 5.11-cm rods is 2 × 5.11 cm, which actually means

$$5.11\ \text{m} + 5.11\ \text{m} = 10.22\ \text{m}$$

The number of significant figures in the answer is therefore obtained from the rules for addition. The same holds for division by an integer, so

$$\frac{10.22\ \text{m}}{2} = 5.11\ \text{m}$$

In other words, integers have no role in restricting the number of significant figures in a calculation.

1.8 MASS PERCENTAGE COMPOSITION

We conclude this chapter with a concept that draws on many of the points we have described earlier and that is widely used in chemistry. We have introduced the concept of mixture; now we shall see how to report the composition of a mixture and how to use the composition in calculations. Ideally, composition should be reported in terms of *intensive* properties, so that the information applies to samples of any size. In contrast, most cooking recipes (which specify the composition of the mixture that should be cooked) are *extensive*, because they are designed to lead to a cake or dish of a predetermined size.

N This section could be postponed until Chapter 10.

A convenient way of reporting the composition of a mixture is to state the "mass percentage" (mass %) of each substance present:

The **mass percentage** of a substance A is the mass of A present in a sample, expressed as a percentage of the sample's total mass:

$$\text{Mass \% A} = \frac{\text{mass of A in a sample}}{\text{total mass of sample}} \times 100\%$$

Mass percentage is an intensive property because the ratio does not change as the size of the sample increases. A list of alloys might state that a sample of brass is 35% zinc by mass. We would then know that a 100-g sample contained 35 g of zinc and 65 g of copper. A 10.0-kg brass candlestick of the same alloy would contain 3.5 kg of zinc and 6.5 kg of copper.

FIGURE 1.24 Nickel-steel was known even in prehistorical times. The head of this Chinese dagger is crafted from a nickel-steel meteorite. The dagger dates from the Chou dynasty, which began in 1027 B.C.

U Exercises 1.95 and 1.96.

▼ EXAMPLE 1.12 Using mass percentage composition

One meteorite is known to have a mass percentage composition of 26% nickel and 74% iron (Fig. 1.24). Calculate the mass of a sample of the meteorite that would contain 55 g of nickel.

STRATEGY We can use the mass percentage composition to set up a conversion factor between the mass of nickel required and the mass of sample that should be used. We know that 26 g of nickel is present in 100 g of meteorite, so we base the conversion factor on the relation

$$26 \text{ g nickel} = 100 \text{ g meteorite}$$

SOLUTION The conversion we require is

$$\underset{\text{Information required}}{\text{Mass of meteorite (g)}} = \underset{\text{Information given}}{55 \text{ g nickel}} \times \underset{\text{Conversion factor}}{\frac{100 \text{ g meteorite}}{26 \text{ g nickel}}}$$

$$= 2.1 \times 10^2 \text{ g meteorite}$$

EXERCISE E1.12 The composition of a commercial cold remedy is 87% aspirin, 7% caffeine, 6% vitamin C. What mass of the mixture should you measure out to obtain 5 g of caffeine.

[*Answer*: 70 g (1 sf)] ▲

SUMMARY

1.1 Chemistry makes progress as a result of careful observation of the properties of **samples** of **matter**, the material of the universe. It deals with **substances**, different pure kinds of matter that are distinguished by their **properties**, their characteristics. Properties are classified as either **physical**, if their measurement does not involve a chemical change, or **chemical**, if it does. Physical properties include the **physical state** of the substance (whether it is **gas**, **liquid**, or **solid**) and the temperature at which **changes of state** (which include melting, freezing, boiling, or vaporization, and condensation) occur. The chemical properties of a substance include its behavior in the presence of other substances or as a result of electrolysis.

1.2 The composition of a sample is determined by **chemical analysis**, the first step of which is to check if a sample is a single **substance** or a **mixture** of several substances. Mixtures may be **homogeneous** or **heterogeneous**, the former including **solutions** obtained when a **solute** dissolves in a **solvent**. The components of mixtures can be separated by techniques that depend on the different physical properties of the components, including **filtration**, **distillation**, **crystallization**, and **chroma-**

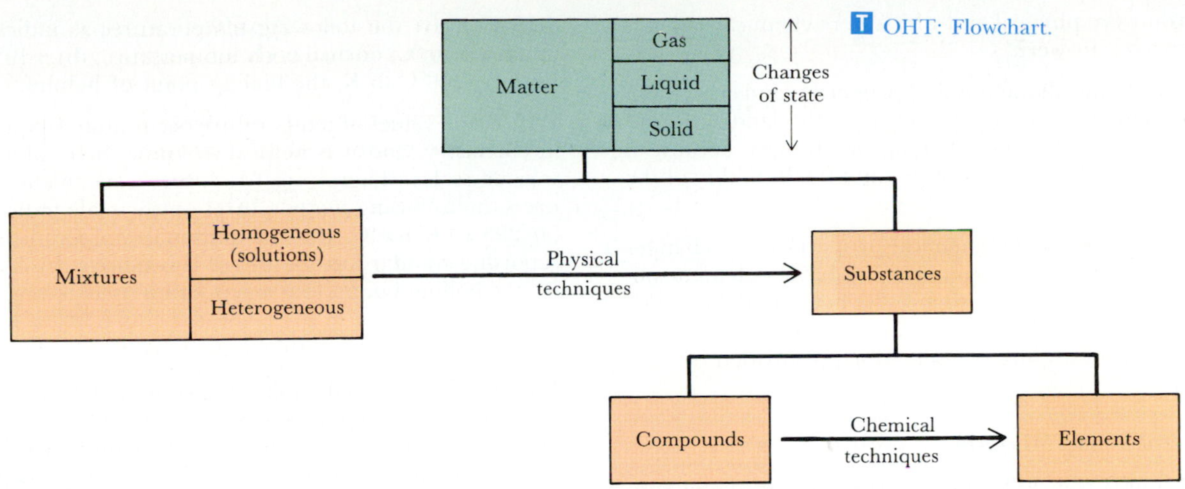

tography. A pure substance may be either an element or a compound. An **element** is a substance that cannot be separated into simpler components using chemical techniques. A **compound** is a specific chemical combination of elements that can be separated by chemical techniques. The relation between the different forms of matter is summarized in the chart.

1.3 Many of the observations made in chemistry are measurements, or observations with a numerical outcome. Four properties frequently measured are **mass**, **volume**, **temperature**, and time. The results of measurements are reported using the **International System of Units** (SI units). The **base units** of the system include the kilogram for mass, the meter for length, the second for time, and the kelvin for temperature. **Derived units** for other properties, such as volume and density, are obtained from these base units. Multiples and subdivisions of units are constructed with prefixes denoting different powers of ten (see Table 1.6).

1.4 A property that depends on the size of the sample is called an **extensive property**; one that is independent of size is called an **intensive property**. A substance is usually identified by some of its intensive properties.

1.5 Measurements are converted from one set of units to another using **conversion factors** derived from the relation between units. Units are multiplied and divided like numbers.

1.6 Great care must be used not to give the impression that a reported value is more reliable than is the case. The convention is adopted that the last digit of a measured or experimental value is taken to be uncertain to the extent of ± 1. The number of digits up to and including the first uncertain digit is the number of **significant figures** in the measurement. If in a series of measurements there is only a small **random error**, then the measurements are said to be **precise**. A measurement is **accurate** if it is close to the true or accepted value, which implies that there is no **systematic error**, an error that does not average to zero.

1.7 When **data**, or supplied information, are used in a calculation, the result must not give a false impression of its reliability. Rules were given for estimating the number of significant figures justified by the data, but they must be used cautiously.

1.8 Measurements of the composition of mixtures are reported intensively as **mass percentages**, the mass of each substance expressed as a percentage of the whole.

CLASSIFIED EXERCISES

Physical and Chemical Properties

1.1 Classify the following as chemical or physical properties: (a) the color of copper(II) sulfate is blue; (b) the melting point of sodium metal is 97.8°C; (c) iron rusts in areas with high humidity.

1.2 A chemist investigates the boiling point, melting point, and flammability of a compound called acetone, a component of fingernail polish remover. Which of these properties are physical properties and which are chemical properties?

1.3 Classify the following changes as chemical or physical: (a) the freezing of water; (b) the vaporization of alcohol; (c) the corrosion of aluminum when exposed to air.

1.4 A general chemistry student was asked to decide which of the changes of melting, rusting, and subliming for iron were physical and which were chemical. What is the correct answer?

1.5 Underline all the physical properties and changes in the statement, "The temperature of the land is an important factor for the ripening of oranges, because it affects the evaporation of water and the humidity of the surrounding air."

1.6 Underline all the chemical properties and changes in the statement, "Copper is a red-brown element obtained from copper sulfide ores by heating in air, which forms copper oxide. Heating the copper oxide with carbon produces impure copper, which is purified by electrolysis."

Substances and Mixtures

1.7 Identify the following as mixtures or single substances: (a) sodium bicarbonate, commonly called baking soda; (b) brass; (c) diamond, a crystalline form of carbon.

1.8 Identify the following substances as elements or compounds: (a) gold; (b) chlorine gas; (c) table salt.

1.9 Identify the following as homogeneous or heterogeneous mixtures and suggest a technique (by referring to Table 1.2) for separating their components: (a) alcohol and water; (b) chalk and table salt; (c) the gases in air; (d) salt water.

1.10 Identify the following as homogeneous or heterogeneous mixtures and suggest a technique (by referring to Table 1.2) for separating their components: (a) gasoline and motor oil; (b) the pigments in a bucket of paint; (c) drinking water; (d) carbonated water.

International System (SI) of Units

1.11 Complete the following equalities, using the information in Table 1.6.
(a) 250 g (3 sf) = ________ kg
(b) 250 μs (3 sf) = ________ ms
(c) 1.49 cm = ________ dm
(d) 2.48 cg = ________ g

1.12 Complete the following equalities, using the information in Table 1.6.
(a) 200 μg (3 sf) = ________ g
(b) 88 nm = ________ pm
(c) 0.0789 mg = ________ μg
(d) 25.0 mL = ________ kL

1.13 Complete the following equations, using scientific notation:
(a) 1 μm = ________ m
(b) 550 nm (3 sf) = ________ mm
(c) 0.10 g = ________ mg

1.14 Express the following measurements in scientific notation: (a) 0.000 000 001 K, the lowest-ever recorded temperature; (b) $\frac{1}{100,000,000}$ m; (c) 0.000 000 535 m, the approximate wavelength of green light.

1.15 Convert the following temperatures as indicated: (a) 98.6°F to °C, normal body temperature; (b) −40°C to °F; (c) −269°C to K, the boiling point of helium.

1.16 Some values of temperature occur quite frequently in chemistry, and it is helpful to know their values on various scales. Therefore, for future convenience, express the following temperatures on the scale indicated: (a) 298.15 K to °C, the conventional temperature for reporting standard properties of substances; (b) 77 K to °C, the boiling point of nitrogen, often used as a moderately cheap coolant in the laboratory; (c) 65°C to K and °F, "hot" water that is still comfortable to the skin.

1.17 Work through the following steps to determine the units of pressure in terms of SI base units: (a) velocity = distance per time; (b) acceleration = change in velocity per time; (c) force = mass × acceleration; (d) pressure = force per area.

1.18 Work through the following steps to determine the units of energy in terms of SI base units: (a) velocity = distance per time; (b) acceleration = change in velocity per time; (c) force = mass × acceleration; (d) energy = force × distance.

1.19 Convert the following derived units as indicated:
(a) 1 cm^3 = ________ m^3
(b) 30 m/s = ________ cm/μs
(c) 25 cm^3 = ________ mL

1.20 Convert the following values to the designated units in brackets: (a) The density of water at 3.98°C [°F] is 1.0 g/mL [mg/L]. (b) The volume of a laboratory test tube is 3.0 mL [dL]. (c) A penny has an approximate mass of 3 g [mg], an approximate diameter of 2 cm [mm], and a thickness of 1 mm. Given that the volume of a cylinder is $\pi r^2 h$, determine the density of a penny in units of mg/mm^3.

1.21 Rewrite the following statement, converting the measurements to the units in brackets: "A sample of tin of area 1.0 cm^2 [mm^2] was set on a small block of lead of volume 10.0 cm^3 [m^3]. The two metals were placed in a 100 mL [L] flask and 25.0 mL [cm^3] of acid was added."

1.22 Make the appropriate conversions in the following statement: "A minute sample of the metal iridium of volume 0.5 mm^3 [μm^3] was recovered from a land area of 1.5 km^2 [m^2]. The chemist, who analyzed a 25-mL [dL] sample (as part of a research project to determine whether there had been a major comet impact on the Earth at the time of the extinction of the dinosaurs), wanted to know the amount of iridium to be found in 1.0 cm^3 [m^3] of soil."

Extensive and Intensive Properties

1.23 State whether the following are extensive or intensive properties: (a) the temperature of boiling liquid oxygen (LOX); (b) the heat content of a beaker of water; (c) the interatomic distance in a hydrogen molecule.

1.24 Some everyday properties are extensive and some are intensive. State which of the following are extensive and which are intensive properties: (a) the air pressure in a tire; (b) the bleaching power of a liquid laundry bleach; (c) the hardness of concrete.

Conversion Factors

1.25 Use the conversion factors in Table 1.6 and inside the back cover to express the following measurements in the designated units: (a) 25 L to m^3; (b) 25 g/L to mg/dL; (c) 4.77 mg to μg; (d) 4.2 L/h to mL/s.

1.26 Use the conversion factors in Table 1.6 and inside the back cover to express the following measurements in the designated units: (a) 4.82 nm to pm; (b) 1.83 mL/min to mm^3/s; (c) 0.044 g/L to mg/cm^3; (d) 4.2°C/s to °C/min.

1.27 On occasion it becomes necessary to convert a measurement from one system of units to another. Make the following conversions: (a) $\frac{4}{5}$ quart bottle of spirits to milliliters; (b) the compacted metal from an automobile salvage yard has a density of about 450 lb/ft^3, into kg/m^3; (c) the density of water, 1.0 g/mL, in lb/ft^3.

1.28 Make the following conversions: (a) The approximate concentration of sodium in seawater is 10.5 g/L. Express this concentration in ounces per gallon. (b) Over 40 billion kilograms of sulfuric acid are produced in the United States annually. Express this production in units of pounds per day. (c) The mass of an electron, one of the subatomic particles that occurs inside atoms, is 9.11×10^{-28} g. What is this mass in pounds?

1.29 Convert the following information into the units in brackets: (a) Gas entered the balloon at a rate of 35 cm^3/min [mm^3/s]. (b) The load on the laboratory floor was 125 kg/ft^2 [lb/in^2]. (c) A certain insecticide can treat 10 hectares/kg [m^2/g], 1 hectare = 10^4 m^2.

1.30 Convert the following information into the units in brackets: (a) An industrial process required sulfuric acid to be supplied at 25 L/h [cm^3/s]. (b) A chemical reaction produced carbon dioxide at a rate of 1.05 cm^3/s [L/h]. (c) The surface area of finely divided talcum powder can be as much as 400 m^2/g [km^2/kg].

1.31 The angstrom unit (1 Å = 10^{-10} m) is still widely used to report measurements of the sizes of atoms and molecules. Express the following data in angstroms: (a) The radius of a sodium atom is 180 pm (2 sf). (b) The wavelength of yellow light is 550 nm (2 sf). (c) Write a (single) conversion factor between angstroms and nanometers.

1.32 Many "old" units of measurement still appear in the literature and must be understood by modern chemists. Make the appropriate conversions to these old units in the following instances: (a) The *calorie* (1 cal = 4.184 J), a unit of energy. The burning of a sample of gasoline produces 400 kJ (2 sf) of heat. Convert this energy to calories. (b) The *atmosphere* (1 atm = 101.3 kPa), a unit of pressure. Normal atmospheric pressure in the high plains is 0.9 atm. What is this pressure in pascals (Pa)? (c) The *barrel* (1 barrel = 42 gal), a unit of volume for crude oil. Express the volume of 1.00 barrel of crude oil in liters.

1.33 Identify the smaller of each of the following pairs of measurements: (a) 1.04 g/L or 1.04 mg/cL; (b) 13.6 g/cm^3 (the density of mercury) or 2.25×10^{-4} mg/nL (the density of platinum); (c) 0.000 000 45 kg or 45 μg.

1.34 Identify the larger of each of the following pairs of measurements: (a) 2698 kg/m^3 or 3.700 g/cm^3; (b) 1.00 cL/min or 450 cL/h; (c) 2.0 kg/cm^2 or 1.0×10^{-10} g/nm^2.

1.35 Express (a) the diameter of the Earth, 1.3×10^4 km, in miles and (b) the diameter of an aluminum atom, 250 pm, in nanometers.

1.36 The normal recommended maximum intake of copper by an adult is 3 mg/day. In the course of a lifetime of 80 years, to how many (a) grams, (b) ounces does this intake correspond?

1.37 Gold ore contains typically 10 g (1 sf) of gold per 1000 kg (4 sf) of ore. Assuming the current price of gold is \$410/oz, what mass of ore in metric tons (1 metric ton = 1000 kg) must be processed to obtain \$1 million worth of gold?

1.38 The source of magnesium for the commercial production of the metal is seawater. The concentration of magnesium in seawater is 1.35 g/L. How many cubic miles of seawater must be processed to obtain 1000 kg (1 metric ton) of magnesium?

Reliability of Measurements and Calculations

1.39 State the number of significant figures in the following quantities: (a) 2.00 g of silver; (b) 0.0200 s; (c) 2.00×10^2 mL of water; (d) 6 thermometers.

1.40 State the number of significant figures in the following quantities: (a) 3.00100 g of sugar; (b) 2.998×10^8 m/s (speed of light); (c) 22 beakers; (d) 0.0001 K.

1.41 Shown below are two pieces of laboratory glassware graduated in milliliters. Report the volumes, using the correct number of significant figures.

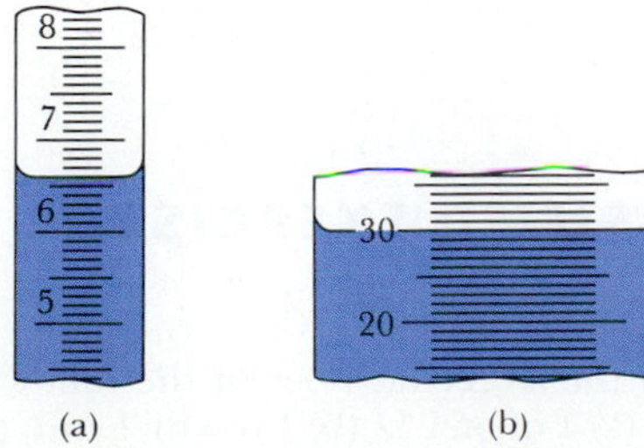

1.42 Record the volumes of solution in milliliters in (a) the buret and (b) the graduated cylinder. Report the volumes, using the correct number of significant figures.

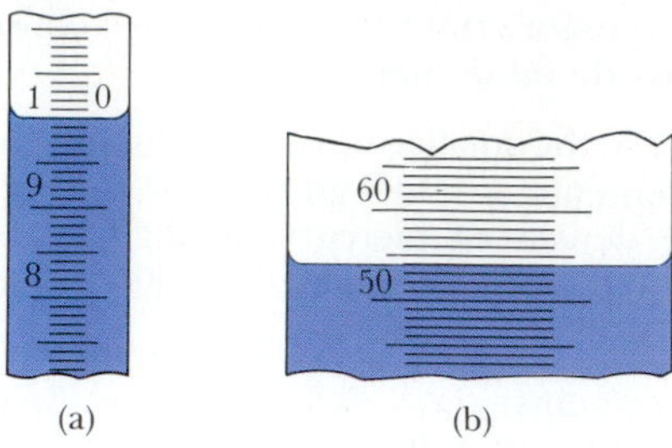

1.43 The masses of copper, zinc, and manganese in a sample of an alloy were measured as 2.011 g, 1.02 g, and 1.4 g, respectively. What should the reported mass of the alloy be?

1.44 A forensic chemist collected three samples from the scene of a crime. Their masses were 0.220 g, 0.03476 g, and 0.0001 g. What is the total mass of the collected sample?

1.45 A pharmacist made up a capsule consisting of 0.21 g of one drug, 0.124 g of another, and 1.311 g of a "filler." What is the total mass of the contents of the capsule?

1.46 A 2.001-μg diamond was removed from a sample of earth of mass 1.78×10^{-4} g. How much of the sample of earth is left?

1.47 The speed of light is 2.998×10^8 m/s. (a) How far can light travel in 5.0 s? (b) The distance from the Earth to the Sun is 9.3×10^7 miles. How long (in seconds) does it take light from the Sun to reach Earth?

1.48 When 0.25 lb of octane (the major component of gasoline) burns, about 5.47×10^6 J of energy is evolved. How much energy is produced in the combustion of 25 g of octane?

1.49 The concentration of gold in seawater is 0.011 μg/L. What mass of gold (in kg) is present in the Atlantic Ocean, of volume 3.23×10^{11} km^3?

1.50 A supersonic transport (SST) airplane consumes about 18,000 L of kerosene per hour of flight. Kerosene has a density of 0.965 g/mL. What mass of kerosene is consumed on a 3-h flight?

Mass Percentage Composition

1.51 Brass is an alloy of copper and zinc. Calculate the mass percentage composition of brass formed by mixing 325 g of zinc with 482 g of copper.

1.52 Bronze is an alloy of copper and tin (and, in specialized bronzes, of other metals too). Calculate the mass percentage composition of a bronze formed by alloying 84 g of tin and 286 g of copper.

1.53 Pure gold is called "24-carat gold," where each carat is $\frac{1}{24}$ part. Thus 18-carat gold is a mixture with 18 parts gold by mass and 6 parts of another metal (normally silver). Calculate the mass percentage composition of (a) 18-carat gold; (b) 14-carat gold.

1.54 Pewter is an alloy of 6% antimony, 1.5% copper, and 92.5% tin. What mass of each element is present in a pewter figurine that has a mass of 47.7 g?

1.55 "Solders" are often alloys of tin and lead. What masses of tin and lead are present in a 2.50-kg sample of solder that is known to be 43% tin?

1.56 An inexpensive lawn fertilizer is 22% nitrogen by mass. How many kilograms of nitrogen are in a 25.0-kg bag of fertilizer?

1.57 Seawater is a very complex mixture, but its main component other than water is sodium chloride (table salt). A scientist investigating the corrosive action of seawater on steel prepared a solution resembling it by dissolving 5.24 g of table salt in 50.0 g of water. What are (a) the mass percentage composition of the solution; (b) the mass of solute per kilogram of solution; (c) the mass of solute per kilogram of solvent?

1.58 "Concentrated" sulfuric acid is 95% sulfuric acid by mass (and the remainder water) and has a density of 1.84 g/mL. (a) What volume of concentrated sulfuric acid solution has 14.9 mg of sulfuric acid? (b) Will a 400-mL beaker hold 500 g of concentrated sulfuric solution?

1.59 Concentrated orange juice contains 40% solids. If one can of orange concentrate is mixed with three cans of water, what mass of solids will be present in an 8-oz (1 fl oz = 28.3 g) glass of orange juice?

1.60 A hospital medical technician needs to prepare 750 g solution that is 5% glucose by mass. What masses of glucose and water must be mixed to prepare the solution?

UNCLASSIFIED EXERCISES

1.61 Identify the larger member of the following pairs of quantities: (a) 1 qt or 1 L; (b) 1 nm or 1 μm; (c) 1°C or 1°F; (d) 1 dL or 1 mL.

1.62 Classify the following as an element, a compound, a homogeneous mixture, or a heterogeneous mixture: (a) iron; (b) table salt; (c) helium; (d) seawater; (e) dirt.

1.63 Classify the following changes as chemical or physical: (a) decay of wood; (b) combustion of wood; (c) sawing wood.

1.64 Classify the following properties as extensive or intensive: (a) melting point; (b) hardness; (c) the ability of iron to become corroded; (d) density.

1.65 Select the correct base unit and from Table 1.6 a convenient unit for recording (a) your mass; (b) the diameter of an atom; (c) the tolerance (of the order of 0.00001 m) on a finely machined tool.

1.66 The density of balsa wood is 0.16 g/cm^3. What mass occupies 1.00 ft^3?

1.67 The student with the highest grade on the first exam was 5 ft 3 in tall and weighed 134 lb. Express these measurements in SI base units.

1.68 An aspirin tablet has a mass of about 325 mg. Express the mass (in kilograms and ounces) of aspirin in a bottle that contains 250 tablets.

1.69 Express the volume in milliliters of a "1.00-cup" measure of milk, given that 2 cups = 1 pint, 2 pints = 1 quart.

1.70 The distance for a marathon is 26 miles and 385 yards. Convert this distance to kilometers, given that 1 mi = 1760 yd.

1.71 Express in nanometers the separation of the centers of a carbon atom and a tungsten atom in a particular compound that is reported to be 1.81 Å, given that 1 Å = 1×10^{-10} m.

1.72 The naked eye can detect an object that is about 0.1 mm in diameter. Express this diameter in inches.

1.73 The diameter of a typical aerosol smog particle is 0.1 μm. Express this diameter in centimeters.

1.74 The reference points on a newly proposed temperature scale expressed in °X are the freezing and boiling points of water, set equal to 50°X and 250°X. (a) Derive a formula for converting temperatures on the Celsius scale to the new scale. (b) Comfortable room temperature is 22°C. What is that temperature in °X?

1.75 An international committee of distinguished scientists met to establish the standard snail's pace, P_{sn}. The average snail covered 1 inch in 2 minutes. Express the average P_{sn} in centimeters per second (cm/s).

1.76 Approximately 45 billion aspirins are consumed annually in the United States. Considering the population to be 250 million (2 sf), determine the number of aspirins taken per person per day.

1.77 An experiment calls for 383 g of toluene. A bottle containing 0.90 lb is available. Does it contain sufficient toluene for the experiment?

1.78 The density of diamond is 3.51 g/cm^3. The international (but non-SI) unit for weighing diamonds is the carat: 1 carat = 200 mg. What is the volume of a 0.300-carat diamond?

1.79 What volume (in cubic centimeters) of lead (of density 11.3 g/cm^3) has the same mass as 100 cm^3 of a piece of redwood (of density 0.38 g/cm^3)?

1.80 Give the units in which the following physical quantities would be expressed if the base units were taken as gram, centimeter, second, and kelvin. Some of the combinations may seem odd, but we shall see later that they play an important role in chemistry: (a) velocity (distance per time); (b) kinetic energy ($\frac{1}{2} \times$ mass $\times$ (velocity)^2); (c) acceleration (change in velocity per time); (d) force (mass $\times$ acceleration); (e) energy (force $\times$ distance); (f) pressure (force per area).

1.81 Rewrite the following statements using the temperature scales indicated in brackets: (a) The melting point of gold is 1063°C [K]; (b) Neon boils at −411°F [K]; (c) The temperature of space is 2.7 K [°C].

1.82 Express the following quantities using scientific notation: (a) 56,000,000 km, which is the closest approach of Mars to Earth; (b) 93,900,000 Hz, which is the frequency of a popular FM radio station; (c) 0.000 000 000 000 000 000 000 020 g, the mass of one carbon atom.

1.83 The "specific gravity" of a substance is the ratio of its density to that of water at the same temperature. Is specific gravity an extensive or an intensive property? Explain your reasoning.

1.84 The average distance between the centers of carbon and hydrogen atoms in methane, the major component of natural gas, is 109 pm. Convert this distance to angstroms, where 1 Å = 10^{-10} m.

1.85 Make the following conversions, using the correct scientific notation: (a) The area occupied by an atom on a surface is 5.0×10^4 pm^2 [nm^2]. (b) The volume of the smallest particle of sediment found in a sample of river water was 100 nm^3 [cm^3]. (c) The volume of a chlorine gas cylinder is 10.0 ft^3 [dm^3].

1.86 One property of light that is used in spectroscopy is its wavenumber. Yellow light has a wavenumber of 18,000 cm^{-1}. Convert this measurement to units of m^{-1}.

1.87 The rates of chemical reactions are reported in terms of quantities called rate constants. The rate constant for the decomposition of nitrous oxide gas is 3.4 s^{-1}. Report this rate constant in units of reciprocal hours (h^{-1}).

1.88 One light-year is the distance light travels in 1 year and is equal to 9.46×10^{15} m. (a) How far is 1.00 light-year in miles? (b) How long (in meters) is 1.00 light-nanosecond? (c) The distance from Earth to the Moon is about 221,000 miles. A mirror, planted on the Moon's surface reflects a laser beam back to the Earth. How long

CASE 1 ▸ PROCESSING AN ORE: A PHYSICAL SEPARATION

Most metals that exist in nature occur in compounds that are mixed with rocks, sand, clay, and other materials. Collectively this mixture is called an *ore*. The metal-bearing compound itself is the *mineral* and the nonessential material is the *gangue*. Several metallurgical steps are used in processing the ore to extract the metal as the element. First, the ore is pretreated to separate the mineral from the gangue, the mineral is then chemically separated into its elements, and finally the metal is purified.

The pretreatment step usually involves a physical separation of the ore: the choice for the process in this step takes advantage of the physical properties of the mineral and the gangue, such as their relative densities, their ability to adsorb materials, and their magnetic properties. In particular, the wetting property—the ability of a liquid to spread over a surface (and not merely to form beads on it, like water on a newly waxed surface)—is used to separate metal sulfide-bearing minerals from the gangue. The technique relies on the fact

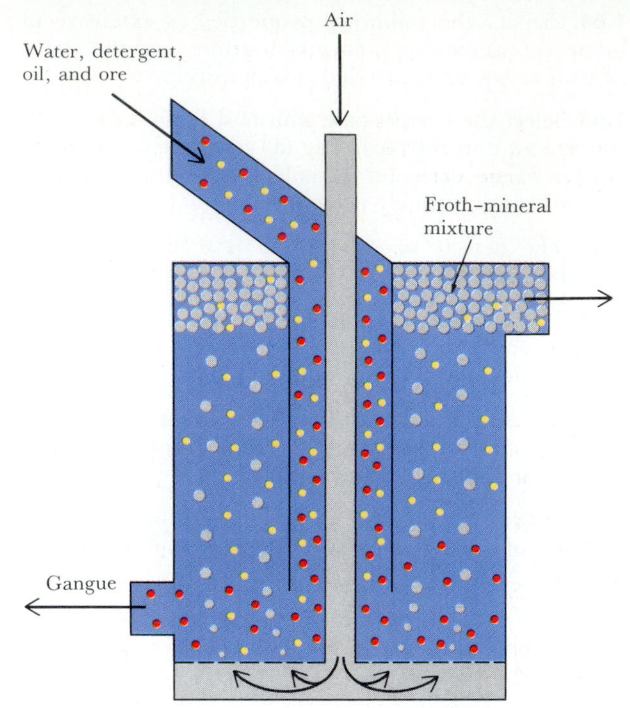

(in seconds) will it take for the laser beam to travel from the Earth to the Moon and back?

1.89 Rewrite the following statement using the designated units: "A typical young adult weighs 140 lb [kg] and needs to consume about 1.8 ounces [mg] of protein per day."

1.90 The temperature on the day side of the lunar surface is 127°C and the temperature on the night side of the lunar surface is −183°C. Convert each temperature to K and °F.

1.91 The density of gold is 19.3 g/cm^3. A gold nugget having a mass of 16.7 mg will displace what volume of water when placed into a graduated cylinder? The density of water is 1.00 g/cm^3.

1.92 The "specific heat capacity," which is discussed in Chapter 6, is defined as the heat needed to change the temperature of 1 g of substance by 1°C, and has units of J/(g·°C). The specific heat capacity of ice is 2.0 J/(g·°C). How much heat is needed to increase the temperature of a 20-g (2 sf) ice cube from −30°C to 0°C?

1.93 A 160-lb adult has about 5.0 qt of blood. If the concentration of cholesterol in the blood is 1.3 g of cholesterol per liter, how many grams of cholesterol are present?

1.94 A chemist determined in a set of four experiments that the density for magnesium metal was 1.68 g/cm^3, 1.67 g/cm^3, 1.69 g/cm^3, 1.69 g/cm^3. The accepted value for its density is 1.74 g/cm^3. What can you conclude about the precision and accuracy of the chemist's data?

1.95 A contraceptive pill contains 0.050% of the active agent ethynodiol diacetate. What mass of pills contains 5.0 g of the compound?

1.96 Many foods are sweetened with sugar, but fears have been expressed that high concentrations of sugar can cause cancer. This possibility has been investigated by exposing living cells to solutions containing high concentrations of sugar. As part of the investigation, a scientist prepared a solution containing 15 g of sugar in 250 g of water. What are (a) the mass percentage composition of the solution; (b) the mass of solute per kilogram of solution; (c) the mass of solute per kilogram of solvent?

that sulfide minerals, such as chalcocite (a compound of copper and sulfur) and galena (a compound of lead and sulfur) are wetted by oil but not by water. In the *flotation process,* ore that has been crushed into a slurry by a ball mill is placed into a large vat containing an oil-water mixture. The oil wets and spreads over the surface of the mineral; the water beads on the surface of the mineral but wets the gangue. The mixture is agitated with air that has been blown in from the bottom of the vat. The air bubbles stick preferentially to the oil, and, if the bubbles are large enough, they carry the mineral upward to the surface and a froth appears. A detergent is generally added to prevent the bubbles from collapsing at the surface. The froth-mineral mixture is then skimmed from the vat and dried to form a concentrate. The gangue sinks to the bottom of the vat and is discarded.

By controlling the kinds and amounts of oil and detergent in the vat, it is possible not only to separate a metal-sulfur mineral from its gangue but also to separate metal-sulfur minerals from one another. For example, chalcocite, galena, and pyrite (an iron mineral) can be selectively separated by the flotation process.

QUESTIONS

1. A certain ore is 4.5% by mass chalcocite. How many metric tons (1 metric ton = 10^3 kg) of ore must pass through the flotation process to obtain 500 kg of chalcocite?

2. An equivalent of about 7.5 million metric tons of chalcocite are mined throughout the world each year. If an average ore is 3% by mass chalcocite, what mass of the earth is moved annually.

3. The density of chalcocite is about twice that of sand (5.6 g/cm^3 and 2.7 g/cm^3, respectively). Explain how the flotation process seems to defy these relative densities in their separation.

This chapter introduces one of the greatest simplifying concepts of chemistry: the periodic table of the elements. We learn about the structure of atoms, which compose the elements, and about how atoms form compounds and how these are named. We also meet the "mole," a unit used throughout chemistry. The design and manufacture of microchips, such as this one, require knowledge of the composition of matter.

2 THE COMPOSITION OF MATTER

The single most important entity in chemistry is the atom:

> An **atom** is the smallest particle of an element that has the *chemical* properties of that element.

An atom of gold has the chemical properties of gold; an atom of lead has the chemical properties of lead; and so on.

Atoms are far too small to be seen with the naked eye, but sophisticated microscopes can now make them out. An image of individual xenon atoms arranged on the surface of a sample of nickel is shown in Fig. 2.1. The existence of atoms is so well established and so fundamental to chemistry that chemists now take them as the starting point of their definition of an element:

> An **element** is a substance that consists of atoms having the same chemical properties.

E This definition of an element is still not final because we cannot yet use the term *atomic number*: that term is introduced in Section 2.3.

All the atoms in a piece of gold have the same chemical properties, as do all the atoms in a piece of lead. The atoms of gold, however, are different from the atoms of lead.

One of the tasks of this chapter is to explain how atoms of different elements differ from one another. We shall see that atoms have a definite structure and that atoms of different elements are built from different combinations of the same **subatomic particles**, particles smaller than atoms. Scientists are beginning to understand that all the rich diversity of the world around us can be traced back to a handful of components. Whatever the material—sugar, chalk, seawater, a rock, a rose, or a piece of a brain—we can follow its composition back to atoms and beyond, back to the subatomic particles. We trace some of the connecting links in this chapter.

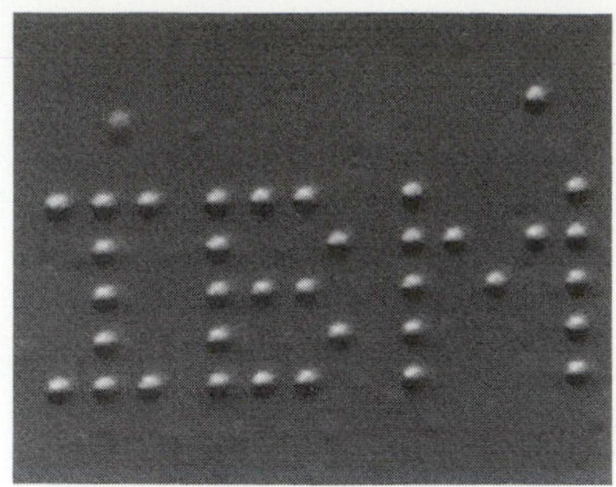

FIGURE 2.1 The small dots are individual xenon atoms that have been arranged into the world's smallest advertisement. Advances in technique have only recently made pictures like this possible.

E Atomic resolution imaging of adsorbates on metal surfaces in air: Iodine adsorption on Pt(111). *Science*, **1989**, *243*, 1050.

V Video 4: Elements.

N Origin of the names of chemical elements. *J. Chem. Ed.*, **1989**, *66*, 731.

N The symbols for the elements that currently have only systematic names may have up to three letters: thus, unnilennium has the symbol Une. Section 22.4 has more information on the nomenclature of elements with atomic numbers greater than 103.

ELEMENTS

One hundred and nine elements have been identified, and each one has been given a name and a symbol. Although 109 may seem like a large number of elements to remember and study, chemists have organized them into a pattern that makes it very easy to learn their properties. Moreover, for the purposes of chemistry, chemists regard all the elements as consisting of only three subatomic particles.

2.1 THE NAMES AND SYMBOLS OF THE ELEMENTS

The names of some elements are very ancient because those elements have been known for a long time. The name *copper* is derived from Cyprus, where the metal was once mined (Fig. 2.2); the name *gold* comes from the Old English word meaning "yellow." Elements identified more recently have been named, in a more or less chaotic fashion, by their discoverers. In some cases the name refers to a characteristic property. Chlorine is a yellow-green gas, and its name is derived from the Greek word meaning "yellow-green." For other names, chemists seem to have turned poets. Vanadium, which forms attractively colored compounds, is named after Vanadis, the Scandinavian goddess of beauty (Fig. 2.3). Several elements are named in honor of people or places; these include americium, berkelium, californium, einsteinium, and curium. The most recently discovered elements have not yet been given their final names. They are known by temporary names, such as unnilennium (for element 109, *un* standing for 1, *nil* for 0, and *enn* for 9), until scientists agree on a permanent name.

We represent each element by a **chemical symbol**, an abbreviation of the element's name. Many of the symbols are the first, or first two, letters of the element's name:

hydrogen H	carbon C	nitrogen N	oxygen O
helium He	aluminum Al	nickel Ni	silicon Si

FIGURE 2.2 Samples of common elements. Clockwise from the red-brown liquid bromine are the silvery liquid mercury and the solids iodine, cadmium, red phosphorus, and copper.

FIGURE 2.3 The element vanadium is named after the Scandinavian goddess of beauty, Vanadis, because some of the compounds of vanadium produce solutions of attractive hues.

V Video 6: Oxidation states of vanadium.

Note that the first letter of a symbol is always uppercase and the second letter always lowercase (for example, Ni, not NI). Some elements have symbols formed from the first letter of the name and a later letter:

magnesium Mg   chlorine Cl   zinc Zn   plutonium Pu

The rest have symbols taken from their Latin or German names (Table 2.1). Appendix 2D lists the names and symbols of all the elements and gives the origins of their names.

2.2 THE PERIODIC TABLE

The known elements are listed inside the front cover of this book in a special arrangement called the **periodic table**. Because this table is very important, it is a good idea to become familiar with it as early as possible. An element's location in the table is an excellent guide to its properties and the kinds of compounds it forms. Therefore, the first time you meet an element you should note its position and which elements are its neighbors, because it is likely to resemble them.

Groups and periods. The vertical columns of the periodic table are called **groups**; the horizontal rows are called **periods** (Fig. 2.4). An element's **congeners**, the other members of its group, are generally very similar and show a gradation in their properties. The properties of sodium (Na), for example, are a good clue to the properties of its congeners in Group I, namely, lithium (Li), potassium (K), rubidium (Rb), cesium (Cs), and the extremely rare and short-lived element francium (Fr). These elements, which collectively are called the **alkali metals**, show a smooth variation in properties from lithium to cesium. All of them are soft, silvery metals that melt at low temperatures: lithium at 181°C, potassium at 64°C, and cesium at 29°C. They all produce hydrogen from water (Fig. 2.5), lithium gently but cesium with explosive violence. All the alkali metals are stored under oil to keep them out of contact with air and moisture.

TABLE 2.1 Elements with symbols taken from Latin and German names*

Element	Symbol	Latin or German name
antimony	Sb	*stibium*
copper	Cu	*cuprum*
gold	Au	*aurum*
iron	Fe	*ferrum*
lead	Pb	*plumbum*
mercury	Hg	*hydrargyrum*
potassium	K	*kalium*
silver	Ag	*argentum*
sodium	Na	*natrium*
tin	Sn	*stannum*
tungsten	W	*wolfram*

*The origins of these names are given in Appendix 2D.

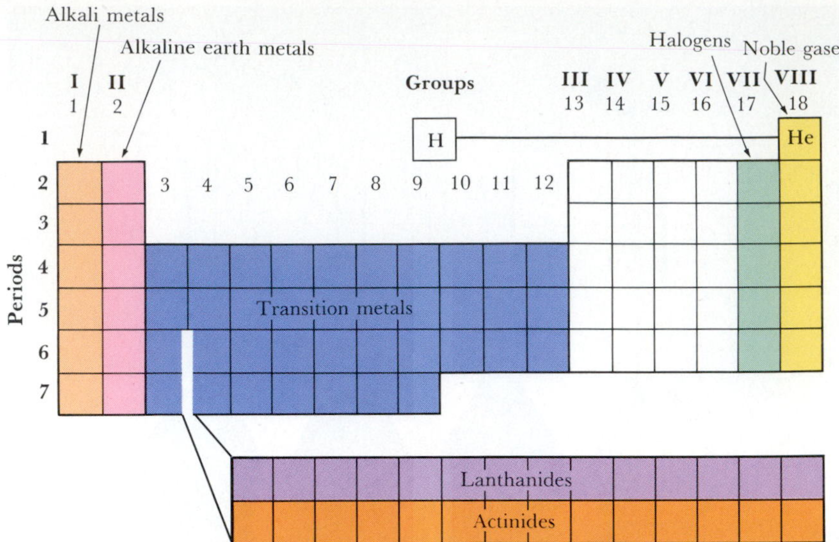

FIGURE 2.4 The structure of the periodic table, showing the names of some of the regions and groups. The periods are denoted by the Arabic numbers 1 through 7 and the groups are denoted by the Roman numerals I through VIII. An international committee has recommended that the groups be numbered from 1 through 18 (numbers shown just above the table). Hydrogen, which heads the entire table but does not belong to any group, and helium are the only members of Period 1.

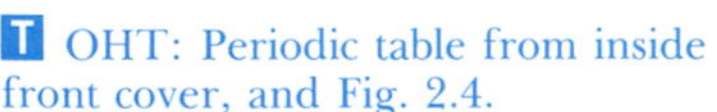
T OHT: Periodic table from inside front cover, and Fig. 2.4.

D Oxidation of sodium. CD1, 83. Reactions of magnesium. TDC, 13.2–13.5 (see the interesting footnote). Burning magnesium in carbon dioxide. CD2, 31.

Next to the alkali metals (Group I) are the **alkaline earth metals**, the elements in Group II. They resemble the alkali metals in several ways but react with water less vigorously. Calcium (Ca), strontium (Sr), and barium (Ba) react strongly enough to release hydrogen from water, but magnesium (Mg) does so only when heated; beryllium (Be) does not react with water even when it is red hot.

On the right side of the table, in Group VIII, are the elements known as the **noble gases**. They are so called because they form so few compounds—they are chemically aloof. Indeed, until the 1960s they were called the inert gases, because it was thought that they formed no compounds at all. Since then, a few dozen compounds of krypton (Kr), xenon (Xe), and radon (Rn) have been prepared. All Group VIII elements are colorless, odorless gases. They are **monatomic** gases, in the sense that they are found as independent, single atoms.

Next to the noble gases in the periodic table are the **halogens** of Group VII: fluorine (F), chlorine (Cl), bromine (Br), iodine (I), and astatine (At). Many of the properties of the halogens show a smooth

FIGURE 2.5 The alkali metals react with water, producing gaseous hydrogen and heat. (a) Lithium reacts quietly. (b) Sodium reacts so vigorously that the accompanying heat melts the unreacted metal. (c) Potassium reacts even more vigorously, producing so much heat that the hydrogen produced from the water is ignited and burns in the air above the water.

(a)

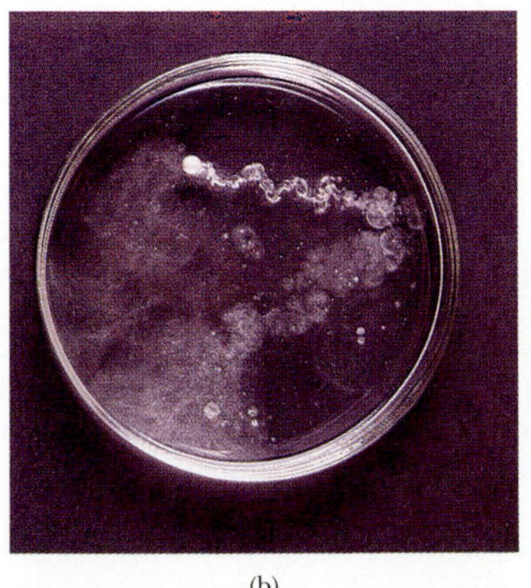
(b)

(c)

FIGURE 2.6 The halogens are colored elements. From left to right, chlorine is yellow-green, bromine is red-brown, and iodine is violet. At room temperature, chlorine is a gas; bromine, a liquid; and iodine, a solid.

D Show samples of these three halogens. Generate chlorine gas by mixing equal volumes of chlorox and dilute hydrochloric acid in a flask (in the hood). See TDC, 20.1s, 20.2s, and 10.3s.

D War gases exhibit. TDC, 20–26. Halogens compete for electrons. CD2, 33.

variation from fluorine to iodine. Fluorine is a very pale yellow, almost colorless gas; chlorine is a yellow-green gas; bromine, a red-brown liquid; and iodine, a purple-black solid (Fig. 2.6). Astatine exists only in amounts that are too minute to be seen.

The elements in the central part of the table, between Groups II and III, are called the **transition metals**. They include the important structural elements titanium (Ti) and iron (Fe) and the coinage metals copper (Cu), silver (Ag), and gold (Au). The transition metals are so called because their chemical and physical properties range from those of the chemically active metals of Groups I and II to those of the much less active metals of Groups III and IV.

Metals, nonmetals, and metalloids. A very useful classification of the elements contains three categories: metals, nonmetals, and metalloids.

> A **metal** is a substance that conducts electricity, has a metallic luster, and is malleable and ductile.

A malleable substance (from the Latin word meaning "hammer") is one that can be hammered into thin sheets (Fig. 2.7). A ductile substance (from the Latin word meaning "drawing out") is one that can be drawn out into wires. Copper, for example, conducts electricity, has a luster when polished, is malleable, and is quite ductile (it is readily drawn out into wire for electric installations). Later we shall see that metallic elements also have characteristic chemical properties, so knowing that copper is a metal immediately enables us to predict some of the reactions of the element and its compounds.

> A **nonmetal** is a substance that does not conduct electricity and is neither malleable nor ductile.

For example, sulfur is a brittle yellow solid that does not conduct electricity, cannot be hammered into thin sheets, and cannot be drawn out

FIGURE 2.7 All metals can be deformed by hammering. Gold can be hammered into a sheet so thin that light can pass through it. Here it is possible to see the light through the sheet of gold.

FIGURE 2.8 The location of the six elements commonly regarded as metalloids; these elements have characteristics of both metals and nonmetals. Other elements (notably beryllium, boron, and bismuth) are sometimes included in the classification.

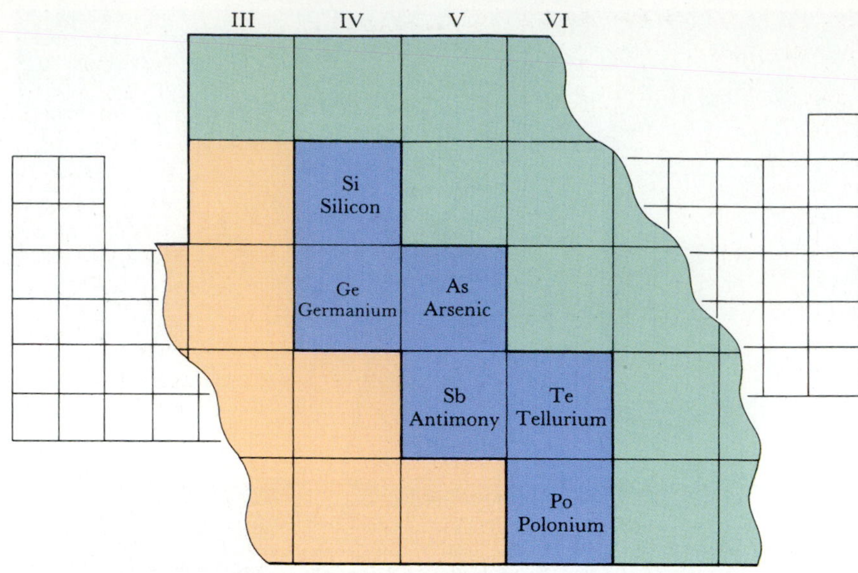

E Metalloids. *J. Chem. Ed.*, **1982**, *59*, 526.

N The term *semimetal* is often encountered as a synonym for metalloid. However, in solid-state physics a semimetal has a different meaning relating to band structure, and it is better to avoid possible confusion.

E Beryllium, boron, and bismuth are also sometimes classified as metalloids.

T OHT: Fig. 2.9.

into wires. All gases are nonmetals. We shall see later that nonmetals and their compounds have characteristic properties.

Some elements have characteristics of both metals and nonmetals; these elements are called metalloids:

> A **metalloid** has the physical appearance and properties of a metal but behaves chemically like a nonmetal.

The distinction between metals and metalloids and between metalloids and nonmetals is not very precise (and not always made), but the metalloids are often taken to be the six elements shown in Fig. 2.8.

A striking feature of the periodic table becomes apparent when we mark the positions of the metals, metalloids, and nonmetals (Fig. 2.9). We find that all the metallic elements occur to the left of the table and all the nonmetallic elements occur to the upper right. The metalloids lie between these two regions. One of the first applications of the periodic table now becomes clear: with a glance at the table we can see whether an element is a metal, a metalloid, or a nonmetal. We may

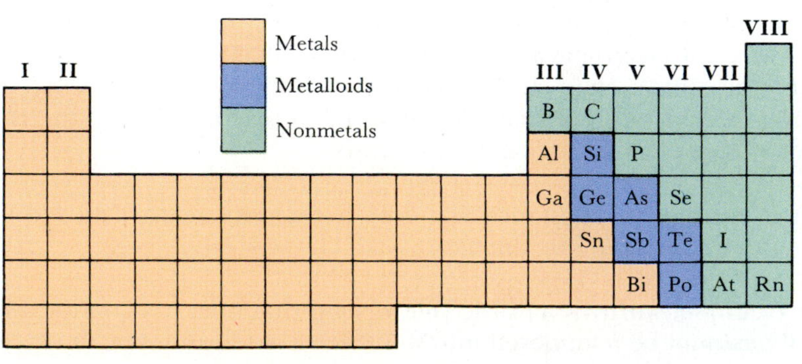

FIGURE 2.9 The distribution of metals, metalloids, and nonmetals in the periodic table. The lanthanides and the actinides, which are all metals, have been omitted, as has the nonmetal hydrogen.

never have heard of indium (In) before, but its position in the periodic table indicates that it is a metal. We can even guess that, because it follows the transition metals, it will not be a very reactive metal.

ATOMS

The first pieces of experimental evidence for the existence of atoms were assembled in 1805 by John Dalton, a scruffy but learned English schoolteacher (Fig. 2.10). His **atomic hypothesis** was that

1. All matter is composed of atoms.
2. All the atoms of a given element are identical.
3. The atoms of different elements have different masses.
4. A compound is a specific combination of atoms of more than one element.
5. In a chemical reaction, atoms are neither created nor destroyed but are merely rearranged to produce new substances.

(A *hypothesis* is a suggestion put forward to account for a series of observations.) Dalton based his proposals on measurements of the masses of separate and combined elements. We do not need to go through the same arguments today because experimental techniques have progressed since Dalton's day and the evidence for atoms is now much more direct (Box 2.1). For instance, as we saw in Fig. 2.1, it is now possible to photograph individual atoms. Other support comes from a newly invented technique that in effect feels the surface of a solid and shows the position of its atoms as bumps (Fig. 2.11).

But what are atoms? Are atoms like very hard billiard balls and truly "uncuttable," as Dalton thought and as their name suggests (the word *atom* comes from the Greek words meaning "not cuttable"), or can they be taken apart into subatomic particles? If the latter is the case and we

FIGURE 2.10 John Dalton (1776–1844), the English schoolteacher who used experimental measurements to argue that matter consisted of atoms. This very early photograph—a daguerreotype—was made in 1842.

N Is an atom of copper malleable? *J. Chem. Ed.*, **1986**, *63*, 64.

E The colors in Fig. 2.11 indicate the electron densities of the two types of atom.

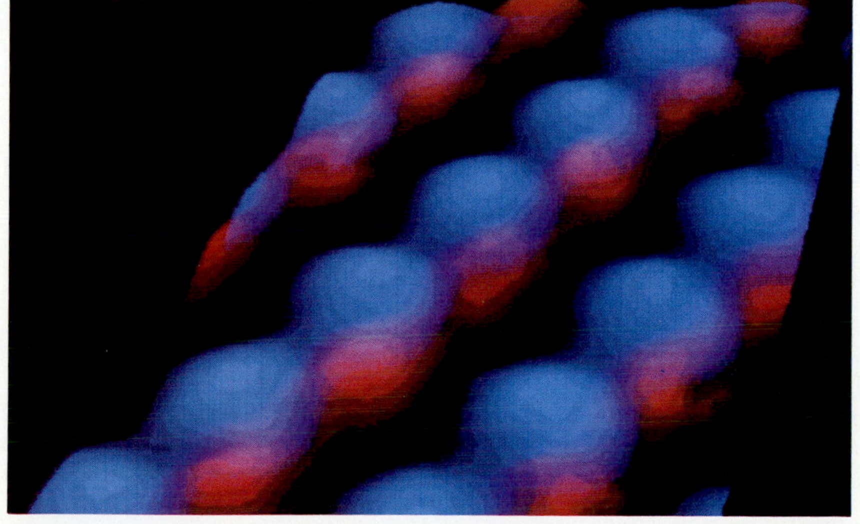

FIGURE 2.11 Atoms can be discerned as bumps on the surface of a solid by the technique called scanning tunneling microscopy (STM). This is an image of the surface of gallium arsenide. The gallium atoms are depicted as blue and the arsenic atoms as red (these are not their actual colors).

BOX 2.1 ▸ THE SCANNING TUNNELING MICROSCOPE

C. F. QUATE, *Ginzton Laboratory, Stanford University*

In March 1981, 300 years after the invention of the optical microscope and 50 years after the development of the electron microscope, a new form of microscope was introduced by Gerd Binnig and Heinrich Röhrer of the IBM Research Laboratory in Zurich, Switzerland. The new microscope used electrons that "tunneled" across a narrow gap between two conducting surfaces. The introduction of this new type of microscope proved worthy of the Nobel Prize in 1986.

In the scanning electron microscope—the SEM—the source of electrons is a sharp tungsten tip from which the electrons emerge as free particles. They form into a beam that threads through a series of plates containing small apertures. These plates, charged to a potential difference of several hundred kilovolts, form a lens that focuses the electron beam onto the surface under investigation. In the scanning tunneling microscope—the STM—the electron source is similar to that in the SEM, but the physical arrangement is much simpler: there is no electron beam, nor is there a focusing lens.

In the STM, the sharp tungsten tip is brought close to the sample and scanned over the surface in a series of closely spaced lines similar to the scanning pattern used to form images on a television screen. The electrons travel across the gap between the tip and the surface, with a potential difference of only about 1 V between tip and sample. This transfer is called a tunneling process because the electrons seem to burrow through a region where they should not be found.

The tunneling current had been expected—indeed, the first experiment was fashioned to exploit the phenomenon—but the low degree of sideways spread of the electron current was a surprise, and the first experiment indicated that the tunneling current was confined to a *single atom* at

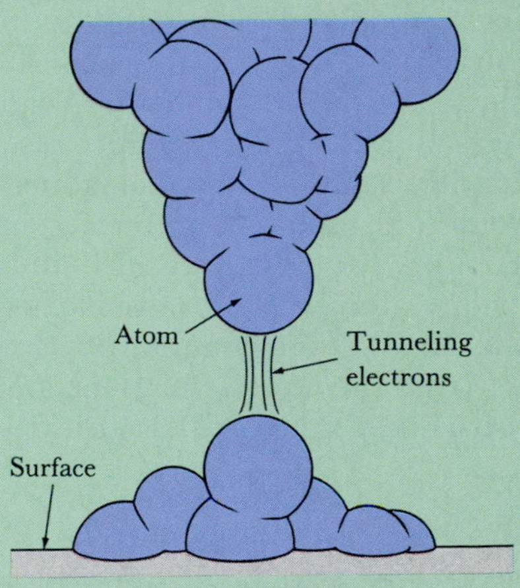

can discover how an atom is built, then we might be able to take a lead atom apart and reassemble it as gold! Even if that turns out to be impossible, we should at least be able to account for the existence of different elements and understand why they have different properties.

2.3 THE NUCLEAR ATOM

Experiments carried out at the end of the nineteenth century and early in the twentieth showed that atoms *are* made up of smaller particles. In fact, they showed that atoms have an internal structure made up of electrons, protons, and neutrons.

the end of the tip. After the fact, the behavior is easily rationalized. Because the tunneling current decreases exponentially as the gap spacing is increased, the protruding atom on the tip closest to the surface totally dominates the tunneling current. As shown in the drawing, the current passes from that single atom into the nearby surface. The strength of the tunneling current is a measure of the density of electrons on the conducting surface. This relation makes it possible to study the atomic structure of the surface because different species of atoms represent different concentrations of electrons (see Figure 2.11). The technique can show images of the surface in striking detail; for example, it shows how atoms are arranged in planes separated by "cliffs" one or two atoms high.

The instruments operate over a wide range of temperatures and in a variety of environments. The original instruments were placed in a vacuum. However, the experimenters soon realized that molecules could not get between the surface and tip anyway. Because the tip is positioned only a few atomic diameters from the sample, the space is not large enough to accommodate molecules from the surrounding atmosphere. It soon became evident that STMs could operate not only in a vacuum but also in gases and in liquids. Vacuum conditions are now reserved for scientific investigations of ultraclean surfaces. Currently, the instruments are placed in oil to monitor the processes of wear in moving bearings, in electrolytic cells to monitor the deposition of metallic layers, and in water to image single molecules in molecular layers formed on smooth substrates.

Scanning tunneling microscopes are used in physics, in chemistry, in biology, and in many areas of technology. In the future, STMs will be used to study the details of chemical and biological processes that occur when atoms come together to form molecules and molecular complexes.

Perspective STM image of a 32 × 36 nm region of the surface of silicon, showing the steps between layers of atoms. The steps are one atom high. (Courtesy of R. Wiesendanger and colleagues, University of Basel, Switzerland.)

Electrons. Just before the end of the nineteenth century (in 1897), the British physicist J. J. Thomson (Fig. 2.12) investigated the "cathode rays" that are emitted when a high potential difference (a high "voltage") is applied between two electrodes (metal contacts) in a glass tube containing a small amount of gas (Fig. 2.13). His observations confirmed earlier work suggesting that a stream of particles moves from one electrode (the cathode) toward the other electrode. These cathode rays caused a spot of light to appear when they hit a specially treated screen, and the spot could be moved by placing electrically charged plates or a magnet near the path of the rays. Thomson found that the properties of the rays were the same no matter what metal he used for the electrodes.

D Set up a cathode-ray tube (or fluorescent light) apparatus using a Tesla coil and show the effects of a magnet.

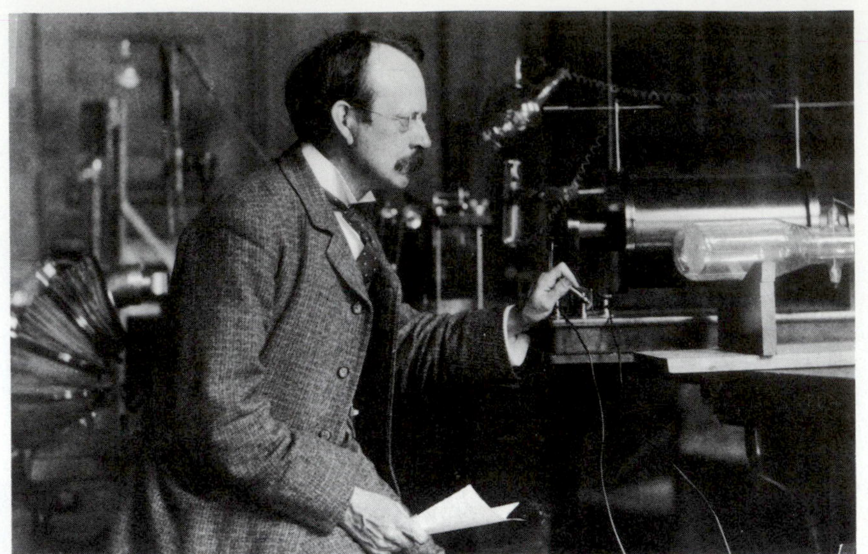

FIGURE 2.12 Joseph John Thomson (1856–1940), who measured the ratio of the charge to the mass of an electron, with the apparatus he used.

? What experimental data have led scientists to the view that electrons are fundamental components of matter?

N The discovery of the electron, proton, and neutron. *J. Chem. Ed.* **1989**, *66*, 73.

Thomson showed that cathode rays are streams of negatively charged particles coming from inside the atoms that make up the cathode. These particles are now called **electrons** and denoted e^-. The fact that electrons are produced regardless of the metal used for the cathode suggests that they are part of the makeup of all atoms. Thomson determined some of the properties of electrons from the way the spot moved when he altered the charge on the charged plates or the strength of the magnet. He measured the value of e/m, which is the ratio of the size of the electron's charge e to its mass m, and found it to be 1.76×10^8 C/g. (The letter C is the abbreviation for the SI unit of charge, the coulomb.) The technological outcome of Thomson's discovery was the cathode-ray tube used in television sets and computer monitors. Each time we see a television picture, we are watching a (sometimes) more entertaining version of Thomson's experiment.

Later workers, most notably the American Robert Millikan, devised experiments to measure the charge of an electron separately. Millikan found that $e = 1.6 \times 10^{-19}$ C. Therefore, by combining this value with the ratio determined by Thomson, we can calculate the mass of an electron:

$$m = e \times \frac{m}{e} = 1.6 \times 10^{-19}\ \cancel{\text{C}} \times \frac{\text{g}}{1.76 \times 10^8\ \cancel{\text{C}}} = 9.1 \times 10^{-28}\ \text{g}$$

This tiny value makes the electron by far the least massive of the subatomic particles of interest in chemistry (see Table 2.2). We shall not use the absolute numerical value of the charge again until much later; instead, we shall refer to the charge of an electron as one unit of negative charge, or -1.

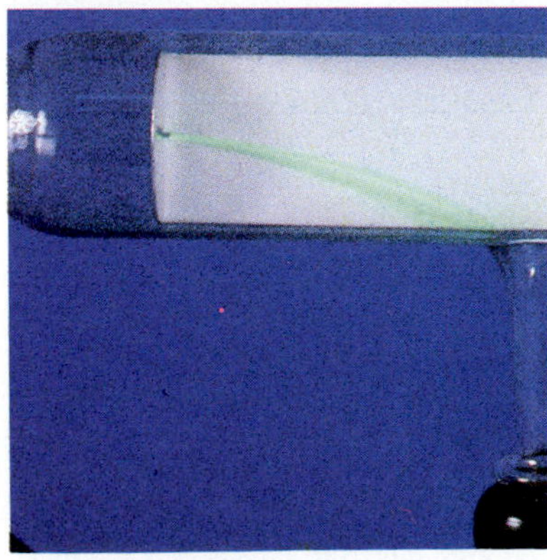

FIGURE 2.13 A close-up of the glow near a cathode in an apparatus like that used by Thomson. Note the cathode ray and the deflection of the ray by a magnetic field.

Nuclei. Ordinary matter is neither attracted to nor repelled by charged electrodes. This observation indicates that atoms do not have an electric charge: they are electrically neutral. Thomson's experiment, however, showed that atoms contain electrons, which are negatively charged particles. Atoms must therefore also contain enough positive charge to cancel the negative charges of the electrons. Thomson suggested a "plum-pudding" model of the atom in which the electrons are

TABLE 2.2 Properties of subatomic particles

Particle	Symbol	Charge*	Mass, g	Mass, u†
electron	e^-	−1	9.109×10^{-28}	0.00055
proton	p	+1	1.673×10^{-24}	1.0078
neutron	n	0	1.675×10^{-24}	1.0087

*Charges are given as multiples of 1.602×10^{-19} C.
†The atomic mass unit, u, is explained in Section 2.4.

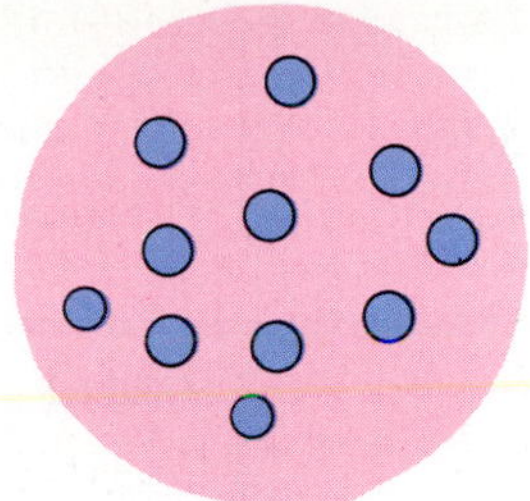

FIGURE 2.14 Thomson's "plum-pudding" model of the atom. Thomson proposed that an atom consists of electrons (blue circles) embedded in a jellylike, positively charged substance.

scattered throughout a jellylike, positively charged substance (Fig. 2.14). However, this view was overthrown by the results of an experiment suggested by the New Zealander Ernest Rutherford (Fig. 2.15).

Rutherford was interested in the observation that some elements, including radium, emit streams of positively charged particles that he called **alpha particles** (α particles). By studying how α particles behave in the presence of electrically charged plates and magnets, he identified these particles as helium atoms that had lost their electrons (as we shall see, a helium atom has only two electrons). According to Thomson's view, α particles should be blobs of positively charged jelly. Two of Rutherford's students, Hans Geiger and Ernest Marsden, carried out an experiment that Rutherford thought up to introduce Marsden to a particular piece of apparatus. They shot α particles toward a very thin gold foil only a few atoms thick (like the one shown in Fig. 2.7) and observed the positions where the particles hit a screen (Fig. 2.16). Because they thought that the atoms in the foil were all spheres of jelly, they expected all the α particles, which had speeds close to 2×10^4 km/s, to pass through the foil, with perhaps a slight deflection from their paths in some cases. What they actually observed astonished them. Although almost all the α particles did pass through and some were indeed deflected only very slightly, about 1 in 20,000 was deflected through more than 90°. Sometimes an α particle bounced straight back in the direction from which it had come. "It was almost as incredible," said Rutherford, "as if you had fired a 15-inch shell at a piece of tissue paper and it had come back and hit you."

The explanation had to be that atoms are not soft blobs of jelly but have pointlike centers of positive charge that contain most of the atom's mass. Rutherford suggested a structure, called the **nuclear atom**, to account for this conclusion. He proposed that all the positive charge and most of the mass of an atom are concentrated in a minute lump, the **atomic nucleus**, and that the electrons occupy the surrounding space. According to Rutherford's model, an atom is almost all empty space. Rutherford's suggestion accounts well for Geiger and Marsden's observations. The positively charged α particles are the minute nuclei of helium atoms. Those that score a direct hit on one of the minute gold nuclei are strongly repelled by the positive charge of the nucleus, and the α particle is deflected through a large angle, like a rebounding billiard ball (Fig. 2.17). The picture of the atom as mostly empty space accounts for the observation that most of the α particles passed through the foil.

Measurement of the deflection angles gives information about the charges and diameters of atomic nuclei. Rutherford concluded that the nucleus of a gold atom carries nearly 100 units of positive charge (the

T OHT: Figs. 2.14, 2.16, and 2.17.

? What are the differences between a Dalton atom, a Thomson atom, and a Rutherford atom? How can they be distinguished experimentally?

D A simulation of the Rutherford experiment. *J. Chem. Ed.*, **1982**, *59*, 973.

FIGURE 2.15 Ernest Rutherford (1871–1937), who was responsible for many discoveries about the structure of the atom and its nucleus.

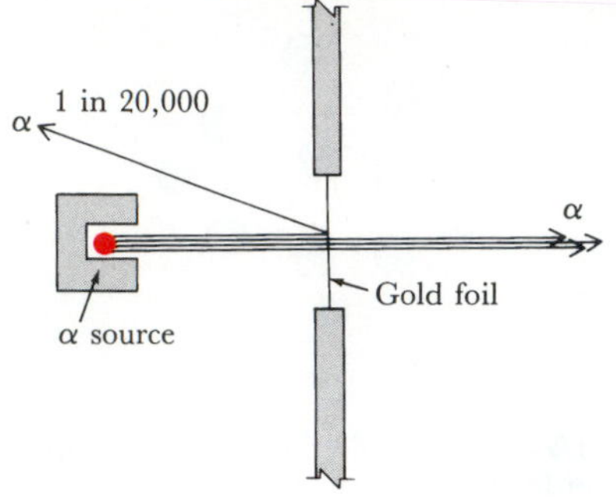

FIGURE 2.16 The experimental arrangement used by Geiger and Marsden. The α particles came from a sample of the radioactive gas radon. Their deflections were measured by observing flashes of light ("scintillations") where they struck a zinc sulfide screen. About 1 in 20,000 α particles were deflected through very large angles; most went straight through the foil with very little deflection.

value is now known to be 79) and that it is a minute, dense sphere about 10^{-4} m in diameter, or around $\frac{1}{10,000}$ the diameter of the atom itself. If we were to think of an atom as magnified to the size of a football stadium, the nucleus would be the size of a fly at the stadium's center.

Protons and the atomic number. Thomson and his contemporaries showed that an atom contains electrons. Rutherford showed that these electrons occupy the space around a minute, positively charged, central nucleus. Nuclear physicists have continued this process of taking atoms apart into simpler components. They have found that atomic nuclei are built from two kinds of particles called **nucleons**. One nucleon is the **proton** (p) and the other is a **neutron** (n). A proton is 1836 times heavier than an electron; it has a positive charge that exactly cancels an electron's negative charge. A neutron has almost the same mass as a proton has, but, as its name suggests, a neutron is electrically neutral (see Table 2.2).

The number of protons in an atomic nucleus is called the **atomic number** Z of the element. Henry Moseley, a young British scientist, was the first to measure atomic numbers accurately (shortly before he was killed in action in World War I). Moseley knew that when elements are bombarded with rapidly moving electrons (the cathode rays of Thomson's experiment) they emit x-rays. These rays are like light rays in some respects, but they can pass through many substances. He found that the energy of the x-rays emitted by an element depends on its atomic number; and by studying the x-rays of many elements, he was able to determine the value of Z for them. We now know the atomic numbers of all the elements. They are shown above the chemical symbols in the periodic table inside the front cover. In many versions of the periodic table, the data are arranged as shown in the margin. (The meaning of molar mass will be explained later in the chapter.)

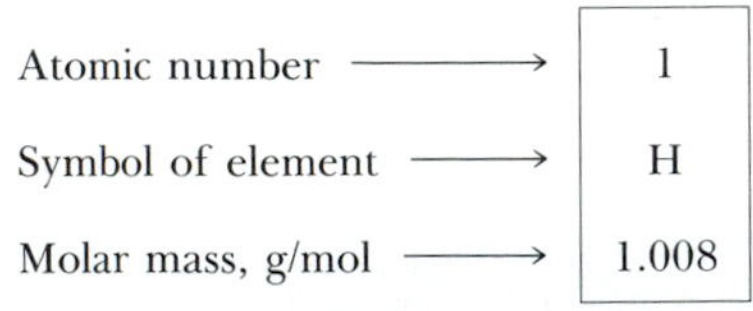

N We now have the modern definition of an element as a substance that consists of atoms with the same atomic number.

Because an atom is electrically neutral, the number of protons in its nucleus must be the same as the number of electrons outside its nucleus. Therefore, counting the number of protons in the nucleus is an indirect way of counting the number of electrons. For hydrogen, $Z = 1$; so we know at once that a hydrogen atom has one electron. A gold atom, with $Z = 79$, has 79 protons in its nucleus and 79 electrons. For uranium, $Z = 92$; so we know that every uranium atom has 92 electrons. The atomic number of an element is sometimes added as a subscript just in front of its chemical symbol, for example, ${}_{1}H$ and ${}_{92}U$.

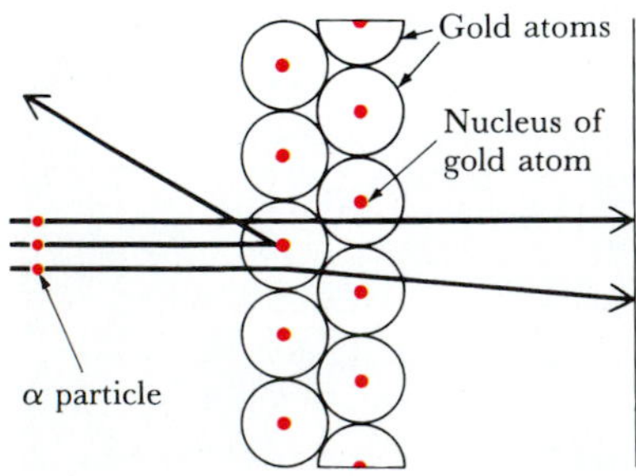

FIGURE 2.17 Rutherford's model of the atom explains why most α particles pass almost straight through, whereas a very few undergo large deflections.

EXAMPLE 2.1 Using the atomic number

How many protons and electrons are present in an atom of fluorine?

STRATEGY We know that the number of protons and the number of electrons are both equal to the atomic number Z. Therefore, all we need to do is look up the atomic number of the element in the periodic table (or the list of elements inside the back cover).

SOLUTION Fluorine (F) has atomic number 9. Therefore, there are nine protons in its nucleus, and there are nine electrons occupying the space around the nucleus.

EXERCISE E2.1 How many protons and electrons are present in an atom of bismuth?

[*Answer*: 83 of each]

We can now see that, in the periodic table *the elements are arranged in order of increasing atomic number*. The elements in the table are arranged to show how the properties of the elements repeat periodically. Therefore, it follows that these properties vary periodically as the number of electrons increases. This **periodicity** of the elements, that is, the regular repetition of properties as Z increases, is one of the most remarkable features of matter. Periodicity is explained in Chapter 7.

2.4 THE MASSES OF ATOMS

When Dalton first proposed his atomic hypothesis, the mass of an individual atom could not be measured directly. He did try to deduce the *relative* masses of atoms in a series of clever experiments based on the masses of elements that combined together. He deduced (in 1803) that an atom of copper is about 60 times heavier than an atom of hydrogen—a correct conclusion. Hence, he knew that a 1-g sample of copper has only $\frac{1}{60}$ the number of atoms that a 1-g sample of hydrogen has. However, he also thought that an atom of oxygen is about 8 times heavier than an atom of hydrogen, whereas it is in fact 16 times heavier.

E Dalton's error about the relative mass of oxygen stemmed from his view that compounds were the simplest combinations of atoms; so he presumed that the formula of water was HO.

Dalton's values were corrected later in the nineteenth century, and chemists have long been able to determine the number of atoms in a sample from its mass. It was not until earlier in this century, however, that they found ways to determine the mass of an individual atom directly, accurately, and precisely. The mass of an atom is now measured by using the technique of **mass spectrometry** (see Box 2.2). Consequently, we now know that the mass of a hydrogen atom is 1.67×10^{-24} g and that of a carbon atom is 1.99×10^{-23} g. Indeed, we now know the masses of the atoms of all the elements, and they are all in the region of 10^{-24} to 10^{-22} g.

Isotopes and mass number. The measurement of precise atomic masses led to a major discovery. Dalton assumed that all the atoms of a given element are identical. However, when an element is studied with a mass spectrometer, atoms with slightly different masses are generally detected, even though a chemically pure sample is being used. In a sample of pure neon, for example, most of the atoms have mass 3.32×10^{-23} g, which is about 20 times greater than that of a hydrogen atom. Some, however, are about 22 times heavier than hydrogen, and a few are about 21 times heavier. All three types of atoms have the chemical properties of neon and are called isotopes of neon:

E One illustration of the classical determination of relative atomic masses (and a useful test question) is to burn a known mass of magnesium in oxygen, and to measure the increase in mass. If it is supposed (correctly) that the formula of the product is MgO, then the relative masses of Mg and O can be determined quite simply.

> **Isotopes** are atoms that have the same atomic number but different atomic masses.

The name *isotope* comes from the Greek words for "equal place"; it signifies that although the atoms have different masses, they belong to an element that occupies one place in the periodic table. The atomic number Z (the number of protons in the nucleus) determines the identity and properties of the element, so atoms of an element can have different masses as long as they have the same number of protons.

E The symbol Z is taken from the German word for number, *Zahlen*.

It is easy to explain the existence of isotopes by supposing that *the atomic nucleus of a given element has a fixed number of protons but a variable number of neutrons*. Neutrons add to the mass of the atom but do not affect the number of electrons it needs to achieve electrical neutrality.

BOX 2.2 ▸ MASS SPECTROMETRY

The masses of atoms and molecules are now measured with a mass spectrometer. Atoms or molecules of the elements—either a gaseous element (neon or oxygen, for instance) or the vapor of a liquid or a solid element (such as mercury or zinc)—are fed into the spectrometer's ionization chamber. There they are exposed to a beam of rapidly moving electrons. When one of the accelerated electrons collides with an atom, it knocks another electron out of it, leaving the atom with a positive charge. The collision creates a positive ion, an electrically charged atom (or group of atoms). The positive ions are accelerated out of the chamber by a strong electric field applied between two metal grids. The speeds reached by the ions depends on their masses, with lighter ions reaching higher speeds than heavy ones.

As the accelerated ion passes through a magnetic field generated by an electromagnet, its path is bent by an amount that depends on its speed (and hence on its mass). The strength of the magnetic field is slowly changed, and a signal is produced when the magnetic field is just strong enough to bend the beam of ions so that they arrive at the detector. The mass of the ion is then calculated from the accelerating voltage and the strength of the magnetic field used to produce a signal. The mass spectrum is a plot of the detector signal against the magnetic field. The positions of the peaks along the abscissa are used to calculate the masses of the accelerated ions, and the heights of the peaks indicate the proportions of ions of each type.

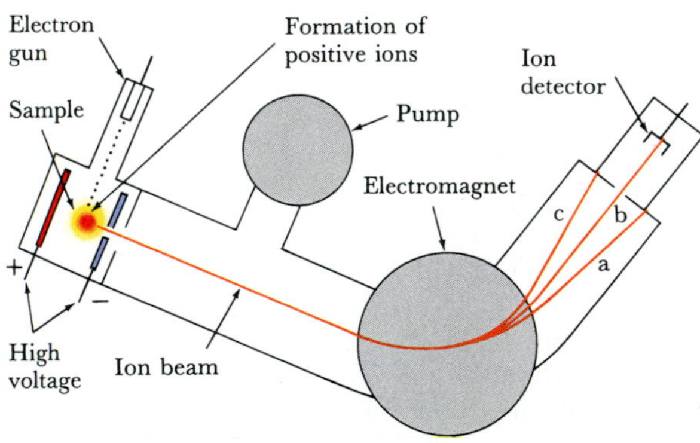

A mass spectrometer is used to measure the mass and abundance of an isotope. As the strength of the magnetic field is changed, the path of the accelerated ions moves from *a* to *c*. When the path is at *b*, the ion detector sends a signal to a recorder. At a fixed magnetic field, the three paths would represent the trajectories of three ions of isotopes of three different masses.

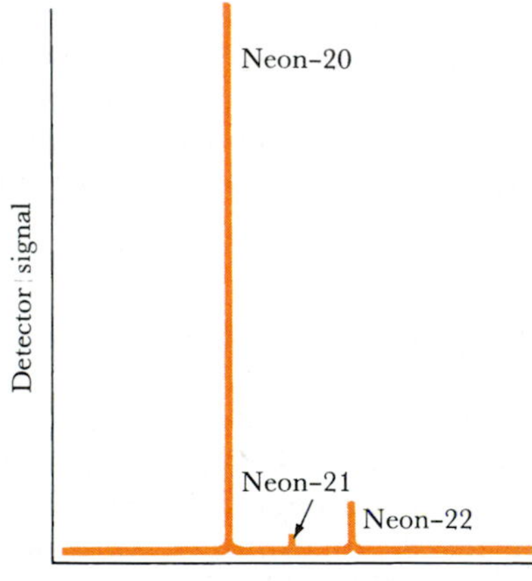

The signal obtained from a mass spectrometer, which is called the mass spectrum of the sample, shows that neon consists of atoms with different masses. Neon-20, an atom of neon about 20 times heavier than hydrogen, is the most abundant.

T OHT: The two illustrations in Box 2.2.

Hence, the different numbers of neutrons in the isotopes of an element significantly affect their masses but not their chemical properties. The total number of protons and neutrons in the nucleus of an isotope determines the value of a property we call mass number:

> The **mass number** *A* of an isotope is the total number of nucleons (protons plus neutrons) in the nucleus of one of its atoms.

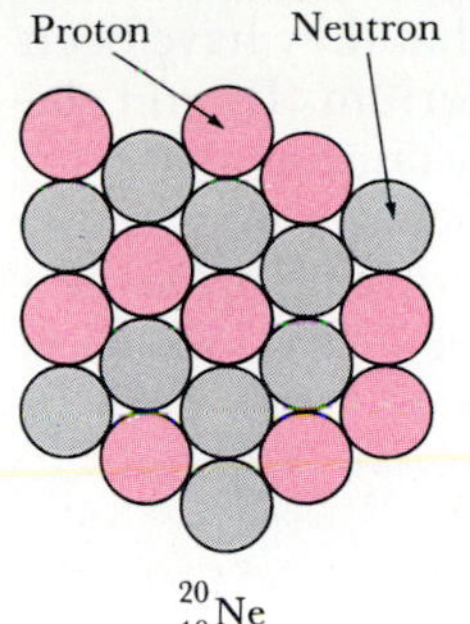

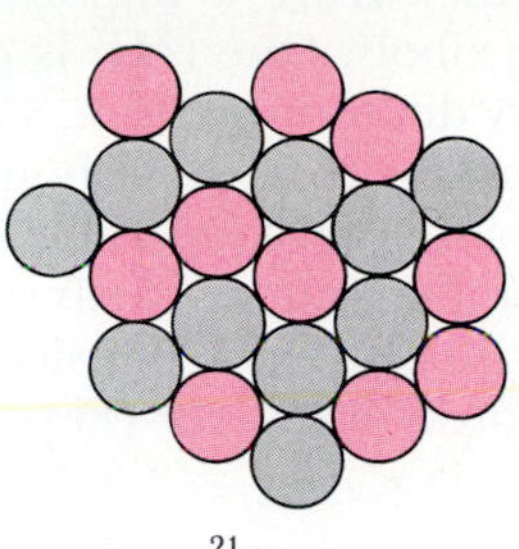

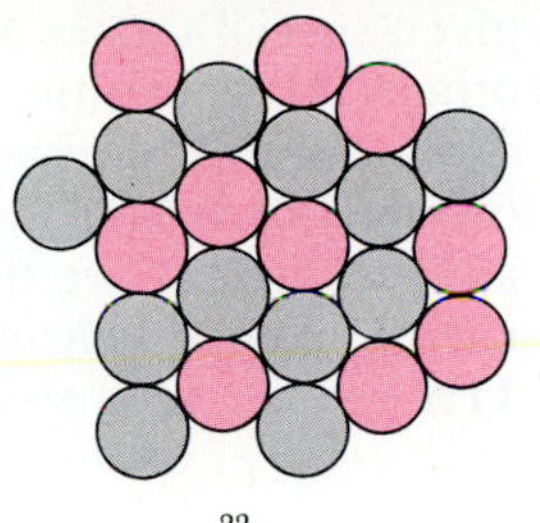

FIGURE 2.18 The nuclei of isotopes have the same numbers of protons (pink circles) but different numbers of neutrons (gray circles). These three diagrams represent three isotopes of neon.

It follows that *isotopes of a given element have the same atomic number but different mass numbers.*

An isotope of mass number A is about A times heavier than a hydrogen atom because a nucleon (a proton or a neutron) and an atom of the common isotope of hydrogen each have almost the same mass. (Recall that the neutral hydrogen atom has one proton and one electron and that the mass of the electron is very much smaller than the mass of the proton. Consequently the mass of a hydrogen atom is essentially identical to the mass of its single proton.) Helium, for example, has a mass number of 4, so it has four nucleons and a mass about four times greater than that of a hydrogen atom.

E A species with a given atomic number and mass number is called a *nuclide*. The term is not used until Chapter 22. Nuclides with the same mass number but different atomic number (such as ^{22}Ne and ^{22}Na) are called *isobars*.

If we know both the mass of an isotope and its atomic number Z, then we can determine the number of neutrons in its nucleus. Mass spectrometry data show that the masses of the isotopes of neon are about 20, 21, and 22 times greater than that of a hydrogen atom, so their mass numbers must be 20, 21, and 22, respectively. Because each nucleus has 10 protons ($Z = 10$ for neon), the isotopes must have 10, 11, and 12 neutrons, respectively (Fig. 2.18).

? What is the difference between an atom of uranium-235 and an atom of uranium-238?

An isotope is named by writing its mass number after the name of the element, as in neon-20, neon-21, and neon-22. By convention, we write the mass number as a superscript to the left of the chemical symbol: ^{20}Ne, ^{21}Ne, and ^{22}Ne. In general, we use the notation shown in the margin. The atomic number of the element may be written as a subscript on the left of the symbol when it is helpful to display Z too: $^{22}_{10}$Ne.

Mass number
↓
^{A}E
↑
Chemical symbol of element

Hydrogen has three isotopes (see Table 2.3). The most common (^{1}H) has no neutrons, so its nucleus is a lone proton. The other two isotopes

TABLE 2.3 Some common isotopes

Name	Atomic number Z	Number of neutrons	Mass number A	Mass, u	Abundance, %*	Symbol
hydrogen	1	0	1	1.008	99.985	^{1}H
deuterium	1	1	2	2.014	0.015	^{2}H or D
tritium	1	2	3	3.016	—†	^{3}H or T
carbon-12	6	6	12	12	98.90	^{12}C
carbon-13	6	7	13	13.003	1.10	^{13}C
oxygen-16	8	8	16	15.995	99.76	^{16}O

*Abundances vary slightly between sources of different samples.
†Radioactive, short-lived.

are less common but nevertheless are so important that they have been given special names and symbols. One (^2H) is deuterium (D) and the other (^3H) is tritium (T). A deuterium atom, with a nucleus that consists of one proton and one neutron, has about twice the mass of an ordinary hydrogen atom (more precisely, 1.998 times greater). Oxygen and deuterium are the components of "heavy water," D_2O, which is used in some nuclear reactors. It has a density of 1.11 g/mL at 25°C, 11% greater than that of ordinary water (Fig. 2.19).

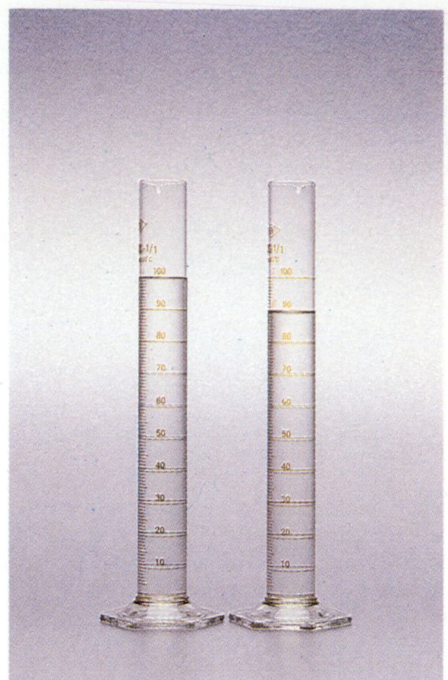

FIGURE 2.19 These two samples, both of which have a mass of 100 g, illustrate the difference in the densities of heavy water (D_2O) and ordinary water. The volume occupied by 100 g of heavy water (right) is 11% less than that occupied by the same mass of ordinary water (left).

D Freezing of H_2O, D_2O. TDC, 3–18.

▼ EXAMPLE 2.2 Using mass numbers

How many neutrons are present in atoms of (a) uranium-238; (b) ^{57}Fe?

STRATEGY The number identifying the isotope is the mass number A, or the total number of protons and neutrons. The atomic number Z (and hence the number of protons) is obtained by reference to the periodic table inside the front cover. The difference $A - Z$ is the number of neutrons.

SOLUTION (a) The mass number A of uranium-238 is 238, so the total number of nucleons is 238. For uranium, $Z = 92$: so its nucleus contains 92 protons. The number of neutrons is therefore $238 - 92 = 146$. (b) For ^{57}Fe (or iron-57), $Z = 26$ and $A = 57$. Therefore, the number of neutrons in its nucleus is $57 - 26 = 31$.

EXERCISE E2.2 How many neutrons are present in an atom of (a) nitrogen-14; (b) ^{35}Cl?

[*Answer*: (a) 7; (b) 18] ▲

Isotopic abundance. A sample of an element in naturally occurring material (rock, soil, air) is typically a mixture of isotopes. The **abundance** of an isotope is the percentage (in terms of the numbers of atoms) of that isotope present in a sample of the element:

$$\text{Abundance of X} = \frac{\text{number of atoms of isotope X}}{\text{total number of atoms in sample}} \times 100\%$$

We can find the number of atoms of each isotope present with a mass spectrometer, because the heights of the peaks on a mass spectrum (as explained in Box 2.2) are proportional to the numbers of atoms of each mass.

E The natural abundance of isotopes varies with the source, and the quoted values are typical values. The variation of isotopic abundance can give valuable information about the history of the source, including climate. For example, the abundance of deuterium in polar ice can be interpreted in terms of the climate at the time the rain clouds that were responsible for the precipitation were formed by evaporation from the oceans.

The **natural abundance** of an isotope is its abundance in a sample of naturally occurring material. The natural abundance of neon-20 is 91%, which means that 91 atoms out of 100 neon atoms in a sample of neon are neon-20. The natural abundance of uranium-235, the uranium isotope widely used as a nuclear fuel, is only 0.7%, so only 7 atoms out of every 1000 uranium atoms in a naturally occurring ore of uranium are uranium-235. The natural abundances of six common isotopes are listed in Table 2.3. As illustrated in Fig. 2.20, some elements have one or very few isotopes, but others are rich mixtures of several.

Atomic mass units. Because it is convenient to use simple numerical values in calculations, atomic masses are often reported as relative masses, that is, as multiples of the **atomic mass unit** (abbreviated as u; formerly denoted as amu).

One **atomic mass unit** is $\frac{1}{12}$ the mass of an atom of carbon-12.

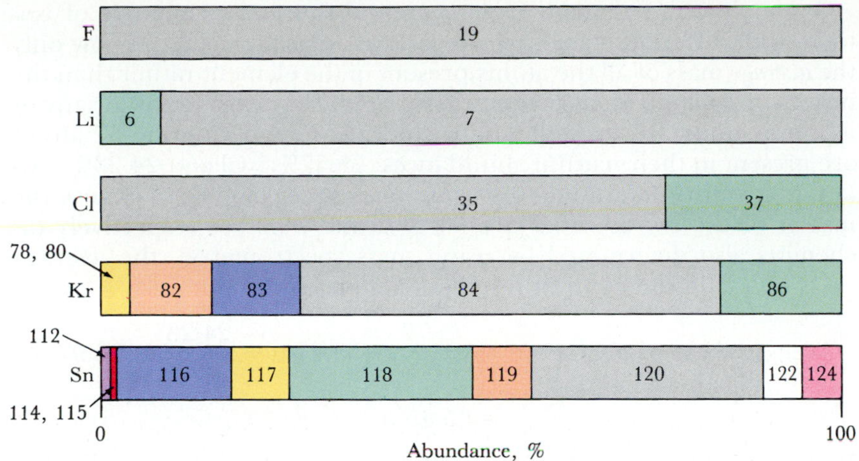

FIGURE 2.20 The abundances of the isotopes of five elements. The widths of the colored segments of the bars are proportional to the percentage of each isotope in a naturally occurring mixture. The numbers within the segments are the mass numbers of the isotopes.

Consequently, the mass of one atom of carbon-12 is exactly 12 u. Thus, an atom with a mass of 24 u (magnesium) is twice as heavy as an atom of carbon-12. (Why carbon-12 has been chosen as a reference mass and not hydrogen has a long history, which we will not discuss here.) Because a carbon atom has 12 nucleons in its nucleus (six protons and six neutrons), each nucleon has a mass of about 1 u. If an atom has 25 nucleons in its nucleus (that is, if its mass number is 25), we know at once that its mass is about 25 u. (We remarked earlier that electrons have such a small mass that they make a negligible contribution to the mass of an atom: the mass of an atom is almost the same as the mass of its nucleus.)

? What is the distinction between mass number and atomic mass?

U Exercise 2.71.

N Relative atomic mass scale. *J. Chem. Ed.*, **1988**, *65*, 16.

The mass of a single carbon-12 atom is 1.9926×10^{-23} g. Because we defined 1 u as $\frac{1}{12}$ that mass, we see that

$$1 \text{ u} = \frac{1.9926 \times 10^{-23} \text{ g}}{12} = 1.6605 \times 10^{-24} \text{ g}$$

The mass of any atom can be expressed in atomic mass units by using this relation as a conversion factor; typical atomic masses in atomic mass units are given in Table 2.3. They range from 1 u (for hydrogen) to about 250 u for the heaviest elements (uranium and the synthetic elements following it in the periodic table).

▼ EXAMPLE 2.3 Expressing mass in terms of atomic mass units

The mass of an atom of oxygen-16 is 15.995 u. What is its mass in grams?

STRATEGY We treat the calculation as a straightforward conversion from atomic mass units to grams, basing the conversion factor on the relation $1 \text{ u} = 1.6605 \times 10^{-24}$ g.

U Exercise 2.72.

SOLUTION

$$\text{Mass (g)} = 15.995 \text{ u} \times \frac{1.6605 \times 10^{-24} \text{ g}}{1 \text{ u}} = 2.6560 \times 10^{-23} \text{ g}$$

EXERCISE E2.3 The mass of a uranium-235 atom is 235.044 u. Express this mass in grams.

[*Answer*: 3.9029×10^{-22} g] ▲

U Exercises 2.67–69.

A sample of an element from a natural source is a mixture of isotopes with different masses. However, we normally need to know only the *average* mass of all the atoms present in the element rather than the masses of the individual isotopes. In a sample of chlorine gas, or any of its compounds, for example, both chlorine-35 and chlorine-37 atoms are present in their natural abundances: 75.77% ^{35}Cl and 24.23% ^{37}Cl (a fact chemists determined by using mass spectrometry). Because the masses of the two isotopes are 34.97 u and 36.97 u, respectively (as chemists also determined by using mass spectrometry), the average mass of a chlorine atom in a typical sample is

$$\text{Average mass of a Cl atom} = \frac{75.77}{100} \times 34.97\text{ u} + \frac{24.23}{100} \times 36.97\text{ u}$$
$$= 35.45\text{ u}$$

In other words, the average mass of the atoms of an element that has two or more isotopes is equal to the sum of the products of the natural abundance of each isotope and its atomic mass.

The number below each element's symbol in the periodic table inside the front cover is the average atomic mass of that element in atomic mass units. When we refer to the atomic mass of an element, we generally mean *average mass* rather than the mass of a single isotope.

2.5 MOLES AND MOLAR MASS

N How large is a mole? *J. Chem. Ed.*, **1990**, *67*, 481.

N How to visualize Avogadro's number. *J. Chem. Ed.*, **1989**, *66*, 762.

N Moles, pennies, and nickels. *J. Chem. Ed.*, **1989**, *66*, 249.

Even a small sample of hydrogen contains a vast number of atoms: 1 g of the gas consists of about 6×10^{23} atoms. When we pick up a penny, we are holding about 10^{22} atoms of various kinds—more atoms than the number of stars in the visible universe. A unit has been invented to express numbers as big as this and to make calculating the numbers of atoms in samples much more convenient. It is the mole, one of the most important units in chemistry.

The mole. The mole (abbreviated mol) is the unit chemists use for expressing large numbers of atoms, ions, and molecules. The name comes—somewhat ironically—from the Latin word for "massive heap."

One **mole** is the number of atoms in exactly 12 g of carbon-12.

N When using the term *mole*, it is best to specify the entities to avoid any ambiguity. Thus, 1 mol of oxygen is ambiguous, but 1 mol of oxygen atoms (1 mol O) or 1 mol of oxygen molecules (1 mol O_2) are not.

Although 1 mol is defined in terms of carbon atoms, the unit applies to any object, just as 1 dozen means 12 of anything. As noted before, the mass of one ^{12}C atom is exactly 12 u; by writing 1 ^{12}C atom = 12 u and $1\text{ u} = 1.6605 \times 10^{-24}$ g we can calculate the number of atoms in exactly 12 g of carbon-12:

$$\text{Number of } ^{12}\text{C atoms} = 12\text{ g } ^{12}\text{C} \times \frac{1\text{ u}}{1.6605 \times 10^{-24}\text{ g}} \times \frac{1\ ^{12}\text{C atom}}{12\text{ u}}$$
$$= 6.022 \times 10^{23}\ ^{12}\text{C atoms}$$

It follows that 1 mol of atoms (of any element) is 6.022×10^{23} atoms of the element. The conclusion also applies to 1 mol of any objects—atoms, ions, or molecules: 1 mol of objects always means 6.022×10^{23} of those objects.

We know that units such as grams and meters are used to express properties such as mass and length, but what property does the unit *mole* express? The official recommendation is that the mole expresses the "amount" of substance present in a sample. Thus, 1 mol of hydrogen (6.022×10^{23} hydrogen atoms) is the amount of hydrogen atoms (expressed in moles) in a sample. Chemists, however, have been reluctant to adopt the term *amount* (partly because it is already used widely in casual conversation), and you will almost always hear them speaking of "the number of moles" rather than the "amount."

The number 6.022×10^{23} is called **Avogadro's number** N_A, in honor of the nineteenth-century Italian scientist Amedeo Avogadro (Fig. 2.21), who argued in favor of the existence of atoms. Several awe-inspiring illustrations can be devised to help visualize the enormous size of Avogadro's number. For instance, an Avogadro's number of soft-drink cans would cover the surface of the Earth to a height of over 200 mi.

Avogadro's number is very useful for converting between the number of atoms or molecules and the amount (in moles), because we can construct a conversion factor based on the relation

$$1 \text{ mol} = 6.022 \times 10^{23} \text{ objects}$$

FIGURE 2.21 Lorenzo Romano Amedeo Carlo Avogadro, count of Quaregna and Cerreto (1776–1856).

E Strictly speaking, N_A is a *constant* not a number: it is a multiple of the unit mol^{-1}. Thus, Avogadro's constant is the number of entities per mole, or $6.022 \times 10^{23}\ \text{mol}^{-1}$.

D Relative weights [masses] of atoms analogy. TDC, 205.

EXAMPLE 2.4 Converting number of atoms to amount in moles

Suppose that in an electrolysis experiment, 1.29×10^{24} H atoms were produced (as hydrogen gas). Express this number as an amount.

STRATEGY Because the number of atoms produced is greater than Avogadro's number, we anticipate that more than 1 mol of atoms was produced. For the actual calculation, set up a conversion factor using the relation given above.

SOLUTION

$$\underset{\substack{\uparrow \\ \text{Information} \\ \text{required}}}{\text{Amount of H (mol)}} = \underset{\substack{\uparrow \\ \text{Information} \\ \text{given}}}{1.29 \times 10^{24}\ \text{H atoms}} \times \underset{\substack{\uparrow \\ \text{Conversion factor} \\ \text{from atoms to moles}}}{\frac{1 \text{ mol H}}{6.022 \times 10^{23}\ \text{H atoms}}}$$

$$= 2.14 \text{ mol H}$$

Notice how much simpler it is to report that 2.14 mol H was formed rather than the actual number of atoms formed.

EXERCISE E2.4 In the same experiment, 6.43×10^{23} atoms oxygen were produced (as oxygen gas). Express this number as an amount in moles.

[*Answer*: 1.07 mol O]

U Exercise 2.70.

Molar mass. One of the most important measures of the mass of the atoms of an element (and, as we shall see in Sections 2.6 and 2.7, of the molecules they form) is the molar mass:

> The **molar mass** of an element is the mass per mole of atoms of the element.

? What is the distinction between mass number and molar mass?

E One exception is the use of molar in *molar concentration* (Section 4.4), where it is an adjective implying that the concentration is expressed in moles per liter. Another exception, which is not used in this text, is *molar conductivity*, which is the electrical conductivity of a solution divided by the molar concentration of the solute in the solution.

Thus, the molar mass of carbon is the mass per mole of carbon atoms, and the molar mass of uranium is the mass per mole of uranium atoms. (The word *molar* is often, but not exclusively, used to mean "per mole.") We shall always consider samples of naturally occurring material, so the molar mass of carbon is actually the *average* molar mass of a sample containing carbon-12 and carbon-13 in their natural abundances. The average molar mass of an element is still widely called the *atomic weight.* You should be aware of the term, but we shall use it only rarely in this text.

The molar mass of a sample of natural carbon is found by converting the average mass per atom to the mass per mole:

$$\text{Molar mass (mass per mole)} = \text{average mass per atom} \times \text{number of atoms per mole } (N_A)$$

For example, we know that the average mass per chlorine atom is 35.45 u per Cl atom, so the mass (in grams) per mole is

$$\underset{\substack{\uparrow\\ \text{Information required}}}{\text{Molar mass (g/mol)}} = \underset{\text{Information given}}{\frac{35.45\ \text{u}}{1\ \text{Cl atom}}} \times \underset{\text{Conversion factor from u to g}}{\frac{1.6605 \times 10^{-24}\ \text{g}}{1\ \text{u}}} \times \underset{\text{Conversion factor from 1/atom to 1/mol}}{\frac{6.022 \times 10^{23}\ \text{Cl atoms}}{1\ \text{mol Cl}}}$$

$$= 35.45\ \text{g/mol Cl}$$

We often see this result reported as "the atomic weight of chlorine is 35.45," sometimes with the units grams per mole (g/mol) included and sometimes omitted.

? What is the distinction between molar mass and atomic mass? What units are used for each one?

An important point is that the value for molar mass in grams per mole is numerically identical to that for the average atomic mass in atomic mass units (u). This relation makes it very easy to go between molar masses and the masses of the individual atoms. Thus, if we are told that the molar mass of hydrogen is 1.008 g/mol, then we know at once that the average mass of a single hydrogen atom in a sample of the element is 1.008 u.

The average molar masses of all the elements (their "atomic weights") are given in the tables inside the back cover. They are so important that the values are also included in the periodic table inside the front cover; so, almost anywhere we look we see the name or symbol of the element accompanied by its molar mass.

N The expression 1 mol Cl = 35.45 g Cl is the first of several cases where the equals sign is used to mean "is equivalent to." Whereas 1 in = 2.54 cm is a true equality (the two lengths are the same), the relation 1 mol Cl = 35.45 g Cl expresses the equivalent of a certain amount of atoms and a certain mass of matter. We have not thought it appropriate to burden the student with this distinction, or to use a different symbol, but it may be helpful to keep the distinction in mind. (Other examples will arise later.)

The number of moles of atoms in a sample. The molar mass is important because it provides a conversion factor between the mass of a sample of an element and the moles of atoms present. For chlorine, with molar mass 35.45 g/mol, we can write

$$1\ \text{mol Cl} = 35.45\ \text{g Cl}$$

and use this relation to construct a conversion factor. Suppose we want to know how many moles of Cl atoms are present in 15.0 g of chlorine.

We write

$$\text{Amount of Cl (mol)} = 15.0 \text{ g Cl} \times \frac{1 \text{ mol Cl}}{35.45 \text{ g Cl}} = 0.423 \text{ mol Cl}$$

↑ Information required ↑ Information given ↑ Conversion from g to mol

Measuring out moles of an element. Another major application of molar mass is in the preparation of a sample of an element (and, as we shall see later, of compounds) that contains a specified number of atoms. For example, we may want a sample of gold that contains a definite number of gold atoms. Because we know (from the periodic table or from inside the back cover) that the molar mass of gold is 196.97 g/mol, we can write

$$1 \text{ mol Au} = 196.97 \text{ g Au}$$

Therefore, to obtain 1.0000 mol Au (corresponding to 6.022×10^{23} gold atoms), we measure out 196.97 g of gold. An analogous statement can be made for any element; for example:

N Gram formula weights [formula unit molar masses] and fruit salad. *J. Chem. Ed.*, **1985**, *62*, 61.

- To obtain 1.000 mol Cu (molar mass, 63.54 g/mol), we measure out 63.54 g of copper.
- To obtain 1.0000 mol Hg (molar mass, 200.59 g/mol), we measure out 200.59 g of mercury.

And so on. The "massive heaps" corresponding to 1 mol of atoms of some elements are shown in Fig. 2.22; they have all been obtained by measuring out a mass of each element in grams equal to its molar mass given in the periodic table.

If we want a sample that contains an amount other than exactly 1 mol of atoms, then we use the molar mass as a conversion factor to determine the mass of the element that we should measure out. This calculation is illustrated in the following example.

FIGURE 2.22 Each sample consists of 1 mol of atoms of the element. Clockwise from the upper right are 32 g of sulfur, 201 g of mercury, 207 g of lead, 64 g of copper, and 12 g of carbon.

▼ **EXAMPLE 2.5 Measuring out a specified amount of a substance**

What mass of iron is needed to obtain a sample that contains 1.345 mol of iron atoms?

STRATEGY We use the molar mass of iron (from the periodic table inside the front cover) as a factor to convert moles of Fe atoms to mass of iron.

SOLUTION The molar mass of iron (55.85 g/mol) is written in the form

$$1 \text{ mol Fe} = 55.85 \text{ g Fe}$$

and then expressed as a conversion factor:

$$\text{Mass of iron (g)} = 1.345 \cancel{\text{ mol Fe}} \times \frac{55.85 \text{ g Fe}}{1 \cancel{\text{ mol Fe}}} = 75.12 \text{ g Fe}$$

That is, if we measure out 75.12 g of iron, then we shall have a sample that contains 1.345 mol of iron atoms.

EXERCISE E2.5 Which mass of tin contains 0.775 mol of tin atoms?
[*Answer*: 92.0 g Sn] ▲

U Exercises 2.73–77 and 2.81.

? Where in chemistry might kilomole quantities of reactants be used?

When we are dealing with very small amounts of matter, it is useful to use one of the SI prefixes introduced in Table 1.6. For example, 10 mg of magnesium corresponds to 4.1×10^{-4} mol Mg, which we can write in millimoles (mmol) as 0.41 mmol Mg, with

$$1 \text{ mmol} = 10^{-3} \text{ mol}$$

When chemists, and particularly biochemists, deal with very tiny amounts of expensive or rare material, they use micromoles (μmol) of material, where

$$1 \text{ } \mu\text{mol} = 10^{-6} \text{ mol}$$

D Mixtures and compounds. TDC, 1–5.

COMPOUNDS

Now we turn our attention from elements to compounds. In Chapter 1 we identified a compound as a substance containing more than one element and having a definite composition. We can now express the definition in terms of atoms:

> A **compound** is a substance consisting of atoms of two or more elements in a definite ratio.

Water, for instance, is invariably made up of hydrogen and oxygen atoms, with two hydrogen atoms for each oxygen atom. It is therefore a compound of hydrogen and oxygen. Common table salt, sodium chloride, is a compound of the elements sodium and chlorine in which there is invariably one sodium atom for each chlorine atom. The different elements in a compound are not just mixed together; their atoms are joined—the technical term is *bonded*—to one another in a specific way (Fig. 2.23). We shall explore in Chapter 8 how that bonding takes place.

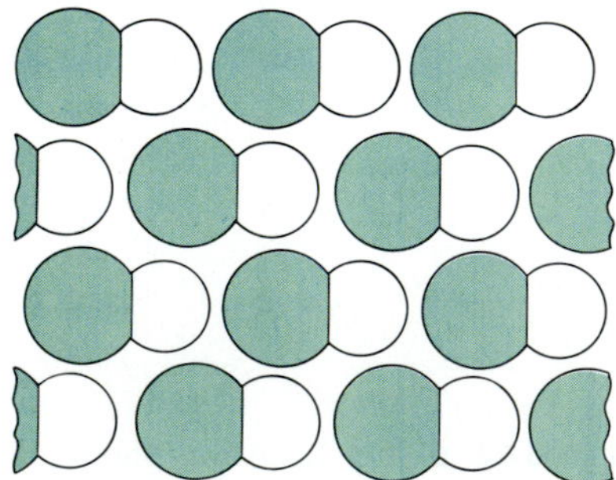

FIGURE 2.23 A compound consists of different atoms bonded together in a strict ratio: here each molecule of a compound consists of one white atom bonded to one green atom.

All we need to know at this stage is that there are two kinds of compounds, one type consisting of molecules and the other of ions:

A **molecule** is a definite and distinct, electrically neutral group of bonded atoms; it is the smallest particle of a compound that possesses the chemical properties of the compound.

An **ion** is a positively or negatively charged atom or group of atoms.

(The name *ion* comes from the Greek word meaning "go"—because of the way charged particles go either toward or away from a charged electrode.) We call a compound **molecular** if it consists of molecules and **ionic** if it consists of ions. We shall discuss the two classes of compounds separately. In general, *molecular compounds are formed by a reaction between nonmetallic elements or between nonmetallic elements and metalloids; one of the ions in an ionic compound is usually an ion of a metallic element.* Note, however, that some compounds of metals are molecular and some compounds that consist entirely of nonmetallic elements are ionic. Molecular compounds may be solid, liquid, or gaseous at room temperature; ionic compounds are solids at room temperature and usually have high melting points.

? Is there any way of telling from a compound's formula whether it is molecular or ionic?

D Reaction of zinc and sulfur. BZS1, 1.21.

N The structures shown in the margin below (and subsequently) are intended to give an early familiarity with the constitution and shapes of molecules. The atoms are color coded consistently, but we always label one atom of each element as a reminder. The double lines (and triple lines in some structures) in structure **3** are, of course, double (and triple) bonds. This concept is developed in Section 8.3; at this stage, students could be told that the double lines indicate the bonding pattern, which will be explained later.

2.6 MOLECULES AND MOLECULAR COMPOUNDS

Examples of molecular compounds are water, methane, ammonia, carbon dioxide, cane sugar (sucrose), benzene, and aspirin. We can speak, therefore, of molecules of water, methane, and carbon dioxide. These three molecules are shown in the models in the margin, where the black spheres represent carbon atoms; the white spheres, hydrogen atoms; and the red spheres, oxygen atoms. A water molecule (**1**) consists of one oxygen atom bonded to two hydrogen atoms, a methane molecule (**2**) consists of one carbon atom bonded to four hydrogen atoms, and a carbon dioxide molecule (**3**) consists of a carbon atom bonded to two oxygen atoms. (The double lines between atoms will be explained later.)

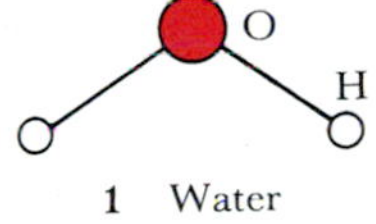

1 Water

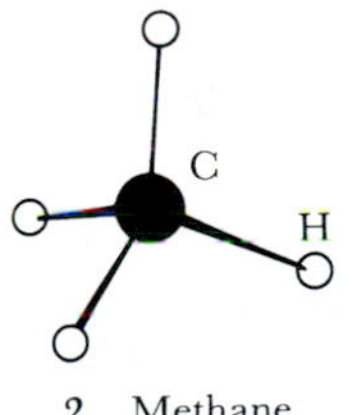

2 Methane

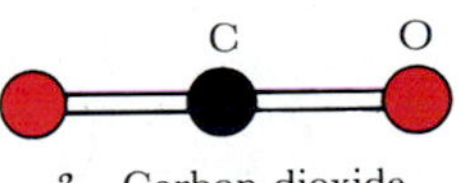

3 Carbon dioxide

Empirical formulas. The mass percentage composition of a compound can be used to derive the ratios of the numbers of atoms of each element in a compound, but not the actual numbers. That is, it tells us the empirical formula of the compound:

An **empirical formula** shows the *relative* numbers of atoms of each element in a compound in terms of the chemical symbols of the elements.

The relative numbers of atoms of each element in the compound are written as subscripts to the chemical symbols of the elements. For example, the empirical formula of phosphorus(V) oxide is P_2O_5, the subscripts indicating that the atoms are present in the ratio 2:5. An empirical formula is always written in terms of whole numbers, so we write P_2O_5, not $PO_{2.5}$.

▼ **EXAMPLE 2.6 Calculating the empirical formula from the mass percentage composition**

A laboratory analysis of vitamin C showed that its mass percentage composition is 40.9% C, 4.57% H, and 54.5% O. What is its empirical formula?

N Perplexed by students puzzled by percent? *J. Chem. Ed.* **1979**, *56*, 45.

STRATEGY The meaning of "mass percentage composition" was explained in Section 1.8. The definition implies that the mass of each element present in a sample of mass 100 g is equal to the mass percentage composition expressed as grams (so 40.9% C implies the presence of 40.9 g of carbon in 100 g of compound). We then calculate the amount of atoms of each element (expressed as moles) in the sample by using the mass percentage composition and the molar masses of the elements. We write the empirical formula by expressing the ratios of the numbers of moles of the components, using the smallest whole numbers we can.

SOLUTION The amount (in moles) of each element present in a 100-g sample is

U Exercise 2.82.

$$\text{Amount of C (mol)} = 40.9\ \cancel{\text{g C}} \times \frac{1\ \text{mol C}}{12.01\ \cancel{\text{g C}}} = 3.41\ \text{mol C}$$

$$\text{Amount of H (mol)} = 4.57\ \cancel{\text{g H}} \times \frac{1\ \text{mol H}}{1.008\ \cancel{\text{g H}}} = 4.53\ \text{mol H}$$

$$\text{Amount of O (mol)} = 54.5\ \cancel{\text{g O}} \times \frac{1\ \text{mol O}}{16.00\ \cancel{\text{g O}}} = 3.41\ \text{mol O}$$

We know that the number of atoms is directly proportional to the number of moles of each element, so the atoms of the three elements are present in the ratio C : H : O = 3.41:4.53:3.41. Dividing through by the smallest number (3.41) gives C : H : O = 1.00:1.33:1.00. Multiplication by 3 to produce whole numbers of atoms gives C : H : O = 3.00:4.00:3.00. Hence, the empirical formula of vitamin C is $C_3H_4O_3$.

EXERCISE E2.6 The mass percentage composition of the compound thionyl difluoride is 18.59% O, 37.25% S, and 44.16% F. Calculate its empirical formula.

[*Answer*: OSF_2]

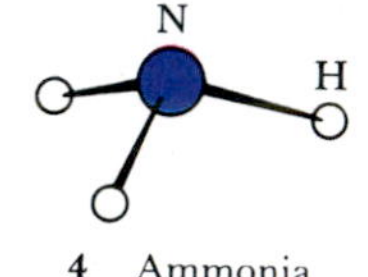

4 Ammonia

In Section 4.3 we shall meet a laboratory technique for determining empirical formulas by analyzing the products formed when a sample burns in oxygen.

Molecular formulas. The number of atoms of each element in a molecule is given as a subscript in the molecular formula of the compound:

> A **molecular formula** shows the actual numbers of atoms of each element present in a molecule.

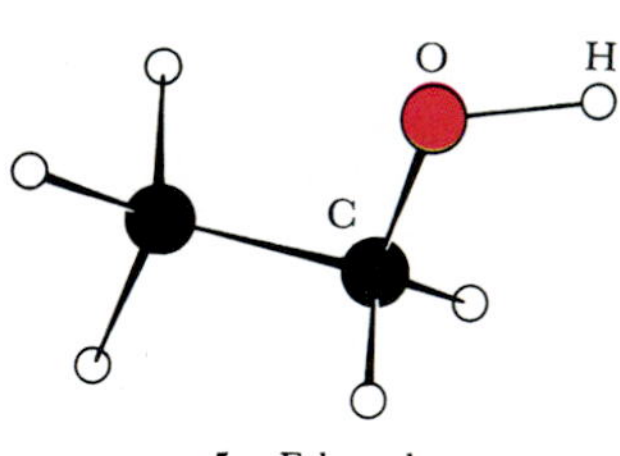

5 Ethanol

6 Glucose

The molecular formulas of water, methane, and carbon dioxide are H_2O, CH_4, and CO_2 respectively. An ammonia molecule (**4**) consists of one nitrogen atom (colored blue in the diagram) and three hydrogen atoms, so its molecular formula is NH_3. The molecular formula of ethanol is C_2H_6O; the formula shows that ethanol consists of two carbon atoms, six hydrogen atoms, and one oxygen atom (**5**). Glucose (**6**) is a much more complicated molecule. Although its *empirical formula* is CH_2O, showing that the atoms of carbon, hydrogen, and oxygen are present in the ratio 1:2:1, its *molecular formula* $C_6H_{12}O_6$ shows that each molecule consists of 6 C atoms, 12 H atoms, and 6 O atoms. The molecular formula of a compound is written in the same way as its empirical formula is written; however, there is little likelihood of confusing them because, unless we explicitly state that we are writing the empirical formula, the formulas given throughout the text will be molecular formulas.

A **diatomic molecule** is a molecule that consists of only two atoms. The lethal gas carbon monoxide (**7**), which is produced when fuels burn in a limited supply of air, consists of diatomic molecules of formula CO: each molecule has one carbon atom and one oxygen atom. A **polyatomic molecule** is one that consists of more than two atoms (*poly* comes from the Greek word for "many"). Carbon dioxide (**3**), the gas produced when fuels burn in plenty of air, consists of polyatomic molecules (in this case, triatomic molecules) of formula CO_2. The female sex hormone estradiol (**8**) is much larger than CO_2: its molecular formula is $C_{18}H_{24}O_2$, which shows that it consists of 18 C atoms, 24 H atoms, and 2 O atoms.

7 Carbon monoxide

8 Estradiol

Elements—not only compounds—may be molecular, but each molecule consists of only one type of atom. Hydrogen gas, for example, consists of diatomic molecules of formula H_2. The nitrogen and oxygen of the air exist as diatomic molecules, so their molecular formulas are N_2 and O_2, respectively. Elemental phosphorus exists in several forms, one of which is a waxlike solid that is known as white phosphorus and consists of P_4 molecules (**9**). Sulfur is made up of S_8 molecules in which eight S atoms form a crownlike ring (**10**).

The molar masses of compounds. The molar mass of a molecular compound is the mass per mole of the molecules. For example, the molar mass of water is the mass per mole of H_2O. Because water molecules each consist of three atoms, 1 mol of H_2O consists of 2 mol of H and 1 mol of O; so the molar mass of H_2O is the sum of the molar masses of its components:

$$\text{Molar mass of } H_2O = \text{molar mass of } (2 \times \text{H} + \text{O})$$
$$= (2 \times 1.008 + 16.00)\ \text{g/mol} = 18.02\ \text{g/mol}$$

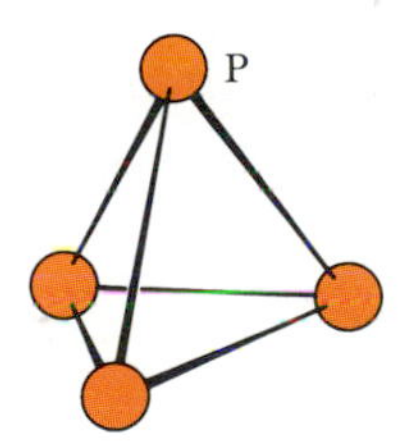

9 Phosphorus

For historical reasons, the molar mass of a molecular compound is still widely called the *molecular weight* of the compound. Thus, you will still often see or hear the expression "the molecular weight of water is 18.02."

The importance of the property molar mass becomes apparent when we measure out 18.02 g of water: we know that we have a sample containing 1.000 mol of H_2O (or 6.022×10^{23} H_2O molecules). It is often much more important for a chemist to know the *amount* of substance (expressed in moles of molecules or atoms) in a sample than to know the *mass* of the substance present. A principal reason for measuring the mass of a sample, in fact, is to determine the number of moles of atoms, ions, or molecules it contains.

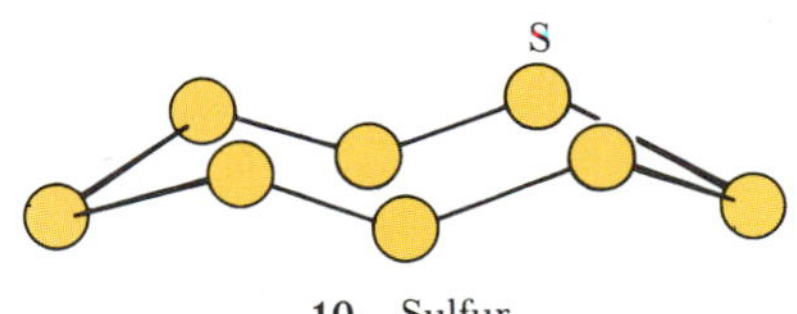

10 Sulfur

The determination of a molecular formula from an empirical formula. Earlier we used the mass percentage composition of vitamin C to deduce that its *empirical* formula is $C_3H_4O_3$, but we could not deduce its molecular formula. The mass percentage composition tells us only that the C, H, and O atoms are present in the ratio 3:4:3. The empirical formula $C_3H_4O_3$ is consistent with the molecular formulas $C_3H_4O_3$ and $C_6H_8O_6$, and any other formula containing whole-number multiples of the empirical formula. To decide on the molecular formula of a compound, we must measure the compound's molar mass and see how many empirical formula units we need to account for it. The most precise method for measuring the molar mass of a simple molecule (one containing fewer than about 30 atoms) is mass spectrometry (see

E The simpler procedures mentioned here include the measurement of the densities of gases (Section 5.3) and the techniques based on the colligative properties of solutions (Sections 11.6 and 11.7).

Box 2.2), in which molecules of the compound rather than atoms are used to form the beam of ions. Other methods that use simpler apparatus are also available. They are sometimes more suitable than mass spectrometry, especially if the compound is not volatile. Some are described in Chapters 5 and 11.

Mass spectrometry shows that the molar mass of ethanol is 46.07 g/mol. The empirical formula of ethanol is C_2H_6O, and the molar mass corresponding to that formula is

$$\begin{aligned}\text{Molar mass of } C_2H_6O &= \text{molar mass of } (2 \times C + 6 \times H + O)\\ &= (2 \times 12.01 + 6 \times 1.008 + 16.00)\ \text{g/mol}\\ &= 46.07\ \text{g/mol}\end{aligned}$$

This value is the same as the measured value, so we conclude that the molecular formula of ethanol is also C_2H_6O and that each molecule contains two C atoms, six H atoms, and one O atom. When we calculate the molar mass from the empirical formula of vitamin C, we find that

U Exercise 2.84.

$$\begin{aligned}\text{Molar mass of } C_3H_4O_3 &= \text{molar mass of } (3 \times C + 4 \times H + 3 \times O)\\ &= 88.06\ \text{g/mol}\end{aligned}$$

whereas its measured value is 176.12 g/mol, a value twice as large. Hence, the molecular formula of vitamin C must be $2 \times C_3H_4O_3$, or $C_6H_8O_6$.

Using the molar masses of compounds. Two points about the molar masses of compounds follow directly from our discussion of the molar masses of elements. The first is that the numerical value of the molar mass of a compound in grams per mole (g/mol) is equal to the numerical value of the mass of one molecule in atomic mass units (u). For example, the molar mass of water is 18.02 g/mol, so the mass of one H_2O molecule is 18.02 u. In a few applications in chemistry we need to know the masses of individual molecules, and this relation gives us a very easy way of determining them from the information in the periodic table.

The second point is much more important. As in the case of atoms, we can use the molar mass of a compound to derive a conversion factor to calculate the amount (in moles) of a compound from its mass, or to calculate the mass we need to measure out to obtain the desired amount of compound (Fig. 2.24). For example, the molar mass of water (18.02 g/mol) implies that

$$1\ \text{mol}\ H_2O = 18.02\ \text{g}\ H_2O$$

and this relation can be used to form a conversion factor.

▼ EXAMPLE 2.7 Using molar mass to calculate the amount of molecules in a sample

The molecular formula of ethanol is C_2H_6O. How many moles of ethanol molecules are present in a sample of mass 100 g (3 sf)?

STRATEGY First, we calculate the molar mass of the compound (from the sum of the molar masses of the atoms present) and write it in the form

$$x\ \text{g}\ C_2H_6O = 1\ \text{mol}\ C_2H_6O$$

FIGURE 2.24 Each sample contains 1 mol of molecules of a molecular compound. From left to right are 18 g of water (H_2O), 46 g of ethanol (C_2H_6O), 180 g of glucose ($C_6H_{12}O_6$), and 342 g of sucrose ($C_{12}H_{22}O_{11}$).

Then we use this relation to set up a conversion factor from grams of compound to moles of compound.

SOLUTION The molar mass of ethanol is

$$\begin{aligned}\text{Molar mass of } C_2H_6O &= \text{molar mass of } (2 \times C + 6 \times H + O) \\ &= (2 \times 12.01 + 6 \times 1.008 + 16.00)\ \text{g/mol} \\ &= 46.07\ \text{g/mol}\end{aligned}$$

Hence,

$$46.07\ \text{g } C_2H_6O = 1\ \text{mol } C_2H_6O$$

The conversion from mass to moles is therefore

$$\begin{aligned}\text{Amount of } C_2H_6O\ (\text{mol}) &= 100\ \text{g } C_2H_6O \times \frac{1\ \text{mol } C_2H_6O}{46.07\ \text{g } C_2H_6O} \\ &= 2.17\ \text{mol } C_2H_6O\end{aligned}$$

EXERCISE E2.7 Calculate the amount of sucrose ($C_{12}H_{22}O_{11}$) in a 3.0-g cube of sugar.

[*Answer*: 8.8×10^{-3} mol $C_{12}H_{22}O_{11}$]

U Exercise 2.78.

T OHT: Fig. 2.25.

2.7 IONS AND IONIC COMPOUNDS

Ionic compounds consist of positive and negative ions held together by the attraction between their opposite charges. An example is sodium chloride, which consists of sodium ions (positively charged sodium atoms, denoted Na^+) and chloride ions (negatively charged chlorine atoms, denoted Cl^-). Each crystal of the compound is a huge collection of Na^+ and Cl^- ions (Fig. 2.25). When you take a pinch of salt, you are picking up crystals consisting of huge numbers of ions held together in this way.

In the presence of charged electrodes, oppositely charged ions move in opposite directions. They are called either cations or anions, depending on their charge:

A **cation** is a positively charged ion.

An **anion** is a negatively charged ion.

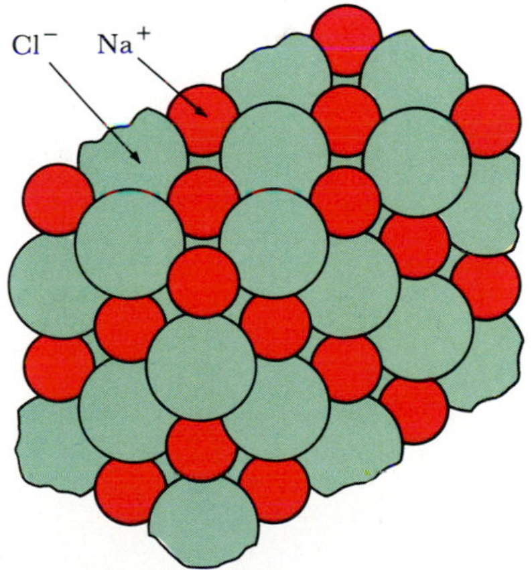

FIGURE 2.25 In solid sodium chloride, an ionic compound, the ions are held together by the attraction of their opposite charges. The red spheres represent sodium ions, the green spheres chloride ions. The ions become free to move when the solid dissolves in water.

FIGURE 2.26 A cation of an element M is formed by the loss of one or more electrons. An anion of an element X is formed by the gain of electrons.

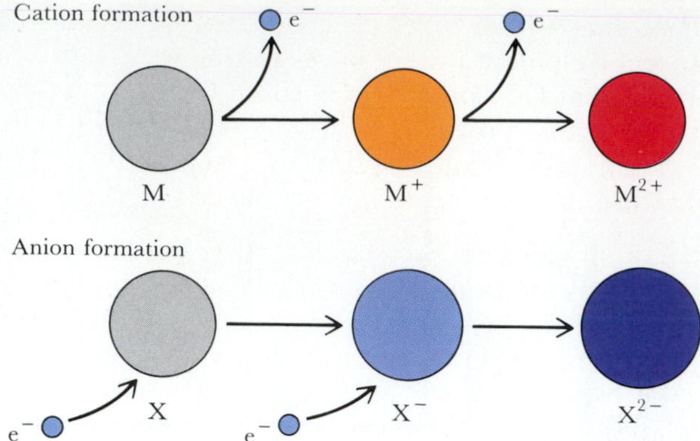

(The prefixes *cat-* and *an-* come from the Greek words for "down" and "up," which reflect two opposite directions of travel.) In sodium chloride, the sodium ions are the cations and the chloride ions are the anions.

Cations. It is easy to explain the existence of ions in terms of the nuclear atom. Because an electron has one unit of negative charge, removing one electron from a neutral atom leaves behind a cation with one unit of positive charge; the charge of the remaining electrons no longer cancels the positive charge of the nucleus (Fig. 2.26). This ability to lose electrons and form cations is characteristic of metal atoms. A sodium cation is a sodium atom that has lost one electron; it is denoted Na^+. Each electron that an atom loses increases its overall positive charge by one unit. When a calcium atom loses two electrons, it becomes the calcium ion, Ca^{2+} (read "calcium two-plus"), an ion of charge number +2. In general, the **charge number** of an ion is the number of positive or negative charges it carries. When an aluminum atom loses three electrons, it becomes the aluminum ion, Al^{3+}, an ion of charge number +3.

N The charge number of an ion is a signed quantity: cations have positive charge numbers, anions have negative charge numbers.

T OHT: Tables 2.4 and 2.6.

N The explanation of why some elements can form a variety of cations is given in Section 7.9 after concepts relating to atomic structure have been introduced.

Cations formed by various metallic elements are listed in Table 2.4. There is a very useful rule for predicting charge number: in Groups I, II, and III the group number is the *maximum* (and sometimes the only) positive charge of the cations. Thus, indium is in Group III and forms In^{3+} (and also In^+); barium is in Group II and forms only Ba^{2+}.

Table 2.4 shows that some elements form two or more ions with different charge numbers. An iron atom, for instance, can lose two

TABLE 2.4 Cations formed by some elements*

Group I	Group II	Transition metals			Group III	Group IV
Li^+	Be^{2+}					
Na^+	Mg^{2+}				Al^{3+}	
K^+	Ca^{2+}	Fe^{2+}, Fe^{3+}	Cu^+, Cu^{2+}	Zn^{2+}	Ga^{3+}	
Rb^+	Sr^{2+}		Ag^+	Cd^{2+}	In^+, In^{3+}	Sn^{2+}, Sn^{4+}
Cs^+	Ba^{2+}		Au^+, Au^{3+}	Hg_2^{2+}, Hg^{2+}	Tl^+, Tl^{3+}	Pb^{2+}, Pb^{4+}

*The most important polyatomic cation is NH_4^+, the ammonium ion.

electrons to become Fe^{2+} or three electrons to become Fe^{3+} (Fig. 2.27). Copper is another element that behaves in this way: it can lose either one electron to form Cu^+ or two to become Cu^{2+}. This variability of charge number is most pronounced among the transition metals, but the heavier elements in Groups III and IV also show it (as we noted earlier for indium).

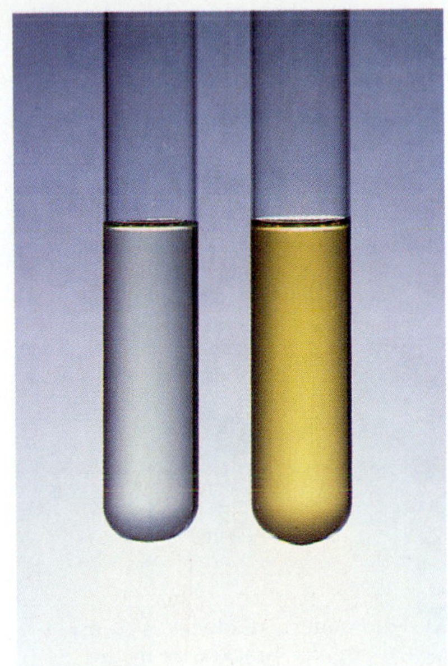

FIGURE 2.27 Iron is an example of an element that can form ions with different charge numbers. Solutions containing Fe^{2+} are usually pale green and solutions containing Fe^{3+} are usually yellow-brown.

Anions. If an atom gains an electron, then it acquires a single negative charge and becomes an anion. This ability to gain electrons is characteristic of atoms of the nonmetallic elements. Each electron that an atom gains increases its overall negative charge by one unit. A chloride ion (the names of ions are discussed in Section 2.8) is a chlorine atom that has gained an additional electron; it is denoted Cl^-. An oxygen atom gains two electrons to become O^{2-}, an ion with charge number -2. A nitrogen atom gains three electrons to become N^{3-}, an ion with charge number -3.

Anions formed by various nonmetallic elements are listed in Table 2.5. There is also a rule for predicting the charge on the ions of elements on the right side of the periodic table: the maximum charge number of the monatomic anion an element forms is equal to the group number minus 8 (why this rule is successful is explained in Chapter 7). Thus, oxygen (in Group VI) forms the oxide ion, O^{2-}, with charge number $6 - 8 = -2$. Phosphorus (in Group V) forms the phosphide ion, P^{3-}, with charge number $5 - 8 = -3$.

▼ EXAMPLE 2.8 Predicting the anion an element is likely to form

What monatomic anion would you expect selenium to form? How many electrons will the anion have?

STRATEGY To determine the charge number of the anion, identify the group to which the element belongs, and then subtract 8. To count the number of electrons, note the atomic number Z of the element (which gives the number of electrons in the neutral atom), and then add the number of electrons needed to form the anion.

SOLUTION Selenium (Se) belongs to Group VI. Because $6 - 8 = -2$, it forms the anion Se^{2-}. The atomic number of selenium is 34, so the neutral atom has 34 electrons. The ion Se^{2-} therefore has $34 + 2 = 36$ electrons.

EXERCISE E2.8 What anion is iodine likely to form? How many electrons will the anion possess?

[*Answer*: I^-; 54] ▲

T OHT: Tables 2.5 and 2.7.

D Color changes in Fe(II) and Fe(III) solutions. CD1, 88.

TABLE 2.5 Some common monatomic and diatomic anions*

Group V	Group VI	Group VII
nitride N^{3-}	oxide O^{2-}	fluoride F^-
phosphide P^{3-}	sulfide S^{2-}	chloride Cl^-
cyanide CN^-†	hydroxide OH^-	bromide Br^-
		iodide I^-

*A more complete list of anions is given in Appendix 3A.
†The cyanide ion, CN^-, is formed from elements of Groups IV and V.

E Triatomic anions include the azide ion, N_3^-, the ozonide ion, O_3^-, and the thiocyanate ion, SCN^-. There are many common triatomic oxoanions, such as the nitrite ion, NO_2^-, and the chlorite ion, ClO_2^-.

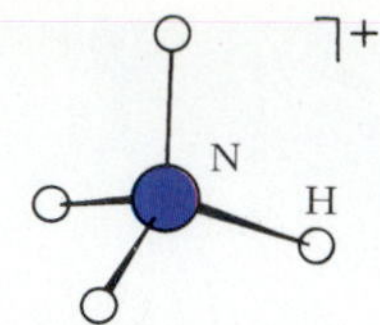

11 Ammonium ion

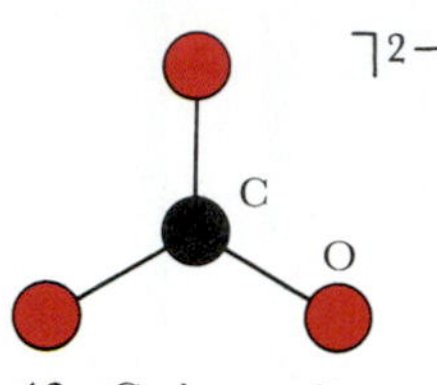

12 Carbonate ion

N The modern tendency is to use a partial angular bracket, as in the structures above, to show that a charge belongs to an entire species. The partial bracket is less typographically intrusive than a pair of braces; however, all structures for which we have used a partial bracket could be written enclosed in braces if that is preferred.

D Ions in slow motion. CD2, 63.

N The formula of an ionic compound represents the ratio of the numbers of cations and anions, not a precise chemical unit; for the latter we use the term *formula unit*; see next page.

U Exercise 2.86.

Polyatomic ions. A **polyatomic ion** is a bonded group of atoms with a net positive or negative charge. A simple example of a polyatomic cation is the ammonium ion, NH_4^+ (**11**), which occurs in ammonium chloride. As its formula shows, the ammonium ion consists of one N atom and four H atoms, but it has one electron less than required for electrical neutrality.

The most common polyatomic anions are the **oxoanions**, polyatomic anions that contain oxygen. They include the anions carbonate (CO_3^{2-}), nitrate (NO_3^-), nitrite (NO_2^-), and sulfate (SO_4^{2-}). The carbonate ion (**12**), for example, has one C atom, three O atoms, and an additional two electrons. The nitrite ion consists of one N atom, two O atoms, and one additional electron. The neutral molecule (nitrogen dioxide, NO_2) is known, but its properties are entirely different from those of the nitrite ion. For instance, nitrogen dioxide is a dark brown gas that contributes to the color of smog, whereas the nitrite ion is colorless and occurs as a component of ionic solids, such as sodium nitrite.

The formulas of ionic compounds. The chemical formula of an ionic compound shows the *ratio* of cations to anions in the compound. In sodium chloride, for instance, there is one Na^+ for each Cl^-, so its chemical formula is NaCl. In sodium carbonate, there are two sodium ions (Na^+) per carbonate ion (CO_3^{2-}), so its formula is Na_2CO_3. When a subscript has to be added to a polyatomic ion, the ion is written within parentheses, as in $(NH_4)_2SO_4$; here the parentheses show that there are two NH_4^+ for each SO_4^{2-} in ammonium sulfate. Note that we omit the charges on the ions when writing the formula of a neutral compound.

The ratio of cations to anions in an ionic compound is easy to determine once we know the charges of the ions present. Because the compound as a whole is electrically neutral, the charges of the cations and the anions must cancel one another. In sodium chloride the positive charge of each Na^+ is canceled if there is one Cl^- present for each Na^+ (as we saw in Fig. 2.25); so NaCl is the correct formula. In magnesium chloride the charge of one Mg^{2+} is canceled by two Cl^-; so its formula is $MgCl_2$. In ammonium sulfate the double charge of SO_4^{2-} is canceled by two NH_4^+, so its formula must be $(NH_4)_2SO_4$.

▼ EXAMPLE 2.9 Working out the formula of an ionic compound

Write the formula of aluminum sulfate, given that an aluminum ion is Al^{3+} and a sulfate ion is SO_4^{2-}.

STRATEGY We try to match the charges by taking one cation with one, two, and three anions in turn; we stop when the overall charge is zero. If we get a net negative charge without getting zero first, then we take two cations with one, two, and three anions—and so on.

SOLUTION The two ions in the compound are Al^{3+} and SO_4^{2-}. The charge of one Al^{3+} cannot be neutralized by one, two, or three SO_4^{2-}. But two Al^{3+} (six positive charges) are neutralized by three SO_4^{2-} ions, which we denote $(SO_4)_3$ in the formula. The formula of aluminum sulfate is therefore $Al_2(SO_4)_3$.

EXERCISE E2.9 Write the formula for magnesium phosphate. The formula of the phosphate ion is PO_4^{3-}.

▲ [*Answer*: $Mg_3(PO_4)_2$]

Formula units and molar masses of ionic compounds. A **formula unit** is the group of ions that matches the formula of an ionic compound. Thus, the formula unit NaCl of sodium chloride consists of one Na^+ and one Cl^-. The formula unit $(NH_4)_2SO_4$ of ammonium sulfate consists of two NH_4^+ and one SO_4^{2-}. Although the formula unit NaCl looks exactly the same as the empirical formula of the compound, there is a distinction. The formula unit denotes the actual group of ions we are considering (one Na^+ and one Cl^-), and hence is the analog of a molecular formula of a molecular compound. The formula NaCl refers to the *entire* compound and simply specifies that we are dealing with a compound in which the ions Na^+ and Cl^- are present in equal numbers.

The molar mass of an ionic compound is the mass per mole of the compound's formula units. (The molar mass of an ionic compound is sometimes called the *formula weight,* by analogy with the terms *atomic weight* and *molecular weight* we mentioned earlier.) For example, the molar mass of sodium chloride is the mass per mole of NaCl. Like the molar mass of molecular compounds, the molar mass of an ionic compound is obtained by adding together the molar masses of the atoms present in the formula unit. Thus, the molar mass of the ionic compound magnesium chloride is

? What is wrong with the statement "the molecular weight of magnesium chloride is 95.21 g/mol"?

$$\text{Molar mass of } MgCl_2 = \text{molar mass of } (Mg + 2 \times Cl)$$
$$= (24.31 + 2 \times 35.45) \text{ g/mol} = 95.21 \text{ g/mol}$$

We now know that if we need 1.000 mol of $MgCl_2$ (actually 1.000 mol of Mg^{2+} and 2.000 mol of Cl^-), we must measure out 95.21 g of magnesium chloride. The "massive heaps" corresponding to 1 mol of formula units of some common ionic compounds are shown in Fig. 2.28.

▼ EXAMPLE 2.10 Calculating the mass of a given amount of an ionic compound

Calculate the mass of magnesium chloride that should be measured out to obtain 0.223 mol of $MgCl_2$.

STRATEGY The calculation involves a conversion from moles of $MgCl_2$ to mass of $MgCl_2$. Therefore, we use the molar mass of the compound as a conversion factor in the usual way.

FIGURE 2.28 Each sample contains 1 mol of formula units of an ionic compound. From left to right are 58 g of sodium chloride (NaCl), 100 g of calcium carbonate ($CaCO_3$), 278 g of iron(II) sulfate heptahydrate ($FeSO_4 \cdot 7H_2O$), and 78 g of sodium peroxide (Na_2O_2).

U Exercises 2.79., 2.80., and 2.83.

SOLUTION We have already calculated the molar mass of $MgCl_2$, namely, 95.21 g/mol. Therefore, we write 1 mol $MgCl_2$ = 95.21 g $MgCl_2$ and use this relation to form a conversion factor:

$$\text{Mass of } MgCl_2 \text{ (g)} = 0.223 \cancel{\text{mol } MgCl_2} \times \frac{95.21 \text{ g } MgCl_2}{1 \cancel{\text{mol } MgCl_2}} = 21.2 \text{ g } MgCl_2$$

That is, we should measure out 21.2 g of magnesium chloride.

EXERCISE E2.10 Calculate the mass of aluminum sulfate that contains 0.223 mol of $Al_2(SO_4)_3$.

[*Answer*: 76.3 g]

2.8 CHEMICAL NOMENCLATURE

E The word *nomenclature* is derived from the Latin words meaning "name calling."

D Bring bottles of reagents into class to show that these names are really used!

Many compounds have at least two names. A **common name** is one that has become familiar from everyday use but may give little or no clue to the compound's composition. Many of them were given before the compound's composition was determined; they include water, salt, sugar, ammonia, and quartz. A **systematic name** reveals which elements are present (and, in some cases, how the atoms are arranged) and is constructed by certain rules. The systematic name of table salt is sodium chloride, which shows at once that it is a compound of sodium and chlorine. A systematic name specifies the components of a compound exactly. This correspondence is useful when the compound is unfamiliar or new and is essential when it has no common name. The systematic naming of compounds—**chemical nomenclature**—follows a set of rules. Like defining SI units, deciding on the rules for naming is an international activity.

E The distinction between organic and inorganic compounds has become more imprecise now that so many organometallic compounds have been prepared. The terms are broadly useful classifications rather than precise demarcations of two distinct regions of chemistry.

A distinction is traditionally made between organic and inorganic compounds. **Organic compounds** are compounds containing the element carbon, and usually hydrogen. They include methane, propane, glucose, and millions of other substances. These compounds are called organic because it was once (incorrectly) believed that they could be formed only by living organisms. **Inorganic compounds** are all the other compounds; these include water, calcium sulfate, ammonia, silica, hydrochloric acid, and many more. In addition, some very simple carbon compounds, particularly carbon dioxide and the carbonates (which include chalk, or calcium carbonate), are treated as inorganic compounds even though they are formed by living organisms. We shall deal with the nomenclature of a few common organic compounds later in this section, but for now we shall concentrate on naming inorganic compounds.

Element	Characteristic charge number cation
Group I (the alkali metals)	+1
Group II (the alkaline earth metals)	+2
zinc and cadmium	+2
aluminum	+3

Naming cations. The names of cations of the metallic elements are formed simply by adding the word *ion* to the name of the element, as in sodium ion. Except in special circumstances, certain elements always form cations of one characteristic charge number. Thus, potassium is always present as K^+ in its compounds, zinc is always present as Zn^{2+}, and aluminum is always present as Al^{3+}. The common elements that behave in this way include those shown in the margin.

When an atom can form more than one kind of cation, such as Cu^+ and Cu^{2+}, we need to distinguish between the different kinds. The most straightforward way of doing so is to use the **Stock number**, a Roman numeral equal to the positive charge on the cation. Then Cu^+ is

TABLE 2.6 Names of cations with variable charge numbers

Element	Cation	Old style name	Modern style name
cobalt	Co^{2+}	cobaltous	cobalt(II)
	Co^{3+}	cobaltic	cobalt(III)
copper	Cu^{+}	cuprous	copper(I)
	Cu^{2+}	cupric	copper(II)
iron	Fe^{2+}	ferrous	iron(II)
	Fe^{3+}	ferric	iron(III)
lead	Pb^{2+}	plumbous	lead(II)
	Pb^{4+}	plumbic	lead(IV)
manganese	Mn^{2+}	manganous	manganese(II)
	Mn^{3+}	manganic	manganese(III)
mercury	$Hg_2{}^{2+}$	mercurous	mercury(I)
	Hg^{2+}	mercuric	mercury(II)
tin	Sn^{2+}	stannous	tin(II)
	Sn^{4+}	stannic	tin(IV)

T OHT: Tables 2.4 and 2.6.

N Select the entries the student should memorize.

D Show the different appearances of Fe^{2+} and Fe^{3+} in aqueous solution.

D The oxidation states of manganese. CD1, 80.

called the copper(I) ion and Cu^{2+} is called the copper(II) ion. Similarly, Fe^{2+} is the iron(II) ion and Fe^{3+} is the iron(III) ion.

The transition metals and some of the elements that follow them form so many different kinds of ions that it is usually necessary to show their charges and to include a Stock number in the names of their compounds. However, the general aim of chemical nomenclature is to be unambiguous but brief. Therefore, when an element does in fact form only one kind of cation, there is no need to give a Stock number. The most common example is silver: silver chloride always means silver(I) chloride, even though some silver(II) compounds are known. Most scandium compounds contain Sc^{3+}, and there is no need to specify them as scandium(III). It is never wrong to include a Stock number; so if you are in doubt, and the element is not one of those in the table opposite, it is not wrong to give the Stock number.

E Stock was the German chemist Alfred Stock (1876–1946), who is chiefly famous for his work on boron hydrides. The rules of inorganic nomenclature are set out in a IUPAC publication known familiarly as "the Red Book." Its official title is *Nomenclature of inorganic chemistry*, ed. G. J. Leigh, Blackwell Scientific (Oxford), 1990.

N Origin and adoption of the Stock system. *J. Chem. Ed.*, **1985**, *62*, 243.

In an older system of nomenclature, the different cations of one element are denoted by the endings *-ous* and *-ic*. In some cases the Latin names of the elements are used. Thus, in this system, iron(II) ions are called ferrous ions and iron(III) ions are called ferric ions. Table 2.6 can be used to translate these older names (which are still often encountered) into modern names.

Naming anions. The names of monatomic anions are formed by adding the suffix *-ide* to the stem of the element, as shown in the margin. There is no need to worry about alternatives because monatomic anions never have more than one charge number. The ions formed by the halogens are collectively called halide ions, and include fluoride (F^-), chloride (Cl^-), bromide (Br^-), and iodide (I^-).

Element	Stem	Ion
fluorine	fluor-	fluoride ion F^-
oxygen	ox-	oxide ion O^{2-}
nitrogen	nitr-	nitride ion N^{3-}

The nomenclature of oxoanions is more complex because they are so numerous. The names of the more common oxoanions are summa-

E Sodium hydrogencarbonate is also known as bicarbonate of soda and baking soda.

T OHT: Tables 2.5 and 2.7; Appendix 3A table.

N A mnemonic for oxyanions [oxoanions]. *J. Chem. Ed.*, **1990**, *67*, 149.

TABLE 2.7 Some common oxoanions and their parent acids*

Oxoanion	Parent oxoacid
hydrogencarbonate (bicarbonate) HCO_3^-	carbonic acid H_2CO_3
carbonate CO_3^{2-}	
nitrate NO_3^-	nitric acid HNO_3
phosphate PO_4^{3-}	phosphoric acid H_3PO_4
hydrogenphosphate HPO_4^{2-}	
dihydrogenphosphate $H_2PO_4^-$	
hydrogensulfate (bisulfate) HSO_4^-	sulfuric acid H_2SO_4
sulfate SO_4^{2-}	

*A more extensive list of oxoanions is given in Appendix 3A.

Element	Stem	Oxoanion
carbon	carbon-	carbonate ion CO_3^{2-}
sulfur	sulf-	sulfate ion SO_4^{2-}
chlorine	chlor-	chlorate ion ClO_3^-

Element	Stem	Oxoanion
nitrogen	nitr-	nitrate ion NO_3^- nitrite ion NO_2^-
sulfur	sulf-	sulfate ion SO_4^{2-} sulfite ion SO_3^{2-}

rized in Table 2.7 and in Appendix 3A. The role of the acids listed in the table will be explained shortly. The basic rule is that the names of oxoanions are formed by adding the suffix *-ate* to the stem of the name of the element that is not oxygen; examples are shown in the margin.

The problem in naming oxoanions is that a given element can often form a variety of oxoanions with different numbers of oxygen atoms; nitrogen, for example, forms NO_2^- and NO_3^-. When an element forms only two oxoanions, the ion with the larger number of oxygen atoms is given the suffix *-ate*, and the one with the smaller number of oxygen atoms is given the suffix *-ite* (think of a mite as being small); examples are included in the lower table in the margin.

Some oxoanions—particularly those of the halogens—have relatives with an even smaller proportion of oxygen. Their names are formed by adding the prefix *hypo-* (from the Greek word meaning "under") to the stem of the *-ite* suffix, for example, the *hypo*chlorite ion, ClO^-. Other oxoanions have a higher proportion of oxygen than the *-ate* oxoanions do and are named with the prefix *per-* added to the *-ate* form of the name (*per* is the Latin word meaning "all over"; this choice of prefixes suggests that the element's ability to combine with oxygen is finally satisfied). An example is the *per*chlor*ate* ion, ClO_4^- (Fig. 2.29). Chlorine forms four oxoanions:

perchlorate ion	ClO_4^-	highest oxygen content
chlorate ion	ClO_3^-	
chlorite ion	ClO_2^-	
hypochlorite ion	ClO^-	lowest oxygen content
chloride ion	Cl^-	no oxygen

Bromine and iodine form similar oxoanions.

Some important oxoanions contain hydrogen. One of the most common is the hydroxide ion OH^-; others include versions of the oxoanions mentioned above (see Table 2.7). The hydrogen-containing oxoanions are named by adding the word *hydrogen* to the name of the oxoanion, for example,

hydrogencarbonate ion HCO_3^- (bicarbonate ion)
hydrogensulfate ion HSO_4^- (bisulfate ion)
hydrogensulfite ion HSO_3^- (bisulfite ion)

(The older names—bicarbonate, bisulfate, and bisulfite, respectively—are still quite widely used.) When two hydrogen atoms are present, the prefix *di-* (from the Greek word meaning "two") is added, for example, dihydrogenphosphate ion, $H_2PO_4^-$. Prefixes representing the number of atoms are listed in Table 2.8.

Naming ionic compounds. The name of an ionic compound is built from the names of the ions present, in the order (cation)(anion), but with the word *"ion"* omitted. Typical names include sodium chloride (a compound containing the ions Na^+ and Cl^-), ammonium nitrate (containing the ions NH_4^+ and NO_3^-), and calcium hydrogencarbonate (Ca^{2+} and HCO_3^-). The copper chloride that contains Cu^+ is called copper(I) chloride and the chloride that contains Cu^{2+} is called copper(II) chloride.

Prefixes are not used in the systematic nomenclature of ionic compounds because the numbers of anions present can be worked out from the charge of the cation, which is known or is given by the Stock number. A compound such as $CuCl_2$ is called simply copper(II) chloride, not copper(II) dichloride, because there is no ambiguity regarding the number of anions (Cl^-) present. The same is true of $CaCl_2$, which is called simply calcium chloride, and of Al_2O_3, aluminum oxide: $CaCl_2$ is the *only* chloride of calcium (because Ca^{2+} is the only cation calcium forms) and Al_2O_3 is the only oxide of aluminum. The *common* names of some ionic compounds do occasionally have prefixes; two examples are MnO_2, manganese(IV) oxide, which is commonly called manganese dioxide, and TiO_2, titanium(IV) oxide, which is commonly called titanium dioxide.

FIGURE 2.29 All three compounds shown here contain the potassium ion K^+. The different colors are due to the presence of different anions. From top to bottom: potassium chromate, K_2CrO_4; potassium chlorate, $KClO_3$; and potassium permanganate, $KMnO_4$.

D Place a selection of different compounds in test tubes and display them so as to establish an association between names and compounds (exhibiting their color, texture, physical state, and so on).

U Exercise 2.85.

TABLE 2.8 Prefixes used for naming compounds

Prefix	Meaning
mono-	1
di-	2
tri-	3
tetra-	4
penta-	5
hexa-	6
hepta-	7
octa-	8
nona-	9
deca-	10
undeca-	11
dodeca-	12

EXAMPLE 2.11 Naming ionic compounds

Give the systematic names of (a) $CrCl_3$ and (b) $Ba(ClO_4)_2$.

STRATEGY In each case, identify the cation and the anion, by reference to Tables 2.4 through 2.7 if necessary. If the cation is of an element that exists in only one charge state (as set out in the table of characteristic charge numbers), omit the Stock number; if not, then give its Stock number (to do so, we need to note the charge number of the anion and decide what cation charge number that implies).

SOLUTION (a) The cation is a chromium ion and the anion is a chloride ion, Cl^-. The cation must be Cr^{3+} (to achieve electrical neutrality), so chromium (a transition metal) is present as chromium(III). The compound is chromium(III) chloride. (b) The cation is a barium ion, Ba^{2+} (the only cation that barium forms). The anion is a perchlorate ion, ClO_4^-. Hence, the compound is barium perchlorate.

EXERCISE E2.11 Name (a) $FeCl_2$; (b) $AlBr_3$; (c) $Cr(ClO_3)_2$.

[*Answer*: (a) iron(II) chloride; (b) aluminum bromide; (c) chromium(II) chlorate]

EXAMPLE 2.12 Writing chemical formulas from names

Write formulas for the compounds (a) calcium sulfate and (b) cobalt(III) oxide.

STRATEGY In each case determine the charge number of the cation either from its Stock number or from the table listing characteristic charge

FIGURE 2.30 Blue copper(II) sulfate crystals ($CuSO_4 \cdot 5H_2O$) lose water above 150°C and form the white anhydrous powder ($CuSO_4$). The color is restored when water is added. A test based on this color change can be used to identify water.

U Exercise 2.86.

V Video 5: The dehydration of copper(II) sulfate pentahydrate.

D The effect of temperature on a hydrate. CD1, 24. The magic handkerchief. CD2, 42.

numbers. Then note the charge number of the anion and select the relative numbers of cations and anions that produce electrical neutrality for the compound.

SOLUTION (a) Calcium forms only Ca^{2+}. The sulfate ion is SO_4^{2-}. The electrically neutral compound therefore has formula $CaSO_4$. (b) Cobalt(III) signifies the presence of Co^{3+}. The oxide ion is O^{2-}. Therefore, to achieve electrical neutrality, the formula must contain $2 \times Co^{3+} + 3 \times O^{2-}$ and be Co_2O_3.

EXERCISE E2.12 Write formulas for the compounds (a) copper(I) bromide; (b) cesium nitrate; (c) lead(IV) oxide.

[*Answer*: (a) CuBr; (b) $CsNO_3$; (c) PbO_2]

Prefixes are also used in naming **hydrates**, which are solid, inorganic compounds containing H_2O molecules. For example, blue copper(II) sulfate crystals contain five H_2O molecules for every pair of ions, Cu^{2+} and SO_4^{2-}. The overall formula unit is therefore written $CuSO_4 \cdot 5H_2O$, in which the 5 is understood to multiply the entire H_2O unit and the raised dot simply separates the parts of the compound. This hydrate is called copper(II) sulfate pentahydrate. The water in the crystal is called **water of hydration**. It can be driven out of the copper(II) sulfate pentahydrate (but not out of all other hydrates) by heating the crystals, which crumble to a white powder of anhydrous copper(II) sulfate (formula, $CuSO_4$). **Anhydrous** means "without water." When water is added to the powder, the blue is restored as the pentahydrate reforms (Fig. 2.30).

Sodium carbonate is another example of a compound that can exist in various hydrated forms. Two of the most important are the decahydrate $Na_2CO_3 \cdot 10H_2O$, which is the compound used as washing soda, and the anhydrous form Na_2CO_3. The latter, which is called soda ash, is used in large amounts for making glass.

Naming molecular compounds. The names of many molecular compounds are formed as though the compound were ionic. This similarity is an advantage in that we do not need to know whether a compound is ionic or molecular before naming it. It has a disadvantage in that we cannot tell from its name whether a compound is ionic or molecular. That knowledge comes with experience; but, as remarked earlier, a good guide is that compounds containing only nonmetals or metalloids (that is, elements on the right of the periodic table or hydrogen) are likely to be molecular; some important exceptions are compounds that contain the ammonium ion. A compound is likely to be ionic if one of the elements is a metal, especially an element from Group I or II.

A **binary compound** is an ionic or molecular compound that consists of only two elements. Binary molecular compounds include HCl, H_2O, and CO_2. Except in a few cases, the molecular formula of a binary compound of hydrogen and a nonmetal has the H atom written first and named first:

hydrogen chloride HCl   hydrogen sulfide H_2S

Some binary compounds of hydrogen are so common that they are always referred to by their common names. These compounds are included in Table 2.9.

TABLE 2.9 Some common names for molecular compounds

Formula*	Common name
H_2O	water
H_2O_2	hydrogen peroxide
NH_3	ammonia
N_2H_4	hydrazine
NH_2OH	hydroxylamine
PH_3	phosphine
NO	nitric oxide
N_2O	nitrous oxide

*For historical reasons, the molecular formulas of binary hydrogen compounds of Group V elements are written with the Group V element first.

Most other common binary molecular compounds have at least one element from Group VI or VII. The elements from these groups are

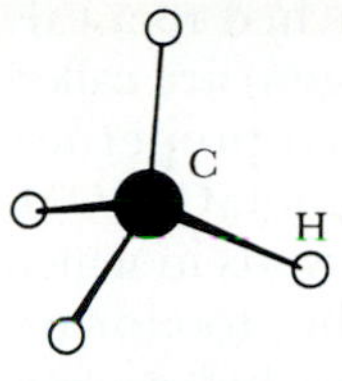

13 Methane

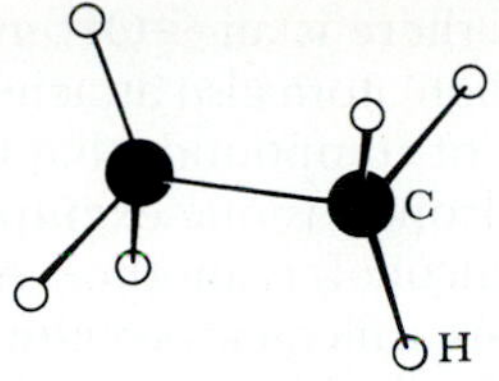

14 Ethane

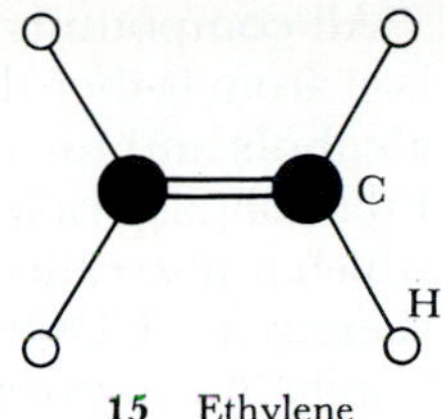

15 Ethylene

named second, and the number of atoms of each type is indicated with a prefix from Table 2.8:

phosphorus trichloride PCl_3	dinitrogen oxide N_2O
sulfur hexafluoride SF_6	dinitrogen pentoxide N_2O_5

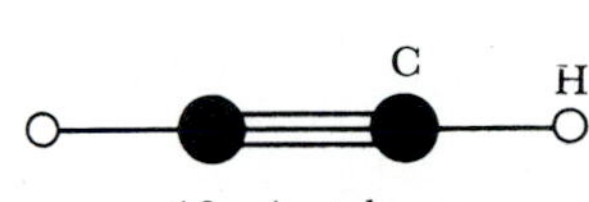

16 Acetylene

The prefix *mono-* is never used with the first element and only very rarely with the second: the only common example of its use is in carbon monoxide. Binary compounds of oxygen are called oxides (unless, for reasons that will become clear in Chapter 8, the other element is fluorine):

chlorine dioxide ClO_2 sulfur trioxide SO_3

Some oxides have common names that are still widely used; two are also included in Table 2.9.

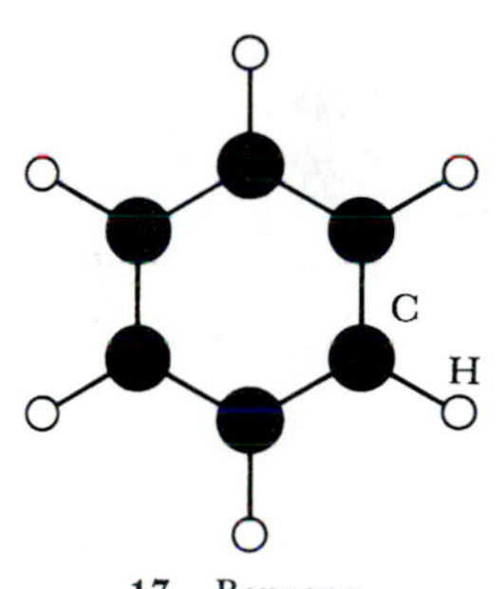

17 Benzene

The nomenclature of simple organic compounds. We shall discuss the systematic naming of organic compounds in Chapters 23 and 24. However, some organic compounds are so common that it is useful to meet their names at this stage. An organic compound is called a **hydrocarbon** if it contains only carbon and hydrogen. Five examples of hydrocarbons are CH_4, methane (**13**); C_2H_6, ethane (**14**); C_2H_4, ethylene (**15**); C_2H_2, acetylene (**16**); and C_6H_6, benzene (**17**).

Two compounds that contain oxygen as well as carbon and hydrogen are C_2H_6O ethanol (**18**) and $C_2H_4O_2$ acetic acid (**19**). Neither of these last two molecular formulas gives much indication of the arrangements of the atoms in the molecules depicted in (**18**) and (**19**). Therefore, chemists often write their formulas in a way that shows how the atoms are grouped together. These alternative formulas are a shortened form of the **structural formula** of the molecule (see Chapter 24). Thus, the structural formula of ethanol is C_2H_5OH (or, in more detail, CH_3CH_2OH), which shows that an OH group exists in the molecule. The structural formula of acetic acid is $CH_3CO(OH)$, or more commonly CH_3COOH, which shows that a CO group, with an OH group attached to it, is present. We now see that chemists have three ways of writing formulas, and which form they choose depends on how much they know or what they want to emphasize:

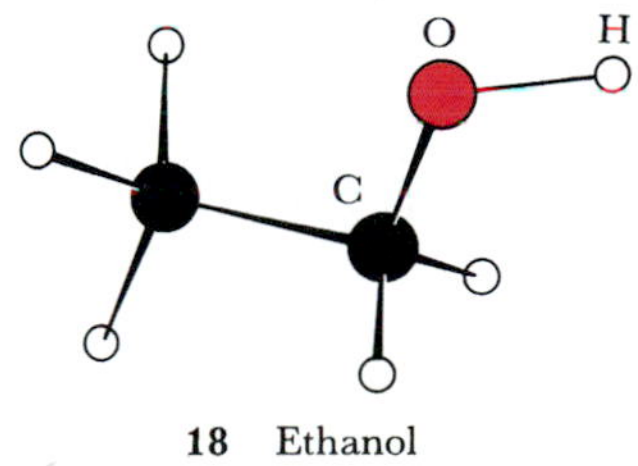

18 Ethanol

empirical formula: CH_2O	when only the ratio of numbers of atoms present in the compound is known
molecular formula: $C_2H_4O_2$	to show the actual numbers of atoms present in each molecule
structural formula: CH_3COOH	to show the groups of atoms

The general name for any one of these formulas is a **chemical formula**.

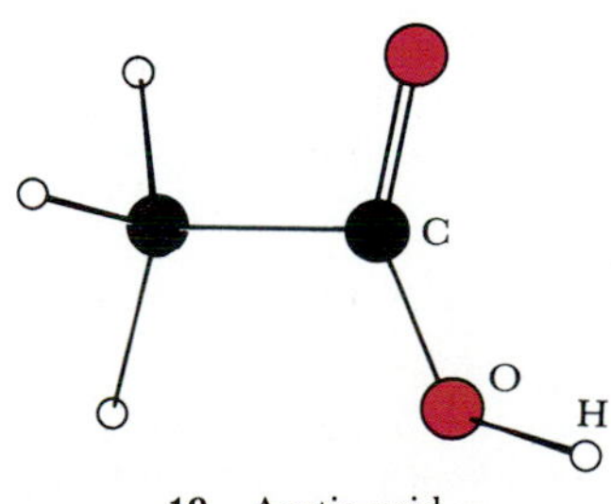

19 Acetic acid

E The rules for organic nomenclature are introduced in Chapters 23 (hydrocarbons) and Chapter 24 (compounds with functional groups). The IUPAC rules are set out in "the Blue Book." Its official title is *Nomenclature of organic chemistry*, ed. J. Rigaudy and S. P. Klesney, Pergamon Press (Oxford), 1979.

? How can you tell from its chemical formula that a substance is an alcohol or a carboxylic acid?

All compounds in which there is an —OH group attached to a carbon atom (other than a carbon atom also attached to oxygen) are called **alcohols** and form a family of compounds that have similar properties. Ethanol (also called ethyl alcohol) is one example; methanol (CH_3OH, which is also called methyl alcohol) is another. All compounds in which there is a —COOH group are called **carboxylic acids**. They too form a family of compounds with similar properties; the family includes acetic acid (the acid responsible for the sharpness of the taste of vinegar) and formic acid, HCOOH, the acid in ant venom.

Acids. Both HCl and H_2S are examples of an important class of compounds called acids. For now we define an acid as a compound that contains hydrogen and can release hydrogen ions, H^+, in water. (Acids are very important in chemistry; and starting in Chapter 3, we will discuss them more fully.) Hydrogen chloride is an acid because it dissolves in water to give a solution of hydrogen cations and chloride anions known as hydrochloric acid. Methane (CH_4) contains hydrogen; but it does not release hydrogen ions in water, so it is not an acid. A clue to whether a substance is an acid is whether or not its formula begins with H, as in HCl and H_2SO_4. However, acetic acid, CH_3COOH (**19**), is an acid because it can release the hydrogen attached to the oxygen atom in the —COOH group. Its molecular formula is often written $HC_2H_3O_2$ to emphasize that it is an acid and can lose *one* of its hydrogen atoms.

D Show the action of acetic acid and hydrochloric acid on zinc or magnesium.

Formally, the hydrogen halides—HF, HCl, HBr, and HI—are named like other binary compounds of nonmetals: hydrogen fluoride, hydrogen chloride, and so on. However, aqueous solutions of these compounds have special names, formed by adding the prefix *hydro-* and the suffix *-ic acid* to the stem of the element. Thus, an aqueous solution of HCl is called *hydro*chlor*ic acid.* These acids are the parent acids of the halides, in the sense that the halide ions are left when the hydrogen ion has been released. All binary acids (that is, compounds that contain hydrogen and one other element) are named with the prefix *hydro-* and the suffix *-ic* acid, as in *hydro*chlor*ic* acid, HCl, and *hydro*iod*ic* acid, HI.

Among the most important acids are the **oxoacids**, acids containing oxygen. These molecular compounds are the parent acids of the oxoanions and consist of molecules in which all the negative charges of an oxoanion are canceled by hydrogens, as in H_2SO_4 and H_3PO_4. The acid is named by replacing the *-ate* suffix of the anion by *-ic acid;* thus, the parent acid of the carbonates is H_2CO_3, carbonic acid. The more important oxoanions and oxoacids are listed in Table 2.7.

When both *-ate* (larger number of O atoms) and *-ite* (smaller number of O atoms) oxoanions exist the parent acids are named as the *-ic acid* and the *-ous acid,* respectively:

- *-ic* oxoacids are the parent acids of *-ate* oxoanions
- *-ous* oxoacids are the parent acids of *-ite* oxoanions

Thus, we have sulfuric acid (H_2SO_4) as the parent of the sulfate anion (SO_4^{2-}), and sulfurous acid (H_2SO_3) as the parent of the sulfite anion (SO_3^{2-}). This rule carries over into the parent acids of the *per-* and *hypo-* oxoanions: $HClO_4$, the parent acid of the perchlor*ate* ion is called perchlor*ic* acid; HClO, the parent acid of the hypochlor*ite* ion is called hypochlor*ous* acid. The complete range of chlorine acids is therefore

perchlorate ion	ClO_4^-	$HClO_4$	perchloric acid (highest oxygen content)
chlorate ion	ClO_3^-	$HClO_3$	chloric acid
chlorite ion	ClO_2^-	$HClO_2$	chlorous acid
hypochlorite ion	ClO^-	$HClO$	hypochlorous acid (lowest oxygen content)
chloride ion	Cl^-	HCl	hydrochloric acid (no oxygen)

A similar range of acids exists for bromine and iodine.

E The systematic names of oxoacids (which we do not use in this text) specify the oxidation number of the "central" atom and the number of oxygen atoms, as in tetraoxo-choric(VII) acid for $HClO_4$.

▼ **EXAMPLE 2.13 Interpreting the names and formulas of acids**

Give the names of (a) HBr in water and (b) H_2S in water; give the formulas of (c) phosphoric acid and (d) perbromic acid.

E Phosphoric acid is added to many cola drinks to provide their tart taste.

STRATEGY Most of the names and formulas can be taken from the tables or the discussion in the text. Binary acids all have names of the form *hydro . . . ic acid.* Remember that the prefix *per-* indicates that a halogen is bonded to its maximum number of oxygen atoms, as in perchloric acid, $HClO_4$.

SOLUTION (a) HBr is a binary acid, and is hydrobromic acid. (b) H_2S is also a binary acid, and is hydrosulfuric acid. (c) Phosphoric acid, an oxoacid, is H_3PO_4 (see Table 2.7). (d) Perbromic acid is $HBrO_4$ (by analogy with $HClO_4$).

EXERCISE E2.13 Give (a) the names of HIO and (b) the formula of selenic acid.

[*Answer*: (a) hypoiodous acid; (b) H_2SeO_4]

▲

In this chapter we have taken matter apart as far as chemists need to go and have encountered elements, compounds, atoms, ions, molecules, electrons, and nuclei. We have now met the building blocks of matter. It is time to see how to use these blocks to build new materials.

T OHT: Chart in Summary.

SUMMARY

2.1 The classification of substances into their fundamental components is summarized in the chart (a continuation of the chart at the end of Chapter 1). Every **element** has a name and a **chemical symbol**.

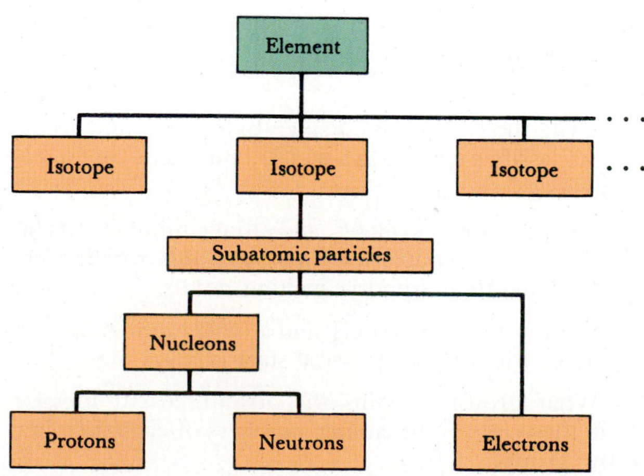

2.2 The elements are listed in an arrangement called the **periodic table**. The columns of the table are called **groups** and the rows are called **periods**. Some groups have special names, including the **alkali metals** of Group I, the **alkaline earth metals** of Group II, the **halogens** of Group VII, and the **noble gases** of Group VIII. The **transition metals** lie between Groups II and III. Elements with similar characteristics are close to each other in the periodic table. All the **metals**, for instance, lie to the left of the **nonmetals**, which are found on the upper right; the **metalloids** lie in a diagonal band between the metals and nonmetals.

2.3 According to Dalton's **atomic hypothesis**, all the atoms of a given element are the same, and compounds are specific combinations of atoms. Atoms are made up of **subatomic particles**, particles smaller than atoms, of which the most important in chemistry are the **electron**, the **proton**, and the **neutron**. An atom consists of a central **nucleus** surrounded by electrons. The **atomic number**, the number of protons in the nucleus, determines the identity of the element and the number of electrons in the neutral atom.

2.4 The **mass number**, the total number of **nucleons** in the nucleus, determines the mass of the atom. Individual atomic masses are measured with a **mass spectrometer** and are often expressed in **atomic mass units**. Most elements have **isotopes**, atoms that differ in mass because they have nuclei with different numbers of neutrons (but the same number of protons). Most samples of elements from naturally occurring material are actually mixtures of isotopes.

2.5 The very large numbers of atoms in typical samples are expressed more conveniently in terms of the **mole**, the unit of the amount of atoms (or other objects) present in the sample. The **molar mass** of an element is the mass per mole of its atoms.

2.6 **Molecular compounds** consist of **molecules**. Molecular compounds are typically combinations of nonmetals or of nonmetals and metalloids. The **molar mass** of a compound is the sum of the molar masses of the elements in the compound. A molecular compound is symbolized by an **empirical formula**, which gives the relative numbers of atoms of each element present, or by a **molecular formula**, which shows the actual numbers of atoms in each molecule.

2.7 Atoms may lose or gain electrons and become **ions**, which may form **ionic compounds**. Ionic compounds are compounds of elements that usually include a metal (but not all compounds of metals are ionic). Positively charged ions are called **cations** and negatively charged ions are called **anions**. Important polyatomic anions include the **oxoanions**, anions that contain oxygen.

2.8 Every compound or molecule also has a name and is represented by a **chemical formula**, which consists of the symbols of the constituent elements. The rules of **chemical nomenclature** describe how to name compounds systematically. The names of cations are listed in Table 2.4 and those of anions are listed in Tables 2.5 and 2.6 and Appendix 3A. The names of common oxoanions are listed in Table 2.7. For transition elements, the **Stock number**, the charge of the cation, must always be given.

CLASSIFIED EXERCISES

The elements are listed in the periodic table inside the front cover and in alphabetical order inside the back cover.

2.1 Name the elements (a) Li; (b) Ga; (c) Xe. List their group numbers in the periodic table. Identify each one as a metal, a nonmetal, or a metalloid.

2.2 Name the elements (a) P; (b) Sb; (c) Ag. List their group numbers in the periodic table. Identify each one as a metal, a nonmetal, or a metalloid.

2.3 Write the symbol of (a) iodine; (b) chromium; (c) aluminum. Classify each one as a metal, a nonmetal, or a metalloid.

2.4 Write the symbol of (a) zinc; (b) barium; (c) germanium. Classify each one as a metal, a nonmetal, or a metalloid.

2.5 List the names, symbols, and atomic numbers of the alkali metals. Characterize their reactions with water, and describe their trend in melting points.

2.6 List the names, symbols, and atomic numbers of the halogens. Identify the physical state of each one.

2.7 What group contains the alkaline earth metals? Write the symbols and atomic numbers of each member of the group.

2.8 What is meant by the term *congener?* Identify the congeners of carbon by name and symbol. Which of the congeners of carbon are malleable?

The Nuclear Atom

2.9 Give the number of protons, neutrons, and electrons in an atom of (a) carbon-13; (b) ^{37}Cl; (c) chlorine-35; (d) ^{235}U.

2.10 Give the number of protons, neutrons, and electrons in an atom of (a) tritium, ^{3}H; (b) ^{60}Co; (c) oxygen-16; (d) ^{204}Pb.

2.11 Using the values in Table 2.2, calculate the masses of neutrons, protons, and electrons in 1.00 g of iron-57. The mass of one atom of iron-57 is 9.45×10^{-23} g.

2.12 Use the values in Table 2.2 to calculate the masses of neutrons, protons, and electrons in 1.00 g of gold-197. The mass of one atom of gold-197 is 3.3×10^{-22} g.

2.13 In chemistry, it is common to make the approximation that the mass of a cation is the same as that of the parent atom. (a) To how many significant figures must the mass of a hydrogen atom be reported before the difference of mass between it and a hydrogen cation is significant? (b) What is the difference in mass of a hydrogen cation, H^+, from that of a hydrogen atom, expressed as a percentage of the atom's mass?

2.14 Confirm that, to the number of significant figures given in the periodic table, there is a negligible error in assuming that the mass of an Fe^{3+} ion is the same as that of the Fe atom.

2.15 (a) Determine the total number of protons, neutrons, and electrons in one water molecule, H_2O. (b) What are the total masses of protons, neutrons, and electrons in one water molecule? (c) What fraction of your own mass is due to the neutrons in your body, assuming you consist primarily of water?

2.16 Determine the fraction of the total mass of an ^{56}Fe atom that is due to (a) neutrons; (b) protons; (c) electrons. (d) What is the mass of protons in a 1000-kg automobile? Assume the total mass of the vehicle is due to ^{56}Fe.

Masses of Atoms

2.17 Calculate the molar mass of bromine in a brine sample that contains 50.54% ^{79}Br (of mass 78.918 u) and 49.46% ^{81}Br (of mass 80.916 u).

2.18 Calculate the molar mass of krypton in a natural sample, which consists of 0.3% ^{78}Kr (of mass 77.92 u), 2.3% ^{80}Kr (of mass 79.91 u), 11.6% ^{82}Kr (of mass 81.91 u), 11.5% ^{83}Kr (of mass 82.92 u), 56.9% ^{84}Kr (of mass 83.91 u), and 17.4% ^{86}Kr (of mass 85.91 u).

2.19 By now, you should realize that atoms are very small and have a very small mass. Considering that 1 u = 1.6605×10^{-24} g, determine the average mass, in grams, of (a) one gold atom, and the number of atoms in 1.0 g of gold; (b) one krypton atom, and the number of atoms in 5.0 L of krypton of mass 18.7 g; (c) one deuterium atom, ^2H (of mass 2.01 u), and the number of atoms in 1.0 mg of deuterium.

2.20 Given that 1 u = 1.6605×10^{-24} g, express the mass, in atomic mass units (u) of (a) one ^{35}Cl atom, of mass 5.81×10^{-23} g; (b) one ^{238}U atom, of mass 3.953×10^{-22} g; (c) one ^{19}F atom, of mass 3.15×10^{-23} g.

Fun with Atoms and Moles

2.21 Rutherford estimated that the volume of a nucleus was about $1/10^{15}$ that of an atom. If the volume of the nucleus were the size of a grain of sand of volume 1 mm^3, what would the volume of the atom be in cubic meters? Compare this volume with that of a cubic football field (300 ft × 300 ft × 300 ft).

2.22 Assume that the entire mass of an atom is concentrated in its nucleus, a sphere of radius 1.5×10^{-5} pm. (a) What is the density of a carbon nucleus? The volume of a sphere is $\frac{4}{3}\pi r^3$, where r is its radius. (b) What would be the radius of the Earth if its matter were compressed to the same density? (Its average radius is 6.4×10^3 km and its average density is 5.5 g/cm^3.)

2.23 (a) The approximate population of the Earth is 5.3 billion people. How many moles of people are on the Earth? (b) If all people were pea pickers and pea counters how long would it take for the Earth's population to count out one mole of peas at the rate of one pea per second, 24 hours per day, 365 days per year?

2.24 (a) About 1000 metric tons of sand contain about a trillion (10^{12}) grains of sand. How many metric tons of sand are needed to provide 1 mol of sand? (b) If we assume that the volume of a grain of sand is 1 mm^3 and the land area of the continental United States is 3,600,000 mi^2, then how deep would the sand pile be over the United States if it were covered with one mole of grains of sand?

Moles and Molar Masses of Elements

2.25 State (with the correct units) the molar masses of (a) iron; (b) xenon; (c) sodium.

2.26 Identify by name and symbol the elements that have the following molar masses: (a) 101.07 g/mol; (b) 69.72 g/mol; (c) 14.01 g/mol.

2.27 Calculate the amount in moles of (a) 4.82×10^{22} atoms of ^{35}Cl (mass 34.97 u); (b) 2.22 g of copper; (c) 8.96 μg of iron.

2.28 Determine the number of atoms in (a) 3.97 mol of Xe; (b) 18.3 μg of Sc; (c) 12.8 pg of Li.

2.29 Which sample in each of the following pairs contains the greater number of moles of atoms? (a) 25 g of carbon or 35 g of silicon; (b) 1.0 g of Au or 1.0 g of Hg; (c) 2.49×10^{22} atoms of Au or 2.49×10^{22} atoms of Hg.

2.30 Determine the mass of aluminum that has the same number of atoms as there are in (a) 6.29 mg of silver; (b) 6.29 mg of gold.

Moles, and the Molar Masses of Compounds

2.31 Determine the empirical formula from the following analyses. (a) The mass composition of cryolite, a compound used in the electrolytic production of aluminum, is 32.79% Na, 13.02% Al, and 54.19% F. (b) A compound used to generate O_2 gas in the laboratory has mass composition 31.91% K, 28.93% Cl, the remainder being oxygen. (c) A mixed fertilizer shows an analysis of 12.2% N, 5.26% H, 26.9% P, and 55.6% O. What is the empirical formula of the fertilizer?

2.32 Determine the empirical formula of each compound from the following data. (a) Talc (used in talcum powder) has mass composition 19.2% Mg, 29.6% Si, 42.2% O, and 9.0% OH. (b) Saccharin, a sweetening agent, has mass composition 45.89% C, 2.75% H, 7.65% N, 26.20% O, and 17.50% S. (c) Salicylic acid, used in the synthesis of aspirin, has mass composition 60.87% C, 4.38% H, and 34.75% O.

2.33 Determine the molar mass of the compounds (a) $CaBr_2$; (b) $NiSO_4 \cdot 6H_2O$; (c) CH_4 (methane, the major component of natural gas).

2.34 Determine the mass of 1.00 mol of (a) sulfur tetrafluoride, SF_4; (b) hydrazine, N_2H_4; (c) sodium cyanide, NaCN; (d) sucrose, $C_{12}H_{22}O_{11}$; (e) $CuCl_2 \cdot 4H_2O$.

2.35 Calculate the amount of molecules (in moles) and the number of molecules (or atoms) in (a) 10.0 g of carbon tetrachloride, CCl_4, a cleaning solvent; (b) 1.65 mg of hydrogen iodide, HI; (c) 3.77 μg of hydrazine, N_2H_4, a fuel for rockets; (d) 500 g of sucrose, $C_{12}H_{22}O_{11}$, cane sugar; (e) 2.33 g of oxygen as O atoms, as found in the upper part of the atmosphere.

2.36 Convert each of the following masses to amount (in moles) and to number of molecules (or atoms): (a) 1.00 kg of H_2O; (b) 1.00 kg of C_2H_5OH (ethanol); (c) 10.0 g of sulfur, as S atoms and as S_8 molecules; (d) 3.0 g of CO_2; (e) 3.0 g of NO_2.

2.37 Calculate the amount (in moles) of (a) Ag^+ in 2.00 g of AgCl; (b) UO_3 in 600 g of UO_3 (3 sf); (c) Cl^- in 4.19 mg of $FeCl_3$; (d) H_2O in 1.00 g of $AuCl_3 \cdot 2H_2O$.

2.38 Calculate the amount (in moles) of (a) CN^- in 1.00 g of KCN; (b) H atoms in 200 mg of H_2O; (c) $CaCO_3$ in 500 g of $CaCO_3$ (3 sf); (d) H_2O in 5.00 g of $La_2(SO_4)_3 \cdot 9H_2O$.

2.39 (a) Calculate the amount (in moles) of testosterone, $C_{19}H_{28}O_2$, a male sex hormone, in a 1-μg sample. (b) What is the mass percentage composition of testosterone?

2.40 (a) Aspartame, $C_{14}H_{18}N_2O_5$, is an artificial sweetener; it is also known as NutraSweet. How many molecules are present in 1.0 mg of aspartame? (b) What is the mass percentage composition of aspartame?

2.41 (a) Calculate the mass, in grams, of a single water molecule. (b) Determine the number of H_2O molecules in 1.00 g of H_2O.

2.42 Octane, C_8H_{18}, is typical of the molecules found in gasoline. (a) Calculate the mass (in grams) of one octane molecule. (b) Determine the number of C_8H_{18} molecules in 1.00 mL of C_8H_{18}, the mass of which is 0.82 g.

2.43 A chemist measured out 5.50 g of copper(II) bromide tetrahydrate, $CuBr_2 \cdot 4H_2O$. (a) How many moles of $CuBr_2 \cdot 4H_2O$ were measured out? (b) How many moles of Br^- are present in the sample? (c) How many water molecules are present in the sample? (d) What fraction of the total mass of the sample was due to copper?

2.44 A chemist prepared an aqueous solution by mixing 2.50 g of ammonium phosphate trihydrate, $(NH_4)_3PO_4 \cdot 3H_2O$, and 1.50 g of potassium phosphate, K_3PO_4, with 500 g water. (a) Determine the number of moles of each compound that was measured. (b) How many moles of PO_4^{3-} are present in solution? (c) Calculate the mass of phosphate ions present in the solution. (d) What is the total mass of the water present in the solution?

2.45 Lindane, used as an insecticide, has mass composition 24.78% C, 2.08% H, and 73.14% Cl and molar mass 290.85 g/mol. What is the molecular formula of lindane?

2.46 Nicotine has mass composition 74.03% C, 8.70% H, and 17.27% N and molar mass 162.23 g/mol. Determine the molecular formula of nicotine.

2.47 Caffeine, a stimulant in coffee and tea, has molar mass 194.19 g/mol and mass composition 49.48% C, 5.19% H, 28.85% N, and 16.48% O. What is the molecular formula of caffeine?

2.48 Cacodyl, which has an intolerable garlicky odor and is used in the manufacture of cacodylic acid, a cotton herbicide, has mass composition 22.88% C, 5.76% H, and 71.36% As and molar mass 209.96 g/mol. What is the molecular formula of cacodyl?

Cations, Anions, and Their Nomenclature

2.49 Determine whether the following elements are most likely to form a cation or anion and list the formula for that ion: (a) sulfur; (b) potassium; (c) silver; (d) chlorine.

2.50 Determine whether the following elements are most likely to form a cation or anion and list the formula for that ion: (a) zinc; (b) magnesium; (c) nitrogen; (d) oxygen.

2.51 How many protons, neutrons, and electrons are present in (a) $^{2}H^+$; (b) $^{9}Be^{2+}$; (c) $^{80}Br^-$?

2.52 How many protons, neutrons, and electrons are present in (a) $^{40}Ca^{2+}$; (b) $^{115}In^{3+}$; (c) $^{127}Te^{2-}$?

2.53 Write formulas for the ionic compounds formed from (a) sodium and oxide ions; (b) potassium and sulfate ions; (c) zinc and nitrate ions.

2.54 Write formulas for the ionic compounds formed from (a) iron(III) and sulfate ions; (b) ammonium and iodide ions; (c) lithium and sulfide ions.

Chemical Nomenclature

2.55 Write the names of the following ions, giving both the old and modern names where appropriate: (a) HSO_4^-; (b) Hg_2^{2+}; (c) CN^-; (d) HCO_3^-.

2.56 Write the formulas of (a) cupric ion; (b) chlorite ion; (c) hydride ion.

2.57 Write the formulas of (a) calcium phosphate (the major inorganic component of bones); (b) aluminum sulfate; (c) calcium nitride.

2.58 The following ionic compounds are commonly found in laboratories. Write both the old and modern names wherever appropriate: (a) $NaHCO_3$ (baking soda); (b) Hg_2Cl_2 (calomel); (c) ZnO (calamine).

2.59 Write the names of (a) $Cu(NO_3) \cdot 6H_2O$; (b) $NdCl_3 \cdot 6H_2O$; (c) $NiF_2 \cdot 4H_2O$.

2.60 Give the chemical formulas of (a) sodium carbonate monohydrate; (b) indium(III) nitrate pentahydrate; (c) copper(II) perchlorate hexahydrate.

2.61 Write the formulas of (a) dinitrogen tetroxide; (b) dichlorine heptoxide; (c) disulfur dichloride.

2.62 The following molecular compounds are often found in chemical laboratories. Name each compound. (a) SiO_2 (silica); (b) SiC (carborundum); (c) N_2O (a general anesthetic).

2.63 Classify each of the following organic compounds as a hydrocarbon, an alcohol, or a carboxylic acid: (a) C_3H_7OH; (b) HCOOH; (c) C_3H_6.

2.64 The names and formulas of several organic compounds will appear throughout the text and lectures. Give the chemical name or formula for each one. (a) CH_3OH (wood alcohol); (b) ethanol (also called grain alcohol); (c) HCOOH.

2.65 Name the following aqueous solutions as acids: (a) H_2SO_4; (b) HNO_3; (c) CH_3COOH; (d) H_3PO_4.

2.66 The following acids are less common than those in the preceding exercise but are reasonably widely used in chemical laboratories. Write their formulas. (a) perchloric acid; (b) hypochlorous acid; (c) hypoiodous acid; (d) hydrofluoric acid.

UNCLASSIFIED EXERCISES

2.67 Two isotopes of an element Q are ^{97}Q (23.4% abundance) and ^{94}Q (76.6% abundance). ^{97}Q is 8.082 times heavier than ^{12}C and ^{94}Q is 7.833 times heavier than ^{12}C. What is the average atomic mass of the element Q?

2.68 The nuclear power industry needs ^{6}Li but not ^{7}Li and is gradually removing ^{6}Li from samples that are sold commercially. The current abundances of the two isotopes are 7.42% and 92.58%, respectively, and their masses are 6.015 u and 7.016 u. (a) What is the molar mass of a sample of naturally occurring lithium? (b) What will be the molar mass when the abundance of ^{6}Li is reduced to 5.67%?

2.69 The molar mass of boron atoms in a sample of naturally occurring ore is 10.81 g/mol. The sample is known to consist of ^{10}B (of mass 10.013 u) and ^{11}B (of mass 11.093 u). What are the abundances of the two isotopes?

2.70 A 1.0-cm^3 cube of uranium (density, 19.05 g/cm^3) is placed next to a 2.0-cm^3 cube of niobium (density, 8.57 g/cm^3). Which cube contains the greater number of atoms?

2.71 Calculate the average density of a single carbon atom by assuming that it is a uniform sphere of radius 77 pm. The volume of a sphere is $\frac{4}{3}\pi r^3$, where r is its radius. Express the density of carbon in grams per cubic centimeter. The density of diamond, a crystalline form of carbon, is 3.5 g/cm^3. What does your answer suggest about the way the atoms are packed together in diamond?

2.72 (a) The current cost of gold bullion is approximately $370 per troy ounce (31.30 g). How much does one gold atom cost? (b) A 2.50-carat (1 carat = 200 mg) diamond costs $7950. What is the price of one carbon atom in this diamond?

2.73 A chemical reaction requires at least 0.683 mol of sulfur atoms to react with 0.683 mol of copper atoms. (a) How many S atoms are required? (b) How many sulfur molecules, S_8, are necessary? (c) What mass of sulfur is needed for the reaction?

2.74 A sample of brass consists of 75% copper and 25% zinc by mass. (a) How many moles of each are present in a 100-g sample (3 sf)? (b) How many atoms of each are present in a 100-g sample (3 sf)?

2.75 The alcohol content in most wines is 12% by volume, meaning that for every 100 mL of wine, 12 mL is alcohol and the remainder is water (not necessarily 88 mL of water, because the volumes of water and alcohol are not additive in a mixture). The formula of ethanol, the fermented alcohol from grains, is C_2H_5OH. How many molecules of ethanol are in 12 mL of wine? The density of ethanol is 0.79 g/mL.

2.76 A paper clip has a mass of 455 mg. Assuming the paper clip to be pure iron (in fact, it is a mixture of elements but is principally iron), calculate the number of atoms it contains.

2.77 A silver dollar weighs about 0.943 ounces. Express this mass in milligrams. Assuming the silver dollar to be pure silver (which it is not), how many silver atoms are present?

2.78 A test rocket is loaded with 3.77 mol of N_2H_4 (hydrazine) and 8.99×10^{24} molecules of N_2O_4 (dinitrogen tetroxide). Calculate the total mass of the mixture loaded on the rocket.

2.79 Epsom salts consists of magnesium sulfate heptahydrate. Write its formula. (a) How many atoms of magnesium are in 2.00 g of Epsom salts? (b) How many moles of H_2O are in 2.00 g of Epsom salts? (c) How many formula units of the compound are present in 2.00 g?

CASE 2 ▸ WHAT DOES 18-18-5 MEAN?

Plants grow in many types of soils, but some varieties grow better when additional specific nutrients have been added. We buy these nutrients as solid fertilizers that contain the three major nutrients for plants: nitrogen, phosphorus, and potassium. In addition, commercial fertilizers also may include conditioners (to improve their physical qualities), fillers (to dilute the nutrients), pesticides (to kill weeds, insects, and fungi), and compounds containing various trace elements, such as the three secondary plant nutrients (calcium, magnesium, and sulfur) and various micronutrients (e.g., copper, zinc, iron, molybdenum).

Usually there are three numbers on the label of the fertilizer pack denoting the mass percentage composition of nitrogen (N), phosphorus (V) oxide (P_4O_{10}), and potassium oxide (K_2O) in the mixture. Thus, a fertilizer label that reads 18-18-5 is interpreted as 18% N, 18% P_4O_{10}, and 5% K_2O by mass. The remaining 59% is conditioner, filler, and other inert materials.

A potentially confusing point is that the potassium is actually present as KCl, but the mass percentage composition is calculated as though the same amount of potassium were present as the compound K_2O. Likewise, the phosphorus is actually present as phosphates and hydrogenphosphates of various kinds. In fact, many fertilizers contain the water-soluble compounds ammonium sulfate, $(NH_4)_2SO_4$, calcium dihydrogenphosphate, $Ca(H_2PO_4)_2$, and potassium chloride, KCl, which are mixed to give the desired percentages denoted by the numbers on the label. Other compounds that are sometimes used include urea, $CO(NH_2)_2$, calcium cyanamide, $CaCN_2$, which is toxic to plants and must be applied weeks in advance of planting, potassium sulfate, K_2SO_4, and ammonium chloride, NH_4Cl.

Pure chemicals also may be sold as fertilizers. For example, ammonium nitrate, NH_4NO_3, is a 35-0-0 fertilizer, because its mass percentage composition is 35% N; potassium nitrate, KNO_3, is a 13-0-38 fertilizer; anhydrous ammonia, $NH_3(g)$, is an 82-0-0 fertilizer.

2.80 Express the following data as suggested: (a) the mass (in grams) of 3.98 mol $Na_2B_4O_7 \cdot 10H_2O$ (known as borax), used as a water softener and for washing clothes; (b) the amount (in moles) in 9.47×10^{19} formula units of $K_2Hg(CN)_4$, used in the manufacturing of mirrors to prevent the yellowing of the silver coating.

2.81 One of the newly discovered materials that is a superconductor at about 90 K is called a "123 compound" because its formula is $YBa_2Cu_3O_{9-x}$. What is the mass ratio of the elements Y, Ba, and Cu in the compound?

2.82 L-Dopa, a drug used for the treatment of Parkinson's disease, is 54.8% carbon, 5.62% hydrogen, 7.10% nitrogen, and 32.6% oxygen. What is the empirical formula of the compound?

2.83 A chemist wants to extract the gold from 15.0 g of gold(III) chloride dihydrate, $AuCl_3 \cdot 2H_2O$, by electrolysis of an aqueous solution (this technique is explained fully in Chapter 17). What mass of gold can be obtained from the sample?

2.84 A chemical analysis of a complex carbohydrate is 40.0% C, 6.72% H, and 53.5% O. Its molar mass is approximately 860 g/mol. (a) What is the empirical formula of the carbohydrate? (b) What is the molecular formula of the carbohydrate?

2.85 Name the compounds: (a) $Co(NO_3)_2 \cdot 6H_2O$; (b) CuCl; (c) $Hg(NO_3)_2$; (d) V_2O_5; (e) $CrPO_4$.

2.86 Write formulas of (a) aluminum phosphate; (b) barium nitrate dihydrate; (c) silver sulfate; (d) sodium acetate; (e) copper(II) oxide.

QUESTIONS

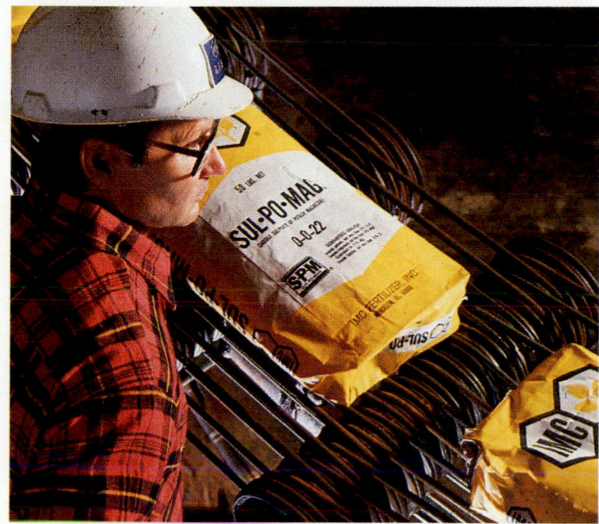

1 (a) What is the mass percentage of phosphorus in P_4O_{10}? (b) What is the mass percentage of potassium in K_2O? (c) What would a 20-10-5 label on a bag of fertilizer read if instead of referring to N, P_4O_{10}, and K_2O it referred to N, P, and K?

2 If a fertilizer label indicates 8% P_4O_{10}, what mass of $Ca(H_2PO_4)_2$ would be present in 100 g of fertilizer?

3 What should the fertilizer label read on a bag of ammonium sulfate, $(NH_4)_2SO_4$?

4 What should the numbers be on the label of a box of fertilizer if 20.0 g of $(NH_4)_2SO_4$, 10.0 g of $Ca(H_2PO_4)_2$, and 10.0 g of KCl are mixed with 100 g of filler and conditioner?

5 A corn crop removes about 35 kg of nitrogen, 5.5 kg of phosphorus, and 6.8 kg of potassium from one acre (0.4 hectare) of soil. Typical barnyard manure is a 0.25-0.25-0.50 fertilizer. What mass of barnyard manure must be applied to the one acre of soil to replenish (a) the nitrogen and (b) the phosphorus? (c) What mass of ammonium nitrate, calcium dihydrogenphosphate, and potassium chloride should be added to replenish the lost nutrients? (d) If the fertilizer used contains 50% filler and conditioner, what should the label for the mixed fertilizer in (c) be?

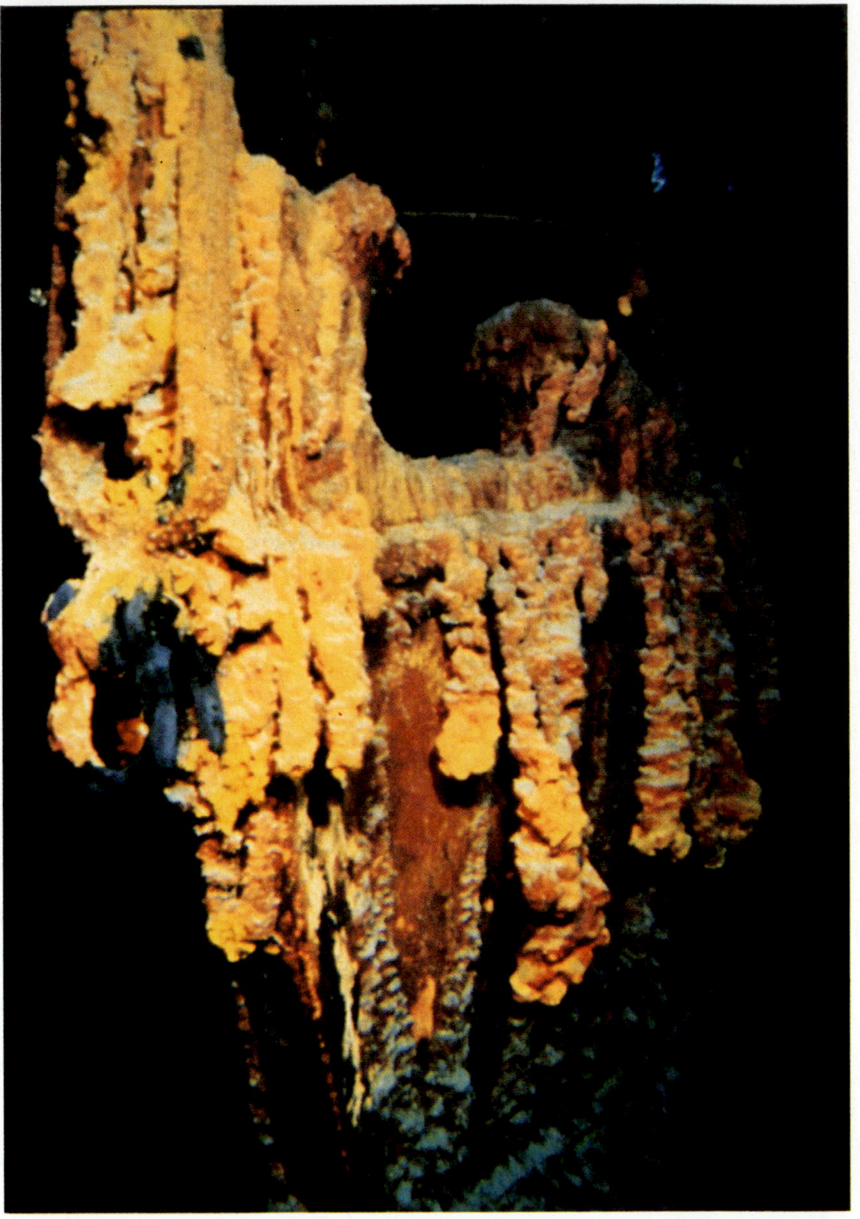

The change of one substance into others by chemical reaction is the essence of chemistry. In this chapter we see how chemists describe reactions symbolically, and how they organize observations on reactions by identifying certain characteristic features. The illustration shows the corrosion that has formed on the Titanic *during its immersion. The reactions responsible for corrosion are among those we encounter in this chapter.*

3 CHEMICAL REACTIONS

Chemists have developed a powerful language for discussing reactions in terms of chemical formulas and equations. The language, which we explore here, will allow us to group reactions into classes, some of which we discuss in this chapter. As so often happens in science, classification helps us recognize patterns that make the subject much easier to understand and use. Moreover, once we have identified the different types of reactions that are likely to occur when substances are mixed or heated, we can begin to make predictions about specific cases.

Several different types of reactions are described in this chapter (Table 3.1). Chemical reactions can be thought of as occurring as the result of certain "driving forces" that seem to "drive" a new substance out of the starting materials. These forces are only a convenient fiction, and we shall not discuss the *true* driving force of reactions until Chapter 16. One apparent driving force is the formation of an insoluble precipitate, which seems to drag a new substance into existence as it tumbles out of solution. Another apparent driving force is the formation of a gas, as occurs when hydrochloric acid is poured on calcium carbonate; the reaction proceeds because the carbon dioxide gas escapes. Two other apparent driving forces that will play a major role in this and later chapters are the formation of water from the reaction of two particular types of substances, acids and bases, and the tendency of one substance to transfer an electron to another substance.

D A simple reaction to produce foam. CD2, 25.

CHEMICAL EQUATIONS

The starting materials in a chemical reaction are called the **reactants**. The substances formed as a result of the reaction are called the **products**. A chemical reaction can therefore be summarized as

$$\text{Reactants} \longrightarrow \text{products}$$

N A unified approach to the study of chemical reactions in freshman chemistry. *J. Chem. Ed.* **1982**, *59*, 377.

Chemists speak of reagents as well as of reactants: a **reagent** is a substance or solution that reacts with other substances and, for that reason,

E Later, we shall see that reactions can be classified in a more fundamental manner, in terms of proton transfer (acid-base reactions), electron transfer (redox reactions), and electron-pair sharing (Lewis acid-base reactions). It is important to emphasize that there are several ways of classifying reactions.

N Are students to memorize any of these reaction types and which types will be covered in the lecture?

D Dancing spaghetti. CD2, 18.

TABLE 3.1 Classifications of reactions

Type of reaction	Distinguishing features*
Classification scheme 1 (based on compositions of reactants and products)	
synthesis	formation of compound from simpler starting materials *Example*: $2H_2(g) + O_2(g) \longrightarrow 2H_2O(l)$
decomposition	formation of simpler substances from the starting material *Example*: $CaCO_3(s) \longrightarrow CaO(s) + CO_2(g)$
double replacement (metathesis)	the exchange of partners *Example*: $2NaCl(aq) + Pb(NO_3)_2(aq) \longrightarrow 2NaNO_3(aq) + PbCl_2(s)$
combustion	reaction with oxygen to form CO_2, H_2O, N_2, and oxides of any other elements present *Example*: $CH_4(g) + 2O_2(g) \longrightarrow CO_2(g) + 2H_2O(g)$
corrosion	reaction of a metal with oxygen to form the metal oxide *Example*: $4Fe(s) + 3O_2(g) \longrightarrow 2Fe_2O_3(s)$
Classification scheme 2 (based on driving force of reaction)	
gas evolution	formation of gas *Example*: $CaCO_3(s) + 2HCl(aq) \longrightarrow CaCl_2(aq) + H_2O(l) + CO_2(g)$ *Driving force*: Escape of gas
precipitation	formation of precipitate when one solution is added to another *Example*: $3CaCl_2(aq) + 2Na_3PO_4(aq) \longrightarrow Ca_3(PO_4)_2(s) + 6NaCl(aq)$ *Driving force*: Formation of insoluble solid
neutralization	reaction between an acid and a base *Example*: $HCl(aq) + NaOH(aq) \longrightarrow NaCl(aq) + H_2O(l)$ *Driving force*: Formation of solvent (water)
redox	transfer of electrons from one species to another (accompanied by atoms in many cases) *Example*: $2Mg(s) + O_2(g) \longrightarrow 2MgO(s)$ *Driving force*: Electron transfer to achieve greater stability

*Key to symbols representing physical states of reactants and products: (*aq*), aqueous solution; (*g*), gas; (*l*), liquid; (*s*), solid.

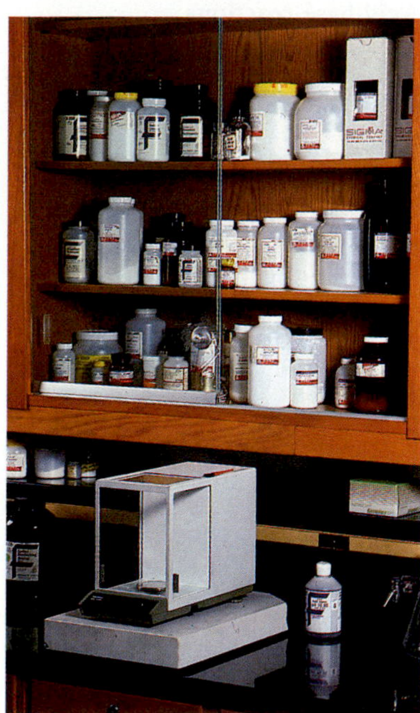

FIGURE 3.1 A typical chemical laboratory is equipped with a wide range of reagents that can be used as reactants.

is regularly stocked in chemistry laboratories (Fig. 3.1). Hydrochloric acid is one example of a reagent. A reactant is a reagent that is taking part in a specified reaction.

3.1 SYMBOLIZING REACTIONS

A simple example of a reaction is the formation of water from hydrogen and oxygen:

$$\text{Hydrogen} + \text{oxygen} \longrightarrow \text{water}$$

This reaction is an example of a **synthesis**, a reaction in which a substance is formed from simpler starting materials (in this case, elements). Another type of reaction is **decomposition**, in which a substance is broken down into simpler substances (Fig. 3.2). An example of a decomposition is the reaction that occurs when chalk (calcium carbonate) is heated to about 800°C:

$$\text{Calcium carbonate} \longrightarrow \text{calcium oxide} + \text{carbon dioxide}$$

This **thermal decomposition**, a decomposition brought about by heat, is typical of most carbonates. The driving force of this thermal decomposition is the escape of the gas from the solid calcium carbonate.

Chemists report a reaction by writing its **chemical equation**, which is an expression showing the chemical formulas of the reactants and products. For the reactions described above, chemists write

$$2H_2 + O_2 \longrightarrow 2H_2O$$

$$CaCO_3 \xrightarrow{\Delta} CaO + CO_2$$

Note that we distinguish a chemical reaction, the actual chemical change, from a chemical equation, its representation in terms of chemical symbols. The symbol Δ (the Greek letter delta) signifies that the reaction occurs at elevated temperatures (in this case, at about 800°C); the symbol is used only when it is desirable to emphasize that a high temperature is needed. The numbers multiplying *entire* chemical formulas are called the **stoichiometric coefficients** of the substances:

$$\underset{\uparrow}{2} \times H_2 + \underset{\uparrow}{1} \times O_2 \longrightarrow \underset{\uparrow}{2} \times H_2O$$

Stoichiometric coefficients

This awkward name comes from the Greek words for "element" and "measure." The stoichiometric coefficients are included to ensure that the same number of atoms of each element appears in both the reactants and the products, because atoms are neither created nor destroyed in a chemical reaction. A result of the preservation of the number of atoms is that mass is conserved (kept constant) in a chemical reaction.

In the chemical equation describing the formation of water from hydrogen and oxygen, there are four hydrogen atoms on the left of the arrow (in the reactants) and four on the right of the arrow (in the products); there are also two oxygen atoms on each side of the arrow. This chemical equation and all others therefore take into account Dalton's view that a chemical change is a *rearrangement* of atoms (Fig. 3.3).

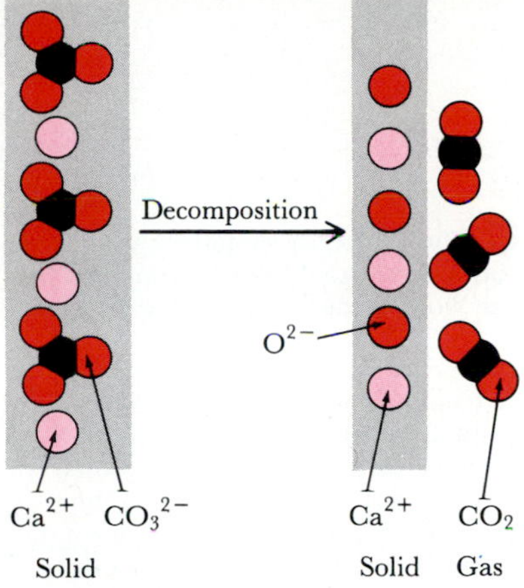

FIGURE 3.2 Solid calcium carbonate (an ionic compound) decomposes into calcium oxide (another ionic compound) and carbon dioxide (a gas). In a decomposition reaction a substance changes into less complex substances, which may be either elements or other compounds. In this case, the solid compound $CaCO_3$ decomposes into two compounds, the solid CaO and the gas CO_2.

T OHT: Figs. 3.2 and 3.3.

E The thermal decomposition of calcium carbonate is a key reaction in the cement industry.

N The illustration below is intended to indicate that a chemical equation shows the net or overall outcome of a reaction. As we shall see later, when we deal with the mechanisms of reactions, it is probable that a pair of H atoms from an H_2 molecule end up in two different H_2O molecules.

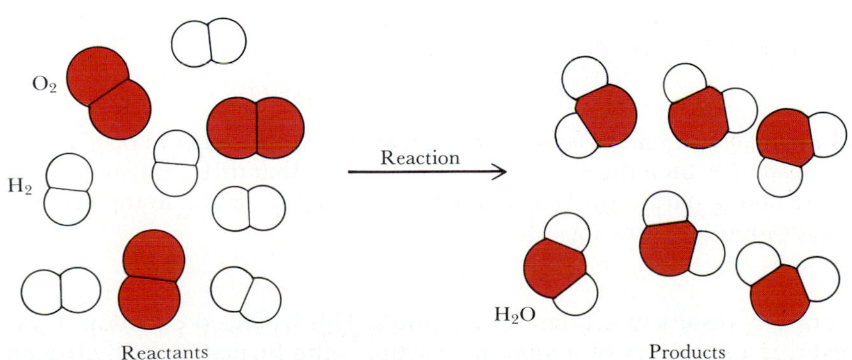

FIGURE 3.3 A chemical reaction is the rearrangement of atoms present in the reactants into the groupings characteristic of the products.

A chemical equation summarizes the identities of the substances (the qualitative information) and shows that the number of each type of atom is preserved (the quantitative information).

A chemical equation conveys even more information if it also shows the physical state (solid, liquid, or gas) of each substance taking part. This information is included by adding one of the following labels:

(*s*) for solid (*l*) for liquid (*g*) for gas

The label (*aq*) is used to show that a substance is in aqueous solution (dissolved in water). The full-dress versions of the equations given above are

$$2H_2(g) + O_2(g) \longrightarrow 2H_2O(l)$$

if liquid water (as distinct from water vapor or ice) is the product of the synthesis, and

$$CaCO_3(s) \xrightarrow{\Delta} CaO(s) + CO_2(g)$$

for the decomposition. The first equation states that hydrogen and oxygen (both of which exist as gases composed of diatomic molecules) form the molecular compound water as the liquid and that, for each O_2 molecule that reacts, two H_2O molecules are formed and two H_2 molecules are consumed. The second equation states that solid calcium carbonate decomposes into solid calcium oxide and carbon dioxide gas and that, for each $CaCO_3$ formula unit that decomposes, one CaO formula unit and one CO_2 molecule are formed.

N The symbol Δ should be used with caution. It should not be replaced by the word "heat" because that will lead to confusion when we come to consider exothermic and endothermic reactions. The symbol should be interpreted as "elevated temperature." How elevated is highly ambiguous, for in some cases it means a little above room temperature; in others it might mean over 1000°C. Many reactions need to be carried out at elevated temperature either for thermodynamic reasons or for kinetic reasons. We have used Δ sparingly, and have shown it only when the conditions are an essential part of the point we are making. The annotations will sometimes be more specific.

D Explosive reaction of hydrogen and oxygen. BZS1, 1.42.

N The chemical equation. Part 1: Simple reactions. *J. Chem. Ed.* **1978**, *55*, 507.

3.2 BALANCING EQUATIONS

N How should equation balancing be taught? *J. Chem. Ed.* **1985**, *62*, 507.

A chemical equation is **balanced** when the same number of atoms of each element appears on both sides of the arrow. (Those discussed so far have been balanced.) Simple reactions can often be balanced almost at a glance; others need more work.

Formulating equations. To formulate—and balance—a chemical equation, we begin by writing the **skeletal equation**, an unbalanced equation that summarizes the qualitative information about the reaction, for example,

$$\underset{\text{Hydrogen}}{H_2} + \underset{\text{oxygen}}{O_2} \underset{\text{forms}}{\longrightarrow} \underset{\text{water}}{H_2O} \qquad ⚠$$

(We shall use the international *hazard* road sign ⚠ to warn that a skeletal equation is not balanced.) Then we find the stoichiometric coefficients that balance all the elements:

$$2 \times (H_2) + O_2 \longrightarrow 2 \times (H_2O)$$

An equation must never be balanced by changing the subscripts in the chemical formulas, for then the equation would indicate that different substances were taking part in the reaction. Changing H_2O to H_2O_2 in the original expression and writing

$$H_2 + O_2 \longrightarrow H_2O_2$$

certainly results in a balanced equation. This balanced equation, however, is a summary of a *different* reaction—the formation of hydrogen peroxide from its elements. Similarly, writing

$$2H + O \longrightarrow H_2O$$

also gives a balanced reaction but one that expresses the reaction between hydrogen and oxygen *atoms,* not the diatomic molecules that are the actual starting materials.

A technique for balancing. Often, when we multiply one formula by a coefficient in order to balance a particular element, the balance of the other elements is upset. This process can be very frustrating when an equation contains a number of different elements. It is therefore wise to reduce the amount of work by proceeding systematically. One procedure is to balance the element that occurs the least number of times first and balance last the element that occurs the most times. With this procedure, subsequent changes affect only the previously balanced elements, which occur less often.

We illustrate this procedure by balancing the equation of a **combustion**, the reaction that occurs when an element or compound burns in oxygen (Fig. 3.4). When organic compounds containing only carbon, hydrogen, and oxygen burn in a plentiful supply of air, the only products are the gas carbon dioxide and water (which usually escapes as a vapor); if in addition the compound contains nitrogen, then N_2 gas is the most likely product in addition to carbon dioxide and water. It follows that if the organic compound is the methane (CH_4) of natural gas, the skeletal equation is

$$CH_4 + O_2 \longrightarrow CO_2 + H_2O \quad ⚠$$

Because C and H each occur in two formulas and O occurs in three, we begin with C and H. The C atoms are already balanced. We balance the H atoms by using a stoichiometric coefficient of 2 for H_2O to give four H atoms on each side:

$$CH_4 + O_2 \longrightarrow CO_2 + 2H_2O \quad ⚠$$

Now only O remains to be balanced. Because there are four O atoms on the right but only two on the left, the O_2 should have a stoichiometric coefficient of 2. The result is

$$CH_4 + 2O_2 \longrightarrow CO_2 + 2H_2O$$

The equation is now balanced, as we can verify by comparing the numbers of each type of element on both sides of the arrow. Now we specify the states. If water vapor is the product, we write

$$CH_4(g) + 2O_2(g) \longrightarrow CO_2(g) + 2H_2O(g)$$

FIGURE 3.4 A methane flame gives out heat as a result of the combustion of methane to carbon dioxide and water. The flame is blue because the heat generated by the reaction excites the C_2 and CH units formed in the course of the combustion; the excited molecules radiate blue light.

N Methane is the major component of natural gas, which is formed by anaerobic decomposition of organic matter. "Sour" natural gas contains hydrogen sulfide; "sweet" gas does not.

N A question of basic chemical literacy. *J. Chem. Ed.* **1989**, *66*, 217.

▼ EXAMPLE 3.1 Balancing equations

The element barium is obtained from its oxide, BaO, by reaction with molten aluminum. When aluminum and barium oxide are heated together, a vigorous reaction begins, and elemental liquid barium and aluminum oxide, Al_2O_3, are formed. Write the chemical equation for the reaction.

STRATEGY We begin by writing the unbalanced, skeletal equation for the conversion of the reactants (Al and BaO) to the products (Al_2O_3 and Ba). Because all three elements occur the same number of times (twice, in each case), we can start the balancing procedure with any one of them. Finally, we identify the physical states of the substances in the reaction.

D Place a small piece of calcium metal beneath an inverted test tube filled with water and collect the hydrogen gas; ignite the gas. Also, place a small piece of calcium metal in a beaker of water containing a few drops of phenolphthalein.

SOLUTION The skeletal equation is

$$Al + BaO \longrightarrow Al_2O_3 + Ba \quad ⚠$$

Balance the Al atoms with a stoichiometric coefficient of 2 for Al:

$$2Al + BaO \longrightarrow Al_2O_3 + Ba \quad ⚠$$

Balance the O atoms with a stoichiometric coefficient of 3 for BaO:

$$2Al + 3BaO \longrightarrow Al_2O_3 + Ba \quad ⚠$$

Finally, balance the Ba atoms by a stoichiometric coefficient of 3 for Ba:

$$2Al + 3BaO \longrightarrow Al_2O_3 + 3Ba$$

The physical states of the substances are added:

$$2Al(l) + 3BaO(s) \xrightarrow{\Delta} Al_2O_3(s) + 3Ba(l)$$

Because BaO and Al_2O_3 are ionic compounds, we predict that they will be solids. The delta is added over the arrow to indicate that the reaction takes place at an elevated temperature.

EXERCISE E3.1 The alkaline earth metal calcium reacts with cold water to form an aqueous solution of calcium hydroxide, $Ca(OH)_2$, and hydrogen gas. Write the balanced equation for the reaction.

▲ [*Answer*: $Ca(s) + 2H_2O(l) \rightarrow Ca(OH)_2(aq) + H_2(g)$]

In some cases, this balancing procedure leads to fractional stoichiometric coefficients. In the equation for the combustion of butane (C_4H_{10}), we get

$$C_4H_{10}(g) + \tfrac{13}{2}O_2(g) \longrightarrow 4CO_2(g) + 5H_2O(g)$$

Because it is much more convenient to deal with whole numbers, it is usually sensible (but not necessary) to clear the fractions by multiplying the *entire* equation by a numerical factor. We can clear the fraction $\tfrac{13}{2}$ by multiplying through by 2, thereby obtaining

$$2C_4H_{10}(g) + 13O_2(g) \longrightarrow 8CO_2(g) + 10H_2O(g)$$

PRECIPITATION REACTIONS

A soluble ionic compound dissolves in water to give a solution that contains cations and anions. An example is a solution of sodium chloride in water, which consists of Na^+ cations and Cl^- anions moving freely and largely independently throughout the solvent. Similarly, an aqueous solution of silver nitrate contains Ag^+ cations and NO_3^- anions. When we mix the two aqueous solutions, we immediately get a white precipitate, which analysis shows is silver chloride. The remaining solution still contains Na^+ cations and NO_3^- anions because sodium nitrate ($NaNO_3$) is soluble in water.

The reaction we have just described is an example of a precipitation reaction (Fig. 3.5):

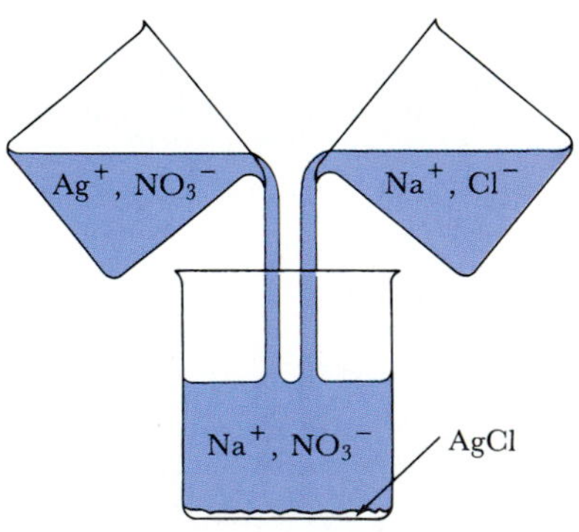

FIGURE 3.5 In a precipitation reaction, two soluble substances react to produce a precipitate of a new substance. A second, soluble product is generally left in solution.

A **precipitation reaction** is a reaction in which an insoluble solid product is formed when two solutions are mixed.

In our case, the reaction is

$$AgNO_3(aq) + NaCl(aq) \longrightarrow AgCl(s) + NaNO_3(aq)$$

A colorful example is the formation of the insoluble deep red solid silver chromate, Ag_2CrO_4, when solutions of potassium chromate and silver nitrate iodide are mixed (Fig. 3.6):

$$K_2CrO_4(aq) + 2AgNO_3(aq) \longrightarrow Ag_2CrO_4(s) + 2KNO_3(aq)$$

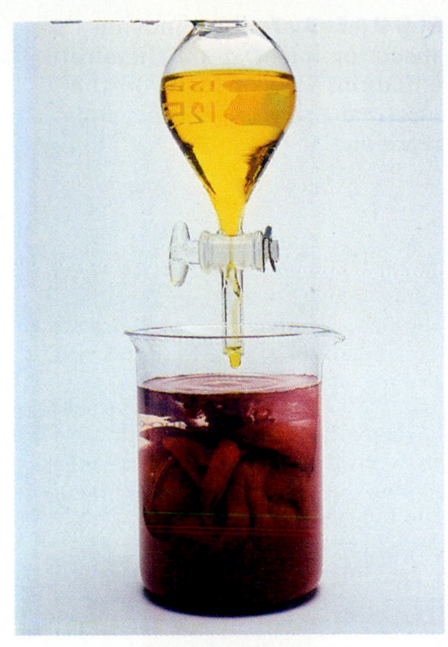

FIGURE 3.6 When potassium chromate solution (in the dropper) is added to a colorless silver nitrate solution (in the flask), a deep red solid, silver chromate, forms as a precipitate.

3.3 NET IONIC EQUATIONS

The essential character of a precipitation reaction becomes clear when we write its **ionic equation**, the chemical equation showing the ions explicitly. The ionic equation for the silver chloride precipitation shown in Fig. 3.5 is

$$Ag^+(aq) + \cancel{NO_3^-(aq)} + \cancel{Na^+(aq)} + Cl^-(aq) \longrightarrow AgCl(s) + \cancel{Na^+(aq)} + \cancel{NO_3^-(aq)}$$

Because the $Na^+(aq)$ and $NO_3^-(aq)$ ions appear as both reactants and products, they must not play a direct role in the reaction. In other words, they are **spectator ions**, ions that are present but remain unchanged. The **net ionic equation**, the equation that shows the net change, is obtained by canceling the spectator ions, a step that leaves

$$Ag^+(aq) + Cl^-(aq) \longrightarrow AgCl(s)$$

This equation focuses on the essential feature of the reaction (Fig. 3.7). It shows that in the precipitation Ag^+ ions in one solution combine with the Cl^- ions in the other, and precipitate as insoluble solid AgCl. As we remarked earlier, the precipitation of the insoluble solid is an example of an apparent driving force of a reaction.

E Silver chloride emulsions are used to produce photographic prints.

Because the hydroxides of Group I are soluble but those of other elements are not, we can predict that a solid hydroxide is obtained as a precipitate in reactions like

E Iron(III) hydroxide has the appearance of rust, $Fe_2O_3 \cdot xH_2O$.

$$FeCl_3(aq) + 3KOH(aq) \longrightarrow Fe(OH)_3(s) + 3KCl(aq)$$

In this case the driving force for the reaction is the precipitation of the insoluble hydroxide. The net ionic equation is readily found. First write the ionic equation:

$$Fe^{3+}(aq) + \cancel{3Cl^-(aq)} + \cancel{3K^+(aq)} + 3OH^-(aq) \longrightarrow Fe(OH)_3(s) + \cancel{3K^+(aq)} + \cancel{3Cl^-(aq)}$$

and then cancel the ions that appear on both sides of the arrow:

$$Fe^{3+}(aq) + 3OH^-(aq) \longrightarrow Fe(OH)_3(s)$$

We see that the net outcome of the reaction is the formation of the solid iron(III) hydroxide from Fe^{3+} and OH^- ions.

▼ EXAMPLE 3.2 Writing a net ionic equation

The chemical equation for the precipitation reaction in which barium iodate, $Ba(IO_3)_2$, is formed when aqueous solutions of barium nitrate, $Ba(NO_3)_2$, and ammonium iodate, NH_4IO_3, are mixed is

$$Ba(NO_3)_2(aq) + 2NH_4IO_3(aq) \longrightarrow Ba(IO_3)_2(s) + 2NH_4NO_3(aq)$$

FIGURE 3.7 By removing the spectator ions from a chemical equation, we can focus on the essential process—the net ionic reaction.

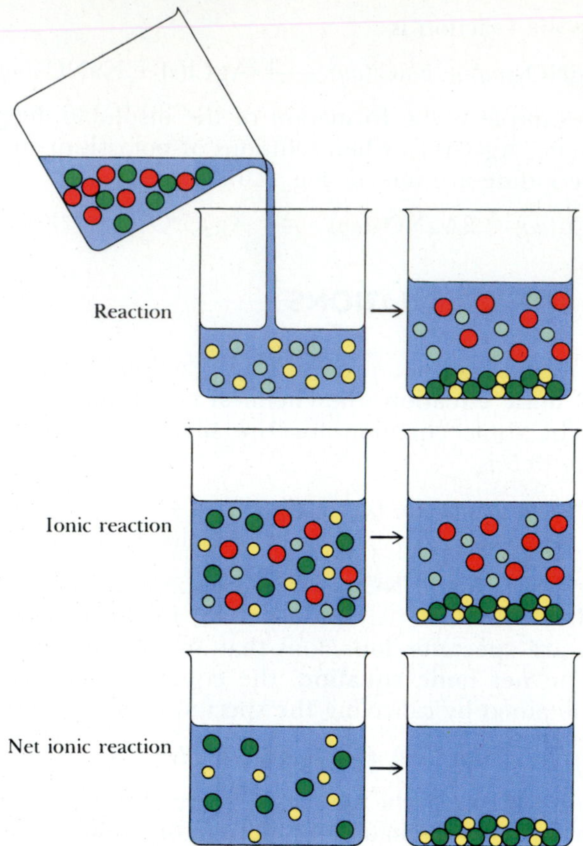

N In a number of cases throughout the text, we juxtapose a schematic diagram (like the one shown here) and a photograph of an actual process (as in Fig. 3.8). We consider it an important pedagogical point to establish a bridge between the observed process and a chemist's interpretation in terms of the behavior of atoms, molecules, and ions.

T OHT: Fig. 3.7.

D Use a medicine dropper to add a precipitating reagent to a salt solution in a glass dish placed on an overhead projector.

V Video 39: Precipitation I. Reactions of a cation.

V Video 40: Precipitation II. Reactions of an anion.

Write the net ionic equation of the reaction.

STRATEGY We write the balanced ionic equation showing all the ions in solution *separately*, as they actually exist in their separate solutions, then cancel the ions that remain in solution on both sides.

SOLUTION The ionic equation, with the ions written as they exist in solution, is

$$Ba^{2+}(aq) + \cancel{2NO_3^-(aq)} + \cancel{2NH_4^+(aq)} + 2IO_3^-(aq) \longrightarrow Ba(IO_3)_2(s) + \cancel{2NH_4^+(aq)} + \cancel{2NO_3^-(aq)}$$

Cancellation of the ions $NH_4^+(aq)$ and $NO_3^-(aq)$ on each side of the arrow leaves

$$Ba^{2+}(aq) + 2IO_3^-(aq) \longrightarrow Ba(IO_3)_2(s)$$

as the net ionic equation.

EXERCISE E3.2 Write the net ionic equation for the reaction in which aqueous solutions of silver nitrate and sodium chromate are mixed and give a precipitate of silver chromate.

[*Answer*: $2Ag^+(aq) + CrO_4^{2-}(aq) \rightarrow Ag_2CrO_4(s)$]

N The principles outlined in this section could be delayed until Section 15.6.

D Precipitate formation: black and white. CD1, 18. Precipitate formation: blue. CD1, 19.

3.4 USING PRECIPITATION REACTIONS IN CHEMISTRY

If a compound MX is insoluble in water, then the net ionic equation for its precipitation is

$$M^+(aq) + X^-(aq) \longrightarrow MX(s)$$

An example is

$$Ag^+(aq) + Cl^-(aq) \longrightarrow AgCl(s)$$

Therefore, if we can find two soluble compounds MA and BX, where A^- and B^+ act as spectator ions, then the general form of the precipitation reaction will be

$$MA(aq) + BX(aq) \longrightarrow MX(s) + BA(aq)$$

For example,

$$AgNO_3(aq) + NaCl(aq) \longrightarrow AgCl(s) + NaNO_3(aq)$$

Precipitation reactions are one kind of **double replacement reaction** (or metathesis, from the Greek words for "transpose"), a reaction in which atoms or ions exchange partners. Thus, in the starting solutions Ag^+ and NO_3^- were partners, as were the ions Na^+ and Cl^- of the second solution. After the reaction, Ag^+ and Cl^- are partners in the solid precipitate and Na^+ and NO_3^- are partners in the remaining solution.

By choosing the starting solutions carefully, we can use a precipitation reaction to make the compound we want and then separate it from the reaction mixture by filtration (Fig. 3.8). However, to choose the starting solutions, we need to know which compounds are soluble and which are not. Once we know that, we should be able to predict what will precipitate (if anything) when we mix any two aqueous solutions.

The empirical (observation-based) **solubility rules** given in Table 3.2 summarize the solubility patterns of a range of common compounds in water. We see, for instance, that PbI_2 is insoluble. Thus, it is formed whenever solutions containing Pb^{2+} and I^- are mixed:

$$Pb^{2+}(aq) + 2I^-(aq) \longrightarrow PbI_2(s)$$

N See Fig. 15.16 for the formation of lead(II) iodide.

Because $Pb(NO_3)_2$ is soluble (like all nitrates) and KI is soluble (like all common Group I compounds), we can predict that mixing solutions of

(a)

(b)

FIGURE 3.8 Two precipitation reactions for preparing compounds. (a) Sodium sulfide solution is added to a lead nitrate solution, giving a precipitate of black lead sulfide. (b) Sodium carbonate solution is added to a cobalt(II) chloride solution, giving a precipitate of blue cobalt(II) carbonate.

OHT: Table 3.2.

TABLE 3.2 Solubility of ionic compounds

Classes of soluble compounds	Classes of insoluble compounds
compounds of Group I elements	carbonates (CO_3^{2-}), phosphates (PO_4^{3-}), oxalates ($C_2O_4^{2-}$), and chromates (CrO_4^{2-}), **EXCEPT** Group I elements and NH_4^+
ammonium (NH_4^+) compounds	sulfides (S^{2-}), **EXCEPT** Group I elements, Group II elements, and NH_4^+
chlorides (Cl^-), bromides (Br^-), and iodides (I^-), **EXCEPT** Ag^+, Hg_2^{2+}, Pb^{2+}	hydroxides (OH^-) and oxides (O^{2-}), **EXCEPT** Group I elements and Group II elements*
nitrates (NO_3^-), chlorates (ClO_3^-), perchlorates (ClO_4^-), acetates ($CH_3CO_2^-$)	
sulfates (SO_4^{2-}), **EXCEPT** Sr^{2+}, Ba^{2+}, Pb^{2+}	

*$Ca(OH)_2$ and $Sr(OH)_2$ are sparingly (slightly) soluble; $Mg(OH)_2$ is only very slightly soluble.

lead(II) nitrate and potassium iodide will result in the precipitation of lead(II) iodide:

$$Pb^{2+}(aq) + 2NO_3^-(aq) + 2K^+(aq) + 2I^-(aq) \longrightarrow PbI_2(s) + 2K^+(aq) + 2NO_3^-(aq)$$

The K^+ and NO_3^- are spectator ions and remain in solution. The same precipitate can be expected when we mix *any* soluble lead(II) compound with *any* soluble iodide, because in every case the lead(II) and iodide ions will form insoluble lead(II) iodide. For example, we could also form lead(II) iodide from aqueous solutions of lead(II) acetate and sodium iodide, both of which are soluble:

$$Pb(CH_3CO_2)_2(aq) + 2NaI(aq) \longrightarrow PbI_2(s) + 2NaCH_3CO_2(aq)$$

Now the Na^+ and $CH_3CO_2^-$ ions are the spectators. The ionic equation is

$$Pb^{2+}(aq) + 2CH_3CO_2^-(aq) + 2Na^+(aq) + 2I^-(aq) \longrightarrow PbI_2(s) + 2Na^+(aq) + 2CH_3CO_2^-(aq)$$

After cancellation of the spectator ions, the net ionic equation is

$$Pb^{2+}(aq) + 2I^-(aq) \longrightarrow PbI_2(s)$$

which is the same as before.

▼ EXAMPLE 3.3 Predicting the outcome of a precipitation reaction

Predict the likely product when aqueous solutions of sodium phosphate, Na_3PO_4, and lead(II) nitrate, $Pb(NO_3)_2$, are mixed, and write the net ionic equation.

STRATEGY All compounds of Group I are soluble, but phosphates of other elements are generally insoluble. Hence, we predict that lead(II) phosphate will precipitate. Refer to the solubility rules in Table 3.2 for further information.

SOLUTION The compounds Na_3PO_4 (like all Group I compounds) and $Pb(NO_3)_2$ (like all nitrates) are soluble. Of the two possible products of a double replacement reaction, $NaNO_3$ and $Pb_3(PO_4)_2$, the former is soluble and the latter is insoluble. We therefore predict that $Pb_3(PO_4)_2$ will be formed. The net ionic equation is

$$3Pb^{2+}(aq) + 2PO_4^{3-}(aq) \longrightarrow Pb_3(PO_4)_2(s)$$

EXERCISE E3.3 Predict the identity of the solid product when aqueous solutions of ammonium sulfide, $(NH_4)_2S$, and copper(II) sulfate, $CuSO_4$, are mixed, and write the net ionic equation of the reaction.

[*Answer*: copper(II) sulfide; $Cu^{2+}(aq) + S^{2-}(aq) \rightarrow CuS(s)$]

We now have enough information to be able to predict when a precipitation reaction is likely to occur. If the solutions contain ions that, when mixed, correspond to one of the insoluble compounds listed in Table 3.2, then that compound will form a solid precipitate that we can separate from the remaining ions by filtration.

ACID-BASE REACTIONS

In Section 2.8 we defined an acid as a compound that contains hydrogen and releases hydrogen ions in water. This definition was proposed by the Swedish chemist Svante Arrhenius toward the end of the nineteenth century. We begin this discussion of acids and bases with the outdated but still quite useful Arrhenius definitions and then turn to a more modern definition.

U Exercises 3.68 and 3.69.

N Acids and bases. *J. Chem. Ed.* **1978**, *55*, 459.

3.5 ARRHENIUS ACIDS AND BASES

Arrhenius defined acids and bases as substances with the following characteristics:

An **Arrhenius acid** is a compound that contains hydrogen and releases hydrogen ions (H^+) in water.

An **Arrhenius base** is a compound that produces hydroxide ions (OH^-) in water.

Arrhenius acids and bases are recognized by their effect on the color of certain dyes known as indicators (Fig. 3.9). (We will learn more about

N The emphasis on *compound* has considerable force in this definition: sodium metal produces OH^- ions in water, but it is not an Arrhenius base.

D Food is usually acidic, cleaners are usually basic. BZS3, 8.6.

D Use several indicators to identify acids and bases. Other common substances that act as indicators include grape juice, tannin (in tea), Ex-Lax (which contains phenolphthalein), and red rose petals.

V Video 15: Indicators.

FIGURE 3.9 The acidities of various household products can be demonstrated by adding an indicator (an extract of red cabbage in this case) and noting the resulting color. Red indicates acidic, blue basic. From left to right, the household products are lemon juice, soda water, 7-Up, vinegar, ammonia, Drāno, milk of magnesia, and detergent in water. Note that ammonia and Drāno are such strong bases that they destroy the red cabbage dye, so no blue color is visible.

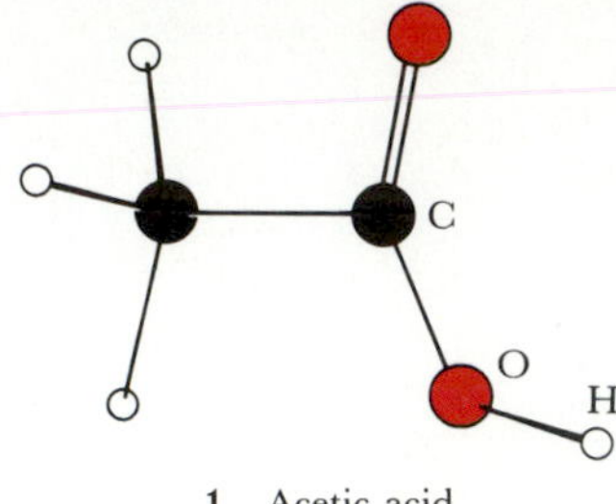

1 Acetic acid

indicators and their uses in Section 15.4.) One of the most famous indicators is litmus, a vegetable dye obtained from lichen. Aqueous solutions of acids turn litmus red; aqueous solutions of bases turn it blue.

Hydrogen chloride, HCl, and nitric acid, HNO_3, are Arrhenius acids because they both contain hydrogen and release hydrogen ions in water. Methane, CH_4, is not an Arrhenius acid because, although it contains hydrogen, it does not release hydrogen ions in water. Acetic acid, CH_3COOH (**1**), releases *one* of its hydrogen atoms as H^+ in water (the one attached to an oxygen atom), so it is an Arrhenius acid. As we saw in Section 2.8, an acidic hydrogen atom (the hydrogen atom that can be released in water) is often written first in a molecular formula. Carboxylic acids, however, are normally written in terms of their (abbreviated) structural formula and contain the group —COOH. It follows that we can generally recognize which compounds are Arrhenius acids by noting whether their molecular formulas begin with H or, if they are organic compounds, whether they contain the group —COOH. Thus, we can immediately recognize that HCl, H_2CO_3, HSO_3^-, and C_6H_5COOH (benzoic acid) are Arrhenius acids and that CH_4, NH_3, and $CH_3CO_2^-$ (the acetate ion) are not.

E Sodium hydroxide is found in Drano and most oven cleaners; it is used for the manufacture of soap from fats.

E Aqueous ammonia mixed with soap is the "sudsy ammonia" used as a household cleaner.

Sodium hydroxide is an Arrhenius base because it dissolves in water to give a solution of hydroxide ions. Ammonia, NH_3, is also an Arrhenius base because, even though it does not contain hydroxide ions, when it dissolves in water it gives rise to them as a result of the reaction

$$NH_3(g) + H_2O(l) \longrightarrow NH_4^+(aq) + OH^-(aq)$$

Aqueous solutions of bases are called **alkalis**: thus, aqueous sodium hydroxide and aqueous ammonia are both alkalis.

Acids and bases are produced in huge amounts by industry (Table 3.3). Sulfuric acid, for example, is manufactured in greater mass than

TABLE 3.3 The top twelve acids and bases (by mass manufactured in the United States)

Substance	Rank*	Production, 10^9 kg†	Class
sulfuric acid H_2SO_4	1	40.3	acid
lime CaO	5	15.8	base
ammonia NH_3	6	15.4	base
phosphoric acid H_3PO_4	7	11.1	acid
sodium hydroxide NaOH	8	10.0	base
sodium carbonate Na_2CO_3	11	9.0	base‡
urea $CO(NH_2)_2$	12	7.2	base
nitric acid HNO_3	13	7.1	acid
terephthalic acid $C_6H_4(COOH)_2$	22	3.50	acid
hydrochloric acid HCl	31	2.13	acid
acetic acid CH_3COOH	33	1.71	acid
adipic acid $C_4H_8(COOH)_2$	46	0.75	acid

*Rank among all manufactured chemicals.
†Data for 1990.
‡For reasons explained in Chapter 15, a solution of sodium carbonate is basic.

E Terephthalic acid is used in the manufacture of polyesters and adipic acid is used in the manufacture of nylon (see Chapter 24).

any other chemical. Phosphoric acid ranks seventh; nitric acid, thirteenth; and hydrochloric acid is thirty-first. Sulfuric acid was until recently so important to the manufacturing industry that its annual production was used as a measure of a nation's degree of industrialization and commercial prosperity. However, most sulfuric acid is now used to make fertilizers from phosphates, so its production is increasingly a measure of a nation's *agricultural* activity (Fig. 3.10). Bases too are produced on an enormous scale. Ammonia ranks sixth in annual mass produced among industrial chemicals; however, it ranks first according to the number of molecules manufactured because each NH_3 molecule is very light. (The molar mass of ammonia is 17 g/mol; sulfuric acid's is 98 g/mol.) Sodium hydroxide ranks eighth in annual production.

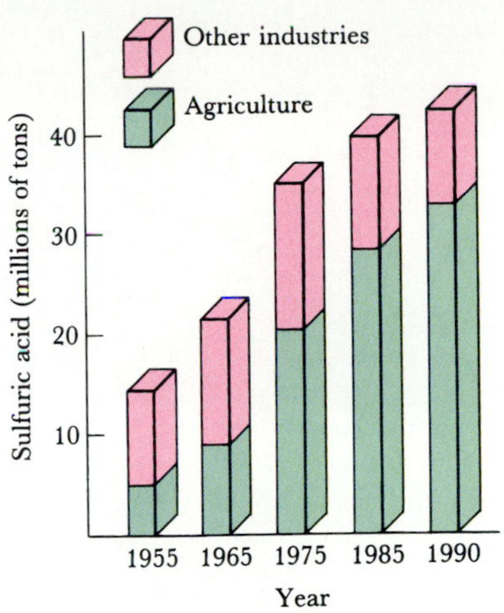

FIGURE 3.10 The uses of sulfuric acid have shifted over the past 35 years from mainly industrial to mainly agricultural.

D The removal of ions by metathesis reactions. TDC, 163. (See also TDC 6.14.)

3.6 NEUTRALIZATION

The reaction between an acid and a base is the **neutralization** of one by the other to form a "neutral" compound, which is neither an acid nor a base, and a molecule of the solvent (water). For example, the reaction between hydrochloric acid and aqueous sodium hydroxide leads to the formation of sodium chloride and water, neither of which is an Arrhenius acid or base:

$$HCl(aq) + NaOH(aq) \longrightarrow NaCl(aq) + H_2O(l)$$

Ionic compounds that can be formed in this way are called **salts**, that name coming from one of the best-known ionic compounds, table salt or sodium chloride. The general form of an Arrhenius neutralization is therefore

$$\text{Acid} + \text{base} \longrightarrow \text{salt} + \text{water}$$

As in the formation of sodium chloride, the cation of the salt is provided by the base and the anion is provided by the acid:

$$\underset{HCl}{\text{Acid}} + \underset{NaOH}{\text{base}} \longrightarrow \underset{NaCl}{[\text{cation}][\text{anion}]} + \underset{H_2O}{\text{water}}$$

The net ionic reaction is obtained by writing out the chemical equation in terms of ions and canceling the spectator ions that appear on both sides of the arrow:

$$H^+(aq) + \cancel{Cl^-(aq)} + \cancel{Na^+(aq)} + OH^-(aq) \longrightarrow \cancel{Na^+(aq)} + \cancel{Cl^-(aq)} + H_2O(l)$$

The resulting equation is

$$H^+(aq) + OH^-(aq) \longrightarrow H_2O(l)$$

That is, the net reaction of a neutralization of an Arrhenius acid and base is the formation of water from hydrogen ions and hydroxide ions. This net water-forming reaction is the apparent driving force of neutralization reactions.

▼ EXAMPLE 3.4 Predicting the outcome of a neutralization reaction

What salt will be obtained in the reaction between aqueous solutions of barium hydroxide and sulfuric acid?

STRATEGY The base (barium hydroxide) provides the cation, Ba^{2+}, and the acid provides the anion (SO_4^{2-}).

SOLUTION The neutralization will result in the formation of insoluble barium sulfate (see Table 3.2) in the reaction

$$Ba(OH)_2(aq) + H_2SO_4(aq) \longrightarrow BaSO_4(s) + 2H_2O(l)$$

EXERCISE E3.4 What acid and base solutions could you use to prepare rubidium nitrate?

▲ [*Answer*: $RbOH(aq) + HNO_3(aq) \rightarrow RbNO_3(aq) + H_2O(l)$]

Note that a neutralization reaction may also be a precipitation reaction. Thus, in the reaction between barium hydroxide and sulfuric acid treated in Example 3.4, the barium sulfate is insoluble and precipitates as soon as it is formed:

precipitation

$$Ba^{2+}(aq) + 2OH^-(aq) + 2H^+(aq) + SO_4^{2-}(aq) \longrightarrow BaSO_4(s) + 2H_2O(l)$$

neutralization

3.7 THE BRØNSTED DEFINITIONS

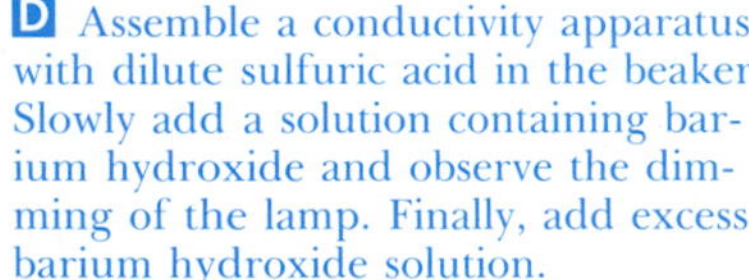

D Assemble a conductivity apparatus with dilute sulfuric acid in the beaker. Slowly add a solution containing barium hydroxide and observe the dimming of the lamp. Finally, add excess barium hydroxide solution.

N The Brønsted-Lowry acid-base concept. *J. Chem. Ed.* **1988**, *65*, 28.

In 1923, the Danish chemist Johannes Brønsted proposed more fundamental definitions of acids and bases:

A **Brønsted acid** is a proton donor.

A **Brønsted base** is a proton acceptor.

The same definitions were proposed independently by the English chemist Thomas Lowry, and the theory based on them is widely called the **Brønsted-Lowry theory** of acids and bases (Fig. 3.11). When we speak of an acid or a base from now on, we will mean a species that fits the Brønsted-Lowry definitions. In the context of Brønsted acids and bases, a proton means a hydrogen ion (H^+). Thus, the definitions focus attention on the transfer of H^+ from a donor molecule or ion (the acid) to an acceptor molecule or ion (the base). These definitions are wider in scope than the Arrhenius definitions, do not make an arbitrary distinction between elements and compounds, and focus on the *essential* feature of acids and bases—that is, their ability to participate in the transfer of protons. We shall see in Chapters 14 and 15 that the Brønsted-Lowry definitions are also more helpful in practice, because they enable us to account for the behavior of solutions of a wide range of substances.

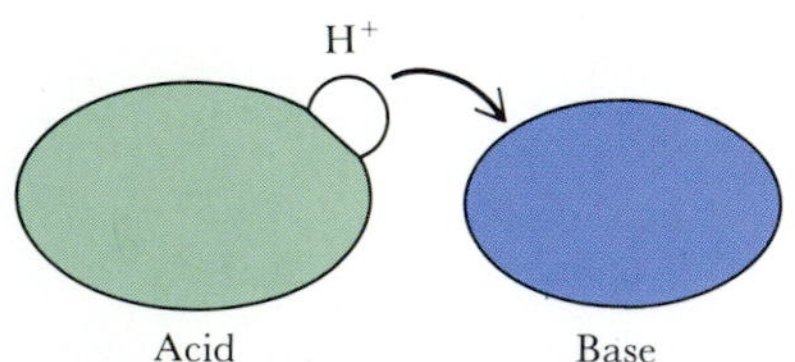

FIGURE 3.11 A Brønsted acid is a proton donor; a Brønsted base is a proton acceptor.

Strong and weak Brønsted acids and bases. The Arrhenius acid HCl is also an acid according to the Brønsted definition: when it dissolves in water it donates a proton to a neighboring water molecule. That is,

$$HCl(aq) + H_2O(l) \longrightarrow H_3O^+(aq) + Cl^-(aq)$$

Because proton donation leads to the formation of ions (HCl, for example, forming H_3O^+ and Cl^-), we call the process **ionization** and say that HCl is ionized when it dissolves in water.

A water molecule that acts as a Brønsted base and accepts a proton from a proton donor becomes a **hydronium ion**, H_3O^+ (**2**). The H^+ ion is so firmly attached to the H_2O molecule that we should think of the hydronium ion as a polyatomic cation (the analog of $NH_4{}^+$), not just a loosely hydrated proton. We shall see in later chapters that the hydronium ion plays a very special role in chemistry.

All the hydrogen chloride molecules that dissolve in water ionize to hydronium ions and chloride ions. As a result, HCl is an example of a strong acid:

A **strong acid** is an acid that is fully ionized in aqueous solution at normal concentrations.

There are very few strong acids, and all of them are listed in Table 3.4. The majority of acids are classified as weak acids:

A **weak acid** is an acid that is only partially ionized in aqueous solution at normal concentrations.

For example, a typical aqueous solution of acetic acid consists of about 95% nonionized acetic acid molecules and about 5% acetate ions and hydronium ions that are the product of its ionization.

The organic compound phenol, C_6H_5OH, which is used to make the phenolic resins for electric plugs, only partially ionizes in water to produce hydronium and phenoxide ions, $C_6H_5O^-$. It is therefore a weak Brønsted acid.

$$C_6H_5OH(aq) + H_2O(l) \longrightarrow H_3O^+(aq) + C_6H_5O^-(aq)$$

Most oxides of nonmetals, such as SO_2 and P_4O_{10}, are **acidic oxides,** because they produce hydronium ions in aqueous solutions:

$$SO_2(g) + 2H_2O(l) \longrightarrow H_3O^+(aq) + HSO_3{}^-(aq)$$

$$P_4O_{10}(s) + 10H_2O(l) \longrightarrow 4H_3O^+(aq) + 4H_2PO_4{}^-(aq)$$

Carbon dioxide dissolved in water produces the slight acidity of many soft drinks. When sodium hydroxide dissolves in water, the hydroxide ions it provides can accept protons from any hydronium ions present:

$$H_3O^+(aq) + OH^-(aq) \longrightarrow 2H_2O(aq)$$

Because OH^- accepts a proton, it is a Brønsted base. Hence, anything that contains OH^- ions (such as sodium hydroxide) behaves as a base.

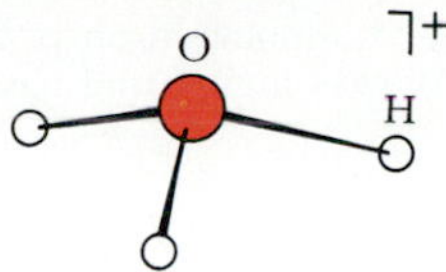

2 Hydronium ion

E The extent of ionization of all acids, including strong acids, does in fact depend on the concentration of the acid. However, at normal concentrations (of the order of 1 M and less), the fraction of acid molecules that are not ionized is negligibly small. Similarly, a weak acid may approach a fully ionized state as the concentration is reduced toward zero, but in the systems we normally encounter (where the molar concentration is 0.1 M or more), a high proportion of the acid molecules remain intact. These points are discussed quantitatively in Chapter 14. The best distinction between weak and strong acids is in terms of their acidity constants (K_a).

N The great fallacy of the H^+ ion and the true nature of H_3O^+. *J. Chem. Ed.* **1979**, *56*, 571.

U Exercise 3.71.

E Phenol is commonly called carbolic acid.

D Differences between acid strength and concentration. BZS3, 8.20.

T OHT: Table 3.4.

N Which strong acids and bases should be memorized?

TABLE 3.4 The strong acids and strong bases (in water)

Strong acids
hydrobromic acid HBr
hydrochloric acid HCl
hydroiodic acid HI
nitric acid HNO_3
perchloric acid $HClO_4$
chloric acid $HClO_3$
sulfuric acid H_2SO_4 (to $HSO_4{}^-$)
Strong bases
Group I hydroxides (soluble): LiOH, NaOH, KOH, RbOH, CsOH
Group II hydroxides (a trend from sparingly soluble to soluble): $Ca(OH)_2$*, $Sr(OH)_2$, $Ba(OH)_2$†

*Sparingly soluble.
†Soluble.

N The first annotation on p. 99 also applies to weak and strong bases.

U Exercise 3.67.

N Strong and weak acids and bases. *J. Chem. Ed.* **1979**, *56*, 814.

Compounds that do not contain OH^- ions may also be bases. An oxide ion, for instance, is a Brønsted base. When an ionic oxide (such as CaO) dissolves in water, the negative charge of the O^{2-} ions attracts protons from water molecules, and the oxide ion is converted to a hydroxide ion:

$$O^{2-}(aq) + H_2O(l) \longrightarrow 2OH^-(aq)$$

The strong ability of O^{2-} to attract protons accounts for the observation that aqueous solutions of ionic oxides do not exist: all soluble oxides immediately form hydroxides when they dissolve. Most oxides of metals, such as Li_2O and CaO, are **basic oxides,** because they produce hydroxide ions in aqueous solution.

$$Li_2O(s) + H_2O(l) \longrightarrow 2Li^+(aq) + 2OH^-(aq)$$

$$CaO(s) + H_2O(l) \longrightarrow Ca^{2+}(aq) + 2OH^-(aq)$$

Calcium oxide (known as lime or quicklime) reacts so vigorously with water that the wooden boats once used to transport it sometimes caught fire. It is more safely transported as "slaked lime," $Ca(OH)_2$, lime with its thirst for water slaked. Substances that are ionic in the solid state, such as sodium hydroxide and barium hydroxide, are also fully ionized in aqueous solution, and are classified as strong bases:

A **strong base** is a base that is fully ionized in aqueous solution at normal concentrations.

All soluble and partially soluble ionic hydroxides are strong bases, which includes all the alkali metal hydroxides and the alkaline earth metal hydroxides. As we have just seen, the oxides of these elements also react with water to produce hydroxide ions; hence, the oxides of Groups I and II metals are strong bases in water (see Table 3.4).

Ammonia is also a proton acceptor and hence a Brønsted base:

$$NH_3(aq) + H_2O(l) \longrightarrow NH_4^+(aq) + OH^-(aq)$$

N With which weak bases should the student be familiar?

(In this example H_2O has provided the proton and thus serves as a Brønsted acid.) This behavior of NH_3 resembles that of the oxide ion. However, whereas all the O^{2-} ions are converted to OH^- ions in water, only a small proportion (typically 1 in 100) of the NH_3 molecules are converted to NH_4^+ ions because an uncharged NH_3 molecule has much less proton-accepting power than a highly charged O^{2-} ion. Because only a small proportion of the ammonia molecules form ions in solution, ammonia is classified as a weak base:

A **weak base** is a base that is only partially ionized in solutions at normal concentrations.

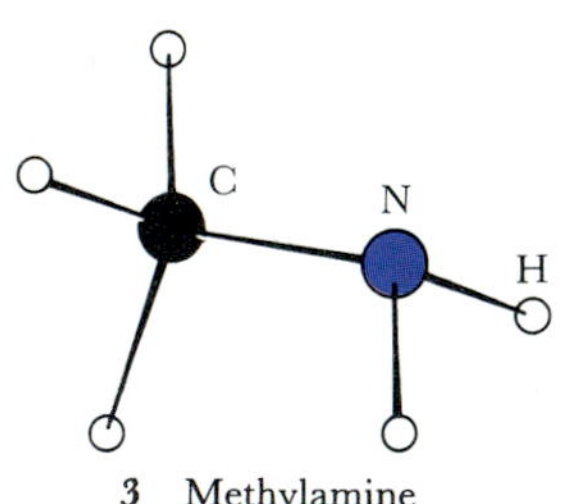

3 Methylamine

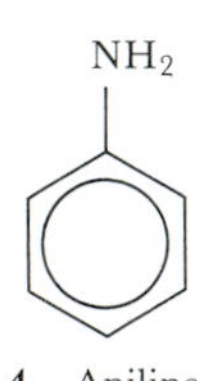

4 Aniline

Many organic compounds that contain a nitrogen atom are weak bases in water. They include methylamine, CH_3NH_2 (**3**), which contributes to the smell of dogs and rotting fish, and aniline, $C_6H_5NH_2$ (**4**), a compound widely used in the manufacture of dyes.

Proton transfer. We saw in Section 3.6 that we can identify the essential feature of a neutralization reaction by writing a typical salt formation reaction, such as

$$HCl(aq) + NaOH(aq) \longrightarrow NaCl(aq) + H_2O(l)$$

as a net ionic reaction. According to the Brønsted-Lowry theory, the hydrochloric acid is present as hydronium ions (not as hydrogen ions) and chloride ions, and the sodium hydroxide is present as sodium ions and hydroxide ions. Therefore, the reaction is

$$H_3O^+(aq) + Cl^-(aq) + Na^+(aq) + OH^-(aq) \rightarrow Na^+(aq) + Cl^-(aq) + 2H_2O(l)$$

As we saw before, the sodium and chloride ions are spectators in this reaction and cancel on each side of the arrow. The net ionic reaction is therefore

$$H_3O^+(aq) + OH^-(aq) \longrightarrow 2H_2O(l)$$

That is, the net ionic reaction is *proton transfer* from a hydronium ion to a hydroxide ion. In a reaction between Brønsted acids and bases in aqueous solution, therefore, the apparent driving force is the transfer of a proton from a hydronium ion to a hydroxide ion and hence the formation of water (Fig. 3.12).

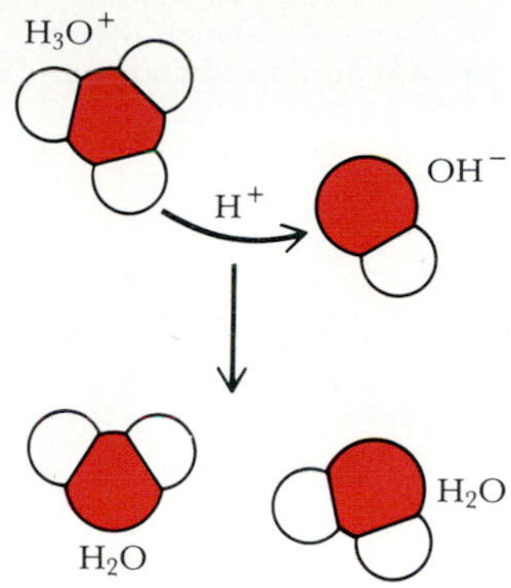

FIGURE 3.12 The net ionic reaction between an acid and a base is the formation of water by the transfer of a proton from a hydronium ion to a hydroxide ion.

▼ EXAMPLE 3.5 Expressing a neutralization reaction in terms of proton transfer

When aqueous ammonia reacts with hydrochloric acid, ammonium chloride is produced. Confirm that the neutralization reaction takes place by proton transfer.

STRATEGY We must show that the reaction involves the transfer of a proton from an acid to a base. Because most of the NH_3 molecules are present in nonionized form and all the HCl molecules are ionized, we should be able to express the reaction as proton transfer from H_3O^+ ions to NH_3 molecules.

SOLUTION The ionic reaction can be written

$$NH_3(aq) + H_3O^+(aq) + Cl^-(aq) \longrightarrow NH_4^+(aq) + Cl^-(aq) + H_2O(l)$$

and the net ionic reaction is

$$NH_3(aq) + H_3O^+(aq) \longrightarrow NH_4^+(aq) + H_2O(l)$$

We see that proton transfer has occurred from H_3O^+ to NH_3.

EXERCISE E3.5 Show that the reaction between hydrochloric acid and silver nitrate is *not* an acid-base proton-transfer reaction.

[*Answer*: net ionic reaction, $Ag^+(aq) + Cl^-(aq) \rightarrow AgCl(s)$; no proton transfer] ▲

U Exercise 3.70.

Proton transfer is characteristic of all acid-base reactions whether or not water is present as solvent. For example, hydrogen chloride and ammonia gases can react directly:

$$NH_3(g) + HCl(g) \longrightarrow NH_4Cl(s)$$

? Why is a slice of lemon usually served with an order of fish in a restaurant?

Proton transfer is again the driving force of the reaction, but in this reaction no water is formed.

We now have enough information to recognize a neutralization reaction when we meet it. We know how to recognize certain acids (they generally have an acidic hydrogen atom written at the beginning of the molecular formula; in the abbreviated structural formulas of organic compounds, an acid is signaled by the presence of a —COOH group). We also know how to recognize a number of bases: they include the oxides and hydroxides of the elements in Groups I and II and certain organic compounds containing nitrogen. We know that when aqueous solutions of acids and bases are mixed, a neutralization reaction occurs

N The chemical equation, Part 2: Oxidation-reduction reactions. *J. Chem. Ed.* **1978**, 55, 326.

because proton transfer occurs from an acid to a base. A by-product of that proton transfer reaction is a salt, formed from the cation supplied by the base and the anion supplied by the acid. That salt might remain in solution, but if it is insoluble, it will form a precipitate that we can separate from the mixture by filtration.

REDOX REACTIONS

D The combustion of a small strip of magnesium is always spectacular.

D Corrosion of an iron nail. CD2, 99. Reversible oxidation-reduction color changes. CD2, 96.

The reactions between acids and bases involve the transfer of protons from one molecule or ion to another. Reactions of another important class involve the transfer of electrons. Reactions of this kind include the combustion of methane,

$$CH_4(g) + 2O_2(g) \longrightarrow CO_2(g) + 2H_2O(l)$$

and of magnesium,

$$2Mg(s) + O_2(g) \longrightarrow 2MgO(s)$$

the reaction between a metal and an acid to produce hydrogen,

$$Zn(s) + 2HCl(aq) \longrightarrow ZnCl_2(aq) + H_2(g)$$

and the reaction used to obtain iron from its ore,

$$Fe_2O_3(s) + 3CO(g) \xrightarrow{\Delta} 2Fe(l) + 3CO_2(g)$$

Although all these reactions look very different, we shall now see that they have a common feature. By recognizing common features in chemistry (such as the proton transfer in the reactions of acids and bases and the electron transfer here) we can make chemistry a much more systematic subject than it might appear at first glance.

FIGURE 3.13 An example of electron transfer: magnesium burning brightly in oxygen. Magnesium also burns in water and carbon dioxide; consequently, magnesium fires are very difficult to extinguish.

3.8 ELECTRON TRANSFER

Magnesium metal burns in air to form magnesium oxide (Fig. 3.13). In the reaction between magnesium and oxygen, the Mg atoms in the solid magnesium have lost electrons to form Mg^{2+} ions and the O atoms in O_2 have each gained electrons to form O^{2-} ions:

$$2Mg(s) + O_2(g) \longrightarrow \underbrace{2Mg^{2+}(s) + 2O^{2-}(s)}_{2MgO(s)} \quad (2 \times 2e^- \text{ from Mg to } O_2)$$

The reaction does not actually take place by a simple electron transfer, because complicated events occur as the metal is converted to oxide. However, the overall outcome is that of electron transfer. Similarly, in the reaction that occurs between magnesium and chlorine, the overall outcome is that of a transfer of electrons, in this case from Mg atoms to Cl atoms:

$$Mg(s) + Cl_2(g) \longrightarrow \underbrace{Mg^{2+}(s) + 2Cl^-(s)}_{MgCl_2(s)} \quad (2e^- \text{ from Mg to } Cl_2)$$

Oxidation and reduction. The reaction of a substance with oxygen is called oxidation. However, the meaning of this term has been extended to cover any reaction that involves loss of electrons:

Oxidation is electron loss.

When magnesium reacts with oxygen, the magnesium loses electrons; and hence it is oxidized. Magnesium also loses electrons when it reacts with chlorine, so it is also oxidized in that reaction—even though no oxygen is involved. In each case we can express the electron loss by writing an oxidation **half-reaction**, an equation that shows the electron loss explicitly:

$$Mg(s) \longrightarrow Mg^{2+}(s) + 2e^-$$

Note that the state of the electrons is not given: they are "in transit" and do not have a definite physical state.

The name *reduction* originally meant the reduction of the oxygen content of compounds and, in particular, the extraction of metals from their ores, for example,

$$Fe_2O_3(s) + 3CO(g) \xrightarrow{\Delta} 2Fe(l) + 3CO_2(g)$$

An essential feature of this reaction is the conversion of Fe^{3+} ions in solid Fe_2O_3 to Fe atoms by electron gain. The modern meaning of the term *reduction* generalizes this feature:

Reduction is electron gain.

The Fe^{3+} ions in the oxide gain electrons when they form Fe atoms, so the conversion of Fe^{3+} to Fe is an example of a reduction in the modern sense. The O atoms in the O_2 molecules that react with magnesium gain electrons when they become O^{2-} ions:

$$O_2(g) + 4e^- \longrightarrow 2O^{2-}(s)$$

so the oxygen is reduced when magnesium burns in oxygen. The Cl atoms in the Cl_2 molecules that react with magnesium gain electrons when they form Cl^- ions:

$$Cl_2(g) + 2e^- \longrightarrow 2Cl^-(s)$$

so the chlorine is reduced when it reacts with magnesium. Many reactions follow the same general pattern as that of reactions with oxygen or with chlorine (Fig. 3.14). The electron transfer leads to products that are more stable than the reactants, and this change in stability is the driving force of these reactions.

FIGURE 3.14 When bromine is poured on red phosphorus, the subsequent reaction is a redox reaction, which follows the same general pattern as a reaction with oxygen or with hydrogen.

E There is a distinction between the proton transfer of acids and bases and the electron transfer of redox reactions. Whereas proton transfer is often the sole process, electron transfer is often accompanied by an atom or groups of atoms.

N Everyday examples of oxidation processes. *J. Chem. Ed.* **1978**, *55*, 332.

▼ EXAMPLE 3.6 Writing half-reactions

Write the equation for the half-reaction in which (a) zinc metal is oxidized to zinc ions, Zn^{2+}, in aqueous solution and (b) hydrogen ions in aqueous solution are reduced to hydrogen gas.

STRATEGY We can proceed in much the same way as we did in Section 3.2 for the chemical equations, the only difference being that "stateless" electrons are included as reactants (in reductions) and as products (in oxidations). We have to ensure that the electric charge is balanced on both sides of each half-reaction and do so by including the appropriate stoichiometric coefficient.

SOLUTION (a) The skeletal equation for the oxidation of zinc is

$$Zn \longrightarrow Zn^{2+} + e^- \quad ⚠$$

The Zn atoms are balanced. We balance the electric charge with a stoichiometric coefficient of 2 for e^-; we then add the physical states:

$$Zn(s) \longrightarrow Zn^{2+}(aq) + 2e^-$$

(b) The skeletal equation for the reduction of H^+ is

$$H^+ + e^- \longrightarrow H_2 \quad ⚠$$

The H atoms are balanced with a stoichiometric coefficient of 2 for H^+:

$$2H^+ + e^- \longrightarrow H_2 \quad ⚠$$

The charge is balanced if we include a stoichiometric coefficient of 2 for the e^-; we can add the states at the same time:

$$2H^+(aq) + 2e^- \longrightarrow H_2(g)$$

EXERCISE E3.6 Write the equation for each half-reaction: (a) the oxidation of iron(II) ions to iron(III) ions in aqueous solution; (b) the reduction of copper(II) ions in aqueous solution to copper metal.

▲ [*Answer*: (a) $Fe^{2+}(aq) \rightarrow Fe^{3+}(aq) + e^-$; (b) $Cu^{2+}(aq) + 2e^- \rightarrow Cu(s)$]

D Add a strip of zinc to hydrochloric acid; do the same with a strip of copper.

U Exercise 3.74.

N The concept of oxidation number is explained more fully in Section 9.4 after the concept of electronegativity has been introduced. The rules in Table 3.5 are essentially a summary of the outcome of the electron drifts that stem from the electronegativities of the element in the compound. At this stage, the interpretation of oxidation numbers in terms of the "exaggerated ionic" picture described on p. 107 is adequate.

Oxidation and reduction do not necessarily occur by electron release from one species and its capture by a species elsewhere, for in many instances atoms are transferred as well and the electrons effectively go along for the ride. Half-reactions are only a *conceptual* way of dividing the overall reaction into two processes. However, in some cases a half-reaction accurately mirrors the actual process. Such a process occurs in photography, and is explained in Box 3.1.

Oxidation numbers. A simple way of judging whether a substance has undergone oxidation or reduction in a reaction is to note whether there is a change in the charge number of an atom or ion. For example, an increase in the charge number of a monatomic ion, as in the conversion of Fe^{2+} to Fe^{3+}, implies an oxidation. A decrease in charge number, as occurs in the conversion of Br to Br^-, implies a reduction. The change in charge number is easy to identify when we are dealing with ions, but chemists have also found a way of assigning to any atom in any kind of compound an *effective* charge number, called the **oxidation number** N_{ox}. The oxidation number is defined such that an increase in its value corresponds to oxidation and a decrease corresponds to reduction.

N There is no conventional symbol for oxidation number. Some texts simply write O.N., and that may be sensible. We have avoided O.N. (in favor of N_{ox}) because it makes expressions (such as those in Example 3.7) quite cumbersome.

N Oxidation numbers and their limitations. *J. Chem. Ed.* **1988**, *65*, 45.

An oxidation number is assigned by applying the rules in Table 3.5. These rules are based on the ideas chemists have developed about the way atoms in molecules share electrons (see Section 9.4). The rules must be applied in the order given, and we must stop as soon as the oxidation number has been obtained (because a later rule might contradict an earlier one). The rules imply the following two simple points:

1. The oxidation number of an elemental substance is zero.
2. The oxidation number of a monatomic ion is equal to the charge number of that ion.

BOX 3.1 ▸ REDOX REACTIONS IN PHOTOGRAPHY

The photographic emulsions used to coat photographic film contain silver bromide and small amounts of silver iodide. Photographic printing papers are coated with silver chloride. Each of these emulsions contains microscopic crystals, or "grains," of the silver halide. These grains are typically about 500 nm in diameter and are held in the gelatin with which the flat film or paper has been coated.

The reaction that records the image is an example of a *photochemical reaction,* a reaction caused by light. When the emulsion is exposed to light, an electron is driven out of a halide ion wherever the light falls:

$$Br^- \xrightarrow{\text{light}} Br + e^-$$

This half-reaction, the loss of an electron from the anion, is an oxidation. The liberated electron wanders through the grain and attaches to a nearby silver cation:

$$Ag^+ + e^- \longrightarrow Ag$$

This half-reaction, the gain of an electron by the cation, is a reduction. The equation for the overall photochemical reaction is the sum of the equations for the half-reactions. It is a redox reaction in which a silver cation is reduced by a bromide anion:

$$Ag^+ + Br^- \xrightarrow{\text{light}} Ag + Br$$

The redox reaction occurs only where the light falls, and it results in the formation of small clusters of silver atoms within the grains. There must be about four or more silver atoms in a cluster for the grain to be developable, and grains that have this minimum number form the latent image.

The film is developed by reducing the silver ions remaining in the grains that contain the latent image but not those in the unexposed grains. This task is achieved with a mild reducing agent, typically the organic compound hydroquinone HOC_6H_4OH, which has the structure

$$HO-\langle\bigcirc\rangle-OH$$

Developing *amplifies* the latent image, because the grains containing four or so silver atoms are converted to microscopic particles of metallic silver containing around 10^{10} atoms. The developing process is completed by dissolving the excess silver halide in aqueous sodium thiosulfate ($Na_2S_2O_3{\cdot}5H_2O$), photographer's hypo (a shortening of its old name). Thiosulfate anions attach to the remaining silver ions, forming the soluble $Ag(S_2O_3)_2^{3-}$ anion, but not to the metallic silver that forms the image. Rinsing with water washes away the soluble silver ion, leaving the metallic silver behind and the film insensitive to light.

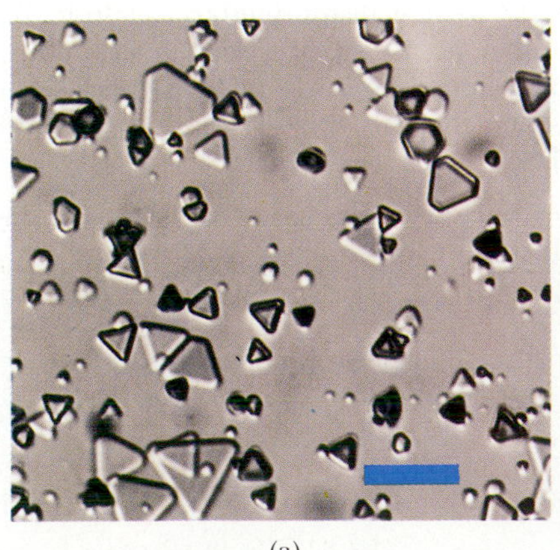

(a)

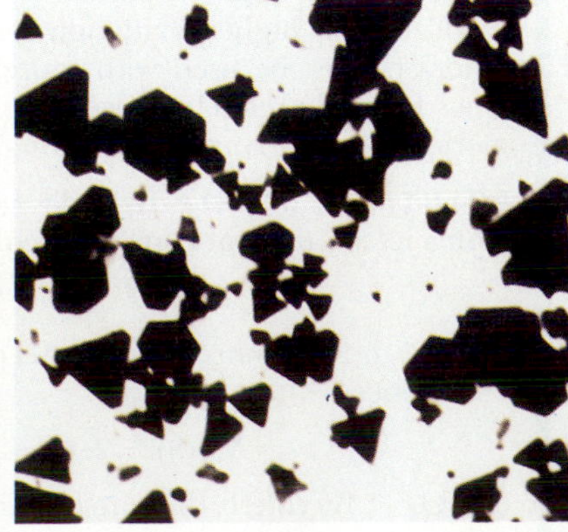

(b)

Micrographs of the silver halide crystals in a photographic emulsion before (a) and after (b) development. The scale bar represents 10 μm.

T OHT: Table 3.5.

N Simple method for determination of atoms in compounds. *J. Chem. Ed.* **1986**, *63*, 474.

TABLE 3.5 Rules for assigning oxidation numbers

Work through the following rules *in the order given*. Stop as soon as the oxidation number has been assigned.

	Oxidation number
1. The sum of the oxidation numbers of all the atoms in the species is equal to its total charge.	
2. For atoms in their elemental form	0
3. For elements of Group I	+1
Group II	+2
Group III (except B)	+3 for M^{3+} +1 for M^{+}
Group IV (except C, Si)	+4 for M^{4+} +2 for M^{2+}
4. For hydrogen	+1 in combination with nonmetals −1 in combination with metals
5. For fluorine	−1 in all its compounds
6. For oxygen	−2 unless combined with F −1 in peroxides (O_2^{2-}) $-\frac{1}{2}$ in superoxides (O_2^{-}) $-\frac{1}{3}$ in ozonides (O_3^{-})

N These rules are not exhaustive: for example, they do not lead to assignments in HCN.

V Video 6: Oxidation states of vanadium.

Thus, hydrogen, oxygen, iron, and all the elements have oxidation numbers of 0 in their elemental forms; the oxidation number of Fe^{3+} is +3 and that of Br^- is −1. The change of Fe to Fe^{3+} is an oxidation (because the oxidation number of iron increases from 0 to +3), and the change of Br to Br^- is a reduction (because the oxidation number decreases from 0 to −1). Note that the definition of oxidation number and its relation to oxidation and reduction are consistent with the definitions of oxidation and reduction in terms of electron loss and gain.

▼ EXAMPLE 3.7 Calculating oxidation numbers

Calculate the oxidation numbers of the elements in (a) SO_2, (b) SO_4^{2-}.

STRATEGY Because the sulfur atom has more oxygen atoms attached to it in the sulfate ion than in sulfur dioxide, we may be tempted to conclude that it will have a higher oxidation number in the sulfate ion. However, this criterion must be used with caution when comparing neutral species and ions. The only reliable procedure—and the only way to obtain precise numerical values—is to work through the rules in Table 3.5.

SOLUTION (a) SO_2: By rule 1, the sum of oxidation numbers of the atoms in the neutral compound must be 0:

$$N_{ox}(S) + 2 \times N_{ox}(O) = 0$$

Rules 2 to 5 are not relevant. According to rule 6, each oxygen has N_{ox} = −2. Hence,

$$N_{ox}(S) + 2 \times (-2) = 0, \text{ or } N_{ox}(S) = +4$$

(b) SO_4^{2-}: By rule 1, the sum of oxidation numbers of the atoms in the ion is −2:

$$N_{ox}(S) + 4 \times N_{ox}(O) = -2$$

Rules 2 to 5 are not relevant. According to rule 6, $N_{ox}(O) = -2$. Hence,

$$N_{ox}(S) + 4 \times (-2) = -2, \text{ or } N_{ox}(S) = +6$$

The sulfur is more highly oxidized in the sulfate ion than in sulfur dioxide.

EXERCISE E3.7 Calculate the oxidation numbers of the elements in (a) H_2S; (b) PO_4^{3-}; (c) NO_3^-.

[*Answer*: (a) $N_{ox}(H) = +1$, $N_{ox}(S) = -2$; (b) $N_{ox}(P) = +5$, $N_{ox}(O) = -2$; (c) $N_{ox}(N) = +5$, $N_{ox}(O) = -2$]

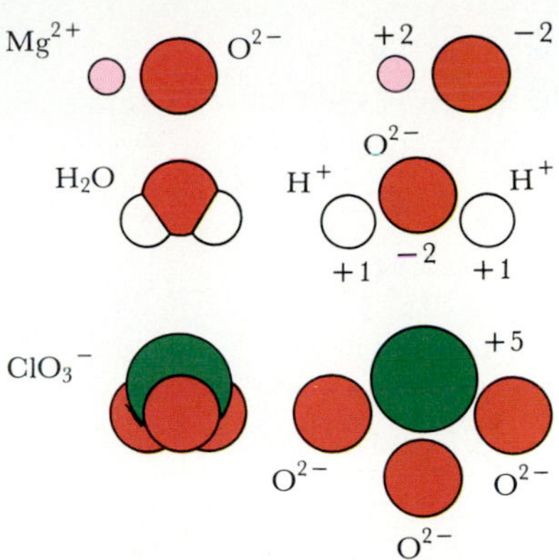

FIGURE 3.15 The oxidation number of an element is the charge it would have if each O atom present in the compound were an O^{2-} ion. In MgO the oxidation number of Mg is +2. The oxidation number of H in H_2O is +1 and that of Cl in ClO_3^- is +5.

Insight into the significance of oxidation numbers for elements in polyatomic species comes from noting that the oxidation number assigned to an element is equal to the charge that the atom would have if every O atom in the compound were present as O^{2-} (Fig. 3.15). For example, if the oxygen atom in H_2O is assumed to be O^{2-}, then, to maintain overall charge neutrality, each H atom must be assumed to be H^+. Hence, the oxidation number of H in H_2O is +1, the value that would be obtained by applying the rules in the table. In the formation of water from its elements, the oxidation number of hydrogen changes from 0 to +1, corresponding to oxidation. The oxidation number of oxygen changes from 0 to −2 in the reaction; therefore, we conclude that it is reduced:

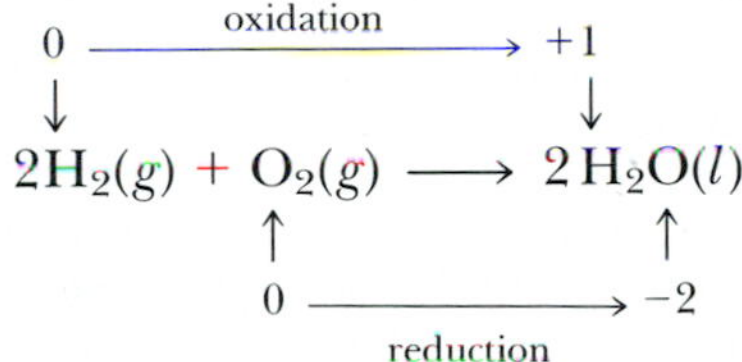

This conclusion is consistent with our understanding of oxidation and reduction in terms of electron loss and gain, respectively.

N When oxidation occurs, N_{ox} increases; when reduction occurs, N_{ox} decreases.

Oxidation-reduction reactions. One application of oxidation numbers is to decide whether a given reaction involves oxidation and reduction. Oxidation (electron loss) must be accompanied by reduction (electron gain) because one ion or molecule cannot lose electrons unless another one gains them. Because oxidation cannot occur without reduction, the overall process is called a redox reaction.

A **redox reaction** is a reaction in which oxidation and reduction take place.

Changes in oxidation numbers occur as a result of the electron transfer, so we can identify a redox reaction by noting whether a reaction is accompanied by changes in the oxidation numbers of the elements in the reactants and products: *if there is a change in the oxidation numbers, then the reaction is redox.* For instance, in the reaction of magnesium with chlorine, the oxidation number of magnesium increases (from 0 to +2, an oxidation) and that of chlorine decreases (from 0 to −2, a reduction); hence, we conclude that the reaction is redox.

The chemical equation of a redox reaction is the sum of the oxidation and reduction half-reactions into which the overall process may be *conceptually* divided. As an example, the chemical equation for the com-

E When proton transfer occurs, only a proton is transferred and the oxidation numbers of all the elements are left unchanged. The transfer of an H atom would result in a reduction in oxidation number of an element in the acceptor molecule (according to the rules in Table 3.5), but the fact that an electron does not accompany the proton means that that reduction does not in fact take place. Thus, proton transfer does not change any oxidation numbers, and hence acid-base reactions are distinct from redox reactions.

bustion of magnesium can be written as the sum of the following half-reactions:

Oxidation half-reaction of magnesium:

$$2Mg(s) \longrightarrow 2Mg^{2+}(s) + 4e^-$$

Reduction half-reaction of oxygen:

$$O_2(g) + 4e^- \longrightarrow 2O^{2-}(s)$$

Redox reaction:

$$2Mg(s) + O_2(g) + \cancel{4e^-} \longrightarrow \underbrace{2Mg^{2+}(s) + 2O^{2-}(s)}_{2MgO(s)} + \cancel{4e^-}$$

which is equivalent to

$$2Mg(s) + O_2(g) \longrightarrow 2MgO(s)$$

▼ EXAMPLE 3.8 Writing the equation for a redox reaction

Use oxidation and reduction half-reactions to write the chemical equation for the oxidation of aluminum to alumina (Al_2O_3). This is a corrosion reaction that occurs when aluminum is exposed to air.

STRATEGY We must first determine the oxidation numbers of each element and note the changes in them. Then we write the skeletal equations for the oxidation and reduction half-reactions separately, balance them individually, and add enough electrons to each half-reaction to balance the charges. We then form the overall equation by adding together the two half-reactions, after first making sure that all the electrons released in the oxidation are taken up by the reduction.

SOLUTION Because the oxidation number of O in Al_2O_3 is -2, the oxidation number of Al is $+3$. Because the oxidation number of Al changes from 0 to $+3$, we balance the charges in the oxidation half-reaction $Al(s) \rightarrow Al^{3+}(s)$ by adding three electrons to the right-hand side, obtaining

$$Al(s) \longrightarrow Al^{3+}(s) + 3e^-$$

The oxidation number of O has changed from 0 in O_2 to -2 in Al_2O_3. Therefore, we balance the charges in the reduction half-reaction $O_2(g) \longrightarrow 2O^{2-}$ by adding four electrons to the left-hand side, getting

$$O_2(g) + 4e^- \longrightarrow 2O^{2-}(s)$$

The numbers of electrons lost and gained will match if we multiply the first equation through by 4 and the second by 3. Then we can form the overall reaction by adding the two half-reactions:

$$4Al(s) \longrightarrow 4Al^{3+}(s) + 12e^-$$

$$3O_2(g) + 12e^- \longrightarrow 6O^{2-}(s)$$

$$4Al(s) + 3O_2(g) + \cancel{12e^-} \longrightarrow \underbrace{4Al^{3+}(s) + 6O^{2-}(s)}_{2Al_2O_3(s)} + \cancel{12e^-}$$

The overall balanced redox equation is therefore

$$4Al(s) + 3O_2(g) \longrightarrow 2Al_2O_3(s)$$

EXERCISE E3.8 Use the oxidation and reduction half-reactions to write the balanced chemical equation for the reaction of iron with oxygen to form Fe_2O_3.

[*Answer*: $4Fe(s) + 3O_2(g) \rightarrow 2Fe_2O_3(s)$]

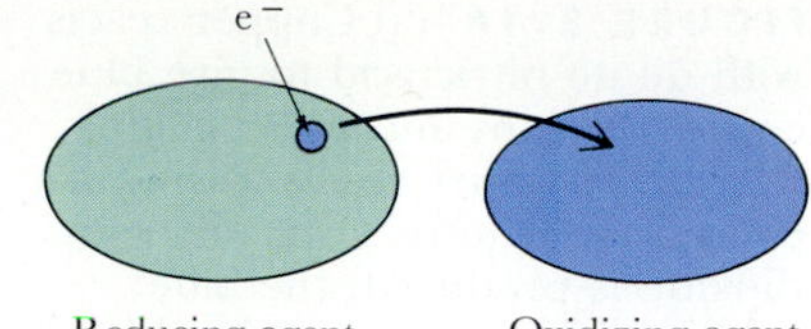

FIGURE 3.16 Oxidation is electron loss and reduction is electron gain. The electron donor is the reducing agent and the acceptor is the oxidizing agent.

Some redox reactions can be difficult to balance, but a systematic procedure for more complicated cases is set out in Section 3.10.

Oxidizing and reducing agents. The species that causes oxidation in a redox reaction is called the oxidizing agent. Oxygen and chlorine are two of the oxidizing agents we have met so far. We have used the word *species* rather than *substance* because ions, such as the permanganate ion and the nitrate ion, can also act as oxidizing agents. In terms of electron transfer, the **oxidizing agent** is the species that removes electrons from the species being oxidized. In terms of oxidation numbers, an element in the oxidizing agent undergoes a *decrease* in oxidation number because it gains electrons when the oxidizing agent reacts. Thus, chlorine removes electrons from magnesium; and, because chlorine accepts those electrons, its oxidation number decreases from 0 to −1: chlorine has been reduced, and hence is the oxidizing agent.

The species that causes reduction is called the reducing agent. In other words, the **reducing agent** supplies electrons to another reactant. In terms of oxidation numbers, an element in the reducing agent undergoes an *increase* in oxidation number because it loses electrons when it reacts. Magnesium metal supplies electrons to the chlorine, a transfer resulting in the reduction of chlorine. In doing so, the magnesium loses electrons, its oxidation number increases (from 0 to +2), and thus it undergoes oxidation: magnesium is the reducing agent.

The characteristics of oxidizing agents and reducing agents are complementary (Fig. 3.16):

- An oxidizing agent removes electrons and becomes reduced in the reaction; an element in the oxidizing agent undergoes a *decrease* in oxidation number.
- A reducing agent supplies electrons and becomes oxidized in the reaction; an element in the reducing agent undergoes an *increase* in oxidation number.

Identification of the oxidizing and reducing agents is very simple when monatomic ions are involved. For example, when a piece of zinc is placed in a copper(II) solution, a film of copper metal is deposited on the surface of the zinc and the blue solution of copper(II) ions fades to a colorless zinc(II) solution (Fig. 3.17). The reaction is

$$Zn(s) + Cu^{2+}(aq) \longrightarrow Zn^{2+}(aq) + Cu(s)$$

We see that the zinc metal is oxidized to Zn^{2+} ions and the electrons released from the zinc are transferred to and reduce the Cu^{2+} ions to copper metal. Therefore, zinc is the reducing agent in this reaction. Conversely, the Cu^{2+} ions withdraw electrons from the zinc metal and hence act as an oxidizing agent.

Some oxoanions (NO_3^- and MnO_4^-, for instance, but not CO_3^{2-} or PO_4^{3-}) are powerful oxidizing agents, particularly in acidic solution. This property is illustrated by the action of nitric acid on copper, for

N When reduction occurs, the oxidation number of *an* element takes place (it is meaningless to speak of the decrease in oxidation number of an entire species, unless it is monatomic). Therefore, the species that is reduced (the oxidizing agent) is the species that *contains* the atom of the element that is reduced. Similar remarks apply to oxidation. It is largely a matter of convention, or of what we are trying to identify, whether a species is identified as a reducing agent or as a species that is oxidized (and likewise for an oxidizing agent and a species that is reduced).

D A rapid redox reaction occurs when a polished strip of zinc is placed in a solution of copper(II) ions.

D Metal trees. CD1, 90.

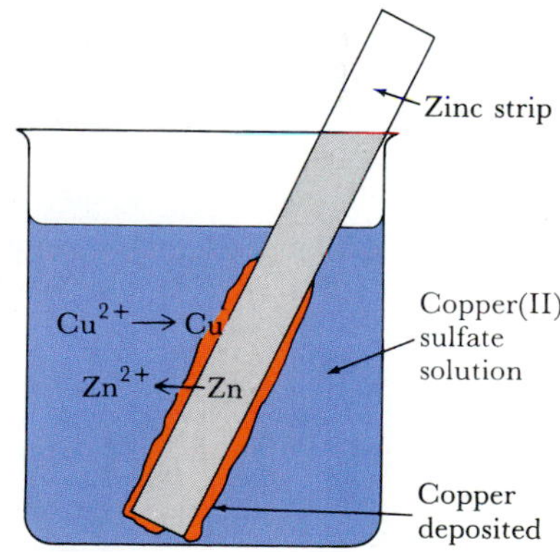

FIGURE 3.17 When a strip of zinc is placed in a solution of Cu^{2+} ions, the blue solution slowly becomes colorless and copper metal is deposited on the zinc. In this redox reaction, the zinc metal is reducing the Cu^{2+} ions to copper and the Cu^{2+} ions are oxidizing the zinc metal to Zn^{2+} ions.

FIGURE 3.18 (a) Copper reacts with dilute nitric acid to give blue copper(II) ions and nitric oxide. (b) When copper reacts with concentrated nitric acid, nitrogen dioxide is produced; the blue solution is turned green by this brown gas.

(a)

(b)

D Place a copper penny in concentrated nitric acid (in a hood, or a well-ventilated area).

V Video 36: Chemical equilibrium in a gaseous system. (The beginning of the sequence shows the preparation of nitrogen dioxide.)

the nitrate ion present in the acid can oxidize the metal. When copper is added to warm, dilute nitric acid, a blue solution of copper(II) ions and the colorless gas nitric oxide are produced (Fig. 3.18a):

$$3Cu(s) + 8H^+(aq) + 2NO_3^-(aq) \longrightarrow 3Cu^{2+}(aq) + 2NO(g) + 4H_2O(l)$$

The oxidation number of the copper is increased from 0 to +2, so it has been oxidized and has acted as a reducing agent. The oxidation number of nitrogen has changed from +5 in NO_3^- to +2 in NO; hence the nitrogen in the nitrate ion has undergone reduction and the nitrate ion has acted as the oxidizing agent. If copper metal is added to concentrated nitric acid instead, then the product includes nitrogen dioxide (Fig. 3.18b) in place of NO:

$$Cu(s) + 4H^+(aq) + 2NO_3^-(aq) \longrightarrow Cu^{2+}(aq) + 2NO_2(g) + 2H_2O(l)$$

Once again, copper is the reducing agent and nitric acid is the oxidizing agent. Sulfuric acid is also an oxidizing agent, and the hot, concentrated acid oxidizes copper and is reduced to sulfur dioxide:

$$Cu(s) + 2H_2SO_4(aq,\ conc) \xrightarrow{\Delta} Cu^{2+}(aq) + SO_4^{2-}(aq) + SO_2(g) + 2H_2O(l)$$

▼ EXAMPLE 3.9 Identifying oxidizing agents and reducing agents

When an acidic solution of potassium dichromate, $K_2Cr_2O_7$, is mixed with an iron(II) chloride solution, iron(III) ions and chromium(III) ions form. Which species is oxidized and which is reduced? Identify the oxidizing agent and the reducing agent.

STRATEGY First, we use Table 3.2 to determine whether the salts are soluble. If they are, then we separate the salts into their ions and determine the oxidation numbers of the elements. We identify the oxidation and reduction processes by looking for a change in the oxidation number of an element in the species.

SOLUTION Potassium dichromate is soluble and forms K^+ and $Cr_2O_7^{2-}$ (**5**) in solution. The oxidation number of oxygen in the $Cr_2O_7^{2-}$ ion is −2, so the oxidation number of chromium must be +6. The problem states that chromium(III) ions are formed, so the oxidation number of chromium decreases from +6 to +3. Therefore, the chromium in the dichromate ion is reduced and the dichromate ion is the oxidizing agent. Iron(II) chloride is soluble, forming Fe^{2+} and Cl^- ions. In the reaction, Fe^{3+} is a product, so the oxidation number of iron increases from +2 to +3. This increase in oxidation number indicates that iron is oxidized in the reaction and hence that Fe^{2+} is the reducing agent.

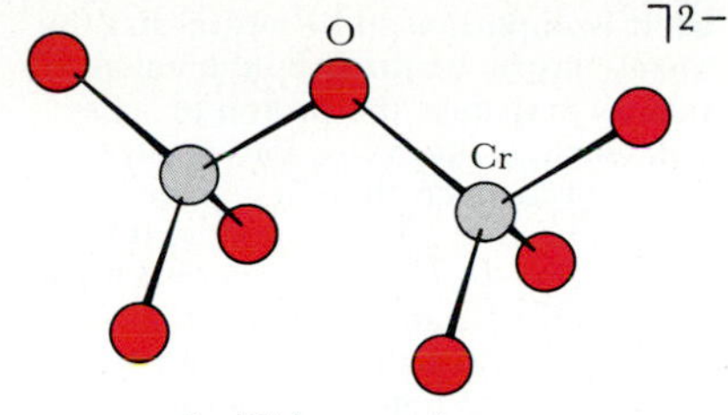

5 Dichromate ion

EXERCISE E3.9 In the Claus process for the recovery of sulfur from natural gas and petroleum, hydrogen sulfide, H_2S, reacts with sulfur dioxide, SO_2, to form elemental sulfur and water: $2H_2S(g) + SO_2(g) \rightarrow 3S(s) + 2H_2O(l)$. What species is oxidized and what is reduced? Identify the oxidizing agent and the reducing agent.

[*Answer*: S atom in H_2S is oxidized and H_2S is the reducing agent; S atom in SO_2 is reduced and SO_2 is the oxidizing agent]

U Exercises 3.72, 3.73, and 3.75.

T OHT: Table 3.6.

D The activity series for some metals. CD1, 89.

In principle, any species can accept or donate electrons, but whether it does so depends on the other species taking part in the reaction. For example, nitrogen can be reduced to ammonia with hydrogen (the first step in the production of fertilizers) and it can be oxidized to nitric oxide with oxygen (the first step in the production of nitric acid from ammonia, and in one of the processes that contributes to pollution and the formation of acid rain). The relative strengths of elements and compounds as oxidizing and reducing agents can be discussed quantitatively, and we shall do so in Chapter 17 after we have assembled more concepts. However, the relative oxidizing and reducing powers of metals can be discussed quite simply, as the following section shows.

3.9 THE ACTIVITY SERIES

We have seen that zinc metal can reduce Cu^{2+} ions to copper metal. However, it is also found that copper metal does not reduce Zn^{2+} ions to zinc metal. These observations suggest that it may be possible to arrange metals in some order that is based on their abilities to reduce the ions of other metals (and perhaps nonmetals too) and that reflects their readiness to release electrons and the readiness of their ions to gain electrons. Thus, the fact that zinc can reduce Cu^{2+} ions, and not vice versa, shows that zinc metal is more willing to give up electrons to Cu^{2+} ions than copper metal is willing to give up electrons to Zn^{2+} ions. In subsequent chapters we explore the reasons for this difference, but at this stage we shall simply use the empirical observation that metals can be arranged into an **activity series** (Table 3.6), a list of metals arranged such that a metal can reduce the cations formed by any of the metals below it in the list.

TABLE 3.6 The metal activity series*

Element	Reduced form	Oxidized form
Most strongly reducing		
potassium	K	K^+
sodium	Na	Na^+
magnesium	Mg	Mg^{2+}
chromium	Cr	Cr^{2+}
zinc	Zn	Zn^{2+}
iron	Fe	Fe^{2+}
nickel	Ni	Ni^{2+}
tin	Sn	Sn^{2+}
lead	Pb	Pb^{2+}
(hydrogen)	H_2	H^+
copper	Cu	Cu^{2+}
mercury	Hg	Hg_2^{2+}
silver	Ag	Ag^+
platinum	Pt	Pt^{2+}
gold	Au	Au^+
Least strongly reducing		

*A more complete version of this table (with additional information) is given in Appendix 2B.

Hydrogen is included in the series to enable us to predict whether a metal can reduce hydrogen ions to hydrogen gas in solution. That is, the relative positions of a metal and hydrogen tell us whether the metal can produce hydrogen gas when acted on by an acid. Metals that lie

N It is important to be aware that the metal may be oxidized by the anion that accompanies the proton in solution (such as the NO_3^- ion in nitric acid). Thus, even though copper is not oxidized by H^+, it is oxidized by NO_3^-, and so copper reacts with nitric acid. Similarly, an accompanying anion may form a complex with a metal ion, and hence an acid may result in reaction with a metal even though the proton alone is impotent. That is the case with the action of aqua regia (a mixture of hydrochloric and nitric acids) on gold, for gold is complexed by Cl^- ions present in the acid. These points are treated in Sections 17.4 and 15.8, respectively.

D Place a polished strip of zinc in a copper(II) sulfate solution and a polished strip of copper in a zinc(II) nitrate solution.

above hydrogen in the activity series can reduce H^+ ions in acids and produce hydrogen gas. Put another way, metals that lie above hydrogen in the activity series can be oxidized by hydrogen ions. Metals that lie below hydrogen in the activity series are not oxidized by hydrogen ions. Copper, for example, lies below hydrogen and is unaffected by hydrochloric acid.

▼ **EXAMPLE 3.10 Deciding whether an acid can oxidize a metal**

Would hydrogen be evolved if hydrochloric acid were poured on a piece of nickel?

STRATEGY The production of hydrogen from an acid takes place when the metal reduces the hydrogen ions provided by the acid. Therefore, check whether the metal lies above hydrogen in the activity series in Table 3.6.

SOLUTION Because nickel lies above hydrogen in the activity series, it can reduce H^+ ions to hydrogen gas, so H_2 will be evolved when an acid acts on nickel. It follows that H^+ ions can oxidize metallic nickel. The equations for the half-reactions are

$$\textit{Reduction half-reaction: } 2H^+(aq) + 2e^- \longrightarrow H_2(g)$$

$$\textit{Oxidation half-reaction: } Ni(s) \longrightarrow Ni^{2+}(aq) + 2e^-$$

and the overall redox reaction is their sum:

$$Ni(s) + 2H^+(aq) \longrightarrow Ni^{2+}(aq) + H_2(g)$$

EXERCISE E3.10 Should we expect silver to produce hydrogen gas from dilute hydrochloric acid?

[*Answer*: no] ▲

3.10 BALANCING REACTIONS BY USING HALF-REACTIONS

N The material in this section could be delayed until the more detailed coverage of redox reactions in Chapter 17.

N Balancing complex redox equations by inspection. *J. Chem. Ed.* **1981**, *58*, 642.

N More on balancing redox equations. *J. Chem. Ed.* **1979**, *56*, 181.

This chapter began with an introduction to chemical reactions; then we saw how to balance them. Now we discuss how to use the information that we have acquired in the course of the chapter to balance some particularly tricky equations. In particular, we shall see that chemical equations for redox reactions, which are often quite difficult to balance, can be balanced *systematically* using half-reactions. The scheme involves balancing the equation for each half-reaction separately and then adding the two half-reactions after making sure that all the electrons released in the oxidation are taken up in the reduction.

Redox reactions in acidic solution. The first example we consider is the oxidation of oxalic acid, $H_2C_2O_4$ (**6**), by permanganate ion, MnO_4^-, in acidic solution (Fig. 3.19). It might be helpful to have a sneak preview of the final equation:

$$2MnO_4^-(aq) + 6H^+(aq) + 5H_2C_2O_4(aq) \longrightarrow 2Mn^{2+}(aq) + 10CO_2(g) + 8H_2O(l)$$

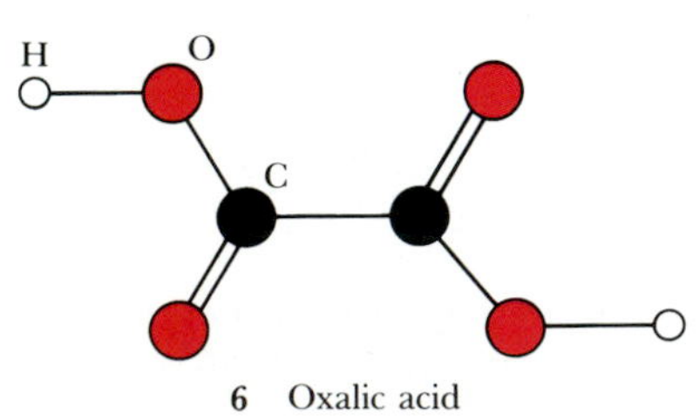

6 Oxalic acid

To balance such a complicated equation by elementary procedures could be very time consuming and not a little frustrating.

First, we identify the oxidized and reduced species from the changes in oxidation numbers. For example, we write the skeletal equation:

$$\overset{+7}{\downarrow}\xrightarrow{\text{reduction}}\overset{+2}{\downarrow}$$

$$MnO_4^-(aq) + H_2C_2O_4(aq) \longrightarrow Mn^{2+}(aq) + CO_2(g)$$

$$\overset{\uparrow}{+3}\xrightarrow[\text{oxidation}]{}\overset{\uparrow}{+4}$$

The oxidation number of manganese decreases from +7 to +2, so we conclude that MnO_4^- is the oxidizing agent. The oxidation number of carbon increases from +3 to +4, so we conclude that oxalic acid is the reducing agent in the reaction.

Next, we use the following procedure:

1. Write the two skeletal equations for the half-reactions.

   Reduction half-reaction: $MnO_4^- \longrightarrow Mn^{2+}$

   Oxidation half-reaction: $H_2C_2O_4 \longrightarrow CO_2$

2. Balance by inspection all elements in the half-reactions except O, H, and the charge.

   Reduction half-reaction: $MnO_4^- \longrightarrow Mn^{2+}$

   Oxidation half-reaction: $H_2C_2O_4 \longrightarrow 2CO_2$

3. Balance the oxygen and hydrogen atoms. For oxoanions in *acidic* solution, we balance H and O by adding H^+ and H_2O. (a) First, balance the O atoms by adding H_2O:

   Reduction half-reaction. Add $4H_2O$ on the right to balance the four O atoms on the left:

   $$MnO_4^- \longrightarrow Mn^{2+} + 4H_2O$$

   Oxidation half-reaction. The O atoms are already balanced:

   $$H_2C_2O_4 \longrightarrow 2CO_2$$

   (b) Next, add H^+ to balance the H atoms on either side of the half-reactions:

   Reduction half-reaction. Add $8H^+$ on the left to balance the eight H atoms on the right:

   $$MnO_4^- + 8H^+ \longrightarrow Mn^{2+} + 4H_2O$$

   Oxidation half-reaction. The equation needs $2H^+$ on the right to balance the H atoms in the $H_2C_2O_4$ molecule on the left:

   $$H_2C_2O_4 \longrightarrow 2CO_2 + 2H^+$$

4. Balance the electric charges by adding electrons.

   Reduction half-reaction. To balance the +2 charge on the right and the +7 charge on the left, we need to add five electrons to the left:

   $$MnO_4^- + 8H^+ + 5e^- \longrightarrow Mn^{2+} + 4H_2O$$

   Oxidation half-reaction. We need to add two electrons on the right to cancel the two positive charges:

   $$H_2C_2O_4 \longrightarrow 2CO_2 + 2H^+ + 2e^-$$

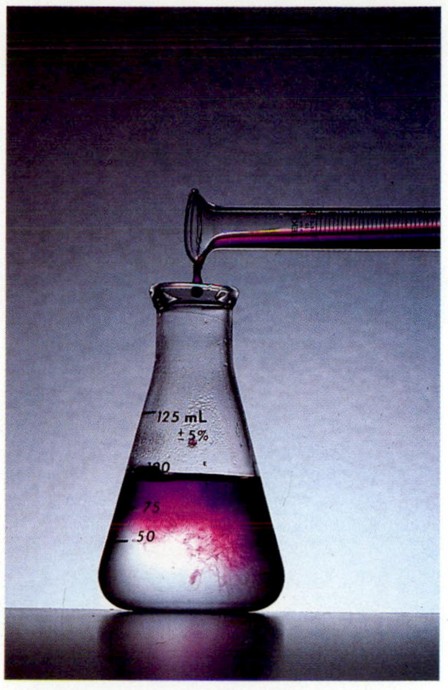

FIGURE 3.19 The addition of a potassium permanganate solution to an oxalic acid solution causes a redox reaction. The oxalic acid that is present reduces the permanganate to very pale pink Mn^{2+} ions. When all the oxalic acid has been oxidized, the addition of more permanganate turns the solution purple (as shown here) because the purple MnO_4^- ions survive.

V Video 7: Oxidation of ethanol by manganese(VII) oxide.

V Video 55: The reaction of potassium permanganate and glycerin.

5. Prepare the two half-reactions for summation by making the number of electrons the same in both. Multiply the reduction half-reaction by 2 and the oxidation half-reaction by 5:

   Reduction half-reaction: $2MnO_4^- + 16H^+ + 10e^- \longrightarrow 2Mn^{2+} + 8H_2O$

   Oxidation half-reaction: $5H_2C_2O_4 \longrightarrow 10CO_2 + 10H^+ + 10e^-$

6. Combine the two half-reactions. (a) First, add the two half-reactions in step 5:

   $$2MnO_4^- + 16H^+ + 10e^- + 5H_2C_2O_4 \longrightarrow 2Mn^{2+} + 8H_2O + 10CO_2 + 10H^+ + 10e^-$$

   (b) Next, simplify the balanced equation by canceling like species on both sides of the equation. In this example, $10e^-$ and $10H^+$ can be canceled:

   $$2MnO_4^- + 5H_2C_2O_4 + 6H^+ \longrightarrow 2Mn^{2+} + 8H_2O + 10CO_2$$

   (c) Finally, insert the state of each species:

   $$2MnO_4^-(aq) + 5H_2C_2O_4(aq) + 6H^+(aq) \longrightarrow 2Mn^{2+}(aq) + 8H_2O(l) + 10CO_2(aq)$$

   This expression is the fully balanced net ionic equation.

▼ EXAMPLE 3.11 Balancing redox equations by using half-reactions

When dilute nitric acid is poured on a piece of copper metal, copper(II) ions and the gas nitric oxide, NO, are formed. Write the balanced equation for the reaction.

STRATEGY The skeletal equation is

$$Cu(s) + NO_3^-(aq) + H^+(aq) \longrightarrow Cu^{2+}(aq) + NO(g)$$ ⚠

and it is easy to see (either by inspection or by calculating the oxidation numbers) that copper has been oxidized and NO_3^- ions have been reduced:

oxidation: $0 \longrightarrow +2$

$$Cu(s) + NO_3^-(aq) \longrightarrow Cu^{2+}(aq) + NO(g)$$ ⚠

reduction: $+5 \longrightarrow +2$

We then work through the procedure given above. The copper oxidation half-reaction is so straightforward that it can be written down at once; the reduction half-reaction will take a little more work.

SOLUTION

1. Write the two skeletal equations for the half-reactions.

   Reduction half-reaction: $NO_3^- \longrightarrow NO$ ⚠

   Oxidation half-reaction: $Cu \longrightarrow Cu^{2+}$ ⚠

2. Balance by inspection all elements in the half-reactions except O, H, and the charge. The equations of both half-reactions are already balanced in this sense.

3. Balance the oxygen and hydrogen atoms. (a) First, balance the oxygen atoms by adding H_2O (this step is relevant only to the reduction half-reaction).

D The copper penny/nitric acid demonstration (p. 110) could be done here.

Reduction half-reaction. Add $2H_2O$ on the right to balance the three O atoms on the left:

$$NO_3^- \longrightarrow NO + 2H_2O$$ ⚠

(b) Next, add H^+ to balance the H atoms on either side of the half-reactions. (Once again, this step is relevant only to the reduction half-reaction.)

Reduction half-reaction. Add $4H^+$ on the left to balance the four H atoms on the right:

$$NO_3^- + 4H^+ \longrightarrow NO + 2H_2O$$ ⚠

4. Balance the electric charges by adding electrons.

Reduction half-reaction. To balance the zero charge on the right and the $-1 + 4 = +3$ charge on the left, we need to add three electrons to the left:

$$NO_3^- + 4H^+ + 3e^- \longrightarrow NO + 2H_2O$$

Oxidation half-reaction. We need to add two electrons on the right to cancel the two positive charges:

$$Cu \longrightarrow Cu^{2+} + 2e^-$$

5. Prepare the two half-equations for summation by making the number of electrons the same in both. Multiply the reduction half-reaction by 2 and the oxidation half-reaction by 3:

Reduction half-reaction: $2NO_3^- + 8H^+ + 6e^- \longrightarrow 2NO + 4H_2O$

Oxidation half-reaction: $3Cu \longrightarrow 3Cu^{2+} + 6e^-$

6. Combine the two half-reactions. (a) The sum is

$$3Cu + 2NO_3^- + 8H^+ + 6e^- \longrightarrow 3Cu^{2+} + 6e^- + 2NO + 4H_2O$$

(b) Simplify the balanced equation:

$$3Cu + 2NO_3^- + 8H^+ \longrightarrow 3Cu^{2+} + 2NO + 4H_2O$$

(c) Finally, insert the state of each species:

$$3Cu(s) + 2NO_3^-(aq) + 8H^+(aq) \longrightarrow 3Cu^{2+}(aq) + 2NO(g) + 4H_2O(l)$$

This expression is the fully balanced net ionic equation.

EXERCISE E3.11 When acidified potassium permanganate is mixed with a solution of sulfurous acid, H_2SO_3, sulfuric acid and manganese(II) ions are produced. Write the chemical equation for the reaction. In acidic aqueous solution, H_2SO_3 is present as the nonionized molecules and sulfuric acid is present as HSO_4^- ions.

[*Answer*: $2MnO_4^-(aq) + 5H_2SO_3(aq) + H^+(aq) \rightarrow 2Mn^{2+}(aq) + 5HSO_4^-(aq) + 3H_2O(l)$]

▲

Redox reactions in basic solution. When the reaction is carried out in basic solution, the solution is rich in hydroxide ions, OH^-, and H_2O molecules, and we use these two species to balance the O and H atoms. The balancing is slightly more involved because both species provide O and H atoms. We shall illustrate the procedure by considering the reaction of permanganate ions with bromide ions in basic solution, with the formation of solid manganese(IV) oxide, MnO_2, and bromate ions, BrO_3^-.

U Exercises 3.76 and 3.77.

In the reaction the oxidation number of manganese changes from +7 in MnO_4^- to +4 in MnO_2, so the manganese in MnO_4^- is reduced (and MnO_4^- acts as the oxidizing agent). The oxidation number of bromine changes from -1 as Br^- to $+5$ in BrO_3^-, so Br^- acts as the reducing agent. The procedure is similar to that for balancing redox equations in acidic solution, but Step 3 is a little more involved.

1. Write the two skeletal equations for the half-reactions.

   Reduction half-reaction: $MnO_4^- \longrightarrow MnO_2$

   Oxidation half-reaction: $Br^- \longrightarrow BrO_3^-$

2. Balance by inspection all elements in the half-reactions except O, H, and the charge. The equations of both half-reactions are already balanced in this sense.

3. Balance the oxygen and hydrogen atoms. For oxoanions in *basic* solution, we balance H and O by adding OH^- and H_2O. (a) First, balance the O atoms by adding H_2O:

   Reduction half-reaction. Add $2H_2O$ on the right to balance the two extra O atoms on the left:

   $$MnO_4^- \longrightarrow MnO_2 + 2H_2O$$

   Oxidation half-reaction. We need to add $3H_2O$ on the left to balance the three O atoms on the right:

   $$Br^- + 3H_2O \longrightarrow BrO_3^-$$

   (b) Next, balance the H atoms by adding H_2O on the side needing the H atoms and then adding the same number of OH^- ions to the other side of the equation. The net result is an addition of an H atom to the side to which we have added H_2O. When we add

   $$\ldots OH^- \ldots \longrightarrow \ldots H_2O \ldots$$

   we are effectively adding one H atom to the right; when we add

   $$\ldots H_2O \ldots \longrightarrow \ldots OH^- \ldots$$

   we are effectively adding an H atom to the left.

   Reduction half-reaction. Add $4H_2O$ on the left and $4OH^-$ on the right to balance the four H atoms on the right:

   $$MnO_4^- + 4H_2O \longrightarrow MnO_2 + 2H_2O + 4OH^-$$

   Oxidation half-reaction. The equation needs $6H_2O$ on the right and $6OH^-$ on the left to balance the six H atoms on the left:

   $$Br^- + 3H_2O + 6OH^- \longrightarrow BrO_3^- + 6H_2O$$

   (c) In the final part of this step, we simplify the two half-reactions by canceling like species on each side:

   Reduction half-reaction: $MnO_4^- + 2H_2O \longrightarrow MnO_2 + 4OH^-$

   Oxidation half-reaction: $Br^- + 6OH^- \longrightarrow BrO_3^- + 3H_2O$

4. Balance the electric charges by adding electrons.

   Reduction half-reaction. To balance the -4 charge on the right and the -1 charge on the left, we need to add three electrons to the left:

   $$MnO_4^- + 2H_2O + 3e^- \longrightarrow MnO_2 + 4OH^-$$

Oxidation half-reaction. To balance the −7 charge on the left and the −1 charge on the right, we need to add six electrons to the right:

$$Br^- + 6OH^- \longrightarrow BrO_3^- + 3H_2O + 6e^-$$

5. Prepare the two half-reactions for summation by making the number of electrons the same in both.

   Multiply the reduction half-reaction by 2:

   Reduction half-reaction: $2MnO_4^- + 4H_2O + 6e^- \longrightarrow 2MnO_2 + 8OH^-$

   Oxidation half-reaction: $Br^- + 6OH^- \longrightarrow BrO_3^- + 3H_2O + 6e^-$

6. Combine the two half-reactions. (a) First, add the equations of the two half-reactions as they are written:

   $$2MnO_4^- + 4H_2O + 6e^- + Br^- + 6OH^- \longrightarrow 2MnO_2 + 8OH^- + BrO_3^- + 3H_2O + 6e^-$$

   (b) Next, simplify the balanced equation by canceling like species on both sides of the equation. In this example, $6e^-$, $3H_2O$, and $6OH^-$ can be canceled:

   $$2MnO_4^- + H_2O + Br^- \longrightarrow 2MnO_2 + 2OH^- + BrO_3^-$$

   (c) Finally, insert the state of each species:

   $$2MnO_4^-(aq) + H_2O(l) + Br^-(aq) \longrightarrow 2MnO_2(s) + 2OH^-(aq) + BrO_3^-(aq)$$

   This expression is the fully balanced net ionic equation. It would have been very difficult to balance such a complicated equation without breaking it down into half-reactions and working systematically on each of them.

E Oxidation by permanganate in basic solution is less commonly used than in acidic solution (for redox titrations, at least), because the formation of the cloudy black precipitate of manganese(IV) oxide obscures the end-point of the titration.

This chapter has concentrated on the qualitative aspects of reactions. The three types of reactions introduced here have key features in common, as we explain in later chapters. Precipitation reactions are dealt with in Chapter 15. Acids and bases are so important that we devote two chapters, Chapters 14 and 15, to their properties. We have not explained how we know what substances can oxidize or reduce a given substance. This point is taken up in Chapter 17, where we also see how redox reactions are used to generate electricity.

SUMMARY

3.1 and 3.2 A chemical reaction is summarized by writing its **chemical equation**. The varieties of chemical reactions commonly encountered in chemistry are summarized in Table 3.1. A chemical equation summarizes qualitative information about the identities of the **reactants**, the starting materials, and the **products**, the materials formed. When the **skeletal equation** is **balanced** by including the **stoichiometric coefficients**, it also expresses the fact that no atoms are created or destroyed in the course of the reaction.

3.3 In a **precipitation reaction** a solid is formed when two solutions are mixed. The essential feature of a precipitation reaction (and other reactions involving ions) is made clear by writing the **net ionic equation**, which is derived by canceling the **spectator ions**, the ions left unchanged by the reaction.

3.4 A precipitation reaction is a special case of a **double-replacement reaction** in which atoms or ions exchange partners. The **solubility rules** (Table 3.2), a summary of the solubilities of common compounds, are useful in deciding what product is likely in a precipitation reaction.

3.5 An **Arrhenius acid** is a compound that contains hydrogen and acts as a source of hydrogen ions in water. An **Arrhenius base** is a compound that acts as a source of hydroxide ions in water. Arrhenius acids and bases are detected by the color changes they bring about in compounds called indicators.

3.6 The product of a **neutralization** of an Arrhenius acid and an Arrhenius base is water and a **salt** in which the cation is supplied by the base and the anion by the acid. The net ionic reaction of a neutralization reaction is

$$H^+(aq) + OH^-(aq) \longrightarrow H_2O(l)$$

3.7 The **Brønsted-Lowry theory** of acids and bases is broader than the Arrhenius definitions and concentrates on the ability of species to participate in **proton transfer**. A **Brønsted acid** donates protons and a **Brønsted base** accepts them. A **strong acid** is an acid that is fully **ionized** in water in the sense that all its molecules have donated their protons to water molecules. A **strong base** is a base that has fully accepted hydrogen ions from water. A **weak acid** is an acid that is only partially ionized in water, in the sense that many acid molecules retain their protons. A **weak base** is a base that is only partially ionized in water in the sense that only a few of its ions or molecules have accepted protons from water molecules.

3.8 Oxidation is electron loss and **reduction** is electron gain; the two processes may be represented in terms of oxidation and reduction **half-reactions**. The species that undergo oxidation and reduction can be identified by calculating the **oxidation number**, the effective charge, of each element. The species containing an element with an oxidation number that is increased in a reaction has been oxidized, and the species containing the element with an oxidation number that is decreased has been reduced. The combination of oxidation and reduction is a **redox reaction**; the chemical equation of a redox reaction is the sum of the oxidation and reduction half-reactions into which it may be *conceptually* divided. A **reducing agent** is a species that transfers electrons to an **oxidizing agent**. The oxidation number of an element in a reducing agent increases in a reaction whereas that of an element in an oxidizing agent decreases.

3.9 The reducing strengths of metals are listed in the order called the **activity series**, a part of which is given in Table 3.6. A metal can reduce the ions of any metal that lies below it in the series.

3.10 Oxidation and reduction half-reactions are useful for balancing redox equations because they break a complicated equation down into two simpler components. If the redox reaction occurs in acidic solution, the equation is balanced with H^+ and H_2O; if it takes place in basic solution, it is balanced with OH^- and H_2O.

CLASSIFIED EXERCISES

Balancing Equations

3.1 Write balanced chemical equations for the following reactions: (a) The reaction of sodium oxide, Na_2O, and water produces sodium hydroxide. (b) Hot lithium metal reacts in a nitrogen atmosphere to produce lithium nitride, Li_3N. (c) The reaction of calcium metal with water leads to the evolution of hydrogen gas and the formation of calcium hydroxide, $Ca(OH)_2$.

3.2 Write balanced equations for the following reactions: (a) One way to remove the air pollutant sulfur dioxide is to allow it to react with hydrogen sulfide, H_2S; this reaction produces elemental sulfur and water. (b) The photosynthesis reaction is one in which carbon dioxide and water combine to form glucose, $C_6H_{12}O_6$, and oxygen gas. (c) Magnesium metal reacts with boron oxide, B_2O_3, to form elemental boron and magnesium oxide.

3.3 In one stage in the commercial production of iron metal in a blast furnace, the iron(III) oxide, Fe_2O_3, reacts with carbon monoxide to form Fe_3O_4 and carbon dioxide. In a second stage, the Fe_3O_4 reacts further with carbon monoxide to produce elemental iron and carbon dioxide. Write a balanced equation for each stage in the reaction.

3.4 One of the roles of stratospheric ozone, O_3, is to remove damaging ultraviolet radiation from sunlight. The ozone absorbs the energy of the harmful rays and dissociates into molecular oxygen and atomic oxygen. Write the balanced equation for the reaction.

3.5 When nitrogen and oxygen react in the cylinder of an automobile engine, nitric oxide (NO) is formed. After it escapes into the atmosphere along with the other exhaust gases, the nitric oxide reacts with oxygen to produce nitrogen dioxide, one of the precursors of acid rain. Write the two balanced equations leading to the formation of nitrogen dioxide.

3.6 Hydrofluoric acid is used to etch glass because a frosted surface is created where the acid reacts with silica, SiO_2, the main component of glass. The products of the reaction are silicon tetrafluoride and water. Write the balanced equation for the reaction.

3.7 Sodium thiosulfate (which as the pentahydrate, $Na_2S_2O_3{\cdot}5H_2O$, forms the large white crystals used as photographer's hypo) can be prepared by bubbling oxygen through a solution of sodium polysulfide, Na_2S_5. Sodium polysulfide is made by the action of hydrogen sulfide on a solution of sodium sulfide, Na_2S, which, in turn, is made by the neutralization of hydrogen sulfide, H_2S, with sodium hydroxide. Write the three chemical equations that show how hypo is prepared from hydrogen sulfide and sodium hydroxide.

3.8 The first stage in the production of nitric acid by the Ostwald process is the reaction of ammonia with oxygen; the products of this reaction are nitric oxide (NO) and water. The nitric oxide further reacts with oxygen to produce nitrogen dioxide, which, when dissolved in water, produces nitric acid and nitric oxide. Write the three balanced equations that lead to the production of nitric acid.

3.9 Write a balanced equation for the complete combustion of octane, C_8H_{18}, the major component of gasoline.

3.10 Energy is obtained in the body by a series of reactions that, overall, is equivalent to the combustion of sucrose (cane sugar), $C_{12}H_{22}O_{11}$. Write a balanced equation for its combustion.

3.11 The psychoactive drug sold as Methadrine (also known as "speed"), $C_{10}H_{15}N$, undergoes a series of reactions in the body, the net outcome of which is the equivalent of combustion. Write the balanced equation for the reaction.

3.12 Write the balanced equation for the combustion of the analgesic (pain killer) sold as Tylenol, $C_8H_9O_2N$.

Precipitation Reactions

3.13 Use the information in Table 3.2 to classify the following salts as soluble or insoluble in water: (a) lead(II) nitrate, $Pb(NO_3)_2$; (b) silver nitrate, $AgNO_3$; (c) sodium sulfate, Na_2SO_4.

3.14 Use the information in Table 3.2 to classify the following compounds as soluble or insoluble in water: (a) zinc acetate, $Zn(CH_3CO_2)_2$; (b) iron(III) chloride, $FeCl_3$; (c) copper(II) hydroxide, $Cu(OH)_2$.

3.15 What are the principal species present in aqueous solutions of (a) NaI; (b) Ag_2CO_3; (c) $(NH_4)_3PO_4$?

3.16 What are the principal species present in aqueous solutions of (a) K_2CO_3; (b) K_2CrO_4; (c) Hg_2Cl_2?

3.17 (a) When aqueous solutions of iron(III) sulfate, $Fe_2(SO_4)_3$, and sodium hydroxide were mixed, a precipitate formed. Write the formula of the precipitate. (b) Does a precipitation form when aqueous solutions of silver nitrate, $AgNO_3$, and potassium carbonate are mixed? If so, write the formula of the precipitate. (c) Aqueous solutions of lead(II) nitrate, $Pb(NO_3)_2$, and sodium acetate are mixed. Does a precipitate form? If so, write the formula of the precipitate.

3.18 (a) Solid ammonium nitrate, NH_4NO_3, and solid calcium chloride were placed in water and the mixture stirred. Is the formation of a precipitate expected? If so, write the formula of the precipitate. (b) Solid magnesium carbonate, $MgCO_3$, and solid sodium nitrate were mixed and placed in water and stirred. What is observed? If a precipitate is present, write its formula. (c) Aqueous solutions of sodium sulfate, Na_2SO_4, and barium chloride are mixed and a precipitate forms. What is the formula of the precipitate?

3.19 When the solution in Beaker 1 (below) is mixed with the solution in Beaker 2, a precipitate forms. Write three equations describing the formation of the precipitate in terms of the overall reaction, the full ionic form of the reaction, and the net ionic reaction, and then identify the spectator ions.

(a) Beaker 1: $FeCl_3(aq)$ Beaker 2: $NaOH(aq)$
(b) Beaker 1: $AgNO_3(aq)$ Beaker 2: $KI(aq)$
(c) Beaker 1: $Pb(NO_3)_2(aq)$ Beaker 2: $K_2SO_4(aq)$

3.20 Write the three chemical equations describing the formation of the precipitate (showing the overall reaction, the full ionic form of the equation, and the net ionic reaction), and then identify the spectator ions. (a) $(NH_4)_2CrO_4(aq)$ is mixed with $BaCl_2(aq)$; (b) $CuSO_4(aq)$ is mixed with $Na_2S(aq)$; (c) $FeCl_2(aq)$ is mixed with $(NH_4)_3PO_4(aq)$.

3.21 Write the full ionic form of the equation and the net ionic equation for each of the following reactions:

(a) $Pb(ClO_4)_2(aq) + 2NaBr(aq) \longrightarrow PbBr_2(s) + 2NaClO_4(aq)$
(b) $Sr(NO_3)_2(aq) + 2NH_4Cl(aq) \longrightarrow SrCl_2(aq) + 2NH_4NO_3(aq)$
(c) $2NaOH(aq) + Cu(NO_3)_2(aq) \longrightarrow Cu(OH)_2(s) + 2NaNO_3(aq)$

3.22 Write the full ionic form of the equation and the net ionic equation for each of the following reactions:

(a) $3BaCl_2(aq) + 2K_3PO_4(aq) \longrightarrow Ba_3(PO_4)_2(s) + 6KCl(aq)$
(b) $Hg_2(NO_3)_2(aq) + 2KCl(aq) \longrightarrow Hg_2Cl_2(s) + 2KNO_3(aq)$
(c) $Mg(CH_3CO_2)_2(aq) + Ba(OH)_2(aq) \longrightarrow Mg(OH)_2(s) + Ba(CH_3CO_2)_2(aq)$

3.23 Suggest two soluble salts that, when mixed together in water, give the product of each of the following net ionic equations: (a) $Ca^{2+}(aq) + CO_3^{2-}(aq) \rightarrow CaCO_3(s)$, the reaction responsible for the deposition of chalk hills and of the formation of sea urchin spines. (b) $Cd^{2+}(aq) + S^{2-}(aq) \rightarrow CdS(s)$, one of the substances used to color glass yellow.

3.24 Suggest two soluble salts that, when mixed together in water, give the product of each of the following net ionic equations: (a) $Mg^{2+}(aq) + 2OH^-(aq) \rightarrow Mg(OH)_2(s)$, the suspension present in milk of magnesia. (b) $3Ca^{2+}(aq) + 2PO_4^{3-}(aq) \rightarrow Ca_3(PO_4)_2(s)$, the major component of phosphate rock.

3.25 Write the net ionic equation for the formation of each salt in aqueous solution: (a) lead(II) sulfate, $PbSO_4$, a precipitate formed in a lead-acid battery; (b) copper(II) sulfide, CuS; (c) cobalt(II) carbonate, $CoCO_3$. (d) Select two soluble salts that, when mixed in solution, form each of the above insoluble salts. Identify the spectator ions.

3.26 Write the net ionic equation for the formation of each of the following compounds in aqueous solution: (a) lead(II) carbonate, $PbCO_3$, the white pigment in putty; (b) aluminum hydroxide, $Al(OH)_3$; (c) zinc chromate, $ZnCrO_4$. (d) Select two soluble salts that, when mixed in solution, form each of the above insoluble substances. Identify the spectator ions.

Acids and Bases

3.27 Identify the following as either an Arrhenius acid or an Arrhenius base: (a) $NH_3(aq)$; (b) $HCl(aq)$; (c) $NaOH(aq)$.

3.28 Classify the following as either an Arrhenius acid or an Arrhenius base: (a) $CH_3NH_2(aq)$, a derivative of ammonia; (b) $CH_3COOH(aq)$; (c) $HClO_4(aq)$.

3.29 Identify all the principal species that exist in aqueous solutions of (a) HCl; (b) NH_3; (c) HNO_3.

3.30 Identify all the principal species that exist in aqueous solutions of (a) CH_3NH_2; (b) HI; (c) $HClO_4$.

3.31 Complete and write the overall equation, the full ionic form of the equation, and the net ionic equation for the acid-base reactions given below. If the substance is a weak acid or weak base, then write it in its molecular form in the equations.

(a) $HCl(aq) + NaOH(aq) \longrightarrow$
(b) $NH_3(aq) + HNO_3(aq) \longrightarrow$
(c) $CH_3NH_2(aq) + HI(aq) \longrightarrow$

3.32 Complete and write the overall equation, the full ionic form of the equation, and the net ionic equation for the acid-base reactions given below. If the substance is a weak acid or weak base, then write it in its molecular form in the equations.

(a) $H_2SO_4(aq) + KOH(aq) \longrightarrow$
(b) $Ba(OH)_2(aq) + HCN(aq) \longrightarrow$
(c) $NH_3(aq) + HClO_4(aq) \longrightarrow$

3.33 Select an acid and a base for a neutralization reaction that results in the formation of (a) potassium bromide, KBr; (b) calcium cyanide, $Ca(CN)_2$; (c) potassium phosphate, K_3PO_4. Write the overall equation for each reaction.

3.34 Determine the salt that is produced from the acid-base neutralization reactions between (a) potassium hydroxide and acetic acid, CH_3COOH; (b) ammonia, NH_3, and hydroiodic acid; (c) barium hydroxide and sulfuric acid, H_2SO_4 (both H atoms react). Write the full ionic form of the equation for each reaction.

3.35 Identify the Brønsted acid and the Brønsted base in the following equations:

(a) $C_2H_5NH_2(l) + HCl(g) \longrightarrow C_2H_5NH_3Cl(s)$
(b) $C_2H_5NH_2(aq) + HCl(aq) \longrightarrow C_2H_5NH_3^+(aq) + Cl^-(aq)$
(c) $HCO_3^-(aq) + HI(aq) \longrightarrow H_2O(l) + CO_2(g) + I^-(aq)$

3.36 Identify the Brønsted acid and the Brønsted base in the following equations:

(a) $HPO_4^{2-}(aq) + HSO_4^-(aq) \longrightarrow H_2PO_4^-(aq) + SO_4^{2-}(aq)$
(b) $SO_2(g) + 2H_2O(l) \longrightarrow HSO_3^-(aq) + H_3O^+(aq)$
(c) $O^{2-}(aq) + H_2O(l) \longrightarrow 2OH^-(aq)$

Redox Reactions, Half-Reactions

3.37 Balance each half-reaction and state whether it is an oxidation or a reduction:

(a) $AgCN(s) \longrightarrow Ag(s) + CN^-(aq)$
(b) $ClO_3^-(aq) + H_2O(l) \longrightarrow ClO_2^-(aq) + OH^-(aq)$
(c) $F^-(aq) + H_2O(l) \longrightarrow OF_2(g) + 2H^+(aq)$

3.38 Balance each half-reaction and state whether it is an oxidation or a reduction:

(a) $Hg_2Cl_2(s) \longrightarrow Hg(l) + 2Cl^-(aq)$
(b) $O_2(g) + H^+(aq) \longrightarrow H_2O(l)$
(c) $Cd(s) + 2OH^-(aq) \longrightarrow Cd(OH)_2(s)$, a half-reaction that occurs in a Nicad cell

3.39 Write the balanced equations for the oxidation and reduction half-reactions occurring in the following redox reactions: (a) The displacement of copper(II) ion from solution by magnesium metal, $Mg(s) + Cu^{2+}(aq) \rightarrow Mg^{2+}(aq) + Cu(s)$. (b) The formation of iron(III) ion by the reaction $Fe^{2+}(aq) + Ce^{4+}(aq) \rightarrow Fe^{3+}(aq) + Ce^{3+}(aq)$. (c) A simplified equation for the formation of rust, $Fe(s) + O_2(g) \rightarrow Fe_2O_3(s)$.

3.40 Write the balanced equations for the oxidation and reduction half-reactions occurring in the following redox equations: (a) The production of titanium metal by magnesium metal, $TiCl_4(g) + Mg(l) \rightarrow MgCl_2(s) + Ti(s)$. (b) The decomposition of hydrogen peroxide, $H_2O_2(aq) \rightarrow 2H_2O(l) + O_2(g)$. (c) The industrial production of elemental bromine from brine, $Cl_2(g) + Br^-(aq) \rightarrow Br_2(l) + Cl^-(aq)$.

Oxidation Numbers

3.41 Use Table 3.5 to determine the oxidation number of the italicized element: (a) $\mathit{I}O_3^-$; (b) $HCl\mathit{O}$; (c) $\mathit{N}O$; (d) $H\mathit{N}O_3$.

3.42 Identify the oxidation number of each italicized element: (a) $\mathit{Mn}O_4^-$; (b) $\mathit{S}_2O_3^{2-}$; (c) $\mathit{S}O_4^{2-}$; (d) $\mathit{Mn}O_4^{2-}$.

3.43 For the following reactions that are redox reactions, identify the substance oxidized and the substance reduced by the change in oxidation numbers: (a) $CH_3Br(aq) + OH^-(aq) \rightarrow CH_3OH(aq) + Br^-(aq)$; (b) $BrO_3^-(aq) + 5Br^-(aq) + 6H^+(aq) \rightarrow 3Br_2(l) + H_2O(l)$; (c) $2F_2(g) + 2H_2O(l) \rightarrow 4HF(aq) + O_2(g)$.

3.44 In each of the following reactions, use oxidation numbers to identify the substance oxidized and the substance reduced: (a) The production of iodine from seawater, $Cl_2(g) + 2I^-(aq) \rightarrow I_2(aq) + 2Cl^-(aq)$. (b) A reaction to prepare "bleach," $Cl_2(g) + 2NaOH(aq) \rightarrow NaCl(aq) + NaClO(aq) + H_2O(l)$. (c) A reaction that destroys ozone in the stratosphere, $NO(g) + O_3(g) \rightarrow NO_2(g) + O_2(g)$.

Oxidizing and Reducing Agents

3.45 Identify the oxidizing agent and the reducing agent in each of the following reactions: (a) $Zn(s) + 2HCl(aq) \rightarrow ZnCl_2(aq) + H_2(g)$, a simple way for preparing H_2 gas in the laboratory. (b) $2H_2S(g) + SO_2(g) \rightarrow 3S(s) + 2H_2O(l)$, a reaction used to produce sulfur from hydrogen sulfide, the "sour gas" in natural gas. (c) $B_2O_3(s) + 3Mg(s) \rightarrow 2B(s) + 3MgO(s)$, a preparation of elemental boron.

3.46 Identify the oxidizing agent and the reducing agent in each of the following reactions: (a) $2Al(s) + Cr_2O_3(s) \rightarrow Al_2O_3(s) + 2Cr(s)$, an example of a thermite reaction used to obtain some metals from their ores. (b) $6Li(s) + N_2(g) \rightarrow 2Li_3N(s)$, a reaction that shows the similarity of lithium and magnesium. (c) $2Ca_3(PO_4)_2(s) + 6SiO_2(s) + 10C(s) \rightarrow P_4(g) + 6CaSiO_3(s) + 10CO(g)$, a reaction for the preparation of elemental phosphorus.

3.47 Decide whether to choose an oxidizing agent or a reducing agent to bring about each of the following changes: (a) $Br^-(aq)$ to $BrO_3^-(aq)$; (b) $NO_3^-(aq)$ to $NO(g)$; (c) HCHO (formaldehyde) to CH_3OH (methanol).

3.48 Would you choose an oxidizing agent or a reducing agent to make the following conversions? (a) $SO_4^{2-}(aq)$ to $S^{2-}(aq)$; (b) $Mn^{2+}(aq)$ to $MnO_2(s)$; (c) HCHO (formaldehyde) to HCOOH (formic acid).

3.49 Use the metal activity series in Table 3.6 to decide which of the following metals can be oxidized by $HCl(aq)$ to produce H_2 gas: zinc; tin; silver.

3.50 Refer to the activity series to predict whether any observable reaction can be expected when the following actions are performed. Explain your conclusions. (a) Copper wire is placed into $HCl(aq)$. (b) Lead pellets are dropped into $CuSO_4(aq)$. (c) Tin foil is placed into $FeSO_4(aq)$.

3.51 Cations can act as oxidizing agents in certain circumstances because they may be able to remove electrons from more readily ionizable species. Use Table 3.6 to determine whether the following reactions can be expected to occur. For those that do, write corresponding oxidation and reduction half-reactions. (a) $Cu^{2+}(aq)$ oxidizes nickel metal. (b) $Cu^{2+}(aq)$ oxidizes platinum metal. (c) $Zn^{2+}(aq)$ oxidizes copper metal.

3.52 From Table 3.6 determine whether the following reactions can occur.

(a) $Cu^{2+}(aq) + 2Au(s) \rightarrow 2Au^+(aq) + Cu(s)$
(b) $Sn^{2+}(aq) + Fe(s) \rightarrow Fe^{2+}(aq) + Sn(s)$
(c) $3Mg(s) + 2Cr^{3+}(aq) \rightarrow 3Mg^{2+}(aq) + 2Cr(s)$

Balancing Redox Equations

3.53 Balance the following skeletal equations for each half-reaction occurring in an acidic solution. Identify each as an oxidation or reduction half-reaction. (a) The conversion of vanadyl ions to vanadium(III) ions, $VO^{2+}(aq) \rightarrow V^{3+}(aq)$. (b) One of the half-reactions that occurs when a lead-acid battery is charged, $PbSO_4(s) \rightarrow PbO_2(s) + SO_4^{2-}(aq)$. (c) The decomposition of hydrogen peroxide to oxygen gas, $H_2O_2(aq) \rightarrow O_2(g)$.

3.54 Balance the following skeletal equations for each half-reaction occurring in an acidic solution. Identify each as an oxidation or reduction half-reaction.
(a) $Cr_2O_7^{2-}(aq) \rightarrow Cr^{3+}(aq)$
(b) $I^-(aq) \rightarrow IO_3^-(aq)$
(c) $NO(g) \rightarrow NO_3^-(aq)$

3.55 Balance the following skeletal equations for each half-reaction occurring in a basic solution. Identify each as an oxidation or reduction half-reaction. (a) The conversion of hypochlorite ions to chloride ions (the "action" of bleach), $ClO^-(aq) \rightarrow Cl^-(aq)$. (b) The conversion of iodate ions to hypoiodite ions, $IO_3^-(aq) \rightarrow IO^-(aq)$. (c) The conversion of sulfite ions to dithionite ions, $SO_3^{2-}(aq) \rightarrow S_2O_4^{2-}(aq)$.

3.56 Balance the following skeletal equations for each half-reaction occurring in a basic solution. Identify each as an oxidation or reduction half-reaction.
(a) $NO_3^-(aq) \rightarrow NO_2^-(aq)$
(b) $CrO_4^{2-}(aq) \rightarrow Cr(OH)_3(s)$
(c) $MnO_4^{2-}(aq) \rightarrow MnO_2(s)$

3.57 When water is electrolyzed (Chapter 17), hydrogen is produced at the electrode called the cathode and oxygen is produced at the electrode called the anode. The skeletal half-reactions for the production of H_2 and O_2 from H_2O are $H_2O(l) \rightarrow H_2(g) + OH^-(aq)$ and $H_2O(l) \rightarrow O_2(g) + H^+(aq)$. Balance the half-reactions. At which electrode does reduction occur and at which electrode does oxidation occur?

3.58 The industrial production of sodium metal and chlorine gas makes use of the Downs process, in which molten sodium chloride is electrolyzed (Chapter 17). Write the balanced half-reactions for the production of each element from molten sodium chloride.

3.59 Balance the following skeletal equations by using oxidation and reduction half-reactions. All reactions occur in an acidic solution. Identify the oxidizing agent and reducing agent in each reaction. (a) The reaction of thiosulfate ion with chlorine gas, $Cl_2(g) + S_2O_3^{2-}(aq) \rightarrow Cl^-(aq) + SO_4^{2-}(g)$. (b) The action of the permanganate ion on sulfurous acid, $MnO_4^-(aq) + H_2SO_3(aq) \rightarrow Mn^{2+}(aq) + HSO_4^-(aq)$. (c) The reaction of hydrosulfuric acid with chlorine, $H_2S(aq) + Cl_2(g) \rightarrow S(s) + Cl^-(aq)$.

CASE 3 ▸ FROM RIVER WATER TO DRINKING WATER

The sources of public drinking water (called raw water) vary from underground aquifers (water-bearing layers) to murky river water laden with, among other pollutants, bacteria, pesticides, and metal ions. It is the responsibility of the water chemists and sanitary engineers to convert this water into a safe, potable (consumable) drinking water supply that is free of color, odor, suspended solids, toxic compounds, and bacteria. In most cases the water treatment process utilizes physical separation techniques, similar to those discussed in Chapter 1 for separating a mixture into its component substances.

Some of the steps used in the treatment process are shown in the illustration. The raw water is aerated to remove foul-smelling dissolved gases, such as H_2S and CH_4, and to add oxygen and nitrogen. Nonaerated water tastes flat. Aeration also oxidizes any Fe^{2+} ions to Fe^{3+} ions. The addition of lime, as CaO or $Ca(OH)_2$, raises the pH and causes the precipitation of Mg^{2+} and Fe^{3+} ions as hydroxides and Ca^{2+} ions as carbonate (the ions responsible for hardness, Mg^{2+} and Ca^{2+}, are present generally as the hydrogencarbonates):

$$Mg^{2+}(aq) + HCO_3^-(aq) + Ca(OH)_2(s) \longrightarrow Mg(OH)_2(s) + Ca^{2+}(aq) + HCO_3^-(aq)$$

$$Ca^{2+}(aq) + 2HCO_3^-(aq) + Ca(OH)_2(aq) \longrightarrow 2CaCO_3(s) + 2H_2O(l)$$

$$2Fe^{3+}(aq) + 3Ca(OH)_2(aq) \longrightarrow 2Fe(OH)_3(s) + 3Ca^{2+}(aq)$$

The precipitates are allowed to settle in a primary settling basin. Because much of the $CaCO_3$ and $MgCO_3$ remains in suspension, the water is separated from the precipitate and treated with a coagulant. A coagulant, the most common being alum, $Al_2(SO_4)_3 \cdot 18H_2O$, is added to form a gelatinous aluminum hydroxide precipitate with the remaining hydrogencarbonate ions. The acidity of the water solution is raised by the addition of

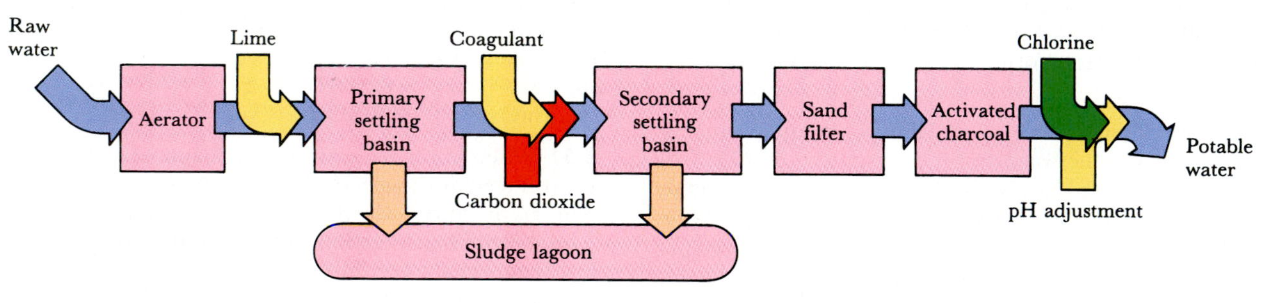

3.60 Balance the following equations by using oxidation and reduction half-reactions. All reactions occur in an acidic solution. Identify the oxidizing agent and reducing agent in each reaction. (a) The conversion of iron(II) to iron(III) by the orange dichromate ion, $Fe^{2+}(aq) + Cr_2O_7^{2-}(aq) \rightarrow Fe^{3+}(aq) + Cr^{3+}(aq)$. (b) The formation of acetic acid from ethanol by the action of purple permanganate ion, $C_2H_5OH(aq) + MnO_4^-(aq) \rightarrow Mn^{2+}(aq) + CH_3COOH(aq)$. (c) The reaction of iodine with nitric acid, $I^-(aq) + NO_3^-(aq) \rightarrow I_2(aq) + NO(g)$.

3.61 Balance the following equations by using oxidation and reduction half-reactions. All reactions occur in a basic solution. Identify the oxidizing agent and reducing agent in each reaction. (a) The action of ozone on bromide ions, $O_3(g) + Br^-(aq) \rightarrow O_2(g) + BrO^-(aq)$. (b) The reaction of bromine in water, $Br_2(l) \rightarrow BrO_3^-(aq) + Br^-(aq)$. (c) The formation of chromate ions from chromium(III) ions, $Cr^{3+}(aq) + MnO_2(s) \rightarrow Mn^{2+}(aq) + CrO_4^{2-}(aq)$.

CO_2, and the lower pH causes the formation of a flocculent (fluffy) precipitate:

$$Al(H_2O)_6^{3+}(aq) + 3HCO_3^-(aq) \longrightarrow Al(OH)_3(s) + 3CO_2(g) + 6H_2O(l)$$

The precipitate settles slowly in a secondary basin and in so doing adsorbs any remaining suspended $CaCO_3$, bacteria, and other suspended particles, such as dirt and algal growth. The precipitates from the primary and secondary basins are combined in a sludge lagoon for disposal. Other flocculating substances are being investigated (partly to avoid the health hazards that may result from the use of aluminum). These include starch and cellulose derivatives and synthetic polymers and polyelectrolytes. Iron(II) sulfate is sometimes used instead of alum as a coagulant.

The clear water is then passed through a sand filter to remove any remaining suspended particles. Another stage currently is being added to many city water-treatment plants to remove synthetic organic compounds, most of which are pesticides that originate from the runoff water from farmland, and natural organic compounds, which may contribute a taste or odor to drinking water. The most common procedure is to pass the water through a bed of activated charcoal (finely divided carbon that has a very large surface area), which adsorbs the organic compounds onto its highly porous surface.

The final step in the process before the water enters the distribution system is to adjust it to a slightly basic pH (to avoid acid corrosion of the pipes) and to add disinfectant, usually chlorine. By law, the chlorine level is required to be greater than 1 g Cl_2 per 10^3 kg of water at the point of usage in the water distribution system. In water the chlorine forms hypochlorous acid:

$$Cl_2(g) + 2H_2O(l) \longrightarrow H_3O^+(aq) + Cl^-(aq) + HClO(aq)$$

The hypochlorous acid attacks bacteria. Fluoride ions, in the form of sodium fluoride, may be added at low concentrations (less than 1 g/10^3 kg of water) to reduce tooth decay in children.

Depending on the source of water and the contaminants that it may contain, other water purification steps may be introduced. Techniques such as ion exchange, reverse osmosis (Chapter 11), and additional precipitation procedures may be introduced.

QUESTIONS

1. Using the solubility rules in Table 3.2, write the full ionic equation and the net ionic equation for the reaction of aluminum sulfate with calcium hydroxide.

2. Write the formulas and name the compounds that are used in the purification of water.

3. Identify (a) the physical processes and (b) the chemical process used for purifying raw water.

4. Classify the reactions that are used in the purification of water.

3.62 Balance the following equations by using oxidation and reduction half-reactions. All reactions occur in a basic solution. Identify the oxidizing agent and reducing agent in each reaction. (a) The production of chlorite ions from chlorine heptoxide, $Cl_2O_7(g) + H_2O_2(aq) \rightarrow ClO_2^-(aq) + O_2(g)$. (b) The action of permanganate ions on sulfide ions, $MnO_4^-(aq) + S^{2-}(aq) \rightarrow S(s) + MnO_2(s)$. (c) The reaction of hydrazine with chlorate ions, $N_2H_4(g) + ClO_3^-(aq) \rightarrow NO(g) + Cl^-(aq)$.

3.63 (a) The hydrogensulfite ion, HSO_3^-, is a moderately strong reducing agent in acidic solutions and, depending upon the conditions, is oxidized to either the hydrogensulfate ion, HSO_4^-, or the dithionate ion, $S_2O_6^{2-}$. Write the equations for each half-reaction. (b) The reaction of the hydrogensulfite ion with iodine to form iodide ions and hydrogensulfate ions is used to determine its concentration in solution in the laboratory. Write the equation for each half-reaction and the overall equation for the reaction.

3.64 The thiosulfate ion, $S_2O_3^{2-}$, is a moderately strong reducing agent in acidic solutions. It is used to determine the concentration of elemental iodine in solution, forming iodide ions and tetrathionate ions, $S_4O_6^{2-}$, in the reaction. Write the equation for each half-reaction and the overall equation for the reaction.

3.65 Potassium permanganate is an excellent oxidizing agent for laboratory use and in sewage treatment. It readily reacts with the organic compounds in sewage to produce carbon dioxide and water. Write the equation for each half-reaction and the overall equation for the oxidation of glucose, $C_6H_{12}O_6$, the skeletal form of which is $MnO_4^-(aq) + C_6H_{12}O_6(aq) \rightarrow Mn^{2+}(aq) + CO_2(g) + H_2O(l)$.

3.66 Nitric oxide gas can be produced from the reaction of nitrite salts with water in a dilute sulfuric acid solution. The skeletal equation for the reaction is $NO_2^-(aq) \rightarrow NO(g) + NO_3^-(aq)$. Write the equation for each half-reaction and the overall equation for the reaction.

UNCLASSIFIED EXERCISES

3.67 Identify the following as a strong acid, a strong base, a soluble ionic compound, or an insoluble ionic compound in water: (a) HNO_3; (b) KOH; (c) NH_3; (d) $CuSO_4$; (e) HCOOH; (f) $Ca_3(PO_4)_2$.

3.68 The contents of Beaker 1 are mixed with those of Beaker 2. If a reaction occurs, write the net ionic equation and indicate the spectator ions.

Beaker 1	Beaker 2
(a) $H_2SO_4(aq)$	$KOH(aq)$
(b) $H_3PO_4(aq)$	$CuCl_2(aq)$
(c) $K_2S(aq)$	$AgNO_3(aq)$
(d) $NiSO_4(aq)$	$(NH_4)_2CO_3(aq)$

3.69 Select two soluble ionic compounds that, when mixed in solution, produce the precipitates (a) $MgCO_3$; (b) Ag_2SO_4; (c) $Zn(OH)_2$; (d) $PbCrO_4$.

3.70 An acid-base neutralization reaction that occurs in liquid ammonia (denoted by *am*) is the reaction of ammonium chloride and potassium amide: $NH_4Cl(am) + KNH_2(am) \rightarrow 2NH_3(l) + KCl(am)$. Identify the Brønsted acid and the Brønsted base in the reaction.

3.71 Select a strong acid and a strong base that, when mixed in solution, produce the salts (a) $Na_2SO_4(aq)$; (b) $KClO_4(aq)$; (c) $NiCl_2(aq)$.

3.72 Classify each of the following reactions as precipitation, acid-base neutralization, or redox. If it is a precipitation reaction, then write a net ionic equation; if it is a neutralization reaction, then identify the acid and the base; if it is a redox reaction, then identify the oxidizing agent and the reducing agent. (a) The formation of magnesium chloride by the action of hydrochloric acid on magnesium hydroxide solution, $Mg(OH)_2(aq) + 2HCl(aq) \rightarrow MgCl_2(aq) + 2H_2O(l)$. (b) The production of sulfur trioxide from sulfur dioxide for the manufacture of sulfuric acid, $2SO_2(g) + O_2(g) \rightarrow 2SO_3(g)$. (c) The reaction used to produce elemental phosphorus from its oxide, $P_4O_{10}(s) + 10C(s) \rightarrow P_4(s) + 10CO(g)$. (d) The formation of silver chromate, Ag_2CrO_4, from sodium chromate, Na_2CrO_4, and silver nitrate aqueous solutions, $2AgNO_3(aq) + Na_2CrO_4(aq) \rightarrow Ag_2CrO_4(s) + 2NaNO_3(aq)$.

3.73 Classify each of the following reactions as precipitation, acid-base neutralization, or redox. If it is a precipitation reaction, then write a net ionic equation; if it is a neutralization reaction, then identify the acid and the base; if it is a redox reaction, then identify the oxidizing agent and the reducing agent. (a) The reaction often used to monitor the amount of iodine in a sample, in which the iodine reacts with thiosulfate ions, $S_2O_3^{2-}$, and forms tetrathionate ions, $S_4O_6^{2-}$: $I_2(aq) + 2S_2O_3^{2-}(aq) \rightarrow 2I^-(aq) + S_4O_6^{2-}(aq)$. (b) The test for bromide ions in solution, in which a dense, creamy precipitate of silver bromide is formed when a drop of the solution is added to silver nitrate solution, $AgNO_3(aq) + Br^-(aq) \rightarrow AgBr(s) + NO_3^-(aq)$. (c) The heating of uranium tetrafluoride with magnesium, one stage in the purification of uranium metal, $UF_4(g) + 2Mg(s) \rightarrow U(s) + 2MgF_2(s)$.

3.74 Balance the skeletal equation for the following half-reactions and state whether the half-reaction is an oxidation or a reduction.
(a) $I^-(aq) \rightarrow I_3^-(aq)$
(b) $NO_2^-(aq) + H_2O(l) \rightarrow N_2O_2^{2-}(aq) + OH^-(aq)$
(c) $SeO_4^{2-}(aq) + H_2O(l) \rightarrow SeO_3^{2-}(aq) + OH^-(aq)$

3.75 Identify the oxidizing agent and the reducing agent for each of the following reactions: (a) The production of tungsten metal from its oxide by the reaction $WO_3(s) + 3H_2(g) \rightarrow W(s) + 3H_2O(l)$. (b) The generation of hydrogen gas in the laboratory, $Zn(s) + 2HCl(aq) \rightarrow H_2(g) + ZnCl_2(aq)$. (c) Hydrazine has a low molar mass and releases a significant amount of energy in its reaction with dinitrogen tetroxide; the combination is used as a rocket propellant, $2N_2H_4(g) + N_2O_4(g) \rightarrow 3N_2(g) + 4H_2O(g)$.

3.76 Nitric acid is a good oxidizing agent; its product depends on the concentration of the acid and the strength of the reducing agent. Write the balanced equations for the half-reactions and for the overall reaction for each of the following cases:

(a) $Cu(s) + NO_3^-(aq) \rightarrow$
$Cu^{2+}(aq) + NO(g)$ (warm, dilute HNO_3)

(b) $Cu(s) + NO_3^-(aq) \rightarrow$
$Cu^{2+}(aq) + NO_2(g)$ (concentrated HNO_3)

(c) $Zn(s) + NO_3^-(aq) \rightarrow$
$Zn^{2+}(aq) + NH_4^+(aq)$ (dilute HNO_3)

(d) $Zn(s) + NO_3^-(aq) \rightarrow$
$Zn^{2+}(aq) + N_2(g)$ (very dilute HNO_3)

3.77 One stage in the extraction of gold from rocks involves dissolving the metal from the rock with a basic solution of sodium cyanide that has been thoroughly aerated. This stage results in the formation of the soluble $Au(CN)_2^-$ ions. The next stage is to precipitate the gold by the addition of zinc dust, forming another ion, $Zn(CN)_4^{2-}$. Write the balanced equations for the half-reactions and the overall redox equation for both stages.

In this chapter we see how to interpret chemical equations quantitatively. The techniques we learn here enable us to predict the amount of product formed in a reaction, calculate the required amounts of reactants, and use measurements of mass and volume to determine the compositions of compounds. Calculations like these are used to decide how much hydrogen and oxygen were needed to launch the space shuttle, seen here at the moment of a nighttime launch.

4 REACTION STOICHIOMETRY

A balanced equation is a quantitative statement about a chemical reaction; hence, it can be used to make numerical predictions about the outcome of the reaction. In this chapter we shall see how to use a chemical equation to predict how much product will result when a given mass of starting material reacts completely and how much of each reactant we must supply to obtain a given mass of product. Calculations like these are exactly the kind a chemist needs to do when designing a chemical plant (Fig. 4.1), analyzing the suitability of a fuel, or working out how much reactant to use in an experiment. We can also use the same kinds of calculations to analyze the compositions of samples and to determine the chemical formulas of compounds.

This chapter is central to the practice of chemistry because it establishes the framework for almost all the calculations that we meet later in the text and when—beyond the course—we need to apply chemical arguments to environmental and other issues. We shall see how central the concept of the mole is in chemistry and, by expressing amounts of substances in moles, how we can make a wide variety of predictions. Indeed, the concepts in this chapter give us power to make quantitative predictions throughout chemistry, and time and again we shall return to the techniques introduced here.

N A constant theme of this chapter is: *to make predictions, convert from the data supplied to moles, and then from moles to the information required.* The same theme will recur throughout the text: we focus all chemical calculations on the mole.

INTERPRETING STOICHIOMETRIC COEFFICIENTS

The chemical equation

$$2H_2(g) + O_2(g) \longrightarrow 2H_2O(l)$$

N Teaching stoichiometry: a two-cycle approach. *J. Chem. Ed.* **1989**, *66*, 57.

tells us that for every two H_2 molecules and one O_2 molecule that react, two H_2O molecules are formed. It follows that for every 2 *moles* of H_2

FIGURE 4.1 The large, round structures are the ammonia storage tanks in a modern plant for producing ammonia by the Haber synthesis. The quantities of hydrogen and nitrogen that must be supplied to produce a given amount of ammonia can be calculated by using the techniques described in this chapter.

N The mole, as the pivot of chemical calculations, is stressed throughout the text.

($2 \times N_A$ H_2 molecules, where N_A is Avogadro's number) that react, 1 mole of O_2 reacts and 2 moles of H_2O are formed:

$$\underset{2\ \text{mol}}{2H_2(g)} + \underset{1\ \text{mol}}{O_2(g)} \longrightarrow \underset{2\ \text{mol}}{2H_2O(l)}$$

N The extent of reaction as a unifying basis for stoichiometry in elementary chemistry. *J. Chem. Ed.* **1974**, *51*, 194.

That is, the stoichiometric coefficients—the numbers multiplying the chemical formulas in a balanced chemical equation—tell us the number of moles of each substance that is consumed or is produced in the reaction.

4.1 MOLE CALCULATIONS

The interpretation of stoichiometric coefficients as numbers of moles opens up the path to a wide range of calculations. We introduce the general strategy here and illustrate two types of approach to making predictions. In later chapters the same strategy will be used in other cases.

N It is emphasized on the following page that this equality is not a true equality because in no sense is 2 mol H_2 "equal" to 2 mol H_2O (or, more apparently, is 2 mol H_2 "equal" to 1 mol O_2). The "equality" is actually another example of an equivalence (see the annotation on p. 58). Moreover, the equivalence is valid only for the reaction under discussion. Some texts employ other symbols for this stoichiometric equivalence, such as $\triangleq$ or $\approx$. We have decided to use the simpler =, and to point out that its interpretation requires some care (in principle, but not in practice!).

The amount of a substance that reacts. We know from the equation for the formation of water that 2 mol of H_2O is formed when 2 mol of H_2 reacts. However, in general we are interested in the amount of product formed when the amount of reactants is not the same as one of the stoichiometric coefficients in the equation. For example, we might be interested in knowing how much water is formed when 0.25 mol of H_2, not 2 mol of H_2, reacts with oxygen.

The calculation we have to do makes use of the conversion factor technique introduced in Chapter 1. According to the chemical equation, 2 mol of H_2 reacts to form 2 mol of H_2O; so we can write

$$2\ \text{mol}\ H_2 = 2\ \text{mol}\ H_2O$$

N Almost all stoichiometric calculations employ a stoichiometric relation.

This expression, which equates the relative amounts of reactants and products that participate in a reaction is called a **stoichiometric relation**. We can use it in exactly the same way we used a relation between units (such as $1\ u = 1.66 \times 10^{-24}$ g) to construct a conversion factor.

So, if we want to know the amount of H_2O formed from 0.25 mol of H_2, we write

$$\text{Amount of } H_2O \text{ (mol)} = 0.25 \cancel{\text{mol } H_2} \times \frac{2 \text{ mol } H_2O}{2 \cancel{\text{mol } H_2}} = 0.25 \text{ mol } H_2O$$

↑ Information required — ↑ Information given — ↑ Conversion from moles of H_2 to moles of H_2O

Note how the substances cancel, just as though they were units.

N All stoichiometric calculations are set up using the same format: Information required = information given × conversion factors.

▼ **EXAMPLE 4.1 Calculating the amount of a substance taking part in a reaction**

What amount of N_2 (in moles) is needed to produce 5.0 mol of NH_3 in the Haber synthesis? The equation for the synthesis is

$$N_2(g) + 3H_2(g) \longrightarrow 2NH_3(g)$$

STRATEGY In all stoichiometry calculations, use the balanced equation to identify the stoichiometric relation that is needed, and obtain the required conversion factor from the stoichiometric coefficients in the equation.

SOLUTION Because 2 mol NH_3 = 1 mol N_2, the conversion we require is

$$\text{Amount of } N_2 \text{ (mol)} = 5.0 \cancel{\text{mol } NH_3} \times \frac{1 \text{ mol } N_2}{2 \cancel{\text{mol } NH_3}} = 2.5 \text{ mol } N_2$$

↑ Information required — ↑ Information given — ↑ Conversion from moles of NH_3 to moles of N_2

EXERCISE E4.1 How many moles of NH_3 can be produced from 2.0 mol H_2 if all the hydrogen reacts?

[*Answer*: 1.3 mol NH_3]

▲

N The information given is also called the *dependent variable* in algebra or the *measured quantity* in an experiment.

U Exercise 4.81.

It should be noted that a stoichiometric relation is a little different from the relation between pairs of units. Thus, 2 mol of H_2 is not *equal* to 2 mol H_2O in the sense that 1 u is the same as 1.66×10^{-24} g. The stoichiometric relation shows that *for the reaction we are considering*, the amounts of each substance consumed and formed are related in the manner shown by the stoichiometry.

The masses taking part in a reaction. We have just seen that the amount (number of moles) of reactants taking part in a reaction can be related to the amount (number of moles) of products. We have already seen (in Section 2.6) how to use molar masses to convert between amounts and masses. By combining the two calculations, we obtain a way of calculating the *masses* of the substances produced in a reaction from the masses of the reactants. The general procedure is

- Convert the masses of the reactants in grams to amounts in moles by using the molar mass of the reactant.
- Use the reaction stoichiometry (the balanced equation) to convert amount of reactant in moles to amount of product in moles.
- Convert from amount of product in moles to the mass of product in grams by using the molar mass of the product.

N This diagram is developed and expanded in the next several chapters. The mole is the pivot of chemical calculation, and the reaction stoichiometry establishes the chemical relationship between the reactants and the products.

T OHT: Diagrams **1**, **2**, and **3** (from p. 141).

N A logic diagram for teaching stoichiometry. *J. Chem. Ed.* **1975**, *52*, 492.

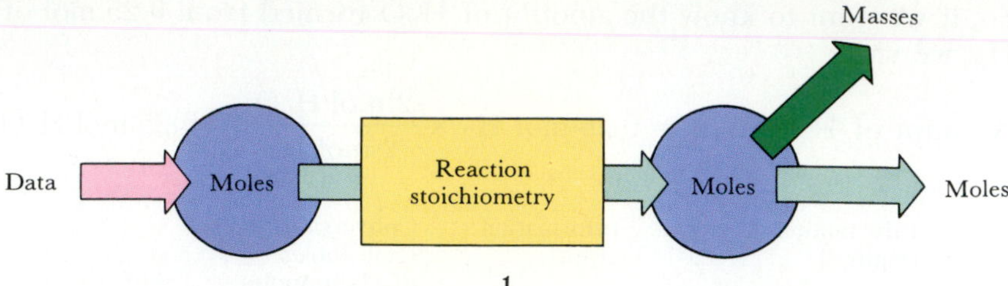

1

This procedure is illustrated in diagram (**1**). As we work through the text, we shall see that an almost identical procedure can be used to calculate other aspects of the reaction, such as the heat a reaction produces and the volume of gas evolved. *The strategy will always be the same*: we convert the data to moles, carry out a stoichiometric conversion, and then convert the new amount of substance to the output required (**2**).

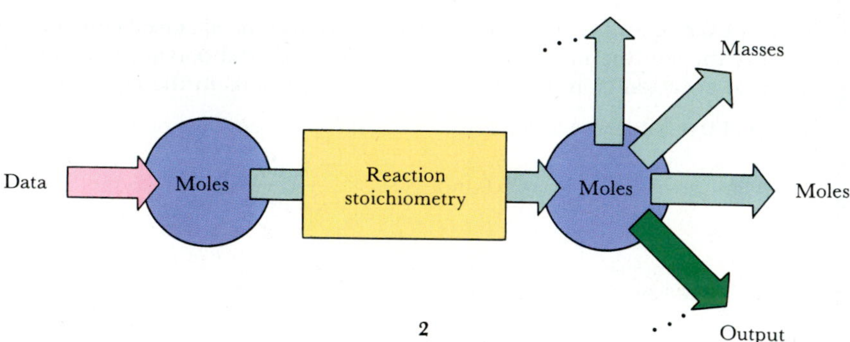

2

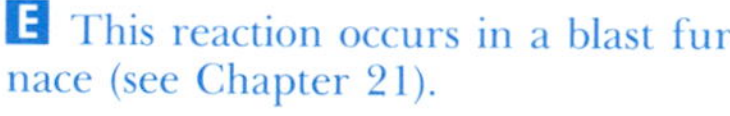

E This reaction occurs in a blast furnace (see Chapter 21).

At this stage we concentrate on calculations dealing with masses. Suppose that we want to know the mass of iron that can be obtained from 10.0 kg of Fe_2O_3 present in iron ore by the reaction

$$Fe_2O_3(s) + 3CO(s) \longrightarrow 2Fe(s) + 3CO_2(s)$$

We can find the mass of iron by converting the mass of Fe_2O_3 to moles of Fe_2O_3, then use the chemical equation to set up the stoichiometric relation between moles of Fe_2O_3 and moles of Fe, and finally convert moles of Fe to mass of iron.

For the conversion of the mass of Fe_2O_3 to amount of Fe_2O_3 we must convert from kilograms to grams by using the relation 1 kg = 10^3 g and then use the molar mass of Fe_2O_3 (159.70 g/mol) to write

$$159.70 \text{ g } Fe_2O_3 = 1 \text{ mol } Fe_2O_3$$

For the mole-to-mole conversion we use the stoichiometric relation

$$1 \text{ mol } Fe_2O_3 = 2 \text{ mol Fe}$$

and for the conversion from moles of Fe to mass of iron we use the molar mass of Fe in the form

$$1 \text{ mol Fe} = 55.85 \text{ g Fe}$$

These relations give

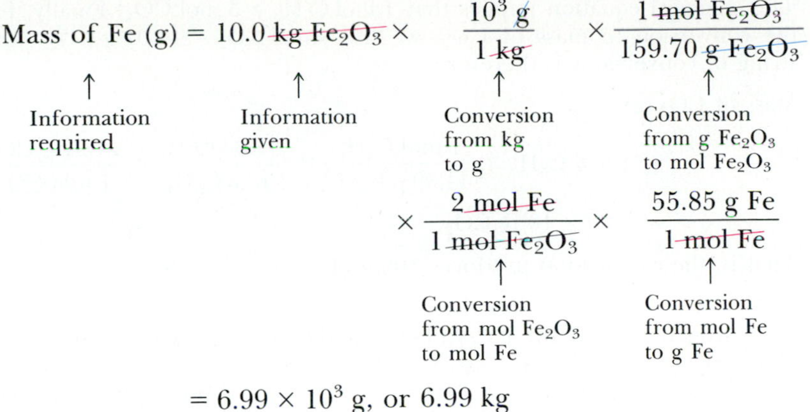

Mass of Fe (g) = $10.0\ \text{kg Fe}_2\text{O}_3 \times \frac{10^3\ \text{g}}{1\ \text{kg}} \times \frac{1\ \text{mol Fe}_2\text{O}_3}{159.70\ \text{g Fe}_2\text{O}_3}$
Information required
Information given
Conversion from kg to g
Conversion from g Fe_2O_3 to mol Fe_2O_3
$\times \frac{2\ \text{mol Fe}}{1\ \text{mol Fe}_2\text{O}_3} \times \frac{55.85\ \text{g Fe}}{1\ \text{mol Fe}}$
Conversion from mol Fe_2O_3 to mol Fe
Conversion from mol Fe to g Fe

$$= 6.99 \times 10^3\ \text{g, or } 6.99\ \text{kg}$$

We can get just under 7 kg of iron from 10.0 kg iron(III) oxide.

▼ EXAMPLE 4.2 Predicting the mass of a product

One environmental problem receiving wide attention is the increase of carbon dioxide in the atmosphere (Fig. 4.2). Calculate the mass of carbon dioxide produced when 100 g of propane (C_3H_8) is burned. The combustion equation is

$$C_3H_8(g) + 5O_2(g) \longrightarrow 3CO_2(g) + 4H_2O(g)$$

STRATEGY Because the six O atoms in the carbon dioxide formed have greater mass than the eight H atoms in the fuel, the mass of CO_2 produced will be greater than that (100 g) of the fuel. Set up the string of conversion factors needed to convert from mass to moles of reactant, from moles of reactant to moles of product, and finally from moles of product to mass of product. Construct the conversion factors from the molar masses and the chemical equation.

SOLUTION The first conversion required is from mass of fuel to amount of C_3H_8 (which has molar mass 44.09 g/mol): 44.09 g C_3H_8 = 1 mol C_3H_8.

E Propane is the bottled gas used for barbeque grills and in camping trailers.

N Follow the strategy through diagram **1**.

? Why does the concentration of carbon dioxide increase at Point Barrow, Alaska while, at the same time, it decreases at the South Pole?

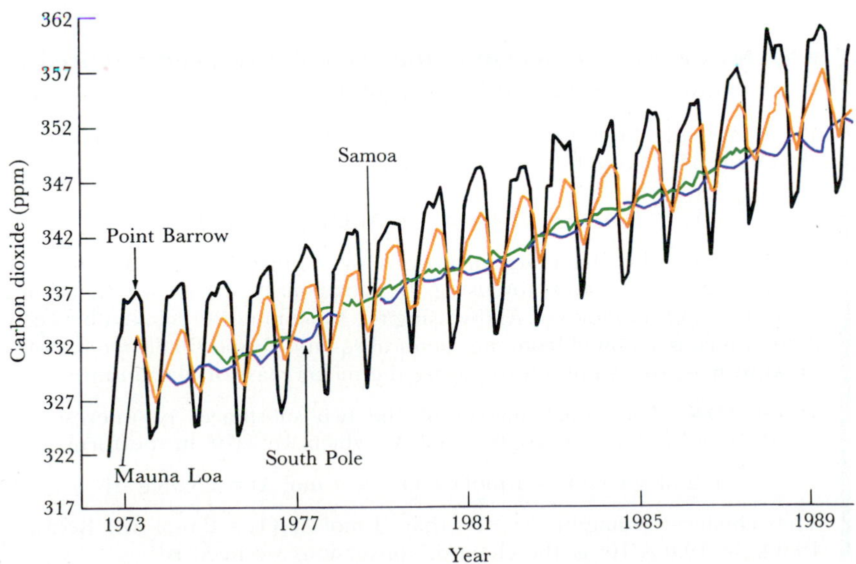

FIGURE 4.2 Atmospheric CO_2 concentrations measured at NOAA's (National Oceanic and Atmospheric Administration) continuous monitoring stations located at Point Barrow, Alaska, 70°N; Mauna Loa, Hawaii, 20°N; Somoa, 14°S; and the South Pole, 90°S. Each data point represents the monthly, mean CO_2 concentration, adjusted for background conditions. The seasonal variations are due to the photosynthesis and respiration of land plants. The average CO_2 growth rate is about 1.5 ppm per year.

U Exercise 4.57.

The chemical equation implies that 1 mol C_3H_8 = 3 mol CO_2. Finally, for the conversion to mass of CO_2, we use 1 mol CO_2 = 44.01 g CO_2. The string of conversions is therefore

Mass of CO_2 (g)

$$= 100 \text{ g } C_3H_8 \times \frac{1 \text{ mol } C_3H_8}{44.09 \text{ g } C_3H_8} \times \frac{3 \text{ mol } CO_2}{1 \text{ mol } C_3H_8} \times \frac{44.01 \text{ g } CO_2}{1 \text{ mol } CO_2}$$

$$= 3.00 \times 10^2 \text{ g } CO_2$$

That is, the combustion produces 300 g of carbon dioxide.

EXERCISE E4.2 Calculate the mass of ammonia produced when 50 kg of hydrogen is used for its synthesis in the reaction $N_2(g) + 3H_2(g) \rightarrow 2NH_3(g)$

▲ [*Answer*: 2.8×10^2 kg]

A very similar analysis can be carried out when we are asked to calculate the amount of a particular product that will be obtained when we started with a specified mass of reactant. The first conversion is from the mass of the reactant to its amount in moles. Then we use the reaction stoichiometry to convert from moles of reactant to moles of product, which is the output we require.

Our next example will show how to calculate the mass of one reactant that is needed to react completely with a given mass of another reactant. For example, we might be interested in knowing the mass of aluminum needed to react completely with a given mass of chromium(III) oxide in the redox reaction used to obtain chromium from its ore:

N Here the Δ indicates that the two solids must be heated initially, but once the reaction has been initiated, so much heat is generated that the reaction continues vigorously even in the absence of any external source of heat. The reaction is a variation of the conventional thermite reaction with iron(III) oxide. See Example 4.5.

V Video 8: Thermite.

$$2Al(l) + Cr_2O_3(s) \xrightarrow{\Delta} Al_2O_3(s) + 2Cr(l)$$

We perform exactly the same kind of calculation as before, but this time we need to use the stoichiometric relation between the aluminum metal and the chromium(III) oxide.

▼ **EXAMPLE 4.3 Calculating the mass of a reactant required to react completely with another reactant**

What mass of aluminum is needed to react completely with 10.0 kg of chromium(III) oxide to produce chromium metal in the reaction

$$2Al(l) + Cr_2O_3(s) \xrightarrow{\Delta} Al_2O_3(s) + 2Cr(l)$$

N Follow this strategy through diagram **1**.

STRATEGY The chain of conversions we require is from the mass of Cr_2O_3 to moles of Cr_2O_3 (by using the molar mass of Cr_2O_3), then from moles of Cr_2O_3 to moles of Al (by using the stoichiometric relation between them, which is obtained from the chemical equation), and finally from moles of Al to mass of aluminum (by using the molar mass of aluminum).

SOLUTION The molar masses of the two substances of interest are 152.00 g/mol Cr_2O_3 and 26.98 g/mol Al, which we write in the form

$$152.00 \text{ g } Cr_2O_3 = 1 \text{ mol } Cr_2O_3 \qquad 1 \text{ mol Al} = 26.98 \text{ g Al}$$

The chemical equation shows that 1 mol Cr_2O_3 = 2 mol Al. Because 10.0 kg = 10.0×10^3 g, the chain of conversions we need is

Mass of Al (g)

$$= 10.0 \times 10^3 \text{ g } Cr_2O_3 \times \frac{1 \text{ mol } Cr_2O_3}{152.00 \text{ g } Cr_2O_3} \times \frac{2 \text{ mol Al}}{1 \text{ mol } Cr_2O_3} \times \frac{26.98 \text{ g Al}}{1 \text{ mol Al}}$$

$$= 3.55 \times 10^3 \text{ g Al}$$

That is, we need to use 3.55 kg of aluminum.

EXERCISE E4.3 Calculate the mass of potassium needed to react with 0.450 g of hydrogen gas to produce potassium hydride in the reaction $2K(l) + H_2(g) \rightarrow 2KH(s)$.

[*Answer*: 17.4 g]

N The potassium is indicated as present as a liquid because, in the reaction, hydrogen is passed over the hot, molten metal.

U Exercises 4.58 and 4.59.

Reaction yield. The chemical equation for the combustion of octane (C_8H_{18}), a representative compound in the mixture sold as gasoline, is

$$2C_8H_{18}(l) + 25O_2(g) \longrightarrow 16CO_2(g) + 18H_2O(g)$$

Using this equation and the procedure illustrated in Example 4.2, we calculate that when 100 g of octane burns in a plentiful supply of oxygen, 308 g of carbon dioxide should be produced. This 308 g is the theoretical yield of CO_2:

The **theoretical yield** is the maximum quantity of product that can be obtained, according to the reaction stoichiometry, from a given quantity of a specified reactant.

By *quantity* we can mean mass, volume of gas, or amount, but we shall usually express theoretical yields in terms of the mass of product formed from a given mass of reactant.

When octane burns in a limited supply of oxygen, carbon monoxide is formed as well as carbon dioxide; so, in addition to the reaction written above, the reaction

$$2C_8H_{18}(l) + 17O_2(g) \longrightarrow 16CO(g) + 18H_2O(g)$$

also takes place. It is not unusual for a number of different reactions to take place at the same time, in which case a single equation is an incomplete description of the changes taking place. That is, when other reactions occur along with the reaction of interest, or when for some reason the reaction does not run its full course, the actual mass of a particular product formed (CO_2, for instance) might be less than the theoretical yield (Fig. 4.3). We then speak of the percentage yield of the product:

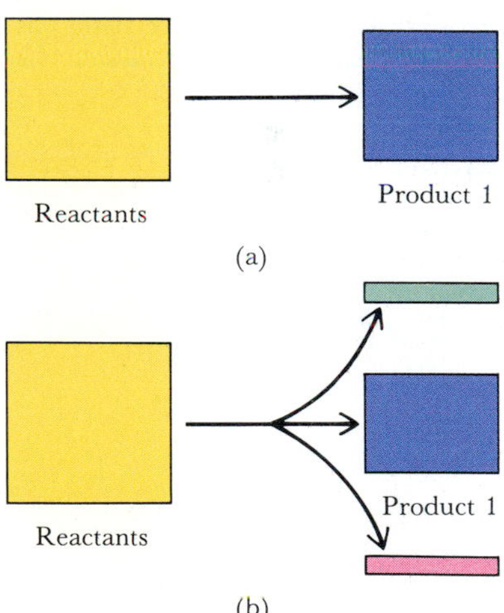

FIGURE 4.3 (a) The yield of a product would be 100% if no competing reactions were taking place. (b) However, if a reactant can take part in several reactions simultaneously, then the yield of a particular product will be less than 100% because other products will also form.

The **percentage yield** of a product is the percentage of its theoretical yield achieved in practice:

$$\text{Percentage yield} = \frac{\text{actual yield}}{\text{theoretical yield}} \times 100\%$$

Suppose we find that in an actual combustion of 100 g of octane, only 92 g of carbon dioxide is produced, then

$$\text{Percentage yield of } CO_2 = \frac{92 \text{ g } CO_2}{308 \text{ g } CO_2} \times 100\% = 30\%$$

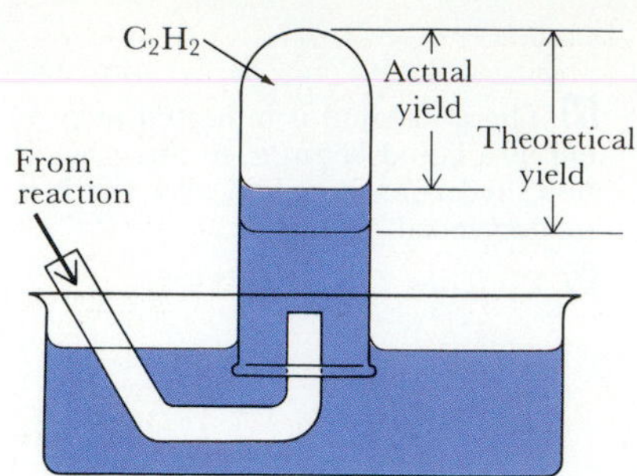

FIGURE 4.4 If calcium carbide reacts to give the stoichiometric amount of acetylene, then we expect the volume indicated as the theoretical yield. However, in practice, a smaller volume is obtained and the actual yield is less than 100%. As we will see in Chapter 5, apparatus like this is often used to collect gases produced by reactions.

T OHT: Fig. 4.4.

N Follow the strategy through diagram **1**; the reaction is used again in Example 4.6.

D Add a few lumps of calcium carbide to water and ignite the gas evolved. (See Fig. 4.6.) Acetylene was used for lighting on early automobiles and bicycles.

U Exercises 4.60–4.64.

If the percentage yield is very close to 100%, then we say that the reaction is complete or that it is quantitative.

▼ EXAMPLE 4.4 Calculating the percentage yield

When a large amount of water was poured on 100 g of calcium carbide, 28.3 g of acetylene was produced (Fig. 4.4). Calculate the percentage yield of acetylene in the reaction

$$CaC_2(s) + 2H_2O(l) \longrightarrow Ca(OH)_2(s) + C_2H_2(g)$$

STRATEGY First, we need to know the theoretical yield. We calculate it by using the relations

$64.10\ g\ CaC_2 = 1\ mol\ CaC_2$	to convert from grams of CaC_2 to moles of CaC_2
$1\ mol\ CaC_2 = 1\ mol\ C_2H_2$	to convert from moles of CaC_2 to moles of C_2H_2
$1\ mol\ C_2H_2 = 26.04\ g\ C_2H_2$	to convert from moles of C_2H_2 to grams of C_2H_2

Then we express the actual yield as a percentage of the theoretical yield.

SOLUTION The theoretical yield of acetylene is

Mass of C_2H_2 (g)

$$= 100\ \cancel{g\ CaC_2} \times \frac{1\ \cancel{mol\ CaC_2}}{64.10\ \cancel{g\ CaC_2}} \times \frac{1\ \cancel{mol\ C_2H_2}}{1\ \cancel{mol\ CaC_2}} \times \frac{26.04\ g\ C_2H_2}{1\ \cancel{mol\ C_2H_2}}$$

$$= 40.6\ g\ C_2H_2$$

The actual yield is 28.3 g C_2H_2, so

$$\text{Percentage yield of } C_2H_2 = \frac{28.3\ g\ C_2H_2}{40.6\ g\ C_2H_2} \times 100\% = 69.7\%$$

EXERCISE E4.4 When 24.0 g of potassium nitrate was heated with lead, 13.8 g of potassium nitrite was formed in the reaction $Pb(s) + KNO_3(s) \rightarrow PbO(s) + KNO_2(s)$. Calculate the percentage yield of potassium nitrite.

▲ [*Answer*: 68.3%]

If the percentage yield of product in each step of a long series of reactions is low, then very little of the final product will be formed from even a large amount of starting material. A low percentage yield in an industrial process may be very significant economically. One of the achievements of chemistry—one that we shall examine in later chapters—has been to find ways of improving percentage yields so that costs can be minimized and useful substances made more generally available.

N Limiting reagents. *J. Chem. Ed.* **1978**, *55*, 726.

N Limiting reagents problems made simple for students. *J. Chem. Ed.* **1985**, *62*, 106.

D Producing hydrogen gas from calcium metal. CD2, 27. (Vary amounts of calcium and water to produce different amounts of hydrogen and different limiting reactants.)

4.2 LIMITING REACTANTS

Example 4.3 illustrates a calculation of the mass of aluminum needed to react with a given mass of chromium(III) oxide. We can use that method to solve similar problems. However, if we supply less than 3.55 kg of aluminum for its reaction with 10.0 kg of Cr_2O_3, then less of the Cr_2O_3 will be reduced and less of the product will form. Because the insufficient amount of aluminum limits the amount of product formed, aluminum is called the limiting reactant in the reaction:

> A **limiting reactant** is the reactant that governs the theoretical yield of product in a given reaction.

An analogy is an automobile factory that has 6000 tires and 1000 car bodies. The maximum number of automobiles it can produce is limited by the number of bodies (because the 1000 bodies require 4000 tires), so the bodies are the limiting "reactant." When there are no more car bodies, 2000 excess tires remain unused (Fig. 4.5).

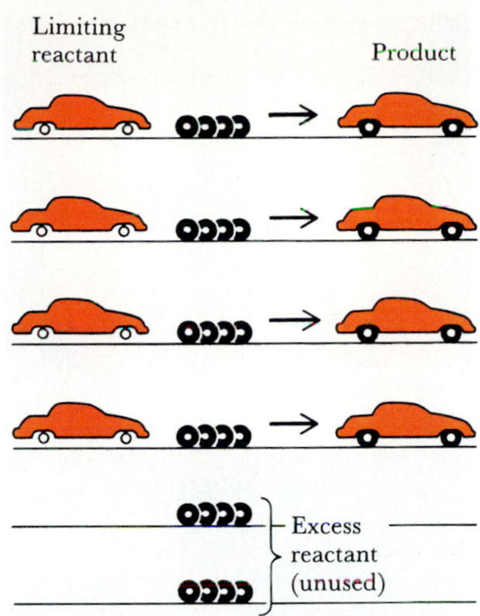

FIGURE 4.5 The role of a limiting reactant: only as much product (the finished cars) can be produced as the limiting reactant (the car bodies) allows.

▼ EXAMPLE 4.5 Calculating the mass of reactant remaining

How much chromium(III) oxide of the original 10.0-kg sample in Example 4.3 would have remained if only 2.54 kg of aluminum had been used for its reduction?

STRATEGY The mass of Cr_2O_3 remaining is the difference between the mass supplied and the mass that is reduced by the available aluminum. We can calculate the mass that is reduced from the mass of aluminum supplied and the stoichiometric relation between aluminum and chromium(III) oxide in the reaction. All the data are in Example 4.3.

SOLUTION From the information in Example 4.3, we calculate the mass of chromium(III) oxide reduced by 2.54 kg of aluminum as

Mass of Cr_2O_3 (g)

$$= 2.54 \times 10^3 \cancel{\text{g Al}} \times \frac{1 \cancel{\text{mol Al}}}{26.98 \cancel{\text{g Al}}} \times \frac{1 \cancel{\text{mol Cr}_2\text{O}_3}}{2 \cancel{\text{mol Al}}} \times \frac{152.00 \text{ g Cr}_2\text{O}_3}{1 \cancel{\text{mol Cr}_2\text{O}_3}}$$

$$= 7.15 \times 10^3 \text{ g Cr}_2\text{O}_3$$

That is, 7.15 kg of chromium(III) oxide can be reduced. The mass remaining after the reaction is therefore

$$\text{Mass remaining} = (10.0 - 7.15) \text{ kg Cr}_2\text{O}_3 = 2.8 \text{ kg Cr}_2\text{O}_3$$

EXERCISE E4.5 Suppose 25.0 g of potassium was supplied in the preparation of potassium hydride mentioned in Exercise E4.3, but the same amount of hydrogen gas was available. What mass of potassium would remain at the end of the reaction?

[*Answer*: 7.6 g potassium] ▲

Unless we supply exactly the right relative amounts of reactants, one of them is always limiting. We can decide which one that is by comparing the amounts of reactants supplied with the stoichiometric coefficients in the chemical equation. The reactant that is present in the smallest amount in relation to the stoichiometric coefficients is the limiting reactant. Thus, the "equation" for the construction of an automobile (A) from a car body (B) and the tires (T) is

$$B + 4T \longrightarrow A$$

We see from the equation that we need four tires for each body. If we have more than four tires per body, then bodies are the limiting "reactant." If we have fewer than four tires per body, then tires are the limiting "reactant."

N All industrial chemical processes have a limiting reactant.

D S'mores: A demonstration of stoichiometric relationships. *J. Chem. Ed.* **1986**, *63*, 1048.

? The methane in a Bunsen burner may burn with a blue or a yellow flame. How does the concept of a limiting reactant account for these observations.

D Explore the above question with an ignited Bunsen burner.

FIGURE 4.6 Calcium carbide is a dirty-white solid formed from limestone and carbon. When, as here, water is poured on it, acetylene, which burns with a brilliant white flame, is produced.

N Try the calculation assuming that H_2O is the limiting reactant. In this case, the amount of CaC_2 needed (2.77 mol) exceeds the supply.

U Exercise 4.65.

▼ EXAMPLE 4.6 Identifying the limiting reactant

As we saw in Example 4.4, calcium carbide (CaC_2) reacts with water to form calcium hydroxide and the flammable gas acetylene (C_2H_2). This reaction was once used in lamps on vehicles because the acetylene it forms burns with a bright white flame (Fig. 4.6). Today the reaction is a potentially important route from inorganic to organic materials. Which is the limiting reactant when 100 g of water reacts with 100 g of calcium carbide? The chemical equation is

$$CaC_2(s) + 2H_2O(l) \longrightarrow Ca(OH)_2(s) + C_2H_2(g)$$

STRATEGY We are given the masses of the reactants, not the amounts in moles. Therefore, we begin by converting masses to moles by using the molar masses of the substances. Then, by using the stoichiometric coefficients, we determine the amount of one reactant A needed for the complete reaction of the specified amount of the other reactant B. If the given amount of A is greater than the amount calculated, then all the B can react and B is the limiting reactant.

SOLUTION First, we need the molar masses of the two reactants (for the conversion of the mass of each reactant to amounts in moles):

$$64.10 \text{ g } CaC_2 = 1 \text{ mol } CaC_2 \qquad 18.02 \text{ g } H_2O = 1 \text{ mol } H_2O$$

The amounts of reactants supplied are therefore

$$\text{Amount of } CaC_2 \text{ (mol)} = 100 \cancel{\text{ g } CaC_2} \times \frac{1 \text{ mol } CaC_2}{64.10 \cancel{\text{ g } CaC_2}} = 1.56 \text{ mol } CaC_2$$

$$\text{Amount of } H_2O \text{ (mol)} = 100 \cancel{\text{ g } H_2O} \times \frac{1 \text{ mol } H_2O}{18.02 \cancel{\text{ g } H_2O}} = 5.55 \text{ mol } H_2O$$

Next, to determine whether CaC_2 is the limiting reactant, we need to know the amount of H_2O that must be present to react with 1.56 mol CaC_2 completely:

$$\text{Amount of } H_2O \text{ required (mol)} = 1.56 \cancel{\text{ mol } CaC_2} \times \frac{2 \text{ mol } H_2O}{1 \cancel{\text{ mol } CaC_2}}$$
$$= 3.12 \text{ mol } H_2O$$

Thus, because only 3.12 mol H_2O is required and 5.55 mol H_2O is supplied, all the CaC_2 can react; so the CaC_2 is the limiting reactant. When the reaction is complete, (5.55 − 3.12) mol = 2.43 mol H_2O remains.

EXERCISE E4.6 Which reactant is limiting when 100 kg of hydrogen reacts with 800 kg of nitrogen in the synthesis of ammonia? The chemical equation is $N_2(g) + 3H_2(g) \rightarrow 2NH_3(g)$. ▲

[*Answer*: hydrogen]

The limiting reactant determines the amounts of the products that can be formed. Therefore, once we have identified the limiting reactant, we can use the information to calculate both the amounts of the products that can be formed and the amount of the excess reagent that remains at the end of the reaction.

▼ EXAMPLE 4.7 Using the limiting reactant to calculate a yield

In the reaction

$$CCl_4(g) + 2HF(g) \longrightarrow CCl_2F_2(g) + 2HCl(g)$$

for the manufacture of the compound CCl_2F_2 (known commercially as Freon-12), 100.0 g of CCl_4 is mixed with 30.00 g of HF. What mass of Freon-12 can be produced and what mass of excess reactant remains at the end of the reaction?

Freon is a DuPont trade name for compounds having the formula $C_xCl_yF_z$. Most Freons are colorless, odorless, nonflammable, and nontoxic. Freon-12 is the refrigerant used in air conditioners and refrigerators.

STRATEGY The limiting reactant must be identified according to the procedures in Example 4.6. Because the limiting reactant determines the theoretical yield of the reaction, we calculate the mass of Freon-12 that can be produced from the limiting reactant by using the molar masses of the compounds and the stoichiometric relation between them. The mass of excess reactant that remains at the end of the reaction is the difference between the mass supplied and the mass that reacts.

SOLUTION The conversion factors we require are obtained from the molar masses of CCl_4 (153.81 g/mol), HF (20.01 g/mol), and Freon-12 (120.91 g/mol). The amounts of reactants are

$$\text{Amount of } CCl_4 \text{ (mol)} = 100.0 \text{ g } CCl_4 \times \frac{1 \text{ mol } CCl_4}{153.81 \text{ g } CCl_4} = 0.6502 \text{ mol } CCl_4$$

$$\text{Amount of HF (mol)} = 30.00 \text{ g HF} \times \frac{1 \text{ mol HF}}{20.01 \text{ g HF}} = 1.499 \text{ mol HF}$$

The amount of HF needed to react with 0.6502 mol CCl_4, given that the reaction stoichiometry implies that 2 mol HF = 1 mol CCl_4, is

$$\text{Amount of HF required (mol)} = 0.6502 \text{ mol } CCl_4 \times \frac{2 \text{ mol HF}}{1 \text{ mol } CCl_4}$$
$$= 1.300 \text{ mol HF}$$

Because 1.300 mol HF is needed to react with all the CCl_4 and 1.499 mol HF is provided, the CCl_4 is the limiting reactant and hence controls the amount of CCl_2F_2 formed. The reaction stoichiometry implies that 1 mol CCl_4 = 1 mol CCl_2F_2, so the mass of product that can be formed from 0.6502 mol CCl_4 is

Exercises 4.66–4.69.

$$\text{Mass of Freon-12 (g)} = 0.6502 \text{ mol } CCl_4 \times \frac{1 \text{ mol } CCl_2F_2}{1 \text{ mol } CCl_4} \times \frac{120.91 \text{ g } CCl_2F_2}{1 \text{ mol } CCl_2F_2}$$
$$= 78.62 \text{ g } CCl_2F_2$$

The mass of excess reactant (HF) is the difference between the mass supplied and the mass that reacts:

$$\text{Mass of HF reacted (g)} = 1.300 \text{ mol HF} \times \frac{20.01 \text{ g HF}}{1 \text{ mol HF}} = 26.01 \text{ g HF}$$

Hence, the mass of HF remaining after the reaction is (30.00 − 26.01) g = 3.99 g HF.

EXERCISE E4.7 (a) Identify the limiting reactant in the reaction $6Na(l) + Al_2O_3(s) \rightarrow 2Al(l) + 3Na_2O(s)$ when 5.52 g of sodium is mixed with 5.10 g of Al_2O_3. (b) What mass of aluminum can be produced, and (c) what mass of excess reactant remains at the end of the reaction?

[*Answer*: (a) Na; (b) 2.16 g Al; (c) 1.02 g Al_2O_3]

4.3 CHEMICAL COMPOSITION FROM MEASUREMENTS OF MASS

The calculations described so far can be adapted to the determination of the composition of compounds and, in particular, to the determination of empirical formulas.

N In Section 2.6, the empirical formula of a compound was determined from mass percentage; here the calculation is based on mass itself.

The determinants of empirical formulas. We saw in Section 2.6 that an empirical formula shows the relative numbers of atoms in a compound and that it can be obtained from the mass percentage composition, as described in Example 2.6. The following example is a review of how an empirical formula can be determined from the masses of substances taking part in a reaction.

V Video 9: Direct formation of a binary compound.

V Video 10: Iron and sulfur. A given mass of iron reacts with an excess of sulfur to give a certain mass of product.

D Determine the mass of a polished 1-cm strip of magnesium metal, burn the strip in a Bunsen flame while holding the metal with tongs, and determine the mass of the product (which should be assumed to be MgO). The experimental data will be only approximate, but the technique can be demonstrated quickly.

? How could the mole ratio of Na_2SO_4 to H_2O in $Na_2SO_4{\cdot}xH_2O$ be determined in the laboratory?

▼ EXAMPLE 4.8 Determining an empirical formula from the masses of reactants and products

It is found that a sample of magnesium of mass 0.450 g reacts with excess nitrogen to form 0.623 g of magnesium nitride. Determine the empirical formula of magnesium nitride.

STRATEGY The number of moles of atoms of each element in the compound can be determined from the molar mass of the element and the mass of element present. We are given the mass of magnesium present in the compound (it is the same as the mass of magnesium used as reactant). The mass of nitrogen present in the nitride is the mass of magnesium nitride less the mass of magnesium present in it.

SOLUTION The molar masses that we need for conversions between masses and moles are

$$24.31 \text{ g Mg} = 1 \text{ mol Mg} \qquad 14.01 \text{ g N} = 1 \text{ mol N}$$

The amount of Mg present is

$$\text{Amount of Mg (mol)} = 0.450 \cancel{\text{g Mg}} \times \frac{1 \text{ mol Mg}}{24.31 \cancel{\text{g Mg}}} = 0.0185 \text{ mol Mg}$$

The mass of nitrogen in the compound is

$$\text{Mass of N (g)} = (0.623 - 0.450) \text{ g} = 0.173 \text{ g}$$

Hence, the amount of N present is

$$\text{Amount of N (mol)} = 0.173 \cancel{\text{g N}} \times \frac{1 \text{ mol N}}{14.01 \cancel{\text{g N}}} = 0.0123 \text{ mol N}$$

The ratio of number of atoms of each element in the compound is equal to the ratio of the number of moles of each type of atom:

$$\underset{\substack{\uparrow \\ \text{Ratio of atoms}}}{\text{Mg:N}} = \underset{\substack{\uparrow \\ \text{Ratio of moles}}}{0.0185:0.0123}$$

Dividing through by the smallest number (0.0123) gives

$$\text{Mg:N} = 1.5:1$$

The corresponding formula is $Mg_{1.5}N$; but the subscripts in empirical formulas are usually expressed as whole numbers, so we multiply through by 2 and report the empirical formula of magnesium nitride as Mg_3N_2.

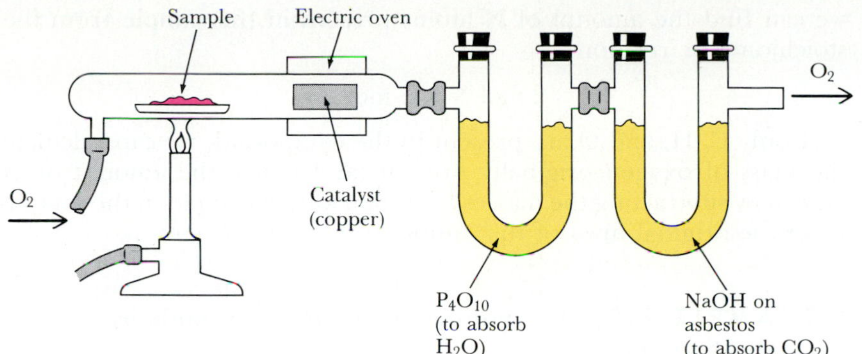

FIGURE 4.7 The apparatus used for a combustion analysis. The masses of CO_2 and H_2O produced are obtained from the differences of the masses of the collecting tubes before and after the experiment. The catalyst ensures that any CO produced is oxidized to CO_2.

EXERCISE E4.8 A sample of bromine of mass 1.546 g reacts with fluorine to form 2.649 g of a bromine fluoride. Determine the empirical formula of the compound.

[*Answer*: BrF_3]

T OHT: Fig. 4.7.

U Exercises 4.70 and 4.71.

Combustion analysis. In the procedure called **combustion analysis** the composition of an organic compound is determined by burning it in an unlimited supply of oxygen and measuring the masses of carbon dioxide and water produced. The apparatus used is shown in Fig. 4.7. The carbon dioxide is trapped in the tube containing sodium hydroxide, and the increase in mass of that tube is equal to the mass of carbon dioxide produced. The mass of water produced is equal to the increase in mass of the tube containing the water-absorbing compound phosphorus(V) oxide, P_4O_{10} (the common name of this compound is phosphorus pentoxide).

Oxygen is in excess, so the carbon in the compound limits the amount of carbon dioxide formed. Moreover, because each atom of carbon ends up in one molecule of carbon dioxide, the combustion has the form

$$\ldots C \ldots + O_2(g) \longrightarrow CO_2(g)$$

for each C atom in the compound (Fig. 4.8). The relation between the carbon in the compound and in the product is

$$1 \text{ mol C} = 1 \text{ mol } CO_2$$

Thus, by measuring the mass of carbon dioxide produced, we can find the amount (in moles) of C atoms in the sample.

Under the same circumstances (that is, presence of excess oxygen), all the hydrogen in a compound is converted into water:

$$4(\ldots H \ldots) + O_2(g) \longrightarrow 2H_2O(g)$$

(see Fig. 4.8.). Therefore, if we measure the mass of water produced when the compound burns, we can find the amount of H atoms in the sample by using the stoichiometric relation 4 mol H = 2 mol H_2O, or, more simply,

$$2 \text{ mol H} = 1 \text{ mol } H_2O$$

So long as the heat is not too intense, any nitrogen in the compound is released as nitrogen gas. Because

$$2(\ldots N \ldots) \longrightarrow N_2(g)$$

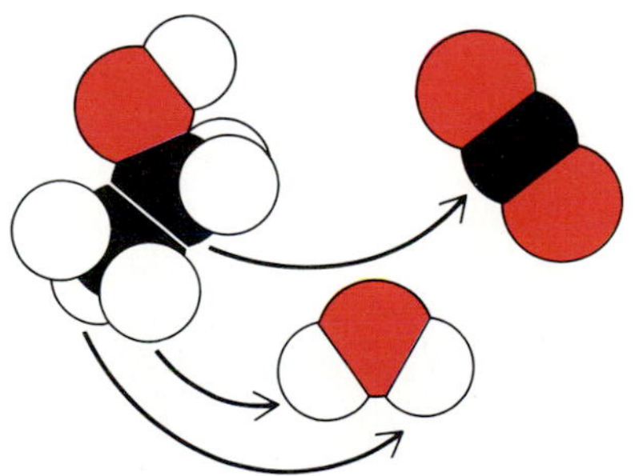

FIGURE 4.8 During combustion in a plentiful supply of oxygen, each C atom produces one CO_2 molecule, and each pair of H atoms produces one H_2O molecule.

E Under very hot conditions, nitrogen present in the compound may be converted to nitrogen oxide, NO. Any oxides of nitrogen that are produced may be reduced to elemental nitrogen by passing the gaseous mixture over hot copper wire.

we can find the amount of N atoms present in the sample from the stoichiometric relation

$$2 \text{ mol N} = 1 \text{ mol } N_2$$

If only C, H, and O are present in the compound, we can calculate the mass of oxygen originally present (and hence the amount of O atoms) by subtracting the masses of carbon and hydrogen in the sample from the original mass of the sample.

▼ EXAMPLE 4.9 Determining an empirical formula by combustion analysis

A combustion analysis was carried out on 1.621 g of ethanol, which is known to contain only C, H, and O. The masses of water and carbon dioxide produced were 1.902 g and 3.095 g, respectively. What is the empirical formula of ethanol?

N Follow the strategy by reference to diagram **1**.

STRATEGY The empirical formula shows the relative numbers of atoms, and therefore the relative amounts (in moles), of the elements present. Therefore, we need to determine the amounts of C, H, and O in the sample and express the result as a formula of the form $C_aH_bO_c$. We first need to find the masses of carbon and hydrogen in the sample, and then convert those masses to amounts of C and H. The mass of oxygen in the sample is obtained by subtraction of the combined masses of carbon and hydrogen from the mass of the original sample. It is then converted to moles of O.

SOLUTION To convert the mass of carbon dioxide to amount of C, we use the molar mass of carbon dioxide (44.01 g/mol CO_2) and the stoichiometric relation between CO_2 and C atoms, 1 mol CO_2 = 1 mol C, given earlier:

$$\text{Amount of C (mol)} = 3.095 \cancel{\text{g } CO_2} \times \frac{1 \cancel{\text{mol } CO_2}}{44.01 \cancel{\text{g } CO_2}} \times \frac{1 \text{ mol C}}{1 \cancel{\text{mol } CO_2}}$$
$$= 0.07032 \text{ mol C}$$

The mass of carbon in the sample is therefore

$$\text{Mass of C (g)} = 0.07032 \cancel{\text{mol C}} \times \frac{12.01 \text{ g C}}{1 \cancel{\text{mol C}}} = 0.8446 \text{ g C}$$

For the conversion of mass of water to moles of H we use the molar mass of water (18.02 g/mol H_2O) and the relation between H_2O and H atoms, 1 mol H_2O = 2 mol H, given earlier:

$$\text{Amount of H (mol)} = 1.902 \cancel{\text{g } H_2O} \times \frac{1 \cancel{\text{mol } H_2O}}{18.02 \cancel{\text{g } H_2O}} \times \frac{2 \text{ mol H}}{1 \cancel{\text{mol } H_2O}}$$
$$= 0.2111 \text{ mol H}$$

The mass of hydrogen in the sample is therefore

$$\text{Mass of hydrogen (g)} = 0.2111 \cancel{\text{mol H}} \times \frac{1.008 \text{ g H}}{1 \cancel{\text{mol H}}} = 0.2128 \text{ g H}$$

The total mass of carbon and hydrogen is (0.8446 + 0.2128) g = 1.0574 g, and hence the mass of oxygen in the sample is

$$\text{Mass of oxygen (g)} = (1.621 - 1.0574) \text{ g} = 0.564 \text{ g}$$

The amount of O (as atoms) is therefore

$$\text{Amount of O (mol)} = 0.564 \cancel{\text{g O}} \times \frac{1 \text{ mol O}}{16.00 \cancel{\text{g O}}} = 0.0352 \text{ mol O}$$

At this stage we know that the relative numbers of atoms are

$$C:H:O = 0.07032:0.2111:0.0352$$

Division by 0.0352, the smallest number in the ratio, gives

$$C:H:O = 2.00:6.00:1.00$$

The ratios show that the empirical formula of ethanol is C_2H_6O.

EXERCISE E4.9 When 0.528 g of sucrose, a compound of carbon, hydrogen, and oxygen, was burned, 0.306 g of water and 0.815 g of carbon dioxide were formed. Deduce the empirical formula of sucrose.

▲ [*Answer*: $C_{12}H_{22}O_{11}$]

THE STOICHIOMETRY OF REACTIONS IN SOLUTION

Chemists very often study reactions in solution, particularly aqueous solution. To do this they need to know how to calculate the *volumes* of solutions that react together. For example, they may want to know the volume of acid that would be required to neutralize a given volume of a solution of a base. In addition, the composition of a solution is variable because a solution is a mixture, not a compound with a fixed composition. Consequently, the *concentration* of the solute—that is, the amount of solute per liter of solution—must be taken into account. Like calculations involving masses that react, calculations involving volumes of solutions that react are conveniently done by converting all the relevant quantities to moles. In other words, we are going to add a new arrow to the calculations diagram (**3**).

N We speak in terms of the concentration of the solute, not the concentration of the solution. We are concerned with the amount of solute in a given volume of solution.

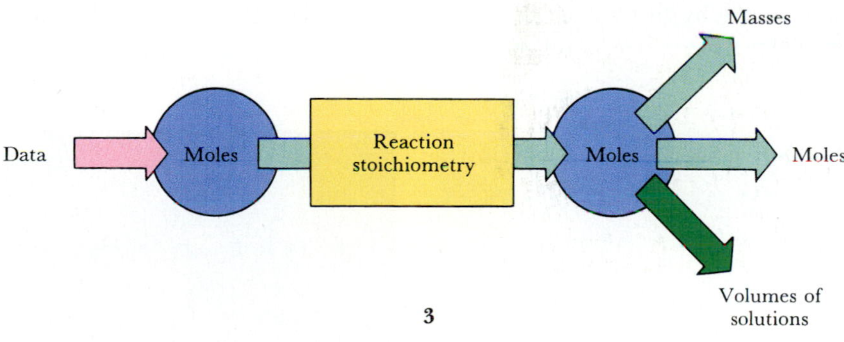

3

T OHT: Diagrams **1**, **2**, and **3**.

4.4 MOLAR CONCENTRATION

The volume of a solution and the amount of solute it contains are linked by a quantity known as molar concentration:

> The **molar concentration** of a solute in a solution is the amount (in moles) of solute per liter of solution.

In other words,

$$\text{Molar concentration} = \frac{\text{moles of solute}}{\text{liters of solution}}$$

(Molar concentration is often called molarity.)

E The IUPAC recommendation is *amount concentration* (or *amount of substance concentration*). This term has not been widely adopted, and we have adopted the term *molar concentration,* which is permitted in "the Green Book" (the IUPAC recommendations for physical chemistry; more formally *Quantities, Units, and Symbols in Physical Chemistry*, I. M. Mills (ed.), Blackwell Scientific, Oxford 1988). We do not use the term *molarity* because of the possibility of confusion with *molality.*

The unit of molar concentration is moles per liter (mol/L). However, because this unit is used so frequently, chemists denote it with the symbol M, which is sometimes read molar. The molar concentration of sucrose in a solution prepared by dissolving 0.010 mol of sucrose molecules ($C_{12}H_{22}O_{11}$) in enough water to give 100 mL of solution would be

$$\text{Molar concentration of } C_{12}H_{22}O_{11} = \frac{0.010 \text{ mol } C_{12}H_{22}O_{11}}{0.100 \text{ L}}$$

$$= 0.10 \text{ M } C_{12}H_{22}O_{11}(aq)$$

Molar concentration is defined in terms of the volume of *solution,* not in terms of the volume of solvent used to prepare the solution. This choice makes sense, because we need to know what volume of the solution to measure out to supply a given amount of solute. The usual way to prepare an aqueous solution of known molar concentration of a solid substance is to transfer a known mass of the solid into a volumetric flask, dissolve it in a little water, and then fill the flask up to the mark with water (Fig. 4.9).

FIGURE 4.9 The steps involved in making up a solution of known concentration (here, a solution of potassium permanganate). (a) A known mass of solute is dispensed into a volumetric flask. (b) Some water is added to dissolve it. (c) Finally, water is added up to the mark. The bottom of the solution's meniscus should be level with the mark.

▼ EXAMPLE 4.10 Calculating the molar concentration of a solute

A student prepared a solution of potassium nitrate, KNO_3, by dissolving 1.345 g of the salt in enough water to prepare 25.00 mL of solution. What is the molar concentration of KNO_3 in the solution.

STRATEGY Because the molar concentration is the number of moles of solute per liter of solution, we begin by converting the mass of solute to moles by using the molar mass. Then we calculate the molar concentration of KNO_3 by dividing the amount of solute by the volume of the solution (in liters).

(a)

(b)

(c)

SOLUTION The molar mass of potassium nitrate is 101.11 g/mol KNO_3, so we use the relation 101.11 g KNO_3 = 1 mol KNO_3. Then, for 25.00 mL = 0.02500 L of solution,

$$\text{Molar concentration of } KNO_3 = \frac{1.345 \text{ g } KNO_3 \times \left(\dfrac{1 \text{ mol } KNO_3}{101.11 \text{ g } KNO_3}\right)}{0.02500 \text{ L}}$$

$$= 0.5321 \text{ mol } KNO_3/\text{L}$$

That is, the solution will be 0.5321 M $KNO_3(aq)$.

EXERCISE E4.10 Calculate the molar concentration of NaCl in a solution made by dissolving 2.357 g of sodium chloride in enough water to make 75.00 mL of solution.

[*Answer*: 0.5378 M NaCl(*aq*)]

N Molar concentration calculations are performed using the units moles per liter not the symbol M.

U Exercises 4.72 and 4.73.

D Measure volume of sodium chloride solution of known molar concentration, evaporate the water, determine the mass of NaCl remaining, and compare the calculated amount in moles with the amount determined by mass.

The relation between volume of solution and amount of solute. The molar concentration of a solute is actually a conversion factor that relates the moles of solute to the volume of solution. For example, suppose we want to know the amount of sucrose molecules in 15 mL (0.015 L) of 0.10 M $C_{12}H_{22}O_{11}(aq)$ solution. We know from the molar concentration that

$$1 \text{ L solution} = 0.10 \text{ mol } C_{12}H_{22}O_{11}$$

so we can obtain the amount of sucrose molecules in 15 mL of solution by setting up the conversion

$$\text{Amount of } C_{12}H_{22}O_{11} \text{ (mol)} = 0.015 \text{ L solution} \times \frac{0.10 \text{ mol } C_{12}H_{22}O_{11}}{1 \text{ L solution}}$$

$$= 1.5 \times 10^{-3} \text{ mol } C_{12}H_{22}O_{11}$$

EXAMPLE 4.11 Preparing a solution of specified concentration

Calculate the mass of potassium permanganate needed to prepare 250 mL of a 0.0380 M $KMnO_4(aq)$ solution.

STRATEGY To calculate the mass of solute, we need to know the amount (in moles) of solute in 250 mL of solution, and then convert this amount to mass of solute by using the molar mass. The first part of the calculation is carried out by using the molar concentration as a conversion factor.

SOLUTION The conversion factors we need are from volume of solution to amount of $KMnO_4$:

$$1 \text{ L solution} = 0.0380 \text{ mol } KMnO_4$$

and from amount of $KMnO_4$ to mass,

$$1 \text{ mol } KMnO_4 = 158.04 \text{ } KMnO_4$$

Hence, the mass of $KMnO_4$ in 250 mL (0.250 L) of solution is

$$\text{Mass of } KMnO_4 \text{ (g)} = 0.250 \text{ L} \times \frac{0.0380 \text{ mol } KMnO_4}{1 \text{ L}} \times \frac{158.04 \text{ g } KMnO_4}{1 \text{ mol } KMnO_4}$$

$$= 1.50 \text{ g } KMnO_4$$

D The preparation of a potassium permanganate solution is colorful.

That is, 1.50 g of potassium permanganate should be dissolved in water in a 250-mL volumetric flask and then diluted to the mark.

EXERCISE E4.11 Calculate the mass of glucose needed to prepare 150 mL of a 0.442 M $C_6H_{12}O_6(aq)$ solution.

[*Answer*: 11.9 g]

Transferring a specified amount of solute. Once we know the concentration of solute, we can decide what volume of solution contains a specific amount of solute. Suppose, for example, we want 0.760 mmol $KMnO_4$ for a certain redox reaction and have available the 0.0380 M $KMnO_4(aq)$ solution prepared in Example 4.11. From the molar concentration of the solute, we have the relation

$$1 \text{ L solution} = 0.0380 \text{ mol } KMnO_4$$

so we can write

$$\text{Volume of solution (L)} = 0.760 \times 10^{-3} \cancel{\text{mol } KMnO_4} \times \frac{1 \text{ L solution}}{0.0380 \cancel{\text{mol } KMnO_4}} = 20.0 \times 10^{-3} \text{ L solution}$$

That is, we should transfer 20.0 mL of the permanganate solution (from a buret) into the flask to give us 0.760 mmol of $KMnO_4$.

This calculation emphasizes one of the advantages of using solutions in chemistry. It shows that we can transfer small amounts of substances (in this case, 0.760 mmol $KMnO_4$) readily and precisely by dissolving them in a manageable amount of solvent and measuring out the volume that corresponds to the desired amount of solute. It is also very much easier to transfer some substances when they are dissolved than when they are not. For instance, it is much easier to transfer HCl and NH_3 as their solutions than as the gases themselves.

Dilution. Another advantage of using solutions is that they may be used to transfer not just small but *very* small amounts of a substance. This capability is important when only a little sample is available, perhaps because of its rarity (for example, Moon dust) or its cost (for example, a new pharmaceutical compound). It can also be important when a sample being examined contains very little of the substance of interest, as in the analysis of an ore for a rare metal or of a contaminated food for traces of poison.

Suppose we were investigating a rare compound and needed to add 0.010 mmol (1.0×10^{-5} mol) of $KMnO_4$ to a sample to oxidize it. One method would be to transfer 0.26 mL of the 0.0380 M $KMnO_4(aq)$ solution. A well-equipped modern research laboratory has an instrument called a micropipet, which can measure out very tiny volumes of liquid accurately, perhaps as little as 10^{-3} mL (1 microliter, 1 μL)—about the volume of a pinhead. However, for many applications, there is a much simpler approach that does not require such expensive equipment. This is simply to **dilute**, or reduce the concentration of, the solute by adding solvent. Then a larger volume of solution contains the same amount of solute and the transfer can be carried out accurately by using an ordinary pipet or buret. If our original 0.0380 M solution were diluted 100-fold, then 0.010 mmol of $KMnO_4$ would be contained in 26 mL of solution, which can be measured out easily and accurately in the normal way.

To use dilution, we need to know how to reduce the concentration of a solute from an initial value to some final target value. Suppose we transfer a certain volume of the original solution to a volumetric flask. The amount of solute we have transferred is obtained by using the initial molar concentration:

$$\text{Amount of solute (mol)} = \text{volume transferred (L)} \times \frac{\text{moles of solute}}{\text{liters of solution}}$$

When we add enough solvent to increase the volume to its final value, the final molar concentration is this same amount of solute divided by the final volume.

U Exercises 4.74 and 4.75.

N The formula $V_1M_1 = V_2M_2$ is not introduced because we wish to develop a uniform approach to calculations based solely on the mole (and to avoid memorized equations).

EXAMPLE 4.12 Calculating the volume of solution to dilute

What volume of 0.0380 M $KMnO_4(aq)$ solution should be used to prepare 250 mL of a 1.50×10^{-3} M $KMnO_4(aq)$ solution?

STRATEGY The volume of the initial, more concentrated solution must be determined. Because the volume of the final, more dilute solution is known, we can calculate the amount of $KMnO_4$ needed in the final solution by using its molar concentration, and then convert that amount to a volume of the *initial* solution by using the molar concentration of $KMnO_4$ in the initial solution as a conversion factor.

SOLUTION The conversions we require are obtained from the following relations:

From the initial concentration, 0.0380 mol $KMnO_4$ = 1 L solution

From the final concentration, 1 L solution = 1.50×10^{-3} mol $KMnO_4$

The conversion we require is therefore

Volume of initial solution (L)

$$= 0.250\ \text{L} \times \frac{1.50 \times 10^{-3}\ \text{mol KMnO}_4}{1\ \text{L}} \times \frac{1\ \text{L}}{0.0380\ \text{mol KMnO}_4}$$

$$= 9.87 \times 10^{-3}\ \text{L}$$

That is, we should measure 9.87 mL of the initial, more concentrated solution into a 250-mL volumetric flask (using a buret) and add water up to the mark (Fig. 4.10).

EXERCISE E4.12 Calculate the volume of 0.0155 M $HCl(aq)$ that should be used to prepare 100 mL of a 5.23×10^{-4} M $HCl(aq)$ solution.

[*Answer*: 3.37 mL]

FIGURE 4.10 Dilution of a concentrated solution enables a chemist to transfer very small amounts of substance conveniently and accurately. The flask on the left contains 9.87 mL of the 0.0380 M $KMnO_4(aq)$ solution prepared as shown in Fig. 4.9. The flask on the right contains another 9.87 mL of the same 0.0380 M $KMnO_4(aq)$ solution to which has been added sufficient water to make up 250 mL of solution. A 25-mL sample of the solution on the right contains only 0.0375 mmol $KMnO_4$. This same amount of $KMnO_4$ is contained in only 0.987 mL of the original solution, a quantity difficult to measure accurately.

The use of solutions and of techniques like dilution give chemists very precise control over the amounts of the substances they are handling, even very small quantities. Pipeting 25.0 mL of the 1.50×10^{-3} M $KMnO_4(aq)$ solution prepared in Example 4.12 corresponds to transferring as little as 37.5 μmol of $KMnO_4$, or only 5.93 mg of the compound. Furthermore, solutions can be stored in a concentrated form to save space (and hence money) and then diluted to whatever concentration is appropriate for their intended use.

N We regard the concept of pH as so much a part of a chemist's (and others') language, that we consider it appropriate to introduce it at this early stage. Here it is introduced simply as a way of reporting the molar concentration by using logarithms. The term inside the logarithm is in fact the *numerical value* of the molar concentration, not the molar concentration itself (a logarithm cannot be taken of a quantity with units). One way round this difficulty is to say that the pH is defined as the negative logarithm of the molar concentration of H^+ ions relative to a standard concentration of 1 mol L^{-1}; then the units moles per liter cancel. In advanced work, the pH is defined in terms of the dimensionless activity.

V Video 15: Indicators.

The pH of solutions. Measures of concentration are introduced in Chapter 11. But one is useful at this stage—because it is so widely used in all branches of chemistry—the **pH** of a solution. This measure of concentration is specific to hydrogen ions and is defined in terms of their molar concentration as

$$\text{pH} = -\log(\text{molar concentration of H}^+)$$

The logarithm in this expression is to the base 10. For example, if the molar concentration of hydrogen ions is 4.2×10^{-4} M, the pH of the solution is

$$\text{pH} = -\log(4.2 \times 10^{-4}) = 3.38$$

It is important to remember that *the higher the pH, the lower the concentration of hydrogen ions*. For example, if the concentration of hydrogen ions is doubled to 8.4×10^{-4} M, the pH falls to

$$\text{pH} = -\log(8.4 \times 10^{-4}) = 3.08$$

The pH is a central concept of chemistry, as we shall see in Chapters 14 and 15. At this stage, however, we use it simply as an alternative way of reporting the concentration of hydrogen ions in aqueous solution.

▼ **EXAMPLE 4.13 Converting between pH and molar concentration**

The pH of acid rain lies in the range between pH 3.0 and pH 5.8. What is the range in terms of the molar concentration of hydrogen ions?

STRATEGY Because the pH is the *logarithm* of the molar concentration, a change in one unit of pH represents a tenfold change in molar concentration. Therefore, because the pH range spans nearly three units, the range of concentration is about 10^3. For the precise calculation, we find the molar concentration by taking the antilogarithm of the pH (see Appendix 1B):

$$\text{Molar concentration of H}^+ = 10^{-\text{pH}}\ \text{M}$$

SOLUTION The two molar concentrations are

For pH 3.0, molar concentration of $H^+ = 10^{-3.0}\ \text{M} = 1.0 \times 10^{-3}$ M

For pH 5.8, molar concentration of $H^+ = 10^{-5.8}\ \text{M} = 1.6 \times 10^{-6}$ M

As we anticipated, the molar concentration spans a 1000-fold increase.

EXERCISE E4.13 Calculate the concentration of hydrogen ions in vinegar, of pH 2.8.

[*Answer*: 1.6×10^{-3} M]
▲

4.5 THE VOLUME OF SOLUTION REQUIRED FOR REACTION

We are now at the stage where we can calculate the volume of one solution that is needed to react with a given volume of another solution. This kind of analysis is important in acid-base reactions, where we may be interested in knowing the volume of acid needed to neutralize a solution of a base. It is also needed in redox chemistry, where we may be interested in knowing how much oxidizing agent we should add to a solution containing a reducing agent to make a particular amount of product. There are other applications as well, including measuring the concentration of a solute, discovering the identity of a solute, and measuring molar masses of solutes.

The general strategy is to calculate the amount of one reactant from the volume of solution and its molar concentration. That amount is then converted to an amount of the other reactant (using the stoichiometric relation from the chemical equation). The final step is to convert the amount of the second reactant to the required volume of its solution, by using its molar concentration in that solution. The chain of conversions is illustrated in diagram **4**:

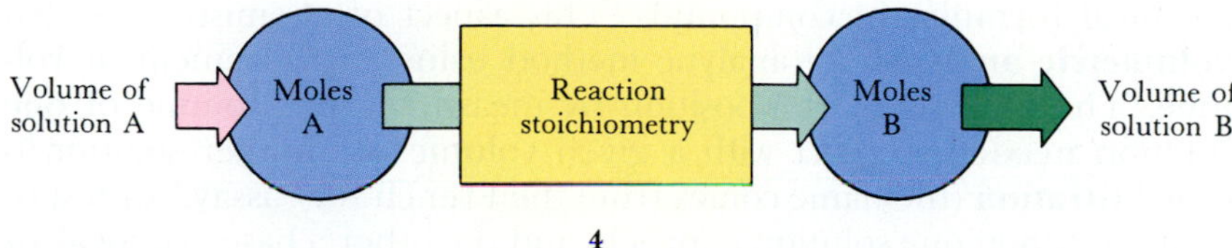

4

▼ EXAMPLE 4.14 Calculating the volume of solution needed for a reaction

Calculate the volume of 0.250 M HCl(*aq*) needed to neutralize completely 25.00 mL of 0.100 M NaOH(*aq*). The chemical equation for the neutralization reaction is

$$\text{HCl}(aq) + \text{NaOH}(aq) \longrightarrow \text{NaCl}(aq) + \text{H}_2\text{O}(l)$$

STRATEGY The pivot of the calculation is the relation between the amount of HCl that reacts with a given amount of NaOH, so we set up the calculation in moles. The amount of NaOH comes from the molar concentration of the NaOH solution and its volume. The amount of HCl to which this corresponds is given by the stoichiometric relation between HCl and NaOH, which we take from the chemical equation for the neutralization reaction. The volume of acid containing that amount of HCl is then found from the molar concentration of the acid. We can estimate the answer to this example by noting that 1 mol HCl is required to neutralize 1 mol NaOH. If both solutions in this example were the same concentration, then we would expect our answer to be 25.00 mL HCl (to neutralize 25.00 mL NaOH). However, the concentrations of the two reactants are different; the acid solution is 2.5 times more concentrated than the base solution. So we estimate that the answer will be $\frac{1}{2.5}$ (= 0.4) of 25.00 mL, or 10 mL HCl.

N Follow the strategy using diagram **4**. Note how the mole remains the pivot of the calculation and how the reaction stoichiometry relates the species taking part in the reaction.

SOLUTION The amount of NaOH in the solution is obtained by using the concentration 1 L solution = 0.100 mol NaOH. The chemical equation for the neutralization reaction shows that the stoichiometric relation between NaOH and HCl is 1 mol NaOH = 1 mol HCl. The volume of solution containing a specified amount of HCl is obtained from the concentration of the acid, 0.250 mol HCl = 1 L solution. The volume of acid required for neutralization is therefore

Volume of HCl solution (L)

$$= 25.00 \times 10^{-3}\ \cancel{\text{L NaOH solution}} \times \frac{0.100\ \cancel{\text{mol NaOH}}}{1\ \cancel{\text{L solution}}}$$

$$\times \frac{1\ \cancel{\text{mol HCl}}}{1\ \cancel{\text{mol NaOH}}} \times \frac{1\ \text{L solution}}{0.250\ \cancel{\text{mol HCl}}}$$

$$= 10.0 \times 10^{-3}\ \text{L solution} = 10.0\ \text{mL solution}$$

U Exercises 4.76, 4.77, and 4.80.

That is, 10.0 mL of 0.250 M HCl neutralizes the OH^- in 25.00 mL of 0.100 M NaOH.

EXERCISE E4.14 Calculate the volume of 0.220 M NaOH(*aq*) that should be added to 25.0 mL of 0.743 M H_2SO_4(*aq*) to react with it completely. [*Answer*: 169 mL]

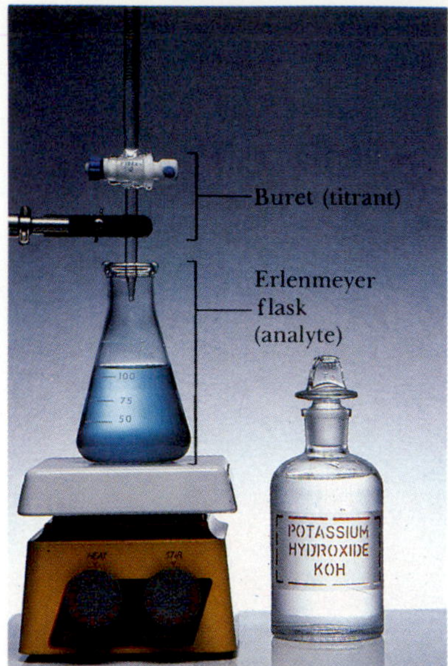

FIGURE 4.11 The apparatus typically used for a titration: magnetic stirrer; flask containing the analyte; clamp and buret containing the titrant, potassium hydroxide.

N The coverage of titrations in Section 4.6 could be delayed until Section 14.6.

N Which terms should the students memorize?

4.6 TITRATIONS

The techniques described in this chapter are widely used to determine the compositions of mixtures, the concentrations of solutes, and the chemical formulas of compounds. This aspect of chemistry is called **volumetric analysis**, an analytic method using measurement of volume. The analysis of composition by measuring the volume of one solution needed to react with a given volume of another solution is called **titration** (the name comes from the French for "assay," or test of quality). When one solution is an acid and the other a base, we speak of an **acid-base titration**. When one solution contains a reducing agent and the other an oxidizing agent, we speak of a **redox titration**.

In a titration, we are typically trying to determine the concentration of a solute in one of two solutions. The solution of unknown concentration is called the **analyte**, and we assume that we have a known volume of that solution in a flask. The solution of known concentration in the buret is the **titrant** (Fig. 4.11). In a titration we slowly add titrant to the analyte and measure the volume of titrant needed to react with all the analyte. (Recall that in Section 4.1 we saw that a reaction is complete when the percentage yield of products is very close to 100%.) The point at which exactly the right volume of titrant has been added to complete the reaction is widely called the equivalence point of the titration. The name **stoichiometric point** is more appropriate, however, because it conveys the sense that at this point in a titration enough titrant has been added to the analyte to complete the reaction according to the chemical equation. To obtain accurate results, it is important to use a reaction that is quantitative (that is, one that has 100% yield).

Acid-base titrations. In an acid-base titration, the stoichiometric point is reached when exactly enough acid has been added to convert all the base to a salt and water. The stoichiometric point can be detected with an **indicator** (such as litmus or phenolphthalein), which is a substance that changes color as the solution changes from basic to acidic (Fig. 4.12). The correct choice of an acid-base indicator from the wide range available is an important aspect of titrations, and we discuss this decision in detail in Chapters 14 and 15.

At the stoichiometric point of the titration, we know the volumes of the base and acid that have reacted and the concentration of solute in one of them (the titrant). Our aim is to find the concentration of solute in the other. To do so, we follow the usual route for stoichiometric calculations—we use the stoichiometric relation between moles of reactants as the pivot of the calculation.

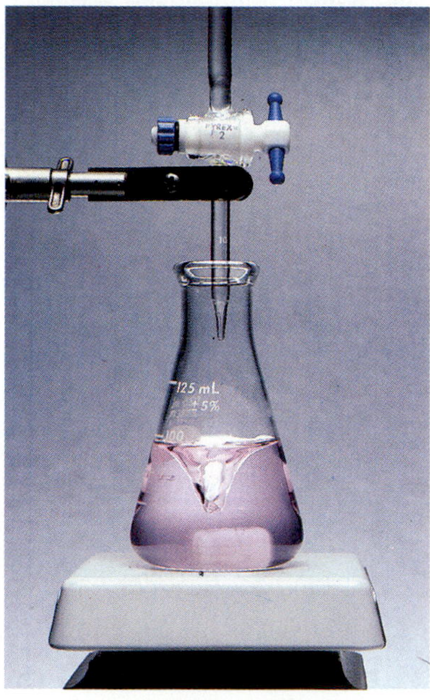

FIGURE 4.12 An acid-base titration in progress. The indicator is phenolphthalein.

EXAMPLE 4.15 Measuring a concentration by titration

A 25.0-mL sample of vinegar (a dilute solution of acetic acid, CH_3COOH) was titrated with 0.500 M NaOH(*aq*). The stoichiometric point was reached when 38.1 mL of the base had been added. Find the molar concentration of acetic acid in the vinegar. The chemical equation of the reaction is

$$CH_3COOH(aq) + NaOH(aq) \longrightarrow NaCH_3CO_2(aq) + H_2O(l)$$

STRATEGY We know the volume and concentration of the titrant, so we can calculate the amount of NaOH used. The stoichiometric coefficients of the chemical equation enable us to convert from the amount of NaOH to the amount of CH_3COOH in the 25.0-mL of analyte. Knowing the amount of CH_3COOH contained in that volume, we can calculate the initial molar concentration of the acetic acid. To estimate the answer, we note that more than 25.0 mL of base is needed to neutralize the 25.0-mL sample of vinegar. Therefore, we know that the concentration of acetic acid in the vinegar sample must be greater than 0.500 M, the concentration of the neutralizing base. So we expect our answer to be greater than 0.500 M.

SOLUTION We need the conversions 1 L solution = 0.500 mol NaOH and, from the stoichiometric coefficients of the chemical equation, 1 mol NaOH = 1 mol CH_3COOH. Hence, the amount of CH_3COOH in the sample is

$$\begin{aligned}\text{Amount of } CH_3COOH \text{ (mol)} \\ &= 38.1 \times 10^{-3}\ \cancel{\text{L solution}} \times \frac{0.500\ \cancel{\text{mol NaOH}}}{1\ \cancel{\text{L solution}}} \times \frac{1\ \text{mol } CH_3COOH}{1\ \cancel{\text{mol NaOH}}} \\ &= 1.90 \times 10^{-2}\ \text{mol } CH_3COOH\end{aligned}$$

The initial concentration of the acid is therefore

$$\text{Molar concentration of } CH_3COOH = \frac{1.90 \times 10^{-2}\ \text{mol } CH_3COOH}{25.0 \times 10^{-3}\ \text{L}}$$

$$= 0.762\ \text{mol/L } CH_3COOH$$

That is, the vinegar is 0.762 M $CH_3COOH(aq)$.

EXERCISE E4.15 A student prepared a sample of hydrochloric acid that contained 0.72 g of hydrogen chloride in 250 mL of solution. This acid solution was used to titrate 25.0 mL of a solution of calcium hydroxide. The stoichiometric point was reached when 15.1 mL of acid had been added. What was the concentration of the calcium hydroxide solution?

[*Answer*: 0.024 M $Ca(OH)_2(aq)$]

E It is common to distinguish three titration procedures. In a *direct titration* (the only type we consider), the analyte is titrated directly with the titrant. In an *indirect titration*, the analyte is reacted with an excess of another reagent (so the analyte is the limiting reactant), and the amount of product is determined by titration and then interpreted in terms of the amount of original analyte. In a *back titration*, the analyte is reacted with a known (and excess) amount of another reagent, and then the amount of unreacted reagent is determined by titration.

D Electrolytic titration. CD2, 70. Acid-base titration. TDC, 4.17.

U Exercises 4.78 and 4.79.

Redox titrations. In a redox titration, the stoichiometric point occurs when sufficient oxidizing agent has been added to oxidize all the reducing agent. It can be detected by watching for the color change of a small amount of a redox indicator, a substance that changes color when it is converted from its oxidized to its reduced state. An example of a redox indicator is ferroin, a complex compound of iron that changes from an almost colorless pale blue in its oxidized form to red in its reduced form. Sometimes, however, the stoichiometric point can be detected without using an indicator—when the reduced and oxidized forms of the reagents have different colors. This is the case when oxalic acid, $(COOH)_2$, is oxidized to carbon dioxide by permanganate ions (Fig. 4.13). The permanganate ion is purple, but in acid solution it is reduced to the pale pink Mn^{2+} ion in the half-reaction

$$MnO_4^-(aq) + 8H^+(aq) + 5e^- \longrightarrow Mn^{2+}(aq) + 4H_2O(l)$$

The oxalic acid is oxidized in the half-reaction

$$(COOH)_2(aq) \longrightarrow 2CO_2(g) + 2H^+(aq) + 2e^-$$

FIGURE 4.13 A redox reaction of oxalic acid with potassium permanganate. Before the stoichiometric point (left), the permanganate is converted to the very pale pink Mn^{2+} ion. The stoichiometric point is recognized by noting when the added titrant survives as permanganate ions (right).

D Add a dilute solution of potassium permanganate slowly to a reducing agent (such as iron(II) ions or oxalic acid).

N Oxalic acid is found in the leaves of rhubarb.

The overall redox reaction is

$$5(COOH)_2(aq) + 2MnO_4^-(aq) + 6H^+(aq) \longrightarrow 10CO_2(g) + 2Mn^{2+}(aq) + 8H_2O(l)$$

The solution becomes purple immediately after the stoichiometric point is reached, when an excess of permanganate titrant has been added.

We can use the same type of calculations for both redox titrations and acid-base titrations. In both cases the stoichiometric coefficients of the chemical equation are used to derive the appropriate conversion factor.

SUMMARY

4.1 The quantitative uses of chemical reactions are based on the interpretation of stoichiometric coefficients as moles of reacting substance. In stoichiometry calculations, the general strategy is to convert from the data to moles of reactants or products, then use the reaction stoichiometry (from the balanced chemical equation) to convert the moles of one substance to moles of another, and finally to convert the moles to the property of interest. The **theoretical yield** is the yield expected on the basis of the chemical equation. However, other reactions taking place may reduce the actual yield below the theoretical yield, and the ratio of actual yield to theoretical yield is reported as the **percentage yield**.

4.2 and 4.3 The **limiting reactant** is the reactant that governs the theoretical amount of product that can be formed in a given reaction. The technique of **combustion analysis**, for the determination of empirical formulas, especially of organic compounds, is based on the principle of limiting reactants.

4.4 The link between volumes of solution and moles of solute is the **molar concentration** (or molarity), the number of moles of solute per liter of solution. Solutions are very useful in chemistry, partly because, through dilution (the reduction of concentration), we can transfer very small amounts of solute. A convenient procedure for reporting the hydrogen ion concentration in solutions is to give the **pH** of the solution (the negative logarithm of the molar concentration of H^+ ions).

4.5 The volume of one solution needed to react with a stated volume of another solution is obtained by converting each volume to moles of solute and relating the solutes by using the reaction stoichiometry as a conversion factor.

4.6 The compositions of mixtures, solutions, and compounds may be determined by **volumetric analysis**, of which the most important technique is the **titration** of an **analyte** with a **titrant**. The normal procedure is to measure one solution into another until the **stoichiometric point** (the equivalence point) is reached. At the stoichiometric point, enough titrant has been added to react with all the analyte. The stoichiometric point is shown by a color change either of the reagents or of an added **indicator**. In **acid-base titrations**, an acid reacts with a base. In **redox titrations**, an oxidizing agent reacts with a reducing agent.

CLASSIFIED EXERCISES

Mole Calculations

4.1 (a) How many moles of H_2 are needed to convert 5.0 mol of O_2 to water? (b) Determine the amount (in moles) of H_2 needed to convert 5.0 mol of O_2 to hydrogen peroxide, H_2O_2.

4.2 (a) Calculate the number of moles of product gases produced from the explosion of 2.0 mol TNT by the reaction

$$4C_7H_5O_6N_3(s) + 21O_2(g) \longrightarrow 28CO_2(g) + 6N_2(g) + 10H_2O(g)$$

(b) Calculate the number of moles of product gases produced from the detonation of 3.2 mol of nitroglycerin molecules by the reaction

$$4C_3H_5O_9N_3(l) \longrightarrow 12CO_2(g) + 6N_2(g) + O_2(g) + 10H_2O(g)$$

4.3 (a) In the commercial manufacture of nitric acid, how many moles of NO_2 produce 7.33 mol of HNO_3 in the reaction $3NO_2(g) + H_2O(l) \rightarrow 2HNO_3(aq) + NO(g)$? (b) The concentration of iodide ions in a solution can be determined by its oxidation to iodine with permanganate ion according to the reaction $2MnO_4^-(aq) + 16H^+(aq) + 10I^-(aq) \rightarrow 2Mn^{2+}(aq) + 5I_2(aq) + 8H_2O(l)$.

How many moles of permanganate ion react with 0.042 mol of I^-?

4.4 (a) The removal of hydrogen sulfide gas from sour natural gas occurs by the reaction

$$2H_2S(g) + SO_2(g) \longrightarrow 3S(s) + 2H_2O(l)$$

How many moles of sulfur form from the reaction of 5.0 mol of H_2S? (b) The neutralization of phosphoric acid with potassium hydroxide occurs by the reaction

$$H_3PO_4(aq) + 3KOH(aq) \longrightarrow K_3PO_4(aq) + 3H_2O(l)$$

Calculate the amount (in moles) of NaOH that reacts with 0.22 mol of H_3PO_4.

4.5 Calculate the amount of CO_2 produced when 1.5 mol of hexane molecules, C_6H_{14}, burn in air by the reaction

$$2C_6H_{14}(l) + 19O_2(g) \longrightarrow 12CO_2(g) + 14H_2O(g)$$

4.6 A method used by the Environmental Protection Agency (EPA) for determining the concentration of ozone in air is to pass the air sample through a "bubbler" containing iodide ion, which removes the ozone according to the equation $O_3(g) + 2I^-(aq) + H_2O(l) \rightarrow O_2(g) + I_2(aq) + 2OH^-(aq)$. What amount of iodide ions is needed to remove 3.5×10^{-5} mol of O_3?

Mass–Mole Relationships

4.7 Ammonia burns in oxygen according to the equation

$$4NH_3(g) + 3O_2(g) \longrightarrow 2N_2(g) + 6H_2O(g)$$

(a) Calculate the number of moles of H_2O produced in the combustion of 1.0 g of NH_3. (b) How many grams of O_2 are required for the complete reaction of 13.7 mol of NH_3?

4.8 Sodium thiosulfate, photographer's hypo, reacts with unexposed silver bromide in the film emulsion to form sodium bromide and a soluble compound of formula $Na_3Ag(S_2O_3)_2$:

$$2Na_2S_2O_3(aq) + AgBr(s) \longrightarrow NaBr(aq) + Na_3Ag(S_2O_3)_2(aq)$$

(a) How many moles of $Na_2S_2O_3$ are needed to dissolve 1.0 mg of AgBr? (b) Calculate the mass of silver bromide that will produce 0.033 mol of $Na_3Ag(S_2O_3)_2$.

4.9 Small bottles of propane gas are sold in hardware stores for convenient, portable heat sources (for soldering, for example). The combustion reaction of propane is

$$C_3H_8(g) + 5O_2(g) \longrightarrow 3CO_2(g) + 4H_2O(l)$$

(a) What mass of CO_2 is produced from the combustion of 1.55 mol of C_2H_8? (b) How many moles of water accompany the production of 4.40 g of CO_2?

4.10 Impure phosphoric acid for use in the manufacture of fertilizers is produced by the reaction of sulfuric acid on phosphate rock, of which a principal component is $Ca_3(PO_4)_2$. The reaction is

$$Ca_3(PO_4)_2(s) + 3H_2SO_4(aq) \longrightarrow 3CaSO_4(s) + 2H_3PO_4(aq)$$

(a) What amount of H_3PO_4 is produced from the reaction of 200 kg of H_2SO_4? (b) Determine the mass of calcium sulfate that is produced as a by-product of the reaction of 200 mol of $Ca_3(PO_4)_2$.

Mass–Mass Relationships

4.11 The surface atoms of aluminum metal corrode in air to form an impervious aluminum oxide coating that prevents further corrosion of the lower layers of atoms. The oxidation reaction is

$$4Al(s) + 3O_2(g) \longrightarrow 2Al_2O_3(s)$$

(a) Determine the mass of aluminum oxide formed from the corrosion of 10.0 g of aluminum. (b) In the reaction of 10.0 g of aluminum, what mass of oxygen is needed?

4.12 A typical problem in the iron industry is to determine the mass of iron that can be obtained from a mixture of iron(III) oxide, the principal component of the ore hematite, and carbon obtained from coal. The iron reduction reaction is

$$Fe_2O_3(s) + 3CO(g) \longrightarrow 2Fe(s) + 3CO_2(g)$$

(a) Calculate the mass of iron that can be produced from 1.0 metric ton of Fe_2O_3 (1 metric ton = 10^3 kg). (b) What mass of carbon monoxide is needed for the reduction of Fe_2O_3 to produce 500 kg of iron?

4.13 The solid fuel in the booster stage of the space shuttle is a mixture of ammonium perchlorate and aluminum powder. Upon ignition, the reaction that occurs is

$$6NH_4ClO_4(s) + 10Al(s) \longrightarrow 5Al_2O_3(s) + 3N_2(g) + 6HCl(g) + 9H_2O(g)$$

(a) What mass of aluminum should be mixed with 1.5×10^4 kg of NH_4ClO_4 for this reaction? (b) Determine the mass of Al_2O_3 (alumina, a finely divided white powder that is produced as billows of white smoke) formed in the reaction of 5000 kg of aluminum.

4.14 The compound diborane, B_2H_6, was at one time considered for use as a rocket fuel. Its combustion reaction is

$$B_2H_6(g) + 3O_2(l) \longrightarrow 2HBO_2(g) + 2H_2O(l)$$

The fact that HBO_2, a reactive compound, was produced rather that the relatively inert B_2O_3 was a factor in the discontinuation of the investigation of diborane as a fuel. (a) What mass of liquid oxygen (LOX) would be needed to burn 50.0 g of B_2H_6? (b) Determine the mass of HBO_2 produced from the combustion of 30.0 g of B_2H_6.

4.15 The camel stores the fat tristearin, $C_{57}H_{110}O_6$, in its hump. As well as being a source of energy, the fat is also a source of water because, when it is used, the reaction

$$2C_{57}H_{110}O_6(s) + 163O_2(g) \longrightarrow 114CO_2(g) + 110H_2O(l)$$

takes place. (a) What mass of water is available from 2.5 kg of this fat? (b) What mass of oxygen is needed to oxidize 2.5 g of tristearin?

4.16 Potassium superoxide, KO_2, is utilized in closed-system breathing apparatus. Exhaled air contains carbon dioxide and water, both of which are removed; and the removal of water generates oxygen for breathing by the reaction

$$4KO_2(s) + 2H_2O(l) \longrightarrow 3O_2(g) + 4KOH(s)$$

The potassium hydroxide removes carbon dioxide from the apparatus by the reaction

$$KOH(s) + CO_2(g) \longrightarrow KHCO_3(s)$$

(a) What mass of potassium superoxide generates 20.0 g of O_2? (b) What mass of CO_2 can be removed from the apparatus by 100 g of KO_2?

4.17 When a hydrocarbon burns, water is produced as well as carbon dioxide. (For this reason, clouds of condensed water droplets are often seen coming from automobile exhausts, especially on a cold day.) The density of gasoline is 0.79 g/mL. Assume gasoline to be pure octane, C_8H_{18}, for which the combustion reaction is

$$2C_8H_{18}(l) + 25O_2(g) \longrightarrow 16CO_2(g) + 18H_2O(l)$$

Calculate the mass of water produced from the combustion of 1.0 L of gasoline.

4.18 Ultrapure silicon used in solid-state electronics is produced by a zone-refining process of an impure silicon, which is produced by the high-temperature reduction of silicon dioxide (SiO_2, the principal component of sand and quartz) with carbon:

$$SiO_2(s) + 2C(s) \xrightarrow{\Delta} Si(l) + 2CO(g)$$

(a) How many Si atoms are produced in conjunction with 5.00 μg of CO? (b) How many C atoms react with 7.33×10^{22} SiO_2 formula units?

Reaction Yield

4.19 The theoretical yield of sodium perxenate, Na_4XeO_4, from a certain reaction is 1.25 mg. In a certain laboratory preparation, only 1.07 mg was obtained. What was the percentage yield?

4.20 The theoretical production yield of ammonia in an industrial synthesis is 95 metric tons per day. The production manager recorded a production of only 62.9 metric tons on a given day. What was the percentage yield for that day?

4.21 When limestone rock, which is principally $CaCO_3$, is heated, carbon dioxide and quicklime, CaO, are produced by the reaction

$$CaCO_3(s) \longrightarrow CaO(s) + CO_2(g)$$

If 11.7 g of CO_2 is produced from the thermal decomposition of 30.7 g $CaCO_3$, what is the percentage yield of the reaction?

4.22 Phosphorus trichloride, PCl_3, is produced from the reaction of white phosphorus, P_4, and chlorine:

$$P_4(s) + 6Cl_2(g) \longrightarrow 4PCl_3(l)$$

A 16.4-g sample of PCl_3 was collected from the reaction of 5.00 g of P_4 with excess chlorine. What is the percentage yield of the reaction?

Limiting Reactants

4.23 The souring of wine occurs when ethanol, C_2H_5OH, is converted by oxidation into acetic acid:

$$C_2H_5OH(aq) + O_2(g) \longrightarrow CH_3COOH(aq) + H_2O(l)$$

If 12.0 g of ethanol and 10.0 g of oxygen were sealed in a wine bottle, which would be the limiting reactant for the oxidation?

4.24 Phosphorus trichloride, PCl_3, reacts with water to form phosphorous acid, H_3PO_3, and hydrochloric acid:

$$PCl_3(l) + 3H_2O(l) \longrightarrow H_3PO_3(aq) + 3HCl(aq)$$

(a) Which is the limiting reactant when 12.4 g of PCl_3 is mixed with 10.0 g of H_2O? (b) What masses of phosphorous acid and hydrochloric acid are formed?

4.25 A mixture of 7.45 g iron(II) oxide and 0.111 mol Al as aluminum metal is placed in a crucible and heated in a high-temperature oven, where a reduction of the oxide occurs:

$$3FeO(s) + 2Al(l) \longrightarrow 3Fe(l) + Al_2O_3(s)$$

(a) Determine the maximum amount of iron (in moles of Fe) that can be produced. (b) Calculate the mass of excess reactant remaining in the crucible.

4.26 A solution containing 3.44 g of $AgNO_3$ is mixed with a solution containing 4.22 g of K_3PO_4. A precipitate of Ag_3PO_4 forms. (a) Write the balanced equation for the reaction. (b) What mass of Ag_3PO_4 can form in the mixture?

4.27 A mixture is prepared that consists of 4.0 L of SO_2 and 4.0 L of H_2S. Upon heating, a reaction occurs, forming elemental sulfur and water:

$$SO_2(g) + 2H_2S(g) \longrightarrow 3S(s) + 2H_2O(g)$$

Under the conditions of the experiment, the density of SO_2 is 2.86 g/L and that of H_2S is 1.52 g/L. (a) What is the limiting reactant in the formation of sulfur? (b) Calculate the mass of excess reactant in the system. (c) What masses of sulfur and of water can form? (d) Confirm that the sum of the masses of SO_2 and H_2S before reaction is the same as the sum of the masses of the excess reactant, sulfur, and water after reaction.

4.28 Lithium metal is unique to Group I elements in that it is the only member that reacts directly with nitrogen to produce a nitride, Li_3N:

$$6Li(s) + N_2(g) \longrightarrow 2Li_3N(s)$$

(a) What mass of lithium nitride can form from 48.0 g of lithium and 1.0×10^{24} molecules of N_2? (b) Determine the mass of excess reactant in the system. (c) Confirm that the sum of the masses of lithium and nitrogen before reaction is the same as the sum of the masses of the excess reactant and lithium nitride after reaction. This confirmation is called a *mass balance* of the reaction.

Chemical Composition from Measurements of Mass

4.29 A chemical analysis of a 4.39-g sample of benzoic acid, a common carboxylic acid used as a food preservative, is 3.03 g C, 1.14 g O, and the remainder, hydrogen. What is the empirical formula of benzoic acid?

4.30 Oxalic acid occurs in rhubarb. Analysis of a sample of oxalic acid of mass 10.0 g showed that it contained 0.22 g H, 2.7 g C, and the remainder oxygen. What is the empirical formula of oxalic acid?

4.31 The analysis of 0.922 g of aniline, a common organic base used in some varnishes, is 0.714 g C, 0.138 g N, and the remainder hydrogen. What is the empirical formula of aniline?

4.32 Urea is used as a commercial fertilizer because of its nitrogen content. An analysis of 25.0 mg of urea showed that it contained 5.0 mg C, 11.68 mg N, 6.65 mg O, and the remainder, hydrogen. What is the empirical formula of urea?

4.33 In a combustion analysis of a 0.152-g sample of the artificial sweetener aspartame, it was found that 0.318 g of carbon dioxide, 0.084 g of water, and 0.0145 g of nitrogen were produced. What is the empirical formula of aspartame? The molar mass of aspartame is 294 g/mol. What is its molecular formula?

4.34 The bitter-tasting compound quinine is a component of tonic water and is used as a protection against malaria. When a sample of mass 0.487 g was burned, 1.321 g of carbon dioxide, 0.325 g of water, and 0.0421 g of nitrogen were produced. The molar mass of quinine is 324 g/mol. Determine the empirical and molecular formulas of quinine.

4.35 The stimulant in coffee and tea is caffeine, a substance of molar mass 194 g/mol. When 0.376 g of caffeine was burned, 0.682 g of carbon dioxide, 0.174 g of water, and 0.110 g of nitrogen were formed. Determine the empirical and molecular formulas of caffeine, and write the equation for its combustion.

4.36 Nicotine, the stimulant in tobacco, causes a very complex set of physiological effects in the body. It is known to have a molar mass of 162 g/mol. When a sample of mass 0.385 g was burned, 1.072 g of carbon dioxide, 0.307 g of water, and 0.068 g of nitrogen were produced. What are the empirical and molecular formulas of nicotine? Write the equation for its combustion.

Molar Concentration

4.37 (a) An aqueous solution was prepared by dissolving 1.567 mol of $AgNO_3$ in enough water to make 250.0 mL of solution. What is the molar concentration of silver nitrate in the solution? (b) A 2.11-g sample of NaCl is placed into a 1500-mL volumetric flask. The sample is dissolved and diluted to the mark on the flask with water. What is the molar concentration of NaCl in the solution?

4.38 (a) A chemist prepared a solution by dissolving 1.230 g KCl in enough water to make 150.0 mL of solution. What molar concentration of potassium chloride should appear on the label? (b) If the chemist had mistakenly used a 500-mL volumetric flask (instead of the 150-mL flask in (a), what molar concentration of potassium chloride should appear on the label?

4.39 An experimental procedure requires 25.0 mL of 0.45 M HCl(*aq*) for an analysis. What mass of HCl must be dispensed from the reagent bottle?

4.40 A river-water sample is analyzed for total alkalinity by its reaction with 0.010 M $H_2SO_4(aq)$. If the chemist needs 50.0 mL of the acid, how many moles of H_2SO_4 are needed for its preparation?

4.41 A chemist studying the properties of photographic emulsions needed to prepare 25.00 mL of 0.155 M $AgNO_3(aq)$. What mass of silver nitrate must be placed into a 25-mL volumetric flask and dissolved and diluted to the mark with water?

4.42 What mass of $Na_2CO_3 \cdot 10H_2O$, a compound used in detergents, must be dissolved and diluted to the mark in a 500-mL volumetric flask to prepare a 0.10 M $Na_2CO_3(aq)$ solution?

4.43 A student prepared a solution of barium hydroxide by adding 2.577 g of the solid to a 250-mL volumetric flask and adding water to the mark. Some of this solution was transferred to a buret. What volume of solution should the student run into a flask to transfer (a) 50.0 mg $Ba(OH)_2$; (b) 3.5 mmol of OH^- ions?

4.44 A student investigating the properties of solutions containing carbonate ions prepared a solution containing 7.112 g of Na_2CO_3 in a 250-mL volumetric flask. Some of the solution was transferred to a buret. What volume of solution should be dispensed from the buret to provide (a) 5.112×10^{-3} mol of Na_2CO_3; (b) 3.451×10^{-3} mol of CO_3^{2-}?

4.45 (a) A 12.56-mL sample of 1.345 M $K_2SO_4(aq)$ is diluted to 250.0 mL. What is the concentration of K_2SO_4 in the diluted solution? (b) A 25-mL sample of 0.366 M HCl(*aq*) is drawn from a reagent bottle with a pipet. The sample is transferred to a 125-mL volumetric flask and diluted to the mark with water. What is the concentration of the dilute hydrochloric acid solution?

4.46 (a) What volume of a 0.778 M $Na_2CO_3(aq)$ solution should be diluted to 150.0 mL with water to reduce its concentration to 0.0234 M $Na_2CO_3(aq)$? (b) An experiment requires the use of 60.0 mL of 0.50 M NaOH(*aq*). The stockroom assistant can only find a reagent bottle of 2.5 M NaOH(*aq*). How is the 0.50 M NaOH(*aq*) solution to be prepared?

4.47 (a) What are the molar concentrations of H^+ and Cl^- in 0.010 M HCl(*aq*)? What is the pH of the solution? (b) What is the pH of a 0.00020 M $HNO_3(aq)$ solution? (c) The pH of a solution of milk is about 6.5. What is the molar concentration of H^+ ions in milk?

4.48 (a) Determine the molar concentrations of H^+ and NO_3^- in a 0.030 M $HNO_3(aq)$ solution. What is the pH of the solution? (b) The pH of stomach acid is about 1.2. What is the molar concentration of H^+ in the stomach? (c) What is the pH of tomato juice, in which the molar concentration of H^+ is 7.9×10^{-5} M?

Acid-Base Titrations

4.49 A 15.00-mL sample of sodium hydroxide was titrated to the stoichiometric point with 17.40 mL of 0.234 M HCl(*aq*). (a) What is the molar concentration of the NaOH solution? (b) Calculate the mass of NaOH in the solution.

4.50 A 15.00-mL sample of oxalic acid, $H_2C_2O_4$ (with two acidic protons), was titrated to the stoichiometric point with 17.02 mL of 0.288 M NaOH(*aq*). (a) What is the molar concentration of the oxalic acid solution? (b) Determine the mass of oxalic acid in the solution.

4.51 A 9.670-g sample of barium hydroxide is dissolved and diluted to the mark in a 250-mL volumetric flask. It was found that 11.56 mL of this solution was needed to reach the stoichiometric point in a titration of 25.0 mL of a nitric acid solution. (a) Calculate the molar concentration of the HNO_3 solution. (b) What is the pH of the HNO_3 solution? (c) What mass of HNO_3 is in solution?

4.52 A 10.0-mL volume of 3.0 M KOH(*aq*) is transferred to a 250-mL volumetric flask and diluted to the mark. It was found that 38.5 mL of this diluted solution was needed to reach the stoichiometric point in a titration of 10.0 mL of a phosphoric acid solution according to the reaction

$$3KOH(aq) + H_3PO_4(aq) \longrightarrow K_3PO_4(aq) + 3H_2O(l)$$

(a) Calculate the molar concentration of the H_3PO_4 solution. (b) What mass of H_3PO_4 is in solution?

Redox Titrations

4.53 A solution of iron(II) ions was titrated with a cerium(IV) sulfate solution:

$$Fe^{2+}(aq) + Ce^{4+}(aq) \longrightarrow Fe^{3+}(aq) + Ce^{3+}(aq)$$

When a 25.00-mL sample of an iron(II) solution was analyzed, 13.45 mL of 1.340 M $Ce(SO_4)_2(aq)$ was needed to reach the stoichiometric point. What is the molar concentration of the iron(II) in the solution?

4.54 A chemist prepared a solution of 6.148 g of sodium thiosulfate pentahydrate, $Na_2S_2O_3{\cdot}5H_2O$ in 250.0 mL of solution. It was found that 14.58 mL of this solution was needed to reach the stoichiometric point with 25.0 mL of a test solution containing iodine, the reaction being

$$2S_2O_3^{2-}(aq) + I_2(aq) \longrightarrow 2I^-(aq) + S_4O_6^{2-}(aq)$$

What mass of iodine was in the test solution?

4.55 The iron content in an ore can be determined by titrating a sample with a $KMnO_4$ solution. The ore is dissolved in hydrochloric acid, forming iron(II) ions; the latter react with MnO_4^-:

$$5Fe^{2+}(aq) + MnO_4^-(aq) + 8H^+(aq) \longrightarrow 5Fe^{3+}(aq) + Mn^{2+}(aq) + 4H_2O(l)$$

A sample of ore of mass 0.202 g needed 16.7 mL of 0.0108 M $KMnO_4(aq)$ to reach the stoichiometric point. (a) What mass of iron(II) ions is present? (b) What is the mass percentage of iron in the ore sample?

4.56 Hydrogen peroxide solutions are used as hair bleach. To analyze for the concentration of H_2O_2 in a commercial product, a sample of $KMnO_4$ of mass 1.872 g was dissolved and diluted to the mark in a 100-mL volumetric flask. A volume of 29.79 mL of this solution was used to titrate 10.00 mL of the acidified H_2O_2 solution to the stoichiometric point (detected by the retention of the purple color of the permanganate ion):

$$5H_2O_2(aq) + 2MnO_4^-(aq) + 6H^+(aq) \longrightarrow 2Mn^{2+}(aq) + 5O_2(g) + 8H_2O(l)$$

(a) What is the molar concentration of H_2O_2 in the hair bleach sample? (b) Express the concentration of hydrogen peroxide in units of grams of H_2O_2 per liter of solution.

UNCLASSIFIED EXERCISES

4.57 Sodium thiosulfate, $Na_2S_2O_3$, is often used in the laboratory to determine the molar concentration of iodine solutions:

$$2S_2O_3^{2-}(aq) + I_2(aq) \longrightarrow 2I^-(aq) + S_4O_6^{2-}(aq)$$

If 1.134 g of $Na_2S_2O_3$ was needed to react with all the iodine in a sample, then how many moles of iodine were present?

4.58 The density of oak wood is 0.72 g/cm^3. Assume oak wood has the formula $C_6H_{12}O_6$ and calculate the mass of water produced when a log 12 cm × 14 cm × 25 cm is burned.

4.59 A mixture of hydrogen peroxide and hydrazine can be used as a rocket propellant:

$$7H_2O_2(g) + N_2H_4(l) \longrightarrow 2HNO_3(aq) + 8H_2O(l)$$

(a) How many moles of H_2O_2 react with 0.477 mol of N_2H_4? (b) How many moles of HNO_3 can be produced in a reaction of 6.77 g of H_2O_2 with excess N_2H_4? (c) What mass of H_2O can be produced in a reaction of 49.6 mg H_2O_2 with excess N_2H_4?

4.60 Nitrogen dioxide contributes to the formation of acid rain as a result of its reaction with water in the air:

$$3NO_2(g) + H_2O(g) \longrightarrow 2HNO_3(aq) + NO(g)$$

(a) What mass (in milligrams) of HNO_3 can be produced from the reaction of 2.93 mg of NO_2 with an excess of H_2O? (b) If, in the reaction, only 1.91 mg of HNO_3 is produced, what is the percentage yield?

4.61 Small amounts of chlorine gas can be generated in the laboratory from the reaction of manganese(IV) oxide with hydrochloric acid:

$$4HCl(aq) + MnO_2(s) \longrightarrow 2H_2O(l) + MnCl_2(s) + Cl_2(g)$$

(a) What mass of Cl_2 will be produced from 42.7 g of MnO_2 with an excess of $HCl(aq)$? (b) What volume of chlorine gas (of density 3.17 g/L) will be produced from the reaction of 300 mL of 0.100 M $HCl(aq)$ with an excess of MnO_2? (c) Suppose only 150 mL of chlorine was produced in the reaction in (b). What is the percentage yield of the reaction?

4.62 The first stage in the production of nitric acid is the oxidation of ammonia to nitric oxide:

$$4NH_3(g) + 5O_2(g) \xrightarrow{Pt} 4NO(g) + 6H_2O(g)$$

Calculate the mass (in grams) of NO that is produced from the reaction of 600 L of $O_2(g)$ with an excess of ammonia, assuming 90% yield. The density of O_2 under the conditions of the reaction is 1.43 g/L.

4.63 Octane, the principal component of gasoline, burns in excess air by the reaction

$$2C_8H_{18}(l) + 25O_2(g) \longrightarrow 16CO_2(g) + 18H_2O(g)$$

(a) Calculate the volume of oxygen gas needed to react with 2.27 mg of C_8H_{18} given that the density of oxygen is 1.43 g/L under the conditions of the experiment. (b) What volume of air is required in (a), given that air is 21% O_2 by volume.

4.64 The equation describing the reaction used for the analysis of ozone in the atmosphere is

$$O_3(g) + 2KI(aq) + H_2O(l) \longrightarrow O_2(g) + I_2(s) + 2KOH(aq)$$

(a) Determine the amount (in moles) of KI needed for the reaction of 0.20 mol of O_3. (b) Calculate the mass of I_2 produced from the reaction of 49.7 μg of O_3. (c) Suppose in (b), 0.118 mg I_2 is produced, what is the percentage yield in the reaction?

4.65 What is the limiting reactant in a reaction mixture of 12.0 kg of SO_2 and 8.0 kg H_2S for the reaction

$$SO_2(g) + 2H_2S(g) \longrightarrow 3S(s) + 2H_2O(l)$$

4.66 Sodium metal can reduce aluminum oxide according to the equation

$$6Na(s) + Al_2O_3(s) \longrightarrow 2Al(s) + 3Na_2O(s)$$

A mixture of 10.0 g of sodium and 10.0 g of Al_2O_3 is heated in an inert atmosphere (so that oxygen does not react with the sodium metal). (a) Which reactant limits the amount of aluminum that can be produced in the reaction? (b) What is the mass of aluminum that can be produced? (c) What is the mass of excess reactant remaining in the system? (d) If 1.77 g of aluminum is produced in the reaction, what is the percentage yield?

4.67 Butane, C_4H_{10}, is used as a relatively cheap, portable heat source (for example, in cigarette lighters). From the combustion of a mixture of 4.66 g of butane and 11.1 L of O_2 (density, 1.43 g/L),

$$2C_4H_{10}(g) + 13O_2(g) \longrightarrow 8CO_2(g) + 10H_2O(g)$$

12.7 g of CO_2 was collected. (a) What is the percentage yield? (b) Calculate the mass of excess reactant.

4.68 A mixture of 4.94 g of 85.0% pure phosphine, PH_3, and 0.110 kg of $CuSO_4 \cdot 5H_2O$ (molar mass, 250 g/mol) is placed in a reaction vessel. Calculate the mass (in grams) of Cu_3P_2 (molar mass, 253 g/mol) with a 6.31% yield that would be produced in the reaction

$$3CuSO_4 \cdot 5H_2O(s) + 2PH_3(g) \longrightarrow Cu_3P_2(s) + 3H_2SO_4(aq) + 15H_2O(l)$$

4.69 What mass of the excess reactant remains in a reaction mixture consisting of 1.54 g of $Cr(NO_3)_3$ dissolved in 120 mL of 0.10 M $H_2S(aq)$? The reaction is

$$2Cr(NO_3)_3(aq) + 3H_2S(aq) \longrightarrow Cr_2S_3(s) + 6HNO_3(aq)$$

4.70 Teflon is the DuPont trade name for a class of synthetic fluorocarbon plastics. (Fluorocarbons are compounds of fluorine and carbon.) A chemist from a rival laboratory analyzed a sample and found it to have mass percentage composition 24% C and 76% F. What is the empirical formula of the plastic?

4.71 Fructose is a type of sugar that occurs in fruit and is the principal compound responsible for the sweetness of honey. Analysis of a 2.0-g sample showed it to contain 0.80 g C, 1.06 g O, and the remainder hydrogen. (a) What is the empirical formula of fructose? (b) Given that the molar mass of fructose is 180 g/mol, determine the molecular formula of the compound.

4.72 (a) Calculate the molar concentration of the solution formed when 12.7 g of Na_2CO_3 is dissolved in water and diluted to 250 mL in a volumetric flask. (b) When

12.7 g $Na_2CO_3 \cdot 10H_2O$ is dissolved in water and diluted to 250 mL in a volumetric flask, what is the molar concentration of the solution?

4.73 (a) Determine the mass of anhydrous copper(II) sulfate that must be used to prepare 250 mL of a 0.20 M $CuSO_4(aq)$ solution. (b) Determine the mass of $CuSO_4 \cdot 5H_2O$ that must be used to prepare 250 mL of a 0.20 M $CuSO_4(aq)$ solution.

4.74 The sulfuric acid solution that is purchased for the stockroom has a molar concentration of 17.8 M; all sulfuric acid solutions for experiments are prepared by dilution of this stock solution. (a) Determine the volume of 17.8 M H_2SO_4 that must be diluted to 250 mL to prepare a 2.0 M $H_2SO_4(aq)$ solution. (b) An experiment requires a 0.50 M $H_2SO_4(aq)$ solution. The stockroom manager estimates that a laboratory section needs 6.0 L of the acid. What volume of 17.8 M $H_2SO_4(aq)$ must be used for the preparation?

4.75 A solution of hydrochloric acid was prepared by measuring 10.00 mL of the concentrated acid into a 1.000-L volumetric flask and adding water up to the mark. Another solution was prepared by adding 0.530 g of anhydrous sodium carbonate to a 100.0-mL volumetric flask and adding water up to the mark. Then, 25.00 mL of the latter solution was pipeted into a flask and titrated with the diluted acid. The stoichiometric point was reached after 26.50 mL of acid had been added. (a) Write the balanced equation for the reaction of $HCl(aq)$ with $Na_2CO_3(aq)$. (b) What is the molar concentration of the original hydrochloric acid?

4.76 The concentration of a potassium permanganate solution was measured by a redox titration with arsenic(III) oxide in an acidic solution. The skeletal equation for the reaction is

$$As_4O_6(s) + MnO_4^-(aq) \longrightarrow Mn^{2+}(aq) + H_3AsO_4(aq)$$

(a) Write a balanced equation for the reaction. (b) When 0.1236 g of As_4O_6 was dissolved in 50.00 mL of water, 18.11 mL of the permanganate solution was needed to reach the stoichiometric point. What is the molar concentration of the permanganate solution?

4.77 A vitamin C tablet was analyzed to determine whether it did in fact contain, as the manufacturer claimed, 1.0 g of the vitamin. A tablet was dissolved in 100.00 mL of water and a 10.0-mL sample was titrated with iodine (as potassium triiodide). It required 10.1 mL of 0.0521 M I_3^- to reach the stoichiometric point in the titration. Given that 1 mol I_3^- = 1 mol vitamin C in the reaction, is the manufacturer's claim correct? The molar mass of vitamin C is 176 g/mol.

4.78 The mass percentage of sulfur in a fuel, such as coal, may be determined by burning the fuel in air and passing the resulting SO_2 and SO_3 gases into dilute H_2O_2, which converts the oxides to sulfuric acid. The amount of the sulfuric acid may be determined by a titration with a base. In one experiment, 6.43 g of coal was burned in excess air and the resulting H_2SO_4 was titrated to the stoichiometric point with 17.4 mL of 0.100 M NaOH. (a) Determine the amount (in moles) of H_2SO_4 that were produced. (b) What is the mass percentage of sulfur in the coal sample?

4.79 A forensic chemist needed to determine the concentration of HCN in the blood of a suspected homicide victim and decided to titrate a diluted sample of the blood with iodine using the reaction

$$HCN(aq) + I_3^-(aq) \longrightarrow ICN(aq) + 2I^-(aq) + H^+(aq)$$

A diluted blood sample of volume 15.00 mL was titrated to the stoichiometric point with 5.21 mL of an I_3^- solution. The molar concentration of the I_3^- solution was determined by titrating it against arsenic(III) oxide, As_4O_6, which in solution forms arsenious acid, H_3AsO_3. A volume of 10.42 mL of the triiodide solution was needed to reach the stoichiometric point on a 0.122-g sample of As_4O_6 in the reaction

$$H_3AsO_3(aq) + I_3^-(aq) + H_2O(l) \longrightarrow H_3AsO_4(aq) + 3I^-(aq) + 2H^+(aq)$$

(a) What is the molar concentration of the triiodide solution? (b) What is the molar concentration of HCN in the blood sample?

4.80 To prepare a very dilute solution, it is advisable to perform successive dilutions of a single prepared reagent solution, rather than weighing a very small mass or measuring a very small volume of stock chemical. A solution was prepared by transferring 0.661 g of $K_2Cr_2O_7$ to a 250-mL volumetric flask and adding water to the mark. A 1.0-mL sample of this solution was transferred to a 500-mL volumetric flask and diluted to the mark with water. (a) What is the final concentration of $K_2Cr_2O_7$ in solution? (b) What mass of $K_2Cr_2O_7$ is in this final solution? (Note that the answer to the latter question gives the mass that would have had to have been measured out if the solution had been prepared directly.)

4.81 Pure ethanol is produced commercially by the reaction of ethylene and water at elevated temperatures:

$$C_2H_4(g) + H_2O(g) \longrightarrow C_2H_5OH(g)$$

Approximately 1 mol C_2H_5OH in 20 reacts to form diethyl ether as a by-product:

$$2C_2H_5OH(g) \longrightarrow (C_2H_5)_2O(g) + H_2O(g)$$

The product mixture is then condensed and the mixture separated by fractional distillation. A mixture having a ratio of 3 mol $C_2H_4(g)$ for every 2 mol $H_2O(g)$ is fed into the reaction vessel. If no water vapor remains in the reaction vessel, what is the composition (in moles) of the system after the reaction?

CASE 4 ▸ ACID INDIGESTION AND ANTACIDS

The normal pH of the contents of the stomach is about 1.5, because of the presence of gastric juice, a mixture of components that is the equivalent of 0.030 M HCl(*aq*). The purpose of the gastric juice, which is secreted by cells in the stomach lining, is to assist in the digestion of food by enzymes. When the stomach is additionally stimulated by the presence of certain foods, an excess of hydrogen ions is secreted and acid indigestion (and heart burn) occurs.

To compensate for excess acid, we take an antacid, a base that neutralizes some of the stomach acid. An example of a simple antacid is milk of magnesia, which is an aqueous suspension of magnesium hydroxide and acts by providing hydroxide ions to neutralize the excess hydrogen ions:

$$Mg(OH)_2(s) + 2H^+(aq) \longrightarrow Mg^{2+}(aq) + H_2O(l)$$

Another simple and cheap antacid is sodium hydrogencarbonate (called either sodium bicarbonate or baking soda in the home):

$$NaHCO_3(aq) + H^+(aq) \longrightarrow Na^+(aq) + H_2O(l) + CO_2(g)$$

In this reaction, the evolution of carbon dioxide causes one to belch.

Most other antacids are formulations of pure crushed limestone (calcium carbonate) or crushed magnesium carbonate. For example, powdered calcium carbonate together with a binder is sold in tablet form as Tums. The reaction that takes place when one swallows this remedy is

$$CaCO_3(s) + 2H^+(aq) \longrightarrow Ca^{2+}(aq) + H_2O(l) + CO_2(g)$$

Calcium carbonate formulations are also sold as a source of calcium for the prevention of osteoporosis, a weakening of bones that generally occurs in elderly women.

The formulation sold as Alka-Seltzer contains citric acid and sodium hydrogencarbonate. The tablets release carbon dioxide immediately upon contact with water, because the citric acid reacts with solid hydrogencarbonate ions in a reaction similar to that of an aqueous solution of baking soda in the stomach.

Aspirin, which is a weak acid (its chemical name is acetylsalicylic acid), can often cause acid indigestion. To minimize this effect, an antacid (typically magnesium carbonate) is added to the aspirin. One such formulation is sold as Bufferin. The table below lists the composition of some commercial antacids.

QUESTIONS

1. Which antacid is more effective on a mass basis, milk of magnesia or Tums?

2. What mass (in milligrams) of Tums neutralizes 5.0 mmol HCl in 500 mL of solution?

3. The volume of an average stomach is about 1.0 L. Suppose that a particular condition of acid indigestion is equivalent to the presence of a 0.5% HCl (by mass) solution. (a) Determine the mass of milk of magnesia required to reduce the stomach acidity to pH = 2.0. (Assume the density of stomach acid is 1.0 g/mL.) (b) If each Tums tablet contains 500 mg of calcium carbonate, how many tablets should be taken to relieve the same acid indigestion?

Commercial antacids and their active ingredients

Commercial antacid	Active ingredient	Formula
Milk of Magnesia	magnesium hydroxide	$Mg(OH)_2$
Tums®	calcium carbonate	$CaCO_3$
Alka-Seltzer®	citric acid and sodium hydrogencarbonate	$C_3H_4(OH)(COOH)_3$, $NaHCO_3$
Rolaids®	dihydroxyaluminum sodium carbonate	$NaAl(OH)_2CO_3$
Bufferin®	aspirin and magnesium carbonate	$MgCO_3$

This chapter introduces the properties of the simplest state of matter—the gaseous state. It shows how the response of gases to changes in pressure and temperature can be predicted, and it extends the stoichiometric calculations of Chapter 4 to the prediction of the volumes of gases consumed in or generated by reactions. In addition, we see how the properties of gases can be understood in terms of their molecules. The illustration shows one of the most important gases of all: the atmosphere of the Earth.

5 THE PROPERTIES OF GASES

The importance of gases to the economy can be judged from the fact that they include half the top ten industrial chemicals (Appendix 2E). Among them are nitrogen, which is extracted from the atmosphere and then reduced to ammonia, and oxygen, which is used in steel making. Oxygen's role in steel making is to oxidize the impurities left in the crude iron produced in the blast furnace; about 1 ton of oxygen is needed to produce 1 ton of steel. Fourth in annual production is the gas ethylene (C_2H_4), much of which is used to make the plastic polyethylene. Air, that most important of all gaseous mixtures, affects everything we do (Table 5.1). Some of the many chemical processes that take place in the atmosphere are surveyed in Box 5.1. Eleven elements are gases under normal conditions, as are hundreds of compounds (Table 5.2 and Fig. 5.1). Most gases are molecular, including hydrogen (H_2), ammonia (NH_3), and sulfur dioxide (SO_2). The exceptions are the six noble gases (helium through radon), which are monatomic.

One goal of this chapter is to explain how to use reaction stoichiometry to predict the volumes of gases that take part in reactions. The first goal, though, is to understand the properties of the gaseous state, the

E The volume percentage composition of the "US standard atmosphere" at sea level is N_2 (78.084), O_2 (20.9476), Ar (0.934), CO_2 (0.0314), Ne (1.818 10^{-3}), He (5.24×10^{-4}), CH_4 (2×10^{-4}), Kr (1.14×10^{-4}), H_2 (5×10^{-5}), N_2O (5×10^{-5}), Xe (8.7×10^{-6}), with trace amounts of other gases.

N The properties of solids and liquids are described in Chapter 10.

T OHT: Table 5.1.

TABLE 5.1 The composition of dry air at sea level

Constituent	Molar mass,* g/mol	Percentage composition	
		Volume	Mass
nitrogen N_2	28.02	78.09	75.52
oxygen O_2	32.00	20.95	23.14
argon Ar	39.95	0.93	1.29
carbon dioxide CO_2	44.01	0.03	0.05

*The average molar mass of molecules in the air, allowing for the different abundances, is 28.97 g/mol. The percentage of water vapor in ordinary air varies with the humidity.

BOX 5.1 ▸ THE ATMOSPHERE

The atmosphere (from the Greek words for "vapor ball") is a relatively thin layer of gas on the surface of the Earth. If the Earth were the size of a basketball, its atmosphere would be only 1 mm thick. The atmosphere is not uniform in pressure, temperature, or composition. This nonuniformity is partly a result of the Sun's radiation, which causes different chemical reactions at different altitudes. It is also partly a result of the compressibility of air. The pressure exerted by the air causes the air at sea level to be most dense, so the atmosphere becomes thinner at higher altitudes. Half the mass of the atmosphere lies below 5.5 km; an airplane at a typical cruising altitude of 10 km is flying above 70% of the mass of the atmosphere.

The atmosphere is divided into several regions. The lowest region, where weather largely occurs, is called the troposphere (from the Greek for "sphere of change"). In this region the temperature decreases with increasing altitude at an average lapse rate of around 6.5°C per kilometer. Between about 11 and 16 km lies the tropopause, a region where the temperature remains constant at about −55°C. The precise thickness of the tropopause depends on the season and latitude. Above about 16 km, the stratosphere begins.

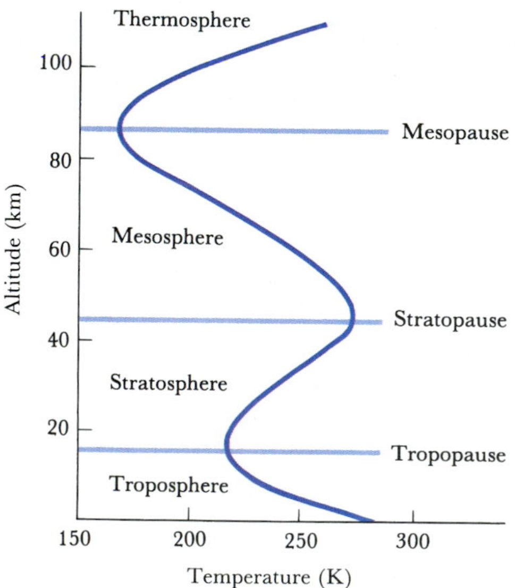

The variation of temperature with altitude in the atmosphere and the various zones into which the atmosphere is divided. The temperature profile represents the consequences of the differing interactions of molecules with solar radiation.

Here the temperature rises and reaches about 0°C in the stratopause, a region lying about 45 km above sea level. Above that height, in the mesosphere, the temperature again decreases as altitude increases, up to the mesopause. After the mesopause, in the thermosphere, the temperature rises, and the atoms, ions, and molecules reach speeds corresponding to over 1000°C.

The increase in temperature throughout the stratosphere is due to chemical reactions caused by the sun's radiation. One of the most important reactions in the stratosphere is the formation of ozone (O_3). However, there are numerous other reactions, some of which may stem from the presence of pollutants generated by human activities.

Stratospheric ozone is formed in two steps. The first is the decomposition of O_2 and other molecules into atoms of oxygen by the absorption of solar radiation:

$$O_2(g) \xrightarrow{\text{sunlight}} 2O(g)$$

This reaction absorbs harmful ultraviolet rays from sunlight. The oxygen atoms react with other oxygen molecules, forming ozone:

$$O(g) + O_2(g) \longrightarrow O_3(g)$$

This reaction releases energy that raises the temperature of the stratosphere.

The ozone produced in the ozone layer has several possible fates. One is decomposition:

$$O_3(g) \xrightarrow{\text{sunlight}} O(g) + O_2(g)$$

Because this reaction also absorbs ultraviolet radiation, it helps to shield the surface of the Earth from damage. Ozone in the troposphere—most of which is formed by electrical storms—is destroyed by reaction with other substances. A little of it reacts with airborne organic compounds leaking from engines and fuel tanks. This reaction converts them to substances that act as the eye irritants in smog.

Tropospheric nitrogen oxides, collectively referred to as NO_x (read "nox"), include nitric oxide (NO), nitrogen dioxide (NO_2), and dinitrogen oxide (N_2O). Dinitrogen oxide occurs naturally as the product of the action of bacteria in the soil on nitrogen. Nitric oxide and nitrogen dioxide are produced by automobiles, by the jet engines of aircraft, and by lightning. At ground level, nitric oxide is oxidized by atmospheric oxygen to the brown gas nitrogen dioxide (see Fig. 5.1), which dissolves in water to form nitric acid:

$$3NO_2(g) + H_2O(g) \longrightarrow 2HNO_3(aq) + NO(g)$$

In dry climates the dioxide may hang in the air and contribute to the color of smog.

Whatever their origin, NO_x molecules may diffuse up through the troposphere and enter the stratosphere through breaks in the tropopause. In both the troposphere and the stratosphere, NO_2 molecules are decomposed by solar radiation:

$$NO_2(g) \xrightarrow{\text{sunlight}} NO(g) + O(g)$$

The nitric oxide reacts with the ozone

$$NO(g) + O_3(g) \longrightarrow NO_2(g) + O_2(g)$$

and the oxygen atoms react with the nitrogen dioxide formed in this reaction:

$$NO_2(g) + O(g) \longrightarrow NO(g) + O_2(g)$$

The net result is the destruction of ozone and the regeneration of nitric oxide, which is then free to destroy more ozone. Because NO is regenerated in this way, a small amount of nitric oxide can eliminate a large amount of ozone.

The normal concentration of gases in the stratosphere is such that ozone formation is balanced by ozone destruction. In the heart of the ozone layer the concentration is maintained at about 0.1 μmol O_3/L. There is currently fear, however, that the balance might be upset by human meddling. This interference includes the generation of large amounts of NO_x by jet engines, by automobile emissions, and by the degradation of nitrogen-based fertilizers. The world's agriculture industry adds 50 billion kilograms of nitrogen fertilizer to the soil annually, and it is not known how much of this reaches the stratosphere as NO_x.

Another threat to the ozone balance in the stratosphere is the presence of chlorofluorocarbons (CFC), a class of compounds that includes the Freons once widely used as aerosol propellants but banned in the United States since 1978. When these compounds reach the stratosphere, they are broken down by ultraviolet solar radiation and chlorine atoms are produced. These atoms behave like nitric oxide molecules: they attack ozone and are regenerated in the process:

$$Cl(g) + O_3(g) \longrightarrow ClO(g) + O_2(g)$$
$$ClO(g) + O(g) \longrightarrow Cl(g) + O_2(g)$$

A single Cl atom survives in the stratosphere for four to ten years, and during that time it can destroy countless O_3 molecules.

While it is tempting to be complacent with the thought that CFCs are now less widely used in aerosols, they are still used in refrigerators and air conditioners, from which there is inevitable leakage. Moreover, it takes years for the CFC molecules to diffuse up into the stratosphere, the location of the ozone layer. Much of the aerosol propellant used during the last decade may still be on its way up.

Sulfur oxides in the troposphere are referred to as SO_x (read "sox") and include sulfur dioxide (SO_2) and sulfur trioxide (SO_3). About 70 billion kilograms of the dioxide result from the decomposition of vegetation and emission from volcanoes. A further 100 billion kilograms of the foul smelling, poisonous gas hydrogen sulfide is produced naturally and reacts with oxygen to form sulfur dioxide:

$$2H_2S(g) + 3O_2(g) \longrightarrow 2SO_2(g) + 2H_2O(g)$$

An SO_2 molecule survives for a few days in the atmosphere before it is oxidized to SO_3. The direct reaction

$$2SO_2(g) + O_2(g) \longrightarrow 2SO_3(g)$$

is very slow, but it is much faster on the walls of buildings and particulate matter (such as that from coal combustion sources) and in the presence of metal cations like Fe^{3+} dissolved in droplets of water. Other routes of conversion from SO_2 to SO_3 include reaction with OH, H_2O_2, and O_3 formed by the effect of sunlight on air and water vapor or by the attack of O atoms (from NO_2 dissociation) on hydrocarbons:

$$CH_4(g) + O(g) \longrightarrow CH_3(g) + OH(g)$$

The sulfur trioxide produced by these processes reacts with atmospheric water to form dilute sulfuric acid:

$$SO_3(g) + H_2O(l) \longrightarrow H_2SO_4(aq)$$

This very dilute acid, together with the nitric acid formed by dissolved NO_x, falls as acid rain. This acidic precipitation increases the acidity of rivers and lakes and leads to substantial damage to forests, perhaps by encouraging infection by viruses.

TABLE 5.2 The preparation and properties of some gases

Gas	Preparation and properties
Monatomic gases	
helium He	fractional distillation of liquefied natural gas colorless; odorless; insoluble; forms no compounds; condenses at −269°C (4 K); used in dirigibles and weather balloons and for low-temperature research
neon Ne	fractional distillation of liquid air colorless; odorless; insoluble; forms no compounds; condenses at −246°C (27 K); used in neon lights
Diatomic gases	
oxygen O_2	*commercial production*: fractional distillation of liquid air *laboratory preparation*: $2KClO_3(s) \xrightarrow{\Delta,\ MnO_2} 2KCl(s) + 3O_2(g)$ colorless; highly reactive; life supporting; not very soluble in water; condenses at −183°C to pale blue liquid
nitrogen N_2	fractional distillation of liquid air colorless; not very reactive; not very soluble in water; condenses at −196°C; used for making ammonia and other fertilizers and nitrogen-containing compounds (including plastics) and for providing inert atmospheres
fluorine F_2	electrolysis of molten KF–HF mixture almost colorless; highly reactive; reacts with water, forming HF; condenses at −188°C; used for isotope separation (UF_6) and for producing fluorocarbons, including Teflon® and Freons®
chlorine Cl_2	electrolysis of molten sodium chloride: $2NaCl(l) \longrightarrow 2Na(l) + Cl_2(g)$ pale yellow-green; reactive; reacts with water forming HClO; condenses at −34°C to yellow-green liquid; used as a water disinfecting agent and in many industrial processes, such as the production of PVC (polyvinyl chloride)
carbon monoxide CO	combustion of carbon or carbon compounds in a restricted supply of oxygen *commercial production*: water-gas reaction: $C(s) + H_2O(g) \longrightarrow H_2(g) + CO(g)$ *laboratory preparation*: action of sulfuric acid on formic acid: $HCOOH(l) \xrightarrow{\Delta,\ conc.H_2SO_4} CO(g) + H_2O(l)$ colorless; flammable; very poisonous; almost insoluble in water; used as a reducing agent in steelmaking
Triatomic gases	
hydrogen sulfide H_2S	*commercial production*: present in natural gas *laboratory preparation*: $FeS(s) + 2HCl(aq) \longrightarrow FeCl_2(aq) + H_2S(g)$ colorless; rotten egg odor; poisonous; soluble in water to give an acidic solution; condenses at −60°C; used in chemical analysis to produce insoluble sulfides; used as source of sulfur and to manufacture sulfuric acid
carbon dioxide CO_2	product of fermentation and combustion *laboratory preparation*: action of acids on carbonates: $CO_3^{2-}(s) + 2H^+(aq) \longrightarrow CO_2(g) + H_2O(l)$ colorless; odorless; soluble in water to give an acidic solution (carbonic acid); condenses directly to a solid at −79°C; used naturally in photosynthesis; solid carbon dioxide (dry ice) is a freezing agent
sulfur dioxide SO_2	*commercial production*: $S(s) + O_2(g) \longrightarrow SO_2(g)$ *laboratory preparation*: action of acids on sulfites: $SO_3^{2-}(aq) + 2H^+(aq) \longrightarrow H_2O(l) + SO_2(g)$ colorless; choking, pungent odor; poisonous; soluble in water to give acidic solution (sulfurous acid, H_2SO_3); condenses at −10°C; used in the production of sulfuric acid
Tetraatomic gases	
ammonia NH_3	*commercial production*: Haber synthesis (Section 12.4), $N_2(g) + 3H_2(g) \xrightarrow{\Delta,\ pressure} 2NH_3(g)$ colorless; pungent; very soluble in water to give a basic solution; condenses at −33°C; used as fertilizer and for production of HNO_3 and other nitrogen-containing compounds.

FIGURE 5.1 All gases are transparent and most gases are colorless, but here we see three with characteristic colors: brown nitrogen dioxide, pale yellow-green chlorine, and purple iodine vapor.

E The turbidity of some "vapors," such as steam, arises from the fact that they consist of droplets of liquid, and hence are not true vapors. The droplets scatter the incident light, and only a diffuse image of the source penetrates the sample. The change in color of cigarette smoke from blue (before it is inhaled) to brown (after it is inhaled) stems from the change in size of the smoke particles as they stick together after becoming moist. Examples of a blue gas are ozone and nitrogen trioxide. Bromine vapor and some interhalogens are red.

N Decide which (if any) of the reactions in Table 5.2 are worthy of special consideration by students.

simplest state of matter. These two topics are interdependent, however, because the volume of a gas changes sharply when the temperature and the pressure change and we cannot make reliable predictions about reacting volumes until we know how to take these factors into account.

? What is the effect of having air in the brake lines of an automobile?

? How do rapidly expanding gases cause propulsion?

THE GAS LAWS

We have defined a gas as the state of matter that fills any container it occupies. In molecular terms, a gas is a collection of widely separated molecules in chaotic, random motion (Fig. 5.2). This picture is reflected in the name *gas,* which is derived from the same root as the word *chaos.* The large distances between molecules in a gas account for the fact that it is highly compressible and can readily be confined to a smaller volume, for instance, by pushing a piston into a cylinder. A gas that is free to expand and contract also responds strongly to changes in temperature, occupying a larger volume as it is heated and a smaller one as it is cooled. These responses are expressed quantitatively by the laws described in this section.

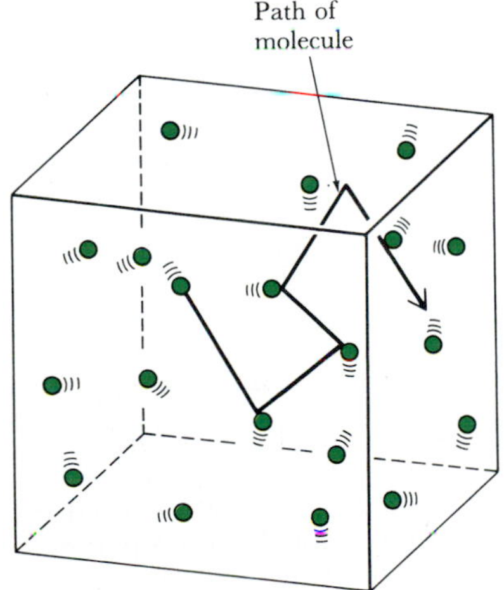

FIGURE 5.2 A gas may be pictured as a collection of widely spaced molecules in continuous, chaotic motion.

5.1 PRESSURE

Everyone who has ever pumped up a bicycle tire or squeezed an inflated balloon has experienced the opposing force arising from the "pressure" of the confined air. The scientific definition of **pressure** P is the force exerted per unit area:

$$P = \frac{\text{force}}{\text{area}}$$

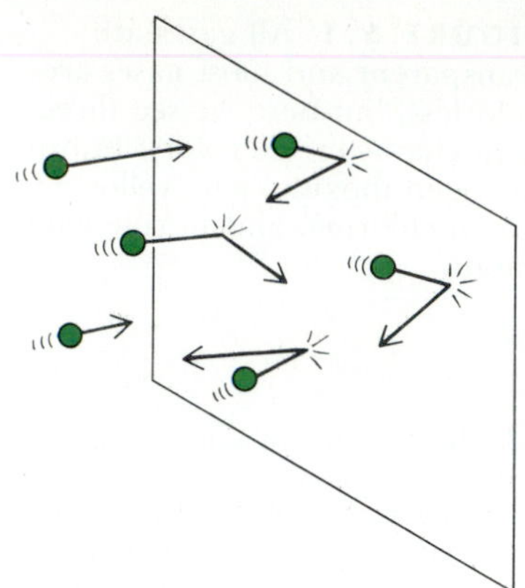

FIGURE 5.3 As the molecules of a gas collide with the walls of their container, they exert a force on it. The average force per unit area is the pressure of the gas.

U Exercises 5.67 and 5.68.

? Is there a pressure in the body of the gas, away from the walls?

A gas exerts a pressure on an object as a result of the collisions that the molecules have with the object's surface (Fig. 5.3). A vigorous, chaotic storm of molecules against a surface produces a strong force and hence a high pressure; a gentle storm of molecules results in a low pressure.

The pressure exerted by the atmosphere is measured with a barometer. This instrument was invented in the seventeenth century by Evangelista Torricelli, a student of Galileo. Torricelli (whose name, incidentally, means "little tower") formed a little tower of mercury by sealing a long glass tube at one end, filling it with mercury (a dense liquid), and inverting it into a beaker (Fig. 5.4). The column of mercury fell until the pressure exerted by the liquid mercury matched the pressure exerted by the atmosphere outside. The final height of the column is proportional to the atmospheric pressure* and hence can be used to measure it.

The height of the mercury column in a barometer on a typical day at sea level is about 760 mm (a value corresponding to the 30 inches referred to in weather forecasts). Hence, a pressure of 760 millimeters of mercury (written 760 mmHg) corresponds to normal atmospheric pressure at sea level. The modern tendency is to replace the symbol mmHg by Torr (named for Torricelli):

$$1 \text{ Torr} = 1 \text{ mmHg}$$

Atmospheric pressure is about 760 Torr at sea level on an average day. A low-pressure gas, such as one being used to study reactions high in the atmosphere, might have a pressure of only a few Torr (a few millimeters of mercury). The atmospheric pressure at the cruising height of a commercial aircraft (10 km) is about 200 Torr.

Chemists often deal with substances at atmospheric pressure, so they also find it convenient to use a unit called the atmosphere (atm):

$$1 \text{ atm} = 760 \text{ Torr}$$

This (exact) relation can be used as a conversion factor between Torr and atmospheres. For instance, a pressure of 500 Torr corresponds to

$$\text{Pressure (atm)} = 500 \cancel{\text{Torr}} \times \frac{1 \text{ atm}}{760 \cancel{\text{Torr}}} = 0.658 \text{ atm}$$

A very low pressure atmospheric region (an "area of low pressure" on a weather chart) typically has a pressure of about 0.98 atm, and the eye of a hurricane might reach as low as 0.90 atm. A typical summer anticyclone, a region of high pressure, is about 1.03 atm.

Pressures are also reported in other units, including the SI units pascal (Pa) and kilopascal (1 kPa = 10^3 Pa). The relations between all these units are given in Table 5.3 (more are given inside the back cover). It is useful to remember that typical atmospheric pressure is close to 100 kPa.

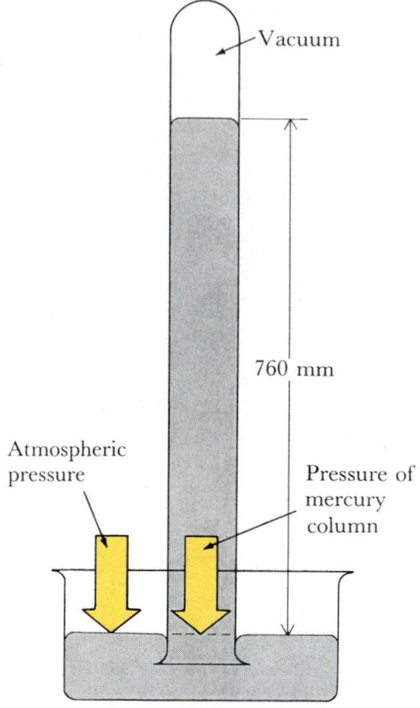

FIGURE 5.4 In a mercury barometer the pressure of a column of mercury is balanced by the pressure of the atmosphere; the height of the column is directly proportional to the atmospheric pressure. The space above the mercury is a vacuum, so it gives rise to no additional pressure inside the barometer tube.

*The precise relation is $P = dgh$, where P is the pressure, d is the density of the liquid, g is the acceleration of free fall (9.81 m/s^2), and h is the height of the column.

5.2 THE IDEAL GAS

The first measurements of the properties of gases were made by the Anglo-Irish scientist Robert Boyle in the seventeenth century. Technological motives for studying gases led to more discoveries when, in the eighteenth century, people began to fly in balloons. Their interest in ballooning stimulated the French scientists Jacques Charles and Joseph-Louis Gay-Lussac to make measurements of how the temperature of a gas affected its pressure, volume, and density. Charles built the first hydrogen balloon, which became known as the *charlière*. Gay-Lussac established a world altitude record of 23,018 feet in 1804.

TABLE 5.3 Pressure units*

SI unit: pascal (Pa)
1 Pa = 1 kg/(m·s^2) = 1 N/m^2
Conventional units
1 bar = 100 kPa
1 atm = 101.3 kPa
1 atm = 760 Torr (exactly)
1 atm = 14.7 lb/in^2

*See inside the back cover for more relations.

E The entry 1 atm = 14.7 lb/in^2 in Table 5.3 is another example of an equivalence rather than an equality (all the other entries are equalities). The dimensions of the unit atmosphere are N/m^2 (like the pascal); to convert pounds per square inch to a quantity with the same dimensions we should include a factor of g, the acceleration of free fall.

? When inflating a bicycle tire with a hand pump, what changes in pressure and volume of the air occur as one "pump" of air is transferred from the pump to the tire?

N Boyle's law, Charles's law, Rudberg's law, and the ideal gas law. *J Chem. Ed.* **1978**, *50*, 692.

T OHT: Figs. 5.5 and 5.6.

Boyle's law. The experiments done by Boyle, Charles, Gay-Lussac, and Avogadro (in the nineteenth century) showed that the pressure P, the volume V, and the temperature T of a fixed mass of gas all depend on one another, and a change in one affects the others. The volume of a gas decreases as the pressure is increased, in a manner summarized by Boyle's law:

> **Boyle's law**: At constant temperature, and for a given sample of gas, the volume is inversely proportional to the pressure:
>
> $$V \propto \frac{1}{P}$$

For example, doubling the pressure reduces the volume by half (Fig. 5.5). Gases respond readily to pressure because there is so much space between the molecules that they can easily be confined to a smaller volume. Boyle's law—which is illustrated in Fig. 5.6—applies to all gases—argon, carbon dioxide, water vapor, or any other.

A proportionality can always be expressed as an equality by introducing a constant factor, the constant of proportionality; so we can write

$$V = \text{constant} \times \frac{1}{P}$$

This expression is the equation of a straight line passing through the origin, because we can write it as

$$\begin{array}{ccccc} V = 0 & + & \text{constant} & \times & \dfrac{1}{P} \\ y = b & + & m & \times & x \\ \uparrow & & \uparrow & & \\ \text{Intercept} & & \text{Slope} & & \end{array}$$

Hence, the validity of Boyle's law for a variety of gases can be tested by plotting V against $1/P$ for each gas and getting a straight line in each case. As Fig. 5.7 shows, gases obey the law reasonably well at and below about 1 atm. For reasons we shall examine in Section 5.7, gases deviate from the law at high pressures and low temperatures.

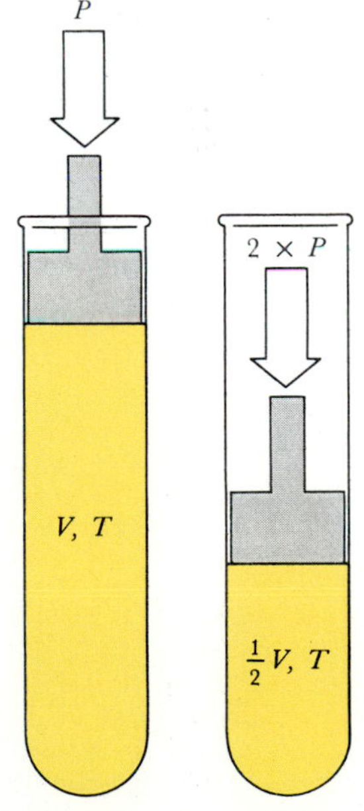

FIGURE 5.5 When pressure is applied to a sample of gas at constant temperature, its volume is decreased. Doubling the pressure reduces the volume of the gas by half.

▼ EXAMPLE 5.1 Using Boyle's law

In the industrial production of ammonia, nitrogen is compressed to about 250 atm before it reacts with hydrogen. If a certain amount of the nitrogen gas occupies 100 L at 100 kPa, what pressure is needed to compress it into 10 L at the same temperature?

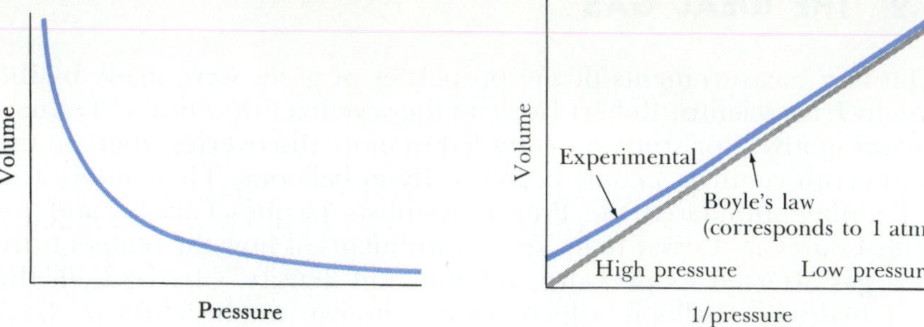

FIGURE 5.6 The variation of volume with pressure predicted by Boyle's law. As the pressure is increased, the volume of the gas is decreased. The curve is for a single temperature and a fixed mass of gas.

FIGURE 5.7 Boyle's law, $V \propto 1/P$, suggests that a plot of the volume of a gas against the inverse of its pressure should be a straight line. Experiments show that gases obey this law well at low pressures.

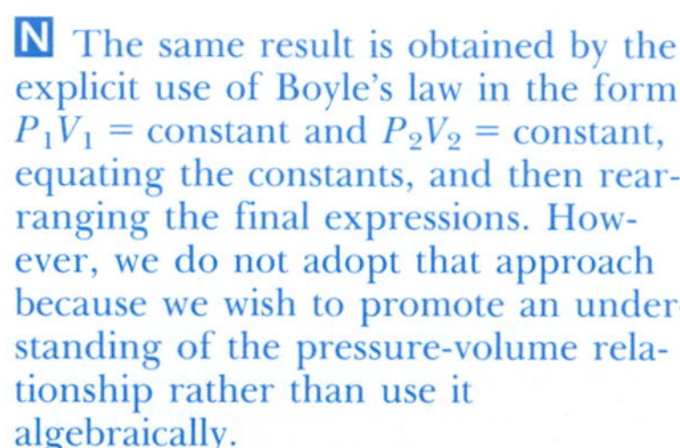

N The same result is obtained by the explicit use of Boyle's law in the form P_1V_1 = constant and P_2V_2 = constant, equating the constants, and then rearranging the final expressions. However, we do not adopt that approach because we wish to promote an understanding of the pressure-volume relationship rather than use it algebraically.

D Boyle's law and the monster marshmallow. *J. Chem. Ed.* **1982**, *59*, 974. Charles's law. BZS2, 5.7. A miniature hot air balloon and Charles's law. *J. Chem. Ed.* **1990**, *67*, 672.

U Exercises 5.69a and b and 5.70.

T OHT: Figs. 5.8 and 5.9.

STRATEGY The pressure of a gas is inversely proportional to the volume it occupies. Because the volume decreases, we need a pressure higher than the initial pressure. Therefore, to find the pressure required, we should multiply the initial pressure by a ratio of volumes that is greater than 1.

SOLUTION Of the two ratios of volumes,

$$\frac{100 \text{ L}}{10 \text{ L}} \quad \text{and} \quad \frac{10 \text{ L}}{100 \text{ L}}$$

that we can form, only the former is greater than 1. Therefore, the pressure required is

$$P = 100 \text{ kPa} \times \frac{100 \text{ L}}{10 \text{ L}} = 1.0 \times 10^3 \text{ kPa}$$

That is, to bring about a tenfold decrease in volume, we must apply a tenfold increase in pressure.

EXERCISE E5.1 What pressure is needed to confine a sample of argon in a 300-mL container if it occupies 500 mL when the pressure is 750 Torr?

[*Answer*: 1250 Torr]

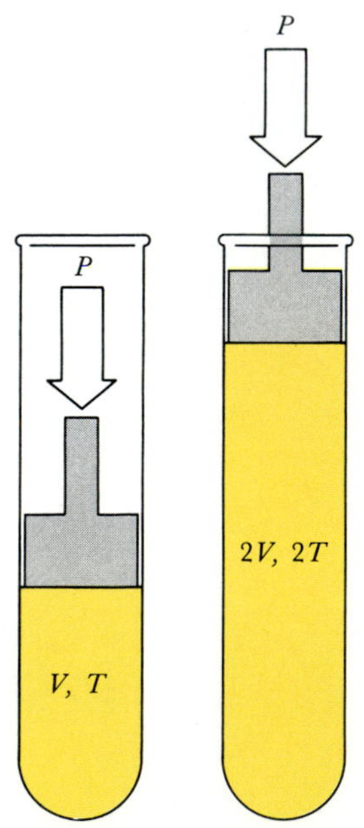

FIGURE 5.8 When a gas is heated at constant pressure, it expands. Doubling the temperature on the Kelvin scale doubles its volume.

Charles's law. The volume of a gas increases when it is heated, in a manner summarized by Charles's law:

> **Charles's law**: The volume of a given sample of gas at constant pressure is proportional to its absolute temperature:
>
> $$V \propto T$$

In agreement with this law, it is found experimentally that doubling the temperature from 293 K (20°C) to 586 K (313°C) doubles the volume of any gas that is free to expand against a constant pressure (Fig. 5.8). Like Boyle's law, Charles's law is obeyed at and below about 1 atm but not at high pressures or low temperatures.

If we introduce a constant of proportionality, Charles's law can be expressed as the equality

$$V = \text{constant} \times T$$

This expression is also the equation of a straight line:

$$V = 0 + \text{constant} \times T$$

$$y = b + m \times x$$

The law can therefore be tested by plotting the volume of a sample of gas against the temperature (Fig. 5.9); the result should be a straight line that, when extrapolated, passes through $V = 0$ at $T = 0$ (at $-273°C$). This is how the absolute zero of temperature was first identified: extrapolated plots of Charles's law passed through zero volume at $-273°C$ no matter what gas was used in the experiment. Because the same value was observed for all gases, and a volume cannot be negative, the graphs suggested that $-273°C$ (more precisely, $-273.15°C$) is the lowest temperature that can be reached by any substance. Lord Kelvin then set the zero of his scale at this lowest possible temperature.

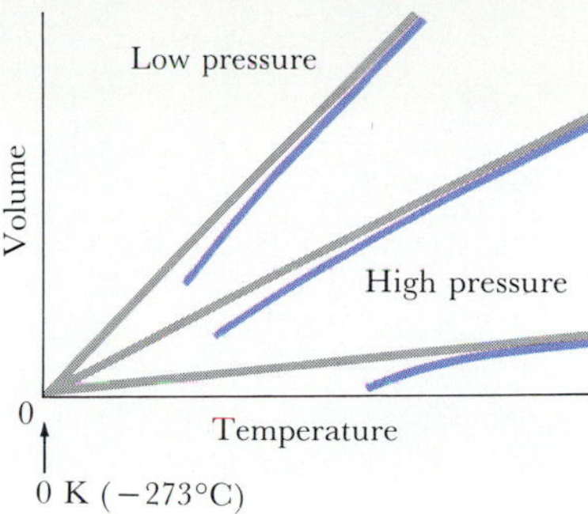

FIGURE 5.9 Charles's law, $V \propto T$, suggests that a plot of the volume of a gas against temperature should be a straight line. Note that all the lines extrapolate to $V = 0$ at $T = 0$ (corresponding to $-273°C$); however, this point is well below the temperature at which Charles's law is valid for real gases.

N It is not necessary to write $T = 0$ K (that is, to display the units: temperature T is expressed on the thermodynamic (absolute) temperature scale, and so the value $T = 0$ is independent of the actual size of the units of that scale. (We do not write $m = 0$ g or $l = 0$ cm, and so on, because if $m = 0$ in grams it is also equal to 0 in any other units, such as kilograms or pounds.)

▼ EXAMPLE 5.2 Using Charles's law

A hot-air balloon rises because the hot air inside the balloon is less dense than the cooler air outside. Calculate the volume that a sample of air would occupy at 40°C if it occupies 1.00 L at 20°C and the same pressure.

STRATEGY The volume of a gas is directly proportional to the temperature in kelvins. We anticipate that, because volume increases with temperature, the volume is greater at 40°C than at 20°C. Therefore, we multiply the initial volume by the ratio of temperatures (on the Kelvin scale) that is greater than 1.

SOLUTION The two temperatures are (20 + 273) K = 293 K and (40 + 273) K = 313 K. Of the two ratios that can be formed,

$$\frac{293\text{ K}}{313\text{ K}} \quad \text{and} \quad \frac{313\text{ K}}{293\text{ K}}$$

only the latter is greater than 1. Therefore, the final volume will be

$$V = 1.00\text{ L} \times \frac{313\text{ K}}{293\text{ K}} = 1.07\text{ L}$$

That is, the sample occupies 1.07 L, an increase of 7%. This increase in volume of a given sample of air implies that its density is 7% less at the higher temperature.

EXERCISE E5.2 To what temperature should a sample of gas be cooled from 25°C to reduce its volume to half its initial value, at constant pressure?
[*Answer*: 149 K, −124°C] ▲

D Cool an inflated balloon with liquid nitrogen.

D Determination of absolute zero. BZS2, 5.8.

Closely related to Charles's law is the experimental observation that, for a given sample of gas held in a container of constant volume, the pressure is proportional to the temperature:

$$P \propto T$$

Hence, doubling the temperature (for example, from 300 K, corresponding to 27°C, to 600 K, corresponding to 327°C) doubles the pressure. Aerosol containers should not be thrown in a fire because the pressure inside them may rise high enough to cause them to explode.

U Exercise 5.71.

? What happens to the air pressure in an automobile tire after it has been driven a distance?

The combined gas law. Boyle's and Charles's laws can be combined into a single expression:

$$\left.\begin{aligned} V &\propto \frac{1}{P} \\ V &\propto T \end{aligned}\right\} V \propto \frac{T}{P} \quad \text{or} \quad V = \text{constant} \times \frac{T}{P}$$

The combined expression can be reorganized into

$$\frac{PV}{T} = \text{constant} \qquad (1)$$

which we call the **combined gas law**. We shall determine the value of the constant at the end of this section.

The combined gas law is widely used to calculate the changes that occur to a sample when any of the conditions P, V, or T change. This relation could be important, for instance, in the design of an ammonia synthesis plant, because the pressure of nitrogen changes when a sample is heated to the temperature required by the process.

▼ **EXAMPLE 5.3 Using the combined gas law**

A deep-sea diver exhales a 15.0-mL bubble of air at a depth where the pressure is 12.0 atm and the temperature is 8.0°C. What is the volume of the bubble at the surface, where the atmospheric pressure is 770 Torr and the temperature is 20.0°C?

N We do not use the explicit approach to this calculation, using $P_1V_1/T_1 = P_2V_2/T_2$ because we want students to understand (rather than memorize) the interdependency of pressure, volume, and temperature.

STRATEGY The increase in temperature increases the volume occupied by the gas, so—according to Charles's law—the initial volume should be multiplied by the ratio of temperatures that is greater than 1. The decrease in pressure also increases the volume, so—according to Boyle's law—we should also multiply the volume by the ratio of pressures that is greater than 1. We begin by collecting the data.

SOLUTION The pressure of 12.0 atm corresponds to 9120 Torr, the initial temperature is 281.2 K, and the final temperature is 293.2 K. The temperature ratio that is greater than 1 is

$$\frac{293.2 \text{ K}}{281.2 \text{ K}}$$

and the pressure ratio that is greater than 1 is

$$\frac{9120 \text{ Torr}}{770 \text{ Torr}}$$

Therefore, the final volume will be

$$V = 15.0 \text{ mL} \times \frac{293.2 \text{ K}}{281.2 \text{ K}} \times \frac{9120 \text{ Torr}}{770 \text{ Torr}} = 185 \text{ mL}$$

U Exercises 5.69c, 5.72, 5.74, and 5.75.

The volume increases from 15.0 to 185 mL, a factor of more than 12.

EXERCISE E5.3 A sample of methane gas is stored in a 227-mL flask at 600 Torr and 10.0°C. The gas is allowed to expand into another flask of volume 1.78 L at 27.0°C. What is the pressure of methane in the second flask?

▲ [*Answer*: 81.1 Torr]

Avogadro's principle. Avogadro's contribution to our understanding of gases was to suggest that the volume occupied by a gas is a measure of the amount of molecules present, independent of their identity:

Avogadro's principle: The volume of a sample of gas at a given temperature and pressure is proportional to the amount of gas molecules in the sample.

$$V \propto n, \text{ or } V = \text{constant} \times n$$

where n is the amount (the number of moles) of molecules present. Hence, doubling the amount of molecules in the sample doubles the volume of the gas if its pressure and temperature are held constant.

We can use the data in Fig. 5.10 to judge the validity of Avogadro's view that the volume occupied by a given amount of gaseous molecules is independent of their chemical identity. The volumes occupied by 1 mol of various kinds of molecules are only approximately the same at normal pressures, but the volumes all approach the same value as the pressure is lowered.

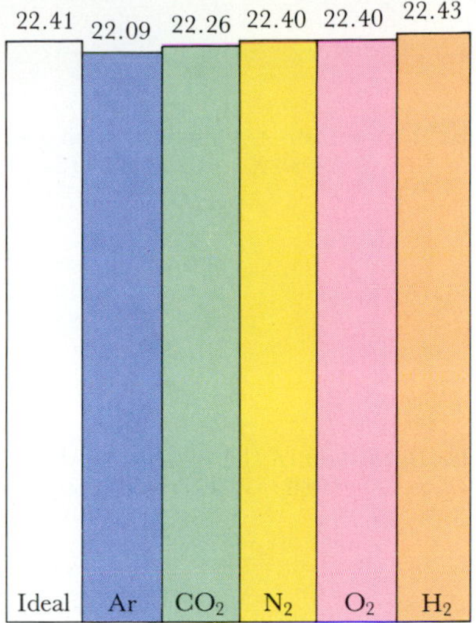

FIGURE 5.10 The volumes occupied by 1 mol of various gases at the same temperature and pressure (at 0°C and 1 atm in this case) are all very similar.

T OHT: Figs. 5.10 and 5.15.

N Avogadro's principle is not a summary of experience (it is a statement that depends on a certain model of a system), and so it is not strictly a scientific law.

D Molar volume of carbon dioxide. CD1, 10.

The **molar volume** V_m of a substance (*any* substance, not only a gas) is the volume it occupies per mole of molecules:

$$V_m = \frac{\text{volume occupied}}{\text{moles of molecules}} = \frac{V}{n}$$

According to Avogadro's principle, a gas consisting of 1 mol of X molecules occupies the same volume as one consisting of 1 mol of Y molecules, whatever the identities of the substances X and Y. Therefore, the molar volume of O_2 should be the same as that of CO_2 or any other gas at the same temperature and pressure. The data in Fig. 5.10 confirm that this is approximately true under normal conditions.

To avoid having to calculate the molar volume of a gas each time a calculation is done, chemists often report properties at an internationally accepted **standard temperature and pressure** (STP) defined as 0°C (273.15 K) and 1 atm. At STP, the molar volume of most gases is found to be approximately 22.4 L (about the volume of a cube of side 1 ft; Fig. 5.11).

The molar volume is used to convert from the *amount* of gas to the volume it occupies. However, sometimes we need the volume that a given *mass* of gas occupies at STP. Once again, the conversion to moles is the key: we convert the mass of the sample to amount in moles, and then use the molar volume at STP to convert this amount to liters.

FIGURE 5.11 The molar volume of an ideal gas at 0°C and 1 atm.

▼ EXAMPLE 5.4 Calculating the volume of a given mass of gas

Calculate the volume occupied by 10 g of CO_2 at STP by using the typical molar volume of 22.4 L/mol.

STRATEGY Once we know the amount of CO_2 in the sample, we can use the molar volume to convert to liters at STP. Therefore, we convert from mass to moles first and then from moles to liters.

SOLUTION We convert from mass to moles by using the molar mass of CO_2 (44.01 g/mol) in the form

$$44.01 \text{ g } CO_2 = 1 \text{ mol } CO_2$$

N 3 basketballs = 1 mole of ideal gas at STP. *J. Chem. Ed.* **1977**, *54*, 112.

and from moles to liters at STP by using the relation

$$1 \text{ mol } CO_2 = 22.4 \text{ L } CO_2$$

$$\text{Volume of } CO_2 \text{ (L)} = 10 \text{ g } CO_2 \times \frac{1 \text{ mol } CO_2}{44.01 \text{ g } CO_2} \times \frac{22.4 \text{ L } CO_2}{1 \text{ mol } CO_2}$$

$$= 5.2 \text{ L } CO_2 \text{ at STP}$$

EXERCISE E5.4 Calculate the volume occupied by 1.0 kg of hydrogen at STP.

[*Answer*: 1.1×10^4 L]

N Derivation of the ideal gas law. *J. Chem. Ed.* **1980**, *57*, 201; **1985**, *62*, 399.

The ideal gas law. We can now extend the combined gas law (Eq. 1) to include Avogadro's principle:

$$\left.\begin{array}{rl} \text{Boyle's law:} & V \propto \dfrac{1}{P} \\ \text{Charles's law:} & V \propto T \\ \text{Avogadro's principle:} & V \propto n \end{array}\right\} \text{combine into } V \propto \frac{nT}{P}$$

When we introduce a constant of proportionality R, the fully combined law can be written

$$V = R \times \frac{nT}{P}$$

This expression can be rearranged into the ideal gas law:

$$\textbf{Ideal gas law}: PV = nRT \qquad (2)$$

? What gases may be expected to behave markedly differently from an ideal gas under normal conditions? (Vapors of volatile liquids, such as water, ethanol, gasoline.)

The ideal gas law is approximately true for all gases under normal laboratory conditions (room temperature and 1 atm), and all gases are found to obey the law more and more closely as the pressure is reduced. A gas that obeys the law is called an **ideal gas**; and although no real gas is ideal at all temperatures and pressures (for reasons explained in Section 5.7), we can usually assume that they are ideal in the calculations we shall do. The *physical* reason why gases behave nearly ideally under normal conditions is that their molecules move largely independently and—because of the very large distances between molecules—interact only weakly with one another. As the pressure is lowered, the interactions between the widely spaced molecules decrease even further and the molecules move more and more independently. In an ideal gas, the molecules move completely independently of one another except during collisions.

It is an experimental fact of the greatest importance that the constant R, which is called the **gas constant**, has the same value for every gas. Once R has been measured for one gas, for instance by measuring n, P, V, and T for a given sample and calculating

$$R = \frac{PV}{nT}$$

the same value may be used for any gas. (In practice, other methods are used to measure R more precisely, including, as we shall see, relating it to the speed of sound in a gas.) It is found that at 0°C (273 K) and

1.000 atm, 1.000 mol of an ideal gas occupies 22.414 L, so

$$R = \frac{1.000 \text{ atm} \times 22.414 \text{ L}}{1.000 \text{ mol} \times 273.15 \text{ K}} = 0.08206 \text{ L·atm/(K·mol)}$$

(The result is read as "0.08206 liter-atmospheres per kelvin-mole.") This value of R should be used only when the pressure is given in atmospheres and the volume is given in liters. Values of R for pressure and volume in different units are given inside the back cover.

N The appearance of some of the calculations illustrated in this chapter would be simplified by using exponential notation, with $R = 0.08206$ $\text{L·atm·K}^{-1}\text{·mol}^{-1}$.

5.3 USING THE IDEAL GAS LAW

The ideal gas law is an example of an equation of state:

> An **equation of state** is a mathematical expression relating the pressure, volume, temperature, and amount of substance present in a sample.

Equations of state are very useful because they make it possible to calculate one property once we know the other three. In the case of the ideal gas equation of state (Eq. 2), for example, if we know the amount of gas, its temperature in kelvins, and its volume in liters, then we can calculate its pressure in atmospheres from

$$P = \frac{nRT}{V} \text{ with } R = 0.08206 \text{ L·atm/(K·mol)}$$

whatever the identity of the gas.

D The ideal gas law at the center of the sun. *J. Chem. Ed.* **1989**, *66*, 826.

▼ EXAMPLE 5.5 Using the ideal gas law to calculate a pressure

Calculate the pressure (in atm) inside a television picture tube, given that its volume is 5.0 L, its temperature is 23°C, and it contains 0.010 mg of nitrogen.

STRATEGY Assemble the data in the form required by the equation. Temperature must be in kelvins and the mass of nitrogen must be converted to moles of N_2 by using the relation 28.02 g N_2 = 1 mol N_2. When the volume is given in liters and we require the pressure in atmospheres, we use $R = 0.08206$ L·atm/(K·mol).

U Exercises 5.73 and 5.90.

SOLUTION The temperature corresponds to $T = (273 + 23) \text{ K} = 296 \text{ K}$. The number of moles of N_2 is

$$n = 0.010 \times 10^{-3} \text{ g } N_2 \times \frac{1 \text{ mol } N_2}{28.02 \text{ g } N_2} = 3.6 \times 10^{-7} \text{ mol } N_2$$

Hence, the pressure is

$$P = \frac{nRT}{V} = \frac{3.6 \times 10^{-7} \text{ mol } N_2 \times \left(0.08206 \frac{\text{L·atm}}{\text{K·mol}}\right) \times 296 \text{ K}}{5.0 \text{ L}}$$

$$= 1.7 \times 10^{-6} \text{ atm}$$

The very low pressure of gas (corresponding to a very small number of molecules striking the walls of the tube) is necessary, so that the electrons in the beam are not deflected by collisions with gas molecules, for that would give a blurred, dim picture.

EXERCISE E5.5 Calculate the pressure (in kilopascals) exerted by 1.0 g of carbon dioxide in a 1.0-L flask at 300°C.

[*Answer*: 1.1×10^2 kPa]

The density of a gas. The ideal gas law is used to determine how other physical properties vary with temperature and pressure. We shall use it to find the density d of a gas—its mass per volume—in terms of its temperature and pressure. As in our considerations of stoichiometry in Chapter 4, the mole is the link between what we are given and what we want to know. The procedure is illustrated in the following example.

U Exercises 5.76a and 5.67.

? Why does a hot-air balloon rise?

? What happens to the density of a gas in a container of constant volume if the gas is heated?

EXAMPLE 5.6 Calculating the density of a gas

Estimate the density of nitrogen at 20°C and 1.0 atm pressure.

STRATEGY Density is mass per volume. The procedure to adopt is

- Calculate n/V, the number of moles of molecules per liter of gas, by using the ideal gas equation.
- Convert the number of moles to mass by using the molar mass of the gas.

After these two conversions, we know the mass per liter of gas at the temperature of interest, which is its density. The molar mass of nitrogen is 28.02 g/mol N_2.

SOLUTION The ideal gas law $PV = nRT$ rearranges to

$$\frac{n}{V} = \frac{P}{RT}$$

Because 20°C corresponds to $T = 293$ K and $P = 1.0$ atm, we find

$$\frac{n}{V} = \frac{1.0 \text{ atm}}{\left(0.08206 \dfrac{\text{L·atm}}{\text{K·mol}}\right) \times 293 \text{ K}} = 4.2 \times 10^{-2} \text{ mol/L}$$

Therefore, the density of nitrogen is

$$\text{Density (g/L)} = 4.2 \times 10^{-2} \frac{\text{mol } N_2}{\text{L}} \times \frac{28.02 \text{ g } N_2}{1 \text{ mol } N_2} = 1.2 \text{ g/L } N_2$$

The density of nitrogen is 1.2 g/L, nearly a thousand times less than the density of water (1.0 kg/L).

EXERCISE E5.6 Calculate the density of carbon dioxide at 1.0 atm and 25°C.

[*Answer*: 1.8 g/L]

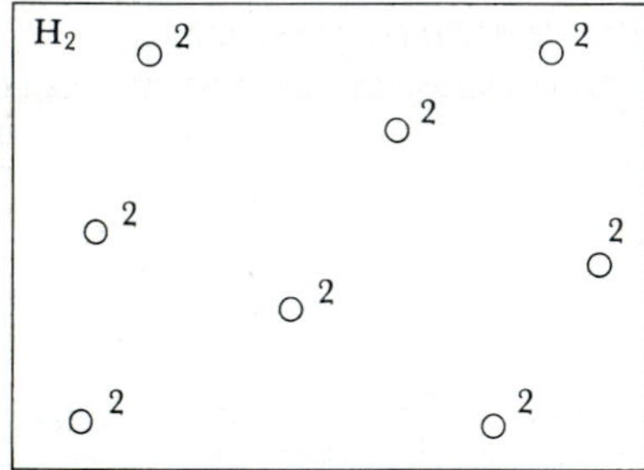

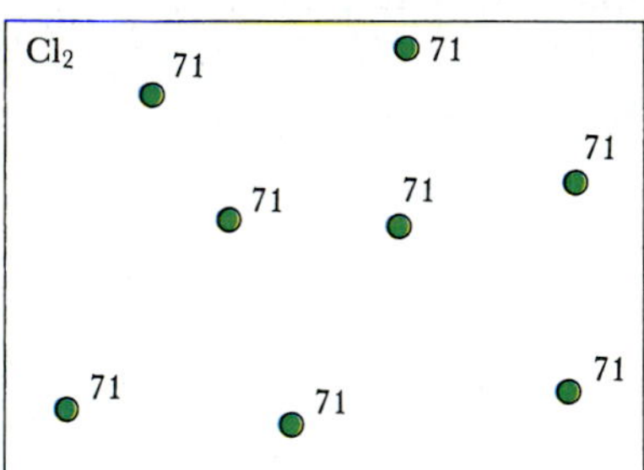

FIGURE 5.12 At the same temperature and pressure, a molecule occupies the same volume in any gas; hence, the greater the mass of each molecule, the greater the density of the gas. The illustration shows samples of hydrogen (top) and chlorine (bottom); the numbers are masses in atomic mass units.

The molar mass of a gas. According to Avogadro's principle, at a given temperature and pressure, 1 L of any gas contains the same number of molecules. However, because the total mass of the gas in that volume is the product of the number of moles of molecules and their molar mass, the *mass* of the sample is proportional to the molar mass. Hence, the greater the molar mass of the gas, the greater its density (at a given temperature and pressure; Fig. 5.12). Thus, chlorine (molar mass, 71 g/mol) is much denser than hydrogen (molar mass, 2 g/mol) under the same conditions of temperature and pressure.

The next example shows how the measurement of density is used to determine the molar mass of a volatile substance.

▼ EXAMPLE 5.7 Determining molar mass from gas density

The volatile organic compound geraniol, a component of oil of roses, is used in perfumery. It was found that the density of the vapor at 260°C is 0.480 g/L when the pressure is 103 Torr. Calculate the molar mass of geraniol.

D Determining the molecular weight of a gas. CD1, 11.

STRATEGY We know that the mass of 1.00 L of the vapor is 0.480 g. If we can find the number of moles of geraniol molecules in this volume, then we can divide the mass by the number of moles to find the molar mass. Therefore, the main part of the calculation is the determination of the number of moles in 1.00 L of gas under the conditions specified. When using the ideal gas law, with $R = 0.08206\ \text{L·atm/(K·mol)}$, we need to convert the pressure in Torr to atmospheres using 1 atm = 760 Torr.

SOLUTION The amount n of geraniol molecules in $V = 1.00$ L at $T = 533$ K (corresponding to 260°C) and $P = 103$ Torr is

$$n = \frac{PV}{RT} = \frac{\left(103\ \text{Torr} \times \dfrac{1\ \text{atm}}{760\ \text{Torr}}\right) \times 1.00\ \text{L}}{\left(0.08206\ \dfrac{\text{L·atm}}{\text{K·mol}}\right) \times 533\ \text{K}} = 3.10 \times 10^{-3}\ \text{mol}$$

U Exercises 5.76b, 5.78, and 5.79.

It follows that the molar mass of geraniol is

$$\text{Molar mass} = \frac{0.480\ \text{g}}{3.10 \times 10^{-3}\ \text{mol}} = 155\ \text{g/mol}$$

EXERCISE E5.7 The oil produced from eucalyptus leaves contains the volatile organic compound eucalyptol. At 190°C, a sample of eucalyptol vapor had a density of 0.400 g/L and a pressure of 60.0 Torr. Calculate the molar mass of eucalyptol.

▲ [*Answer*: 193 g/mol]

THE STOICHIOMETRY OF REACTING GASES

We saw in Section 4.1 how to calculate the amount of product from a given amount or mass of reactant. If the product is a gas, then from its molar volume we can go on to predict the volume of gas produced at STP by using the relation

$$1\ \text{mol X} = 22.4\ \text{L X (at STP)}$$

N This relation is another example of an equivalence (between an amount and a volume). It is valid only at STP. The corresponding relation at 25°C is 1 atm = 24.47 L.

Then, if desired, we can use the ideal gas law to convert the volume of gas to any other temperature and pressure. Thus, we can now add another arrow to the stoichiometry diagram (**1**) and calculate reacting volumes for any reaction involving gases.

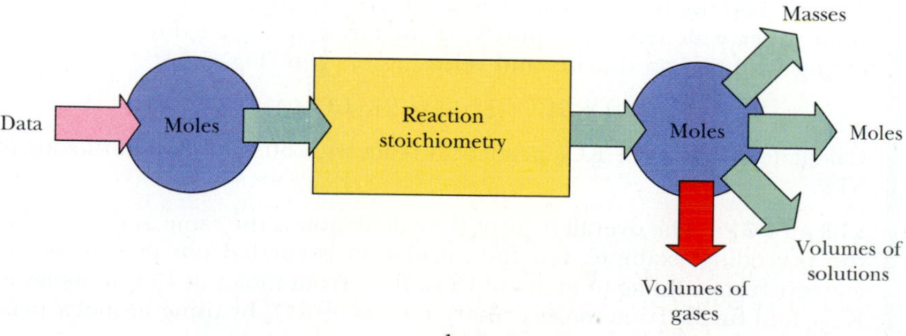

1

5.4 GAS VOLUMES AND REACTION STOICHIOMETRY

FIGURE 5.13 Sulfur burns with a blue flame and produces the dense gas sulfur dioxide.

E The blue emission in Fig. 5.13 is from electronically excited S_2 molecules formed in the flame when S_8 rings are fragmented.

U Exercises 5.80–5.85, 5.93.

As an example of a chemical problem that we can now analyze, we shall find the volume of sulfur dioxide produced by burning a specified mass of sulfur in the reaction

$$S_8(s) + 8O_2(g) \longrightarrow 8SO_2(g)$$

(see Fig. 5.13). As we did in the stoichiometry calculations of Chapter 4, we convert the given mass of sulfur to moles of S_8, then use the stoichiometry relation to convert moles of S_8 to moles of SO_2, and finally (the new step in this chapter), convert moles of SO_2 to liters of the gas. This procedure is illustrated in the following example.

▼ EXAMPLE 5.8 Calculating the volume of gas produced by a given mass of reactant

What volume of sulfur dioxide is produced at STP by the combustion of 10 g of sulfur according to the reaction

$$S_8(s) + 8O_2(g) \longrightarrow 8SO_2(g)$$

STRATEGY Follow the strategy set out above, converting from mass to moles of S_8, from moles of S_8 to moles of SO_2, and finally from moles of SO_2 to volume at STP.

SOLUTION The conversion from mass of sulfur to moles of S_8 (molar mass, 256.48 g/mol) is based on the relation 256.48 g S_8 = 1 mol S_8. The chemical equation shows that 1 mol S_8 = 8 mol SO_2 and we know that at STP, 1 mol SO_2 = 22.4 L SO_2. It follows that, at STP,

$$\text{Volume of } SO_2 \text{ (L)} = 10 \text{ g } S_8 \times \frac{1 \text{ mol } S_8}{256.48 \text{ g } S_8} \times \frac{8 \text{ mol } SO_2}{1 \text{ mol } S_8} \times \frac{22.4 \text{ L } SO_2}{1 \text{ mol } SO_2}$$

$$= 7.0 \text{ L } SO_2$$

EXERCISE E5.8 Calculate the volume of acetylene (C_2H_2) produced at STP when 10 g of calcium carbide reacts completely with water in the reaction $CaC_2(s) + 2H_2O(l) \longrightarrow Ca(OH)_2(s) + C_2H_2(g)$.

[*Answer*: 3.5 L C_2H_2 at STP]

▲

FIGURE 5.14 When carbon dioxide is passed over yellow potassium superoxide, they react to form colorless potassium carbonate and oxygen gas. The reaction is used to remove CO_2 from air in a closed-system breathing apparatus.

▼ EXAMPLE 5.9 Calculating the mass of reagent needed to react with a specified volume of gas

A team working on a submarine design was investigating the use of potassium superoxide (KO_2) for purifying the air. The superoxide acts by combining with carbon dioxide and releasing oxygen (Fig. 5.14):

$$4KO_2(s) + 2CO_2(g) \longrightarrow 2K_2CO_3(s) + 3O_2(g)$$

Calculate the mass of KO_2 needed to react with 50 L of carbon dioxide at STP.

STRATEGY The overall logic of the calculation is the same as that used in the preceding example, but the calculation is carried out in reverse: we convert from volume to moles of CO_2, then from moles of CO_2 to moles of KO_2, and finally from moles of KO_2 to mass of KO_2 by using its molar mass (71.1 g/mol).

SOLUTION The three conversions are based on three relations:

$22.4\ \text{L CO}_2 = 1\ \text{mol CO}_2$, from the molar volume at STP

$2\ \text{mol CO}_2 = 4\ \text{mol KO}_2$, from the chemical equation

$1\ \text{mol KO}_2 = 71.1\ \text{g KO}_2$, from the molar mass of KO_2

The result we require is therefore

$$\text{Mass of KO}_2\ (\text{g}) = 50\ \text{L CO}_2 \times \frac{1\ \text{mol CO}_2}{22.4\ \text{L CO}_2} \times \frac{4\ \text{mol KO}_2}{2\ \text{mol CO}_2} \times \frac{71.1\ \text{g KO}_2}{1\ \text{mol KO}_2}$$

$$= 3.2 \times 10^2\ \text{g KO}_2$$

EXERCISE E5.9 Calculate the volume of carbon dioxide at STP needed to make 1.00 g of glucose ($C_6H_{12}O_6$) by photosynthesis in the reaction $6CO_2(g) + 6H_2O(g) \longrightarrow C_6H_{12}O_6(s) + 6O_2(g)$.

[*Answer*: 0.75 L]

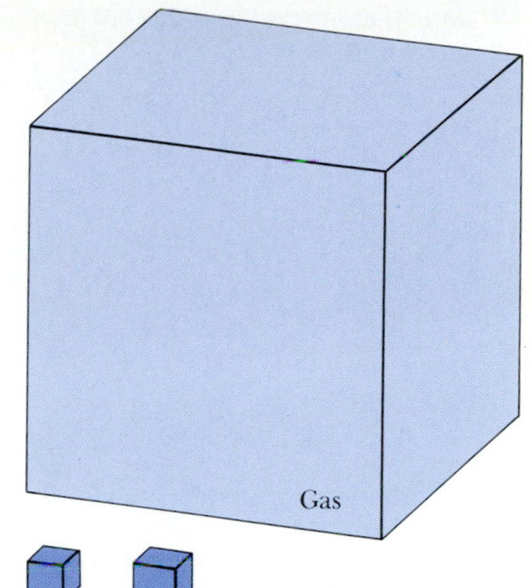

FIGURE 5.15 The molar volumes of solid, liquid, and gaseous substances at STP.

The molar volumes of gases, which are close to 22 L/mol at STP, are much larger than those of liquids and solids, which are more typically a few tens of milliliters per mole. The molar volume of liquid water, for instance, is 18 mL/mol. In other words, a mole of gas molecules occupies a volume about a thousand times greater than the volume of a mole of molecules in a liquid or solid (Fig. 5.15). It follows that when liquids or solids react to form a gas, there may be a thousandfold increase in volume. The increase may be even greater if several gas molecules are produced from each reactant molecule (Fig. 5.16).

One example of the consequence of the increase in volume is the explosive action of nitroglycerin, $C_3H_5(NO_3)_3$ (**2**). When subjected to a shock wave from a detonator, it decomposes into many small gaseous molecules:

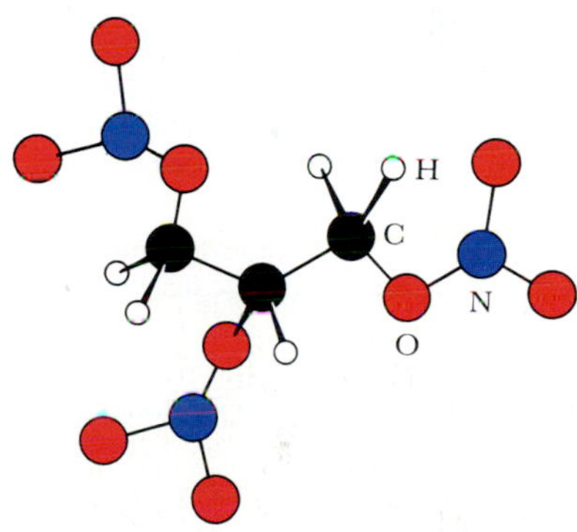

2 Nitroglycerin

$$4C_3H_5(NO_3)_3(l) \longrightarrow 6N_2(g) + O_2(g) + 12CO_2(g) + 10H_2O(g)$$

In this reaction, 4 mol $C_3H_5(NO_3)_3$, corresponding to a little over 500 mL of the liquid, produces 29 mol of gas molecules of various kinds, or a total of about 600 L of gas at STP. The pressure wave from the sudden 1200-fold expansion is the destructive shock of the explosion. The detonator, which is typically lead azide, $Pb(N_3)_2$, works on a similar principle; it releases a large volume of nitrogen gas when it is struck:

$$Pb(N_3)_2(s) \longrightarrow Pb(s) + 3N_2(g)$$

The sudden shock of expansion of the detonator stimulates the explosive itself to react. For example, a 1.0-g sample of lead azide produces 1.6×10^3 L of nitrogen gas at STP.

FIGURE 5.16 An explosion caused by the ignition of coal dust. A shock wave is created by the tremendous expansion of volume as large numbers of gas molecules form.

5.5 GASEOUS MIXTURES

What we know about the properties of pure gases can be adapted very easily to describe the behavior of mixtures of nonreactive gases. This extension of the gas laws should not be surprising. We have seen that the response to changes in the pressure, volume, and temperature is largely independent of the identity of the gas. Hence, it is unimportant whether or not all the molecules in a sample are the same, and we can expect a mixture of gases that do not react with one another (such as air) to behave like a single pure gas.

E John Dalton was led to his law by his interest in meteorology.

Dalton's law. Suppose that when a certain amount of oxygen is introduced into a container, it results in a pressure of 10 kPa. Suppose also that when some nitrogen at the same temperature is introduced instead, it results in a pressure of 20 kPa. John Dalton—whose contribution to atomic theory is described in Chapter 2—wondered what the total pressure would be if the two gases were present simultaneously. He made some fairly crude measurements and concluded that the total pressure when both gases were in the same container would be 30 kPa, the sum of the individual pressures.

Dalton summarized his observations in terms of the partial pressures of gases:

E The modern, precise definition of the partial pressure of a gas is the product of the total pressure and the mole fraction of the gas in question. Thus, $P_A = x_A P$: this definition applies to real gases as well as to ideal gases, and Dalton's law then becomes strictly true (because the sum of mole fractions is 1 for any mixture). The definition used here, which gives the partial pressure a physical significance, is valid only for ideal gases.

? What molecules (by name) keep an automobile tire inflated? What would happen if all the N_2 molecules were removed from the tire?

? How does the removal of moisture affect the pressure inside a laboratory dry box?

D Dalton's law of partial pressures. BZS2, 5.10.

> The **partial pressure** of a gas in a mixture is the pressure it would exert if it alone occupied the container.

In our example, the partial pressures of oxygen and nitrogen in the mixture are 10 kPa and 20 kPa, respectively, because those are the pressures the gases exert when each one is alone in the container. Dalton then described the behavior of gaseous mixtures by his law of partial pressures:

> **Dalton's law of partial pressures**: The total pressure of a mixture of gases is the sum of the partial pressures of its components.

If we write the partial pressures of the gases, A, B, . . . as P_A, P_B, . . . and the total pressure of the mixture as P, Dalton's law can be written

$$P = P_A + P_B + \ldots$$

The law is illustrated in Fig. 5.17; it is exactly true only for gases that behave ideally.

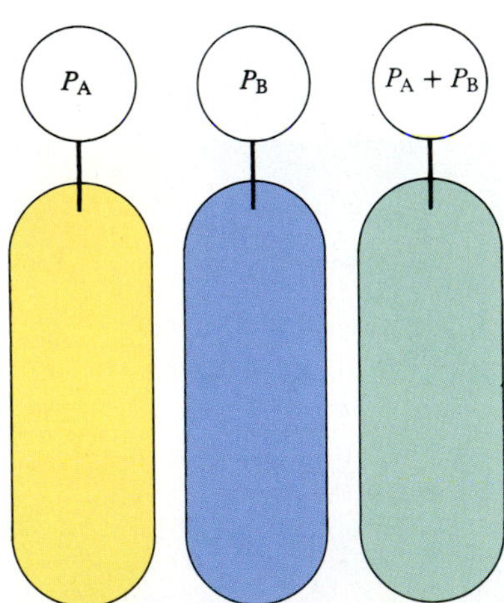

FIGURE 5.17 According to Dalton's law, the total pressure P of a mixture of gases is the sum of the partial pressures P_A and P_B of the components, the pressures they exert when they each occupy the container alone (at the same temperature).

The partial pressure P_A of a gas A depends on the amount n_A of A present, the temperature, and the total volume the mixture occupies. It is calculated as though no other gases were present and hence can be found from the ideal gas law in the form

$$P_A = \frac{n_A RT}{V} \qquad (3)$$

▼ EXAMPLE 5.10 Calculating partial pressures

A 1.00-g sample of air consists of approximately 0.76 g of nitrogen and 0.24 g of oxygen. Calculate the partial pressures and the total pressure when this sample occupies a 1.00-L vessel at 20°C.

STRATEGY Collect the data and decide whether there is enough information to calculate the total pressure from the equation defining the partial pressure (Eq. 3). It may be necessary to convert another item of data or to obtain additional information from tables.

SOLUTION The data needed for use in Eq. 3 are

$$P_{N_2} = ? \quad n_{N_2} = ? \quad V = 1.00\text{ L} \quad T = 293\text{ K (20°C)}$$

$$P_{O_2} = ? \quad n_{O_2} = ? \quad V = 1.00\text{ L} \quad T = 293\text{ K (20°C)}$$

The values of n_{N_2} and n_{O_2} are obtained from the masses and molar masses (28.02 g/mol for N_2 and 32.00 g/mol for O_2):

$$\text{Amount of } N_2 \text{ (mol)} = 0.76 \text{ g } N_2 \times \frac{1 \text{ mol } N_2}{28.02 \text{ g } N_2} = 0.027 \text{ mol } N_2$$

$$\text{Amount of } O_2 \text{ (mol)} = 0.24 \text{ g } O_2 \times \frac{1 \text{ mol } O_2}{32.00 \text{ g } O_2} = 0.0075 \text{ mol } O_2$$

Then, from Eq. 3,

$$P_{N_2} = \frac{0.027 \text{ mol } N_2 \times \left(0.08206 \dfrac{\text{L·atm}}{\text{K·mol}}\right) \times 293 \text{ K}}{1.00 \text{ L}} = 0.65 \text{ atm } N_2$$

$$P_{O_2} = \frac{0.0075 \text{ mol } O_2 \times \left(0.08206 \dfrac{\text{L·atm}}{\text{K·mol}}\right) \times 293 \text{ K}}{1.00 \text{ L}} = 0.18 \text{ atm } O_2$$

The total pressure is the sum of the partial pressures of nitrogen and oxygen:

$$P = P_{N_2} + P_{O_2} = (0.65 + 0.18) \text{ atm} = 0.83 \text{ atm}$$

EXERCISE E5.10 The composition of dry air is given more precisely in Table 5.1. Calculate the partial pressures of all the components listed there, and the total pressure of the sample, when 1.00 g of dry air is confined to a 500-mL container (such as a bicycle tire) at 25°C.

[*Answer:* $P_{N_2} = 1.32$ atm; $P_{O_2} = 0.354$ atm; $P_{Ar} = 0.0158$ atm; $P_{CO_2} = 5.56 \times 10^{-4}$ atm; $P = 1.69$ atm]

Measuring the amount of gas collected over water. To study chemical reactions that produce gases, chemists often collect the gas over water. This technique is commonly used to collect a sample of a gas that has been prepared in the laboratory and is not very soluble in water (for example, oxygen). However, because the gas inside the inverted collecting bottle is a mixture of water vapor and oxygen, to find the amount of O_2 produced we need to correct the total pressure for the amount of water vapor present. We might, for example, be interested in the amount of O_2 that can be obtained by heating a sample of potassium chlorate:

T OHT: Fig. 5.18.

$$2KClO_3(s) \xrightarrow{\Delta,\ MnO_2} 2KCl(s) + 3O_2(g)$$

(In practice this reaction is carried out in the presence of a little manganese dioxide, which acts as a catalyst, a substance that helps the reaction to run quickly and smoothly.)

To find the amount of O_2 from the volume of gas collected, we could use

$$n_{O_2} = \frac{P_{O_2}V}{RT}$$

where P_{O_2} is the partial pressure of oxygen in the collecting bottle. The problem is to determine P_{O_2} even though we can measure only the *total* pressure, which we know is the sum of the partial pressures of the oxygen and the water vapor. In practice, we measure P by adjusting the height of the bottle so that the levels of the water inside and outside the bottle are equal (Fig. 5.18). Then we know that the total pressure inside the bottle is equal to the atmospheric pressure, which we measure with a barometer.

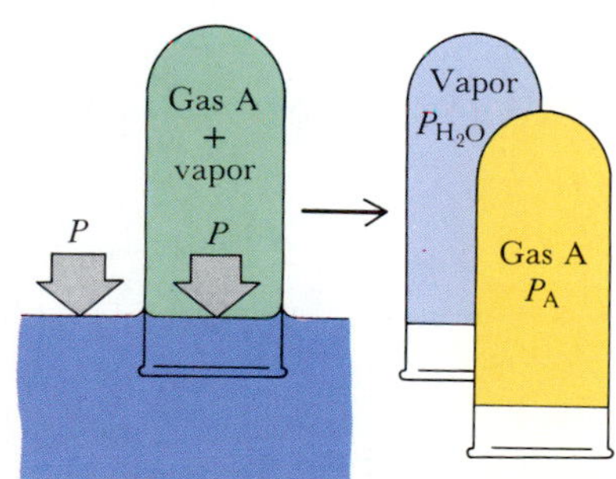

FIGURE 5.18 The pressure of gas A can be measured by equalizing the heights of water inside and outside the bottle; at that height the total pressure inside is equal to the atmospheric pressure. If the vapor pressure of the water is subtracted from the atmospheric pressure, then the difference is the partial pressure of the gas itself.

TABLE 5.4 The vapor pressure of water

Temperature, °C	Vapor pressure, Torr
0	4.58
10	9.21
20	17.54
21	18.65
22	19.83
23	21.07
24	22.38
25	23.76
30	31.83
37*	47.08
40	55.34
60	149.44
80	355.26
100	760.00

*Body temperature.

T OHT: Table 5.4.

N Plot the data in Table 5.4.

U Exercises 5.86–5.88.

D Collect a gas over water (if it is not done in the laboratory).

Because the total pressure (P) of the gas inside the bottle is the sum of the partial pressures of the oxygen (P_{O_2}) and the water vapor (P_{H_2O}),

$$P_{O_2} = P - P_{H_2O}$$

The partial pressure of water vapor above liquid water is called its vapor pressure (a property that we explore in much more detail in Section 10.4). For the present all we need know is that the vapor pressure of water depends on the temperature (see Table 5.4).

▼ **EXAMPLE 5.11 Measuring the amount of gas produced by a reaction**

A sample of potassium chlorate was heated with a little manganese dioxide as catalyst on a day when the atmospheric pressure was 755.2 Torr, and 370 mL of oxygen was collected in a gas bottle over water at 20°C. How many moles of O_2 were produced?

STRATEGY We can calculate the number of moles of O_2 from the ideal gas law once we know the partial pressure, the volume, and the temperature of the gaseous oxygen. The volume and temperature are given. According to Dalton's law, the partial pressure of oxygen is the difference between the total pressure and the vapor pressure of water, which can be obtained from Table 5.4.

SOLUTION At 20°C, the vapor pressure of water is 17.5 Torr. The partial pressure of oxygen in the bottle is therefore

$$P_{O_2} = (755.2 - 17.5) \text{ Torr} = 737.7 \text{ Torr}$$

Because we are using $R = 0.08206$ L·atm/(K·mol), we need this pressure in atmospheres:

$$P_{O_2} = 737.7 \text{ Torr} \times \frac{1 \text{ atm}}{760 \text{ Torr}} = 0.971 \text{ atm}$$

The volume occupied by the gases is

$$V_{O_2} = 0.370 \text{ L}$$

and the temperature is

$$T = (20 + 273) \text{ K} = 293 \text{ K}$$

The amount of O_2 produced is therefore

$$n_{O_2} = \frac{P_{O_2}V}{RT} = \frac{0.971 \text{ atm} \times 0.370 \text{ L}}{\left(0.08206 \frac{\text{L·atm}}{\text{K·mol}}\right) \times 293 \text{ K}} = 1.49 \times 10^{-2} \text{ mol}$$

That is, 14.9 mmol of O_2 is produced.

EXERCISE E5.11 When a small piece of zinc was dissolved in dilute hydrochloric acid, 446 mL of hydrogen was collected over water at 24°C and 760 Torr. Calculate the amount of H_2 produced.

[*Answer*: 17.8 mmol H_2]

▲

THE KINETIC THEORY OF GASES

A gas is a collection of molecules that are in continuous, chaotic motion and, except during collisions, are widely separated from one another. A molecule may be traveling through space one moment, but a fraction of a second later it collides with another. After the collision, it travels off in a different direction, only to collide almost at once with yet another molecule. This description is the basis of the **kinetic theory of gases** (*kinetic* is from the Greek word for "move").

The kinetic theory is based on three assumptions:

1. A gas consists of a collection of molecules in continuous random motion.
2. Molecules are infinitely small pointlike particles that move in straight lines until they collide.
3. Molecules do not influence one another except during collisions.

The third assumption means that there are no attractive or repulsive forces between ideal-gas molecules. The average distance that a molecule travels between collisions is called the **mean free path**. For a gas like air at 1 atm, the mean free path is about 70 nm, or about 200 molecular diameters. For a molecule magnified to the size of a tennis ball, the mean free path would be about 15 m (nearly as long as a tennis court).

Despite the simplicity of the model, the theory leads to a number of important relations between the properties of the gas as a whole and the properties of its molecules. We shall concentrate on the relation between the speeds of molecules and the temperature of the gas, for we shall need this information when (in Chapter 12) we discuss the rates at which reactions occur in gases, including reactions that take place in the atmosphere.

? Are molecules in fact pointlike particles?

? Do real molecules influence each other when they are at short distances from one another?

? How does the kinetic theory of gases account for Boyle's and Charles's laws?

N Do the particles of an ideal gas collide. *J. Chem. Ed.* **1974**, *51*, 141.

N Boyle's law—a different view (and the kinetic theory of gases). *J. Chem. Ed.* **1982**, *59*, 827.

5.6 MOLECULAR SPEEDS

Gas molecules travel at about the speed of sound. The reasoning behind this assertion is that sound is a wave of alternating high and low pressure carried through a substance by its molecules (Fig. 5.19). Because the motion of the pressure wave depends on the speed at which molecules can move into new locations, we can expect the speed of sound in a gas to be close to the average speed of its molecules. The speed of sound in air is about 300 m/s (about 700 mi/h), so we can expect average molecular speeds to be close to that value.

N Velocity and energy distributions in gases. *J. Chem. Ed.* **1982**, *59*, 193.

Speed and temperature. The kinetic theory of gases leads to the conclusion that the pressure and volume of a gas are related by the equation

$$PV = \tfrac{1}{3}nMv^2$$

where n is the amount of gas molecules, M is their molar mass, and v is their average speed.* However, we also know that $PV = nRT$. Hence, we can equate the two expressions, obtaining

$$\tfrac{1}{3}nMv^2 = nRT$$

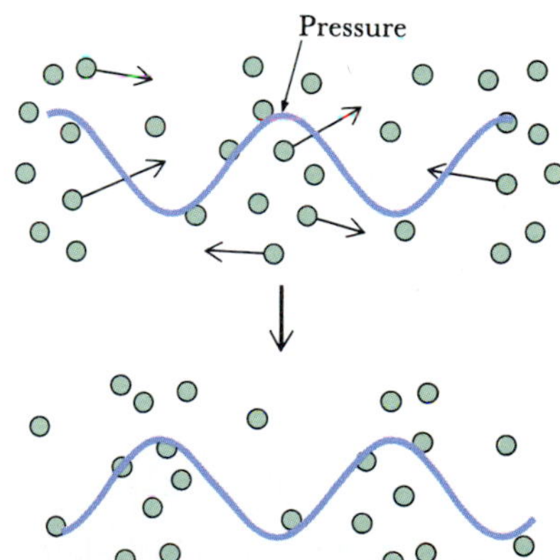

FIGURE 5.19 A sound wave is a wave (blue) of alternating high- and low-pressure regions traveling through a gas. The speed at which the wave travels—the speed of sound in the gas—is greater if the individual molecules can move quickly.

*More accurately, v is the root mean square speed (the rms speed). This quantity is defined as the square root of the average value of the squares of the speeds of all the molecules in the sample: $v = \sqrt{(v_1^2 + v_2^2 + \cdots + v_N^2)/N}$, where N is the number of molecules, v_1 is the speed of molecule 1, v_2 is the speed of molecule 2, and so on.

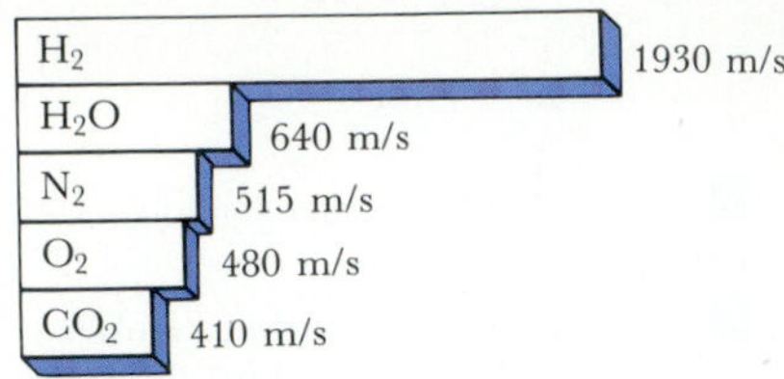

FIGURE 5.20 The average speeds of H_2 and of the molecules in air at 25°C. Hydrogen is included to emphasize how much faster light molecules travel than heavy ones.

T OHT: Fig. 5.20.

U Exercise 5.91.

D Heat water to boiling in a can (for example, a duplicating fluid can), seal the can, and then remove the heat source.

Solving the equation for the average speed v gives

$$v = \sqrt{\frac{3RT}{M}} \tag{4}$$

To calculate the average speed of nitrogen molecules in air at 25°C we substitute the appropriate temperature and molar mass (28.02 g/mol N_2) in the formula. However, it is important to use R in units that result in a speed in meters per second, so we take the value in SI units, $R = 8.314$ J/(K·mol), from inside the back cover, where J stands for kg·m^2/s^2 (as we explain in more detail in Chapter 6). When we substitute the data into Eq. 4, we find $v = 515$ m/s.

Equation 4 shows that the average speed of gas molecules increases as the temperature is raised. In particular, it shows that their average speed is proportional to the square root of the absolute temperature: this relation we write as

$$v \propto \sqrt{T}$$

Doubling the temperature of *any* gas (from 200 to 400 K, for example) increases the average speed of its molecules by a factor of $\sqrt{2} = 1.4$.

Speed and molar mass. Equation 4 also shows that, for a given temperature, the greater the molar mass of the gas molecules, the lower their average speed. Thus, the average speed of carbon dioxide molecules in the atmosphere is slower than the average speeds of the oxygen and nitrogen molecules in air. Specifically, the equation shows that the average speed of molecules in a gas is *inversely* proportional to the *square root* of the molar mass. This relation we write as

$$v \propto \frac{1}{\sqrt{M}}$$

For example, the ratio of the speeds of carbon dioxide and nitrogen molecules at the same temperature are

$$\frac{v(CO_2)}{v(H_2O)} = \sqrt{\frac{\text{molar mass of } H_2O}{\text{molar mass of } CO_2}} = \sqrt{\frac{18.02 \text{ g/mol}}{44.01 \text{ g/mol}}} = 0.6399$$

At any given temperature, CO_2 molecules have an average speed only 64% that of H_2O molecules. A sample of damp air at 25°C can therefore be pictured as a storm of molecules, with the CO_2 molecules, the heaviest molecules present, lumbering along at about 410 m/s and H_2O molecules, the lightest molecules present, zipping around at 640 m/s or so (Fig. 5.20).

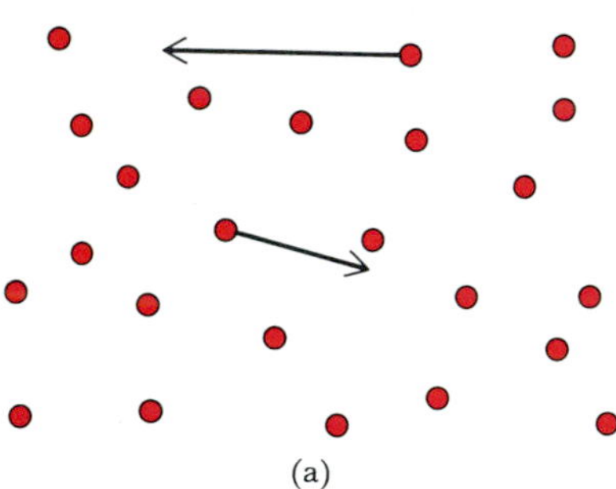

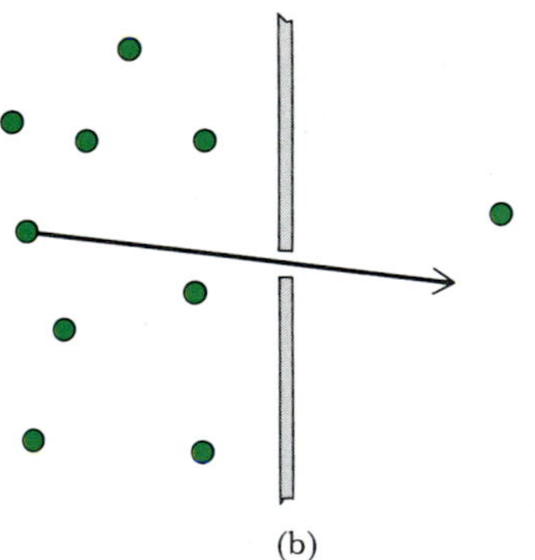

FIGURE 5.21 (a) In diffusion the molecules of one substance spread into the region occupied by the other. (b) In effusion the molecules of gas escape through a small hole in a barrier. In both cases the rate of the phenomenon increases with increasing temperature and decreases with increasing molar mass.

Diffusion and effusion. The spreading of one substance through another substance is called **diffusion** (Fig. 5.21a). The escape of one substance (particularly a gas) through a small hole is called **effusion** (Fig. 5.21b).

A solid diffuses into a liquid as it dissolves, and one solid might also diffuse slowly into another where they are in contact. Here we shall concentrate on the diffusion of one gas into another because all gases mix freely with one another and the molecules travel so quickly that diffusion is relatively fast. Diffusion occurs whenever two or more gases are introduced into the same region of space. It takes place because both types of molecules move chaotically and quickly mingle together. Effusion occurs whenever a gas is separated from a region of lower pressure by a porous barrier, which contains microscopic holes.

The diffusion of one gas into another is important for the transport of pheromones (chemical signals between animals) and perfumes through air. It helps to keep the composition of the atmosphere approximately constant, because abnormally high concentrations of one gas diffuse away and disperse. Normally wind is a bigger factor, but even in the absence of wind, diffusion helps to disperse the substances responsible for the aromas of food, flowers, and perfumes.

D Kinetic molecular theory and the movement of molecules and ions. TDC, 153.

N Graham's law: defining gas velocities. *J. Chem. Ed.* **1990**, *67*, 871.

The rates of diffusion and effusion of a gas depend on the speed of its molecules, for the faster the molecules move, the more quickly the gases mingle or escape through a small hole. We have seen that, at the same temperature, molecules with small molar mass travel most rapidly. Therefore, the rates of diffusion and effusion should show the same dependence. The dependence on molar mass was confirmed for effusion by Thomas Graham, a nineteenth-century Scottish chemist, and is now known as Graham's law:

> **Graham's law of effusion**: The rate of effusion of a gas is inversely proportional to the square root of its molar mass.

The diffusion of a gas also follows an *approximate* inverse square root dependence of the same kind, with heavy molecules diffusing more slowly than light molecules (Fig. 5.22).

D Diffusion of gases. CD1, 8.

Graham's law has been used to measure the molar masses of volatile compounds, but more accurate methods (particularly mass spectrometry; see Box 2.2) are now available. Because the rate of effusion is inversely proportional to the time a given amount of gas takes to escape, Graham's law implies that the time t required for diffusion is *directly* proportional to the square root of the molar mass:

$$t \propto \sqrt{M}$$

Heavy molecules move more slowly than light molecules do, so the heavier ones take longer to escape through a hole. The ratio of the times it takes two different gases A and B to effuse when they are at the same temperature and pressure is therefore

$$\frac{t_A}{t_B} = \sqrt{\frac{\text{molar mass of A}}{\text{molar mass of B}}} \qquad (5)$$

E The mutual diffusion of gases is a more complex problem than effusion because the two types of molecule move with different average speeds and each gas spreads into the other gas.

V Video 16: The rate of diffusion of gases.

? Would you expect ideal or real gases to effuse and diffuse more rapidly?

▼ EXAMPLE 5.12 Measuring molar mass by using Graham's law

Chlorofluorocarbon (abbreviated CFC) gases and volatile liquids sold commercially as Freons are still used in refrigeration units. A chemist found that a certain amount of a CFC gas required 186 s to effuse through a porous

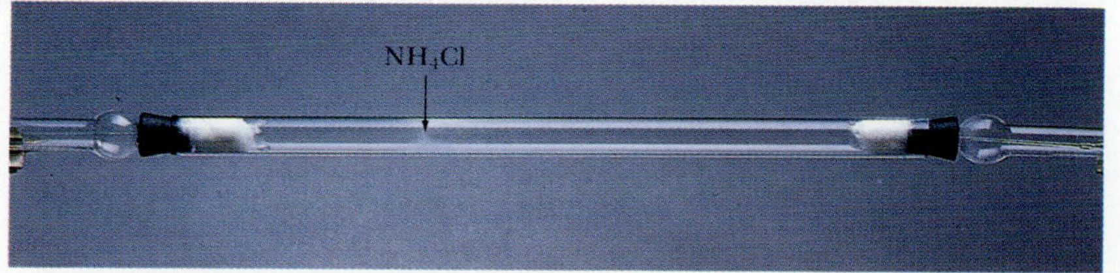

FIGURE 5.22 The plug on the left is soaked in hydrochloric acid and that on the right in aqueous ammonia. Formation of NH_4Cl occurs where gaseous HCl and NH_3 meet. The reaction occurs closer to the HCl plug because HCl has the greater molar mass and thus has more slowly diffusing molecules.

FIGURE 5.23 The individual diffusion stages in the original uranium-235 diffusion plant at Oak Ridge, Tennessee. There are thousands of stages in the entire plant. Note the size of the components relative to that of the technician.

plug when the pressure was held constant. The same amount of carbon dioxide required 112 s. Calculate the molar mass of the chlorofluorocarbon.

STRATEGY Because the CFC gas takes longer to effuse than the same amount of carbon dioxide does, we can guess that its molar mass is greater than that of CO_2 (44 g/mol). The data gives the time for carbon dioxide to effuse, and its molar mass is known; so the only unknown is the molar mass of the chlorofluorocarbon. This quantity is obtained by rearranging Eq. 5 to

$$\text{Molar mass of CFC} = \text{molar mass of } CO_2 \times \left(\frac{t_{CFC}}{t_{CO_2}}\right)^2$$

and substituting the data.

SOLUTION The rearranged equation gives

$$\text{Molar mass of CFC} = 44.01 \text{ g/mol} \times \left(\frac{186 \text{ s}}{112 \text{ s}}\right)^2 = 121 \text{ g/mol}$$

The value is consistent with the molecular formula CCl_2F_2 for this chlorofluorocarbon, which is sold as Freon-12.

EXERCISE E5.12 Carvone is the flavor component of spearmint. The time taken for a certain amount of carvone molecules to effuse through a porous plug was 186 s. Under the same conditions, the same amount of argon atoms effused in 96 s. Calculate the molar mass of carvone.

[*Answer*: 150 g/mol]

U Exercise 5.89.

E There are several advantages in using UF_6: they include the volatility of the substance (a consequence of the low polarizability of the F atoms and the nonpolar character of the octahedral molecule), and the fact that ^{19}F occurs in 100% natural abundance, so the isotopic composition of the F atoms does not interfere with the separation procedure. The large-scale use of fluorine in this process has led to fluorine's wider availability and hence, indirectly, to the availability of chlorofluorocarbons and of polytetrafluoroethylene plastics.

T OHT: Fig. 5.24.

N Visualizing Boltzmann-like distributions. *J. Chem. Ed.* **1988**, *65*, 24.

Graham's law explains why some isotope separation plants are so large and difficult to hide (a factor that is important when foreign powers want to monitor each other's nuclear capabilities). Nuclear power generation depends on our ability to separate uranium-235 from the much more abundant uranium-238. One process uses a series of reactions to convert uranium ore into the volatile solid uranium(VI) fluoride. The UF_6 vapor is then allowed to effuse through a series of porous barriers. The UF_6 molecules containing uranium-235, which are lighter than those containing uranium-238, effuse more quickly and thus may be separated from the rest. However, the ratio of the times that the same amounts of $^{235}UF_6$ and $^{238}UF_6$ take to effuse is only 1.004, so very little separation occurs at each stage. To improve the separation, the vapor is passed through many effusion stages; consequently, the plants are very large. The original plant in Oak Ridge, Tennessee (called, somewhat loosely, the gaseous diffusion plant) uses 4000 stages and covers an area of 43 acres (Fig. 5.23).

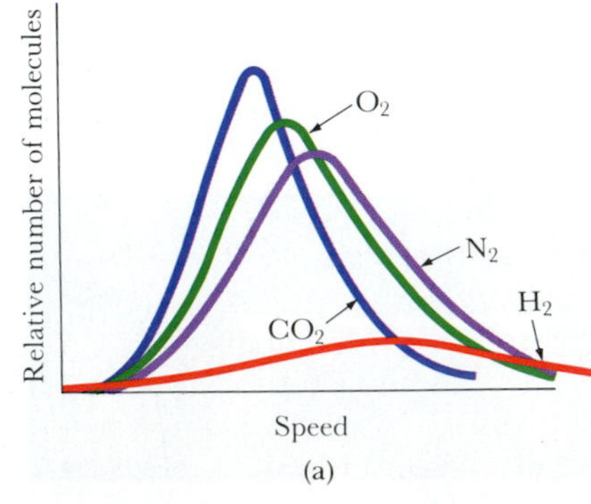

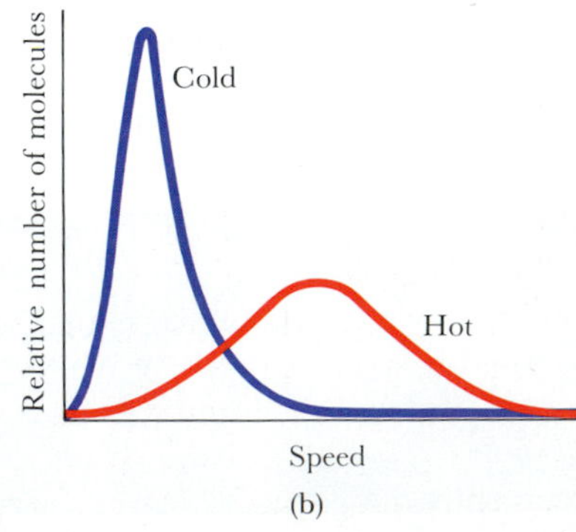

FIGURE 5.24 (a) The range of molecular speeds for several gases, as given by the Maxwell distribution. All the curves correspond to the same temperature. The greater the molar mass, the narrower the spread of speeds. (b) The Maxwell distribution again, but now the curves correspond to the speeds of a single substance at different temperatures. The higher the temperature, the broader the spread of speeds.

The Maxwell distribution of speeds. Only the *average* speeds of gas molecules are given by Eq. 4. Like cars in traffic, individual molecules have speeds that vary over a wide range. Moreover, as in a traffic collision, a molecule might be brought almost to a standstill when it collides with another. Then, in the next instant, it might be struck to one side by another and move off at the speed of sound. An individual molecule undergoes several billion changes of speed each second.

The percentage of gas molecules moving at each speed at any instant is called the **distribution of molecular speeds**. The general formula was first calculated by the Scottish scientist James Maxwell, whose conclusions are summarized in Fig. 5.24. The curves in Fig. 5.24a show that most of the heaviest molecules (CO_2) travel with speeds close to their average value. Light molecules (H_2) not only have a higher average speed, but the speeds of many of them are very different from their average speed. This wide range of speeds implies that many light molecules are likely to have such high speeds that they escape from the gravitational pull of small planets and go off into space. Consequently, hydrogen molecules and helium atoms, which are both very light, are very rare in the Earth's atmosphere.

The curves in Fig. 5.24b show that the spread of speeds widens as the temperature increases. At low temperatures most molecules have speeds close to their average speed. At high temperatures a high proportion have speeds widely different from their average speed. This observation has important consequences for the rates of chemical reactions, as we shall see in Chapter 12.

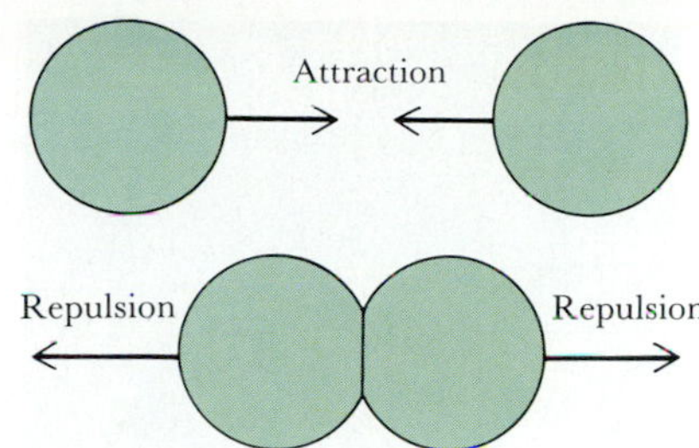

FIGURE 5.25 Molecules attract each other over several molecular diameters. However, as soon as they come into contact, they repel each other strongly.

? What are the differences (in molecular terms, and in behavior) of ideal and real gases?

? Under what conditions does a real gas approach ideal behavior?

5.7 REAL GASES

We remarked in Section 5.2 that real gases do not behave exactly as predicted by the ideal gas law. The deviations arise from **intermolecular forces**, the attractions and repulsions between molecules (Fig. 5.25). These interactions become important at high pressures, when the molecules are closer together. The evidence for attraction is the existence of liquids and solids, which consist of molecules held together by the attractions between them. The evidence for repulsion between molecules that are very close together is the difficulty with which liquids and solids are compressed, which suggests that their molecules resist being pressed together. The effect of intermolecular forces on the properties of substances will be covered more fully in Chapter 10.

Many suggestions have been proposed for changing the ideal gas law to accommodate the effect of intermolecular forces. One of the earliest and most useful of these improved equations of state was proposed by Johannes van der Waals, a nineteenth-century Dutch scientist:

$$\textbf{Van der Waals equation: } \left(P + \frac{an^2}{V^2}\right) \times (V - nb) = nRT \qquad (6)$$

The constant a represents the effect of attractions, and the constant b is a measure of the volume taken up by the gas molecules themselves; it is zero in an ideal gas, in which the pointlike molecules have zero volume. A few experimental values for a and b are listed in Table 5.5. The two constants are measured by adjusting their values until the van der Waals equation fits the observed dependence of the pressure

TABLE 5.5 Van der Waals constants for common gases

	a, $L^2 \cdot atm/mol^2$	b, L/mol
air	1.4	0.039
ammonia	4.17	0.037
argon	1.35	0.032
carbon dioxide	3.59	0.043
ethylene	4.47	0.057
helium	0.034	0.024
hydrogen	0.244	0.027
nitrogen	1.39	0.039
oxygen	1.36	0.032

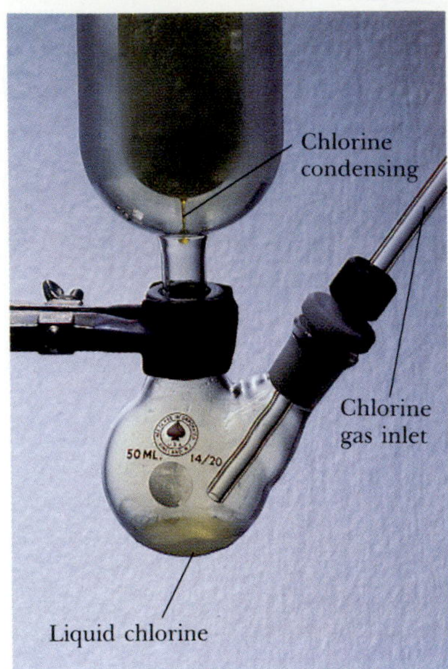

FIGURE 5.26 Chlorine can be liquefied at atmospheric pressure by cooling it to −35°C or lower.

U Exercise 5.92.

D Discharge a CO_2 fire extinguisher and account for the formation of dry ice.

D Simple demonstrations of the liquefaction of gases. *J. Chem. Ed.* **1986**, *63*, 436.

E Athlete trainers spray Freon on to a strain or sprain.

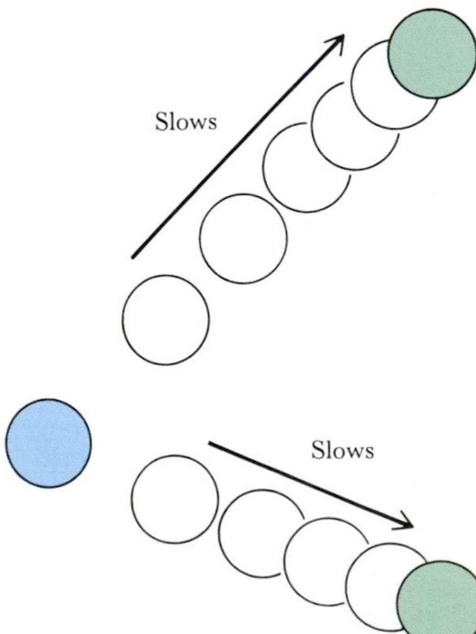

FIGURE 5.27 Cooling by the Joule-Thomson effect can be visualized as a slowing of the molecules as they climb away from each others' attractions.

on the volume, temperature, and amount of the gas present. The values of a and b differ from gas to gas because the sizes of molecules differ (which affects b), and their ability to attract each other differs too (which affects a).

The volume occupied by a gas is very large at very low pressures. Then $V - nb$ is virtually identical to V, and an^2/V is so much smaller than P that it can be neglected in the first term in parentheses in Eq. 5. That is, at very low pressures, a good approximation to the van der Waals equation is

$$P \times V = nRT$$

which is the ideal gas law. In other words, a real gas that obeys the van der Waals equation at high pressures, increasingly obeys the ideal gas law as the pressure is reduced to low values. Real gases behave as ideal gases at very low pressures. They also behave ideally at high temperatures, when the molecules move so rapidly that intermolecular forces have little effect on their motion.

5.8 THE LIQUEFACTION OF GASES

At low temperatures, gas molecules move so slowly that intermolecular attractions may cause one molecule to be captured by another and stick to it instead of moving freely. When the temperature is reduced to below the boiling point of the substance, the gas condenses to a liquid: the molecules move too slowly to escape from one another, and the entire sample condenses to a jostling crowd of molecules held together by the attractions between them.

The simplest method of liquefying a gas is to immerse a sample in a bath kept at a temperature lower than the boiling point of the substance. Chlorine (boiling point −35°C) condenses to a liquid when the gas comes into contact with a surface cooled by solid carbon dioxide (Fig. 5.26). Temperatures as low as about −77°C can be reached by adding chips of solid carbon dioxide to a liquid with a low freezing point, such as acetone.

Gases can also be liquefied by making use of the relation between temperature and molecular speed. Because lower average speed corresponds to lower temperature, slowing the molecules is equivalent to cooling the gas. Molecules can be slowed by making use of the attractions between them exactly as the gravitational pull of the Earth slows a ball that has been thrown up into the air. An expanding gas is like a vast number of balls separating from a vast number of Earths; and although the attractions are not gravitational, they have the same effect: increasing the average separation of molecules in a gas lowers their average speed (Fig. 5.27). In other words, as a gas expands it cools. This is called the **Joule-Thomson effect** in honor of James Joule and William Thomson (later to become Lord Kelvin and known for his temperature scale), who first studied it. (There are some exceptions: under some circumstances a gas might become hotter as it expands.)

The Joule-Thomson effect is used in some commercial refrigerators to liquefy gases. The gas to be liquefied is compressed and then allowed to expand through a small hole, called the throttle. The gas cools as it expands, and the cooled gas circulates past the incoming compressed gas. This contact cools the incoming gas before it expands and cools still further. As the gas is continually recompressed and recirculated, its temperature progressively falls until finally it condenses to liquid. If the gas is a mixture, like air, then the liquid it forms can be fractionally distilled to separate its components. This technique is used for harvesting nitrogen, oxygen, neon, argon, krypton, and xenon from the atmosphere.

E Joule-Thomson cooling occurs below the *inversion temperature* of a gas. Above that temperature, expansion results in heating. Some gases (particularly hydrogen and helium) must first be cooled to below the inversion temperature by some other technique before Joule-Thomson cooling can occur.

SUMMARY

5.1 Gases exert a **pressure**, or force per unit area, which is measured by using a barometer. Gas pressure arises from the impact of gas molecules on surfaces.

5.2 **Boyle's law** ($V \propto 1/P$) summarizes the effect of pressure on the volume of a gas and **Charles's law** ($V \propto T$) summarizes the effect of temperature on the volume. Both laws are approximations at normal temperatures and pressures but are obeyed increasingly well as the pressure is reduced. The relation between the volume and the amount of molecules in a gaseous sample is summarized by **Avogadro's principle** ($V \propto n$). The gas laws are summarized by the **ideal gas law**, which is an example of an **equation of state**, a relation between pressure, volume, temperature, and amount of substance. As the pressure approaches zero, all gases obey the ideal gas law. A gas that obeys the ideal gas law is called an **ideal gas**. The **standard temperature and pressure** for reporting data is 0°C (273.15 K) and 1 atm (101.325 kPa).

5.3 Applications of the ideal gas law include the calculation of the pressure of a gas from the amount of molecules present and of the density of a gas, and the determination of molar mass.

5.4 At the standard temperature and pressure, all gases have very nearly the same **molar volume**, the volume per mole. The molar volume of an ideal gas can be used to relate the number of moles of gas molecules in a sample to the volume it occupies. It can also be used to extend the stoichiometry calculations of Chapter 4 to the calculation of the volumes of gases taking part in reactions. When volumes at other temperatures and pressures are needed, they are obtained by using the combined gas law or the ideal gas law.

5.5 Mixtures of nonreacting gases are described by **Dalton's law of partial pressures**, which treats each gas in a mixture as though it alone were present in the container. The **partial pressure** of a gas in a mixture is the pressure that a gas would exert if it alone occupied the container.

5.6 The **kinetic theory of gases** supposes that the molecules of a gas are in chaotic random motion, have negligible size, and move independently in straight lines except during collisions. The theory leads to the conclusion that the average speed of molecules in a gas is directly proportional to the square root of the temperature and inversely proportional to the square root of their molar mass. The mobility of gas molecules accounts for the rapidity with which they **diffuse**, or spread, through substances and the rate at which they **effuse**, or escape through a small hole. The kinetic theory accounts for **Graham's law**, that the rate of effusion of a gas is inversely proportional to the square root of the molar mass. The proportion of molecules having a particular speed at any temperature is given by the **Maxwell distribution**, which is summarized in Fig. 5.24.

5.7 Real gases behave differently from an ideal gas, especially at high pressures. The behavior of a real gas is described by the **van der Waals equation**, which takes into account the effects of the **intermolecular forces**—the attractions and repulsions between molecules.

5.8 Gases are liquefied either by cooling in a cold bath or by making use of the **Joule-Thomson effect**, the cooling caused by expansion.

CLASSIFIED EXERCISES

Pressure

5.1 Gas pressures are expressed in different units, depending on the application. Make the following conversions:
(a) 1.00 bar = ________ atm
(b) 1.00 mmHg = ________ Pa
(c) 1.00 kPa = ________ atm

5.2 Several different units of pressure will be used in these exercises. Complete the following conversions:
(a) 725 Torr = ________ atm
(b) 150 kPa = ________ atm
(c) 10.7 mmHg = ________ Torr

5.3 A chemist noted that the pressure exerted by a sample of water vapor at 200°C was 1268 Torr. What is this pressure in kilopascals?

5.4 The pressure needed to make synthetic diamonds from graphite is 80,000 atm (1 sf). Express this pressure in (a) kilobars; (b) pascals.

5.5 Suppose you were marooned on a tropical island and had to make a primitive barometer using seawater (density, 1.10 g/mL). What height would the water reach in your barometer when a mercury barometer would reach 77.5 cm? The density of mercury is 13.6 g/cm^3.

5.6 Assume that the width of your body (across your shoulders) is 20 inches and the depth of your body (chest to back) is 10 inches. If atmospheric pressure is 14.7 lb/in^2, what mass of air does your body support when you are in an upright position?

The Ideal Gas

5.7 Calculate the final volume when (a) the pressure on 1.0 L H_2 changes from 2.20 kPa to 3.00 atm; (b) the pressure on 25.0 mL of CO_2 changes from 200 Torr to 0.500 atm. Assume that the temperature is constant.

5.8 (a) A helium balloon had a volume of 22 L at sea level where the atmospheric pressure is 0.951 atm. When the balloon rose to a height at which the atmospheric pressure is 73.3 kPa, the balloon burst. What was the volume of the balloon just before it burst? Assume a constant temperature. (b) A 1.0-mm^3 bubble at the bottom of a lake where the pressure is 3.5 atm rises to the surface. What is the volume of the bubble when it hits the top where the atmospheric pressure is 780 Torr?

5.9 (a) A 300-m^3 storage vessel for hydrogen gas is at 1.5 atm and 10°C at 2 A.M. By 2 P.M., the temperature has risen to 30°C. What is the new pressure of the hydrogen gas in the vessel? (b) A chemist prepared a sample of hydrogen bromide and found that it occupied 255 mL at 85°C and 600 Torr. What volume would it occupy at 0°C at the same pressure?

5.10 (a) A 250-mL aerosol can at 25°C and 1.10 atm was thrown into an incinerator where the temperature was 600°C. What pressure does the aerosol in the can withstand before it explodes? (b) A helium balloon has a volume of 22 L at sea level where the pressure is 0.951 atm and the temperature is 18°C. The balloon rises at a constant pressure until the temperature is −15°C. What is the volume of the balloon at this stage?

5.11 A sample of 150 mL of nitrogen was stored at 50 kPa and 16°C. If the volume of the sample is to be reduced to 65 mL, what must be the final temperature, assuming no pressure change?

5.12 A 50-mL glass vial of xenon gas at 23°C and 130 kPa was placed in the refrigerator, where the temperature was 3.6°C. What is the pressure of the chilled xenon gas in the vial?

5.13 The pressure in an automobile tire is 30 lb/in^2 at 20°C. The automobile is driven for several hundred kilometers and pressure in the tire then reads 34 lb/in^2. Assuming a constant volume and no leaks in the tire, calculate the new temperature of the air in the tire?

5.14 A gas relief valve on an industrial storage tank operates if the pressure of the carbon dioxide gas exceeds 115 atm. In December, the tank was filled with carbon dioxide at 100 atm pressure when the ambient temperature was −10°C. On a hot summer day, the temperature rose to 35°C. Should the gas relief valve have operated?

5.15 In a certain chemical process, 2500 L of sulfur dioxide gas is fed from a storage tank at 2000 kPa and 20°C into a 6000-L chemical reactor at 150°C. What is the pressure of the SO_2 just prior to reaction?

5.16 A lung-full of air (350 cm^3) is exhaled into a machine that measures lung capacity. If the air is exhaled from the lungs at a pressure of 1.08 atm at 37°C, but the machine is at ambient conditions of 0.958 atm and 23°C, what is the volume of air measured by the machine?

5.17 A 20-mL sample of xenon exerts a pressure of 0.48 atm at −15°C. (a) What volume does the sample occupy at STP? (b) What pressure would it exert if it were transferred to a 12-mL flask at 20°C? (c) Calculate the temperature needed for the xenon to exert a pressure of 500 Torr in the 12-mL flask.

5.18 The volume of a sample of sulfur hexafluoride gas at STP is 150 μL. (a) What volume does the gas occupy at 180 Torr and 17°C? (b) At what temperature will it occupy 217 μL at 25 kPa? (c) If the sample is stored at a temperature of 9°C in the 217-μL vessel, what will be its pressure (in kilopascals)?

Molar Volumes

5.19 Assume ideal gas behavior and calculate the volume at STP occupied by (a) 33.9 μg of SiF_4; (b) 0.572 mg of Xe; (c) 3.55 g of air (average molar mass, 28.97 g/mol).

5.20 Calculate the number of molecules present in (a) 6.99 nL of $TiCl_4$; (b) a spherical flask of ammonia with a diameter of 10 m ($V = \frac{4}{3}\pi r^3$); (c) an ampoule of krypton with a diameter of 1.4 mm and a length of 16 mm ($V = \pi r^2 l$). All measurements are at STP.

The Ideal Gas Law

5.21 (a) A 120-mL flask contains 2.7 μg of O_2 at 17°C. What is the pressure (in Torr)? (b) A 16.7-g sample of krypton exerts a pressure of 100 mTorr at 44°C. What is the volume of the container (in liters)? (c) A 2.6-μL ampoule of xenon has a pressure of 2.00 Torr at 15°C. How many Xe atoms are present?

5.22 (a) A 23.9-mg sample of bromine trifluoride exerts a pressure of 10 Torr at 100°C. What is the volume of the container (in milliliters)? (b) A 100-mL flask contains sulfur dioxide at 0.77 atm and 30°C. What mass of gas is present? (c) A 6000-m^3 storage tank contains methane at 129 kPa and 15°C. What amount of CH_4 is present (in moles)?

5.23 What mass of ammonia will exert the same pressure as 12 mg of hydrogen sulfide, H_2S, in the same container under the same conditions?

5.24 A 2.00-mg sample of argon is confined to a 50-mL vial at 20°C; 2.00 mg of krypton is also confined to a 50-mL vial. What must the temperature of the krypton be if it is to have the same pressure as the argon?

5.25 A sample of helium gas occupies a spherical tank measuring 12.0 inches in diameter and exerts a pressure of 100.0 lb/in^2 at 15.0°C. What mass of helium is in the tank? The volume of a sphere of radius r is $V = \frac{4}{3}\pi r^3$.

5.26 A neon sign consists of a piece of glass tubing that is 10.0 m long and 1.0 cm in diameter and filled with neon gas at a pressure of 1.2 Torr and 20°C. What mass of neon is in the tube. The volume of a tube of radius r and length l is $V = \pi r^2 l$.

Density and Molar Mass

5.27 What is the density (in g/L) of chloroform ($CHCl_3$) vapor at (a) STP; (b) 100.0°C and 1.00 atm?

5.28 The density of xenon is 5.84 g/L at STP. What is its density at −150°C and 0.50 atm?

5.29 A sample of 21.3 g of a gas is confined to a vessel of volume 7.73 L at 0.880 atm and 30°C. (a) What is the density of the gas at STP? (b) What is the molar mass of the gas?

5.30 A sample of halogen gas has a mass of 0.239 g and exerts a pressure of 600 Torr at 14°C in a 100-mL flask. (a) Calculate the density of the gas at STP. (b) What is the molar mass of the halogen? (c) Identify the halogen.

5.31 The density of a gaseous compound of phosphorus is 0.360 g/L at 420 K when its pressure is 96.9 kPa. (a) What is the molar mass of the compound? (b) What is the density of the gas at STP?

5.32 A gaseous fluorinated methane compound has a density of 8.0 g/L at 2.81 atm and 300 K. (a) What is the molar mass of the compound? (b) What is the density of the gas at STP?

5.33 The elemental analysis of a hydrocarbon revealed that it was 85.7% C and 14.3% H by mass. When 1.77 g of the gas was stored in a 1.500-L flask at 17°C, it exerted a pressure of 508 Torr. What is the molecular formula of the hydrocarbon?

5.34 A compound used in the manufacture of saran is 24.7% C, 2.1% H, and 73.2% Cl by mass. The storage of 3.557 g of the compound in a 750-mL vessel at 0°C results in a pressure of 1.10 atm. What is the molecular formula of the compound?

Stoichiometry of Reacting Gases

5.35 Oxygen gas is generated in the laboratory by the thermal decomposition of potassium chlorate. What volume of O_2 (at STP) is generated from 1.00 g of potassium chlorate in the reaction

$$2KClO_3(s) \xrightarrow{\Delta} 2KCl(s) + 3O_2(g)$$

5.36 Calculate the mass of ammonium nitrate that should be heated to obtain 100 mL of dinitrogen oxide, N_2O, at STP in the reaction

$$NH_4NO_3(s) \xrightarrow{\Delta} N_2O(g) + 2H_2O(g)$$

5.37 One industrial process for the removal of hydrogen sulfide from natural gas is its reaction with sulfur dioxide: $2H_2S(g) + SO_2(g) \rightarrow 3S(s) + 2H_2O(g)$. (a) What volume of sulfur dioxide at STP is needed to produce 1.00 kg of sulfur? (b) What volume of sulfur dioxide is needed in (a) if it is supplied at 5.0 atm and 250°C?

5.38 The Haber process for the synthesis of ammonia is one of the most significant industrial processes for the well-being of humanity. (a) What volume of hydrogen at STP must be supplied to produce 1.0 metric ton (1 metric ton = 10^3 kg) of NH_3? (b) What volume of hydrogen is needed in (a) if it is supplied at 200 atm and 400°C?

5.39 Urea, $CO(NH_2)_2$, is used as a fertilizer and is made by the reaction of carbon dioxide and ammonia: $CO_2(g) + 2NH_3(g) \rightarrow CO(NH_2)_2(s) + H_2O(g)$. What volumes of CO_2 and NH_3 at 200 atm and 450°C are needed to produce 2.50 kg urea?

5.40 The first stage in the production of nitric acid by the Ostwald process is the oxidation of ammonia:

$$4NH_3(g) + 5O_2(g) \longrightarrow 4NO(g) + 6H_2O(g)$$

(a) Calculate the mass of nitric oxide that can be produced from the reaction of 150 L of ammonia at 15.0 atm and 200°C with an excess of oxygen. (b) If the water produced in (a) is condensed to a liquid with a density of 1.0 g/mL, what volume will it occupy?

5.41 Nitroglycerin is a shock-sensitive liquid that detonates by the reaction

$$4C_3H_5(NO_3)_3(l) \longrightarrow 6N_2(g) + 10H_2O(l) + 12CO_2(g) + O_2(g)$$

Calculate the total volume of product gases at 150 kPa and 100°C from the detonation of 1.0 g of nitroglycerin.

5.42 Xenon and fluorine react at 350°C to produce a mixture of XeF_2, XeF_4, and XeF_6. What volumes of xenon and fluorine at 200 kPa and 350°C are needed to produce 1.00 μg of XeF_4, assuming a 100% yield in the reaction $Xe(g) + 2F_2(g) \rightarrow XeF_4(g)$.

Gaseous Mixtures

5.43 Calculate the partial pressure of each gas and the total pressure of the following mixtures, each of which occupies a 500-mL vessel at 0°C: (a) 0.020 mol of N_2 and

2.33 g of O_2; (b) 0.015 mol of H_2, 4.22 mg of He, and 0.030 mol of NH_3.

5.44 The following gases were placed in a 3500-mL flask at 25°C; (a) 0.0195 g and CH_4 and 0.0195 g of CO_2; (b) 1.00 mg of Ar, 2.00 mg of Kr, and 3.00 mg of Xe. Calculate the partial pressure of each gas and the total pressure.

5.45 A sample of damp air in a 1.00-L container exerts a pressure of 762.0 Torr at 20°C; but when it is cooled to −10.0°C, the pressure falls to 607.1 Torr as the water condenses. What mass of water was present?

5.46 A 2.00-L flask contains carbon dioxide at 20°C and 606.0 Torr. After 1.0 g of nitrogen is added to the flask and heated to 200°C, what is the expected pressure in the flask?

5.47 An apparatus consists of a 4.0-L flask containing nitrogen gas at 25°C and 80.3 kPa, joined by a tube to a 10.0-L flask containing argon gas at 25°C and 47.2 kPa. The stopcock on the connecting tube is opened and the gases mix. (a) What is the partial pressure of each gas after mixing? (b) What is the total pressure of the gas mixture?

5.48 The total pressure of a mixture of sulfur dioxide and nitrogen gases at 25°C in a 500-mL vessel is 1.09 atm. The mixture is passed over a warm calcium oxide powder, which removes the sulfur dioxide by the reaction $CaO(s) + SO_2(g) \longrightarrow CaSO_3(s)$, and is then transferred to a 150-mL vessel where the pressure is 1.09 atm at 50°C. (a) What was the partial pressure of the SO_2 in the initial mixture? (b) What is the mass of SO_2 in the initial mixture?

5.49 In the course of an electrolysis of water, 220 mL of hydrogen was collected over water at 20°C when the external pressure was 756.7 Torr. (a) What is the partial pressure of hydrogen? (b) What mass of oxygen was also produced in the electrolysis?

5.50 When a potassium chlorate sample was heated in the presence of MnO_2 (a catalyst for this reaction), 25.7 mL of $O_2(g)$ was collected over water at 14°C and 97.6 kPa. What was the mass of $KClO_3$ in the sample? The decomposition reaction is

$$2KClO_3(s) \xrightarrow{\Delta,\ MnO_2} 2KCl(s) + 3O_2(g)$$

5.51 A 37.6-mL sample of dry hydrogen gas was collected at 18°C over a nonvolatile liquid when the atmospheric pressure was 107 kPa. What volume would the hydrogen gas have occupied had it been collected over water at 18°C?

5.52 Dinitrogen oxide, N_2O, gas was generated from the thermal decomposition of ammonium nitrate and collected over water. The wet gas occupied 126 mL at 21°C when the atmospheric pressure was 755 Torr. What volume would the same amount of dry dinitrogen oxide have occupied if collected at 755 Torr and 21°C?

Molecular Speeds

5.53 Calculate the average speeds of (a) H_2 molecules at 0°C; (b) Xe atoms at 25°C; (c) He atoms at 25°C.

5.54 Calculate the average speeds of CO_2 molecules at 263 K, 273 K, 298 K, 323 K, 348 K, and 373 K, and then plot the speed against the temperature.

5.55 (a) By what factor does the speed of UF_6 molecules increase when the temperature is increased from 25°C to 100°C? (b) Calculate the relative speed of air molecules at room temperature (at 26°C) relative to their average speed in a freezer (at −20°C). The average molar mass of air in the troposphere is 28.97 g/mol.

5.56 (a) Determine the factor by which the speed of helium atoms increases when the temperature increases from 10K to 20K. (b) Calculate the ratio of the average speed of helium atoms on the Earth's surface (at 25°C) to their speed on the surface of the sun (at 6500°C).

5.57 A sample of argon gas effuses through a porous plug in 147 s. Calculate the time required for the same amount of the following gases to effuse under the same conditions of pressure and temperature: (a) CO_2; (b) C_2H_4; (c) SO_2.

5.58 A sample of CO_2 gas effused through a porous plug in 13.7 min. Determine the molar mass of the gases that required the following periods of time for the same amount of gas to effuse at the same conditions of temperature and pressure: (a) 23.1 min; (b) 7.33 min; (c) 3000 s.

5.59 What is the molar mass of a compound that takes 2.7 times longer to effuse through a porous plug than it did for the same amount of XeF_2 at the same temperature and pressure.

5.60 What is the molecular formula of a compound of empirical formula CH that diffuses 3.22 times more slowly than krypton at the same temperature and pressure.

5.61 The vapor of an oxide of phosphorus took 302 s to diffuse, whereas the same amount of CO_2 took only 135 s at the same conditions of temperature and pressure. Is the compound phosphorus(III) oxide or phosphorus(V) oxide?

5.62 One of the components of pineapple flavor is ethyl butyrate, which has the empirical formula C_3H_6O. Its vapor took 162 s to effuse through a porous plug; the same amount of CO_2 molecules effused through the plug in 100 s under the same conditions. What is the molecular formula of the compound?

5.63 A compound used to make polyvinyl chloride (PVC) has composition 38.4% C, 4.82% H, and 56.8% Cl by mass. It took 7.73 min for a given mass of the compound to effuse through a porous plug, but it took only 6.18 min for the same amount of Ar to diffuse at the same temperature and pressure. What is the molecular formula of the compound?

5.64 A compound that contains sulfur and has been used in chemical warfare has a composition 23.76% S, 23.71% O, and 52.54% Cl by mass. If it takes 46.0 s for this compound to effuse through the same porous plug through which the same amount of argon passed in 25.0 s, what is the molecular formula of the compound?

Real Gases

5.65 Use the van der Waals equation and Table 5.5 to calculate the pressure of (a) 10.0 g of carbon dioxide in a 500-mL container at 10°C; (b) 10.0 g of helium in a 250-mL vessel at 100°C. (c) Repeat (a) and (b), using the ideal gas law equation. (d) Determine the percentage error in assuming that carbon dioxide and helium behave as ideal gases under the conditions given in (a) and (b).

5.66 Use the van der Waals equation and Table 5.5 to calculate the pressure of (a) 10.0 g of ammonia in a 500-mL container at 25°C; (b) 10.0 g of hydrogen in a 250-mL vessel at 100°C. (c) Repeat (a) and (b), using the ideal gas law equation. (d) Determine the percentage error in assuming that ammonia and hydrogen behave as ideal gases under the same conditions given in (a) and (b).

UNCLASSIFIED EXERCISES

5.67 (a) The weather report states that the current atmospheric pressure is 28.92 inches of mercury. Convert this pressure to Torr and atmospheres. (b) The pressure of nitrogen gas in a steel cylinder is 59.5 kPa. What is the pressure in Torr?

5.68 Low-pressure gauges in research laboratories are occasionally calibrated in inches of water (inH_2O). Assuming that the density of mercury is 13.6 g/cm^3 and the density of water is 1.0 g/cm^3, calculate the pressure (in Torr) inside a gas cylinder that reads 1.5 inH_2O?

5.69 A certain four-cylinder automobile engine is said to be a 2.2-L engine. (a) What is the maximum volume (the displacement) of each cylinder? (b) If a gas-fuel mixture at 0.70 atm is injected into the cylinder at its maximum displacement and then compressed to a volume of 80 mL, what is the pressure of the mixture prior to ignition? Assume no temperature change. (c) Temperatures inside a cylinder can approach 2000°C. If a gas-fuel mixture at 50°C and 5.3 atm ignites in one of the cylinders, thereby causing an expansion of the gases in the cylinder by a factor of 7.5 while the temperature reaches 2000°C, what is the pressure of the mixture? Assume no change in the amount of gas from reactant to product (although that is not actually the case).

5.70 How much additional pressure would you need to exert on a sample of helium at 765 Torr to compress it from 555 mL to 125 mL at constant temperature?

5.71 A sample of chlorine was kept inside a container fitted with a fragile glass membrane that would break if the internal pressure rose above 225 kPa. If the pressure of the gas at 20°C was 755 Torr, what is the maximum temperature to which the chlorine can be heated before the membrane will break?

5.72 A natural gas well with an approximate volume of 2.0×10^9 L has a reservoir pressure of 1.20 atm and 40°C. If all the gas is to be transferred to a steel tank where the pressure is not to exceed 6.00 atm on a very cold day (−25°C), what must the volume of the tank be?

5.73 The Goodyear blimp has a volume of about 25×10^6 L. (a) What mass of helium is needed to fill the blimp with helium to a pressure of 780 Torr at 20°C? (b) The "lift" of the blimp is the difference between the mass of helium and the mass of air (molar mass, 28.97 g/mol) used to fill the blimp. What is the lift, in kilograms, of the blimp?

5.74 At the top of the troposphere, the temperature is about −50°C and the pressure is about 0.25 atm. If a balloon that is filled with helium to a volume of 10.0 L at 2.00 atm pressure and 27.0°C is released at ground level, what will the volume of the balloon be at the top of the troposphere?

5.75 A 2.10-mL bubble of methane gas forms at the bottom of a Louisiana swamp, where the temperature is 8.1°C and the pressure is 6.4 atm, and rises to the top of the swamp, where the temperature is 25°C and the pressure is 1.00 atm. What is the volume of the methane bubble before it bursts?

5.76 A 0.297-g sample of an unidentified gas occupies 250 mL at 300 K and 670 Torr. (a) What is the density of the gas at STP? (b) What is the molar mass of the gas?

5.77 Carbon monoxide gas is purchased in a 425-mL bottle at a pressure of 5.00 atm and 23°C. (a) What mass, in grams, of the gas has been purchased? (b) What is the density of the gas in the bottle?

5.78 A 0.473-g sample of a gas that occupies 200 mL at 1.81 atm and 25°C was analyzed and found to contain 0.414 g of N and 0.0591 g of H. What is the molecular formula of the compound?

5.79 A 1.509-g sample of an osmium oxide (which melts at 40°C and boils at 130°C) is placed in a cylinder with a movable piston that enables the cylinder to expand against the atmospheric pressure of 745 Torr. When the sample is heated to 200°C, it is completely vaporized and the volume of the cylinder expands by 235 mL. What is the molar mass of the oxide? Assuming that the formula of the oxide is OsO_x, what is the value of x?

5.80 When titanium tetrachloride is mixed with water, voluminous clouds of a white smoke of titanium dioxide particles are produced by the reaction

$$TiCl_4(g) + 2H_2O(l) \longrightarrow TiO_2(s) + 4HCl(g)$$

(a) What mass of TiO_2 is produced from the reaction of 200 L of $TiCl_4$, measured at 500 kPa and 30°C? (b) Determine the volume of hydrogen chloride gas produced at STP.

5.81 A 5-gal propane tank used for an outdoor barbecue grill is filled with 20 lb of liquid propane (density, 0.57 g/mL). All the liquid is burned in an excess of oxygen; the equation for the combustion is

$$C_3H_8(l) + 5O_2(g) \longrightarrow 3CO_2(g) + 4H_2O(g)$$

(a) Calculate the mass of carbon dioxide released to the atmosphere. (b) What pressure would the carbon dioxide exert if it were collected and stored in a second 5-gal tank at 16°C?

5.82 A 200-mL sample of hydrogen chloride at 690 Torr and 20°C is dissolved in 100 mL of water. The solution was titrated to the stoichiometric point with 15.7 mL of a sodium hydroxide solution. What is the molar concentration of the NaOH in the solution?

5.83 Through a series of enzymatic steps, carbon dioxide and water combine during photosynthesis to produce glucose and oxygen according to the equation

$$6CO_2(g) + 6H_2O(l) \longrightarrow C_6H_{12}O_6(s) + 6O_2(g)$$

Given that the partial pressure of carbon dioxide in the troposphere is 0.26 Torr and that the temperature is 25°C, calculate the volume of air that is needed to support the production of 10.0 g of glucose.

5.84 A 15-mL sample of ammonia gas at 100 Torr and 30°C is mixed with 25 mL of hydrogen chloride at 150 Torr and 25°C, and the reaction $NH_3(g) + HCl(g) \rightarrow NH_4Cl(s)$ occurs. (a) Calculate the mass of NH_4Cl that forms. (b) Identify the gas in excess and determine its pressure (in the combined volume of the original two flasks) at 27°C.

5.85 Iron pyrite, FeS_2, is the form in which much of the sulfur occurs in coal. In the combustion of the coal, oxygen reacts with iron pyrite to produce iron(III) oxide and sulfur dioxide:

$$4FeS_2(s) + 11O_2(g) \xrightarrow{\Delta} 2Fe_2O_3(s) + 8SO_2(g)$$

(a) Calculate the mass of Fe_2O_3 that is produced from the reaction of 75.0 L of oxygen at 2.33 atm and 150°C with an excess of iron pyrite. (b) If the sulfur dioxide that is generated in (a) is dissolved to form 5.00 L of aqueous solution, what is the molar concentration of the resulting sulfurous acid, H_2SO_3, solution?

5.86 Small quantities of hydrogen gas in the laboratory can be generated by the action of dilute hydrochloric acid on zinc metal. When 0.40 g of impure zinc was reacted with an excess of hydrochloric acid, 127 mL of hydrogen was collected over water at 17°C when the external pressure was 737.7 Torr. (a) What amount (in moles) of H_2 was collected? (b) What is the percentage purity of the zinc?

5.87 An adult takes about 15 breaths per minute, with each breath having a volume of 500 mL. If the air that is inhaled is "dry," but the exhaled air at 1 atm pressure is saturated with water vapor at 37°C (body temperature), what mass of water is lost from the body in 1 h? The vapor pressure of water at 37°C is 47.1 Torr.

5.88 The fermentation of the sugars in grain results in the formation of ethanol by the reaction $C_6H_{12}O_6(aq) \rightarrow 2C_2H_5OH(aq) + 2CO_2(g)$. If 1.8 g of ethanol is produced in a bottle of beer and all the carbon dioxide is collected in the 25-mL space above the liquid at 10°C, what is pressure of the gas? (Assume that carbon dioxide is insoluble in water, even though it is not. In practice, not all the fermentation occurs in the bottle, so the pressure calculated is not the actual pressure of carbon dioxide in the bottle.)

5.89 When 2.36 g of phosphorus was burned in chlorine, 10.5 g of a phosphorus chloride was produced. Its vapor took 1.8 times longer to diffuse than the same amount of CO_2 under the same conditions of temperature and pressure. What is the molar mass and molecular formula of the phosphorus chloride?

5.90 At 180.0°C, 27.20 mg of the vapor of the antimalarial drug quinine in a 250.0-mL container gave rise to a pressure of 9.48 Torr. In a combustion experiment, 17.2 mg of quinine produced 46.7 mg of carbon dioxide, 11.5 mg of water, and 1.48 mg of nitrogen. What is the molecular formula of quinine?

5.91 Compare the average speed of air molecules at the top of the troposphere (at −50°C) and at ground level (at 20°C). The average molar mass of air in the troposphere is 28.97 g/mol.

5.92 A 1.0-L vessel contains 1.0 mol of ideal gas atoms at 273 K. Another 1.0-L vessel contains 1.0 mol C_2H_4 at 273 K (treated as a van der Waals gas). (a) Which vessel has the greater pressure when the volumes are large? (b) Which vessel has more free space between molecules? (c) Which molecular system exerts greater pressure when the volumes are small?

5.93 Portland cement is made in rotary kilns at temperatures approaching 1500°C. The composition of the initial mixture placed into the kiln is 20% clay and 80% limestone (90% $CaCO_3$ and 10% impurities) by mass. At 900°C the calcium carbonate in the limestone decomposes to lime, CaO, and carbon dioxide. If carbon dioxide is evolved from the kiln at the rate of 1000 m^3/h at 1000°C and 1 atm, what mass of limestone and what mass of the "charge" must be being added to the kiln per hour?

CASE 5 ▸ SKIING, MOUNTAIN CLIMBING, AND THE GAS LAWS

Atmospheric pressure is the force per unit area exerted by the mixture of gases in the atmosphere. According to Dalton's law, each gas contributes its partial pressure to the total, and oxygen exerts about 160 Torr at sea level. We inhale about 0.5 L of air (about 0.02 mol O_2) with each breath, and the partial pressure of oxygen maintains our hemoglobin about 97% saturated with oxygen. This concentration of oxygen allows normal metabolic processes to occur.

At higher altitudes, the composition of the air remains almost constant; however, the density of the air decreases and, as a result, the atmospheric pressure decreases too. At 1500 m (5000 ft), atmospheric pressure is about 630 Torr and the partial pressure of oxygen is 130 Torr; at 7000 m (20,000 ft), atmospheric pressure decreases to 350 Torr and the oxygen partial pressure is only 73 Torr. These pressures can be calculated from the equation

$$P = P_0 e^{-Mgh/RT}$$

where M is the average molar mass of air (29 g/mol at sea level), g is the acceleration of free fall (9.81 m/s^2), h is the altitude in meters, R is the gas constant (8.314 J/(K·mol)), and P_0 is the pressure at sea level.

Suppose you travel by air to the ski slopes at 3000 m (10,000 ft) and arrive in a few hours. At this altitude, atmospheric pressure is about 520 Torr and the oxygen partial pressure is 110 Torr—about two-thirds that at sea level. As a result, your lungs (and consequently, your arterial blood) receive only two-thirds the oxygen they normally receive with each breath. To make up for this deficiency, you breath more deeply (lungs have a capacity of up to about 5 L) or more often. In addition, your heart pumps more vigorously to deliver the normal oxygen supply where it is needed, your blood pressure increases, and your arteries expand to provide for the increased capacity of blood flow through the circulatory system. If you ascend or exercise beyond the point at which your lungs and heart can maintain the normal oxygen flow to the cells through the circulatory system, high-altitude sickness results. This sickness is characterized by nausea, dizziness, and sluggishness, none of which is immediately relieved by resting. To avoid such problems, you could proceed more cautiously and acclimatize your cell functions to the higher altitude, or, in preparation for the skiing weekend, you could exercise to increase lung capacity.

A rapid and lengthy ascent or descent in an elevator or airplane produces similar effects; however, it is the immediate effects that are more obvious. Whereas a human body is mostly solid and liquid, there are air cavities, such as the ear canals and sinus cavities. When a sudden ascent occurs, the pressure in the air cavities cannot adjust quickly to the decreased atmospheric pressure; as a result, the pressure in the air cavity is greater than atmospheric pressure and a "sinus headache" may be experienced; to relieve the increased pressure in the ear canals, your ears may "pop."

QUESTIONS

1. A sky diver jumps from an airplane at 1000 m and falls to sea level. If the sky diver assumes a full breath of air (5 L) at the start of the jump and (because of anxiety) does not breath during the jump, what volume of air would be exhaled at the end of the jump where the atmospheric pressure is 770 Torr? Assume that the temperature remains constant.

2. Assume that the temperature of air decreases by 6.5°C per kilometer of altitude and that the temperature at sea level is 22°C. What is the volume change of the breath of air in Question 1?

3. You are on the ground floor of a skyscraper. You take an express elevator to the 120th floor, approximately 500 m higher. By what factor will the pressure on your eardrum change? Assume that there is no change in temperature and that your ear canal has a constant volume.

4. What is the partial pressure of oxygen at the summit of Mt. Everest, elevation 8.9 km, where the temperature is −60°C?

This chapter explores one of the roles of energy in chemistry. We concentrate here on the heat given out or absorbed during physical and chemical changes, and how it is described, measured, and predicted. We also see how to extend stoichiometry calculations to deal with heat outputs and requirements of reactions. The illustration shows the vigorous conflagration of a burning forest. Calculations like the ones we encounter in this chapter let us assess the energy output during such combustions and analogous processes under more controlled conditions.

6 ENERGY, HEAT, AND THERMOCHEMISTRY

Many chemical reactions release heat, which may be allowed to go to waste or is used to provide warmth or to power an engine. Some chemical reactions absorb heat as they occur. One of the most striking heat-absorbing reactions occurs when the solids barium hydroxide octahydrate, $Ba(OH)_2{\cdot}8H_2O$, and ammonium thiocyanate, NH_4SCN, are mixed together in a flask:

$$Ba(OH)_2{\cdot}8H_2O(s) + 2NH_4SCN(s) \longrightarrow Ba(SCN)_2(aq) + 2NH_3(g) + 10H_2O(s)$$

So much heat is absorbed as the reaction occurs that frost (condensed moisture from the air) appears on the outside of the flask (Fig. 6.1). Examples such as this one show that heat is a product of many reactions but is used—like a reactant—in others. It follows that we ought to be able to add one more arrow to the mole diagram (**1**) and use relations between moles to make predictions about the heat that a reaction can release or absorb. This aspect of reaction stoichiometry is called **thermochemistry**.

D Endothermic reaction: Two solids. CD1, 40.

N The first law of thermodynamics is discussed in Sections 16.1 and 16.2; that material could be introduced here.

In thermochemistry we typically begin with given masses of reactant as data, convert masses to moles, and then predict the quantity of heat the reaction releases or absorbs. By analogy with the discussions in Chapters 4 and 5, we can anticipate that calculations of this kind depend on knowing how much heat is produced from a known amount of reactant (or product), writing a relation of the form

$$\text{Amount of reactant (or product)} \simeq \text{quantity of heat}$$

and then setting up the appropriate conversion factor. Once we know the conversion factor, we can calculate the heat released (or absorbed) when any amount of reactant is consumed:

FIGURE 6.1 The reaction between ammonium thiocyanate and barium hydroxide octahydrate absorbs a lot of heat and can cause water vapor in the air to freeze on the outside of the flask.

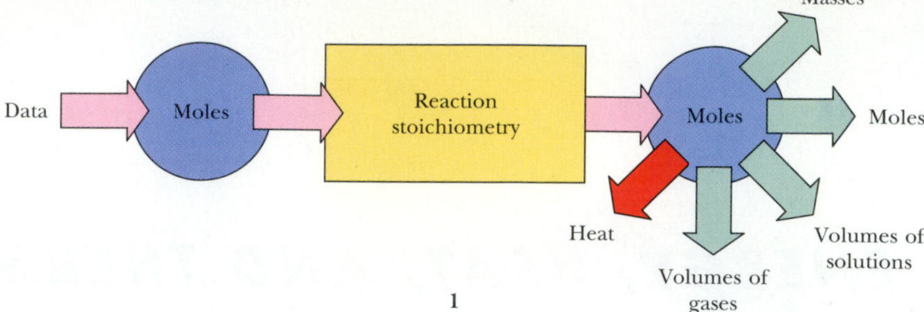

$$\text{Heat released} = \text{amount of reactant consumed} \times \frac{\text{quantity of heat}}{\text{amount of reactant}}$$

A similar expression can be written for the relation between the heat released or absorbed and the amount of product formed.

ENTHALPY AND CALORIMETRY

Throughout this chapter we shall refer to a system and its surroundings. For our purposes, a **system** is the reaction vessel and its contents, the reactants and products in which we have an interest (Fig. 6.2). A typical system might be a flask and its contents. The **surroundings** consist of everything outside the system. People are systems in this sense, and the rooms they inhabit, with all their furniture, walls, and the world outside, are the surroundings.

N The definition of heat. *J. Chem. Ed.* **1976**, *53*, 782.

6.1 ENERGY AND HEAT

A term that we shall use throughout the remainder of the book is *energy:*

> The **energy** of a system is a measure of the system's capacity to do work or to supply heat.

A wound-up spring possesses energy, because as it unwinds it can do work by raising a weight against the pull of gravity. A fuel possesses energy because as it burns it supplies heat to its surroundings. A fuel can also be used to do the work of propelling a vehicle along a highway.

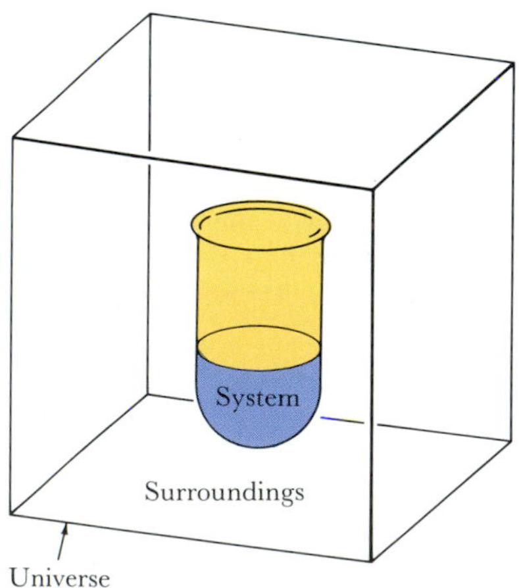

FIGURE 6.2 The system is the sample or reaction mixture that we are interested in and the reaction vessel. Outside the system are the surroundings. The system plus its surroundings is sometimes called the universe.

Heat. Two familiar terms, work and heat, are used in the definition of energy. By **work** we mean the energy expended during the act of moving an object against an opposing force, such as raising a weight on the surface of the Earth. Little will be said about work in this chapter but we shall deal with it in some detail in Chapter 16. By **heat** we mean the energy that is transferred as a result of a temperature difference between the system and its surroundings. Most of this chapter will deal with the energy that reactions release or absorb as heat.

If a system has a higher temperature than its surroundings, then energy flows out of the system as heat and the system's energy decreases. The escaping energy increases the **thermal motion** (the random, chaotic motion) of the atoms in the surroundings (Fig. 6.3). If the system has a lower temperature than that of its surroundings, then energy flows in as heat and the system's energy rises. This flow of energy into the system decreases the thermal motion of the atoms in the surroundings.

The form—as heat or as work—in which energy is released to the surroundings depends on how the reaction takes place. Consider the reaction between hydrogen and oxygen, which normally gives out a great deal of energy as heat:

$$2H_2(g) + O_2(g) \longrightarrow 2H_2O(g) + \text{energy as heat}$$

However, we could carry out the same reaction in a fuel cell, a device that is used to generate electricity from a chemical reaction and has been used on some space missions. The reaction releases the same amount of energy, but the energy is released as electrical power and can be used to run an electric motor:

$$2H_2(g) + O_2(g) \longrightarrow 2H_2O(g) + \text{energy as work}$$

It is worth noting the difference in the complexity of the equipment needed to obtain the energy as heat and as work. Heat is a more primitive way of releasing energy (Fig. 6.4); harnessing a reaction to obtain the energy as work usually requires more sophisticated equipment. There was a gap of thousands of years between the discovery of fire (energy obtained from fuel as heat) and the invention of the steam engine (energy obtained from fuel as work).

Reactions that release heat when they take place are called **exothermic reactions** (Fig. 6.5); reactions that absorb heat are **endothermic reactions** (Fig. 6.1). All combustions are exothermic. The **thermite reaction**, the reduction of a metal oxide by aluminum—for example,

$$2Al(s) + Fe_2O_3(s) \longrightarrow Al_2O_3(s) + 2Fe(s)$$

is spectacularly exothermic (Fig. 6.6); once started, it produces enough heat to melt the iron. The reaction cited earlier between barium hydroxide octahydrate and ammonium thiocyanate is endothermic. Heat

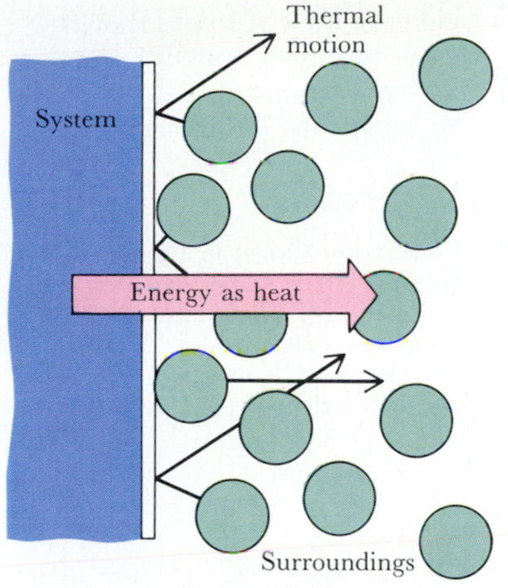

FIGURE 6.3 When energy leaves a system as a result of a temperature difference between the system and the surroundings (that is, as heat), it stimulates random thermal motion of the atoms in the surroundings.

E The transfer of energy as work involves the stimulation of *orderly* motion of atoms in the surroundings, as when a weight is raised (and all its atoms move in the same direction). It is relatively easy to extract energy as disorderly motion; more complex machines are required to extract energy as orderly motion.

E Iron melts at 1540°C. In earlier times, the thermite reaction was used to weld rails together.

D Thermite reaction. BZS1, 1.36.

FIGURE 6.4 Uncontrolled combustion takes place during a forest fire. The cellulose of the vegetation is oxidized back to the carbon dioxide and water from which it was originally formed by photosynthesis. A considerable amount of carbon monoxide is also formed.

N Neither of these reactions should be used as classroom demonstrations.

V Video 20: Exothermic reactions, III. An ammonium dichromate volcano.

V Video 8: Thermite.

E The reaction shown in Fig. 6.5 is $2(NH_4)_2Cr_2O_7(s) \rightarrow 2Cr_2O_3(s) + 2N_2(g) + 8H_2O(g)$. The volcano is formed by the gases produced in the reaction; the "lava" consists of the incandescent powdered reactants and products.

FIGURE 6.5 The decomposition of ammonium dichromate to chromium(III) oxide is so exothermic that, once ignited, it produces a miniature volcano.

FIGURE 6.6 The thermite reaction is another highly exothermic reaction—one that can melt the metal it produces. In this reaction, aluminum metal is reacting with iron(III) oxide, causing a shower of molten iron sparks.

must be applied to vaporize water: therefore, the system (water) absorbs heat and vaporization is endothermic. All vaporizations are endothermic because heat is absorbed from the surroundings as the liquid changes to vapor. Dissolving may be either slightly exothermic or endothermic, depending on the solute and the solvent. When ammonium nitrate dissolves in water, the process is quite strongly endothermic. This property is the basis of the instant cooling packs in some first-aid kits. These packs absorb heat (from the skin) when a tube of the salt is broken inside the bag of water and dissolves (see Case 16).

The units of energy. The quantity of energy released as heat can be determined by measuring the increase in temperature when the reaction takes place in an insulated container (Fig. 6.7). Such a container, used for measuring heat released or absorbed, is called a **calorimeter**, and the process of measurement is called **calorimetry**. A more complicated calorimeter, a bomb calorimeter, is often used to study combustions (Fig. 6.8). The mass of a sample is measured in a small crucible and sealed into the central container, the bomb. Next, the bomb is filled with compressed oxygen and immersed in the water bath. The sample is then ignited electrically and the change in temperature of the water in the bath is measured. The increase in temperature of the calorimeter is proportional to the heat released by the reaction.

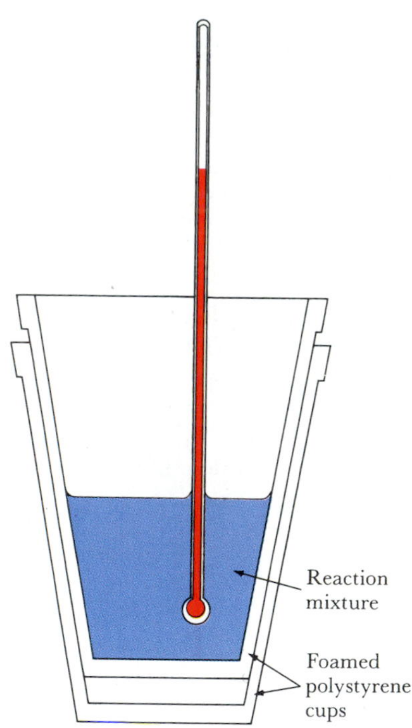

FIGURE 6.7 The quantity of heat released or absorbed by a reaction can be measured in this primitive version of a calorimeter. The outer polystyrene cup acts as an extra layer of insulation to ensure that no heat leaves or enters the inner cup. The quantity of heat released or absorbed is proportional to the change in temperature of the reaction mixture.

The heat released is expressed in **joules** (J); this unit is defined such that 1 J is the heat required to increase the temperature of 0.2390 g of water by 1°C. This statement can be written

$$1\ \text{J} = (0.2390\ \text{g water}) \times 1°\text{C} = 0.2390\ \text{g}\cdot°\text{C water}$$

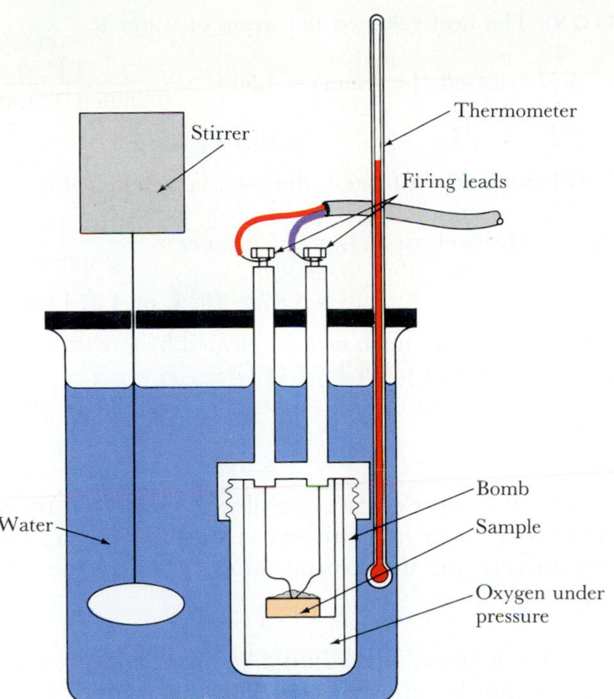

FIGURE 6.8 A bomb calorimeter. The combustion is started with an electrically ignited fuse, and the temperature of the entire assembly is monitored.

E In one type of calorimeter, the *adiabatic bomb calorimeter*, the entire assembly is immersed in a second water bath that has a temperature readjusted continuously to the temperature of the calorimeter's water bath. Because the temperatures of the two baths are equal at all times, there is no heat loss from the calorimeter (the entire assembly shown in the illustration), so the rise in temperature can be ascribed to the heat released in the reaction.

(This relation between joules and temperature change is a consequence of a more fundamental definition of the joule in the International System of Units.*) In thermochemistry, energies are normally expressed in kilojoules (kJ), where

$$1 \text{ kJ} = 10^3 \text{ J}$$

and occasionally in megajoules (MJ):

$$1 \text{ MJ} = 10^6 \text{ J} \ (= 10^3 \text{ kJ})$$

T OHT: Figs. 6.7 and 6.8.

D A bomb calorimeter is an intriguing item of apparatus for students to see.

E The "energy content" of food is measured with a bomb calorimeter.

▼ EXAMPLE 6.1 Assessing the heat released by a system

A certain reaction increased the temperature of 255 g of water by 4.56°C. How many joules of heat did the reaction release?

STRATEGY We know that 1 J of heat raises the temperature of 0.2390 g of water by 1°C. A larger mass of water heated through a larger temperature change will require more heat. Specifically, because we know that 1 J = 0.2390 g·°C water, we can use this relation to convert the data to heat in joules.

E Each beat of a human heart uses about 1 J of energy.

*In the International System of Units, the definition of the joule, a derived unit, is $1 \text{ J} = 1 \text{ kg·m}^2/\text{s}^2$. The name honors the nineteenth-century English scientist James Joule, a student of Dalton, who made a major contribution to our understanding of energy. Each beat of a human heart uses about 1 J of energy as it drives the blood through the body. To lift this book from the floor to the table, we need to use about 15 J of energy. To throw it across the room at 5 m/s requires nearly 20 J.

SOLUTION The heat released per gram of water is

$$\text{Heat released (J/g water)} = 4.56\text{°C} \times \frac{1\text{ J}}{0.2390\text{ g·°C water}}$$

$$= 19.1\text{ J/g water}$$

Because 255 g of water is heated, the total heat released is

$$\text{Heat released (J)} = 255\text{ g water} \times \frac{19.1\text{ J}}{1\text{ g water}}$$

$$= 4.87 \times 10^3\text{ J, or } 4.87\text{ kJ}$$

EXERCISE E6.1 How much heat is released by a reaction that raises the temperature of 359 g of water by 4.11°C?

[*Answer*: 6.17 kJ]

U Exercises 6.59, 6.62, and 6.64.

E This relation is now the *definition* of a calorie in terms of the SI unit, the joule. An earlier, now outdated, and less fundamental definition was expressed in terms of the heat required to raise the temperature of 1 g of water from 14.5°C to 15.5°C.

T OHT: Table 6.1.

Another unit, the **calorie** (cal), is still widely used in thermochemistry. It is defined as the heat required to raise 1 g of water through 1°C. It is easy to convert one unit to the other by using the relation

$$1\text{ cal} = 4.184\text{ J exactly}$$

Note that 1 cal is a larger unit than 1 J. For example, if the heat released by a reaction is reported as 155 kcal (where 1 kcal = 10^3 cal), then the heat released in kilojoules is

$$\text{Heat released (J)} = 155 \times 10^3\text{ cal} \times \frac{4.184\text{ J}}{1\text{ cal}} = 649 \times 10^3\text{ J} = 649\text{ kJ}$$

In this text we shall always use joules (and kilojoules) because that unit is simpler to use in later calculations (and its use is the modern trend).

Heat capacity and the calorimeter constant. The quantity of heat released or absorbed in a reaction is measured by monitoring the change in temperature of the system. One way of converting the observed change in temperature to the quantity of heat released is to make use of the property called **heat capacity**. The heat capacity is a characteristic property of a substance that is defined as

$$\text{Heat capacity} = \frac{\text{heat supplied}}{\text{temperature rise produced}}$$

If we observe a temperature increase of 2.6°C when we supply 98 kJ of heat to a substance, then we conclude that the heat capacity of the substance is

$$\text{Heat capacity} = \frac{98\text{ kJ}}{2.0\text{°C}} = 49\text{ kJ/°C}$$

Note that the units of heat capacity that we shall use are joules per degree Celsius (J/°C). The heat capacity of a substance is an extensive property, so the larger the sample, the greater its heat capacity. It is therefore common to report the **specific heat capacity** (which is often called just the specific heat), the heat capacity per gram of substance. The specific heat capacities of a few common substances are given in Table 6.1.

TABLE 6.1 Specific heat capacities of common materials

Substance	Specific heat capacity, J/(g·°C)
air	1.01
benzene	1.05
brass	0.37
copper	0.38
ethanol	2.42
glass (Pyrex)	0.78
granite	0.80
marble	0.84
polyethylene	2.3
stainless steel	0.51
water	2.03 solid 4.18 liquid 2.01 vapor

Once we know the specific heat capacity and the mass of a substance, we can interpret changes in its temperature in terms of the quantity of heat supplied to or lost from it. If we measure a change in temperature of a particular sample of a substance, then

Heat supplied
$$= \text{temperature change} \times \text{heat capacity}$$
$$= \text{temperature change} \times \text{mass of sample} \times \text{specific heat capacity}$$

This relation is illustrated in the following example.

N Heat capacity and the equipartition theorem. *J. Chem. Ed.* **1972**, *49*, 798.

▼ EXAMPLE 6.2 Using the specific heat capacity

How much heat is required to heat 500 g of water from 20.0°C to boiling (100.0°C) when it is contained in a glass beaker of mass 150 g?

STRATEGY We need the *total* quantity of heat required to raise the temperature of both the water and the beaker:

$$\text{Heat required} = \text{heat(water)} + \text{heat(beaker)}$$

Each quantity can be calculated from the specific heat capacity and mass of the substance (water and glass, respectively), and the change in temperature.

SOLUTION Because the specific heat capacity of water is 4.184 J/(g·°C), to raise the temperature of 500 g of water by 80.0°C requires

$$\text{Heat(water)} = 80.0°\text{C} \times 500\ \text{g} \times 4.184\ \frac{\text{J}}{\text{g}\cdot°\text{C}} = 1.67 \times 10^5\ \text{J, or } 167\ \text{kJ}$$

N The solution focuses on the relation "Information required = information given."

The specific heat capacity of glass is 0.78 J/(g·°C); so to increase the temperature of 150 g of glass by 80°C requires

$$\text{Heat(glass)} = 80.0°\text{C} \times 150\ \text{g} \times 0.78\ \frac{\text{J}}{\text{g}\cdot°\text{C}} = 9.4 \times 10^3\ \text{J, or } 9.4\ \text{kJ}$$

Therefore, the total quantity of heat required is

$$\text{Heat required} = 167\ \text{kJ} + 9.4\ \text{kJ} = 176\ \text{kJ}$$

EXERCISE E6.2 Calculate the quantity of heat required to increase the temperature of a 1.0-kg brass statue standing on a 500-g granite block from 20° to 100°C. See Table 6.1 for data.

[*Answer*: 62 kJ] ▲

In calorimetry we must take into account the fact that the heat released by a reaction increases the temperature of the entire calorimeter—the reaction vessel, its contents, and (for a bomb calorimeter) the water bath (Fig. 6.9). Although it would be possible to use the heat capacity of the entire calorimeter to convert the observed change in temperature to heat released, it proves more accurate in practice to compare the observed increase with that caused by a reaction of known heat output and to express the relation as a **calorimeter constant**:

$$\text{Calorimeter constant} = \frac{\text{heat supplied to calorimeter}}{\text{temperature rise observed}}$$

Once the calorimeter constant has been determined, we can interpret an observed temperature rise caused by a second chemical reaction in the same calorimeter with the same contents (in practice, the same mass of solution) by using the relation

$$\text{Heat released} = \text{temperature rise} \times \text{calorimeter constant}$$

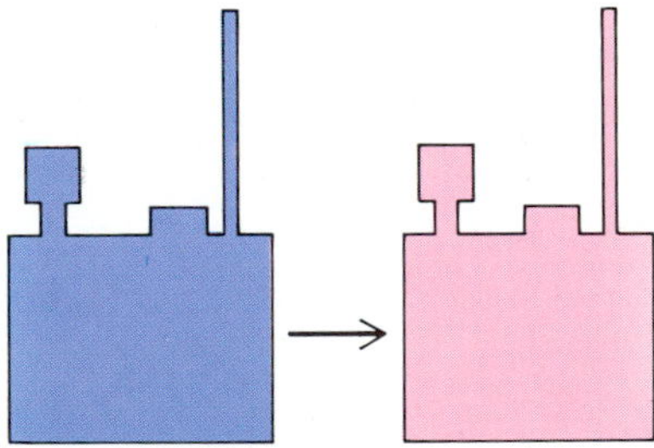

FIGURE 6.9 When a reaction takes place in a calorimeter, the temperature of the entire apparatus increases (shown here by a change of color). Therefore, to interpret the temperature increase, we need to know the heat capacity (in this context, the calorimeter constant) of the entire apparatus or to calibrate it by using a reaction of known heat output.

This procedure is an example of **calibration**, the interpretation of an observation by comparison with known information.

For example, suppose a reaction that releases 80.0 kJ of heat increases the temperature of the calorimeter by 8.3°C, then we can calculate the calorimeter constant as

$$\text{Calorimeter constant} = \frac{80.0\ \text{kJ}}{8.3°\text{C}} = 9.6\ \text{kJ/°C}$$

If under identical conditions a reaction of interest produces a temperature increase of 5.2°C, then we can conclude that the heat released by the reaction is

$$\text{Heat output (kJ)} = 5.2°\text{C} \times 9.6\ \frac{\text{kJ}}{°\text{C}} = 50\ \text{kJ}$$

▼ **EXAMPLE 6.3 Interpreting a temperature change as a heat transfer**

When a reaction that was known to release 17.55 kJ of heat took place in a calorimeter like that shown in Fig. 6.8 and containing 100 mL of water, the temperature rose by 3.65°C. When 50 mL of hydrochloric acid was mixed with 50 mL of aqueous sodium hydroxide solution in the same container, the temperature rose by 1.15°C. How much heat was released by the neutralization reaction?

STRATEGY We are given the calibration data and can use it to determine the calorimeter constant. Then we use this constant to convert the observed temperature change to the heat released by the neutralization reaction. (Note that in the neutralization reaction, there is the same amount of solution in the calorimeter as there was during the calibration experiment.)

U Exercises 6.57, 6.58, and 6.66.

SOLUTION The calibration data (a temperature rise of 3.65°C when 17.55 kJ of heat is supplied) gives the following calorimeter constant:

$$\text{Calorimeter constant} = \frac{17.55\ \text{kJ}}{3.65°\text{C}} = 4.81\ \text{kJ/°C}$$

Therefore, the heat output of the neutralization is

$$\text{Heat output (J)} = 1.15°\text{C} \times 4.81\ \frac{\text{kJ}}{°\text{C}} = 5.53\ \text{kJ}$$

EXERCISE E6.3 A small piece of calcium carbonate was placed in the same calorimeter used in Example 6.3 and 100 mL of dilute hydrochloric acid was poured onto it. The temperature of the calorimeter rose by 3.25°C. What is the heat output of the reaction?

▲ [*Answer*: 15.6 kJ]

We can also calibrate a calorimeter by passing an electric current through a heater inside the calorimeter (and the appropriate amount of solution) for a known time and measuring the temperature rise. The energy supplied by the heater is

$$\text{Heat supplied} = \text{time} \times \text{power}$$

We use this value and the temperature rise observed to calibrate the calorimeter. **Power**, the rate of supply of energy, is expressed in watts

(W), where 1 W is 1 J/s. For example, if a 10.0-W heater is on for 155 s, then the quantity of heat it supplies is

$$\text{Heat supplied} = 155 \text{ s} \times 10.0 \text{ J/s} = 1.55 \times 10^3 \text{ J, or } 1.55 \text{ kJ}$$

E The power (in watts) of an electric current is given by the product of the current (in amperes) and the potential (in volts) of the supply: $P = IV$. It follows that the heat supplied when a current I from a supply of potential V for a time t is equal to IVt.

▼ EXAMPLE 6.4 Measuring an enthalpy change calorimetrically

When a heater rated at 27.0 W was on for 182 s, the temperature of a calorimeter rose by 3.461°C (this precision can be attained with an electronic thermometer). A small sample of sulfur was burned in the same calorimeter, and the temperature rose by 2.245°C. How much heat is released during the oxidation of sulfur to sulfur dioxide?

STRATEGY First, we determine the calorimeter constant. To do so, we find the quantity of heat supplied, by using the expression above and the data on the electrical heating. Then we use this calorimeter constant to convert the observed temperature increase to the quantity of heat released by the reaction. Again we use the relation 1 W = 1 J/s.

SOLUTION The quantity of heat supplied by the electric heater is

$$\text{Heat supplied} = 182 \text{ s} \times 27.0 \text{ J/s} = 4.91 \times 10^3 \text{ J} = 4.91 \text{ kJ}$$

Therefore, the calorimeter constant is

$$\text{Calorimeter constant} = \frac{4.91 \text{ kJ}}{3.461°\text{C}} = 1.42 \text{ kJ/°C}$$

The quantity of heat released in the combustion of sulfur, which caused an increase in temperature of 2.245°C, is therefore

$$\text{Heat released} = 2.245°\text{C} \times 1.42 \frac{\text{kJ}}{°\text{C}} = 3.18 \text{ kJ}$$

EXERCISE E6.4 When current was passed for 136 s through a 12.5-W heater in a calorimeter, the temperature rose by 2.115°C. A small piece of phosphorus was burned in the same calorimeter, and the temperature rose from 18.15° to 20.43°C. What is the heat output of the oxidation of phosphorus?

[*Answer*: 1.83 kJ] ▲

D A chemical hand warmer. CD2, 54.

6.2 ENTHALPY

Most of the chemical reactions that we shall consider will be carried out in a container that is open to the atmosphere, as a simple calorimeter is. That is, these reactions occur at constant pressure. In thermochemistry we identify the release or absorption of heat by a system at constant pressure as a change in a property called the enthalpy H of the system:

> The change in the **enthalpy** of a system is equal to the heat released or absorbed at constant pressure.

E Enthalpy is not stored as heat. Heat is the manner in which the energy is transferred between the system and the surroundings. The enthalpy is a measure of the capacity of the system to supply energy as heat when a certain change occurs.

The name comes from the Greek words for "heat inside" and reflects the manner in which energy is lost or is supplied during the change. *The enthalpy is like a reservoir of energy that can be obtained as heat.*

The enthalpy of a sample of a substance depends on its physical state (whether it is a gas, a liquid, or a solid), the temperature, and the pres-

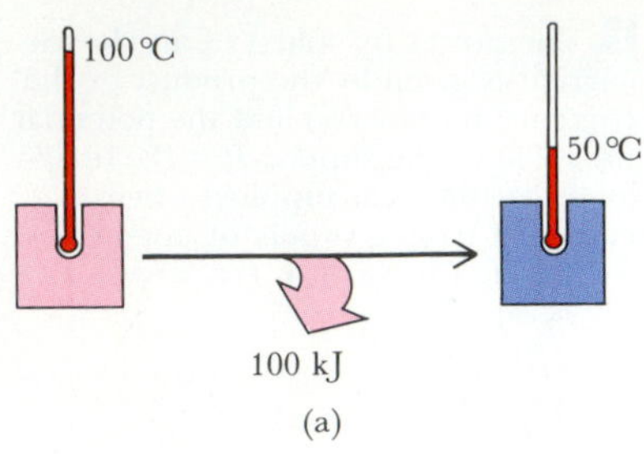

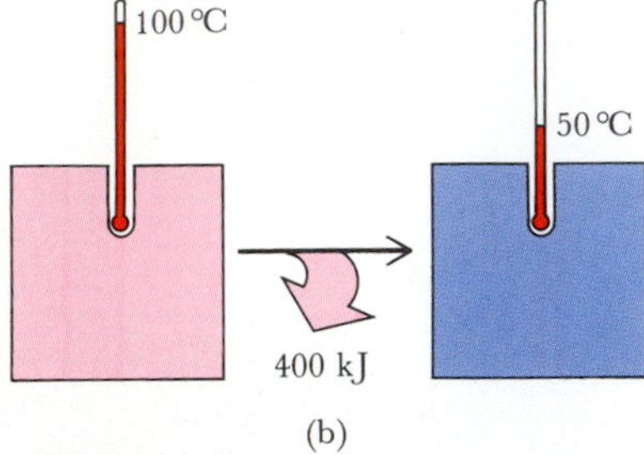

FIGURE 6.10 The heat released when (a) a small sample undergoes a change in temperature is less than the heat released when (b) a large sample of the same substance undergoes the same change in temperature. The larger sample is a bigger reservoir of energy.

sure. It also depends on the size of the sample. In other words, the enthalpy is an *extensive* property (Section 1.4). For example, 200 g of iron has four times the enthalpy of 50 g of iron at the same temperature and pressure. That is, compared with a 50-g block of iron, 200 g of iron is four times as big a reservoir of heat (Fig. 6.10).

Like the level of water in a reservoir, the enthalpy of a system falls as heat is released (at constant pressure) and it increases when heat is absorbed (once again, at constant pressure). Changes in the enthalpy from an initial value H_i to a final value H_f are denoted by the Greek uppercase delta (Δ); so

$$\Delta H = H_f - H_i$$

If heat is absorbed in a particular process, then we write

$$\Delta H = +(\text{heat absorbed at constant pressure})$$

The plus sign is present because heat is absorbed by the system, H_f is greater than H_i, and ΔH is positive (Fig. 6.11a). Because a process that absorbs heat is classified as endothermic, we see that

An **endothermic process** is a change in which ΔH is positive.

On the other hand, if heat is released, then

$$\Delta H = -(\text{heat released at constant pressure})$$

The minus sign is included because in this case H_f is less than H_i, so ΔH is negative (Fig. 6.11b). Because a process that releases heat is classified as exothermic,

An **exothermic process** is a change in which ΔH is negative.

T OHT: Figs. 6.10 and 6.11.

For example, measurements on boiling water show that 55 kJ of heat is needed to vaporize a sample at its boiling point in an open container, so we write

$$\Delta H = +55 \text{ kJ}$$

because 55 kJ of heat is absorbed, and the enthalpy of the water increases as it is vaporized.

We must always remember that the enthalpy change can be equated to the quantity of heat released or absorbed only if the pressure is constant while the change takes place. The pressure is constant when a reaction takes place in a simple calorimeter like that shown in Fig. 6.7. However, the volume of a bomb calorimeter is constant; so if gases are produced in the reaction, the pressure inside the bomb changes. Hence, the heat released or absorbed by a reaction that takes place in a bomb calorimeter is not quite equal to the change in enthalpy, and the measurement must be treated slightly differently (see Chapter 16).

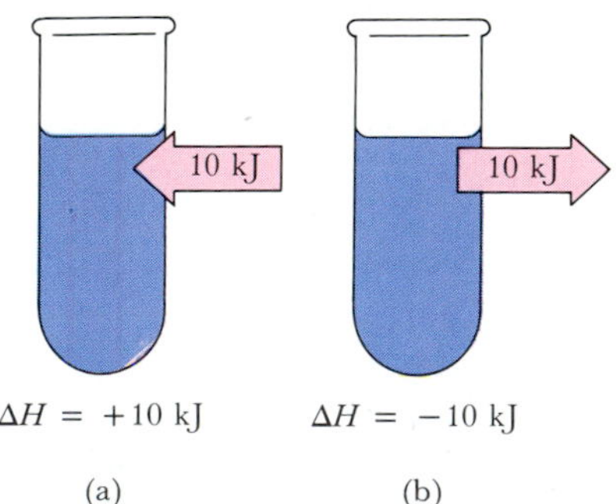

FIGURE 6.11 This system is open to the atmosphere and, hence, free to expand or contract. (a) When 10 kJ of energy enters the system as heat, the enthalpy increases by 10 kJ. (b) When 10 kJ leaves the system, the enthalpy decreases by the same amount.

Enthalpy as a state property. A very important property of the enthalpy of a substance is that it is the same at a given temperature and pressure, whatever may have happened to the sample in the past and whatever its origin. A 100-g sample of water at 25°C and 1 atm has exactly the same enthalpy whether it has been freshly synthesized from hydrogen and oxygen or obtained by distillation from a solution. The enthalpy of the sample is also the same whether it has been kept at constant temperature or heated and cooled back to 25°C. These remarks are summarized by saying that enthalpy is a state property:

A **state property** is a property of a substance that is independent of how the sample was prepared.

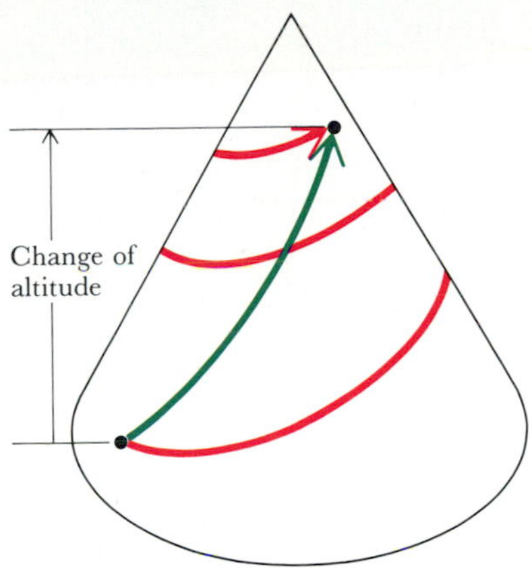

FIGURE 6.12 The altitude of someone on a mountain is a kind of state property: the same change of altitude results, whatever the route between two points on the mountain.

An everyday analogue of a state property is altitude, or height above sea level: once we arrive at the top of a mountain, our altitude is the same no matter how we got there (Fig. 6.12). In other words, the same *change* in altitude occurs whatever route we take between our starting and finishing points. Other examples of state properties we have already met include pressure, volume, and temperature, because all three properties depend only on the current state of the system, not on its previous history.

Because enthalpy is a state property, the change in its value when one substance changes into another is independent of how the change comes about (Fig. 6.13). The starting substance has a certain enthalpy; so does the final substance. The change is equal to the difference. This change is exactly like the difference of altitude between two points on a mountain, which is independent of the path taken to go from one to the other. For example, we might measure the change in enthalpy when 100 g of water is vaporized at a certain temperature and pressure and find that it is 55 kJ. If we then measured out another 100 g of water, electrolyzed it, and then discharged a spark through the resulting hydrogen and oxygen to produce water vapor at the stated temperature and pressure, we would find that the overall change in enthalpy for the second sample is also 55 kJ.

N Another analogue of a state property is the distance between two cities.

T OHT: Figs. 6.12 and 6.13.

The enthalpy of physical change. The enthalpy of any substance increases when it is heated. We can normally detect this increase as a rise in temperature. For example, if 10 kJ of heat is supplied to 100 g of water, the enthalpy of the water increases by 10 kJ. Because the heat capacity of 100 g of water is 0.418 kJ/°C, this increase in enthalpy appears as a temperature rise of about 24°C.

If we continue to heat the sample, then its temperature continues to rise. At 100°C the water starts to boil (if the external pressure is 1 atm). However, as we continue to supply heat to the sample (and, hence, continue to increase the enthalpy of the system), the temperature of the water stops rising. The energy we are supplying is now being used to overcome the attractions between the water molecules, that is, to vaporize the water. As we noted in Section 1.1, this behavior is characteristic of all changes of state: whenever a substance freezes or boils, its temperature remains unchanged until the transition is complete. Because heat has been supplied to boil away the water, the enthalpy of the vapor at 100°C must be higher than that of the same amount of liquid at the same temperature. Steam can cause severe scalding when it comes in contact with the skin, because a large *reduction* of enthalpy occurs as it condenses back to water (still at 100°C), and the heat released by the condensing steam (about 2 kJ per gram of water) is absorbed by and damages the living tissue.

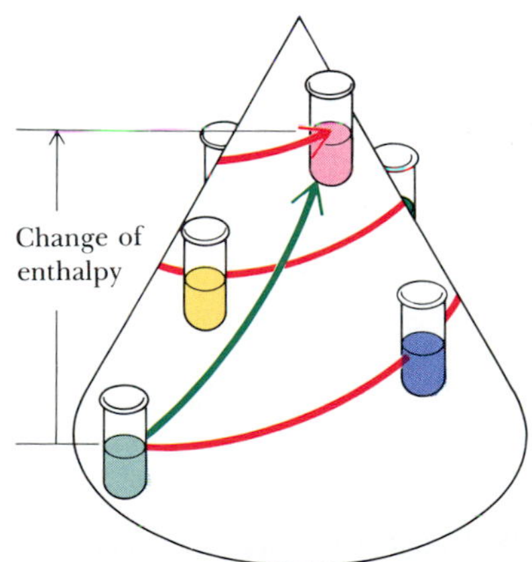

FIGURE 6.13 The overall change in enthalpy is independent of the path taken from the initial state to the final state. The two different paths represented here (one a direct change and one a series of changes) lead to the same change in enthalpy.

The difference in enthalpy *per mole* between the vapor and liquid states of a substance is called the **enthalpy of vaporization**, ΔH_{vap}, of that substance. For water at 100°C,

$$\Delta H_{vap} = H_{vapor} - H_{liquid} = +40.7 \text{ kJ/mol}$$

Values for other substances are given in Table 6.2. The absorption of heat from the skin when perspiration evaporates is one of the body's strategies for keeping us cool. It also accounts for the action of cosmetic "cold creams," which are mixtures of oil and water. They feel cool because the water they contain evaporates after the mixture is applied to the skin.

▼ EXAMPLE 6.5 Measuring the enthalpy of vaporization

An electric heater was used to heat a sample of water to its boiling point in a simple calorimeter. Heating was then continued until 35 g of water had vaporized. From the power rating of the heater and the time it was on, it was calculated that the vaporization required 79 kJ of heat. Calculate the enthalpy of vaporization of water at 100°C.

STRATEGY We can anticipate an *increase* in enthalpy (a positive ΔH), because heat is absorbed. Because vaporization occurs at constant pressure, the heat absorbed by 35 g of water is equal to the increase in its enthalpy. We can convert this enthalpy change to an enthalpy change *per mole* of H_2O by dividing the observed enthalpy change by the number of moles that have been vaporized.

SOLUTION From the data, we know that the change in enthalpy of the water is +79 kJ when 35 g is vaporized. The number of moles of H_2O in 35 g of water is

$$\text{Amount of } H_2O \text{ (mol)} = 35 \text{ g } H_2O \times \frac{1 \text{ mol } H_2O}{18.02 \text{ mol } H_2O} = 1.94 \text{ mol } H_2O$$

The enthalpy change per mole of H_2O is therefore

$$\Delta H_{\text{vap}} = \frac{+79 \text{ kJ}}{1.94 \text{ mol}} = +41 \text{ kJ/mol}$$

EXERCISE E6.5 The same heater was used to heat a sample of benzene, C_6H_6, to 80°C, its boiling point. The heating was continued, during which time 71 g of boiling benzene was evaporated (in a hood, because benzene is carcinogenic and toxic) and 28 kJ of heat was supplied. Calculate the enthalpy of vaporization of benzene at its boiling point.

▲ [*Answer*: +31 kJ/mol]

The enthalpy change that accompanies melting (fusion), expressed as kilojoules per mole of molecules, is called the **enthalpy of melting**:

$$\Delta H_{\text{melt}} = H_{\text{liquid}} - H_{\text{solid}}$$

(The enthalpy of melting is also called the enthalpy of fusion.) The enthalpy of melting of water is +6 kJ/mol, a value implying that 6 kJ of energy must be supplied as heat to melt 18 g of ice at 0°C. Melting is endothermic, because heat is absorbed when a substance melts and its molecules break away from their neighbors. Some enthalpies of melting are given in Table 6.2.

E Enthalpies of melting are generally smaller than enthalpies of vaporization because the particles of the solid are only loosened when a solid melts, and the distance between them (and hence the strength of their interaction) does not change very much. In contrast, when a liquid vaporizes, the particles become widely separated and the intermolecular forces are reduced to zero.

The **enthalpy of freezing** is the negative of the enthalpy of melting: the same quantity of heat is released when a liquid freezes as is absorbed when it melts. For water at 0°C, the enthalpy of freezing is −6 kJ/mol. Hence, to freeze 18 g of water at 0°C in a refrigerator, 6 kJ of energy must be removed as heat. The enthalpy change for the reverse of *any* process (any chemical reaction or physical change) is the

TABLE 6.2 Standard enthalpies of physical change*

Substance	Formula	Freezing point, K	ΔH°_{melt}, kJ/mol	Boiling point, K	ΔH°_{vap}, kJ/mol
acetone	CH_3COCH_3	177.8	5.72	329.4	29.1
ammonia	NH_3	195.3	5.65	239.7	23.4
argon	Ar	83.8	1.2	87.3	6.5
benzene	C_6H_6	278.7	9.87	353.3	30.8
ethanol	C_2H_5OH	158.7	4.60	351.5	43.5
helium	He	3.5	0.02	4.22	0.08
mercury	Hg	234.3	2.292	629.7	59.30
methane	CH_4	90.7	0.94	111.7	8.2
methanol	CH_3OH	175.5	3.16	337.2	35.3
water	H_2O	273.2	6.01	373.2	40.7 (44.0 at 25°C)

*Values correspond to the transition temperature. The superscript ° signifies that the change takes place at 1 atm and that the substance is pure. Note that, generally, the higher the freezing point, the higher the enthalpy of melting.

negative of the enthalpy change for the original process:

$$\Delta H(\text{reverse change}) = -\Delta H(\text{forward change})$$

The **enthalpy of sublimation** is the enthalpy change per mole of molecules when a solid changes directly into a vapor:

$$\Delta H_{sub} = H_{vapor} - H_{solid}$$

Sublimation is always endothermic (ΔH is positive), because heat is absorbed when a substance changes from the solid state, in which the molecules are packed close together, to the gaseous state, in which they move freely. The disappearance of ice on a cold, dry morning is the result of the direct sublimation of ice into water vapor. Solid carbon dioxide is called dry ice because it sublimes directly to the gas and does not form an intermediate liquid phase.

Because enthalpy is a state property, the enthalpy change for the conversion from solid to vapor is independent of the path taken. It follows that the enthalpy change for the *direct* conversion of a solid to a vapor (its sublimation) is equal to the sum of the enthalpy changes for melting followed by vaporization (**2**):

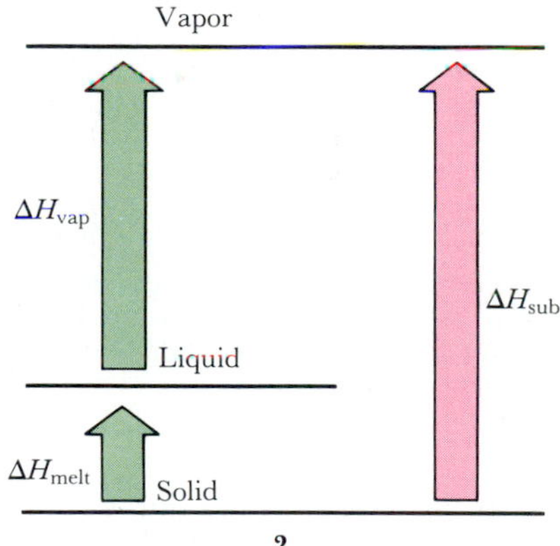

2

$$\Delta H_{sub} = \Delta H_{melt} + \Delta H_{vap}$$

For example, at 25°C the enthalpy of melting of sodium metal is +2.6 kJ/mol and the enthalpy of vaporization of liquid sodium is +98 kJ/mol, so the enthalpy of sublimation of solid sodium at 25°C is

$$\Delta H_{sub} = +2.6 \text{ kJ/mol} + 98 \text{ kJ/mol} = +101 \text{ kJ/mol}$$

The existence of enthalpies of fusion and vaporization affects the appearance of the **heating curve** of a substance, the graph showing the variation of the temperature of a sample as it is heated. As we see in Fig. 6.14, the temperature of the solid rises steadily up to its melting point as it is heated (the slope of the graph is determined by the specific heat capacity and the mass of the solid sample). The temperature remains constant at the melting point as heat supplies the energy re-

FIGURE 6.14 The temperature of a solid rises as heat is supplied. At the melting (freezing) point, the temperature remains constant and the heat is used to melt the sample. When enough heat has been supplied to melt all the solid, the temperature begins to rise again. A similar pause occurs at the boiling point.

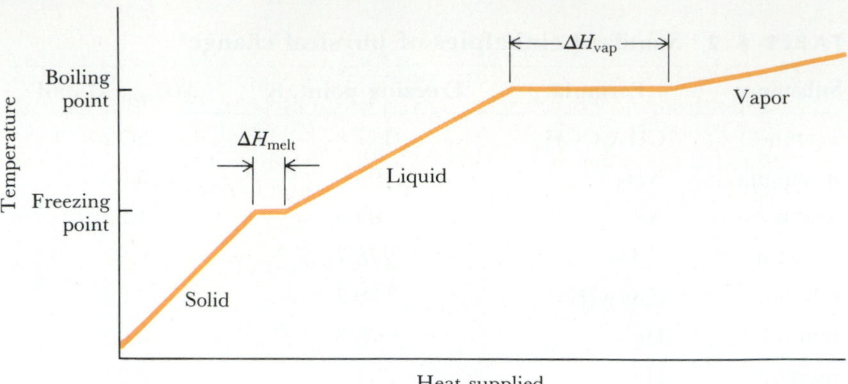

T OHT: Fig. 6.14.

U Exercises 6.60 and 6.61.

quired to form the liquid (the enthalpy of melting). Then it rises again (with a slope characteristic of the specific heat capacity and the mass of the liquid) up to the boiling point. The temperature now remains constant as heat supplies the energy required to form the vapor (the enthalpy of vaporization). After the sample has evaporated, the temperature rises again as heating continues; and now the slope of the graph is determined by the specific heat capacity and the mass of the vapor. This topic is developed further in Section 10.5.

THE ENTHALPY OF CHEMICAL CHANGE

When 1 mol of CH_4 burns in air, 890 kJ of heat is produced. That is, when the reaction

$$CH_4(g) + 2O_2(g) \longrightarrow CO_2(g) + 2H_2O(l)$$

takes place, the total enthalpy of the system (in this case, the reaction mixture alone) decreases by 890 kJ for each mole of CH_4 that react. This decrease of 890 kJ is the **reaction enthalpy** ΔH, the change of enthalpy for the reaction *exactly as the chemical equation is written.* We report this value by writing the **thermochemical equation**, which consists of the chemical equation and the reaction enthalpy:

N Whenever we write a thermodynamic quantity (such as ΔH, but later also for other thermodynamic functions) next to a chemical reaction, the equation should be interpreted as a thermochemical equation.

$$CH_4(g) + 2O_2(g) \longrightarrow CO_2(g) + 2H_2O(l) \qquad \Delta H = -890 \text{ kJ}$$

A thermochemical equation shows the enthalpy change for the reaction in which the stoichiometric coefficients are interpreted as the *actual amounts* of reactants that have been consumed (in this case, 1 mol of CH_4 and 2 mol of O_2). If the same reaction is written with the coefficients all multiplied by 2 (the new equation implying that 2 mol CH_4 molecules have been burned), the reaction enthalpy would be twice as great, and the thermochemical equation would be

$$2CH_4(g) + 4O_2(g) \longrightarrow 2CO_2(g) + 4H_2O(l) \qquad \Delta H = -1780 \text{ kJ}$$

D Exothermic reaction: sodium sulfite and bleach. CD1, 42.

V Video 4: Iron and sulfur.

▼ EXAMPLE 6.6 Calculating the reaction enthalpy

When 1.25 g of iron reacts with sulfur to form iron(II) sulfide, 2.24 kJ of heat is released. Calculate the enthalpy of the reaction

$$Fe(s) + S(s) \longrightarrow FeS(s)$$

and write the thermochemical equation for the reaction of 1 mol of Fe.

STRATEGY We need the enthalpy change that accompanies the reaction of 1 mol Fe. First, we convert the given mass of iron to the amount in moles, and then we express the heat output as an enthalpy change per mole of Fe. Because the reaction is exothermic, ΔH is negative. To write the thermochemical equation, we write the chemical equation and include the reaction enthalpy for amounts of reactants equal to the stoichiometric coefficients.

SOLUTION The molar mass of iron is 55.85 g/mol Fe, so

$$\text{Amount of Fe (mol)} = 1.25\text{ g Fe} \times \frac{1\text{ mol Fe}}{55.85\text{ g Fe}} = 2.24 \times 10^{-2}\text{ mol Fe}$$

The enthalpy change per mole of Fe that reacts is therefore

$$\Delta H = \frac{-2.24\text{ kJ}}{2.24 \times 10^{-2}\text{ mol Fe}} = -100\text{ kJ/mol Fe}$$

For the reaction as written, in which 1 mol of Fe reacts with 1 mol of S, the thermochemical equation is

$$Fe(s) + S(s) \longrightarrow FeS(s) \qquad \Delta H = -100\text{ kJ}$$

EXERCISE E6.6 When 2.31 g of phosphorus reacts with chlorine to form phosphorus trichloride, 23.9 kJ of heat is released. Write the thermochemical equation for the reaction $2P(s) + 3Cl_2(g) \rightarrow 2PCl_3(l)$ in which 2 mol P reacts.

[*Answer*: $2P(s) + 3Cl_2(g) \rightarrow 2PCl_3(l)$, $\Delta H = -638$ kJ]

6.3 REACTION ENTHALPIES

Once we know a reaction enthalpy, we can use the methods described in Chapter 4 to calculate the enthalpy change when any amount of reactant is consumed or product formed. For example, suppose we want to know the heat output from the combustion of 150 g of methane. We begin by converting the given mass to moles of CH_4 by using the relation 16.04 g CH_4 = 1 mol CH_4 and then using the reaction enthalpy (from the thermochemical equation) in the form 1 mol CH_4 = −890 kJ:

$$\text{Enthalpy change (kJ)} = 150\text{ g } CH_4 \times \frac{1\text{ mol } CH_4}{16.04\text{ g } CH_4} \times \frac{-890\text{ kJ}}{1\text{ mol } CH_4}$$
$$= -8.32 \times 10^3\text{ kJ}$$

That is, the enthalpy of the system decreases by 8.32×10^3 kJ; or, in other words, the combustion releases 8.32×10^3 kJ of heat.

EXAMPLE 6.7 Calculating the heat output of a fuel

The combustion of propane is described by the thermochemical equation

$$C_3H_8(g) + 5O_2(g) \longrightarrow 3CO_2(g) + 4H_2O(l) \qquad \Delta H = -2220\text{ kJ}$$

Calculate the mass of propane that must be burned to obtain 350 kJ of heat, which is just enough energy to heat 1 L of water from room temperature (20°C) to boiling.

STRATEGY As in all stoichiometry calculations, we convert to moles and use the mole relations given by the chemical equation. Because we are given the heat output, the first step is to convert heat output to moles of propane by using the thermochemical equation. Then we convert from moles of propane to grams by using the molar mass of propane (44.09 g/mol).

N The strategy illustrates an extension of diagram **2** in Chapter 4 (p. 130).

SOLUTION The two conversion factors are obtained from

$$-2220\text{ kJ} = 1\text{ mol } C_3H_8$$

$$1\text{ mol } C_3H_8 = 44.09\text{ g } C_3H_8$$

Therefore,

$$\text{Mass of } C_3H_8\text{ (g)} = -350\text{ kJ} \times \frac{1\text{ mol } C_3H_8}{-2220\text{ kJ}} \times \frac{44.09\text{ g } C_3H_8}{1\text{ mol } C_3H_8}$$

$$= 6.95\text{ g } C_3H_8$$

EXERCISE E6.7 The thermochemical equation for the combustion of butane is

$$2C_4H_{10}(g) + 13O_2(g) \longrightarrow 8CO_2(g) + 10H_2O(l) \qquad \Delta H = -5756\text{ kJ}$$

What mass of butane supplies 350 kJ of heat?

[*Answer*: 7.07 g]

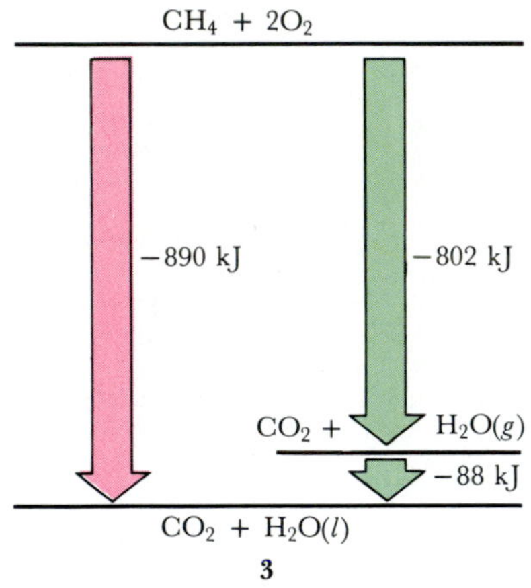

3

Standard reaction enthalpies. The heat given out or taken in by a reaction depends on the physical states of the reactants and products. Less heat is produced by the combustion of methane if the water is formed as a vapor rather than as a liquid:

$$CH_4(g) + 2O_2(g) \longrightarrow CO_2(g) + 2H_2O(g) \qquad \Delta H = -802\text{ kJ}$$

$$CH_4(g) + 2O_2(g) \longrightarrow CO_2(g) + 2H_2O(l) \qquad \Delta H = -890\text{ kJ}$$

The enthalpy of water vapor is 44 kJ/mol higher than that of liquid water at 25°C (Table 6.2); hence, 88 kJ less heat (for 2 mol H_2O) is released in the reaction if the vapor is formed rather than the liquid (**3**).

Because enthalpy changes depend on temperature and pressure, chemists find it convenient to report reaction enthalpies under an internationally accepted set of standard conditions. Chemists report values for reactions in which the reactants and products are in their standard states:

The **standard state** of a substance is its pure form at 1 atm.

The standard state of liquid water at some specified temperature is pure water at that temperature and 1 atm. The standard state of ice at some temperature is pure ice at that temperature and 1 atm. A reaction enthalpy based on such standard states is called a standard reaction enthalpy ($\Delta H°$, the superscript ° always denotes a "standard" value) and is defined as follows:

E The modern definition of standard states has replaced 1 atm by 1 bar (where 1 bar = 10^5 Pa; 1.00 bar is equal to 1.01 atm). There are few significant changes stemming from this definition in elementary work, and we have retained the more familiar definition. Note, however, that all the thermodynamic data in the text (including those in Appendix 2A) are actually for 1 bar.

A **standard reaction enthalpy** is the reaction enthalpy when reactants in their standard states change to products in their standard states.

Standard reaction enthalpies can be reported for any temperature, but data are usually for 25°C (more precisely, for 298.15 K). All the values used in this text are for this temperature unless noted otherwise.

T OHT: Fig. 6.15.

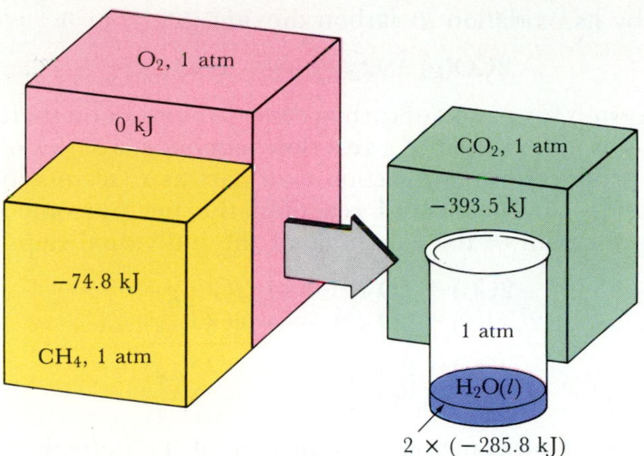

FIGURE 6.15 The standard reaction enthalpy is the difference in enthalpy between the *pure* products each at 1 atm and the pure reactants at the same pressure and the specified temperature (typically but not necessarily 25°C). The relative values are expressed on a scale defined such that O_2 is zero (see Section 6.4).

The standard reaction enthalpy for the combustion of methane is the change in enthalpy when pure methane and pure oxygen gases, each at 1 atm, react to form pure carbon dioxide and pure water, each at that pressure (Fig. 6.15). If the product is liquid water, then we have

$$CH_4(g) + 2O_2(g) \longrightarrow CO_2(g) + 2H_2O(l) \qquad \Delta H° = -890 \text{ kJ}$$

If the product is water vapor, then

$$CH_4(g) + 2O_2(g) \longrightarrow CO_2(g) + 2H_2O(g) \qquad \Delta H° = -802 \text{ kJ}$$

Because the enthalpy change of any reverse process is the negative of the change for the forward process, the standard reaction enthalpy of the reverse reaction is obtained simply by reversing the sign of the standard reaction enthalpy of the forward reaction (**4**):

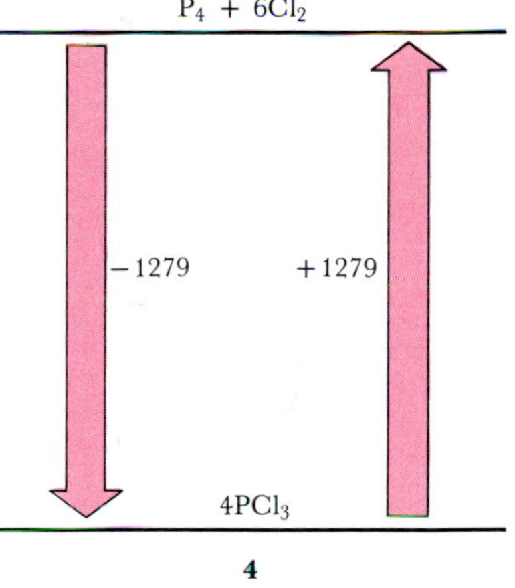

4

$$P_4(s) + 6Cl_2(g) \longrightarrow 4PCl_3(l) \qquad \Delta H° = -1279 \text{ kJ}$$

$$4PCl_3(l) \longrightarrow P_4(s) + 6Cl_2(g) \qquad \Delta H° = +1279 \text{ kJ}$$

A reaction that is exothermic in one direction is endothermic in the reverse direction.

Hess's law. Because enthalpy is a state property, the change in its value is the same whatever the path from given reactants to specified products. We used this property in determining the enthalpy of sublimation of sodium in Section 6.2. Now we apply the same argument to chemical changes.

N Hess's law is a consequence of the fact that enthalpy is a state property; it has no special status in thermodynamics, and it is merely for historical reasons that it is commonly honored as a law. We would have left it in the anonymity it deserves, except for the fact that the name is so widely encountered.

For example, consider the oxidation of carbon to carbon dioxide

$$2C(s) + 2O_2(g) \longrightarrow 2CO_2(g) \qquad \text{(a)}$$

The enthalpy change is the same (provided the conditions of temperature and pressure are the same) as the total enthalpy change in the formation of carbon monoxide

$$2C(s) + O_2(g) \longrightarrow 2CO(g) \qquad \text{(b)}$$

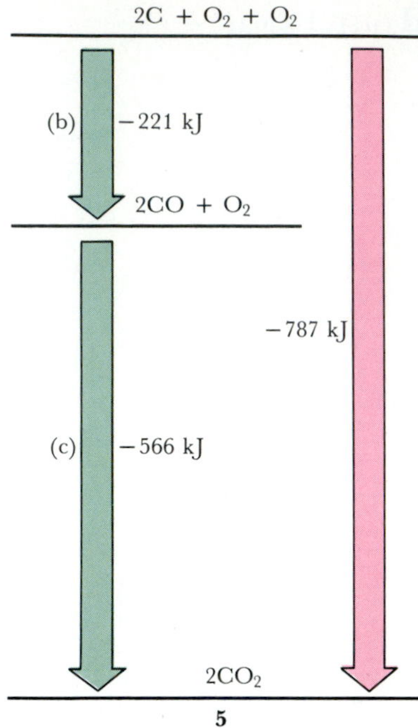

T Diagram **5** and Fig. 6.16.

followed by its oxidation to carbon dioxide (**5**):

$$2CO(g) + O_2(g) \longrightarrow 2CO_2(g) \qquad \text{(c)}$$

The two-step formation of carbon dioxide by reaction (b) followed by reaction (c) is an example of a **reaction sequence**, a series of reactions in which products of one reaction take part as reactants in the next. The equation for the **overall reaction**, the net outcome of the sequence, is the sum of the equations of the individual steps:

$$\begin{array}{rcll} 2C(s) + O_2(g) & \longrightarrow & 2CO(g) & \text{(b)} \\ 2CO(g) + O_2(g) & \longrightarrow & 2CO_2(g) & \text{(c)} \\ \hline 2C(s) + 2O_2(g) & \longrightarrow & 2CO_2(g) & \text{(b + c)} \\ 2C(s) + 2O_2(g) & \longrightarrow & 2CO_2(g) & \text{(a)} \end{array}$$

Because the overall equation (a) is the sum of the individual equations (b) and (c), the reaction enthalpy of (a) is the sum of the reaction enthalpies of (b) and (c):

$$\begin{array}{rclll} & \text{Add equations} & & \text{Add reaction enthalpies} & \\ & \downarrow & & \downarrow & \\ 2C(s) + O_2(g) & \longrightarrow & 2CO(g) & \Delta H^\circ = -221.0\ \text{kJ} & \text{(b)} \\ 2CO(g) + O_2(g) & \longrightarrow & 2CO_2(g) & \Delta H^\circ = -566.0\ \text{kJ} & \text{(c)} \\ \hline 2C(s) + 2O_2(g) & \longrightarrow & 2CO_2(g) & \Delta H^\circ = -787.0\ \text{kJ} & \text{(a)} \end{array}$$

The same conclusion—that reaction enthalpies can be combined like individual chemical equations—applies to any reaction that can be considered to take place in several steps. This concept is summarized as follows (Fig. 6.16):

> **Hess's law**: A reaction enthalpy is the sum of the enthalpies of any sequence of reactions (at the same temperature and pressure) into which the overall reaction may be divided.

Germain Hess was a Swiss chemist who proposed the law in 1840. We now know that it is a consequence of the fact that enthalpy changes are independent of the path by which the specified reactants are transformed into the specified products.

The importance of Hess's law is that it enables us to predict the enthalpies of reactions that we cannot measure directly in the laboratory. For example, the reaction

$$3C(s) + 4H_2(g) \longrightarrow C_3H_8(g)$$

cannot be carried out in practice, but it can be expressed as the sum of reactions that can be studied (as we shall shortly see). Moreover, the intermediate reactions in a sequence need not be ones that can actually be carried out in the laboratory. That is, the intermediate reactions may be "hypothetical." For instance, the reaction above could be used in a reaction sequence to determine the enthalpy of an overall reaction. The only requirement is that the individual chemical equations in the sequence must balance and must add up to the equation for the reaction of interest. Subject to these conditions, a reaction enthalpy can be calculated from any convenient sequence.

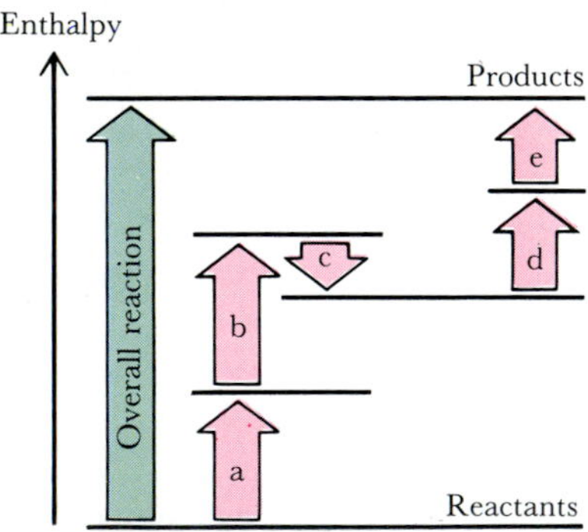

FIGURE 6.16 If the overall reaction can be broken down into a series of steps, then the corresponding overall reaction enthalpy is the sum of the reaction enthalpies of those steps. None of the steps need be a reaction that can actually be carried out in the laboratory.

When we use Hess's law, we add together a series of chemical equations that results in the desired overall chemical equation. A good plan is set out below; it illustrates each step with an equation for a reaction that contributes to the overall reaction for the (hypothetical) synthesis of propane

$$3C(s) + 4H_2(g) \longrightarrow C_3H_8(g)$$

The thermochemical equations we shall use are

$$C_3H_8(g) + 5O_2(g) \longrightarrow 3CO_2(g) + 4H_2O(l) \qquad \Delta H° = -2220\ \text{kJ} \qquad \text{(a)}$$

$$C(s) + O_2(g) \longrightarrow CO_2(g) \qquad \Delta H° = -394\ \text{kJ} \qquad \text{(b)}$$

$$H_2(g) + \tfrac{1}{2}O_2(g) \longrightarrow H_2O(l) \qquad \Delta H° = -286\ \text{kJ} \qquad \text{(c)}$$

Remember that if we reverse a thermochemical equation, we must change the sign of the reaction enthalpy. The overall reaction enthalpy is the sum of the individual reaction enthalpies combined in the same way we combined the individual chemical equations.

1. Start with a thermochemical equation that has at least one of the reactants on the correct side of the arrow in the overall chemical equation. A sensible choice would be either (b) or (c). We select (b) and multiply it through by 3:

$$3C(s) + 3O_2(g) \longrightarrow 3CO_2(g) \qquad \Delta H° = 3 \times (-394\ \text{kJ}) = -1182\ \text{kJ}$$

N Hess's law is summarized by Figs. 6.12 and 6.13.

2. Add to it an equation that results in a desired product on the right of the arrow. To obtain 1 mol of C_3H_8 on the right, we reverse (a), change the sign of its reaction enthalpy, and add it to this equation:

$$\begin{array}{lll} 3C(s) + 3O_2(g) \longrightarrow 3CO_2(g) & & \Delta H° = -1182\ \text{kJ} \\ 3CO_2(g) + 4H_2O(l) \longrightarrow C_3H_8(g) + 5O_2(g) & & \Delta H° = -(-2220\ \text{kJ}) \\ \hline 3C(s) + 3O_2(g) + 3CO_2(g) + 4H_2O(l) \longrightarrow C_3H_8(g) + 5O_2(g) + 3CO_2(g) & & \end{array}$$

which simplifies to

$$3C(s) + 4H_2O(l) \longrightarrow C_3H_8(g) + 2O_2(g) \qquad \Delta H° = +1038\ \text{kJ}$$

3. To cancel an unwanted reactant, we add an equation that has that substance on the *right* of the arrow. In this case, we want to cancel the $4H_2O$, so we add equation (c) after multiplying it by 4:

$$\begin{array}{ll} 3C(s) + 4H_2O(l) \longrightarrow C_3H_8(g) + 2O_2(g) & \Delta H° = +1038\ \text{kJ} \\ 4H_2(g) + 2O_2(g) \longrightarrow 4H_2O(l) & \Delta H° = 4 \times (-286\ \text{kJ}) \\ & \qquad = -1144\ \text{kJ} \\ \hline 3C(s) + 4H_2(g) + 4H_2O(l) + 2O_2(g) \longrightarrow C_3H_8(g) + 2O_2(g) + 4H_2O(l) & \end{array}$$

which simplifies to

$$3C(s) + 4H_2(g) \longrightarrow C_3H_8(g) \qquad \Delta H° = -106\ \text{kJ}$$

4. To cancel a product, we add an equation that has that substance on the *left* of the arrow. In our example, this step has already been taken into account by the addition of 4 times equation (c).

▼ EXAMPLE 6.8 Combining reaction enthalpies

Calculate the standard reaction enthalpy for the incomplete combustion of octane to carbon monoxide and water:

$$2C_8H_{18}(l) + 17O_2(g) \longrightarrow 16CO(g) + 18H_2O(l)$$

given the standard reaction enthalpies for the complete combustion of 2 mol C_8H_{18}:

$$2C_8H_{18}(l) + 25O_2(g) \longrightarrow 16CO_2(g) + 18H_2O(l) \quad \Delta H^\circ = -10{,}942 \text{ kJ} \quad \text{(a)}$$

and for the oxidation of 2 mol CO:

$$2CO(g) + O_2(g) \longrightarrow 2CO_2(g) \quad \Delta H^\circ = -566.0 \text{ kJ} \quad \text{(b)}$$

STRATEGY We can anticipate that the enthalpy of partial combustion of octane to carbon monoxide is smaller (less negative) than the enthalpy of its complete combustion to carbon dioxide: less heat is given out when the oxidation terminates at a less fully oxidized product. We use the same steps described above.

SOLUTION Equation (a) has 2 mol C_8H_{18} on the left of the arrow, the same amount as in the overall equation; so we write it first:

$$2C_8H_{18}(l) + 25O_2(g) \longrightarrow 16CO_2(g) + 18H_2O(l) \quad \Delta H^\circ = -10{,}942 \text{ kJ}$$

To get 16 mol CO on the right of the arrow, we multiply (b) by 8, reverse it, and change the sign of the reaction enthalpy:

$$16CO_2(g) \longrightarrow 16CO(g) + 8O_2(g)$$

$$\Delta H^\circ = -8 \times (-566.0 \text{ kJ}) = +4528.0 \text{ kJ}$$

Summing the two equations (and simplifying) gives

$$\begin{array}{rll} 2C_8H_{18}(l) + 25O_2(g) \longrightarrow & 16CO_2(g) + 18H_2O(l) & \Delta H^\circ = -10{,}942 \text{ kJ} \\ 16CO_2(g) \longrightarrow & 16CO(g) + 8O_2(g) & \Delta H^\circ = +4{,}528.0 \text{ kJ} \\ \hline 2C_8H_{18}(l) + 17O_2(g) \longrightarrow & 16CO(g) + 18H_2O(l) & \Delta H^\circ = -6{,}414 \text{ kJ} \end{array}$$

U Exercises 6.70, 6.71, and 6.81.

This thermochemical equation describes the desired overall reaction, so we do not need to proceed any further. It shows that the partial combustion releases only about 6 MJ of energy as heat, whereas the complete combustion releases about 11 MJ.

EXERCISE E6.8 Calculate the standard reaction enthalpy for the incomplete combustion of propane to carbon monoxide and water in the reaction

$$2C_3H_8(g) + 7O_2(g) \longrightarrow 6CO(g) + 8H_2O(l)$$

given that

$$C_3H_8(g) + 5O_2(g) \longrightarrow 3CO_2(g) + 4H_2O(l) \quad \Delta H^\circ = -2220 \text{ kJ}$$

[*Answer*: −2742 kJ] ▲

We can use the same procedure to calculate enthalpies of a wide range of reactions. In all such calculations the skill we need is the ability to find a reaction sequence that leads to the overall reaction required.

Standard enthalpy of combustion. The reaction enthalpies of combustions are important for judging the suitability of fuels and for measuring the enthalpies of other kinds of reactions. Their values are usually reported as standard enthalpies of combustion ΔH_c°:

> The **standard enthalpy of combustion** of a substance is the change in enthalpy per mole of the substance when it burns (reacts with oxygen) completely under standard conditions.

TABLE 6.3 Standard enthalpies of combustion at 25°C

Substance*	Formula	ΔH_c°, kJ/mol
acetylene	$C_2H_2(g)$	−1300
benzene	$C_6H_6(l)$	−3268
carbon	C(*s*, graphite)	−394
ethanol	$C_2H_5OH(l)$	−1368
glucose	$C_6H_{12}O_6(s)$	−2808
hydrogen	$H_2(g)$	−286
methane	$CH_4(g)$	−890
octane	$C_8H_{18}(l)$	−5471
propane	$C_3H_8(g)$	−2220
urea	$CO(NH_2)_2(s)$	−632

*C is converted to $CO_2(g)$, H to $H_2O(l)$, and N to $N_2(g)$. More values are given in Appendix 2A(2).

The complete combustion of hydrocarbons produces carbon dioxide and water. An example is the combustion of propane,

$$C_3H_8(g) + 5O_2(g) \longrightarrow 3CO_2(g) + 4H_2O(l) \qquad \Delta H^\circ = -2220 \text{ kJ}$$

which shows that the standard enthalpy of combustion of propane is

$$\Delta H_c^\circ = \frac{-2220 \text{ kJ}}{1 \text{ mol } C_3H_8} = -2220 \text{ kJ/mol } C_3H_8$$

Many standard enthalpies of combustion have been measured, and a few are listed in Table 6.3 (see also Appendix 2A(2)). All enthalpies of combustion of organic compounds are negative because all such combustions are exothermic.

▼ EXAMPLE 6.9 Using enthalpy of combustion to calculate heat released

Gasoline, although a mixture, is thermochemically similar to pure octane. Calculate the quantity of heat released when 1.0 L of gasoline (density, 0.70 g/mL) burns completely under standard conditions at 25°C. Use data from Table 6.3.

U Exercise 6.65.

STRATEGY To calculate the quantity of heat released, we need to know the number of moles of C_8H_{18} burned, and then we can use the enthalpy of combustion (Table 6.3). The amount is obtained from the mass of the octane, which we can calculate from the volume and the density, both of which are given.

SOLUTION The three conversion factors we need are derived from the relations

N All three of these "equalities" are in fact equivalences.

$$1.0 \text{ mL octane} = 0.70 \text{ g } C_8H_{18}$$

$$114.2 \text{ g } C_8H_{18} = 1 \text{ mol } C_8H_{18}$$

$$1 \text{ mol } C_8H_{18} = -5471 \text{ kJ}$$

The enthalpy change is therefore

U Exercises 6.78 and 6.79.

Enthalpy change (kJ)

$$= 1.0 \times 10^3 \text{ mL } C_8H_{18} \times 0.70 \frac{\text{g}}{\text{mL}} \times \frac{1 \text{ mol } C_8H_{18}}{114.2 \text{ g } C_8H_{18}} \times \frac{-5471 \text{ kJ}}{1 \text{ mol } C_8H_{18}}$$
$$= -3.4 \times 10^4 \text{ kJ}$$

The combustion is exothermic and releases 34 MJ of heat into the surroundings, which is enough to heat more than 120 L of water from room temperature to boiling.

EXERCISE E6.9 The density of ethanol is 0.79 g/mL. Calculate the quantity of heat produced when 1.0 L burns under standard conditions. [*Answer*: 23 MJ]

N Nitrogen is converted to nitrogen oxides if the combustion results in very high temperatures.

Combustion reactions are a useful source of data because they all result in CO_2 and H_2O (in the case of hydrocarbons; but also N_2 when nitrogen is present and SO_2 from any sulfur), and these products act as a pool of atoms that can be used to rebuild other compounds. Their usefulness is illustrated by the following example.

EXAMPLE 6.10 Using combustion reactions to reproduce a synthesis reaction

Use standard enthalpies of combustion to calculate the standard reaction enthalpy for the synthesis of urea, $CO(NH_2)_2$, from its elements:

$$C(s) + 4H_2(g) + \tfrac{1}{2}O_2(g) + N_2(g) \longrightarrow CO(NH_2)_2(s)$$

STRATEGY We must decide how to reproduce the overall equation as sums and differences of combustion equations. We write all the combustion reactions we shall need (of C to CO_2, H_2 to H_2O, and $CO(NH_2)_2$ to CO_2, H_2O, and N_2), and then combine them (by using the strategy described previously) to reproduce the overall equation. Standard enthalpies of combustion are listed in Table 6.3.

SOLUTION The three thermochemical equations we shall need are

$$C(s) + O_2(g) \longrightarrow CO_2(g) \qquad \Delta H^\circ = -394 \text{ kJ} \quad \text{(a)}$$
$$H_2(g) + \tfrac{1}{2}O_2(g) \longrightarrow H_2O(l) \qquad \Delta H^\circ = -286 \text{ kJ} \quad \text{(b)}$$
$$CO(NH_2)_2(s) + \tfrac{3}{2}O_2(g) \longrightarrow CO_2(g) + 2H_2O(l) + N_2(g) \qquad \Delta H^\circ = -632 \text{ kJ} \quad \text{(c)}$$

The first equation has the correct number of C atoms on the left, so we start with it:

$$C(s) + O_2(g) \longrightarrow CO_2(g) \qquad \Delta H^\circ = -394 \text{ kJ}$$

We need to reverse (c) to obtain urea as a product:

$$CO_2(g) + 2H_2O(l) + N_2(g) \longrightarrow CO(NH_2)_2(s) + \tfrac{3}{2}O_2(g) \qquad \Delta H^\circ = -(-632 \text{ kJ})$$

The sum of these two thermochemical equations is

$$C(s) + O_2(g) \longrightarrow CO_2(g) \qquad \Delta H^\circ = -394 \text{ kJ}$$
$$CO_2(g) + 2H_2O(l) + N_2(g) \longrightarrow CO(NH_2)_2(s) + \tfrac{3}{2}O_2(g) \qquad \Delta H^\circ = +632 \text{ kJ}$$
$$C(s) + 2H_2O(l) + N_2(g) \longrightarrow CO(NH_2)_2(s) + \tfrac{1}{2}O_2(g) \qquad \Delta H^\circ = +238 \text{ kJ}$$

We now need to cancel the 2 mol of H_2O by adding reaction (b) multiplied by 2:

$$\begin{array}{rll} C(s) + 2H_2O(l) + N_2(g) \longrightarrow & CO(NH_2)_2(s) + \frac{1}{2}O_2(g) & \\ & & \Delta H^\circ = +238 \text{ kJ} \\ 2H_2(g) + O_2(g) \longrightarrow & 2H_2O(l) & \Delta H^\circ = 2 \times (-286) \text{ kJ} \\ \hline C(s) + 4H_2(g) + \frac{1}{2}O_2(g) + N_2(g) \longrightarrow & CO(NH_2)_2(s) & \Delta H^\circ = -334 \text{ kJ} \end{array}$$

This result shows that the synthesis reaction is exothermic and releases 334 kJ of heat for each mole of urea molecules formed.

EXERCISE E6.10 Use the standard enthalpy of combustion of benzene and other enthalpies of combustion from Table 6.3 to calculate the standard reaction enthalpy for the synthesis of 1 mol of $C_6H_6(l)$ from its elements.

[*Answer*: +46 kJ]

U Exercise 6.72.

6.4 ENTHALPIES OF FORMATION

In everyday conversation we speak of altitudes above sea level, without knowing (or caring about) actual distances from the center of the Earth. Analogously, in chemistry we report the enthalpies of compounds as a kind of "thermochemical altitude" above a "thermochemical sea level."

Standard enthalpies of formation. We let the elements define thermochemical sea level. Then any compound is at a thermochemical altitude given by the enthalpy change that occurs when it is formed from its elements (Fig. 6.17). We therefore define the standard enthalpy of formation ΔH_f° as follows:

E As remarked above, the modern definition refers to a pressure at 1 bar.

> The **standard enthalpy of formation** of a compound is the standard reaction enthalpy per mole of compound for the compound's synthesis from its elements in their most stable form at 1 atm and the specified temperature.

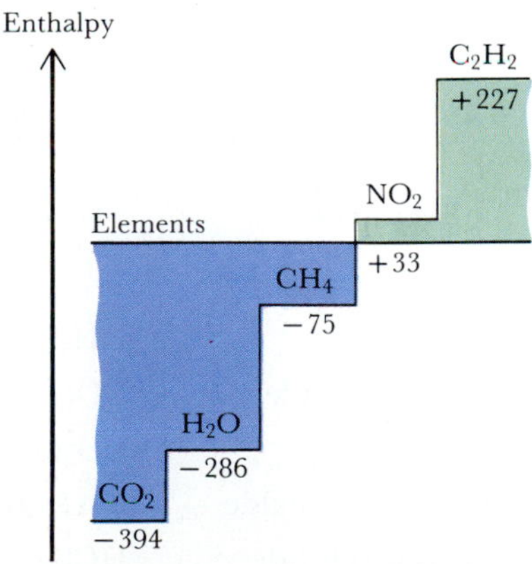

FIGURE 6.17 The elements define a thermochemical "sea level" for reporting the enthalpies of compounds. Some compounds have enthalpies of formation that are positive; others have negative enthalpies of formation.

EXAMPLE 6.11 Writing equations for enthalpies of formation

Write the chemical equation (and describe the states of the elements) to which the standard enthalpy of formation applies in the case of (a) ammonia gas and (b) ethanol, C_2H_5OH, vapor at 298 K.

STRATEGY In each case we write the thermochemical equation for the synthesis of the compound from its elements and adjust the stoichiometric coefficients so that 1 mol of the compound appears as the product. The states of the elements that we should take are the most stable ones at 1 atm and the stated temperature (298 K, or 25°C). The compound is in its specified state at 1 atm and the specified temperature.

SOLUTION (a) The balanced equation for the synthesis of ammonia as a gas is

$$N_2(g) + 3H_2(g) \longrightarrow 2NH_3(g)$$

We divide through by 2, so that we have NH_3 in place of $2NH_3$ on the right:

$$\tfrac{1}{2}N_2(g) + \tfrac{3}{2}H_2(g) \longrightarrow NH_3(g)$$

The nitrogen and hydrogen are in their gaseous states at 1 atm and 298 K. The product is ammonia gas at 1 atm and 298 K. (b) The balanced equation for the formation of ethanol vapor is

$$4C(s) + 6H_2(g) + O_2(g) \longrightarrow 2C_2H_5OH(g)$$

and division by 2 gives

$$2C(s) + 3H_2(g) + \tfrac{1}{2}O_2(g) \longrightarrow C_2H_5OH(g)$$

The most stable form of carbon at 1 atm and 298 K is graphite; the hydrogen and oxygen are the pure gases under the same conditions. The product is the *vapor* despite the fact that ethanol is a liquid at 1 atm and 298 K (we can take the product to be in *any* specified state; only the reactants must be in their most stable forms). Note also, that the synthesis reaction may be hypothetical (as in the case of ethanol).

EXERCISE E6.11 Repeat the example for the synthesis of (a) ice; (b) solid glucose.

[*Answer*: (a) $H_2(g) + \tfrac{1}{2}O_2(g) \rightarrow H_2O(s)$; (b) $6C(s, gr) + 6H_2(g) + 3O_2(g) \rightarrow C_6H_{12}O_6(s)$]

▲

U Exercises 6.68 and 6.69.

E There is one exception to the rule about using the most stable form of the element as the reference state: although white phosphorus is not the most stable form of phosphorus at 25°C, it is selected as the reference state: the other allotropes (the red and black forms of phosphorus) are less well characterized than white phosphorus.

T OHT: Table 6.4.

The standard enthalpy of formation ΔH_f° of an *element* in its most stable form is zero (by definition). However, the standard enthalpy of formation of an element in some other state is not zero. For example, the standard enthalpy of formation of carbon as graphite is zero, because it is the enthalpy of the null reaction

$$C(s, gr) \longrightarrow C(s, gr) \qquad \Delta H^\circ = 0$$

However, the enthalpy of formation of carbon as diamond is different from zero:

$$C(s, gr) \longrightarrow C(s, diamond) \qquad \Delta H^\circ = +1.9 \text{ kJ}$$

Values of enthalpies of formation for some compounds are listed in Table 6.4, which is extracted from a longer table in Appendix 2A. All the tabulated data in this text are for 25°C. The most stable form of

TABLE 6.4 Standard enthalpies of formation at 25°C

Substance*	Formula	ΔH_f°, kJ/mol	Substance*	Formula	ΔH_f°, kJ/mol
Inorganic compounds					
ammonia	$NH_3(g)$	−46.11	nitric oxide	$NO(g)$	+90.25
carbon monoxide	$CO(g)$	−110.53	nitrogen dioxide	$NO_2(g)$	+33.18
carbon dioxide	$CO_2(g)$	−393.51	sodium chloride	$NaCl(s)$	−411.15
dinitrogen tetroxide	$N_2O_4(g)$	+9.16	water	$H_2O(l)$	−285.83
hydrogen chloride	$HCl(g)$	−92.31		$H_2O(g)$	−241.82
hydrogen fluoride	$HF(g)$	−271.1			
Organic compounds					
acetylene	$C_2H_2(g)$	+226.73	glucose	$C_6H_{12}O_6(s)$	−1268
benzene	$C_6H_6(l)$	+49.0	methane	$CH_4(g)$	−74.81
ethanol	$C_2H_5OH(l)$	−277.69			

*A much longer list is given in Appendix 2A.

hydrogen at this temperature and 1 atm is the gas, that of bromine is the liquid, and that of iron is the solid. Because graphite is the most stable form of carbon at 25°C and 1 atm, enthalpies of formation of organic compounds refer to their synthesis starting from graphite.

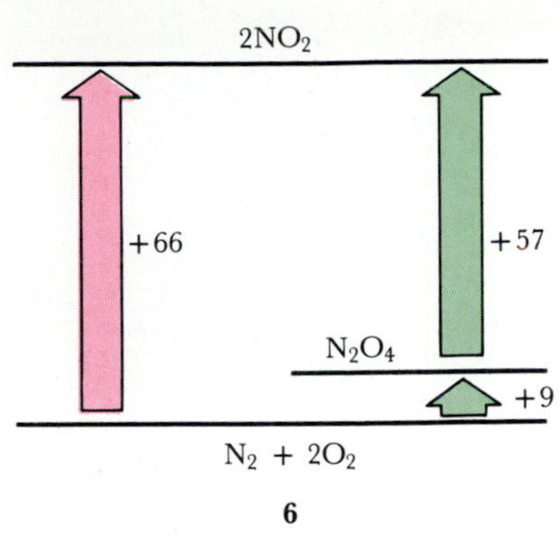

6

Standard enthalpies of formation are reported as kilojoules per mole of molecules (or formula units) of the compound. As an example, in the synthesis of liquid water,

$$2H_2(g) + O_2(g) \longrightarrow 2H_2O(l)$$

$\Delta H° = -571.6$ kJ and 2 mol H_2O molecules are formed; hence, the standard enthalpy of formation of liquid water is

$$\Delta H_f^\circ = \frac{-571.6 \text{ kJ}}{2 \text{ mol } H_2O(l)} = -285.8 \text{ kJ/mol } H_2O(l)$$

A common way of determining an enthalpy of formation when the compound cannot be synthesized directly is to use combustion enthalpies and Hess's law. The procedure was illustrated for urea in Example 6.10.

Combining standard enthalpies of formation. Standard enthalpies of formation may be used to calculate the standard enthalpy of *any* reaction. Suppose, for example, that we are interested in the reaction in which dinitrogen tetroxide decomposes into nitrogen dioxide:

$$N_2O_4(g) \longrightarrow 2NO_2(g) \qquad \text{(a)}$$

This equation can be expressed as the *difference* between the formation of the product and the formation of the reactant:

$$\begin{array}{lll} & N_2(g) + 2O_2(g) \longrightarrow 2NO_2(g) & \text{(b)} \\ \text{Subtract:} & N_2(g) + 2O_2(g) \longrightarrow N_2O_4(g) & \text{(c)} \\ \hline & — \quad\quad — \quad \longrightarrow 2NO_2(g) - N_2O_4(g) & \end{array}$$

which with N_2O_4 added to both sides is the same as (a). It follows (**6**) that the standard reaction enthalpy of (a) is the difference of the standard reaction enthalpies of (b) and (c). We can obtain this difference from the enthalpies of formation of the two compounds (Table 6.4):

$$\Delta H°(\text{b}) = 2 \text{ mol } NO_2 \times (+33.18 \text{ kJ/mol } NO_2) = +66.36 \text{ kJ}$$

$$\Delta H°(\text{c}) = 1 \text{ mol } N_2O_4 \times (+9.16 \text{ kJ/mol } N_2O_4) = +9.16 \text{ kJ}$$

The reaction enthalpy for reaction (a) is therefore the difference:

$$\Delta H°(\text{a}) = 66.36 \text{ kJ} - 9.16 \text{ kJ} = +57.20 \text{ kJ}$$

This same approach can be used for any reaction. For a reaction

$$\text{Reactants} \longrightarrow \text{products}$$

the standard reaction enthalpy is the difference between the reaction enthalpies for the formation of the products and the standard reaction enthalpies for the formation of the reactants (**7**). This statement is often written

$$\Delta H° = \sum n\Delta H_f^\circ\,(\text{products}) - \sum n\Delta H_f^\circ\,(\text{reactants}) \qquad (2)$$

↑ Sum of the enthalpies of formation of each product multiplied by its stoichiometric number of moles in the thermochemical equation

↑ Sum of the enthalpies of formation of each reactant multiplied by its stoichiometric number of moles in the thermochemical equation

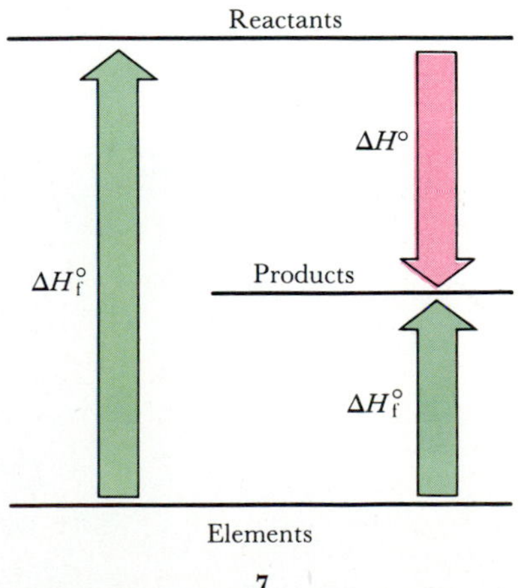

7

▼ **EXAMPLE 6.12 Using enthalpies of formation**

Use the information in Table 6.4 to calculate the standard enthalpy of combustion of benzene:

$$2C_6H_6(l) + 15O_2(g) \longrightarrow 12CO_2(g) + 6H_2O(l)$$

STRATEGY We can expect a negative value, because all combustions of hydrocarbons are exothermic. We can see from the chemical equation that the standard reaction enthalpy is given by the difference of the enthalpies of formation of the two products (12 mol CO_2 and 6 mol H_2O) and the two reactants (2 mol C_6H_6 and 15 mol O_2). The enthalpy of formation of an element is zero; the other values can be taken from Table 6.4. To find the enthalpy of combustion, we must divide the reaction enthalpy by the number of moles of C_6H_6 taking part.

U Exercises 6.63, 6.67, and 6.73–6.77.

SOLUTION From Table 6.4, we find that the total enthalpy change for the formation of products is

$$\begin{aligned}\sum n\Delta H_f^\circ \text{ (products)} &= 12 \text{ mol } CO_2 \times (-393.51 \text{ kJ/mol } CO_2) \\ &\quad + 6 \text{ mol } H_2O \times (-285.83 \text{ kJ/mol } H_2O) \\ &= -4722.12 \text{ kJ} + (-1714.98 \text{ kJ}) = -6437.10 \text{ kJ}\end{aligned}$$

The total enthalpy change for the formation of reactants is

$$\sum n\Delta H_f^\circ \text{ (reactants)} = 2 \text{ mol } C_6H_6 \times (+49.0 \text{ kJ/mol } C_6H_6) = +98.0 \text{ kJ}$$

The difference is

$$\Delta H^\circ = -6437.10 \text{ kJ} - (+98.0 \text{ kJ}) = -6535.1 \text{ kJ}$$

Because 2 mol C_6H_6 molecules are burned in the reaction, the standard enthalpy of combustion of benzene is

$$\Delta H_c^\circ = \frac{-6535.1 \text{ kJ}}{2 \text{ mol } C_6H_6} = -3267.6 \text{ kJ/mol } C_6H_6$$

EXERCISE E6.12 Calculate the standard enthalpy of combustion of glucose from the information in Table 6.4.

▲ [*Answer*: -2808 kJ/mol $C_6H_{12}O_6$]

6.5 THE ENTHALPY OF FUELS

Many factors affect the decision as to which fuel is best for a particular job. They include its costs, the ease with which it burns, the pollution its use may cause, and the amount of fuel that must be supplied to achieve a particular effect. We consider only the last.

The **specific enthalpy** of a fuel is its enthalpy of combustion per gram (expressed without the negative sign). Its **enthalpy density** is the enthalpy of combustion per liter (similarly, without the negative sign). Fuels with a high specific enthalpy release a lot of heat per gram when they burn. Hence, specific enthalpy is an important criterion when mass is of concern, as in airplanes and rockets. A fuel with a high enthalpy density releases a lot of heat per liter when it burns. Hence, enthalpy density is an important criterion when fuel storage volume must be considered. In some cases, both volume and mass are important; for example, increasing the size of an airplane to include more tanks for a less dense fuel may increase the retarding forces the aircraft experiences to an uneconomical level.

TABLE 6.5 **Thermochemical properties of four fuels**

Fuel	Combustion equation	ΔH_c°, kJ/mol	Specific enthalpy, kJ/g	Enthalpy density,* kJ/L
hydrogen	$2H_2(g) + O_2(g) \longrightarrow 2H_2O(l)$	−286	142	13
methane	$CH_4(g) + 2O_2(g) \longrightarrow CO_2(g) + 2H_2O(l)$	−890	55	40
octane	$2C_8H_{18}(l) + 25O_2(g) \longrightarrow 16CO_2(g) + 18H_2O(l)$	−5471	48	3.8×10^4
methanol	$2CH_3OH(l) + 3O_2(g) \longrightarrow 2CO_2(g) + 4H_2O(l)$	−726	23	1.8×10^4

*At atmospheric pressure and room temperature.

Three examples of readily available fuels are hydrogen, methane, and octane (Table 6.5). Methane is the major component of natural gas. Although octane is only one of a large number of components of gasoline, most of the molecules in a gallon are hydrocarbons with about eight carbon atoms, so octane is representative of the mixture. The data in Table 6.5 show that one advantage of using gasoline in mobile engines arises from its high enthalpy density of 38 MJ/L. Because this value is high, a fuel tank need not be large to carry a large store of energy. Where mass is important, as in a rocket, hydrogen may be an attractive fuel because of its high specific enthalpy. Mass is rarely a consideration in the supply of energy for domestic use. The higher enthalpy density of methane compared with that of hydrogen means that, for the same heating effect, a smaller volume of methane must be pumped through the pipes or supplied to the home in cylinders.

A great advantage of hydrogen over hydrocarbon fuels is that it produces no carbon dioxide when it burns. (Carbon dioxide is potentially harmful, because a higher concentration of CO_2 in the atmosphere might lead to a worldwide rise in temperature through the insulating effect known as the greenhouse effect.) The widespread adoption of hydrogen as fuel—perhaps by using solar-powered decomposition or electrolysis of water to produce it—would lead to what is called a hydrogen economy. However, one difficulty with hydrogen gas is its low enthalpy density. A solution might be to supply liquid hydrogen, which has a higher enthalpy density because the same volume contains many more molecules. Unfortunately, hydrogen condenses to a liquid only at low temperatures (at 20 K and below), and it is costly to refrigerate and store. Furthermore, even liquid hydrogen has such a low density (0.070 g/mL) that its enthalpy density is only 10 MJ/L, about a fourth that of gasoline, so fuel tanks for hydrogen would have to be large.

Chemists are currently exploring the feasibility of synthesizing compounds that can be decomposed to release hydrogen as it is needed (Fig. 6.18). Candidates include the hydrides formed when some metals (including titanium, copper, and their alloys with other metals) are heated in hydrogen. These compounds carry hydrogen in a smaller volume than that of liquid hydrogen itself and release it when they are heated or treated with acid. One example is the iron titanium hydride of approximate formula $FeTiH_2$. However, because of its iron and titanium content, the compound is relatively dense and its specific enthalpy is therefore small.

U Exercises 6.80 and 6.82.

N Exact thermodynamics of propane combustion. *J. Chem. Ed.* **1974**, *51*, 505.

N Applying thermodynamics to fuels: heats of combustion from elemental composition. *J. Chem. Ed.* **1980**, *57*, 56.

FIGURE 6.18 Calcium hydride releases hydrogen when it reacts with an acid; the reaction is exothermic, and produces clouds of water vapor (which condenses to a white mist).

6.6 THE ENTHALPY OF FOOD

Over 50% by mass of our food is in the form of carbohydrates. One kind of carbohydrate is the cellulose of wood, stalks, leaves, the outer parts of seeds, and the walls of plant cells. Cellulose is ingested but not digested, and it helps move material through the intestines. The digestible carbohydrates are the starches and sugars. The first stage of their digestion is a breakdown into glucose, which is soluble in water and is carried throughout the body in the bloodstream. Glucose is therefore the carbohydrate actually used as fuel in the body.

When glucose is used in animal cells, it is oxidized to carbon dioxide and water. The overall reaction of glucose in the body is thus the same as the combustion of glucose. The enthalpy of combustion of glucose is −2.8 MJ/mol, which corresponds to a specific enthalpy of 16 kJ/g: oxidizing 1 g of glucose produces 16 kJ of energy, enough to heat 1 L of water by about 4°C. The *average* specific enthalpy of all types of digestible carbohydrates, including starches, is about 17 kJ/g. Nutritional scientists often report this value in units called food Calories (Cal)

$$1 \text{ Calorie} = 4.184 \text{ kJ}$$

(Note the uppercase C, which distinguishes these Calories from the calories introduced in Section 6.1.) The average specific enthalpy of carbohydrates is therefore about 4 Cal/g.

A second major energy source is fats, which are compounds with long $—CH_2—CH_2—CH_2—$ chains, for example, tristearin ($C_{57}H_{110}O_6$). The enthalpy of combustion of tristearin is −75 MJ/mol, which corresponds to a specific enthalpy of about 84 kJ/g or 20 Cal/g. This value is significantly higher than the value for carbohydrates and similar to the value for gasoline (48 kJ/g). The difference arises from the higher proportion of oxygen already present in carbohydrate molecules. In other words carbohydrate molecules are already partially oxidized, so a smaller enthalpy change occurs during complete oxidation to carbon dioxide and water. Fats are like gasoline in the sense that their molecules contain very little oxygen. That is, they are initially only very slightly oxidized and release a great deal of heat when complete oxidation finally takes place. It is interesting to note that our bodies store energy in compounds that resemble the compounds found so convenient for storing energy in automobiles. Animals, airplanes, and automobiles all need to maximize their efficiency by storing energy in lightweight materials.

The third type of food is proteins, which are complex compounds of carbon, hydrogen, oxygen, and nitrogen made up of small units called amino acids (these are examined in Chapter 24). Proteins are large molecules having molar masses that sometimes reach 10^6 g/mol. They carry out many of the functions of a living cell and are too important to be used merely as fuel; nevertheless, in due course they are oxidized and, in mammals, form urea, $CO(NH_2)_2$ (**8**). This change corresponds to a specific enthalpy of about 17 kJ/g, similar to that of carbohydrates.

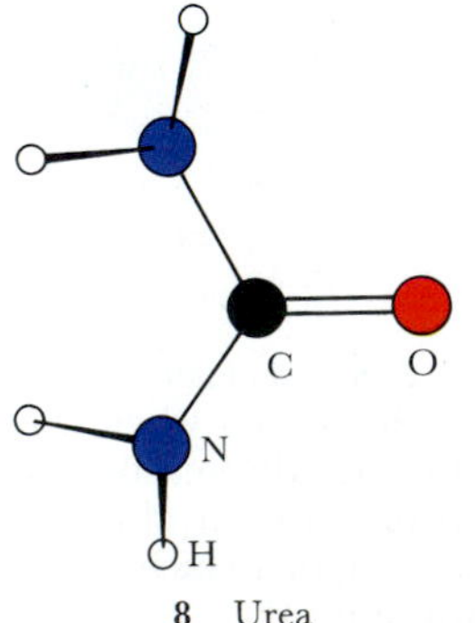

8 Urea

The composition and specific enthalpy resources of some typical foods are given in Table 6.6. The recommended consumption for 18- to 20-year-old males is 12 MJ/day; for females of the same age, it is 9 MJ/day. That energy intake is not used merely to produce warmth; nor is it always simply stored as fat (the analogue of the Earth's deposits

TABLE 6.6 Thermochemical properties of some foods

Food	Percentage composition				Specific enthalpy, kJ/g
	Water	Protein	Fat	Carbohydrate	
apples	84.3	0.3	0	11.9	2.5
beef	54.3	23.6	21.1	0	13.1
bread	39.0	7.8	1.7	49.7	12.6
cheese	37.0	26.0	33.5	0	17.0
cod	76.6	21.4	1.2	0	3.1
hamburger	40.9	15.8	14.2	29.1	17.3
milk	87.6	3.3	3.8	4.7	2.6
potatoes	80.5	1.4	0.1	19.7	3.5

of fossil fuel). Food in animals, like fuel in engines, is used to do work. This work may be mechanical, and used to move objects, including the animal itself. It may also be electrical or chemical, as in the propagation of nerve signals and in growth. We discuss this broader aspect of thermochemistry in Chapters 16 and 17.

N Heat, work, and metabolism. *J. Chem. Ed.* **1985**, *62*, 571.

SUMMARY

6.1 A **system** is the reaction vessel and its contents in which we have a particular interest. The **surroundings** consist of everything outside the system. The **energy** of a system is a measure of its capacity to do work or supply heat. The **heat capacity** of a substance is the proportionality between heat input and temperature rise. Reactions that release heat are called **exothermic reactions**, and reactions that absorb heat are called **endothermic reactions**. Transfers of energy as heat are measured with a **calorimeter**, and the observed change in temperature is interpreted in terms of the heat released or absorbed using the **calorimeter constant**.

6.2 The heat transferred at constant pressure is equal to the change of **enthalpy** H of the system. For an exothermic change, ΔH is negative; for an endothermic change, ΔH is positive. The enthalpy of a system is a **state property**, which means that it depends only on the current state of the system and not on how that state was prepared. Enthalpy changes are measured by calorimetry. Enthalpies of physical change include the **enthalpies of vaporization**, of **melting**, of **freezing**, and of **sublimation**. The enthalpy change for the reverse of any process is the negative of the enthalpy change for the original process.

6.3 A **reaction enthalpy** is the change in enthalpy that accompanies a reaction and is equal to the enthalpy of the products minus the enthalpy of the reactants. Reaction enthalpies are usually reported for substances in their **standard states**, their pure form at 1 atm pressure. A **thermochemical equation** is a chemical equation combined with a reaction enthalpy, and the latter refers to the enthalpy change that occurs for amounts of reactants and products in moles equal to the stoichiometric coefficients. Thermochemical equations are used to give the conversion factor from moles of reactant to the heat released or absorbed in a reaction. According to **Hess's law**, reaction enthalpies may be calculated from the most convenient **reaction sequence**, a series of (possibly hypothetical) reactions with the same net outcome. A special case of a reaction enthalpy is the **enthalpy of combustion**; an application of enthalpies of combustion is the determination of standard enthalpies of formation.

6.4 Standard reaction enthalpies may be calculated by combining **standard enthalpies of formation**, the standard enthalpy change when compounds are formed from their elements in their most stable forms.

6.5 and 6.6 Fuels are evaluated according to several criteria, including **specific enthalpy**, or the enthalpy of combustion per gram, and the **enthalpy density**, or the enthalpy of combustion per liter.

CLASSIFIED EXERCISES

Energy and Heat

6.1 One tablespoon of a salad dressing is reported as having an "energy content" of 16 kcal. What is this value in kilojoules?

6.2 It requires about 100 kJ of energy to heat water from room temperature to 100°C to make a cup of coffee. Express this energy requirement in kilocalories.

6.3 A 2.0-W electric heater is used for 5 min. Calculate the number of kilojoules of energy provided.

6.4 A 100-W light bulb burns for 3.5 h in a living room. Determine the energy released into the room (in kilojoules).

6.5 (a) Near room temperature, the specific heat capacity of benzene is 1.05 J/(g·°C). Calculate the heat needed to raise the temperature of 50.0 g of benzene from 25.3° to 37.2°C. (b) A 1.0-kg block of aluminum is supplied with 490 kJ of heat. What is the temperature change of the aluminum? The specific heat capacity of aluminum is 0.90 J/(g·°C).

6.6 (a) Near room temperature, the specific heat capacity of ethanol is 2.42 J/(g·°C). Calculate the heat that must be removed to reduce the temperature of 150.0 g C_2H_5OH from 50.0° to 16.6°C. (b) What mass of copper can be heated from 10° to 200°C when 400 kJ of energy is available?

6.7 The specific heat capacity of water is 4.18 J/(g·°C) and that of stainless steel is 0.51 J/(g·°C). Calculate the heat that must be supplied to a 500.0-g stainless steel vessel containing 450.0 g of water to raise its temperature from 25°C to the boiling point of water, 100°C. What percentage of the heat is used to raise the temperature of the water?

6.8 (a) Calculate the amount of energy needed to raise the temperature of 10.0 g of iron (specific heat capacity, 0.45 J/(g·°C)) from 25° to 500°C. (b) What mass of gold (specific heat capacity, 0.13 J/(g·°C)) can be heated through the same temperature difference when supplied with the same amount of energy as in (a)?

6.9 A 20.0-g piece of metal at 100°C is placed in a calorimeter containing 50.7 g of water at 22.0°C. The final temperature of the mixture is 25.7°C. What is the specific heat capacity of the metal? (*Hint:* Use the law of the conservation of energy to write a relation in which the energy lost by the metal is equal to the energy gained by the water.)

6.10 A 20.0-g piece of copper at 100°C is placed in a vessel of negligible heat capacity but containing 50.7 g of water at 22.0°C. Calculate the final temperature of the water. (*Hint:* Use the law of the conservation of energy to write a relation in which the energy lost by the copper is equal to the energy gained by the water.)

Enthalpy Change

6.11 Classify the following processes as exothermic or endothermic. (a) The formation of acetylene, which is used for oxyacetylene welding:

$$2C(s) + 2H_2(g) \longrightarrow C_2H_2(g) \qquad \Delta H° = +227\text{ kJ}$$

(b) The freezing of water:

$$H_2O(l) \longrightarrow H_2O(s) \qquad \Delta H° = -6.0\text{ kJ}$$

6.12 State whether the temperature will rise or fall when the following reactions are carried out in an insulated calorimeter. (a) The reaction which, in industry, is used to produce carbon disulfide from natural gas:

$$CH_4(g) + 4S(s) \longrightarrow CS_2(l) + 2H_2S(g) \quad \Delta H° = -106\text{ kJ}$$

(b) The dissolution of table salt in water:

$$NaCl(s) \longrightarrow NaCl(aq) \qquad \Delta H° = +3.9\text{ kJ}$$

6.13 A 100.0-g sample of water at 62.5°C was poured into a calorimeter containing 100.0 g of water at 19.8°C. The final temperature was 40.1°C. What is the calorimeter constant of the calorimeter containing 100.0 g of water?

6.14 The enthalpy of combustion of benzoic acid, C_6H_5COOH, which is often used to calibrate calorimeters, is −3227 kJ/mol. When 1.236 g of benzoic acid was burned in a calorimeter, the temperature increased by 2.345 K. What is the value of the calorimeter constant?

6.15 A 250-W electric heater increased the temperature of a calorimeter by 4.22°C in 55 s. When the oxidation of a methanol sample was carried out, the temperature rose from 22.49° to 26.77°C. What is the enthalpy change for the oxidation?

6.16 A heater was rated at 124 W. When it was turned on for 258 s, the temperature of a calorimeter rose from 18.23° to 22.48°C. Calculate the increase in temperature that would occur if a reaction released 12.5 kJ of heat in the same calorimeter.

6.17 A calorimeter constant was measured as 5.24 kJ/°C. With the combustion of a piece of asparagus, the temperature rose from 22.45° to 23.17°C. What is the enthalpy change for the combustion?

6.18 A 50.0-mL sample of 0.500 M $NaOH(aq)$ and 50.0 mL of 0.500 M $HNO_3(aq)$, both initially at 18.6°C, were mixed and stirred in a calorimeter having a calorimeter constant equal to 525.0 J/°C. The temperature of the mixture rose to 21.3°C. (a) What is the change in enthalpy for the neutralization reaction? (b) What is the change in enthalpy for the neutralization in kilojoules per mole of HNO_3?

Enthalpy of Physical Change

6.19 (a) The vaporization of 0.235 mol of liquid CH_4 requires 1.93 kJ of heat. What is the enthalpy of vaporization of methane? (b) An electric heater was immersed in a flask of boiling ethanol, C_2H_5OH, and 22.45 g of ethanol was vaporized when 21.2 kJ of energy was supplied. What is the enthalpy of vaporization of ethanol?

6.20 (a) When 25.23 g of methanol, CH_3OH, froze, 2.49 kJ of heat was released. What is the enthalpy of fusion of methanol? (b) A sample of benzene was vaporized at a reduced pressure at 25°C by using a 75.0-W heater for 500 s; 95 g of the liquid benzene vaporized. What is the enthalpy of vaporization of benzene at 25°C?

6.21 Use the information in Table 6.2 to calculate the enthalpy change for (a) the vaporization of 100.0 g of water at 373.2 K; (b) the melting of 600 g of solid ammonia at its freezing point (195.3 K).

6.22 Use the information in Table 6.2 to calculate the enthalpy change that occurs when (a) 200 g of methanol condenses at its boiling point (337.2 K); (b) 17.7 g of acetone, CH_3COCH_3, freezes at its freezing point (177.8 K).

6.23 How much heat is needed to melt 50.0 g of ice at 0°C and then heat the water to 25°C (Tables 6.1 and 6.2).

6.24 If we start with 150 g of water at 30°C, then how much heat must we add to convert it to steam at 100°C? (See Tables 6.1 and 6.2.)

6.25 Determine the time needed to melt 15.0 g of benzene at its melting point using a 1.0-kW heater.

6.26 Strong sunshine delivers about 1 kW/m^2. Calculate the maximum mass of pure ethanol that can be vaporized in 10 min from a beaker left in strong sunshine, assuming the surface area of the ethanol to be 50 cm^2. Assume all heat results in vaporization, not an increase in temperature.

Enthalpy of Chemical Change

6.27 The industrial preparation of carbon disulfide from coke and elemental sulfur occurs from a direct combination of the elements:

$$4C(s) + S_8(s) \xrightarrow{\Delta} 4CS_2(l) \qquad \Delta H° = +358.8 \text{ kJ}$$

(a) How much heat is absorbed in the reaction of 0.20 mol S_8? (b) Calculate the heat absorbed in the reaction of 20.0 g of carbon with an excess of sulfur.

6.28 The oxidation of nitrogen in the hot exhaust of jet engines and automobiles occurs by the reaction

$$N_2(g) + O_2(g) \longrightarrow 2NO(g) \qquad \Delta H° = +180.5 \text{ kJ}$$

(a) How much heat is absorbed in the formation of 0.70 mol NO? (b) How much heat is absorbed in the oxidation of 17.4 L of nitrogen measured at STP?

6.29 The combustion of octane is expressed by the thermochemical equation

$$2C_8H_{18}(l) + 25O_2(g) \longrightarrow 16CO_2(g) + 18H_2O(l) \qquad \Delta H° = -10{,}942 \text{ kJ}$$

(a) Calculate the mass of octane that must be burned to produce 12 MJ of heat. (b) How much heat will be evolved from the combustion of 1.0 gal of gasoline (assumed to be exclusively octane)? The density of octane is 0.70 g/mL.

6.30 Suppose that coal, of density 1.5 g/cm^3, is carbon (it is in fact much more complicated, but this is a reasonable first approximation). The combustion of carbon is described by the equation

$$C(s) + O_2(g) \longrightarrow CO_2(g) \qquad \Delta H° = -394 \text{ kJ}$$

(a) Calculate the heat produced when a lump of coal of size 7 cm × 6 cm × 5 cm is burned. (b) Estimate the mass of water that can be heated from 15° to 100°C with this piece of coal.

6.31 How much heat can be produced from a reaction mixture of 50.0 g of iron(III) oxide and 25.0 g of aluminum in the reaction

$$Fe_2O_3(s) + 2Al(s) \longrightarrow Al_2O_3(s) + 2Fe(s) \qquad \Delta H° = -851.5 \text{ kJ}$$

6.32 Calculate the amount of heat evolved from a reaction mixture of 13.4 L of sulfur dioxide at STP and 15.0 g oxygen in the reaction

$$2SO_2(g) + O_2(g) \longrightarrow 2SO_3(g) \qquad \Delta H° = -198 \text{ kJ}$$

Hess's Law

6.33 The standard enthalpies of combustion of graphite and diamond are −393.51 and −395.41 kJ/mol, respectively. Calculate the change in enthalpy for the graphite to diamond transition.

6.34 Elemental sulfur occurs in several forms, with rhombic sulfur the most stable under normal conditions and monoclinic sulfur somewhat less stable. The standard enthalpies of combustion of the two forms to sulfur dioxide are −296.83 and −297.16 kJ/mol, respectively. Calculate the change in enthalpy for the rhombic to monoclinic transition.

6.35 Two successive stages in the industrial manufacture of sulfuric acid are the combustion of sulfur and the oxidation of sulfur dioxide to sulfur trioxide. From the standard reaction enthalpies

$$S(s) + O_2(g) \longrightarrow SO_2(g) \qquad \Delta H° = -296.83 \text{ kJ}$$

$$2S(s) + 3O_2(g) \longrightarrow 2SO_3(g) \qquad \Delta H° = -791.44 \text{ kJ}$$

calculate the reaction enthalpy for the oxidation of sulfur dioxide to sulfur trioxide in the reaction

$$2SO_2(g) + O_2(g) \longrightarrow 2SO_3(g)$$

6.36 In the manufacture of nitric acid by the oxidation of ammonia, the first product is nitric oxide, which is then oxidized to nitrogen dioxide. From the standard reaction enthalpies

$$N_2(g) + O_2(g) \longrightarrow 2NO(g) \qquad \Delta H° = +180.5 \text{ kJ}$$

$$N_2(g) + 2O_2(g) \longrightarrow 2NO_2(g) \qquad \Delta H° = +66.4 \text{ kJ}$$

calculate the standard reaction enthalpy for the oxidation of nitric oxide to nitrogen dioxide:

$$2NO(g) + O_2(g) \longrightarrow 2NO_2(g)$$

6.37 Determine the enthalpy of reaction for the hydrogenation of acetylene to form ethane,

$$C_2H_2(g) + 2H_2(g) \longrightarrow C_2H_6(g)$$

from the following data:

$$2C_2H_2(g) + 5O_2(g) \longrightarrow 4CO_2(g) + 2H_2O(l) \qquad \Delta H^\circ = -2600 \text{ kJ}$$

$$2C_2H_6(g) + 7O_2(g) \longrightarrow 4CO_2(g) + 6H_2O(l) \qquad \Delta H^\circ = -3120 \text{ kJ}$$

$$H_2(g) + \tfrac{1}{2}O_2(g) \longrightarrow H_2O(l) \qquad \Delta H^\circ = -286 \text{ kJ}$$

6.38 Determine the enthalpy of reaction for the combustion of methane to carbon monoxide:

$$2CH_4(g) + 3O_2(g) \longrightarrow 2CO(g) + 4H_2O(l)$$

Use the following data:

$$CH_4(g) + 2O_2(g) \longrightarrow CO_2(g) + 2H_2O(l) \qquad \Delta H^\circ = -890 \text{ kJ}$$

$$2CO(g) + O_2(g) \longrightarrow 2CO_2(g) \qquad \Delta H^\circ = -566.0 \text{ kJ}$$

6.39 Calculate the enthalpy of reaction for the synthesis of hydrogen chloride gas

$$H_2(g) + Cl_2(g) \longrightarrow 2HCl(g)$$

from the following data:

$$NH_3(g) + HCl(g) \longrightarrow NH_4Cl(s) \qquad \Delta H^\circ = -176.0 \text{ kJ}$$

$$N_2(g) + 3H_2(g) \longrightarrow 2NH_3(g) \qquad \Delta H^\circ = -92.22 \text{ kJ}$$

$$N_2(g) + 4H_2(g) + Cl_2(g) \longrightarrow 2NH_4Cl(s) \qquad \Delta H^\circ = -628.86 \text{ kJ}$$

6.40 Calculate the reaction enthalpy for the formation of anhydrous aluminum chloride

$$2Al(s) + 3Cl_2(g) \longrightarrow 2AlCl_3(s)$$

from the following data:

$$2Al(s) + 6HCl(aq) \longrightarrow 2AlCl_3(aq) + 3H_2(g) \qquad \Delta H^\circ = -1049 \text{ kJ}$$

$$HCl(g) \longrightarrow HCl(aq) \qquad \Delta H^\circ = -73.5 \text{ kJ}$$

$$H_2(g) + Cl_2(g) \longrightarrow 2HCl(g) \qquad \Delta H^\circ = -185 \text{ kJ}$$

$$AlCl_3(s) \longrightarrow AlCl_3(aq) \qquad \Delta H^\circ = -323 \text{ kJ}$$

Enthalpy of Formation

6.41 Write chemical equations to which the values of the standard enthalpy of formation refer for (a) $KClO_3(s)$, potassium chlorate; (b) $H_2NCH_2COOH(s)$, glycine; (c) $Al_2O_3(s)$, alumina.

6.42 Write chemical equations to which the values of the standard enthalpy of formation refer for (a) $CH_3COOH(l)$; (b) $SO_3(g)$; (c) $CO_2(g)$.

6.43 Calculate the standard enthalpy of formation of dinitrogen pentoxide from the data

$$2NO(g) + O_2(g) \longrightarrow 2NO_2(g) \qquad \Delta H^\circ = -114.1 \text{ kJ}$$

$$4NO_2(g) + O_2(g) \longrightarrow 2N_2O_5(g) \qquad \Delta H^\circ = -110.2 \text{ kJ}$$

and the standard enthalpy of formation of nitric oxide.

6.44 An important reaction that occurs in the atmosphere is $NO_2(g) \rightarrow NO(g) + O(g)$, which is brought about by sunlight. How much energy must be supplied by the sun to cause it? Calculate the standard enthalpy of the reaction from the following information

$$O_2(g) \longrightarrow 2O(g) \qquad \Delta H^\circ = +498.4 \text{ kJ}$$

$$NO(g) + O_3(g) \longrightarrow NO_2(g) + O_2(g) \qquad \Delta H^\circ = -200 \text{ kJ}$$

and additional information from Appendix 2A.

6.45 Calculate the standard enthalpy of formation of $PCl_5(s)$ from the enthalpy of formation of $PCl_3(l)$ (Appendix 2A) and

$$PCl_3(l) + Cl_2(g) \longrightarrow PCl_5(s) \qquad \Delta H^\circ = -124 \text{ kJ}$$

6.46 When 1.92 g of magnesium reacts with nitrogen to form magnesium nitride, the heat evolved is 12.2 kJ. Calculate the standard enthalpy of formation of Mg_3N_2.

6.47 Use the information in Appendix 2A to calculate the enthalpy of reaction of (a) the oxidation of 10.0 g of sulfur dioxide,

$$2SO_2(g) + O_2 \longrightarrow 2SO_3(g)$$

(b) the reduction of 1.00 mol of $CuO(s)$ with hydrogen,

$$CuO(s) + H_2(g) \longrightarrow Cu(s) + H_2O(l)$$

6.48 Use the enthalpies of formation from Appendix 2A to determine the enthalpy of reaction of (a) the hydrogenation of 50.0 g of benzene to form cyclohexane,

$$C_6H_6(l) + 3H_2(g) \longrightarrow C_6H_{12}(l)$$

(b) the hydrogenation of 50.0 g of ethene to produce ethane,

$$C_2H_4(g) + H_2(g) \longrightarrow C_2H_6(g)$$

6.49 Calculate the standard reaction enthalpy of the following reactions. Use the enthalpies of formation in Appendix 2A. (a) The replacement of deuterium by ordinary hydrogen in heavy water:

$$H_2(g) + D_2O(l) \longrightarrow H_2O(l) + D_2(g)$$

(b) The formation of carbon disulfide from natural gas:

$$CH_4(g) + 4S(s) \longrightarrow CS_2(l) + 2H_2S(g)$$

(c) The oxidation of ammonia:

$$4NH_3(g) + 5O_2(g) \longrightarrow 4NO(g) + 6H_2O(g)$$

6.50 Calculate the standard reaction enthalpy for each reaction. Use the standard enthalpies of formation in Appendix 2A. (a) The final stage in the production of nitric acid, when nitrogen dioxide dissolves in and reacts with water:

$$3NO_2(g) + H_2O(l) \longrightarrow 2HNO_3(aq) + NO(g)$$

(b) The formation of boron trifluoride, which is widely used as a catalyst in the chemical industry:

$$B_2O_3(s) + 3CaF_2(s) \longrightarrow 2BF_3(g) + 3CaO(s)$$

(c) The formation of a sulfide by the action of hydrogen sulfide on an aqueous solution of a base:

$$H_2S(aq) + 2KOH(aq) \longrightarrow K_2S(aq) + 2H_2O(l)$$

Fuels and Foods

6.51 A minor component of gasoline is heptane (C_7H_{16}), which has a standard enthalpy of combustion of −4854 kJ/mol and a density of 0.68 g/mL. Calculate the specific enthalpy of heptane and its enthalpy density.

6.52 Another minor component of gasoline is toluene (C_7H_8), with a standard enthalpy of combustion of −3910 kJ/mol and a density of 0.867 g/mL. Calculate the specific enthalpy of toluene and its enthalpy density.

6.53 Calculate the specific enthalpy of magnesium from its enthalpy of combustion to magnesium oxide. Use the standard enthalpy of formation in Appendix 2A. Would aluminum, which burns to aluminum oxide, be a better fuel if mass were the only consideration?

6.54 Calculate the specific enthalpy of phosphorus from its enthalpy of combustion in oxygen to P_4O_{10}. Would sulfur burned to either SO_2 or SO_3 be a more efficient fuel if mass were the only consideration?

6.55 One problem with fuels containing carbon is that they produce carbon dioxide when they burn, so one consideration governing the selection of a fuel could be the heat per mole of CO_2 produced. (a) Calculate this quantity for methane and octane. (b) Calculate the heat produced per mole of CO_2 from the combustion of glucose.

6.56 The booster rockets of the space shuttle use a mixture of powdered aluminum and ammonium perchlorate in the exothermic redox reaction

$$2Al(s) + 2NH_4ClO_4(s) \longrightarrow Al_2O_3(s) + 2HCl(g) + 2NO(g) + 3H_2O(g)$$

(a) Calculate the specific enthalpy of a stoichiometric mixture of aluminum and ammonium perchlorate. (b) Would it be better to use magnesium in place of aluminum? Explain your conclusion.

UNCLASSIFIED EXERCISES

6.57 A calorimeter constant is 8.92 kJ/°C. The combustion of 1.00 oz (28.3 g) of cheese produces about 460 kJ. What mass of cheese will produce a temperature change in the calorimeter of 2.37°C?

6.58 A 100-g serving of shrimp provides 91 Calories. If a 20.0-g sample is burned in a calorimeter with a calorimeter constant of 4.66 kJ/°C, what is the expected temperature change?

6.59 A slice of bread has a mass of about 30 g. If there are 24 slices of bread in one loaf, and each slice supplies 85 Calories, how much energy (in kilojoules) can be obtained from a loaf of bread?

6.60 A 20.0-g ice cube at −14°C is heated until it becomes steam at 110°C. Using Tables 6.1 and 6.2, determine the total heat that must have been supplied. Use Fig. 6.14 as a guide to the calculation.

6.61 Calculate the heat required to melt 10 g of solid ethanol at its freezing point at 158.7 K and vaporize it at its boiling point of 351.5 K. Use Tables 6.1 and 6.2. Sketch the heating curve.

6.62 A 50.0-g ice cube at 0°C is added to a glass containing 400 g of water at 45°C. What is the final temperature of the system? Assume that no heat is lost to the glass.

6.63 (a) The thermite reaction can be used for the production of manganese:

$$3MnO_2(s) + 4Al(s) \longrightarrow 2Al_2O_3(s) + 3Mn(s)$$

Use the information in Appendix 2A and the enthalpy of formation of $MnO_2(s)$, which is −521 kJ/mol, to calculate the standard enthalpy of reaction. (b) What is the enthalpy change in the production of 10.0 g of manganese?

6.64 The enthalpy of combustion of methanol, CH_3OH, is −726 kJ/mol. What mass of methanol must be burned to heat 200 g of water in a 50-g Pyrex beaker from 20° to 100°C? (See Table 6.1.)

6.65 An important quantity in thermochemistry on account of its role in the analysis of biochemical processes is the enthalpy of combustion of glucose, $C_6H_{12}O_6$. In an experiment, 1.49 g of glucose was burned in a calorimeter with a calorimeter constant of 7.66 kJ/°C, and the temperature rose from 21.53° to 24.56°C. Calculate the enthalpy of combustion of glucose.

6.66 A 0.922-g sample of naphthalene, $C_{10}H_8$, a major component of moth balls, is burned in a calorimeter that has a calorimeter constant of 9.44 kJ/°C. The temperature of the calorimeter rose from 15.73° to 19.66°C. Calculate the enthalpy of combustion for naphthalene.

6.67 Would the designer of an industrial plant need to supply heat or remove it in the final step in the industrial preparation of urea, a commercial fertilizer? The reaction is

$$CO_2(g) + 2NH_3(g) \longrightarrow H_2O(g) + CO(NH_2)_2(s)$$

6.68 When 3.245 g of lead(IV) oxide is formed from lead metal and oxygen, 3.76 kJ of heat is released. What is the enthalpy of formation of $PbO_2(s)$?

6.69 The standard enthalpy of combustion of sulfur to sulfur dioxide is −2374.4 kJ/mol S_8. What is the standard enthalpy of formation of $SO_2(g)$?

6.70 Calculate the standard reaction enthalpy of the reduction of iron(II) oxide by carbon monoxide, a step in the production of iron,

$$FeO(s) + CO(g) \longrightarrow Fe(s) + CO_2(g)$$

given the following thermochemical equations:

$$3Fe_2O_3(s) + CO(g) \longrightarrow 2Fe_3O_4(s) + CO_2(g) \qquad \Delta H° = -47.2\ kJ$$

$$Fe_2O_3(s) + 3CO(g) \longrightarrow 2Fe(s) + 3CO_2(g) \qquad \Delta H° = -24.7\ kJ$$

$$Fe_3O_4(s) + CO(g) \longrightarrow 3FeO(s) + CO_2(g) \qquad \Delta H° = +35.9\ kJ$$

6.71 Calculate the enthalpy of formation of acetylene,

$$2C(s) + H_2(g) \longrightarrow C_2H_2(g)$$

from the following information:

$$2C_2H_2(g) + 5O_2(g) \longrightarrow 4CO_2(g) + 2H_2O(l) \qquad \Delta H° = -2600\ kJ$$

$$C(s) + O_2(g) \longrightarrow CO_2(g) \qquad \Delta H° = -393\ kJ$$

$$2H_2(g) + O_2(g) \longrightarrow 2H_2O(g) \qquad \Delta H° = -483.6\ kJ$$

$$H_2O(l) \longrightarrow H_2O(g) \qquad \Delta H° = +44\ kJ$$

6.72 Calculate the standard reaction enthalpy for the hydrogenation of acetylene to ethylene:

$$C_2H_2(g) + H_2(g) \longrightarrow C_2H_4(g)$$

This is a reaction that cannot be easily performed in the laboratory because of the formation of many by-products in the reaction. Use the following data:

$$2C_2H_2(g) + 5O_2(g) \longrightarrow 4CO_2(g) + 2H_2O(l) \qquad \Delta H° = -2600\ kJ$$

$$2C_2H_4(g) + 6O_2(g) \longrightarrow 4CO_2(g) + 4H_2O(l) \qquad \Delta H° = -2822\ kJ$$

$$2H_2(g) + O_2(g) \longrightarrow 2H_2O(l) \qquad \Delta H° = -572\ kJ$$

6.73 (a) Is the production of water gas (a cheap, low-grade industrial fuel) exothermic or endothermic? The reaction is

$$C(s) + H_2O(g) \longrightarrow CO(g) + H_2(g)$$

(b) Calculate the enthalpy change in the production of 200 L (at 500 Torr and 65°C) of hydrogen by this reaction.

6.74 (a) Acetic acid can be produced by the reaction of carbon monoxide with methanol with a nickel iodide catalyst:

$$CO(g) + CH_3OH(l) \longrightarrow CH_3COOH(l)$$

Use the information in Appendix 2A to determine whether this reaction is exothermic or endothermic by calculating its standard reaction enthalpy. (b) Acetic acid can also be formed by the oxidation of ethanol

$$C_2H_5OH(l) + O_2(g) \longrightarrow CH_3COOH(l) + H_2O(l)$$

This reaction occurs when wine goes sour. Decide whether this reaction is exothermic or endothermic by calculating its standard reaction enthalpy.

6.75 How much heat is evolved in the dissolution of 20.0 g of sodium hydroxide? The process is

$$NaOH(s) \longrightarrow Na^+(aq) + OH^-(aq)$$

6.76 (a) Calculate the standard reaction enthalpy for the reaction of calcium carbonate (limestone) with hydrochloric acid:

$$CaCO_3(s) + 2HCl(aq) \longrightarrow CaCl_2(aq) + H_2O(l) + CO_2(g)$$

6.77 A "silver tree" can be made in the laboratory by cutting a piece of copper metal in the shape of a tree and placing it into a silver nitrate solution, where the reaction

$$2Ag^+(aq) + Cu(s) \longrightarrow 2Ag(s) + Cu^{2+}(aq)$$

takes place. What is the change in enthalpy for the formation of 1.88 g of silver? Is the reaction an endothermic or exothermic process?

6.78 A natural gas mixture is burned in a furnace at a power-generating station at a rate of 13.0 mol per minute. If the fuel consists of 9.3 mol methane, 3.1 mol ethane, 0.40 mol propane, and 0.20 mol butane, what mass of CO_2 is produced per minute? How much heat is released per minute?

6.79 Suppose that of the heat produced by the combustion of 100 g of propane, C_3H_8, only 70% is useful in heating. What will the resulting temperature change be if the fuel is used to heat 250.0 g of ethanol?

6.80 The enthalpy of formation of trinitrotoluene (TNT) is −67 kJ/mol and its density is 1.65 g/cm^3. In principle, it could be used as a fuel, with the gases resulting from its decomposition (nitrogen, carbon dioxide, oxygen, and water vapor) streaming out of the rear of the rocket to give the required thrust. In practice, of course, it would be extremely dangerous as a fuel because it is sensitive to shock. Explore its potential as a rocket fuel by calculating its enthalpy density for the reaction

$$4C_7H_5N_3O_6(s) + 21O_2(g) \longrightarrow 28CO_2(g) + 10O_2(g) + 10H_2O(g) + 6N_2(g)$$

6.81 Calculate the enthalpy of vaporization of solid potassium bromide to a gas of ions, the process $KBr(s) \longrightarrow K^+(g) + Br^-(g)$ from the following information and the enthalpy of formation of KBr(*s*), which is −394 kJ/mol.

Atomization of potassium:

$$K(s) \longrightarrow K(g) \qquad \Delta H° = +89.2\ kJ$$

Ionization of potassium:

$$K(g) \longrightarrow K^+(g) + e^-(g) \qquad \Delta H° = +425.0\ kJ$$

Vaporization of bromine:

$$Br_2(l) \longrightarrow Br_2(g) \qquad \Delta H° = +30.9\ kJ$$

Dissociation of bromine:

$$Br_2(g) \longrightarrow 2Br(g) \qquad \Delta H° = +192.9\ kJ$$

Electron attachment to bromine:

$$Br(g) + e^-(g) \longrightarrow Br^-(g) \qquad \Delta H° = -331.0\ kJ$$

6.82 If you had to choose between carrying propane or butane on a camping expedition, which would you choose on the basis of its specific enthalpy?

CASE 6 ▸ ROCKET PROPELLANTS

Liquid and solid fuels are used in several stages to propel the space shuttles into orbit. Liquid hydrogen is stored at 20 K in the fuel tank; the oxidizing agent is oxygen, which is stored as a liquid at 90 K. For a typical shuttle mission, about 1.4×10^6 L (1.4 ML) of liquid hydrogen and 0.54 ML of liquid oxygen are used. For combustion, the hydrogen and oxygen mix as gases; the rapid expansion of the water vapor at the elevated temperature provides the thrust for the main engines. The combustion reaction is

$$2H_2(g) + O_2(g) \longrightarrow 2H_2O(g) \qquad \Delta H° = -483\ kJ$$

The reaction mixture for the main engines is adjusted to be fuel-rich—the excess hydrogen is used as a coolant for the main engine nozzles during combustion and lift-off.

The solid fuel in the booster rockets consists of ammonium perchlorate (the oxidizing agent), aluminum powder (the fuel), iron(III) oxide (the catalyst), and a binding agent. The skeletal equation of the combustion reaction is

$$NH_4ClO_4(s) + Al(s) \xrightarrow{\text{binder, } Fe_2O_3} Al_2O_3(s) + HCl(g) + H_2O(g) + NO_x(g) \quad \triangle$$

This skeletal equation is an idealized form of the overall reaction: in the turbulent conditions of the rocket exhaust, the initial products are all kinds of molecular fragments.

QUESTIONS

1. Assuming a stoichiometric burn of liquid hydrogen and liquid oxygen, calculate the heat evolved from the reaction of 1.4 ML of liquid hydrogen and 0.54 ML of liquid oxygen. The density of liquid hydrogen at 20 K is 0.070 g/cm^3 and that of liquid oxygen at 90 K is 1.15 g/cm^3.

2. Estimate (a) the enthalpy of reaction for the solid fuel in the booster rocket and (b) the specific enthalpy (the enthalpy of reaction per gram) of the aluminum in this reaction. Assume NO_x to be a 50%-by-volume mixture of NO and NO_2. ($\Delta H_f°$ of NH_4ClO_4 is −295 kJ/mol.)

3. The solid propellant in the booster rocket is about 16% aluminum and 70% ammonium perchlorate by mass. Binding agent, the iron(III) oxide catalyst, and other materials make up the remainder of the mixture. If the propellant for each booster rocket has a mass of 5.0×10^5 kg, what amounts (in moles) of fuel and oxidizing agent are mixed? Which is the limiting reactant?

4. Use the information in Questions 2 and 3 to calculate how much heat is released from one solid-fuel booster rocket.

This chapter introduces the main features of the structures of atoms and accounts for the form of the periodic table in terms of them. We begin to see how some of the properties of an element are related to its location in the table. One of the principal methods for investigating the structures of atoms is spectroscopy, in which the light emitted by atoms is analyzed. A prism, like the one shown here, is used to split up the light into its component colors, and these colors are then interpreted in terms of the arrangement of the electrons in atoms.

7 ATOMIC STRUCTURE AND THE PERIODIC TABLE

The experiments described in Section 2.3 showed that an atom of atomic number Z consists of a central, minute, dense, positively charged nucleus surrounded by Z electrons. When Ernest Rutherford proposed his model of the nuclear atom, he expected to be able to describe the motion of these electrons in terms of **classical mechanics**, the laws of motion proposed by Isaac Newton in the late seventeenth century. However, it quickly became clear that these laws failed when they were applied to electrons in atoms. Their replacement by the laws of quantum mechanics caused an intellectual earthquake that shook science to its foundations.

We shall see a little of that earthquake in this chapter. We shall also see, in this and the next two chapters, the importance of understanding the **electronic structure** of atoms, the details of the distribution of the electrons that surround the central nucleus of an atom. The electronic structure of atoms is the key to understanding the properties of the elements, the compounds they can form, the reactions they undergo, and the shapes of the molecules they form.

LIGHT AND SPECTROSCOPY

Much of our current understanding of atomic structure has come from **spectroscopy**, the analysis of the light emitted or absorbed by substances. (The name *spectroscopy* comes from the Latin word for "appearance.") Some elements emit light of a distinctive color when their compounds are heated in a flame (Fig. 7.1) or their vapors are exposed to an electric discharge (a storm of electrons and ions passing between two electrodes). Sodium atoms emit the yellow light characteristic of some highway and city streetlights. Potassium atoms emit violet light. Rubidium gives a red-violet flame (hence its name, from the Latin word for "red"), and cesium a blue one (hence its name, too, from the Latin word for "sky-blue"). These colors can often be seen in fireworks dis-

V Video 22: Flame tests of metals.

D Spectacular classroom demonstration of the flame test for metal ions. *J. Chem. Ed.* **1990**, *67*, 791. Vivid flame tests. *J. Chem. Ed.* **1988**, *65*, 545.

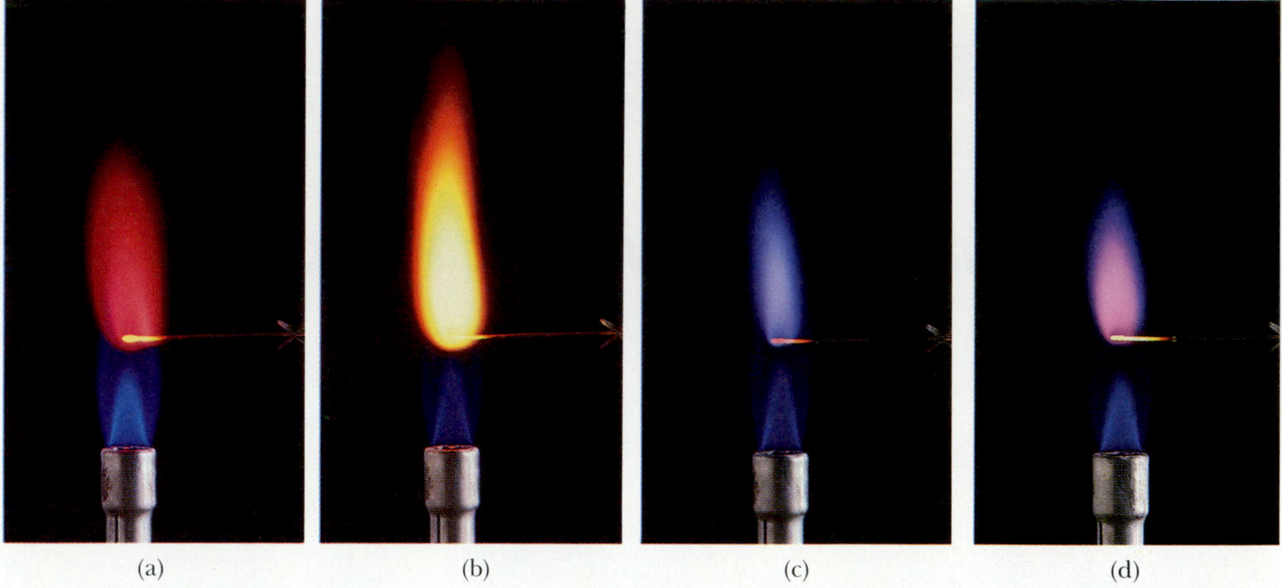

FIGURE 7.1 Flame tests are used to identify the elements in a compound. In particular, they provide an easy way of distinguishing the alkali metals. In each case except lithium, the colors come from energetically excited atoms. In lithium's case, LiOH molecules are responsible for the emission. (a) lithium; (b) sodium; (c) potassium; (d) rubidium.

plays (Box 7.1). The emitted light is in general a mixture of colors that can be separated by passing it through a prism, as sunlight is separated by raindrops to give a rainbow. The separated colors are recorded photographically as a **spectrum** (Fig. 7.2). Each ray of color gives an image of the slit the light passed through initially, so each individual color is recorded as one of a series of **spectral lines**, a record of each color in a spectrum.

7.1 THE CHARACTERISTICS OF LIGHT

To interpret the information provided by spectroscopy, we need to understand some of the properties of light. Light is **electromagnetic radiation**, a wave of oscillating electric and magnetic fields. A magnetic field is an influence that exerts a force on a moving charged particle. An electric field, the only field we consider here because it has the

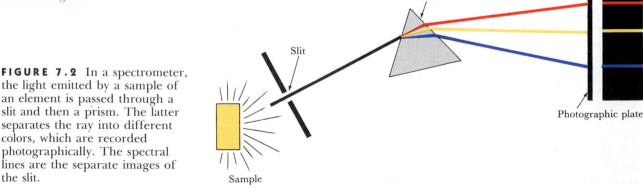

D Place a spatula amount of each salt in a Petri dish, add 10 mL of methanol, and ignite: LiCl will be red, NaI yellow, $CaCl_2$ orange red, $CuBr_2$ green with blue edges, RbCl faint violet, $SrCl_2$ crimson, $BaCl_2$ yellow green.

T OHT: Fig. 7.2

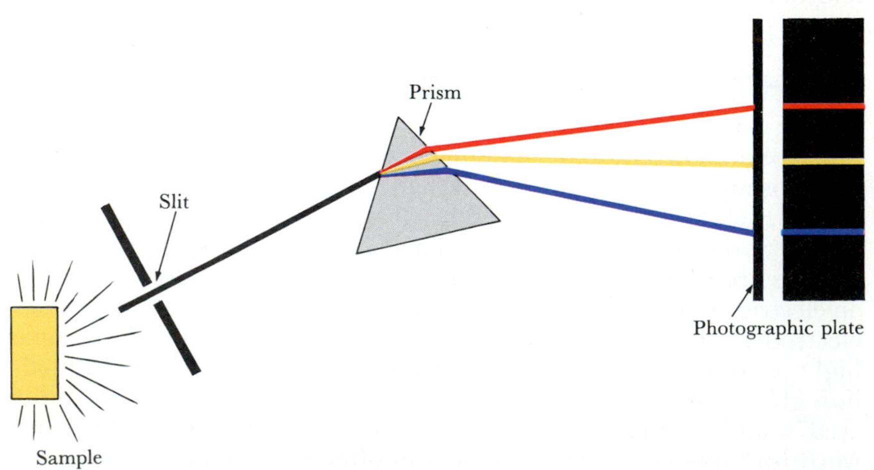

FIGURE 7.2 In a spectrometer, the light emitted by a sample of an element is passed through a slit and then a prism. The latter separates the ray into different colors, which are recorded photographically. The spectral lines are the separate images of the slit.

stronger effect on matter, is an influence that exerts a force on any charged particle (such as an electron), moving or not. The oscillation of the electric field of a ray of light indicates that the force the field exerts acts alternately in one direction and then in the opposite direction (Fig. 7.3). The wave is like a ripple in space that travels away from its source—an atom that emits the radiation, or the hot filament of an electric lamp—at 3.00×10^8 m/s (more precisely, 2.99792×10^8 m/s, a speed corresponding to over 670 million miles per hour). The speed of light is always denoted c and, in exponential notation (which will prove convenient in calculations), is

$$c = 3.00 \times 10^8 \text{ m·s}^{-1}$$

All electromagnetic radiation travels through empty space at this speed.

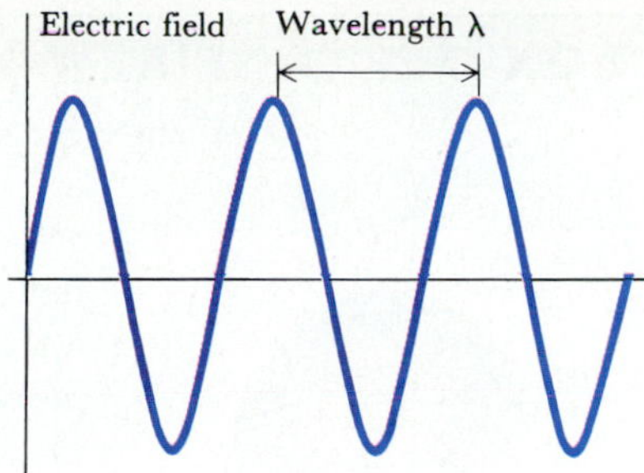

FIGURE 7.3 The electric field of electromagnetic radiation oscillates in space and time: this diagram represents a snapshot of the wave at a given instant. The distance between peaks is the wavelength λ. The crests of the wave move through space at a speed c, approximately 3.00×10^8 m/s. The number of oscillations at a given point that the field makes each second is the frequency ν of the radiation.

Frequency and wavelength. One complete reversal of the direction of the field (the direction of the force that a charged particle experiences as the radiation passes by) and its return to the original direction is called a **cycle**. The number of cycles per second is the **frequency** ν of the light (ν is the Greek letter nu). The unit of frequency is the **hertz** (Hz), and it is defined as 1 cycle per second (1 s^{-1}):

$$1 \text{ Hz} = 1 \text{ s}^{-1}$$

The unit honors Heinrich Hertz, one of the pioneers of the study of electromagnetic radiation.

The frequency of light determines its color (Table 7.1). If the radiation oscillates at 6.4×10^{14} Hz, for example, we see it as blue light. The light from a traffic signal changes frequency from 5.7×10^{14} to 5.2×10^{14} and then to 4.3×10^{14} Hz as it changes from green to yellow and then to red.

E Light of different frequencies excites different molecules in the retina of the eye, and we perceive the stimulation as different colors. There is only one type of retinal molecule (the photosensitive species), but it is attached to three different protein molecules. The three species of retinal-protein complex are sensitive to light of wavelengths close to 455 nm (blue), 530 nm (green), and 625 nm (red), respectively.

E Electromagnetic radiation of the shortest wavelength and highest frequency yet detected is that in cosmic radiation. The origin of cosmic radiation is still not fully settled. Typical x-ray wavelengths are about 100 pm. The wavelength of a 100-MHz radio transmitter is 3 m.

N Light and color. *J. Chem. Ed.* **1980**, *57*, 256.

TABLE 7.1 Color, frequency, and wavelength of electromagnetic radiation*

Electromagnetic radiation	Frequency, 10^{14} Hz	Wavelength, nm	Energy per photon, 10^{-19} J/photon
x-rays and γ-rays	$\geq 10^3$	≤ 3	≥ 660
ultraviolet	10	350	6.6
visible light			
violet	7.1	420	4.7
blue	6.4	470	4.2
green	5.7	530	3.7
yellow	5.2	580	3.4
orange	4.8	620	3.2
red	4.3	700	2.8
infrared	3.0	1000	1.9
microwaves and radiowaves	$\leq 3 \times 10^{-11}$ Hz	$\geq 3 \times 10^6$	$\leq 2.0 \times 10^{-22}$ J

*The values given are approximate but typical.

T OHT: Table 7.1.

U Exercise 7.81.

D Demonstrating color-wavelength relations. *J. Chem. Ed.* **1982**, *59*, 383.

BOX 7.1 ▸ FIREWORKS AND COLORED FLAMES

JOHN A. CONKLING, *Washington College; Executive Director, American Pyrotechnics Association*

The dazzling colors observed when a fireworks canister explodes high in the night sky are the product of a sequence of high-temperature chemical processes. The use of chemical mixtures for such purposes dates back over 1000 years to China and the discovery of black powder; an intimate mixture of 75% potassium nitrate, 15% charcoal, and 10% sulfur). Knowledge of the reactions involved, however, has been acquired mainly during this century.

A charge of black powder is used to launch the fireworks canister several hundred feet into the air, and a second charge of black powder bursts the cardboard casing and ignites dozens of marble-sized pellets, called stars. Each star contains a mixture of chemicals that burns to produce specific flame colors or that emits light when heated to high temperatures. The formulas for these pyrotechnics mixtures are closely guarded by the fireworks manufacturers.

The reactions that take place in each star during the combustion of the stars are electron transfer (redox) reactions. Potassium perchlorate ($KClO_4$) and potassium chlorate ($KClO_3$) are typical pyrotechnic oxidizing agents. Fuels are generally organic compounds that are typically natural products such as charcoal, tree resins, or cornstarch derivatives or synthetic products, such as polyvinylchloride or chlorinated rubber. The pyrotechnics mixture also contains elements such as strontium, barium, and sodium, the compounds of which emit light when heated to the high temperatures produced during the energetic redox reaction:

$$KClO_4 + \text{fuel} \xrightarrow{\Delta} KCl + \text{combustion products} + \text{heat}$$

Metallic fuels, such as aluminum, magnesium, and titanium, react exothermically to produce brilliant white flames. The particle size of the metal determines the duration of the flash. These mixtures are less frequently used to generate specific flame colors because their high flame tem-

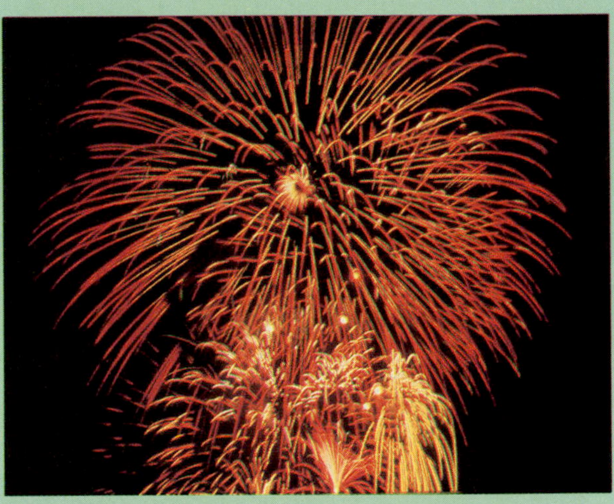

A sphere of red color is expelled from a bursting aerial fireworks device. The formulas for the pyrotechnics mixtures are closely guarded secrets.

A wave is also characterized by its **wavelength** λ (the Greek letter lambda), the peak-to-peak distance. Wavelength and frequency are related by the expression

$$\lambda \times \nu = c \qquad (1)$$

U Exercises 7.82–7.84.

Because the product of λ and ν is a constant, Eq. 1 shows that the higher the frequency, the shorter the wavelength (Fig. 7.4). Hence, blue light has a shorter wavelength (of about 470 nm, where 1 nm = 10^{-9} m) than red light (700 nm). The light from the three lamps of a

peratures (close to 3500°C) tend to destroy the other more fragile molecules that also produce specific colors. "Sparklers" use aluminum as the fuel.

The emission of colored starbursts occurs as follows. First, a molecule such as SrCl is generated in the combustion reaction by the combination of strontium and chlorine atoms. The electrons of the molecule are excited to a high energy level by the high temperature of the redox reaction:

$$\text{SrCl} \longrightarrow \text{SrCl}^*$$

The excited molecule then returns to a lower energy level and, in so doing, emits a photon of red light of specific wavelength corresponding to the energy difference between the excited and ground states.

$$\text{SrCl}^* \longrightarrow \text{SrCl} + h\nu \qquad (\lambda = 620\text{–}750 \text{ nm})$$

Modern instrumentation has enabled scientists to identify many of the principal species responsible for the other colors produced. Green flame color (light in the 490- to 560-nm wavelength range) is emitted by BaCl molecules, and CuCl molecules are the best blue emitters (435–440 nm). Atomic sodium, formed by the vaporization of a sodium compound, produces the yellow-orange light (589 nm) seen in some displays.

The chemical mixtures developed by fireworks manufacturers produce beautiful red, white, green, and orange flames. But a deep blue has remained elusive. The best blue emitter, CuCl, is unstable at the high flame temperatures (above 1500°C) needed to produce intense light emissions. In addition, blue light is less efficiently perceived by humans than other colors are. So there is still a need—even after 1000 years of experimentation—for a chemical mixture that burns with a deep blue flame. Perhaps someone is developing the right mixture at this moment. Look for a deep blue color next time you see a fireworks display.

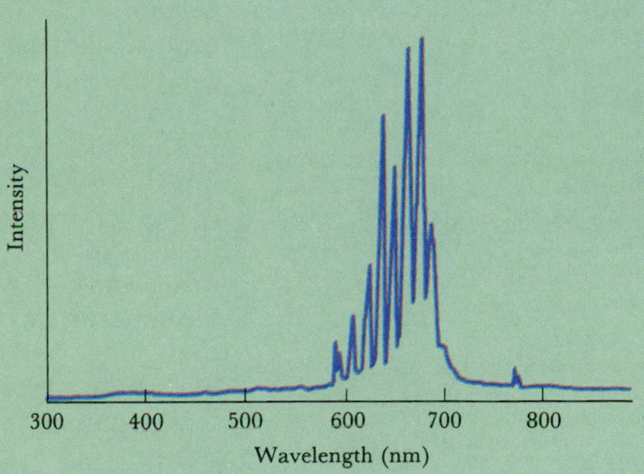

The light emission from pyrotechnic mixtures can be carefully studied by using modern instrumental methods. This diagram shows the intensity of light from a burning red-flame composition consisting of ammonium perchlorate (NH_4ClO_4, an oxidizer), strontium carbonate (a source of strontium atoms), and chlorinated rubber (a fuel and a source of chlorine atoms). In the flame, strontium and chlorine atoms combine to generate SrCl, which emits light in the red region (620–750 nm) of the visible spectrum when exposed to high temperature.

traffic signal goes from 530 to 580 and then to 700 nm, as they change from green through yellow to red (Fig. 7.5).

Our eyes can detect electromagnetic radiation with wavelengths in the range 700 nm (red light) to 400 nm (violet light). The radiation in this range is called **visible light**. White light, which includes sunlight, is a mixture of all wavelengths of visible light. Electromagnetic radiation also occurs with wavelengths ranging from less than a picometer up to wavelengths of more than several kilometers, but our eyes do not detect it—either because it is absorbed as it passes through the lens of the

E White light can be simulated by mixing three colors. One selection (used for TV screens) is red, green, and blue. In color printing, colors are created by mixing cyan (green-blue), magenta (red-purple), and yellow.

N The causes of color. *Sci. Amer.* **1980**, October, 124.

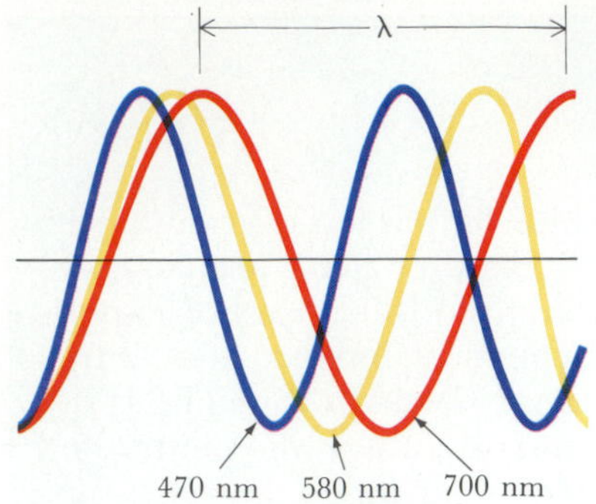

FIGURE 7.4 The color of radiation is determined by its wavelength (and frequency). Red light has a longer wavelength than blue light has. For example, light with a wavelength of 700 nm is perceived as red, whereas light of 470 nm is seen as blue.

T OHT: Fig. 7.4.

N How a photon is created or destroyed. *J. Chem. Ed.* **1979**, *56*, 631.

eye, or because the molecules in the retina do not respond to it. **Ultraviolet radiation** is radiation with a higher frequency (shorter wavelength) than that of violet light (*ultra* is Latin for "beyond"). **Infrared radiation**, the radiation we experience as heat, is radiation with a lower frequency (longer wavelength) than that of red light (*infra* is Latin for "below"). **Microwaves**, which are used in radar and microwave cookers, have wavelengths in the range millimeters to centimeters.

7.2 QUANTIZATION AND PHOTONS

One of the principal features of the theory known as quantum mechanics, which we meet in stages in the course of this chapter, is that it treats the transfer of energy as taking place in small packets. A packet of energy is called a **quantum** of energy (from the Latin for "amount").

Quanta. The existence of quanta is completely contrary to the concepts of classical mechanics. According to classical mechanics, an object can have any total energy. A pendulum, for instance, loses energy continuously by friction and can be given any quantity of energy simply by pushing it harder. Macroscopic (everyday-sized) objects like pendulums *seem* to us to be able to take on any energy. However, when we make very careful observations, we find that they can accept energy only in discrete amounts. Transferring energy is like pouring water into a bucket. Water *seems* to be a continuous fluid, and it seems that any amount can be transferred. However, the smallest amount of water that can be added is one H_2O molecule, which could be thought of as one quantum of water.

We summarize the existence of quanta by saying that energy is quantized:

> The **quantization** of a property is the restriction of that property to discrete values.

FIGURE 7.5 Each lamp in a traffic signal generates white light, a mixture of all colors of light, but the screens over the lamps allow only certain frequencies and wavelengths to pass.

The modern theory of light illustrates the concept of quantization very clearly. First, we note that electromagnetic radiation is a form of energy and hence that a light ray is a stream of energy emitted by the source. Then we take the viewpoint of quantum mechanics, and regard this stream of energy as made up of discrete energy packets. These particlelike packets of electromagnetic energy, or light quanta, are called **photons**. The more intense the light, the greater the number of photons in the ray. A dim source of light emits relatively few photons; a bright source of light emits a dense stream of photons. We can feel the energy of the photons in the infrared radiation emitted by the sun as we stand in warm sunlight.

The energy per photon is proportional to the frequency of the radiation. We write this proportionality as

$$E = h \times \nu \qquad (2)$$

where h is **Planck's constant**, a fundamental constant of nature with the value

$$h = 6.63 \times 10^{-34} \text{ J·s/photon}$$

The constant is named for Max Planck, the German physicist who first introduced the idea that energy is transferred in packets. For example, blue light of frequency 6.4×10^{14} Hz (or $6.4 \times 10^{14}\ s^{-1}$) consists of a stream of photons, the energy per photon being

$$E = 6.63 \times 10^{-34} \frac{\text{J·s}}{\text{photon}} \times 6.4 \times 10^{14}\ \text{s}^{-1}$$
$$= 4.2 \times 10^{-19}\ \text{J/photon}$$

A ray of red light also consists of a stream of photons, but because the frequency of the light is lower, each of its photons has less energy (2.8×10^{-19} J/photon). The photon energies of light of various colors are included in Table 7.1.

The combination of Eqs. 1 and 2 gives the energy of a photon in terms of its wavelength:

$$E = h \times \frac{c}{\lambda} \tag{3}$$

E A photon can be regarded as a measure of the extent of excitation of one particular mode of the electromagnetic field. Each mode (of a particular frequency) can be excited to its first, second, third, . . . , energy levels. We interpret the first state of oscillation as the presence of one photon, the second as the presence of two photons, and so on.

▼ EXAMPLE 7.1 Interpreting a light ray in terms of photons

In 1.0 s, a certain lamp gives out 25 J of energy in the form of yellow light of wavelength 580 nm (the rest of the energy is given out as light of different colors and as infrared radiation). How many photons of yellow light does the lamp generate in 1.0 s?

STRATEGY We should expect a large number of photons because the energy of a single photon is very small. We can calculate the energy per photon from Eq. 3. Because we know the total energy emitted as yellow light in 1.0 s, the number of photons is that total energy divided by the energy per photon.

SOLUTION It follows from Eq. 3 that the energy per photon of 580-nm light (for which $\lambda = 5.8 \times 10^{-7}$ m) is

$$E = 6.63 \times 10^{-34} \frac{\text{J·s}}{\text{photon}} \times \frac{3.00 \times 10^{8}\ \text{m·s}^{-1}}{5.8 \times 10^{-7}\ \text{m}}$$
$$= 3.4 \times 10^{-19}\ \text{J/photon}$$

Therefore, the number of photons that accounts for 25 J of energy is

$$\text{Number of photons} = 25\ \text{J} \times \frac{1\ \text{photon}}{3.4 \times 10^{-19}\ \text{J}} = 7.4 \times 10^{19}\ \text{photons}$$

This result means that when you turn on an electric lamp, it generates about 10^{20} photons of yellow light each second.

EXERCISE E7.1 How long should a lamp be left on in order to generate 1.0×10^{20} photons of blue (470 nm) light if it produces 5.0 J of energy per second in that region of the spectrum?

[*Answer*: 8.5 s] ▲

U Exercise 7.88.

E The relation between the frequency ν of the incident radiation and the speed v of the electrons ejected in the photoelectric effect (see below) is given by the *Einstein relation* $h\nu = \Phi + \frac{1}{2}m_e v^2$, where Φ is the work function of the metal, the minimum energy required to remove an electron from the metal to infinity. This relation follows from the conservation of energy. It implies that $v \propto \sqrt{\nu}$.

The photoelectric effect. Evidence for the existence of photons comes from the **photoelectric effect**, the emission of electrons from the surface of metals when electromagnetic radiation strikes it (Fig. 7.6). It is found that no electrons are emitted when the frequency of the radiation is below a certain threshold value characteristic of the metal. This observation is evidence for the existence of packets of energy, for if a photon has too low an energy, then it cannot eject an electron.

The photoelectric effect also confirms the relation between the energy of a photon and the frequency of the radiation expressed by Eq. 1.

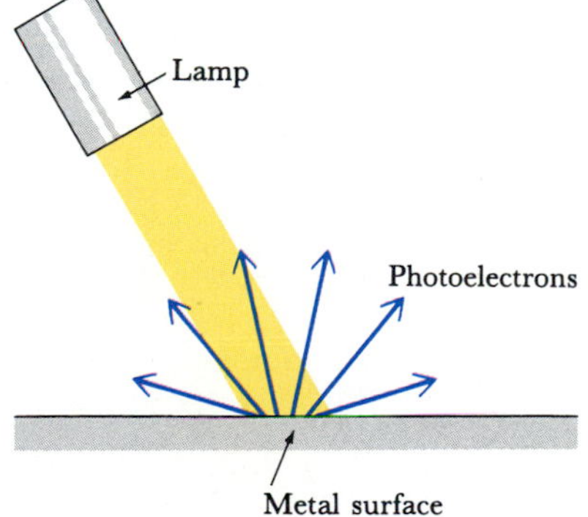

FIGURE 7.6 In the photoelectric effect, ultraviolet radiation from a lamp falls on the surface of a metal and ejects electrons. The ejected electrons are called photoelectrons.

Thus, it is found that the energy of the ejected electron is proportional to the frequency of the radiation (as long as the frequency is greater than the threshold value for emission). The more energy the photon brings, the more energy the electron will carry away.

THE STRUCTURE OF THE HYDROGEN ATOM

The energy emitted as radiation from an atom acts like a messenger that tells us about the atom's structure. We shall now see how to interpret this important information for the simplest atom, hydrogen.

7.3 THE SPECTRUM OF ATOMIC HYDROGEN

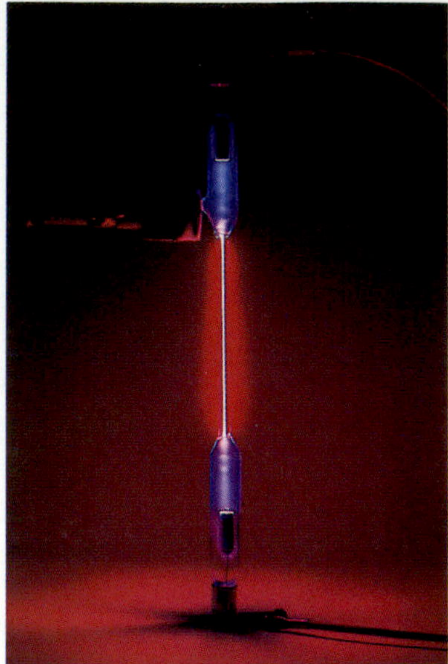

FIGURE 7.7 The red glow from this hydrogen discharge lamp comes from excited hydrogen atoms that are returning to a lower energy state and emitting the excess energy as visible radiation.

The spectrum of atomic hydrogen is observed when an electric current is passed through a low-pressure sample of hydrogen gas. The current—which is like a storm of electrons—breaks up the H_2 molecules and produces hydrogen atoms in energetically excited states. These atoms subsequently lose their excess energy as electromagnetic radiation, which we detect as the spectrum, and then combine to form H_2 molecules again. In the visible region, the spectrum of atomic hydrogen consists of four lines. The brightest line (at 656 nm) is red, and the excited sample glows with this red light (Fig. 7.7). Energetically excited hydrogen atoms also emit ultraviolet and infrared radiation, which can be detected electronically and photographically. The complete spectrum is shown diagrammatically in Fig. 7.8.

U Exercises 7.89 and 7.101.

T OHT: Fig. 7.8.

E The *Lyman series* occurs in the ultraviolet region of the spectrum; the *Paschen, Brackett,* and *Pfund series* occur in the infrared region; and the *Humphreys series* occurs in the very far infrared. The series involve transitions that terminate in the level with $n = 1$ (Lyman), 2 (Balmer), 3 (Paschen), 4 (Brackett), 5 (Pfund), and 6 (Humphreys).

N The spectrum of atomic hydrogen. *Sci. Amer.*, **1979**, March, 94.

Energy quantization. Although the complete spectrum looks like a jungle of lines, it actually follows a precise pattern, and the lines fall into several families called **series**. A part of the pattern was recognized by a Swiss schoolteacher, Joseph Balmer, in 1885; the **Balmer series** is the family that includes the lines in the visible region of the spectrum. The full pattern was recognized by the Swedish spectroscopist Johannes Rydberg. He discovered that the frequencies of all the lines in the spectrum are given by the formula

$$\nu = \mathcal{R} \times \left(\frac{1}{n_l^2} - \frac{1}{n_u^2}\right) \tag{4}$$

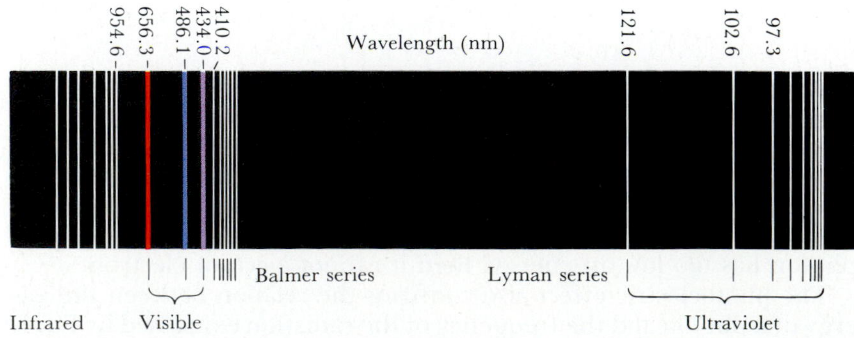

FIGURE 7.8 The complete spectrum of atomic hydrogen. The spectral lines have been assigned to various groups called series, two of which are shown.

where the constant $\mathcal{R}$, which is called the **Rydberg constant**, has the value 3.29×10^{15} Hz and n_l and n_u are whole numbers (for reasons that will be explained shortly, l stands for lower and u for upper). Each series is given by a different value of $n_l = 1, 2, \ldots$ and each individual line of a series is obtained by setting $n_u = n_l + 1, n_l + 2$, and so on, in turn.

The key idea that connects the specific frequencies of the light emitted by any kind of atom with its structure is that *each photon is emitted by one atom and carries away energy from the atom that emitted it.* Heating or passing an electrical discharge through the sample provides energy that changes the structure of the atoms. As that distorted structure readjusts, all or some of the excess energy is lost as a photon of light. If the energy of an atom decreases by ΔE, that amount of energy is carried away as a photon of light. Because the energy of a photon is $h\nu$, the frequency of the light generated by the atom is given by rearranging $h\nu = \Delta E$ into

$$\nu = \frac{\Delta E}{h} \quad (5)$$

This relation shows that if the decrease in the energy of the atom is large, then a photon of high-frequency (short-wavelength) radiation is generated. In that case we may detect blue or ultraviolet radiation coming from the sample. If the decrease in energy is small, then we may detect lower-frequency (longer-wavelength) red or infrared radiation (Fig. 7.9).

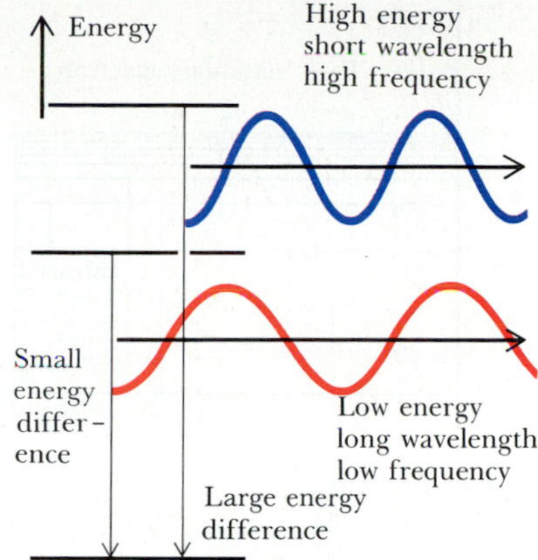

FIGURE 7.9 When an atom undergoes a transition from a state of higher energy to one of lower energy, it loses energy that is carried away as a photon. The greater the energy loss, the higher the frequency (and the shorter the wavelength) of the radiation emitted.

N Presenting the Bohr atom. *J. Chem. Ed.* **1982**, *59*, 372.

E Bohr's principal assertion was that only those electron orbits could exist for which the angular momentum of the electron was an integral multiple of $h/2\pi$. He then found the stable orbits by balancing the centrifugal force against the Coulombic attraction.

N Niels Bohr. *J. Chem. Ed.* **1982**, *59*, 303.

The Bohr model. The Danish physicist Niels Bohr (Fig. 7.10) attempted to account for the spectrum of atomic hydrogen by proposing what we now call the **Bohr model** of the atom. We shall see in Section 7.4 that the model is wrong in one significant respect; nevertheless its proposal was a very important stepping stone in the development of the modern description of atoms.

Bohr proposed that the single electron in the H atom travels in definite circular orbits around the central nucleus. He asserted that only certain of these orbits were possible, and he deduced that the energies of the allowed orbits are given by

$$E = -h \times \frac{\mathcal{R}}{n^2} \quad \text{with } n = 1, 2, \ldots \quad (6)$$

This expression gives the energy of the atom relative to the widely separated electron and nucleus. The negative sign means that the energy of the atom is lower when the electron is in one of the allowed orbits than when the electron and nucleus are so far apart that they do not interact. The energy is lowest when $n = 1$; the state of lowest energy is called the **ground state** of the atom.

The allowed energy levels predicted by Bohr's theory are shown in Fig. 7.11. The quantum number n is an integer (whole number) that labels the allowed orbits, with $n = 1$ for the lowest energy orbit, $n = 2$ for the next higher, and so on to infinity. The orbit with $n = 1$ lies closest to the nucleus and, as n increases, the orbits of the electron occur progressively farther from the nucleus.

Bohr's formula gives the energy of each orbit. His final job was to calculate the change in energy when an electron fell from a high-energy orbit with quantum number n_u (the upper orbit) into an orbit of

FIGURE 7.10 Niels Bohr (1885–1962).

lower energy with quantum number n_l (the lower orbit). The change in energy when this **transition**, or change of electronic state, occurs is

$$\Delta E = h \times \mathcal{R} \times \left(\frac{1}{n_l^2} - \frac{1}{n_u^2}\right)$$

When Bohr set ΔE equal to the energy of a photon $h\nu$ and canceled the h from both sides of the result, he obtained Eq. 4, which we know describes the spectrum. He also obtained a formula for $\mathcal{R}$ in terms of Planck's constant and the mass and charge of the electron.* With this formula he calculated that $\mathcal{R} = 3.29 \times 10^{15}$ Hz, a value in excellent agreement with Rydberg's experimental value. It is easy to imagine the thrill Bohr must have felt at this point in his calculations.

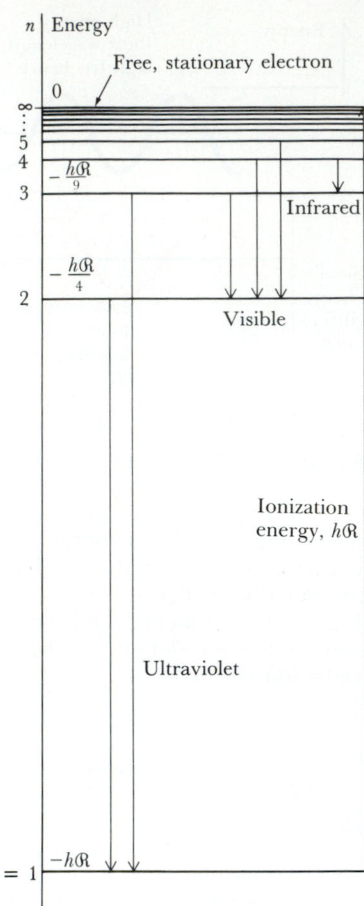

FIGURE 7.11 The energy levels predicted by Bohr, and some of the transitions responsible for the observed spectrum. The zero of energy corresponds to the widely separated nucleus and electron.

T OHT: Fig. 7.11.

U Exercises 7.86 and 7.87.

▼ EXAMPLE 7.2 Identifying a line in the spectrum of atomic hydrogen

Calculate the energies of the states of the hydrogen atom with $n = 2$ and $n = 3$, and calculate the wavelength of a photon emitted by the atom when an electron makes a transition between these two states. Identify the spectral line in Fig. 7.8.

STRATEGY The energy of each state is given by Eq. 6 (it is sensible to express the energy as a multiple of h because that constant will cancel later in the calculation). Then calculate the difference in energy between the two states and use Eq. 5 to express the difference as a frequency. Finally, convert that frequency to a wavelength by using Eq. 1.

SOLUTION The energies of the two states are

$$\text{For } n = 2, \quad E = -h \times \frac{\mathcal{R}}{n^2} = -h \times \frac{3.29 \times 10^{15}\text{ Hz}}{4}$$
$$= -h \times 8.23 \times 10^{14}\text{ Hz}$$

$$\text{For } n = 3, \quad E = -h \times \frac{\mathcal{R}}{n^2} = -h \times \frac{3.29 \times 10^{15}\text{ Hz}}{9}$$
$$= -h \times 3.66 \times 10^{14}\text{ Hz}$$

The difference is

$$\Delta E = -h \times (3.66 - 8.23) \times 10^{14}\text{ Hz} = h \times 4.57 \times 10^{14}\text{ Hz}$$

Then, from Eq. 5,

$$\Delta E = h\nu = h \times 4.57 \times 10^{14}\text{ Hz}$$

The h cancels on each side, to give $\nu = 4.57 \times 10^{14}$ Hz. Therefore, the wavelength of the radiation is

$$\lambda = \frac{c}{\nu} = \frac{3.00 \times 10^8\text{ m}\cdot\text{s}^{-1}}{4.57 \times 10^{14}\text{ s}^{-1}} = 6.56 \times 10^{-7}\text{ m}, \quad \text{or } 656\text{ nm}$$

The transition gives the red line in the spectrum shown in Fig. 7.8.

EXERCISE E7.2 Repeat the calculation for the transition from the state with $n = 4$ to $n = 2$ and identify the spectral line in Fig. 7.8.

[*Answer*: 486 nm; blue line] ▲

*Bohr's formula is $\mathcal{R} = m_e e^4/8h^3\epsilon_0^2$, where m_e and e are the mass and charge of the electron, h is Planck's constant, and ϵ_0 is a fundamental constant with the value $8.85 \times 10^{-12}\text{ C}^2/(\text{J}\cdot\text{m})$.

7.4 PARTICLES AND WAVES

Bohr's calculation was a spectacular numerical success. However, when all attempts to extend it to more complicated atoms failed, people began to suspect that there was something wrong with it. That view was confirmed when further experiments were carried out on the behavior of matter. It turns out that Bohr was right in his assumption that electrons exist in certain energy states (that is, that their energy is quantized), and he was right in ascribing the emission of radiation to transitions of electrons between these definite energy states. However, he was wrong in a very surprising and fundamental way about the motion of electrons in orbits. We shall now begin to introduce the modern description of the motion of electrons, which we shall see *cannot be thought of as movement along definite circular paths.*

E Bohr's theory is inferior to modern quantum mechanics theory not only because it cannot be extended to many-electron atoms: it is fundamentally flawed conceptually. One indication of the flaw is that it predicts that an electron has nonzero orbital angular momentum in the ground state: in fact, the orbital angular momentum is zero. Moreover, the Bohr model does not allow the electron to be found at the nucleus, whereas in fact it has a nonzero probability of being found there. For these reasons, we consider it best not to draw Bohrlike planetary atoms, for they are hard to eradicate.

The de Broglie relation. We have seen that a light ray, which classically is treated as a wave, should in fact also be thought of as a stream of particlelike photons. The French scientist Louis de Broglie had the curious idea that the same **wave-particle duality**, or combined wavelike and particlelike character of light, should apply to matter too. In 1924 he suggested that we should also think of an electron as having the properties of a wave. He proposed that every particle has wavelike properties and that its wavelength is related to its mass by the **de Broglie relation**:

N The de Broglie relation is usually written $\lambda = h/p$, where p denotes the linear momentum ($p = mv$).

U Exercise 7.85.

$$\lambda = \frac{h}{\text{mass} \times \text{velocity}} \qquad (7)$$

According to this relation, a heavy particle traveling rapidly has a short wavelength (small λ). A particle of small mass that is traveling slowly has a relatively long wavelength.

One of the earliest experiments to confirm the wavelike character of electrons was done by the American scientists Clinton Davisson and Lester Germer in 1927. They knew that when waves of any kind pass through a grid with a spacing comparable to the wavelength, characteristic diffraction patterns appear. The **Davisson-Germer experiment** showed that electrons gave the expected pattern when they were reflected from a crystal, where the layers of atoms act as the grid. They also found that the pattern corresponds exactly to that expected for electrons with a wavelength given by the de Broglie relation (Fig. 7.12).

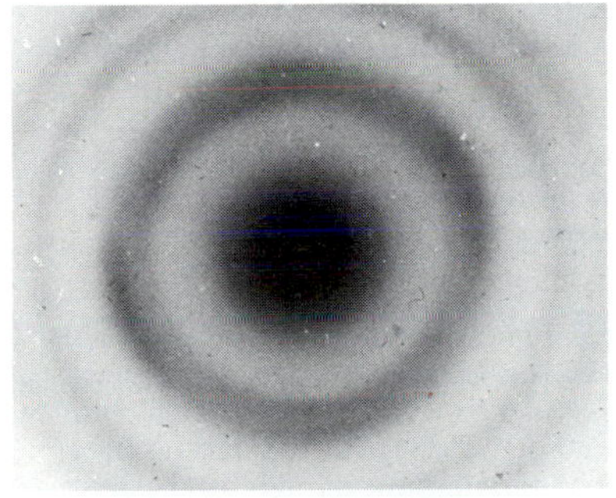

FIGURE 7.12 Davisson and Germer showed that electrons give a diffraction pattern when reflected from a crystal. G. P. Thomson working in Aberdeen, Scotland showed that they also give a diffraction pattern when they pass through a very thin gold foil. The latter is shown here. [G. P. Thomson was the son of J. J. Thomson, who identified the electron (Section 2.3). Both received Nobel prizes—J. J. for showing that the electron is a particle and G. P. for showing that it is a wave.]

We cannot detect the wavelike character of ordinary objects because their wavelengths are so short. The wavelength of a 100-g tennis ball traveling at 65 km/h (40 mi/h) is less than 10^{-33} m, which is far smaller than the diameter of an atomic nucleus. The scientists who developed classical mechanics got excellent agreement with their observations because the wavelike character of matter, of which they knew nothing, was completely undetectable for the objects they could observe (such as balls and planets). However, the wavelength of an electron moving at 2000 km/s is 360 pm, which is not very much bigger than the diameter of an atom (about 106 pm). When trying to account for the properties of electrons in atoms, we must take their wavelike character into account.

The uncertainty principle. The wavelike character of an electron has an important consequence: just as we cannot say where a wave in water

FIGURE 7.13 According to the uncertainty principle, the more precisely we know the position of a particle, the less precisely we can know its speed, and vice versa.

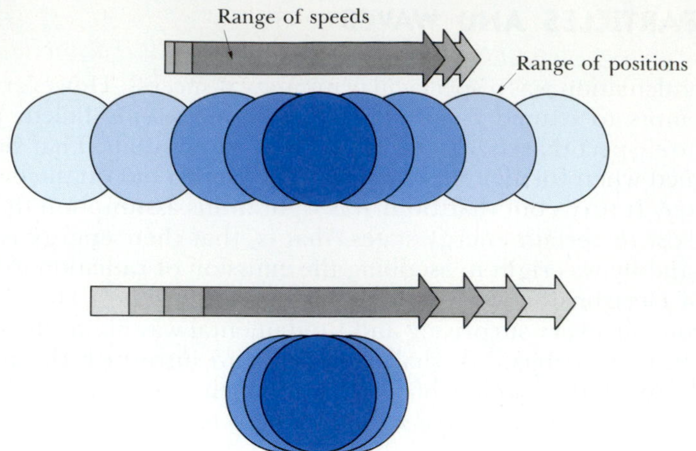

actually *is,* so we cannot say precisely where an electron is. Our fundamental inability to know the position of an electron is expressed by the **uncertainty principle** proposed by the German Werner Heisenberg in 1927. Heisenberg deduced that the more precisely we know the position of a particle (any particle, not only an electron), the less we can say about its speed, and vice versa (Fig. 7.13).* If we know the position of a particle precisely, then it does not have a definite speed. If we know the speed of a particle precisely, then it does not have a definite position. We have to choose how to describe the particle. If we choose to describe it in terms of a definite position, we cannot say anything about its speed. If, on the other hand, we choose to describe the particle in terms of how fast and in which direction it is traveling, then we can say nothing about its present or future position. It is not simply a case of not being able to *measure* speed and position simultaneously: a particle cannot *possess* the two properties simultaneously.

The relevance of the uncertainty principle to the description of the electronic structures of atoms (and hence to chemistry) is that it implies that Bohr's picture of an electron traveling in a precise orbit cannot be valid. An electron in a definite orbit would have a definite position and speed at every instant, but that is forbidden by the uncertainty principle. Bohr's theory must be wrong, because electrons cannot simultaneously be at a particular point in an orbit and be traveling with a definite speed. His theory is not just a little bit wrong; it is deeply and fundamentally wrong. An electron simply does not behave in the way he supposed—it does not travel around the central nucleus like a planet around the sun. To see how it does behave, we have to take our next step, which takes us into the heart of quantum mechanics.

The wavefunction of an electron. In quantum mechanics, we do not speak of the precise position of a particle; instead, we speak of the *probability* that the particle will be at a given location. For example, we

*More specifically, if we know the position of a particle of mass m to within a range Δx, then its speed must be uncertain by at least an amount Δv, where $\Delta x \times (m \times \Delta v) \geq h/4\pi$. If $\Delta x = 0$ (perfect knowledge about the location), then the only way this inequality can be satisfied is for Δv to be infinite (total ignorance about the speed). Similarly, if the speed is certain ($\Delta v = 0$), then the position must be completely uncertain (Δx infinite).

E Heisenberg devised a number of "thought experiments" to demonstrate that the uncertainty principle is consistent with observation. These experiments are often used to argue that we are limited in our ability to measure complementary observables simultaneously. We are adopting the stronger interpretation that the state of a system can be specified *either* by stating position coordinates *or* by stating the components of momentum. That is, if we specify position, then it is meaningless to try to specify momentum. The inability to measure these observables simultaneously is then implied and is supported by the analysis of the thought experiments. Classical physics is an overcomplete specification of nature.

E The uncertainty restricts the specification of position and momentum parallel to the position axis (such as x and p_x); it does not restrict position and momentum parallel to a perpendicular axis (such as x and p_y).

D Illustrating the problem described by Heisenberg's uncertainty principle. *J. Chem. Ed.* **1982**, *59*, 300.

N Perspectives on the uncertainty principle and quantum reality. *J. Chem. Ed.* **1985**, *62*, 192.

N The wave-particle duality. Teaching via a visual metaphor. *J. Chem. Ed.* **1988**, *65*, 339.

N Particles, waves, and the interpretation of quantum mechanics. *J. Chem. Ed.* **1975**, *52*, 573.

might be able to say that the probability of finding an electron close to the hydrogen nucleus is 0.01, which means that in 100 observations of the atom, the electron would be found close to the nucleus on one occasion.

The mathematical expression that summarizes where an electron is likely to be found is called its **wavefunction** and is denoted by ψ (the Greek letter psi). This wavefunction has high values in some regions and small values in others, just like the height of a wave in water. The interpretation of the wavefunction that we shall use was proposed by the German scientist Max Born. According to the **Born interpretation**,

E The wavefunction is a *probability amplitude* and its square is a *probability density*. The product of a probability density and a volume is a probability (see the note below).

> The probability of finding an electron at a given location is proportional to the square of ψ at that location.

E The precise statement of the Born interpretation is that the probability of finding a particle in an infinitesimal region of space δV where the wavefunction has the value ψ is proportional to $\psi^2 \times \delta V$.

For instance, if $\psi = 0.1$ at one location and $\psi = -0.2$ at another location, then the square of the wavefunction is 0.01 and 0.04, respectively, and there is a *fourfold* greater chance of finding the electron at the second location than at the first. If we know the wavefunction of an electron in an atom, then we can use the Born interpretation to predict where the electron is most likely to be found. We shall explore this feature of electrons in the rest of the chapter and shall use it throughout the text when looking for explanations of chemical properties in terms of the arrangement of electrons in atoms.

The wavefunction of an electron in an atom is called an **atomic orbital**. (The name suggests something similar to, but less definite than, the orbits of the Bohr model.) We can think of an atomic orbital as defining a region of space in which there is a high probability of finding the electron. We shall see shortly that for most purposes it is possible to draw diagrams of atomic orbitals and to discuss their characteristic shapes without going into their mathematical details. A hydrogen atom has an infinite number of different orbitals. Each of these orbitals corresponds to a different region of space, as we shall shortly see. Which orbital the electron occupies determines the energy of the atom; and when the electron makes a transition from one orbital to another, the change in energy is emitted as a photon, just as in the Bohr model.

E Each orbital corresponds to a particular spatial distribution of the electron around the nucleus (and therefore to a particular potential energy) and to a particular state of motion (and therefore to a particular kinetic energy). Each orbital therefore corresponds to a certain total energy.

D A lecture demonstration model of the quantum mechanical atom. *J. Chem. Ed.* **1981**, *58*, 801.

The atomic orbitals of hydrogen. The equation that must be solved to find the atomic orbitals of the hydrogen atom was discovered by the Austrian Erwin Schrödinger in 1926 (Fig. 7.14). When he solved the equation, he found that atomic orbitals could exist only for certain energies. Hence, whereas Bohr had to *assume* that only certain orbits existed, Schrödinger *deduced* from his equation that the energy of the atom is quantized and that electrons could exist only in discrete energy states. By a remarkable coincidence, the permitted energy levels calculated from the **Schrödinger equation** turned out to be exactly the same as those obtained by Bohr (see Eq. 6). Because Bohr's calculations agree with the spectroscopic data for hydrogen, Schrödinger's model of the atom also agrees with experiment. However, Schrödinger's model is more securely based in quantum mechanics than Bohr's model and can be extended to more complex atoms.

FIGURE 7.14 Erwin Schrödinger (1887–1961).

Schrödinger found that each atomic orbital is identified by three quantum numbers:

> A **quantum number** is a number that labels the state of an electron and specifies the value of a property.

TABLE 7.2 Quantum numbers of the hydrogen atom

Name of quantum number	Symbol	Values	Meaning
principal	n	1, 2, . . .	labels shell, specifies energy
azimuthal*	l	0, 1, . . . ,$n - 1$	labels subshell: $l = 0\ 1\ 2\ 3\ 4\ \ldots$ $s\ p\ d\ f\ g\ \ldots$
magnetic	m_l	$l, l - 1, \ldots, -l$	labels orbitals of subshell
spin-magnetic	m_s	$+\frac{1}{2}, -\frac{1}{2}$	labels spin state

*Also called the *orbital angular momentum quantum number.*

T OHT: Table 7.2.

We have already seen an example: in the Bohr theory each orbit is labeled by the quantum number n and its value gives the energy of the orbit. Schrödinger's three quantum numbers (Table 7.2; the fourth quantum number given there will be explained later) are the principal quantum number n, the azimuthal quantum number l, and the magnetic quantum number m_l.

E Quantum numbers arise from the requirement that the wavefunction, which may be pictured as a wave, behaves properly (just as a violin vibration must fit the length of the string). The principal quantum number stems from the requirement that the wavefunction behaves properly along a radius, and guarantees that the wavefunction approaches zero at large distances.

The **principal quantum number** n specifies the energy of an electron in a hydrogen atom exactly the way it did in the Bohr model (Eq. 6); and it can take the values 1, 2, 3, . . . up to infinity. We shall see shortly that there is only one orbital with $n = 1$ but that there are four orbitals with $n = 2$, nine orbitals with $n = 3$, and so on—in general, there are n^2 orbitals for each value of n. All orbitals with the same value of n are said to belong to the same **shell** of the atom. Thus, the four orbitals with $n = 2$ belong to the same shell, and all nine orbitals with $n = 3$ belong to another shell, and so on.

The principal quantum number also governs the average distance of the electron from the nucleus. (Remember that we cannot be definite about the position of the electron, but we can speak in terms of its average distance from a nucleus.) The important point is that the average distance of an electron increases as n increases. Thus, when the electron is in the orbital of the lowest energy shell ($n = 1$; when the atom is in its ground state), its average distance from the nucleus is smaller than when it occupies any of the orbitals of the shell with $n = 2$.

E The azimuthal and magnetic quantum numbers guarantee that the wavefunction behaves properly as the two angular coordinates (the latitude and longitude) of the electron are changed.

The **azimuthal quantum number** l is a label that groups the orbitals of a given shell into different sets called **subshells**: all the orbitals of a given shell that have the same value of l belong to the same subshell. For a shell of principal quantum number n, the azimuthal quantum number can take the values 0, 1, . . . up to $n - 1$, giving n values in all. This range of values for l means that there is only one subshell of the $n = 1$ shell (the one with $l = 0$), two subshells of the shell with $n = 2$ ($l = 0$ and 1), and so on. It is common practice to refer to the subshells by letters rather than numbers, using the following correspondence:

$$\begin{matrix} l = 0 & 1 & 2 & 3 & 4 \ldots \\ s & p & d & f & g \ldots \end{matrix}$$

The **magnetic quantum number** m_l specifies the individual orbital in a subshell. A subshell with azimuthal quantum number l is made up of $2l + 1$ individual orbitals, each orbital corresponding to one of the permitted values of m_l, namely, $l, l-1, l-2, \ldots$ down to $-l$. For example, because the orbitals in the p subshell of any shell have $l = 1$, there are three such orbitals and they have the magnetic quantum numbers $+1$, 0, and -1. It is often more convenient to use a different set of labels for these three orbitals and to name them p_x, p_y, and p_z. These alternative labels specify the shapes of the orbitals more directly, as we shall see.

E Orbitals with a specific value of orbital angular momentum about an axis (and having a definite value of m_l) correspond to running waves (the sense in which they run is the direction of the electron's angular momentum). The orbitals labeled p_x, etc., are linear combinations of running waves and correspond to standing waves.

▼ EXAMPLE 7.3 Identifying the number of orbitals in a shell

How many orbitals are there in the shell with $n = 4$?

STRATEGY We have to decide which subshells the $n = 4$ shell has, write down the number of orbitals in each one, and then add these numbers together. We know from the discussion above that l has values from 0 up to $n - 1$. We also know that the number of orbitals in a subshell of given value of l is $2l + 1$. We must combine the two pieces of information.

SOLUTION For $n = 4$, there are four subshells with $l = 0, 1, 2, 3$ designated s, p, d, f. Now, we draw up the following table:

l	Subshell	m_l values		Number of orbitals ($2l + 1$)
0	s	0		1
1	p	+1,0,−1		3
2	d	+2,+1,0,−1,−2		5
3	f	+3,+2,+1,0,−1,−2,−3		7
			Total:	16

That is, there are 16 orbitals in the shell with $n = 4$.

EXERCISE E7.3 Calculate the total number of orbitals in the shell with $n = 6$. What is the general formula?

[*Answer*: 36; number of orbitals = n^2] ▲

E The formal proof that the number of orbitals is n^2 is to sum $2l + 1$ from $l = 0$ to $l = n - 1$ using the formula for summing finite series.

U Exercise 7.91.

T OHT: Fig. 7.15.

The relation between shells, subshells, and orbitals is summarized in the chart in Fig. 7.15.

The s orbitals. The lowest-energy orbital of the hydrogen atom is the **1*s* orbital** (the orbital with $n = 1, l = 0$, and $m_l = 0$). It is the only orbital permitted when $n = 1$. An electron with a probability distribution given by a 1*s* orbital is said to **occupy** a 1*s* orbital and to be a **1*s* electron**.

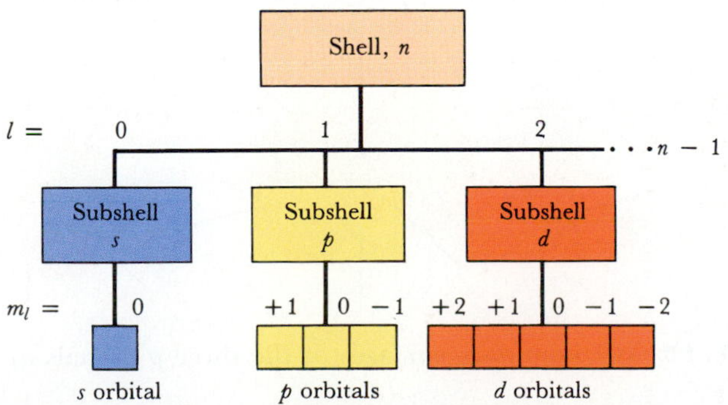

FIGURE 7.15 A chart showing the hierarchy of shells, subshells, and orbitals in an atom. A shell is designated by the value of the principal quantum number n. A given shell has n subshells, each one of which is designated by the quantum number l. Each subshell consists of $2l + 1$ individual orbitals that can be designated by the quantum number m_l. The color code shown here will be used throughout.

FIGURE 7.16 (a) The 1s orbital of hydrogen, drawn so that the density of shading represents the probability of finding the electron at any point. (b) The graph shows that the probability of finding the electron at any point along a given radius is greatest at the nucleus and decreases sharply with distance.

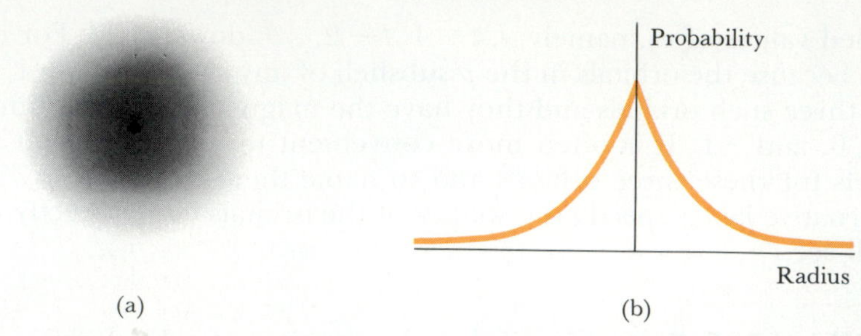

T OHT: Fig. 7.16.

E The amplitude of a 1s orbital decays exponentially from a nonzero, finite value at the nucleus to zero at infinity. It is proportional to $e^{-r/2a}$, where a is the Bohr radius (53 pm).

We can visualize the shape of the 1s orbital in several ways. Figure 7.16 uses shading to show the probability of finding the electron in the region surrounding the nucleus. The shading is darkest close to the nucleus, because the electron is most likely to be found near there. The shading becomes lighter with increasing distance from the nucleus, thus showing that there is a decreasing probability of finding the electron the farther we go from the nucleus. The shaded region is sometimes called the **electron cloud**. The illustration also shows a plot of the probability of the electron being found at any point along a given radius.

E That s orbitals are spherically symmetrical implies that they correspond to zero orbital angular momentum around the nucleus. That lack of orbital momentum allows the electron to be found at the nucleus.

N Order out of chaos: shapes of hydrogen orbitals. *J. Chem. Ed.* **1988**, *65*, 31.

N Orbital shape presentations. *J. Chem. Ed.* **1985**, *62*, 206.

All s orbitals are spherical. That is, the probability of finding the electron at a given distance from the nucleus is the same in every direction. The orbital is therefore often drawn as a sphere called a **boundary surface** (Fig. 7.17). The surface of the sphere is the boundary within which there is about 90% probability of finding the electron. For a hydrogen atom, the radius of the boundary surface is 140 pm.

The 2s orbital, the s orbital belonging to the shell with $n = 2$, is similar to the 1s orbital but spreads over a larger volume. Its boundary surface, like that of the 1s orbital and all s orbitals, is spherical, and the 2s orbital also is normally drawn as a sphere.

FIGURE 7.17 The simplest way of drawing any atomic orbital is as a boundary surface, a surface within which there is a high probability (typically 90%) of finding the electron. This sphere represents the boundary surface of an s orbital.

The p orbitals. Three orbitals with $l = 1$ ($m_l = +1,0,-1$) can occur for shells with $n \geq 2$. These orbitals are called ***p* orbitals** and have the dumbbell shapes shown in Fig. 7.18. All three have the same shape, but each lies along one of three perpendicular axes. They are labeled with the name of the axis they lie along, which accounts for the notation p_x, p_y, and p_z. A **p_x electron**, an electron in a p_x orbital, is most likely to be found somewhere near or along the x axis.

A p orbital has a **nodal plane** running through the nucleus, a plane on which a p electron will never be found. Whereas an s electron may be found at the nucleus, a p electron will never be found there. The node can be seen clearly in Fig. 7.19, which shows the probability of finding the electron at different distances from the nucleus.

N A p orbital is dumbbell shaped.

T OHT: Fig. 7.18.

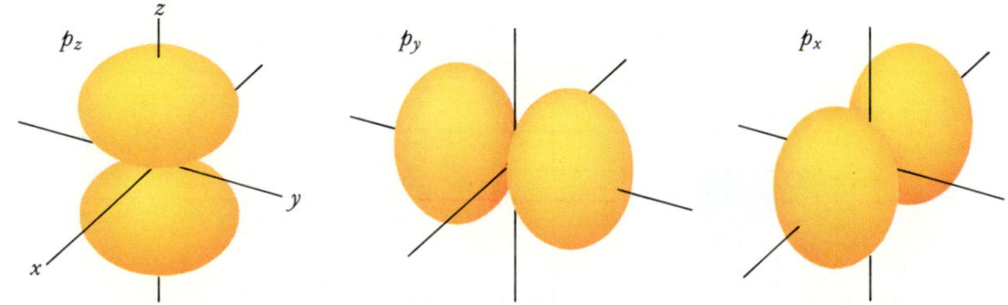

FIGURE 7.18 The boundary surfaces of the three p orbitals of a shell with $n \geq 2$.

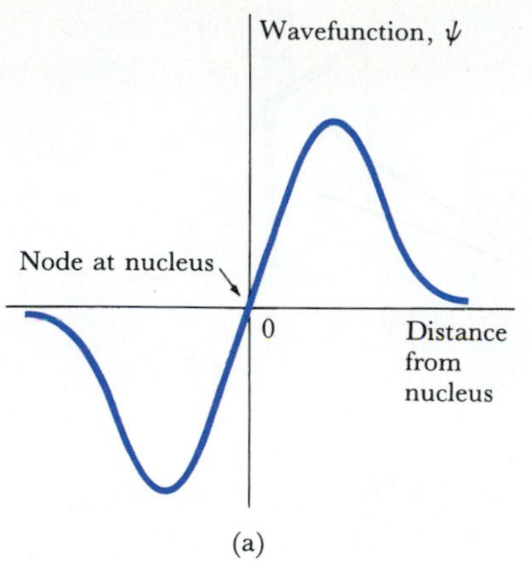

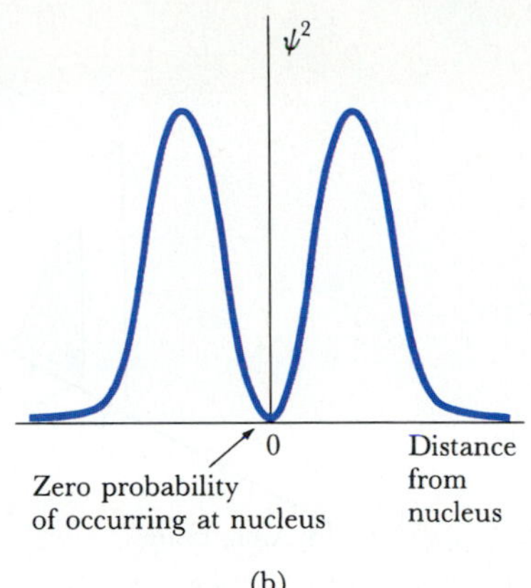

FIGURE 7.19 The variation of (a) the wavefunction and (b) the square of a wavefunction (which gives the probability of finding the electron at each location) for a $2p$ orbital of the hydrogen atom. Note that the wavefunction is zero at the nucleus, so there is zero probability of finding the electron there. The corresponding graph for an s orbital is nonzero at the nucleus, so an s electron can be found right at the nucleus.

The d and f orbitals. Five orbitals with $l = 2$ ($m_l = +2, +1, 0, -1, -2$), the ***d* orbitals**, can occur for shells with $n \geq 3$. Their boundary surfaces and labels are shown in Fig. 7.20. Four of the d orbitals have a double dumbbell shape; the d_{z^2} orbital is different.

For $n \geq 4$, ***f* orbitals** can occur. There are seven f orbitals, but their shapes are complicated and there is no need to try to remember them.

N The shapes of d orbitals will play an important role in Chapter 21, but they need not be emphasized until then.

N Five equivalent d orbitals. *J. Chem. Ed.* **1970**, *47*, 15.

E Orbitals with higher values of l are called $g, h, j, k, \ldots$, orbitals (for $l = 4, 5, 6, 7, \ldots$).

Electron spin. A careful analysis of the spectrum of atomic hydrogen shows that the spectral lines do not have exactly the frequencies predicted by Schrödinger's calculation. An explanation of the small differences was offered by the Dutch-American physicists Samuel Goudsmit and George Uhlenbeck. They proposed that an electron behaves in some respects like a sphere rotating on its axis, something like a gyroscope. This property is called the **spin** of the electron.

An electron has only two spin states, represented by the arrows ↑ (up) and ↓ (down). A helpful analogy is to think of an electron as being able to spin clockwise (the ↑ state) or counterclockwise (the ↓ state). The two electron spin states are distinguished by a fourth quantum number, the **spin magnetic quantum number** m_s. This quantum number can have only two values, $+\frac{1}{2}$ for an ↑ electron and $-\frac{1}{2}$ for a ↓ electron.

T OHT: Fig. 7.20.

The existence of only two spin states explained an experiment that had been carried out by Otto Stern and Walter Gerlach in 1920. They passed a beam of silver atoms between the poles of a powerful magnet that produced an inhomogeneous field (one that varies from place to

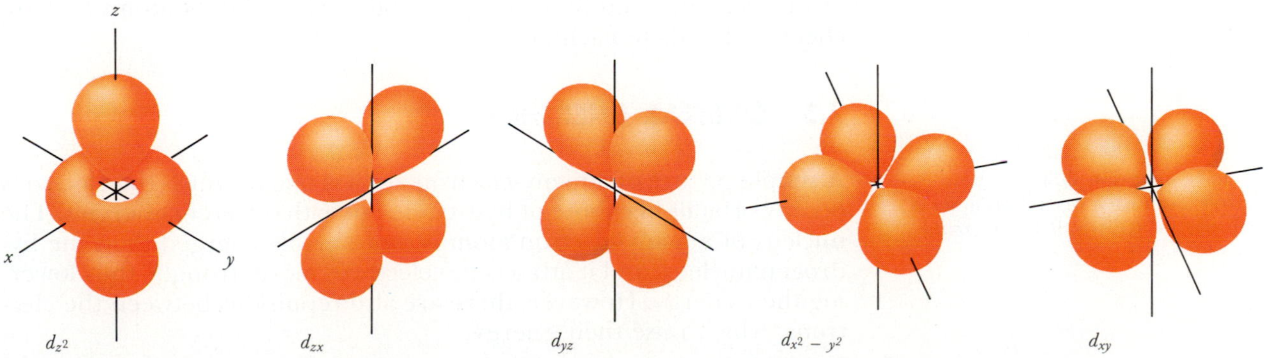

FIGURE 7.20 The boundary surfaces and labels of the five d orbitals of a shell with $n \geq 3$.

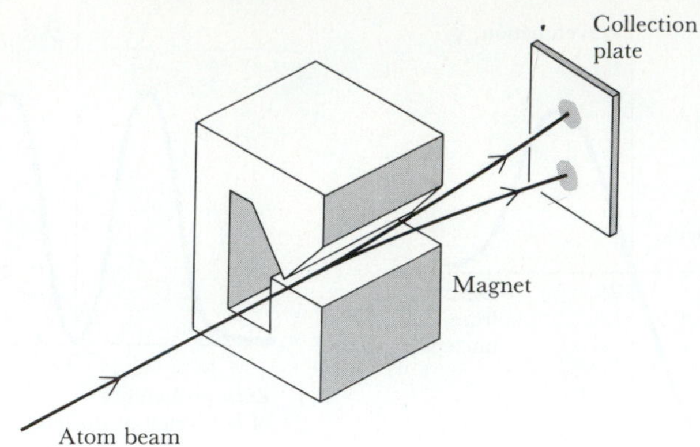

FIGURE 7.21 The quantization of electron spin is confirmed by the Stern-Gerlach experiment, in which a stream of atoms splits into two as it passes between the poles of a magnet. The atoms in one stream have an odd ↑ electron, and those in the other an odd ↓ electron.

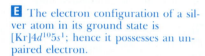

E The beam must be inhomogeneous in order to give a net force on a magnetic dipole.

place) and found that the beam split into two (Fig. 7.21). The explanation is based on the fact that a silver atom has an odd number of electrons ($Z = 47$) and, in the **Stern-Gerlach experiment**, behaves like a hydrogen atom with its one electron. As a result of the spin of the odd electron, the atom acts like a tiny magnet and its path through the apparatus is bent by the laboratory magnet. The initial beam is split into two beams because atoms with an odd ↑ electron are pushed in one direction by the field and those having an odd ↓ electron are pushed in the other direction.

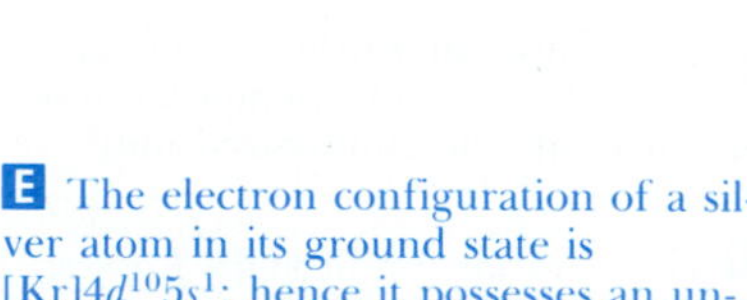

E The electron configuration of a silver atom in its ground state is $[Kr]4d^{10}5s^1$; hence it possesses an unpaired electron.

THE STRUCTURES OF MANY-ELECTRON ATOMS

All neutral atoms other than hydrogen have more than one electron. The helium atom ($Z = 2$) has two, the lithium atom ($Z = 3$) has three, and an atom of an element with atomic number Z has Z electrons. All these are examples of **many-electron atoms**, atoms with more than one electron. Our task now is to see how we can use the concepts already introduced to describe the structure of the hydrogen atom—principally the concepts of atomic orbitals and their energies—to describe the electronic structures of atoms of other elements. With the structures established, we shall be able to account for the form of the periodic table, the periodicity of the chemical and physical properties of the elements, and (in the next chapter) the ability of atoms to form chemical bonds to each other.

7.5 ORBITAL ENERGIES

E The assumption that in a many-electron atom each orbital occupies a particular orbital is called the *orbital approximation.*

A simple picture of a many-electron atom is one in which the electrons occupy orbitals like those of hydrogen, but with different energies. The nucleus of a many-electron atom is more highly charged than the hydrogen nucleus, and it attracts the electrons more strongly, thus lowering their energy. However, there are also repulsions between the electrons, which raise their energy.

The effective nuclear charge. In the hydrogen atom, where there are no electron-electron repulsions, all the orbitals of a given shell have the

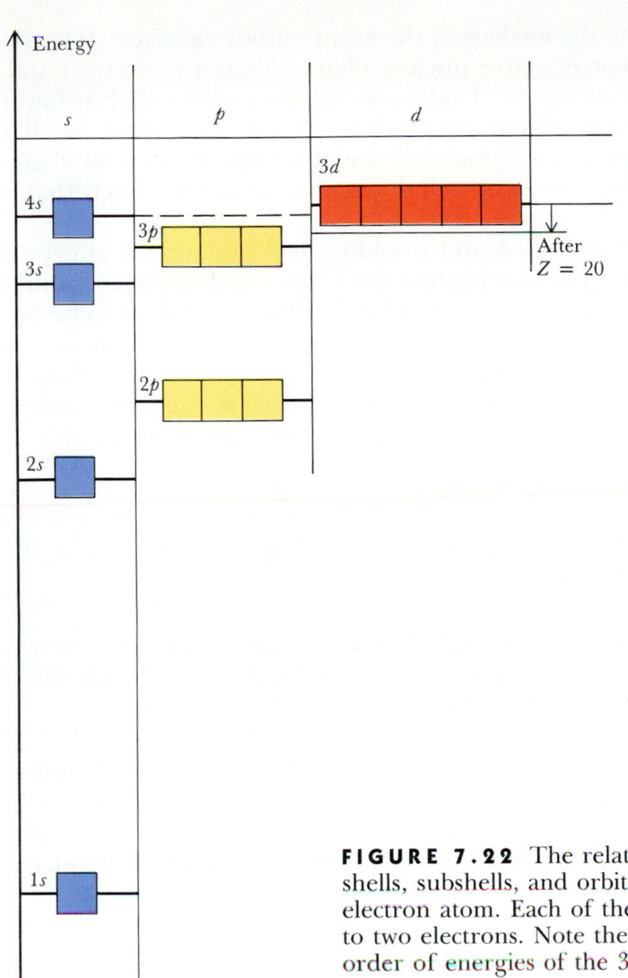

FIGURE 7.22 The relative energies of the shells, subshells, and orbitals in a many-electron atom. Each of the boxes can hold up to two electrons. Note the change in the order of energies of the 3*d* orbitals after $Z = 20$.

T OHT: Fig. 7.22.

N Figure 7.22 is a generic diagram, and specific atoms exhibit various modifications of the energy level scheme.

same energy. In many-electron atoms, electron-electron repulsions cause the energy of a *p* subshell to be higher than that of an *s* subshell of the same shell, and the energy of a *d* subshell to be higher than a *p* subshell of the same shell. The orbitals of a given subshell are, however, equal in energy to one another. Each of the three 2*p* orbitals, for instance, has the same energy. This order of energy levels is shown in Fig. 7.22 and will be used for discussing atomic structure throughout the rest of the chapter.

The difference in energies of subshells of the same shell can be traced to the shapes of their orbitals. There are two factors to take into account. First, as noted above, an *s* electron may be found very close to the nucleus, but a *p* electron may not. We say that an *s* electron **penetrates** closer to the nucleus than a *p* electron. Second, each electron is repelled by the other electrons in the atom and, hence, is less tightly bound to the nucleus than would be the case if those other electrons were absent. We say that each electron is **shielded** from the full attraction of the nucleus by the other electrons in the atom and that the **effective nuclear charge** is less than the actual charge (Fig. 7.23).

Now we bring the two factors together. Because an *s* electron penetrates closer to the nucleus than a *p* electron of the same shell does, it

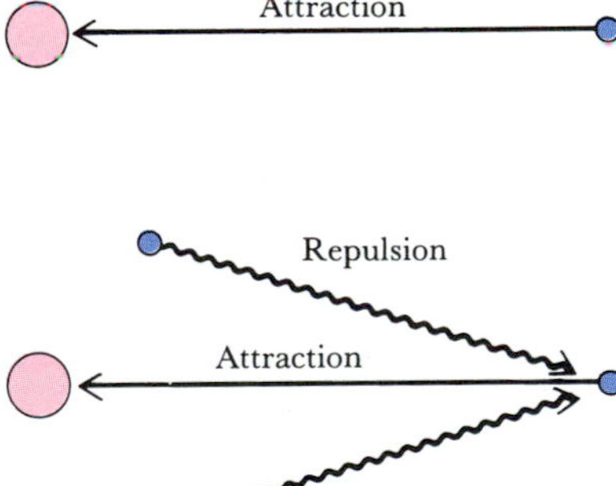

FIGURE 7.23 An electron near a bare nucleus experiences the full charge of the nucleus. In a many-electron atom, the electron is repelled by the other electrons present and experiences a smaller attraction. The net effect is that the effective nuclear charge is less than the atomic number.

is less shielded from the nucleus by the atom's other electrons. It therefore feels a stronger effective nuclear charge than a p electron and, hence, is bound more tightly. That is, an s electron has a slightly lower (more negative) energy than a p electron of the same shell. Similar differences between penetration and shielding apply to p and d subshells, because d electrons approach the nucleus even less closely than p electrons.

The effects of penetration and shielding can be large: a $4s$ orbital may be so much lower in energy than the $4p$ and $4d$ orbitals that it may also be lower in energy than the $3d$ orbital of the same atom. Whether or not this is in fact the case depends on the number of electrons in the atom. In some atoms the $4s$ orbital is lower in energy than the $3d$ orbitals, but in others it is higher in energy. We return to this point shortly when we deal with the atomic structures of the transition metals.

U Exercise 7.92.

N An entirely fabulous account of the origin of the Pauli exclusion principle. *J. Chem. Ed.* **1989**, *66*, 983.

The exclusion principle. The lowest total energy of an atom is not obtained by letting all its electrons occupy the 1s orbital. That arrangement is forbidden by a fundamental feature of nature identified by the Austrian Wolfgang Pauli in 1925:

> The **Pauli exclusion principle**: No more than two electrons may occupy any given orbital; and when two electrons do occupy one orbital, their spins must be paired.

The spins of two electrons are said to be **paired** if one is ↑ and the other is ↓. Paired spins are denoted ↑↓. Because an atomic orbital is designated by three quantum numbers (n, l, and m_l), and the two spin states are specified by a fourth quantum number (m_s), another way of expressing the principle *for atoms* is

> No two electrons in an atom can have the same set of four quantum numbers.

In practice, the exclusion principle means that no more than two electrons can enter each orbital in the energy level diagram in Fig. 7.22. For $Z > 2$, the atom's electrons cannot all enter the 1s orbital, some must occupy higher-energy orbitals.

The configuration of hydrogen through lithium. The list of occupied orbitals in an atom is a statement of its **electron configuration**. The hydrogen atom in its ground state (its lowest energy state) has one electron in a 1s orbital. We denote this by using a single arrow in the 1s orbital in a diagram like that of Fig. 7.22, and we report its configuration as $1s^1$ ("one s one"). The ground-state electron configuration of helium ($Z = 2$) is that in which both electrons are in a 1s orbital, which is reported as $1s^2$ ("one s two"). We show only the occupied orbital:

N In these box diagrams, energy increases from left to right. There is no line extending to the left of the box because a 1s orbital is the orbital of lowest energy. The color codes are blue for s orbitals, yellow for p orbitals, and (later) orange for d orbitals, pink for f orbitals.

He [↑↓]—
$1s^2$

The full $n = 1$ shell of helium is an example of a closed shell:

> A **closed shell** (or subshell) is a shell (or subshell) containing the maximum number of electrons allowed by the exclusion principle.

The shell with $n = 2$ has two subshells ($l = 0,1$); it consists of one $2s$ orbital and three $2p$ orbitals and, hence, four orbitals in all. The greatest number of electrons that can be accommodated in the $n = 2$ shell is therefore eight, and a shell with $n = 2$ is closed when it contains eight electrons.

Lithium ($Z = 3$) has three electrons. Two electrons can occupy the $1s$ orbital and complete the $n = 1$ shell. The third electron occupies an orbital of the shell with $n = 2$. That shell has two subshells, with the $2s$ subshell lower in energy than the $2p$. The third electron therefore enters the $2s$ orbital to give the electron configuration $1s^22s^1$:

Li ⇅ ↑

$1s^2$ $2s^1$

(The energy increases from left to right in this diagram, a convention that is widely used when depicting electron configurations.) We can think of the atom as consisting of a **core** made up of inner closed shells (in this case, the single $n = 1$ shell) and, around that core, a single electron in a $2s$ orbital. Because they are so close to the nucleus, lithium's core electrons are much more tightly bound than the outer $2s$ electron, and they can be removed from the atom only if a lot of energy is supplied. In general, **core electrons**, the electrons that belong to the core, are much more inert than outer electrons and are not involved in an atom's chemical reactions.

E Core electrons are also more compact: the $1s$ electrons experience an almost unshielded nuclear charge, and are attracted strongly to the nucleus.

D A low-cost classroom demonstration of the *Aufbau* principle. *J. Chem. Ed.* **1979**, *56*, 747.

▼ EXAMPLE 7.4 Building up an electron configuration

Predict the ground-state electron configuration of boron.

STRATEGY Although we have not yet discussed all the rules for obtaining an electron configuration, we have enough information to deal with this atom. First, decide how many electrons are present (from the atomic number). Then add arrows to boxes arranged like those in Fig. 7.22, starting at the lowest orbital and filling each orbital before moving to the next higher orbital. Finally, list the occupied orbitals as an electron configuration.

SOLUTION A boron atom has five electrons. Two enter the $1s$ orbital and complete the $n = 1$ shell. The $2s$ orbital is the next orbital to be occupied. It can accommodate two of the remaining electrons, so the fifth electron must occupy an orbital of the next available subshell, which Fig. 7.22 shows is a $2p$ subshell, to give the configuration $1s^22s^22p^1$:

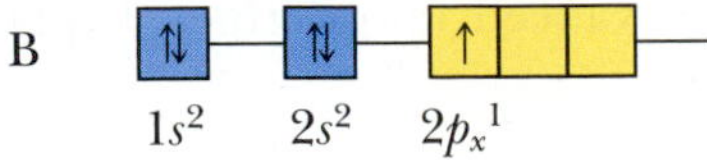

EXERCISE E7.4 Predict the lowest-energy electron configuration of neon.

[*Answer*: $1s^22s^22p^6$] ▲

We have seen in Example 7.4 that the ground-state configuration of a boron atom is $1s^22s^22p^1$. However, there are three $2p$ orbitals; so when we want to emphasize that electrons occupy different p orbitals, we write the configuration in more detail. For example, if in a certain atom there are two electrons in different $2p$ orbitals, then we write the configuration as $1s^22s^22p_x^{\ 1}2p_y^{\ 1}$. If we do not need to emphasize that two different orbitals are occupied, then we write the configuration more simply as $1s^22s^22p^2$.

7.6 THE BUILDING-UP PRINCIPLE

The lowest-energy configuration of any atom can be predicted by a generalization of the approach we used above for H, He, and Li. The procedure is called the **building-up principle** (some call it by its name in German, the *Aufbau* principle). The principle leads to the electron configuration with the lowest *total* energy of the atom, taking into account the attraction of the electrons to the nucleus and their repulsion of one another. It does this by specifying the order in which the orbitals are to be occupied as one electron after another is added, until all Z electrons are present. A clue to the order is given by the order of energy levels in Fig. 7.22:

$$1s < 2s < 2p < 3s < 3p < 4s < 3d$$

for $Z = 1$ through 20. However, there is no need to make a special effort to memorize this order, because we can arrive at the same configuration if we add electrons in the order shown in Fig. 7.24, which matches the structure of the periodic table. It is important to remember that the sequence of occupation summarized by Fig. 7.24 usually leads to the correct ground-state electron configuration—the configuration of lowest *total* energy.

N A number of mnemonics have been proposed for remembering the order in which orbitals are occupied. However, we consider it healthier to relate the order directly to the structure of the periodic table. This diagram is also quicker, more versatile, and more reliable than a mnemonic.

T OHT: Fig. 7.24.

To assign a configuration to an element with atomic number Z, we proceed as follows:

1. Add Z electrons, one after the other, to the orbitals in the order shown in Fig. 7.24, but with no more than two electrons in any one orbital.
2. If more than one orbital in a subshell is available, add electrons with parallel spins to *different* orbitals of that subshell.

The second rule is called **Hund's rule**, for the German spectroscopist Fritz Hund, who first proposed it. Electrons have **parallel spins** (denoted ↑↑) if they spin in the same direction. The explanation of why electrons occupy *different* orbitals of the same subshell can be traced to the effect of electron repulsion. If electrons occupy different orbitals, they generally stay farther apart than if they are in the same orbital. As a result, they repel each other less and the energy of the atom is lower. The explanation of why they should also have parallel spins is complex and requires more advanced knowledge of quantum mechanics than we have covered here.

FIGURE 7.24 The order in which atomic orbitals are filled according to the building-up principle. Each time an electron is added, we move one place to the right until all the electrons are accommodated.

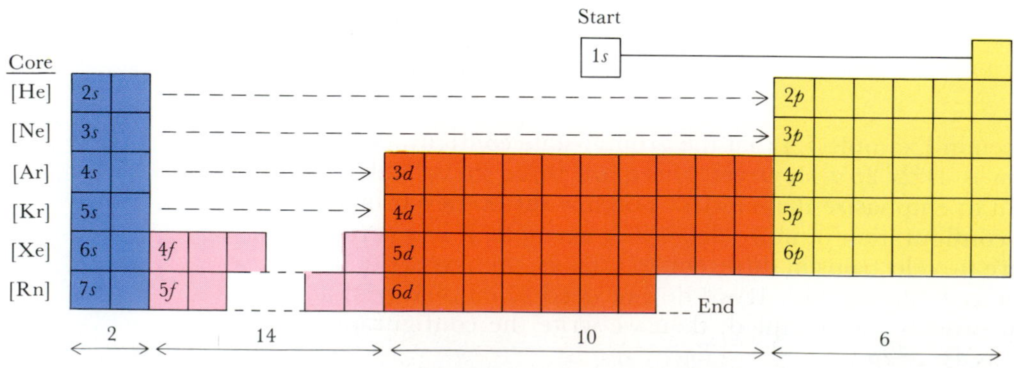

Lithium through sodium. Lithium, $1s^22s^1$, was described earlier. Because it consists of a single $2s$ electron outside a heliumlike $1s^2$ core, its configuration is more simply written $[He]2s^1$. It is often useful to think of an atom as a noble-gas core surrounded by electrons in the outermost shell (the shell with the largest value of n). The outermost shell is called the **valence shell** of the atom, because its orbitals are the ones mainly responsible for compound formation (and the number of bonds an atom can form is called its *valence*); the electrons that occupy the valence shell are called **valence electrons**. All the atoms in a given period have the same valence shell, and the principal quantum number n of the valence shell is equal to the period number. For example, the valence shell of the elements in Period 2 is the shell with $n = 2$. All the atoms in a given period also have the same core. All the atoms of Period 2, for example, have a heliumlike $1s^2$ core, written [He], and all those of Period 3 have a neonlike $1s^22s^22p^6$ core, written [Ne].

The element with $Z = 4$ is beryllium, Be. The first three electrons form $[He]2s^1$, like Li. The fourth electron pairs with the $2s$ electron, thereby giving $[He]2s^2$. A Be atom therefore has a heliumlike core surrounded by a valence shell containing two paired electrons. The next element, boron, has $Z = 5$. Its additional electron cannot enter the full $2s$ subshell; instead, it occupies any one of the three $2p$ orbitals, thus giving $[He]2s^22p^1$ (Example 7.4):

E Although the cores have the same designation (such as $1s^22s^22p^6$), the orbitals become more compact as the atomic number of the element increases, because the greater nuclear charge draws the electrons inward.

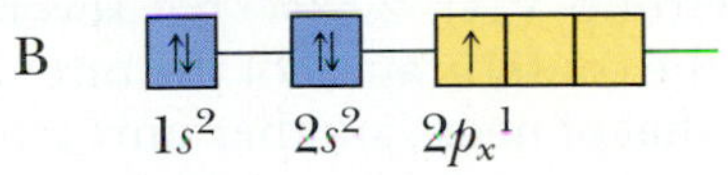

B [⇅]—[⇅]—[↑| |]—
$1s^2$ $2s^2$ $2p_x^{\,1}$

(Boxes joined together and, in this case, representing the three $2p$ orbitals, all have the same energy.) The last electron can occupy any one of the three $2p$ orbitals, because they all have the same energy. Carbon, with $Z = 6$, has one more electron than boron and the configuration $[He]2s^22p^2$. According to Hund's rule, the fifth and sixth electrons occupy different $2p$ orbitals and do so with parallel spins:

U Exercise 7.96.

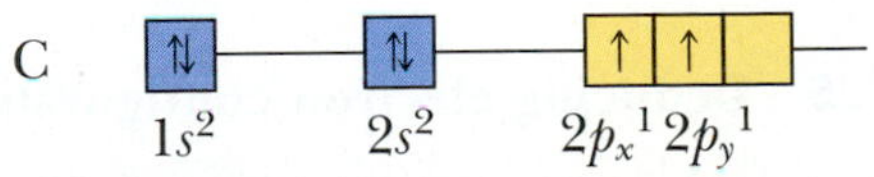

C [⇅]—[⇅]—[↑|↑|]—
$1s^2$ $2s^2$ $2p_x^{\,1}2p_y^{\,1}$

Nitrogen has $Z = 7$ and one more electron than carbon, so nitrogen's electron configuration is $[He]2s^22p^3$. Each p electron occupies a different orbital, and the three have parallel spins:

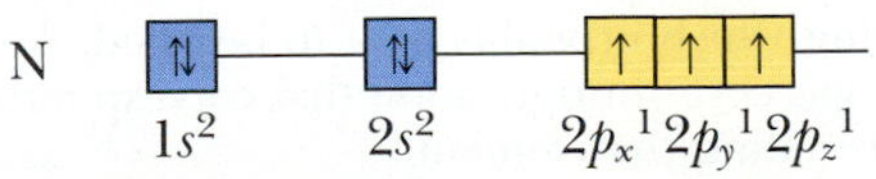

N [⇅]—[⇅]—[↑|↑|↑]—
$1s^2$ $2s^2$ $2p_x^{\,1}2p_y^{\,1}2p_z^{\,1}$

Oxygen has $Z = 8$, and one more electron than N. The eighth electron must pair with one already present, thereby giving the configuration $[He]2s^22p^4$:

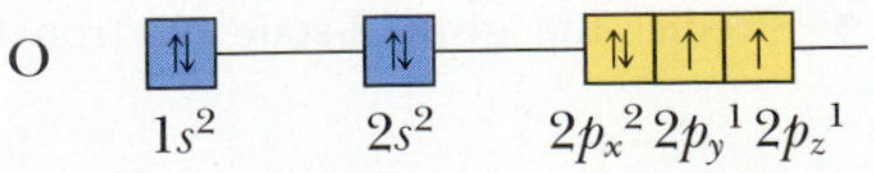

O [⇅]—[⇅]—[⇅|↑|↑]—
$1s^2$ $2s^2$ $2p_x^{\,2}2p_y^{\,1}2p_z^{\,1}$

Similarly, fluorine, with $Z = 9$ and one more electron than O, has the configuration [He]$2s^22p^5$:

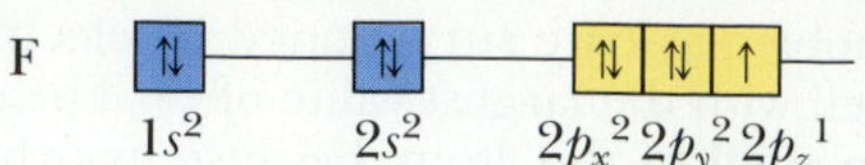

The fluorine atom can be pictured as having a heliumlike core surrounded by a valence shell that is complete except for one p electron. Neon, with $Z = 10$, has one more electron than fluorine has. This electron completes the $2p$ subshell, thus giving [He]$2s^22p^6$:

Ne

$1s^2$ $2s^2$ $2p_x^{\,2}\,2p_y^{\,2}\,2p_z^{\,2}$

According to Fig. 7.24, the next electron enters the $3s$ orbital, the lowest-energy orbital of the next shell. The configuration of sodium is therefore [He]$2s^22p^63s^1$, or more briefly [Ne]$3s^1$, where [Ne] represents the neonlike core.

A feature that is revealed by the preceding examples is that *the number of valence electrons in the ground state of an atom is equal to the group number.* For example, carbon (Group IV), $2s^22p^2$, has four valence electrons and fluorine (Group VII), $2s^22p^5$, has seven.

We have seen that the configuration of the unreactive gas helium is a closed shell. So too is that of neon, another unreactive gas. We have also seen that the configuration of lithium, which is a reactive metal, consists of a single electron outside a core. So too is that of sodium, another reactive metal. Thus, with a few simple ideas about atomic structure, we have begun to account for the periodicity of the elements: *the building-up principle leads periodically to analogous electron configurations and, hence, to similar chemical properties.* The electron configurations of all the elements are listed in Appendix 2C.

U Exercises 7.92 and 7.94.

E It should be stressed that the electron configurations given in Appendix 2C are the ones that are found experimentally (from spectroscopic and magnetic measurements). The building-up principle is a largely successful attempt to predict and to rationalize these experimentally determined configurations.

▼ EXAMPLE 7.5 Deducing electron configurations

Predict the ground-state electron configuration of silicon. Which element already mentioned has a similar valence electron configuration?

STRATEGY Because silicon is in the same group as carbon, we can anticipate that the two elements have analogous ns^2np^2 configurations. The precise configuration is deduced with the building-up principle, beginning with the $1s$ orbital. Each time an electron is added, we move one place to the right in Fig. 7.24, noting which orbital is next to be filled. A shortcut is to make use of the noble-gas core configuration that corresponds to the completion of the period preceding the element.

SOLUTION For Si, $Z = 14$. The first ten electrons give a neonlike $1s^22s^22p^6$ core, [Ne]. Now we can start at the left of the period in Fig. 7.24 that has [Ne] as a core, and add four electrons. Two electrons complete the $3s$ subshell. The remaining two occupy $3p$ orbitals. Hence, the lowest-energy configuration is [Ne]$3s^23p^2$.

EXERCISE E7.5 Predict the ground-state electron configuration of a chlorine atom.

▲ [*Answer*: [Ne]$3s^23p^5$]

The filling of d orbitals. The *s* and *p* subshells of the $n = 3$ shell are full in an atom of argon, which is a colorless, odorless, unreactive gas resembling neon. As a result of penetration and shielding, the 4*s* orbitals have lower energy than the 3*d* orbitals and are occupied next. Hence, the next two electron configurations are $[Ar]4s^1$ for potassium and $[Ar]4s^2$ for calcium. Now the 3*d* orbitals come into line for occupation, and there is a change in the rhythm of the periodic table.

According to Fig. 7.24, the next ten electrons (for scandium, with $Z = 21$, through Zn, with $Z = 30$) enter the 3*d* subshell. The ground-state electron configuration of scandium, for example, is $[Ar]3d^14s^2$, and that of its neighbor titanium is $[Ar]3d^24s^2$. Note that we write the 4*s* electrons after the 3*d* electrons. This arrangement reflects the change in the order of the energies of the orbitals, with 3*d* below 4*s*, a change that begins at scandium. The building-up principle leads to the correct overall configuration, even though the order in which electrons are added according to the principle is not precisely the same as the actual order of energies in the atom.

The ground-state electron configurations are now not quite as straightforward as before. The *basic* rule to remember is that the configurations of the first series of transition metals are $[Ar]3d^14s^2$ for scandium, $[Ar]3d^34s^2$ for vanadium, and so on. However, there are *two* exceptions, because the half-complete subshell configuration d^5 and the complete subshell configuration d^{10} turn out experimentally to be more stable than simple theory suggests. In some cases, the neutral atom has a lower total energy if the 3*d* subshell is half-completed (to d^5) or completed (to d^{10}) by transferring a 4*s* electron into it. For example, the experimental electron configuration of chromium is $[Ar]3d^54s^1$ and that of copper is $[Ar]3d^{10}4s^1$. Other exceptions to the building-up principle can be found in Appendix 2C.

After the 3*d* subshell is full, starting with gallium, the 4*p* orbitals are occupied. The configuration of germanium (in Group IV), for example, is obtained by adding two electrons to the 4*p* subshell outside the completed 3*d* subshell and is $[Ar]3d^{10}4s^24p^2$. The fourth period of the periodic table contains 18 elements, because the 4*s* and 4*p* orbitals can accommodate eight electrons and the 3*d* orbitals can accommodate ten. It is the first **long period** of the periodic table.

Next in line for occupation is the 5*s* orbital, followed by the 4*d* orbitals. As in Period 4, the energy of the 4*d* orbitals falls below that of the 5*s* orbital after two electrons have been accommodated in the 5*s* orbital. A similar change happens in Period 6, but now another complication arises, because the 4*f* orbitals come into line for occupation. Neodymium, for example, has the configuration $[Xe]4f^46s^2$. Electrons then continue to occupy the seven 4*f* orbitals, which are complete after 14 electrons have been added, at ytterbium, $[Xe]4f^{14}6s^2$. Next the 5*d* orbitals are occupied. The 6*p* orbitals are occupied only after mercury completes the 5*d* orbitals; thallium, for example, has the configuration $[Xe]4f^{14}5d^{10}6s^26p^1$.

N The building-up principle becomes more ad hoc in the *d* block (and even more in the *f* block). The central idea that should be stressed is that the ground-state configuration is the state of lowest *total* energy for the atom as a whole, taking into account all the electron-electron interactions.

E A complication is that the orbital energies vary quite considerably, and whereas the 3*d* orbitals lie close to the 4*s* orbitals in the region of calcium and scandium, the 3*d* orbitals lie much lower in energy than the 4*s* (and 4*p*) orbitals at gallium, and for that element and the ones that follow it in the period, the 3*d* orbitals can be largely ignored, and the chemical properties of the atoms are largely explained by the 4*s* and 4*p* orbitals alone.

E The identity of the members of the *f* block remains controversial (at this level, it is not particularly important). We have adopted the suggestion that the *f* block should extend from lanthanum to ytterbium. Some texts start one element later, and run from cerium to lutetium.

▼ EXAMPLE 7.6 Writing the electron configuration of a heavy atom

Write the ground-state electron configuration of a lead atom.

STRATEGY We note the group number, which tells us the number of valence electrons in the ground state of the atom, and the period number, which tells us the value of the principal quantum number of the valence shell. The core consists of the preceding noble-gas configuration together with any completed d and f subshells.

SOLUTION Lead, Pb, belongs to Group IV and Period 6. It therefore has four electrons in its valence shell, two in a 6s orbital and two in different 6p orbitals. The period has complete 5d and 4f subshells, and the preceding noble gas is xenon. The electron configuration of lead is therefore $[Xe]4f^{14}5d^{10}6s^{2}6p^{2}$.

EXERCISE E7.6 Write the ground-state electron configuration of a bismuth atom.

▲ [*Answer*: $[Xe]4f^{14}5d^{10}6s^{2}6p^{3}$]

U Exercises 7.95 and 7.97.

The configurations of ions. Cations are formed by the removal of electrons from the configuration predicted for the neutral atom. If the principal quantum number of the valence shell is n, then we remove electrons in the order np first (because that subshell has the highest energy), then ns, and finally $(n-1)d$, until the appropriate number of electrons has been removed. Thus, for the Fe^{3+} ion, we first work out the configuration of the Fe atom, which is $[Ar]3d^{6}4s^{2}$, and then remove three electrons from it. There are no 4p electrons, so the first two electrons removed are 4s electrons. The third electron comes from the 3d subshell, thus giving $[Ar]3d^{5}$.

▼ EXAMPLE 7.7 Predicting the configurations of cations

Predict the configurations of the In^{+} and In^{3+} ions.

STRATEGY From the group and period of the element, we determine the configuration of the neutral atom. One electron is removed to produce In^{+}, and then two more, to produce In^{3+}. We remove electrons from valence shell p orbitals first, then from the s orbitals, and finally, if necessary, from the d orbitals in the next lowest shell.

SOLUTION Indium, in Group III, has three valence electrons. It is in Period 5 (so its valence shell has $n = 5$) and is preceded in Period 4 by the noble gas krypton (so its core is [Kr]). Its ground-state configuration is therefore $[Kr]4d^{10}5s^{2}5p^{1}$. When one electron is lost from the 5p orbital, we obtain In^{+} as $[Kr]4d^{10}5s^{2}$:

In [⇅]—[↑][ ][ ]— ⟶ In^{+} [⇅]—[ ][ ][ ]—

$5s^{2}$ $5p_x{}^{1}$ $5s^{2}$

When the next two electrons are lost from the 5s orbital, we obtain In^{3+} as $[Kr]4d^{10}$:

In [⇅]—[↑][ ][ ]— ⟶ In^{3+} [ ]—[ ][ ][ ]—

$5s^{2}$ $5p_x{}^{1}$

EXERCISE E7.7 Predict the electron configurations of the copper(I) and copper(II) ions.

▲ [*Answer*: $[Ar]3d^{10}$; $[Ar]3d^{9}$]

Monatomic anions are formed by adding enough electrons to the vacant orbitals of the valence shell to achieve the configuration of the next noble-gas atom. Thus, to predict the electron configuration of the anion formed by a nitrogen atom, we first note that, because nitrogen is in Group V (and, hence, has five valence electrons), three electrons are needed to reach a noble-gas configuration; therefore, the ion will be N^{3-}:

$$\text{N}\quad [\uparrow\downarrow]\ 2s^2 \quad [\uparrow|\uparrow|\uparrow]\ 2p_x^{\,1}2p_y^{\,1}2p_z^{\,1} \longrightarrow \text{N}^{3-}\quad [\uparrow\downarrow]\ 2s^2 \quad [\uparrow\downarrow|\uparrow\downarrow|\uparrow\downarrow]\ 2p_x^{\,2}2p_y^{\,2}2p_z^{\,2}$$

The configuration of the nitrogen atom is $[He]2s^22p^3$, with room for three more electrons in the $2p$ subshell. Thus, the ion N^{3-} has the configuration $[He]2s^22p^6$, the same as that of neon.

▼ EXAMPLE 7.8 Predicting the configurations of anions

Predict the electron configuration of the oxide ion.

STRATEGY First, we establish the ground configuration of the neutral O atom by using the building-up principle. Then we add enough electrons to complete the vacant orbitals and achieve the closed-shell configuration of the next noble gas atom (Ne).

SOLUTION We have already established that the electron configuration of an O atom is $1s^22s^22p^4$. The $2p$ subshell can accommodate two more electrons, so we expect the oxide ion to be O^{2-} with configuration $1s^22s^22p^6$, the same configuration as Ne has:

$$\text{O}\quad [\uparrow\downarrow]\ 2s^2 \quad [\uparrow\downarrow|\uparrow|\uparrow]\ 2p_x^{\,2}2p_y^{\,1}2p_z^{\,1} \longrightarrow \text{O}^{2-}\quad [\uparrow\downarrow]\ 2s^2 \quad [\uparrow\downarrow|\uparrow\downarrow|\uparrow\downarrow]\ 2p_x^{\,2}2p_y^{\,2}2p_z^{\,2}$$

EXERCISE E7.8 Predict the electron configuration of the phosphide ion.
[*Answer*: P^{3-}; $[Ne]3s^23p^6$]
▲

The magnetic properties of atoms. In our discussion of the Stern-Gerlach experiment, we remarked that, because an electron is charged and is spinning, it behaves like a tiny magnet. We can think of an electron with ↑ spin as behaving like a minute bar magnet in one orientation and of an electron with ↓ spin as behaving like a bar magnet in the opposite orientation. A closed subshell of electrons has equal numbers of ↑ and ↓ electrons, so it does not respond to an applied magnetic field: the influence of the applied field on the ↑ electrons cancels the effect of the field on the ↓ electrons. However, if the valence shell of an atom has unequal numbers of ↑ and ↓ electrons, then the effect of an applied magnetic field on the ↑ electrons does not completely cancel its effect on the ↓ electrons.

It follows that we can verify whether an atom or ion has unpaired electrons by examining the effect of a magnetic field on a sample. If the atom is drawn into a magnetic field, then it has unpaired electrons and is said to be **paramagnetic**. If it is pushed out of a magnetic field it has no unpaired electrons and is said to be **diamagnetic**. For example, the electron configuration of an Ag atom is $[Kr]4d^{10}5s^1$: the [Kr] core and the full $4d$ subshell have no unpaired electrons. But there is one unpaired electron in the valence shell, an arrangement consistent with the results of the Stern-Gerlach experiment.

E Diamagnetism arises from the ability of the applied magnetic field to stir up electron currents in a molecule. These currents give rise to a magnetic field that opposes the applied field. All species have a diamagnetic component in their response to a magnetic field, but if unpaired electron spins are present the paramagnetism they cause greatly outweighs the diamagnetism induced by the applied field, and the species is paramagnetic. The names come from the Greek words for along (*para*) and across (*dia*), indicating the orientation a cylindrical sample of material will adopt in an applied field.

U Exercise 7.90.

D Magnetic susceptibilities—Iron(II) ion. TDC, 208. Paramagnetic properties of Fe(II) and Fe(III). *J. Chem. Ed.* **1977**, *54*, 431.

FIGURE 7.25 Dmitri Ivanovich Mendeleev (1834–1907).

A SURVEY OF PERIODIC PROPERTIES

The overall pattern of the periodic table was discovered by the Russian chemist Dmitri Mendeleev (Fig. 7.25) during a single day of furious thought, according to legend, on February 17, 1869. Mendeleev arranged the elements in groups and periods in order of increasing atomic mass, not atomic number, for in his time the latter concept was unknown. However, because atomic mass almost always increases with atomic number, and hence with the number of electrons in the atom, Mendeleev was unwittingly arranging the atoms in order of increasing number of electrons. Because the periodicity of the elements reflects the periodicity of their electron configurations, he stumbled on the pattern of configurations without knowing anything about atomic structure. Mendeleev's success also sprang from his chemical insight, which led him to leave gaps for elements that seemed to be needed to complete the pattern but were then unknown. He was even able to predict the properties of some of these missing elements. When later they were discovered, he turned out to be strikingly correct.

7.7 BLOCKS, PERIODS, AND GROUPS

The periodic table is divided into **blocks** named for the last subshell that is occupied according to the building-up principle (Fig. 7.26). Sodium, [Ne]$3s^1$, and calcium, [Ar]$4s^2$, both belong to the ***s* block**. Nitrogen, [He]$2s^22p^3$, and neon, [He]$2s^22p^6$, belong to the ***p* block**. Iron, [Ar]$3d^64s^2$, belongs to the ***d* block** (the transition metals). Hydrogen is separated from the main table because, although it has an s^1 configuration, like lithium and sodium, it has quite different properties. The closed-shell noble gas helium is so similar to the noble gases neon and argon that it is treated as a member of the *p* block even though it has no *p* electrons. The members of the *s* and *p* blocks are called the **main-group elements**. Calcium is a main-group element; iron is not.

As we have seen, the **period** number is the same as the principal quantum number of the valence shell of its atoms. Period 1 consists of two elements (H and He) in which the single 1*s* orbital of the $n = 1$ shell is being filled with electrons. Period 2 consists of the eight elements (Li through Ne) in which the *s* and *p* orbitals of the $n = 2$ shell are being filled with electrons. In Period 3 (Na through Ar) the *s* and *p* orbitals of the $n = 3$ shell are being occupied, and so on. The members of Periods 2 and 3 are called the **representative elements** of their respective groups.

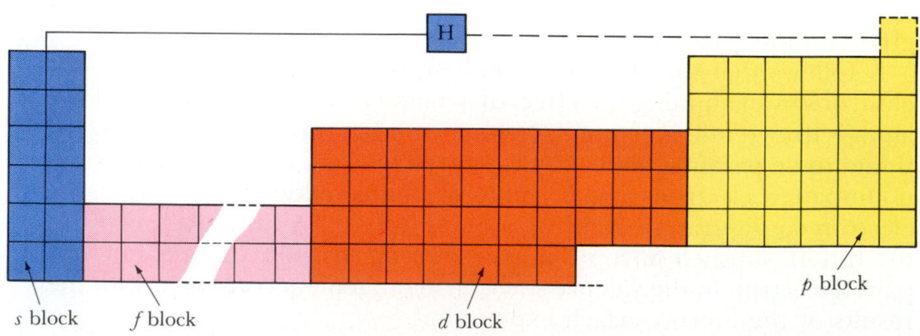

FIGURE 7.26 The block structure of the periodic table is based on the last subshell occupied in an element according to the building-up principle.

With the exception of helium, the group number is equal to the number of s and p electrons in the valence shell. Lithium (Group I) has one valence-shell s electron, and bromine (Group VII) has seven valence-shell electrons (two s electrons and five p electrons). All members of the same group have the same valence electron configuration (with different values of n).

▼ **EXAMPLE 7.9 Writing an electron configuration from a group number**

Write the ground-state electron configuration of a halogen atom.

STRATEGY We know that the (Roman) group number of an element equals the number of s and p electrons in its valence shell. We can write the configuration by allowing two electrons to occupy an s orbital, thereby leaving the rest to occupy p orbitals. The core configuration is that of the preceding noble gas.

SOLUTION The halogens are in Group VII and, hence, have seven valence electrons. Two enter an s orbital, thus leaving five to enter p orbitals. Therefore, their characteristic valence-shell electron configuration is ns^2np^5. For Br, a Period 4 halogen, the core is [Ar] and the configuration is $[Ar]3d^{10}4s^24p^5$.

U Exercise 7.93.

EXERCISE E7.9 Give the general valence-shell configuration of the atoms of the group that contains tin.

[*Answer*: ns^2np^2]

▲

7.8 PERIODICITY OF PHYSICAL PROPERTIES

The physical properties of the elements show a striking periodicity. This periodicity is particularly clear when we examine the variation of atomic sizes and the energies needed to remove electrons from atoms. We shall see later in this chapter, and again in Chapter 8, that the readiness with which atoms lose and gain electrons and the radii of atoms and ions are among the most useful properties for accounting for the chemical behavior of the elements. Therefore, if we can account for the periodicity of these physical properties of the atoms, then we shall be well on the way to accounting for the periodicity of the chemical properties of the elements.

Atomic radii. Because the electron clouds of atoms and ions do not have sharp edges, it is necessary to define what we mean by their radii.

The **atomic radius** of an element is half the distance between the centers of neighboring atoms (**1**). If the element is a metal, then the appropriate distance is that between the nuclei of neighboring atoms in a solid sample. If the element is a nonmetal, then the appropriate distance is that between the nuclei of two atoms of that element joined by a single chemical bond. Because the distance between neighboring nuclei in solid copper is 270 pm, its atomic radius is 135 pm (remember that 1 pm = 10^{-12} m). The distance between the nuclei in a Cl_2 molecule is 198 pm, so the atomic radius of chlorine is 99 pm.

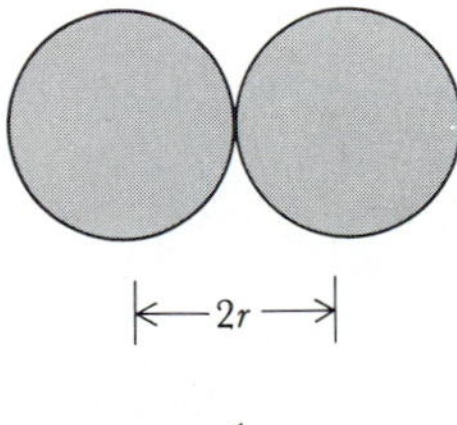

1

E Atomic radii increase with the coordination number of the atom. To adjust for this variation, the atomic radii of metallic elements are usually adjusted to correspond to a coordination number of 12.

D Electronegativity, atomic diameter, and ionization energy. CD1, 1.

Atomic radii are illustrated in Fig. 7.27; their sawtooth periodicity is illustrated in Fig. 7.28. (We shall see how the values are measured in Section 10.6.) The illustration shows that atomic radius generally decreases from left to right across a period and increases down a group.

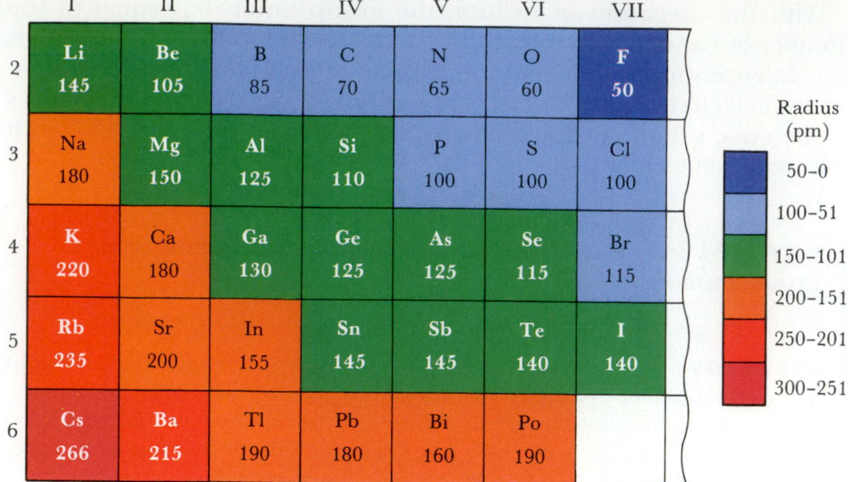

FIGURE 7.27 The atomic radii (in picometers: 1 pm = 10^{-12} m) of the main-group elements. The radii decrease from left to right in a period and increase down a group.

T OHT: Figs. 7.27 and 7.28.

The decrease across a period, like that from Li to Ne, is a result of increasing effective nuclear charge. The increase down a group, like that from Li to Cs, is a result of the fact that the outermost electrons occupy shells that lie farther and farther from the nucleus.

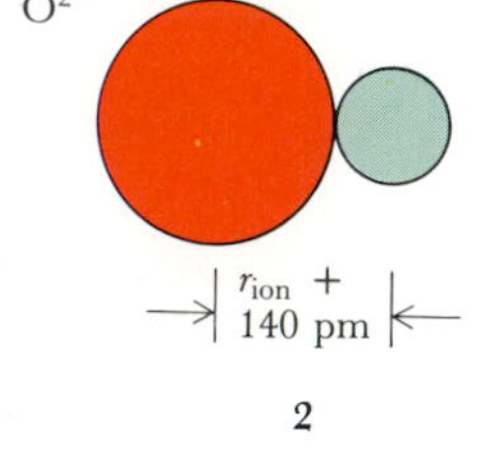

2

Ionic radii. The **ionic radius** of an element is its contribution to the distance between neighboring ions in a solid ionic compound (**2**). The distance between the nuclei of a cation and its neighboring anion is the sum of the two ionic radii. In practice, we take the radius of the oxide ion (O^{2-}) as 140 pm and use this value to calculate the radii of other ions. For example, because the separation between Mg and O nuclei in magnesium oxide is 205 pm, the radius of the ion Mg^{2+} is taken as (205 − 140) pm = 65 pm. This value for Mg^{2+} can then be used to deduce the ionic radius of Cl^- from measurements on magnesium chloride, and so on.

U Exercise 7.100.

N Periodic contractions among the elements, or on being the right size. *J. Chem. Ed.* **1988**, *65*, 17.

Ionic radii obtained in this way are given in Fig. 7.29 and relative sizes of ions and parent atoms are illustrated in Fig. 7.30. The illustration shows that all cations are smaller than their parent atoms. Because a cation is formed when an atom loses one or more electrons, only its

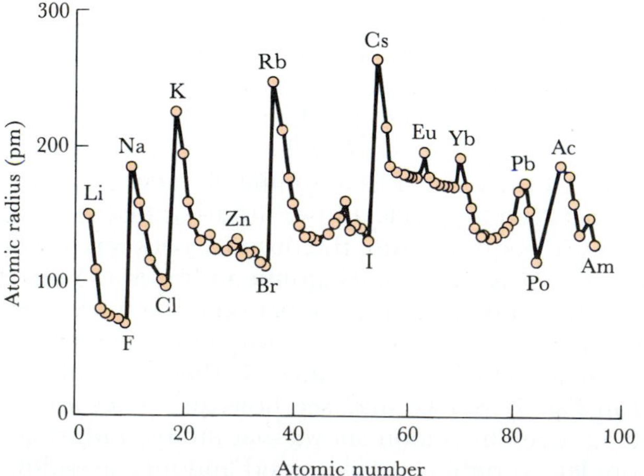

FIGURE 7.28 The periodic variation of the atomic radii of the elements. This periodic variation can be explained in terms of the effect of increasing atomic number and the occupation of shells with increasing principal quantum number.

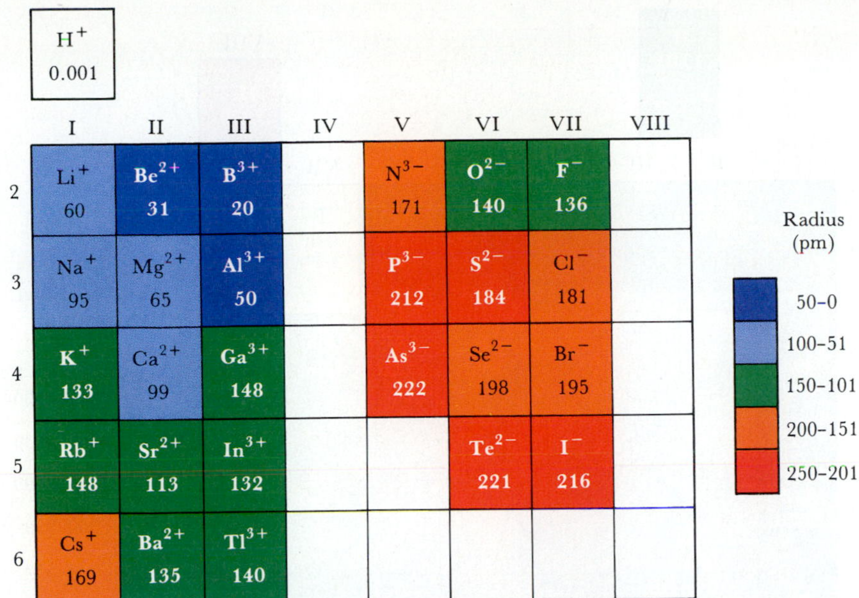

H+
0.001
I II III IV V VI VII VIII
2
Li+ 60
Be2+ 31
B3+ 20
N3− 171
O2− 140
F− 136
3
Na+ 95
Mg2+ 65
Al3+ 50
P3− 212
S2− 184
Cl− 181
4
K+ 133
Ca2+ 99
Ga3+ 148
As3− 222
Se2− 198
Br− 195
5
Rb+ 148
Sr2+ 113
In3+ 132
Te2− 221
I− 216
6
Cs+ 169
Ba2+ 135
Tl3+ 140
Radius (pm)
50–0
100–51
150–101
200–151
250–201

FIGURE 7.29 The ionic radii of ions (in picometers) with noble-gas or pseudo–noble-gas configurations. The radii of cations are smaller than the atomic radii of their parent atoms, whereas the radii of anions are larger than the radii of their parent atoms. The radius of H^+ is the radius of a proton—about 0.001 pm.

core of tightly bound, closed-shell electrons remain. The difference in radius can be large; the core is generally much smaller than the parent atom because the core electrons are so tightly bound to the nucleus. The atomic radius of Li is 145 pm, but the ionic radius of Li^+, the bare heliumlike $1s^2$ core, is only 60 pm. This difference in size is comparable to the difference between a cherry and its pit. Like atomic radii, and for the same reasons, cation radii decrease across a period (as a result of increasing effective nuclear charge) and increase down each group (because shells with higher principal quantum numbers are being occupied).

Figure 7.30 also shows that anions are larger than their parent atoms. The reason can be traced to the increased number of electrons in the valence shell of the anion and the repulsive effects they exert on one another. The variation in the radii of anions is the same as that for atoms and cations, with the smallest at the upper right of the periodic table, close to fluorine.

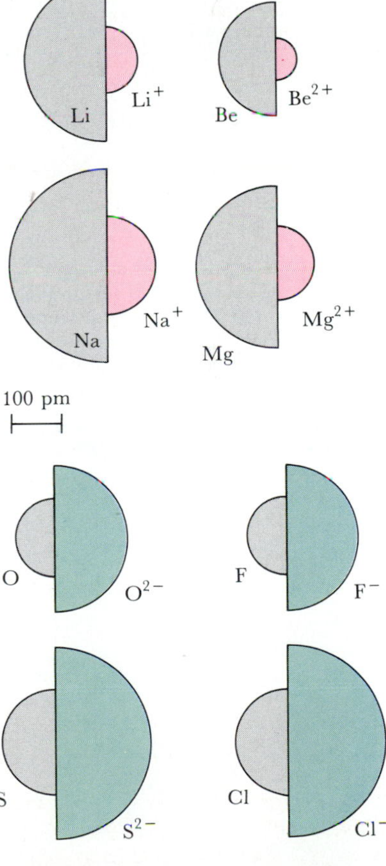

FIGURE 7.30 Relative atomic and ionic radii of eight elements. Note that cations (pink) are smaller than their parent atoms, but anions (green) are larger.

▼ EXAMPLE 7.10 Predicting the relative sizes of ions

Arrange (a) Mg^{2+} and Ca^{2+} and (b) O^{2-} and F^- in order of increasing ionic radius in each case.

STRATEGY Ions follow the same patterns that atoms follow, so the smaller member of the pair will be an ion of an element that lies either further to the right in a period or higher in a group.

SOLUTION (a) Because Mg lies above Ca in Group II, it will have the smaller ionic radius: $r(Mg^{2+}) < r(Ca^{2+})$. The actual values are 65 pm and 99 pm, respectively. (b) Because F lies to the right of O in Period 2, it will have the smaller ionic radius: $r(F^-) < r(O^{2-})$. The actual values are 136 pm and 140 pm, respectively.

EXERCISE E7.10 Arrange (a) Mg^{2+} and Al^{3+} and (b) O^{2-} and S^{2-} in order of increasing ionic radius in each case.

[*Answer*: $r(Al^{3+}) < r(Mg^{2+})$; $r(O^{2-}) < r(S^{2-})$] ▲

FIGURE 7.31 First (and higher) ionization energies of the elements, in kilojoules per mole (kJ/mol). The colors indicate the value of the first ionization energy.

T OHT: Figs. 7.31 and 7.32.

	I	II	III	IV	V	VI	VII	VIII
1		H 1310						He 2370 5250
2	Li 519 7300	Be 900 1760	B 799 2420 14,800	C 1090 2350 3660 25,000	N 1400 2860	O 1310 3390	F 1680 3370	Ne 2080 3950
3	Na 494 4560	Mg 736 1450 7740	Al 577 1820 2740 11,600	Si 786	P 1060	S 1000	Cl 1260	Ar 1520
4	K 418 3070	Ca 590 1150 4940	Ga 577	Ge 762	As 966	Se 941	Br 1140	Kr 1350
5	Rb 402 2650	Sr 548 1060 4120	In 556	Sn 707	Sb 833	Te 870	I 1010	Xe 1170
6	Cs 376 2420 3300	Ba 502 966 3390	Tl 812	Pb 920	Bi 1040	Po 812	At 920	Rn 1040

Ionization energy (kJ/mol)

2000–2500

1500–1999

1000–1499

500–999

0–499

E An electron can be removed if more energy is supplied, but then the electron would acquire excess kinetic energy. When the minimum energy is supplied, the electron is removed to infinity, and is stationary there.

N Ionization potentials for isoelectronic series. *J. Chem. Ed.* **1988**, *65*, 42.

E Ionization energies are often reported in *electronvolts* (eV), where 1 eV = 96.485 kJ/mol.

Ionization energies. The ease with which an electron can be removed from a gas-phase atom is measured by its ionization energy I:

> The **ionization energy** of an element is the minimum energy required to remove an electron from the ground state of a gaseous atom.

More precisely, the first ionization energy I_1 of an element E is the minimum energy needed to remove an electron from a neutral, gas-phase E atom:*

$$E(g) \longrightarrow E^+(g) + e^-(g) \qquad \text{energy required} = I_1$$

The second ionization energy I_2 of the element is the minimum energy needed to remove an electron from the singly charged, gas-phase cation:

$$E^+(g) \longrightarrow E^{2+}(g) + e^-(g) \qquad \text{energy required} = I_2$$

The energy needed to ionize a gaseous sample of atoms can be supplied as heat. At constant pressure, the heat required is equal to the

*Unlike the electrons that appear in the half-reactions used to discuss redox reactions in Chapter 3, the electrons removed during the ionization of a gas-phase atom are also in the gas phase. This ionization is an actual process, not a hypothetical half-reaction, so the electron is assigned a definite state, namely, (g).

enthalpy of ionization ΔH_{ion}, and the thermochemical equation is

$$E(g) \longrightarrow E^{+}(g) + e^{-}(g) \qquad \Delta H_{\text{ion}}$$

The first (and some second, third, and fourth) ionization energies of the main-group elements are listed in Fig. 7.31. Most first ionization energies lie in the range 500 to 1000 kJ/mol. The second ionization energy of an element is always higher than its first, because more energy is needed to remove an electron from a positively charged ion than from a neutral atom. In some cases (see the Group I elements) the second ionization energy is considerably larger than the first, but in others (see Group II) they are relatively close in value. The energy needed to remove an electron from the noble-gas core of an atom is always much larger than that needed to remove a valence electron. The core electrons have lower principal quantum numbers and are so close to the nucleus that they are strongly attracted to it. Because the Group I elements have an ns^1 valence-shell electron configuration, although the removal of the first electron requires only a small amount of energy, a second electron must come from the noble-gas core; consequently, its removal requires much more energy. The Group II elements have an ns^2 valence-shell electron configuration; and although two electrons are readily removed, a third is very difficult to remove because it must come from the core.

E Atoms with large radii have low ionization energies.

Ionization energy varies periodically with atomic number. This pattern is shown in Fig. 7.32 and is summarized for the first ionization energies of the main-group elements in Fig. 7.33. With a few exceptions, the ionization energy increases from left to right across a period and then falls back to a lower value at the start of the succeeding period. The lowest values occur at the bottom left of the periodic table (near Cs) and the highest at the upper right (near He). In other words, less energy is needed to remove an electron from atoms of elements near Cs and more energy is needed to remove an electron from atoms of elements near He. This finding is consistent with the fact that cesium is so highly reactive that it reduces water explosively.

The increase in first ionization energy across each period follows the trend in atomic radius. We have seen that atoms become smaller as the effective nuclear charge increases from left to right across the period. As a result of their smaller size, electrons of atoms on the right are closer to a more highly charged nucleus, are attracted more strongly to it, and, hence, are more difficult to remove. The small departures from these trends can usually be traced to the effects of repulsions between electrons, particularly electrons occupying the same orbital.

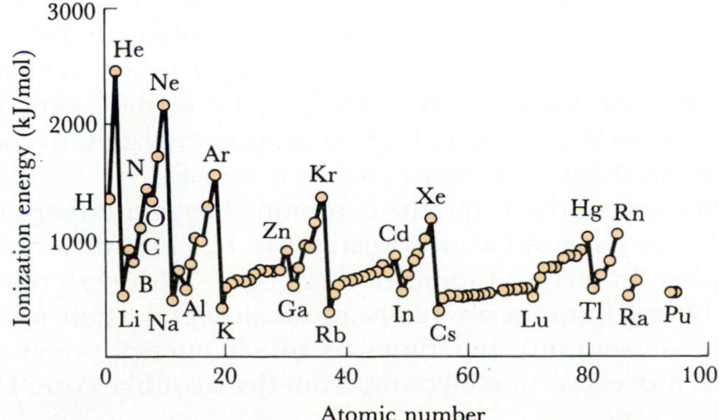

FIGURE 7.32 The periodic variation of the ionization energies of the elements. This periodic variation is like that depicted in Fig. 7.28.

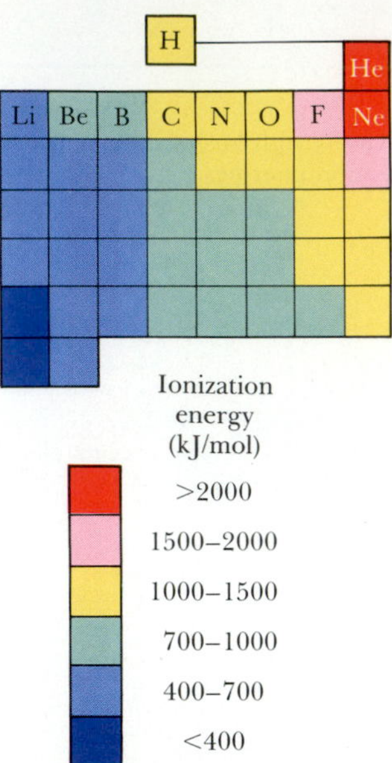

FIGURE 7.33 The variation of the first ionization energy within the main groups of the periodic table. Red indicates highest, blue lowest.

? Why are the second ionization energies of the elements in Group I larger than those of the elements in Group II?

▼ EXAMPLE 7.11 Accounting for the variation of ionization energies within a period

Suggest a reason for the small decrease of ionization energy between nitrogen (1400 kJ/mol) and oxygen (1310 kJ/mol).

STRATEGY Because the outermost electron is closer to the nucleus in O than in N and the O nucleus is more strongly charged, we expect oxygen to have a higher (not lower) ionization energy. When faced with a conflict between a prediction and an observation, we should look for an influence that has been ignored. Here, if there were a reason why the outermost electron would experience a strong repulsion from the other electrons, that would explain why less than the expected energy is needed to remove it. Therefore, we should examine the electron configurations of the two atoms to see whether they suggest a greater repulsion for the outermost electron in O than in N.

SOLUTION The outermost three electrons in N occupy three different $2p$ orbitals:

N [⇅] [↑|↑|↑]

$2s^2$ $2p_x^{\,1}2p_y^{\,1}2p_z^{\,1}$

One of the $2p$ orbitals in O, however, is full:

O [⇅] [⇅|↑|↑]

$2s^2$ $2p_x^{\,2}2p_y^{\,1}2p_z^{\,1}$

The electrons in the doubly occupied $2p$ orbital repel one another strongly, and less energy is needed to remove one of them. Hence, the ionization energy of O is less than that of N.

EXERCISE E7.11 Account for the decrease in ionization energy between beryllium and boron.

[*Answer*: new, higher-energy subshell being occupied] ▲

The decrease in ionization energy on descending a group results from the fact that the valence electron occupies an increasingly high-energy shell, with a higher principal quantum number, and, hence, in each case is bound less tightly.

The low ionization energies of elements at the lower left of the periodic table account for their metallic character. A metal consists of a collection of cations of the element surrounded by a sea of valence electrons that the atoms have lost. For example, a piece of silver consists of a stack of Ag^+ ions held together by a sea of electrons formed from one electron supplied by each of the atoms in the sample. Only elements with low ionization energies—the members of the *s* block, the *d* block, and the lower left of the *p* block—can form such solids, because only they can lose their electrons easily. The elements in the upper right of the periodic table have high ionization energies, so they do not readily lose electrons and, hence, are not metals.

U Exercise 7.99.

We noted in Section 2.7 that the common charge number of Group I cations is +1 and that of Group II cations is +2. This difference is now easy to explain in terms of ionization energies. Although one electron can be removed quite easily from an alkali-metal atom (at a cost of 494 kJ/mol for sodium), ten times as much energy (4560 kJ/mol) is needed to remove a second electron from the neonlike core. Hence, E^+

is the typical charge type of Group I cations. There is a much smaller difference between the first two ionization energies of the elements in Group II (for magnesium, they are 736 kJ/mol and 1450 kJ/mol), and it is energetically feasible to remove both electrons. However, a huge energy (7740 kJ/mol) is needed to remove a third electron from magnesium, because that electron would have to come from the neonlike core. Hence, in Group II the expected charge type of the ions is E^{2+}.

Inert pairs. When we consider the ions that *p*-block elements can form, we need to take into account the difference in ionization energies between the *s* and *p* electrons in the valence shell. The *p* electrons are lost first; but because the *s* electrons have a higher ionization energy, they might not be lost at all. It is found that the *difference* between the ionization energies for *s* and *p* electrons increases down a group. This experimental fact is expressed by saying that the two valence-shell *s* electrons of a heavy element can act as an **inert pair**. ("Lazy pair" would be a better name, because the pair of electrons can participate in reactions if conditions are right.) Thus, when aluminum forms an ion, it gives up all three of its valence electrons and forms only Al^{3+} ions, but its heavier congener indium loses three electrons in some compounds, to form In^{3+} ions, and only one in others, to form In^{+} ions. The elements for which inert pairs are important are shown in Fig. 7.34.

E The inert pair effect is largely a relativistic effect, and can be interpreted as stemming from the increase in mass of an electron when it is moving very rapidly (which inner electrons, with their high kinetic energies, do).

Electron affinity. The electron affinity E_{ea} of an element is a measure of the change in energy that occurs when an electron is added to an atom or ion:

> The **electron affinity** is the energy released when an electron is added to a gas-phase atom or ion of the element.

A high electron affinity means that a lot of energy is released when an electron attaches to the atom. A negative electron affinity means that energy must be supplied to attach an electron to an atom. We can measure the enthalpy change that occurs when an electron attaches to an atom; it is called the **electron-gain enthalpy** ΔH_{gain}. The thermochemical equation for electron gain is

$$E(g) + e^-(g) \longrightarrow E^-(g) \qquad \Delta H_{gain}$$

An element has a high electron affinity if its electron-gain enthalpy is strongly negative (and electron gain is exothermic). Some values of ΔH_{gain} are given in Fig. 7.35. Broadly speaking, electron gain becomes more exothermic (and electron affinities increase) toward the upper right of the period table and is most exothermic close to fluorine. In these atoms the incoming electron occupies a *p* orbital close to a highly charged nucleus.

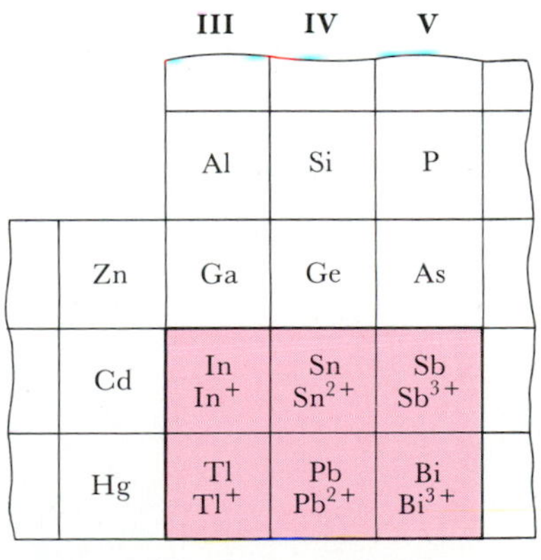

	III	IV	V
	Al	Si	P
Zn	Ga	Ge	As
Cd	In In^{+}	Sn Sn^{2+}	Sb Sb^{3+}
Hg	Tl Tl^{+}	Pb Pb^{2+}	Bi Bi^{3+}

FIGURE 7.34 The shaded boxes show elements that can lose *s* and *p* electrons in stages. The ions listed here are the ones that are formed when the *s* electrons are retained by the atom.

▼ EXAMPLE 7.12 Accounting for the variation in electron affinity

Suggest a reason for the decrease in electron affinity between carbon and nitrogen.

STRATEGY The basic theory suggests that we should expect more energy to be released when an electron enters the N atom, because it is smaller than the C atom and its nucleus is more highly charged. That the opposite is observed suggests that we may have ignored the effects of electron repul-

FIGURE 7.35 Electron-gain enthalpies (in kilojoules per mole, kJ/mol) of the main-group elements. Electron affinities are the negative of electron-gain enthalpies. For example, the electron affinity for hydrogen is 72 kJ/mol. Two values are given for O and S: the first refers to the formation of the ion X^- from the neutral atom X; the second, to the formation of X^{2-} from X^-.

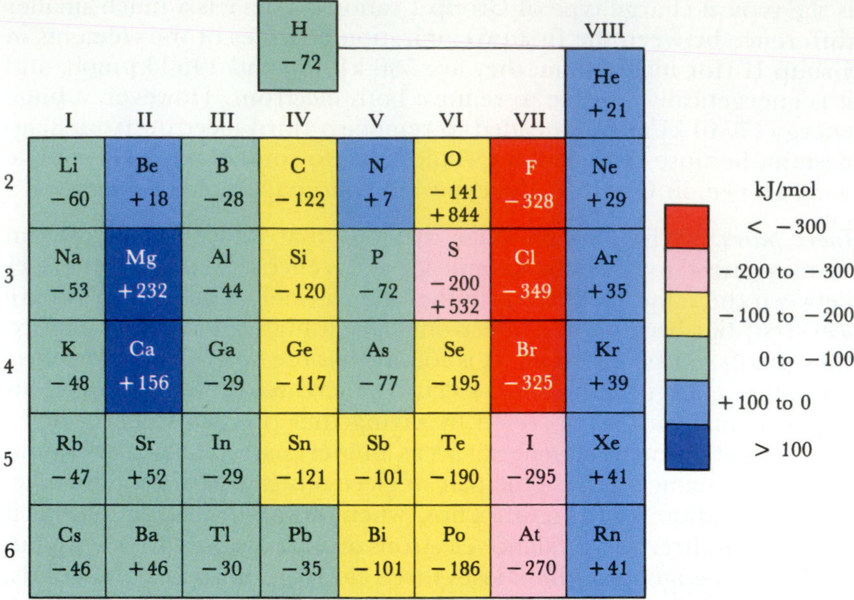

T OHT: Fig. 7.35.

N Atoms with high ionization energies often (but not always) have high electron affinities.

N A point to stress is that the individual energy required to make an ion is only partly relevant to deciding whether or not an ionic compound can form: the deciding factor is the *overall* change in energy that arises from the formation of the ions and the strengths of the interactions between them. This point is discussed in Section 8.1.

N The experimental values of atomic electron affinities. Their selection and periodic behavior. *J. Chem. Ed.* **1975**, *52*, 486.

N The periodicity of electron affinity. *J. Chem. Ed.* **1990**, *67*, 307.

sion in the resulting anions. We should therefore examine the configurations of the anions to see whether the effect of electron repulsion is greater in N^- than in C^-. If it is, then less energy will be released when the incoming electron attaches to the N atom than is released when the electron attaches to the C atom.

SOLUTION On forming C^- from C, the additional electron occupies an empty 2*p* orbital:

C^- [⇅]—[↑|↑|↑]—

On forming N^- from N, the additional electron must occupy a 2*p* orbital that is already half full:

N^- [⇅]—[⇅|↑|↑]—

Although the nuclear charge of nitrogen is greater than that of carbon, its effective nuclear charge is reduced by repulsion from the electron already in the 2*p* orbital; hence, less energy is released when N^- is formed, and the electron affinity of nitrogen is lower than that of carbon.

EXERCISE E7.12 Account for the large decrease in electron affinity between lithium and beryllium.

[*Answer*: the additional electron enters 2*s* in Li, but 2*p* in Be; and a 2*s* electron is more tightly bound than a 2*p* electron]

Once an electron has entered the single vacancy in the valence shell of a Group VII atom, the shell is complete and any additional electron would have to begin a new shell. In that shell it would not only be farther from the nucleus but also would feel the repulsion of the negative charge already present. The addition of a second electron to a halide ion is therefore strongly *endo*thermic. A consequence is that the ionic compounds of the halogens are built from singly charged ions such as F^- and never from doubly charged ions such as F^{2-}.

A Group VI atom, such as O or S, has two vacancies in its valence-shell *p* orbitals and can accommodate two additional electrons. The first electron gain is exothermic. However, the second electron gain is endothermic, because energy must be supplied to overcome repulsion by the negative charge already present. Because 141 kJ/mol of energy is released when the first electron joins the neutral atom to form O^- and 844 kJ/mol is needed to add a second electron to form O^{2-}, the net energy required is the difference, 703 kJ/mol.

Electronegativity. The **electronegativity** χ (the Greek letter chi) of an element is a measure of the electron-attracting power of its atoms when they are part of a compound. An element with a high electronegativity has a strong electron-attracting power; an element with a low electronegativity is likely to lose an electron to a more electronegative atom in a compound. The concept of electronegativity was originally devised by the American chemist Linus Pauling, who won a Nobel prize for his work (and, for his other activities, the prize for peace).

E There have been many versions of the definition of electronegativity. The first was Pauling's, which was in terms of the strengths of bonds. The definition in terms of ionization and electron affinity is essentially the Mulliken definition. Electronegativity is discussed more fully in Section 9.3.

N Principles of electronegativity. Part 1. General nature. *J. Chem. Ed.* **1988**, *65*, 112.

The values of χ for the main-group elements are summarized in Fig. 7.36. The variation of electronegativity through the periodic table is depicted in Fig. 7.37. The electronegativity is highest for elements at the top right of the periodic table, close to fluorine, and such elements are said to be **electronegative**. The electronegativity is lowest for elements at the bottom left of the table, close to cesium. An element with a low electronegativity (one that is likely to give up electrons to another element on compound formation) is called **electropositive**. The members of the *s* block, especially cesium, are the most electropositive elements.

We can find an explanation of the variation in electronegativity by noting that an atom is likely to form a cation if it has a low ionization energy and a low electron affinity. On the other hand, it is likely to form an anion if it has a high ionization energy and a high electron affinity. These observations suggest the pattern of behavior shown in the first table in the margin, which is summarized neatly if we write

Ionization energy	Electron affinity	Behavior
low	low	forms cation
high	high	forms anion

$$\chi \propto I + E_{\text{ea}}$$

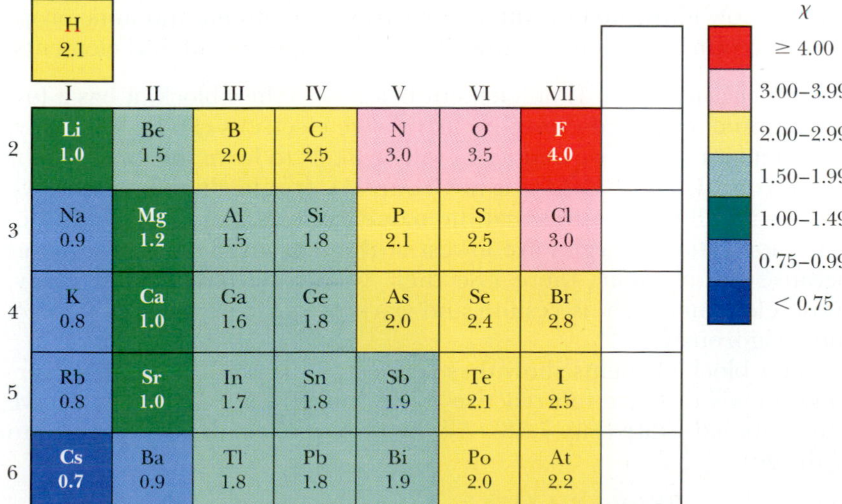

FIGURE 7.36 Electronegativities of the main-group elements.

FIGURE 7.37 The periodic variation of the electronegativities of the elements. The periodic variation is like that depicted in Fig. 7.28.

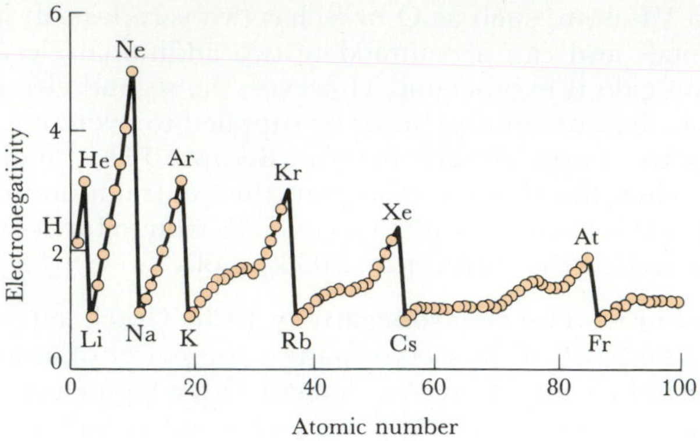

T OHT: Figs. 7.36 and 7.37.

Electronegativity	Behavior
low	forms cations
high	forms anions

When both ionization and electron affinity are low, χ is low; when both are high, χ is high. In terms of χ, the table may be written as shown in the lower table in the margin. This table is consistent with the definition of electronegativity as the ability of an atom to attract electrons when it is part of a compound. It follows that we should expect a high electronegativity for an element that has a high ionization energy *and* a high electron affinity, which is the case for fluorine and its neighbors in Groups VI and VII. Conversely, we expect a low electronegativity for elements that have a low ionization energy and a low electron affinity, which is the case for cesium and its neighbors in the *s* block.

The high electronegativities of the halogens and the high electropositivities of the alkali metals account for the formation of ionic compounds when they react together. Oxygen also has a very high electronegativity, so we would expect it to form the oxide ion when it reacts with the *s*-block elements. We shall see in more detail in Chapter 8 that the electronegativity is indeed a useful guide to the type of compound an element forms.

U Exercise 7.98.

7.9 TRENDS IN CHEMICAL PROPERTIES

We can now illustrate how the various trends in atomic and ionic properties account for some of the chemical properties of the elements.

The s-block elements. If an element belongs to the *s* block, it has a low ionization energy and therefore its valence electrons can be lost easily. The element (which we denote E) will be likely to form the ions E^+ (if it is in Group I) and E^{2+} (if it is in Group II). It is likely to be a reactive metal with all the features that the name *metal* implies (Table 7.3). Because ionization energies are lowest at the bottom of each group and because the elements there lose their valence electrons most easily, those elements such as cesium and barium can be expected to react most vigorously.

N More properties of the *s*-block elements are described in Chapter 18.

D Reactivity of sodium and potassium. TDC, 12.4 (and accompanying safety note 1.3).

N Table 7.3 is very important, because it summarizes many of the correlations between the location of an element in the periodic table and its chemical and physical properties.

T OHT: Table 7.3.

The *s*-block elements show these expected properties. All are silver-gray metals that are too reactive to be found naturally in the native (uncombined) state (Fig. 7.38). All the Group I metals reduce water to hydrogen:

$$2K(s) + 2H_2O(l) \longrightarrow 2KOH(aq) + H_2(g)$$

TABLE 7.3 Characteristics of metals and nonmetals

Metals	Nonmetals
Physical properties	
good conductors of electricity	poor conductors of electricity
malleable	not malleable
ductile	not ductile
lustrous	not lustrous
typically:	
solid	solid, liquid, or gas
high melting point	low melting point
good conductors of heat	poor conductors of heat
Chemical properties	
react with acids	do not react with acids
form basic oxides (which react with acids)	form acidic oxides (which react with bases)
form cations	form anions
form ionic halides	form covalent halides

(a)

(b)

FIGURE 7.38 All the alkali metals are soft, silvery metals. Sodium is kept under paraffin oil to protect it from air, and a freshly cut surface (a) soon becomes covered with oxides (b).

The vigor of this reaction increases down the group. Lithium reacts gently, sodium moderately vigorously, and cesium with explosive violence. All the Group II metals except beryllium also reduce water to hydrogen. However, the reaction is less violent than that involving the alkali metals, and magnesium reacts only with hot water.

All the *s*-block elements have **basic oxides** that react with water to form hydroxides:

$$CaO(s) + H_2O(l) \longrightarrow Ca(OH)_2(aq)$$

Beryllium, at the head of Group II, has the highest ionization energy of the block. It therefore loses its valence electrons less readily than its congeners and has the least pronounced metallic character. The compounds of all the *s*-block elements (with the exception of beryllium) are ionic.

Diagonal relationships. Another feature of the periodic table becomes apparent when we compare neighbors on diagonal lines running from upper left to lower right, such as Li and Mg or Be and Al (Fig. 7.39). Such elements show a diagonal relationship:

> A **diagonal relationship** is a similarity in properties between diagonal neighbors in the periodic table, especially for elements in Periods 2 and 3 at the left of the table.

Diagonal relationships show up in many different ways and are very helpful for making predictions about the properties of elements and their compounds. The diagonal band of metalloids dividing the metals from the nonmetals is one example we have already met. A diagonal relationship shows up in the *s* block as a close similarity between the

N More on diagonal relationships and the representative elements appears in Chapters 18 and 19.

D Reaction of calcium oxide and water (slaking lime). BZS1, 1.7.

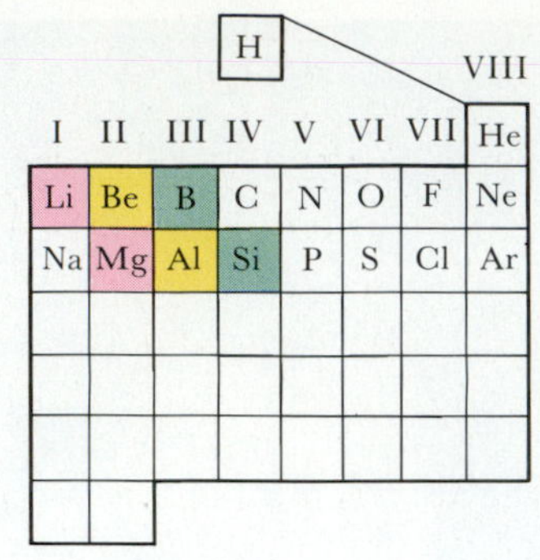

FIGURE 7.39 The pairs of elements represented by colored boxes show a strong diagonal relationship to each other.

T OHT: Figures 7.39 and 7.40.

N Magnesium metal reacts vigorously and exothermically with oxygen, nitrogen, carbon dioxide, sulfur dioxide, and hot water.

? How may a magnesium fire be extinguished?

E Solid Drano is a mixture of sodium hydroxide and small pieces of aluminum.

D Amphoteric properties of the hydroxides of aluminum, zinc, chromium, and lead. BZS3, 8.19.

N More on the properties of the *p*-block elements appear in Chapters 19 and 20.

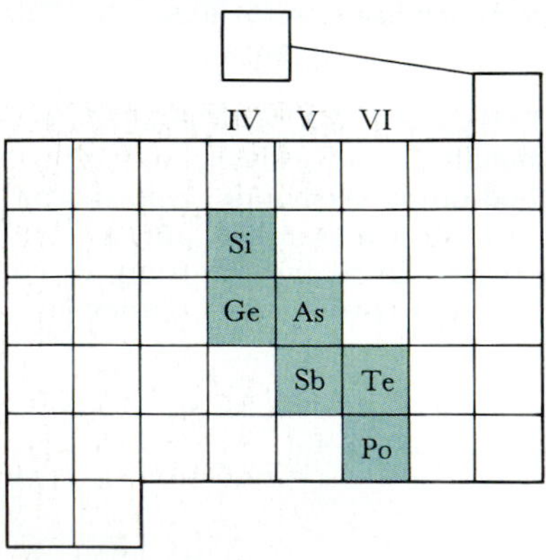

FIGURE 7.40 The metalloids fall in a diagonal band across the periodic table, between the metals and the nonmetals. The location of the metalloids reflects the diagonal relationship between elements.

chemical properties of lithium and magnesium. Lithium is the only member of Group I that burns in nitrogen and, like magnesium, forms the nitride:

$$6Li(s) + N_2(g) \longrightarrow 2Li_3N(s)$$

$$3Mg(s) + N_2(g) \longrightarrow Mg_3N_2(s)$$

The latter reaction is one reason why magnesium burns so brightly in air: it combines exothermically with both the oxygen and the nitrogen.

The similarity of beryllium to its *p*-block diagonal neighbor aluminum is another example of a diagonal relationship. Like aluminum, beryllium is **amphoteric**, that is, able to react with both acids and bases. The two elements react with acids:

$$Be(s) + 2H^+(aq) \longrightarrow Be^{2+}(aq) + H_2(g)$$

$$2Al(s) + 6H^+(aq) \longrightarrow 2Al^{3+}(aq) + 3H_2(g)$$

(Aluminum appears not to react with nitric acid and, in fact, is often used to transport it. However, the acid oxidizes the surface of the metal to a protective film of aluminum oxide, which prevents further reaction.) They also react with alkalis:

$$Be(s) + 2OH^-(aq) + 2H_2O(l) \longrightarrow Be(OH)_4{}^{2-}(aq) + H_2(g)$$

$$2Al(s) + 2OH^-(aq) + 6H_2O(l) \longrightarrow 2Al(OH)_4{}^-(aq) + 3H_2(g)$$

The reaction between aluminum and sodium hydroxide is partly responsible for the action of domestic drain cleaners, which are mixtures of sodium hydroxide and aluminum shavings. The reaction produces heat, which melts the fats and grease, and hydrogen, which helps to break up the mass. The sodium hydroxide also reacts with the fat and breaks it down into simpler soaplike substances that can be washed away. Beryllium is not used in drain cleaners because it is much more expensive than aluminum and because its salts are very poisonous.

The p-block elements. An element on the left of the *p* block has a low enough ionization energy to possess some of the metallic properties of the *s*-block elements, especially if it is in a late period. The frontier between *p* block metals and nonmetals—the metalloids shown in Fig. 7.40—runs diagonally across the table. However, the ionization energies of all the *p*-block metals are quite high, and they are less reactive than the members of the *s* block.

In Group IV, lead and tin are metals. They are lustrous and malleable and conduct electricity, but they are not nearly as reactive as the *s*-block elements and many *d*-block elements. As a practical example, steel cans are tin-plated—covered in tin by dipping them into the molten metal—to protect them against corrosion. Lead, too, was used for about 2000 years to make pipes for domestic water supplies, for it is very resistant to corrosion. It is no longer used, because lead compounds are now known to be toxic. Lead and tin also show more than a hint of nonmetallic character because they form amphoteric oxides:

$$SnO(s) + 2H^+(aq) \longrightarrow Sn^{2+}(aq) + H_2O(l)$$

$$SnO(s) + OH^-(aq) + H_2O(l) \longrightarrow Sn(OH)_3{}^-(aq)$$

The inert-pair is most important at the bottom of a *p*-block group. We have seen that aluminum exists as Al^{3+} ions but that indium, near the bottom of the same group, forms both In^{3+} and In^{+} ions. Tin, in Group IV, forms tin(IV) oxide when heated in air. In contrast, its heavier congener lead loses only two *p* electrons and forms lead(II) oxide:

$$Sn(s) + O_2(g) \longrightarrow SnO_2(g)$$

$$2Pb(s) + O_2(g) \longrightarrow 2PbO(s)$$

Tin(II) oxide can be formed as a blue-black solid by adding alkali to a solution of tin(II) salt and then heating the precipitate in the absence of air. When it is heated to 300°C in air, it forms tin(IV) oxide (the inert pair now being lost) with the evolution of so much heat that the solid becomes incandescent (Fig. 7.41).

FIGURE 7.41 When tin(II) oxide is heated in air, it becomes incandescent as it reacts to form tin(IV) oxide. Even without being heated, it smolders and can ignite.

The characteristic property of elements at the right of the *p* block is their high electron affinity, and in particular their ability to gain electrons and thus complete closed-shell configurations. The elements on the upper right of the block, such as oxygen, sulfur, and the halogens, are therefore nonmetals and typically occur in ionic compounds as anions. Fluorine forms ionic or molecular compounds with every element except helium, neon, and argon.

The d-block elements. All *d*-block elements are metals. Elements that lie toward the left of the block, near scandium, have lower ionization energies (see Fig. 7.32) and therefore are reactive, resembling the *s*-block metals. Those that lie to the right of the block, as copper and gold do, have higher ionization energies and are less reactive, resembling the more chemically sluggish metals of the *p* block. The coinage metals can be used as such because they are so unreactive. The properties of the *d*-block elements are transitional between the *s*- and *p*-block elements; this character led to the origin of their alternative name, the transition metals.

One very characteristic feature of the *d*-block elements is that many of them form a variety of cations with different oxidation numbers. Iron, we have seen, forms iron(II) and iron(III) ions, Fe^{2+} and Fe^{3+}. Copper forms copper(I) and copper(II), Cu^{+} and Cu^{2+}. Potassium, an *s*-block metal, has a $4s^1$ valence configuration like copper, but forms only one type of ion, K^{+}. The reason for the difference between copper and potassium can be seen by comparing their second ionization energies, which are 1960 kJ/mol and 3070 kJ/mol, respectively. To form Cu^{2+}, an electron is removed from the *d* subshell of $[Ar]3d^{10}$; but to form K^{2+}, the electron would have to be removed from potassium's argonlike core. Because such huge amounts of energy are not readily available in chemical reactions, potassium can lose only its 4*s* electron.

We are beginning to see why many chemists claim that the periodic table is the single most important discovery in their field. In the remainder of the text, beginning with the ideas we meet in the next chapter, we shall see that it is indeed an enormously powerful way of organizing knowledge.

N The properties of the *d*-block elements are covered more fully in Chapter 21.

? The common ion of zinc is Zn^{2+}: why are Zn^{+} and Zn^{3+} highly unlikely to be prepared?

SUMMARY

7.1 and 7.2 Light is **electromagnetic radiation**; the **frequency** and **wavelength** of its waves determine its color. According to **quantum mechanics**, energy is **quantized** and can be transferred in **quanta**, or packets of energy. Light rays consist of streams of **photons**, packets of energy of magnitude $h\nu$. Support for this view comes from the **photoelectric effect**, the ejection of electrons from the surfaces of meals by high-frequency electromagnetic radiation. When electrons in an atom change into a spatial arrangement of lower energy, the difference in energy is lost as a photon. The existence of **spectral lines** at discrete frequencies in atomic spectra indicates that the energies of electrons in atoms are **quantized**, or confined to certain values.

7.3 The **Bohr model** was an attempt to explain the **Rydberg formula**, which summarizes the spectrum of atomic hydrogen. In the Bohr model, the electrons are presumed to travel in definite orbits with a certain energy.

7.4 The Bohr model had to be rejected because electrons have a **wave-particle duality**, a dual wavelike and particlelike character, with a wavelength given by the **de Broglie relation**. The wave character of electrons is confirmed by the diffraction pattern formed in the **Davisson-Germer experiment**. Because they have wavelike properties, electrons do not travel in precisely determined paths; the precision with which the position and speed can be known is limited by Heisenberg's **uncertainty principle**. The wavelike character of electrons is taken into account by the **Schrödinger equation**, which gives the **orbital**, the region in which the electron is most likely to be found in an atom. The atomic orbitals of hydrogen are classified as $s, p, d, f, \ldots$ and lie in a series of **shells** and **subshells**. Orbitals are labeled with the **quantum numbers** n, l, and m_l. In the hydrogen atom, all orbitals of a given shell have the same energy. The **ground state**, the state of lowest energy, of the hydrogen atom consists of an electron in the 1s orbital. The **Stern-Gerlach experiment**, in which a beam of atoms passes through an inhomogeneous magnetic field, shows that an electron may also have one of two **spin** states (↑ or ↓).

7.5 The electronic structures of **many-electron atoms** are based on the orbitals obtained by solving Schrödinger's equation for the hydrogen atom. As a result of **penetration** (the approach of an electron to the nucleus through inner shells) and **shielding** (the repulsion by other electrons), the **effective nuclear charges** experienced by electrons in different subshells are different. Hence, in many-electron atoms, subshells have different energies. The electrons of an atom occupy the orbitals that result in the lowest total energy subject to the requirements of Pauli's **exclusion principle**—that no more than two electrons can occupy a given orbital.

7.6 The ground-state **electron configuration**, or list of occupied orbitals, of any atom is obtained by using the **building-up principle**, in which electrons are imagined as entering the orbitals of the atom one after the other in a certain order and in accordance with the exclusion principle and **Hund's rule** concerning the occupation of different orbitals. The availability of d orbitals in Period 4 and later periods accounts for the greater length of those periods. An atom is **paramagnetic** (pulled into a magnetic field) if it has unpaired electrons and **diamagnetic** (expelled from a magnetic field) if it does not.

7.7 The periodic table reflects the periodicity of the electron configurations of the elements. Its **groups** contain elements with analogous valence electron configurations and its **periods** correspond to the occupation of different shells. The **blocks** of the periodic table correspond to the occupation of subshells by the electrons added last according to the building-up principle.

7.8 For a review of periodicity, see the table below. **Atomic radii** and **ionic radii** vary periodically, with the largest radii at the lower left in the periodic table and smallest at the upper right. **Ionization energy** and the **electron affinity** (as measured by the **electron-gain enthalpy**) also vary periodically, with smallest values at the lower left and greatest at the upper right. **Core electrons**, or electrons in completed inner shells and subshells, have much higher ionization energies than do the **valence electrons**, the outermost electrons. The difference is less pronounced for a d subshell, so d electrons may be lost reasonably easily from elements of the d block. The ability of metals in the lower regions of the p block to form cations with different charge numbers reflects the **inert-pair effect** (the relative inertness of a pair of valence-shell s electrons), which depends on differences between s and p electron ionization energies. **Electronegativity**, a measure of the electron-attracting power of an atom in a compound, is greatest at the upper right of the periodic table.

Summary of periodic trends*

Property	Trend
atomic radius	decreases L → R across period; increases down group
ionic radius	decreases L → R across period; increases down group; in the same period, anions generally have larger radii than cations
ionization energy	increases L → R across period; decreases down group
electron affinity	highest values at upper right; lowest values at lower left
electronegativity	increases L → R across period; decreases down group

*In each case there are exceptions; these are the general trends.

7.9 As well as being closely related to their congeners in a group, a number of elements (especially those in Periods 2 and 3 toward the left of the periodic table) show a strong **diagonal relationship** to each other.

CLASSIFIED EXERCISES

Frequency and Wavelength

7.1 (a) The approximate wavelength range of the visible spectrum is 400 to 700 nm. Express this range in hertz (Hz). (b) The wavelength of an AM radio band is about 250 m. What is the frequency of this radio band?

7.2 (a) Infrared radiation has wavelengths greater than about 800 nm. What is the frequency of 800-nm infrared radiation? (b) Microwaves, such as those used for radar and to heat food in a microwave oven, have wavelengths greater than 3 mm. What is the frequency of this 3-mm radiation?

7.3 (a) Violet light has a frequency of 7.1×10^{14} Hz. What is the wavelength (in nm) of violet light? (b) When an electron beam strikes a block of copper, x-rays with a frequency of 2.0×10^{18} Hz are emitted. What is the wavelength of these x-rays?

7.4 (a) Radio waves for the FM station "Rock 99" at "99.3" on the FM dial are emitted with a frequency of 99.3 MHz. What is wavelength of this radiation? (b) Radioastronomers use 1420 MHz waves to look at interstellar clouds of hydrogen atoms. What is the wavelength of this radiation?

Photons

Note: Planck's constant, $h = 6.63 \times 10^{-34}$ J·s/photon, may also be written $h = 6.63 \times 10^{-34}$ J/(Hz·photon).

7.5 Sodium vapor lamps, used for public lighting, emit 589-nm yellow light. What energy is emitted by (a) an excited sodium atom when it generates a photon and (b) 1 mol of excited sodium atoms at this wavelength?

7.6 When an electron beam strikes a block of copper, x-rays with a frequency of 2.0×10^{18} Hz are emitted. What energy is emitted by (a) an excited copper atom when it generates an x-ray photon and (b) 1 mol of excited copper atoms at this wavelength?

7.7 (a) Gamma radiation, which is emitted when the protons and neutrons inside the nucleus of an iron-57 atom adjust their positions, has a wavelength of 86 pm. Calculate the energy of one photon of gamma radiation. (b) A mixture of argon and mercury vapor used in blue advertising signs emits 470-nm light. Calculate the energy change resulting from the emission on 1.00 mol of photons at this wavelength.

7.8 (a) The frequency for the FM radio station "Z-95" is 95.5 MHz. Calculate the energy produced in the transmission of 1.00 mol of photons at this frequency. (b) Ultraviolet radiation has wavelengths less than about 350 nm. What is the energy accompanying the emission of 1.00 mol of photons at this wavelength?

7.9 A high-power 100-MW carbon dioxide laser emits a pulse of 1.05-μm infrared radiation lasting for 50 ns. How many photons does it generate?

7.10 A certain laser has an average power output of 7.5 W with a 100-ps pulse at 1064 nm. How many photons does it generate per pulse?

Spectrum of Atomic Hydrogen

7.11 (a) Use the Rydberg formula for atomic hydrogen to calculate the wavelength for the transition from $n = 6$ to $n = 2$. (b) Using Table 7.1, determine to what color in the visible spectrum this transition corresponds.

7.12 The violet line in the hydrogen spectrum is at 434.0 nm. What is the principal quantum number of the orbital corresponding to the upper energy state of the electron that produces a photon of this wavelength? (*Hint:* For the visible spectrum of hydrogen, the principal quantum number of the lower state is $n = 2$.)

7.13 What is the energy change of a sodium atom that emits a yellow photon with a wavelength of 589 nm? This is commonly called the sodium "D line."

7.14 A transition in a mercury atom emits a blue photon with a wavelength of 435.8 nm. What is the energy change of the atom?

Particles and Waves

7.15 Calculate the wavelengths of (a) an electron and (b) a neutron, both with a velocity of 1.5×10^{8} m/s (one-half the speed of light).

7.16 (a) A baseball must weigh between 5 and $5\frac{1}{4}$ ounces (1 oz = 28.3 g). What is the wavelength of a 5-oz baseball thrown at 92 mi/h? (b) The velocity of an electron that is emitted from a metallic surface by a photon in the photoelectric effect is 2200 km/s. What is the wavelength of the electron?

7.17 How many subshells are there for n equal to (a) 2 and (b) 3. (c) What are the permitted values of l when $n = 3$?

7.18 Designate the subshell as *s*, *p*, and so on for l equal to (a) 0; (b) 2; (c) 1; (d) 3.

7.19 How many orbitals are in subshells with l equal to (a) 0; (b) 2; (c) 1; (d) 3?

7.20 State the number of (a) orbitals in a 4*s* subshell; (b) subshells with $n = 1$; (c) orbitals in a 5*p* subshell; (d) orbitals in a 6*d* subshell.

7.21 Write the subshell notation (in the form 3*d*, for instance) and the number of orbitals having these quantum numbers in an atom: (a) $n = 3, l = 2$; (b) $n = 1, l = 0$; (c) $n = 6, l = 3$; (d) $n = 2, l = 1$.

7.22 Identify the values of the principal quantum number n and the azimuthal quantum number l for the following subshells: (a) $2p$; (b) $5f$; (c) $3s$; (d) $4d$.

7.23 How many electrons can have the following quantum numbers in an atom: (a) $n = 2, l = 1$; (b) $n = 4, l = 2, m_l = -2$; (c) $n = 2$; (d) $n = 3, l = 2, m_l = +1$?

7.24 List the numbers of values (and the values themselves) for the magnetic quantum number for the following subshells: (a) $l = 0$; (b) $5s$; (c) $l = 3$; (d) $4f$.

7.25 Which subshells with the following designations cannot exist in an atom? (a) $2d$; (b) $4d$; (c) $4g$; (d) $6f$.

7.26 Which subshells with the following designations cannot exist in an atom? (a) $1p$; (b) $3f$; (c) $7p$; (d) $5d$.

7.27 Identify the sets of four quantum numbers $\{n, l, m_l, m_s\}$ that cannot exist for an electron in an atom and explain why not: (a) $\{4, 2, -1, +\frac{1}{2}\}$; (b) $\{5, 0, -1, +\frac{1}{2}\}$; (c) $\{4, 4, -1, -\frac{1}{2}\}$.

7.28 Identify the sets of four quantum numbers $\{n,l,m_l,m_s\}$ that cannot exist for an electron in an atom and explain why not: (a) $\{2, 2, -1, +\frac{1}{2}\}$; (b) $\{6, 0, 0, +\frac{1}{2}\}$; (c) $\{3, 2, +3, +\frac{1}{2}\}$.

7.29 Identify the sets of four quantum numbers $\{n, l, m_l, m_s\}$ for the bold electron in the diagrams below. Select the values of m_l by numbering from $+l$ to $-l$ from left to right.

(a) $2p$ [⇅|⇅|↑]

(b) $5d$ [⇅|⇅|⇅|↑|]

7.30 Identify the sets of four quantum numbers $\{n, l, m_l, m_s\}$ for the bold electron in the diagrams below. Select the values of m_l by numbering from $+l$ to $-l$ from left to right.

(a) $6p$ [⇅|⇅|↑]

(b) $5f$ [⇅|⇅|⇅|⇅|⇅|⇅|]

Penetration and Shielding

7.31 (a) List the shells with n ranging from 1 to 4 in order of increasing energy. (b) For a given shell, list the subshells with l ranging from 0 to 4 in order of increasing energy.

7.32 For electrons in the s, p, d, and f subshells, list them in order of (a) the probability of being found close to the nucleus and (b) the effectiveness with which they can shield electrons in higher shells.

7.33 Atomic sodium emits light at 389 nm when an excited electron moves from a $4s$ orbital to a $3s$ orbital (this emission is, in fact, very weak), and at 300 nm when an electron moves from a $4p$ orbital to the same $3s$ orbital. What is the energy separation (in joules and kilojoules per mole) between the $4s$ and $4p$ orbitals?

7.34 Atomic potassium emits light at 365 nm when an excited electron moves from a $4d$ orbital to a $4s$ orbital (this emission is very weak) and at 689 nm when an electron moves from a $4d$ orbital to a $4p$ orbital. What is the energy separation (in joules and kilojoules per mole) between the $4s$ and $4p$ orbitals? Why is the separation larger than that for the same two orbitals in sodium (Exercise 7.33).

Electron Configurations and the Building-Up Principle

7.35 Refer to the periodic table and list the subshells $3s$, $5d$, $1s$, $2p$, $3d$ in order of increasing energy.

7.36 Refer to the periodic table and list the subshells $2s$, $4d$, $2p$, $3p$, $1s$ in order of increasing energy.

7.37 Use the periodic table and the Pauli exclusion principle to determine the set of quantum numbers $\{n, l, m_l, m_s\}$ that, according to the building-up principle, correspond to the (a) first; (b) 9th; (c) 20th electron added to an atom.

7.38 Use the periodic table and the Pauli exclusion principle to determine the set of quantum numbers $\{n, l, m_l, m_s\}$ that, according to the building-up principle, correspond to the (a) 6th; (b) 33rd; (c) 54th electron added to an atom.

7.39 Predict the ground-state electron configuration of an atom of (a) calcium; (b) nitrogen; (c) bromine; (d) uranium.

7.40 Starting with the previous noble gas, write the ground-state electron configuration of an atom of (a) vanadium; (b) osmium; (c) tellurium; (d) mercury.

7.41 The general electron configuration for Group III elements is ns^2np^1. Write the general electron configuration for the elements of (a) Group II; (b) Group VIII.

7.42 The general electron configuration for Group III elements is ns^2np^1. Write the general electron configuration for the elements of (a) Group V; (b) the manganese group.

7.43 Starting with the previous noble gas, write the electron configurations of the following ions: (a) Fe^{2+}; (b) Cl^-; (c) Tl^+.

7.44 Starting with the previous noble gas, write the electron configurations of the following ions: (a) Ni^{2+}; (b) Pb^{2+}; (c) N^{3-}.

7.45 Identify the following atoms as paramagnetic or diamagnetic. For the paramagnetic atoms, list the number of unpaired electrons: (a) iron; (b) magnesium; (c) zirconium.

7.46 Identify these ions as paramagnetic or diamagnetic. For the paramagnetic ions, list the number of unpaired electrons: (a) Fe^{2+}; (b) Pb^{2+}; (c) Zn^{2+}.

7.47 Which of the following pairs of atoms or ions is more strongly attracted to a magnetic field? (a) Fe or Mn; (b) Na or Mg; (c) V or N.

7.48 Which of the following pairs of atoms or ions is more strongly attracted to a magnetic field? (a) Cr or Cr^{3+}; (b) Mn^{2+} or Fe^{2+}; (c) Cu or Ni.

7.49 Use the periodic table to determine the number of electrons in each of the *s*, *p*, and *d* orbitals outside the previous noble-gas core for the following elements: (a) silicon; (b) chlorine; (c) manganese; (d) cobalt.

7.50 Use the periodic table to determine the number of electrons in the outermost shell of atoms of the following elements: (a) tin; (b) iron; (c) sulfur; (d) bismuth.

Atomic and Ionic Radii

7.51 (a) From Fig. 7.27, determine which group of elements has the largest atomic radius. (b) From Figs. 7.27 and 7.29, what general statement can be made regarding the ionic radius of a cation relative to its neutral atom? (c) What general statement can be made regarding the ionic radii of cations having the same number of electrons, but a different number of protons (for example, Al^{3+} versus Na^{+})?

7.52 (a) From Fig. 7.28, determine which group of elements has the smallest atomic radii. (b) How may the relative sizes of the ionic radius of an anion and its parent neutral atom be summarized? (c) Summarize the relative ionic radii of anions having the same number of electrons, but different number of protons (for example N^{3-} versus F^{-}).

7.53 Identify the atom or ion with the larger radius in each of the following pairs: (a) Cl or S; (b) Cl^{-} or S^{2-}; (c) Na or Mg; (d) Mg^{2+} or Al^{3+}.

7.54 Which atom or ion has the smaller radius: (a) Li^{+} or Na^{+}; (b) Cl or Cl^{-}; (c) Al or Al^{3+}; (d) N^{3-} or O^{2-}?

Ionization Energy

7.55 (a) Which group of elements has the highest ionization energies? (b) Which group of elements has the lowest ionization energies?

7.56 From Fig. 7.32, identify the groups of elements that are exceptions to the trend in ionization energies across Periods 2 and 3.

7.57 Predict which atom of each pair has the greater first ionization energy (be alert for any exceptions to the general trend): (a) Na or Mg; (b) C or N; (c) P or S.

7.58 Predict which atom of each pair has the greater first ionization energy (be alert for any exceptions to the general trend): (a) Ba or Ca; (b) Be or B; (c) Ar or Xe.

7.59 In terms of electron configurations obtained from the building-up principle, explain why the ionization energies of Group VI elements are smaller than those of Group V elements.

7.60 In terms of electron configurations obtained from the building-up principle, explain why the ionization energies of Group III elements are smaller than those of Group II elements.

7.61 Explain why the second ionization energy of sodium is significantly higher than the second ionization energy of magnesium, even though the first ionization energy of sodium is less than that of magnesium. (See Fig. 7.31.)

7.62 Explain why the second ionization energy of sodium is even larger than the third ionization energy of aluminum, even though the first ionization energy of sodium is low. (See Fig. 7.31.)

Electron Affinity

7.63 (a) Which group of elements tends to have high electron affinities? (b) What is the general trend in electrons affinities across a period?

7.64 (a) Explain why the electron affinity of chlorine is greater than that of bromine. (b) Explain why the electron affinity of the S atom is greater than that of the S^{-} ion.

7.65 Select the atom or ion that has the greater electron-gain enthalpy in the following pairs (be alert for any exceptions to the general trend): (a) S or Cl; (b) C or N; (c) Cl or Br.

7.66 Select the atom or ion that has the greater electron-gain enthalpy (be alert for any exceptions to the general trend): (a) S or Se; (b) Si or P; (c) N or O.

7.67 Describe the correlation in the periodic trends of atomic radii and ionization energies, both across a period and down a group.

7.68 Describe the correlation in the periodic trends of electron affinities and ionization energies, both across a period and down a group.

Electronegativity

7.69 Arrange the elements N, Al, C, and B in order of increasing electronegativity.

7.70 Select the element that has the greatest electronegativity in each set: (a) Cl, P, or S; (b) F, O, or S; (c) Sr, In, or Ge.

7.71 Identify the element in each of the following pairs that tends to acquire electrons from the other when forming a bond: (a) H and C; (b) O and F; (c) S and P.

7.72 Identify the element in each of the following pairs that tends to acquire electrons from the other when forming a bond: (a) Si and Al; (b) S and As; (c) Cl and Br.

7.73 Describe the correlation in the periodic trends of ionization energies and electronegativities, both across a period and down a group.

CASE 7 ▸ USING LIGHT FOR CHEMICAL ANALYSIS

A substance exhibits a color as a result of its absorption of visible light from white light (which consists of all wavelengths of visible light, from 400 to 700 nm): the substance absorbs some of the light and transmits or reflects the rest. The perceived color for various absorptions from incident white light is shown in the table below. For instance, if a substance absorbs blue light (in the range 500 to 435 nm), it will appear orange. Orange is called the *complementary color* of the absorbed blue light, a term referring to the color perceived when the original color is removed from white light.

Color and wavelength in the visible region of the electromagnetic spectrum

Color absorbed	Wavelength, nm*	Color perceived
red	750 to 620	green-blue
orange	620 to 580	blue
yellow	580 to 560	blue-violet
green	560 to 490	red-violet (purple)
blue	490 to 430	orange
violet	430 to 380	yellow

*The wavelength range for each color is only approximate.

Because a substance absorbs a characteristic wavelength and because the intensity of its transmitted light is proportional to the number of absorbing species in a solution, the absorption of light can be used to identify and determine the concentration of a substance in a solution. The instrument used to make the measurement is called a *spectrophotometer,* in which a selected wavelength of radiation from a white-light source is passed through a sample and the intensity of the absorption monitored.

The ratio of the intensity of the transmitted light, I, to that of the incident light, I_0, expressed as a percentage, is called the *percentage transmittance, T,* of the sample:

$$\%T = \frac{I}{I_0} \times 100\%$$

The graph of the percentage transmittance against the wavelength of the incident radiation is called the *spectrum* of the species.

It is sometimes more convenient to express the intensity of light absorbed in terms of the *absorbance A,* which is defined as

$$A = \log\left(\frac{I_0}{I}\right) = \log\left(\frac{100}{\%T}\right)$$

The advantage of using the absorbance is that, according to *Beer's law,* the absorbance is proportional to the molar concentration:

7.74 Describe the correlation in the periodic trends of atomic radii and electronegativities, both across a period and down a group.

Trends in Chemical Properties

7.75 Complete and balance each of the following equations:
(a) $Na + H_2O \rightarrow$
(b) $BaO + H_2O \rightarrow$

7.76 Write a balanced equation for the reactions that occur (a) between lithium metal and water and (b) between magnesium oxide and water.

7.77 Identify the following elements as metals, nonmetals, or metalloids: (a) aluminum; (b) carbon; (c) germanium; (d) arsenic.

7.78 Which of the following pairs of elements do *not* exhibit the diagonal relationship: (a) B and Si; (b) Be and Al; (c) Al and Ge; (d) Na and Ca.

7.79 According to Table 7.3, metals form basic oxides and nonmetals form acidic oxides. Write balanced equations for the following reactions (assuming that none is a redox reaction):
(a) $Na_2O + H_2O \rightarrow$
(b) $SO_3 + H_2O \rightarrow$

$$A = \text{constant} \times \text{sample length} \times \text{molar concentration}$$

The constant in this expression depends on the absorbing species and is called the *molar absorption coefficient.* Therefore, when the spectrophotometer is calibrated by measuring the absorbance of a sample of known concentration, the concentration of an unknown solution can be determined by measuring its absorbance at the same wavelength.

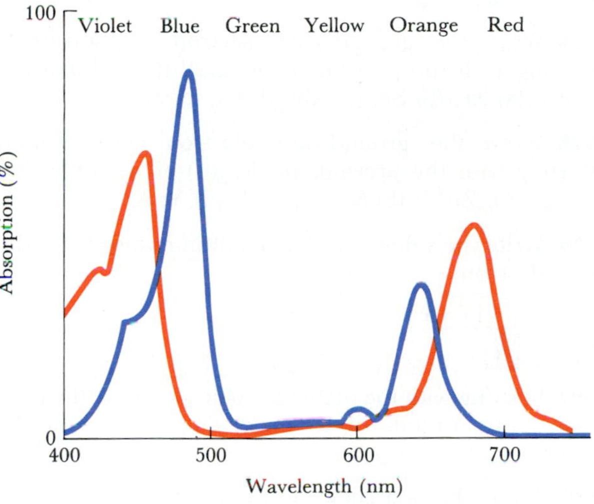

The optical absorption spectra of chlorophyll as a plot of percentage absorption against wavelength. Chlorophyll *a* is shown in red, chlorophyll *b* in blue.

QUESTIONS

1. The $Cu(H_2O)_4^{2+}$ ion is light blue and the $Cu(NH_3)_4^{2+}$ ion is blue-violet. (a) Estimate the wavelength range over which each ion absorbs visible radiation. (b) Which ion absorbs photons of higher energy?

2. If a sample gives a 22% transmittance, what is the absorbance of the species in the solution?

3. (a) Plot the following data obtained from a spectrophotometer set at 532 nm and equipped with a 1.0-cm cell, and from the slope of the plot determine the molar absorption coefficient of the absorbing species:

A	Molar concentration, mol/L
0.12	3.2×10^{-3}
0.27	7.2×10^{-3}
0.43	1.1×10^{-2}
0.66	1.8×10^{-2}
0.94	2.5×10^{-2}

(b) A solution containing the same absorbing species but of unknown concentration has an absorbance of 0.78. What is its molar concentration in solution?

7.80 Refer to the previous exercise, and write balanced equations for the following reactions (assuming that none is a redox reaction):

(a) $CO_2 + H_2O \rightarrow$

(b) $BaO + H_2O \rightarrow$

UNCLASSIFIED EXERCISES

7.81 Name the regions of the electromagnetic spectrum adjacent to the visible spectrum at the (a) high energy and (b) low energy ends.

7.82 (a) Identify the product of the velocity of light and the inverse of the wavelength. (b) Name the SI unit that is convenient for reporting the wavelength of visible light.

7.83 The wavelength of radar is about 3 cm. What is the frequency of radar waves?

7.84 X-rays of wavelength 149 pm are generated from a particle accelerator called a synchrotron. What is the frequency of these x-rays?

7.85 The average speed of a hydrogen molecule at 20°C is 1930 m/s. What is the wavelength of the H_2 molecule at this temperature? (Remember to calculate the mass of one H_2 molecule.)

7.86 How much energy in kilojoules per mole is released when an electron makes a transition from $n = 5$ to $n = 2$ in a hydrogen atom? Is this energy sufficient to break the H—H bond (436 kJ/mol is needed to break this bond)?

7.87 The energy required to break C—C bonds in a molecule is 348 kJ/mol. Will violet light of wavelength 420 nm be able to break the bond?

7.88 Many fireworks mixtures depend on the highly exothermic combustion of magnesium to magnesium oxide, in which the heat causes the oxide to become incandescent and to give out a bright white light. The color of the light can be changed by including nitrates and chlorides of elements that have spectra in the visible region. Barium nitrate is often added to produce a yellow-green color. The excited barium ions generate 487-nm, 524-nm, 543-nm, 553-nm, and 578-nm light. Calculate the change in energy of the ions in each case, and the change in energy per mole of ions.

7.89 The photoelectric effect for mercury is observed when the energy of a photon is not less than 7.25×10^{-19} J. (a) What is the maximum wavelength of light that causes mercury to exhibit the photoelectric effect? (b) If UV radiation of wavelength 250 nm shines on the surface of the mercury metal, what will the kinetic energy of the ejected photon be?

7.90 (a) How many unpaired electrons are there in the ground state of an iron atom? (b) How many *p* electrons (allowing for all shells) are there in the ground state of a phosphorus atom? (c) What is the maximum number of electrons that can be accommodated in a shell with $n = 3$? (d) Which group of elements has a [noble gas]ns^2 electron configuration?

7.91 Fill in the blanks with the correct response:

(a) The number of orbitals with the quantum numbers $n = 3$, $l = 2$, and $m_l = 0$ is ________.

(b) The number of valence electrons in the outermost *p* subshell of an S atom is ________.

(c) The number of unpaired electrons in a Mn^{2+} ion is ________.

(d) The subshell with the quantum numbers $n = 4$, $l = 2$ is ________.

(e) The m_l values allowed for a *d* orbital are ________.

(f) The allowed values of l for the shell with $n = 4$ are ________.

(g) The number of unpaired electrons in the cobalt atom is ________.

(h) The number of orbitals in a shell with $n = 3$ is ________.

(i) The number of orbitals with $n = 3$ and $l = 1$ is ________.

(j) The maximum number of electrons with quantum numbers $n = 3$ and $l = 2$ is ________.

(k) When $n = 2$, l can be ________.

(l) When $n = 2$, the possible values of m_l are ________.

(m) The number of electrons with $n = 4$, $l = 1$ is ________.

(n) The quantum number that characterizes the angular shape of an atomic orbital is ________.

(o) A subshell with $n = 3$ and $l = 1$ is designated as the ________ subshell.

(p) The lowest value of n for which a *d* subshell can occur is ________.

7.92 Which sets of quantum numbers are unacceptable?

(a) $n = 3$, $l = -2$, $m_l = 0$, $m_s = +\frac{1}{2}$

(b) $n = 2$, $l = 2$, $m_l = -1$, $m_s = -\frac{1}{2}$

(c) $n = 6$, $l = 2$, $m_l = -2$, $m_s = +\frac{1}{2}$

7.93 Identify the group in the periodic table the members of which have the following ground-state electron configurations: (a) ns^2np^3; (b) ns^1; (c) ns^2np^6; (d) ns^2.

7.94 Write the ground-state electron configurations (starting with the previous noble gas) of the following atoms: (a) Zr; (b) Se; (c) Rb; (d) Cl; (e) Sb.

7.95 Write the ground-state electron configuration (starting with the previous noble gas) of the following species: (a) Zn^{2+}; (b) Se^{2-}; (c) I^-; (d) Y; (e) Po.

7.96 Write the values for the quantum numbers for the bold electron:

(a) 3*p* [⇅][⇅][↑]

(b) 3*s* [⇅]

7.97 Starting with the previous noble gas, write the electron configurations of the following species: (a) V; (b) Cl^-; (c) Sn^{2+}; (d) Ni; (e) N^{3-}.

7.98 Circle the best choice in the list:

(a)	highest first ionization energy	Se, S, Te
(b)	smallest radius	Cl^-, Br^-, F^-
(c)	lowest electron affinity	Ba, Sr, Cs
(d)	largest ionization energy	O, S, F
(e)	lowest second ionization energy	Ar, K, Ca
(f)	most paramagnetic	Fe, Co, Ni
(g)	largest ionic radius	Ca^{2+}, Mg^{2+}, Ba^{2+}
(h)	largest radius	S^{2-}, Cl^-, Cl
(i)	highest first ionization energy	C, N, O
(j)	highest electron affinity	P, S, Cl
(k)	highest electronegativity	As, Sn, S
(l)	smallest atom	Sn, I, Bi
(m)	lowest ionization energy	K, Na, Ca
(n)	impossible subshell designation	4*g*, 5*d*, 4*p*
(o)	number of orbitals for $n = 2$	2, 4, 8
(p)	number of 5*f* orbitals	14, 7, 9

7.99 Explain why the elements zinc, cadmium, and mercury have higher ionization energies than the coinage metals (copper, silver, and gold). (See Fig. 7.32.)

7.100 The German physicist Lothar Mayer observed a periodicity in the physical properties of the elements at about the same time as Mendeleev was working on their chemical properties. Some of his observations can be reproduced by plotting the molar volume (the volume occupied per mole of atoms) for the solid forms of the elements against atomic number. Do this for Periods 2 and 3, given the following densities of the elements in their solid forms:

Element	Li	Be	B	C	N	O	F	Ne
Density, g/cm^3	0.53	1.85	2.47	2.27	1.04	1.46	1.14	1.44

Element	Na	Mg	Al	Si	P	S	Cl	Ar
Density, g/cm^3	0.97	1.74	2.70	2.33	1.82	2.09	2.03	1.66

Suggest a reason for the sharp change in molar volume between the *s*-block and *p*-block elements.

7.101 In the spectroscopic technique known as photoelectron spectroscopy (PES), ultraviolet radiation is directed at an atom or molecule. Electrons are ejected from the valence shell, and their kinetic energies are measured. Because the energy of the incoming ultraviolet photon is known and the kinetic energy of the outgoing electron is measured, the ionization energy can be deduced from the fact that the total energy is conserved. (a) Show that the speed v of the electron and the frequency (ν) of the radiation are related by

$$h\nu = I + \tfrac{1}{2}m_e v^2$$

(b) Use this relation to calculate the ionization energy of a rubidium atom, given that light of wavelength 58.4 nm produces electrons with a speed of 2450 km/s.

This chapter explains, in terms of the properties of electrons and nuclei, how atoms bond together to form compounds. It shows how to predict the types of compounds that an element can form, the number of bonds it can make with other atoms, and the geometrical shapes of the resulting molecules. As well as symbolizing bonding, the illustration shows a consequence of bonding: an adhesive forms chemical bonds with the two surfaces it joins.

8 THE CHEMICAL BOND

The concepts discussed in this chapter account for the formation and strengths of chemical bonds in terms of the ideas about atomic structure developed in Chapter 7. In particular, they show the role of the valence electrons in bonding. The importance of these electrons for the strengths of bonds is the origin of the name *valence,* which comes from the Latin word for "strength." *"Vale!"*—"Be strong!"—was what the Romans said on parting.

It will be useful to keep in mind the general plan of this chapter and to have a preview of the concepts that we shall meet. A chemist generally thinks of compounds as held together by ionic or covalent bonds. As we shall see, an ionic bond arises from the attractions between cations and anions and a covalent bond is formed when atoms share an electron pair. Ionic bonds give rise to ionic compounds (of which sodium chloride and silver sulfate are examples); these substances are solids at normal temperatures and pressures. Covalent bonds give rise to substances that are molecular solids, liquids, or gases (glucose, water, and ammonia, for example) or solids that consist of extended arrays of linked atoms (as in diamond, boron nitride, and silica). Ionic bonds are generally formed between elements that differ strongly in their electronegativity. Covalent bonds are generally formed by elements that have similar electronegativities.

A chemical bond forms if, as a result, the energy of the bonded atoms is lower than that of the separate atoms. If the lowest energy can be achieved by ion formation, then the bonding will be ionic. If the lowest energy can be reached by electron sharing, then the bonding will be covalent. To come to any conclusion about the type of bonding present in a compound, we must consider the *energy changes* that accompany bond formation. As a result, the energetics of ion formation and electron sharing will be a central focus of this chapter. We shall see that considerations of energy enable us not only to account for the bonds that atoms can form but also to predict the shapes of polyatomic molecules and ions. One of the considerable abilities you will acquire by the

N This chapter is essentially a Lewis chapter. Depending on the level of your course, Chapter 8 could be your only chapter on bonding without affecting your presentation of the material in the chapters following Chapter 9.

D Making sodium chloride from sodium and chlorine. BZS3, 8.19.

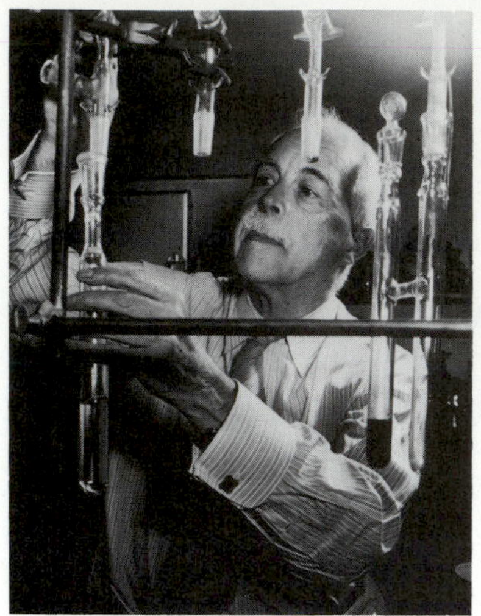

FIGURE 8.1 Gilbert Newton Lewis (1875–1946).

end of this chapter will be the ability to predict the shapes of simple molecules, starting only from a knowledge of the numbers of electrons that each atom brings to the molecule.

A considerable contribution to our understanding of bond formation was made by G. N. Lewis (Fig. 8.1), one of the greatest of all chemists. We shall see that his views (essentially, that a covalent bond is a shared electron pair and that atoms transfer or share electrons until they have reached a noble-gas electron configuration) are of enormous help in understanding the valence of elements and in rationalizing much of chemistry. If there were a single individual to whom this chapter should be dedicated, it would be Lewis. We shall see several of his contributions in this chapter: they range from the basic concepts, through the formulation of Lewis structures, which show how atoms are bonded together, to a theory of acids and bases that is even more widely applicable than the Brønsted theory outlined in Section 3.7.

IONIC BONDS

Ionic compounds are electrically neutral assemblies of cations and anions held together by the attraction between ions of opposite charge. This attraction is called an ionic bond:

> An **ionic bond** is the attraction between the opposite charges of cations and anions.

No bond is purely ionic (as we shall see in Section 9.3), but the **ionic model**, the description of bonding in terms of ions, is a good starting point for a discussion of the bonding in many compounds, particularly compounds containing elements from the *s* block, such as sodium chloride and magnesium oxide.

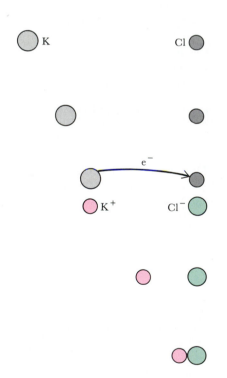

FIGURE 8.2 As a potassium atom and a chlorine atom approach, there comes a stage when it is energetically favorable for an electron to move from the K atom to the Cl atom.

8.1 THE ENERGETICS OF IONIC BOND FORMATION

In this section we shall consider the energy changes that occur when an ionic bond is formed. We begin by considering a very simple case in which a single potassium atom and a single chlorine atom come together to form an **ion pair**, one cation next to one anion. This example allows us to identify several contributions to the energy, even though the formation of a bond between one K atom and one Cl atom is unimportant in practice. We shall see that energy is released when an electropositive atom (like K) transfers an electron to an electronegative atom (like Cl) nearby. Energy release also occurs when a solid sample containing enormous numbers of ions is formed from atoms, so the ideas we develop concerning the formation of a single ion pair also apply to bulk compounds.

The formation of an ion pair. As a gaseous K atom and a gaseous Cl atom approach each other, at some point (when the process leads to a lower overall energy) the K atom loses an electron, becoming a K^+ cation, and the Cl atom gains one, becoming a Cl^- anion (Fig. 8.2). The overall energy change accompanying this transfer is the sum of three

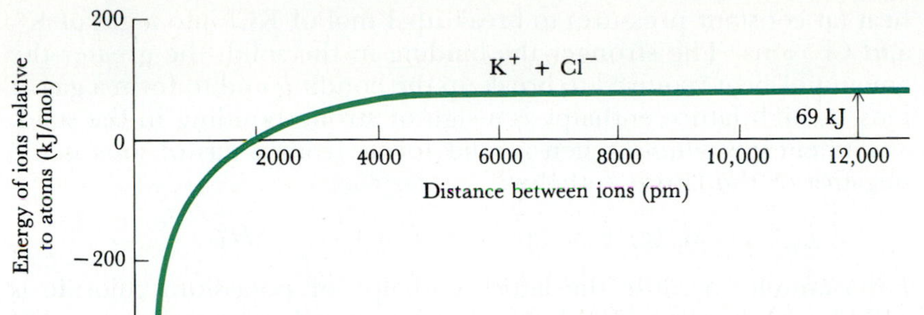

FIGURE 8.3 The energy of two ions, separated by a distance d, relative to that of the parent atoms. When the energy falls below zero, it is energetically advantageous for the ion pair to form.

contributions. One is the energy needed to form the K^+ cation from a K atom in the gas phase; this is the ionization energy of potassium. The second is the energy released when the Cl^- anion forms from a Cl atom in the gas phase; this is the electron affinity of chlorine. The thermochemical equations for the two processes and their sum are

D Set up a conductivity demonstration to show the virtual absence of ions in pure water and the presence of ions when a salt is dissolved in water.

$$\begin{array}{lr} & \Delta H \\ K(g) \longrightarrow K^+(g) + e^-(g) & +418\text{ kJ} \\ Cl(g) + e^-(g) \longrightarrow Cl^-(g) & -349\text{ kJ} \\ \hline K(g) + Cl(g) \longrightarrow K^+(g) + Cl^-(g) & +69\text{ kJ} \end{array}$$

We see that 69 kJ of energy is *needed* to form 1 mol of each of the ions. At this point of the discussion there seems to be no reason why the ions should form.

The third contribution, the energy of the attraction between the ions, is critical to ionic bond formation. This energy depends on the distance between the centers of the two neighboring ions (Fig. 8.3), and the closer they are, the stronger their interaction. The interaction energy of K^+ and Cl^- would be zero if the two ions were formed when they were very far apart. It becomes stronger than 69 kJ when the two ions are separated by 2000 pm, about six ionic diameters. This attractive interaction *lowers* the overall energy of the ion pair; so as soon as the atoms are closer than 2000 pm, the net energy of formation of the ion pair is negative and energy is released if the ion pair forms.

T OHT: Fig. 8.4.

It should be easy to appreciate that there is also a net reduction in energy when a large number of gas-phase K^+ and Cl^- ions come together to form a bulk ionic solid. The decrease in overall energy is the net outcome of the attraction of every cation for every anion in the solid, less the repulsions that each cation has for the other cations and each anion has for the other anions (Fig. 8.4).

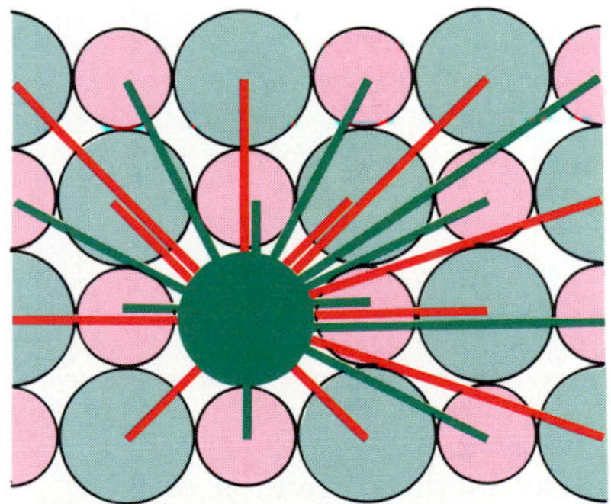

FIGURE 8.4 In an ionic solid, each ion interacts with all the others. Here we show some of the attractions (green lines) and repulsions (red lines) experienced by one anion. The net effect is attraction, partly because the anion is closest to ions of opposite charge.

The energy changes that accompany ionic bonding. A measure of the strength of ionic bonding in a bulk ionic solid is the lattice enthalpy ΔH_L°:

The **lattice enthalpy** of an ionic solid is the standard enthalpy change for the conversion of the solid to a gas of ions:

$$MX(s) \longrightarrow M^+(g) + X^-(g) \qquad \Delta H_L^\circ$$

The process always absorbs energy (is endothermic), so all lattice enthalpies are positive and we do not need to show the sign explicitly. For example, we shall see that the lattice enthalpy of solid potassium chloride is 717 kJ, which signifies that 717 kJ of energy must be supplied as

E The *Born-Meyer equation* may be used to estimate the lattice enthalpy. Its essential content is that the lattice enthalpy is proportional to the product of the charge numbers of the ions and inversely proportional to the separation of nearest neighbors: $\Delta H_L \propto z_A z_B/d$. Thus, highly charged, small ions result in high lattice enthalpies. (Al_2O_3 is an example.)

T OHT: Figs. 8.5 and 8.6.

N Madelung constants and other lattice sums. *J. Chem. Ed.* **1975**, *52*, 58.

heat (at constant pressure) to break up 1 mol of KCl into a gas of K^+ and Cl^- ions. The stronger the binding in the solid, the greater the amount of heat required to break up the bonding and to form a gas of ions. A high lattice enthalpy is a sign of strong bonding in the solid.

The energy *released* when a solid forms from a gas of ions is the negative of the lattice enthalpy:

$$M^+(g) + X^-(g) \longrightarrow MX(s) \qquad -\Delta H_L^\circ$$

For example, because the lattice enthalpy of potassium chloride is 717 kJ, we know that 717 kJ of energy is released as heat when 1 mol of KCl is formed from a gas of ions.

A knowledge of the lattice enthalpy is important for judging whether an ionic compound is likely to form. We have seen that the formation of an ion pair depends on the *overall* change in energy when ions are formed. A net input of energy is required to form the gas-phase ions from the atoms, and the ion pair forms only because the interaction between two neighboring ions is so strong that the *net* outcome is a reduction in energy. Much the same argument applies to the formation of an ionic solid. We must invest energy to form the gas of ions from the gas of atoms, and only if the lattice enthalpy is large will there be a net reduction in energy. We can anticipate that an ionic compound will form only if

- The energy required to form the gaseous cations and anions is not too great.
- The interaction between ions results in a large lattice enthalpy.

Our immediate task is to see how the lattice enthalpy may be measured.

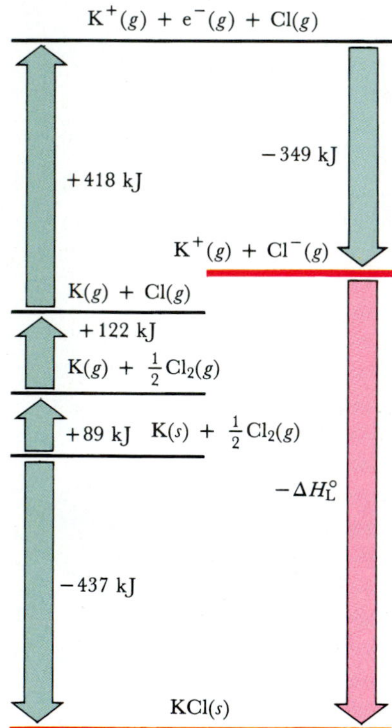

FIGURE 8.5 A Born-Haber cycle for the formation of KCl from potassium and chlorine. The sum of the enthalpy changes along the long route (green) is equal to the enthalpy change for the direct route (red). Upward pointing arrows denote endothermic processes, and downward pointing arrows denote exothermic processes.

The Born-Haber cycle. The formation of solid potassium chloride in the reaction

$$K(s) + \tfrac{1}{2}Cl_2(g) \longrightarrow KCl(s) \qquad \Delta H^\circ = -437 \text{ kJ}$$

is found experimentally to be exothermic. If we suppose that the overall reaction takes place in a series of steps, then the *overall* change in enthalpy will be equal to the sum of the enthalpy changes accompanying the individual steps. These steps are shown in Fig. 8.5, which we shall interpret below. The diagram is called a **Born-Haber cycle** in recognition of the work of Max Born and Fritz Haber, who first introduced it. It is called a cycle because we can travel from any starting point, pass through a series of changes, and return to the starting point. The enthalpy changes accompanying all the individual steps are known experimentally (and are listed in the tables in Chapter 7 and Appendix 2A) with the exception of the change accompanying the formation of the solid from the gas of ions. Therefore, we can use the cycle to calculate the missing enthalpy change, which is the lattice enthalpy we require.

The first step is the formation of a gas of K atoms from solid potassium:

$$K(s) \longrightarrow K(g) \qquad \Delta H^\circ = \Delta H^\circ_{sub} = +89 \text{ kJ}$$

The enthalpy change is the standard enthalpy of formation of potassium atoms in the gas phase, which has been measured and is listed in Appendix 2A. It is the same as the enthalpy of sublimation of solid potassium. This step is shown with an upward-pointing arrow in the Born-Haber cycle, showing that it is an endothermic step and that the enthalpy of the system increases when it takes place. The second step is the formation of Cl atoms from molecular chlorine:

$$\tfrac{1}{2}Cl_2(g) \longrightarrow Cl(g) \qquad \Delta H^\circ = \Delta H_f^\circ(Cl, g) = +122 \text{ kJ}$$

This step, the formation of chlorine atoms by the dissociation of molecular chlorine, is also endothermic and is represented by the upward-pointing arrow in Fig. 8.5.

At this stage the system consists of a gas of potassium and chlorine atoms. The next step is the ionization of potassium to form K^+ ions, which requires an input of energy equal to the ionization energy of potassium:

$$K(g) \longrightarrow K^+(g) + e^-(g) \qquad \Delta H_{ion}^\circ = +418 \text{ kJ}$$

The following step is the formation of Cl^- by electron gain:

$$Cl(g) + e^-(g) \longrightarrow Cl^-(g) \qquad \Delta H_{gain}^\circ = -349 \text{ kJ}$$

This step releases an energy equal to the electron affinity of chlorine. The arrow representing this process points downward because the process is exothermic and is accompanied by a reduction in enthalpy.

At this stage in the hypothetical process, we have a gas of K^+ and Cl^- ions. Now we allow the ions to come together to form a solid:

$$K^+(g) + Cl^-(g) \longrightarrow KCl(s) \qquad \Delta H^\circ = -\Delta H_L^\circ$$

The change in enthalpy is the negative of the lattice enthalpy, the unknown quantity we are trying to calculate. Because lattice formation is exothermic, this step is represented by a downward pointing arrow that ends at the level marked "KCl(*s*)," the product of the overall reaction. The overall change is the sum of these individual steps

$$\begin{array}{rcll} & & & \Delta H^\circ \\ K(s) & \longrightarrow & K(g) & +89 \text{ kJ} \\ \tfrac{1}{2}Cl_2(g) & \longrightarrow & Cl(g) & +122 \text{ kJ} \\ K(g) & \longrightarrow & K^+(g) + e^-(g) & +418 \text{ kJ} \\ Cl(g) + e^-(g) & \longrightarrow & Cl^-(g) & -349 \text{ kJ} \\ K^+(g) + Cl^-(g) & \longrightarrow & KCl(s) & -\Delta H_L^\circ \\ \hline K(s) + \tfrac{1}{2}Cl_2(g) & \longrightarrow & KCl(s) & 280 \text{ kJ} - \Delta H_L^\circ \end{array}$$

However, we also know (from the information in Appendix 2A) the standard enthalpy of formation of potassium chloride:

$$K(s) + \tfrac{1}{2}Cl_2(g) \longrightarrow KCl(s) \qquad \Delta H^\circ = -437 \text{ kJ}$$

Because the two enthalpy changes must be the same, we can write

$$280 \text{ kJ} - \Delta H_L^\circ = -437 \text{ kJ}$$

which solves to

$$\Delta H_L^\circ = 280 \text{ kJ} + 437 \text{ kJ} = 717 \text{ kJ}$$

That is, the lattice enthalpy of potassium chloride is 717 kJ. The lattice enthalpies of other substances may be obtained in the same way, and a few are listed in Table 8.1.

N Thermochemistry, dilithium crystals, and Star Trek. *J. Chem. Ed.* **1987**, *64*, 1039.

N It is instructive to use a Born-Haber cycle to estimate the enthalpy of formation of a hypothetical compound, such as $NaCl_2$, and (in this case) to show that the formation is highly endothermic. See Exercise 8.9.

N The cycle is similar to the use of Hess's law in Chapter 6.

T OHT: Table 8.1.

T OHT: Figs. 8.5 and 8.6.

N Lattice enthalpies of ionic halides, hydrides, oxides, and sulfides. *J. Chem. Ed.* **1990**, *67*, 304.

TABLE 8.1 Lattice enthalpies at 25°C in kilojoules per mole

Halides				
	LiF 1046	LiCl 861	LiBr 818	LiI 759
	NaF 929	NaCl 787	NaBr 751	NaI 700
	KF 826	KCl 717	KBr 689	KI 645
	AgF 971	AgCl 916	AgBr 903	AgI 887
	$BeCl_2$ 3017	$MgCl_2$ 2524	$CaCl_2$ 2260	$SrCl_2$ 2153
		MgF_2 2961	$CaBr_2$ 1984	
Oxides				
	MgO 3850	CaO 3461	SrO 3283	BaO 3114
Sulfides				
	MgS 3406	CaS 3119	SrS 2974	BaS 2832

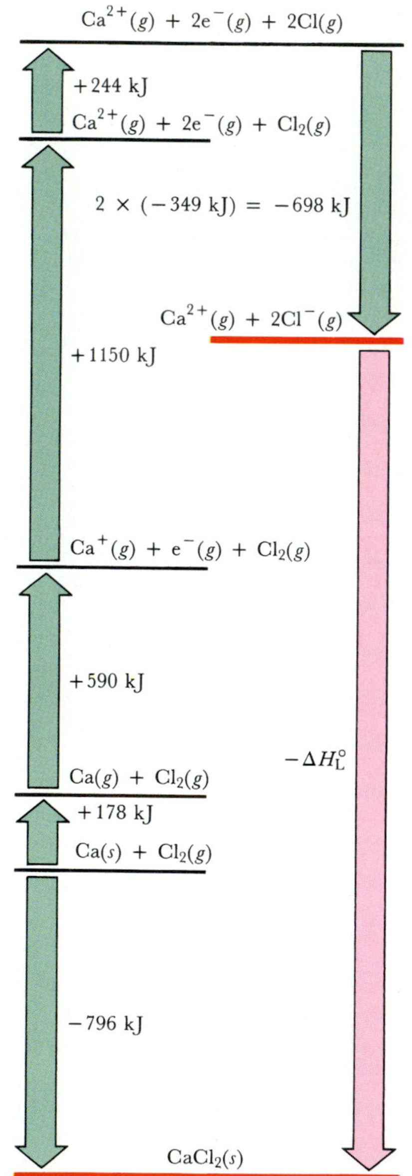

FIGURE 8.6 The Born-Haber cycle for the formation of $CaCl_2$.

▼ EXAMPLE 8.1 Determining a lattice enthalpy

Use the information in Appendix 2A and Figs. 7.31 and 7.35 to determine the lattice enthalpy of calcium chloride, $CaCl_2$.

STRATEGY We require the enthalpy of the process

$$CaCl_2(s) \longrightarrow Ca^{2+}(g) + 2Cl^-(g) \qquad \Delta H_L^\circ$$

To find it, we set up a Born-Haber cycle that includes this process (or its reverse) as one of the steps. The cycle should start from solid calcium and gaseous chlorine and lead to the formation of solid calcium chloride. When carrying out the addition to find the unknown enthalpy change, a useful rule is that the sum of the enthalpy changes on the left of the diagram must equal the sum of the changes on the right.

SOLUTION The Born-Haber cycle we shall use is shown in Fig. 8.6, and the data have been taken from Appendix 2A with the addition of ionization energies from Fig. 7.31 and electron-gain enthalpies from Fig. 7.35. The individual steps are as follows:

		ΔH°
Formation of Ca atoms:	$Ca(s) \longrightarrow Ca(g)$	+178 kJ
First ionization of Ca:	$Ca(g) \longrightarrow Ca^+(g) + e^-(g)$	+590 kJ
Second ionization of Ca:	$Ca^+(g) \longrightarrow Ca^{2+}(g) + e^-(g)$	+1150 kJ
Dissociation of Cl_2:	$Cl_2(g) \longrightarrow 2Cl(g)$	+244 kJ
Electron gain by Cl:	$2Cl(g) + 2e^-(g) \longrightarrow 2Cl^-(g)$	2 × (−349 kJ)
Lattice formation:	$Ca^{2+}(g) + 2Cl^-(g) \longrightarrow CaCl_2(s)$	$-\Delta H_L^\circ$
Overall:	$Ca(s) + Cl_2(g) \longrightarrow CaCl_2(s)$	$1464\text{ kJ} - \Delta H_L^\circ$

From Appendix 2A we also know that

$$Ca(s) + Cl_2(g) \longrightarrow CaCl_2(s) \qquad \Delta H° = -796 \text{ kJ}$$

Hence, we can write 1464 kJ − $\Delta H_L°$ = −796 kJ and conclude that $\Delta H_L°$ = 2260 kJ. Note that this lattice enthalpy is considerably greater than that of potassium chloride, largely because the Ca^{2+} ion has a higher charge than the K^+ ion and attracts the Cl^- ions more strongly.

EXERCISE E8.1 Calculate the lattice enthalpy of magnesium bromide, $MgBr_2$; use data from the same sources.

[*Answer*: 2432 kJ]

N Another instructive example is to show why calcium forms $CaCl_2$ rather than CaCl (both substances are exothermic): one approach is to show that the reaction $2CaCl(s) \rightarrow CaCl_2(s) + Ca(s)$ is highly exothermic. See Exercise 8.10 for a similar question.

U Exercises 8.63 and 8.64.

8.2 IONIC BONDS AND THE PERIODIC TABLE

Ionic bonds can be expected to form only if the energy released in one hypothetical step—the formation of the solid from the gas of ions—exceeds the energy investment in the other hypothetical step, the formation of the ions. The enthalpy of ion formation is quite small as long as the ionization energy of the cation-forming element is not very high and the electron affinity of the anion-forming element is not too low. These conditions effectively confine ionic bonds to compounds formed between strongly electropositive and strongly electronegative elements.

D Ionic versus covalent bonding. *J. Chem. Ed.* **1969**, *46*, A497.

The energy released in the hypothetical step in which the ionic solid forms from the gas of ions is the negative of the lattice enthalpy. Small, highly charged ions, which have strong interactions (Fig. 8.7), lead to high lattice enthalpies. The loss of all valence electrons from an *s*-block atom is therefore favored energetically, for in that way the atom reaches its highest charge (such as Na^+ and Ca^{2+}) without using the electrons from the core (which have *very* high ionization energies). The completion of valence shells of atoms to the right of the *p* block is also favored energetically, for in that way the atoms also reach the highest charge (such as Cl^- and O^{2-}) and hence attain the highest lattice enthalpy possible.

T OHT: The periodic table.

N A justification of this remark is the Born-Meyer equation; see the annotation on p. 282.

It is found, as we noted in Section 2.7, that *ionic compounds are formed between elements toward the left of the periodic table and elements on the right of the table.* That is, *ionic compounds typically form between the metallic and nonmetallic elements.* The elements on the left have low ionization energies and provide the cations; the elements on the right have high electron affinities and provide the anions. As shown in Fig. 8.8, *s*-block elements tend to form cations with an oxidation number equal to their (Roman) group number (Na^+ for Group I and Mg^{2+} for Group II). Elements to the lower left of the *p* block form cations with an oxidation number equal to the group number or to the group number minus two (In^{3+} and In^+, Sn^{4+} and Sn^{2+}). Elements to the upper right of the *p* block form anions by acquiring enough electrons to complete the valence shell: they form anions with an oxidation number equal to the group number minus 8 (as in O^{2-} in Group VI and Cl^- in Group VII).

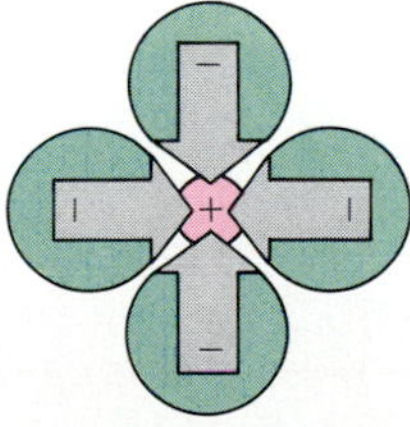

FIGURE 8.7 Small, highly charged ions (pink) attract ions of the opposite charge (green) very strongly. The ion in the center is a cation and the surrounding ions are anions. The same attraction exists between small anions and large cations.

Pseudonoble-gas configurations. The elements near the left of the *p* block and the right of the *d* block include copper, zinc, and indium. Their electron configurations are

$$\text{Cu: [Ar]}3d^{10}4s^1 \qquad \text{Zn: [Ar]}3d^{10}4s^2 \qquad \text{In: [Kr]}4d^{10}5s^25p^1$$

FIGURE 8.8 The typical ions of the main-group elements. Cations, pink; anions, green.

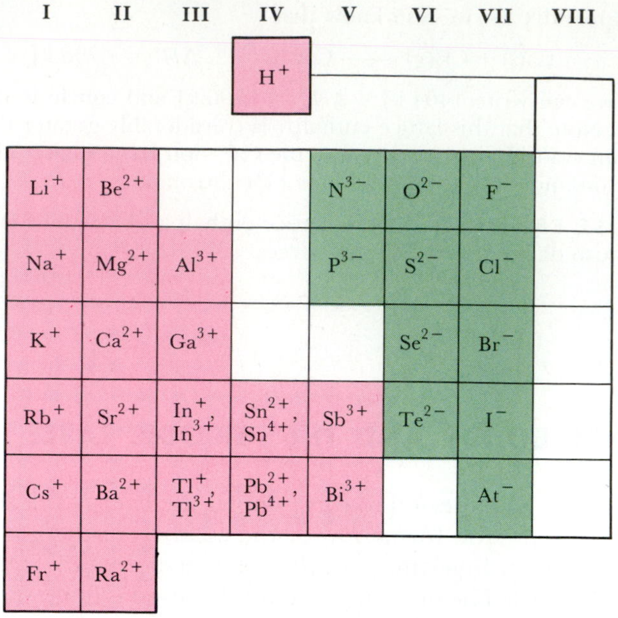

I	II	III	IV	V	VI	VII	VIII
H^+							
Li^+	Be^{2+}			N^{3-}	O^{2-}	F^-	
Na^+	Mg^{2+}	Al^{3+}		P^{3-}	S^{2-}	Cl^-	
K^+	Ca^{2+}	Ga^{3+}			Se^{2-}	Br^-	
Rb^+	Sr^{2+}	In^+, In^{3+}	Sn^{2+}, Sn^{4+}	Sb^{3+}	Te^{2-}	I^-	
Cs^+	Ba^{2+}	Tl^+, Tl^{3+}	Pb^{2+}, Pb^{4+}	Bi^{3+}		At^-	
Fr^+	Ra^{2+}						

T OHT: Fig. 8.8.

E It is sometimes said that a noble-gas electron configuration is "stable" and that elements tend to acquire it. That remark is a highly abbreviated summary: what is really meant is that electron loss and gain results in a net lowering of energy so long as a noble-gas configuration is not passed. Electron loss from a closed shell is highly endothermic. Electron gain by a closed shell may be weakly exothermic but results in an anion that readily gives up its electron or is so large that the lattice enthalpy is too small to compensate for the ionization energy of the cationic species.

Too much energy is required for these atoms to reach a noble-gas configuration by losing all their outer electrons. In fact, once the *s* and *p* electrons have been lost, the electrons in the complete *d* subshell have relatively high ionization energies and the $[Ar]3d^{10}$ and $[Kr]4d^{10}$ configurations act like stable closed shells. The configuration [noble gas]nd^{10} of these ions is an example of a **pseudonoble-gas configuration**. Elements that form cations with pseudonoble-gas configurations are shown in Fig. 8.9.

Although some elements discard electrons to achieve a pseudonoble-gas configuration for their ions, that configuration is more easily broken up than a true noble-gas core. For example, elements in the *d* block lose all their valence-shell *s* electrons when they form ions, but most of them can also lose electrons from their *d* subshell and, as a result, can form a range of cations with different charges. The subtlety of the energy balances that are involved in breaking into the pseudonoble-gas core are shown by copper's congeners silver and gold and its neighbor zinc, which give up *d* electrons to strikingly different extents even though they are close to one another in the periodic table. Zinc ($[Ar]3d^{10}4s^2$) forms no compounds in which the *d* subshell is used. Silver ($[Kr]4d^{10}5s^1$) is much more common as silver(I) than as silver(II), and its chemistry is dominated by silver(I) compounds. However, a strong oxidizing agent can break into its *d* subshell; fluorine reacts with silver to form silver(II) fluoride, a brown solid. Gold ($[Xe]4f^{14}5d^{10}6s^1$) has oxidation numbers +1 and +3. In gold(III), two electrons have been lost from the *d* subshell. Gold resembles copper in that gold(I) readily forms gold metal and gold(III) on contact with water:

$$3Au^+(aq) \longrightarrow 2Au(s) + Au^{3+}(aq)$$

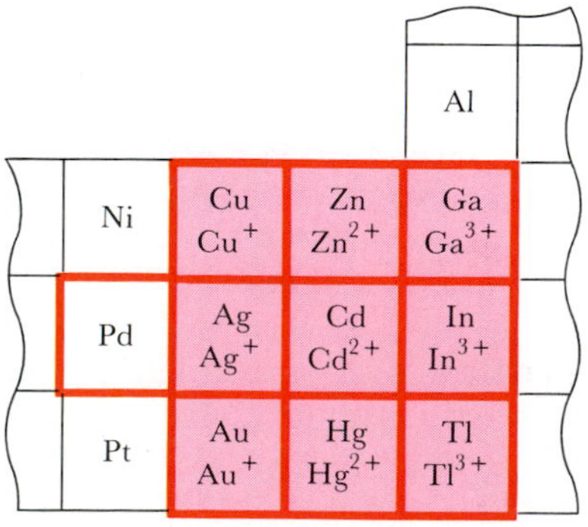

			Al
Ni	Cu Cu^+	Zn Zn^{2+}	Ga Ga^{3+}
Pd	Ag Ag^+	Cd Cd^{2+}	In In^{3+}
Pt	Au Au^+	Hg Hg^{2+}	Tl Tl^{3+}

FIGURE 8.9 Elements that form ions with a pseudonoble-gas electron configuration.

Lewis diagrams for atoms and ions. A simple way of keeping track of the valence electrons of an atom is to write a **Lewis diagram**. A Lewis diagram consists of the chemical symbol and a dot for each valence electron. Some examples are

$$\mathrm{H}\cdot \qquad \mathrm{He}: \qquad :\dot{\underset{\cdot}{\mathrm{N}}}\cdot \qquad :\dot{\underset{\cdot}{\mathrm{O}}}: \qquad :\ddot{\underset{\cdot\cdot}{\mathrm{Cl}}}\cdot \qquad \mathrm{K}\cdot \qquad \mathrm{Mg}:$$

A single dot represents an electron alone in an orbital. The dots grouped in pairs represent the two electrons that occupy the same orbital. The Lewis diagram of nitrogen, for example, represents the valence electron configuration

N [⇅] — [↑|↑|↑] —

$2s^2 \qquad 2p_x{}^1 2p_y{}^1 2p_z{}^1$

We can represent the formation of potassium chloride as the transfer of the single dot from a K atom into the single vacancy of a Cl atom:

$$\mathrm{K}\cdot + :\ddot{\underset{\cdot\cdot}{\mathrm{Cl}}}\cdot \longrightarrow \mathrm{K}^+, [:\ddot{\underset{\cdot\cdot}{\mathrm{Cl}}}:]^-$$

$\mathrm{K}^+, [:\ddot{\underset{\cdot\cdot}{\mathrm{Cl}}}:]^-$ is the **Lewis formula** of potassium chloride. The ns^2np^6 valence electron configuration reached by each atom is called an **octet** of electrons. A Lewis formula is a neat summary of the ability of some atoms to form ions by completing a noble-gas configuration either by electron loss (for elements in the *s* block) or by electron gain (for elements on the right of the *p* block). We shall see later that Lewis diagrams are also a powerful device for visualizing covalent bonding.

N A Lewis formula emphasizes the nature of the ionic bond.

U Exercise 8.65.

▼ EXAMPLE 8.2 Writing the Lewis formula of a compound

Write the Lewis formula of aluminum oxide.

STRATEGY We write the Lewis diagrams of Al and O atoms by noting that, as main-group elements, they have the same number of valence electrons (and hence dots) as their group numbers. The Lewis formula of the compound is then formed by allowing the valence electrons lost from two Al atoms to fill the vacancies in the valence shell of the three O atoms.

SOLUTION Aluminum belongs to Group III and has three valence electrons ($:\mathrm{Al}\cdot$). Oxygen belongs to Group VI and has six valence electrons ($:\dot{\underset{\cdot}{\mathrm{O}}}:$). The $2 \times 3 = 6$ dots released by the two Al atoms can be accommodated in the $3 \times 2 = 6$ gaps of the three O atoms:

$$2:\mathrm{Al}\cdot + 3:\dot{\underset{\cdot}{\mathrm{O}}}: \longrightarrow 2\mathrm{Al}^{3+}, 3[:\ddot{\underset{\cdot\cdot}{\mathrm{O}}}:]^{2-}$$

Hence, the Lewis formula of aluminum oxide is $2\mathrm{Al}^{3+}, 3[:\ddot{\underset{\cdot\cdot}{\mathrm{O}}}:]^{2-}$.

EXERCISE E8.2 Write the Lewis formulas of (a) calcium nitride and (b) sodium telluride.

[*Answer*: (a) $3\mathrm{Ca}^{2+}, 2[:\ddot{\underset{\cdot\cdot}{\mathrm{N}}}:]^{3-}$; (b) $2\mathrm{Na}^+, [:\ddot{\underset{\cdot\cdot}{\mathrm{Te}}}:]^{2-}$] ▲

D Reaction of aluminum with oxygen. TDC, 14.9.

One of the points we must be alert to has already been mentioned: elements that belong to the *d* block may lose various numbers of *d* electrons after they have lost their valence *s* electrons and, hence, may exhibit **variable valence**, the ability to form ions with different charge numbers. For example, copper forms a series of copper(I) and copper(II) compounds, and iron forms iron(II) and iron(III) compounds.

▼ **EXAMPLE 8.3 Writing the Lewis formula of a *d*-block compound**

Write the Lewis formula of iron(III) oxide.

STRATEGY We need to know the electron configuration of the *d*-block element, which can be worked out from the building-up principle (taking care to allow for the extra stability of half-filled and full *d* subshells). We need to show the electrons in the five orbitals of the 3*d* subshell as well as the electrons in the 4*s* subshell. Then we proceed as we did in Example 8.2.

SOLUTION The electron configuration of Fe is $[Ar]3d^6 4s^2$, so we depict it as

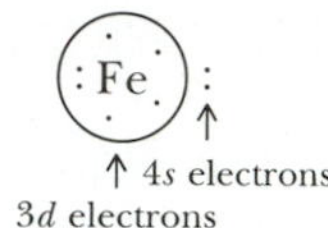

with the inner circle of dots representing the *d* electrons. Oxygen is represented in the usual way as $:\dot{O}:$. The three electrons (two 4*s* and one 3*d*) released by each of two Fe atoms can be accommodated in the two gaps in each of three O atoms:

$$2(:\text{Fe}\cdot): + 3:\dot{O}: \longrightarrow 2[(:\text{Fe}\cdot)]^{3+}, 3[:\ddot{O}:]^{2-}$$

The only difference between this example and that of Al_2O_3 is that we have had to use electrons from the *d* subshell.

EXERCISE E8.3 Write the Lewis formula of manganese(IV) oxide, MnO_2.

[*Answer*: $[(\cdot\text{Mn})]^{4+}, 2[:\ddot{O}:]^{2-}$]

▲

N There are five 3*d* electrons inside the circle drawn round the Fe^{3+} ion. Lewis diagrams for *d*-block species are much less useful than for others. At best, they give some indication of the numbers of *d* electrons that remain on the cation.

U Exercise 8.66.

COVALENT BONDS

We introduced ionic bonds by showing that considerable energy lowering occurs when an electropositive atom (such as K) transfers an electron to an electronegative atom (such as Cl) nearby. The same transfer does not lead to a lower energy when one of the elements is not electropositive. Because bonds between electronegative atoms do exist (for example, in PCl_3, SO_2, and ClO_3^-), we have to conclude that a second type of bond is needed to account for the existence of molecules and polyatomic ions.

N Grade 12 misconceptions of covalent bonding and structure. *J. Chem. Ed.* **1989**, *66*, 459.

8.3 THE ELECTRON-PAIR BOND

In 1916 Lewis proposed an explanation of bonding in molecular compounds (Section 2.6). With brilliant insight, and before anyone knew about quantum mechanics, Lewis identified the essential feature of a covalent bond, the bond responsible for the formation of molecules from atoms:

N The faithful couple: The electron pair. *J. Chem. Ed.* **1978**, *55*, 344.

A **covalent bond** is a pair of electrons shared between two atoms.

In most cases, each atom contributes one electron to the pair it shares with its neighbor. However, as we shall see in Section 8.5, in some cases one atom contributes both electrons. In either case, the shared pair of electrons lies between the two neighboring atoms.

Lewis thought that, because the two electrons of a shared pair can attract the nuclei they lie between, they would pull the two atoms together (**1**). He had no way of knowing why it had to be a *pair* of electrons—two, and not some other number. The explanation came only with the development of quantum mechanics in the 1920s and the formulation of the Pauli exclusion principle. We shall explore this point in Section 9.7.

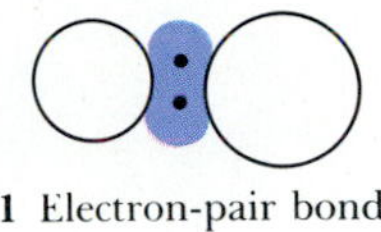

1 Electron-pair bond

The octet rule. We have already seen that when an ionic bond is formed, one atom loses electrons and the other gains them until both atoms have reached a noble-gas electron configuration (a duplet in the valence shell for elements close to helium and an octet for all other elements). Lewis suggested that atoms could also *share* electrons until they had reached a noble-gas configuration. He expressed this as the octet rule:

N Lewis structures and the octet rule. An automatic procedure for writing canonical forms. *J. Chem. Ed.* **1972**, *49*, 819.

> **Octet rule**: Atoms proceed as far as possible toward completing their octets by sharing electron pairs.

Nitrogen ($:\dot{\underset{\cdot}{N}}\cdot$) has five valence electrons and needs three more to complete its octet. Chlorine ($:\ddot{\underset{\cdot\cdot}{Cl}}\cdot$) has seven valence electrons and needs a share in one more to complete its octet. Argon ($:\ddot{\underset{\cdot\cdot}{Ar}}:$) already has a complete octet and has no tendency to share any more electrons. Hydrogen ($H\cdot$) completes the heliumlike configuration $1s^2$, a duplet (rather than an octet) of electrons.

As long as no *d* orbitals are available for occupation (as is the case for the nonmetals in Period 2), octet completion is the end of the line for any energy advantages that may come from sharing electron pairs. However, when empty *d* orbitals are available (as is the case for the elements in Period 3 and later), more than eight electrons can be accommodated around an atom. Thus, although the octet rule works well for the Period 2 elements C, N, O, and F, it often fails for elements in later periods that have empty *d* orbitals available. The five bonds between P and Cl in PCl_5, for instance, require phosphorus to use its *d* orbitals to accommodate an additional pair of electrons beyond the octet. This molecule is said to have an **expanded octet**, a valence shell that has more than eight electrons. The octet rule also fails for boron, which often forms compounds in which it has only six valence electrons. Boron trifluoride (BF_3) is an example of a molecule with an **incomplete octet**, one in which the valence shell has fewer than eight electrons. (We address these compounds more fully in Section 8.7.)

E The availability of *d* orbitals may be only a correlation not a reason why octet expansion can occur. The actual reason why elements of Period 3 and later can form more bonds may be a consequence of their larger sizes. The additional electrons are accommodated in molecular orbitals. This point is explained later (Section 9.7); for the present, it is probably best to pretend that *d* orbitals can house the additional electrons.

E Boron trifluoride (b.p. −100°C) is a common industrial catalyst (Section 8.7) and is used as a flux for soldering magnesium.

At first sight it may appear that the Lewis octet rule has so many exceptions that it is useless. However, the existence of expanded and incomplete octets can be explained quite simply, and we can usually anticipate when they are likely to occur. We shall concentrate first on molecules that satisfy the octet rule.

Lewis structures of diatomic molecules. The number of covalent bonds an atom can form depends on how many electrons it needs to complete its octet (or duplet). This number is most easily found by using Lewis diagrams.

Consider molecular hydrogen (H_2) as an example. The Lewis diagram of each hydrogen atom is H · . Each atom completes its helium-like duplet by sharing its electron with the other:

$$\text{H}\cdot + \cdot\text{H forms H:H, or H—H}$$

H—H, in which the line represents the electron pair, is the simplest example of a **Lewis structure**, a diagram showing how electron pairs are shared between atoms in a molecule.

A fluorine atom has seven valence electrons and needs one more to complete its octet. It can achieve this by sharing a pair, perhaps with another fluorine atom:

$$:\ddot{\underset{..}{\text{F}}}\cdot + \cdot\ddot{\underset{..}{\text{F}}}: \text{ forms } (:\ddot{\underset{..}{\text{F}}}(:)\ddot{\underset{..}{\text{F}}}:), \text{ denoted } :\ddot{\underset{..}{\text{F}}}\text{—}\ddot{\underset{..}{\text{F}}}:$$

The two circles have been drawn around each F atom to show how each one obtains an octet of electrons by sharing one electron pair. One difference between H_2 and F_2 is that the latter possesses lone pairs (or nonbonding pairs) of electrons:

> A **lone pair** of electrons is a pair of valence electrons that is not involved in bonding.

The lone pairs on neighboring F atoms repel one another and almost overcome the favorable attractions that hold the F_2 molecule together. This repelling effect is one of the reasons why fluorine gas is so reactive: the atoms are bound together as F_2 molecules only very weakly. Many other molecules (not only diatomic molecules) have lone pairs of electrons; among the numerous diatomic molecules, only H_2 has no lone pairs.

E The explanation of the low dissociation energy of fluorine in terms of molecular orbitals is that the antibonding orbitals are almost fully occupied and its bond order is low.

▼ EXAMPLE 8.4 Writing the Lewis structure of a diatomic molecule

Write the Lewis structure for the hydrogen chloride molecule and state how many lone pairs there are on each atom.

STRATEGY Write the Lewis diagrams for the H and Cl atoms, then combine the two atoms and share the electrons so that H has a complete duplet of electrons and Cl has a complete octet.

SOLUTION An H atom has one electron and a Cl atom has seven electrons in its valence shell. The formation of an HCl molecule can be written

$$\text{H}\cdot + \cdot\ddot{\underset{..}{\text{Cl}}}: \text{ forms H}:\ddot{\underset{..}{\text{Cl}}}:, \text{ or H—}\ddot{\underset{..}{\text{Cl}}}:$$

The H atom has no lone pairs; the Cl atom has three.

EXERCISE E8.4 Write the Lewis structure for the "interhalogen" compound chlorine monofluoride, ClF, and state how many lone pairs each atom possesses.

[*Answer*: $:\ddot{\underset{..}{\text{Cl}}}\text{—}\ddot{\underset{..}{\text{F}}}:$; three on each atom] ▲

Multiple bonds. Two atoms can bond together by sharing more than one pair of electrons:

A **double bond** consists of two shared electron pairs.

A **triple bond** consists of three shared electron pairs.

For the present, we shall call the number of shared pairs the **bond order**. For instance, O_2 is formed when the two oxygen atoms ($:\dot{\underset{\cdot}{O}}:$) share two pairs of electrons:

$$:\dot{\underset{\cdot}{O}}: + :\dot{\underset{\cdot}{O}}: \text{ forms } \ddot{\underset{..}{O}}::\ddot{\underset{..}{O}}, \text{ or } \ddot{\underset{..}{O}}{=}\ddot{\underset{..}{O}}$$

Each of the two atoms achieves a neonlike octet, and the bond order is 2. The Lewis structure of N_2 is obtained when the two atoms share three electron pairs, and the bond order of the N_2 molecule is 3:

$$:\dot{\underset{\cdot}{N}}\cdot + \cdot\dot{\underset{\cdot}{N}}: \text{ forms } :N:::N:, \text{ or } :N{\equiv}N:$$

The two atoms in N_2 are bound together very strongly because the three electron pairs draw the nuclei together tightly.

Writing Lewis structures. The Lewis structure of a diatomic molecule is obtained by writing the Lewis diagrams for the two atoms. The sum of the numbers of valence electrons in the diagrams is the number of dots that must appear in the final Lewis structure of the molecule. In carbon monoxide, for instance, the C atom has four valence electrons and the O atom has six, so ten dots must appear in the Lewis structure. The dots must be added in pairs to the structure so that each atom (other than hydrogen) ends up with an octet and all the dots are used. The same procedure is used to draw the Lewis structures of diatomic anions (such as the cyanide ion, CN^-), but an additional electron is added for each negative charge. For example, the CN^- ion has 4 + 5 + 1 = 10 electrons (4 from the C atom, 5 from the N atom, and 1 for the additional negative charge). Therefore, the Lewis structure is formed as follows:

$$[:\dot{\underset{\cdot}{C}} + \cdot\dot{\underset{\cdot}{N}}: + \cdot]^- \text{ forms } [:C:::N:]^-, \text{ denoted } [:C{\equiv}N:]^-$$

and we see that there is a triple bond between the two atoms.

8.4 LEWIS STRUCTURES OF POLYATOMIC MOLECULES

Lewis's ideas about the structure of diatomic molecules also apply to polyatomic molecules. Each atom of a polyatomic molecule completes its octet by sharing pairs of electrons with its immediate neighbors. Each pair of electrons shared by two neighbors is a covalent bond, just like a covalent bond in diatomic molecules. An example is methane:

$$\begin{array}{c}H\\ \cdot\\ \cdot\\ H\cdot\ \cdot C\cdot\ \cdot H\\ \cdot\\ \cdot\\ H\end{array} \text{ forms } \begin{array}{c}H\\ ..\\ H:C:H\\ ..\\ H\end{array}, \text{ or } \begin{array}{c}H\\ |\\ H{-}C{-}H\\ |\\ H\end{array}$$

This Lewis structure shows that the molecule consists of four carbon-hydrogen single bonds. However, what a Lewis structure does *not* show is the three-dimensional arrangement of atoms in space. (We shall see in Section 8.8 that a methane molecule consists of a carbon atom at the center of a regular tetrahedron of hydrogen atoms.) It is very important to remember that a Lewis structure is a two-dimensional diagram showing the links between atoms and not, except for special cases (such as a diatomic molecule), the shape of the molecule.

N A Lewis structure is a topological map of the bonding.

The procedure for writing Lewis structures of polyatomic molecules is essentially the same as the procedure used with diatomic molecules. We use all the dots representing the valence electrons, and we arrange them so that each atom has the same number of electrons as a noble-gas atom and hence has either a duplet (for H) or an octet of electrons. The only complication is that we now need to know which atoms are linked to which other atoms. For instance, we need to know that the arrangement of atoms in carbon dioxide is OCO and not COO. Like the atoms of carbon dioxide, the atoms of other polyatomic molecules are often symmetrically arranged with the less electronegative atom at the center. For instance, SO_2 is OSO and not SOO. In chemical formulas, especially simple ones, the central atom is often written first, followed by the terminal atoms (the atoms attached to the central atom). For example, OF_2 is FOF, not OFF; and in the PO_4^{3-} ion, the P atom lies at the center of four O atoms. An exception to the symmetrical arrangement of atoms is dinitrogen oxide, N_2O, which has the structure NNO. The convention of writing the formulas of acids with the acidic hydrogens first does not reflect the arrangement of the atoms. If the species is an oxoacid, then the acidic hydrogen atoms are attached to oxygen atoms, which in turn are attached to the central atom. Two examples are H_2SO_4, which has the structure $(HO)_2SO_2$, and HClO, which has the structure (HO)Cl.

N Molecular structures are determined experimentally; these rules are summaries of what is found typically. We shall see in Chapter 9 that the concept of formal charge helps to rationalize the structure adopted in each case.

The procedure for writing Lewis structures of polyatomic molecules and ions can be summarized as follows:

1. Determine the total number of electron dots to be used by adding up the numbers of valence electrons on each atom. Each hydrogen atom supplies one electron. Each main-group element supplies its (Roman) group number of electrons. For a cation, subtract one dot for each positive charge. For an anion, add one dot for each negative charge. Divide this total number by 2 to get the number of electron pairs. For example, the HCN molecule has $1 + 4 + 5 = 10$ valence electrons and hence five electron pairs.

2. Write the chemical symbols of the atoms to show which are neighbors (and bonded together). The HCN molecule is written

   H C N

   because the H atom is linked to the C atom and the C atom is linked to the N atom.

3. Use electron pairs to form single bonds linking each atom to its neighbor. For HCN, we write

   H : C : N

   which leaves three electron pairs unused.

4. Try to place any remaining electron pairs around the atoms to complete their octets (or duplet in the case of H). If there are not

enough electron pairs, use one or more of the pairs to form multiple bonds. The H atom has already completed its duplet, so we try putting all three remaining pairs on the N atom in HCN:

H : C : N :

However, this arrangement does not complete carbon's octet, but if we do the same for C, then we do not complete nitrogen's octet:

H : C : N :

Therefore, we rearrange the electron pairs to form multiple bonds between carbon and nitrogen:

H : C : : : N :, written H—C≡N :

In this way, both the nitrogen atom and the carbon atom achieve octets by sharing electron pairs. It is useful to remember that terminal halogen atoms always have three lone pairs and form only single bonds:

—F :   —Cl :   —Br :   —I :

Terminal oxygen and sulfur atoms form either one or two bonds and have either three or two lone pairs:

—O : or =O   —S : or =S

Terminal nitrogen atoms may have up to three bonds:

—N :   =N   ≡N :

▼ EXAMPLE 8.5 Writing Lewis structures for molecules

Write the Lewis structure for hypochlorous acid, HClO.

STRATEGY We follow the procedure set out above, bearing in mind that the molecule is an acid with the hydrogen atom attached to the oxygen atom: the *structural* formula of hypochlorous acid is (HO)Cl, as noted earlier.

N Hypochlorous acid is formed when chlorine dissolves in water; it forms when chlorine is used to disinfect drinking water, swimming pools, and sewage effluent.

SOLUTION The total number of valence electrons is

$$\begin{array}{ccccc} \text{H} & & \text{O} & & \text{Cl} \\ 1 & + & 6 & + & 7 = 14 \end{array}$$

so seven electron pairs must be accommodated. The atomic arrangement is

H O Cl

and we form single bonds between each neighboring pair of atoms:

H : O : Cl

There are five more electron pairs to accommodate. Three can be used to complete the Cl octet, leaving two to complete the O octet:

$$\mathrm{H:\ddot{\underset{..}{O}}:\ddot{\underset{..}{Cl}}:} \quad \text{or} \quad \mathrm{H{-}\ddot{\underset{..}{O}}{-}\ddot{\underset{..}{Cl}}:}$$

? What other species are isolectronic with CO_2? (Hint: identify the congeners of oxygen.)

EXERCISE E8.5 Write the Lewis structure for carbon dioxide.

[*Answer*: $\mathrm{\ddot{O}{=}C{=}\ddot{O}}$]

In the Lewis structure of an ionic compound containing a polyatomic ion, the cation and the anion should be treated separately, to show that they are individual ions and not linked by shared pairs. The structure of sodium sulfate, Na_2SO_4, for instance, is written

N This example serves to illustrate that a compound may be both ionic and covalent: the ions of which it is composed may be a covalently bonded molecular species.

$$2\mathrm{Na}^+,\ \left[\begin{array}{c} :\ddot{\mathrm{O}}: \\ | \\ :\ddot{\mathrm{O}}{-}\mathrm{S}{-}\ddot{\mathrm{O}}: \\ | \\ :\underset{..}{\mathrm{O}}: \end{array}\right]^{2-} \qquad \textit{not}\ \ \mathrm{Na{-}\ddot{O}{-}\overset{\displaystyle :\ddot{O}:}{\underset{\displaystyle :\underset{..}{O}:}{S}}{-}\ddot{O}{-}Na}$$

Note that the charge belongs to the entire anion (this fact is signified by the symbol $]^{2-}$) and is not localized on a particular atom. This rule applies to all polyatomic ions.

8.5 LEWIS ACIDS AND BASES

N We consider that the introduction of Lewis acids and bases at this stage is the most natural location, because it flows naturally from electron pair formation. An alternative position is in a chapter on acids and bases, as a generalization of Brønsted theory (Chapter 14). The section introduces some descriptive chemistry into the chapter; furthermore, it helps to relate concepts of structure to chemical reactions and reaction mechanism.

Lewis's familiarity with the importance of electron pairs in bond formation led him to propose a new definition of acids and bases. We have already seen that Brønsted's definition of acids and bases is more general than Arrhenius's definition. Brønsted's definition (that an acid is a proton donor and a base is a proton acceptor) is more general because it applies to more substances and unifies their behavior into a single description. Lewis went one stage further. His definition of acids and bases in terms of electron pairs included not only all the substances that Brønsted's definition included but also other substances that showed similar behavior. Once again we see an aspect of the way that chemistry develops: one chemist proposes a primitive idea and then another chemist refines it into a more general, widely embracing theory, and then perhaps a third chemist repeats the process until we can understand the behavior of many apparently different substances in terms of a single theory.

The Lewis definitions of acids and bases are as follows:

A **Lewis acid** is an electron pair acceptor.

A **Lewis base** is an electron pair donor.

The simplest example of a Lewis acid is a proton, H^+, which accepts a pair of electrons when it bonds to an ammonia molecule to form an ammonium ion:

$$\mathrm{H^+ + :\overset{\displaystyle H}{\underset{\displaystyle H}{N}}{-}H \longrightarrow \left[H{-}\overset{\displaystyle H}{\underset{\displaystyle H}{N}}{-}H\right]^+}$$

The ammonia molecule has supplied the electron pair and hence, according to the Lewis definition, is a base, just as it is in the Brønsted theory. The result of the combination of a Lewis base and a Lewis acid (the ammonium ion in this example) is called a **complex**. Note that the link between the acid and the base stems from the formation of a covalent bond in which the electron pair provided by the base is *shared* with the acid.

The advantage of the Lewis definition over the Brønsted definition is that we can identify substances as acids and bases even if there are no protons transferred (the role of the proton is essential in the Brønsted definition, so the classification of a substance as an acid or a base depends on the involvement of protons). For example, the formation of a hydrated aluminum ion when an aluminum salt dissolves in water is a Lewis acid-base reaction:

$$\underset{\text{Acid}}{Al^{3+}} + \underset{\text{Base}}{6H_2O} \longrightarrow \underset{\text{Complex}}{Al(H_2O)_6{}^{3+}}$$

The attachment of one water molecule to the aluminum ion occurs by the donation of an electron pair on the oxygen atom to the aluminum ion:

$$\underset{\text{Acid}}{Al^{3+}} + \underset{\text{Base}}{:\ddot{O}(H)—H} \longrightarrow \underset{\text{Complex}}{[Al—\ddot{O}(H)—H]^{3+}}$$

The remaining five water molecules attach in a similar way to form the fully hydrated ion.

Another example of a Lewis acid-base reaction is the reaction in which a proton is transferred from a hydronium ion to a hydroxide ion:

$$H_3O^+(aq) + OH^-(aq) \longrightarrow 2H_2O(l)$$

The Lewis base is the hydroxide ion, $:\ddot{\underset{..}{O}}—H^-$, with its three lone pairs, any one of which can be used to form a covalent bond. We have to look harder to find the Lewis acid. It is a hydrogen ion that is already bonded to a Lewis base, H_2O, and is present as a hydronium ion:

$$\underset{\text{Acid}}{H^+} + \underset{\text{Base}}{:\ddot{O}(H)—H} \longrightarrow \underset{\text{Complex}}{[H—\ddot{O}(H)—H]^{+}}$$

In the Brønsted neutralization reaction, one Lewis base (the OH^- ion) uses its lone pair to drive away the other Lewis base, H_2O, and to form a new covalent bond with the Lewis acid H^+. We can show the way the electron pairs move by drawing a curved arrow:

$$\underset{\text{Complex}}{[H—\ddot{O}(H)—H]^{+}} \curvearrowleft \underset{\text{Base}}{[:\ddot{\underset{..}{O}}(H):]^{-}} \longrightarrow H—\ddot{O}(H): + H—\ddot{O}(H):$$

Therefore, this neutralization reaction can be considered as a reaction in which a proton (a Lewis acid) transfers from one Lewis base (H_2O) to

N We consider it inappropriate to use the old-fashioned terms *dative bond* or *coordinate bond* for a link in which the two electrons are provided by the same atom.

N The term *adduct* is still quite widely used for a complex. IUPAC favors the elimination of the term *complex* (because it has other connotations), and its replacement by the term *coordination compound*. We consider it useful, however, to distinguish a complex from a coordination compound (see Section 21.4).

N The structures of many complexes are discussed in Chapter 21.

U Exercise 8.74.

N Once a Lewis base has been used to form a bond to a Lewis acid, its other lone pairs, if any, are usually pointing in the wrong direction to form multiple bonds.

another (OH^-). The reaction occurs because the hydroxide ion, with its negative charge, is a stronger electron pair donor than the water molecule. It is a good idea always to keep in mind the ability of a hydronium ion to act as a source of the Lewis acid H^+ and to transfer H^+ to a Lewis base that can attract it more strongly than can an electrically neutral water molecule. In the same way, a water molecule can also act as a source of hydrogen ions:

$$\text{H–}\ddot{\text{O}}\text{–H (Complex)} \longrightarrow [\text{H–}\ddot{\text{O}}\text{:}]^- \text{ (Base)} + H^+ \text{ (Acid)}$$

This reaction will take place if a species that is present is an even stronger Lewis base than the hydroxide ion and can drive out the hydrogen ion from the complex, the water molecule.

Exercises 8.77 and 8.78.

▼ EXAMPLE 8.6 Identifying a Lewis acid and a Lewis base

When water is added to calcium oxide (quick lime), a vigorous reaction takes place in which calcium hydroxide (slaked lime) is formed. Express the reaction in terms of the formation of a Lewis acid-base complex.

STRATEGY We need to look for the species with a lone pair of electrons that it can donate to form a covalent bond; such a species is a Lewis base. The species to which the lone pair is donated is the Lewis acid. We should bear in mind (as stated in the text) that the water molecule can act as a source of the Lewis acid H^+ if a very strong Lewis base is present.

SOLUTION The oxide ion, O^{2-}, is a very strong Lewis base because of its high charge and small radius. We can suspect that it is a stronger base than the singly charged hydroxide ion. In that case, we can expect the reaction

$$\text{H–}\ddot{\text{O}}\text{–H} + :\ddot{\text{O}}:^{2-} \longrightarrow [\text{H–}\ddot{\text{O}}\text{:}]^- + :\text{H–}\ddot{\text{O}}:^-$$

That is, when water is added to calcium oxide, the oxide ions extract protons from the water molecules and form hydroxide ions. An oxide ion is such a strong Lewis base that it never exists as such in water but always reacts to form a hydroxide.

EXERCISE E8.6 Account for the formation of ammonium and hydroxide ions when ammonia dissolves in water.

[*Answer*: $:NH_3$ acts as Lewis base in a competition with OH^- ions for protons, the Lewis acid:

$$\text{H–}\ddot{\text{O}}\text{–H} + :NH_3 \longrightarrow [\text{H–}\ddot{\text{O}}\text{:}]^- + [NH_4]^+ \;]$$

▲

Exercises 8.75, 8.76, and 8.79.

All Brønsted bases are Lewis bases and all Lewis bases are Brønsted bases, because an electron pair that makes a molecule a Lewis base can form a bond to a proton, thus ensuring that the molecule can act as a proton acceptor (the definition of a Brønsted base). We saw this pattern in connection with the attachment of a proton to ammonia. All Brønsted acids are Lewis acids in the sense that all Brønsted acids are sources of hydrogen ions, which (as we have seen) are Lewis acids. However, not all Lewis acids are Brønsted acids, because species other than hydrogen ions can act as electron pair acceptors. We saw that the Al^{3+} ion is an electron pair acceptor, but it is not a proton donor. Aluminum ions in water *are* proton donors, but only because the complex $Al(H_2O)_6{}^{3+}$ has many protons and can supply some of them to a base.

E A Brønsted acid is not itself a Lewis acid: it is a complex of a Lewis acid (a proton) and a Lewis base (the conjugate base of the acid), and can supply the Lewis acid (the proton) to another Lewis base.

8.6 RESONANCE

One of the problems that arises when writing the Lewis structures of certain molecules is that the multiple bonds may be written in several equivalent locations. For example, three possible Lewis structures for the nitrate ion ($NO_3{}^-$) are

$$\left[:\ddot{O}-N(-\ddot{O}:)=\ddot{O}\right]^- \qquad \left[:\ddot{O}-N(=\ddot{O})-\ddot{O}:\right]^- \qquad \left[\ddot{O}=N(-\ddot{O}:)-\ddot{O}:\right]^-$$

The three structures differ only in the position of the double bond and therefore correspond to exactly the same energy. Because experiments on nitrates have shown that all three NO bonds are identical (for instance, all three bonds have the same length), the display of only one of the structures would give a false impression. In fact, the ion is taken to be a blend of all three Lewis structures. This blending of structures, called **resonance**, is depicted as follows:

N Where does resonance energy come from? *J. Chem. Ed.* **1977**, *54*, 217.

D A visual aid for teaching the resonance concept. *J. Chem. Ed.* **1989**, *66*, 461.

$$\left[:\ddot{O}-N(-\ddot{O}:)=\ddot{O}\right]^- \longleftrightarrow \left[:\ddot{O}-N(=\ddot{O})-\ddot{O}:\right]^- \longleftrightarrow \left[\ddot{O}=N(-\ddot{O}:)-\ddot{O}:\right]^-$$

The structure resulting from resonance is a **resonance hybrid** of the contributing Lewis structures. Resonance should not be thought of as a flickering of a molecule between Lewis structures but as a *blending* of the individual structures. This concept is a little like the fact that a mule is a hybrid of a horse and a donkey, not a creature that flickers between the two.

E In quantum mechanical terms, resonance is the representation of a structure as a superposition of the wavefunctions of the individual structures. Thus, if the three structures of $NO_3{}^-$ are denoted χ_A, χ_B, and χ_C, then the actual structure is the wavefunction $\chi_A + \chi_B + \chi_C$.

▼ EXAMPLE 8.7 Writing resonance structures

Suggest two resonance structures for the SO_2 molecule. Experimental data show that the S atom lies between two O atoms and that the SO bond lengths are the same.

STRATEGY Write a Lewis structure for the molecule by using the method in Example 8.5. Decide whether there is another equivalent structure that results from the interchange of a single bond and a double bond. Write the actual structure as a resonance hybrid of these Lewis structures.

SOLUTION All three atoms are members of Group VI, so the total number of valence electrons in the molecule is $3 \times 6 = 18$. One structure is therefore $\ddot{O}=\ddot{S}-\ddot{O}:$. Interchanging the bonds gives $:\ddot{O}-\ddot{S}=\ddot{O}$. The overall structure is the resonance hybrid

$$\ddot{O}=\ddot{S}-\ddot{O}: \longleftrightarrow :\ddot{O}-\ddot{S}=\ddot{O}$$

EXERCISE E8.7 Write resonance hybrids for the acetate ion, $CH_3CO_2^-$.

[*Answer*: resonance structures of acetate ion]

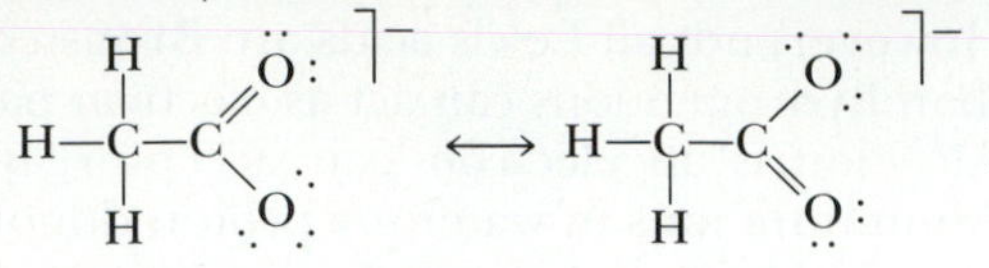

U Exercises 8.71, 8.73, and 8.83.

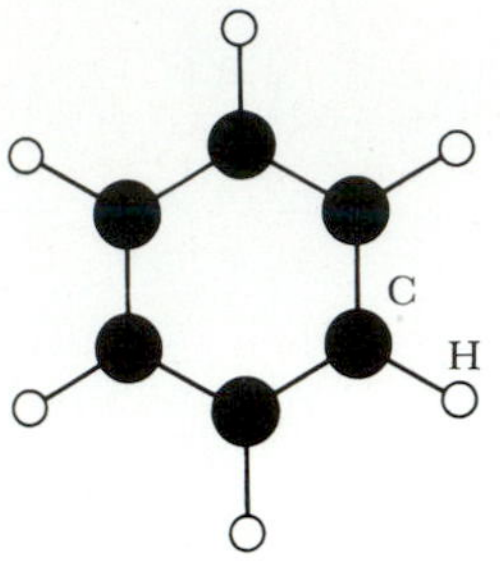

Benzene (C_6H_6) is another example of a molecule with a resonance hybrid structure. A benzene molecule consists of a hexagonal ring of six carbon atoms with a hydrogen atom attached to each C (**2**). Two structures that contribute to the resonance hybrid are

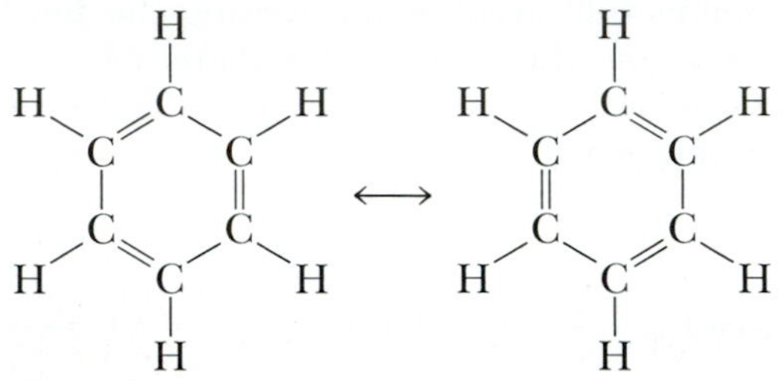

2 Benzene

These are called **Kekulé structures** for the German chemist Friedrich Kekulé, who first proposed (in 1865) that benzene had a cyclic structure with alternating single and double bonds. Compounds derived from benzene occur so widely in chemistry that the Kekulé structures are normally abbreviated to hexagons:

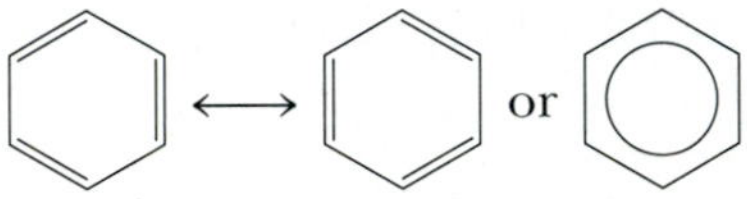

N Resonance theory and the enumeration of Kekulé structures. *J. Chem. Ed.* **1974**, *51*, 10.

The circle-in-a-hexagon representation of the resonance hybrid of the two Kekulé structures conveys the idea that all the C—C bonds in the benzene molecule are equivalent. For example, all of them are found to have the same length (134 pm). Moreover, all six C atoms are the same; the Kekulé structures, on the other hand, suggest that there ought to be two dichlorobenzenes in which the chlorine atoms are neighbors:

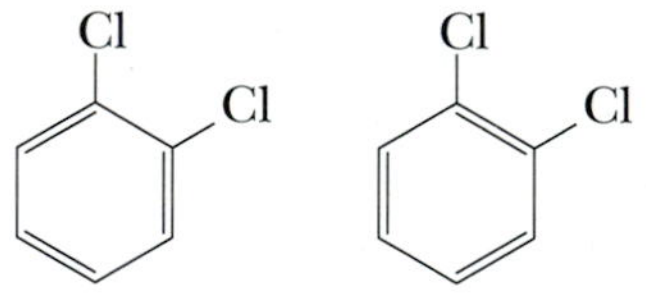

N Dewar resonance energy. *J. Chem. Ed.* **1971**, *48*, 509.

However, only one such compound is known, and it is written

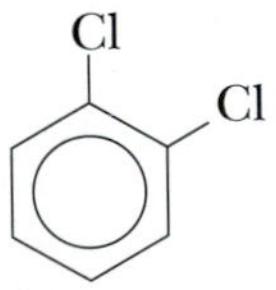

The resonance hybrid of the two Kekulé structures of benzene has a significantly lower energy than either of the individual structures. This energy lowering, which can be explained only in terms of quantum mechanics, plays an important role in the chemical properties of benzene. For example, benzene is more stable and less reactive than expected for a molecule with three carbon-carbon double bonds. The

N The lowering of energy is difficult to explain at this level, but it can be likened to the beating of two or more wavemotions. The beating results in a motion with frequencies that are the sums and differences of the individual motions, and one of the composite motions has a lower frequency than the components. In quantum mechanical terms, the lower frequency corresponds to the lower energy.

point to remember is that whenever resonance occurs, the energy of the resonance hybrid is lower than expected on the basis of any one of the contributing structures alone.

Resonance produces the greatest lowering of energy when the contributing structures have equal energies. All the structures described so far fulfill this condition because they differ only in the positions of their double bonds. However, the best description for other molecules is obtained by considering them to be a blend of all their possible Lewis structures, including those having different energies. For these molecules, the general rule is that the lowest-energy structures contribute most strongly to the overall structure.

A final point to note is that *resonance occurs between structures with the same arrangement of atoms but different arrangements of electrons.* For example, although we might be able to write two hypothetical structures for the dinitrogen oxide molecule, NNO and NON, there is no resonance between them because the atoms lie in different locations.

N For example, the resonance structure of benzene includes contributions from the Dewar structures, the three structures of the form

However, these three structures have a higher energy than the two Kekulé structures, and contribute less to the overall resonance hybrid (about 11% in all).

8.7 EXCEPTIONS TO THE LEWIS OCTET RULE

The octet rule helps us to predict the ability of an element to form bonds and to write Lewis structures in terms of shared electron pairs. However, as noted in Section 8.3, it is not rigorously obeyed, particularly by elements that belong to Period 3 and later (such as phosphorus, sulfur, chlorine, and their heavier congeners). These elements are able to accommodate more than an octet of electrons. A second type of deviation occurs with some elements of Period 2, particularly boron, which form reasonably stable molecules in which there are fewer than eight electrons in the valence shell. Finally, a third class of compounds that break the octet rule comprises molecules with an odd number of electrons.

E As remarked in a previous annotation (p. 289), it is highly likely that the principal reason for the difference between elements of the second period and those of later periods stems from the larger radii of the latter. There are enough molecular orbitals that can be formed from the *s* and *p* orbitals of the central atom and the orbitals of the ligands without needing to call on the *d* orbitals. Calculations on SF_6 show that *d* orbitals play only a minor role in the bonding in this molecule.

D Reaction of white phosphorus and chlorine. BZS1, 1.29.

Expanded octets. The octet rule is based on the idea that eight electrons fill a valence shell consisting of one *s* orbital and three *p* orbitals. However, if an atom has empty *d* orbitals available, then it may be able to use them to accommodate more than eight electrons and, hence, to "expand its octet" to 10, 12, or even more electrons. That capability may allow the central atom to form additional multiple bonds to the atoms attached to it, or to form additional single bonds to more atoms. The formation of additional bonds is common for the Period 3 elements of the *p* block, where the 3*d* orbitals are only slightly higher in energy than the 3*s* and 3*p* orbitals; it is also true of the elements in later periods. However, because 2*d* orbitals do not exist, Period 2 elements, such as N, O, and F, cannot expand their octets. Another factor—and possibly the main factor—in determining whether more atoms can bond to a central atom than are allowed by the octet rule is the size of the central atom. A P atom is big enough for five (and even six) Cl atoms to fit comfortably around it. An N atom, however, is too small, and NCl_5 is unknown (Fig. 8.10).

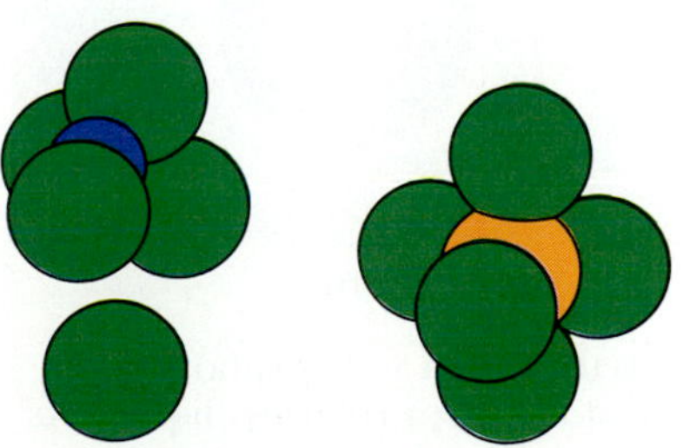

FIGURE 8.10 An N atom is so small, five Cl atoms cannot fit around it, but a P atom is big enough. Thus, NCl_5, is unknown, but PCl_5 exists.

Elements that can expand their octets show **variable covalence**, the ability to form different numbers of covalent bonds. Phosphorus is a good example. It reacts directly with a limited supply of chlorine to form the colorless liquid phosphorus trichloride:

$$P_4(g) + 6Cl_2(g) \longrightarrow 4PCl_3(l)$$

N The sublimation of PCl_5 can be used as an interesting example of a Lewis acid-base reaction, in which one species (PCl_6^-) is a complex (of the acid PCl_5 and the base Cl^-) and the other (PCl_4^+) is a Lewis acid: when sublimation occurs, the Lewis base (Cl^-) is transferred from one acid to the other.

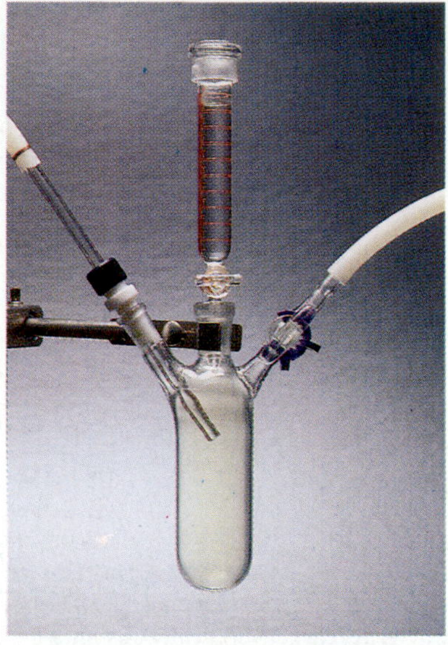

(a)

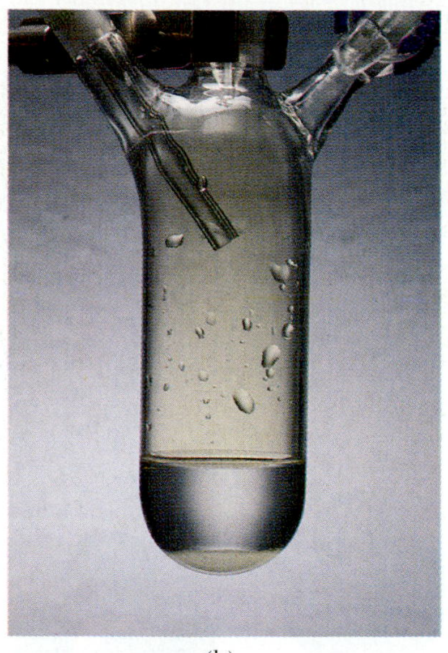

(b)

FIGURE 8.11 Phosphorus trichloride is a colorless liquid. As it drips into the flask (a), it reacts with yellow-green chlorine gas to produce the very pale yellow solid phosphorus pentachloride (b), which is ionic and consists of PCl_4^+ cations and PCl_6^- anions.

The Lewis structure of the PCl_3 molecule is

:Cl:P:Cl: (with :Cl: above P), denoted :Cl—P—Cl: (with :Cl: bonded above P)

However, when excess chlorine is present or when the trichloride reacts with more chlorine (Fig. 8.11), the pentachloride, a pale yellow crystalline solid, is produced:

$$P_4(s) + 10Cl_2(g) \longrightarrow 4PCl_5(s)$$

$$PCl_3(l) + Cl_2(g) \longrightarrow PCl_5(s)$$

Phosphorus pentachloride is an ionic solid that consists of PCl_4^+ cations and PCl_6^- anions and sublimes at 160°C to a gas of PCl_5 molecules. The Lewis structures of the polyatomic ions and the molecule are as follows:

[:Cl—P—Cl: with :Cl: above and below P]$^+$ [P bonded to six :Cl: atoms]$^-$ [P bonded to five :Cl: atoms]

Although PCl_4^+ is a polyatomic ion in which the P atom does not need to expand its octet, in PCl_6^- the P atom has expanded its valence shell to 12 electrons—the four extra electrons are in two of its 3*d* orbitals. In PCl_5 the P atom has expanded its valence shell to ten electrons by using one 3*d* orbital.

▼ EXAMPLE 8.8 Writing Lewis structures with expanded octets

A fluoride of composition SF_4 is formed when fluorine diluted with nitrogen is passed over a film of sulfur at −75°C in the absence of oxygen and moisture. Write the Lewis structure of sulfur tetrafluoride.

STRATEGY A fluorine atom forms only single bonds, so we anticipate that the Lewis structure consists of a shared pair of electrons between the central S atom and each of the four surrounding F atoms. However, if each F supplies one bonding electron and S already has six electrons in its valence shell, then there are ten electrons to accommodate. This octet expansion is possible because sulfur is in Period 3 and has empty 3*d* orbitals available. With that octet expansion in mind, we write the Lewis structure in the usual way, using the procedure given above.

SOLUTION Sulfur (:S:) supplies six valence electrons and each fluorine atom (:F·) supplies seven. Hence, there are $6 + (4 \times 7) = 34$ electrons, or 17 electron pairs, to accommodate. We write each terminal F atom with three lone pairs and a bonding pair shared with the central S atom:

:F—S—F: (with two :F: atoms bonded below S)

All 34 electrons are accounted for. The S atom has ten electrons in its valence shell, which require at least five orbitals, so one 3*d* orbital must be used in addition to the four 3*s* and 3*p* orbitals.

EXERCISE E8.8 Write the Lewis structure for xenon tetrafluoride, XeF_4, and give the number of electrons in the expanded octet.

[*Answer:* (Lewis structure of XeF_4: Xe with two lone pairs bonded to four F atoms, each with three lone pairs); 12]

A molecule or ion may be a resonance hybrid of octet and expanded-octet Lewis structures. An example is the sulfate ion, SO_4^{2-}. Two of the numerous Lewis structures that may be written for the ion are

$$\left[\begin{array}{c} :\ddot{O}: \\ | \\ :\ddot{O}-S-\ddot{O}: \\ | \\ :\ddot{O}: \end{array}\right]^{2-} \qquad \left[\begin{array}{c} :\ddot{O}: \\ | \\ \dot{\dot{O}}=S=\dot{\dot{O}} \\ | \\ :\ddot{O}: \end{array}\right]^{2-}$$

The sulfur octet is not expanded in the first, but it is expanded (to 12 electrons) in the second. Because the two structures differ by more than the location of double bonds, they can be expected to have different energies. The actual structure of the ion is a resonance hybrid of the two shown and several similar structures in which the double bonds are in different locations. It turns out (from calculations based on the Schrödinger equation) that the expanded-octet structures have lower energy than the others. Hence, the best description of the molecule is as a resonance hybrid of all these structures, but the expanded octet structures make the greatest contribution to the hybrid.

N In Section 9.4 we shall see that the analysis of the formal charges on the atoms will predict a lower energy for the expanded octet structure of SO_4^{2-} and other polyatomic oxoanions.

Incomplete octets. Some elements on the left of the *p* block, most notably boron, form compounds in which their atoms have incomplete octets. For example, the colorless gas boron trifluoride (BF_3) has the Lewis structure

$$\begin{array}{c} :\ddot{F}: \\ :\ddot{F}:B:\ddot{F}: \end{array} \quad \text{or} \quad \begin{array}{c} :\ddot{F}: \\ | \\ :\ddot{F}-B-\ddot{F}: \end{array}$$

The central boron atom has only six electrons.

Boron trifluoride is far from being a laboratory curiosity, since it is produced in large quantities by heating together boron oxide and calcium fluoride:

$$B_2O_3(s) + 3CaF_2(s) \xrightarrow{\Delta} 2BF_3(g) + 3CaO(s)$$

The boron oxide is obtained from the minerals borax and kernite, which are both forms of sodium borate; the calcium fluoride is obtained from the mineral fluorspar. So much BF_3 is manufactured because it is a useful Lewis acid: it can fill its incomplete octet by attaching to lone pairs of electrons on other molecules and ions, thereby bringing about a variety of reactions. For example, when the gas is passed over a metal fluoride, it forms the tetrafluoroborate anion, BF_4^-:

$$\begin{array}{c} :\ddot{F}: \\ | \\ :\ddot{F}-B \\ | \\ :\ddot{F}: \end{array} \;\; :\ddot{F}:^- \longrightarrow \left[\begin{array}{c} :\ddot{F}: \\ | \\ :\ddot{F}-B-\ddot{F}: \\ | \\ :\ddot{F}: \end{array}\right]^-$$

N It is possible to draw Lewis structures with one boron-fluorine double bond, but then the fluorine atom would carry a formal positive charge (see Section 9.4), which is energetically more unfavorable than leaving boron's octet incomplete.

In this reaction the fluoride ion is acting as a Lewis base and the product is a complex ion. Boron trifluoride also acts as a Lewis acid when it reacts with ammonia:

$$BF_3 + :NH_3 \longrightarrow F_3B{-}NH_3$$

In this reaction the two gases combine to form a white solid. Through analogous reactions, particularly with organic compounds, boron trifluoride acts as an important industrial and laboratory catalyst.

N Both borax and kernite have the formula $Na_2B_4O_7{\cdot}xH_2O$, with $x = 10$ for borax and $x = 4$ for kernite.

N The structure of solid aluminum chloride. *J. Chem. Ed.* **1969**, *46*, 495.

Boron trichloride is another molecule with an incomplete octet. It is made industrially (for use as a catalyst, like BF_3) by heating a mixture of boron oxide and carbon in chlorine:

$$B_2O_3(s) + 3C(s) + 3Cl_2(g) \xrightarrow{\Delta} 2BCl_3(g) + 3CO(g)$$

Boron trichloride is a reactive gas. The trichloride of boron's congener aluminum, however, is not a gas but a volatile white solid that sublimes at 180°C to a gas consisting of Al_2Cl_6 molecules. These molecules survive in the gas up to about 200°C but then fall apart into $AlCl_3$ molecules. The Al_2Cl_6 molecule exists because a Cl atom of one $AlCl_3$ molecule acts as a Lewis base by forming a bond to the Al atom, a Lewis acid, of a neighboring $AlCl_3$ molecule:

N Catalysts are discussed in more detail in Chapter 12.

E Boron trifluoride is used in the manufacture of epoxy resins.

E Aluminum trichloride is used in the Friedel-Crafts alkylation and acylation reactions (see Section 23.9). In industry, it is used as a catalyst in organic reactions, such as hydrocarbon reforming reactions, and in metallurgical and metal-finishing processes.

$$Cl_2Al(\mu\text{-}Cl)_2AlCl_2 \quad \text{or} \quad Cl_2Al(\mu\text{-}Cl)_2AlCl_2$$

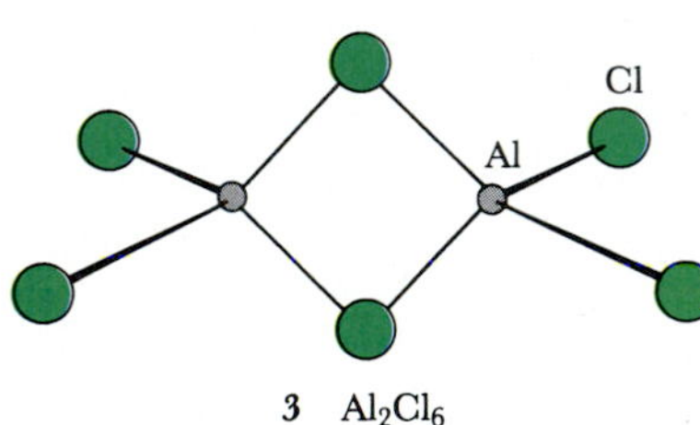

3 Al_2Cl_6

The two Cl atoms that act as bridges between the $AlCl_3$ units lie above and below a plane defined by the two Al and the four other Cl atoms (**3**). An interesting feature of aluminum trichloride is that at its melting point (which is 192°C when the compound is under slightly greater than atmospheric pressure, to prevent sublimation and allow the liquid to form), it nearly doubles in volume. This expansion occurs because each Al^{3+} ion in the solid is surrounded by six Cl^- ions, but on melting, this compact arrangement breaks up as the liquid of Al_2Cl_6 molecules is formed.

Odd-electron molecules. Some molecules have an odd number of valence electrons, so at least one atom cannot have an octet. Nitric oxide is such a molecule; it is prepared industrially as an intermediate in the production of nitric acid by the oxidation of ammonia:

$$4NH_3(g) + 5O_2(g) \xrightarrow{1000°C,\ Pt} 4NO(g) + 6H_2O(g)$$

It is also formed by the direct reaction of nitrogen and oxygen in the hot gases of automobile exhausts and jet engines:

$$N_2(g) + O_2(g) \xrightarrow{\Delta} 2NO(g)$$

Because the N atom supplies 5 valence electrons and the O atom supplies 6, the NO molecule has 11 valence electrons. One possible Lewis structure is

$$\cdot\ddot{N}::\ddot{O}\cdot \quad \text{or} \quad \cdot\dot{N}=\ddot{O}\cdot$$

with an unpaired electron on the nitrogen atom. These odd-electron molecules are called radicals:

> A **radical** is an atom, molecule, or ion with at least one unpaired electron.

Other examples of radicals include the fragments obtained when molecules break up during a reaction, such as the methyl radicals formed by breaking the carbon-carbon bond in ethane (C_2H_6):

$$\begin{array}{c} \text{H}\ \ \text{H} \\ |\ \ \ | \\ \text{H—C—C—H} \\ |\ \ \ | \\ \text{H}\ \ \text{H} \end{array} \longrightarrow \begin{array}{c} \text{H} \\ | \\ \text{H—C}\cdot \\ | \\ \text{H} \end{array} + \begin{array}{c} \text{H} \\ | \\ \cdot\text{C—H} \\ | \\ \text{H} \end{array}$$

A methyl radical is denoted $\cdot CH_3$, the dot signifying the unpaired electron. Another example is the hydroxyl radical, $\cdot\ddot{O}:H$ or, more simply, $\cdot OH$. This radical is formed briefly when a mixture of hydrogen and oxygen is ignited by a spark. It is also present in the upper atmosphere (as a result of the action of the sun's radiation on water molecules) and in the troposphere in smog-laden areas.

Like most radicals, $\cdot CH_3$ and $\cdot OH$ are highly reactive and survive only for very short times under normal conditions. The reactivity of the hydroxyl radical is partly responsible for the explosive violence with which hydrogen and oxygen gases combine with each other. The methyl radical is involved in the explosion that occurs when ethane is ignited in air. Under abnormal conditions, however, a radical can survive indefinitely. Hydroxyl radicals have been detected in interstellar gas clouds, where they can survive for millions of years. In that environment they collide very rarely with other molecules, so they do not have an opportunity to react.

Radicals also play a very important role in the chemistry of the upper atmosphere, where they contribute to the formation and decomposition of ozone (see Box 5.1). Another function of radicals is in the series of reactions that are responsible for the spoilage of foods and the degradation of plastics in sunlight. Both processes can be delayed by including an **antioxidant**, which is a substance composed of molecules that react rapidly with radicals and remove them before they have a chance to react with the molecules of the substance the antioxidant is present to protect.

N Free radicals generated by radiolysis of aqueous solutions. *J. Chem. Ed.* **1981**, *58*, 101.

N The term *free radical* is now obsolescent (but still widely used).

E Much of the chemistry of smog formation results from the formation of radicals by photodissociation. The initiation reaction is $NO_2 \rightarrow NO + O$ under the influence of radiation of wavelength shorter than 350 nm.

N The structure and evolution of interstellar grains. *Sci. Amer.* **1984**, June, 124.

E Antioxidants are radical scavengers.

D Equilibrium between nitrogen dioxide and dinitrogen tetroxide. BZS2, 6.17.

▼ EXAMPLE 8.9 Writing the Lewis structure of an odd-electron molecule

Write the Lewis structure of nitrogen dioxide, NO_2. A gas of these molecules, when cooled, forms dinitrogen tetroxide, N_2O_4. Suggest a reason.

STRATEGY The first step is to establish a Lewis structure for the NO_2 molecule. We proceed in the usual way, the only difference being that the molecule has an odd number of electrons, so one of the atoms must have an incomplete octet. We must account for the formation of the dimer N_2O_4 by the formation of an electron-pair covalent bond.

SOLUTION The number of valence electrons present in the molecule is

$$\begin{matrix} \text{O} & \text{N} & \text{O} & \\ 6 & +5 & +6 & = 17; \quad 8 \text{ pairs} + 1 \end{matrix}$$

The less electronegative nitrogen atom is the central atom, so the arrangement of atoms is

O N O

Writing single bonds gives

O : N : O

and 6 pairs + 1 electron remain to be accommodated. One Lewis structure that accounts for the six pairs is

O : : N : O :

but the nitrogen atom has an incomplete octet. It can accommodate the remaining one electron:

O : : N : O : ⟷ : O : N : : O or

O=N—O : ⟷ : O—N=O

The formation of N_2O_4 can be ascribed to the sharing of the two unpaired electrons to form a single covalent bond between the two NO_2 radicals:

O=N· + ·N=O (with O's) ⟶ N : N or N—N

(This structure is only one of four similar resonance structures.)

EXERCISE E8.9 Write a Lewis structure for the peroxyl radical, HOO, which plays an important role in atmospheric chemistry and has been implicated in the degeneration of neurons. (Think of the radical as the molecule remaining after the loss of an H atom from H_2O_2.)

[*Answer*: H on : O—O ·]

▲

N Another interesting opportunity for a discussion is a comparison of N_2O_4 and the oxalate ion, $C_2O_4^{2-}$: the two species are isolectronic, but the carbon-carbon bond in the oxalate ion is appreciably stronger than the nitrogen-nitrogen bond in dinitrogen tetroxide, and the oxalate ion survives, even in solution (where hydration should favor its dissociation).

THE SHAPES OF MOLECULES

A molecule's shape is often an essential part of the explanation of its properties. The shape of an H_2O molecule, for instance, is a factor in water's ability to act as a good solvent for ionic compounds. In addition, the shapes of many organic molecules affect their odors, their tastes, their action as drugs, as well as many of the reactions going on inside our bodies as we live, think, and work. Molecular shapes can be accounted for in terms of the ideas we have developed about atomic and molecular structure.

In this section we deal mainly with the shapes of simple molecules—those containing no more than about a dozen atoms. Such molecules can be classified by their shapes (see the rightmost column in Table 8.2; ignore the rest of the information in the table for now). We identify the shape of a molecule by noting the arrangement of the atoms, not the locations of any lone pairs that may be present. We classify the ammonia molecule, NH_3, with its one lone pair, for example, as trigonal pyramidal, and the ammonium ion, NH_4^+, as tetrahedral. Later we shall see that the four electron pairs in a water molecule are arranged tetrahedrally around the oxygen atom. However, because only two of the pairs form bonds between atoms, we classify the H_2O molecule as angular, not tetrahedral.

When we want to report shapes precisely, we report the bond angles. The **bond angle** in an A—B—C molecule or part of a molecule is the angle between the A—B and B—C bonds. The bond angle in H_2O, for instance, is the angle (104.5°) between the two O—H bonds.

N More precisely, the A—B—C bond angle is the angle between the A—B and B—C internuclear directions.

N To what extent are students to memorize Table 8.2?

T OHT: Both pages of Table 8.2.

N The VSEPR theory is reliable only for main group elements; it can give very misleading predictions for compounds containing *d*- and *f*-block elements.

N A visual aid to demonstrate VSEPR theory. *J. Chem. Ed.* **1980**, *57*, 668.

8.8 ELECTRON-PAIR REPULSIONS

To account for molecular shapes in terms of their Lewis structures, we need the following simple addition to Lewis's ideas: *electron pairs repel one another.* As a result of this repulsion, the valence electron pairs on an atom (which include the lone pairs as well as the bonding pairs) take up positions as far from one another as possible, for then the repulsion forces are smallest (Fig. 8.12). The positions of the atoms, and hence the shape of the molecule, are then dictated by the positions of the bonding electron pairs. This idea was first explored by the British chemists Nevil Sidgwick and Herbert Powell and was developed to a considerable extent by the Canadian chemist Ronald Gillespie. It is called the **valence-shell electron-pair repulsion theory**, or VSEPR theory.

The VSEPR theory focuses on the positions taken up by a given number of electron pairs in the valence shell of the central atom of a molecule (the S atom in SF_6 and the C atom in CO_2). These positions are decided by assuming that all the pairs (the lone pairs as well as the bonding pairs) move as far apart as possible on the surface of an imaginary sphere surrounding the central atom. The "most-distant" arrangements of two through seven valence-shell electron pairs (arrangements found by using trigonometry) are shown in the second column of Table 8.2. These arrangements minimize the electron repulsions and thus result in molecular structures of lowest energy.

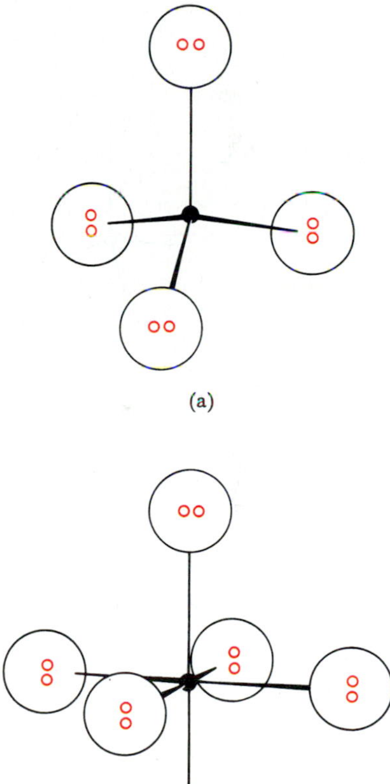

FIGURE 8.12 Electron pairs (represented by circles) repel each other, and according to VSEPR theory, take up positions as far apart as possible. (a) Atom with four electron pairs; (b) atom with six electron pairs.

Lewis structures with single bonds. The first step in using VSEPR theory to predict the shape of a molecule is to identify its bonding arrangement and its lone pairs by writing the Lewis structure of the molecule. For the methane molecule (CH_4) and the sulfite ion (SO_3^{2-}), these are

```
                         ..
     H                  :O:      ]2-
     |             ..    |    ..
   H—C—H          :O—S—O:
     |             ..   ..   ..
     H
```

Methane, CH_4, is an example of an AX_4 molecule and SO_3^{2-} an example of an AX_3E polyatomic ion, where A is the central atom, X an attached atom, and E a lone pair of electrons. In both these structures, the central atom is surrounded by four electron pairs. (At this stage we do not distinguish between bonding and lone pairs.) Table 8.2 indicates that the four pairs adopt a tetrahedral arrangement.

TABLE 8.2 The shapes of simple molecules

Number of electron pairs	Arrangement of electron pairs	Number of bonding electron pairs	Number of nonbonding electron pairs	VSEPR formula	Shape
2	Linear 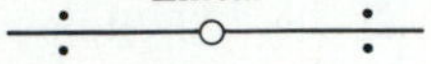	2	0	AX_2	Linear 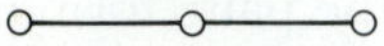
3	Trigonal planar 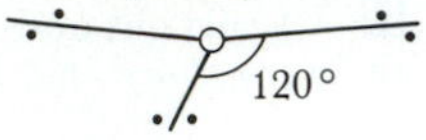 (120°)	3	0	AX_3	Trigonal planar 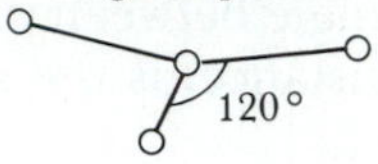 (120°)
		2	1	AX_2E	Angular 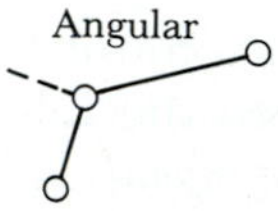
4	Tetrahedral 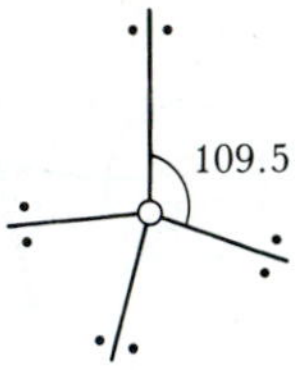 (109.5°)	4	0	AX_4	Tetrahedral 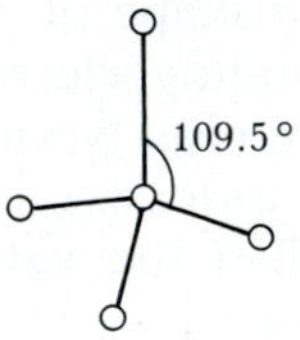 (109.5°)
		3	1	AX_3E	Trigonal pyramidal 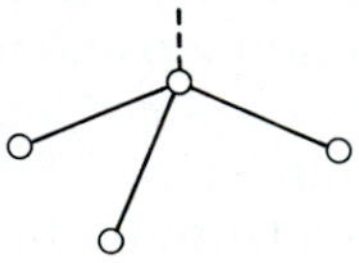
		2	2	AX_2E_2	Angular 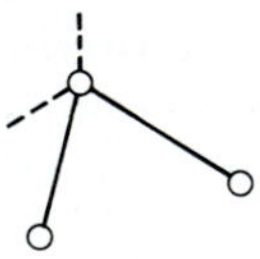
5	Trigonal bipyramidal 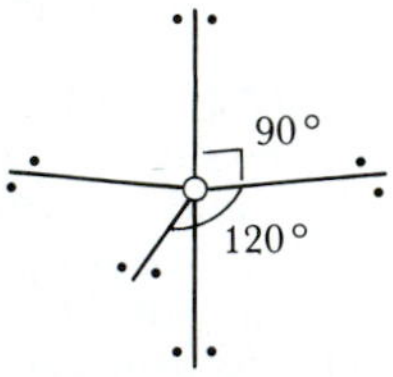 (90°, 120°)	5	0	AX_5	Trigonal bipyramidal 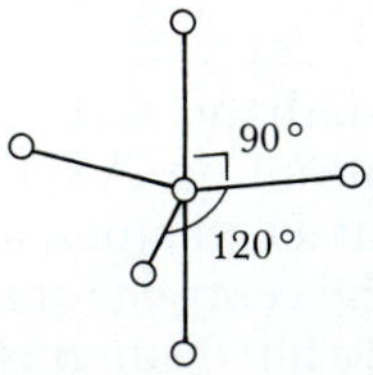 (90°, 120°)

Number of electron pairs	Arrangement of electron pairs	Number of bonding electron pairs	Number of nonbonding electron pairs	VSEPR formula	Shape
		4	1	AX_4E	Seesaw
		3	2	AX_3E_2	T-shaped
6	Octahedral (90°, 90°)	6	0	AX_6	Octahedral (90°, 90°)
		5	1	AX_5E	Square pyramidal
		4	2	AX_4E_2	Square planar
7	Pentagonal bipyramidal (90°, 72°)	7	0	AX_7	Pentagonal bipyramidal (90°, 72°)

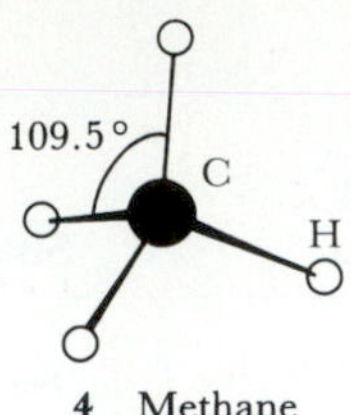

4 Methane

The next stage is to decide which electrons are bonding pairs and which are lone pairs. Then, by noting the arrangement of the bonded atoms but ignoring the lone pairs, we identify the molecule's shape by referring to the right-hand half of Table 8.2. Because in CH_4 each pair bonds an H atom, we predict that the molecule is tetrahedral (**4**), with an HCH angle of 109.5°C, the tetrahedral angle. Because in SO_3^{2-} three electron pairs bond the three O atoms, we predict that the ion is trigonal pyramidal with OSO angles of 109.5°. Experimental data show that the methane molecule is tetrahedral with HCH angles equal to 109.5°. However, although the sulfite ion is trigonal pyramidal, the experimental bond angle is only 106°, not the 109.5° predicted. Clearly, although VSEPR theory is broadly correct, it must be refined if it is to be used to make accurate predictions.

N The formulas AX_4, AX_3E, and so on in Table 8.2 are examples of *VSEPR formulas*. In general, a VSEPR formula has the form AX_mE_n and shows the number of atoms (m) and lone pairs (n) on the central atom (A).

EXAMPLE 8.10 Predicting the shape of a simple molecule

Predict the shape of the nitrogen trifluoride molecule, NF_3.

STRATEGY We suspect that NF_3, an AX_3E molecule, has the same shape as NH_3, another AX_3E molecule (a trigonal pyramid), because the only difference between them is the replacement of hydrogen atoms by fluorine atoms. First, we draw the Lewis structure, then determine from Table 8.2 how the electron pairs are arranged around the central (N) atom. Finally, we decide which (if any) of the pairs are lone pairs and identify the molecule's shape by referring to Table 8.2.

SOLUTION The Lewis structure of nitrogen trifluoride is

```
       ..
      :F:
  ..   |   ..
 :F—N—F:
  ..  ..  ..
```

The four electron pairs around the N atom adopt a tetrahedral arrangement. Because one of the pairs is a lone pair, we expect the molecule to be trigonal pyramidal (**5**), like ammonia. That is in fact the case.

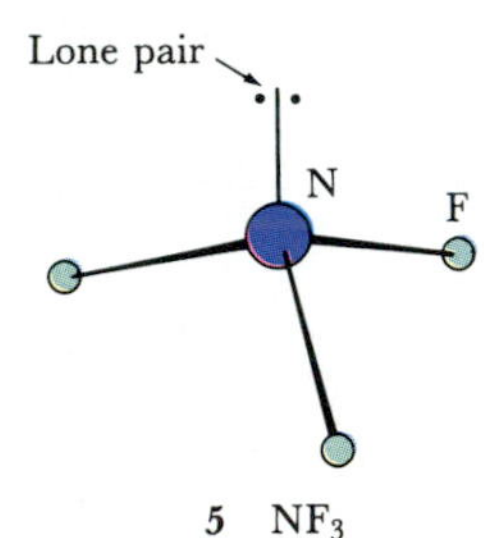

5 NF_3

EXERCISE E8.10 Predict the shape of the PCl_5 molecule.

[*Answer*: trigonal bipyramidal]

U Exercise 8.81.

The repelling effect of lone pairs. The first refinement of VSEPR theory takes into account the greater repelling effect of a lone pair than of a bonding pair. The difference is sometimes explained by arguing that a lone pair is spread over a greater volume than a bonding pair is, because the latter is confined to the space between two atoms (Fig. 8.13). Then, because a lone pair occupies more space, the argument continues, it repels other electron pairs from a greater region of space. Whatever the true reason (no one really knows), VSEPR theory asserts the following order for the strengths of electron pair repulsions:

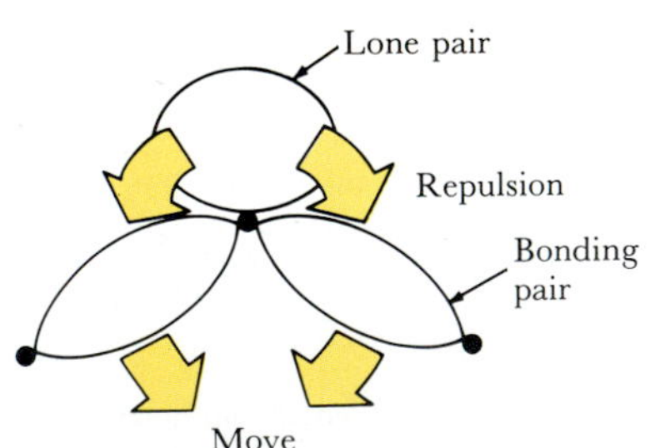

FIGURE 8.13 Lone pairs appear to exclude other electron pairs from a greater region of space than bonding pairs do, so the latter seem to move away from them.

- Lone pair–lone pair repulsions are stronger than lone pair–bonding pair repulsions.
- Lone pair–bonding pair repulsions are stronger than bonding pair–bonding pair repulsions.
- Bonding pair–bonding pair repulsions are least strong.

It is therefore energetically most favorable for lone pairs to be as far from one another as possible. It is also energetically most favorable for bonding pairs to be far from lone pairs even though that behavior might bring them closer to other bonding pairs.

This additional feature accounts for the bond angle of the AX_3E sulfite ion. The four electron pairs adopt a tetrahedral arrangement. However, one of them is a lone pair (**6**) and exerts a strong repulsion on the three bonding pairs, forcing them to move together slightly. Hence, the OSO angle is reduced from the 109.5° of the regular tetrahedron toward the 106° observed experimentally. Note that although VSEPR theory can predict the *direction* of the distortion, it cannot predict its extent: we can predict that the angle will be less than 109.5°, but we cannot say it will be 3° less.

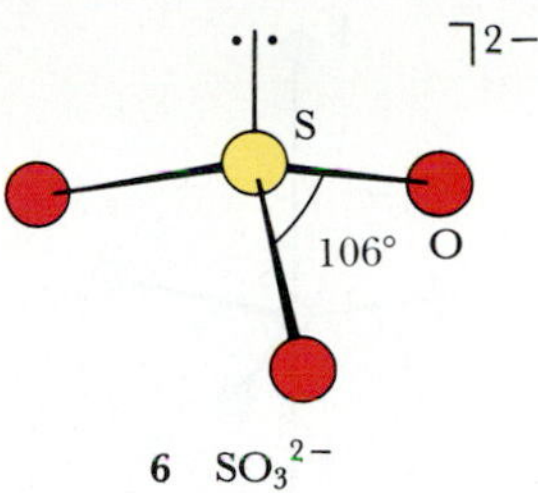

6 SO_3^{2-}

▼ EXAMPLE 8.11 Using VSEPR theory to predict shapes: I

Account for the shape of the water molecule (and of AX_2E_2 molecules in general).

STRATEGY According to VSEPR theory, we must count the number of lone pairs on the central atom and identify the arrangement they adopt by using Table 8.2. After that, we identify the bonding pairs and, hence, identify the shape of the molecule. Finally, we allow the lone pairs to move apart, at the expense of the distance between the bonding pairs, which move together slightly.

SOLUTION The central O atom has four pairs of electrons in a tetrahedral arrangement. In H_2O the central oxygen atom has two lone pairs and two bonding pairs, a configuration giving an angular molecule (**7**). Because the lone pairs move apart slightly and the bonding pairs move together to avoid them, we predict an HOH bond angle of less than 109.5°. This prediction is in agreement with the experimental value of 104.5°.

7 H_2O

EXERCISE E8.11 Predict the shape of the NH_3 molecule (and of AX_3E molecules in general).

▲ [*Answer*: trigonal pyramidal (**8**); HNH angle less than 109.5°]

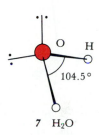

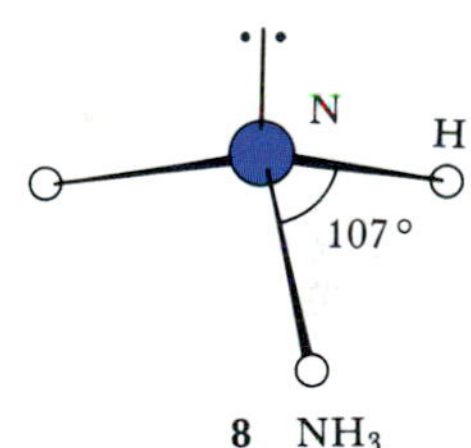

8 NH_3

In trigonal bipyramidal (and pentagonal bipyramidal) molecules we must distinguish between axial and equatorial electron pairs. An **axial pair** lies on the long axis of the molecule whereas an **equatorial pair** lies in one of the positions around the "equator" of the molecule (**9**). A pair in an axial position repels three other pairs strongly, whereas in the equatorial position it repels only two other pairs strongly. Therefore, if one of the pairs is a lone pair, it is energetically more favorable for it to be in an equatorial position.

N Sulfur tetrafluoride (b.p. −38°C) reacts violently with water and attacks glass.

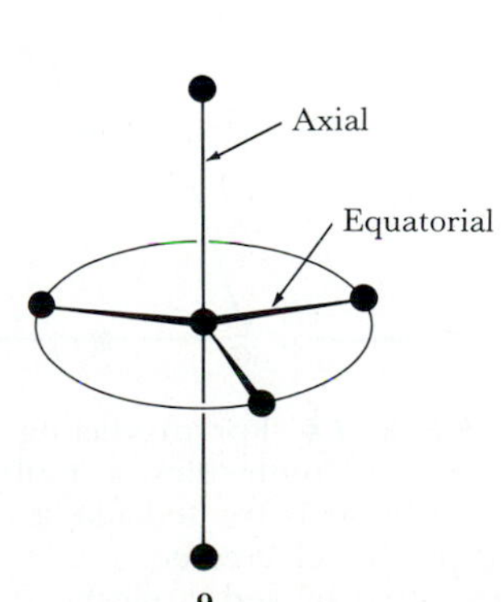

9

▼ EXAMPLE 8.12 Using VSEPR theory to predict shapes: II

Predict the shape of a sulfur tetrafluoride molecule.

STRATEGY First, we decide how many electron pairs are present by writing a Lewis structure of the molecule (as we did in Example 8.11) and identify them as lone pairs and bonding pairs. We then identify the arrangement of electron pairs by referring to Table 8.2. After that, we identify the bonding pairs and allow the molecule to distort so that lone pairs are as far from one another and from bonding pairs as possible. Finally, we identify the molecular shape by using Table 8.2.

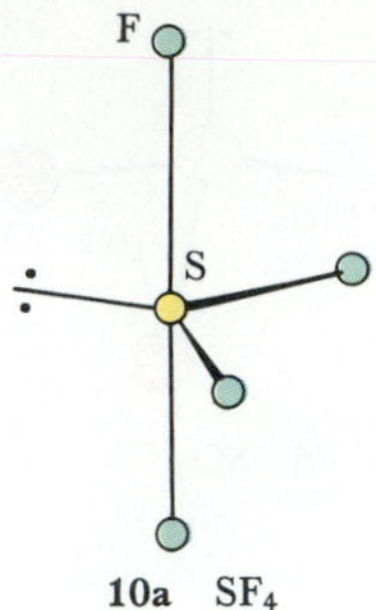

10a SF_4

SOLUTION The Lewis structure of SF_4 was found in Example 8.8 to be

$$:\ddot{F}—\ddot{S}—\ddot{F}: \quad \text{with two further } :\ddot{F}: \text{ atoms bonded to S below}$$

The S atom has five electron pairs and the molecule is AX_4E. The pairs adopt a trigonal bipyramidal arrangement. The lone pair can be either axial or equatorial, but the latter arrangement (**10a**) is likely to have lower energy. The energy is lowered still further if the four bonds bend away from the lone pair (**10b**). This seesaw shape is the one found experimentally.

EXERCISE E8.12 Predict the shape of a xenon difluoride molecule.

[*Answer*: linear]

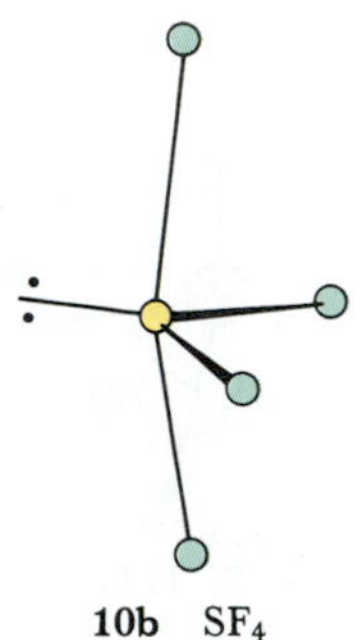
10b SF_4

As this example shows, it is important to be alert to the presence of lone pairs in molecules and not to conclude that every AB_3 molecule is trigonal planar and that every AB_4 molecule is tetrahedral. In each case one or more lone pairs may be present and affect the shape of the molecule. However, all AX_3 molecules (with no lone pairs) can be expected to have the same shape, as can all AX_3E molecules (with one lone pair) and all AX_3E_2 molecules (with two lone pairs).

N If it is desired to account for the planarity of $CH_2{=}CH_2$ in VSEPR terms (see below), one approach is to regard each CH_2 group as an AX_2E_2 group, which results in a tetrahedral disposition of electron pairs; however, two of the pairs are shared with the neighboring (tetrahedral) unit, so the two tetrahedral arrangements are locked together in such a way that the carbon-hydrogen bonding pairs lie in a plane.

U Exercises 8.80 and 8.82.

8.9 MOLECULES WITH MULTIPLE BONDS

The VSEPR theory also includes rules about how to deal with multiple bonds. In an $X{=}AX_2$ molecule in which one X atom is joined to the central A atom by a double bond, the two electron pairs of the double bond hold the X atom in nearly the same place as it would be in an $X{-}AX_2$ molecule (Fig. 8.14). Similarly, in an $X{\equiv}AX$ molecule, the three electron pairs of the triple bond hold the X atom in the same place a single electron pair would hold it. These remarks suggest that when applying VSEPR theory to predict molecular shapes, we should treat a multiple bond as though it were a *single* electron pair.

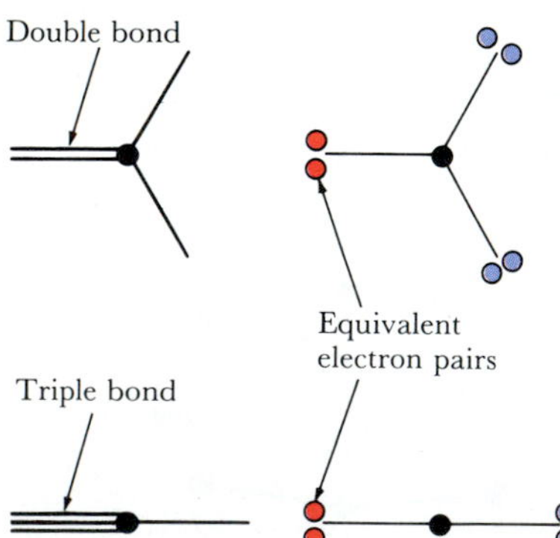

FIGURE 8.14 For predicting the shapes of molecules, a double or triple bond is treated like a single pair of electrons (represented by red circles).

Suppose we wanted to predict the shape of an ethylene molecule, $CH_2{=}CH_2$. The first step, as always, is to write its Lewis structure:

$$\begin{matrix} H & & & & H \\ & \diagdown & & \diagup & \\ & & C{=}C & & \\ & \diagup & & \diagdown & \\ H & & & & H \end{matrix}$$

Each carbon is treated as having *three* electron pairs because the two pairs of the double bond are treated as one. The arrangement around each carbon atom is therefore trigonal planar. We predict that the molecule will have the shape shown in (**11**) with the HCH and HCC angles both 120°, as is found experimentally. A feature not brought out very clearly by VSEPR theory, however, is that the two CH_2 groups lie in the same plane. Double bonds between carbon atoms always result in a planar arrangement of atoms, as we shall see in Chapter 9.

▼ EXAMPLE 8.13 Predicting the shape of a molecule with multiple bonds

Suggest a shape for the acetylene molecule, HC≡CH.

STRATEGY The shape is given by VSEPR theory, with the triple bond treated as a single electron pair. Write the Lewis structure, count the *effective* number of electron pairs around each C atom, and predict the arrangement of the atoms around each one.

SOLUTION The Lewis structure of the molecule is H—C≡C—H. Because the triple bond behaves as one electron pair, each C atom should be treated as having two electron pairs: one shared with the other C atom and the other bonding an H atom. Their arrangement on each atom is linear, with one pair diametrically opposite the other. Therefore the molecule is also linear (**12**).

EXERCISE E8.13 Predict the shape of a carbon suboxide molecule, C_3O_2, in which the atoms lie in the order OCCCO.

[*Answer*: linear] ▲

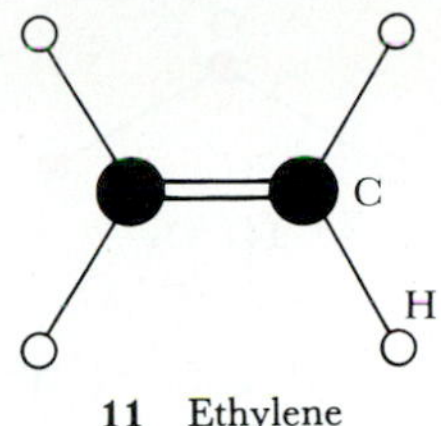

11 Ethylene

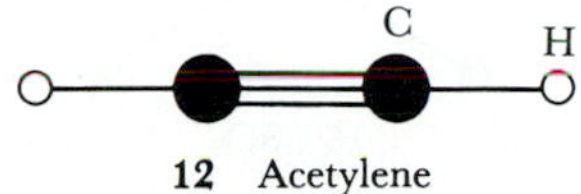

12 Acetylene

Treating multiple bonds as single pairs does away with any worries about which of several resonance structures to choose. This point can be illustrated by the sulfate ion and the following two contributions to its resonance hybrid:

$$\left[\begin{array}{ccc} & :\ddot{O}: & \\ :\ddot{O}- & S & -\ddot{O}: \\ & :\ddot{O}: & \end{array}\right]^{2-} \qquad \left[\begin{array}{ccc} & :\ddot{O}: & \\ \ddot{O}= & S & =\ddot{O} \\ & :\ddot{O}: & \end{array}\right]^{2-}$$

E Acetylene is the fuel in oxyacetylene torches.

U Exercises 8.67–8.70.

In the structure on the left, the S atom has four bonding pairs, and we predict an AX_4 tetrahedral species. In the other structure, the S atom is to be treated as though it had four electron pairs, so again we predict that an AX_4 ion is tetrahedral (**13**).

The same shape is obtained whatever the resonance structure of a polyatomic molecule or ion is, so all we need to know to predict its shape is the number of lone pairs and atoms linked to the central atom. We do not need to know whether the atoms are linked by single or multiple bonds. This simplification is the basis of a shortcut for predicting shapes.

First, count the total number of lone pairs and *atoms* (not bonds) attached to the central atom. Next, determine the arrangement of lone pairs and atoms by using Table 8.2 and then obtain the molecular shape by deciding on the locations of the atoms. Finally, take minor distortions into account by allowing for the greater repelling power of lone pairs.

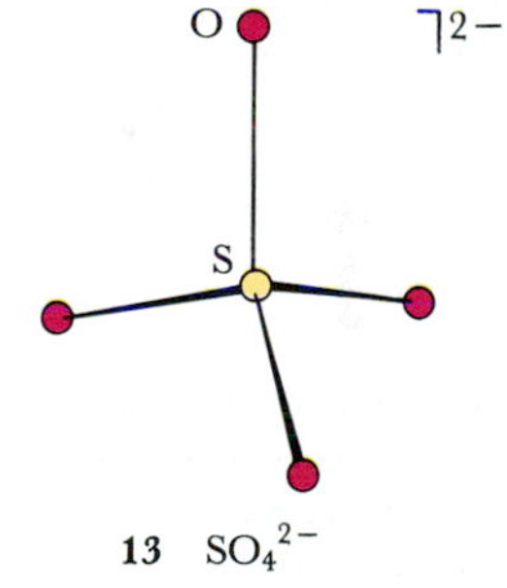

13 SO_4^{2-}

E Ozone is generated in the troposphere by the reaction $O_2 + O \rightarrow O_3$; the O atom is produced by the photodissociation of NO_2.

▼ EXAMPLE 8.14 Predicting the shape of an AX_mE_n molecule

Predict the shape of an ozone molecule, O_3.

STRATEGY After writing the Lewis structure, we follow the steps outlined in the previous paragraph.

SOLUTION The Lewis structure $\ddot{O}=\ddot{O}-\ddot{O}:$ shows that the effective number of lone pairs and attached atoms on the central oxygen atom is three: O_3 is an AX_2E molecule (in this special case, the A and X atoms are all

U Exercise 8.72.

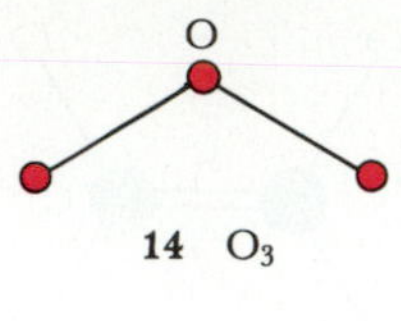

14 O_3

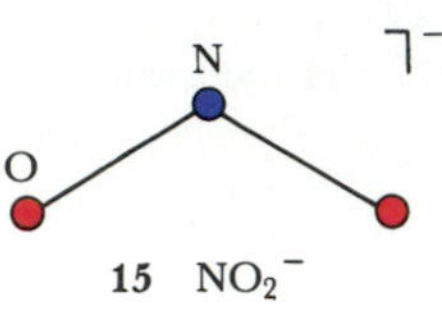

15 NO_2^-

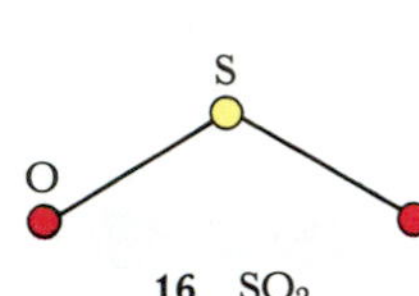

16 SO_2

O atoms). Hence, the arrangement of the three effective pairs is expected to be trigonal planar. Because two atoms are joined to the central atom, the molecule is angular (**14**). The two outer O atoms are bent away from the lone pair, giving an expected bond angle slightly smaller than 120°. Experimentally, the O_3 molecule is angular, and the OOO bond angle is 117°.

EXERCISE E8.14 Predict the shape of a sulfur trioxide molecule.

[*Answer*: trigonal planar]

A helpful point to remember when thinking about molecular shapes is to note whether the molecule of interest is **isoelectronic** with another molecule with a known shape, that is, whether the two molecules have the same number of valence electrons, for their shapes will be the same. For instance, O_3 is isoelectronic with NO_2^- and SO_2. If we know that O_3 is angular, then we can predict that NO_2^- and SO_2 are angular too (**15** and **16**).

SUMMARY

8.1 An **ionic bond**, the attraction between oppositely charged ions, is formed between elements with widely different electronegativities. Whether or not an ionic bond can form depends on whether the energy needed to form the ions is recovered from the strength of their interaction in the ionic solid. The latter is measured by the **lattice enthalpy**, which is the enthalpy change for the formation of a gas of ions from the solid and is large for small, highly charged ions. The lattice enthalpy may be determined by using a **Born-Haber cycle** and thermochemical data.

8.2 Ionic bonds are formed between elements toward the left of the periodic table and elements on the right of the table. That is, ionic bonds typically form between the metallic and nonmetallic elements. Electron gain or loss generally proceeds until an ion has the same number of electrons as a nearby noble gas. Electron loss and gain are conveniently depicted in terms of a **Lewis formula** constructed from the **Lewis diagrams** of the contributing atoms, in which valence electrons are shown as dots. In the *p* block, cations are formed by losing electrons until they achieve a **pseudonoble-gas configuration**, a closed core and a complete *d* subshell.

8.3 and 8.4 A single **covalent bond** consists of a shared pair of electrons. Covalent bonds are formed between elements that do not differ greatly in electronegativity. **Double bonds** and **triple bonds** consist of two and three shared pairs, respectively. A guide to the number of pairs that an atom can share, especially for Period 2 elements, is the **octet rule**, that an atom tends to complete its octet (more precisely, the nearest noble-gas electron configuration). The bonding arrangement in a molecule is shown by its **Lewis structure**, with lines representing shared electron pairs.

8.5 A **Lewis acid** is an electron pair acceptor, and a **Lewis base** is an electron pair donor. The combination of a Lewis acid and a Lewis base is called a **complex**. The Lewis definitions of acids and bases are more general than the Brønsted definitions because they do not require the involvement of a proton.

8.6 When several Lewis structures can be written for one molecule, the actual structure is a **resonance hybrid**, a blend of all Lewis structures with the atoms in the same positions, with the lowest-energy structures dominant. In benzene, the two most important resonance structures are the **Kekulé structures**. Resonance between single and double bonds equalizes bond lengths and averages bond strengths to intermediate values. The energy of a resonance hybrid is lower than the energy of any of the individual contributing structures.

8.7 Elements in Period 3 and later periods can form compounds with expanded octets by using their vacant *d* orbitals to accommodate more than eight electrons. This possibility gives rise to **variable covalence**, the ability of elements to form different numbers of covalent bonds. Octet formation is not universal: some elements (notably boron) can form compounds in which they have an incomplete octet and some species are **radicals**, with one or more unpaired electrons.

8.8 and 8.9 The shape of a molecule is classified on the basis of the relative locations of its atoms. **VSEPR theory** explains molecular shape in terms of the repulsions between electron pairs in the valence shell of the central atom. Repulsion strengths are in the order (LP, LP) > (LP, BP) > (BP, BP) where LP = lone pair and BP = bonding pair. A multiple bond is treated as the equivalent of a single electron pair.

CLASSIFIED EXERCISES

Formation of Ionic Bonds

8.1 Decide which of the following compounds are likely to be ionic, and explain why or why not: (a) magnesium oxide; (b) nitrogen triiodide; (c) iron(II) oxide.

8.2 Decide which of the following compounds are likely to be ionic, and explain why or why not: (a) sulfur tetrachloride; (b) cobalt(III) oxide; (c) oxygen difluoride.

8.3 Write the most likely oxidation number for each element: (a) Li; (b) S; (c) Ca; (d) Al.

8.4 Write the most likely oxidation number for each element: (a) F; (b) Ba; (c) Se; (d) O.

8.5 Explain why the lattice enthalpy of magnesium oxide (3850 kJ/mol) is greater than that of barium oxide (3114 kJ/mol).

8.6 Explain why the lattice enthalpy of magnesium oxide (3850 kJ/mol) is greater than that of magnesium sulfide (3405 kJ/mol).

8.7 Calculate the lattice enthalpy of silver fluoride from the data in Appendix 2A and the following information: enthalpy of formation of Ag(*g*), +284 kJ/mol; enthalpy of ionization of Ag(*g*), +731 kJ/mol; enthalpy of formation of F(*g*), +79 kJ/mol; enthalpy of formation of AgF(*s*), −205 kJ/mol.

8.8 Calculate the lattice enthalpy of calcium sulfide from the data in the chapter and the following information: enthalpy of formation of Ca(*g*), +178 kJ/mol; enthalpy of formation of S(*g*), +279 kJ/mol; enthalpy of formation of CaS(*s*), −482 kJ/mol.

8.9 By assuming that the lattice enthalpy of $NaCl_2$ would be the same as that of $MgCl_2$, use enthalpy arguments to explain why $NaCl_2$ is an unlikely compound. The enthalpy of formation of Na(*g*) is +107 kJ/mol.

8.10 By assuming that the lattice enthalpy of MgCl would be the same as that of KCl, use enthalpy arguments to explain why MgCl is an unlikely compound.

Lewis Diagrams and Lewis Formulas

8.11 Write the Lewis diagrams of (a) calcium; (b) sulfur; (c) oxide ion; (d) nitride ion.

8.12 Write the Lewis diagrams of (a) chlorine; (b) arsenic; (c) carbon; (d) chloride ion.

8.13 Assuming that only *s* and *p* valence electrons are used for writing Lewis diagrams for an element, write the Lewis diagrams of (a) iron; (b) manganese; (c) cobalt.

8.14 Assuming that only *s* and *p* valence electrons are used for writing Lewis diagrams for an element, write the Lewis diagrams of (a) titanium; (b) nickel; (c) cadmium.

8.15 Write the Lewis formulas (for the ionic model of the compounds) of (a) potassium fluoride; (b) aluminum sulfide; (c) calcium nitride; (d) cobalt(II) chloride.

8.16 Write the Lewis formulas (for the ionic model of the compounds) of (a) cesium boride; (b) sodium oxide; (c) calcium phosphide; (d) chromium(III) oxide.

8.17 Assuming that only *s* and *p* valence electrons are involved in their ionic bond formation, write the Lewis formulas of (a) iron chloride; (b) manganese sulfide; (c) zinc oxide.

8.18 Assuming that only *s* and *p* valence electrons are involved in their ionic bond formation, write the Lewis formulas of (a) nickel bromide; (b) cobalt oxide; (c) vanadium oxide.

Lewis Structures

8.19 Determine the number of valence electrons that must be gained by covalent bond formation if each element is to obey the octet rule: (a) nitrogen; (b) chlorine; (c) silicon.

8.20 How many electron pairs must be shared with another element to obey the octet rule in forming a compound with all single bonds to the elements (a) iodine; (b) oxygen; (c) carbon.

8.21 Write the Lewis structures of (a) hydrogen fluoride; (b) ammonia; (c) methane.

8.22 Write the Lewis structures of (a) nitronium ion, ONO^+; (b) chlorite ion, ClO_2^-; (c) peroxide ion, O_2^{2-}.

8.23 Write the complete Lewis structures for each of the following salts: (a) ammonium chloride; (b) potassium phosphate; (c) sodium hypochlorite.

8.24 Write the complete Lewis structures for each of the following compounds: (a) zinc cyanide; (b) potassium tetrafluoroborate; (c) barium peroxide (the peroxide ion is O_2^{2-}).

8.25 Potassium ozonide, KO_3, is formed by the reaction of ozone with potassium hydroxide. What is the oxidation number of oxygen in the ozonide ion, O_3^-? Write the Lewis structure for the ozonide ion.

8.26 A decomposition product of sulfur monoxide is disulfur oxide, S_2O. Write the Lewis structure of S_2O.

8.27 Write the Lewis structures of the following organic compounds: (a) formaldehyde, H_2CO, which, as its aqueous solution called formalin, is used to preserve biological specimens; (b) acetone, CH_3COCH_3, a common organic solvent used in fingernail polish removers; (c) glycine, $CH_2(NH_2)COOH$, the simplest example of an amino acid, the building blocks of proteins.

8.28 Write the Lewis structures of the following organic compounds: (a) ethanol, CH_3CH_2OH, which is also called grain alcohol; (b) methylamine, CH_3NH_2, a putrid-smelling substance formed when living matter, including fish flesh, decays; (c) formic acid, HCOOH, a component of the venom injected by ants.

Lewis Acids and Bases

8.29 Identify the following species as Lewis acids or bases: (a) NH_3; (b) BF_3; (c) Na^+; (d) F^-.

8.30 Identify the following species as Lewis acids or bases: (a) H^+; (b) Al^{3+}; (c) CN^-; (d) NO_2^-.

8.31 Write the Lewis structure of the methoxide ion, CH_3O^-. Would you expect it to be a Lewis acid or a Lewis base? Explain your answer.

8.32 Write the Lewis structure of the hydride ion, Would you expect it to be a Lewis acid or a Lewis base? Explain your answer.

8.33 Write the Lewis structures of each reactant, identify the Lewis acid and the Lewis base, and then write the product (a complex) for the following Lewis acid-base reactions:
(a) $PF_5 + F^- \rightarrow$ (b) $SO_2 + Cl^- \rightarrow$
(c) $Cu^{2+} + 4NH_3 \rightarrow$

8.34 Write the Lewis structures of each reactant, identify the Lewis acid and the Lewis base, and then write the product (a complex) for the following Lewis acid-base reactions:
(a) $GaCl_3 + Cl^- \rightarrow$ (b) $SF_4 + F^- \rightarrow$
(c) $Ag^+ + 2CN^- \rightarrow$

Resonance

8.35 (a) Write the resonance structures for the nitrite ion, NO_2^-. (b) Write the resonance structures for nitryl chloride, $ClNO_2$.

8.36 (a) Write the resonance structures for ozone, O_3. (b) Write the resonance structures for the formate ion, HCO_2^-.

8.37 Write Lewis structures, including typical resonance structures where appropriate (for example, allowing for the possibility of octet expansion as well as double bonds in different locations) for (a) sulfate ion; (b) sulfite ion; (c) chlorate ion; (d) nitrate ion.

8.38 Write Lewis structures, including typical resonance structures where appropriate (for example, allowing for the possibility of octet expansion as well as double bonds in different locations) for (a) phosphate ion; (b) hydrogensulfite ion; (c) perchlorate ion; (d) nitrite ion.

Exceptions to the Octet Rule

8.39 Write the Lewis structure, and note the number of lone pairs on the central atom for (a) sulfur tetrachloride; (b) iodine trichloride; (c) IF_4^-.

8.40 Write the Lewis structure, and note the number of lone pairs on the central atom for (a) PF_4^-; (b) ICl_4^+; (c) xenon tetrachloride.

8.41 Determine the numbers of electron pairs (both bonding and lone pairs) on the iodine atom in (a) ICl_2^+; (b) ICl_4^-; (c) ICl_3; (d) ICl_5.

8.42 Determine the numbers of electron pairs (both bonding and lone pairs) on the phosphorus atom in (a) PCl_3; (b) PCl_5; (c) PCl_4^+; (d) PCl_4^-.

8.43 Write the Lewis structures of the following components of smog: (a) NO; (b) OH (hydroxyl radical); (c) NO_2.

8.44 Write Lewis structures of (a) the superoxide ion, O_2^-; (b) the methoxy radical, CH_3O; (c) the methyl radical, CH_3.

8.45 Write Lewis structures and state the number of lone pairs on the central atom of the following compounds of xenon: (a) $XeOF_4$; (b) XeF_2; (c) XeO_4; (d) $HXeO_4^-$.

8.46 Write Lewis structures and state the number of lone pairs on the central atom of (a) ClF_3; (b) AsF_5; (c) BCl_3.

The Shapes of Molecules and Ions

8.47 What is the expected shape and bond angle in a molecule of VSEPR formula (a) AX_5; (b) AX_2; (c) AX_3E; (d) AX_2E_2?

8.48 What is the expected shape and bond angle in a molecule of VSEPR formula (a) AX_4E; (b) AX_3; (c) AX_6; (d) AX_3E_2?

8.49 Using Lewis structures and VSEPR theory, predict the shape of each of the following molecules: (a) CO_2; (b) SO_2; (c) O_3.

8.50 Using Lewis structures and VSEPR theory, predict the shape of each of the following molecules: (a) CS_2; (b) N_2O; (c) Cl_2O.

8.51 Using Lewis structures and VSEPR theory, predict the shape of each of the following ions: (a) $CH_3NH_3^+$; (b) ClO_3^-; (c) PF_4^+.

8.52 Using Lewis structures and VSEPR theory, predict the shape of each of the following species: (a) PF_4^-; (b) ICl_4^+; (c) xenon tetrafluoride.

8.53 Predict the bond angles in each of the following: (a) I_3^-; (b) ICl_3; (c) IO_4^-.

8.54 Predict the value of the ClICl bond angle in (a) ICl_2^-; (b) ICl_2^+; (c) ICl_4^-; (d) ICl_4^+.

8.55 Write the Lewis structure, write the VSEPR formula, list the shape, and predict the approximate bond angles in (a) CF_3Cl; (b) GaI_3; (c) CH_3^-.

8.56 Write the Lewis structure, write the VSEPR formula, list the shape, and predict the approximate bond angles in (a) SnF_4; (b) CH_3^+; (c) SnF_6^{2-}.

8.57 Trifluoroamine oxide, NF_3O, reacts with antimony pentafluoride, SbF_5, to produce the ionic salt, $NF_2O^+SbF_6^-$, in anhydrous hydrogen fluoride at

−95°C. Write the Lewis structure and predict the shape of each ion in the salt.

8.58 A gaseous mixture of nitrogen trifluoride, krypton difluoride, and arsenic pentafluoride at 80°C and 75 atm react, producing tetrafluoroammonium hexafluoroarsenate: $NF_3(g) + KrF_2(g) + AsF_5(g) \rightarrow NF_4AsF_6(s) + Kr(g)$. Write the Lewis structure and predict the shape of each ion in the salt.

8.59 Estimate the bond angles marked with arcs and lowercase letters in acrolein, an eye irritant in smoke with a pungent smell:

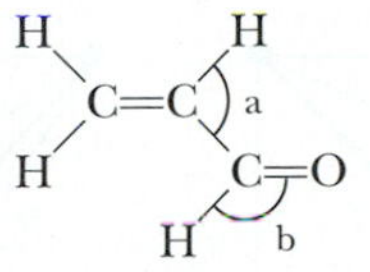

8.60 Estimate the bond angles marked with arcs and lowercase letters in peroxyacetyl nitrate, an eye irritant in smog:

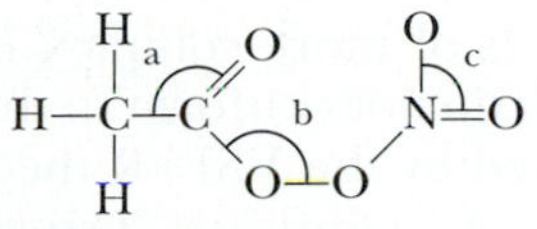

8.61 When BF_3 reacts with NH_3, a white solid of molecular formula $F_3B—NH_3$ is formed. Draw Lewis structures of the reactants and the product, and estimate the bond angles.

8.62 Hydrogen fluoride reacts with antimony pentafluoride to give a salt in which the cation is H_2F^+ and the anion is SbF_6^-. Draw Lewis structures of the reactants and the product, and estimate the bond angles.

UNCLASSIFIED EXERCISES

8.63 Complete the following table (all values are in kJ/mol; M denotes the metallic element; X denotes the halide):

Compound (MX)	ΔH_f° (M, *g*)	ΔH_i° (M)	ΔH_f° (X, *g*)	ΔH_{gain}° (X)	ΔH_L° (MX)	ΔH_f° (MX, *s*)
(a) NaCl	108	494	122	−349	787	?
(b) KBr	89	418	112	−325	?	−394
(c) RbF	?	402	79	−328	785	−558

8.64 Use the data in Exercise 8.63, Figs. 7.31 and 7.35, Appendix 2A, and the information given below to calculate the lattice enthalpy of (a) sodium oxide; (b) aluminum chloride: $\Delta H_f^\circ(O, g) = +249$ kJ/mol; $\Delta H_f^\circ(Na_2O, s) = -409$ kJ/mol; $\Delta H_f^\circ(Al, g) = +342$ kJ/mol.

8.65 Write the Lewis diagrams of (a) N^{3-}; (b) O^{2-}; (c) Sn^{2+}.

8.66 Write the Lewis formulas of the following compounds (assuming an ionic model of their bonding): (a) lithium hydride; (b) copper(II) chloride; (c) barium nitride; (d) cobalt(III) oxide.

8.67 Write the Lewis structures and predict the shapes of (a) TeF_4; (b) NH_2^-; (c) H_3O^+; (d) OCS.

8.68 Write the Lewis structures and give the approximate bond angles of (a) ClCN; (b) $OPCl_3$; (c) N_2H_4.

8.69 Write the Lewis structures and predict the shapes of (a) $OCCl_2$; (b) $OSCl_2$; (c) $OSbCl_3$. In each case the formula expresses the atomic arrangement.

8.70 Phosgene has been used in chemical warfare as a poisonous gas. A chemical analysis of phosgene is 12.1% carbon, 16.2% oxygen, and 71.7% chlorine by mass. Write the Lewis structure of phosgene.

8.71 Fluorine azide, N_3F, is prepared by the reaction of hydrogen azide, HN_3, with fluorine in the gas phase. Fluorine azide has a normal boiling point of −30°C and is extremely unstable in the liquid and solid phases. It decomposes slowly to difluorodiazine, N_2F_2. Write the Lewis structures (including resonance structures) of fluorine azide (atom arrangement, NNNF) and difluorodiazine (atom arrangement, FNNF) and estimate all bond angles.

8.72 The first krypton-nitrogen bond has been reported in the compound $[HC{\equiv}N{—}KrF]^+$, $[AsF_6]^-$, which is stable at temperatures below −50°C. Write the Lewis structure and predict the shape of each ion of the salt.

8.73 Write the (noted) number of resonance structures for the following compounds: (a) CN_2^{2-} (two); (b) N_2O_3 (three, for the arrangement $ONNO_2$); (c) OCN^- (three); (d) C_3O_2 (three, for the arrangement OCCCO).

8.74 Identify each of the following as a Lewis acid or a Lewis base: (a) BCl_3; (b) Fe^{3+}; (c) GaI_3.

8.75 Write the Lewis structure for each reactant, identify the Lewis acid and Lewis base, and then write the formula of the product (a complex) for the following reactions:

(a) $AlCl_3 + Cl^- \rightarrow$
(b) $I_2 + I^- \rightarrow$
(c) $Cr^{3+} + 6NH_3 \rightarrow$
(d) $SnCl_4 + 2Cl^- \rightarrow$

8.76 In the reaction of sulfur dioxide with water, SO_2 functions as a Lewis acid. Write the chemical equation for the reaction in terms of Lewis structures.

CASE 8 ▸ STRUCTURES BEYOND VSEPR

The structures of more complex inorganic compounds often do not conform to the simple structures predicted by the VSEPR theory outlined in this chapter. A significant distortion from the VSEPR prediction, or in some cases an entirely different structure, may result from the effect of the size or the partial electric charges of the atoms (or groups of atoms) bonded to the central atom. The shape may also be affected by the properties of the electron distribution in the valence shell of the central atom itself. These "non-VSEPR" structures occur most often where five or more atoms (or groups of atoms) are bonded to the central atom and when the central atom is a *d*-metal cation.

As an example, the electron configuration of the Ni^{2+} ion in $Ni(CN)_5{}^{3-}$ is $[Ar]3d^8$, with no *s* and *p* valence electrons. The CN^- ion is a Lewis base (an electron-pair donor) and forms a complex with the Ni^{2+} ion (the Lewis acid). Because each CN^- ion provides two bonding electrons, we can visualize the structure as a central metal atom with five electron pairs in its valence shell. On those grounds, VSEPR would predict a trigonal bipyramidal structure. In fact, a structural analysis of $Ni(CN)_5{}^{3-}$ shows it to be a square pyramid:

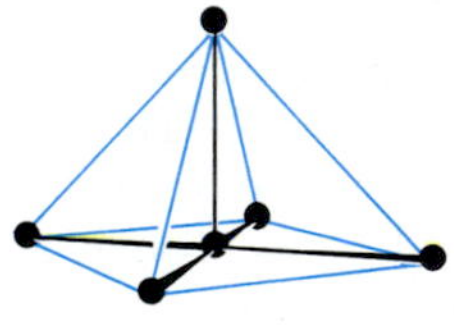

Square pyramid

Compounds of formula AB_6 are usually octahedral; however, a few are trigonal prismatic or trigonal antiprismatic. In thorium(II) iodide, ThI_2, the I^- ions act as Lewis bases to more than one Th^{2+} ion and both trigonal prismatic (left) and trigonal antiprismatic (right) structures are present in the solid:

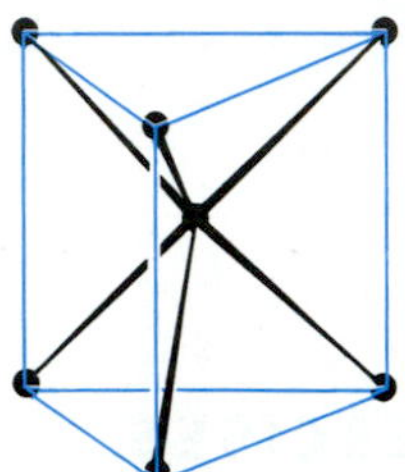

Trigonal prism

Trigonal antiprism

Structures with more than six electron pairs in the valence shell occur when the central atom is large enough to accommodate more atoms. For example, iodine heptafluoride, IF_7, is pentagonal bipyramidal:

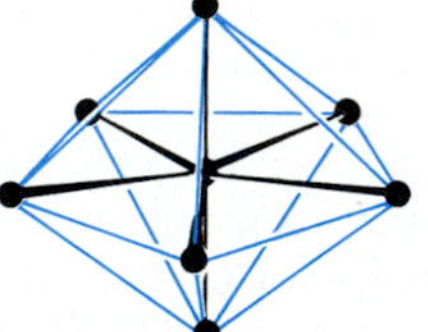

Pentagonal bipyramid

8.77 Which of the following pairs of species is the stronger Lewis acid: (a) BF_3 or GaI_3; (b) Al^{3+} or Fe^{3+}? Explain your reasoning.

8.78 Which of the following pairs of species is the stronger Lewis base: (a) $CH_3{}^-$ or Cl^-; (b) $NO_2{}^-$ or $NO_3{}^-$? Explain your reasoning.

8.79 Limestone (calcium carbonate) is used to remove impurities from iron ore in a blast furnace. When limestone is heated, it decomposes into calcium oxide and carbon dioxide, and the calcium oxide combines with impurities such as silica, SiO_2, to form calcium silicate, $CaSiO_3$. The latter is commonly called slag. Write the chemical equation for the Lewis acid-base decomposition of carbonate ions (the complex) and a Lewis acid-base reaction for the formation of calcium silicate. Identify the Lewis acid and base in each reaction.

8.80 For each molecule or ion, write the Lewis structure, list the number of lone pairs on the central atom,

but BrF_7 is unknown. Similar structures occur with some Periods 5 and 6 *f*-block Lewis acid-base complexes. For example, UF_7^{2-} is pentagonal bipyramidal but NbF_7^{2-} is a capped trigonal prism:

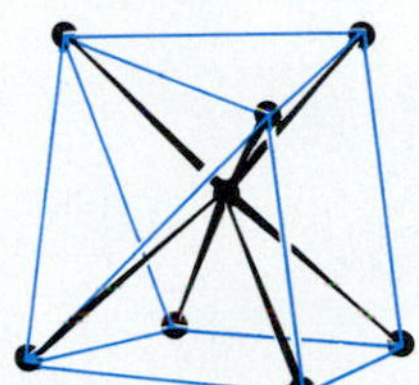

Capped trigonal prism

An example of an AB_8 compound is $Mo(CN)_8^{4-}$, which is a dodecahedron (left), and TaF_8, which is a tetragonal antiprism (a cube with the top twisted through 45°; right):

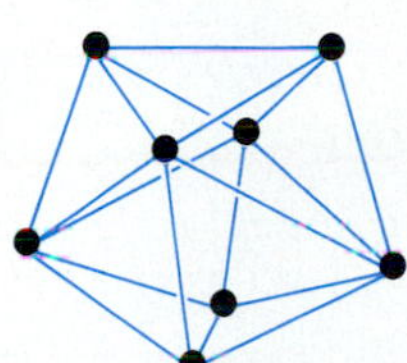

Dodecahedron

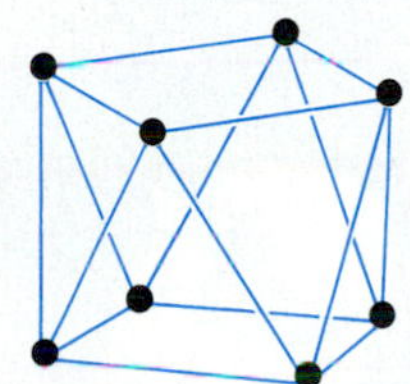

Tetragonal antiprism

Only ligands are illustrated in these two models.

QUESTIONS

1. (a) What is the F—I—F bond angle in the plane of the pentagonal bipyramid structure in IF_7? (b) How many faces are present in a pentagonal bipyramid structure?

2. (a) Draw diagrams to show the relationship between a trigonal prism, a trigonal antiprism, and an octahedron. (b) Predict the effect of temperature on the rate of interconversion of one structure to another.

3. How many *different* B—B distances are there in a dodecahedral AB_8 structure? Which distances would need to be increased to create a tetragonal antiprism?

4. In which of the following AX_n structures are *all* bond angles the same: (a) square pyramid; (b) trigonal antiprism; (c) pentagonal bipyramid; (d) dodecahedron; (e) trigonal prism; (f) tetragonal antiprism?

identify the shape, and estimate the bond angle: (a) SF_5^+; (b) TeH_2; (c) BrO_3^-; (d) IF_5.

8.81 Predict the shapes and estimate the bond angles of (a) the thiosulfate ion, $S_2O_3^{2-}$; (b) BH_2^-; (c) $SnCl_2$.

8.82 Hexamethyltellurium, $Te(CH_3)_6$, has just recently been synthesized by a research group at the University of Illinois, Chicago. The starting material was tetramethyldifluorotellurium, $TeF_2(CH_3)_4$. Write the Lewis structure of each compound and predict its shape.

8.83 It has been mentioned in the text, and we shall see it again on several more occasions, that if resonance is possible, then the molecule or ion has a lower energy than any of the contributing structures. Show how resonance may occur in the following species: (a) the acetate ion, $CH_3CO_2^-$; (b) an enolate ion, $CH_2(CO)CH_3^-$, a resonance form of an "ene" type ion containing a C=C double bond and the anion formed from an "ol" type alcohol containing an —OH group (converted to $—O^-$ in the ion); (c) an allyl cation, $CH_2CHCH_2^+$.

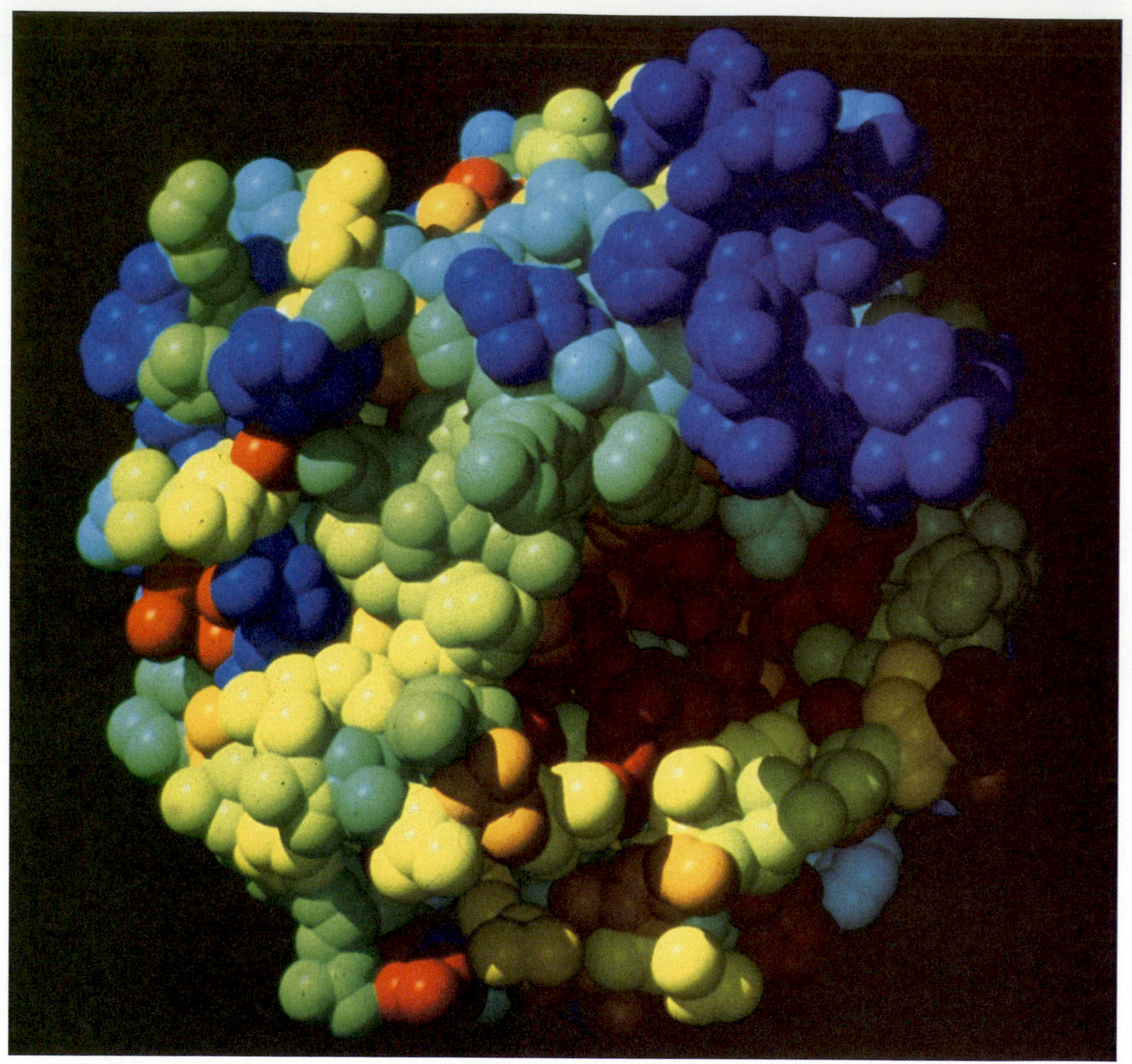

This chapter explains how electrons are distributed in molecules. It explores how electrons respond to the presence of atoms of different electronegativity, and then develops two descriptions of bonding that take into account the quantum mechanical properties of electrons. We see that electrons occupy orbitals that may be understood by extending the description of atomic orbitals to orbitals that spread between atoms and over the entire molecule. The properties of a complex molecule, such as the enzyme molecule shown in this computer-generated model, can be understood only by taking into account the details of the locations of the electrons in the molecule.

9 THE STRUCTURES OF MOLECULES

In this chapter we take three further steps forward in the description of covalent bonds. First, we look at bonds more closely and see to what extent they can be discussed quantitatively. We shall see that we can make reasonably reliable predictions about the strengths and lengths of bonds. Next, we look more closely at ionic and covalent bonding and see that they are best thought of as two extreme versions of a single type of bond, with all actual bonds lying somewhere between the two. In particular, we shall see how electrons are distributed in molecules in response to the electronegativities—the electron-attracting power—of the atoms present.

The third step we take involves a major change of attitude. Throughout Chapter 8 we were concerned with Lewis structures. However, we know from our study of atomic structure in Chapter 7 that electrons occupy orbitals in atoms and that their locations cannot be specified precisely. So, what does it mean to say that a covalent bond consists of an electron pair shared by two neighboring atoms: where are the electrons? Later in the chapter we shall see how the Lewis account of chemical bonding can be reconciled with the existence of atomic orbitals. That work will bring us right up to date with the way that chemists currently think about chemical bonds and the distribution of electrons in molecules.

BOND PARAMETERS

A covalent bond between two specific atoms has certain properties that are approximately the same in all the molecules in which that bond occurs. The C—H bond, for example, has about the same strength and the same length in all organic compounds, and all O—N bonds are similar, no matter in which compounds they occur. These characteristic features are called **bond parameters**.

N The sections on valence bond and molecular orbital theories can be excluded from the discussion without damaging the flow of the text. Valence bond theory could be included as part of Chapter 23.

E G. N. Lewis and the chemical bond. *J. Chem. Educ.*, **1984**, *61*, 201.

TABLE 9.1 Bond enthalpies of diatomic molecules

Molecule	ΔH_B, kJ/mol
H_2	436
N_2	944
O_2	496
CO	1074
F_2	158
Cl_2	242
Br_2	193
I_2	151
HF	565
HCl	431
HBr	366
HI	299

T OHT: Tables 9.1 and 9.2.

E A photon of wavelength no greater than 317 nm is needed to break the C—O bond in CH_3OH.

E Strength of chemical bonds. *J. Chem. Educ.*, **1973**, *50*, 176; 871. Bond free energies. *J. Chem. Educ.*, **1979**, *56*, 453.

Diatomic molecule	Lewis structure	Bond enthalpy, kJ/mol
N_2	:N≡N:	944
O_2	O=O (with two lone pairs on each atom)	496
F_2	:F—F: (with three lone pairs on each atom)	158

N Bond enthalpies can be used to estimate standard enthalpies of formation, as in the reaction $\frac{1}{2}H_2(g) + \frac{1}{2}Cl_2(g) \rightarrow HCl(g)$.

Molecule	Lewis structure	Bond enthalpy, kJ/mol
HF	H—F: (three lone pairs on F)	565
HCl	H—Cl: (three lone pairs on Cl)	431
HBr	H—Br: (three lone pairs on Br)	366
HI	H—I: (three lone pairs on I)	299

9.1 BOND STRENGTH

The strength of a covalent bond is measured by the **bond enthalpy**, ΔH_B, the change in enthalpy when a bond is broken. For example, the bond enthalpy of H_2 is the change in enthalpy for the process

$$H_2(g) \longrightarrow 2H(g) \qquad \Delta H° = +436 \text{ kJ}$$

The breaking of a bond is called **dissociation**. We write $\Delta H_B(\text{H—H}) = 436$ kJ/mol to report this bond enthalpy. (The value implies that the heat needed to dissociate 2 g of hydrogen molecules is about the same as that needed to heat 1 kg of water from 0°C to its boiling point.) In general, we need to specify which bond is undergoing dissociation, particularly in polyatomic molecules. The value $\Delta H_B(CH_3\text{—OH}) = 377$ kJ/mol, for instance, refers to the dissociation of the C—O bond in methanol:

$$CH_3OH(g) \longrightarrow CH_3(g) + OH(g) \qquad \Delta H° = +377 \text{ kJ}$$

whereas $\Delta H_B(CH_3O\text{—H}) = 437$ kJ/mol refers to the dissociation of the O—H bond:

$$CH_3OH(g) \longrightarrow CH_3O(g) + H(g) \qquad \Delta H° = +437 \text{ kJ}$$

All bond dissociations are endothermic, because energy must be supplied to break bonds; therefore, all bond enthalpies are positive.

Bond enthalpies in diatomic molecules. The bond enthalpies of some diatomic molecules are listed in Table 9.1. They range from 151 kJ/mol for I_2 up to 1074 kJ/mol for CO, the highest value for any diatomic molecule.

The trends in the values shown in Table 9.1 are quite easy to explain in terms of the Lewis structures of the molecules. Consider, for example, nitrogen, oxygen, and fluorine, as shown in the margin. One reason for the decrease in bond enthalpy is the decrease in the number of bonds between the atoms, from three in N_2, to two in O_2, to one in F_2. Multiple bonds are stronger than single bonds: this statement is always true for bonds between the same pair of elements and is almost always true whatever the elements (for instance, the triple bond in N_2 is stronger than the double bond in O_2). A second reason for the decrease in bond enthalpy can be found by examining the Lewis structures more closely. We note that the number of lone pairs on the neighboring atoms increases from N_2 (which has one lone pair on each atom), to two in O_2, to three in F_2. Lone pairs on neighboring atoms repel one another, so there is a stronger force pushing the atoms apart in F_2 than in N_2.

In general, when considering bond strengths we need to consider:

1. The existence of multiple bonds;
2. The repulsions between lone pairs on neighboring atoms; and
3. Effects related to the sizes of atoms.

Statement 3 means that if the nuclei of atoms cannot come very close together, then the shared electron pair will not be very effective in bonding them together (because it is a long way from the nuclei), and the bond will be weak. For example, the bond enthalpies of the hydrogen halides decrease from HF to HI, as shown in the margin. All the structures are singly bonded, so multiple bonding plays no role in their relative strengths. Because there are no lone pairs on hydrogen atoms, the repulsion between lone pairs on different atoms is also irrelevant. The atomic radii of the halogens, however, increase from fluorine to iodine (see Fig. 7.27), a trend suggesting that the hydrogen nucleus cannot get as close to the iodine nucleus as it can to the fluorine nucleus. The bond enthalpy decreases along the series because the hydrogen atom is progressively farther away from the nucleus of the halogen atom.

E The relative strengths of carbon-carbon bonds are not always reflected in the relative strengths of bonds between other elements. Thus, an N≡N bond is *more* than three times the strength of an N—N bond. A contribution to the reason is the change in bond length, and the σ contribution is strengthened when the formation of π bonds draws the atoms together.

Average bond enthalpies in polyatomic molecules. The bond enthalpy of any particular bond in a polyatomic molecule depends on the other atoms present. For instance, the values for the HO—H bond in water, 492 kJ/mol, and the CH_3O—H bond in methanol, 435 kJ/mol, are not quite the same. As this example shows, however, variations in strength for bonds of the same order between the same two elements are not very great, so it is useful to list the **average bond enthalpy** ΔH_B(A—B), which is the average of A—B bond enthalpies for a number of different molecules containing the A—B bond. The values of several average bond enthalpies are given in Table 9.2.

T OHT: Tables 9.1 and 9.2.

N Average bond enthalpies are enthalpies of last resort: they are only approximately applicable to a specific molecule. Whenever possible, use the bond enthalpy for the species in question.

When using average bond enthalpies, however, we do need to distinguish between single and multiple bonds. A multiple bond is stronger than a single bond between the same pair of elements, as is illustrated by the values for carbon-carbon bonds in the shaded box in Table 9.2. As we see from these values, a C=C bond is not twice as strong as a C—C bond, and a C≡C bond is a lot less than three times as strong:

$$\Delta H_B(\text{C=C}) = 612 \text{ kJ/mol} \qquad \Delta H_B(\text{C≡C}) = 837 \text{ kJ/mol}$$
$$2 \times \Delta H_B(\text{C—C}) = 696 \text{ kJ/mol} \qquad 3 \times \Delta H_B(\text{C—C}) = 1044 \text{ kJ/mol}$$

This weakness of a carbon-carbon multiple bond relative to the same number of single carbon-carbon bonds plays a very important role in organic chemistry and is a factor that contributes to the high reactivity of compounds containing carbon-carbon multiple bonds (such as ethylene, H_2C=CH_2, and acetylene, HC≡CH). One contribution to the weakness of a multiple carbon-carbon bond relative to that of two or three single bonds is that the electron pairs that lie between the atoms and form the multiple bond repel each other slightly; thus they are not quite as effective in bonding as a single bonding pair.

Another feature illustrated by the data in Table 9.2 is the role of resonance. We saw in Section 8.6 that resonance averages out the bond characteristics. Hence, resonance between structures with single and double bonds should result in bonds of an intermediate character. For example, compare the entries for carbon-carbon single and double bonds with the value for the bonds between carbon atoms in benzene (which are denoted C┄C). We see that the bond enthalpy of the bonds in benzene is intermediate between those of a single and a double bond (but closer to that of a double bond).

TABLE 9.2 Average bond enthalpies

Bond	Average bond enthalpy, kJ/mol
C—H	412
C—C	348
C=C	612
C┄C	518 (benzene)
C≡C	837
C—O	360
C=O	743
C—N	305
N—H	388
N—N	163
N=N	409
N—O	210
N=O	630
N—F	195
N—Cl	381
O—H	463
O—O	157

▼ **EXAMPLE 9.1 Using bond enthalpies to assess the effect of resonance**

Use the average bond enthalpies in Table 9.2 to calculate the lowering in energy that occurs when resonance is allowed between the Kekulé structures of benzene.

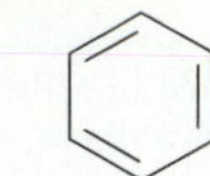

1 Benzene, C_6H_6

2 Benzene

STRATEGY The Kekulé structures of benzene were introduced in Section 8.6; one is shown as (**1**) and the resonance hybrid of the two structures is shown as (**2**). The only differences between the structures are in the nature of the bonding between the C atoms, so we concentrate on their bond enthalpies and ignore the C—H bonds (which are unchanged in the two structures). We can calculate the energy of a Kekulé structure by finding the sum of three single and three double carbon-carbon bond enthalpies. For the resonance hybrid, we find the sum of six carbon-carbon "benzene" bonds. The difference in the two total bond enthalpies is the energy lowering that we can ascribe to resonance.

SOLUTION The total carbon-carbon bond enthalpy in a Kekulé structure is

$$\text{Bond enthalpy (Kekulé)} = 3 \times \Delta H_B(\text{C—C}) + 3 \times \Delta H_B(\text{C=C})$$
$$= (3 \times 348 + 3 \times 612)\ \text{kJ/mol} = 2880\ \text{kJ/mol}$$

The corresponding quantity in the resonance hybrid is

$$\text{Bond enthalpy (hybrid)} = 6 \times \Delta H_B(\text{C}\text{⋯}\text{C})$$
$$= 6 \times 518\ \text{kJ/mol} = 3108\ \text{kJ/mol}$$

The difference is 228 kJ/mol. Because the hybrid has the greater total bond enthalpy (that is, more energy must be supplied to dissociate it into atoms), the hybrid lies 228 kJ/mol lower in energy than either of the Kekulé structures.

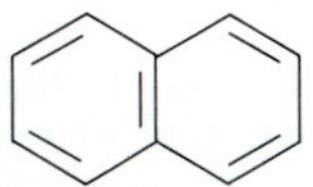

3 Naphthalene, $C_{10}H_6$

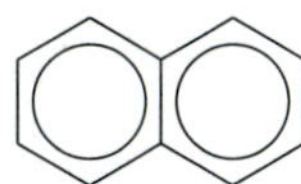

4 Naphthalene

EXERCISE E9.1 Estimate the lowering in energy between the structure of the compound naphthalene (**3**) and the resonance hybrid (**4**) in which all carbon-carbon bonds are taken to be the same.

▲ [*Answer*: 550 kJ/mol]

Average bond enthalpies can be used to explain chemical properties. For example, the enthalpies of E—H bonds, where E represents a Group IV element, decrease down the group from carbon (412 kJ/mol) to lead (205 kJ/mol) (Fig. 9.1). This trend parallels a decrease in the stability of the hydrides of the heavier members of the group: methane (CH_4) can be kept indefinitely in air at room temperature, stannane

E Naphthalene and *p*-dichlorobenzene are sublimable solids and are components of mothballs.

U Exercises 9.62–9.64.

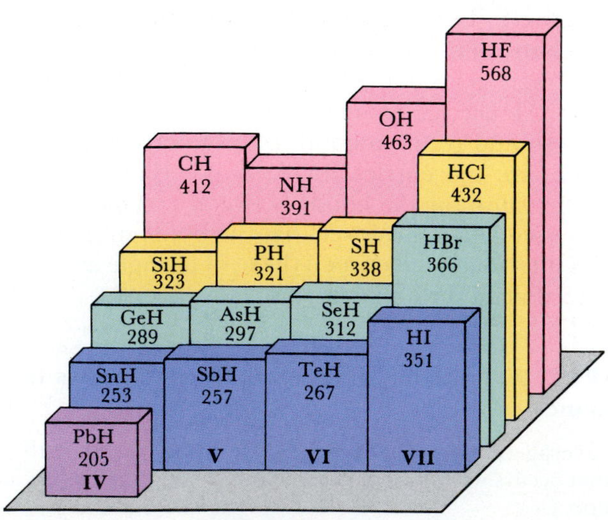

FIGURE 9.1 The bond enthalpies (in kilojoules per mole) for the bonds between hydrogen and several main-group elements.

(SnH_4) decomposes into tin and hydrogen, and plumbane (PbH_4) has never been prepared except perhaps in trace amounts. Average bond enthalpies can also be used to estimate reaction enthalpies when accurate data are not available. This use of bond enthalpies is shown in the next example.

▼ EXAMPLE 9.2 Estimating reaction enthalpy from bond enthalpies

Estimate the enthalpy of combustion of ethanol vapor. Ethanol is CH_3CH_2OH and the reaction is

$$C_2H_5OH(g) + 3O_2(g) \longrightarrow 2CO_2(g) + 3H_2O(g)$$

STRATEGY We expect a negative value because all combustions are exothermic. To determine the overall change in enthalpy, we need to know the change in enthalpy that accompanies dissociation of all the bonds in the reactants and the change that accompanies the formation of new bonds in the products. We identify the bonds that are broken and formed by writing the reaction in terms of Lewis structures. Then we use the bond enthalpies in Tables 9.1 and 9.2 to calculate the heat required to break all the bonds in the reactants and the heat released when the products are formed from the atoms. The enthalpy of combustion is the difference between the two.

SOLUTION In terms of Lewis structures the combustion reaction is

$$\mathrm{H{-}\overset{\overset{H}{|}}{\underset{\underset{H}{|}}{C}}{-}\overset{\overset{H}{|}}{\underset{\underset{H}{|}}{C}}{-}\ddot{\underset{\cdot\cdot}{O}}{-}H} + 3\ \mathrm{\ddot{O}{=}\ddot{O}} \longrightarrow 2\ \mathrm{\ddot{O}{=}C{=}\ddot{O}} + 3\ \mathrm{H{-}\ddot{\underset{\cdot\cdot}{O}}{-}H}$$

The change in enthalpy when all bonds are broken to form a gas of atoms (Fig. 9.2) is

ΔH(dissociation of reactants)

$$= 5\ \Delta H_B(\text{C—H}) + \Delta H_B(\text{C—C}) + \Delta H_B(\text{C—O}) + \Delta H_B(\text{O—H}) + 3\ \Delta H_B(\text{O=O})$$

$$= 5 \times 412\ \text{kJ} + 348\ \text{kJ} + 360\ \text{kJ} + 463\ \text{kJ} + 3 \times 496\ \text{kJ} = +4719\ \text{kJ}$$

The enthalpy change when the products form from the gas of atoms is

$$\Delta H(\text{formation of products}) = -\{2 \times 2\ \Delta H_B(\text{C=O}) + 3 \times 2\ \Delta H_B(\text{O—H})\}$$
$$= -(2 \times 2 \times 743\ \text{kJ} + 3 \times 2 \times 463)\ \text{kJ}$$
$$= -5750\ \text{kJ}$$

The overall change, the sum of the two contributions, is

$$\Delta H = \Delta H(\text{dissociation of reactants}) + \Delta H(\text{formation of products})$$
$$= 4719\ \text{kJ} - 5750\ \text{kJ} = -1031\ \text{kJ}$$

Because the thermochemical equation refers to 1 mol C_2H_5OH, we estimate the enthalpy of combustion of ethanol as −1031 kJ/mol. The experimental enthalpy of combustion is −1236 kJ/mol. That the estimated value is different from the experimental value is a consequence of using *average* bond enthalpies rather than the true enthalpies of atomization of the compounds.

EXERCISE E9.2 Estimate the standard enthalpy of formation of gaseous ammonia.

[*Answer*: −38 kJ/mol] ▲

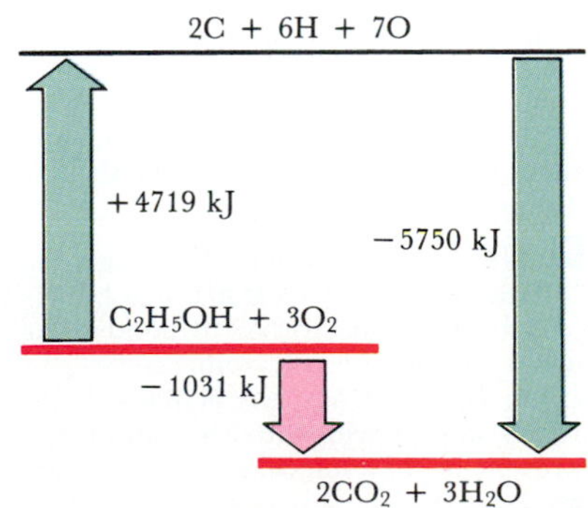

FIGURE 9.2 The enthalpy changes for the combustion of ethanol estimated by using average bond enthalpies.

9.2 BOND LENGTHS

Another measurable property of covalent bonds is the **bond length**, the distance between the nuclei of two atoms joined by a chemical bond. Several values are given in Table 9.3: typical values for bonds between the elements of Period 2 lie in the range 100 to 150 pm (where 1 pm = 10^{-12} m).

Molecule	Bond length, pm
F—F	142
Cl—Cl	199
Br—Br	228
I—I	268

E Experimental bond lengths are averages over the internuclear distance, taking into account the vibrational motion of the atoms. The *equilibrium bond length* is the internuclear distance at the minimum of the molecular potential energy curve.

E Bond length-bond energy term correlations for bonds in Al_2X_6 compounds. *J. Chem. Educ.*, **1972**, *49*, 502.

Bonds between heavy atoms tend to be longer than those between light atoms because heavy atoms have more electrons and hence greater atomic radii. This feature is illustrated in the margin by the trend in bond lengths for the halogen molecules. Multiple bonds are shorter than single bonds between the same pair of elements because the additional bonding electrons bind the atoms together more closely. This feature is illustrated for carbon-carbon bonds in Table 9.3. Once again, we can see the averaging effect of resonance by comparing the lengths of single and double carbon-carbon bonds with the values for bonds in benzene. The length of the bond in benzene is intermediate between the lengths of the single and double bonds of a Kekulé structure (but closer to that of a double bond). Another useful feature revealed in Tables 9.2 and 9.3 is that, for bonds between the same pairs of atoms, the *stronger* the bond, the *shorter* it is. Thus, a C≡C triple bond is both stronger and shorter than a C—C single bond. Similarly, a C=O double bond is both stronger and shorter than a C—O single bond.

A bond length is approximately the sum of the **covalent radii** of the two bonded atoms (**5**). The O—H bond length in ethanol, C_2H_5OH, for example, is the sum of the covalent radius of H and that of O, or 111 pm. Some covalent radii are given in Table 9.4.

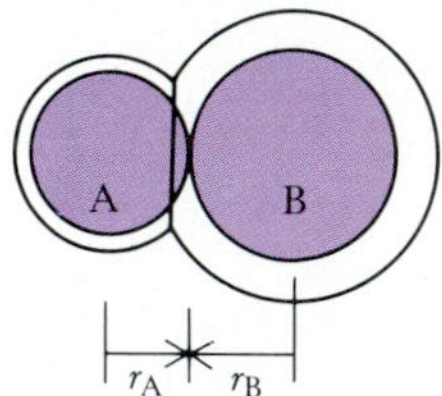

5 Covalent radii

Like atomic radii in Section 7.8 (and for the same reason), covalent radii typically decrease from left to right across a period: the increasing nuclear charge draws in the electrons and makes the atom more compact. Again like atomic radii, covalent radii increase down a group, because in successive periods the valence electrons occupy shells that are more distant from the nucleus. The new information in Table 9.4 is

E The covalent radius should be distinguished from the *van der Waals radius* of an atom, which is its contribution to the separation of the nuclei of two closed-shell atoms when the completed valence shells of the atoms are in contact.

N The actual bond lengths are given for diatomic molecules; average bond lengths are given for A—B bonds that occur as parts of polyatomic molecules. For the latter, the actual value varies slightly with the other atoms present in the molecule.

T OHT: Table 9.3.

E Bent bonds and multiple bonds. *J. Chem. Educ.*, **1980**, *57*, 329.

TABLE 9.3 Average and actual bond lengths

Bond	Average bond length, pm	Molecule	Bond length, pm
C—H	109	H_2	74
C—C	154	N_2	110
C=C	134	O_2	121
C┄C	139 (benzene)	F_2	142
C≡C	120	Cl_2	199
C—O	143	Br_2	228
C=O	112	I_2	268
O—H	96		
N—H	101		

TABLE 9.4 Covalent radii* in picometers

		H 37			
Be 90	B 82	C 77 67 60	N 75 60 55	O 74 60	F 72
	Al 118	Si 111	P 120	S 102	Cl 98
	Ga 126	Ge 122	As 119	Se 117	Br 114
	In 144	Sn 141	Sb 138	Te 135	I 134

*Alternative radii refer to single, double, and triple bonds, respectively.

the variation of covalent radius with the number of bonds. As we have already noted, the greater the number of bonds between a pair of atoms, the shorter the bond and therefore the smaller the covalent radius.

▼ EXAMPLE 9.3 Estimating the bond lengths in a molecule

Estimate the values of all the bond lengths in a molecule of acetic acid, CH_3COOH.

STRATEGY The first step is to write down the Lewis structure of the molecule so that we can identify the single and multiple bonds. Then we express each bond length as a sum of the appropriate covalent radii taken from Table 9.4. We must be careful to use the radius corresponding to the correct bond order (single, double, or triple) for each bond.

SOLUTION The Lewis structure of acetic acid is

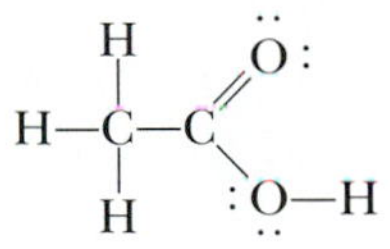

```
      H    ..
      |    O:
      |   //
  H—C—C
      |   \
      H    :O—H
           ..
```

The bond lengths can be analyzed into the following contributions taken from Table 9.4:

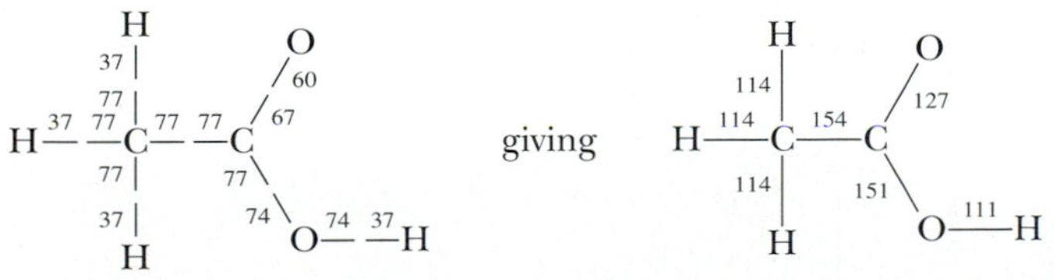

```
       H                                   H
    37 |        O                          |       O
    77 |       / 60                    114 |      /
  37 77|  77 77 / 67                   114 |  154/ 127
H—— ——C—— ——C          giving      H——————C——————C
    77 |     77\                       114 |  151 \
    37 |     74 \ 74 37                    |       \  111
       H         O—— ——H                   H        O——————H
```

In this molecule it would probably be more accurate to use the bond lengths shown in Table 9.3: those values are actual bond lengths of the atom pairs shown. Table 9.4 gives *average* values calculated using data from a variety of different molecules.

EXERCISE E9.3 Estimate the bond lengths in the hydrogen cyanide molecule, HCN.

[*Answer*: $H\overset{114}{—}C\overset{115}{\equiv}N$] ▲

N As for average bond enthalpies, covalent radii are data of last resort: they are (quite good) average values, and actual bond lengths should be used whenever possible.

T OHT: Table 9.4.

E A continuous quantitative relationship between bond length, bond order, and electronegativity for homo- and heteronuclear bonds. *J. Chem. Educ.*, **1986**, *63*, 123.

U Exercises 9.61, 9.65, and 9.66.

CHARGE DISTRIBUTIONS IN COMPOUNDS

Electron distributions in molecules have far-reaching consequences in chemistry, and in life in general. For example, with a knowledge of electron distributions, we can explain why water is such a good solvent for ionic compounds whereas benzene, for instance, is not (see Section 11.3). The chemical properties of elements depend on the properties of the electrons in atoms, and it should not be surprising that a knowledge of how electrons are distributed in the molecules these atoms form is a crucial additional piece of information. The properties of molecules, like the properties of atoms, are essentially the properties of the electrons they contain.

9.3 IONIC VERSUS COVALENT BONDING

E The experimental basis of the validity of the ionic description is a comparison of the calculated and observed values of the lattice enthalpy. The calculated value is obtained by proposing a purely ionic structure and evaluating the lattice enthalpy. The experimental value is obtained by using a Born-Haber cycle, as explained in Section 8.1. If the two values are in good agreement, the ionic model is valid. If not, then there is significant covalent character in the bonding.

E The chemical bond as an atomic tug-of-war. *J. Chem. Educ.*, **1984**, *61*, 677.

E Convenient relationships for the estimation of bond ionicity in A—B type compounds. *J. Chem. Educ.*, **1983**, *60*, 640.

The concepts of ionic and covalent bonding, although very useful, are *extreme* descriptions of actual bonds: they are ideal models. No actual bond is purely ionic or purely covalent. We have noted that whether two elements form an ionic or a covalent compound depends on the difference in their electronegativities (Section 7.8), their power to attract electrons to themselves when part of a compound. If the electronegativity difference is large (as it is between an alkali metal and a halogen), then we expect the more electronegative element to gain control of electrons from the electropositive element, and we expect the bonding to be ionic. If the electronegativities of the elements are similar (if they both come from the *p* block, for instance), then neither will win the tug-of-war for the electrons, and we expect the bond to be largely covalent. However, in neither case will there be complete victory or complete stalemate. Even a very electropositive element, such as cesium, will not lose *complete* control of its valence electrons, so no com-

N Fig. 9.3 indicates that there is no sharp distinction between ionic and covalent bonding: the two classes of bonding are analogous to left- and right-wing politicians.

T OHT: Fig. 9.3.

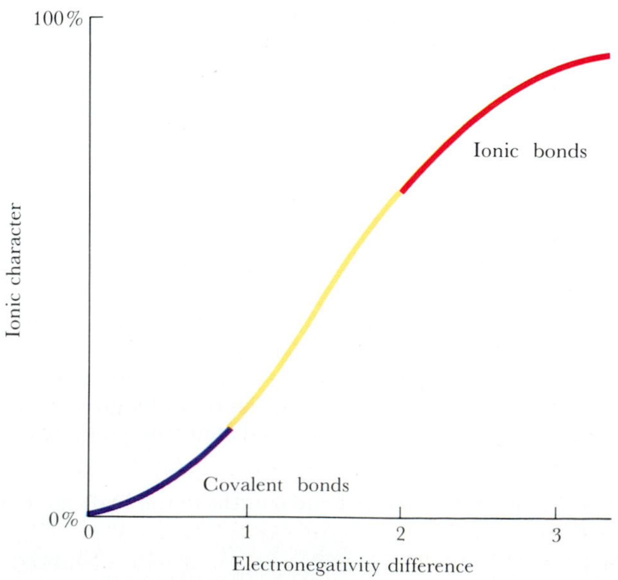

FIGURE 9.3 The ionic character of bonds increases with the difference in electronegativity between the two elements. Even in CsF (which has the largest electronegativity difference), the bond is only 95% ionic. No part of the curve is very precise, and the yellow area is particularly ambiguous.

pound can be expected to be purely ionic. Similarly, no compound can be expected to be purely covalent.

The variation of ionic character with electronegativity. Figure 9.3 shows how the ionic character of bonds varies with the electronegativity difference of the two linked atoms. In very broad terms we can refer to a bond between atoms that differ in electronegativity by less than about 1 as covalent (the bond between hydrogen and chlorine, for instance, for which the difference is 0.9). Similarly, when we refer to an ionic bond, we mean a bond between atoms that differ in electronegativity by more than about 2 (for example, the bond between sodium and chlorine, for which the difference is 2.1). However, even for electronegativity differences less than 1, a covalent bond has appreciable ionic character. Likewise, even for electronegativity differences greater than 2, an ionic bond has appreciable covalent character. *When we speak of an ionic compound, we mean a compound in which ionic bonding is a good first approximation.* Likewise, *when we speak of a covalent compound, we mean a compound in which covalent bonding is a good first approximation.* When the electronegativity difference lies between 1 and 2, the bonding is neither clearly ionic nor clearly covalent.

Polarizability and polarizing power. Consider a monatomic anion that is next to a cation. Because of the attraction of the cation's charge for the anion's electrons, the spherical electron cloud of the anion becomes distorted. We can think of this distortion as the tendency of an electron pair to move into the bonding region between the nuclei (**6**) and to form a covalent bond. Highly distorted ions are well on the way to becoming covalently bonded, and we can expect bonds to become more covalent as the distortion of the electron cloud increases.

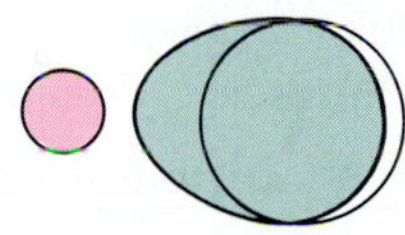

6 Polarization

Atoms and ions that readily undergo a large distortion are said to be highly **polarizable**. Ions that can cause large distortions are said to have a high **polarizing power**. An anion can be expected to be highly polarizable if it is large, like the I^- ion. In such an ion the nucleus exerts little control over its outermost electrons because they are farther away. As a result, the electron cloud of the ion is easily distorted. A cation has a high polarizing power if it is small and highly charged, like the Al^{3+} ion.

U Exercise 9.67.

E Ions having a high polarizing power are those with a high charge density (a large charge and small volume).

Compounds composed of a small, highly charged cation and a large, polarizable anion will tend to have bonds with a significant covalent character. In Group II the very small beryllium atom gives the element strikingly different properties from those of its large congeners. The Group II chlorides are largely ionic; however, beryllium chloride, which is made by the action of carbon tetrachloride on beryllium oxide,

$$2BeO(s) + CCl_4(g) \xrightarrow{800°C} 2BeCl_2(g) + CO_2(g)$$

is not ionic. It consists of individual $BeCl_2$ molecules in the vapor and long chains of covalently bonded $BeCl_2$ units in the solid, as shown in the margin.

E The colors of the silver halides are due to charge-transfer transitions in which an electron migrates from a halide ion to a silver ion: $Ag^+X^- \rightarrow AgX$. As the covalent character of the bonding increases, the energy of the AgX form is lowered relative to that of the Ag^+X^- form and the transition moves into the visible region of the spectrum.

Cl Cl Cl Cl
Be Be Be
Cl Cl Cl Cl

The covalent character of the silver halides increases from AgCl to AgI as the anion becomes more polarizable. One consequence is their decreasing solubility in water from AgCl to AgI. Silver fluoride is freely soluble in water and is predominantly ionic. Another consequence (for more complicated reasons) is their darkening of color from white for AgCl through pale yellow for AgBr to yellow for AgI (Fig. 9.4).

FIGURE 9.4 The silver halides, which are formed by precipitation from silver nitrate and sodium halide solutions, become increasingly insoluble and deeply colored down the group from Cl to I.

N A *dipole* is the pair of charges; a *dipole moment* is a property of that pair.

T OHT: Table 9.5.

TABLE 9.5 Dipole moments of selected molecules

Molecule	μ, debye
HF	1.91
HCl	1.08
HBr	0.80
HI	0.42
CO	0.12
ClF	0.88
NaCl	9.00
CsCl	10.42
H_2O	1.85
NH_3	1.47
PH_3	0.58
AsH_3	0.20
SbH_3	0.12
O_3	0.53
CO_2	0
BF_3	0
CH_4	0
cis-CHCl=CHCl	1.90
trans-CHCl=CHCl	0

Polarizability and polarizing power vary periodically through the periodic table and help to account for the diagonal relationships in it (see Section 7.9 and Fig. 7.39). Cations become smaller, more highly charged, and hence more strongly polarizing from left to right across a period. Thus Be^{2+} is more strongly polarizing than Li^+, Mg^{2+} more than Na^+, and Al^{3+} still more than Mg^{2+}. Cations become larger and hence less strongly polarizing down a group. Thus Na^+ is less strongly polarizing than Li^+, and K^+ is less polarizing still. Likewise, Mg^{2+} is less strongly polarizing than Be^{2+}. Because polarizing power increases from Li^+ to Be^{2+} but decreases from Be^{2+} to Mg^{2+}, it follows that the polarizing power of the diagonal neighbors Li^+ and Mg^{2+} should be similar. Likewise, because B^{3+} is more strongly polarizing than Be^{2+} but Al^{3+} is less strongly polarizing than B^{3+}, the polarizing power of the diagonal neighbors Be^{2+} and Al^{3+} should also be similar. Hence, we expect similarities in the properties of diagonally related neighbors.

Dipole moments. In an HCl molecule, the electronegative Cl atom has a small negative charge attributable to its partial takeover of the bonding electron pair, and the more electropositive H atom has a small positive charge. Such **partial charges**, or charges arising from small shifts in the distributions of electrons, are reported by writing $^{\delta+}\text{H—Cl}^{\delta-}$, where $\delta+$ signifies a partial positive charge and $\delta-$ a partial negative charge. (The Greek lowercase letter *delta*, δ, is often used to indicate a small quantity.) The HCl molecule is said to have an electric dipole:

> An **electric dipole** is a positive charge next to an equal but opposite negative charge.

The electric dipole in the HCl molecule is denoted $\overset{+\!\!\longrightarrow}{\text{H—Cl}}$ where + marks the site of the partial positive charge.

The magnitude of an electric dipole is reported as the **electric dipole moment** μ (the Greek letter *mu*) in units called **debye** (D). The debye is defined so that a single negative charge (an electron) separated by 100 pm from a single positive charge (a proton) has a dipole moment of 4.80 D.* The unit is named for the Dutch chemist Peter Debye, who made important studies of dipole moments. The dipole moment of HCl, for example, is 1.08 D (Table 9.5).

A **polar bond** is a covalent bond between atoms that have partial electric charges. We can visualize a polar bond (Fig. 9.5) as a resonance hybrid of a pure covalent bond (in which the electron pair is exactly equally shared) and a pure ionic bond (in which an electron has been transferred completely):

$$\text{H—}\ddot{\underset{\cdot\cdot}{\text{Cl}}}\text{:} \longleftrightarrow \text{H}^+ \ \ \text{:}\ddot{\underset{\cdot\cdot}{\text{Cl}}}\text{:}^-$$

The ionic structure contributes more to the resonance hybrid as the electronegativity difference between the two atoms increases. As that difference increases, the partial charges increase, so the dipole moment increases too.

*The 4.8 in the definition of the debye arises from an earlier system of units used for electric charge. The SI unit of dipole moment is 1 C·m (1 coulomb-meter). It is the dipole moment of a charge of 1 C separated from a charge of −1 C by a distance of 1 m. The relation between debyes and coulomb-meters is 1 D = 3.336×10^{-30} C·m.

An *approximate* relation between the electric dipole moment of a molecule AB and the electronegativities χ of the two atoms is

$$\mu\ (\mathrm{D}) = \chi_A - \chi_B$$

with the more electronegative atom forming the negative end of the dipole (because that atom tends to attract the electron pair). For example, because the electronegativities of H and Cl are 2.1 and 3.0, we predict that

$$\mu = (3.0 - 2.1)\ \mathrm{D} = 0.9\ \mathrm{D}$$

in moderate agreement with the experimental value of 1.08 D. We predict that a C—H bond will be much less polar (only about 0.4 D), with the positive end of the dipole located on the H atom. This value is so low that in most cases C—H bonds can be treated as almost **nonpolar**, that is, as a covalent bond between atoms with zero net partial charges, and hence with zero electric dipole moment.

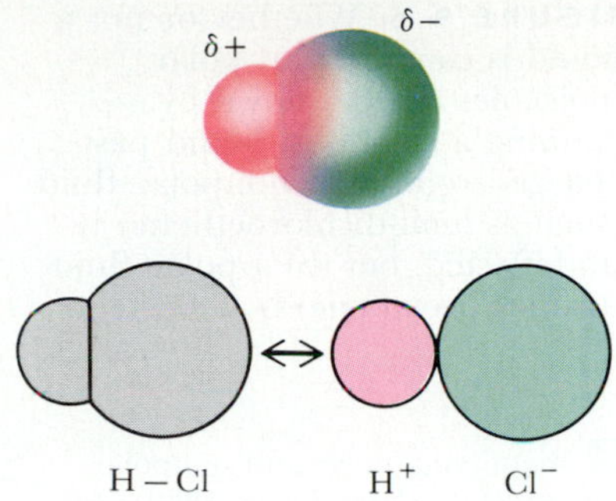

FIGURE 9.5 The HCl molecule has an electric dipole arising from the partial positive charge, δ^+ (red), on the H atom and the partial negative charge, δ^- (green), on the Cl atom. We can think of the charges as arising from resonance, in which a small proportion of ionic structure blends into the pure covalent structure.

Even a homonuclear diatomic molecule (one consisting of two identical atoms, like Cl_2) can be regarded as a resonance hybrid of covalent and ionic structures:

$$:\ddot{\mathrm{Cl}}^+\ \ :\ddot{\mathrm{Cl}}:^- \longleftrightarrow :\ddot{\mathrm{Cl}}\text{—}\ddot{\mathrm{Cl}}: \longleftrightarrow :\ddot{\mathrm{Cl}}:^-\ \ :\ddot{\mathrm{Cl}}^+$$

However, because the Cl^+Cl^- and Cl^-Cl^+ forms contribute equally (because they have the same energy), there is no net electric dipole moment and the bond is nonpolar. We can think of the Cl_2 molecule as held together by a bond that is largely covalent but in which fluctuations in the positions of the bonding pair occur and result in them sometimes being entirely on one Cl atom (giving Cl^+Cl^-) and for an equal length of time being entirely on the other atom (giving Cl^-Cl^+).

E In terms of wavefunctions, ionic-covalent resonance corresponds to the existence of a wavefunction of the form $\psi = \psi_{\text{covalent}} + \lambda\psi_{\text{ionic}}$. An ionic bond has $\lambda \approx 1$; a covalent bond has $\lambda \approx 0$.

The extreme case of a polar bond theoretically occurs when one atom is so electronegative that it gains complete control of the bonding electron pair and the other atom is so electropositive that it gives up any control of the bonding electron pair. Then only the ionic structure would contribute to the resonance, and the bond would be purely ionic, with the electron pair entirely on the more electronegative atom. However, this limit is never completely reached, so there is no such thing as a *purely* ionic bond. Even in CsF, where the electronegativity difference of 3.3 is the largest occurring between any pair of elements, the fluorine atom wins only a 95% share of the electron pair.

Polar molecules. The concept of polar bonds can be generalized to include polar molecules:

> A **polar molecule** is a molecule with a nonzero electric dipole moment. Conversely, a **nonpolar molecule** has a zero electric dipole moment.

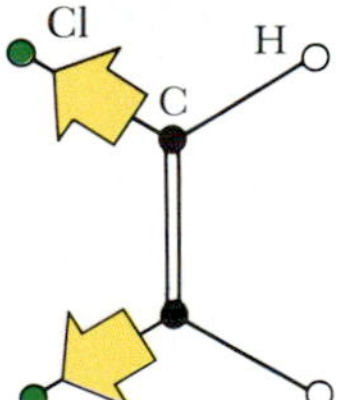

7 *cis*-Dichloroethene

It is very important to distinguish between a polar *molecule* and a polar *bond*. An individual bond is polar if the electrons are shared unequally between the two atoms in the bond. However, although each bond in a polyatomic molecule may be polar, the molecule as a whole may be nonpolar if its shape is such that the dipoles of the individual bonds point in opposite directions and cancel each other. An illustration of the importance of shape is the difference in polarity between *cis*-dichloroethene (**7**) and *trans*-dichloroethene (**8**): the bonds are the same in

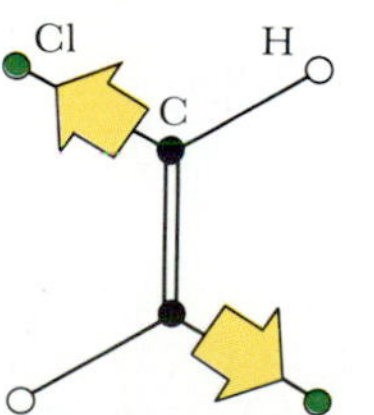

8 *trans*-Dichloroethene

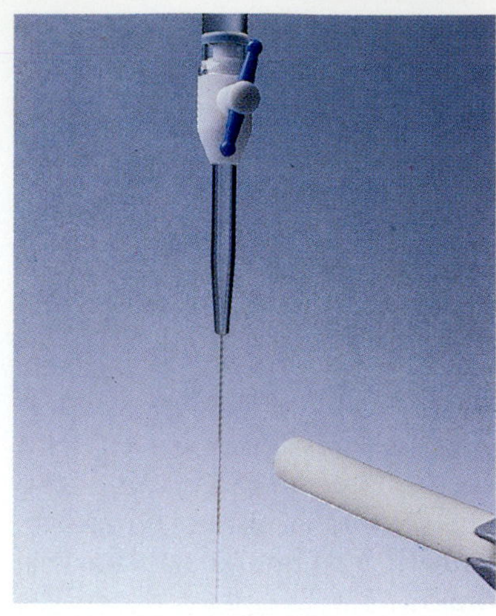

FIGURE 9.6 Whether or not a liquid is composed of polar molecules can be shown by running a stream of liquid past a charged rod. (a) A nonpolar fluid (such as *trans*-dichloroethene) is undeflected, but (b) a polar fluid (*cis*-dichloroethene) is deflected.

E Identifying polar and nonpolar molecules. *J. Chem. Educ.*, **1989**, *66*, 1007. Heating in microwave ovens. *J. Chem. Educ.*, **1983**, *60*, 1043.

D Bending a stream of water. CD2, 48.

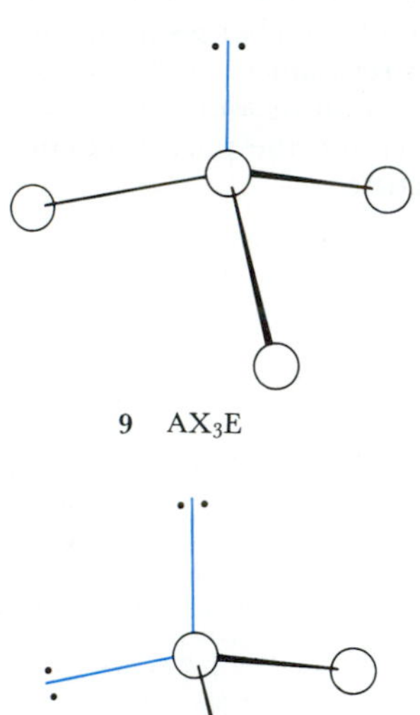

9 AX_3E

10 AX_2E_2

E The symmetries of molecules provide the best criteria for the possession of a nonzero dipole moment. A molecule is nonpolar if it has a rotation axis perpendicular to a mirror plane. If a polar molecule has a rotation axis, then the dipole moment must be parallel to the axis of symmetry.

T OHT: Table 9.6.

both molecules, but in the latter the dipoles (represented by the arrows) cancel each other and the compound is nonpolar. A simple practical test for polarity of molecules in a liquid is whether a stream of the liquid is deflected by an electrically charged rod (Fig. 9.6).

Tetrahedral AB_4 molecules with the same atom at each corner (like CCl_4) are nonpolar because the dipoles of the four bonds cancel. If one atom is replaced by a different atom (for example, $CHCl_3$) or by a lone pair (for example, NH_3 and PH_3, or—in the notation used in Chapter 8—any other AX_3E molecule, (**9**), then the cancellation does not occur and the molecule is polar. Similarly, we can think of an AX_2E_2 (**10**) molecule (such as H_2O and O_3) as being derived from a nonpolar tetrahedral molecule by the replacement of two atoms by lone pairs. The bond dipoles no longer cancel, and the molecule is polar. The H_2O molecule is thus polar; and because oxygen is more electronegative than hydrogen, the O atom is the negative end of the dipole.

Whether or not a molecule with polar bonds is polar depends on the symmetry of the arrangement of its individual bonds: highly symmetrical arrangements (linear, trigonal planar, and tetrahedral among them) are nonpolar when all the atoms X joined to the central atom A are identical. Table 9.6 shows which molecular shapes have canceling bond dipoles. Whether a given molecule is polar can often be decided by seeing whether it has one of these formulas or can be derived from one of them by loss or replacement of an atom.

▼ EXAMPLE 9.4 Predicting whether a molecule is polar or nonpolar

Predict whether (a) a boron trifluoride molecule and (b) an ozone molecule are polar or nonpolar.

STRATEGY In each case we must identify (from VSEPR theory) the class of the molecule and decide whether it has any lone pairs on the central atom. Then, if the molecule has one of the nonpolar shapes in Table 9.6 and the atoms joined to the central atom are identical, it is nonpolar. If the shape is not shown in the table, then we must decide whether it can be obtained from one in the table by removal or replacement of an atom. If it can, and if the symmetry of the molecule is upset so that the bond dipoles do not cancel, then the molecule is polar.

TABLE 9.6 Nonpolar and polar molecules

VSEPR class	Nonpolar shape	Polar shape
AX_2	 CO_2	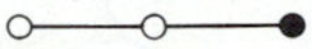 N_2O
AX_3	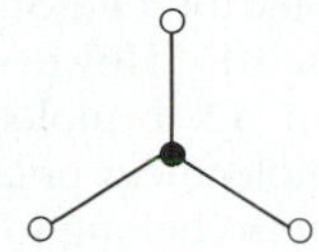 BF_3	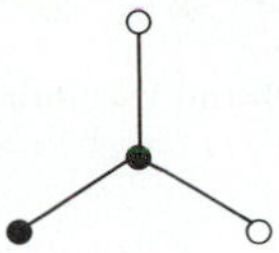 $COCl_2$
AX_2E	none	 SO_2
AX_4	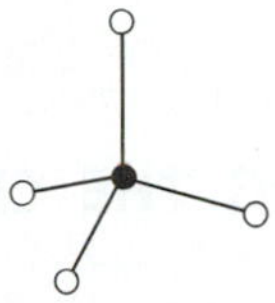 CH_4	 CH_3Cl
AX_3E	none	 NH_3, PCl_3
AX_2E_2	none	 H_2O
AX_5	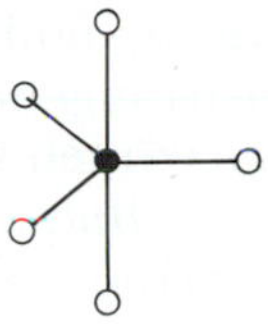 PCl_5	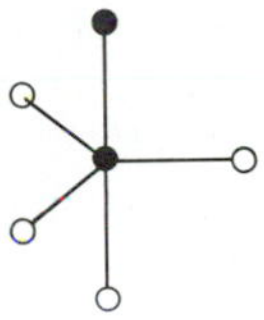 PCl_4F
AX_4E	none	 SF_4
AX_6	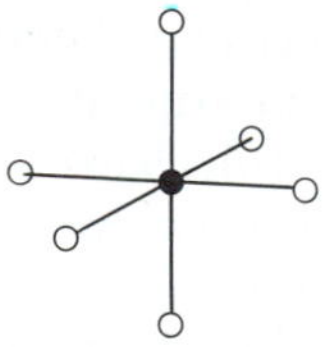 SF_6	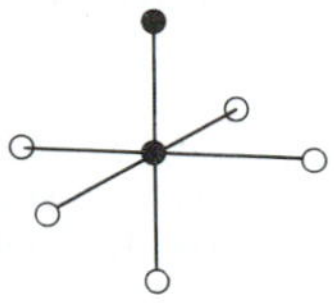
AX_5E	none	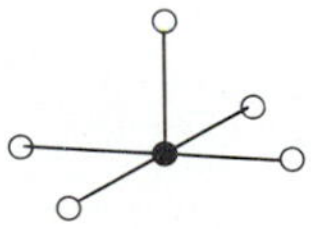 IF_5
AX_4E_2	 XeF_4	

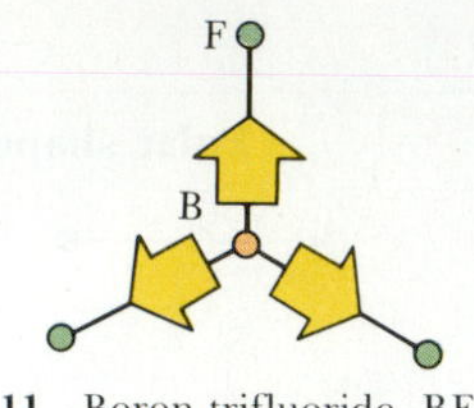

11 Boron trifluoride, BF_3

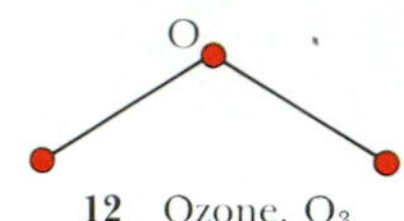

12 Ozone, O_3

SOLUTION (a) According to VSEPR theory, BF_3 is a trigonal planar AX_3 molecule. This shape is one of the nonpolar shapes shown in Table 9.6. Because it has the same atom at each vertex, we predict that it is nonpolar (**11**). (b) According to VSEPR theory, the O_3 molecule is angular, with one lone pair on the central O atom (**12**). If all three positions around that O atom were occupied by O atoms, then the molecule would be a nonpolar AX_3 molecule, like BF_3. However, because one position is occupied by a lone pair, making it an AX_2E molecule, the dipole cancellation is upset; so we predict that the molecule is polar. This example shows that even a molecule in which all the atoms belong to the same element may be polar: in this case, the central oxygen atom is not identical to the other two.

EXERCISE E9.4 Predict whether (a) SF_6 and (b) SF_4 are polar or nonpolar.

[*Answer*: (a) nonpolar; (b) polar]

▲

U Exercises 9.68, 9.70, 9.71, 9.73, and 9.74.

9.4 ASSESSING THE CHARGE DISTRIBUTION

In a pure ionic bond, electrons are transferred completely from one atom to another. In a pure covalent bond, each atom shares electrons equally with a neighbor. In a polar bond, there is unequal sharing of electrons. How can we determine the degree to which each atom controls the electrons in the molecule? We know that when two atoms have similar electronegativities, as do those that bond covalently, there is unlikely to be a significant lowering of energy when an electron transfers from one atom to another. This observation suggests that if we keep track of the rearrangement of electrons that occurs when various Lewis structures are written for molecules built from atoms of similar electronegativities, *the structures involving the least transfer of electrons between atoms are likely to correspond to the lowest energy*. Such structures will correspond to the most accurate description of the molecule.

E The concept of formal charge can be helpful in a number of instances. For example, it is consistent with the choice of the lowest energy Lewis structure for SO_4^{2-} in which there are two double bonds (see Example 9.5); it is consistent with the rejection of a completed octet structure for BF_3 in favor of an incomplete octet representation; it is consistent with the existence of a dipole on CO in which the positive partial charge resides on the *oxygen* atom. It is also useful for distinguishing between different atomic arrangements (see p. 334).

Formal charge. A simple way of assigning ownership of electrons to the atoms in a molecule is to assume that each atom has an equal share in a bonding electron pair but "owns" its lone pairs completely (Fig. 9.7). This approach corresponds to a "perfectly covalent" model of the molecule in which bonding electron pairs are equally shared. If this assignment results in an atom having more electrons in the molecule than it has when it is a free, neutral atom, then we say that the atom has a negative **formal charge** in the Lewis structure (like an anion). If the assignment of electrons leaves the atom with fewer electrons than it has when it is neutral, then we say that the atom has a positive formal charge (like a cation).

The formal charge on an atom in a specific Lewis structure can be evaluated very simply by drawing a line round all the electrons it owns, including all its lone pairs and one electron from each pair that it shares. Then the difference between this number and the number of valence electrons in the free atom is equal to the atom's formal charge.*

*The same result is obtained from the formula

$$\text{Formal charge} = V - (L + \tfrac{1}{2}S)$$

where V is the number of valence electrons in the free atom, L the number of lone pair electrons, and S is the number of shared electrons.

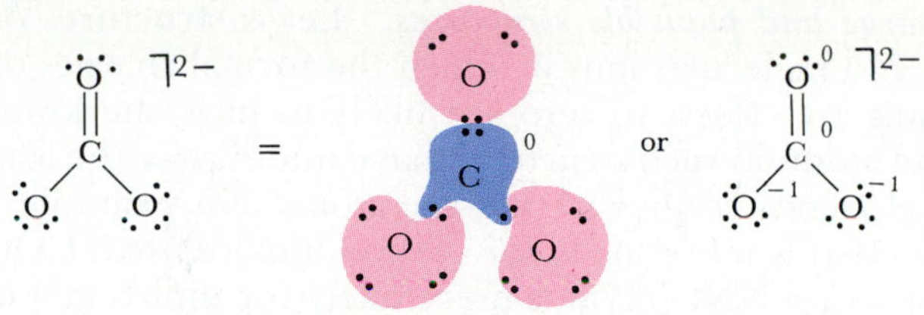

FIGURE 9.7 The formal charge on an atom in a particular Lewis structure is calculated by supposing that each atom owns the lone pairs on it and one-half the electrons that are shared with other atoms. The lines show the ownership of the electrons in one of the Lewis structures of the CO_3^{2-} ion.

▼ EXAMPLE 9.5 Calculating the formal charges on a structure

Calculate the formal charges on the atoms in the three Lewis structures of a sulfate ion shown in (**13**).

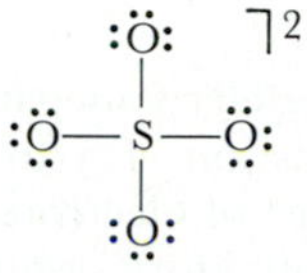

13a 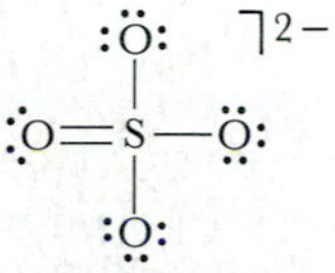 **13b** 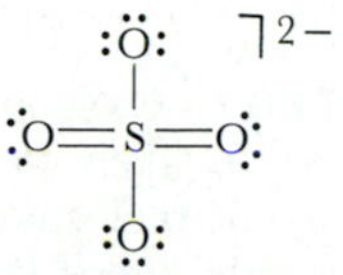 **13c**

STRATEGY Each individual atom has its own formal charge. To find it, we draw the Lewis structures showing each electron pair as dots (not as a bar), then draw a line that captures the electrons that the atom owns, as specified above, and finally subtract this number from the number of valence electrons in the free atom. All equivalent oxygen atoms have the same formal charge, so we need show only one charge in each case. Both O and S belong to Group VI, so the free atoms each have six valence electrons.

SOLUTION The enclosures that capture the electrons are shown in (**14**).

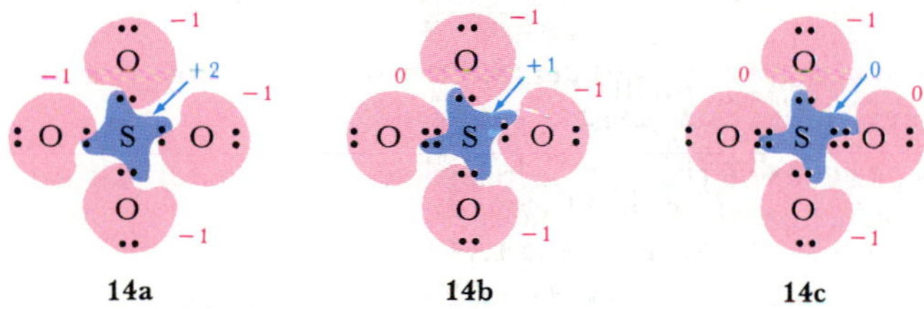

To calculate the formal charges shown in (**14**), draw up the following table:

	Number of electrons		
Lewis structure	Owned by bonded atom (A)	On free atom (B)	Formal charge (B − A)
S in **14a**	4	6	+2
S in **14b**	5	6	+1
S in **14c**	6	6	0
—Ö:	7	6	−1
=O:	6	6	0

U Exercises 9.72 and 9.75.

EXERCISE E9.5 Calculate the formal charges for the two Lewis structures of the phosphate ion shown in (**15**).

[*Answer*: the formal charges are shown on the structures] ▲

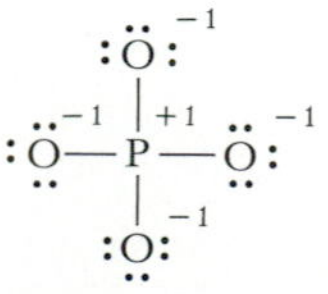

15a 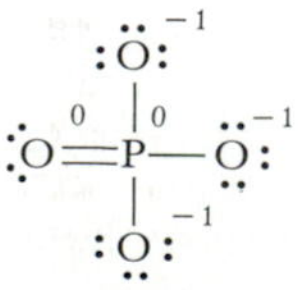 **15b**

$\overset{0}{:\ddot{O}}=\overset{0}{C}=\overset{0}{\ddot{O}:}$

16a

$\overset{0}{:\ddot{O}}=\overset{+2}{O}=\overset{-2}{\ddot{C}:}$

16b

$\overset{-1}{:N}=\overset{+1}{N}=\overset{0}{\ddot{O}:}$

17a

$\overset{-1}{:N}=\overset{+2}{O}=\overset{-1}{N:}$

17b

Formal charge and plausible structures. Lewis structures of covalent molecules and molecular ions in which the formal charges of the individual atoms are closest to zero are likely to have the lowest energy, because the atoms in such structures have undergone the least redistribution of electrons relative to the free atoms. For example, the structure OCO (**16a**) is more likely for carbon dioxide than COO (**16b**) is, and the structure NNO (**17a**) is more likely for dinitrogen oxide than NON (**17b**) is.

▼ **EXAMPLE 9.6 Judging the plausibility of a structure**

Write three Lewis structures with different atomic arrangements for the cyanate ion, NCO^-, and suggest which is likely to be the most plausible structure.

STRATEGY Oxygen and nitrogen are both more electronegative than carbon, so C is likely to be the central atom of this ion. Try to think of an isoelectronic molecule (one with the same number of electrons), because it may provide a good clue to the structure of this ion. More systematically, we should calculate the formal charges on the three possible arrangements of atoms and select the one with formal charges closest to zero.

🅴 Sodium cyanate is used in the treatment of sickle-cell anemia.

SOLUTION The NCO^- ion has the same number of atoms and valence electrons (**16**) as a CO_2 molecule, which we know to be OCO. This similarity suggests that the atomic arrangement in the ion is NCO. The formal charges of the Lewis structures corresponding to the three possible arrangements are

$$\overset{-1}{\ddot{N}}=\overset{0}{C}=\overset{0}{\ddot{O}} \qquad \overset{-2}{\ddot{C}}=\overset{+1}{N}=\overset{0}{\ddot{O}} \qquad \overset{-2}{\ddot{C}}=\overset{+2}{O}=\overset{-1}{\ddot{N}}$$

The individual formal charges are closest to zero in the first. That structure is therefore the likely one, as anticipated.

EXERCISE E9.6 Suggest a plausible structure for ClF_3, and write its Lewis structure and formal charges.

[*Answer*: $:\ddot{F}-\ddot{Cl}^{0}-\ddot{F}:$ with a third $:\ddot{F}:^{0}$ bonded above Cl]

▲

🅽 Formal charge is a charge assessment based on an exaggerated covalent model; oxidation number is a charge assessment based on an exaggerated ionic model of the species.

Oxidation numbers. We can now see the origin of the rules for calculating the oxidation number (N_{ox}) of an element in a compound, which were introduced in Section 3.8. As we saw there, the oxidation number is an *effective* ionic charge obtained by exaggerating the drift of electrons in a covalent bond and supposing that the transfer is complete. In contrast to the formal charge, which supposes that the bonding is purely covalent, the oxidation number exaggerates shifts of electrons stemming from differences in electronegativity and treats the compound as purely ionic.

In HCl, for example, the oxidation numbers of the two elements are calculated by treating the more electronegative chlorine atom as owning *both* electrons of the bond:

$$H:\ddot{Cl}: \text{ exaggerated to } H^+ \; :\ddot{Cl}:^-$$

In the gaseous interhalogen compound chlorine monofluoride (ClF), the bonding pair is assigned to the F atom because that atom is more electronegative than the Cl atom:

$$:\ddot{\underset{..}{Cl}}:\ddot{\underset{..}{F}}: \quad \text{exaggerated to} \quad :\ddot{\underset{..}{Cl}}^{+} \quad :\ddot{\underset{..}{F}}:^{-}$$

U Exercise 9.77.

The oxidation number of fluorine is -1 in this compound and that of chlorine is $+1$.

When there are several atoms in a molecule, tracing the ownership of electrons on the basis of the electronegativities of the atoms can be quite confusing. It turns out to be much simpler to use the rules set out in Table 3.5, which have been designed to lead to the same result as would be obtained by exaggerating the drift of shared electron pairs to the more electronegative atoms in a compound.

THE VALENCE-BOND MODEL OF BONDING

The Lewis theory of the chemical bond and its extension, VSEPR theory, assume that each bonding electron pair is located between the two bonded atoms. However, we know from Chapter 7 that electrons are not located at precise positions but are spread over a region defined by the atomic orbitals they occupy. Our task in this section is to see how to combine the Lewis approach with the description of a molecule in terms of orbitals. This extension, which is called valence-bond theory, also introduces concepts and a terminology that are used throughout chemistry. In brief, **valence-bond theory** is the description of bond formation in terms of the merging of atomic orbitals in the valence shells of neighboring atoms and of the pairing of the spins of the electrons that occupy these orbitals.

N The coverage of valence-bond theory could be delayed until Chapter 23 (on the hydrocarbons).

E Valence-bond (VB) theory originated in the work of Heitler, London, Slater, and Pauling in the 1930s. It is motivated by the Lewis concept of a shared electron pair, and conforms to the chemist's concept of a localized electron-pair bond.

E Principles of chemical bonding. *J. Chem. Educ.*, **1961**, *38*, 382. The chemical bond. *J. Chem. Educ.*, **1987**, *64*, 934.

9.5 BONDING IN DIATOMIC MOLECULES

The simplest molecule of all is the hydrogen molecule, H_2. We shall consider the valence-bond description of the bonding in this molecule and then see how the same ideas can be extended to more complex molecules.

***σ* bonds.** The structure of a single hydrogen atom was explained in Chapter 7. It consists of a single electron in a spherical 1*s* orbital that surrounds the nucleus. When two H atoms come together to form H_2, their 1*s* electrons pair and occupy a single orbital that spreads over both atoms (Fig. 9.8). The resulting sausage-shaped distribution of electrons is called a σ bond.

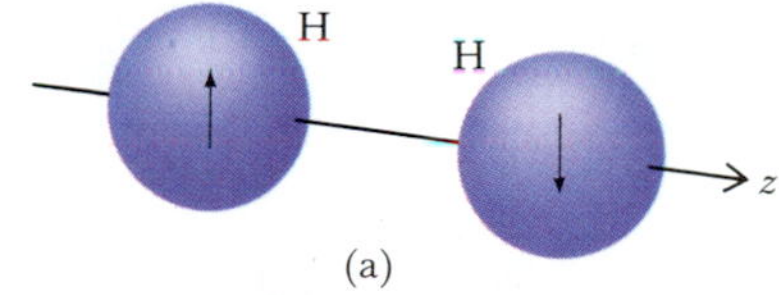

(a)

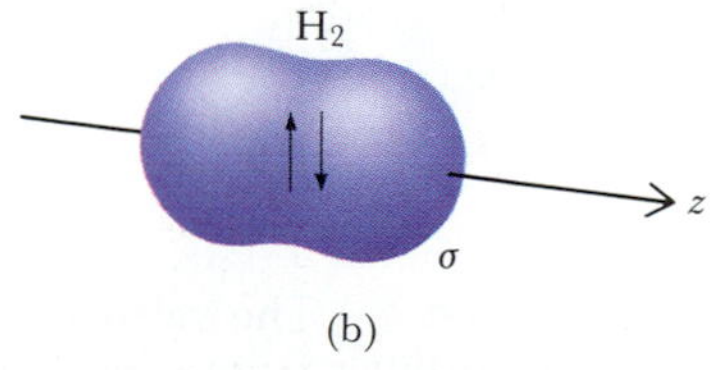

(b)

FIGURE 9.8 When electrons (depicted as ↑ and ↓) in two hydrogen 1*s* orbitals pair and the orbitals overlap, they form a σ bond, which is depicted as an electron cloud with cylindrical symmetry along the internuclear axis that spreads over both nuclei.

> A ***σ* bond** consists of two electrons in an orbital that has cylindrical symmetry about the internuclear axis.

(The Greek letter *sigma*, σ, is chosen because, looking along the internuclear axis, the electron distribution resembles that of an *s* orbital.) The hydrogen molecule is held together by one such σ bond. The merging of the two 1*s* orbitals that gives rise to the single orbital occupied by the paired electrons is called the **overlap** of atomic orbitals.

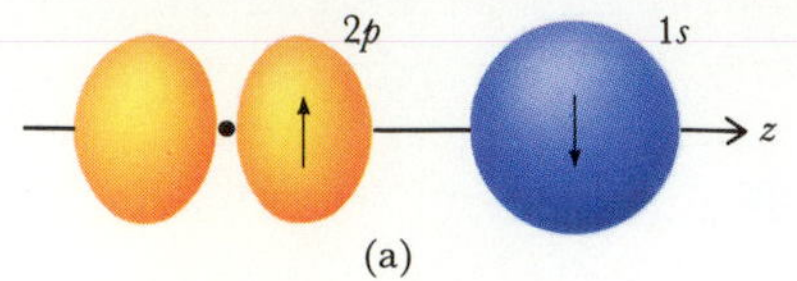

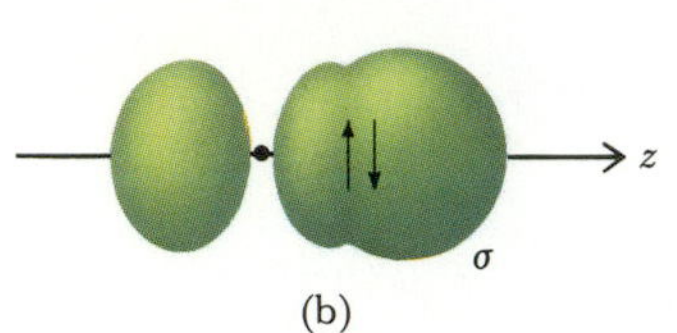

FIGURE 9.9 (a) A σ bond is formed when electrons in $2p$ and $1s$ orbitals pair. (b) The orbitals overlap to give an electron cloud of cylindrical symmetry. (Although the paired electrons, ↑↓, are shown inside one lobe of the bond, they actually are spread over both green regions.)

T OHT: Figs. 9.8 and 9.10.

U Exercise 9.78.

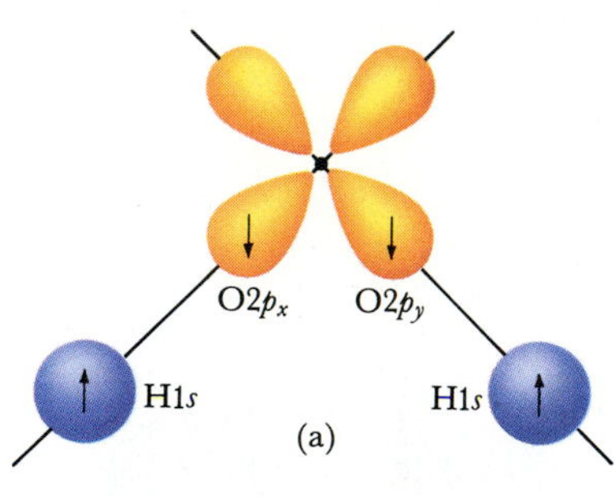

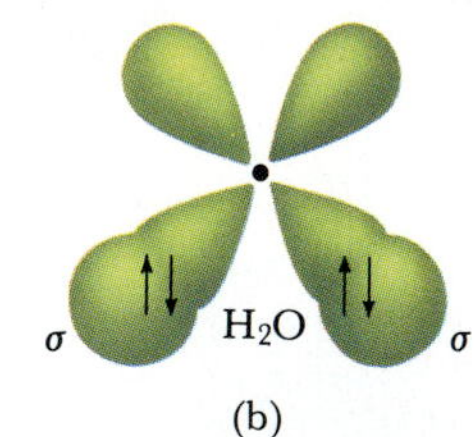

FIGURE 9.10 (a) The valence-shell atomic orbitals from which the valence-bond description of the H_2O molecule is formed. (b) The two σ(O—H) bonds formed when the electrons pair and the orbitals overlap. Note that this model implies a bond angle of 90° in H_2O, whereas the experimental value is 104.5°.

Much the same kind of σ bonding occurs in the hydrogen halides. First, we identify the electrons that will pair to form a bond and note the atomic orbitals they occupy. The Lewis diagrams and electron configurations of the atoms are

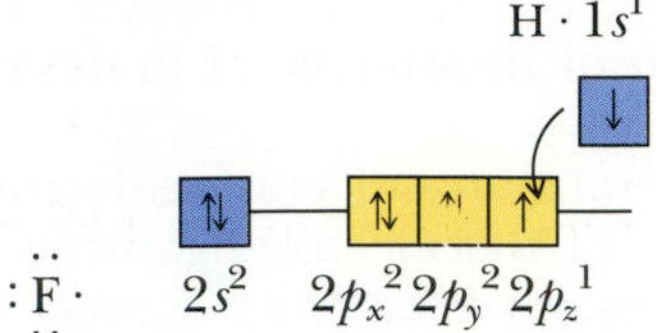

The unpaired electron on the fluorine atom occupies a $2p_z$ orbital, and the unpaired electron on the hydrogen atom occupies a $1s$ orbital. We identify these two electrons as the ones that will pair to form a bond and then think of the bond as formed from the overlap of the $2p_z$ and $1s$ atomic orbitals (Fig. 9.9). Although the resulting bond has a more complicated shape than the σ bond in H_2 when viewed from the side, it looks like one when viewed along the internuclear axis; hence, it is designated a σ bond.

▼ EXAMPLE 9.7 Identifying the structure of the bonds in a molecule

Suggest how σ bonds might be formed in the water molecule, and indicate whether you expect the molecule to be linear or angular.

STRATEGY We know that σ bonds may form by the pairing of electrons in orbitals formed by the overlap of s and p atomic orbitals, so we should inspect the electron configurations of the H and O atoms and see which atomic orbitals contain one electron. Then we imagine these electrons as pairing and forming a bond as the corresponding atomic orbitals overlap. As to the shape, we know from VSEPR theory that the H_2O molecule is angular, so we have a clue to the answer. We should examine the p orbitals on the oxygen atom that we use to form the σ bonds and see whether they indicate an angular shape too.

SOLUTION The electron configurations of the atoms are

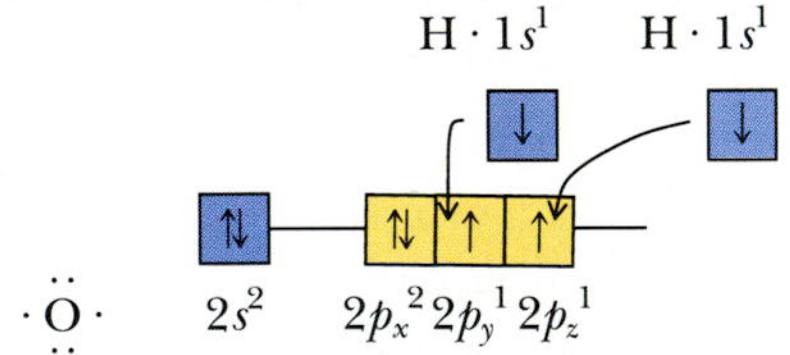

The configurations suggest that we should pair the hydrogen atom electrons with each of the two unpaired $2p$ electrons of the O atom. The two half-full $2p$ orbitals can each *separately* overlap with a hydrogen $1s$ orbital to form a σ bond, as shown in Fig. 9.10. Because the p orbitals that are used point at 90° to each other, we anticipate that the H_2O molecule will be angular, as depicted in the illustration. However, we have more work to do, because although we have predicted that the bond angle will be 90°, it is actually observed to be 104.5°.

EXERCISE E9.7 Identify the structure of the bonds in the ammonia molecule, and suggest a shape for the molecule.

[*Answer*: three σ bonds formed from H1s,N2p-orbital overlap; trigonal pyramidal] ▲

All *single* covalent bonds consist of a σ bond in which two electrons spread over the two bonded atoms. A σ bond can be formed by the pairing of electrons in two *s* orbitals (as in H_2), an *s* orbital and a *p* orbital (as in the hydrogen halides), and two *p* orbitals. We shall see an example of this last possibility in a moment. Valence-bond theory gets its name from this emphasis on the use of electrons from the valence shell of atoms to form bonds.

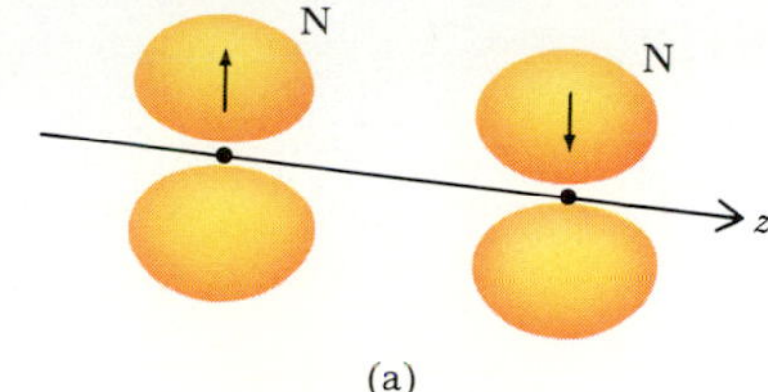

(a)

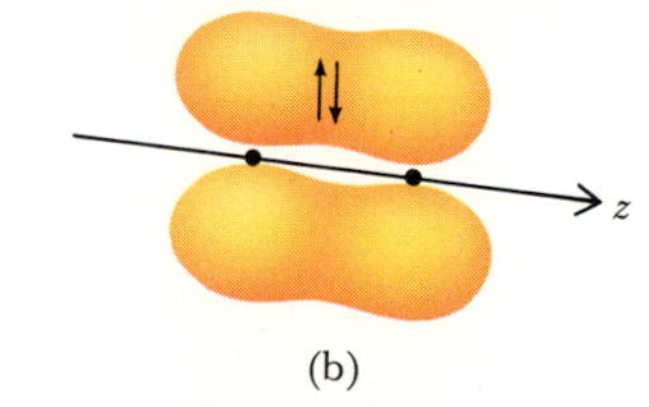

(b)

FIGURE 9.11 (a) When there are unpaired electrons in two neighboring *p* orbitals that are aligned side-by-side as shown here, they may pair and the orbitals overlap to give a double-lobed π bond. (b) We shall always call the internuclear axis *z*, so π bonds form from a side-by-side overlap of p_x or p_y orbitals on neighboring atoms.

π bonds. We meet a different type of bond when we consider the structure of a nitrogen molecule. Suppose we follow the same procedure as above. We write the electron configurations as

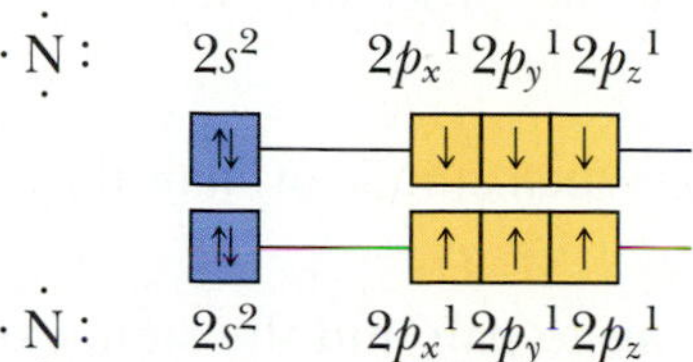

We appear to be on the right track, because there is an electron in each of the three $2p$ orbitals on each atom, an arrangement suggesting that we should be able to pair them and form three bonds. However, we now meet a new concept: although one $2p_z$ electron on each nitrogen atom can pair with a $2p_z$ electron on the other atom to form a σ bond, the other two $2p$ orbitals ($2p_x$ and $2p_y$) cannot overlap end-to-end to form σ bonds (Fig. 9.11) because they are perpendicular to the internuclear axis. The remaining *p* electrons do pair, but their orbitals overlap in a side-by-side arrangement and form π bonds.

> A **π bond** consists of two electrons that occupy an orbital that has two lobes, one each side of the internuclear axis.

A π bond is so called because the Greek letter *pi,* π, is the equivalent of our letter *p*, and when we imagine looking along the internuclear axis, a π bond resembles a pair of electrons in a *p* orbital. The fact that a π bond has electron clouds on each side of the internuclear axis does not mean that it consists of two distinct bonds: a π bond is *one* bond with *two* lobes.

N Note that at this stage we are being careful to use the terms σ and π *bonds*, not orbitals. The terms σ and π *orbitals* will emerge once we turn to molecular orbital theory.

T OHT: Figs. 9.11 and 9.12.

N In all hydrocarbons, carbon atoms use only their *s* and *p* orbitals for bond formation.

Now we can describe the structure of the nitrogen molecule. It consists of a σ bond (of two electrons) formed from the end-to-end overlap of two *p* atomic orbitals directed along the internuclear axis, and two π bonds (of two electrons each) formed from the side-by-side overlap of the remaining *p* atomic orbitals on the two atoms. This approach leads to the conclusion that there are three bonds in all (one σ and two π; Fig. 9.12), in agreement with the Lewis structure :N≡N:.

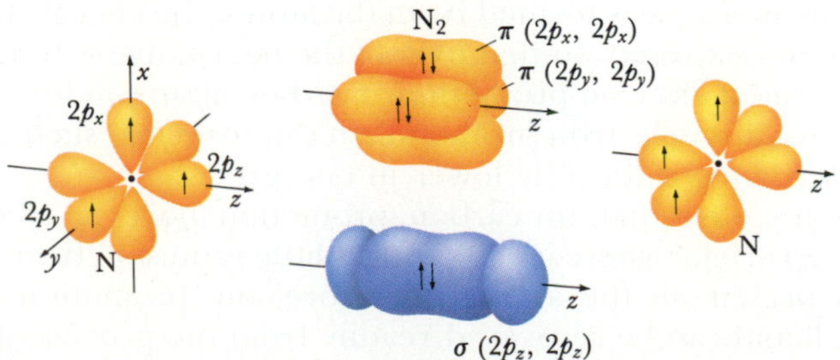

FIGURE 9.12 The structure of the N_2 molecule in terms of electron pairing and orbital overlap. The two atoms are bonded together by one σ bond and two π bonds.

We can generalize from the conclusion we have just drawn concerning the role of π bonds in the description of a multiply bonded species:

> A **single bond** is one shared electron pair that forms a σ bond.
>
> A **double bond** is two shared electron pairs that form one σ bond and one π bond.
>
> A **triple bond** is three shared electron pairs that form one σ bond and two π bonds.

In short, the valence-bond procedure for describing bonding can be expressed as follows:

1. Identify the valence-shell atomic orbitals that contain unpaired electrons.
2. Allow these electrons to pair and the atomic orbitals they occupy to overlap to form σ bonds (by end-to-end overlap) or π bonds (by side-by-side overlap).

9.6 HYBRIDIZATION

E Hybrid orbitals in molecular orbital theory. *J. Chem. Educ.*, **1969**, *46*, 487.

When we deal with polyatomic molecules, we need to make use of a new concept. The problem is readily identified by considering the bonding in methane, CH_4. We start by writing down the electron configuration of the carbon atom and looking for the half-filled orbitals that we can use for bond formation (remember that, in these diagrams, energy increases from left to right, except for boxes that are touching each other):

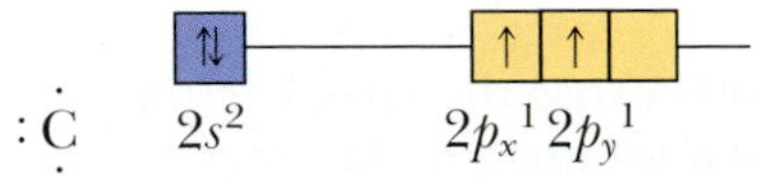

[⇅] —— [↑][↑][ ]

:C̈ $2s^2$ $2p_x^{\,1}2p_y^{\,1}$

However, it looks as though we cannot form a molecule with four σ bonds because there are only two half-filled $2p$ orbitals on the carbon atom.

N Whenever we consider the origin of bonding, we must assess whether the overall outcome of a series of changes is a net lowering of energy. In carbon, not only is the promotion energy small and four bonds are formed in place of the two bonds of the unpromoted atom, but the σ bonds are stronger than in the absence of hybridization because the hybrid orbitals are strongly directional and have a large overlap with the orbitals of the neighboring atom.

E Bond angle relationships in some AX_nY_m molecules. *J. Chem. Educ.*, **1974**, *51*, 571.

A step toward a solution is to imagine that we can invest some energy into the carbon atom to **promote** a $2s$ electron into a higher-energy $2p$ orbital to give the configuration

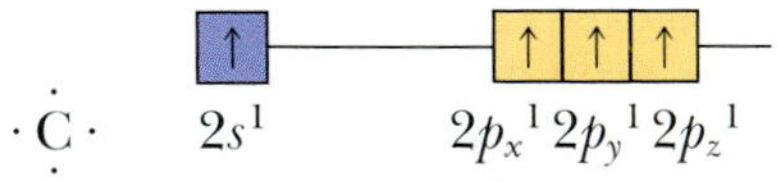

[↑] —— [↑][↑][↑]

·Ċ· $2s^1$ $2p_x^{\,1}2p_y^{\,1}2p_z^{\,1}$

This promotion of an electron from one orbital to another is legitimate if, when the molecule is formed from the atoms, the energy investment is more than recovered, so that overall the net change is toward lower energy. Without electron promotion, a carbon atom can form only two C—H bonds; after electron promotion, it can form *four* such bonds and thereby become significantly lower in energy.

A point to note is that, for carbon, promotion moves an electron into an empty $2p$ orbital where it experiences little repulsion from the other electrons present in the atom. Therefore, the promotion energy is quite small and can be recovered readily from the process of forming four bonds. One consequence is the enormous number of compounds in organic chemistry in which carbon has four bonds.

However, our troubles are not yet over. Even after promotion of the 2*s* electron, it appears that when we pair the electrons in the hydrogen atoms with those of the carbon atom, we form two *different* types of σ bond: one from the overlap of an H1*s* orbital with a C2*s* orbital, and three other σ bonds from the overlap of H1*s* orbitals with C2*p* orbitals at 90° to one another. However, experimentally, the four C—H bonds are equivalent and at 109.5° to one another.

FIGURE 9.13 One *s* orbital and three *p* orbitals blend into four sp^3 hybrid orbitals pointing toward the corners of a tetrahedron.

***s*- and *p*-orbital hybridization.** The way out of this difficulty lies in the fact that *s* and *p* orbitals are like waves centered on the nucleus. Like waves in water, the four orbitals produce patterns where they mix. The result of this mixing is the formation of four identical **hybrid orbitals**, or mixed orbitals, that point toward the corners of a tetrahedron (Fig. 9.13). The four orbitals shown there are called sp^3 **hybrids** because they are formed from one *s* orbital and three *p* orbitals. In an orbital-energy diagram we can represent the hybridization as the formation of four orbitals that have equal energy and are intermediate in energy between those of the *s* and *p* orbitals from which they are constructed:

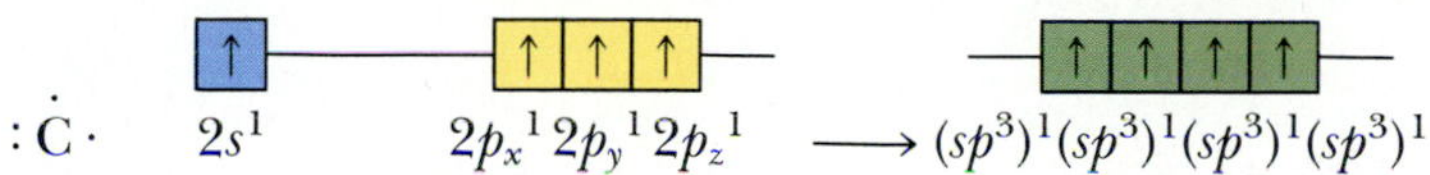

A C atom with a promoted electron is *equivalent* to an atom in which one electron occupies each of four sp^3 hybrid orbitals directed toward the corners of a tetrahedron. Put another way, the overall electron cloud of the atom with the promoted electron is the same as the electron cloud of an atom in which the electrons occupy the four hybrid orbitals.

It is now simple to account for the bonding in the methane molecule. The electron in each sp^3 hybrid orbital can pair with an electron in a hydrogen 1*s* orbital and form a σ bond:

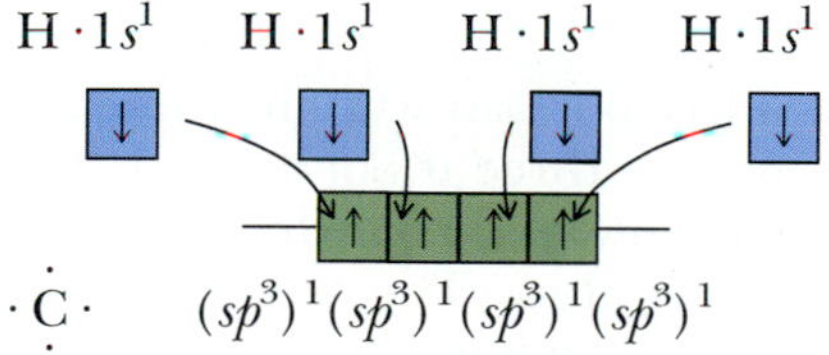

Because the four hybrid orbitals point toward the corners of a tetrahedron, the σ bonds also point in those directions. Moreover, all four bonds are identical. This description is in accord with VSEPR theory and with experiment.

N We are denoting all hybridized orbitals by the color green, signifying a blend of blue (*s*) and yellow (*p*) character.

E The valence-bond interpretation of molecular geometry. *J. Chem. Educ.*, **1980**, *57*, 107.

▼ EXAMPLE 9.8 Accounting for the structure of a hydrocarbon molecule

Describe the structure of the ethane molecule, C_2H_6, in terms of hybrid orbitals and σ bonds.

STRATEGY We start with Lewis structures to decide on the arrangement of electron pairs around each C atom in the molecule. Next we decide on the

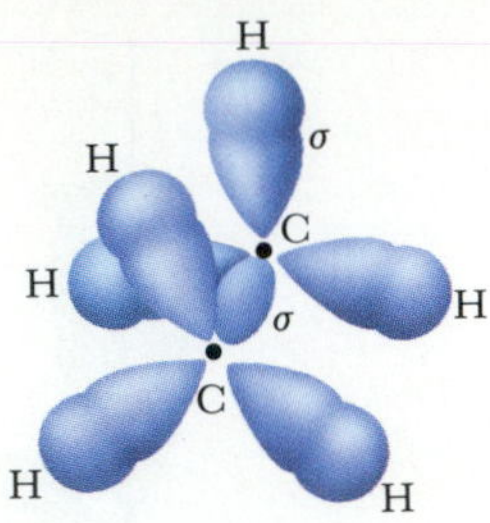

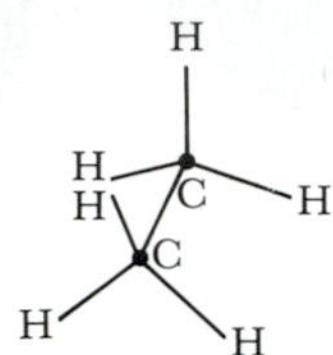

FIGURE 9.14 The valence-bond description of the ethane molecule, C_2H_6. (The bottom diagram shows the bonds as lines.) Each pair of neighboring atoms is linked by a σ bond formed by the pairing of electrons in either H1*s* orbitals or Csp^3 hybrid orbitals, and all bond angles are close to 109.5° (the tetrahedral angle).

U Exercises 9.79–9.81, 9.83, and 9.84.

N Hydridization is more commonly invoked for *s* and *p* orbitals than for *d* orbitals.

E Hybrid orbitals and atomic configuration. *J. Chem. Educ.*, **1971**, *48*, 603.

hybridization of the C atom that reproduces these locations. Then we describe the bonding in the molecule in terms of the formation of σ bonds arising from the pairing of electrons that occupy sp^3 hybrids on each atom.

SOLUTION The Lewis structure of ethane is

```
   H  H
   |  |
H—C—C—H
   |  |
   H  H
```

Each C atom forms four bonds pointing toward the corners of a regular tetrahedron; therefore, its four valence electrons must be available for bonding and occupy a set of four sp^3 hybrid orbitals, each orbital having a single electron. The pairing of two electrons, each in a hybrid orbital on a C atom, produces a σ(C—C) bond between the two C atoms, which we can denote as $\sigma(Csp^3,Csp^3)$. The remaining electrons in the three hybrid orbitals on each C atom can be used to form σ(C—H) bonds by pairing with the six H atoms; we denote these bonds as $\sigma(Csp^3,H1s)$. The structure of the ethane molecule is illustrated in Fig. 9.14.

EXERCISE E9.8 Suggest a bonding structure for the ion CH_3^+.

[*Answer*: three $\sigma(Csp^2,H1s)$ bonds; trigonal planar]

A similar description accounts for the bonding in ammonia, but now we see how to describe the bonding in a molecule with a lone pair. We treat the $2s^22p^3$ ground-state configuration of the nitrogen atom as the equivalent of a configuration in which the five valence electrons occupy four sp^3 hybrid orbitals (one of which is doubly occupied):

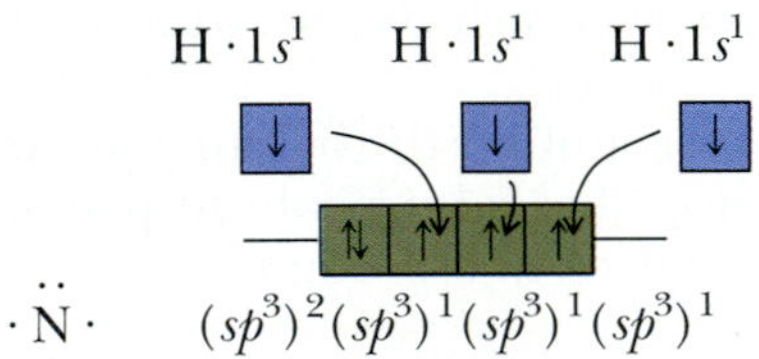

The three hydrogen electrons pair with the three electrons in the three sp^3 hybrid orbitals, to give three σ bonds, and the other two valence electrons remain in a **nonbonding orbital**, an orbital that has not formed a bond to another atom. The two electrons in the nonbonding orbital on the N atom are the lone pair of the NH_3 molecule. In general, in valence-bond theory, all the valence orbitals of an atom are considered to be hybridized, and lone pairs of electrons occupy hybrid orbitals just as the bonding pairs do.

Other hybridization types. Different patterns of hybrid orbitals result from other mixtures of atomic orbitals (Table 9.7). For example, if the electron pairing involves the hybridization of one *s* orbital and two *p* orbitals, the result is three sp^2 **hybrids** that point toward the corners of an equilateral triangle (Fig. 9.15). The three most important types of hybridizations in chemistry are tetrahedral hybridization (sp^3 hybridization, bond angle 109.5°), trigonal planar hybridization (sp^2 hybridization, bond angle 120°), and linear hybridization (sp hybridization, bond angle 180°).

TABLE 9.7 Hybridization and molecular shape

Number of orbitals blended	Pattern of hybrid orbitals	Hybridization type
2	linear	*sp*
3	trigonal planar	sp^2
4	tetrahedral	sp^3
5	trigonal bipyramidal	sp^3d
6	octahedral	sp^3d^2

T OHT: Table 9.7.

E A simple method for generating sets of orthonormal hybrid orbitals. *J. Chem. Educ.*, **1973**, *50*, 114.

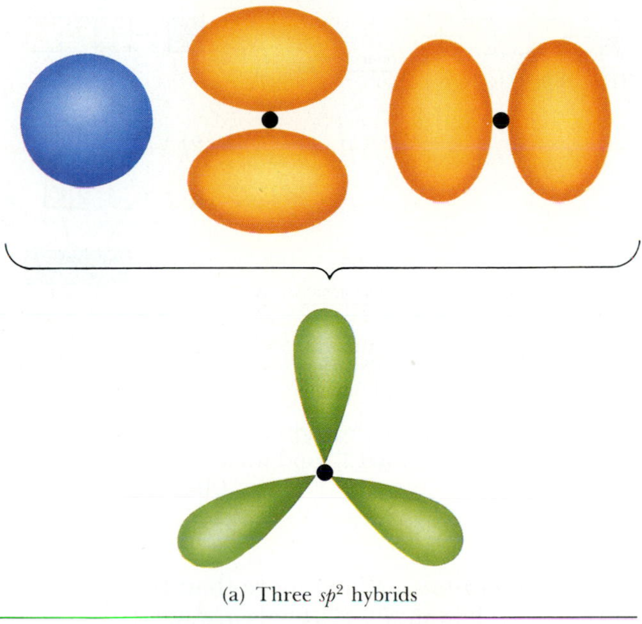

(a) Three sp^2 hybrids

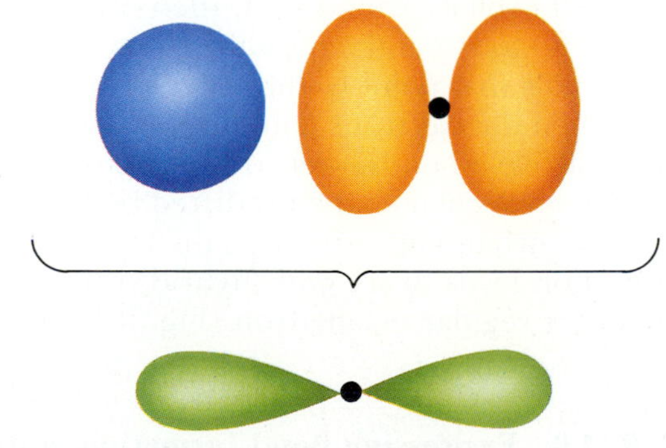

(b) Two *sp* hybrids

FIGURE 9.15 (a) One *s* orbital (blue) and two *p* orbitals (yellow) form three sp^2 hybrid orbitals (green) pointing to the corners of an equilateral triangle. (b) One *s* orbital and one *p* orbital form two linear *sp* hybrid orbitals.

T OHT: Fig. 9.15.

E A more detailed picture of an sp^3 hybrid is given by the following contour diagram:

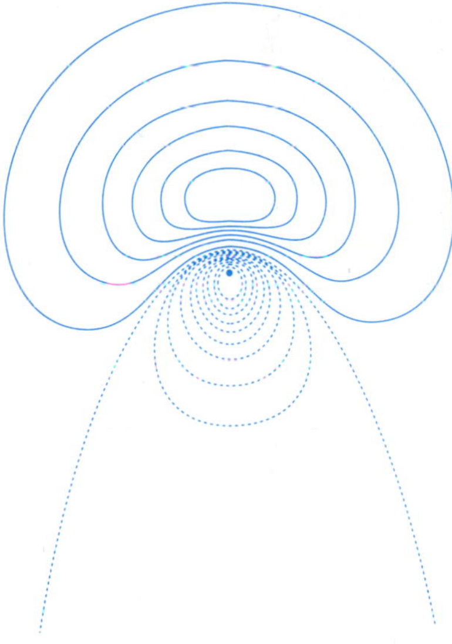

Note that the nucleus no longer lies in a nodal plane, because the hydbrid orbital contains some *s* character.

▼ EXAMPLE 9.9 Selecting a hybridization scheme

What is the hybridization of the B atom in BF_3, in which there are three equivalent bonds?

STRATEGY First, we write the valence electron configuration of the ground-state atom and then consider what promotion scheme, if any, is needed to obtain the stated pattern of bonding. We then determine the hybridization type of the atom that ensures that the bonds are equivalent and identify the pattern of bonds by referring to Table 9.7.

SOLUTION The ground-state valence-shell electron configuration of fluorine is $2s^22p^5$ and of boron it is $2s^22p^1$. Three unpaired electrons can be obtained in boron by promoting one $2s$ electron into a $2p$ orbital; then from one $2s$ orbital and two $2p$ orbitals, we can form three identical bonds by sp^2 hybridization and overlap with the $2p_z$ orbital on each of the F atoms (the orbital that contains the unpaired electron on the F atoms):

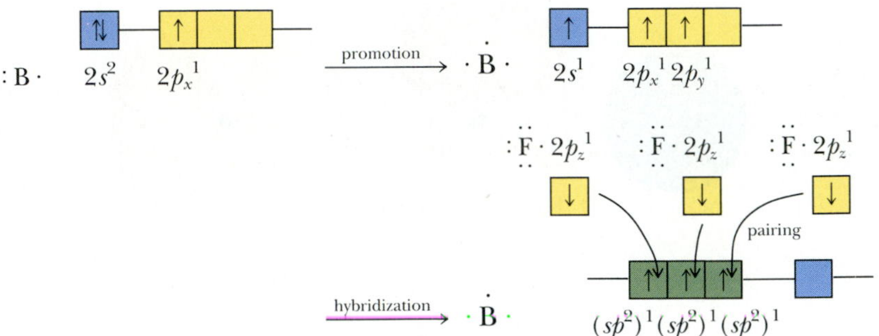

The three sp^2 hybrids point toward the corners of an equilateral triangle, and the molecule is trigonal planar.

EXERCISE E9.9 What is the hybridization of the beryllium atom in the beryllium chloride molecule, $BeCl_2$, and what is the shape of the molecule?

[*Answer*: *sp* hybridized; linear] ▲

Hybrids including d orbitals. When more than four electron pairs are present on the central atom, its *d* orbitals are needed to accommodate the expanded octet. If five pairs of valence electrons are present, so that five valence-shell orbitals are needed, then one *d* orbital must be used in addition to the four *s* and *p* orbitals. Then hybridization gives five sp^3d hybrid orbitals that point to the corners of a trigonal bipyramid (Fig. 9.16a). When the electrons pair and the hybrid orbitals overlap to form a total of five σ bonds, their trigonal bipyramidal arrangement reproduces the molecular shape predicted by VSEPR theory. Six σ bonds require two *d* orbitals in addition to the four *s* and *p* orbitals; in this case, hybridization leads to six sp^3d^2 hybrid orbitals that point toward the corners of a regular octahedron (Fig. 9.16b).

▼ EXAMPLE 9.10 Expressing bond formation with *d* orbitals

Use valence-shell atomic diagrams to show how promotion and hybridization may be used to describe the bonding in phosphorus pentafluoride, PF_5.

STRATEGY First, we draw the valence-electron configuration of the ground-state phosphorus atom, using boxes to represent orbitals; because we know that we shall need to use phosphorus *d* orbitals, we show those

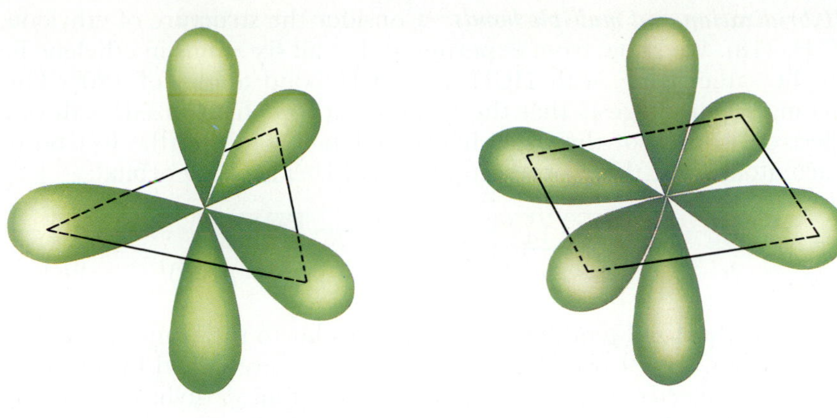

FIGURE 9.16 (a) The five sp^3d and (b) the six sp^3d^2 hybrid orbitals in which d orbitals are used as well as s and p orbitals.

T OHT: Fig. 9.16.

(empty) orbitals too. Then we promote an electron so that the phosphorus atom has five unpaired electrons (because we need to form five bonds). We hybridize all the *occupied* orbitals to give hybrid orbitals of intermediate energy. Finally, we pair the electrons in each hybridized orbital with an electron in a $2p$ orbital on an F atom; the orbitals overlap to give σ bonds.

SOLUTION The valence-shell ground-state electron configuration of the phosphorus atom is

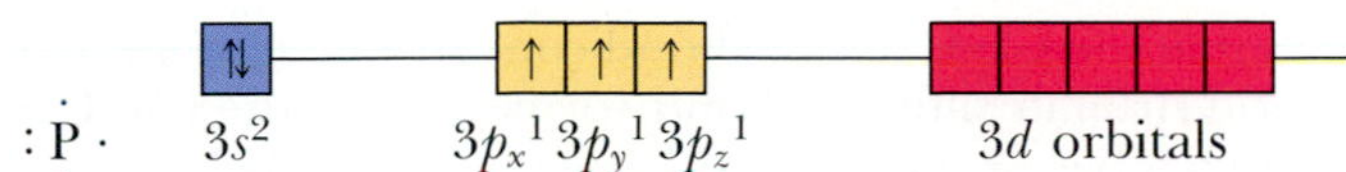

The atom with the promoted electron (with five unpaired electrons) is

·P· $3s^1$ $3p_x{}^1 3p_y{}^1 3p_z{}^1$ $3d^1$

Next, we hybridize the occupied orbitals:

·P· $(sp^3d)^1(sp^3d)^1(sp^3d)^1(sp^3d)^1(sp^3d)^1$

Finally, we pair an F2p electron with each unpaired electron; the five F2p orbitals overlap with the five hybrids to give five σ(Psp^3d,F2p) bonds. The resulting structure is trigonal bipyramidal.

EXERCISE E9.10 Repeat the exercise for the SF_6 molecule.

[*Answer*: six σ(Ssp^3d^2,F2p) bonds] ▲

N The number of boxes represents the numbers of orbitals and the groupings of the boxes represents their relative energies. In the unhybridized atom the boxes form three distinct groups, corresponding to the different energies of the s, p, and d orbitals. After promotion and hybridization, the boxes form two groups: one consists of five hybrid orbitals at an energy that is broadly the average of the contributing orbitals, and the four unused d orbitals remain unchanged. Unoccupied orbitals are unlabeled but their color denotes their type. Even though the hybrids contain d character (orange), we continue to denote them by the color green (to signify a mixture).

In Exercise 9.10 we treated the terminal F atom in PF_5 as though it were pure $2p$ in character. It is very likely, however, that terminal atoms also undergo hybridization in a manner that reflects the arrangement of their electron pairs. Thus, the four electron pairs around each F atom are arranged tetrahedrally (Fig. 9.17), an arrangement indicating that it is sp^3 hybridized. Each bond between P and F in PF_5 is therefore best described as arising from the overlap of a sp^3d hybrid orbital on the P atom and an sp^3 hybrid on the F atom—denoted σ(Psp^3d,Fsp^3).

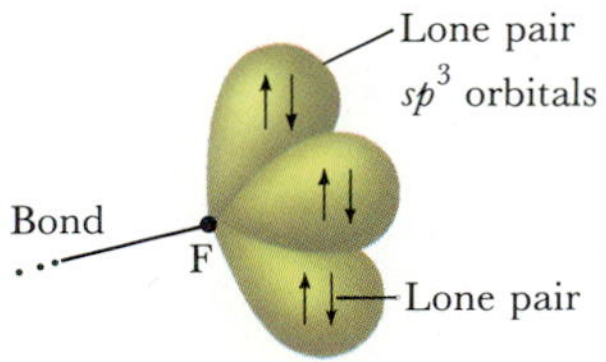

FIGURE 9.17 A terminal F atom with three lone pairs in sp^3 hybrid orbitals and a bond formed with the fourth sp^3 hybrid orbital.

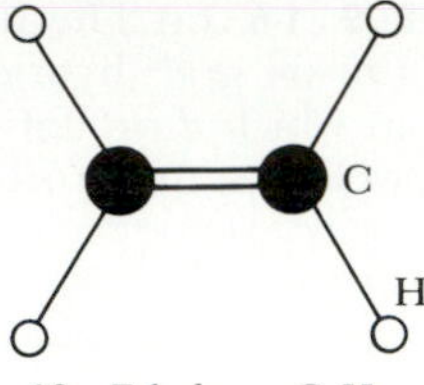

18 Ethylene, C_2H_4

Hybridization and multiple bonds. Consider the structure of ethylene, C_2H_4 (**18**). We know from experiment that all six atoms in ethylene lie in the same plane, with HCH and CCH bond angles of 120°. This arrangement suggests that the C atoms are sp^2 hybridized, with one electron in each of the three hybrid orbitals. The fourth electron of each atom must therefore occupy an unhybridized $2p$ orbital:

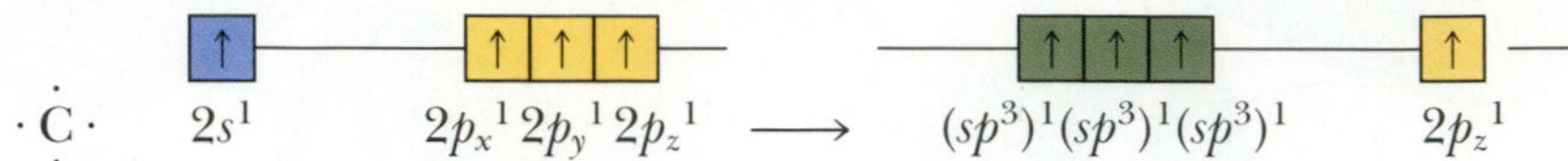

The unhybridized p orbital lies perpendicular to the plane formed by the hybrids (Fig. 9.18). The two carbon atoms are linked by a σ bond formed by electron pairing and the overlap of an sp^2 hybrid orbital on each C atom. The H atoms form σ bonds with the remaining lobes of the sp^2 hybrids. The electrons in the two unhybridized $2p$ orbitals pair to form a π bond:

N The π bond is weak because of the small degree of overlap of the two p orbitals that contribute to it.

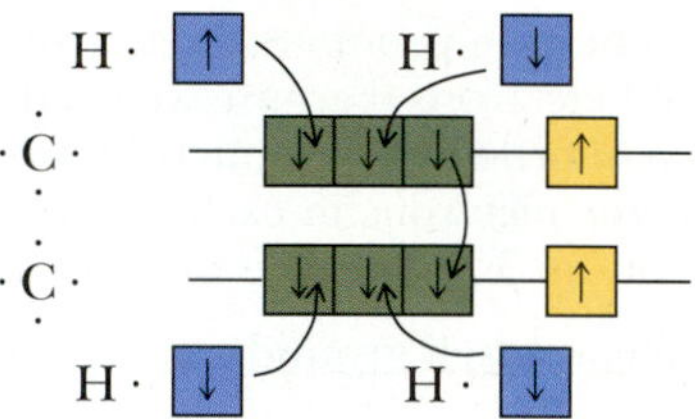

It follows that the C═C double bond consists of a σ bond formed from Csp^2 hybrid orbitals plus a π bond formed from the two C$2p$ orbitals, with the two lobes of the π bond lying above and below the molecular plane.

The structure of one of the Kekulé structures of benzene can be expressed in very similar terms. Each carbon atom is sp^2 hybridized (as in ethylene), with one electron in each hybrid orbital and one electron in a $2p$ orbital perpendicular to the plane of the hybrids. The electrons in sp^2 orbitals on neighboring atoms pair and form six σ(C—C) bonds, and an H$1s$ electron pairs with the electron in the third hybrid orbital on each C atom, thereby forming six σ(C—H) bonds (Fig. 9.19). The ring so formed brings the carbon $2p$ electrons close to each other; so each one can pair and the orbitals can overlap side-by-side with *one* of its neighbors to form a π bond (Fig. 9.20). Depending on which neigh-

T OHT: Figs. 9.18 and 9.19.

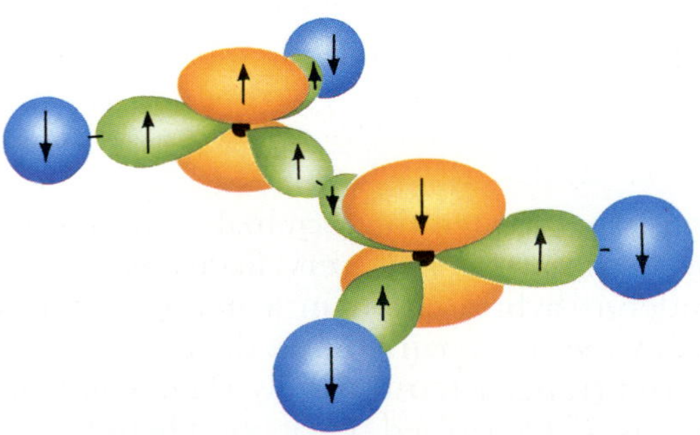

FIGURE 9.18 An exploded view of the bonding structure in ethylene, showing the σ framework and the π bond formed by the side-by-side overlap of unhybridized C$2p$ orbitals.

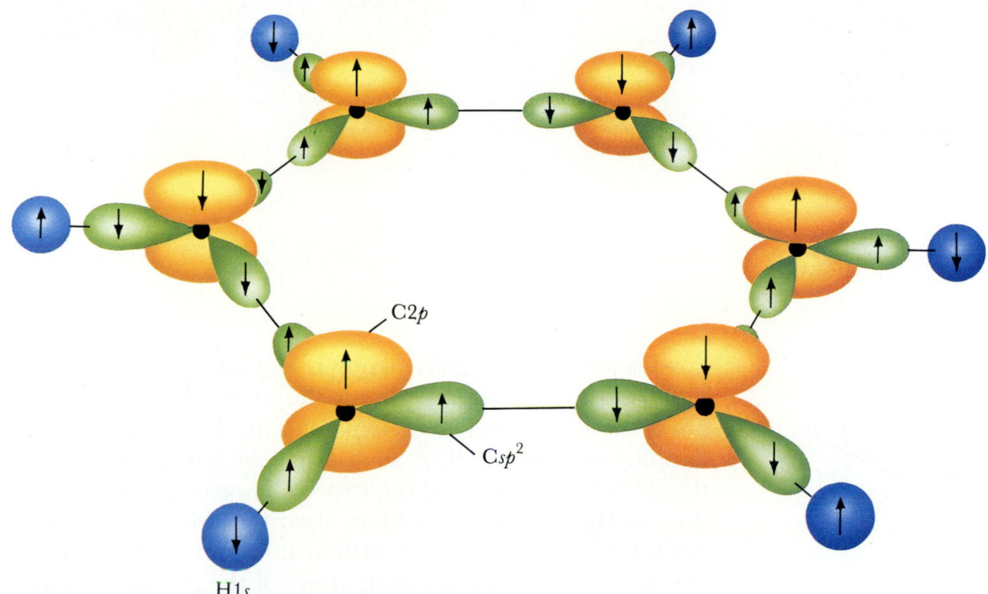

bors bond together in this way, we get either one of the two Kekulé structures. The actual structure of the benzene molecule is then taken to be a resonance hybrid of the two alternatives. (Note that the term *hybrid* occurring in the description of benzene has two slightly different meanings. In their bonds the carbon atoms use hybrid *orbitals,* a mixture of *atomic orbitals;* the resulting overall orbital structures are then used to form a *resonance* hybrid, a mixture of *structures.*)

The structure of acetylene can be described similarly. The Lewis structure is H—C≡C—H, and to achieve a linear structure, the C atoms must be *sp* hybridized. There is one electron in each of the *sp* hybrid orbitals and one electron in each of the two perpendicular unhybridized $2p$ orbitals. The electrons in an *sp* hybrid orbital on each carbon atom pair and form a σ(C—C) bond, and the electrons in the two remaining *sp* hybrid orbitals pair with H1*s* electrons to form two σ(C—H) bonds. The electrons in the perpendicular $2p$ orbitals pair, and the side-by-side overlap of the orbitals results in the formation of *two* π bonds at 90° to each other. The resulting structure of acetylene is as shown in Fig. 9.21.

Properties of double bonds. The data in Table 9.2 show that a C=C double bond is stronger than a C—C single bond but weaker than the sum of two C—C single bonds. (This comparison is not, however, true of all A=B and A—B bonds.) We also saw that a C≡C triple bond is weaker than the sum of three single bonds. We can now see another part of the reason: whereas a single bond is a σ bond, the extra bonding in a multiple bond is a result of π bonding. In carbon (but not in general) π bonds are about 84 kJ/mol weaker than σ bonds because their electron pairs are in less favorable locations for pulling the atoms together.

A further point about multiple bonds: in Section 8.9 we remarked that a double bond prevents one part of a molecule from rotating relative to the other. This constraint is summarized by saying that a double

FIGURE 9.19 An exploded view of the valence orbitals that are used to describe the structure of benzene. The sp^2 hybrid orbitals each contain an unpaired electron, as do the remaining six $2p$ orbitals and the six hydrogen atoms.

T OHT: Figs. 9.18 and 9.19.

N See Case 9, on electrical conduction by covalent compounds.

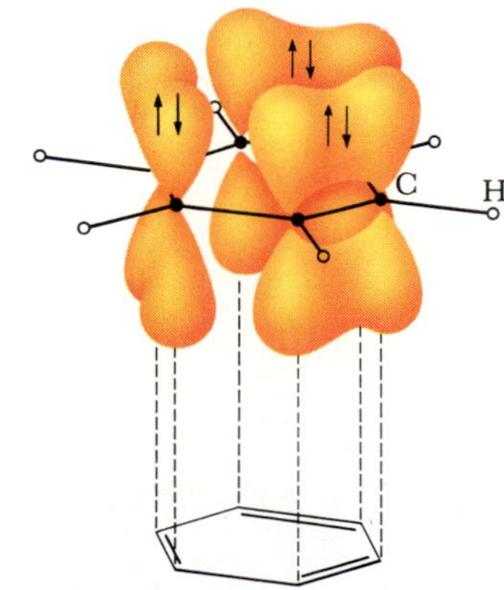

FIGURE 9.20 One valence-bond structure of benzene and the corresponding Kekulé structure. Electrons in the $2p$ orbitals of neighboring C atoms pair and the orbitals overlap side-by-side to give three π bonds.

FIGURE 9.21 (a) The hybridized atomic orbitals that are used to form the bonds in acetylene, HC≡CH. (b) When the electrons pair, they form σ bonds between pairs of neighboring atoms (by end-to-end overlap) and two π bonds (by side-by-side overlap).

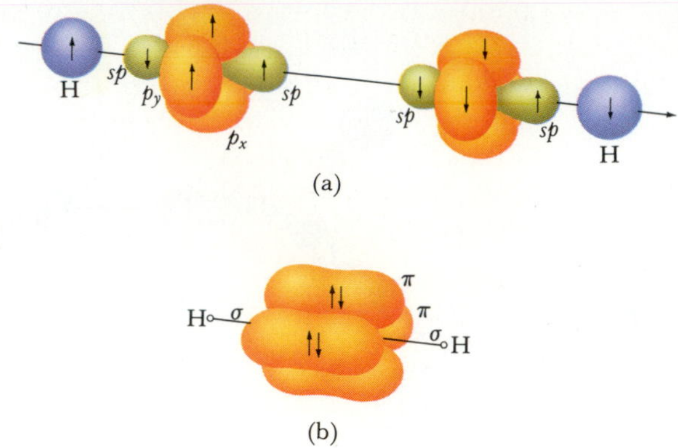

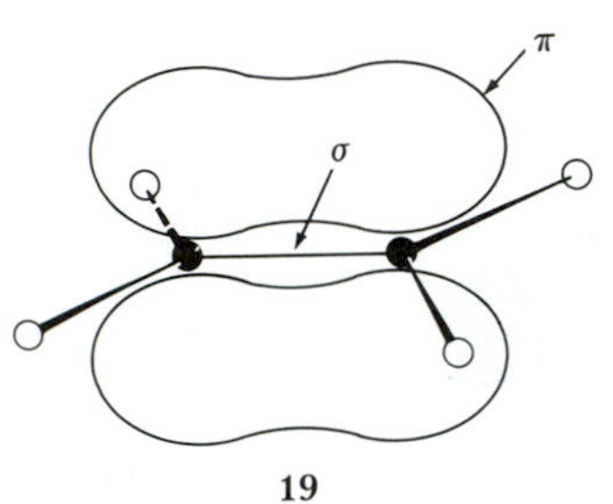

19

20

21

bond is torsionally rigid (torsion is a twisting motion). The double bond of ethylene, for example, holds the molecule flat (**19**). Again, valence-bond theory provides an explanation. The drawing of the ethylene double bond in Fig. 9.18 shows that the two $2p$ orbitals overlap best if the sp^2 hybrid orbitals of the two $—CH_2$ groups lie in the same plane. If the $—CH_2$ groups are twisted, then the side-by-side overlap of the $2p$ orbitals lessens and the π orbital no longer forms. This lessening of overlap removes the bonding effect of the two electrons in that orbital, and the energy of the molecule rises. In other words, the planar arrangement is rigid because energy must be supplied to twist one $—CH_2$ group relative to the other.

One consequence of the torsional rigidity of double bonds is your ability to read these words. Vision depends on the shape of the molecule retinal, which is found in the retina of the eye, the image-receiving screen of cells. In its normal form (**20**), retinal is held rigid by its double bonds. When a photon of visible light enters the eye the photon is absorbed by a retinal molecule. The photon's energy excites an electron out of the π orbital of the double bond marked by the arrow, thereby reducing retinal's torsional rigidity. Before the excited electron has time to fall back, the molecule is free to rotate around the remaining σ bond. When the double bond does reform, the molecule is trapped in its new shape (**21**). This change of shape triggers a signal along the optic nerve and is interpreted by the brain as the sensation of vision.

A final feature clarified by the orbital picture of bonding is the rarity of multiple bonds between elements of Period 3 and later: although C=C and P=O bonds are common, Si=Si and P=S bonds are rare. The reason can be traced to the absence of π overlap that occurs between p orbitals on large atoms (**22**); Period 3 atoms are so large that they cannot come close enough for their p orbitals to overlap side-by-side and give significant multiple bonding.

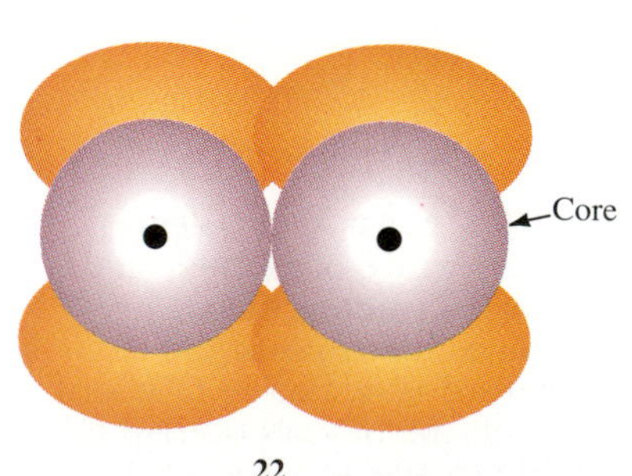

22

MOLECULAR ORBITAL THEORY

Lewis's original ideas were brilliant, and their extension as in VSEPR theory made them even more powerful. However, they were little more than inspired guesswork. Lewis had no way of knowing why the electron pair—the essential feature of his approach—was so important. In addition, his approach fails for some molecules.

An example of the failure of the Lewis approach is the compound diborane (B_2H_6, **23**), a colorless gas that bursts into flame on contact with air. Diborane was prepared by the German chemist Alfred Stock as a part of his pioneering work on boron chemistry, which he began in 1912. Stock produced it by the action of acid on magnesium boride (Mg_3B_2). The modern method of diborane production makes use of the action of boron trifluoride on lithium tetrahydroborate ($LiBH_4$) dissolved in ether:

$$3LiBH_4 + 4BF_3 \longrightarrow 2B_2H_6 + 3LiBF_4$$

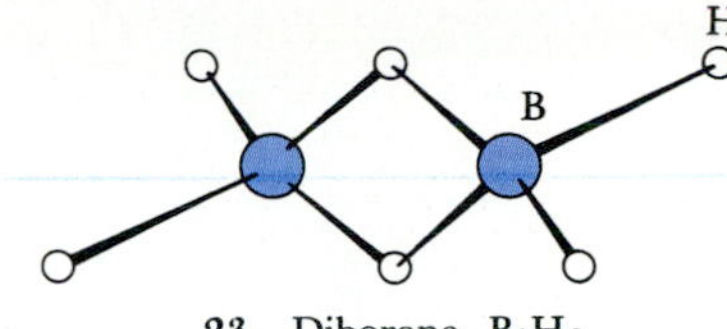

23 Diborane, B_2H_6

The problem with diborane, from the point of view of Lewis structures, is that it has only 12 valence electrons (three from each B atom, one from each H atom) but it needs at least 7 bonds, and therefore 14 electrons, to bind the eight atoms together! Diborane is an example of an electron-deficient compound:

An **electron-deficient compound** is a compound with too few valence electrons for it to be assigned a Lewis structure.

Another puzzle of a different kind arose in connection with the Lewis description of O_2 as $\ddot{O}{=}\ddot{O}$, because oxygen is a paramagnetic substance (Fig. 9.22). Paramagnetism (Section 7.6) is a property of *unpaired* electrons, each of which acts as a tiny magnet and gives rise to a magnetic field. The magnetism of O_2 therefore contradicts the Lewis structure (and the valence-bond description) of the molecule, which requires all the electrons to be paired.

The modern theory of chemical bonding, which was formulated when quantum theory was established in the late 1920s, overcomes all these difficulties. It shows very naturally why and when the electron pair is so important. It accommodates the boron hydrides just as naturally as it deals with methane and water. It also predicts that oxygen is

E The molecular orbital (MO) theory of the chemical bond was formulated by Hund and Mulliken shortly after the VB theory. It identifies an orbital as belonging to the molecule as a whole, and thus expresses its structure in terms of delocalized orbitals.

E Oxygen gas is colorless; liquid and solid oxygen are pale blue. The color of the latter is also a consequence of the unpaired electron configuration, for the transition depends on the existence of a weak coupling between neighboring molecules in the condensed phase. Isolated molecules (as in the gas phase) do not interact and do not undergo the transition.

D Observation of a paramagnetic property of oxygen by simple method: A simple experiment for college chemistry and physics courses. *J. Chem. Educ.*, **1990**, *67*, 63.

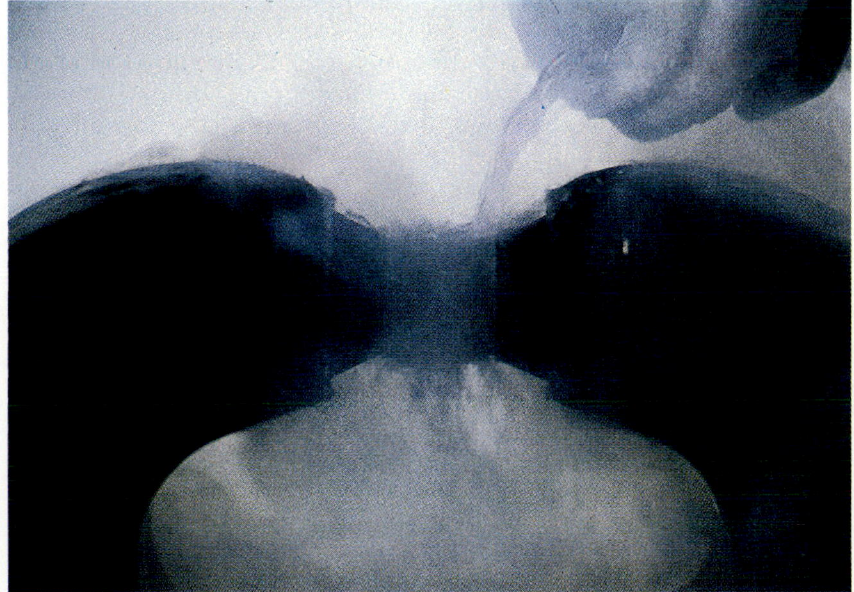

FIGURE 9.22 Oxygen is a paramagnetic substance; liquid and solid oxygen (as shown here) sticks to the poles of a magnet.

paramagnetic. However, in doing all this it shows how close Lewis was to the truth and that his concepts can still be used with only minor refinement.

9.7 MOLECULAR ORBITALS

N The discussion of molecular orbitals could be delayed until metallic bonding is discussed in Section 10.6.

The σ and π bonds we have been discussing represent an intermediate stage between Lewis's concept of a localized electron pair and the modern theory of bonding. So far, we have treated bonding pairs of electrons as an electron cloud that spreads over two neighboring atoms, but we have not been very definite about what it means for the atomic orbitals to merge together when they overlap. In **molecular orbital theory**, electrons occupy precisely defined orbitals that spread throughout the molecule. That is, electrons occupy molecular orbitals:

> A **molecular orbital** is an orbital that spreads over all the atoms in a molecule.

N When a molecular orbital is constructed, we use *all* the orbitals available on all the atoms, and do not seek to construct localized orbitals by invoking hybridization. Thus, a general molecular orbital is composed of *s* and *p* contributions from each atom in the molecule. In practice, symmetry considerations exclude the contribution of some atomic orbitals to some molecular orbitals.

E Molecular orbitals may also be nonbonding, and neither raise nor lower the energy of the molecule if they are occupied. A nonbonding orbital may be a single atomic orbital on one atom, or a molecular orbital in which the contributing atomic orbitals are not on neighboring atoms.

E Loosely bound diatomic molecules. *J. Chem. Educ.*, **1979**, *56*, 452.

Once we know the form of a molecular orbital, we can say where an electron that occupies it is likely to be found.

The extension of the concept of orbitals that are centered on one nucleus to orbitals that spread over several nuclei immediately explains a central concept of bonding theory. We saw in Chapter 7 that the Pauli exclusion principle restricts the number of electrons that can occupy an orbital to two; and the two electrons must have paired spins. Exactly the same is true of orbitals in molecules. Because an orbital can accommodate only two electrons, we see at once the importance of the electron pair in bonding theory: *a covalent bond consists of two paired electrons, because that is the greatest number that can occupy a molecular orbital.* Lewis had no idea why an electron pair was so important. Today, we can see it as a direct consequence of the Pauli exclusion principle.

Molecular orbitals of diatomic molecules. We can think of a molecular orbital as formed from the overlap of atomic orbitals. Two *s* atomic orbitals, and also two *p* orbitals lying end-to-end can overlap to form a **σ orbital**, a sausage-shaped orbital that, when occupied by two electrons, forms a σ bond. Two *p* atomic orbitals lying side-by-side can overlap to give a **π orbital**, which when occupied by two electrons forms a π bond. Molecular orbital theory goes on to introduce the important concept of bonding orbitals and antibonding orbitals:

> A **bonding orbital** is a molecular orbital that, if occupied, lowers the energy of a molecule relative to that of the separated atoms.
>
> An **antibonding orbital** is a molecular orbital that, if occupied, raises the energy of a molecule relative to that of the separated atoms.

According to molecular orbital theory, when two atomic orbitals overlap, they give rise to *two* molecular orbitals. An occupied σ bonding orbital is the same as the σ bond we have already described in connection with valence-bond theory. We can think of the orbital as being formed from two atomic orbitals acting like waves that add together where they overlap (Fig. 9.23a). If an electron occupies a σ orbital, then it is likely to be found between the nuclei and can interact with them

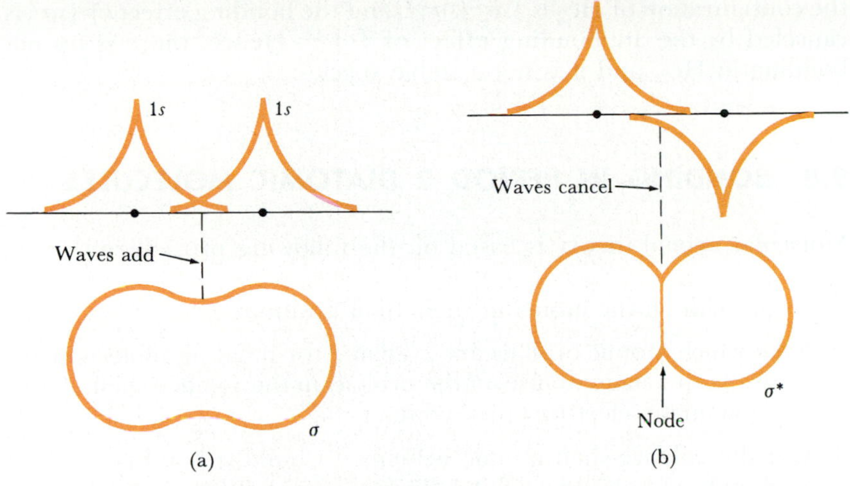

FIGURE 9.23 (a) The formation of a bonding molecular σ orbital from the overlap of two $1s$ orbitals (the upper diagram shows a section through the orbitals, and the lower diagram is a boundary surface). The addition of wave amplitudes occurs where the orbitals overlap (in the internuclear region), and the probability of finding the electron is high in that region. (b) The formation of an antibonding σ orbital as a result of the cancellation of the orbitals where they overlap. There is zero probability of finding the electrons on the internuclear axis (where the orbitals cancel exactly). Each orbital can hold up to two electrons.

both. As a result of the attraction, an electron in a σ orbital helps to lower the energy of a molecule and contributes to its bonding.

Another possibility is that when two atomic orbitals spread into the same region of space, they behave like waves that subtract from each other. This subtraction gives rise to a molecular orbital like the one labeled σ^* in Fig. 9.23b. We see that there is a plane of zero probability half way between the nuclei, showing that the overlap of atomic orbitals has canceled completely. That is, in a σ^* orbital there is *zero* probability of finding an electron midway between the nuclei. An electron in a σ^* orbital is excluded from the internuclear regions and is in a poor position for binding the two atoms together, and there is a net repulsion between the two nuclei. As a result, an electron in a σ^* orbital has an antibonding influence on the molecule.

The relative energies of the original atomic orbitals and the bonding orbitals and the antibonding molecular orbitals are shown in a **molecular-orbital energy-level diagram** (Fig. 9.24). The destabilization caused by an antibonding orbital is about equal to or a little greater than the stabilization caused by the bonding orbital.

E Sum of two waves (computer program). *J. Chem. Educ.*, **1988**, *65*, 389.

N When N atomic orbitals are available, they can be used to form N molecular orbitals. About $\frac{1}{2}N$ of the molecular orbitals will be bonding and about $\frac{1}{2}N$ will be antibonding. A few may be essentially nonbonding.

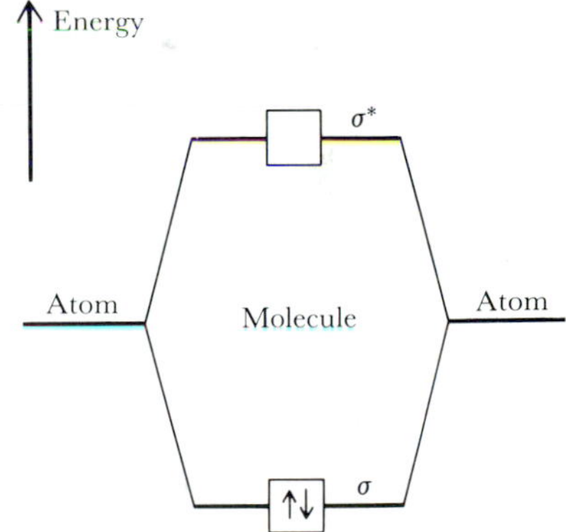

FIGURE 9.24 Two electrons in a bonding orbital (σ) lower the energy of a molecule below that of the separated atoms (left and right). However, because electrons in an antibonding orbital are likely to be found outside the bonding region, they tend to break the bond by pulling the atoms apart. Therefore, the antibonding orbital (σ^*) corresponds to an energy for the molecule that is higher than the energy of the separated atoms.

The role of the electron pair. We can use the building-up principle to fill molecular orbitals, just as we used it in Section 7.5 for filling atomic orbitals. In H_2, two atomic $1s$ orbitals (one on each atom) merge to form two molecular orbitals: we call the bonding orbital $1s\sigma$ and the antibonding orbital $1s\sigma^*$, the $1s$ showing the atomic orbitals from which the molecular orbitals are formed. Two electrons, one from each H atom, are available. Both occupy the bonding orbital (the lowest-energy orbital), producing the configuration $1s\sigma^2$. Because only the bonding orbital is occupied, the energy of the molecule is lower than that of the separate atoms, and H_2 exists as a molecule.

The energy-level diagram for He_2 will resemble that for H_2 because both are built from atoms that have their valence electrons in $1s$ orbitals. This similarity suggests that He_2 can be discussed in terms of the diagram in Fig. 9.24, but four electrons are involved instead of two. We can use the building-up principle and examine the net bonding effect of the resulting electron configuration. The first two electrons enter $1s\sigma$, which is then full. The remaining two must enter $1s\sigma^*$. Therefore,

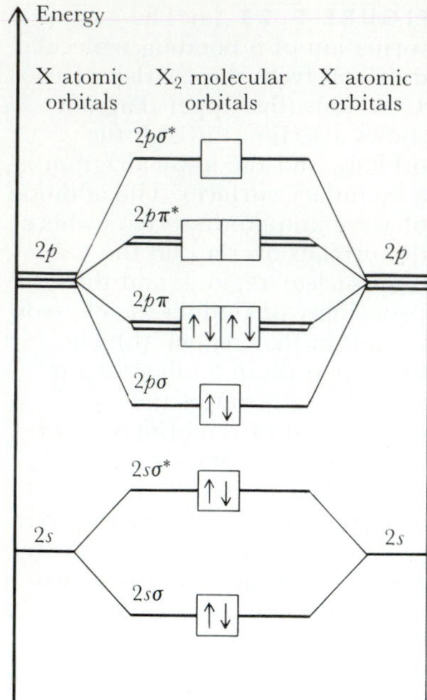

FIGURE 9.25 The molecular-orbital energy-level diagram for the Period 2 diatomic molecules. Each box can hold up to two electrons; the electron configuration for N_2 is shown.

T OHT: Fig. 9.25.

E Orbital energy levels of molecular hydrogen: a simple approach. *J. Chem. Educ.*, **1988**, *65*, 418.

the configuration of He_2 is $1s\sigma^2 1s\sigma^{*2}$, and the bonding effect of $1s\sigma^2$ is canceled by the antibonding effect of $1s\sigma^{*2}$. Hence, there is no net bonding in He_2, and it is not a stable species.

9.8 BONDING IN PERIOD 2 DIATOMIC MOLECULES

Molecular orbital theory is based on the following procedure:

1. Begin with all the atoms in their final positions.
2. Note which atomic orbitals are available for building molecular orbitals. At this stage we use *all* the orbitals in the valence shell, ignoring how many electrons they contain.
3. Use the valence-shell atomic orbitals, to build a bonding and an antibonding molecular orbital (and still ignore the electrons).
4. Finally, note how many valence electrons are to be accommodated, and then add them to the molecular orbitals according to the building-up principle.

Constructing the orbitals. The second and third steps involve building up the molecular-orbital energy-level diagram from the valence-shell atomic orbitals. Period 2 atoms have $2s$ and $2p$ orbitals in their valence shells, so we form molecular orbitals from the overlap of these atomic orbitals. The two $2s$ orbitals overlap to form two sausage-shaped molecular orbitals, one bonding (the $2s\sigma$ orbital) and the other antibonding (the $2s\sigma^*$ orbital). The three $2p$ orbitals on each neighboring atom can overlap in two distinct ways. The two $2p$ orbitals that are directed toward each other along the internuclear axis form a bonding σ orbital ($2p\sigma$) and an antibonding σ^* orbital ($2p\sigma^*$) where they overlap. Two $2p$ orbitals that are perpendicular to the internuclear axis overlap side-by-side to form the bonding and antibonding π orbitals. Because there are two perpendicular $2p$ orbitals on each atom, two $2p\pi$ orbitals and two $2p\pi^*$ orbitals are formed by their overlap. The resulting energy-level diagram is shown in Fig. 9.25.

Electron configurations and bond orders. Now that we know what molecular orbitals are available, we can obtain the ground-state electron configurations of the molecules by using the building-up principle.

Consider N_2. Nitrogen belongs to Group V, so each atom supplies five valence electrons. A total of ten electrons must therefore be assigned to the molecular orbitals. Two fill the $2s\sigma$ orbital. The next two fill the $2s\sigma^*$ orbital. Next in line for occupation is the $2p\sigma$ orbital, which can hold two electrons. The last four electrons then enter the two $2p\pi$ orbitals. The ground configuration is therefore

$$N_2 \quad (2s\sigma)^2(2s\sigma^*)^2(2p\sigma)^2(2p\pi_x)^2(2p\pi_y)^2$$

The energy-level diagram, where now the boxes represent *molecular* orbitals, is

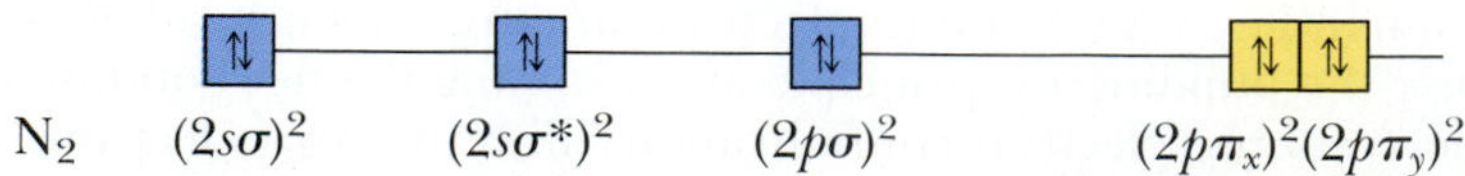

The molecular-orbital description of N_2 looks quite different from :N≡N:, the Lewis description. However, it is in fact very closely related. We can see this by calculating the **bond order**, the net number of bonds existing after the cancellation of bonds by antibonds:

$$\text{Bond order} = \tfrac{1}{2} \times (B - A)$$

where B is the number of electrons in bonding molecular orbitals and A is the number of electrons in antibonding molecular orbitals. This definition is the source of the generalized definition given in Section 8.3, where we simply counted shared pairs. In N_2 there are eight electrons in bonding orbitals and two in antibonding orbitals, so

$$\text{Bond order} = \tfrac{1}{2} \times (8 - 2) = 3$$

Because its bond order is 3, N_2 effectively has three bonds between the N atoms, just as the Lewis structure suggests.

E Figure 9.25 is a simplification. It applies best to the diatomics N_2, O_2, and F_2. For the elements earlier in Period 2, the bonding $2p\sigma$ and $2p\pi$ orbitals lie in the reverse order:

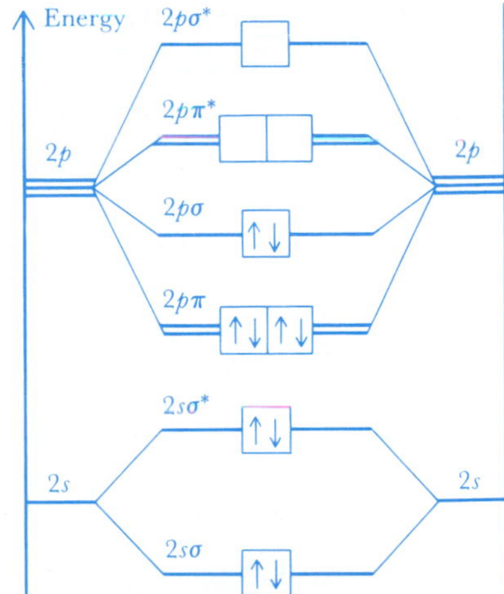

The four σ orbitals are built from all available orbitals of the correct symmetry (not by taking the $2s$ orbitals pairwise and the $2p$ orbitals pairwise). Then the most bonding and most antibonding orbitals are in much the same location as in Fig. 9.25, but—depending on atomic energy levels and orbital overlap—the two other σ orbitals lie between them and may adopt the positions shown either in Fig. 9.25 or here.

E Explanation of the permutation of the σ_p and π levels in homonuclear diatomic molecules. *J. Chem. Educ.*, **1978**, 55, 571.

▼ **EXAMPLE 9.11 Deducing the electron configurations of Period 2 diatomic molecules**

Deduce the electron configuration of the fluorine molecule, and calculate its bond order.

STRATEGY The Lewis structure of F_2 is :F̤̈—F̤̈:, so we anticipate that the bond order is 1. To calculate it formally, we need to know the numbers of electrons in bonding and antibonding orbitals in the molecule. Therefore, we first use the building-up principle to accommodate the 2 × 7 = 14 valence electrons in the energy-level diagram. Then we calculate the bond order from the resulting configuration.

SOLUTION The first ten electrons repeat the N_2 configuration. The remaining four can enter the two antibonding $2p\pi^*$ orbitals. The configuration is therefore

$$F_2 \quad (2s\sigma)^2(2s\sigma^*)^2(2p\sigma)^2(2p\pi_x)^2(2p\pi_y)^2(2p\pi_x^*)^2(2p\pi_x^*)^2$$

and in terms of orbital energy levels:

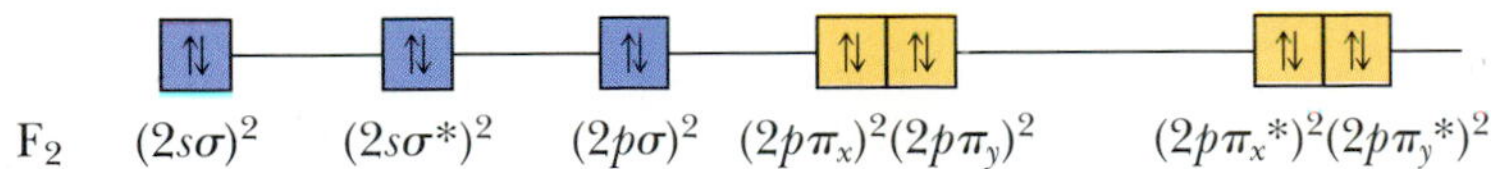

F_2 $(2s\sigma)^2$ $(2s\sigma^*)^2$ $(2p\sigma)^2$ $(2p\pi_x)^2(2p\pi_y)^2$ $(2p\pi_x^*)^2(2p\pi_y^*)^2$

The bond order is

$$\text{Bond order} = \tfrac{1}{2} \times \{(2 + 2 + 4) - (2 + 4)\} = 1$$

Hence, F_2 is a singly bonded molecule, in agreement with the Lewis structure :F̤̈—F̤̈:.

EXERCISE E9.11 Deduce the electronic configuration and calculate the bond order of the carbide ion (C_2^{2-}).

[*Answer*: $(2s\sigma)^2(2s\sigma^*)^2(2p\sigma)^2(2p\pi_x)^2(2p\pi_y)^2$; 3] ▲

U Exercises 9.86–9.89.

Earlier in this section we remarked that Lewis theory cannot account for the paramagnetism of O_2. In molecular orbital theory the configuration of O_2 is obtained by feeding its 12 valence electrons into the molecular orbitals shown in Fig. 9.25. The first ten repeat the N_2 configuration. According to Hund's rule (Chapter 7), the last two electrons occupy the two separate $2p\pi^*$ orbitals with parallel spins. The configuration is therefore

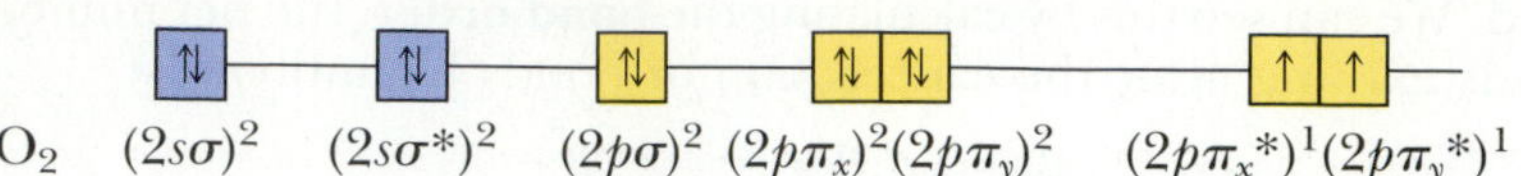

This outcome is a minor triumph. Because the last two spins are not paired, their magnetic fields do not cancel, and the molecule is predicted to be paramagnetic—exactly as observed. The bond order of O_2 is

$$\text{Bond order} = \tfrac{1}{2} \times \{(2 + 2 + 4) - (2 + 1 + 1)\} = 2$$

This value is consistent with the Lewis structure O=O. However, the Lewis structure conceals the fact that the double bond is actually a σ bond plus two *half* π bonds, a bonding pattern that leaves the molecule with two unpaired electrons.

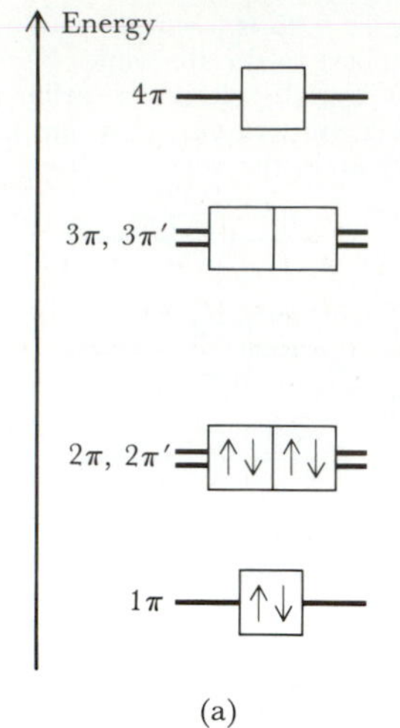

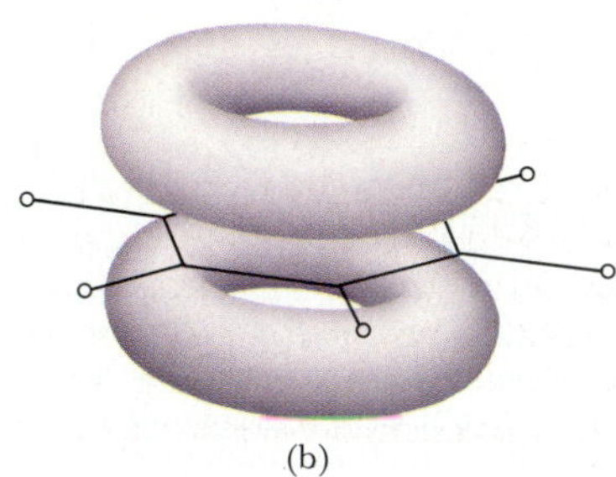

FIGURE 9.26 (a) The energies of the molecular orbitals of benzene and the ground-state electron configuration. Note that only the bonding orbitals are occupied. (b) The net outcome of occupying the three orbitals of lowest energy (1π, 2π, and $2\pi'$) is an electronic structure of benzene in which six electrons occupy two donut-shaped regions, one above the plane of the molecule and one below.

E The "semidelocalized" picture of bonding is a mongrel form of MO theory, but is widely used to show how the electronic structure of a molecule can be constructed from fragments. The MO diagram for the B—H—B fragment, consists of three levels: one bonding, one nonbonding, and the third antibonding. Two electrons enter the most bonding orbital and bind the B—H—B fragment together.

N Molecular orbital theory can be used to account for molecular shapes (and can be used to predict actual bond values), but the procedures are much less simple than in VSEPR and valence-bond theory.

E Pictorial approach to teaching MO concepts in polyatomic molecules. *J. Chem. Educ.*, **1977**, *54*, 590.

9.9 ORBITALS IN POLYATOMIC MOLECULES

We will not go into the molecular orbital theory of polyatomic molecules except to say that it follows the same principles used for diatomic molecules. The only elaboration is that the molecular orbitals spread over *all* the atoms in the molecule. Therefore, an electron pair that occupies a bonding orbital helps to bind together the *whole* molecule, not just an individual pair of atoms. This picture of **delocalized orbitals**, molecular orbitals that spread over the entire molecule, accounts very neatly for both the existence of conventional molecules and the existence of electron-deficient compounds.

In the molecular-orbital description of benzene, *all* the unhybridized carbon $2p$ orbitals perpendicular to the ring contribute to π molecular orbitals that spread all the way around the ring. From the six $2p$ orbitals, we form six delocalized π molecular orbitals: their energies are shown in Fig. 9.26. Three of the π orbitals are bonding and three are antibonding.

Each carbon atom provides one electron for the π orbitals. Two electrons occupy the lowest orbital, the remaining four occupy the next two orbitals (which both have the same energy). We immediately see one of the reasons for the considerable stability of benzene: the π electrons occupy only orbitals with a net bonding effect; and there are no occupied destabilizing antibonding orbitals. The electronic structure of benzene resembles that of N_2 (compare Figs. 9.25 and 9.26), which we know is also a very stable molecule.

The existence of delocalized bonds also accounts for the occurrence of electron-deficient molecules. Electron delocalization implies that the bonding influence of an electron pair is spread over all the atoms in the molecule. Therefore, there is no need to provide one pair of electrons for each pair of atoms that are bonded together because a smaller number of pairs of electrons spread throughout the molecule is able to bind all the atoms together, particularly if the nuclei are not highly charged and so do not repel each other strongly. This arrangement occurs in diborane, B_2H_6, where six electron pairs hold the eight nuclei together.

SUMMARY

9.1 The characteristics of bonds known as **bond parameters** include the bond enthalpy and the bond length. The strength of a covalent bond is measured by its **bond enthalpy**, the enthalpy change for the **dissociation**, or breaking, of the specified bond. Reaction enthalpies can be estimated from **average bond enthalpies**, which are the average enthalpies of A—B bonds found in a range of compounds.

9.2 In general, the stronger the bond between a given pair of atoms, the shorter the **bond length**, the distance between the nuclei of the bonded atoms. Bond lengths may be estimated by taking the sum of the **covalent radii** of the atoms involved.

9.3 Ions are distorted by **polarization** effects. Small, highly charged cations are highly polarizing, and large anions are highly **polarizable**. The distortion of ions represents a tendency toward covalent bonding. Covalent bonds between different atoms are **polar** in the sense that one atom (the more electronegative element) has a partial negative charge and the other a partial positive charge: this effect gives rise to an **electric dipole**, a pair of equal and opposite charges with a strength measured by the **dipole moment**. The periodicity of **polarizing power** is part of the explanation of diagonal relationships in the periodic table. There is no such thing as a *pure* ionic bond or a *pure* covalent bond. Depending on its shape and the polar character of its bonds, a molecule may be **polar**; that is, it may possess a nonzero dipole moment.

9.4 The reorganization of electrons that takes place as a result of bond formation may be analyzed in terms of the **formal charge** of an atom, the charge it carries if it is assumed to possess half of each shared electron pair. The loss or gain of electrons that occurs when atoms form bonds is assessed from the oxidation number of each element in the compound, its effective ionic charge. Formal charge is a "pure covalent" assessment of charge; oxidation number is a "pure ionic" assessment of charge.

9.5 According to the **valence-bond theory** of molecular electronic structure, bonds form when electrons pair and orbitals on one atom can **overlap** (merge with) those of a neighbor. The result of electron pairing is either a σ **bond**, a bond with cylindrical symmetry about the z-axis, or a π **bond**, a bond with double-lobed electron distribution. A **single bond** is a σ bond, a **double bond** is a σ bond plus a π bond, and a **triple bond** is a σ bond plus two π bonds. Double bonds are resistant to twisting.

9.6 The location of valence electron pairs can be expressed in terms of atomic orbitals by assuming that the latter **hybridize**, or mix together, to match the shape of the molecule. In general, N atomic orbitals give N **hybrid orbitals**.

9.7 Chemical bonds can be expressed in terms of **molecular orbitals**, which are orbitals that spread throughout the molecule, formed by the overlap of atomic orbitals. They may be **bonding** (or energy lowering) or they may be **antibonding** (or energy raising). Depending on their shape, they are either σ **orbitals** or π **orbitals**. Electrons occupy the available orbitals in accord with the building-up principle. The importance of the electron pair in bond formation can be traced to the exclusion principle, which limits the occupation of a bonding orbital to two paired electrons.

9.8 A **molecular-orbital energy-level diagram** is constructed from the available valence-shell atomic orbitals by allowing those orbitals to overlap to give either σ or π orbitals. From the viewpoint of molecular orbital theory, a Lewis diagram is a representation of **bond order**, the net number of bonds existing after the cancellation of bonds by antibonds. Molecular orbital theory accounts for the paramagnetism of oxygen.

9.9 A molecular orbital is **delocalized** and spreads the bonding influence of its electrons over several atoms. Delocalization accounts for the existence of **electron-deficient compounds**, those for which Lewis structures cannot be written and in which two electrons bind groups of atoms together.

CLASSIFIED EXERCISES

Bond Strengths and Bond Lengths

Refer to Tables 9.1 and 9.2 for information to help with the following exercises.

9.1 Estimate the enthalpy change that occurs when each molecule is dissociated into its gaseous atoms: (a) CO_2; (b) CH_3COOH.

9.2 Estimate the standard enthalpy of formation, ΔH_f°, of the following molecules in the gas phase: (a) N_2O (NNO); (b) NH_2OH.

9.3 Use the information in Tables 6.2 and 9.2 to estimate the enthalpy of formation of the following compounds in the liquid state, given that the standard enthalpy of sublimation of carbon is +717 kJ/mol: (a) CH_3OH (methanol); (b) C_6H_6 (assume *no* resonance), (c) C_6H_6 (assume resonance).

9.4 Use the information in Tables 6.2 and 9.2 to estimate the enthalpy of formation of the following compounds in the liquid state, given that the standard enthalpy of sublimation of carbon is +717 kJ/mol: (a) NH_3; (b) C_2H_5OH (ethanol); (c) CH_3COCH_3 (acetone).

9.5 Estimate the reaction enthalpy of
(a) $HCl(g) + F_2(g) \rightarrow HF(g) + ClF(g)$, given that $\Delta H_B(\text{Cl—F}) = 256$ kJ/mol
(b) $C_2H_4(g) + HCl(g) \rightarrow C_2H_5Cl(g)$
(c) $C_2H_2(g) + 2H_2(g) \rightarrow CH_3CH_3(g)$

9.6 Estimate the reaction enthalpy of
(a) $N_2(g) + 3F_2(g) \rightarrow 2NF_3(g)$
(b) $CH_3CH{=}CH_2(g) + H_2O(g) \rightarrow CH_3{-}CHOH{-}CH_3(g)$
(c) $CH_4(g) + Cl_2(g) \xrightarrow{h\nu} CH_3Cl(g) + HCl(g)$

9.7 List the carbon-oxygen bonds in the following compounds in order of increasing length: (a) CH_3CH_2OH; (b) H_2CO; (c) $CH_3CO_2^-$.

9.8 List the nitrogen-nitrogen bonds in the following compounds in order of increasing length: (a) NH_2NH_2 (hydrazine); (b) N_2; (c) HNNH (nitrogen hydride).

9.9 Compare the carbon-carbon double bond lengths in $[CH_2{=}CH{-}CH_2]^+$ and $CH_2{=}CH{-}CH_3$.

9.10 Compare the carbon-oxygen bond lengths in $[CH_2{-}CO{-}CH_3]^+$ and $CH_3{-}CO{-}CH_3$.

9.11 Use the information in Table 9.3 to estimate bond lengths in (a) H_2NNH_2 (hydrazine); (b) CO_2; (c) $OC(NH_2)_2$ (urea).

9.12 Use the information in Table 9.3 to estimate bond lengths in (a) H_2CO (formaldehyde); (b) CH_3OCH_3 (dimethyl ether); (c) CH_3SH.

Ionic and Covalent Bonding

9.13 Use Figs. 7.35 and 9.3 to label the following bonds as ionic, covalent, or a significant mix of covalent and ionic character: (a) Ba—Cl; (b) Bi—I; (c) Si—H.

9.14 Use Figs. 7.35 and 9.3 to label the following bonds as ionic, covalent, or a significant mix of covalent and ionic character: (a) N—H; (b) C—S; (c) P—F.

9.15 Arrange the cations Li^+, Be^{2+}, Sr^{2+}, H^+ in order of increasing polarizing power. Give an explanation of your arrangement.

9.16 Arrange the cations K^+, Mg^{2+}, Al^{3+}, Cs^+ in order of increasing polarizing power. Give an explanation of your arrangement.

9.17 Arrange the anions Cl^-, Br^-, N^{3-}, O^{2-} in order of increasing polarizability. Give reasons for your decisions.

9.18 Arrange the anions N^{3-}, P^{3-}, I^-, At^- in order of increasing polarizability. Give reasons for your decisions.

9.19 Classify the bonding in the following compounds as mainly ionic or significantly covalent: (a) AgF; (b) AgI; (c) $AlCl_3$; (d) AlF_3.

9.20 Classify the bonding in the following compounds as mainly ionic or significantly covalent: (a) $BeCl_2$; (b) $CaCl_2$; (c) TlCl; (d) $InCl_3$.

Polarity, Polarizability, and Polar Molecules

9.21 Indicate with a dipole arrow (+⟶ or ⟵+) the electric dipole in the bonds and, from the information in Fig. 7.35, estimate their dipole moments: (a) O—H; (b) F—Cl; (c) O—S.

9.22 Indicate with a dipole arrow (+⟶ or ⟵+) the electric dipole in the bonds and, from the information in Fig. 7.35, estimate their dipole moments: (a) N—O; (b) C—O; (c) C—N.

9.23 Identify in the following molecules the bonds that are polar and nonpolar: (a) Br_2; (b) CH_4; (c) O_3.

9.24 Classify the following molecules as polar or nonpolar (use VSEPR theory to determine their shapes). Assume that all Xs are atoms of the same element. (a) AX_3E_2; (b) AX_4E; (c) AX_5.

9.25 Write the Lewis structures, and predict whether each molecule is polar or nonpolar: (a) CCl_4; (b) PCl_5; (c) XeF_4.

9.26 Write the Lewis structures, and predict whether each molecule is polar or nonpolar: (a) H_2S; (b) PCl_3; (c) SF_4.

9.27 Organic molecules are generally nonpolar or only slightly polar. Predict whether the following molecules are likely to be either polar or nonpolar: (a) C_6H_6 (benzene); (b) CH_3OH; (c) H_2CO (formaldehyde, used to preserve biological specimens).

9.28 Organic molecules are generally nonpolar or only slightly polar. Predict whether the following molecules are likely to be either polar or nonpolar: (a) C_6H_5Cl (chlorobenzene); (b) $CH_3{-}CHOH{-}CH_3$ (2-propanol, rubbing alcohol); (c) CCl_2F_2 (Freon-12, a refrigerant).

9.29 There are three different dichlorobenzenes, $C_6H_4Cl_2$, which differ in the relative positions of the Cl atoms in the benzene ring (**a, b, c**). (a) Which of the three forms are polar and which are nonpolar? (b) Which has the largest dipole moment?

9.30 There are three different dichloroethenes, $C_2H_2Cl_2$, which differ in the locations of the chlorine atoms (**a, b, c**). (a) Which of the forms are polar and which are nonpolar? (b) Which has the largest dipole moment?

a	b	c
Cl(Cl)C=C(H)H	Cl(H)C=C(H)Cl	Cl(H)C=C(Cl)H

Charge Distributions in Molecules

9.31 Determine the formal charge of each atom in the following Lewis structures:

(a) hydrazine, $H_2\ddot{N}{-}\ddot{N}H_2$

(b) carbon dioxide, $\ddot{O}{=}C{=}\ddot{O}$

9.32 Determine the formal charges of each atom in the following Lewis structures:

(a) hydroxylamine, $H_2\ddot{N}{-}\ddot{O}{-}H$

(b) a Kekulé structure of benzene

9.33 Given below are some resonance structures of a number of species. Determine the formal charge on each atom and then, if possible, identify the resonance structure of lowest energy for each species.

(a) $:\ddot{O}{-}\ddot{S}{=}\ddot{O}$ $\quad$ $\ddot{O}{=}\ddot{S}{=}\ddot{O}$

(b) $O{=}S(=O){=}O$ $\quad$ $O{=}S(-\ddot{O}:){=}O$

9.34 Given below are some resonance structures of a number of species. Determine the formal charge on each atom and then, if possible, select the resonance structure of lowest energy for each species.

(a) $[O{=}\ddot{S}(-\ddot{O}:){-}\ddot{O}{-}H]^-$ $\quad$ $[:\ddot{O}{-}\ddot{S}(-\ddot{O}:){-}\ddot{O}{-}H]^-$

(b) $[O{=}S(-\ddot{O}:)_2{-}\ddot{O}{-}H]^-$ $\quad$ $[O{=}S(=O)(=O){-}\ddot{O}{-}H]^-$

9.35 Use formal charge arguments to predict the resonance structure of lowest energy of sulfurous acid:

(a) $H{-}\ddot{O}{-}\ddot{S}(-\ddot{O}:){-}\ddot{O}{-}H$ $\quad$ (b) $H{-}\ddot{O}{-}\ddot{S}(=O){-}\ddot{O}{-}H$

9.36 Use formal charge arguments to identify the resonance structure of lowest energy of sulfuric acid:

(a) $H{-}\ddot{O}{-}S(-\ddot{O}:)_2{-}\ddot{O}{-}H$ $\quad$ (b) $H{-}\ddot{O}{-}S(=O)(=O){-}\ddot{O}{-}H$

9.37 Select from each of the following pairs of Lewis structures the one that is likely to make a dominant contribution to a resonance hybrid.

(a) $:\ddot{F}{-}\ddot{O}{-}\ddot{F}:$ or $F{=}O{=}F$

(b) $\ddot{O}{=}C{=}\ddot{O}$ or $:\ddot{O}{-}C{\equiv}O:$

9.38 Select from each of the following pairs of Lewis structures the one that is likely to make a dominant contribution to a resonance hybrid.

(a) $\ddot{N}{=}N{=}\ddot{O}$ or $:N{\equiv}N{-}\ddot{O}:$

(b) $[\ddot{O}{=}P(-\ddot{O}:)_2{-}\ddot{O}:]^{3-}$ or $[\ddot{O}{=}P(-\ddot{O}:)_2{=}\ddot{O}]^{3-}$

Bonds and Hybridization

9.39 State whether σ bonds, π bonds, or neither are formed by the overlap of the given orbitals on neighboring atoms, where the internuclear distance lies along the z axis. (a) ($1s$, $1s$); (b) ($2p_x$,$2p_x$); (c) ($2s$,$2p_y$); (d) ($2p_z$,$2p_z$).

9.40 State whether σ bonds, π bonds, or neither are formed by the overlap of the given orbitals on neighboring atoms, where the internuclear distance lies along the z axis. (a) ($1s$, $2p_z$); (b) ($2p_y$,$2p_y$); (c) ($3s$,$4p_z$); (d) ($2p_x$,$2p_z$).

9.41 State the relative orientations of the following hybrid orbitals: (a) sp^3; (b) sp; (c) sp^3d^2; (d) sp^2.

9.42 The relative orientation of bonds on a central atom of a molecule that has no lone pairs is given below. What is the hybridization of the orbitals used by the central atom for its bonds: (a) trigonal bipyramidal; (b) octahedral; (c) linear?

9.43 State the hybridization of the atom in bold type in the following molecules: (a) **S**F_4; (b) **B**Cl_3; (c) $(CH_3)_2$**Be**. (Hint: Lone-pair electrons occupy hybrid orbitals that are very similar to those occupied by bonding electrons.)

9.44 State the hybridization of the atom in bold type in the following molecules: (a) **S**F_6; (b) O_3Cl—**O**—ClO_3; (c) **O**C$(NH_2)_2$ (urea). (See the hint in Exercise 9.43.)

9.45 Use an orbital energy-level diagram to identify the hybrid orbitals used by the central atoms for bonding in the following species: (a) CH_3^+; (b) $AlCl_4^-$; (c) Cl_2O.

9.46 Use an orbital energy-level diagram to identify the hybrid orbitals used by the central atoms for bonding in the following species: (a) CH_3^-; (b) BiI_4^-; (c) NH_2^-.

9.47 Give the composition of the indicated bonds—in the form (C*sp*, H1*s*), for instance—in the following molecules: (a) C—H in C_2H_6; (b) O—H in H_2O_2; (c) P—Cl in PCl_5.

9.48 Give the composition of the indicated bonds—in the form (C*sp*, H1*s*), for instance—in the following molecules: (a) N—H in N_2H_4; (b) O—H in CH_3OH; (c) I—F in IF_4^+.

9.49 State the hybridization of the boldface atoms:
(a) acetic acid,

$CH_3\mathbf{C}(=O)OH$

(b) acetylsalicylic acid, the active component of aspirin,

$H_3\mathbf{C}—\mathbf{C}(=O)—\mathbf{O}—C_6H_4—C(=O)OH$

9.50 State the hybridization of the boldface atoms in (a) propylene, $\mathbf{C}H_3—\mathbf{C}H=\mathbf{C}H_2$, the hydrocarbon used to make polypropylene; (b) *β*-ionone, a molecule partly responsible for the odor of freshly cut hay:

H₃C CH₃ … CH=**C**H—**C**(=O)—CH₃ (ring with CH₃)

Molecular Orbitals

9.51 Write the ground-state electron configurations and bond orders of (a) N_2; (b) F_2; (c) Be_2; (d) Ne_2.

9.52 Write the ground-state electron configurations and bond orders of (a) C_2^{2-} (carbide ion); (b) O_2^{2-} (peroxide ion); (c) C_2^{2+}; (d) O_2^{2+}.

9.53 Which of the following species are paramagnetic: (a) O_2; (b) O_2^- (superoxide ion); (c) O_3^- (ozonide ion); (d) O_2^+?

9.54 Which of the following species are diamagnetic: (a) N_2^-; (b) O_2^{2-}; (c) F_2^+?

9.55 Determine the bond orders and use them to predict which species of each pair has the stronger bond: (a) F_2 or F_2^{2-}; (b) B_2 or B_2^+.

9.56 Determine the bond orders and use them to predict which species of each pair has the stronger bond: (a) C_2 or C_2^{2+}; (b) O_2 or O_2^{2+}.

9.57 Predict the electron configurations and bond orders of (a) CN^-; (b) NO. (Fig. 9.25 is approximately valid for diatomic molecules composed of dissimilar atoms.)

9.58 Predict the electron configurations and bond orders of each of the following species. On that basis, predict which member of each pair will have the greater bond enthalpy. (a) CO or CO^+; (b) CN or CN^-; (c) O_2 or O_2^+. (Fig. 9.25 is approximately valid for diatomic molecules composed of dissimilar atoms.)

9.59 State the numbers of bonding and antibonding π orbitals that can be constructed in the following species. How many are occupied in the ground states of the molecules: (a) $H_2C=CH—CH=CH_2$; (b) $HC\equiv C—C\equiv CH$?

9.60 State the numbers of bonding and antibonding π orbitals that can be constructed in the following species. How many are occupied in the ground states of the molecules: (a) $H_2C=CH—CH_2—C\equiv C—CH=CH_2$; (b) $C_{10}H_8$(naphthalene)?

UNCLASSIFIED EXERCISES

9.61 Arrange the following compounds in order of increasing carbon-nitrogen bond length: (a) CH_3CN; (b) $CH_3CH_2NH_2$; (c) CH_3CHNH.

9.62 The production of Freon-12, CCl_2F_2, the coolant medium used in air conditioners, makes use of the reaction between carbon tetrachloride and hydrogen fluoride:

$$CCl_4(g) + 2HF(g) \longrightarrow CCl_2F_2(g) + 2HCl(g)$$

Use bond energies to estimate the standard enthalpy of this reaction. (See Table 23.4.)

9.63 A two-step reaction in the stratosphere that may be responsible for the depletion of the ozone layer is:

Step 1. $Cl + O_3 \rightarrow ClO + O_2$
Step 2. $ClO + O \rightarrow Cl + O_2$

Given that the enthalpy of the Cl—O bond is 203 kJ/mol, (a) determine the enthalpy of reaction for each step; (b) determine the enthalpy of the overall reaction.

9.64 Investigate whether the replacement of a carbon-carbon double bond by single bonds is energetically favored by calculating the reaction enthalpy for the conversion of ethylene, C_2H_4, to ethane, C_2H_6. The reaction is $H_2C{=}CH_2 + H_2 \rightarrow CH_3{-}CH_3$.

9.65 Compare the carbon-nitrogen bond lengths in $N{\equiv}C{-}C{\equiv}N$ and $N{\equiv}C{-}N^{2-}$.

9.66 Use covalent radii to calculate the bond lengths in (a) CF_4; (b) SiF_4; (c) SnF_4. Account for the trend in the values you calculate.

9.67 The difference in electronegativity between silicon and fluorine is 2.2. Explain why silicon tetrafluoride exhibits the properties of covalent bonding.

9.68 Write the Lewis structure of each of the following compounds, predict its shape, and state whether it is polar or nonpolar: (a) O_2SCl_2; (b) $AsCl_5$; (c) SiF_4.

9.69 Predict the types of hybrid orbitals used in bonding by the central atom, the shape, and the bond angles in (a) PCl_3; (b) SiF_6^{2-}; (c) CH_3^+.

9.70 Polar molecules attract other polar molecules through dipole-dipole intermolecular forces. If a liquid composed of polar molecules functions as a solvent, then polar solutes tend to have higher solubilities than nonpolar solutes. Which of the following compounds will have the higher solubility in water? (a) SiF_4 or PF_3; (b) SF_6 or SF_4.

9.71 The halogens form compounds among themselves (called interhalogens). These have the formulas XX', XX'_3, XX'_5, and XX'_7, where X is the heavier halogen. Predict their structures and bond angles. Which of them are polar?

9.72 Sulfur trioxide melts at 16.9°C and boils at 44.6°C. In the liquid state, sulfur trioxide is a cyclic trimer, a molecule having alternating S and O atoms in a six-membered ring with the formula $(SO_3)_3$. Write the Lewis structure for the trimer, minimizing the formal charge on each atom.

9.73 Confirm, by using trigonometry, that the dipoles of the three bonds in a trigonal planar AB_3 molecule cancel and that the molecule is nonpolar. Go on to show that the dipoles of the four bonds in a tetrahedral AB_4 molecule cancel and that the molecule is nonpolar.

9.74 Based on the arguments in Exercise 9.70, suggest which solute is likely to have the greater solubility in benzene: (a) H_2CO (formaldehyde) or C_8H_{18} (octane); (b) the three dichlorobenzenes, **a**, **b**, and **c**, shown in Exercise 9.29.

9.75 Determine the formal charge of each atom in the following species. Where two Lewis structures are given, identify the structure of lower energy.

(a) :Ö=C̈l—Ö—H (with =O below Cl)   :Ö—C̈l—Ö—H (with —:Ö: below Cl)

(b) :Ö=C=S̈:

(c) H—C≡N:

(d) $[\ddot{N}{=}C{=}\ddot{N}]^{2-}$ $[{:}N{\equiv}C{-}\ddot{N}{:}]^{3-}$

(e) $[AsO_4]^{3-}$: As bonded to four :Ö: by single bonds $[AsO_4]^{3-}$: As bonded to two :Ö: by single bonds and two O by double bonds

9.76 Hexamethyltellurium, $Te(CH_3)_6$, has recently been synthesized by a research group at the University of Illinois, Chicago. The starting material was tetramethyldifluorotellurium, $TeF_2(CH_3)_4$. Predict the hybridized orbitals on tellurium used in the bonding of the hexamethyltellurium, $Te(CH_3)_6$, and tetramethyldifluorotellurium, $TeF_2(CH_3)_4$, molecules.

9.77 Determine the maximum oxidation number that each of the following elements may have in compounds: (a) lead; (b) vanadium; (c) sulfur; (d) chlorine.

9.78 Oxygen difluoride is the only compound in which oxygen is assigned a positive oxidation number. Write its Lewis structure and then determine the composition of the hybrid orbitals it uses to form bonds to the fluorine atoms. Predict the bond angle.

9.79 The molecules (**a**) and (**b**) are two of the bases that are part of the nucleic acids involved in the genetic code. Identify the hybridizations of the carbon and nitrogen atoms, the composition of the bonds, and the orbitals occupied by the lone pairs.

(Structures: **a** — purine ring with NH_2 substituent, ring atoms N, HC, N, C, C, N, CH, N—H; **b** — purine ring with C=O, HN, C, N—H_2, N, C, N, CH, N—H)

a   **b**

9.80 Write three (equal energy) resonance structures of the carbonate ion, CO_3^{2-}. Determine the formal charge on each atom of the structure. What are the hybrid orbitals that the carbon atom uses to form bonds to the oxygen atoms?

9.81 Acrylonitrile, CH_2CHCN, is used in the synthesis of acrylic fibers (polyacrylonitriles), such as Orlon. Write the Lewis structure of acrylonitrile, and describe the hybrid orbitals on each carbon atom. What is the approximate value of the CCC bond angle?

9.82 Write the Lewis structure of ClF_3, and identify the orbitals used for bonding by the Cl and F atoms. What is the shape of the molecule? Estimate the value of the FClF bond angle.

9.83 Borazine, $B_3N_3H_6$, a compound that has been called "inorganic benzene" because of its similar hexagonal structure (but with alternating B and N atoms in place of C atoms), is the basis of a large class of boron-

CASE 9 ▸ ELECTRICAL CONDUCTION BY COVALENT COMPOUNDS

The movement of charge from one point to another commonly occurs through a metal, such as copper, or through a molten salt or aqueous salt solution, such as aqueous sodium chloride; most covalently bonded solids are nonconductors of electricity. Graphite, with its planar sheets of sp^2-hybridized C atoms (and a virtually infinite extension of benzenelike rings in each plane), is a familiar exception, for its unhybridized *p* orbitals provide a pathway by which an electron can move from one C atom to another when a potential difference is applied.

The recent emergence of covalently bonded *conducting polymers* (which are sometimes called synthetic metals), however, has changed the way scientists think about moving charge. Conducting polymers have properties that are very different from those of metals; but one day conducting polymers may be used in place of metals in electrical wiring because they do not corrode. Already conducting polymers are being used for some rechargeable batteries, thus eliminating the need for cadmium in these units (see Chapter 17).

All conducting polymers have a common bonding feature: a linear chain of carbon-carbon bonds that are sp^2-hybridized, with N or S atoms as a component of the chains. The first conducting polymer was polyacetylene, $(CH)_n$, a long chain of atoms in which each C atom is sp^2-hybridized:

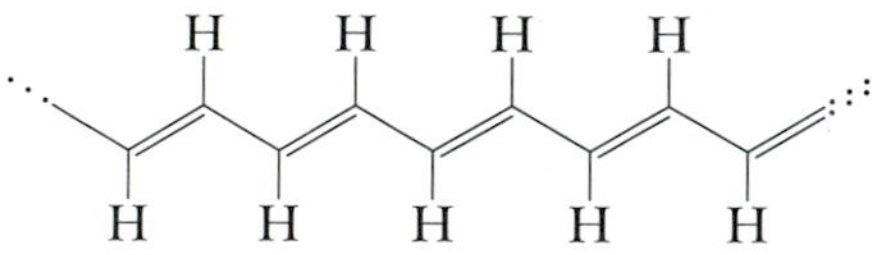

When an electron from an external source enters an unhybridized *p* orbital of a C atom in the polyacetylene molecule, it moves down the chain of C—C bonds as shown in the illustration:

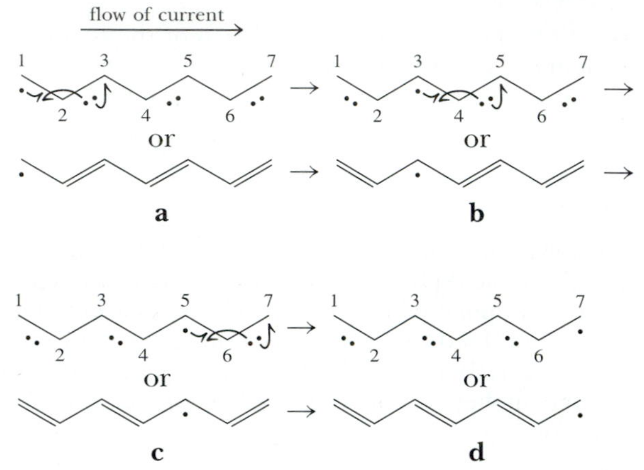

flow of current

a b c d

When an electron enters the chain at carbon atom 1, we can visualize the breakage of the π bond between carbon atoms 2 and 3 (**a**); then one of the broken π bond's electrons pairs with the lone electron to form a new π bond between C atoms 1 and 2, thereby leaving an unpaired electron on C atom 3 (**b**). Next, the π bond between C atoms 4 and 5 is broken, one electron pairing with the lone electron on C atom 3 and forming a new π bond between C atoms 3 and 4. The lone electron now appears on C atom 5 (**c**). Then the additional electron can move to C atom 7 by the same sequence of events (**d**). The process can continue down the chain of the molecule, thereby turning it into a conducting path.

nitrogen compounds. Write its Lewis structure, and predict the composition of the hybrid orbitals used by each B and N atom.

9.84 Nitrogen, phosphorus, oxygen, and sulfur exist as N_2, tetrahedral P_4, O_2, and ringlike S_8 molecules. Rationalize this fact in terms of the abilities of the atoms to form different types of bonds with one another. In each case, describe the bonding in terms of the hybridization of the atoms.

9.85 Xenon forms XeO_3, XeO_4^{2-}, and XeO_6^{4-}, all of which are powerful oxidizing agents. Give their Lewis structures, and account for the bonding in terms of hybridization of the xenon atom.

9.86 Explain (in terms of molecular orbitals) why the CF^+ bond is shorter than the CF bond.

Other conducting polymers that have been extensively researched are polypyrrole

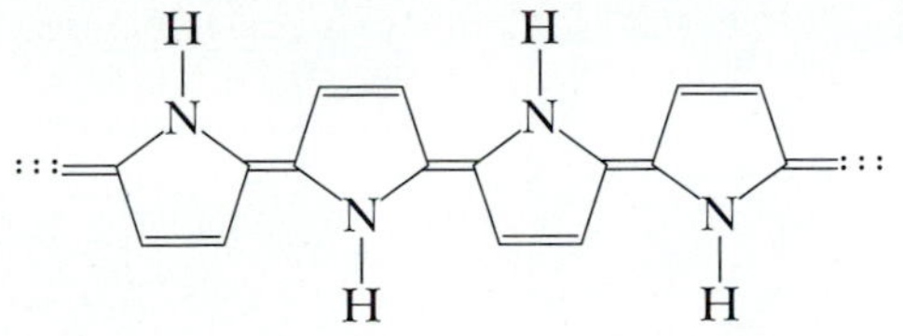

and polyaniline

Polypyrrole has been used in "smart" windows, windows that change color (from a transparent yellow-green to blue-black) in response to changes in temperature or amount of sunlight. Polyaniline has been used to design rechargeable batteries made as flat buttons or as laminated, rolled films.

QUESTIONS

1. (a) Write the Lewis structure of acetylene. (b) Sketch the valence-bond picture for six C atoms in a polyacetylene of formula —$(CH)_6$—, showing all σ and π bonds. (c) Explain how an electron added to the first C atom in the carbon-carbon chain migrates to C atom 6. (d) What is the hybridization of C atom 5 shown in part (c) of the illustration on the facing page?

2. (a) What is the hybridization of the N atom in polypyrrole? (b) Sketch the valence-bond structure of polypyrrole. (c) Starting with the diagram, show how polypyrrole conducts electricity by the migration of the position of the double bonds as the electron migrates from A to B to C to D.

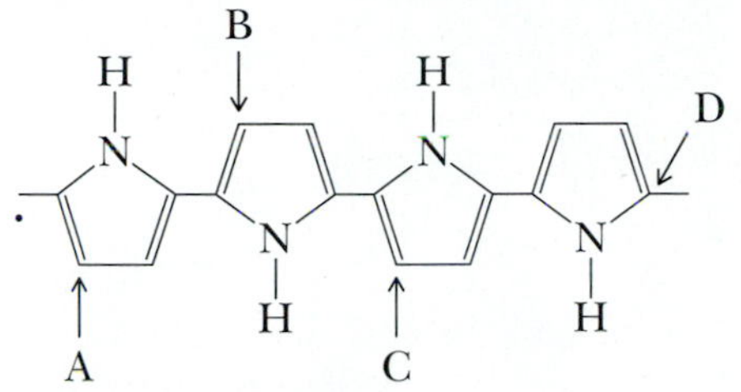

3. Using molecular orbital theory, account for the stability of polyacetylene by sketching the molecular-orbital energy-level diagram for the electrons in the π system; use —$(CH)_6$— as a model.

9.87 Would you expect the HeH^- ion to exist? What would be its bond order? Compare the stability of HeH^- with that of HeH^+.

9.88 Molecular orbital theory predicts that O_2 is paramagnetic. What other Period 2 homonuclear diatomic molecules are also paramagnetic?

9.89 The electron configurations of molecules derived by using Fig. 9.25 depend on the Pauli exclusion principle. Suppose we lived in a universe in which *three* electrons could occupy one orbital. What would be the electronic structure of O_2? Could Ar_2 exist? Suggest what rules for writing Lewis structures you would have proposed if you had been the G. N. Lewis who lived in that universe.

This chapter explains the origin of the forces between molecules, and how those forces account for the condensation of gases to liquids and solids. We see that forces between atoms, molecules, and ions account for the physical properties of liquids and the structures of solids. The illustration shows the condensation of water on the interior of a flask: that water vapor condenses to a liquid and also adheres to glass is a sign that there are forces acting between molecules.

10 LIQUIDS AND SOLIDS

Social revolutions have followed from the discovery of new liquids and solids. The discovery of metals raised civilization out of the Stone Age. The development of steels yielded materials strong enough to make possible the Industrial Revolution. The discovery of natural sources of liquid hydrocarbons—petroleum—transformed society by providing fuel for light and mobile power sources in vehicles. The organic solids we call plastics—including polyethylene and nylon—have transformed people's lives during this century. We are still in the middle of a revolution that started with the discovery of semiconductors. We are at the start of the revolution that many hope will follow from the discovery, in the late 1980s, of high-temperature superconductors, substances that conduct electricity without resistance not far below room temperature and which may contribute to faster transportation, smaller and faster computers, and better communications.

In this chapter we consider some of the properties of the liquid and solid states of matter and the forces that bind them together. The only liquids we consider are those made up of molecules, which include water, ethanol, benzene, and liquefied gases. These molecular liquids are of the greatest importance in chemistry and the everyday world, because they include common solvents and natural waters. We do not consider molten salts, which are liquid mixtures of ions, nor do we consider molten metals.

The only substance that has more than one liquid form is helium, but many have several solid forms, which collectively are called **polymorphs**. Diamond and graphite, for instance, are polymorphs of carbon; and calcite and aragonite are polymorphs of calcium carbonate. The term **allotrope** denotes forms of *elements* that differ in the way the atoms are bonded. Thus, carbon and diamond are allotropes of carbon but calcite and aragonite are not allotropes (they are compounds, not elements). Oxygen, O_2, and ozone, O_3, on the other hand are allotropes but not polymorphs (because they are both gases).

D Allotropic forms exhibit. TDC, 19-5.

N Ice forms a number of solid phases with different crystal structures. Which one is stable depends on the temperature and the pressure; some phases melt only at 100°C, but that is under great pressure. See Case 10.

The term used to distinguish the different physical states and the alternative forms of each state is **phase** (from the Greek word for "appearance"). Thus we can speak of the gas, liquid, and solid phases of water. We can also speak of the diamond and graphite phases of solid carbon and of the calcite and aragonite phases of calcium carbonate (Fig. 10.1). The properties of the gas phase of matter were discussed in Chapter 5.

Different phases of solids exist because the atoms, ions, or molecules may stack in several ways, the different arrangements depending on the temperature and pressure at which the solid forms. Even when the conditions change after the solid has formed (as when diamonds are moved upward by geological forces from the site where they were formed deep inside the Earth's crust), the particles cannot rearrange themselves and the phase survives. Some solid phases, however, do convert to the most stable form at room temperature and pressure if they are left under those conditions for long periods. For example, aragonite forms from warm, tropical oceans and slowly converts to calcite over geological periods of time. By measuring the proportion of calcite and aragonite in a rock sample it is possible to estimate the approximate date at which the rock was deposited.

FORCES BETWEEN ATOMS, IONS, AND MOLECULES

E Teaching ion-ion, ion-dipole, and dipole-dipole interactions. *J. Chem. Educ.*, **1977**, *54*, 402.

The bulk of this section describes the origins of **intermolecular forces**, the attractions and repulsions (collectively, the interactions) between neutral closed-shell atoms and molecules (we first encountered these interactions in Chapter 5). Intermolecular forces are also called **van der Waals forces** after Johannes van der Waals, whose contribution to the study of the effects of intermolecular forces in real gases was described in Section 5.7. We shall concentrate mainly on the attractive forces.

FIGURE 10.1 The calcite (left) and aragonite phases (right) of calcium carbonate.

TABLE 10.1 Interionic and intermolecular forces

Type of interaction	Typical energy,* kJ/mol	Interacting species
ion-ion	250	ions only
ion-dipole	15	ions and polar molecules
dipole-dipole	2	stationary polar molecules
	0.3	rotating polar molecules
London (dispersion)	2	all types of molecules
hydrogen bonding	20	N,O,F; the link is a shared H atom

*Typical strengths are for a distance of 500 pm.

E The distance dependence of the potential energy of interaction varies as $1/r$ for ion-ion interactions, as $1/r^3$ for ion-dipole interactions and between stationary polar molecules, as $1/r^6$ for interactions between rotating polar molecules and for the London interaction. The hydrogen bond occurs only for species in contact.

10.1 ION AND DIPOLE FORCES

The various types of interionic and intermolecular forces are summarized in Table 10.1. We have already met the electrostatic forces between ions of opposite charge in an ionic solid (Section 8.1). Only slightly more complex is the force between an ion and the *partial* charges of the electric dipole of a polar molecule or between the partial charges of two dipoles. More problematic is the origin of the interaction between nonpolar molecules, but we shall see shortly how they arise.

The stronger the force, the more likely it is to play a dominating role in determining the structure and properties of a substance. However, not all substances possess all types of interactions. The strongest interaction—that between ions—is present only when the interacting species are ions; dipole-dipole interactions exist only when both species are polar. The strong hydrogen bonding interaction exists only when the molecules fulfil the criteria for hydrogen bonding (see Section 10.2). The only interaction that exists between *all* types of species—ions, polar molecules, and nonpolar molecules—is the London interaction.

N When discussing the distance-dependence, care should be taken to distinguish between the *force* of an interaction and the *potential energy* of that interaction. If the potential energy varies as $1/r^n$, then the force varies as $1/r^{n+1}$. Thus, although the potential energy of the London interaction varies as $1/r^6$, the force of the interaction varies as $1/r^7$. We have used the term *force* when that is appropriate, and the term *interaction* when there is no need to distinguish between the potential energy and the force.

T OHT: Table 10.2 and Fig. 10.2.

Ion-ion interactions. As we saw in Section 8.1, the strength of the interaction between ions (more precisely, their potential energy) decreases as the ions separate and increases as the magnitude of their charge increases. The strength of the **ion-ion interaction** decreases quite slowly with increasing distance. For example, it falls to only half its value when the distance between the ions doubles. As a result, it binds together ions separated by long distances, not just the closest ions.

The strong, long-range attraction between ions of opposite charge is responsible for the high lattice enthalpies and generally high melting and boiling points of ionic compounds (Table 10.2). Melting points and lattice enthalpies are particularly high for solids composed of small, highly charged ions, because the electrostatic forces can grip such ions together strongly. This strength of the interaction is shown by the fact that the aluminum oxide melts only at 2015°C and that magnesium oxide sublimes at 2800°C. The trend to lower melting points with increasing ion size is illustrated by the alkali chlorides, which have melting points that fall from 801°C for NaCl to 645°C for CsCl (Fig. 10.2).

When the melting point of a specific compound conflicts with this trend, for example, when a small, highly charged ion has an unusually

TABLE 10.2 Normal melting and boiling points of ionic solids

Ionic solid	Melting point, °C	Boiling point, °C
LiF	842	1676
LiCl	614	1382
NaCl	801	1413
KCl	776	1500*s**
$MgCl_2$	708	1412
MgO	2800*s*	3600
$CaCl_2$	782	2000
Al_2O_3	2015	2980

*The *s* indicates that the solid sublimes.

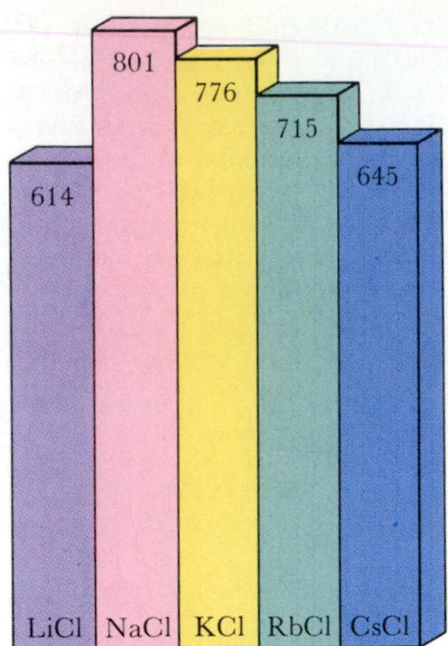

FIGURE 10.2 The melting points of the alkali metal halides show a variation that can be attributed in large part to the strength of the interactions between the ions. The values given here are in degrees Celsius.

low melting point, we take it as evidence that there is an appreciable amount of covalent character in the bonding of the crystal. Lithium chloride is such a solid; in LiCl the small Li^+ cation polarizes the Cl^- ions that surround it and hence introduces covalent character into the bonding. A similar remark applies to the oxides of aluminum and magnesium. As noted earlier, the solid with the *smaller* cation (Al_2O_3) has the *lower* melting point (2015°C compared with 2800°C for MgO). We conclude that the polarizing power of the highly charged, small cation Al^{3+} polarizes the neighboring oxide ions, introduces some covalent character into the bonding, and hence lowers the melting point of the solid.

Ion-dipole interactions. The **ion-dipole interaction** is the attraction between a cation and the partial negative charge of a polar molecule (**1**) or between an anion and the partial positive charge (**2**). The strength of the ion-dipole interaction decreases more rapidly with increasing distance than the ion-ion interaction does because the partial charge at one end of the polar molecule tends to cancel the effect of the opposite partial charge at the other end. Not only does the strength of the effect of *each* charge diminish with distance, but also the two partial charges of the dipole appear to cancel each other at large distances; so their net effect decreases rapidly with distance. To be precise, the energy of the interaction decreases as the inverse square of the distance d between them—as $1/d^2$ rather than $1/d$ (the effect of a single charge in the ion-ion interaction).

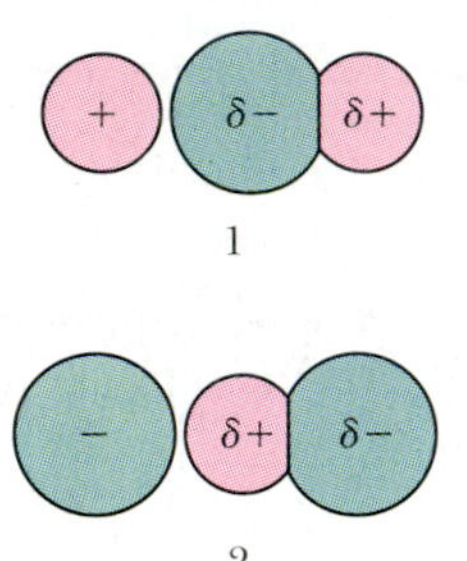

1

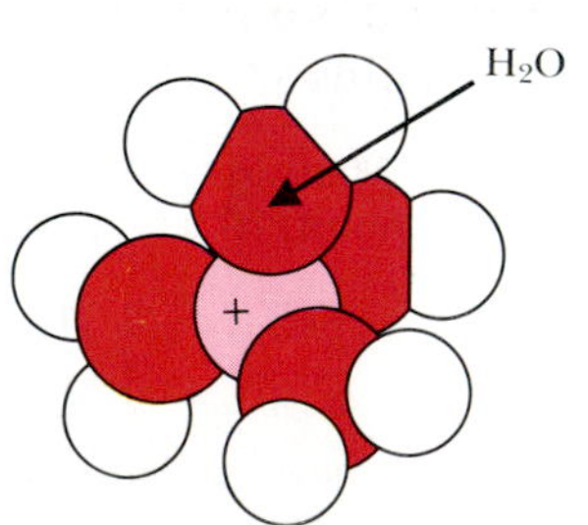

2

The ion-dipole interaction is very important for the **hydration** of cations in solution (the strong attraction of the polar water molecules to an ion), because the partial negative charge on the O atom of H_2O is attracted to the positive charge of the cation (Fig. 10.3). Hydration often persists in the solid, where it accounts for the existence of salt hydrates such as $Na_2CO_3 \cdot 10H_2O$ and $CuSO_4 \cdot 5H_2O$. Because the strength of the interaction is greater the smaller the distance between the ion and the dipole, we expect small cations to be more strongly hydrated than large ones. This expectation is consistent with the fact that lithium and sodium commonly form hydrated salts, whereas their larger congeners (potassium, rubidium, and cesium) do not. Ammonium salts are usually anhydrous for a similar reason, because the NH_4^+ ion has about the same radius (143 pm) as a Rb^+ ion has (148 pm). The effect of charge is seen by comparing barium and potassium. Their cations have similar radii (135 pm for Ba^{2+} and 133 pm for K^+), but barium salts are often hydrated because of the barium ion's higher charge (for example, barium chloride is found as $BaCl_2 \cdot 2H_2O$ but potassium chloride exists as KCl). Lanthanum, barium's neighbor, is both smaller (115 pm) and more highly charged (La^{3+}), and its salts include $La(NO_3)_3 \cdot 6H_2O$ and $La_2(SO_4)_3 \cdot 9H_2O$.

FIGURE 10.3 The ion-dipole interaction makes an important contribution to the hydration of ions, particularly of cations. Here we see how several H_2O molecules are attached to a central cation as a result of the interaction.

EXAMPLE 10.1 Judging the relative strengths of ion-dipole interactions

Which has the stronger interaction with a water molecule, a Na^+ ion or a K^+ ion?

STRATEGY Because the energy of the interaction depends on the distance between the cation and the electric dipole, we need to decide which ion can attract the water molecule more strongly. The strength of attraction increases as charge increases and distance between them decreases. Both ions have the same charge, so their size must be the feature to compare. (Recall from Chapter 7 that atomic and ionic radii vary.)

SOLUTION Ionic radii increase down groups, so we expect a K^+ ion to be larger than a Na^+ ion. That being so, a K^+ ion will have the weaker interaction with a water molecule. In practice, the energy of the interaction of K^+ is only about half that of Na^+.

EXERCISE E10.1 Which has the stronger attraction to a water molecule, Na^+ or Mg^{2+}?

[*Answer*: Mg^{2+}; because of its greater charge]

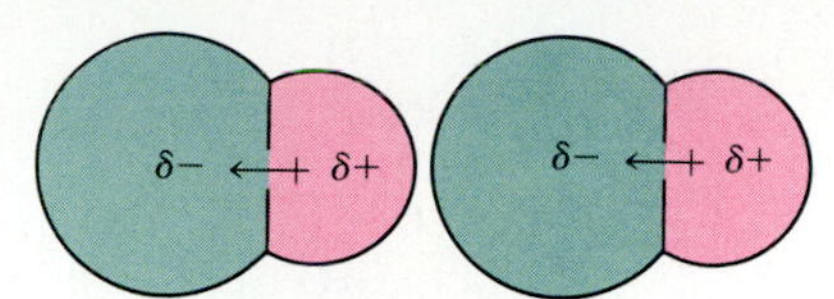

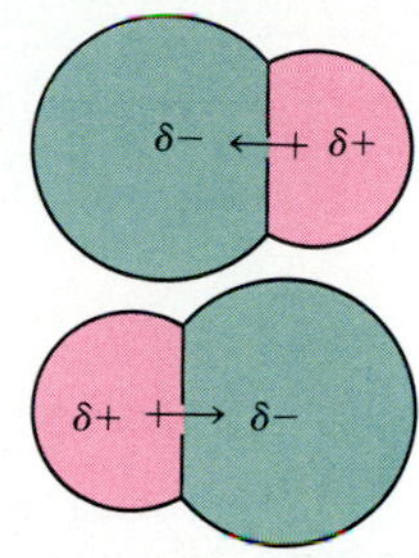

FIGURE 10.4 Polar molecules attract each other by the interaction between the partial charges of their electric dipoles (represented by the arrows). Either of these orientations (end-to-end or side-by-side) is energetically favorable.

Dipole-dipole interactions. A **dipole-dipole interaction** is the attraction between the electric dipoles of polar molecules. When two neighboring polar molecules have either of the orientations shown in Fig. 10.4, the partial charges of opposite sign are closer together than those of like sign and there is a net attraction between the molecules. Because the dipole-dipole force arises from the interaction between *partial* charges, it is weaker than the attraction between oppositely charged ions (which have a full charge). The relative strengths of dipole-dipole and ion-ion interactions can be seen by comparing the enthalpy of vaporization with the lattice enthalpy. The enthalpy of vaporization, ΔH°_{vap}, is the enthalpy change per mole for the process

$$A(s \text{ or } l) \longrightarrow A(g)$$

and the lattice enthalpy is the enthalpy change per mole for the process

$$MX(s) \longrightarrow M^+(g) + X^-(g)$$

so both quantities measure the energy needed to rip the liquid or solid apart into its components. The enthalpy of vaporization of solid hydrogen chloride, which is 18 kJ/mol, refers to the process in which the molecules are separated into a gas:

$$HCl(s) \longrightarrow HCl(g)$$

Much of the energy requirement is used to overcome the dipole-dipole interactions between the molecules when they are close together in the solid. The lattice enthalpy of sodium chloride, 787 kJ/mol, refers to the process

$$NaCl(s) \longrightarrow Na^+(g) + Cl^-(g)$$

It is much larger because ion-ion interactions are much stronger than dipole-dipole interactions.

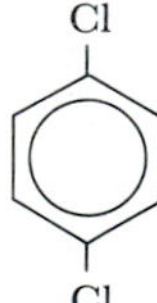

3 *p*-Dichlorobenzene

Cl
Cl

4 *o*-Dichlorobenzene

▼ EXAMPLE 10.2 Predicting boiling points on the basis of dipole-dipole interactions

Which would you expect to have the greater boiling point, *p*-dichlorobenzene (**3**) or *o*-dichlorobenzene (**4**)?

STRATEGY The two molecules have the same numbers of atoms of each element, so a clue may be found in the arrangement of atoms in the molecule. If the two molecules have different dipole moments, then we could ascribe the higher boiling point to the more strongly polar molecule because we expect polar molecules to interact more strongly than otherwise similar

Cl H
C
C
Cl H

5 *cis*-Dichloroethene

Cl H
C
C
H Cl

6 *trans*-Dichloroethene

E The *London formula* states that the potential energy of the interaction is proportional to $\alpha^2 I/r^6$, where α is the polarizability of the molecules, I is their ionization energy, and r their separation. We see that the more polarizable the molecules, the greater the strength of the London interactions. Highly polarizable molecules are those for which the nuclei have little control over the valence electrons, and therefore the fluctuations in the electron cloud can be large.

N A flickering instantaneous dipole could be likened to the wobbling of a cube of jello on a plate: its shape is momentarily distorted, and the larger the cube the greater the amplitude of the wobbles. The London interaction is responsible for sulfur being a solid (see front cover).

T OHT: Figs. 10.5 and 10.6.

molecules. We saw in Section 9.3 how to decide whether a molecule is polar or not.

SOLUTION The *p*-dichlorobenzene molecule is nonpolar because the dipole moment of one C—Cl bond is canceled by that of the other C—Cl bond diametrically opposite across the ring. The *o*-dichlorobenzene molecule is polar, because the two C—Cl dipole moments do not cancel. We therefore predict that *o*-dichlorobenzene will have a higher boiling point than *p*-dichlorobenzene. The experimental values are 180.4°C for *o*-dichlorobenzene and 174.1°C for *p*-dichlorobenzene.

EXERCISE E10.2 Which will have the higher boiling point, *cis*-dichloroethene (**5**) or *trans*-dichloroethene (**6**)?

▲ [*Answer*: *cis*-dichloroethene (experimental: cis, 60.7°C; trans, 47.7°C)]

The London interaction. The **London interaction** is an attractive interaction that occurs even between nonpolar species, such as atoms and nonpolar molecules. It arises from the fluctuations in the electron distributions in the species. We should not think of a molecular electron cloud as a frozen, static distribution but as the average position of the continuously moving electrons. As the electrons move about, some of their momentary locations may produce an instantaneous electric dipole in the molecule. This instantaneous dipole occurs even if the *average* positions of the electrons are those of a nonpolar molecule. In fact, even a noble-gas atom can have an instantaneous dipole of this kind. The flickering instantaneous dipoles on two neighboring molecules tend to align and come into step with each other, because the energy is lower when the partial positive charge of one is near the partial negative charge of the other (Fig. 10.5). As a result, the two instantaneous dipoles attract each other and give rise to the London interaction.

The London interaction is named for the German-American physicist Fritz London, who first explained it. It acts between *all* types of molecules, polar as well as nonpolar (and between noble-gas atoms); it is always attractive. The London interaction is responsible for the existence of benzene as a liquid at normal temperatures, for the condensation of carbon dioxide to a solid, and for the condensation of oxygen, nitrogen, and the noble gases to liquids at low temperatures.

The strength of the London interaction increases as molecular size (both molar mass and molecular radius) increases. Larger molecules have more electrons, which produce greater fluctuations in charge as they flicker between different positions, and these, in turn, lead to bigger instantaneous dipoles. The strength of the interaction depends on the polarizabilities (the ease of distortion, which was discussed in

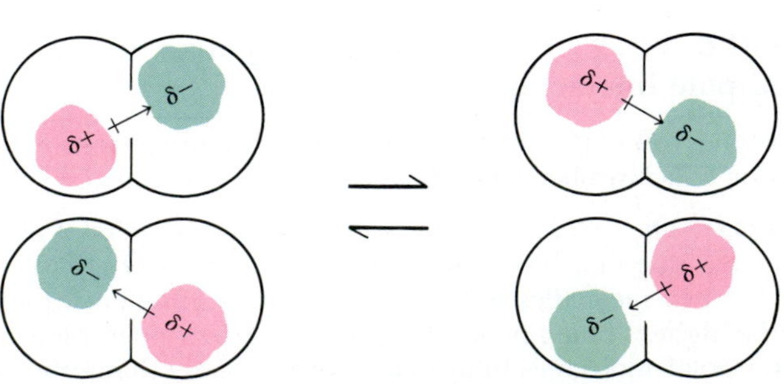

FIGURE 10.5 The London force arises from the attraction between two instantaneous dipoles. Although they are continuously changing direction, they remain in step in the most favorable orientation.

Section 9.3) of the molecules and is largest when both are highly polarizable. This feature fits the flickering-dipole picture of the interaction, because a highly polarizable molecule is one in which the nucleus has only weak control over its outer electrons and the fluctuations in electron distribution can be large. The potential energy of the London interaction decreases very rapidly as the distance between the two molecules increases (as the sixth power of the distance, or as $1/d^6$).

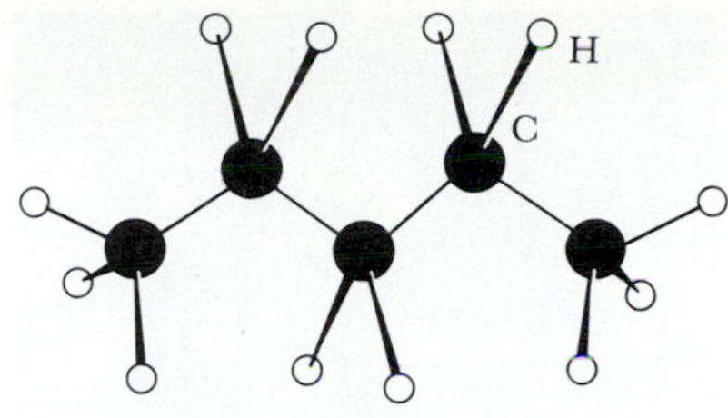

7 Pentane

Because an H_2 molecule has only two electrons, its instantaneous dipoles are very small. As a result, it interacts so weakly with neighboring H_2 molecules that hydrogen gas condenses to a liquid only if the temperature is lowered to −253°C (20 K) at 1 atm pressure. This effect is general: *small nonpolar molecules, with few electrons, have very weak London interactions and hence low melting and boiling points.* However, it is very difficult to make precise predictions about boiling points, because the strength of the London interaction is also determined in part by the shape of the molecules and how they pack together in the liquid phase. Both pentane (**7**) and neopentane (**8**), for instance, have the molecular formula C_5H_{12}, but they have markedly different boiling points (36°C and 10°C, respectively). The instantaneous dipoles on neighboring cylindrical molecules can approach one another closely, whereas those belonging to neighboring spherical molecules cannot (Fig. 10.6). Hence the London interaction is stronger between cylindrical molecules than between spherical molecules of the same mass.

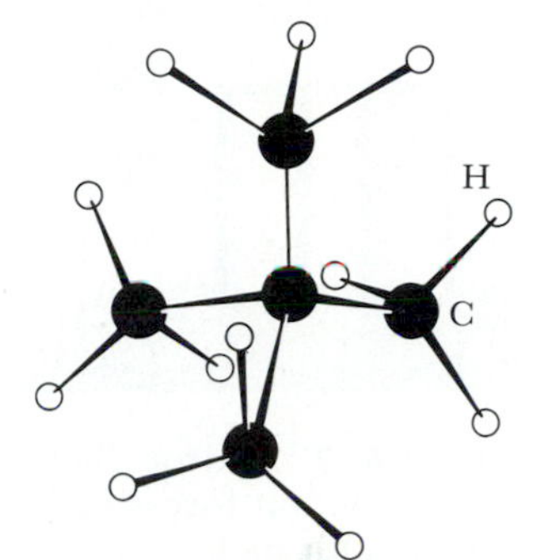

8 Neopentane

Large differences in polarizability, however, can be more important than molecular shape. This characteristic is one of the reasons why hydrocarbons with up to 4 C atoms are gases, hydrocarbons with 5 through 17 C atoms (such as benzene and octane) are liquids, and the heavier hydrocarbons (those containing 18 or more C atoms) are waxy solids (Fig. 10.7). In each case, the compound with the greater molar mass—and thus the greater number of electrons—has the stronger London interactions with its neighbors.

The increase in strength of the London force is striking when heavier atoms are substituted for hydrogen atoms in a molecule. Methane boils at −162°C but carbon tetrachloride, with its much more polarizable Cl atoms, is a liquid at room temperature and boils at 77°C. Carbon tetrabromide is even more polarizable. It is a solid at room temperature, melting at 90°C and boiling at 190°C. Covalent fluorides are generally more volatile than covalent chlorides, for fluorides have fewer electrons. Compare, for example, the values for CF_4 and CCl_4 in Table 10.3. The high nuclear charge of the fluorine nucleus compared with the proton and its firm control over the surrounding electrons also means that fluorocarbons are significantly less polarizable than hydrocarbons of a similar shape. Thus, although CF_4 and neopentane (**8**) have similar molar masses (88 g/mol and 72 g/mol, respectively), the boiling point of CF_4 is 139°C lower (−129°C for tetrafluoromethane, 10°C for neopentane). One of the reasons why chlorofluorocarbons (which are hydrocarbons with some or all of their hydrogen atoms replaced by chlorine and fluorine atoms) have been used in refrigerating and air conditioning units is that they have convenient boiling points for such applications, are not flammable, and are not toxic. They are highly reactive in the upper atmosphere, however, as explained in Box 5.1, and are enemies of the ozone layer.

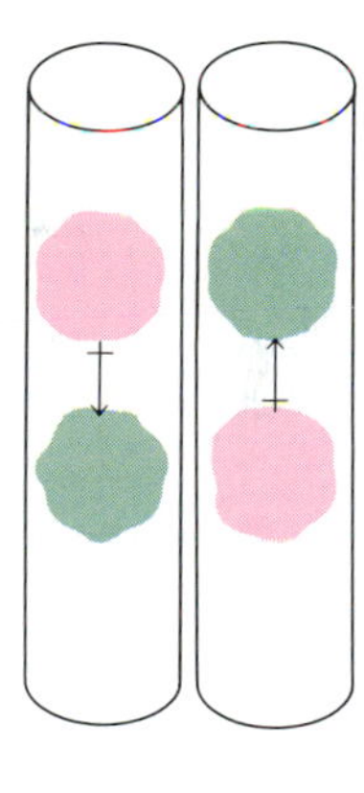

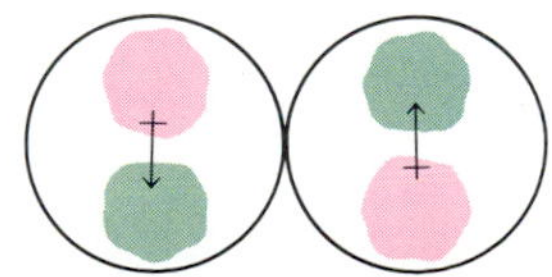

FIGURE 10.6 The instantaneous dipole moments in two neighboring cylindrical molecules tend to be close together and interact strongly. Those in neighboring spherical molecules (such as neopentane, $(CH_3)_4C$) tend to be far apart and interact weakly.

The London interaction between polar molecules is usually stronger than their dipole-dipole interaction. As the following example shows,

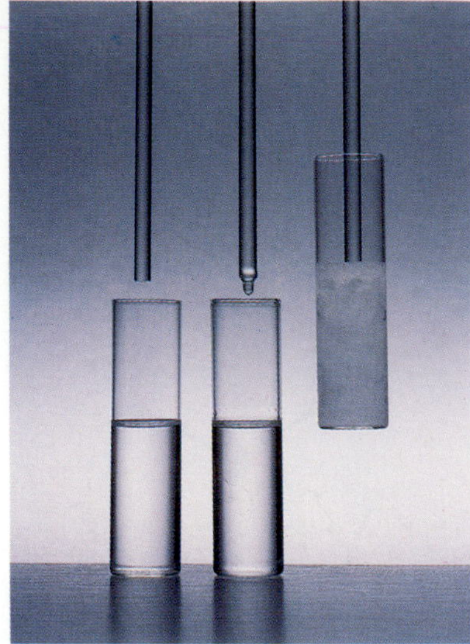

FIGURE 10.7 Hydrocarbons show how London forces increase in strength with molar mass. Pentane is a mobile fluid (left); pentadecane ($C_{15}H_{32}$), a viscous liquid; and octadecane ($C_{18}H_{38}$), a waxy solid (right).

TABLE 10.3 Normal melting and boiling points of substances

Substance	Melting point, °C	Boiling point, °C	Substance	Melting point, °C	Boiling point, °C
Noble gases			*Small inorganic species*		
He	−270	−269	H_2	−259	−253
	(3.5 K)	(4.2 K)	N_2	−210	−196
Ne	−249	−246	O_2	−218	−183
Ar	−189	−186	H_2O	0	100
Kr	−157	−153	H_2	−86	−60
Xe	−112	−108	NH_3	−78	−33
Halogens			CO_2	—	−78*s**
F_2	−220	−188	SO_2	−76	−10
Cl_2	−101	−34	*Organic compounds*		
Br_2	−7	59	CH_4	−183	−162
I_2	114	184	CF_4	−150	−129
Hydrogen halides			CCl_4	−23	77
HF	−83	20	C_6H_6	6	80
HCl	−114	−85	CH_3OH	−94	65
HBr	−89	−67	glucose	142	*d**
HI	−51	−35	sucrose	184d	—

*An *s* indicates that the solid sublimes, a *d* that it decomposes.

N It is probably worth emphasizing (as the following example helps to do) that the London interaction is the strongest interaction (other than hydrogen bonding) between closed-shell neutral molecules.

T OHT: Table 10.3.

U Exercise 10.65.

E On the boiling points of the alkyl halides. *J. Chem. Educ.*, **1988**, *65*, 62.

D Flow of liquids through pipes: liquid viscosities. BZS3, 9.27.

we can infer from trends in boiling points that the condensation of hydrogen chloride to a liquid is due more to the London interaction between the molecules than to their dipolar interaction.

▼ EXAMPLE 10.3 Accounting for a trend in boiling points

Explain the trend in the boiling points of the hydrogen halides, bearing in mind that the electronegativity difference between hydrogen and the halogen *decreases* from HCl to HI: HCl, −85°C; HBr, −67°C; HI, −35°C.

STRATEGY We need to consider the strengths of the dipole-dipole and London interactions and their likely effects on boiling points. Stronger attractive interactions suggest higher boiling points. The size of the dipoles, and hence of the dipole-dipole interaction, increases as the difference in electronegativity between hydrogen and the halogen increases. The strength of the London interaction increases as the number of electrons increases.

SOLUTION The electronegativity differences decrease from HCl to HI. Hence the molecules become less polar from HCl to HI, and the dipolar interactions decrease too, a trend suggesting that the boiling points should decrease. This prediction conflicts with the data. The number of electrons in the molecules increases from HCl to HI, so we expect the London interaction to increase too. Hence, the boiling points should increase from HCl to

HI, in accord with the data. This analysis suggests that the London interaction dominates the dipolar interactions for these molecules.

EXERCISE E10.3 Account for the trend in boiling points of the noble gases, which increase from helium to radon.

[*Answer*: London interactions increase as the number of electrons increases]

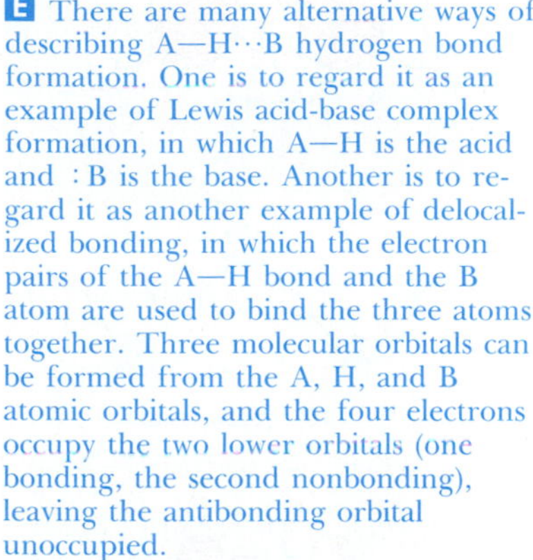

E There are many alternative ways of describing A—H···B hydrogen bond formation. One is to regard it as an example of Lewis acid-base complex formation, in which A—H is the acid and :B is the base. Another is to regard it as another example of delocalized bonding, in which the electron pairs of the A—H bond and the B atom are used to bind the three atoms together. Three molecular orbitals can be formed from the A, H, and B atomic orbitals, and the four electrons occupy the two lower orbitals (one bonding, the second nonbonding), leaving the antibonding orbital unoccupied.

E A model for hydrogen bonding. *J. Chem. Educ.*, **1986**, *63*, 503. Dielectric aspects of the hydrogen bond interaction. *J. Chem. Educ.*, **1969**, *46*, 17.

10.2 HYDROGEN BONDING

The existence of another type of intermolecular force becomes apparent when we examine the trends in boiling points of a range of compounds containing hydrogen (Fig. 10.8). Water boils at a much higher temperature (100°C) than hydrogen sulfide (−60°C). In fact, hydrogen sulfide is a gas at room temperature even though an H_2S molecule has many more electrons—and hence stronger London forces—than an H_2O molecule has. Ammonia and hydrogen fluoride also have higher boiling points than the analogous hydrogen compounds of their congeners. The unusually high boiling points of water, ammonia, and hydrogen fluoride suggest that there are unusually strong forces between their molecules.

The strong intermolecular forces between NH_3, H_2O, and HF molecules are the result of hydrogen bonding:

A **hydrogen bond** is a link formed by a hydrogen atom lying between two strongly electronegative atoms.

N Hydrogen bonding is similarly important in substituted NH_3 and H_2O molecules, such as amines and alcohols (but not ethers).

D Hydrogen bonding in liquids. TDC, 143.

The other intermolecular forces we have already considered act between compounds built from any elements. Hydrogen bonding is exceptional because it requires the joint presence of hydrogen and two electronegative atoms. Only N, O, and F are sufficiently electronegative to take part in hydrogen bonding in neutral molecules.

The key to the formation of a hydrogen bond is the strongly polar nature of the A—H bond when A is strongly electronegative. The resulting partial positive charge on H can attract the lone-pair electrons

T OHT: Fig. 10.8.

E Hydrogen bonding and proton transfer. *J. Chem. Educ.*, **1982**, *59*, 362.

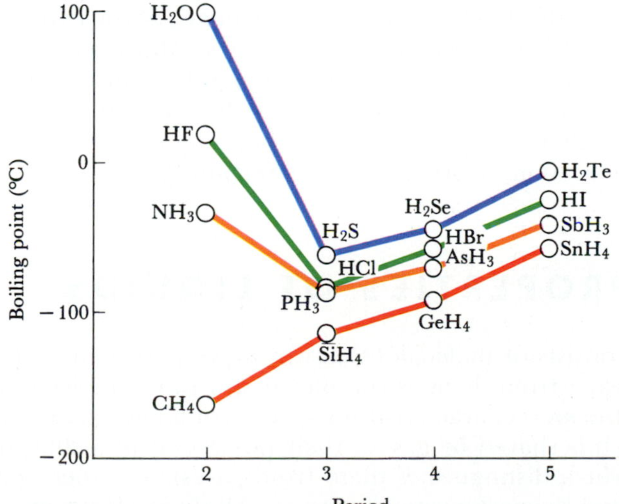

FIGURE 10.8 The boiling points of the hydrogen compounds of the *p*-block elements show a smooth variation, except those of nitrogen (NH_3), oxygen (H_2O), and fluorine (HF), which are strikingly out of line.

of a neighboring atom B, especially if that atom is so electronegative that it also has a strong partial negative charge. Hydrogen is unique in being able to form this kind of bond because it is so small that the atom B can approach very close to the partial charge of the hydrogen atom and interact with it strongly. The only important cases are

N—H···N	O—H···N	F—H···N
N—H···O	O—H···O	F—H···O
N—H···F	O—H···F	F—H···F

9 O—H···O

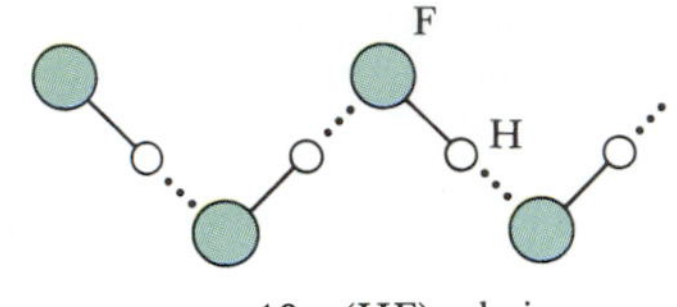

10 $(HF)_4$ chain

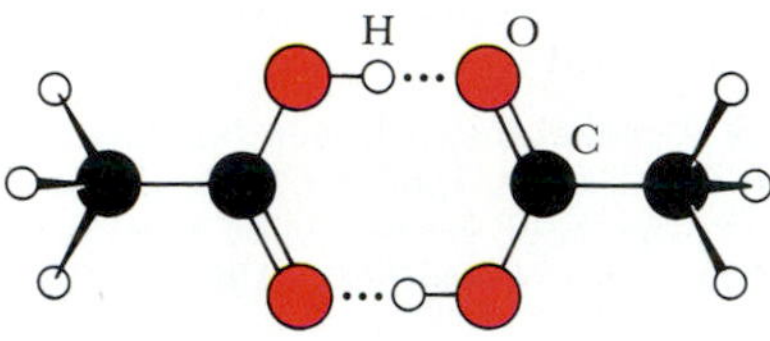

11 Acetic acid dimer

D Using silica to demonstrate hydrogen bonding. *J. Chem. Educ.*, **1972**, *49*, 419.

E Hydrogen bonding is responsible for the linking of the two strands of a DNA double helix: it is strong enough for stability but not so strong that enzymatic activity cannot overcome the linking between the strands.

E The loss of structure of a protein is called its *denaturation*. The milkiness stems from the presence of denatured molecules that form groups with dimensions comparable to that of the wavelength of light.

E The description of liquid structure. *J. Chem. Educ.*, **1973**, *50*, 119.

Hydrogen bonds may form to the anions of other elements, for example, the O—H···Cl^- bond between a chloride ion and a hydrating water molecule. In such cases, the negative charge of the anion assists the bond formation.

The strength of an O—H···O hydrogen bond (**9**), about 20 kJ/mol, is only a fraction of the strength of a normal O—H bond, which is 463 kJ/mol. When it exists, however, a hydrogen bond is strong enough to dominate all the other types of intermolecular interactions. Hydrogen bonding is so strong that it can survive even in a vapor. Liquid hydrogen fluoride, for instance, contains zigzag chains of HF molecules (**10**), and the vapor contains short fragments of the chains and $(HF)_6$ rings. Acetic acid vapor contains **dimers**—pairs of molecules (**11**)—linked by two hydrogen bonds.

Hydrogen bonding, in conjunction with ion-dipole forces, is responsible for the hydration of oxoanions in solution and in salt hydrates: it contributes to the high degree of hydration in $Ba(OH)_2 \cdot 8H_2O$, for example. We shall also see that it has a profound influence on the properties of water itself, including its existence as a liquid at room temperature. Hydrogen bonding also plays a very important role in determining the shapes and properties of biological molecules. The shapes of protein molecules, which govern almost all the chemical reactions in living cells, are determined largely by hydrogen bonds; and when the bonds are broken, the protein loses its function. When we cook an egg, the clear albumin becomes milky white because the heat breaks the hydrogen bonds in its molecules, which collapse into a random jumble. Cellulose molecules, which have many —OH groups, can form many hydrogen bonds with one another, and the strength of wood can be traced in large part to the strength of the hydrogen bonds between neighboring ribbonlike molecules. Many wood and paper glues are substances that form hydrogen bonds to the two surfaces they join. Hydrogen bonding is also the force that binds the two chains of a DNA molecule together, so hydrogen bonding is a key to understanding the replication of organisms (see Section 24.10).

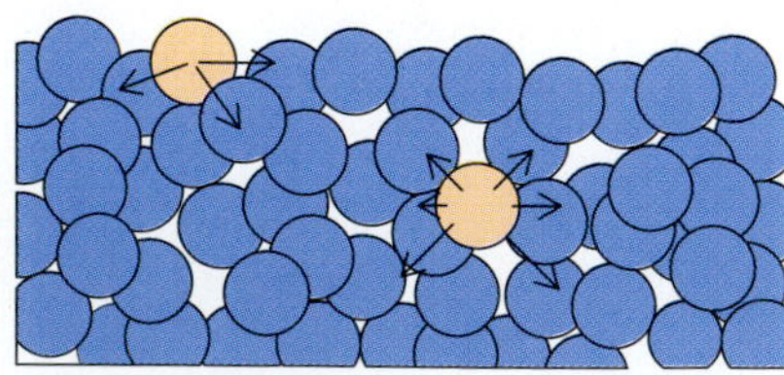

FIGURE 10.9 Surface tension exists because of attractive forces between the molecules in the bulk liquid and the molecules at the surface. A molecule at the surface experiences a net inward force.

THE PROPERTIES OF LIQUIDS

A liquid consists of molecules that can move past their neighbors but cannot escape from them as completely as can the molecules in a gas. Three of the most characteristic properties of liquids are their ability to flow (which is shared by gases), their possession of a sharply defined surface (which distinguishes them from gases), and their tendency to vaporize and to exert a vapor pressure. All three effects are related to the strengths of the intermolecular forces. Just as important is the ability of a liquid to act as a solvent for a solid, a liquid, or a gas. We shall see in Chapter 11 that the presence of a solute affects the physical properties of liquids as well as providing an environment in which chemical reactions may occur readily.

10.3 SURFACE TENSION

The effect of intermolecular forces is to draw the surface molecules inward from the surface of a liquid, thereby packing them together and forming a smooth surface. When a small amount of water is poured onto a waxed surface, it forms nearly spherical beads. This beading occurs as a result of the effects of two intermolecular forces, the forces between the water molecules and the forces between the water and wax molecules. Because the attractive forces between the water molecules are greater than those between the water and wax molecules, there is a net inward attractive force on the water molecules at the surface toward the bulk of the liquid (Fig. 10.9). The geometric figure with the least surface area and greatest volume is a sphere, so a drop of water tends to adopt that shape.

The unbalanced forces acting at the surface of a liquid are responsible for the liquid's surface tension (γ, the Greek letter gamma):

> The **surface tension** of a liquid is a measure of the force that must be applied to surface molecules so that they experience the same force as molecules in the interior of the liquid.

Some values are listed in Table 10.4; note that the surface tension decreases with increasing temperature. The greater the surface tension, the greater the force that must be applied to equalize the forces experienced by all the molecules of the sample. The surface tension of water is about three times higher than that of most other common liquids; water's high surface tension is another consequence of its strong hydrogen bonds. The very high surface tension of mercury, more than six times that of water, can be traced to the strength of the bonding between its atoms.

In the absence of gravity and other external influences (such as the air resistance encountered by a falling raindrop), a sample of water would form a sphere, for then the maximum number of molecules are in the interior where they are surrounded by neighbors. However, in the presence of gravity, the lowest energy is achieved when the droplet is slightly flattened. For example, the droplets of water on the waxy surface of vegetation are nearly spherical but are slightly distorted by gravity (Fig. 10.10). In a gravity-free environment, as in an orbiting space shuttle, the shape of liquid droplets is governed by surface tension alone. Tiny (0.01-mm diameter), perfect spheres that have been formed in space are now commercially available and are used to calibrate particle sizes for powdered pharmaceuticals (Fig. 10.11). Raindrops would be spherical in the absence of air resistance, but their fall through the air results in a teardrop shape.

Another observable effect of intermolecular forces is **capillary action**, the rise of liquids up narrow tubes (the name comes from the Latin word for "hair"). The liquid rises up the tube because there are

TABLE 10.4 Surface tensions of common liquids

Substance	Temperature, °C	Relative surface tension*
water	18	1.02
	25	1
	100	0.806
ethanol	0	0.334
	10	0.328
	20	0.316
	30	0.304
methanol	20	0.314
benzene	20	0.401
mercury	20	6.56

*Values are relative to that of water at 25°C, which is defined as 1.

T OHT: Table 10.4.

V Video 26: Surface tension.

D Surface tension of water: The floating needle. CD2, 9.

FIGURE 10.10 The nearly spherical shape of these "beads" of water on a waxed surface arises from the effect of surface tension. The beads are flattened slightly by the attraction of the Earth's gravity.

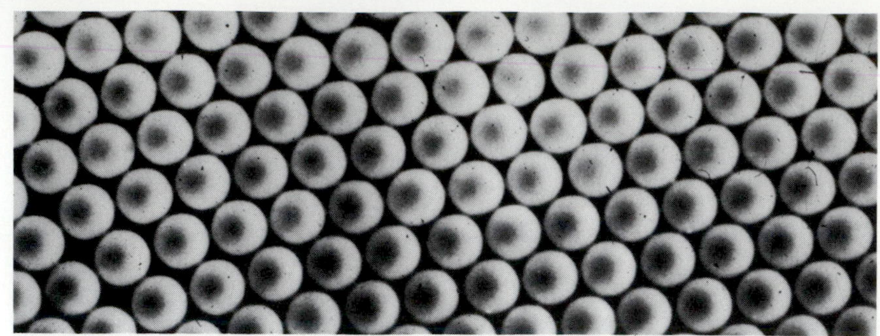

FIGURE 10.11 Latex spheres produced on one of the space shuttle missions. Each one is a perfect sphere of diameter 0.01 mm, formed by surface tension in the absence of gravity.

D At the water's edge: Surface spreading and surface tension. BZS3, 9.23.

U Exercises 10.67 and 10.64.

favorable attractions between its molecules and the tube's inner surface. These interactions are forces of **adhesion**, which are forces that bind a substance to a surface. Adhesive forces should be distinguished from the forces of **cohesion**, which are forces that bind the molecules of a substance together to form a bulk material and which are responsible for condensation. The liquid can climb only so high, however, because its potential energy increases as it climbs.* Narrow tubes and high surface tensions result in tall columns of liquid. Capillary action is partly responsible for the ability of a paper towel to mop up a water spill (because the water spreads between the fibers as a result of the strong adhesive forces) and for the use of a paper support in paper chromatography (Section 1.2).

The **meniscus** of a liquid (from the Greek word for "crescent") is the curved surface the liquid forms in a narrow tube (Fig. 10.12). The meniscus of water in a glass capillary is curved upward (forming a ∪ shape) because the forces between water molecules and the oxygen atoms and hydroxide groups on the surface of the glass are stronger than the forces of cohesion between the water molecules. The water therefore tends to cover the greatest possible area of glass. The meniscus of mercury curves downward in glass (forming a ∩ shape). This shape is a sign that the cohesive forces between the mercury atoms are stronger than their adhesion to the glass, for in this case the liquid tends to reduce its contact with the glass. The same effect accounts for the difference in the shape of the meniscus of water in a glass and a plastic graduated cylinder. It also explains why water droplets do not spread on the surface of a freshly waxed car but do flatten out and spread over the surface of an unwaxed car (which has many polar groups and ionic sites on it).

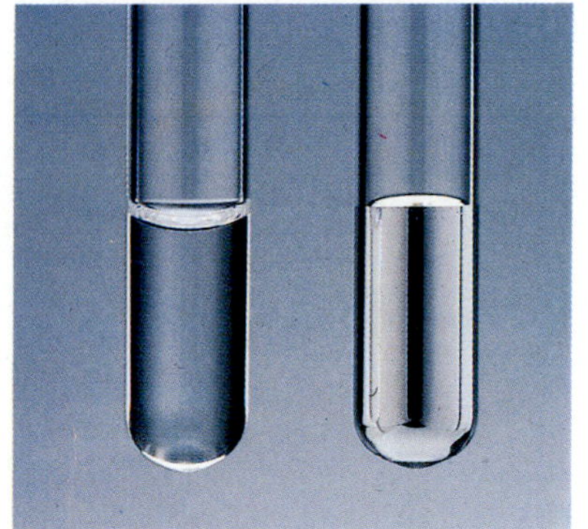

FIGURE 10.12 When the adhesive forces between a liquid and glass are stronger than the cohesive forces within the liquid, the liquid forms the meniscus shown here for water in glass (left). When the cohesive forces are stronger than the adhesive forces (as they are for mercury in glass), the meniscus is curved downward (right).

10.4 VAPOR PRESSURE

We first encountered vapor pressure in Section 5.5, in connection with collecting gases over water. There we saw that it is a measure of the volatility of substances. In this section we discuss the concept of vapor pressure more precisely and examine its relationship to intermolecular forces.

*The final height h of the liquid is proportional to the surface tension γ and inversely proportional to the density d (because more energy is needed to raise a dense liquid): $h = 2\gamma/dgr$. In this expression, r is the radius of the tube and g is the acceleration of free fall (9.81 m/s^2).

The measurement of vapor pressure. Suppose we introduce a little water (or any other liquid) into the vacuum above the mercury in a barometer tube. Water molecules immediately evaporate and fill the space with vapor. As a result, the pressure rises, thereby depressing the column of mercury a few millimeters. If we supply only a trace of water, then it all evaporates and the pressure exerted by the vapor depends on the amount of water added. However, if we add so much water that some remains on the surface of the mercury in the tube, then the pressure exerted by the vapor becomes constant (Fig. 10.13). That is, as long as some water is present, at a fixed temperature its vapor exerts the same, characteristic pressure. For example, at 40°C the mercury falls 55 mm, so the pressure exerted by the vapor is 55 Torr; the outcome is the same whether there is 0.1 mL or 1 mL (or any other volume) of liquid water present. This characteristic pressure is the vapor pressure of the liquid. Values of the vapor pressures of some common liquids at room temperature are given in Table 10.5.

Some solids vaporize (sublime) and fill the space in a closed container with their vapor; they therefore exert a characteristic **sublimation vapor pressure**. The compound *p*-dichlorobenzene, a component of mothballs, sublimes and creates a vapor pressure (of 0.40 Torr at 20°C). However, the vapor pressures of most solids at room temperature are much smaller than those of liquids, and the rate at which the solid evaporates is often so low that its vapor never reaches its final pressure.

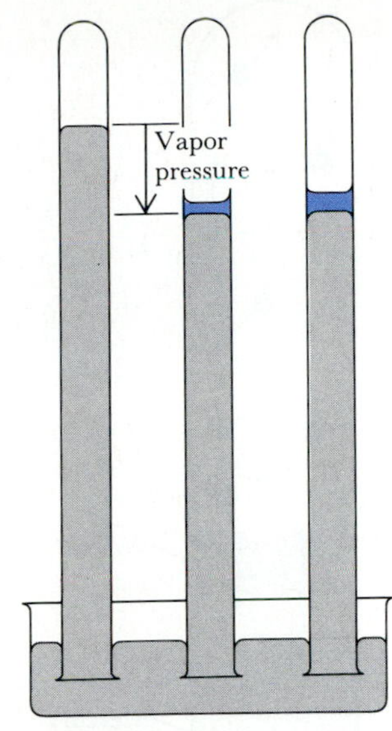

FIGURE 10.13 An apparatus for demonstrating the pressure exerted by the vapor of a liquid. The vapor pressure exerted by the liquid water (in Torr) is equal to the distance by which the column of mercury is lowered. The vapor pressure is the same however much liquid water is present.

N In Fig. 10.13, the vapor pressure is slightly less than the length of the arrow because the condensed (but not very dense) liquid also pushes the mercury down very slightly.

T OHT: Fig. 10.13 and Table 10.5.

U Exercise 10.76.

D Vapor pressures of pure liquids and solutions. BZS3, 9.5A.

E Evaporation of liquids. A kinetic approach. *J. Chem. Educ.*, **1974**, *51*, 276.

D A permanent demonstration of vapor pressure of solids, liquids, and mixtures. *J. Chem. Educ.*, **1981**, *58*, 725.

Vapor pressure as a dynamic process. We can build up a precise definition of vapor pressure by thinking in molecular terms about what is happening at the surface of the water shown in Fig. 10.13. When water is first introduced into the vacuum above the mercury, some of its molecules leave the liquid and form a vapor. When molecules that have escaped strike the surface of the remaining liquid, they may be recaptured by the attractive intermolecular forces. As the number of molecules in the vapor increases, more of them strike the surface, and eventually a stage is reached at which the number of molecules returning to the liquid exactly matches the number escaping from it. At this stage, the vapor is condensing as fast as the liquid is vaporizing. That is, the rate of vaporization, in moles of H_2O per second, is equal to the rate of condensation. The concentration of molecules in the vapor, and hence its pressure, remains constant and the liquid and vapor are in dynamic equilibrium (Fig. 10.14):

> **Dynamic equilibrium** is a condition in which a forward process and its reverse are occurring simultaneously at equal rates.

Dynamic is a key word in this definition, and it implies continuous activity. Except at the beginning and end of the day, a busy store is at dynamic equilibrium, with the number of customers arriving matching the number leaving. Dynamic equilibrium is quite unlike the static equilibrium of a ball at rest at the foot of a hill. A store is at static equilibrium before it opens for business.

In a closed container, a liquid and its vapor reach the dynamic equilibrium,

$$\text{Rate of evaporation} = \text{rate of condensation}$$

TABLE 10.5 Vapor pressures at 25°C

Substance	Vapor pressure, Torr
benzene	94.6
ethanol	58.9
mercury	0.0017
methanol	122.7
water*	23.8

*For values at other temperatures, see Table 5.4.

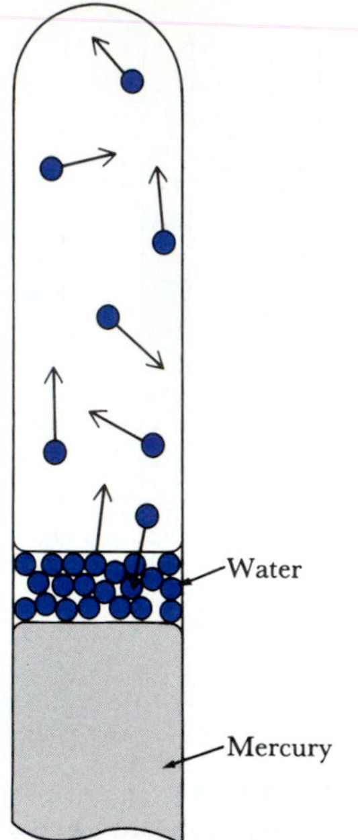

FIGURE 10.14 When a liquid and its vapor are in dynamic equilibrium inside a closed container, the rate at which molecules leave the liquid is equal to the rate at which they return. The arrows indicate the random movement of water molecules in the vapor.

T OHT: Figs. 10.15 and 10.17.

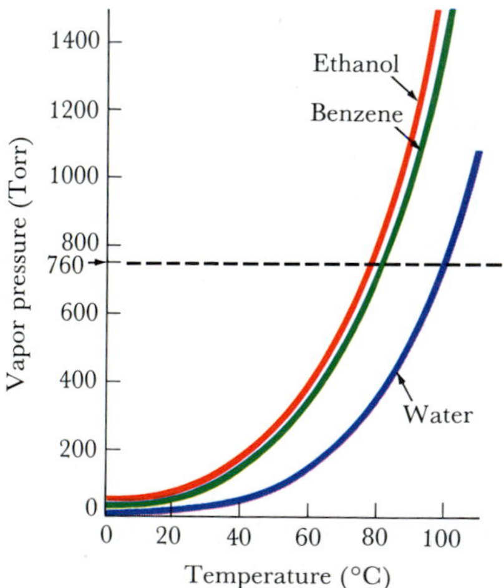

FIGURE 10.15 The dependence of the vapor pressures on the temperature for ethanol (red), benzene (green), and water (blue). The normal boiling point is the temperature at which the vapor pressure is 1 atm (760 Torr).

when the pressure of the vapor has risen to a definite value (which depends on the liquid and the temperature). If the liquid is water, then we denote this dynamic equilibrium

$$H_2O(l) \rightleftharpoons H_2O(g)$$

where the double-headed arrow signifies the dynamic equilibrium between the forward and reverse processes.

The existence of dynamic equilibrium suggests the following definition of vapor pressure:

> The **vapor pressure** of a liquid (or a solid) is the pressure exerted by its vapor when the vapor and the liquid (or solid) are in dynamic equilibrium.

A low vapor pressure (like that of a solid or cold water) is a sign that the rate of condensation matches the rate of vaporization when the number of molecules in the vapor is low. On the other hand, when the vapor pressure is high (like that of ether or hot water), it is a sign that the rate of condensation matches the rate of vaporization only when there are many gas molecules present.

In an open container, such as a beaker of water, the vapor that is formed disperses away from the liquid. Little, if any, is recaptured at the surface of the liquid, the rate of condensation never increases to the point at which it matches the rate of vaporization, and a condition of dynamic equilibrium is never reached. Given enough time, the liquid **evaporates**, or vaporizes completely. Substances with high vapor pressures (like ether, gasoline, and hot water) vaporize more readily than substances with low vapor pressures.

Vapor pressure increases with temperature because the molecules in the heated liquid move more energetically and can escape (to form the vapor) more readily from their neighbors (Fig. 10.15). Practically all evaporation takes place from the surface of the liquid because the molecules there are least strongly bound and can escape to the vapor more easily than those in the bulk liquid.

Vapor pressure and molecular structure. A liquid has a low vapor pressure at room temperature if the intermolecular forces are strong. We therefore expect molecules capable of forming hydrogen bonds to exist as liquids that are much less volatile than others (Fig. 10.16). This characteristic is shown strikingly by dimethyl ether (**12**) and ethanol (**13**), which both have the molecular formula C_2H_6O and so might be expected to have similar vapor pressures. However, ethanol molecules each have an —OH group and can link together by forming hydrogen

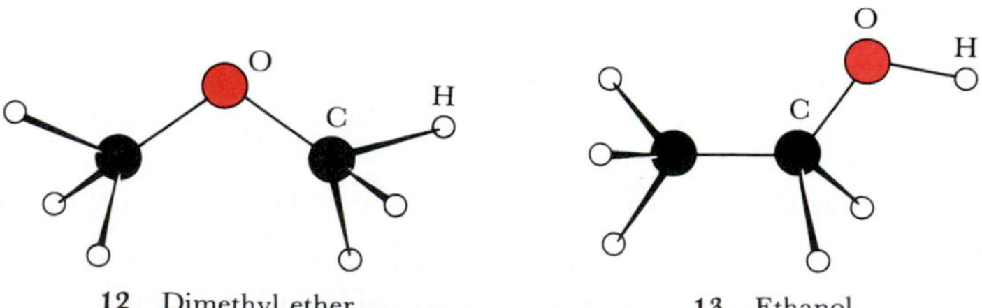

12 Dimethyl ether

13 Ethanol

bonds. The ether molecules cannot form hydrogen bonds and interact only through London and dipole-dipole forces. As a result, ethanol is a liquid at room temperature whereas dimethyl ether is a gas.

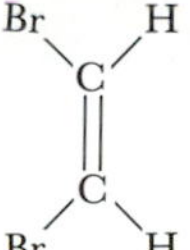

14 *cis*-Dibromoethene

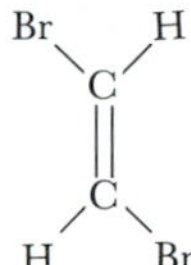

15 *trans*-Dibromoethene

▼ **EXAMPLE 10.4 Predicting the relative vapor pressures of substances**

Which liquid do you expect to have higher vapor pressure, *cis*-dibromoethene (**14**) or *trans*-dibromoethene (**15**)?

STRATEGY The two molecules have the same molar masses and differ only in the arrangement of their atoms. We should identify which intermolecular interactions they can have and then decide which has the stronger interactions. That substance will be likely to have the lower vapor pressure and hence be less volatile.

SOLUTION Both substances can interact by London forces, but the strengths will be about the same in each case because they are composed of the same atoms. Only the cis compound is polar, so only it can have dipole-dipole interactions. Therefore, we can expect *cis*-dibromoethene to have stronger intermolecular forces than *trans*-dibromoethene. It follows that *trans*-dibromoethene should have a higher vapor pressure than *cis*-dibromoethene. (The boiling point of *trans*-dibromoethene is 108°C and that of *cis*-dibromoethene is 113°C, in agreement with this conclusion.)

EXERCISE E10.4 Which do you expect to be more volatile, carbon tetrabromide, CBr_4, or carbon tetrachloride, CCl_4? Give reasons.

[*Answer*: CCl_4; weaker London forces] ▲

U Exercises 10.62 and 10.63.

D Boiling at reduced pressure. CD2, 13.

Vapor pressure and boiling point. Now consider what happens when we heat water in an open container and the vapor pressure of the water increases. When the temperature of the liquid is raised to the point at which its vapor pressure matches the pressure exerted by the atmosphere, vaporization can occur throughout the liquid because any vapor formed can drive back the atmosphere and make room for itself. Thus, bubbles of vapor form in the liquid and rise to the surface. We call this condition "boiling," and the temperature at which it occurs is the **boiling temperature**.

The boiling temperature increases with increasing external pressure because a higher temperature must be reached before the vapor pressure matches the external pressure. Because we are normally interested in the properties of liquids under atmospheric pressure, it is convenient to report boiling temperatures at 1 atm pressure:

> The **normal boiling point** T_b of a liquid is the temperature at which its vapor pressure is equal to 1 atm.

In other words, the normal boiling point is the boiling temperature when the external pressure is 1 atm. The normal boiling points of a number of substances are given in Table 10.3. Reporting the normal boiling point of water as 100°C ($T_b = 373$ K) means that its vapor pressure rises to 1 atm when it is heated to 100°C and that it boils at 100°C when the atmospheric pressure is 1 atm. The boiling temperature is higher when the external pressure is increased, as in a pressure cooker, and it is lower when the external pressure is lower. This phenomenon is familiar to mountain climbers and to inhabitants of high-altitude

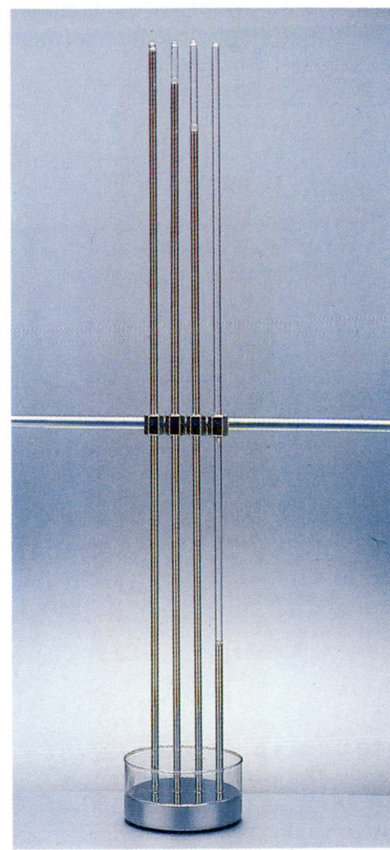

FIGURE 10.16 The height of the column of mercury (far left) is depressed by the vapor pressures of water, ethanol, and ether (left to right). Ether has a high vapor pressure because its molecules cannot form hydrogen bonds with one another.

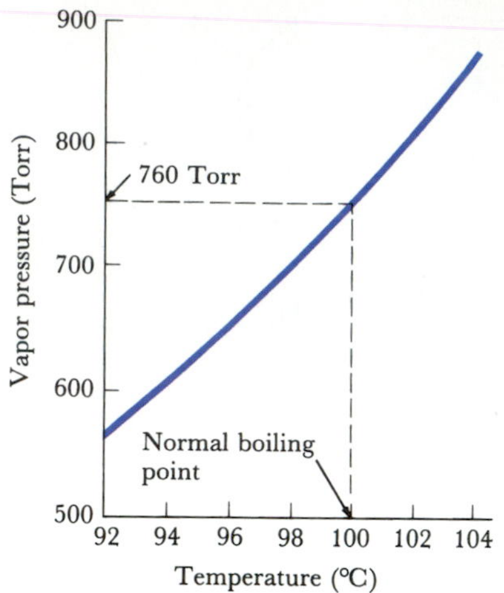

FIGURE 10.17 The temperature variation of the vapor pressure of water in the region close to its normal boiling point (marked by the intersection of the dashed lines).

U Exercises 10.61, 10.68, 10.72–10.75, and 10.77.

? Account for the trends in boiling point of water (100°C), methanol (65°C), and ethanol (78°C).

U Exercises 10.59, 10.60, and 10.66.

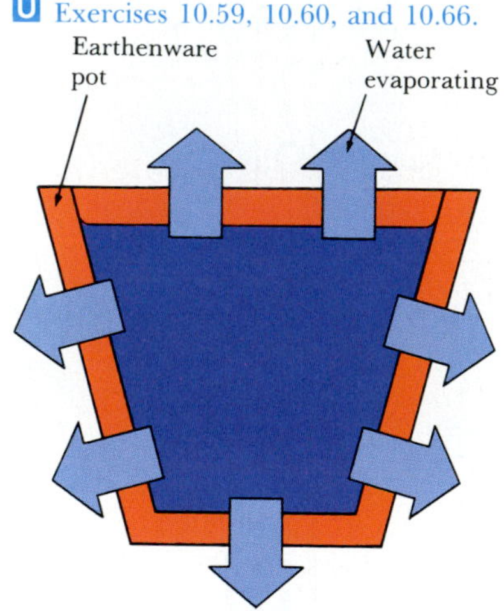

FIGURE 10.18 The water molecules with enough energy escape from the liquid and leave the less energetic molecules behind. Because molecules can escape from the top surface and through the porous walls of the pot, vaporization is rapid and heat does not flow in rapidly from the surroundings. As a result, the liquid cools to below the temperature of the surroundings.

cities; the boiling temperature of water at the summit of Mt. Everest—where the pressure is about 240 Torr—is only 71°C, which is nearly cool enough to place your hand in (the maximum tolerable temperature is about 60°C).

▼ EXAMPLE 10.5 Predicting the boiling temperature at a given pressure

Use Fig. 10.17 to predict the boiling temperature of water at the atmospheric pressure of Denver, Colorado, which on a typical day is 650 Torr.

STRATEGY Because the atmospheric pressure (650 Torr) is lower than 760 Torr, we expect a boiling temperature lower than the normal boiling point. To predict the boiling temperature, we can use Fig. 10.17 to find the temperature at which the vapor pressure of water reaches 650 Torr.

SOLUTION From Fig. 10.17, the vapor pressure of water is 650 Torr at 96°C. Hence its boiling temperature at that pressure is 96°C.

EXERCISE E10.5 Predict the boiling temperature of water on a day when the atmospheric pressure is 770 Torr.

▲ [*Answer*: 100.3°C]

Because boiling temperatures are related to vapor pressure, they also are related to the strengths of intermolecular forces. In general, the normal boiling point is high when the intermolecular forces are strong, because then high temperatures are needed to raise the pressure of the vapor to 1 atm. This relation accounts for the anomalously high boiling point of water relative to that of hydrogen sulfide: hydrogen bonds are strong in water but absent in hydrogen sulfide.

Enthalpy of vaporization. A quantitative measure of the strength of intermolecular forces is the enthalpy of vaporization, ΔH_{vap}, introduced in Section 6.2 and listed in Table 6.2. The enthalpy of vaporization is the enthalpy change per mole that accompanies the process

$$\text{Liquid} \longrightarrow \text{vapor}$$

It corresponds to the heat (per mole of molecules) that must be supplied to the liquid (under conditions of constant pressure) to overcome the intermolecular forces binding the molecules together and to liberate them as a vapor. Hence, the greater the strength of the intermolecular forces, the higher the enthalpy of vaporization.

Because vaporization is an endothermic process, heat is always absorbed when a liquid vaporizes. This property is the basis of the cooling effect of rubbing alcohol (isopropanol, $(CH_3)_2CHOH$) and why we feel cool when leaving the water after a swim. In the laboratory we notice different enthalpies of vaporization because we find that to vaporize some liquids we need to turn an electric heater to a higher setting than for others. Furthermore, water left in an unglazed pot is slightly cooler than its surroundings because when the liquid vaporizes, which it does rapidly because the walls of the pot are porous, the enthalpy of vaporization is withdrawn from the liquid more rapidly than heat can flow in from the surroundings; consequently the bulk liquid cools (Fig. 10.18).

The critical temperature. A liquid does not boil when it is heated in a closed, constant-volume container. To see what happens instead, suppose initially that the tube shown in Fig. 10.19 contains liquid water and water vapor at 24 Torr (its vapor pressure at 25°C). As the water and vapor are heated, the vapor pressure increases, and at 100°C the water is in dynamic equilibrium with vapor at a pressure of 760 Torr. However, the water does not now boil because there is no space for the water to vaporize freely. At 200°C, the vapor pressure has risen to 11,700 Torr (15.4 atm); liquid and vapor are still in dynamic equilibrium, but now the vapor is very dense because it is at such a high pressure.

At 374°C, with the vapor pressure at 218 atm—the container must be very strong!—the density of the vapor is so great that it is equal to that of the remaining liquid. Now there is no surface separating liquid from vapor, the liquid and vapor cannot be distinguished, and a single uniform substance fills the container. Because a substance that fills any container it occupies is by definition a gas, we conclude that we have reached a temperature above which the liquid phase does not exist. We have reached the critical temperature (T_c) of water:

> The **critical temperature** of a substance is the temperature above which it cannot exist as a liquid.

If a substance is *at* its critical temperature, then we can still liquefy it by applying pressure, and the minimum pressure required to convert the gas to a liquid is the **critical pressure** P_c of the substance. At temperatures higher than the critical temperature, a substance cannot be liquified at any pressure. For example, the critical pressure of carbon dioxide is 73 atm, and a sample of the gas at 31°C (its critical temperature) may be condensed to a liquid if the pressure is greater than 73 atm. However, the nitrogen and oxygen in air can never be liquefied by compression at room temperature, for both gases are then well above their critical temperatures.

The critical temperatures and pressures of some substances are listed in Table 10.6. Their values are important in practice because it is useless to try to liquefy a gas by applying pressure if it is above its critical temperature. For example, the critical temperature of carbon dioxide is 31°C, so we know that it cannot be compressed to a liquid if its temperature is higher than 31°C. Because the critical temperature of helium is 5.2 K, the gas must be cooled almost to absolute zero before it can be liquefied by pressure.

A substance at a temperature higher than its critical temperature may be so dense that, although a gas, it can act as a solvent like a liquid. Supercritical carbon dioxide—carbon dioxide at high pressure but above its critical temperature—can dissolve organic compounds. It is used to remove caffeine from coffee beans and to extract perfumes from flowers. Supercritical hydrocarbons are used to dissolve coal and separate it from ash, and it has been proposed that they could be used to extract oil from oil-rich tar sands.

10.5 SOLIDIFICATION

A liquid solidifies when it is cooled to its freezing point. At this temperature and below, the molecules have such low energies that they can move past their neighbors only with great difficulty. The molecules oscillate about their average positions and only rarely migrate from site to site.

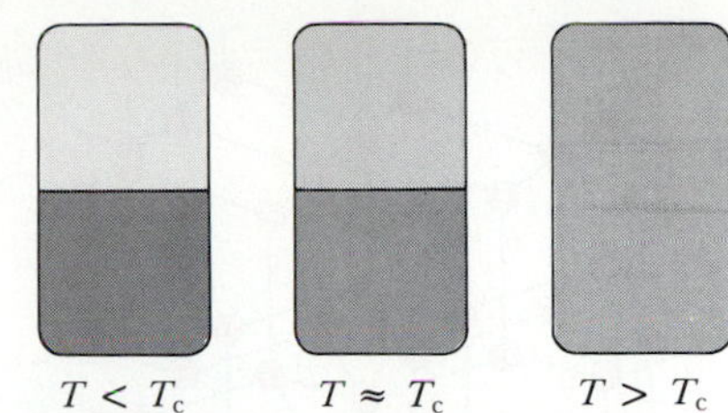

FIGURE 10.19 When the temperature of a liquid in a sealed, constant-volume container is raised, the density of the liquid decreases and the pressure and density of the vapor increase. At and above the critical temperature T_c, the density of the vapor is the same as the density of the liquid and a single uniform phase fills the container.

D Place some dry ice into a pipet, clamp tightly with pliers, and submerge in a beaker of water. Liquid carbon dioxide can be observed momentarily (just before the pipet bursts).

D Critical phenomena. TDC, 5-31s.

T OHT: Table 10.6.

TABLE 10.6 Critical temperatures and pressures of substances

Substance	Critical temperature, °C	Critical pressure, atm
He	−268 (5.2 K)	2.3
Ne	−229	27
Ar	−123	48
Kr	−64	54
Xe	17	58
H_2	−240	13
O_2	−118	50
H_2O	374	218
N_2	−147	34
NH_3	132	111
CO_2	31	73
CH_4	−83	46
C_6H_6	289	49

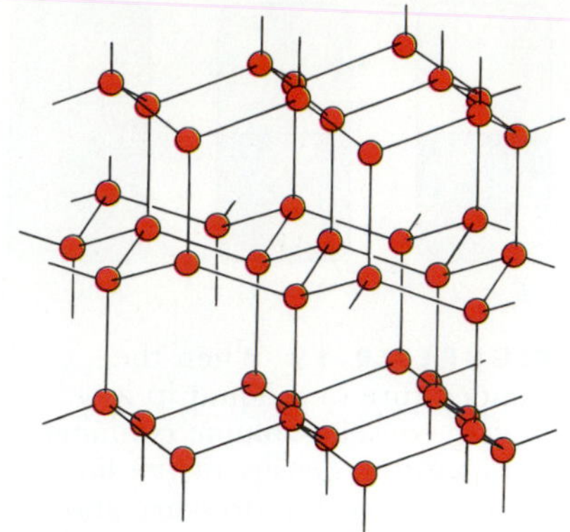

FIGURE 10.20 The structure of ice. Only the oxygen atoms are shown here. The hydrogen atoms lie at about the midpoints of the lines that join neighboring O atoms. (Compare this structure with that of diamond shown later in Fig. 10.42.) A view of the structure of ice from a different perspective is shown in Fig. 10.44.

T OHT: Figs. 10.20 and 10.44 (p. 396).

E Subtleties of phenomena involving ice-water equilibria. *J. Chem. Educ.*, **1986**, *63*, 115.

Freezing point and pressure. The **normal freezing point** T_f of a liquid is the temperature at which it freezes under 1 atm pressure. Under greater pressure, most liquids freeze at higher temperature: pressure pushes the molecules together, thereby reducing the distance between them. Consequently, they attract each other more strongly. The effect of pressure is usually quite small unless the pressure is large. Iron, for example, melts at 1800 K under 1 atm pressure; but it melts at a temperature only a few degrees higher when the pressure is a thousand times as great. However, at the very center of the earth, the pressure is high enough for iron to be solid despite the high temperatures there, so the earth's innermost core is solid.

Water is an exception to the general rule that freezing points increase with pressure. Water freezes at a *lower* temperature under pressure: at 1000 atm it freezes at −5°C. The reason can be traced to the effect of the hydrogen bonds, which hold the H_2O molecules in a structure that is more open in ice than in liquid water: the open structure of ice partially collapses when the solid melts (Fig. 10.20). The effect of this collapse can be seen in the way the density of water varies with temperature—a maximum at 4°C and a sharp decrease on freezing (Fig. 10.21). When pressure is applied to a sample, the water tends to remain liquid, for then it occupies a smaller volume than it would if it froze. Conversely, ice tends to melt under pressure, for the more compact structure characteristic of the liquid is then favored.

The melting of ice under pressure is thought to contribute to the advance of glaciers. The weight of ice pressing on the edges of rocks deep under the glacier results in very high local pressures. The liquid forms despite the low temperatures, and the glacier slides slowly downhill on a film of water.

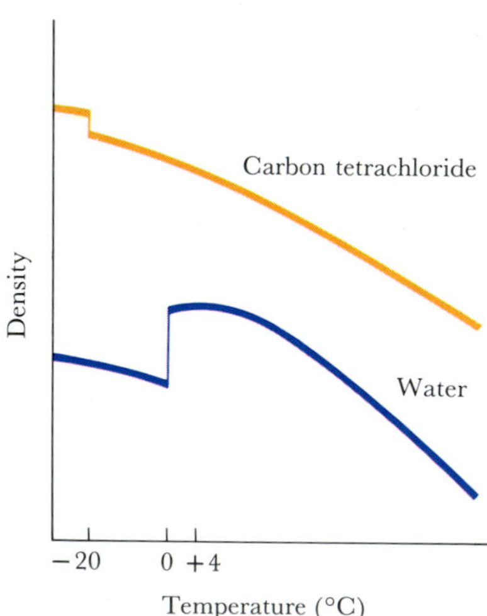

FIGURE 10.21 The variation of the densities of water and carbon tetrachloride with temperature. Note that ice is less dense than water at its freezing point and that water itself has its maximum density at 4°C.

Phase diagrams. A **phase diagram** is a summary in graphical form of the conditions of temperature and pressure at which the various solid, liquid, and gaseous phases of a substance exist. The phase diagram for water is shown in Fig. 10.22. Any sample of water at a temperature and pressure that place it within the area labeled "ice" exists as ice. Similarly, a sample represented by any point in the "liquid" area is liquid. A sample at any point in the "vapor" region is a vapor. From phase diagrams we can predict, almost at a glance, the phase of a substance at any temperature and pressure.

The lines separating the areas in the diagram are called **phase boundaries**. The points *on* a phase boundary (such as point C at 0°C and 760 Torr on the phase diagram of water) show the conditions under which two phases (ice and liquid water in this case) coexist in dynamic equilibrium. The phase boundary between liquid and vapor shows the pressures at which the vapor and liquid coexist in dynamic equilibrium and is therefore a plot of the vapor pressure of the liquid against the temperature. The phase boundary between the solid and the vapor shows the pressures at which the solid is in equilibrium with its vapor, and hence is a plot of the sublimation vapor pressure of the solid.

▼ EXAMPLE 10.6 Interpreting a phase diagram

Use the phase diagram in Fig. 10.22 to decide the physical state of water at 70°C and pressures of (a) 5 Torr and (b) 800 Torr.

STRATEGY Locate the point corresponding to the conditions on the phase diagram. The region in which it lies shows the state of the sample under those conditions. If the point lies on one of the curves, then both phases are present in equilibrium.

SOLUTION (a) This is point A, and it lies in the vapor region. (b) Increasing the pressure to 1.1 atm takes the system to point B, which lies in the liquid region. The sample is a liquid because the applied pressure is greater than the vapor pressure and all the molecules have been pushed into the condensed phase.

EXERCISE E10.6 The phase diagram for carbon dioxide is shown in Fig. 10.23. What is the phase of a sample at 10 atm and −15°C?

[*Answer*: liquid]

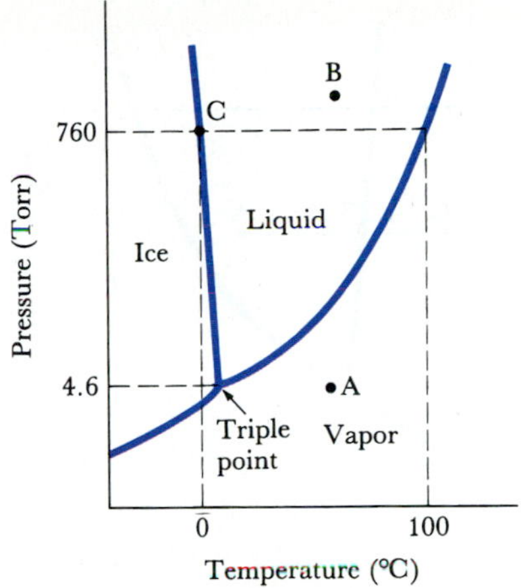

FIGURE 10.22 The phase diagram for water. The solid lines define the regions of pressure and temperature in which each phase is most stable. Note that the freezing point decreases with increasing pressure. The letters A, B, and C refer to Example 10.6 and the text. The phase diagram is not precisely to scale.

The **triple point** is the point where the three phase boundaries meet. For water, it occurs at 4.6 Torr and 0.01°C. Under these conditions (and only these) all three phases (ice, liquid, and vapor) coexist in dynamic equilibrium. At the triple point, the vapor pressure of liquid water is equal to that of ice. Because the triple point of water is a characteristic, fixed property, it is a better choice for defining a temperature scale than the freezing or boiling temperatures, which depend on the applied pressure. The triple point is in fact used to define the size of the kelvin: by definition, there are exactly 273.16 kelvins between absolute zero and the triple point of water.

Liquid water can exist only if the pressure of the vapor is greater than 4.6 Torr. On a cold, dry morning on Earth, the partial pressure of water in the air may be lower than 4.6 Torr and frost may appear and disappear, like solid carbon dioxide on Mars, without turning to liquid water first.

Cooling curves. We can use a phase diagram to predict how the temperature of a liquid changes as it cools (that is, as energy is withdrawn as heat from the sample). For example, what would we observe if we placed 100 mL of water in a refrigerator and monitored its temperature?

D Sublimation pressure visualized. TDC, 151.

N The effect of a solute on the phase diagram in Fig. 10.22 is explained in Chapter 11.

T OHT: Figs. 10.22; 10.23.

U Exercises 10.69 and 10.70.

E The temperature of the triple point is defined as 273.16 K; the melting point of water is then found to lie 0.01 K below the triple point; so the melting point of water is 273.15 K.

E Freezing points, triple points, and phase equilibria. *J. Chem. Educ.*, **1974**, *51*, 658. Relation of melting point to triple point. *J. Chem. Educ.*, **1975**, *52*, 276.

U Exercises 10.71 and 10.78.

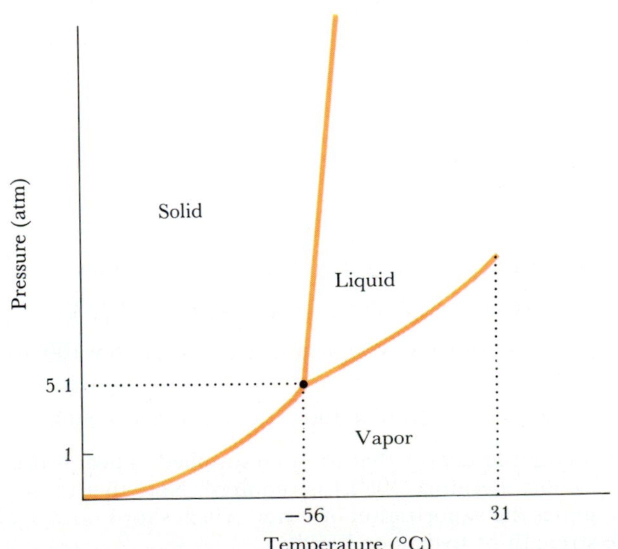

FIGURE 10.23 The phase diagram for carbon dioxide. Note the slope of the boundary between the solid and liquid phases; it shows that the freezing point rises as pressure is applied.

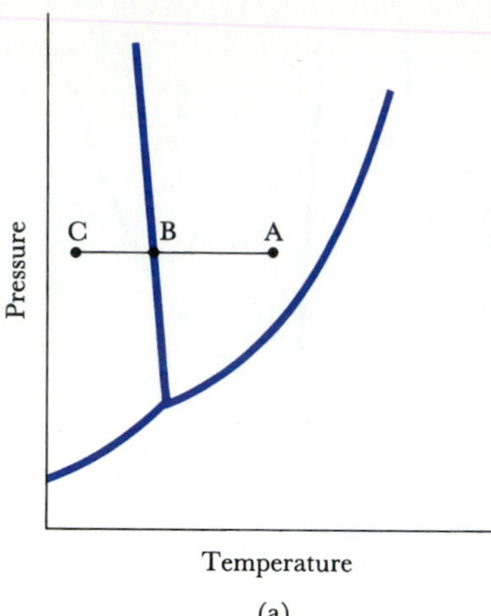

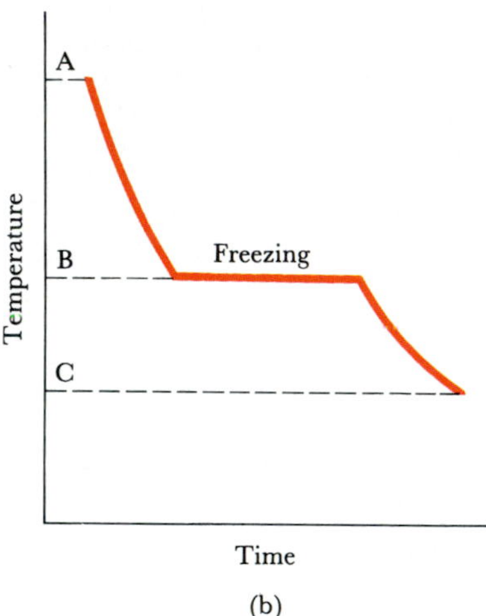

FIGURE 10.24 (a) The phase diagram of water (like that in Fig. 10.22). (b) The cooling curve when a liquid sample at point A cools at constant pressure through B to ice at C. The pause in the decline of the cooling curve at B is due to the release of heat when the liquid freezes.

E The slope of a cooling curve is determined by the heat capacity of the sample (and factors related to the insulation of the sample). The greater the heat capacity, the slower the cooling (because more energy must be lost for a given change in temperature).

The initial state of the sample might be point A in Fig. 10.24a. The cooling occurs at constant pressure, so the state of the sample is represented by the points on the horizontal line starting at A. The temperature of the water falls steadily (Fig. 10.24b) as it loses energy to its surroundings, until point B is reached on the solid-liquid phase boundary. Then, instead of the temperature falling as energy is lost, the liquid begins to freeze. As it does so, heat is released from the enthalpy of freezing to the surroundings and the temperature of the sample remains constant. This stage is represented by the horizontal portion of the cooling curve in Fig. 10.24b. As soon as all the liquid has frozen, the temperature begins to fall again, because now the heat is removed from the solid ice.

▼ EXAMPLE 10.7 Calculating the heat required to achieve a temperature increase

Calculate the heat required to convert 100 g of ice at −20°C to water vapor at 110°C by using the following data:

Specific heat capacity of $H_2O(s)$:	2.1 J/(g·°C)
Specific heat capacity of $H_2O(l)$:	4.2 J/(g·°C)
Specific heat capacity of $H_2O(g)$:	2.0 J/(g·°C)
Enthalpy of melting of $H_2O(s)$:	6.01 kJ/mol
Enthalpy of vaporization of $H_2O(l)$:	40.7 kJ/mol

STRATEGY The steps involved are illustrated in Fig. 10.25, where it is assumed that heat is being added to the sample at a constant rate. Heat is required to raise the temperature of ice to 0°C, to melt the ice at that temperature, to raise the temperature of water to 100°C, to vaporize the water at that temperature, and then to heat the vapor to 110°C. The overall heat required is the sum of the individual contributions. For the conversion from grams of water to moles of H_2O, use the molar mass of H_2O, 18.02 g/mol.

SOLUTION (1) To heat the ice through 20°C (from −20° to 0°C), we require

$$\text{Heat (J)} = 20°\text{C} \times 100\ \text{g} \times 2.1\ \text{J/(g·°C)} = 4.2\ \text{kJ}$$

(2) To melt 100 g of ice at 0°C (which corresponds to 5.55 mol H_2O), we require

$$\text{Heat (J)} = 5.55\ \text{mol} \times 6.01\ \text{kJ/mol} = 33.4\ \text{kJ}$$

(3) To heat 100 g of water through 100°C (from 0° to 100°C), we must supply

$$\text{Heat (J)} = 100°\text{C} \times 100\ \text{g} \times 4.2\ \text{J/(g·°C)} = 42\ \text{kJ}$$

(4) To vaporize 100 g of water at 100°C, we must supply

$$\text{Heat (J)} = 5.55\ \text{mol} \times 40.7\ \text{kJ/mol} = 226\ \text{kJ}$$

(5) To heat the resulting water vapor through 10°C (from 100° to 110°C), we must supply

$$\text{Heat (J)} = 10°\text{C} \times 100\ \text{g} \times 2.0\ \text{J/(g·°C)} = 2.0\ \text{kJ}$$

The total amount of energy that must be supplied as heat is the sum of the individual steps; therefore 308 kJ is required. Note that the most energy-intensive step is the vaporization of water, which shows once again the considerable strength of hydrogen bonds.

EXERCISE E10.7 What amount of heat is required to convert 15 g of ice at −50°C into vapor at 200°C?

[*Answer*: 50 kJ]

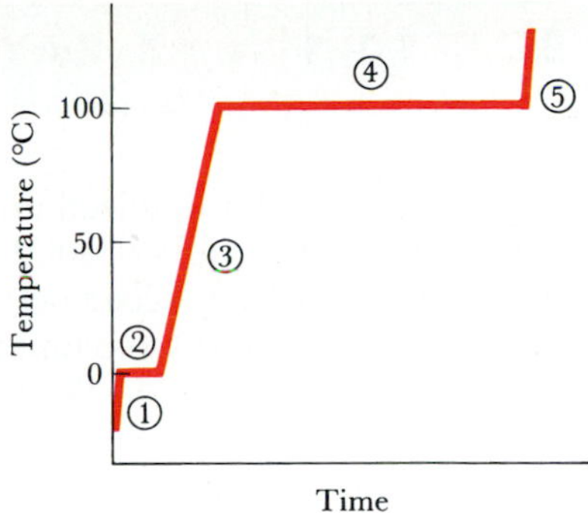

FIGURE 10.25 The variation in the temperature with time of heating described in Example 10.7. In this plot, heat is supplied at a constant rate, so the horizontal axis represents the quantity of heat that has been supplied.

Cooling curves are used to construct phase diagrams. For example, we can trace the ice-liquid curve in the phase diagram of water by subjecting the sample of a liquid to a series of different pressures and monitoring the cooling curve at each one. The temperature of the flat portions in the cooling curves enables us to identify the transition temperature at the corresponding pressure, so we can draw the entire phase boundary.

THE STRUCTURES AND PROPERTIES OF SOLIDS

The solids we consider fall into four classes (Fig. 10.26):

Metals consist of cations held together by a sea of electrons.

Ionic solids are built from cations and anions.

Network solids consist of atoms bonded together covalently throughout the solid.

Molecular solids are collections of individual molecules.

Some characteristics and examples of each type of solid are given in Table 10.7; one of the best methods for determining the structures of solids is **x-ray diffraction**, which is described in Box 10.1.

Solids may also be classified as crystalline or amorphous. A **crystalline solid** is a solid in which the atoms, ions, or molecules lie in an

N Notice in Fig. 10.25 the greater heat required to overcome hydrogen bonding at the boiling point than at the melting point: in the latter, the bonding pattern is readjusted, but in the former the bonds are severed completely as the molecules separate.

E An introduction to the principles of the solid state. *J. Chem. Educ.*, **1970**, *47*, 501. The chemistry of solids. *J. Chem. Educ.*, **1982**, *59*, 100.

FIGURE 10.26 Examples of the four classes of solids. Clockwise from upper left: an ionic solid, nickel(II) sulfate (blue-green); a network solid, boron nitride (white); a molecular solid, carotene, the compound responsible for the color of carrots (red); and a metal, titanium (silvery gray).

BOX 10.1 ▸ X-RAY DIFFRACTION

To see why von Laue seized on x-rays as a way of exploring the interiors of solids, we need to know that *interference* may occur between waves. Imagine two waves of electromagnetic radiation in the same region of space. Where the peaks and troughs of the two waves coincide, they add together to give a wave of greater displacement, as shown in the figure on the left below. This enhancement of displacement is called *constructive interference.* When the combined wave is detected photographically, the spot obtained is brighter than that obtained with either ray alone. However, when the peaks of one of the waves coincide with the troughs of the other, they partly cancel and give a wave of lower amplitude. This cancellation is called *destructive interference.* When this combined wave is detected photographically, a dimmer spot is obtained. No spot at all is detected when the peaks and troughs match exactly and cancellation is complete.

The phenomenon of *diffraction* is the interference between waves that is caused by an object in their path. The resulting pattern of bright spots against a dark background is called a *diffraction pattern.* A crystal can cause diffraction in a beam of x-rays, and a bright spot of constructive interference is obtained when the crystal is held at a certain angle relative to the beam. The angle θ (the Greek letter theta) at which constructive interference occurs is related to the wavelength λ of the x-rays and the distance d between the atoms (below right) by the Bragg equation:

$$2d \sin \theta = \lambda$$

For example, if we find a spot at 17.5° when we use x-rays of wavelength 154 pm, we can conclude that there are layers of atoms separated by a distance

$$d = \frac{\lambda}{2 \sin \theta} = \frac{154 \text{ pm}}{2 \sin 17.5°} = 256 \text{ pm}$$

Therefore, from the measurement of the angle at which the spot is obtained and from the wavelength of the radiation, the distance between the atoms can be calculated. Because $\sin \theta$ cannot be greater than 1, the smallest distance d that can be measured in this way is $\frac{1}{2}\lambda$. Von Laue realized that x-rays provide a way of exploring inside crystals because their wavelengths are so short: they can be used to measure distances comparable to the distances between atoms in a molecule.

(a)

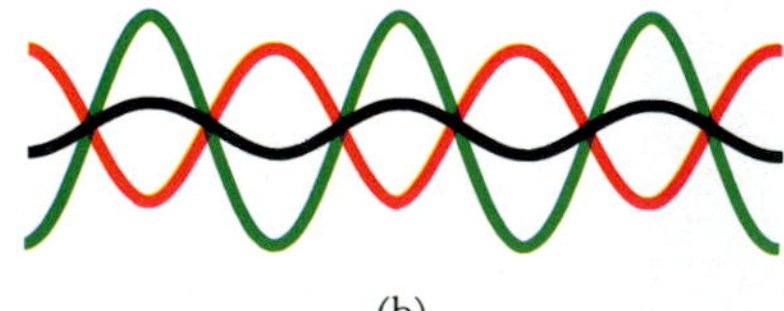

(b)

Interference between waves (the red and green lines) may be either (a) constructive, to give a wave of greater amplitude (the solid line), or (b) destructive, to give a wave of smaller amplitude.

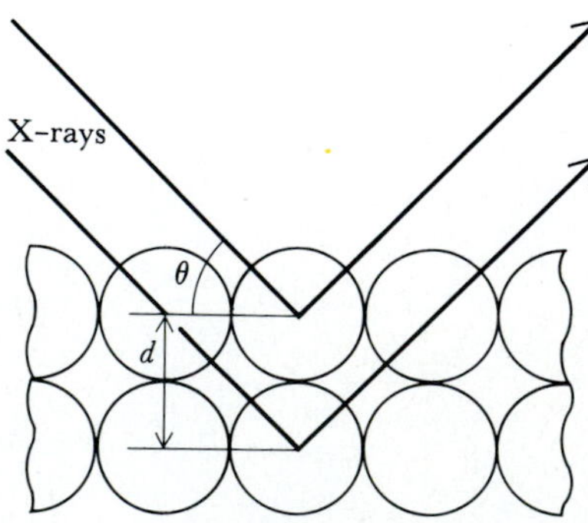

The definitions of the distance d between atoms and the angle θ that occur in the Bragg equation.

TABLE 10.7 Examples and characteristics of solids

Class	Examples	Characteristics
metals	*s*- and *d*-block elements	malleable, ductile, lustrous; electrically and thermally conducting
ionic	NaCl, KNO_3, $CuSO_4{\cdot}5H_2O$	hard, rigid, brittle; high melting and boiling points; those soluble in water give conducting solutions
network	B, C, black P, BN, SiO_2	hard, rigid, brittle; very high melting and boiling points; insoluble in water
molecular	$BeCl_2$, S_8, P_4, I_2, ice, glucose, naphthalene	relatively low melting and boiling points; brittle if pure

orderly array (Fig. 10.27). Crystalline solids include a very wide range of substances. They include metallic elements and alloys (such as copper, iron, and brass), in which the atoms of the elements are stacked together in ways that resemble that shown in Fig. 10.27b. They also include the nonmetallic solid elements, such as sulfur, phosphorus, iodine, and gases and liquids that have been solidified by freezing, such as solid argon. In all of these, the atoms are ordered in a regular pattern. Crystalline solids include ionic compounds, such as sodium chloride and potassium nitrate, where the cations and anions are stacked together, sometimes with water molecules of hydration (as in $CuSO_4{\cdot}5H_2O$). They also include network solids, such as carbon (as diamond and graphite) and boron nitride, where the atoms are linked by covalent bonds, and solids, such as silver chloride, in which there is a mixture of ionic and covalent bonding. Finally, crystalline solids include molecular solids, such as sucrose (cane sugar) and aspirin, where the entities that are stacked together are individual molecules. Crystalline solids usually have flat, well-defined surfaces—called **faces**—that make definite angles with their neighbors. These faces are the edges of orderly stacks of particles.

U Exercise 10.79.

E Most consumer products (such as fibers, plastics, paper, and solid pharmaceuticals) are molecular solids. Consumer products that are considered to be ionic solids include sodium chloride, garden fertilizers, baking soda, and drain cleaners.

D Growing ammonium oxalate crystals. CD2, 73.

E Antique windowpanes and the flow of supercooled liquids. *J. Chem. Educ.*, **1989**, *66*, 994. Glass formation and crystal structure. *J. Chem. Educ.*, **1974**, *51*, 28.

An **amorphous solid** (from the Greek words for "without form") is one in which the atoms, ions, or molecules lie in a random jumble, as

(a)

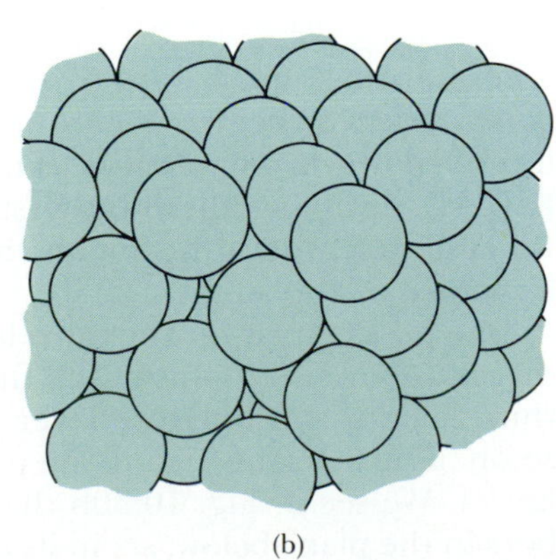

(b)

FIGURE 10.27 Crystals have well-defined faces and an orderly internal structure. Each face is the edge of a stack of atoms, molecules, or ions. (a) An electron micrograph of sodium chloride crystals; (b) the stacking of spheres responsible for three of the faces.

FIGURE 10.28 (a) Quartz is a crystalline form of silica with the atoms in an orderly network, represented here in two dimensions. (b) When molten silica solidifies, it becomes glass, in which the atoms form a disorderly network.

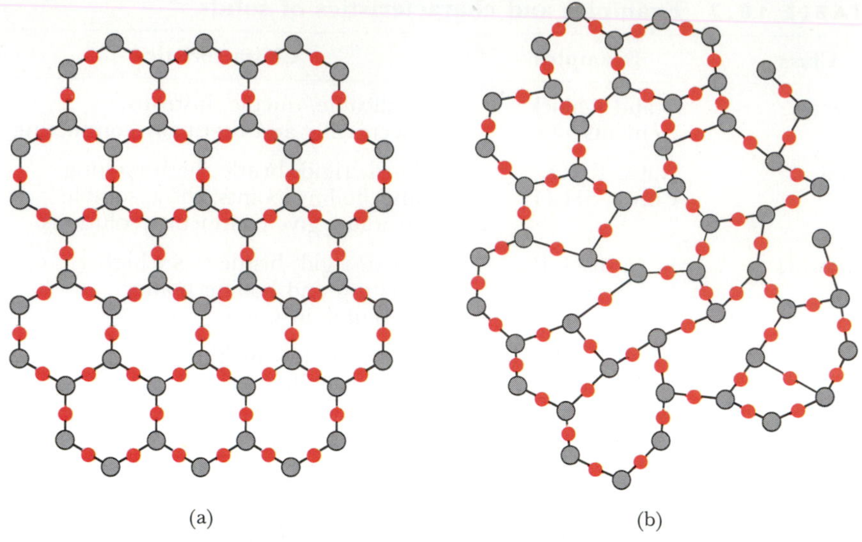

E Other amorphous solids are tar, (cold) molasses, and tree sap (resin).

E Qualitative observations concerning packing densities for liquids, solutions, and random assemblies of spheres. *J. Chem. Educ.*, **1977**, *54*, 139.

E Grapefruit and oranges are packed for shipping according to the patterns shown in Fig. 10.30. Incidentally, no one has yet *proved* that the closest packing of identical spheres is that shown in the illustration.

they do in butter, rubber, and glass (Fig. 10.28). These solids do not have well-defined faces (unless they have been molded or formed into sheets, like a sheet of glass).

10.6 METALS AND SEMICONDUCTORS

FIGURE 10.29 The stacks of apples, oranges, and other produce in a grocery store form faces with different slopes. They illustrate how atoms stack together in metals to give single crystals with flat faces.

The structures of solid metallic elements are quite easy to describe because all the atoms are the same size. Consequently, we can model a crystal of a pure metal, whether aluminum, gold, or iron, by using identical spheres to represent the atoms and stacking them together like fruit in a grocery display. Indeed, if you look at such a stack of fruit, you can see how stacked spheres can form crystal faces (Fig. 10.29).

Close-packing. One kind of structure often found in metals is said to be close-packed:

> A **close-packed structure** is one in which atoms occupy the smallest total volume with the least empty space.

There are two main ways of stacking identical atoms together into a close-packed structure. We can picture both as beginning with the two layers shown in Fig. 10.30a. The atoms of the second (upper) layer (layer B) lie in the dips of the first layer (layer A). A third layer can take up either of two arrangements, depending on the details of electronic structure of the atoms.

One possibility is for the third layer of atoms to lie in the dips that are directly above the atoms of the first layer, as shown in Fig. 10.30b. The third layer then duplicates layer A, the fourth layer duplicates B, and so on. This stacking arrangement results in an ABABA . . . pattern of layers. We see in Fig. 10.30b that each atom has three nearest neighbors in the plane below, six in its own plane and three in the one above,

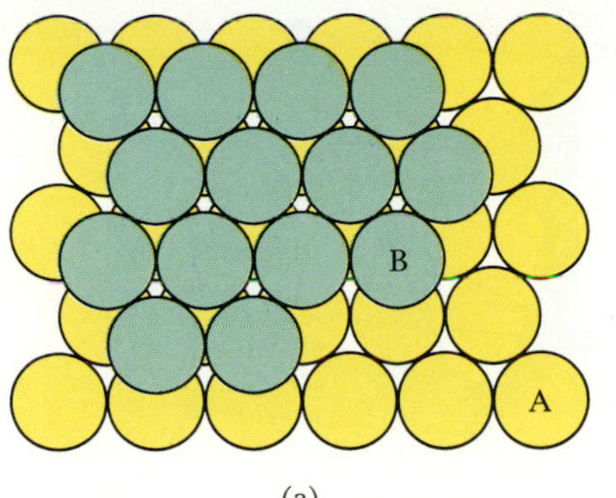

(a)

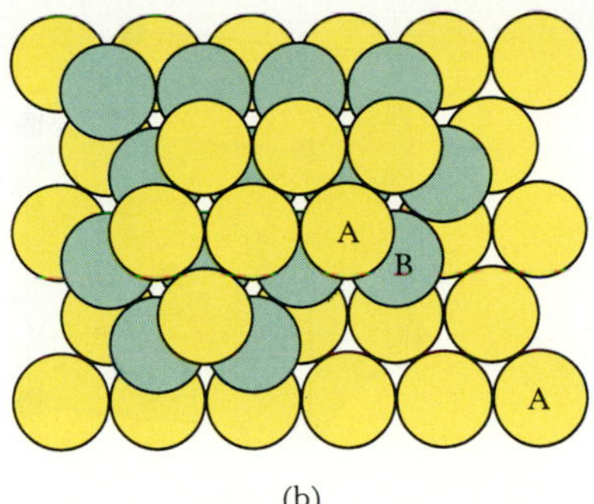

(b)

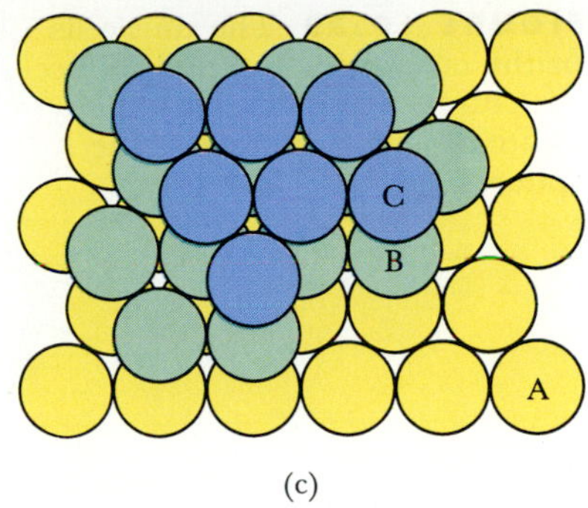

(c)

FIGURE 10.30 A close-packed structure can be built up in stages. (a) The first layer (A) is laid down with minimum waste of space, and the second layer (B) lies in the dips, or depressions, of the first. (b) The third layer can lie above the spheres of the first layer to give a hexagonal close-packed (hcp) structure (ABABAB . . .). (c) Alternatively, the third layer can lie above the dips in the first layer to give a cubic close-packed (ccp) structure (ABCABC . . .).

or twelve in all. This value is reported by saying that the coordination number of each atom in the solid is 12:

> The **coordination number** of an atom is the number of nearest neighbors it has in a solid.

(For ionic solids, the coordination number of an ion is the number of nearest neighbors of opposite charge.) The close-packed structure just described is called the **hexagonal close-packed** (hcp) structure, because it is possible to make out a hexagonal pattern in the arrangement of atoms (Fig. 10.31a). The metals magnesium and zinc crystallize in this way.

The second possibility is for the third layer of atoms to lie in the dips of the second layer that are not directly over the atoms of the first (Fig. 10.30c). If we call this third layer C, then the resulting structure has an ABCABC . . . pattern of layers. The coordination number here is also 12, because each atom has three nearest neighbors in the layer below, six in its own layer, and three in the layer above. It is called the **cubic close-packed** (ccp) structure because it exhibits a cubic pattern (Fig. 10.31b). Aluminum, copper, silver, and gold crystallize in this way.

Some metals do not have close-packed structures. The electronic structures of their atoms are such that the solid has a lower energy if the neighboring atoms lie in positions other than those corresponding to close-packing. One common form is shown in Fig. 10.32. This structure has a central atom lying at the center of a cube formed by eight others, so it is called the **body-centered cubic** (bcc) structure. Iron, sodium, and potassium crystallize as body-centered cubic solids.

T OHT: Fig. 10.30.

E Some monatomic gases also crystallize in close-packed structures. Thus, solid helium (which forms under pressure), is hcp and solid neon, argon, and xenon are ccp. Alloys (Table 10.9, p. 388) often assume one of these packing patterns because the atomic radii of the metals present are very similar.

E Physical and chemical properties and bonding of metallic elements. *J. Chem. Educ.*, **1979**, *56*, 712.

D Models for simple close-packed crystal structures. *J. Chem. Educ.*, **1973**, *50*, 652.

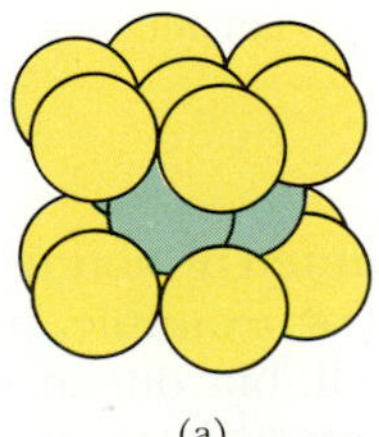
(a)

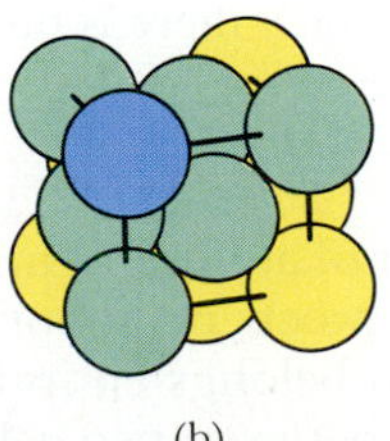
(b)

FIGURE 10.31 The fragments of (a) the hcp structure and (b) the ccp structure give these structures the names hexagonal and cubic. The layers have the same color coding used in Fig. 10.30.

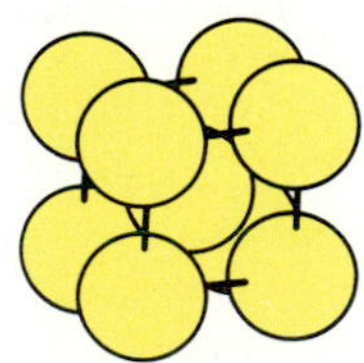

FIGURE 10.32 The body-centered cubic (bcc) structure.

FIGURE 10.33 The unit cells of the (a) ccp (or fcc) and (b) bcc structures. The dots mark the locations of the centers of the spheres.

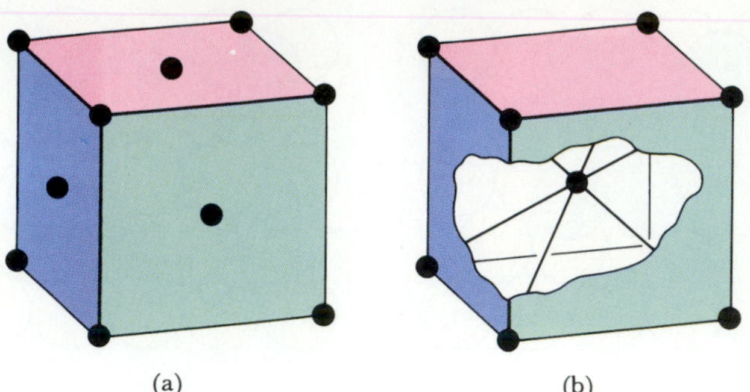

E The unit cells of a crystal are related by pure translations. If it is found that a cell must be rotated in order to reproduce the stacking pattern, then the cell is not a unit cell and a larger cell must be selected.

T OHT: Fig. 10.33.

D Unit cells. *J. Chem. Educ.*, **1975**, *52*, 509.

N A child's building block could be used as an example of a unit cell.

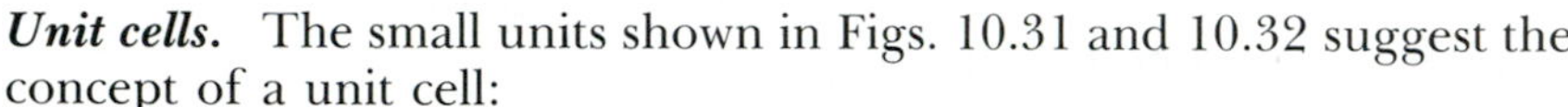

Unit cells. The small units shown in Figs. 10.31 and 10.32 suggest the concept of a unit cell:

> A **unit cell** is the smallest unit that, when stacked together repeatedly without any gaps, can reproduce the entire crystal.

A unit cell is to a crystal as a brick is to a wall. Unit cells are sometimes drawn by representing each atom by a dot that marks the location of its center (Fig. 10.33). Notice that the ccp unit cell has a dot at the center of each face. For this reason, it is also called a **face-centered cubic** (fcc) unit cell (Fig. 10.33a).

▼ **EXAMPLE 10.8 Calculating the coordination number of a crystal structure**

What is the coordination number of a body-centered cubic (bcc) structure?

STRATEGY All we need do is examine the neighborhood of a single atom in the structure. Then we count the number of nearest neighbors. We must make sure that if we select an outside atom of a unit cell, we take into account the atoms in the neighboring cells, because *all* nearest neighbors must be counted, not just those in a single unit cell.

SOLUTION The atom at the center of the bcc unit cell in Fig. 10.32 has eight *nearest* neighbors (one at each corner of the cell), so the coordination number of the central atom is 8. If we focus on one of the corner atoms, then we imagine eight unit cells stacked together, with the corner atom at their center. Now we count eight *nearest* neighbors, which are the center atoms of eight different unit cells.

EXERCISE E10.8 What is the coordination number of a primitive cubic structure in which there is one atom at each corner of a cube (**16**)?

[*Answer*: 6] ▲

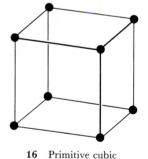

16 Primitive cubic

The number of atoms in a unit cell is counted by noting how the atoms are shared between neighboring cells. For example, an atom at the center of a cell belongs entirely to that cell, but one at the center of a face is shared between two cells and counts as $\frac{1}{2}$ (an atom). For an fcc structure, each of the eight corner atoms is shared by eight cells; so collectively they contribute 1 to the cell. Each atom at the center of a face contributes $\frac{1}{2}$; so jointly the atoms at the centers of all six faces contribute $6 \times \frac{1}{2} = 3$. The net number of atoms in the fcc unit cell is therefore $1 + 3 = 4$; so the mass of the unit cell is four times the mass of one atom.

E Common misconceptions about crystal lattices and crystal symmetry. *J. Chem. Educ.*, **1980**, *57*, 552.

▼ EXAMPLE 10.9 Counting the number of atoms in a unit cell

How many atoms are there in the bcc cubic cell shown in Fig. 10.32?

STRATEGY As explained above, each atom that is not shared by other cells counts as 1, and each atom on a face counts as $\frac{1}{2}$. When an atom is on an edge, it counts as $\frac{1}{4}$ (because it is shared by four cells); and each atom at a corner (where it is shared by eight cells) counts as $\frac{1}{8}$.

SOLUTION Inspection of Fig. 10.32 shows that there is one central atom (count as 1) and eight corner atoms (count as $8 \times \frac{1}{8} = 1$). Therefore, the net content of each bcc unit cell is two atoms.

EXERCISE E10.9 How many atoms are there in a primitive cubic cell (as specified in Exercise E10.8)?

[*Answer*: 1] ▲

The characteristics of five unit cells are listed in Table 10.8. One result of knowing the structure of a unit cell and the number of atoms it contains is the ability to predict the density of a metal from its atomic radius. We saw in Chapter 7 how the atomic radius varies throughout the periodic table: the calculation we shall now illustrate is the link between that periodicity and the variation of the densities of metals.

▼ EXAMPLE 10.10 Calculating the density of a metal

The atomic radius of copper is 128 pm, and it crystallizes with an fcc structure. Calculate the density of the metal.

STRATEGY We know that density is mass per volume and that it is an intensive property. Because it is intensive, we can calculate the density for a sample of any size, including one as small as a unit cell. The volume of the unit cell (which is cubic) is the cube of its edge. That can be obtained from the radius of the copper atom and the fact that the diagonal of one face of the cube is four times the radius (**17**). The mass of a unit cell is the sum of the masses of the atoms it contains, which is the net number of atoms (converted to moles) times the molar mass. It is helpful to bear in mind that typical metal densities are about 10 g/cm^3.

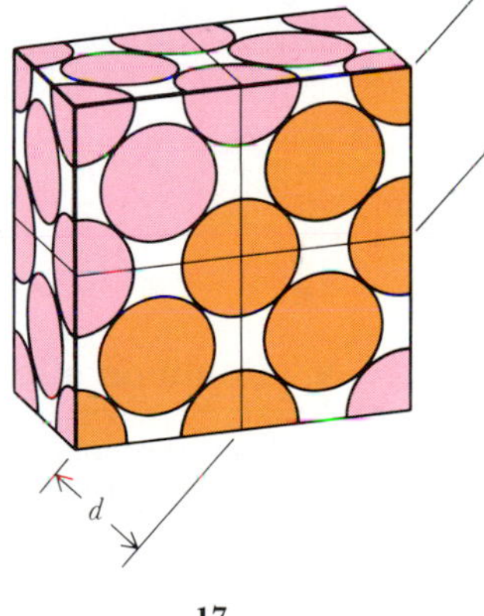

17

TABLE 10.8 Characteristics of unit cells

Type of cells	Number of atoms or ions per cell	Coordination number
primitive cubic	1	6
body-centered cubic, bcc	2	8
face-centered cubic, fcc*	4	12
rock-salt	4 cations, 4 anions	(6,6)
cesium chloride	1 cation, 1 anion	(8,8)

*Also called a ccp (cubic close-packed) unit cell.

T OHT: Table 10.8.

N For a bcc unit cell, the atoms touch along the diagonal through the center of the cube. The Pythagorean theorem for this cube diagonal is

$$\text{Edge}^2 + (\text{face diagonal})^2 = (\text{cube diagonal})^2$$

which is the same as

$$\text{Edge}^2 + (\text{edge}^2 + \text{edge}^2)^2 = (\text{cube diagonal})^2$$

SOLUTION From (**17**), the Pythagorean theorem ($a^2 + b^2 = c^2$), and r = 128 pm, we have

$$\text{Edge}^2 + \text{edge}^2 = (4 \times 128 \text{ pm})^2$$

so

$$\text{Edge}^2 = \tfrac{1}{2} \times (4 \times 128 \text{ pm})^2 = 1.31 \times 10^5 \text{ pm}^2$$

and

$$\text{Edge} = 362 \text{ pm}$$

The volume of the unit cell is therefore

$$V = (362 \times 10^{-12} \text{ m})^3 = 4.75 \times 10^{-29} \text{ m}^3 = 4.75 \times 10^{-23} \text{ cm}^3$$

Each fcc unit cell contains $8 \times \frac{1}{8} + 6 \times \frac{1}{2} = 4$ atoms. The molar mass of copper is 63.54 g/mol, so the total mass of the unit cell is the mass of four Cu atoms:

$$\text{Mass (g)} = 4 \text{ Cu atoms} \times \frac{1 \text{ mol Cu}}{6.022 \times 10^{23} \text{ Cu atoms}} \times \frac{63.54 \text{ g}}{1 \text{ mol Cu}}$$
$$= 4.221 \times 10^{-22} \text{ g}$$

The density of the unit cell, and hence of the solid itself, is therefore

$$\text{Density} = \frac{4.221 \times 10^{-22} \text{ g}}{4.74 \times 10^{-23} \text{ cm}^3} = 8.89 \text{ g/cm}^3$$

The experimental value is 8.92 g/cm^3.

EXERCISE E10.10 Calculate the atomic radius of silver, given that its density is 10.5 g/cm^3 and that it has an fcc (ccp) unit cell.

[*Answer*: 145 pm]

U Exercises 10.80–10.86.

Alloys. An **alloy** is a homogeneous mixture of two or more metals. Common alloys are listed in Table 10.9. Their structures are more complicated than those of pure metals because the two or more types of metal atoms have different radii. The packing problem is now akin to that of a storekeeper trying to stack oranges and grapefruit in the same pile. However, because the metallic radii of the *d*-block elements are all quite similar, for them the problem is not so acute as for some other metals. They form an extensive range of alloys with one another because one type of atom can take the place of another with little distortion of the crystal structure.

Some alloys are stronger than their component metals alone. Copper containing 2% by mass of beryllium is much harder than pure copper, for the copper atoms are pinned together by the small beryllium atoms lying between them. This alloy has a useful feature: the high electrical conductivity of the copper allows charge to leak away rapidly, so a sample of this alloy does not spark when struck. It is therefore used for tools in oil refineries, where a spark could be disastrous. In contrast, a nickel-barium alloy produces sparks readily because barium has so many electrons that they can form a conducting path through air; it is used for the electrodes in spark plugs.

Some alloys are softer than the component metals. The presence of big bismuth atoms helps to soften a metal and lower its melting point, much as melons destabilize a stack of oranges. A low melting point alloy of lead, tin, and bismuth is used to control water sprinklers employed in certain fire-extinguishing systems.

TABLE 10.9 Common alloys

Alloy	Mass percentage composition
brass	up to 40% zinc in copper (yellow brass: 35% Zn)
bronze	a metal other than zinc or nickel in copper (casting bronze: 10% Sn and 5% Pb)
cupro-nickel	nickel in copper (coinage cupronickel: 25% Ni)
pewter	6% antimony and 1.5% copper in tin
solder	tin and lead
stainless steel	over 12% chromium in iron

Electrical conduction. One of the most characteristic properties of a metal is its ability to conduct an electric current, the flow of electric charge. In **electronic conduction**—the type of **electrical conduction** that occurs in metals—the charge is carried by electrons. In **ionic conduction**, the charge is carried by ions. The latter is the mechanism of electrical conduction in a molten salt or an aqueous salt solution. Because ions are too bulky to travel easily through most solids, the flow of charge through solids is almost always a result of electronic conduction.

The ability of a substance to conduct electricity is measured by its electrical resistance: the lower the resistance, the better it conducts.* Substances may be classified according to the resistance they show to the passage of a current (Fig. 10.34):

An **insulator** is a substance that does not conduct electricity.

A **metallic conductor** is an electronic conductor with a resistance that increases as the temperature is raised.

A **semiconductor** is an electronic conductor with a resistance that decreases as the temperature is raised.

A **superconductor** is an electronic conductor that conducts electricity with zero resistance.

Insulators include gases, most solid ionic compounds, almost all organic compounds, all molecular liquids, and all molecular and network solids. Metallic conductors include all metals and some other solids. An example of a semiconductor is a crystal of pure silicon containing a tiny amount of arsenic or indium. Until 1987 most superconductors were metals (such as lead) or compounds cooled to almost absolute zero. In 1987, however, the first high-temperature superconductors were reported (see Box 10.2); these substances could be used at about 100 K (Fig. 10.35). Their discovery was important because liquid nitrogen

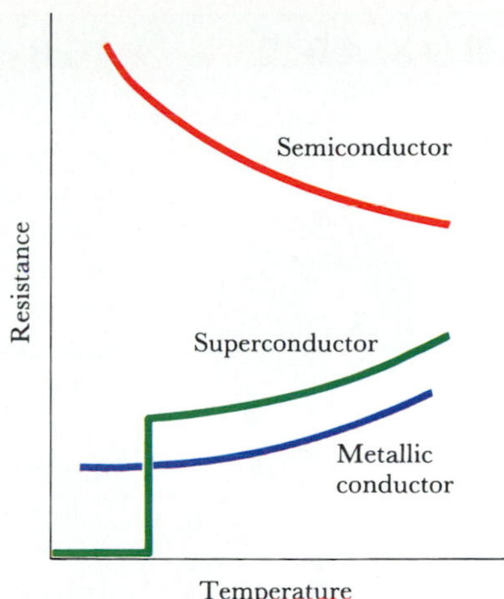

FIGURE 10.34 The variation of the resistance of a substance with temperature is the basis of its classification as a metallic conductor, a semiconductor, or a superconductor. An insulator is a semiconductor with a very high resistance.

E All solid "insulators" are in fact semiconductors, for their resistances, although very high, decrease as the temperature is raised.

E Chemistry and physics of amorphous semiconductors. *J. Chem. Educ.*, **1969**, *46*, 80. Fundamental principles of semiconductors. *J. Chem. Educ.*, **1969**, *46*, 80.

FIGURE 10.35 A sample of a high-temperature superconductor, one of the materials first produced in 1987. This sample is repelled strongly by a magnetic field. When the material becomes warm, superconductivity will be lost and the sample will no longer hover above the magnet.

*Specifically, according to Ohm's law, the current I in amperes is related to the potential difference V in volts by $I = V/R$, where R is the resistance in ohms. The higher the resistance, the smaller the current for a given potential difference.

BOX 10.2 ▸ HIGH-TEMPERATURE SUPERCONDUCTIVITY

C. W. CHU, *Texas Center for Superconductivity, University of Texas*

Superconductivity is the loss of all electrical resistance when a substance is cooled below a certain characteristic transition temperature (T_s). It was first observed in 1911 in mercury, for which $T_s = 4$ K (also the boiling point of liquid helium). In the next few decades, superconductivity was observed in many other metallic and intermetallic compounds; but the transition temperatures for these substances were never higher than 23 K.

Research chemists found that the conducting properties of a crystalline substance can be modified by adding an impurity, such as a few atoms of another element. This process is known as doping. For instance, the silicon used in computer chips is doped with phosphorus, antimony, boron, or indium. In 1986 a record-high T_s of 35 K was observed for a lanthanum-copper oxide doped with barium. Early in 1987 a new record T_s of 93 K was set with yttrium-barium-copper and a series of related oxides. In 1988 two more oxide series of bismuth-strontium-calcium-copper and thallium-barium-calcium-copper exhibited transition temperatures of 110 and 125 K, respectively. Since then, intensive research has uncovered twelve families of high-temperature superconductors that have transition temperatures above 20 K.

Almost all the high-temperature superconductors have sheets of copper and oxygen atoms sandwiched between layers of either cations or a combination of cations and oxygen, and all are derived from their respective parent insulators by doping. Because of their layered structure, their electrical and magnetic properties depend strongly on the orientation of the crystal. The observations of superconductivity above 77 K in such unusual classes of materials defy the predictions of earlier theories, a situation that makes research on these substances intriguing. But these materials are also intriguing because they behave in other unusual ways at temperatures above their transition temperatures. The causes for and consequences of these observations pose great challenges to physicists, chemists, and materials scientists.

Many applications for superconductors—from ultrasensitive detectors to extremely powerful magnets (which might be used in suspended trains)—have been proposed and a few have been demonstrated. Already superconductors are being used in magnetic neural imaging (MNI), magnetic resonance imaging (MRI), and atom smashers (such as the superconducting supercollider, SSC). But the full-scale commercial application of superconductors has been hampered by the need to use liquid helium (a rare, expensive, and inefficient coolant that is difficult to handle) to cool the conventional superconducting components below their transition temperatures. So when superconductivity was observed above the boiling point of liquid nitrogen (77 K), it seemed that applications of superconductors would soon multiply dramatically. Liquid nitrogen is an abundant, inexpensive, and efficient coolant and is easy to handle. Nevertheless, despite some successes, many of the great promises of superconductivity technology have yet to be realized. The difficulties encountered can be attributed to the many material and engineering problems of high-temperature superconductors, for example, making these materials into long wires that can carry high current without excessive heat loss and that can retain their superconducting properties over a long period without chemical or physical degradation.

After the discovery of the transistor in 1947, almost 40 years passed before the introduction of the one-megabyte memory chip that is vital to today's powerful computers. If history serves as a guide, then the wonderland of high-temperature superconductor applications is destined to become reality in the foreseeable future, given determination, persistence, and imaginative experimentation by scientists with vision.

(which boils at 77 K) can be used to cool them rather than the much more expensive liquid helium. These high-temperature superconductors are complicated ionic oxides, such as $YBa_2Cu_3O_7$.

Conduction by metals. Electronic conduction in metals can be explained in terms of molecular orbitals that spread throughout the solid. As we saw in Section 9.7, two atomic orbitals overlap to form two molecular orbitals. In general, when N atomic orbitals merge together in a molecule, they form N molecular orbitals. The same is true in a metal, but there N is enormous (around 10^{23} for 10 g of copper, for instance). Instead of the few molecular orbitals with their widely different energies, which is typical of small molecules, the huge number of molecular orbitals in a metal are so close together in energy that they form a nearly continuous band (Fig. 10.36).

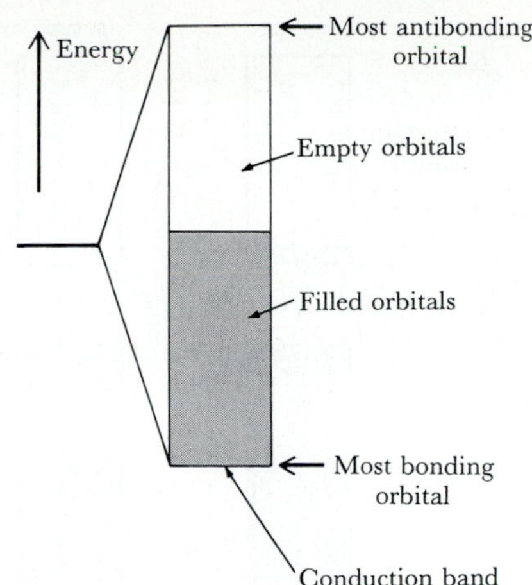

FIGURE 10.36 When a large number of atomic orbitals overlap, they form the same large number of molecular orbitals (but only the most strongly bonding and antibonding orbitals are shown here). Their energies lie in an almost continuous band. Each orbital can hold up to two electrons.

U Exercise 10.87.

E An introduction to principles of solid-state extrinsic semiconductors. *J. Chem. Educ.*, **1971**, *48*, 831.

Consider a sample of an alkali metal, such as sodium. Each atom contributes one valence orbital (the 3*s* orbital in this case) and one valence electron. If there are N atoms in the sample, then the N 3*s* orbitals merge to give a band of N molecular orbitals, of which half are bonding and half antibonding. Because each of the N atoms supplies one valence electron, a total of N electrons must be accommodated in the orbitals. These N electrons occupy the orbitals according to the building-up principle, filling the lower $\frac{1}{2}N$ bonding orbitals (because two electrons can occupy each orbital). An incompletely filled band of molecular orbitals is called a **conduction band**. The bonding orbitals in the band lie so close together in energy that it takes very little additional energy to excite the electrons from the topmost filled molecular orbitals to the empty bonding orbitals just above. These electrons move freely throughout the solid, which is like one big molecule, and hence they can carry an electric current through the solid.

The resistance of a metal increases with temperature because, when a metal is heated, its atoms vibrate more vigorously. As they do so, the atoms collide with any electrons traveling past them. These collisions reduce the flow of electrons, so the solid is a poorer conductor at high temperatures than at low.

Solid insulators, such as diamond and sodium chloride, can also be described in terms of band theory, but in these substances the valence electrons fill the entire band of molecular orbitals. The full band is called a **valence band** to distinguish it from an incompletely filled conduction band. There is a substantial **band gap**, a range of energies for which there are no orbitals, before the next band of empty orbitals begins (Fig. 10.37). The electrons in the valence band can be excited into the upper empty band only by a large injection of energy. The upper band is formed by the overlap of higher atomic orbitals. Hence the electrons of the valence band effectively are not mobile because that band is full, and the solid does not conduct electricity.

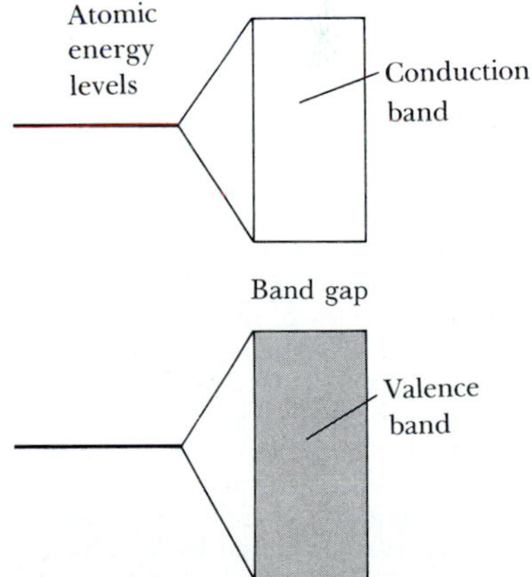

FIGURE 10.37 In an insulator, there is a substantial energy gap between the uppermost filled molecular orbital (of the valence band) and the first empty molecular orbital (of the conduction band) and the electrons are relatively immobile.

Semiconductors. One type of semiconductor consists of very pure silicon to which a minute amount of arsenic impurity has been added in a process called **doping**. The arsenic increases the number of electrons in the solid: an Si atom (Group IV) has four valence electrons, and an As atom (Group V) has five. The additional electrons enter the upper, normally empty conduction band of silicon and hence enable the solid to conduct (Fig. 10.38a). This type of material is called an ***n*-type semiconductor**, the *n* indicating the presence of excess *n*egatively charged electrons.

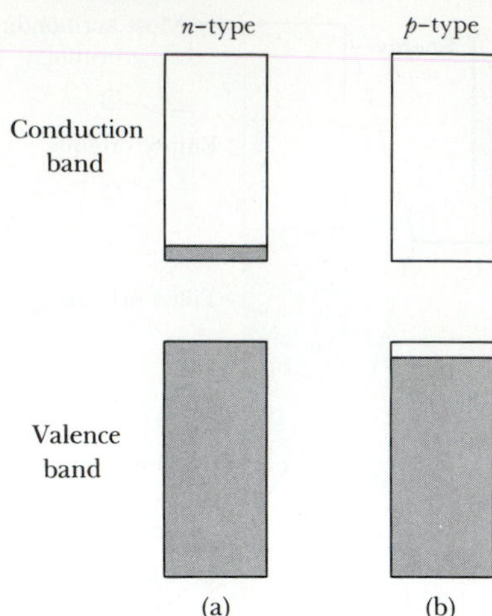

FIGURE 10.38 (a) In an n-type semiconductor the additional atoms provide extra electrons that occupy the conduction band and carry the electric current. (b) In a p-type semiconductor the additional atoms remove electrons from the valence band, and the electrons near the top of the filled orbitals are free to conduct.

When silicon (Group IV) is doped with indium (Group III) instead of arsenic, the resulting solid has fewer valence electrons than the pure element, so the lower band (which in pure silicon is full) is now not completely full (Fig. 10.38b). The solid is now electrically conducting because the uppermost occupied molecular orbital has unoccupied neighbors. This type of material is called a ***p*-type semiconductor**. The *p* indicates that, because the bands of silicon contain fewer electrons, they have a less negative charge and, hence—in a sense—greater *p*ositive charge. Solid-state electronic devices (transistors and integrated circuits) are formed from so-called *p-n* junctions in which a *p*-type semiconductor is in contact with an *n*-type semiconductor.

10.7 IONIC SOLIDS

As in metals, the ions making up an ionic solid may be pictured as spheres; and the crystal structure of an ionic solid is built up by stacking them together. However, because cations and anions have different charges as well as different sizes, the packing problem is more complicated. Each unit cell is made up of ions with different radii and charges, yet it must be electrically neutral overall.

The rock-salt structure. One of the solutions that nature has found to the ionic packing problem is the **rock-salt structure**, the structure of the mineral form of sodium chloride ("rock salt") and of several other substances. In the rock-salt structure the anions (the green spheres in the illustration, such as the Cl^- ions in NaCl itself) form a face-centered cubic (fcc) array (Fig. 10.39). The smaller cations (the pink spheres in the illustration, such as the Na^+ ions in NaCl itself) also form an fcc array, one that fits snugly among the anions. The rock-salt structure therefore consists of two interpenetrating fcc arrays of ions. Each cation is surrounded by six anions, giving a coordination number of 6. Each anion is surrounded by six cations, so the anions also have a coordination number of 6. The structure is therefore described as having (6,6)-coordination; in this notation, the first number is the cation coordination number and the second is that of the anion.

E Many crystal structures can be regarded as consisting of a basic structure (such as fcc) of one type of ion with the second type of ion occupying the gaps in the array. An example is provided by the rock-salt structure, for the Cl^- anions can be considered to form an fcc array, and the Na^+ occupy the gaps between the anions. Each gap is surrounded by six anions arranged to form an octahedron, so each Na^+ ion is in an *octahedral hole*. Equivalently, the Na^+ ions form an fcc array with the Cl^- occupying the octahedral holes between them.

E Crystal systems and general chemistry. *J. Chem. Educ.*, **1982**, *59*, 742. Some simple AX and AX_2 structures. *J. Chem. Educ.*, **1982**, *59*, 630.

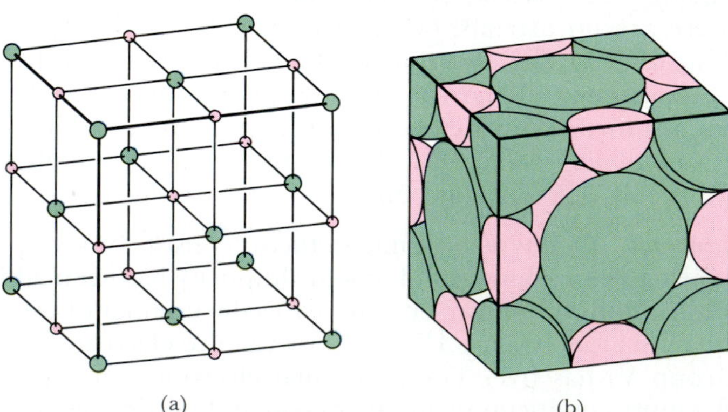

FIGURE 10.39 The arrangement of the ions in (a) the rock-salt structure. The pink dots show the centers of the cations, and the green dots the centers of the anions. (b) The unit cell of the rock-salt structure.

▼ EXAMPLE 10.11 Counting the ions in a unit cell

How many Na^+ and Cl^- ions are there in the unit cell shown in Fig. 10.39. Confirm that the cell is electrically neutral.

STRATEGY An ion at a corner or on a face or an edge is shared by neighboring unit cells, so in a crystal only a fraction of it (and its charge) belongs to the unit cell in question. An ion on a face is shared by two cells and counts as $\frac{1}{2}$ (of an ion). An ion on an edge is shared by four neighboring cells, so it contributes $\frac{1}{4}$. An ion at a corner is shared by eight neighboring cells and contributes $\frac{1}{8}$ to each cell. To confirm electrical neutrality, we must count the number of cations and anions and make sure that their charges cancel.

SOLUTION Inspection of Fig. 10.39 gives the following count:

	Number of ions	
	Na^+	Cl^-
at center of cell	1	0
on 6 faces	0	$6 \times \frac{1}{2} = 3$
on 12 edges	$12 \times \frac{1}{4} = 3$	0
at 8 corners	0	$8 \times \frac{1}{8} = 1$
total	4	4

That is, the unit cell contains four cations and four anions. The total charge is therefore zero because the charge of $4Na^+$ is canceled by that of $4Cl^-$.

EXERCISE E10.11 Repeat the calculation for the unit cell of the cesium chloride structure shown in Fig. 10.40.

[*Answer*: one cation, one anion] ▲

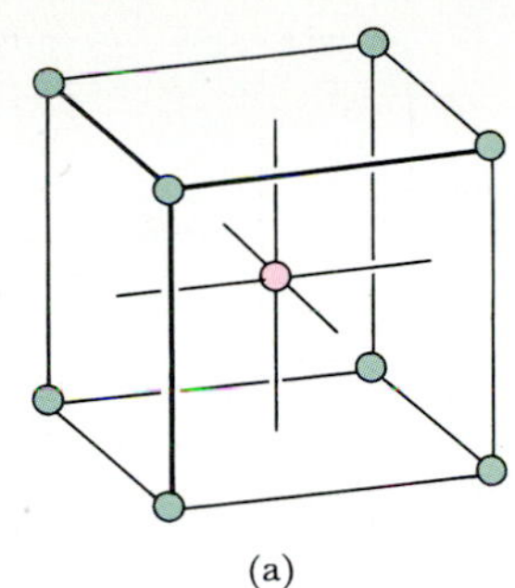

(a)

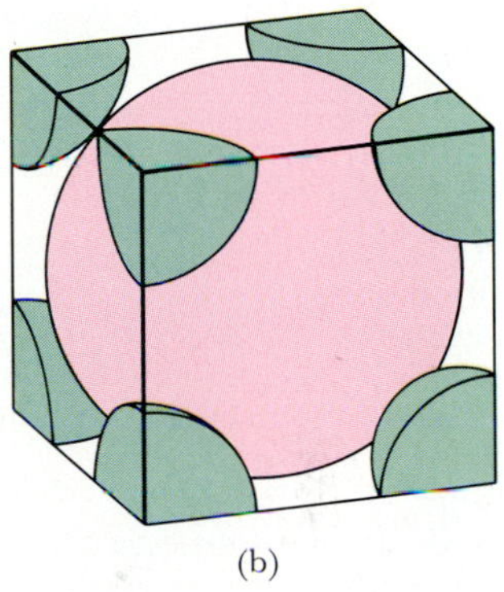

(b)

FIGURE 10.40 The structure of (a) cesium chloride and (b) its unit cell, with dots showing the locations of the ions.

The cesium chloride structure. The rock-salt structure is not the lowest-energy arrangement when the radii of the cations and anions are very different. In this case a lower energy is obtained if the ions adopt the **cesium chloride structure** (Fig. 10.40). The bulky cation (Cs^+ in CsCl itself) is surrounded by eight anions (Cl^-), and each anion ion is surrounded by eight cations, an arrangement with (8,8) coordination. It is much less common than the rock-salt structure, but it illustrates how different structures arise from ions of different size. Further examples of crystal structures are listed in Table 10.10.

N An alterative description of the cesium-chloride structure is to regard the Cl^- ions as forming a primitive cubic array, and the Cs^+ ion in the hole between them. An equivalent description is that the Cs^+ ions define a primitive cubic lattice and the Cl^- ions inhabit the holes between them.

TABLE 10.10 Crystal structures and examples

Crystal structure	Examples*
rock-salt	**NaCl**, LiCl, KBr, RbI, MgO, CaO, AgCl
cesium chloride	**CsCl**, CsI, TlSb
fluorite	$\mathbf{CaF_2}$, Na_2O, UO_2
rutile	$\mathbf{TiO_2}$, MnO_2, SnO_2, MgF_2
perovskite	$\mathbf{CaTiO_3}$, $BaTiO_3$, $SrFeO_3$, $LaMnO_3$

*The substance in bold type is the one that gives its name to the structure. For the fluorite, rutile, and perovskite structures, see the illustrations for Exercises 10.45, 10.46, and 10.47.

Using the unit cell. Once we know the structure of the unit cell of a compound, we can use it to deduce other properties of the solid. An application of particular chemical significance is the determination of the value of Avogadro's number from the measurement of the density of a solid.

▼ EXAMPLE 10.12 Determining Avogadro's number from a crystal structure

The density of sodium chloride is 2.163 g/cm^3 and the length of the edge of the unit cell in rock salt is 564.1 pm. Use this information to deduce a value for Avogadro's number.

STRATEGY Avogadro's number, N_A, is the number of objects per mole, in this case, the number of NaCl formula units per mole of NaCl. Because we know (from Example 10.11) that there are four NaCl formula units per unit cell, we can calculate the number of formula units per cubic meter. Then, because we are given the density, we can calculate the number of formula units per gram. Finally, using the molar mass, we can calculate the number of formula units per mole of NaCl.

SOLUTION The conversions we need are obtained from the volume of the unit cell:

$$1 \text{ unit cell} = (564.1 \text{ pm})^3$$

The conversion from picometers to meters is 1 pm = 10^{-12} m; the conversion from meters to centimeters, 10^{-2} m = 1 cm; the conversion from volume to mass (using the density), 1 cm^3 NaCl = 2.163 g NaCl; and the conversion from mass to moles (using the molar mass of NaCl) is 58.44 g NaCl = 1 mol NaCl. The overall conversion is therefore

Number of formula units per mole

$$= \frac{4 \text{ NaCl}}{\text{unit cell}} \times \frac{1 \text{ unit cell}}{(564.1 \text{ pm})^3} \times \left(\frac{1 \text{ pm}}{10^{-12} \text{ m}}\right)^3 \times \left(\frac{10^{-2} \text{ m}}{1 \text{ cm}}\right)^3$$

$$\times \frac{1 \text{ cm}^3 \text{ NaCl}}{2.163 \text{ g NaCl}} \times \frac{58.44 \text{ g NaCl}}{1 \text{ mol NaCl}}$$

$$= 6.021 \times 10^{23} \text{ NaCl/mol}$$

EXERCISE E10.12 Repeat the determination of Avogadro's number, using the facts that the density of cesium chloride is 3.988 g/cm^3 and its unit cell is a cube of edge 411 pm.

▲ [*Answer*: 6.08 × 10^{23} CsCl/mol]

10.8 OTHER TYPES OF SOLIDS

The atoms in network solids are joined to their neighbors by covalent bonds that form a network extending throughout the crystal. Network solids show the strength of the covalent bonds that bind them by being very hard, rigid materials with high melting and boiling temperatures. In contrast, molecular solids—solids composed of individual molecules—often have low melting points, because only a small amount of thermal motion is needed to disrupt the relatively weak intermolecular forces that hold the molecules in place. Titanium tetrachloride ($TiCl_4$), for instance, is a liquid (and often yellow because of iron impurities) that boils at 136°C and freezes at −25°C to a molecular solid. (It is used

FIGURE 10.41 Tiny synthetic diamonds about 350 μm in diameter.

to generate smoke screens, because it reacts with water to give dense clouds of titanium dioxide.) The solid is composed of tetrahedral $TiCl_4$ units, held together largely by the London forces between the electron-rich chlorine atoms. Not all molecular solids are soft: sucrose, a brittle solid, is a molecular solid in which the $C_{12}H_{22}O_{11}$ molecules are held together by hydrogen bonds between their numerous —OH groups. The bonding between sucrose molecules is so strong that by the time the melting temperature has been reached (at 184°C), the molecules themselves have started to decompose. The half-decomposed mixture of products is called caramel.

E Finely divided titanium dioxide acts as a smokescreen because it has a high refractive index and the tiny crystals that are formed scatter the light strongly. It is used as a brilliant white paint for the same reason.

Network solids: Diamond and graphite. Two of the most important examples of network solids are diamond and graphite, two allotropes of carbon. Diamond is found in nature as a crystalline form of elemental carbon embedded in a soft rock called kimberlite. This rock forms volcanic pipes that raise the diamonds from deep in the earth where they have been formed under intense pressure. The erosion of kimberlite releases diamonds into the sediments of some rivers. One method for making synthetic diamonds recreates the geological conditions that produce natural diamonds (Fig. 10.41). This process involves compressing graphite at over 80,000 atm at above 1500°C in the presence of small amounts of metals such as chromium and iron. It is thought that the molten metals dissolve the graphite and then, as they cool, deposit crystals of diamond (which are less soluble than graphite) in the molten metal.

Each C atom in diamond is covalently bonded to four neighbors through sp^3-hybrid σ bonds (Fig. 10.42). The rigid three-dimensional structure is analogous to the steel framework of a large building, and its rigidity accounts for the great hardness of the solid. Diamond is one of the best conductors of heat, so it is used as a base for some integrated circuits to prevent overheating. This good thermal conductivity is also a result of the crystal structure and the rigidly linked atoms, for the vigorous vibration of an atom in a hot part of the crystal is rapidly transmitted to distant, cooler parts.

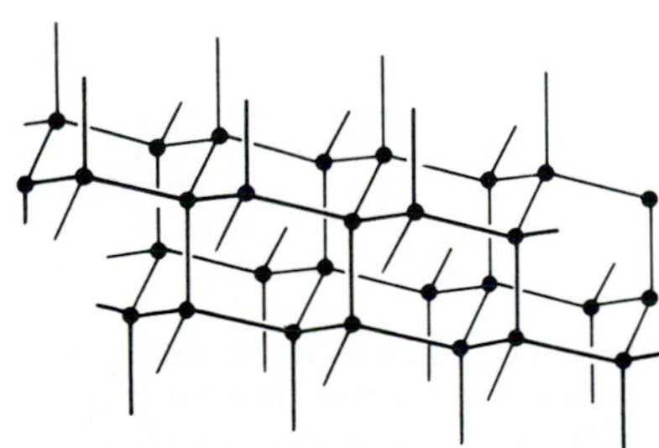

FIGURE 10.42 The structure of diamond. Each dot indicates the location of the center of a carbon atom. Each atom forms an sp^3-hybrid covalent bond to each of its four neighbors.

Graphite is produced naturally as a result of changes brought about on ancient organic remains or on carbonates. Most commercial graphite is produced by heating carbon rods to a high temperature in an electric furnace for several days. Graphite is a black, lustrous, electrically conducting, slippery solid that sublimes at 3700°C. It consists of

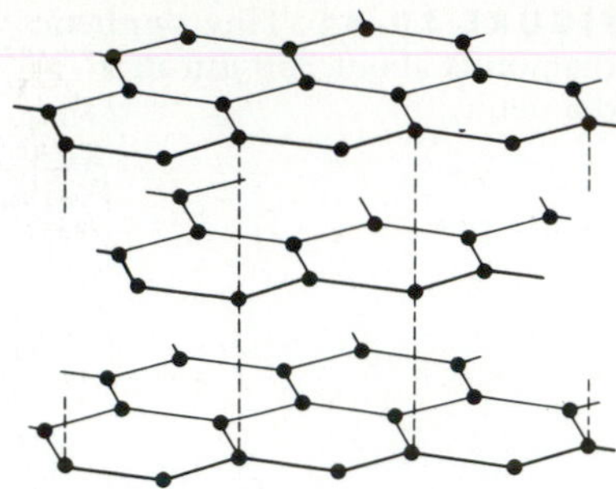

FIGURE 10.43 Graphite consists of layers of hexagonal rings of sp^2-hybridized carbon atoms. The slipperiness of graphite is due to the ease with which the layers can slide over one another when there are impurity atoms lying between the planes.

flat sheets of sp^2-hybridized carbon atoms bonded covalently into hexagons, and the structure of each sheet resembles chicken wire (Fig. 10.43). There are also covalent bonds between the sheets. However, because these bonds are so long and weak, they resemble London forces and the sheets can slide over one another. Electrons can move within the sheets in the delocalized π orbitals formed from the unhybridized p orbitals on the C atoms, but they move much less readily from one sheet to another. Hence, graphite conducts electricity better parallel to the sheets than perpendicular to them.

Ice, a molecular solid. Because molecules have such widely differing shapes, they stack together in a wide variety of ways. We consider only one example here, namely, ice (Fig. 10.44). In ice, each O atom is surrounded by four H atoms. Two of these four H atoms are linked to the O atom through σ bonds. The other two belong to neighboring H_2O molecules and are linked to the O atom by hydrogen bonds. As a result, the structure of ice is an open network of H_2O molecules held together by hydrogen bonds. Some of the hydrogen bonds break, so the open structure collapses when the solid melts and the molecules pack less uniformly but more densely. The openness of the network in ice compared with that in the liquid explains why ice has a lower density than liquid water has (0.92 g/cm^3 and 1.00 g/cm^3, respectively, at 0°C). This feature was mentioned in Section 10.5, where we saw that it accounts for the unusual property that ice melts when subjected to pressure. Solid benzene and solid carbon dioxide, in contrast, have higher densities than their liquids. Their molecules are held together by London forces and pack together more closely in the solid than in the liquid (Fig. 10.45).

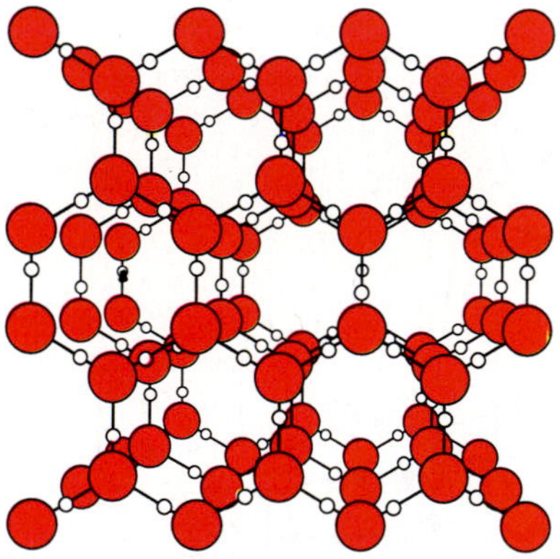

E Graphite is an example of a "dry" or "greaseless" lubricant. The impurities trapped between the layers of carbon atoms appear to be essential to graphite's slipperiness.

V Video 27: Benzenebergs and icebergs.

E In this view, the honeycomblike structure of ice can readily be seen and it is easy to appreciate that the density of ice is low relative to that of the liquid.

D The ice bomb: Expansion of water as it freezes. BZS3, 9.26.

FIGURE 10.44 Ice is made up of water molecules that are held together by hydrogen bonds in a relatively open structure. Each O atom is surrounded tetrahedrally by four H atoms, two of which are σ-bonded to it and two of which are hydrogen-bonded to it. This view of ice is different from that shown in Fig. 10.20.

FIGURE 10.45 As a result of its open structure, ice is less dense than water and floats in it (left). Solid benzene is denser than liquid benzene, and "benzenebergs" sink in liquid benzene (right).

SUMMARY

10.1 Forces between ions and polar molecules are called **intermolecular forces** and include the **ion-dipole interaction**, which is responsible for the **hydration** of cations—the attachment of water molecules to an ion. The partial charges of the dipoles of polar molecules attract each other by the **dipole-dipole interaction**. All molecules, polar or nonpolar, attract each other by the **London interaction**, the attraction between instantaneous electric dipoles.

10.2 The **hydrogen bond** is a link formed by a hydrogen atom lying between two strongly electronegative atoms (N, O, or F). It dominates the other intermolecular interactions when it is present.

10.3 The smooth surface of liquids is due to the **surface tension**, a measure of the force required to balance the force acting on the surface molecules relative to that acting on the molecules in the interior of a sample. The **adhesion** of a liquid to the walls of its container accounts for **capillary action**, the tendency of a liquid to climb up inside a narrow tube, and the formation of a **meniscus**, a crescent-shaped surface. The interactions that hold atoms, ions, and molecules together are called **cohesive forces**.

10.4 A liquid in a closed container reaches **dynamic equilibrium** with its vapor, a condition in which forward and reverse processes (in this case, of vaporization and condensation) continue at equal rates. At a given temperature, each liquid has a characteristic **vapor pressure**, the pressure of the vapor in dynamic equilibrium with its liquid (or solid) phase. Vapor pressures increase with temperature. A liquid in an open container boils when the vapor pressure is equal to the external pressure. A gas cannot be liquefied by the application of pressure if its temperature is above its **critical temperature**.

10.5 Freezing temperatures depend weakly on the pressure and normally increase as the pressure is raised. Water is unusual, because it melts at a lower temperature under pressure. The conditions of pressure and temperature under which solids, liquids, and gases are stable are shown on the **phase diagram** of the substance. Three phases are in dynamic equilibrium with each other at a **triple point**. The location of boundaries on a phase diagram can be determined by plotting the cooling curve of a substance and noting the temperature at which the temperature change halts while the phase change is in progress.

10.6 Metals are held together by electrons that spread among the cations of the element. **Ionic solids** are held together by **ion-ion interactions**. In **network solids**, atoms are held together by covalent bonding that extends throughout the solid. In **molecular solids**, the individual molecules are held together by intermolecular forces. A **crystalline solid** is an orderly array of atoms, ions, or molecules. The structures of crystals are determined by **x-ray diffraction**. The structures of metals can be described in terms of the stacking of spheres; in many cases a **close-packed structure**, a structure with the least waste of space, is adopted. Two such structures are most common, namely, **hexagonal** and **cubic close-packed**. The structure of the crystal can be summarized by drawing its **unit cell**, the unit that can reproduce the entire crystal when repeatedly stacked together. **Alloys**, which are homogeneous mixtures of metals, can be discussed similarly; but the spheres of the constituents have different radii. **Electrical conduction** may be electronic or ionic. The temperature dependence of electronic conduction distinguishes **metallic conductors** from **semiconductors**. Electronic conduction may be discussed in terms of **bands** of molecular orbitals. A metallic conductor is a substance in which the bands of orbitals are incompletely filled. **Insulators** are substances in which the bands are completely filled. In an ***n*-type semiconductor**, the **doping** increases the number of valence electrons in the solid; in a ***p*-type semiconductor**, it reduces them.

10.7 Two simple structures characteristic of several ionic compounds are the **rock-salt** and **cesium chloride** structures.

10.8 Network solids are often very hard (diamond, for example). Other forms of network solids may be slippery (graphite, for example). Ice, a molecular solid, has an open structure because of the hydrogen bonds between its molecules.

CLASSIFIED EXERCISES

Intermolecular Forces

10.1 Identify the kinds of intermolecular forces that might arise between pairs of the following molecules: (a) Cl_2; (b) HCl; (c) C_6H_5Cl.

10.2 Identify the kinds of intermolecular forces that might arise between pairs of the following molecules: (a) N_2; (b) $CH_2{=}CH_2$; (c) $CHCl{=}CHCl$ (both isomers).

10.3 Write the Lewis structure and indicate the dominant type of intermolecular forces for (a) CO; (b) NF_3; (c) CH_4.

10.4 Write the Lewis structure and indicate the dominant type of intermolecular forces for (a) CH_3Cl; (b) OF_2; (c) Xe.

10.5 Which of the following molecules are likely to form hydrogen bonds: (a) CH_4; (b) NH_3; (c) CH_3OH?

10.6 Which of the following molecules is likely to form hydrogen bonds? Suggest the consequences in each case. (a) D_2O; (b) CH_3CH_2OH (ethanol); (c) H_3PO_4.

10.7 Which of the following ions has the stronger interaction with a hydrating water molecule: (a) Mg^{2+} or Ca^{2+}; (b) Fe^{3+} or Fe^{2+}?

10.8 Which of the following ions has the stronger interaction with a hydrating water molecule: (a) Na^+ or Li^+; (b) Al^{3+} or Mg^{2+}?

10.9 Suggest, giving reasons, which substance in each pair is likely to have the higher normal boiling point (Lewis structures may help your arguments): (a) CH_4 or SiH_4; (b) HF or HCl; (c) H_2O (to compare, write as HOH) or CH_3OH (methanol).

10.10 Suggest, giving reasons, which substance in each pair is likely to have the higher normal melting point (Lewis structures may help your arguments): (a) H_2S or Na_2S; (b) NH_3 or PH_3; (c) $C_2H_5{-}O{-}C_2H_5$ (diethyl ether) or C_4H_9OH (butanol), both molecules have the same molar mass.

10.11 Account for the following observations in terms of the type and strength of intermolecular forces. (a) The melting point of xenon is −112°C and that of argon is −189°C. (b) The critical temperature of HI is 151°C and that of HCl is 52°C. (c) The vapor pressure of diethyl ether ($C_2H_5OC_2H_5$) is greater than that of water.

10.12 Account for the following observations in terms of the type and strength of intermolecular forces. (a) The boiling point of NH_3 is −33°C and that of PH_3 is −87°C. (b) The critical temperature of H_2O is 374°C and that of HF is 188°C. (c) The vapor pressure of CH_3OH is greater than that of CH_3SH.

10.13 Give the shape (from VSEPR considerations) of each of the following molecules and predict which substance of each pair has the higher boiling point: (a) SF_4 or SF_6; (b) SF_4 or CF_4; (c) *cis*-CHCl=CHCl or *trans*-CHCl=CHCl.

10.14 Give the shape (from VSEPR considerations) of each of the following molecules and predict which substance of each pair has the higher boiling point: (a) PF_3 or PCl_3; (b) SO_2 or CO_2; (c) $AsCl_3$ or $AsCl_5$.

Vapor Pressure

10.15 Suppose you were to collect 1.0 L of air by slowly passing it through water at 20°C and into a container. Estimate the mass of water vapor in the collected air, assuming that the air is saturated with water. At 20°C, the vapor pressure of water is 17.5 Torr.

10.16 Suppose you were to collect 500 mL of nitrogen by slowly passing it through ethanol at 25°C, at which temperature the vapor pressure of ethanol is 60 Torr. Estimate the mass of ethanol vapor in the collected gas, assuming that the nitrogen is saturated with it.

10.17 What mass of water vapor can you expect to find in the air of a bathroom of dimensions 4 × 3 × 3 m when water has been left in the bathtub at 40°C? The vapor pressure at that temperature is 7.4 kPa. Assume that the air is saturated with water vapor and that its temperature is 25°C.

10.18 A bottle of mercury at 25°C was left unstoppered in a chemical supply room measuring 3 × 2 × 2 m. What mass of mercury vapor would be present if the air became saturated with mercury? The vapor pressure of mercury at 25°C is 0.224 kPa.

10.19 Use the vapor-pressure curve in Fig. 10.17 to estimate the boiling point of water when the atmospheric pressure is (a) 700 Torr; (b) 770 Torr; (c) 600 Torr.

10.20 Use the vapor-pressure curve in Fig. 10.15 to estimate the boiling point of ethanol when the atmospheric pressure is (a) 700 Torr; (b) 770 Torr; (c) 100 Torr.

Phase Diagrams

10.21 Use Fig. 10.22 (the phase diagram for water) to predict the state of a sample of water under the following sets of conditions: (a) 2 atm, 200°C; (b) 3 Torr, 0°C; (c) 218 atm, 374°C.

10.22 Using Fig. 10.23 (the phase diagram for carbon dioxide) to predict the state of a sample of CO_2 under the following sets of conditions: (a) 1 atm, −80°C; (b) 1 atm, −78°C; (c) 5.1 atm, −56°C.

10.23 Use the phase diagram for helium to answer the following questions: (a) What is the maximum temperature at which superfluid helium-II can exist? (b) What is the minimum pressure at which solid helium can exist? (c) Can solid helium sublime?

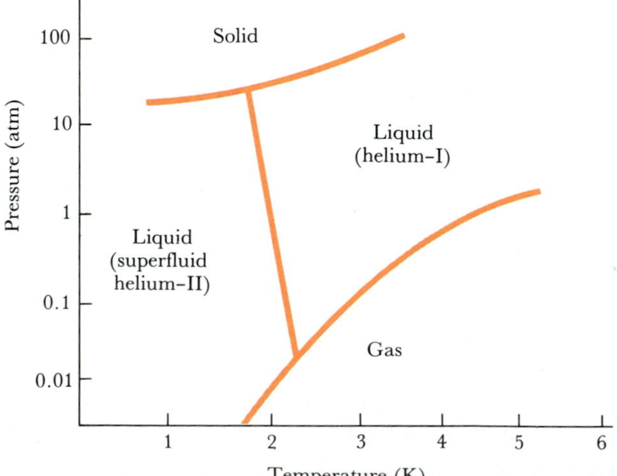

10.24 The phase diagram for carbon indicates the extreme conditions that are needed to form diamonds from graphite. Use the diagram to answer the following questions. (a) At 2000 K, what is the minimum pressure needed before graphite changes to diamond? (b) What is the minimum temperature at which liquid carbon can exist at pressures below 10,000 atm? (c) At what pressure does graphite melt at 4000 K?

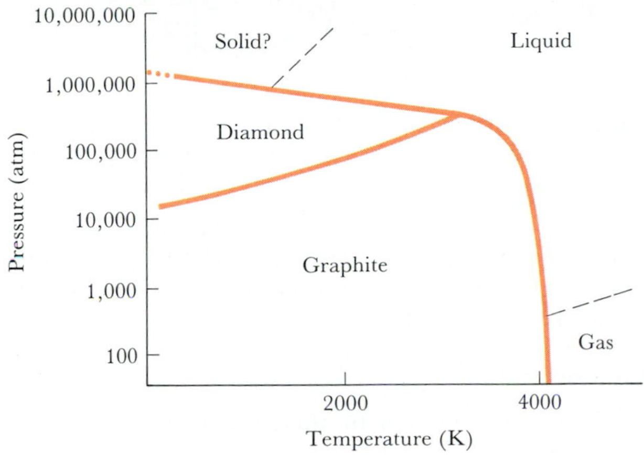

10.25 Naphthalene sublimes at room temperature and pressure. What can be said about the triple point of naphthalene?

10.26 Can liquid carbon dioxide exist at room temperature? Explain your conclusion.

Classification of Solids

10.27 Use Table 10.7 to classify each of the following solids according to its bonding type: (a) sodium chloride; (b) solid nitrogen; (c) sugar (sucrose); (d) copper.

10.28 Use Table 10.7 to classify each of the following solids according to its bonding type: (a) quartz; (b) sulfur; (c) iron(II) sulfate; (d) brass.

10.29 Calcium crystallizes in a ccp (fcc) structure; its atomic radius is 180 pm. (a) What is the length of the edge of the unit cell? (b) How many unit cells are there in 1 cm^3 of calcium?

10.30 The metal polonium (which was named by Marie Curie after her native Poland) crystallizes in a primitive cubic structure, with an atom at each corner of a cubic unit cell. The atomic radius of polonium is 190 pm. Sketch the unit cell and determine (a) the number of atoms per unit cell; (b) the coordination number; (c) the length of the edge of the unit cell.

10.31 Potassium crystallizes in a bcc structure. The atomic radius of potassium is 220 pm. Calculate (a) the number of atoms per unit cell; (b) the coordination number of the lattice; (c) the length of the edge of the unit cell.

10.32 Calculate the density of each of the following metals from the data: (a) platinum (fcc structure), atomic radius 138 pm; (b) cesium (bcc structure), atomic radius 266 pm.

10.33 Calculate the atomic radius of the following elements from the data: (a) gold (fcc structure), density 19.3 g/cm^3; (b) vanadium (bcc structure), density 5.96 g/cm^3.

10.34 Calculate the atomic radius of each of the following elements from the data: (a) iridium (fcc structure), density 22.5 g/cm^3, the densest of all elements; (b) molybdenum (bcc structure), density 10.2 g/cm^3.

10.35 Calculate the relative number of atoms of each element contained in each alloy: (a) yellow brass, which is 35% Zn by mass in copper; (b) casting bronze, which is 10% Sn and 5% Pb by mass in copper; (c) a stainless steel that is 19% Cr and 9% Ni by mass in iron.

10.36 Calculate the relative number of atoms of each element contained in each alloy: (a) coinage cupronickel, which is 25% Ni by mass in copper; (b) pewter, which is about 6% Sb and 3% Cu by mass in tin; (c) Wood's metal, a low melting point alloy that is used to trigger automatic sprinkler systems and is 12.5% Sn, 12.5% Cd, and 25% Pb by mass in bismuth.

Electrical Conduction

10.37 Distinguish between an electrical conductor and an electrical insulator in terms of molecular orbitals.

10.38 Gallium arsenide, GaAs, can be used as a semiconductor for solar cells. Explain how GaAs can conduct current in the presence of sunlight.

10.39 The electrical conductivity of graphite parallel to its planes is different from that perpendicular to them. Parallel to the planes, the conductivity decreases as the temperature is raised, but perpendicular to them it rises. In what sense is graphite a metallic conductor or a semiconductor?

10.40 "Graphite bisulfates" are formed by heating graphite with a mixture of sulfuric and nitric acids. In the reaction the graphite planes are partially oxidized (so there is an average of one positive charge shared among 24 carbon atoms) and the HSO_4^- anions are distributed between the planes. What effects should this have on the electrical conductivity?

10.41 State whether each of the following materials is an *n*- or a *p*-type semiconductor: (a) Si doped with P; (b) Si doped with In; (c) Ge doped with Sb.

10.42 State whether each of the following materials is an *n*- or a *p*-type conductor: (a) Si doped with Al; (b) Ge doped with Ga; (c) Ge doped with As.

10.43 Silicon is a semiconductor even without being doped (diamond is too). Account for this property in terms of bands, using as a clue the fact that at absolute zero ($T = 0$) the conductivities of silicon and carbon are zero.

10.44 Zinc oxide is a semiconductor. Its conductivity increases when it is heated in a vacuum but decreases when it is heated in oxygen. Account for these observations.

Ionic Solids

10.45 Calculate the number of cations, anions, and formula units per unit cell in the following solids: (a) the rock-salt unit cell shown in Fig. 10.39b; (b) the fluorite, CaF_2, unit cell shown below. (c) What are the coordination numbers of the ions in fluorite?

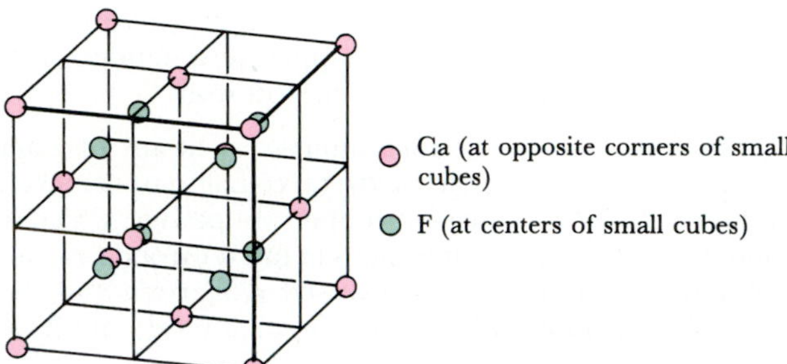

10.46 Calculate the number of cations, anions, and formula units per unit cell in the following solids: (a) the cesium chloride unit cell shown in Fig. 10.40b; (b) the rutile, TiO_2, unit cell shown below. (c) What are the coordination numbers of the ions in rutile?

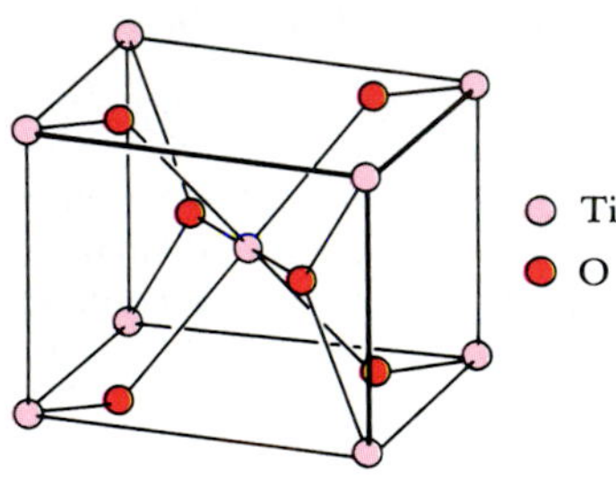

10.47 A unit cell of the mineral perovskite is drawn below. What is its formula?

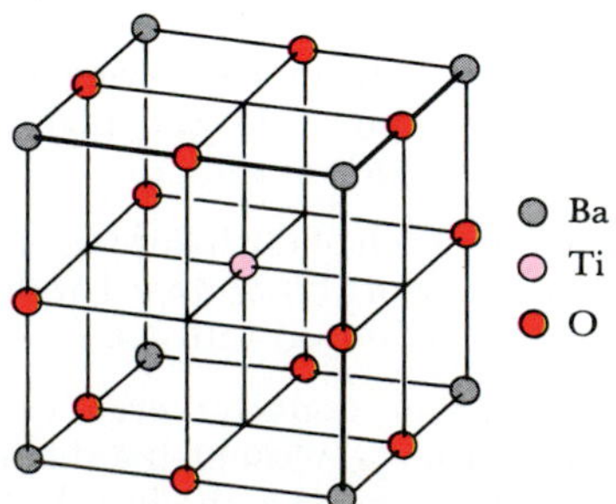

10.48 A unit cell of one of the new high-temperature superconductors is shown below. What is its formula?

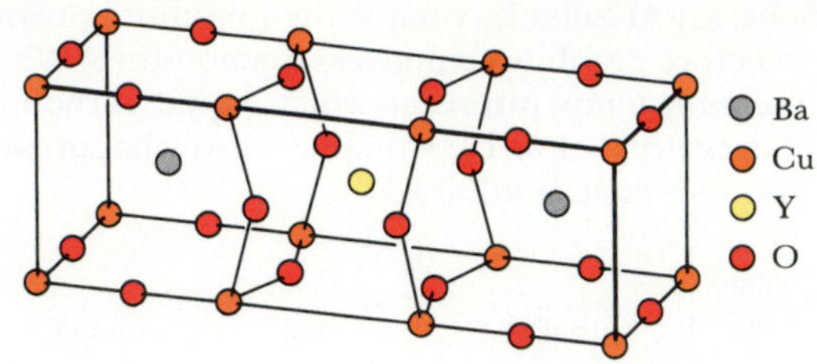

10.49 How many unit cells of the kind shown in Fig. 10.39 are there in a 1.00-mm^3 grain of potassium bromide? The density of potassium bromide is 2.75 g/cm^3.

10.50 Calculate the density of the following solids from the data in Fig. 7.29: (a) calcium oxide (rock-salt structure; Fig. 10.39); (b) cesium bromide (cesium chloride structure; Fig. 10.40).

Network Solids

10.51 Name four elements that, in at least one allotropic form, are network solids.

10.52 Name two compounds that exist as network solids at room temperature and pressure.

10.53 The enthalpy of sublimation of diamond is 713 kJ/mol. What is the C—C bond enthalpy in the solid?

10.54 The enthalpy of formation of BN(*s*) is −254 kJ/mol and that of BN(*g*) is +647 kJ/mol. Given that the enthalpies of formation of gaseous B and N atoms are +563 kJ/mol and +473 kJ/mol, respectively, calculate the B—N bond enthalpy in the solid.

Molecular Solids

10.55 Identify the type of intermolecular forces that are responsible for the existence of each of the following molecular solids: (a) solid argon; (b) solid HCl; (c) glucose.

10.56 Identify the type of intermolecular forces that are responsible for the existence of each of the following molecular solids: (a) iodine; (b) oxalic acid, $H_2C_2O_4$; (c) solid benzene.

10.57 The figure shows a unit cell of a molecular compound. What is its molecular formula?

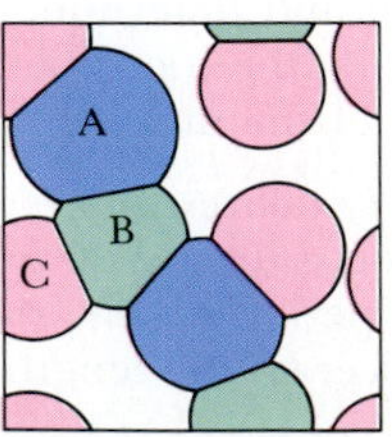

10.58 State the net number of A atoms in the unit cell shown on the right.

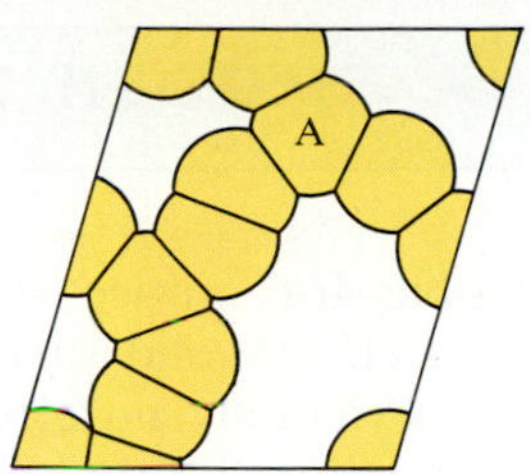

UNCLASSIFIED EXERCISES

10.59 Complete the following statements about the effects of intermolecular forces on the physical properties of a substance: (a) The higher the boiling point of a liquid, the (stronger, weaker) its intermolecular forces. (b) Substances with strong intermolecular forces have (high, low) vapor pressures. (c) Substances with strong intermolecular forces typically have (high, low) surface tensions. (d) The higher the vapor pressure of a liquid, the (stronger, weaker) its intermolecular forces. (e) Because nitrogen, N_2, has (strong, weak) intermolecular forces, it has a (high, low) critical temperature. (f) Substances with high vapor pressures have correspondingly (high, low) boiling points. (g) Because water has a relatively high boiling point, it must have (strong, weak) intermolecular forces and a correspondingly (high, low) enthalpy of vaporization.

10.60 Select the one substance that has the corresponding property. Justify your answer in each case. (a) The strongest hydrogen bonding: H_2O, H_2S, CH_3OH (as liquids). (b) The greatest surface tension: CH_3OH, C_2H_5OH, C_3H_7OH (as liquids). (c) The highest vapor pressure: CO_2, SO_2, SiO_2 (as solids). (d) The lowest enthalpy of vaporization: H_2O, H_2S, H_2Te (as liquids). (e) The highest boiling point: AsH_3, PH_3, NH_3 (as liquids). (f) The strongest dipole-dipole forces: H_2S, SCl_2, SF_2. (g) The strongest London forces: CH_4, SiF_4, GeF_4.

10.61 Hydrogen peroxide, H_2O_2, is a syrupy liquid with a vapor pressure lower than that of water and a boiling point of 152.2°C. Account for the differences between these properties and those of water.

10.62 What effect does an increase in temperature have on the following properties? Explain your answer. (a) surface tension; (b) vapor pressure; (c) evaporation rate.

10.63 Explain how the vapor pressure of a liquid is affected by each of the following changes in conditions: (a) an increase in temperature; (b) an increase in surface area of the liquid; (c) an increase in volume above the liquid.

10.64 Use the data in Table 10.4 to list the following substances in order of increasing values of surface tension: water, benzene, methanol, ethanol.

10.65 Use Table 10.3 to list the following in order of increasing strength of their intermolecular forces: bromine, oxygen, carbon tetrachloride, ammonia.

10.66 Explain the observation that, in hot weather, water is cooled relative to ambient temperature, by placing it in an unglazed urn.

10.67 Paper towels are effective in cleaning up water spills. Synthetic towels from a synthetic fiber (such as a polyester) may be more economical to produce. Explain why paper towels are effective. Could a synthetic towel be more effective? If so, what properties should the synthetic fiber exhibit?

10.68 The normal boiling temperature of water is 100°C. Suppose a cyclonic region (a region of low pressure) moves into the area. State and explain what happens to the boiling point of the water.

10.69 A vacuum pump is attached to a flask of water at 0°C and 1 atm pressure. As the pressure above the liquid is decreased, explain what would be observed, based on the phase diagram of water, Fig. 10.22.

10.70 Explain what would be observed if the water in Exercise 10.69 were at 50°C instead of 0°C.

10.71 Predict what would happen to a sample of carbon dioxide gas at 20°C if its pressure were suddenly increased to 10 atm. In a second step, what would be the physical state of the carbon dioxide if the temperature were decreased to (a) −10°C; (b) to −60°C?

10.72 An air mass is saturated with water vapor at 10°C. Given that at 10°C the vapor pressure of water is 9.21 Torr, explain what happens to the water in the air mass if it meets (a) a warm weather front; (b) a cold front.

10.73 Relative humidity is defined as

$$\text{Relative humidity (\%)} = \frac{\text{partial pressure of water}}{\text{vapor pressure of water}} \times 100\%$$

The vapor pressure of water at various temperatures is given in Table 5.4. (a) What is the relative humidity at 30°C when the partial pressure of water is 25.0 Torr? (b) Explain what would be observed if the temperature of the air in (a) fell to 25°C.

10.74 The 5:00 P.M. weather report states that the ambient temperature is 94°F (34°C) and the relative humidity

CASE 10 ▸ SMOOTH SKATING

How is it possible for an ice skater to glide smoothly across a clean, solid surface on a cold strip of steel with virtually no effort? How can two solids slide across each other with seemingly little or no friction? A look at the phase diagram for water (see Fig. 10.22) gives us a clue. If at point C, the normal freezing point of water, only ice is present, an applied pressure causes the ice to convert to the liquid. The backward slope of the solid-liquid equilibrium continues up to 2045 atm and −22.0°C (see the illustration on the right). This phase diagram therefore indicates that liquid water cannot exist below −22°C, whatever the pressure; conversely, a sufficiently great pressure can cause ice to melt at any temperature above −22°C. As a consequence, a liquid film can form and act as a lubricant between the ice and the ice skate.

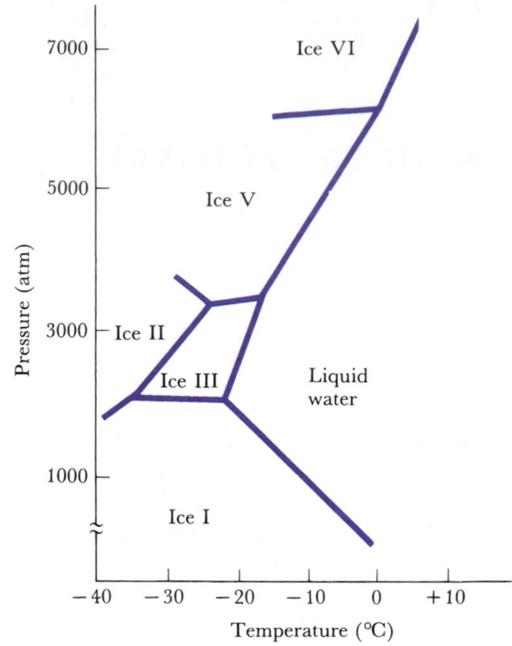

The surface in ice rinks is normally maintained at about −8°C. According to the phase diagram, the formation of water at this temperature requires the application of about 1000 atm. The blade of a figure skate is about 30 cm long and the width of its knife edge is about 0.001 cm, so the area in contact with the ice is about 0.03 cm^2. A 70-kg person on such a skate would therefore exert about 2300 atm, which easily exceeds the pressure needed to melt the ice. This pressure is so high, in fact, that fracturing of the ice may occur; this effect is evident on the ice surface after an ice skater has passed.

Another significant factor in the formation of the water film is the heat generated by friction, for friction converts the kinetic energy of the

is 65% (see Exercise 10.73); an overnight low of 77°F (25°C) is expected. Can fog or rain showers be predicted? Explain your reasoning.

10.75 The vapor pressure of ethanol at 25°C is 58.9 Torr. A sample of ethanol vapor at 25°C in equilibrium with its liquid is mixed with dry air in a 10-L container at a total pressure of 750 Torr. The volume of the container is then reduced to 5.0 L. (a) What is the partial pressure of ethanol in the smaller volume? Explain your reasoning. (b) What is the total pressure of the mixture?

10.76 Calculate the vapor pressure of water, given that when 25.0 L of argon is bubbled slowly through water at 80°C, 7.3 g of water is vaporized.

10.77 A small beaker of water was heated in a sealed room of volume 30 m^3 that was filled with 2.0 kg of nitrogen at 20°C. At what temperature did the water boil?

10.78 A new substance developed in a laboratory has the following properties: Normal melting point, 83.7°C; normal boiling point, 177°C; triple point, 200 Torr and 38.6°C. (a) Sketch the phase diagram and label the solid, liquid, and gaseous phases and the solid/liquid, liquid/gas, and solid/gas phase boundaries. (b) Sketch a warming curve for a sample at constant pressure, beginning at 500 Torr and 25°C and ending at 200°C.

10.79 Classify the following solids as ionic, network, or molecular: (a) limestone, $CaCO_3$; (b) dry ice, CO_2; (c) sucrose, $C_{12}H_{22}O_{11}$; (d) polyethylene, a polymer of repeating —CH_2CH_2— units.

10.80 Sodium metal has a density of 0.97 g/cm^3 and forms a body-centered lattice. (a) Calculate the length (in pm) of the edge of the unit cell. (b) What is the atomic radius of sodium? (c) What is the mass of one unit cell of sodium?

10.81 Lead metal has a density of 11.34 g/cm^3 and crystallizes in a face-centered lattice. (a) Using the value of Avogadro's number given inside the back cover, deter-

skater into heat that can be absorbed by the ice. For example, a 70-kg skater moving at 5 m/s has a kinetic energy of 875 J; if all this energy were transferred to ice at −8°C (raising its temperature to 0°C), then about 2.5 g of ice could melt (because the specific heat capacity of ice is 2.1 J/(g·°C) and ΔH_{melt} = 6.01 kJ/mol). For a 5-m/s glide on a skate having a width of 0.001 cm and travelling for 20 m, the average thickness of the water film between the knife edge of the skate and the ice would be over 1 cm if all the frictional heat were transferred to the ice, which is certainly enough lubricant for the skate. In practice, much less heat is transferred to the ice, but even the small fraction transferred produces an adequate film.

In summary, the combination of the very high pressures exerted by the ice skate on the ice surface and the friction between the skate and ice serve to provide a thin water film that serves as the lubricant for ice skaters.

QUESTIONS

1. (a) Calculate the pressure (in kilopascals and atmospheres) exerted by a 50-kg person on a 25-cm skate of edge 0.01 mm. (The force exerted by a mass m is mg, where g is the acceleration of free fall, 9.81 m/s^2.) (b) Use the phase diagram for water to decide whether the pressure will cause the ice to melt at −10°C.

2. (a) Determine the kinetic energy of a 50-kg ice skater moving at 10 m/s. (The kinetic energy of an object of mass m and speed v is $\frac{1}{2}mv^2$.) (b) If all that energy is transferred to ice at −10°C, how much ice will melt? (c) If a glide of 40 m takes place on the skate in question 1, what will the thickness of the water film between the skate and the ice be?

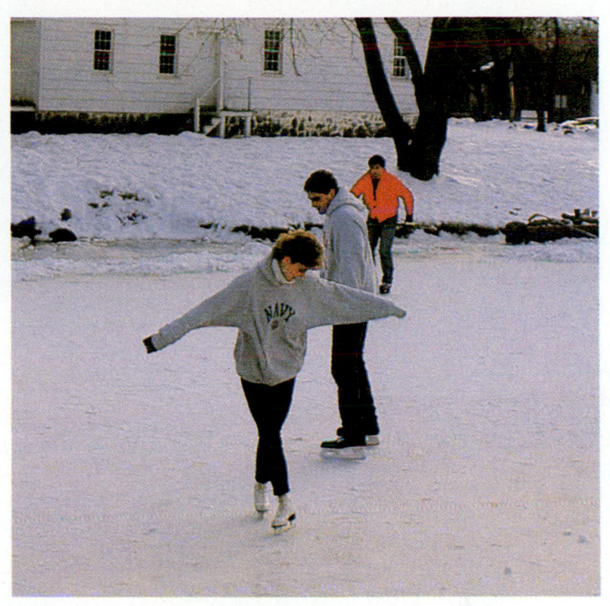

mine the length of the edge of the unit cell. (b) Determine the volume of one unit cell. (c) Calculate the atomic radius of lead.

10.82 Krypton crystallizes with a face-centered cubic unit cell of edge 559 pm. (a) What is the density of solid krypton? (b) What is the atomic radius of krypton? (c) What is the volume of one krypton atom? (The volume of a sphere of radius r is $\frac{4}{3}\pi r^3$.) (d) What percentage of the unit cell is empty space?

10.83 Aluminum metal has a density of 2.72 g/cm^3 and crystallizes in a lattice with an edge of 404 pm. (a) What type of cubic unit cell is formed by aluminum? (b) What is the coordination number of aluminum?

10.84 Metals with bcc structures are not close-packed. Therefore, their densities would be greater if they were to change to a ccp structure (under pressure, for instance). What would the density of tungsten be if it had a ccp structure rather than bcc? Its actual density is 19.4 g/cm^3.

10.85 Calculate the percentage of space occupied by the spheres in a cubic close-packed (fcc) structure.

10.86 All the noble gases except helium crystallize with fcc structures. Find an equation relating the atomic radius to the density of an fcc solid of given molar mass and apply it to deduce the atomic radii of the noble gases, given the following densities (in g/cm^3): Ne, 1.20; Ar, 1.40; Kr, 2.16; Xe, 2.83; Rn, 4.4.

10.87 (a) Suggest, in terms of energy bands, why the electrical conductivity of some substances increases when they are exposed to light. (b) The band gap in amorphous selenium, which is used in the xerographic photocopying process, is 2.9×10^{-19} J (1.8 eV). What is the longest wavelength of light that can make it conduct? (Recall from Chapter 7 that $E_{photon} = h\nu$.) (c) Can amorphous selenium be used in infrared burglar alarms?

In this chapter we explore the solubilities of substances in different solvents and under different conditions. We also see how the solute affects the physical properties of a solvent. One very important effect of a solute is the generation of an osmotic pressure as a pure solvent flows into a solution through a membrane. An adaptation of the effect, reverse osmosis, can be used to obtain pure water from seawater. The illustration shows water dripping out of a filter in a commercial desalination plant that uses reverse osmosis.

11 THE PROPERTIES OF SOLUTIONS

Solutions are homogeneous mixtures of one substance in another—the solute in the solvent. Aqueous solutions, those in which the solvent is water, are so important that we shall concentrate almost exclusively on them in this chapter. Some aqueous solutions occur on a vast scale. The oceans, for instance, account for 1.4×10^{18} kg of the Earth's surface water. Although the main solutes in seawater are Na^+ and Cl^- ions, the sea contains at least a trace of every naturally occurring element and huge amounts of some (Table 11.1). The rivers that feed the oceans are also solutions on a grand scale. Samples from the Columbia River, Washington, taken at a point about 50 mi inland from its mouth, show that each year about 1.8 trillion liters (1.8×10^{12} L) of water flow past and carry 83 million moles of PO_4^{3-} ions, 2.1 billion moles of NO_3^- ions, 27 billion moles of Si atoms (as various silicates), and 190 billion moles of HCO_3^- ions.

Chemists carry out many of their reactions in solution (usually on a much smaller scale), partly because the reactants are then mobile and can come together and react. We shall see numerous examples of these chemical aspects of solutions in Part III of the text. In this chapter we shall see how some of the *physical* properties of solutions are explained in terms of the ions and molecules they contain.

We shall focus primarily on aqueous solutions, but many of our remarks apply equally to nonaqueous solutions. Substances that dissolve to give solutions that conduct electricity (for example, sodium chloride and acetic acid) because of the ions they contain are called **electrolytes**. Substances that give solutions that do not conduct electricity (for example, glucose and ethanol), because the solute remains molecular, are called **nonelectrolytes**.

MEASURES OF CONCENTRATION

To consider the stoichiometry of reactions taking place in solution, we need to know the amount of solute in a given volume of solution. For instance, to interpret an acid-base titration, we may need to know that

TABLE 11.1 The principal ions found in seawater

Element	Principal form	Concentration g/L
Cl	Cl^-	19.0
Na	Na^+	10.5
Mg	Mg^{2+}	1.35
S	SO_4^{2-}	0.89
Ca	Ca^{2+}	0.40
K	K^+	0.38
Br	Br^-	0.065
C	HCO_3^-, H_2CO_3, CO_3^{2-}	0.028

N We have grouped the discussion of concentration units into two types: one in which chemists are primarily interested in the amount of solute in a certain volume, and another where the relative amounts of solute and solvent are important.

E The IUPAC recommendation for molar concentration is *amount concentration* as an acceptable abbreviation of amount-of-substance concentration. A solution of molar concentration 1 mol/L is often called a 1-molar solution, denoted 1. However, IUPAC advises against using M as a symbol for the unit 1 mol L^{-1} in the sense that it should not be used with SI prefixes or in conjunction with other units. That is, M is more of an adjective (as in 1 M HCl(*aq*)) than a unit (as in a concentration of 1 mol HCl L^{-1}).

25.0 mL of a certain solution contains 0.100 mol OH^- ions. To predict the effect of solutes on the physical properties of solutions, however, we often need to know only the *relative numbers* of solute and solvent molecules; the absolute numbers of each type of molecule is less important (Fig. 11.1). For instance, if we know that one out of every hundred molecules in an aqueous solution is a solute molecule (the other 99 being water), then we can predict the boiling and freezing points of the solution, no matter how large or how small the sample is. The different ways of measuring composition in order to emphasize the amount of solute or the relative numbers of solute and solvent molecules are summarized in Table 11.2 and described below.

11.1 EMPHASIZING THE AMOUNT OF SOLUTE IN SOLUTION

In this section, we review three measures of composition that are used when it is important to know how much solute is present in a sample of known total volume.

Molar concentration. When we want to focus on the *amount* of solute present in a given volume of solution, we use the molar concentration introduced in Section 4.4:

$$\text{Molar concentration} = \frac{\text{amount (mol) of solute}}{\text{volume (L) of solution}}$$

The units of molar concentration, which is also widely called molarity, are moles per liter (mol/L), written M. The molar concentration acts as a conversion factor between volume of solution and amount of solute in moles:

$$\text{Amount of solute} = \text{liters of solution} \times \frac{\text{moles of solute}}{\text{liters of solution}}$$

TABLE 11.2 Concentration units

Measure	Units	Note
molar concentration (molarity)	moles per liter, mol/L, M	moles of solute per liter of solution
volume percentage		volume of component expressed as a percentage of total volume
parts per million by volume		volume in milliliters per 10^3 liters of sample; abbreviated as ppm
mass percentage		mass of component expressed as a percentage of the total mass
parts per million by mass		mass of solute in milligrams per kilogram of sample; abbreviated as ppm
mole fraction (x)		number of moles as a fraction of the total number of moles of solute and solvent molecules: $x_A + x_B + \cdots = 1$
molality	moles per kilogram, mol/kg, m	moles of solute per kilogram of solvent

T OHT: Table 11.2.

For example, if we have 25.00 mL of a 0.150 M NaOH(*aq*) solution, then the amount of NaOH present is

$$\text{Amount of NaOH (mol)} = (25.00 \times 10^{-3}\ \text{L solution}) \times \frac{0.150\ \text{mol NaOH}}{1\ \text{L solution}} = 3.75 \times 10^{-3}\ \text{mol NaOH}$$

Volume percentage composition. Many solutions are prepared by measuring the volume of a solute, instead of its mass, and mixing it with the solvent. A convenient expression of the resulting composition is in terms of the **volume percentage composition**, the volume of a substance present in a solution expressed as a percentage of the total volume. Volume percentage is the basis of the measurement of alcohol (ethanol) content of beers, wines, and spirits. A typical wine has about 15% by volume of alcohol (so 100 mL of wine contains 15 mL of ethanol) and a beer is typically about 5% by volume of alcohol (so 100 mL contains 5 mL of ethanol). The proof scale equates 100 proof with 50% by volume of ethanol. So a strong vodka reported as 110 proof contains 55% by volume of ethanol, and 100 mL of the spirit contains 55 mL of pure ethanol.

When there is only a small quantity of solute in a solution, and particularly when the solution is gaseous (for example, a pollutant in air), the composition is often expressed in **parts per million by volume** (ppm by volume), which is 10^6 times the ratio of the volume of the solute to the volume of the sample. It follows that parts per million by volume is numerically equal to the volume percentage composition multiplied by 10^4 (Fig. 11.2). Thus, an SO_2 concentration expressed as 42 ppm by volume would correspond to 42 mL of sulfur dioxide in a sample of volume 10^3 L.

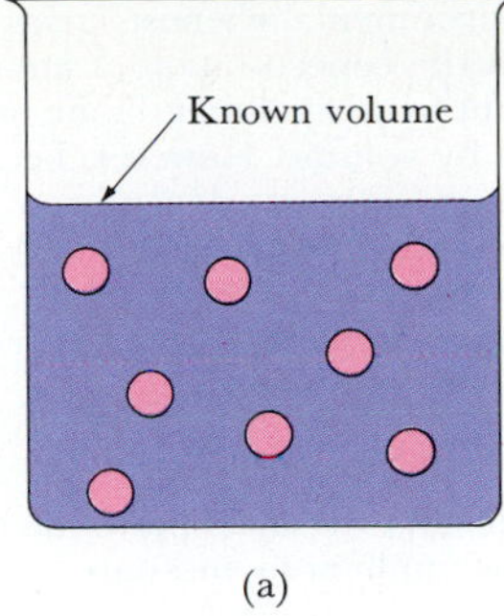

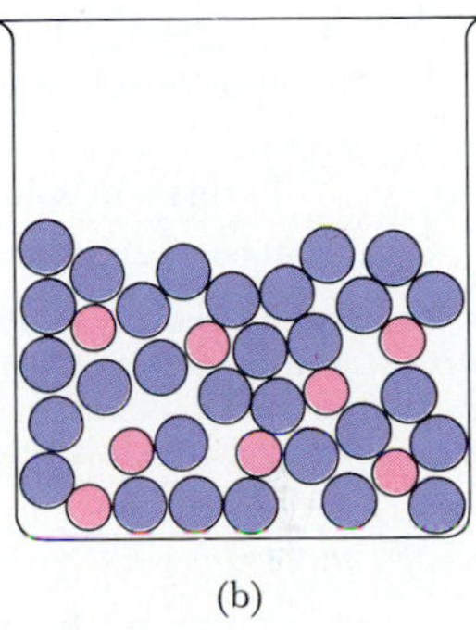

FIGURE 11.1 (a) In some applications we need to emphasize the amount of solute (pink) molecules or ions in a known volume, so we use the molar concentration. (b) In others, we need to emphasize the relative numbers of solute and solvent (blue) molecules, so we use either the molality or the mole fraction.

E Compositions may also be expressed in parts per billion (ppb), 1 part in 10^9.

D Nonadditivity of volumes. CD2, 6.

N The volume percentage and parts per million by volume are defined as

$$\text{Volume percentage (\%)} = \frac{\text{volume of solute}}{\text{volume of solution}} \times 10^2$$

$$\text{ppm by volume (ppm)} = \frac{\text{volume of solute}}{\text{volume of solution}} \times 10^6$$

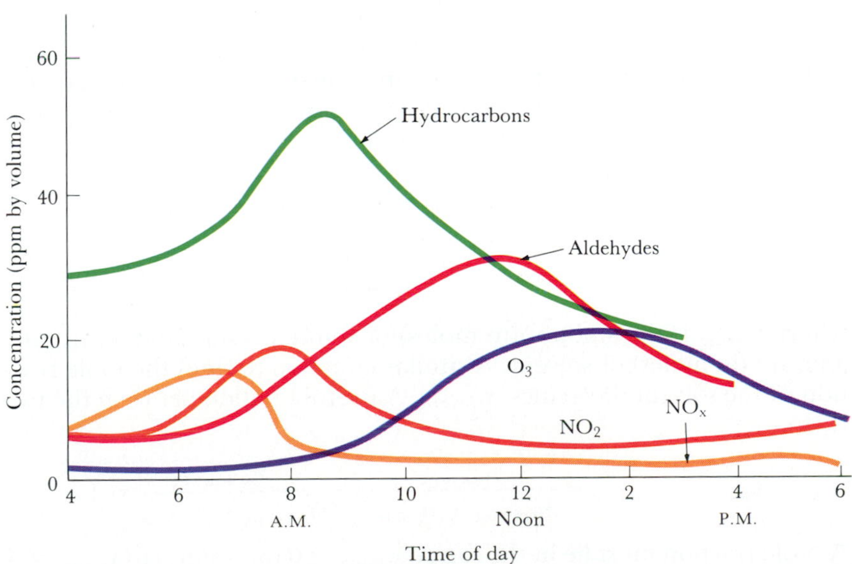

FIGURE 11.2 The concentration (in parts per million by volume) of some components of the atmosphere at different times during the day in downtown Los Angeles before emission controls were increased.

E Environmental chemists generally express the concentration of an air pollutant as μL pollutant/L air, which is ppm by volume. However, because the volume of a gas depends on the temperature and pressure, atmospheric chemists currently report concentration in milligrams per cubic meter at 1 atm and 25°C.

U Exercises 11.70 and 11.72.

N Note that the mass percentage and parts per million by mass are defined as

Mass percentage

$$= \frac{\text{mass of solute}}{\text{mass of solution} \times 10^2}$$

ppm by mass (ppm)

$$= \frac{\text{mass of solute}}{\text{mass of solution} \times 10^6}$$

E Environmental chemists express the concentration of a pollutant in water in milligrams of pollutant per kilogram of sample, which is parts per million by mass.

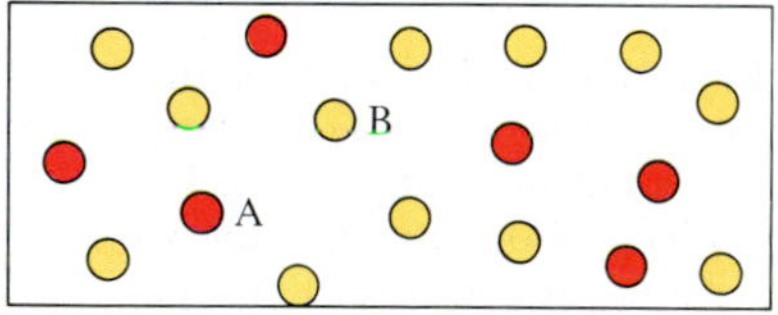

$x_A = 0.33$

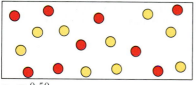

$x_A = 0.50$

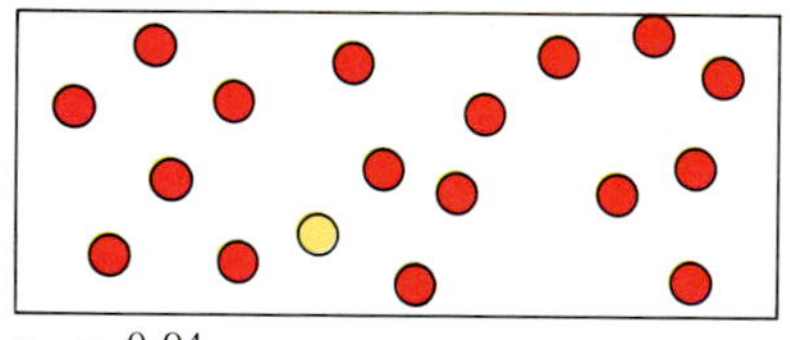

$x_A = 0.94$

FIGURE 11.3 The mole fraction x_A of a substance A in a mixture is the number of moles of A expressed as a fraction of the total number of moles in the mixture. Because the number of moles is proportional to the number of molecules, the mole fraction also gives the fraction of molecules that are A molecules.

11.2 EMPHASIZING RELATIVE AMOUNTS OF SOLUTE AND SOLVENT MOLECULES

The three principal measures of the relative numbers of solute and solvent molecules are the mass percentage composition, the mole fraction, and the molality of the solution.

Mass percentage composition. We met the mass percentage composition in Section 1.8 and saw that it is the mass of one substance present in a mixture expressed as a percentage of the total mass:

$$\text{Mass percentage of A} = \frac{\text{mass of A in the solution}}{\text{total mass of solution}} \times 100\%$$

$$= \frac{m_A}{m_A + m_B + \cdots} \times 100\%$$

This measure of composition is applicable to a solution (which, after all, is a homogeneous mixture) and is often used when reporting the composition of a solution of a solid solute. For example, we might report the composition of a solution of sucrose in water as being 22.1% by mass of sucrose.

When the mass of solute in the solution is very small, a widely used measure of composition is **parts per million by mass** (ppm by mass). For the analysis of pollutants in water, parts per million is 10^6 times the ratio of the mass of solute to the mass of the sample. (Therefore, the concentration in parts per million is equal to the mass percentage composition multiplied by 10^4.) In practice, it is the mass of solute in milligrams per kilogram of sample. For example, the pollution of a lake or stream expressed as 22 ppm would imply that 22 mg of pollutant species was present in 1 kg of a sample.

Mole fraction. The mole fraction is closely related to the mass percentage composition, but is expressed in terms of the amounts of species present in place of their masses (Fig. 11.3):

> The **mole fraction** x of a species in a mixture is the number of moles of that species expressed as a fraction of the total number of moles of ions and molecules in the mixture.

For solute molecules in a nonelectrolyte solution, the mole fraction is

$$x_{\text{solute}} = \frac{n_{\text{solute}}}{n_{\text{solvent}} + n_{\text{solute}}}$$

where n_{solute} is the amount (in moles) of solute present and n_{solvent} the amount (in moles) of solvent. A similar equation defines the mole fraction of the solvent molecules, x_{solvent}. A useful relation between the two is

$$x_{\text{solvent}} + x_{\text{solute}} = \frac{n_{\text{solvent}}}{n_{\text{solvent}} + n_{\text{solute}}} + \frac{n_{\text{solute}}}{n_{\text{solvent}} + n_{\text{solute}}} = 1$$

A mole fraction must lie in the range $x_{\text{solute}} = 0$ (no solute) to $x_{\text{solute}} = 1$ (pure solute).

For an electrolyte solution, the mole fraction is calculated by treating the cations and anions as individual particles; hence

$$x_{\text{cations}} = \frac{n_{\text{cations}}}{n_{\text{solvent}} + n_{\text{cations}} + n_{\text{anions}}}$$

A similar expression can be written for the anions, and

$$x_{\text{solvent}} + x_{\text{cations}} + x_{\text{anions}} = 1$$

▼ **EXAMPLE 11.1 Calculating a mole fraction**

Calculate the mole fractions of the nonelectrolyte sucrose ($C_{12}H_{22}O_{11}$) and water in a solution prepared by dissolving 5.00 g of sucrose in 100.0 g of water.

STRATEGY We need to know the numbers of moles of solute and solvent molecules in the solution. Therefore, we begin by converting the masses of solute and solvent to moles, using molar masses, and substituting the amounts in moles into the definitions above. The answer can be checked by verifying that $x_{C_{12}H_{22}O_{11}} + x_{H_2O} = 1$.

SOLUTION From the molar masses of sucrose (342.3 g/mol) and water (18.02 g/mol), the conversions from grams to moles are

$$\text{Amount of } C_{12}H_{22}O_{11} \text{ (mol)} = 5.00 \text{ g } C_{12}H_{22}O_{11} \times \frac{1 \text{ mol } C_{12}H_{22}O_{11}}{342.3 \text{ g } C_{12}H_{22}O_{11}}$$

$$= 0.0146 \text{ mol } C_{12}H_{22}O_{11}$$

$$\text{Amount of } H_2O \text{ (mol)} = 100.0 \text{ g } H_2O \times \frac{1 \text{ mol } H_2O}{18.02 \text{ g } H_2O}$$

$$= 5.549 \text{ mol } H_2O$$

The total amount of both compounds in the solution is

$$n = 0.0146 \text{ mol} + 5.549 \text{ mol} = 5.564 \text{ mol}$$

The two mole fractions are therefore

$$x_{C_{12}H_{22}O_{11}} = \frac{0.0146 \text{ mol}}{5.564 \text{ mol}} = 0.00262$$

$$x_{H_2O} = \frac{5.549 \text{ mol}}{5.564 \text{ mol}} = 0.9973$$

Note that 0.00262 + 0.9973 = 0.9999, or virtually 1.

EXERCISE E11.1 Calculate the mole fractions of H_2O and CH_3CH_2OH in a mixture of equal masses of water and ethanol.

[*Answer*: $x_{H_2O} = 0.719$; $x_{C_2H_5OH} = 0.281$] ▲

N An alternative introduction to the mole fraction. *J. Chem. Educ.*, **1982**, *59*, 153.

E Industrial chemists often express components of a reaction mixture in mole percentages, especially if the reaction takes place in an nonaqueous system.

Molality. The molality of a solution also emphasizes the relative numbers of solute and solvent molecules:

The **molality** of a solution is the amount of solute per kilogram of solvent:

$$\text{Molality} = \frac{\text{amount (mol) of solute}}{\text{mass (kg) of solvent}}$$

The units of molality, moles per kilogram of solvent (mol/kg), are often abbreviated as *m* and read "molal." The emphasis on *solvent* in the

N One reason why we favor molar concentration rather than the more colloquial "molarity" is to avoid the risk of confusion between molarity and molality.

E A practical application of molality. *J. Chem. Educ.*, **1983**, *60*, 63.

definition should be noted. It means that to prepare a 1 *m* $NiSO_4(aq)$ solution, 1 mol $NiSO_4$ is dissolved in 1 kg of water. Because 1 kg of solvent consists of a definite number of moles of molecules (55.5 mol H_2O per kilogram of water), the higher the molality of a given solution, the higher the proportion of solute molecules.

The molality is a conversion factor from the mass of solvent in a sample to the number of moles of solute in it:

$$\text{Amount of solute (mol)} = \text{mass of solvent (kg)} \times \frac{\text{amount (mol) of solute}}{\text{mass (kg) of solvent}}$$

▼ EXAMPLE 11.2 Preparing a solution of given molality

What mass of potassium nitrate should be added to 250 g of water to prepare a 0.200 *m* $KNO_3(aq)$ solution?

STRATEGY If we knew how many moles of KNO_3 were needed, we could convert that to grams by using the molar mass of KNO_3. To find the number of moles we can use the molality as a conversion factor for the mass of solute, as set out above.

SOLUTION The number of moles of KNO_3 in 0.250 kg of water in an 0.200 *m* solution is

$$\text{Amount of KNO}_3\text{ (mol)} = 0.250\text{ kg H}_2\text{O} \times \frac{0.200\text{ mol KNO}_3}{1\text{ kg H}_2\text{O}}$$
$$= 0.0500\text{ mol KNO}_3$$

The molar mass of KNO_3 is 101.1 g/mol, so

$$\text{Mass of KNO}_3\text{ (g)} = 0.0500\text{ mol KNO}_3 \times \frac{101.1\text{ g KNO}_3}{1\text{ mol KNO}_3}$$
$$= 5.06\text{ g KNO}_3$$

U Exercise 11.74.

EXERCISE E11.2 What mass of potassium permanganate is needed to prepare a 0.150 *m* $KMnO_4(aq)$ solution with 500 g of water?

[*Answer*: 11.6 g] ▲

▼ EXAMPLE 11.3 Converting between mole fraction and molality

Express as a molality the composition of a solution of benzene (C_6H_6) in toluene ($C_6H_5CH_3$) reported as $x_{\text{benzene}} = 0.150$.

STRATEGY To work out a molality, we need to know the number of moles of solute (benzene) molecules and the mass of solvent (toluene). The mole fraction tells us that for each mole of molecules in a sample, there are x_{benzene} mol of benzene molecules and $(1 - x_{\text{benzene}})$ mol of toluene molecules. The number of moles of toluene molecules can be converted to a mass by using the molar mass of toluene (92.13 g/mol). Then we will know both the number of moles of solute and the mass of solvent.

SOLUTION We set the number of moles of benzene molecules in the sample at 0.150 mol, thereby implying that the number of moles of toluene molecules is 0.850 mol. The total mass of toluene in the sample is

$$\text{Mass of toluene (g)} = 0.850\text{ mol C}_6\text{H}_5\text{CH}_3 \times \frac{92.13\text{ g C}_6\text{H}_5\text{CH}_3}{1\text{ mol C}_6\text{H}_5\text{CH}_3}$$
$$= 78.3\text{ g C}_6\text{H}_5\text{CH}_3$$

Therefore,

$$\text{Molality} = \frac{0.150 \text{ mol } C_6H_6}{78.3 \times 10^{-3} \text{ kg } C_6H_5CH_3} = 1.92 \text{ mol } C_6H_6/(\text{kg } C_6H_5CH_3)$$

The molality of benzene in the solution is therefore 1.92 m C_6H_6.

EXERCISE E11.3 Calculate the molality of a solution of toluene in benzene, given that the mole fraction of toluene is 0.150.

[*Answer*: 2.26 m]

The molality of a solution must be carefully distinguished from its molarity (its molar concentration). The molality is the moles of solute per kilogram of *solvent*. The molar concentration is the moles of solute per liter of *solution*. There are two good reasons for introducing molality in addition to molar concentration. One is that the molality makes it very easy to calculate the relative numbers of solute and solvent molecules in a solution. The second is that the molality is independent of temperature, whereas the molar concentration is not (because the volume of the solution increases as the temperature is raised).

The relation between the volume of a solution and the mass of solvent present in it is obtained by using the density of the solution (to obtain the total mass of the solution), the mass of solute (which is known from the molar concentration and the molar mass of the solute), and the relation

$$\text{Mass of solution} = \text{mass(solvent)} + \text{mass(solute)}$$

to obtain the mass of solvent. The following example illustrates the procedure.

EXAMPLE 11.4 Converting between molality and molar concentration

A 1.06 M $C_{12}H_{22}O_{11}(aq)$ solution has a density of 1.14 g/mL. Calculate the molality of the sucrose solution.

STRATEGY Because molality is defined as moles of solute per kilogram of solvent, we need to know the mass of water in the solution. If we know the molar concentration, then we can obtain both the amount (in moles) and the mass (in grams) of solute in 1.00 L of the solution. The mass of solute can then be subtracted from the mass of the solution to determine the mass of solvent. At this stage we will know the amount of solute and the mass of the solvent, and from those values, we can calculate the molality.

SOLUTION The molar concentration implies that 1 L of solution contains 1.06 mol $C_{12}H_{22}O_{11}$, so the amount of sucrose in 1.00 L of solution is

$$\text{Amount of } C_{12}H_{22}O_{11} \text{ (mol)} = 1.00 \text{ L solution} \times \frac{1.06 \text{ mol } C_{12}H_{22}O_{11}}{1 \text{ L solution}}$$

$$= 1.06 \text{ mol } C_{12}H_{22}O_{11}$$

The mass of sucrose in the solution is obtained by using the molar mass of sucrose (342.3 g/mol):

$$\text{Mass of } C_{12}H_{22}O_{11} \text{ (g)} = 1.06 \text{ mol } C_{12}H_{22}O_{11} \times \frac{342.3 \text{ g } C_{12}H_{22}O_{11}}{1 \text{ mol } C_{12}H_{22}O_{11}}$$

$$= 363 \text{ g } C_{12}H_{22}O_{11}$$

N For very dilute *aqueous* solutions, the molar concentration and the molality are numerically equal because the volume of an aqueous solution in liters is equal to its mass in kilograms.

Now we use the density of the solution to convert its volume (1.00 L) to a mass (a mass, that is, of the solvent plus the solute):

$$\text{Mass of solution} = 1.00\ \text{L} \times \frac{1\ \text{mL}}{10^{-3}\ \text{L}} \times \frac{1.14\ \text{g}}{\text{mL}} = 1.14 \times 10^{3}\ \text{g}$$

Therefore, because the mass of solute is 363 g,

$$\text{Mass of water (g)} = 1.14 \times 10^{3}\ \text{g} - 363\ \text{g} = 777\ \text{g}$$

which corresponds to 0.777 kg of water. Therefore, the molality of the solution is

$$\text{Molality} = \frac{1.06\ \text{mol}\ C_{12}H_{22}O_{11}}{0.777\ \text{kg}\ H_2O} = 1.36\ \text{mol}\ C_{12}H_{22}O_{11}/(\text{kg}\ H_2O)$$

Hence, the solution is 1.36 *m* $C_{12}H_{22}O_{11}(aq)$

EXERCISE E11.4 Battery acid is 4.27 M $H_2SO_4(aq)$ and has a density of 1.25 g/mL. What is the molality of the solution?

▲ [*Answer*: 5.14 *m* $H_2SO_4(aq)$]

SOLUBILITY

U Exercise 11.71.

E Solutions: Something new from the past. *J. Chem. Educ.*, **1983**, *60*, 64. A simple model to demonstrate solubility principles. *J. Chem. Educ.*, **1971**, *48*, 668.

In this part of the chapter we focus on what happens when one substance dissolves in another and see how to discuss the outcome quantitatively. This discussion is an essential preparation for the later parts of the chapter, where we see how the presence of a solute affects the physical properties of a solution.

First we shall consider the *molecular* events that occur when a substance dissolves in a solvent. When a solid is added to a solvent, events take place at the interface between the solid and the liquid. If the solid is a crystal of sucrose, then there is a competition for the formation of hydrogen bonds between the sucrose molecules themselves and between them and the water molecules (Fig. 11.4). When a sucrose molecule forms hydrogen bonds with the water molecules, it breaks away from its sucrose neighbors and, surrounded by water molecules, drifts off into the solution. If the solid is ionic, then the partially negatively charged oxygen atoms of the polar water molecules surround the cations and pry them away from the crystal lattice (Fig. 11.5). The water

T OHT: Figs. 11.4 and 11.5.

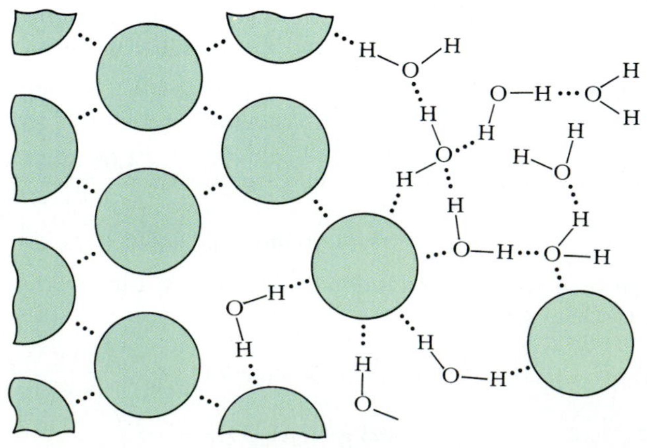

FIGURE 11.4 In the dissolution of sucrose, the hydrogen bonds between the sucrose molecules (green) in the solid are replaced by hydrogen bonds between the sucrose molecules and the water molecules of the solvent.

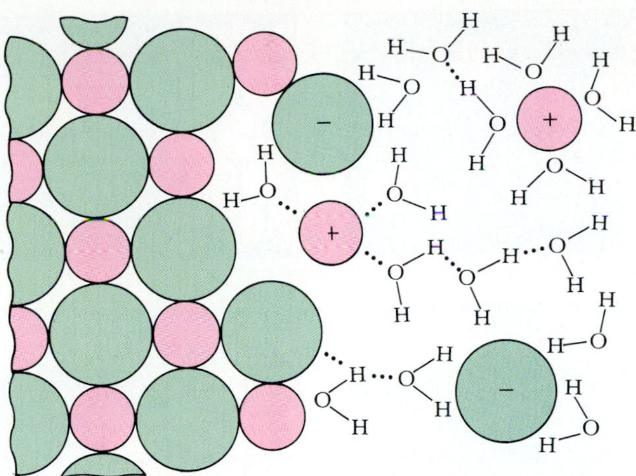

FIGURE 11.5 When an ionic solid dissolves in water, the water molecules hydrate the ions: the cations (pink) are bound to the water molecules by their attraction for the partial negative charge of the O atoms and the anions (green) are bound to the water molecules by their attraction for the partial positive charge on the H atoms and by hydrogen bonding.

molecules also attack the anions at the interface: they can attach to them by hydrogen bonds or by the attraction between the negative charge of the anion and the partial positive charge of the hydrogen atoms in the water molecules. As the attack continues, molecules or ions are carried away from the newly exposed surface of the solid, and in due course we report that it has dissolved.

? When iron(III) chloride crystallizes from an aqueous solution, the composition of the crystal is $FeCl_3{\cdot}6H_2O$ and not $FeCl_3$. Suggest an explanation.

11.3 SATURATION AND SOLUBILITY

When we add 20 g of sucrose (cane sugar) to 100 mL of water at room temperature, all the sucrose dissolves. However, when we add 300 g, although most dissolves, some does not (Fig. 11.6). When the solvent has dissolved all the solute it can and some undissolved solute remains, the solution is said to be saturated. We shall now see why the solute behaves like this.

D Solubility of some silver compounds. CD2, 45.

The definition of solubility. If we could follow a single sucrose molecule in a saturated solution, we might find that at some instant it is part of

FIGURE 11.6 When a little sucrose is stirred with 100 mL of water, it all dissolves (left). However, when a large amount is added, some undissolved sucrose remains (right).

FIGURE 11.7 The solute in a saturated solution is in dynamic equilibrium with the undissolved solute. If we could follow the history of a single solute particle (the orange circle), we would sometimes find it in solution and sometimes in the solid.

? Why, if large crystals are required, should the solvent be allowed to evaporate very slowly.

E A nonequilibrium solution that contains less solute than required for saturation is an *unsaturated solution;* one that contains more than is required for saturation is called a *supersaturated solution.* Supersaturated solutions are thermodynamically unstable with respect to the saturated solution, but may exist in a metastable state for considerable periods; precipitation of excess solid occurs when a nucleation center is introduced. Similarly, rain might not form in a cloud unless a nucleation center is present that can induce the formation of a raindrop. Finely divided silver iodide has been used to seed clouds.

D Solutions containing solids. TDC, 5-17.

the surface layer of a sucrose crystal (Fig. 11.7). Shortly thereafter, the molecule might be found in solution. Still later, it might be buried more deeply in a crystal, under many layers of molecules that have settled on top of it. There it will remain until it is exposed again and is able to return to the solution. In other words, a saturated solution is another example of a dynamic equilibrium (Section 10.4), one in which a forward process and its reverse occur at equal rates. In this case, the solute continues to dissolve at a rate that exactly matches the rate of the reverse process, the return of the solute to the solid. This discussion suggests the following definition:

> A **saturated solution** is a solution in which the dissolved and undissolved solute are in dynamic equilibrium.

We represent this equilibrium with a double-headed arrow ($\rightleftharpoons$, exactly as in the dynamic equilibrium that is central to the discussion of vapor pressure):

$$\text{Solute}(s) \rightleftharpoons \text{solute}(aq)$$

A saturated solution represents the limit of a solute's ability to dissolve in a given quantity of solvent. It is therefore a natural measure of the solute's solubility:

> The **solubility** of a substance is its concentration in the saturated solution.

The solubilities of substances depend on the solvent, the temperature, and—for gases—the pressure.

The dependence of solubility on the solute. Some substances are soluble in water, others sparingly (slightly) soluble, and others almost insoluble. We can know which behavior to expect by referring to the solubility rules, which are given in Table 3.2. We used the rules in Chapter 3 to choose reagents for precipitation reactions; these rules also help us understand the behavior of many common substances and the properties of minerals. For instance, because most nitrates are very soluble, they are rarely found in mineral deposits, for they are usually carried away by the water that trickles through the ground. An exception is the large deposit of impure sodium nitrate in the arid coastal region of Chile, where groundwater is absent (Fig. 11.8). This Chile saltpeter was the main source of nitrates for fertilizers and explosives until the Haber process for ammonia was developed at the start of this century.

The low solubility of most phosphates is an advantage for animals with skeletons, because bone consists largely of calcium phosphate (much of the rest is the protein collagen). However, this insolubility is inconvenient for agriculture, because it means that phosphorus is slow to circulate through the ecosystem. One of chemistry's achievements has been the development of manufacturing processes to speed phosphates on their way as fertilizers. The phosphates used for this purpose are obtained from phosphate rocks, principally the apatites—hydroxyapatite, $Ca_5(PO_4)_3OH$, and fluorapatite, $Ca_5(PO_4)_3F$—by treating them with concentrated sulfuric acid:

$$Ca_5(PO_4)_3OH(s) + 2H_2SO_4(aq) \longrightarrow 3CaHPO_4(aq) + 2CaSO_4(s) + H_2O(l)$$

The phosphate rocks themselves were once alive, for they are the crushed and compressed remains of the skeletons of prehistoric animals. Calcium hydrogenphosphate ($CaHPO_4$) is more soluble than calcium phosphate and is included in commercial phosphate fertilizers.

Just as hydrogenphosphates are more soluble than phosphates, so hydrogencarbonates (bicarbonates, HCO_3^-) are more soluble than carbonates. This difference is responsible for the behavior of hard water, which is water that contains dissolved calcium and magnesium salts. In particular, the difference accounts for the deposit of scale inside hot water pipes and for the formation of a soap scum in hard water. The behavior of hard water stems from the fact that rain water contains dissolved carbon dioxide and hence some carbonic acid from the reaction

$$CO_2(g) + H_2O(l) \longrightarrow H_2CO_3(aq)$$

As the water runs along and through the ground, the carbonic acid reacts with the calcium carbonate of limestone or chalk and forms the more soluble hydrogencarbonate:

$$CaCO_3(s) + H_2CO_3(aq) \longrightarrow Ca^{2+}(aq) + 2HCO_3^-(aq)$$

These reactions are reversed when the water is heated in a kettle or furnace:

$$Ca^{2+}(aq) + 2HCO_3^-(aq) \xrightarrow{\Delta} CaCO_3(s) + CO_2(g) + H_2O(l)$$

The carbon dioxide is driven off and the almost insoluble calcium carbonate is deposited as scale.

FIGURE 11.8 This saltpeter has survived in the arid region of Chile where it is mined because there is too little groundwater and rain water to dissolve it and wash it away.

D Solubility. TDC, 211.

E The lattice enthalpies of hydrogencarbonates are smaller than those of carbonates because of the lower charge of the HCO_3^- ion compared with that of the CO_3^{2-} ion. The hydrogencarbonate ion has the same charge, and a similar size, as that of the NO_3^- ion, and we have remarked that all nitrates are soluble.

N See Case 13.

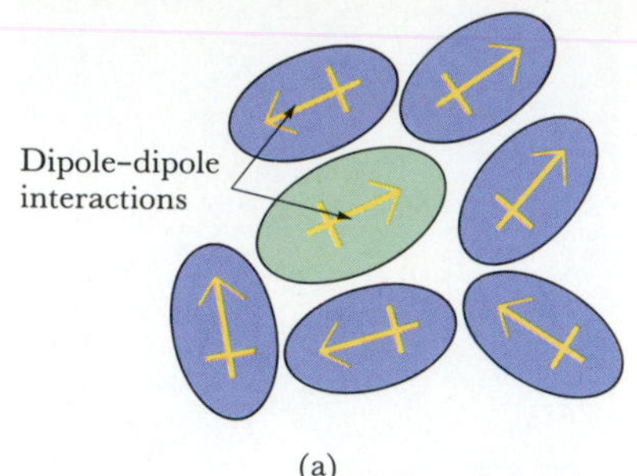

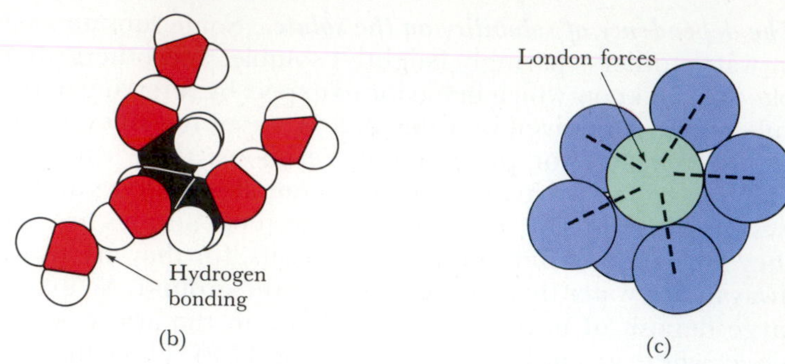

FIGURE 11.9 Like often dissolves like. Intermolecular interactions help (a) a polar solvent to dissolve other polar substances, (b) a hydrogen-bonding solvent to dissolve substances held together by hydrogen bonds, and (c) a solvent with strong London forces to dissolve nonpolar molecular solutes.

T OHT: Fig. 11.9.

D Polar properties and solubility. CD1, 26.

V Video 31: Distribution of a solute between immiscible liquids.

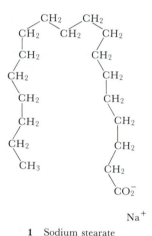

CH_2 CH_2 CH_2 CH_2 CH_2 CH_2 CH_2 CH_2 CH_2 CH_2 CH_2 CH_2 CH_2 CH_2 CH_2 CH_2 CH_3 CO_2^- Na^+

1 Sodium stearate

The dependence of solubility on the solvent. In many instances, the dependence of the solubility of a substance on the identity of the solvent can be summarized by the rule that *like dissolves like.* That is, chemists have found that a polar liquid (such as water) is generally a much better solvent for ionic and polar compounds than a nonpolar one (such as benzene) is. Conversely, nonpolar liquids such as benzene and the tetrachloroethene (C_2Cl_4) used for dry cleaning are often better solvents for nonpolar compounds than for polar compounds (Fig. 11.9). The reason that like dissolves like is that the energy of a solute molecule in the solution is similar to what it was in the original solid if the intermolecular forces in solution and solid are similar.

If the principal cohesive forces in a solute are hydrogen bonds, the like-dissolves-like rule implies that the solute is more likely to dissolve in a hydrogen-bonding solvent than in other solvents. Sucrose, for example, dissolves readily in water but not in benzene. Similarly, if the principal cohesive forces are London forces, then the best solvent is likely to be one held together by the same kind of forces. For example, carbon disulfide is a far better solvent for sulfur than water is because solid sulfur is a molecular solid of S_8 molecules held together by London forces (Fig. 11.10).

FIGURE 11.10 The molecular solid sulfur does not dissolve in water (left) but it does dissolve in carbon disulfide (right), with which its molecules have favorable London interactions.

Soaps and detergents. Modern soaps and detergents represent a practical application of the principle of like dissolves like. Soaps are the sodium salts of organic acids with long hydrocarbon chains, including sodium stearate (**1**); we shall denote the soap as NaA and the parent organic acid as HA. The anions (A^-) each have a polar group (the negatively charged segment called the head group) at one end of a nonpolar hydrocarbon chain. Their long, nonpolar, and thus **hydrophobic** (water-repelling) hydrocarbon tails sink into a blob of grease. The **hydrophilic** (water-attracting) head groups remain on the surface of the grease blob, thus coating it with a skin of polar, hydrogen-bonding groups (Fig. 11.11). This arrangement enables the whole unit to dissolve in water and be washed away.

A problem with soaps is that they form a scum in hard water. The scum is the product of a precipitation reaction that occurs because calcium salts are less soluble than sodium salts:

$$Ca^{2+}(aq) + 2A^-(aq) \longrightarrow CaA_2(s)$$

One way of avoiding the formation of scum is to use another precipitation reaction to remove the Ca^{2+} ions from the water before the soap is used. This result can be achieved by adding sodium carbonate (washing soda) to the water and precipitating the Ca^{2+} ions as calcium carbonate:

$$Ca(HCO_3)_2(aq) + Na_2CO_3(aq) \longrightarrow CaCO_3(s) + 2NaHCO_3(aq)$$

Another way to avoid soap scum is to include polyphosphate ions in the detergent. Polyphosphate ions (**2**) are formed when phosphates are heated; they consist of chains and rings of PO_4 groups. The first step in their formation is

$$\mathrm{HO{-}\underset{\underset{\textstyle OH}{|}}{\overset{\overset{\textstyle O}{\|}}{P}}{-}OH} + \mathrm{HO{-}\underset{\underset{\textstyle OH}{|}}{\overset{\overset{\textstyle O}{\|}}{P}}{-}OH} + \mathrm{HO{-}\underset{\underset{\textstyle OH}{|}}{\overset{\overset{\textstyle O}{\|}}{P}}{-}OH} \xrightarrow{\Delta}$$

$$\mathrm{HO{-}\underset{\underset{\textstyle OH}{|}}{\overset{\overset{\textstyle O}{\|}}{P}}{-}O{-}\underset{\underset{\textstyle OH}{|}}{\overset{\overset{\textstyle O}{\|}}{P}}{-}O{-}\underset{\underset{\textstyle OH}{|}}{\overset{\overset{\textstyle O}{\|}}{P}}{-}OH} + 2H_2O$$

In basic solution the product of this reaction is present as the polyphosphate anion, $P_3O_{10}{}^{5-}$:

$$\mathrm{{}^-O{-}\underset{\underset{\textstyle O^-}{|}}{\overset{\overset{\textstyle O}{\|}}{P}}{-}O{-}\underset{\underset{\textstyle O^-}{|}}{\overset{\overset{\textstyle O}{\|}}{P}}{-}O{-}\underset{\underset{\textstyle O^-}{|}}{\overset{\overset{\textstyle O}{\|}}{P}}{-}O^-}$$

Polyphosphate ions are big; and, when added to hard water, they wrap around the calcium cations and hide them away from other anions with which they would normally precipitate. This wrapping up of one ion by another is called **sequestration** of the ion (from the Latin word for "hiding away"), and the polyphosphates are called sequestering agents.

Modern commercial detergents are mixtures of compounds, of which the most important is the surface-active agent, or **surfactant**, that takes the place of the soap molecule. Surfactant molecules are synthetic organic compounds that resemble the one shown as (**3**). Like the stearate ion, they have a hydrophilic head group and a hydrophobic tail; and they act similarly. Detergents also contain polyphosphates to sequester calcium ions and to raise the pH of the solution. Other additives in the detergent mixture fluoresce (absorb ultraviolet radiation and then give out visible light) and give the impression of greater cleanliness.

$$^{-}O_3S{-}C_6H_4{-}CH_2{-}CH_2{-}CH_2{-}CH_2{-}CH_2{-}CH_2{-}CH_2{-}CH_2{-}CH_2{-}CH_2{-}CH_2{-}CH_3$$

3 A typical surfactant molecule

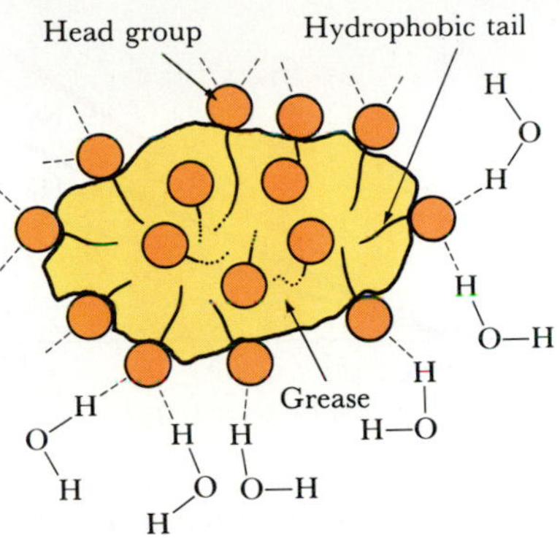

FIGURE 11.11 The hydrophobic groups of a soap or surfactant molecule dissolve in grease, leaving the hydrophilic polar head groups on the surface.

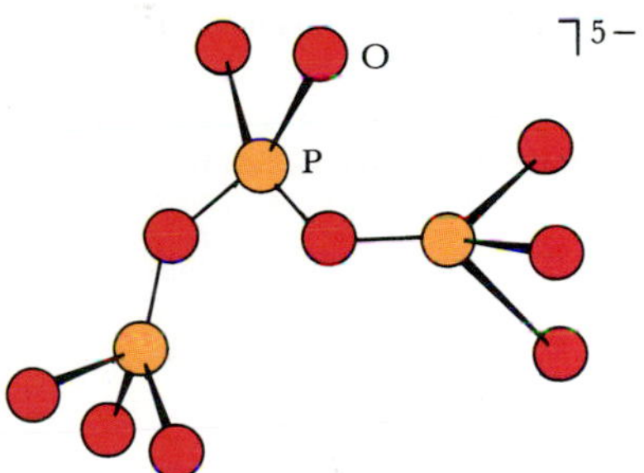

2 Polyphosphate ion, $P_3O_{10}{}^{5-}$

E The early detergents were formulated with linear alkyl chain substituents on the benzene ring, as in (**3**), and were indigestible by bacteria. Biodegradable detergents are made using branched-chain substituents, which can be digested by bacteria and hence removed from the environment. A difficulty with phosphates in fertilizers is that they can encourage the growth of organisms; in newer formulations of detergents borates have replaced phosphates.

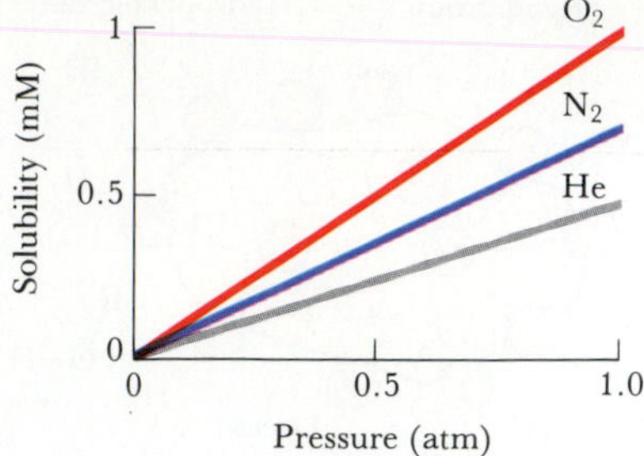

FIGURE 11.12 The variation of the solubilities of oxygen, nitrogen, and helium with partial pressure. Note that solubility of each gas is doubled when the pressure is doubled.

11.4 THE EFFECT OF PRESSURE ON GAS SOLUBILITY

The solubility of a gas in a liquid increases as the partial pressure of the gas increases (Fig. 11.12). An application of this pressure dependence is the production of soft drinks and champagne. In each case carbon dioxide is dissolved in the liquid under pressure (in champagne, as a result of fermentation that continues in the sealed bottle). When the bottle is opened, the pressure above the solution is released, the solubility of the gas is greatly reduced, and the gas effervesces (bubbles out of solution), pushing out the cork with a festive pop. A more serious consequence of the dependence of gas solubility on pressure is its effect on the nitrogen that dissolves in the blood of deep-sea divers when they are working at great depths. The dissolved nitrogen comes out of solution when the diver returns to the surface, thereby forming numerous small bubbles in the bloodstream (Fig. 11.13). These bubbles can block the capillaries (the narrow vessels that distribute the blood) and starve the tissues of oxygen; this painful condition is known as the bends and can be fatal. The risk of the bends is reduced if helium is used to dilute the diver's oxygen supply instead of nitrogen; helium is much less soluble than nitrogen and, being smaller, can pass through cell walls more readily.

E William Henry (1775–1836) was a close friend of John Dalton. He also established the molecular formula of ammonia.

D Effect of pressure on the solubility of gases in liquids. BZS3, 9.18. Solutions of gases, TDC, 5-13.

T OHT: Fig. 11.2 and Table 11.3.

Henry's law. The dependence of the solubility of a gas on its partial pressure was summarized in 1801 by the English chemist William Henry:

> **Henry's law**: The solubility of a gas in a liquid is proportional to its partial pressure above the liquid.

The partial pressure of a gas was introduced in Section 5.5: it is the contribution the particular gas makes to the total pressure of the sample. The partial pressure of oxygen in the atmosphere at sea level, for instance, is about 0.2 atm; in other words, the oxygen molecules bombard the surface of a liquid at a rate that corresponds to a pressure of 0.2 atm. The other molecules in the air also bombard the surface at a

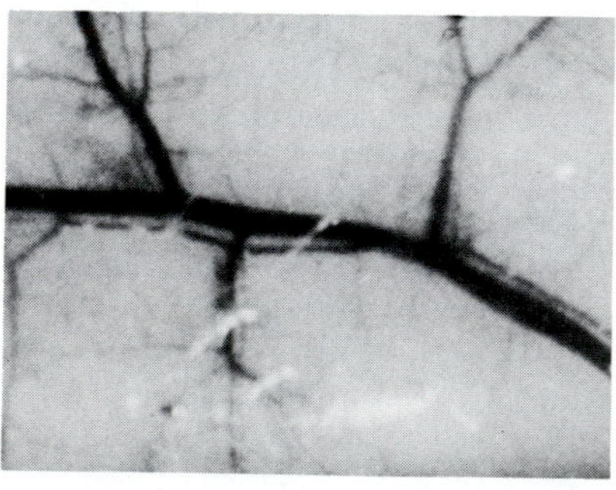

(a)

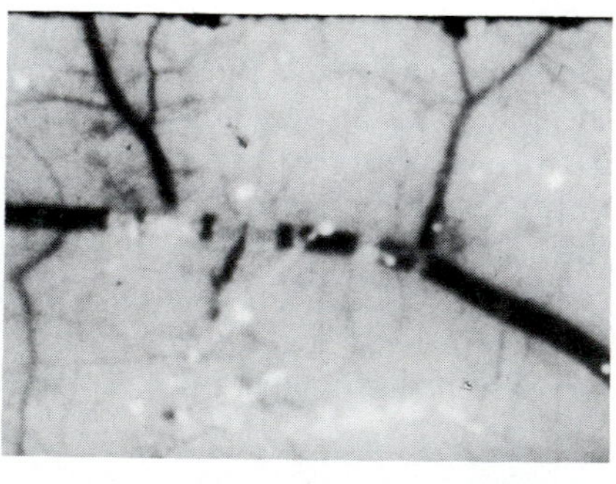

(b)

FIGURE 11.13 Small bubbles of air are responsible for the bends. (a) Normal blood vessels; (b) catastrophic collapse as bubbles escape from solution in the blood plasma.

rate corresponding to their partial pressures. However, when we are considering the solubility of oxygen, we need consider only the partial pressure of oxygen, for it is the partial pressure of each gas that determines the rate at which its molecules strike a surface.

Henry's law is normally written

$$\text{Solubility} = k_H \times \text{partial pressure}$$

where k_H, which is called **Henry's constant**, depends on the gas, the solvent, and the temperature (see Table 11.3). The law implies that, at constant temperature, doubling the partial pressure of the gas doubles its solubility.

As an illustration, consider oxygen in water, for which $k_H = 1.3 \times 10^{-3}$ mol/(L·atm). At sea level, where the partial pressure of oxygen is 0.21 atm, the solubility of oxygen is

$$\text{Solubility (mol/L)} = 0.21 \text{ atm} \times 1.3 \times 10^{-3} \frac{\text{mol}}{\text{L·atm}} = 2.7 \times 10^{-4} \text{ mol/L}$$

which corresponds to 8.6 mg of oxygen per liter of water. Under normal conditions, the lowest concentration of O_2 that can support aquatic life is about 1.3×10^{-4} mol/L (about 4 mg/L).

The increase in the solubility of a gas with pressure can be explained in terms of the dynamic equilibrium between the gas molecules in the solution and those in the space above it (Fig. 11.14). When a solvent is saturated with dissolved gas, the rate at which gas molecules enter the solvent matches the rate at which they leave it. When the partial pressure increases, the rate at which molecules enter the solution increases, because they strike its surface more often. As a result, the concentration of gas in the solution rises. However, the increased concentration of dissolved gas molecules means that there are more dissolved molecules available to escape back to the gas, so their rate of escape increases. A new equilibrium is reached when the two rates are equal, and at this new equilibrium there is a higher concentration of gas in the solution than before; that is, the gas has a higher solubility.

TABLE 11.3 Henry's constants for gases in water at 20°C

Gas	k_H, mol/(L·atm)
air	7.9×10^{-4}
argon	1.5×10^{-3}
carbon dioxide	2.3×10^{-2}
helium	3.7×10^{-4}
neon	5.0×10^{-4}
hydrogen	8.5×10^{-4}
nitrogen	7.0×10^{-4}
oxygen	1.3×10^{-3}

E The solubility of gases arising from the anaerobic decay of plants in swamps increases with the depth of water in the swamp because of the higher pressure they experience at greater depths.

? Why is the solubility of oxygen less in salt water than in fresh water?

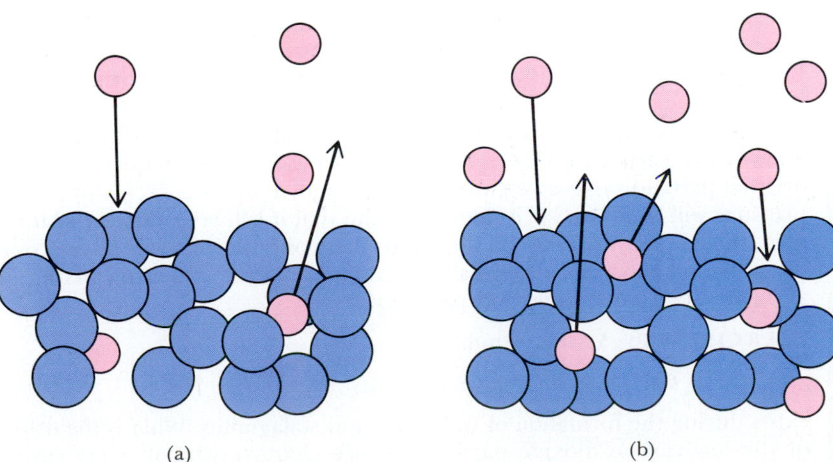

FIGURE 11.14 The effect of increased pressure, represented by the change from (a) to (b), is to increase the rate at which gas molecules dissolve. The concentration of gas in solution rises until the rate at which gas molecules leave the solution matches the new rate at which they enter it.

E Henri Louis Le Chatelier (1850–1936) also invented the platinum-rhodium thermocouple for the measurement of high temperatures and proposed the use of acetylene for welding.

Le Chatelier's principle. The increase in gas solubility with pressure is an instance of a general characteristic of dynamic equilibria. This characteristic was first described by the French chemist Henri Le Chatelier in 1884:

> **Le Chatelier's principle**: When a stress is applied to a system in dynamic equilibrium, the equilibrium adjusts to minimize the effect of the stress.

Accordingly, when the partial pressure of a gas is increased, the dynamic equilibrium adjusts by increasing the concentration of dissolved gas in the solution.

$$\text{Gas}(g) \xrightarrow{\text{encouraged by increased partial pressure}} \text{gas}(aq)$$

That response tends to decrease the number of molecules in the gas in the space above the solution and hence to minimize the increase in pressure.

Le Chatelier's principle applies only to *dynamic* equilibria, not to static equilibria. A dynamic equilibrium is in a sense a "living" equilibrium: it can adjust because both forward and reverse processes are taking place simultaneously and make the system responsive to changes. A pencil balanced on its point is in a condition of static equilibrium: it has no ability to recover its balance when a force pushes it to the side. We shall see in this and the next few chapters that Le Chatelier's principle is very helpful for predicting the effects of changes in pressure, temperature, and composition on a chemical system at equilibrium.

EXAMPLE 11.5 Using Le Chatelier's principle

The carbon dioxide in the atmosphere is in dynamic equilibrium with the carbon dioxide dissolved in the oceans. What does the principle suggest will happen as the amount of carbon dioxide gas in the atmosphere increases?

STRATEGY The equilibrium is dynamic, so Le Chatelier's principle can be used to discuss how it responds to changes. We have to decide what change the increased amount of carbon dioxide represents and then apply the principle.

SOLUTION An increase in the amount of carbon dioxide in the atmosphere results in an increase in the partial pressure of the gas. According to Le Chatelier's principle, the solubility equilibrium will respond by tending to minimize the increase, which it can do by dissolving more of the gas. However, the net outcome will not necessarily be a stabilization of the amount of atmospheric carbon dioxide, for the increase in solubility will not restore the original partial pressure: the principle says only that the increase in partial pressure will be *minimized,* it does not claim that it will be eliminated. Moreover, the equilibrium between the atmosphere and the oceans is a very complex phenomenon and depends, for example, on the extent of mixing of surface waters caused by wind patterns.

U Exercise 11.75.

EXERCISE E11.5 The equilibrium

$$Ca(HCO_3)_2(aq) \rightleftharpoons CaCO_3(s) + CO_2(g) + H_2O(l)$$

exists during the formation of stalactites and stalagmites. What is the effect of the low carbon dioxide partial pressure characteristic of many caves?

[*Answer*: deposition of calcium carbonate]

11.5 THE EFFECT OF TEMPERATURE ON SOLUBILITY

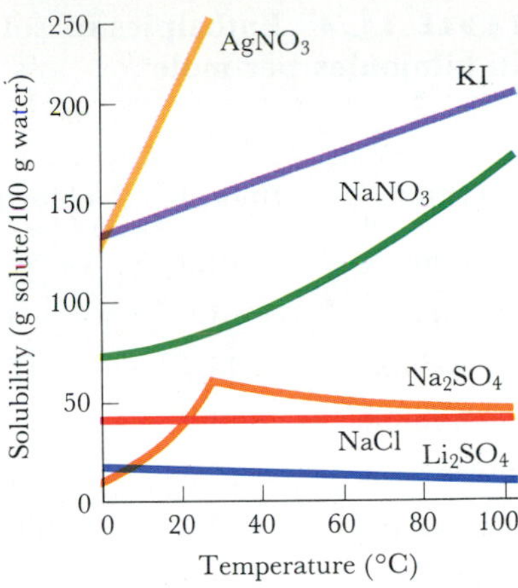

FIGURE 11.15 The variation with temperature of the solubilities of six substances in water.

In a discussion of the effect of temperature on solubility, we must distinguish between the *rate* at which a substance dissolves and the *concentration* it finally reaches. Although many substances dissolve more quickly at higher temperatures, their saturated solutions may have a lower concentration at the higher temperature.

All gases become *less* soluble as the temperature is raised. In contrast, most ionic and molecular solids are *more* soluble in hot water than in cold (Fig. 11.15). The magnitude of the increase varies sharply from one substance to another. The solubility of sodium chloride in water increases by only about 10%, from 35.7 g per 100 g of water (written 35.7 g/100 g) to 39.8 g/100 g when the temperature is raised from 0° to 100°C. The solubility of silver nitrate, however, increases by over 700%, from 122 to 951 g/100 g over the same temperature range. A few solids are less soluble at high temperatures than at low. The solubility of lithium sulfate, for instance, decreases by about 10% from 25.3 g/100 g at 0°C to 23.1 g/100 g at 100°C. Some compounds show a mixed behavior: the solubility of sodium sulfate increases up to 32°C but then decreases as the temperature is raised further.

The lower solubility of gases at higher temperatures is responsible for the tiny bubbles that appear when cool water from the faucet is left to stand in a warm room: the gas is air that dissolved when the water was cooler. A more serious effect is **thermal pollution**, the damage caused to the environment by the waste heat of an industrial process. One form this effect takes is a reduced oxygen concentration in rivers, resulting from the hot water discharged from power stations. The problem is aggravated by the lower density of the warm water, which causes it to rise to the top. Once there, it blocks the penetration of oxygen to the cooler water below. As a result, marine life (principally fish) may be killed.

Predicting the effect of temperature. According to Le Chatelier's principle, when a saturated solution is cooled or heated, the dynamic equilibrium adjusts to minimize the change in temperature. If heat is absorbed when the substance dissolves, then we can expect more of the substance to dissolve as the temperature is raised. On the other hand, if dissolving releases heat, then we can expect the solubility to increase when the temperature is lowered. That is, a substance that dissolves endothermically becomes more soluble as the temperature is raised. Conversely, a substance that dissolves exothermically becomes less soluble as the temperature is raised.

Because dissolving normally occurs at constant pressure, the heat given out or absorbed is equal to an enthalpy change, called the **enthalpy of solution**, ΔH_{sol}. The enthalpy of solution can be measured by calorimetry (although there are more accurate methods). Some values for very dilute solutions are given in Table 11.4; like the values in that table, enthalpies of solution are normally reported in kilojoules per mole of the solute.

All gases dissolve exothermically and hence are less soluble in hot solvent than in cold. As a result, some gases can be collected over hot water (as described in Section 5.5) even though they are too soluble in cold water for this to be feasible. Dinitrogen oxide is an example of a gas that can be collected in this way.

T OHT: Fig. 11.15.

V Video 30: Effect of temperature on solubility.

? Why might a major fish-kill occur as a result of a sudden increase in temperature of the water the fish inhabit?

? Why might a carbonated beverage go "flat" more quickly at room temperature than inside a refrigerator?

D A glittering shower of lead iodide crystals. CD2, 68. Solutions containing solids. TDC, 5-18.

E Le Chatelier's principle and the prediction of the effect of temperatures on solubilities. *J. Chem. Educ.*, **1982**, *59*, 550. Le Chatelier's principle applied to the temperature dependence of solubility. *J. Chem. Educ.*, **1984**, *61*, 499.

N The effect of temperature is the most difficult application of Le Chatelier's principle to explain at this stage, and is probably best left until later. In Section 12.3 we shall see that it can be traced to the different sensitivities of the forward and reverse reaction rates (or rates of a physical process) to changes in temperature.

E The dissolution of a gas is accompanied by a reduction in entropy. Therefore, for dissolving to be spontaneous, the entropy of the surroundings must increase by more than the entropy of the system decreases, so heat must be released into the surroundings. Hence, the dissolution of a gas must be exothermic.

TABLE 11.4 **Enthalpies of solution, ΔH_{sol}, at 25°C for dilute aqueous solution, in kilojoules per mole***

Cation	Anion							
	fluoride	chloride	bromide	iodide	hydroxide	carbonate	nitrate	sulfate
lithium	+4.9	−37.0	−48.8	−63.3	−23.6	−18.2	−2.7	−29.8
sodium	+1.9	+3.9	−0.6	−7.5	−44.5	−26.7	+20.4	−2.4
potassium	−17.7	+17.2	+19.9	+20.3	−57.1	−30.9	+34.9	+23.8
ammonium	−1.2	+14.8	+16.0	+13.7	—	—	+25.7	+6.6
silver	−22.5	+65.5	+84.4	+112.2	—	+41.8	+22.6	+17.8
magnesium	−17.7	−160.0	−185.6	−213.2	+2.3	−25.3	−90.9	−91.2
calcium	+11.5	−81.3	−103.1	−119.7	−16.7	−13.1	−19.2	−18.0
aluminum	−27	−329	−368	−385	—	—	—	−350

*The value for silver iodide, for example, is the entry found where the row labeled "silver" intersects the column labeled "iodide." A positive value for ΔH_{sol} denotes an endothermic process.

T OHT: Table 11.4.

▼ EXAMPLE 11.6 Predicting the temperature dependence of solubility

Is silver bromide likely to be less soluble in hot water than in cold?

STRATEGY The data in Table 11.4 are for very dilute solutions; they can be used in connection with Le Chatelier's principle only if the saturated solution is also very dilute. This is the case for silver bromide, which is a very sparingly soluble salt (Table 3.2). We have to determine whether the dissolution is endothermic or exothermic. If it is endothermic, then the solubility will be greater in hot water than in cold. If it is exothermic, then the solubility will be lower in hot water than in cold.

SOLUTION The enthalpy of solution is +84.4 kJ/mol, a value indicating an endothermic process. Heating therefore favors dissolving, and silver bromide should be more soluble in hot water. (It is.)

EXERCISE E11.6 Is magnesium fluoride likely to be more or less soluble in hot water than in cold?

[*Answer*: less] ▲

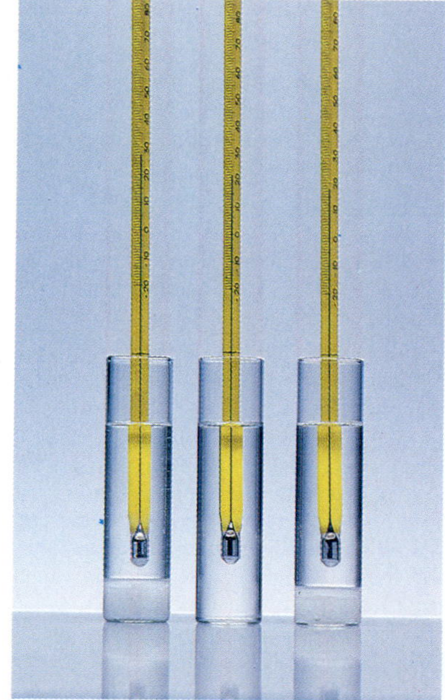

FIGURE 11.16 The exothermic dissolution of lithium chloride (left) is shown by the rise in temperature above that of the original water (center); in contrast, ammonium nitrate dissolves endothermically (right).

Contributions to the enthalpy of solution. Table 11.4 shows that some solids dissolve exothermically ($MgCl_2$ for example) and others endothermically (K_2SO_4, for example). Lithium chloride has a negative enthalpy of solution; when added to water, the solution becomes noticeably warm (Fig. 11.16). In contrast, when ammonium nitrate is added to water, the temperature falls, a change indicating that the dissolving process is endothermic. We can understand the sign and magnitude of the enthalpy of solution by thinking of the overall dissolving process as taking place in two imaginary steps (Fig. 11.17). The first is the breakup of the solid, and the other the interaction of the separated molecules or ions with the solvent.

In the first step, we imagine that the solid is vaporized to a gas of ions or molecules. As we saw in Section 8.1, the standard change of enthalpy in this endothermic step is the lattice enthalpy, ΔH_L, of the solid. For

NaCl, it is the enthalpy change for the process

$$\text{NaCl}(s) \longrightarrow \text{Na}^+(g) + \text{Cl}^-(g) \qquad \Delta H_L$$

Typical values are given in Table 8.1. As we noted, ionic compounds formed from small, highly charged ions (such as Mg^{2+} and O^{2-}) have high lattice enthalpies.

In the second step, we imagine that the gas of ions enters the solvent and that heat is released as the ions are **solvated**, or surrounded by solvent molecules. **Hydration** is a special case of solvation and occurs when the solvent is water. In aqueous solutions, water molecules hydrate cations as a result of ion-dipole interactions; they hydrate anions by ion-dipole interactions and by hydrogen bonding. The change of enthalpy in this second imaginary step is called the **enthalpy of hydration**, ΔH_H:

$$\text{Na}^+(g) + \text{Cl}^-(g) \longrightarrow \text{Na}^+(aq) + \text{Cl}^-(aq) \qquad \Delta H_H$$

Hydration is always exothermic for ions (Table 11.5). It is also exothermic for solutes that can form hydrogen bonds with the water. Examples include sucrose, glucose, acetone, and ethanol.

The enthalpy of solution of sodium chloride is the sum of the lattice enthalpy and the enthalpy of hydration. For NaCl, we have

$$\begin{array}{rcll} \text{NaCl}(s) & \longrightarrow & \text{Na}^+(g) + \text{Cl}^-(g) & \Delta H_L = +787 \text{ kJ} \\ \text{Na}^+(g) + \text{Cl}^-(g) & \longrightarrow & \text{Na}^+(aq) + \text{Cl}^-(aq) & \Delta H_H = -784 \text{ kJ} \\ \hline \text{NaCl}(s) & \longrightarrow & \text{Na}^+(aq) + \text{Cl}^-(aq) & \Delta H = \quad +3 \text{ kJ} \end{array}$$

We see that the dissolving of NaCl is slightly endothermic. However, we also see that whether the overall process turns out to be endothermic or exothermic depends on a very delicate balance between the lattice enthalpy and the enthalpy of hydration. If sodium chloride had only a 0.5% smaller lattice enthalpy, +783 in place of +787 kJ/mol, then it would dissolve exothermically instead of endothermically.

The prediction of the enthalpy of solution from the difference of tabulated lattice enthalpies and enthalpies of hydration is very unreliable, because a small inaccuracy in either quantity might lead to an answer with the wrong sign. It is like trying to calculate the mass of the

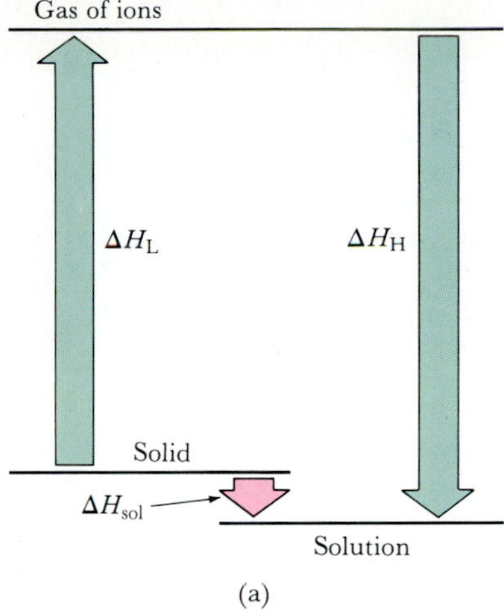

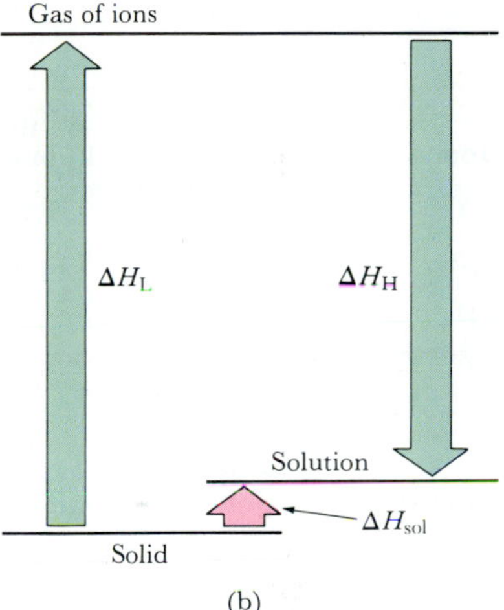

FIGURE 11.17 The enthalpy of solution, ΔH_{sol}, is the sum of the enthalpy changes required to separate the ions, ΔH_L, and the enthalpy change required for their hydration, ΔH_H. The outcome is finely balanced: (a) in some cases it is exothermic; (b) in others it is endothermic. For gaseous solutes, the lattice enthalpy, ΔH_L, is zero.

T OHT: Fig. 11.17; Table 11.5.

U Exercises 11.76 and 11.77.

? Why are the enthalpies of solution of lithium halides significantly more negative than those of the corresponding sodium, potassium, ammonium, and silver halides.

E Ionic hydration energies. *J. Chem. Educ.*, **1977**, *54*, 540. Empirical correlations for the evaluation of the free energies of solvation of some gaseous monovalent ions. *J. Chem. Educ.*, **1989**, *66*, 903.

TABLE 11.5 Enthalpies of hydration, ΔH_H, of some halides, in kilojoules per mole*

	Anion			
Cation	F^-	Cl^-	Br^-	I^-
H^+	−1613	−1470	−1439	−1426
Li^+	−1041	−898	−867	−854
Na^+	−927	−784	−753	−740
K^+	−844	−701	−670	−657
Ag^+	−993	−850	−819	−806
Ca^{2+}	—	−1387	—	—

*The entry where the row labeled Na^+ intersects the column labeled Cl^-, for instance, is the enthalpy change, −784 kJ/mol, for $Na^+(g) + Cl^-(g) \rightarrow Na^+(aq) + Cl^-(aq)$; the values here apply only when the resulting solution is very dilute.

captain of a ship by measuring the mass of the ship with and without the captain aboard. However, it is reasonably safe to expect an endothermic value when the lattice enthalpy is very high. Similarly, we can predict an exothermic value when the lattice enthalpy is low and the enthalpy of hydration is large.

Individual ion hydration enthalpies. The enthalpies of hydration in Table 11.5 are the enthalpy changes that occur when widely separated cations and anions in a gas form a very dilute solution. It would be useful to know the values for the cations and anions separately, for then we could predict the value for NaCl(*g*) by adding together the values for $Na^+(g)$ and $Cl^-(g)$. But the enthalpies of hydration of cations generally cannot be measured without anions being present too. However, the value for the hydrogen ion has been measured by a special technique that makes use of a mass spectrometer. It is

$$H^+(g) \longrightarrow H^+(aq) \qquad \Delta H_H = -1130 \text{ kJ}$$

This value can be combined with the experimentally determined values in Table 11.5 to obtain the values for other ions. Some typical values are given in Table 11.6 on the facing page.

Ion	ΔH_H, kJ/mol
Li^+	−588
Be^{2+}	−1435
Al^{3+}	−2537

Ion	Radius *r*, pm	ΔH_H, kJ/mol
Cations		
Li^+	60	−558
Na^+	95	−444
K^+	133	−361
Anions		
Cl^-	181	−340
Br^-	195	−309
I^-	216	−296

Ion hydration enthalpies show patterns. First, they are more strongly exothermic for ions with greater ionic charges (see values in the upper table in the margin). The change in ΔH_H to more negative values (corresponding to move exothermic hydration) reflects the much greater strength of the interaction between a highly charged ion and the polar water molecules. Second, for ions of the same charge, hydration enthalpies are more strongly exothermic for ions with smaller radii (see values in the lower table in the margin). We observe this pattern because water molecules can approach a small ion more closely and hence interact with it more strongly. There are exceptions: Ag^+ is bigger than Na^+ but its hydration enthalpy is more exothermic. This unusually large hydration enthalpy may be due to the ability of the Ag^+ ion to form covalent bonds with the hydrating water molecules.

▼ EXAMPLE 11.7 Predicting relative ion hydration enthalpies

Which ion can be expected to have a more negative hydration enthalpy, Ca^{2+} or Sr^{2+}?

STRATEGY We need to consider two factors, the charges of the ions (the higher the charge, the more exothermic their hydration), and their radii (the smaller the ion, the more exothermic their hydration). The latter can be judged from the trends in ionic radii in the periodic table, as explained in Chapter 7.

SOLUTION The ions have the same charge. Strontium is further down Group II than calcium, so we expect it to be larger. Therefore, we predict that Ca^{2+} will be hydrated more exothermically and therefore have a more negative hydration enthalpy. This prediction is confirmed by the data in Table 11.6.

EXERCISE E11.7 Why does Cl^- have a more negative hydration enthalpy than I^-?

[*Answer*: Cl^- has smaller radius] ▲

T OHT: Table 11.6.

At the beginning of this section, we saw that the solubilities of gases decrease with increasing temperature. We are now in a position to relate this observation to the enthalpy changes that occur when gases dissolve. We can be confident that the enthalpies of solution of all gases are negative, because we do not have to take into account the lattice enthalpy: the molecules are already widely separated and do not interact with one another. Hence, the enthalpy of solution is equal to the enthalpy of hydration, which is always negative. Consequently, according to Le Chatelier's principle, because dissolving is always exothermic, the dynamic equilibrium between gas and dissolved gas shifts toward dissolved gas when the temperature is lowered:

$$\text{Gas}(g) \xrightarrow{\text{lower temperature}} \text{gas}(aq)$$

Enthalpy, entropy, and solubility. It may seem obvious that substances with exothermic enthalpies of solvation should be soluble, because the solid loses energy as it dissolves. But why does ammonium nitrate dissolve in water, when in doing so it draws in heat from the surroundings and *gains* energy? In fact, a moment's thought shows that there is a similar problem even with exothermic dissolving: because energy cannot be created or destroyed, the total energy of the solution and its surroundings remains unchanged. Therefore, even for exothermic dissolving, although the energy of the solution has gone down, that of its surroundings has gone up by the same amount. During endothermic dissolving, the energy of the solution goes up, but the energy of the surroundings decreases to the same extent. In each case there is no net change of energy. Why, then, does *either* process occur?

The answer lies in a very simple and natural idea: *energy and matter tend to disperse*. By disperse, we mean spread out in a disorderly way (Fig. 11.18). Compared with a solid or a liquid, molecules are more dispersed if they are spread out as a gas or as a solute in a solvent. Energy is more dispersed if it spreads as chaotic thermal motion (Fig. 11.19). In this and the next five chapters, we shall increasingly see that the idea of energy and matter tending to disperse is a major principle of science and an explanation of all kinds of change. In particular, the idea accounts for the occurrence of **spontaneous changes**, changes that can occur without our having to make them happen: *all spontaneous changes are accompanied by an increase in disorder.*

The technical term for the degree of disorder is **entropy**. We shall see in more detail in Chapter 16 how this property is expressed and used; all we need know at this stage is that, for our present purposes, *when disorder increases, the entropy increases; when disorder decreases, entropy decreases.* It follows that the statement about spontaneous changes being accompanied by an increase in disorder can now be stated in the form:

Spontaneous changes tend to occur in the direction of increasing entropy.

TABLE 11.6 Ion hydration enthalpies, ΔH_H

Ion	Hydration enthalpy, kJ/mol*
Cations	
H^+	−1130
Li^+	−558
Na^+	−444
K^+	−361
Ag^+	−510
Mg^{2+}	−2003
Ca^{2+}	−1657
Sr^{2+}	−1524
Al^{3+}	−4797
Anions	
F^-	−483
Cl^-	−340
Br^-	−309
I^-	−296

*The enthalpy change for the reaction ion(g) → ion(aq) at 25°C in dilute aqueous solution.

D Endothermic reaction: Ammonium nitrate. CD1, 39. Exothermic reaction: Calcium chloride. CD1, 41.

N The purpose of this discussion is to prepare the student for the emergence of the entropy as the determinant of spontaneous change (see Chapter 16). In Chapter 3 we introduced some of the empirical driving forces of reactions (see Table 3.1); in Chapter 6 we encountered one contribution (the reaction enthalpy) to the actual driving force; here we see a second contribution (the reaction entropy). The two are combined in Chapter 16 where we see the true origin of the driving force of a reaction.

T OHT: Figs. 11.18 and 11.19.

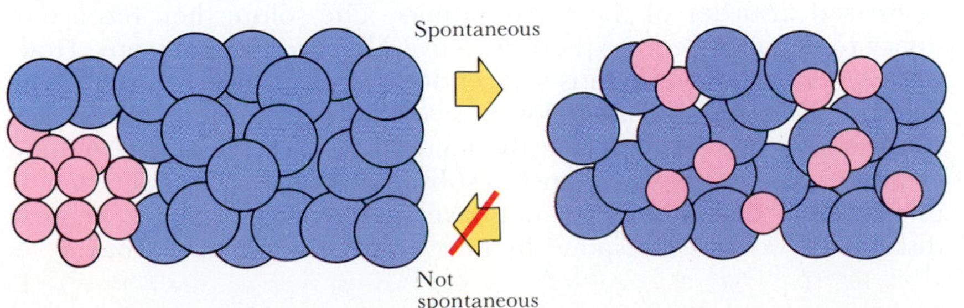

FIGURE 11.18 Matter has a universal tendency to disperse in a disorderly way. Here we see matter spreading out spontaneously. The reverse process is not spontaneous.

FIGURE 11.19 Energy also tends to spread out and to become more disordered. The reverse process, in which dispersed energy collects in a small region, is not spontaneous.

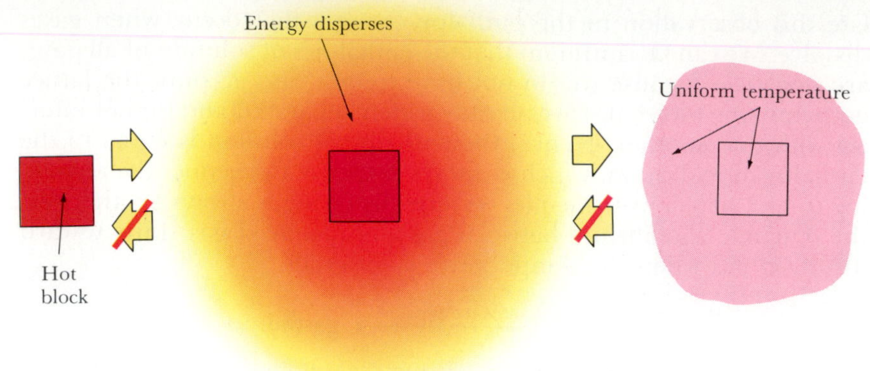

All we have done is to change the words, not the concepts. However, when in Chapter 16 we meet entropy again, we shall see that it is a quantitative concept of considerable power and importance.

When considering entropy, we have to make sure that we consider the *total* change that accompanies a process: the same is true of energy—we can confirm that energy is conserved only if we consider *all* the changes that occur. As an illustration of an entropy argument, we shall consider the dissolution of a solid. The solution becomes more disordered as the solute spreads through the solvent, because the solute molecules or ions are much more disorderly in the solution than in the original solid. However, energy flows into or out of the solution as the solid dissolves, and its contribution to the entropy must be considered too. If energy flows out of the solution, then it stirs up the molecules in the surroundings into more vigorous chaotic motion, and their entropy is increased; similarly, if energy flows into the solution, the entropy of the surroundings is reduced. The solid will tend to dissolve only if *overall* (the solution and its surroundings) the entropy increases.

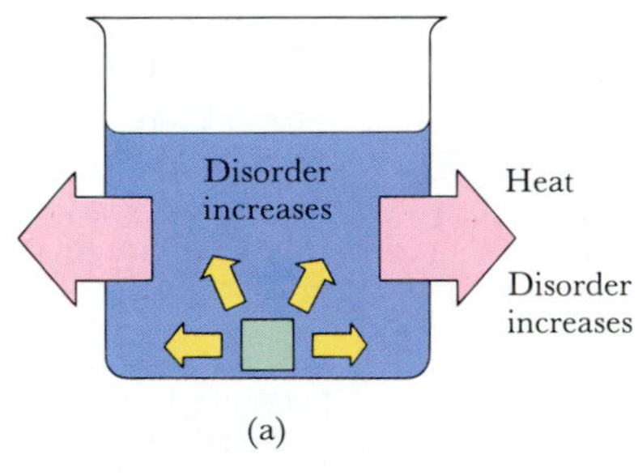

(a)

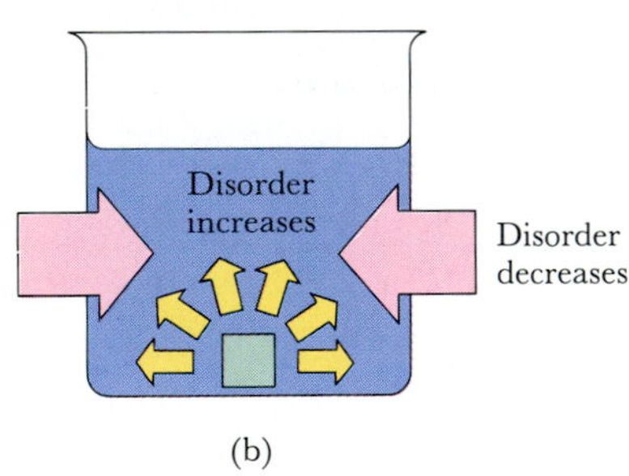

(b)

FIGURE 11.20 (a) When a solute dissolves exothermically, the system and the surroundings become more disordered, so the process is spontaneous. (b) When a substance dissolves endothermically, energy must be concentrated into the system. Hence, only if the disorder caused by the solute dispersing is great enough will there be an overall increase in disorder and a spontaneous process.

When a substance dissolves exothermically, not only does the entropy of the solution increase (because the ions or molecules become more disordered) but also the entropy of the surroundings increases as a result of the flow of energy into them. Hence, the overall entropy increases and exothermic dissolution is spontaneous (Fig. 11.20a). When a substance dissolves endothermically, the ions or molecules also become more disordered, but because energy flows in, the entropy of the surroundings decreases (Fig. 11.20b). Whether or not dissolving is spontaneous depends on the balance of these two effects. When the dissolution is only slightly endothermic (as it is for NaCl or even NH_4NO_3), the increased disorder of the solution is greater than the decreased disorder of the surroundings. The solute then has a tendency to dissolve, because there is an overall increase in entropy. However, when the dissolution is very endothermic (as it is for MgO), the amount of energy that must flow in from the surroundings and collect in the solution is so great that the increase in entropy of the solution cannot compensate for it. Hence, a substance with a strongly endothermic enthalpy of solution does not have a tendency to dissolve, because dissolution would correspond to an overall decrease in entropy.

COLLIGATIVE PROPERTIES

A solute affects the physical properties of a solvent. We shall consider four related effects:

- The lowering of the vapor pressure of the solvent
- The elevation of the boiling point of the solvent
- The depression of the freezing point of the solvent
- Osmosis

(The last property, osmosis, will be explained in Section 11.7.) An important piece of experimental information is that *all four effects depend only on the number of solute particles present in a solution and not on their chemical composition.* For example, an aqueous glucose solution in which the glucose is present at a mole fraction of 0.01—where 1 out of every 100 molecules is a glucose molecule—has the same vapor pressure, boiling point, freezing point, and osmotic pressure as an aqueous sucrose solution in which the sucrose mole fraction is also 0.01. These observations are summarized by saying that all four properties are colligative properties:

> A **colligative property** is a property that depends only on the number of solute particles present in a solution, and not on their chemical composition.

(*Colligative* means "depending on the collection.")

Because a colligative property depends only on the number of solute particles present, cations and anions in an electrolyte solution contribute separately and equally to the total. Therefore, when 1 mol NaCl is added to a solvent, the solute consists of 2 mol of ions (1 mol Na^+ ions and 1 mol Cl^- ions). When 1 mol $CaCl_2$ is added, the total number of moles of solute is 3 mol of ions (1 mol Ca^{2+} ions and 2 mol Cl^- ions). On the other hand, in a nonelectrolyte solution, each solute molecule is present as a single unit. Therefore, when 1 mole of glucose molecules is dissolved, the total amount of solute present is also 1 mol.

E Colligative properties can all be presented as though they arose from the solute particles "getting in the way" of the escape of the solvent molecules from the solution. The actual thermodynamic reason is the effect of the solute on the entropy of the solution: the presence of the solute renders the solution more chaotic, and hence more stable (so that the liquid has a lessened tendency to form the vapor or to freeze). A solute also affects the surface tension, viscosity, rate of evaporation, the enthalpy of vaporization, etc., but these are not colligative properties because their magnitudes vary with the solvent.

D Vapor pressure of pure liquids and solutions. BZS3, 9.5B,C.

E Molecular size and Raoult's law. *J. Chem Educ.*, **1985**, *62*, 1090.

11.6 CHANGES IN VAPOR PRESSURE, BOILING POINTS, AND FREEZING POINTS

A solvent has a lower vapor pressure when a nonvolatile solute is present (Fig. 11.21). For example, the vapor pressure of pure water at 40°C is 55 Torr, but that of a 0.1 mol/kg aqueous sodium chloride solution is only 44 Torr at the same temperature. Because a lower vapor pressure results in a higher boiling point (Section 10.4), we conclude that the presence of a nonvolatile solute raises the boiling point of the solvent.

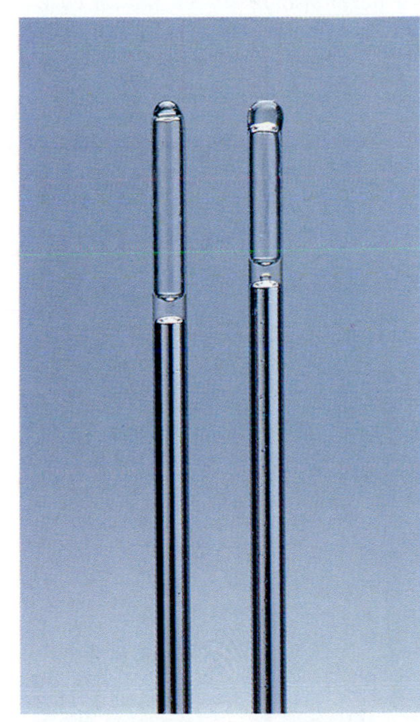

FIGURE 11.21 The vapor pressure of a solvent is lowered by a nonvolatile solute. The barometer tube on the left has a small amount of pure water floating on the mercury. That on the right has a small amount of 0.1 *m* NaCl(*aq*) solution and has a lower vapor pressure.

Raoult's law. The French scientist François-Marie Raoult spent much of his life measuring vapor pressures. He found that the effect of a solute could be summarized as follows:

> **Raoult's law**: The vapor pressure of a solvent in a solution of a nonvolatile solute is directly proportional to the mole fraction of the *solvent* in the solution.

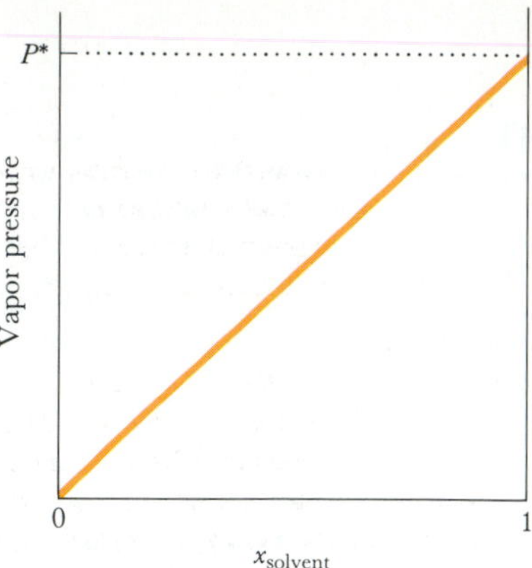

FIGURE 11.22 Raoult's law predicts that the vapor pressure of a solvent in a solution should be proportional to the mole fraction of the solvent molecules ($x_{solvent}$). P^* is the vapor pressure of the pure solvent.

E Relationship between rate of evaporation and the vapor pressure of binary systems. *J. Chem. Educ.*, **1975**, *52*, 439.

N The expressions we give for the colligative properties are valid only for very dilute solution.

E Phase change formula $\Delta T = K \cdot m$. *J. Chem. Educ.*, **1979**, *56*, 259.

N The kinetic explanation is that the solute particles block the surface to vaporization. The thermodynamic explanation is the increase in entropy, and hence increase in stability, of the liquid phase.

T OHT: Figs. 11.23, 11.28, and 11.30.

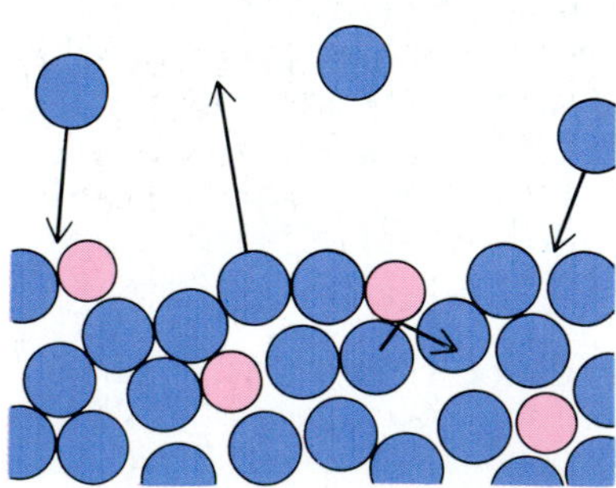

FIGURE 11.23 A nonvolatile solute particle (pink) can block the escape of the solvent particles (blue) but has no effect on the return of the solvent particles from the vapor to the solution.

This law is normally written

$$P = x_{solvent} \times P^*$$

where P^* is the vapor pressure of the pure solvent, $x_{solvent}$ is the mole fraction of the solvent, and P is the vapor pressure of the solvent in the solution when the nonvolatile solute is present (Fig. 11.22).

▼ EXAMPLE 11.8 Using Raoult's law

Calculate the vapor pressure of water at 100°C in a solution prepared by dissolving 5.00 g of sucrose ($C_{12}H_{22}O_{11}$) in 100 g of water.

STRATEGY This calculation becomes a straightforward application of Raoult's law once we know the mole fraction of the solvent (water) in the solution. (That mole fraction was in fact calculated in Example 11.1.) We need the vapor pressure of the pure solvent. We could get the value from Table 5.4; but 100°C is the normal boiling point of water and we already know that its vapor pressure at that temperature is 760 Torr.

SOLUTION From Example 11.1 we know that $x_{H_2O} = 0.997$. Therefore, because $P^* = 760$ Torr,

$$P = 0.997 \times 760 \text{ Torr} = 758 \text{ Torr}$$

EXERCISE E11.8 Calculate the vapor pressure of water at 90°C in a solution prepared by dissolving 5.00 g of glucose ($C_6H_{12}O_6$) in 100 g of water. The vapor pressure of water at 90°C is 525.8 Torr.

▲ [*Answer*: 523 Torr]

The vapor pressure of a solvent is lowered by a nonvolatile solute because the solute blocks part of the surface and hence reduces the rate at which the solvent molecules leave the solution (Fig. 11.23). However, the solute has no effect on the rate at which the solvent molecules return, because a returning molecule can stick to any point of the surface. Because the rate of escape is reduced but the rate of return is unaffected, there is a net flow of molecules back to the solution. Therefore, the vapor pressure is lowered to a new equilibrium value.

Modern experiments have shown that Raoult's law is reliable only at low concentrations, when there is a large separation between the solute molecules. However, many solutions behave approximately like an **ideal solution**, one that obeys Raoult's law at any concentration. Real solutions resemble ideal solutions more closely the lower their concentration: the agreement is quite good below about 0.1 mol/kg for nonelectrolyte solutions and 0.01 mol/kg for electrolyte solutions (in the latter, interactions between ions have a marked effect). We shall assume that all the solutions we meet are ideal.

The boiling-point elevation. Because the vapor pressure of a solvent in a solution is lower than that of the pure solvent, its normal boiling point is higher (Fig. 11.24). The increase, which is called the **boiling-point elevation**, is usually quite small. A 0.1 mol/kg aqueous sucrose solution, for instance, boils at 100.05°C. The boiling-point elevation is directly proportional to the molality m of the solute in the solution. For a molecular solute, it is normally written

$$\text{Elevation of boiling point: } \Delta T_b = k_b \times m$$

where k_b is the boiling-point constant of the solvent. Values for various solvents are given in Table 11.7.

TABLE 11.7 Boiling-point and freezing-point constants

Solvent	Freezing point, °C	k_f, °C·kg/mol	Boiling point, °C	k_b, °C·kg/mol
acetone	−95.35	2.40	56.2	1.71
benzene	5.5	5.12	80.1	2.53
camphor	178.8	39.7	5.61	—
carbon tetrachloride	−23	29.8	76.5	4.95
cyclohexane	6.5	20.1	80.7	2.79
naphthalene	80.5	6.94	5.80	—
phenol	43	7.27	3.04	—
water	0	1.86	100.0	0.51

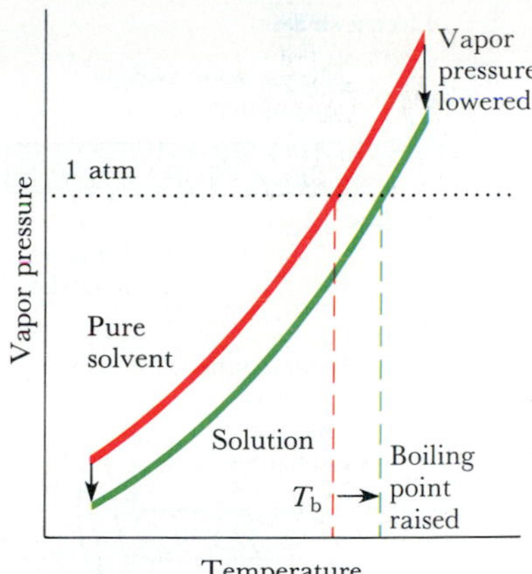

FIGURE 11.24 The lowering of vapor pressure of a solution leads to an increase in its boiling point (T_b).

D Getting hotter: Boiling-point elevation by nonvolatile solutes. BZS3, 9.21. Getting colder: Freezing point depression. BZS3, 9.21.

N The van't Hoff *i* factor appears again on p. 431.

When sodium chloride dissolves, each formula unit gives two ions. For very dilute solutions, we can treat the cations and anions as independent contributors, so the total solute molality is twice the molality in terms of NaCl formula units. When the solution is not extremely dilute, ions of opposite charge do not move independently but instead form ion pairs (Fig. 11.25). In this case, the effective molality of particles—individual ions or ions pairs—is different from the molality of NaCl treated as a collection of independent ions. The ions need not actually come into contact for their effective number to be reduced. All that is required is for cations to spend a significant proportion of their time near an anion, and vice versa.

The difference between the effective molality and the actual molality of ions is taken into account by introducing the **van't Hoff *i* factor** for electrolyte solutions:

$$\text{Elevation of boiling point: } \Delta T_b = i \times k_b \times m$$

In a very dilute solution, when all ions are independent, $i = 2$ for MX salts and 3 for MX_2 salts such as $CaCl_2$, and so on. The *i* factor is named for Jacobus van't Hoff, a Dutch chemist who studied the properties of solutions and in 1901 was awarded the first Nobel prize for chemistry.

The freezing-point depression. A solute lowers the freezing point of the solvent. This effect is called the **freezing-point depression**. For example, seawater, an aqueous solution rich in Na^+ and Cl^- ions, freezes at lower temperatures than fresh water does. In winter, salt spread on highways in northern latitudes also makes use of the effect, for the salt lowers the freezing point of water and prevents ice formation. Salt spread on ice that is already formed melts the ice (so long as the temperature is not too low) because a few of the ions can attract water molecules from the surfaces of the ice crystals and gradually form liquid water. Similarly, chemists can judge the purity of a compound by checking its melting point, which will be lower than normal if impurities are present.

The freezing-point depression caused by a solute also accounts for the fact that mixtures freeze over a range of temperatures rather than at one precise melting point. When the solution begins to freeze, *only the solvent solidifies* and solute is left behind in the solution. As a result, the concentration of the remaining solution rises and the freezing

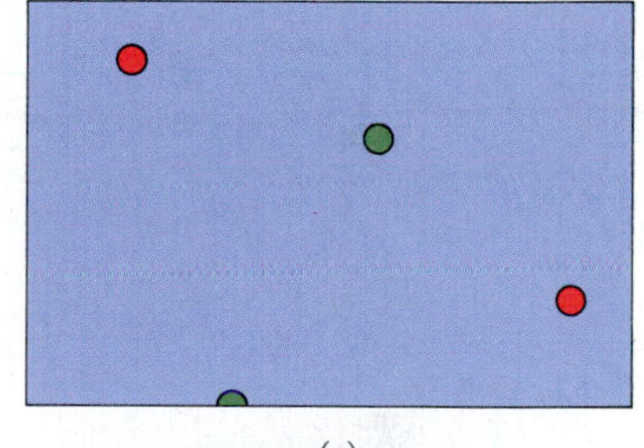

(a)

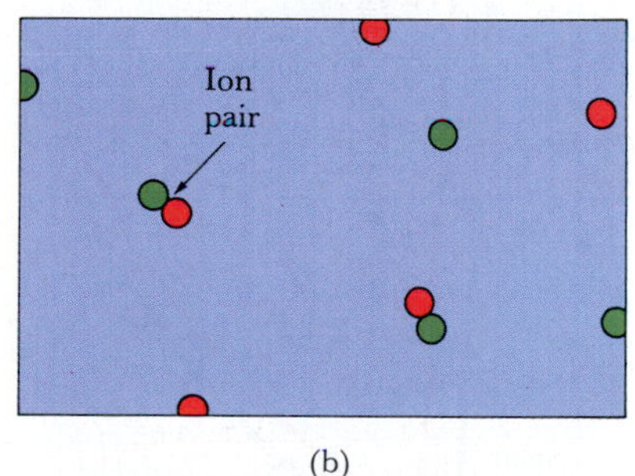

(b)

FIGURE 11.25 (a) In a very dilute solution, most ions move independently of one another and contribute separately to the molality. (b) As the ionic concentration increases, more ion pairs are formed, and the effective molality is lower than the actual molality of ions. For example, ten ions behave as eight effective "particles."

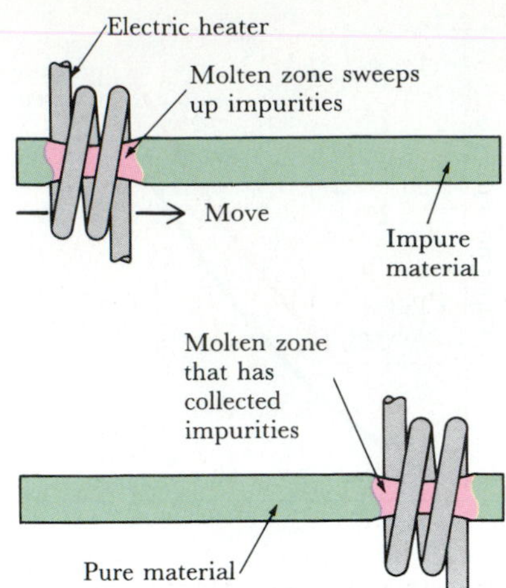

FIGURE 11.26 In the technique of zone refining, a molten zone is passed repeatedly from one end of the solid sample to the other. The impurities collect in the zone.

N Teaching freezing-point lowering. *J. Chem. Educ.*, **1990**, *67*, 676. The freezing-point depression law in physical chemistry. *J. Chem. Educ.*, **1988**, *65*, 1077.

U Exercises 11.73, 11.83, 11.90, and 11.91.

T OHT: Fig. 11.27.

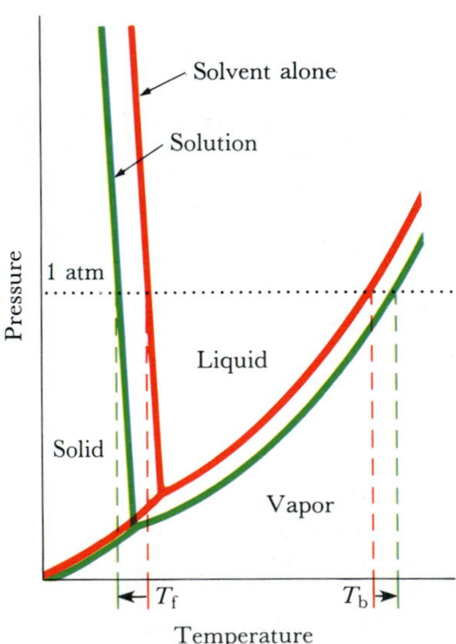

FIGURE 11.27 The effect of a solute on the freezing point (T_f) of water. The effect on vapor pressure and boiling point (T_b) are also shown.

point of the solvent becomes even lower. The concentration of the solution continues to rise, and the freezing point continues to fall, until all the solvent has frozen. This phenomenon has been used to purify solvents. It is also the basis of the proposal to drag icebergs from polar regions as sources of fresh water, for the ice was largely salt-free when it formed. The technique called **zone refining**, which is used to prepare very pure samples of substances (such as silicon for use in the manufacture of semiconductors) also makes use of the effect. In zone refining, a cylindrical heater is moved several times from one end of a cylindrical solid sample to the other, so that a molten zone passes repeatedly through the solid (Fig. 11.26). Impurities dissolve in the molten zone as it passes, and the purer material freezes. The moving molten zone sweeps the impurities to one end of the sample, some of which can be discarded.

The effect of a solute on the freezing point of water is shown in (Fig. 11.27): the solid-liquid boundary is shifted to lower temperatures as the solute concentration is increased. The figure also shows that the vapor pressure of the solution is lowered and the boiling point is increased (as we have already seen).

The explanation of the freezing-point depression lies in the effect of the solute on the rates at which solvent molecules form the solid and leave it to return to the liquid. At the freezing point of the pure solvent, these two rates are equal. When a solute is present, fewer solvent molecules in the solution are in contact with the surface of the solid solvent, because some of the places they might occupy are taken by solute particles (Fig. 11.28). As a result, they attach to the surface more slowly. However, the rate at which molecules leave the solid, which is pure solid solvent, is unchanged because a molecule can break away from the solid even though a solute particle is next to it in the solution. Therefore, when solute is added to the pure solvent, there is a net flow of molecules away from the solid, and the solid melts. Only if the temperature is lowered will flow be stopped and the equilibrium restored.

The depression of freezing point of an ideal solution is directly proportional to the molality m. For a molecular solute,

$$\text{Depression of freezing point: } \Delta T_f = k_f \times m$$

where k_f is the freezing-point constant of the solvent (see Table 11.7). For an electrolyte solution, we include the van't Hoff i factor on the right of the equation. The freezing-point constant is almost always larger than the boiling-point constant for the same solvent, which implies that freezing-point depressions are larger than boiling-point elevations. A 0.1 mol/kg aqueous sucrose solution, for instance, has

$$\Delta T_f = 1.86 \frac{°\text{C}\cdot\text{kg}}{\text{mol}} \times 0.1 \frac{\text{mol}}{\text{kg}} = 0.2°\text{C}$$

and hence freezes at -0.2°C.

The freezing-point depression can be used to measure the molar mass of the solute. This is more reliable than using boiling-point elevations because a freezing-point depression is generally larger and hence easier to measure (and avoids the problem of the variation of boiling point with pressure). Moreover, heat-sensitive molecules are less likely to be damaged by freezing than by boiling, and the problems caused by decomposition products and evaporation are avoided. Because the organic compound camphor has a large freezing-point constant, solutes depress its freezing point significantly, and it is often used as a solvent in this kind of measurement.

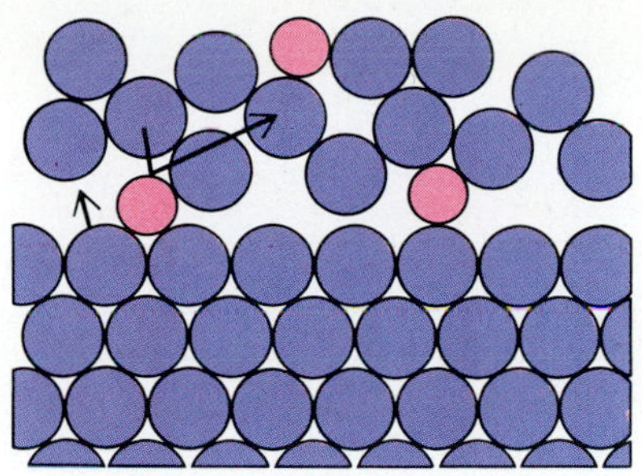

FIGURE 11.28 The rate at which solvent molecules (blue) leave the pure solid solvent is unaffected by the presence of solute particles (pink) nearby in the solution, but their rate of return to the solid is reduced.

▼ EXAMPLE 11.9 Using freezing-point depression to measure molar mass

Solid sulfur is a *molecular* solid of formula S_x. The addition of 0.24 g of sulfur to 100 g of carbon tetrachloride lowered its freezing point by 0.28°C. Find the value of x in the molecular formula of solid sulfur.

STRATEGY The value of x in the molecular formula S_x can be found by measuring the molar mass of the molecules and comparing it with the molar mass of its atoms. Because the freezing-point constant for carbon tetrachloride (see Table 11.7) can be used to convert a freezing-point depression to a molality, we can find the number of moles of sulfur molecules in the solution. We are given their total mass, therefore we can combine the two to obtain the molar mass. Then x can be found by dividing the molar mass of the sulfur molecules by the molar mass of sulfur atoms.

SOLUTION The molality of the solution is obtained by rearranging the freezing point depression equation into

$$m = \frac{\Delta T_f}{k_f}$$

and inserting the data:

$$m = 0.28°\text{C} \times \frac{\text{mol}}{29.8°\text{C}\cdot\text{kg}} = 0.0094 \text{ mol/kg}$$

Therefore, the amount of S_x molecules in 100 g of CCl_4 is

$$\text{Amount of } S_x \text{ (mol)} = 0.100 \text{ kg } CCl_4 \times 0.0094 \frac{\text{mol } S_x}{1 \text{ kg } CCl_4}$$

$$= 9.4 \times 10^{-4} \text{ mol } S_x$$

The total mass of sulfur present is 0.24 g. Therefore, the molar mass of sulfur is

$$\frac{0.24 \text{ g } S_x}{9.4 \times 10^{-4} \text{ mol } S_x} = 260 \text{ g/mol}$$

The molar mass of S atoms is 32.1 g/mol, so

$$x = \frac{260 \text{ g/mol}}{32.1 \text{ g/mol}} = 8$$

Elemental sulfur is therefore composed of S_8 molecules.

EXERCISE E11.9 The addition of 250 mg of eugenol, the compound responsible for the odor of oil of cloves, lowered the freezing point of 100 g of camphor by 0.62°C. Calculate the molar mass of eugenol molecules. ▲

[*Answer*: 160 g/mol (actual: 164.2 g/mol)]

U Exercise 11.78.

E The van't Hoff factor can be defined as

$$i = \frac{\text{actual amount (in moles) of solute particles in solution}}{\text{amount of solute particles added to solvent}}$$

Thus, for a *very* weak dilute aqueous solution of $CaCl_2$, $i = 3$. For a weak electrolyte such as CH_3COOH,

$$i = \frac{(C - x) + x + x}{C} = \frac{C + x}{C}$$

where C is the initial amount (in moles) of CH_3COOH and x is the amount of CH_3COOH that undergoes deprotonation and hence gives rise to H_3O^+ and $CH_3CO_2^-$ ions. The percentage ionization of the weak electrolyte would be

$$\text{Percentage ionized (\%)} = \frac{x}{C} \times 100\%$$

U Exercises 11.79 to 11.81.

The use of antifreeze, typically the organic compound ethylene glycol (**4**), is often given as an illustration of a colligative property. Indeed, a solution of antifreeze does have a lower freezing point than water does. However, antifreeze is used in much greater concentrations than are typical of the solutions we have been considering. It is more realistic to think of the contents of a car radiator as a mixture that remains fluid down to very low temperatures because the two kinds of molecules,

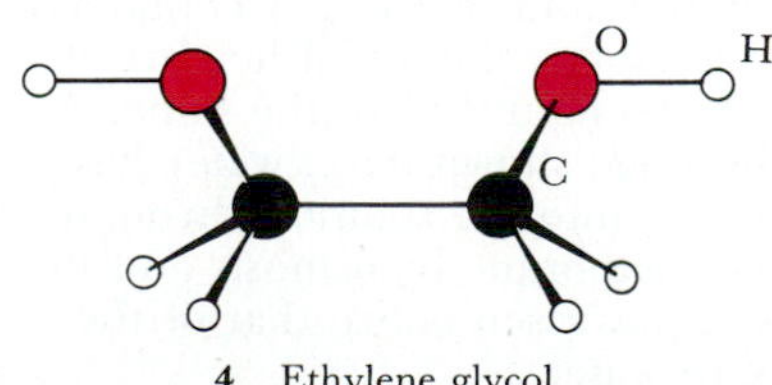

4 Ethylene glycol

water and ethylene glycol, pack together so badly. In other words, the hydrocarbon part of the glycol molecule acts as a kind of lubricant for the water molecules in the mixture.

E Chemical potential and osmotic pressure. *J. Chem. Educ.*, **1979**, *56*, 579.

D Osmosis and the egg membrane. CD2, 72. Osmotic pressure of a sugar solution. BZS3, 9.20.

11.7 OSMOSIS

The process of osmosis is a very important phenomenon both for the function of life and in laboratory measurements. It is illustrated by the experiment shown in Fig. 11.29. A solution and a pure solvent are separated by a sheet of cellulose acetate (a material that is widely used as a transparent wrapper on candy boxes). Initially the heights of the solution inside the tube and the pure solvent outside the tube are the same, but the solution inside rises because the pure solvent pushes through the membrane into the solution, thereby increasing the latter's volume. Equilibrium, when the flow of solvent into the tube occurs at the same rate as the flow of solvent out of it, is reached when the pressure exerted by the additional height of solution, which is called the **hydrostatic pressure** of the column of liquid, can push solvent molecules back through the membrane at a matching rate.

The special characteristic of a cellulose acetate membrane is its **semipermeability**; that is, only certain types of molecules or ions can pass through it. Cellulose acetate, for example, allows water molecules to pass but not solute molecules or ions, with their bulky coating of hydrating water molecules. The result, shown in Fig. 11.29, is called osmosis:

FIGURE 11.29 An experiment to illustrate osmosis. Initially, the inverted tube contained a sucrose solution and the beaker contained pure water; the initial heights of the two liquids were the same. At the stage shown here, water has passed into the solution through the membrane by osmosis and its level has risen above that of the pure water.

Osmosis is the tendency of a solvent to flow through a semipermeable membrane into a more concentrated solution.

(The name *osmosis* comes from the Greek word for "push.") The hydrostatic pressure inside the tube needed to balance the flow of solvent across the semipermeable membrane is called the **osmotic pressure** (denoted by Π, the Greek uppercase letter pi). At this pressure, a dynamic equilibrium exists across the membrane:

$$\text{Solvent(solution)} \rightleftharpoons \text{solvent(pure)}$$

The greater the osmotic pressure, the greater the height of the solution above that of the pure solvent.

The flow occurs because the solvent molecules can pass readily through the membrane from the pure solvent into the solution, but the presence of solute molecules blocks the return of some solvent molecules from the solution into the pure solvent (Fig. 11.30). As a result, the flow from the pure solvent into the solution is faster than the return flow. The return flow can be increased by applying pressure to the solution, which forces the solvent molecules to pass through the membrane more quickly.

Van't Hoff showed that the osmotic pressure is related to the molar concentration of the solution by

$$\Pi = i \times RT \times \text{molar concentration}$$

where i is the van't Hoff factor, R is the gas constant, and T is the temperature. This expression is now known as the van't Hoff equation.

The osmotic pressure of a 0.010 M solution of any nonelectrolyte at 25° (298 K) is therefore

$$\Pi = 1 \times 0.0821 \frac{\text{L·atm}}{\text{K·mol}} \times 298\ \text{K} \times 0.010 \frac{\text{mol}}{\text{L}} = 0.24\ \text{atm}$$

This pressure is enough to push a column of water to a height of over 2 m (the height needed to exert a hydrostatic pressure of 0.24 atm). Osmotic pressure is a colligative property in the sense that its value depends only on the number of solute molecules present (as given by their molar concentration) and not on their chemical identity.

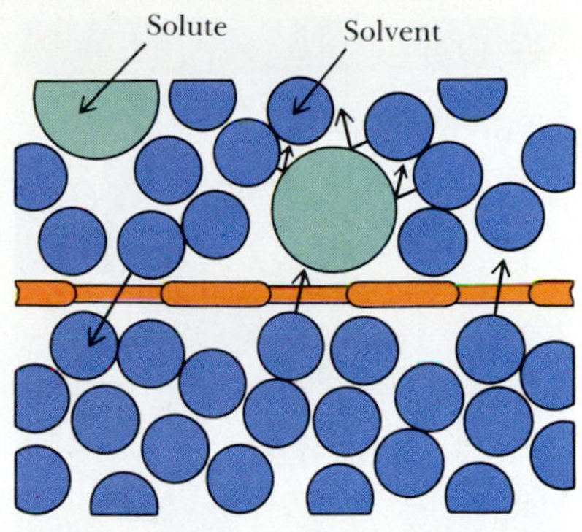

FIGURE 11.30 During osmosis, the presence of solute (green) on one side of the membrane hinders solvent molecules (blue) from passing through the membrane into the side containing pure solvent.

Osmometry. Like the other colligative properties, osmotic pressure can be used to measure the molar mass of the solute. The technique is called **osmometry** and involves the measurements of the osmotic pressure of a solution containing a known mass of the solute.

▼ EXAMPLE 11.10 Measuring molar mass by osmometry

In an experiment to measure the molar mass of polyethylene (which we shall denote PE and which consists of long —CH_2CH_2— chains), 2.20 g of the PE plastic was dissolved in enough of the solvent toluene to produce 100 mL of solution. Its osmotic pressure at 25°C was measured as 1.10×10^{-2} atm (which corresponds to a 13-cm column of the toluene solution). Calculate the molar mass of the polyethylene.

STRATEGY We know the mass of PE in the solution, so we can find its molar mass if we know the number of moles of PE present. Because we know the volume of the solution, the amount of PE present can be found from the molar concentration of PE. That in turn can be found from the osmotic pressure by using the van't Hoff equation.

SOLUTION From the van't Hoff equation, with $i = 1$,

$$\text{Molar concentration of PE} = \frac{\Pi}{RT} = \frac{1.10 \times 10^{-2}\ \text{atm}}{(0.0821\ \text{L·atm/K·mol}) \times 298\ \text{K}}$$
$$= 4.6 \times 10^{-4}\ \text{mol/L}$$

U Exercise 11.85.

That is, 1 L solution contains 4.6×10^{-4} mol PE. The amount of PE in 100 mL of the solution is therefore

$$\text{Amount of PE (mol)} = 0.100\ \text{L solution} \times \frac{4.6 \times 10^{-4}\ \text{mol PE}}{1\ \text{L solution}}$$
$$= 4.6 \times 10^{-5}\ \text{mol PE}$$

Because the total mass of PE in the solution is 2.20 g,

$$\text{Molar mass of PE (g/mol)} = \frac{2.20\ \text{g PE}}{4.6 \times 10^{-5}\ \text{mol PE}} = 4.8 \times 10^{4}\ \text{g/mol}$$

This molar mass corresponds to a chain of 1700 —CH_2CH_2— units. The molar mass measured in this experiment is an *average* molar mass, for the molecules present in the sample do not all have the same chain length.

EXERCISE E11.10 The osmotic pressure of 3.0 g of polystyrene in enough benzene to produce 150 mL of solution was 1.21 kPa at 25°C. Calculate the molar mass of the sample of polystyrene.

[*Answer*: 40 kg/mol] ▲

E Solvent tension or solvent dilution. Which gives a simpler measure of osmotic pressure? *J. Chem. Educ.*, **1978**, *55*, 21. Molecular weights of drugs and diluents by osmometry. *J. Chem. Educ.*, **1982**, *59*, 672.

U Exercises 11.84 and 11.86–11.88.

The advantage of determining molar mass by osmometry rather than by using the other colligative methods is its very high sensitivity.

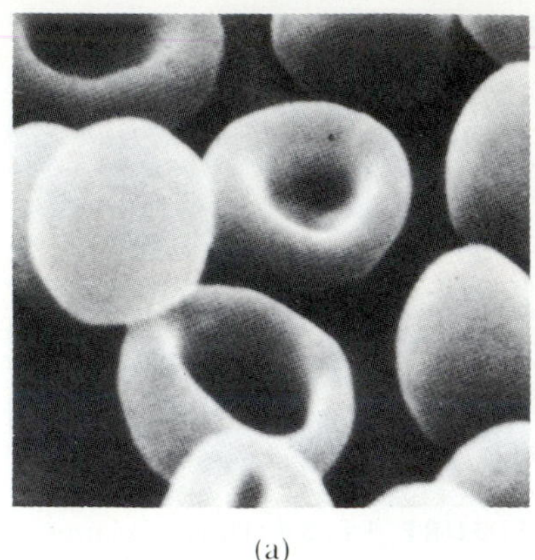

(a)

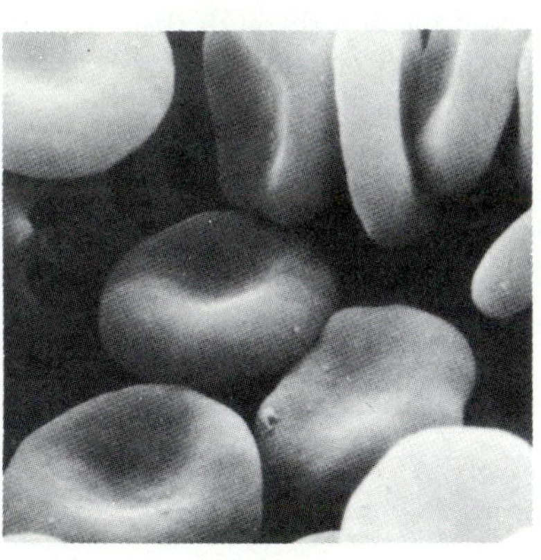

(b)

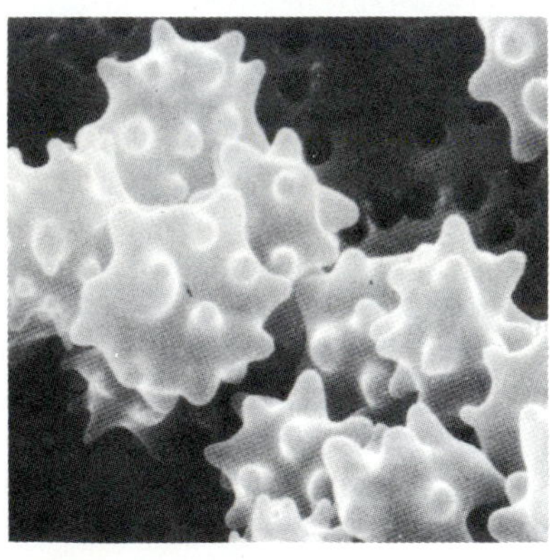

(c)

FIGURE 11.31 (a) Red blood corpuscles need to be in solutions of the correct solute concentration if they are to function properly. (b) When the solution is too dilute, water passes into them and they burst. (c) When it is too concentrated, water flows out of them and they shrivel up.

Whereas a 0.01 M aqueous sucrose solution shows a boiling-point elevation of 0.005°C and a freezing-point depression of 0.02°C, it has an osmotic pressure equivalent to a column of water about 2.5 m high. The two small temperature changes are difficult to measure, but the large difference in heights can be measured easily and accurately. This sensitivity is particularly important when the solute has a very high molar mass, as do enzymes, proteins, and synthetic polymers. Even though the mass of these solutes in the solution may be appreciable, their molar masses are so high (of the order of 10,000 g/mol and more) that the actual *number* of molecules present may be very small and the colligative properties correspondingly minute. The concentration of a solution of 0.10 g of hemoglobin (which has a molar mass of 66,500 g/mol) in 100 mL of solution is only 1.6×10^{-5} M. It has an unmeasurably small freezing-point depression but an osmotic pressure equivalent to about 4 mm of water, which is measurable.

Some applications of osmosis. Biological cell walls act as semipermeable membranes that allow water, small molecules, and hydrated ions to pass. However, they block the passage of the enzymes and proteins that have been synthesized within the cell. The difference in concentrations of solute inside and outside a plant cell gives rise to osmosis, and water passes into the more concentrated solution in the interior of the cell, carrying small nutrient molecules with it. This influx of water also keeps the cell turgid, or swollen. When the supply is cut off, the turgidity is lost and the plant wilts. Salted meat is preserved from bacterial attack by osmosis; the concentrated salt dehydrates the bacteria by causing water to flow out of them.

Osmosis maintains the turgidity of the red corpuscles in blood (Fig. 11.31). The walls of the red blood cells are impermeable to sodium ions, so the presence of these ions on one side or the other affects the direction of osmotic flow. If the concentration of Na^+ ions in the surrounding blood plasma is too low, then water flows into the blood cells; if the flow continues, they burst. Because the walls of the blood capillaries are permeable to all species except the big protein molecules, the latter's presence in blood plasma governs the direction of the flow of water through the capillary walls. If the concentration of protein in the blood plasma is reduced, as it is in cases of extreme hunger, water flows out of the plasma. This condition can be corrected, and a life saved, by intravenous injections of protein-enriched blood plasma.

Reverse osmosis. A modification of osmosis called **reverse osmosis** is used to produce fresh water for drinking and irrigation by removing salts from seawater. In reverse osmosis, a pressure greater than the osmotic pressure is applied to the solution side of the semipermeable membrane. This application of pressure increases the rate at which water molecules leave the solution. The water is almost literally squeezed out of the salt solution through the membrane. The technological challenge is to fabricate membranes into tiny tubes that are strong enough to withstand the high pressures needed yet are permeable enough to permit a good flow of water without becoming clogged. Commercial plants use cellulose acetate as the membrane and pressures of up to 70 atm. The membrane is bundled into containers and each cubic meter of membrane can produce about 250,000 L of pure water a day.

MIXTURES OF LIQUIDS

Some of the ideas we have met in this chapter are also applicable to mixtures of liquids, such as a solution of ethanol in water or of benzene in another liquid hydrocarbon. One difference is that *both* components of the solution are now volatile, and the vapor above it is a mixture of the vapors of both substances. This feature has important consequences for the fractional distillation of mixtures, such as liquid air (to obtain oxygen, nitrogen, and the noble gases) and of petroleum for gasoline and other fractions.

E Two solutions having the same osmotic pressure are *isotonic;* if one has a greater osmotic pressure than another, then it is called *hypertonic;* if less, then it is *hypotonic*. These terms are widely used in the description of blood plasma.

E Reverse osmosis water purification units are used in many arid regions of the world where salt water is readily available: two examples are Saudi Arabia and Southern California.

11.8 RAOULT'S LAW FOR MIXTURES OF LIQUIDS

A number of binary liquid mixtures, such as benzene and toluene, form ideal solutions in which *both* components in the solution individually obey Raoult's law. More precisely, the total vapor pressure of an ideal solution is the sum of two partial pressures, one being the vapor pressure of one of the liquids and the other the vapor pressure of the second liquid:

$$\text{Total vapor pressure} = P_A + P_B$$

These two individual vapor pressures each obey Raoult's law:

$$P_A = x_A \times P_A^* \quad \text{and} \quad P_B = x_B \times P_B^*$$

where x_A and x_B are the mole fractions of the two liquids in the mixture and P_A^* and P_B^* are the vapor pressures of the pure liquids. The graph in Fig. 11.32 shows how the vapor pressure of the mixture varies with composition.

▼ **EXAMPLE 11.11 Calculating the vapor pressure of a mixture of liquids**

The vapor pressure of benzene at 25°C is 94.6 Torr and that of toluene is 29.1 Torr. What is the vapor pressure of a liquid in which one-third of the molecules are benzene and two-thirds are toluene?

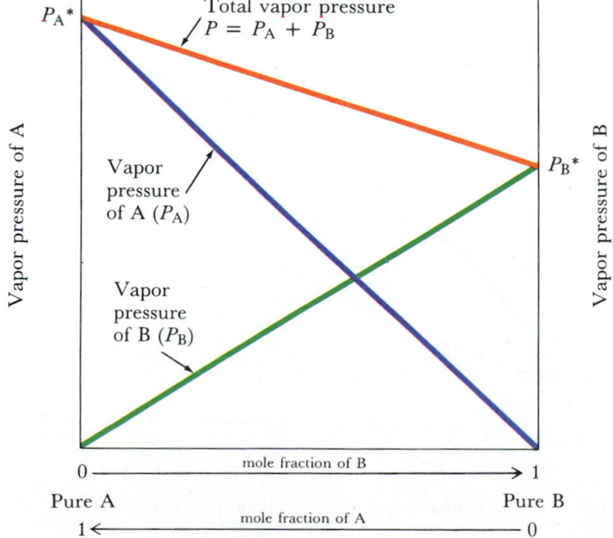

FIGURE 11.32 The total vapor pressure of a mixture of two liquids is the sum of the vapor pressure that each one exerts in accord with the prediction of Raoult's law. The blue and green lines are the individual vapor pressures of the liquids in the mixture, and the orange line is their sum.

STRATEGY Benzene and toluene are sufficiently alike to form a nearly ideal solution with each other. Therefore, Raoult's law should be applicable. To use the law, we need the mole fractions of each component, but that information is provided in the question.

SOLUTION The mole fraction of benzene is $\frac{1}{3}$ and that of toluene is $\frac{2}{3}$, so according to Raoult's law,

$$\begin{aligned}\text{Vapor pressure} &= P(\text{benzene}) + P(\text{toluene}) \\ &= x(\text{benzene}) \times P^*(\text{benzene}) + x(\text{toluene}) \times P^*(\text{toluene}) \\ &= \tfrac{1}{3} \times 94.6 \text{ Torr} + \tfrac{2}{3} \times 29.1 \text{ Torr} = 50.9 \text{ Torr}\end{aligned}$$

EXERCISE E11.11 Calculate the vapor pressure of a mixture in which one in ten of the molecules is toluene and the rest are benzene (at 25°C). [*Answer*: 88.1 Torr]

11.9 THE DISTILLATION OF MIXTURES OF LIQUIDS

U Exercise 11.89.

N The technique of fractional distillation was introduced in Chapter 1; here we see that the separation is not complete for certain mixtures of liquids, even when the boiling points are well separated.

E A deviation from Raoult's law. The mercury-water system. *J. Chem. Educ.*, **1975**, *52*, 117. *J. Chem. Educ.*, **1975**, *52*, 815.

We saw in Example 11.11 that the vapor pressure of an ideal solution containing two kinds of molecules (benzene and toluene) is the sum of the vapor pressures of the two components. The same will be true as the mixture is heated right up to its boiling point. Because boiling occurs when the vapor pressure is equal to the atmospheric pressure, the mixture will boil when the total vapor pressure is equal to that pressure. However, the vapor itself will be richer in the more volatile component, because its molecules leave the liquid phase more easily than the molecules of the less volatile component. A 1:1 mixture of benzene and toluene, for instance, boils to give a vapor that is significantly richer in benzene (the more volatile component). The compositions of the liquid and the vapor in equilibrium with it are shown by the points marked on Fig. 11.33.

If we were to remove some of the vapor and condense it to a liquid, we would find it richer in benzene than in toluene. Because the vapor is richer in benzene than in toluene, the remaining liquid is richer in

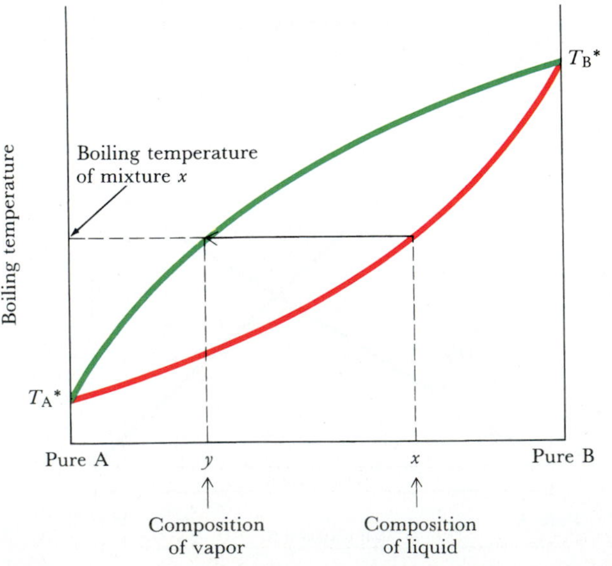

FIGURE 11.33 The red curve shows the boiling point of a mixture of two liquids as the composition of the mixture is changed from pure A on the left to pure B on the right. For example, the liquid of composition x boils at the temperature shown. The horizontal line at the boiling point cuts the green line at a composition y, which is the composition of the *vapor* that is in equilibrium with the boiling liquid. Note that it is richer in the lower-boiling-point component A than in the liquid mixture (that is, y is closer to the left-hand side than x is).

toluene than in benzene. The result is a partial separation of the two components.

Suppose we boil the condensed vapor again. Because it is richer in (the more volatile) benzene than in toluene, it vaporizes at a lower temperature. The vapor is now even richer in the more volatile component; and if we were to condense it, we would have a liquid that is significantly richer in benzene than in the original liquid. Indeed, we could continue the process of vaporization and condensation and end up with a sample of nearly pure benzene.

The process just described should remind you of the procedure for fractional distillation described in Chapter 1. Indeed, fractional distillation is exactly this sequence of many vaporizations and condensations. However, the steps all take place inside one piece of apparatus, the fractionating column. Fractional distillation is a widely used laboratory procedure; it is also used in industry to separate mixtures of volatile liquids, particularly liquid air and petroleum.

SUMMARY

11.1 An **electrolyte** is a substance that dissolves to give a solution capable of conducting electricity; a **nonelectrolyte** is a substance that dissolves to give a nonconducting solution. The measure of concentration that emphasizes the amount of solute in a given volume of solution is the **molar concentration** (molarity, moles per liter of solution). The **volume percentage** and **parts per million by volume** are also sometimes used.

11.2 The measures of concentration that emphasize the relative numbers of solute and solvent particles are the **mass percentage composition**, the **mole fraction** (the number of moles of solute molecules as a fraction of the total), and the **molality** (moles per kilogram of solvent). In certain instances the expression **parts per million by mass** is also used.

11.3 A **saturated solution** is one in which the dissolved and undissolved solute are in dynamic equilibrium. The concept is used to define the **solubility** of a substance, which is its concentration in a saturated solution. Solubility depends on the identity of the solute and the solvent ("like dissolves like"), on the pressure (for gases), and on the temperature. Soaps and detergents contain ions and molecules with **hydrophobic** (water-repelling) and **hydrophilic** (water-attracting) groups that dissolve in oil and water, respectively.

11.4 The increase in solubility of a gas with pressure is described by **Henry's law**, which states that the solubility of a gas is proportional to its partial pressure above the liquid. This law is an example of **Le Chatelier's principle**, which states that a system at equilibrium tends to minimize any change in the conditions. Le Chatelier's principle applies only to dynamic equilibria.

11.5 The energetics of dissolving are discussed in terms of the **enthalpy of solution**, which is equal to the sum of the lattice enthalpy and the enthalpy of hydration of the solute. An enthalpy of hydration can be expressed as the sum of the individual **ion hydration enthalpies** for each type of ion. The enthalpy of solution can be used to predict the effect of temperature on the solubility of a sparingly soluble compound. The direction of the dissolving process is the one that leads to an increase in **entropy**, and increased **dispersal** of matter and energy. Substances with large enthalpies of solution are unlikely to be soluble because, were they to dissolve, a large quantity of energy would need to collect in the solution.

11.6 Colligative properties depend on the number of particles in a solution and not on their chemical identity. Colligative properties include **lowering of vapor pressure**, **elevation of boiling point**, and **depression of freezing point**, all of which are directly proportional to the molality of the solute in the solution. The lowering of vapor pressure is described by **Raoult's law**, which states that the vapor pressure is directly proportional to the mole fraction of the solvent. Measurements of boiling-point elevation and of freezing-point depression can be used to measure molar masses of solutes.

11.7 Osmosis, another colligative property, is the passage of a solvent through a **semipermeable membrane**, one that permits the passage of some types of particles but not all. **Osmometry**, the measurement of osmotic pressure, is used to measure molar masses, especially those of large molecules.

11.8 The total vapor pressure of a mixture of liquids is equal to the sum of the vapor pressures of the components as given by Raoult's law.

11.9 In the process of **fractional distillation**, successive vaporizations and condensations result in the separation of a mixture of volatile liquids.

CLASSIFIED EXERCISES

Measures of Concentration

11.1 Calculate the molar concentration of (a) 50.0 g of KNO_3 dissolved in 1.50 L of solution and (b) 10.0 g of glucose, $C_6H_{12}O_6$, dissolved in 250 mL of solution.

11.2 The stockroom person was asked to prepare the following solutions: (a) 1.63 g anhydrous sodium carbonate, Na_2CO_3, dissolved in water and diluted to 200 mL and (b) 6.29 g anhydrous calcium chloride, $CaCl_2$, dissolved in water and diluted to 750 mL. Calculate the molar concentrations of the two solutions.

11.3 What mass (in grams) of anhydrous solute is needed to prepare each of the following solutions: (a) 1.0 L of 0.10 M NaCl(*aq*); (b) 250 mL of 0.10 M $CaCl_2$(*aq*); (c) 500 mL of 0.63 M $C_6H_{12}O_6$(*aq*)?

11.4 Describe the preparation of each solution, starting with the anhydrous solute and water and using the corresponding volumetric flask: (a) 25.0 mL of 6.0 M NaOH(*aq*); (b) 1.0 L of 0.10 M $BaCl_2$(*aq*); (c) 500 mL of 0.0010 M $AgNO_3$(*aq*).

11.5 Determine the mass percentage of each solute in the following solutions: (a) 4.0 g of NaCl dissolved in 100 g of aqueous solution; (b) 4.0 g of NaCl dissolved in 100 g of water; (c) 1.66 g of $C_{12}H_{22}O_{11}$ dissolved in 200 g of water.

11.6 What mass (in grams) of each solute should be added in the following solutions: (a) to 200 g water to form a 3.0% by mass KBr solution; (b) to 25.0 g water to form a 6.0% by mass $AgNO_3$ solution; (c) to 500 g water to form a 10.5% by mass C_2H_5OH solution?

11.7 Calculate the mass percentage compositions and the mole fractions of each component in the following solutions: (a) 25.0 g of water and 50.0 g of ethanol, C_2H_5OH; (b) 25.0 g of water and 50.0 g of methanol, CH_3OH; (c) a glucose solution that is 0.10 *m* $C_6H_{12}O_6$(*aq*).

11.8 Calculate the mass percentage compositions and the mole fractions of each component in the following solutions: (a) 25.0 g of benzene, C_6H_6, and 50.0 g of toluene, $C_6H_5CH_3$; (b) 25.0 g of benzene, 10.0 g of carbon tetrachloride, CCl_4, and 50.0 g of naphthalene, $C_{10}H_8$; (c) a sucrose solution that is 0.020 *m* $C_{12}H_{22}O_{11}$(*aq*).

11.9 In a laboratory exercise, a student mixed 25.0 g of ethanol, C_2H_5OH, with 150 g of water. (a) What is the mass percentage of ethanol in the solution? (b) What is the mole fraction of ethanol in the solution? (c) What is the molality of ethanol in the solution?

11.10 In the laboratories of an oil company, a research chemist dissolved 1.0 g of octane, C_8H_{18}, in 50.0 g of benzene, C_6H_6. (a) What is the mass percentage of octane in the solution? (b) What is the mole fraction of octane in the solution? (c) What is the molality of octane in the solution?

11.11 Calculate the molality of each of the following solutions: (a) 10.0 g of NaCl dissolved in 250 g water; (b) 0.48 mol of KOH dissolved in 50.0 g water; (c) 1.94 g of urea, $CO(NH_2)_2$, dissolved in 200 g water.

11.12 (a) What mass (in grams) of NaOH must be mixed with 250 g of water to prepare a 0.22 *m* NaOH solution? (b) Calculate the amount (in mol) of ethylene glycol, HOC_2H_4OH, that should be added to 2.0 kg of water to prepare a 0.44 *m* HOC_2H_4OH(*aq*) solution. (c) Determine the amount (in mol) of HCl that must be dissolved in 500 g water to prepare a 0.0010 *m* HCl(*aq*) solution.

11.13 Calculate the mass of solute required: (a) $ZnCl_2$ to dissolve in 150 g of water to prepare a 0.200 *m* $ZnCl_2$(*aq*) solution; (b) $KClO_3$ to dissolve in 20.0 g of water to prepare a 3.0% by mass $KClO_3$(*aq*) solution; (c) $KClO_3$ to dissolve in 20.0 g of water to prepare a 3.0 *m* $KClO_3$(*aq*) solution.

11.14 Calculate the mass of solute required: (a) $CuSO_4$ to dissolve in 2.0 kg of water to prepare a 0.066 *m* $CuSO_4$(*aq*) solution; (b) $Na_2Cr_2O_7$ to dissolve in 200 g of water to prepare a 1.5% by mass $Na_2Cr_2O_7$(*aq*) solution; (c) $Na_2Cr_2O_7$ to dissolve in 200 g water to prepare a 1.5 *m* $Na_2Cr_2O_7$(*aq*) solution.

11.15 What is the molality of (a) 5.0% NaCl(*aq*); (b) 10.0% HCl(*aq*); (c) 50% C_2H_5OH(*aq*)? The compositions are mass percentages.

11.16 A student helper was asked by the stockroom director to label the concentration of each aqueous solution as a mass percentage. What labels should the student have prepared for (a) 0.33 *m* $CO(NH_2)_2$(*aq*); (b) $x(C_2H_5OH) = 0.28$; (c) 1.04 *m* K_2SO_4(*aq*)?

11.17 Calculate the mole fractions of the cations, anions, and water in (a) 0.10 *m* NaCl(*aq*); (b) 0.20 *m* Na_2CO_3(*aq*); (c) 10.0% KNO_3(*aq*).

11.18 What are the mole fractions of the cations and anions in (a) 0.10 *m* $MgSO_4$(*aq*); (b) 0.55% $MgBr_2$(*aq*); (c) 0.72 *m* $Al_2(SO_4)_3$(*aq*)?

11.19 (a) Calculate the mass of $CaCl_2{\cdot}6H_2O$ needed to prepare a 0.10 *m* $CaCl_2$(*aq*) solution, using 250 g water. (b) What mass of $NiSO_4{\cdot}6H_2O$ must be dissolved in 500 g water to produce a 0.22 *m* $NiSO_4$(*aq*) solution?

11.20 (a) Determine the mass of $Na_2CO_3{\cdot}10H_2O$ needed to prepare a 1.0 *m* Na_2CO_3(*aq*) solution, using 250 g water. (b) Find a general expression for the mass of solute needed to prepare a solution of the solute B of molality *m* when the hydrated solute has the formula $B{\cdot}xH_2O$.

11.21 The density of a 0.35 M $(NH_4)_2SO_4$ aqueous solution is 1.027 g/mL. Determine (a) the molality and (b) the mass percentage of ammonium sulfate in the solution.

11.22 A 5.00% K_3PO_4 aqueous solution has a density of 1.043 g/mL. Determine (a) the molality and (b) the molar concentration of potassium phosphate in solution.

Solubility and Saturation

11.23 The following solubilities are expressed as mass of solute per mass of solvent. Convert the solubility of each salt to molar concentration and molality of solute: (a) KCl: solubility = 28.1 g KCl per 100 g H_2O, density = 1.15 g/mL; (b) NaCl: solubility = 35.7 g NaCl per 100 g H_2O, density = 1.20 g/mL.

11.24 The following solubilities are expressed as mass of solute per mass of solvent. Convert the solubility of each salt to molar concentration, molality, and mass percentage of solute: (a) $BaCl_2$: solubility = 30.5 g $BaCl_2$ per 100 g H_2O, density = 1.24 g/mL; (b) MgF_2: solubility = 8 mg per 100 g H_2O, density = 1.00 g/mL.

Gas Solubility

11.25 State the solubility (in moles per liter) of (a) O_2 at 50 kPa; (b) CO_2 at 500 Torr; (c) CO_2 at 0.1 atm. The temperature in each case is 20°C and the pressures are partial pressures of the gas. Use the information in Table 11.3.

11.26 Calculate the solubility (in milligrams per liter) of (a) air at 1 atm; (b) He at 1 atm; (c) He at 25 kPa. The temperature is 20°C in each case and the pressures are partial pressures of the gas. Use the information in Table 11.3.

11.27 The minimum concentration of oxygen required for fish life is 4 ppm by mass. (a) Assuming the density of the solution to be 1.00 g/mL, express this concentration in moles of O_2 per liter. (b) What is the minimum partial pressure of O_2 that would supply this concentration in water at 20°C?

11.28 A soft drink is made by dissolving CO_2 at 3 atm in a flavored solution and sealed in an aluminum can. What volume of CO_2 is released when a 355-mL can is opened to 1 atm of pressure? The temperature is 20°C throughout.

11.29 The solubility equilibrium for CO_2 is $CO_2(g) \rightleftharpoons CO_2(aq)$, and the dissolution is exothermic. Use Le Chatelier's principle to predict what happens to the solubility of CO_2 when (a) the pressure of CO_2 is increased and (b) the temperature is raised.

11.30 The solubility of oxygen in rivers and streams depends on various factors, such as temperature, salt concentration, depth, and biological activity. Its solubility equilibrium is $O_2(g) \rightleftharpoons O_2(aq)$. Use Le Chatelier's principle to compare the dissolved oxygen concentration at 1 atm total pressure (0.2 atm partial pressure) with its concentration when the partial pressure of oxygen is increased to 0.5 atm.

Enthalpy of Solution

11.31 The equilibrium for a saturated solution of Li_2SO_4 is $Li_2SO_4(s) \rightleftharpoons 2Li^+(aq) + SO_4^{2-}(aq)$, and the dissolution is exothermic. (a) Which is larger for lithium sulfate, the lattice enthalpy or the enthalpy of hydration? (b) Use Le Chatelier's principle to predict what happens to the solubility of Li_2SO_4 when the temperature of the system is raised.

11.32 The enthalpy of solution of ammonium nitrate is positive. (a) Write an equilibrium to represent the dissolving process. (b) Use Le Chatelier's principle to predict the effect of temperature on the solubility of ammonium nitrate. (c) Which is larger for NH_4NO_3, its lattice enthalpy or the enthalpy of hydration?

11.33 Calculate the heat evolved or absorbed when 10.0 g of (a) NaCl; (b) NaBr; (c) $AlCl_3$; (d) NH_4NO_3 is dissolved in water. Assume that the enthalpies of solution in Table 11.4 are applicable.

11.34 Determine the temperature change when 10.0 g of (a) KCl; (b) $MgBr_2$; (c) KNO_3; (d) NaOH is dissolved in 100.0 g of water. Assume that the specific heat capacity of the solution is 4.18 J/(g·°C) and that the enthalpies of solution in Table 11.4 are applicable.

11.35 From the information in Table 11.4, decide whether the solubility of (a) AgCl; (b) Li_2CO_3; (c) AgF increases or decreases when the temperature is raised. For each salt determine which value is larger, ΔH_L or ΔH_H.

11.36 State how the solubilities of (a) $Ca(OH)_2$; (b) $Al_2(SO_4)_3$; (c) KI are affected by a decrease in temperature. Base your answer on the data in Table 11.4. For each salt, determine which value is larger, ΔH_L or ΔH_H.

11.37 Estimate the enthalpy of solution of (a) LiCl; (b) NaCl; (c) AgCl from ΔH_L and ΔH_H in Tables 8.1 and 11.5. Identify and explain any trend that may exist.

11.38 Estimate the enthalpy of solution of (a) NaF; (b) NaBr; (c) NaI from ΔH_L and ΔH_H in Tables 8.1 and 11.5. Identify and explain the trend in the enthalpy of solution on proceeding down the group of halides. Compare this trend with values in Table 11.4.

Lowering of Vapor Pressure

11.39 Calculate the vapor pressure of the solvent in each solution; use the data in Table 5.4 for the vapor pressures of water at various temperatures: (a) an aqueous solution at 100°C in which the mole fraction of sucrose is 0.10; (b) an aqueous solution at 100°C in which the molality of sucrose is 0.10 *m*.

11.40 What is the vapor pressure of the solvent in each of the solutions listed below? Use the data in Table 5.4 for the vapor pressures of water at various temperatures. (a) The mole fraction of glucose is 0.050 in an aqueous solution at 80°C. (b) An aqueous solution at 25°C that is 0.10 *m* urea, $CO(NH_2)_2$, a nonelectrolyte.

11.41 (a) Calculate the vapor pressure of a 1.0% ethylene glycol, $C_2H_4(OH)_2$, solution at 0°C. (b) What is the vapor pressure of 0.10 *m* NaOH solution at 80°C? (c) Determine the *change* in the vapor pressure when 6.6 g of

$CO(NH_2)_2$ is dissolved in 100 g water at 10°C. For data, see Table 5.4.

11.42 (a) Calculate the *change* in vapor pressure for an aqueous solution at 40°C in which the mole fraction of fructose is 0.22. (b) Determine the vapor pressure of a saturated magnesium fluoride solution at 20°C. The solubility of MgF_2 at 20°C is 8 mg/100 g water. (c) What is the vapor pressure of a 0.010 *m* $Fe(NO_3)_3$ solution at 0°C? For data, see Table 5.4.

11.43 When 8.05 g of an unknown compound X was dissolved in 100 g of benzene, C_6H_6, the vapor pressure of the benzene decreased from 100 to 94.8 Torr at 26°C. What is (a) the mole fraction and (b) the molar mass of X?

11.44 The normal boiling point of ethanol, C_2H_5OH, is 78.4°C. When 9.15 g of a soluble nonelectrolyte was dissolved in 100 g ethanol, the vapor pressure of the solution was 740 Torr. (a) What are the mole fractions of ethanol and solute? (b) What is the molar mass of the solute?

Boiling-Point Elevation

11.45 Estimate the boiling-point elevation and the normal boiling points of (a) 0.10 *m* $C_{12}H_{22}O_{11}(aq)$; (b) 0.22 *m* $NaCl(aq)$; (c) a saturated solution of LiF, of solubility 230 mg/100 g water at 100°C.

11.46 Estimate the boiling-point elevation and the normal boiling points of (a) 0.22 *m* $CaCl_2(aq)$; (b) a saturated solution of Li_2CO_3, of solubility 0.72 g/100 g water at 100°C; (c) 1.7% $CO(NH_2)_2(aq)$.

11.47 (a) What is the normal boiling point of an aqueous solution that has a vapor pressure of 751 Torr? (b) Determine the normal boiling point of a benzene solution that has a vapor pressure of 740 Torr. The normal boiling point of benzene is 80.1°C.

11.48 (a) What is the normal boiling point of an aqueous solution that has a freezing point of −1.04°C? (b) The freezing point of a benzene solution is 2.0°C. The normal freezing point of benzene is 5.5°C. What is the expected normal boiling point of the solution? The normal boiling point of benzene is 80.1°C.

11.49 A 1.05-g sample of a molecular compound is dissolved in 100 g of carbon tetrachloride, CCl_4. The normal boiling point of the solution is 61.51°C; the normal boiling point of CCl_4 is 61.20°C. What is the molar mass of the compound?

11.50 When 2.25 g of an unknown compound is dissolved in 150 g cyclohexane, the boiling point increased by 0.481°C. Determine the molar mass of the compound.

Freezing-Point Depression

11.51 Estimate the freezing-point depression and the freezing point of each of the following aqueous solutions: (a) 0.10 *m* $C_{12}H_{22}O_{11}(aq)$; (b) 0.22 *m* $NaCl(aq)$; (c) a saturated solution of LiF, of solubility 120 mg/100 g water at 0°C.

11.52 Estimate the freezing point of the following aqueous solutions: (a) 0.10 *m* $C_6H_{12}O_6(aq)$; (b) 0.22 *m* $CaCl_2(aq)$; (c) a saturated solution of Li_2CO_3, of solubility 1.54 mg/100 g water at 0°C.

11.53 (a) What is the freezing point of a benzene solution that boils at 82.0°C, rather than 80.1°C, the normal boiling point of pure benzene? The freezing point of benzene is 5.5°C. (b) An aqueous solution freezes at −3.04°C. What is the molality of the solution? (c) The freezing point of an aqueous solution containing a nonelectrolyte dissolved in 200 g water is −1.94°C. How many moles of the solute are present?

11.54 (a) A benzene solution has a vapor pressure of 740 Torr at 80.1°C, the normal boiling point of pure benzene. What is the expected freezing point of the solution? The freezing point of benzene is 5.5°C. (b) How many moles of $CO(NH_2)_2$ are present in 1200 g water, given that the freezing point of the solution is −4.02°C? (c) A 1.0-g sample of a protein (of molar mass 100,000 g/mol) is dissolved in 1.0 kg of water. What is the expected freezing point of the solution?

11.55 A 1.14-g sample of a molecular substance dissolved in 100 g camphor (freezing point 179.8°C) freezes at 177.3°C. What is the molar mass of the substance?

11.56 When 2.11 g of a nonpolar solute is dissolved in 50.0 g phenol, the latter's freezing point was lowered by 1.753°C. Calculate the molar mass of the solute.

11.57 A 1.00% $NaCl(aq)$ solution has a freezing point of −0.593°C. (a) Estimate the van't Hoff *i* factor from the data. (b) Determine the actual ionic molality of the NaCl. (c) Calculate the percentage dissociation of NaCl in this solution. (*Hint:* Consider the effective molality of the solute; see Fig. 11.25.)

11.58 A 1.00% $MgSO_4(aq)$ solution has a freezing point of −0.192°C. (a) Estimate the van't Hoff *i* factor for the solution. (b) Determine the actual ionic molality of the $MgSO_4$. (c) Calculate the percentage dissociation of the $MgSO_4$ in this solution. (*Hint:* Consider the effective molality of the solute; see Fig. 11.25.)

Osmosis and Osmometry

11.59 What is the osmotic pressure of (a) 0.010 M $C_{12}H_{22}O_{11}(aq)$; (b) 1.0 M $HCl(aq)$; (c) 0.010 M $CaCl_2(aq)$ at 20°C?

11.60 Which of the following solutions has the highest osmotic pressure at 50°C? Justify your answer by calculating the osmotic pressure of each solution. (a) 0.10 M $KCl(aq)$; (b) 0.60 M $CO(NH_2)_2(aq)$; (c) 0.30 M $K_2SO_4(aq)$.

11.61 What is the osmotic pressure of the following aqueous solutions? Assume the density of the solutions is the same as the density of water at the indicated temperature: (a) a saturated solution of Li_2CO_3 of solubility of

1.54 g/100 g H_2O at 0°C; (b) a solution that has a vapor pressure of 751 Torr at 100°C; (c) a solution that has a boiling point of 101°C.

11.62 (a) A benzene solution has vapor pressure of 740 Torr at 80.1°C, the normal boiling point of benzene. What is the osmotic pressure of this solution at 20°C? The density of benzene is 0.88 g/mL. (b) The freezing point of a benzene solution was 5.4°C; the normal freezing point of benzene is 5.5°C. Determine the osmotic pressure of the solution at 10°C. (c) Determine the height to which osmosis will force the solution in (b) to rise in an arrangement like that shown in Fig. 11.29. The density of mercury is 13.6 g/mL.

11.63 Calculate the osmotic pressure of each solution listed below and then determine the height to which osmosis will force the solution to rise in an arrangement like that shown in Fig. 11.29. The density of mercury is 13.6 g/mL. Assume the density of the solution to be equal to the density of water (0.998 g/mL) at 20°C: (a) 0.050 M $C_{12}H_{22}O_{11}(aq)$; (b) 0.0010 M $NaCl(aq)$; (c) a saturated solution of AgCN of solubility 2.3×10^{-5} g/100 g water.

11.64 To what height will the solution column rise in the apparatus shown in Fig. 11.29 at 20°C for the solutions specified below? (See Exercise 11.63 for additional data.) (a) 3.0×10^{-3} M $C_6H_{12}O_6(aq)$; (b) 2.0×10^{-3} M $CaCl_2(aq)$; (c) 0.010 M $K_2SO_4(aq)$.

11.65 A 0.10-g sample of a polymer, dissolved in 100 mL of toluene, has an osmotic pressure of 5.4 Torr at 20°C. What is the molar mass of the polymer?

11.66 A solution of toluene (of density 0.867 g/mL) containing 0.50 g of a polymer in 200 mL toluene showed a 9.14-mm rise of the solution on an osmometer at 20°C. What is the molar mass of the polymer? The density of mercury is 13.6 g/mL.

Mixtures of Liquids

11.67 At 25°C, the vapor pressure of benzene is 95 Torr and the vapor pressure of toluene is 28 Torr. Assuming the solution to be ideal, what is the vapor pressure of a solution that is 35% C_6H_6 and 65% $C_6H_5CH_3$ by mass?

11.68 A mixture of 20 g of water and 80 g ethanol is prepared. Assuming the solution to be ideal, what is the vapor pressure of the solution at 19°C? At 19°C, the vapor pressure of ethanol is 40 Torr; that of water is 16.6 Torr.

UNCLASSIFIED EXERCISES

11.69 A bottle of spirits is labeled "80 proof." (a) What is the percentage by volume of ethanol in the solution? (b) The density of ethanol is 0.79 g/mL. What is the molar concentration of a solution labeled "80 proof"?

11.70 The solubility of iron(III) ammonium sulfate dodecahydrate, $FeNH_4(SO_4)_2 \cdot 12H_2O$, is 124 g/100 g water. What is the mass percentage of the resulting solution? (Remember to consider the mass of the water in the hydrate.)

11.71 A saturated magnesium chloride solution is 34.6% $MgCl_2(aq)$ and has a density of 1.27 g/mL. Calculate (a) the molality of $MgCl_2$; (b) the mole fraction of water; and (c) the molar concentration of $MgCl_2$ in the solution.

11.72 A 10.0% $H_2SO_4(aq)$ solution has a density of 1.07 g/mL. (a) How many milliliters of solution contain 6.32 g of H_2SO_4? (b) What is the molality of H_2SO_4 in solution? (c) What mass (in grams) of H_2SO_4 is in 300 mL of solution?

11.73 The density of a 16.0% $C_{12}H_{22}O_{11}(aq)$ solution is 1.0635 g/mL at 20°C. (a) What is the molar concentration of the solution? (b) What is the vapor pressure of the solution at 20°C? (c) Predict the boiling point of the solution.

11.74 Nitric acid is purchased from chemical suppliers as a solution that is 70% HNO_3 by mass. What mass (in grams) of a 70% $HNO_3(aq)$ solution is necessary to prepare 250 mL of a 2.0 *m* $HNO_3(aq)$ solution? The density of 70% $HNO_3(aq)$ is 1.42 g/mL.

11.75 The carbon dioxide dissolved in a certain sample of water has reached equilibrium with its partial pressure above the solution. Air also occupies the space above the liquid. Explain what happens to the solubility of the $CO_2(aq)$ if (a) the partial pressure of the CO_2 is increased by the addition of more CO_2; (b) the total pressure of the gas above the liquid is increased by the addition of nitrogen.

11.76 (a) Calculate the enthalpy of hydration of Br^-, using the appropriate data from Tables 8.1, 11.4, and 11.6. (b) Use the value obtained in (a) to determine the enthalpy of hydration of Rb^+, given that the enthalpy of solution of RbBr is +22 kJ/mol and its lattice enthalpy is −651 kJ/mol.

11.77 Estimate the enthalpy of solution of $SrCl_2$ from the data for the enthalpy of hydration and the lattice enthalpy in Tables 8.1 and 11.5.

11.78 Water temperatures in the Gulf of Mexico can be as high as 30°C along the coast during the summer. Estimate the vapor pressure at that temperature, at the boiling point, and at the freezing point of seawater, assuming that a 0.50 *m* $NaCl(aq)$ solution simulates seawater.

CASE 11 ▸ A BOTTLE OF CHEMISTRY

Soda water was invented by the English chemist and priest Joseph Priestley, the discoverer of oxygen and several other gases. He was politically active, and his revolutionary views led to him fleeing from England and settling in Pennsylvania, where he died in 1804. Soda water is a solution of carbon dioxide in water. The solution is sealed in a can or bottle at a CO_2 pressure of 2 to 3 atm, and equilibrium between the $CO_2(g)$ and the dissolved $CO_2(aq)$ is established:

$$CO_2(g) \rightleftharpoons CO_2(aq)$$

A series of events begin when the seal is broken and the pressure in the bottle falls immediately to 1 atm: the rapid expansion of the pressurized CO_2 gas at the mouth of the bottle causes a rapid cooling of the gaseous mixture by the Joule-Thomson effect (Section 5.8), and the temperature can go as low as −30°C for an instant. The resulting condensation of the water vapor in the air produces the fine, white mist that appears at the mouth of the bottle.

In addition, the release of carbon dioxide shifts the equilibrium in favor of the undissolved gas; immediately, some of the dissolved carbon dioxide begins to escape from the solution as bubbles. When the pressure at the surface is reduced, the carbon dioxide molecules form tiny microbubbles from which visible bubbles shortly appear. Microbubbles form at *nucleation sites,* small regions where a large number of CO_2 molecules accumulate. As the microbubble escapes from the nucleation site and rises, other CO_2 molecules join the molecules already in the bubble and the bubble increases in size. Nucleation sites may be at a scratch on the surface of the glass or on minute floating particles.

The escape of carbon dioxide from the drink to form a frothy head can be accelerated by several methods. If a small amount of salt is added, then the grains provide an enormous number of pits and cracks for nucleation sites and cause a rapid evolution of gas. In addition, a *salting-out effect* contributes to the rapid formation of a foamy head. Ions of the salt are more strongly hydrated than the CO_2 molecules in solution are; therefore, when salt is added, fewer H_2O molecules are free to hydrate the CO_2 molecules, and the carbon dioxide bubbles out of solution. The agitation of the liquid by stirring or pouring the drink rapidly also accelerates the escape of CO_2 molecules. The agitation increases the probability of more microbubbles being formed on nucleation sites and, consequently, of more CO_2 molecules collecting on the microbubbles to form many large, visible bubbles.

11.79 Organic chemists sometimes use freezing-point and boiling-point measurements to determine the molar mass of compounds they have synthesized. If 0.30 g of a nonvolatile solute is dissolved in 30.0 g CCl_4, the boiling point increased from 76.80°C (the normal boiling point of CCl_4) to 77.19°C. What is the molar mass of the compound?

11.80 A 10.0-g sample of *p*-dichlorobenzene, a component of mothballs, dissolves in 80.0 g benzene, C_6H_6. The freezing point of the solution is 1.20°C. The freezing point of pure benzene is 5.48°C. (a) What is an approximate molar mass of *p*-dichlorobenzene? (b) An elemental analysis of *p*-dichlorobenzene indicated that the empirical formula is C_3H_2Cl. What is the molecular formula for *p*-dichlorobenzene?

11.81 An elemental analysis of adrenaline is 59.0% carbon, 26.2% oxygen, 7.10% hydrogen, and 7.65% nitrogen. When 0.64 g of adrenaline is dissolved in 36.0 g benzene, the freezing point decreased by 0.50°C. (a) Determine the empirical formula of adrenaline. (b) What is the molar mass of adrenaline? (c) Deduce the molecular formula of adrenaline.

11.82 The freezing point of a 5.00% $CH_3COOH(aq)$ solution is −1.576°C. Determine the experimental van't Hoff *i* factor for this solution. On the basis of your understanding of intermolecular forces, account for its value.

11.83 A 0.020 M $C_6H_{12}O_6(aq)$ solution of glucose is separated from a 0.050 M $CO(NH_2)_2(aq)$ solution of urea by a semipermeable membrane. (a) Which solution has the higher osmotic pressure? (b) Which solution becomes more dilute with the passage of H_2O molecules through the membrane? (c) To which solution should an external pressure be applied in order to maintain an equilibrium flow of H_2O molecules across the membrane? (d) What external pressure (in atm) should be applied in (c)?

QUESTIONS

1. The smallest bubble visible to the naked eye has a diameter of about 0.1 mm. Assume a spherical bubble (of volume $\frac{4}{3}\pi r^3$) in which the temperature of the soda water is 4.5°C and the CO_2 gas pressure is 1.1 atm. (a) Calculate the number of CO_2 molecules in the bubble. (b) Determine the mass of CO_2 in the bubble.

2. Explain why the concentration of dissolved oxygen is lower in seawater than in fresh water (at the same temperature).

3. The external pressure of a bubble at a distance h below the surface in the bulk of the soda water is greater than that at the surface, P(atmosphere), because of the additional pressure caused by the column of liquid, P(liquid):

$$P = P(\text{atmosphere}) + P(\text{liquid}) \qquad P(\text{liquid}) = gdh$$

where d is the density of the liquid and g is the acceleration of free fall (9.81 m/s^2). If a CO_2 gas bubble forms 5 cm below the surface and rises, by what fraction will the volume of the bubble change when it reaches the surface?

4. Determine the mass of dissolved CO_2 in 355 mL of soda water (the volume of a typical can), assuming the temperature of the liquid to be 20°C and the pressure of $CO_2(g)$ above the surface to be 2.5 atm.

5. What is the mass of dissolved CO_2 in the liquid after the seal has been broken and the $CO_2(g) \rightleftharpoons CO_2(aq)$ equilibrium is reestablished with dry air at 1.0 atm? The percentage of CO_2 in the atmosphere is 0.034% by volume.

11.84 A 0.40-g sample of a polypeptide dissolved in 1.0-L of an aqueous solution at 27°C has an osmotic pressure of 3.74 Torr. What is the molar mass of the polypeptide?

11.85 Intravenous solutions are 5% glucose, $C_6H_{12}O_6$, aqueous solutions. What is the osmotic pressure of such solutions at 37°C (body temperature)?

11.86 When 0.1 g of insulin is dissolved in 200 mL of water, the osmotic pressure is 2.30 Torr at 20°C. What is the molar mass of insulin?

11.87 Catalase, a liver enzyme, dissolves in water. A 10-mL solution containing 0.166 g of catalase exhibits an osmotic pressure of 1.2 Torr at 20°C. What is the molar mass of catalase?

11.88 An aqueous solution of 0.010 g of a protein in 10 mL of water at 20°C shows a 5.22-cm rise in the apparatus shown in Fig. 11.29. Assume the density of the solution to be 0.998 g/mL and the density of mercury to be 13.6 g/cm^3. (a) What is the molar mass of the protein? (b) What is the freezing point of the solution?

11.89 Acetone, CH_3COCH_3, is a solvent both for nonpolar compounds and, to a more limited extent, for polar compounds. The vapor pressures of acetone and water at 7.7°C are 100 Torr and 8.0 Torr, respectively. (a) What is the vapor pressure of an aqueous solution that is 10.0% acetone at 7.7°C? (b) What is the vapor pressure of an aqueous solution that is 10.0% water at 7.7°C?

11.90 Determine the freezing point of a 0.10 m aqueous solution of a weak electrolyte that is 7.5% dissociated into two ions.

11.91 A 0.124 m trichloroacetic acid, $CCl_3COOH(aq)$, solution has a freezing point of −0.423°C. What is the percentage ionization of the acid?

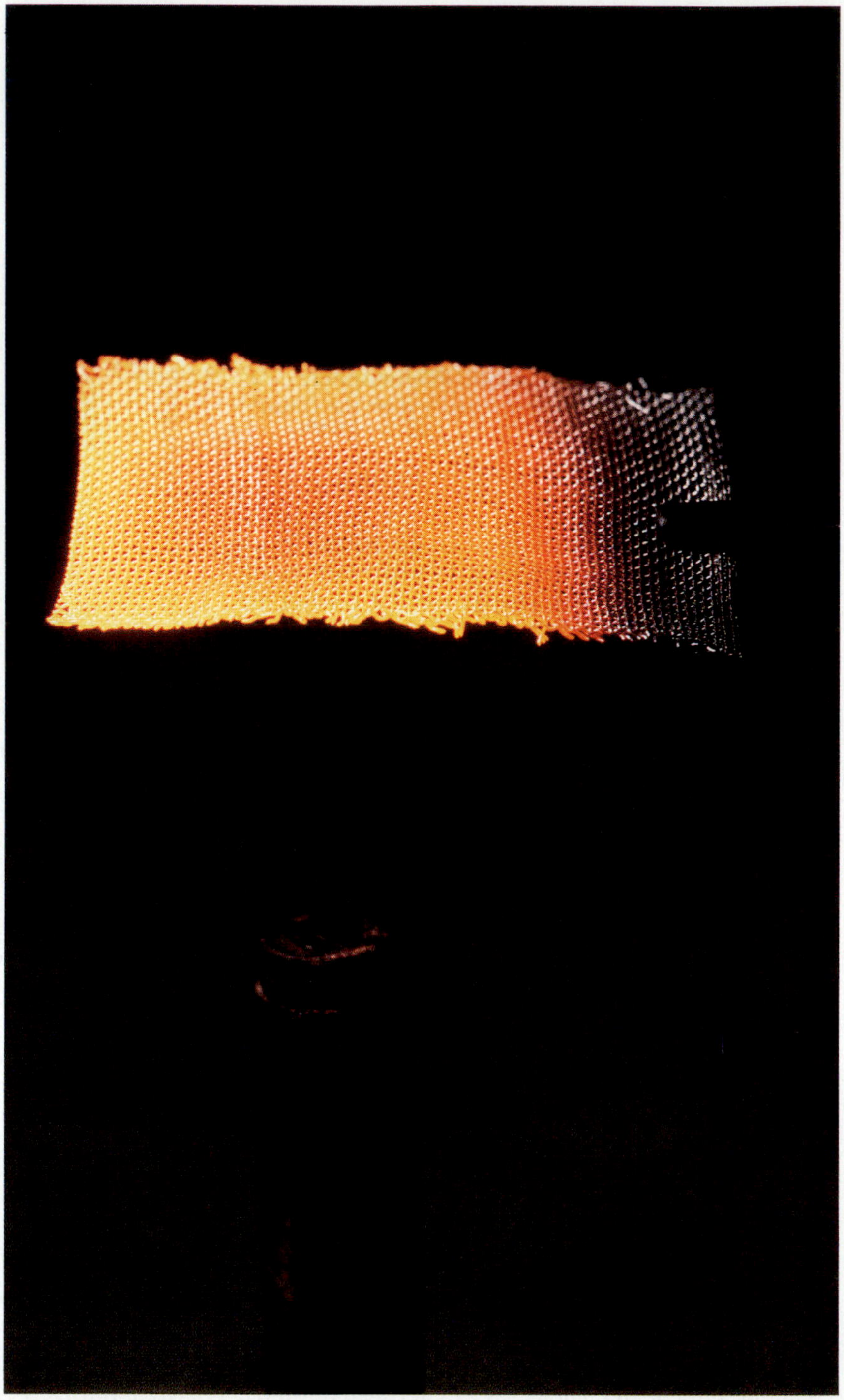

One important feature of a chemical reaction is the rate at which it takes place. In this chapter we see how to define reaction rate and express it quantitatively. In doing so, we also see how to classify reactions according to their rates and how to use experimental information to predict the composition of a reaction mixture at any time after the reaction begins. We also see why most reactions go faster when they are heated or when catalysts are added to the mixture. The illustration shows platinum acting as a catalyst for the combustion of hydrogen: when the platinum is held in a stream of hydrogen gas, it glows with a brilliant incandescence as a result of the heat released in the reaction.

12 THE RATES OF REACTIONS

The study of the rates of chemical reactions and the steps by which they occur is called **chemical kinetics**. The subject is important in the chemical industry because the designer of a chemical plant must know how the rates of the reactions taking place in it are affected by changes of temperature, pressure, and concentration. It is important in biology and medicine, because health represents a balance between a large number of simultaneous, interdependent reactions. Illness is often a sign that the rates of biologically important reactions have changed too much. Chemical kinetics is also important in the study of reactions in the atmosphere. Among other things, it deals with the reactions that generate and destroy the ozone layer and with the reactions that lead to the formation and removal of pollutants.

Chemical kinetics also provides a link between the topics described in this part and those treated in Part II. We saw in Part II that some of the physical properties of solutions, such as vapor pressure and osmosis, can be explained in terms of dynamic equilibria in which the rates of forward and reverse *physical* processes are balanced. In this chapter we shall see how to describe the rates of forward and reverse *chemical* processes. Once we have done that, we shall see how the discussion of dynamic equilibria is extended to include the enormously important subject of chemical equilibrium, the main subject of Part III.

E Chemical kinetics: Reaction rates. *J. Chem. Educ.*, **1980**, *57*, 659. On chemical kinetics. *J. Chem. Educ.*, **1969**, *46*, 308. Some common oversimplifications in teaching chemical kinetics. *J. Chem. Educ.*, **1978**, *55*, 84. The definition of the rate of a chemical reaction. *J. Chem. Educ.*, **1973**, *50*, 200.

THE DESCRIPTION OF REACTION RATES

What is a reaction rate and how is it measured? In this section we shall learn the answers to these questions and also see that reaction rates show patterns of behavior that enable us to classify reactions into several types.

D The starch-iodine clock reaction. CD1, 63. Burning of magnesium (as example of fast reaction). BZS1, 1.15.

E Individual molecular processes are very fast. The rate of change refers to the *overall* outcome of the changes taking place; for example, although a reaction event may be very fast when it occurs, the reactants might meet only very rarely, and hence the perceived rate will be slow.

N Reaction rates may be determined by monitoring, among other properties, the change in color, the change in amount of precipitation, change in pH, change in the amount of gas evolved, and the change in pressure of a system.

12.1 REACTION RATES

Rates in chemistry are defined like rates in other fields: a **rate** is the change in a property per unit time (for instance, per second). An everyday example is *speed,* the rate of change of position, which is distance traveled divided by the travel time. Likewise, the rate of a chemical reaction is the change in concentration of a substance divided by the time it takes for that change to occur. The concentration of an acid changes very rapidly when a base is added, so the rate of the neutralization reaction is very high. In a fermentation, the concentration of alcohol rises slowly, so the rate of the reaction is low.

The definition of rate. Although we shall soon see that we must refine the definition, initially we can define **reaction rate** as

$$\text{Rate} = \frac{\text{change in concentration}}{\text{time for the change to occur}}$$

If we let $\Delta[X]$ represent the change in molar concentration of a substance X, where [X] denotes the molar concentration (in moles per liter) of X, and Δt is the time it takes for that change to occur, then we have

$$\text{Rate} = \frac{\Delta[X]}{\Delta t}$$

Suppose, for example, that we are studying the reaction

$$2HI(g) \longrightarrow H_2(g) + I_2(g)$$

and find that during an interval of 100 s the concentration of HI decreases by 0.50 mol/L. Then the reaction rate would be

$$\text{Rate of decomposition of HI} = \frac{0.50 \text{ mol HI/L}}{100 \text{ s}}$$
$$= 5.0 \times 10^{-3} \text{ mol HI/(L·s)}$$

The units of reaction rate are mol/(L·s), which is read "moles per liter-second." More convenient values are sometimes obtained if rates are expressed in mmol/(L·s), mol/(L·min), or mol/(L·h). In this chapter we shall sometimes find it convenient to use exponential notation (Section 1.3) and to write the units of a rate as mol L^{-1} s^{-1} instead of mol/(L·s).

It is important to specify the substance for which the reaction rate is defined. In the example above, the rate of decomposition of HI is not the same as the rate of formation of H_2 or I_2: only one H_2 molecule is formed for every two HI molecules that react, so in a given time the concentration of H_2 changes only half as much as that of HI; consequently, the rate of formation of H_2 is half the rate of reaction of HI. However, it is quite easy to use the reaction stoichiometry to convert a rate expressed in terms of one substance to a rate expressed in terms of another, as the following example shows.

FIGURE 12.1 The gradual appearance of color signals the formation of NO_2 (a brown gas) as N_2O_5 (a colorless vapor) decomposes at 65°C. The molar concentration of N_2O_5 can be calculated from the initial concentration, the measured concentration of NO_2, and the stoichiometric relation between them.

▼ EXAMPLE 12.1 Expressing the reaction rate for different substances

If the rate of decomposition of HI is 5.0×10^{-3} mol/(L·s), what is the rate of formation of hydrogen in the same reaction?

STRATEGY We always begin by writing the chemical equation for the reaction, for then it is quite easy to see the relation between the rates at which the various substances change. In this case, the reaction is

$$2HI(g) \longrightarrow H_2(g) + I_2(g)$$

and we see that we should expect a lower rate of change of H_2 concentration, because only one molecule of H_2 is formed for every two HI molecules that decompose. Because we know the change in molar concentration of the HI, we can calculate the change in concentration of H_2 by using the stoichiometric relation 2 mol HI = 1 mol H_2.

SOLUTION We carry out the following conversion:

$$\underset{\substack{\uparrow \\ \text{Information} \\ \text{required}}}{\text{Rate}\left(\frac{\text{mol H}_2}{\text{L·s}}\right)} = \underset{\substack{\uparrow \\ \text{Information} \\ \text{given}}}{5.0 \times 10^{-3}\,\frac{\text{mol HI}}{\text{L·s}}} \times \underset{\substack{\uparrow \\ \text{Conversion factor} \\ \text{from mol HI to} \\ \text{mol H}_2}}{\frac{1\text{ mol H}_2}{2\text{ mol HI}}}$$

$$= 2.5 \times 10^{-3}\,\frac{\text{mol H}_2}{\text{L·s}}$$

The H_2 formation rate, 2.5×10^{-3} mol H_2/(L·s), is half the rate of decomposition of HI.

EXERCISE E12.1 The rate of formation of ammonia from nitrogen and hydrogen in the reaction $N_2(g) + 3H_2(g) \rightarrow 2NH_3(g)$ is reported as 1.15 mol NH_3 L^{-1} h^{-1}. What is the rate at which hydrogen is used?

[*Answer*: 1.72 mol H_2 L^{-1} h^{-1}]

N The determination of reaction rate should be rapid, accurate, and reproducible; moreover, it should not interfere with the progress of the reaction.

U Exercises 12.57–12.59.

N When discussing the illustration below, it may be helpful to develop successive approximations to the tangent, by drawing a series of straight lines centered on 1000 s, and calculating their slopes. Then the true instantaneous rate is obtained in the limit when the two points converge on 1000 s and the slope is that of the tangent to the curve.

The instantaneous reaction rate. The rates of most reactions change as the reactants are consumed. It follows that we cannot speak of *the* rate of a reaction any more than we can speak of *the* speed of an automobile over an entire journey. To see how to take this changing rate into account, we consider the gas-phase decomposition of dinitrogen pentoxide, N_2O_5, vapor (Fig. 12.1):

$$2N_2O_5(g) \longrightarrow 4NO_2(g) + O_2(g)$$

The graph in Fig. 12.2 is a plot of the concentration of N_2O_5 against time for a sample kept at 65°C. At any instant,

$$\text{Rate of decomposition of N}_2\text{O}_5 = \frac{\text{decrease in concentration of N}_2\text{O}_5}{\text{time for the change to occur}}$$

Suppose we wanted to find the rate at 1000 s after the start of the reaction. By monitoring the pressure of the gas in the reaction vessel, we can measure the concentration of N_2O_5 at each instant. But how do we measure the *change* in the concentration at a specified instant? As explained in Appendix 1D, we draw the *tangent* to the concentration curve at the time of interest and calculate its slope. The rate calculated in this way is called the **instantaneous rate** of the reaction at the specified time. In this case, the instantaneous rate at 1000 s is found to be 2.06×10^{-5} mol N_2O_5/(L·s). In general, *the instantaneous rate of any reaction at any time is obtained from the slope of the tangent to the concentration curve at that time.* From now on, when we speak of a reaction rate, we will always mean an *instantaneous* rate at a specified time.

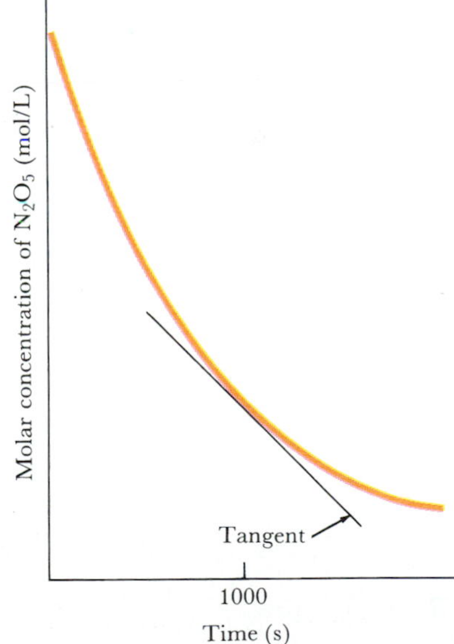

FIGURE 12.2 The time-dependence of the concentration of N_2O_5 at 65°C as it decomposes into NO_2 and O_2. The reaction rate at any time after the start of the reaction is given by the tangent to the curve at that time.

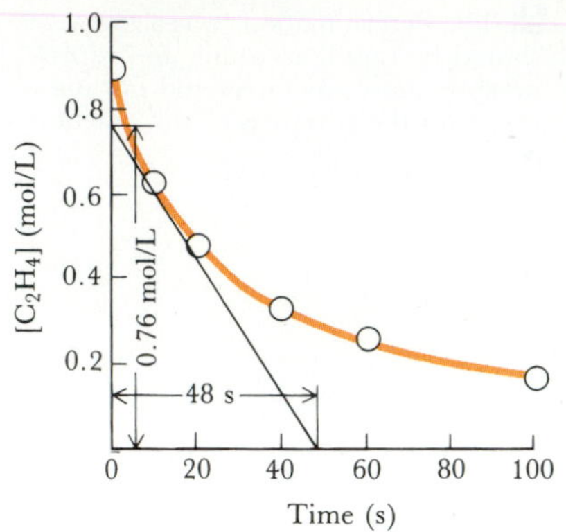

FIGURE 12.3 The determination of the rate of consumption of ethylene at 10 s after the start of the reaction, as described in Example 12.2.

T OHT: Figs. 12.3 and 12.4.

E Rate constants from initial concentration data. *J. Chem. Educ.*, **1976**, *53*, 493.

U Exercise 12.60.

N The initial rate of a reaction is the most reproducible. The accuracy with which an initial reaction rate is determined by the procedure shown in Fig. 12.4 is limited by the precision with which the tangent can be drawn; shortly, we present a much more accurate procedure (in terms of integrated rate laws).

▼ EXAMPLE 12.2 Determining the rate of a reaction

The concentration of ethylene, C_2H_4, in the reaction

$$2C_2H_4(g) \longrightarrow C_4H_8(g)$$

was measured at 900 K and varied with time as follows:

time, s	0	10	20	40	60	100
$[C_2H_4]$, mol/L	0.884	0.621	0.479	0.328	0.250	0.169

What is the rate of change of the ethylene concentration 10 s after the start of the reaction?

STRATEGY We can *estimate* the reaction rate by noting the change in concentration of ethylene between $t = 0$ and $t = 20$ s, an interval that spans 20 s and is centered on the time of interest. A more accurate procedure is to plot the concentration against time, draw the tangent to the graph at the specified time, and calculate its slope.

SOLUTION The *estimate* of the reaction rate is

$$\text{Rate} = \frac{\text{decrease in concentration of } C_2H_4}{\text{time for the change to occur}}$$

Concentrations: at $t = 0$; at $t = 20$ s

$$= \frac{(0.884 - 0.479) \text{ mol } C_2H_4/L}{20 \text{ s}} = \frac{0.405 \text{ mol } C_2H_4/L}{20 \text{ s}}$$

$$= 0.020 \text{ mol } C_2H_4/(L{\cdot}s)$$

To obtain an accurate value, we plot the data and draw the tangent to the curve at $t = 10$ s. As shown in Fig. 12.3, the slope of the tangent is

$$\text{Rate} = \frac{0.76 \text{ mol } C_2H_4/L}{48 \text{ s}} = 0.016 \text{ mol } C_2H_4/(L{\cdot}s)$$

EXERCISE E12.2 Determine the reaction rate (with respect to the concentration of ethylene) at $t = 40$ s from the data above.

[*Answer*: 0.0051 mol $C_2H_4/(L{\cdot}s)$] ▲

Initial rates. Once a product has been formed in a reaction, it may react with another substance, perhaps with the original reactants. Because this secondary reaction can make the analysis of reaction rates quite difficult, chemists have developed an important technique: they measure the reaction's **initial rate**, that is, the rate at the start of the reaction (at $t = 0$) when no products are present. Initial rates are measured just like other instantaneous rates; consequently, the tangent to the concentration curve is drawn at the very start of the reaction (at time $t = 0$).

▼ EXAMPLE 12.3 Calculating initial rates

What is the initial rate of the reaction described in Example 12.2?

STRATEGY The best procedure is to draw the tangent to the curve in Fig. 12.3 at $t = 0$ and then determine its slope.

SOLUTION As shown by the red line in Fig. 12.4 (which is a copy of the graph in Fig. 12.3), the slope of the tangent at $t = 0$ is

$$\text{Rate} = \frac{0.88 \text{ mol } C_2H_4/L}{14 \text{ s}} = 0.063 \text{ mol } C_2H_4/(L\cdot s)$$

That is, the initial rate is 6.3×10^{-2} mol $C_2H_4/(L\cdot s)$. Notice that the initial rate is higher than the rate 10 s later (see Example 12.2). We explore this feature in the following paragraphs.

EXERCISE E12.3 The blue line in Fig. 12.4 shows how the concentration of ethylene changes at a higher temperature. What is the initial rate of the reaction?

[*Answer*: 0.15 mol $C_2H_4/(L\cdot s)$]

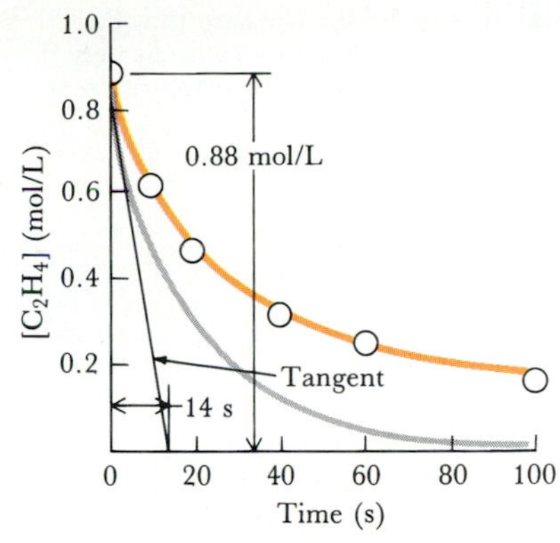

FIGURE 12.4 The determination of the initial rate of consumption of ethylene as described in Example 12.3. The concentration curve is the same as that shown in Fig. 12.3, and the tangent is taken at the start of the curve (at $t = 0$).

The initial rates of most reactions depend on the initial concentrations of the reactants; and, usually, the higher the concentration of the reactants, the greater the initial rate (Fig. 12.5). As an example, suppose we were to measure different masses of N_2O_5 into different flasks, immerse them all in a water bath at 65°C, and measure the initial rates at which the vapor in each flask decomposed. We would find greater initial rates of decomposition in the flasks with greater initial concentrations of N_2O_5 (Fig. 12.5a). The pattern in such data becomes clear when we look at actual values: chemists have found that the initial rate for an initial concentration of 0.08 mol/L of N_2O_5 vapor is twice the initial rate for an initial concentration of 0.04 mol/L. This doubling of the rate when the concentration is doubled suggests that the initial rate is directly proportional to the initial molar concentration of vapor. That is,

$$\text{Initial rate} = k \times [N_2O_5]_0$$

T OHT: Fig. 12.5.

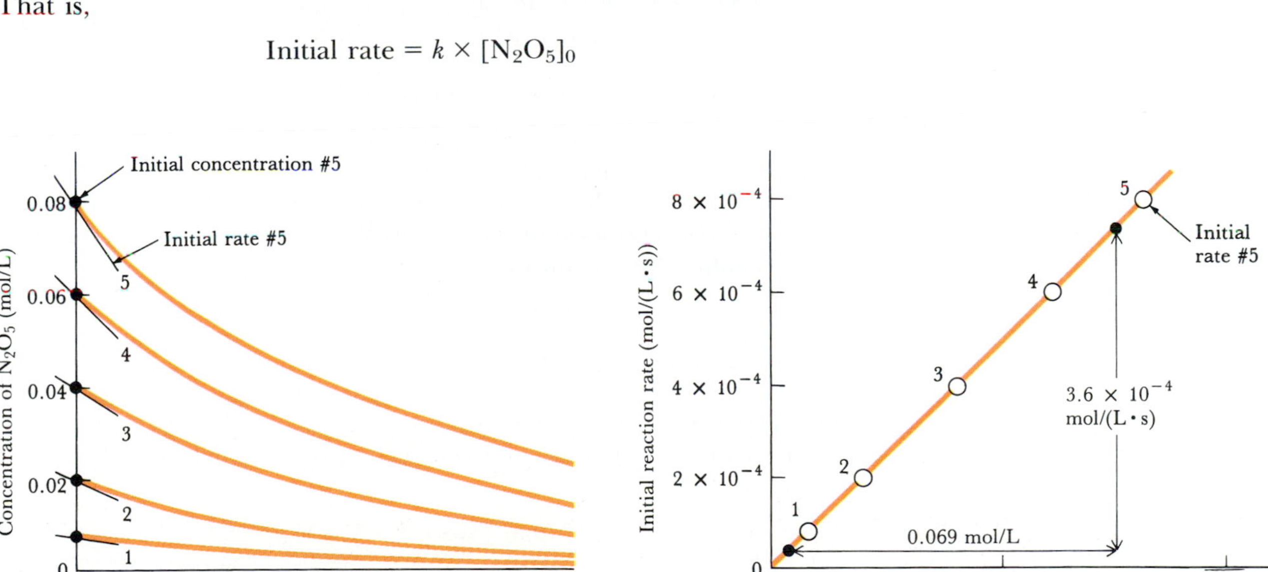

FIGURE 12.5 (a) The initial rate of decomposition of N_2O_5, as given by the slope of the tangents at $t = 0$, is greater when the initial concentration of N_2O_5 is higher. (b) The initial rate is proportional to the initial concentration, as shown by the straight line obtained when that rate is plotted against that concentration for the five samples in (a).

N It is generally the case that data can be interpreted more accurately if a scheme can be devised that gives a linear plot.

where k is a constant, called the **rate constant** for the reaction, and the subscript 0 means "initial value." This expression is the equation of a straight line with intercept 0 and slope k:

$$\text{Initial rate} = 0 + k \times [N_2O_5]_0$$

$$y \quad = \underset{\substack{\uparrow \\ \text{Intercept}}}{b} + \underset{\substack{\uparrow \\ \text{Slope}}}{m} \times \quad x$$

We can therefore confirm the proportionality between the rate and the concentration by plotting the initial rate against the initial concentration of N_2O_5; we hope to obtain a straight line. As Fig. 12.5b shows, a straight line is obtained and confirms the proportionality. Moreover, we can go on to obtain the value of the rate constant k by determining the slope of the straight line. For this calculation, it is more convenient to use exponential notation for the initial rates (and express them in mol L^{-1} s^{-1}). So, from the information in Fig. 12.5b,

N The convenience of exponential notation arises from the ease with which cancellations can be carried out: the notation avoids having denominators in denominators. A further point is that first-order rate constants look slightly odd if written, for example, 1×10^5/s, in place of $1 \times 10^5\ s^{-1}$.

$$\text{Slope} = k = \frac{3.6 \times 10^{-4}\ \text{mol L}^{-1}\ \text{s}^{-1}}{0.069\ \text{mol L}^{-1}} = 5.2 \times 10^{-3}\ \text{s}^{-1}$$

(Notice the units of k in this case: s^{-1}, read "per second.")

Now that we know the value of k for this reaction at 65°C, we can use it to predict the initial rate of the *same reaction* at the *same temperature* for any initial concentration. For instance, if $[N_2O_5]_0 = 5.0 \times 10^{-2}$ mol L^{-1} and the temperature is 65°C, the initial rate of decomposition is

$$\begin{aligned}\text{Initial rate} &= k \times [N_2O_5]_0 \\ &= (5.2 \times 10^{-3}\ \text{s}^{-1}) \times (5.0 \times 10^{-2}\ \text{mol N}_2\text{O}_5\ \text{L}^{-1}) \\ &= 2.6 \times 10^{-4}\ \text{mol N}_2\text{O}_5\ \text{L}^{-1}\ \text{s}^{-1}\end{aligned}$$

That is, the initial rate of disappearance of N_2O_5 is 2.6×10^{-4} mol N_2O_5/(L·s) when the initial concentration of N_2O_5 is 5.0×10^{-2} mol/L.

Not all reactions have initial rates that are directly proportional to the initial concentration of a reactant. When we plot the initial rate of the decomposition of nitrogen dioxide at 300°C,

$$2NO_2(g) \longrightarrow 2NO(g) + O_2(g)$$

against the initial NO_2 concentration, we do not obtain a straight line (Fig. 12.6a). However, a plot of the initial rate against the *square* of the initial concentration, is a straight line (Fig. 12.6b). This plot shows that the initial rate of this reaction is directly proportional to $[NO_2]_0{}^2$, and therefore that

$$\text{Initial rate} = k \times [NO_2]_0{}^2$$

where k is this reaction's rate constant (not the same as that for N_2O_5 found above). The units of this k are L/(mol·h) (or L mol^{-1} h^{-1}), as can be verified by dividing the units for the initial rate (mol L^{-1} h^{-1}) by the square of the initial concentration (mol L^{-1}):

N Put another way, the units of k must be such that when k multiplies the concentration raised to the appropriate power, then the dimensions molar concentration per time are obtained.

$$\begin{aligned}\text{Units of } k &= \frac{\text{units of initial rate}}{\text{units of (initial concentration)}^2} = \frac{\text{mol L}^{-1}\ \text{h}^{-1}}{\text{mol}^2\ \text{L}^{-2}} \\ &= \text{L mol}^{-1}\ \text{h}^{-1}\end{aligned}$$

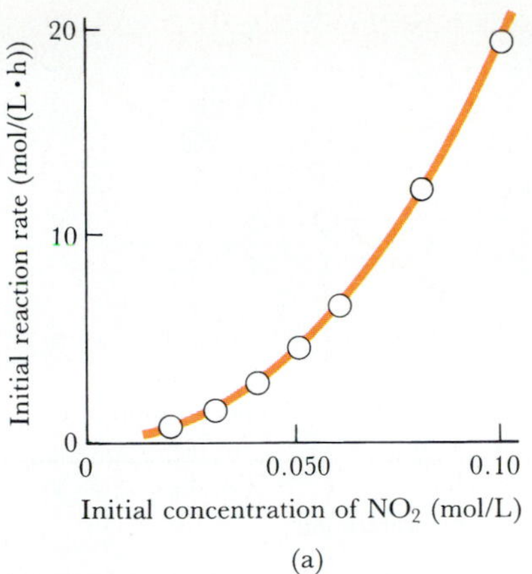

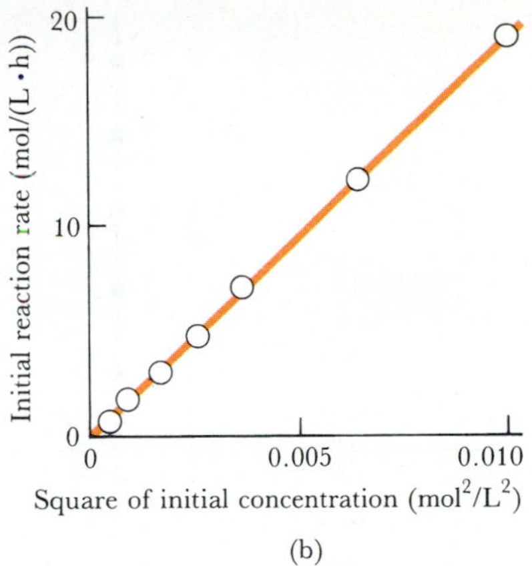

FIGURE 12.6 (a) When the initial rate of decomposition of NO_2 into NO and O_2 at 300°C is plotted against the concentration of NO_2, a straight line is not obtained. (b) A straight line is obtained, however, when that rate is plotted against the square of the concentration of NO_2.

The numerical value of k is obtained from the slope of the graph and, for this reaction, is 0.54 L/(mol·h) at 300°C.

12.2 RATE LAWS

The expressions we have derived show how initial rates depend on the initial concentrations of the reactants. Because these are initial values, they do not take into account any reactions that may occur between the reactants and the products—initially, there are no products. In many cases, however, the products do not take part in the reaction, so the expression for the initial rate is also applicable at later stages of the reaction when products are present. For reactions of this kind, the expression for the reaction rate applies at all stages of the reaction. The only difference is that at each stage we must use the *current* value of the concentration to calculate the rate. In the N_2O_5 decomposition at 65°C, for example, when the N_2O_5 concentration has fallen to 1.0×10^{-2} mol L^{-1} from an initial concentration of 5.0×10^{-2} mol L^{-1}, the reaction rate falls to

$$\text{Rate} = k \times [N_2O_5] = (5.2 \times 10^{-3}\ s^{-1}) \times (1.0 \times 10^{-2}\ \text{mol}\ N_2O_5\ L^{-1})$$
$$= 5.2 \times 10^{-5}\ \text{mol}\ N_2O_5\ L^{-1}\ s^{-1}$$

There is a lower concentration of N_2O_5 at this stage of the reaction, and the reaction is correspondingly slower.

Figure 12.7a shows that the reaction rate (the tangent to the concentration curve) decreases as the reaction proceeds and N_2O_5 is removed. The rate of reaction is less at time e than at time d, the rate at time d is less than at time c, and so on. We can confirm that the reaction rate is directly proportional to the remaining concentration of N_2O_5 at any stage of the reaction by plotting the rate at different times after the

V Video 34: The iodine clock reaction.

D A variation of the starch-iodine clock reaction (vary concentration). CD2, 78.

E Some reactions may proceed more rapidly at a later stage despite the fact that the concentration of reactants is lower than initially. Such reactions are *autocatalytic* in the sense that the products participate in the reaction, so the greater the concentration of products the more rapid the reaction. (See, e.g., Video 32: Autocatalysis.)

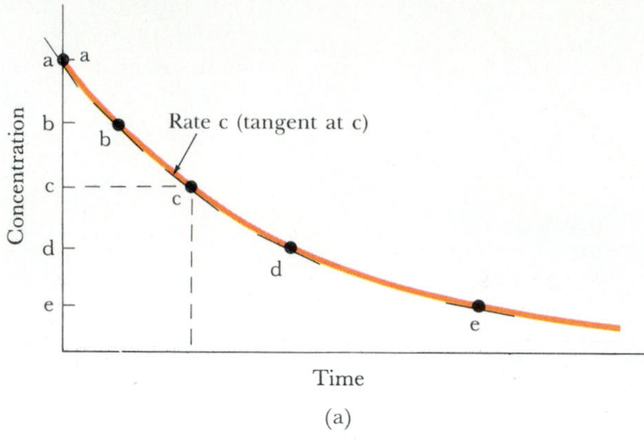

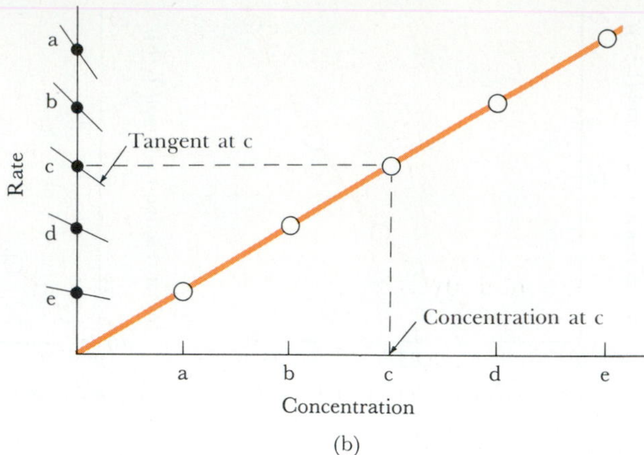

FIGURE 12.7 (a) The instantaneous reaction rates of the N_2O_5 decomposition at different times during a reaction are obtained as the slopes of the tangents. (b) When these rates are plotted against the concentration of N_2O_5 remaining, a straight line is obtained.

start of the reaction against the concentration of N_2O_5 that remains at each time (Fig. 12.7b). We obtain a straight line, which shows that *at each stage of the reaction*

$$\text{Rate} = k \times [N_2O_5]$$

where $[N_2O_5]$ is the concentration at the time of interest. Furthermore, the value of k (the slope of the line in Fig. 12.5b) is exactly the same as the slope we found earlier when considering only the initial rates, and at 65°C is $5.2 \times 10^{-3}\ s^{-1}$.

The equation for the rate in terms of the current concentration of N_2O_5 is an example of a rate law:

E The overall rate law of a reaction is expressed in terms of the concentrations of the reactants and the products only: intermediates (species that do not appear in the chemical equation for the overall process) should not appear in the rate law.

> A **rate law** is an equation expressing the instantaneous reaction rate in terms of the concentrations, at that instant, of the substances taking part in the reaction.

Each reaction has its own characteristic rate law, and we see later how these laws are determined and what they signify. The rate constant k that appears in the rate law is independent of the concentration of the reactants but depends on the reaction and the temperature. That is, not only does each reaction have its characteristic rate equation, but it also has a characteristic rate constant, with different values of the constant at different temperatures. The rate law for the NO_2 decomposition, for example, is

T OHT: Fig. 12.7.

? Comment on the plot of the data in Fig. 12.5, for which a series of initial concentrations were needed, in relation to the plot in Fig. 12.7, for which only one initial concentration is needed. Both sets of data provide a value for k.

$$\text{Rate of decomposition} = k \times [NO_2]^2$$

with $k = 0.54\ L\ mol^{-1}\ h^{-1}$ at 300°C.

We shall see that the concentrations of substances taking part in reactions with similar rate laws change with time in a similar way. A useful first step in classifying reactions is therefore to group them according to their rate laws. The two rate laws given so far differ in the power to which the concentration is raised. Both are summarized by the expression

$$\text{Rate} = k \times [\text{reactant}]^a$$

with $a = 1$ for the N_2O_5 reaction and $a = 2$ for the NO_2 reaction.

The N_2O_5 reaction, for which the rate is proportional to $[N_2O_5]$, is an example of a **first-order reaction**, a reaction in which the rate is proportional to the first power of the concentration of a substance (that is, $a = 1$). Doubling the concentration of a reactant in such a first-order reaction doubles the reaction rate. The NO_2 decomposition, for which the rate is proportional to $[NO_2]^2$, is an example of a **second-order reaction,** one for which the rate is proportional to the second power (the square) of the concentration of a substance (that is, $a = 2$). Doubling the concentration of the reactant in such a second-order reaction increases the reaction rate by a factor of $2^2 = 4$.

It is important to appreciate that we cannot tell in advance of an experimental investigation whether a reaction will turn out to be first order, second order, or some other order. *The rate law is an experimentally determined characteristic of the reaction; in general, it cannot be written down from the stoichiometry of the chemical equation for the reaction.* The advantage of knowing the order of a reaction is that all reactions with the same order have certain similar characteristics: the rates of first-order reactions have similar characteristics, as do those of second-order reactions. Therefore, we do not need to study all reactions individually; we need study only the characteristics of classes of reactions instead. Once again, we see how chemists classify reactions in order to identify similarities in behavior and enable them to make general statements.

▼ EXAMPLE 12.4 Identifying the order of a reaction

The rate of the reaction

$$S_2O_8^{2-}(aq) + 3I^-(aq) \longrightarrow 2SO_4^{2-}(aq) + I_3^-(aq)$$

in which iodide ions are oxidized by persulfate ions, $S_2O_8^{2-}$, doubles when the concentration of iodide ions is doubled and the concentration of persulfate ions remains constant. What is the order of the reaction with respect to the iodide ion?

STRATEGY We know that in a first-order reaction the rate is directly proportional to the concentration, and in a second-order reaction the rate is directly proportional to the square of the concentration. We have to decide which of these two possibilities fits the data. Note that we *cannot* infer the rate law from the fact that $3I^-$ occurs in the chemical equation.

SOLUTION When $[I^-]$ doubles, the rate doubles. Therefore, the reaction rate is directly proportional to $[I^-]$. We can conclude that the reaction is first order with respect to the iodide ion.

EXERCISE E12.4 The rate of decomposition of dinitrogen oxide in the reaction $2N_2O(g) \longrightarrow 2N_2(g) + O_2(g)$ falls by a factor of 3 when the concentration of dinitrogen oxide is reduced from 6×10^{-3} to 2×10^{-3} mol/L. What is the reaction order?

▲ [*Answer*: first order]

Some reactions have more complicated rate laws than those discussed so far. For example, the redox reaction between persulfate and iodide ions

$$S_2O_8^{2-}(aq) + 3I^-(aq) \longrightarrow 2SO_4^{2-}(aq) + I_3^-(aq)$$

obeys the experimental rate law

$$\text{Rate} = k[S_2O_8^{2-}][I^-]$$

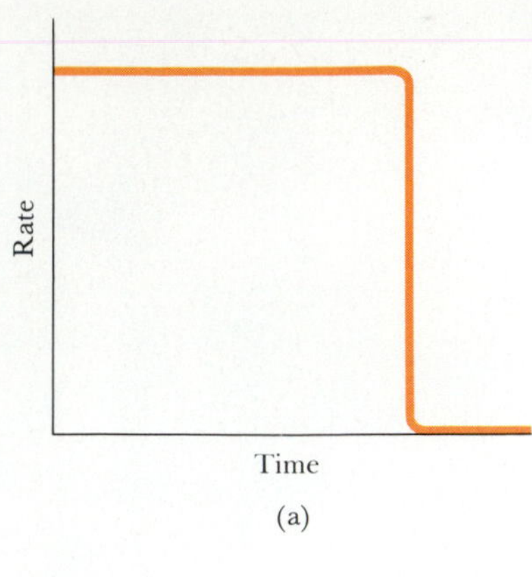

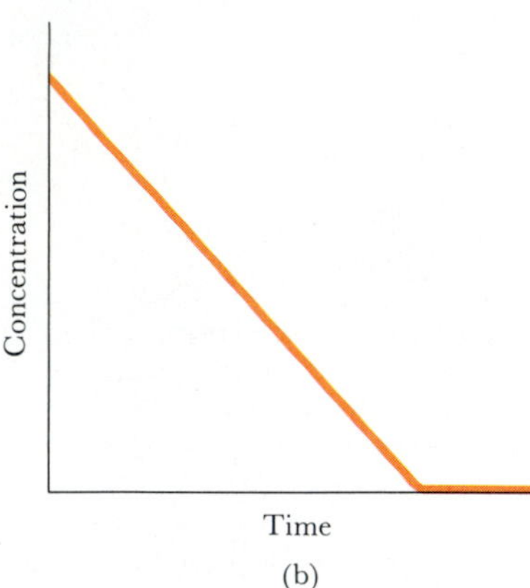

FIGURE 12.8 (a) The reaction rate of a zero-order reaction is independent of the concentration of the reactant and remains constant until all the reactant has been consumed. (b) In such a reaction, the concentration of the reactant falls at a constant rate until it is exhausted, at which time the reaction comes to an abrupt halt.

E Example of a constant rate reaction. *J. Chem. Educ.*, **1987**, *64*, 534.

E In terms to be introduced later, the rate determining step of the ammonia decomposition reaction occurs to a species that has been adsorbed on to the platinum surface, so the rate at which ammonia molecules arrive at the surface (which is proportional to their concentration) has no effect on the overall rate of the reaction until the concentration is so low that the rate of arrival is less than the surface reaction. In the final stages of the reaction the rate determining step becomes the process of adsorbing on the surface, and then the rate is first order in NH_3.

We say that the reaction is first order with respect to $S_2O_8^{2-}$ (or "in" $S_2O_8^{2-}$) and first order in I^-, where the **order** with respect to a single substance is the power to which its concentration is raised in the rate law. Doubling either the $S_2O_8^{2-}$ ion concentration or the I^- ion concentration doubles the reaction rate. When, as in this example, the rate depends on the concentrations of more than one substance, we can also speak of its overall order:

> The **overall order** of a reaction is the sum of the powers to which the individual concentrations are raised in its rate law.

In the persulfate reaction the overall order is 2. In general, if

$$\text{Rate} = k[\text{A}]^a[\text{B}]^b \ldots$$

the overall order is $a + b + \ldots$. For example, the reaction

$$2\text{NO}(g) + \text{O}_2(g) \longrightarrow 2\text{NO}_2(g) \qquad \text{Rate} = k[\text{NO}]^2[\text{O}_2]$$

is second order in NO, first order in O_2, and third order overall.

Reactions can be orders other than first, second, or third: some reactions are zero order ($a = 0$), others are fractional orders (such as $a = \frac{1}{2}$), and yet others are negative orders (such as $a = -1$). A reaction is **zero order** if its rate is independent of the concentration. The name *zero order* comes from the mathematical result that anything raised to the zero power is equal to one ($x^0 = 1$). In particular, $[\text{reactant}]^0 = 1$, so the rate law for a zero-order reaction is

$$\text{Rate} = k \times [\text{reactant}]^0 = k, \text{ a constant}$$

An example of a zero-order reaction is the decomposition of ammonia on a hot platinum wire:

$$2\text{NH}_3(g) \longrightarrow \text{N}_2(g) + 3\text{H}_2(g)$$

The ammonia decomposes at a constant rate until it has all disappeared. Then the reaction stops abruptly (Fig. 12.8). The reduction of nitrogen dioxide by carbon monoxide

$$\text{NO}_2(g) + \text{CO}(g) \longrightarrow \text{NO}(g) + \text{CO}_2(g)$$

obeys the rate law

$$\text{Rate} = k[\text{NO}_2]^2$$

It is second order in NO_2, zero order in CO, and second order overall. By zero order in CO we mean that, although some CO must be present for the reaction to occur at all, the rate is independent of its concentration.

▼ EXAMPLE 12.5 Classifying reactions by order

The oxidation of sulfur dioxide to sulfur trioxide in the presence of platinum (Fig. 12.9) in the reaction $2\text{SO}_2(aq) + \text{O}_2(g) \rightarrow 2\text{SO}_3(g)$ has the rate law

$$\text{Rate} = k\frac{[\text{SO}_2]}{[\text{SO}_3]^{1/2}}$$

What is the overall order of the reaction?

FIGURE 12.9 When sulfur dioxide and oxygen are bubbled through concentrated sulfuric acid to remove moisture and then passed over hot platinum (left), they combine to form sulfur trioxide. This compound forms dense white fumes when it comes into contact with moisture in the atmosphere (right). The rate law for its formation, given in Example 12.5, shows that the rate of formation of SO_3 decreases as its concentration increases.

STRATEGY We have to identify the order with respect to each substance and then add these individual orders to find the overall order. Because $1/x^a = x^{-a}$, if a concentration occurs in a denominator, it has a negative order.

SOLUTION The rate law has the form

$$\text{Rate} = k[SO_2][SO_3]^{-1/2}$$

so it is first order in SO_2 and minus one-half order in SO_3. Because $1 + (-\frac{1}{2}) = \frac{1}{2}$, the reaction is one-half order overall. Because the reaction has a negative order with respect to SO_3, the rate of the reaction *decreases* as the concentration of SO_3 increases.

EXERCISE E12.5 The reduction of bromate ions (BrO_3^-) by bromide ions in acid solution,

$$BrO_3^-(aq) + 5Br^-(aq) + 6H^+(aq) \longrightarrow 3Br_2(aq) + 3H_2O(l)$$

has a rate law

$$\text{Rate} = k[BrO_3^-][Br^-][H^+]^2$$

What are the orders with respect to the reactants and the overall order?
[*Answer*: first order in BrO_3^- and Br^-; second order in H^+; fourth order overall]

▲

E One reason why a reaction might go more slowly in the presence of product (as for the reaction in Example 12.5) is that the product reacts with an intermediate and decomposes back into reactant.

? Why does a fuel burn more rapidly in pure oxygen than in air?

U Exercise 12.61.

Typical rate laws and rate constants are given in Table 12.1. Recall that, except in some special cases that we shall describe later, *a rate law cannot be written down from a chemical equation; it must be determined experimentally*. The N_2O_5 and the NO_2 reactions both have equations of the form $2A \rightarrow$ products, but one is first order and the other is second order. Another example of a rate law that shows little relation to its chemical equation is the iodination of acetone,

$$CH_3COCH_3(aq) + I_2(aq) \xrightarrow{\text{acid}} (CH_2I)COCH_3(aq) + HI(aq)$$
$$\text{Rate} = k[CH_3COCH_3][H^+]$$

in which $[I_2]$ does not appear in the rate law but $[H^+]$ does, even though H^+ does not appear as a reactant in the chemical equation.

TABLE 12.1 Rate laws and rate constants

Reaction	Rate law*	Temperature, K	k
Gas phase			
$H_2 + I_2 \rightarrow 2\mathbf{HI}$	$k[H_2][I_2]$	500	4.3×10^{-7} L mol^{-1} s^{-1}
		600	4.4×10^{-4}
		700	6.3×10^{-2}
		800	2.6
$2\mathbf{HI} \rightarrow H_2 + I_2$	$k[HI]^2$	500	6.4×10^{-9} L mol^{-1} s^{-1}
		600	9.7×10^{-6}
		700	1.8×10^{-3}
		800	9.2×10^{4}
$2\mathbf{N_2O_5} \rightarrow 4NO_2 + O_2$	$k[N_2O_5]$	298	3.7×10^{-5} s^{-1}
		318	5.1×10^{-4}
		328	1.7×10^{-3}
		338	5.2×10^{-3}
$2\mathbf{N_2O} \rightarrow 2N_2 + O_2$	$k[N_2O]$	1000	0.76 s^{-1}
		1050	3.4
$2\mathbf{NO_2} \rightarrow 2NO + O_2$	$k[NO_2]^2$	573	0.54 L mol^{-1} s^{-1}
$\mathbf{C_2H_6} \rightarrow 2CH_3$	$k[C_2H_6]$	973	5.5×10^{-4} s^{-1}
cyclopropane → propene	k[cyclopropane]	773	6.7×10^{-4} s^{-1}
Aqueous solution			
$\mathbf{H^+} + OH^- \rightarrow H_2O$	$k[H^+][OH^-]$	298	1.5×10^{11} L mol^{-1} s^{-1}
$\mathbf{CH_3Br} + OH^- \rightarrow CH_3OH + Br^-$	$k[CH_3Br][OH^-]$	298	2.8×10^{-4} L mol^{-1} s^{-1}
$\mathbf{C_{12}H_{22}O_{11}} + H_2O \xrightarrow{\text{acid}} 2C_6H_{12}O_6$	$k[C_{12}O_{22}O_{11}][H^+]$	298	1.8×10^{-4} L mol^{-1} s^{-1}

*For the rate of formation or reaction of the substance in bold type in the reaction column.

E One of the fastest reactions known is the recombination of a proton and a hydroxide ion to form H_2O in water (and in ice): see the entry in Table 12.1.

D Appearing red. CD2, 76.

▼ EXAMPLE 12.6 Determining reaction order from experimental data

The data that follow show how the initial rate of the reaction

$$BrO_3^-(aq) + 5Br^-(aq) + 6H^+(aq) \longrightarrow 3Br_2(aq) + 3H_2O(l)$$

varies as the concentrations of the reactants are changed. What is the order of the reaction with respect to each reactant? Write the rate law for the reaction and state the overall order of the reaction.

	Initial concentration, mol/L			
Experiment	$\mathbf{BrO_3^-}$	$\mathbf{Br^-}$	$\mathbf{H^+}$	**Initial rate mol BrO_3^-/(L·s)**
1	0.10	0.10	0.10	1.2×10^{-3}
2	0.20	0.10	0.10	2.4×10^{-3}
3	0.10	0.30	0.10	3.5×10^{-3}
4	0.20	0.10	0.15	5.4×10^{-3}

STRATEGY Suppose the concentration of a substance is increased by a certain factor f. If the rate increases by f^a, then the reaction has order a in that substance. Therefore, we have to inspect the data to see how the rate changes when the concentration of each substance is changed. To isolate the effect of each substance, we compare experiments that differ in the concentration of only *one* substance at a time.

SOLUTION When the concentration of BrO_3^- is doubled from 0.10 to 0.20 mol/L (from experiment 1 to experiment 2, in which the concentrations of H^+ and Br^- remain unchanged), the rate also doubles. Therefore, the reaction is first order in BrO_3^-. When the concentration of Br^- is changed from 0.10 to 0.30 mol/L (from experiment 1 to experiment 3, in which the concentrations of H^+ and BrO_3^- are constant), that is, by a factor of 3.0, the rate changes from 1.2×10^{-3} to 3.5×10^{-3} mol/(L·s), that is, by a factor of $3.5/1.2 = 2.9$. Therefore, allowing for experimental error, we can deduce that the reaction is also first order in Br^-. When the concentration of H^+ ions is increased from 0.10 to 0.15 mol/L (from experiment 2 to experiment 4, when the concentrations of Br^- and BrO_3^- remain constant), that is, by a factor of 1.5, the rate increases by a factor of $5.4/2.4 = 2.3$. Thus we need to solve $1.5^a = 2.3$ for a. Because $1.5^2 = 2.3$, the reaction is second order in H^+. The rate law is therefore

$$\text{Rate} = k[BrO_3^-][Br^-][H^+]^2$$

and is fourth order overall.

EXERCISE E12.6 Repeat this example for the reaction between persulfate ions and iodide ions,

$$S_2O_8^{2-}(aq) + 3I^-(aq) \longrightarrow 2SO_4^{2-}(aq) + I_3^-(aq)$$

given the following data:

Experiment	Initial concentration, mol/L $S_2O_8^{2-}$	I^-	Initial rate mol $S_2O_8^{2-}$/(L·s)
1	0.15	0.21	1.14
2	0.22	0.21	1.70
3	0.22	0.12	0.98

[*Answer*: rate = $k[S_2O_8^{2-}][I^-]$; second order overall]

N It is usually very simple to solve expressions of the form $1.5^a = 2.3$ that arise in calculations of this kind, because a is usually either 1 or 2. In general, one could solve for a by taking logarithms, obtaining $a \log 1.5 = \log 2.3$ and then solving for a.

U Exercises 12.64 and 12.65.

As we have seen, a reaction order need be neither a whole number nor positive. For instance, the decomposition of ozone (O_3),

$$2O_3(g) \longrightarrow 3O_2(g) \qquad \text{Rate} = \frac{k[O_3]^2}{[O_2]}$$

is another example in which the concentration of a reaction product appears in the rate law and the reaction is of order -1 in O_2 (because $1/x = x^{-1}$). A negative order signifies that the reaction rate *decreases* as the concentration of a substance, in this case oxygen, increases.

E We shall see later (in Section 12.5) why the presence of oxygen slows the rate at which ozone decomposes. The reason lies in the ability of O_2 molecules to react with O atoms formed in the decomposition of O_3, and hence to re-form O_3 molecules.

Some reactions do not have an order with respect to each substance taking part in the reaction, and some do not have an overall order. The reaction

$$H_2(g) + Br_2(g) \longrightarrow 2HBr(g)$$

E A simple method for analyzing first-order kinetics. *J. Chem. Educ.*, **1990**, *67*, 459.

is found to obey

$$\text{Rate} = \frac{k[H_2][Br_2]^{3/2}}{[Br_2] + k'[HBr]}$$

The reaction is first order in H_2; but, because the rate law does not have the form Rate = $k[H_2]^a[Br_2]^b[HBr]^c$ . . . , it does not have a specific order with respect to either Br_2 or HBr, nor does it have an overall order. The rate law is perfectly well defined, however, so we can discuss the kinetics of the reaction. All we cannot do is to slot it into one of the simple classes of reactions that we shall now study.

N We shall see in Chapter 22 that all radioactive decay processes are first-order.

The time dependence of first-order reactions. A major practical advantage of knowing the rate law and the rate constant is that they can be used to predict the concentrations of substances at any stage of a reaction. Conversely, knowledge of the time variation of the concentrations allows us to identify the rate law. Applications of this material include the time variation of gases in the atmosphere—such as the rate of disappearance of ozone or the rate of formation of smog-forming pollutants—and the rate at which a particular product is formed in an industrial process. Whenever chemists need to know how fast a particular product is formed or a reactant is used in a process, they need information of the kind we are about to meet.

We begin with first-order reactions. Consider a reaction of a substance A for which it has been found that

$$\text{Rate} = k[A]$$

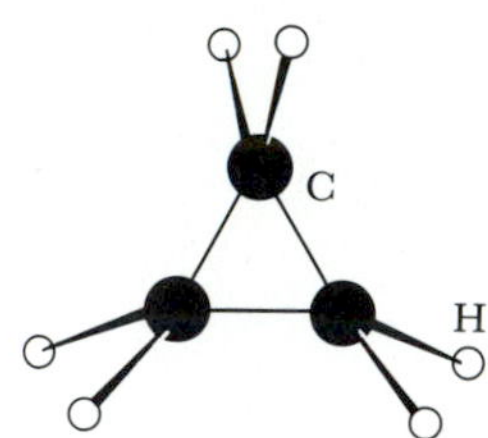

1 Cyclopropane, C_3H_6

An example is the rearrangement of the bonds in cyclopropane (C_3H_6, **1**) to form propene ($CH_3CH{=}CH_2$, **2**):

$$C_3H_6(g) \xrightarrow{\Delta} CH_3{-}CH_2{=}CH_3(g) \qquad \text{Rate} = k[C_3H_6]$$

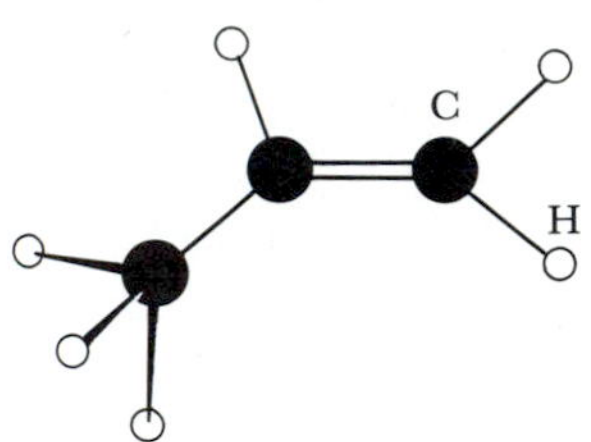

2 Propene, C_3H_6

Another example is the decomposition of N_2O_5 discussed earlier.

We often know the concentration of a reactant at the start of a reaction ($[A]_0$) but want to know the concentration [A] at some later time t. The expression required is called the **integrated rate law** and is derived (by using calculus) from the rate law itself. For a first-order reaction, the integrated rate law is

D First-order reaction: Inversion of sugar. TDC, 8-2s.

$$\ln\left(\frac{[A]_0}{[A]}\right) = kt \qquad (1)$$

(The logarithm, ln, is the *natural* logarithm; it is usually labeled "ln" on calculators.) The variation of [A] with time predicted by this equation is plotted in Fig. 12.10.* The graph, which is called an exponential decay, reproduces the initially rapid but then slower change of concentration observed experimentally (compare with Fig. 12.2). Notice that the larger the rate constant, the more rapid the initial decay.

N Natural logarithms emerge naturally in calculations of this kind, and result in simpler expressions than those obtained using common logarithms; electronic calculators have made natural logarithms as easy to handle as common logarithms, and we see no reason for resorting to common logarithms. If needed, the appropriate expression would be

$$\log\left(\frac{[A]_0}{[A]}\right) = \frac{kt}{2.303}$$

E Kinetic rate equations. *J. Chem. Educ.*, **1969**, *46*, 226.

*The dependence of the concentration on the time is obtained by taking the natural antilogarithm (e^x on a calculator) of both sides of the equation and rearranging the result to $[A] = [A]_0e^{-kt}$. This exponential dependence on the time is the origin of the name *exponential decay* for the shape of the curves in Fig. 12.10.

▼ EXAMPLE 12.7 Calculating a concentration from a rate law

Calculate the concentration of N_2O_5 remaining 600 s (10 min) after the start of its decomposition at 65°C when its concentration was 0.040 mol/L. The reaction $2N_2O_5(g) \rightarrow 4NO_2(g) + O_2(g)$ and its rate law is

$$\text{Rate of decomposition of } N_2O_5 = k[N_2O_5]$$

with $k = 5.2 \times 10^{-3}\ s^{-1}$.

STRATEGY We can anticipate that the concentration will be smaller than 0.040 mol/L, because some of the reactant decomposes as the reaction proceeds. We also know that, because the reaction is first order, we can use Eq. 1 to predict the concentration at any time after the reaction begins. Equation 1 is an expression for $\ln([A]_0/[A])$, but we need [A] itself. We can find it if we first take the natural antilogarithm of kt (in other words, enter e^x on your calculator, with x equal to the value of kt).

SOLUTION Substituting for k, t, and $[N_2O_5]_0$ in Eq. 1. gives

$$\ln\left(\frac{0.040\ \text{mol/L}}{[N_2O_5]}\right) = 5.2 \times 10^{-3}\ s^{-1} \times 600\ s = 3.1$$

The natural antilogarithm of 3.1 is 22 (that is, on a calculator $e^{3.1} = 22$), so

$$\frac{0.040\ \text{mol/L}}{[N_2O_5]} = 22$$

which can be rearranged to

$$[N_2O_5] = \frac{0.040\ \text{mol/L}}{22} = 0.0018\ \text{mol/L}$$

That is, after 600 s, the concentration of N_2O_5 will have fallen from 0.040 to 0.0018 mol/L.

EXERCISE E12.7 Calculate the concentration of N_2O remaining after its decomposition has continued at 780°C for 100 ms; the initial concentration of N_2O was 0.20 mol/L and $k = 3.4\ s^{-1}$. The reaction is $2N_2O(g) \rightarrow 2N_2(g) + O_2(g)$ and its rate law is

$$\text{Rate of decomposition of } N_2O = k[N_2O]$$

▲ [*Answer*: 0.14 mol/L]

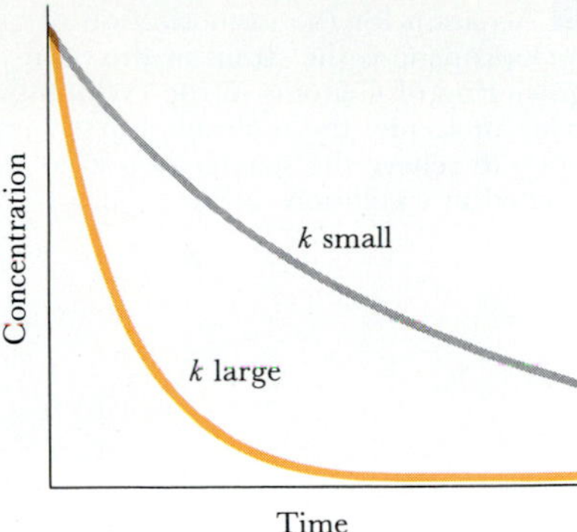

FIGURE 12.10 The characteristic shape of the time dependence of the concentration of a reactant in a first-order reaction is this exponential decay. The larger the rate constant, the faster the decay from the same initial concentration.

Equation 1 is also used to determine whether a reaction is first order and, if it is, to calculate its rate constant. We should note that if we rearrange Eq. 1, we obtain the equation of a straight line:

$$\ln[A] = \ln[A]_0 + (-k) \times t$$
$$y = b + m \times x$$

Therefore, when we suspect that a reaction is first order in A, we plot ln [A] against t. If the line is straight, then we have confirmed that the reaction is first order. Moreover, when the line is straight, its slope is equal to $-k$, so we have a route to the measurement of the reaction's rate constant too. Once we know the value of k, we can start making predictions about the time dependence of the reactants and products, as we illustrated in Example 12.7. Soon we shall see that we can also use the value of k to suggest the details of how the reaction takes place.

U Exercises 12.67, 12.69, and 12.72.

E A reason for the isomerization of cyclopropane is the strain in the triangular ring of C atoms in the cyclopropane molecule: the molecule bursts open to relieve the strain when it is excited in a collision.

U Exercises 12.62 and 12.85.

T OHT: Figs. 12.11 and 12.13.

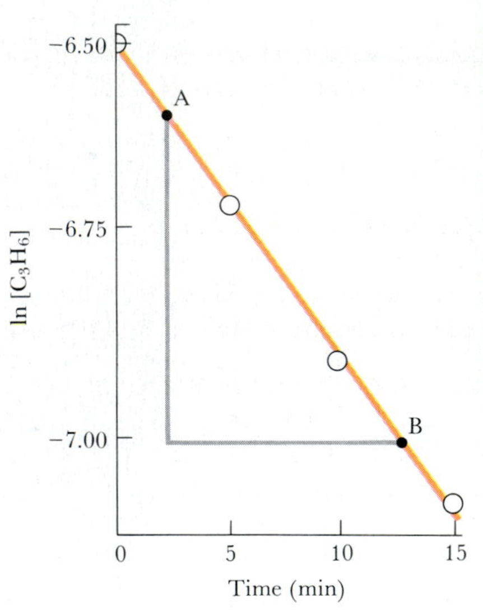

FIGURE 12.11 We test for a first-order reaction by plotting the natural logarithm of the reactant concentration against the time. This method is simpler and more accurate than calculating the rate at each moment and plotting it against concentration. The slope of the line gives the (negative of the) rate constant.

▼ **EXAMPLE 12.8 Measuring a rate constant**

When cyclopropane (**1**) is heated to 500°C, it changes into propene (**2**). The following data were obtained in one experiment:

t, min	0	5	10	15
$[C_3H_6]$, mol/L	1.5×10^{-3}	1.24×10^{-3}	1.00×10^{-3}	0.83×10^{-3}

Confirm that the reaction is first order in C_3H_6 and calculate the rate constant.

STRATEGY As outlined above, the strategy in problems of this kind is to plot ln [A] against t and to see whether we obtain a straight line. If we do, then the reaction is first order and the slope of the graph has the value $-k$.

SOLUTION Begin by drawing up the data table:

t, min	0	5	10	15
$\ln [C_3H_6]$	−6.50	−6.69	−6.91	−7.09

The points are plotted in Fig. 12.11: the graph is a straight line, thus confirming that the reaction is first order in cyclopropane. The slope of the line is

$$\text{Slope} = \frac{\ln [C_3H_6]_B - \ln [C_3H_6]_A}{\text{time B} - \text{time A}}$$

$$= \frac{(-7.02) - (-6.56)}{13.3 \text{ min} - 1.7 \text{ min}} = -0.040 \text{ min}^{-1}$$

Therefore, because $k = -\text{slope}$,

$$k = -(-0.040 \text{ min}^{-1}) = 0.040 \text{ min}^{-1}$$

This value is equivalent to $k = 6.7 \times 10^{-4}\ \text{s}^{-1}$, the value in Table 12.1

EXERCISE E12.8 Data on the decomposition of N_2O_5 at 25°C are as follows:

t, min	0	200	400	600	800	1000
$[N_2O_5]$, mol/L	1.50 $\times 10^{-2}$	9.6 $\times 10^{-3}$	6.2 $\times 10^{-3}$	4.0 $\times 10^{-3}$	2.5 $\times 10^{-3}$	1.6 $\times 10^{-3}$

Confirm that the reaction is first order and find the value of k.

▲ [*Answer*: $3.7 \times 10^{-5}\ \text{s}^{-1}$]

The half-life of first-order reactions. A widely used measure of the rate of a reaction is the half-life, $t_{1/2}$, of a substance:

> The **half-life** of a substance is the time needed for its concentration to fall to one-half the initial concentration.

If the substance is denoted A, then, after one half-life, the concentration of A falls from $[A]_0$ to $\frac{1}{2}[A]_0$ (Fig. 12.12). For a first-order reaction, from Eq. 1,

$$\ln\left(\frac{[A]_0}{\frac{1}{2}[A]_0}\right) = kt_{1/2}$$

or, because the $[A]_0$ cancel and $\ln(1/\frac{1}{2}) = \ln 2$,

$$\ln 2 = kt_{1/2}$$

Then, because $\ln 2 = 0.693$, we get

$$t_{1/2} = \frac{0.693}{k} \quad (2)$$

The initial concentration does not appear in this expression, so the half-life of a first-order reaction is independent of the initial concentration. As we shall see, this equation is true only of first-order reactions.

For the N_2O_5 decomposition at 65°C ($k = 5.2 \times 10^{-3}\ s^{-1}$),

$$t_{1/2} = \frac{0.693}{5.2 \times 10^{-3}\ s^{-1}} = 130\ s$$

Then it follows that if $[N_2O_5] = 1.00 \times 10^{-2}$ mol/L, the concentration decreases to 0.50×10^{-2} mol/L after 130 s. Because the half-life is independent of the concentration at the start of the period of interest, the concentration falls to 0.25×10^{-2} mol/L during the next 130 s, to half that value in another 130 s, and so on.

N We consider it easier to remember the relation between k and $t_{1/2}$ in terms of ln 2 than in terms of the value of ln 2 as given in Eq. 2. In general, if an equation is going to be remembered (which in some cases is helpful), then we consider that it is best (because the room for error is least) if the simplest form of the expression is remembered.

N The larger the rate constant, the shorter the half-life (see Fig. 12.10).

▼ EXAMPLE 12.9 Using half-lives

Calculate the time needed for the concentration of cyclopropane to fall to (a) one-half and (b) one-fourth its initial value at 500°C in the reaction in which it forms propene. From Table 12.1, we know that $k = 6.7 \times 10^{-4}\ s^{-1}$.

STRATEGY We first decide whether the reaction is first order. If it is, then the time it takes for the concentration to fall to half its initial value is $t_{1/2} = 0.693/k$. The total time required for the concentration to fall to one-fourth its initial value is the sum of two successive half-lives. For a first-order reaction, the two half-lives are equal, so the time required is $2t_{1/2}$.

SOLUTION For this reaction $k = 6.7 \times 10^{-4}\ s^{-1}$, so

$$t_{1/2} = \frac{0.693}{6.7 \times 10^{-4}\ s^{-1}} = 1.0 \times 10^3\ s$$

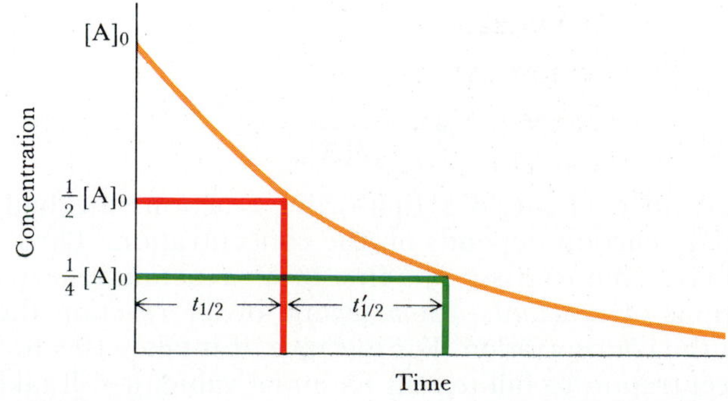

FIGURE 12.12 The half-life of a substance is the time needed for it to fall to one-half its initial value. For first-order reactions, the half-life is the same whatever the concentration at the start of the period; that is, in terms of the illustration, $t'_{1/2} = t_{1/2}$.

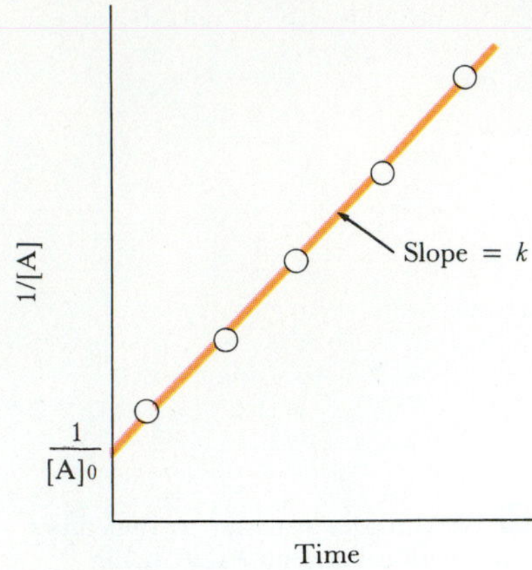

FIGURE 12.13 We test for a second-order reaction by plotting 1/[A], the reciprocal of the concentration of a reactant, against time. If we obtain a straight line, then the reaction is second order. The slope of the line is the rate constant, k.

The time required for part (a) is therefore 1.0×10^3 s, or just under 17 min. The time required for part (b) is twice this, or 33 min.

EXERCISE E12.9 Calculate the time needed for the concentration of N_2O to fall to (a) one-half and (b) one-eighth of its initial value at 1000 K.

[*Answer*: (a) 0.91 s; (b) 2.7 s]

The half-life provides a quick way of detecting a first-order reaction. For any two different initial concentrations, we measure the time it takes each to fall by one-half. If both times are the same, then the reaction is first order.

The time dependence of second-order reactions. The integrated rate law for second-order reactions of the form

$$\text{Rate} = k[\text{A}]^2$$

is (from calculus)

$$\frac{1}{[\text{A}]} = \frac{1}{[\text{A}]_0} + kt \qquad (3)$$

Comparison of this expression with the equation for a straight line,

$$y = b + mx$$

shows that a plot of 1/[A] against time, t, gives a straight line for a second-order reaction (Fig. 12.13) and that the slope of the graph is equal to the rate constant, k.

U Exercises 12.63 and 12.68.

The variation of concentration with time as predicted by Eq. 3 is shown by the orange lines in Fig. 12.14. The concentration decreases rapidly at first, but then changes more slowly than it does in a first-order reaction with the same initial rate (shown by gray lines). Comparison of the two sets of curves in the illustration shows one difficulty with determining rate laws: the time dependences of the concentrations in first- and second-order reactions are not very different until well after the start of the reaction. Measurements must be made over several half-lives before the two reaction orders can be distinguished.

The half-life of a second-order reaction is much less useful than that of a first-order reaction. The reason can be seen by setting $[\text{A}] = \frac{1}{2}[\text{A}]_0$ in Eq. 3, which gives

$$\frac{1}{\frac{1}{2}[\text{A}]_0} = \frac{1}{[\text{A}]_0} + kt_{1/2}$$

which rearranges to

$$t_{1/2} = \frac{1}{k[\text{A}]_0} \qquad (4)$$

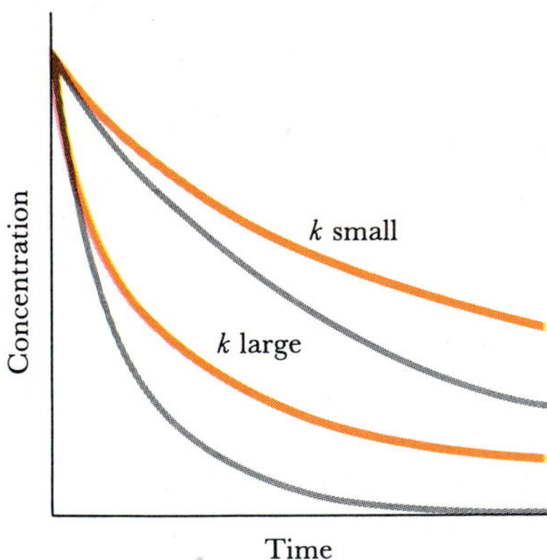

FIGURE 12.14 The characteristic shape (orange lines) of the time dependence of the concentration of the reactant during a second-order reaction. The gray lines are the curves for first-order reactions with the same initial rates.

We see that, in contrast to a first-order reaction, the half-life of a second-order reaction depends on the concentration. Thus, although the half-life is a characteristic of a first-order reaction regardless of the concentrations of reactants, for a second-order reaction the half-life varies with the concentration. For instance, if it takes 10 s for a certain initial concentration to fall to half its initial value, it will take 20 s for that new concentration to fall to half that new value. The doubling of the half-life after each preceding half-life has passed is reflected in the flattening of the orange curves in Fig. 12.14.

Some reactions of the form A + B → products have the overall second-order rate law

$$\text{Rate} = k[\text{A}][\text{B}]$$

An example is the reaction

$$CH_3Br(aq) + OH^-(aq) \longrightarrow CH_3OH(aq) + Br^-(aq)$$
$$\text{Rate} = k[CH_3Br][OH^-]$$

There are two cases in which a rate law of this kind can be simplified to a type we have already met. One is when a reactant has such a high concentration that it barely changes during the reaction. For instance, suppose the molar concentration of OH^- is 1.00 mol/L and the methyl bromide concentration is only 0.01 mol/L; then at the end of the reaction, the OH^- concentration will have fallen to 1.00 mol/L − 0.01 mol/L = 0.99 mol/L, or by only 1%. Under these conditions, the concentration $[OH^-]$ in this reaction, and the concentration [B] in general, can be treated as constant. Then the rate law can be simplified to

$$\text{Rate} = k'[\text{A}] \qquad \text{with } k' = k \times [\text{B}]$$

A reaction such as this one, with a rate law that is *effectively* first order because one substance has a nearly constant concentration, is called a **pseudo–first-order reaction**. The time dependence of the concentration of A is given by Eq. 1, with a rate constant equal to k[B]. Thus, if $[OH^-]$ = 1.0 mol/L in the methyl bromide reaction, then

$$k' = 2.8 \times 10^{-4}\ \text{L mol}^{-1}\ \text{s}^{-1} \times 1.0\ \text{mol L}^{-1} = 2.8 \times 10^{-4}\ \text{s}^{-1}$$

and the half-life of the methyl bromide, as given by Eq. 2, is 2500 s at 25°C.

N An application of a half-life for a third-order reaction is in Exercise 12.82.

E The order of a reaction with respect to each of its participants can be determined by *isolating* each one in turn by ensuring that the concentrations of all the others are in such large excess that they are effectively constant. Then the rate of the reaction of each of the isolated species can be studied in turn. If it is found that the rate is pseudo–first-order in A and pseudo–second-order in B, then we would know that the rate was proportional to $[A][B]^2$ when both species were present with variable concentrations.

U Exercise 12.82.

? Explain why sticks rather than logs of wood are used to start a fire.

CONTROLLING RATES OF REACTIONS

The rates of chemical reactions are affected by concentration, the exposed surface area of the reactants, the temperature, and the presence of catalysts. We have already discussed the effect of concentration, which is taken into account in the rate laws. A striking illustration of the importance of the surface area exposed to a reagent, especially in reactions involving solids, is shown in Fig. 12.15: a dust of very finely divided iron particles ignites spontaneously in air. Dangerous explosions can occur in dusty environments, such as grain elevators and coal mines, for the minute particles have a large surface area exposed to the air and burn very rapidly.

FIGURE 12.15 A solid rod of iron can be heated in a flame without catching fire. However, a powder of very finely divided iron oxidizes rapidly in air to form Fe_2O_3, because the powder presents a much greater surface area for reaction with oxygen.

12.3 THE TEMPERATURE DEPENDENCE OF REACTION RATES

Reaction rates almost always increase when the temperature is raised (Fig. 12.16). An increase of 10°C from room temperature approximately doubles the rate. (In reality, the factor is usually in the range from 1.5 to 4.) In this section we shall explore the reasons for this dependence of rate on temperature.

FIGURE 12.16 Reaction rates almost always increase with temperature. The beaker on the left contains magnesium in cold water and that on the right contains magnesium in hot water. An indicator has been added to show the formation of an alkaline solution as the magnesium reacts.

E Svante Arrhenius (1859–1927), who was encountered in Chapter 3 in connection with a definition of acids and bases, was awarded the lowest class of doctorate, for his examiners did not recognize the importance of his work on solutions of electrolytes. Nevertheless, he went on to win a Nobel prize (1903). His wide interests included an investigation of the greenhouse effect (Section 6.5).

N We consider that there is little advantage in writing the Arrhenius expression in terms of common logarithms. If desired, it would be

$$\log k = \log [A] - \frac{E_a}{2.303R} \times \frac{1}{T}$$

? Comment on the growth rate of tadpoles in water at 15°C and in water at 25°C.

V Video 33: Chemical kinetics and temperature.

E Dependence of reaction rate on temperature. *J. Chem. Educ.*, **1969**, *46*, 674. The origin and status of the Arrhenius equation. *J. Chem. Educ.*, **1982**, *59*, 279. The development of the Arrhenius equation. *J. Chem. Educ.*, **1984**, *61*, 494.

D A variation of the starch-iodine clock reaction (vary temperature). CD2, 78. An organic clock reaction, CD2, 84.

Arrhenius behavior. The fact that a reaction rate increases when the temperature is raised implies that the rate constant of the reaction has increased. The increase in k is often found to obey an equation suggested by Svante Arrhenius in 1889:

$$\ln k = \ln A - \frac{E_a}{RT} \tag{5}$$

The **Arrhenius parameters** A and E_a are constants that depend on the reaction; R is the gas constant, 8.31 J/(K·mol); and T is the temperature (in kelvins). The constant A is called the **frequency factor**. The constant E_a is called the **activation energy** (we shall see its significance and the reason for this name shortly). When we compare the Arrhenius equation with the equation of a straight line

$$\ln k = \ln [A] - \frac{E_a}{R} \times \frac{1}{T}$$

$$y \quad = \quad b \quad + m \times x$$

we see that we can determine whether a reaction rate constant obeys Eq. 5 by drawing an **Arrhenius plot**, a graph of ln k against $1/T$. If the Arrhenius plot is a straight line, then the reaction rate constant obeys Eq. 5 and we can use the graph to determine the values of the two Arrhenius parameters. The intercept with the vertical axis at $1/T = 0$ is equal to ln A, and the slope of the line is equal to $-E_a/R$ (Fig. 12.17).

As Fig. 12.18 shows, the greater the activation energy E_a, the stronger the temperature dependence of the reaction rate. Reactions with small activation energies (about 10 kJ/mol) have rates that increase only slightly with temperature. Reactions with large activation energies (above about 60 kJ/mol, with steep Arrhenius plots) have rates that depend strongly on the temperature.

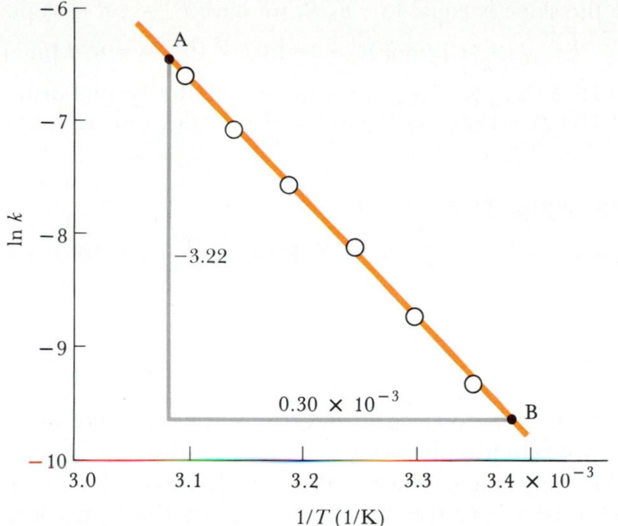

FIGURE 12.17 An Arrhenius plot is a graph of ln k against $1/T$. If, as here, the line is straight, then the reaction is said to show Arrhenius behavior in the temperature range studied, and the activation energy can be obtained from the slope.

T OHT: Fig. 12.17.

? Rate constants are obtained from data plots like those in Fig. 12.7. How would the plots in Figs. 12.7a and 12.7b differ if data were provided for a higher temperature?

D Temperature and reduction of potassium permanganate. CD2, 74.

▼ EXAMPLE 12.10 Measuring an activation energy

The second-order rate constant for the reaction between ethyl bromide and hydroxide ions in water,

$$C_2H_5Br(aq) + OH^-(aq) \longrightarrow C_2H_5OH(aq) + Br^-(aq)$$

was measured at several temperatures, with the following results:

Temperature, °C	25	30	35	40	45	50
k, L mol^{-1} s^{-1}	8.8×10^{-5}	1.6×10^{-4}	2.8×10^{-4}	5.0×10^{-4}	8.5×10^{-4}	1.40×10^{-3}

Find the activation energy of the reaction.

STRATEGY Activation energies are found by plotting ln k against $1/T$ and measuring the slope of the line.

SOLUTION We begin by converting the temperature to the Kelvin scale and drawing up a table of $1/T$ and ln k:

Temperature, °C	T, K	$1/T$, K^{-1}	ln k
25	298	3.35×10^{-3}	−9.34
30	303	3.30×10^{-3}	−8.74
35	308	3.25×10^{-3}	−8.18
40	313	3.19×10^{-3}	−7.60
45	318	3.14×10^{-3}	−7.07
50	323	3.10×10^{-3}	−6.57

The points are plotted in Fig. 12.17. Using points A and B, the slope is

$$\text{Slope} = \frac{-3.22}{0.30 \times 10^{-3}\ \text{K}^{-1}} = -1.07 \times 10^4\ \text{K}$$

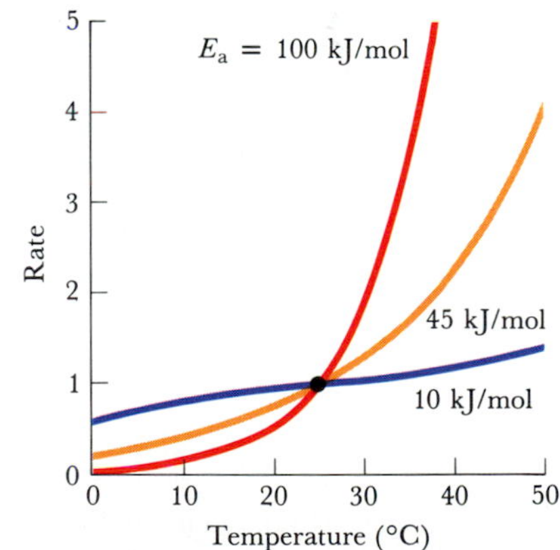

FIGURE 12.18 Reaction rate depends on temperature and activation energy. The upper and lower curves represent large and small activation energies and the middle curve a typical activation energy (45 kJ/mol). All values have been adjusted so that the reaction rate is 1 at 25°C.

? How can rate constants be obtained at different temperatures? (Refer to Fig. 12.7.)

U Exercises 12.77–12.79.

Because the slope is equal to $-E_a/R$, we have $E_a = -R \times$ slope:

$$E_a = -8.31 \text{ J/(mol·K)} \times (-1.07 \times 10^4 \text{ K}) = 89 \text{ kJ/mol}$$

EXERCISE E12.10 The rate constant of the second-order gas-phase reaction $HO(g) + H_2(g) \rightarrow H_2O(g) + H(g)$ varies with temperature as follows:

Temperature, °C	100	200	300	400
k, L mol^{-1} s^{-1}	1.1×10^{-9}	1.8×10^{-8}	1.2×10^{-7}	4.4×10^{-7}

Calculate the activation energy.

[*Answer*: 42 kJ/mol]

Reactions that give a straight line when ln k is plotted against $1/T$ are said to show **Arrhenius behavior**. In fact, a wide range of reactions—not only those in the gas phase—show Arrhenius behavior. Even tropical fireflies flash more quickly on hot nights than on cold, and their rate of flashing is Arrhenius-like over a small range of temperatures. We can conclude that the biochemical reactions responsible for the flashing have rate constants that increase exponentially with temperature in accord with Eq. 5. Some Arrhenius parameters are listed in Table 12.2. Once the Arrhenius parameters have been measured for a reaction, it is very easy to predict the rate constant at any temperature that lies within the range of the original measurements (we cannot be sure that the reaction is Arrhenius-like outside that range).

The activation energy is the more useful of the two Arrhenius parameters, because we can use it to calculate the rate constant at one temperature if its value is already known at another temperature. For example, we might know the rate constant for the synthesis of ammo-

? Why is the activation energy for the recombination of CH_3 radicals zero? Would you expect it to be exactly zero?

? Why is the activation energy for the reaction between O and N_2 so large?

TABLE 12.2 Arrhenius parameters for various reactions

Reaction	A	E_a, kJ/mol
First order, gas phase		
cyclopropane → propene	2.0×10^{15} s^{-1}	272
CH_3—CN → CH_3—NC	4.0×10^{15}	160
$C_2H_6 \rightarrow 2CH_3$	2.5×10^{17}	384
$N_2O \rightarrow N_2 + O$	8.0×10^{11}	250
$2N_2O_5 \rightarrow 4NO_2 + O_2$	6.3×10^{14}	88
Second order, gas phase		
$O + N_2 \rightarrow NO + N$	1×10^{11} L mol^{-1} s^{-1}	315
$OH + H_2 \rightarrow H_2O + H$	8×10^{10}	42
$2CH_3 \rightarrow C_2H_6$	2×10^{10}	0
Second order, in aqueous solution		
$C_2H_5Br + OH^- \rightarrow C_2H_5OH + Br^-$	4.3×10^{11} L mol^{-1} s^{-1}	90
$CO_2 + OH^- \rightarrow HCO_3^-$	1.5×10^{10}	38
acid hydrolysis of sucrose	1.5×10^{15}	108

nia at 300°C but decide that we want to run the reaction at 400°C in an industrial plant. To do this kind of calculation, we need to know only E_a, as the following example illustrates.

▼ EXAMPLE 12.11 Predicting the rate constant at one temperature from its value at another temperature

Calculate the rate constant at 35°C for the hydrolysis of sucrose, given that it is equal to 1.0×10^{-3} L/(mol·s) at 37°C and that the activation energy of the reaction is 108 kJ/mol (see Table 12.2).

STRATEGY Because the temperature is lower, we expect a smaller rate constant at 35°C than at 37°C. We are given only E_a, so we cannot use Eq. 5 directly. However, we can write Eq. 5 for two temperatures and rearrange the equation to eliminate A:

$$\text{At temperature } T': \quad \ln k' = \ln A - \frac{E_a}{RT'}$$

$$\text{At temperature } T: \quad \ln k = \ln A - \frac{E_a}{RT}$$

Subtraction of the second equation from the first gives

$$\ln k' - \ln k = -\frac{E_a}{RT'} + \frac{E_a}{RT}$$

which can be rearranged to

$$\ln \frac{k'}{k} = \frac{E_a}{R}\left(\frac{1}{T} - \frac{1}{T'}\right)$$

Now all we need to do is substitute the data into this expression (and remember to express temperature on the Kelvin scale).

SOLUTION Because $E_a = 108$ kJ/mol, $R = 8.314$ J/(K·mol) or 8.314×10^{-3} kJ/(K·mol), $T = 310$ K, and $T' = 308$ K,

$$\ln \frac{k'}{k} = \frac{108 \text{ kJ/mol}}{8.314 \times 10^{-3} \text{ kJ/mol}} \times \left\{\frac{1}{310} - \frac{1}{308}\right\} = -0.272$$

Hence,

$$\frac{k'}{k} = 0.762$$

(The K in the temperature has been canceled by the 1/K in the units of R.) Because $k = 1.0 \times 10^{-3}$ L/(mol·s) at 37°C,

$$k' = 0.762 \times [1.0 \times 10^{-3} \text{ L/(mol·s)}] = 6.9 \times 10^{-4} \text{ L/(mol·s) at 35°C}$$

The very high activation energy of the reaction makes its rate very sensitive to temperature.

EXERCISE E12.11 The rate constant of the second-order reaction between CH_3Br and OH^- ions is 2.8×10^{-4} L/(mol·s) at 25°C. What is it at 50°C? See Table 12.2.

[*Answer*: 4.7×10^{-3} L/(mol·s)] ▲

E A really attentive student might wonder why the gas constant plays a role in the description of reaction rates in solution. The answer, of course, is that R is a modification of the Boltzmann constant k_B ($R = N_A k$), and the Boltzmann constant occurs wherever the distribution of molecules over available energy levels is in question (as it is when discussing the rates of reactions). Really attentive *instructors* might therefore wonder, instead, why R occurs in the ideal gas equation!

Collision theory. Up to this point, we have treated the Arrhenius parameters as quantities that summarize the experimental observations on the temperature dependence of reaction rates. The *explanation* of Arrhenius behavior and an interpretation of the parameters comes from the **collision theory** of elementary gas-phase bimolecular reac-

U Exercise 12.76.

E Collision theory of chemical reactions. *J. Chem. Educ.*, **1974**, *51*, 790. Collision probabilities and rate of chemical reaction. *J. Chem. Educ.*, **1971**, *48*, 390.

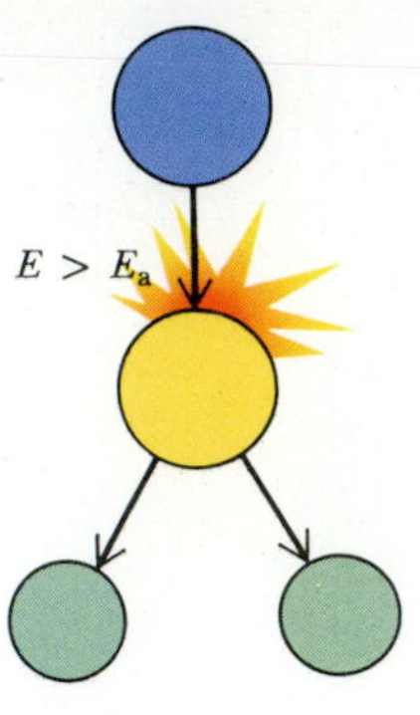

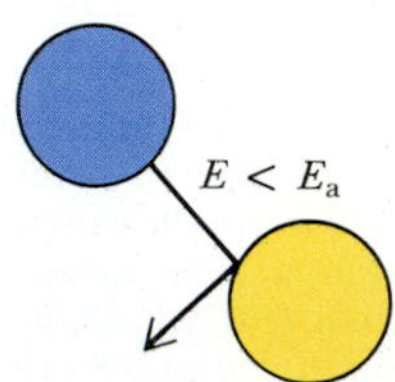

FIGURE 12.19 In the collision theory of chemical reactions, reaction occurs only when two molecules collide with a certain minimum kinetic energy, the activation energy of the reaction (top). Otherwise they simply bounce apart (bottom).

N The factors that determine the rate constant for a reaction are the probability of encounter, the orientation of the colliding species, and the energy required for reaction when a collision occurs.

E The high activation energy required for the reaction of N_2 (see Table 12.2) is one reason why nitrogen is a useful, inert diluent of the atmosphere. Although N_2 molecules undergo vast numbers of collisions every second, hardly any of them results in reaction because the activation energy is so high.

D Hydrolysis of CO_2 and SO_2 (dependence of rate on structure). TDC, 8–9s(c).

T OHT: Fig. 12.20.

tions. This theory supposes that molecules react only if they smash together with at least enough kinetic energy for bonds to be broken (Fig. 12.19). In this respect, molecules behave like billiard balls; they bounce apart if they collide at low speed, but smash into pieces when the impact is really powerful. The essential content of collision theory is that the frequency factor A is proportional to the frequency with which molecules collide, and the activation energy E_a determines whether a collision is likely to lead to the formation of products.

Consider a reaction between two substances A and B. Because the rate at which collisions occur between A and B molecules is directly proportional to their concentrations, we can write

$$\text{Rate of collisions} = \text{constant} \times [\text{A}][\text{B}]$$

If every collision were successful, we would conclude that the constant was the rate constant itself. However, it is easy to see that not every collision can be successful. The kinetic theory of gases lets us calculate that, in a gas at 25°C and 1 atm pressure, each molecule collides with another within about 10^{-10} s. Therefore, if the reaction rate were equal to the rate at which molecules met, most gas-phase reactions would have half-lives of about 10^{-9} to 10^{-10} s. Because as we have seen, some half-lives are minutes and even hours long, we must conclude that only a very tiny fraction of the collisions are successful.

In collision theory, it is assumed that *a collision is successful only if the molecules collide with at least the activation energy of the reaction.* In other words, the **activation energy** is the minimum energy required for reaction. The fraction of collisions with at least the energy E_a is given by the Maxwell distribution of speeds (Section 5.6 and Fig. 12.20). As shown by the shaded area under the blue curve in the illustration, at room temperature only a very tiny fraction of molecules have enough energy to react. At higher temperatures, a much larger fraction of molecules can react, as represented by the shaded area under the red curve. The precise value of the fraction of collisions that occur with at least the energy E_a can be calculated; for $E_a = 45$ kJ/mol (a typical value of the activation energy), that fraction is about 10^{-8}. This result shows that fewer than 1 in 100 million collisions can be expected to lead to reaction.

The energy is not the only factor determining whether a collision is successful. In many cases, the molecules need to collide in the correct relative orientation, and this additional requirement is also taken into account by the value of the frequency factor A. For example, in the reaction

$$\text{Cl} + \text{HI} \longrightarrow \text{HCl} + \text{I}$$

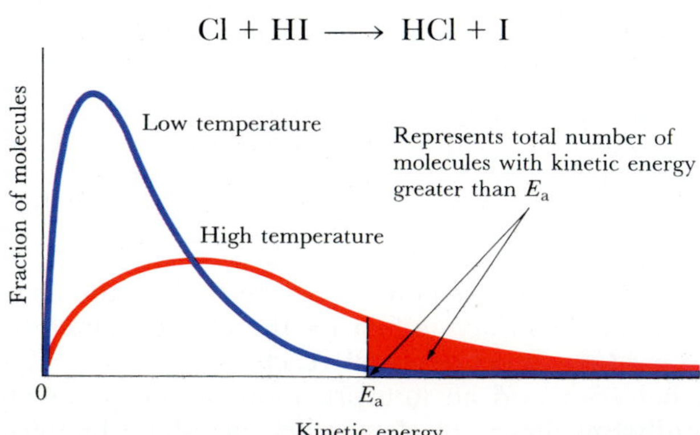

FIGURE 12.20 The fraction of molecules that collide with at least a kinetic energy E_a is given by the shaded areas under each curve. This fraction increases as the temperature is raised.

the Cl atom reacts with the HI molecule only if its line of approach is inside a cone of half-angle 30° surrounding the hydrogen atom (Fig. 12.21). For collisions between more complex species, the orientation criterion may be even more severe and the reaction rate considerably lower than the rate at which sufficiently energetic collisions occur.

Activation barriers. Collision theory can be extended to explain why the Arrhenius equation also applies to reactions in solution. This more general theory is called **activated complex theory**. In this theory, two molecules approach and distort as they meet and form an **activated complex**, a combination of the two molecules that either can go on to form products or can fall apart into the unchanged reactants. When we follow the potential energy of the two molecules, we see that it rises (at the expense of the kinetic energy that they had initially) as they merge to form the activated complex and then falls again as the product molecules form and separate (Fig. 12.22). The entire curve is called the **reaction profile**, and the hump between reactants and products is the **activation barrier**. Because only reactant molecules that have enough kinetic energy and the correct orientation in the activated complex to cross the barrier can turn into products (the rest just departing unchanged from the encounter), the height of the barrier is the activation energy E_a of the reaction.

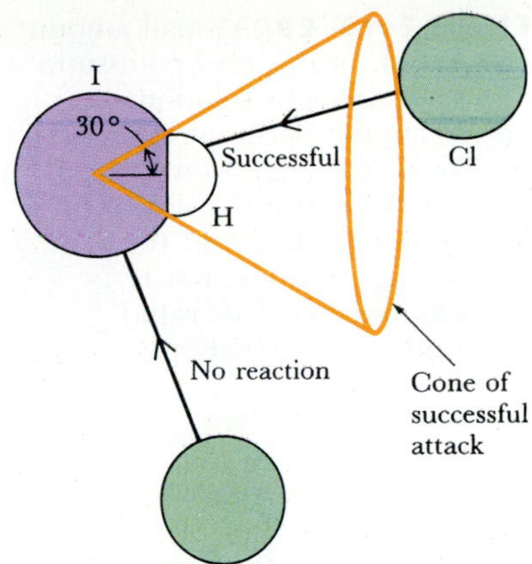

FIGURE 12.21 Whether or not a reaction occurs when two species collide also depends on their relative orientation. In the reaction between a Cl atom and an HI molecule, for example, only those collisions in which the Cl atom approaches the HI molecule along a line of approach that lies inside the brown cone lead to reaction, even though the energy of other collisions may exceed the activation energy.

T OHT: Fig. 12.21.

E Just what is a transition state? *J. Chem. Educ.*, **1988**, *65*, 540. Interpretation of activation energy. *J. Chem. Educ.*, **1978**, *55*, 309. Changing conceptions of activation energy. *J. Chem. Educ.*, **1981**, *58*, 612.

U Exercises 12.70 and 12.71.

V Video 32: Autocatalysis.

T OHT: Figs. 12.22 and 12.24.

D Acetone oxidation with copper catalyst. TDC, 8–7s(c). Catalytic copper. CD2, 79.

12.4 CATALYSIS

The rates of some reactions are increased by the addition of small amounts of certain substances that can often be recovered at the end of the reaction (Fig. 12.23). Such substances are called catalysts (from the Greek words meaning roughly "breaking down by coming together"):

> A **catalyst** is a substance that increases the rate of a reaction without being consumed in the reaction.

The Chinese name for catalyst, "marriage broker," captures the sense quite well. An example of a reaction that can be catalyzed is the thermal decomposition of potassium chlorate:

$$2KClO_3(s) \xrightarrow{\Delta} 2KCl(s) + 3O_2(g)$$

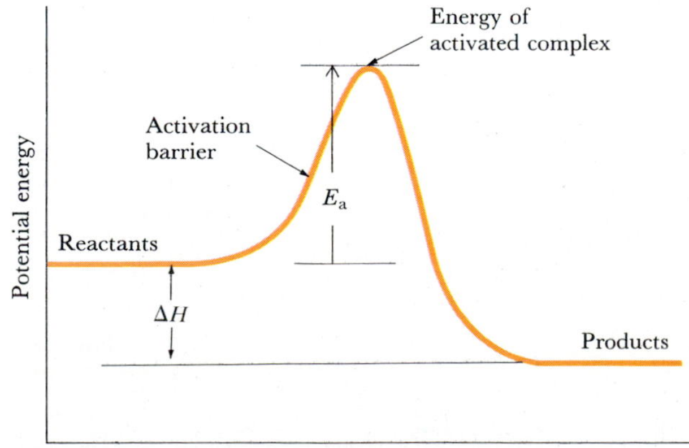

FIGURE 12.22 A reaction profile for an exothermic reaction. In the activated complex theory of chemical reactions, it is supposed that the potential energy of the reactant molecules rises as they approach each other, reaches a maximum as they form an activated complex, and then decreases as the product molecules form and separate. Only molecules with enough energy and the correct relative orientation can cross the barrier and react when they meet.

FIGURE 12.23 A small amount of catalyst, in this case potassium iodide in aqueous solution, can accelerate the decomposition of hydrogen peroxide to water and oxygen. This effect is shown by (a) the slow inflation of the balloon when no catalyst is present and by (b) its rapid inflation when a catalyst is present.

(a)

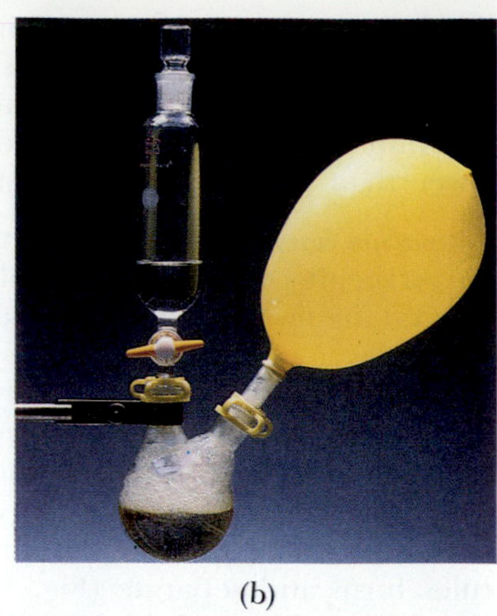

(b)

The reaction is very slow even when the solid is heated to a high temperature. However, when a small amount of manganese dioxide is added to the potassium chlorate, oxygen is evolved briskly at the salt's melting point (356°C), and the manganese dioxide can be recovered at the end of the reaction. Another example is the use of finely divided iron to catalyze the synthesis of ammonia:

$$N_2(g) + 3H_2(g) \xrightarrow{\text{Fe}} 2NH_3(g)$$

(Note how the catalyst is written above the arrow in the chemical equation.) This catalyst was discovered by the German chemist Fritz Haber shortly before World War I, and it has made available enormous quantities of ammonia. At first the ammonia was used largely for explosives, but now its principal destinations are fertilizers and plastics.

A catalyst speeds up a reaction by providing an alternative pathway from reactants to products. This new pathway has a lower activation energy than that of the original pathway (Fig. 12.24). At the same temperature, a greater fraction of reactant molecules can cross the barrier and turn into products.

E Homogeneous, heterogeneous, and enzymatic catalysis. *J. Chem. Educ.*, **1988**, *65*, 765.

E The new path for the catalyzed reaction may involve several additional steps, each with a lower activation energy than that for the uncatalyzed reaction. The reaction profile may change from one with a single activation energy to one having several lower activation energies.

D Oxidation of manganese(II) sulfate by a catalyst (homogeneous catalysis). CD2, 83.

Homogeneous and heterogeneous catalysis. A catalyst is **homogeneous** if it is present in the same phase as the reactants. For reactants that are

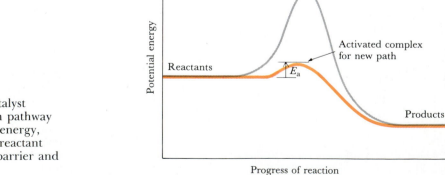

FIGURE 12.24 A catalyst provides a new reaction pathway with a lower activation energy, thereby allowing more reactant molecules to cross the barrier and form products.

gases, a homogeneous catalyst is also a gas. When the reactants are liquids, a homogeneous catalyst is a dissolved liquid or solid.

Dissolved bromine is a homogeneous liquid-phase catalyst for the decomposition of aqueous hydrogen peroxide:

$$2H_2O_2(aq) \xrightarrow{Br_2} 2H_2O(l) + O_2(g)$$

In the absence of bromine, a solution of hydrogen peroxide can be stored for a long time at room temperature; however, bubbles of oxygen form as soon as a drop of bromine is added. Bromine's role is believed to be its reduction to Br^- in one step followed by oxidation back to Br_2 in a second step:

$$Br_2(aq) + H_2O_2(aq) \longrightarrow 2Br^-(aq) + 2H^+(aq) + O_2(g)$$

$$2Br^-(aq) + H_2O_2(aq) + 2H^+(aq) \longrightarrow Br_2(aq) + 2H_2O(l)$$

When we add the two steps, the Br_2 and $2Br^-$ cancel, leaving the overall reaction as written above. Hence, although Br_2 molecules have participated in the reaction, they are not consumed.

D Place 3% $H_2O_2(q)$ in a glass dish on an overhead projector. Then add a pinch of manganese(IV) oxide or a drop of bromine.

A catalyst is **heterogeneous** if it is in a phase different from that of the reactants. The most common heterogeneous catalysts are finely divided solids used in gas-phase or liquid-phase reactions; these are often called **contact catalysts**. One example is the iron catalyst used in the Haber process. Another is finely divided vanadium pentoxide, V_2O_5, which is used in the contact process for the production of sulfuric acid:

E Vanadium pentoxide is the most common compound of vanadium used industrially.

V Video 35: Catalytic oxidation of ammonia.

$$2SO_2(g) + O_2(g) \xrightarrow{V_2O_5} 2SO_3(g)$$

The rhodium catalyst used in the Ostwald process for the production of nitric acid from ammonia brings about a large increase in the oxidation number of nitrogen (from −3 to +2) and avoids the formation of molecular nitrogen, a less highly oxidized form (Fig. 12.25):

$$4NH_3(g) + 5O_2(g) \xrightarrow{\Delta,Rh} 4NO(g) + 6H_2O(g)$$

(a)

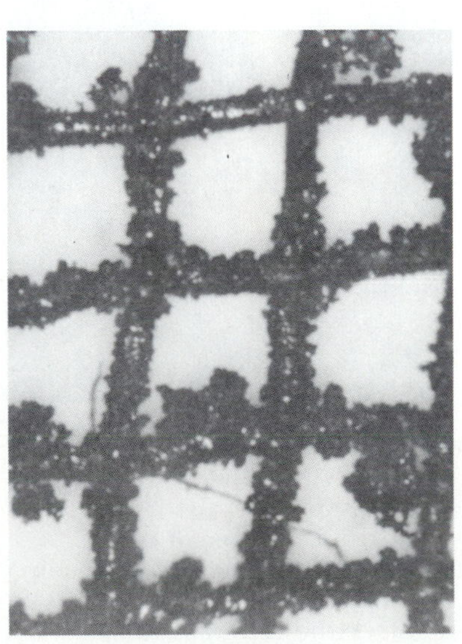

(b)

FIGURE 12.25 (a) Installing a rhodium gauze catalyst for the production of nitric acid by the oxidation of ammonia. (b) An enlarged view of the catalyst after use.

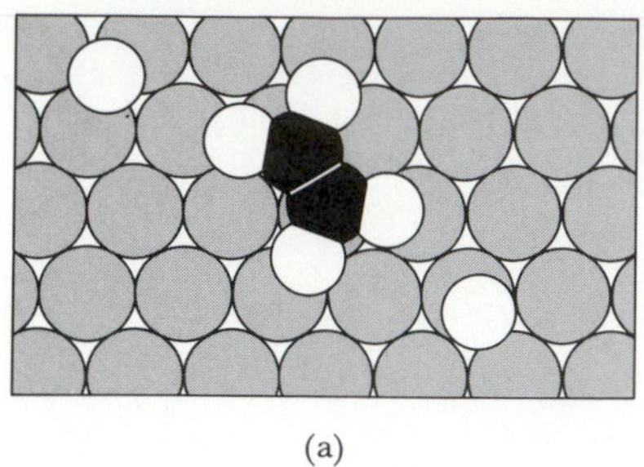
(a)

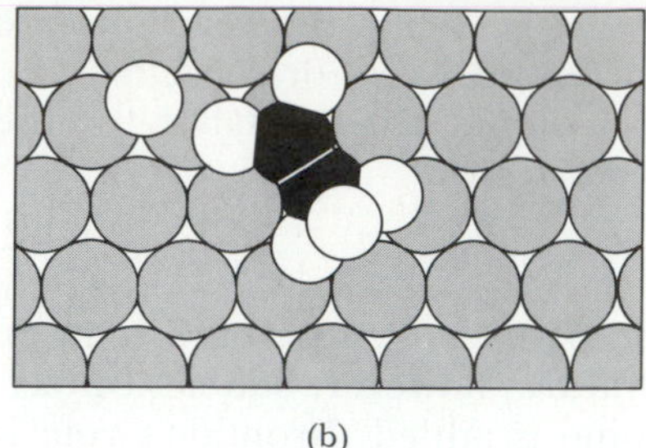
(b)

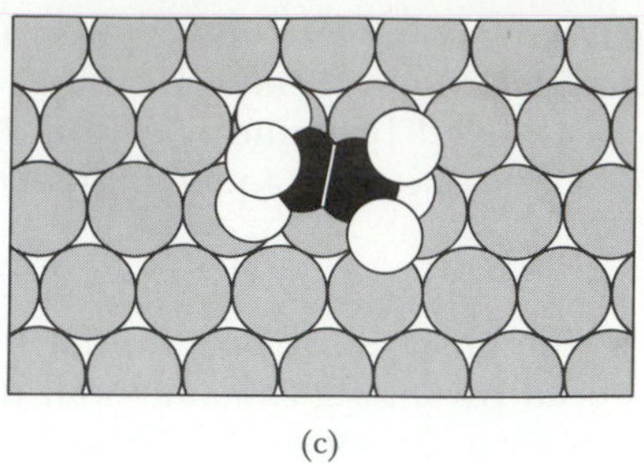
(c)

FIGURE 12.26 The reaction between ethylene, C_2H_4, and hydrogen on a metal surface. (a) The surface adsorbs the reactant molecules and the H_2 molecules dissociate into atoms. (b) A hydrogen atom has moved across the surface, collided with a C_2H_4 molecule, and formed a C—H bond. (c) A second hydrogen atom has collided with the C_2H_5— and formed another C—H bond; the resulting C_2H_6 molecule leaves the surface.

The new reaction pathway opened by a heterogeneous catalyst generally depends on the catalyst's ability to adsorb the reactant onto its surface. This adsorption often results in the dissociation of the reactant molecule—or at least the weakening of its bonds—and the reaction can then proceed more quickly (Fig. 12.26). The reaction pathway for the ammonia synthesis is still not fully understood, but one important step may be the adsorption of N_2 molecules by the iron surface and the weakening of the strong N≡N triple bond.

Catalysts are used in the catalytic converters of automobiles to facilitate the complete and rapid combustion of unburned fuel. Some fuel always remains as unburned gases in the exhaust, a mixture that includes carbon monoxide, the unburned hydrocarbons of gasoline, and the nitrogen oxides collectively referred to as NO_x. The pollution these substances can cause is lessened if the carbon compounds are oxidized to carbon dioxide and the NO_x reduced, by another catalyst, to nitrogen. However, although it is easy to find metal and metal oxide catalysts to speed up these reactions, selecting practical catalysts is very difficult. For instance, a catalyst that causes the reduction of NO_x to proceed beyond N_2 to NH_3 is undesirable, because ammonia will then be released into the atmosphere, where it will be oxidized back to NO_x. Furthermore, the catalyst must work even at the relatively low temperatures encountered when the engine is first started, for that is when the emission problem is worst. It is for this reason that cheap metals such as copper are less commonly used than the expensive noble metals, especially platinum: the latter are active at low temperatures. A further problem arises from the presence of sulfur in fuel, for the catalyst used to oxidize the unburned hydrocarbons may also catalyze the oxidation of SO_2 to SO_3, thus causing the vehicle to lay a trail of sulfuric acid in an urban, mobile version of the contact process.

Catalysts can be **poisoned**, or inactivated. A common cause of such poisoning is the adsorption of a molecule so tightly by the catalyst that it seals the catalyst's surface against further reaction. Some heavy metals, especially lead, are very potent catalyst poisons, which is why lead-free gasoline must be used in engines fitted with catalytic converters. The elimination of lead in gasoline has the further benefit of decreasing the amount of lead in the environment: lead is also a poison for people, who as living organisms depend on their own systems of catalysts to survive and grow.

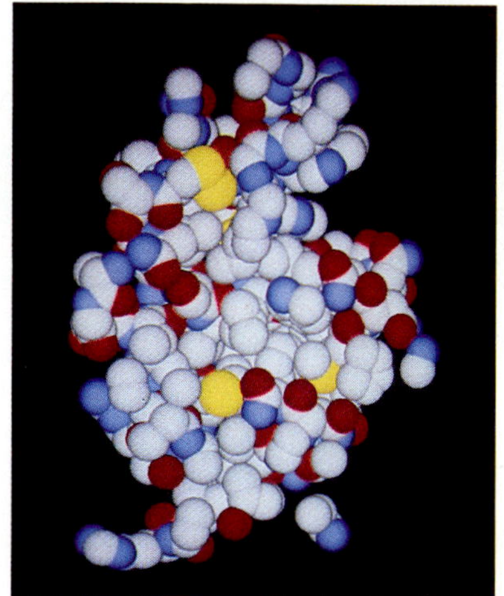

FIGURE 12.27 The lysozyme molecule shown here is a typical enzyme molecule. Lysozyme occurs in a number of places, including tears and the mucus in the nose. One of its functions is to attack the cell walls of bacteria and destroy them.

Enzymes. Biological catalysts are called **enzymes** (from the Greek words for "in yeast"; Fig. 12.27). They are proteins with a slotlike **active site**, where reaction takes place. The **substrate**, the molecule on which the enzyme acts, is recognized by its ability to fit into the slot, as a key fits into a lock (Fig. 12.28). Once in the slot, the substrate undergoes reaction, which in many cases takes the form of an acid-base neutralization in which hydrogen ions are removed from one group and added to another. The modified substrate molecule is then released for use in the next stage, which is controlled by another enzyme, and the original enzyme molecule is free to receive the next substrate molecule.

One form of biological poisoning mirrors the effect of lead on a catalytic converter. The activity of enzymes is destroyed when an alien substrate such as lead attaches too strongly to the reactive site, for then the site is blocked and made unavailable to the substrate (Fig. 12.29). As a result, the chain of biochemical reactions in the cell stops, and the cell dies. The action of nerve gases is believed to stem from their ability to block the enzyme-controlled reactions that enable impulses to travel through nerves. Arsenic, that favorite of fictional poisoners, acts in a similar way. After ingestion as As(V) in the form of arsenate ions (AsO_4^{3-}), it is reduced to As(III), which binds to the —SH groups of enzymes and inhibits their action.

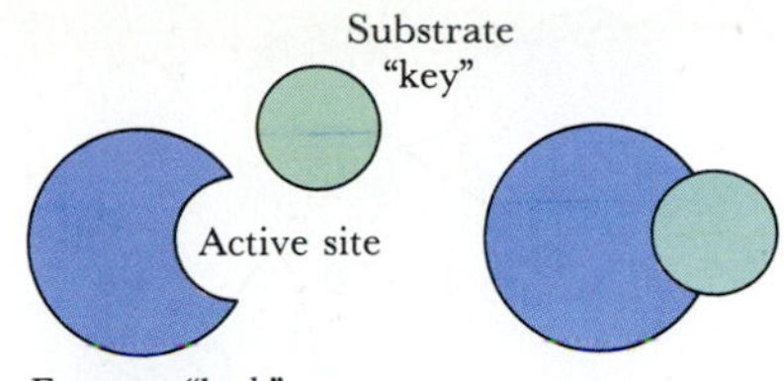

FIGURE 12.28 In the lock-and-key model of enzyme action, the correct substrate is recognized by its ability to fit into the active site.

REACTION MECHANISMS

Until now, we have used chemical reactions to show the stoichiometry of reactions. For instance, the equation

$$H_2(g) + I_2(g) \longrightarrow 2HI(g)$$

is interpreted as meaning that 1 mol H_2 reacts with 1 mol I_2 to give 2 mol HI: the equation does not mean that one particular H_2 molecule reacts with one particular I_2 molecule to produce two HI molecules. The question we now explore is how the rearrangement of atoms in the sample actually takes place. Questions of this type are of great importance in chemistry, and many reactions can be understood, and their products predicted, once we know the details of events taking place between molecules. For example, how is it some drugs are effective in combating viral infections and others are not? How does a DNA molecule replicate itself? Many questions like these can be answered only after years of research into reaction mechanisms.

T OHT: Figs. 12.28 and 12.29.

E A link between kinetics and bonding theory. Frontier molecular orbitals. *J. Chem. Educ.*, **1973**, *50*, 463. Arthur Lapworth: The genesis of reaction mechanism. *J. Chem. Educ.*, **1972**, *49*, 750. Rate laws for elementary chemical reactions. *J. Chem. Educ.*, **1974**, *51*, 254.

12.5 ELEMENTARY REACTIONS

All but the simplest reactions are in fact the outcome of several, and sometimes many, steps, each of which is called an **elementary reaction**. In the formation of HI, these steps include the dissociation of the I_2 molecules into atoms and their attack on H_2 molecules. The probable identities of the elementary reactions involved in a reaction are discovered by determining the rate law for the reaction experimentally and then trying to account for it in terms of a mechanism:

> A **reaction mechanism** is the pathway that is proposed for the overall reaction and accounts for the experimental rate law.

Elementary reactions are expressed by writing chemical equations in the usual way. However, these equations have a different significance: they show how *individual* atoms and molecules take part in the reaction

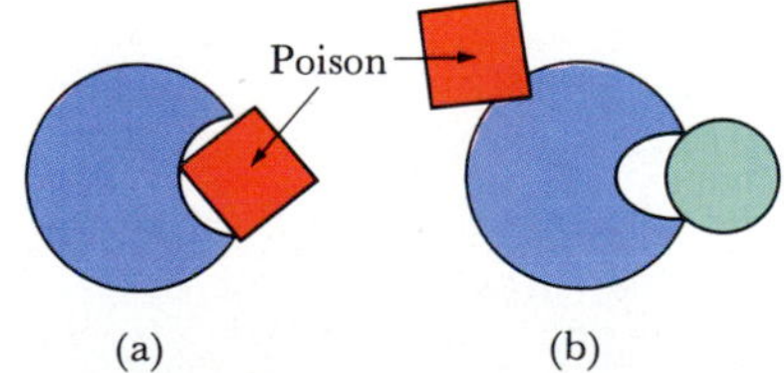

FIGURE 12.29 (a) An enzyme poison (represented by the orange square) can act by attaching so strongly to the active site that it blocks the site, thereby taking the enzyme out of operation. (b) Alternatively, the poison may attach elsewhere, thereby distorting the active site so that the substrate no longer fits.

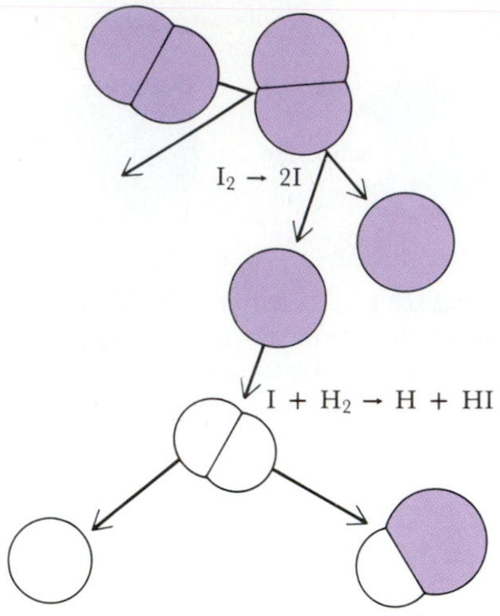

FIGURE 12.30 The chemical equations for elementary reactions show the *individual* events that are taking place between atoms and molecules that encounter one another. This figure illustrates two of the steps in the production of HI. In one ($I_2 \rightarrow 2I$), a collision results in the dissociation of an I_2 molecule. In the other, one of the I atoms so produced attacks an H_2 molecule ($I + H_2 \rightarrow HI + H$).

U Exercise 12.82.

E Three-body collisions are important for removing the excess energy of two atoms that collide: without the third body, the newly formed molecule would immediately dissociate, but that is avoided if some energy is removed by a third molecule.

(Fig. 12.30). For example, one step in the formation of HI is the dissociation of an I_2 molecule. This elementary reaction is written in the usual way:

$$I_2 \longrightarrow 2I$$

Because we are dealing with an elementary reaction, the equation is taken to mean that a *particular* I_2 molecule gives two particular I atoms. To emphasize this difference from the interpretation of chemical equations we have used up to now, we shall always write elementary reactions without showing the physical state (*s*, *l*, and so on). Thus, if we believe that the dissociation of I_2 is followed by the attack of one of these I atoms on an H_2 molecule, then we write

$$I + H_2 \longrightarrow HI + H$$

When there is a risk of confusion between the equations for an elementary reaction and those of the overall outcome of reactions, the intended interpretation will be stated.

The two examples of elementary reactions given above differ in their molecularity:

> The **molecularity** of an elementary reaction is the number of reactant molecules (or atoms) taking part in it.

Thus, the dissociation $I_2 \rightarrow 2I$ is a **unimolecular reaction**, an elementary reaction in which a single reactant molecule (I_2) changes into products. The reaction between an iodine atom and a hydrogen molecule, $I + H_2 \rightarrow HI + H$, is a **bimolecular reaction**, an elementary reaction in which two species (atoms, molecules, or ions) take part. It is possible to think of a termolecular reaction in which three atoms or molecules collide simultaneously. However, the chance of this happening is very slight, and most reaction mechanisms ignore the possibility; in some reactions, however, a termolecular collision is the only way of forming products, and they play a significant role in the reactions in the upper atmosphere.

EXAMPLE 12.12 Judging the molecularity of elementary reactions

What is the molecularity of the elementary reactions (a) $2NO_2 \rightarrow NO + NO_3$ and (b) $Ar + 2O \rightarrow Ar + O_2$?

STRATEGY All we need to do is count the number of species (molecules or atoms) on the left of the arrow and quote that as the molecularity of the reaction.

SOLUTION (a) Two NO_2 molecules collide to give the products, so the reaction is bimolecular. (b) Three atoms collide to give the products (one unchanged Ar atom and an O_2 molecule), so the reaction is termolecular.

EXERCISE E12.12 What is the molecularity of the elementary reactions (a) $C_2N_2 \rightarrow 2CN$ and (b) $Ar + N + O \rightarrow Ar + NO$?

[*Answer*: (a) unimolecular; (b) termolecular]

We will now show that once we know the molecularity of an elementary reaction, we can write down the rate law of that elementary reaction from its chemical equation. It must be stressed again, however, that this is possible *only* for elementary reactions. We shall soon see that an ability to write down rate laws for elementary reactions is the first step in accounting for the rate laws of more complicated, multistep reactions.

Unimolecular reactions. As we saw earlier, the elementary reaction $I_2 \rightarrow 2I$ is a unimolecular reaction. Another unimolecular reaction is the reaction in which an energetically excited ozone molecule vibrates so vigorously that it shakes off an oxygen atom:

$$O_3 \longrightarrow O_2 + O$$

The unimolecular decomposition of ozone is one of the elementary reactions that contribute to the very complex reactions taking place in the stratosphere. Ozone molecules in the stratosphere are subjected to intense ultraviolet radiation from the sun. Upon absorbing the highly energetic photons, they dissociate into fragments that take part in further reactions.

N An analogy between unimolecular and bimolecular collisions can be made using one-car explosions and two-car collisions.

E Mechanistic implications and ambiguities of rate laws. *J. Chem. Educ.*, **1970**, *47*, 805; Activated complex theory of bimolecular gas reactions. *J. Chem. Educ.*, **1972**, *49*, 480.

In a unimolecular reaction, the excited molecules (for instance, the vigorously vibrating O_3 molecules in the stratosphere) fall apart at random. Because the number dissociating is directly proportional to the total number of molecules present, the reaction rate for the decomposition is directly proportional to the concentration. In other words, an elementary unimolecular reaction has a first-order rate law:

$$A \longrightarrow \text{products} \qquad \text{Rate} = k[A]$$

For the elementary ozone reaction, we can write

$$\text{Rate of } O_3 \text{ decomposition} = k[O_3]$$

Bimolecular reactions. As noted earlier, the reaction between an iodine atom and a hydrogen molecule, $I + H_2 \rightarrow HI + H$, is a bimolecular reaction. In this reaction, an I atom collides with an H_2 molecule and carries off one of the H atoms. Another bimolecular reaction is the reaction that occurs in the atmosphere when an O atom collides with an O_3 molecule. The force of the impact breaks one of the O—O bonds in the O_3 molecule, and the incoming atom carries off an O atom from the O_3 molecule:

$$O + O_3 \longrightarrow 2O_2$$

A bimolecular reaction occurs only if two species meet. This requirement suggests that the reaction rate should be directly proportional to the concentrations of both species, because the chance of two molecules meeting increases if the concentration of either is increased. In other words, a bimolecular reaction has an overall second-order rate law:

$$A + B \longrightarrow \text{products} \qquad \text{Rate} = k[A][B]$$

For the elementary reaction above,

$$\text{Rate of } O_2 \text{ formation} = k[O][O_3]$$

The overall reaction. Our task now is to see how a series of unimolecular and bimolecular elementary reactions combine to account for the overall reaction. We first propose a mechanism, writing down a series of elementary reactions that are likely to play a role in the overall reaction. (These reactions are often suggested by additional experimental evidence.) The sum of the elementary reactions must be the overall

The term *free radical* is now obsolescent (but still widely used).

reaction. Finally, we test the proposed mechanism by seeing whether the rate laws for the elementary reactions combine to give the experimental rate law for the overall reaction.

Some reactions are believed to be single-step processes in which one reactant molecule attacks another. The reaction between methyl bromide and hydroxide ions discussed earlier is such a reaction. The observed second-order rate law is consistent with a single bimolecular reaction in which an OH^- ion attacks a CH_3Br molecule. This case therefore involves the single elementary reaction

$$CH_3Br + OH^- \longrightarrow CH_3OH + Br^-$$

which has the same appearance as the overall reaction

$$CH_3Br(aq) + OH^-(aq) \longrightarrow CH_3OH(aq) + Br^-(aq)$$

Because the elementary step is a bimolecular reaction, its rate is directly proportional to the concentrations of the two reactants, and we can write

$$\text{Rate} = k[CH_3Br][OH^-]$$

E The role of the solvent greatly complicates the interpretation of reactions in solution. In some cases, the *diffusion-controlled reactions*, the rate of diffusion through the solvent determines the reaction rate; in others, the *activation-controlled reactions*, the rate-determining step is the rate at which the two reactants acquire enough energy to react.

This expression is in agreement with the observed second-order rate law for the overall reaction.

In building a multistep reaction mechanism, we must take into account the possibility that the products of one step will be the reactants in the next. Because the reverse reactions may also occur, we include them in the mechanism unless we know from the size of the rate constants that they are so slow that they can be ignored. These points can be illustrated with the reaction between nitrogen dioxide and carbon monoxide, which is second order in NO_2:

$$NO_2(g) + CO(g) \longrightarrow NO(g) + CO_2(g) \qquad \text{Rate} = k[NO_2]^2$$

The fact that CO does not appear in the rate law suggests that the reaction is not an elementary one, but has at least two steps. A mechanism has been proposed in which, in the first step, two NO_2 molecules collide to give an NO_3 molecule and the NO molecule:

1. $NO_2 + NO_2 \rightarrow NO_3 + NO$

$$\text{Rate of formation of } NO_3 = k_1[NO_2]^2$$

E The existence of a reaction intermediate is often confirmed by the addition of a scavenger species. For example, to remove oxygen atoms, a radical scavenger for oxygen may be introduced into the reaction system. Nitrogen monoxide, NO, is a good radical scavenger, and radical reaction rates slow dramatically if it is introduced.

The NO_3 molecule does not appear in the overall reaction because it is not present at the beginning or at the end. However, it does take part in the elementary reactions that lead to the final product: it is therefore an example of a **reaction intermediate**, a species that is produced and consumed during the reaction but does not appear in the overall equation. As in this case, a reaction intermediate is often a **radical**, a molecular fragment with one or more unpaired electrons. When a proposed mechanism involves a reaction intermediate, one test of the mechanism is to determine experimentally whether the intermediate is present while the reaction is in progress. This determination can often be done spectroscopically, because the intermediate is likely to absorb light of a characteristic wavelength.

In the second step, the NO_3 molecule collides with a CO molecule and gives up one of its O atoms:

U Exercise 12.75.

2. $NO_3 + CO \rightarrow NO_2 + CO_2$

$$\text{Rate of formation of } CO_2 = k_2[NO_3][CO]$$

The overall reaction is the sum of these two steps:

$$2NO_2 + NO_3 + CO \longrightarrow NO_3 + NO_2 + NO + CO_2$$

which simplifies to

$$NO_2 + CO \longrightarrow NO + CO_2$$

as required.

E Rate-controlling step: A necessary or useful concept? *J. Chem. Educ.*, **1988**, *65*, 250. What is the rate-limiting step of a multistep reaction? *J. Chem. Educ.*, **1981**, *58*, 32.

Rate-determining steps. In general, the overall rate is determined by both steps. However, if the first step is much slower than the second, so that the products of the first step immediately take part in the second step, the overall rate of the reaction will be equal to the rate of the first step. In our example, as soon as the NO_3 radical is formed, it reacts with a CO molecule; hence, overall,

$$\text{Rate of formation of } CO_2 = \text{rate of formation of } NO_3 = k[NO_2]^2$$

in agreement with the observed rate law.

Step 1 in the reaction between NO_2 and CO is an example of a rate-determining step:

> The **rate-determining step** is the slowest step in a multistep reaction and therefore the step that governs the rate of the overall reaction.

A rate-determining step is like a slow ferry on the route between two cities (Fig. 12.31). The rate at which the traffic arrives at its destination is governed by the rate at which it is ferried across the river, because this step is much slower than the rate at which the traffic travels along the roads to the ferry. The action of a catalyst is like opening a new highway that avoids the ferry crossing.

N A catalyst does not lower the activation energy of an existing pathway: it provides a new pathway with a lower activation energy.

? Could a substance act as an *anticatalyst* to retard a reaction that in the absence of a catalyst runs smoothly and rapidly?

Another example of a mechanism with a rate-determining step is the one proposed for the decomposition of ozone in the upper atmosphere. The overall reaction is

$$2O_3(g) \longrightarrow 3O_2(g)$$

and the experimental rate law, which was mentioned before, is

$$\text{Rate of decomposition of } O_3 = \frac{k[O_3]^2}{[O_2]}$$

If the rate law had been simply $k[O_3]^2$, we might have been tempted to suppose that the mechanism involved one bimolecular step:

$$O_3 + O_3 \longrightarrow 3O_2 \qquad \text{Rate} = k[O_3]^2$$

However, the fact that O_2 appears in the rate law shows that a simple bimolecular elementary reaction between two O_3 molecules cannot be

T OHT: Fig. 12.31.

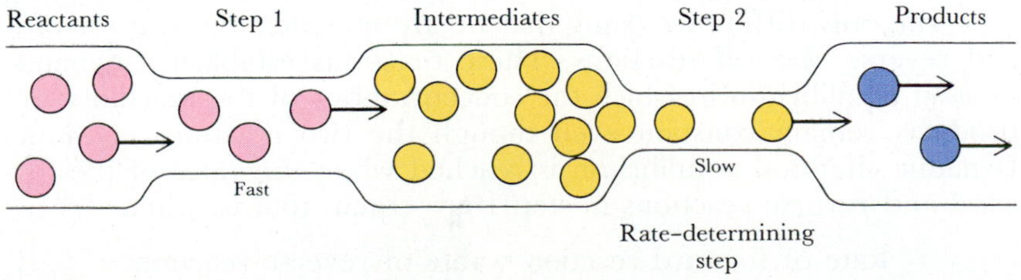

FIGURE 12.31 The rate-determining step in a reaction is the step that governs the rate at which products are formed in a multistep reaction.

the complete mechanism. We need a more complex mechanism to account for the decrease in rate as the oxygen concentration increases.

One possible mechanism is

1. Unimolecular dissociation of excited O_3:

$$O_3 \longrightarrow O_2 + O \qquad \text{Rate} = k_1[O_3]$$

The O atoms are reaction intermediates. The presence of atomic oxygen in a sample of ozone irradiated with ultraviolet light is evidence in favor of this step (Fig. 12.32). The reverse reaction may also occur:

$$O_2 + O \longrightarrow O_3 \qquad \text{Rate} = k_1'[O_2][O]$$

(We label the rate constants with the number of the step and use a prime to denote the reverse reaction.)

2. Bimolecular attack of an O atom on another O_3 molecule:

$$O_3 + O \longrightarrow 2O_2 \qquad \text{Rate} = k_2[O_3][O]$$

The reverse reaction

$$2O_2 \longrightarrow O_3 + O \qquad \text{Rate} = k_2'[O_2]^2$$

is so slow that it can be ignored. The overall forward reaction is the sum of the two forward elementary reactions:

$$\begin{array}{r} O_3 \longrightarrow O_2 + O \\ O_3 + O \longrightarrow 2O_2 \\ \hline 2O_3 \longrightarrow 3O_2 \end{array}$$

Measurements of the rates of the elementary reactions show that the slowest one is step 2. We therefore select it as the rate-determining step and equate the overall rate to the rate of this step:

$$\text{Overall rate} = k_2[O_3][O]$$

Because the O atoms are intermediates and are not part of the experimental rate law, [O] must not appear in the overall rate law derived from the proposed mechanism. We shall now see how the concentration of O atoms is related to the concentrations of O_3 and O_2. To do so, we consider the relatively fast forward and reverse reactions in step 1, which have a pronounced effect on the concentration of O atoms because they can build it up quickly or drain it away, and ignore the much slower reaction in step 2, which has a much smaller effect.

As we saw in the discussion of solubility in Section 11.3, a system reaches a state of dynamic equilibrium when forward and reverse processes both occur at the same rate. There we were concerned with an equilibrium that produced a steady concentration of solute in a saturated solution. We are concerned with almost exactly the same thing here, the only difference being that we are now dealing with forward and reverse *chemical* reactions. These reactions establish a dynamic *chemical* equilibrium in which the concentrations of the reactants and products remain constant even though the two reactions continue. Dynamic chemical equilibrium is reached when the rates of the forward and reverse reactions in step 1 are equal; that is, when

$$\text{Rate of forward reaction} = \text{rate of reverse reaction}$$

FIGURE 12.32 The presence of O atoms in the upper atmosphere is partly responsible for the formation of auroras, the "northern lights" and "southern lights" that appear as colored bands of lights in northern and southern latitudes. Excited oxygen atoms emit crimson and whitish green light. Excited N_2 molecules give a pink light, and N_2^+ ions give out violet and blue light when they combine with electrons.

We are dealing only with step 1 here, because only that step has an appreciable reverse reaction. It follows from this equality and the rate laws for the forward and reverse reactions that

$$k_1[O_3] = k_1'[O_2][O]$$

which we can rearrange to

$$[O] = \frac{k_1[O_3]}{k_1'[O_2]}$$

This expression is the relation we required between the concentration of O atoms and the concentrations of O_2 and O_3 molecules. When it is inserted into the rate law for step 2, the rate-determining step, we get

$$\text{Rate} = k_2[O_3][O] = k_2[O_3] \times \frac{k_1[O_3]}{k_1'[O_2]}$$

$$= \frac{k_1k_2}{k_1'} \times \frac{[O_3]^2}{[O_2]}$$

This result, which has been derived from the proposed mechanism, has exactly the same form as the experimental rate law,

$$\text{Rate} = k \times \frac{[O_3]^2}{[O_2]}$$

and we can identify the observed rate constant k with the combination k_1k_2/k_1' of the rate constants for the elementary steps.

Although the calculated rate law (from the proposed mechanism) and the experimental rate law may be the same, this does not *prove* that the proposed mechanism is correct: some other mechanism might also lead to the same rate law. Kinetic information can only *support* a proposed mechanism; it can never *prove* that a mechanism is correct. The acceptance of a suggested mechanism is more like the process of proof in a court of law than in mathematics, with evidence being assembled to give a convincing, consistent picture. In the case of the ozone decomposition, for instance, support for a mechanism should include measurement of the concentration of oxygen atoms during the reaction, to see whether it is proportional to $[O_3]/[O_2]$ as predicted by the equilibrium expression. The individual rate constants should also be measured to see whether k is indeed equal to k_1k_2/k_1'.

E The O_2 molecules inhibit the decay of ozone because they provide a route for the removal of O atoms. That removal has two effects: the reduction in the concentration of O atoms reduces the rate of the second step, in which an O atom attacks an O_3 molecule (in step 2); the reaction between an O atom and an O_2 molecule restores some of the O_3 molecules that have decomposed, and so reduces the net rate of decomposition of ozone.

▼ EXAMPLE 12.13 Deducing a rate law from a mechanism

The reaction $2NO(g) + O_2(g) \rightarrow 2NO_2(g)$ is thought to have the following mechanism:

1. $2NO \rightarrow N_2O_2$, and its reverse, $N_2O_2 \rightarrow 2NO$ (both fast)
2. $N_2O_2 + O_2 \rightarrow 2NO_2$ (slow)

Derive the rate law implied by this proposed mechanism and find the relation between the rate constant k for the overall reaction and the rate constants for the elementary reactions.

STRATEGY If we knew the rate law for the rate-determining step, then we could identify the overall rate as equal to that rate. Therefore, we begin by identifying the rate-determining step (the slowest step). The rate law obtained at this stage often includes the concentrations of one or more of the reaction intermediates. These can be expressed in terms of the concentrations of reactants and products by assuming that the earlier fast forward and reverse steps in the proposed mechanism have reached equilibrium and that their rates have become equal. We write first-order rate laws for unimolecular reactions and second-order rate laws for bimolecular reactions.

SOLUTION The rate-determining elementary reaction is the bimolecular reaction in step 2, so

$$\text{Rate of formation of } NO_2 = k_2[N_2O_2][O_2]$$

We need to replace the concentration of the intermediate N_2O_2 by an expression that involves only the concentrations of reactants and products from the overall reaction. When the rates of the forward and reverse reactions in step 1 are equal,

$$k_1[NO]^2 = k_1'[N_2O_2]$$

This expression may be rearranged to

$$[N_2O_2] = \frac{k_1}{k_1'} \times [NO]^2$$

Substituting this expression into the rate law for step 2 gives

$$\text{Rate of formation of } NO_2 = \frac{k_1k_2}{k_1'} \times [NO]^2[O_2]$$

which has the form

$$\text{Rate of formation of } NO_2 = k[NO]^2[O_2]$$

with $k = k_1k_2/k_1'$.

EXERCISE E12.13 The proposed mechanism for a reaction is $AH + B \rightarrow BH^+ + A^-$ and its reverse $BH^+ + A^- \rightarrow AH + B$, both of which are fast, followed by $A^- + AH \rightarrow$ products, which is slow. Find the rate law with A^- treated as the intermediate.

[*Answer*: rate = $(k_1k_2/k_1')[AH]^2[B]/[BH^+]$]

U Exercises 12.74, 12.80, and 12.86.

E The reaction in Example 12.13 goes more *slowly* as the temperature is raised: it shows *anti-Arrhenius behavior*. The explanation is that each rate constant in the mechanism increases with temperature, but k_1' increases more rapidly than k_1 and k_2, so the observed rate constant $k = k_1k_2/k_1'$ decreases with temperature. Anti-Arrhenius behavior is always a sign of a complex mechanism with a composite rate constant.

12.6 CHAIN REACTIONS

The rate law for the reaction between hydrogen and bromine is quite complicated:

$$H_2(g) + Br_2(g) \longrightarrow 2HBr(g) \qquad \text{Rate} = \frac{k[H_2][Br_2]^{3/2}}{[Br_2] + k'[HBr]}$$

This rate law suggests that the reaction takes place by a complicated mechanism. A chain reaction has been proposed:

> A **chain reaction** is a reaction in which an intermediate reacts to produce another intermediate.

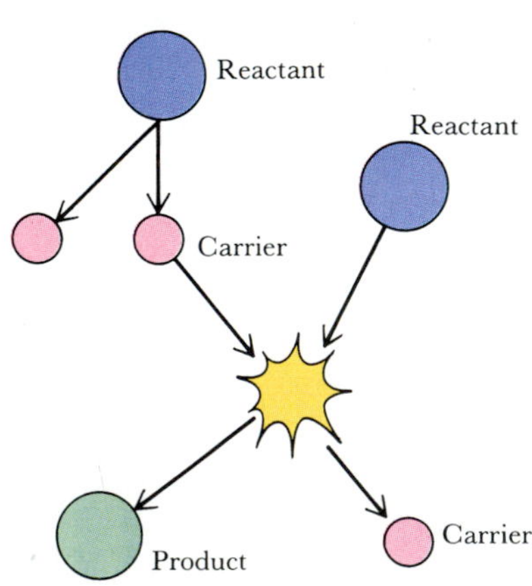

FIGURE 12.33 In a chain reaction, the product (represented by a pink circle) of one step is a reactant in another that in turn produces substances that can take part in more reactions.

In many cases, including the HBr synthesis, the reaction intermediate—which in this context is often called a **chain carrier**—is a radical (Section 8.7), and the reaction is called a **radical chain reaction**. In such a reaction, one radical reacts with a molecule to produce another radical, which goes on to attack another molecule (Fig. 12.33). In the HBr synthesis, the chain carriers are the hydrogen atom (H ·) and the bro-

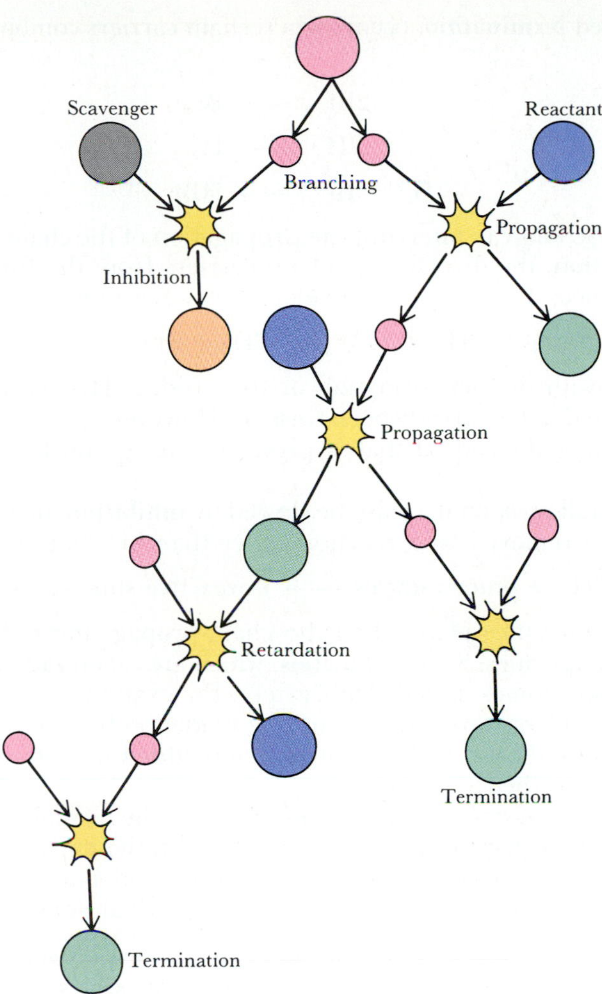

FIGURE 12.34 The characteristic steps of a chain reaction. The chain carriers, which are radicals in a radical chain reaction, are pink, the reactants blue, and the products green. If the chain includes a branching step, the concentration of intermediates rises very rapidly and the reaction may become explosively fast.

T OHT: Fig. 12.34.

D Reaction of hydrogen and chlorine. TDC, 154.

mine atom (Br·). (Recall from Section 8.7, that the unpaired electron of a radical is represented by a dot.)

U Exercise 12.81.

E The development of free radical chemistry. *J. Chem. Educ.*, **1986**, *63*, 99.

Radical chains. We can use the HBr synthesis to introduce the characteristic steps of a chain reaction (Fig. 12.34). The first step in any chain reaction is **initiation**, the formation of chain carriers from a reactant:

$$Br_2 \xrightarrow{\Delta \text{ or } h\nu} 2Br\cdot$$

Once chain carriers have been formed, the chain can **propagate** in a series of reactions in which one carrier reacts with a reactant molecule to produce another carrier. The elementary reactions for the propagation of the chain are

$$Br\cdot + H_2 \longrightarrow HBr + H\cdot$$

$$H\cdot + Br_2 \longrightarrow HBr + Br\cdot$$

The chain carriers—radicals here—produced in these reactions can go on to attack other reactant (H_2 and Br_2) molecules, thus allowing the chain reaction to continue. The elementary reaction that ends the chain, called **termination**, occurs when chain carriers combine to form products:

$$2Br\cdot \longrightarrow Br_2$$

$$2H\cdot \longrightarrow H_2$$

$$H\cdot + Br\cdot \longrightarrow HBr$$

Some reactions can interrupt the propagation of the chain and result in **retardation**, the diversion of chain carriers from the formation of final products. For example, the elementary reaction

$$H\cdot + HBr \longrightarrow H_2 + Br\cdot$$

interferes with the net formation of the product HBr because it destroys some that has already been formed. However, it does not end the chain because the radical that is produced can go on to form more product.

E Radical reactions are implicated in degenerative nerve diseases. A particularly virulent radical that occurs naturally in the brain and the central nervous system is the superoxide ion, O_2^-; nitrogen monoxide is also present in the brain, and is involved in degenerative disorders. One approach to chemical control of disease is to trap radicals before they have a chance to attack cell membranes.

The overall reaction may also be slowed by **inhibition**, in which chain carriers are removed by a reaction other than by chain termination:

$$H\cdot + \text{other radicals} \longrightarrow \text{unreactive substances}$$

Inhibition may take place when the chain-propagating radicals combine with impurities. Such impurities, which are called **radical scavengers**, are sometimes added deliberately; for example, food additives called antioxidants are added to inhibit oxidation by removing chain-propagating radicals. Antioxidants are also added to plastics and rubber to prevent their degradation.

U Exercises 12.83 and 12.84.

When all these steps are taken into account, the overall rate law implied by the mechanism is found to agree with the experimental rate law given previously, and the rate constants k and k' are found to be combinations of the rate constants for the individual steps in the chain.

Explosions. In some cases a chain can propagate explosively. An explosion is likely to happen when chain **branching** occurs, that is, when more than one chain carrier is formed in a propagation step. The characteristic pop that occurs when a mixture of hydrogen and oxygen is ignited is a consequence of chain branching. The two gases combine in a radical chain reaction in which the initiation step may be the formation of hydrogen atoms:

$$\textit{Initiation}:\ H_2 \xrightarrow{\text{spark}} 2H\cdot$$

Two new radicals are formed when one hydrogen atom attacks an oxygen molecule:

$$\textit{Branching}:\ H\cdot + O_2 \longrightarrow HO\cdot + \cdot O\cdot$$

Note that the oxygen atom, with valence electron configuration $2s^2 2p_x^{\ 2} 2p_y^{\ 1} 2p_z^{\ 1}$, has two electrons with unpaired spins (its complete Lewis dot diagram is $\cdot\ddot{\underset{\cdot\cdot}{O}}\cdot$). Two radicals are also produced when the oxygen atom attacks a hydrogen molecule:

$$\textit{Branching}:\ \cdot O\cdot + H_2 \longrightarrow HO\cdot + H\cdot$$

As a result of these branching processes, the chain produces a large number of radicals that can take part in even more branching steps. The reaction rate increases rapidly, and an explosion may occur (Fig. 12.35).

FIGURE 12.35 This flame front inside an internal combustion engine cylinder is a region where a radical chain reaction, in this case a combustion, is in progress.

SUMMARY

12.1 Chemical kinetics is the study of the rates of chemical reactions. The **instantaneous rate** of a reaction is the tangent to the concentration-versus-time graph for the reaction. The **initial rate** is the rate at the start of the reaction.

12.2 The **rate law** is the relation between the rate and the concentrations of substances present in the reaction mixture. The rate is reported in terms of the **rate constant** of the reaction; and reactions are classified by their **order**, the power to which the concentrations of the reactants (and products) are raised in the rate law. The rate law must be determined experimentally and cannot be deduced from the stoichiometry of the chemical equation for the reaction. Once the **integrated rate law**, the expression for the concentration in terms of the time, is known, it may be used to predict the concentration at any stage of the reaction. The **half-life** is the time taken for the concentration of a substance to fall to half its initial value. The half-life of a first-order reaction is independent of the initial concentration.

12.3 The rate of a reaction is affected by the surface area of the reactants, the presence of catalysts, the temperature, and (in most cases) the concentrations of the substances in the reaction mixture. The temperature-dependence of the rate constant is expressed by the **Arrhenius equation** and the two **Arrhenius parameters**, the **frequency factor** and the **activation energy**. Their role is explained by **collision theory**, in which A is proportional to the frequency of collisions and the relative orientation of the reactants during the encounter and E_a is the minimum energy needed for reaction. This theory is generalized by the **activated complex theory** in which it is proposed that an **activated complex**, an energetically excited combination of reactant molecules, is formed and that the reactants must cross an **activation barrier**, an energy barrier to reaction, to form products.

12.4 A **catalyst**, a substance that speeds a reaction but is not itself consumed, may be either **homogeneous** (if it is in the same phase as the reactants) or **heterogeneous** (if it is in a different phase). Catalysts act by providing a path with a lower activation energy. **Enzymes** are biological catalysts.

12.5 A **reaction mechanism** consists of a sequence of **elementary reactions**, or reaction steps involving individual molecules. Elementary reactions can be classified according to their **molecularity**, the number of molecules taking part. **Unimolecular reactions**, in which a single molecule forms a product, obey first-order kinetics. **Bimolecular reactions**, elementary reactions between two molecules, obey second-order kinetics. A proposed mechanism must agree with the experimental rate law. The rate law implied by a mechanism is found by identifying the **rate-determining step**, the slowest step, and finding an expression for the concentration of any **reaction intermediates**, species that are produced and consumed in the reaction and taking part in it but are not present as initial reactants or final products.

12.6 Chain reactions are a class of reactions in which one intermediate reacts to produce another. In **radical chain reactions**, the intermediates, or **chain carriers**, are radicals. These reactions may include characteristic **initiation**, **propagation**, **retardation**, **inhibition**, and **termination** steps. Chain **branching**, in which more than one radical is produced in an elementary step, may lead to an explosion.

CLASSIFIED EXERCISES

Reaction Rates

12.1 (a) The rate of formation of O_2 is 1.5×10^{-3} mol/(L·s) in the reaction $2O_3(g) \rightarrow 3O_2(g)$. What is the rate of decomposition of ozone? (b) The rate of formation of the dichromate ions is 0.14 mol/(L·s) in the reaction $2CrO_4^{2-}(aq) + 2H^+(aq) \rightarrow Cr_2O_7^{2-}(aq) + H_2O(l)$. What is the rate of reaction of the chromate ion in the reaction?

12.2 (a) Nitrogen dioxide decomposes at the rate of 6.5×10^{-3} mol/(L·s) by the reaction $2NO_2(g) \rightarrow N_2O_2(g) + O_2(g)$. Determine the rate of formation of O_2. (b) Manganate ions, MnO_4^{2-}, form permanganate ions and manganese(IV) oxide in an acidic solution at a rate of 2.0 mol/(L·min): $3MnO_4^{2-}(aq) + 4H^+(aq) \rightarrow 2MnO_4^-(aq) + MnO_2(l) + 2H_2O(l)$. What is the rate of formation of permanganate ions? What is the rate of reaction of $H^+(aq)$?

Instantaneous Reaction Rate

12.3 The decomposition of gaseous dinitrogen pentoxide in the reaction $2N_2O_5(g) \rightarrow 4NO_2(g) + O_2(g)$ gives the following data at 298 K.

Time, h	$[N_2O_5]$, mol/L
0	2.50×10^{-3}
1.11	1.88×10^{-3}
2.22	1.64×10^{-3}
3.33	1.43×10^{-3}
4.44	1.25×10^{-3}

(a) Plot the concentration of N_2O_5 against the time. (b) Estimate the rate of decomposition of N_2O_5 at each time. (c) Plot the concentrations of NO_2 and O_2 against the time on the same graph.

12.4 The decomposition of gaseous hydrogen iodide, $2HI(g) \rightarrow H_2(g) + I_2(g)$, gives the following data at 700 K.

Time, s	0	1000	2000	3000	4000	5000
[HI], mmol/L	10	4.4	2.8	2.1	1.6	1.3

(a) Plot the concentration of HI against the time. (b) Estimate the rate of decomposition of HI at each time. (c) Plot the concentrations of H_2 and I_2 against the time on the same graph.

12.5 Express the units for rate constants when the concentrations are in moles per liter and time is in seconds for (a) zero-order reactions; (b) first-order reactions; (c) second-order reactions.

12.6 If $k = 1.35\ \mathrm{L^2/(mol^2 \cdot h)}$, what is the order of the reaction?

12.7 What is the reaction rate for the decomposition of N_2O_5 when 2.0 g of N_2O_5 is confined in a 1.0-L container and heated to 65°C? From Table 12.1, $k = 5.2 \times 10^{-3}\ \mathrm{s^{-1}}$ at 65°C.

12.8 Ethane, C_2H_6, decomposes to methyl radicals at 700°C by a first-order reaction. If a 100-mg sample of ethane is confined to a 250-mL reaction vessel and heated to 700°C, what is the reaction rate of ethane decomposition? From Table 12.1, $k = 5.5 \times 10^{-4}\ \mathrm{s^{-1}}$ at 700°C.

12.9 A 0.15-g sample of H_2 and a 0.32-g sample of I_2 are confined to a 500-mL reaction vessel and heated to 700 K, when $k = 0.063\ \mathrm{L/(mol \cdot s)}$. (a) What is the initial reaction rate? (b) By what factor does the reaction rate increase if the mass of H_2 present in the mixture is doubled?

12.10 A 100-mg sample of NO_2, confined to a 200-mL reaction vessel, is heated to 300°C, when $k = 0.54\ \mathrm{L/(mol \cdot s)}$. (a) What is the initial reaction rate? (b) How does the reaction rate change (and by what factor) if the mass of NO_2 present in the container is increased to 200 mg?

12.11 Write the rate law for the following reactions. (a) The reaction $X + 2Y \rightarrow Z_2$ is second order in X and one-half order in Y. (b) The reaction $2A + B \rightarrow C$ is found to be first order in A, first order in B, and three-halves order overall.

12.12 (a) The reaction $A + 2B + C \rightarrow D + 4E$ is found to be first order in A, first order in B, and zero order in C. What is the rate law for the reaction? (b) The reaction $2A + B \rightarrow C + D$ is found to be first order in A, one-half order in B, three-halves order in D, and zero order overall. What is the rate law for the reaction?

12.13 State the order of the reaction with respect to each reactant and the overall order of each of the following reactions:

(a) $2SO_2(g) + O_2(g) \xrightarrow{\text{Pt}} 2SO_3(g)$
$$\text{Rate} = k[SO_2]/[SO_3]^{1/2}$$
(b) $A(g) + 2B(g) + C(g) \rightarrow \text{products}$
$$\text{Rate} = k[A]^2[B]^{-2}$$

12.14 Determine the units for the rate constants for each of the following reactions, assuming that the concentration is expressed in moles per liter and time is expressed in seconds.

(a) $S_2O_8{}^{2-}(aq) + 3I^-(aq) \rightarrow 2SO_4{}^{2-}(aq) + I_3{}^-(aq)$
$$\text{Rate} = k[S_2O_8{}^{2-}][I^-]$$
(b) $A(g) + 3B(g) \rightarrow \text{products}$ $\quad \text{Rate} = k[A]^2[B]^{3/2}$

12.15 In the reaction $CH_3Br(aq) + OH^-(aq) \rightarrow CH_3OH(aq) + Br^-(aq)$, when the OH^- concentration was doubled, the rate doubled; when the CH_3Br concentration was increased by a factor of 1.2, the rate increased by 1.2. Write the rate law for the reaction.

12.16 In the reaction $2NO(g) + O_2(g) \rightarrow 2NO_2(g)$, when the NO concentration was doubled, the rate increased by a factor of 4; when both the NO and O_2 concentrations were increased by a factor of 2, the rate increased by a factor of 8. What is the rate law for the reaction?

12.17 The following kinetic data were obtained for the reaction $A_2(g) + 2B_3(g) \rightarrow$ product:

	Initial concentration, mol/L		
Experiment	**A_2**	**B_3**	**Initial rate, mol/(L·s)**
1	0.60	0.30	12.6
2	0.20	0.30	1.4
3	0.60	0.10	4.2
4	0.17	0.25	?

(a) What is the order with respect to each reactant, and the overall order of the reaction? (b) Write the rate law for the reaction. (c) From the data, determine the value of the rate constant. (d) Use the data to predict the reaction rate for experiment 4.

12.18 The reaction $2ICl(g) + H_2(g) \rightarrow I_2(g) + 2HCl(g)$ produced the following data:

Experiment	Initial concentration, mol/L ICl	H_2	Initial rate, mol HCl/(L·s)
1	1.5×10^{-3}	1.5×10^{-3}	3.7×10^{-7}
2	3.0×10^{-3}	1.5×10^{-3}	7.4×10^{-7}
3	3.0×10^{-3}	4.5×10^{-3}	$22 \quad \times 10^{-7}$
4	4.7×10^{-3}	2.7×10^{-3}	?

(a) Write the rate law for the reaction. (b) From the data, determine the value of the rate constant. (c) Use the data to predict the reaction rate for experiment 4.

12.19 The following kinetic data were obtained for the reaction $NO_2(g) + O_3(g) \rightarrow NO_3(g) + O_2(g)$:

Experiment	Initial concentration, mol/L NO_2	O_3	Initial rate, mol O_2/(L·s)
1	0.21×10^{-3}	0.70×10^{-3}	6.3
2	0.21×10^{-3}	1.39×10^{-3}	12.5
3	0.38×10^{-3}	0.70×10^{-3}	11.4
4	0.66×10^{-3}	0.18×10^{-3}	?

(a) Write the rate law for the reaction. (b) What is the order of the reaction? (c) From the data, determine the value of the rate constant. (d) Use the data to predict the reaction rate for experiment 4.

12.20 The following data were obtained for the reaction A + B + C → products:

Experiment	Initial concentration, mmol/L A	B	C	Initial rate, mmol D/(L·s)
1	1.25	1.25	1.25	0.87×10^{-2}
2	2.50	1.25	1.25	1.74×10^{-2}
3	1.25	3.02	1.25	5.08×10^{-2}
4	1.25	3.02	3.75	0.457
5	3.01	1.00	1.15	?

(a) Write the rate law for the reaction. (b) What is the order of the reaction? (c) Determine the value of the rate constant. (d) Use the data to predict the reaction rate for experiment 5.

Time Dependence of Reactions

12.21 Determine the rate constants for the following first-order reactions: (a) A → B, given that the concentration of A decreases to one-half its initial value in 1000 s. (b) 2A → B + C, given that $[A]_0 = 0.050$ mol/L and that after 120 s the concentration of B rises to 0.015 mol/L.

12.22 Determine the rate constants for the following first-order reactions: (a) 2A → B + C, given that the concentration of A decreases to one-third its initial value in 25 min. (b) 2A → 3B + C, given that $[A]_0 = 0.050$ mol/L and that after 7.7 min the concentration of B rises to 0.050 mol/L.

12.23 Dinitrogen pentoxide, N_2O_5, decomposes by first-order kinetics with a rate constant of $3.7 \times 10^{-5}\ s^{-1}$ at 298 K. (a) What is the half-life (in hours) for the decomposition of N_2O_5 at 298 K? (b) If $[N_2O_5]_0 = 2.33 \times 10^{-2}$ mol/L, what will be the concentration of N_2O_5 after 2.0 h? (c) How much time in minutes will elapse before the N_2O_5 concentration decreases from 2.33×10^{-2} to 1.76×10^{-2} mol/L?

12.24 The first-order rate constant for the photodissociation of A is $0.173\ s^{-1}$. Calculate the time needed for the concentration of A to decrease to (a) one-fourth; (b) one-thirty-second; (c) one-fifth of its initial concentration.

12.25 The half-life is 50.5 s for a second-order reaction when $[A]_0 = 0.84$ mol/L in the reaction of a substance A. Calculate the time needed for the concentration of A to decrease to (a) one-sixteenth; (b) one-fourth; (c) one-fifth of its original value.

12.26 The second-order rate constant for the decomposition of NO_2 (to NO and O_2) at 573 K is 0.54 L/(mol·s). Calculate the time for an initial NO_2 concentration of 0.20 mol/L to decrease to (a) one-half; (b) one-sixteenth; (c) one-ninth of its initial concentration.

12.27 Determine the rate constant for the following second-order reactions: (a) 2A → B + 2C, given that the concentration of A decreases from 2.5×10^{-3} to 1.25×10^{-3} mol/L in 100 s. (b) 3A → 2D + C, given that $[A]_0 =$ 0.30 mol/L and that the concentration of C increases to 0.010 mol/L in 200 s.

12.28 Determine the time required for each of the following second-order reactions to occur: (a) 2A → B + C, for the concentration of A to decrease from 0.10 to 0.080 mol/L, given that $k = 0.010$ L/(mol·min). (b) A → 2B + C, when $[A]_0 = 0.45$ mol/L, for the concentration of B to increase to 0.45 mol/L, given that $k =$ 0.0045 L/(mol·min).

12.29 Sulfuryl chloride, SO_2Cl_2, decomposes by first-order kinetics, and at $k = 2.81 \times 10^{-3}\ min^{-1}$ at a certain temperature. (a) Write the rate law for the reaction. (b) Determine the half-life for the reaction. (c) If a 14.0-g sample of SO_2Cl_2 is sealed in a 2500-L reaction vessel and heated to the specified temperature, what mass will remain after 1.5 h?

12.30 Ethane, C_2H_6, forms $\cdot CH_3$ radicals at 700°C in a first-order reaction, for which $k = 1.98\ h^{-1}$. (a) What is the half-life for the reaction? (b) Calculate the time needed for the amount of ethane to fall from 1.15×10^{-3} to 2.35×10^{-4} mol in a 500-mL reaction vessel at 700°C. (c) How much of a 6.88-mg sample of ethane in a 500-mL reaction vessel at 700°C will remain after 45 min?

12.31 A substance of chemical formula A_2 forms B in the first-order reaction $A_2 \rightarrow 2B$, in which the concentration of A_2 falls to 20% of its original concentration in 120 s. (a) What is the rate constant for the reaction? (b) Determine the time in which the concentration of A_2 falls to 10% of its original concentration.

12.32 For the first-order reaction $A_3 \rightarrow 3B_2 + C$, when $[A_3]_0 = 0.015$ mol/L the concentration of B_2 increases to 0.020 mol/L in 3.0 min. (a) What is the rate constant for the reaction? (b) How much more time would be needed for the concentration of B_2 to increase to 0.040 mol/L?

12.33 The following data were collected for the reaction $C_2H_6(g) \rightarrow 2 \cdot CH_3(g)$ at 700°C:

Time, s	0	1000	2000	3000	4000
$[C_2H_6]$, mmol/L	1.59	0.92	0.53	0.31	0.18

(a) Plot the data to confirm that the reaction is first order. (b) From the graph, determine the reaction rate constant.

12.34 The following data were collected for the reaction $2HI(g) \rightarrow H_2(g) + I_2(g)$ at 580 K.

Time, min	0	16.7	33.3	50.0	66.7
[HI], mmol/L	300	120	61	41	31

(a) Plot the data to confirm that the rate law is rate = $k[HI]^2$. (b) From the graph, determine the reaction rate constant.

12.35 The following data were obtained on the reaction $2A \rightarrow B$:

Time, s	0	5	10	15	20
[A], mmol/L	100	14.1	7.8	5.3	4.0

(a) Plot the data to determine the order of the reaction. (b) Determine the rate constant.

12.36 The following data were obtained on the reaction $2A_2 \rightarrow B$:

Time, s	0	100	200	300	400
$[A_2]$, mmol/L	250	143	81	45	25

(a) Plot the data to determine the order of the reaction. (b) Determine the rate constant.

Temperature Dependence of Reaction Rates

12.37 (a) Calculate the activation energy for the conversion of cyclopropane to propene from an Arrhenius plot of the following data:

T, K	750	800	850	900
k, s^{-1}	1.8×10^{-4}	2.7×10^{-3}	3.0×10^{-2}	0.26

(b) What is the value of the reaction rate constant at 600°C?

12.38 (a) Determine the activation energy for $C_2H_5I(g) \rightarrow C_2H_4(g) + HI(g)$ from an Arrhenius plot of the following data:

T, K	660	680	720	760
k, s^{-1}	7.2×10^{-4}	2.2×10^{-3}	1.7×10^{-2}	0.11

(b) What is the value of the reaction rate constant at 400°C?

12.39 The rate constant of the first-order reaction $2N_2O(g) \rightarrow 2N_2(g) + O_2(g)$ is $0.38\ s^{-1}$ at 1000 K and $0.87\ s^{-1}$ at 1030 K. What is the activation energy of the reaction?

12.40 The rate constant of the second-order reaction $2HI(g) \rightarrow H_2(g) + I_2(g)$ at 575 K is 2.4×10^{-6} L/(mol·s); and at 630 K, it is 6.0×10^{-5} L/(mol·s). Calculate the activation energy of the reaction.

12.41 The rate constant of the reaction $O(g) + N_2(g) \rightarrow NO(g) + N(g)$, which occurs in the atmosphere, is 9.7×10^{10} L/(mol·s) at 800°C. The activation energy of the reaction is 315 kJ/mol. Determine the reaction rate constant at 700°C.

12.42 The rate constant of the reaction between CO_2 and OH^- in aqueous solution, to give the HCO_3^- ion is 1.5×10^{10} L/(mol·s) at 25°C. What is the reaction rate constant at blood temperature (37°C), given that the activation energy for the reaction is 38 kJ/mol?

12.43 Draw a reaction profile for an endothermic reaction. Label the activation energy, E_a, for the forward reaction and the enthalpy of reaction, ΔH. Mark the location of the activated complex. What is the relation of the activation energies of the reverse reaction, E'_a, to that of the forward reaction?

12.44 The activation energy for a certain reaction is 125 kJ/mol and its enthalpy of reaction is −30 kJ/mol.

Draw a reaction profile for the reaction. Indicate the location of the activated complex. What is the activation energy, E'_a, of the reverse reaction?

Catalysis

12.45 The contribution to the destruction of the ozone layer caused by high flying aircraft has been attributed to the following mechanism:

Step 1. $O_3 + NO \rightarrow NO_2 + O_2$
Step 2. $NO_2 + O \rightarrow NO + O_2$

(a) What is the catalyst in the reaction? (b) What is the reaction intermediate?

12.46 A reaction was believed to occur by the following mechanism.

Step 1. $A_2 \rightarrow 2A$
Step 2. $2A + B \rightarrow C + 2D$
Step 3. $D + D \rightarrow A_2 + E$

(a) What is the overall reaction? (b) What is the catalyst in the reaction? (c) Which species are the reaction intermediates?

12.47 The presence of a catalyst reduces the activation energy of a certain reaction from 100 to 50 kJ/mol. By what factor does the rate of the reaction increase at 400 K, all other factors being equal?

12.48 The presence of a catalyst reduces the activation energy of a certain reaction from 88 to 62 kJ/mol. By what factor does the rate of the reaction increase at 300 K, all other factors being equal?

12.49 A reaction rate increases by a factor of 1000 in the presence of a catalyst at 25°C. The activation energy of the original pathway is 98 kJ/mol. What is the activation energy of the new pathway, all other factors being equal?

12.50 A reaction rate increases by a factor of 500 in the presence of a catalyst at 37°C. The activation energy of the original pathway is 106 kJ/mol. What is the activation energy of the new pathway, all other factors being equal?

Reaction Mechanisms

12.51 Each of the following steps is an elementary reaction. Write its rate law and indicate its molecularity:

(a) $2NO \rightarrow N_2O_2$
(b) $Cl_2 \rightarrow 2Cl$
(c) $2NO_2 \rightarrow NO + NO_3$
(d) Which of these reactions might be radical chain initiating?

12.52 Each of the following is an elementary reaction. Write its rate law and indicate its molecularity:
(a) $O + CF_2Cl_2 \rightarrow ClO + CF_2Cl$
(b) $OH + NO_2 + N_2 \rightarrow HNO_3 + N_2$
(c) $ClO^- + H_2O \rightarrow HClO + OH^-$
(d) Which of these reactions might be radical chain propagating?

12.53 Write the overall reaction for the mechanism proposed below and identify the reaction intermediates:

Step 1. $ICl + H_2 \rightarrow HI + HCl$
Step 2. $HI + ICl \rightarrow HCl + I_2$

12.54 Write the overall reaction for the mechanism proposed below and identify the reaction intermediates:

Step 1. $Cl_2 \rightarrow 2Cl$
Step 2. $Cl + CO \rightarrow COCl$
Step 3. $COCl + Cl_2 \rightarrow COCl_2 + Cl$

12.55 The production of phosgene, $COCl_2$, from carbon monoxide and chlorine is believed to take place by the following mechanism:

Step 1. $Cl_2 \rightarrow 2Cl$, and its reverse, $2Cl \rightarrow Cl_2$ (both fast, equilibrium)
Step 2. $Cl + CO \rightarrow COCl$, and its reverse, $COCl \rightarrow CO + Cl$ (both fast, equilibrium)
Step 3. $COCl + Cl_2 \rightarrow COCl_2 + Cl$ (slow)

Write the rate law implied by this mechanism.

12.56 The mechanism proposed for the oxidation of iodide ion by the hypochlorite ion in aqueous solution is as follows:

Step 1. $ClO^- + H_2O \rightarrow HClO + OH^-$, and its reverse, $HClO + OH^- \rightarrow ClO^- + H_2O$ (both fast, equilibrium)
Step 2. $I^- + HClO \rightarrow HIO + Cl^-$ (slow)
Step 3. $HIO + OH^- \rightarrow IO^- + H_2O$ (fast)

Write the rate law implied by this mechanism.

UNCLASSIFIED EXERCISES

12.57 (a) The rate of formation of Cl^- in the reaction $3ClO^-(aq) \rightarrow Cl^-(aq) + ClO_3^-(aq)$, is 3.6 mol/(L·min). What is the rate of reaction of ClO^-? (b) In the Haber process for the industrial production of ammonia, $N_2(g) + 3H_2(g) \rightarrow 2NH_3(g)$, the rate of ammonia production is 2.7×10^{-3} mol/(L·s). What is the rate of reaction of H_2?

12.58 Indicate two ways of expressing the reaction rate for each of the following reactions, using, in each case, the concentration of the species in bold print. Assume that the concentrations are in moles per liter and that time is expressed in seconds.
(a) $4\mathbf{NH_3}(g) + 3O_2(g) \rightarrow 2\mathbf{N_2}(g) + 6H_2O(g)$
(b) $\mathbf{CH_3CHO}(g) \rightarrow \mathbf{CH_4}(g) + CO(g)$

12.59 Complete the following statements relating to the production of ammonia by the Haber process, for which the overall reaction is $N_2(g) + 3H_2(g) \rightarrow 2NH_3(g)$. (a) The rate of decomposition of N_2 is ______ times the rate of decomposition of H_2. (b) The rate of formation of NH_3 is ______ times the rate of decomposition of H_2. (c) The rate of formation of NH_3 is ______ times the rate of decomposition of N_2.

12.60 The following data were collected for the reaction $2N_2O_5(g) \rightarrow 4NO_2(g) + O_2(g)$ at 308 K:

Time, 10^3 s	$[N_2O_5]$, mol/L
0	2.57×10^{-3}
4	1.50×10^{-3}
8	0.87×10^{-3}
12	0.51×10^{-3}
16	0.30×10^{-3}

(a) Plot the $[N_2O_5]$ concentration against the time. (b) Estimate the rate of decomposition of N_2O_5 at each time. (c) Plot the $[NO_2]$ and $[O_2]$ concentrations against the time on the same graph made for (a).

12.61 The rate law for the reaction $H_2SeO_3(aq) + 6I^-(aq) + 4H^+(aq) \rightarrow Se(s) + 2I_3^-(aq) + 3H_2O(l)$ is rate $= k[H_2SeO_3][I^-]^3[H^+]^2$ with $k = 5.0 \times 10^5$ $L^5/(mol^5 \cdot s)$. What is the initial reaction rate when $[H_2SeO_3]_0 = [I^-]_0 = 0.020$ mol/L and $[H^+]_0 = 0.010$ mol/L?

12.62 From a set of kinetic data, a plot of the logarithm of the concentration of a reactant against the time yielded a straight line with a negative slope. What was the order of the reaction?

12.63 From a set of kinetic data, a plot of the reciprocal concentration of a reactant against the time yielded a straight line with a positive slope. What was the order of the reaction?

12.64 The following rate data were collected for the decomposition of acetaldehyde, CH_3CHO, in the reaction $CH_3CHO(g) \rightarrow CH_4(g) + CO(g)$ at a certain temperature:

Experiment	Initial concentration, mol/L: CH_3CHO	Initial rate, mol CO/(L · s)
1	0.1	9.0×10^{-7}
2	0.2	36×10^{-7}
3	0.3	81×10^{-7}
4	0.4	144×10^{-7}
5	0.73	?

(a) What is the order of the reaction? (b) Write the rate law for the reaction. (c) Determine the reaction rate constant. (d) Predict the reaction rate for experiment 5.

12.65 The following rate data were collected for the reaction $2A(g) + 2B(g) + C(g) \rightarrow 3G(g) + 4F(g)$:

Experiment	Initial concentration, mmol/L			Initial rate, mmol G/(L·s)
	A	B	C	
1	10	100	700	2.0
2	20	100	300	4.0
3	20	200	200	16
4	10	100	400	2.0
5	4.62	177	124	?

(a) What is the order for each reactant and the overall order of the reaction? (b) Write the rate law for the reaction. (c) Determine the reaction rate constant. (d) Predict the reaction rate for experiment 5.

12.66 (a) The rate law for the thermal decomposition of acetaldehyde, $CH_3CHO(g) \rightarrow CH_4(g) + CO(g)$, under certain conditions is rate $= k[CH_3CHO]^{3/2}$. What is the order of the reaction? (b) The rate law for the "hot wire" decomposition of ammonia, $2NH_3(g) \rightarrow N_2(g) + 3H_2(g)$, is rate $= k$. What is the order of the reaction?

12.67 The decomposition of hydrogen peroxide, $2H_2O_2(aq) \rightarrow 2H_2O(l) + O_2(g)$, follows first-order kinetics with respect to H_2O_2 and has $k = 0.0410$ min^{-1}. (a) If the initial concentration of H_2O_2 is 0.20 mol/L, what is its concentration after 10 min? (b) How much time will it take for the H_2O_2 concentration to decrease from 0.50 to 0.10 mol/L? (c) Calculate the time needed for the H_2O_2 concentration to decrease by 75%.

12.68 (a) For the second-order reaction, $A_2 \rightarrow B + C_2$, the concentration of the species A_2 falls from 0.040 to 0.0050 mol/L in 12 h. What is the reaction rate constant? (b) In the second-order reaction $CX_2 \rightarrow C + 2X$, the concentration of X increases to 0.070 mol/L in 15 h when $[CX_2]_0 = 0.040$ mol/L. What is the reaction rate constant for the decomposition of CX_2?

12.69 The half-life for the (first-order) decomposition of azomethane, $CH_3N{=}NCH_3$, in the reaction $CH_3N{=}NCH_3(g) \rightarrow N_2(g) + C_2H_6(g)$ is 1.02 s at 300°C. A 45.0-mg sample of azomethane is placed in a 300-mL reaction vessel and heated to 300°C. (a) What mass (in milligrams) of azomethane remains after 10 s? (b) Determine the partial pressure exerted by the $N_2(g)$ in the reaction vessel after 3.0 s.

12.70 The mechanism of the reaction $A \rightarrow B$, consists of two steps. Overall, the reaction is exothermic. (a) Sketch the reaction profile, labeling the activation energies for each step and the overall enthalpy of reaction. (b) Indicate on the same diagram the effect of a catalyst on the first step of the reaction.

12.71 An exothermic reaction for which $\Delta H = -200$ kJ/mol has an activation energy of 100 kJ/mol. Estimate the activation energy for the reverse reaction.

12.72 The concentration of a species A was initially 0.20 mol/L, and decreased by the reaction $2A \rightarrow B + C$ to 0.10 mol/L in 100 s by first-order kinetics. Calculate the time needed for the concentration of A to fall to (a) one-eighth and (b) one-thirty-second of its initial concentration.

12.73 Determine the molecularity of the following elementary reactions: (a) $CH_3Br + OH^- \rightarrow CH_3OH + Br^-$; (b) $C_2N_2 \rightarrow 2CN\cdot$; (c) $Ar + 2O \rightarrow Ar + O_2$. Suggest a reason why an argon atom is needed in the last reaction.

12.74 The rate law of the reaction $2NO(g) + 2H_2(g) \rightarrow N_2(g) + 2H_2O(g)$ is Rate $= k[NO]^2[H_2]$, and the mechanism that has been proposed is

Step 1. $2NO \rightarrow N_2O_2$
Step 2. $N_2O_2 + H_2 \rightarrow N_2O + H_2O$
Step 3. $N_2O + H_2 \rightarrow N_2 + H_2O$

(a) Which step in the mechanism is likely to be rate determining? Explain your answer. (b) Sketch a reaction profile for the (exothermic) overall reaction. Label the activation energies of each step and the overall reaction enthalpy.

12.75 The old industrial process for the production of sulfuric acid was called the lead chamber process. A greatly oversimplified mechanism of the formation of sulfur trioxide, SO_3, which dissolved in water produces sulfuric acid, is:

Step 1. $O_2 + 2NO \rightarrow 2NO_2$
Step 2. $2NO_2 + SO_2 \rightarrow SO_3 + 2NO$

What is (a) the catalyst in the reaction and (b) the reaction intermediate?

12.76 The activation energy of the reaction $H_2(g) + I_2(g) \rightarrow 2HI(g)$ is reduced from 184 to 59 kJ/mol in the presence of a platinum catalyst. By what factor will the reaction rate be increased by the platinum at 600 K, all other factors being equal?

12.77 (a) Calculate the activation energy for the acid hydrolysis of sucrose from an Arrhenius plot of the following data:

Temperature, °C	k, L/(mol·s)
24	4.8×10^{-3}
28	7.8×10^{-3}
32	13×10^{-3}
36	20×10^{-3}
40	32×10^{-3}

(b) Calculate the rate constant at 37°C (body temperature).

12.78 (a) Calculate the activation energy for the reaction between ethyl bromide, C_2H_5Br, and hydroxide ions in water from an Arrhenius plot of the following data:

Temperature, °C	k, L/(mol·s)
24	1.3×10^{-3}
28	2.0×10^{-3}
32	3.0×10^{-3}
36	4.4×10^{-3}
40	6.4×10^{-3}

(b) Calculate the rate constant at 25°C.

12.79 Raw milk sours in about 4 h at 28°C, but in about 48 h in a refrigerator at 5°C. What is the activation energy for the souring of milk?

12.80 Under certain conditions, the reaction $H_2(g) + Br_2(g) \rightarrow 2HBr(g)$ obeys the rate law

$$\text{Rate} = k[H_2][Br_2]^{1/2}$$

However, the reaction hardly proceeds at all if another substance is added that rapidly removes hydrogen and bromine atoms. Suggest a mechanism for the reaction.

12.81 The contribution to the destruction of the ozone layer by the presence of chlorofluorocarbons has been explained by the following mechanism:

Step 1. $O_3 + Cl \rightarrow ClO + O_2$
Step 2. $ClO + O \rightarrow Cl + O_2$

(a) What is the catalyst in the reaction? (b) What is the reaction intermediate? (c) Identify the radicals in the mechanism. (d) Identify the steps as initiating, propagating, or terminating. (e) Write a chain-terminating step for the reaction.

12.82 The half-life of a substance taking part in a third-order reaction $A \rightarrow$ products is inversely proportional to the square of the initial concentration of A. How may this half-life be used to predict the time needed for the concentration to fall to (a) one-half; (b) one-fourth; (c) one-sixteenth of its initial value?

12.83 When a 1:1 mole ratio mixture of hydrogen and chlorine is exposed to sunlight, the reaction $H_2(g) + Cl_2(g) \rightarrow 2HCl(g)$ occurs explosively. The proposed chain-reaction mechanism is thought to be:

Step 1. $Cl_2 \rightarrow 2Cl$, a chain-initiation step caused by light
Step 2. a chain-propagation step that results in the formation of a hydrogen atom
Step 3. a chain-propagation step in which the hydrogen atom reacts with Cl_2
Steps 4, 5, 6. three chain-termination steps that occur between the two radicals

CASE 12 ▸ SMOG FORMATION

The use of fossil fuels for the generation of energy has both a positive and a negative effect on the quality of life. The positive side includes all the processes of civilization. On the negative side, the high-temperature combustion and incomplete combustion of fossil fuels, primarily in automobiles, results in the emission of a mixture of primary pollutants into the atmosphere. Many of these primary pollutants are then converted by further photochemical reactions into secondary pollutants, which have various concentrations, compositions, and structures. Smog is the outcome of the presence of both primary and secondary pollutants.

The photochemical mechanism for the production of secondary pollutants varies according to the intensity of the sunlight, the meteorological conditions, and the concentration of the primary pollutants. The first step is the production of nitric oxide, NO, in the high-temperature conditions inside the combustion cylinder of an automobile:

$$N_2(g) + O_2(g) \longrightarrow 2NO(g)$$

Once vented from the automobile into the atmosphere, the nitric oxide is readily oxidized to nitrogen dioxide:

$$2NO(g) + O_2(g) \longrightarrow 2NO_2(g)$$

Sunlight at wavelengths less than 400 nm dissociates nitrogen dioxide into NO and an O atom:

$$NO_2(g) + \xrightarrow{h\nu,\ \lambda < 400\text{ nm}} NO(g) + O(g)$$

Because 400-nm light is short-wavelength light in the visible spectrum (the violet end), the long-wavelength light (the red-yellow end) is transmitted. This color is visible in the brown haze over a city; another contributor to the brown color is light scattered from small dust particles in the atmosphere, which also imparts a red-brown hue.

An oxygen atom, $\cdot O \cdot$, is a highly reactive diradical (a radical with two unpaired electrons). It is a strong oxidizing agent and can bring about the oxidation of many of the primary pollutants; it even oxidizes oxygen (to ozone). Ozone and lachrymators (eye irritants) commonly associated with smog, called peroxyacylnitrates (PAN), are the most evident products of the oxidation. The mechanism for PAN formation is quite complex and open-ended, in the sense that many pathways can lead to a PAN. One proposed mechanism for PAN formation follows.

One of the first reactions of an O atom is the abstraction of a H atom from some of the unburned fuel to form the hydroxyl radical, $\cdot OH$, the more common oxidizing agent in smog; for example, with methane:

$$CH_4 + \cdot O \cdot \longrightarrow H_3C\cdot + \cdot OH$$

The methyl radical, $H_3C\cdot$, is also highly reactive. Some of the unburned fuel in an automobile's exhaust may be only partially oxidized (either in the combustion chamber or in the atmosphere containing the oxygen atoms) to form compounds such as acetaldehyde, CH_3CHO. Acetal-

Write the equations for the elementary reactions in steps 2 to 6.

12.84 The reaction of methane, CH_4, with chlorine gas occurs by a mechanism similar to that described in Exercise 12.83. Write the mechanism. (a) Identify the initiation, propagation, and termination steps. (b) What products are predicted from the mechanism?

12.85 At 328 K, the total pressure of the dinitrogen pentoxide decomposition to NO_2 and O_2 varied with time as shown by the following data. Use the data to find the rate in L/(mol·min) at each time.

Time, min	0	5	10	15	20	30
Pressure, kPa	27.3	43.7	53.6	59.4	63.0	66.3

dehyde reacts with the highly reactive hydroxyl radical, thereby producing yet another radical:

$$CH_3CHO + \cdot OH \longrightarrow CH_3CO\cdot + H_2O$$

The $CH_3CO\cdot$ radical is also highly reactive and combines with molecular oxygen to form a peroxyl radical:

$$CH_3CO\cdot + O_2 \longrightarrow CH_3\overset{\overset{\displaystyle O}{\|}}{C}OO\cdot$$

which combines with NO_2 (an odd-electron molecule, see Chapter 8, and therefore also a radical) to form peroxyacetylnitrate:

$$CH_3\overset{\overset{\displaystyle O}{\|}}{C}OO\cdot + NO_2 \longrightarrow CH_3C(=O){-}O{-}O{-}NO_2$$

Peroxyacylnitrates, like ozone, are highly toxic, even at concentrations of less than 0.1 ppm; and exposure to them at concentrations of 0.5 ppm for 5 min can cause eye irritation.

QUESTIONS

1. Write the chemical equation for the reaction of the hydroxyl radical with (a) ethylene, $CH_2{=}CH_2$, and (b) ethane, CH_3CH_3.

2. Write the chemical equation for (a) the reaction that is a possible source of ozone in a smoggy atmosphere and (b) the reaction of a hydroxyl radical with benzene.

3. (a) Suggest an equation for a reaction that results in the formation of nitrogen trioxide, NO_3, in smog. (b) What is the most likely product from the reaction of NO_3 with NO_2? Draw Lewis structures to assist in your prediction. (c) What is the resulting solution if the substance formed in (b) is dissolved in water?

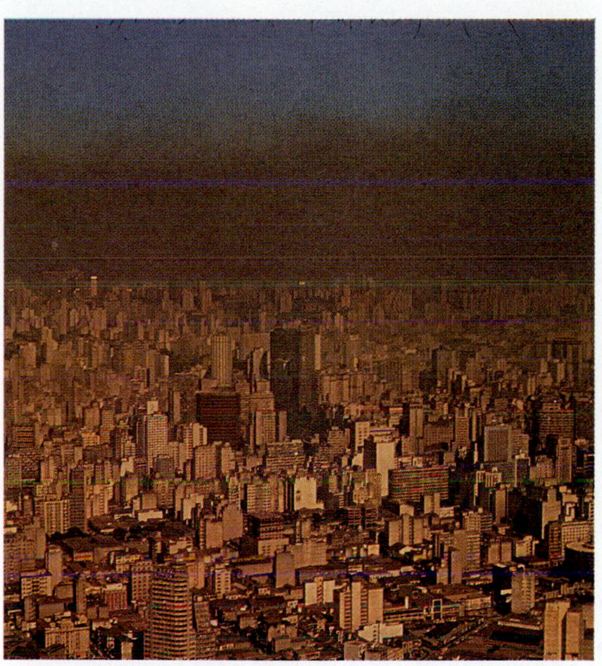

12.86 When the rate of the reaction $2NO(g) + O_2(g) \rightarrow 2NO_2(g)$ was studied, it was found that the rate doubled when the O_2 concentration was doubled, but quadrupled when the NO concentration was doubled. Which of the following mechanisms accounts for these observations?

(a) Step 1. $NO + O_2 \rightarrow NO_3$, and its reverse (both fast, equilibrium)
Step 2. $NO + NO_3 \rightarrow 2NO_2$ (slow)

(b) Step 1. $2NO \rightarrow N_2O_2$ (slow)
Step 2. $O_2 + N_2O_2 \rightarrow N_2O_4$ (fast)
Step 3. $N_2O_4 \rightarrow 2NO_2$ (fast)

In this chapter we see how to describe the dynamic equilibria reached by chemical reactions and how to predict the composition of the reaction mixture when equilibrium has been reached. We also see how reaction mixtures respond to changes in the pressure and temperature at which the reaction is being carried out. The illustration shows a chemical garden grown by adding salts to a solution of sodium silicate: when the growth of the garden has ceased, the solutions and the solid growths are in dynamic equilibrium.

CHEMICAL EQUILIBRIUM

We have seen several examples of physical processes—including vaporization and dissolution—that reach dynamic equilibrium, the state in which the forward and reverse processes are continuing at equal rates. For example, when water is present in a closed container, the equilibrium

$$H_2O(l) \rightleftharpoons H_2O(g)$$

is established between the liquid and the vapor; and for a saturated solution of sodium chloride, the equilibrium

$$NaCl(s) \rightleftharpoons Na^+(aq) + Cl^-(aq)$$

exists between the solid solute and the solute in solution. As we saw in Chapter 11, the symbol $\rightleftharpoons$ indicates that the system is at dynamic equilibrium. Now we shall see that exactly the same ideas apply to chemical reactions: they also reach a dynamic equilibrium, which we write

$$\text{Reactants} \rightleftharpoons \text{products}$$

where, once again, the symbol $\rightleftharpoons$ indicates that the forward and reverse reaction rates are equal.

That chemical reactions tend toward (and often reach) dynamic equilibrium is of the greatest practical importance. The existence of equilibrium means that, like the vaporization of a liquid in a closed container, a chemical reaction does not always "go to completion"; instead, some reactants and products coexist in the system. However, because equilibrium amounts respond to changes in pressure (or concentration) and temperature, we can often encourage the formation of a product (just as heating increases the vapor pressure of a liquid or changes the solubility of a solute) and hence increase the yield of a reaction.

E The solution used in a crystal garden is saturated with sodium silicate, Na_2SiO_3 (which is also called water glass).

E Using the equilibrium concept. *J. Chem. Educ.*, **1981**, *58*, 56. Chemical equilibrium. *J. Chem. Educ.*, **1988**, *65*, 618. Chemical equilibrium. *J. Chem. Educ.*, **1980**, *57*, 801.

THE DESCRIPTION OF CHEMICAL EQUILIBRIUM

N A summary of the Haber process begins on p. 519. References to the process occur throughout the text.

The **Haber process**, the synthesis of ammonia from nitrogen and hydrogen, provides a good example of chemical equilibrium. This process is the source of over 1.5×10^{10} kg of ammonia in the United States each year. In the process, nitrogen and hydrogen react to produce ammonia in the reaction

$$N_2(g) + 3H_2(g) \longrightarrow 2NH_3(g)$$

The rate of this reaction decreases as the concentrations of nitrogen and hydrogen decrease. The reverse reaction

$$2NH_3(g) \longrightarrow N_2(g) + 3H_2(g)$$

V Video 36: Chemical equilibrium in a gaseous system.

D Chromate-dichromate equilibrium. TDC. 182. Secular equilibrium. TDC, 185.

E On the dynamic nature of chemical equilibrium. *J. Chem. Educ.*, **1983**, *60*, 930.

occurs at an increasing rate as the concentration of ammonia increases and the newly formed ammonia decomposes back into nitrogen and hydrogen. The dynamic equilibrium

$$N_2(g) + 3H_2(g) \rightleftharpoons 2NH_3(g)$$

is reached when the rate of the reverse reaction has increased to equal that of the forward reaction.

13.1 REACTIONS AT EQUILIBRIUM

N The first part of the curve—the approach to equilibrium—was the focus of the discussion of chemical kinetics in Chapter 12 (although we did not consider there the effect of the reverse reaction).

T OHT: Fig. 13.1.

Figure 13.1 shows how the molar concentrations of N_2, H_2, and NH_3 change with time in a mixture kept at 500°C and 250 atm. After a period of time, the composition of the mixture remains the same even though some nitrogen and hydrogen are still present. The absence of apparent change is exactly like the formation of a saturated solution of sodium chloride, when some undissolved solid remains, and like the vaporization of water in a closed container, when some liquid water remains.

In each case a dynamic equilibrium exists. To confirm that the N_2, H_2, NH_3 equilibrium is dynamic, we can carry out two ammonia syntheses with exactly the same starting conditions, except that in one of the syntheses we use D_2 in place of H_2. The two reaction mixtures reach equilibrium with the same composition, apart from the presence of D_2 and ND_3 in one of the systems instead of H_2 and NH_3. Now consider what happens when the two systems are combined (Fig. 13.2). Titration of a cooled sample shows that the concentration of ammonia remains constant in the mixture.* However, when the sample is analyzed with a mass spectrometer, all isotopic forms of ammonia (NH_3, NH_2D, NHD_2, and ND_3) and all isotopic forms of hydrogen (H_2, HD, and D_2) are found. This scrambling of H and D must be a result of a continuation of the forward and reverse reactions even though the reaction mixture has reached equilibrium.

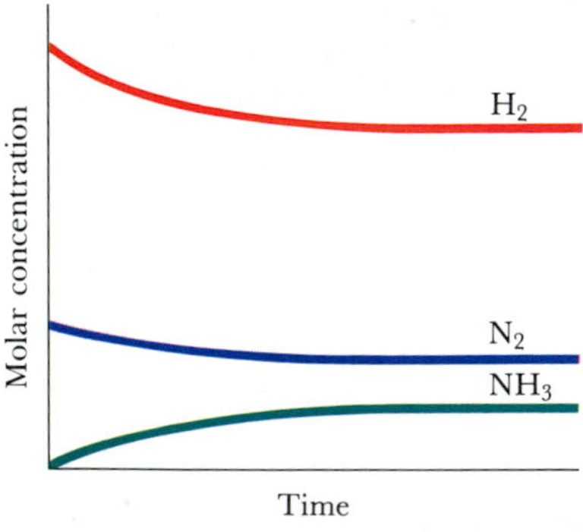

FIGURE 13.1 In the synthesis of ammonia, the concentrations of N_2, H_2, and NH_3 change with time until they finally settle into a state in which all three are present and there is no further net change in concentration.

*The equilibrium composition is slightly affected by the replacement of hydrogen by deuterium, but not enough to affect this argument.

13.2 THE EQUILIBRIUM CONSTANT

The key concept for discussing chemical equilibria quantitatively was identified by the Norwegians Cato Guldberg (a mathematician) and Peter Waage (a chemist) in 1864. They noticed that the molar concentrations of the reactants and products in a reaction mixture at equilibrium always satisfied a certain relation (which will be specified shortly). The existence of this relation can be illustrated with the data in Table 13.1 for the reaction between SO_2 and O_2 to produce SO_3, for which the dynamic equilibrium is

$$2SO_2(g) + O_2(g) \rightleftharpoons 2SO_3(g)$$

This reaction is used in the production of sulfuric acid—the SO_3 is subsequently dissolved in concentrated sulfuric acid and then diluted with water to produce more sulfuric acid.

The law of mass action and the reaction quotient. In the experiments reported in Table 13.1, several different mixtures with different initial compositions were prepared at 1000 K and allowed to reach equilibrium. Analysis of the equilibrium mixtures showed that the reactants and products were present in a variety of different concentrations. However, these values have a pattern that becomes clear when we calculate for each experiment the value of the **reaction quotient**, which for this reaction is

$$Q_c = \frac{[SO_3]^2}{[SO_2]^2[O_2]}$$

where [X] denotes the molar concentration of the species X *relative to a standard molar concentration.* For a species in a mixture, we take the standard molar concentration to be 1 mol/L:

$$[X] = \frac{\text{molar concentration of X}}{1\text{ mol/L}}$$

The subscript c reminds us that the reaction quotient is defined in terms of concentrations. For example, if the molar concentration of SO_2 happens to be 2.3×10^{-3} mol/L, then

$$[SO_2] = \frac{2.3 \times 10^{-3}\text{ mol/L}}{1\text{ mol/L}} = 2.3 \times 10^{-3}$$

TABLE 13.1 Equilibrium data for the reaction $2SO_2(g) + O_2(g) \rightleftharpoons 2SO_3(g)$ at 1000 K

$[SO_2]$, mol/L	$[O_2]$, mol/L	$[SO_3]$, mol/L	K_c
0.660	0.390	0.0840	0.0415
0.0380	0.220	0.00360	0.0409
0.110	0.110	0.00750	0.0423
0.950	0.880	0.180	0.0408
1.44	1.98	0.410	0.0409
		Average:	0.0413

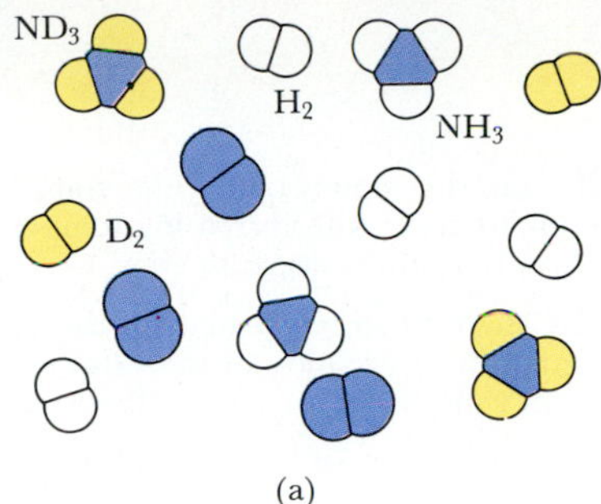

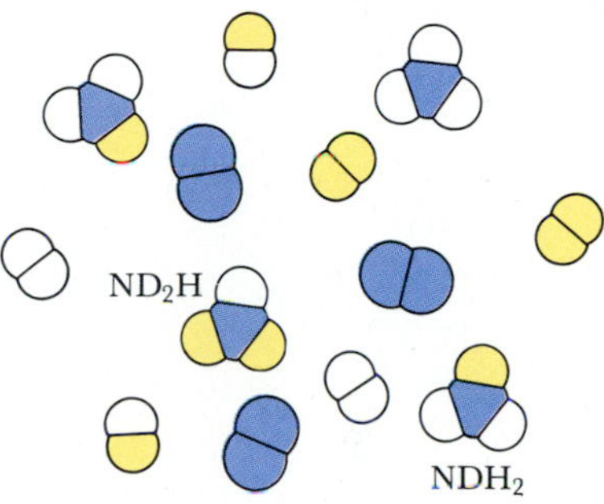

FIGURE 13.2 In an experiment showing that equilibrium is dynamic, two ammonia-synthesis reaction mixtures, one with H_2 and the other with D_2, are allowed to reach equilibrium. (a) When they are mixed, the concentration of ammonia does not change, but (b) in due course NH_2D and NHD_2 appear in the products.

N The reaction quotient Q_c is also called the *mass-action expression.*

N The advantage of dividing the molar concentration by a standard molar concentration is that Q_c (and, later, K_c) turn out to be dimensionless, which is essential when we come to take their logarithms (in Chapters 14, 15, and 16). A simpler approach would be to interpret [X] as the molar concentration in moles per liter, but then the description of equilibria involving pure liquids and solids (see below) would be more cumbersome.

U Exercise 13.64.

T OHT: Table 13.1.

E Clarifying the concept of equilibrium in chemically reacting systems. *J. Chem. Educ.*, **1982**, *59*, 1034. An elementary discussion of chemical equilibrium. *J. Chem. Educ.*, **1988**, *65*, 407. Units of an equilibrium constant. *J. Chem. Educ.*, **1990**, *67*, 88.

N Note that some equilibrium constants increase with increasing temperature, but others decrease.

? Can you detect a pattern in the temperature variation of the values of K_c in Table 13.2?

E A better way of dealing with chemical equilibrium. *J. Chem. Educ.*, **1986**, *63*, 582.

U Exercise 13.61.

TABLE 13.2 Equilibrium constants, K_c, for various reactions

Reaction	Temperature, K	K_c
$H_2(g) + Cl_2(g) \rightleftharpoons 2HCl(g)$	300	4.0×10^{31}
	500	4.0×10^{18}
	1000	5.1×10^{8}
$H_2(g) + Br_2(g) \rightleftharpoons 2HBr(g)$	300	1.9×10^{17}
	500	1.3×10^{10}
	1000	3.8×10^{4}
$H_2(g) + I_2(g) \rightleftharpoons 2HI(g)$	298	794
	500	160
	700	54
$2BrCl(g) \rightleftharpoons Br_2(g) + Cl_2(g)$	300	377
	500	32
	1000	5
$2HD(g) \rightleftharpoons H_2(g) + D_2(g)$	100	0.52
	500	0.28
	1000	0.26
$F_2(g) \rightleftharpoons 2F(g)$	500	7.3×10^{-13}
	1000	1.2×10^{-4}
	1200	2.7×10^{-3}
$Cl_2(g) \rightleftharpoons 2Cl(g)$	1000	1.2×10^{-7}
	1200	1.7×10^{-5}
$Br_2(g) \rightleftharpoons 2Br(g)$	1000	4.1×10^{-7}
	1200	1.7×10^{-5}
$I_2(g) \rightleftharpoons 2I(g)$	800	3.1×10^{-5}
	1000	3.1×10^{-3}
	1200	6.8×10^{-2}

The striking result is that, within experimental error, when the molar concentrations of the various equilibrium mixtures are used to calculate the reaction quotient, *the same value of Q_c is obtained in each case.* This common value is called the **equilibrium constant**, K_c, of the reaction.

As we see in Table 13.1, equilibrium constants defined in terms of concentrations are not strictly constant but vary within a small range, typically of about 5%. Hence, the calculations we shall encounter in this chapter are accurate to within about 5%. In more advanced work, equilibrium constants are defined in a manner that takes into account the effect of the molecules on one another, and their values are then strictly constant.

Guldberg and Waage studied a variety of reactions and found that, in every case, the equilibrium composition of a particular reaction could be expressed in a similar way. They proposed the law of mass action to summarize their conclusions:

Law of mass action: For an equilibrium of the form, $a\text{A} + b\text{B} \rightleftharpoons c\text{C} + d\text{D}$ the reaction quotient

$$Q_c = \frac{[\text{C}]^c[\text{D}]^d}{[\text{A}]^a[\text{B}]^b}$$

evaluated by using the equilibrium molar concentrations of the reactants and products, is equal to a constant, K_c, which has a specific value for a given reaction and temperature.

Notice that the products (C and D) occur in the numerator and the reactants (A and B) in the denominator. Each [X] is raised to a power equal to the stoichiometric coefficient in the balanced equation. For example, the reaction quotient for

$$4\text{NH}_3(g) + 5\text{O}_2(g) \rightleftharpoons 4\text{NO}(g) + 6\text{H}_2\text{O}(g)$$

is

$$Q_c = \frac{[\text{NO}]^4[\text{H}_2\text{O}]^6}{[\text{NH}_3]^4[\text{O}_2]^5}$$

To obtain the value of K_c, we use the molar concentrations at equilibrium in this expression.

Each reaction has its own characteristic equilibrium constant, but that value is dependent on the temperature (see Table 13.2). Whatever the initial composition of the reaction mixture, at a given temperature its equilibrium composition will always correspond (in practice, to within about 5%) to the value of K_c for that reaction. Therefore, to measure an equilibrium constant, we can take any convenient initial mixture of reagents, allow the reaction to reach equilibrium at the temperature of interest, measure the concentration of the reactants and products, and insert them into the expression for Q_c. Because all the molar concentrations in the reaction quotient are divided by a standard molar concentration (1 mol/L), both Q_c and K_c are unitless.

N We think it is helpful to consider the reaction quotient as the primary object of discussion, because it is applicable at any stage of the reaction; then K_c is the numerical value of Q_c at a particular stage, namely equilibrium.

The magnitude of K_c. The value of an equilibrium constant is an indication of the extent to which the reaction favors the reactants or products at equilibrium (Fig. 13.3):

- Large values of K_c (larger than 10^3): The equilibrium favors the products strongly.
- Intermediate values of K_c (in the range 10^{-3} to 10^3): Reactants and products are present in similar amounts at equilibrium.
- Small values of K_c (smaller than 10^{-3}): The equilibrium favors the reactants strongly.

For example, consider the reaction

$$\text{H}_2(g) + \text{Cl}_2(g) \rightleftharpoons 2\text{HCl}(g)$$

$$Q_c = \frac{[\text{HCl}]^2}{[\text{H}_2][\text{Cl}_2]}, \quad K_c = 4.0 \times 10^{31} \text{ at } 300°\text{C}$$

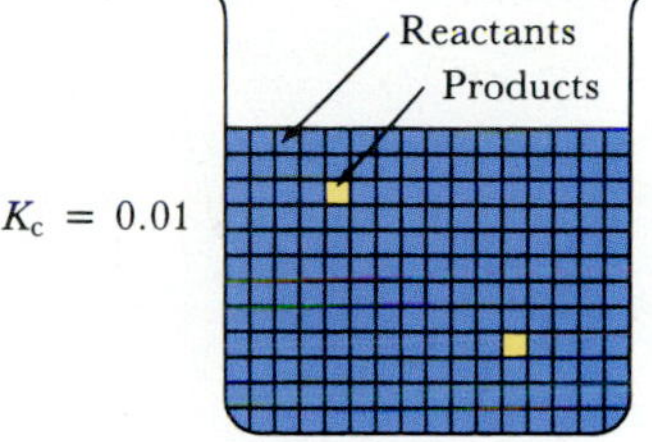

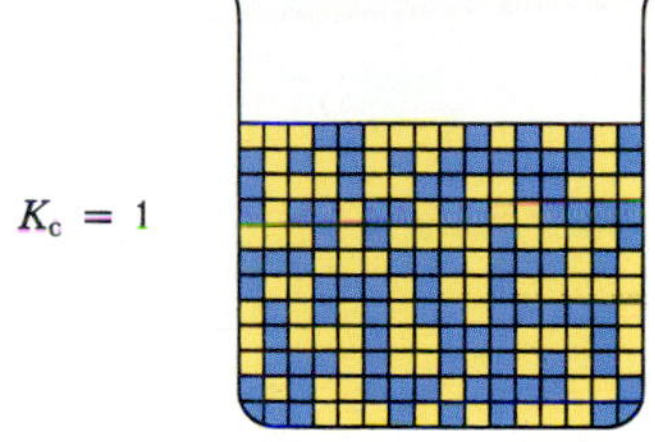

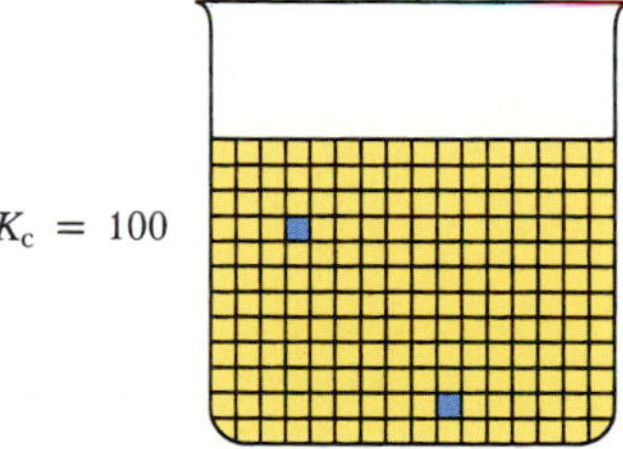

FIGURE 13.3 The size of the equilibrium constant shows whether the reactants or the products are favored. The reactants are represented by blue squares and the products by yellow squares. (a) K_c very small (reactants favored); (b) K_c near 1 (reactants and products present in similar concentrations); (c) K_c large (products favored).

N It is instructive to repeat this type of calculation for very small values of K_c and for values close to 1.

With such a large value for K_c, the equilibrium lies in favor of HCl. For instance, suppose that in this system the equilibrium concentrations of H_2 and Cl_2 are both extremely small, at 1.0×10^{-16} mol/L—only about 6×10^7 molecules of each species per liter—then the equilibrium molar concentration of HCl is given by the solution of the expression

$$4.0 \times 10^{31} = \frac{[HCl]^2}{1.0 \times 10^{-16} \times 1.0 \times 10^{-16}}$$

which rearranges to

$$[HCl] = \sqrt{4.0 \times 10^{31} \times (1.0 \times 10^{-16})^2} = 0.63$$

That is, the molar concentration of HCl at equilibrium is 0.63×1 mol/L $= 0.63$ mol/L. We see that at equilibrium there is an overwhelming amount of product in the system.

The equilibrium constant and the chemical equation. The reaction quotient and its value at equilibrium depend on how the chemical equation for the reaction is written. For example, earlier we wrote

$$4NH_3(g) + 5O_2(g) \rightleftharpoons 4NO(g) + 6H_2O(g) \qquad Q_c = \frac{[NO]^4[H_2O]^6}{[NH_3]^4[O_2]^5}$$

However, if we had written the reaction as

$$4NO(g) + 6H_2O(g) \rightleftharpoons 4NH_3(g) + 5O_2(g)$$

we would have written the reaction quotient as

$$Q_c = \frac{[NH_3]^4[O_2]^5}{[NO]^4[H_2O]^6}$$

which is the *reciprocal* of the earlier expression. In general, *if we reverse the way in which the reaction is written, the new reaction quotient (and the equilibrium constant) is the reciprocal of the old.* Similarly, had we written the reaction in the form

$$2NH_3(g) + \tfrac{5}{2}O_2(g) \rightleftharpoons 2NO(g) + 3H_2O(g)$$

(with the stoichiometric coefficients all divided by 2), we would write the reaction quotient

$$Q_c = \frac{[NO]^2[H_2O]^3}{[NH_3]^2[O_2]^{5/2}} = \sqrt{\frac{[NO]^4[H_2O]^6}{[NH_3]^4[O_2]^5}}$$

and the new reaction quotient and the equilibrium constant are now the *square root* of the original ones.

The direction of the reaction. We shall now see how to decide whether a reaction mixture that is not at equilibrium has a tendency to form more product or to decompose into reactant in order to reach equilibrium. To do so, we calculate the reaction quotient by using the specified molar concentrations of each substance in the reaction mixture. Recall that only when the reaction mixture is at equilibrium is the value of the reaction quotient equal to the equilibrium constant of the reaction.

N We are adopting the procedure that a reaction quotient is an algebraic expression that has a value that depends on the values of the molar concentrations that are substituted into it. If the *equilibrium* values of the molar concentrations are used, then Q_c takes a particular numerical value, which we call the equilibrium constant. That is, K_c is not an algebraic quantity: it is a particular numerical value of an algebraic quantity Q_c.

The direction in which the mixture tends to react (toward more product or back toward reactants) can be predicted by comparing the value of the reaction quotient with the equilibrium constant. Suppose an analysis of a reaction mixture reveals that $Q_c < K_c$; then we know that the concentrations of the products are too low (or the concentrations of the reactants are too high) for equilibrium. Hence, the reaction has a tendency to proceed right toward products, for in that way the mixture will become poorer in reactants and richer in products. If, on the other hand, the analysis of the system reveals that $Q_c > K_c$, then the concentrations of the products are too high (or the concentrations of the reactants are too low) and the reaction tends to proceed to the left, toward reactants, to reach equilibrium. If Q_c turns out to be *equal* to K_c, then the mixture is at equilibrium and has no tendency to change in either direction.

The tendency of the reaction mixture to change toward its equilibrium composition, and hence for Q_c to become equal to K_c, may be summarized as follows (Fig. 13.4):

When $Q_c < K_c$, the reaction tends to proceed right, toward products.
When $Q_c = K_c$, the reaction is at equilibrium.
When $Q_c > K_c$, the reaction tends to proceed left, toward reactants.

▼ EXAMPLE 13.1 Predicting the direction of reaction

A mixture of hydrogen, iodine, and hydrogen iodide, each at a concentration of 0.0020 mol/L, was introduced into a container heated to 490°C. At this temperature, $K_c = 46$ for the reaction $H_2(g) + I_2(g) \rightleftharpoons 2HI(g)$. Predict whether or not more HI has a tendency to form.

STRATEGY We calculate Q_c and compare it with K_c. Because K_c has an "intermediate" value, we expect the concentrations of the reactants and products to be similar to one another. We can therefore anticipate that, although there may be some shift toward products or reactants, the extent of the reaction will be quite small.

SOLUTION The reaction quotient is

$$Q_c = \frac{[HI]^2}{[H_2][I_2]} = \frac{(0.0020)^2}{0.0020 \times 0.0020} = 1.0$$

Because $Q_c < K_c$ (because $K_c = 46$), we know that the numerator—the concentration of the product—is too small for the composition of the system to correspond to equilibrium. Therefore, the reaction will tend to proceed right to form more product and consequently to consume reactant.

EXERCISE E13.1 A mixture of composition 3.0×10^{-3} mol H_2/L, 1.0×10^{-3} mol N_2/L, and 2.0×10^{-3} mol NH_3/L was prepared and heated to 500°C, at which temperature $K_c = 0.11$ for $N_2(g) + 3H_2(g) \rightleftharpoons 2NH_3(g)$. Calculate Q_c and predict whether ammonia will tend to form or decompose.

▲ [*Answer*: $Q_c = 1.5 \times 10^5$; tend to decompose]

? What is the likelihood that, when a solution is prepared in the laboratory, it will be at equilibrium?

U Exercises 13.65 and 13.66.

T OHT: Fig. 13.4.

N The values of Q_c and K_c could be likened to the heights of water in two connected vessels: equilibrium will be reached if the water flows in the direction that equalizes the heights.

Notice that when Q_c and K_c are not equal, we say that there is a *tendency* to form either reactants or products. In reality, the reaction may be so slow in both directions that an equilibrium is never reached. A mixture of hydrogen and oxygen has a tendency to form water; nevertheless, at room temperature and in the absence of an initiating spark, the reaction is almost infinitely slow. The tendency, although present, is not realized, and the mixture of gases can be kept indefinitely. In dealing with equilibria, we have to bear in mind that the *rates* of the forward and reverse reactions may be too slow for our prediction to be realized in practice. *A tendency toward a particular substance does not mean that the substance will actually be formed under the conditions of the reaction.*

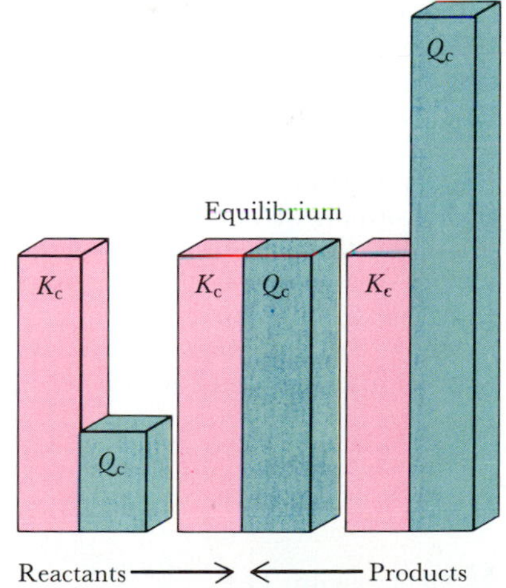

FIGURE 13.4 The relative sizes of the reaction quotient Q_c and the equilibrium constant K_c indicate the direction in which a reaction mixture tends to change. The progress toward equilibrium may, however, be unobservably slow.

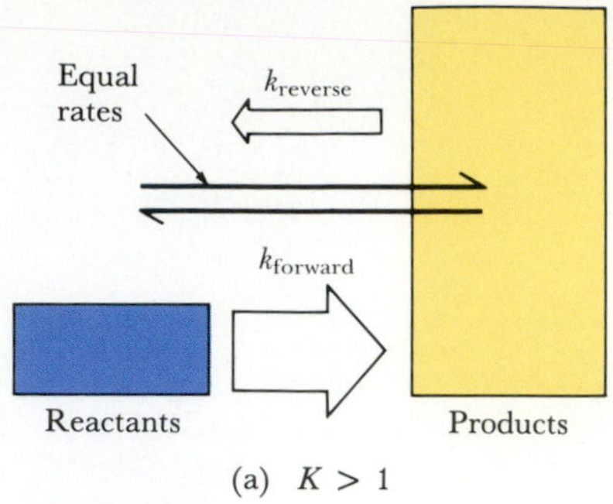

(a) $K > 1$

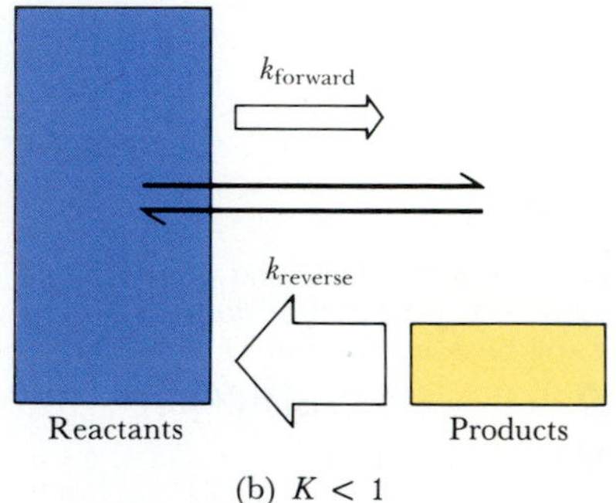

(b) $K < 1$

FIGURE 13.5 The equilibrium constant K expresses the concentration condition under which the rates of the forward and reverse reactions are equal. (a) When the rate constant for the forward reaction is large compared with that for the reverse reaction, only a low concentration of reactants is needed to match the rate of decomposition of the products; the equilibrium constant is large, and products are favored. (b) When the rate constant for the reverse reaction is large compared with that for the forward reaction, the equilibrium constant is small and reactants are favored.

T OHT: Figure 13.5.

N To be strictly consistent with the thermodynamic data in Appendix 2A, the partial pressure should be relative to 1 bar; however, the difference between 1 atm and 1 bar is insignificant at this level.

D Demonstrating the relation between rate constants and the equilibrium constant. *J. Chem. Educ.*, **1970**, *47*, 646.

The relation of equilibrium composition to reaction rate. Although the concept of the equilibrium constant was discovered by analyzing experimental data, the reason for its existence is easy to understand. We know that chemical equilibria are dynamic and occur when the forward and reverse reactions have the same rates. Because reaction rates change with concentration, there will always be a set of reactant and product concentrations that correspond to equal forward and reverse rates. The equilibrium constant expresses the relationship between the concentrations that guarantees this equality of rates.

To see that there is such a set of concentrations, consider again the reaction for the synthesis of hydrogen iodide treated in Example 13.1. Experiments have shown that the rate laws for the forward and reverse reactions are both second order overall:

Forward: $H_2(g) + I_2(g) \longrightarrow 2HI(g)$ $\quad$ $\text{rate}_{\text{forward}} = k_{\text{forward}}[H_2][I_2]$

Reverse: $2HI(g) \longrightarrow H_2(g) + I_2(g)$ $\quad$ $\text{rate}_{\text{reverse}} = k_{\text{reverse}}[HI]^2$

At equilibrium, the forward and reverse rates are equal, so

$$k_{\text{forward}}[H_2][I_2] = k_{\text{reverse}}[HI]^2 \text{ at equilibrium}$$

This expression may be rearranged into

$$\frac{k_{\text{forward}}}{k_{\text{reverse}}} = \frac{[HI]^2}{[H_2][I_2]} = K_c \text{ at equilibrium}$$

That is, *the equilibrium constant for the reaction is equal to the ratio of the forward and reverse rate constants.* If the rate constant for the forward reaction is large compared with the rate constant for the reverse reaction, then K_c is large and products are favored. If the opposite is true, then K_c is small and reactants are favored (Fig. 13.5).

The equilibrium constant in terms of partial pressures. The molar concentration of a gas X is directly proportional to its partial pressure P_x (Section 5.6). It is therefore often convenient to express the composition of a gaseous reaction mixture in terms of the partial pressures of the components rather than in terms of their molar concentrations.

When writing reaction quotients in terms of partial pressures, we express each partial pressure relative to a standard value, which is taken to be 1 atm. Therefore, when writing a reaction quotient, we take P_x to mean the ratio

$$P_x = \frac{\text{partial pressure of X}}{1\text{ atm}}$$

The reaction quotient is written Q_P and has the form

$$Q_P = \frac{\text{partial pressures of products}}{\text{partial pressures of reactants}}$$

with each partial pressure raised to a power equal to the stoichiometric coefficient in the chemical equation. Thus, for the reaction

$$4NO(g) + 6H_2O(g) \rightleftharpoons 4NH_3(g) + 5O_2(g)$$

we write

$$Q_P = \frac{(P_{NH_3})^4(P_{O_2})^5}{(P_{NO})^4(P_{H_2O})^6}$$

TABLE 13.3 Equilibrium constants, K_P, for various reactions

Reaction	Temperature, K	K_P
$N_2(g) + 3H_2(g) \rightleftharpoons 2NH_3(g)$	298	6.8×10^5
	400	41
	500	3.5×10^{-2}
$H_2(g) + I_2(g) \rightleftharpoons 2HI(g)$	298	794
	500	160
	700	54
$2SO_2(g) + O_2(g) \rightleftharpoons 2SO_3(g)$	298	4.0×10^{24}
	500	2.5×10^{10}
	700	3.0×10^4
$N_2O_4(g) \rightleftharpoons 2NO_2(g)$	298	0.98
	400	47.9
	500	1700

? Suggest a reason why the value of K_P for the reaction between hydrogen and iodine is numerically the same as the value for K_c for the same reaction (Table 13.2).

When the reaction is at equilibrium and the partial pressures have the corresponding equilibrium values, Q_P is equal to the equilibrium constant for the reaction, which is denoted K_P. In the ammonia synthesis, for instance,

$$N_2(g) + 3H_2(g) \rightleftharpoons 2NH_3(g) \qquad Q_P = \frac{(P_{NH_3})^2}{P_{N_2}(P_{H_2})^3}$$

and when the system is at equilibrium, the value of Q_P is the equilibrium constant K_P. For example, suppose that, in a typical experiment at 500 K, it was found that the partial pressures of ammonia, nitrogen, and hydrogen are 0.15 atm, 1.2 atm, and 0.81 atm, respectively, when the reaction reaches equilibrium. Then at that temperature

$$K_P = \frac{(0.15)^2}{1.2 \times (0.81)^3} = 0.035$$

Values of K_P for other reactions are given in Table 13.3.

N The formula for converting between K_c and K_P is in the footnote on p. 502.

N Note that K_P is a dimensionless quantity.

▼ EXAMPLE 13.2 Using K_P to determine a partial pressure

The equilibrium constant for $PCl_5(g) \rightleftharpoons PCl_3(g) + Cl_2(g)$ is $K_P = 490$ at 340°C. The partial pressures of PCl_5 and Cl_2 are 0.0021 atm and 0.48 atm, respectively, in a certain reaction system at equilibrium. What is the equilibrium partial pressure of PCl_3?

STRATEGY Because the composition is given in terms of partial pressures, we first write the reaction quotient Q_P in terms of the partial pressures of the reactants and products and then substitute the known values. The resulting expression is then solved for the one unknown (in this case, for the partial pressure of phosphorus trichloride vapor).

SOLUTION The reaction quotient is

$$Q_P = \frac{P_{PCl_3}P_{Cl_2}}{P_{PCl_5}}$$

Insertion of the equilibrium data into the expression gives

$$490 = \frac{P_{PCl_3} \times 0.48}{0.0021}$$

which rearranges to

$$P_{PCl_3} = \frac{0.0021 \times 490}{0.48} = 2.1$$

That is, the partial pressure of phosphorus trichloride at equilibrium in the mixture is 2.1 atm.

U Exercises 13.67 and 13.68.

E The expression quoted in the footnote is obtained by substituting the expression $P_X = n_X RT/V$ for the partial pressure of each substance in the expression for Q_P and recognizing that n_X/V is the molar concentration [X] of that species.

EXERCISE E13.2 Nitrosyl chloride, NOCl, decomposes into NO and Cl_2 according to the equation $2NOCl(g) \rightleftharpoons 2NO(g) + Cl_2(g)$, with $K_P = 1.8 \times 10^{-2}$ at 500 K. An analysis of a reaction mixture at equilibrium indicates that the partial pressures of NO and Cl_2 are 0.11 atm and 0.84 atm, respectively. What is the equilibrium partial pressure of NOCl?

[*Answer*: 0.75 atm]

U Exercises 13.62 and 13.63.

The values of K_P and K_c for a reaction at a given temperature are generally different. However, should the need ever arise, it is reasonably straightforward to convert one into the other.*

13.3 HETEROGENEOUS EQUILIBRIA

E A heterogeneous equilibrium exists between the polar ice caps and the local oceans (but not the global ocean system).

D The common-ion effect: Ammonium hydroxide and ammonium acetate. CD1, 57.

E Equilibrium constants of chemical reactions involving condensed phases: Pressure dependence and choice of standard state. *J. Chem. Educ.*, **1984**, *61*, 782.

Chemical equilibria in which all the substances taking part are in the same phase are called **homogeneous equilibria**. All the equilibria mentioned in this chapter so far have been homogeneous. Many other reactions, however, lead to **heterogeneous equilibria**, equilibria in which at least one substance is in a phase different from that of the rest. We encountered several examples of heterogeneous physical equilibria in earlier chapters. The vapor pressure of a liquid was described as being the pressure exerted by a vapor when it is in a state of dynamic equilibrium with its liquid (liquid $\rightleftharpoons$ vapor). In that case, two phases coexist in the system. For example, the vapor pressure of water can be represented as a dynamic equilibrium between liquid water and water vapor:

$$H_2O(l) \rightleftharpoons H_2O(g)$$

The existence of a saturated salt solution is another example of a heterogeneous equilibrium, because the solid salt coexists with its ions in aqueous solution:

$$CaCl_2(s) \rightleftharpoons Ca^{2+}(aq) + 2Cl^-(aq)$$

? What other examples of heterogeneous equilibria can you think of?

The consequences of heterogeneous solubility equilibria are discussed in Chapter 15.

An example of a heterogeneous chemical equilibrium is the initial step in the Mond process for the purification of nickel, in which nickel

*For a reaction $aA(g) + bB(g) \rightleftharpoons cC(g) + dD(g)$, the relation between K_c and K_P is $K_P = K_c \times (RT)^{\Delta n}$, where $\Delta n = (c + d) - (a + b)$ is the difference in the number of gas molecules on the right and left of the chemical equation.

metal reacts with carbon monoxide at 50°C to form gaseous nickel tetracarbonyl, $Ni(CO)_4$:

$$Ni(s) + 4CO(g) \rightleftharpoons Ni(CO)_4(g)$$

E This reaction is reversed (that is, K_c falls well below 1) when the temperature is raised above 200°C; then pure nickel is deposited.

Experiments have shown that the amount of solid nickel present in the system has no affect on the amounts of CO and $Ni(CO)_4$ at equilibrium and that only the temperature affects the relative amounts of CO and $Ni(CO)_4$. Similarly, in the thermal decomposition of limestone, $CaCO_3$,

$$CaCO_3(s) \rightleftharpoons CaO(s) + CO_2(g)$$

the concentration of CO_2 is dependent only on the temperature and not on the amounts of $CaCO_3$ and CaO.

The experimental observations that the equilibrium composition is independent of the amounts of pure solid (or liquid) present is explicable as soon as we note that the molar concentration of a pure solid or liquid (in moles per liter) is proportional to the ratio of the mass of the substance and its volume (not the volume of the vessel, as for gases). As a result, the molar concentration is proportional to the density of the pure solid or liquid (independent of the volume of the vessel), with its molar mass being the conversion factor between the two. Because density is an intensive property (a property with a value that is independent of the size of the sample; Section 1.4), it follows that the molar concentration of pure solids and liquids is intensive too. Therefore, wherever the molar concentration of a pure solid or liquid appears in a reaction quotient, it has the same value however much of the substance is present. Because molar concentrations that appear in a reaction quotient are always expressed relative to a standard value, it is sensible to take the standard as the molar concentration of the pure substance, for then at any stage of the reaction we can write

N This procedure is admissible because the concentration of a solid (which is proportional to its density) is an intensive property, and hence independent of the quantity of material present.

$$[X] = \frac{\text{molar concentration of X}}{\text{standard molar concentration of X}} = 1$$

where X is any pure solid or liquid. For example, for the two heterogeneous equilibria mentioned above, we write

$$Ni(s) + 4CO(g) \rightleftharpoons Ni(CO)_4(g) \qquad Q_c = \frac{[Ni(CO)_4]}{[CO]^4}$$

and

$$CaCO_3(s) \rightleftharpoons CaO(s) + CO_2(g) \qquad Q_c = [CO_2]$$

The relative molar concentrations of the pure solids do not appear (being equal to 1). The pure substances must be present for the equilibrium to exist, but they do not appear in the reaction quotient.

Similar remarks apply to the reaction quotient in terms of partial pressures. For the two equilibria above, we write

$$Q_P = \frac{P_{Ni(CO)_4}}{(P_{CO})^4} \quad \text{and} \quad Q_P = P_{CO_2}$$

(with the partial pressure relative to 1 atm in each case). At equilibrium, the reaction quotients are equal to the equilibrium constants K_P for the reactions. Thus, for the decomposition of calcium carbonate, the equilibrium partial pressure of the CO_2 in the system is equal to K_P for the decomposition reaction. Because at 800°C that pressure is 0.22 atm, $K_P = 0.22$ at 800°C.

E The equilibrium partial pressure of carbon dioxide in the presence of calcium carbonate and calcium oxide, which is equal to K_P, is called the *decomposition vapor pressure* of calcium carbonate.

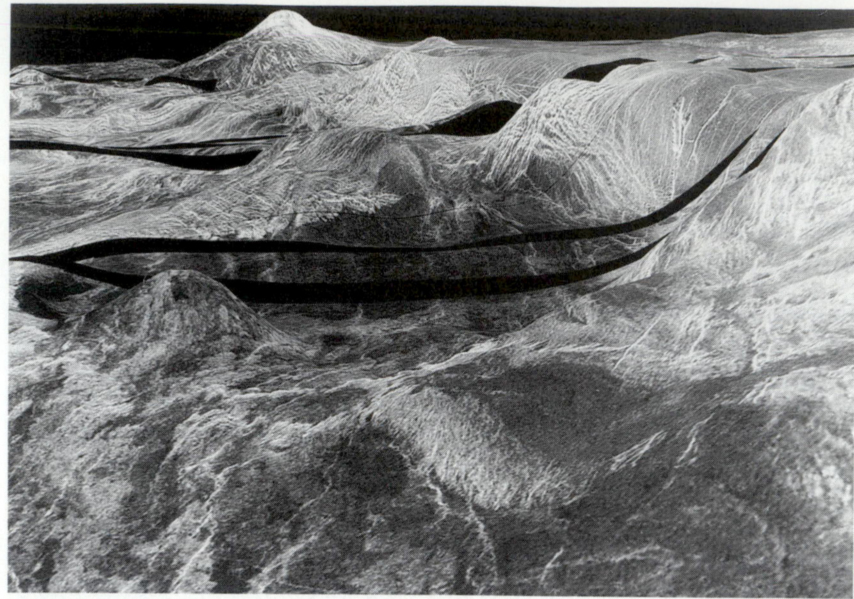

FIGURE 13.6 A radar image of the surface of Venus. Although the rocks are very hot, the partial pressure of carbon dioxide in the atmosphere is so great that carbonates may be abundant. (Black stripes reflect areas where information was not collected.)

When calcium carbonate is heated to 800°C in an *open* container, like a lime kiln or a blast furnace, the gas escapes into the surroundings, the partial pressure of carbon dioxide does not rise to 0.22 atm, equilibrium is never reached, and all the calcium carbonate decomposes. However, when the surroundings are already so rich in carbon dioxide that its partial pressure exceeds 0.22 atm, no net decomposition occurs: for every CO_2 molecule that is formed, one is converted back to carbonate. This tendency to form solid carbonate is probably what occurs on the very hot surface of Venus (Fig. 13.6), where the partial pressure of carbon dioxide is 87 atm. This high value of the pressure has led to speculation that the planet's surface is rich in carbonates, in spite of its high temperature (about 550°C).

EQUILIBRIUM CALCULATIONS

N A Q without a subscript can be either Q_c or Q_P (and likewise for K).

A knowledge of the equilibrium constant of a reaction lets us predict and interpret many aspects of the system's composition at equilibrium. We have already seen that the magnitude of K indicates the "position" of an equilibrium—whether the reactants or the products are favored at equilibrium. In addition, comparison of the values of Q and K indicates the direction in which the reaction has a tendency to occur. As we shall see, once we know K, we can predict the equilibrium composition given the original composition of the reaction mixture. In addition, we shall see that a knowledge of the properties of K enables us to interpret the change in composition that results from a change in reaction conditions, such as the temperature and pressure. These applications are important throughout chemistry, for they are used to discuss solubilities, the behavior of acids, bases, and salts, and the outcome of redox reactions. An indication of their importance is the fact that this chapter and the next four chapters are devoted to an exploration of the properties of equilibrium constants.

In this section we shall meet the ideas that serve as a foundation for the following chapters. We shall see that there is a single unifying theme that, once grasped, accounts for a very wide range of behavior: *at equilibrium, although the individual concentrations of substances in a reaction mixture may vary, collectively the concentrations are such that the reaction quotient is equal to the equilibrium constant.* Only when $Q = K$ are all the substances taking part in the reaction in dynamic equilibrium with one another.

The same format for discussing the properties of a system at equilibrium will be used here and throughout Chapters 14 and 15. Unless stated otherwise, we shall always begin our analysis of equilibrium systems by using the following steps. First, we shall write the balanced chemical equation for the equilibrium. Then

1. Set up an equilibrium table, a table showing the initial molar concentrations of each substance taking part in the reaction.

This step shows how the chemist prepares the reaction system, that is, what and how much of each substance is placed into the reaction vessel. As always, we express the concentrations relative to standard values:

- For mixtures, molar concentrations are relative to 1 mol/L and partial pressures are relative to 1 atm.
- For pure solids and liquids, the relative molar concentrations are all set equal to 1.

2. Write the changes in the molar concentrations that are needed for the reaction to reach equilibrium.

It is often the case that we do not know the changes, so we write *one* of them as x and then use the reaction stoichiometry to express the other changes in terms of that x.

3. Write the equilibrium molar concentrations by adding the change in concentration (from step 2) to the initial concentration of each substance (from step 1).

We should always remember that although a *change* in concentration may be positive (an increase) or negative (a decrease), the *value* of the concentration must always be positive.

4. Use the reaction quotient and the equilibrium constant to determine the value of the unknown molar concentration at equilibrium.

In this step, the equilibrium concentrations that were determined in step 3 are substituted into the reaction quotient. Because the value of the reaction quotient at equilibrium is the equilibrium constant K_c, the resulting expression can be solved for x. The same procedure can be used to calculate compositions in terms of partial pressures.

N Dynamic equilibrium is a universal principle of chemistry. When applying the principle to the reactions, acids and bases, buffers, salts, and solubilities in this and the next few chapters, we consistently use an *equilibrium table* to formulate the quantitative description of the system. The format of the table, together with the approach and attitude, is the same in Chapters 13 to 15; only the system of interest is different.

E Are the equilibrium conditions for a chemical reaction always determined by the initial conditions? *J. Chem. Educ.*, **1990**, *67*, 548.

13.4 SPECIFIC INITIAL CONCENTRATIONS

As a first illustration, we consider how an equilibrium constant is determined from experimental data. The procedure is straightforward and is best shown by example.

N In this and the following sections we present several reaction types, but the analysis of their composition at equilibrium is the same in each case.

T OHT: A blank equilibrium table.

N A consistent use of the equilibrium table greatly simplifies the approach to solving equilibrium problems; note, however, that it is important to present the material that keeps the chemical process in focus, and not as a repetitive exercise in algebra.

N Step 2 is most critical: here the composition of the system changes in accord with the reaction stoichiometry.

N Review in advance Examples 13.3 to 13.9 and note the similarities of their solutions. Only a slight adjustment of focus on the nature of the system is made from one example to the next.

U Exercises 13.69, 13.70, 13.73, 13.74, and 13.82.

N We are using the term *species* because we shall also want to use the same format for atoms and ions, and "species" is a convenient catch-all term.

N We emphasize once again that Q_c is an algebraic expression that may have any value (depending on the molar concentrations used as data); the equilibrium constant is a quantity with a specific numerical value, here 0.278, to which Q_c is equal when the molar concentrations are those at equilibrium.

▼ EXAMPLE 13.3 Calculating an equilibrium constant

A mixture consisting of 0.500 mol N_2/L and 0.800 mol H_2/L in a reaction vessel reacts and reaches equilibrium. At equilibrium, the concentration of ammonia is 0.150 mol/L. Calculate the value of the equilibrium constant for

$$N_2(g) + 3H_2(g) \rightleftharpoons 2NH_3(g)$$

STRATEGY We need to know the equilibrium concentrations of each substance in the reaction mixture and then substitute those values into the reaction quotient. Because the initial molar concentrations of each reactant are known (step 1; initially there is no ammonia present) and the increase in the equilibrium molar concentration of the product is known (step 2), the decrease in the molar concentration of each reactant can be calculated from the reaction stoichiometry. More formally, we work through the procedure set out above.

SOLUTION First we set up the equilibrium table:

Equilibrium equation:	N_2	+	$3H_2$	$\rightleftharpoons$	$2NH_3$
Species:	N_2		H_2		NH_3
Step 1. Initial concentration, mol/L	0.500		0.800		0
Step 2. Change in concentration, mol/L	$-\frac{1}{2}(0.150)$		$-\frac{3}{2}(0.150)$		+0.150

(The increase in the molar concentration of NH_3 from the reaction of N_2 and H_2 is given, and the reaction stoichiometry then implies the decrease in molar concentrations of the nitrogen and hydrogen.)

	N_2	H_2	NH_3
Step 3. Equilibrium concentration, mol/L	0.425	0.575	0.150

(These values are the result of adding the changes, step 2, to the initial values, step 1.)

Step 4. To obtain the equilibrium constant, the equilibrium values from step 3 are now inserted in Q_c:

$$Q_c = \frac{[NH_3]^2}{[N_2][H_2]^3} = K_c \text{ at equilibrium}$$

which gives

$$\frac{(0.150)^2}{0.425 \times (0.575)^3} = 0.278$$

EXERCISE E13.3 A 0.0540-mol sample of dinitrogen tetroxide (N_2O_4) was allowed to vaporize in a 500-mL flask and reach equilibrium with its decomposition product, nitrogen dioxide. At equilibrium, the sample contained 0.0480 mol NO_2. Calculate the equilibrium constant K_c for the decomposition.

▲ [*Answer*: 0.160]

One unknown concentration. Suppose we know the equilibrium concentrations of all but one of the substances taking part in a reaction and want to find the unknown concentration. For the reaction $H_2(g) + I_2(g) \rightleftharpoons 2\,HI(g)$, for instance, we might know the concentrations of iodine vapor and hydrogen iodide vapor at equilibrium and want to determine the hydrogen concentration. In such a case, if we know the equilibrium constant, we can simply rearrange the expression for K_c to calculate the unknown concentration.

▼ EXAMPLE 13.4 Calculating an unknown concentration

A mixture of hydrogen gas and iodine vapor was heated to 490°C and allowed to reach equilibrium. At that temperature, $K_c = 46$ for $H_2(g) + I_2(g) \rightleftharpoons 2HI(g)$. The equilibrium concentrations were measured spectroscopically as 0.0031 mol I_2/L and 0.0027 mol HI/L. Calculate the equilibrium molar concentration of H_2.

STRATEGY Because only one concentration in the reaction quotient is unknown, it can be calculated from K_c and the molar concentrations of the remaining substances. To do so, we work through the sequence of steps set out above. However, because we are already given the equilibrium molar concentrations of two of the substances, most of the work has already been done, and we can take up the general procedure at step 3, writing x for the unknown molar concentration.

SOLUTION We set up the only part of the equilibrium table that we need:

Equilibrium equation:	H_2	+	I_2	$\rightleftharpoons$	2HI
Species:	H_2		I_2		HI
Step 3. Equilibrium concentration, mol/L	x		0.0031		0.0027

Step 4. Substitute the equilibrium concentrations into the reaction quotient:

$$Q_c = \frac{[HI]^2}{[H_2][I_2]} = K_c \text{ at equilibrium}$$

Because $K_c = 46$, we obtain:

$$\frac{(0.0027)^2}{x \times 0.0031} = 46$$

This expression rearranges to

$$x = \frac{(0.0027)^2}{46 \times 0.0031} = 5.1 \times 10^{-5}$$

Therefore, from step 3, the equilibrium concentration of hydrogen is 5.1×10^{-5} mol/L.

EXERCISE E13.4 For the reaction $H_2(g) + CO_2(g) \rightleftharpoons CO(g) + H_2O(g)$, $K_c = 0.18$ at 500°C. Calculate the concentration of H_2 in an equilibrium mixture in which the molar concentrations are 0.011 mol CO_2/L, 0.0030 mol H_2O/L, and 0.041 mol CO/L.

[*Answer*: 0.062 mol/L]

▲

? What would the form of Q_c be for this reaction if iodine were present as a solid?

N The formulation of equilibrium calculations is well-suited to programming on a spreadsheet. Students with access to computers could be invited to prepare a spreadsheet for the examples treated here, and then to explore the consequences of using different initial conditions. (It would be best to program the calculation without making the approximations employed here, and to avoid problems that lead to cubic and higher equations.)

Decomposition reaction. A decomposition reaction is often studied by placing a known amount of a compound into a reaction vessel and heating it. A typical question is how much decomposes and how much of each substance (the original and the decomposition products) remains when equilibrium is established. For example, if a known amount of PCl_5 is placed in a reaction vessel and heated, how much decomposes in the reaction

$$PCl_5(g) \longrightarrow PCl_3(g) + Cl_2(g)$$

and how much PCl_3 and Cl_2 are present at equilibrium? Questions like this can be answered quite easily from the equilibrium constant, even though there are several unknowns (the concentrations of PCl_5, PCl_3, and Cl_2), because the concentrations are linked by the stoichiometry of the reaction. The calculation is illustrated in the next example, which also shows how to calculate the percentage of PCl_5 molecules that have decomposed:

$$\text{Percentage decomposed} = \frac{\text{amount decomposed}}{\text{initial amount}} \times 100\%$$

N There are basically two types of equilibrium problems: those in which all the concentrations in the table are known and K_c is to be determined, and those in which one or more unknown concentrations occur in the table and are to be determined from the given value of K_c.

▼ EXAMPLE 13.5 Determining the equilibrium composition of a decomposition reaction

When a 1.50-mol sample of PCl_5 is placed into a 500-mL reaction vessel and heated to 250°C, the decomposition

$$PCl_5(g) \rightleftharpoons PCl_3(g) + Cl_2(g)$$

occurs and reaches equilibrium with $K_c = 1.80$. What is the composition of the equilibrium mixture? What percentage of PCl_5 has decomposed at 250°C?

STRATEGY First, we set up the equilibrium table, with x denoting the change in molar concentration of PCl_5: the resulting changes in the other two components can then be found by using the stoichiometry of the reaction. Finally, we express Q_c in terms of x, set the reaction quotient equal to K_c, and solve the resulting equation to find the value of x.

SOLUTION The data let us conclude that the initial molar concentration of PCl_5 vapor is

$$[PCl_5] = \frac{1.50 \text{ mol } PCl_5}{0.500 \text{ L}} = 3.00 \text{ mol } PCl_5/\text{L}$$

The initial concentrations of PCl_3 and Cl_2 are zero because neither substance is present initially. Hence, the equilibrium table is

Equilibrium equation:	$PCl_5(g) \rightleftharpoons$	$PCl_3(g)$ +	$Cl_2(g)$
Species:	PCl_5	PCl_3	Cl_2
Step 1. Initial concentration, mol/L	3.00	0	0
Step 2. Change in concentration, mol/L	$-x$	$+x$	$+x$

(If the decomposition of PCl_5 reduces its molar concentration by x, then from the reaction stoichiometry the molar concentrations of PCl_3 and Cl_2 will both increase by x.)

Step 3. Equilibrium concentration, mol/L	$3.00 - x$	x	x

(These values are the sums of the initial concentrations, step 1, and the changes in concentration brought about by reaction, step 2.)

Step 4. Substitution of these equilibrium values into the reaction quotient

$$Q_c = \frac{[PCl_3][Cl_2]}{[PCl_5]} = K_c \text{ at equilibrium}$$

with $K_c = 1.80$ then gives

$$\frac{x \times x}{3.00 - x} = 1.80$$

This expression rearranges to the quadratic equation

$$x^2 + 1.80x - 5.60 = 0$$

The solutions of this quadratic equation (see Appendix 1C) are

$$x = \frac{-1.80 \pm \sqrt{(1.80)^2 - 4(1)(-5.60)}}{2} = 1.59 \text{ and } -3.39$$

Because concentrations must be positive and because (from step 3) x is the concentration of PCl_3 and Cl_2, we select the solution 1.59. It follows that at equilibrium

$$[PCl_5] = 3.00 - x = 3.00 - 1.59 = 1.41$$

$$[PCl_3] = x = 1.59 \quad [Cl_2] = x = 1.59$$

and the equilibrium concentrations are 1.41 mol PCl_5/L, 1.59 mol PCl_3/L, and 1.59 mol Cl_2/L.

The initial amount of PCl_5 present was 1.50 mol, and the amount that has decomposed is 0.500 L × 1.59 mol/L = 0.795 mol, the percentage of PCl_5 that has decomposed is

$$\text{Percentage } PCl_5 \text{ decomposed} = \frac{\text{amount decomposed}}{\text{initial amount}} \times 100\%$$

$$= \frac{0.795 \text{ mol}}{1.50 \text{ mol}} \times 100\% = 53.0\%$$

EXERCISE E13.5 The gas bromine monochloride, BrCl, decomposes into bromine and chlorine and reaches the equilibrium $2BrCl(g) \rightleftharpoons Br_2(g) + Cl_2(g)$, for which $K_c = 32$ at 500 K. If initially pure BrCl is present at a concentration of 3.30×10^{-3} mol/L, then what is its concentration in the mixture at equilibrium?

[*Answer:* 2.7×10^{-4} mol/L]

U Exercises 13.75, 13.77, 13.80, and 13.86.

Synthesis reactions. A reaction chemists often encounter is one in which two or more products are formed from one or more reactants, and they may need to know the concentrations of the reactants and products when the reaction has reached equilibrium. We shall now see that we can use the equilibrium constant and the reaction stoichiometry to analyze the composition of the equilibrium mixture.

▼ EXAMPLE 13.6 Calculating the equilibrium composition of a synthesis reaction

Nitric oxide, NO, is an air pollutant generated in the internal combustion engine. This product comes from the reaction between nitrogen and oxygen at the high temperatures typical of the combustion reaction. The equilibrium constant of the reaction $N_2(g) + O_2(g) \rightleftharpoons 2NO(g)$ is $K_c = 1.00 \times 10^{-5}$ at 1200°C. Calculate the equilibrium composition in a reaction vessel that initially held 0.800 mol N_2/L and 0.200 mol O_2/L.

STRATEGY Set up the equilibrium table by writing the change in the molar concentration of N_2 as x and using the reaction stoichiometry to determine the corresponding changes in the molar concentrations of the other substances. The value of x is determined from the equilibrium constant, which is given as data.

SOLUTION The equilibrium table is set up by noting that the reaction vessel initially contains only nitrogen and oxygen:

Equilibrium equation:	$N_2(g)$ +	$O_2(g)$ $\rightleftharpoons$	$2NO(g)$
Species:	N_2	O_2	NO
Step 1. Initial concentration, mol/L	0.800	0.200	0
Step 2. Change in concentration, mol/L	$-x$	$-x$	$+2x$
Step 3. Equilibrium concentration, mol/L	$0.800 - x$	$0.200 - x$	$2x$

N Note the similarity of Example 13.6 to Examples 13.5 and 13.7–13.9.

? In Step 2, why is the change in concentration of NO listed as $+2x$ whereas those of N_2 and O_2 are listed as $-x$?

Step 4. Now we use the reaction quotient (and its value at equilibrium, which is given as data) to find the value of x:

$$Q_c = \frac{[NO]^2}{[N_2][O_2]} = K_c \text{ at equilibrium}$$

which, with $K_c = 1.00 \times 10^{-5}$, becomes

$$\frac{(2x)^2}{(0.800 - x) \times (0.200 - x)} = 1.00 \times 10^{-5}$$

The expression can be arranged first into

$$\frac{4x^2}{0.160 - x + x^2} = 1.00 \times 10^{-5}$$

which rearranges into the quadratic equation

$$4x^2 - (1.00 \times 10^{-5})x^2 + (1.00 \times 10^{-5})x - 1.60 \times 10^{-6} = 0$$

Because there are only three significant figures in the data, we can set $4 - 1.00 \times 10^{-5} = 4.00$; so the expression simplifies to

$$4.00x^2 + (1.00 \times 10^{-5})x - 1.60 \times 10^{-6} = 0$$

The quadratic formula (see Appendix 1C) then gives

$$x = \frac{-1.00 \times 10^{-5} \pm \sqrt{(1.00 \times 10^{-5})^2 - 4 \times 4.00 \times (-1.60 \times 10^{-6})}}{2 \times 4.00}$$

$$= 6.31 \times 10^{-4} \text{ and } -6.34 \times 10^{-4}$$

We must adopt the positive value of x because the molar concentration of NO (which is equal to $2x$) must be positive. It follows that

$$[N_2] = 0.800 - x = 0.800 - 6.31 \times 10^{-4} = 0.799$$

$$[O_2] = 0.200 - x = 0.200 - 6.31 \times 10^{-4} = 0.199$$

$$[NO] = 2x = 2 \times (6.31 \times 10^{-4}) = 1.26 \times 10^{-3}$$

and hence the equilibrium concentrations are 0.799 mol N_2/L, 0.199 mol O_2/L, and 1.26×10^{-3} mol NO/L.

E Solving quadratic equations. *J. Chem. Educ.*, **1990**, *67*, 409.

EXERCISE E13.6 Hydrogen gas and iodine vapor are mixed in a reaction vessel at concentrations of 0.54 mol/L and 0.71 mol/L, respectively. The system is heated to 490°C, at which temperature $K_c = 46$ for the equilibrium

$H_2(g) + I_2 \rightleftharpoons 2HI(g)$. Calculate the equilibrium concentrations of H_2, I_2, and HI.

[*Answer*: 0.08 mol H_2/L; 0.25 mol I_2/L; 0.92 mol HI/L]

U Exercises 13.72 and 13.85.

13.5 ARBITRARY INITIAL CONCENTRATIONS

Up to this point we have considered reactions that are run using "ideal" initial concentrations. In industry, however, it may be the case that a reaction starts with the components present in proportions that are quite different from that given by the stoichiometric coefficients. The analysis of such mixtures is a little more complex than the ones we have considered so far, but the approach to the calculation is very similar.

Analysis of a general reaction mixture. Although we might not be told whether a reaction is at equilibrium, we might be asked to decide whether additional product has a tendency to form and, if so, how much. The procedure for making this decision uses the equilibrium table, as the following example illustrates.

EXAMPLE 13.7 Analyzing a general reaction mixture

An analysis of a chemical reaction system at 340°C shows that PCl_5, PCl_3, and Cl_2 are present at equal concentrations of 0.120 mol/L. Is the reaction mixture at equilibrium? If not, determine the equilibrium concentration of each component at 340°C, given that $K_c = 0.800$ for the equilibrium $PCl_5(g) \rightleftharpoons PCl_3(g) + Cl_2(g)$.

STRATEGY We can determine whether the system is at equilibrium by calculating the value of Q_c and comparing it with K_c. If the reaction is not at equilibrium (and we can suspect that it is not, because few equilibria correspond to a system in which all the species are present at the same concentration), then we set up the equilibrium table for the reaction.

SOLUTION The expression for the reaction quotient is

$$Q_c = \frac{[PCl_3][Cl_2]}{[PCl_5]}$$

and its value at the stage in the reaction specified is

$$Q_c = \frac{0.120 \times 0.120}{0.120} = 0.120$$

Because $Q_c < K_c$, the reaction tends to produce more product at the expense of the PCl_5. For the second part of the problem, we suppose that equilibrium is reached when the molar concentration of PCl_5 has been reduced by x and (from the reaction stoichiometry) that the molar concentrations of both PCl_3 and Cl_2 have been increased by x. We draw up the following table:

Equilibrium equation:	$PCl_5(g)$ $\rightleftharpoons$	$PCl_3(g)$ +	$Cl_2(g)$
Species:	PCl_5	PCl_3	Cl_2
Step 1. Initial concentration, mol/L	0.120	0.120	0.120
Step 2. Change in concentration, mol/L	$-x$	$+x$	$+x$
Step 3. Equilibrium concentration, mol/L	$0.120 - x$	$0.120 + x$	$0.120 + x$

Step 4. The reaction quotient is

$$Q_c = \frac{[PCl_3][Cl_2]}{[PCl_5]} = K_c \text{ at equilibrium}$$

Because $K_c = 0.800$, at equilibrium

$$\frac{(0.120 + x) \times (0.120 + x)}{0.120 - x} = 0.800$$

This expression rearranges to

$$0.800 \times (0.120 - x) = 0.0144 + 0.240x + x^2$$

and then to

$$x^2 + 1.040x - 0.0816 = 0$$

The solutions of this quadratic equation are given by the quadratic formula:

$$x = \frac{-1.040 \pm \sqrt{(1.040)^2 - 4(-0.0816)}}{2} = 0.0733 \text{ and } -1.113$$

Reference to step 3 in the equilibrium table shows that only the first solution is acceptable. Therefore, at equilibrium

$$[PCl_5] = 0.120 - x = 0.120 - 0.0733 = 0.047$$
$$[PCl_3] = 0.120 + x = 0.120 + 0.0733 = 0.193$$
$$[Cl_2] = 0.120 + x = 0.120 + 0.0733 = 0.193$$

and the molar concentrations at equilibrium are 0.047 mol PCl_5/L, 0.193 mol PCl_3/L, and 0.193 mol Cl_2/L.

EXERCISE E13.7 Suppose the initial concentrations in this example are 0.012 mol PCl_5/L, 0.024 mol PCl_3/L, and 0.036 mol Cl_2/L. Calculate the equilibrium composition at 340°C.

▲ [*Answer*: 0.0020 mol PCl_5/L; 0.034 mol PCl_3/L; 0.046 mol Cl_2/L]

N It is always important to refer to Step 3 when selecting one of the various alternative values of x.

N We consider the process of simplification and approximation to be a central component of the analysis of equilibrium problems, and it will be deployed extensively in the following two chapters. This material is the first introduction of a technique that will be used frequently (and, used circumspectly, is an important scientific skill). Appreciating the physical motivation of an approximation is an aid to understanding the chemical system as much as it is an aid to finding a mathematical solution. In each case we consider it important to verify at the end of the calculation that the approximation is appropriate.

U Exercises 13.71, 13.76, 13.81, and 13.97.

Simplification by approximation. Even when the expression for x in terms of K_c cannot be solved easily, we may still be able to obtain an approximate solution, especially when the equilibrium constant is small. An example is the formation of N_2O in the reaction of N_2 and O_2:

$$2N_2(g) + O_2(g) \rightleftharpoons 2N_2O(g) \qquad Q_c = \frac{[N_2O]^2}{[N_2]^2[O_2]}, \quad K_c = 2.0 \times 10^{-37}$$

If the concentration of O_2 decreases by x, we shall find (as we see below) that we end up with an equation in x^3, which is difficult to solve. However, we can adopt a simplification if the change in concentration is less than about 5% of the initial concentration. If

$$\frac{\text{Change in concentration of X}}{\text{Initial concentration of X}} \times 100\%$$

is less than about 5%, then the error in the approximate calculation is less than about 5%. That is an acceptable margin of error within the limitations of the experimental data and the imprecision of the equilibrium constant. When we meet a complicated expression, we first assume that the change x is much smaller than any nonzero initial concentration and replace all expressions like $0.100 - x$ or $0.100 - 2x$ by 0.100, for example. At the end of the calculation, we must verify that the value of x that we calculate is indeed smaller than about 5% of the initial values (see the following example). If it is not, then the only valid procedure is to try to solve the equation without approximation.

EXAMPLE 13.8 Finding an equilibrium concentration by approximation

A mixture of 0.482 mol N_2 and 0.933 mol O_2 is placed in a reaction vessel of volume 10.0 L. Calculate the equilibrium composition of the system at a temperature at which $K_c = 2.0 \times 10^{-37}$ for the equilibrium $2N_2(g) + O_2(g) \rightleftharpoons 2N_2O(g)$.

STRATEGY As usual, we set up the equilibrium table in terms of the unknown change in molar concentration x. We shall see that the equation for x is complicated, so we make the approximations described above. At the end of the calculation, we must verify the validity of the approximation. Because K_c is so small, we expect only a very low molar concentration of N_2O at equilibrium.

SOLUTION The initial concentrations of nitrogen and oxygen are

$$\text{Molar concentration of } N_2 = \frac{0.482 \text{ mol } N_2}{10.0 \text{ L}} = 0.0482 \text{ mol } N_2/\text{L}$$

$$\text{Molar concentration of } O_2 = \frac{0.933 \text{ mol } O_2}{10.0 \text{ L}} = 0.0933 \text{ mol } O_2/\text{L}$$

Because there is no N_2O initially, the equilibrium table is as follows:

Equilibrium equation:	$2N_2(g)$	+	$O_2(g)$	$\rightleftharpoons$	$2N_2O(g)$
Species:	N_2		O_2		N_2O
Step 1. Initial concentration, mol/L	0.0482		0.0933		0
Step 2. Change in concentration, mol/L	$-2x$		$-x$		$+2x$
Step 3. Equilibrium concentration, mol/L	$0.0482 - 2x$		$0.0933 - x$		$2x$

Step 4. The reaction quotient is

$$Q_c = \frac{[N_2O]^2}{[N_2]^2[O_2]} = K_c \text{ at equilibrium}$$

It follows from step 3 that

$$\frac{(2x)^2}{(0.0482 - 2x)^2 \times (0.0933 - x)} = 2.0 \times 10^{-37}$$

When rearranged, this equation is a cubic equation (an equation in x^3). It is difficult to solve cubic equations exactly (there is a formula like the quadratic formula, but it is complicated). However, because K_c is very small, the initial concentrations adjust only very slightly to reach equilibrium. Therefore, we can ignore the changes in the concentrations of N_2 and O_2 relative to their starting concentrations and replace $0.0482 - 2x$ by 0.0482 and $0.0933 - x$ by 0.0933. These approximations simplify the equation above to

$$\frac{(2x)^2}{(0.0482)^2 \times 0.0933} \approx 2.0 \times 10^{-37}$$

which is easily solved. (The sign $\approx$ means "approximately equal to.") Because $K_c = 2.0 \times 10^{-37}$, we obtain

$$x \approx \tfrac{1}{2} \times \sqrt{(0.0482)^2(0.0933) \times 2.0 \times 10^{-37}} = 3.3 \times 10^{-21}$$

E This reaction occurs during electrical storms and is the source of naturally occurring dinitrogen oxide, N_2O.

? Why in Step 2 are $-2x$ and $-x$ listed for the reactants, but $+2x$ listed for the product?

N The solution of cubic equations (by successive numerical estimation) is outlined in Appendix 1C. Analytical solutions do exist, but they are very cumbersome.

N If students are not sure whether to use the approximation, they should be encouraged to study the chemical system: they should inspect the value of K_c and the molar concentrations of the species present. It is often (but not invariably) the case that if the molar concentrations are about three or more orders of magnitude greater than K_c, then the approximation will be valid.

N Review Example 13.5 to see if an approximation would have been valid.

The value of x is *very* small compared with 0.0482:

$$\frac{\text{Change in concentration of } N_2}{\text{Initial concentration of } N_2} = \frac{2(3.3 \times 10^{-21})}{0.0482} \times 100\%$$

$$= 1.4 \times 10^{-17}\%$$

The error in the approximate calculation is only $1.4 \times 10^{-17}\%$, which is well inside the 5% limit for the approximation to be valid. Therefore, at equilibrium,

$$[N_2] = 0.0482 - 2x \approx 0.0482$$

$$[O_2] = 0.0933 - x \approx 0.0933$$

$$[N_2O] = 2x \approx 6.6 \times 10^{-21}$$

The equilibrium concentrations are therefore 0.0482 mol N_2/L, 0.0933 mol O_2/L, and 6.6×10^{-21} mol N_2O/L.

EXERCISE E13.8 The initial concentrations of nitrogen and hydrogen were 0.010 mol/L and 0.020 mol/L, respectively. The mixture is heated to 770 K, at which temperature $K_c = 0.11$ for the reaction $N_2(g) + 3H_2 \rightleftharpoons 2NH_3(g)$. What is the equilibrium composition of the mixture?

[*Answer*: $[N_2] \approx 0.010$ mol/L; $[H_2] \approx 0.020$ mol/L; $[NH_3] \approx 9.4 \times 10^{-5}$ mol/L]

In some cases, when chemists fill a reaction vessel with reactant gases, the amount of each gas introduced is measured by using a pressure gauge. Therefore, partial pressures, in units of atmospheres, are usually recorded for the data.

U Exercise 13.79.

EXAMPLE 13.9 Determining equilibrium partial pressures by approximation

The partial pressures of carbon monoxide and hydrogen in a reaction vessel at 225°C are 0.37 atm and 0.72 atm, respectively. The equilibrium constant for the reaction $CO(g) + 2H_2(g) \rightleftharpoons CH_3OH(g)$ is $K_P = 0.049$. What are the partial pressures of the gases in the reaction mixture at equilibrium?

N Remember that the partial pressure of a gas is proportional to its molar concentration: $P_X = nRT/V$, with $[X] = n/V$ gives $P_X = [X]RT$, and the partial pressure is proportional to the molar concentration.

STRATEGY The analysis of this system is much like that of Example 13.8, except that the amounts of reactants are expressed as partial pressures rather than as concentrations. Because partial pressure and molar concentration are directly proportional to each other, the procedure used earlier can be applied to the partial pressures.

SOLUTION Initially, only carbon monoxide and hydrogen are present, so the equilibrium table is

Equilibrium equation:	$CO(g)$	+	$2H_2(g)$	$\rightleftharpoons$	$CH_3OH(g)$
Species:	**CO**		H_2		CH_3OH
Step 1. Initial partial pressure, atm	0.37		0.72		0
Step 2. Change in partial pressure, atm	$-x$		$-2x$		$+x$
Step 3. Equilibrium partial pressure, atm	$0.37 - x$		$0.72 - 2x$		x

N In Example 13.9, K_P is *not* more than three orders of magnitude different from the initial partial pressures; therefore, it is not clear in advance that the usual approximation procedure is valid.

Step 4. We insert the equilibrium partial pressures into the reaction quotient:

$$Q_P = \frac{P_{CH_3OH}}{(P_{CO})(P_{H_2})^2} = K_P \text{ at equilibrium}$$

which gives

$$\frac{x}{(0.37 - x) \times (0.72 - 2x)^2} = 0.049$$

Because K_P is small, as a first approximation we assume that x is much smaller than 0.37 and $2x$ much smaller than 0.72 and write

$$\frac{x}{0.37 \times (0.72)^2} \approx 0.049$$

which solves to $x \approx 9.4 \times 10^{-3}$. To verify the validity of the approximation, we calculate the value of the expression

$$\frac{\text{Change in partial pressure of CO}}{\text{Initial partial pressure of CO}} \times 100\% = \frac{9.4 \times 10^{-3}}{0.37} \times 100\% = 2.5\%$$

which is smaller than 5%. Similarly, the partial pressure of hydrogen decreases by 2.6%. Hence, the approximations are valid. The equilibrium partial pressures are therefore

$$P_{CO} = 0.37 - 9.4 \times 10^{-3} \approx 0.36$$

$$P_{H_2} = 0.72 - 2(9.4 \times 10^{-3}) \approx 0.70$$

$$P_{CH_3OH} \approx 9.4 \times 10^{-3}$$

and the partial pressures are 0.36 atm, 0.70 atm, and 9.4×10^{-3} atm, respectively.

EXERCISE E13.9 Hydrogen chloride gas is added to a reaction vessel containing solid iodine until its partial pressure reaches 0.367 atm. At the temperature of the experiment, $K_P = 3.9 \times 10^{-33}$ for the reaction $2HCl(g) + I_2(s) \rightleftharpoons 2HI(g) + Cl_2(g)$. What are the equilibrium partial pressures of the gases?

[*Answer:* 0.367 atm HCl; 1.0×10^{-11} atm HI; 5.1×10^{-12} atm Cl_2]

U Exercise 13.78.

THE RESPONSE OF EQUILIBRIA TO THE REACTION CONDITIONS

In the industrial production of chemicals such as NH_3, it is necessary to ensure that the equilibrium concentration of product is high enough to be economical. How can that concentration be raised? By raising the temperature or by lowering it? By raising the pressure or lowering it?

Chemical equilibria, being dynamic, are responsive to changes in the conditions under which the reaction takes place. If a change in conditions increases the rate at which reactants change into products, then the equilibrium composition adjusts until the rate of the reverse reaction has risen to match the new forward rate. If the change reduces the rate of the forward reaction, then products decompose into reactants until the two rates are equal again. Because a catalyst affects the rates of both the forward and the reverse reaction equally, it has no effect on the composition of the mixture at equilibrium.

E Only rarely are industrial processes allowed to settle into equilibrium: in many plants, product is removed as soon as it is formed, so the process undergoes a continuous hunt for equilibrium. Nevertheless, equilibrium considerations are a good guide to the viability of a synthetic procedure, because a reaction with a small equilibrium constant will form so little product that its removal to encourage further reaction will be a slow and difficult (and therefore uneconomical) process.

In Section 11.4 we saw that the response of dynamic equilibria to changes in the conditions can often be predicted using Le Chatelier's

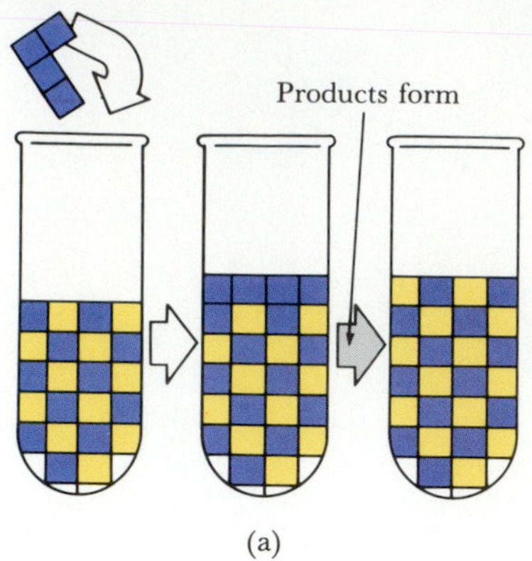

(a)

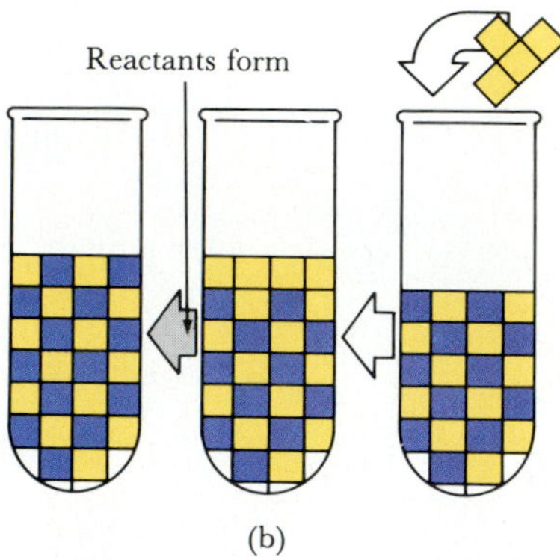

(b)

FIGURE 13.7 (a) When a reactant (blue) is added to a reaction mixture at equilibrium, the products have a tendency to form. (b) When a product (yellow) is added instead, the reactants tend to be formed. (For this reaction, $K_c = 1$.)

V Video 38: Effect of change of concentration on chemical equilibrium.

D The dissociation [ionization] of acetic acid (common ion effect). CD2, 85. Equilibrium and Le Chatelier's principle. CD1, 46.

principle, that a dynamic equilibrium tends to minimize any change in the conditions. In this section, we apply the principle to chemical equilibria. However, it is important to realize that Le Chatelier's principle only suggests an outcome; the principle does not provide an explanation or produce numerical values. For an *explanation* of the effect of changes in the conditions, we need to examine their effect on the rates of the forward and reverse reactions. For a *quantitative prediction* of their effect, we have to carry out a calculation based on the equilibrium constant.

13.6 THE EFFECT OF ADDED REAGENTS

Suppose we add hydrogen to an equilibrium mixture in the Haber synthesis reaction for ammonia, for which the reaction is

$$N_2(g) + 3H_2(g) \rightleftharpoons 2NH_3(g)$$

According to Le Chatelier's principle, the equilibrium tends to adjust to minimize the increase in the number of hydrogen molecules (Fig. 13.7). This adjustment is achieved when the reaction produces additional ammonia, with a consequent decrease in the concentrations of N_2 and H_2:

$$N_2(g) + 3H_2(g) \longrightarrow 2NH_3(g)$$

Conversely, if instead we add some ammonia, then the equilibrium would adjust to minimize the effect of the added ammonia and thus the equilibrium composition would be shifted in favor of the reactants:

$$N_2(g) + 3H_2(g) \longleftarrow 2NH_3(g)$$

The explanation of the effect can be found by thinking about the rates of the forward and reverse reactions and the effect on them of changes in concentration. The addition of H_2 increases its concentration in the reaction mixture and hence increases the rate of the forward reaction. The reaction therefore forms more product until the reverse rate (the decomposition of NH_3) rises to match the forward rate and the dynamic equilibrium is reestablished. Similarly, the addition of ammonia to the equilibrium mixture increases the reverse rate and forms nitrogen and hydrogen. As a result, the rate of the forward reaction increases until it matches the rate of the reverse reaction and reestablishes the dynamic equilibrium.

▼ EXAMPLE 13.10 Predicting the effect of added reagents on a chemical equilibrium

Consider the equilibrium

$$4NH_3(g) + 3O_2(g) \rightleftharpoons 2N_2(g) + 6H_2O(g)$$

Predict the effect on the equilibrium of (a) the addition of N_2; (b) the removal of NH_3; and (c) the removal of H_2O.

STRATEGY Le Chatelier's principle states that an equilibrium mixture tends to minimize any change. Therefore, if a substance is added, then the equilibrium tends to lower the concentration of that substance. If a substance is removed, then the equilibrium tends to replace that substance.

SOLUTION (a) The addition of N_2 to the equilibrium mixture causes the reaction to shift in the direction that minimizes the increase in N_2. Therefore, the equilibrium shifts left, thereby increasing the amounts of NH_3 and O_2 while decreasing the amount of H_2O. (b) When NH_3 is removed from the system at equilibrium, the reaction adjusts to minimize the reduction in its concentration. So the reaction shifts left, thereby increasing the amount of O_2 and decreasing the amounts of N_2 and H_2O. (c) The removal of H_2O causes the equilibrium to shift right in favor of products to restore (partially) the amount removed. This shift increases the amount of N_2 while decreasing the amounts of NH_3 and O_2.

EXERCISE E13.10 Consider the equilibrium $SO_3(g) + NO(g) \rightleftharpoons SO_2(g) + NO_2(g)$. Predict the effect on the equilibrium of (a) the addition of NO; (b) the removal of SO_2; and (c) the addition of NO_2.

▲ [*Answer*: the equilibrium shifts (a) right; (b) right; (c) left]

13.7 THE EFFECT OF PRESSURE

N A change in pressure corresponds to a change in molar concentration only if it is brought about by a change in the volume occupied by the system; no change in volume results from the addition of an inert gas to the system.

T OHT: Fig. 13.8.

D Effect of pressure on equilibrium. CD1, 52.

Equilibria respond to changes in pressure. We shall consider gas-phase reactions here because they are most affected by pressure.

Predicting the effect of pressure from Le Chatelier's principle. According to Le Chatelier's principle, a gas-phase equilibrium responds to an increase in pressure by tending to shift in the direction that minimizes the increase. Because the formation of NH_3 from N_2 and H_2 decreases the number of gas molecules in the container (from 4 to 2 mol) and hence the pressure the mixture exerts, the equilibrium composition will tend to shift in favor of product. Therefore, to increase the yield of ammonia in the Haber process, the synthesis should be carried out at high pressure. The actual industrial process uses pressures of 250 atm and more. Some other equilibria respond similarly: when the pressure is increased, a reaction at equilibrium adjusts to reduce the number of molecules in the gas phase (Fig. 13.8).

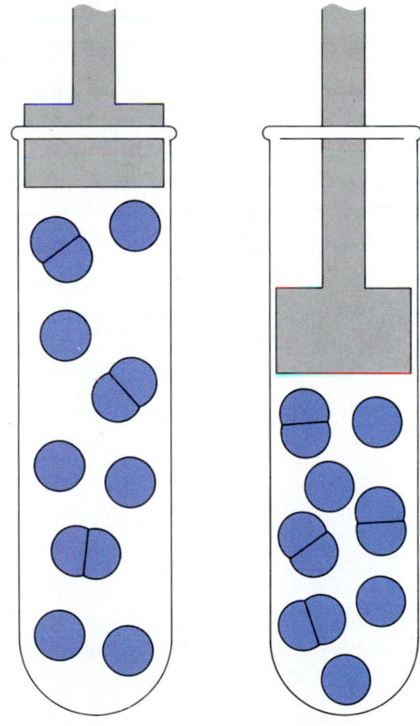

FIGURE 13.8 Le Chatelier's principle predicts that when pressure is applied to a reaction at equilibrium, the composition tends to shift in the direction that corresponds to the smaller number of gas-phase molecules.

▼ **EXAMPLE 13.11 Predicting the effect of pressure on an equilibrium**

Predict the effect of an increase in pressure on the equilibrium composition of the reactions (a) $N_2O_4(g) \rightleftharpoons 2NO_2(g)$ and (b) $H_2(g) + I_2(g) \rightleftharpoons 2HI(g)$.

STRATEGY A glance at the chemical equation is normally enough to show which direction corresponds to a decrease in the number of gas-phase molecules. The composition of the equilibrium mixture will tend to shift in that direction when the pressure is increased.

SOLUTION (a) In the reverse reaction, two NO_2 molecules combine to form one N_2O_4 molecule. Hence, an increase in pressure favors the formation of N_2O_4. (b) Because neither direction corresponds to a reduction of gas-phase molecules, increasing the pressure should have little effect on the composition of the equilibrium mixture.

EXERCISE E13.11 Predict the effect of an increase in pressure on the equilibrium compositions of (a) $CH_4(g) + H_2O(g) \rightleftharpoons CO(g) + 3H_2(g)$ and (b) C(diamond) $\rightleftharpoons$ C(graphite). For the latter, consider the densities of the two solids, which are 3.5 g/cm^3 for diamond and 2.0 g/cm^3 for graphite.

▲ [*Answer*: (a) reactants favored; (b) diamond favored]

N When predicting the effect of pressure on a reaction at equilibrium, consider the changes in the partial pressures of each gas.

***K_c* and the effect of pressure.** When the pressure of a gas mixture is increased by pushing in a piston, the volume of the sample is decreased. This reduction in volume raises the concentrations of all the substances in the sample. We have already seen that the value of K_c is independent of the individual concentrations, so a change in pressure must leave K_c unchanged. However, although the overall equilibrium constant is unchanged, the amounts of each substance do in general adjust with the added pressure.

Suppose we want to discover the effect of a reduction in volume on the equilibrium

$$N_2O_4(g) \rightleftharpoons 2NO_2(g) \qquad Q_c = \frac{[NO_2]^2}{[N_2O_4]}$$

? In Example 13.11, how does the addition of neon gas affect the composition of the hydrogen iodide synthesis reaction? Would the addition of chlorine affect the equilibrium; if so, how?

First, we express the reaction quotient in terms of the volume of the system. The molar concentration of each gas is the amount n of that gas divided by the total volume V of the reaction vessel:

$$[NO_2] = \frac{n_{NO_2}}{V} \qquad [N_2O_4] = \frac{n_{N_2O_4}}{V}$$

Hence, the reaction quotient is

$$Q_c = \frac{n_{NO_2}{}^2 \times (1/V)^2}{n_{N_2O_4} \times (1/V)} = \frac{n_{NO_2}{}^2}{n_{N_2O_4}} \times (1/V) = K_c \text{ at equilibrium}$$

For this expression to remain constant when the pressure on the system is increased (and, as a result, the volume of the system is reduced), the ratio $n_{NO_2}{}^2/n_{N_2O_4}$ must decrease. That is, the amount of NO_2 must decrease and the amount of N_2O_4 must increase. Therefore, as we have seen before, as the volume of the system is decreased, the equilibrium shifts to a smaller total number of gas molecules even though K_c remains unchanged.

? How could we show that a catalyst has no effect on the equilibrium composition of a system?

V Video 37: Effect of a change in temperature on chemical equilibrium.

U Exercises 13.83 and 13.84.

D Equilibrium: Temperature and the ammonium hydroxide [aqueous ammonia]-ammonia system. CD1, 51.

Suppose, however, that at equilibrium we increase the pressure within a reaction vessel by pumping in argon or some other inert gas. In this case the equilibrium composition is unaffected because the reacting gases themselves continue to occupy the same volume, so their individual concentrations and partial pressures remain unchanged.

13.8 THE EFFECT OF TEMPERATURE

Le Chatelier's principle can also be used to predict how a chemical equilibrium will respond to a change in temperature. If a reaction is exothermic, like the Haber process for ammonia production, then lowering the temperature will favor ammonia production because the heat generated in the reaction will tend to minimize the lowering of the temperature. In an endothermic reaction, such as the decomposition of PCl_5, heat must be supplied to shift the equilibrium toward products.

D Effects of concentration and temperature on equilibrium: Copper complex. CD1, 49.

▼ EXAMPLE 13.12 Predicting the effect of temperature on an equilibrium

One stage in the manufacture of sulfuric acid is the formation of sulfur trioxide by the reaction of SO_2 with O_2 in the presence of a V_2O_5 catalyst.

Predict how the equilibrium composition for the sulfur trioxide synthesis will tend to change when the temperature is raised. The thermochemical reaction is

$$2SO_2(g) + O_2(g) \rightleftharpoons 2SO_3(g) \qquad \Delta H^\circ = -198\ \text{kJ}$$

STRATEGY When the temperature is raised, the equilibrium will tend to shift in the direction that corresponds to the endothermic reaction. For a negative reaction enthalpy, the reaction is exothermic in the forward direction; therefore heat is released in the formation of SO_3.

SOLUTION Because the synthesis is exothermic, the reverse reaction is endothermic. Hence, heating the equilibrium mixture favors the formation of SO_2 and O_2 from the decomposition of SO_3.

EXERCISE E13.12 Predict the effect of raising the temperature on the equilibria (a) $N_2O_4(g) \rightleftharpoons 2NO_2(g)$, $\Delta H^\circ = +60.2$ kJ, and (b) $2CO(g) + O_2(g) \rightleftharpoons 2CO_2$, $\Delta H^\circ = -566$ kJ.

[*Answer*: (a) NO_2 favored; (b) CO and O_2 favored]

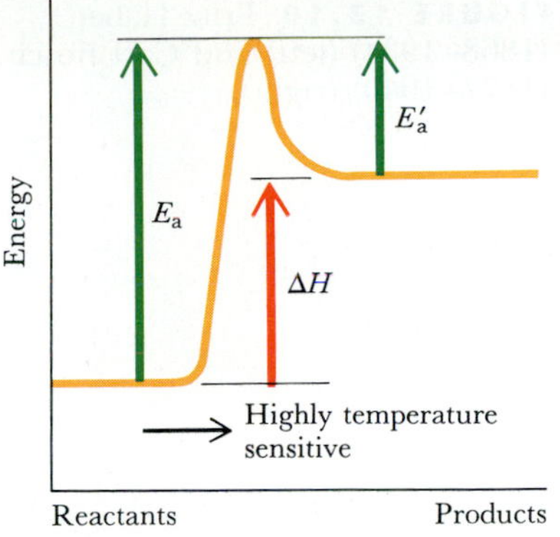

(a) Endothermic

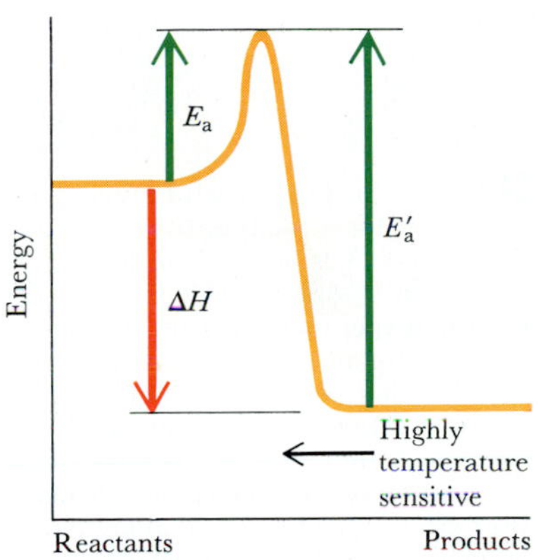

(b) Exothermic

FIGURE 13.9 (a) The activation energy for an endothermic reaction is larger in the forward direction than in the reverse, so the rate of the forward reaction is more sensitive to temperature. (b) The opposite is true for an exothermic reaction, and the reverse reaction is more sensitive to temperature.

Explaining the effect of temperature. The explanation of the effect of temperature on equilibria is found in its effect on the forward and reverse reaction rates. As explained in Section 12.5, the higher the activation energy of a reaction, the more sensitive is its rate to changes of temperature. Figure 13.9a shows that the activation energy of an endothermic reaction is larger than that of its reverse reaction. Therefore, the forward reaction rate increases more rapidly with temperature than the reverse reaction rate. As a result, when the temperature of the equilibrium mixture is raised, more reactants are converted into products until the concentration of products has risen enough for the reverse reaction rate to match the forward rate. The same argument applies to an exothermic reaction, but now the reverse reaction is more sensitive to temperature and generates more reactants when the temperature is raised (Fig. 13.9b).

The numerical calculation of the effect of temperature is quite straightforward: we simply use the value of K_c (or K_P) for the new temperature in calculations like those described already.* Values at various temperatures are given in Tables 13.2 and 13.3.

Haber's achievement. We can now comprehend Fritz Haber's problem, approach, and achievement in designing a process for the commercial production of ammonia (Section 12.7). Because the synthesis of ammonia is exothermic, low temperatures favor the product. This shift toward product is shown by the large increase in K_P: from 7.8×10^{-5} at 450°C to 6.8×10^5 at 25°C, a change of 10 orders of magnitude. However, the rate at which nitrogen and hydrogen combine is virtually zero near room temperature, and at that temperature the reaction proceeds so slowly—in effect, infinitely slowly—that it never reaches equilib-

N We saw in Chapter 11 how temperature affects the solubilities of salts.

T OHT: Fig. 13.9.

? Review the $H_2(g) + X_2(g) \rightleftharpoons 2HX(g)$ series of reactions in Table 13.2. How and why does K_c vary with temperature?

U Exercises 13.88 and 13.89.

*It is best to use the measured value of the equilibrium constant at the new temperature. However, it is also possible to estimate its value from an equation deduced by van't Hoff:

$$\ln\left(\frac{K'_P}{K_P}\right) = \frac{\Delta H^\circ}{R} \times \left(\frac{1}{T} - \frac{1}{T'}\right)$$

where K'_P is the equilibrium constant at the temperature T', K_P the constant at T, and ΔH° the standard reaction enthalpy at the temperature T.

FIGURE 13.10 Fritz Haber (1868–1934) (left) and Carl Bosch (1874–1940) (right).

E Haber was unsuccessful initially in his search for a catalyst, for the activity of the metals he used, including iron, soon deteriorated. It was not until he experimented with iron obtained from an ore from Gallivàre, Sweden, that he was successful. Analysis showed that the iron contained some aluminum and potassium oxides. We now know that the catalyst is not very effective unless small amounts of other metals are present, which was the case for the Swedish sample. The oxides act as *promotors* of the catalytic activity: the alumina prevents the iron particles from fusing together, and the potassium oxide modifies the electronic structure of the metal. A modern promoted iron catalyst contains 0.8% K_2O, 2.0% CaO, 2.5% Al_2O_3, 0.4% SiO_2, 0.3% MgO, and traces of titanium, vanadium, and zirconium oxides.

rium. Haber was therefore faced with a dilemma. He had to use high temperatures to achieve an acceptable *rate* of conversion, but if he did, the *extent* of conversion would be very low.

A part of the solution, as we saw in Section 12.7, was to use a catalyst. Haber found that iron oxide was reduced to porous iron in the hydrogen atmosphere within the reaction vessel and that the iron lowered the activation energy for the reaction. As a result, the synthesis took place more quickly at lower temperatures. However, the presence of a catalyst speeds the reverse reaction as well as the forward reaction, leaving the equilibrium constant unchanged. In other words, although Haber achieved a faster rate of formation of ammonia, he simultaneously achieved a faster rate of decomposition! Although the equilibrium composition was reached more quickly, it contained very little ammonia.

FIGURE 13.11 One of the high-pressure containers inside which the catalytic synthesis of ammonia takes place.

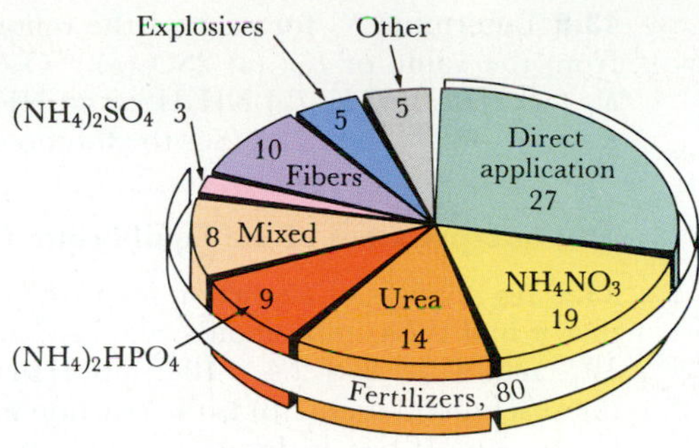

FIGURE 13.12 The Haber process is still used to produce almost all the ammonia manufactured in the world. This chart shows how the ammonia is used (the figures are percentages); 80% is used as fertilizer, either directly or after conversion to another compound.

Haber's solution to the equilibrium problem was to increase the pressure at which the synthesis took place and to remove the ammonia as it was formed. As we have seen, an increase in pressure shifts the equilibrium composition in favor of the products and makes the process economical; removing the product encourages more to be formed. This is where Carl Bosch (Fig. 13.10), Haber's chemical engineer colleague, earned his Nobel prize. He designed the first high-temperature, high-pressure, catalytic industrial process (Fig. 13.11). Though he was working at the limits of technology, he succeeded; and the Haber-Bosch process, a process that led to two Nobel prizes, is still the sole source of the ammonia produced throughout the world (Fig. 13.12).

SUMMARY

13.1 A chemical equilibrium is a dynamic equilibrium in which the rate of the forward reaction for the formation of products equals the rate of the reverse reaction for the re-formation of reactants.

13.2 The value of the **reaction quotient** is a constant, the **equilibrium constant**, for a given reaction at equilibrium at a specified temperature. The reaction quotient and the equilibrium constant may be expressed either in molar concentrations relative to a standard molar concentration (which is 1 mol/L for mixtures), which gives K_c, or in terms of partial pressures relative to a standard pressure of 1 atm (which gives K_P). The equilibrium constant summarizes the combination of concentrations (or partial pressures) that guarantees that the rates of the forward and reverse reactions are equal. At equilibrium, although individual concentrations (and partial pressures) may vary, collectively they must satisfy the expression that defines K_c or K_P. A large equilibrium constant indicates that products dominate the equilibrium, and a small equilibrium constant indicates that reactants dominate the equilibrium. When $Q > K$, reactants tends to form; when $Q < K$, products tend to form; when $Q = K$, the reaction is at equilibrium.

13.3 An equilibrium is **homogeneous** when all the reactants and products are in the same phase; it is **heterogeneous** when the reactants and products are not all in the same phase. Reaction quotients for heterogeneous equilibria are written by expressing the composition relative to the molar concentration of a pure solid or liquid. Examples of heterogeneous equilibria include vapor pressure and solubility.

13.4 and 13.5 Equilibrium calculations are based on the fact that although individual concentrations (and partial pressures) may vary, collectively they combine to give a value equal to the equilibrium constant for the reaction at the specified temperature. Most equilibrium calculations can be carried out systematically by using an equilibrium table that identifies the initial concentrations, the changes in concentrations, and the equilibrium concentrations. The equilibrium concentrations are then substituted into the reaction quotient, which is set equal to the equilibrium constant; the expression is solved for any unknown quantity. When the calculation leads to a complicated mathematical equation, the simplifying assumption is made that the change in concentration of a substance is small compared with its initial concentration. This procedure is valid when the change in concentration is no greater than 5% of the starting concentrations. The validity of the approximation must be verified at the end of the calculation.

13.6 to 13.8 Equilibria are unaffected by the presence of a catalyst. They respond to concentration, pressure, and temperature changes in accordance with Le Chatelier's principle. Reactions that reduce the number of gas-phase molecules are favored by an increase in pressure. Endothermic reactions are favored by an increase in temperature.

CLASSIFIED EXERCISES

Reaction Quotients and Equilibrium Constants

13.1 Write the reaction quotients (in terms of molar concentrations) for the following reactions: (a) $CO(g) + Cl_2(g) \rightleftharpoons COCl(g) + Cl(g)$; (b) $H_2(g) + Br_2(g) \rightleftharpoons 2HBr(g)$; (c) $2O_3(g) \rightleftharpoons 3O_2(g)$.

13.2 Write the reaction quotients (in terms of partial pressures) for the following reactions: (a) $2NO(g) + O_2(g) \rightleftharpoons 2NO_2(g)$; (b) $SbCl_5(g) \rightleftharpoons SbCl_3(g) + Cl_2(g)$; (c) $N_2(g) + 2H_2(g) \rightleftharpoons N_2H_4(g)$.

13.3 For the reaction $N_2(g) + 3H_2(g) \rightleftharpoons 2NH_3(g)$ at 400 K, $K_P = 41$. Find the values of the equilibrium constants (at the same temperature) of the following reactions: (a) $2NH_3(g) \rightleftharpoons N_2(g) + 3H_2(g)$; (b) $\frac{1}{2}N_2(g) + \frac{3}{2}H_2(g) \rightleftharpoons NH_3(g)$.

13.4 The equilibrium constant for the reaction $2SO_2(g) + O_2(g) \rightleftharpoons 2SO_3(g)$ has the value $K_P = 2.5 \times 10^{10}$ at 500 K. Find the values of the equilibrium constants of the following reactions at the same temperature: (a) $SO_2(g) + \frac{1}{2}O_2(g) \rightleftharpoons SO_3(g)$; (b) $SO_3(g) \rightleftharpoons SO_2(g) + \frac{1}{2}O_2(g)$.

Equilibrium Constants from Equilibrium Amounts

13.5 Use the following data, which were collected at 460°C and are equilibrium molar concentrations, to determine K_c for the reaction $H_2(g) + I_2(g) \rightleftharpoons 2HI(g)$.

$[H_2]$, mol/L	$[I_2]$, mol/L	$[HI]$, mol/L
6.47×10^{-3}	0.594×10^{-3}	0.0137
3.84×10^{-3}	1.52×10^{-3}	0.0169
1.43×10^{-3}	1.43×10^{-3}	0.0100

13.6 From the following equilibrium data, which were collected at 1000 K, determine K_P for the equilibrium $2SO_2(g) + O_2(g) \rightleftharpoons 2SO_3(g)$.

P_{SO_2}, atm	P_{O_2}, atm	P_{SO_3}, atm
0.309	0.353	0.338
0.456	0.180	0.364
0.564	0.102	0.333

13.7 Determine K_c for each of the following equilibria from the value of K_P: (a) $N_2O(g) \rightleftharpoons 2NO_2(g)$, $K_P = 0.62$ at 45°C; (b) $CaCO_3(s) \rightleftharpoons CaO(s) + CO_2(g)$, $K_P = 167$ at 1073 K. (See the footnote on page 502.)

13.8 Determine K_c for each of the following equilibria from the value of K_P: (a) $2SO_2(g) + O_2(g) \rightleftharpoons 2SO_3(g)$, $K_P = 3.4$ at 1000 K; (b) $NH_4HS(s) \rightleftharpoons NH_3(g) + H_2S(g)$, $K_P = 9.4 \times 10^{-2}$ at 24°C. (See the footnote on page 502.)

The Interpretation of the Equilibrium Constant

13.9 An analysis of a reaction mixture at 500°C showed that it had the composition 4.8×10^{-3} mol H_2/L, 2.4×10^{-3} mol I_2/L, and 2.4×10^{-3} mol HI/L. (a) Calculate the reaction quotient. (b) Is the reaction mixture at equilibrium? (c) If not, is there a tendency to form more reactants or more products? Use data from Table 13.2.

13.10 Analysis of a reaction mixture showed that it had the composition 0.417 mol N_2/L, 0.524 mol H_2/L, and 0.122 mol NH_3/L at a temperature at which $K_c = 0.278$ for $N_2(g) + 3H_2(g) \rightleftharpoons 2NH_3(g)$. (a) Calculate the reaction quotient. (b) Is the reaction mixture at equilibrium? (c) If not, is there a tendency to form more reactants or more products?

13.11 A 500-mL reaction vessel at 700 K contains 1.20×10^{-3} mol $SO_2(g)$, 5.0×10^{-4} mol $O_2(g)$, and 1.0×10^{-4} mol $SO_3(g)$. At 700 K, $K_P = 3.0 \times 10^4$ for the equilibrium $2SO_2(g) + O_2(g) \rightleftharpoons 2SO_3(g)$. (a) Calculate the reaction quotients Q_c and Q_P. (b) Will more $SO_3(g)$ tend to form?

13.12 Given that $K_P = 3.6 \times 10^{-2}$ for the reaction $N_2(g) + 3H_2(g) \rightleftharpoons 2NH_3(g)$ at 500 K, calculate whether more ammonia will tend to form when a mixture of composition 2.23×10^{-3} mol N_2/L, 1.24×10^{-3} mol H_2/L, and 1.12×10^{-4} mol NH_3/L is present in a container at 500 K.

Heterogeneous Equilibria

13.13 Explain why the following equilibria are heterogeneous and write the reaction quotient Q_P for each one:
(a) $NH_4HS(s) \rightleftharpoons NH_3(g) + H_2S(g)$
(b) $NH_4(NH_2CO_2)(s) \rightleftharpoons 2NH_3(g) + CO_2(g)$

13.14 Explain why the following equilibria are heterogeneous and write the reaction quotient Q_P for each one:
(a) $NH_4Cl(s) \rightleftharpoons NH_3(g) + HCl(g)$
(b) $Na_2CO_3 \cdot 10H_2O(s) \rightleftharpoons Na_2CO_3(s) + 10H_2O(g)$

13.15 Write the reaction quotients Q_c for (a) $Cu(s) + Cl_2(g) \rightleftharpoons CuCl_2(s)$; (b) $NH_4NO_3(s) \rightleftharpoons N_2O(g) + 2H_2O(g)$.

13.16 Write the reaction quotients Q_c for:
(a) $NH_4HS(s) \rightleftharpoons NH_3(g) + H_2S(g)$
(b) $Na_2CO_3 \cdot 10H_2O(s) \rightleftharpoons Na_2CO_3(s) + 10H_2O(g)$

Equilibrium Constants from Initial Amounts

13.17 When 0.0172 mol of HI is heated to 500 K in a 2.00-L sealed container, the resulting equilibrium mixture contains 1.90 g of HI. Calculate K_c for the decomposition reaction $2HI(g) \rightleftharpoons H_2(g) + I_2(g)$.

13.18 When 1.00 g of I_2 is heated to 1000 K in a 1.00-L sealed container, the resulting equilibrium mixture contains 0.830 g of I_2. Calculate K_c for the dissociation equilibrium $I_2(g) \rightleftharpoons 2I(g)$.

13.19 A 25.0-g sample of ammonium carbamate, $NH_4(NH_2CO_2)$, was placed in an evacuated 250-mL flask and kept at 25°C. At equilibrium, 17.4 mg of CO_2 was present. What is the value of K_c for the decomposition of ammonium carbamate into ammonia and carbon dioxide? The reaction is $NH_4(NH_2CO_2)(s) \rightleftharpoons 2NH_3(g) + CO_2(g)$.

13.20 Carbon monoxide and water vapor, each at 200 Torr pressure were introduced into a 250-mL container. When the mixture had reached equilibrium at 700°C, the partial pressure of CO_2 was 88 Torr. Calculate the value of K_P for the equilibrium $CO(g) + H_2O(g) \rightleftharpoons CO_2(g) + H_2(g)$.

13.21 In a gas-phase equilibrium mixture of H_2, I_2, and HI at 500 K, $[HI] = 2.21 \times 10^{-3}$ mol/L and $[I_2] = 1.46 \times 10^{-3}$ mol/L. Given the value of the equilibrium constant in Table 13.2, calculate the concentration of H_2.

13.22 In a gas-phase equilibrium mixture of H_2, Cl_2, and HCl at 1000 K, $[HCl] = 1.45 \times 10^{-3}$ mol/L and $[Cl_2] = 2.45 \times 10^{-3}$ mol/L. Use the information in Table 13.2 to calculate the concentration of H_2.

13.23 In a gas-phase equilibrium mixture of PCl_5, PCl_3, and Cl_2 at 500 K, $P_{PCl_5} = 0.15$ atm and $P_{Cl_2} = 0.20$ atm. What is the partial pressure of PCl_3 given that $K_P = 25$ for the reaction $PCl_5(g) \rightleftharpoons PCl_3(g) + Cl_2(g)$?

13.24 In a gas-phase equilibrium mixture of $SbCl_5$, $SbCl_3$, and Cl_2 at 500 K, $P_{SbCl_5} = 0.15$ atm and $P_{SbCl_3} = 0.20$ atm. Calculate the equilibrium partial pressure of Cl_2, given that $K_P = 3.5 \times 10^{-4}$ for the reaction $SbCl_5(g) \rightleftharpoons SbCl_3(g) + Cl_2(g)$.

Decomposition Reactions

13.25 (a) A sample of 2.0 mmol of Cl_2 was sealed into a 2.0-L reaction vessel and heated to 1000 K to study its dissociation into Cl atoms. Use the information in Table 13.2 to calculate the equilibrium composition of the mixture. What is the percentage decomposition of the Cl_2? (b) If 2.0 mmol of F_2 were placed into the reaction vessel instead of the chlorine, what would be its equilibrium composition at 1000 K? Account for the difference in percentage decomposition of fluorine and chlorine at 1000 K.

13.26 (a) A sample of 5.0 mmol of Cl_2 was sealed into a 2.0-L reaction vessel and heated to 1200 K, and the dissociation equilibrium was established. What is the equilibrium composition of the mixture? What is the percentage decomposition of the Cl_2? Use the information in Table 13.2. (b) If 5.0 mmol of Br_2 were placed into the reaction vessel instead of the chlorine, what would be the equilibrium composition at 1200 K? (c) Account for the difference between the percentage decompositions of bromine and chlorine at 1200 K.

13.27 The initial concentration of HBr in a reaction vessel is 1.2×10^{-3} mol/L. If the vessel is heated to 500 K, what is the percentage decomposition of the HBr and the equilibrium composition of the mixture? For $2HBr(g) \rightleftharpoons H_2(g) + Br_2(g)$, $K_c = 7.7 \times 10^{-11}$.

13.28 The initial concentration of BrCl in a reaction vessel is 1.4×10^{-3} mol/L. If the vessel is heated to 500 K, what is the percentage decomposition of the BrCl and what is the equilibrium composition of the mixture? See Table 13.2 for data on the reaction.

13.29 The equilibrium constant $K_P = 0.36$ for the reaction $PCl_5(g) \rightleftharpoons PCl_3(g) + Cl_2(g)$ at 400 K. (a) Given that 1.0 g of PCl_5 was initially placed in a 250-mL reaction vessel, determine the molar concentrations in the mixture at equilibrium. (b) What percentage of PCl_5 is decomposed at 400 K?

13.30 For the reaction $PCl_5(g) \rightleftharpoons PCl_3(g) + Cl_2(g)$, $K_P = 25$ at 500 K. (a) Calculate the equilibrium molar concentrations of the components in the mixture when 2.0 g of PCl_5 is placed in a 300-mL reaction vessel and allowed to come to equilibrium. (b) What is the percentage of PCl_5 decomposed at 500 K?

13.31 When solid NH_4HS and 0.400 mol of gaseous NH_3 were placed into a 2.0-L vessel at 24°C, the equilibrium $NH_4HS(s) \rightleftharpoons NH_3(g) + H_2S(g)$, for which $K_c = 1.6 \times 10^{-4}$, was reached. What are the equilibrium molar concentrations of NH_3 and H_2S?

13.32 When solid NH_4HS and 0.200 mol of gaseous NH_3 were placed into a 2.0-L vessel at 24°C, the equilibrium $NH_4HS(s) \rightleftharpoons NH_3(g) + H_2S(g)$, for which $K_c = 1.6 \times 10^{-4}$, was reached. What are the equilibrium molar concentrations of NH_3 and H_2S?

13.33 At 760°C, $K_c = 33.3$ for $PCl_5(g) \rightleftharpoons PCl_3(g) + Cl_2(g)$. If a mixture that consists of 0.200 mol PCl_5 and 0.600 mol PCl_3 is placed in a 4.00-L reaction vessel and heated to 760°C, what is the equilibrium composition of the system?

13.34 At 760°C, $K_c = 33.3$ for $PCl_5(g) \rightleftharpoons PCl_3(g) + Cl_2(g)$. If a mixture that consists of 0.200 mol PCl_3 and 0.600 mol Cl_2 is placed in a 4.00-L reaction vessel and heated to 760°C, what is the equilibrium composition of the system?

Synthesis Reactions

13.35 The equilibrium constant K_c for the reaction $N_2(g) + O_2(g) \rightleftharpoons 2NO(g)$ at 1200°C is 1.00×10^{-5}. Calculate the equilibrium molar concentrations of NO, N_2, and O_2 in a 1.00-L reaction vessel that initially held 0.114 mol N_2 and 0.114 mol O_2.

13.36 The equilibrium constant K_c for the reaction $N_2(g) + O_2(g) \rightleftharpoons 2NO(g)$ at 1200°C is 1.00×10^{-5}. Calculate the equilibrium molar concentrations of NO, N_2, and O_2 in a 10.0-L reaction vessel that initially held 0.014 mol N_2 and 0.214 mol O_2.

13.37 A reaction mixture that consisted of 0.400 mol of H_2 and 1.60 mol of I_2 was placed into a 3.00-L flask and heated. At equilibrium, 60.0% of the hydrogen gas had reacted. What is the equilibrium constant for the reaction $H_2(g) + I_2(g) \rightleftharpoons 2HI(g)$ at this temperature?

13.38 A reaction mixture that consisted of 0.20 mol N_2 and 0.20 mol H_2 was placed into a 25.0-L reactor and heated. At equilibrium 5.0% of the nitrogen gas had reacted. What is the value of the equilibrium constant K_c for the reaction $N_2(g) + 3H_2(g) \rightleftharpoons 2NH(g)$ at this temperature?

13.39 The equilibrium constant K_c of the reaction $2CO(g) + O_2(g) \rightleftharpoons 2CO_2(g)$ is 0.66 at 2000°C. If 0.28 g of CO and 0.032 g of O_2 are placed in a 2.0-L reaction vessel and heated to 2000°C, what will the equilibrium composition of the system be?

13.40 In the Haber process for ammonia synthesis, $K_P = 0.036$ for the reaction $N_2(g) + 3H_2(g) \rightleftharpoons 2NH_3(g)$ at 500 K. If a reactor is charged with partial pressures of 0.010 atm of N_2 and 0.010 atm of H_2, what will the equilibrium partial pressure of the components be?

13.41 An "ester" is formed in the reaction of an organic acid with an alcohol. For example, in the reaction of acetic acid, CH_3COOH, with ethanol, C_2H_5OH, the ester ethyl acetate, $CH_3COOC_2H_5$, and water form, the reaction being

$$CH_3COOH(l) + C_2H_5OH(l) \rightleftharpoons CH_3COOC_2H_5(l) + H_2O(l) \quad K_c = 4.0 \text{ at } 100°C$$

If the initial concentrations of CH_3COOH and C_2H_5OH are 0.32 mol/L and 6.3 mol/L, respectively, what will the equilibrium molar concentration of the ester be?

13.42 In the esterification reaction specified in Exercise 13.41, if the initial concentrations of CH_3COOH and C_2H_5OH are 0.024 mol/L and 0.059 mol/L, respectively, but in addition the concentration of the water in the initial mixture is 0.015 mol/L, what will the equilibrium molar concentration of the ester be?

13.43 Let the equilibrium constants for the reactions $2H_2O(g) \rightleftharpoons 2H_2(g) + O_2(g)$ and $2CO_2(g) \rightleftharpoons 2CO(g) + O_2(g)$ be K_{P1} and K_{P2}, respectively. Show that the equilibrium constant for the reaction $CO_2(g) + H_2(g) \rightleftharpoons H_2O(g) + CO(g)$ is $K_{P3} = \sqrt{K_{P2}/K_{P1}}$ and evaluate it at 1565 K, at which temperature $K_{P1} = 1.6 \times 10^{-11}$ and $K_{P2} = 1.3 \times 10^{-10}$.

13.44 Suppose that in an esterification reaction in which an organic acid reacts with an alcohol to produce an ester, the amounts A mol of acid and B mol of alcohol are mixed and heated to 100°C. Find an expression for the amount (in moles) of ester that is present at equilibrium, in terms of A, B, and K_c. Evaluate the expression for $A = 1.0$, $B = 0.50$, and $K_c = 3.5$.

Effect of Added Reagents

13.45 Consider the equilibrium $CO(g) + H_2O(g) \rightleftharpoons CO_2(g) + H_2(g)$. (a) If the partial pressure of CO_2 is increased, what happens to the partial pressure of H_2? (b) If the partial pressure of CO is decreased, what happens to the partial pressure of CO_2? (c) If the concentration of CO is increased, what happens to the concentration of H_2?

13.46 Consider the equilibrium $CH_4(g) + 2O_2(g) \rightleftharpoons CO_2(g) + 2H_2O(g)$. (a) If the partial pressure of CO_2 is increased, what happens to the partial pressure of CH_4? (b) If the partial pressure of CH_4 is decreased, what happens to the partial pressure of CO_2? (c) If the concentration of CH_4 is increased, what happens to the equilibrium constant for the reaction?

13.47 The four gases NH_3, O_2, NO, and H_2O are mixed in a reaction vessel and allowed to reach equilibrium in the reaction $4NH_3(g) + 5O_2(g) \rightleftharpoons 4NO(g) + 6H_2O(g)$. Certain changes (see the table below) are then made to this mixture. Considering each change separately, state the effect (increase, i; decrease, d; no change, nc) that the change has on the original equilibrium value of the quantity in the second column. The temperature and volume are constant unless otherwise noted.

Change	Quantity	Effect
add NO	amount of H_2O	i d nc
add NO	amount of O_2	i d nc
remove H_2O	amount of NO	i d nc
remove O_2	amount of NH_3	i d nc
remove NH_3	K_c	i d nc
remove NO	amount of NH_3	i d nc
add NH_3	amount of O_2	i d nc

13.48 The four substances HCl, I_2, HI, and Cl_2 are mixed in a reaction vessel and allowed to reach equilibrium in the reaction $2HCl(g) + I_2(s) \rightleftharpoons 2HI(g) + Cl_2(g)$. Certain changes (which are specified in the first column in the table below) are then made to this mixture. Considering each change separately, state the effect (increase, i; decrease, d; no change, nc) that the change has on the original equilibrium value of the quantity in the second column. The temperature and volume are constant unless otherwise noted.

Change	Quantity	Effect
add HCl	amount of HI	i d nc
add I_2	amount of Cl_2	i d nc
remove HI	amount of Cl_2	i d nc
remove Cl_2	amount of HCl	i d nc
add HCl	K_c	i d nc
remove HCl	amount of I_2	i d nc
add I_2	K_c	i d nc

Response to Pressure

13.49 State whether reactants or products will be favored by an increase in the total pressure on each equilibrium listed below. If no change occurs, explain why that is so. (a) $2O_3(g) \rightleftharpoons 3O_2(g)$; (b) $H_2O(g) + C(s) \rightleftharpoons H_2(g) + CO(g)$; (c) $4NH_3(g) + 5O_2(g) \rightleftharpoons 4NO(g) + 6H_2O(g)$.

13.50 State what happens to the concentration of the indicated substance when the total pressure on each equilibrium is increased. (a) $NO_2(g)$ in $2Pb(NO_3)_2(s) \rightleftharpoons 2PbO(s) + 4NO_2(g) + O_2(g)$; (b) $NO(g)$ in $3NO_2(g) + H_2O(l) \rightleftharpoons 2HNO_3(aq) + NO(g)$; (c) $HI(g)$ in $2HCl(g) + I_2(s) \rightleftharpoons 2HI(g) + Cl_2(g)$.

13.51 Consider the equilibrium $4NH_3(g) + 5O_2(g) \rightleftharpoons 4NO(g) + 6H_2O(g)$. (a) What happens to the partial pressure of NH_3 when the partial pressure of NO is increased? (b) Does the partial pressure of O_2 increase or decrease when the partial pressure of NH_3 decreases?

13.52 Consider the equilibrium $2SO_2(g) + O_2(g) \rightleftharpoons 2SO_3(g)$. (a) What happens to the partial pressure of SO_3 when the partial pressure of SO_2 is decreased? (b) If the partial pressure of SO_2 increases, what happens to the partial pressure of O_2?

13.53 The density of quartz (SiO_2) is greater than that of the glassy form of silica (also SiO_2). Would glass or quartz be favored under pressure?

13.54 The density of red phosphorus is 2.34 g/cm^3, and that of its white allotrope is 1.82 g/cm^3. Would you expect the red or the white allotrope to be favored under pressure?

13.55 Let α be the fraction of PCl_5 molecules that have decomposed to PCl_3 and Cl_2 in the reaction $PCl_5(g) \rightleftharpoons PCl_3(g) + Cl_2(g)$, so that the amount of PCl_5 at equilibrium is $n(1 - \alpha)$, where n is the amount present initially. Derive an equation for K_P in terms of α and the total pressure P and solve it for α in terms of P. Calculate the fraction decomposed at 556 K, at which temperature $K_P = 4.96$, and pressures of (a) 0.50 atm and (b) 1.0 atm.

13.56 Express K_P for the ammonia synthesis equilibrium $N_2(g) + 3H_2(g) \rightleftharpoons 2NH_3(g)$ in terms of the total pressure P and the fraction α of nitrogen that has reacted in an initially stoichiometric mixture of nitrogen and hydrogen. Show that when α is small ($\alpha \ll 1$) it is proportional to the total pressure ($\alpha \propto P$).

The Response to Temperature

13.57 Predict whether each equilibrium will shift toward products or reactants with a temperature increase: (a) $N_2O_4(g) \rightleftharpoons 2NO_2(g)$, $\Delta H° = +57$ kJ; (b) $X_2(g) \rightleftharpoons 2X(g)$, where X is a halogen; (c) $Ni(s) + 4CO(g) \rightleftharpoons Ni(CO)_4(g)$, $\Delta H° = -161$ kJ.

13.58 Predict whether each equilibrium will shift toward products or reactants with a temperature decrease: (a) $CH_4(g) + H_2O(g) \rightleftharpoons CO(g) + 3H_2(g)$, $\Delta H° = +206$ kJ; (b) $CO(g) + H_2O(g) \rightleftharpoons CO_2(g) + H_2(g)$, $\Delta H° = -41$ kJ; (c) $2SO_2(g) + O_2(g) \rightleftharpoons 2SO_3(g)$, $\Delta H° = -198$ kJ.

13.59 A mixture consisting of 2.23×10^{-3} mol of N_2 and 6.69×10^{-3} mol of H_2 in a 500-mL container was heated to 600 K and allowed to reach equilibrium. Will more ammonia be formed if that equilibrium mixture is then heated to 700 K? For $N_2(g) + 3H_2(g) \rightleftharpoons 2NH_3(g)$, $K_P = 1.7 \times 10^{-3}$ at 600 K and $K_P = 7.8 \times 10^{-5}$ at 700 K.

13.60 A mixture consisting of 1.1 mmol of SO_2 and 2.2 mmol of O_2 in a 250-mL container was heated to 500 K and allowed to reach equilibrium. Will more sulfur trioxide be formed if that equilibrium mixture is cooled to 25°C? For the reaction $2SO_2(g) + O_2(g) \rightleftharpoons 2SO_3(g)$, $K_P = 2.5 \times 10^{10}$ at 500 K and $K_P = 4.0 \times 10^{24}$ at 25°C.

UNCLASSIFIED EXERCISES

13.61 Write the reaction quotients Q_c and Q_P for the following reactions: (a) $S(s) + O_2(g) \rightleftharpoons SO_2(g)$; (b) $SO_3(g) + H_2(g) \rightleftharpoons SO_2(g) + H_2O(g)$; (c) $W(s) + 6HCl(g) \rightleftharpoons WCl_6(g) + 3H_2(g)$.

13.62 The value of the equilibrium constant K_c for the reaction $F_2(g) \rightleftharpoons 2F(g)$ is 2.7×10^{-3} at 1200 K. Determine the value of K_P for the reactions (a) $F_2(g) \rightleftharpoons 2F(g)$ and (b) $2F(g) \rightleftharpoons F_2(g)$. (See the footnote on page 502.)

13.63 The value of the equilibrium constant K_P for the reaction $2SO_2(g) + O_2(g) \rightleftharpoons 2SO_3(g)$ is 3.0×10^4 at 700 K. Determine the value of K_c for the reactions (a) $2SO_2(g) + O_2(g) \rightleftharpoons 2SO_3(g)$ and (b) $SO_3(g) \rightleftharpoons SO_2(g) + \frac{1}{2}O_2(g)$. (See the footnote on page 502.)

CASE 13 ▸ STALACTITES AND STALAGMITES

Natural caves with stalactites (iciclelike conical formations that hang from the roof) and stalagmites (formations that rise from the floor) are often found in regions where high concentrations of calcium carbonate (limestone) deposits exist. Calcium carbonate is a sparingly soluble salt, and in water it is in equilibrium with its ions:

$$CaCO_3(s) \rightleftharpoons Ca^{2+}(aq) + CO_3^{2-}(aq) \qquad K_c = 8.7 \times 10^{-9}$$

Any effect that shifts this equilibrium to the right increases the solubility of $CaCO_3$. The CO_3^{2-} ion is a Brønsted base and can be removed by proton transfer from an acid, as occurs in the reaction

$$CO_3^{2-}(aq) + H_3O^+(aq) \rightleftharpoons H_2O(l) + HCO_3^-(aq) \qquad K_c = 1.8 \times 10^{10}$$

The natural water percolating through the limestone is rich in acid. For instance, the pH of natural rainfall is 5.7 as a result of the presence of dissolved carbon dioxide

$$CO_2(g) + 2H_2O(l) \rightleftharpoons H_3O^+(aq) + HCO_3^-(aq) \qquad K_c = 4.3 \times 10^{-7}$$

In addition, soils may be acidic as a result of humic acids in the soil or the decay of organic matter to CO_2. The acidic groundwater protonates the carbonate ions in equilibrium with the solid calcium carbonate, removes them from the solubility equilibrium, and hence encourages more calcium carbonate to dissolve to restore the equilibrium. As a result, holes are formed in the limestone, and over geological time, the cavity grows to the size of a cave.

Once the cave has formed, the formation of the stalactites and stalagmites begins. Consider the following sum of the first and third equations above:

$$CaCO_3(s) + CO_2(g) + H_2O(l) \rightleftharpoons Ca^{2+}(aq) + 2HCO_3^-(aq)$$

When a drop of the percolating groundwater containing Ca^{2+} and HCO_3^- ions reaches the ceiling of the cave, it encounters new atmospheric conditions because the partial pressure of carbon dioxide in the cave is low. As a result, CO_2 escapes from the solution, the equilibrium shifts toward the left, and that shift results in the deposition of calcium carbonate: stalactite formation occurs. Contrary to common thought, it is not the evaporation of the water that is the principal cause of the deposition—the concentration of water vapor in the cave is high, and most caves are damp because of the high relative humidity. If the drop falls to the floor, then the same shift in equilib-

13.64 Determine K_P from the following equilibrium data collected at 24°C for the reaction $NH_4HS(s) \rightleftharpoons NH_3(g) + H_2S(g)$.

P_{NH_3}, atm	P_{H_2S}, atm
0.309	0.309
0.364	0.258
0.539	0.174

13.65 At 500°C, $K_c = 0.060$ for $N_2(g) + 3H_2 \rightleftharpoons 2NH_3(g)$. When analysis shows that the composition is 3.00 mol N_2/L, 2.00 mol H_2/L, and 0.500 mol NH_3/L, is the reaction at equilibrium? If not, then in which direction does the reaction tend to proceed to reach equilibrium?

13.66 At 2500 K, the equilibrium constant is $K_c = 20$ for the reaction $Cl_2(g) + F_2(g) \rightleftharpoons 2ClF(g)$. An analysis of a reaction vessel at 2500 K reveals the presence of 0.18 mol Cl_2/L, 0.31 mol F_2/L, and 0.92 mol ClF. Will ClF tend to form or to decompose as the reaction proceeds toward equilibrium?

13.67 At 500 K, the equilibrium constant is $K_c = 0.031$ for the reaction $Cl_2(g) + Br_2(g) \rightleftharpoons 2BrCl(g)$. If the equilibrium composition is 0.22 mol Cl_2/L and 0.097

rium occurs, the CO_2 leaves the equilibrium and enters the atmosphere of the cave: stalagmite formation occurs. The rate of stalactite and stalagmite formations is typically 0.2 mm/y.

Stalactites and stalagmites have different colors as a result of the presence of *d*-block ion impurities found in the soil solution, most commonly iron(III) ions.

QUESTIONS

1. The length of a certain stalactite is 6.0 m. Assuming an average growth of 0.2 mm/y, calculate the time required for its formation.

2. (a) Write the reaction quotient for $CaCO_3(s) + CO_2(g) + H_2O(l) \rightleftharpoons Ca^{2+}(aq) + 2HCO_3^-(aq)$ and (b) determine the equilibrium constant of the reaction at 25°C from the information given above.

3. Determine whether the formation of stalactites and stalagmites is favored in each of the following cases: (a) the partial pressure of carbon dixoide in the cave increases; (b) the hardness of the groundwater increases; (c) the temperature of the cave increases; (d) the pH of the soil solution increases; (e) the relative humidity of the cave decreases.

4. (a) For a normal partial pressure of CO_2 in the atmosphere in equilibrium with groundwater, the concentration of dissolved CO_2 is 1.03×10^{-5} mol/L. Under these conditions, what are the concentrations of the Ca^{2+} and HCO_3^- ions? Use the equilibrium constant determined in Question 2. (b) If the partial pressure of the carbon dioxide falls (as it does in a cave) to the point at which the dissolved CO_2 concentration decreases to 5.0×10^{-6} mol/L, what are the concentrations of the Ca^{2+} and HCO_3^- ions? (c) As a result of the changes from the conditions in (a) to those in (b), what mass of $CaCO_3$ is deposited per drop, where the volume of one drop is 0.05 mL?

mol BrCl/L, what is the equilibrium molar concentration of Br_2?

13.68 The equilibrium constant is $K_P = 3.5 \times 10^4$ for reaction $PCl_3(g) + Cl_2(g) \rightleftharpoons PCl_5(g)$ at 760°C. At equilibrium, the partial pressure of PCl_5 was 2.2×10^{-4} atm and that of PCl_3 was 1.33 atm. What is the equilibrium partial pressure of Cl_2?

13.69 A reaction mixture consisting of 2.00 mol CO and 3.00 mol H_2 are placed into a 10.0-L reaction vessel and heated to 1200 K. At equilibrium, 0.478 mol of CH_4 was present in the system. Determine the values of K_c and K_P for the reaction $CO(g) + 3H_2(g) \rightleftharpoons CH_4(g) + H_2O(g)$.

13.70 A mixture consisting of 1.0 mol $H_2O(g)$ and 1.0 mol $CO(g)$ is placed in a 10.0-L reaction vessel at 800 K. At equilibrium, 0.665 mol $CO_2(g)$ is present as a result of the reaction $CO(g) + H_2O(g) \rightleftharpoons CO_2(g) + H_2(g)$. What are (a) the equilibrium molar concentrations for all substances and (b) the values of K_c?

13.71 A reaction mixture was prepared by mixing 0.100 mol SO_2, 0.200 mol NO_2, 0.100 mol NO, and 0.150 mol SO_3 in a 5.00-L reaction vessel. The reaction $SO_2(g) + NO_2(g) \rightleftharpoons NO(g) + SO_3(g)$ is allowed to reach equilibrium at 460°C, when $K_c = 85.0$. What is the equilibrium molar concentration of each substance?

13.72 For the equilibrium $SO_2(g) + Cl_2(g) \rightleftharpoons SO_2Cl_2(g)$, $K_P = 0.654$ at 300 K. (a) Calculate the value of K_c for the equilibrium at 300 K. (b) Suppose that a mixture is prepared that consists of 1.47 g of SO_2 and 0.082 mol Cl_2 in a 4.0-L reaction vessel and heated to 300 K. What is the equilibrium composition in the mixture?

13.73 A 0.100-mol sample of H_2S is placed in a 10.0-L reaction vessel and heated to 1132°C. At equilibrium, 0.0285 mol of H_2 is present. Calculate the value of K_c for the reaction $2H_2S(g) \rightleftharpoons 2H_2(g) + S_2(g)$.

13.74 A mixture of 0.0560 mol O_2 and 0.0200 mol N_2O is placed in a 1.00-L reaction vessel at 25°C. When the reaction $2N_2O(g) + 3O_2(g) \rightleftharpoons 4NO_2(g)$ is at equilibrium, 0.0200 mol NO_2 is present. (a) What are the equilibrium molar concentrations of O_2 and N_2O? (b) What is the value of K_c?

13.75 At 500 K, 1.0 mol NOCl is 9.0% dissociated in a 1.0-L vessel. Calculate the value of K_c for the reaction $2NOCl(g) \rightleftharpoons 2NO(g) + Cl_2(g)$.

13.76 The equilibrium constant $K_c = 0.56$ for the reaction $PCl_3(g) + Cl_2(g) \rightleftharpoons PCl_5(g)$ at 250°C. Upon analysis, it was found that 1.5 mol PCl_5, 3.0 mol PCl_3, and 0.50 mol Cl_2 were present in a 500-mL reaction vessel. (a) Is the reaction at equilibrium? (b) If not, in which direction does it tend to proceed and (c) what is the equilibrium composition of the reaction system?

13.77 A 1.50-mol sample of PCl_5 is placed into a 500-mL reaction vessel. What is the molar concentration of each substance present in the system when the reaction $PCl_5(g) \rightleftharpoons PCl_3(g) + Cl_2(g)$ has reached equilibrium at 250°C (when $K_c = 1.80$).

13.78 At 25°C, $K_P = 3.2 \times 10^{-34}$ for the reaction $2HCl(g) \rightleftharpoons H_2(g) + Cl_2(g)$. If a 1.0-L reaction vessel is filled with HCl at a pressure of 0.22 atm, (a) what are the equilibrium partial pressures of HCl, H_2, and Cl_2; (b) what is the total pressure in the vessel?

13.79 If 4.00 L of HCl(*g*) at STP and 26.0 g of $I_2(s)$ are transferred to a 12.00-L reaction vessel and heated to 25°C, what will be the equilibrium molar concentrations of HCl, HI, and Cl_2? $K_c = 1.6 \times 10^{-34}$ at 25°C for $2HCl(g) + I_2(s) \rightleftharpoons 2HI(g) + Cl_2(g)$.

13.80 A 30.1-g sample of NOCl is placed into a 200-mL reaction vessel and heated to 500 K. The value of K_P for the decomposition of NOCl at 500 K in the reaction $2NOCl(g) \rightleftharpoons 2NO(g) + Cl_2(g)$ is 1.13×10^{-3}. (a) What are the equilibrium partial pressures of NOCl, NO, and Cl_2? (b) What is the percentage decomposition of NOCl at this temperature?

13.81 At 25°C, $K_c = 4.66 \times 10^{-3}$ for $N_2O_4(g) \rightleftharpoons 2NO_2(g)$. If 2.50 g N_2O_4 and 0.33 g NO_2 are placed in a 2.0-L reaction vessel, what are the equilibrium molar concentrations of N_2O_4 and NO_2?

13.82 A 16-mol sample of SO_3 is placed in a 2.0-L reaction vessel at 600°C. When the reaction $2SO_2(g) + O_2(g) \rightleftharpoons 2SO_3(g)$ is at equilibrium, the system contains 2.0 mol O_2. (a) What is the percentage decomposition of the SO_3 in the system? (b) Calculate the value of K_c for the reaction.

13.83 The photosynthesis reaction is $6CO_2(g) + 6H_2O(l) \rightarrow C_6H_{12}O_6(s) + 6O_2(g)$, $\Delta H° = +2802$ kJ. Suppose that the reaction is at equilibrium and state the effect that each of the following changes will have on the equilibrium composition (tend to shift toward the formation of reactants, toward the formation products, or have no effect). (a) The partial pressure of O_2 is increased. (b) The atmospheric pressure increases. (c) The amount of CO_2 is increased. (d) The temperature is increased. (e) Some of the $C_6H_{12}O_6$ is removed. (f) Water is added. (g) The partial pressure of CO_2 is decreased.

13.84 Use Le Chatelier's principle to predict the effect (increase, i; decrease, d; no change, nc) that the change given in the first column of the table below has on the quantity in the second column for the following equilibrium system:

$$5CO(g) + I_2O_5(s) \rightleftharpoons I_2(g) + 5CO_2(g) \quad \Delta H° = -1175 \text{ kJ}$$

Each change is applied separately to the system.

Change	Quantity	Effect
decrease volume	K_c	i d nc
increase volume	amount of CO	i d nc
raise temperature	K_c	i d nc
add I_2	amount of CO_2	i d nc
add I_2O_5	amount of I_2	i d nc
remove CO_2	amount of I_2	i d nc
increase pressure	amount of CO	i d nc
lower pressure	amount of CO_2	i d nc
add CO_2	concentration of I_2O_5	i d nc

13.85 A 3.00-L reaction vessel is filled with 0.150 mol CO, 0.0900 mol H_2, and 0.180 mol CH_3OH. Equilibrium is reached in the presence of a zinc oxide/chromium(III) oxide catalyst, and at 300°C, $K_c = 1.1 \times 10^{-2}$ for the reaction $CO(g) + 2H_2(g) \rightleftharpoons CH_3OH(g)$. (a) As the reaction approaches equilibrium, will the molar concentration of CH_3OH increase, decrease, or remain unchanged? Explain your answer. (b) What is the equilibrium composition of the mixture?

13.86 The equilibrium constant is $K_c = 0.395$ at 350°C for the reaction $2NH_3(g) \rightleftharpoons N_2(g) + 3H_2(g)$. A 15.0-g sample of NH_3 is placed in a 5.00-L reaction vessel and heated to 350°C. What are the equilibrium molar concentrations of NH_3, N_2, and H_2?

13.87 At 25°C, $K_c = 4.66 \times 10^{-3}$ for the reaction $N_2O_4(g) \rightleftharpoons 2NO_2(g)$. If a 2.50-g N_2O_4 sample is placed in a 2.00-L reaction flask, what will the composition of the equilibrium mixture be?

13.88 The van't Hoff equation relates the equilibrium constant K'_P at a temperature T' to its value K_P at T:

$$\ln\left(\frac{K'_P}{K_P}\right) = \frac{\Delta H^\circ}{R} \times \left(\frac{1}{T} - \frac{1}{T'}\right)$$

where ΔH° is the enthalpy of the forward process. The temperature dependence of the equilibrium constant of the reaction $N_2(g) + O_2(g) \rightleftharpoons 2NO(g)$, which makes an important contribution to atmospheric nitrogen oxides, can be expressed as $\ln K_P = 2.5 - 21{,}700/T$. What is the standard enthalpy of the forward reaction?

13.89 The dissociation vapor pressure of ammonium hydrogen sulfide (NH_4HS) is 501 Torr at 298.3 K and 919 Torr at 308.8 K. Using the information in Exercise 13.88, estimate its enthalpy of dissociation and the temperature at which it would "boil" when the external pressure is 1 atm.

Two of the most important concepts in chemistry are those of "acid" and "base," which were first introduced in Chapter 3. Here we discuss these substances in terms of the transfer of a proton from one species to another. We also see how to describe the behavior of acids and bases in terms of equilibrium constants, and we examine the molecular features that affect their strengths. Acids and bases have a profound influence on the everyday world: for example, the difference in the acidities of the sap of hydrangeas causes them to be either blue (acidic) or red (basic).

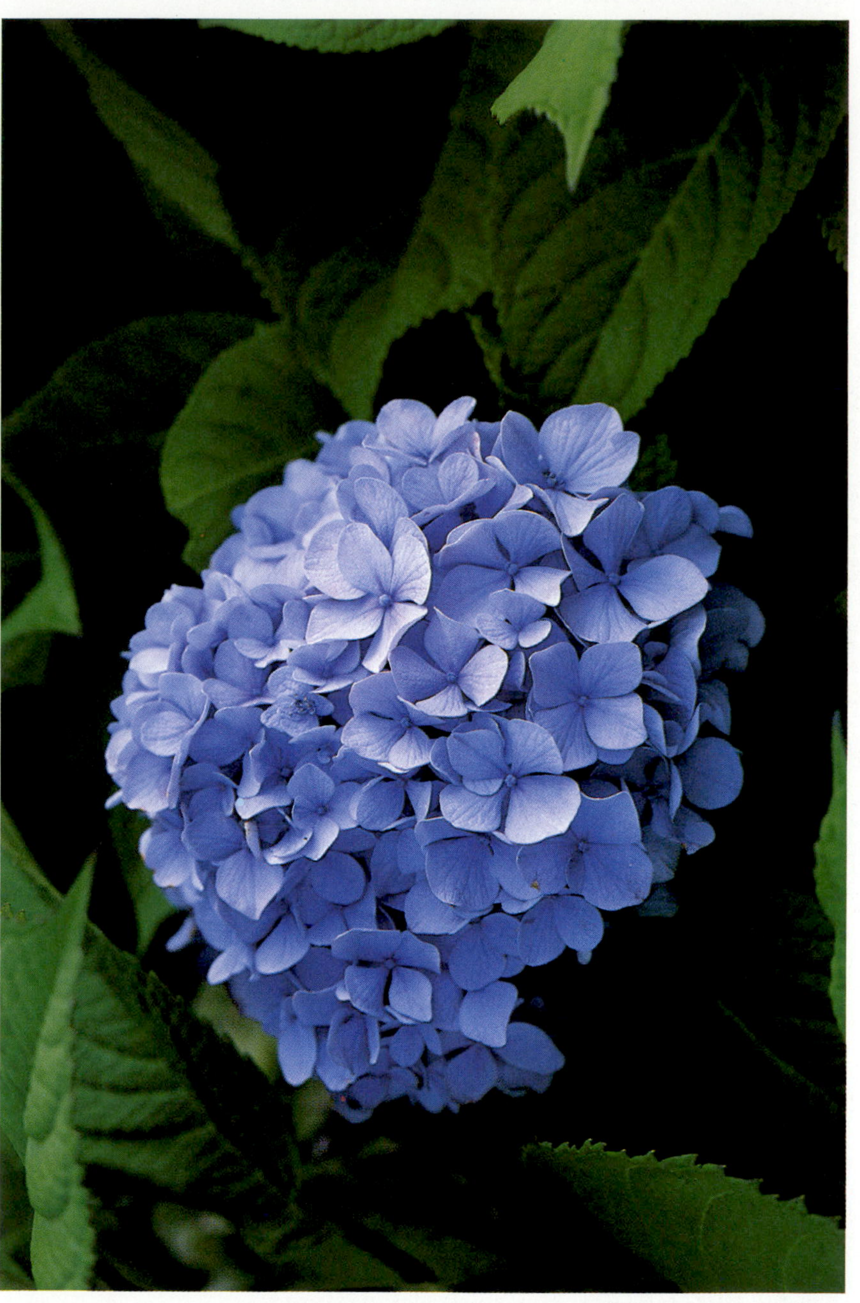

14 ACIDS AND BASES

This chapter builds on Chapter 3, where we first encountered acids and bases in terms of the Arrhenius definition: an acid is a compound that contains hydrogen and releases hydrogen ions in water; a base is a compound that produces hydroxide ions in water. We saw there, however, that a broader view of acids and bases is obtained if we adopt the Brønsted-Lowry definitions, in which an acid is a proton donor and a base is a proton acceptor.

Our understanding of acids and bases can be enriched considerably by combining the Brønsted-Lowry definitions with what we have learned about chemical equilibria in Chapter 13. In particular, we shall see that we can set up a scale of the relative strengths of acids and bases by using equilibrium constants and that we can make quantitative predictions about reactions involving acids and bases.

Table 14.1 summarizes some of the techniques for preparing and manufacturing common acids and bases.

THE DEFINITIONS OF ACIDS AND BASES

The Brønsted-Lowry definitions of acids and bases were introduced in Section 3.7. Those definitions focus attention on the transfer of a proton (H^+) from one molecule (the acid) to another (the base). Here we briefly review the central points of the theory and then extend it.

14.1 BRØNSTED ACIDS AND BASES

The Brønsted definitions of acids and bases are as follows (Fig. 14.1):

A **Brønsted acid** is a proton donor.

A **Brønsted base** is a proton acceptor.

E Johannes Brønsted (1879–1947) was a Danish chemist whose specialty was electrochemistry and reaction kinetics. Thomas Lowry (1874–1936) was the first professor of physical chemistry at the University of Cambridge, England.

D Acid base indicators: A voice activated chemical reaction. CD1, 36. Acids and bases with litmus, TDC, 4-13.

E See annotations on pp. 95, 98.

TABLE 14.1 The preparation and manufacture of some common acids and bases

Acid or base	Method of preparation or manufacture
hydrohalic acids	1. action of a nonvolatile acid (H_2SO_4 or H_3PO_4) on a metal halide: $CaF_2(s) + 2H_2SO_4(l) \longrightarrow Ca(HSO_4)_2(s) + 2HF(g)$ (Use phosphoric acid if oxidation is a danger, as it is with HBr and HI.) 2. action of water on nonmetal halide: $PCl_3(l) + 3H_2O(l) \longrightarrow H_3PO_3(aq) + 3HCl(aq)$
hypohalous acids	action of halogen on water: $Cl_2(g) + H_2O(l) \longrightarrow HClO(aq) + HCl(aq)$
sulfuric acid	contact process (catalytic oxidation of SO_2, followed by reaction with water): $2SO_2(g) + O_2(g) \xrightarrow{V_2O_5} 2SO_3(g)$; $SO_3(g) + H_2SO_4(aq) \longrightarrow H_2S_2O_7(aq) \xrightarrow{\text{water}} 2H_2SO_4(aq)$
nitric acid	Ostwald process (catalytic oxidation of ammonia, followed by reaction with water): $4NH_3(g) + 7O_2(g) \xrightarrow{Pt} 4NO_2(g) + 6H_2O(g)$; $3NO_2(g) + H_2O(l) \longrightarrow 2HNO_3(aq) + NO(g)$
acetic acid	oxidation of ethanol: $C_2H_5OH(l) \xrightarrow{\text{oxidation}} CH_3COOH(l)$
sodium hydroxide	electrolysis of brine: $2NaCl(aq) + 2H_2O(l) \longrightarrow 2Na^+(aq) + 2OH^-(aq) + H_2(g) + Cl_2(g)$
calcium hydroxide	thermal decomposition of limestone, followed by reaction with water: $CaCO_3(s) \xrightarrow{\Delta} CaO(s) + CO_2(g)$; $CaO(s) + H_2O(l) \longrightarrow Ca(OH)_2(s)$
ammonia	Haber process (catalytic high-pressure synthesis): $N_2(g) + 3H_2(g) \xrightarrow{Pt} 2NH_3(g)$

E In this chapter and the next we see that a major part of chemistry can be ascribed to the transfer of one fundamental particle (a proton) from one species to another. Later, in Chapter 17, we shall see that the transfer of another fundamental particle (an electron) is responsible for another wide range of reactions.

We shall normally denote a Brønsted acid as HA and a Brønsted base as B. Then proton donation from acid to water is

$$HA(aq) + H_2O(l) \longrightarrow H_3O^+(aq) + A^-(aq)$$

An H_2O molecule that has accepted a proton from a proton donor becomes a hydronium ion, H_3O^+. (As we saw in Section 3.7, the formula H_3O^+ for the hydronium ion may be a considerable simplification of the actual entity in solution.) Similarly, proton donation from water to a base in aqueous solution is

$$H_2O(l) + B(aq) \longrightarrow HB^+(aq) + OH^-(aq)$$

The H_2O molecule that has donated the proton becomes a hydroxide ion, OH^-. Note that in its reaction with the acid HA, the H_2O molecule accepts the proton from HA and hence functions as a Brønsted base. However, in its reaction with the base B, the water molecule donates a proton to B and hence functions as a Brønsted acid.

We saw in Section 3.7 that an acid or base is classified as "strong" if it is fully ionized in solution (as are nitric acid, HNO_3, and sodium hydroxide, NaOH). An acid or base is classified as "weak" if only a small fraction (usually a very tiny fraction) is ionized in solution. Examples of a weak acid and a weak base are acetic acid, CH_3COOH, and ammonia, NH_3, respectively. In other words, the strength of an acid or a base refers to the extent to which it is ionized (deprotonated in the case of an acid, and protonated in the case of a base) in solution.

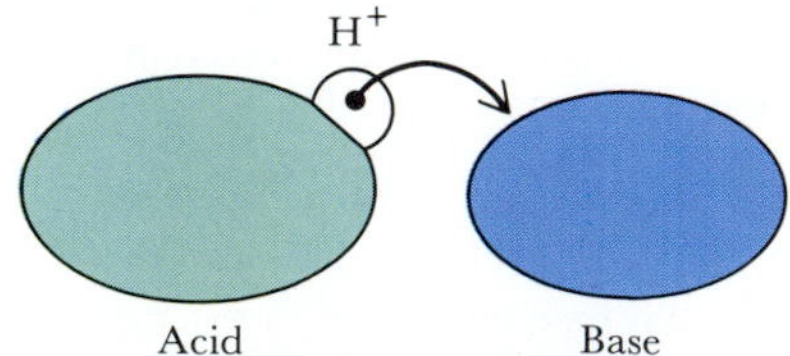

FIGURE 14.1 A Brønsted acid is a proton donor and a Brønsted base is a proton acceptor.

The Brønsted-Lowry theory summarizes all strong acid-base **neutralization reactions** in aqueous solution by a single chemical equation (Fig. 14.2):

$$H_3O^+(aq) + OH^-(aq) \longrightarrow 2H_2O(l)$$

The same net ionic equation applies to *all* strong acid-strong base neutralizations in water, because the acid present always produces H_3O^+ by donating a proton to an H_2O molecule and the base present always produces OH^- by accepting a proton from another H_2O molecule.

Support for the view that all neutralizations in water are essentially the same comes from thermochemistry. When the reaction enthalpies of neutralizations are measured in very dilute solutions (so that the ions do not interact with one another), approximately the same value, −57 kJ/mol, is obtained whatever the acids and bases used (Table 14.2). This observation suggests that all strong acid-strong base neutralization reactions are actually the same and that the enthalpy change is that of the same proton transfer reaction in every case.

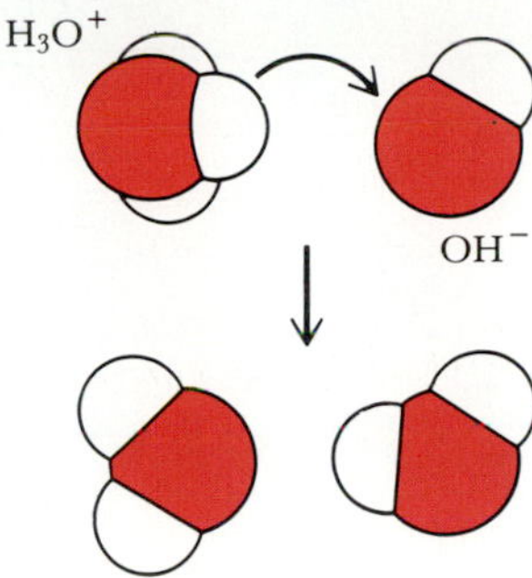

FIGURE 14.2 In the Brønsted-Lowry theory of acids and bases, the neutralization of a strong acid and a strong base is the transfer of protons from hydronium ions to hydroxide ions. Oxygen atoms, red; hydrogen atoms, white.

14.2 BRØNSTED ACID-BASE EQUILIBRIA

Proton donating and accepting are dynamic processes for all acids and bases. For example, in aqueous solution, acetic acid molecules donate protons to water molecules:

$$CH_3COOH(aq) + H_2O(l) \longrightarrow H_3O^+(aq) + CH_3CO_2^-(aq)$$

At the same time, acetate ions may accept protons from hydronium ions and become acetic acid molecules:

$$H_3O^+(aq) + CH_3CO_2^-(aq) \longrightarrow CH_3COOH(aq) + H_2O(l)$$

When these two reactions occur at equal rates, the species are in dynamic equilibrium:

$$CH_3COOH(aq) + H_2O(l) \rightleftharpoons H_3O^+(aq) + CH_3CO_2^-(aq)$$

The same is true of a solution of a base, where a proton transfer equilibrium such as

$$H_2O(l) + NH_3(aq) \rightleftharpoons NH_4^+(aq) + OH^-(aq)$$

is quickly established, with ammonia molecules accepting protons from water molecules at the same rate as ammonium ions donate protons to hydroxide ions.

D Acid neutralizing properties of lake beds. BZS3, 8.18.

E Emphasis on acids and bases. *J. Chem. Educ.*, **1977**, *54*, 626. The conjugate acid-base chart. *J. Chem. Educ.*, **1986**, *63*, 939.

E Proton transfer is so fast that equilibrium is reached almost the instant an acid or base is added to a solution. We may therefore always presume that the concentrations are those at equilibrium, and assume that a reaction quotient for proton transfer reaction is equal to the appropriate equilibrium constant.

Conjugate acids and bases. In the proton transfer reaction between CH_3COOH and H_2O, a CH_3COOH molecule donates a proton and

TABLE 14.2 Standard enthalpies of neutralization at 25°C

Acid	Base	Salt	$\Delta H°$, kJ/mol
HCl	NaOH	NaCl	−57.1
HCl	KOH	KCl	−57.2
HNO_3	NaOH	$NaNO_3$	−57.3
2HCl	$Ba(OH)_2$	$BaCl_2$	2 × (−58.2)

hence is a Brønsted acid; an H_2O molecule accepts a proton and hence is a Brønsted base. In the reverse reaction, the acetate ion, $CH_3CO_2^-$ accepts a proton from the hydronium ion. Hence, $CH_3CO_2^-$ is a Brønsted base and, in particular, is the conjugate base of the acid CH_3COOH:

The **conjugate base** of a Brønsted acid is the base that forms when the acid has donated a proton.

$$\text{Brønsted acid} \xrightarrow{-H^+} \text{conjugate Brønsted base}$$

(*Conjugate* in this context means "related.") Likewise, in the forward reaction, H_2O acts as a Brønsted base and accepts a proton from CH_3COOH, becoming H_3O^+ in the process. This H_3O^+ ion acts as an acid in the reverse reaction by donating the proton to a $CH_3CO_2^-$ ion. Hence, H_3O^+ is the conjugate acid of the base H_2O:

N The acetate ion is written $CH_3CO_2^-$, and not CH_3COO^-, because we wish to emphasize the equivalence of the two carboxyl O atoms in the ion (but not in the acid).

The **conjugate acid** of a Brønsted base is the acid that forms when the base has accepted a proton.

$$\text{Brønsted base} \xrightarrow{+H^+} \text{conjugate Brønsted acid}$$

Every Brønsted base has its conjugate acid and every Brønsted acid its conjugate base (Table 14.3). The acid and its conjugate base (or the base and its conjugate acid) are called a **conjugate acid-base pair**. Note that a conjugate acid is a Brønsted acid, just like any other acid, and a conjugate base is a Brønsted base. They are called conjugate to emphasize the fact that they are related to each other, *not* because they show any difference in behavior from other acids or bases.

▼ EXAMPLE 14.1 Writing the formulas for conjugate acids and bases

Write the formula for (a) the conjugate acid of NH_2NH_2; (b) the conjugate base of OH^-; (c) the conjugate acid of HPO_4^{2-}; (d) the conjugate base of HPO_4^{2-}.

STRATEGY From the definitions, we know that a conjugate acid has one more proton than the base has and that a conjugate base has one less proton than the acid has. We must remember that *one proton* means "an H^+ ion," so we must add (or remove) one H from the formula of the starting material *and* one positive charge.

SOLUTION (a) The formula of the conjugate acid of NH_2NH_2 must have one more proton; hence it is $NH_2NH_3^+$. (b) The formula of the conjugate base of OH^- must have one less proton, so we remove one H and reduce the charge by one unit, obtaining the formula O^{2-}. (c) The formula of the conjugate acid of HPO_4^{2-} must have one more H^+, so we add an H and increase the charge number by one unit, obtaining $H_2PO_4^-$. (d) The formula of the conjugate base of HPO_4^{2-} must have one less H^+, so we remove an H and reduce the charge by one unit, obtaining PO_4^{3-}.

U Exercise 14.71.

EXERCISE E14.1 Write chemical formulas for (a) the conjugate acids of CH_3NH_2 and CN^- and (b) the conjugate bases of HSO_4^- and HI.
[*Answer*: (a) $CH_3NH_3^+$ and HCN; (b) SO_4^{2-} and I^-] ▲

TABLE 14.3 Conjugate acid-base pairs arranged by strength

Acid		Base	
Name	**Formula**	**Formula**	**Name**
perchloric acid	$HClO_4$	ClO_4^-	perchlorate ion
sulfuric acid	H_2SO_4	HSO_4^-	hydrogensulfate ion
hydroiodic acid	HI	I^-	iodide ion
hydrobromic acid	HBr	Br^-	bromide ion
hydrochloric acid	HCl	Cl^-	chloride ion
nitric acid	HNO_3	NO_2^-	nitrate ion
hydronium ion	H_3O^+	H_2O	*water*
hydrogensulfate ion	HSO_4^-	SO_4^{2-}	sulfate ion
hydrofluoric acid	HF	F^-	fluoride ion
nitrous acid	HNO_2	NO_2^-	nitrite ion
acetic acid	CH_3COOH	$CH_3CO_2^-$	acetate ion
carbonic acid	H_2CO_3	HCO_3^-	hydrogencarbonate ion
hydrosulfuric acid	H_2S	HS^-	hydrogensulfide ion
ammonium ion	NH_4^+	NH_3	ammonia
hydrocyanic acid	HCN	CN^-	cyanide ion
hydrogencarbonate ion	HCO_3^-	CO_3^{2-}	carbonate ion
methylammonium ion	$CH_3NH_3^+$	CH_3NH_2	methylamine
water	H_2O	OH^-	*hydroxide ion*
ammonia	NH_3	NH_2^-	amide ion
hydrogen	H_2	H^-	hydride ion
methane	CH_4	CH_3^-	methide ion
hydroxide ion	OH^-	O^{2-}	oxide ion

Strong acids (perchloric acid to nitric acid); Increasing acid strength in water (upward, hydroxide ion to hydrogensulfate ion); Increasing base strength in water (downward, water to hydroxide ion); Strong bases (amide ion to oxide ion).

All the equilibria discussed so far in this section have the form

$$\text{Acid}_1 + \text{base}_2 \rightleftharpoons \text{acid}_2 + \text{base}_1$$

(Conjugate: base_2 and acid_2; Conjugate: Acid_1 and base_1)

where acid_2 is the conjugate acid of base_2 and base_1 is the conjugate base of acid_1. It follows that the composition of the solutions are described by the reaction quotient

$$Q_c = \frac{[\text{acid}_2][\text{base}_1]}{[\text{acid}_1][\text{base}_2]}$$

Proton transfer is so fast that a system containing acid and base is always at equilibrium, so the reaction quotient always has its equilibrium value K_c. For aqueous acetic acid, we have

$$\underset{\text{Acid}_1}{CH_3COOH(aq)} + \underset{\text{Base}_2}{H_2O(l)} \rightleftharpoons \underset{\text{Acid}_2}{H_3O^+(aq)} + \underset{\text{Base}_1}{CH_3CO_2^-(aq)}$$

(Conjugate: Base_2 and Acid_2; Conjugate: Acid_1 and Base_1)

T OHT: Table 14.3.

N The conjugate bases of strong acids (upper right in Table 14.3) are all approximately equally weak, and the conjugate acids of strong bases (lower left) are all approximately equally weak. They are so weak that neither group is considered a base or an acid, respectively, in aqueous solution. Strong acids can be ranked in strength by considering their proton donating powers in solvents other than water.

N All Brønsted equilibria are consistently written according to this format (with the acid first on each side of the arrows).

U Exercises 14.72 and 14.73.

and the concentrations are related by

$$\frac{[H_3O^+][CH_3CO_2^-]}{[CH_3COOH][H_2O]} = K_c$$

where $CH_3COOH/CH_3CO_2^-$ and H_3O^+/H_2O are the two conjugate acid-base pairs. Similarly, for aqueous ammonia,

$$\underset{\text{Acid}_1}{H_2O(l)} + \underset{\text{Base}_2}{NH_3(aq)} \rightleftharpoons \underset{\text{Acid}_2}{NH_4^+(aq)} + \underset{\text{Base}_1}{OH^-(aq)}$$

Conjugate (Base₂ ↔ Acid₂)
Conjugate (Acid₁ ↔ Base₁)

where NH_4^+/NH_3 and H_2O/OH^- are now the two conjugate acid-base pairs. The concentrations satisfy

$$\frac{[NH_4^+][OH^-]}{[H_2O][NH_3]} = K_c$$

Now we have the link we need with the ideas of Chapter 13: the properties of acids and bases in solution can be described in terms of equilibrium constants.

Not only molecules but also ions may function as Brønsted acids and bases (Fig. 14.3). The ions of a salt in water may function either as proton donors to water, thereby forming H_3O^+ and producing an acidic solution, or as proton acceptors from water, thereby producing OH^- and a basic solution. An acetate ion is a Brønsted base (because it can accept a proton), so an aqueous solution of acetate ions is a solution of a base. A solution of sodium acetate, for example, is basic; and the equilibrium

N We shall establish in Chapter 15 that an $Na^+(aq)$ ion is an extremely weak acid, so the ability of sodium acetate to act as a base stems from the more powerful base character of the acetate ion. Similar remarks apply to NH_4^+ and Cl^-, the latter being an extremely weak base.

$$H_2O(l) + CH_3CO_2^-(aq) \rightleftharpoons CH_3COOH(aq) + OH^-(aq)$$

is established when the salt is dissolved in water. An ammonium ion is a Brønsted acid (because it can donate protons), so an aqueous solution of an ammonium salt is a solution of an acid; and the proton transfer equilibrium

$$NH_4^+(aq) + H_2O(l) \rightleftharpoons H_3O^+(aq) + NH_3(aq)$$

FIGURE 14.3 Cations and anions that are Brønsted acids and bases in water are shown by the colors of the indicators for four aqueous solutions: from left to right, NH_4^+ (pH = 5.1); $CH_3CO_2^-$ (pH = 8.9); CO_3^{2-} (pH = 11.6); PO_4^{3-} (pH = 12.7). All concentrations are 0.10 mol/L.

will be established. This feature of salt solutions will be covered in detail in Section 15.1.

A summary of the remarks made so far is that *acids are molecules or ions that can donate protons to other molecules;* in aqueous solution they generate H_3O^+ ions by proton transfer to water molecules. *Bases are molecules or ions that can accept protons from other molecules;* in aqueous solution they generate OH^- ions by accepting protons from water molecules.

It is helpful to note that the ions that most commonly produce H_3O^+ are cations, such as NH_4^+, because their positive charge aids the loss of a proton. Ions that most commonly produce OH^- are anions, such as $CH_3CO_2^-$, because their negative charge aids their role as proton acceptors. The strengths with which molecules and ions can donate or accept protons are discussed in more detail in Section 15.1.

Autoionization. According to the Brønsted-Lowry theory, the H_2O molecule can act both as a base (when it accepts a proton from an acid) and as an acid (when it donates a proton to a base). Because it can act as both a proton donor and a proton acceptor, water is said to be **amphiprotic**.

E The name amphiprotic comes from the Greek prefix *amphi-*, which means "of both kinds."

E Many species are amphiprotic; for example, the hydrogensulfate ion, HSO_4^-, can donate a proton to form SO_4^{2-}, and accept one, to form H_2SO_4.

? Can you think of other amphiprotic species?

E Other protic solvents also undergo autoionization to a small extent: they include ammonia (to NH_4^+ and NH_2^-) and anhydrous acetic acid (to $CH_3C(OH)_2^+$ and $CH_3CO_2^-$).

The amphiprotic character of water means that it can take part in proton transfer even in the absence of another acid or base. Thus, even in pure water one H_2O molecule (acting as an acid) can donate a proton to another H_2O molecule (acting as a base):

$$\underset{\text{Acid}_1}{H_2O(l)} + \underset{\text{Base}_2}{H_2O(l)} \rightleftharpoons \underset{\text{Acid}_2}{H_3O^+(aq)} + \underset{\text{Base}_1}{OH^-(aq)}$$

This type of proton transfer equilibrium, in which a substance ionizes itself, is called **autoionization** (Fig. 14.4). We shall see that autoionization plays a special role in the properties of acids and bases in water.

The reaction quotient for the autoionization of water is

$$Q_c = \frac{[H_3O^+][OH^-]}{[H_2O]^2}$$

As explained in Section 13.2, [X] is the molar concentration relative to a standard molar concentration. For solutes, that standard is 1 mol/L. In dilute aqueous solutions (the only ones we consider), the solvent water is very nearly pure, so it is convenient to take the standard as the molar concentration of pure water. Then, to a very good approximation, we can write

$$[H_2O] = \frac{\text{molar concentration of } H_2O}{\text{molar concentration of pure water}} \approx 1$$

in all reaction quotients. It follows that the autoionization reaction quotient given above simplifies to $[H_3O][OH^-]$ and that, at equilibrium,

$$[H_3O^+][OH^-] = K_w$$

where K_w is the **water autoionization constant** (or the ion-product constant for water). *At equilibrium, the product of the concentrations of the H_3O^+ and OH^- ions must always be equal to K_w.* This relation is true even if either ion concentration is modified by adding acid (which increases $[H_3O^+]$) or by adding base (which increases $[OH^-]$).

At 25°C, the equilibrium concentration of hydronium ions in pure water is 1.0×10^{-7} mol/L. From the stoichiometry of the equilibrium,

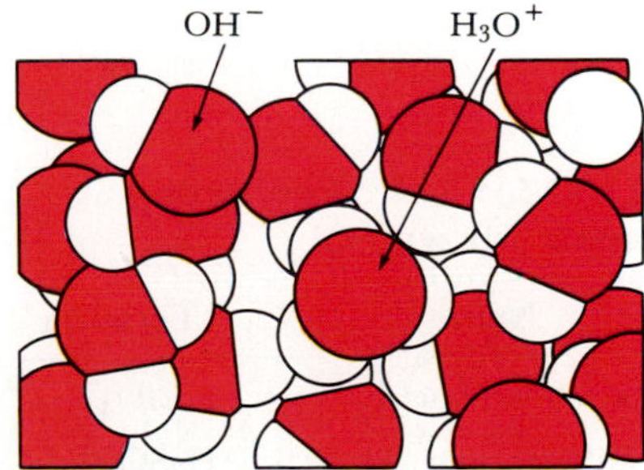

FIGURE 14.4 A sample of water is not a purely molecular liquid but also contains hydronium ions and hydroxide ions that result from its autoionization. Oxygen atoms, red; hydrogen atoms, white.

each H_3O^+ ion that is formed is accompanied by the formation of an OH^- ion, so $[H_3O^+] = [OH^-]$ in pure water and $[OH^-] = 1.0 \times 10^{-7}$ mol/L too. Therefore, at 25°C,

$$K_w = (1.0 \times 10^{-7}) \times (1.0 \times 10^{-7}) = 1.0 \times 10^{-14}$$

The value $K_w = 1.0 \times 10^{-14}$ applies to any aqueous solution at equilibrium at 25°C, whatever the individual concentrations of H_3O^+ and OH^-. Proton transfer is so fast that we can always be confident that it has reached equilibrium. That the solution is at equilibrium means that if the concentration of hydronium ions is increased by the addition of an acid, then the concentration of hydroxide ions must decrease so that the product $[H_3O^+] \times [OH^-]$ continues to equal 1.0×10^{-14}.

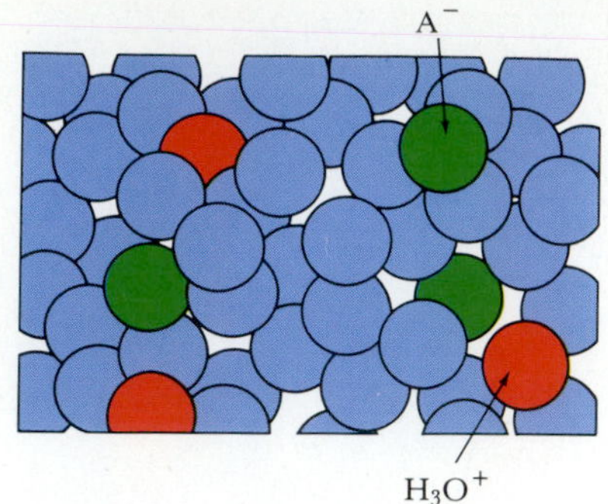

FIGURE 14.5 In a solution of a strong acid, all the ionizable hydrogen atoms are present as hydronium ions (red) and the solution contains no nonionized acid molecules. Water molecules, blue; anion of strong acid, green.

N The value of K_w depends on the temperature; see Exercises 14.7 to 14.9 and 14.74.

? Would you expect K_w to increase or decrease when the temperature is increased?

U Exercise 14.74.

N The properties of solutions of strong acids, strong bases, and mixtures of them are covered in detail in this chapter before the quantitative aspects of weak acids and bases are covered later in this chapter and more thoroughly in Chapter 15.

T OHT: Tables 14.4 and 14.5.

D Color acid-base indicators. BZS3, 8.1.

EXAMPLE 14.2 Calculating the hydroxide concentration from a hydronium ion concentration

The hydronium ion concentration in raw carrot juice is 1.0×10^{-5} mol/L. What is the hydroxide ion concentration?

STRATEGY The value of $[H_3O^+]$ is given and we know the value of K_w of water at 25°C, so we can calculate the value of $[OH^-]$ by rearranging the expression $[H_3O^+][OH^-] = K_w$. When calculations are carried out with K_w, molar concentrations are always taken to be in moles per liter.

SOLUTION The expression for K_w is rearranged to

$$[OH^-] = \frac{K_w}{[H_3O^+]} = \frac{1.0 \times 10^{-14}}{1.0 \times 10^{-5}} = 1.0 \times 10^{-9}$$

That is, the concentration of hydroxide ions in the juice is 1.0×10^{-9} mol/L.

EXERCISE E14.2 The concentration of hydroxide ions in household ammonia is about 3.2×10^{-3} mol/L. What is the molar concentration of hydronium ions in the solution?

[*Answer*: 3.2×10^{-12} mol/L]

SOLUTIONS OF STRONG ACIDS AND BASES

Solutions of strong acids and bases, which include the acids hydrochloric acid, nitric acid, and sulfuric acid and the bases sodium hydroxide and potassium hydroxide, are used in many laboratory procedures. Sulfuric acid is used extensively in industry and ranks first in worldwide production by a considerable margin. In 1990, U.S. production of sulfuric acid exceeded 40 billion kilograms; the production of the second ranked chemical, nitrogen, was more than 26 billion kilograms. Sodium hydroxide, a common strong base, ranked eighth; its total production exceeded 10 billion kilograms.

TABLE 14.4 The strong acids (in water)

Name	Formula
perchloric acid	$HClO_4$
sulfuric acid	H_2SO_4 (to HSO_4^-)
hydroiodic acid	HI
hydrochloric acid	HCl
hydrobromic acid	HBr
nitric acid	HNO_3

14.3 STRONG ACIDS AND BASES

In Brønsted-Lowry terms, a strong acid HA transfers its proton to water to form H_3O^+ and A^-, and no molecular HA remains in solution (Fig. 14.5). Essentially the proton transfer equilibrium

$$HA(aq) + H_2O(l) \rightleftharpoons H_3O^+(aq) + A^-(aq)$$

lies completely in favor of the products. It follows that a 0.1 M $HA(aq)$ solution consists of a solution of H_3O^+ ions and A^- ions, with each kind of ion present at a concentration of 0.1 mol/L and virtually no HA molecules present. Only a few acids are strong; they are all listed in Table 14.4.

It is important to distinguish between a *concentrated acid* and a *strong acid. Concentration* refers to the amount of acid per liter; *strength* refers to the proton-donating power of the acidic species. A strong acid, such as hydrochloric acid or perchloric acid, may be concentrated (for example, 10 M) or dilute (for example, 0.01 M). Whichever the case, the acid will be almost fully ionized and, hence, strong.

For bases, *strength* refers to the proton-accepting power of the basic species. Strong bases are substances for which the proton transfer equilibrium lies completely in favor of OH^- ions in solution. All the molecules or ions of a strong base B accept a proton from water to form OH^-, and the equilibrium

$$H_2O(l) + B(aq) \rightleftharpoons HB^+(aq) + OH^-(aq)$$

lies almost fully in favor of the products (Fig. 14.6). For example, almost all amide ions, NH_2^-, when added to water (as sodium amide, for instance) accept protons and produce OH^-:

$$H_2O(l) + NH_2^-(aq) \longrightarrow NH_3(aq) + OH^-(aq)$$

and there are very few amide ions present in solution. The addition of sodium oxide, Na_2O, to water produces a strongly basic solution because the oxide ion accepts a proton from water to form OH^- ions; the concentration of oxide ions in solution is very close to zero. Several soluble strong bases are listed in Table 14.5.

The concentration of OH^- ions in a solution of a strong base is known once we know the concentration of the base added to the solution and the number of OH^- ions each formula unit supplies. For example, because each formula unit of $Ba(OH)_2$ provides two OH^- ions, the hydroxide ion concentration in 0.020 M $Ba(OH)_2(aq)$ will be 2×0.020 mol/L = 0.040 mol/L.

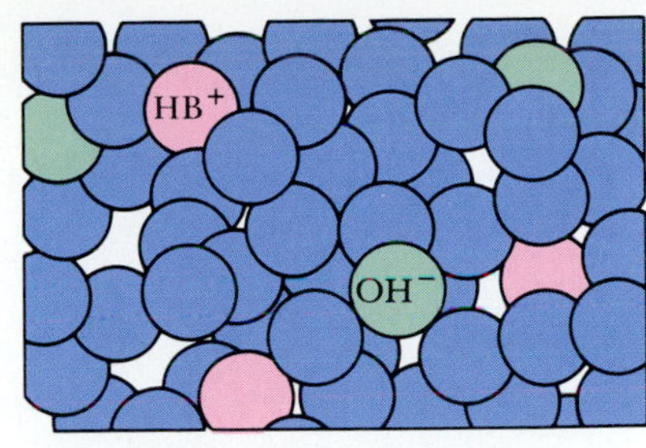

FIGURE 14.6 In a solution of a strong base, all the base molecules have accepted protons from water (blue) and are present as HB^+ ions (pink). Hydroxide ion, green.

N Most metal hydroxides—for example, $Mg(OH)_2$ and $Fe(OH)_3$—are fully ionized in aqueous solution and thus are properly considered strong bases (more precisely, are considered the *source* of the strong base OH^-). However, because they are only sparingly soluble they are not listed in Table 14.5.

▼ EXAMPLE 14.3 Calculating the concentrations of ions in a solution of a strong acid

What are the molar concentrations of H_3O^+, OH^-, and NO_3^- in a 0.0030 M $HNO_3(aq)$ solution at 25°C?

STRATEGY First we decide whether nitric acid is a strong acid by referring to Table 14.4. If it is, then we know that in aqueous solution it is present only as H_3O^+ and NO_3^-. That being the case, the concentrations of H_3O^+ and NO_3^- will be the same as the original concentration of HNO_3. Then, to find the concentration of OH^- ions, we use the water autoionization constant and set $[H_3O^+][OH^-] = K_w$.

SOLUTION Nitric acid is a strong acid (Table 14.4), so it is fully ionized in aqueous solution. The concentration of the acid is 0.0030 M, and the ionization is

$$HNO_3(aq) + H_2O(l) \longrightarrow H_3O^+(aq) + NO_3^-(aq)$$

so we know that in solution $[H_3O^+] = [NO_3^-] = 0.0030$ mol/L. Then, for

TABLE 14.5 The soluble strong bases (in water)

Group	Compounds	Examples
I	hydroxides	NaOH, KOH
	oxides*	Na_2O
	amides	KNH_2
II	hydroxides (except Be)	$Sr(OH)_2$, $Ba(OH)_2$
	oxides*	CaO, BaO
	amides	$Ca(NH_2)_2$

*In water, the oxide ion in the compound is converted to OH^-.

the concentration of hydroxide ions, we write

$$[OH^-] = \frac{K_w}{[H_3O^+]} = \frac{1.0 \times 10^{-14}}{0.0030} = 3.3 \times 10^{-12}$$

That is, the concentration of OH^- ions is only 3.3×10^{-12} mol/L.

EXERCISE E14.3 Determine the molar concentrations of H_3O^+, OH^-, and I^- in a 6.0×10^{-5} M HI(*aq*) solution.

▲ [*Answer*: $[H_3O^+] = [I^-] = 6.0 \times 10^{-5}$ mol/L; $[OH^-] = 1.7 \times 10^{-10}$ mol/L]

▼ **EXAMPLE 14.4 Calculating the concentrations of ions in a solution of a strong base**

Calculate the concentrations of H_3O^+, OH^-, NH_3, and K^+ in 0.010 M $KNH_2(aq)$ at 25°C. The amide ion, NH_2^-, is a strong base in water.

STRATEGY First, we note that, because KNH_2 is a salt, its solution consists of K^+ and NH_2^- ions, each at a concentration of 0.010 mol/L. However, NH_2^- is a strong base, so almost all the NH_2^- ions accept protons from water to form NH_3 molecules and OH^- ions. To find the concentration of hydronium ions, we use the water autoionization constant $[H_3O^+][OH^-] = K_w$.

SOLUTION The K^+ ion is neither a proton donor nor a proton acceptor, so its concentration remains unchanged at 0.010 mol/L. Because NH_2^- is a strong base (Table 14.5), in solution the equilibrium

$$H_2O(l) + NH_2^-(aq) \rightleftharpoons NH_3(aq) + OH^-(aq)$$

favors products almost completely. Therefore, the concentrations of the various species in solution are

$$[OH^-] = [NH_3] = 0.010 \text{ mol/L}$$

$$[H_3O^+] = \frac{K_w}{[OH^-]} = \frac{1.0 \times 10^{-14}}{0.010} = 1.0 \times 10^{-12}$$

That is, the concentration of H_3O^+ ions is 1.0×10^{-12} mol/L.

EXERCISE E14.4 Determine the concentrations of H_3O^+, OH^-, NH_3, and K^+ in a 6.0×10^{-5} M $KNH_2(aq)$ solution.

[*Answer*: $[K^+] = [OH^-] = [NH_3] = 6.0 \times 10^{-5}$ mol/L;
▲ $[H_3O^+] = 1.7 \times 10^{-10}$ mol/L]

N Note that NH_3 is the conjugate acid of NH_2^-, and hence, as an acid, appears first on the right of the arrows.

D Rainbow colors with mixed acid base indicators. BZS3, 8.2.

14.4 HYDROGEN ION CONCENTRATION AND pH

A knowledge of the molar concentration of hydrogen ions (more specifically, hydronium ions) in aqueous solutions is of the greatest importance in chemistry, biology, and medicine because it helps to explain the behavior of many reactions. In this section we shall see how to calculate this concentration for solutions of an acid or of a base. In Sections 14.5 and Section 15.2 we shall see how to do similar calculations for mixtures of the two. All these calculations are applications of the principles described in Chapter 13.

D Acid-base indicators and pH. CD1, 35.

The concentrations of H_3O^+ and OH^- ions vary over several orders of magnitude for the solutions we encounter in everyday life and in our laboratories. The concentrations of hydronium and hydroxide ions are usually no larger than 1 mol/L, but as we have seen in some of the examples, they may be as low as 10^{-12} mol/L, and in fact can be even smaller. As we saw in Chapter 3, chemists usually report hydrogen ion concentrations in terms of the pH of the solution, which is defined as

$$pH = -\log [H_3O^+]$$

where $[H_3O^+]$ is interpreted as the molar concentration of the hydronium ions relative to the standard concentration of 1 mol/L. (In Chapter 3, the pH was defined as $-\log [H^+]$, but we now know that all protons in aqueous solution are present as hydronium ions.) This definition indicates that *the higher the hydronium concentration (the more acidic the solution), the lower the pH.* Conversely, the lower the hydronium concentration (or, from $[H_3O^+][OH^-] = K_w$, the higher the concentration of hydroxide ions), the higher the pH. The definition also indicates that a change of one pH unit represents a change in the concentration of H_3O^+ ions by a factor of 10. For example, when the pH decreases from 5 to 4, the H_3O^+ concentration increases by a factor of 10, from 10^{-5} to 10^{-4} mol/L.

The pH scale was introduced by the Danish chemist Søren Sørensen in 1909 in the course of his work on quality control in the brewing of beer. It is now used throughout chemistry, biochemistry, environmental chemistry, geology, industrial chemistry, medicine, and agriculture. The pH of our blood must be maintained at a particular value to ensure the proper functioning of many biological processes.

N The *relative* molar concentration is used in the definition of pH because a logarithm cannot be taken of a quantity with units.

N The molar concentrations of H_3O^+ and OH^- are connected like two people on a see-saw: if one goes up, the other must go down.

EXAMPLE 14.5 Calculating a pH

What is the pH of (a) human blood, in which the hydronium ion concentration is 4.0×10^{-8} mol/L; (b) 0.020 M HCl(*aq*); (c) 0.040 M KOH(*aq*)?

STRATEGY The definition of pH is $-\log [H_3O^+]$; so once we know the H_3O^+ concentration (in moles per liter), we can calculate the pH simply by taking the logarithm (to the base 10) and changing the sign. For strong monoprotic (one proton) acids, the molar concentration of hydronium ions is equal to the molar concentration of the acid. For strong bases, the molar concentration of hydroxide ions is equal to the molar concentration of the base multiplied by the number of OH^- ions per formula unit. The conversion to $[H_3O^+]$ from $[OH^-]$ is made by using the relation $[H_3O^+][OH^-] = K_w$.

SOLUTION (a) For a solution in which $[H_3O^+] = 4.0 \times 10^{-8}$ mol/L,

$$pH = -\log (4.0 \times 10^{-8}) = 7.40$$

(b) Because HCl is a strong acid, $[H_3O^+] = 0.020$ mol/L. Hence,

$$pH = -\log 0.020 = 1.70$$

(c) Each formula unit of KOH (a strong base) provides one OH^- ion; therefore $[OH^-] = 0.040$ mol/L and

$$[H_3O^+] = \frac{K_w}{[OH^-]} = \frac{1.0 \times 10^{-14}}{0.040} = 2.5 \times 10^{-13}$$

Hence

$$pH = -\log (2.5 \times 10^{-13}) = 12.60$$

(When a logarithm is calculated, the number of significant figures in the data—two in this case—determines the number of significant digits that *follow* the decimal point; see Appendix 1B.)

EXERCISE E14.5 Calculate the pH of (a) household ammonia, for which the hydronium ion concentration is about 3×10^{-12} mol/L; (b) 6.0×10^{-5} M $HClO_4$; (c) 0.077 M NaOH(*aq*).

[*Answer:* (a) 11.5; (b) 4.22; (c) 12.89]

E Electrodes are now available that measure pCl, pNa, pNH_4, etc. Bioelectrodes can be used to monitor the concentrations of nonelectrolytes, such as glucose in blood. See the article in *Sci. Amer.*, **1991**, August, 48.

T OHT: Fig. 14.7.

U Exercises 14.75–14.77.

? Why is the pH of milk less than 7? Which would you expect to taste more sour, tomato juice or soda?

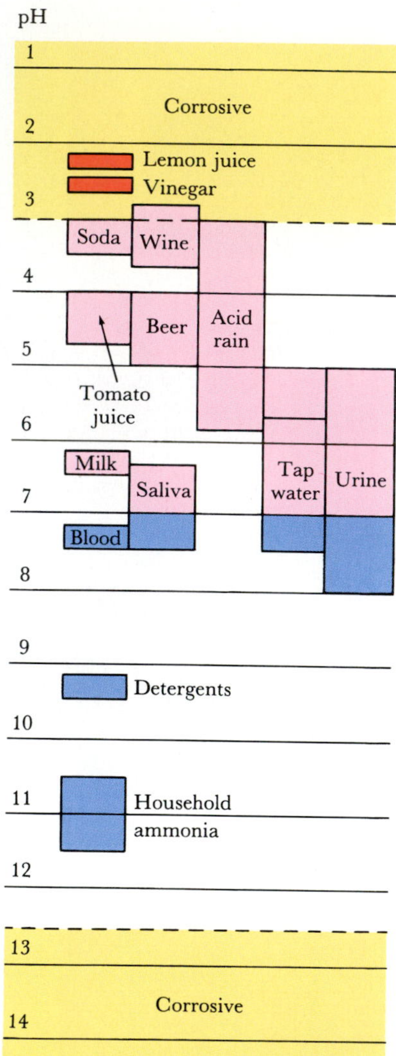

FIGURE 14.7 Typical pH values of common aqueous solutions. The dotted lines indicate the pHs beyond which solutions are regarded as corrosive.

Many equations involving acids and bases are greatly simplified by expressing $[H_3O^+]$,$[OH^-]$, and K_w in terms of their logarithms. That is, chemists have introduced the quantity pX as a generalization of pH, taking it to signify the negative logarithm of X:

$$\text{pX} = -\log \text{X}$$

For example, by pK_w we mean

$$pK_w = -\log K_w = -\log(1.0 \times 10^{-14}) = 14.00$$

As an illustration of the use of pX, if we take logarithms of both sides of the expression $[H_3O^+][OH^-] = K_w$ and use the fact that $\log ab = \log a + \log b$, we obtain

$$\log [H_3O^+] + \log [OH^-] = \log K_w$$

Then, multiplication of both sides of the equation by -1 gives

$$-\log [H_3O^+] + (-\log [OH^-]) = -\log K_w$$

which is the same as

$$\text{pH} + \text{pOH} = pK_w$$

Because $pK_w = 14.00$ at 25°C, at that temperature

$$\text{pH} + \text{pOH} = 14.00$$

This equation could have been used in place of the procedure we adopted in Example 14.5(c): because $[OH^-] = 0.040$ mol/L for the solution treated there, pOH $= -\log(0.040) = 1.40$, and hence pH $= 14.00 - 1.40 = 12.60$.

EXAMPLE 14.6 Calculating the pH of a solution of a strong base

Calculate the pH of a 0.010 M $Sr(OH)_2(aq)$ solution at 25°C.

STRATEGY First, we confirm that the base is strong by referring to Table 14.5. For a strong base, the value of $[OH^-]$ is equal to the molar concentration of the base in solution multiplied by the number of OH^- ions per formula unit. Then we calculate $[OH^-]$; and from that value, we calculate pOH by taking the negative logarithm. Once pOH is known, we calculate the pH from pH + pOH = 14.00.

SOLUTION Table 14.5 confirms that strontium hydroxide is a strong base in water. Each $Sr(OH)_2$ formula unit produces two OH^- ions, so $[OH^-] = 2 \times 0.010$ mol/L $= 0.020$ mol/L. Hence, we can conclude that pOH $= -\log(0.020) = 1.70$ and pH $= 14.00 - 1.70 = 12.30$.

EXERCISE E14.6 Calculate the pH of a 1.5×10^{-3} M $Ca(OH)_2$ aqueous solution.

[*Answer*: 11.48]

The pH of a solution can be estimated with a strip of universal indicator paper, which turns different colors at different pH values. It is instructive to test a selection of common liquids and beverages with the paper to assess their relative acidities (Fig. 14.7). More precise measurements are made with a pH meter (Fig. 14.8). This instrument consists of a voltmeter connected to two electrodes that dip into the solution. The potential difference across the electrodes is proportional to the pH (as explained in Chapter 17); so once the scale on the meter has been calibrated, the pH of a solution can be read directly.

We can convert a measured pH to a hydronium ion concentration by taking antilogarithms of the negative value of the pH (because pH = $-\log [H_3O^+]$):

$$[H_3O^+] = 10^{-pH} \text{ mol/L}$$

The pH of fresh orange juice is about 3.8, so

$$[H_3O^+] = 10^{-3.8} \text{ mol/L} = 2 \times 10^{-4} \text{ mol/L}$$

Fresh lemon juice has a pH of 2.5, corresponding to a hydrogen ion concentration of 3×10^{-3} mol/L; so lemon juice is more than ten times more acidic than orange juice. Lemon juice also tastes sharper because hydrogen ions stimulate the "sour" taste receptors on the surface of the tongue. The pH of pure water is 7.00 at 25°C, an observation implying that

$$[H_3O^+] = 10^{-7.00} \text{ mol/L} = 1.0 \times 10^{-7} \text{ mol/L}$$

We conclude that, of the 56 mol of H_2O molecules in 1 L of water, only 10^{-7} mol, or 1 in 560 million, is ionized. This low concentration of ions explains why pure water is a very poor conductor of electricity. (Its reputation as a conductor stems from the ions present in impure water.)

Water marks the dividing line between acids and bases. Because its pH is 7 at 25°C and because, by definition, pH increases as the hydrogen ion concentration decreases, pH enables us to classify solutions as acidic, neutral, or basic (see margin). The Environmental Protection Agency (EPA) defines waste as **corrosive** if its pH is either lower than 3.0 (highly acidic) or higher than 12.5 (highly basic).

(a)

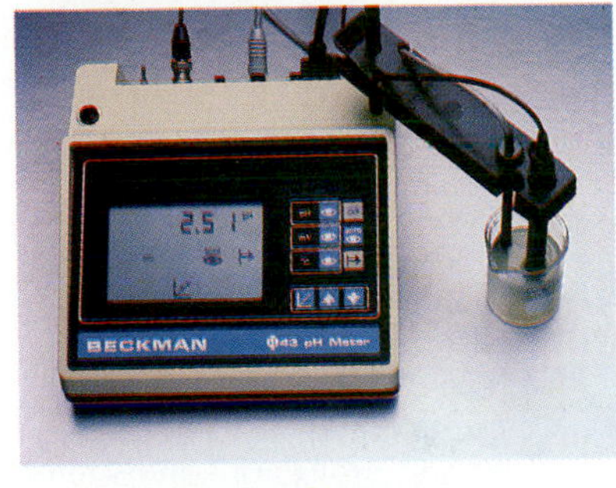

(b)

FIGURE 14.8 A pH meter is a voltmeter that measures the potential difference between the two electrodes immersed in the solution. The display gives the pH directly. The samples are (a) orange juice and (b) lemon juice.

14.5 MIXTURES OF STRONG ACIDS AND BASES

We saw in Section 14.1 that the Brønsted-Lowry theory summarizes all acid-base neutralizations in aqueous solution by the net ionic equation

$$H_3O^+(aq) + OH^-(aq) \longrightarrow 2H_2O(l)$$

We now know that all strong acids produce H_3O^+ ions and that all strong bases produce OH^- ions. When a solution that contains H_3O^+ ions (an acid) is mixed with a second solution containing an equal amount of OH^-, complete neutralization occurs, water is the product, and the pH of the mixed solution is equal to 7. When solutions containing unequal amounts of H_3O^+ and OH^- ions are mixed, neutralization is incomplete, excess H_3O^+ or OH^- ions remain, and the pH is less than or greater than 7, respectively.

In the case of incomplete neutralization of a strong acid and strong base, we calculate the pH (or pOH) as follows:

1. Determine the amount (in moles) of the excess H_3O^+ (or OH^-) ions that remains in the solution.
2. Divide that amount by the *total* volume of the combined solutions, thereby obtaining the molar concentration of the H_3O^+ (or OH^-) ions in the solution.

Solution's pH	Solution type
>7	basic
7	neutral
<7	acidic

? The autoionization constant of liquid ammonia at −33°C is $pK_{am} = 33$. What corresponds to "acidic," "neutral," and "basic" in liquid ammonia?

D Set up a calibrated pH meter and perform this example as a demonstration.

3. Take the negative logarithm of the relative molar concentration of the excess H_3O^+ or OH^- ions to find the pH or the pOH. The pOH can be converted to pH by using the relation pH + pOH = 14.00.

▼ EXAMPLE 14.7 Calculating the pH of a strong acid-strong base mixture

Calculate the pH of a solution that results from the addition of 15.0 mL of 0.340 M NaOH(*aq*) to 25.0 mL of 0.250 M HCl(*aq*).

STRATEGY The reaction is between a strong base, NaOH, and a strong acid, HCl, both of which are fully ionized in solution, producing Na^+, OH^-, H_3O^+, and Cl^- ions. The hydronium ions provided by the acid react with the hydroxide ions provided by the base, and the reaction

$$H_3O^+(aq) + OH^-(aq) \longrightarrow 2H_2O(l)$$

occurs. This equation implies that 1 mol H_3O^+ reacts with 1 mol OH^-. Hence, we need to determine the amounts of H_3O^+ and OH^- ions supplied by each solution, decide which is the limiting reagent, and calculate the amount of the ion in excess that remains after the reaction. Then we divide that amount by the total volume of the solution and convert the resulting molar concentration to pH.

N The analysis of a strong acid-strong base mixture can be considered a limiting reactant problem.

N We can effectively ignore equilibria in solutions of strong acids and strong bases because proton transfer is virtually complete (equilibrium does exist, of course, but it is strongly in favor of proton transfer); moreover, the concentrations of H_3O^+ and OH^- ions formed by the added acid or base far exceed those arising from the autoionization of water.

SOLUTION The amounts of ions supplied by each solution are

Amount of H_3O^+ (from the acid, in moles)

$$= 0.0250 \text{ L} \times \frac{0.250 \text{ mol } H_3O^+}{1 \text{ L}} = 6.25 \times 10^{-3} \text{ mol } H_3O^+$$

Amount of OH^- (from the base, in moles)

$$= 0.0150 \text{ L} \times \frac{0.340 \text{ mol } OH^-}{1 \text{ L}} = 5.10 \times 10^{-3} \text{ mol } OH^-$$

We see that the H_3O^+ ions are in excess; and after the reaction of all the OH^- ions, the amount of H_3O^+ remaining is $(6.25 \times 10^{-3} - 5.10 \times 10^{-3})$ mol $H_3O^+ = 1.15 \times 10^{-3}$ mol H_3O^+. Because the total volume of the solution is 40.0 mL, or 0.0400 L, the molar concentration of hydronium ions after the reaction is

$$[H_3O^+] = \frac{1.15 \times 10^{-3} \text{ mol } H_3O^+}{0.0400 \text{ L}} = 2.88 \times 10^{-2} \text{ mol/L}$$

Hence,

$$\text{pH} = -\log (2.88 \times 10^{-2}) = 1.541$$

Calculations of pH are rarely sufficiently accurate to justify more than one figure after the decimal point (because interactions between ions affect the result), so we should report the answer as pH = 1.5.

U Exercises 14.78 and 14.79.

EXERCISE E14.7 What is the pH of a solution that results from the addition of 20.0 mL of 0.340 M NaOH(*aq*) to 25.0 mL of 0.250 M HCl(*aq*)?

[*Answer*: 12.1] ▲

In summary, to calculate the pH of a strong acid, a strong base, or a mixture of a strong acid and a strong base, we determine the amounts (in moles) of H_3O^+ or OH^- ions that exist in solution. In the case of a mixture of the two, we first allow for the neutralization reaction $H_3O^+(aq) + OH^-(aq) \rightarrow 2H_2O(l)$. The amount of H_3O^+ (or OH^-) is divided by the total volume of the solution to obtain the molar concentration of H_3O^+ (or OH^-), which is then converted to pH (if necessary, by using the relation pH + pOH = 14.00).

14.6 pH CURVES FOR STRONG ACID-STRONG BASE TITRATIONS

N We consider the term *stoichiometric point* to be more in the spirit of the discussion so far and more evocative of the type of calculation involved; the term also distances the text from the obsolete term "equivalence."

In Section 4.6 we introduced the volumetric analysis technique of titration for determining the compositions of mixtures and the concentrations of solutions. A titration makes use of the addition of a **titrant** (the solution in the buret) to an **analyte** (the solution being analyzed, which is contained in the receiving flask) until an **end point** (a color change of the indicator) is reached. The indicator is selected so that the end point occurs very close to the **stoichiometric point**, the point at which the reaction under investigation is complete.*

A **pH curve** is the plot of pH of the reaction mixture against the volume of titrant added to the analyte in an acid-base titration. The coordinates of the graph are pH (on the vertical axis) and volume of titrant (on the horizontal axis). The stoichiometric point is reached when stoichiometric amounts of acid and base are present in solution; for a strong acid and a strong base, this point occurs at pH = 7. To be specific, we shall consider the titration of aqueous sodium hydroxide with hydrochloric acid, but the discussion can be applied to a titration of any strong base with any strong acid. The titration of weak acids and weak bases is considered in Chapter 15.

▼ EXAMPLE 14.8 Calculating the pH before the stoichiometric point in a strong base-strong acid titration

N Proton transfer equilibria are taken into account explicitly only when the acid or base is weak.

Calculate the initial pH of 25.00 mL of 0.150 M NaOH(*aq*) and the pH after the addition of 5.00 mL of 0.100 M HCl(*aq*).

STRATEGY Because the sodium hydroxide is in excess in both cases, both solutions are basic; so we expect pH > 7. Because OH^- ions are in excess, a convenient approach is to calculate pOH first and then convert it to pH by using the relation pH + pOH = pK_w. The concentration of OH^- ions depends on the amount of OH^- present and the total volume of the solution (the sum of the initial volume of the basic solution and the volume of acid added). We can find the amount of OH^- ions in the solution after the addition of acid from the stoichiometry of the neutralization reaction and the volume of acid added.

SOLUTION The initial concentration of OH^- ions is 0.150 mol/L, a value corresponding to pOH = −log 0.150 = 0.824. The pH is therefore

$$\text{pH} = 14.00 - 0.824 = 13.18$$

This pH is point A in Fig. 14.9. The original 25.00 mL of 0.150 M NaOH(*aq*) contains

$$\text{Amount of OH}^- \text{ (mol)} = 0.025 \text{ L} \times \frac{0.150 \text{ mol OH}^-}{1 \text{ L}}$$
$$= 3.75 \times 10^{-3} \text{ mol OH}^-$$

The added 5.00 mL of 0.100 M HCl(*aq*) contains

$$\text{Amount of H}_3\text{O}^+ \text{ (mol)} = 5.00 \times 10^{-3} \text{ L} \times \frac{0.100 \text{ mol H}_3\text{O}^+}{1 \text{ L}}$$
$$= 5.00 \times 10^{-4} \text{ mol H}_3\text{O}^+$$

*The stoichiometric point is also widely called the equivalence point.

FIGURE 14.9 The variation of pH during the titration of a strong base, 25.00 mL of 0.150 M NaOH(*aq*), with a strong acid, 0.100 M HCl(*aq*). The stoichiometric point (S) occurs at pH = 7. See text for explanation of points A through F.

D Instrumental recording of a titration curve. BZS3, 8.27 (HCl as acid).

T OHT: Fig. 14.9.

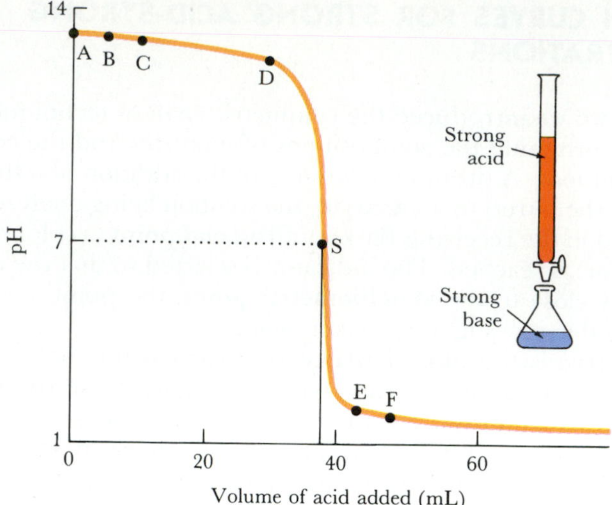

We know that 1 mol of H_3O^+ reacts with 1 mol of OH^-, so 5.00×10^{-4} mol H_3O^+ reacts with 5.00×10^{-4} mol OH^-, leaving

$$(3.75 \times 10^{-3} - 5.00 \times 10^{-4}) \text{ mol } OH^- = 3.25 \times 10^{-3} \text{ mol } OH^-$$

in a total volume of (25.00 + 5.00) mL = 30.0 mL. The molar concentration of OH^- ions is therefore

$$[OH^-] = \frac{3.25 \times 10^{-3} \text{ mol}}{30.0 \times 10^{-3} \text{ L}} = 0.108 \text{ mol/L}$$

This concentration corresponds to pOH = 0.967, so

$$pH = 14.00 - 0.967 = 13.03$$

or about 13.0. This pH is point B in Fig. 14.9.

EXERCISE E14.8 Calculate the pH after the addition of another 5.00 mL of acid.

[*Answer*: 12.90; point C in Fig. 14.9]

The pH is greater than 7 when the analyte is a solution of a strong base. In the titration treated in the example and illustrated in Fig. 14.9, we see that when some of the strong base is neutralized with the strong acid, the pH of the analyte is still greater than 7 (Fig. 14.9, point D), but less than its original pH. The pH continues to decrease as strong acid is added; and at the stoichiometric point (S), it has reached pH = 7. In the titration treated in the example, this point occurs after the addition of 37.5 mL of acid. Recall that the amount of H_3O^+ added must equal the amount of OH^- initially present; so we must add 3.75×10^{-3} mol H_3O^+, which is present in

$$3.75 \times 10^{-3} \text{ mol } H_3O^+ \times \frac{1 \text{ L}}{0.100 \text{ mol } H_3O^+} \times \frac{1 \text{ mL}}{10^{-3} \text{ L}} = 37.5 \text{ mL}$$

In practice, most titrations are halted at the stoichiometric point. However, it is instructive to plot the pH curve as more acid is added.

N Consider using a spreadsheet to calculate and plot the pH after 1-mL additions of 0.100 M HCl(*aq*) are added to 0.150 M NaOH(*aq*). This task could be assigned as a student project.

U Exercises 14.80 and 14.81 parallel Exercise E14.9.

N The principles involved in the selection of indicators (see Fig. 14.10) are described in Chapter 15.

After the stoichiometric point, all the added H_3O^+ ions survive as further acid is supplied, and the pH falls below 7. All the hydronium ions supplied contribute to the pH, but their concentration is reduced because they are present in a larger volume. The actual pH depends on the amount of excess acid added and the total volume of the solution.

(a) Basic

(b) Neutral

(c) Acidic

FIGURE 14.10 The stoichiometric point of a titration of a strong base (the analyte) and a strong acid (the titrant) can be readily detected with an indicator. Three stages are shown here: (a) close to, (b) at, and (c) just after the stoichiometric point. Thymol blue is the indicator.

▼ EXAMPLE 14.9 Calculating the pH after the stoichiometric point of a strong base-strong acid titration

Calculate the pH of the solution in Example 14.8 after 5.00 mL of the acid has been added beyond the stoichiometric point.

STRATEGY We need to find the molar concentration of H_3O^+ ions in the solution after the stoichiometric point has been passed and then convert that concentration to pH. The H_3O^+ ions are in a total volume consisting of the volume of the analyte plus the volume of added titrant. The calculation of the concentration requires two steps: the determination of the amount of H_3O^+ ions in solution after the stoichiometric point and the determination of the total volume of the solution.

SOLUTION The amount of H_3O^+ ions in the 5.00 mL of 0.100 M HCl(*aq*) added beyond the stoichiometric point is

$$\text{Amount of } H_3O^+ \text{ (mol)} = 5.00 \times 10^{-3}\text{ L} \times \frac{0.100 \text{ mol } H_3O^+}{1 \text{ L}}$$

$$= 5.00 \times 10^{-4} \text{ mol } H_3O^+$$

The total volume of the solution is the sum of the initial volume of the analyte (25.00 mL), the volume of titrant to reach the stoichiometric point (37.5 mL), and the volume of titrant beyond the stoichiometric point (5.00 mL): these sum to 67.5 mL. The hydronium ion concentration is therefore

$$[H_3O^+] = \frac{5.00 \times 10^{-4} \text{ mol}}{67.5 \times 10^{-3} \text{ L}} = 7.41 \times 10^{-3} \text{ mol/L}$$

It follows that the pH of the solution is

$$\text{pH} = -\log(7.41 \times 10^{-3}) = 2.130, \text{ or about } 2.1$$

This pH is point E in Fig. 14.9.

EXERCISE E14.9 Calculate the pH of the solution after the addition of another 5.00 mL of 0.100 M HCl(*aq*).

[*Answer*: 1.86; point F in Fig. 14.9] ▲

The shape of the pH curve in Fig. 14.9 is typical of all titrations in which a strong acid is added to a strong base. Initially the pH falls slowly. As the stoichiometric point is passed, there is a sudden fall through pH = 7 as the hydronium ion concentration increases sharply to a value typical of a diluted solution of the added acid. The pH then falls slowly toward the value of the acid itself as the dilution caused by the original analyte solution becomes less and less important. The sudden decrease in pH thus marks the stoichiometric point; it may be detected with either a pH meter or an indicator (Fig. 14.10).

The pH curve for the titration of a strong acid (the analyte) with a strong base (the titrant) mirrors that of a strong base with a strong acid.

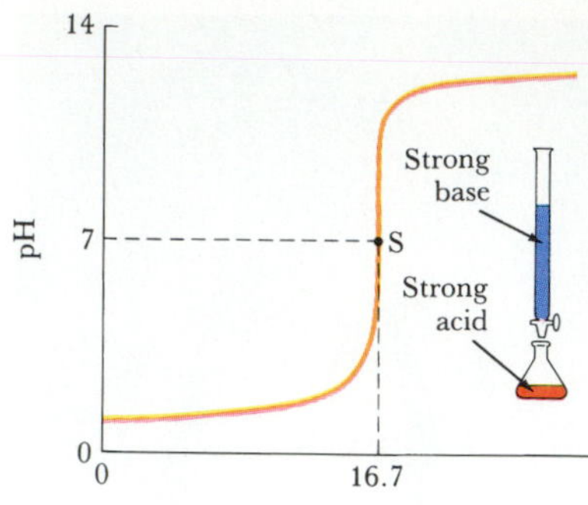

FIGURE 14.11 The variation of pH during the titration of a strong acid, 25 mL of 0.10 M HCl(*aq*), with a strong base, 0.15 M NaOH(*aq*). The stoichiometric point (S) occurs at pH = 7.

T OHT: Fig. 14.11.

N The same equilibrium table as was introduced in Chapter 13 is used here (see OHT of blank table).

E Protein molecules are very elaborate concatenations of weak acids and weak bases: they respond sensitively to the pH of their environment in a living cell. A part of the biochemical basis of life is sensitivity toward and sensitive control of pH.

As shown in (Fig. 14.11), the initial pH is low; it increases only slightly until just before the stoichiometric point; and then it increases sharply. The pH continues to increase after the stoichiometric point; but then it levels off on account of the presence of excess strong base in the solution, and gradually approaches the pH of the original strong base.

EQUILIBRIA IN SOLUTIONS OF ACIDS AND BASES

Strong acids and bases are fully ionized, so we have not needed to consider the existence of any equilibrium between conjugate acid-base pairs. However, we saw in Chapter 3 that most acids and bases are weak and that only a small proportion of their molecules are ionized in solution. For weak acids and bases, it is essential to consider the role of the dynamic equilibrium between the ionized and nonionized forms in solution. That is, to understand the behavior of these substances in solution, we need to combine the concepts of the Brønsted-Lowry theory of acids and bases with the concepts of equilibria described in Chapter 13. As in Chapter 13, we will be able to predict the composition of solutions of weak acids and bases by writing down the chemical equation for the appropriate Brønsted equilibrium, setting up the equilibrium table, and then using the reaction quotient and the equilibrium constant to calculate the quantity of interest (typically the pH but also the degree of ionization).

A key feature of the Brønsted-Lowry theory is its emphasis on the equilibrium between conjugate acids and bases. As noted earlier, proton transfer is so fast that a system containing an acid and a base is always at equilibrium; so as soon as a solution of a weak acid HA is prepared, the solution contains HA and its conjugate base A^- in their equilibrium concentrations. The stronger the acid, the greater the relative proportion of A^- ions (and the H_3O^+ ions that are formed by proton transfer to the water); however, as we shall see, the actual degree of ionization depends on the concentration of the solution. Similarly, a solution of a weak base B contains equilibrium concentrations of the base and its conjugate acid HB^+ virtually the instant it is prepared. The stronger the base, the greater the relative proportion of HB^+ ions (and the OH^- ions that are formed by proton transfer from the water); however, the actual degree of ionization of weak bases, like that of weak acids, depends on the concentration of the solution.

Most acids and bases that exist in nature are weak. For example, the natural acidity of a stream is generally due to the presence of carbonic acid (H_2CO_3, from dissolved CO_2), hydrogenphosphate ions (such as $H_2PO_4^-$, from fertilizer runoff), or acids arising from the natural degradation of plant tissue. Similarly, most naturally occurring bases are weak. They often arise from the decomposition of nitrogen-containing compounds in the absence of air. For example, the odor of dead fish is due to amines (compounds of the form R—NH_2, where R is an organic group of atoms, as in methylamine, CH_3NH_2), which produce basic solutions. (The use of a few drops of acidic lemon juice to a fish dish helps to eliminate the smell of these amines.)

Some weak bases are also important commercially, the most significant being ammonia, ranked sixth in U.S. production in 1990, with over 15 billion kilograms produced in that year. Urea (**1**), another base used as a fertilizer, ranks twelfth, with more than 7 billion kilograms produced.

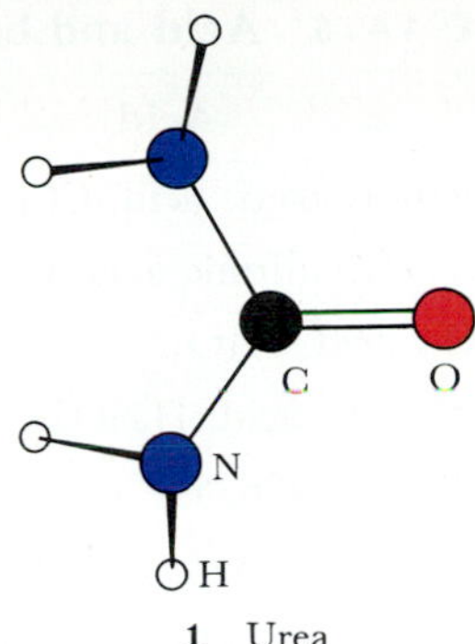

1 Urea

14.7 IONIZATION CONSTANTS

Many chemical reactions can be understood in terms of a competition between acids and bases of different strengths to donate or accept protons. Because so many reactions take place in water, an important aspect of this competition is the strength with which various acids and bases transfer protons to and from water.

Acid ionization constants. The proton transfer equilibrium for a weak acid HA in water is

$$HA(aq) + H_2O(l) \rightleftharpoons H_3O^+(aq) + A^-(aq) \qquad Q_c = \frac{[H_3O^+][A^-]}{[HA]}$$

For reasons discussed in Section 14.2, $[H_2O] = 1$ in all reaction quotients for dilute aqueous solutions. The equilibrium value of the reaction quotient is called the **acid ionization constant** K_a:

$$\frac{[H_3O^+][A^-]}{[HA]} = K_a \text{ at equilibrium}$$

N The term *dissociation constant* is still in widespread use as a synonym for acid ionization constant, but we consider it to be inappropriate because it suggests a kind of homolysis. An alternative name for K_a, which is gaining acceptance, is *acidity constant.*

For acetic acid, for example,

$$CH_3COOH(aq) + H_2O(l) \rightleftharpoons H_3O^+(aq) + CH_3CO_2^-(aq)$$

$$K_a = \frac{[H_3O^+][CH_3CO_2^-]}{[CH_3COOH]}$$

Similarly, for trichloroacetic acid,

$$CCl_3COOH(aq) + H_2O(l) \rightleftharpoons H_3O^+(aq) + CCl_3CO_2^-(aq)$$

$$K_a = \frac{[H_3O^+][CCl_3CO_2^-]}{[CCl_3COOH]}$$

The significance of K_a. The ionization constants of a number of weak acids are listed in Table 14.6. The point to note is that the larger the value of K_a, the stronger the acid; at a given concentration, the equilibrium lies more in favor of the ionized form the larger the value of K_a. Thus trichloroacetic acid, CCl_3COOH, which has $K_a = 0.30$, is stronger than acetic acid, for which $K_a = 1.8 \times 10^{-5}$. Acid ionization constants are commonly reported as their negative logarithms, by defining

$$pK_a = -\log K_a$$

Thus the pK_a of acetic acid is $-\log(1.8 \times 10^{-5}) = 4.74$ and that of trichloroacetic acid is 0.52. As these two examples show, the larger the acid ionization constant, the smaller the value of pK_a. Hence, the lower the value of pK_a, the stronger the acid.

Because CCl_3COOH is the stronger acid, $CCl_3CO_2^-$ must be the weaker conjugate base in the sense that it has a weaker tendency to accept a proton from the H_3O^+ and re-form the acid. Conversely, that CH_3COOH is the weaker acid indicates that the $CH_3CO_2^-$ ion (the conjugate base of CH_3COOH) has the greater tendency to accept a proton from H_3O^+ and re-form its conjugate acid. We can conclude

E Acid ionization constants can be used to discuss proton transfer equilibria in solutions of strong acids; K_a for such acids are greater than 1, and pK_a values for strong acids are negative.

TABLE 14.6 Acid and base ionization constants at 25°C*

Acid	K_a	pK_a	Base	K_b	pK_b
trichloroacetic acid, CCl_3COOH	3.0×10^{-1}	0.52	urea, NH_2CONH_2	1.3×10^{-14}	13.90
benzenesulfonic acid, $C_6H_5SO_3H$	2×10^{-1}	0.70	aniline, $C_6H_5NH_2$	4.3×10^{-10}	9.37
iodic acid, HIO_3	1.7×10^{-1}	0.77	pyridine, C_5H_5N	1.8×10^{-9}	8.75
sulfurous acid, H_2SO_3	1.6×10^{-2}	1.81	hydroxylamine, NH_2OH	1.1×10^{-8}	7.97
chlorous acid, $HClO_2$	1.0×10^{-2}	2.00	nicotine, $C_{10}H_{14}N_2$	1.0×10^{-6}	5.98
phosphoric acid, H_3PO_4	7.6×10^{-3}	2.12	morphine, $C_{17}H_{19}O_3N$	1.6×10^{-6}	5.79
chloroacetic acid, $CH_2ClCOOH$	1.4×10^{-3}	2.85	hydrazine, NH_2NH_2	1.7×10^{-6}	5.77
lactic acid, $CH_3CH(OH)COOH$	8.4×10^{-4}	3.08	ammonia, NH_3	1.8×10^{-5}	4.74
nitrous acid, HNO_2	4.3×10^{-4}	3.37	trimethylamine, $(CH_3)_3N$	6.5×10^{-5}	4.19
hydrofluoric acid, HF	3.5×10^{-4}	3.45	methylamine, CH_3NH_2	3.6×10^{-4}	3.44
formic acid, HCOOH	1.8×10^{-4}	3.75	dimethylamine, $(CH_3)_2NH$	5.4×10^{-4}	3.27
benzoic acid, C_6H_5COOH	6.5×10^{-5}	4.19	ethylamine, $C_2H_5NH_2$	6.5×10^{-4}	3.19
acetic acid, CH_3COOH	1.8×10^{-5}	4.74	triethylamine, $(C_2H_5)_3N$	1.0×10^{-3}	2.99
carbonic acid, H_2CO_3	4.3×10^{-7}	6.37			
hypochlorous acid, HClO	3.0×10^{-8}	7.53			
hypobromous acid, HBrO	2.0×10^{-9}	8.69			
boric acid, $B(OH)_3$†	7.2×10^{-10}	9.14			
hydrocyanic acid, HCN	4.9×10^{-10}	9.31			
phenol, C_6H_5OH	1.3×10^{-10}	9.89			
hypoiodous acid, HIO	2.3×10^{-11}	10.64			

(Increasing acid strength: upward arrow beside acid column. Increasing base strength: downward arrow beside base column.)

*Values for polyprotic acids—those capable of donating more than one proton—refer to the first ionization.
†The proton transfer equilibrium is $B(OH)_3(aq) + 2H_2O(l) \rightleftharpoons H_3O^+(aq) + B(OH)_4^-(aq)$.

T OHT: Table 14.6.

E Most weak molecular bases are organic derivatives of ammonia, such as amines and derivatives of pyridine; all accept a proton at the N: site.

that *the stronger the Brønsted acid, the weaker its conjugate base.* Thus, hydrochloric acid is a strong acid and its conjugate base, Cl^-, has very little tendency to accept protons from H_3O^+; on the other hand, acetic acid is a weak acid and its conjugate base, the acetate ion, $CH_3CO_2^-$, is a good proton acceptor.

Base ionization constants. The proton transfer equilibrium and the corresponding equilibrium constant for a Brønsted base in aqueous solution is

$$H_2O(l) + B(aq) \rightleftharpoons HB^+(aq) + OH^-(aq)$$

$$\frac{[HB^+][OH^-]}{[B]} = K_b \text{ at equilibrium}$$

The constant K_b is called the **base ionization constant**. We always assume that a solution of a base is at equilibrium. If B is the organic base methylamine, CH_3NH_2 (**2**), then HB^+ is the methylammonium ion, $CH_3NH_3^+$; and the proton transfer equilibrium is

$$H_2O(l) + CH_3NH_2(aq) \rightleftharpoons CH_3NH_3^+(aq) + OH^-(aq)$$

$$\frac{[CH_3NH_3^+][OH^-]}{[CH_3NH_2]} = K_b$$

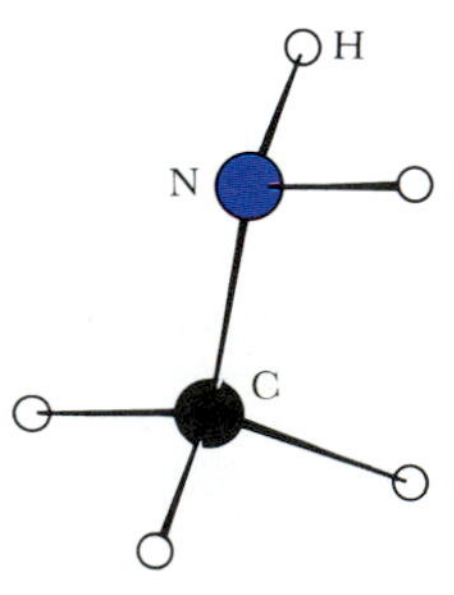

2 Methylamine

Similarly, when B is urea, NH_2CONH_2, the equilibrium is

$$H_2O(l) + NH_2CONH_2(aq) \rightleftharpoons NH_2CONH_3^+(aq) + OH^-(aq)$$

$$\frac{[NH_2CONH_3^+][OH^-]}{[NH_2CONH_2]} = K_b$$

The values of K_b for a variety of weak bases are listed in Table 14.6. We see, for instance, that K_b for CH_3NH_2 is 3.6×10^{-4} and that for NH_2CONH_2 is 1.3×10^{-14}. We can conclude that methylamine is the stronger base, because in water its equilibrium lies more in favor of the conjugate acid than does that of urea. Alternatively, we can express the strengths in terms of pK_b, where

$$pK_b = -\log K_b$$

Thus, for methylamine,

$$pK_b = -\log (3.6 \times 10^{-4}) = 3.44$$

Similarly, for urea, $pK_b = 13.90$. Again we should bear in mind that the stronger the base, the larger the value of K_b and the smaller the value of pK_b.

The relative strengths of the conjugate acids of the weak bases can also be determined from their K_b values. Because the K_b of methylamine is larger than that of urea, methylamine is a stronger base. That is, the conjugate acid of methylamine—the methylammonium ion, $CH_3NH_3^+$—has a weaker tendency to donate a proton to H_2O to reform CH_3NH_2. Because urea is the weaker of the two bases, the conjugate acid of urea, $NH_2CONH_3^+$, must be a better proton donor to H_2O. The $NH_2CONH_3^+$ ion is a relatively strong acid. So, weak base equilibria are like weak acid equilibria: *the stronger the Brønsted base, the weaker its conjugate acid.*

E Urea is used as a commercial fertilizer. It is also a component of mammalian urine. Urea was also the first organic material to be synthesized from inorganic precursors: Freidrich Wöhler's classic synthesis in 1828, in which he produced urea by heating ammonium cyanate, NH_4CNO, demonstrated that organic compounds were not the manifestation of some "vital force."

E Most molecular bases, the amines, have a putrid odor. This odor is also associated with the presence of lone pairs on the N atom, which enable the molecules to interact with sites on the olfactory epithelium in the nose. These sites are less sensitive to shape than the principal sites for olfaction, so most amines have a similar pungent smell.

The relation between* pK_a*,* pK_b*, and* pK_w*. A very useful relation exists between the pK_b of a base (such as NH_3) and the pK_a of its conjugate acid (here NH_4^+). We can derive it by considering the equilibrium of the base NH_3 in water,

$$H_2O(l) + NH_3(aq) \rightleftharpoons NH_4^+(aq) + OH^-(aq) \quad K_b = \frac{[NH_4^+][OH^-]}{[NH_3]}$$

and the equilibrium of ammonia's conjugate acid, NH_4^+, in water,

$$NH_4^+(aq) + H_2O(l) \rightleftharpoons H_3O^+(aq) + NH_3(aq) \quad K_a = \frac{[H_3O^+][NH_3]}{[NH_4^+]}$$

When we multiply the two ionization constants together, we get

$$K_a \times K_b = \frac{[H_3O^+][NH_3]}{[NH_4^+]} \times \frac{[NH_4^+][OH^-]}{[NH_3]}$$

Cancellation of like terms on the right-hand side of the equation gives

$$K_a \times K_b = [H_3O^+][OH^-] = K_w$$

If K_b increases, then K_a must decrease to keep the product $K_a \times K_b$ equal to K_w. This relation confirms what we have already seen, that a stronger base must have a weaker conjugate acid. Similarly, the stronger an acid, the weaker its conjugate base. We have already seen

N The values of K_a and K_b are also related like the positions of two people on a see-saw: as one goes up, the other must go down.

some examples of this relation:

- CH_3COOH is a weak proton donor and its conjugate base $CH_3CO_2^-$ is a strong proton acceptor; CCl_3COOH is a stronger proton donor and its conjugate base $CCl_3CO_2^-$ is a weaker proton acceptor; HCl is a *very* strong proton donor and its conjugate base Cl^- a *very* weak proton acceptor.
- NH_2CONH_2, urea, is a very weak proton acceptor and its conjugate acid $NH_2CONH_3^+$ is a strong proton donor; CH_3NH_2, methylamine, is a stronger proton acceptor and its conjugate acid $CH_3NH_3^+$ is a weaker proton donor; O^{2-} is a *very* strong proton acceptor and its conjugate acid OH^- is a *very* weak proton donor.

If we take logarithms of both sides of the relation $K_a \times K_b = K_w$ we obtain

$$\log K_a + \log K_b = \log K_w$$

After a change of signs, this expression becomes

$$pK_a + pK_b = pK_w$$

As an example, the pK_b of NH_3 is 4.74, so the pK_a of NH_4^+ at 25°C is

$$pK_a = pK_w - pK_b = 14.00 - 4.74 = 9.26$$

This value shows that NH_4^+ is a weaker proton donor than acetic acid ($pK_a = 4.74$) but stronger than hypoiodous acid (HIO, $pK_a = 10.64$). The conjugate acid of urea, $NH_2CONH_3^+$, has $pK_a = 0.10$, a value showing that it is a stronger proton donor than any of the acids listed in Table 14.6.

EXAMPLE 14.10 Predicting relative strengths of acids and bases

U Exercises 14.82 and 14.83.

Use Table 14.6 to decide which member of each of the following pairs of species is the stronger acid or base in water: (a) HF or HIO_3; (b) NO_2^- or CN^-.

STRATEGY In determining the relative strengths of Brønsted acids and bases, we should bear in mind that the larger the K_a of a weak acid, the stronger the acid and the weaker its conjugate base; and that the larger the K_b of a weak base, the stronger the base and the weaker its conjugate acid.

SOLUTION (a) Because $K_a(HIO_3) > K_a(HF)$, it follows that HIO_3 is the stronger acid. (b) Because $K_a(HNO_2) > K_a(HCN)$, and because the stronger acid has the weaker conjugate base, it follows that NO_2^- is weaker as a base than CN^-. Hence, CN^- is the stronger base.

EXERCISE E14.10 Use Table 14.6 to decide which species of each pair is the stronger acid or base. (a) HF or HIO; (b) $C_6H_5CO_2^-$ or $CH_2ClCO_2^-$; (c) $C_6H_5NH_2$ or $(CH_3)_3N$; (d) $C_6H_5NH_3^+$ or $(CH_3)_3NH^+$.

[*Answer*: (a) HF; (b) $C_6H_5CO_2^-$; (c) $(CH_3)_3N$; (d) $C_6H_5NH_3^+$]

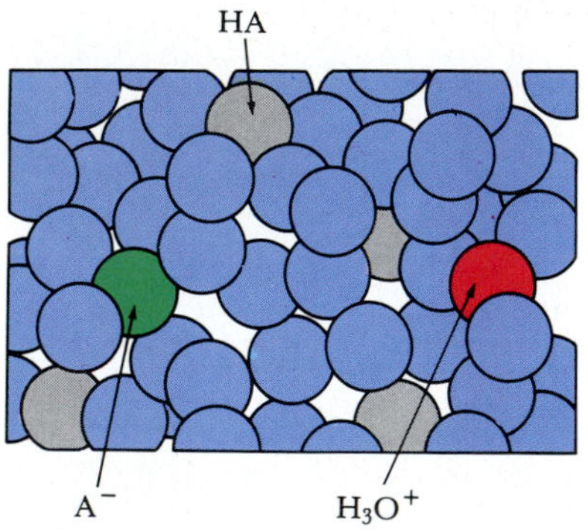

FIGURE 14.12 In a solution of a weak acid, only some of the ionizable hydrogen atoms are present as hydronium ions (red), and the solution contains a high proportion of the original acid molecules (HA, gray). Water molecules, blue; anion of weak acid, green.

14.8 WEAK ACIDS AND BASES

Once we know the ionization constants of a number of weak acids and bases (Table 14.6), we can list them according to their relative strengths and seek reasons for their behavior. Again we must remember that *strength* means the extent to which the acid or base ionizes (that is, the extent of proton transfer), not its concentration. We continue to limit our discussion to aqueous solutions.

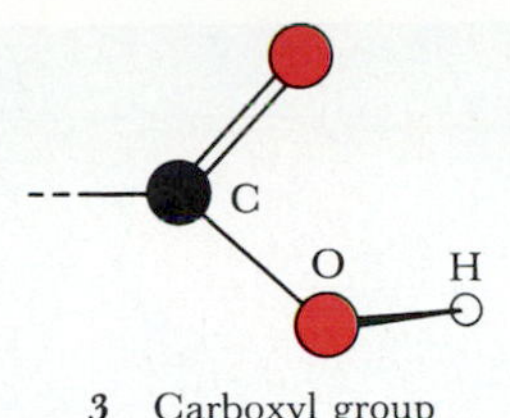

3 Carboxyl group

Weak acids. Most acids are weak acids in the sense that only a small percentage (usually less than 10%, and often very much less) of the acid molecules in solution ionize to produce the hydronium ion and the conjugate base. This very small extent of ionization is indicated by the ionization constants in Table 14.6, most of which are much smaller than 1. In a solution of a weak acid at ordinary concentrations (neither concentrated nor very dilute), the proton transfer equilibrium favors the nonionized HA (Fig. 14.12). All carboxylic acids, which are organic acids containing the —COOH group (**3**), are weak acids in water. For example, acetic acid has $pK_a = 4.74$, and although a 1 M $CH_3COOH(aq)$ solution contains some H_3O^+ and $CH_3CO_2^-$ ions, most of the solute is present as CH_3COOH.

Like the distinction between strong acids and concentrated acids, a distinction must be made between a *weak acid* and a *dilute acid:* a weak acid may be concentrated or dilute. Weakness, like strength, refers to the extent of ionization and therefore the proton-donating power of the acid molecules, not to the number of moles of acid per liter.

One practical difference between strong and weak acids is shown by dropping pieces of magnesium into 0.1 M $HCl(aq)$ and 0.1 M $CH_3COOH(aq)$. Hydrogen is evolved much more rapidly from the hydrochloric acid, which suggests that hydrogen ions are much more numerous in that acid than in acetic acid, even though the concentrations of the acids are the same (Fig. 14.13). Another sign of the weakness of acetic acid is its electrical conductivity: 0.1 M $CH_3COOH(aq)$ is a much poorer conductor of electricity than 0.1 M $HCl(aq)$ is. The concentration of ions, which carry the current through the solution, is much lower in acetic acid than in hydrochloric acid (Fig. 14.14).

Weak bases. Ammonia, a weak base, has $pK_b = 4.74$ and is largely present in aqueous solution as nonionized NH_3 molecules (represented as B in Fig. 14.15). Only a small percentage (usually less than 10%, and often very much less) of the molecules of a weak base ionize at ordinary concentrations to produce the conjugate acid and hydroxide ions. As we see from Table 14.6, most K_b values are very much smaller than 1, which indicates that their proton transfer equilibria lie in favor of the nonionized molecules or ions. All amines (substances containing the $—NH_2$ group) are weak bases in water.

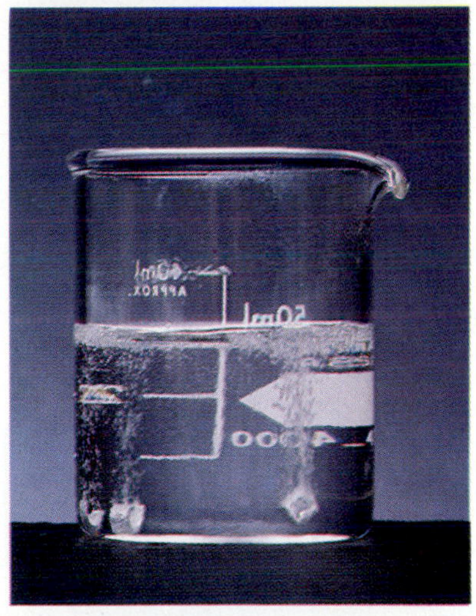

FIGURE 14.13 The effects of a strong acid (top) and a weak acid (bottom) on magnesium. Although the acids have the same concentrations, the rate of hydrogen evolution, which depends on the presence of hydronium ions, is much greater in the strong acid.

14.9 THE STRUCTURES AND STRENGTHS OF ACIDS

Now we shall consider the structural features of molecules and ions that determine their strengths as acids and bases.

The strengths of binary acids. The K_a values for acids (and K_b values for bases) are difficult to predict because they depend on a number of factors. These factors include the strength of the H—A bond, the strength of the O—H bond in H_3O^+, and the extent to which the conjugate base, A^-, of the acid is hydrated. The ability of the water molecules to become organized around the hydronium ion when it is formed also plays an important, sometimes dominant, role. However,

(a)

(b)

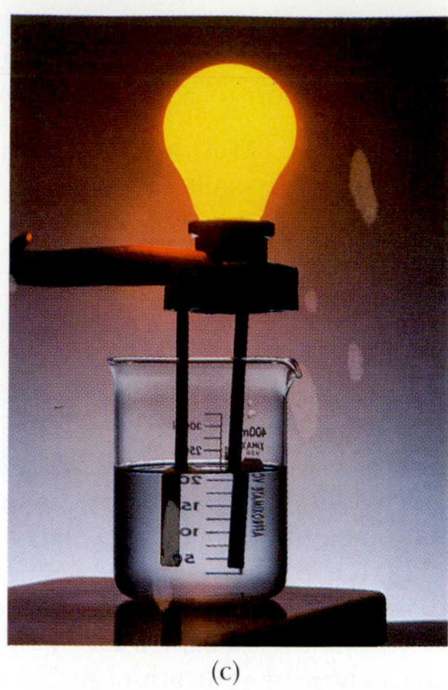

(c)

FIGURE 14.14 For the same reason demonstrated in Fig. 14.13, a solution of hydrochloric acid is a much better electrical conductor than an acetic acid solution of the same concentration is. The samples are (a) deionized water, (b) 0.1 M $CH_3COOH(aq)$, and (c) 0.1 M $HCl(aq)$.

D Set up a conductivity apparatus, like that shown in Fig. 14.14, and test various strong and weak acids and bases.

N It is extremely difficult to predict the strengths of acids and bases because the values depend on the properties of the solvent (the changes in its organization) as well as properties intrinsic to the species.

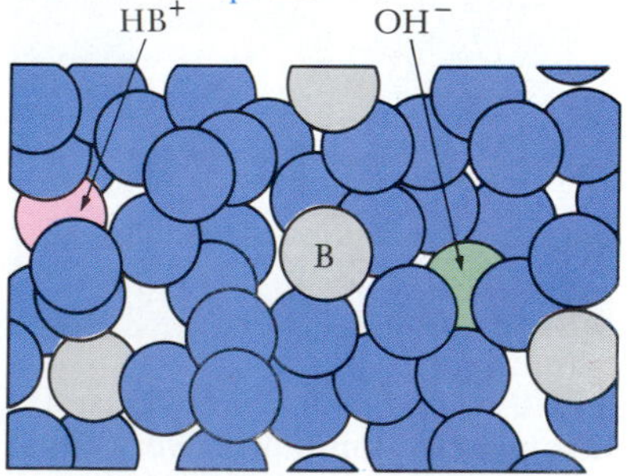

FIGURE 14.15 In a solution of a weak base, only a small proportion of the base molecules (gray) have accepted protons from water (blue) to form HB^+ (pink). Hydroxide ion, green.

trends among series of compounds with similar structures can sometimes be explained by focusing on the features we now consider.

The ease with which an acid molecule HA transfers a proton depends on the strength of the hydrogen bond it forms with the O atom of the H_2O molecule:

$$H_2O\cdots H—A \longrightarrow H_2O—H^+ + A^-$$

A strong hydrogen bond weakens the H—A bond and aids in the formation of the H_3O^+ ion. Generally, the more polar the H—A bond, the stronger the O···H—A hydrogen bond. Therefore, we expect an acid HA with a polar H—A bond to be stronger than one with a less polar H—A bond. Because the polarity of the H—A bond increases with the electronegativity of A, we can also predict that the greater the electronegativity of A, the stronger the acid HA. An example of an increase in acidity with electronegativity is provided by the binary hydrogen compounds in Period 2, namely, NH_3, H_2O, and HF. The electronegativity difference in the N—H bond is 0.9; for O—H, it is 1.4; and for F—H, it is 1.9 (Fig. 14.16; for individual electronegativities, see Fig. 7.38). Therefore, the H—F bond is markedly more polar than the N—H bond. Experimentally, HF is acidic, H_2O is neutral, and NH_3 is basic in aqueous solutions. For these three compounds at least, the bond polarity dominates the trend of acid strengths.

Even though an H—A bond may be polar and able to form a hydrogen bond, it may be too strong to be broken. Because the H—A bond can be broken more readily if it is weak, we can also expect that the weaker the H—A bond, the stronger the acid HA. An example is the anomalous position of HF among the hydrohalic acids. Even though the H—F bond is the most polar of the group, HF is a weak acid in water whereas all other hydrohalic acids are strong. A part of the reason is the great strength of the H—F bond (562 kJ/mol; compare this value with 431 kJ/mol for HCl, 366 kJ/mol for HBr, and 299 kJ/mol for HI); consequently the proton is lost from HF only with difficulty. In a solvent that is a poorer proton acceptor than water (such as pure acetic acid) in which it is possible to measure the relative strengths of the hydrohalic acids, the order of strengths is $HF < HCl < HBr < HI$, which is consistent with the weakening of the H—A bond down the group but not with the decrease in bond polarity. The same trend is found for the Group VI acids in aqueous solution, where the acidities are in the order $H_2O < H_2S < H_2Se < H_2Te$, and the bond strengths and bond polarities decrease down the group. Therefore it appears that bond strength, but not bond polarity, makes an important contribution to the trend in acid strength in these two sets of binary acids.

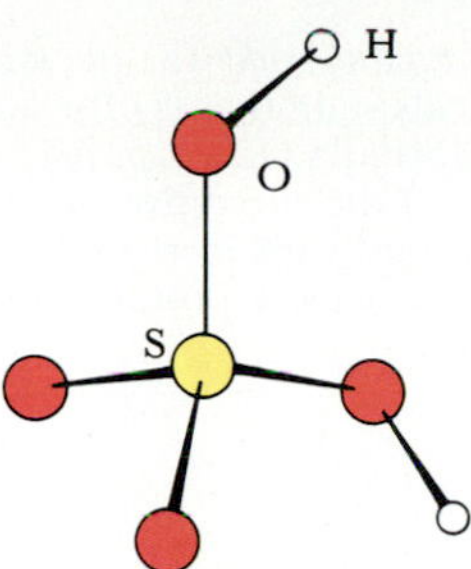

4 Sulfuric acid

The strengths of oxoacids. When considering the strengths of oxoacids, we need to be aware that the acidic protons (the ones that can be donated) are always attached to an oxygen atom. Thus, although we always write the formula of sulfuric acid as H_2SO_4, its structure is actually $(HO)_2SO_2$ (**4**). The high polarity of the O—H bond is one reason why the proton of an —OH group in an oxoacid molecule is acidic. This observation is shown strikingly by phosphorous acid H_3PO_3, which has the structure $(HO)_2PHO$ (**5**), for it can donate the protons from its two —OH groups but not the proton attached directly to the phosphorus atom.

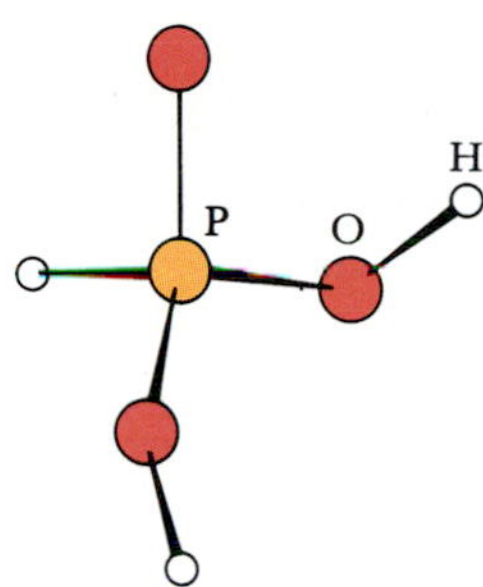

5 Phosphorous acid

The oxoacids form two structurally related series. In one, the number of O atoms bonded to the atom X is variable, as in the chlorine oxoacids HClO, $HClO_2$, $HClO_3$, and $HClO_4$ (for which X = Cl) or as in the sulfur oxoacids H_2SO_3 and H_2SO_4 (for which X = S). In these families of acids, it is found that the greater the number of oxygen atoms attached to the central atom, the stronger the acid. Because the oxidation number of X also increases as the number of O atoms increases, we can also conclude that the greater the oxidation number of X, the stronger the acid. The effect is demonstrated by the following acids:

T OHT: Text tables, pp. 555 and 556.

Acid	Model	Oxidation number	pK_a
HClO (hypochlorous acid)	:Cl—O—H	+1	7.52
$HClO_2$ (chlorous acid)	O=Cl—O—H	+3	2.00
$HClO_3$ (chloric acid)	O=Cl(=O)—O—H	+5	strong
$HClO_4$ (perchloric acid)	O=Cl(=O)(=O)—O—H	+7	strong

The arrows indicate the direction and magnitude of the shift of electron density away from the O—H bond.

					He
B −0.1	C 0.4	N 0.9	O 1.4	F 1.9	Ne
Al −0.6	Si −0.3	P 0	S 0.4	Cl 0.9	Ar
Ga −0.5	Ge −0.3	As −0.1	Se 0.3	Br 0.7	Kr
In −0.4	Sn −0.3	Sb −0.2	Te 0	I 0.4	Xe
Tl −0.3	Pb −0.3	Bi −0.2	Po −0.1	At 0.1	Rn

FIGURE 14.16 The differences of electronegativities between hydrogen (2.1) and the *p*-block elements. Note the high values for oxygen and fluorine. (Blue squares mark elements that are less electronegative than hydrogen.)

E The *Pauling rules* for the strengths of oxoacids state that (1) for an oxoacid of formula $O_{p}E(OH)_{q}$, $pK_a \approx 8 - 5p$, and (2) the successive pK_a values of polyprotic acids increase by 5 units for each successive proton transfer.

In the second structurally related series of oxoacids, the identity of the atom X is variable and the number of O atoms is constant, as in the hypohalous acids, HClO, HBrO, and HIO. In this series, we observe that the greater the electronegativity of the central atom X, the stronger the oxoacid. This trend is demonstrated by the following acids:

Acid (HXO)	Model	Electronegativity of atom X	pK_a
HClO	:Cl—O—H	3.0	7.52
HBrO	:Br—O—H	2.8	8.70
HIO	:I—O—H	2.5	10.64

A partial explanation is that electrons are withdrawn slightly from the O—H bond as the electronegativity of the halogen atom increases (Fig. 14.17). As these bonding electrons move toward the O atom, the O—H bond becomes more polar; the more polar the bond, the more readily the O—H proton is donated to H_2O and the stronger the acid. In addition, the greater the electronegativity of the atom X of the acid, the greater the stability of the negatively charged conjugate base.

U Exercise 14.84.

EXAMPLE 14.11 Predicting the relative strengths of acids

In the following pairs, predict which acid is stronger and explain why: (a) H_2S and H_2Se; (b) H_2SO_4 and H_2SO_3; (c) H_2SO_4 and H_3PO_4.

STRATEGY We recall that for binary acids:

1. The more polar the H—A bond, the stronger the acid. This effect is dominant for acids of the same period.
2. The weaker the H—A bond, the stronger the acid. This effect is dominant for acids of the same group.

For oxoacids:

3. The greater the number of O atoms attached to the central atom (or the greater the oxidation number of the central atom), the stronger the acid.
4. If the same number of O atoms are attached to the central atom, then the greater the electronegativity of the central atom, the stronger the acid.

SOLUTION (a) Sulfur and selenium are in the same group, and we expect the H—Se bond to be weaker than the H—S bond. For that reason, we predict that H_2Se is the stronger acid. (b) H_2SO_4 has the greater number of O atoms bonded to the S atom and the oxidation number of sulfur is +6, whereas in H_2SO_3 the sulfur has an oxidation number of only +4. Hence H_2SO_4 is the stronger acid. (c) Both acids have four O atoms bonded to the central atom; but the electronegativity of sulfur is greater than that of phosphorus, so H_2SO_4 is the stronger acid.

EXERCISE E14.11 In the following pairs, predict which acid is stronger and explain why: (a) H_2S and HCl; (b) HNO_2 and HNO_3; (c) H_2SO_3 and $HClO_3$

[*Answer*: (a) HCl, rule 1; (b) HNO_3, rule 3; (c) $HClO_3$, rule 4]

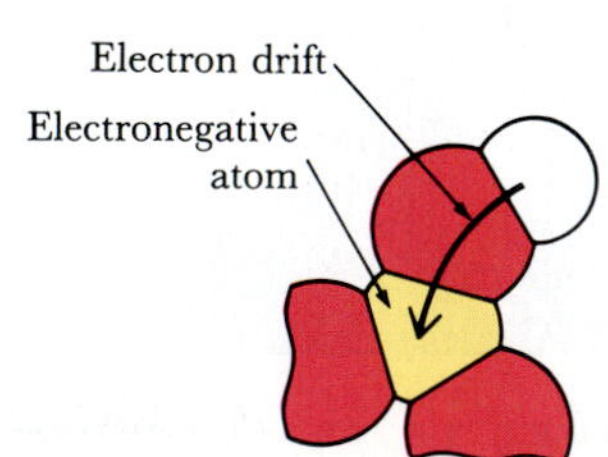

FIGURE 14.17 The greater the electronegativity of the atom X, the more the electrons in the O—H bond drift toward it, thereby producing a stronger acid.

The strengths of organic acids. The effect on acid strength of the electronegativity of the atom or group of atoms attached to an —OH group can be seen in organic compounds. Alcohols are organic compounds that contain a hydroxyl group, —OH, covalently bonded to a carbon atom, which in turn is covalently bonded to hydrogen atoms or carbon atoms. An example is ethanol, CH_3CH_2—OH (**6**), the alcohol produced from the fermentation of grains such as wheat, barley, and rye. If we compare the structure of alcohols with that of hypochlorous acid, Cl—OH (commonly written HClO), we can see that because the Cl atom has a far higher electronegativity than the CH_3CH_2— group, HClO should be a stronger acid. Indeed, the pK_a of CH_3CH_2—OH is 16 whereas that of HClO is 7.52. Alcohols have such weak proton-donating power that they are not usually regarded as oxoacids.

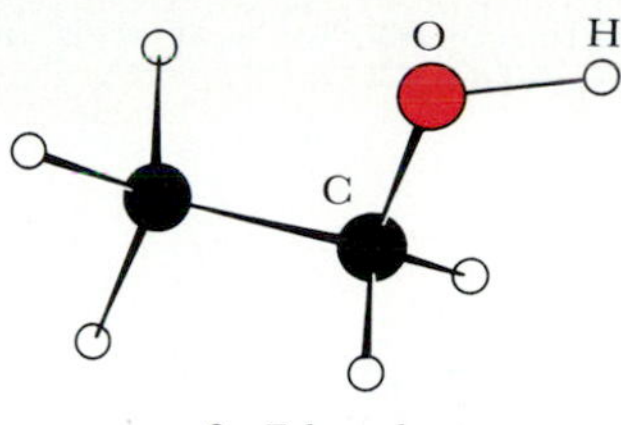

6 Ethanol

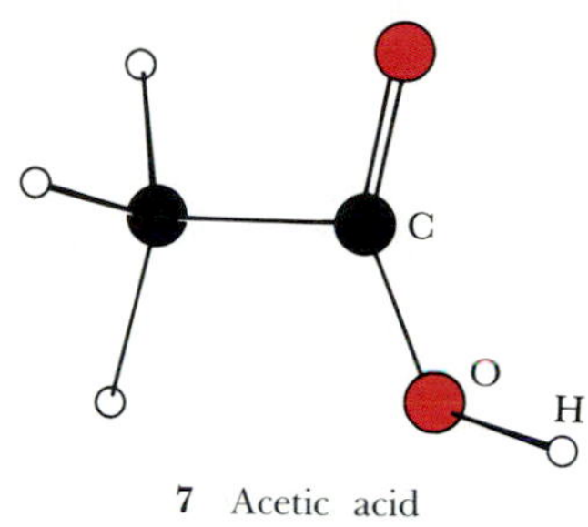

7 Acetic acid

The effect of the number of O atoms on the acidities of organic compounds is shown by the difference between alcohols and carboxylic acids. Carboxylic acids have an additional O atom bonded to the carbon atom to which the —OH group is attached, to form the carboxyl group —COOH, as in acetic acid (**7**). Although carboxylic acids are weak acids, they are much stronger than alcohols.

The strength of an acid is also increased by the stabilizing effect of the extra oxygens on the conjugate base. The second O atom of the carboxyl group, for instance, provides an additional electronegative atom over which the negative charge of the conjugate base can spread: this arrangement stabilizes the anion by resonance:

$$CH_3\text{—}C(\text{—}\ddot{\underset{..}{O}}{:}^-)(=\ddot{O}{:}) \longleftrightarrow CH_3\text{—}C(=\ddot{O}{:})(\text{—}\ddot{\underset{..}{O}}{:}^-)$$

The lower energy of the conjugate base as a result of this resonance contributes to the lower tendency of the base to accept a proton. With the base weakened in this way, the conjugate acid is stronger.

E Entropy and equilibria. A reassessment of ionization data for substituted acetic acids. *J. Chem. Educ.*, **1982**, *59*, 354.

The strengths of carboxylic acids also vary directly with total electron-withdrawing power of the atoms bonded to the carboxyl group. For example, because hydrogen is less electronegative than chlorine, the CH_3— group bonded to —COOH in acetic acid is less electron withdrawing than the CCl_3— group in trichloroacetic acid. Therefore, we expect CCl_3COOH to be a stronger acid than CH_3COOH. In agreement with this prediction, the pK_a of acetic acid is 4.74 whereas that of trichloroacetic acid is 0.52.

▼ EXAMPLE 14.12 Predicting the trend in strengths of carboxylic acids

List the following carboxylic acids in order of increasing strength: CH_3COOH, CH_2FCOOH, CHF_2COOH, and CF_3COOH.

STRATEGY As we saw in the text, the greater the electron-withdrawing power of the group attached to the carboxyl group, the stronger the acid. Therefore, we need to judge the electron-withdrawing power of that group by comparing the electronegativities of the atoms.

SOLUTION Because fluorine is more electronegative than hydrogen, the groups attached to —COOH increase in electron-attracting power in the order $CH_3\text{—} < CH_2F\text{—} < CHF_2\text{—} < CF_3\text{—}$. Therefore the acid strengths increase accordingly, and we predict that they lie in the order $CH_3COOH <$

D Different pHs (of weak acids and bases). TDC, 4-15.

E Approach to acid-base equilibrium calculations. *J. Chem. Educ.*, **1978**, *55*, 576.

N The principles of chemical equilibrium are applied in a uniform manner in Chapters 13 through 15; only the chemical system changes. The consistent use of an equilibrium table makes equilibrium calculations much more understandable.

N As remarked before, we distinguish between ionization and dissociation: ionization involves the formation of at least one ion, dissociation is homolytic cleavage of a bond or bonds. In the context of Brønsted theory, *deprotonation* would be a better term, and we would calculate the *percentage deprotonated*.

$CH_2FCOOH < CHF_2COOH < CF_3COOH$.

EXERCISE E14.12 List the following carboxylic acids in order of increasing strength: CH_2ClOOH, $CH_2BrCOOH$, and CH_2FCOOH.
[*Answer*: $CH_2BrCOOH < CH_2ClCOOH < CH_2FCOOH$]

14.10 THE pH OF WEAK ACIDS AND BASES

Weak acids produce a lower concentration of H_3O^+ ions in aqueous solution than do strong acids prepared with the same concentration. For example, a 0.010 M HCl(*aq*) solution has a pH of 2.00; however, a 0.010 M CH_3COOH(*aq*) solution has a much lower concentration of H_3O^+ ions, and its pH is 3.4. (Remember that the pH increases as the hydronium ion concentration decreases, and an increase in pH by 1 corresponds to a tenfold decrease in concentration.) Similarly, weak bases produce a lower concentration of OH^- ions than do strong bases of the same (base) concentration; therefore, the pH of the weak base will be less than that of the strong base of the same molar concentration. A 0.010 M NaOH(*aq*) solution has a pH of about 12, but the pH of a 0.010 M NH_3(*aq*) solution is about 10.6. In this section we shall see how to calculate the pH of solutions of weak acids and weak bases.

Weak acids. The concentration of hydronium ions in a solution of a weak acid is calculated by taking into account the equilibrium between the acid HA and its conjugate base A^-:

$$HA(aq) + H_2O(l) \rightleftharpoons H_3O^+(aq) + A^-(aq) \qquad K_a = \frac{[H_3O^+][A^-]}{[HA]}$$

Because a dynamic equilibrium is established, we have the same conditions we met in the discussion of equilibria in Chapter 13, and we can use the same method to calculate the equilibrium composition of the solutions. Therefore, to calculate the pH of a solution of a weak acid, we use the equilibrium table to calculate the molar concentration of hydronium ions for a given molar concentration of acid.

For the initial concentration in step 1 of the scheme set out in Chapter 13, we use the molar concentration of the acid molecules on the assumption that none is ionized. We assume in step 2 that, to reach equilibrium, the acid ionizes and its concentration decreases by x mol/L. The resulting concentrations of H_3O^+ and A^- ions can then be calculated in step 3, and the unknown concentration x found by using K_a. It may be possible to simplify the calculation with an approximation like those illustrated in Section 13.5, by neglecting changes of less than 5% of the initial concentration of the acid.

The percentage of HA molecules that are ionized is another item of information that gives some insight into the proton transfer equilibrium of weak acids in solution:

$$\text{Percentage ionized (\%)} = \frac{\text{concentration of HA ionized}}{\text{original concentration of HA}} \times 100\%$$

A small percentage of ionized molecules indicates that the solution consists primarily of the acid HA with very low concentrations of hydronium and conjugate base ions.

▼ EXAMPLE 14.13 Calculating the pH of a solution of a weak acid

Calculate the pH and percentage of acid ionized in 0.10 M $CH_3COOH(aq)$, given that K_a for acetic acid is 1.8×10^{-5}.

STRATEGY Acetic acid is a weak acid, so we expect the molar concentration of H_3O^+ ions to be less than 0.10 mol/L; therefore its pH should be greater than 1.0 (because $-\log 0.10 = 1.0$). To find the actual value, we set up an equilibrium table in the usual way, with the initial concentration of acid equal to 0.10 mol/L, and allow the concentration of acid to decrease by x mol/L to achieve equilibrium. If x is less than about 5% of the initial concentration of acid, then the expression for the equilibrium constant can be simplified by ignoring x relative to the initial concentration of the acid. However, at the end of the calculation, the value of x must be checked to ensure that it is consistent with any approximation that has been made. If it is not, then the exact expression for K_a must be solved for x; the calculation often involves solving a quadratic equation (Appendix 1C).

SOLUTION The equilibrium table is

Equilibrium equation:	$CH_3COOH(aq) + H_2O(l) \rightleftharpoons H_3O^+(aq) + CH_3CO_2^-(aq)$		
Species:	CH_3COOH	H_3O^+	$CH_3CO_2^-$
Step 1. Initial concentration, mol/L	0.10	0	0
(The initial molar concentration of CH_3COOH is given and the initial concentrations of H_3O^+ and $CH_3CO_2^-$ are zero.)			
Step 2. Change in concentration, mol/L	$-x$	$+x$	$+x$
(We assume that the concentration of CH_3COOH decreases by x mol/L as it approaches equilibrium, and the reaction stoichiometry then implies that the concentrations of H_3O^+ and $CH_3CO_2^-$ must increase by the same amount.)			
Step 3. Equilibrium concentration, mol/L	$0.10 - x$	x	x

Step 4. Because

$$\frac{[H_3O^+][CH_3CO_2^-]}{[CH_3COOH]} = K_a \text{ at equilibrium}$$

with $K_a = 1.8 \times 10^{-5}$ and the concentrations from step 3, we obtain

$$\frac{x \times x}{0.10 - x} = 1.8 \times 10^{-5}$$

Anticipating that $x \ll 0.10$, we approximate this expression as

$$\frac{x^2}{0.10} \approx 1.8 \times 10^{-5}$$

Solving for x gives

$$x \approx \sqrt{0.10 \times 1.8 \times 10^{-5}} = 1.3 \times 10^{-3}$$

From step 3, $x = [H_3O^+] = 1.3 \times 10^{-3}$ mol/L, so

$$\text{pH} = -\log(1.3 \times 10^{-3}) = 2.89, \text{ or about } 2.9$$

E An efficient method for the treatment of weak acid/base equilibria. *J. Chem. Educ.*, **1990**, *67*, 224. Dissociation of a weak acid. *J. Chem. Educ.*, **1980**, *57*, 669.

T OHT: Generic equilibrium table from Chapter 13.

N This analysis of a chemical equilibrium is exactly like that presented in Example 13.5; only the chemical system is different. Note that *Step 3 = step 1 + step 2*. In Step 4, insert Step 3 values into the reaction quotient and set the quotient equal to the equilibrium constant.

E A criterion for the simple approximation in dissociation [ionization] equilibria. *J. Chem. Educ.*, **1976**, *53*, 460.

N The negative root is an alternative solution of the equation for x, but it is excluded by the requirement that x be positive (because it is equal to the molar concentration of H_3O^+).

We have assumed that x is less than about 5% of 0.10; in fact, it is

$$\frac{1.3 \times 10^{-3}}{0.10} \times 100\% = 1.3\%$$

and the approximation is valid. The percentage of acid ionized is

$$\text{Percentage ionized (\%)} = \frac{\text{concentration of } CH_3COOH \text{ ionized}}{\text{original concentration of } CH_3COOH} \times 100\%$$

$$= \frac{1.3 \times 10^{-3}}{0.10} \times 100\% = 1.3\%$$

That is, 1.3% of the acetic acid molecules have ionized.

EXERCISE E14.13 Calculate the pH and percentage of ionization of 0.20 M lactic acid. See Table 14.6 for K_a. Be sure to check your approximation to see whether it is valid.

[*Answer*: 1.90; 6.3%; must use exact solution]

N It is relatively easy, but somewhat tedious, to incorporate the hydronium ion concentration that stems from the autoionization of water: in Step 1 of the equilibrium table, the initial concentration of hydronium ions is set equal to 1.0×10^{-7}, and the resulting quadratic equation is solved without approximation.

U Exercises 14.85, 14.86, and 14.88.

One limitation of this approach to pH calculation is that it ignores the concentration of hydrogen ions that result from the autoionization of water. When the acid is in such a low concentration that the calculation predicts a hydronium ion concentration of less than 10^{-7} mol/L, we must not report its pH as being higher than 7, because the autoionization already provides hydrogen ions at a concentration of 10^{-7} mol/L. For example, the pH of 1.0×10^{-8} M HCl solution is not 8, because the autoionization of water leads to $[H_3O^+] = 1.0 \times 10^{-7}$ mol/L, which is ten times greater than the concentration produced by the acid, and gives rise to a pH of 7. We can safely ignore the contribution of the autoionization of water only when the calculated hydronium ion concentration is substantially (about three times) higher than 10^{-7} mol/L.

? How could you add 1.0×10^{-8} mol HCl to 1 L of water?

E $[A^-]$ in weak acid solution. *J. Chem. Educ.*, **1986**, *63*, 597.

Weak bases. A weak base has a lower pH than a strong base of the same molar concentration: a weak base in solution establishes a dynamic equilibrium with its conjugate acid and therefore produces less OH^- than a comparable strong base. We can calculate the pH of solutions of weak bases in the same way that we calculated the pH of solutions of weak acids—by using the equilibrium table technique. The equilibrium concentrations of ions in the solution always satisfy

$$H_2O(l) + B(aq) \rightleftharpoons HB^+(aq) + OH^-(aq) \qquad \frac{[HB^+][OH^-]}{[B]} = K_b$$

To calculate the pH of the solution, we first calculate the molar concentration of OH^- ions at equilibrium, express that as pOH, and then calculate pH from pH + pOH = 14.00 at 25°C.

EXAMPLE 14.14 Calculating the pH of a solution of a weak base

Calculate the pH and percentage ionization of 0.20 M aqueous methylamine, CH_3NH_2. The K_b for CH_3NH_2 is 3.6×10^{-4}.

STRATEGY We expect pH > 7 because the solution is basic. We can calculate the equilibrium molar concentration of OH^- ions by using the equilibrium table technique explained in Chapter 13 and used again in Example 14.13, but we use K_b instead of K_a. Once we know $[OH^-]$, we can calculate pOH and convert it to pH by using the relation pH + pOH = 14.00.

SOLUTION The equilibrium table is

Equilibrium equation:	$H_2O(l) + CH_3NH_2(aq) \rightleftharpoons CH_3NH_3^+(aq) + OH^-(aq)$		
Species:	CH_3NH_2	$CH_3NH_3^+$	OH^-
Step 1. Initial concentration, mol/L	0.20	0	0
Step 2. Change in concentration, mol/L	$-x$	$+x$	$+x$
Step 3. Equilibrium concentration, mol/L	$0.20 - x$	x	x

N The initial concentration of OH^- should be set equal to 1.0×10^{-7} if the protonation of the base leads to a comparable concentration of OH^- ions to that arising from the autoionization of water.

Step 4. Because

$$\frac{[CH_3NH_3^+][OH^-]}{[CH_3NH_2]} = K_b \text{ at equilibrium}$$

and $K_b = 3.6 \times 10^{-4}$, we obtain

$$\frac{x \times x}{0.20 - x} = 3.6 \times 10^{-4}$$

N Only the positive square root is taken, because x, which is equal to a concentration, must be positive.

We now anticipate that x is less than 5% of 0.20 and approximate this expression by

$$\frac{x^2}{0.20} \approx 3.6 \times 10^{-4}$$

N Yet again, the equilibrium analysis is the same as that presented for the weak acid described in Example 14.13; only the chemical system is different.

Therefore,

$$x \approx \sqrt{0.20 \times 3.6 \times 10^{-4}} = 8.5 \times 10^{-3}$$

The value of x differs from 0.20 by

$$\frac{8.5 \times 10^{-3}}{0.20} \times 100\% = 4.3\%$$

so the approximation is valid (just). According to step 3, the concentration of hydroxide ions is x mol/L, or 8.5×10^{-3} mol/L; so

$$\text{pOH} = -\log(8.5 \times 10^{-3}) = 2.07$$

and hence

$$\text{pH} = 14.00 - 2.07 = 11.93, \text{ or about } 11.9$$

The percentage of methylamine ionized is

$$\text{Percentage ionized (\%)} = \frac{\text{concentration of ionized } CH_3NH_2}{\text{original concentration of } CH_3NH_2} \times 100\%$$

$$= \frac{8.5 \times 10^{-3}}{0.20} \times 100\% = 4.3\%$$

That is, 4.3% of the methylamine is present as the protonated form, $CH_3NH_3^+$.

U Exercise 14.87.

EXERCISE E14.14 Calculate the pH and percentage of base ionized in 0.15 M NH_2OH (hydroxylamine).

[*Answer*: 9.61; 0.027%]

Determining K_a and K_b. So far, we have listed and used acid and base ionization constants without explaining how they are measured. In fact, they are measured by determining the pH of a solution of known acid or base concentration and carrying out a calculation like those we have already illustrated. The procedure is illustrated in the following example.

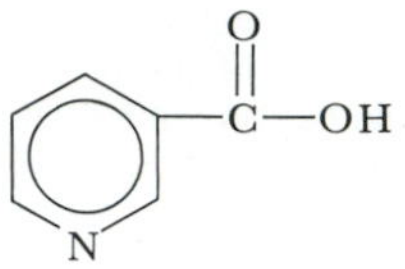

▼ EXAMPLE 14.15 Determining the K_a of an acid

8 Nicotinic acid

Nicotinic acid (C_5H_4NCOOH, **8**) is prepared by the oxidation of nicotine (a byproduct of the tobacco industry) with concentrated nitric acid. The pH of a 0.030 M nicotinic acid solution is 3.19. What is the K_a and the percentage ionization of nicotinic acid?

STRATEGY We know that nicotinic acid is weak because it does not appear in the list of strong acids (Table 14.4). Moreover, if it were a strong acid, the pH of the solution would be $-\log(0.030) = 1.52$. Because the pH of the solution is greater than this value, the acid must be weak. To calculate K_a, we need to know the molar concentration of H_3O^+, which we are given (in the form of the pH). We also need to know the molar concentration of the conjugate base—but that must be the same as the concentration of hydronium ions because one A^- ion is formed for each H_3O^+ ion that is formed. Finally, we need to know the molar concentration of nonionized acid, which is the initial molar concentration minus that which ionizes. With the three concentrations known (step 3), all we need to do is substitute their values into the expression for K_a. For the percentage ionized, we need the molar concentrations of the ionized acid and the nonionized acid.

SOLUTION The molar concentration of the hydronium ion (and also that of the conjugate base) is obtained from the pH:

$$[H_3O^+] = 10^{-3.19}\ \text{mol/L} = 6.5 \times 10^{-4}\ \text{mol/L}$$

$$[C_5H_4NCO_2^-] = 6.5 \times 10^{-4}\ \text{mol/L}$$

The equilibrium table is

Equilibrium equation:	$C_5H_4NCOOH(aq) + H_2O(l) \rightleftharpoons H_3O^+(aq) + C_5H_4NCO_2^-(aq)$		
Species:	C_5H_4NCOOH	H_3O^+	$C_5H_4NCO_2^-$
Step 1. Initial concentration, mol/L	0.030	0	0
Step 2. Change in concentration, mol/L	-6.5×10^{-4}	$+6.5 \times 10^{-4}$	$+6.5 \times 10^{-4}$
(The reaction stoichiometry indicates that the molar concentrations of H_3O^+ and $C_5H_4NCO_2^-$ are produced from the ionization of C_5H_4NCOOH.)			
Step 3. Equilibrium concentration, mol/L	$0.030 - 6.5 \times 10^{-4}$	6.5×10^{-4}	6.5×10^{-4}

Step 4. To obtain the equilibrium constant, the equilibrium values from step 3 are inserted into the reaction quotient, which at equilibrium gives

$$\frac{(6.5 \times 10^{-4})(6.5 \times 10^{-4})}{(0.030 - 6.5 \times 10^{-4})} \approx \frac{(6.5 \times 10^{-4})(6.5 \times 10^{-4})}{(0.030)} = K_a = 1.4 \times 10^{-5}$$

Hence, $pK_a = 4.85$. The percentage of the nicotinic acid ionized is

$$\text{Percentage ionized (\%)} = \frac{\text{concentration of } C_5H_4NCOOH \text{ ionized}}{\text{original concentration of } C_5H_4NCOOH} \times 100\%$$

$$= \frac{6.5 \times 10^{-4}}{0.030} \times 100\% = 2.2\%$$

EXERCISE E14.15 The organic base codeine is present at 0.7 to 2.5% levels in opium, the precise value depending on the source. The pH of a 0.010 M aqueous solution of the codeine is 10.10. What is the value of K_b for codeine?

[*Answer*: 1.6×10^{-6}]

14.11 POLYPROTIC ACIDS AND BASES

Brønsted acids that can donate more than one proton are called **polyprotic acids**. The ionization constants for the loss of successive protons from a triprotic acid of formula H_3A are denoted K_{a1} for the loss of the first proton, K_{a2} for the loss of the second proton, and K_{a3} for the loss of the third proton:

E To be strictly correct, there is no such thing as a "polyprotic acid" in Brønsted theory: H_3A, H_2A^-, and HA^{2-} are each acids in their own right, since each one can donate a proton to give the conjugate bases H_2A^-, HA^{2-}, and A^{3-}, respectively.

U Exercises 14.88 and 14.89.

$$H_3A(aq) + H_2O(l) \rightleftharpoons H_3O^+(aq) + H_2A^-(aq)$$

$$\frac{[H_3O^+][H_2A^-]}{[H_3A]} = K_{a1}$$

$$H_2A^-(aq) + H_2O(l) \rightleftharpoons H_3O^+(aq) + HA^{2-}(aq)$$

$$\frac{[H_3O^+][HA^{2-}]}{[H_2A^-]} = K_{a2}$$

$$HA^{2-}(aq) + H_2O(l) \rightleftharpoons H_3O^+(aq) + A^{3-}(aq) \qquad \frac{[H_3O^+][A^{3-}]}{[HA^{2-}]} = K_{a3}$$

Common examples of diprotic acids are sulfuric acid, H_2SO_4, and carbonic acid, H_2CO_3, each of which can donate two protons. Phosphoric acid, H_3PO_4, is an example of a triprotic acid. The values of the ionization constants of some polyprotic acids are listed in Table 14.7. **Polyprotic bases** are species that can accept more than one proton. Examples include the carbonate anion, $CO_3{}^{2-}$, and the oxalate anion,

T OHT: Table 14.7.

TABLE 14.7 Ionization constants of polyprotic acids

Acid	K_{a1}	pK_{a1}	K_{a2}	pK_{a2}	K_{a3}	pK_{a3}
sulfuric acid, H_2SO_4	strong	—	1.2×10^{-2}	1.92		
oxalic acid, $(COOH)_2$	5.9×10^{-2}	1.23	6.5×10^{-5}	4.19		
sulfurous acid, H_2SO_3	1.5×10^{-2}	1.81	1.2×10^{-7}	6.91		
phosphorous acid, H_2PO_3	1.0×10^{-2}	2.00	2.6×10^{-7}	6.59		
phosphoric acid, H_3PO_4	7.6×10^{-3}	2.12	6.2×10^{-8}	7.21	2.1×10^{-13}	12.67
tartaric acid, $C_2H_4O_2(COOH)_2$	6.0×10^{-4}	3.22	1.5×10^{-5}	4.82		
carbonic acid, H_2CO_3	4.3×10^{-7}	6.37	5.6×10^{-11}	10.25		
hydrosulfuric acid, H_2S	1.3×10^{-7}	6.88	7.1×10^{-15}	14.15		

E Other examples of polyprotic bases are diamines, such as $H_2NCH_2CH_2NH_2$, and protein molecules.

$C_2O_4^{2-}$, both of which can accept two protons, and the phosphate anion, PO_4^{3-}, which can accept three protons.

Carbonic acid has the following ionization equilibria:

$$H_2CO_3(aq) + H_2O(l) \rightleftharpoons H_3O^+(aq) + HCO_3^-(aq)$$

$$\frac{[H_3O^+][HCO_3^-]}{[H_2CO_3]} = K_{a1}$$

$$HCO_3^-(aq) + H_2O(l) \rightleftharpoons H_3O^+(aq) + CO_3^{2-}(aq)$$

$$\frac{[H_3O^+][CO_3^{2-}]}{[HCO_3^-]} = K_{a2}$$

U Exercise 14.90.

The conjugate base of H_2CO_3 in the first equilibrium, HCO_3^-, acts as an acid in the second equilibrium, producing, in turn, its conjugate base, CO_3^{2-}. The equation for the overall ionization is the sum of the two individual equations:

$$H_2CO_3(aq) + 2H_2O(l) \rightleftharpoons 2H_3O^+(aq) + CO_3^{2-}(aq)$$

For the overall ionization,

$$\frac{[H_3O^+]^2[CO_3^{2-}]}{[H_2CO_3]} = K_a \text{ at equilibrium}$$

We can show that this K_a is the product of the two individual ionization constants,

$$K_a = K_{a1} \times K_{a2}$$

by multiplying the expressions for the two constants together and canceling $[HCO_3^-]$, which appears in both:

$$K_{a1} \times K_{a2} = \frac{[H_3O^+][HCO_3^-]}{[H_2CO_3]} \times \frac{[H_3O^+][CO_3^{2-}]}{[HCO_3^-]}$$

$$= \frac{[H_3O^+]^2[CO_3^{2-}]}{[H_2CO_3]} = K_a$$

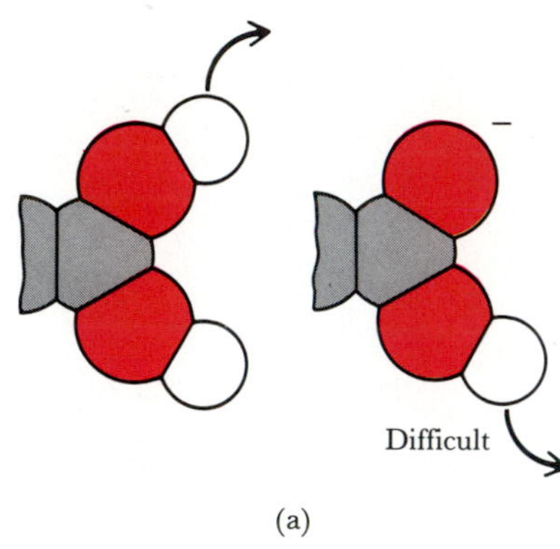

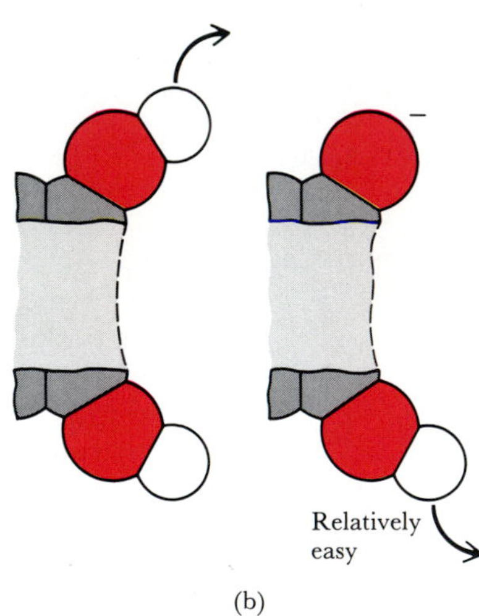

FIGURE 14.18 (a) The loss of a proton from a small molecule can have a strong influence on the loss of a second proton. (b) In a protein molecule, however, the second proton may come from a site far away from the first, and its loss is largely unaffected by the loss of the first.

The strengths of polyprotic acids. The entries in Table 14.7 show that the strengths of polyprotic acids decrease as protons are lost. That is,

$$K_{a1} > K_{a2} > \ldots$$

The decrease is largely due to the greater difficulty of losing a positively charged proton from a negatively charged ion (such as HCO_3^-) than from the original neutral molecule (H_2CO_3). Sulfuric acid shows this decrease in strength very well. The H_2SO_4 molecule is a strong acid and loses its first proton to give its conjugate base, the hydrogensulfate ion, HSO_4^-; this ion, however, is a weak acid:

$$H_2SO_4(aq) + H_2O(l) \rightleftharpoons H_3O^+(aq) + HSO_4^-(aq) \qquad \text{Strong}$$

$$HSO_4^-(aq) + H_2O(l) \rightleftharpoons H_3O^+(aq) + SO_4^{2-}(aq) \qquad pK_a = 1.92$$

Some proteins are complicated polyprotic acids that can donate dozens of protons in reactions. The decrease in acid strength is less pronounced for these species because the protons are usually lost from widely separated sites. The loss of a proton from one site has little influence on a distant site (Fig. 14.18).

▼ EXAMPLE 14.16 Calculating the pH of a polyprotic acid

Calculate the pH of a 0.010 M $H_2SO_4(aq)$ at 25°C. Use the appropriate information from Table 14.7.

STRATEGY The pH depends on the total hydrogen ion concentration, taking both ionization steps into account. The first ionization is complete and yields a hydronium ion concentration equal to the original concentration of the acid, 0.010 mol/L. This concentration corresponds to pH = 2.0. The second ionization adds to the hydrogen ion concentration slightly, so we should expect an overall pH of slightly below 2.0. The contribution of the second ionization is obtained by treating the HSO_4^- formed in the first ionization as a weak acid of concentration 0.010 mol/L and setting up an equilibrium table in the usual way. However, the initial concentration of hydrogen ions in the second ionization is not zero but 0.010 mol/L. Because $K_{a2} = 0.012$ is not very small, it will be necessary to solve a quadratic equation without taking simplifying short cuts.

SOLUTION The equilibrium table is

Equilibrium equation:	$HSO_4^-(aq) + H_2O(l) \rightleftharpoons H_3O^+(aq) + SO_4^{2-}(aq)$		
Species:	HSO_4^-	H_3O^+	SO_4^{2-}
Step 1. Initial concentration, mol/L	0.010	0.010	0
(The initial concentrations of HSO_4^- and H_3O^+ are 0.010 as a result of the ionization of H_2SO_4, a strong acid.)			
Step 2. Change in concentration, mol/L	$-x$	$+x$	$+x$
(When the HSO_4^- ionizes, its concentration is reduced by x and that of H_3O^+ and SO_4^{2-} both increase by x.)			
Step 3. Equilibrium concentration, mol/L	$0.010 - x$	$0.010 + x$	x

Step 4. Now we substitute the concentrations in step 3 into

$$\frac{[H_3O^+][SO_4^{2-}]}{[HSO_4^-]} = K_{a2} \text{ at equilibrium}$$

Because $K_{a2} = 0.012$, this expression becomes

$$\frac{(0.010 + x) \times x}{0.010 - x} = 0.012$$

We have to find x. First, we rearrange this expression into a quadratic equation:

$$x^2 + 0.022x - 1.2 \times 10^{-4} = 0$$

We solve this equation by using the quadratic formula (Appendix 1C):

$$x = \frac{-0.022 \pm \sqrt{(0.022)^2 - 4(-1.2 \times 10^{-4})}}{2} = 4.5 \times 10^{-3} \text{ and } -2.7 \times 10^{-2}$$

Because $x = [SO_4^{2-}]$ (step 3), the value of x cannot be negative, so we select the solution $x = 4.5 \times 10^{-3}$. The total concentration of H_3O^+ ions is the sum of the concentration supplied by the ionization of H_2SO_4 and the additional contribution from the ionization of HSO_4^-. Thus,

$$[H_3O^+] = 0.010 + x = 0.010 + 4.5 \times 10^{-3} = 1.5 \times 10^{-2}$$

so

$$pH = -\log(1.5 \times 10^{-2}) = 1.84, \text{ or about } 1.8$$

D pH of phosphates. TDC, 18-14.

N For *weak* polyprotic acids, the pH is generally due only to the first ionization. Generally, analysis of the second (or last) ionization is used only to determine the concentration of the fully unprotonated conjugate base of the polyprotic acid.

U Exercise 14.91.

The pH of the solution is less than 2.0, as we predicted.

EXERCISE E14.16 Calculate the pH of a 0.10 M oxalic acid solution. Refer to Table 14.7 for K_a values. *Hint:* In this diprotic acid, the first ionization is not 100% as it is for sulfuric acid—the first ionization must be treated as a weak acid, as well as the second ionization.

[*Answer*: 1.28]

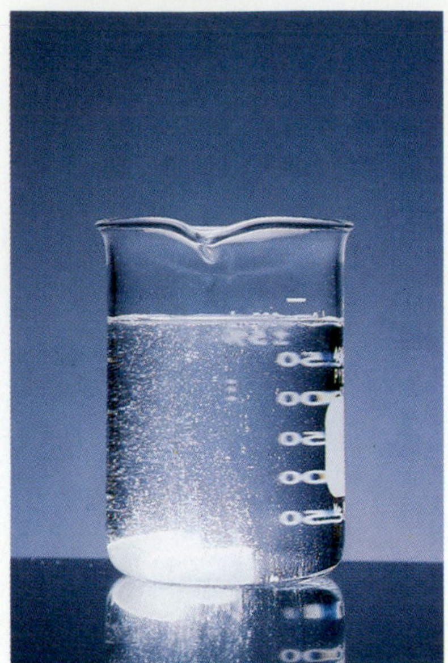

FIGURE 14.19 The effervescence that occurs when an antacid tablet containing hydrogencarbonate ions is dissolved in acid is the result of a disturbance that is passed from one equilibrium to another.

In calculating the pH of a polyprotic acid like carbonic acid for which all the ionization stages are weak, it is normally sufficient (except at very low concentrations) to take only the first ionization into account: the second ionization is much less important than the first. In the case of H_2CO_3, for instance, $K_{a2} = 5.6 \times 10^{-11}$, which is much smaller than $K_{a1} = 4.3 \times 10^{-7}$. That being so, we can ignore the contribution of the second ionization to the pH and treat the acid as being a monoprotic acid with $K_{a1} = 4.3 \times 10^{-7}$. The pH of the solution is calculated by the same method used in Example 14.13.

The second ionization of carbonic acid is important when we want to calculate the concentration of CO_3^{2-} ions at equilibrium. The first ionization does not produce these ions directly, but they are one of the products of the second ionization.

The effervescence of carbon dioxide that occurs when an acid is added to a carbonate or a hydrogencarbonate can be discussed in terms of proton transfer equilibria (Fig. 14.19). According to Le Chatelier's principle, the added hydrogen ions shift the equilibrium composition to the left, thereby favoring the formation of H_2CO_3:

$$H_2CO_3(aq) + H_2O(l) \rightleftharpoons H_3O^+(aq) + HCO_3^-(aq)$$

Carbonic acid in solution is also in equilibrium with dissolved CO_2 molecules according to

$$H_2CO_3(aq) \rightleftharpoons H_2O(l) + CO_2(aq)$$

However, when the H_2CO_3 concentration increases, the CO_2 concentration increases in response. The latter increase is so great that the CO_2 bubbles out of solution with the familiar fizz. As in this everyday example, we often see chemical effects being transmitted along chains of equilibria, with the disturbance of one equilibrium affecting another equilibrium and perhaps still others. This coupling of equilibria is particularly important when acids and bases are both present in the same solution, as we shall see in the next chapter.

N The environmental carbonic acid equilibrium is very complex because of the solubility equilibria that involve CO_3^{2-} ions with Ca^{2+} and Mg^{2+} ions. Carbonic acid exists only at very low concentrations in aqueous solutions; at higher concentrations the equilibrium between dissolved CO_2 and H_2CO_3 molecules moves in favor of the dissolved CO_2.

U Exercises 14.92–14.94.

D Reactions of carbon dioxide in aqueous solution. BZS2, 6.2.

SUMMARY

14.1 A **Brønsted acid** is a proton donor and a **Brønsted base** is a proton acceptor. In a **neutralization reaction**, a proton is transferred from a Brønsted acid (which in solutions of a strong acid is H_3O^+) to a Brønsted base (OH^- in a solution of a strong base).

14.2 Acids and bases exist in solution in a state of dynamic equilibrium. The **conjugate base** of a Brønsted acid is the base that is formed when the acid has donated a proton; the **conjugate acid** of a Brønsted base is the acid that is formed when the base has accepted a proton. In solution, the dynamic equilibrium has the form $acid_1 + base_2 \rightleftharpoons acid_2 + base_1$. Water is **amphiprotic**, being both a proton donor and a proton acceptor; and in solution it undergoes **autoionization**, the transfer of a proton from one molecule to another of the same kind. The autoionization equilibrium is described by the **water autoionization constant** K_w.

14.3 A **strong acid** is a Brønsted acid in which the proton transfer equilibrium lies completely in favor of H_3O^+ and the acid's conjugate base. A **strong base** is a Brønsted base in which the proton transfer equilibrium lies completely in favor of OH^- and the base's conjugate acid. The hydrogen ion concentration in a solution of a strong base is calculated by determining the concentration of OH^- ions and then taking into account the water autoionization equilibrium.

14.4 Hydronium ion concentrations are often reported in terms of the **pH** of the solution: the higher the hydronium ion concentration (the more acidic the solution), the lower the pH. The pH of an acidic aqueous solution is less than 7, that of pure water is 7, and the pH of a basic aqueous solution is greater than 7.

14.5 The pH of a strong acid, a strong base, and a mixture of the two is calculated by determining the amount of hydronium ions in solution and dividing by the total volume of the solution. When the solution is basic, the OH^- ion concentration is calculated and then converted to hydronium ion concentration by using the water autoionization constant.

14.6 A **pH curve** is the plot of the changing pH of a solution as **titrant** (the solution in a buret) is added to **analyte** (the solution being analyzed) in an acid-base titration. (A typical pH curve is shown in Fig. 14.9.) The pH at the **stoichiometric point** of a strong acid-strong base titration is 7.

14.7 A key feature of the Brønsted-Lowry theory of acids and bases is its emphasis on the equilibrium between conjugate acids and bases in solution. The proton-transferring ability of an acid is expressed in terms of the **acid ionization constant** K_a and that of a base in terms of the **base ionization constant** K_b. The stronger the Brønsted base, the weaker its conjugate acid.

14.8 A **weak acid** is a proton donor that is only partially ionized in water. A **weak base** is a proton acceptor that is partially ionized in water.

14.9 The variation of acid and base strength is very difficult to explain because it depends on the properties of the solvent as well as those of the solute. However, in some cases it is possible to account for trends. For binary acids, the more polar and the weaker the H—A bond, the stronger the acid. For oxoacids, the greater the number of O atoms attached to the central atom (or the greater the oxidation number of the central atom), the stronger the acid. When the same number of O atoms are attached to the central atom, then the greater the electronegativity of the central atom, the stronger the acid.

14.10 The pH of a solution of a weak acid is calculated by taking into account the equilibrium between the acid HA and its conjugate base by using the equilibrium table technique established in Chapter 13. A similar procedure is used for solutions of a weak base, but pOH is calculated first and then converted to pH.

14.11 A **polyprotic acid** is a Brønsted acid that can donate more than one proton, and its successive ionizations are described by successive ionization constants. A **polyprotic base** is a Brønsted base that can accept more than one proton. The strengths of polyprotic acids decrease as protons are lost. In calculating the pH of a polyprotic acid for which all ionization stages are weak, it is normally sufficient (except at very low concentrations) to take only the first ionization into account. The second ionization of a weak acid is important when calculating the concentration of ions that the first ionization does not produce directly.

CLASSIFIED EXERCISES

Unless stated otherwise, assume that all solutions are aqueous and that the temperature is 25°C.

Brønsted Acids and Bases

14.1 Write the formulas for the conjugate acids of (a) CH_3NH_2 (methylamine) and (b) HCO_3^- and the conjugate bases of (c) HCO_3^- and (d) C_6H_5OH (phenol).

14.2 Write the formulas for the conjugate acids of (a) H_2O and (b) $C_6H_5NH_2$ (aniline) and the conjugate bases of (c) H_2S and (d) HPO_4^{2-} (hydrogenphosphate ion).

14.3 Write the proton transfer equilibria for the following acids in aqueous solution and identify the conjugate acid-base pairs in each one: (a) H_2SO_4; (b) $C_6H_5NH_3^+$ (anilinium ion); (c) $H_2PO_4^-$ (dihydrogenphosphate ion); (d) $NH_2NH_3^+$ (hydrazinium ion).

14.4 Write the proton transfer equilibria for the following bases in aqueous solution and identify the conjugate acid-base pairs in each one: (a) CN^-; (b) NH_2NH_2 (hydrazine); (c) CO_3^{2-}; (d) NH_2CONH_2 (urea).

14.5 Write the two proton transfer equilibria that demonstrate the amphiprotic character of (a) HCO_3^- and (b) HPO_4^{2-} and identify the conjugate acid-base pairs in each equilibrium.

14.6 Write the two proton transfer equilibria that show the amphiprotic character of (a) $H_2PO_4^-$ and (b) $HC_2O_4^-$ (hydrogenoxalate ion).

Autoionization

14.7 The value of K_w for water at body temperature (37°C) is 2.5×10^{-14}. (a) What is the value of $[H_3O^+]$ and the pH of neutral water at 37°C? (b) What is the molar concentration of OH^- ions and the pOH of neutral water at 37°C?

14.8 The concentration of H_3O^+ ions at the freezing point of water is 3.9×10^{-8} mol/L. (a) Calculate K_w and pK_w at 0°C. (b) What are the values of pH and pOH for neutral water at 0°C?

14.9 Heavy water, D_2O, is used in some nuclear reactors (see Chapter 22). The K_w for heavy water at 25°C is 1.35×10^{-15}. (a) Write the chemical equation for the autoionization of D_2O. (b) Evaluate pK_w for D_2O at 25°C. (c) Calculate the molar concentrations of D_3O^+ and OD^- in neutral heavy water at 25°C. (d) Evaluate the pD and pOD of neutral heavy water at 25°C? (e) Find the relation between pD, pOD, and pK_w.

14.10 Although many chemical reactions take place in water, it is often necessary to use other solvents instead, and liquid ammonia (b.p. −33°C) has been used extensively. Many of the reactions that occur in water have analogous reactions in liquid ammonia. (a) Write the chemical equation for the autoionization of NH_3. (b) What are the formulas of the acid and base species that result from the autoionization of liquid ammonia? (c) The autoionization constant, K_{am}, of liquid ammonia has the value 1×10^{-33} at −35°C. What is the value of pK_{am} at that temperature? (d) What is the molar concentration of NH_4^+ ions in neutral liquid ammonia? (e) Evaluate pNH_4 and pNH_2 in neutral liquid ammonia at −35°C. (f) Determine the relation between pNH_4, pNH_2, and pK_{am}.

Strong Acids and Bases

14.11 Calculate the molar concentrations of HCl, H_3O^+, Cl^-, and OH^- in an aqueous solution that contains 0.48 mol of HCl in 500 mL of solution.

14.12 Calculate the molar concentrations of KNH_2, K^+, NH_2^-, OH^-, and H_3O^+ in an aqueous solution that contains 1.0 g of KNH_2 in 250 mL of solution.

14.13 The pH of several solutions was measured in the research laboratories of a food company; convert each of the following pH values to the molar concentration of H_3O^+ ions: (a) 6.7 (the pH of a saliva sample); (b) 4.4 (the pH of beer); (c) 5.3 (the pH of a urine sample).

14.14 The pH of several solutions was measured in a hospital laboratory; convert each of the following pH values to the molar concentration of H_3O^+ ions: (a) 5 (the pH of coffee); (b) 7.4 (the pH of blood); (c) 10.5 (the pH of milk of magnesia).

14.15 The molar concentration of H_3O^+ ions in the following solutions was measured at 25°C. Calculate the pH and pOH of the solutions: (a) 2.0×10^{-5} mol/L (sample of rain water); (b) 1.0 mol/L; (c) 5.02×10^{-5} mol/L.

14.16 The molar concentration of OH^- ions in the following solutions was measured at 25°C. Calculate the pH and pOH of the solutions: (a) 6.7×10^{-8} mol/L (a sample of milk); (b) 1.00×10^{-8} mol/L (a sample of tap water); (c) 7.09×10^{-4} mol/L.

14.17 Calculate the pH and pOH of each of the following aqueous solutions of strong acid or base: (a) 0.010 M HNO_3; (b) 0.22 M HCl; (c) 10.0 mL of 0.022 M KOH after dilution to 250 mL; (d) 50.0 mL of 0.00043 M HBr after dilution to 250 mL.

14.18 Calculate the pH and pOH of each of the following aqueous solutions of strong acid or base: (a) 0.0149 M HI; (b) 0.0602 M HCl; (c) 1.73×10^{-3} M $Ba(OH)_2$; (d) 10.0 mL of 0.0022 M NaOH after dilution to 250 mL.

Titrations

14.19 Calculate the molar concentration of H_3O^+ ions and the pH of the following solutions: (a) 25.0 mL of 0.30 M HCl was added to 25.0 mL of 0.20 M NaOH; (b) 25.0 mL of 0.15 M HCl was added to 50.0 mL of 0.15 M KOH.

14.20 Calculate the molar concentration of H_3O^+ ions and the pH of the following solutions: (a) 17.3 mL of 0.25 M HCl was added to 15.0 mL of 0.33 M NaOH; (b) 15.94 mL of 0.101 M NaOH was added to 25.0 mL of 0.094 M HNO_3.

14.21 A 14.0-g sample of NaOH was dissolved in 250 mL of solution, and then 25.0 mL was pipeted into 50.0 mL of 0.20 M HBr. What is the pH of the resulting solution?

14.22 A 0.150-g sample of $Ba(OH)_2$ was dissolved in 50.0 mL of solution, and then 25.0 mL was pipeted into 100 mL of 0.0010 M HCl. What is the pH of the resulting solution?

14.23 What volume of 0.0631 M HCl is required to neutralize 25.0 mL of a 0.0497 M KOH solution?

14.24 It was found that 24.7 mL of 0.184 M HI solution was required to neutralize 20.0 mL of a $Ba(OH)_2$ solution. What is the molar concentration of the $Ba(OH)_2$ solution?

14.25 Sketch reasonably accurately the pH curve for the titration of 20.0 mL of 0.10 M HCl with 0.20 M KOH. Mark on the curve (a) the initial pH and (b) the pH at the stoichiometric point.

14.26 Sketch reasonably accurately the pH curve for the titration of 20.0 mL of 0.10 M $Ba(OH)_2$ with 0.20 M HCl. Mark on the curve (a) the initial pH and (b) the pH at the stoichiometric point.

14.27 Calculate the volume of 0.150 M HCl required to neutralize (a) one-half and (b) all the hydroxide ions in 25.0 mL of 0.110 M NaOH. (c) What is the concentration of Na^+ ions at the stoichiometric point? (d) Calculate the pH of the solution after the addition of 20.0 mL of 0.150 M HCl to 25.0 mL of 0.110 M NaOH.

14.28 Calculate the volume of 0.116 M HCl required to neutralize (a) one-half and (b) all the hydroxide ions in 25.0 mL of 0.215 M KOH. (c) What is the concentration of Cl^- ions at the stoichiometric point? (d) Calculate the pH of the solution after the addition of 40.0 mL of 0.116 M HCl to 25.0 mL of 0.215 M KOH.

14.29 (a) A 2.54-g sample of NaOH was dissolved in 25.0 mL of solution. What volume (in milliliters) of 0.150 M HCl is required to reach the stoichiometric point? (b) What is the molar concentration of Cl^- ions at the stoichiometric point?

14.30 (a) A 2.88-g sample of KOH was dissolved in 25.0 mL of solution. What volume of 0.200 M HNO_3 is required to reach the stoichiometric point? (b) What is the molar concentration of NO_3^- ions at the stoichiometric point?

14.31 A 0.968-g sample of impure sodium hydroxide was dissolved in 200 mL of aqueous solution. A 20.0-mL portion of this solution was titrated to the stoichiometric point with 15.8 mL of 0.107 M HCl. What is the percentage purity of the original sample?

14.32 A 1.331-g sample of impure barium hydroxide was dissolved in 250 mL of aqueous solution. A 35.0-mL portion of this solution was titrated to the stoichiometric point with 17.6 mL of 0.0935 M HCl. What is the percentage purity of the original sample?

14.33 Calculate the pH at each stage in the titration for the addition of 0.150 M HCl to 25.0 mL of 0.110 M NaOH (a) initially; (b) after the addition of 5.0 mL of acid; (c) after the addition of a further 5.0 mL; (d) at the stoichiometric point; (e) after the addition of 5.0 mL of acid beyond the stoichiometric point; (f) after the addition of 10 mL of acid beyond the stoichiometric point.

14.34 Calculate the pH at each stage in the titration in which 0.116 M HCl is added to 25.0 mL of 0.215 M KOH (a) initially; (b) after the addition of 5.0 mL of acid; (c) after the addition of a further 5.0 mL; (d) at the stoichiometric point; (e) after the addition of 5.0 mL of acid beyond the stoichiometric point; (f) after the addition of 10 mL of acid beyond the stoichiometric point.

14.35 For the titration specified in Exercise 14.33, calculate the pH at (a) 0.1 mL and (b) 0.01 mL before the stoichiometric point and (c) 0.01 mL and (d) 0.1 mL after the stoichiometric point. Sketch the pH curve.

14.36 For the titration specified in Exercise 14.34, calculate the pH at (a) 0.1 mL and (b) 0.01 mL before the stoichiometric point and (c) 0.01 mL and (d) 0.1 mL after the stoichiometric point. Sketch the pH curve.

Ionization Constants

14.37 Refer to Table 14.6. Name the following acids, write the K_a and pK_a values, and list the acids in order of increasing strength: (a) HCOOH; (b) CH_3COOH; (c) CCl_3COOH; (d) C_6H_5COOH.

14.38 Refer to Table 14.6. Name the following acids, write the K_a and pK_a values, and list the acids in order of increasing strength: (a) HCN; (b) HIO_3; (c) HNO_2; (d) HF.

14.39 Give the K_a values and list the following acids in order of increasing strength: (a) phosphoric acid, H_3PO_4, $pK_{a1} = 2.12$; (b) phosphorous acid, H_3PO_3, $pK_{a1} = 2.00$; (c) selenous acid, H_2SeO_3, $pK_{a1} = 2.46$; (d) selenic acid, H_2SeO_4, $pK_{a1} = 1.92$.

14.40 Give the pK_b values and list the following bases in order of increasing strength: (a) ammonia, NH_3, $K_b = 1.8 \times 10^{-5}$; (b) deuterated ammonia, ND_3, $K_b = 1.1 \times 10^{-5}$; (c) hydrazine, NH_2NH_2, $K_b = 1.7 \times 10^{-6}$; (d) hydroxylamine, NH_2OH, $K_b = 1.1 \times 10^{-8}$.

14.41 (a) Write the names and formulas for the strongest and weakest conjugate acids of the bases listed in Exercise 14.40. (b) What are the K_a values for the two acids? (c) Which acid, dissolved in water to a given concentration, would produce a solution with the highest pH?

14.42 (a) Write the names and formulas for the strongest and weakest conjugate bases of the acids listed in Exercise 14.39. (b) What are the K_b values for the two bases? (c) Which base, dissolved in water to a given concentration, would produce a solution with the highest pH?

Structures and Strengths of Acids

14.43 Determine (and explain) which acid is stronger: (a) HClO or $HClO_2$; (b) HBrO or HClO; (c) H_2SO_4 or H_3PO_4; (d) HNO_3 or HNO_2.

14.44 Determine (and explain) which acid is stronger: (a) $HBrO_3$ or HBrO; (b) H_3PO_4 or H_3PO_3; (c) H_2S or H_2Se; (d) HClO or HIO.

14.45 The values of K_a for phenol (left) and 2,4,6-trichlorophenol (right) are 1.3×10^{-10} and 1.0×10^{-6}, respectively. Which is the stronger acid? Account for the differences in acid strength.

Phenol   2,4,6-Trichlorophenol

14.46 The values of pK_b for aniline (left) is 9.37 and that for 4-chloroaniline (right) is 9.85. Which is the stronger base? Account for the differences in base strength.

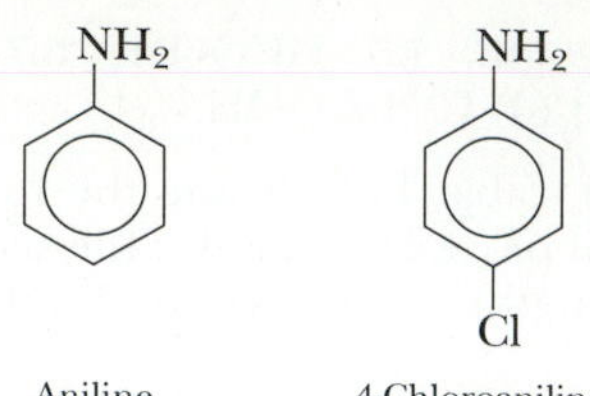

Aniline 4-Chloroaniline

14.47 Arrange the following bases in order of increasing strength by using the pK_a values of their conjugate acids, which are given in parentheses: (a) ammonia (9.25); (b) methylamine (10.56); (c) ethylamine (10.72); (d) aniline (4.63). Is there a simple pattern of strengths?

14.48 Arrange the following bases in order of increasing strength by using the pK_a values of their conjugate acids, which are given in parentheses: (a) aniline (4.63); (b) 2-hydroxyaniline (left, 4.72); (c) 3-hydroxyaniline (middle, 4.17); (d) 4-hydroxyaniline (right, 5.47). Is there a simple pattern of strengths?

2-Hydroxyaniline 3-Hydroxyaniline 4-Hydroxyaniline

Weak Acids and Bases

Refer to Tables 14.6 and 14.7 for the appropriate K_a and K_b values for the following exercises.

14.49 Determine the percentage ionization of the solute in (a) 0.20 M C_6H_5COOH and (b) 0.20 M NH_2NH_2 (hydrazine).

14.50 Determine the percentage ionization of the solute in (a) 0.15 M HCOOH and (b) 0.10 M NH_3.

14.51 Assume the molar concentration of water is 55.5 M. What fraction of H_2O molecules accept protons to form H_3O^+ ions in a 0.20 M C_6H_5COOH solution?

14.52 Assume the molar concentration of water is 55.5 M. What fraction of H_2O molecules donate protons to form OH^- ions in a 0.10 M NH_3 solution?

14.53 Calculate the pH, pOH, and percentage ionization of the solute in the following aqueous solutions: (a) 0.15 M CH_3COOH; (b) 0.15 M CCl_3COOH.

14.54 Calculate the pH, pOH, and percentage ionization of solute in the following aqueous solutions: (a) 0.20 M $CH_3CH(OH)COOH$ (lactic acid); (b) 0.10 M $C_6H_5SO_3H$ (benzenesulfonic acid).

14.55 Calculate the pH, pOH, and percentage ionization of solute in the following aqueous solutions: (a) 0.017 M NH_2OH; (b) 0.020 M codeine, given that the pK_a of its conjugate acid is 8.21. Codeine, a cough suppressant, is extracted from opium.

14.56 Calculate the pOH, pH, and percentage ionization of solute in the following aqueous solutions: (a) 0.0058 M $C_{10}H_{14}N_2$ (nicotine); (b) 0.020 M quinine, given that the pK_a of its conjugate acid is 8.52; (c) 0.011 M strychnine, given that the K_a of its conjugate acid is 5.49×10^{-9}.

14.57 (a) When the pH of a 0.10 M $HClO_2$ aqueous solution was measured, it was found to be 1.57. What are the values of K_a and pK_a of chlorous acid? (b) The pH of a 0.10 M propylamine, $C_3H_7NH_2$, was measured as 11.86. What are the values of K_b and pK_b of propylamine?

14.58 (a) The pH of a 0.015 M HNO_2 aqueous solution was measured as 2.63. What are the values of K_a and pK_a of nitrous acid? (b) The pH of a 0.10 M butylamine, $C_4H_9NH_2$, aqueous solution was measured as 12.04. What are the percentage ionization and the values of K_b and pK_b of butylamine?

14.59 Cacodylic acid is used as a cotton defoliant. A 0.011 M cacodylic acid solution is 0.77% ionized in water. What is the pH of the solution and the K_a of cacodylic acid?

14.60 The percentage ionization of veronal (diethylbarbituric acid) in a 0.020 M aqueous solution is 0.14%. What is the pH of the solution and the K_a of veronal?

14.61 The percentage ionization of octylamine (an organic base) in a 0.10 M aqueous solution is 6.7%. What is the pH of the solution and the K_b of octylamine?

14.62 When 150 mg of an organic base of molar mass 31.06 g/mol is dissolved in 50.0 mL of water, the pH is found to be 10.05. What is the percentage ionization of the base? Calculate the pK_b of the base and the pK_a of its conjugate acid.

Polyprotic Acids

Refer to Tables 14.6 and 14.7 for the appropriate K_a and K_b values for the following exercises.

14.63 Write the stepwise proton transfer equilibria for the ionization of (a) sulfuric acid, H_2SO_4; (b) arsenic acid, H_3AsO_4; (c) phthalic acid, $C_6H_4(COOH)_2$.

14.64 Write the stepwise proton transfer equilibria for the ionization of (a) phosphoric acid, H_3PO_4; (b) adipic acid, $(CH_2)_4(COOH)_2$; (c) succinic acid, $(CH_2)_2(COOH)_2$.

14.65 Calculate the pH of 0.15 M $H_2SO_4(aq)$ at 25°C.

14.66 Calculate the pH of 0.010 M $H_2SeO_4(aq)$ solution given that K_{a1} is very large and $K_{a2} = 1.2 \times 10^{-2}$.

14.67 Calculate the pH of the following diprotic acid solutions at 25°C; ignore second ionizations only when that approximation is justified: (a) 0.10 M H_2S; (b) 1.1×10^{-3} M H_2TeO_4 (telluric acid, for which $K_{a1} = 2.1 \times 10^{-8}$ and $K_{a2} = 6.5 \times 10^{-12}$).

14.68 Calculate the pH of the following acid solutions at 25°C; ignore second ionizations only when that approximation is justified: (a) 1.0×10^{-4} M H_3BO_3 (boric acid acts as a monoprotic acid); (b) 0.015 M H_3PO_4; (c) 0.10 M H_2SO_3.

14.69 (a) Calculate the molar concentrations of $(COOH)_2$, $HOOCCO_2^-$, $(CO_2)_2^{2-}$, H_3O^+, and OH^- in a 0.10 M $(COOH)_2$ solution. (b) Calculate the molar concentrations of H_2S, HS^-, S^{2-}, H_3O^+, and OH^- in a 0.050 M H_2S solution.

14.70 Calculate the molar concentrations of H_2SO_3, HSO_3^-, SO_3^{2-}, H_3O^+, and OH^- in a 0.10 M H_2SO_3 solution.

UNCLASSIFIED EXERCISES

14.71 Write the formula for the conjugate acids of (a) the carbonate ion, CO_3^{2-}, and (b) acetic acid, CH_3COOH, and the conjugate bases of (c) the hydrogenphosphate ion, $H_2PO_4^-$, and (d) the hydroxide ion, OH^-.

14.72 Identify the conjugate acid-base pairs and write the proton transfer equilibria for (a) propionic acid, C_2H_5COOH; (b) acetylsalicylic acid (aspirin), $C_8H_7O_2COOH$; (c) caffeine (a base), $C_8H_{10}N_4O_2$.

14.73 Classify each of the following species as an acid, a base, or amphiprotic substance in aqueous solution (refer to Tables 14.3 and 14.6): (a) H_2O; (b) PO_4^{3-}; (c) CH_3NH_2; (d) $C_6H_5NH_3^+$.

14.74 The value of K_w at 40°C is 3.8×10^{-14}. What is the pH of water at 40°C?

14.75 Calculate the pH and pOH of (a) 0.026 M HNO_3 and (b) 3.19×10^{-3} M KOH.

14.76 Calculate the molar concentration of H_3O^+ ions in a solution having a pH of (a) 7.95; (b) 0.01; (c) 1.99; (d) 11.95.

14.77 A 1.00-g sample of NaOH was dissolved in 100 mL of solution and then diluted to 500 mL. What is the pH of the resulting solution?

14.78 In the determination of the heat evolved in the neutralization of a strong acid by a strong base, 21.0 mL of a 3.0 M HNO_3 solution was mixed with 25.0 mL of a 2.5 M NaOH solution. What is the pH of the resulting solution?

14.79 A 25.0-mL sample of 6.0 M HCl is diluted to 1.0 L, and 15.7 mL of this diluted solution was required to reach the stoichiometric point in the titration of 25.0 mL of a KOH solution. What is the molar concentration of the KOH solution?

14.80 A 20-mL sample of 0.020 M HCl solution was titrated with 0.035 M KOH. Calculate the pH at the following points in the titration and sketch the pH curve: (a) no KOH added; (b) 5.00 mL KOH added; (c) an additional 5.00 mL KOH (for a total of 10.0 mL) added; (d) another 5.0 mL KOH added; (e) another 5.00 mL KOH added. (f) Determine the volume of KOH required to reach the stoichiometric point.

14.81 An old bottle labeled "Standardized 6.0 M NaOH" was found on the back of a shelf in the stockroom. Over time, some of the NaOH had reacted with the glass and the solution was no longer 6.0 M. To determine its purity, 5.0 mL of the solution was diluted to 100 mL and titrated to the stoichiometric point with 11.8 mL of 2.05 M HCl. What is the molar concentration of the sodium hydroxide solution?

14.82 The K_a of phenol (C_6H_5OH, which is also called carbolic acid) is 1.3×10^{-10} and that for the ammonium ion is 5.6×10^{-10}. (a) Write the proton transfer equilibria for each in aqueous solution. (b) Calculate the pK_a for each. (c) Which acid is the stronger acid?

14.83 Determine the value of K_b for the following anions (all of which are the conjugate bases of acids) and list them in order of increasing base strength (refer to Tables 14.4, 14.6, and 14.7 for information): F^-, $CH_2ClCO_2^-$; CO_3^{2-}; IO_3^-; Cl^-.

14.84 Identify the stronger acid in each of the following pairs and give reasons for your choice: (a) $HBrO_4$ or HIO_4; (b) HIO_2 or HIO_3; (c) H_3AsO_4 or H_2SeO_4.

14.85 Calculate the percentage ionization of the solute in (a) 0.15 M C_6H_5COOH (benzoic acid); (b) 0.0058 M $C_{10}H_{14}N_2$ (nicotine, a base).

14.86 Calculate the pH and the percentage ionization of the solute in (a) 0.0477 M HCN (hydrocyanic acid) and (b) 0.023 M HBrO.

14.87 Calculate the pOH and pH of each of the following aqueous solutions: (a) 0.10 M $C_6H_{11}NH_2$ (cyclohexamine, for which $pK_b = 3.36$); (b) 0.0194 M $C_{17}H_{19}O_3N$ (morphine, a base); (c) 0.015 M $NH_2CH_2CH_2NH_2$ (ethylenediamine), given that the K_a of its conjugate acid is 1.9×10^{-11}.

14.88 (a) The percentage ionization of benzoic acid in a 0.110 M solution is 2.4%. What is the pH of the solution and the K_a of benzoic acid? (b) The percentage ionization of barbituric acid in a 0.20 M solution is 2.2%. What is the pH of the solution and the K_a of barbituric acid?

14.89 (a) The pH of a 0.025 M aqueous solution of a base was 11.6. What is the pK_b of the base and the pK_a of its conjugate acid? (b) The percentage ionization of thiazole (an organic base) in a 0.0010 M solution is 5.2 ×

CASE 14 ▸ RAIN IS NATURALLY ACIDIC, SO WHY THE CONCERN?

Pure water has a pH of 7.0, but the observed pH of rainfall varies according to the components present in the atmosphere when precipitation occurs. The major component affecting the acidity of rainfall is the presence of CO_2. Carbon dioxide is an acidic oxide, which when dissolved in water produces an acidic solution:

$$CO_2(g) + 2H_2O(l) \longrightarrow H_3O^+(aq) + HCO_3^-(aq)$$

The partial pressure of carbon dioxide in air that is saturated with water vapor at 25°C and at an overall pressure of 1 atm is 3.04×10^{-4} atm. Because Henry's law constant for carbon dioxide in water is 2.3×10^{-2} mol/(L·atm) and for carbonic acid, H_2CO_3, $pK_{a1} = 6.37$, the pH of "normal" rainwater is 5.7 (see Question 1).

The concern about acid rain centers on the fact that rainwater often has a pH of less than 5.7. The northeastern part of the United States, the neighboring provinces of Canada, and the Scandinavian countries have all experienced rainfall with a pH as low as 4. In most heavily industrial and populated areas where fossil fuel combustion is necessary either as an industrial energy source or for transportation, the pH of rainfall is lower than normal. This low pH is caused by the acidic oxides sulfur dioxide, SO_2, and nitrogen oxides (denoted NO_x, signifying both NO_2 and NO). In fact, all nonmetallic oxides, being acidic oxides, contribute to the acidity of rain. In the western and southwestern United States, the pH of rainfall is often near 7. This value reflects the upsweep of minerals from the dry soils in the region into the atmosphere. Because these minerals often contain carbonates, the normally acidic rain is neutralized.

One major source of atmospheric SO_2 is the oxidation of sulfur during the combustion of fossil fuel, for all fossil fuels contain various amounts of sulfur (from a trace to about 4%). There are two possible fates for the sulfur dioxide in the atmosphere. It may combine with water to form sulfurous acid:

$$SO_2(g) + 2H_2O(l) \longrightarrow H_3O^+(aq) + HSO_3^-(aq)$$

Alternatively, in the presence of particulate matter and aerosols, it may react with oxygen to form sulfur trioxide, the acidic oxide of sulfuric acid:

$$2SO_2(g) + O_2(g) \xrightarrow{\text{particulates}} 2SO_3(g)$$

$$SO_3(g) + 2H_2O(l) \longrightarrow H_3O^+(aq) + HSO_4^-(aq)$$

Both mechanisms of SO_2 removal generate acidic conditions.

An important source of nitrogen oxides is the reaction of nitrogen and oxygen in the high-temperature combustion chambers of automobile engines and electrical power generating plants. Nitric oxide, NO, reacts with atmospheric oxygen to form nitrogen dioxide, NO_2:

$$2NO(g) + O_2(g) \longrightarrow 2NO_2(g)$$

The NO_2 reacts with water, forming nitric acid and nitric oxide:

$$3NO_2(g) + 3H_2O(l) \longrightarrow 2H_3O^+(aq) + 2NO_3^-(aq) + NO(g)$$

Again, acid is formed in the raindrops.

The effects of acid rain are far-reaching. A decrease in the pH of lakes and streams can affect the quality, quantity, and type of aquatic life. A decrease in the pH of soils affects the type of plants and crops that can be grown effectively, and the attack of acid on limestone materials in monuments, statues, and buildings and on structural steel can accelerate their deterioration.

10^{-3}%. What is the pH of the solution and K_b of thiazole?

14.90 Write the stepwise proton transfer equilibria for the ionization of (a) hydrosulfuric acid, H_2S; (b) tartaric acid, $C_2H_4O_2(COOH)_2$; (c) malonic acid, $CH_2(COOH)_2$.

14.91 Calculate the pH of the following diprotic acid solutions at 25°C, ignoring second ionizations only when that approximation is justified: (a) 0.0010 M H_2CO_3; (b) 0.10 M $(COOH)_2$.

14.92 Estimate the enthalpy of ionization of formic acid at 25°C given that K_a is 1.765×10^{-4} at 20°C and 1.768×10^{-4} at 30°C. (*Hint:* Use the van't Hoff equation, Exercise 13.88.)

14.93 Convert the van't Hoff equation (stated in Exercise 13.88) for the temperature dependence of an equilibrium constant to an expression for the temperature dependence of pK_w. Estimate the value of pK_w at blood temperature (37°C) from the enthalpy of the water autoionization reaction, which is +57 kJ for the reaction

One consequence of acid rain is vividly illustrated by these two photographs of a forest site in Germany. The one on the left was taken in 1970 and that on the right in 1986.

However, we should keep the problem in perspective, for the implicated detrimental effects associated with acid rain may be less than is currently widely believed. There has been a lack of consistent monitoring (attributable to variations in technique and equipment) and of experimental data to confirm the occurrence of a significant change in pH since the onset of the Industrial Revolution. Moreover, it still remains difficult to understand the relation between the location of the source and the location of the effect.

QUESTIONS

1. Verify by calculation that the pH of "normal" rainwater is about 5.7.

2. One metric ton (1000 kg) of coal that contains 2.5% of sulfur is burned in a coal-fired power plant. (a) What mass of SO_2 is produced? (b) What is the pH of rainwater when the SO_2 dissolves in a volume of water equivalent to 2.0 cm of rainfall over a 2.6 km^2 area? (The pK_{a1} of sulfurous acid is 1.82. Consider the water to be initially pure and at a pH of 7.) (c) If the SO_2 is first oxidized to SO_3 before the rainfall occurs, what would the pH of the rainwater be?

3. One cleanup process installed in coal-fired plants is the passage of the stack gases from the combustion through a dry bed of pulverized calcium carbonate or a wet calcium carbonate slurry. In both cases the reaction

$$CaCO_3(s) + SO_2(g) + \tfrac{1}{2}O_2(g) \longrightarrow CaSO_4(s) + CO_2(g)$$

takes place. (a) What mass of limestone is needed to remove the sulfur dioxide produced in Question 2 if the removal process is 90% efficient? (b) Discuss the costs and trade-offs in the prevention of SO_2 from entering the atmosphere from a coal-fired power plant.

$2H_2O(l) \rightarrow H_3O^+(aq) + OH^-(aq)$. What is the pH of pure water at that temperature?

14.94 Use the van't Hoff equation (see Exercise 13.88) to estimate the enthalpy of the ionization of heavy water, D_2O, given the following data on its autoionization:

Temperature, °C	10	20	25	30	40	50
pK_w	15.44	15.05	14.87	14.70	14.39	14.10

What is the pD (pD = $-\log[D_3O^+]$) of pure heavy water at 25°C? What equations in this chapter would be changed if fully deuterated compounds and heavy water were used throughout?

The goal of this chapter is to explain how acids, bases, and their salts affect the pH of solutions. We also see how mixtures can be used to stabilize solutions against changes in pH, how to interpret changes in pH during acid-base titrations, and how to choose an indicator to detect a stoichiometric point accurately. We shall see that the same techniques can be used to discuss the solubilities of solids in solutions of various kinds. The illustration shows a tiny section of an ocean: the pH and composition of natural waters are maintained by some of the processes that we describe in this chapter.

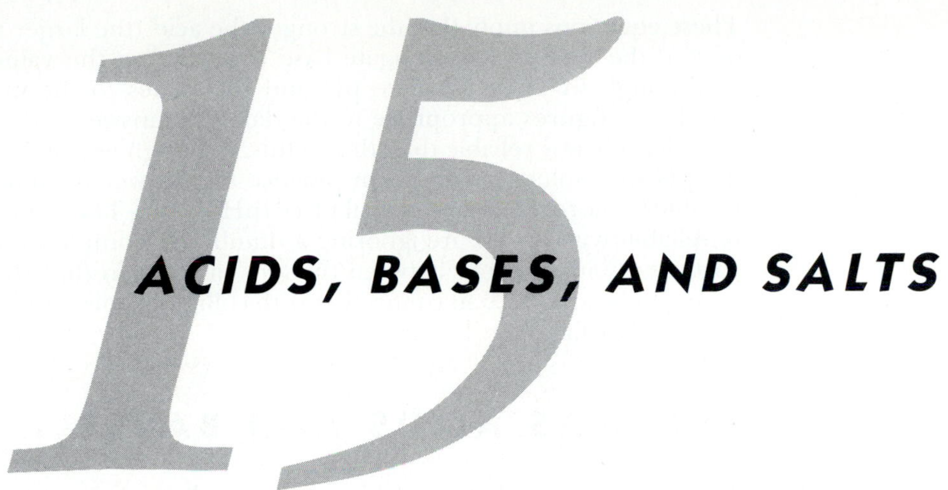

ACIDS, BASES, AND SALTS

A striking illustration of the importance of pH is that we are likely to die if the pH of our blood plasma falls by more than 0.4 from its normal value of 7.4. This change in pH can happen as a result of disease and shock, both of which generate acidic conditions in the body. Death is also likely if the pH of the blood plasma reaches 7.8, as sometimes happens during the early stages of recovery from severe burns. To survive, the body must control its own pH, perhaps with the aid of physicians.

The pH of aqueous solutions—not only blood plasma, but also seawater, detergents, sap, and reaction mixtures—is controlled by the proton transfer that takes place between ions and water molecules. The presence of NH_4^+ ions, which are Brønsted acids, for example, makes a solution acidic; the presence of $CH_3CO_2^-$ ions, which are Brønsted bases, makes it basic. In this chapter we shall see how these ions affect the pH of aqueous solutions. In doing so, we shall make use of two of the equations that were derived in Chapter 14. The first expresses the autoionization equilibrium for water:

$$[H_3O^+][OH^-] = K_w$$

The same relation in logarithmic form is

$$pH + pOH = pK_w$$

According to this equation, if the concentration of either H_3O^+ or OH^- increases, then the concentration of the other must decrease to preserve the value of K_w. At 25°C (the only temperature we shall consider), $pK_w = 14.00$. The other equation expresses the relation between the strength of an acid and its conjugate base:

$$K_a \times K_b = K_w$$

This relation in logarithmic form is

$$pK_a + pK_b = pK_w$$

N This chapter raises the level of the discussion of acid-base equilibria to include contributions to the pH from cations and anions present in the solution. It also presents a discussion of mixed solutions in which one component is a weak electrolyte. Special consideration is given to the common-ion effect, titrations, and buffers. However, in each case we emphasize that each system is merely a particular realization of the basic principles of chemical equilibrium presented in Chapter 13.

N The relation between pH and pOH is like the two sides of a see-saw: if one goes down the other must rise. The same may be said of pK_a and pK_b.

These equations imply that the stronger the acid (the larger the value of K_a), the weaker its conjugate base (the smaller the value of K_b).

Although we shall calculate pH and pK values to the number of significant figures appropriate to the data, the answers may often be considerably less reliable than that. Thus, although we might calculate the pH of a solution as 8.82, in practice the answer is unlikely to be reliable to more than one decimal place (pH = 8.8). The reason for this unreliability is that we are ignoring a number of complicating effects, particularly interactions between the ions in solution (like those mentioned in Section 11.6 in connection with colligative properties and the van't Hoff i factor).

SALTS AS ACIDS AND BASES

D Hydrolysis: Acidic and basic properties of salts. BZS3, 8.13.

N We refrain from using the term *hydrolysis* in this context, and regard it simplest to recognize Brønsted acids and bases directly from their proton-donating or -accepting abilities. The term *hydrolysis* is reserved to mean a reaction with water in which a new element-oxygen bond is formed.

D The puzzling case of ammonium sulfate. CD2, 95. Precipitates and complexes of iron. BZS1, 4.11.

E There are no basic cations because the positive charge repels an incoming proton.

E Cations in the general chemistry course. *J. Chem. Educ.*, **1981**, *58*, 349.

The question we address in this section is what happens when a salt containing an ion that is a Brønsted acid or base is dissolved in water. For example, sodium acetate, $NaCH_3CO_2$, provides the acetate ion, a base; hence, the presence of the dissolved salt in water raises the pH of the solution. Similarly, NH_4Cl provides the NH_4^+ ion, an acid; hence, the addition of ammonium chloride to water lowers the pH of the solution. A salt such as sodium chloride, NaCl, provides neither a Brønsted acid nor a Brønsted base, so its solutions are neutral. These observations may not sound very interesting at first, but the effect of salts on the pH of solutions turns out to be fundamental to chemistry and essential to much of biochemistry.

15.1 IONS AS ACIDS AND BASES

It is relatively easy to decide whether the ions of a salt will affect the pH of a solution. First, we note that, because a salt is a strong electrolyte, it is present only as ions in solution. We can always use the salt's formula to decide which ions are present (Section 2.8). For example, sodium acetate, $NaCH_3CO_2$, is present in solution as Na^+ ions and $CH_3CO_2^-$ ions; potassium carbonate, K_2CO_3, is present in solution as K^+ ions and

TABLE 15.1 Acidic character and K_a values of some common cations in water

Character	Examples	K_a	pK_a
(a) *Acidic cations.* These cations are conjugate acids of weak bases. (See Table 14.6 for a more complete list.)	anilinium ion, $C_6H_5NH_3^+$	2.3×10^{-5}	4.64
	pyridinium ion, $C_5H_5NH^+$	5.6×10^{-6}	5.24
	ammonium ion, NH_4^+	5.6×10^{-10}	9.25
	methylammonium ion, $CH_3NH_3^+$	2.8×10^{-11}	10.56
(b) *Acidic metal cations.* All metal cations produce acidic solutions except those described as neutral cations.	Fe^{3+}	3.5×10^{-3}	2.46
	Cr^{3+}	1.3×10^{-4}	3.89
	Al^{3+}	1.4×10^{-5}	4.85
	Fe^{2+}	1.3×10^{-6}	5.89
	Cu^{2+}	3.2×10^{-8}	7.49
	Ni^{2+}	9.3×10^{-10}	9.03
(c) *Neutral cations.* All Groups I and II metal cations and all other metal cations with oxidation number +1.	Li^+, Na^+, K^+, Mg^{2+}, Ca^{2+}, Ag^+		
(d) *Basic cations.* None	none		

CO_3^{2-} ions. Then, to decide whether a salt solution is acidic or basic, we must decide which ions can affect the acidity of a solution and, for those that do, determine their strength.

Cations as acids. For a cation to be a Brønsted acid, it must have a proton to donate. All cations that are the conjugate acids of weak bases function as Brønsted acids. Table 14.6 lists common weak bases—the cations of all of these bases produce acidic solutions (Table 15.1a). For example, NH_3 is a weak base, so any salt containing the NH_4^+ ion (the conjugate acid of NH_3), such as NH_4Cl, is acidic. The NH_4^+ ion is a proton donor that produces hydronium ions in solution and lowers the pH of the solution:

$$NH_4^+(aq) + H_2O(l) \rightleftharpoons H_3O^+(aq) + NH_3(aq)$$

The general equation for the equilibrium that leads to an acidic solution when a salt containing the conjugate acid ion HB^+ is present is

$$HB^+(aq) + H_2O(l) \rightleftharpoons H_3O^+(aq) + B(aq) \qquad \frac{[H_3O^+][B]}{[HB^+]} = K_a$$

The K_a of an acid HB^+ can always be calculated from the K_b of the parent base B (Table 14.6) by using the relation $K_a \times K_b = K_w$ quoted above.

Some small, highly charged metal cations, such as Al^{3+} and Fe^{3+}, also produce acidic solutions. Even though the cations themselves have no protons to donate, they are strongly hydrated in solution, and the resulting hydrated ion can act as a Brønsted acid if an O—H bond in one of the hydrating H_2O molecules is weakened enough by the electron-withdrawing influence of the cation (Fig. 15.1). For Al^{3+}, which exists in solution as $Al(H_2O)_6^{3+}$, the small, highly charged Al^{3+} ion polarizes the O—H bonds of the hydrating H_2O molecules and makes it possible for at least one or two protons to be lost:

$$Al(H_2O)_6^{3+}(aq) + H_2O(l) \rightleftharpoons H_3O^+(aq) + Al(H_2O)_5OH^{2+}(aq)$$

Some cations that act as acids in this way are listed in Table 15.1b. An example of hydrated metal-cation equilibria of this kind is

$$Fe(H_2O)_6^{3+}(aq) + H_2O(l) \rightleftharpoons H_3O^+(aq) + Fe(H_2O)_5OH^{2+}(aq)$$

$$\frac{[H_3O^+][Fe(H_2O)_5OH^{2+}]}{[Fe(H_2O)_6^{3+}]} = K_a$$

Solutions of the salts of small, highly charged metal cations such as Al^{3+} and Fe^{3+} are often about as acidic as dilute acetic acid (Fig. 15.2). An aqueous solution of titanium(III) sulfate, which contains the Brønsted acid $Ti(H_2O)_6^{3+}$, is so acidic that, like hydrochloric acid, it produces hydrogen sulfide when poured on sodium sulfide (Fig. 15.3). On the other hand, hydrated Na^+ ions, like other metal cations of Groups I and II and those of charge +1 from other groups, are so weak as Brønsted acids that they are considered neutral (Table 15.1c). Any cation that does not affect the pH of a solution is called a **spectator ion**. Such a cation is too large to have an appreciable polarizing effect on the hydrating water molecules that surround it, so these molecules do not readily release their protons. We can state the general rule that *all metal cations produce an acidic solution except those of Groups I and II and those with charge +1.*

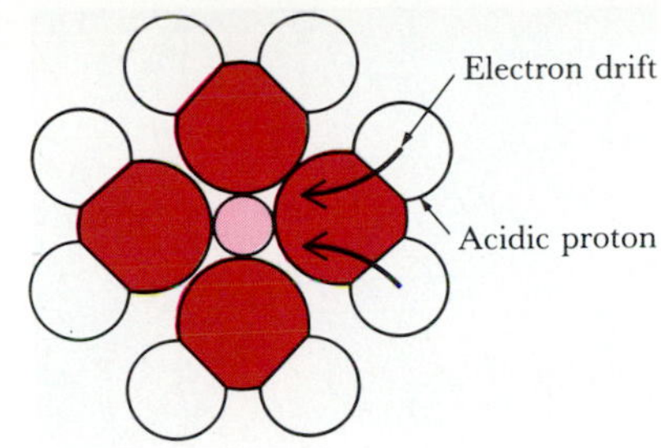

FIGURE 15.1 In water, Al^{3+} cations exist as hydrated ions that can act as Brønsted acids. Although for clarity only four H_2O molecules are shown here, metal cations typically have six H_2O molecules bonded to them.

N Although it is conventional to enclose a complex ion between brackets, we do not introduce that concept until Chapter 21; moreover, its use here would lead to confusion with the use of square brackets in the expressions for the acid and base ionization constants.

T OHT: Tables 15.1 and 15.2.

N To what extent should students memorize Tables 15.1 and 15.2?

E Solutions of Fe^{3+} and Cr^{3+} are more acidic than an equal molar concentration of CH_3COOH.

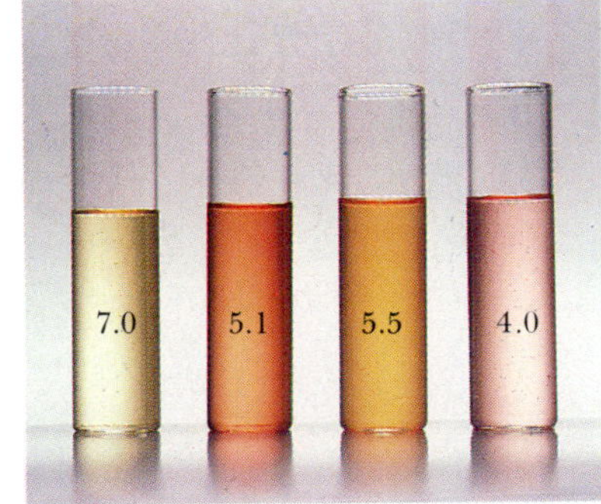

FIGURE 15.2 These four solutions show that hydrated cations can be significantly acidic. The test tubes contain (from left to right) deionized water, 0.1 M aluminum sulfate, 0.1 M titanium(III) sulfate, and 0.1 M acetic acid. All four test tubes contain the same universal indicator. The superimposed numbers are the pH of the solution.

FIGURE 15.3 A solution of titanium(III) sulfate is so acidic that it can release H_2S from some sulfides.

N We use HCO_2^- for the chemical formula of the formate ion to emphasize the equivalence of the O atoms (as in CO_3^{2-} and $CH_3CO_2^-$).

Anions as bases. For an anion to be a Brønsted base, it must be able to accept a proton; all anions that are the conjugate bases of weak acids may act in this way. It follows that all the anions of the acids listed in Table 14.6 produce basic solutions when they are present in water. For example, because formic acid, HCOOH, is a weak acid, any formate salt, such as $NaHCO_2$, provides the formate anion, HCO_2^-, which acts as a base and increases the pH of the solution:

$$H_2O(l) + HCO_2^-(aq) \rightleftharpoons HCOOH(aq) + OH^-(aq)$$

Other anions that produce basic solutions are listed in Table 15.2c.

Anions that are the conjugate bases of weak polyprotic acids (Table 14.7) also accept protons and produce a basic solution. For example, the phosphate ion, PO_4^{3-} (the conjugate base of HPO_4^{2-}), and the carbonate ion, CO_3^{2-} (the conjugate base of HCO_3^-), both produce basic aqueous solutions:

$$H_2O(l) + PO_4^{3-}(aq) \rightleftharpoons HPO_4^{2-}(aq) + OH^-(aq)$$

$$H_2O(l) + CO_3^{2-}(aq) \rightleftharpoons HCO_3^-(aq) + OH^-(aq)$$

The equilibrium established by any anionic base A^- in water is

$$H_2O(l) + A^-(aq) \rightleftharpoons HA(aq) + OH^-(aq) \qquad \frac{[HA][OH^-]}{[A^-]} = K_b$$

The K_b value for any anion that is not listed can be calculated from the K_a value of the parent acid HA (from Tables 14.6 and 14.7) by using the relation $K_a \times K_b = K_w$.

The anions of strong acids are *very* weak Brønsted bases and therefore are assumed not to affect the pH of a solution (Table 15.2b). These neutral, spectator anions include Cl^-, Br^-, I^-, NO_3^-, and ClO_4^-. Only a few anions are acids: they are generally the conjugate bases of polyprotic acids, such as HSO_4^- and $H_2PO_4^-$, which take part in equilibria such as

$$HSO_4^-(aq) + H_2O(l) \rightleftharpoons H_3O^+(aq) + SO_4^{2-}(aq)$$

Examples of these anions are listed in Table 15.2a.

▼ EXAMPLE 15.1 Estimating the pH of a salt solution

Decide whether aqueous solutions of the salts (a) $Ba(NO_2)_2$ and (b) $CrCl_3$ have a pH equal to 7, greater than 7, or less than 7. If the pH is other than 7,

TABLE 15.2 Acidic and basic character of some common anions in water*

Character	Examples
(a) *Acidic anions.* Very few	HSO_4^-, $H_2PO_4^-$
(b) *Neutral anions.* The conjugate bases of strong acids	Cl^-, Br^-, I^-, NO_3^-, ClO_4^-
(c) *Basic anions.* The conjugate bases of weak acids (see Tables 14.6 and 14.7)	F^-, O^{2-}, OH^-, S^{2-}, HS^-, CN^-, CO_3^{2-}, PO_4^{3-}, NO_2^-, $CH_3CO_2^-$, other carboxylate ions

*For K_a and K_b values, see Tables 14.6 and 14.7.

then write a chemical equation that supports your conclusion. For information, refer to Tables 14.6, 14.7, 15.1, and 15.2.

STRATEGY We should first decide what ions are present in the solution and then decide whether either ion is an acid or base, in which case the solution will be acidic (pH < 7) or basic (pH > 7), respectively. If either ion is an acid, then we write the corresponding proton transfer equilibrium as $HB^+(aq) + H_2O(l) \rightleftharpoons H_3O^+(aq) + B(aq)$, where HB^+ is the cation or the hydrated metal cation. If either ion is a base, then we write the equilibrium as $H_2O(l) + A^-(aq) \rightleftharpoons HA(aq) + OH^-(aq)$, where A^- is the anion.

SOLUTION (a) The salt $Ba(NO_2)_2$ is present as Ba^{2+} and NO_2^- ions in water; Ba^{2+} is neutral (it is a Group II ion); NO_2^- is the conjugate base of a weak acid (HNO_2), so the solution will be basic, with pH > 7. The basicity arises from the proton transfer equilibrium

$$H_2O(l) + NO_2^-(aq) \rightleftharpoons HNO_2(aq) + OH^-(aq)$$

(b) The salt $CrCl_3$ is present as Cr^{3+} and Cl^- ions. The small, highly charged Cr^{3+} ion is hydrated in solution and is an acid; the Cl^- ion (the conjugate base of the strong acid HCl) is neutral. Therefore, the resulting salt solution will be acidic, with pH < 7. The proton transfer equilibrium is

$$Cr(H_2O)_6^{3+}(aq) + H_2O(l) \rightleftharpoons H_3O^+(aq) + Cr(H_2O)_5OH^{2+}(aq)$$

EXERCISE E15.1 State whether aqueous solutions of (a) Na_2CO_3, (b) $AlCl_3$, and (c) KNO_3 have a pH less than, equal to, or greater than 7. If the pH of the solution is not 7, then write an equation to illustrate your conclusion.

[*Answer*: (a) pH > 7, $H_2O(l) + CO_3^{2-}(aq) \rightleftharpoons HCO_3^-(aq) + OH^-(aq)$; (b) pH < 7, $Al(H_2O)_6^{3+}(aq) + H_2O(l) \rightleftharpoons H_3O^+(aq) + Al(H_2O)_5OH^{2+}(aq)$; (c) pH $= 7$]

The pH of a salt solution. We can calculate the pH of a salt solution containing ions that are acids or bases by treating it as a problem in chemical equilibrium and setting up an equilibrium table like those used in Chapter 13. We assume that the solution consists of only the cations and anions of the salt in water at the instant the solution is prepared; but the composition immediately adjusts and reaches equilibrium. Some idea of the changes in composition of a solution that occur as it approaches equilibrium can be obtained by calculating the percentage of acid ions HB^+ that have donated a proton to a water molecule (that is, that have been deprotonated):

N We consider it important to show that the same type of procedure may be used in the context of salt solutions as has been used in the preceding chapters on equilibrium.

U Exercises 15.73 and 15.74.

$$\text{Percentage deprotonated (\%)} = \frac{\text{concentration of HB}^+\text{ deprotonated}}{\text{original concentration of HB}^+} \times 100\%$$

As we shall see, the percentage of deprotonated acid ions is usually very small.

EXAMPLE 15.2 Calculating the pH of a salt solution containing an acidic cation

Calculate the pH of 0.15 M $NH_4Cl(aq)$. What is the percentage of the NH_4^+ ions that have donated a proton?

STRATEGY The NH_4^+ cations are acids and the Cl^- anions are neutral, so we expect an acidic solution with a pH of less than 7. For the calculation, we set up an equilibrium table, taking the initial molar concentration of the acid cation (NH_4^+) to be the concentration of salt stated in the question and setting the initial concentrations of H_3O^+ and NH_3 equal to zero. Chloride ion is a neutral spectator ion, so we ignore its presence when calculating pH. We also ignore the autoionization of water, an approximation that is valid as long as the pH of the equilibrium solution is significantly different from 7. Then, in step 2 of the table, we set the decrease in concentration of NH_4^+ as the system adjusts to equilibrium equal to x mol/L and use the reaction stoichiometry to write the changes in the concentrations of H_3O^+ and NH_3 that are implied. Finally, we express the reaction quotient in terms of x, set it equal to the acid ionization constant, and solve the equation for x. The value of K_a for NH_4^+ is not given in Table 14.6 but is obtained from the value of K_b for NH_3 by using $K_a = K_w/K_b$. Because x mol/L is the hydronium ion concentration, the pH of the solution is $-\log x$. The percentage of NH_4^+ ions that have donated a proton is calculated from the expression in the text, with the concentration of deprotonated NH_4^+ ions equal to x mol/L.

N Once again we are using the format of the equilibrium table established in Chapter 13, but applying it to a particular system.

SOLUTION The equilibrium table is

Equilibrium equation:	$NH_4^+(aq) + H_2O(l) \rightleftharpoons$	$H_3O^+(aq) +$	$NH_3(aq)$
Species:	NH_4^+	H_3O^+	NH_3
Step 1. Initial concentration, mol/L	0.15	0	0
Step 2. Change in concentration, mol/L	$-x$	$+x$	$+x$
Step 3. Equilibrium concentration, mol/L	$0.15 - x$	x	x

From Table 14.6, the value of K_b for NH_3 is 1.8×10^{-5}. Therefore, for NH_4^+,

$$K_a = \frac{K_w}{K_b} = \frac{1.0 \times 10^{-14}}{1.8 \times 10^{-5}} = 5.6 \times 10^{-10}$$

Step 4. Inserting the information from step 3 into

$$\frac{[H_3O^+][NH_3]}{[NH_4^+]} = K_a \text{ at equilibrium}$$

gives

$$\frac{x \times x}{0.15 - x} = 5.6 \times 10^{-10}$$

We assume that x is less than 5% of 0.15 and simplify this expression to

$$\frac{x^2}{0.15} \approx 5.6 \times 10^{-10}$$

The solution is

$$x \approx \sqrt{0.15 \times 5.6 \times 10^{-10}} = 9.2 \times 10^{-6}$$

? How should the calculation be carried out if the approximation is invalid?

The approximation that x is less than 5% of 0.15 is valid:

$$\frac{9.2 \times 10^{-6}}{0.15} \times 100\% = 6.1 \times 10^{-3}\%$$

The concentration of hydronium ions (9.2×10^{-6} mol/L, from step 3) is also much larger than that generated by the autoionization of water (1.0×10^{-7} mol/L), so the neglect of the latter contribution is also valid. Because the molar concentration of H_3O^+ ions is 9.2×10^{-6} mol/L, the pH of the solution is

$$\text{pH} = -\log(9.2 \times 10^{-6}) = 5.04, \text{ or about } 5.0$$

The percentage of NH_4^+ ions that have donated protons is

$$\text{Percentage deprotonated (\%)} = \frac{\text{concentration of } NH_4^+ \text{ deprotonated}}{\text{original concentration of } NH_4^+} \times 100\%$$

$$= \frac{9.2 \times 10^{-6}}{0.15} \times 100\% = 6.1 \times 10^{-3}\%$$

Thus, only a very small proportion (about 1 in 16,000) of the NH_4^+ ions in solution have lost a proton.

EXERCISE E15.2 Calculate the pH and the percentage of acid cations that have donated a proton in 0.10 M $CH_3NH_3Cl(aq)$, where CH_3NH_3Cl is methylammonium chloride. See Table 14.6 for the K_b of methylamine.

▲ [*Answer*: pH = 5.8; $1.6 \times 10^{-3}\%$]

U Exercises 15.76 and 15.78.

E The hydrolysis of salts derived from a weak monoprotic acid and a weak monoprotic base. *J. Chem. Educ.*, **1990**, *67*, 226.

The percentage of anions A^- that accept a proton in a solution of a salt that provides basic anions can also be calculated:

$$\text{Percentage protonated (\%)} = \frac{\text{concentration of } A^- \text{ protonated}}{\text{original concentration of } A^- \text{ added}} \times 100\%$$

▼ **EXAMPLE 15.3 Calculating the pH of a solution containing a basic anion**

Calculate the pH and the percentage of anions that are protonated in 0.15 M $NaCN(aq)$.

STRATEGY The CN^- anions are bases and the Na^+ cations are neutral, so the solution will be basic, and we expect pH > 7. We follow the same procedure we used in the previous example. To find the base ionization constant of the CN^- ion from the acid ionization constant of its conjugate acid, hydrocyanic acid, we use $K_b = K_w/K_a$. As before, we assume that the concentration of OH^- ions produced by proton transfer is much greater than that produced by the autoionization of water; so we neglect the latter. The calculation leads to pOH; conversion to pH is made by using the relation pH + pOH = pK_w. For the calculation of the percentage of anions that become protonated, we use the expression given in the text above.

N Always write the proton-transfer equilibrium for the species that is responsible for modifying the pH of the solution (the weak base or weak acid).

SOLUTION The equilibrium table is

Equilibrium equation:	$H_2O(l) + CN^-(aq) \rightleftharpoons HCN(aq) + OH^-(aq)$		
Species:	CN^-	HCN	OH^-
Step 1. Initial concentration, mol/L	0.15	0	0
Step 2. Change in concentration, mol/L	$-x$	$+x$	$+x$
Step 3. Equilibrium concentration, mol/L	$0.15 - x$	x	x

Step 4. Table 14.6 gives the K_a of HCN as 4.9×10^{-10}, the K_b of its conjugate base, the CN^- ion, is

$$K_b = \frac{K_w}{K_a} = \frac{1.0 \times 10^{-14}}{4.9 \times 10^{-10}} = 2.0 \times 10^{-5}$$

Inserting the values from step 3 and the value of K_b into

$$\frac{[HCN][OH^-]}{[CN^-]} = K_b \text{ at equilibrium}$$

gives

$$\frac{x \times x}{0.15 - x} = 2.0 \times 10^{-5}$$

Because K_b is so small, we can anticipate that x is less than 5% of 0.15; hence we simplify this expression to

$$\frac{x^2}{0.15} \approx 2.0 \times 10^{-5}$$

It follows that

$$x \approx \sqrt{0.15 \times 2.0 \times 10^{-5}} = 1.7 \times 10^{-3}$$

We must verify that x is less than 5% of 0.15:

$$\frac{1.7 \times 10^{-3}}{0.15} \times 100\% = 1.1\%$$

The concentration of OH^- ions (1.7×10^{-3} mol/L, from step 3) arising from the proton transfer equilibrium is much larger than that arising from the autoionization of water (1.0×10^{-7} mol/L), so the neglect of the latter is also valid. Because the equilibrium concentration of OH^- ions is 1.7×10^{-3} mol/L, it follows that

$$\text{pOH} = -\log(1.7 \times 10^{-3}) = 2.77$$

U Exercises 15.75 and 15.77.

and hence that

$$\text{pH} = 14.00 - 2.77 = 11.23$$

The solution is basic, as we expected. Because the equilibrium concentration of HCN molecules is 1.7×10^{-3} mol/L,

$$\text{Percentage protonated (\%)} = \frac{\text{concentration CN}^- \text{ protonated}}{\text{original concentration of CN}^-} \times 100\%$$

$$= \frac{1.7 \times 10^{-3}}{0.15} \times 100\% = 1.1\%$$

Therefore, only a very small proportion (about 1 in 100) of the cyanide ions becomes protonated in solution.

EXERCISE E15.3 Calculate the pH and percentage of anions protonated in a 0.10 M potassium benzoate solution. See Table 14.6 for the K_a of benzoic acid, C_6H_5COOH.

[*Answer*: 8.6; 4.0×10^{-3}%]

15.2 MIXED SOLUTIONS

D Effect of acetate ion on the acidity of acetic acid: Common-ion effect. BZS3, 8.24.

So far, we have been considering aqueous equilibria involving only a single solute (a weak acid, a weak base, or a salt) that affects the pH of the solution. However, most solutions contain more (in some cases, many more) than one solute. For instance, river water may contain dissolved carbonates from limestone deposits, dissolved nitrates and phosphates from fertilizers, and chlorides dissolved out of the soil and rocks; calcium, sodium, ammonium, iron, aluminum, and magnesium cations may also be present. Each of these ions affects the quantity and quality of the aquatic life and determines the kind of treatment necessary to make the natural water suitable for drinking.

The common-ion effect. In this section we consider what happens to the pH of a solution when a salt is added to a solution that already contains some of the parent weak acid or weak base. Such a salt is said to provide a **common ion**, in the sense that it has an ion in common with the solute already present. The effect of such a salt on the pH of the solution is called the **common-ion effect**. The addition of sodium acetate to a solution of acetic acid provides an example: a mixed solution of this kind is present during the titration of acetic acid with sodium hydroxide. Because the $CH_3CO_2^-$ ion is a base, we expect the solution to become more basic and hence the pH of the solution to increase toward 7. Likewise, if we add sodium nitrite, $NaNO_2$, to aqueous nitrous acid, HNO_2, then the pH will increase because NO_2^- is also a base (Fig. 15.4). Once again, we can express the effect in terms of the proton transfer equilibrium, which in the latter case is

$$HNO_2(aq) + H_2O(l) \rightleftharpoons H_3O^+(aq) + NO_2^-(aq) \qquad K_a = 4.3 \times 10^{-4}$$

According to Le Chatelier's principle, the addition of NO_2^- (as the salt) shifts the equilibrium toward the HNO_2 and water, thereby decreasing the concentration of H_3O^+ ions and increasing the pH. Such an increase in pH is expected whenever a base is added to any solution.

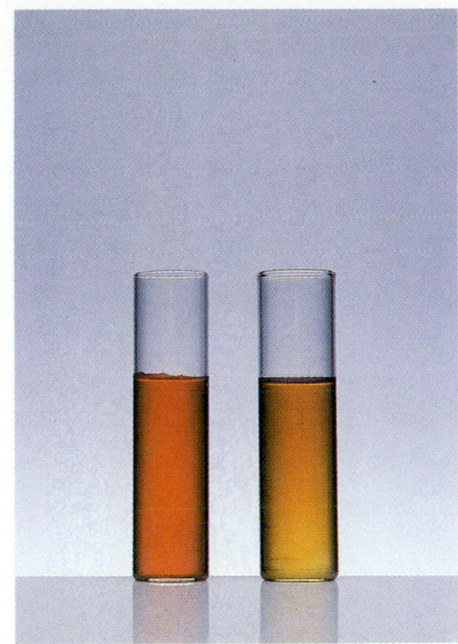

FIGURE 15.4 A 0.1 M $HNO_2(aq)$ solution has a pH of about 3 (left). When nitrite ions, which are bases, are added, the pH rises toward 7. The solution on the right is 0.1 M $HNO_2(aq)$ plus 0.1 M $NaNO_2(aq)$. A universal indicator is present in both solutions.

The pH of a mixed solution. We can treat mixed solutions that contain a common ion in the same way we addressed single-solute solutions. That is, we set up an equilibrium table and use the ionization constants to calculate the unknown concentration. The neutral spectator ions are ignored because they have no effect on the pH of the solution.

As an example, suppose we want to know the pH of a solution that contains both acetic acid and sodium acetate. The salt provides Na^+ ions, which are spectator ions, and $CH_3CO_2^-$ ions, which are bases. The latter take part in the proton transfer equilibrium that already exists in the acetic acid solution:

$$CH_3COOH(aq) + H_2O(l) \rightleftharpoons H_3O^+(aq) + \underset{\substack{\uparrow \\ \text{Provided by the acid} \\ \textit{and}\text{ the added salt}}}{CH_3CO_2^-(aq)}$$

The following example shows how to handle this kind of problem.

U Exercise 15.79.

N When difficulties are encountered in writing Brønsted equilibria for mixed solutions, follow these guidelines:

- If the solution contains a molecular weak acid or base, write the proton-transfer equilibrium for it and use the corresponding K_a or K_b from the tables. Use the chemical equation for the equilibrium to set up the equilibrium table.
- Conjugate acids and bases always appear on opposite sides of the chemical equation for the proton-transfer equilibrium.
- Omit all spectator ions from the chemical equation.

▼ EXAMPLE 15.4 Calculating the pH of a solution of a weak acid and its salt

Calculate the pH of a solution that is 0.500 M $HNO_2(aq)$ and 0.100 M $KNO_2(aq)$. Table 14.6 lists K_a for HNO_2 as 4.3×10^{-4}.

STRATEGY The solution contains a common ion (the NO_2^- ion, a base), so we expect the pH to be higher than that of nitrous acid alone. The K^+ ion supplied by the salt is a neutral spectator ion and has no effect on the pH of the solution. To calculate the pH of the mixed solution, we set up an equilibrium table in the usual way and consider the initial concentration of the acid to be 0.500 mol HNO_2/L; but instead of setting the concentration of the anion to zero, we take it to be equal to the concentration of salt, because each KNO_2 formula unit supplies one NO_2^- anion. Then we suppose that the concentration of acid molecules decreases by x mol/L as the system approaches equilibrium. The concentration of the conjugate base increases by x mol/L, as does that of the H_3O^+ ions. The value of x is then found by using the acid ionization constant. Because the acid supplies so many hydronium ions, we can ignore those arising from the autoionization of water.

D Measure the pH of a weak acid solution (either with a pH meter or with universal indicator); slowly add a salt containing the conjugate base of the acid, and observe the change in pH.

SOLUTION The equilibrium table is

Equilibrium equation:	$HNO_2(aq) + H_2O(l) \rightleftharpoons H_3O^+(aq) + NO_2^-(aq)$		
Species:	HNO_2	H_3O^+	NO_2^-
Step 1. Initial concentration, mol/L	0.500	0	0.100
Step 2. Change in concentration, mol/L	$-x$	$+x$	$+x$
Step 3. Equilibrium concentration, mol/L	$0.500 - x$	x	$0.100 + x$

Step 4. Inserting the values from step 3 and $K_a = 4.3 \times 10^{-4}$ into

$$\frac{[H_3O^+][NO_2^-]}{[HNO_2]} = K_a \text{ at equilibrium}$$

gives

$$\frac{x \times (0.100 + x)}{0.500 - x} = 4.3 \times 10^{-4}$$

Assuming that x is less than 5% of 0.100 (and therefore also less than 5% of 0.500) lets us write

$$\frac{x \times 0.100}{0.500} \approx 4.3 \times 10^{-4}$$

which solves to $x \approx 2.2 \times 10^{-3}$. To verify the validity of the approximation, we evaluate

$$\frac{2.2 \times 10^{-3}}{0.100} \times 100\% = 2.2\%$$

U Exercises 15.80 and 15.81.

Because the value of x is 2.2% of 0.100, it is only 0.44% of 0.500, so the approximations are valid. It follows that the equilibrium concentration of hydronium ions is 2.2×10^{-3} mol/L. This concentration is much larger than the concentration arising from the autoionization of water, so the neglect of autoionization is also valid. The pH of the solution is therefore

$$pH = -\log(2.2 \times 10^{-3}) = 2.67, \text{ or about } 2.7$$

The pH of 0.500 M $HNO_2(aq)$ is 1.8, so, as anticipated, the pH of the mixed solution is higher.

EXERCISE E15.4 Calculate the pH of a solution that is 0.300 M $CH_3NH_2(aq)$ and 0.146 M $CH_3NH_3Cl(aq)$. The K_b of CH_3NH_2 is 3.6×10^{-4}.

▲ [*Answer*: 10.87]

Dilution of mixed solutions. The dilution of an aqueous solution affects the concentration of the various solutes and hence influences the pH of the solution. For example, if 50.0 mL of 1.0 M HCl(aq), which has a pH close to 0, is diluted to 100 mL, then the concentration decreases to 0.50 M HCl(aq) and the pH rises to 0.3. Similarly, when a solution of one solute is mixed with a solution of another solute, the concentrations of both solutes are reduced; and there is a corresponding effect on the pH of the combined solution that results solely from the effect of dilution.

N When dilution takes place, amounts of substance remain unchanged, but the volume they occupy becomes larger.

E In practice, dilution has two effects: it reduces the concentration of the solute and it reduces the interactions between the ions. Here we consider only the former effect, but in real solutions the effect on ion-ion interactions may be significant too.

To determine the concentration of a solute in a diluted, mixed solution, we determine the amount of solute present in the initial solution and divide that amount by the *total volume* of the mixed solution.

Molar concentration of solute in diluted, mixed solution $= \frac{\text{amount (mol) of solute in initial solution}}{\text{total volume (L) of combined solutions}}$

TITRATIONS AND pH CURVES

The success of an acid-base titration as an analytical technique depends on our ability to recognize the stoichiometric point, that is, the point at which the amount of acid added to a base, or vice versa, brings about complete neutralization (with moles of acid added equal to moles of base initially present). Although the pH of the solution is 7 at the stoichiometric point of the titration of a strong acid and a strong base (see Section 14.6), that is not the case in the titration of a weak acid (or base). At the stoichiometric point of such a titration, the solution contains the salt and there is no excess acid or base. As we have seen, however, the ions of the salt may themselves be acids or bases and hence give rise to a solution that is acidic (if the ions present are acids) or basic (if the ions are bases).

At the stoichiometric point in the titration of a weak acid with a strong base, the solution contains the anions of the acid; these anions are bases, so the pH of the solution is greater than 7. For example, in a titration of formic acid (HCOOH) with sodium hydroxide, at the stoichiometric point the solution contains sodium formate ($NaHCO_2$) and water. A formate ion (HCO_2^-) is a base, so the solution is basic (the Na^+ ions are neutral) and the pH is greater than 7. Similarly, at the stoichiometric point in the titration of a weak base with a strong acid, the solution contains cations of the base; these cations are acids, so the solution is acidic and its pH is less than 7. For example, in a titration of aqueous ammonia with hydrochloric acid, ammonium chloride is present at the stoichiometric point; but, because an NH_4^+ ion is an acid and Cl^- ions are neutral, the solution is acidic and its pH is less than 7.

N As remarked before, the name *stoichiometric point* more accurately represents the principle that a stoichiometric amount, according to the chemical equation for the reaction, of acid has been added to a base.

U Exercise 15.83.

D Comparison of strong acid and weak acid titration curves. *J. Chem. Educ.*, **1979**, *56*, 194.

15.3 THE VARIATION OF pH DURING A TITRATION

If we are to detect the stoichiometric points accurately in different types of acid-base titrations, then we must know how to interpret the pH of the solution at different stages of the titration. In this section we shall see how to construct the pH curve for two common cases, the titration of a weak acid with a strong base and the titration of a weak base with a strong acid. (The titration of a weak acid with a weak base is so inaccurate that it is never used, and we shall not consider it.)

D Instrumental recording of a titration curve. BZS3, 8.27.

E Another approach to titration curves. Which is the time dependent variable? *J. Chem. Educ.*, **1981**, *58*, 659. Titration behavior of monoprotic and diprotic acids. *J. Chem. Educ.*, **1970**, *47*, 658.

Titration of a weak acid with a strong base. Suppose we titrate 25.00 mL of 0.100 M HCOOH(*aq*) (formic acid, the acid present in ant venom) with 0.150 M NaOH(*aq*). Initially, the pH of the analyte (the acid solution) is that of 0.100 M HCOOH(*aq*). The pH of this solution can be found by using the technique described for acetic acid in Example 14.13; it is 2.37. This pH is point A in Fig. 15.5. To calculate the rest of the curve, we need to take into account the neutralization of the acid as base is added *and* the increase in volume of the solution as the titrant is added to the analyte. Each stage of the curve can be dealt with by setting up the appropriate equilibrium table, as the following sequence of examples illustrates.

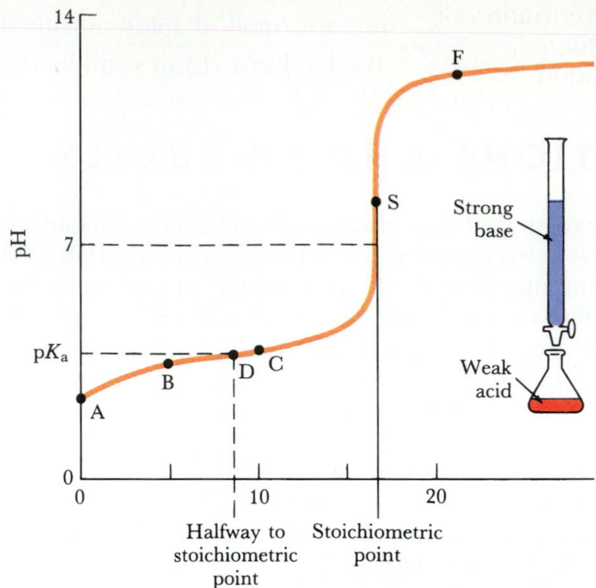

FIGURE 15.5 The pH curve for the titration of a weak acid (25 mL of 0.100 M HCOOH(*aq*)) with a strong base (0.150 M NaOH(*aq*)). The stoichiometric point (point S) occurs on the basic side of pH = 7 because the anion HCO_2^- is a base.

T OHT: Fig. 15.5.

N Because pH is proportional to the logarithm of the molar concentration of hydronium ions, a moderately slow increase in pH (as from B to C) still corresponds to a very rapid decrease in concentration (by about an order of magnitude). A pH curve compresses many orders of magnitude into a small numerical range.

D A simple, effective demonstration of titration curves and indicator selection. *J. Chem. Educ.*, **1973**, *50*, 262.

EXAMPLE 15.5 Calculating the pH before the stoichiometric point is reached in a weak acid-strong base titration

Calculate the pH of a solution of 25.00 mL of 0.100 M HCOOH(*aq*) after the addition of 5.00 mL of 0.150 M NaOH(*aq*). The K_a for HCOOH is 1.8×10^{-4}.

STRATEGY The pH of the analyte is initially 2.37 (see text above), and we expect the pH to be higher than 2.37 after some base has been added. The pH of the mixed solution of HCOOH molecules and HCO_2^- ions can be calculated as explained in Section 15.2. We must first calculate the initial amount of HCOOH and the amount of OH^- added as titrant. Next, we calculate the amount of HCOOH remaining and the amount of HCO_2^- that forms by using the relation

$$1 \text{ mol } OH^- \text{ added} = 1 \text{ mol HCOOH reacted}$$

(which follows from the reaction stoichiometry). Then we calculate the concentrations of HCOOH and HCO_2^- in the combined solution, by dividing the amounts of each by the total volume of the solution. Finally, we set up the equilibrium table in the usual way and use it to calculate the pH of the solution as we illustrated in Example 15.4.

? Why is the presence of Na^+ ignored in the calculation?

SOLUTION The initial amount of HCOOH in the analyte is

$$\text{Amount of HCOOH (mol)} = 25.00 \times 10^{-3}\text{ L} \times \frac{0.100 \text{ mol HCOOH}}{1 \text{ L}}$$

$$= 2.50 \times 10^{-3} \text{ mol HCOOH}$$

The amount of OH^- in 5.00 mL of the titrant is

$$\text{Amount of } OH^- \text{ (mol)} = 5.00 \times 10^{-3}\text{ L} \times \frac{0.150 \text{ mol } OH^-}{1 \text{ L}}$$

$$= 7.50 \times 10^{-4} \text{ mol } OH^-$$

The addition of 7.50×10^{-4} mol OH^- removes 7.50×10^{-4} mol of HCOOH, leaving

N Note how we identify the species (HCOOH) that affects the pH, write the chemical equation for its proton-transfer equilibrium, and then use that equation as a guide to formulation of the equilibrium table.

$$\text{Amount of HCOOH (mol)} = (2.50 \times 10^{-3} - 7.50 \times 10^{-4}) \text{ mol HCOOH}$$
$$= 1.75 \times 10^{-3} \text{ mol HCOOH}$$

and forming 7.50×10^{-4} mol HCO_2^- ions. The total volume of solution at this stage is (25.00 + 5.00) mL = 30.00 mL, so the concentrations of acid and conjugate base are

$$[\text{HCOOH}] = \frac{1.75 \times 10^{-3} \text{ mol}}{30.00 \times 10^{-3} \text{ L}} = 5.83 \times 10^{-2} \text{ mol/L}$$

$$[\text{HCO}_2^-] = \frac{7.50 \times 10^{-4} \text{ mol}}{30.00 \times 10^{-3} \text{ L}} = 2.50 \times 10^{-2} \text{ mol/L}$$

We are now ready to take into account the proton transfer equilibria in the solution by following the procedure described in Example 15.4. The equilibrium table is

Equilibrium equation:	**$HCOOH(aq) + H_3O(l) \rightleftharpoons H_2O^+(aq) + HCO_2^-(aq)$**		
Species:	**HCOOH**	**H_3O^+**	**HCO_2^-**
Step 1. Initial concentration, mol/L	5.83×10^{-2}	0	2.50×10^{-2}
Step 2. Change in concentration, mol/L	$-x$	$+x$	$+x$
Step 3. Equilibrium concentration, mol/L	$5.83 \times 10^{-2} - x$	x	$2.50 \times 10^{-2} + x$

Step 4. Inserting the concentrations from step 3 and $K_a = 1.8 \times 10^{-4}$ into

$$\frac{[H_3O^+][HCO_2^-]}{[HCOOH]} = K_a \text{ at equilibrium}$$

gives

$$\frac{x \times (2.50 \times 10^{-2} + x)}{5.83 \times 10^{-2} - x} = 1.8 \times 10^{-4}$$

We assume that x is less than 5% of 2.50×10^{-2} (which would automatically make it less than 5% of 5.83×10^{-2}) and simplify this expression to

$$\frac{x \times (2.50 \times 10^{-2})}{5.83 \times 10^{-2}} \approx 1.8 \times 10^{-4}$$

This equation solves to $x \approx 4.2 \times 10^{-4}$. This value is small enough for the approximations to be valid. We conclude that $[H_3O^+] = 4.2 \times 10^{-4}$ mol/L (so the neglect of autoionization is also valid), and

$$\text{pH} = -\log(4.2 \times 10^{-4}) = 3.38, \text{ or about } 3.4$$

This pH is point B in Fig. 15.5.

EXERCISE E15.5 Calculate the pH of the solution after the addition of another 5.00 mL of 0.150 M NaOH(aq).

▲ [*Answer*: 3.93; point C in Fig. 15.5]

A special stage in a titration occurs when the concentrations of the weak acid and its conjugate base (the anion of the salt) are equal in the mixed solution. This stage is reached when half the stoichiometric amount of base has been added to a weak acid. For instance, at this stage in the titration of formic acid with sodium hydroxide, $[HCO_2^-] =$

[HCOOH], and the expression for the acid ionization constant simplifies to

$$K_a = \frac{[H_3O^+]\cancel{[HCO_2^-]}}{\cancel{[HCOOH]}} = [H_3O^+]$$

E It follows from this equality, that a good procedure for determining the pK_a of an acid (or conjugate acid of a base, and hence the pK_b of that base) is to plot a pH curve, and identify the pH at the halfway point. In practice, a glass electrode is used and the experimental pH is monitored electronically.

Therefore, on taking the negative logarithm of both sides, we see that *at the halfway point of any weak acid-strong base titration,*

$$pH = pK_a$$

For the formic acid titration, $pK_a = 3.75$; so halfway to the stoichiometric point, pH = 3.75. This pH is point D in Fig. 15.5.

The stoichiometric point. At the stoichiometric point, all the HCOOH has been neutralized with an equal amount of OH^-; therefore, only the salt, sodium formate, is present in the solution. The HCO_2^- ion is a base, so the pH of the solution is greater than 7. To calculate its value, we use the same technique we used in Example 15.3.

▼ **EXAMPLE 15.6 Calculating the pH at the stoichiometric point of a weak acid-strong base titration**

Calculate the pH of the solution of Example 15.5 at the stoichiometric point of the titration.

STRATEGY The solution contains the base HCO_2^-, so we expect the pH to be greater than 7. Because all the HCOOH is neutralized at the stoichiometric point, the amount of HCO_2^- in the solution is equal to the initial amount of HCOOH in the analyte. The concentration of HCO_2^- ions is this amount divided by the total volume of the solution at the stoichiometric point. Once we have determined the molar concentration of HCO_2^-, we can calculate the pH with the method we used in Example 15.3.

E The pH at the stoichiometric point of the titration of a weak acid with a strong base is greater than 7 not because of the strong base but because of the presence of the conjugate base of the neutralized weak acid.

U Exercise 15.82.

SOLUTION The amount of HCO_2^- present in the solution at the stoichiometric point is equal to the initial amount of HCOOH, which from Example 15.5 is 2.50×10^{-3} mol. The volume of 0.150 M NaOH(*aq*) that contains the same amount of OH^- ions is

$$\text{Volume (L)} = 2.50 \times 10^{-3}\text{ mol OH}^- \times \frac{1\text{ L}}{0.150\text{ mol OH}^-} = 16.7 \times 10^{-3}\text{ L}$$

or 16.7 mL; therefore, the total volume of the combined solutions at the stoichiometric point is (25.00 + 16.7) mL = 41.7 mL. It follows that at the stoichiometric point

$$[HCO_2^-] = \frac{2.5 \times 10^{-3}\text{ mol}}{41.7 \times 10^{-3}\text{ L}} = 0.060\text{ mol/L}$$

Then following exactly the same procedure we used in Example 15.3, with $[HCO_2^-] = 0.060$ mol/L and $K_a = 1.8 \times 10^{-4}$, we find that pH = 8.26, or about 8.3. This pH is point S in Fig. 15.5.

EXERCISE E15.6 Calculate the pH at the stoichiometric point of the titration of 25.00 mL of 0.010 M HClO(*aq*) with 0.020 M KOH(*aq*).

▲ [*Answer*: 9.7]

N One point that might be helpful in considering the shape of the pH curve after the stoichiometric point has been passed is that if a huge amount of base is added, then the initial volume of analyte is irrelevant, and so the pH will be that of the titrant.

Beyond the stoichiometric point. Beyond the stoichiometric point, the pH depends only on the concentration of excess strong base. If another 5.00 mL of base (for a total of 21.7 mL of base, and another 7.50 ×

10^{-4} mol OH^-) is added, then the concentration of OH^- ions rises to

$$[OH^-] = \frac{7.50 \times 10^{-4}\ \text{mol}}{(25.00 + 16.7 + 5.00) \times 10^{-3}\ \text{L}} = 0.0161\ \text{mol/L}$$

This concentration corresponds to pOH = 1.79 and hence to pH = 12.21, which is point F in Fig. 15.6. The addition of more base results in the gradual rise of the pH toward 13.2, the pH of the 0.150 M $NaOH(aq)$ solution itself.

The shape of the pH curve shown in Fig. 15.5 is typical of a titration of a weak acid with a strong base. The pH of the weak acid solution initially rises slowly. It then passes rapidly through the pH at the stoichiometric point, which is *higher* than 7 because the salt anion is a base. Finally, it climbs slowly toward a strongly basic value.

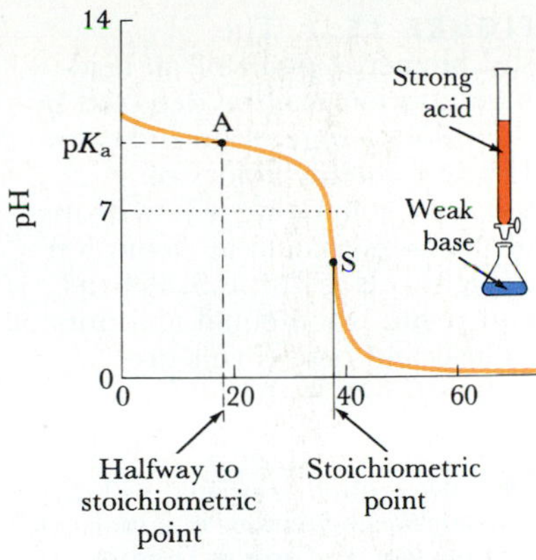

FIGURE 15.6 The variation of pH during the titration of a weak base (25 mL of 0.15 M $NH_3(aq)$) with a strong acid (0.10 M $HCl(aq)$). The stoichiometric point (point S) occurs on the acidic side of pH = 7 because the cation NH_4^+ is an acid.

Titration of a weak base with a strong acid. Titration of a weak base with hydrochloric acid gives a solution of the conjugate acid of the base at the stoichiometric point. Thus, if ammonia is the base, the titration results in the production of NH_4^+ ions. Because the NH_4^+ ion is a Brønsted acid, the solution is acidic and the pH is less than 7. The pH curve is shown in Fig. 15.6: it is calculated in exactly the same way we calculated the pH curve for the titration of a weak acid with a strong base. The pH begins above 7 and falls slowly at first, passes rapidly through the stoichiometric point on the acid side of neutrality, and then decreases slowly toward the pH of the strong acid titrant.

It is worth noting that when the concentrations of the weak base and its conjugate acid (the cation of the weak base) are equal, so that $[NH_3] = [NH_4^+]$, the expression for K_a simplifies to

$$K_a = \frac{[\cancel{NH_3}][H_3O^+]}{[\cancel{NH_4^+}]} = [H_3O^+]$$

Hence, halfway to the stoichiometric point S, the pH of the solution is equal to the pK_a of the acid cation. This pH is point A in Fig. 15.6.

T OHT: Fig. 15.6.

? Why is the pH at the stoichiometric point of a titration of a weak base with a strong acid less than 7?

? What salt is present at the stoichiometric point of a titration of aqueous ammonia with hydrochloric acid?

U Exercises 15.84 and 15.85.

15.4 INDICATORS AS WEAK ACIDS

Automatic titrators monitor the pH of the analyte solution and detect the stoichiometric point by indicating the characteristic rapid change in pH. When automatic equipment is not available, a stoichiometric point may be detected with a pH meter or by the change in color of an **acid-base indicator**. Such an indicator is a weak Brønsted acid that has one color when it is present as its acid form (HIn, where In stands for indicator), and another when it is present as its conjugate base form (In^-). In solution, an indicator takes part in the proton transfer equilibrium

D Add universal indicator to a solution in a large graduated cylinder made strongly basic with sodium hydroxide. Add dry ice and observe the color change. Repeat the experiment with hydrochloric acid instead of dry ice.

$$HIn(aq) + H_2O(l) \rightleftharpoons H_3O^+(aq) + In^-(aq) \qquad \frac{[H_3O^+][In^-]}{[HIn]} = K_{In}$$

If the concentration of HIn is much greater than that of In^-, then the solution is the color of the acid form of the indicator; if the concentration of HIn is much less than that of In^-, then the solution is the color of the base form of the indicator. The **end point** of the titration, the stage at which the indicator changes color, occurs when $[HIn] \approx [In^-]$, or at $[H_3O^+] = K_{In}$ (substituting $[HIn] = [In^-]$ into the expression for

D Acid-base indicators extracted from plants. BZS3, 8.4.

N It might be helpful to rearrange the expression for K_{In} into an expression for the ratio of concentrations of the In^- and HIn forms of the indicator, and to show how the ratio changes as the pH passes through the value of pK_{In}.

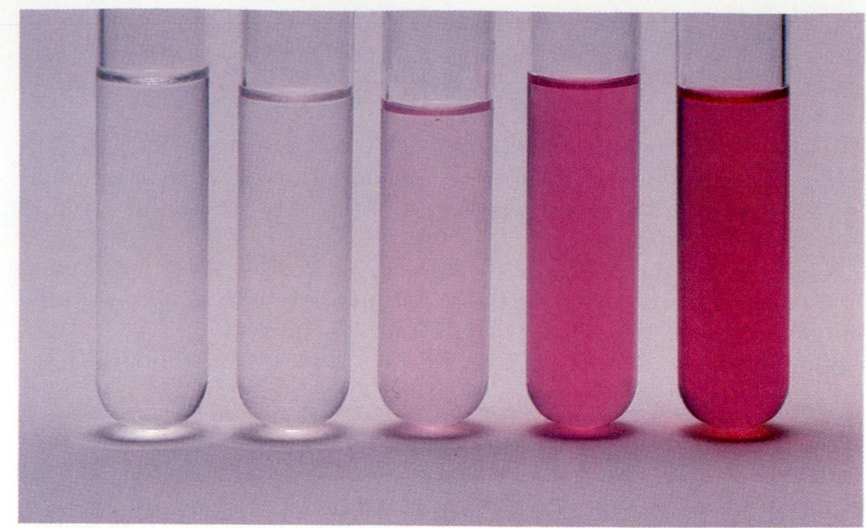

FIGURE 15.7 The stoichiometric point of an acid-base titration may be detected by the color change of an indicator. Here we see the colors of solutions containing a few drops of phenolphthalein at (from left to right) pH = 7.0; 8.5; 9.4 (its end point, when equal amounts of its acid and base forms are present); 9.8; and 12.0.

E Color is often associated with a quinonelike component in a molecule, and the loss of a proton from structure (**1**) results in the formation of this chromophore.

1 Phenolphthalein (Acid form, colorless)

2 Phenolphthalein (Base form, red)

K_{In} above). That is, the color change occurs when pH = pK_{In}. The addition of a strong acid or a strong base to an analyte affects the relative amounts of HIn and In^- and hence modifies the color of the solution. If, as is the case in practice, only one or two drops of indicator have been added to the solution, then only a small amount of the titrant reacts with the indicator; most of the titrant reacts with the analyte itself, and the accuracy of the titration is not upset by the presence of the indicator.

Phenolphthalein (**1**) is a common indicator for acid-base titrations. This large organic acid molecule is colorless in an acidic solution (in which it is present as its acid form, HIn); its conjugate base anion (**2**) is pink (Fig. 15.7). The transition from colorless to pink starts at pH = 8.2 and is complete by pH = 10 (notice that the color change occurs on the basic side of neutrality: we shall see the importance of that fact shortly). Another well-known, naturally occurring indicator is litmus, which is red for pH < 5 and blue for pH > 8. Other organic dyes that change color with the pH of the medium include the anthocyanidins. One of these compounds is responsible for the colors of both red poppies and blue cornflowers (Fig. 15.8).

Because litmus changes color close to pH = 7, it is suitable for titrating strong bases with strong acids, which have stoichiometric points

E The acidity of the sap of poppies and cornflowers is genetically controlled and not subject to change. The red and blue forms of hydrangeas (see the chapter opener for Chapter 14) is another illustration of the effect of pH on color: in this case, the sap responds to the pH of the soil, and so the color can be changed from red in basic environments to blue in acidic environments.

(a)

(b)

FIGURE 15.8 The same dye is responsible for the red of poppies (a) and the blue of cornflowers (b). The color difference is a consequence of the more acidic sap of poppies.

close to pH = 7. However, the pH of the solution sweeps quickly through a wide range of values near the stoichiometric point, so any indicator that changes color within about 2 units of pH = 7 can also be used, as long as the change occurs within a narrow pH range (Fig. 15.9). Phenolphthalein is often used because its color change (at pH = 9) is sharp and easy to detect. In contrast, one color change of thymol blue (from yellow to red) occurs in the pH range 2.8 to 1.2 and would give inaccurate results. Ideally, it is best to select an indicator that has a color change close to the stoichiometric point of the titration: the pK_{In} of the indicator should be within about one pH unit of the stoichiometric point of the titration:

$$pK_{In} \approx pH(\text{stoichiometric point}) \pm 1$$

Phenolphthalein, which changes color in the pH range 8.2 to 10.0, can be used for titrations with a stoichiometric point near pH = 9; methyl orange should not be used because it changes color between pH = 3.2 and pH = 4.4 (Fig. 15.10). However, in the titration of a weak base with a strong acid, the stoichiometric point occurs at pH < 7; hence methyl orange is suitable for this titration but phenolphthalein is not. The color change that methyl orange undergoes is shown in Fig. 15.11; some other indicators and their color changes are listed in Table 15.3.

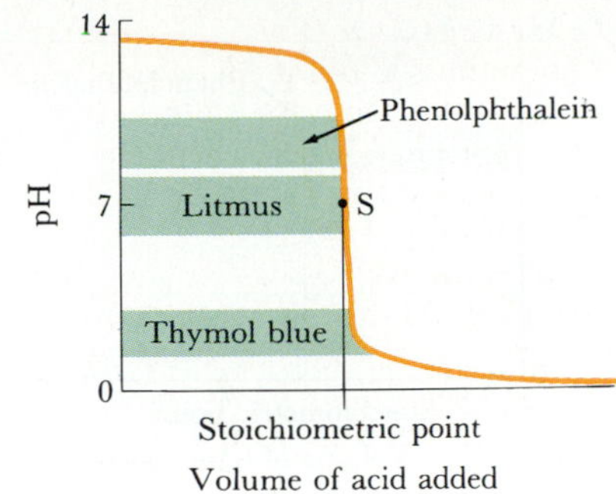

FIGURE 15.9 Ideally, an indicator should have a sharp color change close to the stoichiometric point (pH = 7 for a strong acid-strong base titration). However, the change of pH is so abrupt that phenolphthalein can be used for a strong acid-strong base titration. The color change from yellow to red (pH range 2.8 to 1.2) of thymol blue would give a less accurate result.

? For most titrations, the protocol requires the addition of one to three drops of indicator. How would the addition of 20 drops of indicator affect the accuracy of the titration?

15.5 BUFFER SOLUTIONS

One feature of the pH curves in Figs. 15.5 and 15.6 is the gradual change in pH halfway to the stoichiometric point, close to $pH = pK_a$. The slowness of this change means that when a small amount of acid or base is added to a solution containing roughly equal concentrations of an ion and its parent weak acid or base, the pH changes much less than when the same amount of acid or base is added to pure water.

E On the preparation of buffer solutions. *J. Chem. Educ.*, **1981**, *58*, 743.

T OHT: Table 15.3; Figs. 15.9 and 15.10.

D Buffering action of Alka-Seltzer. BZS3, 8.29.

T OHT: Figs. 15.9 and 15.10.

TABLE 15.3 Indicator color changes

Indicator	Color of acidic form	pH range of color change	pK_{In}	Color of basic form
thymol blue	red	1.2 to 2.8	1.7	yellow
methyl orange	red	3.2 to 4.4	3.4	yellow
bromophenol blue	yellow	3.0 to 4.6	3.9	blue
bromocresol green	yellow	3.8 to 5.4	4.7	blue
methyl red	red	4.8 to 6.0	5.0	yellow
bromothymol blue	yellow	6.0 to 7.6	7.1	blue
litmus	red	5.0 to 8.0	6.5	blue
phenol red	yellow	6.6 to 8.0	7.9	red
thymol blue	yellow	8.0 to 9.6	8.9	blue
phenolphthalein	colorless	8.2 to 10.0	9.4	pink
alizarin yellow R	yellow	10.1 to 12.0	11.2	red
alizarin	red	11.0 to 12.4	11.7	purple

T OHT: Table 15.3.

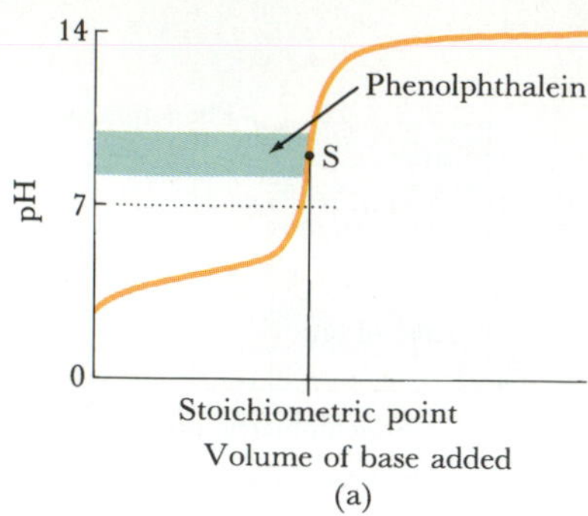

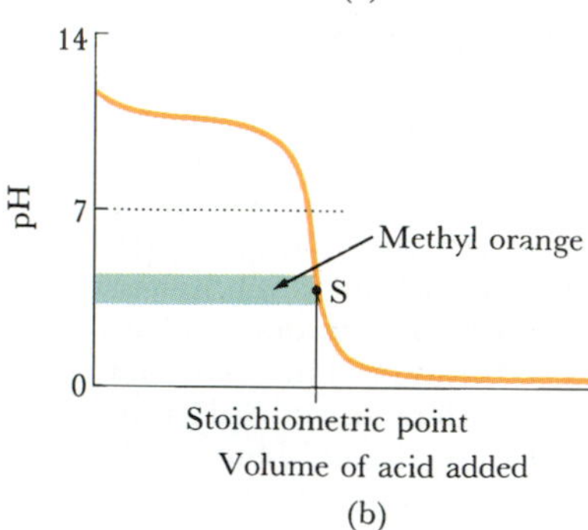

FIGURE 15.10
(a) Phenolphthalein can be used to detect the stoichiometric point of a weak acid-strong base titration. (b) Methyl orange can be used for a weak base-strong acid titration. The pH curves are superimposed on approximations to the color of the indicator at each pH value.

E As well as the physical processes that occur in the solution (the supply and removal of protons), another contribution to the flatness of the pH curve in the buffer region is that pH is given by the logarithm of a concentration, and logarithms of numbers change much more slowly than numbers themselves. For instance, if x changes from 10 to 1000, $\log x$ changes from 1 to only 3. So, even if the hydrogen in concentration changes a little, the pH will change much less.

Solutions that resist any change in pH when small amounts of acid or base are added are called **buffers** (Fig. 15.12). Blood is maintained at pH = 7.4 by buffer action, as are other cell fluids too. The oceans are maintained at about pH = 8.4 by a complex buffering process that depends on the concentrations of the hydrogencarbonates and silicates. Buffer solutions are particularly important in biological systems. For example, the growth of bacteria can be sustained only in buffered solutions; in unbuffered solutions, the acidity or basicity of the wastes from the bacteria causes so great a change in the pH of the solution that the bacteria die. Similarly, the many buffering systems in our bodies are extremely important to our health.

Buffer action. The action of buffers can be explained by considering how the acetic acid equilibrium

$$CH_3COOH(aq) + H_2O(l) \rightleftharpoons H_3O^+(aq) + CH_3CO_2^-(aq)$$

(or some other similar Brønsted equilibrium) responds to the addition of acid or base. Suppose a solution contains large numbers of $CH_3CO_2^-$ ions and CH_3COOH molecules, in about equal concentrations. When a small amount of acid is added, the newly arrived H_3O^+ ions transfer protons to the $CH_3CO_2^-$ ions to form CH_3COOH molecules, thereby leaving the pH nearly unchanged. When a small amount of base is added, the new OH^- ions remove protons from the CH_3COOH molecules. As a result, the OH^- concentration remains nearly unchanged and the H_3O^+ concentration (and the pH) is also left nearly unchanged. In other words, the weak acid acts as a source of hydrogen ions to react with OH^- (from added strong base), and its conjugate base (the anion of the weak acid) acts as a proton acceptor from H_3O^+ (from added strong acid) (Fig. 15.13). It follows that a buffer solution must consist of a weak acid (to neutralize strong base) and its conjugate base (to neutralize strong acid); alternatively, it may also consist of a weak base (to neutralize strong acid) and its conjugate acid (to neutralize strong base).

Figure 15.5 shows that the flattest part of the pH curve of a solution of a weak acid and its conjugate base occurs where their concentrations are equal (halfway to the stoichiometric point). The pH is on the acid side of neutral and equal to the pK_a of the weak acid. The mixture is an

FIGURE 15.11 The color change shown by methyl orange. From left to right, the pH of the solution is 3.0, 3.5, 4.0, 4.5, and 5.0.

acid buffer and stabilizes the solution at pH $\approx$ pK_a. Similarly, Fig. 15.6 shows that a mixture of a weak base and its conjugate acid stabilizes the pH on the basic side of neutrality, at pH $\approx$ pK_a, where K_a is the ionization constant of the conjugate acid of the base. These mixtures are called **base buffers**. Mixtures in which the common ion and acid (or base) have unequal concentrations are also buffers but are less effective than those in which the concentrations are nearly equal. Some typical buffer systems are listed in Table 15.4.

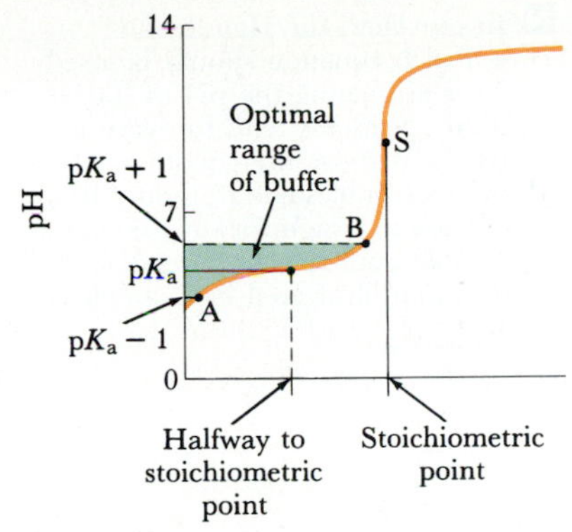

FIGURE 15.12 Although the ratio of anion to acid is increased considerably from A to B, the pH changes very little. In other words, the solution acts as a buffer in that region.

The analysis of buffer solutions. Because buffer solutions are mixed solutions similar to those described in Section 15.2, we can calculate their pH as we did in Example 15.4. The proton transfer equilibrium we must consider for a solution containing a weak acid HA and a salt that provides the conjugate base anions A^- is

$$HA(aq) + H_2O(l) \rightleftharpoons H_3O^+(aq) + A^-(aq) \qquad \frac{[H_3O^+][A^-]}{[HA]} = K_a$$

We can obtain the pH for a mixed solution by rearranging the equilibrium condition to

$$[H_3O^+] = K_a \times \frac{[HA]}{[A^-]}$$

Then, taking the negative logarithms of both sides gives

$$-\log [H_3O^+] = -\log K_a - \log\left(\frac{[HA]}{[A^-]}\right)$$

which is equivalent to

$$pH = pK_a + \log\left(\frac{[A^-]}{[HA]}\right)$$

T OHT: Fig. 15.12.

U Exercise 15.86.

E Thermal buffers: An illustrative analogy. *J. Chem. Educ.*, **1979**, *56*, 710.

At this stage we have made no approximation, and [X] represents the actual concentrations of acid and of base. However, Example 15.4 shows that the percentage ionization of HA and protonation of A^- are usually both very low in a mixed solution. That being so, we can take their concentrations to be approximately equal to those of the acid and the salt, respectively, that were added initially. Then

$$pH \approx pK_a + \log\left(\frac{[\text{base added initially}]}{[\text{acid added initially}]}\right)$$

This equation, which is called the **Henderson-Hasselbalch equation**, is commonly used by biochemists in their research into biochemical systems. Because so many chemical reactions in our bodies occur in buffered environments, biochemists use the equation for quick estimates

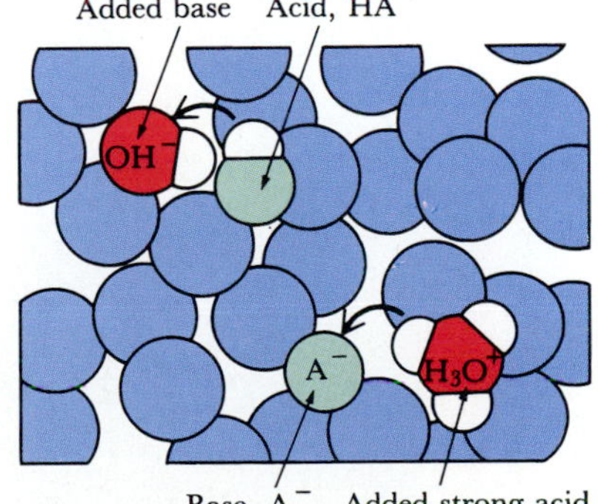

FIGURE 15.13 Buffer action depends on the donation of protons by the acid molecules when a base is added and the acceptance of protons by the base anions when an acid is added.

TABLE 15.4 Typical buffer systems

Composition	pK_a	Composition	pK_a
Acid buffers		*Base buffers*	
CH_3COOH and $CH_3CO_2^-$	4.74	NH_3 and NH_4^+	9.26
HNO_2 and NO_2^-	3.37	$(CH_3)_3N$ and $(CH_3)_3NH^+$	9.81
$HClO_2$ and ClO_2^-	2.00	$H_2PO_4^-$ and HPO_4^{2-}	7.21

N In our view, the Henderson-Hasselbalch equation should be used only for estimating the pH of buffer solutions. Students tend to overuse, misuse, and abuse the equation. It is often less troublesome to ignore it, and to use a straightforward equilibrium table approach to a calculation, such as that illustrated in Example 15.5.

of the pH of the medium while keeping in mind the assumptions on which it is based.

The Henderson-Hasselbalch equation is essentially the expression for K_a written in terms of logarithms, so it is usually just as quick to derive it as to remember it. As an example of its use, consider a solution that is 0.10 M $NaF(aq)$ and 0.20 M $HF(aq)$. The pH of this solution, given that pK_a of HF is 3.45, is approximately

$$pH \approx 3.45 + \log\left(\frac{0.10}{0.20}\right) = 3.15$$

or, more realistically, pH = 3.2. The pH of a solution that is 0.10 M $NH_3(aq)$ and 0.20 M $NH_4Cl(aq)$, given that pK_a of NH_4^+ is 9.25, is

$$pH \approx 9.25 + \log\left(\frac{0.10}{0.20}\right) = 8.95$$

or, more realistically, pH = 9.0. (In this calculation, we identify the acid HA as NH_4^+; hence, its conjugate base A^- must be NH_3.)

N We consider that it would be courting disaster to attempt to introduce more varieties of the Henderson-Hasselbalch equation, such as one for pOH in terms of pK_b.

▼ EXAMPLE 15.7 Calculating the pH of a buffer solution

Calculate the pH of a buffer solution that is 0.040 M $Na_2HPO_4(aq)$ and 0.080 M $KH_2PO_4(aq)$ at 25°C.

STRATEGY This system is similar to that of Example 15.4 in that a weak acid and its conjugate base are present. Because buffer action depends on the simultaneous presence of a weak acid and base, we should begin by identifying the weak acid and its conjugate base (the acid has one more hydrogen atom than the base). Once they have been identified, we can write the proton transfer equilibrium between them, rearrange the expression for K_a to give $[H_3O^+]$, and from that calculate the pH.

SOLUTION The acid is the dihydrogenphosphate ion, $H_2PO_4^-$; its conjugate base is the hydrogenphosphate ion, HPO_4^{2-}. The equilibrium to consider is

$$H_2PO_4^-(aq) + H_2O(l) \rightleftharpoons H_3O^+(aq) + HPO_4^{2-}(aq)$$

$$\frac{[H_3O^+][HPO_4^{2-}]}{[H_2PO_4^-]} = K_{a2}$$

E The mixtures $H_2PO_4^-/HPO_4^{2-}$, CO_2/HCO_3^-, and HCO_3^-/CO_3^{2-} make an important contribution to the very complex buffering system in blood.

where K_{a2} is the second ionization constant of phosphoric acid (from Table 14.7, $pK_{a2} = 7.21$). Rearranging the expression for K_{a2} gives first

$$[H_3O^+] = K_{a2} \times \frac{[H_2PO_4^-]}{[HPO_4^{2-}]}$$

and then, taking logarithms and multiplying by -1,

$$pH = pK_{a2} + \log\left(\frac{[HPO_4^{2-}]}{[H_2PO_4^-]}\right)$$

Inserting the initial concentrations of acid ($H_2PO_4^-$) and base (HPO_4^{2-}), given as data, leads to

$$pH = 7.21 + \log\left(\frac{0.040}{0.080}\right) = 6.91$$

That is, the solution acts as a buffer close to pH = 7.

EXERCISE E15.7 Calculate the pH of a buffer solution that is 0.040 M $NH_4Cl(aq)$ and 0.030 M $NH_3(aq)$.

▲ [*Answer*: 9.1]

▼ EXAMPLE 15.8 Calculating the pH change when acid is added to a buffer solution

Suppose that 1.2 g (0.030 mol) of NaOH is dissolved in 500 mL of the buffer solution prepared as described in Example 15.7. Calculate the pH of the resulting solution and the change in pH. Assume there are no volume or temperature changes.

STRATEGY Once we know the molar concentrations of the acid and its conjugate base, we can rearrange the expression for K_a (into the form of the Henderson-Hasselbalch equation) to obtain the pH of the solution. The 0.030 mol of OH^- that is added to the buffer solution reacts with the acid of the buffer system, $H_2PO_4^-$, thereby reducing its amount by 0.030 mol and increasing the amount of its conjugate base, HPO_4^{2-}, by 0.030 mol.

SOLUTION The amount of the acid $H_2PO_4^-$ in the solution initially is

$$\text{Amount of } H_2PO_4^- \text{ (mol)} = 0.500 \text{ L} \times \frac{0.080 \text{ mol } H_2PO_4^-}{1 \text{ L}} = 0.040 \text{ mol } H_2PO_4^-$$

This amount is reduced by 0.030 mol to 0.010 mol $H_2PO_4^-$ by the addition of sodium hydroxide; hence its concentration is reduced to

$$[H_2PO_4^-] = \frac{0.010 \text{ mol}}{0.500 \text{ L}} = 0.020 \text{ mol/L}$$

Similarly, the amount of HPO_4^{2-} in the 500 mL of solution initially is

$$\text{Amount of } HPO_4^{2-} \text{ (mol)} = 0.500 \text{ L} \times \frac{0.040 \text{ mol}}{1 \text{ L}} = 0.020 \text{ mol } HPO_4^{2-}$$

The addition of 0.030 mol OH^- ions increases this amount to (0.020 + 0.030) mol = 0.050 mol and hence increases the concentration of the base to

$$[HPO_4^{2-}] = \frac{0.050 \text{ mol}}{0.500 \text{ L}} = 0.10 \text{ mol/L}$$

The proton transfer equilibrium is

$$H_2PO_4^-(aq) + H_2O(l) \rightleftharpoons H_3O^+(aq) + HPO_4^{2-}(aq) \qquad \frac{[H_3O^+][HPO_4^{2-}]}{[H_2PO_4^-]} = K_{a2}$$

so the hydronium ion concentration is

$$[H_3O^+] = K_{a2} \times \frac{[H_2PO_4^-]}{[HPO_4^{2-}]}$$

Therefore,

$$pH = pK_{a2} + \log\left(\frac{[HPO_4^{2-}]}{[H_2PO_4^-]}\right) = 7.21 + \log\left(\frac{0.10}{0.020}\right) = 7.91, \text{ or about } 7.9$$

That is, the pH of the solution changes from about 6.9 to about 7.9, an increase of 1.0. If the 0.030 mol of strong base had been added to 500 mL of water, the pH would have changed from 7.0 to 12.8, a change in pH of 5.8.

EXERCISE E15.8 Suppose that 0.010 mol of HCl gas is dissolved in 500 mL of the buffer solution in Example 15.7. Calculate the pH of the resulting solution and the change in pH. Assume that the volume and temperature do not change.

[*Answer*: 6.5; a decrease of 0.4] ▲

D Place equal volumes of a buffer solution and water in a beaker and measure the pH of each. Add a few drops of a strong acid to each one, and record the pH.

U Exercises 15.87 and 15.88.

E The following buffer solutions are often employed:

Composition	pH at 25°C
saturated potassium hydrogentartrate	3.56
0.05 mol/kg potassium hydrogenphthalate	4.01
0.025 mol/kg KH_2PO_4 + 0.025 mol/kg Na_2HPO_4	6.87
0.0087 mol/kg KH_2PO_4 + 0.030 mol/kg Na_2HPO_4	7.41
0.01 mol/kg $Na_2B_4O_7$	9.18

E Buffer capacities can be increased somewhat by increasing the amounts of acid and base in the buffer system.

? What happens to the change in pH of a solution if the buffer capacity is exceeded by the addition of a strong acid? Sketch a pH curve for the process, and explain its shape (see Fig. 15.6).

D Buffering action and capacity. BZS3, 8.28.

U Exercise 15.92

E Many bacterial cultures are grown in buffered media, and the selection and pH range of the buffer are critical to their growth and survival.

Commercially available buffer solutions can be purchased for virtually any desired pH and are often used to calibrate instruments. For instance, pH meters are often calibrated by using a 0.025 M $Na_2HPO_4(aq)$ and 0.025 M $KH_2PO_4(aq)$ mixed solution, which has pH = 6.87 at 25°C. The method demonstrated in Example 15.7 predicts pH = 7.2 for this solution; however, as noted at the start of the chapter, these calculations ignore ion-ion interactions, which modify the pH slightly.

Buffer capacity. The **buffer capacity** of a solution is the amount (in moles) of hydronium or hydroxide ions that needs to be added to the solution to change its pH by 1 unit. The concept of buffer capacity applies to any solution, but it is most useful when applied to an actual buffer solution. It is then an indication of the amount of acid or base that can be added before the buffer loses its ability to resist the change in pH: the bigger the capacity, the more the solution can absorb the addition of acid or base without a significant change in pH. This exhaustion of the buffer may occur because most of the base (which may be HPO_4^{2-} ions) has been converted to acid or because most of the acid (the $H_2PO_4^-$ ions) has been converted to base (see Example 15.8).

As we have seen, the pH of a buffer solution varies least with added acid or base in the region of pH = pK_a, so the solution has its maximum buffer capacity when its composition corresponds to this pH. It is quite easy to show that if the buffer is to be in the region of maximum capacity, the ratio of the base concentration to the acid concentration should not be less than 0.1 or more than 10. This conclusion follows from the Henderson-Hasselbalch equation, for a concentration ratio ranging from 1:10 to 10:1 implies a pH ranging from

$$\text{pH} = \text{p}K_a + \log\left(\frac{1}{10}\right) = \text{p}K_a - 1$$

to

$$\text{pH} = \text{p}K_a + \log\left(\frac{10}{1}\right) = \text{p}K_a + 1$$

For instance, the $H_2PO_4^-/HPO_4^{2-}$ phosphate buffer, for which $pK_{a2} = 7.21$, undergoes a change in pH from only 6.2 to 8.2 as the concentration ratio changes by a factor of 100, from 1:10 to 10:1.

Selection and preparation of a buffer solution. In the preparation of a buffer solution with a given pH, we select a weak acid (or weak base) with a pK_a (or pK_b) that is within one pH unit of the desired pH. Thus, from Table 14.6 or Table 15.4, we can conclude that an acetic acid/acetate buffer system is good over a pH range of 4.74 ± 1; and similarly, from Table 14.7, that a hydrogencarbonate/carbonate buffer is good over a pH range of 10.25 ± 1. The precise buffering range depends on the ratio of the concentrations of the acid and base.

▼ EXAMPLE 15.9 Determining the ratio of concentrations for a buffer solution with a given pH

Calculate the ratio of the concentrations of CO_3^{2-} and HCO_3^- ions needed to achieve buffering at pH 9.50. The pK_{a2} of H_2CO_3 is 10.25.

STRATEGY The pH of an HA/A$^-$ buffer is determined by the Henderson-Hasselbalch equation in the form

$$pH = pK_a + \log\left(\frac{[A^-]}{[HA]}\right)$$

So, all we need to do is identify the acid and base and rearrange this expression to give us the desired concentration ratio.

SOLUTION The base is CO_3^{2-} and the acid is HCO_3^-; hence,

$$\log\left(\frac{[CO_3^{2-}]}{[HCO_3^-]}\right) = pH - pK_a = 9.50 - 10.25 = -0.75$$

The antilogarithm of -0.75 is 0.18, so the ratio of concentrations is

$$\frac{[CO_3^{2-}]}{[HCO_3^-]} = 0.18$$

Therefore, the solution will buffer a solution at a pH close to 9.50 when it is prepared by mixing a ratio of 0.18 mol CO_3^{2-} to 1.0 mol HCO_3^-.

U Exercise 15.89.

EXERCISE E15.9 Calculate the ratio of the concentrations of acetate ions and acetic acid needed to buffer a solution at pH = 5.25. The pK_a of CH_3COOH is 4.74.

[*Answer*: 3.24]

▼ EXAMPLE 15.10 Preparing a buffer solution starting with a solid

What mass of sodium carbonate must be added to 500 mL of 0.40 M $NaHCO_3(aq)$ to prepare the buffer solution in Example 15.9? Assume that the volume of the solution does not change when the solid is added.

STRATEGY The Henderson-Hasselbalch equation tells us the ratio of concentrations of CO_3^{2-} and HCO_3^- ions that are needed (as explained in Example 15.9). Next, we note that, because the volume of the solution cancels, the ratio of concentrations of acid and base is equal to the ratio of amounts (in moles) of the acid and base. Then, from the amount of acid (HCO_3^-) already in the solution, we can obtain the amount of base (CO_3^{2-}) that is needed to produce the ratio of CO_3^{2-} and HCO_3^- ion concentrations to achieve the stated pH. Finally, we use the molar mass to calculate the mass of the salt that must be added to provide the calculated amount of basic anions.

? If you wanted to prepare a buffer solution for pH = 9.5 for a biology experiment, would you select HCN/CN$^-$ or HCO_3^-/CO_3^{2-}? Explain your reasoning.

SOLUTION We know from Example 15.9 that

$$\frac{[CO_3^{2-}]}{[HCO_3^-]} = 0.18$$

The amount of HCO_3^- ions already present in the solution is

$$\text{Amount of } HCO_3^- \text{ (mol)} = 0.500\text{ L} \times \frac{0.40\text{ mol } HCO_3^-}{1\text{ L}}$$
$$= 0.20\text{ mol } HCO_3^-$$

Therefore, the amount of CO_3^{2-} that must be added is

$$\text{Amount of } CO_3^{2-} = 0.20\text{ mol } HCO_3^- \times \frac{0.18\text{ mol } CO_3^{2-}}{1\text{ mol } HCO_3^-}$$
$$= 0.036\text{ mol } CO_3^{2-}$$

Because 1 mol CO_3^{2-} = 1 mol Na_2CO_3 and 1 mol Na_2CO_3 = 105.99 g Na_2CO_3,

BOX 15.1 ▸ pH AND ELECTROLYTE CHANGES AFTER BURN INJURY

BASIL A. PRUITT, JR., M.D.,
Commander and Director,
U.S. Army Institute of Surgical Research

ARTHUR D. MASON, JR., M.D.,
Chief, Laboratory Division,
U.S. Army Institute of Surgical Research

Basil A. Pruitt

Arthur D. Mason

The pH of human blood is maintained within a narrow range (7.35-7.45) by several buffer systems, among which the carbonic acid (H_2CO_3)/hydrogencarbonate (bicarbonate) ion (HCO_3^-) system is of greatest importance. Maintenance of normal blood pH is crucial to many life-sustaining functions.

The body controls the concentrations of the two components of the H_2CO_3/HCO_3^- buffer system by different mechanisms. Blood carbonic acid concentration is controlled by varying respiration, the amount of CO_2 exhaled being proportional to the rate and depth of breathing. Breathing faster and more deeply increases the amount of CO_2 exhaled and lowers the carbonic acid concentration in the blood. The concentration of hydrogencarbonate ion is controlled by varying the rate of excretion of this ion in the urine.

The normal ratio of HCO_3^- to H_2CO_3 in the blood is 20:1. When the concentration of HCO_3^- increases relative to that of H_2CO_3, blood pH rises. If this increase is sufficient to cause blood pH to exceed the normal range, the condition is called *alkalosis*. Conversely, blood pH decreases when the ratio decreases, and when blood pH falls below the normal range the condition is called *acidosis*.

When a person is burned, blood plasma leaks from the circulatory system into the injured area, producing swelling (edema) and reducing the blood volume. If the burned area is large, this loss of blood volume may be sufficient to reduce blood flow and oxygen supply to all the body's tissues. Lack of oxygen, in turn, causes the tissues to produce an excessive amount of lactic acid, which enters the blood stream, reacts with bicarbonate ion to produce H_2CO_3, and shifts the ratio of HCO_3^- to H_2CO_3 to a lower value. To minimize the consequent decrease in pH (called *metabolic* acidosis), the injured individual breathes harder, increasing the respiratory excretion of CO_2. If blood volume loss and tissue acid production exceed levels for which the body can compensate, a vicious circle ensues in which blood flow decreases still further, blood pressure falls, CO_2 excretion diminishes, and acidosis becomes more severe. Individuals in this state are said to be in *shock*, and will die if not treated promptly.

The dangerous condition of shock is avoided or treated by intravenous infusion of large vol-

Mass of Na_2CO_3 (g)

$$= 0.036 \text{ mol } CO_3^{2-} \times \frac{1 \text{ mol } Na_2CO_3}{1 \text{ mol } CO_3^{2-}} \times \frac{105.99 \text{ g } Na_2CO_3}{1 \text{ mol } Na_2CO_3}$$

$$= 3.82 \text{ g } Na_2CO_3$$

U Exercises 15.90 and 15.91.

That is, to prepare a hydrogencarbonate/carbonate buffer solution with a pH close to 9.5, starting with solid sodium carbonate and 500 mL of 0.40 M $NaHCO_3$, we should add 3.82 g of Na_2CO_3 to the solution.

EXERCISE E15.10 What mass of sodium acetate should be added to 100 mL of 0.100 M $CH_3COOH(aq)$ to obtain a solution that buffers close to pH = 5.25? See Exercise E15.9.

[*Answer*: 2.66 g]

umes of a salt-containing solution, usually one known as *lactated Ringer's solution.* With such infusion, blood volume and blood flow increase, oxygen delivery is improved, the HCO_3^-/H_2CO_3 ratio increases toward normal, and the severely injured person can survive.

When smoke is inhaled at the time of burn injury, the chemicals contained in the smoke cause an inflammatory response in the lungs that may impair breathing sufficiently to interfere with the normal excretion of CO_2. The consequent accumulation of H_2CO_3 in the blood decreases the HCO_3^-/H_2CO_3 ratio and blood pH. Such a pH abnormality, called *respiratory acidosis,* is treated by assisting the person's breathing with a mechanical ventilator. With such assistance, CO_2 excretion is increased and blood pH is corrected.

Other components of the blood are also involved in the body's response to burn injury; among these sodium and potassium ions are the blood electrolytes most profoundly affected. The plasma lost from the blood stream into the burn wound contains sodium ions, a loss that is replaced by the same intravenous infusion of salt-containing solution used to combat shock. Even though this infusion increases the total amount of sodium ion in the body, mild *hyponatremia* (reduced plasma sodium concentration) often ensues because the sodium concentration of the lactated Ringer's solution is somewhat lower than that of normal blood plasma. This hyponatremia may be worsened if the patient receives salt-free solutions. When hyponatremia is severe, it produces clinical effects ranging from disorientation to coma.

To guard against severe hyponatremia, medical personnel monitor plasma sodium concentrations frequently in burned patients and make daily measurements of body weight and urinary sodium excretion. This information permits the calculation of appropriate doses of salt and water to avoid or treat hyponatremia. Hyponatremia may be corrected by restricting water administration, by inducing increased urine output, or by infusing concentrated salt solutions.

Burned patients also tend to develop *hypernatremia* (increased blood sodium concentration) as a result of both extensive evaporative losses of salt-free water through the burn wound and hormonally driven retention of sodium by the kidneys. The careful monitoring routine noted above and calculation of appropriate salt and water doses permit control of hypernatremia.

Most of the body's potassium resides within cells. After burn injury, heat-damaged cells release potassium ion into the blood stream. The resulting increase in potassium concentration in the blood plasma is called *hyperkalemia.* In most patients, this is mild and of little consequence.

Later in the hospital course of treatment, urinary excretion of potassium and the uptake of potassium by healing tissue may produce a reduced plasma potassium concentration, or *hypokalemia.* If necessary, this deficit can be corrected by intravenous administration of solutions containing potassium salts.

The composition of blood plasma, with the concentration of HCO_3^- ions about 20 times that of H_2CO_3, seems to be outside the range for optimum buffering. However, this ratio is advantageous because the principal waste products of living cells are carboxylic acids, such as lactic acid. With its relatively high concentration of HCO_3^- ions, the plasma can absorb a significant surge of hydrogen ions from the carboxylic acids. Another consequence of the high proportion of HCO_3^- is that bodies are better able to withstand disturbances leading to excess acid (disease and shock) than those leading to excess base (burns); see Box 15.1.

FIGURE 15.14 The formation of a silver chloride precipitate occurs immediately after a small amount of Ag^+ ions is added to a solution containing Cl^- ions.

N We use the term *solubility constant* to match the similar terms that have already been encountered (such as equilibrium constant, acid ionization constant, etc.). All compounds have a solubility constant, even those that are freely soluble, but the concept is useful only when the compound is so sparingly soluble that K_s can be interpreted in terms of molar concentrations.

N The "more accurate procedure" is to use standard electrode potentials and their relation to equilibrium constants. See Section 17.5.

D Place an aqueous solution of sodium chloride in a glass dish on an overhead projector. Slowly add several drops of silver nitrate solution. Save this solution for p. 609 of the text, and then add repeatedly aqueous ammonia and nitric acid solutions.

V Video 39: Precipitation I. Reactions of a cation.

V Video 40: Precipitation II. Reactions of an anion.

D Solubility product: Effect of concentration. CD1, 54. The solubility of some silver compounds. CD2, 45.

E A different look at the solubility product principle. *J. Chem. Educ.,* **1985**, *62*, 645.

SOLUBILITY EQUILIBRIA

We saw in Section 11.3 that in a saturated solution of a salt the solid salt is in dynamic equilibrium with its ions in solution. We can now extend the discussion that we began there and use the techniques we have been developing to discuss the response of saturated solutions to changes in conditions. One such change is a modification of the pH of the solution by the addition of an acid or base; another is the effect of a common ion. We shall confine our attention to sparingly soluble salts because their solutions are very dilute and we do not have to worry about the effects of the interactions between the ions in the solution even when that solution is saturated.

15.6 THE SOLUBILITY CONSTANT

Examples of sparingly soluble salts were given in Table 3.2. One such salt is silver chloride, and Fig. 15.14 shows that a small volume of a solution containing Ag^+ ions (from a silver nitrate solution) added to a solution containing Cl^- ions (a sodium chloride solution) produces an immediate precipitate. The precipitate of silver chloride quickly establishes a heterogeneous dynamic equilibrium with its ions in the solution, and we can write

$$AgCl(s) \rightleftharpoons Ag^+(aq) + Cl^-(aq) \qquad \frac{[Ag^+][Cl^-]}{[AgCl]} = K_c \text{ at equilibrium}$$

However, we have seen that the *relative* molar concentration of a pure solid is 1, so the condition for equilibrium can be expressed in terms of the **solubility constant K_s**, where

$$[Ag^+][Cl^-] = K_s$$

(K_s is often called the solubility product constant and denoted K_{sp}.) The molar concentrations appearing in this expression are those at equilibrium (in the saturated solution) relative to the standard molar concentration of 1 mol/L. A more complicated example of an expression for the solubility constant is

$$Ag_3PO_4(s) \rightleftharpoons 3Ag^+(aq) + PO_4{}^{3-}(aq) \qquad [Ag^+]^3[PO_4{}^{3-}] = K_s$$

The solubility constants of a number of compounds are listed in Table 15.5. They can be obtained directly from the **molar solubility** S of the compound, which is the molar concentration of the compound in a saturated solution. For example, the molar solubility of silver chloride in water at 25°C is 1.3×10^{-5} mol/L. Therefore, because each formula unit of AgCl provides one Ag^+ ion and one Cl^- ion, a molar solubility of 1.3×10^{-5} mol AgCl/L corresponds to concentrations of 1.3×10^{-5} mol Ag^+/L and 1.3×10^{-5} mol Cl^-/L. Hence

$$K_s = [Ag^+][Cl^-] = (1.3 \times 10^{-5}) \times (1.3 \times 10^{-5}) = 1.7 \times 10^{-10}$$

The K_s listed in Table 15.5 differs slightly from this value; it was found by a more accurate procedure that is discussed in Chapter 17.

TABLE 15.5 Solubility constants at 25°C

Compound	Formula	K_s	Compound	Formula	K_s
aluminum hydroxide	$Al(OH)_3$	1.0×10^{-33}	fluoride	PbF_2	3.7×10^{-8}
antimony sulfide	Sb_2S_3	1.7×10^{-93}	iodate	$Pb(IO_3)_2$	2.6×10^{-13}
barium carbonate	$BaCO_3$	8.1×10^{-9}	iodide	PbI_2	1.4×10^{-8}
fluoride	BaF_2	1.7×10^{-6}	sulfate	$PbSO_4$	1.6×10^{-8}
sulfate	$BaSO_4$	1.1×10^{-10}	sulfide	PbS	8.8×10^{-29}
bismuth sulfide	Bi_2S_3	1.0×10^{-97}	magnesium ammonium phosphate	$MgNH_4PO_4$	2.5×10^{-13}
calcium carbonate	$CaCO_3$	8.7×10^{-9}	carbonate	$MgCO_3$	1.0×10^{-5}
fluoride	CaF_2	4.0×10^{-11}	fluoride	MgF_2	6.4×10^{-9}
hydroxide	$Ca(OH)_2$	5.5×10^{-6}	hydroxide	$Mg(OH)_2$	1.1×10^{-11}
sulfate	$CaSO_4$	2.4×10^{-5}	mercury(I) chloride	Hg_2Cl_2	1.3×10^{-18}
copper(I) bromide	$CuBr$	4.2×10^{-8}	iodide	Hg_2I_2	1.2×10^{-28}
chloride	$CuCl$	1.0×10^{-6}	mercury(II) sulfide	HgS, black	1.6×10^{-52}
iodide	CuI	5.1×10^{-12}		HgS, red	1.4×10^{-53}
sulfide	Cu_2S	2.0×10^{-47}	nickel(II) hydroxide	$Ni(OH)_2$	6.5×10^{-18}
copper(II) iodate	$Cu(IO_3)_2$	1.4×10^{-7}	silver bromide	$AgBr$	7.7×10^{-13}
oxalate	CuC_2O_4	2.9×10^{-8}	carbonate	Ag_2CO_3	6.2×10^{-12}
sulfide	CuS	1.3×10^{-36}	chloride	$AgCl$	1.6×10^{-10}
iron(II) hydroxide	$Fe(OH)_2$	1.6×10^{-14}	hydroxide	$AgOH$	1.5×10^{-8}
sulfide	FeS	6.3×10^{-18}	iodide	AgI	1.5×10^{-16}
iron(III) hydroxide	$Fe(OH)_3$	2.0×10^{-39}	sulfide	Ag_2S	6.3×10^{-51}
lead(II) bromide	$PbBr_2$	7.9×10^{-5}	zinc hydroxide	$Zn(OH)_2$	2.0×10^{-17}
chloride	$PbCl_2$	1.6×10^{-5}	sulfide	ZnS	1.6×10^{-24}

▼ EXAMPLE 15.11 Determining the solubility constant from the molar solubility

The molar solubility of silver chromate, Ag_2CrO_4, is 6.5×10^{-5} mol/L. Calculate the value of K_s for Ag_2CrO_4.

STRATEGY First, we write the chemical equation for the dynamic equilibrium that exists between the solid salt and its ions in solution and write the condition for equilibrium in terms of the solubility constant. The molar concentration of CrO_4^{2-} ions is equal to that of the salt; the molar concentration of Ag^+ ions can be calculated by noting that each Ag_2CrO_4 formula unit gives rise to two Ag^+ ions.

SOLUTION The solubility equilibrium is

$$Ag_2CrO_4(s) \rightleftharpoons 2Ag^+(aq) + CrO_4^{2-}(aq)$$

T OHT: Table 15.5.

? What additional information do you need if you want to predict how K_s for the compounds in Table 15.5 change with temperature?

E Relation between solubility and solubility product. *J. Chem. Educ.*, **1978**, *55*, 452. Solubility products and solubility: *Plus ça change. J. Chem. Educ.*, **1989**, *66*, 184.

U Exercise 15.93.

Because $[CrO_4^{2-}] = 6.5 \times 10^{-5}$ mol/L, we know that $[Ag^+] = 2 \times 6.5 \times 10^{-5}$ mol/L. Therefore,

$$K_s = [Ag^+]^2[CrO_4^{2-}] = (2 \times 6.5 \times 10^{-5})^2 \times (6.5 \times 10^{-5})$$
$$= 1.1 \times 10^{-12}$$

EXERCISE E15.11 The molar solubility of $Pb(IO_3)_2$ is 4.0×10^{-5} mol/L. What is the K_s of lead iodate?

[*Answer*: 2.6×10^{-13}]

One of the applications of solubility constants is the calculation of the molar solubility of the salt. (This calculation is not just a reversal of the determination of K_s, because we can determine the solubility constant from measurements other than solubilities.) The procedure is illustrated in the following example.

N The calculation is valid only if the solubility is so low that ion-ion interactions can be ignored (for sparingly soluble solids). Note that we continue to use an equilibrium table as a unifying factor for all equilibria.

E Ion association, solubilities, and reduction potentials in aqueous solution. *J. Chem. Educ.*, **1989**, *66*, 148.

EXAMPLE 15.12 Determining the molar solubility of a salt from its solubility constant

The K_s of PbI_2 is 1.4×10^{-8}. Calculate the molar solubility of lead(II) iodide and the mass of salt that can dissolve in 1.0 L of water at 25°C.

STRATEGY We can do the problem the same way we have done other equilibrium calculations, by drawing up the equilibrium table and then solving for the unknown concentration by using the solubility constant. The molar solubility is converted to mass dissolving in the stated volume by using the molar mass of the solute.

SOLUTION The equilibrium table is

Equilibrium equation:	$PbI_2(s) \rightleftharpoons$ $Pb^{2+}(aq)$ +	$2I^-(aq)$
Species:	Pb^{2+}	I^-
Step 1. Initial concentration, mol/L	0	0
Step 2. Change in concentration, mol/L	$+x$	$+2x$
Step 3. Equilibrium concentration, mol/L	x	$2x$

Step 4. Inserting the values from step 3 and $K_s = 1.4 \times 10^{-8}$ into

$$[Pb^{2+}][I^-]^2 = K_s \text{ at equilibrium}$$

gives

$$x \times (2x)^2 = 1.4 \times 10^{-8}$$

which solves to $x = 1.5 \times 10^{-3}$. From step 3, the Pb^{2+} ion concentration at equilibrium (in a saturated solution) is 1.5×10^{-3} mol/L. Because one formula unit of PbI_2 results in one Pb^{2+} ion, the molar solubility of PbI_2 is 1.5×10^{-3} mol/L. It follows that 1.0 L of the saturated solution contains 1.5×10^{-3} mol PbI_2, which corresponds to

$$\text{Mass } PbI_2 \text{ (g)} = 1.5 \times 10^{-3} \text{ mol } PbI_2 \times \frac{460.99 \text{ g } PbI_2}{1 \text{ mol } PbI_2} = 0.69 \text{ g } PbI_2$$

EXERCISE E15.12 The K_s of Ag_2SO_4 is 1.4×10^{-5}. Calculate the molar solubility of Ag_2SO_4.

[*Answer*: 1.5×10^{-2} mol/L]

U Exercises 15.94 and 15.97–99.

(a)

(b)

FIGURE 15.15 (a) A saturated solution of zinc acetate. (b) When acetate ions are added (as solid sodium acetate in the spatula shown in (a)), the solubility of the zinc acetate is significantly reduced, and zinc acetate precipitates.

D Solubility product effect of concentration. CD1, 54.

In Section 15.2 we saw how the common-ion effect reduces the extent of ionization of a weak acid when a salt with a common ion is added to the solution. Similarly, when a soluble salt that has an ion in common with a sparingly soluble salt is also present in solution, the solubility of the sparingly soluble salt is reduced even further (Fig. 15.15).

To see how this effect comes about, consider the dynamic equilibrium of the sparingly soluble salt AgCl in water:

$$AgCl(s) \rightleftharpoons Ag^+(aq) + Cl^-(aq) \qquad [Ag^+][Cl^-] = K_s$$

with $K_s = 1.6 \times 10^{-10}$ at 25°C. The molar solubility of AgCl in pure water is 1.3×10^{-5} mol/L. The addition of sodium chloride to a saturated solution of silver chloride increases the concentration of Cl^- ions. The system responds according to Le Chatelier's principle, and the added Cl^- ions shift the equilibrium toward AgCl(s), thereby reducing the Ag^+ concentration by precipitating AgCl. That is, the solubility of AgCl in a solution with added chloride ions is lower than in pure water.

We can make a quantitative prediction of the effect of a common ion by noting the fact that K_s is constant even when the additional ions are added. In any solution of silver and chloride ions, for example,

$$[Ag^+] = \frac{K_s}{[Cl^-]}$$

where K_s is a constant. Suppose that we try to dissolve silver chloride in 0.10 M NaCl(aq), in which $[Cl^-] = 0.10$ mol/L. The silver chloride dissolves until the concentration of Ag^+ ions reaches

$$[Ag^+] = \frac{1.6 \times 10^{-10}}{0.10} \text{ mol/L} = 1.6 \times 10^{-11} \text{ mol/L}$$

This solubility is a 10,000-fold decrease from its solubility in pure water.

E This calculation is highly unreliable because the addition of Cl^- causes a sharp rise in ion-ion interactions (that is, it is no longer valid to ignore activity coefficients), and K_s cannot be interpreted in terms of molar concentrations. The calculation, however, does capture the strong decrease in solubility even though the actual numbers obtained are inaccurate.

EXAMPLE 15.13 Determining the molar solubility of a salt in the presence of a common ion

In Exercise E15.11, the molar solubility of a saturated $Pb(IO_3)_2$ solution (in pure water) was given as 4.0×10^{-5} mol/L. What is its molar solubility in 0.10 M $Pb(NO_3)_2(aq)$? The K_s of $Pb(IO_3)_2$ is 2.6×10^{-13}.

STRATEGY When enough sparingly soluble salt has dissolved to saturate the solution, it has reached a dynamic equilibrium with its ions in solution. The solution also contains a common cation (Pb^{2+}) from $Pb(NO_3)_2$, which lowers the solubility of the sparingly soluble salt by causing the solubility equilibrium to shift toward reactants (the undissolved salt). Because the problem concerns the composition of a system at dynamic equilibrium, we should set up an equilibrium table, using the concentration of the lead nitrate solution as the starting composition of the system. Then we should solve for the unknown concentration in terms of the equilibrium constant—in this application, the solubility constant.

N The same comment applies to this calculation as in the previous annotation: the qualitative change is correct; the quantitative result is inaccurate.

SOLUTION The equilibrium table is

Equilibrium equation:	$Pb(IO_3)_2(s) \rightleftharpoons Pb^{2+}(aq) + 2IO_3^-(aq)$	
Species:	Pb^{2+}	IO_3^-
Step 1. Initial concentration, mol/L	0.10	0
(The initial Pb^{2+} ion concentration is 0.10 mol/L, from the 0.10 M $Pb(NO_3)_2$ solution, and the initial IO_3^- ion concentration is zero. The NO_3^- ion is a spectator, and the solid salt does not appear in the calculation.)		
Step 2. Change in concentration, mol/L	$+x$	$+2x$
Step 3. Equilibrium concentration, mol/L	$0.10 + x$	$2x$

Step 4. Inserting the values in step 3 and $K_s = 2.6 \times 10^{-13}$ into $[Pb^{2+}][IO_3^-]^2 = K_s$ at equilibrium gives

$$(0.10 + x)(2x)^2 = 2.6 \times 10^{-13}$$

Because the initial molar concentration of Pb^{2+} is much greater than the increase (x mol/L) in these ions when the sparingly soluble salt dissolves, we can neglect x in the first factor and solve

$$0.10 \times (2x)^2 \approx 2.6 \times 10^{-13}$$

The solution is

$$x \approx \sqrt{\frac{2.6 \times 10^{-13}}{0.40}} = 8.1 \times 10^{-7}$$

Because the concentration of Pb^{2+} ions arising from the sparingly soluble salt is x (from step 3) and each $Pb(IO_3)_2$ formula unit gives rise to one Pb^{2+} ion, the molar solubility of lead(II) iodate is 8.1×10^{-7} mol/L. The solubility of lead(II) iodate in pure water is 4.0×10^{-5} mol/L, so the common ion reduces its solubility by a factor of about 50.

EXERCISE E15.13 Calculate the molar solubility of Ag_2SO_4 in 0.20 M $Na_2SO_4(aq)$, given that the K_s of Ag_2SO_4 is 1.4×10^{-5}.

[*Answer*: 4.2×10^{-3} mol/L]

U Exercise 15.95.

15.7 PRECIPITATION REACTIONS AND QUALITATIVE ANALYSIS

FIGURE 15.16 When a few drops of lead(II) nitrate solution are added to a solution of potassium iodide, yellow lead(II) iodide immediately precipitates.

For a precipitation reaction to be successful, the concentrations of the two solutions that are mixed, one containing the cation and the other containing the anion of a sparingly soluble salt, must be high enough to result in the formation of precipitate. The question of whether a precipitate will form can be discussed in terms of the solubility constant of the sparingly soluble salt and the techniques we have already described.

Consider the reaction that occurs when drops of lead nitrate solution are added to a potassium iodide solution (Fig. 15.16). The overall equation for the reaction is

$$Pb(NO_3)_2(aq) + 2KI(aq) \longrightarrow 2KNO_3(aq) + PbI_2(s)$$

and the net ionic reaction (Section 3.3) is

$$Pb^{2+}(aq) + 2I^-(aq) \longrightarrow PbI_2(s)$$

The instant the solutions are mixed, the concentrations of the Pb^{2+} and I^- ions are very high (typically 0.1 mol/L) and a precipitate forms. As is customary, the solubility equilibrium for the sparingly soluble salt is written

$$PbI_2(s) \rightleftharpoons Pb^{2+}(aq) + 2I^-(aq) \qquad [Pb^{2+}][I^-]^2 = K_s \text{ at equilibrium}$$

with $K_s = 1.4 \times 10^{-8}$ at 25°C. The **solubility quotient** Q_s, the analogue of the reaction quotient considered in Section 13.2, is

$$Q_s = [Pb^{2+}][I^-]^2$$

where the concentrations are not necessarily those at equilibrium.

Just as we used the reaction quotient in Section 13.2 to determine the position of a system relative to its equilibrium composition, we can use the relative values of Q_s and K_s to judge whether a precipitate has a tendency to form, a solid has a tendency to dissolve, or a solid and its ions are in equilibrium (saturated):

E The rate of a precipitation reaction is determined largely by the rate at which the ions can move through the solvent, for as soon as they meet they clump together. The only activation energy required (other than that required to move through the solvent) is the energy to remove the hydrating H_2O molecules that they must discard as they form ionic bonds with each other.

- When $Q_s > K_s$, the concentrations of the ions are too high for equilibrium and the compound precipitates.
- When $Q_s = K_s$, the solution is saturated; the system is at equilibrium.
- When $Q_s < K_s$, the concentrations of the ions are too low for equilibrium and no precipitate forms.

To bring about a precipitation reaction, we must contrive to produce ion concentrations that result in a value of Q_s that is greater than that of the K_s of the salt. In our lead iodide example, a precipitate formed when the solutions were first mixed, so $Q_s > K_s$. The precipitate continues to form until $Q_s = K_s$.

For PbI_2 to precipitate, the ion concentrations must be such that $Q_s > 1.4 \times 10^{-8}$. For example, the instant after equal volumes of 0.2 M $Pb(NO_3)_2(aq)$ and 0.4 M $KI(aq)$ solutions are mixed, the ion concentrations are

$$[Pb^{2+}] = 0.1 \text{ mol/L}, \qquad [I^-] = 0.2 \text{ mol/L}$$

(The concentration of each ion is reduced by one-half because the combined volume of the two solutions has doubled.) Hence,

$$Q_s = [Pb^{2+}][I^-]^2 = (0.1) \times (0.2)^2 = 4 \times 10^{-3}$$

which is considerably greater than $K_s = 1.4 \times 10^{-8}$ (at 25°C). Consequently, a precipitate will form.

▼ EXAMPLE 15.14 Predicting whether a salt will precipitate

Suppose we mix 25.0 mL of 0.0010 M $AgNO_3(aq)$ with 75.0 mL of 0.0010 M $Na_2CO_3(aq)$. Does a precipitate of Ag_2CO_3 form? The K_s of Ag_2CO_3 is 6.2×10^{-12} at 25°C.

STRATEGY The outcome of the reaction can be predicted by comparing the values of Q_s and K_s for Ag_2CO_3. We must first write the expression for Q_s, then substitute the concentrations (not forgetting to correct for dilution) of each ion and compare the result with K_s.

SOLUTION The equilibrium to consider is

$$Ag_2CO_3(s) \rightleftharpoons 2Ag^+(aq) + CO_3^{2-}(aq) \qquad Q_s = [Ag^+]^2[CO_3^{2-}]$$

The total volume of the mixture is 100 mL, or 0.100 L. Therefore, immediately after mixing, the concentrations of the two ions in the combined solution are

$$[Ag^+] = \frac{0.025\text{ L} \times 0.0010\text{ mol/L}}{0.100\text{ L}} = 2.5 \times 10^{-4}\text{ mol/L}$$

$$[CO_3^{2-}] = \frac{0.075\text{ L} \times 0.0010\text{ mol/L}}{0.100\text{ L}} = 7.5 \times 10^{-4}\text{ mol/L}$$

U Exercises 15.96, 15.100, and 15.101.

Inserting these concentrations into Q_s gives

$$Q_s = [Ag^+]^2[CO_3^{2-}] = (2.5 \times 10^{-4})^2 \times (7.5 \times 10^{-4}) = 4.7 \times 10^{-11}$$

Because $K_s = 6.2 \times 10^{-12}$, we see that $Q_s > K_s$ and conclude that a precipitate of Ag_2CO_3 will form.

EXAMPLE E15.14 Does a precipitate of AgCl form when 200 mL of 1.0×10^{-4} M $AgNO_3(aq)$ and 900 mL of 1.0×10^{-6} M $KCl(aq)$ are mixed? The K_s of AgCl is 1.6×10^{-10}.

▲ [*Answer*: $Q_s = 1.5 \times 10^{-11} < K_s$; no precipitate forms]

The relationship between the solubility quotient and the solubility constant explains some of the techniques that have been developed for **qualitative analysis**, the analysis of an unknown sample to discover what cations it contains. There are many different schemes of analysis, but all have features in common. One such feature is the formation of a precipitate when certain reagents are added to a solution of the sample, often after it has been made acidic or basic. We shall work through a single procedure in the order actually adopted in the laboratory, interpreting each step in terms of the solubility equilibrium involved. The scheme described here is only a fragment of the systematic procedure actually employed, but it is enough to illustrate the general idea.

The scheme is set out in Table 15.6 and illustrated in Fig. 15.17. Certain chlorides (Table 15.6, step 1) are precipitated when hydrochloric acid is added to a solution of the sample. The idea behind this step is that silver, mercury(I), and lead(II) chlorides have such small values of K_s that the addition of Cl^- ions to solutions of those metal ions raises Q_s above K_s, with the result that the chlorides precipitate. The hydrogen

U Exercises 15.102–15.105 and 15.107.

ions provided by the acid play no role in this step; they simply accompany the chloride ions.

At a later stage in a typical analysis scheme, certain sulfides (Table 15.6, step 2) are precipitated by the addition of S^{2-} ions. The goal at this stage is to precipitate and identify *some* but not all the metal cations that may still be in solution; that is, to achieve **selective precipitation** by making use of the widely different solubilities (and solubility constants) of different compounds. Some metal sulfides (such as CuS, HgS, and Sb_2S_3) have extremely small solubility constants and are precipitated if there is the merest trace of S^{2-} ions in the solution. Such a very low concentration of S^{2-} ions is achieved by adding hydrogen sulfide, H_2S, to an acidified solution. A higher hydronium ion concentration ensures that the equilibrium

$$H_2S(aq) + H_2O(l) \rightleftharpoons 2H_3O^+(aq) + S^{2-}(aq)$$

is shifted to the left and that almost all the H_2S is present in its nonionized form. Hence, very little S^{2-} will be present. Nevertheless, that small amount will result in the precipitation of highly insoluble solids if the appropriate cations are present.

Next, ammonia is added to the solution. The base removes the hydronium ion from the H_2S equilibrium, thereby shifting the equilibrium in favor of S^{2-} ions. The higher concentration of S^{2-} ions increases the Q_s of any remaining metal sulfides above their K_s values, and they precipitate. This step detects the presence of metal sulfides that have larger solubility constants than those in the preceding step (Table 15.6, step 3), including ZnS and MnS.

(a)

(b)

(c)

FIGURE 15.17 The sequence for the analysis of cations by selective precipitation. The original solution contains Ag^+, Cu^{2+}, and Zn^{2+} ions. (a) Addition of HCl precipitates AgCl, which can be removed by filtration. (b) Addition of H_2S to the remaining solution precipitates CuS, which can also be removed. (c) Making the solution basic precipitates ZnS.

15.8 DISSOLUTION OF PRECIPITATES

After a precipitate of a sparingly soluble salt has formed, it is often necessary to dissolve the precipitate so that the cations or anions can be identified or used in another reaction. For example, if a silver salt precipitates, it may be necessary to dissolve it again so that it can be electroplated as metallic silver (Chapter 17). The strategy for dissolving a precipitate is to remove one of the ions from the solubility equilibrium, so that the precipitate continues to dissolve in order to reestablish

TABLE 15.6 Part of a qualitative analysis scheme

Step	Possible precipitate	K_s
(1) Add HCl(*aq*)	AgCl	1.6×10^{-10}
	Hg_2Cl_2	1.3×10^{-18}
	$PbCl_2$	1.6×10^{-5}
(2) Add $H_2S(aq)$ (in acidic solution, low S^{2-} concentration)	Bi_2S_3	1.0×10^{-97}
	CdS	4.0×10^{-29}
	CuS	1.3×10^{-36}
	HgS	1.6×10^{-52}
	Sb_2S_3	1.7×10^{-93}
(3) $H_2S(aq)$ (in basic solution, higher S^{2-} concentration)	FeS	6.3×10^{-18}
	MnS	1.3×10^{-15}
	NiS	1.3×10^{-24}
	ZnS	1.6×10^{-24}

N The dissolution of a precipitate requires an equilibrium shift toward products: according to Le Chatelier's principle, that implies that at least one of the ionic species must be removed.

D Precipitates and complexes of iron(III). BZS1, 4.11.

equilibrium. The removal of an ion can be done by one of several procedures.

The effect of pH. Solubilities of certain sparingly soluble compounds can be controlled by varying the pH. Consider, for example, nickel hydroxide in water:

$$Ni(OH)_2(s) \rightleftharpoons Ni^{2+}(aq) + 2OH^-(aq)$$
$$[Ni^{2+}][OH^-]^2 = K_s \text{ at equilibrium}$$

Because the solubility equilibrium involves OH^- ions, the addition of hydroxide ions to the solution will decrease the solubility of $Ni(OH)_2$. However, because the addition of hydronium ions removes OH^- ions, Le Chatelier's principle leads us to expect a shift in the equilibrium toward product and therefore an increase in the solubility of $Ni(OH)_2$ when H_3O^+ ions are added to the solution.

? Rust, $Fe_2O_3{\cdot}xH_2O$, can be removed from a steel furnace with acid. Explain the process in terms of iron(III) oxide being a sparingly soluble compound.

A compound need not be a hydroxide for its solubility to depend on pH: any Brønsted acid and base responds to the addition of an acid. Suppose we have a saturated solution of calcium fluoride, CaF_2. Because the F^- ion is the conjugate base of the weak acid hydrofluoric acid, the F^- ion is a proton acceptor. With the addition of a strong acid, the F^- ion accepts a proton from H_3O^+ to form the conjugate acid HF, thereby reducing the concentration of F^- ions in the solution. As a result, the solubility equilibrium

$$CaF_2(s) \rightleftharpoons Ca^{2+}(aq) + 2F^-(aq) \qquad [Ca^{2+}][F^-]^2 = K_s$$

also shifts right, more CaF_2 dissolves, and we report that its solubility has increased. Any Brønsted base is affected in the same manner when acid is added: the strong acid protonates the base and hence encourages the solubility equilibrium to shift in favor of dissolution. This behavior is also shown, for example, by sparingly soluble carbonates, sulfides, oxalates, and chromates, which contain the basic anions CO_3^{2-}, S^{2-}, $C_2O_4^{2-}$, and CrO_4^{2-}, respectively. In all cases their sparingly soluble salts are more soluble in acid than in pure water.

D Perform the following series of tests (in order), and account for each observation in terms of solubility, dissolution of precipitates, and Le Chatelier's principle: Start with $AgNO_3$ and add $Na_2CO_3(aq)$, form $Ag_2CO_3(s)$; add $HNO_3(aq)$, form CO_2; add $HCl(aq)$, form $AgCl(s)$; add $NH_3(aq)$, form $Ag(NH_3)_2^+(aq)$; add $HNO_3(aq)$, form $AgCl(s)$; add $KI(aq)$, form $AgI(s)$; add $Na_2S(aq)$, form $Ag_2S(s)$.

E Calculation of solubilities of carbonates and phosphates in water as influenced by competitive acid-base reactions. *J. Chem. Educ.*, **1990**, *67*, 934.

Carbonate, sulfite, and sulfide precipitates can often be dissolved by the addition of acid because the anions react to form a gas. For example, in a saturated solution of zinc carbonate, $ZnCO_3$, in equilibrium with its ions

$$ZnCO_3(s) \rightleftharpoons Zn^{2+}(aq) + CO_3^{2-}(aq)$$

the CO_3^{2-} reacts with the H_3O^+ from the acid that is added and forms CO_2:

$$CO_3^{2-}(aq) + 2H_3O^+(aq) \longrightarrow CO_2(g) + 3H_2O(l)$$

The removal of the carbonate ion from the solution causes the zinc carbonate solubility equilibrium to adjust in favor of the dissolved ions. Similarly, a zinc sulfite, $ZnSO_3$, precipitate is in equilibrium with its ions:

$$ZnSO_3(s) \rightleftharpoons Zn^{2+}(aq) + SO_3^{2-}(aq)$$

When acid is added, the hydronium ions react with the dissolved sulfite ions to produce sulfur dioxide:

$$SO_3^{2-}(aq) + 2H_3O^+(aq) \longrightarrow SO_2(g) + 3H_2O(l)$$

Because the sulfite ion is removed, the sulfite precipitate dissolves in an attempt to regain equilibrium. Much the same happens when acid is added to a zinc sulfide, ZnS, precipitate; the solubility equilibrium

$$ZnS(s) \rightleftharpoons Zn^{2+}(aq) + S^{2-}(aq)$$

adjusts by forming more sulfide ion when the S^{2-} reacts with H_3O^+ to form H_2S:

$$S^{2-}(aq) + 2H_3O^+(aq) \longrightarrow H_2S(g) + 2H_2O(l)$$

The effect of oxidizing agents. When the use of an acid does not modify the solubility equilibrium of a sparingly soluble salt, then an alternative procedure is tried. We can remove an ion in solution by changing its oxidation number (for that change turns it into a different ion). For example, the very insoluble sulfide precipitates can be dissolved by the addition of an oxidizing agent that oxidizes the sulfide ion to elemental sulfur. Copper(II) sulfide, CuS, takes part in the equilibrium

N Complex ion formation is discussed further in Chapter 21.

D Precipitates and complexes of copper(II). BZS1, 4.8.

$$CuS(s) \rightleftharpoons Cu^{2+}(aq) + S^{2-}(aq)$$

Unlike zinc sulfide, copper(II) sulfide does not dissolve when hydrochloric acid is added. When nitric acid is added, however, the S^{2-} ions are oxidized to elemental sulfur in the reaction

$$3S^{2-}(aq) + 8H^+(aq) + 2NO_3^-(aq) \longrightarrow 3S(s) + 2NO(g) + 4H_2O(l)$$

This oxidation removes sulfide ions from the equilibrium and the solid CuS dissolves.

The effect of Lewis bases. The formation of a complex ion (the product of a Lewis acid-base reaction; Section 8.5) can also remove an ion from its participation in a solubility equilibrium. For example, when aqueous ammonia is added to a saturated solution of silver chloride, the complex ion diamminesilver(I) forms:

D Precipitates and complexes of silver(I). BZS1, 4.6.

$$Ag^+(aq) + 2NH_3(aq) \rightleftharpoons Ag(NH_3)_2^+(aq)$$

The solid silver chloride present dissolves because the Ag^+ ions are removed from the equilibrium

$$AgCl(s) \rightleftharpoons Ag^+(aq) + Cl^-(aq)$$

But the product $[Ag^+][Cl^-]$ must remain equal to the value of K_s for silver chloride, so more silver chloride dissolves to make up for the loss of Ag^+ ions. Each new Ag^+ ion that goes into solution is immediately converted into $Ag(NH_3)_2^+$ by the ammonia present in the solution. If enough ammonia is present, then all the precipitate dissolves. A similar procedure is used to remove the silver halide emulsion from an exposed photographic film after it has been developed (see Box 3.1). In this case, the reagent used to form the complex ion is the thiosulfate ion, $S_2O_3^{2-}$, and the equilibrium is

$$Ag^+(aq) + 2S_2O_3^{2-}(aq) \rightleftharpoons Ag(S_2O_3)_2^{3-}(aq)$$

Formation constants of complex ions. An Ag^+ ion is a Lewis acid (an electron pair acceptor) and an NH_3 molecule is a Lewis base (an electron pair donor). The reaction between them is a Lewis acid-base reaction and the product is a complex ion. (The same is true of the reaction between Ag^+ ions and thiosulfate ions.) Here, then, is yet another kind of equilibrium that can affect the properties of a solution. The equilibrium constant for complex formation is called a **formation constant** (or

U Exercise 15.106.

D The "lemonade" reaction. CD1, 29.

TABLE 15.7 Formation constants in water at 25°C

Equilibrium	K_f
$Ag^+(aq) + 2Cl^-(aq) \rightleftharpoons AgCl_2^-(aq)$	2.5×10^5
$Ag^+(aq) + 2Br^-(aq) \rightleftharpoons AgBr_2^-(aq)$	1.3×10^7
$Ag^+(aq) + 2CN^-(aq) \rightleftharpoons Ag(CN)_2^-(aq)$	5.6×10^8
$Ag^+(aq) + 2NH_3(aq) \rightleftharpoons Ag(NH_3)_2^+(aq)$	1.6×10^7
$Au^+(aq) + 2CN^-(aq) \rightleftharpoons Au(CN)_2^+(aq)$	2.0×10^{38}
$Cu^{2+}(aq) + 4NH_3(aq) \rightleftharpoons Cu(NH_3)_4^{2+}(aq)$	1.2×10^{13}
$Hg^{2+}(aq) + 4Cl^-(aq) \rightleftharpoons HgCl_4^{2-}(aq)$	1.2×10^5
$Fe^{2+}(aq) + 6CN^-(aq) \rightleftharpoons Fe(CN)_6^{4-}(aq)$	7.7×10^{36}
$Ni^{2+}(aq) + 6NH_3(aq) \rightleftharpoons Ni(NH_3)_6^{2+}(aq)$	5.6×10^8

stability constant) K_f and is defined in the usual way:

$$Ag^+(aq) + 2NH_3(aq) \rightleftharpoons Ag(NH_3)^{2+}(aq)$$

$$\frac{[Ag(NH_3)_2^+]}{[Ag^+][NH_3]^2} = K_f, \text{ at equilibrium}$$

At 25°C, $K_f = 1.6 \times 10^7$ for this reaction. Values for other equilibria are given in Table 15.7.

We calculate the solubility of a salt in the presence of a complex ion by ensuring that both the formation and solubility constants are satisfied. For example, for AgCl in the presence of ammonia, we must consider the two equilibria

$$AgCl(s) \rightleftharpoons Ag^+(aq) + Cl^-(aq) \qquad [Ag^+][Cl^-] = K_s$$

$$Ag^+(aq) + 2NH_3(aq) \rightleftharpoons Ag(NH_3)_2^+(aq) \qquad \frac{[Ag(NH_3)_2^+]}{[Ag^+][NH_3]^2} = K_f$$

The sum of the two chemical equations is

$$AgCl(s) + 2NH_3(aq) \rightleftharpoons Ag(NH_3)_2^+(aq) + Cl^-(aq)$$

and the equilibrium condition is

$$\frac{[Ag(NH_3)_2^+][Cl^-]}{[NH_3]^2} = K$$

where (as it is easy to verify)

$$K = K_s \times K_f$$

D Disappearing orange reaction. Now you see it, now you don't. CD1, 65.

This relation can be used to calculate the molar solubility of a sparingly soluble salt in the presence of complex formation, as the following example shows.

EXAMPLE 15.15 Calculating molar solubility in the presence of complex formation

Calculate the molar solubility of silver chloride in 0.10 M $NH_3(aq)$, given that $K_s = 1.6 \times 10^{-10}$ for silver chloride and $K_f = 1.6 \times 10^7$ for the ammonia complex of Ag^+ ions, $Ag(NH_3)_2^+$.

STRATEGY Because 1 mol AgCl = 1 mol $Ag(NH_3)_2^+$, the molar solubility of AgCl is equal to the molar concentration of $Ag(NH_3)_2^+$ in the saturated solution. We can use the same strategy we used in all previous equilibrium calculations: we set up an equilibrium table, express the equilibrium constant for the solubility in terms of the concentrations in the table, and then solve the resulting expression for the unknown concentration. As remarked above, K for the combined reaction is the product of K_s and K_f.

SOLUTION The equilibrium table is

Equilibrium equation:	$AgCl(s) + 2NH_3(aq) \rightleftharpoons$	$Ag(NH_3)_2^+(aq) +$	$Cl^-(aq)$
Species:	NH_3	$Ag(NH_3)_2^+$	Cl^-
Step 1. Initial concentration, mol/L	0.10	0	0
Step 2. Change in concentration, mol/L	$-2x$	$+x$	$+x$
Step 3. Equilibrium concentration, mol/L	$0.10 - 2x$	x	x

Step 4. Inserting the equilibrium concentrations in step 3 into

$$\frac{[Ag(NH_3)_2^+][Cl^-]}{[NH_3]^2} = K = K_s \times K_f \text{ at equilibrium}$$

with $K_s \times K_f = 2.6 \times 10^{-3}$, gives

$$\frac{x \times x}{(0.10 - 2x)^2} = 2.6 \times 10^{-3}$$

Taking the square root of both sides then gives

$$\frac{x}{0.10 - 2x} = 5.1 \times 10^{-2}$$

Rearranging and solving for x gives $x = 4.6 \times 10^{-3}$. Therefore, from step 3, $[Ag(NH_3)_2^+] = 4.6 \times 10^{-3}$ mol/L, and the molar solubility of silver chloride is 4.6×10^{-3} mol/L. The molar solubility of silver chloride in pure water is 1.3×10^{-5} mol/L, different by a factor of over 100 from the molar solubility of silver chloride in an aqueous solution of ammonia.

EXERCISE E15.15 Calculate the molar solubility of silver bromide in 1.0 M $NH_3(aq)$.

[*Answer*: 3.5×10^{-3} mol/L]

▲

Complex ion formation is used in the extraction of gold from low-grade gold-containing rock. The formation constant for the complex $Au(CN)_2^-$ is very large, and as soon as any Au^+ ions are formed (through oxidation) in the presence of CN^- ions, the complex is formed. Complex formation removes Au^+ ions from the solution, so more of them are formed. Therefore, even though gold is not oxidized by air normally, bubbling air through a suspension of gold-containing ore in the presence of cyanide ions leads to the formation of a solution of the complex ion:

$$4Au(s) + 8CN^-(aq) + O_2(g) + 2H_2O(l) \longrightarrow 4Au(CN)_2^-(aq) + 4OH^-(aq)$$

The unwanted material is removed by filtration, and zinc powder is added to the solution. The zinc reduces the gold(I) ion in the complex to metallic gold:

$$2Au(CN)_2^-(aq) + Zn(s) \longrightarrow 2Au(s) + Zn(CN)_4^{2-}(aq)$$

which can then be removed by filtration.

Aqua regia ("kingly water"), a 3:1 mixture by volume of concentrated hydrochloric acid and nitric acid, can also dissolve gold. In this process, the nitric acid is the oxidizing agent and the chloride ions form the complex ion $AuCl_4^-$ with Au^{3+} ions. As the formation of the complex ion removes Au^{3+} ions from the solution, the nitric acid oxidizes more gold metal until it has all disappeared.

SUMMARY

15.1 The presence of a salt in water can affect the pH of a solution because the cation or the anion of the salt may react with water and produce an imbalance of H_3O^+ or OH^- ions. Cations that are conjugate acids of weak bases or metal cations that are not from Groups I or II or do not have an oxidation number +1 produce acidic solutions. Anions that are conjugate bases of weak acids produce basic solutions. The pH of a salt solution is determined in the same manner as the pH of a solution of a weak acid or weak base is determined (see Chapter 14).

15.2 The addition of a **common ion**, an ion that is also a component of a weak acid, to a solution suppresses the ionization of the parent weak acid. The addition of an ion that is common to a weak base in solution suppresses the ionization of the parent weak base. The pH of a mixed solution is determined in the same manner as the pH of a weak acid or weak base solution is determined (Chapter 14).

15.3 The stoichiometric point of the titration of a strong base with a strong acid is at pH = 7 because neither the cation nor the anion of the resulting salt is an acid or base. The pH at the stoichiometric point of the titration of a weak acid with a strong base is higher than 7 because the anion of the weak acid is a base. The pH at the stoichiometric point of the titration of a weak base with a strong acid is less than 7 because the cation of the weak base is an acid. The pH at the halfway point is equal to the pK_a of the weak acid.

15.4 The stoichiometric point in an acid-base titration is detected with an **acid-base indicator**, a weak acid that undergoes a color change when it changes from its acid form, HIn, to its conjugate base form, In^-. The indicator must be chosen so that its end point (the pH of its color change) occurs over a range close to the stoichiometric point of the titration. Specifically, the indicator should have a pK_{In} in the range pH ± 1 around the stoichiometric point.

15.5 The pH of a solution can be stabilized by a **buffer**, a mixture of a weak acid and its conjugate base or a weak base and its conjugate acid, in approximately equal concentrations. A buffer solution acts by providing a base that reacts with hydronium ions and an acid that reacts with hydroxide ions. The pH of a buffer solution can be determined from the **Henderson-Hasselbalch equation**. A buffer system for a desired pH range is selected such that the pK_a of the weak acid lies in the range pH ± 1 around the desired pH. The buffer system for a desired pH is prepared by determining the correct [base]:[acid] ratio. The **buffer capacity** of a solution is the amount of acid or base needed to change its pH by 1 unit.

15.6 The solubility equilibrium of a sparingly soluble solid is described by the equilibrium constant, the **solubility constant** K_s. The molar solubility of the substance can be determined from the value of K_s. Le Chatelier's principle predicts a decrease in solubility in the presence of a common ion: this change in solubility is the **common-ion effect**.

15.7 The precipitation of a sparingly soluble salt may be discussed in terms of the **solubility quotient**, Q_s: precipitation occurs when $Q_s > K_s$. In one scheme of **qualitative analysis**, a procedure for identifying the cations present in a sample, the magnitude of the solubility quotient is modified to bring about the **selective precipitation** of various metal cations.

15.8 There are several procedures for dissolving a precipitate: the general procedure is to remove one of the ions in the solubility equilibrium. Solids that contain a basic anion, especially the OH^- ion, dissolve in an acidic solution because of the formation of the conjugate acid of the anion. Salts that contain an anion that in an acidic solution readily reacts to form a gas, especially the CO_3^{2-}, SO_3^{2-}, or S^{2-} ions, dissolve because the anion is removed from the solubility equilibrium. A change in the oxidation number of an ion also removes the ions from the equilibrium. Some cations act as Lewis acids, forming complex ions with anions or neutral molecules; both act as Lewis bases. Such complex ions are in equilibrium in solution, and the composition of the equilibrium mixture is described by the **formation constant** K_f. Complex ion formation can increase the solubility of a salt by removing the cation from the equilibrium.

CLASSIFIED EXERCISES

The values for the ionization constants of weak acids and bases are listed in Tables 14.6, 14.7, and 15.1. You may need to refer to these and other tables in the text for these exercises. Unless stated otherwise, assume that all solutions are aqueous and at 25°C.

Ions as Acids and Bases

15.1 Determine whether aqueous solutions of the following salts have a pH equal to, greater than, or less than 7. If pH > 7 or pH < 7, then write a chemical equation to justify your answer. (a) NH_4Br; (b) Na_2CO_3; (c) KF; (d) $Co(NO_3)_2$.

15.2 Determine whether aqueous solutions of the following salts have a pH equal to, greater than, or less than 7. If pH > 7 or pH < 7, then write a chemical equation to justify your answer. (a) $K_2C_2O_4$ (potassium oxalate); (b) $Ca(NO_3)_2$; (c) CH_3NH_3Cl; (d) K_3PO_4.

15.3 Determine the value of the acid or base ionization constants of the following ions: (a) NH_4^+; (b) CO_3^{2-}; (c) F^-; (d) HCO_3^-.

15.4 Determine the value of the acid or base ionization constants of the following ions: (a) $N_2H_5^+$; (b) PO_4^{3-}; (c) NO_2^-; (d) $C_5H_5NH^+$.

15.5 Calculate the pH of the solution and the percentage of base or acid protonated or deprotonated, respectively, of the following solutions: (a) 0.10 M $NH_4Cl(aq)$; (b) 0.10 M $AlCl_3(aq)$; (c) 0.15 M $KCN(aq)$.

15.6 Calculate the pH of the solution and the percentage of base or acid protonated or deprotonated, respectively, of the following solutions: (a) 0.15 M $CH_3NH_3Cl(aq)$; (b) 0.20 M $Na_2SO_3(aq)$; (c) 0.30 M $FeCl_3(aq)$.

15.7 (a) A 10.0-g sample of potassium acetate, KCH_3CO_2, is dissolved in 250 mL of solution. What is the pH of the solution? (b) What is the pH of a solution resulting from the dissolution of 5.75 g of ammonium bromide, NH_4Br, in 100 mL of solution?

15.8 (a) A 1.00-g sample of sodium carbonate, Na_2CO_3, is dissolved in 50.0 mL of solution. What is the pH of the solution? (b) An aqueous solution filled to "the mark" in a 1.0-L volumetric flask contains 5.00 mg of methylammonium bromide, CH_3NH_3Br. What percentage of methylammonium ions are deprotonated?

15.9 (a) A 200-mL sample of 0.200 M $NaCH_3CO_2(aq)$ is diluted to 500 mL. What is the concentration of acetic acid at equilibrium? (b) What is the pH of a solution resulting from the dissolution of 5.75 g of ammonium bromide, NH_4Br, in 400 mL of solution?

15.10 (a) A 50.0-mL sample of 0.630 M $KCN(aq)$ is diluted to 125 mL. What is the concentration of hydrocyanic acid present at equilibrium? (b) A 1.00-g sample of sodium carbonate, Na_2CO_3, is dissolved in 150.0 mL of solution. What is the pH of the solution?

Mixed Solutions

15.11 Explain what happens to (a) the concentration of H_3O^+ ions in an acetic acid solution when solid sodium acetate is added and (b) the percentage ionization of benzoic acid in a benzoic acid solution when hydrochloric acid is added.

15.12 Explain what happens to (a) the pH of a phosphoric acid solution on the addition of solid sodium dihydrogenphosphate and (b) the concentration of H_3O^+ ions when pyridinium chloride is added to a pyridine solution.

15.13 A solution of equal concentrations of lactic acid and sodium lactate was found to have pH = 3.08. (a) What are the values of pK_a and K_a of lactic acid? (b) What would the pH be if the acid had twice the concentration of the salt?

15.14 A solution containing equal concentrations of saccharin and its sodium salt was found to have pH = 11.68. (a) What are the values of pK_a and K_a of saccharin? (b) What would the pH be if the salt had twice the concentration of the acid?

15.15 Calculate the concentration of hydronium ions in (a) a solution that is 0.20 M $HBrO(aq)$ and 0.10 M $KBrO(aq)$; (b) a solution that is 0.010 M $(CH_3)_2NH(aq)$ and 0.150 M $(CH_3)_2NH_2Cl(aq)$; (c) a solution that is 0.020 M $(CH_3)_2NH(aq)$ and 0.030 M $(CH_3)_2NH_2Cl(aq)$.

15.16 What is the concentration of hydronium ions in (a) a solution that is 0.050 M $HCN(aq)$ and 0.030 M $NaCN(aq)$; (b) a solution that is 0.10 M $NH_2NH_2(aq)$ and 0.50 M $NaCl(aq)$; (c) a solution that is 0.15 M $NH_2NH_2(aq)$ and 0.15 M $NH_2NH_3Br(aq)$?

15.17 The pH of 0.40 M $HF(aq)$ is 1.93. Calculate the change in pH when 1.0 g of sodium fluoride is added to 25.0 mL of the solution. Ignore any volume change.

15.18 The pH of 0.50 M $Na_2CO_3(aq)$ is 11.98. Calculate the change in pH when 10.0 g of sodium hydrogen carbonate is added to 250 mL of the solution. Ignore any volume change.

15.19 Calculate the pH of the solution that results from mixing (a) 20.0 mL of 0.050 M $HCN(aq)$ with 80.0 mL of 0.030 M $NaCN(aq)$; (b) 25.0 mL of 0.105 M $HCN(aq)$ with 25.0 mL of 0.105 M $NaCl(aq)$.

15.20 Calculate the pH of the solution that results from mixing (a) 100 mL of 0.020 M $(CH_3)_2NH(aq)$ with 300 mL of 0.030 M $(CH_3)_2NH_2Cl(aq)$; (b) 65.0 mL of 0.010 M $(CH_3)_2NH(aq)$ with 10.0 mL of 0.150 M $(CH_3)_2NH_2Cl(aq)$.

Weak Acid-Strong Base and Weak Base-Strong Acid Titrations

15.21 A 25.0-mL sample of 0.10 M $CH_3COOH(aq)$ is titrated with 0.10 M $NaOH(aq)$. The K_a of CH_3COOH is 1.8×10^{-5}. (a) What is the initial pH of the 0.10 M $CH_3COOH(aq)$ solution? (b) What is the pH after the addition of 10.0 mL of 0.10 M $NaOH(aq)$? (c) What volume of 0.10 M $NaOH(aq)$ is required to reach halfway to the stoichiometric point? (d) Calculate the pH at that halfway point. (e) What volume of 0.10 M $NaOH(aq)$ is required to reach the stoichiometric point? (f) Calculate the pH at the stoichiometric point. (g) Select a suitable indicator from Table 15.3.

15.22 A 30.0-mL sample of 0.20 M $C_6H_5COOH(aq)$ solution is titrated with 0.30 M $KOH(aq)$. The K_a of C_6H_5COOH is 6.5×10^{-5}. (a) What is the initial pH of the 0.20 M $C_6H_5COOH(aq)$ solution? (b) What is the pH after the addition of 15.0 mL of 0.30 M $KOH(aq)$? (c) What volume of 0.30 M $KOH(aq)$ is required to reach halfway to the stoichiometric point? (d) Calculate the pH at the halfway point. (e) What volume of 0.30 M $KOH(aq)$ is required to reach the stoichiometric point? (f) Calculate the pH at the stoichiometric point. (g) Select a suitable indicator from Table 15.3.

15.23 A 15.0-mL sample of 0.15 M $NH_3(aq)$ solution is titrated with 0.10 M $HCl(aq)$. The K_b of NH_3 is 1.8×10^{-5}. (a) What is the initial pH of the 0.15 M $NH_3(aq)$ solution? (b) What is the pH after the addition of 15.0 mL of 0.10 M $HCl(aq)$? (c) What volume of 0.10 M $HCl(aq)$ is required to reach halfway to the stoichiometric point? (d) Calculate the pH at the halfway point. (e) What volume of 0.10 M $HCl(aq)$ is required to reach the stoichiometric point? (f) Calculate the pH at the stoichiometric point. (g) Select a suitable indicator from Table 15.3.

15.24 A 50.0-mL sample of 0.25 M $CH_3NH_2(aq)$ solution is titrated with 0.35 M $HCl(aq)$. The K_b of CH_3NH_2

is 3.6×10^{-4}. (a) What is the initial pH of the 0.25 M $CH_3NH_2(aq)$ solution? (b) What is the pH after the addition of 15.0 mL of 0.35 M $HCl(aq)$? (c) What volume of 0.35 M $HCl(aq)$ is required to reach halfway to the stoichiometric point? (d) Calculate the pH at the halfway point. (e) What volume of 0.35 M $HCl(aq)$ is required to reach the stoichiometric point? (f) Calculate the pH at the stoichiometric point. (g) Select a suitable indicator from Table 15.3.

15.25 Calculate the pH of 25.0 mL of 0.110 M aqueous lactic acid being titrated with 0.150 M $NaOH(aq)$ (a) initially; (b) after the addition of 5.0 mL of base; (c) after the addition of a further 5.0 mL of base; (d) at the stoichiometric point; (e) after the addition of 5.0 mL of base beyond the stoichiometric point; (f) after the addition of 10 mL of base beyond the stoichiometric point. (g) Pick a suitable indicator from Table 15.3.

15.26 Calculate the pH of 25.0 mL of 0.215 M chloroacetic acid being titrated with 0.116 M $NaOH(aq)$ (a) initially; (b) after the addition of 5.0 mL of base; (c) after the addition of a further 5.0 mL of base; (d) at the stoichiometric point; (e) after the addition of 5.0 mL of base beyond the stoichiometric point; (f) after the addition of 10 mL of base beyond the stoichiometric point. (g) Pick a suitable indicator from Table 15.3.

Indicators

15.27 Over what pH range can each of the following indicators be used for detecting the stoichiometric point in a titration: (a) methyl orange; (b) litmus; (c) methyl red; (d) phenolphthalein?

15.28 Over what pH range can each of the following indicators be used for detecting the stoichiometric point in a titration: (a) thymol blue; (b) phenol red; (c) bromphenol blue; (d) alizarin.

15.29 For each indicator calculate the ratio of base form to conjugate acid form present in solution as the pH changes in unit steps over the designated pH range: (a) methyl orange (3.0–5.0); (b) litmus (5.0–8.0); (c) methyl red (4.0–6.0); (d) phenolphthalein (8.0–10.0).

15.30 For each indicator calculate the ratio of base form to conjugate acid form present in solution as the pH changes in unit steps over the designated pH range: (a) thymol blue (8.0–10.0); (b) phenol red (6.0–8.0); (c) bromphenol blue (6.0–8.0).

Buffers

15.31 Identify which of the following mixed systems can function as a buffer solution and write an equilibrium equation for each buffer system: (a) equal volumes of 0.10 M $HCl(aq)$ and 0.10 M $NaCl(aq)$; (b) a solution that is 0.10 M $HClO(aq)$ and 0.10 M $NaClO(aq)$; (c) equal volumes of 0.20 M $CH_3COOH(aq)$ and 0.10 M $NaOH(aq)$.

15.32 Identify which of the following mixed systems can function as a buffer solution and write an equilibrium equation for each buffer system: (a) equal volumes of 0.10 M $C_6H_5COOH(aq)$ and 0.10 M $Na(C_6H_5CO_2)(aq)$; (b) a solution that is 0.10 M $HNO_3(aq)$ and 0.10 M $NaNO_3(aq)$; (c) a solution that is 0.10 M $C_5H_5N(aq)$ and 0.10 M $C_5H_5NHCl(aq)$.

15.33 Aspirin is a derivative of salicylic acid, for which $K_a = 1.1 \times 10^{-3}$. Calculate the ratio of the concentrations of the salicylate ion (its conjugate base) to salicylic acid in a solution having a pH adjusted to 2.50.

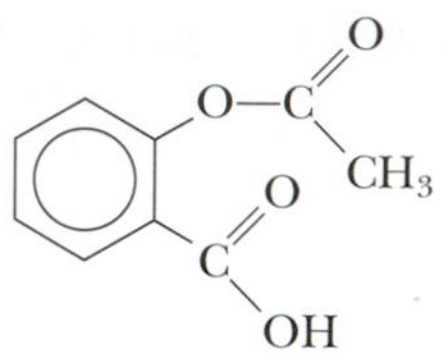

Acetylsalicylic acid (aspirin)

15.34 The narcotic cocaine is a weak base with $pK_b = 5.59$. Calculate the ratio of the concentration of cocaine and its conjugate acid in a solution of pH = 7.00.

15.35 A 100-mL buffer solution consists of 0.10 M $CH_3COOH(aq)$ and 0.10 M $NaCH_3CO_2(aq)$. (a) What is the pH of the buffer solution? (b) What is the pH and pH change resulting from the addition of 3.0 mmol NaOH to the buffer solution? (c) What is the pH and pH change resulting from the addition of 6.0 mmol of HNO_3 to the initial buffer solution?

15.36 A 100-mL buffer solution is 0.15 M $Na_2HPO_4(aq)$ and 0.10 M $KH_2PO_4(aq)$. From Table 14.7, the K_{a2} for phosphoric acid is 6.2×10^{-8}. (a) What is the pH of the buffer solution? (b) What is the pH and pH change resulting from the addition of 8.0 mmol NaOH to the buffer solution? (c) What is the pH and pH change resulting from the addition of 10.0 mmol of HNO_3 to the initial buffer solution?

15.37 A 100-mL buffer solution consists of 0.10 M $CH_3COOH(aq)$ and 0.10 M $NaCH_3CO_2(aq)$. (a) What is the pH and pH change resulting from the addition of 10.0 mL of 0.950 M $NaOH(aq)$ to the buffer solution? (b) What is the pH and pH change resulting from the addition of 20.0 mL of 0.10 M $HNO_3(aq)$ to the initial buffer solution? (*Hint*: Consider the dilution stemming from the addition of strong base or acid.)

15.38 A 100-mL buffer solution is 0.15 M $Na_2HPO_4(aq)$ and 0.10 M $KH_2PO_4(aq)$. (a) What is the pH and pH change resulting from the addition of 80.0 mL of 0.010 M $NaOH(aq)$ to the buffer solution? (b) What is the pH and pH change resulting from the addition of 10.0 mL of 1.0 M $HNO_3(aq)$ to the initial buffer solution? (*Hint*: Consider the dilution stemming from the addition of strong base or acid.)

15.39 Predict the pH region in which each of the following buffers will be effective, assuming equal molar concentrations of the acid and its conjugate base (refer to

Tables 14.6 and 14.7): (a) sodium lactate and lactic acid; (b) sodium benzoate and benzoic acid; (c) potassium hydrogenphosphate and potassium dihydrogenphosphate.

15.40 Refer to Tables 14.6 and 14.7 to suggest a buffer that would be effective at a pH close to (a) 2; (b) 7; (c) 3; (d) 12.

15.41 (a) What must be the ratio of the concentrations of CO_3^{2-} and HCO_3^- ions in a buffer solution having a pH of 11.0? (b) What mass of K_2CO_3 must be added to 1.00 L of 0.100 M $KHCO_3(aq)$ to prepare a buffer solution with a pH of 11.0. (c) What mass of $KHCO_3$ must be added to 1.00 L of 0.100 M $K_2CO_3(aq)$ to prepare a buffer solution with a pH of 11.0. (d) What volume of 0.200 M K_2CO_3 must be added to 100 mL of 0.100 M $KHCO_3(aq)$ to prepare a buffer solution with a pH of 11.0?

15.42 (a) What must be the ratio of the concentrations of PO_4^{3-} and HPO_4^{2-} ions in a buffer solution having a pH of 12.0? (b) What mass of K_3PO_4 must be added to 1.00 L of 0.100 M $K_2HPO_4(aq)$ to prepare a buffer solution with a pH of 12.0? (c) What mass of K_2HPO_4 must be added to 1.00 L of 0.100 M $K_3PO_4(aq)$ to prepare a buffer solution with a pH of 12.0? (d) What volume of 0.150 M K_3PO_4 must be added to 50.0 mL of 0.100 M $K_2HPO_4(aq)$ to prepare a buffer solution with a pH of 12.0?

15.43 Tris is an organic buffer that is commonly used in biochemistry experiments. In its acidic form (HB^+) it has a pK_a of 8.08. Calculate the pH of 1.00 L of solution that contains 0.050 mol of B and 0.100 mol of HB^+.

15.44 Valine is an amino acid that acts like a diprotic acid (H_2A) in its acid form, with $pK_{a1} = 2.29$ and $pK_{a2} = 9.72$. (a) Calculate the two pH values at which valine solutions act as a buffer. (b) Calculate the pH of 1.00 L of solution containing 1.00 mol of H_2A and 1.00 mol of HA^-.

Solubility Constants and Solubilities

15.45 Write the expression for the solubility constant of the following substances: (a) AgBr; (b) Ag_2S; (c) Ag_2CrO_4.

15.46 Write the expression for the solubility of the following substances: (a) AgSCN; (b) Sb_2S_3; (c) $Mg_3(PO_4)_2$.

15.47 Determine the K_s for the following sparingly soluble substances given their molar solubility: (a) $PbCrO_4$, 1.3×10^{-7} mol/L; (b) $Ba(OH)_2$, 0.11 mol/L; (c) MgF_2, 1.2×10^{-3} mol/L.

15.48 Determine the K_s for the following sparingly soluble salts, given their molar solubility: (a) AgI, 9.1×10^{-9} mol/L; (b) Ag_3PO_4, 2.7×10^{-6} mol/L; (c) Hg_2Cl_2, 5.2×10^{-7} mol/L.

15.49 Use the data in Table 15.5 to calculate the molar solubility of (a) Ag_2S; (b) CuS; (c) $CaCO_3$.

15.50 Use the data in Table 15.5 to determine the molar solubility of (a) $PbSO_4$; (b) Ag_2CO_3; (c) $Fe(OH)_2$.

15.51 The concentration of Mg^{2+} ions in seawater is about 1.3 μg/L. In a commercial recovery (Dow Chemical) process, the magnesium is precipitated as the hydroxide. At what pH does magnesium hydroxide precipitate?

15.52 Limestone is composed primarily of calcium carbonate. A 1-mm^3 pebble of limestone was accidentally dropped into a swimming pool, measuring 10 m $\times$ 7 m $\times$ 2 m and filled with water. Assuming that the carbonate ion does not function as a Brønsted base, will the pebble dissolve entirely? The density of calcium carbonate is 2.71 g/cm^3.

15.53 What volume (in liters) of a saturated mercury(II) sulfide, HgS (black) solution contains an average of one mercury(II) ion, Hg^{2+}?

15.54 What volume (in liters) of a saturated Bi_2S_3 solution contains an average of one bismuth ion, Bi^{3+}?

Common-Ion Effect

15.55 Use the data in Table 15.5 to calculate the molar solubility of each sparingly soluble substance in its respective solution: (a) silver chloride in a 0.20 M NaCl solution; (b) mercury(I) chloride in a 0.10 M NaCl(aq) solution; (c) iron(II) hydroxide in a 1.0×10^{-4} M $FeCl_2(aq)$ solution.

15.56 Use the data in Table 15.5 to calculate the molar solubility of each sparingly soluble substance in its respective solution: (a) silver bromide in a 1.0×10^{-3} M NaBr(aq) solution; (b) magnesium carbonate in a 4.2×10^{-5} M $Na_2CO_3(aq)$ solution; (c) nickel hydroxide in a 3.7×10^{-5} M $NiSO_4(aq)$ solution.

15.57 (a) What molar concentration of Ag^+ ions is required for the formation of a precipitate when added to a 1.0×10^{-5} M NaCl(aq) solution. (b) What mass (in micrograms) of $AgNO_3$ must be added to 200 mL of solution to trigger the onset of precipitation in (a)?

15.58 It is necessary to add iodide ions to precipitate lead(II) ion from a 0.0020 M $Pb(NO_3)_2(aq)$ solution. (a) What (minimum) iodide ion concentration is required for the onset of PbI_2 precipitation? (b) What mass (in grams) of KI must be added to 25.0 mL of solution to initiate PbI_2 formation?

15.59 Determine the pH required for the onset of precipitation of $Ni(OH)_2$ from a 0.010 M $NiSO_4(aq)$ solution.

15.60 At what minimum molar concentration of strontium ion will the ion precipitate as $Sr(OH)_2$ from a solution with pH = 12.0?

Precipitation Reactions and Qualitative Analysis

15.61 Decide whether a precipitate will form when the following solutions are mixed: (a) 27.0 mL of 0.0010 M

NaCl(*aq*) and 73.0 mL of 0.0040 M $AgNO_3$(*aq*); (b) 1.0 mL of 1.0 M K_2SO_4(*aq*), 10.0 mL of 0.0030 M $CaCl_2$(*aq*), and 100 mL of water.

15.62 Decide whether a precipitate will form when the following solutions are mixed: (a) 5.0 mL of 0.10 M K_2CrO_4(*aq*) and 1.00 L of 0.010 M $AgNO_3$(*aq*); (b) 3.3 mL of 1.0 M H_3PO_4(*aq*), 4.9 mL of 0.0030 M $AgNO_3$(*aq*), and enough water to dilute the solution to 50.0 mL.

15.63 Suppose that there are typically 20 average-sized drops in 1 mL of an aqueous solution. Will a precipitate form when 1 drop of 0.010 M NaCl(*aq*) is added to 10.0 mL of (a) 0.0040 M $AgNO_3$(*aq*) solution or (b) 0.0040 M $Pb(NO_3)_2$(*aq*) solution?

15.64 Assuming 20 drops per milliliter, will a precipitate form if (a) 7 drops of 0.0029 M K_2CO_3(*aq*) are added to 25.0 mL of a 0.0018 M $CaCl_2$(*aq*) solution or (b) 10 drops of 0.010 M Na_2CrO_4 are added to 10.0 mL of 0.0040 M $AgNO_3$(*aq*) solution?

15.65 The concentrations of calcium and iron(II) ions in an aqueous solution are 0.0010 mol/L. (a) In what order do they precipitate when a KOH solution is added? (b) Determine the pH at which each salt precipitates.

15.66 We want to separate by means of sulfide precipitation the copper(II) ions from the manganese(II) ions in a solution that is 0.20 mol Cu^{2+}/L and 0.20 mol Mn^{2+}/L. The source of the sulfide is from H_2S, which, when saturated, has a molar concentration of 0.1 mol/L. Considering the equilibrium

$$H_2S(aq) + 2H_2O(l) \rightleftharpoons 2H_3O^+(aq) + S^{2-}(aq) \qquad K_{a1}K_{a2} = 9.3 \times 10^{-22}$$

determine a pH that will result in the precipitation of one cation (identify the cation) but not the other.

Dissolution of Precipitates

15.67 Use the data in Table 15.5 to calculate the molar solubility of each sparingly soluble substance in its respective solution: (a) aluminum hydroxide at pH = 7.0; (b) aluminum hydroxide at pH = 4.5; (c) zinc hydroxide at pH = 7.0; (d) zinc hydroxide at pH = 6.0.

15.68 Use the data in Table 15.5 to calculate the solubility of each sparingly soluble compound in its respective solution: iron(III) hydroxide at (a) pH = 11.0 and (b) pH = 3.0; iron(II) hydroxide at (c) pH = 8.0 and (d) pH = 6.0.

15.69 Consider the two equilibria

$$CaF_2(s) \rightleftharpoons Ca^{2+}(aq) + 2F^-(aq) \qquad K_s = 4.0 \times 10^{-11}$$

$$F^-(aq) + H_2O(l) \rightleftharpoons HF(aq) + OH^-(aq) \qquad K_a(HF) = 3.5 \times 10^{-4}$$

(a) Write the chemical equation for the overall equilibrium and determine the corresponding equilibrium constant. (b) Determine the solubility of CaF_2 at pH = 7.0. (c) Determine the solubility of CaF_2 at a pH = 5.0.

15.70 Consider the two equilibria

$$MnS(s) \rightleftharpoons Mn^{2+}(aq) + S^{2-}(aq) \qquad K_s = 1.4 \times 10^{-15}$$

$$S^{2-}(aq) + 2H_2O(l) \rightleftharpoons H_2S(aq) + 2OH^-(aq) \qquad K_{a1}K_{a2}(H_2S) = 9.3 \times 10^{-22}$$

(a) Write the chemical equation for the overall equilibrium and determine the corresponding equilibrium constant. (b) Determine the molar solubility of MnS in a saturated H_2S (0.1 M H_2S(*aq*)) solution adjusted to pH = 7.0. (c) Determine the molar solubility of MnS in a saturated H_2S (0.1 M H_2S(*aq*)) solution adjusted to pH = 10.0.

15.71 Precipitated silver chloride dissolves in hydrochloric acid as a result of the formation of the $AgCl_2^-$ ion. What is the molar solubility of silver chloride in 1.0 M HCl(*aq*)?

15.72 Silver bromide is precipitated when a bromide solution is added to silver nitrate. The precipitate dissolves in the presence of KCN. Suppose the initial concentration of the cyanide ion is 0.10 mol/L. Calculate the final concentration of CN^- in the solution and the molar solubility of AgBr in the solution.

UNCLASSIFIED EXERCISES

15.73 Predict whether aqueous solutions of the following salts will be acidic, basic, or neutral (and justify your prediction): (a) KI; (b) CsF; (c) CrI_3; (d) $C_6H_5NH_3Cl$.

15.74 Write an equilibrium that shows that (a) a $CrCl_3$ solution is acidic; (b) a $(CH_3)_3NCl$ (trimethylammonium chloride) solution is acidic; (c) a Na_3PO_4 solution is basic.

15.75 *p*-Chlorophenylacetic acid has $pK_a = 4.19$. (a) What is the pH of a 0.020 M sodium *p*-chlorophenylacetate solution? (b) Determine the percentage protonation of *p*-chlorophenylacetate ion in the solution.

p-Chlorophenylacetic acid

15.76 *p*-Aminobenzoic acid (PABA) is used as a sunscreen in tanning lotion and has $pK_a = 4.00$. Determine the pH of a 0.0010 M sodium *p*-aminobenzoate solution.

p-Aminobenzoic acid

15.77 When 10.0 mg of sodium barbituate is dissolved in 250 mL of solution, the resulting pH is 7.71. The molar mass of sodium barbituate is 150 g/mol. Determine (a) the percentage protonation of barbituate ions and (b) the K_a of barbituric acid.

15.78 Determine (a) the pH of a 0.0240 M hydroxylammonium chloride (more commonly called hydroxylamine hydrochloride) and (b) the percentage deprotonation of hydroxylammonium ions in the solution.

15.79 A solution is prepared by mixing 200 mL of 0.27 M $Na_3PO_4(aq)$ and 150 mL of 0.62 M $KCl(aq)$. What is the pH of the solution?

15.80 (a) What is the pH of 0.037 M $NaCH_3CO_2(aq)$? (b) If 200 mL of 0.020 M $CH_3COOH(aq)$ is added to 150 mL of the solution in (a), what is the pH of the mixed solution?

15.81 A 60.0-mL sample of 0.10 M $NaHCO_2(aq)$ is mixed with 40.0 mL of 0.070 M $HCl(aq)$. Calculate the pH and the molar concentration of HCOOH in the mixed solution.

15.82 (a) What volume of 0.0400 M $NaOH(aq)$ is required to reach the stoichiometric point in the titration of 10.00 mL of 0.0633 M $HBrO(aq)$? (b) What is the pH at the stoichiometric point? (c) Pick a suitable indicator from Table 15.3.

15.83 What is the pH at each stoichiometric point in the titration of 25.0 mL of 0.20 M $H_2SO_4(aq)$ with 0.20 M $NaOH(aq)$?

15.84 A 25.0-mL sample of 0.20 M $(COOH)_2(aq)$ is titrated with 0.20 M $NaOH(aq)$. From Table 14.7, for oxalic acid, $K_{a1} = 5.9 \times 10^{-2}$ and $K_{a2} = 6.5 \times 10^{-5}$. (a) What volume of 0.20 M $NaOH(aq)$ is required to reach the first stoichiometric point? (b) What is the salt present at that point? (c) Calculate the pH at the first stoichiometric point. (d) What (total) volume of 0.20 M $NaOH(aq)$ is required to reach the second stoichiometric point? What is the salt present at that point? (d) Calculate the pH at the second stoichiometric point. (e) Select a suitable indicator to detect the first stoichiometric point and a second indicator for the second stoichiometric point from Table 15.3.

15.85 A 25.0-mL sample of tartaric acid solution is titrated with 0.100 M $KOH(aq)$ and a pH curve was constructed from the data. A sharp rise in the pH occurred on the addition of 17.0 mL of titrant and a second sharp rise occurred on the addition of 34.0 mL. (a) Explain why there were two rapid increases in pH. (b) What is the molar concentration of the tartaric acid solution? What is the pH of the solution after the addition of (c) 17.0 mL or (d) 34.0 mL of 0.100 M $NaOH(aq)$. (e) What is the tartrate ion concentration after the addition of 34.0 mL of titrant? (f) What is the pH of the solution after the addition of 8.5 mL of titrant?

15.86 Novocaine, which is used by dentists as a local anesthetic, is a weak base with $pK_b = 5.05$. Because blood has a pH of 7.4, what is the ratio of concentrations of novocaine to its conjugate acid in the bloodstream?

15.87 A buffer solution is prepared by mixing 50.0 mL of 0.022 M $C_6H_5COOH(aq)$ and 20.0 mL of 0.032 M $NaC_6H_5CO_2(aq)$. (a) What is the pH of the buffer solution? (b) What is the pH and the change in pH upon the addition of 0.054 mmol HCl to the buffer solution? (c) What would the pH change be if the 0.054 mmol HCl had been added to pure water instead of the buffer solution? (d) What is the pH and the change in pH after 10.0 mL of 0.054 M HCl is added to the original buffer solution?

15.88 For 100 mL of a buffer solution that is 0.150 M $CH_3COOH(aq)$ and 0.50 M $NaCH_3CO_2(aq)$, what is the pH before and after adding (a) 10.0 mL of 1.2 M $HCl(aq)$ or (b) 50.0 mL of 0.094 M $NaOH(aq)$?

15.89 A chemist decides to try to prepare a buffer solution from acetic acid, CH_3COOH, and hydrazine, N_2H_4. The equilibrium representing this buffer could be

$$CH_3COOH(aq) + N_2H_4(aq) \rightleftharpoons N_2H_5^+(aq) + CH_3CO_2^-(aq)$$

(a) Write the expression for the equilibrium constant and use the information in Table 14.6 to determine its value. (b) Derive a Henderson-Hasselbalch equation for this buffer system.

15.90 To simulate blood conditions, a phosphate buffer system with a pH = 7.40 is desired. (a) What must the ratio of the concentrations of HPO_4^{2-} to $H_2PO_4^-$ ions be? (b) What mass of Na_2HPO_4 must be added to 500 mL of 0.10 M $NaH_2PO_4(aq)$ in the preparation of the buffer?

15.91 Describe, with accompanying calculations, the procedure for preparing a buffer solution for pH = 10.0, starting with solid Na_2CO_3 and solid $NaHCO_3$.

15.92 What is the ideal pH range for a buffer solution that uses HBrO and NaBrO as the acid-base pair?

15.93 The molar concentration of CrO_4^{2-} in a saturated Tl_2CrO_4 solution is 6.3×10^{-5} mol/L. What is the K_s of Tl_2CrO_4?

CASE 15 ▸ SWIMMING POOLS AND CHEMISTRY

For health reasons and aesthetic effects, all swimming pools require a certain amount of care: suspended solids (leaves, dirt, and hair) must be removed; the pH of the water must not irritate the skin, eyes, and ears; disinfectants must be present to minimize the growth of bacteria and algae; and stabilizers must be added to maintain the appropriate concentration of disinfectant.

The effectiveness of the disinfectant is closely related to the pH of the water. The disinfectant used in public swimming pools is chlorine gas. Chlorine gas reacts with water:

$$Cl_2(g) + H_2O(l) \rightleftharpoons HClO(aq) + H^+(aq) + Cl^-(aq)$$

The presence of the oxidizing hypochlorous acid maintains low levels of bacteria (and algae) in the water. From the equilibrium, it is evident that an extensive use of chlorine causes the water to become acidic, and the more acidic the water, the less soluble the chlorine. To maintain disinfectant in the water and pH levels that are between 7.2 and 7.8 (ideally, 7.5), the pH is adjusted by adding base, usually sodium carbonate (called soda ash).

Because of the toxic and corrosive nature of chlorine, many home swimming pools are treated with solid calcium hypochlorite, $Ca(ClO)_2$,

$$Ca(ClO)_2(s) + 2H_2O(l) \longrightarrow 2HClO(aq) + Ca(OH)_2(aq)$$

Calcium hydroxide is sparingly soluble but does produce a sufficient supply of hydroxide ions. Consequently, the use of $Ca(ClO)_2$ increases the pH. To offset the increased pH, the pH must occasionally be lowered by adding hydrochloric acid ("muriatic acid") or solid sodium hydrogensulfate, $NaHSO_4$. If the pH becomes too high, a cloudiness appears because of the presence of the insoluble calcium hydroxide.

Alternatively, chlorine may be supplied by adding chlorinated isocyanurates, such as trichloroisocyanuric acid, which is marketed as Tri-Chlor and is the sodium salt of trichloroisocyanuric acid. (Di-Chlor, dichloroisocyanuric acid, is also used.) Trichloroisocyanuric acid produces hypochlorous acid and cyanuric acid in water by the reaction

15.94 (a) What is the molar solubility of Ag_2S, the black tarnish on silverware? (b) What is its molar solubility in 2.0×10^{-4} M $AgNO_3(aq)$? (c) What mass of Ag_2S will dissolve in 10.0 L of an aqueous solution?

15.95 What is the molar solubility of $Al(OH)_3$ in a solution with pH = 6.0?

15.96 Fluoridation of city water supplies produces a fluoride ion concentration close to 5×10^{-5} mol/L. Is it possible that CaF_2 will precipitate in hard water in which the Ca^{2+} ion concentration is 2×10^{-4} mol/L?

15.97 The fluoride ions in drinking water convert the hydroxyapatite, $Ca_5(PO_4)_3OH$, of teeth into fluoroapatite, $Ca_5(PO_4)_3F$. The K_s of the two compounds are 10^{-36} and 10^{-60}, respectively. What are the molar solubilities of each substance? The solubility equilibria to consider are

$$Ca_5(PO_4)_3OH(s) \rightleftharpoons 5Ca^{2+}(aq) + 3PO_4^{3-}(aq) + OH^-(aq)$$

$$Ca_5(PO_4)_3F(s) \rightleftharpoons 5Ca^{2+}(aq) + 3PO_4^{3-}(aq) + F^-(aq)$$

15.98 Milk of magnesia, taken internally for acid indigestion, is a saturated solution of magnesium hydroxide. What is the pH of milk of magnesia?

15.99 Limewater is a saturated aqueous calcium hydroxide solution. (a) What is the pH of limewater? (b) What volume of 0.010 M $HCl(aq)$ is required to titrate 25.0 mL of limewater to the phenolphthalein end point?

15.100 Will Ag_2CO_3 precipitate from a solution formed from a mixture of 100 mL of 1.0×10^{-4} M $AgNO_3(aq)$ and 100 mL of 1.0×10^{-4} M $Na_2CO_3(aq)$?

15.101 To 500 mL of an aqueous solution adjusted to a pH of 8.00, a student adds 1.36 mg of $ZnCl_2$. Does a precipitate of $Zn(OH)_2$ form?

15.102 Which sulfide precipitates first when sulfide ions are added to a solution containing equal amounts of Co^{2+}, Cu^{2+}, and Cd^{2+}? Explain your conclusions.

15.103 In the process of separating Cd^{2+} ions from Fe^{2+} ions as sparingly soluble sulfides, what is the Cd^{2+}

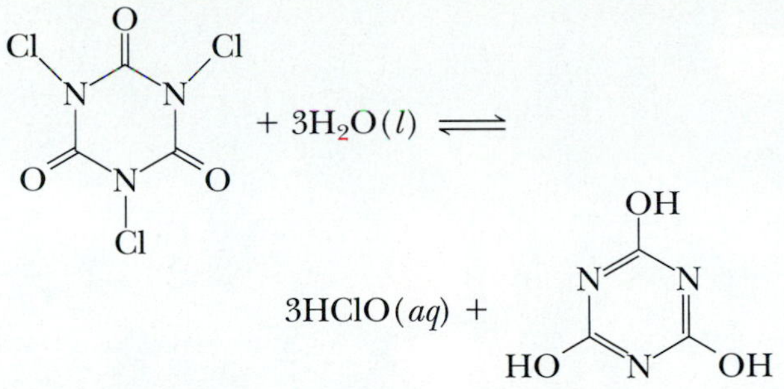

$$+\ 3H_2O(l) \rightleftharpoons$$

$$3HClO(aq)\ +$$

The pK_{a1} of cyanuric acid is 7.20, and so it does not contribute appreciably to the pH of the water. These compounds also are desirable because their release of HClO is gradual; so as the HClO is consumed in the course of its destruction of bacteria additional HClO is produced to maintain equilibrium. In a properly maintained pool, about one-half the total chlorine should be in the form of HClO.

The test for levels of chlorine in swimming pools utilizes an organic compound, *ortho*-toluidine, which forms a yellow compound in its oxidized form. The more intense the yellow color, the higher the concentration of chlorine (as Cl_2, HClO, or ClO^-). Chlorine concentrations in public swimming pools should be no less than 1 ppm (that is, 1 g of Cl_2 per 10^6 g of water).

QUESTIONS

1. (a) Determine the ratio $[ClO^-]/[HClO]$ at pH values of 6.0, 6.5, 7.0, 7.5, 8.0, and 8.5. The pK_a of HClO is 7.52. (b) Plot the concentration ratios in (a) against pH. (c) What is the ratio of concentrations in the pH range that is ideal for swimming pools? (d) Use the plot to describe the effect of pH on the concentration of HClO in an aqueous system.

2. A certain swimming pool has a capacity of 3.5×10^4 L. (a) What amount (in moles) of hydrochloric acid must be added to change the pH of the water from 8.0 to 7.5? (b) Assuming a 10 M HCl(*aq*) solution, what volume of acid should be added?

3. The chlorine level in the swimming pool in Question 2 should be 1 ppm. The density of chlorine gas at STP is 3.17 g/L and that of swimming pool water is 1.0 g/mL. (a) What volume of chlorine gas must dissolve in the swimming pool to provide a concentration of 1 ppm? (b) What mass of calcium hypochlorite, $Ca(ClO)_2$, would produce the same concentration of chlorine?

concentration when Fe^{2+} just begins to precipitate from 0.0010 M $FeCl_2(aq)$?

15.104 An attempt is made to separate barium ion from lead ion by using the sulfate ion as a precipitating agent. (a) What sulfate ion concentrations are required for the precipitation of $BaSO_4$ and $PbSO_4$ from a solution containing 0.010 mol Ba^{2+}/L and 0.010 mol Pb^{2+}/L? (b) What is the concentration of barium ions when the lead sulfate begins to precipitate?

15.105 Consider the two equilibria

$$ZnS(s) \rightleftharpoons Zn^{2+}(aq) + S^{2-}(aq) \qquad K_s = 1.6 \times 10^{-24}$$

$$S^{2-}(aq) + 2H_2O(l) \rightleftharpoons H_2S(aq) + 2OH^-(aq) \qquad K_{a1}K_{a2}(H_2S) = 9.3 \times 10^{-22}$$

(a) Write the chemical equation for the overall equilibrium and determine the corresponding equilibrium constant. (b) Determine the molar solubility of ZnS in a saturated H_2S (0.1 M $H_2S(aq)$) solution adjusted to pH = 7.0. (c) Determine the molar solubility of ZnS in a saturated H_2S (0.1 M $H_2S(aq)$) solution adjusted to pH = 10.0.

15.106 Use appropriate examples to write balanced equations for the dissolution of (a) an insoluble carbonate salt with hydrochloric acid; (b) an insoluble hydroxide salt with hydrochloric acid; (c) an insoluble silver salt with aqueous ammonia; (d) an insoluble sulfide salt with nitric acid; (e) an insoluble oxalate salt with hydrochloric acid.

15.107 It is often useful to know whether two ions can be separated by selective precipitation from a solution. Generally a 99% separation is considered "separated." A solution has concentrations of 0.010 mol Pb^{2+}/L and 0.010 mol Ag^+/L. Chloride ions are added from a sodium chloride solution. (a) Determine the chloride ion concentration required for the precipitation of each cation. (b) Which cation precipitates first? (c) What is the molar concentration of the first cation that precipitates when the second cation begins to precipitate? (d) Determine the percentage of the first cation that remains in solution when the second cation begins to precipitate and the percentage of the first cation that has precipitated when the second cation begins to precipitate.

In this chapter we develop the concepts of energy and entropy as they are used in chemistry. In particular, we see how entropy accounts for the tendency of reaction mixtures to change until they reach equilibrium. We see, too, that substances have certain properties that determine the composition of a reaction mixture at equilibrium. The illustration shows hot copper in a refining plant. The dimming color of the copper depicts the tendency of copper to cool as it emerges from the source of heat: the same principle underlies the approach to chemical equilibrium.

16 THERMODYNAMICS AND EQUILIBRIUM

The branch of chemistry that deals with the transformations of energy is called **thermodynamics**. Thermochemistry, the study of the heat released and absorbed by reactions, is a branch of thermodynamics. So is equilibrium, the subject we have been studying for the past three chapters. We have already dealt with thermochemistry in some detail (in Chapter 6), and all we shall do in this chapter is review some of its features in the context of the broader subject of thermodynamics. In particular, we shall describe the role of the *first law* of thermodynamics, which is essentially a statement about the conservation of energy.

The major part of the chapter is concerned with another aspect of the transformation of energy: the entropy. This property enables us to predict whether a change has a tendency to occur, such as whether a solid has a tendency to dissolve in a given solvent. We have seen that systems tend toward equilibrium and that the composition of the system at equilibrium can be expressed in terms of an equilibrium constant. The question we shall address is *why* this is so. What determines, for instance, whether a liquid has a particular tendency to vaporize or to condense? What determines whether a particular reaction mixture has a tendency to form more products or more reactants? We shall see in this chapter that questions like these are answered by a single unifying law: the *second law* of thermodynamics.

The second law, which is essentially a statement about the increase in entropy, underlies the tendency toward equilibrium of any type of system undergoing any type of change. It accounts for the tendency of a system to reach equilibrium in vaporizing, dissolving, and precipitating; it explains all the colligative properties; it lies behind the equilibrium properties of Brønsted acids and bases. We shall see in Chapter 17 that it also accounts for the tendency of redox reactions to approach equilibrium. It is not far from the truth to say that the second law summarizes chemistry.

The second law also has great practical importance: it can be used to *predict* the values of equilibrium constants. In particular, we shall see how to use tables of data like those in Appendix 2A to predict the value

N Students often have difficulty with thermodynamics on their first exposure to it. Pace is very important, and we consider it better to cover less but to cover it well, rather than rush through the chapter. We consider this a centrally important chapter, for it brings together all the preceding chapters on equilibrium into a single conceptual framework, and acts as a bridge to the remaining chapter in this part (on electrochemistry).

E The first law. For scientists, citizens, poets, and philosophers. *J. Chem. Ed.*, **1973**, *50*, 323.

of the equilibrium constants for a wide range of reactions. We shall see how to predict solubility constants, vapor pressures, pK_a values, and the values of the equilibrium constants for reactions in general.

THE FIRST LAW OF THERMODYNAMICS

E The system is defined by the boundary that separates it from the surroundings. All observations on the properties of the system, and the distinction between the transfer of energy as heat or as work, are made in the surroundings.

We need to recall that in thermochemistry—and in thermodynamics itself—we distinguish between the system, the reaction vessel and contents in which we are interested, and its surroundings, everything else outside the system. We shall make use of both these terms throughout the chapter. We should also recall that energy is defined as a measure of the capacity of a system to do work (Section 6.1). There are two ways of changing the energy of a system (other than adding more material to it). One is by heating the system—the transfer of energy by making use of a temperature difference between the system and its surroundings. The other is by doing work on the system—by pushing against an opposing force; for example, winding a spring.

The other major property that we introduced in Chapter 6—one we shall use extensively in this chapter—was the enthalpy of a system. The change in enthalpy is equal to the energy transferred as heat to a system that is kept at constant pressure, for example, a reaction taking place in a container that is open to the atmosphere.

16.1 HEAT, WORK, AND ENERGY

N There are several advantages in the use of U for the internal energy rather than E (which is also acceptable): one is its distinction from the symbol for a cell potential (E).

E General definitions of work and heat in thermodynamic processes. *J. Chem. Ed.*, **1987**, *64*, 660. A gas kinetic explanation of simple thermodynamic processes. *J. Chem. Ed.*, **1985**, *62*, 224.

In thermodynamics, the total energy of a system is called the **internal energy** U; we shall use that term from now on. The internal energy is the sum of the kinetic energies of all the particles and the potential energy arising from their interactions with one another. Like the enthalpy, the internal energy is a state property, which (as we saw in Section 6.2) means that its value depends only on the current state of the system and is independent of how the system was prepared.

The internal energy of a system can be changed by doing work and by heating. We do work when we wind a spring or compress a gas into a smaller volume; and both winding the spring and compressing the gas increase the internal energy of the spring and the gas, respectively. When we do an amount of work w (for instance, 100 kJ of work) on the system, the internal energy rises by the same amount, and we can write $\Delta U = w$. Alternatively, we can supply energy as heat to the system, which is what happens when we stand a beaker of water (the system) over a flame (a part of the surroundings). When we supply an amount of heat q to the system, the internal energy increases by that amount, and we can write $\Delta U = q$.

E Heat and work are not different forms of energy: they are different modes by which energy may be transferred to or from a system. When energy is transferred as work, the process could (in principle) have been used to raise or lower a weight in the surroundings. Energy is transferred as heat when use is made of a temperature difference between the system and its surroundings. The change in internal energy is the result of transfers of energy by doing work or by heating: it is not a store of work or heat.

When we supply energy to the system both by doing work and by heating, the total change in internal energy is

$$\underbrace{\Delta U}_{\text{Change in internal energy}} = \underbrace{q}_{\text{Energy supplied as heat}} + \underbrace{w}_{\text{work}} \qquad (1)$$

For example, when we supply 50 kJ of heat and do 25 kJ of work on the system, the total change in internal energy is

$$\Delta U = 50 \text{ kJ} + 25 \text{ kJ} = +75 \text{ kJ}$$

an increase of 75 kJ in internal energy. Had 50 kJ of heat leaked out of the system while we were doing 25 kJ of work on it, we would use $q = -50$ kJ, the minus sign indicating that heat has *left* the system. In this case,

$$\Delta U = -50 \text{ kJ} + 25 \text{ kJ} = -25 \text{ kJ}$$

Now there is a net decrease of 25 kJ in the internal energy of the system: the capacity of the system to do work has been reduced by 25 kJ.

N Note that we always affix the sign to ΔU (and any other similar quantity), even if it is positive, for that emphasizes that it is a change in a property.

▼ EXAMPLE 16.1 Calculating the change in internal energy

Suppose that an electric battery drives an electric motor and that the battery plus the motor are the system. During a certain time period, the motor does 555 kJ of work, and the motor and the battery release 124 kJ of heat into the surroundings (perhaps as a result of friction). What is the change in internal energy of the system?

STRATEGY All losses of energy as heat or work reduce the internal energy of the system, so they occur in Eq. 1 with negative signs. All gains of energy increase the internal energy, so they occur in Eq. 1 with positive signs.

SOLUTION Both energy transfers are losses, so we write $q = -124$ kJ and $w = -555$ kJ. Therefore, from Eq. 1,

$$\Delta U = -124 \text{ kJ} - 555 \text{ kJ} = -679 \text{ kJ}$$

That is, the internal energy of the system decreases by 679 kJ during the period when the motor is running.

EXERCISE E16.1 A certain system gains 250 kJ of energy as heat while it is doing 500 kJ of work. By what amount does the internal energy change?
[*Answer*: -250 kJ]
▲

N The work in Example 16.1 may be needed to drive a toy or a cassette player.

U Exercise 16.57.

There are two important features to note about Eq. 1. The first is that any change in internal energy can be brought about either by heat or by work—the system stores the energy as internal energy however that energy is supplied. The internal energy is rather like the reserves of a bank: the bank accepts deposits and withdrawals in either of two currencies (heat or work) but stores them as a common fund, the internal energy. *Heat and work are equivalent ways of changing the energy of a system.*

The second important consequence of Eq. 1 is that, if energy *cannot* be supplied as heat or as work, then the internal energy does not change: if $q = 0$ and $w = 0$, then it follows that $\Delta U = 0$. In other words, *we cannot create or destroy internal energy*—all we can do is change the internal energy by importing energy from, or exporting it to, the surroundings. This conservation of energy is the essential content of the first law of thermodynamics:

First law of thermodynamics: The internal energy of an isolated system is constant.

By an isolated system, we mean one that can do no work and is insulated against transfers of energy as heat.

N Work can be done only if something moves (for example, a piston, the atmosphere, or a winch), for work is defined as force × distance. The units of work are those of energy, as can be seen by writing

$$\begin{aligned}[\text{Work}] &= [\text{force}] \times [\text{distance}]\\ &= [\text{mass}] \times [\text{acceleration}] \times [\text{distance}]\\ &= \text{kg} \times (\text{m}\cdot\text{s}^{-2}) \times \text{m} = \text{kg}\cdot\text{m}^2\cdot\text{s}^{-2} = \text{J}\end{aligned}$$

E Quantities of work in thermodynamic equations. *J. Chem. Ed.*, **1969**, *46*, 380.

E The first law of thermodynamics is a richer statement than the law of the conservation of energy, because it implies the equivalence of heat and work. Work is a concept of mechanics, but "heat" is a new concept, and distinguishes thermodynamics from mechanics.

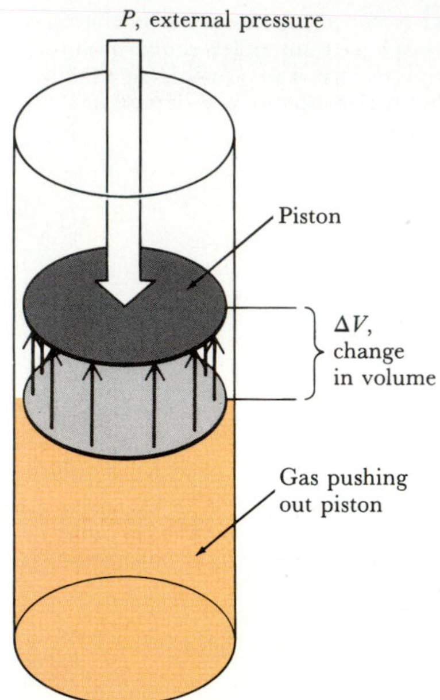

FIGURE 16.1 The work done by a system that expands against an external pressure P is given by the product of the pressure and the volume ΔV that is swept out by the piston. The higher the external pressure, the greater the work the system must do to push out the piston through a given volume.

N The exact relation is 1 L·atm = 101.325 J (because 1 atm = 101.325 kPa exactly).

T OHT: Fig. 16.1; Figs. 16.2 and 16.3.

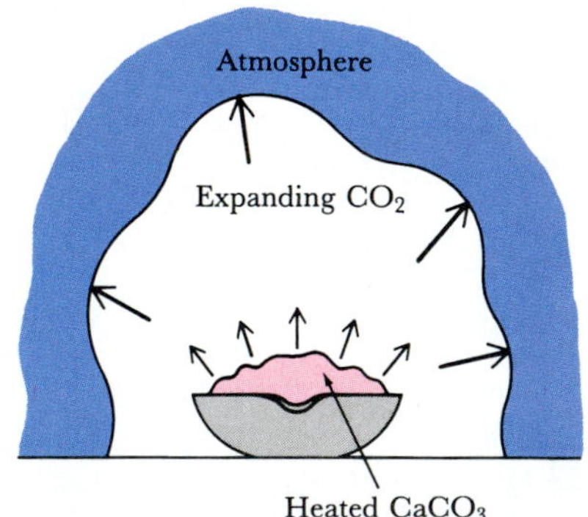

FIGURE 16.2 A system also does work when it generates a gas (for example, when calcium carbonate is heated and decomposes). Because (at constant pressure) the gas must expand as it is formed, it must drive back the surrounding atmosphere.

Work of expansion against the atmosphere. One important way in which a system can do work is by pushing back the atmosphere, as happens when a gas expands inside a cylinder and pushes out a piston (Fig. 16.1). However, the piston is not really necessary (it just makes it easier to harness the work to drive wheels), because any reaction that produces gases must push back the surrounding atmosphere to make room for the products and thereby does work. For example, when calcium carbonate is heated and decomposes in the reaction

$$CaCO_3(s) \longrightarrow CaO(s) + CO_2(g)$$

for every mole of $CaCO_3$ that decomposes, about 30 L of carbon dioxide is produced. Thus, if the reaction is to proceed, then the carbon dioxide must push back the atmosphere by that volume (Fig. 16.2).

When a system expands by ΔV against a pressure P, the change in internal energy of the system can be calculated from the expression

Work of expansion

$$\Delta U = w \qquad \text{with } w = -P \times \Delta V \qquad (2)$$

Decrease in internal energy

In this expression, P is the *external* pressure (which opposes the expansion). The negative sign in w indicates a decrease in internal energy when the system expands. For example, when the volume of a system increases by 30 L by pushing back against a pressure of 1.0 atm, the internal energy of the system decreases by

$$w = -1.0 \text{ atm} \times 30 \text{ L} = -30 \text{ L·atm}$$

The reduction in internal energy can be converted to joules by using the relation*

$$1 \text{ L·atm} = 101 \text{ J}$$

▼ EXAMPLE 16.2 Calculating the work of expansion

Calculate the work that must be done by the reaction at STP (1 atm and 0°C; Section 5.4) to make room for the products of the octane combustion

$$2C_8H_{18}(l) + 25O_2(g) \longrightarrow 16CO_2(g) + 18H_2O(g)$$

when 2.00 mol of octane is burned.

STRATEGY The reaction shows that 25 mol of gas molecules is replaced by 34 mol of gas molecules, a net increase of 9 mol. Because the molar volume of an ideal gas at STP is 22.4 L/mol (see Section 5.2), we can calculate the change in volume and hence the work needed for the expansion. For the conversion from liter-atmospheres to joules, we use 1 L·atm = 101 J, as given in the text above.

SOLUTION The change in volume is

$$\Delta V = 9 \text{ mol} \times 22.4 \text{ L/mol} = 202 \text{ L}$$

*This relation follows from the definition of 1 atm (in Table 5.3) in 1 L·atm = 10^{-3} m^3 × [101 × 10^3 kg/(m·s^2)] = 101 kg·m^2/s^2 and from the definition of the joule as 1 J = 1 kg·m^2/s^2.

Because the external pressure (arising from the atmosphere) is 1.0 atm, the work required is

$$w = -1.0 \text{ atm} \times 202 \text{ L} = -2.0 \times 10^2 \text{ L·atm}$$

which converts to

$$w\text{ (J)} = -2.0 \times 10^2 \text{ L·atm} \times \frac{101 \text{ J}}{1 \text{ L·atm}}$$

$$= -2.0 \times 10^4 \text{ J}$$

That is, when the expansion occurs, the internal energy decreases by 20 kJ.

EXERCISE E16.2 Calculate the work needed to make room for the products in the combustion of 1 mol of glucose molecules, $C_6H_{12}O_6$, to carbon dioxide and water vapor at STP.

[*Answer*: −14 kJ]

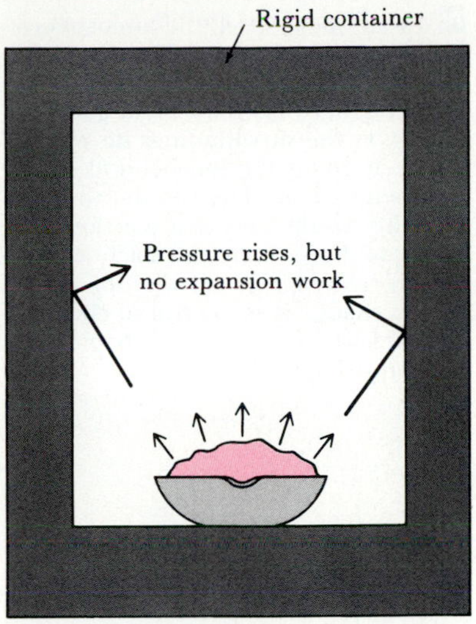

Rigid container

Pressure rises, but no expansion work

FIGURE 16.3 When a reaction (such as the thermal decomposition of calcium carbonate) takes place in a closed, constant-volume container, the pressure rises inside the vessel; but because the volume of the system does not change, it does no work on the surroundings.

Measuring a change in internal energy. When a system is sealed—and cannot change its volume—it cannot push back the atmosphere, and any gases that are produced remain confined to the same volume (Fig. 16.3). This condition exists inside a bomb calorimeter (Section 6.1), where the reaction takes place inside a sealed container. Because $\Delta V = 0$, there is no change in internal energy arising from expansion, so we can write $w = 0$, which implies that $\Delta U = q$. This relation is normally written

$$\Delta U = q_V$$

The subscript V reminds us that the expression is valid only when the heat is supplied at constant volume. It follows that, to measure the change in internal energy of a system, we should measure the heat released (a reduction in internal energy) or the heat absorbed (a gain in internal energy) when the system is held at constant volume. For example, if a reaction that takes place inside a bomb calorimeter releases 100 kJ to the surroundings, then we know that $\Delta U = -100$ kJ. On the other hand, if 100 kJ of heat flows into the calorimeter, then we know that $\Delta U = +100$ kJ.

U Exercise 16.58.

16.2 ENTHALPY

The thermodynamic property at the center of our attention in Chapter 6 and several following chapters was the enthalpy H. Like the internal energy, the enthalpy is a state property. The change in enthalpy of a system, ΔH, is equal to the heat transferred to it at constant pressure:

$$\Delta H = q_P$$

D Endothermic reaction: Ammonium nitrate. CD1, 39. Heat of dilution of sulfuric acid. BZS1, 1.6.

As a consequence of this relation, we can measure the enthalpy change of a system by measuring the heat lost or gained by the system at constant pressure. An example of such a system is a reaction occurring in a vessel that is open to the atmosphere. For example, if in a combustion reaction taking place in an open vessel 100 kJ of energy is released into the surroundings as heat, then we know that the enthalpy of the system has decreased by 100 kJ, and we write $\Delta H = -100$ kJ. On the other hand, if 100 kJ is absorbed by an endothermic process (such as a vaporization or some chemical reactions), then the enthalpy of the system increases by 100 kJ, and we write $\Delta H = +100$ kJ.

In an open vessel, the atmosphere is either pushed outward (so the system loses energy by doing work) or it pushes inward (so the system gains energy as the surroundings do work on it). In many thermochemical experiments we are blind to the work that the system does as a reaction proceeds, and are interested solely in the heat it releases. That is why the enthalpy change is so central to chemistry, and why it occupied a chapter of its own (Chapter 6).

When the system cannot undergo a change in volume, all the heat supplied to it is used to raise the internal energy of the system, and the change in internal energy is equal to the energy supplied as heat ($\Delta U = q_V$). That is, when 100 kJ of heat is supplied at constant volume, the internal energy increases by 100 kJ. However, when the system can change its size (when the reaction takes place in an open vessel at constant pressure), some of the energy supplied as heat is used to drive back the atmosphere as the system expands. Therefore, at constant pressure, only some of the energy supplied is used to increase the internal energy of the system—the rest leaks away into the surroundings as work. For instance, suppose we supply 100 kJ of energy as heat and 10 kJ of that energy leaks away as the system expands and does work by pushing back the atmosphere. We conclude that the enthalpy change is $\Delta H = +100$ kJ (because $\Delta H = q_P$) but that the change in internal energy of the system is only

$$\Delta U = 100 \text{ kJ} - 10 \text{ kJ} = +90 \text{ kJ}$$

In general, when heat is supplied to a system at constant pressure, the resulting change in internal energy is

$$\Delta U = q_P + w$$

We have seen that q_P is equal to the change in enthalpy, ΔH, that accompanies the change and that the energy transferred as work when the system expands through a volume ΔV against a pressure P is $-P\Delta V$ (the -10 kJ in the example). Therefore, writing $q_P = \Delta H$ and $w = -P\Delta V$ in the above expression gives

$$\Delta U = \Delta H - P\Delta V$$

for the change in internal energy of a system at constant pressure. Now we see how to measure the internal energy of a system from measurements of the heat transferred between the system and its surroundings: When the transfer occurs at constant volume, we use $\Delta U = q_V$. When the transfer occurs at constant pressure, we use the formula given above, with $\Delta H = q_P$.

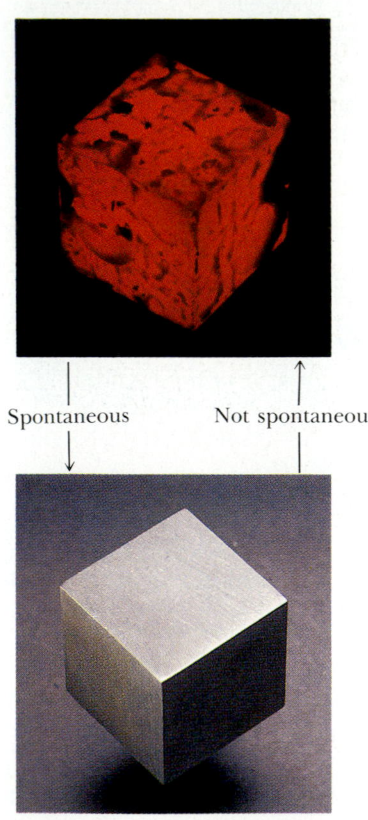

FIGURE 16.4 The direction of spontaneous change is for a hot block of metal (top) to cool to the temperature of its surroundings (bottom). A block at the same temperature as its surroundings does not spontaneously become hotter.

THE DIRECTION OF SPONTANEOUS CHANGE

In principle, changes can run in either direction. For instance, depending on the pressure and temperature, water can either vaporize to a gas or water vapor can condense to a liquid. In Section 11.5 we met the idea that the actual direction of change—such as water's change from liquid to vapor if the pressure is 1 atm and the temperature is 100°C or more—is in the direction that corresponds to the *dispersal* of energy and matter, their spreading in disorder. That is, when a change occurs, it leaves the system and its surroundings in a more disordered state. We met the term *entropy* as a measure of disorder and saw that the natural direction of change is in the direction of increasing entropy.

The tendency to disperse—the tendency for entropy to increase—accounts for the expansion of a gas into a vacuum, because a gas becomes more disordered as its molecules spread. It also accounts for the tendency of substances to dissolve. We saw in Chapter 11 how the entropy of a sample increases as ions and molecules break away from a crystal and spread throughout the solvent. We shall now make these ideas precise and show how the entropy may be defined, measured, and used.

16.3 ENTROPY AND SPONTANEOUS CHANGE

In this chapter we shall be concerned with spontaneous changes:

> A **spontaneous change** is a change that has a natural tendency to occur without needing to be driven by an external influence.

One simple example of a spontaneous physical change is the cooling of a block of hot metal to the temperature of its surroundings (Fig. 16.4). The reverse change, a block of metal growing hotter than its surroundings, is not spontaneous. We can drive that reverse (nonspontaneous) change in a number of ways, for example, by forcing an electric current through the metal (which would heat it); but the reverse change has no tendency to occur unless it is driven. The expansion of a gas into a vacuum is also spontaneous (Fig. 16.5); but a gas has no tendency to contract spontaneously into one part of a container. We can drive a gas into a smaller volume, however, by pushing in a piston.

Although spontaneous changes are sometimes fast (such as the expansion of a gas), that is not necessarily the case. A large block of metal cools spontaneously but slowly. Viscous oil has a spontaneous tendency to flow out of an overturned can, but that flow may be very slow. Throughout this chapter and the next, we must be very careful *not* to assume that, because a process may have a *tendency* to occur, it will take place at a significant rate. Thermodynamics tells us about *tendencies*; it is silent about rates.

In the following sections we shall see that *spontaneous changes are changes that result in an increase in entropy*. That is, if the change would result in a net increase in disorder of the system *and its surroundings*, then the change is spontaneous and has a natural tendency to occur.

Entropy as a measure of disorder. We have already remarked that entropy is a measure of disorder. Now is the time to make this notion precise. We need to understand what it means to say that the entropy of one substance is greater than that of another; and we need to know how to determine the entropy change that occurs in the surroundings when heat spreads into them. We must never forget that if we want to predict whether a change is likely to be spontaneous, we have to assess the *total* entropy change of both the system and the surroundings: only if this *total* disorder of the system and its surroundings increases is the change spontaneous.

It is often quite easy to know when the entropy of one substance is likely to be greater than that of another. Thus, the entropy of a system usually increases when a gas (a disorderly state of matter) is formed from solid or liquid reactants in a reaction. The entropy of a system generally decreases when a gas is converted to a solid or a liquid (both of which are relatively orderly states—solids more so than liquids). When chlorine reacts with PCl_3 in the reaction

$$PCl_3(l) + Cl_2(g) \longrightarrow PCl_5(s)$$

N All viable chemical reactions are spontaneous, and they continue to be spontaneous until they have reached equilibrium. At that stage, although the reaction is continuing (in the forward and reverse directions), there is no further tendency to undergo net change in either direction.

N Entropy and its role in introductory chemistry. *J. Chem. Ed.*, **1982**, *59*, 317.

D The acid in water puzzle. CD2, 55.

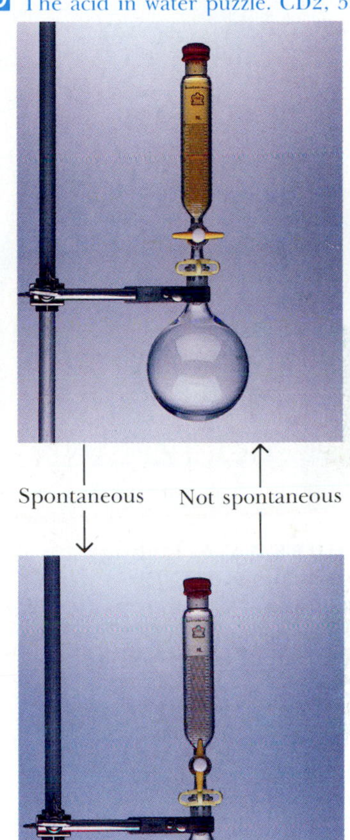

FIGURE 16.5 The direction of spontaneous change is for a gas to fill its container. A gas that already fills its container does not spontaneously collect in a small region of the container. A glass cylinder containing yellow gas (upper piece of glassware, top) is attached to a flask containing a vacuum (lower flask, top). When the stopcock between them is opened, the yellow gas fills both upper and lower containers (bottom).

FIGURE 16.6 Ludwig Boltzmann (1844–1906). His formula for the entropy (using an earlier notation for natural logarithms) became his epitaph.

T OHT: Figs. 16.7 and 16.8.

E Indistinguishable state and $S = k \ln W$. *J. Chem. Ed.*, **1969**, *46*, 719.

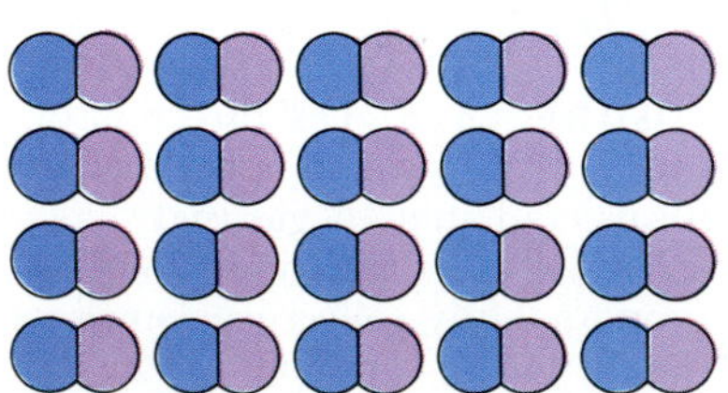

FIGURE 16.7 A sample of 20 heteronuclear diatomic molecules in a perfectly ordered arrangement at $T = 0$ has zero disorder and zero entropy ($S = 0$). This sample represents a perfect crystal.

a liquid and a gas are converted to a solid, so we can anticipate that the entropy of the reaction system decreases. (Do not be puzzled by the fact that this reaction runs in the direction of decreasing disorder—decreasing entropy—of the system; we have not yet taken into account the change in entropy of the surroundings.) In the photosynthesis reaction

$$6CO_2(g) + 6H_2O(l) \longrightarrow C_6H_{12}O_6(s) + 6O_2(g)$$

there is a net reduction in disorder within the reaction system. In this case, many simple reactant molecules are used to build a highly organized glucose molecule. Hence, we can expect the entropy *of the reaction system* to decrease when this reaction occurs. Conversely, the entropy *of the system* increases when glucose burns in oxygen:

$$C_6H_{12}O_6(s) + 6O_2(g) \longrightarrow 6CO_2(g) + 6H_2O(l)$$

In this reaction, which is the reverse of the photosynthesis reaction, a large number of small triatomic molecules are formed from the relatively highly structured glucose molecules.

The formula for calculating the entropy S of a substance at any temperature was proposed by the Austrian Ludwig Boltzmann (Fig. 16.6):

$$S = k \ln W \qquad (3)$$

The factor k, which is called **Boltzmann's constant**, stands for R/N_A, where R is the gas constant and N_A is Avogadro's constant. Its value, $k = 1.38 \times 10^{-23}$ J/K, shows that the units of entropy are joules per kelvin (J/K). The logarithm (ln) is the natural logarithm. The disorder of the system is expressed by W, which is the number of ways that the atoms or molecules in the sample can be arranged yet have the same total energy. Before going any further, we must see what this relation means in practice.

The entropies of simple solids. Suppose we wanted to know the entropy of a solid made up of 20 heteronuclear diatomic molecules. (A real example might be a block of solid carbon monoxide or hydrogen chloride, but the number of molecules would then be closer to 10^{23} than to 20.) Suppose that the 20 molecules form a perfectly ordered crystal (Fig. 16.7). Suppose also that the temperature is zero ($T = 0$), so all motion has been quenched. We expect such a sample to have zero entropy because there is no disorder. This expectation is confirmed by the Boltzmann formula: because there is only one way of arranging the molecules in the perfect crystal, $W = 1$ and (because $\ln 1 = 0$)

$$S = k \ln 1 = 0$$

Now suppose that each molecule can point in either of two directions in the solid (Fig. 16.8). Because each of the 20 molecules can have two orientations, the total number of ways of arranging them is

$$W = 2 \times 2 \times 2 \times \cdots = 2^{20}$$

The entropy of this disorderly solid is therefore

$$S = k \ln 2^{20} = (1.38 \times 10^{-23}\ \text{J/K}) \times (20 \ln 2) = 1.9 \times 10^{-22}\ \text{J/K}$$

(We have used the relation $\ln x^a = a \ln x$; Appendix 1B.) The entropy is now higher than that of the perfectly ordered solid. If the solid contained 1 mol of CO molecules (that is, 6.02×10^{23} of them), then its entropy would be

$$S = (1.38 \times 10^{-23}\ \text{J/K}) \times (6.02 \times 10^{23} \times \ln 2) = 5.76\ \text{J/K}$$

When the entropy of 1 mol of CO is actually measured* close to $T = 0$, the value obtained is 4.6 J/K. This value is close enough to 5.8 J/K to suggest that in the crystal the molecules are indeed arranged nearly randomly, as depicted in Fig. 16.8. The physical reason for this randomness is that the electric dipole moment of a CO molecule is very small, so there is little energy advantage in the molecules lying head-to-tail (as depicted in Fig. 16.7), and the molecules lie in either direction at random. For solid HCl, the same experimental measurements give $S = 0$, showing that at $T = 0$ the molecules are arranged in an orderly way: under the influence of their bigger dipole moments, the HCl molecules lie strictly head to tail at $T = 0$, as depicted in Fig. 16.7.

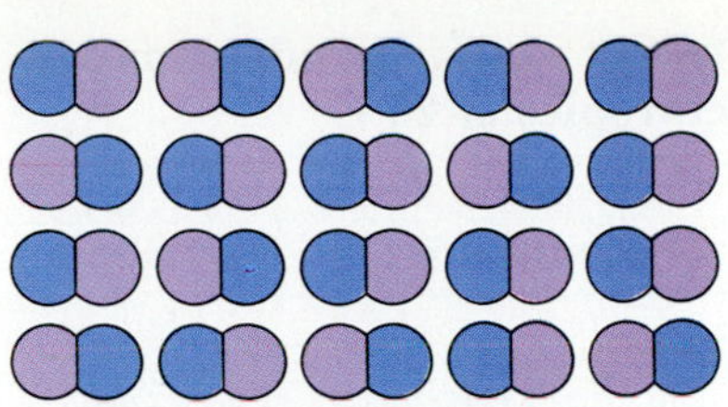

FIGURE 16.8 When each of the 20 molecules can take up either of two orientations without affecting the energy, there are $2 \times 2 \times \cdots = 2^{20}$ different possible arrangements. This illustration shows just one of them. The entropy of this sample is higher than that of the sample in Fig. 16.7.

▼ EXAMPLE 16.3 Using the Boltzmann formula for entropy

The entropy of 1 mol of solid chloryl fluoride, ClO_3F, at absolute zero is 10.1 J/K. Suggest an interpretation.

STRATEGY The existence of nonzero entropy at $T = 0$ suggests that the molecules are disordered. From the shape of the molecule (which can be obtained from VSEPR theory), we need to determine how many orientations (W) it is likely to be able to adopt in a crystal; then we can use the Boltzmann formula to see whether that leads to the observed value of S.

SOLUTION Chloryl fluoride is a tetrahedral molecule, so we can expect it to be able to take up any of four orientations in a crystal (Fig. 16.9). The total number of ways of arranging the ClO_3F molecules in a crystal in which there are N of them is therefore

$$W = 4 \times 4 \times 4 \times \cdots \times 4 = 4^N$$

The entropy is then

$$S = k \ln 4^N = k \times (N \times \ln 4)$$

Because in 1 mol of ClO_3F there are 6.02×10^{23} molecules,

$$S = 1.38 \times 10^{-23}\ \text{J/K} \times (6.02 \times 10^{23} \times \ln 4) = 11.5\ \text{J/K}$$

This value is reasonably close to the experimental value of 10.1 J/K, which suggests that at absolute zero the molecules are arranged in almost random orientations.

EXERCISE E16.3 Explain the observation that at absolute zero the entropy of 1 mol of solid N_2O (NNO) is 6 J/K.

▲ [*Answer*: in the crystal, the orientations NNO and ONN are equally likely]

E Ice is another example of a substance with a nonzero entropy at $T = 0$. In its case, the randomness of the structure arises from the hydrogen bonds that link the H_2O molecules together in a locally tetrahedral array: each O atom is joined to two H atoms by ordinary short O—H bonds, and two others by long O···H—O hydrogen bonds. The randomness is whether any particular oxygen-hydrogen link is short or long. The experimental molar entropy at $T = 0$ is close to the theoretical value of $R \ln \frac{3}{2}$.

To summarize, at absolute zero in a perfect crystal, as in Fig. 16.7, there is perfect order and the entropy is zero. As the crystal is heated, the molecules start to move and become more disordered as their thermal motion increases. The value of W, and hence of $\ln W$, increases because, although there is only one way of being ordered, there are

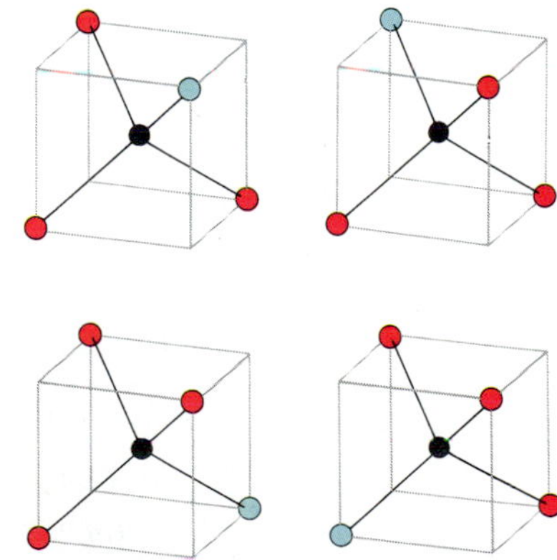

FIGURE 16.9 Each chloryl fluoride molecule can take up one of four orientations at random at each site in the solid, so the entropy of solid ClO_3F is nonzero at $T = 0$. Oxygen, red; fluorine, black; chlorine, green.

*We shall not describe how entropies of substances are measured experimentally other than to say that it depends on measuring heat capacities, C, down to very low temperatures and measuring the area under the curve of a graph of C/T against T.

TABLE 16.1 Standard molar entropies at 25°C

Substance	$S°$, J/(K·mol)
Gases	
ammonia, NH_3	192.4
carbon dioxide, CO_2	213.7
hydrogen, H_2	130.7
nitrogen, N_2	191.6
oxygen, O_2	205.1
Liquids	
benzene, C_6H_6	173.3
ethanol, CH_3CH_2OH	160.7
water, H_2O	69.9
Solids	
calcium oxide, CaO	39.8
calcium carbonate, $CaCO_3$	92.9
diamond, C	2.4
graphite, C	5.7
lead, Pb	64.8

T OHT: Tables 16.1 and 16.2.

U Exercises 16.59 and 16.60.

E All substances at $T > 0$ have a positive entropy because there is always some disorder of the atomic arrangements. The low entropy of diamond compared with that of lead reflects the difficulty with which the C atoms can be excited into vibration around their lattice sites.

TABLE 16.2 The standard molar entropy of water at various temperatures

Phase	Temperature, °C	$S°$, J/(K·mol)
solid	−273 (0 K)	3.4
	0	43.2
liquid	0	65.2
	20	69.6
	50	75.3
	100	86.8
vapor	100	196.9
	200	204.1

N Entropy for the beginner. *J. Chem. Ed.*, **1969**, *46*, 36.

many ways of being disordered (think of the number of disorderly arrangements that are possible for a deck of cards). It follows that the entropy of a substance increases as it is heated and becomes disordered. Hence, we can expect the entropy of any substance at room temperature to be greater than zero and, indeed, to be large if the substance has a very disorderly arrangement of molecules and a lot of thermal motion.

Standard molar entropies. The entropies of many substances at different temperatures have been obtained either by calculation with Boltzmann's formula or by measurement of their heat capacities down to very low temperatures. Some of these standard molar entropies are given in Table 16.1; a longer list can be found in Appendix 2A.

The **standard molar entropy** $S°$ of a substance is the entropy per mole of the pure substance at 1 atm pressure.

The units of molar entropy are joules per kelvin-mole, J/(K·mol). Table 16.1 gives standard molar entropies at 25°C; values for water at other temperatures are shown in Table 16.2 for comparison. Note that all values are positive and that water molecules are more disordered at 25°C than at absolute zero.

The entropies of pure substances. The differences in the standard molar entropies of the substances in Tables 16.1 and 16.2 can be explained in terms of disorder. For example, the molar entropy of diamond, 2.4 J/(K·mol), is much lower than that of lead, 64.8 J/(K·mol). Diamond's low entropy is consistent with the fact that diamond has a much more rigid, orderly structure than lead has. Also, the entropy of water increases as the sample is heated: this trend reflects the increasing disorder of the liquid as it gets hotter and the molecules undergo more vigorous thermal motion. The entropy of any substance increases as its temperature is raised.

The large entropy increase at the boiling point of water—from 87 J/(K·mol) for the liquid to 197 J/(K·mol) for the vapor—shows the effect on the disorder when a liquid changes to a much more chaotic gas. A similar, but usually smaller, increase occurs when a solid melts, because a liquid is more disordered than a solid (Fig. 16.10).

▼ EXAMPLE 16.4 Calculating the entropy of a physical change

Calculate the change in the entropy when 100 g of water, initially at 0°C, freezes at 0°C in a refrigerator ice tray.

STRATEGY Because ice is a more orderly substance than liquid water, in which the molecules are free to move, we expect ice to have a lower entropy; hence the change should be negative. We know the molar entropy of water and ice at 0°C (Table 16.2); the change in entropy is the difference between them. The entropy change for 100 g of water is the product of the molar entropy change and the number of moles of H_2O molecules in 100 g of water, which is obtained from the molar mass of H_2O (18.02 g/mol).

SOLUTION From the standard molar entropies of water and ice at 0°C (Table 16.2), we can calculate the change in standard molar entropy at 0°C.

$$\text{Change in molar entropy} = S°(\text{ice}) - S°(\text{water})$$
$$= (43.2 - 65.2)\ \text{J/(K·mol)} = -22.0\ \text{J/(K·mol)}$$

Change in entropy
= amount of H_2O (mol) × entropy change per mole of H_2O

and for 100 g H_2O

$$\text{Amount of } H_2O \text{ (mol)} = 100 \text{ g } H_2O \times \frac{1 \text{ mol } H_2O}{18.02 \text{ g } H_2O} = 5.55 \text{ mol } H_2O$$

The change in entropy when 100 g H_2O freezes is therefore

$$\text{Change in entropy} = 5.55 \text{ mol } H_2O \times \frac{-22.0 \text{ J}}{\text{K·mol } H_2O} = -122 \text{ J/K}$$

The negative sign indicates that freezing decreases the entropy and that ice is more ordered than water at 0°C.

EXERCISE E16.4 Use the information in Appendix 2A to calculate the change in molar entropy when white tin changes to gray tin at 25°C (Fig. 16.11). Which is the more ordered form?
[*Answer:* −7.5 J/(K·mol); gray]

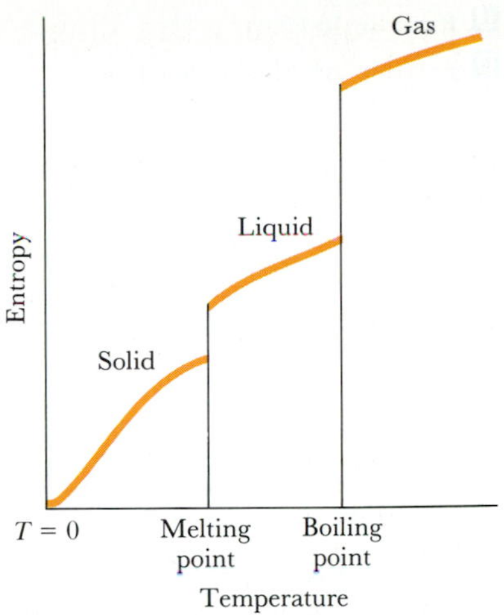

FIGURE 16.10 The entropy of a solid increases when the solid is heated (that is, as its temperature is raised). The entropy increases sharply when the solid melts to the more disordered liquid and then increases steadily up to the boiling point. A second, larger increase in entropy occurs when the liquid boils.

E It is said (we do not know how reliably), that the phase transition depicted in Fig. 16.11 contributed to Napoleon's defeat before Moscow: the tin buttons his soldiers used crumbled in the intense cold.

U Exercise 16.62.

Standard reaction entropies. Example 16.4 shows that the information in Tables 16.1 and 16.2 can be used to calculate the entropy change that occurs during a physical change. It can also be used to obtain the entropy change in a chemical reaction, when one substance changes into another. The entropy change is called the standard reaction entropy $\Delta S°$; and its definition is analogous to that of the standard reaction enthalpy (Section 6.3):

The **standard reaction entropy** of any reaction is the difference in entropy between the products in their standard states and the entropy of the reactants in their standard states:

$$\Delta S° = \Sigma nS°(\text{products}) - \Sigma nS°(\text{reactants})$$

where $\Sigma nS°$(products) is the total standard entropy of the products and $\Sigma nS°$(reactants) is that of the reactants; the n are the stoichiometric coefficients in the thermochemical equation.

EXAMPLE 16.5 Calculating the standard reaction entropy

Calculate the standard reaction entropy for the synthesis of 2 mol of $NH_3(g)$ at 25°C in the reaction

$$N_2(g) + 3H_2(g) \longrightarrow 2NH_3(g)$$

STRATEGY We expect a decrease in entropy because 4 mol of reactant gas molecules occupy a larger volume than 2 mol of product gas molecules at the same pressure. To find the numerical value, we use the chemical equation to write an expression for $\Delta S°$ and then insert values from Table 16.1.

SOLUTION The reaction is

$$N_2(g) + 3H_2(g) \longrightarrow 2NH_3(g)$$

The standard reaction entropy is therefore

$$\Delta S° = \{2 \text{ mol } NH_3 \times S°(NH_3,g)\} - \{1 \text{ mol } N_2 \times S°(N_2,g) + 3 \text{ mol } H_2 \times S°(H_2,g)\}$$
$$= \{2 \times 192.5 \text{ J/K}\} - \{(191.6 + 3 \times 130.7) \text{ J/K}\} = -198.7 \text{ J/K}$$

FIGURE 16.11 Gray tin and white tin are two allotropes of tin. The denser white metallic form is the more stable phase above 13°C. The powdery gray allotrope forms on the metallic white allotrope below 13°C.

U Exercise 16.63.

D Entropy and the rubber band. CD1, 45.

Because $\Delta S°$ is negative, we conclude that the product is less disordered than the reactants (as we expected).

EXERCISE E16.5 Calculate the standard reaction entropy of the reaction $N_2O_4(g) \rightarrow 2NO_2(g)$ at 25°C. Use information from Appendix 2A. [*Answer*: +175.8 J/K]

16.4 THE ENTROPY CHANGE IN THE SURROUNDINGS

N It is an absolutely central component of the teaching of the second law, that attention must be paid to entropy changes in both the system and the surroundings: only if there is a net increase will the change be spontaneous. Orderly structures may form, but only if elsewhere there is a greater production of disorder. The growth (which involves a reduction in disorder as food or carbon dioxide is converted into more complex molecules) of animals and vegetation on Earth can be traced ultimately to the increase in disorder that occurs when nuclear fusion occurs in the Sun.

E With Clausius from energy to entropy. *J. Chem. Ed.*, **1989**, *66*, 1001. Order, chaos, and all that! *J. Chem. Ed.*, **1989**, *66*, 997.

V Video 29: Crystallization.

T OHT: Figs. 16.12, 16.14, and 16.15.

We have seen (in Table 16.2) that the entropy of water is 22.0 J/(K·mol) higher than the entropy of ice at 0°C, which shows that liquid water is more disordered than ice. Freezing therefore corresponds to a *decrease* in disorder as liquid water forms ice. Because a change to a less disorderly state is never spontaneous, some additional factor must account for the fact that the freezing of water is spontaneous below 0°C. Likewise, we know that nitrogen and hydrogen have a tendency to form ammonia, but we have just calculated that NH_3 is *less* disordered than the reactants! Again, there must be an explanation for why this reaction is spontaneous.

The additional factor is the change in the disorder of the surroundings:

$$\Delta S_{tot} = \Delta S + \Delta S_{surr}$$

where ΔS is the entropy change of the system and ΔS_{surr} that of the surroundings. We have shown that the entropy of the *system* (the substance undergoing the physical change, or the reactants and products in a reaction) decreases when water freezes or ammonia forms. However, the freezing of water and the synthesis of ammonia are exothermic processes (for both of them, $\Delta H°$ is negative); so heat passes into the surroundings when they take place. The heat released in each case stimulates chaotic thermal motion of the atoms in the surroundings, thereby increasing their disorder (Fig. 16.12a). When the increase in disorder of the surroundings is greater than the decrease in the disorder of the system, there is an *overall* increase in the total disorder of the system plus its surroundings, and the change is spontaneous. Many exothermic reactions do release enough heat into the surroundings for the increase in their entropy to overcome a possible decrease in entropy of the system, and therefore many exothermic reactions are spontaneous.

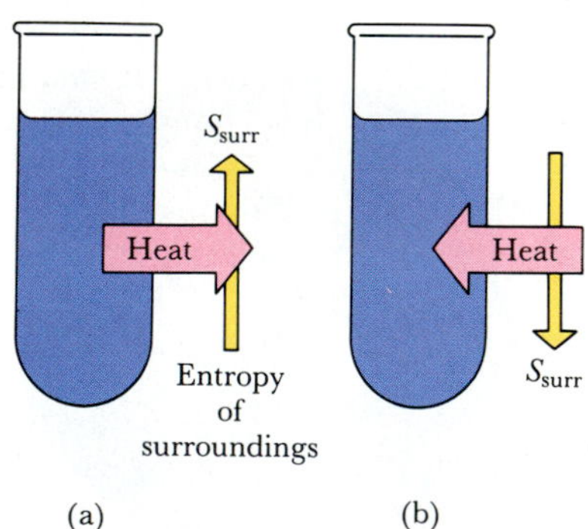

FIGURE 16.12 (a) In an exothermic process, heat escapes into the surroundings and increases their entropy. (b) In an endothermic process, the entropy of the surroundings decreases. An increase in entropy is designated by an upward-pointing arrow; a decrease in entropy is designated by a downward-pointing arrow.

Calculating the entropy change from the enthalpy change. At first sight, calculating the entropy of the surroundings from Boltzmann's formula might look like an impossible task. How is it possible to calculate the number of ways of arranging all the atoms in the water bath, the laboratory, the country, and the planet, all of which make up the surroundings? Fortunately, this cumbersome calculation can be sidestepped by focusing on the entropy *changes* that occur when heat flows into the surroundings. When we deal only with changes, it turns out that we can use another much simpler formula for calculating the change in entropy of the surroundings. Although Boltzmann's formula is still in the background (in the sense that the new formula can be derived from it), we do not need to use it directly. In what follows, we use the fact that the change in the enthalpy of a system is equal to the heat transferred to or from the system at constant pressure.

To derive a formula for ΔS_{surr}, the change in the entropy of the surroundings, we suppose that the process occurring in the system is accompanied by a change in enthalpy ΔH, which might be the enthalpy change of -6 kJ when 1 mol H_2O freezes exothermically or the -92 kJ when 2 mol of NH_3 forms from 1 mol N_2 and 3 mol H_2. When water freezes at constant pressure in an open container, 6 kJ of heat flows into the surroundings and stirs up thermal motion there. In general, if the enthalpy change of the system is ΔH (for instance, -6 kJ), then the amount of heat that flows into the surroundings is $-\Delta H$ (that is, 6 kJ). We assume that the amount of disorder stirred up is proportional to the heat transferred:

$$\Delta S_{surr} \propto -\Delta H$$

Note that when the process is exothermic (with ΔH negative because heat is released), the entropy of the surroundings increases (ΔS_{surr} is positive) (Fig. 16.12a). When the process is endothermic (and ΔH is positive), heat leaves the surroundings, thereby reducing their disorder. Hence, their entropy decreases (ΔS_{surr} is negative) (Fig. 16.12b).

The change in entropy of the surroundings is not *equal* to the change in enthalpy, because the change in entropy caused by a given supply of heat depends on the temperature. If the surroundings are hot, then the molecules in the surroundings are already very chaotic and a small inflow of heat from the system has very little impact (Fig. 16.13a). If the surroundings are cool, however, then the molecules in the surroundings are relatively ordered and the same amount of heat can cause a great deal of disorder (Fig. 16.13b). The difference is like that between sneezing in a crowded street, which may pass unnoticed, and sneezing in a quiet library, which will not. This analogy suggests that the entropy change caused by the transfer of a given amount of heat at constant pressure is inversely proportional to the temperature at which the transfer takes place:

$$\Delta S_{surr} = \frac{-\Delta H}{T} \qquad (4)$$

This formula, which can be derived from the laws of thermodynamics, shows that the entropy of the surroundings increases when heat escapes into them and that the increase is largest when the temperature is low.

For the synthesis of 2 mol of NH_3, for which $\Delta H° = -92.22$ kJ, we find that at 25°C

$$\Delta S_{surr} = \frac{-(-92.22 \times 10^3 \text{ J})}{298 \text{ K}} = +309 \text{ J/K}$$

This increase in the entropy of the surroundings is greater than the decrease in the entropy of the system (in Example 16.5, we calculated the latter to be -199 J/K), so overall there is an increase in disorder when ammonia forms. Hence, the synthesis of ammonia is spontaneous at 25°C:

$$\Delta S_{tot} = -199 \text{ J/K} + (+309 \text{ J/K}) = +110 \text{ J/K}$$

The difference between the two entropy changes is much more pronounced for the combustion of magnesium in the reaction

$$2Mg(s) + O_2(g) \longrightarrow 2MgO(s) \qquad \Delta H° = -1202 \text{ kJ}, \quad \Delta S° = -217 \text{ J/K}$$

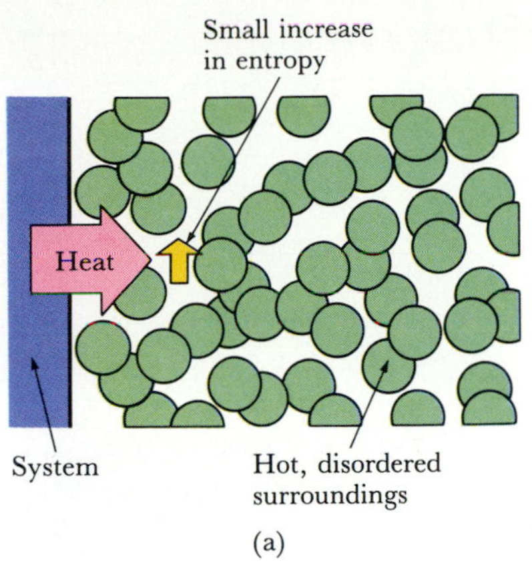

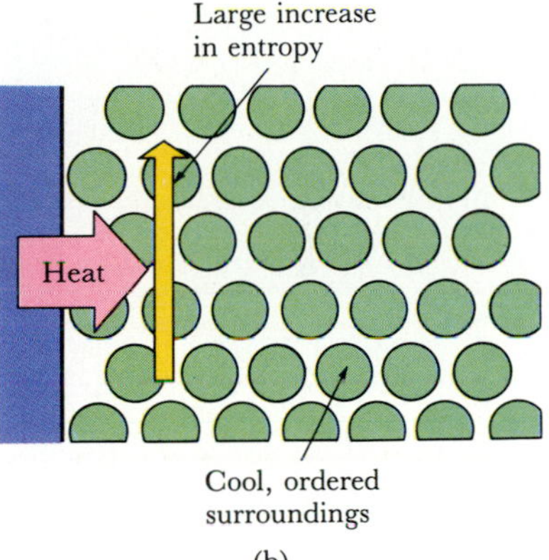

FIGURE 16.13 (a) When a given amount of heat flows into hot surroundings, it produces very little additional chaos. (b) When the surroundings are cool, however, the same quantity of heat can make a considerable difference in the disorder.

N This material on the relationship of ΔS_{tot} to ΔS (of the system) and ΔS_{surr} must be read slowly so as to obtain a clear understanding or otherwise the rest of the chapter is (almost) meaningless.

In this case, the entropy change in the surroundings at 25°C is

$$\Delta S_{\text{surr}} = \frac{-(-1202 \times 10^3\ \text{J})}{298\ \text{K}} = +4.03 \times 10^3\ \text{J/K}$$

This large increase overwhelms the relatively small decrease in the entropy of the system. Hence, as we know from experience, the combustion of magnesium is spontaneous.

Systems at equilibrium. Now consider the same calculation for the freezing of water at 0°C, for which, as we have seen, the entropy of the system decreases by 22.0 J/K. Freezing is exothermic, and $\Delta H° = -6.00$ kJ; hence, the entropy change of the surroundings is

$$\Delta S_{\text{surr}} = \frac{-(-6.00 \times 10^3\ \text{J})}{273\ \text{K}} = +22.0\ \text{J/K}$$

In this process, the increase in entropy of the surroundings cancels the decrease in entropy of the system: overall, the entropy change is zero:

$$\Delta S_{\text{tot}} = -22.0\ \text{J/K} + (+22.0\ \text{J/K}) = 0$$

Because spontaneous changes are accompanied by an *increase* in entropy, we have to conclude that water does not freeze spontaneously at 0°C!

This conclusion is correct. Water does *not* freeze spontaneously at 0°C: water and ice are in *equilibrium* at that temperature. Our calculation shows that, because the *total* entropy of the system and its surroundings neither increases nor decreases when water becomes ice at 0°C, at that temperature water has no spontaneous tendency to freeze and ice has no spontaneous tendency to melt. In kinetic terms, at the melting point of ice (0°C), the rates at which ice melts and water freezes are exactly equal.

E The observation that the enthalpy of vaporization of a liquid divided by its boiling temperature is close to 85 J/(K·mol) for all substances is called *Trouton's law*. We can now see that Trouton's constant is the molar entropy of vaporization, and the explanation of his law is that most substances undergo a similar increase in disorder when they vaporize. The exceptions are liquids with an abnormally high degree of structure, such as water (on account of hydrogen bonds) and mercury (on account of the metallic bonding in the liquid).

▼ EXAMPLE 16.6 Predicting the boiling point of a substance

Predict the normal boiling point of liquid sodium, given that the entropy of 1 mol of Na(*l*) changes by +84.8 J/K when it forms a vapor at 1 atm pressure and that its enthalpy of vaporization is +98.0 kJ/mol.

STRATEGY We know that at the normal boiling point a liquid and its vapor are in equilibrium at 1 atm pressure. In other words, there is no change in total entropy when one phase changes into the other. We are given the entropy change of the system and we can find the entropy change of the surroundings from the enthalpy of vaporization and the temperature. Our task is therefore to find the temperature at which ΔS_{tot}, the sum of the two entropy changes, is zero.

SOLUTION For 1 mol Na(*l*), the entropy change of the system is $\Delta S = +84.8$ J/K and that of the surroundings is

$$\Delta S_{\text{surr}} = \frac{-\Delta H}{T} = \frac{-98.0 \times 10^3\ \text{J}}{T}$$

Hence, the total entropy change is

$$\Delta S_{\text{tot}} = +84.8\ \text{J/K} + \left(\frac{-98.0 \times 10^3\ \text{J}}{T}\right)$$

This change is zero when $T = T_b$, that is, at the boiling point. Hence, we have

$$84.8\ \text{J/K} - \frac{98.0 \times 10^3\ \text{J}}{T_b} = 0$$

from which we find that

$$T_b = \frac{98.0 \times 10^3\ \text{K}}{84.8} = 1160\ \text{K, or about } 890°\text{C}$$

EXERCISE E16.6 Calculate the melting point of chlorine, given that its enthalpy of melting is +6.41 kJ/mol and its entropy of melting is +37.3 J/(K·mol).

[*Answer*: 172 K]

U Exercises 16.77 and 16.78.

E An apparent violation of the second law of thermodynamics in biological systems. *J. Chem. Ed.*, **1979**, *56*, 314.

E There are many equivalent statements of the second law; some are based more closely on empirical experience. For example, one statement is that a given quantity of heat cannot be converted completely into work without other changes elsewhere. Another statement is that heat cannot flow spontaneously from a cold body to a hot one (that is, refrigerators must be driven). The statement given here is a succinct summary and amalgamation of the observations: science sophisticates so as to unify, provide insight, and render succinct.

E Evolution of the second law of thermodynamics. *J. Chem. Ed.*, **1970**, *47*, 331. Entropy: Conceptual disorder. *J. Chem. Ed.*, **1988**, *65*, 403. On the teaching of entropy. *J. Chem. Ed.*, **1973**, *50*, 763.

D Prepare a saturated solution of $Na_2S_2O_3 \cdot 3H_2O$ by dissolving 580 g in 10 to 15 mL of warm water. Allow to cool, and add a seed crystal. (The mixture can be reused.)

E The second law applies to physical processes too. Thus, an automobile is propelled along a highway by the disorder created by the combustion of fuel, and an elevator rises in a building because it is linked to a distant source of disorder, such as a hydroelectric plant.

16.5 THE SECOND LAW

The calculations of the previous section show that when we want to express the direction of spontaneous change in terms of the entropy, we must consider the *total* entropy of the system plus its surroundings. This criterion is expressed by the second law of thermodynamics:

Second law of thermodynamics: A spontaneous change is accompanied by an increase in the total entropy of the system and its surroundings.

At one level this law is just a restatement of the point that spontaneous changes lead to an increase in overall disorder. However, at another level it goes much further, for soon we shall see that we can make numerical predictions: from the information in Table 16.1, we can calculate the entropy change in the system; and by using Eq. 4, we can calculate the entropy change for the surroundings.

Exothermic reactions. The second law explains why exothermic reactions may be spontaneous. These are reactions in which the heat released by the reaction increases the disorder of the surroundings. Even though the disorder of the system may decrease (as it does in the combustion of magnesium where a gas is converted to a solid), the increased disorder of the surroundings is so great that the total entropy increases. In many exothermic reactions the disorder of the system also increases; for example, in the formation of hydrogen fluoride,

$$H_2(g) + F_2(g) \longrightarrow 2HF(g) \qquad \Delta H° = -546.4\ \text{kJ}, \quad \Delta S° = +14.1\ \text{J/K}$$

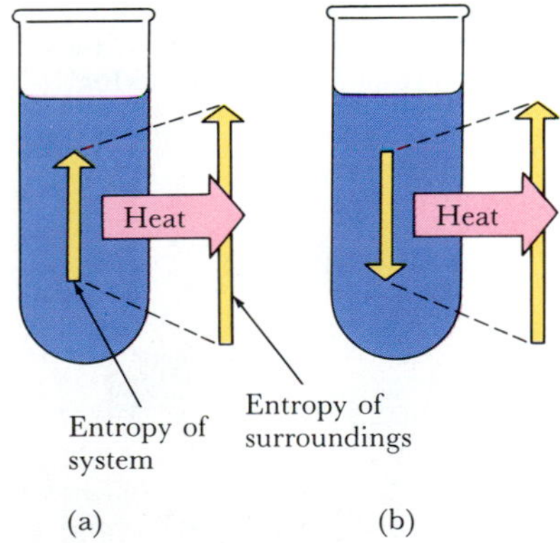

FIGURE 16.14 In an exothermic reaction, (a) the overall entropy change is certainly positive when the entropy of the system increases; (b) the overall change may also be positive when the entropy of the system decreases.

This reaction increases the overall disorder through both the change in the system and the change in its surroundings. As long as ΔH is reasonably large, a reaction with ΔS either positive *or* negative may be spontaneous (Fig. 16.14). In fact, exothermic reactions are common because entropy changes of the system are usually quite small compared with the accompanying increase in the entropy of the surroundings.

Endothermic reactions. The second law also shows why endothermic reactions can occur. These were a puzzle for chemists, who once believed that reactions ran only in the direction of decreasing enthalpy, because an endothermic reaction is one in which the products have a

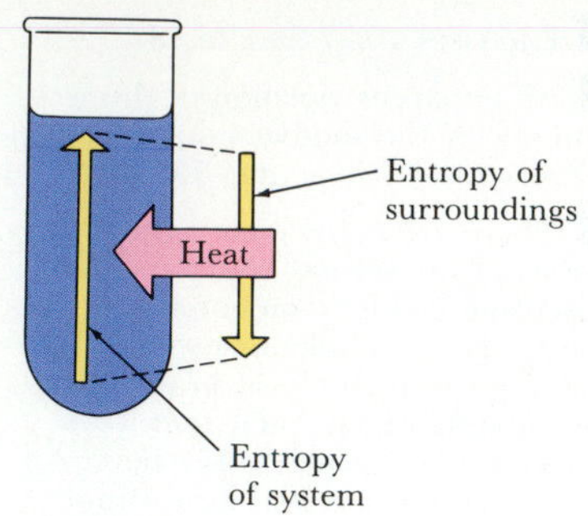

FIGURE 16.15 An endothermic reaction is spontaneous only if the entropy of the system increases enough to overcome the decrease in the entropy of the surroundings.

U Exercise 16.61.

D (1) Mix solid $Ba(OH)_2 \cdot 8H_2O$ with solid NH_4SCN in an Erlenmeyer flask. The endothermicity of the reaction will freeze a wet wooden block to the flask. (2) Heat of solution and colligative properties: An illustration of enthalpy and entropy. *J. Chem. Ed.*, **1990**, *67*, 426.

E Entropy analyses of four familiar processes. *J. Chem. Ed.*, **1988**, *65*, 761. Entropy and heat: fall. *J. Chem. Ed.*, **1982**, *59*, 353.

E Josiah Gibbs spent most of his working life at Yale. He was a poor teacher, but is regarded by some as the greatest native-born American scientist. His work brought thermodynamics into chemistry. He also published on electromagnetism and invented the modern notation for vectors.

FIGURE 16.16 Josiah Willard Gibbs (1839–1903).

higher enthalpy than the reactants. One very striking example of an endothermic reaction—the reaction between barium hydroxide and ammonium thiocyanate—has already been mentioned (in the introduction to Chapter 6), but there are plenty of others, including the decomposition of calcium carbonate at elevated temperatures.

As in exothermic reactions, the driving force of an endothermic reaction comes from the *total* entropy change caused by the reaction. Because heat flows out of the surroundings, their entropy decreases. However, there may still be an overall increase in entropy if the disorder within the system increases enough to compensate for the reduction of disorder in the surroundings. For example, in the reaction between barium hydroxide and ammonium thiocyanate,

$$Ba(OH)_2 \cdot 8H_2O(s) + 2NH_4SCN(s) \longrightarrow Ba(SCN)_2(aq) + 2NH_3(g) + 10H_2O(l)$$

the two solids react to form a solution and a gas. As a result, there is a large increase in entropy of the system and hence an increase in the total entropy of the system and the surroundings. The decomposition of calcium carbonate is also accompanied by a large increase in the entropy of the system as a result of the gas that is produced. Every endothermic reaction must be accompanied by increased disorder *within* the system if it is to be spontaneous (Fig. 16.15), but this constraint does not apply to exothermic reactions. Hence, endothermic reactions are less common than exothermic reactions.

FREE ENERGY

At this point it seems that to predict the direction of spontaneous change we have to carry out two separate entropy-change calculations, one for the system and one for the surroundings. However, there is a convenient way of combining the two calculations so that a *single* piece of information is enough.

16.6 FOCUSING ON THE SYSTEM

As we have seen, the total entropy change ΔS_{tot}, the sum of the changes in the system ΔS and its surroundings ΔS_{surr}, is

$$\Delta S_{tot} = \Delta S + \Delta S_{surr}$$

As long as the pressure is constant, the change in the entropy of the surroundings is given by Eq. 4. Therefore,

$$\Delta S_{tot} = \Delta S - \frac{\Delta H}{T}$$

This equation tells us that we can calculate the total entropy change from information about the system alone (its temperature and the entropy and enthalpy changes it undergoes). That does not mean we are ignoring the surroundings. Their entropy change is simply expressed in terms of the ΔH for the system.

It is common practice to rearrange the last equation into

$$-T\Delta S_{tot} = \Delta H - T\Delta S$$

and to introduce a combination of H and S called the **Gibbs free energy** G:

$$G = H - TS \tag{5}$$

Josiah Gibbs, for whom this state property is named, was a professor at Yale from 1869 until 1903 (Fig. 16.16). He was responsible for turning thermodynamics from an abstract theory into a subject of the greatest usefulness. Some insight into the origin of the name "*free* energy" can be obtained by thinking of G as the total enthalpy H of the system minus a quantity $T \times S$, which represents the system's energy that is already disorderly. Only the orderly part of the enthalpy, the difference $H - TS$, is free to become more disordered and hence to cause change.

For a process that takes place at constant temperature, the resulting changes in enthalpy and entropy produce a change in free energy of

$$\Delta G = \Delta H - T\Delta S \tag{6}$$

By comparing this equation with the expression for ΔS_{tot}, we see that

$$-T\Delta S_{tot} = \Delta G \tag{7}$$

Up to this point we have done nothing except reorganize the expression for the total entropy change. However, this rearrangement has several important consequences.

Free energy and the direction of spontaneous change. The minus sign in Eq. 7 means that an increase in total entropy corresponds to a *decrease* in free energy. Therefore, in terms of G, *the direction of spontaneous change (at constant pressure and temperature) is the direction of decreasing free energy.*

When we plot the free energy of a system against its changing composition as the reaction proceeds, we get a U-shaped curve with a minimum at a certain composition (Fig. 16.17). The reaction will tend to proceed toward the composition at the lowest point of the curve because that is the direction of decreasing free energy. The composition at the lowest point of the curve—the point of minimum free energy—corresponds to equilibrium. For a system at equilibrium, neither the forward nor the reverse change is spontaneous, because both would lead to an increase in free energy (a decrease of total entropy). If the free-energy minimum lies very close to the pure products, then the equilibrium composition strongly favors products (and "goes almost to completion"; Fig. 16.17b). If the free-energy minimum lies very close to the pure reactants, then the equilibrium composition strongly favors the reactants (and the reaction "does not go"; Fig. 16.17c).

Equation 6 shows very clearly the factors that govern the direction of spontaneous change. They are summarized in Table 16.3, which we can examine for factors that make ΔG negative. One condition that tends to do this is large negative ΔH, as for a strongly exothermic process like combustion. Because large negative ΔH corresponds to a large increase in the entropy of the surroundings, combustions and other exothermic processes are spontaneous on account of the disorder they create in the surroundings.

A negative ΔG can occur even when ΔH is positive (an endothermic reaction) if $T\Delta S$ is large and positive. The size and sign of $T\Delta S$ depends on two factors. One is the size and sign of ΔS itself, the change of

T OHT: Fig 16.17.

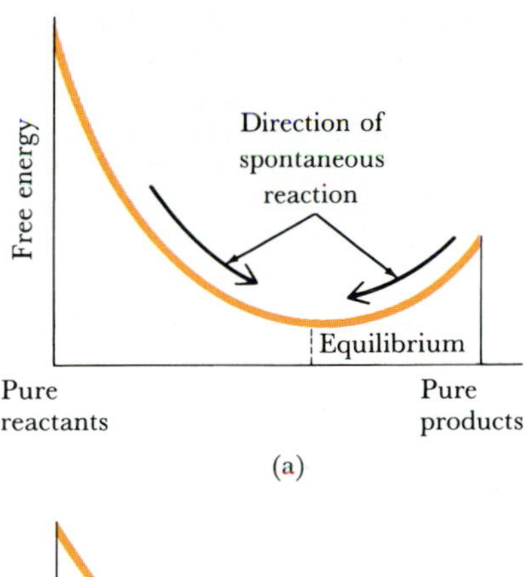

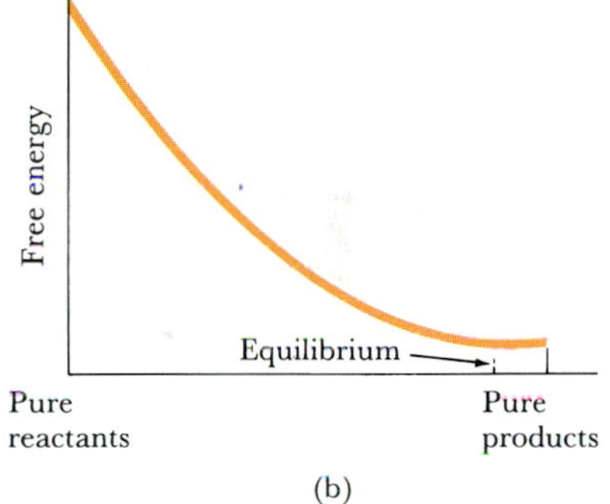

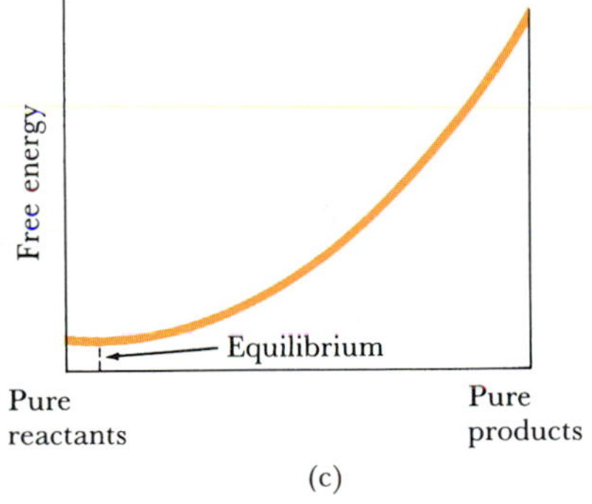

FIGURE 16.17 At constant temperature and pressure, the direction of spontaneous change is toward lower free energy. The equilibrium composition of a reaction mixture corresponds to the lowest point on the curve. (a) A reaction in which reactants and products are present in similar abundance at equilibrium (with $K \approx 1$). (b) A reaction in which the equilibrium favors the formation of products ($K \gg 1$). (c) A reaction in which there is very little tendency for the reactants to form products ($K \ll 1$).

T OHT: Table 16.3.

E The free energy prediction of reaction direction and the principle of Le Chatelier. *J. Chem. Ed.*, **1989**, *66*, 401.

N We consider it important that the impression should not be given that there are *two* criteria for spontaneous change, a decrease in enthalpy and an increase in entropy. The sole criterion is the increase in total entropy, and it just happens that the contribution in the surroundings is given by the change in enthalpy of the system.

TABLE 16.3 Factors that favor reaction

Enthalpy change	Entropy change	Spontaneous reaction?
exothermic ($\Delta H < 0$)	increase ($\Delta S > 0$)	yes, $\Delta G < 0$
exothermic ($\Delta H < 0$)	decrease ($\Delta S < 0$)	yes, if $\lvert T\Delta S\rvert < \lvert\Delta H\rvert$*
endothermic ($\Delta H > 0$)	increase ($\Delta S > 0$)	yes, if $T\Delta S > \Delta H$
endothermic ($\Delta H > 0$)	decrease ($\Delta S < 0$)	no, $\Delta G > 0$

*$\lvert x\rvert$ means the numerical value of x (its value disregarding sign).

entropy of the system; the other is the temperature T. If disorder is created in the system (as it is in vaporization or the decomposition of calcium carbonate), then ΔS is positive and the $-T\Delta S$ term in Eq. 6 contributes a negative term to ΔG. However, the size of its contribution depends on the temperature, and even a large increase in disorder within the system can have a negligible influence if the temperature is low. The role of the entropy of the system becomes more important the higher the temperature. In other words, for reactions in which the entropy change is large (in either direction), the temperature may play an important role in determining whether or not the reaction is spontaneous.

N Free energy is the maximum energy that is available from a system for work other than the work of expansion: $\Delta G = -w(\text{max})$.

▼ EXAMPLE 16.7 Calculating the minimum decomposition temperature

Assume that the reaction enthalpy and reaction entropy are independent of temperature. Estimate the temperature at which calcium carbonate is likely to decompose under 1 atm pressure.

STRATEGY Decomposition is likely when the change in free energy for the reaction

$$CaCO_3(s) \longrightarrow CaO(s) + CO_2(g)$$

is negative. We can calculate the reaction enthalpy from the enthalpies of formation listed in Appendix 2A. We can also calculate the reaction entropy from the entropies listed there. Finally, we can look for the temperature that makes $\Delta H^\circ - T\Delta S^\circ$ negative. The minimum temperature for decomposition is therefore $T = \Delta H^\circ/\Delta S^\circ$, for at all higher temperatures $T\Delta S^\circ$ is larger than ΔH°, and $\Delta H^\circ - T\Delta S^\circ$ is negative.

SOLUTION From the enthalpies of formation in Appendix 2A,

$$\Delta H^\circ = \{1 \text{ mol CaO} \times \Delta H_f^\circ(\text{CaO},s) + 1 \text{ mol CO}_2 \times \Delta H_f^\circ(\text{CO}_2,g)\} - \{1 \text{ mol CaCO}_3 \times \Delta H_f^\circ(\text{CaCO}_3,s)\}$$

$$= \{(-635.09 - 393.51) \text{ kJ}\} - \{-1206.9 \text{ kJ}\} = +178.3 \text{ kJ}$$

From the entropies of the substances,

$$\Delta S^\circ = \{1 \text{ mol CaO} \times S^\circ(\text{CaO},s) + 1 \text{ mol CO}_2 \times S^\circ(\text{CO}_2,g)\} - \{1 \text{ mol CaCO}_3 \times S^\circ(\text{CaCO}_3,s)\}$$

$$= \{(39.75 + 213.74) \text{ J/K}\} - \{92.9 \text{ J/K}\} = +160.6 \text{ J/K}$$

U Exercises 16.70 and 16.74.

E The enthalpy and entropy of reaction both vary with temperature, but the variation is usually quite small and can be neglected in elementary work.

The minimum temperature at which decomposition can occur is therefore

$$T = \frac{\Delta H^\circ}{\Delta S^\circ} = \frac{178.3 \times 10^3 \text{ J}}{160.6 \text{ J/K}} = 1110 \text{ K}$$

That is, we expect the solid to decompose at temperatures higher than 1110 K (about 840°C).

EXERCISE E16.7 Estimate the temperature at which magnesium carbonate can be expected to decompose when heated under 1 atm pressure.
[*Answer*: 575 K]

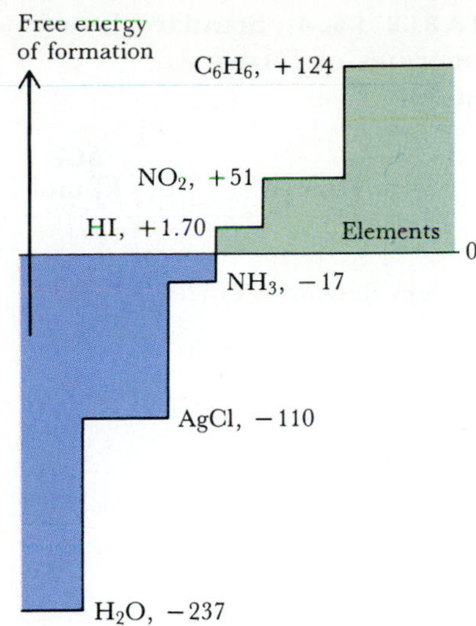

FIGURE 16.18 The standard free energies of formation of compounds are defined as the standard free energy changes that accompany their formation from their elements. Standard free energies of formation represent a "thermodynamic altitude" with respect to the elements at "sea level." The values are in kilojoules per mole.

If a system happens to be at equilibrium, then there is no tendency for spontaneous change in either direction: $\Delta G = -T\Delta S_{tot} = 0$ for a system at equilibrium at constant temperature and pressure. The condition $\Delta G = 0$ applies to any phase change (for example, a liquid in equilibrium with its vapor, or a solid in equilibrium with its liquid) and for a solute in equilibrium with its saturated solution. The same condition applies to a chemical reaction at equilibrium.

Standard free energies of formation and reaction. When nitric oxide reacts with oxygen to form nitrogen dioxide in the reaction

$$2NO(g) + O_2(g) \longrightarrow 2NO_2(g)$$

the entropy, enthalpy, and free energy of the system change. At 25°C, when 2 mol NO at 1 atm pressure reacts with 1 mol O_2 at 1 atm and produces 2 mol NO_2 at the same pressure, the entropy changes by −147 J/K (−0.147 kJ/K) and the enthalpy changes by −114 kJ. The free energy therefore changes by

$$\begin{aligned}\Delta G° &= \Delta H° - T\Delta S° \\ &= -114 \text{ kJ} - 298 \text{ K} \times (-0.147 \text{ kJ/K}) = -70 \text{ kJ}\end{aligned}$$

The symbol $\Delta G°$ is called the standard reaction free energy and is defined like the analogous standard reaction enthalpy and entropy:

The **standard reaction free energy** $\Delta G°$ is the difference between the free energies of the products in their standard states and the reactants in their standard states.

$$\Delta G° = \Sigma nG°(\text{products}) - \Sigma nG°(\text{reactants})$$

As before, the standard state of a substance is its pure form at 1 atm pressure.

E The modern standard state is 1 bar, and the data in Appendix 2A are for that value. The discrepancy is negligible at this level.

U Exercise 16.66.

In Section 6.4 we let the elements define a thermochemical "sea level," and we reported the enthalpy of a compound in terms of the reaction enthalpy for its formation from its elements. The same can be done with free energies. We let the elements define the "sea level" of free energy and take the standard free energy of formation, $\Delta G_f°$, of any compound from its elements to be its height above or below "sea level" (Fig. 16.18):

The **standard free energy of formation** $\Delta G_f°$ of a compound is the standard reaction free energy per mole for its formation from its elements in their most stable forms.

The most stable form of an element is the state with the lowest free energy (see examples in margin). For example, the standard free energy of formation of hydrogen iodide gas at 25°C is the standard reac-

Element	Most stable form at 25°C, 1 atm
H_2, O_2, Cl_2, Xe	gas
Br_2, Hg	liquid
C	graphite
Na, Fe, I_2	solid

TABLE 16.4 Standard free energies of formation at 25°C*

Substance	ΔG_f°, kJ/mol
Gases	
carbon dioxide, CO_2	−394.4
nitrogen dioxide, NO_2	+51.3
water, H_2O	−228.6
Liquids	
benzene, C_6H_6	+124.3
ethanol, CH_3CH_2OH	−174.8
water, H_2O	−237.1
Solids	
calcium carbonate, $CaCO_3$	−1128.8
silver chloride, AgCl	−109.8

*Additional values are given in Appendix 2A.

T OHT: Table 16.4.

U Exercises 16.67–16.69.

tion free energy for

$$\tfrac{1}{2}H_2(g) + \tfrac{1}{2}I_2(s) \longrightarrow HI(g) \qquad \Delta G^\circ = +1.70\text{ kJ}$$

in which hydrogen at 1 atm pressure reacts with solid iodine to give hydrogen iodide gas at 1 atm pressure. It is expressed as an energy per mole of HI molecules:

$$\Delta G_f^\circ = \frac{+1.70\text{ kJ}}{1\text{ mol HI}} = +1.70\text{ kJ/mol HI}$$

Like enthalpies of formation, the standard free energies of formation of elements in their most stable forms are zero: they are at "sea level."

Standard free energies of formation can be measured in a variety of ways, which include combining the enthalpy and entropy data in Tables 6.4 and 16.1. Lists of standard free energies of formation are given in Table 16.4 and Appendix 2A.

▼ **EXAMPLE 16.8 Calculating a standard free energy of formation**

Calculate the standard free energy of formation of HI(g) at 25°C from its standard entropy and its standard enthalpy of formation.

STRATEGY We can calculate ΔG° from the standard enthalpy and entropy of reaction by combining them according to Eq. 6:

$$\Delta G^\circ = \Delta H^\circ - T\Delta S^\circ$$

The standard reaction enthalpy is found from the enthalpies of formation by using data from Appendix 2A. We calculate the standard reaction entropy as we did in Example 16.5, using the data from Appendix 2A. Convert the standard free energy of reaction to a free energy per mole of HI.

SOLUTION The chemical equation is

$$H_2(g) + I_2(s) \longrightarrow 2HI(g)$$

From Appendix 2A, with $\Delta H_f^\circ = 0$ for each of the elements,

$$\begin{aligned}\Delta H^\circ &= \{2\text{ mol HI} \times \Delta H_f^\circ(\text{HI},g)\} \\ &\qquad - \{1\text{ mol H}_2 \times \Delta H_f^\circ(\text{H}_2,g) + 1\text{ mol I}_2 \times \Delta H_f^\circ(\text{I}_2,s)\} \\ &= \{2 \times 26.48\text{ kJ}\} - \{0 + 0\} = +52.96\text{ kJ}\end{aligned}$$

From Appendix 2A,

$$\begin{aligned}\Delta S^\circ &= \{2\text{ mol HI} \times S^\circ(\text{HI},g)\} - \{1\text{ mol H}_2 \times S^\circ(\text{H}_2,g) + 1\text{ mol I}_2 \times S^\circ(\text{I}_2,s)\} \\ &= \{2 \times 206.6\text{ J/K}\} - \{(130.7 + 116.1)\text{ J/K}\} \\ &= +166.4\text{ J/K, or } 0.1664\text{ kJ/K}\end{aligned}$$

Therefore, because $T = 298$ K (corresponding to 25°C),

$$\Delta G^\circ = \Delta H^\circ - T\Delta S^\circ = +52.96\text{ kJ} - 298\text{ K} \times 0.1664\text{ kJ/K} = +3.4\text{ kJ}$$

The reaction produces 2 mol of HI(g), so

$$\Delta G_f^\circ = \frac{+3.4\text{ kJ}}{2\text{ mol HI}} = +1.7\text{ kJ/mol HI}$$

EXERCISE E16.8 Calculate the standard free energy of formation of $NH_3(g)$ at 25°C.

▲ [*Answer*: −16.5 kJ/mol NH_3]

16.7 SPONTANEOUS REACTIONS

The free energies of formation of compounds can be used to determine whether reactions involving those compounds are spontaneous. The simplest reaction of a compound is its decomposition into its elements (the reverse of its synthesis). We begin by considering that kind of reaction and then go on to consider more complicated changes.

Thermodynamic stability. The standard free energy of formation of a compound is a measure of its stability as a compound, with respect to decomposition into its elements, *under standard conditions.* When ΔG_f° is negative at a certain temperature, the elements have a spontaneous tendency to form the compound, and—under standard conditions—the compound is more stable than its elements at that temperature (Fig. 16.19a). When ΔG_f° is positive, the reverse reaction is spontaneous, and—under standard conditions—the compound has a tendency to decompose into its elements (Fig. 16.19b). In the latter case there is no point in trying to synthesize the compound from its elements at that temperature (under standard conditions, at least).

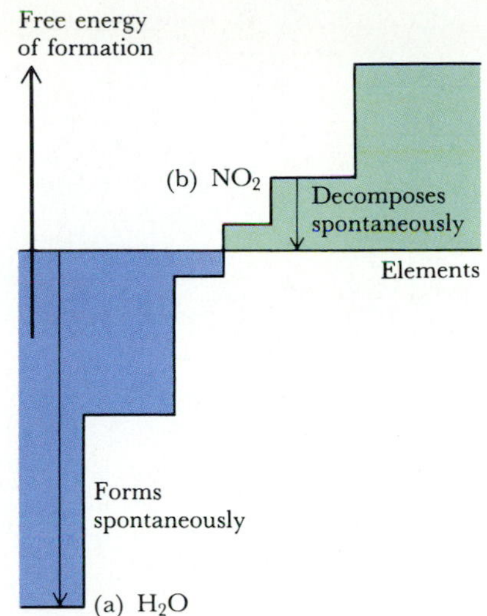

FIGURE 16.19 (a) A compound like H_2O is thermodynamically stable with respect to its elements (under standard conditions) because its standard free energy of formation is negative. (b) A compound like NO_2 is unstable (under standard conditions) and tends to decompose into its elements because its standard free energy of formation is positive.

As an example, the standard free energy of formation of benzene is +124 kJ/mol at 25°C, which means that the synthesis

$$6C(s) + 3H_2(g) \rightarrow C_6H_6(l) \qquad \Delta G^\circ = +124 \text{ kJ}$$

is not spontaneous under standard conditions. There is no point in trying to make benzene by exposing carbon to hydrogen gas at 25°C and 1 atm, even in the presence of a catalyst, because the product has no tendency to form. However, the reverse reaction

$$C_6H_6(l) \rightarrow 6C(s) + 3H_2(g) \qquad \Delta G^\circ = -124 \text{ kJ}$$

is spontaneous at 25°C. The system and its surroundings become more disordered if benzene decomposes into carbon and hydrogen. We say, therefore, that benzene is thermodynamically unstable with respect to its elements under standard conditions.

In general, a **thermodynamically unstable compound** is a compound with a positive standard free energy of formation; such a compound has a thermodynamic tendency to decompose into its elements. However, the *tendency* may not be realized in practice because the kinetics of the decomposition may be very slow. In the language of Section 12.3, the activation energy may be very high. Benzene can, in fact, be kept indefinitely without decomposing at all.

E Whether or not a compound is *endoergic* ($\Delta G_f^\circ > 0$, unstable with respect to its elements) or *exoergic* ($\Delta G_f^\circ < 0$) depends on the temperature, so it may be possible to synthesize an endoergic compound from its elements by changing the temperature to a value at which it becomes exoergic. Although ΔH and ΔS vary only weakly with temperature, $\Delta G = \Delta H - T\Delta S$ may vary quite appreciably on account of the factor T. A thermodynamically unstable compound that survives for long periods (like benzene) is *nonlabile*.

? Comment on the thermodynamic stability and the lability of trinitroglycerin.

▼ EXAMPLE 16.9 Judging the stability of a compound

Would it be worth looking for a catalyst to synthesize glucose from its elements at 25°C and standard conditions?

STRATEGY A synthesis reaction has a spontaneous tendency to proceed in the required direction only if the free energy of formation of the compound is negative. (This statement is true whether or not a catalyst is present, for a catalyst affects only the rate of a reaction, not its direction.) Therefore, we need only inspect a table of data (such as Table 16.4 or Appendix 2A) to see whether ΔG_f° for glucose is negative.

SOLUTION From Appendix 2A, $\Delta G_f^\circ = -910$ kJ/mol for glucose, $C_6H_{12}O_6(s)$. So the synthesis has a spontaneous tendency to proceed in the required direction because ΔG_f° is negative. Hence, on thermodynamic grounds at least, it would not be foolish to look for a catalyst that converts carbon, hydrogen, and oxygen directly into glucose.

EXERCISE E16.9 Is it worth looking for a catalyst to prepare methylamine, $CH_3NH_2(g)$, from its elements at 25°C and standard conditions?

[*Answer*: no]

The spontaneous direction of reactions in general. We use standard free energies of formation like enthalpies of formation to calculate standard reaction free energies:

$$\Delta G^\circ = \Sigma n\Delta G_f^\circ(\text{products}) - \Sigma n\Delta G_f^\circ(\text{reactants})$$

That is, the standard reaction free energy is the difference between the total standard free energy for the formation of the products and that of the reactants. For the oxidation of ammonia,

$$4NH_3(g) + 5O_2(g) \longrightarrow 4NO(g) + 6H_2O(g)$$

we have (from Appendix 2A)

$$\begin{aligned}\Delta G^\circ &= \{4 \text{ mol NO} \times \Delta G_f^\circ(\text{NO},g) + 6 \text{ mol } H_2O \times \Delta G_f^\circ(H_2O,g)\} \\ &\qquad - \{4 \text{ mol } NH_3 \times \Delta G_f^\circ(NH_3,g) + 0\} \\ &= \{(4 \times 86.55 + 6 \times (-228.57))\text{ kJ}\} - \{(4 \times (-16.45) + 0)\text{ kJ}\} \\ &= -959.42 \text{ kJ}\end{aligned}$$

The result shows that the oxidation of ammonia has a strong tendency to proceed under standard conditions. In the Ostwald process for producing nitric acid by the oxidation of ammonia, this tendency is made a reality by passing the gases over a rhodium-platinum catalyst (Fig. 12.25).

The same kind of calculation can be used to decide whether or not a compound has a tendency to decompose into simpler compounds (as distinct from its elements). The decomposition of hydrogen peroxide to water at 25°C is an example:

$$2H_2O_2(l) \longrightarrow 2H_2O(l) + O_2(g) \qquad \Delta G^\circ = -233.6 \text{ kJ}$$

The negative ΔG° shows that pure liquid hydrogen peroxide has a tendency to decompose into water and oxygen under standard conditions at 25°C. This decomposition is rapid in the presence of MnO_2 as a catalyst.

EQUILIBRIA

U Exercises 16.64, 16.65, and 16.71.

We have arrived at the stage at which we can calculate equilibrium constants from tabulated data. Suppose we have a very large reaction mixture—a swimming pool full, for instance—and we consider the change in free energy when 1 mol of a certain reactant changes into products. The amount of reaction is so small compared with the huge amount of material present that we can think of the change as occurring at a definite, constant composition of reactants and products. For instance, we may be interested in the reaction of nitrogen and hydrogen to form ammonia, when 10,000 mol of each substance is present in a large container. The reaction free energy at this fixed composition will be denoted by ΔG. Note that ΔG and ΔG° are quite different properties, and it is important to have the distinction clearly in mind:

- ΔG is the reaction free energy at a definite, *fixed* composition.

In practice, ΔG is the change in free energy when 1 mol of a reactant reacts in such a large amount of reaction mixture that the composition remains effectively constant.

- ΔG° is the difference in free energies between the products and the reactants in their standard states.

In practice, ΔG° is calculated from the standard free energies of formation of the pure reactants and products.

If the composition of the reaction mixture contains more reactants than at equilibrium, then ΔG at that composition will be negative, and the reaction will have a tendency to form more products. If the reaction mixture happens to contain a higher proportion of products than at equilibrium, then ΔG will be positive, and the *reverse* reaction, the formation of reactants, will be spontaneous. However, if the reaction mixture is at equilibrium, then ΔG will be zero, because there will be no tendency for the reaction to run in either direction. That is,

$$\text{At equilibrium: } \Delta G = 0$$

Our task is to discover how ΔG varies with composition. Then, we must solve the resulting equation to find the composition at which $\Delta G = 0$. That composition is the equilibrium composition of the reaction mixture. As we shall see, this strategy leads directly to a formula that relates the equilibrium constant of the reaction to the *standard* free energy of the reaction. Because we have seen how to calculate ΔG° from the information in Appendix 2A, we shall be brought to the point at which we can calculate the equilibrium constant of almost any reaction, including those that have not been studied experimentally.

N This distinction between ΔG and ΔG° might seem labored, but it is crucial to an understanding of the role of free energy in chemical equilibrium. The value of ΔG changes in the course of the reaction, but ΔG° is fixed by the identities of the reactants and products, and has one particular value for a given reaction (at a given temperature). The sign of ΔG reveals whether, at the prevailing composition of the mixture, the reaction has a tendency to proceed in one direction or another. The value of ΔG° is used to predict (as we see below) the composition at which ΔG is zero and hence the composition corresponding to equilibrium.

N An elementary view of thermodynamics for calculating equilibrium. A review for teachers. *J. Chem. Ed.*, **1976**, *53*, 626. On the minimum of the Gibbs free energy involved in chemical equilibrium. *J. Chem. Ed.*, **1987**, *64*, 201. Equilibria and ΔG°. *J. Chem. Ed.*, **1990**, *67*, 745.

16.8 FREE ENERGY AND COMPOSITION

The reaction free energy is related to the composition of a reaction mixture by the formula

$$\Delta G = -RT \ln \left(\frac{K}{Q}\right) \qquad (8)$$

where R is the gas constant, Q is the reaction quotient (which depends on the composition), and K is the equilibrium constant of the reaction. At 25°C (corresponding to 298 K), $RT = 2.48$ kJ/mol. However, because a thermochemical equation always refers to the reaction of a stoichiometric number of moles of each substance, we use $RT = 2.48$ kJ for all calculations at 25°C that refer to thermochemical equations. Equation 8 applies for K, K_P, K_c, and all the other equilibrium constants that we have met.

E A qualitative justification of Eq. 8 will be found in the first edition of this text. Briefly, the fact that ΔG depends logarithmically on K/Q reflects the fact that a reaction has a tendency to form products (a) if $Q < K$, or $K/Q > 1$ (as was discussed in Chapter 13), and (b) if $\Delta G < 0$ (as discussed in this chapter). These two criteria are consistent if ΔG is proportional to $\ln (K/Q)$.

E Error in the minimum free energy curve. *J. Chem. Ed.*, **1984**, *61*, 173.

▼ **EXAMPLE 16.10 Calculating the reaction free energy from the reaction quotient**

The equilibrium constant for the reaction

$$N_2(g) + 3H_2(g) \rightleftharpoons 2NH_3(g)$$

is $K_P = 6.0 \times 10^5$ at 25°C. What is the reaction free energy when the partial pressure of each gas is 100 atm? What is the spontaneous direction of the reaction under these conditions?

STRATEGY All we need to do is calculate the reaction quotient and substitute it and the data into Eq. 8. If ΔG turns out to be negative, then the formation of ammonia is spontaneous; if ΔG is positive, then ammonia has a spontaneous tendency to decompose into its elements.

SOLUTION The reaction quotient is

$$Q_P = \frac{(P_{NH_3})^2}{P_{N_2} \times (P_{H_2})^3} = \frac{(100)^2}{100 \times (100)^3} = 1.00 \times 10^{-4}$$

Therefore, the reaction free energy at this composition is

$$\Delta G = -RT \ln\left(\frac{K_P}{Q_P}\right) = -2.48 \text{ kJ} \times \ln\left(\frac{6.0 \times 10^5}{1.00 \times 10^{-4}}\right) = -56 \text{ kJ}$$

Because the reaction free energy is negative, the formation of products is spontaneous at this composition and temperature.

D Concerning equilibrium, free energy changes, Le Chatelier's principle. TDC, 221 (I), 222 (II), 223 (III).

EXERCISE E16.10 The equilibrium constant for the reaction $H_2(g) + I_2(g) \rightleftharpoons 2HI(g)$ is $K_P = 160$ at 500 K (when $RT = 4.16$ kJ). What is the value of ΔG when the partial pressures of the gases are $P_{H_2} = 1.5$ atm, $P_{I_2} = 0.88$ atm, and $P_{HI} = 0.065$ atm. What is the spontaneous direction of the reaction?

[*Answer*: −45 kJ; toward products]

16.9 THE EQUILIBRIUM CONSTANT

E Free energy and equilibrium: The basis of $\Delta G° = -RT \ln K$ for reactions in solution. *J. Chem. Ed.*, **1983**, *60*, 648.

We are now very close to an equation for K in terms of free energy. Suppose we were considering a gas-phase reaction, such as the synthesis of ammonia or the oxidation of nitric oxide, and were dealing with partial pressures. We would have

$$2NO(g) + O_2(g) \rightleftharpoons 2NO_2(g) \qquad Q_P = \frac{(P_{NO_2})^2}{(P_{NO})^2 P_{O_2}}$$

If each substance were pure and at 1 atm pressure, then ΔG would have its *standard* value, $\Delta G°$. Under these conditions, $Q_P = 1$, because all the partial pressures are equal to 1 atm. Therefore, Eq. 8 would become (with $K = K_P$),

$$\Delta G° = -RT \ln K_P$$

The same equation applies to any kind of equilibrium constant—for K_c, K_s, K_a, K_b, or K_w. Therefore, for any kind of equilibrium,

$$\Delta G° = -RT \ln K \qquad (9)$$

With this result established, we can go back to Eq. 8 written in the form

$$\Delta G = -RT \ln K + RT \ln Q$$

N Equation 10 is needed for the derivation of the Nernst equation in Chapter 17.

and express the relation between ΔG and $\Delta G°$ as

$$\Delta G = \Delta G° + RT \ln Q \qquad (10)$$

Equation 9 is one of the most useful expressions in the whole of thermodynamics. It is the link between the equilibrium constant for *any* change—physical or chemical—and the standard reaction free energy. Because the latter can be found in tables, such as those in

Appendix 2A, Eq. 9 enables us to calculate K for any reaction for which free-energy data are available.

▼ EXAMPLE 16.11 Calculating an equilibrium constant from free energies

Calculate K_P at 25°C for the equilibrium $N_2O_4(g) \rightleftharpoons 2NO_2(g)$.

STRATEGY The required value can be found by solving Eq. 9 for K_P. First, however, the standard reaction free energy must be calculated from the standard free energies of formation in Appendix 2A.

SOLUTION From the data in Appendix 2A:

$$\Delta G^\circ = \{2 \text{ mol } NO_2 \times \Delta G_f^\circ(NO_2,g)\} - \{1 \text{ mol } N_2O_4 \times \Delta G_f^\circ(N_2O_4,g)\}$$
$$= \{2 \times 51.3 \text{ kJ}\} - \{97.89 \text{ kJ}\} = +4.73 \text{ kJ}$$

After rearranging Eq. 9 and substituting this value and (at 25°C) RT = 2.48 kJ, we have

$$\ln K_P = \frac{-\Delta G^\circ}{RT} = \frac{-4.73 \text{ kJ}}{2.48 \text{ kJ}} = -1.91$$

The natural antilogarithm of −1.91 is 0.15, so for this reaction $K_P = 0.15$. If necessary, the expression in Section 13.2 could now be used to find K_c.

EXERCISE E16.11 Calculate the equilibrium constant K_P for the reaction $N_2(g) + 3H_2(g) \rightleftharpoons 2NH_3(g)$ at 25°C.

[*Answer*: 1.7×10^{-6}] ▲

U Exercises 16.72–16.76, 16.79.

A feature of Chapters 14 and 15 on acids, bases, and salts can now be brought into focus. We saw there that pK is often more convenient to use than K itself. Now we can see why: because ln K is proportional to ΔG°, pK is actually a disguised form of ΔG°.* In other words, *pK manipulations are in fact calculations of the free energies of acid and base proton transfer reactions.* A very similar replacement enables us to analyze the equilibria of redox reactions, but in them ΔG° is expressed in volts. We consider this subject in the next chapter.

*More specifically, because $\Delta G^\circ = -RT \ln K$ and $\ln K = 2.303 \times \log K = -2.303 \text{ p}K$, it follows that $\Delta G^\circ = 2.303RT \times \text{p}K$.

SUMMARY

16.1 The **internal energy** U of a system is its total energy. It is a state property and may be changed by heating the system or by doing work. The **first law of thermodynamics** states that the internal energy of an isolated system is constant. The internal energy change accompanying a process is equal to the energy it absorbs as heat at constant volume.

16.2 The change in **enthalpy** ΔH of a system is equal to the heat it absorbs at constant pressure. The enthalpy is a state property and differs from the internal energy by the amount of work that a system must do in the course of a change of state.

16.3 A major question in chemistry is how to decide which reactions and processes are **spontaneous**, or tending to occur without being driven. Changes are spontaneous if they correspond to an increase in the overall disorder of the system and its surroundings. The disorder can be expressed as **entropy**: the greater the disorder, the higher the entropy. The entropy may be measured or calculated from the **Boltzmann formula**. The entropy of any perfect crystal is zero at $T = 0$. Entropy increases as the temperature increases and when the substance melts and vaporizes (the change on vaporization is larger than the change on melting). The change in entropy that accompanies the transformation of reactants into products, the **standard reaction entropy**, is calculated from the differences of their **standard molar entropies**, their entropy per mole under standard conditions.

16.4 The change in entropy of the surroundings is calculated from the temperature and the enthalpy change of the system. Heat released to the surroundings increases the entropy; heat absorbed from the surroundings decreases their entropy. The total entropy change is zero for processes at equilibrium.

16.5 According to the **second law of thermodynamics**, the direction of spontaneous change is that for which the *total* entropy (of the system plus its surroundings) increases. An exothermic reaction may be spontaneous because it releases heat to the surroundings and hence increases their entropy even though the entropy of the reaction system might decrease. Endothermic reactions are less common. They are spontaneous only if the increase in entropy of the system is great enough to offset the decrease in entropy of the surroundings.

16.6 The total entropy change is normally expressed in terms of the **Gibbs free energy**. The direction of spontaneous change (at constant pressure and temperature) is in the direction of decreasing free energy. For a system at equilibrium at constant temperature and pressure, $\Delta G = 0$. Whether or not a reaction is spontaneous can be judged from the **standard reaction free energy** at the temperature of interest: a reaction is spontaneous if the standard reaction free energy is negative. The standard reaction free energy can be calculated from **standard free energies of formation**, the standard reaction free energy for the formation of the compound from its elements.

16.7 The standard free energy of formation of a compound can be used to judge whether it is **thermodynamically unstable** with respect to decomposition into its elements or into other compounds (under standard conditions). The temperature at which the standard reaction free energy becomes negative is the temperature above which a reaction can occur.

16.8 and 16.9 The reaction free energy depends on the composition of the reaction system. From the condition that at equilibrium the reaction free energy is zero, we can deduce the relation between the equilibrium constant and the standard reaction free energy.

CLASSIFIED EXERCISES

Heat, Work, and Energy

16.1 A gas sample is heated in a cylinder, using 550 kJ of heat. A piston compresses the gas, using 700 kJ of work. What is the change in internal energy of the system?

16.2 Calculate the work for a system that absorbs 150 kJ of heat and the change in internal energy is 120 kJ. Is work done on or by the system to cause the resultant change in internal energy?

16.3 A 100-W electric heater (1 W = 1 J/s) operates for 20 min to heat the gas in a cell. The gas expands from 2.0 to 2.5 L against an atmospheric pressure of 1.0 atm. What is the change in internal energy of the gas?

16.4 A gas in a cylinder was placed in a heater and gained 7000 kJ of heat. If the cylinder increased in volume from 700 mL to 1450 mL against an atmospheric pressure of 750 Torr, what is the change in internal energy of the gas in the cylinder?

16.5 The change in internal energy for the combustion of 1 mol CH_4 in a cylinder according to the reaction $CH_4(g) + 2O_2(g) \rightarrow CO_2(g) + 2H_2O(g)$ is -892.4 kJ. If a piston connected to the cylinder performs 492 kJ of expansion work, how much heat is lost from the system?

16.6 In a combustion cylinder the total internal energy change produced from the burning of a fuel is -1740 kJ. The cooling system that surrounds the cylinder absorbs 470 kJ. What is the amount of work that can be done by the fuel in the cylinder?

Entropy

16.7 Without performing a calculation, state whether the entropy of the following systems increases or decreases: (a) the oxidation of nitrogen, $N_2(g) + 2O_2(g) \rightarrow 2NO_2(g)$; (b) the sublimation of dry ice, $CO_2(s) \rightarrow CO_2(g)$; (c) the cooling of water from 50°C to 4°C.

16.8 In the following processes, does the entropy of the system increase or decrease? (Do not attempt to perform any calculations.) (a) The dissolution of table salt, $NaCl(s) \rightarrow NaCl(aq)$; (b) the photosynthesis of glucose, $6CO_2(g) + 6H_2O(l) \rightarrow C_6H_{12}O_6(s) + 6O_2(g)$; (c) the evaporation of water from damp clothes.

16.9 Use the Boltzmann formula, $S = k \ln W$, to calculate the entropy of (a) a stack of 10 pennies arranged as heads up and (b) a sample of solid CO containing 1 mol CO in which the molecules are in either of two orientations with equal probability.

16.10 Use the Boltzmann formula, $S = k \ln W$, to calculate the entropy of (a) a stack of 10 pennies arranged as randomly heads up or down and (b) a sample of a solid containing 1 mol of molecules that can occupy any one of three orientations.

16.11 Use data from Table 6.2 or Appendix 2A to calculate the entropy change (a) for the freezing of 1 mol

H_2O at 0°C, (b) for the vaporization of 50.0 g ethanol, C_2H_5OH, at 351.5 K and 1 atm.

16.12 Use data from Table 6.2 or Appendix 2A to calculate the entropy change (a) for the vaporization of 1 mol H_2O at 100°C and 1 atm, (b) for the freezing of 3.33 g NH_3 at 195.3 K.

Standard Reaction Entropies

16.13 Use data from Table 16.1 or Appendix 2A to calculate the standard reaction entropy for each of the following reactions at 25°C. For each reaction, interpret the sign and magnitude of the reaction entropy. (a) The oxidation of carbon monoxide, $2CO(g) + O_2(g) \rightarrow 2CO_2(g)$; (b) the decomposition of limestone, $CaCO_3(s) \rightarrow CaO(s) + CO_2(g)$; (c) the decomposition of potassium chlorate, $4KClO_3(s) \rightarrow 3KClO_4(s) + KCl(s)$.

16.14 Use data from Table 16.1 or Appendix 2A to calculate the standard reaction entropy for each of the following reactions at 25°C. For each reaction, interpret the sign and magnitude of the reaction entropy. (a) The synthesis of carbon disulfide from natural gas (methane), $2CH_4(g) + S_8(s) \rightarrow 2CS_2(l) + 4H_2S(g)$; (b) the production of acetylene from calcium carbide, $CaC_2(s) + H_2O(l) \rightarrow Ca(OH)_2(s) + C_2H_2(g)$; (c) the industrial synthesis of urea, a common fertilizer, $CO_2(g) + 2NH_3(g) \rightarrow CO(NH_2)_2(s) + H_2O(l)$.

Entropy Change in the Surroundings

16.15 Calculate the change in the entropy of the surroundings when (a) 120 kJ is released to the surroundings at 25°C; (b) 120 kJ is released to the surroundings at 100°C; (c) 100 J is absorbed from the surroundings at 50°C.

16.16 Calculate the change in the entropy of the surroundings when (a) 1 mJ is released to the surroundings at 2×10^{-9} K, the current world record low temperature; (b) 1 J, the energy of a single heartbeat, is released to the surroundings at 37°C; (c) 20 J, the energy released when 1 mol He freezes and enters the surroundings at 3.5 K.

16.17 A human body generates heat at the rate of about 100 W (1 W = 1 J/s). (a) At what rate do you generate entropy in your surroundings, taken to be at 20°C? (b) How much entropy do you generate each day?

16.18 An electric heater is rated at 2 kW (1 W = 1 J/s). (a) At what rate does it generate entropy in a room maintained at 28°C? (b) How much entropy does it generate in the course of a day? (c) Would the entropy be greater or less if the room were maintained at 25°C?

16.19 (a) Calculate the change in entropy of a block of copper at 25°C that absorbs 5 J of energy from a heater. (b) If the block of copper is at 100°C and it absorbs 5 J of energy from the heater, what is its entropy change? (c) Is there a greater entropy change when the block of copper is at 25°C or at 100°C? Explain your conclusions.

16.20 (a) Calculate the change in entropy of 1 L of water at 0°C when it absorbs 500 J of energy from a heater. (b) If the 1 L of water is at 99°C, what is its entropy change? (c) Is there a greater entropy change when the water is at 0°C or at 99°C? Explain your conclusions.

16.21 Consider the reaction for the production of formaldehyde:

$$H_2(g) + CO(g) \longrightarrow HCHO(g)$$
$$\Delta H° = 1.96 \text{ kJ}, \quad \Delta S° = -109.58 \text{ J/K}$$

Calculate the change in entropy of the surroundings and then comment on the spontaneity of the reaction at 25°C.

16.22 Consider the reaction in which deuterium exchanges with ordinary hydrogen in water:

$$D_2(g) + H_2O(l) \longrightarrow H_2(g) + D_2O(l)$$
$$\Delta H° = -7.38 \text{ kJ}, \quad \Delta S° = +114.15 \text{ J/K}$$

Calculate the change in entropy of the surroundings and comment on the spontaneity of the reaction at 25°C.

Entropy and Equilibrium

16.23 Use data from Table 6.2 to calculate the change in entropy of the surroundings and the system, and the overall entropy change for (a) the vaporization of methane at its normal boiling point; (b) the melting of ethanol at its normal melting point; (c) the freezing of ethanol at its normal freezing point. Express your answer as an entropy change per mole of substance.

16.24 Use data from Table 6.2 to calculate the change in entropy of the surroundings and the system, and the overall entropy change for (a) the vaporization of ammonia at its normal melting point; (b) the freezing of methanol at its normal freezing point; (c) the vaporization of water at its normal boiling point. Express your answer as an entropy change per mole of substance.

16.25 The entropy of vaporization of liquid chlorine is 85.4 J/(K·mol) and its enthalpy of vaporization is 20.4 kJ/mol. What is its boiling point?

16.26 Estimate the boiling point of fluorine by assuming that its enthalpy of vaporization is similar to that of argon and that the entropy of vaporization of fluorine is approximately the same as that of chlorine (Exercise 16.25).

16.27 (a) Estimate the enthalpy of vaporization of benzene at its normal boiling point of 80°C. (b) What is the entropy change of the surroundings when 10 g of benzene, C_6H_6, vaporizes at its normal boiling point? The entropy of vaporization of benzene is approximately 85 J/(K·mol).

16.28 (a) Estimate the enthalpy of vaporization of acetone at its normal boiling point of 56.4°C. (b) What is the entropy change of the surroundings when 10 g of acetone, CH_3COCH_3, condenses at its normal boiling point? The entropy of vaporization of acetone is approximately 85 J/(K·mol).

Free Energy

16.29 Predict the sign of ΔG for a reaction that is (a) exothermic, and accompanied by an increase in entropy; (b) endothermic, and accompanied by an increase in entropy. (c) Can a temperature change affect the sign in ΔG in (a) and (b); if so, how?

16.30 Predict the sign of ΔG for a reaction that is (a) exothermic, and accompanied by a decrease in entropy; (b) endothermic, and accompanied by a decrease in entropy. (c) Can a temperature change affect the sign of ΔG in (a) and (b); if so, how?

16.31 Calculate the standard free energy for each reaction by using the equation $\Delta G° = \Delta H° - T\Delta S°$ and data from Appendix 2A. (It may be helpful to note that $\Delta S°$ for each reaction was calculated in Exercise 16.13.) For each reaction, interpret the sign and magnitude of the free energy. (a) The oxidation of carbon monoxide, $2CO(g) + O_2(g) \rightarrow 2CO_2(g)$; (b) the decomposition of limestone, $CaCO_3(s) \rightarrow CaO(s) + CO_2(g)$; (c) the decomposition of potassium chlorate, $4KClO_3(s) \rightarrow 3KClO_4(s) + KCl(s)$.

16.32 Calculate the standard free energy for each reaction by using the equation $\Delta G° = \Delta H° - T\Delta S°$ and data from Appendix 2A. (It may be helpful to note that $\Delta S°$ for each reaction was calculated in Exercise 16.14.) For each reaction, interpret the sign and magnitude of the free energy. (a) The synthesis of carbon disulfide from natural gas (methane), $2CH_4(g) + S_8(s) \rightarrow 2CS_2(l) + 4H_2S(g)$; (b) the production of acetylene from calcium carbide, $CaC_2(s) + H_2O(l) \rightarrow Ca(OH)_2(s) + C_2H_2(g)$; (c) the industrial synthesis of urea, a common fertilizer, $CO_2(g) + 2NH_3(g) \rightarrow CO(NH_2)_2(s) + H_2O(l)$.

16.33 Write a chemical equation for the formation reaction and then calculate the standard free energy of formation of each of the following compounds from the enthalpies of formation and the standard entropies using $\Delta G° = \Delta H° - T\Delta S°$: (a) $NH_3(g)$; (b) $CO(g)$; (c) $NO_2(g)$.

16.34 Write a chemical equation for the formation reaction and then calculate the standard free energy of formation of each of the following compounds from the enthalpies of formation and the standard entropies using $\Delta G° = \Delta H° - T\Delta S°$: (a) $SO_3(g)$; (b) $C_2H_5OH(l)$; (c) $CaCO_3(s)$.

16.35 Use the standard free energies of formation in Appendix 2A to calculate the standard free energy at 25°C of the following reactions. Comment on the spontaneity of the reaction at 25°C.
(a) $2SO_2(g) + O_2(g) \rightarrow 2SO_3(g)$
(b) $SbCl_5(g) \rightarrow SbCl_3(g) + Cl_2(g)$
(c) $2C_8H_{18}(l) + 25O_2(g) \rightarrow 16CO_2(g) + 18H_2O(l)$

16.36 Use the standard free energies of formation in Appendix 2A to calculate the standard free energy of each of the following reactions at 25°C. Comment on the spontaneity of the reaction at 25°C.
(a) $H_2(g) + D_2O(l) \rightarrow D_2(g) + H_2O(l)$
(b) $2NO_2(g) \rightarrow N_2O_4(g)$
(c) $2CH_3OH(g) + 3O_2(g) \rightarrow 2CO_2(g) + 4H_2O(l)$

Thermodynamic Stability

16.37 Determine which of the following compounds are stable with respect to decomposition into their elements at 25°C (see Appendix 2A): (a) HCN; (b) NO; (c) SO_2.

16.38 Determine which of the following compounds are stable with respect to decomposition into their elements at 25°C (see Appendix 2A): (a) CuO; (b) $SbCl_3$; (c) N_2H_4.

16.39 On the basis of the equation $\Delta G° = \Delta H° - T\Delta S°$, which of the following compounds becomes more unstable with respect to its elements as the temperature is raised: (a) HCN; (b) NO; (c) SO_2?

16.40 On the basis of the equation $\Delta G° = \Delta H° - T\Delta S°$, which of the following compounds becomes more unstable with respect to its elements as the temperature is raised: (a) CuO; (b) $SbCl_3$; (c) N_2H_4? $S°(Sb, s) =$ 45.69 J/(K·mol).

16.41 Does potassium chlorate have a thermodynamic tendency to form potassium perchlorate and potassium chloride at 25°C? Might it do so at a higher temperature? *Hint:* Refer to Exercise 16.31c.

16.42 Does methanol have a thermodynamic tendency to decompose into carbon monoxide and water at 25°C and 1 atm? Is the tendency stronger at a higher temperature?

16.43 Determine whether iron(III) oxide can be reduced by carbon at 1000 K in the reactions (a) $Fe_2O_3(s) + 3C(s) \rightarrow 2Fe(s) + 3CO(g)$ and (b) $2Fe_2O_3(s) + 3C(s) \rightarrow 4Fe(s) + 3CO_2(g)$, given that, at 1000 K, $\Delta G_f°(CO,g) = -200$ kJ/mol, $\Delta G_f°(CO_2,g) = -396$ kJ/mol, and $\Delta G_f°(Fe_2O_3,s) = -562$ kJ/mol.

16.44 Determine whether manganese(IV) oxide can be reduced by carbon at 1000 K in the reactions (a) $MnO_2(s) + 2C(s) \rightarrow Mn(s) + 2CO(g)$ and (b) $MnO_2(s) + C(s) \rightarrow Mn(s) + CO_2(g)$, given that, at 1000 K, $\Delta G_f°(CO,g) = -200$ kJ/mol, $\Delta G_f°(CO_2,g) = -396$ kJ/mol, and $\Delta G_f°(MnO_2,s) = -405$ kJ/mol.

16.45 Calculate the standard reaction free energy of the water autoionization $2H_2O(l) \rightarrow H_3O^+(aq) + OH^-(aq)$ at 25°C.

16.46 Do $Cu^+(aq)$ ions have a thermodynamic tendency to form $Cu^{2+}(aq)$ ions and copper metal in water at 25°C?

Free Energy and Equilibrium

16.47 Calculate the equilibrium constant at 25°C for each of the following reactions. (It may be helpful to note that the standard free energy of each reaction was determined in Exercise 16.31.) (a) The oxidation of carbon monoxide, $2CO(g) + O_2(g) \rightarrow 2CO_2(g)$; (b) the decomposition of limestone, $CaCO_3(s) \rightarrow CaO(s) + CO_2(g)$; (c) the decomposition of potassium chlorate, $4KClO_3(s) \rightarrow 3KClO_4(s) + KCl(s)$.

16.48 Calculate the equilibrium constant at 25°C for each reaction. (It may be helpful to note that the standard free energy of each reaction was determined in Exercise 16.32.) (a) The synthesis of carbon disulfide from natural gas (methane), $2CH_4(g) + S_8(s) \rightarrow 2CS_2(l) + 4H_2S(g)$; (b) the production of acetylene from calcium carbide, $CaC_2(s) + H_2O(l) \rightarrow Ca(OH)_2(s) + C_2H_2(g)$; (c) the industrial synthesis of urea, a common fertilizer, $CO_2(g) + 2NH_3(g) \rightarrow CO(NH_2)_2(s) + H_2O(l)$.

16.49 If $Q_P = 1.0$ for the reaction $N_2O_4(g) \rightleftharpoons 2NO_2(g)$ at 25°C, will the reaction have a tendency to form products or reactants, or will it be at equilibrium?

16.50 If $Q_P = 1.0 \times 10^{50}$ for the reaction $C(s) + O_2(g) \rightleftharpoons CO_2(g)$ at 25°C, will the reaction have a tendency to form products or reactants, or will it be at equilibrium?

16.51 Calculate the standard free energy of each of the following reactions:
(a) $N_2(g) + 3H_2(g) \rightleftharpoons 2NH_3(g)$ $\quad K_P = 41$ at 400 K
(b) $2SO_2(g) + O_2(g) \rightleftharpoons 2SO_3(g) K_P = 3.0 \times 10^4$ at 700 K

16.52 Calculate the standard free energy for each of the following reactions:
(a) $H_2(g) + I_2(g) \rightleftharpoons 2HI(g)$ $\quad K_P = 160$ at 500 K
(b) $N_2O_4(g) \rightleftharpoons 2NO_2(g)$ $\quad K_P = 47.9$ at 400 K

16.53 Calculate the reaction free energy of $N_2(g) + 3H_2(g) \rightleftharpoons 2NH_3(g)$ when the partial pressures of N_2, H_2, and NH_3 are 1.0 atm, 4.2 atm, and 63 atm, respectively, and the temperature is 400 K. For this reaction $K_P = 41$ at 400 K. Interpret the meaning of your calculations.

16.54 Calculate the reaction free energy of $H_2(g) + I_2(g) \rightleftharpoons 2HI(g)$ when the concentrations are 0.026 mol H_2/L, 0.33 mol I_2/L, and 1.84 mol HI/L and the temperature is 700 K. For this reaction, $K_c = 54$ at 700 K. Interpret the meaning of your calculations.

16.55 Use data from Appendix 2A to calculate the K_s for each of the following compounds: (a) AgI; (b) $CaCO_3$, given that $\Delta G_f^\circ(CO_3^{2-}, aq) = -527.9$ kJ/mol; (c) FeS_2, given that $\Delta G_f^\circ(S_2^{2-}, aq) = +79.50$ kJ/mol.

16.56 Use data from Appendix 2A to calculate the K_s for each of the following compounds: (a) AgCl; (b) $PbBr_2$, given that $\Delta G_f^\circ(Pb^{2+}, aq) = -24.39$ kJ/mol; (c) $MgCO_3$, given that $\Delta G_f^\circ(CO_3^{2-}, aq) = -527.9$ kJ/mol.

UNCLASSIFIED EXERCISES

16.57 The internal energy of a system increased by 400 J when it absorbed 600 J of heat. (a) Was work done by or on the system? (b) How much work was done?

16.58 (a) Calculate the work that must be done at 25°C and standard pressure against the atmosphere for the expansion of the gaseous products in the combustion $2C_6H_6(l) + 15O_2(g) \rightarrow 12CO_2(g) + 6H_2O(g)$. (b) Using data from Appendix 2A, calculate the standard enthalpy of the reaction. (c) Hence, calculate the change in internal energy, ΔU, of the system.

16.59 List the following in order of increasing standard molar entropy at 25°C: $H_2O(l)$, $H_2O(g)$, $H_2O(s)$, $NaCl(s)$.

16.60 The standard molar entropy of $Cl(g)$ is 165.3 J/(K·mol) and that of $Cl_2(g)$ is 233.07 J/(K·mol). Suggest a reason why the standard molar entropy of the diatomic molecules is greater than that of the atoms at 25°C.

16.61 Potassium nitrate readily dissolves in water and its enthalpy of solution is +34.9 kJ/mol. (a) How does the enthalpy of solution favor the dissolving process? (b) Is the entropy change of the system positive or negative when the salt dissolves? (c) What is the driving force for the dissolution of KNO_3?

16.62 Without attempting to perform any calculations, state whether an increase or decrease in entropy occurs for each of the following processes:
(a) $Cl_2(g) + H_2O(l) \rightarrow HCl(aq) + HClO(aq)$
(b) $Cu_3(PO_4)_2(s) \rightarrow 3Cu^{2+}(aq) + 2PO_4^{3-}(aq)$
(c) $SO_2(g) + Br_2(g) + 2H_2(g) \rightarrow H_2SO_4(aq) + 2HBr(g)$

16.63 Use data from Table 16.1 and Appendix 2A to calculate the standard reaction entropy for each of the following reactions and interpret the answer: (a) the preparation of hydrofluoric acid from fluorine and water, $2F_2(g) + 2H_2O(l) \rightarrow 4HF(aq) + O_2(g)$; (b) the production of "synthesis gas," a low-grade industrial fuel, $CH_4(g) + H_2O(g) \rightarrow CO(g) + 3H_2(g)$; (c) the thermal decomposition of ammonium nitrate at 170°C, $NH_4NO_3(s) \rightarrow N_2O(g) + 2H_2O(g)$.

16.64 Determine whether titanium dioxide can be reduced by carbon at 1000 K in the reactions (a) $TiO_2(s) + 2C(s) \rightarrow Ti(s) + 2CO(g)$ and (b) $TiO_2(s) + C(s) \rightarrow Ti(s) + CO_2(g)$, given that, at 1000 K, $\Delta G_f^\circ(CO,g) = -200$ kJ/mol, $\Delta G_f^\circ(CO_2,g) = -396$ kJ/mol, and $\Delta G_f^\circ(TiO_2,s) = -762$ kJ/mol.

16.65 Is the reaction $4Fe_3O_4(s) + O_2(g) \rightarrow 6Fe_2O_3(s)$ spontaneous at 25°C?

16.66 For the reaction $PCl_3(g) + Cl_2(g) \rightarrow PCl_5(g)$, $\Delta G^\circ = -37.2$ kJ and $\Delta H^\circ = -87.9$ kJ. (a) Calculate ΔS°

CASE 16 ▸ COLD AND HOT PACKS

A twisted ankle or a wrenched knee requires immediate first aid. Ice should be applied to shrink the broken blood vessels around the sprain and minimize any internal bleeding. Ice, however, is often not immediately available, is not easy to store, and is not very portable. Instead, athletic trainers generally use a cold pack, which consists of a divided plastic bag containing a white solid separated from water that has been dyed blue. The white solid is ammonium nitrate, NH_4NO_3, which has a positive enthalpy of solution. When the cold pack is needed, the partition between the ammonium nitrate and water is broken by squeezing the bag; this mixing causes some of the ammonium nitrate to dissolve. In the endothermic dissolution

$$NH_4NO_3(s) \xrightarrow{H_2O} NH_4^+(aq) + NO_3^-(aq) \qquad \Delta H = +26 \text{ kJ}$$

energy is absorbed from the surroundings (the region of the sprained ankle), and they are cooled.

The blue dye in the water enables the user to determine whether the partition has been broken. The blue color also gives the aesthetic effect of coolness; a red dye would not be as effective psychologically. The process of dissolution is spontaneous even though it is strongly endothermic because the disorder of the solution increases more than the disorder of the surroundings (the ankle) decreases. The large increase in entropy of the solution can be traced to the strongly disruptive effect of the NH_4^+ and NO_3^- ions on the arrangement of H_2O molecules of the pure solvent.

Hikers camping at high altitudes or during the autumn and winter months often carry hot packs among their supplies. Two designs of hot packs are known, one containing a solid salt and another containing a supersaturated solution. The solid-salt hot packs are similar to cold packs. The salts, which have negative enthalpies of solution, are generally either anhydrous calcium chloride or magnesium sulfate:

$$CaCl_2(s) \xrightarrow{H_2O} Ca^{2+}(aq) + 2Cl^-(aq) \qquad \Delta H = -81 \text{ kJ}$$

$$MgSO_4(s) \xrightarrow{H_2O} Mg^{2+}(aq) + SO_4^{2-}(aq) \qquad \Delta H = -91 \text{ kJ}$$

Both salts dissolve spontaneously because the dissolution is accompanied by a large increase in the disorder of the surroundings as heat escapes into them; there is also an increase in the disorder of the solution as it forms. Other salts dissolve more endothermically than ammonium nitrate and more exothermically than calcium chloride (see Table 11.4) and could also be used in cold and hot

for the reaction at 25°C. (b) Does the sign of the entropy change agree with your expectations?

16.67 For the reaction, $2NO(g) + O_2(g) \rightleftharpoons 2NO_2(g)$, $\Delta H° = -114.1$ kJ and $\Delta S° = -146.54$ J/K. (a) What is the standard reaction free energy at 25°C? (b) What is the reaction free energy at 700°C? (Assume that $\Delta H°$ and $\Delta S°$ are unaffected by temperature changes.) (c) Temperatures in an internal combustion cylinder approach 2500°C. In the presence of an ample supply of oxygen, which nitrogen oxide is favored at that temperature? (d) Which nitrogen oxide is favored at the temperature of the exhaust gases, approximately 50°C?

16.68 (a) Calculate $\Delta H°$ and $\Delta S°$ at 25°C for the reaction $C_2H_2(g) + 2H_2(g) \rightarrow C_2H_6(g)$. (b) Calculate standard free energy for the reaction from the equation $\Delta G° = \Delta H° - T\Delta S°$. (c) Interpret the calculated values for $\Delta H°$ and $\Delta S°$. (d) Determine the temperature at which $\Delta G° = 0$. What is the significance of that value? (Assume that $\Delta H°$ and $\Delta S°$ are unaffected by temperature changes.)

16.69 (a) Use values from Appendix 2A to calculate $\Delta H°$ and $\Delta G°$ at 25°C for the combustion of methane: $CH_4(g) + 2O_2(g) \rightarrow CO_2(g) + 2H_2O(g)$. (b) Calculate standard reaction entropy from the equation $\Delta G° = \Delta H° - T\Delta S°$. (c) Interpret the calculated values of $\Delta H°$ and $\Delta S°$.

16.70 For the phase transition $M(s_1) \rightarrow M(s_2)$ of a metal M from solid(1) to solid(2), $\Delta H° = -14.07$ J and $\Delta S° = -0.0480$ J/K at -84°C. (a) What is the value of $\Delta G°$ for the transition? (b) At what temperature does $\Delta G° = 0$ and what is the significance of that value? (c) Which phase is favored at 0°C?

16.71 For the reaction $CO(g) + Cl_2(g) \rightleftharpoons COCl(g) + Cl(g)$, $K_P = 9.1 \times 10^{-30}$ at 25°C. Calculate the standard free energy of the reaction.

16.72 Calculate the equilibrium constant at 25°C for each of the following reactions (the standard free energy of each reaction was determined in Exercise 16.36).

packs, but these salts are selected because they are not toxic and will not harm the user if the packs break accidentally.

A supersaturated solution of sodium thiosulfate, $Na_2S_2O_3$, is also used in hot packs. When crystals of solid sodium thiosulfate are added to the supersaturated solution, immediate crystallization occurs, thereby releasing energy as heat. The addition of the seed crystals provides nucleation centers that enable the spontaneous crystallization to occur. This process occurs in two steps. First, the sodium and thiosulfate ions form anhydrous sodium thiosulfate,

$$2Na^+(aq) + S_2O_3^{2-}(aq) \longrightarrow Na_2S_2O_3(s) \qquad \Delta H = +8 \text{ kJ}$$

and then the solid sodium thiosulfate becomes hydrated:

$$Na_2S_2O_3(s) + 5H_2O(l) \longrightarrow Na_2S_2O_3{\cdot}5H_2O(s) \qquad \Delta H = -56 \text{ kJ}$$

The overall crystallization process is exothermic by 48 kJ.

QUESTIONS

1. (a) What mass of ammonium nitrate added to 100 g of water will result in a temperature change of 10°C? (b) What mass of calcium chloride added to 100 g of water will result in a temperature change of 10°C? (In each case, assume that the specific heat capacity of the solution is the same as that of water.)

2. (a) Determine the change in enthalpy when 20.0 g of NH_4NO_3 (about one-tenth the mass in a cold pack) dissolves in water. (b) If this heat is removed from 200 g of water at 37°C (body temperature), what would the final temperature of the water be? The specific heat capacity of water is 4.184 J/(g·°C).

3. (a) Determine the change in enthalpy when 20.0 g of $CaCl_2$ (about one-tenth the mass in a hot pack) dissolves in water. (b) If this heat is transferred to 200 g of water at 20°C, what would the final temperature of the water be?

4. (a) Determine the mass of sodium thiosulfate pentahydrate that must form from crystallization to produce the same energy as the dissolution of 20.0 g $CaCl_2$ (see Question 3a). (b) Explain how solid sodium thiosulfate pentahydrate could be used as a cold pack.

5. The enthalpy of solution of sodium acetate trihydrate, $NaCH_3CO_2{\cdot}3H_2O$ is +19.7 kJ/mol. Because the enthalpy of solution is opposite in sign to the enthalpy of crystallization, what mass of sodium acetate trihydrate must crystallize to produce the same energy as 20.0 g of $CaCl_2$?

(a) $H_2(g) + D_2O(l) \rightarrow D_2(g) + H_2O(l)$
(b) $2NO_2(g) \rightarrow N_2O_4(g)$
(c) $2CH_3OH(g) + 3O_2(g) \rightarrow 2CO_2(g) + 4H_2O(l)$

16.73 For the reaction $H_2O(l) \rightleftharpoons H^+(aq) + OH^-(aq)$, calculate (a) the standard free energy of the reaction at 25°C and (b) the value of K_w.

16.74 Assuming that $\Delta H°$ and $\Delta S°$ are unaffected by temperature changes, use data from Appendix 2A to estimate (a) the normal boiling point of $PCl_3(l)$; (b) the transition temperature of the phase change S(*s*, rhombic) → S(*s*, monoclinic); (c) the normal boiling point of heavy water, D_2O.

16.75 The depletion of the ozone in the stratosphere can be summarized by the net equation, $2O_3(g) \rightarrow 3O_2(g)$. (a) Use data from Appendix 2A to determine the standard free energy and the entropy change of the reaction. (b) What is the equilibrium constant of the reaction? (c) What is the significance of your answers with regard to ozone depletion?

16.76 Determine the equilibrium constant of the reaction $SbCl_5(g) \rightleftharpoons SbCl_3(g) + Cl_2(g)$ at 25°C.

16.77 Calculate the entropy change for (a) the vaporization of 1.0 g of water at its normal boiling point and (b) the melting of 1.0 g of ice at its normal melting point. (c) Why is the entropy change for the vaporization so much larger?

16.78 The enthalpy of vaporization of heavy water is 41.6 kJ/mol at its normal boiling point of 101.4°C. (a) Calculate the entropy change of the surroundings when 10.0 g of heavy water vaporizes at its normal boiling point. (b) Use data from Appendix 2A to determine the vapor pressure of heavy water at 25°C. *Hint:* For $D_2O(l) \rightleftharpoons D_2O(g)$, $K_P = P_{D_2O}$.

16.79 Calculate the reaction free energy of $I_2(g) \rightarrow 2I(g)$ at 1200 K ($K_c = 6.8 \times 10^{-2}$) when the concentrations of I_2 and I are 0.026 mol/L and 0.0084 mol/L, respectively. What is the spontaneous direction of the reaction?

This chapter explains how chemical reactions are used to generate electric currents and how electric currents are used to bring about chemical reactions. It also extends the range of reactions that we can discuss in terms of equilibrium constants, and explains how a scale of oxidizing and reducing strengths is established and used. The illustration shows the interior of one compartment of a lead-acid battery while charging is in progress. The cell stores energy when it is charged and releases it when it is discharged.

17 ELECTROCHEMISTRY

Redox reactions were first introduced in Chapter 3. Our purpose in returning to them now is to examine them with the powerful techniques we have been developing since that early introduction. As reminders of the widespread occurrence and importance of these reactions, Tables 17.1 and 17.2 list some of the redox reactions used to produce elements and compounds from natural sources. Redox reactions are also used in the chemical generation and storage of electric power and in the reverse of these processes, the use of electricity to bring about chemical change. The latter application—electrolysis—was a type of chemical change we met in Chapter 1, so our journey has now come full circle. The branch of chemistry that deals with the use of chemical reactions to produce electricity, the relative strengths of oxidizing and reducing agents, and the use of electricity to produce chemical change is called **electrochemistry**.

The language of redox reactions was introduced in Section 3.8. We learned there that oxidation is electron loss and results in an increase in oxidation number; for example, zinc metal loses two electrons as it is oxidized to Zn^{2+}. The half-reaction for this oxidation is

$$\textit{Oxidation}: Zn(s) \longrightarrow Zn^{2+}(aq) + 2e^-$$

Reduction is electron gain and results in a decrease in oxidation number; for example, Cu^{2+} gains two electrons as it is reduced to copper metal. The half-reaction for this reaction is

$$\textit{Reduction}: Cu^{2+}(aq) + 2e^- \longrightarrow Cu(s)$$

The sum of an oxidation half-reaction and a reduction half-reaction is the overall redox reaction:

$$Cu^{2+}(aq) + Zn(s) \longrightarrow Zn^{2+}(aq) + Cu(s)$$

In this reaction, Cu^{2+} oxidizes the zinc metal (to Zn^{2+}) and is therefore an oxidizing agent. Similarly, zinc metal reduces the Cu^{2+} ions (to Cu) and is therefore the reducing agent.

N A review of the basic principles, Sections 3.8 and 3.10, may be appropriate at this time.

N Goals in teaching electrochemistry. *J. Chem. Ed.*, **1985**, *62*, 1018. Electrochemical errors. *J. Chem. Ed.*, **1985**, *62*, 424.

TABLE 17.1 Elements prepared by reduction

Element	Source	Process
*Easy**		
H_2	H_2O	synthesis gas reaction: $CH_4(g) + H_2O(g) \xrightarrow{800°C,\ Ni} CO(g) + 3H_2(g)$
		shift reaction: $CO(g) + H_2O(g) \xrightarrow{400°C,\ Fe/Cu} CO_2(g) + H_2(g)$
Cu	CuS	copper smelting: $CuS(s) + O_2(g) \xrightarrow{\Delta} Cu(s) + SO_2(g)$
		hydrometallurgy: $Cu^{2+}(aq) + H_2(g) \longrightarrow Cu(s) + 2H^+(aq)$
Moderately difficult		
P_4	PO_4^{3-}	heat with carbon and sand in an electric furnace:
		$2Ca_3(PO_4)_2(l) + 6SiO_2(l) + 10C(s) \xrightarrow{1500°C} P_4(g) + 6CaSiO_3(l) + 10CO(g)$
Fe	Fe_2O_3	blast furnace: $Fe_2O_3(s) + 3CO(g) \xrightarrow{900°} 2Fe(l) + 3CO_2(g)$
Difficult		
Na	NaCl	Downs process: $2NaCl(l) \xrightarrow{\text{electrolysis at } 600°C} 2Na(l) + Cl_2(g)$
K	KCl	reduction with sodium vapor: $KCl(l) + Na(g) \xrightarrow{700°C} K(l) + NaCl(s)$
Si	SiO_2	reduction in electric furnace: $SiO_2(l) + 2C(s) \xrightarrow{1500°C} Si(l) + 2CO(g)$
Al	Al_2O_3	Hall process: $2Al_2O_3(\text{in cryolite}) + 3C(s) \xrightarrow{\text{electrolysis at } 900°C} 4Al(l) + 3CO_2(g)$
Ti	$TiCl_4$	Kroll process: $TiCl_4(g) + 2Mg(l) \xrightarrow{1000°C} Ti(s) + 2MgCl_2(l)$

*The methods are designated easy, moderate, and difficult according to the strength of the reducing agent required. More difficult reductions require stronger agents.

TABLE 17.2 Elements prepared by oxidation

Element	Source	Process
*Easy**		
S	H_2S	Claus process: $2H_2S(g) + 3O_2(g) \longrightarrow 2SO_2(g) + 2H_2O(g)$
		$2H_2S(g) + SO_2(g) \xrightarrow{300°C,\ Fe_2O_3} 3S(g) + 2H_2O(g)$
Moderately difficult		
Cl_2	NaCl	Downs process: $2NaCl(l) \xrightarrow{\text{electrolysis at } 600°C,} 2Na(l) + Cl_2(g)$
Br_2, I_2	Br^-, I^- in brine	oxidation and degassing: $Cl_2(g) + 2Br^-(aq) \longrightarrow 2Cl^-(aq) + Br_2(aq)$
Difficult		
F_2	F^-	Moisson's method: $HF(\text{with dissolved KF}) \xrightarrow{\text{electrolysis near } 100°C} F_2(g) + H_2(g)$
Au	Au	cyanide process:
		$4Au(s, \text{impure}) + 8CN^-(aq) + O_2(g) + 2H_2O(l) \longrightarrow 4Au(CN)_2^-(aq) + 4OH^-(aq)$
		$2Au(CN)_2^-(aq) + Zn(s) \longrightarrow 2Au(s, \text{pure}) + Zn(CN)_4^{2-}(aq)$

*The methods are designated easy, moderate, and difficult according to the strength of the oxidizing agent required. More difficult oxidations require stronger agents.

ELECTROCHEMICAL CELLS

An **electrochemical cell** consists of two metallic conductors called **electrodes** in contact with an electrically conducting medium called an **electrolyte**. There are two types of electrochemical cells. A **galvanic cell** is an electrochemical cell in which a spontaneous chemical reaction is used to generate an electric current. (It was named after the Italian physiologist Luigi Galvani, who first observed the effect of electricity on live muscle.) An **electrolytic cell** is an electrochemical cell in which an electric current is used to bring about an otherwise nonspontaneous chemical reaction. We shall look closely at galvanic cells first: their everyday versions include flashlight batteries, the lead-acid batteries in automobiles, and the lithium cells in heart pacemakers. Later we shall look at electrolytic cells that are used in the commercial production of aluminum and the refining of copper metal.

When an ion or molecule that comes into contact with an electrode is oxidized, it releases electrons into the electrode at that electrode-solution interface. These electrons then travel through an external circuit to the other electrode, where another species of ion or molecule is reduced; that species accepts electrons from the electrode at the electrode-solution interface. The electrode at which oxidation occurs is called the **anode** and the electrode at which reduction occurs is called the **cathode**.

The electrodes and the external metal wire that joins them conduct current by **metallic conduction**, the movement of electrons. Electrical neutrality is maintained in the external wire—for every electron that enters the wire at one electrode (the anode), another electron leaves the wire at the other electrode (the cathode). The electrodes may be either chemically inert (typically platinum or graphite) or actively involved in the reactions at the electrode-solution interface.

The solution conducts current by the process of **ionic conduction**, the movement of cations and anions from one electrode to another. Cations travel toward the cathode and anions travel toward the anode. If a cation forms as a result of electron loss to the anode, then either a negative ion migrates toward the anode or a positive ion migrates away from the anode and toward the cathode. Similarly if a negative ion (or a less positive ion) forms as a result of electron gain from the cathode, then either a positive ion migrates toward or a negative ion moves away from the cathode. Therefore, in all electrochemical cells

1. Electrons flow from the anode to the cathode in the external circuit.
2. Oxidation occurs at the anode, and anions flow toward the anode within the cell.
3. Reduction occurs at the cathode, and cations flow toward the cathode within the cell.

N An effective approach to teaching electrochemistry. *J. Chem. Ed.*, **1990**, *67*, 403.

D A homemade lemon battery. *J. Chem. Ed.*, **1988**, *65*, 158. Electricity generated. TDC, 7-5.

E A metallic conductor need not be a metal: graphite, for example, is a metallic conductor, and is used as an electrode material. The electrolyte need not be a liquid, and currently considerable effort is going into the development of solid electrolytes.

E An electric shock occurs because the body is an ionically conducting medium, and ions in the body conduct the current from the wire to the ground.

V Video 41: Electrical conductivity in solutions.

D Shape a piece of copper wire in the form of a Christmas tree and place it into a silver nitrate solution.

E Thermodynamic and kinetic properties of the electrochemical cell. *J. Chem. Ed.*, **1983**, *60*, 299.

17.1 GALVANIC CELLS

When we place a piece of zinc in an aqueous copper(II) sulfate solution, some of the zinc reacts. As it does so, metallic copper is deposited on the surface of the zinc (Fig. 17.1). This observation shows that the redox reaction

$$Zn(s) + Cu^{2+}(aq) \longrightarrow Zn^{2+}(aq) + Cu(s)$$

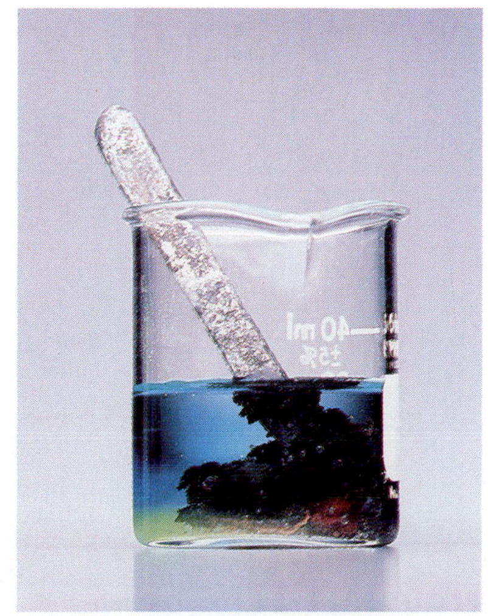

FIGURE 17.1 When a strip of zinc is placed in a beaker of copper(II) sulfate solution, copper is deposited on the zinc and the blue copper(II) ions are gradually replaced by colorless zinc ions.

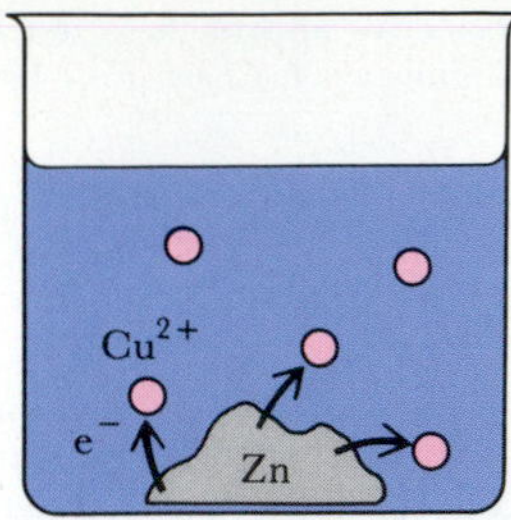

FIGURE 17.2 The reaction shown in Fig. 17.1 takes place at random all over the surface of the zinc as electrons (represented by arrows) are transferred to the Cu^{2+} ions in the solution.

E A generating plant uses a *physical* procedure (the rotation of a magnet or of coils inside a magnet) to produce the uniform flow of electrons; a galvanic cell utilizes a *chemical* procedure.

V Video 43: The Daniell cell.

E The original Daniell cell consisted of a concentrated $CuSO_4$ solution (having a relatively high density) at the bottom of a jar overlaid with a less dense, dilute $ZnSO_4$ solution. The differences in densities of the two solutions was enough to inhibit ion flow directly to the electrodes and therefore the interface served as the porous barrier.

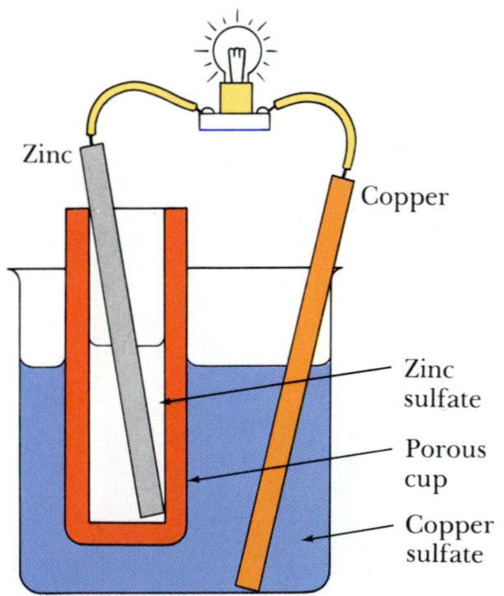

FIGURE 17.3 The Daniell cell consists of copper and zinc electrodes dipping into solutions of copper(II) sulfate and zinc sulfate. The two solutions make contact through the porous cup, which allows ions to pass through to complete the electric circuit.

is spontaneous in the direction of the arrow. As the reaction takes place, electrons are transferred from the zinc to Cu^{2+} ions nearby in the solution (Fig. 17.2). The electrons reduce the Cu^{2+} ions to Cu atoms, which stick to the surface of the zinc or form a finely divided solid deposit in the flask. The piece of zinc slowly dissolves as its atoms give up electrons and form Zn^{2+} ions. The transfer of electrons from the zinc to the copper ions occurs at random all over the surface of the metal and the reaction enthalpy is released as heat.

However, suppose the zinc is separated from the copper solution, as in Fig. 17.3. The same redox reaction takes place, but now the electrons can reach the Cu^{2+} ions only by traveling through the external circuit (the wire and the light bulb). As Cu^{2+} ions are converted to neutral Cu atoms in one compartment and Zn atoms are converted to Zn^{2+} ions in the other, ions—sulfate ions if the solution is copper(II) sulfate—must move between the two compartments (through the porous cup) to preserve electrical neutrality and complete the electric circuit. The reaction therefore produces an orderly flow of electrons—an electric current—through the external wire. This process is the essence of all electrical power generation by chemical reactions. Whenever we use a battery-driven device, we make use of a reaction in which reduction and oxidation are taking place in separate regions and electrons flow between them through an external circuit.

The apparatus in Fig. 17.3 is called a **Daniell cell** for the British chemist John Daniell. He invented the cell in 1836 when, in response to the growth in telegraphy, there was an urgent need for a steady source of electric current. A Daniell cell is a specific case of a galvanic cell. In general, a galvanic cell consists of a container fitted with two electrodes and filled with an electrolyte. The electrodes are usually metal, but graphite is sometimes used. The electrolyte is usually an aqueous solution of ions, but it can be a molten salt or even a solid. The two electrodes occasionally share the same electrolyte in some cells, but most often they are placed in different electrolytes, each in its own compartment. The compartments are joined by a **salt bridge**, a bridge-shaped tube containing a concentrated salt (potassium chloride or nitrate) in a jelly that acts as an electrolyte and allows ions to travel between the two compartments and complete the circuit (Fig. 17.4). In practical cells, such as those sold for use in the home, contact between the two solutions is often made through a porous membrane that allows ions to pass between the compartments. A battery is really a collection of cells connected together, but the word is often used to mean a single cell.

As we have seen, the source of electrons in the Daniell cell is the oxidation of zinc in the anode compartment:

$$\textit{Anode reaction: } Zn(s) \longrightarrow Zn^{2+}(aq) + 2e^-$$

and the electrons are used in the cathode compartment for the reduction of Cu^{2+}:

$$\textit{Cathode reaction: } Cu^{2+}(aq) + 2e^- \longrightarrow Cu(s)$$

Because the electrons leave the cell at the anode, it is defined as the negative (−) terminal of the galvanic cell. The electrons enter the cell at the cathode after passing through the external circuit, so that electrode is defined as the positive (+) terminal. This sign convention is also used on batteries (cells) sold in stores. A complete but schematic diagram of the Daniell cell is shown in Fig. 17.5.

Notation of redox couples. The oxidized and reduced states of each substance taking part in a half-reaction (and thus present in an electrode compartment of a galvanic cell) form a **redox couple**. A single half-reaction is conventionally written as a reduction

Redox couple half-reaction: Oxidized state + ne^- $\longrightarrow$ reduced state

and denoted by the convention

Redox couple notation: Oxidized species/reduced species

For the Daniell cell, the two redox half-reactions and redox couples are written as shown in the margin below right.

The H^+/H_2 redox couple plays a special role in electrochemistry, as we shall see. In this couple, H^+ (more precisely, H_3O^+) is the oxidized species and H_2 is the reduced species. When this couple serves as the reduction (cathodic) reaction, H^+ is reduced:

$$2H^+(aq) + 2e^- \longrightarrow H_2(g) \qquad H^+/H_2 \text{ couple undergoing reduction}$$

However, when the same couple serves as the oxidation (anodic) reaction, H_2 is oxidized:

$$H_2(g) \longrightarrow 2H^+(aq) + 2e^- \qquad H^+/H_2 \text{ couple undergoing oxidation}$$

Whether a couple is oxidized or reduced depends on the other redox couple that makes up the galvanic cell, as we shall see.

To use the H^+/H_2 redox couple in a galvanic cell, a chemically inert metal is needed to accept or remove electrons from the compartment. Platinum is customarily used for the electrode, and hydrogen gas is bubbled over it as it dips into a solution that contains hydrogen ions. This combination of platinum metal and the H^+/H_2 redox couple is commonly called the **hydrogen electrode**. An inert metal must always be included to make electrical contact with the electrolyte if the redox couple does not include a solid metal. Another example of a redox couple of this kind is an Fe^{3+}/Fe^{2+} couple in aqueous solution.

Notation for electrodes. When describing an electrode in an electrochemical cell, chemists use a short vertical line (|) to represent the junc-

FIGURE 17.4 A typical electrochemical cell used in the laboratory. The two electrodes are connected by an external circuit and a salt bridge, which completes the circuit within the cell.

E Platinum is a good material for a hydrogen electrode, because hydrogen chemisorbs (as atoms) to its surface, so facilitating the redox process.

D Set up the Daniell cell as described in the annotation on p. 656: lay the copper electrode in the $CuSO_4$ solution and connect it to a voltmeter (or light bulb) with a coated wire that passes through the $ZnSO_4$ solution. Be sure the Zn electrode does not come into contact with the $CuSO_4$ solution. Electron flow in redox reactions. TDC, 151.

Half-reaction	Redox couple
$Cu^{2+}(aq) + 2e^- \rightarrow Cu(s)$	Cu^{2+}/Cu
$Zn^{2+}(aq) + 2e^- \rightarrow Zn(s)$	Zn^{2+}/Zn

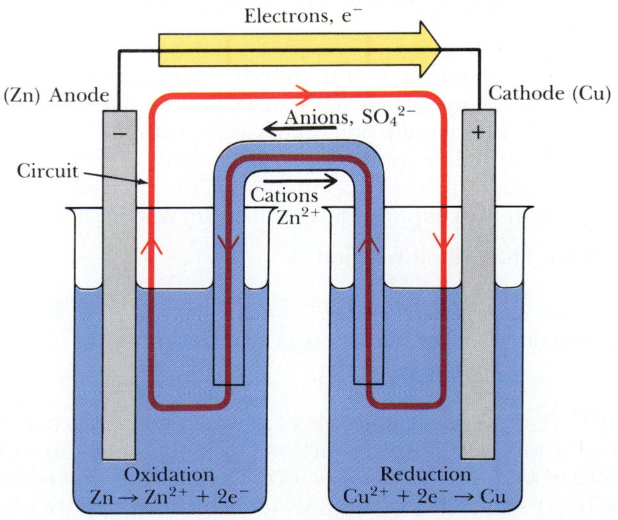

T OHT: Figs. 17.3 and 17.5.

D Do frozen solutions conduct electricity? CD2, 71.

N Another common inert electrode, besides platinum, is graphite, C(gr).

FIGURE 17.5 Electrons leave a galvanic cell at the anode (−), travel through the external circuit, and reenter the cell at the cathode (+). The source of the electrons is oxidation at the anode; the electrons passing through the external circuit cause reduction at the cathode.

tion between the metal and the solution. They use the same vertical line to denote the junction between any two phases (such as that between a gas and a metal or that between an undissolved solid and a metal). By convention, the metal is written first if it is the anode in the galvanic cell but written last if it is the cathode in the cell:

$$\textit{Anode}: \mathrm{Pt}|\mathrm{H_2}(g)|\mathrm{H^+}(aq) \qquad \textit{Cathode}: \mathrm{H^+}(aq)|\mathrm{H_2}(g)|\mathrm{Pt}$$

It is common practice to refer to the entire electrode compartment simply as the electrode, in this case, the hydrogen electrode.

If the oxidized and reduced states of the redox couple are in the same phase, then a comma is used to separate them in the electrode notation. For example, when the Fe^{3+}/Fe^{2+} redox couple undergoes reduction in a galvanic cell with platinum as the chemically inert cathode, the electrode notation is $Fe^{3+}(aq),Fe^{2+}(aq)|Pt$.

Cell diagrams. The arrangement of electrodes in a galvanic cell is summarized by a **cell diagram**, which is built up from the electrode notation just described; the cathode is placed on the right and the anode on the left. The cell diagram for a laboratory version of the Daniell cell is therefore

$$\underset{\text{Anode}}{\mathrm{Zn}(s)|\mathrm{Zn^{2+}}(aq)}||\underset{\text{Cathode}}{\mathrm{Cu^{2+}}(aq)|\mathrm{Cu}(s)}$$

The double vertical bars separating the anode compartment from the cathode compartment represent the salt bridge (or a porous cup like that in Fig. 17.3). We know that the Cu^{2+}/Cu redox couple undergoes reduction (at the cathode) and the Zn^{2+}/Zn redox couple undergoes oxidation (at the anode) because of our observations of the chemical changes taking place. Shortly we shall learn how to make an electrical measurement to identify the cathode, and later we shall see how to predict which electrode is the cathode and which the anode.

▼ EXAMPLE 17.1 Writing the cell diagram for a redox reaction

Write the diagram for a cell that consists of the H^+/H_2 and Fe^{3+}/Fe^{2+} redox couples and in which hydrogen, $H_2(g)$, reduces $Fe^{3+}(aq)$ ions to $Fe^{2+}(aq)$ ions.

STRATEGY We should write the overall reaction and then break it into oxidation and reduction half-reactions (see Section 3.8). We can then identify the electrode where reduction occurs as the cathode; oxidation occurs at the anode. If the redox couple does not include a solid metal, then we should include platinum in the specification of the electrode (to provide electrical contact with the couple). The cell diagram is written anode | | cathode.

SOLUTION The overall reaction is

$$2\mathrm{Fe^{3+}}(aq) + \mathrm{H_2}(g) \longrightarrow 2\mathrm{Fe^{2+}}(aq) + 2\mathrm{H^+}(aq)$$

The half-reaction for the reduction of iron(III) is

$$\mathrm{Fe^{3+}}(aq) + \mathrm{e^-} \longrightarrow \mathrm{Fe^{2+}}(aq)$$

and the Fe^{3+}/Fe^{2+} redox couple serves as the cathodic reaction. Because no solid metal is involved in the reduction, an inert contact must be present, and the electrode is $Fe^{3+}(aq),Fe^{2+}(aq)|Pt$. The half-reaction occurring at the anode is $H_2(g) \rightarrow 2H^+(aq) + 2e^-$. An inert metal is required for this electrode too, which is written $Pt|H_2(g)|H^+(aq)$. The overall cell diagram is

$$Pt|H_2(g)|H^+(aq)||Fe^{3+}(aq),Fe^{2+}(aq)|Pt$$

EXERCISE E17.1 Write the cell diagram for a redox reaction in which the $Cu^{2+}(aq)$ is reduced to $Cu^+(aq)$ by magnesium metal, forming $Mg^{2+}(aq)$ as the oxidized product.

[*Answer*: $Mg(s)|Mg^{2+}(aq)||Cu^{2+}(aq),Cu^+(aq)|Pt$]

N Electrodes are normally written in the form Metal|gas (if any)|solid (if any)|electrolyte. Two examples are $Pt|H_2(g)|H^+(aq)$ and $Ag|AgCl(s)|Cl^-(aq)$. The phase of the metal is not usually written. There is no established convention for the ordering of the species in the electrolyte, but for redox electrodes we adopt the order Metal|ox, red, as in $Pt|Fe^{3+}(aq)$, $Fe^{2+}(aq)$.

Cell potential. The energy released by a redox reaction as electrons spontaneously travel through the external wire can be used either to heat the surroundings if the circuit includes a heater or to do work (such as winding a cassette tape) if the circuit includes an electric motor. The energy available from a cell depends on the electron "pushing" and "pulling" power of the redox reaction. If the oxidation releases electrons readily and the reduction accepts them readily, then electrons will be pushed and pulled through the circuit vigorously and the reaction can provide a great deal of energy.

The pushing and pulling power of a cell reaction is measured as its **cell potential** E. (The cell potential is often called the electromotive force of the cell, or its voltage.) The greater the cell potential, the greater the energy that a given number of electrons can release as they travel between the electrodes. That is, a high cell potential signifies that the cell reaction has a strong tendency to generate a current of electrons.

The SI unit of potential is the **volt** (V). It is defined so that one joule of energy is released when one coulomb of charge* travels between two electrodes that differ in potential by one volt:

$$1 \text{ volt (V)} = \frac{1 \text{ joule (J)}}{1 \text{ coulomb (C)}}$$

(More briefly: $1 \text{ V} = 1 \text{ J/C}$.) Therefore, the energy released when a certain charge (which is determined by the number of electrons) travels between two electrodes having potentials measured in volts is

$$\text{Energy (J)} = \text{charge (C)} \times \text{cell potential (V)}$$

The cell potential is, by definition, a positive quantity. Typical cells used in the home are rated at about 1.5 V, a value meaning that when 1.0 C of charge travels from one electrode to another, the energy released is $1.0 \text{ C} \times 1.5 \text{ V} = 1.5 \text{ J}$. The potential of a Daniell cell is about 1.1 V.

Cell potentials are measured with electronic voltmeters (Fig. 17.6). These are designed to give a positive reading (in volts) when the + terminal of the meter is connected to the + terminal (the cathode) of the cell and the − terminal of the meter is connected to the − terminal of the cell (the anode). Therefore, we can determine *experimentally* which electrode of a galvanic cell is the cathode (the + electrode) by finding the connection that gives a positive reading; the electrode joined to the + terminal of the voltmeter is then the cathode and the other electrode is the anode.

N The + sign on the cathode of a galvanic cell marks the electrode through which electrons enter the cell; the − sign on the anode marks the point of exit from the cell.

In summary, a diagram of a galvanic cell is always written with the positive electrode (the cathode, the port of entry of the electrons into

*A charge of 1 C corresponds to about 6.2×10^{18} electrons, or 1.0×10^{-5} mol of electrons, because each electron has a charge of magnitude 1.602×10^{-19} C.

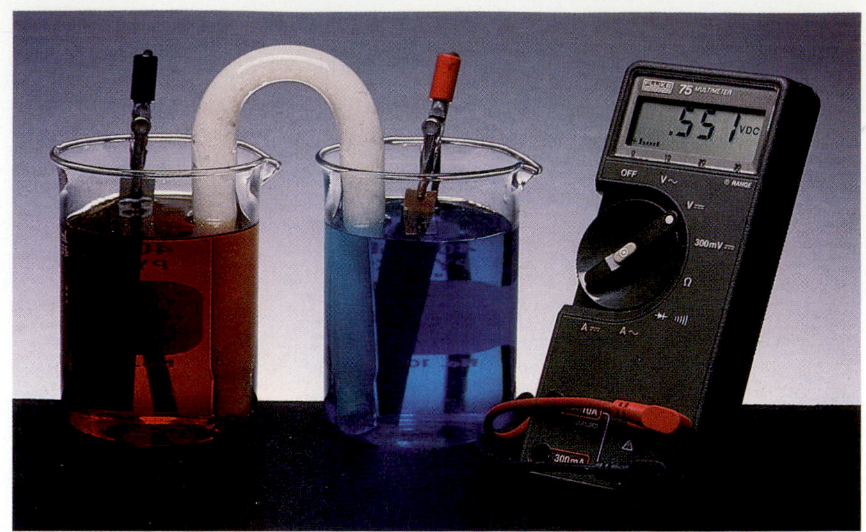

FIGURE 17.6 Cell potential is measured with an electronic voltmeter, a device that draws negligible current so that the composition of the cell does not change during the measurement. The display gives a positive potential when the + terminal of the meter is connected to the (positive) cathode.

the cell, the site of reduction) on the right. We then always know that electrons are tending to flow through the external circuit from the electrode on the left of the diagram to the electrode on the right (in the direction we normally read). We therefore also know that the spontaneous direction of the cell reaction is oxidation in the left-hand, or anode, compartment and reduction in the right-hand, or cathode, compartment (remember: *r* for right and reduction).

▼ EXAMPLE 17.2 Deducing a cell reaction from a cell diagram

The galvanic cell $Pt|H_2(g)|OH^-(aq)||OH^-(aq)|O_2(g)|Pt$, which produces about 1.2 V at 25°C, has been used to produce electric power on some space missions. What is the cell reaction? Refer to Section 3.10 to see how to balance half-reactions in basic media.

STRATEGY We should write the anode half-reaction as an oxidation of hydrogen to water and the cathode half-reaction as a reduction of oxygen to hydroxide ions. As explained in Section 3.10, each half-reaction should be balanced by using only H_2O, OH^- (the electrolyte is basic), and e^-. We can then add the two half-reactions (with electrons balanced) to obtain the overall reaction.

SOLUTION The anode half-reaction is

$$H_2(g) + 2OH^-(aq) \longrightarrow 2H_2O(l) + 2e^-$$

The cathode half-reaction is

$$O_2(g) + 2H_2O(l) + 4e^- \longrightarrow 4OH^-(aq)$$

To balance the electrons, we add two of the first half-reaction equations to one of the second and obtain

$$2H_2(g) + O_2(g) \longrightarrow 2H_2O(l)$$

Hence the cell reaction is the formation of water from hydrogen and oxygen.

EXERCISE E17.2 Identify the anode and cathode half-reactions and then write the reaction for the cell $Cd(s)|Cd^{2+}(aq)||H^+(aq)|O_2(g)|Pt$.

[*Answer*: $2Cd(s) + O_2(g) + 4H^+(aq) \rightarrow 2H_2O(l) + 2Cd^{2+}(aq)$] ▲

17.2 PRACTICAL CELLS

A **primary cell** is a galvanic cell that produces electricity from chemicals that are sealed into it when it is made. This type of cell cannot be recharged: once the cell reaction has reached equilibrium, the cell is discarded. A **secondary cell** is one that must be charged from some other electrical supply before it can be used; this type of cell is normally rechargeable (like an automobile battery). In the charging process, a nonequilibrium mixture of reactants is produced by an external source of electricity. When the cell is in use, it produces electricity as the reaction approaches equilibrium again. A **fuel cell** is a primary cell, but the reactants (the fuel) are supplied from outside while the cell is in use.

D The construction and use of commercial voltaic displays in freshman chemistry. *J. Chem. Ed.*, **1990**, *67*, 158. Electrode potentials—hydrogen and chlorine (construction of a secondary cell). TDC, 160.

Primary cells. The workhorse of primary cells is the **dry cell** (Fig. 17.7), which is widely used to power portable electric equipment. It is also called the Leclanché cell for Georges Leclanché, the French engineer who invented it in about 1866. The cell produces about 1.5 V initially; but this potential falls to about 0.8 V as reaction products accumulate inside it. If the cell is left unused for a day, then its potential may climb back to about 1.3 V as the products disperse throughout the electrolyte.

N The dry cell is not completely dry; it is slightly moist to allow ionic conduction to occur.

T OHT: Figs. 17.7 and 17.8.

E Because the zinc canister is not of uniform thickness and density, "hot" spots develop where the zinc metal preferentially oxidizes. As a result, a hole may form and the cell leak.

E Electrochemical reactions in batteries. Emphasizing the MnO_2 cathode of dry cells. *J. Chem. Ed.*, **1972**, *49*, 587.

About a billion "flashlight batteries" are produced annually in the United States. These dry cells utilize a zinc container lined on the inside with paper (for the porous barrier). The zinc metal functions as the anode for the cell and supplies electrons to the external circuit by the anode reaction:

$$Zn(s) \longrightarrow Zn^{2+}(aq) + 2e^-$$

The exterior of the container usually has a covering (with the company name on it) and is sealed at the top with wax. The container is filled with a moist paste that consists of ammonium chloride, manganese(IV) oxide, finely divided carbon granules, and an inert filler (usually starch).

The cathode is a carbon rod at the center of the container. The cathode reaction is a complicated reduction of manganese(IV) oxide, approximately

$$MnO_2(s) + H_2O(l) + e^- \longrightarrow MnO(OH)(s) + OH^-(aq)$$

The OH^- ions migrate toward the zinc anode and combine with the NH_4^+ ions, thereby forming NH_3 by the Brønsted acid-base proton transfer reaction

$$NH_4^+(aq) + OH^-(aq) \longrightarrow H_2O(l) + NH_3(aq)$$

As the Zn^{2+} ions migrate away from the anode, they combine with ammonia by means of a Lewis acid-base reaction to form the complex ion $Zn(NH_3)_4^{2+}$:

$$Zn^{2+}(aq) + 4NH_3(aq) \longrightarrow Zn(NH_3)_4^{2+}(aq)$$

Thus the Zn^{2+} ion concentration is kept low and diminishes the possibility that local equilibrium concentrations of ions will be established at the anode. At the end of the cell's life, so much $Zn(NH_3)_4^{2+}$ is present that its chloride salt crystallizes and reduces the efficiency of the ionic conduction of current. The salt can be encouraged to diffuse away from the anode by gently warming an exhausted cell for some hours; this treatment partially restores the cell's potential.

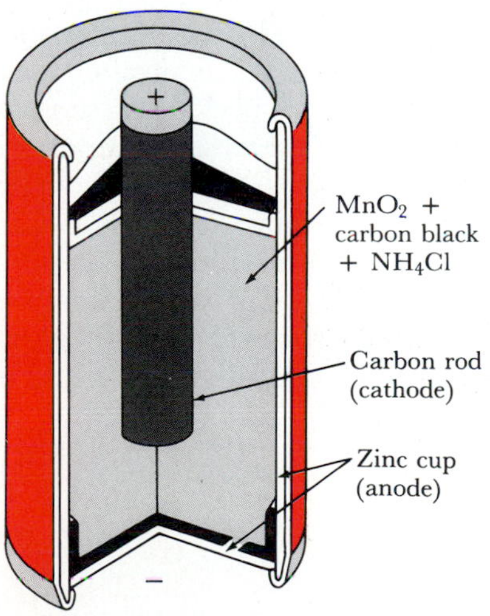

FIGURE 17.7 A commercial dry cell consists of a graphite cathode in a zinc container that acts as the anode. The other components and the cell reaction are described in the text.

BOX 17.1 ▸ PHOTOELECTROCHEMISTRY

MARY ARCHER, *Department of Chemistry, Imperial College, London*

Renewable energy sources have an indispensable part to play in our future energy economy. By definition, they cannot be depleted; and they do not give rise to hazardous by-products. Of the renewable energy sources—sun, wind, waves, tide, biomass and geothermal phenomena—the sun is by far the most important.

The sunlight striking the Earth in 1 h is equivalent to the energy obtained by burning 2×10^{13} kg of coal. Sunlight is readily turned into heat, but we can also use it to produce high-grade energy such as electricity and chemical fuels. The processes we use to bring about this transformation are called photoconversion methods, and they all rely on the absorption of solar photons by appropriate materials. Photosynthesis is a photoconversion method that plants use to generate energy for sugar synthesis.

Absorption of a photon of energy $h\nu$ excites the electron to a higher-energy upper band, leaving a "hole" in the lower band as shown in the drawing on the right. As the excited electron falls from the upper band back into the hole in the ground band, 70% of its stored energy can be captured as work (the rest is dissipated as heat) if it separates (route *a*) from its parent atom rather than directly recombining (route *b*) with the parent. Thus, photoconversion methods require materials that efficiently absorb photons from sunlight to produce long-lived excited states during which the separation can occur. Molecular compounds such as organic dyes and inorganic complex ions can be used as photoconversion materials, but semiconductors are currently the best photoconversion materials available because the lifetime of their hole-electron pairs is much longer than that of such pairs in electronically excited molecules or ions.

Photoconversion methods can produce electricity or fuels. For example, we use semiconductors in solid-state devices called photovoltaic cells, which convert sunlight into electric current. Some small hand-held calculators use photovoltaic cells rather than batteries as a source of

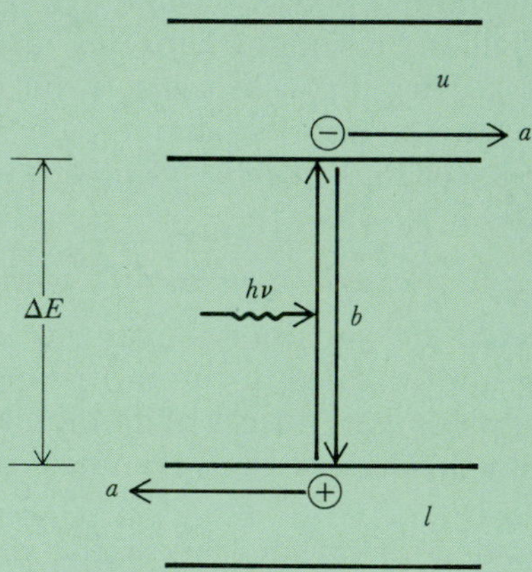

An excited electron, ⊖, occupying a high-energy orbital (*u*) returns to a hole in the ground state (*l*) by one of two paths. (a) It can be separated from its parent atom and pass through several intermediates before falling back to the ground state, or (b) it can fall directly back into the ground state in its parent atom, ⊕.

E Extended use of a dry cell causes its potential to decrease because of the buildup of reaction products at the electrodes. Letting the cell rest allows the ions to diffuse away from the centers of accumulation, so that when the cell is used later it seems to have acquired a new life.

Long-life, "high power" **alkaline cells** are similar to the Leclanché cell, but the ammonium chloride is replaced by sodium or potassium hydroxide. They have a longer life because the zinc is no longer exposed to the acid environment caused by the NH_4^+ ions in the conventional cell. Because the ions move more easily through the electrolyte, alkaline cells produce more power and a steadier current. Their higher cost arises largely from the cost of the extra sealing materials used to prevent hydroxide leakage. The two half-reactions in an alkaline cell are

power. The calculators are powered simply by exposure to light. Certain areas of California use electric power generated by acres of solar-thermal systems.

Electrochemical processes can also be driven by the electrons produced in photoconversion. When semiconductors are used as electrodes, the movement of excited holes or electrons across the electrode-electrolyte interface drives the electrochemical process. In 1972 the Japanese chemists Akira Fujishima and Ken-ichi Honda showed that water can be split by such a process to produce hydrogen, a valuable fuel, and oxygen, a useful by-product (the illustration on the left, below). Similar substances such as hydrogen sulfide and hydrogen bromide also can be split by this photoelectrochemical process to produce hydrogen.

Carbon dioxide seems to pose an environmental hazard in the modern industrial world. Photoelectrochemical cells could be used to reduce carbon dioxide to methanol or methane, both potentially useful as fuels. The technology to carry out this recycling of carbon dioxide is still remote, but the possibility is very attractive.

Semiconductor particles that are coated with a film of metal and suspended in aqueous solutions of various redox couples can act as little electrochemical cells when exposed to sunlight (the illustration on the right, below). The oxidative destruction of hazardous materials such as polychlorinated biphenyls (PCBs), contaminants present in very low concentrations in some industrial effluent streams, can be accomplished in this way.

Over the last 20 years, much progress has been made in these areas of research, and the interdisciplinary field of semiconductor photoelectrochemistry has been firmly established. Although an early adoption of this form of energy production is unlikely, the elegance and environmental harmlessness of the technology makes me hopeful for its long-term use.

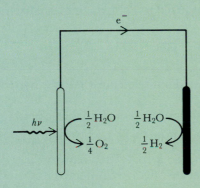

A photon-driven electrochemical cell can produce hydrogen and oxygen from water by electrolysis.

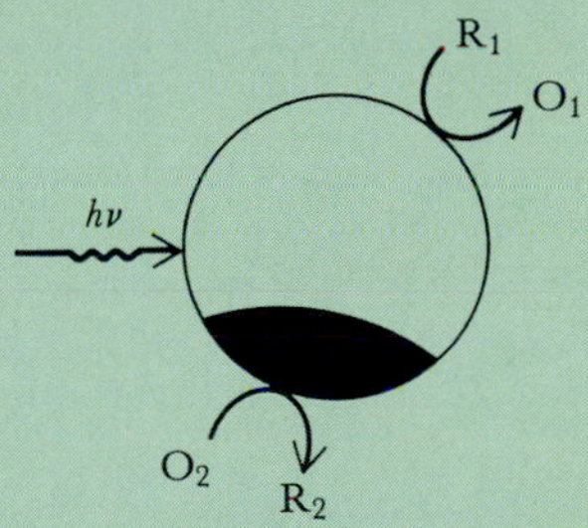

Some day, oxidative destruction of pollutants and hazardous materials (R_1 and R_2) may be carried out by photoelectrochemical cells made of small semiconductor particles coated with metal.

Anode reaction: $Zn(s) + 2OH^-(aq) \longrightarrow Zn(OH)_2(s) + 2e^-$

Cathode reaction: $2MnO_2(s) + 2H_2O(l) + 2e^- \longrightarrow 2MnO(OH)(s) + 2OH^-(aq)$

The cell potential is about 1.54 V. Notice that the OH^- ion that is generated at the cathode is consumed at the anode; this process maintains the concentrations of the ions in the electrolyte and stabilizes the current flow.

Modern research work is directed toward the utilization of solar radiation to generate electrical power (Box 17.1).

E Solar voltaic cells. *J. Chem. Ed.*, **1981**, *58*, 418.

E The *mercury cell*, which is used in hearing aids and cameras, has the same anode reaction, but the cathode half-reaction is
$HgO(s) + H_2O(l) + 2e^- \rightarrow Hg(l) + 2OH^-(aq)$
The *silver cell*, used in watches and some calculators, is similar to the mercury cell but the cathode half-reaction is
$Ag_2O(s) + H_2O(l) + 2e^- \rightarrow 2Ag(s) + 2OH^-(aq)$

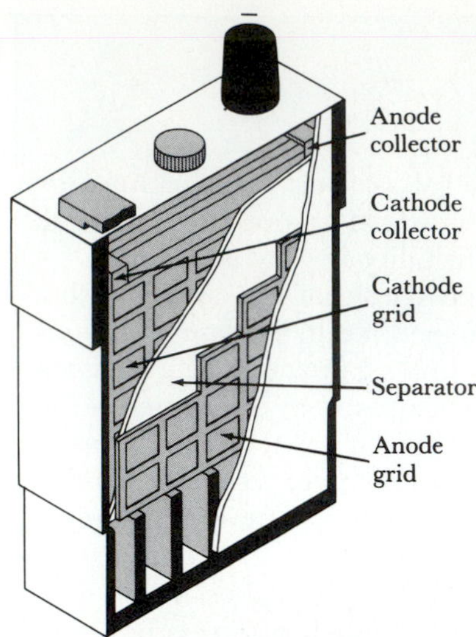

FIGURE 17.8 One cell of a lead-acid battery like those used in automobiles.

? Should water or sulfuric acid be added to a cell that is low in electrolyte? (*Hint*: See the equation for the discharge reaction.)

E Car won't start? *J. Chem. Ed.*, **1970**, *47*, 382.

E The life of the battery is reduced when tap water is added. The Ca^{2+} ions of the tap water precipitate as a bulky $CaSO_4$ on the electrodes, so filling the grids. The $PbSO_4$ then forms on the surface of the $CaSO_4$. Vibration dislodges the $CaSO_4$ from the grids, carrying with it the $PbSO_4$.

E The lead-acid cell produces about 2 V, which is a helpfully high potential. A part of the reason for the size of the potential can be traced to the inert pair effect, which favors Pb(II) (in the product of the cell reaction) relative to Pb(IV) (in the reactants), so there is a strongly negative free energy for the cell reaction. It is interesting to note that a detailed analysis of the inert-pair effect shows that a major contribution to it is the relativistic increase in mass of electrons that move at high speeds; therefore, in this sense, starting a car engine is a relativistic phenomenon.

Secondary cells. Secondary cells are galvanic cells during discharge but electrolytic cells while they are being charged. The electrodes of a secondary cell must be chosen and designed very carefully. In the discharge reaction, the products must be insoluble so that they adhere to the electrodes and are not lost to the electrolyte.

One of the most common secondary cells is the **lead-acid cell** of an automobile battery. Each cell contains several grids that act as electrodes (Fig. 17.8). Because the total surface area of these grids is large, the battery can generate large currents on demand—at least for short periods, as when starting an engine. The electrodes are initially a hard lead-antimony alloy covered with a paste of lead(II) sulfate. The electrolyte is sulfuric acid. During the first charging, some of the lead(II) sulfate is reduced to lead on one of the electrodes (which will act as the anode during discharge) and is oxidized to lead(IV) oxide on the other (which will act as the cathode). Therefore, a fully charged lead-acid cell has lead metal electroplated on a set of lead-antimony grid electrodes (which will function as the anode in the discharge reaction) and lead(IV) oxide deposited on a second set of lead-antimony electrodes (which will function as the cathode in the discharge reaction), submerged in sulfuric acid (approximately 30 to 40% by mass). The set of grids within the cell are alternately anode, cathode, anode, cathode, and so on and are separated from one another by a fibrous material.

The discharge reaction at the anode is the oxidation of the lead metal to Pb^{2+} in the presence of hydrogensulfate ions from the sulfuric acid electrolyte:

$$\textit{Anode reaction}:\ \mathrm{Pb}(s) + \mathrm{HSO_4}^-(aq) \longrightarrow \mathrm{PbSO_4}(s) + \mathrm{H}^+(aq) + 2e^-$$

At the cathode, the PbO_2 is reduced to Pb^{2+} during the discharge reaction:

$$\textit{Cathode reaction}:\ \mathrm{PbO_2}(s) + 3\mathrm{H}^+(aq) + \mathrm{HSO_4}^-(aq) \longrightarrow \mathrm{PbSO_4}(s) + 2\mathrm{H_2O}(l)$$

These Pb^{2+} ions precipitate as $PbSO_4$ within the grid network of the electrodes.

The equation for the overall cell reaction during discharge is

$$\mathrm{PbO_2}(s) + \mathrm{Pb}(s) + 2\mathrm{H}^+(aq) + 2\mathrm{HSO_4}^-(aq) \longrightarrow 2\mathrm{PbSO_4}(s) + 2\mathrm{H_2O}(l)$$

The cell potential is about 2 V. To form a 12-V battery, as is customary for most automobiles, six separate cells are connected in series. The cell reaction shows that sulfuric acid is used up during discharge. When the cell is recharged, the cell reaction is driven in reverse by the external supply, and sulfuric acid is produced. The state of charge of the cell can therefore be judged from the concentration of the sulfuric acid electrolyte and that in turn can be judged from its density. The density of the sulfuric acid in a fully charged cell is 1.25 to 1.30 g/mL. If the density falls below 1.20 g/mL, the cell must be charged. Water must occasionally be added to each cell because in the charging process some of the water is electrolyzed to hydrogen and oxygen, which escape.

The rechargeable **nickel-cadmium cell** is widely used in portable electronic equipment and power tools (Fig. 17.9). The electrodes are stainless steel grids and the electrolyte is a solution of potassium hydroxide. The electron supply to the external circuit is from the oxidation of cadmium metal:

$$\textit{Anode reaction}:\ \mathrm{Cd}(s) + 2\mathrm{OH}^-(aq) \longrightarrow \mathrm{Cd(OH)_2}(s) + 2e^-$$

Because cadmium hydroxide is insoluble, it adheres to the grid as the oxidation occurs. The cathode reaction is the reduction of nickel(III) hydroxide:

$$\textit{Cathode reaction}: 2Ni(OH)_3(s) + 2e^- \longrightarrow 2Ni(OH)_2(s) + 2OH^-(aq)$$

The insoluble nickel(II) hydroxide also adheres to the stainless steel grid and is thus readily available when the cell is charged (when the reactions are reversed). The cell reaction provides about 1.4 V. Because no gases are produced in either the charging or the discharging processes, the cells can be sealed. This feature makes nickel-cadmium cells ideal for transportable equipment.

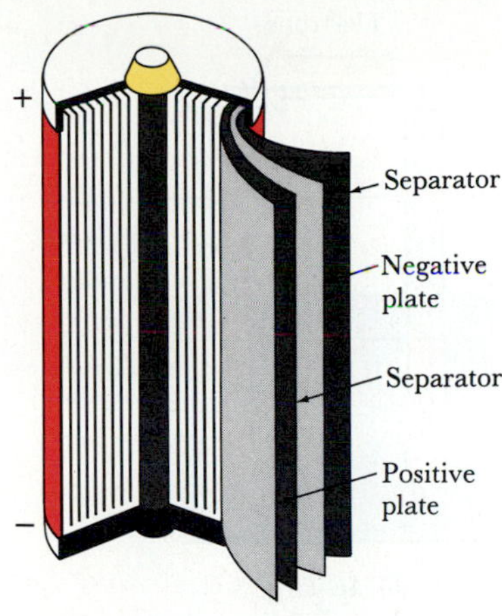

FIGURE 17.9 A rechargeable nickel-cadmium cell. The electrodes are assembled in a jellyroll arrangement and separated by a layer soaked in moist sodium or potassium hydroxide.

D The interconversion of electrical and chemical energy: The electrolysis of water and the hydrogen-oxygen fuel cell. *J. Chem. Ed.*, **1988**, *65*, 725.

E The cells that make up an electric organ are adaptations of muscle cells that generate charge rather than movement.

E Electrochemical principles involved in a fuel cell. *J. Chem. Ed.*, **1970**, *47*, 680. Fuel cells and electrochemical energy. *J. Chem. Ed.*, **1983**, *60*, 320.

Fuel cells. In a simple version of a fuel cell, hydrogen gas—the fuel—is passed over one electrode, oxygen gas is passed over the other, and the electrolyte is aqueous potassium hydroxide. The electron supply is the oxidation of hydrogen:

$$\textit{Anode reaction}: H_2(g) + 2OH^-(aq) \longrightarrow 2H_2O(l) + 2e^-$$

The electron sink—the reaction in which the electrons are used—is the reduction of oxygen at the cathode:

$$\textit{Cathode reaction}: O_2(g) + 2H_2O(l) + 4e^- \longrightarrow 4OH^-(aq)$$

As we saw in Example 17.2, the overall cell reaction is the formation of water. A version of this type of cell is used on the space shuttle, one advantage being that the crew can drink the product of the cell reaction.

Electric eels are mobile, natural fuel cells (Fig. 17.10). They generate their electric charge in an electric organ—a battery of biological electrochemical cells fueled by food. Each cell provides about 0.15 V. It is an incidental feature of nature that the eel's head is its cathode and its tail its anode. The electric catfish has the opposite polarity.

ELECTROCHEMISTRY AND THERMODYNAMICS

A galvanic cell is always written so that it has a positive potential. The corresponding cell reaction is therefore one that has a spontaneous tendency to occur and drive electrons from the anode to the cathode through the external circuit. Therefore, in the language of thermodynamics, such a reaction has a negative free energy. This observation is the link between electrochemistry and thermodynamics, and the topic of this section.

FIGURE 17.10 The electric eel (*Electrophorus electricus*) lives in the Amazon River. The average potential difference it produces along its 1-m length is about 350 V.

17.3 CELL POTENTIAL AND REACTION FREE ENERGY

If ΔG for the cell reaction is large and negative, then the reaction has a strong tendency to occur; hence we expect a large cell potential E (Fig. 17.11). If ΔG is negative but small, then we expect only a small cell potential. If the reaction in the cell is at equilibrium, a condition corresponding to $\Delta G = 0$, then the reaction has no tendency to drive electrons through the circuit and the cell potential is zero. We can conclude

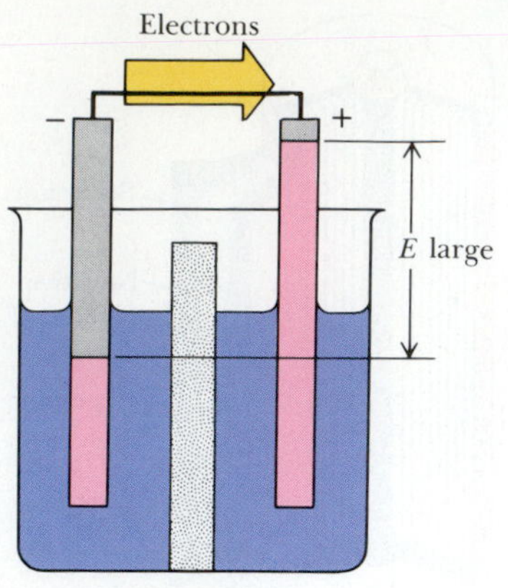

(a) ΔG large

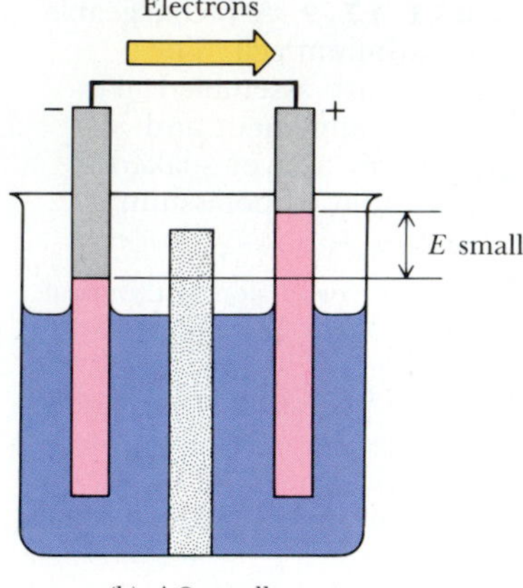

(b) ΔG small

FIGURE 17.11 The more negative the reaction free energy ΔG, the greater the cell potential E.

D Demonstrate the reaction of iodine with zinc powder (with water and in a well-ventilated area), and then show how the energy from this reaction can be harnessed, by setting up a galvanic cell using a KI/I_2 solution (with one graphite electrode and one zinc electrode).

D Energy, heat, or electricity production. TDC, 154. Visible oxidation reduction in electrochemical cells. CD2, 107.

E The definition of the standard state of ions in solution should take into account the interactions between them.

E Electrochemical cell conversions in general chemistry. *J. Chem. Ed.*, **1986**, *63*, 609.

T OHT: Fig. 17.13.

that the cell potential and the reaction free energy are proportional to each other:

$$\Delta G \propto -E$$

The negative sign is included on the right because a cell potential, which is always positive for a galvanic cell, indicates that the cell reaction is spontaneous and hence that ΔG is negative. If ΔG is positive for a reaction, then the *reverse* reaction is spontaneous and we must reverse the cell diagram so that the cathode remains on the right.

The precise relation between E and ΔG for a reaction in which n moles of electrons move from the anode to the cathode is

$$\Delta G = -nFE \tag{1}$$

The constant F, **Faraday's constant**, is the magnitude of the charge per mole of electrons:

$$F = \text{charge of one electron} \times \text{number of electrons per mole}$$
$$= 1.602 \times 10^{-19}\ \text{C} \times 6.022 \times 10^{23}/(\text{mol e}^-) = 9.647 \times 10^4\ \text{C}/(\text{mol e}^-)$$

A more precise value is

$$F = 9.6485 \times 10^4\ \text{C}/(\text{mol e}^-),\ \text{or}\ 96.485\ \text{kC}/(\text{mol e}^-)$$

This constant F is named for the English scientist Michael Faraday, who is regarded as one of the greatest experimental scientists of the nineteenth century (Fig. 17.12).

For the reaction in the Daniell cell, $n = 2$ mol because 2 mol of electrons migrate from 1 mol Zn to 1 mol Cu in the reaction

$$Zn(s) + Cu^{2+}(aq) \longrightarrow Zn^{2+}(aq) + Cu(s)$$

When the concentrations of the Cu^{2+} and Zn^{2+} ions are both 1 mol/L, the cell potential is 1.10 V. Hence, for this reaction, noting that 1.10 V = 1.10 J/C,

$$\Delta G = -nFE = -(2\ \cancel{\text{mol e}^-}) \times 9.65 \times 10^4\ \frac{\cancel{\text{C}}}{\cancel{\text{mol e}}} \times 1.10\ \frac{\text{J}}{\cancel{\text{C}}}$$
$$= -2.12 \times 10^5\ \text{J} = -212\ \text{kJ}$$

We saw in Chapter 16 that a special role is played by the *standard* reaction free energy ΔG°. We can measure this quantity for a cell reaction simply by arranging for the reagents in the cell to be in their standard states, measuring the **standard cell potential** E°, and using the following special case of Eq. 1:

$$\Delta G^\circ = -nFE^\circ \tag{2}$$

We have seen that the standard states of solids and gases are their pure forms at a pressure of 1 atm. The standard state of an ion in solution is one in which its concentration is 1 mol/L. Therefore,

> The **standard cell potential** E° of a galvanic cell is the cell potential measured when the concentration of each type of ion taking part in the cell reaction is 1 mol/L and all the gases are at 1 atm pressure.

As an example, the standard potential of the Daniell cell at 25°C is its potential when the concentrations of the Zn^{2+} ions and the Cu^{2+} ions in their own compartments are each 1 mol/L. This potential is 1.10 V; so $\Delta G^\circ = -212$ kJ, as we calculated above.

There are thousands of possible cells and therefore thousands of possible standard cell potentials that could be tabulated. We can greatly simplify this task by considering each electrode of a cell as making a characteristic contribution to the cell potential. Then the cell potential is the sum of the potentials of its two electrodes (Fig. 17.13). For instance, we can think of the potential of the cell

$$Fe(s)|Fe^{2+}(aq)||Ag^{+}(aq)|Ag(s) \qquad E^\circ = 1.24\ V$$

as the sum of two contributions, one contribution from the silver electrode and another from the iron electrode. These individual contributions to the cell potential are called **electrode potentials**. When the cell is prepared in its standard state, they are called **standard electrode potentials** E°. The standard potential of a cell is the sum of its two standard electrode potentials:

$$E^\circ(\text{cell}) = E^\circ(\text{cathode}) + E^\circ(\text{anode})$$

That is, the standard cell potential is the sum of the electrode potential for the reduction half-reaction (under standard conditions) and the electrode potential for the oxidation half-reaction (also under standard conditions). In the next section we see how the standard electrode potentials are measured.

FIGURE 17.12 Michael Faraday (1791–1867).

E Michael Faraday, born the son of a blacksmith, came to London looking for work. Faraday became the assistant of Humphry Davy (the discoverer of sodium and potassium and the inventor of the miner's safety lamp), and in due course rose to be the director of Davy's laboratory. It has been claimed that he deserved at least six Nobel prizes (for the discovery of electromagnetic induction, the laws of electrolysis, the magnetic properties of matter, the isolation of benzene, the Faraday effect, and the introduction of the concept of an electric and magnetic "field").

D Electrode potentials: Hydrogen electrode. TDC, 141.

The standard hydrogen electrode. A voltmeter placed between the two electrodes of a galvanic cell measures the overall cell potential, not the individual contributions to the potential of the cathode and the anode. However, if by mutual agreement we assign a zero electrode potential, $E^\circ = 0$, to one particular redox couple (and call it the reference redox couple), then the cell potential can be attributed entirely to the second redox couple. In other words the standard potential of the latter electrode becomes equal to E°(cell). Suppose we form another galvanic cell in which our reference redox couple (for which $E^\circ = 0$) serves as one electrode; again the cell potential would equal the electrode potential of the new redox couple. This process can be extended to additional galvanic cells that are similarly constructed until we have as many standard electrode potentials as we want.

We take as the reference redox couple a **standard hydrogen electrode** (SHE), which is the H^+/H_2 redox couple in its standard state (aqueous H^+ ions present at 1 mol/L and hydrogen gas at 1 atm) in the presence of platinum:

$$2H^{+}(aq) + 2e^{-} \longrightarrow H_2(g) \qquad E^\circ = 0$$

Then, to measure the standard electrode potential of, for instance, the Zn^{2+}/Zn redox couple, we use a voltmeter to measure the standard potential of the cell in which the SHE is in contact with a zinc electrode. When this is actually done, we observe that the hydrogen electrode is the cathode and the zinc electrode is the anode. Therefore, the cell diagram and experimental cell potential are

$$Zn|Zn^{2+}(aq)||H^{+}(aq)|H_2(g)|Pt \qquad E^\circ(\text{cell}) = 0.76\ V$$

According to our convention, the zinc electrode contributes the entire 0.76 V to E°(cell), so we write

Cathode reaction: $2H^{+}(aq) + 2e^{-} \longrightarrow H_2(g) \qquad E^\circ(\text{cathode}) = 0$

Anode reaction: $Zn(s) \longrightarrow Zn^{2+}(aq) + 2e^{-} \qquad E^\circ(\text{anode}) = +0.76\ V$

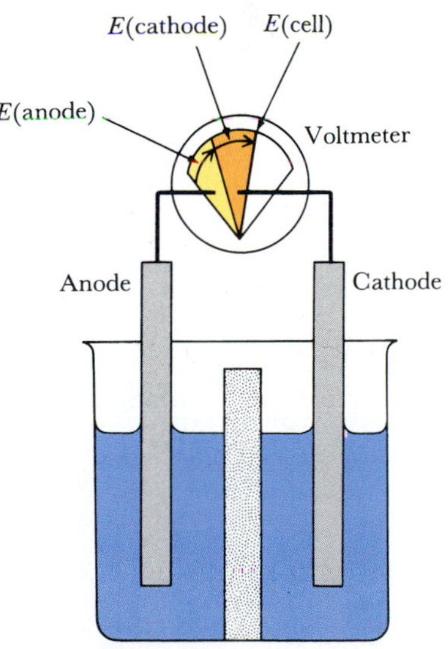

FIGURE 17.13 The cell potential can be thought of as the sum of the voltages of the anode and the cathode.

N There are two pedagogical reasons why we have chosen this format for arriving at cell potentials (rather than using a difference, as is sometimes done): (1) Because the anode half-reaction must be the reverse of the reduction half-reaction in Table 17.3, the sign of the standard potential must be changed. (2) In the discussion of Hess's law in Chapter 6, the sign of ΔH was changed when the reaction was reversed; therefore, the cell reaction is treated like the thermochemical equations in Chapter 6, and the cell potential is the analogue of ΔH.

For reasons that will shortly become clear, we always include the sign of the electrode potential for a redox couple (but not for the overall cell, which is always positive for a galvanic cell).

We can draw two conclusions from the observation that the standard hydrogen electrode is a cathode (the site of reduction) when it is coupled to a zinc electrode:

1. The H^+/H_2 redox couple has a greater tendency to undergo reduction than the Zn^{2+}/Zn redox couple does.
2. Zinc is a stronger reducing agent than hydrogen is; or, conversely, the H^+ ion is a stronger oxidizing agent than the Zn^{2+} ion is.

When a galvanic cell is formed with a SHE and a standard Cu^{2+}/Cu electrode, the cell diagram and cell potential are

$$\mathrm{Pt}|\mathrm{H_2}(g)|\mathrm{H^+}(aq)||\mathrm{Cu^{2+}}(aq)|\mathrm{Cu} \qquad E°(\mathrm{cell}) = 0.34\ \mathrm{V}$$

By convention, the Cu^{2+}/Cu redox couple contributes the entire 0.34 V to the standard cell potential, so we write

$$\textit{Cathode reaction: } \mathrm{Cu^{2+}}(aq) + 2\mathrm{e^-} \longrightarrow \mathrm{Cu} \qquad E°(\mathrm{cathode}) = +0.34\ \mathrm{V}$$

$$\textit{Anode reaction: } \mathrm{H_2}(g) \longrightarrow 2\mathrm{H^+}(aq) + 2\mathrm{e^-} \qquad E°(\mathrm{anode}) = 0$$

Moreover, because the Cu^{2+}/Cu electrode is the cathode, we can conclude that

1. The Cu^{2+}/Cu redox couple has a greater tendency to undergo reduction than the H^+/H_2 redox couple does.
2. Hydrogen is a stronger reducing agent than copper is; conversely, a Cu^{2+} ion is a stronger oxidizing agent than an H^+ ion is.

Because zinc is a stronger reducing agent than hydrogen, and hydrogen is a stronger reducing agent than copper, we can now *predict* that zinc is a stronger reducing agent than copper. That is, when the concentrations of both species of ions are 1 mol/L, the spontaneous direction of reaction between zinc and copper is

$$\mathrm{Zn}(s) + \mathrm{Cu^{2+}}(aq) \longrightarrow \mathrm{Zn^{2+}}(aq) + \mathrm{Cu}(s)$$

and not the reverse. We can also use the two individual standard electrode potentials to predict the standard potential of the corresponding cell,

$$\mathrm{Zn}(s)|\mathrm{Zn^{2+}}(aq)||\mathrm{Cu^{2+}}(aq)|\mathrm{Cu}(s)$$

To do so, we simply add their standard electrode potentials:

$$\textit{Cathode reaction: } \mathrm{Cu^{2+}}(aq) + 2\mathrm{e^-} \longrightarrow \mathrm{Cu}(s) \qquad E°(\mathrm{cathode}) = +0.34\ \mathrm{V}$$

$$\textit{Anode reaction: } \mathrm{Zn}(s) \longrightarrow \mathrm{Zn^{2+}}(aq) + 2\mathrm{e^-} \qquad E°(\mathrm{anode}) = +0.76\ \mathrm{V}$$

$$\textit{Cell reaction: } \mathrm{Cu^{2+}}(aq) + \mathrm{Zn}(s) \longrightarrow \mathrm{Zn^{2+}}(aq) + \mathrm{Cu}(s)$$

$$E°(\mathrm{cell}) = E°(\mathrm{cathode}) + E°(\mathrm{anode}) = +0.34\ \mathrm{V} + (+0.76\ \mathrm{V}) = 1.10\ \mathrm{V}$$

Oxidation and reduction potentials. Of the measured 1.10 V for the cell potential, +0.76 V is due to the oxidation of the Zn^{2+}/Zn couple; this contribution to the cell potential is the **standard oxidation potential** of zinc. If we were to write the half-reaction in reverse, which would imply that we were considering the reduction of the Zn^{2+}/Zn

U Exercises 17.67 and 17.73.

redox couple, there would be a corresponding change in sign of its electrode potential. The value $E° = -0.76$ V associated with the reduction of the Zn^{2+}/Zn redox couple is called the **standard reduction potential** of the Zn^{2+}/Zn couple, and we write

$$Zn^{2+}(aq) + 2e^- \longrightarrow Zn(s) \qquad E° = -0.76 \text{ V}$$

The +0.34 V for the Cu^{2+}/Cu couple quoted above is already a standard reduction potential: the standard oxidation potential of the Cu^{2+}/Cu couple would be −0.34 V. Therefore, an oxidation potential is the negative of the reduction potential.

By convention, *all standard electrode potentials are listed as reduction potentials*. Hence, all the $E°$ values we quote from now on refer to reduction half-reactions unless we state otherwise. A selection of values (at 25°C) is given in Table 17.3 and a longer table can be found in Appendix 2B. Note that the first column in the table gives the oxidized member of the couple. The third column gives the reduced member of the couple. The last column gives the standard reduction potential. Reduction potentials vary in a complicated way throughout the periodic table (Fig. 17.14); generally, however, the most negative are found toward the left of the periodic table and the most positive are found toward the upper right.

17.4 THE ELECTROCHEMICAL SERIES

Table 17.3 provides a wealth of data on redox reactions. We shall now see that we can use the information it contains to predict (among other properties) the relative strengths of oxidizing and reducing agents, whether or not a redox reaction is spontaneous, the amount of energy that we can extract from a redox reaction, and the activity of metals in air or water.

E A thermochemical cycle can be a useful aid in the understanding of the periodic trends (or lack of them) illustrated in Fig. 17.14. Thus, the free energy for the reduction $Li^+(aq) + e^- \rightarrow Li(s)$ can be expressed in terms of the free energy that accompanies vaporization of solid lithium, ionization of Li atoms, and hydration of Li^+ ions. Even if we were to ignore the entropy contribution to the free energy, it is still difficult to predict trends because the contributions vary in a complex manner, and in some cases a trend in one contribution is nearly canceled by the trend in another.

V Video 45: Electrochemical series.

D Activity series. TDC, 7-6. The activity series for some metals. CD1, 89.

E Definition of standard EMF. *J. Chem. Ed.*, **1971**, *48*, 737.

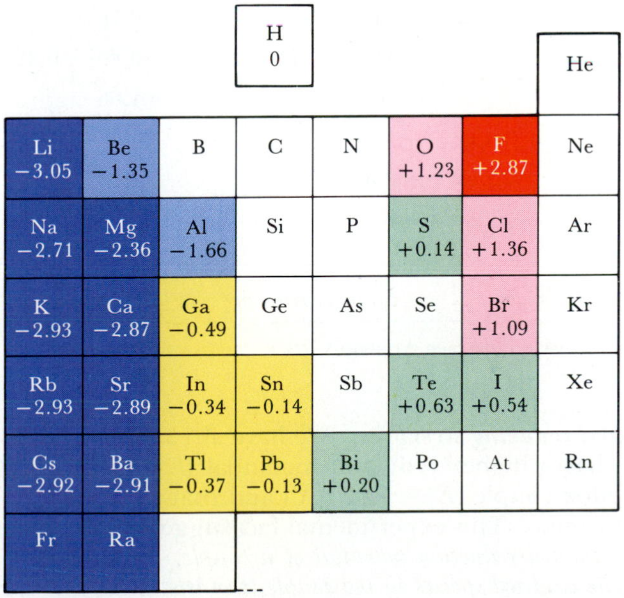

			H 0				He
Li −3.05	Be −1.35	B	C	N	O +1.23	F +2.87	Ne
Na −2.71	Mg −2.36	Al −1.66	Si	P	S +0.14	Cl +1.36	Ar
K −2.93	Ca −2.87	Ga −0.49	Ge	As	Se	Br +1.09	Kr
Rb −2.93	Sr −2.89	In −0.34	Sn −0.14	Sb	Te +0.63	I +0.54	Xe
Cs −2.92	Ba −2.91	Tl −0.37	Pb −0.13	Bi +0.20	Po	At	Rn
Fr	Ra						

FIGURE 17.14 The variation of standard reduction potential throughout the main groups of the periodic table.

T OHT: Table 17.3.

TABLE 17.3 Standard reduction potentials at 25°C*

Oxidizing agent	Reduction half-reaction	Reducing agent	$E°$, V
Strongly oxidizing			
F_2	$+2e^- \rightarrow$	$2F^-$	+2.87
Au^+	$+e^- \rightarrow$	Au	+1.69
Ce^{4+}	$+e^- \rightarrow$	Ce^{3+}	+1.61
$MnO_4^- + 8H^+$	$+5e^- \rightarrow$	$Mn^{2+} + 4H_2O$	+1.51
Cl_2	$+2e^- \rightarrow$	$2Cl^-$	+1.36
$Cr_2O_7^{2-} + 14H^+$	$+6e^- \rightarrow$	$2Cr^{3+} + 7H_2O$	+1.33
$O_2 + 4H^+$	$+4e^- \rightarrow$	$2H_2O$	+1.23, +0.81 at pH = 7
Br_2	$+2e^- \rightarrow$	$2Br^-$	+1.09
$NO_3^- + 4H^+$	$+3e^- \rightarrow$	$NO + 2H_2O$	+0.96
Ag^+	$+e^- \rightarrow$	Ag	+0.80
Fe^{3+}	$+e^- \rightarrow$	Fe^{2+}	+0.77
I_2	$+2e^- \rightarrow$	$2I^-$	+0.54
$O_2 + 2H_2O$	$+4e^- \rightarrow$	$4OH^-$	+0.40, +0.81 at pH = 7
Cu^{2+}	$+2e^- \rightarrow$	Cu	+0.34
$AgCl$	$+e^- \rightarrow$	$Ag + Cl^-$	+0.22
$2H^+$	$+2e^- \rightarrow$	H_2	0, by definition
Fe^{3+}	$+3e^- \rightarrow$	Fe	−0.04
$O_2 + H_2O$	$+2e^- \rightarrow$	$HO_2^- + OH^-$	−0.08
Pb^{2+}	$+2e^- \rightarrow$	Pb	−0.13
Sn^{2+}	$+2e^- \rightarrow$	Sn	−0.14
Fe^{2+}	$+2e^- \rightarrow$	Fe	−0.44
Zn^{2+}	$+2e^- \rightarrow$	Zn	−0.76
$2H_2O$	$+2e^- \rightarrow$	$H_2 + 2OH^-$	−0.83, −0.42 at pH = 7
Al^{3+}	$+3e^- \rightarrow$	Al	−1.66
Mg^{2+}	$+2e^- \rightarrow$	Mg	−2.36
Na^+	$+e^- \rightarrow$	Na	−2.71
K^+	$+e^- \rightarrow$	K	−2.93
Li^+	$+e^- \rightarrow$	Li	−3.05
		Strongly reducing	

*For a more extensive table, see Appendix 2B.

D Place a piece of sodium in water and explain the observations using Table 17.3, or use Video 21 (The reaction of sodium metal with water).

Oxidizing and reducing strengths. We have already seen that the Cu^{2+}/Cu couple has a higher reduction potential ($E° = +0.34$ V) than the Zn^{2+}/Zn redox couple ($E° = -0.76$ V) and that the Cu^{2+} ion can oxidize the zinc metal. This experimental fact suggests that, in general, *the higher the standard reduction potential of a couple, the greater the oxidizing strength of the oxidized species in the couple* (for instance, the Cu^{2+} in the

Cu^{2+}/Cu redox couple). Stated differently, *the higher the position of a couple in Table 17.3, the greater the oxidizing strength of the substance in the first column.* For example, F_2 is a strong oxidizing agent, stronger than Cl_2; Li^+ is a very, very poor oxidizing agent.

We have also seen that Zn ($E° = -0.76$ V) is a stronger reducing agent than H_2 ($E° = 0$). In general, *the lower the standard reduction potential, the greater the reducing strength of the reduced species in the couple* (for instance, the Zn in the Zn^{2+}/Zn redox couple). Put another way: *the lower the position of a couple in Table 17.3, the greater the reducing strength of the substance in the third column.* It follows that the strongest reducing agents are at the bottom of the third column in Table 17.3. For example, lithium metal is a strong reducing agent, stronger than sodium; on the other hand, F^- is a very, very poor reducing agent.

When Table 17.3 is viewed as a list of oxidizing and reducing agents ordered on the basis of strength, it is called the **electrochemical series**. The oxidized species in a redox couple in the list has a tendency to oxidize the reduced member of any couple that lies below it: the free energy for that reaction is negative and the reaction is spontaneous. The reduced species in a couple has a tendency to reduce the oxidized member of any couple that lies above it. Hence, we can see at a glance the direction in which a particular redox reaction will tend to run (under standard conditions).

To judge the oxidizing strength of a substance, we note its location in the first column of the electrochemical series. The higher it is, the greater its oxidizing strength. This process is illustrated by the ability of acidified MnO_4^- solution to oxidize Fe^{2+}: Table 17.3 shows that MnO_4^- is a stronger oxidizing agent than Fe^{3+}, so the oxidation of Fe^{2+} to Fe^{3+} is spontaneous (Fig. 17.15).

V Video 28: The mercury heart.

(a)

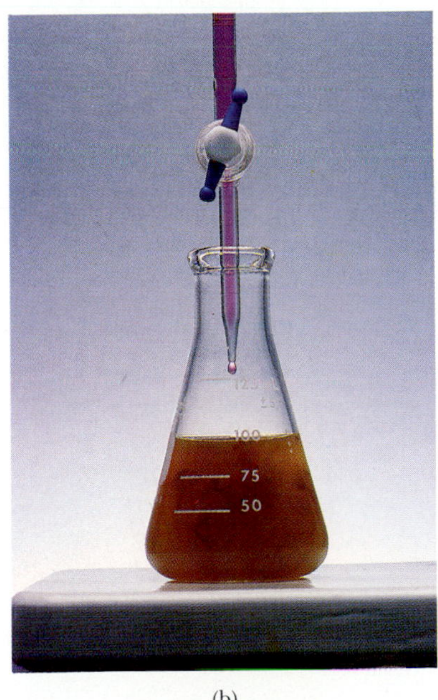

(b)

FIGURE 17.15 (a) In this redox titration, the purple potassium permanganate in the buret is being used to oxidize the pale green Fe^{2+} solution. The stoichiometric point is detected by noting when the violet color of the permanganate ions persists (b).

▼ EXAMPLE 17.3 Predicting relative oxidizing strengths

Is an acidified permanganate solution a more powerful oxidizing agent than an acidified dichromate solution under standard conditions? Confirm your answer numerically.

STRATEGY We inspect Table 17.3 to see whether the permanganate ion MnO_4^- lies above the dichromate ion $Cr_2O_7^{2-}$. We confirm the answer by calculating the cell potential. The half-reaction positioned higher in Table 17.3 is the reduction half-reaction and has the listed reduction potential. The half-reaction lower in Table 17.3 is reversed for oxidation and its (oxidation) potential is the negative of the reduction potential. The cell potential is the sum of the reduction potential and the oxidation potential.

SOLUTION Because the $MnO_4^-,H^+/Mn^{2+}$ couple (+1.51 V) lies above the $Cr_2O_7^{2-},H^+/Cr^{3+}$ couple (+1.33 V), MnO_4^- is the stronger oxidizing agent. We confirm this as follows:

Reduction: $$MnO_4^-(aq) + 8H^+(aq) + 5e^-(aq) \longrightarrow Mn^{2+}(aq) + 4H_2O(l) \qquad E° = +1.51 \text{ V}$$

Oxidation: $$2Cr^{3+}(aq) + 7H_2O(l) \longrightarrow Cr_2O_7^{2-}(aq) + 14H^+(aq) + 6e^- \qquad E° = -1.33 \text{ V}$$

The equation for the overall reaction is obtained by adding six times the reduction half-reaction to five times the oxidation half-reaction (so that the electron supplies match). The cell potential is the sum of the two electrode potentials:

$$E°(\text{cell}) = 1.51 \text{ V} + (-1.33 \text{ V}) = 0.18 \text{ V}$$

U Exercises 17.69, 17.70, and 17.75.

This value indicates that the reaction free energy is negative (because $\Delta G° \propto -E°$) and hence that the reaction is spontaneous. That is, permanganate ions can oxidize Cr^{3+} ions in acid solution.

EXERCISE E17.3 Which is the stronger reducing agent, zinc metal or tin metal?

▲ [*Answer*: zinc]

The arrangement of a galvanic cell. It follows from the discussion above that, when two couples form a galvanic cell, under standard conditions *the couple higher in the table forms the cathode of the cell* (because the oxidized species in the higher couple gets reduced as the reaction proceeds), and *the couple lower in the table forms the anode* (because the reduced member of the lower couple gets oxidized). We always write the half-reaction for the couple higher in the table as a reduction (as it appears in the table) and that for the couple lower in the table as an oxidation (reversing it and changing the sign of the electrode potential). This procedure may be summarized as follows:

Lower couple	Upper couple
Anode compartment|	|Cathode compartment
Red → ox + e^-	Ox + e^- → red
$E°$(anode)	$E°$(cathode)

The $E°$(cathode) for the reduction half-reaction is obtained directly from the table of standard reduction potentials (Table 17.3); the $E°$(anode) for the oxidation half-reaction is the *negative* of the standard reduction potential appearing in the table. This arrangement guarantees that the standard cell potential is positive and therefore that the reaction has a negative free energy; that is, the reaction is spontaneous under standard conditions.

E In the industrial process for the production of bromine from brines that contain bromide ions, chlorine and air are bubbled through the liquid. The chlorine oxidizes the bromide ions to bromine (as illustrated in this example), and the air causes the bromine to evaporate from the solution.

▼ **EXAMPLE 17.4 Writing a cell diagram by using the electrochemical series**

Write the diagram for a galvanic cell in which one redox couple is Cl_2/Cl^- and the other is Br_2/Br^-. Give the cell reaction and the standard cell potential. Oxidation by chlorine gas is the process by which liquid bromine is produced from brine that contains aqueous bromide ions.

STRATEGY From its position in the periodic table and Table 17.3, we expect Cl_2 to be a stronger oxidizing agent than Br_2 and anticipate that Cl_2 will form the cathode (because it will be reduced). More formally, the strategy in such problems is to find the two redox couples in the electrochemical series and then write the cell diagram with the electrode for the upper couple as the cathode (on the right) and that of the lower couple as the anode (on the left). To write the cell reaction, reverse the lower half-reaction (thereby converting a reduction into an oxidation and the reduction potential into an oxidation potential) and add it to the upper half-reaction.

SOLUTION The Cl_2/Cl^- couple lies above the Br_2/Br^- couple, so the cell diagram is

$$\mathrm{Pt}|\mathrm{Br_2}(aq)|\mathrm{Br^-}(l)||\mathrm{Cl^-}(aq)|\mathrm{Cl_2}(g)|\mathrm{Pt}$$

The two half-reactions are

Cathode half-reaction: $Cl_2(g) + 2e^- \longrightarrow Cl^-(aq)$ $E°(\text{cathode}) = +1.36\text{ V}$

Anode half-reaction: $2Br^-(aq) \longrightarrow Br_2(l) + 2e^-$ $E°(\text{anode}) = -1.09\text{ V}$

N Graphite could be substituted for platinum in the cell.

Therefore, the overall reaction is

$$Cl_2(g) + 2Br^-(aq) \longrightarrow 2Cl^-(aq) + Br_2(l)$$

$$E°(\text{cell}) = 1.36 \text{ V} + (-1.09 \text{ V}) = 0.27 \text{ V}$$

EXERCISE E17.4 Write the cell diagram, the standard cell potential, and the equation for the cell reaction for the couples Ag^+/Ag and Fe^{3+}/Fe^{2+}.

[*Answer*: $Pt|Fe^{3+}(aq),Fe^{2+}(aq)||Ag^+(aq)|Ag(s)$; $E°(\text{cell}) = 0.03$ V; $Ag^+(aq) + Fe^{2+}(aq) \rightarrow Ag(s) + Fe^{3+}(aq)$]

EXAMPLE 17.5 Predicting the direction of a reaction from the electrochemical series

Write the equation for the chemical reaction that occurs when tin metal is placed in a solution of $Fe^{2+}(aq)$ and $Fe^{3+}(aq)$ ions. Write the cell diagram for a galvanic cell that could make use of this reaction. What is the standard free energy of the cell reaction?

STRATEGY We need to inspect Table 17.3 for the two redox couples (Sn^{2+}/Sn and Fe^{3+}/Fe^{2+}). The oxidized species (first column) of the couple higher in the table has a tendency to oxidize the reduced species (third column) of the lower couple. The cell diagram is formed with the higher redox couple in the table on the right (as the cathode). We must remember that $E°(\text{anode})$ is the negative of the reduction potential appearing in Table 17.3. The standard free energy of reaction is calculated from the equation, $\Delta G° = -nFE°$.

SOLUTION The Fe^{3+}/Fe^{2+} couple lies above the Sn^{2+}/Sn couple in Table 17.3, so the Fe^{3+} ions oxidize the tin metal to $Sn^{2+}(aq)$. The half-reactions and standard electrode potentials are

Cathode reaction: $Fe^{3+}(aq) + e^- \longrightarrow Fe^{2+}(aq)$ $\qquad E°(\text{cathode}) = +0.77$ V

Anode reaction: $Sn(s) \longrightarrow Sn^{2+}(aq) + 2e^-$ $\qquad E°(\text{anode}) = +0.14$ V

Cell reaction: $Sn(s) + 2Fe^{3+}(aq) \longrightarrow Sn^{2+}(aq) + 2Fe^{2+}(aq)$

$$E°(\text{cell}) = E°(\text{cathode}) + E°(\text{anode}) = +0.77 \text{ V} + (+0.14 \text{ V}) = 0.91 \text{ V}$$

The diagram for a galvanic cell constructed from these two redox couples is

$$Sn(s)|Sn^{2+}(aq)||Fe^{3+}(aq),Fe^{2+}(aq)|Pt$$

The standard free energy of the cell reaction is (using 1 V = 1 J/C)

$$\Delta G° = -nFE° = -(2 \text{ mol e}^-) \times \frac{9.65 \times 10^4 \text{ C}}{\text{mol e}^-} \times 0.91 \text{ J/C} = -1.8 \times 10^5 \text{ J}$$

EXERCISE E17.5 Write the cell reaction, determine the standard cell potential, calculate the standard free energy, and write the cell diagram for the reaction that can occur when copper metal is added to a solution of $Co^{2+}(aq)$ and $Co^{3+}(aq)$ ions. See Appendix 2B for data. What are the oxidizing and the reducing agents in the cell reaction?

[*Answer*: $Cu(s) + 2Co^{3+}(aq) \rightarrow Cu^{2+}(aq) + 2Co^{2+}(aq)$; $E°(\text{cell}) = 1.47$ V; $\Delta G° = -284$ kJ; $Cu(s)|Cu^{2+}(aq)||Co^{3+}(aq),Co^{2+}(aq)|Pt$; Co^{3+} is the oxidizing agent; Cu is the reducing agent]

The reaction of metals with acids. The production of hydrogen gas by the action of an acid on a metal is a redox reaction in which the hydrogen ions of the acid are reduced to H_2. Because the standard potential of the H^+/H_2 couple is defined as zero, only substances with negative standard reduction potentials (those below hydrogen in the third column of the electrochemical series in Table 17.3) can bring this reduc-

U Exercises 17.68, 17.71, and 17.77.

N The activity series in Section 3.9 is generally presented in the order opposite to that adopted for the electrochemical series in Table 17.3. Clarify this convention with students.

D Electrochemical energy in a flash. CD2, 59.

tion about. We can therefore predict that magnesium, iron, indium, tin, and lead have a thermodynamic tendency to produce hydrogen gas. (However, as always, thermodynamics is silent about rates; and although the tendency might exist, the rate of the process might be very slow.) On the other hand, because metals with positive standard reduction potentials (those above hydrogen) cannot reduce hydrogen ions, we know that they cannot produce hydrogen gas when acted on by 1 M acid. For example, copper and the noble metals silver, platinum, and gold are not oxidized by hydrogen ions. These metals, however, may be able to reduce the anions of oxoacids, such as ClO_4^- or NO_3^-, which are often more strongly oxidizing than the hydrogen ion.

FIGURE 17.16 Iron nails stored in oxygen-free water (left) do not rust because the oxidizing power of water is weak. When oxygen is present (right), however, the oxidation is favored and rust soon forms.

T OHT: Figs. 17.17 and 17.20.

Passivation. When a thermodynamic tendency is not realized in practice, there is a good chance that the explanation will be based on kinetics, that is, related to the rates of reactions. A glance at the standard reduction potential of aluminum (−1.66 V) suggests that, like copper, it should react with nitric acid (+0.96 V). However, it does not. Aluminum does not react with the acid because any Al^{3+} ions that are produced immediately form a hard, unreactive, almost impenetrable layer of aluminum oxide on the surface of the metal. This layer prevents further reaction, and we say that the metal has been **passivated**, or protected from further reaction by a surface film. The passivation of aluminum is of great commercial importance because it enables the metal to be used, among many other things, for airplanes and window frames in buildings. Aluminum containers are used to transport nitric acid, for once the surface is passivated, no further reaction occurs.

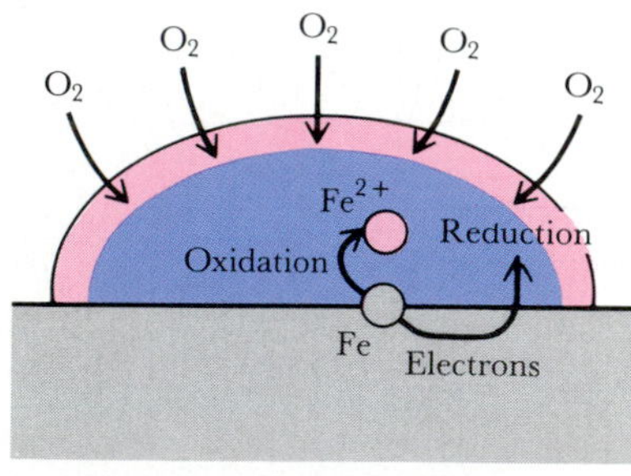

(a)

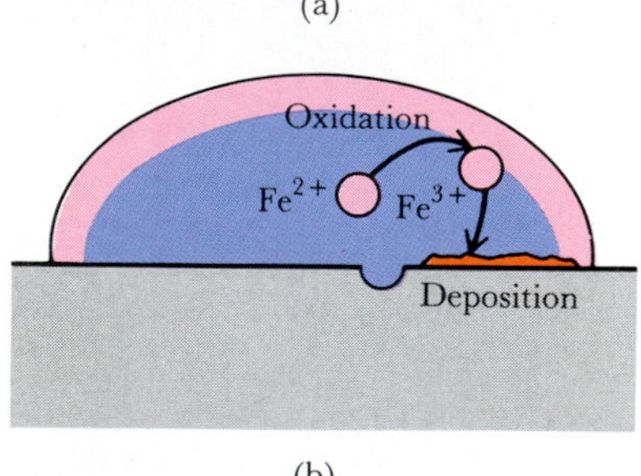

(b)

FIGURE 17.17 The mechanism of rust formation. (a) Oxidation of the iron occurs at a point out of contact with the oxygen of the air, where the metal surface behaves as an anode in a tiny electrochemical cell. (b) Further oxidation of Fe^{2+} to Fe^{3+} results in the deposition of rust on the surface.

Corrosion. The electrochemical series gives us some insight into that often depressing feature of everyday life, corrosion. Any element lower than the $H_2O/H_2,OH^-$ couple in the electrochemical series has a tendency to be oxidized by water as a result of the half-reaction

$$2H_2O(l) + 2e^- \longrightarrow H_2(g) + 2OH^-(aq) \qquad E° = -0.83\ V$$

This standard potential is for an OH^- concentration of 1 mol/L, which corresponds to pH = 14, a strongly basic solution. At pH = 7, this couple has the potential $E = -0.42$ V. Because the Fe^{2+}/Fe couple is almost the same as this potential ($E° = -0.44$ V), iron has only a very slight tendency to be oxidized by pure water. For this reason, iron can be used for making pipes in water supply systems and can be stored in oxygen-free water without rusting (Fig. 17.16). However, when the iron is exposed to damp air, with both oxygen and water present, the half-reaction

$$O_2(g) + 4H^+(aq) + 4e^- \longrightarrow 2H_2O(l) \qquad E° = +1.23\ V$$

must be taken into account. The potential of this couple at pH = 7 is +0.82 V, which lies well above the value for the Fe^{2+}/Fe couple and the Fe^{3+}/Fe^{2+} couple. Hence, oxygen and water can jointly oxidize iron to Fe^{2+}, which can subsequently be oxidized to Fe^{3+} (the iron ion in rust).

The mechanism of rusting is interesting. A drop of water on the surface of iron acts as the electrolyte in a tiny electrochemical cell (Fig. 17.17). At the edge of the drop, where there is dissolved oxygen close to the metal, the oxygen has a tendency to oxidize the iron by the reaction given above. However, the electrons withdrawn from the metal by this oxidation can be restored from another part of the con-

FIGURE 17.18 Metal girders are galvanized by immersion in a bath of molten zinc.

E When your car rusts out. *J. Chem. Ed.*, **1972**, *49*, 29.

ducting metal—in particular, from iron lying beneath the oxygen-poor region of the drop. Iron atoms there give up their electrons, form Fe^{2+} ions, and drift away into the surrounding water. There they meet more oxygen, and become oxidized to Fe^{3+} by the oxidizing agent in any couple above Fe^{3+}/Fe^{2+} in the electrochemical series, including oxygen. As they do so, they precipitate as a hydrated iron(III) oxide, $Fe_2O_3{\cdot}xH_2O$, the brown, insoluble substance we call rust. The water is more highly conducting if it carries dissolved ions, and rusting can occur more rapidly. That is one of the reasons why the salt in the air of coastal cities (and in that of inland cities after salt has been used for deicing highways) can be so damaging.

The prevention of corrosion. The simplest way to prevent corrosion is to ensure that the surface of the metal is not exposed to air and water. This protection can be achieved by painting. A more sophisticated method is to **galvanize** the metal, or coat it with an unbroken film of zinc, either by dipping it into molten zinc (as is done for automobiles and girders; Fig. 17.18) or by electroplating it (a process that is described later). Zinc lies below iron in the electrochemical series, so when a scratch exposes the metal beneath, the more strongly reducing zinc releases electrons to the iron. Hence the zinc, not the iron, is oxidized. The zinc itself survives exposure on an unbroken surface because, like aluminum, it is passivated by a protective film of zinc oxide.

Zinc plating is better than tin plating because the Sn^{2+}/Sn couple, unlike the Zn^{2+}/Zn couple, is more positive than the iron couple. As soon as a tin-plated can is scratched, the more strongly reducing iron supplies electrons to the tin and the iron is oxidized rapidly (Fig. 17.19). For this reason, tin-plated steel cans, which were more common until aluminum began to replace them, corrode rapidly once they have been damaged.

It is not possible to galvanize structures as large as ships, underground pipelines or gasoline storage tanks, and bridges, but a similar measure, called **cathodic protection**, can be used. Instead of the entire surface of the underground pipeline being covered with a more strongly reducing metal (one that lies lower than iron in the third col-

(a)

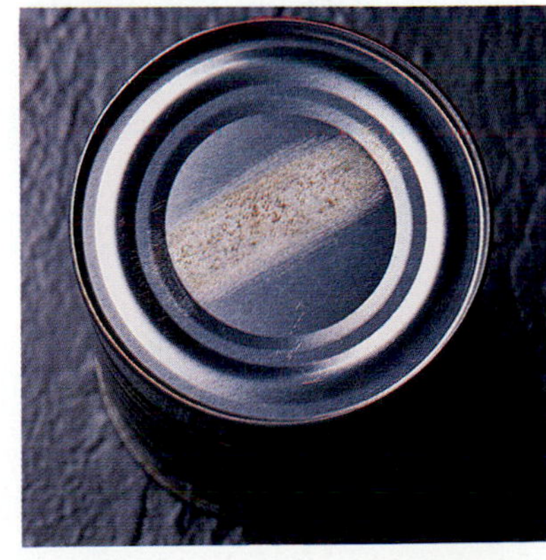

(b)

FIGURE 17.19 The Sn^{2+}/Sn couple lies above the Fe^{2+}/Fe couple in the electrochemical series. When a tin-plated iron can is scratched (a), the iron is rapidly oxidized by Sn^{2+} ions, especially in a damp environment (b).

umn of Table 17.3), a block of the more strongly reducing metal, such as zinc or magnesium, is connected to the underground pipeline (Fig. 17.20). Then the block of magnesium, and not the pipeline, is oxidized and supplies electrons for the reduction of oxygen. The surrounding soil must be damp so that the magnesium cations can migrate away from this anode and anions from the soil can migrate toward the magnesium block. The block of metal, which is called a **sacrificial anode**, protects the iron pipeline (which now serves as the cathode in the galvanic cell) and is cheaper to replace than the pipeline itself. For similar reasons, automobiles generally have "negative ground systems" (recall that the anode is the negative electrode in the galvanic cell) as part of their electric circuitry; that is, the body of the car is connected to the anode of the battery. The decaying anode of the battery is the sacrifice that helps preserve the vehicle itself.

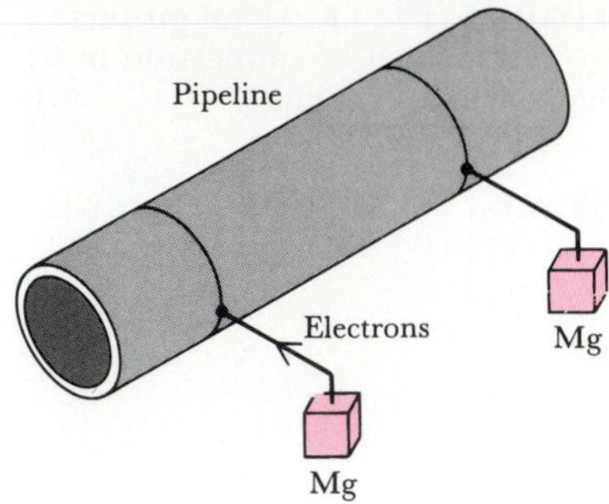

FIGURE 17.20 In the cathodic protection of a buried pipeline, or other large metal construction, the artifact is connected to a number of buried blocks of metal, such as magnesium or zinc. The sacrificial anodes (the magnesium blocks) supply electrons to the pipeline (the cathode of the cell), thereby preserving it from oxidation.

17.5 THE DEPENDENCE OF CELL POTENTIAL ON CONCENTRATION

So far, we have used standard electrode potentials to discuss redox reactions quantitatively and the electrochemical series to discuss them qualitatively. We can now take our discussion of electrochemistry and thermodynamics one step further and see how to determine the equilibrium constants of reactions and the cell potentials when the conditions are not standard.

D Concentration cells: Hydrogen ion. TDC, 144.

E Chemical potential and activities: An electrochemical introduction. *J. Chem. Ed.*, **1986**, *63*, 493.

E Concentration cells with liquid junction. *J. Chem. Ed.*, **1971**, *48*, 741. Thermodynamic parameters from an electrochemical cell. *J. Chem. Ed.*, **1970**, *47*, 365.

Calculating the equilibrium constant. Standard reduction potentials can be used to calculate the equilibrium constant of any reaction that can be expressed as the sum of a reduction half-reaction and an oxidation half-reaction. We saw in Section 16.9 that the standard reaction free energy ΔG° is related to the equilibrium constant of the reaction by

$$\Delta G^\circ = -RT \ln K$$

We also saw in Section 17.3 that the standard reaction free energy also is related to the standard cell potential of a galvanic cell by

$$\Delta G^\circ = -nFE^\circ$$

Combining the two equations gives

$$-nFE^\circ = -RT \ln K$$

This expression rearranges to

$$\ln K = \frac{nFE^\circ}{RT}$$

The combination RT/F occurs frequently in electrochemistry; at 25°C (298.15 K) it has the value

$$\frac{RT}{F} = \frac{8.314 \text{ J/(K·mol)} \times 298.15 \text{ K}}{9.6485 \times 10^4 \text{ C/mol}} = 0.02569 \text{ J/C}$$

That is, at 25°C,

$$\frac{RT}{F} = 0.02569 \text{ V}$$

because 1 J/C = 1 V.

We can calculate $E°$(cell) from standard electrode potentials, so now we can also calculate equilibrium constants. For example, the cell potential for the overall reaction

$$Zn(s) + Cu^{2+}(aq) \rightleftharpoons Zn^{2+}(aq) + Cu(s) \qquad \frac{[Zn^{2+}]}{[Cu^{2+}]} = K_c \text{ at equilibrium}$$

is 1.10 V and $n = 2$ for the reaction as written, so

$$\ln K_c = \frac{2 \times 1.10 \text{ V}}{0.02569 \text{ V}} = 85.6$$

Taking the natural antilogarithm (e^x on a calculator) gives $K_c = 1.6 \times 10^{37}$. Now we know not only that the reaction is spontaneous as written but also that equilibrium is reached only when the concentration of Zn^{2+} ions is over 10^{37} times greater than that of Cu^{2+} ions. For all practical purposes, the reaction goes to completion. Therefore, the equilibrium constant of any reaction that can be derived from two redox couples listed either in Table 17.3 or in Appendix 2B can be calculated.

▼ EXAMPLE 17.6 Calculating an equilibrium constant

Calculate the equilibrium constant of the reaction $AgCl(s) \rightleftharpoons Ag^+(aq) + Cl^-(aq)$.

STRATEGY We should recognize that the equilibrium constant for the reaction is actually the solubility constant, K_s, for AgCl (Section 15.6). We need to find from Table 17.3 or Appendix 2B the two redox couples that, when the reduction half-reaction (the cathode reaction) is added to the oxidation half-reaction (the anode reaction), gives the solubility equilibrium. The sum of the two standard potentials for the half-reactions (the reduction reaction as given in the table and the oxidation reaction with the sign of the potential reversed) gives $E°$ for the overall reaction, from which we can calculate the equilibrium constant by using the formula given in the text.

SOLUTION Table 17.3 lists

$$Ag^+(aq) + e^- \longrightarrow Ag(s) \qquad E° = +0.80 \text{ V}$$

$$AgCl(s) + e^- \longrightarrow Ag(s) + Cl^-(aq) \qquad E° = +0.22 \text{ V}$$

To obtain the solubility equilibrium, we reverse the first half-reaction and add it to the second half-reaction:

Cathode reaction: $AgCl(s) + e^- \longrightarrow Ag(s) + Cl^-(aq)$ $E°(\text{cathode}) = +0.22 \text{ V}$

Anode reaction: $Ag(s) \longrightarrow Ag^+(aq) + e^-$ $E°(\text{anode}) = -0.80 \text{ V}$

Cell reaction: $AgCl(s) \longrightarrow Ag^+(aq) + Cl^-(aq)$ $E°(\text{cell}) = -0.58 \text{ V}$

We see that $n = 1$. Note that $E°$ of the cell is negative; this means that $\Delta G°$ is positive and therefore that the reaction is not thermodynamically spontaneous under standard conditions; it follows that the equilibrium constant is expected to be less than 1. In fact,

$$\ln K = \frac{nFE°}{RT} = \frac{1 \times (-0.58 \text{ V})}{0.02569 \text{ V}} = -22.6$$

Taking the antilogarithm (e^x) gives $K_s = 1.6 \times 10^{-10}$, as in Table 15.5.

EXERCISE E17.6 Calculate the solubility constant of mercury(I) chloride.

[*Answer*: 1.3×10^{-18}] ▲

U Exercises 17.72, 17.74, and 17.76.

E Principles of electrochemical energy conversion. *J. Chem. Ed.*, **1971**, *48*, 732.

E Walther Nernst (1864–1941) was also a chemist of many parts: he proposed the concept of solubility product (Chapter 15) and the use of buffer solutions (in 1889 and 1903, respectively). He proposed a version of the *third law* of thermodynamics and made a fortune by inventing a type of incandescent lamp.

? What happens to the cell potential relative to E°(cell) in Example 17.7 for the conditions $[Cu^{2+}] = 0.010$ mol/L and $[Zn^{2+}] = 1.0$ mol/L?

U Exercises 17.65, 17.66, 17.78, and 17.79.

Example 17.6 shows that we now have an electrochemical method of measuring the solubility constant, and hence the solubility, of a sparingly soluble salt. This procedure is a much more accurate method for obtaining these quantities than trying to measure the minute amount of solid that dissolves in a liter of water.

The Nernst equation. To see how cell potentials depend on concentration and pressure, we make use of the equation relating ΔG to composition (Eq. 10 of Section 16.8):

$$\Delta G = \Delta G^\circ + RT \ln Q$$

Because $\Delta G = -nFE$ and $\Delta G^\circ = -nFE^\circ$, it follows that

$$-nFE = -nFE^\circ + RT \ln Q$$

Rearrangement of this equation produces the **Nernst equation**:

$$E = E^\circ - \frac{RT}{nF} \ln Q$$

The equation is named for Walther Nernst, the German chemist who first derived it. At 25°C,

$$E = E^\circ - \frac{0.02569 \text{ V}}{n} \times \ln Q$$

EXAMPLE 17.7 Using the Nernst equation to determine a cell potential under nonstandard conditions

Calculate the cell potential at 25°C of a Daniell cell in which the concentration of Zn^{2+} ions is 0.10 mol/L and that of the Cu^{2+} ions is 0.0010 mol/L.

STRATEGY First, we determine E° from the standard potentials in Table 17.3 (or Appendix 2B) and then calculate the value of Q_c for the stated conditions. After that, the cell potential is calculated by substituting these values into the Nernst equation.

SOLUTION We have already determined that $E^\circ = 1.10$ V for the Daniell cell. The cell reaction is

$$Cu^{2+}(aq) + Zn(s) \longrightarrow Zn^{2+}(aq) + Cu(s) \qquad n = 2$$

so the reaction quotient is

$$Q_c = \frac{[Zn^{2+}]}{[Cu^{2+}]} = \frac{0.10}{0.0010} = 100$$

Therefore, the Nernst equation is

$$E = E^\circ - \frac{0.02569 \text{ V}}{n} \times \ln Q = 1.10 \text{ V} - \frac{0.02569 \text{ V}}{2} \times \ln 100$$

$$= 1.10 \text{ V} - 0.059 \text{ V} = 1.04 \text{ V}$$

EXERCISE E17.7 Calculate the potential of the galvanic cell $Zn|Zn^{2+}(1.50 \text{ mol/L})||Fe^{2+}(0.10 \text{ mol/L})|Fe$

[*Answer*: 0.28 V]

Measuring the concentrations of ions. It is easy to measure the potential of a galvanic cell with a simple voltmeter; then the Nernst equation can be used to convert that measurement to the concentrations of ions in solution. For example, if the cell potential of a Daniell cell is 1.10 V, then we know from the Nernst equation that the concentrations of the Cu^{2+} and Zn^{2+} ions are equal (because then $Q_c = 1$ and $\ln Q_c = 0$). If the concentration of the Cu^{2+} ions is 0.0010 mol/L, then we know from the voltmeter reading that the concentration of Zn^{2+} ions also is 0.0010 mol/L. If the voltmeter displays the value of 1.04 V (as in Example 17.7), then we know that $Q_c = 100$, and hence that the concentration of Zn^{2+} ions is 100 times that of the Cu^{2+} ions, and hence 0.10 mol/L.

E On the Nernst equations of mercury-mercuric oxide half-cell equations. *J. Chem. Ed.*, **1983**, *60*, 133.

▼ EXAMPLE 17.8 Determining the concentration of an ion in a redox couple

The potential of a cell consisting of a standard $AgCl/Ag,Cl^-$ electrode and a Cu^{2+}/Cu electrode is 0.070 V. What is the molar concentration of Cu^{2+} ions in the cell?

STRATEGY We need to write the diagram for the cell and determine the standard cell potential, and we also need to write the cell reaction and the expression for the reaction quotient. Next, we write the Nernst equation for the cell, solve it for the reaction quotient, and then rearrange Q_c to obtain the value of $[Cu^{2+}]$.

SOLUTION We see in Table 17.3 that the Cu^{2+}/Cu redox couple has a higher reduction potential ($E° = +0.34$ V) than the $AgCl/Ag,Cl^-$ redox couple has ($E° = +0.22$ V). Hence the Cu^{2+}/Cu redox couple serves as the cathode and the $AgCl/Ag,Cl^-$ redox couple serves as the anode in the galvanic cell. The cell diagram is therefore

$$Ag|AgCl(s)|Cl^-(aq,\ 1.0\ mol/L)||Cu^{2+}(aq,\ ?\ mol/L)|Cu$$

Cathode reaction: $Cu^{2+}(aq) + 2e^- \longrightarrow Cu(s)$ $\quad E°(\text{cathode}) = +0.34$ V

Anode reaction: $Ag(s) + Cl^-(aq) \longrightarrow AgCl(s) + e^-$ $\quad E°(\text{anode}) = -0.22$ V

Cell reaction: $Cu^{2+}(aq) + 2Ag(s) + 2Cl^-(aq) \longrightarrow 2AgCl(s) + Cu(s)$ $\quad n = 2$

$$E°(\text{cell}) = E°(\text{cathode}) + E°(\text{anode}) = 0.12\ \text{V}$$

The reaction quotient is

$$Q_c = \frac{1}{[Cu^{2+}][Cl^-]^2}$$

It follows from the Nernst equation that

$$0.070\ \text{V} = 0.12\ \text{V} - \frac{0.02569\ \text{V}}{2} \times \ln\left(\frac{1}{[Cu^{2+}] \times (1.0)^2}\right)$$

This expression rearranges to

$$\ln\left(\frac{1}{[Cu^{2+}]}\right) = \frac{-2 \times (0.070\ \text{V} - 0.12\ \text{V})}{0.02569\ \text{V}} = 3.89$$

and therefore

$$\frac{1}{[Cu^{2+}]} = 48.9, \qquad \text{giving } [Cu^{2+}] = 0.020\ \text{mol/L}$$

EXERCISE E17.8 The cell potential of $Cd|Cd^{2+}(aq, 0.010\ mol/L)||Fe^{2+}(aq,\ ?\ mol/L), Fe^{3+}(aq, 1.0\ mol/L)|Pt$ was measured as 1.53 V. What is the molar concentration of the Fe^{2+} ions in the cell? (For data, see Appendix 2B.)

U Exercises 17.80 and 17.81.

▲ [*Answer*: 8.3×10^{-6} mol/L]

Electrical measurement of pH. One important application of the Nernst equation is the measurement of pH (and, through pH, of the ionization constants of acids and bases, as explained in Section 14.10). The technique makes use of a cell in which one electrode that is sensitive to the H_3O^+ concentration is combined with another electrode. One combination is a hydrogen electrode and a calomel electrode, for which the redox couple is $Hg_2Cl_2/Hg,Cl^-$ (calomel is the common name of mercury(I) chloride); the two electrode compartments are connected through a salt bridge. The reduction half-reaction for the calomel electrode is

N The saturated calomel electrode, in which the electrolyte is a saturated KCl solution, is commonly used as the reference electrode. The saturated calomel electrode has $E = 0.242$ V.

$$Hg_2Cl_2(s) + 2e^- \longrightarrow 2Hg(l) + 2Cl^-(aq) \qquad E° = +0.27 \text{ V}$$

and the complete cell and the cell reaction are

$$Pt|H_2(g)|H^+(aq)||Cl^-(aq)|Hg_2Cl_2(s)|Hg(l)$$

$$Hg_2Cl_2(s) + H_2(g) \longrightarrow 2H^+(aq) + 2Hg(l) + 2Cl^-(aq) \qquad Q = [H^+]^2[Cl^-]^2$$

(We are assuming that the pressure of hydrogen gas is 1 atm.) We want to know the concentration of hydrogen ions in the anode compartment; from the Nernst equation, we can write

$$\begin{aligned} E &= E° - \frac{0.02569 \text{ V}}{2} \times \ln [H^+]^2[Cl^-]^2 \\ &= 0.27 \text{ V} - \frac{0.02569 \text{ V}}{2} \times \ln [Cl^-]^2 - \frac{0.02569 \text{ V}}{2} \times [\ln H^+]^2 \\ &= 0.27 \text{ V} - 0.02569 \text{ V} \times \ln [Cl^-] - 0.02569 \text{ V} \times \ln [H^+] \end{aligned}$$

The concentrations of H^+ and Cl^- ions are unrelated because they occur in different electrode compartments. Moreover, the Cl^- concentration is fixed for a given electrode (it is fixed when the calomel electrode is manufactured), so we take it to be a constant. Therefore, the first two terms on the right are constants and we combine them into a single symbol E', the **cell constant**. Because the relation between a natural logarithm and a common logarithm is $\ln x = 2.303 \log x$, it follows that

$$\begin{aligned} E &= E' - 0.02569 \text{ V} \times 2.303 \times \log [H^+] \\ &= E' + 0.0592 \text{ V} \times \text{pH} \end{aligned}$$

That is, the cell potential is proportional to the pH; so by measuring the potential, we can determine pH.

The hydrogen electrode is difficult to use in practice because it is awkward to set up and because it settles down to a stable reading only very sluggishly. The **glass electrode**, a thin-walled glass bulb containing an electrolyte, is very much easier to use and has a potential that is proportional to the pH (Fig. 17.21). It is normally used in conjunction with a calomel electrode that makes contact with the test solution through a salt bridge. The cell potential is measured and the pH of the solution is displayed directly. The meter is calibrated with a phosphate buffer of known pH, such as that described in Section 15.5

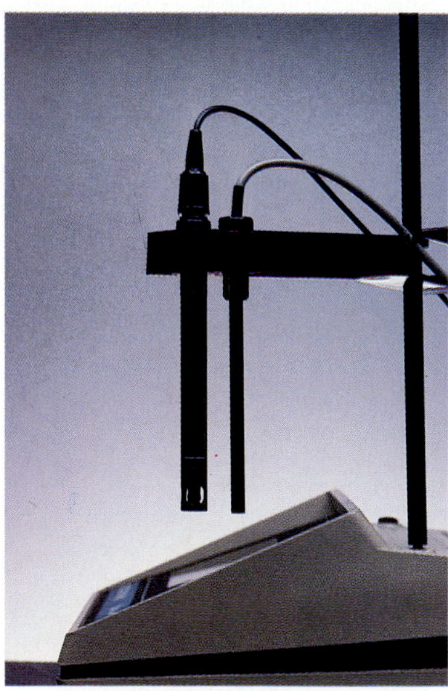

FIGURE 17.21 A glass electrode (left) is used to measure pH. It is generally used in conjunction with a calomel electrode (right).

Because other commercially available electrodes are sensitive to other ions, "pX meters" can be used to measure the concentrations of various ions X. These electrodes are useful for monitoring industrial processes and in pollution control. Electrodes are available for measuring the concentrations of the cations Na^+, K^+, Ca^{2+}, Ag^+, and NH_4^+ and the anions F^-, Cl^-, Br^-, CN^-, and S^{2-}, and many others.

▼ **EXAMPLE 17.9 Measuring the pH of a solution by using a reference calomel electrode**

A voltmeter connected to a glass electrode and a saturated calomel electrode was calibrated with a phosphate buffer having a pH of 7.20; the cell potential was 0.15 V at this pH. When the buffer solution was replaced by a solution of unknown pH, the voltmeter read 0.23 V. What is the pH of the unknown solution?

STRATEGY The first task is to determine the cell constant E' by using the relation between the measured potential and the known pH of the buffer solution, $E = E' + 0.0592\ \text{V} \times \text{pH}$. Once E' has been determined, it may be used to determine the pH of the unknown solution.

D Demonstration of electrochemical cell properties by a simple, colorful oxidation-reduction experiment. *J. Chem. Ed.*, **1982**, *59*, 586.

SOLUTION We can deduce from the calibration measurement that

$$E' = E - 0.0592\ \text{V} \times \text{pH} = 0.15\ \text{V} - 0.0592\ \text{V} \times 7.20 = -0.28\ \text{V}$$

Therefore, for solutions of unknown pH,

$$E = -0.28\ \text{V} + 0.0592\ \text{V} \times \text{pH}$$

If the voltmeter reads 0.23 V for the unknown solution, then the pH is

$$\text{pH} = \frac{E + 0.28\ \text{V}}{0.0592\ \text{V}} = \frac{0.23\ \text{V} + 0.28\ \text{V}}{0.0592\ \text{V}} = 8.61$$

U Exercise 17.82.

T OHT: Fig. 17.22.

EXERCISE E17.9 With the same equipment, the cell potential for another solution is 0.088 V. What is the pH of this solution?

[*Answer*: 6.21] ▲

ELECTROLYSIS

We now turn to redox reactions that are not spontaneous (because they have a positive reaction free energy) and that therefore need to be driven to form products. The process of driving a reaction in a nonspontaneous direction by passing an electric current through a solution is called **electrolysis**, and it is carried out in an electrochemical cell acting as an electrolytic cell. Much of the discussion of galvanic cells also applies to electrolytic cells, but there are three major differences:

1. Two electrodes share the same compartment in an electrolytic cell.
2. An electrolytic cell has a single electrolyte.
3. The conditions are usually far from standard: gas pressures are rarely close to 1 atm and ion concentrations are often not 1 mol/L.

A schematic diagram of an electrolytic cell for the commercial production of magnesium metal (the Dow process) is shown in Fig. 17.22. It has two electrodes that dip into molten magnesium chloride, which melts at 710°C. As in a galvanic cell, oxidation occurs at the anode and reduction occurs at the cathode, electrons travel through the external wire from anode to cathode, cations move through the electrolyte toward the cathode, and anions move toward the anode. Unlike a galvanic cell, however, a current supply drives electrons through the external wire in a predetermined direction, forcing oxidation to occur at one electrode and reduction at the other.

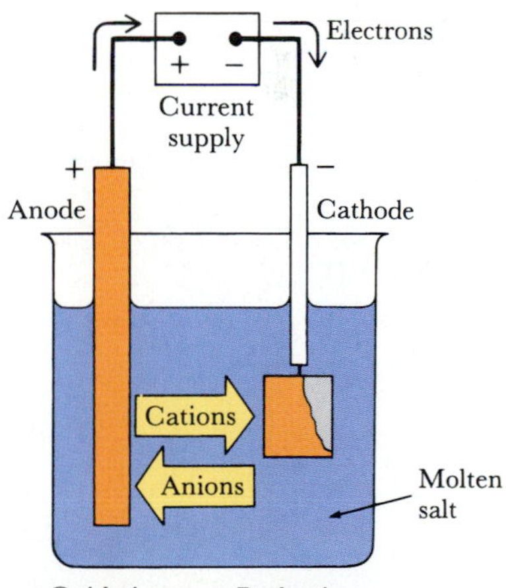

FIGURE 17.22 A simple electrolytic cell. The current enters the cell at the cathode, which is connected to the − terminal of the external supply. It leaves the cell through the anode, which is connected to the + terminal.

N The Dow process is discussed again in Chapter 18.

E Electrolysis of potassium iodide. CD2, 104.

The anode of an electrolytic cell is connected to the positive terminal of the current supply (where electrons enter the supply), and the cathode is connected to the negative terminal of the supply (where electrons leave the supply). Therefore, because the anode is an extension of the positive terminal of the supply, it too is positive; consequently, the anions (the Cl^- ions) are attracted to the anode and undergo oxidation there:

$$\textit{Anode reaction: } 2Cl^-(l) \longrightarrow Cl_2(g) + 2e^-$$

As a result, chlorine gas is evolved at the anode. The cathode, an extension of the negative terminal of the current supply, attracts the cations, which are forced to accept the electrons and hence undergo reduction:

$$\textit{Cathode reaction: } Mg^{2+}(l) + 2e^- \longrightarrow Mg(l)$$

Magnesium metal is produced at the cathode. The overall reaction, the sum of the oxidation and reduction half-reactions, for the commercial production of magnesium is therefore

$$Mg^{2+}(l) + 2Cl^-(l) \xrightarrow{710°C} Mg(l) + Cl_2(g)$$

The current supplied to an electrolytic cell forces the flow of electrons in one direction; therefore, it must come from a direct current source, such as (in laboratory applications) a 9-V battery, a galvanic cell. The anode of the galvanic cell is the source of electrons that enter the electrolytic cell and bring about reduction at the latter's cathode. The cathode of the galvanic cell needs electrons; it acquires them by drawing them out of the electrolytic cell, thereby causing oxidation to occur at the latter's anode (Fig. 17.23).

17.6 THE POTENTIAL NEEDED FOR ELECTROLYSIS

E The coupling of an electrolytic cell to a galvanic cell is an example of the spontaneity of an *overall* process even though one of its component processes (the electrolysis reaction) is not spontaneous.

D Water electrolysis in yellow, green, and blue. CD2, 106.

We mentioned the electrolysis of water in Chapter 2 and are now in a position to discuss it and the electrolysis of various aqueous solutions more fully. As we have seen, the general idea behind electrolysis is to use an electric current to drive a reaction in the reverse of its spontaneous direction. This goal can be achieved by connecting the electrolytic cell to a current supply that generates a potential greater than the reaction would generate if it were free to run in its spontaneous direction. The nonspontaneous reaction

$$2H_2O(l) \longrightarrow 2H_2(g) + O_2(g)$$

illustrates this point. It is the reverse of the cell reaction discussed in Example 17.2:

$$2H_2(g) + O_2(g) \longrightarrow 2H_2O(l)$$

$$Pt|H_2(g)|OH^-(aq)||OH^-(aq)|O_2(g)|Pt \qquad E = 1.23 \text{ V at pH} = 7$$

To reverse this (spontaneous) cell reaction, we must force oxidation (to produce O_2) on the right side of the cell diagram and reduction (to produce H_2) on the left side. The reversal of the spontaneous reaction requires a supply of at least 1.23 V from the external source to overcome the reaction's natural "pushing power."

The driving power that must be supplied can be apportioned between the two electrodes by considering their individual potentials. In

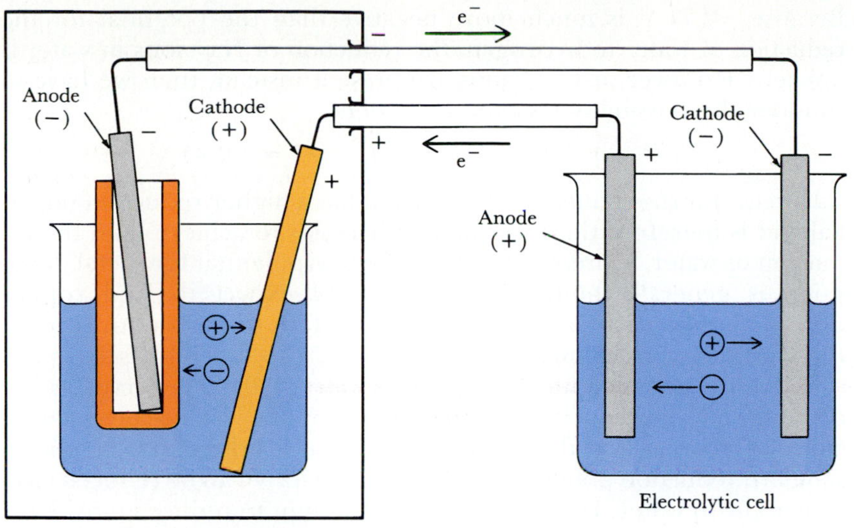

FIGURE 17.23 In this schematic diagram of an electrolysis experiment, the electrons emerge from the galvanic cell at its anode (−) and enter the electrolytic cell at its cathode (−), where they bring about reduction. Electrons are drawn out of the electrolytic cell through its anode (+) and into the galvanic cell at its cathode (+). If the cell reaction in the galvanic cell is more strongly spontaneous than the reaction in the electrolytic cell is nonspontaneous, then the overall process is spontaneous. This experiment is an example of one reaction driving another to which it is coupled.

neutral water, where pH = 7, the two half-reactions we want to drive by electrolysis are

Cathode reaction: $2H_2O(l) + 2e^- \longrightarrow H_2(g) + 2OH^-(aq) E = -0.42\ V$

Anode reaction: $2H_2O(l) \longrightarrow O_2(g) + 4H^+(aq) + 4e^- \quad E = -0.81\ V$

Therefore, of the 1.23 V from the current supply, 0.42 V is needed for the reduction to H_2 and 0.81 V is needed for the oxidation to O_2.

T OHT: Fig. 17.23.

E Overpotential. *J. Chem. Ed.*, **1971**, *48*, 352.

Overpotential. In practice, appreciable product formation from the reversal of a spontaneous reaction by electrolysis is obtained only if the applied potential is significantly greater than the cell potential. The additional potential that must be applied beyond the cell potential is called the **overpotential**. The overpotential for the production of hydrogen and oxygen, using platinum electrodes, is about 0.6 V, so about 1.8 V (and not 1.23 V) must be applied to the electrolytic cell before hydrogen and oxygen are evolved at an appreciable rate. Of this 0.6 V of cell overpotential, 0.5 V can be ascribed to the oxygen evolution, and only 0.1 V is needed for the hydrogen evolution. The overpotential required when other electrode materials are used may be quite different. Hydrogen evolution on a lead electrode needs an overpotential as high as 0.6 V, whereas oxygen evolution requires 0.3 V.

E The high overpotential of hydrogen production on lead is the reason why so little hydrogen is produced while the battery is being recharged. The chapter-opening photograph on p. 652 shows hydrogen evolution only because a high current density is being used and the overpotential is exceeded.

N Because of overpotentials and ionic effects in solution, the predictions of anode and cathode products discussed here are only intelligent guesses.

D Reduction potentials and hydrogen overvoltage: An overhead projector demonstration. *J. Chem. Ed.*, **1982**, *59*, 866. A simple and dramatic demonstration of overvoltage [overpotential]. *J. Chem. Ed.*, **1983**, *60*, 674.

Competing reduction reactions. In the electrolysis of aqueous salt solutions, it is possible not only for the water to be oxidized and reduced but also for the ions of the salt to be oxidized and reduced at the respective electrodes. At the cathode, the reduction of H_2O at pH = 7 requires 0.42 V:

$$2H_2O(l) + 2e^- \longrightarrow H_2(g) + 2OH^-(aq) \qquad E = -0.42\ V$$

Any oxidized species having a reduction potential that is more positive than −0.42 V will be preferentially reduced at the cathode. We can use Table 17.3 as a guide. The standard reduction potential of Na^+ ions, for example, is

$$Na^+(aq) + e^- \longrightarrow Na(s) \qquad E° = -2.71\ V$$

V Video 42: Electrolysis of water.

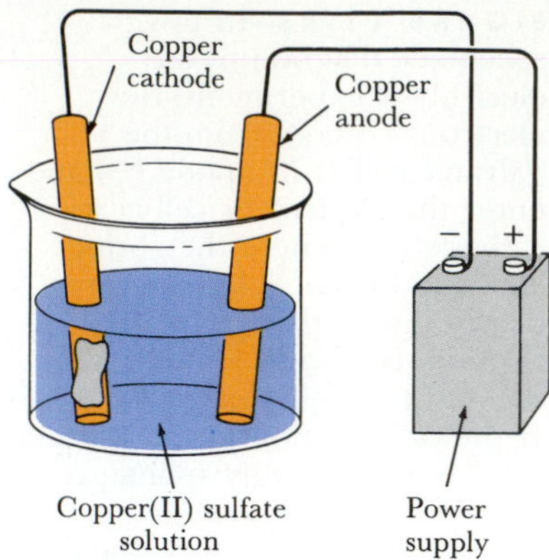

FIGURE 17.24 Copper is refined electrolytically before it is used in electrical wiring. The impurities include precious metals, which, when recovered, help pay for the refining.

Because −2.71 V is much more negative than the potential for the reduction of water to hydrogen, the reduction of Na^+ ions in water is unlikely. However, if Cu^{2+} ions are present instead, then we have to consider the possibility that the reduction

$$Cu^{2+}(aq) + 2e^- \longrightarrow Cu(s) \qquad E° = +0.34 \text{ V}$$

will occur, for the reduction of Cu^{2+} ions has a higher reduction potential and is therefore thermodynamically easier to achieve than the reduction of water. Moreover, because the overpotential for metal deposition is generally quite small, we would expect to find copper deposited rather than hydrogen evolved. In fact, this procedure is used for purifying copper metal (Fig. 17.24).

The reduction potential for neutral water, −0.42 V, serves as the dividing line between couples that can be reduced electrolytically in aqueous solution and those that cannot. Any oxidized species with a standard reduction potential above about −0.42 V can be reduced electrolytically in water. However, when we attempt to reduce an oxidized species lying below about −0.42 V, hydrogen will be produced instead. Aluminum ions (−1.66 V), for example, cannot be reduced to the metal in water; however, silver ions (+0.80 V) can be reduced to silver.

Competing oxidation reactions. We must always consider the possibility that ions present in solution, as well as the water, can be oxidized. At the anode, the oxidation of H_2O at pH = 7 requires at least 0.81 V:

$$2H_2O(l) \longrightarrow O_2(g) + 4H^+(aq) + 4e^- \qquad E = -0.81 \text{ V}$$

The kind of question we need to answer is whether any of the ions present in solution is more or less readily oxidized than water is. That is, does the ion have a higher or lower potential for oxidation? For instance, when Cl^- ions are present (at 1 mol/L) in water, is it possible that they, and not the water, will be oxidized? From Table 17.3, we see that the oxidation of Cl^- requires at least 1.36 V:

$$2Cl^-(aq) \longrightarrow Cl_2(g) + 2e^- \qquad E° = -1.36 \text{ V}$$

Because only 0.81 V is needed to force the oxidation of water but 1.36 V is needed to force the oxidation of Cl^-, it appears that oxygen should be the product at the anode. However, the overpotential for oxygen production is very high, and in practice chlorine may also be produced.

If the solution contains I^- ions (at 1 mol/L), then we need to consider whether they will be oxidized. From Table 17.3, we see that the oxidation of I^- ions requires −0.54 V:

$$2I^-(aq) \longrightarrow I_2(s) + 2e^- \qquad E° = -0.54 \text{ V}$$

Because any reduced species having a more positive oxidation potential than −0.81 V will be preferentially oxidized at the anode in an aqueous solution, the I^- ion will be oxidized.

17.7 THE EXTENT OF ELECTROLYSIS

We now focus on the amount of substance that can be electrolyzed by a given amount of electricity. This is where we add the final arrow to the mole diagram (**1**). The data might be the amount of electricity supplied

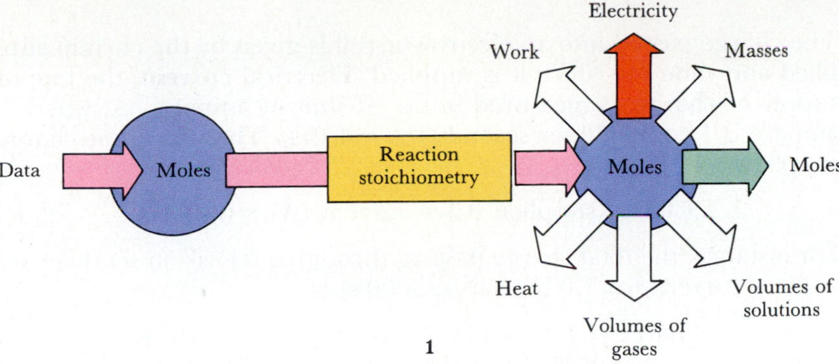

1

and the desired outcome of the calculation the amount, mass, or volume of product that the electrolysis forms.

T OHT: Diagram **1**.

E Edison's chemical meter. *J. Chem. Ed.*, **1972**, *49*, 626.

Faraday's laws. Copper metal is refined electrolytically by using an impure form of copper metal (called blister copper) as the anode in an electrolytic cell. The current supply causes the oxidation of the blister copper to copper(II) ions, and because Cu^{2+} has a higher reduction potential than water and any other ion in the electrolyte, it is reduced at the cathode:

E The blister copper contains the impurities silver and gold, which survive the oxidation and merely fall from the anode to the bottom of the cell to form the *anode mud*. Other impurities include iron and nickel, which are oxidized prior to the copper from the anode, but are not as readily reduced as copper at the cathode and therefore remain in solution.

$$\textit{Cathode reaction: } Cu^{2+}(aq) + 2e^- \longrightarrow Cu(s,\text{pure})$$

$$\textit{Anode reaction: } Cu(s,\text{blister}) \longrightarrow Cu^{2+}(aq) + 2e^-$$

$$\textit{Cell reaction: } Cu(s,\text{blister}) \longrightarrow Cu(s,\text{pure})$$

For the electrolytic deposition of Cu^{2+} at the cathode, the balanced cathodic reaction states that 2 mol e^- = 1 mol Cu. We can use this relation as a conversion factor for predicting the amount (in moles) of copper metal produced when a given amount of electrons is supplied through the cathode. For example, if 4.0 mol of electrons are supplied, then, according to the reaction stoichiometry, the amount of copper produced is

$$\text{Amount of Cu (mol)} = 4 \text{ mol e}^- \times \frac{1 \text{ mol Cu}}{2 \text{ mol e}^-} = 2.0 \text{ mol Cu}$$

This simple calculation is sometimes summarized by the following statement, which is a combined, modern version of Faraday's laws:

Faraday's law of electrolysis: The amount (in moles) of product formed by an electric current is chemically equivalent to the amount (in moles) of electrons supplied.

E The original forms of Faraday's two laws of electrolysis were essentially: (1) The amount of chemical decomposition produced by a current is proportional to the quantity of electricity passing through the electrolytic solution. (2) The amounts of different substances liberated by the same quantity of electricity are proportional to their chemical equivalent weights [molar mass divided by charge number].

From the number of moles of product formed, we can calculate the masses of the products or, if they are gases, their volumes.

Measuring the amount of electricity. The final link between the amount of product formed at an electrode and the quantity of electricity (in coulombs) passed through the electrolytic cell is the determination of the number of moles of electrons supplied by the current. Faraday's constant, the charge per mole of electrons, 9.65×10^4 C/(mol e^-), can be used as a conversion factor to convert charge to moles of electrons:

$$\text{Amount of e}^- \text{ (mol)} = \text{charge supplied (C)} \times \frac{1 \text{ mol e}^-}{9.65 \times 10^4 \text{ C}}$$

U Exercises 17.83 and 17.84.

N Some useful relations are:

$1\ W = 1\ J/s$
$1\ J = 1\ A{\cdot}V{\cdot}s = 1\ C{\cdot}V$
$1\ C = 1\ A{\cdot}s$

The charge passed into an electrolytic cell is given by the current supplied and time for which it is supplied. Electrical **current**, the rate of supply of charge, is measured in the SI unit of amperes, A, where 1 ampere is 1 coulomb per second (1 A = 1 C/s). Therefore, the charge supplied is

$$\text{Charge supplied (C)} = \text{current (A)} \times \text{time (s)}$$

For instance, the total charge passing through a television set that uses a 2.0-A current for 1.0 h (that is, 3600 s) is

$$\text{Charge (C)} = 2.0\ \text{C}{\cdot}\text{s}^{-1} \times 3600\ \text{s} = 7.2 \times 10^3\ \text{C}$$

Hence, by measuring the current and the time for which it flows, we can determine the amount of electrons supplied. The number of moles of electrons passing through the television set, for instance, is

$$\text{Amount of e}^- \text{(mol)} = 7.2 \times 10^3\ \text{C} \times \frac{1\ \text{mol e}^-}{9.65 \times 10^4\ \text{C}} = 7.5 \times 10^{-2}\ \text{mol e}^-$$

Note that even appreciable currents, flowing for long periods, supply very few moles of electrons.

EXAMPLE 17.10 Calculating the amount of product produced from electrolysis

Aluminum is produced by electrolysis of its oxide, Al_2O_3, dissolved in molten cryolite, Na_3AlF_6. Calculate the mass of aluminum that can be produced in one day in an electrolytic cell operating continuously at 1.00×10^5 A.

STRATEGY The focus of calculations like this is the relation between the amount of electrons supplied and the amount of Al produced. We work from the data toward the answer. The charge supplied is the product of the current and the time for which it flows. This charge is converted to the amount of electrons by using the Faraday constant. The amount of electrons is converted to amount of Al by using the stoichiometry of the half-reaction. Finally, the amount of Al is converted to mass of aluminum by using the molar mass of Al (26.98 g/mol).

N The most important step in the calculation is the writing of the appropriate, balanced half-reaction.

SOLUTION The aluminum metal is produced from the reduction of Al^{3+}, which is present in the molten Al_2O_3. The balanced cathode half-reaction is therefore

$$Al^{3+}(l) + 3e^- \longrightarrow Al(l) \qquad 3\ \text{mol e}^- = 1\ \text{mol Al}$$

The amount of electrons passed through the cell when a current of 1.00×10^5 A (1.00×10^5 C/s) flows for 24 h (8.64×10^4 s) is

$$\text{Amount of e}^- \text{(mol)} = 8.64 \times 10^4\ \text{s} \times 1.00 \times 10^5\ \frac{\text{C}}{\text{s}} \times \frac{1\ \text{mol e}^-}{9.65 \times 10^4\ \text{C}}$$
$$= 8.95 \times 10^4\ \text{mol e}^-$$

The mass of aluminum can now be determined from the reaction stoichiometry at the cathode:

$$\text{Mass of Al (g)} = 8.95 \times 10^4\ \text{mol e}^- \times \frac{1\ \text{mol Al}}{3\ \text{mol e}^-} \times \frac{26.98\ \text{g Al}}{1\ \text{mol Al}}$$
$$= 8.05 \times 10^5\ \text{g Al} = 805\ \text{kg Al}$$

The fact that the production of 1 mol Al requires 3 mol e^- accounts for the very high consumption of electricity that is characteristic of aluminum production plants.

U Exercises 17.85–17.88, 17.91, and 17.92.

EXERCISE E17.10 Determine the time, in hours, required to electroplate 7.00 g of magnesium metal from molten magnesium chloride by using a current of 7.30 A. What volume of chlorine gas at STP will be produced at the anode?

[*Answer*: 2.12 h; 6.46 L]

17.8 APPLICATIONS OF ELECTROLYSIS

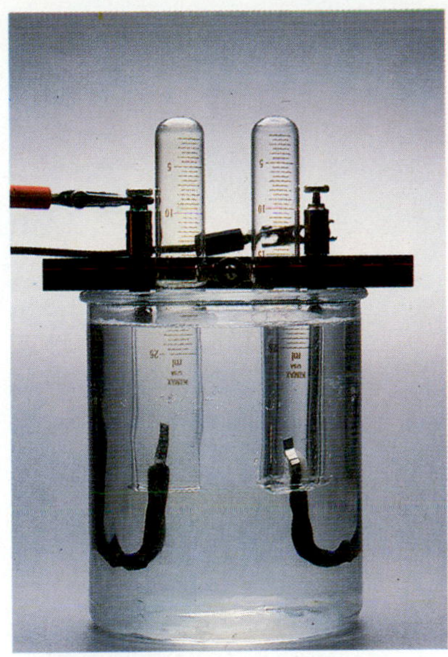

FIGURE 17.25 The electrochemical production of chlorine. An anode can be made a strong enough oxidizing agent to remove electrons from chloride ions in aqueous solution.

A cathode acts as a very powerful reducing agent. Increasing the potential between the electrodes forces electrons into the electrolyte and can bring about a reduction such as $Cu^{2+}(aq) + 2e^- \rightarrow Cu(s)$. An anode acts as a very powerful oxidizing agent. When the applied potential drags a current out of the cell, the electrons are stripped from compounds and ions near the anode, as happens in the oxidation $2Cl^-(aq) \rightarrow Cl_2(g) + 2e^-$ (Fig. 17.25). The oxidizing power of an anode can be made so great (by increasing the voltage) that even fluoride ions, which can be oxidized by no chemical reagent, can be oxidized electrolytically to the element. This was how the French chemist Henri Moisson first isolated fluorine (in 1886) from an anhydrous molten mixture of potassium fluoride and hydrogen fluoride. This method is still used for the commercial preparation of fluorine.

Sodium ion electrolysis. Another example of the industrial application of electrolysis is the production of sodium metal by the **Downs process**:

U Exercise 17.90.

T OHT: Fig. 17.26.

$$\textit{Cathode reaction: } 2Na^+(l) + 2e^- \longrightarrow 2Na(l)$$

$$\textit{Anode reaction: } 2Cl^-(l) \longrightarrow Cl_2(g) + 2e^-$$

The electrolyte is molten sodium chloride, mined as rock salt, with a little calcium chloride added to lower its melting point from 830°C to a more economical 600°C. The principal design requirement for the cell is to keep the sodium and chlorine produced by the electrolysis from contacting each other and air (Fig. 17.26).

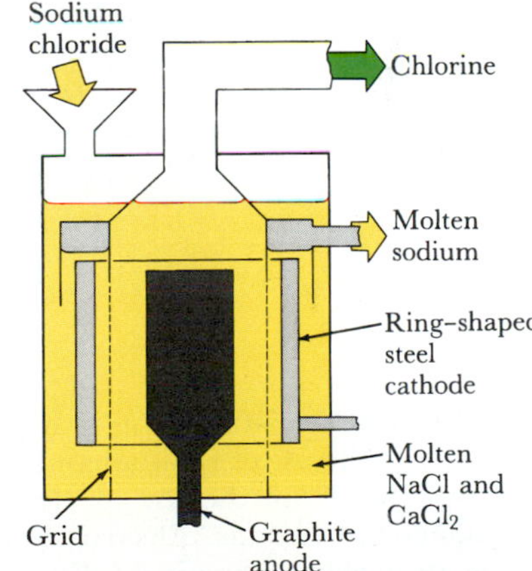

FIGURE 17.26 In the Downs process, molten sodium chloride is electrolyzed with a graphite anode (which oxidizes the Cl^- ions to chlorine) and a steel cathode (which reduces the Na^+ ions to sodium). The sodium and chlorine are kept apart by the hoods surrounding the electrodes.

We have already seen that if aqueous rather than molten sodium chloride is used as the electrolyte, then hydrogen gas and hydroxide ions are produced at the cathode, not sodium:

$$\textit{Cathode reaction: } 2H_2O(l) + 2e^- \longrightarrow H_2(g) + 2OH^-(aq)$$

It follows that the product in the cathode chamber is sodium hydroxide. Because the overpotential of oxygen is so high, chlorine is the principal oxidation product in the anode compartment:

$$\textit{Anode reaction: } 2Cl^-(l) \longrightarrow Cl_2(g) + 2e^-$$

However, when the chlorine is allowed to mix with the sodium hydroxide produced at the cathode, it is oxidized to hypochlorite ions:

$$Cl_2(g) + 2OH^-(aq) \longrightarrow ClO^-(aq) + Cl^-(aq) + H_2O(l)$$

Sodium hypochlorite is widely used as household bleach. Bleaches act by oxidizing colored compounds, and they kill bacteria by oxidation as well. This process has been suggested as a means of treating sewage in coastal areas: the sewage would be mixed with seawater (brine) and electrolyzed; the hypochlorite ions so produced would oxidize the organic matter, thereby making it harmless.

Electroplating. A common industrial application of electrolysis is **electroplating**, the deposition of a thin film of metal on an object by electrolysis. The object to be electroplated is made the cathode. Electroplating and electrolytic refining processes are generally used industrially for those metals that are easily reduced (have high reduction potentials). Many metals can be electroplated.

Metal is deposited on the cathode from the cations in the electrolyte solution. These cations originate either from a salt or from oxidation of the anode, which is made of the metal that is to be deposited. The plated object is usually metal, but plastic objects can also be plated if they are first coated with graphite.

One of the most familiar examples of electroplating is "chrome plating," in which a thin layer of chromium is deposited on another metal. The electrolyte is prepared by dissolving CrO_3 in dilute sulfuric acid. The electrolysis reduces chromium(VI) first to chromium(III) and then to chromium metal, the overall reduction being

$$CrO_3(aq) + 6H^+(aq) + 6e^- \longrightarrow Cr(s) + 3H_2O(l)$$

The chromium deposits on the cathode as a hard protective film. Because *six* electrons must be supplied for each atom of chromium deposited, a large amount of electricity is needed. Attempts to halve the cost of the electricity by starting from a chromium(III) salt have failed because the hydrated ion $Cr(H_2O)_6^{3+}$ is so stable that it cannot be reduced. When chromium(VI) is used, the chromium(III) produced in the first reduction stage is already bound to the cathode surface and can be reduced quite easily.

E Durable chrome plating. *J. Chem. Ed.*, **1972**, *49*, 626.

V Video 44: Electroplating.

U Exercise 17.89.

N The Hall process for the electrolytic production of aluminum is discussed in Chapter 19.

D Electroplating copper. CD2, 103. Electroplating silver plating from an iodide complex. TDC, 193.

We began our study of chemistry, in Chapter 1, with an elementary account of electrolysis. In the intervening chapters, we have progressed to a much deeper understanding of that process, and we can now make both qualitative and quantitative predictions about it. We have encountered other types of reactions along the way and have achieved a similar depth of understanding and capability with them. We have, in fact, built up a considerable store of knowledge about the methods and principles of chemistry. In Part IV, we shall put them to use to explain the characteristics of individual elements.

SUMMARY

17.1 Electrochemistry is the branch of chemistry concerned with the use of spontaneous reactions to generate electricity and with the use of electricity to drive nonspontaneous reactions. Electrochemical measurements are made in **electrochemical cells** of which there are two types: **galvanic cells**, in which a chemical reaction is used to generate electricity, and **electrolytic cells**, in which electricity is used to drive a chemical reaction. A galvanic cell consists of two metallic conductors, known as **electrodes**, dipping into their respective **electrolytes**, or ionically conducting media, usually in separate compartments. The compartments are connected by a **salt bridge**, an ionic conductor that completes the electrical circuit. Electrons conduct the current through an external circuit by the process of **metallic conduction**, and ions conduct the current through the electrolyte in the process of **ionic conduction**. Reduction occurs at the **cathode** and oxidation occurs at the **anode**. Electrons flow from the anode to the cathode in the external circuit; cations flow toward the cathode and anions toward the anode in the electrolyte. The half-reaction at either electrode defines a **redox couple**, denoted Ox/Red, which consists of the oxidized and reduced species taking part in the half-reaction. A galvanic cell is defined by a **cell diagram**, in which the cathode redox couple appears on the right and the anode redox couple on the left. The potential difference of a galvanic cell is reported as the (positive) **cell potential**. A positive cell potential shows that the reducing agent (at the anode) has a natural tendency to reduce the oxidizing agent (at the cathode).

17.2 A **primary cell** is one that produces electricity from reagents built into it when it is manufactured. A **secondary cell** is a storage cell and must be charged. A **fuel cell** produces electricity from reagents that are supplied to it while it is operating. An example of a primary cell is the Leclanché **dry cell**. An example of a collection of secondary cells is the **lead-acid battery** used in automobiles.

17.3 The cell potential (in volts) is proportional to (the negative of) the Gibbs free energy of the cell reaction. The **standard cell potential** is the cell potential when all ions participating in the reaction are at a concentration of 1 mol/L and any gases are at 1 atm pressure. The standard cell potential is the sum of the contributions of the potentials of the redox couples at the cathode and anode. The **standard electrode potentials** of the redox couples are defined relative to the **standard hydrogen electrode** and are reported as **standard reduction potentials**.

17.4 The **electrochemical series** is a list of redox couples arranged in order of standard reduction potentials. The upper redox couple of a pair undergoes reduction and forms the cathode of a galvanic cell; the lower couple undergoes oxidation and functions as the anode. The oxidized species of the redox couples listed near the top of the series are strong oxidizing agents; the reduced species of the redox couples near the bottom of the series are strong reducing agents. Only metals with negative reduction potentials can reduce acids to hydrogen gas. Kinetic factors may interfere with these conclusions, and a strongly reducing metal is oxidized very slowly if it is **passivated** by an inert oxide film. Electrochemical processes include **corrosion** and such corrosion-prevention methods as **galvanizing** and the use of a **sacrificial anode**.

17.5 Redox couples and their standard reduction potentials are used to calculate equilibrium constants of cell reactions. The relation between the actual cell potential and the concentrations or pressures of the reactants and products of a cell reaction under arbitrary conditions is given by the **Nernst equation**. Of special interest is the Nernst equation for a cell containing a hydrogen electrode, in which the cell potential is proportional to pH. A **glass electrode** is a more convenient device for measuring pH.

17.6 An **electrolytic cell** consists of two electrodes in a single electrolyte: reduction occurs at the cathode and oxidation occurs at the anode. The electric current used to bring about the chemical reaction is supplied from a source with a potential that must exceed the potential of the reverse reaction by an amount called the **overpotential**.

17.7 The amount of reaction caused by a given current is described by **Faraday's law of electrolysis** in terms of moles of electrons and substances. The number of moles of electrons supplied is measured by noting the time for which a known current flows through the cell.

17.8 A cathode is a very powerful reducing agent and an anode is a very powerful oxidizing agent. Electrolysis is used commercially for producing aluminum, sodium, chlorine, sodium hydroxide, sodium hypochlorite, and fluorine (among other substances). It is also used for **electroplating**, the electrochemical deposition of a metal on an object.

CLASSIFIED EXERCISES

Assume a temperature of 25°C (298 K) for the following exercises, unless instructed otherwise.

Electrochemical Cells and Cell Reactions

17.1 Complete the following statements: (a) In a galvanic cell, oxidation occurs at the (anode, cathode). (b) The cathode is the (positive, negative) electrode.

17.2 Complete the following statements: (a) In an electrolytic cell, oxidation occurs at the (anode, cathode). (b) The anode is the (positive, negative) electrode.

17.3 Complete the following statement: In an electrolytic cell, the anions migrate toward the (anode, cathode) and the electrons flow from the (anode, cathode) to the (anode, cathode).

17.4 Complete the following statement: In a galvanic cell, the anions migrate toward the (anode, cathode) and the electrons flow from the (anode, cathode) to the (anode, cathode).

17.5 Write the conventional redox couple notation, the reduction half-reaction, and the cathode notation for (a) zinc metal in contact with Zn^{2+} ions; (b) platinum metal dipping into a solution of iron(II) and iron(III) salts; (c) chlorine gas in contact with Cl^- ions; (d) the calomel electrode.

17.6 Write the conventional redox couple notation and the half-reaction that corresponds to the process occurring at each of the following electrodes: (a) $Zn|Zn^{2+}(aq)$; (b) $Br^-(aq)|AgBr(s)|Ag$; (c) $OH^-(aq)|Cd(OH)_2(s)|Cd$; (d) $Pt|O_2(g)|H^+(aq)$.

17.7 Write the cell diagram for each of the following galvanic cells: (a) a H^+/H_2 redox couple in combination with a Ag^+/Ag redox couple; (b) a Cl_2/Cl^- redox couple in combination with a $AgCl/Ag,Cl^-$ redox couple; (c) a calomel electrode in combination with a hydrogen electrode.

17.8 Write the cell diagram for each galvanic cell: (a) an $O_2,H^+/H_2O$ redox couple in combination with a H^+/H_2 redox couple; (b) a Mn^{3+}/Mn^{2+} redox couple in combination with a Cr^{3+}/Cr^{2+} redox couple; (c) a Pb^{2+}/Pb redox couple in combination with a $Hg_2{}^{2+}/Hg$ redox couple.

17.9 Write the cathode and anode half-reactions and the cell reaction for each of the following galvanic cells:
(a) $C(gr)|H_2(g)|H^+(aq)||Cl^-(aq)|Cl_2(g)|Pt$
(b) $U|U^{3+}(aq)||V^{2+}(aq)|V$
(c) $Pt|Sn^{4+}(aq),Sn^{2+}(aq)||Cl^-(aq)|Hg_2Cl_2(s)|Hg$

17.10 Write the cathode and anode half-reactions and the cell reaction for each of the following galvanic cells:
(a) $Cu|Cu^{2+}(aq)||Cu^+(aq)|Cu$
(b) $Pt|I_3{}^-(aq),I^-(aq)||Br^-(aq)|Br_2(l)|C(gr)$, where $I_3{}^-$ is equivalent to $(I_2{\cdot}I)^-$
(c) $Ag|AgI(s)|I^-(aq)||Cl^-(aq)|AgCl(s)|Ag$

17.11 Write the cathode and anode half-reactions, the balanced cell reaction, and the cell diagram for the following skeletal reactions:
(a) $Ni^{2+}(aq) + Zn(s) \rightarrow Ni(s) + Zn^{2+}(aq)$
(b) $Ce^{4+}(aq) + I^-(aq) \rightarrow I_2(s) + Ce^{3+}(aq)$
(c) $Au^+(aq) \rightarrow Au(s) + Au^{3+}(aq)$

17.12 Devise a galvanic cell (write a cell diagram) to study each of the following reactions:
(a) $AgBr(s) \rightleftharpoons Ag^+(aq) + Br^-(aq)$, a solubility equilibrium
(b) $H_3O^+(aq) + OH^-(aq) \rightarrow H_2O(l)$, the Brønsted neutralization reaction
(c) $Cd(s) + Ni(OH)_3(s) \rightarrow Cd(OH)_2(s) + Ni(OH)_2(s)$, the reaction in the nickel-cadmium cell

Practical Cells

17.13 What is (a) the electrolyte and (b) the oxidizing agent in a dry cell? (c) Write the half-reaction for the reducing agent in a dry cell.

17.14 What is (a) the electrolyte and (b) the oxidizing agent during discharge in a lead-acid battery? (c) Write the reaction that occurs at the cathode during the charging of the lead-acid battery.

17.15 (a) What is the electrolyte in a nickel-cadmium cell? (b) Write the reaction that occurs at the anode when the cell is being charged. (c) What is the cell potential for the cell under standard conditions?

17.16 (a) Why are lead-antimony grids used as electrodes in the lead-acid battery rather than smooth plates? (b) The lead-acid cell potential is about 2 V. How, then, does a car battery produce 12 V for its electrical system?

Standard Electrode Potentials

17.17 Write (a) the reduction half-reaction and (b) the cathode notation for the standard hydrogen electrode.

17.18 Suppose the reference electrode for Table 17.3 was the standard calomel electrode, $Hg_2Cl_2/Hg,Cl^-$, with $E°$ for it set equal to 0. Under this system, what would be the standard reduction potential for (a) the standard hydrogen electrode and (b) a Cu^{2+}/Cu redox couple?

17.19 Predict the standard cell potential of each of the following galvanic cells:
(a) $Pt|Cr^{3+}(aq),Cr^{2+}(aq)||Cu^{2+}(aq)|Cu$
(b) $Hg|Hg_2Cl_2(s)|Cl^-(aq)||Hg_2{}^{2+}(aq)|Hg$
(c) $C(gr)|Sn^{4+}(aq),Sn^{2+}(aq)||Pb^{4+}(aq),Pb^{2+}(aq)|Pt$

17.20 Predict the standard cell potential for each of the following galvanic cells:
(a) $Pt|Fe^{3+}(aq),Fe^{2+}(aq)||Ag^+(aq)|Ag$
(b) $Sn|Sn^{2+}(aq)||Sn^{4+}(aq),Sn^{2+}(aq)|Pt$
(c) $Cu|Cu^{2+}(aq)||Au^+(aq)|Au$

Cell Potential and Free Energy

17.21 Predict the standard cell potential and calculate the standard free energy for galvanic cells having the following cell reactions:
(a) $Zn(s) + Fe^{2+}(aq) \rightarrow Zn^{2+}(aq) + Fe(s)$
(b) $Ag^+(aq) + Cl^-(aq) \rightarrow AgCl(s)$
(c) $3Au^+(aq) \rightarrow 2Au(s) + Au^{3+}(aq)$

17.22 Predict the standard cell potential and calculate the standard free energy for the following galvanic cells (it may help to refer to Exercise 17.19, where the standard potentials of these cells were obtained):
(a) $Pt|Cr^{3+}(aq),Cr^{2+}(aq)||Cu^{2+}(aq)|Cu$
(b) $Hg|Hg_2Cl_2(s)|Cl^-(aq)||Hg^{2+}(aq)|Hg$
(c) $C(gr)|Sn^{4+}(aq),Sn^{2+}(aq)||Pb^{4+}(aq),Pb^{2+}(aq)|Pt$

The Electrochemical Series

17.23 Arrange the following species in order of increasing strength as oxidizing agents:
(a) Co^{2+}, Cl_2, Ce^{4+}, In^{3+}
(b) $NO_3{}^-$, $ClO_4{}^-$, HBrO, $Cr_2O_7{}^{2-}$; all in acidic solution
(c) H_2O_2, O_2, $MnO_4{}^-$, HClO; all in acidic solution

17.24 Suppose that the following redox couples are joined to form a galvanic cell under standard conditions, identify the oxidizing agent and the reducing agent and write a cell diagram: (a) Co^{2+}/Co and Ti^{3+}/Ti^{2+}; (b) La^{3+}/La and U^{3+}/U; (c) $O_3/O_2,OH^-$ and Ag^+/Ag.

17.25 Answer the following questions, and for each "yes" response, write a balanced cell reaction and write a cell diagram. (a) Can H_2 reduce Ni^{2+} ions to nickel metal? (b) Can chromium metal reduce Pb^{2+} ions to lead metal? (c) Can permanganate ions oxidize copper metal to Cu^{2+} ions in an acidic solution?

17.26 Identify the spontaneous reactions in the following list and, for each spontaneous reaction, identify the

oxidizing agent and write a cell diagram.
(a) $Cl_2(g) + 2Br^-(aq) \rightarrow 2Cl^-(aq) + Br_2(l)$
(b) $MnO_4^-(aq) + 8H^+(aq) + 5Ce^{3+}(aq) \rightarrow 5Ce^{4+}(aq) + Mn^{2+}(aq) + 4H_2O(l)$
(c) $2Pb^{2+}(aq) \rightarrow Pb(s) + Pb^{4+}(aq)$

17.27 Identify the spontaneous reactions among the following reactions and, for the spontaneous reactions, write balanced reduction and oxidation half-reactions. Show that the reaction is spontaneous by calculating the standard free energy of the reaction. (a) $I_2 + H_2 \rightarrow$?; (b) $Mg^{2+} + Cu \rightarrow$?; (c) $Al + Pb^{2+} \rightarrow$?.

17.28 Identify the spontaneous reactions among the following reactions and, for the spontaneous reactions, write balanced reduction and oxidation half-reactions. Show that the reaction is spontaneous by calculating the standard free energy of the reaction. (a) $Hg_2^{2+} + Ce^{3+} \rightarrow$?; (b) $Zn + Sn^{2+} \rightarrow$?; (c) $O_2 + H^+ + Hg \rightarrow$?.

17.29 Chlorine is used to displace bromine from brine. Could oxygen in an acidified solution be used instead? If so, why is it not used?

17.30 A chemist is interested in the compounds formed by the *d*-block element manganese and wants to find a way for preparing Mn^{3+} from Mn^{2+}. Would an acidified solution of sodium dichromate be suitable? Explain your conclusion.

17.31 A chromium-plated steel (which should be treated as iron for the sake of this exercise) bicycle handlebar is scratched. Will rusting be encouraged or retarded by the chromium?

17.32 A solution consists of the ions Cu^{2+}, Ni^{2+}, and Ag^+. What will happen if a strip of tin metal is placed in the solution?

17.33 (a) What is the approximate chemical formula of "rust"? (b) What is the oxidizing agent in the formation of rust? (c) How does the presence of salt accelerate the rusting process? (d) What is the electrolyte in the formation of rust?

17.34 (a) What is the cathode used in the cathodic protection of an underground storage container? (b) Can aluminum be used for the cathodic protection of an underground storage container? (c) Which of the metals zinc, silver, copper, and magnesium cannot be used as a sacrificial anode in the protection of a buried pipeline? Explain your answer. (d) What is the electrolyte for the cathodic protection of an underground pipeline by a sacrificial anode?

Equilibrium Constants

17.35 Write the expression for the equilibrium constants of the reactions in the following cells:
(a) $Pt|H_2(g)|H^+(aq)||Cl^-(aq)|AgCl(s)|Ag$
(b) $Pt|Fe^{3+}(aq),Fe^{2+}(aq)||NO_3^-(aq),H^+(aq)|NO(g)|Pt$

17.36 Write the expression for the equilibrium constant of the reactions in the following cells:
(a) $Cr|Cr^{3+}(aq)||Br^-(aq)|AgBr(s)|Ag$
(b) $Bi|Bi^{3+}(aq)||OH^-(aq)|O_2(g)|Pt$

17.37 Determine the equilibrium constants for the following reactions and cell reactions:
(a) $Pt|Cr^{3+}(aq),Cr^{2+}(aq)||Cu^{2+}(aq)|Cu$
(b) $Mn(s) + Ti^{2+}(aq) \rightarrow Mn^{2+}(aq) + Ti(s)$
(c) A Pb^{2+}/Pb redox couple in combination with a Hg_2^{2+}/Hg redox couple

17.38 Determine the equilibrium constants for the following reactions and cell reactions:
(a) The $AgI/Ag,I^-$ redox couple in combination with the I_2/I^- redox couple.
(b) $Pt|Sn^{4+}(aq),Sn^{2+}(aq)||Cl^-(aq)|Hg_2Cl_2(s)|Hg$
(c) $Cr(s) + Zn^{2+}(aq) \rightarrow Cr^{2+}(aq) + Zn(s)$

17.39 A chemist wants to make a range of silver(II) compounds. Could aqueous sodium persulfate be used to oxidize a silver(I) compound to silver(II)? If so, what would the equilibrium constant for the reaction be?

17.40 A chemist suspects that manganese(III) might be involved in an unusual biochemical reaction and wants to prepare some of its compounds. Could aqueous potassium permanganate be used to oxidize manganese(II) to manganese(III)? If so, what would the equilibrium constant of the reaction be?

Nernst Equation

17.41 Calculate the reaction quotient Q_c for the cell reaction, given the measured values of the cell potential:
(a) $Pt|Sn^{4+}(aq),Sn^{2+}(aq)||Pb^{4+}(aq),Pb^{2+}(aq)|C(gr)$, $E = 1.33$ V
(b) $Pt|O_2(g)|H^+(aq)||Cr_2O_7^{2-}(aq),H^+(aq),Cr^{3+}(aq)|Pt$, $E = 0.10$ V

17.42 Calculate the reaction quotient Q_c for the cell reaction, given the measured values of the cell potential:
(a) $Ag|Ag^+(aq)||ClO_4^-(aq),H^+(aq),ClO_3^-(aq)|Pt$, $E = 0.40$ V
(b) $C(gr)|Cl^-(aq)|Cl_2(g)||Au^{3+}(aq)|Au$, $E = 0.00$ V

17.43 Determine the potential of the following galvanic cells:
(a) $Pt|H_2(g,1.0\ atm)|HCl(aq,0.0010\ mol/L)||HCl(aq,1.0\ mol/L)|H_2(g,1.0\ atm)|Pt$
(b) $Zn|Zn^{2+}(aq,0.10\ mol/L)||Ni^{2+}(aq,0.0010\ mol/L)|Ni$
(c) $Pt|Cl_2(g,100\ Torr)|HCl(aq,1.0\ mol/L)||HCl(aq,0.010\ mol/L)|H_2(g,450\ Torr)|Pt$

17.44 Determine the potential of the following cells:
(a) $Cr|Cr^{3+}(aq,0.10\ mol/L)||Pb^{2+}(aq,1.00 \times 10^{-5}\ mol/L)|Pb$
(b) $C(gr)|Sn^{2+}(aq,0.10\ mol/L),Sn^{4+}(aq,0.0030\ mol/L)||Fe^{3+}(aq,1.0 \times 10^{-4}\ mol/L),Fe^{2+}(aq,0.40\ mol/L)|Pt$
(c) $Ag|AgI(s)|I^-(aq,0.010\ mol/L)||Cl^-(aq,1.0 \times 10^{-6}\ mol/L)|AgCl(s)|Ag$

17.45 What must be the value of $[Fe^{2+}]/[Fe^{3+}]$ in the cell $Pt|Fe^{3+}(aq),Fe^{2+}(aq)||Hg_2^{2+}(aq,1.0\ mol/L)|Hg$ for it to have a potential of 0.060 V?

17.46 What must be the value of $[In^+]/[In^{3+}]$ in the cell $C(gr)|In^{3+}(aq),In^+(aq)||H^+(aq,pH = 1.0)|H_2(g,1\ atm)|Pt$ for it to have a potential of 0.44 V?

17.47 Determine the unknown in the following cells:
(a) $Pt|H_2(g,1.0\ atm)|H^+(pH = ?)||$
$Cl^-(aq,1.0\ mol/L)|Hg_2Cl_2(s)|Hg$, $E = 0.33$ V
(b) $C(gr)|Cl_2(g,1.0\ atm)|Cl^-(aq,?\ mol/L)||$
$MnO_4^-(aq,0.010\ mol/L)$, $H^+(aq,pH = 4.0)$,
$Mn^{2+}(aq,0.10\ mol/L)|Pt$, $E = -0.30$ V

17.48 Determine the unknown in the following cells:
(a) $Pt|H_2(g,1.0\ atm)|H^+(aq,pH = ?)||$
$Cl^-(aq,1.0\ mol/L)|AgCl(s)|Ag$, $E = 0.30$ V
(b) $Pb|Pb^{2+}(aq,\ ?\ mol/L)||Ni^{2+}(aq,0.10\ mol/L)|Ni$,
$E = 0.040$ V

17.49 Show how a silver-silver chloride electrode and a hydrogen electrode can be used to measure (a) pH, (b) pOH.

17.50 In a neuron (a nerve cell), the concentration of K^+ ions inside the cell is about 20 to 30 times that outside. What potential difference between the inside and the outside of the cell would you expect to measure if the difference is due only to the imbalance of potassium ions? Which solution potential would be more positive, inside or outside?

Electrolysis

For the exercises in this section, base your answers on the potentials listed in Table 17.3 or Appendix 2B, with the exception of the reduction and oxidation of water at a pH = 7:

Reduction: $2H_2O(l) + 2e^- \longrightarrow H_2(g) + 2OH^-(aq)$
$E = -0.42$ V at pH = 7

Oxidation: $2H_2O(l) \longrightarrow O_2(g) + 4H^+(aq) + 4e^-$
$E = -0.81$ V at pH = 7

Do not consider other factors such as passivation or overpotential.

17.51 Aqueous solutions of (a) Mn^{2+}, (b) Al^{3+}, (c) Ni^{2+}, (d) Au^{3+} were electrolyzed. Determine whether the metal ion or water will be reduced at the cathode.

17.52 A 1 M $CsI(aq)$ solution was electrolyzed in a cell with inert electrodes. Write (a) the cathode reaction and (b) the anode reaction. (c) Assuming no overpotential or passivity at the electrodes, what is the minimum potential that must be supplied to the cell for the onset of electrolysis?

17.53 Write the half-reaction and specify the electrode at which each one occurs in an electrolytic cell, for the following processes: (a) the deposition of copper; (b) the production of sodium metal in the Downs cell; (c) the production of chlorine gas in the Downs cell.

17.54 Write the half-reaction and specify the electrode at which each one occurs in an electrolytic cell, for the following processes: (a) the production of aluminum by the Hall process; (b) the production of oxygen gas from an acidic aqueous solution; (c) the production of hypochlorite ions from brine (concentrated aqueous sodium chloride solution).

17.55 Determine the amount (in moles) of electrons needed to produce the indicated substance in an electrolytic cell: (a) 5.12 g of copper from a copper(II) sulfate solution; (b) 200 g of aluminum from molten aluminum oxide dissolved in cryolite; (c) 200 L of oxygen gas at STP from an aqueous sodium sulfate solution.

17.56 A total charge of 9.65×10^4 C is passed through an electrolytic cell. Determine the quantity of substance produced in each case: (a) the mass (in grams) of silver metal from a silver nitrate solution; (b) the volume (in liters at STP) of chlorine gas from a brine solution (concentrated aqueous sodium chloride solution); (c) the mass of copper (in grams) from a copper(II) chloride solution.

17.57 (a) How much time is required to electroplate 4.4 mg of silver from a silver nitrate solution, using a current of 0.50 A? (b) When the same current is used for the same length of time, what mass of copper can be electroplated from a copper(II) sulfate solution?

17.58 (a) When a current of 150 mA is used for 8.0 h, what volume (in liters at STP) of fluorine gas is produced from a molten mixture of potassium and hydrogen fluorides? (b) With the same current and time period, what volume of oxygen gas at STP is produced from the electrolysis of water?

17.59 What current is required to electroplate 6.66 μg of gold in 30.0 min from a gold(III) chloride aqueous solution? (b) How much time is required to electroplate 6.66 μg of chromium from a potassium dichromate solution with a current of 100 mA?

17.60 An aqueous solution of Na_2SO_4 was electrolyzed for 30 min. During the process, 25.0 mL of O_2 gas was collected at the anode over water at 22°C at a total pressure of 722 Torr. Determine the current that was used to produce the oxygen gas.

17.61 When a titanium chloride solution was electrolyzed for 500 s with a 120-mA current, 15.0 mg of titanium was deposited. What is the oxidation number of the titanium in the titanium chloride?

17.62 A 0.26-g sample of mercury was produced from a mercury nitrate aqueous solution when a current of 210 mA was applied for 1200 s. What is the oxidation number of the mercury in the mercury nitrate?

17.63 Thomas Edison was faced with the problem of measuring the electricity that each of his customers had used. His first solution was to use a zinc "coulometer," an electrolytic cell in which the quantity of electricity is determined by measuring the mass of zinc deposited. Only some of the current used by the customer passed through the coulometer. What mass of zinc would be deposited in one month (of 31 days) if 1.0 mA of current passed through the cell continuously?

17.64 An alternative solution to the problem described in Exercise 17.63 is to collect the hydrogen produced by electrolysis and measure its volume. What volume would be collected at STP under the same conditions?

UNCLASSIFIED EXERCISES

17.65 A certain galvanic cell consists of the $O_2,H^+/H_2O$ and Fe^{2+}/Fe redox couples. (a) Write the cathode and anode half-reactions. (b) Write a cell diagram for the galvanic cell. (c) What is the standard potential of the cell? (d) What is the cell potential at a pH = 6.00? Assume all other conditions are standard.

17.66 Consider the galvanic cell

$$Hg|Hg_2Cl_2(s)|Cl^-(aq)||Hg^{2+}(aq),Hg_2^{2+}(aq)|C(gr)$$

(a) Which electrode is the cathode? (b) Write the cell reaction. (c) What is the oxidizing agent and the reducing agent in the cell? (d) What is the standard cell potential? (e) What is the $[Hg_2^{2+}]/[Hg^{2+}]$ ratio when the cell potential is 0.60 V and $[Cl^-] = 1.0$ mol/L?

17.67 A galvanic cell has the cell reaction, $M(s) + 2Zn^{2+}(aq) \rightarrow 2Zn(s) + M^{4+}(aq)$. The standard potential of the cell is 0.16 V. What is the standard reduction potential of the M^{4+}/M redox couple?

17.68 Calculate the standard free energy of the reaction $2MnO_4^-(aq) + 16H^+(aq) + 5Sn^{2+}(aq) \rightarrow 5Sn^{4+}(aq) + 2Mn^{2+}(aq) + 8H_2O(l)$.

17.69 Gold can be oxidized by permanganate ions but not by dichromate ions in an acidic solution. Explain this observation.

17.70 What is the standard reduction potential for the reduction of oxygen to water in (a) an acidic solution and (b) a basic solution? (c) Is MnO_4^{2-} more stable in an acidic or a basic oxygenated solution (a solution saturated with oxygen gas at 1 atm)? Explain your conclusion.

17.71 For each reaction that has a negative standard free energy, write a cell diagram, determine the standard cell potential, and calculate the $\Delta G°$ for the reaction:

(a) $2NO_3^-(aq) + 8H^+(aq) + 6Hg(l) \rightarrow 3Hg_2^{2+}(aq) + 2NO(g) + 4H_2O(l)$

(b) $Cr_2O_7^{2-}(aq) + 14H^+(aq) + 6Pu^{3+}(aq) \rightarrow 6Pu^{4+}(aq) + 2Cr^{3+}(aq) + 7H_2O(l)$

17.72 Determine the equilibrium constant for each of the following reactions:

(a) $PbSO_4(s) \rightarrow Pb^{2+}(aq) + SO_4^{2-}(aq)$

(b) $2Pb^{2+}(aq) \rightarrow Pb(s) + Pb^{4+}(aq)$

(c) $Hg_2Cl_2(s) \rightarrow Hg_2^{2+}(aq) + 2Cl^-(aq)$

17.73 Calculate the standard reduction potential of (a) the Cu^+/Cu couple from those of the Cu^{2+}/Cu^+ and Cu^{2+}/Cu couples; (b) the Fe^{3+}/Fe couple from the values for the Fe^{3+}/Fe^{2+} and Fe^{2+}/Fe couples. Base your calculation on the relation $\Delta G° = -nFE°$ and the fact that $\Delta G°$ of an overall reaction is the sum of the $\Delta G°$ of each reaction into which it may be divided.

17.74 Use standard electrode-potential data to calculate the molar solubility of (a) $AgCl(s)$; (b) $Hg_2Cl_2(s)$; (c) $PbSO_4(s)$.

17.75 Dental amalgam, a solid solution of silver and tin in mercury, is used for filling tooth cavities. Two of the reduction half-reactions that the filling can undergo are

$$3Hg_2^{2+}(aq) + 4Ag(s) + 6e^- \longrightarrow 2Ag_2Hg_3(s) \qquad E° = +0.85 \text{ V}$$

$$Sn^{2+}(aq) + 3Ag(s) + 2e^- \longrightarrow Ag_3Sn(s) \qquad E° = -0.05 \text{ V}$$

Suggest a reason why, when you accidentally bite on a piece of aluminum foil, you may feel pain.

17.76 Given that for the half-reaction $F_2(g) + 2H^+(aq) + 2e^- \rightarrow 2HF(aq)$, $E° = +3.03$ V, calculate the value of K_a for HF.

17.77 The following items are obtained from the stockroom for the construction of a galvanic cell: two 250-mL beakers and a salt bridge, a voltmeter with attached wires and clips, 200 mL of 0.010 M $CrCl_3(aq)$ solution, 400 mL of 0.16 M $CuSO_4(aq)$ solution, a piece of copper wire, and a chrome-plated piece of metal. (a) Describe the construction of the galvanic cell. (b) Write the anode and cathode half-reactions. (c) Write the cell reaction. (d) Write the cell diagram for the galvanic cell. (e) What is the expected cell potential?

17.78 Calculate the standard potential and actual potential of the galvanic cell $Pt|Fe^{3+}(aq,1.0 \text{ mol/L}),Fe^{2+}(aq,0.0010 \text{ mol/L})||Ag^+(aq,0.010 \text{ mol/L})|Ag$. Compare and comment on your answers.

17.79 Consider the galvanic cell $Pt|Fe^{3+}(aq,2.00 \text{ mol/L}),Fe^{2+}(aq,1.00 \times 10^{-5} \text{ mol/L})||Ag^+(aq,2.00 \times 10^{-4} \text{ mol/L})|Ag$. (a) What is the standard cell potential? (b) Write the cell reaction. (c) Calculate the cell potential. (d) Determine the equilibrium constant of the cell reaction. (e) Calculate the free energy of the cell reaction.

17.80 Consider the galvanic cell $Pt|Sn^{4+}(aq,0.010 \text{ mol/L}), Sn^{2+}(aq, 0.10 \text{ mol/L})||O_2(g, 1 \text{ atm})|H^+(aq, \text{pH} = 4)|C(gr)$. (a) What is the standard cell potential? (b) Write the cell reaction. (c) Calculate the cell potential. (d) Determine the equilibrium constant of the cell reaction. (e) Calculate the free energy of the cell reaction? (f) Suppose the cell potential measures 0.89 V. Assuming all other concentrations are precise, what is a more accurate pH of the solution?

CASE 17 ▸ PHOTOCHROMIC GLASS

The colors and the properties of glass are altered by the addition of a variety of compounds to molten silica (sand); for example, cobalt(II) oxide produces a blue glass, boron oxide produces a heat-resistant glass, cadmium selenide yields a red glass, and lead(II) oxide results in lead crystal, a glass with a high refractive index.

Whereas most glasses merely allow light to pass through, certain "new" glasses are built to respond to light. Compounds that are added to the glasses respond photochemically to the light by forming crystalline precipitates or aggregates that absorb visible light. Depending on the wavelengths of light absorbed, the glass can assume a variety of colors. One of these new glasses is *photochromic glass,* which changes color in the presence of high-energy radiation (usually ultraviolet radiation) but reverts back to its original appearance in the absence of this radiation.

For the preparation of photochromic glass, silver and copper nitrates and a metal halide are mixed with the base composition for borosilicate glass and heated until the solids melt, at approximately 1200°C. As the glass cools, small crystallites of silver halide form, along with a small portion of copper halide. The crystallites, which are approximately 10 nm in diameter and 100 nm apart, are too small to scatter or absorb visible light and therefore the glass appears transparent. However, when the glass is exposed to ultraviolet radiation (UV), an electron is removed from the Cu^+ ion and accepted by the Ag^+ ion:

$$Cu^+ + Ag^+ \xrightarrow{UV} Cu^{2+} + Ag$$

Tiny clusters of various dimensions containing the neutral silver atoms absorb most wavelengths from the near ultraviolet (approximately 300 nm), through the visible region and into the near infrared, depending on the halide. When the ha-

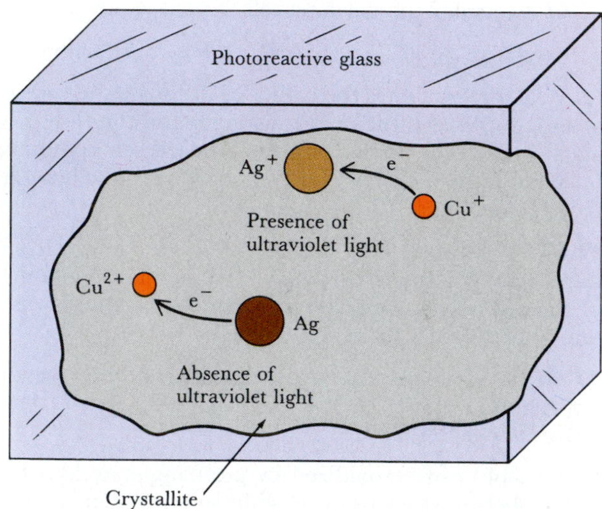

Radiation causes the reduction of Ag^+ to Ag metal, causing the graying of the photoreactive glass. Without ultraviolet radiation, the Ag metal is oxidized, causing the reappearance of colorless glass.

17.81 The measured potential of the galvanic cell $Zn|Zn^{2+}(aq,?\ mol/L)||Pb^{2+}(aq,0.10\ mol/L)|Pb$ is 0.66 V. What is the molar concentration of Zn^{2+} ions?

17.82 When a pH meter was standardized with a boric acid-borate buffer with a pH of 9.40, the cell potential was 0.060 V. When the buffer was replaced with a solution of unknown hydronium ion concentration, the cell potential was 0.22 V. What is the pH in the solution?

17.83 A portable cassette player uses 150 mA of current. Calculate the number of moles of electrons its batteries need to generate when it is on for 1 hour.

17.84 The reaction that powers our body is the oxidation of glucose:

$$C_6H_{12}O_6(aq) + 6O_2(g) \longrightarrow 6CO_2(g) + 6H_2O(l)$$

During normal activity, a person uses the equivalent of about 10 MJ of energy a day. Estimate the average current through your body in the course of a day, assuming that all the energy we use arises from the reduction of O_2 in the glucose oxidation reaction. Also estimate your own power in watts.

17.85 Determine the current required to produce 15.0 L Cl_2 at STP from molten sodium chloride in 1.0 h.

17.86 In the Dow process for the production of magnesium, a current of approximately 10 kA is used. (a) What mass (in kilograms) of magnesium can be produced in 24 h? (b) What volume of chlorine gas at STP is produced during the same period?

17.87 A current of 15.0 A electroplated 50.0 g of haf-

lide is the chloride, the absorption is especially effective in the 300- to 400-nm region; with the bromide, this range extends up to 550 nm and results in a gray appearance. When used in eyeglasses, these bromide glasses are called *photogray lenses.* Maximum absorption occurs within about 1 min. In the absence of ultraviolet radiation, the silver atoms quickly revert to silver ions, thereby causing the reduction of the Cu^{2+} ions and restoring the transparency of the original glass.

$$Cu^{2+} + Ag \xrightarrow{\text{visible light}} Cu^{+} + Ag^{+}$$

Photochromic glass that darkens outdoors. Here only one of the lenses has been exposed to ultraviolet light.

This apparent reversibility means that there is no loss or migration of the reaction products away from the reaction zone. Control of the sizes of silver atom clusters by varying the manufacturing conditions can result in various colors in the presence of ultraviolet radiation, the rate of cluster formation, and the rate of restoration of the original glass. Additional research in the production of photochromic glass may result in its use for automobile safety glass, automobile sunroofs, light control devices, and window glass in homes (for better climate control).

QUESTIONS

1. (a) Write the two half-reactions that cause the graying of photochromic glass. (b) What is the oxidizing agent and what is the reducing agent?

2. How would you expect the properties of photochromic glass to be affected by high temperatures?

3. Comment on how the future of photochromic glass may include its use as an insulation material for homes.

4. Sketch a plot of ultraviolet absorbance by photochromic glass against time at the following stages: (a) before ultraviolet radiation; (b) from the time the photochromic glass is exposed to ultraviolet radiation until the time the light is turned off; (c) from the time the light is turned off until the glass is restored to its initial transparency.

nium metal from an aqueous solution in 2.00 h. What was the oxidation number of hafnium in the solution?

17.88 (a) How many moles of hydronium ion are produced at the anode in the electrolysis of 200 mL of a $CuSO_4$ solution by using a current of 4.00 A for 30 min. (b) If the pH of the solution was initially 7.0, what will the pH of the solution after the electrolysis be? Assume no volume change.

17.89 In the electrolytic refining of copper, blister copper is placed at the anode and oxidized. The copper(II) ion that is produced from its oxidation is then reduced at the cathode to give a metal with a much higher purity. The impurities in the blister copper include iron, nickel, silver, gold, cobalt, and trace amounts of other metals. The material that is not oxidized at the anode falls to the bottom of the electrolytic cell and is called anode mud. What are some of the components of the anode mud?

17.90 In the commercial production of aluminum, alumina (Al_2O_3, m.p. 2000°C) is mixed with cryolite (Na_3AlF_6) to form a melt at about 950°C. Is the aluminum metal produced at the anode or the cathode? Explain your answer.

17.91 A piece of copper metal is to be electroplated with silver to a thickness of 1.0 μm. If the metal strip measures 50 mm × 10 mm × 1 mm, how long must the solution, which contains $Ag(CN)_2^-$ ions, be electrolyzed using current of 100 mA? The density of silver metal is 10.5 g/cm^3.

17.92 A metal forms the salt MCl_3. Electrolysis of the molten salt with a current of 0.70 A for 6.63 h produced 3.00 g of the metal. What is the molar mass of the metal?

In Part IV, we survey the properties of individual elements and families of elements and see how they illustrate the concepts developed in previous chapters. We begin by discussing the properties of hydrogen, the alkali metals, and the alkaline earth metals. The illustration shows a piece of sodium in water that contains a few drops of indicator: the exothermic reaction releases hydrogen and melts the remaining sodium, and the indicator shows that a basic solution is formed.

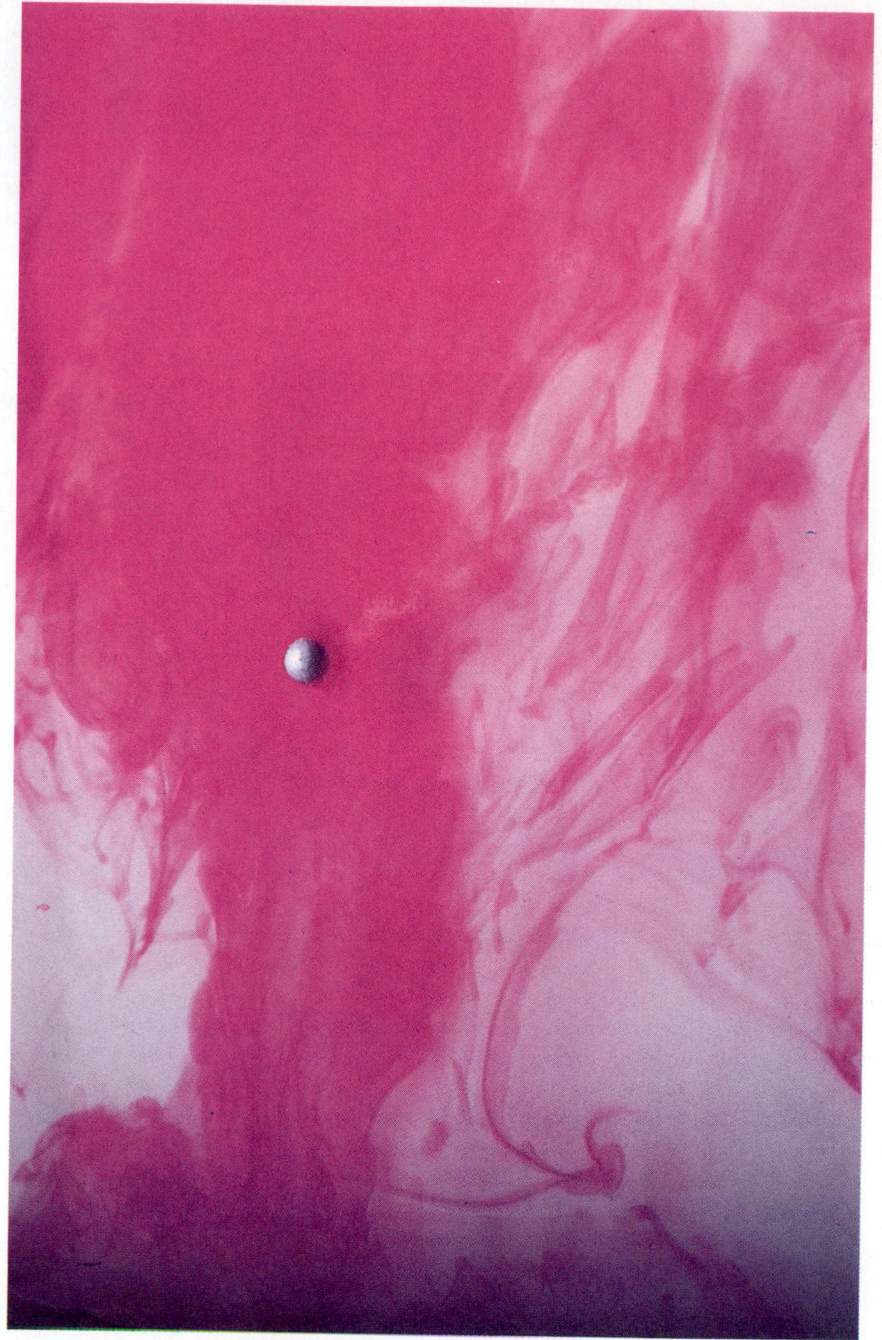

18 HYDROGEN AND THE s-BLOCK ELEMENTS

This chapter is the first of four chapters in which we work systematically but selectively through the periodic table, illustrating earlier statements about how the table helps to organize the properties of the elements and their compounds. Three points should be kept in mind as we do so:

1. The distinctive character of the element at the head of the group
2. The existence of diagonal relationships
3. The trends in metallic and nonmetallic character

Mainly as a result of the small size of its atoms, the element at the head of each group (for example, lithium in Group I and beryllium in Group II) differs significantly from its congeners. Hydrogen, at the head of the entire periodic table, is particularly distinctive; for that reason we treat it separately and not as a member of a group or even a block. Diagonal relationships are particularly important between Period 2 and Period 3 elements toward the left of the periodic table. Thus we can expect lithium to resemble its diagonal neighbor magnesium, and beryllium to resemble its *p*-block diagonal neighbor aluminum. Elements in the same group tend to have similar chemical properties and become more metallike down the group. Elements in the same period tend to become more nonmetallike on progressing from left to right across the period.

N The chemistry described in the next four chapters is made interesting to students by a combination of demonstrations and videos. The text seeks to relate the properties to the principles that have been introduced earlier.

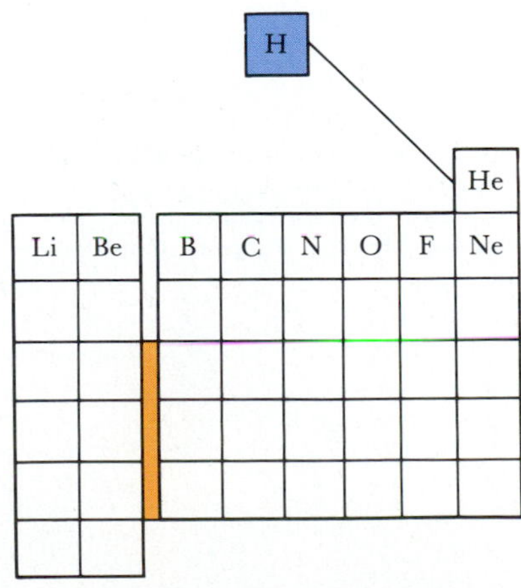

HYDROGEN

There are three isotopes of hydrogen. The most abundant is ^{1}H, with a single proton for its nucleus. Deuterium (^{2}H or D) has one neutron in addition to the proton in the nucleus. The natural abundance of deuterium is 0.02%, so we are likely to find only two ^{2}H atoms in any 10,000 atoms of hydrogen. The presence of the additional neutron doubles the mass of the atom without changing its electronic structure, and for

D Hydrogen: Historical preparation by Cavendish in 1781. TDC, 3-1. Preparation and properties of hydrogen. BZS2, 6.6.

E Deuterium also differs from ordinary hydrogen (protium) in the spin of its nucleus: the proton is a spin-$\frac{1}{2}$ particle, whereas the deuteron is a spin-1 particle. This difference has consequences for their detection by NMR.

E Deuterium may also be used as a tracer: its presence is detected by mass spectrometry.

E Hydrogen is unable to escape from the very massive planets, and is abundant in their atmospheres. It is possible that the hydrogen at the center of Jupiter is so highly compressed that the H_2 molecules have lost their identity and the bulk hydrogen is a metal.

no other element is there such a significant difference between the masses of its isotopes. This difference makes deuterium very valuable for keeping track of hydrogen atoms in a series of reactions: we can synthesize a molecule with a higher than normal proportion of deuterium and then use mass spectrometry to identify which product molecules contain the extra deuterium.

The third isotope, tritium (^3H or T), is radioactive with a half-life of 12.3 d. Tritium is used as a radioactive **tracer**, for it is possible to follow its progress through a complex series of reactions by noting which compounds are radioactive.

Hydrogen is the most abundant element in the universe. However, there is little free hydrogen on earth because H_2 molecules, having such a small mass, move at high average speeds and escape from the atmosphere. The molecular hydrogen that does survive was originally formed by the action of bacteria on ancient vegetable and animal remains and then trapped beneath rock formations.

18.1 THE ELEMENT HYDROGEN

About 3×10^8 kg of hydrogen are used in the United States each year. Although its role as a rocket fuel in the space shuttle is one of its more spectacular uses, that consumption (about 3×10^5 kg per flight) accounts for less than 1% of the total. About 50% of the hydrogen used in industry is converted into ammonia by the Haber process (the high-pressure, high-temperature catalytic synthesis of ammonia from nitrogen and hydrogen). Through the reactions of ammonia, hydrogen finds its way into the numerous important nitrogen compounds for which ammonia is the starting point. Hydrogen also enters the economy through the production of methanol:

$$2H_2(g) + CO(aq) \xrightarrow{\Delta,\ \text{pressure, catalyst}} CH_3OH(l)$$

E A chemical property of hydrogen. *J. Chem. Ed.*, **1982**, *59*, 920.

This process takes place at about 300°C and 250 atm, over a catalyst of zinc, chromium, manganese, and aluminum oxides. Through methanol, hydrogen finds its way into formaldehyde (HCHO), which is used in the manufacture of plastics, and into acetic acid, which is used for the manufacture of synthetic fibers. Methanol is also used as an additive in gasoline, where it promotes smoother combustion and hence helps to improve the octane rating of the fuel (see Case 23).

? Which metal ions in the periodic table cannot be reduced by molecular hydrogen?

N Hydrometallurgical extraction is environmentally more benign than the alternative pyrometallurgical process, in which smelting results in the production of large amounts of sulfur dioxide.

Hydrogen is also used in the **hydrometallurgical extraction** of copper (Section 21.3), that is, extraction of copper from its ores by reduction in aqueous solution:

$$Cu^{2+}(aq) + H_2(g) \longrightarrow Cu(s) + 2H^+(aq)$$

In this process, ores containing copper oxide and sulfide are dissolved in sulfuric acid, and then hydrogen is bubbled through the solution. About a third of the hydrogen that is manufactured is used for this and other similar hydrometallurgical reductions.

EXAMPLE 18.1 Judging the feasibility of reduction with hydrogen

Is the hydrometallurgical extraction of copper thermodynamically favorable under standard conditions at 25°C?

STRATEGY We can judge whether H_2 can reduce Cu^{2+} by checking to see whether the H^+/H_2 couple lies below the Cu^{2+}/Cu couple in the electrochemical series (Appendix 2B).

SOLUTION The redox couple H^+/H_2 ($E° = 0$) lies below Cu^{2+}/Cu ($E° = +0.34$ V) in the electrochemical series. Hence the reduction is thermodynamically feasible under standard conditions.

EXERCISE E18.1 Can zinc be extracted by hydrometallurgical reduction of Zn^{2+} ions?

[*Answer*: no]

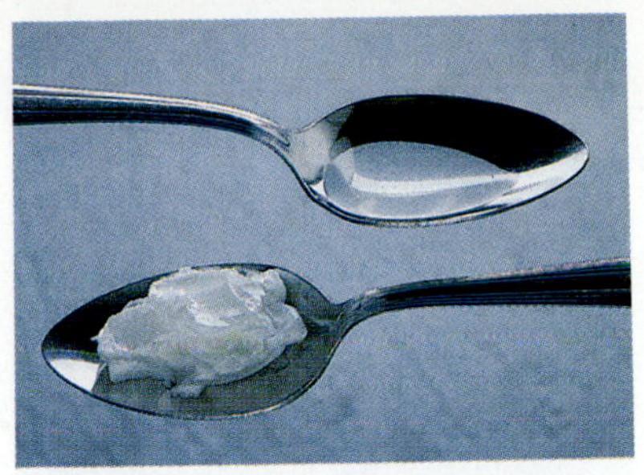

FIGURE 18.1 When the runny oil (top) is hydrogenated, it is converted into a solid—a fat (bottom). At the molecular level, the hydrogen converts carbon-carbon double bonds to single bonds and hence increases the flexibility of the molecules. This enhanced flexibility enables them to pack together more closely.

Hydrogen is used as a reducing agent in other metallurgical processes. Any metal ion having a reduction potential higher than that of H^+/H_2 can be reduced by hydrogen gas. For example, the standard reduction potential of the Ag^+/Ag couple is +0.80 V, so silver ions can be reduced by hydrogen gas:

$$2Ag^+(aq) + H_2(g) \longrightarrow 2Ag(s) + 2H^+(aq)$$

A number of metal oxides can also be reduced by hydrogen. Thus, tungsten is produced industrially from the reduction of its oxide in the reaction

$$WO_3(s) + 3H_2(g) \longrightarrow W(s) + 3H_2O(l)$$

Hydrogen is used in the food industry for converting vegetable oils to shortening (Fig. 18.1). It is added to carbon-carbon double bonds in the reaction called **hydrogenation**:

$$H_2(g) + \ \ldots\ C{=}C\ \ldots \xrightarrow{200°C,\ 30\ atm,\ Ni\ or\ Pt} \ldots\ CH{-}CH\ \ldots$$

D Hydrogenation exhibit. TDC, 3-16.

E Arctic fish have oils that contain more double bonds than those of fish in temperate waters: the double bonds help to ensure that the oils remain fluid down to lower temperatures.

This reaction converts a C=C double bond into a C—C single bond. Oil and fat molecules have long hydrocarbon chains; however, the molecules in oils have more C=C double bonds than the molecules in fats. Because double bonds resist twisting, chains that contain them are stiff and do not pack together well, so the result is a liquid. When they are hydrogenated, the double bonds are replaced by single bonds and the chains become much more flexible; therefore the molecules pack together better and form a solid.

D Hydrogenation of oils. TDC, 3-18s(c).

Manufacture. Natural supplies of hydrogen gas are far too small to satisfy the needs of industry, so most commercial hydrogen is obtained from low-molar-mass hydrocarbons. Hydrogen is produced on an industrial scale as a by-product of the processing and refining of the hydrocarbons in petroleum. For example, the "building" of octane, C_8H_{18}, from smaller hydrocarbon molecules results in the production of hydrogen:

$$2C_2H_6(g) + C_4H_8(g) \longrightarrow C_8H_{18}(l) + H_2(g)$$

The **cracking** of large hydrocarbon molecules also produces hydrogen:

$$C_{12}H_{26}(g) \longrightarrow C_5H_{10}(g) + C_4H_8(g) + C_3H_6(g) + H_2(g)$$

Another process depends on two catalyzed reactions. The first is a **reforming reaction**, a reaction in which a hydrocarbon and steam are converted to carbon monoxide and hydrogen over a nickel catalyst:

$$CH_4(g) + H_2O(g) \xrightarrow{800°C,\ Ni} CO(g) + 3H_2(g)$$

FIGURE 18.2 When water is added to calcium hydride, it oxidizes the hydride ion to molecular hydrogen, which forms the bubbles we see here.

D (1) Slowly bubble hydrogen gas through soapy water and ignite the bubbles. (2) Collect hydrogen gas in a test tube by air displacement and ignite. (3) Inflate two small balloons, one with helium and the other with hydrogen, and ignite (using a long stick).

D Rising moth balls, TDC, 3-21s.

E Hydrogen: The ultimate fuel and energy carrier. *J. Chem. Ed.*, **1988**, *65*, 688.

The mixture of products, which is called **synthesis gas**, is the starting point for the manufacture of numerous compounds, including methanol. In a second reaction, the carbon monoxide in the synthesis gas reacts with more water in the **shift reaction**.

$$CO(g) + H_2O(g) \xrightarrow{400°C,\ Fe/Cu} CO_2(g) + H_2(g)$$

where Fe/Cu denotes a catalyst made of iron and copper.

Hydrogen is prepared in the laboratory by reducing hydrogen ions with a metal that has a negative reduction potential; for example,

$$Zn(s) + 2HCl(aq) \longrightarrow ZnCl_2(aq) + H_2(g)$$

Another useful source is the oxidation of a hydride with water (Fig. 18.2):

$$CaH_2(s) + 2H_2O(l) \longrightarrow Ca(OH)_2(aq) + 2H_2(g)$$

This reaction is a convenient source of hydrogen for filling weather balloons.

Physical properties. The physical properties of hydrogen are summarized in Table 18.1. It is a colorless, odorless, tasteless, almost insoluble gas that condenses to a colorless liquid at 20 K. Its low boiling point and insolubility stem from its very weak intermolecular interactions. Being nonpolar, H_2 molecules can attract one another only by London forces. However, even these forces are weak, because each molecule has only two electrons and hence only a very small instantaneous electric dipole. One striking physical property of the liquid is its very low density (0.09 g/cm^3), less than one-tenth that of water (Fig. 18.3). This low density accounts for the fact that hydrogen occupies over two-thirds the total volume of the space shuttle's main fuel tank (Fig. 18.4). As we saw in Section 6.5, one consequence of its low mass density is that liquid hydrogen has a low enthalpy density. Because hydrogen gas is less dense than air (0.090 g/L compared with 1.3 g/L at STP), it can be collected in the laboratory by downward displacement of air (Fig. 18.5). Its low density results in good lifting power when it is used in balloons; however, helium is normally used to avoid the danger of explosion.

The special characteristics of hydrogen. The smallness of hydrogen atoms allows them to take part in one of the most important of all types

TABLE 18.1 Physical properties of hydrogen

Valence configuration: $1s^1$

*Normal form**: colorless, odorless gas

Z	Name	Symbol	Molar mass, g/mol	Percentage abundance, %	Melting point, °C	Boiling point, °C	Density, g/L
1	hydrogen†	H	1.008	99.88	−259 (14 K)	−253 (20 K)	0.090
1	deuterium	^{2}H or D	2.014	0.02	−254 (19 K)	−249 (24 K)	0.18
1	tritium	^{3}H or T	3.016	radioactive	−252 (21 K)	−248 (25 K)	0.27

**Normal form* means the state and appearance of the element at 25°C and 1 atm pressure.
†The isotope ^{1}H is sometimes called protium.

of intermolecular interactions—hydrogen bonding. As we saw in Section 10.2, hydrogen bonding has far-reaching consequences, among them the low vapor pressure of water, ammonia, and hydrogen fluoride. It also accounts for the expansion of water when it freezes and hence the fact that ice floats on water: in the solid, hydrogen bonds hold the molecules apart in an open structure. Hydrogen bonding is responsible for the solubility in water of many organic molecules that contain —OH and $—NH_2$ groups. It is also responsible for the rigidity of cellulose, where it forms cross-links between neighboring chains of glucoselike carbohydrate molecules. In addition, hydrogen bonding plays a crucial part in controlling the shapes of protein and DNA molecules: it is strong enough to give rise to recognizable molecular structures but weak enough for those structures to be modified easily.

In many cases, particularly when the hydrogen atom is attached to an O atom, a halogen atom, or a positively charged N atom (as in NH_4^+), it can be transferred (as a proton) to another molecule. This proton transfer leads to the rich chemistry of the proton donors that we call Brønsted acids and of the acceptors we call Brønsted bases. Because proton transfer is very fast, acid-base equilibria respond immediately to changes in the conditions, particularly changes in the concentrations of other acids and bases. (See Chapters 14 and 15.)

FIGURE 18.3 The two measuring cylinders contain the same mass of liquid: the liquid on the left is water, that on the right is liquid hydrogen, which is one-tenth as dense.

18.2 IMPORTANT COMPOUNDS OF HYDROGEN

Hydrogen forms a huge number of compounds, and some of its chemical properties are listed in Table 18.2. However, most of its compounds are best treated as compounds of the other elements they contain. Sulfuric acid, for instance, is best treated as a compound of sulfur or as a representative of the oxoacids. Here we consider only the binary compounds of hydrogen, its compounds that contain only one other element. All the main-group elements, with the exception of the noble gases and (possibly) indium and thallium, form binary compounds with hydrogen. So do most of the elements on the left and right of the *d* block (but not those at the center of the block: there is no iron hydride, for instance).

The binary compounds of hydrogen are traditionally divided into three classes called saline (saltlike), molecular, and metallic (Fig. 18.6). However, like most classifications in chemistry, this one is by no means rigid, and the characteristics and structures of the compounds blend from one class to another.

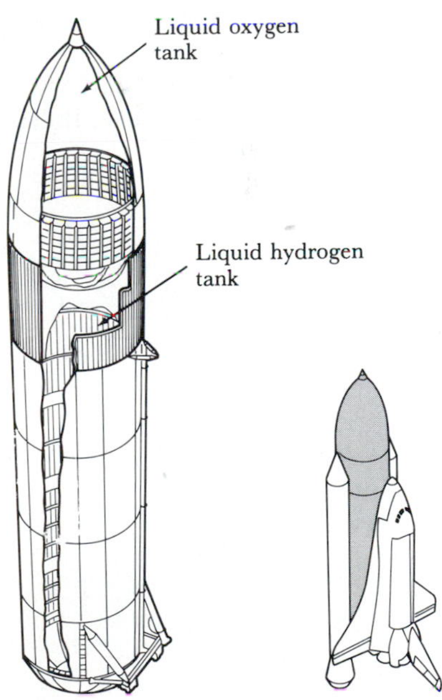

FIGURE 18.4 The arrangement of fuel tanks in the space shuttle. Note the very large size of the hydrogen tank compared with that of the oxygen tank. Liquid hydrogen has a very low density.

Saline hydrides. The **saline hydrides** are compounds of hydrogen and a strongly electropositive metal (any member of the *s* block, with the exception of Be). They are formed by heating the metal in hydrogen:

$$2K(s) + H_2(aq) \xrightarrow{\Delta} 2KH(s)$$

They are white, high-melting-point solids that contain the hydride ion H^-. Their crystalline structures resemble those of the metal halides; the alkali metal hydrides, for instance, have the rock-salt structure.

The single positive charge of the hydrogen atomic nucleus has only a weak control over the two electrons in the H^- ion. The ionic radius of H^- depends on the cations present but lies between the values for F^-

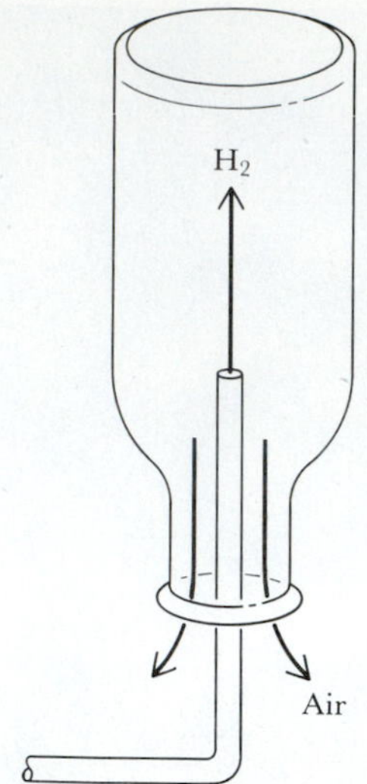

FIGURE 18.5 Because hydrogen is less dense than air, it can be collected by the downward displacement of air from an inverted gas bottle. The hydrogen rises to the top of the bottle and pushes the air out through the open mouth.

U Exercise 18.58.

E Lone-pair orbital energies in Group VI and VII hydrides. *J. Chem. Ed.*, **1973**, *50*, 322.

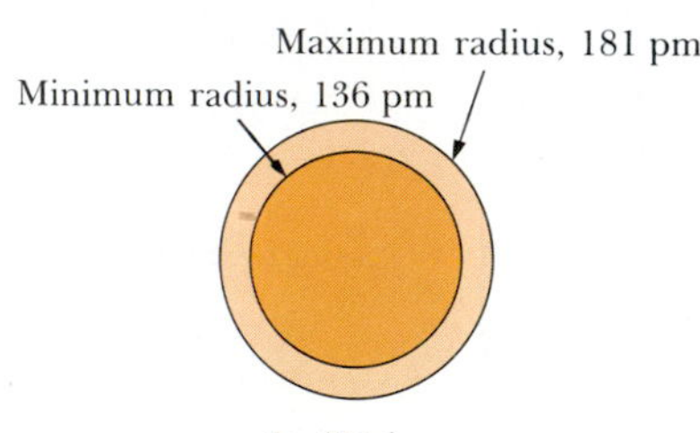

1 H^- ion

TABLE 18.2 Chemical properties of hydrogen

Reactant	Reaction with hydrogen
Group I metals, M	$2M(s) + H_2(g) \rightarrow 2MH(s)$
Group II metals, M	$M(s) + H_2(g) \rightarrow MH_2(s)$; not Be
some *d*-block metals, M	$2M(s) + xH_2(g) \rightarrow 2MH_x(s)$
oxygen, O_2	$O_2(g) + 2H_2(g) \rightarrow 2H_2O(l)$
nitrogen, N_2	$N_2(g) + 3H_2(g) \rightarrow 2NH_3(g)$
halogen, X_2	$X_2(g,l,s) + H_2(g) \rightarrow 2HX(g)$

(136 pm) and Cl^- (181 pm), as indicated in (**1**). The readiness with which one of the electrons can be lost results in ionic hydrides being powerful reducing agents:

$$H_2(g) + 2e^- \longrightarrow 2H^-(aq) \qquad E^\circ = -2.25\text{ V}$$

The reducing power of the hydride ion is similar to that of sodium metal, for the H_2/H^- couple is not very far above Na^+/Na in the electrochemical series.

Saline hydrides reduce water as soon as they come into contact with it:

$$NaH(s) + H_2O(l) \longrightarrow NaOH(aq) + H_2(g)$$

Because this reaction produces hydrogen, saline hydrides are potentially useful as transportable sources of hydrogen.

Molecular compounds of hydrogen. The molecular compounds of hydrogen consist of discrete molecules. They are formed by some metals (tin, for example, forms stannane, SnH_4), the metalloids, and the nonmetals. In most cases they are gases such as silane (SiH_4), ammonia, the hydrogen halides (HF, HCl, HBr, HI), and the numerous hydrocarbons, including methane, ethylene, and acetylene or liquids such as water, ethanol, and benzene. This class of hydrogen compounds includes the electron-deficient boron hydrides, but these are best treated under boron.

The molecular compounds of hydrogen are often prepared by direct action of the elements, as in the Haber process for producing ammonia

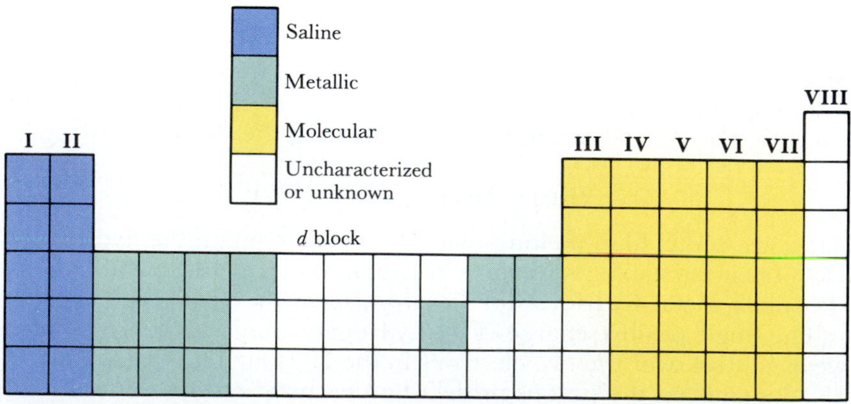

FIGURE 18.6 The different classes of binary hydrogen compounds.

or the synthesis of HCl:

$$N_2(g) + 3H_2(g) \xrightarrow{400\text{–}600°C,\ 150\text{ to }600\text{ atm, Fe}} 2NH_3(g)$$

$$H_2(g) + Cl_2(g) \xrightarrow{\text{light}} 2HCl(g)$$

The vigor of these and similar reactions varies widely: the synthesis of ammonia needs a catalyst, but the reaction of hydrogen and chlorine is explosively violent once it has been initiated. Hydrogen reacts explosively with fluorine as soon as the two gases are mixed. Because the reaction of hydrogen with oxygen is strongly exothermic ($\Delta H° = -286$ kJ/mol H_2), temperatures in excess of 2500°C can be quickly and efficiently reached. Because many metals are molten at this temperature, the reaction is used in oxyhydrogen cutting torches.

The binary molecular compounds of hydrogen may be prepared in a less dramatic way by protonation, or proton transfer to a Brønsted base, such as S^{2-}:

$$FeS(s) + 2HCl(aq) \longrightarrow FeCl_2(aq) + H_2S(g)$$

Another example is the protonation of the carbide ion $C_2^{\ 2-}$ to produce acetylene:

$$:C{\equiv}C:^{2-} + 2H_2O \longrightarrow H{-}C{\equiv}C{-}H + 2OH^-$$

which occurs when water acts on calcium carbide. Volatile binary acids can be prepared in this way, by using a less volatile acid as the proton donor:

$$CaF_2(g) + H_2SO_4(l) \longrightarrow CaSO_4(s) + 2HF(g)$$

The reaction proceeds to the right as the volatile product is removed. When oxidation of the binary acid by sulfuric acid is a danger, an acid that is less strongly oxidizing than sulfuric acid (such as phosphoric acid) should be used, as in the production of HBr:

$$KBr(s) + H_3PO_4(aq) \longrightarrow KH_2PO_4(aq) + HBr(g)$$

Metallic hydrides. The metallic hydrides are electrically conducting. They have a variable composition because the hydrogen atoms occupy gaps (interstices) between metal atoms in the solid (Fig. 18.7). For this reason, metallic hydrides are also called **interstitial hydrides**. These black, powdery solids are prepared by heating certain *d*-block metals in hydrogen; for example,

$$2Cu(s) + H_2(g) \xrightarrow{\Delta} 2CuH(s)$$

Because the metallic hydrides release their hydrogen (as H_2) when heated or treated with acid, they are also currently being considered for use in storing and transporting hydrogen intended for use as a fuel.

GROUP I: ALKALI METALS

The members of Group I, the alkali metals, are shown in the margin; and a few of their physical properties are listed in Table 18.3. Their valence electron configurations are ns^1, where n is the period number. The physical and chemical properties of these elements are dominated by the ease with which the single valence electron can be removed.

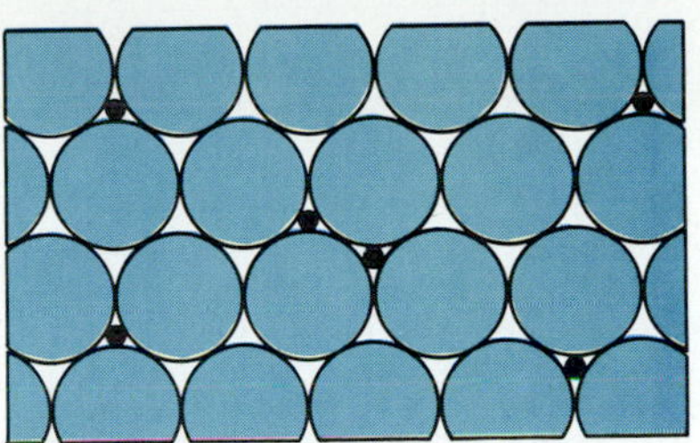

FIGURE 18.7 In a metallic hydride, the small hydrogen atoms (black) occupy gaps—called interstices—between the larger metal atoms (blue).

E Proton power: An intuitive approach to the electronic structures of molecular hydrides. *J. Chem. Ed.*, **1988**, *65*, 976.

D (1) Place solid CaC_2 in a balloon, add water, and tie shut. Using tongs, place the inflated balloon over a Bunsen flame. Be careful, its messy! (2) A calcium carbide cannon can be purchased from Edmund Scientific Co. Spontaneous combustion: acetylene. TDC, 159.

N The carbide (acetylide) ion, $C_2^{\ 2-}$, is an example of a strong Brønsted base which is fully protonated in water. It is isoelectronic with the N_2 molecule.

I	II						He
Li	Be	B	C	N	O	F	Ne
Na	Mg						
K	Ca						
Rb	Sr						
Cs	Ba						
Fr	Ra						

E Lithium is not particularly soft: it is similar in hardness to calcium. Francium has not been isolated in bulk quantities. It has been estimated that there are on average only about 17 naturally occurring atoms of the element throughout the body of the Earth: they are formed by radioactive decay, but decay quickly themselves, and establish a steady state of very low abundance. Its longest lived isotope has a half-life of only 22 min.

E Lithium story. *J. Chem. Ed.*, **1976**, *53*, 291.

D Conductivity and ductility of sodium. TDC, 12-3.

TABLE 18.3 Group I elements: The alkali metals

Valence configuration: ns^1

*Normal form**: soft, silver-gray metals

Z	Name	Symbol	Molar mass, g/mol	Melting point, °C	Boiling point, °C	Density, g/cm³
3	lithium	Li	6.94	181	1347	0.53
11	sodium	Na	22.99	98	883	0.97
19	potassium	K	39.10	64	774	0.86
37	rubidium	Rb	85.47	39	688	1.53
55	cesium	Cs	132.91	28	678	1.87
87	francium	Fr	223	27	677	—

**Normal form* means the state and appearance of the element at 25°C and 1 atm pressure.

18.3 GROUP I ELEMENTS

All the Group I elements are too easily oxidized to be found in the free state in nature, and common reducing agents are unable to extract them from their compounds. The pure metals are obtained by electrolysis of their molten salts (as in the Downs process, Section 17.8) or, in the case of potassium, by exposing molten potassium chloride to sodium vapor:

$$KCl(l) + Na(g) \xrightarrow{750°C} NaCl(s) + K(g)$$

Although the equilibrium constant for this reaction is not particularly favorable, the reaction runs to the right because potassium is more volatile than sodium and is driven off by the heat.

Physical properties. All the Group I elements are soft, silver-gray metals (Fig. 18.8). Lithium is the hardest, but even so it is softer than lead.

(a)

(b)

(c)

(d)

FIGURE 18.8 The alkali metals of Group I: (a) lithium, (b) sodium, (c) potassium, and (d) rubidium (left) and cesium (right). Francium has never been isolated in visible quantities. The first three corrode rapidly in moist air; rubidium and cesium are even more reactive and have to be stored (and photographed) in sealed, airless containers.

The melting points decrease down the group (Fig. 18.9). Cesium, which melts at 28°C, is barely solid at room temperature. Some alloys of sodium and potassium are liquid at room temperature because their atoms pack together poorly and hence produce a fluid structure (Fig. 18.10). They are used as coolants in some nuclear reactors because being metals, they conduct heat very well and are not decomposed by radiation. Lithium was an element with few applications until thermonuclear weapons (which use lithium-6; see Section 22.9) were developed after World War II.

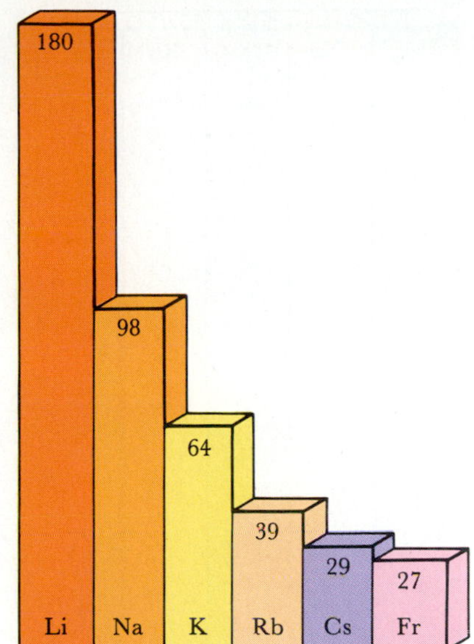

FIGURE 18.9 The melting points of the alkali metals decrease down the group. (Values are given in degrees Celsius.)

Chemical properties. The chemical properties of the alkali metals are summarized in Table 18.4. Because the first ionization energies of the alkali metals are so low, they are most commonly found as singly charged cations. Consequently, most of their compounds are ionic, with the metal present as the M^+ ion. Because of the tendency to exist as M^+ ions, the alkali metals have low reduction potentials and thus are excellent reducing agents. Sodium metal is used to produce zirconium and titanium from their chlorides:

$$TiCl_4(g) + 4Na(l) \longrightarrow 4NaCl(s) + Ti(s)$$

Reduction potentials become more negative down the group, but lithium has an anomalously low reduction potential (−3.05 V; compare with −2.71 V for sodium; Fig. 18.11). This anomaly is due to the small size of the Li atom, which becomes strongly hydrated as the ion—the very exothermic hydration process encourages the atom to lose an electron and hence to act as a strong reducing agent.

N The origin of the standard reduction potential could be demonstrated by constructing a thermodynamic cycle in which the steps are the atomization of the metal, the ionization of its atoms, and the hydration of the resulting cations. Then the influence of atomic and ionic radius on the three terms could be discussed.

V Video 21: The reaction of sodium and water.

T OHT: Tables 18.4 and 18.7.

Reaction with water. Because of their low reduction potentials, the Group I elements can reduce water. For example, from Table 17.3, we can write

$$2H_2O(l) + 2e^- \longrightarrow H_2(g) + 2OH^-(aq) \qquad E = -0.42\text{ V at pH} = 7$$

$$Na(s) \longrightarrow Na^+(aq) + e^- \qquad E° = +2.71\text{ V}$$

$$\textit{Overall}:\ 2Na(s) + 2H_2O(l) \longrightarrow 2NaOH(aq) + H_2(g) \qquad E = 2.29\text{ V}$$

The vigor of this reaction increases uniformly down the group (as illustrated in Fig. 2.5). However, the reaction between water and lithium is gentle, even though lithium's reduction potential is strongly negative. The reaction is vigorous enough with potassium to ignite the hydrogen and dangerously explosive with rubidium and cesium. The danger

TABLE 18.4 Chemical properties of alkali metals

Reactant	Reaction with alkali metal, M
hydrogen, H_2	$2M(s) + H_2(g) \rightarrow 2MH(s)$
oxygen, O_2	$4Li(s) + O_2(g) \rightarrow 2Li_2O(s)$
	$2Na(s) + O_2(g) \rightarrow Na_2O_2(s)$
	$M(s) + O_2(g) \rightarrow MO_2(s)$; M = K, Rb, Cs
nitrogen, N_2	$6Li(s) + N_2(g) \rightarrow 2Li_3N(s)$
halogen, X_2	$2M(s) + X_2(g,l,s) \rightarrow 2MX(s)$
water	$2M(s) + H_2O(l) \rightarrow 2MOH(aq) + H_2(g)$

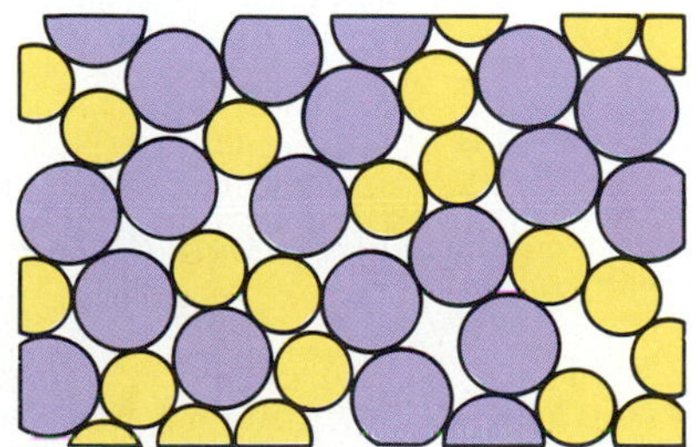

FIGURE 18.10 The atomic radii of sodium (yellow) and potassium (violet) are such that they pack together poorly, and an alloy of the two is a liquid at room temperature.

FIGURE 18.11 The standard reduction potentials of the alkali metals. Note the anomalous value for lithium.

? Are the alkali metals strong oxidizing or strong reducing agents?

E The alkali metals are also oxidized when they dissolve in liquid ammonia: when they do so, the electrons released occupy cavities formed by the NH_3 molecules to give dark blue *metal-ammonia solutions.* At high concentrations the color changes to bronze and the liquids conduct electricity like molten metals.

E The solid KO_2 is paramagnetic on account of the presence of the O_2^- ion, which has one unpaired electron (and a valence configuration $2s\sigma^2 2s\sigma^{*2} 2p\sigma^2 2p\pi^4 2p\pi^3$).

D Oxidation of sodium. CD1, 83.

E Cesium neonide: Molecule or thermochemical exercise? *J. Chem. Ed.*, **1982**, *59*, 637.

with the last two elements stems from the fact that they are denser than water, so they sink and react beneath the surface; the rapidly evolved hydrogen gas then forms a shock wave that can shatter the vessel.

Reaction with nonmetals. All alkali metals react directly with most nonmetals (other than the noble gases). However, only lithium reacts with nitrogen, in which it is oxidized (like its diagonal neighbor magnesium) to lithium nitride:

$$6Li(s) + N_2(g) \xrightarrow{\Delta} 2Li_3N(s)$$

The principal product of the reaction of the alkali metals with oxygen varies down the group (Fig. 18.12). The smaller cations form compounds with small anions (the O^{2-} ion) predominantly, and the larger cations form compounds with progressively larger anions (peroxide, O_2^{2-}, and superoxide, O_2^-). Lithium, for example, forms mainly the oxide:

$$4Li(s) + O_2(g) \longrightarrow 2Li_2O(s)$$

The oxides of the other elements are produced by decomposition of their carbonates:

$$K_2CO_3(s) \xrightarrow{\Delta} K_2O(s) + CO_2(g)$$

The O^{2-} ion is such a strong Brønsted base that a Group I oxide immediately forms the hydroxide when it dissolves in water:

$$O^{2-}(aq) + H_2O(l) \longrightarrow 2OH^-(aq)$$

Sodium reacts with oxygen to form mainly the pale yellow peroxide:

$$2Na(s) + O_2(g) \longrightarrow Na_2O_2(s)$$

The remaining metals form mainly the superoxide when they react with oxygen:

$$K(s) + O_2(g) \longrightarrow KO_2(s)$$

Potassium superoxide, a paramagnetic yellow solid, is used in closed-system breathing apparatus (gas masks). The KO_2 removes the exhaled

FIGURE 18.12 Although the alkali metals give a mixture of products when they react with oxygen, lithium gives mainly the oxide (left); sodium, the very pale yellow peroxide (center); and potassium, the yellow superoxide (right).

water vapor and generates oxygen gas:

$$4KO_2(s) + 2H_2O(g) \longrightarrow 4KOH(s) + 3O_2(g)$$

The KOH produced in this reaction removes the exhaled CO_2:

$$KOH(s) + CO_2(g) \longrightarrow KHCO_3(s)$$

The overall reaction is

$$4KO_2(s) + 2H_2O(g) + 4CO_2(g) \longrightarrow 4KHCO_3(s) + 3O_2(g)$$

Much of the potassium currently produced is used to manufacture potassium superoxide for this purpose. A similar overall reaction occurs with the Group I peroxides:

$$2Na_2O_2(s) + 2H_2O(g) + 4CO_2(g) \longrightarrow 4NaHCO_3(s) + O_2(g)$$

Lithium peroxide reacts similarly. Because it has a lower molar mass, a smaller mass of Li_2O_2 can increase the oxygen content of a given amount of air, so it is favored for use in spacecraft.

18.4 IMPORTANT GROUP I COMPOUNDS

The properties of some of the principal compounds of the Group I elements are summarized in Table 18.5. The presence of the elements in compounds can be detected easily by the characteristic colors they give to flames. As we saw at the beginning of Chapter 7, lithium gives deep red; sodium, bright yellow; potassium, lilac; rubidium, red; and cesium, blue. The color characteristic of lithium is emitted by LiOH molecules formed briefly in the flame, not by Li atoms themselves.

V Video 22: Flame tests of metals.

D Place a spatula-full of LiCl, NaI, KI, RbCl, and CsCl in separate glass dishes containing 5 to 10 mL methanol and ignite.

E A lithium powered heart. *J. Chem. Ed.*, **1986**, *63*, 844.

Lithium compounds. Lithium is typical of an element at the head of its group, in that it differs significantly from its congeners and has a diagonal relationship with a neighbor—in this case, magnesium. The differences stem in part from the small size of the Li^+ cation. This small size gives it a strong polarizing power and hence a tendency toward covalent bonding. A further consequence of the small size of a Li^+ ion is strong ion-dipole interactions, with the result that lithium salts are often hydrated.

Commercial availability of lithium has increased during the past few decades and so has the variety of the metal's applications. Lithium compounds are used in batteries, ceramics, and lubricants. They are also

TABLE 18.5 Properties of Group I compounds

Compounds	Formula*	Comment
oxides	M_2O	formed by decomposition of carbonates; strong bases; react with water to form hydroxides
hydroxides	MOH	formed by reduction of water with the metal or from the oxide; strong bases
carbonates	M_2CO_3	soluble in water; weak bases in water; most decompose into oxides when heated
hydrogencarbonates	$MHCO_3$	weak bases in water; can be obtained as solids
nitrates	MNO_3	decompose to nitrite and evolve oxygen when strongly heated

*M = Group I metal.

E Lithium and mental health. *J. Chem. Ed.*, **1973**, *50*, 343. Sodium. What's the use? *J. Chem. Ed.*, **1990**, *67*, 1046.

used in medicine, and small daily doses of lithium carbonate have been used to treat the mental disorder known as manic depression; the mode of action is still not understood despite a great deal of work. Lithium soaps—the lithium salts of long-chain fatty acids—are used as thickeners in lubricating greases. Potassium and sodium salts of the same fatty acids melt at low temperatures, but the lithium salts do not. Because bearings must remain lubricated at high temperatures, the presence of lithium soap helps to minimize the loss of the lubricant.

E Lithium greases are used at service centers to grease cars.

E The Great Salt Lake in Utah is 15.3% NaCl by mass and is inhabited by only small brine shrimp.

? Vast oil reserves in the United States are stored in underground salt mines in Louisiana. Will the salt dissolve in the oil? Why not?

N The salinity of the oceans is an interesting example of the influence of ionic radius on the environment.

Sodium chloride. The mass of sodium chloride used annually by industry actually exceeds that of the traditional chart-topping sulfuric acid. However, sodium chloride is not normally included in the top 50 chemicals because it is available naturally and does not have to be manufactured. Sodium chloride is mined as rock salt or obtained from the evaporation of brine.

The vastness of both resources raises the interesting question of the origin of the saltiness of the oceans. Why are the oceans nearly 30 times richer in Na^+ ions than in K^+ ions even though these two elements occur with similar overall abundance on earth (Table 11.1)? Three factors contribute to the predominance of sodium in the oceans. One is that K^+ ions, being larger than Na^+ ions, have weaker ion-dipole interactions with H_2O molecules and therefore are slightly less soluble. Another is that Na^+ ions in minerals are less strongly bound than K^+ ions, so they are washed out of the minerals (and, eventually, into the oceans) more easily. Third, K^+ ions are more essential to plant growth than Na^+ ions are, so plant roots have efficient methods for collecting the scarcer K^+ ions from the water that trickles past them in the soil. In effect, plants selectively filter out the K^+ ions, leaving the abundant Na^+ ions to flow on to the sea. The major deposits of potassium chloride (as the mineral sylvite) were probably left by bodies of water that formed and later dried out in environments devoid of vegetation.

Sodium chloride is the starting material for the production of sodium metal (and chlorine, in the Downs process), sodium hydroxide, sodium sulfate, sodium carbonate, and sodium hydrogencarbonate.

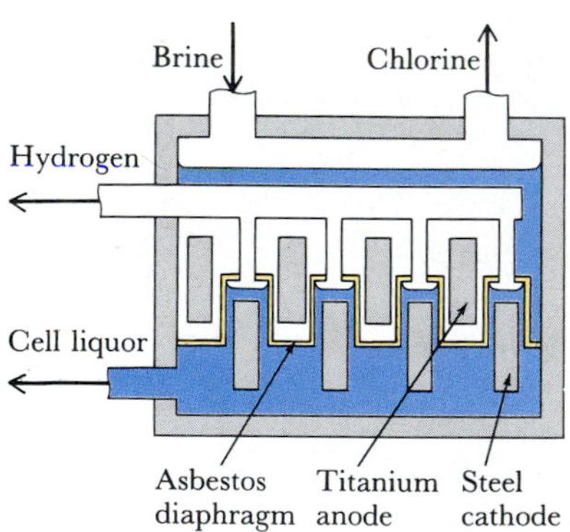

FIGURE 18.13 A diaphragm cell for the electrolytic production of sodium hydroxide from brine. The diaphragm prevents the chlorine from mixing with the hydrogen and the sodium hydroxide. The liquid is drawn off and the water is partially evaporated. The unconverted sodium chloride crystallizes, leaving the sodium hydroxide in solution.

Sodium hydroxide. Sodium chloride is used in large quantities in the electrolytic production of chlorine and of sodium hydroxide (caustic soda) from brine. The hydroxide is a soft, waxy, white, corrosive solid. It is an industrial chemical of considerable importance, being a cheap base for the production of other sodium salts, and it ranks eighth in annual U.S. production.

Brine electrolysis is second only to aluminum extraction in the use of electricity for chemical production. Most modern production uses a diaphragm cell, in which compartments containing steel and titanium electrodes are separated by porous asbestos diaphragms (Fig. 18.13). In the electrolysis

$$2NaCl(aq) + 2H_2O(l) \longrightarrow 2NaOH(aq) + Cl_2(g) + H_2(g)$$

1 mol of Cl_2 is produced for every 2 mol of NaOH; hence the industry must coordinate its hydroxide production with the demand for chlorine. This coordination is difficult, because a rise in demand for one product is not always accompanied by a rise in demand for the other.

Sodium sulfate. Anhydrous sodium sulfate is known as salt cake and the decahydrate ($Na_2SO_4 \cdot 10H_2O$) as Glauber's salt. Much of the sulfate used in industry (for papermaking and as a substitute for phosphates in detergents) now comes from natural sources, particularly the sulfate-rich underground brines found in Texas. In countries lacking natural supplies, it is produced by the action of concentrated sulfuric acid on sodium chloride:

$$H_2SO_4(l) + 2NaCl(s) \longrightarrow Na_2SO_4(s) + 2HCl(g)$$

This reaction is another one that proceeds in the desired direction because a volatile product escapes. It is used to generate hydrogen chloride in the laboratory.

E The sulfates themselves are the excrement of sulfur-oxidizing bacteria.

Sodium hydrogencarbonate and sodium carbonate. Sodium hydrogencarbonate, $NaHCO_3$ (sodium bicarbonate), is commonly called bicarbonate of soda or baking soda. The action of lactic acid, which is found in sour milk or buttermilk, of citric acid from a lemon, of acetic acid in vinegar, or of the hydrated Al^{3+} ions in alum, $NaAl(SO_4)_2 \cdot 12H_2O$, found in baking powder, results in the release of carbon dioxide in the reaction

$$NaHCO_3(aq) + H_3O^+(aq) \longrightarrow Na^+(aq) + 2H_2O(l) + CO_2(g)$$

The release of gas causes dough to rise.

Sodium carbonate decahydrate ($Na_2CO_3 \cdot 10H_2O$) was widely used earlier in this century as washing soda. It was added to water to precipitate Mg^{2+} and Ca^{2+} ions as carbonates,

$$Ca^{2+}(aq) + CO_3{}^{2-}(aq) \longrightarrow CaCO_3(s)$$

and to provide an alkaline environment ($CO_3{}^{2-}$ is a Brønsted base) that helps to remove grease from fabrics. Anhydrous sodium carbonate, or soda-ash, is used in large amounts in the glass industry as a source of sodium oxide into which it decomposes when heated (Section 19.5). This reaction is a Lewis acid-base decomposition, with the Lewis acid CO_2 leaving the Lewis base O^{2-}:

$$CO_3{}^{2-}(s) \xrightarrow{\Delta} O^{2-}(s) + CO_2(g)$$

E Soda ash manufacture: An example of what? *J. Chem. Ed.*, **1973**, *50*, 64.

E The major source of sodium carbonate, before large natural deposits of the mineral trona,

$$Na_2CO_3 \cdot NaHCO_3 \cdot 2H_2O$$

were discovered beneath Wyoming in the 1940s, was the *Solvay process.* The starting materials were sodium chloride and calcium carbonate, and the overall reaction is $2NaCl(aq) + CaCO_3(s) \rightarrow Na_2CO_3(s) + CaCl_2(s)$. However, this simple reaction requires plenty of chemical cunning to be brought about economically, and in 1861 the Belgian industrial chemist Ernest Solvay devised a process that used ammonia at an intermediate stage. For the details, and a flow chart, see the first edition, p. 674. Solvay's achievement was to take a procedure that had been demonstrated by others on a laboratory scale and overcome the engineering difficulties that had foiled others. He became wealthy from his patents, and founded a series of conferences that brought together leading figures of physics and chemistry.

U Exercise 18.68.

E Some baking powders contain $Ca(H_2PO_4)_2$ or sodium tartrate (cream of tartar) instead of alum. (The action of double-acting baking powders is described on p. 738.)

Potassium compounds. The principal mineral sources of potassium are sylvite (KCl) and carnallite ($KCl \cdot MgCl_2 \cdot 6H_2O$). Potassium chloride is incorporated directly into fertilizers as a source of essential potassium; but for crops such as potatoes and tobacco, which cannot tolerate high Cl^- ion concentrations, potassium nitrate is used instead. The latter is formed by fractional crystallization of sodium chloride from a mixed solution of potassium chloride and sodium nitrate:

$$NaNO_3(aq) + KCl(aq) \longrightarrow NaCl(s) + KNO_3(aq)$$

The success of this process depends on the low solubility of sodium chloride relative to that of potassium nitrate.

Potassium compounds are generally more expensive than the corresponding sodium compounds, but sometimes their advantages outweigh the expense. Potassium nitrate releases oxygen when heated, in the reaction

$$2KNO_3(s) \xrightarrow{\Delta} 2KNO_2(s) + O_2(g)$$

and is used to facilitate the ignition of matches. It is more expensive, in common with other potassium compounds, but less hygroscopic (water

N See Fig. 11.15; nitrate salts tend to have higher solubilities than chlorides.

absorbing) than the corresponding sodium compounds because the K^+ cation is larger and has weaker ion-dipole interactions with H_2O molecules. Potassium nitrate is also the oxidizing agent in old-fashioned black gunpowder, a mixture of 75% potassium nitrate, 15% charcoal, and 10% sulfur. It reacts as follows:

$$2KNO_3(s) + 4C(s) \longrightarrow K_2CO_3(s) + 3CO(g) + N_2(g)$$

$$2KNO_3(s) + 2S(s) \longrightarrow K_2SO_4(s) + SO_2(g) + N_2(g)$$

The two reactions result in the rapid formation of 6 mol of gas molecules from 4 mol of solid KNO_3, thus causing a sudden explosive expansion. Potassium carbonate is used as a source of potassium oxide in glassmaking and in the manufacture of soft (liquid) soaps; sodium produces hard (solid) soaps.

GROUP II: ALKALINE EARTH METALS

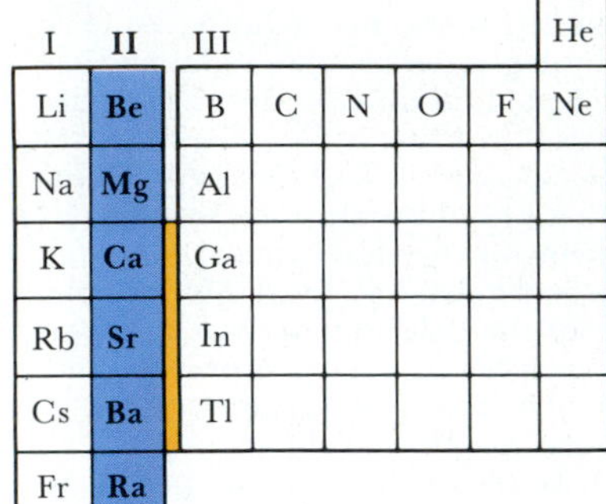

The members of Group II (shown in the margin) and some of their physical properties are listed in Table 18.6. Calcium, strontium, and barium are called the alkaline earth metals because their "earths"—the old name for oxides—are basic. The name *alkaline earth metals* is often extended loosely to all the members of the group.

18.5 GROUP II ELEMENTS

All the elements of the group are metals and are too reactive to occur in the uncombined state in nature (Fig. 18.14). Beryllium occurs mainly as beryl ($3BeO \cdot Al_2O_3 \cdot 6SiO_2$), sometimes in crystals so big that they weigh several tons. Emerald is a form of beryl; its green color is caused by the Cr^{3+} ions present as impurities (Fig. 18.15). Magnesium occurs in seawater and as the mineral dolomite $CaCO_3 \cdot MgCO_3$. Calcium also occurs as $CaCO_3$ in limestone deposits, as calcite, and as chalk. Barium is found as the mineral barite, $BaSO_4$.

E Limestone and chalk are the compressed remains of the shells and skeletons of invertebrates (corals, bivalves, planktonic animals, and so on) deposited on seabeds. Calcium is also deposited as phosphate from the bony skeletons of fish. Some calcium is also precipitated as calcium carbonate directly. Calcium carbonate is initially deposited from seawater as aragonite, but that polymorph changes to calcite in the course of time.

TABLE 18.6 Group II elements

Valence configuration: ns^2

*Normal form**: soft, silver-gray metals

Z	Name	Symbol	Molar mass, g/mol	Melting point, °C	Boiling point, °C	Density, g/cm³
4	beryllium	Be	9.01	1285	2470	1.85
12	magnesium	Mg	24.31	650	1100	1.74
20	calcium†	Ca	40.08	840	1490	1.53
38	strontium†	Sr	87.62	770	1380	2.58
56	barium†	Ba	137.34	710	1640	3.59
88	radium	Ra	226.03	700	1500	5.00

**Normal form* means the state and appearance of the element at 25°C and 1 atm pressure.
†The alkaline earth metals.

E Note on beryllium. *J. Chem. Ed.*, **1969**, *46*, 276.

(a)

(b)

(c)

(d)

(e)

FIGURE 18.14 The elements of Group II: (a) beryllium, (b) magnesium, (c) calcium, (d) strontium, and (e) barium. The four central elements of the group (magnesium through barium) were discovered by Humphry Davy in a single year (1808). The two outer elements were discovered later: beryllium in 1828 (by Friedrich Wöhler) and radium in 1898 (by Pierre and Marie Curie).

Physical properties. Beryllium has grown in importance in recent years, both because of its low density, which makes it suitable for the construction of missiles and satellites, and because of its useful nuclear properties (which are discussed in Chapter 22). Because Be atoms have so few electrons, thin sheets of the metal are transparent to x-rays and can be used as windows for x-ray tubes (to allow the rays to escape). The metal is obtained by electrolytic reduction of molten beryllium chloride. Much of the metal that is produced is added in small amounts to copper; the small Be atoms pin the Cu atoms together in a structure that is more rigid than pure copper is, but they do not destroy copper's excellent ability to conduct electricity. The hard, electrically conducting alloy is formed into nonsparking tools for use in oil refineries.

Magnesium is a silver-white metal that is protected in air by a film of white oxide and hence often looks dull gray. It has a low density (two-thirds that of aluminum) and is widely used as a component of alloys in applications where lightness and toughness are needed, as in airplanes. Metallic magnesium is produced by reduction, either electrolytically or with a chemical reducing agent. In the chemical reduction of magnesium oxide (obtained from the decomposition of dolomite), ferrosilicon, an alloy of iron and silicon, is used as the reducing agent at about

FIGURE 18.15 An emerald is a crystal of beryl with some Cr^{3+} ions, which are responsible for its color.

FIGURE 18.16 This facility is a magnesium extraction plant at Freeport, Texas.

U Exercise 18.62.

T OHT: Fig. 18.17.

E Dolomite, $CaCO_3 \cdot MgCO_3$, is being substituted as the supply of oyster shells has diminished.

1200°C. At this temperature, the magnesium produced is immediately vaporized; so even though the equilibrium does not favor the reduction, the process continues because the product is removed as soon as it is formed.

The electrolytic method uses seawater as its principal raw material (Fig. 18.16). The first stage is the precipitation of magnesium hydroxide ($K_s = 1.1 \times 10^{-11}$) with lime (calcium hydroxide):

$$Mg^{2+}(aq) + 2OH^-(aq) \longrightarrow Mg(OH)_2(s)$$

The lime is produced by the thermal decomposition of the calcium carbonate in shells dredged up from the ocean floor. The precipitated magnesium hydroxide is filtered off and treated with hydrochloric acid:

$$Mg(OH)_2(s) + 2HCl(aq) \longrightarrow MgCl_2(aq) + 2H_2O(l)$$

The acid is obtained by the action of chlorine (a by-product of the electrolysis) on natural gas:

$$CH_4(g) + Cl_2(g) \longrightarrow CH_3Cl(g) + HCl(g)$$

Finally, the magnesium chloride is dried, added to an electrolytic cell, and melted. In the electrolysis, magnesium is produced at the cathode and chlorine at the anode:

$$MgCl_2(l) \xrightarrow{\text{electrolysis}} Mg(s) + Cl_2(g)$$

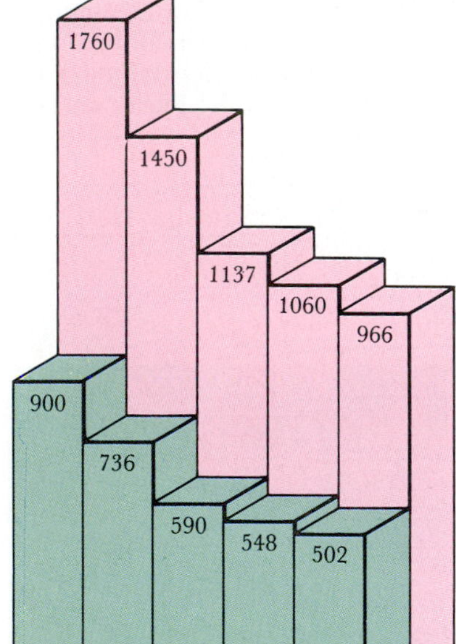

FIGURE 18.17 The variation of the first and second ionization energies (in kJ/mol) of the Group II elements. Although the second ionization energies are larger than the first, they are not enormous, and both valence electrons are lost from each atom in all the compounds of these elements.

The true alkaline earth metals—calcium, strontium, and barium—are obtained either by electrolysis or by reduction with aluminum in a version of the thermite process. For example, for barium,

$$3BaO(s) + 2Al(s) \xrightarrow{\Delta} Al_2O_3(s) + 3Ba(s)$$

Chemical properties. The valence electron configuration of the atoms of the Group II elements is ns^2, where n is the period number. The second ionization energy is low enough that the energy required to form the doubly charged cation can be recovered from the increased lattice enthalpy (Fig. 18.17). Hence, the Group II elements occur with an oxidation number of +2 in all their compounds—as the cation M^{2+}. All are low in the electrochemical series, with strongly negative reduction potentials (Fig. 18.18). Their reducing power increases down the group, beryllium having the least reducing power and radium the greatest. A summary of the chemical properties of the Group II elements is given in Table 18.7.

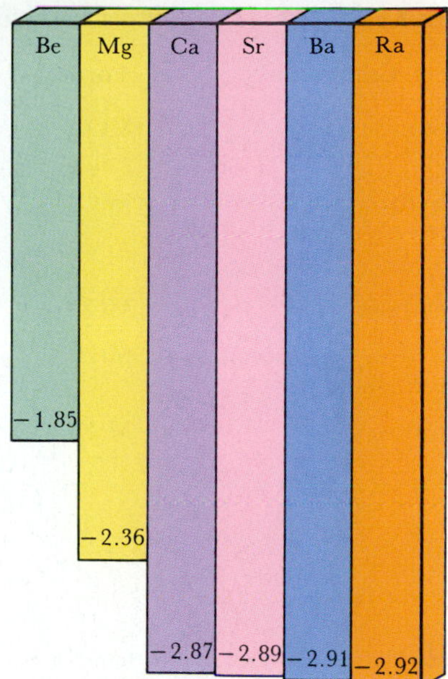

FIGURE 18.18 The standard reduction potentials of the Group II elements.

V Video 46: Reactivity of magnesium.

U Exercises 18.57, 18.59, and 18.66.

D Reduction of sand with magnesium. CD2, 102.

D Producing hydrogen gas from calcium metal. CD2, 27.

Beryllium, at the head of Group II, resembles its diagonal neighbor aluminum in its chemical properties. It is the least metallic element of the group and many of its compounds have properties commonly attributed to covalent bonding. Beryllium shows its amphoteric character (its ability to react with both acids and bases) by being the only member of the group that dissolves (like aluminum) in aqueous sodium hydroxide. In doing so, it forms a beryllate ion by a Lewis acid-base reaction:

$$Be(s) + 2OH^-(aq) + 2H_2O(l) \longrightarrow Be(OH)_4{}^{2-}(aq) + H_2(g)$$

The formation of a tetrahedral BeX_4 covalent unit (**2**) is a typical feature of beryllium's chemistry.

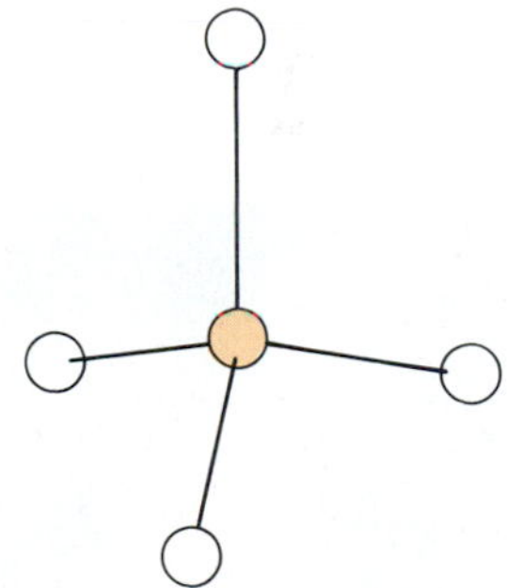

2 BeX_4 unit

Magnesium burns vigorously in air (it reacts with the nitrogen and the carbon dioxide as well as the oxygen, especially when it is sprayed with water); hence, magnesium is used in fireworks and in incendiary devices. Calcium, strontium, and barium, however, are partially passivated in air by a protective surface layer of oxide. Barium reacts particularly quickly and in moist air may ignite. Although this reactivity is a difficult problem for the barium industry, like most problems it can be turned to advantage in particular contexts. Barium is in fact used as a "getter" in vacuum tubes, because it combines with any traces of oxygen and air that remain after manufacture and leaves a better vacuum. Calcium is used on a bigger scale for much the same type of reaction: it is added to some steels in order to remove remaining traces of oxides.

All the Group II elements except beryllium reduce water; for example,

$$Ca(s) + 2H_2O(l) \longrightarrow Ca(OH)_2(aq) + H_2(g)$$

Beryllium does not react with water even when red hot: it is passivated (like aluminum) by an oxide film that survives even at high temperatures. Magnesium reacts when heated, and calcium reacts with cold water. The metals reduce hydrogen ions to hydrogen, but neither beryllium nor magnesium dissolves in nitric acid because they become passivated by a film of oxide.

TABLE 18.7 Chemical properties of alkaline earth metals

Reactant	Reaction with alkaline earth metal, M
hydrogen, H_2	$M(s) + H_2(g) \rightarrow MH_2(s)$; not Be or Mg
oxygen, O_2	$2M(s) + O_2(g) \rightarrow 2MO(s)$
	$Ba(s) + O_2(g) \rightarrow BaO_2(s)$
nitrogen, N_2	$3M(s) + N_2(g) \rightarrow M_3N_2(s)$
halogen, X_2	$M(s) + X_2(g,l,s) \rightarrow MX_2(s)$
water	$M(s) + 2H_2O(l) \rightarrow M(OH)_2(aq) + H_2(g)$; not Be

U Exercise 18.61.

TABLE 18.8 Properties of Group II compounds

Compounds	Formula*	Comment
oxides	MO	formed by decomposition of carbonates; give strongly basic solutions when they react with water (BeO is amphoteric); withstand high temperatures
hydroxides	$M(OH)_2$	formed by action of water on oxides or precipitation from salt solution; sparingly soluble in water (except Ba); strong bases (except Be, which is amphoteric)
carbonates	MCO_3	most decompose on heating; only very slightly soluble in water
hydrogencarbonates (bicarbonates)	$M(HCO_3)_2$	unstable as solids; more soluble than carbonates
nitrates	$M(NO_3)_2$	decompose into NO_2 and O_2 when heated; soluble in water

*M = Group II metal.

18.6 IMPORTANT GROUP II COMPOUNDS

D Place 5 to 10 mL of methanol in three glass dishes, add a spatula full of $CaCl_2$, $SrCl_2$, and $BaCl_2$ to the separate dishes and ignite.

U Exercise 18.64.

The properties of compounds of the Group II elements are summarized in Table 18.8. Like the alkali metals, the compounds of these elements are best treated in terms of their anions. However, certain features are worth emphasizing. Apart from a hint of nonmetallic character in beryllium, these elements have all the chemical characteristics of metals, such as forming cations (M^{2+}) and having basic oxides and hydroxides (see Table 7.3).

The alkaline earth metals can be detected in compounds by the colors they give to flames. Calcium gives orange-red, strontium gives crimson, and barium gives yellow-green (Fig. 18.19). Fireworks are often made from the salts of alkaline earth metals (typically nitrates and chlorates, for the anions then provide an additional supply of oxygen) and magnesium powder. Sparklers are made of aluminum powder in place of magnesium so that they will burn more slowly.

(a)

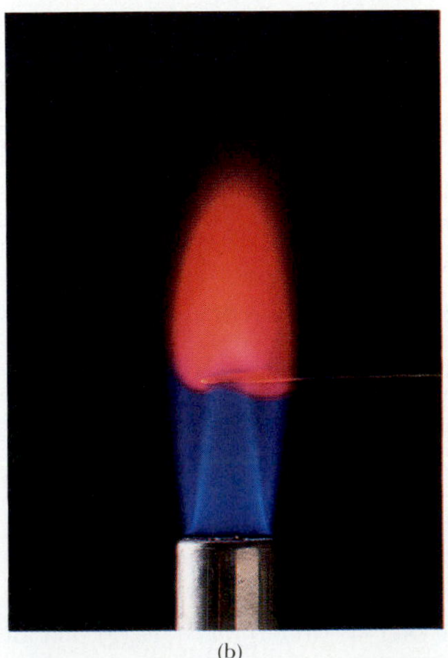

(b)

(c)

FIGURE 18.19 (a) Calcium, (b) strontium, and (c) barium compounds give distinctive colors to flames.

Beryllium compounds. Beryllium compounds are very toxic and must be handled with great caution. Their properties are dominated by the highly polarizing character of the Be^{2+} ion and its small size. The ion's strong polarizing power results in moderately covalent compounds, and its small size limits to four the number of groups that can attach to the ion. These two features together are responsible for the prominence of the tetrahedral BeX_4 unit. This unit is found not only in the oxoanions of beryllium but also in the chloride and the hydride (Fig. 18.20). The chloride is made by the action of chlorine on the oxide in the presence of carbon,

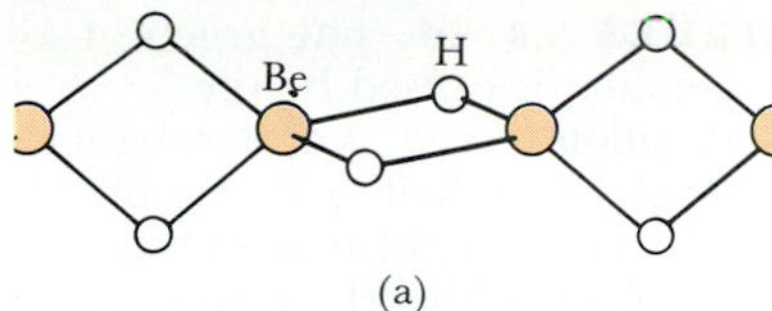

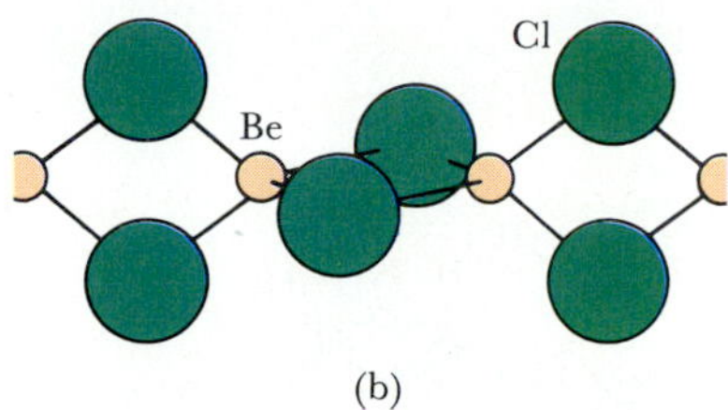

FIGURE 18.20 The structures of (a) BeH_2 and (b) $BeCl_2$.

$$BeO(s) + C(s) + Cl_2(g) \xrightarrow{600° \text{ to } 800°C} BeCl_2(g) + CO(g)$$

the Be atoms act as Lewis acids and accept electron pairs from the Cl atoms of the neighboring $BeCl_2$ groups, forming a chain of overlapping tetrahedral $BeCl_4$ units.

▼ EXAMPLE 18.2 Accounting for the bonding in beryllium hydride

Show, in terms of delocalized molecular orbitals, that beryllium hydride can be expected to be stable.

STRATEGY We need to show that all the valence electrons can be accommodated in orbitals that have a predominantly bonding character, for then the compound will have a lower energy than the separated atoms have. Given N atomic orbitals, we can form N molecular orbitals, of which about half will be bonding (see the discussion of delocalized bonding in benzene in Section 9.9 for which $N = 6$). Therefore, we begin by counting the atomic orbitals, then we count the electrons, and finally we see whether there are enough orbitals to accommodate the electrons without using too many antibonding orbitals.

SOLUTION In a chain of N BeH_2 units, there are N $2s$ orbitals on the Be atoms and $2N$ $1s$ orbitals on the H atoms, thus making $3N$ atomic orbitals in all. In each chain there are therefore $3N$ delocalized molecular orbitals. Of these $3N$ orbitals, about $\frac{3}{2}N$ are bonding and $\frac{3}{2}N$ are antibonding. Each Be atom contributes two electrons and each H atom one electron, so overall there are $2N + N + N = 4N$ electrons to accommodate. Because two electrons can occupy each orbital, these electrons require $2N$ orbitals. Therefore, the bonding orbitals are filled, and only half the antibonding orbitals need be used. This conclusion suggests that, because there are more bonding electrons than antibonding electrons, the chain will be stable. ▲

U Exercise 18.60.

The Be^{2+} ion is hydrated to $Be(H_2O)_4{}^{2+}$ in aqueous solution, and the strong polarizing power of the small ion results in an acidic complex (like $Al(H_2O)_6{}^{3+}$):

$$Be(H_2O)_4{}^{2+}(aq) + H_2O(l) \rightleftharpoons Be(H_2O)_3OH^+(aq) + H_3O^+(aq)$$

Ions more complex than this are also formed, and precipitation of a solid hydroxide begins when sufficient base has been added to bring the average number of OH^- groups per Be atom to 2, giving an empirical formula of $Be(OH)_2$. As in the case of the analogously amphoteric aluminum hydroxide, the beryllium hydroxide precipitate dissolves when more alkali is added.

Magnesium compounds. Magnesium oxide is formed by thermal decomposition of the hydroxide or the carbonate (burning the metal in

FIGURE 18.21 The green of vegetation is caused by the absorption of red and blue light by the chlorophyll molecule in leaves. Green light is not absorbed but reflected, thereby giving the leaves their green color.

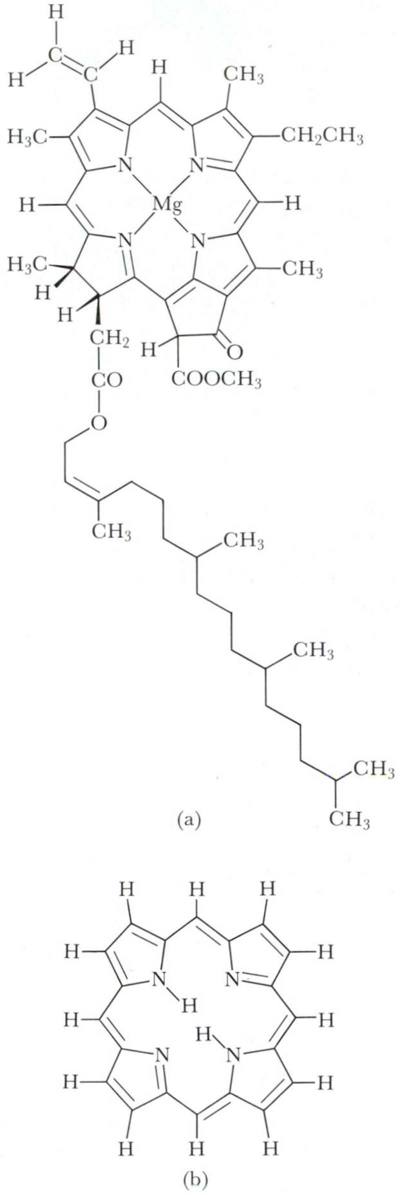

air gives the nitride as well as the oxide). It dissolves only very slowly and sparingly in water. One of its most striking properties is its ability to withstand high temperatures, for it melts only at 2800°C. This high stability (and its corresponding very high lattice enthalpy of 3850 kJ/mol) can be traced to the small ionic radii of the Mg^{2+} and O^{2-} ions and hence to their very strong electrostatic interaction with each other. The oxide has two other characteristics that make it useful: it conducts heat very well and it conducts electricity poorly. All three properties lead to its use as an insulator in electric heaters.

Magnesium hydroxide is a mild base. It is not very soluble in water but instead forms a white colloidal suspension—a mist of small particles dispersed through a liquid—that is known as milk of magnesia and is used as a stomach antacid. Because this base is relatively insoluble, it is not absorbed from the stomach but remains to act on whatever acids are present. The advantage hydroxides have over hydrogencarbonates (which are also used as antacids) is that their neutralization does not lead to the formation of carbon dioxide and its consequence, belching. The disadvantage of milk of magnesia is that the product of neutralization is magnesium chloride, which can act as a purgative. Magnesium sulfate (Epsom salts) is also a popular purgative. Its action, and that of the chloride, appears to be to inhibit the reabsorption of water from the intestine. The resulting increased flow of water triggers the mechanism that results in defecation.

Arguably the most important compound of magnesium is chlorophyll (Fig. 18.21; the name comes from the Greek words for "green leaf"). This green compound captures light from the sun and channels its energy into photosynthesis:

$$6CO_2(g) + 6H_2O(l) \xrightarrow{\text{light, chlorophyll}} C_6H_{12}O_6(aq) + 6O_2(g)$$

In the chlorophyll molecule (Fig. 18.22), the Mg^{2+} acts as a Lewis acid with the N atoms of the large ringlike porphyrin organic group acting as Lewis bases. One function of the Mg^{2+} ion (which lies just above the plane of the ring) appears to be to keep the porphyrin ring rigid. This rigidity helps to ensure that the energy captured from the incoming photon does not spread away as thermal motion before it has been used to bring about a chemical reaction.

(a)

(b)

FIGURE 18.22 (a) In the chlorophyll molecule, a magnesium ion acts as a Lewis acid at the center of a ring of Lewis bases. (b) The ring is a version of a porphyrin molecule.

Calcium carbonate. Calcium occurs naturally as chalk (calcium carbonate, $CaCO_3$) and as limestone (a harder form of the carbonate). Marble is a dense form of calcium carbonate that can be given a high polish; it is often colored by impurities, most commonly iron ions (Fig. 18.23). The two most common forms of pure calcium carbonate are calcite and aragonite (see Fig. 10.1). All these carbonates are the fossilized remains of marine life. As explained in Section 11.3, the hardness of hard water is due to the presence of calcium and magnesium salts (particularly their hydrogencarbonates), dissolved out of the hills through which the water has trickled.

FIGURE 18.23 Marble is a dense form of calcium carbonate and is often colored by ionic impurities.

N The decomposition of calcium carbonate can be regarded as a Lewis acid-base reaction in which the heat induces vibration in the CO_3^{2-} ion (a complex) and the CO_2 molecule (a Lewis acid) is shaken off the O^{2-} ion (a Lewis base).

E Calcium oxide is a major component of Portland cement.

D Decomposition of calcium carbonate. TDC, 13-9. Reactions between carbon dioxide and limewater. BZS1, 4.10.

▼ EXAMPLE 18.3 Accounting for the effect of heat on hard water

Suggest a reason why calcium carbonate is precipitated when hard water, which contains Ca^{2+} and HCO_3^- ions and dissolved CO_2, is boiled.

STRATEGY We should look for a chain of equilibria that connects the heating to the precipitation, for then the effect can be transmitted along the chain. We can draw on the fact that gases are less soluble in hot water than in cold, which suggests that carbon dioxide will be driven out of solution by heating. Then we should think of the effect of decreased CO_2 concentration on other equilibria.

SOLUTION The removal of CO_2 from the solution by heating encourages the decomposition

$$H_2CO_3(aq) \longrightarrow H_2O(l) + CO_2(g)$$

to proceed as written. The resulting lower concentration of H_2CO_3 concentration encourages the Brønsted acid-base reaction

$$\underset{\text{Acid}_1}{HCO_3^-(aq)} + \underset{\text{Base}_2}{HCO_3^-(aq)} \longrightarrow \underset{\text{Acid}_2}{H_2CO_3(aq)} + \underset{\text{Base}_1}{CO_3^{2-}(aq)}$$

Because this reaction increases the CO_3^{2-} ion concentration, it causes the precipitation

$$Ca^{2+}(aq) + CO_3^{2-}(aq) \longrightarrow CaCO_3(s)$$

▲ The net result is the precipitation of calcium carbonate.

Calcium oxide and hydroxide. Calcium carbonate decomposes to the oxide (quicklime) when heated:

$$CaCO_3(s) \xrightarrow{\Delta} CaO(s) + CO_2(g)$$

The decomposition of $CaCO_3$ occurs at a higher temperature (about 800°C) than does that of $MgCO_3$. The difference can be explained by recognizing that the larger Ca^{2+} ion is a weaker Lewis acid than the small Mg^{2+} ion is and hence is less effective at removing an O^{2-} ion from a neighboring CO_3^{2-} ion in the solid. Addition of water to quicklime, CaO, produces slaked lime, $Ca(OH)_2$, with a considerable release of heat. Slaked lime is the form in which lime is normally sold, for quicklime can set fire to moist wood and paper.

A solution of calcium hydroxide, which is sparingly soluble in water, is called lime water and is used as a test for carbon dioxide, which reacts to form a suspension of the even less soluble calcium carbonate:

$$Ca(OH)_2(aq) + CO_2(g) \longrightarrow CaCO_3(s) + H_2O(l)$$

Lime is produced in enormous quantities throughout the world and ranks fifth by mass in the United States. About 40% of this output is

U Exercise 18.65.

N Now, in a continuation of the previous annotation, the SiO_2 acts as a Lewis acid and attaches to the O^{2-} ion, the Lewis base, left behind by the escape of the CO_2.

E Slaked lime is used to neutralize acidic soils; acidity is a characteristic of soils in regions having a high rainfall.

N CaO is a basic anhydride and SiO_2 is an acidic anhydride; the reaction product is a salt (without water).

used in metallurgy. In ironmaking (Section 21.2) it is used as a Lewis base; its O^{2-} ions react in the blast furnace with the silica impurities in the ore to form a liquid slag:

$$CaO(s) + SiO_2(s) \longrightarrow CaSiO_3(l)$$

The oxide SiO_2 is a Lewis acid in this reaction. About 50 kg of lime is needed to produce a ton of iron.

Lime is used as an inexpensive base in industry as well as to adjust the pH of soils in agriculture. Perhaps surprisingly, it is also used to *remove* Ca^{2+} ions from water. Its role here is to convert HCO_3^- to CO_3^{2-} by providing OH^- ions:

$$HCO_3^-(aq) + OH^-(aq) \longrightarrow CO_3^{2-}(aq) + H_2O(l)$$

The increase in the concentration of CO_3^{2-} ions encourages the precipitation of Ca^{2+} ions in the reaction

$$Ca^{2+}(aq) + CO_3^{2-}(aq) \longrightarrow CaCO_3(aq)$$

This reaction removes the Ca^{2+} ions that were present initially and the Ca^{2+} that were added as lime; so overall, the Ca^{2+} ion concentration is reduced.

The reaction that converts calcium oxide into calcium carbide,

$$CaO(s) + 3C(s) \xrightarrow{2200°C} CaC_2(s) + CO(g)$$

is a bridge between inorganic and organic compounds. The carbide ion ($:C{\equiv}C:^{2-}$) is a Brønsted base and is readily protonated by water, a reaction resulting in the formation of the organic compound acetylene:

$$C_2^{2-}(s) + 2H_2O(l) \longrightarrow C_2H_2(g) + 2OH^-(aq)$$

U Exercise 18.63.

Structural calcium. Calcium compounds are often used in (or as) structural components; their rigidity stems from the strength with which the small, highly charged Ca^{2+} cation interacts with its neighbors. Mortar consists of about one part slaked lime and three parts sand (largely silica, SiO_2). It sets to a hard mass as the lime reacts with the carbon dioxide of the air to form the carbonate (Fig. 18.24). Calcium is also found in the rigid structural components of living things, either as the calcium carbonate of the shells of shellfish or the calcium phosphate of bone. About a kilogram of calcium is present in an adult human body, mostly in the form of insoluble calcium phosphate, but also as Ca^{2+} ions in other fluids inside our cells. The calcium in newly formed bone is in dynamic equilibrium with the calcium ions in the body fluids, and about 20 g of calcium is exchanged between the two each day.

FIGURE 18.24 An electron micrograph of the surface of mortar, showing the growth of tiny interlocking crystals as carbon dioxide reacts with calcium oxide and silica.

Tooth enamel is a hydroxyapatite, a mineral of composition $Ca_5(PO_4)_3OH$. Tooth decay can occur when acids attack the enamel:

$$Ca_5(PO_4)_3OH(s) + 4H_3O^+(aq) \longrightarrow 5Ca^{2+}(aq) + 3HPO_4^{2-}(aq) + 5H_2O(l)$$

The principal agents of tooth decay are the carboxylic acids (acids with a —COOH group) produced when bacteria act on the remains of food. A more resistant coating is obtained when the OH^- ions in the hydroxyapatite are replaced by F^- ions to give a fluorapatite:

$$Ca_5(PO_4)_3OH(s) + F^-(aq) \longrightarrow Ca_5(PO_4)_3F(s) + OH^-(aq)$$

The addition of fluoride ions to domestic water supplies (by addition of NaF) is now widespread. Another strategy for strengthening the protective coating of teeth is the use of fluoridated toothpastes, containing either tin(II) fluoride or sodium monofluorophosphate (MFP, Na_2FPO_3).

Calcium sulfate is a component of several rigid materials that are utilized in all kinds of rigid structures. Alabaster is a compact form of gypsum ($CaSO_4{\cdot}2H_2O$); it resembles marble but is softer and is used for ornamental work. When gypsum is heated, it loses some of the water of hydration and becomes plaster of Paris, $CaSO_4{\cdot}\frac{1}{2}H_2O$. This substance is so called because it was originally obtained from the gypsum mined in the Montmarte district of Paris. Plaster of Paris sets to a rigid mass when water is added and the hydration is restored. Because rehydration causes the solid to swell, the plaster takes a good impression from a mold. Calcium sulfate is also a component of cements and concrete, which we describe in the next chapter.

Because strontium produces a very bright red color in a flame, strontium salts are used in fireworks and signal flares. Barium sulfate is used as a drilling mud in oil exploration and is consumed as an aqueous suspension for the purpose of taking x-ray photographs of the gastrointestinal tract.

E The first use of barium sulfate stems from its high density, so that columns of suspensions of barium sulfate exert high pressures at their bases. The second use stems from the possession of each barium ion of a large number of electrons (so it is not a direct consequence of the ion's mass): these electrons scatter the incident x-rays and the suspension appears opaque.

U Exercise 18.67.

SUMMARY

18.1 Hydrogen is the most abundant element in the universe. It is manufactured by the **reforming reaction**, which produces **synthesis gas**, a mixture of H_2 and CO, and by a subsequent **shift reaction**. Alternative procedures are electrolysis and the reduction of $H^+(aq)$ ions by metals. The principal uses of hydrogen are for the synthesis of ammonia and methanol, the reduction of ores, and the hydrogenation of fats. It is also used as a rocket fuel. Hydrogen can take part in hydrogen bonding between electronegative elements. The transfer of the hydrogen ion (the proton) is the central feature of Brønsted acid-base behavior.

18.2 Hydrogen forms three classes of binary compounds. The **saline hydrides**, which contain the hydride ion H^-, are formed by *s*-block elements and are oxidized readily by water. The **molecular compounds** of hydrogen are covalently bonded species formed by elements in the *p* block, and most are gases or volatile liquids. Some of the *d*-block elements form **metallic hydrides**, in which the hydrogen atoms occupy gaps between the metal atoms.

18.3 The elements of Group I are the **alkali metals**. They are all strongly reducing metals that react with water to form hydroxides and hydrogen gas. Lithium is diagonally related to magnesium and forms compounds that are more markedly covalent than those of its congeners. All the elements react directly with the nonmetals (other than the noble gases) and burn in air to form an oxide, a peroxide, or a superoxide.

18.4 **Sodium chloride** is a major industrial chemical. It is electrolyzed to chlorine and **sodium hydroxide** in a diaphragm cell, converted to **sodium sulfate** with sulfuric acid, and used for the production of **sodium hydrogencarbonate** and **sodium carbonate**. Potassium compounds tend to be less hygroscopic than sodium compounds.

18.5 The elements of Group II include the **alkaline earth metals**. A rich source of magnesium is seawater, from which the metal is extracted by precipitation and then electrolysis. Metallic character increases down the group. Beryllium has a strong diagonal resemblance to aluminum, forming amphoteric oxides and compounds that are markedly covalent. Structurally, its compounds are dominated by the tetrahedral BX_4 group. All the Group II metals other than beryllium reduce water with a vigor that increases down the group; all form compounds that contain the M^{2+} cation.

18.6 **Calcium carbonate** is a major source of **lime** and **slaked lime**, which is widely used as a cheap industrial base. Calcium appears in a number of structurally rigid compounds, including the **calcium phosphate** of bone and teeth and the **calcium sulfate** of alabaster, gypsum, and plaster of Paris. **Calcium carbide** is an important link between inorganic and organic compounds.

CLASSIFIED EXERCISES

This chapter includes applications of the chemical principles that have been addressed in earlier chapters. Therefore, a number of the exercises will require you to review and collect data from this earlier material.

Hydrogen

18.1 Write the chemical equations for (a) the shift reaction and for the reactions between (b) lithium and water and between (c) magnesium and hot water.

18.2 Write the chemical equations for (a) the reaction between sodium hydride and water; (b) the formation of synthesis gas; (c) the reaction of magnesium with hydrochloric acid.

18.3 Write the chemical equations for the synthesis, from their elements, of (a) LiH; (b) HI; (c) PH_3.

18.4 A number of binary hydrogen-containing compounds are discussed in this chapter. Suggest, with a chemical equation, a method for preparing each of the following compounds: (a) RbH; (b) CaH_2; (c) HCl.

18.5 Suggest, with chemical equations, methods for the preparation of (a) hydrogen sulfide and (b) barium hydride.

18.6 Write a chemical equation for the preparation of (a) hydrogen iodide and (b) magnesium hydride.

18.7 Classify each of the following compounds as a saline, molecular, or metallic hydride: (a) KH; (b) NH_3; (c) UH_3.

18.8 Classify each of the following compounds as a saline, molecular, or metallic hydride: (a) SiH_4; (b) CaH_2; (c) Pd_2H.

18.9 Use Appendix 2A to determine the standard reaction enthalpy at 25°C for (a) the synthesis gas reaction; (b) the shift reaction; (c) the overall reaction of these two processes.

18.10 Use Appendix 2A to determine the standard free energy at 25°C of (a) the synthesis gas reaction; (b) the shift reaction; (c) the overall reaction of these two processes.

18.11 Calcium hydride is used as a portable source of hydrogen on account of its reaction with water:

$$CaH_2(s) + 2H_2O(l) \longrightarrow Ca(OH)_2(s) + 2H_2(g)$$

(a) What volume of H_2 gas (at STP) can be produced from 500 g of CaH_2? (b) How many milliliters of water should be supplied for the reaction? Assume the density of water to be 1.0 g/mL.

18.12 What volume of hydrogen gas (at STP) is required for the reduction of 20.0 g of WO_3 in the reaction $WO_3(s) + 3H_2(g) \rightarrow W(s) + 3H_2O(l)$.

18.13 About 3×10^8 kg of hydrogen is produced each year in the United States. If 1.5×10^9 kg of ammonia is produced annually by the Haber process, what fraction of the H_2 is used for this purpose?

18.14 Calculate the volume that the annual United States production of hydrogen (3×10^8 kg) would occupy if it were stored as liquid hydrogen. The density of liquid hydrogen is 0.09 g/cm^3.

18.15 Identify the products and indicate the conditions under which hydrogen reacts with (a) chlorine; (b) sodium; (c) ethylene.

18.16 Identify the products and indicate the conditions under which hydrogen reacts with (a) nitrogen; (b) fluorine; (c) potassium.

18.17 Write the Lewis structures of (a) LiH; (b) SiH_4 (silane); (c) SbH_3.

18.18 Fluorine forms one binary hydride, oxygen forms two binary hydrides, and nitrogen forms three binary hydrides (NH_3, N_2H_4, and HN_3). Write the Lewis structures for all six hydrides.

18.19 Hydrogen can be produced from the electrolysis of water. (a) Write the equation for the half-reaction for the production of hydrogen (refer to Chapter 17 if necessary). (b) Is the hydrogen produced at the anode or the cathode? (c) Calculate the volume of hydrogen (at STP) that is produced if a current of 10.0 A is passed through an electrolytic cell for 30 min.

18.20 Hydrogen burns in an atmosphere of bromine to give hydrogen bromide. If 120 mL of H_2 gas at STP combines with a stoichiometric amount of bromine and the resulting hydrogen bromide dissolves to form 150 mL of an aqueous solution, what is the molar concentration of the resulting hydrobromic acid solution?

Group I: The Alkali Metals

18.21 Refer to Fig. 7.1 and describe the color of the flame test for (a) lithium; (b) potassium; (c) sodium; (d) rubidium; (e) cesium.

18.22 (a) Write the valence electron configuration for the alkali metal atoms. (b) Explain why the alkali metals are strong reducing agents in terms of electron configurations and ionization energies.

18.23 Name the minerals that are the principal sources of sodium and potassium.

18.24 Name two industrial processes in which sodium chloride is one of the starting materials.

18.25 Write the chemical equations for the reactions between (a) lithium and nitrogen; (b) sodium and water; (c) potassium superoxide and water.

18.26 Write the chemical equations for the reactions between (a) potassium and oxygen; (b) sodium oxide and water; (c) lithium and hydrochloric acid.

18.27 Write the chemical equations for the thermal decomposition of (a) sodium carbonate and (b) potassium nitrate.

18.28 Complete and balance the following skeletal equations for the reactions:
(a) $Ca(s) + H_2(g) \rightarrow$
(b) $Na(s) + H_2(g) \rightarrow$
(c) $K(s) + N_2(g) \rightarrow$

18.29 Give the chemical names and formulas of the minerals (a) rock salt; (b) sylvite; (c) carnallite.

18.30 Write the chemical formulas for the compounds (a) caustic soda; (b) Glauber's salt; (c) bicarbonate of soda.

18.31 Write the balanced equation for the industrial production of each of the following compounds, starting with sodium chloride: (a) NaOH; (b) Na_2SO_4; (c) sodium metal.

18.32 Sodium metal is produced from the electrolysis of molten sodium chloride in the Downs process (Chapter 17). Determine (a) the standard free energy of the reaction (see Appendix 2A) and (b) the current needed to produce 1.00 kg of sodium in 10.0 h.

Group II: The Alkaline Earth Metals

18.33 Refer to Fig. 18.19 and describe the color of the flame test for (a) calcium; (b) strontium; (c) barium.

18.34 Refer to Chapter 7 and Appendix 2B for data and plot, on the same graph, ionization energies and standard reduction potentials of the Group I and Group II elements against their atomic numbers. What conclusions can be drawn?

18.35 Name the mineral forms used as sources of (a) beryllium; (b) calcium; (c) magnesium.

18.36 Write the chemical equations that describe the production of slaked lime from limestone.

18.37 Give the chemical names and write the formulas of (a) Epsom salts; (b) gypsum; (c) barite; (d) milk of magnesia. (See Appendix 3B.)

18.38 Give the chemical names and write the formulas of (a) dolomite; (b) quicklime; (c) limestone.

18.39 Write the chemical equations for the reactions of (a) beryllium with aqueous sodium hydroxide and (b) aluminum with aqueous sodium hydroxide. (c) Account for the similarities in the reactions.

18.40 Write the chemical equations for (a) the industrial preparation of beryllium; (b) the action of hydrochloric acid on calcium metal; (c) the action of water on calcium metal.

18.41 Write the chemical equations for the reactions between (a) calcium and oxygen; (b) magnesium and hydrochloric acid; (c) strontium and nitrogen.

18.42 Write the chemical equations for the reactions between (a) barium and oxygen; (b) calcium carbonate and hydrochloric acid; (c) magnesium hydroxide and carbon dioxide.

18.43 Explain, with equations, how (a) $Ca(OH)_2$ and (b) $Ca(NO_3)_2$ may be prepared from calcium carbonate.

18.44 Limestone, the major source of calcium in nature, is used as a starting material for many calcium compounds. Use chemical equations to describe the preparation, starting from limestone, of (a) lime; (b) acetylene, C_2H_2; (c) calcium sulfate.

18.45 Predict the products of the following reactions and then balance the equations:
(a) $Mg(OH)_2(s) + HCl(aq) \rightarrow$
(b) $Ca(s) + H_2O(l) \rightarrow$
(c) $BaCO_3(s) \xrightarrow{\Delta}$

18.46 Predict the products of the following reactions and then balance the equations:
(a) $Mg(s) + Br_2(l) \rightarrow$
(b) $Ca(NO_3)_2(s) \xrightarrow{\Delta}$
(c) $CaO(s) + SiO_2(s) \xrightarrow{\Delta}$

18.47 (a) Write the Lewis structure for $BeCl_2$ (see Chapter 8) and (b) predict the Cl—Be—Cl bond angle. (c) What hybrid orbitals are used in the bonding?

18.48 (a) Write the Lewis structure of $MgCl_2$ (see Chapter 8). (b) How does its structure differ from that of $BeCl_2$?

18.49 Magnesium is produced from the electrolysis of molten magnesium chloride. (a) Calculate the mass of magnesium that can be produced in 1.5 h by using a current of 100 A (see Chapter 17). (b) Calculate the volume (at STP) of chlorine gas produced when 1000 kg of magnesium metal is obtained from the process.

18.50 What mass of calcium oxide can be produced from the thermal decomposition of 200 g calcium carbonate?

18.51 (a) Calculate the enthalpy and entropy of reaction for the thermal decomposition of calcium carbonate as calcite (see Appendix 2A for data). (b) Determine the temperature at which the thermal decomposition becomes spontaneous.

18.52 (a) Calculate the standard reaction enthalpy and entropy of the thermal decomposition of magnesium carbonate (see Appendix 2A for data). (b) Estimate the temperature at which the thermal decomposition becomes spontaneous.

18.53 What is the mass percentage of water in Epsom salt? (See Appendix 3B.)

18.54 The concentration of magnesium in seawater is about 1.35 g/L. What volume of water must be processed to collect 1.0 kg of magnesium, assuming an 80% removal.

CASE 18 ▸ LIVING ABOARD THE SPACE SHUTTLE

A large number of factors must be considered when living in a closed environment. Our Earth is so large (compared with an individual) that we often do not consider our personal contributions to the quality of the environment. For example, when we exhale air, we do not concern ourselves about what happens to the CO_2; we know that at some later time photosynthesis will convert the carbon dioxide to carbohydrates. But what considerations are necessary when living aboard the space shuttle or, in the future, a space station? We need to determine not only the sources of food, water, and oxygen but also how to handle all wastes. Rather than being discarded, these wastes must be recycled and used to regenerate the substances that are necessary for survival.

Consider the air quality on board a space shuttle. Only so much air can be taken aboard; the rest must be chemically generated or regenerated while in flight. Contaminated air contains carbon dioxide and substances responsible for odors. To cleanse the contaminated air, it is first passed through an activated charcoal (carbon) filter to remove odor-producing substances. These substances, which have a relatively nonpolar molecular structure, adsorb onto the charcoal surface. During short flights, the filtered air is passed through a lithium hydroxide canister to remove CO_2 by the reaction

$$2LiOH(s) + CO_2(g) \longrightarrow Li_2CO_3(s) + H_2O(l)$$

Lithium hydroxide is preferred to sodium hydroxide (or potassium hydroxide) because of mass considerations. A by-product of this reaction is the valuable commodity water. Other methods of CO_2 removal have been considered, but lithium hydroxide is the one currently used on the shuttle.

During extended flights, or on board a space station, the CO_2 can be removed and used to regenerate oxygen through a series of reactions: the first step is the passage of a mixture of CO_2 and H_2 over a nickel catalyst at 200° to 250°C:

$$CO_2(g) + 4H_2(g) \longrightarrow CH_4(g) + 2H_2O(g)$$

The methane is heated in the absence of air (pyrolyzed) to produce carbon,

$$CH_4(g) \xrightarrow{\Delta} C(s) + 2H_2(g)$$

which can be used to filter odors from the air.

18.55 (a) What is hard water (see Section 11.3)? (b) Write chemical equations for the softening of hardness due to HCO_3^- ions in water that can be achieved by using lime.

18.56 Explain how the existence of the equilibrium $Ca^{2+}(aq) + 2HCO_3^-(aq) \rightleftharpoons CaCO_3(s) + CO_2(g) + H_2O(l)$ can account for the formation of stalactites and stalagmites in caves. (See Case 13.)

UNCLASSIFIED EXERCISES

18.57 Arrange hydrogen and the elements of Groups I and II in order of increasing strength as reducing agents (with the strongest one last). Suggest reasons for the order.

18.58 (a) Write equations for the reactions of hydrogen gas with the halogens, from fluorine to iodine. (b) Comment on the relative vigor of the reactions. (c) Name the products when they are dissolved in aqueous solutions.

18.59 Use data from Chapter 7 and Appendix 2B to plot ionization energy against standard reduction potential for the elements of Groups I and II. What generalizations can be drawn from the graph?

The hydrogen is recovered for reuse in its reaction with CO_2. The water produced in the reaction of CO_2 and H_2 is electrolyzed to produce oxygen for breathing and additional hydrogen for reaction with CO_2:

$$2H_2O(l) \xrightarrow{\text{electrolysis}} 2H_2(g) + O_2(g)$$

The stoichiometry of these reactions indicates that 1 mol O_2 is generated for 1 mol of CO_2 exhaled. The only input, therefore, is an external energy source for the CO_2/H_2 reaction, the pyrolysis of methane, and the electrolysis of the water, all of which can be provided by solar cells on extended solar panels.

QUESTIONS

1. Calculate the CO_2 removal capacity of lithium hydroxide, sodium hydroxide, and potassium hydroxide (expressing your answer as the mass of CO_2 per gram of base).

2. Is energy an additional product of the process or must energy be supplied? Starting with CO_2 and H_2, determine the total enthalpy for the reactions leading to the generation of oxygen.

3. The average adult human requires about 12 MJ of energy per day to function normally. Assuming this energy comes from the combustion of glucose, $C_6H_{12}O_6$, calculate how much CO_2 is produced. (See Appendix 2A for additional data.)

4. Calculate the mass of lithium hydroxide that must be placed aboard the space shuttle to remove the CO_2 generated by five adults during a seven-day flight. Use the information in Question 3.

18.60 (a) Write a chemical equation that illustrates the acidic character of the beryllium(II) ion in aqueous solution. (b) Write a chemical equation showing the effects of the addition of 2 mol OH^- and 4 mol OH^- per mole of beryllium(II) ion in aqueous solution.

18.61 Write the Lewis structures of (a) BaO_2; (b) BeH_2; (c) Na_2O_2; (d) $Be(OH)_4{}^{2-}$.

18.62 Calculate the molar solubility of $Mg(OH)_2$, milk of magnesia, at (a) pH = 8.0 and (b) pH = 9.0. (See Table 15.5 for data.)

18.63 Explain why calcium fluoride is more soluble in an acidic solution than in water.

18.64 Suggest compounds that may be added to fireworks to produce the following colors (more than one compound may be needed for a particular color): (a) red; (b) orange; (c) blue; (d) lilac; (e) yellow; (f) green.

18.65 Heat is generated in the formation of slaked lime from the reaction $CaO(s) + H_2O(l) \rightarrow Ca(OH)_2(s)$ (a) Calculate the standard enthalpy of reaction. (b) If all this heat could be used to heat 250 g of water, what would be the resulting temperature change?

18.66 It has been proposed that the ability of a cation to polarize anions is proportional to its charge divided by its radius (see Chapter 7). (a) Use this criterion to arrange the *s*-block elements in order of increasing polarizing power. (b) Do the resulting values support the diagonal relationships within the block?

18.67 What mass of plaster of Paris can be produced from 500 g of gypsum?

18.68 (a) Describe the taste of pancakes in which sour milk and baking soda are used. (b) If baking powder were substituted for baking soda when making sour milk pancakes, what would be the result? (c) If sweet milk and baking soda were used in the making of pancakes, what would be the result?

This chapter is the first of two that describe the p-block elements and their rich chemical properties. In it we see the transition from the predominantly metallic character of the Group III elements to the predominantly nonmetallic character of the Group V elements. The illustration shows the instant of ignition of a match in a process that epitomizes p-block chemistry: the head consists of a paste of potassium chlorate, antimony sulfide, sulfur, and powdered glass; the striking strip contains red phosphorus; and the wood is a complex mixture of compounds of carbon.

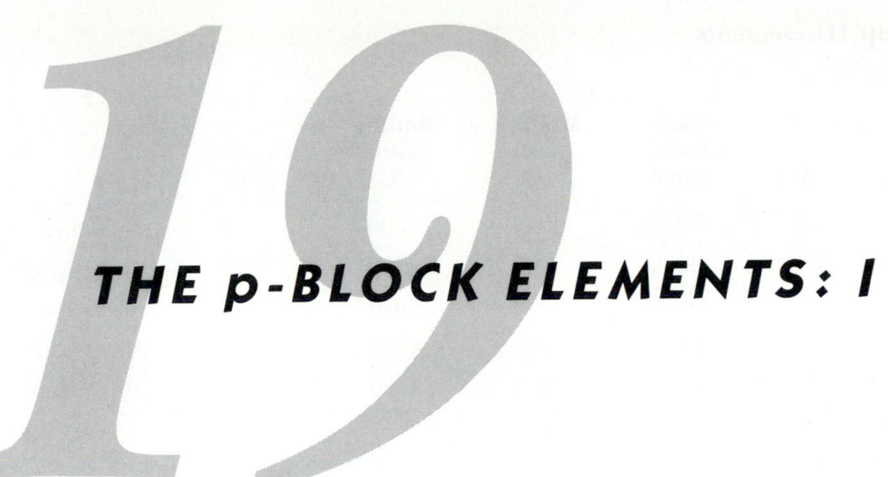

THE p-BLOCK ELEMENTS: I

The *p* block of the periodic table consists of Groups III through VIII. The characteristics of its members range from almost typically metallic (as shown by aluminum) to typically nonmetallic (as we see in the lighter halogens and the noble gases). The *p* block includes carbon and oxygen, the elements central to life and natural intelligence, and silicon and germanium, the elements central to modern technology and artificial intelligence.

Because the *p*-block elements form such a wide and important range of compounds (including the oxides, the oxoacids, the halides, and the millions of organic compounds), we shall discuss them in three stages. This chapter deals with some of the elements of Groups III through V (from boron to nitrogen). Chapter 20 deals with Groups VI through VIII (from oxygen to the noble gases). In Part V we return to carbon and describe the compounds that traditionally are treated as organic.

GROUP III: BORON AND ALUMINUM

The elements in Group III and some of their physical properties are listed Table 19.1; some of their chemical properties are listed in Table 19.2. We shall concentrate on the two most important members of the group, boron and aluminum.

	II	**III**	IV				He
Li	Be	**B**	C	N	O	F	Ne
	Mg	**Al**	Si				
	Ca	**Ga**	Ge				
	Sr	**In**	Sn				
	Ba	**Tl**	Pb				
	Ra						

19.1 GROUP III ELEMENTS

Boron, at the head of Group III, is best regarded as a nonmetal in most of its chemical properties. It has acidic oxides and forms an interesting and extensive range of binary molecular compounds with hydrogen. However, metallic character increases down the group (see Table 7.3) and even boron's immediate neighbor aluminum is a metal. Nonethe-

TABLE 19.1 Group III elements

Valence configuration: ns^2np^1

Z	Name	Symbol	Molar mass, g/mol	Melting point, °C	Boiling point, °C	Density, g/cm^3	Normal form
5	boron	B	10.81	2030	3700	2.47	brown nonmetallic powder
13	aluminum	Al	26.98	660	2350	2.70	silver-white metal
31	gallium	Ga	69.72	30	2070	5.91	silver metal
49	indium	In	114.82	157	2050	7.29	silver-white metal
81	thallium	Tl	204.37	304	1460	11.87	soft metal

E′ All aluminum beverage cans have a plastic liner to prevent this reaction.

D Make a scratch on the inside of an aluminum beverage can, fill with a dilute (0.01 M) $CuSO_4$ solution for about 24 hours, and then empty it. In class, "rip" the can in half by twisting. Amphoteric properties of aluminum. TDC, 14-6.

T OHT: Tables 19.2, 19.4, and 19.6.

D Chemical milling of aluminum. TDC, 162.

less, aluminum is sufficiently far to the right in the periodic table to show a hint of nonmetallic character. For instance, it is amphoteric (that is, reacts with both acids and bases); it reacts both with nonoxidizing acids (such as hydrochloric acid) to form aluminum ions:

$$2Al(s) + 6H^+(aq) \longrightarrow 2Al^{3+}(aq) + 3H_2(g)$$

and with hot aqueous alkali to form aluminate ions, $Al(OH)_4^-$:

$$2Al(s) + 2OH^-(aq) + 6H_2O(l) \longrightarrow 2Al(OH)_4^-(aq) + 3H_2(g)$$

▼ EXAMPLE 19.1 Predicting group trends

Predict the trends in oxidation numbers that can be expected for the elements in Group III.

STRATEGY The group number tells us the maximum oxidation number, but we should also consider the influence of the inert-pair effect (Section 7.8) and how it is likely to vary throughout the group.

SOLUTION In Group III we expect a maximum oxidation number of +3. The inert-pair effect is unimportant at the top of the group, so we expect the oxidation states of boron and aluminum to be +3 in all their compounds. (This prediction is true, except for some complex solids we mention later.) The oxidation number +1 becomes increasingly important on going

TABLE 19.2 Chemical properties of the Group III elements

Reactant	Reaction with Group III elements, E
oxygen	$4E(s) + 3O_2(g) \rightarrow 2E_2O_3(s)$
nitrogen	$2E(s) + N_2(g) \rightarrow 2EN(s)$ E = B, Al
halogen	$2B(s) + 3X_2(g,l,s) \rightarrow 2BX_3(g)$
	$2E(s) + 3X_2(g,l,s) \rightarrow E_2X_6(g)$ E = Al, Ga, In
	$2Tl(s) + X_2(g,l,s) \rightarrow 2TlX(s)$
water	$2Tl(s) + 2H_2O(l) \rightarrow 2TlOH(aq) + H_2(g)$
acid	$2E(s) + 6H^+(aq) \rightarrow 2E^{3+}(aq) + 3H_2(g)$ E = Al, Ga, Tl
base	$2E(s) + 2OH^-(aq) + 6H_2O(l) \rightarrow 2E(OH)_4^-(aq) + 3H_2(g)$ E = Al, Ga

down the group, and we can expect it to be common in thallium. In fact, Tl(I) compounds are as common as Tl(III) compounds.

EXERCISE E19.1 Predict the trend in ionization energies for the elements of Group III.

[*Answer*: decrease down group]

Boron. Boron is mined as borax and kernite ($Na_2B_4O_4{\cdot}xH_2O$, with $x = 10$ and 4, respectively). Large deposits are found in volcanic regions, such as the Mohave Desert region of California. In the extraction process, the minerals are converted to boron oxide with acid and then reduced with magnesium to an impure, brown, amorphous form of boron:

E Boron use. *J. Chem. Ed.*, **1991**, *67*, 14.

D Boron exhibit. TDC, 15-1.

$$B_2O_3(s) + 3Mg(s) \xrightarrow{\Delta} 2B(s) + 3MgO(s)$$

A purer product is obtained by reducing a volatile boron compound, such as BCl_3 or BBr_3, with hydrogen on a heated filament (tantalum is used because it has a very high melting point):

$$2BBr_3(g) + 3H_2(g) \xrightarrow{\Delta} 2B(s) + 6HBr(g)$$

Boron production remains quite low despite the element's desirable properties of hardness and lightness. Here is an opportunity for another young chemist like Charles Hall (the inventor of the process used to extract aluminum) to transform the picture as Hall did for aluminum (see below).

Elemental boron has several allotropes. It is typically either a gray-black nonmetallic, high-melting-point solid or a dark brown powder with a structure based on clusters of 12 atoms (Fig. 19.1). When boron fibers are incorporated in plastics, the result is a very tough material that is stiffer than steel yet lighter than aluminum and potentially useful for aircraft, missiles, and body armor. The element is very inert and is attacked by only the strongest oxidizing agents.

E Many of the new tennis and racquetball rackets are boron composites.

E Aluminum is dispersed throughout the landscape, because many rocks are aluminosilicates (see p. 741), which are silicates with some Si atoms replaced by Al atoms.

Aluminum. Aluminum is the most abundant metallic element in the Earth's crust and, following oxygen and silicon, the third most abundant element. However, the aluminum content in most minerals is low, and the commercial source of aluminum (bauxite) is a hydrated, impure oxide, $Al_2O_3{\cdot}xH_2O$, where x is variable. Bauxite deposits are found in Australia, Jamaica, and Sumatra. The principal impurities in the ore are the oxides of iron, titanium, and silicon.

The bauxite ore, which is red as a result of the iron oxides it contains, is processed to obtain alumina, Al_2O_3, by the **Bayer process**. The crude bauxite is first treated with aqueous sodium hydroxide, which dissolves the amphoteric alumina and the acidic silica, leaving the iron and titanium hydroxides as a precipitate called red mud:

$$Al_2O_3{\cdot}xH_2O(s) + 2OH^-(aq) + 3H_2O(l) \xrightarrow{170°C} 2Al(OH)_4^-(aq) + xH_2O(l)$$

$$Fe_2O_3{\cdot}xH_2O(s) + 3H_2O(l) \xrightarrow{170°C} 2Fe(OH)_3(s) + xH_2O(l)$$

Carbon dioxide, an acidic oxide, is then bubbled through the aluminate solution. It lowers the OH^- concentration, thereby causing a partial decomposition of the aluminate ions and the precipitation of aluminum hydroxide:

$$Al(OH)_4^-(aq) + CO_2(g) \longrightarrow Al(OH)_3(s) + HCO_3^-(aq)$$

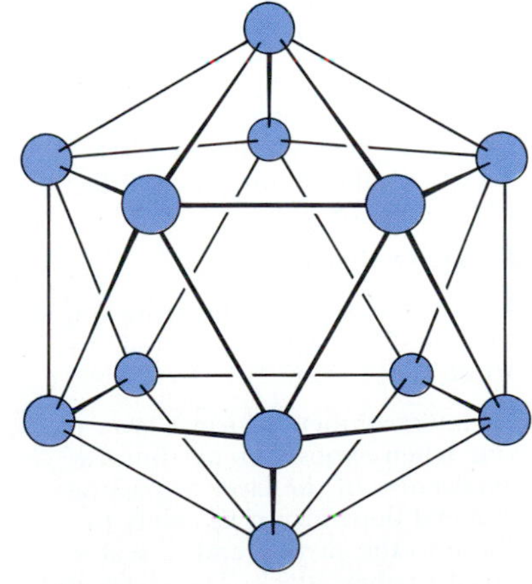

FIGURE 19.1 The structure of boron is based on linked 12-atom units. The unit has 20 faces, so it is called an icosahedron (from the Greek words for "twenty-faced").

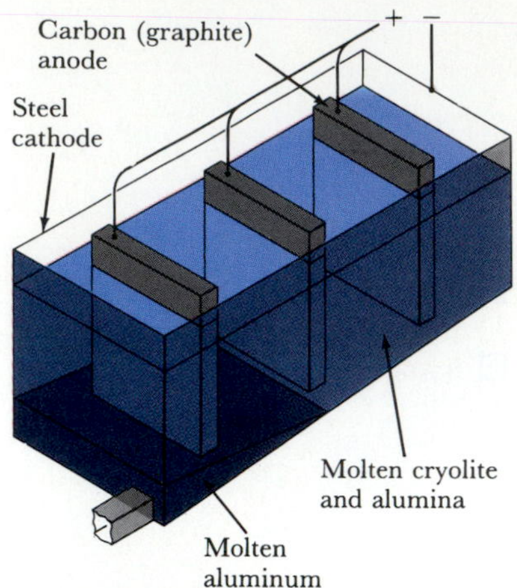

FIGURE 19.2 In the Hall process, aluminum oxide is dissolved in molten cryolite and the mixture is electrolyzed in a cell with carbon anodes and a steel cathode.

T OHT: Fig. 19.2.

E The graphite anodes must be periodically replaced. Other electrolytic and chemical processes that are more economical are currently being investigated for the production of aluminum.

E Hall and the great aluminum revolution. *J. Chem. Ed.*, **1987**, *64*, 690. Charles Morton Hall: The young man, his mentor, and his metal. *J. Chem. Ed.*, **1986**, *63*, 557.

? Why can an aluminum vessel be used to store nitric acid but not hydrochloric acid?

D Aluminum exhibit. TDC, 14-1–14-4. Recycling aluminum. CD2, 29. Passivity of aluminum. *J. Chem. Ed.*, **1974**, *51*, A178.

V Video 8: Thermite.

E Two examples are the reduction of Cr_2O_3 to chromium and of BaO to barium.

E Almost all metals form a layer of oxide when exposed to air, but the effectiveness of the layer to passivate the metal depends on its ability to adhere to the surface and to seal it against further attack. The ability to adhere depends, to some extent, on the matching of the metal-oxide distance in the oxide to the metal-metal distance in the metal so that their structures match.

The aluminum hydroxide is removed and dehydrated to the oxide by heating to 1200°C:

$$2Al(OH)_3(s) \xrightarrow{1200°C} Al_2O_3(s) + 3H_2O(g)$$

Aluminum metal is obtained by the **Hall process**, which was developed by Charles Hall of Oberlin College, Ohio, in 1886, when he was 22 years old, and independently by Paul Héroult in France, who was also 22 years old at the time. They discovered that a mixture of alumina and the mineral cryolite, Na_3AlF_6, melts at 950°C (pure alumina melts at 2050°C). The melt is electrolyzed in a unit consisting of graphite (or carbonized petroleum) anodes and a carbonized steel-lined vat that serves as the cathode (Fig. 19.2). The electrolysis reactions are

$$\textit{Cathode reaction:}\quad Al^{3+}(\textit{melt}) + 3e^- \longrightarrow Al(l)$$

$$\textit{Anode reaction:}\quad 2O^{2-}(\textit{melt}) + C(s,\text{gr}) \longrightarrow CO_2(g) + 4e^-$$

The overall reaction is

$$4Al^{3+}(\textit{melt}) + 6O^{2-}(\textit{melt}) + 3C(s,\text{gr}) \longrightarrow 4Al(l) + 3CO_2(g)$$

From the cathode reaction, we can deduce that a current of 1 A must flow for 80 h to produce 1 mol of Al (27 g of aluminum), about enough for one soft-drink can. The very high electricity consumption—the supply must provide three electrons for every Al atom the industry produces—means that the cost of electricity is of overriding importance in aluminum production. Before the recycling of aluminum was widespread, the aluminum industry in the United States consumed in one day the electricity used by 100,000 people in one year!

Aluminum is a light, strong metal and an excellent electrical conductor. Although it is strongly reducing ($E° = -1.66$ V for the Al^{3+}/Al couple in water), it is resistant to corrosion because it is passivated by a stable oxide film that forms on its surface. The thickness of the oxide layer can be increased by making aluminum the anode of an electrolytic cell; the result is called anodized aluminum. Dyes may be added to the dilute sulfuric acid electrolyte to produce surface layers with different colors. Brown and bronze anodized aluminums produced in this way are widely used in modern architecture.

Aluminum's low density, wide availability, and corrosion resistance makes it ideal for construction. For use in airplanes, it is alloyed with copper and silicon. Its lightness and good electrical conductivity lead to its use for overhead power lines. Another of its applications depends on its highly exothermic **thermite reaction** with metal oxides:

$$Fe_2O_3(s) + 2Al(s) \longrightarrow 2Fe(l) + Al_2O_3(s) \qquad \Delta H° = -852 \text{ kJ}$$

A large part of the reaction enthalpy can be ascribed to the very high enthalpy of formation of Al_2O_3, which is −1676 kJ/mol. This reaction is used to weld rails or large pipes and to reduce oxide ores for which carbon is too weak a reducing agent.

19.2 GROUP III OXIDES

Boron, a nonmetal, has acidic oxides. Aluminum, its metallic neighbor, has amphoteric oxides (like its diagonal relative, beryllium). The oxides of both elements are important in their own right, as sources of the elements and as the starting point for the manufacture of other compounds.

Boron oxide and boric acid. Boric acid, $B(OH)_3$, is obtained commercially by the action of sulfuric acid on borax:

$$Na_2B_4O_7(aq) + H_2SO_4(aq) + 5H_2O(l) \longrightarrow Na_2SO_4(aq) + 4B(OH)_3(s)$$

Boric acid crystallizes as a white solid that melts at 171°C. It is toxic to bacteria and many insects (including roaches) and has long been used as a mild antiseptic. The boron atom in the $B(OH)_3$ molecule has an incomplete octet, so it can act as a Lewis acid by accepting a lone pair of electrons from an H_2O molecule. The complex so formed is a Brønsted acid:

$$\underset{\text{Lewis acid}}{(OH)_3B} + :OH_2 \longrightarrow \underset{\text{Brønsted acid}}{(OH)_3B{-}OH_2}, \qquad \text{written } HB(OH)_4$$

$$HB(OH)_4(aq) + H_2O(l) \rightleftharpoons H_3O^+(aq) + B(OH)_4^-(aq) \qquad pK_a = 9.14$$

The formula $HB(OH)_4$ conveys the sense that this hydrated form of boric acid is a weak *mono*protic acid. Boric acid also retards the spread of flames in cellulosic materials, particularly paper. The scrap paper used to manufacture home insulation contains about 5% boric acid, to retard the spread of flames.

Some of the chemical properties of the *p*-block elements can be rationalized by noting whether a compound is the **anhydride** of an oxoacid, a compound that forms the acid when it reacts with water. It is also helpful to note whether an oxide is the true anhydride or only the **formal anhydride**, a compound that has the formula of an acid less H_2O but does not produce the acid when it dissolves in water. An example is CO, which is the formal anhydride of formic acid, HCOOH. A number of anhydrides can be formed by simply heating the oxoacid. For example, when heated, boric acid dehydrates to its anhydride, boron oxide (B_2O_3):

$$2B(OH)_3(s) \xrightarrow{\Delta} B_2O_3(s) + 3H_2O(g)$$

Because it melts (at 450°C) to a liquid that dissolves many oxides, boron oxide (often as the acid) is used as a flux to remove the oxide coating of metals before they are soldered or welded.

The major use of boric acid is for the production of boron oxide, which in turn is also used as a flame retardant. In addition, boron oxide is used to produce fiberglass and borosilicate glass (Section 19.5), the latter being a glass with a very low thermal expansion.

Alumina. Alumina, Al_2O_3, exists as a variety of solids with different crystal structures. As α-alumina, it is the very hard substance corundum; impure microcrystalline corundum is the purple-black abrasive known as emery. Some impure forms of α-alumina are highly prized (Fig. 19.3). A less dense and more reactive form of the oxide is γ-alumina. This form absorbs water and is used as a support in chromatography.

γ-Alumina is produced by heating aluminum hydroxide in a reaction shown earlier. It is quite reactive and is amphoteric, reacting readily with bases to produce the aluminate ion and with acids to produce the hydrated Al^{3+} ion:

$$Al_2O_3(s) + 2OH^-(aq) + 3H_2O(l) \longrightarrow 2Al(OH)_4^-(aq)$$
$$Al_2O_3(s) + 6H_3O^+(aq) + 3H_2O(l) \longrightarrow 2Al(H_2O)_6^{3+}(aq)$$

D Preparation of boric acid. TDC, 15-2.

E A dilute boric acid/sodium borate solution is used in common eyewashes, such as Visine.

E Flame retardation depends on the ability of the species to produce radicals, which inhibit (by combining with radicals such as OH and CH_3) the propagation of the radical chain reaction that occurs when organic compounds burn. Bromine compounds are also used, for they yield Br atoms when heated, and these atoms inhibit the chain.

U Exercise 19.64.

E Textbook error: Industrial production of alumina. *J. Chem. Ed.*, **1989**, *66*, 313.

E Corundum is the abrasive of emery cloth, which is used for smoothing and cleaning metals. Natural corundum is found in India, Brazil, and South Africa, as well as in the United States (Georgia and the Carolinas). The name corundum is the Tamil word for ruby. Emery is a fine-grained impure variety of corundum that originally came from Greece and Turkey; it takes its name from Cape Emery on the island of Naxos.

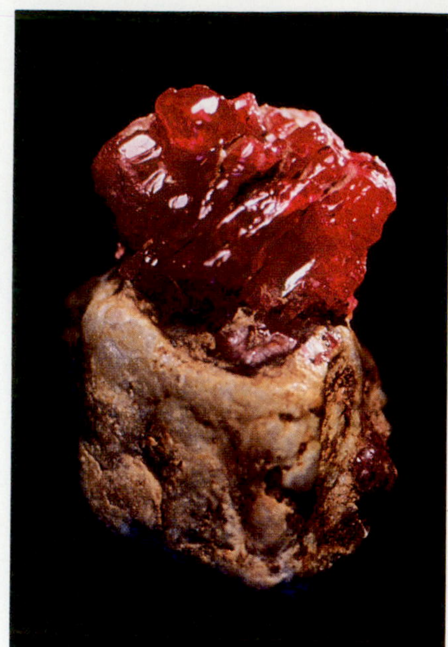

FIGURE 19.3 Some of the impure forms of α-alumina are prized as gems. (a) Alumina with Cr^{3+} ion impurities is ruby. (b) With Fe^{3+} and Ti^{4+} ions, it is sapphire. (c) With Fe^{3+} ions, it is topaz.

N See Case 6 for the use of aluminum as a fuel. The overall reaction is approximately $10Al(s) + 6NH_4ClO_4(s) \rightarrow 5Al_2O_3(s) + 3N_2(g) + 6HCl(g) + 9H_2O(g)$ but the conditions in the rocket exhaust are so intense that many other species and fragments of molecules are also formed.

E Hydrated cations in the general chemistry course. *J. Chem. Ed.*, **1981**, *58*, 349.

D Production of a foam. CD1, 98. Another foam. CD1, 99.

Aluminum hydroxide, which is formed when aluminum oxide reacts with water, also undergoes similar reactions:

$$Al(OH)_3(s) + OH^-(aq) \longrightarrow Al(OH)_4^-(aq)$$

$$Al(OH)_3(s) + 3H_3O^+(aq) \longrightarrow Al(H_2O)_6^{3+}(aq)$$

Because of the strong polarizing power of the small, highly charged Al^{3+} ion, the $Al(H_2O)_6^{3+}$ ion is a Brønsted acid. Solutions of aluminum salts are therefore acidic:

$$Al(H_2O)_6^{3+}(aq) + H_2O(l) \rightleftharpoons H_3O^+(aq) + Al(H_2O)_5OH^{2+}(aq) \quad pK_a = 5.14$$

α-Alumina is produced by heating γ-alumina to about 2000°C. As noted earlier, α-alumina is very hard and is used to synthesize gems such as sapphire, ruby, and topaz. These synthetic gems are often used for bearings in watches and other high-quality instruments. The very high enthalpy of formation ($\Delta H° = -1676$ kJ/mol) of alumina gives it some unique chemical properties. Powdered aluminum is also used as a component of the booster rocket fuel for propelling the space shuttle into orbit. The aluminum is oxidized by ammonium perchlorate, thereby producing a tremendous amount of heat that drives the gases out of the combustion chamber at very high velocity and propels the shuttle upward.

Alums and aluminates. One of the most important aluminum salts prepared by the action of an acid on alumina is aluminum sulfate:

$$Al_2O_3(s) + 3H_2SO_4(aq) \longrightarrow Al_2(SO_4)_3(aq) + 3H_2O(l)$$

Aluminum sulfate is called papermaker's alum and is used in the paper industry to coagulate cellulose fibers into a hard, nonabsorbent surface. True **alums** (from which the element aluminum takes its name) are sulfates of formula $M^+M'^{3+}(SO_4)_2 \cdot 12H_2O$; they include potassium alum, $KAl(SO_4)_2 \cdot 12H_2O$ (which is used in water and sewage treatment), and ammonium alum, $NH_4Al(SO_4)_2 \cdot 12H_2O$ (which is used for pickling cucumbers). Sodium alum provides the Brønsted acid in many baking powders; the hydrated Al^{3+} ion reacts with the HCO_3^- ions and releases carbon dioxide (Section 18.4). Alums are used for waterproofing fabrics and as a mordant in dying and printing textiles. (A mordant is a compound that helps to attach the dye to the fabric.)

E The alums. Interchangeable elements in a versatile crystal structure. *J. Chem. Ed.*, **1970**, *47*, 465.

N See Case 3 for the use of alum in water treatment.

N Again we see the diagonal relationship of beryllium and aluminum.

γ-Alumina dissolves in alkalis to give solutions of the aluminate ion:

$$Al_2O_3(s) + 2OH^-(aq) + 3H_2O(l) \longrightarrow 2Al(OH)_4^-(aq)$$

Sodium aluminate, $NaAl(OH)_4$, in conjunction with aluminum sulfate is used in water purification. When mixed with aluminate ions, the acidic hydrated Al^{3+} cation from the aluminum sulfate produces aluminum hydroxide:

$$Al^{3+}(aq) + 3Al(OH)_4^-(aq) \longrightarrow 4Al(OH)_3(s)$$

The aluminum hydroxide precipitates as a flocculent substance (a fluffy, gelatinous network) that entraps impurities as it settles and then can be filtered off. Using aluminum for both the cation and the anion gives the greatest possible bulk of impurity-collecting aluminum hydroxide because the reaction has no by-products.

19.3 OTHER IMPORTANT GROUP III COMPOUNDS

Both boron and aluminum form a range of compounds that are useful either because they have unusual physical properties or because they are chemically unique.

Carbides and nitrides. When boron is heated to high temperatures with carbon, it forms boron carbide, $B_{12}C_3$, a high-melting-point solid that is almost as hard as diamond. The solid consists of B_{12} groups that are pinned together by the C atoms.

When boron is heated to white heat in ammonia, boron nitride, BN, is formed as a fluffy, slippery powder:

$$2B(s) + 2NH_3(g) \xrightarrow{\Delta} 2BN(s) + 3H_2(g)$$

Boron nitride has a graphitelike structure, with flat planes of hexagons of alternating B and N atoms (Fig. 19.4). Unlike graphite, however, it is white and does not conduct electricity. Under high pressure, boron nitride is converted to a very hard, diamondlike, crystalline form called borazon.

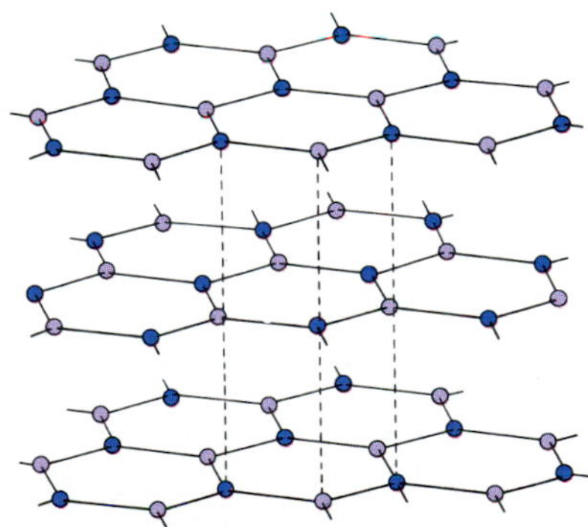

FIGURE 19.4 The structure of boron nitride resembles that of graphite, consisting of flat planes of hexagons. In boron nitride, however, the hexagons consist of alternating B and N atoms (in place of C atoms) and are stacked differently. (Compare with Fig. 10.42.)

Boron halides. The boron halides are made either by direct reaction of the elements at high temperature or from the oxide. The most important is boron trifluoride, BF_3, a widely used industrial catalyst produced by the reaction between boron oxide, calcium fluoride, and sulfuric acid:

$$B_2O_3(s) + 3CaF_2(s) + 3H_2SO_4(l) \xrightarrow{\Delta} 2BF_3(g) + 3CaSO_4(s) + 3H_2O(l)$$

Boron trichloride, which is also widely used as a catalyst, is produced commercially by the action of the halogen on the oxide in the presence of carbon:

$$B_2O_3(s) + 3C(s) + 3Cl_2(g) \xrightarrow{500°C} 2BCl_3(g) + 3CO(g)$$

E The formation of Al_2Cl_6 from two $AlCl_3$ molecules is an example of Lewis acid-base complex formation between identical molecules: the Al atoms are the acids and a Cl atom on the neighboring Al atom provides a lone pair and acts as a Lewis base. The dimers Al_2Br_6 and Al_2I_6 molecules survive in the solid as well as in the vapor because the halogen atoms of aluminum bromide and iodide are more polarizable than the Cl atoms of Al_2Cl_6.

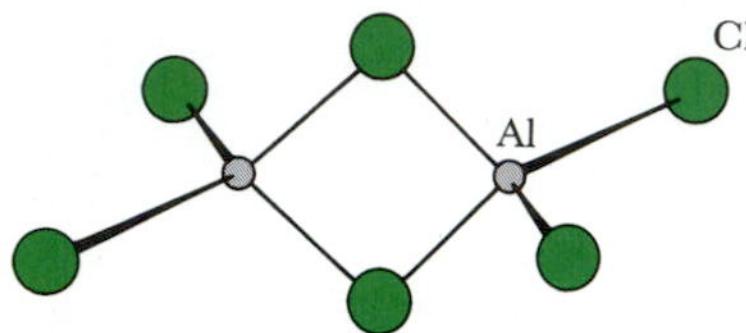

1 Aluminum chloride dimer

In all its trihalides, the B atom has an incomplete octet; the compounds consist of trigonal planar molecules with an empty $2p$ orbital perpendicular to the molecular plane. The empty orbital allows the boron trihalides to act as Lewis acids, and the fluoride and the chloride are widely used as catalysts on this account.

Aluminum halides. Aluminum chloride, another major industrial catalyst, is made by the action of chlorine on aluminum or alumina in the presence of carbon:

$$Al_2O_3(s) + 3C(s) + 3Cl_2(g) \xrightarrow{\Delta} 2AlCl_3(s) + 3CO(g)$$

Aluminum chloride is an ionic solid in which each Al^{3+} ion is surrounded by six Cl^- ions, but it sublimes at 192°C to a vapor of Al_2Cl_6 molecules. The Al_2Cl_6 molecule (**1**) is an example of a dimer, the union of two identical molecules.

The reactions between the aluminum halides and water produce a considerable quantity of heat. When the chloride is exposed to moist air, it produces fumes of hydrochloric acid (Fig. 19.5). The ionic aluminum chloride hexahydrate ($AlCl_3{\cdot}6H_2O$), the chloride salt of $Al(H_2O)_6{}^{3+}$, is prepared by dissolving alumina in concentrated hydrochloric acid. It is used as a deodorant and antiperspirant, its role being to kill the bacteria that convert perspiration into unpleasant-smelling compounds. The hexahydrate does not form the anhydrous chloride when it is heated, because the Al—O bonds are too strong to be replaced by Al—Cl bonds. Instead, heating produces the oxide:

$$2AlCl_3{\cdot}6H_2O(s) \longrightarrow Al_2O_3(s) + 6HCl(g) + 9H_2O(g)$$

Boranes and borohydrides. Boron forms a remarkable series of binary compounds with hydrogen—the **boranes**. These include diborane, B_2H_6, and more complex compounds such as decaborane, $B_{10}H_{14}$. Anionic versions of these compounds, the **borohydrides**, are also known; the most important is BH_4^- as sodium borohydride, $NaBH_4$.

N The high reactivity of many of the boranes have caused many disastrous laboratory accidents.

Sodium borohydride is a white crystalline solid produced from the reaction between sodium hydride and boron trichloride in a nonaqueous solvent:

$$4NaH + BCl_3 \longrightarrow NaBH_4 + 3NaCl$$

N The question is not really why boron forms electron deficient compounds, but why other elements (with one or two exceptions) do not. The answer lies in the small size and low nuclear charge of B atoms, for they can come close enough together for the molecular orbitals that they form with H atoms to act as a bridge between them.

The Na^+ ion is a spectator in the reaction. The compound is prepared industrially by a chemical onslaught on borax that involves sodium metal and hydrogen in the presence of silica at 500°C and under pressure. The product is extracted from the mixture by dissolving it in liquid ammonia. Sodium borohydride is widely used as a reducing agent.

The starting point for the production of the boranes is the reaction (in an organic solvent) of boron trifluoride with sodium borohydride:

$$4BF_3 + 3BH_4^- \longrightarrow 3BF_4^- + 2B_2H_6$$

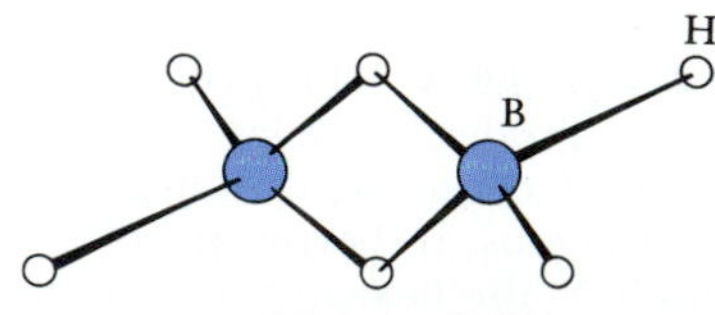

2 Diborane, B_2H_6

The product B_2H_6 is diborane (**2**), a colorless gas that bursts into flame in air. On contact with water, it is immediately oxidized to boric acid and hydrogen:

$$B_2H_6(g) + 6H_2O(l) \longrightarrow 2B(OH)_3(aq) + 6H_2(g)$$

When diborane is heated to a high temperature, it decomposes into hydrogen and pure boron:

$$B_2H_6(g) \xrightarrow{\Delta} 2B(s) + 3H_2(g)$$

This series of reactions is a useful route to the production of the pure element, but more complex boranes are formed when the heating is less intense. When diborane is heated to 100°C, for instance, it forms decaborane, $B_{10}H_{14}$, a solid that melts at 100°C. Decaborane is stable in air, is oxidized by water only slowly, and is an example of the general rule that heavier boranes are less flammable than boranes of low molar mass.

The boranes are electron-deficient compounds. As mentioned just before Section 9.7, these are compounds for which Lewis structures cannot be written because too few electrons are available. However, as we also saw in Section 9.9, there is no difficulty in accounting for their structure in terms of a delocalized bond in which the bonding power of one electron pair is shared by more than two atoms (**2**).

FIGURE 19.5 When anhydrous aluminum chloride is left exposed to moist air, it reacts to form hydrochloric acid. Here we see white fumes of ammonium chloride forming as the hydrochloric acid reacts with ammonia.

Aluminum hydrides. The most important compound of aluminum and hydrogen is lithium aluminum hydride, $LiAlH_4$. This white solid is prepared by the reaction of lithium hydride with aluminum trichloride in an organic solvent:

$$4LiH + AlCl_3 \xrightarrow{\text{diethyl ether}} LiAlH_4 + 3LiCl$$

Its importance lies in its use as a reducing agent, especially in organic chemistry. Lithium aluminum hydride reacts with aluminum trichloride to form aluminum hydride, AlH_3, which is best regarded as a saline hydride:

$$3AlH_4^- + AlCl_3 \longrightarrow 4AlH_3 + 3Cl^-$$

The aluminum analogues of the higher boranes are unknown.

U Exercises 19.74 and 19.79.

N An example is the reduction of a ketone to a secondary alcohol: $CH_3COCH_3 \xrightarrow{LiAlH_5} CH_3CH(OH)CH_3$.

GROUP IV: CARBON AND SILICON

		III	IV	V			He
Li	Be	B	C	N	O	F	Ne
		Al	Si	P			
		Ga	Ge	As			
		In	Sn	Sb			
		Tl	Pb	Bi			

The elements of Group IV and some of their physical properties are listed in Table 19.3. Carbon, at the top of the group, forms a very wide range of compounds and gives rise to a special branch of chemistry—organic chemistry—that we introduce in Part V. However, the oxides and oxoanions of carbon are typically "inorganic" in that they form salts and are therefore included here.

TABLE 19.3 Group IV elements

Valence configuration: ns^2np^2

Z	Name	Symbol	Molar mass, g/mol	Melting point, °C	Boiling point, °C	Density, g/cm³	Normal form
6	carbon	C	12.01	3370s*		1.9 to 2.3	black nonmetal (graphite)
						3.2 to 3.5	transparent nonmetal (diamond)
14	silicon	Si	28.09	1410	2620	2.33	gray metalloid
32	germanium	Ge	72.59	937	2830	5.32	gray-white metalloid
50	tin	Sn	118.69	232	2720	7.29	white lustrous metal
82	lead	Pb	207.19	328	1760	11.34	blue-white lustrous metal

*The symbol s denotes that the element sublimes.

E At pH = 14, the potential of the half-reaction $H_2BO_3^-(aq) + 5H_2O(l) + 8e^- \rightarrow BH_4^-(aq) + 8OH^-(aq)$ is -1.24 V. Because this value is well below the potential of the Ni^{2+}/Ni couple (-0.23 V) in the electrochemical series, the borohydride ion can be used to reduce Ni^{2+} ions to metallic nickel in the process called *chemical plating*. The advantage of chemical plating over electroplating is that the item being plated does not have to be an electrical conductor.

TABLE 19.4 Chemical properties of the Group IV elements

Reactant	Reaction with Group IV elements, E
hydrogen	$C(s) + 2H_2(g) \rightarrow CH_4(g)$ and other hydrocarbons
oxygen	$C(s) + O_2(g) \rightarrow CO(g)$ and $CO_2(g)$
	$E(s) + O_2(g) \rightarrow EO_2(s)$ E = Si, Ge, Sn
	$2Pb(s) + O_2(g) \rightarrow 2PbO(s)$
water	$C(s) + H_2O(g) \xrightarrow{\Delta} CO(g) + H_2(g)$
	$Si(s) + 2H_2O(g) \xrightarrow{\Delta} SiO_2(s) + 2H_2(g)$
halogen	$E(s) + 2X_2(s,l,g) \rightarrow EX_4(s,l,g)$ E = C, Si, Ge, Sn
	$Pb(s) + X_2(s,l,g) \rightarrow PbX_2(s)$
acid	$E(s) + 2H^+(aq) \rightarrow E^{2+}(aq) + H_2(g)$ E = Sn, Pb
base	$E(s) + 2OH^-(aq) + 2H_2O(l) \rightarrow E(OH)_4^{2-}(aq) + H_2(g)$ E = Sn, Pb

E Tetramethylsilane, $(CH_3)_4Si$, is a liquid that is used to calibrate nuclear magnetic resonance (NMR) spectra.

The general chemical properties of the elements of Group IV, which are summarized in Table 19.4, are as expected for their position in the periodic table. Carbon shows definite nonmetallic properties in that it forms covalent compounds with nonmetals and ionic compounds with metals. Silicon and germanium are typical metalloids: they have a metallic appearance, but they form covalent compounds with nonmetals (including the gas silicon tetrafluoride, SiF_4, and the liquid germanium tetrachloride, $GeCl_4$). Tin and, even more so, lead have definite metallic properties. Even though tin is classified as a metal, it is not far from the metalloids in the periodic table, and it does have some amphoteric properties: it reacts with both hot concentrated hydrochloric acid and hot alkali:

$$Sn(s) + 2HCl(aq) \longrightarrow SnCl_2(aq) + H_2(g)$$
$$Sn(s) + 2OH^-(aq) + 2H_2O(l) \longrightarrow Sn(OH)_4^{2-}(aq) + H_2(g)$$

19.4 GROUP IV ELEMENTS

We expect carbon, because it is at the head of its group, to be different from its congeners. However, the differences are more pronounced in Group IV than anywhere else in the periodic table. Some of these differences stem from the wide occurrence of C=C and C=O double bonds, compared with the rarity of Si=Si and Si=O double bonds. Carbon dioxide, which consists of discrete O=C=O molecules, is a gas that we can exhale. Silicon dioxide (silica), which consists of networks of —O—Si—O— groups, is a mineral we can stand on.

The valence electron configuration is ns^2np^2 for all members of the group. All four electrons are approximately equally available for bonding in the lighter elements, and carbon and silicon are characterized by their ability to form four covalent bonds. However, on descending the group, the energy separation between the *s* and *p* orbitals increases and the *s* electrons become progressively less available for bonding; in fact, the lead compounds are most commonly salts in which the oxidation number of lead is +2.

N The decreased availability of the valence ns^2 electrons of these main group metals is the inert-pair effect (Chapter 7); its origin is largely the relativistic increase in mass of electrons that have a high velocity.

Carbon. Solid carbon exists in three crystalline allotropic forms, graphite, diamond, and buckminsterfullerene, with graphite the thermodynamically most stable allotrope under normal conditions. Buckminsterfullerene has only very recently been prepared, and its structure is described in the case at the end of the chapter. Graphite is produced by passing a heavy electric current through coke rods for several days. Synthetic diamonds are made by using high pressure and high temperature (Section 10.9). A more recently developed technique makes use of the thermal decomposition of methane on a hot wire. The carbon atoms that are formed settle on a cooler surface as both graphite and diamond. However, the hydrogen atoms also produced react more quickly with the graphite to form volatile hydrocarbons, so more diamond than graphite survives.

E Graphite is marketed as a dry or greaseless lubricant. It is a metallic conductor parallel to the hexagonal planes and a semiconductor perpendicular to the planes.

? Why should graphite not be used as a lubricant for electric motors or other electrical equipment?

E Diamond films grown by methane decomposition are of considerable (but as yet unrealized) technological significance, because diamond is the best thermal conductor known, so the films could form a base for semiconductor circuits (which need to dissipate heat). Because diamond is also a semiconductor, the actual electronics might even one day be incorporated into the diamond substrate.

Soot and carbon black consist of very small crystals of graphite. Carbon black, which is produced by heating gaseous hydrocarbons to nearly 1000°C in the absence of air, is used to reinforce rubber, in pigments, and in printing inks. Activated carbon (also called activated charcoal) consists of granules of microcrystalline graphite that is obtained by heating waste organic matter in the absence of air and then processed to increase their porosity. Its very high surface area (up to about 2000 m^2/g) enables it to remove organic impurities from liquids and gases by adsorption. It is used in air purifiers, gas masks, and aquarium water filters. On a larger scale, activated carbon is used in emission control canisters of automobiles to minimize loss of gasoline from gas tanks by vaporization and in water purification treatment plants to remove organic compounds from the drinking water supply.

The properties and uses of the two major allotropes of carbon reflect their crystal structures, which were described in Section 10.9. Graphite consists of planar sheets of sp^2-hybridized carbon atoms in a hexagonal network (Fig. 10.42). Because electrons are free to move from one unhybridized p orbital on one carbon atom to another through an endless delocalized π network that spreads across the plane, graphite is a black, lustrous, electrically conducting solid; indeed, graphite is the only major nonmetal used specifically as an electrical conductor. One of its uses is as electrodes of electrolytic cells (in the Hall process, for example). Its slipperiness, which results from the ease with which the flat planes move past one another when impurities are present, leads to its use as a dry lubricant. It is also used as the "lead" in pencils; layers of graphite are rubbed off by friction as the pencil is moved across the paper. In diamond, each carbon atom is sp^3-hybridized and linked tetrahedrally to its four neighbors, with all electrons localized in C—C σ bonds. Diamond is a rigid, transparent, electrically insulating solid. It is the hardest substance known and the best conductor of heat (about five times better than copper). These last two properties make it an ideal abrasive, for it can scratch all other substances yet the heat generated by friction is rapidly conducted away.

FIGURE 19.6 Three of the common forms of silica (SiO_2): quartz (top), quartzite (middle), and cristobalite (bottom) in which the black parts are obsidian, a volcanic rock that also contains silica.

Silicon. The structure of silicon is similar to that of diamond in that the Si atoms are tetrahedrally bonded. It occurs widely, as the second most abundant element in the earth's crust, in the silicates (compounds containing the silicate ion, SiO_3^{2-}) of rocks and as the silica (SiO_2) of sand (Fig. 19.6). It is obtained from quartzite, a granular form of quartz, by reduction with high-purity carbon in an electric arc furnace. The crude product is then exposed to chlorine to form silicon tetrachloride, which is distilled and reduced by hydrogen to a purer form of the element:

$$SiCl_4(l) + 2H_2(g) \longrightarrow Si(s) + 4HCl(g)$$

For its use in semiconductors, more purification is needed. In one process, a large single crystal is grown by pulling a solid rod of the element slowly from the melt (Fig. 19.7). Then the crystals are zone refined to a concentration of less than one impurity atom per billion Si atoms. This "ultrapure" silicon is used for semiconductor chips. Alternative techniques include the thermal decomposition of silane (SiH_4), a method producing a highly crystalline form of the element as the atoms settle onto a cool surface, and the decomposition of silane by an electric discharge. The latter method produces an amorphous form of silicon with a significant hydrogen content. Amorphous silicon is used in photovoltaic devices, which produce electricity from sunlight.

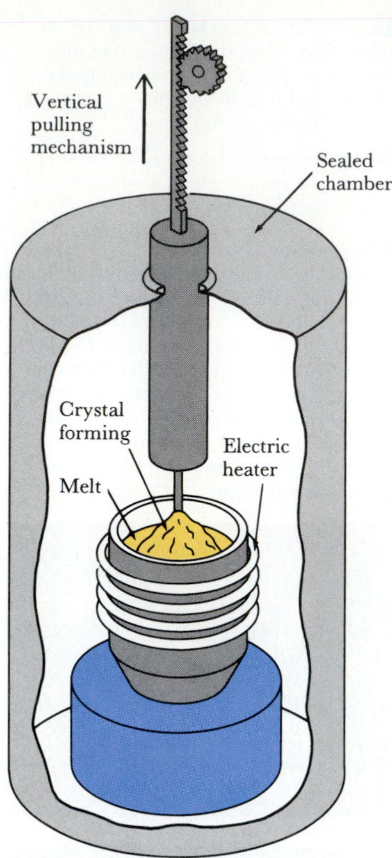

FIGURE 19.7 Pulling a single crystal of silicon from molten silicon. The crystal is withdrawn very slowly from the melt. The technique is called the Czochralski method of crystal growth.

? Pure silicon has a structure like that of diamond. Is pure silicon an electrical conductor?

U Exercises 19.71 and 19.76.

E Until recently, a major use of lead was in the production of tetraethyllead, $(C_2H_5)_4Pb$, an antiknock agent present in gasoline. Whereas its production is on the decline (partly because it poisons people, and partly because it poisons catalysts in pollution control catalytic converters), many countries, states, and provinces have leaded gasoline available at most service centers.

E Lead uses. *J. Chem. Ed.*, **1981**, *58*, 722.

E The PanAm building in Manhattan is based on thick lead foundations to insulate it from the noise of Grand Central Station below.

Germanium, tin, and lead. Germanium is recovered from the flue dust of industrial plants that process zinc ores, in which it occurs as an impurity. It is used mainly—and increasingly—in the semiconductor industry for integrated circuits in very fast computers.

Tin and lead are obtained very easily from their ores and have been known since antiquity. Tin occurs mainly as the mineral cassiterite (SnO_2) and is obtained from it by reduction with carbon at 1200°C:

$$SnO_2(s) + C(s) \xrightarrow{1200°C} Sn(l) + CO_2(g)$$

The main problem with this apparently simple process is ensuring that the iron impurities are reoxidized without oxidizing the tin as well. This delicate balance is achieved by passing oxygen through vigorously stirred molten tin.

The principal lead ore is galena (PbS). It is roasted in air, a process that converts it to PbO, and then reduced with coke:

$$2PbS(s) + 3O_2(g) \longrightarrow 2PbO(s) + 2SO_2(g)$$

$$PbO(s) + C(s) \longrightarrow Pb(s) + CO(g)$$

Tin is expensive and not very strong, but it is resistant to corrosion. Its main use is in tinplating, which accounts for about 40% of its consumption. However, the Sn^{2+}/Sn couple (−0.14 V) lies above the Fe^{2+}/Fe couple (−0.44 V) in the electrochemical series, so the tin-plate encourages the oxidation of iron if the surface is damaged. Tin is also used for the production of alloys (Table 10.10).

One very important property of lead is its durability (its chemical inertness), which makes it useful in the construction industry. The inertness of lead under normal conditions can be traced to the passivation of its surface by oxides, chlorides, and sulfates. This passivation allows lead containers to be used for transporting hot concentrated sulfuric acid; nitric acid cannot be transported in lead containers, however, for lead nitrate is soluble. Another important property of lead is its density, which makes it useful not only as a radiation shield, because its numerous electrons absorb high-energy radiation, but also as soundproofing material, because its heavy atoms move sluggishly and transmit sound waves poorly. Currently, the main use of lead is in electrodes of lead-acid storage batteries.

19.5 GROUP IV OXIDES

Carbon forms several oxides, but the two most important ones are carbon monoxide and carbon dioxide. In striking contrast to these two carbon oxides, which are gases, silicon dioxide is a solid. Derivatives of the oxides and oxoacids of both carbon and silicon are very important components of landscapes and are a rich source of raw materials.

Carbon monoxide. Carbon monoxide is produced when carbon or organic compounds burn in a limited supply of air, as happens when gasoline burns in automobile engines or when tobacco burns in a cigarette. Commercially carbon monoxide is produced as synthesis gas by the reforming reaction (Section 18.1). Carbon monoxide is the formal anhydride of formic acid, HCOOH, and can be produced in the laboratory by the dehydration of that acid with hot, concentrated sulfuric acid:

$$HCOOH(l) \xrightarrow{150°C,\ H_2SO_4} CO(g) + H_2O(l)$$

The reverse of this reaction cannot be carried out directly, but carbon monoxide does react with hydroxide ions in hot alkali to produce formate ions:

$$CO(g) + OH^-(aq) \xrightarrow{\Delta} HCO_2^-(aq)$$

Carbon monoxide is a colorless, flammable, almost insoluble, very toxic gas that condenses to a colorless liquid at −90°C. It is not very reactive, partly on account of its high bond enthalpy (1074 kJ/mol, the highest for any molecule). However, it is a Lewis base, and the lone pair on the carbon atom forms covalent bonds with *d*-block atoms and ions. An example of this type of complex formation is its reaction with nickel to give the intensely poisonous liquid nickel tetracarbonyl:

$$Ni(s) + 4CO(g) \xrightarrow{50°C,\ 1\ atm} Ni(CO)_4(l)$$

Complex formation is also responsible for carbon monoxide's toxicity. It attaches more strongly than oxygen to the iron in hemoglobin, so a person who breathes in carbon monoxide is liable to suffocate because their blood is unable to transport sufficient oxygen.

Carbon monoxide is an industrial reducing agent. It takes this role in the production of a number of metals, most notably the production of iron in the blast furnace (Section 21.2):

$$Fe_2O_3(s) + 3CO(g) \xrightarrow{800°C} 2Fe(l) + 3CO_2(g)$$

Carbon dioxide and the carbonates. The colorless gas carbon dioxide has already featured extensively in earlier chapters. It is formed whenever carbon or organic compounds burn in a liberal supply of air, in fermentation (the conversion of carbohydrates to ethanol by yeast), and by the action of acids on carbonates. Because its triple point is at 5.1 atm (Fig. 10.22), it does not exist as a liquid at ordinary atmospheric pressure. Solid carbon dioxide, which is sold as "dry ice," therefore sublimes directly to the gas—a property that makes it convenient as a refrigerant and a cold pack.

E The reforming reaction is $CH_4(g) + H_2O(g) \rightarrow CO(g) + 3H_2(g)$. One application of the synthesis gas (the mixture of carbon monoxide and hydrogen produced in this way) is the production of methanol: $CO(g) + 2H_2(g) \rightarrow CH_3OH(l)$.

E The formation and subsequent thermal decomposition of nickel tetracarbonyl is used in the Mond process for the production of nickel (Section 21.2).

E Just as significant as its ability to form σ bonds to a *d*-metal atom, is the ability of carbon monoxide to form π bonds with the metal atom by accepting two of the latter's electrons into its empty π^* orbital. The strength of the bonding arises because CO is both a Lewis base (by virtue of its σ lone pair) and a Lewis acid (by virtue of its empty π^* orbitals). A similar remark applies to the isoelectronic CN^- ion.

E The poisonous nature of carbon dioxide (128 years ago!). *J. Chem. Ed.*, **1975**, *52*, 791.

FIGURE 19.8 Double-acting baking powder first forms small cavities in dough when it is moistened. These are later inflated by a second release of carbon dioxide during baking.

Carbon dioxide is the acid anhydride of carbonic acid, H_2CO_3, and in water it reaches the equilibrium

$$CO_2(g) + H_2O(l) \rightleftharpoons H_2CO_3(aq)$$

This equilibrium is utilized in the production of carbonated beverages. Carbonic acid is the parent acid of the carbonates and the hydrogencarbonates (the bicarbonates). Most carbonates decompose to the oxide and carbon dioxide when heated:

$$CO_3^{2-} \xrightarrow{\Delta} CO_2 + O^{2-}$$

An example is the action of heat on calcium carbonate (limestone) in the production of lime:

$$CaCO_3(s) \xrightarrow{\Delta} CO_2(g) + CaO(s)$$

Solid hydrogencarbonates can be isolated only for the alkali metals; others decompose to carbon dioxide and the carbonate when the water is removed.

Carbonates and hydrogencarbonates evolve carbon dioxide when they are treated with acid. The CO_3^{2-} and HCO_3^- ions are Brønsted bases that accept protons to form H_2CO_3 molecules, which at low partial pressures of carbon dioxide dissociate into CO_2 and water:

$$HCO_3^-(aq \text{ or } s) + H_3O^+(aq) \longrightarrow H_2CO_3(aq) + H_2O(l)$$

$$H_2CO_3(aq) \longrightarrow H_2O(l) + CO_2(g)$$

This property is responsible for the action of baking powder, a mixture of sodium hydrogencarbonate and an acid, which may be either sodium alum (where the hydrated aluminum cation is the acid) or calcium dihydrogenphosphate, $Ca(H_2PO_4)_2$. The carbon dioxide evolved is trapped as bubbles in the dough and causes it to rise. However, to get a good dough, the CO_2 should be released in two stages (hence double-acting baking powder). In the first, at about room temperature, a small release makes small cavities (Fig. 19.8). In the second, at baking temperature, the cavities are inflated by a general release of the gas. The first release is achieved by including some tartaric acid in the mixture, which acts on the HCO_3^- ions as soon as the dough is moistened.

D A chemical pop gun. CD1, 14.

U Exercise 19.72.

? Should sour-milk pancakes be made with baking powder or baking soda?

FIGURE 19.9 Impure forms of silica: amethyst (left, in which the color is due to Fe^{3+} impurities), onyx (center), and agate (right).

Silica and glass. Silica, SiO_2, occurs naturally as quartz and as sand, which is an impure form of quartz colored golden brown by iron oxide impurities. Some precious and semiprecious stones are impure silica (Fig. 19.9). Flint is silica colored black by carbon impurities.

Sense can be made of the bewildering array of structures exhibited by silica and the silicates by picturing them as arrangements of SiO_4 tetrahedra. In silica itself, each corner O atom is shared by two Si atoms. Hence, each tetrahedron contributes one Si atom and $4 \times \frac{1}{2} = 2$ O atoms to the solid, an arrangement resulting in the empirical formula SiO_2 (Fig. 19.10). The structure of quartz is complicated, for it is built from helical chains of SiO_4 units wound around one another. When it is heated to about 1500°C, it changes to another arrangement, that of the mineral cristobalite (Fig. 19.11). This structure is simpler to describe: its Si atoms are arranged like the C atoms in diamond, except there is an O atom between each pair of neighboring Si atoms.

Silica resists attack by Brønsted acids but gives way to attack by the strong Lewis base F^- from hydrofluoric acid, with which it forms fluorosilicate ions:

$$SiO_2(s) + 6HF(aq) \longrightarrow SiF_6^{2-}(aq) + 2H_3O^+(l)$$

It is also attacked by the Lewis base OH^- in hot, molten sodium hydroxide and by O^{2-} in the carbonate anion of hot molten sodium carbonate:

$$SiO_2(s) + Na_2CO_3(l) \xrightarrow{1400°C} Na_2SiO_3(s) + CO_2(g)$$

Thus, F^- (from HF), OH^-, and CO_3^{2-} ions all attack and dissolve the silica in glass. This chemical attack is called the etching of glass (Fig. 19.12). The sodium silicate produced by the reaction between silica and sodium carbonate is soluble, and it is sold commercially as water glass. It is used in detergents, partly as an alkaline buffer (SiO_3^{2-} is a Brønsted base) and partly to stop dirt from settling back onto the fabric. The SiO_3^{2-} ions act by attaching to dirt particles and hence giving them a negative charge and preventing them from merging with others into larger, insoluble particles (Fig. 19.13).

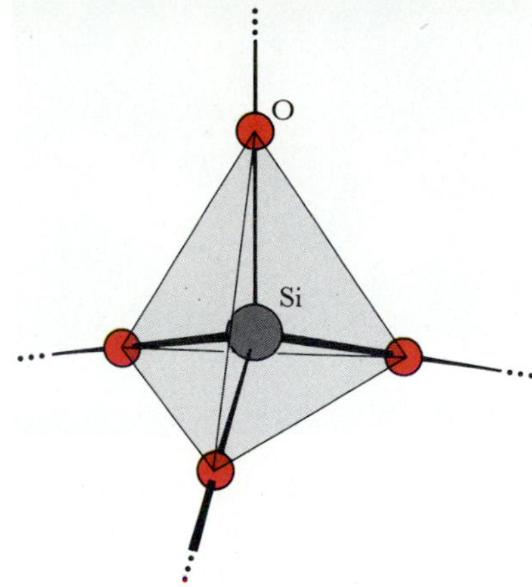

FIGURE 19.10 The structures of silicates are built up from SiO_4 tetrahedra. In different silicates, different numbers of O atoms are shared; in some cases neighboring tetrahedra share one O atom; in others they share two O atoms.

T OHT: Figs. 19.10 and 19.11.

D Show the action of sodium hydroxide on glass by obtaining a bottle from the stockroom that was once used to store NaOH(*aq*).

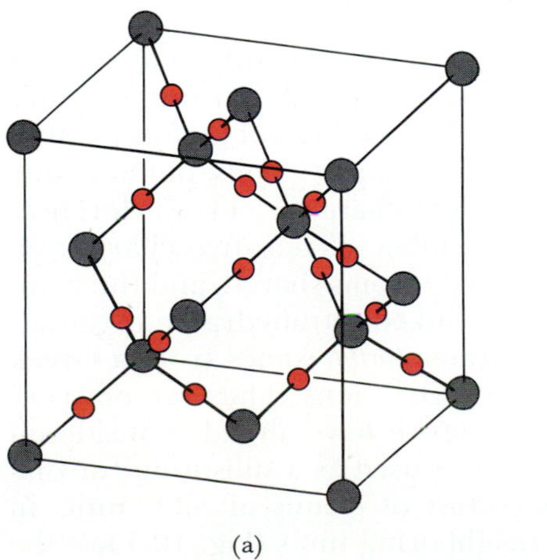

(a)

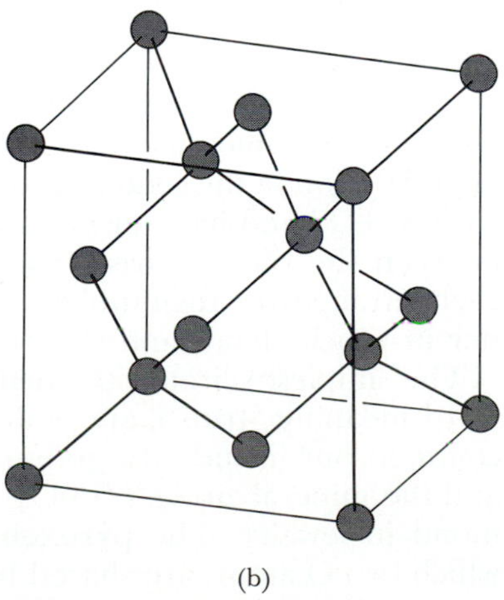

(b)

FIGURE 19.11 The structure of cristobalite (a) is like that of diamond (b) except that an O atom (red) lies between each Si atom (gray) in cristobalite.

FIGURE 19.12 The reaction between hydrofluoric acid and glass is used to etch glass. The surface of the glass is covered with wax, a design is scratched on the wax, and acid is poured over it. This etched glass bowl was designed by the artist Frederick Carder in the 1920s.

E Boron glasses. TDC, 15-14–15-24.

? Why is the procedure for cutting a diamond different from that for cutting glass?

E The brilliance of expensive lead crystal is from the presence of lead oxides (up to as much as 33% by mass). The brilliance is a consequence of the high refractive index of the glass, which in turn stems from the large number of electrons on each lead ion (which therefore interact strongly with the incident light, and refract it). Lead crystal is also much denser than soda-lime glass.

When a solution of sodium silicate is acidified, a gelatinous precipitate of silica is produced:

$$2H_3O^-(aq) + SiO_3^{2-}(aq) + xH_2O(l) \longrightarrow SiO_2{\cdot}xH_2O(s) + 3H_2O(l)$$

After it is washed and dried, this silica gel has a very large surface area (about 700 m^2/g) and is useful as a drying agent, a support for catalysts, a packing for chromatography columns, and a thermal insulator.

Many of the Si—O bonds break when silica (usually in the form of sand) is heated to around 1600°C. The orderly structure of the crystals is not regained when the Si—O bonds reform as the melt cools, and an amorphous, glassy material, called fused silica, results. The presence of the metal oxides MO in the melt leads to **glass**. In the cooling process, the Si atoms of silica form bonds with O^{2-} ions of the metal oxide to give —Si—O^{2-}—M^{2+} groups in place of some of the —Si—O—Si— links present in pure silica.

The addition of sodium oxide alone to silica does not give a very durable glass. Greater durability is obtained by including Ca^{2+} ions too. Almost 90% of all glass now manufactured is **soda-lime glass**. This glass, which is used for windows and bottles, contains about 12% Na_2O (prepared by the action of heat on sodium carbonate—the soda) and 12% CaO (prepared by heating calcium carbonate—the lime). When the proportions of soda and lime are reduced and 16% B_2O_3 is added, a **borosilicate glass**, such as Pyrex, is produced. Because borosilicate glasses do not expand much when heated, they survive rapid heating and cooling and are used for ovenware and laboratory glassware.

Colored glass is produced by adding other substances; cadmium sulfide and selenide, for instance, give ruby glass. Ordinary soda-lime glass is often very pale green as a result of the iron impurities it contains. Cobalt blue glass is colored by Co^{2+} ions. Brown beer-bottle glass is colored by iron sulfides. Photochromic sunglasses, which darken in sunlight, contain silver and copper ions (see Case 17). Like the silver ions in the photographic process, the Ag^+ ions in the lenses are reduced to silver metal by sunlight (see Box 3.1):

$$Ag^+(s) + Cu^+(s) \xrightarrow{\text{light}} Ag(s) + Cu^{2+}(s)$$

Unlike the reducing agent in a photographic emulsion, however, the oxidized reducing agent (in this case Cu^{2+}) cannot escape and, as soon as the light is removed, it oxidizes the silver again.

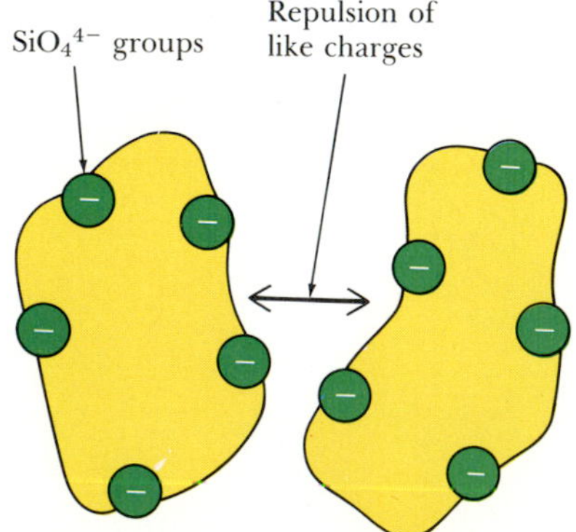

FIGURE 19.13 When ions attach to dirt particles, neighboring particles repel each other and do not collect into larger, insoluble particles.

The silicates. As noted above, the structures of the different silicates can be pictured in terms of SiO_4 tetrahedra with various negative charges. For some purposes it is convenient to think of an SiO_4^{4-} unit as an Si^{4+} ion in the center of four surrounding O^{2-} ions, but in reality each Si—O bond has considerable covalent character. The differences between the silicates arise from the number of negative charges on each tetrahedron, the number of corner O atoms shared, and the manner in which chains and sheets of the linked tetrahedra lie together.

The simplest silicates, the **orthosilicates** (*ortho* comes from a Greek word meaning "true"), are built from SiO_4^{4-} ions. They are not very common but include the sodium silicate we have already considered and the mineral zircon ($ZrSiO_4$), which is used as a substitute for diamond in jewelry. The **pyroxenes** consist of chains of SiO_4 units in which two O atoms are shared by neighboring units (Fig. 19.14a); the

FIGURE 19.14 (a) The basic structural unit of the pyroxenes. (b) The silicates called amphiboles have a ladderlike structure. Each tetrahedron represents an SiO_4 unit (like that in Fig. 19.10), and a shared corner represents a shared O atom (red). The silicon atom (gray) is located at the center of each SiO_4 tetrahedron. The two unshared O atoms each have a negative charge.

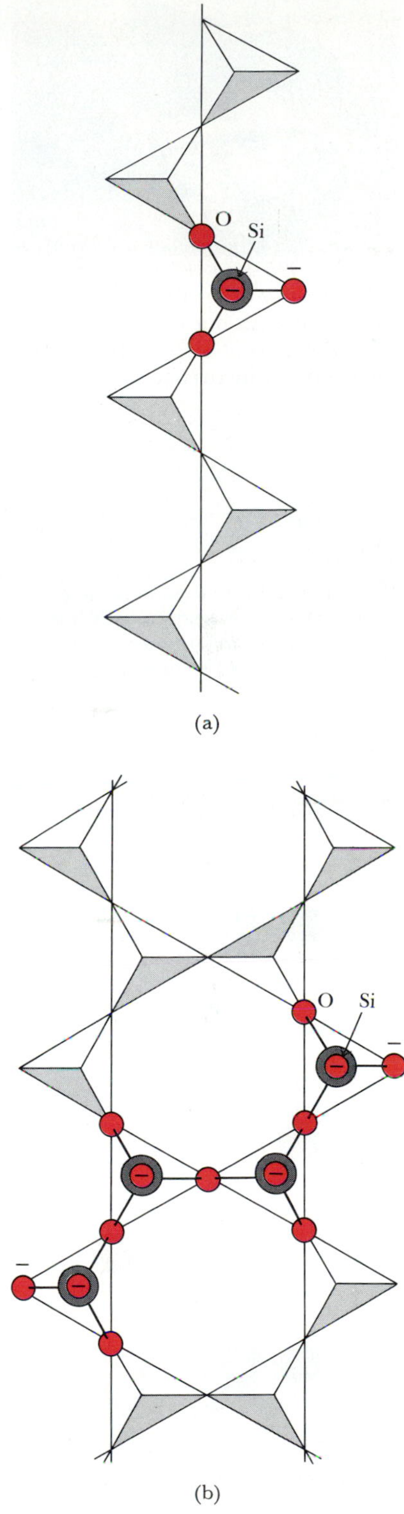

repeating unit has the formula SiO_3^{2-}. Electrical neutrality is provided by cations spaced out along the chain. The pyroxenes include jade, $NaAl(SiO_3)_2$.

The chains of units can link together to form the ladderlike structures which include tremolite, $Ca_2Mg_5(Si_4O_{11})_2(OH)_2$ (Fig. 19.14b). Tremolite is one of the minerals called asbestos, which are characterized by a fibrous structure and an ability to withstand heat. Their fibrous quality reflects the way the ladders lie together but can easily be torn apart. The SiO_4 tetrahedra can also link together to form sheets. An example is talc, a hydrated magnesium silicate, $Mg_3(Si_2O_5)_2(OH)_2$. Talc is soft and slippery because the silicate sheets slide over one another.

The aluminosilicates and ceramics. More complex (and more common) structures are obtained when some of the Si^{4+} ions in silicates are replaced by Al^{3+} ions to form the **aluminosilicates**. The missing positive charge is made up by extra cations, which account for the differences in the properties of the silicate talc and the aluminosilicate mica (Fig. 19.15). One form of mica is $KMg_3(Si_3AlO_{10})(OH)_2$, in which the sheets of tetrahedra are held together by extra K^+ ions. Although it cleaves neatly into layers when the sheets are torn apart, mica is not slippery like talc.

The **feldspars** are aluminosilicates in which more than half the Si^{4+} ions have been replaced by Al^{3+} ions. They are the most abundant silicate materials on earth and include granite, a compressed mixture of mica, quartz, and feldspar. When some of the cations between the crystal layers are washed away as these rocks weather, the structure crumbles to clay, a major component of "dirt." A typical feldspar has the formula $KAlSi_3O_8$. Its weathering by carbon dioxide and water can be described by the equation

$$2KAlSi_3O_8(s) + 2H_2O(l) + CO_2(g) \xrightarrow{\Delta} K_2CO_3(aq) + Al_2Si_2O_5(OH)_4(s) + 4SiO_2(s)$$

The potassium carbonate is soluble and washes away, but the aluminosilicate remains as the clay. **Ceramics** are created by heating clays to drive out the water between the sheets of tetrahedra. What is left is a rigid mass of tiny interlocking crystals bound together by glassy silica. China clay, which is used to make porcelain and china, is a form of aluminum aluminosilicate that can be obtained reasonably free from the iron impurities that make many clays look reddish brown. It is used in large amounts to coat paper (such as this page) to give a smooth, nonabsorbent surface.

When aluminosilicates are melted and then allowed to solidify, various **cements** are obtained. The most widely used is Portland cement, which is made by heating a mixture of silica, clay, and limestone to about 1500°C. The cooled mass is then crushed and some gypsum, $CaSO_4{\cdot}2H_2O$, is added. The main components of the complex mixture

D Colorful stalagmites. The silicate garden. BZS3, 9.49.

FIGURE 19.15 The aluminosilicate mica cleaves into thin transparent sheets. It is used for windows in furnaces.

are various calcium silicates and aluminates. When water is added, complex reactions occur and the mass sets to a solid. High-alumina cement, which is resistant to corrosion in harsh environments, is made by fusing alumina, lime, and silica.

Silicones. The **silicones**, like pyroxenes, consist of long —O—Si—O— chains of tetrahedral units; but the silicon bonding positions that are not used by O atoms are occupied by organic groups, such as the methyl group, $—CH_3$ (Fig. 19.16). Silicones are used to waterproof fabrics because the O atoms that have not been replaced by the organic groups attach to the fabric, thereby leaving the hydrophobic (water-repelling) methyl groups like tiny, inside-out umbrellas sticking up out of the fabric's surface. For similar reasons, these methyl silicones are biologically inert and survive intact when exposed to body fluids. Because they do not coagulate blood or stick to body tissues, they are used for surgical and cosmetic implants.

D Silicones. TDC, 15-25–15-28.

U Exercise 19.77.

E Carborundum (a name coined from carbon + corundum, which it resembles in abrasive power) is an abrasive used for the more expensive grade of sandpaper.

19.6 OTHER IMPORTANT GROUP IV COMPOUNDS

In this section we mention a few of the numerous compounds formed by the Group IV elements and concentrate on those that are of commercial importance.

Carbides. There are three classes of carbides: **saline** (or saltlike), **covalent**, and **interstitial**. The Group I and II metals form saline carbides when their oxides are heated with carbon. The saline carbides are formed most commonly from the Group I and II metals, aluminum, and a few other metals and contain either C_2^{2-} or C^{4-} as anions. All the C^{4-} carbides, which are sometimes called **methides**, produce methane and the corresponding hydroxide in water:

D Production of acetylene. CD1, 9.

$$Al_4C_3(s) + 12H_2O(l) \longrightarrow 4Al(OH)_3(s) + 3CH_4(g)$$

C_2^{2-} is the **acetylide** ion; the carbides that contain it react with water to produce acetylene (the conjugate acid of the acetylide ion) and the corresponding hydroxide in water.

$$CaC_2(s) + H_2O(l) \longrightarrow Ca(OH)_2(s) + HC{\equiv}CH(g)$$

Calcium carbide, CaC_2, is the most common saline carbide.

The covalent carbides include silicon carbide (SiC), which is sold as carborundum:

$$SiO_2(s) + 3C(s) \xrightarrow{2000°C} SiC(s) + 2CO(g)$$

The pure material is colorless, but iron impurities normally produce almost black crystals. Carborundum is used as an abrasive because it is very hard, has a structure like that of diamond, and fractures in a way that leaves its crystals with sharp edges (Fig. 19.17).

The interstitial carbides are formed by the direct reaction of a *d*-block metal and carbon at temperatures above 2000°C. In these compounds the C atoms, like the H atoms in metallic hydrides, occupy the gaps between the metal atoms (Fig. 19.18). Here, however, the C atoms pin the metal atoms together into a rigid structure. This arrangement results in extremely hard substances with melting points often well above 3000°C. Tungsten carbide (W_2C) is used for the cutting surfaces of drills, and iron carbide (Fe_3C) is an important component of steel.

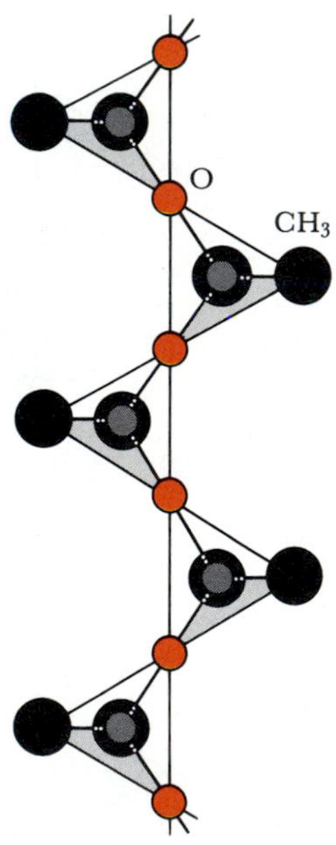

FIGURE 19.16 A typical silicone structure. The hydrocarbon groups give the substance its water-repelling quality. The silicon atom (gray) is located at the center of the $SiO_2(CH_3)_2$ tetrahedron.

Halides. All the elements in the group form liquid molecular tetrachlorides. The least stable is $PbCl_4$, which decomposes to the solid $PbCl_2$ when it is warmed to about 50°C. Carbon tetrachloride, although toxic, was once widely used as an industrial solvent; now, however, it is used primarily as the starting point for the manufacture of chlorofluorocarbons. It is formed by the action of chlorine on methane:

$$CH_4(l) + 4Cl_2(g) \xrightarrow{650°C} CCl_4(l) + 4HCl(g)$$

Chlorofluorocarbons are made from CCl_4 by successive replacement of Cl atoms by F atoms, using hydrogen fluoride in the presence of the catalyst SbF_5:

$$CCl_4(l) + HF(l) \xrightarrow{SbF_5} CFCl_3(g) + HCl(g)$$

Chlorofluorocarbons, which are sold as Freons, are used in refrigeration systems, and in many countries they are still used as aerosol propellants (see Box 5.1).

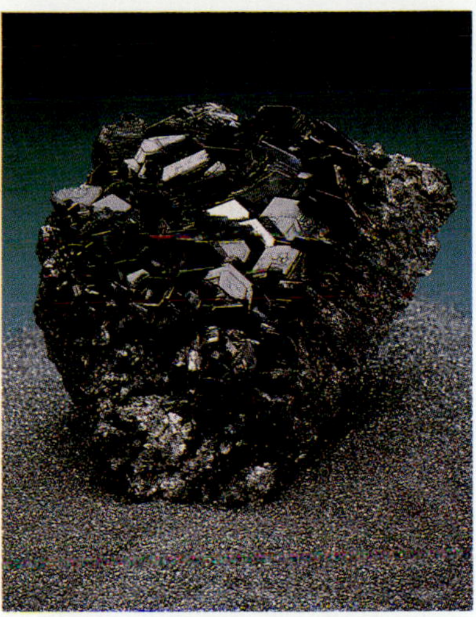

FIGURE 19.17 Carborundum crystals, showing the sharp fractured edges that give the substance its abrasive power.

Cyanides. The cyanide ion is CN^-. The parent acid from which the cyanides are derived is hydrogen cyanide, HCN, which is made by heating ammonia, methane, and air in the presence of a platinum catalyst:

$$2CH_4(g) + 2NH_3(g) + 3O_2(g) \xrightarrow{1100°C,\ Pt} 2HCN(g) + 6H_2O(g)$$

Hydrogen cyanide reacts with lime, CaO, to produce calcium cyanimid, $CaCN_2$:

$$CaO(s) + 2HCN(g) \longrightarrow CaCN_2(s) + CO(g) + H_2(g)$$

Calcium cyanimid is a nitrogen fertilizer and is used in the production of urea (another fertilizer) and as a starting material for the synthesis of organic compounds, most notably acrylic fibers, plastics, and paints. When HCN reacts with acetylene, HC≡CH, in the presence of a copper(I) chloride catalyst, acrylonitrile, $H_2C{=}CHCN$, is produced. This compound is the starting point for the production of the synthetic fibers sold as Acrilan and Orlon.

Cyanides are strong Lewis bases that form a range of complexes with *d*-metal ions. They are also famous as poisons. When they are ingested, they combine with enzymes that regulate the transfer of oxygen to cellular tissues; death results from suffocation.

U Exercises 19.73 and 19.75.

E Freon is a duPont trade name. Freon-12, CCl_2F_2, is used in air conditioners and refrigerators.

E Cyanide combines with the Fe^{3+} ion of cytochrome molecules and thereby blocks the process of oxidative phosphorylation. The oxidation of some hemoglobin by the addition of amyl nitrate creates an iron(III) complex that can withdraw some CN^- ions from the cytochromes, allowing them to function again at the expense of a reduction in oxygen carrying capacity of the blood (which in a short term is less important). The injection of sodium thiosulfate accelerates the conversion of CN^- to the less toxic SCN^- ion: $S_2O_3^{2-}(aq) + CN^-(aq) \rightarrow SO_3^{2-}(aq) + SCN^-(aq)$.

Hydrides. The hydrocarbons are so extensive that they have a chapter (Chapter 23) to themselves. Silicon forms a much smaller number of compounds with hydrogen, the simplest being silane (SiH_4), the analogue of methane. Silane is formed by the action of lithium aluminum hydride on silicon halides dissolved in diethyl ether:

$$SiCl_4 + LiAlH_4 \xrightarrow{\text{diethyl ether}} SiH_4 + LiCl + AlCl_3$$

Silane is much more reactive than methane and bursts into flames on contact with air. Although it is stable in pure water, it forms SiO_2 when a trace of alkali is present:

$$SiH_4(g) + 2H_2O(l) \xrightarrow{OH^-} SiO_2(s) + 4H_2(g)$$

The more complicated **silanes** such as $SiH_3{-}SiH_2{-}SiH_3$, the analogue of propane, decompose on standing.

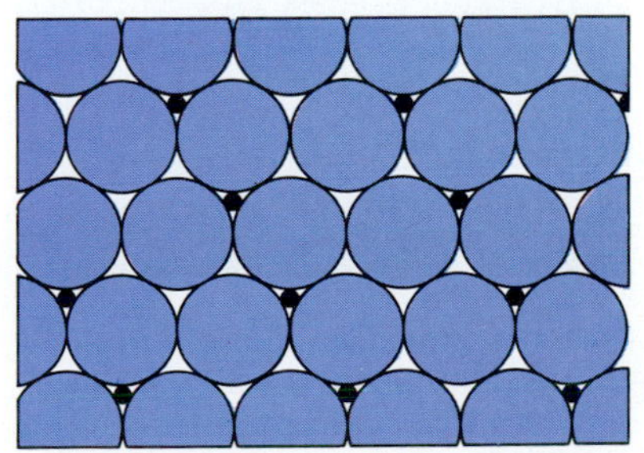

FIGURE 19.18 The structure of an interstitial carbide, in which carbon atoms lie between metal atoms, thereby producing a rigid structure.

TABLE 19.5 Group V elements

Valence configuration: ns^2np^3

Z	Name	Symbol	Molar mass, g/mol	Melting point, °C	Boiling point, °C	Density, g/cm³	Normal form
7	nitrogen	N	14.01	−210	−196	—	colorless gas
15	phosphorus	P	30.97	44	280	1.82	white nonmetal
33	arsenic	As	74.92	613s*	—	5.78	gray metalloid
51	antimony	Sb	121.75	631	1750	6.69	blue-white lustrous metalloid
83	bismuth	Bi	208.98	271	1650	8.90	white-pink metal

*The symbol s denotes that the element sublimes.

			IV	V	VI		He
Li	Be	B	C	N	O	F	Ne
			Si	P	S		
			Ge	As	Se		
			Sn	Sb	Te		
			Pb	Bi	Po		

GROUP V: NITROGEN AND PHOSPHORUS

The elements in Group V are shown in the margin and some of their properties are listed in Table 19.5. The most important members of the group are nitrogen and phosphorus, and we concentrate on them.

19.7 GROUP V ELEMENTS

The elements range in character from the nonmetals nitrogen and phosphorus to the largely metallic bismuth. This range of behavior is represented in their chemical properties: all the oxides of nitrogen and phosphorus are acidic and bismuth's oxide is basic. Chemical properties of the Group V elements are listed in Table 19.6.

Nitrogen. Nitrogen, the principal component of air (of which it forms 78% by volume), is a remarkable element. The considerable bond strength of the N_2 molecule makes it almost as inert as the noble gases. In combination with other elements, however, it has a rich and interesting chemistry. Nitrogen is obtained by the fractional distillation of liquid air. Air is cooled to below −196°C by a series of expansion and

D (1) Show the effects of liquid nitrogen on (hollow) rubber balls, rubber hose, flowers, bananas, etc. (2) Collect liquid oxygen from the tip of a test tube containing liquid nitrogen.

U Exercise 19.62.

E Nitrogen use. *J. Chem. Ed.*, **1990**, *67*, 215.

FIGURE 19.19 These pipes contain liquid nitrogen. The low temperature of liquid nitrogen causes the moisture from the atmosphere to form a thick ice covering, which serves as an insulation. The pipes appear shiny from the condensation of liquid oxygen.

TABLE 19.6 Chemical properties of the Group V elements

Reactant	Reaction with Group V elements, E
hydrogen	$2E(s,g) + 3H_2(g) \rightarrow 2EH_3(g)$ E = N, P
oxygen	$N_2(g) + xO_2(g) \rightarrow 2NO_x$
	$P_4(s) + (3 \text{ or } 5)O_2(g) \rightarrow P_4O_6(s)$ or $P_4O_{10}(s)$
	$4As(s) + 3O_2(g) \rightarrow As_4O_6(s)$
	$4E(s) + 3O_2(g) \rightarrow 2E_2O_3$ E = Sb, Bi
water	no reaction
halogen	$2E(s) + 3X_2(s,l,g) \rightarrow 2EX_3(s,l)$ E = P, As, Sb, Bi
	$2E(s) + 5X_2(s,l,g) \rightarrow 2EX_5(s)$ E = P, As, Sb

compression steps (Fig. 19.19). The liquid mixture is then warmed, and the nitrogen (b.p. −196°C) boils off. This step leaves the higher-boiling-point oxygen (b.p. −183°C). Any oxygen that is mixed with the nitrogen gas is removed by passing the gas over hot copper, which reacts to form copper(II) oxide.

Nitrogen's main use is as the raw material for the Haber synthesis of ammonia. This process is the principal means of **fixing** the element, that is, of combining it with other elements (with hydrogen, for example, in NH_3). Once fixed, nitrogen can be converted to other compounds, such as nitric acid, fertilizers, explosives, and plastics and can be used by plants for their metabolism. Natural fixation reactions can be initiated by lightning, which converts some nitrogen to its oxides; these compounds are then washed into the soil by rain. Nitrogen is also fixed by bacteria found in nodules on the roots of leguminous plants, which include clover, beans, peas, and alfalfa (Fig. 19.20). An intensely active field of research is the search for catalysts that can mimic the bacteria and fix nitrogen at normal temperatures.

FIGURE 19.20 The bacteria that inhabit these nodules on the root of a pea plant are responsible for fixing atmospheric nitrogen and making it available to the plant.

U Exercise 19.61.

Nitrogen differs sharply from its congeners. It is highly electronegative ($\chi = 3.0$, the same as that of chlorine); and, because its atoms are small, it can form multiple bonds by using its *p* orbitals. However, a nitrogen atom is too small to bond to more than four atoms, so analogues of PCl_5 do not exist. It is also unusual in having one of the widest ranges of oxidation numbers of any element: nitrogen compounds are known for each whole-number oxidation number from −3 (in NH_3) to +5 (in nitric acid and the nitrates) and for some fractional oxidation numbers too (such as the $-\frac{1}{3}$ in the azide ion, N_3^-).

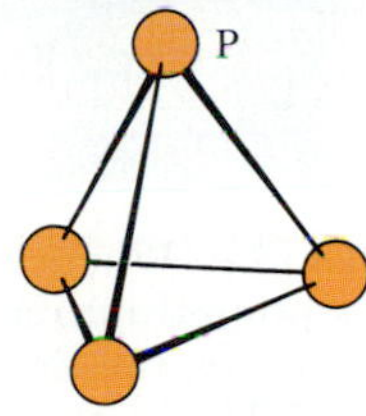

3 Phosphorus, P_4

V Video 13: Barking dogs.

Phosphorus. Phosphorus occurs widely as various kinds of phosphate rocks—particularly the apatites, which are forms of calcium phosphate, $Ca_3(PO_4)_2$. The element is obtained by heating the rocks with carbon in an electric furnace in the presence of sand. The latter removes the calcium by forming a molten slag of calcium silicate:

$$2Ca_3(PO_4)_2(s) + 6SiO_2(s) + 10C(s) \xrightarrow{\Delta} P_4(g) + 6CaSiO_3(l) + 10CO(g)$$

The phosphorus vapor is condensed and forms white phosphorus, a soft, white, reactive, poisonous molecular solid consisting of tetrahedral P_4 molecules (**3**). The reactivity of this allotrope is partly due to the strain associated with the acute angles between the bonds. White phosphorus bursts into flame on contact with air (the name phosphorus comes from the Greek words for "the light bringer") and is normally stored under water.

When it is heated in the absence of air, white phosphorus changes into red phosphorus (Fig. 19.21). This denser allotrope is less reactive; but it can be ignited by friction, so it is used in matches. Its structure is uncertain, but it may consist of chains of linked P_4 tetrahedra.

FIGURE 19.21 The white (top) and red (bottom dish) allotropes of phosphorus. White phosphorus is stored under water because it ignites in air.

19.8 COMPOUNDS WITH HYDROGEN AND THE HALOGENS

The Group V elements have negative oxidation numbers in their binary compounds with hydrogen (−3 for N in NH_3, for instance). Hence they are formed by reduction of the elements, and they tend to be reducing agents themselves. By far the most important is ammonia.

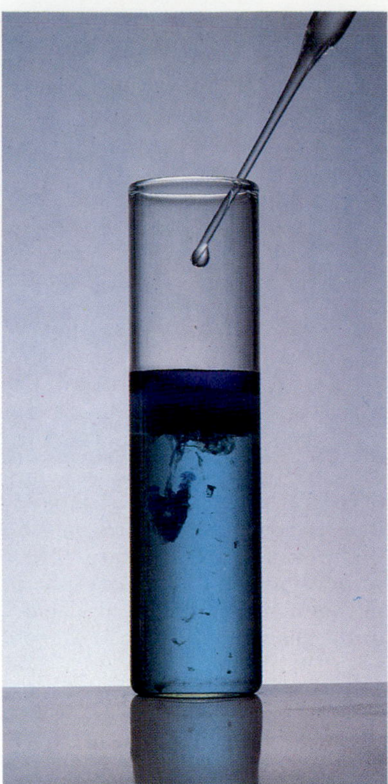

FIGURE 19.22 When aqueous ammonia is poured into aqueous copper(II) sulfate, the dark blue complex $Cu(NH_3)_4^{2+}$ is formed by a Lewis acid-base reaction.

E Small quantities of ammonia are present naturally in the atmosphere as a result of the bacterial decomposition of organic matter in the absence of air, which typically occurs on lake and river beds and in swamps. The pungency of heated ammonium chloride (*sal ammoniac*) was known to the Ammonians, the worshippers of the Egyptian god Amun (hence the name, ammonia):

$$NH_4Cl(s) \xrightarrow{\Delta} NH_3(g) + HCl(g)$$

U Exercise 19.68.

? Which of the following compounds has the strongest smell of ammonia: $(NH_4)_2CO_3$, $NH_4CH_3CO_2$, or NH_4Cl? Explain. (*Hint*: Which salt has the strongest Brønsted base?)

D Colorful complex ions in ammonia. CD2, 41. Ammonia as a reducing agent. CD1, 72.

Ammonia. The **Haber-Bosch process** for the synthesis of ammonia (Section 13.8) stands at the head of the industrial nitrogen chain:

$$N_2(g) + 3H_2(g) \xrightarrow{400° \text{ to } 600°\text{C, } 150 \text{ to } 600 \text{ atm, Fe}} 2NH_3(g)$$

Nitrogen for the synthesis is obtained from the fractional distillation of air, and the hydrogen is obtained from the reforming and shift reactions that start with methane and water (these reactions were described in Section 18.1).

Ammonia is a pungent, toxic gas that condenses to a colorless liquid at −33°C. The liquid resembles water in its physical properties, including its ability to act as a solvent. Gaseous ammonia is very soluble in water because the NH_3 molecules are polar and can form hydrogen bonds to H_2O molecules. Although the aqueous solution is sometimes called ammonium hydroxide, the compound NH_4OH has not been isolated. Ammonia is a weak Brønsted base ($pK_b = 4.75$) in water and a reasonably strong Lewis base, particularly toward *d*-block elements. An example of its ability to form complexes with the latter is its reaction with $Cu^{2+}(aq)$ ions to give a deep blue complex (Fig. 19.22):

$$Cu^{2+}(aq) + 4NH_3(aq) \longrightarrow Cu(NH_3)_4{}^{2+}(aq)$$

We shall see many of these complexes in Chapter 21.

The neutralization of aqueous ammonia with a Brønsted acid gives the corresponding ammonium salt, in which the cation is the ammonium ion NH_4^+. Solutions of these salts are acidic, because NH_4^+ is a weak Brønsted acid. Ammonium salts decompose when heated:

$$(NH_4)_2CO_3 \xrightarrow{\Delta} 2NH_3(g) + CO_2(g) + H_2O(g)$$

The pungent smell of decomposing ammonium carbonate is the reason for its use in smelling salts. If the anion of an ammonium salt is oxidizing, then the ammonium cation may be oxidized:

$$NH_4NO_3(s) \xrightarrow{250°\text{C}} N_2O(g) + 2H_2O(g)$$

$$2NH_4NO_3(s) \xrightarrow{300°\text{C}} 2N_2(g) + O_2(g) + 4H_2O(g)$$

This second reaction can be explosively violent and accounts for the use of ammonium nitrate as a component of dynamite. Ammonium nitrate has a high nitrogen content (33.5% by mass) and is very soluble in water. These two characteristics make it attractive as a fertilizer, and that is now its principal use.

Hydrazine. Hydrazine (NH_2NH_2) is an oily, colorless liquid in which nitrogen has an oxidation number of −2. It is prepared by the gentle oxidation of ammonia with alkaline hypochlorite solution:

$$2NH_3(aq) + OCl^-(aq) \longrightarrow N_2H_4(aq) + Cl^-(aq) + H_2O(l)$$

Its physical properties are very similar to those of water (its melting point is 1.5°C, its boiling point 113°C). However, it is dangerously explosive and, for this reason, is normally used in aqueous solution. The exothermicity of its combustion makes it a valuable rocket fuel: the Apollo lunar missions used a mixture of liquid N_2O_4 and derivatives of hydrazine (such as methyl hydrazine, CH_3NHNH_2) when landing on and leaving the moon. These two liquids—with the nitrogen in one having oxidation number −2 and in the other +4—ignite as soon as they mix, thereby producing a large volume of gas:

$$4CH_3NHNH_2(l) + 5N_2O_4(l) \longrightarrow 9N_2(g) + 12H_2O(g) + 4CO_2(g)$$

A valuable application of hydrazine back on Earth is to eliminate dissolved oxygen from the water used in high-pressure, high-temperature steam furnaces (to prevent corrosion):

$$N_2H_4(aq) + O_2(g) \longrightarrow N_2(aq) + 2H_2O(l)$$

E Phosphine can be prepared by proton transfer to its conjugate base, the phosphide ion, P^{3-}. In practice, the action of water (the proton donor) on calcium phosphide may be used. Alternatively, it may be prepared by proton transfer from a phosphonium ion, the conjugate acid of phosphine. In practice, the action of aqueous potassium hydroxide on phosphonium iodide may be used. Organophosphine compounds have odors that make the corresponding amines resemble fragrant perfumes!

Phosphine. The hydrogen compounds of other members of Group V are much less stable than ammonia, and they decrease in stability down the group. Phosphine, PH_3, is a poisonous gas that smells slightly of garlic and bursts into flames in air if it is slightly impure. It is much less soluble than ammonia in water, a fact that again points to the important role of hydrogen bonding in hydration: NH_3 can form hydrogen bonds to water, but PH_3 does not. Aqueous solutions of phosphine are neutral, for PH_3 has an extremely weak tendency to accept a proton ($pK_b = 27.4$).

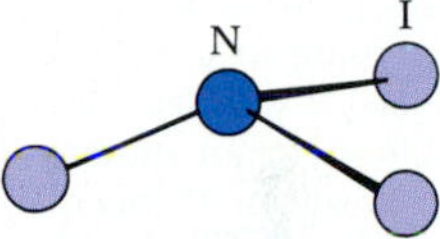

4 Nitrogen triiodide

Nitrogen halides. None of the nitrogen halides is prepared by a direction reaction of nitrogen gas with the halogen. Nitrogen trifluoride, NF_3, is the most stable halide of the series; nitrogen triiodide, NI_3, (**4**), which is formed as the compound $NI_3{\cdot}NH_3$ in the reaction of concentrated aqueous ammonia and iodine, is shock sensitive and decomposes explosively when lightly touched (Fig. 19.23). Nitrogen trifluoride does not react with water; however, nitrogen trichloride, NCl_3, does react with water, forming ammonia and hypochlorous acid.

U Exercises 19.65 and 19.78.

? Why is the fixing of nitrogen by burning magnesium followed by reaction with water not economically viable?

D Explosive decomposition of nitrogen triiodide. BZS1, 1.39.

Nitrides and azides. Lithium and the Group II elements form ionic nitrides by direct combination. Magnesium nitride is formed together with the oxide when magnesium is burned in air:

$$3Mg(s) + N_2(g) \longrightarrow Mg_3N_2(s)$$

Magnesium nitride, like all nitrides, dissolves in water to produce ammonia and the corresponding hydroxide:

$$Mg_3N_2(s) + 6H_2O(l) \longrightarrow 3Mg(OH)_2(s) + 2NH_3(g)$$

The nitride ion is a strong base, accepting three protons from water to form ammonia.

The azide ion is a polyatomic anion of nitrogen, N_3^-. Its most common salt, sodium azide, NaN_3, is prepared from nitrous oxide and molten sodium amide:

$$N_2O(g) + 2NaNH_2(l) \xrightarrow{175°C} NaN_3(l) + NaOH(l) + NH_3(g)$$

FIGURE 19.23 The compound $NI_3{\cdot}NH_3$ explodes when it is lightly touched. Nitrogen triiodide, NI_3, is so unstable that it has never been isolated.

The azide ion is a Brønsted base, accepting protons to form its conjugate acid, hydrazoic acid, HN_3. Hydrazoic acid is unstable and readily detonates; it is a weak acid ($pK_a = 4.77$), similar in strength to acetic acid ($pK_a = 4.74$). Lead azide, $Pb(N_3)_2$, like most azide salts, is shock sensitive and is used as a detonator for some explosives. Sodium azide is used in airbags that, increasingly, are being fitted as a safety feature in automobiles. When the vehicle decelerates suddenly (as a result of a collision), an electric device detonates the azide, and the sudden surge of nitrogen inflates the bag to cushion the impact (Fig. 19.24).

Phosphorus halides. Phosphorus trichloride, PCl_3, and phosphorus pentachloride, PCl_5, are the two most important halides of phosphorus.

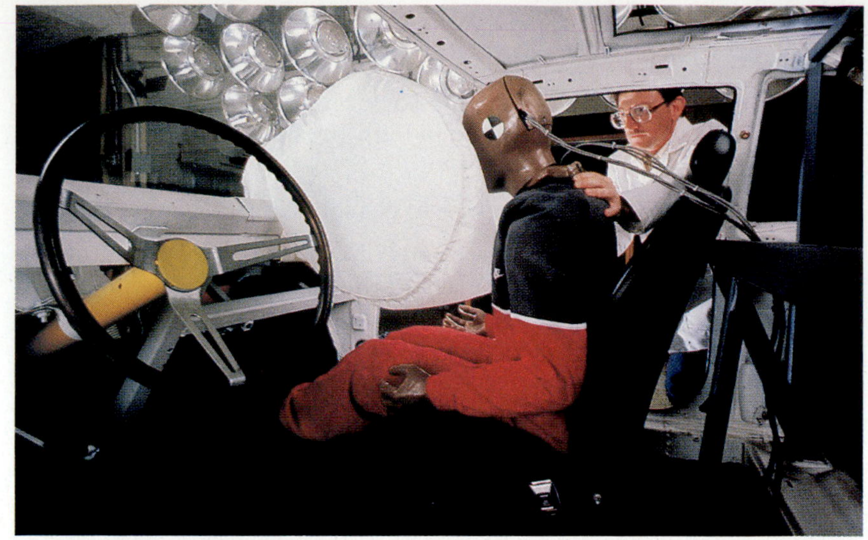

FIGURE 19.24 The rapid decomposition of sodium azide results in the formation of nitrogen gas. A reaction triggered electrically by a switch that responds to sudden deceleration is used in airbags fitted to a safety device in automobiles.

D Reaction of white phosphorus and chlorine. BZS1, 1.29.

E An air-bag system consists of a series of detectors and a "squib." The detector is activated electronically and discriminates between a collision and an incidental or mischievous blow to the front end of the automobile. When a true collision occurs, the detector activates the squib, which produces a shower of red-hot sparks. The heat causes the thermal decomposition of sodium azide to sodium metal and nitrogen gas: $2NaN_3(s) \rightarrow 2Na(s) + 3N_2(g)$. The detector initiates the decomposition within 10 ms of collision, and the air bag is filled with nitrogen within 30 ms. For large air bags, an aspirator is added that draws in air during the rapid expansion of the air bag. One design feature is that the air bag should be porous; whereas it must inflate in 20 to 40 ms, it should deflate in 100 to 200 ms so as to avoid a backlash injury from the air bag itself! The sodium azide, along with other components present for safety and design, is retained in a canister with a filter. The filter avoids solid particles from entering the air bag; only nitrogen gas enters the air bag. Another component inside the canister is a metal oxide, such as iron(III) oxide, that reacts with sodium to produce sodium oxide.

E Chemistry of air bags. *J. Chem. Ed.*, **1990**, *67*, 61.

? What products will form when phosphorus pentachloride reacts with water?

N See Appendix 2E for a list of the top 50 chemicals.

The trichloride is prepared by direct chlorination of phosphorus, and the liquid is distilled out of the reaction mixture before it can react further. It fumes in moist air and reacts readily with water to form phosphorous acid:

$$PCl_3(l) + 3H_2O(l) \longrightarrow H_3PO_3(s) + 3HCl(g)$$

This reaction exhibits a common feature of the reactions of the nonmetal halides: they react with water to give an oxoacid without undergoing a change in oxidation number.

Phosphorus trichloride also takes part in numerous reactions with other substances. It is a major intermediate for the production of phosphorus compounds for use in pesticides, oil additives, and flame retardants. Phosphorus pentachloride is made by allowing the trichloride to react with more chlorine.

An interesting feature of phosphorus pentachloride is its structure: it is an *ionic* solid of tetrahedral PCl_4^+ cations and octahedral PCl_6^- anions, but it vaporizes to a gas of trigonal bipyramidal PCl_5 molecules. Phosphorus pentabromide is also molecular in the vapor and ionic as the solid, but in the solid the anions are simply Br^- ions, presumably because of the difficulty of fitting six bulky Br atoms around a central P atom.

19.9 GROUP V OXIDES AND OXOACIDS

Although the nitrogen oxides can be confusing at first sight, we shall see that their properties and interconversions can be understood by keeping track of oxidation numbers, which range from +1 to +5 (Table 19.7). In atmospheric chemistry, where the oxides play an important role in both maintaining and polluting the atmosphere, they are referred to as NO_x. All nitrogen oxides are acidic. As we work through the various oxidation states of nitrogen, we shall also meet the nitrogen oxoacids (Table 19.7).

Phosphorus compounds make their appearance in the top 50 chemicals by virtue of phosphoric acid, which ranks seventh in U.S. produc-

TABLE 19.7 Oxides and oxoacids of nitrogen

Oxidation number	Oxide: Formula	Oxide: Name	Corresponding acid: Formula	Corresponding acid: Name
5	N_2O_5	dinitrogen pentoxide	HNO_3	nitric acid
4	NO_2*	nitrogen dioxide		
	N_2O_4	dinitrogen tetroxide		
3	N_2O_3	dinitrogen trioxide	HNO_2	nitrous acid
2	NO	nitrogen oxide; nitric oxide		
1	N_2O	dinitrogen oxide; nitrous oxide	$H_2N_2O_2$	hyponitrous acid

*$2NO_2 \rightleftharpoons N_2O_4$.

tion. Its principal use is in the manufacture of fertilizers, which accounts for 85% of its production. Phosphate (fertilizer) production consumes two-thirds of all the sulfuric acid produced. A structural theme in the phosphorus oxides is the tetrahedral PO_4 unit, similar to the unit that occurs in the oxides of its neighbor silicon.

E Almost all the N_2O in the atmosphere is a result of natural processes. Almost all the NO and NO_2 in the atmosphere result from the combustion processes required to sustain our technological society. Nitrous oxide was once used to sedate pregnant women during labor; however, because of the reduced levels of oxygen in the bloodstream, and possible oxygen deficiencies in the baby, the practice has ceased.

? Why is N_2O so soluble in water?

D Preparation and properties of nitrogen(II) oxide. BZS2, 6.15.

E The oxides of nitrogen and their detection in automotive exhaust. *J. Chem. Ed.*, **1972**, *49*, 21.

Dinitrogen oxide. Dinitrogen oxide (or nitrous oxide), N_2O, is the oxide of nitrogen with the lowest oxidation number (+1). It is formed by gently heating ammonium nitrate:

$$NH_4NO_3(s) \xrightarrow{240^\circ C} N_2O(g) + 2H_2O(g)$$

A colorless, very soluble gas with a slight odor, it is used as an anesthetic. Because it is tasteless, nontoxic in small amounts, and dissolves readily in fats, it is also used as a foaming agent and propellant for whipped cream. Dinitrogen oxide is quite unreactive; it does not react at ordinary temperatures with the halogens, the alkali metals, or ozone. However, organic matter burns in it after being ignited.

Nitrogen oxide. Nitrogen oxide, which is more commonly called nitric oxide, NO, is an oxide in which nitrogen's oxidation number is +2. It is prepared industrially by the catalytic oxidation of ammonia:

$$4NH_3(g) + 5O_2(g) \xrightarrow{1000^\circ C,\ Pt} 4NO(g) + 6H_2O(g)$$

Its formation from the elements is endothermic

$$N_2(g) + O_2(g) \longrightarrow 2NO(g) \qquad \Delta H^\circ = +180 \text{ kJ}$$

partly as a result of the high dissociation enthalpy of nitrogen. Hence, by Le Chatelier's principle, the equilibrium composition shifts toward NO as the temperature is raised. As a result, NO is formed in the hot exhausts of airplane and automobile engines.

In the laboratory, nitric oxide can be prepared by reducing a nitrite with a mild reducing agent such as I^-:

$$2NO_2^-(aq) + 2I^-(aq) + 4H^+(aq) \longrightarrow 2NO(g) + I_2(aq) + 2H_2O(l)$$

Nitric oxide is a colorless gas that reacts rapidly on exposure to air,

FIGURE 19.25 Nitric oxide is a colorless gas; but when it mixes with the oxygen of the air, it forms the brown gas NO_2.

D Place several copper pellets in a tall Erlenmeyer flask and add a small quantity of concentrated nitric acid. When the reaction has ceased, stopper the flask or transfer to a fume hood.

E Effects of water and nitrogen dioxide on the stratospheric ozone shield. *J. Chem. Ed.*, **1972**, *49*, 722.

when it is oxidized to nitrogen dioxide (Fig. 19.25):

$$2NO(g) + O_2(g) \longrightarrow 2NO_2(g)$$

Both NO and NO_2 are odd-electron, paramagnetic molecules. Nitric oxide contributes to the problem of acid rain and the formation of smog. It also contributes, like the chlorofluorocarbons, to the destruction of the ozone layer (Box 5.1).

Nitrogen dioxide. Nitrogen dioxide, in which nitrogen has the oxidation number +4, is a choking, poisonous, brown gas that contributes to the color of smog. In the gas phase it exists in equilibrium with its colorless dimer N_2O_4; in the solid only the dimer exists. When it dissolves in water, it reacts to form nitric acid:

$$3NO_2(g) + H_2O(l) \longrightarrow 2HNO_3(aq) + NO(g)$$

Nitrogen dioxide in the atmosphere undergoes the same reaction and contributes to the formation of acid rain (Section 20.3). Industrially the dioxide is made by oxidation of nitric oxide obtained from the oxidation of ammonia. In the laboratory it is prepared either by the action of concentrated nitric acid on copper metal (Fig. 3.18) or by the thermal decomposition of lead nitrate:

$$Cu(s) + 4HNO_3(aq, conc) \longrightarrow Cu(NO_3)_2(aq) + 2NO_2(g) + 2H_2O(l)$$

$$2Pb(NO_3)_2(l) \xrightarrow{400°C} 4NO_2(g) + 2PbO(s) + O_2(g)$$

Nitrogen dioxide initiates a complex sequence of photochemical reactions in the atmosphere. Solar radiation of wavelength less than 350 nm dissociates NO_2 molecules into NO molecules and oxygen atoms:

$$NO_2 \xrightarrow{\text{ultraviolet radiation}} NO + O$$

The highly reactive O atom reacts with hydrocarbons to produce hydroxyl radicals:

$$CH_4 + O \longrightarrow CH_3 + OH$$

or with oxygen to form ozone, a very strong, irritating oxidizing agent:

$$O + O_2 \longrightarrow O_3$$

Nitrous acid. When a mixture of the odd-electron molecules NO and NO_2 is cooled to −20°C, they combine to form dinitrogen trioxide, N_2O_3 (**5**). This compound, in which the oxidation number of the nitrogen is +3, is unstable as a gas but can be collected as a dark blue liquid (Fig. 19.26). It is the anhydride of nitrous acid (HNO_2) and forms that acid when it dissolves in water:

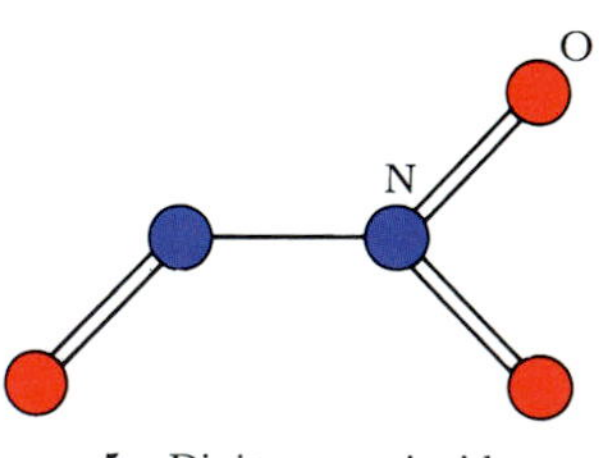

5 Dinitrogen trioxide

$$N_2O_3(g) + H_2O(l) \longrightarrow 2HNO_2(aq)$$

Nitrous acid has not been isolated in the pure form but is quite widely used as an aqueous solution.

Nitrous acid is a weak acid ($pK_a = 3.37$) and the parent of the nitrites, which contain the angular nitrite ion (NO_2^-). However, in practice these salts are more easily produced by the reduction of nitrates with hot metal; for example,

$$KNO_3(s) + Pb(s) \xrightarrow{350°C} KNO_2(s) + PbO(s)$$

Most nitrites are soluble in water and mildly toxic. Despite their toxicity, nitrites have been used for cosmetic treatment of meat because they inhibit the oxidation of the blood (which turns the meat brown) and form a pink complex with the hemoglobin. This compound is responsible for the pink color of ham and other cured meat.

Nitric acid. Nitric acid (HNO_3) ranks thirteenth in annual U.S. production and is used extensively in the production of fertilizers and explosives. It is produced by the three-stage **Ostwald process**:

1. Oxidation of ammonia, from oxidation number −3 to +2:

$$4NH_3(g) + 5O_2(g) \xrightarrow{850^\circ C,\ 5\ atm,\ Pt/Rh} 4NO(g) + 6H_2O(g)$$

2. Oxidation of nitric oxide, from oxidation number +2 to +4:

$$2NO(g) + O_2(g) \longrightarrow 2NO_2(g)$$

3. Reaction with water, from oxidation number +4 to +5 and +2:

$$3NO_2(g) + H_2O(l) \longrightarrow 2HNO_3(aq) + NO(g)$$

Nitric acid is a colorless liquid that boils at 86°C and is normally used in aqueous solution. Concentrated nitric acid is often pale yellow as a result of partial decomposition of the acid to NO_2.

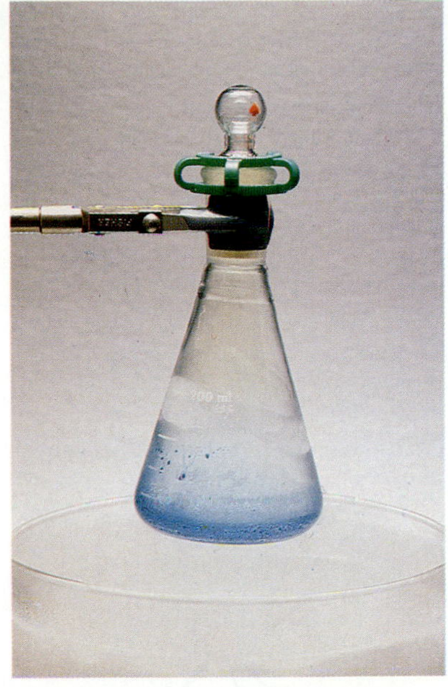

FIGURE 19.26 Dinitrogen trioxide (N_2O_3) condenses to a deep blue liquid that freezes at −100°C to a pale blue solid. On standing, it turns green as a result of partial decomposition into nitrogen dioxide (not shown).

D Catalytic oxidation of ammonia. BZS2, 6.26.

E Nitric acid, chemical of the month. *J. Chem. Ed.*, **1984**, *61*, 174.

D The darker brown a bottle of concentrated nitric acid appears, the older it is. Show a new and old bottle of concentrated nitric acid if your stockroom has them available. Acidic properties of nitrogen(V) oxide. BZS3, 8.16.

U Exercise 19.82.

▼ **EXAMPLE 19.2 Suggesting a preparation of a nitrogen oxide**

Suggest a method of preparing dinitrogen pentoxide, N_2O_5.

STRATEGY We know that the oxides of nitrogen are acidic, so we can expect N_2O_5 to be the anhydride of an oxoacid. If we could identify which one, then we could think of producing the oxide by dehydration of the acid. Therefore, identify the acid of which N_2O_5 is the anhydride (by adding one or more H_2O units to N_2O_5). Alternatively, identify the acid by determining the oxidation number of the nitrogen and identifying the oxoacid with the same oxidation number.

SOLUTION Adding H_2O to N_2O_5 gives $H_2N_2O_6$, which is two HNO_3 units. The oxidation number of nitrogen in N_2O_5 is +5, the same as in HNO_3. Hence, we can expect to produce the oxide by dehydration of nitric acid. The reaction that is used in practice is

$$2HNO_3(l) \xrightarrow{\Delta,\ P_4O_{10}} N_2O_5(s) + H_2O(l)$$

EXERCISE E19.2 What other property, other than Brønsted acidity, can be anticipated for nitric acid?

[*Answer*: oxidizing agent] ▲

Because nitrogen has its highest oxidation number (+5) in HNO_3, nitric acid is an oxidizing agent as well as a Brønsted acid. Some metals (particularly aluminum) become protected by an oxidized layer when they are attacked by nitric acid. The noble metals gold, platinum, iridium, and rhodium are not attacked at all. However, they are attacked if Cl^- ions are also present (as they are in aqua regia, Section 15.8) as a result of the formation of the complex $AuCl_4^-$.

U Exercise 19.81.

E The solubility of nitrates stems from the low charge of the large anion (which results in low lattice enthalpies) and the excellent ability of NO_3^- to form hydrogen bonds with H_2O molecules (which results in a high hydration enthalpy).

E Sodium nitrate could also be used, but the smaller cation renders it more hygroscopic than potassium nitrate.

? What substance could be used to detonate nitroglycerin? (See p. 747.)

E Nobel used the porous granulated siliceous rock *kieselguhr* in his formulation of dynamite. Modern formulations use a mixture of wood flour, ammonium nitrate, sulfur, and sodium nitrate.

V Video 47: The oxidation of phosphorus. Video 48: A rapid oxidation of phosphorus.

D Spontaneous combustion of white phosphorus. BZS1, 1.31.

Nitric acid is the parent of the nitrates, which contain the trigonal planar nitrate ion (NO_3^-). Most nitrates are soluble in water. When heated, nitrates of heavy metals decompose (much like the carbonates) into nitrogen dioxide and the metal oxide; for example,

$$2Pb(NO_3)_2(s) \xrightarrow{\Delta} 2PbO(s) + 4NO_2(g) + O_2(g)$$

The nitrates of the lighter metals decompose into the nitrite and oxygen; for example,

$$2KNO_3(s) \xrightarrow{\Delta} 2KNO_2(s) + O_2(g)$$

This reaction makes nitrates useful as an additional supply of oxygen in matches, flares, and explosives.

Nitric acid is used in manufacturing explosives on account of its reaction with glycerol, $C_3H_5(OH)_3$, and toluene, $C_6H_5CH_3$. The former produces nitroglycerine, the explosive in dynamite:

$$C_3H_5(OH)_3(l) + 3HNO_3(l) \xrightarrow{H_2SO_4} C_3H_5(NO_2)_3(s) + 3H_2O(l)$$

Dynamite was invented by Alfred Nobel (1833–1896), the founder of the Nobel prizes, and consists of nitroglycerin absorbed in a porous material. When detonated with a blasting cap, it decomposes exothermically into gaseous products:

$$4C_3H_5(NO_2)_3(s) \longrightarrow 6N_2(g) + 12CO(g) + 10H_2O(g) + O_2(g) \quad \Delta H^\circ = -5.7\ \text{MJ}$$

This large, rapid increase in the number of gaseous molecules accounts for the formation of a shock wave. Trinitrotoluene (TNT) is produced by the reaction

$$C_6H_5CH_3(l) + 3HNO_3(l) \xrightarrow{H_2SO_4} C_6H_2(NO_2)_3CH_3(s) + 3H_2O(l)$$

The basis of its destructive power is the same as that of nitroglycerin.

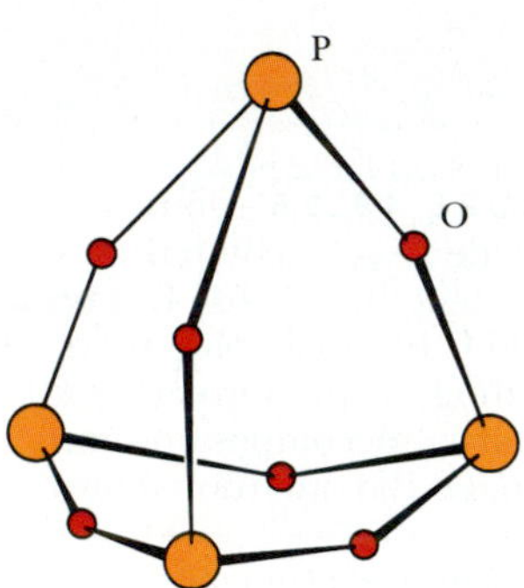

6 Phosphorus(III) oxide

Phosphorus(III) oxide and phosphorous acid. White phosphorus burns in a limited supply of air to form phosphorus(III) oxide, P_4O_6 (**6**):

$$P_4(s, \text{white}) + 3O_2(g) \longrightarrow P_4O_6(s)$$

In structure, its molecules are closely related to the P_4 molecules of phosphorus, but an O atom lies between each pair of P atoms. Phosphorus(III) oxide is the anhydride of phosphorous acid, H_3PO_3 (**7**) and is converted to that acid by cold water.

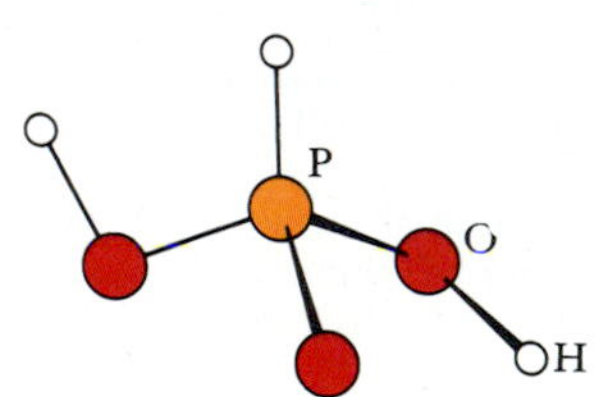

7 Phosphorous acid

$$P_4O_6(s) + 6H_2O(l) \longrightarrow 4H_3PO_3(aq)$$

Although its formula suggests that it should be triprotic, H_3PO_3 is in fact *di*protic because one of the H atoms is attached directly to the P atom.

Phosphorus(V) oxide and phosphoric acid. When phosphorus burns in an ample supply of air, it is oxidized to phosphorus(V) oxide, P_4O_{10} (**8**):

$$P_4(s, \text{white}) + 5O_2(g) \longrightarrow P_4O_{10}(s)$$

The oxide is a white solid that has a greater affinity for water, so it is widely used in the laboratory as a drying agent (to maintain a moisture-free atmosphere inside a container) and as a dehydrating agent (to remove water from compounds; for example, in the preparation of acid anhydrides).

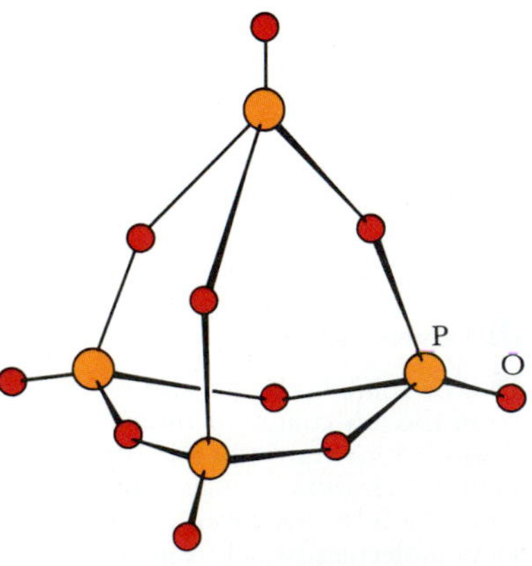

8 Phosphorus(V) oxide

Phosphorus(V) oxide is the anhydride of phosphoric acid, H_3PO_4 (**9**), a triprotic Brønsted acid:

$$P_4O_{10}(s) + 6H_2O(l) \longrightarrow 4H_3PO_4(aq)$$

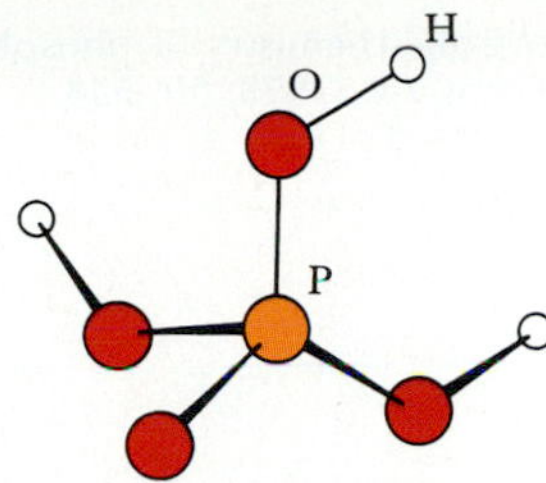

9 Phosphoric acid

Phosphoric acid is used primarily for the production of fertilizers, as a food additive, and in detergents. The tart taste of many soft drinks is due to the presence of small amounts of phosphoric acid. Pure phosphoric acid is a colorless solid of melting point 42°C, but in the laboratory it is normally a syrupy liquid as a result of the water it has absorbed. Phosphoric acid is usually purchased as 85% H_3PO_4, with a density of about 1.7 g/mL. Its high viscosity can be traced to extensive hydrogen bonding between neighboring H_3PO_4 molecules. Although phosphorus has a high oxidation number (+5), the acid shows appreciable oxidizing power only at temperatures above about 350°C; hence it may be used where nitric acid and sulfuric acid would be too oxidizing.

E Most phosphoric acid is manufactured by the action of sulfuric acid on phosphate rock: $Ca_3(PO_4)_2(s) + 3H_2SO_4(l) \rightarrow 2H_3PO_4(l) + 3CaSO_4(s)$. Because fluorine is also commonly present in the rock (as fluorapatite), a more realistic version of the process is $Ca_5F(PO_4)_3(s) + 5H_2SO_4(l) + 10H_2O(l) \rightarrow 3H_3PO_4(l) + 5CaSO_4 \cdot 2H_2O(s) + HF(g)$. The HF is pumped off, mixed with silicon tetrafluoride, and the resulting H_2SiF_6 used for the fluoridation of water supplies.

U Exercises 19.66 and 19.80.

E Phosphorus(V) oxide is used as a drying agent for organic liquids.

? What product does phosphorus(V) oxide form when it is used as a drying agent for organic solvents? Is the product soluble?

U Exercise 19.67.

Phosphoric acid is the parent of the phosphates, which contain the tetrahedral PO_4^{3-} anion and are of great biological and commercial importance. Phosphate rock is mined in huge quantities (especially in Florida and Morocco). After being crushed, it is treated with sulfuric acid to give a mixture of sulfates and phosphates called superphosphate, a major fertilizer:

$$Ca_3(PO_4)_2(s) + 2H_2SO_4(l) \longrightarrow \underbrace{2CaSO_4(s) + Ca(H_2PO_4)_2(s)}_{\text{Superphosphate}}$$

The phosphate rock may be treated with phosphoric acid rather than with sulfuric acid to produce a mixture with a higher phosphate content:

$$Ca_3(PO_4)_2(s) + 4H_3PO_4(l) \longrightarrow 3Ca(H_2PO_4)_2(s)$$

The resulting mixture of calcium phosphates is sold as triple superphosphate. We have already described in Section 18.6 the role of calcium phosphate in skeletons and teeth.

U Exercise 19.70.

Polyphosphates. When phosphoric acid is heated, it undergoes a condensation reaction:

$$\text{HO—}\overset{\overset{\displaystyle O}{\|}}{\underset{\underset{\displaystyle OH}{|}}{P}}\text{—O—H} + \text{HO—}\overset{\overset{\displaystyle O}{\|}}{\underset{\underset{\displaystyle OH}{|}}{P}}\text{—OH} \xrightarrow{\Delta} \text{HO—}\overset{\overset{\displaystyle O}{\|}}{\underset{\underset{\displaystyle OH}{|}}{P}}\text{—O—}\overset{\overset{\displaystyle O}{\|}}{\underset{\underset{\displaystyle OH}{|}}{P}}\text{—OH} + H_2O$$

The product, $H_4P_2O_7$, is pyrophosphoric acid. Further heating gives even more complicated products that have chains and rings of PO_4 groups and are called **polyphosphoric acids**. In this regard, phosphorus is showing a resemblance to the glass-forming power of its Group IV neighbor silicon.

U Exercise 19.69.

Polyphosphates, the salts of the polyphosphoric acids, are added to detergents to sequester the cations responsible for hardness and to prevent their precipitation (see Section 11.3). The most common salt added to detergents is sodium tripolyphosphate, $Na_5P_3O_{10}$ (**10**); it is the principal component of softeners such as Calgon. This polyphosphate anion is a relatively strong Brønsted base, buffering the solution of the wash water to a pH at which the detergent molecules can act most effectively. In addition, the ions attach to the surfaces of dirt

$$Na^+\ ^-\text{O—}\overset{\overset{\displaystyle O}{\|}}{\underset{\underset{\displaystyle O^-}{|}}{P}}\text{—O—}\overset{\overset{\displaystyle O}{\|}}{\underset{\underset{\displaystyle O^-}{|}}{P}}\text{—O—}\overset{\overset{\displaystyle O}{\|}}{\underset{\underset{\displaystyle O^-}{|}}{P}}\text{—O}^-\ Na^+ \quad (Na^+\ Na^+\ Na^+)$$

10 Sodium tripolyphosphate, $Na_5P_3O_{10}$

E Bioorganic chemistry of phosphorus. *J. Chem. Ed.*, **1973**, *50*, 538.

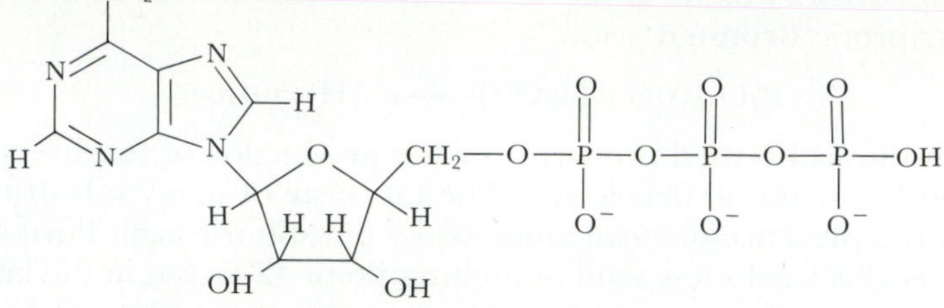

11 Adenosine triphosphate

particles and give them a negative charge. Earlier we saw what happens when sodium silicate is used in this way: neighboring particles repel each other and disperse into the surrounding water.

The most important polyphosphate is adenosine triphosphate (ATP, **11**), for it is found in every living cell. The triphosphate part of this molecule is a chain of three phosphate groups. Its conversion to the diphosphate (ADP) is

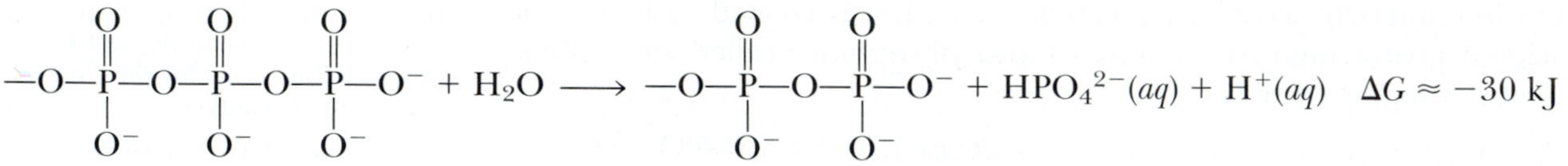

$$-O-\overset{\overset{O}{\|}}{\underset{\underset{O^-}{|}}{P}}-O-\overset{\overset{O}{\|}}{\underset{\underset{O^-}{|}}{P}}-O-\overset{\overset{O}{\|}}{\underset{\underset{O^-}{|}}{P}}-O^- + H_2O \longrightarrow -O-\overset{\overset{O}{\|}}{\underset{\underset{O^-}{|}}{P}}-O-\overset{\overset{O}{\|}}{\underset{\underset{O^-}{|}}{P}}-O^- + HPO_4^{2-}(aq) + H^+(aq) \quad \Delta G \approx -30\text{ kJ}$$

ATP is the fuel of life, for its reactions keep plants and animals alive, enabling them to grow and the ones with brains to think. The death of the organism releases the phosphate to the ecosystem, where it may lie as phosphate rock until nature or industry sends it on its way again.

SUMMARY

19.1 The oxidation number +3 is dominant in the lighter elements of Group III, and metallic character increases down the group. Boron is a hard nonmetal with several allotropic forms; it shows a diagonal relationship to silicon. Aluminum, obtained from bauxite through a combination of the **Bayer process** and the electrolytic **Hall process**, is a metal but has amphoteric oxides. Although low in the electrochemical series (strongly reducing), it is passivated by a tough layer of oxide that protects it from further reaction. Its reducing power is shown in the **thermite reaction**.

19.2 Boron has an acidic oxide, a Lewis acid that in water forms boric acid (a Brønsted acid). Alumina reacts with acids to form the acidic $Al(H_2O)_6^{3+}$ ion, which occurs in the **alums**. It also reacts with alkalis to give **aluminates**.

19.3 Boron and aluminum halides are Lewis acids, with the aluminum halides forming Lewis acid-base dimers (the chloride only in the vapor state). Boron forms an extensive series of electron-deficient **boranes** with structures that depend on delocalized bonds. The **borohydride anion**, BH_4^-, is an important reducing agent. The most important hydride of aluminum is $LiAlH_4$, which is a useful reducing agent.

19.4 Metallic character also increases down Group IV but is less pronounced than in Group III; the inert-pair effect also plays a more important role in this group, particularly in tin and lead. Carbon is a nonmetal. Silicon and germanium are metalloids. Tin and lead are metals but show amphoteric properties.

19.5 Carbon dioxide is the anhydride of carbonic acid and the **carbonates**. Carbon monoxide is a Lewis base and an important reducing agent. Silicon also has an acidic oxide, silica, and forms an extensive and important collection of silicates and **aluminosilicates**, in which Al atoms take the place of some Si atoms. Their structures can be understood in terms of linked SiO_4 tetrahedra. Molten silica with added *s*-block oxides sets to a **glass**.

19.6 Carbides are classified as **saline**, **covalent**, or **interstitial**. The halides of Group IV are progressively less covalent down the group and show the trend from nonmetallic to metallic character in the elements. The **silanes** (such as SiH_4) are much less stable than the hydrocarbons. Part of this instability arises from silicon's ability to expand its octet and accommodate the lone pair of an attacking Lewis base.

19.7 The elements in Group V are mainly nonmetals and metalloids, with metallic character becoming pronounced only at the bottom of the group (in bismuth). Nitrogen has a wide range of oxidation numbers in its compounds (from -3 to $+5$).

19.8 The industrial **fixation** of nitrogen (its combining with other elements) is achieved by the **Haber process** for the synthesis of ammonia. Ammonia is the most important of the binary hydrides of the group; it is more stable and a much stronger base than phosphine PH_3.

19.9 Ammonia is the industrial source of the oxoacids and oxides of nitrogen, particularly its oxidation to nitric acid, the parent of the soluble nitrates, in the **Ostwald process**. The oxides of nitrogen span oxidation numbers $+1$ to $+5$. Nitric acid is used for making fertilizers and explosives. The oxides of phosphorus are also acidic, being the anhydrides of phosphorous acid and the much more important phosphoric acid. The structural theme of the oxoanions of phosphorus is the PO_4 tetrahedron, just as SiO_4 is the structural theme of the silicates. When heated, phosphoric acid condenses to various **polyphosphates** that are used for detergents; the most significant biologically is the molecule ATP, adenosine triphosphate.

CLASSIFIED EXERCISES

This chapter includes applications of the chemical principles that have been addressed in earlier chapters. Therefore, a number of the exercises will require you to review and collect data from this earlier material.

Block Properties

19.1 Write the valence electron configuration of (a) Al; (b) Si; (c) Pb.

19.2 Write the valence electron configuration of (a) C; (b) In; (c) As.

19.3 Write the symbol and name of the following elements and write their valence electron configurations: (a) [4,IV]; (b) [6,III]; (c) [5,V]. The notation [*x*,*y*] denotes Period *x* and Group *y*.

19.4 Write the symbol and name of the following elements and write their valence electron configurations: (a) [3,V]; (b) [5,III]; (c) [2,IV]. The notation [*x*,*y*] denotes Period *x* and Group *y*.

19.5 Arrange the elements aluminum, gallium, indium, thallium, tin, and germanium in order of increasing electropositive character (see Fig. 7.36) and increasing reducing strength.

19.6 Arrange the elements carbon, silicon, hydrogen, nitrogen, and phosphorus in order of increasing electronegativity (see Fig. 7.36).

Group III Elements

19.7 Write a balanced equation for the industrial preparation of aluminum.

19.8 Write a balanced equation for the industrial preparation of high-purity boron.

19.9 Write the formulas of (a) boric acid; (b) alumina; (c) borax; (d) boron oxide.

19.10 Write the formulas of (a) potassium alum; (b) corundum; (c) diborane; (d) lithium aluminum hydride.

19.11 Complete and balance the following equations:
(a) $B_2O_3(s) + Mg(l) \xrightarrow{\Delta}$
(b) $Al(l) + Cr_2O_3(s) \rightarrow$

19.12 Complete and balance the following equations:
(a) $Al_2O_3(s) + OH^-(aq) \rightarrow$
(b) $Al_2O_3(s) + H_3O^+(aq) \rightarrow$

19.13 Identify a use for (a) $B_{12}C_3$; (b) BF_3; (c) α-alumina; (d) $B(OH)_3$.

19.14 Identify a use for (a) NH_3; (b) $NaBH_4$; (c) $LiAlH_4$; (d) $Al_2(SO_4)_3$.

19.15 Balance the following skeletal equations:
(a) $B_2H_6 + H_2O \rightarrow B(OH)_3 + H_2$
(b) $LiH + AlCl_3 \rightarrow LiAlH_4 + LiCl$

19.16 Balance the following skeletal equations:
(a) $B_2H_6 + NaBH_4 \rightarrow Na_2B_{12}H_{12} + H_2$
(b) $LiAlH_4 + H_2SO_4 \rightarrow AlH_3 + Li_2SO_4 + H_2$

19.17 Calculate the mass of aluminum that is needed for the reduction of 100 g of chromium(III) oxide to chromium by the thermite reaction.

19.18 What volume of hydrogen gas (at STP) is needed for the production of 50.0 g of boron from the reduction of boron trichloride?

19.19 Determine the standard reaction enthalpy for the reduction of chromium(III) oxide with aluminum by the thermite reaction; ΔH_f° for $Cr_2O_3(s)$ is -1140 kJ/mol.

19.20 (a) Determine the ΔH° and ΔS° for the reaction $2Al(s) + 3Cl_2(g) \rightarrow 2AlCl_3(s)$. (b) Calculate the standard free energy of the reaction.

19.21 (a) Write a chemical equation showing that $Al^{3+}(aq)$ is an acid. (b) Write equations to show the effect of adding 1, 2, and 3 mol of OH^- to 1 mol of Al^{3+}.

19.22 (a) Write an equation showing that $B(OH)_3$ is an acid. (b) What is the formula of the conjugate base of boric acid?

19.23 What mass of aluminum can be produced by the Hall process in a period of 8.0 h, using a current of 1.0×10^5 A?

19.24 In the production of 1.0×10^3 kg of aluminum by the Hall process, what mass of carbon is lost at the anode?

19.25 The standard free energy of formation of $Tl^{3+}(aq)$ is +215 kJ/mol at 25°C. Calculate the standard reduction potential of the Tl^{3+}/Tl redox couple.

19.26 The standard reduction potential of the Al^{3+}/Al redox couple is −1.66 V. Calculate the standard free energy of formation for $Al^{3+}(aq)$. Account for any differences between the standard free energy of formation of $Tl^{3+}(aq)$ (see Exercise 19.25) and for $Al^{3+}(aq)$.

Group IV Elements

19.27 Write a balanced equation for the industrial preparation of silicon.

19.28 Describe the preparation of high-purity graphite.

19.29 Germanium metal is produced by the reduction of germanium dioxide with hydrogen gas. Write the chemical equation for the process.

19.30 Calcium carbide is produced by the reduction of calcium carbonate with carbon in an electric furnace. Write the chemical equation for the process.

19.31 Write the formulas of (a) carborundum; (b) silica; (c) water glass; (d) calcium cyanide.

19.32 Write formulas for (a) silane; (b) tungsten carbide; (c) a Freon; (d) silica gel.

19.33 Balance the following skeletal equations and classify them as acid-base or redox:
(a) $CH_4 + S \rightarrow CS_2 + H_2S$
(b) $Si_2Cl_6 \rightarrow Si_6Cl_{14} + SiCl_4$
(c) $Sn + KOH + H_2O \rightarrow K_2Sn(OH)_6 + H_2$

19.34 Complete and balance the equations for the following reactions:
(a) $Sn^{2+}(aq) + \text{excess } OH^-(aq) \rightarrow$
(b) $SiH_4(g) + H_2O(l) \xrightarrow{OH^-}$
(c) $MgC_2(s) + H_2O(l) \rightarrow$

19.35 Write a Lewis structure for the orthosilicate unit, SiO_4^{4-}, and deduce the formal charges and oxidation numbers of the atoms. Use VSEPR theory to predict the shape of the unit.

19.36 Use VSEPR theory to estimate the Si—O—Si bond angle in silica.

19.37 Calculate the percentage by mass of silicon in silica.

19.38 Calculate the percentage by mass of silicon in feldspar, $KAlSi_3O_8$.

19.39 Determine the values of $\Delta H°$, $\Delta S°$, and $\Delta G°$ for the production of high-purity silicon by the reaction $SiO_2(s) + 2C(s, gr) \rightarrow Si(s) + 2CO(g)$ and estimate the temperature at which the reaction becomes spontaneous.

19.40 Determine the values of $\Delta H°$, $\Delta S°$, and $\Delta G°$ for the reaction $2CO(g) + O_2(g) \rightarrow 2CO_2(g)$ and estimate the temperature at which the reaction ceases being spontaneous.

19.41 The reduction of tin(IV) oxide to (white) tin metal by carbon proceeds at moderately low temperatures, but the reaction is normally carried out at over 980 K with an excess of carbon monoxide. (a) Write the two chemical equations for the reduction of SnO_2. (b) Determine the value of $\Delta G°$ for each reaction at 25°C. (c) How is the spontaneity affected by temperature in each case?

19.42 Determine the mass of HF in hydrofluoric acid that is required to etch 2.00 mg of SiO_2 from a glass plate in the reaction $SiO_2(s) + 6HF(aq) \rightarrow 2H_3O^+(aq) + SiF_6^{2-}(aq)$.

19.43 What mass of coke containing 98% carbon is needed to reduce the silicon in 1.0 kg of 88.5% pure silica?

19.44 Determine the surface area in square meters of 1.00 mol of activated carbon if the surface area of 1.0 g is 2000 m^2.

19.45 Describe the structures of two silicates in which the silicate tetrahedra share (a) one O atom or (b) two O atoms.

19.46 What is the empirical formula of a potassium silicate in which the silicate tetrahedra share (a) one O atom or (b) two O atoms and form a chain? In each case there are single negative charges on the unshared O atoms.

Group V Elements

19.47 Name the following compounds: (a) HNO_2; (b) H_3PO_4; (c) N_2O_3.

19.48 Name the following compounds: (a) N_2O; (b) H_3PO_3; (c) N_2O_5.

19.49 Write the chemical formula of (a) magnesium nitride; (b) calcium phosphide; (c) hydrazine.

19.50 Write the chemical formula of (a) sodium azide; (b) dinitrogen trioxide; (c) ammonium dihydrogenphosphate.

19.51 Balance the following skeletal equations:
(a) $NH_3 + CuO \rightarrow N_2 + Cu + H_2O$
(b) $NH_3 + F_2 \rightarrow NF_3 + NH_4F$
(c) $NH_3 + O_2 \rightarrow NO + H_2O$

19.52 Balance the following skeletal equations:
(a) $P_4 + KOH + H_2O \rightarrow PH_3 + KH_2PO_2$
(b) $Ca_3(PO_4)_2 + SiO_2 + C \rightarrow P_4 + CaSiO_3 + CO$
(c) $Ca_3(PO_4)_2 + C \rightarrow Ca_3P_2 + CO$

19.53 Determine the oxidation number of nitrogen in

(a) NO; (b) N_2O; (c) HNO_2; (d) N^{3-} and suggest (with equations) methods for transforming them from one to another.

19.54 Determine the oxidation number of phosphorus in (a) P_4; (b) PH_3; (c) H_3PO_4; (d) P_4O_6 and suggest (with equations) methods for transforming them from one to another.

19.55 Urea, $CO(NH_2)_2$, reacts with water to form ammonium carbonate. Write the chemical equation and calculate the mass of ammonium carbonate that can be obtained from 5.0 kg of urea.

19.56 (a) Nitrous acid reacts with hydrazine in acidic solution to form hydrazoic acid, HN_3. Write the chemical equation and determine the mass of hydrazoic acid that can be produced from 20.0 g of hydrazine. (b) Suggest a method of preparation of sodium azide, NaN_3.

19.57 Lead azide, $Pb(N_3)_2$, is used as a detonator. (a) What volume of nitrogen at STP does 1.0 g of the lead azide produce when it decomposes to lead metal and nitrogen gas? (b) Would 1.0 g of mercury(II) azide, $Hg(N_3)_2$, which is also used as a detonator, produce a larger or smaller volume, given that its decomposition products are mercury metal and nitrogen gas?

19.58 Sodium azide is used to inflate protective air bags in automobiles. What mass of solid sodium azide is needed to provide 100 L of N_2 at 1.5 atm and 20°C?

19.59 The common acid anhydrides of nitrogen are N_2O, N_2O_3, and N_2O_5. Write the formulas of their corresponding acids and chemical equations for the formation of the acids.

19.60 The common acid anhydrides of phosphorus are P_4O_6 and P_4O_{10}. Write the formulas of their corresponding acids and chemical equations for the formation of the acids.

UNCLASSIFIED EXERCISES

19.61 List two uses of (a) boron; (b) aluminum; (c) phosphorus; (d) nitrogen; (e) silicon; (f) tin.

19.62 What is meant by the fixation of nitrogen? How is nitrogen fixed naturally and synthetically?

19.63 (a) State the trends in first ionization energies and atomic radii down Groups III, IV, and V. (b) Account for the trends. (c) How do the trends correlate with the properties of the elements?

19.64 (a) Write Lewis structures for boric acid $B(OH)_3$ and boron trifluoride. (b) Using VSEPR theory, predict the structure and bond angles of boric acid and boron trifluoride. (c) What type of hybridization can be ascribed to the boron atom in the bonding in boric acid and boron trifluoride?

19.65 The azide ion has an ionic radius of 148 pm and forms many ionic and covalent compounds that are similar to those of the halides. (a) Write the Lewis formula for the azide ion and predict the N—N—N bond angle. (b) On the basis of its ionic radius, where in Group VII would you place the azide ion? (c) Write the formulas of several ionic and covalent azides.

19.66 Write the formula of the compound obtained when (a) NO_2; (b) CO_2; (c) P_4O_{10}; (d) N_2O_3 dissolve in water.

19.67 What is the mass percentage of phosphorus in (a) superphosphate and (b) triple superphosphate?

19.68 What is the mass percentage of nitrogen in the fertilizers (a) ammonia, NH_3; (b) ammonium nitrate, NH_4NO_3; (c) urea, $CO(NH_2)_2$?

19.69 (a) Distinguish, by drawing their structures, phosphoric acid from tripolyphosphoric acid. (b) Explain why it is preferential to use the sodium salt of tripolyphosphoric acid in detergents rather than the sodium salt of phosphoric acid.

19.70 Determine the volume (in liters) of concentrated sulfuric acid (density, 1.84 g/mL) that is required for the production of 1000 kg of phosphoric acid from phosphate rock by the reaction $Ca_3(PO_4)_2(s) + 3H_2SO_4(l) \rightarrow 2H_3PO_4(l) + 3CaSO_4(s)$.

19.71 What is the formula and use of (a) kernite; (b) quartzite; (c) bauxite; (d) cassiterite; (e) galena?

19.72 When the mineral dolomite, $CaCO_3{\cdot}MgCO_3$, is heated, it gives off carbon dioxide and forms a mixture of a metal oxide and a metal carbonate. Which oxide is formed, CaO or MgO?

19.73 (a) Suggest a reason for the observations that methane is stable in an aqueous alkaline solution, but silane reacts rapidly in the same solution. (b) Write a chemical equation for the reaction of silane in an aqueous alkaline solution.

19.74 Complete and balance the following equations:
(a) $AlCl_3(s) + H_2O(l) \rightarrow$
(b) $B_2H_6(g) \xrightarrow{\text{high temperatures}}$
(c) $B_2H_6(g) \xrightarrow{100°C}$

19.75 Complete and balance the following equations:
(a) $Na_2CO_3(s) + HCl(aq) \rightarrow$
(b) $SiCl_4(l) + H_2O(l) \rightarrow$
(c) $SnO_2(s) + C(s) \rightarrow$

19.76 Tin(II) hydroxide is insoluble in water. Write equations showing its reaction in (a) an acidic solution and (b) a basic solution.

CASE 19 ▸ BUCKMINSTERFULLERENE

A third allotrope of solid carbon, in addition to graphite and diamond, has recently been identified in several research laboratories around the world. Whereas graphite and diamond consist of extended carbon-carbon bonding, this new allotrope is a molecular solid of composition C_{60} (the analogue of P_4 and S_8, but on a larger scale). The C_{60} molecules form a stable solid at room temperature, are soluble in nonpolar solvents such as benzene and hexane, and can be vaporized while retaining the C_{60} structure (see the photograph below).

The C_{60} molecule was first identified by mass spectrometry. The arrangement of carbon atoms has been determined by a variety of spectroscopic techniques, and it is known that all the carbon atoms in the molecule have the same pattern of bonds. The allotrope was first detected in trace amounts in 1985 at Rice University, Texas, in an experiment in which the surface of pure graphite was vaporized with a laser. The vapor was analyzed by mass spectrometry and the presence of a C_{60} molecule was confirmed. The molecule, which is quite stable at high temperatures in the vapor state, was assumed to have a soccer ball structure. As shown in the drawing, the structure of the molecule is reminiscent of the geodesic domes designed by the American architect Buckminster Fuller, so the Rice research group dubbed C_{60} "buckminsterfullerene." The name, and the shorter "fullerene," has remained in the literature since then.

100 μm

A simpler procedure for producing buckminsterfullerene has recently been reported: the soot formed from pure graphite electrodes in a helium atmosphere is 10 to 20% buckminsterfullerene. The buckminsterfullerene is soluble in benzene (graphitic soot is not), forming a wine-red to brown solution. With the evaporation of the benzene, a dark brown to black crystalline solid remains. With this procedure, gram quantities of buckminsterfullerene can be produced quickly, which has facilitated further research and characterization of the compound to be done. The current interest in C_{60} concerns its structure. All carbon atoms are sp^2 hybridized, forming two single bonds and one double bond. The diameter of the cluster is approximately

19.77 Calcium carbide is produced by the reaction of lime with carbon in an electric furnace. Write the chemical equation for the process? What is the function of the carbon?

19.78 Complete and balance the following equations:
(a) $Li_3N(s) + H_2O(l) \rightarrow$
(b) $Ba(s) + N_2(g) \rightarrow$

19.79 The standard enthalpies of formation of $BH_3(g)$ and diborane are +100 kJ/mol and +36 kJ/mol, respectively, and the enthalpies of formation of B(g) and H(g) are +563 kJ/mol and +218 kJ/mol, respectively. Use these values to calculate the mean bond enthalpies of the B—H bonds in each case and to estimate the bond enthalpies of the terminal B—H and B—H—B bonds. Which bonds would you expect to be longer?

19.80 Arsenic(III) sulfide is oxidized by an acidic hydrogen peroxide solution to the arsenate ion AsO_4^{3-}. Write the chemical equation and the reduction and oxidation half-reactions for the reaction.

0.71 nm and the density of the allotrope is 1.65 g/cm^3.

The interest in buckminsterfullerene came from research in the origins of interstellar dust. A portion of the infrared spectra of interstellar matter was thought to be due to graphitic soot, or various carbon clusters. In addition, buckminsterfullerene will soon be available from chemical suppliers and can then be used as a starting material for new chemical syntheses. There are hopes that the nearly spherical shape of the molecule will make it an excellent lubricant (perhaps after the double bonds have been replaced by carbon-fluorine bonds). Because electrons can easily be added to and removed from C_{60}, buckminsterfullerene may turn out to be the basis of an entirely new range of rechargeable batteries. A number of analogues of buckminsterfullerene, which are known as *fullerenes,* also have the carbon-cage structure. The most stable of them have more than 60 carbon atoms, and C_{70}, C_{84}, C_{90}, and C_{94} have been isolated. Compounds of C_{60} were reported in 1991. They include $C_{60}F_{60}$, which may perhaps be useful as a lubricant, and K_3C_{60}, which is superconducting below 18 K.

The current excitement surrounding the existence of buckminsterfullerene and its analogues is being compared to the beginning of aromatic chemistry and the enormously important class of compounds based on benzene.

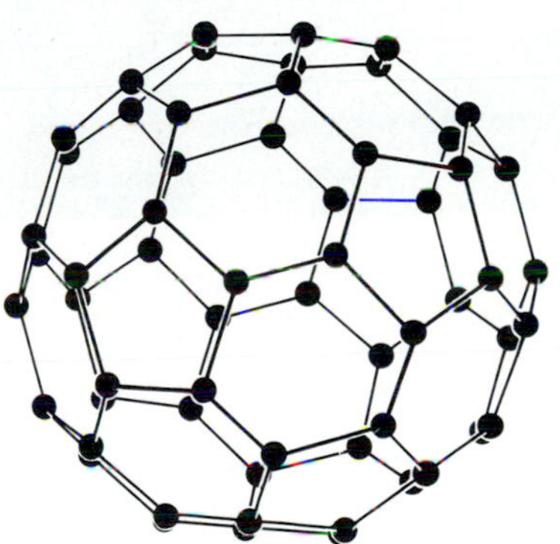

QUESTIONS

1. What is the molar mass of buckminsterfullerene?

2. How many five- and six-membered carbon rings exist in buckminsterfullerene?

3. What is the molecular radius and the molecular volume of C_{60} (the volume of a sphere of radius r is $\frac{4}{3}\pi r^3$).

4. When trace amounts of metals such as potassium, lanthanum, and uranium are a part of the graphite, a metal atom or ion can become trapped inside the C_{60} molecule, thereby forming $C_{60}M^{n+}$. Predict some properties of compounds formed from this shrink-wrapped ion.

5. Would the application of pressure favor the formation of buckminsterfullerene from graphite? (The density of graphite is 2.3 g/cm^3.)

19.81 The concentration of nitrate ion in a basic solution can be determined by the following sequence of steps. (a) Zinc metal can reduce nitrate ions to ammonia in a basic aqueous solution. Write a balanced equation for the reaction. (b) The ammonia is passed into a solution containing a known, but excess, amount of $HCl(aq)$; the unreacted $HCl(aq)$ is titrated with a standard $NaOH(aq)$ solution. A 25.00-mL sample containing $NO_3^-(aq)$ was treated with an excess of zinc metal. The evolved ammonia gas was passed into 50.00 mL of 0.250 M $HCl(aq)$. The unreacted $HCl(aq)$ was titrated to the stoichiometric point with 28.22 mL of 0.150 M $NaOH(aq)$. What is the molar concentration of nitrate ion in the original solution?

19.82 Determine the values of $\Delta H°$ and $\Delta S°$ for each stage of the Ostwald process for the production of nitric acid. Predict the conditions that favor the formation of the products in each case.

Stage 1: $4NH_3(g) + 5O_2(g) \rightarrow 4NO(g) + 6H_2O(g)$
Stage 2: $2NO(g) + O_2(g) \rightarrow 2NO_2(g)$
Stage 3: $3NO_2(g) + H_2O(l) \rightarrow 2HNO_3(aq) + NO(aq)$

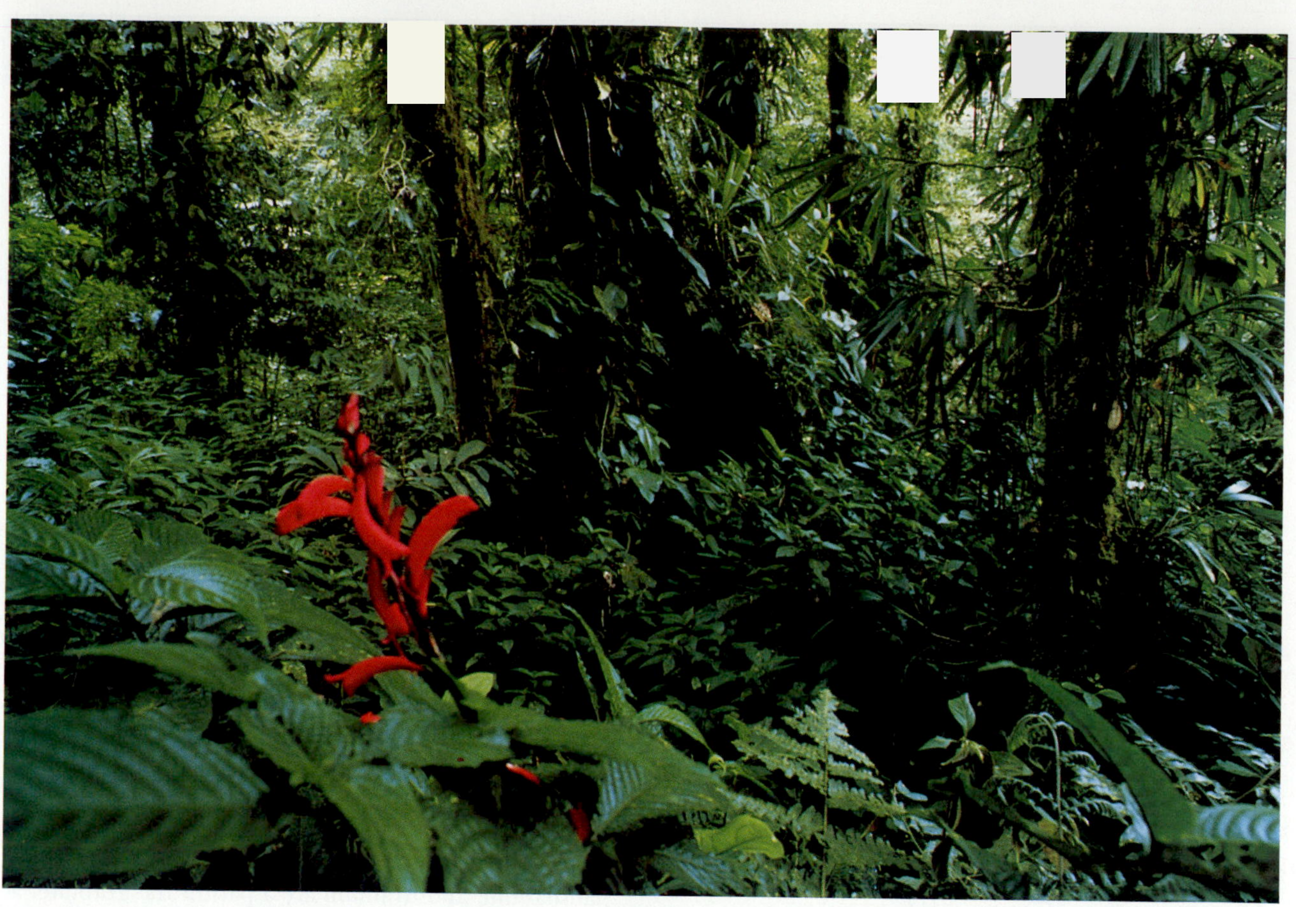

This chapter continues the description of the p-block elements. Here we meet the elements of Group VI, including oxygen, the most abundant element on Earth and one that occurs in numerous compounds, and the vigorously reactive halogens of Group VII. We also discuss the noble gases, a group of elements that until a few years ago were thought to form no compounds at all. Almost all the oxygen present in the atmosphere of the Earth was formed by green vegetation like that in the illustration, as a byproduct of the conversion of carbon dioxide and water to glucose: oxygen is still formed on a massive scale in this way.

20

THE p-BLOCK ELEMENTS: II

E Scheele's priority for the discovery of oxygen. *J. Chem. Ed.*, **1975**, *52*, 442. Oxygen: What's the use? *J. Chem. Ed.*, **1990**, *67*, 298. Oxygen use in sewage treatment. *J. Chem. Ed.*, **1980**, *57*, 137.

The elements in Groups VI through VIII of the *p* block are almost all nonmetals. Metallic character increases down each group, but the elements are now so far to the right in the periodic table that even the most metallic element of the group—polonium—is best regarded as a metalloid. A notable feature of this half of the *p* block is the sharp difference between the reactive halogens in Group VII and the noble gases of Group VIII. The latter are almost chemically inert, and with them we seem to reach the end of chemistry (but not of this text).

GROUP VI: OXYGEN AND SULFUR

The elements in Group VI and some of their properties are summarized in Table 20.1. Oxygen, sulfur, selenium, and tellurium are collectively called the **chalcogens** from the Greek words for "brass giver," because they are found in copper ores and copper is a major component of brass.

D The blue bottle reaction. CD1, 76.

? Name some of the rocks and minerals in which oxygen is found. (Answers could include feldspars, silica, aluminosilicates, limestone, quartz, hematite.)

20.1 GROUP VI ELEMENTS

The two most important members of the group are oxygen and sulfur, and we shall confine our attention principally to their properties.

Oxygen. Oxygen is the most abundant element in the Earth's crust and accounts for 23% by mass of the atmosphere. Some of our present atmospheric oxygen has been produced by the photochemical action of sunlight on water that steams up from the hot interior of the Earth. Most of the rest has been produced by photosynthesis:

$$6CO_2(g) + 6H_2O(l) \longrightarrow C_6H_{12}O_6(s) + 6O_2(g)$$

Nearly 2×10^{10} kg of liquid oxygen are produced each year in the United States (about 80 kg per inhabitant) by fractional distillation of liquid air, and it ranks third by mass produced. The biggest consumer

				V	VI	VII	He
Li	Be	B	C	N	**O**	F	Ne
				P	**S**	Cl	
				As	**Se**	Br	
				Sb	**Te**	I	
				Bi	**Po**	At	

TABLE 20.1 Group VI elements

Valence configuration: ns^2np^4

Z	Name	Symbol	Molar mass, g/mol	Melting point, °C	Boiling point, °C	Density, g/cm^3	Normal form
8	oxygen	O	16.00	−218	−183	1.43×10^{-3} at 0°C	colorless paramagnetic gas (O_2)
				−192	−111	2.14×10^{-3} at 0°C	blue gas (ozone, O_3)
16	sulfur	S	32.06	115	445	2.09	yellow nonmetallic solid (S_8)
34	selenium	Se	78.96	220	685	4.81	gray nonmetallic solid
52	tellurium	Te	127.60	450	990	6.25	silver-white metalloid
84	polonium	Po	210	254	960	9.40	gray metalloid

N More information about the steel industry appears in Chapter 21.

E Atmospheric ozone is also a contributor to the degradation of rubber goods that are exposed to the atmosphere.

D Preparation and properties of oxygen. BZS2, 6.8. Preparation and properties of liquid oxygen. BZS2, 6.10.

E The three forms of molecular oxygen. *J. Chem. Ed.*, **1989**, *66*, 453. Ozone: Properties, toxicity, and applications. *J. Chem. Ed.*, **1973**, *50*, 404. Industrial used of ozone. *J. Chem. Ed.*, **1982**, *59*, 392.

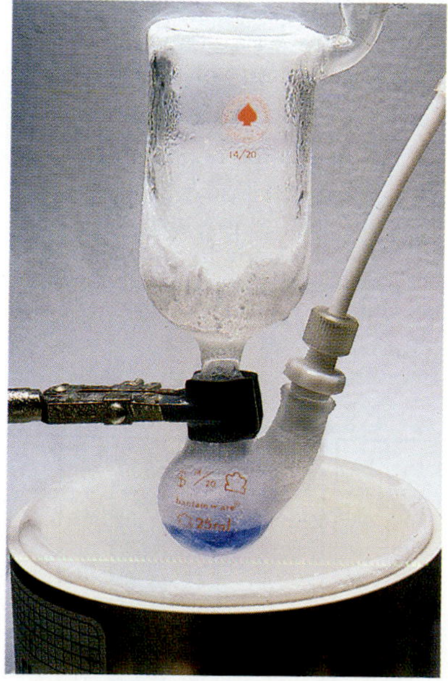

FIGURE 20.1 Liquid ozone is a dark blue, highly unstable liquid.

is the steel industry, which needs about 1 ton of oxygen to produce 1 ton of steel. In this application, oxygen is blown into molten iron to oxidize any impurities, particularly carbon (see Section 21.2). The gas is used in much smaller quantities in medicine, where it is administered to relieve strain on the heart and to act as a stimulant, and in oxyacetylene welding:

$$C_2H_2(g) + \tfrac{5}{2}O_2(g) \longrightarrow 2CO_2(g) + H_2O(g) \qquad \Delta H° = -1300 \text{ kJ}$$

Oxygen is a colorless, tasteless, odorless gas of O_2 molecules that condenses to a pale blue liquid at −183°C. Although O_2 has an even number of electrons, two of them are unpaired and the molecule is paramagnetic (see Fig. 9.22). Molecular orbital theory readily accounts for this unusual feature by showing that the outermost two electrons occupy two different π orbitals, which they can do without pairing their spins. This paramagnetism has a practical application, for the oxygen content of incubators and other life-support systems can be monitored by measuring the magnetism of the gas they contain.

Elemental oxygen also occurs as the allotrope ozone (O_3) in a layer high in the atmosphere, where it is formed by the effect of solar radiation. Its total abundance in the atmosphere is equivalent to a layer that, at normal temperature and pressure, would cover the Earth to a thickness of 3 mm. Ozone can be made in the laboratory by passing an electric discharge through oxygen. It is a blue gas that condenses at −111°C to an explosive liquid that looks like dark blue ink (Fig. 20.1). Its pungent smell (its name comes from the Greek word for "smell") can often be detected near electrical equipment and after lightning. The presence of ozone is partly responsible for the "freshness" of air after an electrical storm accompanied by rain. Ozone is also present in smog-laden areas, where it is produced by the reaction of oxygen molecules with oxygen atoms:

$$O(g) + O_2(g) \longrightarrow O_3(g)$$

The oxygen atoms are produced by the photochemical decomposition of NO_2, a car-emission pollutant:

$$NO_2(g) \xrightarrow{\text{light}} NO(g) + O(g)$$

Sulfur. Sulfur is widely distributed as sulfide ores, which include galena (PbS), chalcocite (Cu_2S), argentite (Ag_2S), and cinnabar (HgS). Iron pyrite, FeS_2, is called fool's gold because of its misleading resemblance to gold metal (Fig. 20.2); it is often found near gold deposits. Sulfur is also found as deposits of the native element (brimstone), which is formed by bacterial action on H_2S. Native sulfur is mined by the ingenious **Frasch process**, which makes use of its low melting point and low density (Fig. 20.3). In this process water at about 165°C and under pressure is pumped down the outermost of three concentric pipes to melt the sulfur trapped beneath deep rock layers. Compressed air is passed down the innermost pipe to force a frothy mixture of sulfur, air, and hot water up the middle pipe. The insoluble sulfur then settles out of the mixture and can be stored in piles like that shown in the cover photograph of this text.

FIGURE 20.2 The mineral pyrite, FeS_2, has a lustrous golden color. It has frequently been mistaken for gold and hence is also known as fool's gold.

Because sulfur is a by-product of a number of metallurgical processes (especially the extraction of copper from its sulfide ores) and must be removed from sulfur-rich petroleum, its recovery from these sources is displacing the mining of the native element. A major source of recovered sulfur is the **Claus process**, in which some of the H_2S that occurs in oil and natural gas wells is first oxidized to sulfur dioxide:

$$2H_2S(g) + 3O_2(g) \longrightarrow 2SO_2(g) + 2H_2O(l)$$

Then the remainder of the hydrogen sulfide is oxidized by reaction with this SO_2:

$$2H_2S(g) + SO_2(g) \xrightarrow{300°C,\ Al_2O_3} 3S(s) + 2H_2O(l)$$

Sulfur is of major industrial importance. Most of the sulfur that is produced is used to make sulfuric acid, but an appreciable amount is used to vulcanize rubber.

Elemental sulfur is a yellow, tasteless, almost odorless, insoluble, nonmetallic, molecular solid of crownlike S_8 rings (**1**). The most stable allotrope under normal conditions is rhombic sulfur (Fig. 20.4). Sulfur vapor has a blue tint from the S_2 molecules present in it. The latter are paramagnetic, like O_2.

E The H_2S that is found in oil wells has been formed by anaerobic bacteria.

E Herman Frasch, sulfur king. *J. Chem. Ed.*, **1981**, *58*, 60.

E Sulfur: Chemical of the month. *J. Chem. Ed.*, **1981**, *58*, 468. The sulfur chemist: The bearer of ill wind. *J. Chem. Ed.*, **1984**, *61*, 372.

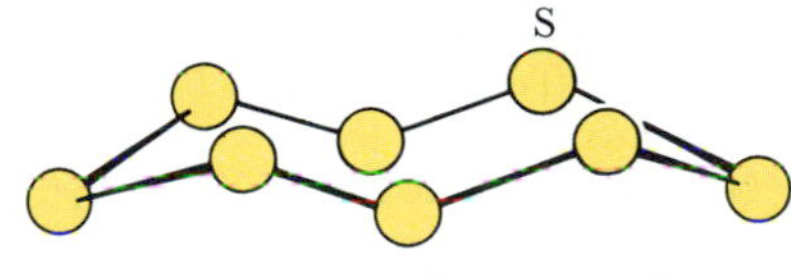

1 S_8

E In the Claus process, the sulfur of H_2S is oxidized (from oxidation number −2 to 0) and the sulfur of SO_2 is reduced (from +4 to 0). In this particular *comproportionation reaction*, two virulent air pollutants react together and annihilate each other. The sulfur shown on the front cover was produced by the Claus process.

V Video 25: Physical properties of sulfur.

T OHT: Fig. 20.3.

D Plastic sulfur. CD2, 28. Preparation and properties of sulfur dioxide. BZS2, 6.18.

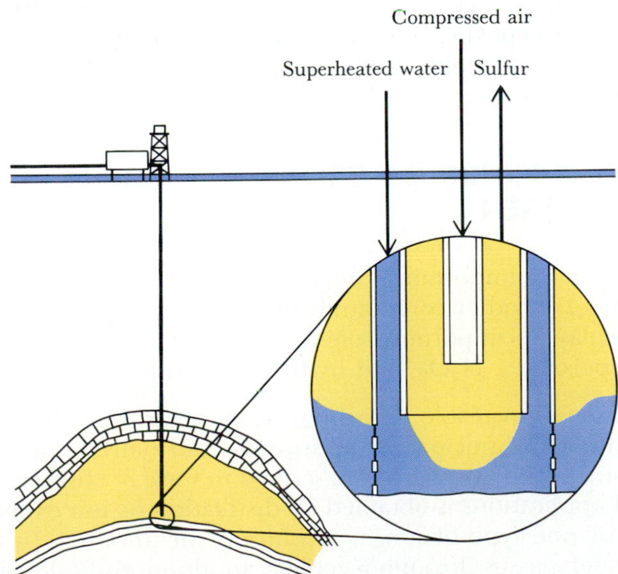

FIGURE 20.3 In the Frasch process, superheated water—water under pressure and heated to above 100°C—is pumped down the outermost of the three concentric pipes; compressed air is passed down the innermost pipe and forces a mixture of liquid sulfur, air, and hot water up the middle pipe.

(a)

(b)

FIGURE 20.4 One of the two most common forms of sulfur is the blocklike rhombic sulfur (a). It differs from the needlelike monoclinic sulfur (b) in the manner in which the S_8 rings stack together.

E The vulcanization of rubber is another process that takes advantage of sulfur's ability to catenate, for short chains of sulfur atoms cross-link the polyisoprene chains (see Section 23.6).

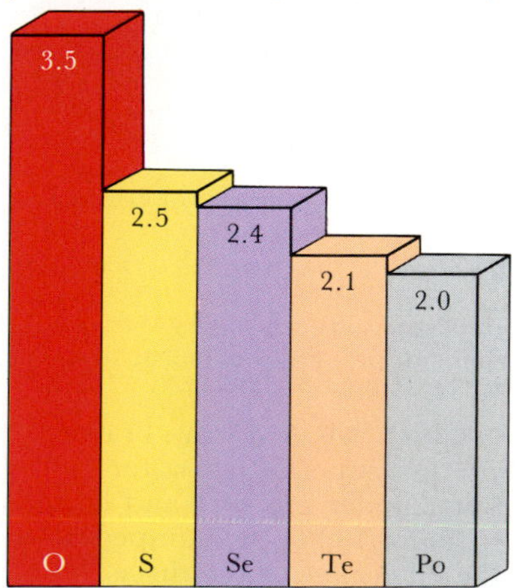

FIGURE 20.5 The electronegativities of the Group VI elements decrease steadily down the group.

Selenium and tellurium. Selenium and tellurium occur in sulfide ores; they are also recovered from the anode sludge formed during the electrolytic refining of copper. Both elements have several allotropes, the most stable consisting of long zigzag chains of atoms. Although these allotropes look like silver-white metals, they are poor electrical conductors. Because the conductivity of selenium is increased in the presence of light, it is used in photoelectric devices and in photocopying machines. Selenium also occurs as a deep red solid composed of Se_8 molecules like those in solid sulfur.

Trends and differences. Oxygen, sulfur, and selenium are typical nonmetals; tellurium and polonium are metalloids. Electron affinities and electronegativities decrease down the group (Fig. 20.5) and atomic and ionic radii increase (Fig. 20.6).

The differences between oxygen and sulfur are emphasized by the latter's striking ability to **catenate**, that is, to form chains of atoms. Oxygen's ability to form chains is very limited, with H_2O_2, O_3, and the anions O_2^-, O_2^{2-}, and O_3^- the only examples. Sulfur's ability is much more developed, for instance, in S_8 rings and their fragments. The —S—S— links in proteins are another example of catenation and make an important contribution to the shapes of proteins containing them.

Oxygen combines directly, often vigorously, and usually exothermically with all but a few elements (principally the noble gases and the noble metals). It forms compounds with all elements except the lighter noble gases (He, Ne, Ar, and Kr). The properties of oxides are one of the chemical criteria for distinguishing between metals (which form basic oxides) and nonmetals (which form acidic oxides). For example, lithium oxide (a metal oxide) dissolves in water to produce a lithium hydroxide solution,

$$Li_2O(s) + H_2O(l) \longrightarrow 2LiOH(aq)$$

whereas sulfur trioxide (a nonmetal oxide) dissolves to produce a sulfuric acid solution,

$$SO_3(g) + H_2O(l) \longrightarrow H_2SO_4(aq)$$

Sulfur is similarly reactive, and it also combines directly with almost all the elements except the noble gases, nitrogen, tellurium, iodine, iridium, platinum, and gold. Compounds of sulfur with these elements (other than the noble gases), however, are known.

20.2 COMPOUNDS OF GROUP VI ELEMENTS AND HYDROGEN

By far the most important compound of oxygen and hydrogen is water, H_2O. Two other compounds of hydrogen and a Group VI element that play an important role in chemistry and the economy are hydrogen peroxide, H_2O_2, and hydrogen sulfide, H_2S.

Water. Water is available on a huge scale worldwide, but in various states of purity. Municipal water supplies normally undergo several stages of purification, which are described in Case 3. High-purity water for special applications is obtained by distillation or **ion exchange**, the exchange of one type of ion in a solution for another. In the latter process, water passes through a zeolite, an aluminosilicate with a very open structure that can capture ions such as Mg^{2+} and Ca^{2+}, often in exchange for H^+ ions. Very pure water is needed industrially for use in high-pressure boilers, because even very small quantities of impurities can lead to the formation of sediment. Industrial boiler water is probably the purest high-tonnage chemical available, for it must have no more than 0.02 ppm of impurity (approximately 2 g of impurity in 100,000 kg of water); this is not far short of the purity of zone-refined silicon. The oxygen concentration of the water is reduced to a very low level by reduction with hydrazine (Section 19.8).

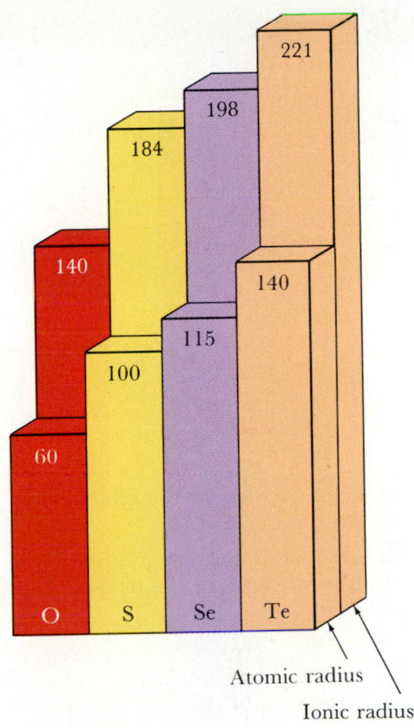

FIGURE 20.6 The atomic and ionic radii of the Group VI elements both increase down the group. The values shown are in picometers.

N Diagonal relationships are less pronounced in Group VI than in previous groups.

N The reduction potential for this half-reaction is +0.81 V at pH = 7 and 25°C.

V Video 52: Reaction of oxygen with the elements.

U Exercise 20.78.

The important chemical properties of water include its high polarity, which together with its ability to form hydrogen bonds to anions gives it a unique ability to act as a solvent for ionic compounds. Water is also amphiprotic and hence can act both as a Brønsted base and as a Brønsted acid (see Section 14.2):

$$CH_3COOH(aq) + H_2O(l) \rightleftharpoons H_3O^+(aq) + CH_3CO_2^-(aq)$$

$$H_2O(l) + NH_3(aq) \rightleftharpoons NH_4^+(aq) + OH^-(aq)$$

The strengths with which H_2O donates and accepts protons makes it the dividing line between weak acids and weak bases (see Table 14.3).

Water is an oxidizing agent:

$$2H_2O(l) + 2e^- \longrightarrow 2OH^-(aq) + H_2(g) \qquad E = -0.42 \text{ V at pH} = 7$$

A simple example is its reaction with the alkali metals, for example,

$$2Na(s) + 2H_2O(l) \longrightarrow 2NaOH(aq) + H_2(g)$$

However, unless the other reagent is a strong reducing agent, water acts as an oxidizing agent only at high temperatures, as it does in the reforming reaction

$$CH_4(g) + H_2O(g) \xrightarrow{\Delta} CO(g) + 3H_2(g)$$

Water is a very mild reducing agent, the half-reaction being

$$2H_2O(l) \longrightarrow 4H^+(aq) + O_2(g) + 4e^-$$

Few oxidizing agents are strong enough to remove the electrons and bring this reaction about, but one that can is fluorine:

$$2H_2O(l) + 2F_2(g) \longrightarrow 4HF(aq) + O_2(g)$$

Water is also a Lewis base, because an H_2O molecule can donate one of its lone pairs to a Lewis acid, as occurs in the formation of complexes such as $Fe(H_2O)_6^{3+}$. The same property is the origin of water's ability to **hydrolyze** other substances, that is, to react with them to form a new element-oxygen bond:

$$PCl_5(s) + 4H_2O(l) \longrightarrow H_3PO_4(aq) + 5HCl(aq)$$

We noted in Section 19.8 that a characteristic reaction of nonmetal halides is their reaction with water to form an oxoacid without change of oxidation number. We can now see that this reaction is a hydrolysis.

Hydrolyses can also occur with change of oxidation number. Such is the case in the reaction between chlorine and water, which can be pictured as a hydrolysis in which the water drives one of the Cl atoms out of the Cl_2 molecule:

$$Cl{-}Cl(g) + H_2O(l) \longrightarrow Cl{-}OH(aq) + HCl(aq)$$

In this reaction, the oxidation number of Cl changes from 0 to +1 (in HClO) and −1 (in HCl). It is an example of a **disproportionation reaction**, which is a reaction in which the oxidation number of an element simultaneously increases and decreases as two different substances are formed from a single substance containing the element.

▼ EXAMPLE 20.1 Accounting for a property of oxides

Suggest a reason why ionic oxides are formed only by elements located on the left of the periodic table.

STRATEGY The discussion of the periodic table in Chapter 7 showed that metals at the left of the table have low ionization energies. We also know from Table 8.1 that oxide ions give rise to relatively high lattice enthalpies.

SOLUTION Elements having low ionization energies, like the metals, tend to combine with elements by forming an ionic bond. The high lattice enthalpies of compounds having ionic bonds enhance the stability of the ionic oxides. Hence, only elements toward the left of the periodic table can be expected to form ionic oxides. ▲

V Video 12: Reaction of oxides with water.

N See Table 14.3 for a comparison of the base strengths of the oxide and hydroxide ions.

The conjugate base of H_2O is the hydroxide ion, OH^-; the conjugate base of OH^- is the oxide ion O^{2-}. Because O^{2-} is such a strong Brønsted base, soluble oxides are immediately and fully protonated in water in the reaction

$$O^{2-}(aq) + H_2O(l) \longrightarrow 2OH^-(aq)$$

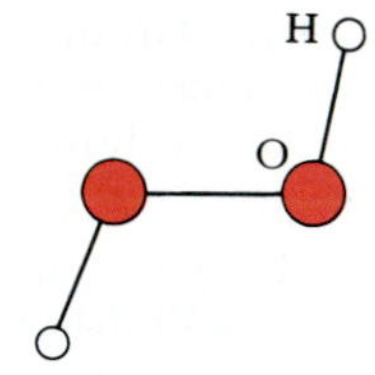

2 Hydrogen peroxide

3 Anthraquinone

The formation of ionic oxides by electropositive elements, together with this proton transfer reaction, accounts for the general rule that the oxides of metals are basic.

Hydrogen peroxide. Hydrogen peroxide, H_2O_2 (**2**), was once made by the electrolysis of dilute sulfuric acid. It is now made by a reaction in which anthraquinone (**3**; abbreviated A) in a hydrocarbon solvent is first allowed to react with hydrogen:

$$A + H_2(g) \xrightarrow{Ni} AH_2$$

Then air is passed through the product:

$$AH_2 + O_2(g) \longrightarrow A + H_2O_2$$

Hydrogen peroxide is a very pale blue liquid that is appreciably denser than water (1.44 g/mL at 25°C). In its other physical properties it is quite similar to water (its melting point is −0.4°C and its boiling point is 150°C). Chemically, however, hydrogen peroxide and water differ greatly. The presence of the second oxygen atom makes H_2O_2 a very weak acid ($pK_{a1} = 11.75$). It can act as an oxidizing agent in both acidic and basic solutions.

Acidic solution: $H_2O_2(aq) + 2H^+(aq) + 2e^- \longrightarrow 2H_2O(l)$ $\quad E° = +1.78$ V

Basic solution: $HO_2^-(aq) + H_2O(l) + 2e^- \longrightarrow 3OH^-(aq)$ $\quad E° = +0.87$ V

E An interesting feature of the oxidation of H_2O_2 is that the O_2 so produced is sometimes in an energetically excited state and emits light as it discards its excess energy. This process is an example of *chemiluminescence,* the emission of light by products formed in energetically excited states when reagents are mixed.

V Video 51: Cool light.

D The "Aladdin's lamp" reaction. CD2, 23.

D Hydrogen peroxide as an oxidizing agent: Black to white. CD1, 87.

For example, its reactions with Fe^{2+} and Mn^{2+} in acidic and basic solutions, respectively, are

$$2Fe^{2+}(aq) + H_2O_2(aq) + 2H^+(aq) \longrightarrow 2Fe^{3+}(aq) + 2H_2O(l)$$
$$Mn^{2+}(aq) + HO_2^-(aq) + H_2O(l) \longrightarrow Mn^{4+}(aq) + 3OH^-(aq)$$

It can also act as a reducing agent in acidic and basic solutions:

Acidic solution: $$O_2(g) + 2H^+(aq) + 2e^- \longrightarrow H_2O_2(aq) \qquad E^\circ = +0.68\text{ V}$$

Basic solution: $$O_2(g) + 2H_2O(l) + 2e^- \longrightarrow H_2O_2(aq) + 2OH^-(aq) \qquad E^\circ = -0.15\text{ V}$$

For example, its reactions with permanganate ions and chlorine in acidic and basic solutions, respectively, are

$$2MnO_4^-(aq) + 5H_2O_2(aq) + 6H^+(aq) \longrightarrow 2Mn^{2+}(aq) + 8H_2O(l) + 5O_2(g)$$

$$Cl_2(g) + H_2O_2(aq) + 2OH^-(aq) \longrightarrow 2Cl^-(aq) + 2H_2O(l) + O_2(g)$$

D Hydrogen peroxide as an oxidizing agent and a reducing agent. CD2, 109.

D Enzyme kinetics: Effects of temperature and an inhibitor on catalase extracted from potato. CD2, 80.

? What is the value of ΔG° for the decomposition of hydrogen peroxide at 25°C?

Hydrogen peroxide is normally sold commercially as a 30% by mass aqueous solution. When used as a hair bleach (as a 6% solution) it acts by oxidizing the pigments in the hair. Because it oxidizes unpleasant effluents without producing any harmful by-products, H_2O_2 increasingly is being used as an oxidizing agent in the control of pollution. A 3% H_2O_2 aqueous solution is used as a mild antiseptic in the home. Contact with blood catalyzes its decomposition to produce oxygen gas,

$$2H_2O_2(aq) \longrightarrow 2H_2O(l) + O_2(g)$$

which cleanses the wound.

The conjugate base of H_2O_2 is the hydroperoxide ion, HO_2^-, and that ion's conjugate base is the peroxide ion, O_2^{2-}. The latter ion occurs in sodium peroxide, which is the product of the oxidation of sodium in air at temperatures above 400°C and is used to bleach textiles and wood pulp for paper. In an alkaline solution, all peroxide salts form hydrogen peroxide:

$$O_2^{2-} + 2H_2O(l) \longrightarrow 2OH^-(aq) + H_2O_2(aq)$$

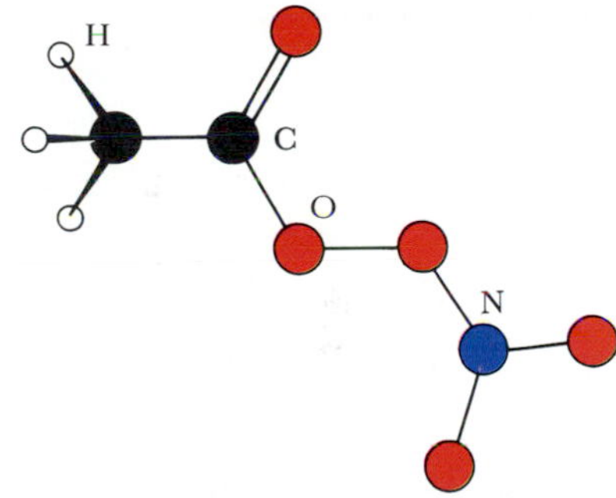

4 Peroxyacetyl nitrate, PAN

Organic peroxides are oxidizing molecular compounds that contain the link —O—O—. They include peroxyacetyl nitrate (PAN, **4**), the eye irritant in smog.

U Exercise 20.71.

N Natural emissions of gaseous sulfur compounds arise primarily from the anaerobic decomposition of organic matter giving rise to the natural abundance of H_2S. Anaerobic decomposition in wet swamps and marshes contributes most, but a significant proportion can be traced to bovine flatulence.

E It was remarked in an annotation on p. 551 that lone pairs stimulate nerves that are different from those usually involved in olfaction. The odors described on p. 768 are aspects of this behavior.

Hydrogen sulfide. The gas hydrogen sulfide, H_2S, is formed either by protonation of the sulfide ion, a Brønsted base, in a reaction such as

$$FeS(s) + 2HCl(aq) \longrightarrow FeCl_2(aq) + H_2S(g)$$

or by direct reaction of the elements at 600°C. When it is being used to precipitate sulfides during qualitative analysis (see Section 15.7), a safer and more convenient method is the hydrolysis of thioacetamide:

$$CH_3\text{—}\overset{\overset{\large S}{\|}}{C}\text{—}NH_2(aq) + 2H_2O(l) \xrightarrow{\Delta} CH_3CO_2^-(aq) + NH_4^+(aq) + H_2S(aq)$$

FIGURE 20.7 The blue stones in this ancient Egyptian ornament are lapis lazuli. This semiprecious stone is an aluminosilicate colored by S_2^- and S_3^- impurities. The blue color is due to S_3^-, and its hint of green to S_2^-.

All the Group VI H_2X compounds other than water are toxic gases with offensive odors. They are insidious poisons because exposure to them paralyzes the olfactory nerve and soon eliminates one's ability to smell them. Rotten eggs smell of hydrogen sulfide because egg proteins contain sulfur and give off the gas when they decompose. Another sign of the formation of sulfides in eggs is the pale green discoloration sometimes seen in cooked eggs where the white meets the yolk: this green material is a deposit of iron(II) sulfide. Hydrogen sulfide is also responsible for the "sourness" of underground natural gas reserves.

The sulfur analog of hydrogen peroxide also exists and is an example of a **polysulfane**, a molecular compound of composition HS—S_n—SH, where n can take on any of the values 0, 1, 2, . . . , 6. In these substances, sulfur is showing its ability to catenate. The polysulfide ions obtained from the polysulfanes include two ions found in lapis lazuli (Fig. 20.7).

20.3 IMPORTANT COMPOUNDS OF SULFUR

The most important oxides and oxoacids of sulfur are the dioxide and trioxide and the corresponding sulfurous and sulfuric acids.

Sulfur dioxide and the sulfites. Sulfur burns in air to form sulfur dioxide, SO_2 (**5**), a colorless, choking, poisonous gas. Sulfur dioxide occurs in the atmosphere as a result of volcanic activity, the combustion of sulfur-rich fuels, and the industrial roasting of sulfide ores.

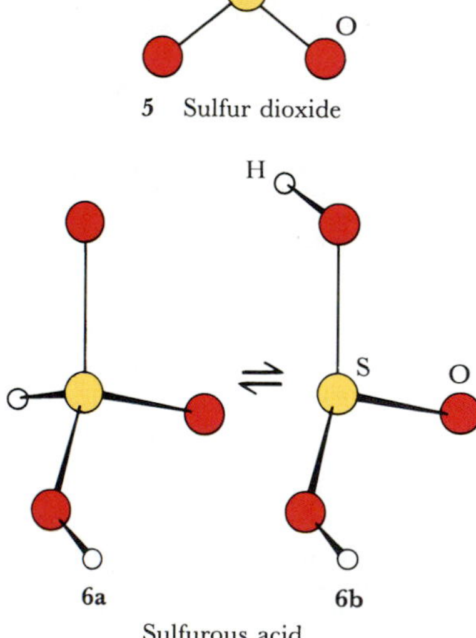

5 Sulfur dioxide

Sulfurous acid

Sulfur oxides in the atmosphere are referred to as SO_x (read "sox"). About 7×10^{10} kg of the dioxide result from the decomposition of vegetation and from volcanic emission. In addition, about 1×10^{11} kg of hydrogen sulfide are formed naturally and then produce the dioxide through oxidation with atmospheric oxygen:

$$2H_2S(g) + 3O_2(g) \longrightarrow 2SO_2(g) + 2H_2O(g)$$

Industry and transport contribute another 1.5×10^{11} kg of the dioxide, of which about 70% comes from oil and coal combustion—mainly in electricity-generating plants. The average atmospheric concentration of SO_2 in rural areas in the northern hemisphere is about 8×10^{-12} mol/L, or 20 ppb.

Sulfur dioxide is an acidic oxide and the anhydride of sulfurous acid (H_2SO_3), which is the parent acid of the hydrogensulfites (or bisulfites) and the sulfites:

$$SO_2(g) + H_2O(l) \longrightarrow H_2SO_3(aq)$$

The H_2SO_3 molecule is an equilibrium mixture of two molecules (**6a** and **6b**); in the former it resembles phosphorous acid, because one of the H atoms is attached directly to the S atom. These sulfurous acid molecules are also in equilibrium with SO_2 molecules, each of which is surrounded by a cage of water molecules. Evidence for this arrangement comes from the observation that crystals of composition $SO_2 \cdot xH_2O$, with x about 7, are obtained when the solution is cooled. Substances like this, in which a molecule sits in a cage, are called **clathrates** (from the Greek word for "cage"). Clathrates are also formed by other gases, including methane, carbon dioxide, and the noble gases. Sulfur dioxide is easily liquified under pressure and can

E Hydrogen sulfide dissolved in water slowly oxidizes to form a cloudy solution consisting of small particles of sulfur, called a *colloidal suspension.*

therefore be used as a refrigerant. It is also used as a preservative for dried fruit and as a bleach for textiles and flour, but its most important use is in the production of sulfuric acid. The sulfite salts are reasonably good reducing agents, being readily oxidized to the sulfate ion.

In sulfur dioxide and the sulfites, the oxidation number of sulfur is +4, an intermediate value in its range from −2 to +6. Hence these compounds can act as either oxidizing agents or reducing agents (Fig. 20.8). For instance, HSO_3^- ions oxidize HS^- ions in a reaction that produces thiosulfate ions, $S_2O_3^{2-}$:

$$4HSO_3^-(aq) + 2HS^-(aq) \longrightarrow 3S_2O_3^{2-}(aq) + 3H_2O(l)$$

Sodium thiosulfate, or photographer's hypo, can also be made by boiling aqueous sodium sulfite with sulfur, which is oxidized:

$$SO_3^{2-}(aq) + S(s) \longrightarrow S_2O_3^{2-}(aq)$$

In Box 3.1 we saw that thiosulfate ions are used to remove any residual silver halide from photographic film emulsions by forming a soluble complex of silver.

FIGURE 20.8 Sulfur dioxide is a reducing agent. When SO_2 is bubbled (center) through an aqueous solution of bromine (left), it reduces the Br_2 to colorless bromide ions (right).

V Video 53: Sulfur dioxide.

D Acid rain. CD2, 87.

Sulfur trioxide and the sulfates. By far the most important reaction of sulfur dioxide is its oxidation to sulfur trioxide, SO_3 (**7**):

$$2SO_2(g) + O_2(g) \longrightarrow 2SO_3(g)$$

The direct reaction is very slow, so an SO_2 molecule survives for a few days in the atmosphere before it is oxidized to SO_3. The oxidation is catalyzed by the metal ions in the minerals on the walls of buildings and in the presence of metal cations like Fe^{3+}, which may be dissolved in droplets of water. Other routes from SO_2 to SO_3 include reactions with $\cdot OH$, H_2O_2, and O_3 formed by the effect of sunlight on air and water vapor. The sulfur trioxide produced by these processes reacts with atmospheric water to form dilute sulfuric acid:

$$SO_3(g) + H_2O(l) \longrightarrow H_2SO_4(aq)$$

This product, together with the nitric acid formed by dissolved NO_x, falls as "acid rain."

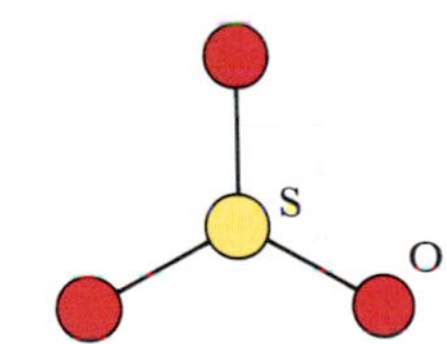

7 Sulfur trioxide

At normal temperatures sulfur trioxide is a volatile liquid (its boiling point is 45°C) composed of trigonal planar SO_3 molecules. In the solid, and to some extent in the liquid, these molecules form trimers (unions of three molecules) of composition S_3O_9 (**8**) and larger combinations.

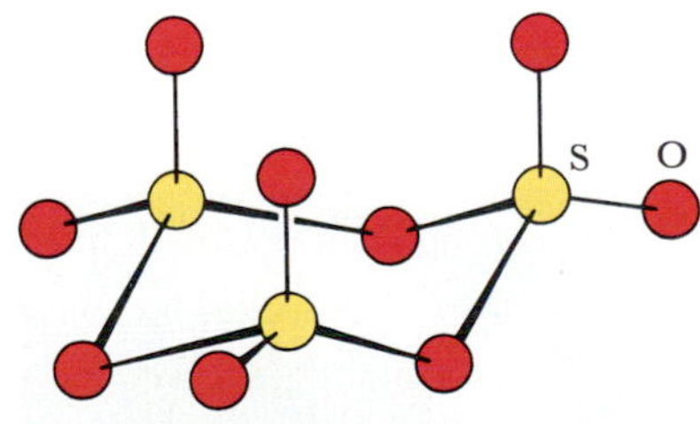

8 Sulfur trioxide trimer

Sulfuric acid is produced commercially in the **contact process**, which is described in Section 12.4 and Case 20. Sulfuric acid is a colorless, corrosive, oily liquid that boils (and decomposes) at about 300°C. It has three chemically important properties: it is a Brønsted acid, a dehydrating agent, and an oxidizing agent (Fig. 20.9).

Sulfuric acid is a strong acid in the sense that its first ionization is almost complete at normal concentrations in water. However, its conjugate base, HSO_4^-, is only a weak acid:

$$H_2SO_4(aq) + H_2O(l) \longrightarrow H_3O^+(aq) + HSO_4^-(aq)$$

$$HSO_4^-(aq) + H_2O(l) \rightleftharpoons H_3O^+(aq) + SO_4^{2-}(aq) \qquad pK_{a2} = 1.92$$

Because sulfuric acid is very cheap, it is widely used in industry, particularly for the production of fertilizers, petrochemicals, dyestuffs, and detergents. About two-thirds of the H_2SO_4 produced is used in the

U Exercises 20.72 and 20.73.

D The self-lighting candle. CD2, 58. Dehydrating action of sulfuric acid. CD1, 37.

D Dehydration of sucrose. CD1, 38.

V Video 19: Exothermic reactions II: The dehydration of sucrose.

manufacture of ammonium sulfate and phosphate fertilizers. Its reaction with phosphate rock to produce superphosphate fertilizer was discussed in Section 19.9.

Sulfuric acid is a powerful dehydrating agent. In a spectacular demonstration of this property, a little of the concentrated acid is poured on sucrose, $C_{12}H_{22}O_{11}$. A black, frothy mass of carbon forms as a result of the extraction of H_2O from sucrose by the acid (Fig. 20.10):

$$C_{12}H_{22}O_{11}(s) \xrightarrow{\text{conc } H_2SO_4} 12C(s) + 11H_2O(l)$$

The froth is caused by the gases CO and CO_2 that are also formed from the partial dehydration of sucrose. A rather more useful reaction of this kind is the production of carbon monoxide from formic acid:

$$HCOOH(l) \xrightarrow{\text{conc } H_2SO_4} CO(g) + H_2O(l)$$

Common dilute acids, such as hydrochloric and sulfuric acids, do not react with copper metal. However, like nitric acid (Section 19.9), concentrated sulfuric acid acts as an oxidizing acid and oxidizes copper metal to produce sulfur dioxide:

$$Cu(s) + 2H_2SO_4(aq, \text{ conc}) \xrightarrow{\Delta} CuSO_4(aq) + SO_2(g) + 2H_2O(l)$$

Some important sulfates and their uses are listed in Table 20.2.

FIGURE 20.9 Sulfuric acid is an oxidizing agent. When some concentrated acid is poured onto solid sodium bromide, the Br^- ions are oxidized to bromine, which colors the solution brown. (This reaction is effectively the reverse of the reaction in Fig. 20.8.)

E Concentrated sulfuric acid is a mild oxidizing agent for Br^- and I^- ions.

Sulfur halides. Sulfur reacts directly with all the halogens except iodine. It ignites in fluorine and burns brightly to give sulfur hexafluoride, SF_6 (**9**), a colorless, tasteless, odorless, nontoxic, thermally stable, insoluble, nonpolar gas. Its stability is, in essence, due to the F atoms that surround the central S atom and protect it from attack. The ionization energy of SF_6 is very high because any electron that is removed must come from the highly electronegative F atoms. Even quite strong electric fields cannot strip off these electrons, so SF_6 is a good gas-phase electrical insulator. It is, in fact, a much better insulator than air and is used in switches on high-voltage power lines.

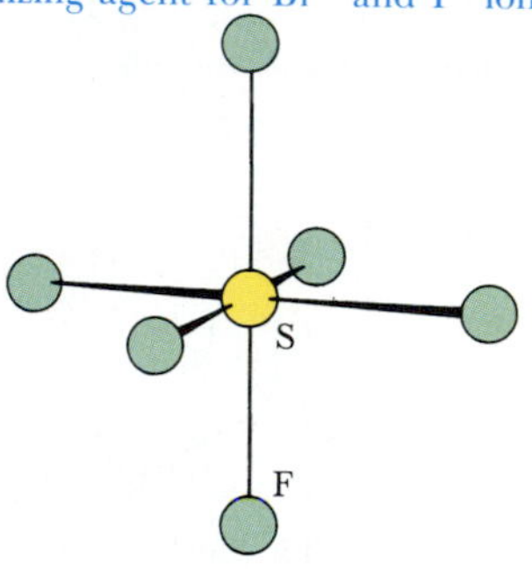

9 Sulfur hexafluoride

D Add a few drops of concentrated sulfuric acid to solid sodium bromide, sodium iodide, and copper(II) sulfate pentahydrate.

Sulfur reacts directly with chlorine. One product is disulfur dichloride, S_2Cl_2, a yellow liquid with a disgusting smell; it is used for the vulcanization of rubber. When disulfur dichloride reacts with more chlorine in the presence of iron(III) chloride as a catalyst, sulfur di-

TABLE 20.2 Some important sulfate salts

Formula	Common name	Application
$CaSO_4 \cdot 2H_2O$	gypsum	soil conditioner, plasterboard
$2CaSO_4 \cdot H_2O$	plaster of Paris	casts for broken limbs
$BaSO_4$	barite	gastrointestinal X-rays, drilling mud in oil fields
$Na_2SO_4 \cdot 10H_2O$	Glauber's salt	glass making
$PbSO_4$	—	formed on discharge of lead-acid battery
$CuSO_4 \cdot 5H_2O$	blue vitriol	fungicide
$FeSO_4 \cdot 7H_2O$	green vitriol	lawn care
$MgSO_4 \cdot 7H_2O$	Epsom salts	laxative

FIGURE 20.10 Sulfuric acid is a dehydrating agent. When concentrated sulfuric acid is poured onto sucrose, the sucrose is dehydrated, leaving a frothy black mass of carbon.

chloride, SCl_2, is produced as an evil-smelling red liquid. Sulfur dichloride reacts with ethylene to give mustard gas (**10**), which has been used for chemical warfare. Mustard gas causes blisters, discharges from the nose, and vomiting; it also destroys the cornea of the eye. All in all, it is easy to see why ancient civilizations associated sulfur with the underworld.

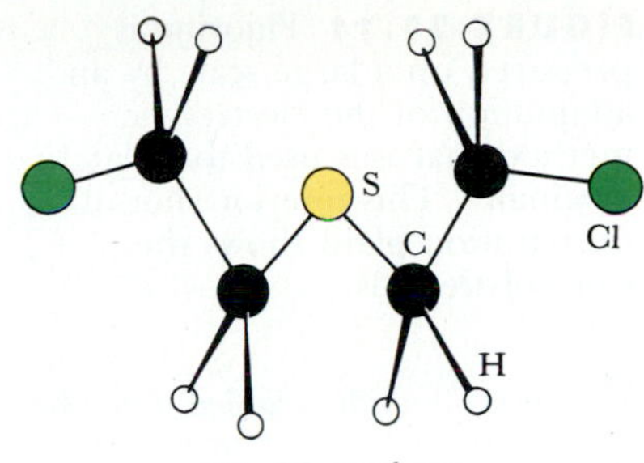

10 Mustard gas

U Exercise 20.70.

GROUP VII: THE HALOGENS

The halogens, a name derived from the Greek words meaning "salt producer," are the elements of Group VII. The elements have high electronegativities, which decrease down the group; small atomic and ionic radii, which increase down the group; and high electron affinities. With metals, the halogens tend to form largely ionic compounds; with nonmetals, their compounds are largely covalent. The elements are listed in the margin and a few of their physical properties are summarized in Table 20.3.

					VI	VII	VIII
							He
Li	Be	B	C	N	O	**F**	Ne
					S	**Cl**	Ar
					Se	**Br**	Kr
					Te	**I**	Xe
					Po	**At**	Rn

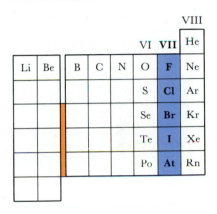

20.4 GROUP VII ELEMENTS

The many and varied chemical properties of the halogens can be traced to the fact that their atoms, which have a valence configuration ns^2np^5, need only one more electron to reach a noble gas configuration. To complete the octet of electrons in the valence shell, all halogen atoms combine to form diatomic molecules, such as F_2 and I_2. Fluorine, at the top of the group, shows some marked differences from its congeners, particularly in the low solubilities of fluoride salts and the fact that hydrofluoric acid is a weak acid.

E The naming of fluorine. *J. Chem. Ed.*, **1976**, *53*, 27. Fluorine with positive oxidation numbers. *J. Chem. Ed.*, **1971**, *48*, 625. Fluorine: What's the use? *J. Chem. Ed.*, **1990**, *67*, 373.

Fluorine. Fluorine occurs widely in many minerals, including fluorspar (CaF_2), cryolite (Na_3AlF_6), and fluorapatite ($Ca_5F(PO_4)_3$). It is so strongly oxidizing ($E°$ for the F_2/F^- couple is +2.87 V) that only an anode can be made more strongly oxidizing (by increasing its positive charge); hence, only in an electrolytic cell can fluorine be driven out of

E Fluorine: Chemical of the month. *J. Chem. Ed.*, **1983**, *60*, 759.

TABLE 20.3 Group VII elements (the halogens)

Valence configuration: ns^2np^5

Z	Name	Symbol	Molar mass, g/mol	Melting point, °C	Boiling point, °C	Density, g/cm³	Normal form
9	fluorine	F	19.00	−220	−188	1.70×10^{-3} at 0°C	almost colorless gas
17	chlorine	Cl	35.45	−101	−34	3.21×10^{-3} at 0°C	yellow-green gas
35	bromine	Br	79.91	−7	59	3.12	red-brown liquid
53	iodine	I	126.90	114	184	4.95	purple-black nonmetallic solid
85	astatine*	At	210	300	350	—	nonmetallic solid

*Radioactive.

FIGURE 20.11 Fluorine is prepared on a large scale by an adaptation of the electrolytic method that was used to isolate it originally. This interior shot of a preparation plant shows the electrolytic cells.

E Tin(II) fluoride is added to toothpaste to convert hydroxyapatite of dental enamel to fluorapatite, which is harder.

U Exercise 20.76.

E The inertness of fluorocarbons stems largely from the greater strength of the C—F bond (484 kJ/mol) than the C—H bond (412 kJ/mol). The larger sizes of the F atoms compared with the H atoms is a kinetic factor as it affords steric protection of the carbon chain against attack by incoming reactants.

N The reaction of iron with chlorine can be performed only in an anhydrous environment. The chlorine gas is first passed through a P_4O_{10} drying tube before it reacts with the iron. Heat is required to initiate the reaction, but the reaction is strongly exothermic once initiated.

D Preparation and properties of chlorine. BZS2, 6.28. Reaction of iron and chlorine. BZS1, 1.27.

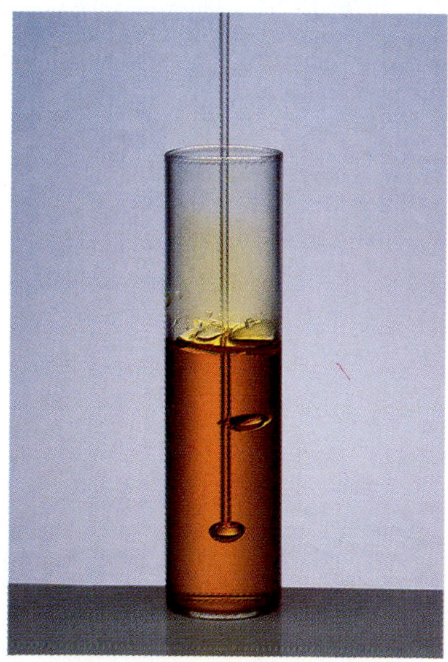

FIGURE 20.12 Chlorine is an oxidizing agent. When chlorine is bubbled through a solution of bromide ions, it oxidizes them to bromine, which colors the solution brown.

its compounds by removing an electron from F^- ions. The normal procedure for producing elemental fluorine is to electrolyze an anhydrous molten mixture of potassium fluoride and hydrogen fluoride at about 75°C with a carbon anode.

Fluorine is a reactive, colorless gas of F_2 molecules. It was little used before the development of the nuclear industry but is now produced on a large scale—at the rate of about 5×10^6 kg a year in the United States (Fig. 20.11). Most of the fluorine produced is used to make the volatile UF_6 as part of the procedure for processing nuclear fuel. Much of the rest is used in the production of SF_6 for electrical equipment and in the production of fluorinated hydrocarbons, such as Teflon and the Freons. Most fluoro-substituted hydrocarbons are relatively inert chemically: they are inert to oxidation by air, hot nitric acid, concentrated sulfuric acid, and other strong oxidizing agents.

Chlorine. Chlorine is ranked tenth in production in the United States, with approximately 1×10^{10} kg being produced annually. It is widely available as sodium chloride and is obtained by electrolysis, either of molten rock salt or of brine (Fig. 17.26). Chlorine is a pale yellow-green gas of Cl_2 molecules that condenses at −34°C. It reacts directly with nearly all the elements (the exceptions being carbon, nitrogen, oxygen, and the noble gases). It is a strong oxidizing agent and oxidizes metals to high oxidation states; for example, anhydrous iron(III) chloride, not iron(II) chloride, is formed when chlorine reacts with iron:

$$2Fe(s) + 3Cl_2(g) \longrightarrow 2FeCl_3(s)$$

Chlorine is used in a number of industrial processes, including the manufacture of plastics, solvents, and organic chemicals in general. For example, chlorine readily extracts hydrogen from many hydrocarbons. Thus, with benzene, chlorine produces chlorobenzene, a slightly polar organic solvent and an intermediate in the manufacture of dyes:

$$C_6H_6(l) + Cl_2(g) \longrightarrow C_6H_5Cl(l) + HCl(g)$$

Chlorine is used as a bleach in the paper and textile industries and as a disinfectant in water treatment—an application that has made large-scale community living possible. The "chlorine" smell of chlorinated water comes largely from the amines (organic compounds containing the $—NH_2$ group) that have become chlorinated (converted to $—NHCl$).

Bromine and iodine. Chlorine is also used to produce bromine from brine wells through the oxidation of Br^- ions (Fig. 20.12).

$$2Br^-(aq) + Cl_2(g) \longrightarrow Br_2(l) + 2Cl^-(aq)$$

Air is bubbled through the solution to vaporize the bromine and drive it out.

Bromine is a corrosive, red-brown fuming liquid of Br_2 molecules with a penetrating odor (its name comes from the Greek word for "stench"). It is increasingly used in industrial chemistry because of the ease with which it can be added to and removed from organic chemicals. Organic bromides are incorporated into textiles as fire retardants and are used as pesticides; inorganic bromides, particularly silver bromide, are used as photographic emulsions. An application of a different kind takes advantage of the very high density of some aqueous bromide solutions. Saturated aqueous zinc bromide, for example, has a density of 2.3 g/mL and is used in the oil industry to control the escape of oil from deep wells. The tall column of solution that results when this solution is poured down a well exerts a very high pressure at its base.

Iodine occurs as iodide ions in brines and as an impurity in Chile saltpeter. It was once obtained from seaweed, which contains high concentrations accumulated from seawater: 2000 kg of seaweed produce about 1 kg of iodine. The best modern source is the brine from oil wells, because the oil itself was produced by the decay of marine organisms that had accumulated the iodine while they were alive. The element is produced by oxidation with chlorine:

$$Cl_2(g) + 2I^-(aq) \longrightarrow I_2(aq) + 2Cl^-(aq)$$

Iodine is only slightly soluble in water, in which it gives a brown solution; in other solvents it dissolves to give a wide variety of colors that arise from the different interactions between the I_2 molecules and the solvent (Fig. 20.13). It is much more soluble if I^- ions are present, because the triiodide ion, I_3^-, is formed:

$$I_2(aq) + I^-(aq) \longrightarrow I_3^-(aq)$$

Iodine is a purple-black, lustrous solid of I_2 molecules. It sublimes to a purple vapor at 184°C. The element itself has few direct uses, but a solution in alcohol is used as a mild oxidizing antiseptic. Iodine is an essential element—in minute quantities—for living systems; a deficiency leads to a swelling of the thyroid gland in the neck. Iodides are added to table salt (to produce "iodized salt") to prevent this deficiency.

Trends and differences. The halogens are a closely related family of elements that (apart from some characteristics of fluorine) show a smooth variation of properties. The melting and boiling points and enthalpies of melting and vaporization all increase steadily down the group, as we would expect when London forces between molecules are dominant (Fig. 20.14). The electronegativities decrease down the

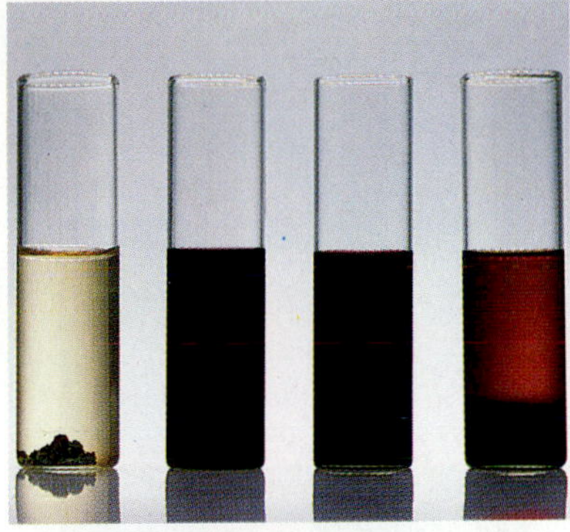

FIGURE 20.13 The color of an iodine solution depends strongly on the identity of the solvent: (from left to right) water, carbon tetrachloride, benzene, ethanol.

E The fire retardant nature of bromine compounds comes from their release of Br atoms when they burn: these odd-electron atoms combine with radicals and inhibit the chain reaction responsible for the combustion.

U Exercise 20.79.

V Video 49: Reactions of a halogen.

D Liquid iodine. CD2, 37. Halogens compete for electrons. CD2, 33. A variation of the starch-iodine clock reaction. CD2, 78.

E Crude iodine production processes. *J. Chem. Ed.*, **1987**, *64*, 152.

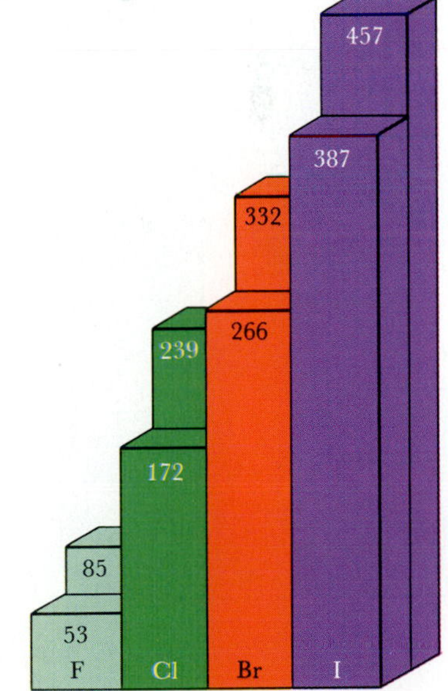

FIGURE 20.14 The normal melting points (front row) and boiling points (back row) of the halogens. The values shown are in kelvins.

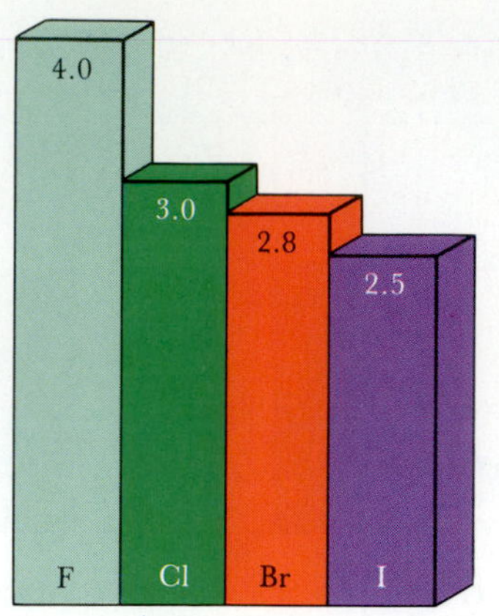

FIGURE 20.15 The electronegativities of the halogens decrease steadily down the group.

group (Fig. 20.15), and the atomic and ionic radii increase smoothly (Fig. 20.16).

▼ EXAMPLE 20.2 Predicting the relative oxidizing abilities of the halogens

Predict the trend in oxidizing character of the halogens in aqueous solution.

STRATEGY Trends in oxidizing ability can be predicted by examining the standard reduction potentials of the X_2/X^- couples, where X is a halogen. The more positive the value of $E°$, the more strongly oxidizing will be the X_2. Data are given in Appendix 2B.

SOLUTION The standard reduction potentials of the X_2/X^- couples change from the highly positive, oxidizing value for fluorine (+2.87 V) to the only slightly positive value for iodine (+0.54 V). Hence, whereas fluorine, chlorine, and bromine are good oxidizing agents, iodine is only weakly oxidizing.

EXERCISE E20.2 Which halide ion is the strongest reducing agent in aqueous solution?

▲ [*Answer*: I^-]

D Reversible oxidation-reduction color changes. CD2, 96.

Fluorine, the first member of the group, has a number of peculiarities that stem from its high electronegativity, small size, and lack of available *d* orbitals. It is the most electronegative element of all and has an oxidation number of −1 in all its compounds. Its high electronegativity and small size contribute to fluorine's ability to induce high oxidation numbers in other elements. Smallness helps because it allows several F atoms to pack around a central atom, as seen in IF_7. Oxygen, however, is sometimes even better in this role than fluorine because an O atom increases the oxidation number of the other element by 2 rather than by 1; hence fewer O atoms are needed to produce the same oxidation number. In XeO_4, for example, Xe has the oxidation number +8, but XeF_8 is unknown.

The small size of the fluoride ion also leads to high lattice enthalpies for its ionic compounds (see Table 8.1). As a consequence of these high lattice enthalpies the solubilities of fluorides are lower than those of the other halides (although there are some exceptions, including AgF). For example, calcium fluoride, CaF_2, and barium fluoride, BaF_2, are sparingly soluble at 25°C, with $K_s = 4.0 \times 10^{-11}$ and 1.7×10^{-6}, respectively, whereas all other calcium and barium halide salts are soluble. This difference in solubility is one of the reasons why the oceans are rich in chlorides rather than fluorides, even though fluorine is more abundant than chlorine in the Earth's crust.

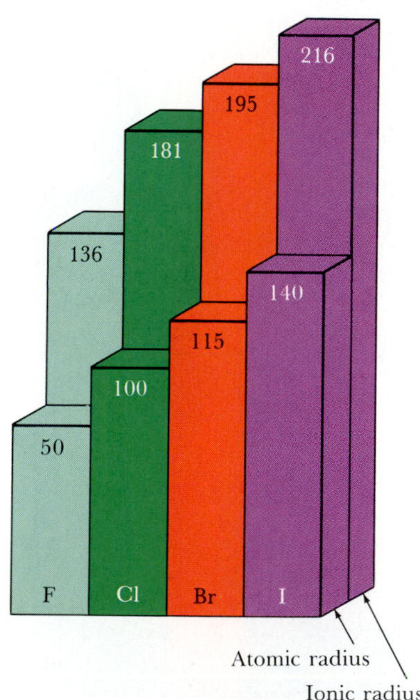

FIGURE 20.16 The atomic and ionic radii of the halogens increase steadily down the group. The values shown are in picometers.

Covalent fluorides are often more volatile than the corresponding covalent chlorides, because the former have fewer electrons, so the London forces between molecules are weaker. However, there are some striking exceptions; for example, hydrogen fluoride is less volatile than hydrogen chloride. This difference is explained by fluorine's ability to take part in hydrogen bonding, with the result that the HF molecules stick together strongly in the liquid.

The interhalogens. The halogens form compounds among themselves. These **interhalogens** have the formulas XX′, XX'_3, XX'_5, and XX'_7, where X is the heavier (and larger) of the two halogens. Only some of the possible combinations have been prepared (Table 20.4), partly for steric reasons. Thus, whereas seven F atoms can fit around an I atom to give IF_7, seven Cl atoms cannot; hence ICl_7 is unknown.

All the interhalogens are prepared by direct reaction of the two halogens, the product formed being determined by the proportions of reactants used. For example,

$$Cl_2(g) + 3F_2(g) \longrightarrow 2ClF_3(g)$$

$$Cl_2(g) + 5F_2(g) \longrightarrow 2ClF_5(g)$$

In general the interhalogens have physical properties that are intermediate between those of their parent halogens. Their chemical properties are dominated by the decreasing X—X′ bond enthalpy as X′ becomes heavier. The fluorides of the heavier halogens are all very reactive. Bromine trifluoride is so reactive that even asbestos burns in it.

TABLE 20.4 Known interhalogens

Interhalogen	Normal form
XF_n	
ClF	colorless gas
ClF_3	colorless gas
ClF_5	colorless gas
BrF	pale brown gas
BrF_3	pale yellow liquid
BrF_5	colorless liquid
IF	unstable
IF_3	yellow solid
IF_5	colorless liquid
IF_7	colorless gas
XCl_n	
$BrCl$	red-brown gas
ICl	red solid
I_2Cl_6	yellow solid
XBr_n	
IBr	black solid

T OHT: Table 20.4.

20.5 HALIDES

The hydrogen halides, HX, can be prepared by the direct reaction of the elements:

$$H_2(g) + X_2(g) \longrightarrow 2HX(g)$$

Fluorine reacts explosively by a radical chain reaction as soon as the gases are mixed. A mixture of hydrogen and chlorine explodes when exposed to light. Bromine and iodine react with hydrogen much more slowly. A less hazardous laboratory source of the hydrogen halides is a mixture of a nonvolatile acid and a metal halide:

$$CaF_2(s) + 2H_2SO_4(aq, \text{conc}) \longrightarrow Ca(HSO_4)_2(aq) + 2HF(g)$$

The product is isolated by distillation. Because Br^- and I^- are oxidized by sulfuric acid, phosphoric acid is used in the preparation of HBr and HI:

$$KI(s) + H_3PO_4(aq) \xrightarrow{\Delta} KH_2PO_4(aq) + HI(g)$$

N See the annotation on p. 770.

All the hydrogen halides are colorless, pungent gases, but hydrogen fluoride is a liquid at temperatures below 20°C (Table 20.5). Its low volatility is a sign of extensive hydrogen bonding, and short zigzag chains of hydrogen-bonded molecules, up to about $(HF)_5$, survive to some extent in the vapor. All the hydrogen halides dissolve in water to give acidic solutions. All except hydrofluoric acid are strong acids, with hydroiodic acid the strongest.

E When organic compounds are analyzed by infrared spectroscopy, the sample cell may have windows cut from a single crystal of sodium chloride (which is transparent down to 700 cm^{-1}) or potassium bromide (down to 400 cm^{-1}). Other ways of preparing the sample include grinding into a paste with a hydrocarbon oil or pressing it into a solid disk, perhaps with powdered potassium bromide.

TABLE 20.5 Hydrogen halides

Compound	Molar mass, g/mol	Melting point, °C	Normal boiling point, °C	pK_a at 25°C	Bond enthalpy, kJ/mol	Bond length, pm
HF	20.01	−83	20	3.45	565	92
HCl	36.46	−115	−85	strong acid	421	127
HBr	80.92	−89	−67	strong acid	366	141
HI	127.91	−51	−35	strong acid	299	161

(a)

(b)

FIGURE 20.17 When a mixture of hydrofluoric acid and ammonium fluoride are swirled inside a flask (a), the reaction with the silica in the cover glass frosts its surface (b).

V Video 50: Etching of glass.

U Exercises 20.77 and 20.80.

Hydrofluoric acid has the distinctive property of attacking glass and silica. The reaction is the attack of the strong Lewis base F^- on the Si atoms, thereby replacing the O atoms:

$$6HF(aq) + SiO_2(s) \longrightarrow H_2SiF_6(aq) + 2H_2O(l)$$

This reaction is used in glass etching (Fig. 19.12), and the interiors of lamp bulbs are frosted with a solution of hydrofluoric acid and ammonium fluoride in a similar reaction (Fig. 20.17).

Anhydrous metal halides may be formed by direct reaction:

$$Cu(s) + Cl_2(g) \longrightarrow CuCl_2(s)$$

They are also formed by reaction of the halogen with an oxide of the metal in the presence of a reducing agent:

$$Cr_2O_3(s) + 3C(s) + 3Cl_2(g) \xrightarrow{\Delta} 2CrCl_3(s) + 3CO(g)$$

Halides—not only anhydrous halides—are also formed by neutralization and by precipitation. An example of the latter is

$$BaCl_2(aq) + 2KF(aq) \longrightarrow BaF_2(s) + 2KCl(aq)$$

Halides of metals tend to be ionic unless the metal has an oxidation number greater than +2. For example, sodium chloride and copper(II) chloride are ionic compounds and have high melting points, but $TiCl_4$ and $FeCl_3$ have properties of covalent compounds in that they sublime as molecules.

20.6 HALOGEN OXIDES AND OXOACIDS

The halogen oxoacids are listed in Table 20.6. It is helpful to keep track of the oxidation number of the halogen in these compounds, as we did for the oxides of nitrogen. By doing so, we can make informed estimates of the relative strengths of the acids as proton donors and,

TABLE 20.6 Halogen oxoacids

Oxidation number	General formula	General acid name	Known examples	pK_a
7	HXO_4	perhalic acid	$HClO_4$	strong
			$HBrO_4$	strong
			HIO_4	1.64
5	HXO_3	halic acid	$HClO_3$	strong
			$HBrO_3$	strong
			HIO_3	0.77
3	HXO_2	halous acid	$HClO_2$	2.00
			$HBrO_2$	unstable
1	HXO	hypohalous acid	HFO	unstable
			HClO	7.53
			HBrO	8.69
			HIO	10.64

because they are *oxo*acids, of their oxidizing strengths (see Section 14.9).

Hypohalous acids. The hypohalous acids, HXO, are prepared by direct reaction of the halogen with water. For example, chlorine gas disproportionates in water to produce hypochlorous acid and hydrochloric acid:

$$Cl_2(g) + H_2O(aq) \longrightarrow HClO(aq) + HCl(aq)$$

Hypofluorous acid is so unstable that it survives only at the freezing point of water: at higher temperatures, fluorine oxidizes water to oxygen:

$$2F_2(g) + 2H_2O(l) \longrightarrow 4HF(aq) + O_2(g)$$

Hypohalite ions, XO^-, are formed when the aqueous solution of a base is used in place of water. Sodium hypochlorite is produced from the electrolysis of brine when the electrolyte is rapidly stirred and the chlorine gas produced at the anode reacts with the hydroxide ion generated at the cathode:

$$\textit{Anode reaction:} \quad 2Cl^-(aq) \longrightarrow Cl_2(g) + 2e^-$$

$$\textit{Cathode reaction:} \quad 2H_2O(l) + 2e^- \longrightarrow H_2(g) + 2OH^-(aq)$$

The chlorine gas disproportionates to produce the hypochlorite ion:

$$Cl_2(g) + 2OH^-(aq) \longrightarrow ClO^-(aq) + Cl^-(aq) + H_2O(l)$$

Calcium hypochlorite is produced by passing chlorine gas over dry calcium oxide (quicklime). Two reactions occur:

$$Cl_2(g) + CaO(s) \longrightarrow CaCl(ClO)(s)$$

$$2Cl_2(g) + 2CaO(s) \longrightarrow Ca(ClO)_2 + CaCl_2(s)$$

Hypochlorites are useful because they can oxidize organic material. Because of this property hypochlorites are used as household bleaches and as disinfectants in swimming pools and other water supplies (see Case 15). Their action as oxidizing agents stems partly from the decomposition of hypochlorous acid in solution:

$$2HClO(aq) \longrightarrow 2H^+(aq) + 2Cl^-(aq) + O_2(g)$$

Chlorates. Chlorate ions, ClO_3^-, in which chlorine has an oxidation number of +5, are formed when chlorine reacts with hot concentrated aqueous alkali:

$$3Cl_2(g) + 6OH^-(aq) \xrightarrow{\Delta} ClO_3^-(aq) + 5Cl^-(aq) + 3H_2O(l)$$

They decompose when heated, to an extent that depends on whether or not a catalyst is present:

$$4KClO_3(s) \xrightarrow{\Delta} 3KClO_4(s) + KCl(s)$$

$$2KClO_3(s) \xrightarrow{\Delta,\ MnO_2} 2KCl(s) + 3O_2(g)$$

The latter reaction is convenient for the preparation of oxygen in the laboratory.

Chlorates have a number of uses that stem from their oxidizing ability. Potassium chlorate is used as an oxygen supply in fireworks and in safety matches. The heads of matches consist of a paste of potassium

N We have been writing the hypohalous acids as HXO throughout the text (to show their relationship to the other halogen oxoacids, HXO_n); their molecular structures, however, are H—O—X.

U Exercise 20.83.

D Preparation of chlorine gas from laundry bleach. CD1, 7.

D Oxidation and reduction of the halides. CD2, 110.

U Exercises 20.69 and 20.81.

D Preparation of oxygen gas from laundry bleach. CD1, 6.

E The oxygen either bubbles out of the solution or attacks oxidizable material. Similar reactions occur when chlorine is used to purify water, the first step being the formation of hypochlorous acid followed by its decomposition. The reverse of the disproportionation reaction, the comproportionation reaction $ClO^-(aq) + Cl^-(aq) + H_2O(l) \rightarrow Cl_2(g) + 2OH^-(aq)$, occurs if the pH of the solution is decreased, and chlorine gas is evolved. Chlorine evolution may occur if swimming pool water becomes acidic, and the odor of chlorine is then detected.

U Exercise 20.74.

V Video 48: A rapid oxidation of phosphorus.

E In strike-anywhere matches, the heat from the friction causes the potassium chlorate to oxidize tetraphorus trisulfide; this subsequent exothermic reaction ignites in turn the wood of the match.

E Thermochemistry of hypochlorite oxidations. *J. Chem. Ed.*, **1969**, *46*, 378.

FIGURE 20.18 A space shuttle booster rocket is being filled with a mixture of ammonium perchlorate and aluminum powder. This mixture provides propellant power for the rocket.

E Special perchloric acid fume hoods are built in some laboratories to avoid any possible reaction of the perchloric acid with organic material which can accumulate in the duct work of the general purpose laboratory fume hoods.

N See Case 6 for details of the "space shuttle" reaction.

chlorate, antimony sulfide, sulfur, and powdered glass; the striking strip contains red phosphorus. Potassium chlorate is used, rather than the cheaper sodium chlorate, because it is less hygroscopic (water-absorbing). Sodium chlorate is widely used as a weedkiller and defoliant in agriculture. It has also been used in warfare. Sodium chlorate is used principally, however, as a source of chlorine dioxide, ClO_2. The chlorine in ClO_2 has oxidation number +4, so a chlorate must be reduced to form it. Sulfur dioxide is a convenient reducing agent:

$$2NaClO_3(aq) + SO_2(g) + H_2SO_4(aq, \text{dilute}) \longrightarrow 2NaHSO_4(aq) + 2ClO_2(g)$$

Chlorine dioxide has an odd number of electrons and is a paramagnetic, yellow-green gas. It is used to bleach paper pulp; it does so by oxidizing various pigments in the pulp without degrading the wood fibers.

Perchlorates. The perchlorates are prepared by electrolytic oxidation of aqueous chlorates:

$$ClO_3^-(aq) + H_2O(l) \longrightarrow ClO_4^-(aq) + 2H^+(aq) + 2e^-$$

Perchloric acid, $HClO_4$, is prepared by the action of concentrated hydrochloric acid on sodium perchlorate, followed by distillation. It is a colorless liquid and the strongest of all common acids. Because chlorine has its highest oxidation number of +7 in the perchlorates, they are also powerful oxidizing agents; contact between perchloric acid and even a small amount of organic material can result in a dangerous explosion. One spectacular example of their oxidizing ability under controlled conditions is the use of a mixture of ammonium perchlorate and aluminum powder in the booster rockets of the space shuttle (Fig. 20.18). The ammonium ions and the aluminum are the "fuel," and the perchlorate ions are the source of oxygen.

GROUP VIII: THE NOBLE GASES

							VIII
						VII	He
Li	Be	B	C	N	O	F	Ne
						Cl	Ar
						Br	Kr
						I	Xe
						At	Rn

N Argon constitutes 0.934% by volume of the atmosphere, whereas carbon dioxide contributes only 0.034%.

D Inert [noble] gases. TDC, 10-7.

The elements in Group VIII, the noble gases, and some of their physical properties are listed in Table 20.7. Until their first compounds were prepared, their closed-shell electron configurations were taken to indicate that these elements are chemically inert; they are still sometimes incorrectly called the inert gases. However, in 1962 the British chemist Neil Bartlett synthesized xenon hexafluoroplatinate, $XePtF_6$, by the reaction of xenon with platinum hexafluoride:

$$Xe(g) + PtF_6(g) \longrightarrow XePtF_6(s)$$

A personal account of these syntheses is presented in Box. 20.1.

20.7 GROUP VIII ELEMENTS

All the Group VIII elements occur in the atmosphere as monatomic gases; the most common is argon, which is thirty times more abundant than carbon dioxide. The noble gases other than helium and radon (for which there are richer sources) are obtained by the fractional distillation of liquid air.

TABLE 20.7 Group VIII elements (the noble gases)

Valence configuration: ns^2np^6

Normal form: colorless monatomic gases

Z	Name	Symbol	Molar mass, g/mol	Melting point, °C	Boiling point, °C
2	helium	He	4.00	—	−269 (4.2 K)
10	neon	Ne	20.18	−249	−246
18	argon	Ar	39.95	−189	−186
36	krypton	Kr	83.80	−157	−153
54	xenon	Xe	131.30	−112	−108
86	radon*	Rn	222	−71	−62

*Radioactive.

E The noble gases and the periodic table. Telling it like it was. *J. Chem. Ed.*, **1969**, *46*, 569.

N Helium constitutes 5.12×10^{-4}% by volume of the atmosphere.

E Liquid helium is shipped in special Dewar containers in which an inner double-walled Dewar vessel is surrounded by another double-walled Dewar vessel.

E Divers who use helium as an oxygen diluent sound like Donald Duck because the speed of sound increases as helium is mixed with oxygen, and the pitch of the voice rises. The speed of sound in oxygen is 316 m/s at 0°C, but in pure helium it is 965 m/s; so vibrations of the same wavelength treble their frequency as one gas is replaced by the other.

D Effect of helium on voice. TDC, 3-8s.

E Neon use. *J. Chem. Ed.*, **1990**, *67*, 588. Light bulbs filled with krypton gas. *J. Chem. Ed.*, **1975**, *52*, 388.

Helium. Helium, the second most abundant element in the universe after hydrogen, is rare on Earth because its atoms are so light that they reach high average speeds and escape from the atmosphere. It is found as a component of natural gases trapped under rock formations in some locations (notably Texas), where it has collected as a result of the emission of α particles by radioactive elements. An α particle is a helium nucleus (He^{2+}), and an atom of the element forms when the particle picks up two electrons from its surroundings:

$$He^{2+} + 2e^- \longrightarrow He$$

Helium gas is twice as dense as hydrogen under the same conditions; however, because it is nonflammable, it is used to provide buoyancy in airships. Helium is also used to dilute oxygen in deep-sea diving and spacecraft atmospheres, as a pressurizing gas for rocket fuels, as a coolant, and as the excited gas in helium-neon lasers, which emit red light of wavelength 633 nm. The element has the lowest boiling point of any substance (4.2 K), and does not freeze to a solid at any temperature unless pressure is applied to hold the light, mobile atoms together. These properties make helium useful for the study of matter at very low temperatures—a field of research known as **cryogenics** (from the Greek word for "frost"). Below 2 K, liquid helium shows the remarkable property of **superfluidity**, the ability to flow without viscosity. The phase diagram in Fig. 20.19 shows the regions of pressure and temperature at which the different phases are stable. Helium is the only substance known to have more than one liquid phase.

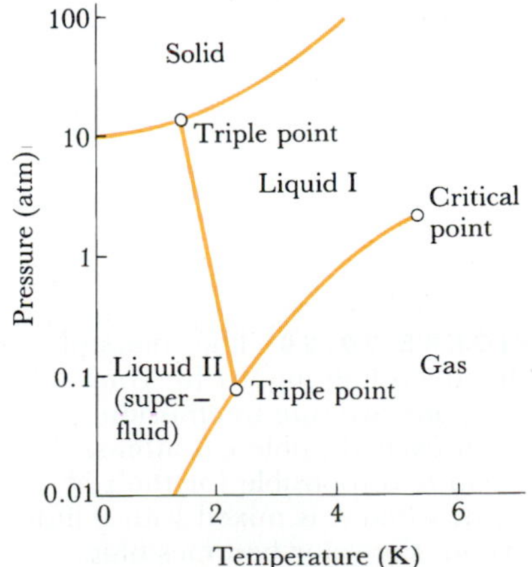

FIGURE 20.19 The phase diagram for helium-4.

Neon through radon. Neon, which emits a red glow when an electric current flows through it, is widely used in advertising signs (Fig. 20.20). Argon is used to provide an inert atmosphere for welding (to prevent oxidation) and to fill some types of light bulbs, where its function is to conduct heat away from the filament. Krypton gives an intense white light when a current is passed through it, so it is used in airport runway lighting. Because it is produced by nuclear fission, its atmospheric abundance is one measure of worldwide nuclear activity. Xenon is used in "halogen" lamps for automobile headlights and in high-speed photographic flash cubes because an electric discharge passing through it—a

BOX 20.1 ▸ NOBLE-GAS CHEMISTRY

NEIL BARTLETT, *Professor of Chemistry, University of California, Berkeley*

Removing an electron from an oxygen molecule requires a large quantity of energy, 1170 kJ. Only an oxidizing agent of extraordinary power can do this. In 1962 Derek Lohmann and I showed that platinum hexafluoride is just such an agent, because it oxidizes O_2 in the reaction

$$O_2(g) + PtF_6(g) \longrightarrow O_2PtF_6(s)$$

While thinking about other ways to test the oxidizing power of PtF_6, I noted that the first ionization energy of the noble gases decreases markedly with the increasing size of the atom, a consequence of the smaller effective nuclear charge experienced by the valence electrons as more and more inner electron shells fill (see Fig. 7.31). This shielding effect is so great in the noble gases that xenon has the same ionization energy as the O_2 molecule; and radon's first ionization energy is even lower than xenon's. Radon, however, is a dangerously radioactive gas, so I chose to use xenon in my tests.

When I mixed PtF_6 (a red gas) with Xe (a colorless gas), an orange-yellow solid ($XePtF_6$) immediately formed. This result was quickly followed by the synthesis in other laboratories of xenon fluorides (XeF_2, XeF_4, and XeF_6), xenon oxofluorides ($XeOF_4$ and XeO_3F_2), xenon oxides (XeO_3 and XeO_4), and a krypton fluoride (KrF_2). Other chemists also demonstrated that radon is more easily oxidized than xenon.

It was soon apparent that the chemically active noble gases are those from which electrons are most easily removed, that is, those with the lowest ionization energies. Indeed, most of our knowledge of the chemistry of noble gases comes from experiments with xenon. It was also evident that noble gases will react only with atoms that are small and highly electronegative (such as fluorine, oxygen, and nitrogen) and that have electronegative groups attached to them (e.g., SO_2F). Larger electronegative atoms, such as chlorine, do not make strong bonds. Consequently, $XeCl_2$ and $XeBr_2$ do not exist at ordinary temperatures and pressures.

We can obtain a more quantitative sense of the possible extent of noble gas chemistry by considering the noble gas difluorides. Both XeF_2 and KrF_2 have been synthesized and both are linear, symmetrical molecules. Xenon difluoride is easily made by placing a mixture of xenon and molecular fluorine in sunlight, whereupon colorless crystals rapidly form. Xenon difluoride is stable, and its dissociation into ground-state atoms requires an input of 272 kJ/mol. Krypton difluoride is also stable, but only 96 kJ/mol are needed to dissociate it into atoms.

Note that the difference in the dissociation energies (176 kJ/mol) of XeF_2 and KrF_2 is almost

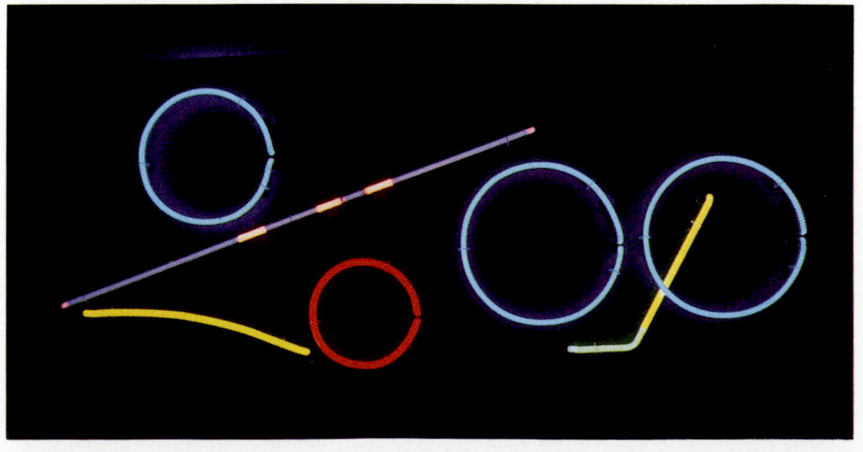

FIGURE 20.20 The colors of this neon lighting art by Tom Anthony are due to emission from excited noble gas atoms. Neon is responsible for the red light; when it is mixed with a little argon, the color becomes blue-green. The yellow color is achieved with appropriately colored glass.

the same as the difference in the first ionization energies of xenon and krypton (180 kJ/mol) (see Fig. 7.31); and recall that both XeF_2 and KrF_2 are linear molecules. In each of these molecules, bond formation requires that the noble gas atom (denoted as G) share an electron with each of the two fluoride atoms. The resulting charge distribution in the molecules is

$$^{-1/2}F—G^{+}—F^{-1/2}$$

The dissociation energy of GF_2 is, roughly, the sum of (1) the ionization energy of G, (2) the electron affinity of the two fluoride atoms, and (3) the Coulomb energy, but only (1) changes markedly, with G.

Argon difluoride, if it existed, would have a similar shape and charge distribution. But the ionization energy of argon is 170 kJ/mol higher than that of krypton. We might therefore expect the dissociation energy of argon difluoride to be that much smaller than the dissociation energy of krypton difluoride. As we have just noted, however, the dissociation energy of krypton is only 96 kJ/mol. It follows that the existence of ArF_2 is unlikely. Does this observation mean that there will be no argon chemistry? Not quite—it might be possible to prepare salts of one species, namely, the cation ArF^+.

The remarkable oxidizing power of the known cation KrF^+ (which has already been used to prepare the previously unknown species BrF_6^+ and ClF_6^+) is due to the high ionization energy of krypton. It follows that ArF^+ should be even more potent. We know, from careful theoretical work, that the cation made from Ar^+ and F is bound together by 205 kJ/mol. But to synthesize this cation, we must first find a suitable anion to stabilize the newly formed cation in an ionic solid. The stabilizing anion must have the following properties: (1) it must be small enough to give a highly favorable lattice enthalpy; (2) it must have a high ionization energy, so that it does not readily lose its electron to the cation; and (3) it must possess such strong bonds that it does not lose a charged ligand (such as F^-) to the cation (because $ArF^+ + F^-$ would give $Ar + F_2$). The best candidate is a hexafluoroanion, MF_6^- (where M denotes a metal), the most likely anion being SbF_6^-. (The structure of the analogous $XeFAsF_6$ salt is illustrated below.)

But how are we to arrange for the simultaneous formation of both ArF^+ and SbF_6^-, when the parent fluoride of the cation, ArF_2, does not exist? This fact poses a formidable (but not necessarily impossible) problem in synthesis. Certainly, ArF^+ salts will be the most powerful of all oxidizing agents—if they can be made.

miniature lightning flash—gives an intense white light. Xenon is also being investigated as a nonpoisonous anesthetic.

The radioactive gas radon seeps out of the ground as a product of radioactive processes deep in the Earth. There is now some concern that its accumulation in buildings can lead to dangerously high levels of radiation (see Case 22).

20.8 COMPOUNDS OF THE NOBLE GASES

The ionization energies of the noble gases are relatively high but decrease down the group (Fig. 20.21); krypton's ionization energy is low enough for electrons to be lost to very electronegative elements, especially fluorine. No compounds of helium, neon, and argon are known (except under very special conditions). Radon is known to react with

E Noble gas compounds: The views of William Ramsey and Giuseppe Oddo in 1902. *J. Chem. Ed.*, **1983**, *60*, 758.

U Exercise 20.82.

E The noble gas compounds: What might have been. *J. Chem. Ed.*, **1984**, *61*, 565.

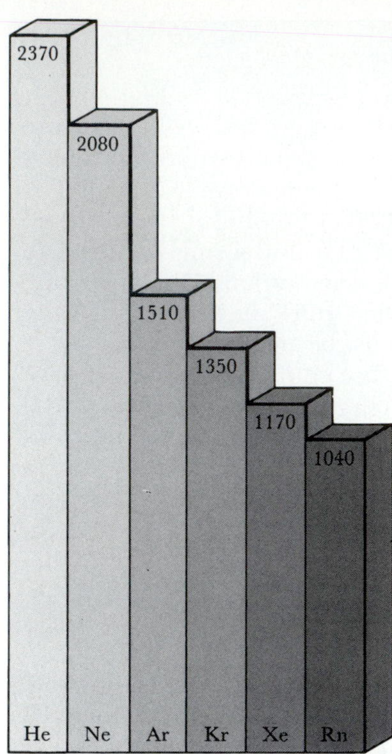

FIGURE 20.21 The ionization energies of the noble gases decrease steadily down the group. The values shown are in kilojoules per mole.

U Exercise 20.75.

E A decade of xenon chemistry. *J. Chem. Ed.*, **1974**, *51*, 628. Horacio Damianovich (1833–1959), Argentine pioneer in noble gas chemistry. *J. Chem. Ed.*, **1982**, *59*, 304. The thermochemical stability of ionic noble gas compounds. *J. Chem. Ed.*, **1988**, *65*, 119.

fluorine, but its radioactivity makes it so dangerous to handle that chemists have been disinclined to study its compounds. Because krypton forms only one known neutral molecule, KrF_2, xenon is the noble gas with the richest chemical properties. It forms several compounds with fluorine and oxygen, and compounds with Xe—N and Xe—C bonds have been reported. In 1988 the first compound with a Kr—N bond was reported, but it is stable only below −50°C.

Xenon fluorides. The starting point for the preparation of xenon compounds is the production of xenon difluoride, XeF_2, and xenon tetrafluoride, XeF_4, by heating a mixture of the elements to between 300° and 400°C. At higher pressures, fluorination proceeds as far as xenon hexafluoride, XeF_6. All three fluorides are crystalline solids. In the gas phase, all are covalent molecular compounds. Solid xenon hexafluoride, however, is ionic; its complex structure consists of XeF_5^+ cations bridged by F^- anions.

The xenon fluorides are used as powerful fluorinating agents (reagents for attaching fluorine atoms to other substances). The tetrafluoride will even fluorinate platinum:

$$Pt(s) + XeF_4(s) \longrightarrow Xe(g) + PtF_4(s)$$

Xenon oxides. The xenon fluorides are used to prepare a series of xenon oxides and oxoacids and, in a series of disproportionations, to bring the oxidation number of xenon up to +8. First, xenon tetrafluoride is hydrolyzed to xenon trioxide, XeO_3:

$$6XeF_4(s) + 12H_2O(l) \longrightarrow 2XeO_3(aq) + 4Xe(g) + 3O_2(g) + 24HF(aq)$$

The trioxide is the anhydride of xenic acid, H_2XeO_4. Xenon trioxide reacts with aqueous alkali to form a hydrogenxenate ion, $HXeO_4^-$. This ion slowly disproportionates into xenon and the octahedral perxenate ion, XeO_6^{4-}, in which the oxidation number of xenon is +8:

$$2HXeO_4^-(aq) + 2OH^-(aq) \longrightarrow XeO_6^{4-}(aq) + Xe(g) + O_2(g) + 2H_2O(l)$$

Perxenate solutions are yellow and, as expected for compounds of elements with such a high oxidation number, very powerful oxidizing agents.

When barium perxenate is treated with sulfuric acid, it is dehydrated to its anhydride, xenon tetroxide, XeO_4. With this compound, which is an explosively unstable gas, our journey through the *p* block of the periodic table comes to an end with a bang.

SUMMARY

20.1 Oxygen, sulfur, selenium, and tellurium are called the **chalcogens**. Oxygen occurs mainly as O_2 but also as its allotrope **ozone**, O_3. It is a highly electronegative element that brings out the high oxidation numbers of other elements. Sulfur is mined by the **Frasch process** and recovered by the **Claus process**, the oxidation of H_2S by SO_2. It is less electronegative than oxygen but more able to **catenate**, or form chains. Oxygen and sulfur react directly with almost all other elements, but sulfur does not react with nitrogen.

20.2 The most important compound of oxygen is **water**, which is amphiprotic (both a Brønsted acid and a Brønsted base), a Lewis base, an oxidizing agent, and a weak reducing agent. It is the parent of the **oxides** and **hydroxides**. Oxides of metals are basic; those of nonmetals are acidic. **Hydrogen peroxide**, a very weak acid, is also an oxidizing and a reducing agent and the parent of the **peroxides**. The sulfur analogue of water is **hydrogen sulfide**, the parent acid of the **sulfides**. Sulfur forms an extensive series of **polysulfanes** of formula H_2S_n.

20.3 Important oxides of sulfur include **sulfur dioxide** and **sulfur trioxide**, the anhydrides of **sulfurous acid** and **sulfuric acid**, respectively. Sulfuric acid is made by the **contact process** and is a Brønsted acid, a dehydrating agent, and an oxidizing agent. Sulfur reacts with the halogens to form the **sulfur halides**, which include the very stable **sulfur hexafluoride**.

20.4 The members of Group VII are called the **halogens**. They include fluorine, the most electronegative element of all. The halogens are characterized by their smoothly varying properties. The early members of the group have high electronegativities and are strongly oxidizing. The iodide ion is a reducing agent. Fluorine is highly reactive, partly because of its low bond enthalpy and the strengths of the bonds it forms with other elements. The character of an **interhalogen**, a binary compound of two different halogens, is intermediate between that of its component elements.

20.5 and 20.6 All the halogens form **hydrogen halides**, which dissolve in water to give the **hydrohalic acids**; except for hydrofluoric acid, these compounds are strong acids in water. The halogens form an important series of oxidizing oxoacids and oxoanions, of which the most important are **hypochlorous**, **chloric**, and **perchloric acids**.

20.7 and 20.8 The **noble gases** of Group VIII all have closed-shell electron configurations. Only xenon forms an extensive series of compounds, of which the most important are the fluorides, oxides, and oxoacids. The fluorides are strong fluorinating agents. They are hydrolyzed to the oxides, which are strong oxidizing agents.

CLASSIFIED EXERCISES

This chapter includes applications of the chemical principles that have been addressed in earlier chapters. A number of the exercises will require you to collect data from and review some of this earlier material.

20.1 Write the ground-state electron configuration of (a) He; (b) O; (c) F; (d) Cl^-.

20.2 Write the ground-state electron configuration of (a) Ar; (b) I; (c) S^{2-}; (d) S.

20.3 Refer to Appendix 2B and arrange the halogens in order of increasing oxidizing strength in water.

20.4 Refer to Appendix 2B and arrange S, O_2 (to H_2O), Cl_2, and H_2O_2 (to H_2O) in order of increasing oxidizing strength in water.

20.5 Arrange the elements Cl, O, Br, S in order of increasing electronegativity.

20.6 Arrange the elements S, Br, O, F in order of increasing atomic radius.

Group VI Elements

20.7 Describe, with chemical equations, two laboratory preparations of oxygen.

20.8 Write the chemical equation that represents the production of oxygen by photosynthesis.

20.9 Write the chemical formula of (a) calcium sulfite; (b) ozone; (c) barium peroxide.

20.10 Write the chemical formula of (a) sulfurous acid; (b) sodium thiosulfate; (c) disulfur dichloride.

20.11 Describe the Frasch process for extracting elemental sulfur from underground sulfur deposits.

20.12 Use chemical equations to describe the Claus process for the production of elemental sulfur from petroleum and natural gas wells.

20.13 Write equations for (a) the burning of lithium in oxygen; (b) the reaction of fluorine gas with water; (c) the oxidation of water at the anode of an electrolytic cell.

20.14 Write equations for the reaction of (a) sodium oxide and water; (b) sodium peroxide and water; (c) sulfur dioxide and water.

20.15 Complete and balance the following equations:
(a) $Na_2SO_3(aq) + S(s) \rightarrow$
(b) $H_2S(aq) + O_2(g) \rightarrow$
(c) $CaO(s) + H_2O(l) \rightarrow$

20.16 Complete and balance the following equations:
(a) $Cl_2(g) + H_2O(l) \rightarrow$
(b) $H_2(g) + S(s) \xrightarrow{\Delta}$
(c) $Cu(s) + H_2SO_4(aq, \text{conc}) \rightarrow$

20.17 Write a Lewis structure of SO_3 in which the formal charge on each atom is zero. What is the shape of the molecule?

20.18 Write the Lewis structure and predict the shape of (a) O_3; (b) SO_2; (c) SO_4^{2-}.

20.19 Hydrogen peroxide is unstable with respect to decomposition into water and oxygen when exposed to light, heat, or a catalyst. If 500 mL of a 3% H_2O_2 aqueous solution (like that sold in drugstores) decomposes, what volume of oxygen at STP will be produced? Assume the density of the solution to be 1.0 g/mL.

20.20 If 2.00 g of sodium peroxide is dissolved to form 200 mL of an aqueous solution, what would be the pH of the solution?

20.21 If you were titrating 0.10 M $H_2S(aq)$ with 0.10 M $NaOH(aq)$, at what pH would you have a 0.050 M $NaHS(aq)$ solution?

20.22 Calculate the volume of pure (concentrated) sulfuric acid (of density 1.84 g/mL) that can be produced from 100 g of sulfur.

20.23 Describe the trend in acidity of the binary acids of Group VI and account for the trend in terms of bond strength.

20.24 Write chemical equations that represent the acid or base character of (a) two metal oxides and (b) two nonmetal oxides.

20.25 (a) Calculate the standard enthalpy and entropy of reaction for the formation of ozone from oxygen. (b) Is the reaction favored at high temperatures or low temperatures? (c) Does the reaction entropy favor the spontaneous formation of ozone? Explain your conclusions.

20.26 When lead(II) sulfide is treated with hydrogen peroxide, the possible products are either (a) lead(IV) oxide and sulfur dioxide or (b) lead(II) sulfate. In terms of the reaction free energy, which product or products are more likely?

20.27 Fool's gold is FeS_2 and is so called because it looks like gold. What physical and chemical tests would distinguish it from gold?

20.28 Explain how to distinguish chemically between a solution containing the SO_3^{2-} ion and one containing the SO_4^{2-} ion.

Group VII Elements

20.29 List the natural sources of fluorine and chlorine.

20.30 List the natural sources of bromine and iodine.

20.31 Write chemical equations that describe the preparation of fluorine and chlorine.

20.32 Write chemical equations that describe the preparation of bromine and iodine.

20.33 Name each of the following compounds: (a) $HBr(aq)$; (b) ClO_2; (c) $NaIO_3$.

20.34 Write the chemical formulas of (a) potassium periodate; (b) ammonium perchlorate; (c) chloric acid.

20.35 Identify the oxidation number of the halogen in (a) ClO_2; (b) dichlorine heptoxide; (c) $NaIO_3$.

20.36 Identify the oxidation number of the halogen in (a) IF_7; (b) sodium periodate; (c) $HClO_2$.

20.37 Write the Lewis structure for each of the following species, write its VSEPR formula, and predict its shape: (a) ClO_4^-; (b) IO_3^-; (c) BrF_3.

20.38 Write the Lewis structure for each of the following species, write its VSEPR formula, and predict its shape: (a) IO_2^-; (b) I_3^-; (c) BrF_4^-.

20.39 Write the balanced equations for (a) the thermal decomposition of potassium chlorate without a catalyst; (b) the reaction of bromine with water; (c) the reaction between sodium chloride and concentrated sulfuric acid.

20.40 Complete the following skeletal equations:
(a) $I_2(s) + F_2(g) \rightarrow$ (any *one* of a number of products)
(b) $I_2(aq) + I^-(aq) \rightarrow$
(c) $Cl_2(g) + H_2O(l) \rightarrow$

20.41 Explain why iodine is much more soluble in a solution of potassium iodide than in pure water.

20.42 Write the chemical equations for the reaction of chlorine in (a) a neutral aqueous solution; (b) a dilute basic solution; (c) a concentrated basic solution.

20.43 (a) Arrange the chlorine oxoacids in order of increasing oxidizing strength. (b) Suggest an interpretation of that order in terms of oxidation numbers.

20.44 (a) Arrange the hypohalous acids in order of increasing acid strength. (b) Suggest an interpretation of that order in terms of electronegativities.

20.45 The booster rockets of the space shuttle are fueled by a mixture of aluminum and ammonium perchlorate. (a) Write the chemical equation for the reaction, assuming that the products are nitrogen, water, hydrochloric acid, and aluminum oxide. (b) Calculate the mass of ammonium perchlorate that should be carried for each kilogram of aluminum.

20.46 Sodium iodate is an impurity in Chile saltpeter, which was once the major source of iodine. The element was obtained by the reduction of an acidic solution of the iodate ion with sodium hydrogensulfite. (a) Write the chemical equation for the reaction, assuming the oxidized product to be HSO_4^-. (b) Calculate the mass of sodium hydrogensulfite needed to produce 50.0 g iodine.

20.47 Anhydrous hydrogen fluoride is made by the action of concentrated sulfuric acid on fluorspar, CaF_2. (a) Write the equation for the process. (b) Why is silica an undesirable impurity in this process? Use chemical equations to support your statement.

20.48 Write equations for reactions used to synthesize gaseous samples of (a) hydrogen chloride; (b) hydrogen bromide; (c) hydrogen iodide.

20.49 Write the Lewis structure for Cl_2O. What is the shape of the Cl_2O molecule and what is the Cl—O—Cl bond angle?

20.50 Write the Lewis structure for BrF_3. What is the hybridization of the bromine atom in the molecule?

20.51 Plot a graph of the standard free energy of formation of the hydrogen halides against the period number of the halogens. What observations can be drawn from the graph?

20.52 Use the data from Table 20.5 to plot the normal boiling point of the hydrogen halides against the period number of the halogens. Account for the trend revealed by the graph.

20.53 Use the data in Appendix 2B to determine whether chlorine gas will oxidize Mn^{2+} to form the permanganate ion in an acidic solution.

20.54 Which is the stronger oxidizing agent, ozone or fluorine? Explain your conclusion.

20.55 The standard free energy of formation of HI is +1.70 kJ/mol at 25°C. Determine the equilibrium constant for the formation of HI from its elements.

20.56 The standard free energy of formation of HCl is −95.3 kJ/mol at 25°C. What is the equilibrium constant for the decomposition of HCl(*g*)?

20.57 The concentration of F^- ions in an aqueous solution can be measured by adding lead(II) chloride solution and weighing the lead(II) chlorofluoride (PbClF) precipitate. Calculate the molar concentration of F^- ions in 25.00 mL of a solution that gave a lead chlorofluoride precipitate of mass 0.765 g.

20.58 Suppose 25.00 mL of an aqueous solution of iodine was titrated with 0.025 M $Na_2S_2O_3(aq)$; starch was the indicator. The blue color of the starch-iodine complex disappeared when 28.45 mL of the thiosulfate solution had been added. What was the molar concentration of I_2 in the original solution? The titration reaction is

$$I_2(aq) + 2S_2O_3^{2-}(aq) \longrightarrow 2I^-(aq) + S_4O_6^{2-}(aq)$$

Group VIII Elements

20.59 Write chemical equations for the preparation of (a) XeF_4 and (b) XeO_3.

20.60 Write chemical equations for the preparation of (a) XeF_6; (b) H_2XeO_4; (c) XeO_4.

20.61 Determine the oxidation number of the noble gases in (a) KrF_2; (b) XeF_2; (c) KrF_4; (d) XeO_4^{2-}.

20.62 Determine the oxidation number of the noble gases in (a) XeO_3; (b) XeO_6^{4-}; (c) XeF_6; (d) $HXeO_4^-$.

20.63 Xenon tetrafluoride is a powerful oxidizing agent. In an acidic solution it is reduced to xenon. Write the corresponding half-reaction.

20.64 Xenon hexafluoride reacts with water to produce xenon trioxide and hydrofluoric acid. Write the chemical equation for the reaction.

20.65 Predict the relative acid strengths of H_2XeO_4 and H_4XeO_6. Explain your conclusions.

20.66 Predict the relative oxidizing strengths of H_2XeO_4 and H_4XeO_6. Explain your conclusions.

20.67 Write the Lewis structure for XeF_4. Estimate the F—Xe—F bond angle.

20.68 Write the Lewis structure for XeO_3. What is the hybridization on the xenon atom in the molecule?

UNCLASSIFIED EXERCISES

20.69 What chemicals are used for disinfecting swimming-pool water and for controlling its pH?

20.70 Complete and balance the following skeletal equations:

(a) $H_2(g) + S_8(s) \xrightarrow{800°C}$
(b) $S_8(s) + F_2(g) \rightarrow$
(c) $N_2O_4(g) + O_2(g) \rightarrow$

20.71 When the enthalpy of vaporization of water is divided by its boiling point (on the Kelvin scale), the result is 110 J/(K·mol). For hydrogen sulfide, the same calculation gives 88 J/(K·mol). (a) What is the significance of these values? (b) Explain why the value for water is greater than that for hydrogen sulfide.

20.72 (a) Describe the sources and production of SO_2 and SO_3 and their corresponding oxoacids. (b) How does their production contribute to acid rain?

20.73 The annual production of sulfuric acid in the United States is approximately 4×10^{10} kg. (a) If the acid were all produced from elemental sulfur, what mass of sulfur would be used for the production of sulfuric acid? (b) What volume of sulfur trioxide (at 25°C and 5 atm) must be produced for the annual production of sulfuric acid?

20.74 Balance each skeletal equation, classify the reaction as acid-base or redox, and describe the chemical property that the equation shows:
(a) $KClO_3 \rightarrow KCl + O_2$
(b) $O_3 + H^+ + e^- \rightarrow O_2 + H_2O$
(c) $H_2S + O_2 \rightarrow SO_2 + H_2O$

20.75 Balance each skeletal equation, classify the reaction as acid-base or redox, and describe the chemical property that the equation shows:
(a) $F_2 + NaOH \rightarrow OF_2 + NaF + H_2O$
(b) $S_2O_3^{2-} + Cl_2 + H_2O \rightarrow HSO_4^- + H^+ + Cl^-$
(c) $XeF_6 + OH^- \rightarrow XeO_6^{4-} + Xe + O_2 + F^- + H_2O$

20.76 (a) Write the chemical equation for the produc-

CASE 20 ▸ THE NUMBER-1 CHEMICAL

Sulfuric acid is the chemical manufactured in the greatest mass: in 1990 the production of sulfuric acid in the United States amounted to 4.03×10^{10} kg. The acid, which was once called *oil of vitriol,* is often referred to as the "old horse of chemistry" because of its versatility and low cost of production; it is used in most industrial processes as an acid, but in concentrated form can be used as an oxidizing agent too. Because of its widespread use, sulfuric acid production is also a business indicator, that is, a measure of the economic vigor of a nation (see illustration opposite).

Sulfur dioxide, SO_2, which is obtained from the combustion of sulfur or recovered from the combustion of fossil fuel that contains sulfur, is the raw material for the production of sulfuric acid in the *contact process.* In this process, a heated mixture of sulfur dioxide and oxygen is passed over a heated vanadium pentoxide catalyst, and the sulfur dioxide is oxidized to sulfur trioxide, SO_3, the acid anhydride of sulfuric acid:

$$2SO_2(g) + O_2(g) \xrightarrow{V_2O_5,\ 500°C} 2SO_3(g) \qquad \Delta H° = -198 \text{ kJ}$$

The reaction yield is 96 to 98%, but modifications to the process can increase the yield to over 99%. The mixture of product gases, which is mostly sulfur trioxide, is cooled to 100°C before entering a tower that is packed with quartz and in which is maintained a continuously circulating concentrated sulfuric acid solution. The SO_3 reacts with the concentrated H_2SO_4 to form oleum, $H_2S_2O_7$ (which is also called fuming sulfuric acid). In the final step of the process, the oleum is slowly diluted with water to produce sulfuric acid:

$$H_2S_2O_7(l) + H_2O(l) \longrightarrow 2H_2SO_4(l)$$

The sulfur trioxide is not added directly to water because the exothermicity of the reaction results in the formation of a fine mist and a substantial loss of the sulfur trioxide.

The principal use of sulfuric acid (nearly 60%) is in the production of fertilizers, most commonly phosphate and ammonium fertilizers (see Fig. 3.10):

$$2NH_3(g) + H_2SO_4(l) \longrightarrow (NH_4)_2SO_4(s)$$

$$Ca_3(PO_4)_2(s, \text{ phosphate rock}) + 2H_2SO_4(l) \longrightarrow 2CaSO_4(s) + Ca(H_2PO_4)_2(s)$$

The combined product is called superphosphate. Ammonium sulfate and superphosphate are garden and lawn fertilizers. Other major uses account for 10% of the output. Only about 0.5% of the sulfuric acid output is used in the lead-acid battery for automobiles: in this application, it is used as a 33.5% H_2SO_4 solution having a density of 1.25 g/mL.

Concentrated sulfuric acid (98.0% by mass and 18.4 M) is a colorless, odorless, dense (1.84 g/mL), oily liquid. Sulfuric acid is a mild oxidizing agent, and its strength in this role increases with temperature and concentration. Concentrated sulfuric acid is a strong dehydrating agent and can extract water molecules from simple sugar molecules such as sucrose (table

tion of chlorine from the electrolysis of an aqueous sodium chloride solution. (b) The annual production of chlorine gas in the United States approaches 1.0×10^{10} kg; how many coulombs of electricity is required for the production of this chlorine? (c) If the chlorine is generated 24 hours a day for 365 days a year, what is the average current (in amperes) that must be used?

20.77 Write the equation for the reaction between hydrofluoric acid and glass.

20.78 Considering (a) Cl_2O and (b) Cl_2O_7 to be acid anhydrides, write the formula of the corresponding acid in aqueous solution.

20.79 Complete and balance the following skeletal equations:
(a) $I_2(aq) + H_2S(aq) \rightarrow S(s) +$ ________
(b) $MnO_2(s) + I^-(aq) + H^+(aq) \rightarrow Mn^{2+}(aq) + H_2O(l) +$ ________

20.80 The concentration of Cl^- ions can be measured gravimetrically by precipitating silver chloride, with silver nitrate as the precipitating reagent in the presence of dilute nitric acid. The white precipitate is filtered off and its mass is determined. Calculate the Cl^- ion concentration in 25.00 mL of a solution that gave a silver chloride precipitate of mass 3.050 g. Why is the method inappropriate for measuring the concentration of fluoride ions?

sugar), $C_{12}H_{22}O_{11}$ (see Fig. 20.10). The dilution of concentrated sulfuric acid should be done with extreme care. Its reaction with water is very exothermic—1 kg (about 550 mL) of 100% sulfuric acid diluted to 20% liberates 700 kJ, enough energy to heat 2 kg of water from room temperature to boiling. Because of this large heat output, concentrated acid should always be added slowly and with stirring to water. The reverse, the addition of water to the concentrated acid, results in the immediate and violent vaporization of water, causing the acid mixture to spit out of the flask. For that reason, *always dilute an acid by adding the acid slowly to the water,* never the reverse.

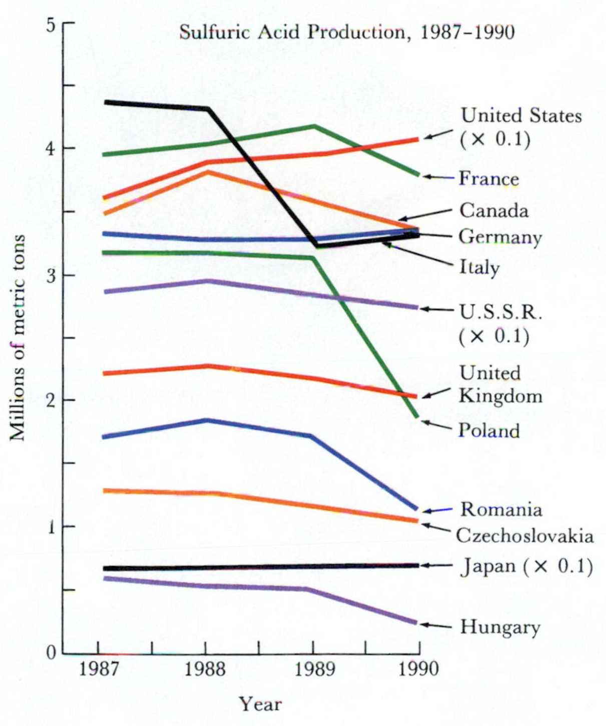

QUESTIONS

1. Determine the mass of sulfur dioxide required to produce 50 metric tons (1 metric ton = 10^3 kg) of 98% (by mass) sulfuric acid. Assume the percentage conversion of SO_2 to SO_3 is 95%.

2. Determine the volume of air at STP needed for the process in the same sulfuric acid plant as in Question 1. Assume air to be 21% O_2 by volume.

3. Suppose a 2.0-L bottle of 98% sulfuric acid was accidentally dropped and broken. Determine the mass of sodium hydrogencarbonate that would be needed to neutralize the sulfuric acid. The reaction is

$$H_2SO_4(aq) + 2NaHCO_3(s) \longrightarrow Na_2SO_4(aq) + 2H_2O(l) + 2CO_2(g)$$

(Use your answer to imagine the magnitude of the cleanup operation if a railroad car of 98% sulfuric acid suddenly burst open!)

4. Show that an aqueous solution of sulfuric acid that is 98.0% H_2SO_4 is 18.4 M $H_2SO_4(aq)$.

T OHT: Sulfuric acid production.

20.81 The concentration of hypochlorite ions in a solution can be determined by adding a known volume to a solution containing excess I^- ions, which are oxidized to iodine:

$$ClO^-(aq) + 2I^-(aq) + H_2O(l) \longrightarrow I_2(aq) + Cl^-(aq) + 2OH^-(aq)$$

The iodine concentration is then measured by titration with sodium thiosulfate:

$$I_2(aq) + 2S_2O_3^{2-}(aq) \longrightarrow 2I^-(aq) + S_4O_6^{2-}(aq)$$

In one experiment, 10.00 mL of ClO^- solution was added to a KI solution, which in turn required 28.34 mL of 0.110 M $Na_2S_2O_3(aq)$ to reach the stoichiometric point. Calculate the molar concentration of ClO^- ions in the original solution.

20.82 Xenon hexafluoride exists as the ionic solid $XeF_5{}^+F^-$. Write the Lewis structure for $XeF_5{}^+$ and, from VSEPR theory, predict its shape.

20.83 (a) Write a Lewis structure for ClO_2 in which the formal charge of each atom is zero. (b) What is the shape of the molecule? (c) Comment on its likely chemical reactivity and account for any similarities it may have with NO_2.

The elements of the d block share some fascinating properties. In particular, they form compounds with a wide variety of oxidation states. They also form an important series of compounds, the coordination compounds, that have colors and magnetic properties that can be explained by a single unified theory. The illustration shows steel girders being welded in a construction project. Steels are alloys of iron with carbon and other members of the d block.

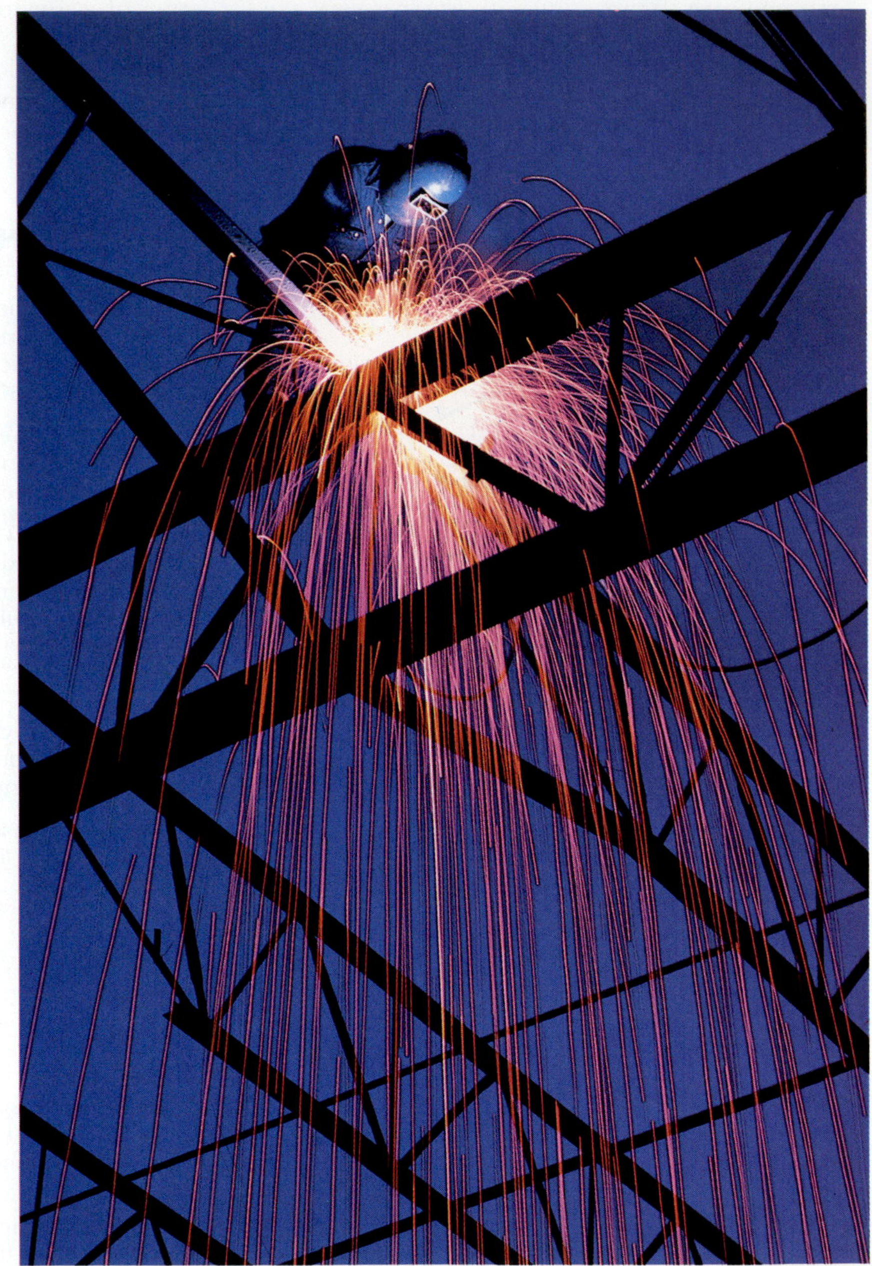

21 THE *d*-BLOCK ELEMENTS

The members of the *d* block, which are also widely called the **transition metals**, are the workhorse elements of the periodic table (Fig. 21.1). They include the construction metals—notably iron and copper—that helped civilization rise from the Stone Age and that are still very widely used in the modern world. They include the metals of the new technologies, such as titanium for the aerospace industry and vanadium for catalysts in the petrochemical industry. They also include the precious metals silver, platinum, and gold, which are prized as much for their appearance, rarity, and durability as for their usefulness.

N Civilization has effectively progressed from right to left across the periodic table, from the Stone Age in the main group elements (particularly silica of flint), through the Bronze Age when materials were made of copper, gold, and brass, to the Iron Age and the Industrial Revolution, and now to extensive use of titanium and its alloys. The widespread use of silicon in computers is in a sense a regression to the main group and the Stone Age.

The *d* block is so called because its members have atoms with electron configurations in which, according to the building-up principle, the *d* orbitals are being occupied. However, as explained in Section 7.6, the building-up principle is only a formal procedure for arriving at an atom's configuration; in fact, in the *d*-block atoms the *d* electrons are bound more strongly than the valence *s* electrons. Hence, when a *d*-block atom becomes a cation, the valence *s* electrons are lost first, after which a variable number of *d* electrons may be lost. The configuration of iron, for example, is $[Ar]3d^6 4s^2$, and its two most common ions are Fe^{2+} with configuration $[Ar]3d^6$ and Fe^{3+} with configuration $[Ar]3d^5$.

Because there are five *d* orbitals in a given shell and because each one can accommodate up to two electrons, there are ten elements in each row of the *d* block. However, the seven 4*f* orbitals begin to be occupied at lanthanum, and the 14 elements of the first row of the *f* block (the lanthanides) interrupt the completion of the *d* block. Although we shall not deal with the *f* block in detail, we should be aware of its presence because it affects the properties of elements that follow it in Period 6.

THE *d*-BLOCK ELEMENTS AND THEIR COMPOUNDS

N The *d*-block elements are characterized by the variability of their oxidation states in compounds and by the formation of complex ions with characteristic colors and magnetic properties.

All the *d*-block elements are metals (and we shall often refer to them as the *d* metals). Most are good electrical conductors, silver being the best of all elements at room temperature. They are usually malleable,

FIGURE 21.1 The elements in the d block of the periodic table. Note that the f block intervenes at the beginning of the block in Periods 6 and 7.

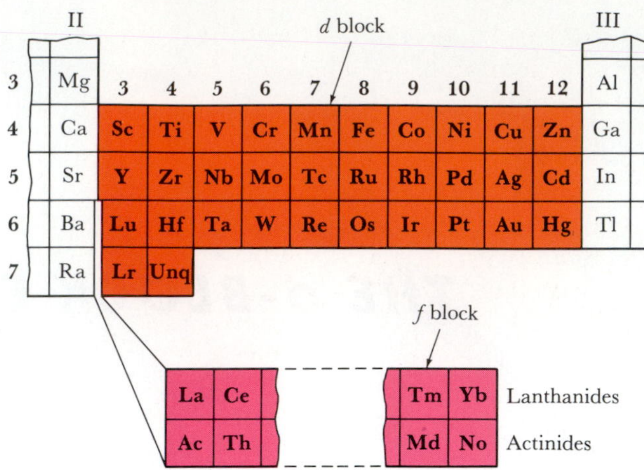

U Exercise 21.76.

N Because of the similarities in atomic radii and ionization energies, the second and third series of d-block elements often coexist in the Earth's crust.

ductile, lustrous, and silver-white in color, and generally they have higher melting and boiling points than the main-group elements. However, there are some notable exceptions: mercury has such a low melting point that it is a liquid at room temperature; copper is red-brown in color, and gold is yellow.

21.1 TRENDS IN PROPERTIES

E The liquid character of mercury and the yellow color of gold are manifestations of the relativistic increase in mass of an electron with its velocity. If relativity is ignored, mercury is not predicted to be a liquid and gold is not predicted to be yellow.

The properties of the d metals can be traced in large measure to the directional character of the lobes of the d orbitals. Because of this directionality, electrons in different d orbitals occupy markedly different regions of space. As a result, they repel each other weakly and two d electrons interact less than two s electrons do. Another consequence of the shape of d orbitals is that the nucleus is only poorly shielded by a d electron: in effect, the nuclear charge is more exposed through the gaps between the lobes of the d orbitals than it is through p orbitals.

Trends in physical properties. Both the atomic number and the number of d electrons increase from left to right across each period. However, because of the weak repulsions between d electrons, the increasing nuclear charge succeeds in drawing not only the d electrons inward but also the electrons in the s orbital of the next higher shell. This attraction has two consequences.

First, the increasing nuclear charge reduces the radius of the atoms. Thus, a $3d^64s^2$ iron atom (with atomic radius 126 pm) is smaller than a $3d^14s^2$ scandium atom (160 pm), even though the iron atom has more d electrons (Fig. 21.2). Nevertheless, the range of d-metal atomic radii is not very great, and some of the atoms of one d metal can be replaced by atoms of another without causing too much strain in a solid. The d metals can therefore be mixed together to form a wide range of alloys, of which the numerous steels are important examples.

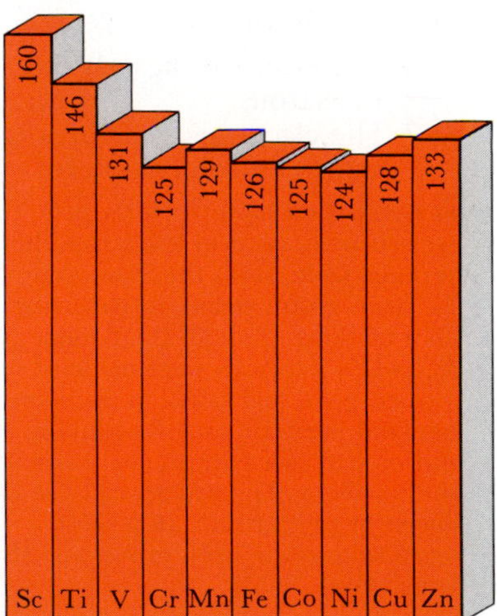

FIGURE 21.2 The atomic radii (in picometers) of the elements of the first row of the d block. The radii decrease from Sc to Ni, because of the increasing nuclear charge and the poor shielding of the d electrons.

Second, as the nuclear charge increases from left to right across the period, the first ionization energies also increase—from 632 kJ/mol for scandium to 762 kJ/mol for iron for the first row of the block (Fig. 21.3). The decrease in ionization energy after iron is due to the double occupation of the $3d$ orbitals at this stage, an arrangement that results in stronger electron-electron repulsions.

The elements in the second and third rows of the *d*-block show the same general trend in radii and ionization energies as those in the first row. However, although there is a general increase in atomic radius from the first row to the second (because a new quantum shell is being occupied), the atomic radii of the elements in the third row are about the same as those in the second row. The failure of the third-row (Period 6) atomic radii to increase as much as expected is a result of the **lanthanide contraction**. It is so called because it stems from the presence of the *f* block (and, in particular, the lanthanides). The *f* electrons that are present in the lanthanides are even poorer at shielding the nucleus than the *d* electrons are. As a result, there is a marked decrease of atomic radius along the *f* block as the nuclear charge increases and pulls the valence electrons inward. When the *d* block resumes (at lutetium) the atomic radius has fallen from 188 pm to 157 pm. Hence, the atoms of all the following elements are smaller than might be expected.

One effect of the lanthanide contraction is the high density of the Period 6 elements (Fig. 21.4). This characteristic is due to the fact that their atomic radii are comparable to those of the elements in Period 5, whereas their atomic masses are about twice as large. The atomic mass of iridium is nearly twice that of rhodium in the row above (192 u and 103 u, respectively), but the atomic radii of iridium and rhodium are almost identical. Iridium is actually one of the two densest elements; its neighbor osmium is the other. Another effect of the contraction is the low reactivity of gold and platinum. This property can be traced to the fact that their valence electrons are relatively close to the nucleus and hence tightly bound.

Most of the *d*-block elements have pronounced paramagnetic properties as a result of the unpaired electrons in the *d* orbitals of their atoms. (Paramagnetism, it should be recalled from Section 7.6, is the tendency of a substance to move into a magnetic field and arises from the presence of unpaired electrons.) In the first row of the *d* block, the paramagnetism increases to chromium with its six unpaired electrons

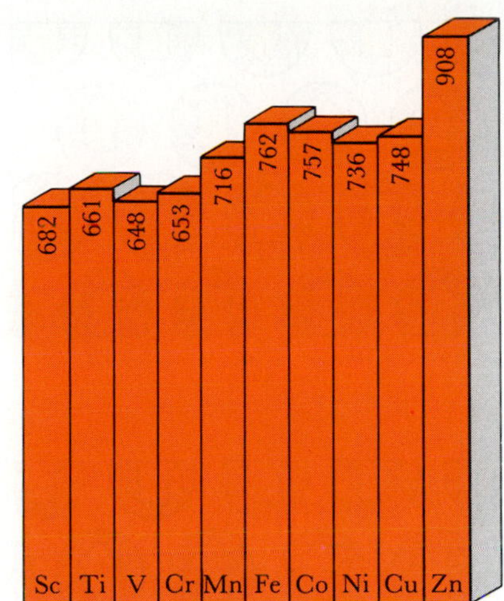

FIGURE 21.3 The first ionization energies (in kilojoules per mole) of the Period 3 *d* metals increase up to Fe.

E Periodic contractions among the elements: Or, on being the right size. *J. Chem. Ed.*, **1988**, *65*, 17.

? Why is the ionization energy of zinc significantly higher than those of the other *d*-block elements in Period 4?

T OHT: Fig. 21.4.

E A detailed analysis suggests that the poor shielding character of the *f* orbitals is a consequence of their radial shape, not their angular shape: On the lanthanide and "scandinide" contractions. *J. Chem. Ed.*, **1986**, *63*, 502.

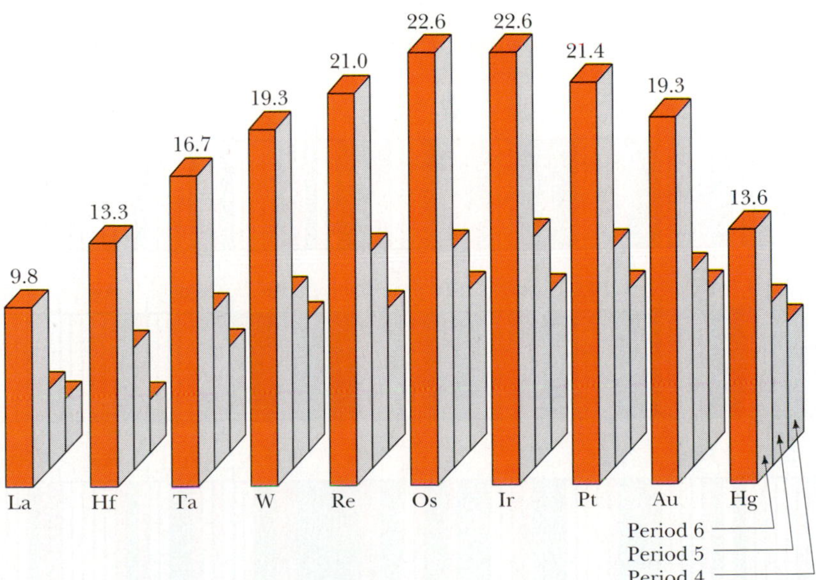

FIGURE 21.4 The densities (in grams per cubic centimeter) of the *d* metals at 25°C. The lanthanide contraction has a pronounced effect on the densities of the elements in Period 6 (front row), which are among the densest of all elements.

(its configuration is [Ar]$4s^1 3d^5$) and then decreases; zinc is diamagnetic (its configuration is [Ar]$4s^2 3d^{10}$, and all its electrons are paired). The property of **ferromagnetism**, the ability to become permanently magnetized, which is shown by iron, cobalt, and nickel, is caused by an alignment of the spins of the electrons in domains of large numbers of atoms or ions (Fig. 21.5). Because large numbers of electrons spin in the same direction throughout a domain, they give rise to a strong magnetic field. The alignment of electron spins can be achieved by exposing the metal to a magnetic field, such as that caused by an electric current; the interaction between the electrons ensures that the alignment is maintained even after the applied field is turned off.

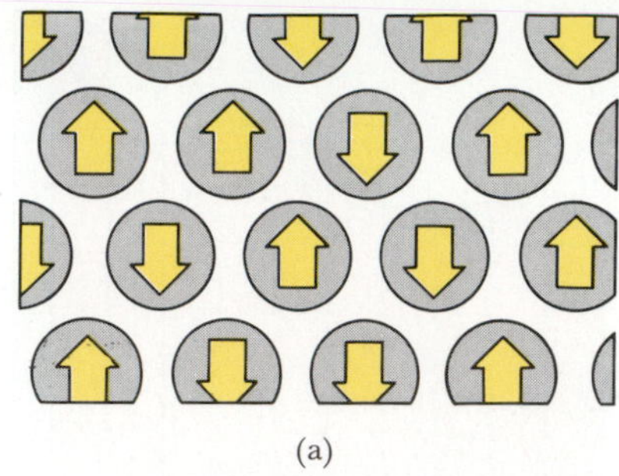

(a)

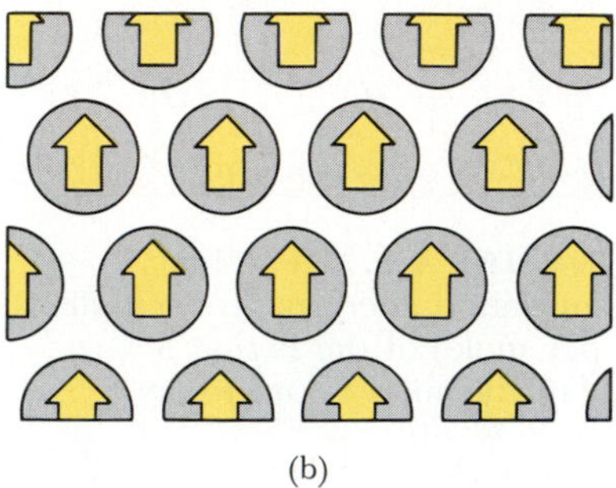

(b)

FIGURE 21.5 (a) A ferromagnetic material, such as iron, magnetite, or cobalt, consists of atoms in which many electrons spin in the same direction and give rise to a strong magnetic field: (a) before magnetization; (b) after magnetization. The magnetization persists after the magnetizing influence has been removed. The yellow arrows represent the aligned spins of the electrons in each atom.

Trends in chemical properties. The principal chemical characteristics of the *d* metals are their ability to form numerous complexes (which is described in Section 21.4), their action as catalysts, and their extensive range of oxidation numbers.

The long list of *d*-metal oxidation numbers might seem daunting at first sight (Fig. 21.6). However, it is helpful to keep in mind that, with the exception of mercury, the elements at the ends of each row of the *d* block have only one oxidation number other than zero. Scandium, for example, occurs only with an oxidation number +3, and zinc only with an oxidation number +2. All the other elements of the row can have at least two oxidation numbers. The most common oxidation numbers of copper, for example, are +1 (as in CuCl) and +2 (as in $CuCl_2$). Elements close to the center of each row have the widest range of oxidation numbers; this property is shown strikingly by manganese, with its seven known oxidation numbers. Another helpful fact is that elements toward the left and in the second and third rows of the *d* block are more likely to form compounds with a higher oxidation number. For example, the Cr atom in the CrO_4^{2-} ion, with oxidation number +6, is easily reduced to a lower oxidation number, whereas the W atom in WO_4^{2-} ion, which also has oxidation number +6, is quite stable.

Oxidation numbers help to rationalize the properties of *d*-metal compounds. A compound in which an element has an oxidation number

T OHT: Fig. 21.6.

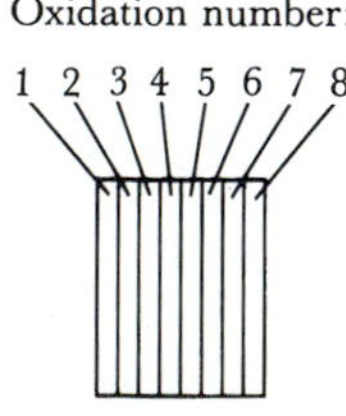

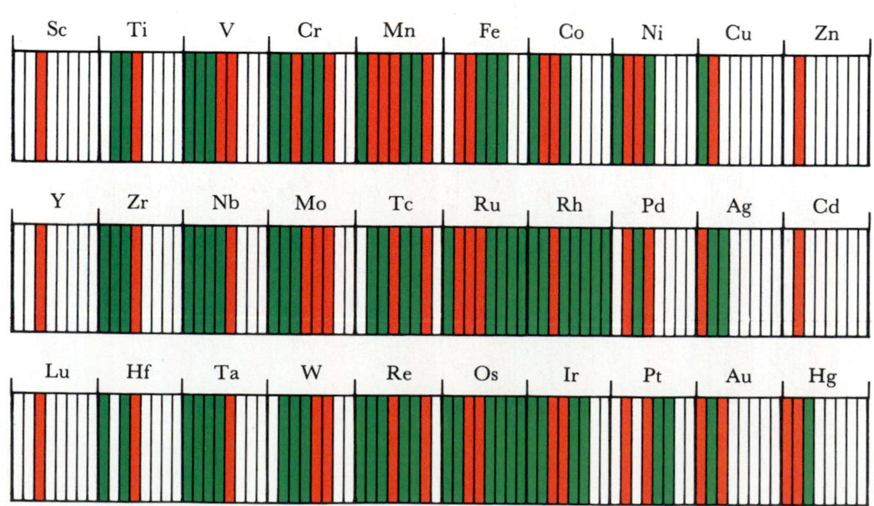

FIGURE 21.6 The oxidation numbers of the *d*-block elements. Known oxidation states are indicated by the colored bars; the orange bars mark the common oxidation states, and the green bars mark the rest.

high in its range tends to be a good oxidizing agent. An example is the strong oxidizing power of the permanganate ion, MnO_4^- (in which manganese has oxidation number +7), in acidic solution:

$$MnO_4^-(aq) + 8H^+(aq) + 5e^- \longrightarrow Mn^{2+}(aq) + 4H_2O(l) \qquad E° = +1.51\ V$$

Compounds in which the oxidation number is intermediate in the element's range often disproportionate, as copper(I) compounds do in water:

$$2Cu^+(aq) \longrightarrow Cu(s) + Cu^{2+}(aq)$$

Compounds in which the element has an oxidation number low in its range, such as Fe^{2+}, are often good reducing agents:

$$Fe^{2+}(aq) + 2e^- \longrightarrow Fe(s) \qquad E° = -0.44\ V$$

In addition, although most *d*-metal oxides are basic, the oxides of a given element show a shift toward acidic character with increasing oxidation number. A good example is provided by the family of chromium oxides (see list in margin). The last oxide, CrO_3, is the anhydride of chromic acid, H_2CrO_4, the parent acid of the chromates.

Among other similarities, the elements on the left side of the *d* block resemble the *s*-block metals in being much more difficult to extract from their ores than the metals on the right side. Indeed, as we move across the *d* block from right to left, we encounter the elements in very roughly the order in which they became available for exploitation. On the far right are copper and zinc, which jointly were responsible for the Bronze Age. That age was succeeded by the Iron Age, as higher temperatures became attainable and iron ore reduction feasible. Finally, at the left of the block, we have metals like titanium that require such extreme conditions for their extraction—including utilization of the thermite process and electrolysis—that they have become widely available only in this century (Fig. 21.7).

N The standard reduction potentials for MnO_4^-, TcO_4^-, and ReO_4^- in acidic solution are +1.51 V, +0.78 V, and +0.51 V, respectively. The permanganate ion is the strongest oxidizing agent.

Oxide	Oxidation number	Property
CrO	+2	basic
Cr_2O_3	+3	amphoteric
CrO_3	+6	acidic

(a)

(b)

(c)

FIGURE 21.7 These three artifacts represent the progress that has been made in the extraction of the *d* metals. (a) An ancient bronze chariot axle cap from China made from a metal that was easy to extract. (b) An early iron steam engine made from a metal that was moderately easy to extract once high temperatures could be achieved. (c) A twentieth-century airplane engine with titanium components that had to await high temperatures and advanced technology before the element became available.

E Scandium iodide is added to mercury vapor lamps to achieve a bright light of daylight quality.

N Scandium has a +3 oxidation state in all its compounds.

U Exercise 21.75.

D Titanium exhibit. TDC, 21-1. Hydrolysis of $TiCl_4$. TDC, 20-16.

E Titanium dioxide: Chemical of the month. *J. Chem. Ed.*, **1982**, *59*, 158.

N For ilmenite, the reaction with coke is

$$FeTiO_3(s) + 3Cl_2(g) \xrightarrow{\Delta} FeCl_2(s) + 3CO(g)$$

E This procedure is called the *Kroll process*. Its development in 1946 converted titanium from a laboratory curiosity to an important constructional metal. It has about half the density of steel, but as strong; although it is 60% denser than aluminum, it is twice as strong.

N Titanium(IV) compounds are colorless but Ti(III) compounds are not. See Section 21.6.

? What would you expect to happen when an aqueous solution of titanium(III) sulfate is poured on solid sodium sulfide?

E Vanadium in the living world. *J. Chem. Ed.*, **1974**, *51*, 503.

21.2 SCANDIUM THROUGH NICKEL

Some of the physical properties of the elements in the first row of the *d* block, from scandium to nickel, are summarized in Table 21.1. Here we consider only their most important chemical properties.

Scandium and titanium. Scandium, which was discovered in 1879, was not isolated until 1937. It is a reactive metal (reacting with water about as vigorously as calcium and tarnishing in air) with few uses. The small, highly charged Sc^{3+} ion is strongly hydrated in water (like Al^{3+}), and the resulting $Sc(H_2O)_6{}^{3+}$ complex is about as strong a Brønsted acid as acetic acid.

Titanium is a light, strong metal and is used where these properties are vital—in airplanes, for example. It is resistant to corrosion because a skin of oxide forms on its surface. The principal sources of the metal are the ores ilmenite, $FeTiO_3$, and rutile, TiO_2. Titanium requires vigorous reducing conditions for extraction from its ores. It was not exploited commercially until quite recently, when the demand from the aerospace industry had grown. The metal is obtained first by treating the ores with chlorine in the presence of coke to form titanium(IV) chloride. The chloride is then reduced with liquid magnesium:

$$TiCl_4(g) + 2Mg(l) \xrightarrow{700°C} Ti(s) + 2MgCl_2(s)$$

The most stable oxidation state of titanium is +4. Its most important compound is the dioxide (TiO_2), a brilliantly white, nontoxic, stable compound that is used as the white pigment in paints and as a paper whitener. Titanium tetrachloride ($TiCl_4$) is a liquid that boils at 136°C and fumes in moist air because it is hydrolyzed to the dioxide:

$$TiCl_4(l) + 2H_2O(l) \longrightarrow TiO_2(s) + 4HCl(aq)$$

This reaction is used for skywriting, either on the grand scale, as in the formation of military smokescreens, or in advertising.

Vanadium, chromium, and manganese. Pure vanadium, which is a soft silver-gray metal, is produced by reducing vanadium pentoxide, V_2O_5, with calcium,

$$V_2O_5(s) + 5Ca(l) \xrightarrow{\Delta} 2V(s) + 5CaO(s)$$

TABLE 21.1 Properties of the *d*-block elements scandium through nickel

Z	Name	Symbol	Electron configuration	Molar mass, g/mol	Melting point, °C	Boiling point, °C	Density, g/cm³
21	scandium	Sc	$3d^14s^2$	44.96	1540	2800	2.99
22	titanium	Ti	$3d^24s^2$	47.90	1660	3300	4.51
23	vanadium	V	$3d^34s^2$	50.94	1920	3400	6.09
24	chromium	Cr	$3d^54s^1$	52.00	1860	2600	7.19
25	manganese	Mn	$3d^54s^2$	54.94	1250	2120	7.47
26	iron	Fe	$3d^64s^2$	55.85	1540	2760	7.87
27	cobalt	Co	$3d^74s^2$	58.93	1494	2900	8.80
28	nickel	Ni	$3d^84s^2$	58.71	1455	2150	8.91

or reducing vanadium(II) chloride with magnesium:

$$VCl_2(s) + Mg(l) \xrightarrow{\Delta} V(s) + MgCl_2(s)$$

The electrolysis of molten vanadium(II) chloride is also used for the commercial production of vanadium metal.

When other metals are added to iron in steelmaking, it is not usually the pure metal that is added—because that practice would be uneconomical. Instead, a **ferroalloy** of the metal is added; these alloys also contain iron and carbon and are less expensive to produce. Such is the case with vanadium, which is normally added to iron as ferrovanadium, a mixture of about 85% V, 12% C, and 2% Fe by mass. It is prepared by reducing V_2O_5 with aluminum in the presence of iron. The final product, vanadium steel, is a tough ferroalloy that is used for automobile and truck springs (Table 21.2).

Vanadium pentoxide is the most important compound of vanadium. This orange-yellow solid is used as an oxidizing agent and as an oxidizing catalyst in the contact process for the manufacture of sulfuric acid. The wide range of colors of vanadium compounds, including the blue of the vanadyl ion, VO^{2+}, accounts for their use as glazes in the ceramics industry.

Chromium is a bright, lustrous, corrosion-resistant metal that got its name from its colorful compounds (*chroma* is the Greek word for "color"). It is obtained from chromite ore, $FeCr_2O_4$, by reduction with carbon in an arc furnace:

$$FeCr_2O_4(s) + 4C(s) \longrightarrow Fe(s) + 2Cr(s) + 4CO(g)$$

Chromium metal is also produced by the thermite reaction:

$$Cr_2O_3(s) + 2Al(s) \longrightarrow Al_2O_3(s) + 2Cr(l)$$

The main product of the former reduction is ferrochromium, which is used in the production of stainless steels (they typically contain about 15% chromium by mass). Chromium(IV) oxide, CrO_2, is used for coating "chrome" recording tapes because it responds better to high-frequency magnetic fields than conventional "ferric" (Fe_2O_3) tapes do. Chrome plating is also a major use for chromium (Section 17.8).

Sodium chromate, Na_2CrO_4, is a yellow solid and one of the most important chromium compounds. It is made commercially by heating

E Oxidation states of vanadium. CD1, 84 and 85.

D Chromium plating. TDC, 21-6s.

V Video 6: Oxidation states of vanadium.

N The chromium quickly reacts with the oxygen in air to form a thin impervious oxide layer of Cr_2O_3, which passivates the surface and minimizes any further corrosion of the iron or other metals of the alloy.

T OHT: Tables 21.2 and 21.3.

TABLE 21.2 Compositions of different steels

Ingredient added to iron	Typical amount	Effect
manganese	0.5 to 1.0%	increases strength and hardness but lowers ductility
	13%	increases wear resistance
nickel	<5%	increases strength and shock resistance
	>5%	increases corrosion resistance (stainless) and heat resistance
chromium	—	increases hardness and wear resistance
	>12%	enhances corrosion resistance (stainless steel, with nickel)
vanadium	—	increases hardness
tungsten	<20%	increases hardness, especially at high temperatures

U Exercise 21.81.

D Add a few drops of concentrated hydrochloric acid to an aqueous solution of Na_2CrO_4 solutions in a shallow dish on an overhead projector; then adjust the equilibrium with aqueous sodium hydroxide solution and back again with hydrochloric acid.

D Preparation of chromate, manganate, and permanganate. TDC, 175. Oxidation states of manganese. CD1, 79 and 80. Oxidation states of manganese. *J. Chem. Ed.*, **1988**, *65*, 1004. Oxidation of glycerin by permanganate. CD1, 73.

E Potassium dichromate in concentrated sulfuric acid is a cleaning agent for preparing very clean glassware for analytical determinations.

E Lead chromate is the pigment *chrome yellow;* strontium chromate has a lemon yellow color. Orange pigments are made by mixing lead chromate, lead molybdate, and lead sulfate. Chrome green is a mixture of chrome yellow and Prussian blue ($Fe_4[Fe(CN)_6]_3$).

E Sulfur dioxide reduces sodium chromate to $Cr(OH)SO_4$, a compound used in the tanning of leather. The compound attaches to the collagen in the animal skins, making it insoluble and protecting it from biological degradation, but leaving it flexible.

E A rich supply of manganese lies in the manganese nodules that litter the ocean floors. These nodules range in diameter from millimeters to meters and are lumps of the oxides of iron, manganese, and other elements. However, because this source is technically difficult to exploit, manganese is currently obtained by the thermite process from pyrolusite, a mineral form of manganese dioxide (MnO_2).

E Because ferromanganese alloys are very hard, tough, and resistant to abrasion, they are used for bulldozer blades and naval armor plate.

N Chromium(II) survives in an aqueous solution but is quickly oxidized by the presence of any dissolved oxygen.

N Manganese(IV) oxide, MnO_2, in which manganese has an intermediate oxidation number of +4, is both a mild oxidizing agent and a mild reducing agent.

an intimate mixture of chromite ore and sodium carbonate in the presence of oxygen:

$$4FeCr_2O_4(s) + 8Na_2CO_3(s) + 7O_2(g) \xrightarrow{\Delta} 8Na_2CrO_4(s) + 2Fe_2O_3(s) + 8CO_2(g)$$

The chromate is converted into the orange dichromate ion, $Cr_2O_7{}^{2-}$, in the presence of acid:

$$2CrO_4{}^{2-}(aq) + 2H^+(aq) \longrightarrow Cr_2O_7{}^{2-}(aq) + H_2O(l)$$

In the laboratory, acidified solutions of dichromates, in which the oxidation number of chromium is +6, are useful oxidizing agents:

$$Cr_2O_7{}^{2-}(aq) + 14H^+(aq) + 6e^- \longrightarrow 2Cr^{3+}(aq) + 7H_2O(l) \qquad E^\circ = +1.33\text{ V}$$

Sodium chromate and sodium dichromate are both starting points for the production of a number of pigments, corrosion inhibitors, fungicides, and ceramics. Chromium(III) oxide is a deep green solid that is used as a paint pigment; most other chromium(III) compounds are violet. Because chromium compounds exhibit a wide variety of colors (which vary with oxidation number), approximately one-third of all chromium compounds are used as pigments.

Manganese is a gray metal that resembles iron. It is much less resistant to corrosion than chromium and becomes coated with a thin brown layer of oxide when exposed to air. The metal is rarely used alone, although it is an important component of alloys. In steel, where it is added as ferromanganese, it helps to remove sulfur by forming a sulfide. It also increases hardness, toughness, and resistance to abrasion (see Table 21.2). Manganese is also alloyed with various nonferrous metals. One such alloy is manganese bronze (39% Zn, 1% Mn, some iron and aluminum, and the rest copper), which is very resistant to corrosion and is used for marine propellers. The increasing use of aluminum cans for beverages has helped to increase the consumption of manganese, because it is alloyed with the aluminum to increase the stiffness of the metal container.

Manganese lies at the center of its period and exhibits a wide variety of oxidation numbers. The most stable is +2, but +4 and +7 also appear in many manganese compounds. Its most important compound is manganese dioxide, a black solid used in dry cells, as a decolorizer to conceal the green tint of glass, and as the starting point for the production of other manganese compounds. As examples of the last use, the green manganate ion, $MnO_4{}^{2-}$, and the purple permanganate ion, $MnO_4{}^-$, are produced from manganese dioxide by oxidation followed by disproportionation in acidic solution:

$$MnO_2(s) + 2KOH(s) \xrightarrow{\Delta} K_2MnO_4(s) + H_2(g)$$

$$3MnO_4{}^{2-}(aq) + 4H^+(aq) \longrightarrow 2MnO_4{}^-(aq) + MnO_2(s) + 2H_2O(l)$$

However, this process is not a good commercial route to permanganate ion because the disproportionation also produces manganese dioxide, the original starting material. To avoid this problem, electrolytic oxidation of the manganate ion is used instead:

$$MnO_4{}^{2-}(aq) \longrightarrow MnO_4{}^-(aq) + e^-$$

Potassium permanganate, in which the manganese has its highest oxidation number (+7), is useful as an oxidizing agent in acidic solution. Its usefulness stems not only from its thermodynamic tendency to oxidize other species but also from its ability to act by a variety of mechanisms and hence to discover a path with low activation energy. Potassium permanganate is used for oxidations in organic chemistry and as a mild disinfectant.

V Video 7: Oxidation of ethanol by manganese(VII) oxide. Video 55: The reaction of $KMnO_4$ and glycerin.

D A chemical hand warmer. CD2, 54.

Iron. Iron, the most widely used of all the *d* metals, is the second most abundant metal in the Earth's crust (aluminum is first). Its principal ores are the oxides hematite, Fe_2O_3, and magnetite, Fe_3O_4. The sulfide mineral pyrite, FeS_2 (fool's gold; see Fig. 20.2), is also widely available; it is not used in steelmaking because the sulfur is difficult to remove.

D Iron metallurgy exhibit. TDC, 21-26. Iron blast furnace reaction. TDC, 21-10s.

E Iron is an essential element in the diet, largely because of its role as the site at which oxygen molecules attach to the oxygen-carrying protein hemoglobin. A healthy adult human body contains around 3 g of iron, mostly as hemoglobin. Around 1 mg is lost daily as sweat, feces, and hair (and about 20 mg is lost during menstruation), so about 1 mg must be ingested daily in order to maintain the balance. Iron deficiency, or anemia, results in reduced transport of oxygen to the brain and muscles, and an early symptom is chronic tiredness.

Iron ore is reduced to iron metal in a **blast furnace** (Fig. 21.8). The furnace, which is approximately 40 m in height, is continuously filled at the top with a "charge" consisting of the ore, coke, and limestone. Each kilogram of iron produced requires about 1.75 kg of ore, 0.75 kg of coke, and 0.25 kg of limestone. The limestone, which is primarily calcium carbonate, undergoes thermal decomposition to calcium oxide (lime) and carbon dioxide. The calcium oxide helps to remove the acid

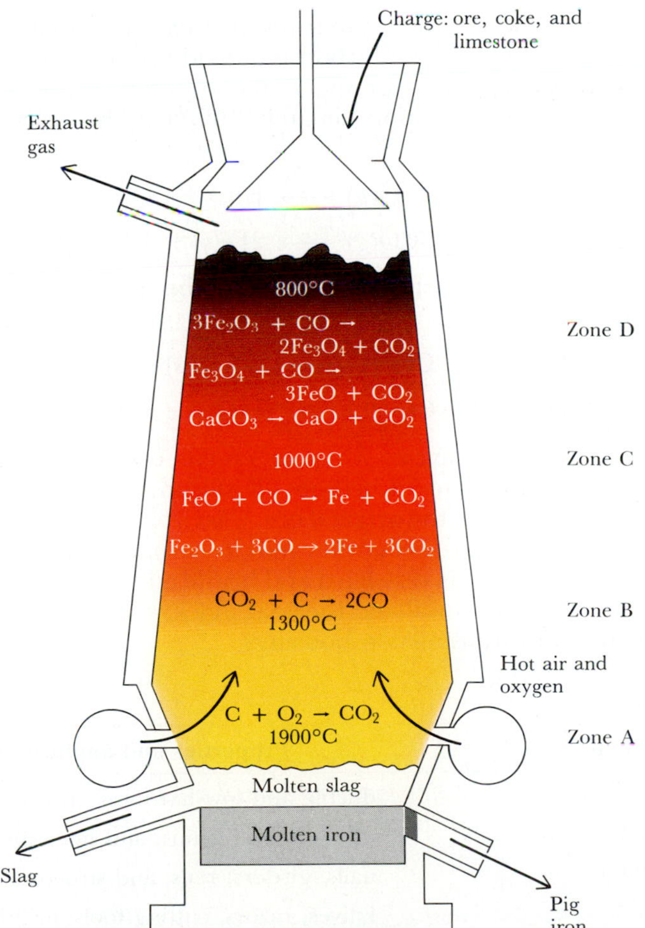

T OHT: Fig. 21.8.

FIGURE 21.8 The reduction of iron ore takes place in a blast furnace containing a charge of the ore with coke and limestone. Different reactions occur in different zones when the blast of air and oxygen is admitted. The ore, an oxide, is reduced to the metal by reaction with the carbon monoxide produced in the furnace.

anhydride (nonmetal oxide) impurities from the ore:

$$CaO(s) + SiO_2(s) \xrightarrow{\Delta} CaSiO_3(l)$$

$$CaO(s) + Al_2O_3(s) \xrightarrow{\Delta} Ca(AlO_2)_2(l)$$

$$6CaO(s) + P_4O_{10}(s) \xrightarrow{\Delta} 2Ca_3(PO_4)_2(l)$$

This mixture, which is known as slag, is molten at the temperatures in the blast furnace and floats on the denser molten iron. It is drawn off and used to make rocklike material for the construction industry.

Molten iron is produced through a series of reactions in four main temperature zones of the blast furnace. In Zone A, preheated air is blown into the furnace where the coke is oxidized. The reaction

$$C(s) + O_2(g) \longrightarrow CO_2(g) \qquad \Delta H° = -394 \text{ kJ}$$

is exothermic and the temperature approaches 1900°C. As the reduced iron moves from Zone C to Zone A, the iron becomes molten (the melting point of pure iron is 1540°C, but it is lowered to 1015°C when 4% of carbon is present). As the carbon dioxide moves upward through the furnace to Zone B, it reacts with some of the coke, producing carbon monoxide:

$$CO_2(g) + C(s) \longrightarrow 2CO(g) \qquad \Delta H° = +173 \text{ kJ}$$

This reaction is endothermic, and it results in a substantial decrease in temperature to 1300°C. The carbon monoxide produced in this reaction is the reducing agent for the reduction of iron ore in Zone C, where temperatures are approximately 1000°C. The reactions there are

$$FeO(s) + CO(g) \longrightarrow Fe(s) + CO_2(g)$$

$$Fe_2O_3(s) + 3CO(g) \longrightarrow 2Fe(s) + 3CO_2(g)$$

Other reductions and partial reductions of the iron ore also occur in Zone D and include

$$3Fe_2O_3(s) + CO(g) \longrightarrow 2Fe_3O_4(s) + CO_2(g)$$

$$Fe_3O_4(s) + CO(g) \longrightarrow 3FeO(s) + CO_2(g)$$

The molten iron (of density 7.9 g/cm^3) is run off as pig iron, consisting of 90 to 95% iron, 3 to 5% carbon, 2% silicon, and trace amounts of other elements found in the original ore. Cast iron is similar to pig iron, but some impurities have been removed; its carbon content is still usually greater than 2%. It is very hard and brittle and is used in parts that experience little mechanical and thermal shock, such as engine blocks, brake drums, and transmission housings.

FIGURE 21.9 In the basic oxygen process, a blast of oxygen and powdered limestone is used to purify the molten iron by, respectively, oxidizing and combining with the impurities in it.

N Each of the reactions at the top of the page is a Lewis acid-base reaction in which O^{2-} is the Lewis base.

U Exercise 21.78.

E The name *pig iron* is derived from the manner in which the molten iron from the blast furnace was cooled and solidified in sand beds that had a main channel with various side pockets. The arrangement looked similar to a sow with its suckling pigs.

E Sir Henry Bessemer (1813–1898) developed his converter to provide the steel he needed for the new type of gun, with a rifled barrel, that he invented during the Crimean war.

TABLE 21.3 The effect of carbon on iron

Type of steel	Carbon content, %	Properties and applications
low-carbon steel	less than 0.15	ductile and low hardness, iron wire
mild-carbon steel	0.15 to 0.25	cables, nails, chains, and horseshoes
medium-carbon steel	0.20 to 0.60	nails, girders, rails, and structural purposes
high-carbon steel	0.61 to 1.5	knives, razors, cutting tools, drill bits

The first stage in steelmaking is to lower the carbon content of the pig iron and to remove the remaining impurities, which include silicon, phosphorus, and sulfur. The modern approach uses a version of the early **Bessemer converter**, which is essentially a big metal pot lined with basic material—typically dolomite, $CaMg(CO_3)_2$—that decomposes to oxides at the operating temperature. Air is blown through the molten impure iron in the converter, the impurities react with the basic lining to form a molten slag, and the carbon is oxidized and removed as the monoxide. In the modern version, called the **basic oxygen process**, oxygen and powdered limestone are forced through the molten metal (Fig. 21.9). In the second stage, steels are produced by adding the appropriate metals, often as ferroalloys, to the molten iron.

Iron is not a very hard metal. Its physical properties are improved by partial reaction with carbon, for the carbides so formed strengthen the solid. These "steels" have various hardnesses, tensile strengths, and ductilities; in general, the higher the carbon content, the harder and more brittle the product (Table 21.3). The corrosion resistance of iron is greatly improved by alloying to form a variety of steels. Iron and steel are ferromagnetic, as are some iron oxides (such as magnetite).

FIGURE 21.10 These solutions show the typical colors of iron ions in aqueous solution. From left to right: $FeSO_4$, $FeCl_2$, $Fe_2(SO_4)_3$, $FeCl_3$.

E Solutions of Fe^{3+} are sufficiently acidic to produce carbon dioxide from sodium carbonate: $Fe(H_2O)_6{}^{3+}(aq) + CO_3{}^{2-}(aq) \rightarrow Fe(H_2O)_4(OH)_2{}^{+}(aq) + CO_2(g) + H_2O(l)$. To preserve iron(III) salts in water, it is necessary to shift the equilibrium in favor of the hydrated ion by adding acid and maintaining a pH close to zero. If the pH rises above 2, a red-brown hydrated iron(III) oxide (a version of rust) precipitates, often as a colloid.

D Corrosion of an iron nail. CD2, 99.

Iron is quite reactive and corrodes (oxidizes) in moist air. It reacts with acids that have nonoxidizing anions, evolving hydrogen and forming iron(II) salts. The colors of these salts vary from pale yellow to dark green-brown; $Fe(H_2O)_6{}^{2+}$ itself is pale green (Fig. 21.10). Iron(II) salts are quite readily oxidized to iron(III) salts. The oxidation is slow in acidic solution but rapid in basic solution, where insoluble iron(III) hydroxide, $Fe(OH)_3$, is precipitated. Although $Fe(H_2O)_6{}^{3+}$ ions are pale purple, the colors of aqueous solutions of iron(III) salts are dominated by the yellow $Fe(H_2O)_5OH^{2+}$ ion, which is the conjugate base of the acidic hydrated ion:

$$Fe(H_2O)_6{}^{3+}(aq) + H_2O(l) \rightleftharpoons H_3O^{+}(aq) + Fe(H_2O)_5OH^{2+}(aq)$$

Compounds in which iron has oxidation number 0 are also known. When iron is heated in carbon monoxide, it reacts to form iron pentacarbonyl, $Fe(CO)_5$, a yellow molecular liquid that boils at 103°C. The CO group is not very electronegative, so the number of electrons on the Fe atom is barely changed from the number on the free atom. Hence the iron is ascribed an oxidation number of 0 in this and similar compounds. Iron(II) chloride reacts with the cyclopentadienide ion, $C_5H_5{}^-$ (**1**), to form the molecular compound ferrocene, $Fe(C_5H_5)_2$, wittily named a "sandwich compound" (**2**). Ferrocene is one of a number of similar compounds that are more formally known as **metallocenes**, compounds in which a metal atom lies between two organic rings (Fig. 21.11). This orange solid, which melts at 174°C, is insoluble in water and thermally stable up to 500°C. Other *d* metals can be used to produce sandwiches with different fillings.

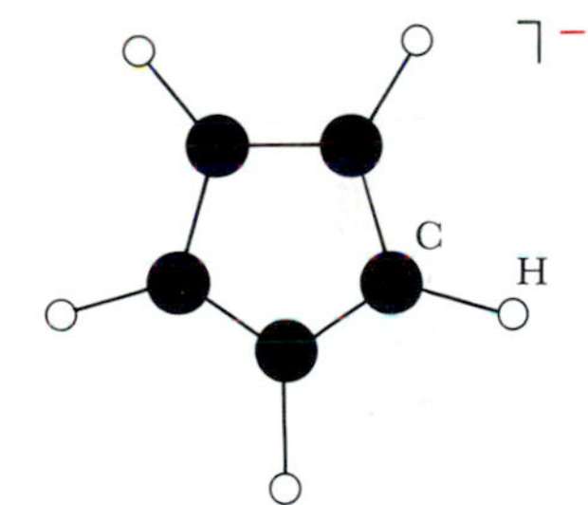

1 Cyclopentadienide ion

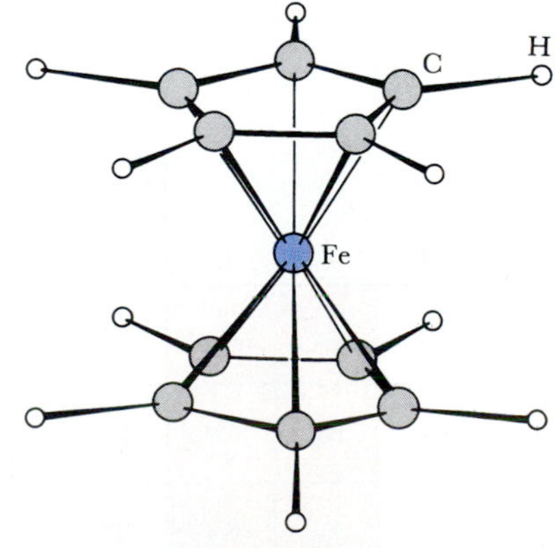

2 Ferrocene

Cobalt and nickel. Economically workable deposits of cobalt are found in association with copper sulfide. Miners called it *Kobold* (which means "evil spirit" in German) because it interfered with the production of copper. Cobalt is a silver-gray metal and is used mainly for alloying with iron. Alnico steel, an alloy of iron, cobalt, nickel, and aluminum, is used to make permanent magnets like those used in loudspeakers. Cobalt steels are also used as surgical steels, drill bits, and lathe tools. Cobalt(II) oxide is a blue salt used to color glass and ceramic glazes.

FIGURE 21.11 A sample of ferrocene, the first and one of the most stable "sandwich compounds" in which a metal atom lies between two planar organic rings.

Most interest in cobalt arises from its ability to form complexes, as described in Section 21.5. Cobalt is an essential element in the body, for it is a component of vitamin B_{12}.

About 70% of the western world's nickel (the origin of its name is "Old Nick," so named for much the same reason as cobalt) comes from iron and nickel sulfide ores that were brought close to the surface by the impact of a meteor at Sudbury, Ontario. The first step in obtaining the metal from its ore (which is in fact a complex mixture of sulfides) is to roast it in air to form the oxide:

$$2NiS(s) + 3O_2(g) \longrightarrow 2NiO(s) + 2SO_2(g)$$

The oxide is then either reduced with carbon and refined electrolytically or reduced and purified by the **Mond process**. In the latter, the reduction is carried out with hydrogen:

$$NiO(s) + H_2(g) \xrightarrow{\Delta} Ni(s) + H_2O(g)$$

Then the impure nickel is exposed to carbon monoxide, when it forms nickel tetracarbonyl, $Ni(CO)_4$:

$$Ni(s) + 4CO(g) \xrightarrow{50°C} Ni(CO)_4(g)$$

N Hydrated cobalt(II) compounds are pink because of the presence of the $Co(H_2O)_6{}^{2+}$ ion; hydrated nickel(II) compounds are green on account of the presence of $Ni(H_2O)_6{}^{2+}$.

D Precipitate formation: Blue. CD1, 19.

E The structure of metal carbonyls. *J. Chem. Ed.*, **1971**, *48*, 372, 813.

Nickel tetracarbonyl is a volatile, poisonous liquid that boils at 43°C; in it, nickel has an oxidation number of 0 (for the same reason iron is 0 in pentacarbonyl). The reaction is reversed by heating nickel tetracarbonyl at about 200°C, and the decomposition gives pure nickel and carbon monoxide:

$$Ni(CO)_4 \xrightarrow{200°C} Ni(s) + 4CO(g)$$

Nickel is a hard, silver-white metal used mainly for the production of stainless steel and for alloying with copper to produce cupronickels, which are the alloys used for coins and consist of about 25% Ni and 75% Cu. Cupronickels are slightly yellow but are whitened by the addition of small amounts of cobalt. Nickel is also used as a catalyst, especially for the addition of hydrogen to organic compounds, as in the manufacture of edible fats from vegetable oils. Its most stable oxidation number is +2, and $Ni(H_2O)_6{}^{2+}$ ions are green. Nickel(III) is reduced to nickel(II) by cadmium at the cathode during the discharge of the rechargeable nickel-cadmium cells (Section 17.2).

N The reduction and oxidation half-reactions are
Cathode: $Ni(OH)_3(s) + e^- \rightarrow Ni(OH)_2(s) + OH^-(aq)$
Anode: $Cd(s) + 2OH^-(aq) \rightarrow Cd(OH)_2(s) + 2e^-$

21.3 COPPER, ZINC, AND THEIR CONGENERS

	10	11	12	III	
				Al	
	Ni	**Cu**	**Zn**	Ga	
	Pd	**Ag**	**Cd**	In	
	Pt	**Au**	**Hg**	Tl	

The six elements in the two groups at the far right of the *d* block are the **coinage metals** copper, silver, and gold, all of which have $(n - 1)d^{10}ns^1$ electron configurations, and their neighbors zinc, cadmium, and mercury, with configurations $(n - 1)d^{10}ns^2$. Some of their physical properties are summarized in Table 21.4.

Copper. Copper is unreactive enough for some to be found native (Fig. 21.12). Most, however, is produced from its sulfides—particularly the ore chalcopyrite, $CuFeS_2$. The crushed and ground ore is separated from excess rock by **froth flotation** (see Case 1).

Processes for extracting metals from their ores are generally classified as **pyrometallurgical**, if high temperatures are used, or **hydrometallurgical**, if aqueous solutions are used. Copper is extracted by both

TABLE 21.4 Properties of copper, zinc, and their congeners

Z	Name	Symbol	Electron configuration	Molar mass, g/mol	Melting point, °C	Boiling point, °C	Density, g/cm³
29	copper	Cu	$3d^{10}4s^1$	63.54	1083	2567	8.93
47	silver	Ag	$4d^{10}5s^1$	107.87	962	2212	10.50
79	gold	Au	$5d^{10}6s^1$	196.97	1064	2807	19.28
30	zinc	Zn	$3d^{10}4s^2$	65.37	420	907	7.14
48	cadmium	Cd	$4d^{10}5s^2$	112.40	321	765	8.65
80	mercury	Hg	$5d^{10}6s^2$	200.59	−39	357	13.55

methods. In the pyrometallurgical process for the extraction of copper, the enriched ore is **roasted** in air:

$$2CuFeS_2(s) + 3O_2(g) \xrightarrow{\Delta} 2CuS(s) + 2FeO(s) + 2SO_2(g)$$

(This step and the next one can contribute an alarming amount of SO_2 to the atmosphere unless precautions are taken to remove it.) The CuS is then **smelted**, or reduced by melting it with a reducing agent. In this step, air is blown through the molten mixture to remove the sulfur as SO_2:

$$CuS(s) + O_2(g) \longrightarrow Cu(l) + SO_2(g)$$

Limestone and sand, which are added to the mixture, form a molten slag that removes many of the impurities. The solidified copper product is known as blister copper because it has entrapped air bubbles.

In the hydrometallurgical processes for extracting copper, Cu^{2+} ions are first formed by the action of sulfuric acid on the ores. Then the metal is obtained by reducing these ions in aqueous solution. In principle, the reduction can be performed with any reducing agent that has a more negative reduction potential than copper. Cheap, and therefore economically viable, reducing agents include hydrogen,

$$Cu^{2+}(aq) + H_2(g) \longrightarrow Cu(s) + 2H^+(aq)$$

and scrap iron,

$$Cu^{2+}(aq) + Fe(s) \longrightarrow Cu(s) + Fe^{2+}(aq)$$

The reduction of copper(II) ion can also be carried out electrically. The impure copper from either process is refined electrolytically: it is made into anodes and plated onto cathodes of pure copper. The rare metals—most notably, platinum, silver, and gold—obtained from the anode sludge help substantially in paying for the electricity used in the electrolysis.

Copper is an excellent electrical conductor when pure and is widely used in the electrical industry. It is also alloyed with zinc to form brass, with tin to form bronze, and with nickel to form cupronickel (Table 10.9). We saw in Section 17.4 that, because copper lies above hydrogen in the electrochemical series, it cannot displace hydrogen from acidic solutions; however, it can be oxidized by oxidizing acids, in reactions like that with dilute nitric acid:

$$3Cu(s) + 8H^+(aq) + 2NO_3^-(aq) \longrightarrow 3Cu^{2+}(aq) + 2NO(g) + 4H_2O(l)$$

E The ore from the Kennecott Bingham Canyon Mine near Salt Lake City, Utah, the largest active open-pit copper mine in the world, is only 0.6% copper. It is concentrated to 28% copper and formed into anodes that are 98% copper before the electrolytic refining step.

U Exercise 21.79.

N For example, the calcium oxide (a basic oxide) from the limestone reacts with the SO_2 (an acidic oxide) to produce calcium sulfite: $CaO(s) + SO_2(g) \rightarrow CaSO_3(s)$.

D Coin-operated red, white, and blue demonstration: Fountain effect with nitric acid and copper. BZS3, 8.9.

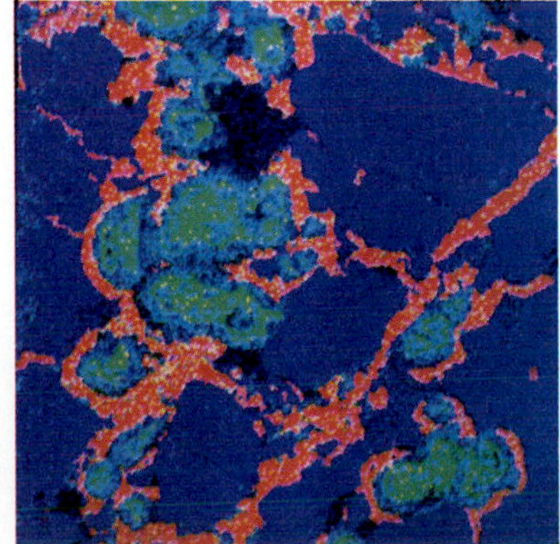

FIGURE 21.12 A color-enhanced x-ray photograph showing copper (red) filling the fractures of a rock. The metal can be recovered from such sources by mining.

Copper corrodes in moist air as a result of oxidation caused by a mixture of water, oxygen, and carbon dioxide:

$$2Cu(s) + H_2O(l) + O_2(g) + CO_2(g) \longrightarrow Cu_2(OH)_2CO_3(s)$$

The pale green product is called basic copper carbonate and is responsible for the green patina of copper and bronze objects (Fig. 21.13). It adheres to the surface, protects the metal, and gives a pleasing appearance.

FIGURE 21.13 Copper corrodes in air to form a pale green layer of basic copper carbonate. This patina, or incrustation, passivates the surface.

Copper, in common with all the coinage metals, forms compounds with oxidation number +1; however, in water, copper(I) salts disproportionate into metallic copper and copper(II) ions. The latter exist as pale blue $Cu(H_2O)_6^{2+}$ ions in water, the color of copper(II) sulfate solutions. The deeper blue of the solid pentahydrate, $CuSO_4{\cdot}5H_2O$, is due to $Cu(H_2O)_4^{2+}$ ions (the fifth H_2O links this ion to the sulfate anion). Copper(II) salts, particularly the acetate, chloride, carbonate, and hydroxide, are used as fungicides because they are toxic to fungi, algae, and bacteria. An intimate mixture of copper(II) carbonate with stoichiometric amounts of Y_2O_3 and $BaCO_3$ that is heated to approximately 1000°C forms a "123 superconductor ceramic" of composition $YBa_2Cu_3O_{6.5-7.0}$ (the numbers 1, 2, and 3 denote the proportions in which the metal atoms Y, Ba, and Cu, respectively, are present in the compound). This solid becomes superconducting at temperatures below 92 K. Because liquid nitrogen has a boiling point of 77 K, the compound is superconducting at liquid nitrogen temperatures (see Box 10.2).

E The color of aqueous copper(II) sulfate is pale because the *d-d* transition in the octahedral complex $Cu(H_2O)_6^{2+}$ is forbidden. That of the ion in the solid is more intense because the complex is tetrahedral, and the *d-d* transition is no longer forbidden as there is no center of symmetry in the complex.

U Exercise 21.80.

Silver and gold. Very little silver occurs native; most is obtained as a by-product of the refining of copper and lead, and a considerable amount is recycled through the photographic industry. Silver lies above hydrogen in the electrochemical series and so does not reduce $H^+(aq)$ to hydrogen. However, silver is oxidized by nitric acid in a reaction resembling copper's, but which leads to the production of Ag^+ ions:

$$3Ag(s) + 4H^+(aq) + NO_3^-(aq) \longrightarrow 3Ag^+(aq) + NO(g) + 2H_2O(l)$$

Silver(I) does not disproportionate in aqueous solution, and almost all silver compounds have oxidation number +1. Apart from silver nitrate and silver fluoride, silver salts are generally only sparingly soluble in water. Silver nitrate, $AgNO_3$, is the most important compound of silver and the starting point for the manufacture of silver halides for use in photography.

Gold is so inert that most of it is found native. Pure gold is called 24-carat gold, and its alloys with silver and copper, which differ in hardness and hue, are classified according to the proportion of gold they contain (Fig. 21.14). For example, 10- and 14-carat golds contain respectively $\frac{10}{24}$ and $\frac{14}{24}$ parts by mass of gold.

FIGURE 21.14 The color of commercial gold depends on its composition: from left to right 8-carat gold, 14-carat gold, white gold, 18-carat gold, and 24-carat gold. The white gold is 6 parts Au and 18 parts Ag.

▼ EXAMPLE 21.1 Predicting the reactions of gold

Should we expect gold to react with acids?

STRATEGY We need to consider (by inspection of the location of gold in the electrochemical series, Appendix 2B) whether it can be oxidized by H^+ ions or by oxoanions. Even if neither reaction is viable, it may still be possible for a reaction to occur if the oxidized gold is effectively removed from solution by complex formation (a point discussed in Section 15.8).

SOLUTION Gold lies well above hydrogen in the electrochemical series and is too noble to react even with nitric acid. Both the following gold couples lie above H^+/H_2 and above $NO_3^-,H^+/NO,N_2O$:

$$Au^+(aq) + e^- \longrightarrow Au(s) \qquad E° = +1.69\text{ V}$$

$$Au^{3+}(aq) + 3e^- \longrightarrow Au(s) \qquad E° = +1.40\text{ V}$$

$$NO_3^-(aq) + 4H^+(aq) + 3e^- \longrightarrow NO(g) + 2H_2O(l) \qquad E° = +0.96\text{ V}$$

However, we saw in Section 15.8 that gold does react with aqua regia (a mixture of concentrated nitric and hydrochloric acids) because the Au^{3+} ions form the complex ion $AuCl_4^-$ with Cl^- ions:

$$Au(s) + 6H^+(aq) + 3NO_3^-(aq) + 4Cl^-(aq) \longrightarrow AuCl_4^-(aq) + 3NO_2(g) + 3H_2O(l)$$

Even though the equilibrium constant for the oxidation of Au to Au^{3+} is very unfavorable, the reaction proceeds because Au^{3+} ions are effectively removed from solution as $AuCl_4^-$. ▲

The most common oxidation number of gold in its compounds is +3. Gold(I) compounds are also known, but in aqueous solution they tend to disproportionate into metallic gold and gold(III) unless they are stabilized by complex formation. This means of stabilization accounts for the ability of air to oxidize gold to gold(I) if cyanide ions are present (Section 15.8).

Zinc, cadmium, and mercury. Zinc is found mainly as the sulfide ZnS in sphalerite, often in association with lead ores. The ore is concentrated by froth flotation, and the metal is extracted by roasting and then smelting with coke:

$$2ZnS(s) + 3O_2(g) \xrightarrow{\Delta} 2ZnO(s) + 2SO_2(g)$$

$$ZnO(s) + C(s) \xrightarrow{\Delta} Zn(l) + CO(g)$$

Cadmium is obtained in a similar manner. Zinc and cadmium are silvery, reactive metals. Zinc is used mainly for galvanizing iron; like copper, it is protected by a hard film of basic carbonate, $Zn_2(OH)_2CO_3$, that forms on contact with air.

Zinc and cadmium are similar to each other but differ sharply from mercury. Zinc is amphoteric (like its main-group neighbor aluminum). It reacts with acids to form Zn^{2+} ions and with alkalis to form the zincate ion, $Zn(OH)_4^{2-}$:

$$Zn(s) + 2OH^-(aq) + 2H_2O(l) \longrightarrow Zn(OH)_4^{2-}(aq) + H_2(g)$$

Galvanized containers should therefore not be used for transporting alkalis. For that purpose, cadmium, which is lower down the group and is more metallic—and hence has a more basic oxide—can be used. However, this application is hazardous, because cadmium salts are toxic. Zinc and cadmium have an oxidation number of +2 in all their compounds.

Mercury occurs mainly as the mineral cinnabar, HgS, from which it is separated by froth flotation and then roasting in air:

$$HgS(s) + O_2(g) \xrightarrow{\Delta} Hg(g) + SO_2(g)$$

V Video 11: Photoeffect and silver chloride.

N See Case 15 for the use of silver in photoactive glass.

D The silver mirror reaction. CD1, 75.

E The recovery of silver from films. *J. Chem. Ed.*, **1976**, *53*, 370. Silver refinement and debasement. *J. Chem. Ed.*, **1988**, *65*, 153.

E Gold and anemia: Teaching the skills of science. *J. Chem. Ed.*, **1988**, *65*, 1000. Reclamation of gold. *J. Chem. Ed.*, **1972**, *49*, 286. The oxidation states of gold. *J. Chem. Ed.*, **1975**, *52*, 731.

U Exercises 21.77 and 21.88.

E Coinage gold in the United States (for commemorative medals) is 90% gold and 10% copper; 18-carat gold is 18 parts gold and 6 parts silver; 15-carat gold is 15 parts gold, 6 parts silver, and 1 part copper; 14-carat gold is 14 parts gold, 8.25 parts silver, and 1.75 parts copper. Gold is the most malleable of metals, and 1 g of gold can be worked into a leaf covering about 1 m^2 or a wire over 2 km in length. Gold leaf is used for decoration, as in cathedrals.

E Zinc enzymes. *J. Chem. Ed.*, **1985**, *62*, 924.

U Exercise 21.82.

D Heating ZnS in a stream of oxygen. TDC, 11-12.

E Mercury: Chemical of the month. *J. Chem. Ed.*, **1982**, *59*, 971. Mercury poisoning. *J. Chem. Ed.*, **1972**, *49*, 28.

D Amphoteric properties of metal hydroxides. CD2, 89.

E What were once thought to be deaths due to zinc poisoning may in fact have been caused by the trace amounts of cadmium, always present in conjunction with zinc (which is almost always present with zinc).

E Mercury metal has many uses, including electrical switches, cathodes in chlor-alkali plants, and pumps to achieve high vacuums. Mercury(II) nitrate was once used for the making and shaping of felt hats: thus the reference to the "mad hatter" in Lewis Carroll's *Alice in Wonderland*.

E The mercury(I) cation, Hg_2^{2+}, is the simplest example of a *cluster* compound in which there is a metal-metal bond. Mercury(I) chloride, Hg_2Cl_2, is called *calomel* and is used in the preparation of the reference electrodes of pH meters.

FIGURE 21.15 When ammonia is added to a silver chloride precipitate, the precipitate dissolves. However, when ammonia is added to a precipitate of mercury(I) chloride, mercury metal is formed by disproportionation and the mass turns gray. Left to right: silver chloride in water; silver chloride in aqueous ammonia; mercury(I) chloride in water; mercury(I) chloride in aqueous ammonia.

D Precipitate formation: Black and white. CD1, 18.

E Thanksgiving dinner and transition metal complexes. *J. Chem. Ed.*, **1971**, *48*, 265.

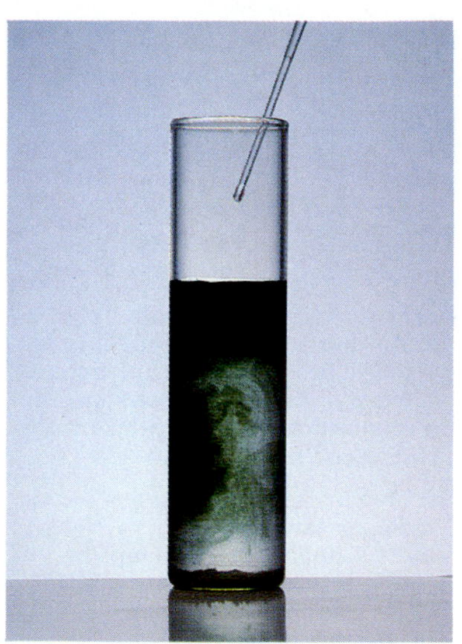

FIGURE 21.16 When cyanide ions (in the form of potassium cyanide) are added to a solution of iron(II) sulfate, they displace the H_2O molecules from the $[Fe(H_2O)_6]^{2+}$ ion and produce a new complex, the more strongly colored hexacyanoferrate(II) ion.

The volatile metal is distilled off and condensed. Mercury is unique in being the only metallic element that is liquid at room temperature; and with the large range of temperatures over which it is liquid (from its melting point of $-39°C$ to its boiling point of 357°C), it is well suited for its use in thermometers.

Mercury lies above hydrogen in the electrochemical series, so it is not oxidized by hydrogen ions. However, it does react with nitric acid:

$$3Hg(l) + 8H^+(aq) + 2NO_3^-(aq) \longrightarrow 3Hg^{2+}(aq) + 2NO(g) + 4H_2O(l)$$

In compounds it has the oxidation number +1 or +2. Compounds with oxidation number +1 are unusual in that the mercury(I) cation is the covalently bonded diatomic ion $(Hg—Hg)^{2+}$, written Hg_2^{2+}. The zinc and cadmium analogues of this cation, Zn_2^{2+} and Cd_2^{2+}, have recently been found in their molten salts.

Mercury(I) chloride is insoluble, and like silver chloride, it too precipitates when chloride solution is added to a solution undergoing qualitative analysis. However, unlike silver chloride, which dissolves when ammonia is added, the solid mercury(I) chloride disproportionates to metallic mercury and mercury(II) ions:

$$Hg_2^{2+}(s) \longrightarrow Hg(l) + Hg^{2+}(aq)$$

The actual reaction is

$$Hg_2Cl_2(s) + 2NH_3(aq) \longrightarrow Hg(l) + HgNH_2Cl(s) + NH_4^+(aq) + Cl^-(aq)$$

Finely divided mercury metal appears black; so in the reaction the white $HgNH_2Cl$ precipitate appears as blackish gray (Fig 21.15). Because its color distinguishes Hg_2Cl_2 from AgCl, the presence of mercury ions can be distinguished from that of silver ions.

COMPLEXES OF THE *d*-BLOCK ELEMENTS

One of the most outstanding properties of the *d* metals is their ability to act as Lewis acids and to form numerous complexes with Lewis bases. A **complex** is the product of a reaction between a Lewis acid and a Lewis base (Section 8.5); a *d*-metal complex consists of a single central metal atom or ion (the Lewis acid), to which are attached several molecules or anions (the Lewis bases) that also have an independent existence. One example is the formation of the hexacyanoferrate(II) ion (formerly called the ferrocyanide ion), $Fe(CN)_6^{4-}$, in which the Lewis acid Fe^{2+} forms bonds by sharing electron pairs provided by the CN^- ions. When giving the chemical formula of a complex, it is common practice to enclose all the chemical symbols in brackets, and hence to write the hexacyanoferrate(II) ion as $[Fe(CN)_6]^{4-}$. Another example of a complex is the neutral complex $Ni(CO)_4$, in which Ni acts as the Lewis acid and the CO molecules act as Lewis bases: although square brackets should also be used for neutral complexes, they are normally omitted. Main-group elements also form complexes—an example is $[Al(H_2O)_6]^{3+}$—but the range of compounds is much wider in the *d* block than in the *s* and *p* blocks.

21.4 THE STRUCTURES OF COMPLEXES

Many complexes are prepared simply by mixing solutions of a *d*-metal ion with the appropriate Lewis base (Fig. 21.16); for example,

$$[Fe(H_2O)_6]^{2+}(aq) + 6CN^-(aq) \longrightarrow [Fe(CN)_6]^{4-}(aq) + 6H_2O(l)$$

Complexes can be formed by **substitution reactions**, in which one Lewis base expels another and takes its place. That process occurs in the reaction above: the CN^- ions drive out H_2O molecules from the $[Fe(H_2O)_6]^{2+}$ complex. A less complete replacement occurs when Cl^- ions are added to an iron(II) solution:

$$[Fe(H_2O)_6]^{2+}(aq) + Cl^-(aq) \longrightarrow [Fe(H_2O)_5Cl]^+(aq) + H_2O(l)$$

The impressive changes of color that often accompany these substitution reactions are one of the features we explore here (Fig. 21.17).

The central metal ion in a complex is a Lewis acid, and the groups attached to it are Lewis bases. These groups are called **ligands** (from the Latin word for "bound"), and some of the common ones are listed in Table 21.5. We say that the ligands **coordinate** to the metal when they form the complex. This type of bonding is the origin of the term **coordination compound**, which means either a neutral complex, such as $Ni(CO)_4$, or an ionic compound in which at least one of the ions is a complex, such as potassium hexacyanoferrate(II), $K_4[Fe(CN)_6]$. The ligands directly attached to the central ion in a complex (and generally enclosed within brackets) constitute the **coordination sphere** of the

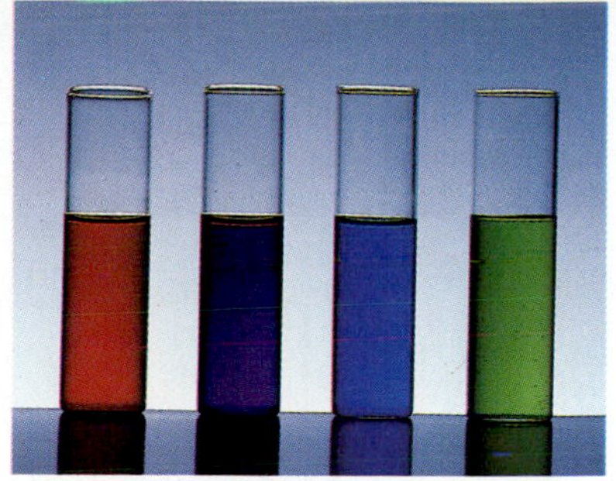

FIGURE 21.17 Some of the highly colored compounds that result when complexes are formed. Left to right: aqueous solutions of $[Fe(SCN)(H_2O)_5]^{2+}$; $[Co(SCN)_4(H_2O)_2]^{2-}$; $[Cu(NH_4)_4(H_2O)_2]^{2+}$; $[CuBr_4]^{2-}$.

D Add a few drops of concentrated aqueous ammonia solution to copper(II) sulfate and nickel(II) sulfate solutions placed in dishes on an overhead projector.

T OHT: Table 21.5.

TABLE 21.5 Common ligands

Formula	Name	Formula	Name
Neutral ligands			
H_2O	aqua	CO	carbonyl
NH_3	ammine	$NH_2CH_2CH_2NH_2$	ethylenediamine (en)*
NO	nitrosyl	$NH_2CH_2CH_2NHCH_2CH_2NH_2$	diethylenetriamine (dien)†
Anionic ligands			
F^-	fluoro	NO_2^-	nitro (as M—NO_2)
Cl^-	chloro	CO_3^{2-}	carbonato
Br^-	bromo	$C_2O_4^{2-}$	oxalato (ox)*
I^-	iodo	$(^-O(O{=})C{-}CH_2)_2N{-}CH_2{-}CH_2{-}N(CH_2{-}C({=}O)O^-)_2$	ethylenediaminetetraacetato‡ (EDTA)
OH^-	hydroxo		
O^{2-}	oxo		
CN^-	cyano (as M—CN)		
NC^-	isocyano (as M—NC)		
SCN^-	thiocyanato (as M—SCN)		
NCS^-	isothiocyanato (as M—NCS)		
NO_2^-	nitrito (as M—ONO)		

*Bidentate.
†Tridentate.
‡Hexadentate.

N This statement is true only if all the ligands are monodentate. The coordination number is actually the number of atoms that are directly linked to the central metal atom.

central ion. The number of atoms directly linked to the metal ion is called the **coordination number** of the complex.

Types of complexes. Many important complexes have a coordination number of either four or six, which means that four or six ligands are attached to the metal ion. In the great majority of cases, complexes with a coordination number of six, such as $[Fe(CN)_6]^{4-}$, have the six ligands at the corners of an octahedron and the metal ion at the center; they are called **octahedral complexes** (Fig. 21.18a). It is conventional to represent these octahedral complexes by a simplified diagram (**3**). In complexes with a coordination number of four, the ligands are either at the corners of a tetrahedron, as in $[Cu(NH_3)_4]^{2+}$, or (most notably for d^8 electron configurations, such as Pt^{2+} and Au^{3+}) at the corners of a square. The two types are called **tetrahedral complexes** and **square-planar complexes**, respectively (Fig. 21.18b,c).

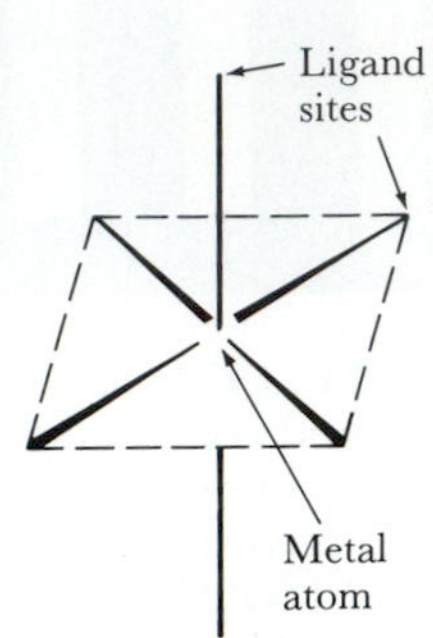

3 Simplified octahedral complex

Many ligands, including H_2O, NH_3, and CN^-, occupy only one site in a coordination sphere and are called **monodentate ligands** (for "one-toothed"). Some ligands can simultaneously occupy more than one site and are called **polydentate ligands** ("many-toothed"). Ethylenediamine (NH_2—CH_2—CH_2—NH_2; **4**) has a tooth (a nitrogen lone pair) at each end and is therefore a **bidentate ligand**. This ligand occurs widely in coordination chemistry and is abbreviated to "en," as in $[Co(en)_3]^{3+}$ (**5**).

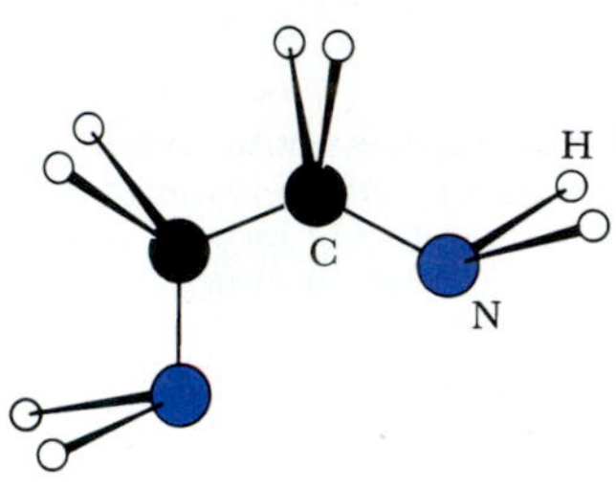

4 Ethylenediamine

N The systematic name of ethylenediamine (**4**) is 1,2-diaminoethane.

Chelates. The metal atom in $[Co(en)_3]^{3+}$ lies at the center of the three ligands as though pinched by three two-pronged molecular claws. It is an example of a **chelate**, which is a complex containing at least one polydentate ligand that forms a ring of atoms including the central metal atom. (The name comes from the Greek word for "claw.") An example of a hexadentate chelating ligand, one that can occupy all six octahedral sites around an ion and grip it in a single six-pronged claw, is the ethylenediaminetetraacetate ion, $EDTA^{4-}$ (Fig. 21.19). This ligand demonstrates the fact that main-group elements can form complexes, for it forms a complex with Ca^{2+} and is used to soften water and to remove scale (which is $CaCO_3$). It also forms a complex with Pb^{2+} and hence acts as an antidote to lead poisoning. Polyphosphate chelating agents are used in detergents to complex the ions Ca^{2+} and Mg^{2+} that cause hardness: the triphosphate ion $P_3O_{10}{}^{5-}$ (**6**), which is included in the formulation of detergents, acts as a pentadentate ligand.

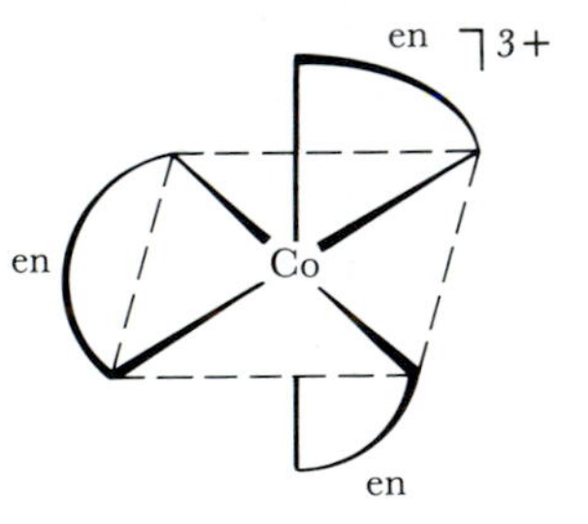

5 $[Co(en)_3]^{3+}$

Chelating ligands are quite common in nature. Mosses and lichens secrete chelating ligands to capture essential metal ions from the rocks they inhabit. Chelate formation also lies behind the body's strategy of running a fever when infected by bacteria. The higher temperature kills bacteria by reducing their ability to synthesize a particular iron-chelating ligand.

E The formal rules are set out in the "Red Book", *Nomenclature of inorganic chemistry*, G. J. Leigh (ed.), Blackwell Scientific Publications, Oxford (1990).

The nomenclature of complexes. The names of the common ligands and their abbreviations used in formulas are given in Table 21.5. General rules for naming complexes are as follows:

1. Neutral ligands have the same name as the molecule, except for H_2O (aqua), NH_3 (ammine), CO (carbonyl), and NO (nitrosyl).
2. Anionic ligands end in *-o*; specifically,

   -ide ⟶ *-o* *-ate* ⟶ *-ato* *-ite* ⟶ *-ito*

   Examples include chloro, cyano, sulfato, carbonato, and sulfito.

$$^-O-\overset{\overset{\displaystyle O}{\|}}{\underset{\underset{\displaystyle O^-}{|}}{P}}-O-\overset{\overset{\displaystyle O}{\|}}{\underset{\underset{\displaystyle O^-}{|}}{P}}-O-\overset{\overset{\displaystyle O}{\|}}{\underset{\underset{\displaystyle O^-}{|}}{P}}-O^-$$

6 Triphosphate ion, $P_3O_{10}{}^{5-}$

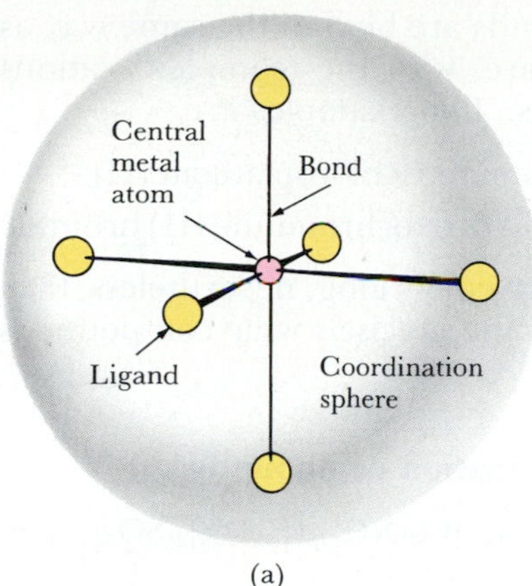

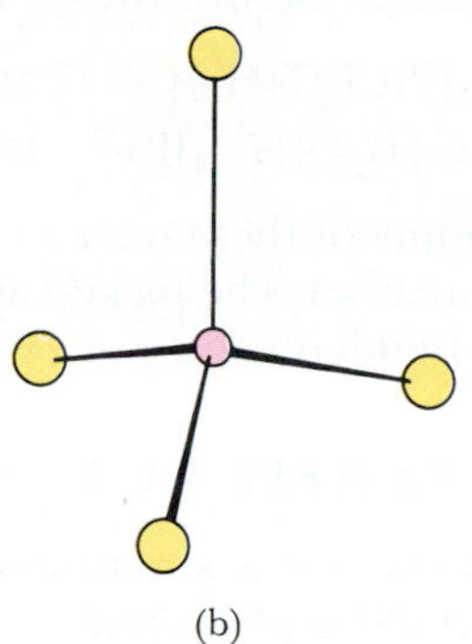

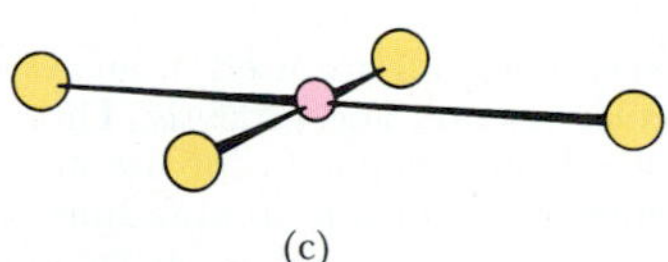

FIGURE 21.18 The various components of a complex may be either neutral or charged. (a) Almost all six-coordinate complexes are octahedral. Four-coordinate complexes are either (b) tetrahedral or (c) square planar.

3. Greek prefixes are used to denote the number of the same ligand in the complex ion:

2	3	4	5	6	...
di-	*tri-*	*tetra-*	*penta-*	*hexa-*	

If the ligand already contains a Greek prefix (as in ethylenediamine) or if it is polydentate, then the italic prefixes

2	3	4	...
bis-	*tris-*	*tetrakis-*	

are used instead. Parentheses are then often used to enclose the name of the ligand.

4. The oxidation number of the central metal ion is given as a Roman numeral after its chemical name.

5. The ligands are named before the metal ion in alphabetical order irrespective of the Greek prefix that denotes the number of each one present. For example,

$[FeCl(H_2O)_5]^+$	pentaaquachloroiron(II) ion
$[CrCl_2(NH_3)_4]^+$	tetraamminedichlorochromium(III) ion
$[Co(en)_3]^{3+}$	*tris*(ethylenediamine)cobalt(III) ion

Notice that the Cl_2 in the coordination sphere of the second complex represents two chloride ligands (named as dichloro) and not a Cl_2 molecular ligand. Also note that the symbols of anionic ligands (such as Cl^-) precede those of neutral ligands (such as H_2O and NH_3) in the chemical formula of the complex (but not necessarily in its name).

6. If the complex itself is anionic, the suffix *-ate* is added to the stem of the metal's name. If the symbol of the metal originates from a Latin name (as listed in Table 2.1), the Latin stem is used. For example, the symbol for iron is Fe, from the Latin *ferrum*. Therefore, any anionic complex of iron ends with *-ferrate* and the oxidation of iron in Roman numerals:

$[Fe(CN)_6]^{4-}$	hexacyanoferrate(II) ion
$[Ni(CN)_4]^{2-}$	tetracyanonickelate(II) ion

E An ingenious impudence: Alfred Werner's coordination theory. *J. Chem. Ed.*, **1976**, *53*, 445. Valency and inorganic metal complexes: Quantum mechanics and chemical bonding in inorganic complexes. *J. Chem. Ed.*, **1979**, *56*, 294.

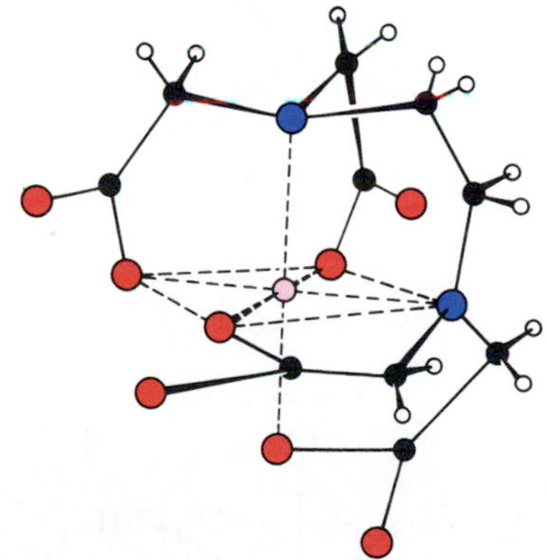

FIGURE 21.19 $EDTA^{4-}$ is a hexadentate chelating ligand. Its two N atoms (blue) and four of its O atoms (red) can occupy all six octahedral bonding positions around a central metal atom (pink). (Carbon, black; hydrogen, white.)

N The main group metals, such as aluminum and magnesium, which have a relatively high charge density and are therefore strong Lewis acids, also form complexes. A great deal of modern chemical research focuses on the structure, properties, and uses of complexes, partly because they take part in many biological reactions. Hemoglobin and vitamin B_{12}, for example, are both complexes—one of iron and the other of cobalt. Complexes are often brightly colored and magnetic, and are used in analysis, color science, and catalysis. They are now being examined for use in solar energy conversion, in atmospheric nitrogen fixation, and as pharmaceuticals. Complexes are also used in catalysis, as dyes, and in the electroplating of metals.

N Many common names are still familiar, such as the ferricyanide and ferrocyanide complex ions.

U Examples 21.83 and 21.84.

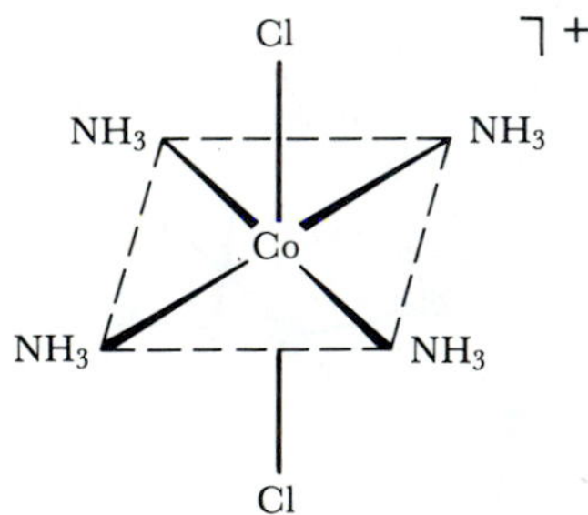

7 *trans*-$[CoCl_2(NH_3)_4]^+$

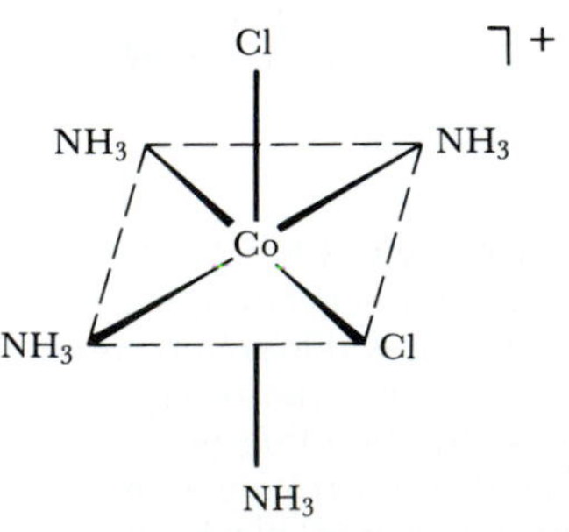

8 *cis*-$[CoCl_2(NH_3)_4]^+$

7. The name of coordination compounds are built in the same way as the names of simple compounds are, with the (complex) cations named before the (complex) anions. Two examples are

$NH_4[PtCl_3(NH_3)]$	ammonium amminetrichloroplatinate(II)
$[Cr(OH)_2(NH_3)_4]Br$	tetraamminedihydroxochromium(III) bromide

Some of these names can become extremely long; nevertheless, they can always be picked apart to determine precisely what compound is intended.

EXAMPLE 21.2 Naming coordination compounds

Name the coordination compounds (a) $[Co(NH_3)_3(H_2O)_3]_2(SO_4)_3$ and (b) $(NH_4)_3[Fe(CN)_6]$.

STRATEGY We need to identify the cation and anion, name each separately, and then combine them. For each complex, we should first note the name and oxidation number of the metal ion; then identify the ligands; and finally, string the names together in alphabetical order with the appropriate prefixes, ending with the name of the metal.

SOLUTION (a) The charge on the complex must be +3 for charge neutrality of the compound (there are three SO_4^{2-} ions for two complex ions, and $2 \times (+3) + 3 \times (-2) = 0$); it follows that the complex cation is $[Co(NH_3)_3(H_2O)_3]^{3+}$, implying cobalt(III). It follows that the name of the cation is triamminetriaquacobalt(III), so the compound is triamminetriaquacobalt(III) sulfate. (b) For charge neutrality, the complex anion must be $[Fe(CN)_6]^{3-}$. This complex must contain iron(III), because each CN^- ligand contributes one negative charge and Fe^{3+} is required to give an overall charge of -3 for the complex. This anion is therefore hexacyanoferrate(III). It follows that the name of the compound is ammonium hexacyanoferrate(III).

EXERCISE E21.2 Name the compounds (a) $[FeOH(H_2O)_5]Cl_2$ and (b) $K_2[Cr(ox)_2(H_2O)_2]$.

[*Answer*: (a) pentaaquahydroxoiron(III) chloride; (b) potassium diaqua*bis*(oxalato)chromate(II)]

21.5 ISOMERISM

A characteristic feature of complexes and of coordination compounds is the existence of isomers:

Isomers are compounds that contain the same numbers of the same atoms in different arrangements.

(The name comes from the Greek for "equal parts.") Isomers are different compounds built from the same kit of parts. The tetraamminedichlorocobalt(III) ions (**7**) and (**8**) differ only in the positions of the Cl^- ligands, but they are distinct species, for they have different physical and chemical properties.

Isomerism—the existence of isomers—can take a number of forms (Fig. 21.20). The two major classes of isomers are **structural isomers**, in which the atoms have different partners, and **stereoisomers**, in which the atoms have the same partners but are arranged differently in space. Structural isomerism can be subdivided into several different types, namely, ionization isomerism, hydrate isomerism, linkage isomerism, and coordination isomerism.

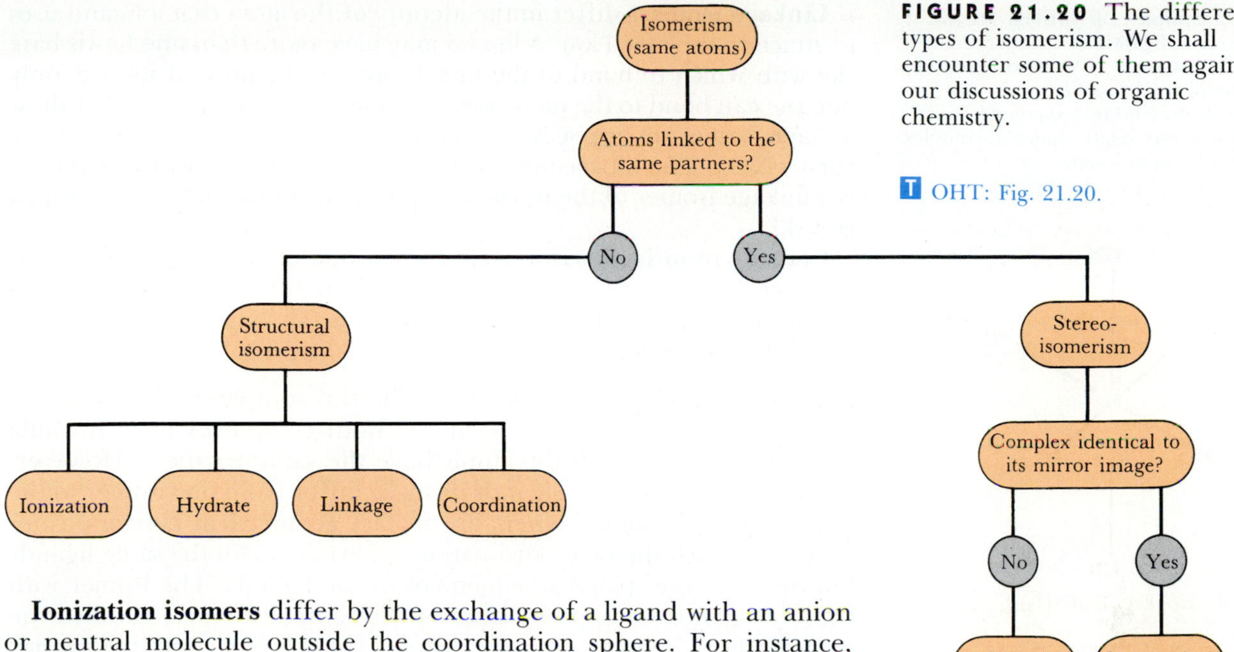

FIGURE 21.20 The different types of isomerism. We shall encounter some of them again in our discussions of organic chemistry.

T OHT: Fig. 21.20.

Ionization isomers differ by the exchange of a ligand with an anion or neutral molecule outside the coordination sphere. For instance, $[CoBr(NH_3)_5]SO_4$ and $[CoSO_4(NH_3)_5]Br$ are ionization isomers because the Br^- ion is a ligand of the cobalt in the former compound but an accompanying anion in the latter compound. The two compounds are physically quite different: the bromo complex is violet and the sulfato complex is red (Fig. 21.21). The chemical properties of the two ionization isomers are also different: the red isomer forms an off-white precipitate of AgBr upon addition of Ag^+ions, but no precipitate upon addition of Ba^{2+}:

E The use of conductivity data for the structure determination of metal complexes. *J. Chem. Ed.*, **1975**, *52*, 649.

N Silver sulfate has marginal solubility, so an excess of the silver nitrate should not be added.

U Exercise 21.85.

N Hydrate isomers can be considered special cases of ionization isomers.

$$[CoSO_4(NH_3)_5]Br(aq, \text{red}) + Ag^+(aq) \longrightarrow AgBr(s) + [CoSO_4(NH_3)_5]^+(aq)$$

$$[CoSO_4(NH_3)_5]Br(aq, \text{red}) + Ba^{2+}(aq) \longrightarrow \text{no reaction}$$

On the other hand, the violet isomer is unaffected by the addition of Ag^+ ions, but it forms a white precipitate of $BaSO_4$ upon addition of Ba^{2+} ions:

$$[CoBr(NH_3)_5]SO_4(aq, \text{violet}) + Ag^+(aq) \longrightarrow \text{no reaction}$$

$$[CoBr(NH_3)_5]SO_4(aq, \text{violet}) + Ba^{2+}(aq) \longrightarrow BaSO_4(s) + [CoBr(NH_3)_5]^{2+}(aq)$$

FIGURE 21.21 Solutions of the two coordination compounds $[CoBr(NH_3)_5]SO_4$ (left) and $[CoSO_4(NH_3)_5]Br$ (right). Although they are built of the same atoms, they are different compounds with their own characteristic properties.

Hydrate isomers differ by an exchange between an H_2O molecule and a ligand in the coordination sphere. For example, the hexahydrate of chromium(III) chloride, $CrCl_3 \cdot 6H_2O$, may be any of the three compounds

$[Cr(H_2O)_6]Cl_3$	violet
$[CrCl(H_2O)_5]Cl_2 \cdot H_2O$	blue-green
$[CrCl_2(H_2O)_4]Cl \cdot 2H_2O$	green

The addition of $AgNO_3$ results in the precipitation of different amounts of AgCl from each of the three salts. The water molecules that are not part of the coordination sphere are easily removed in a drying oven.

E In a complex of formula ML_3X_3, if the three L ligands lie at the corners of a triangular face of the octahedron, the complex is classified as *fac* (for facial); if they lie in a line (one equatorial and two axial), then the complex is *mer* (for meridional).

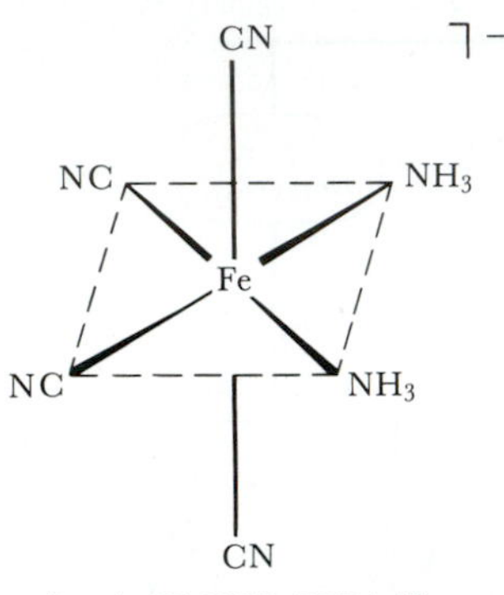

9 *cis*-$[Fe(CN)_4(NH_3)_2]^-$

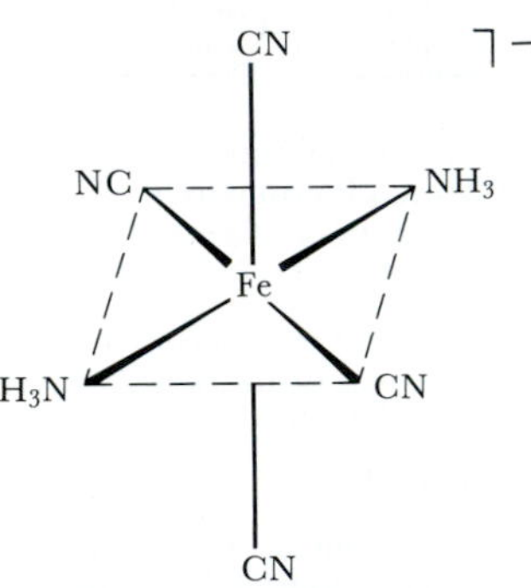

10 *trans*-$[Fe(CN)_4(NH_3)_2]^-$

U Exercise 21.92.

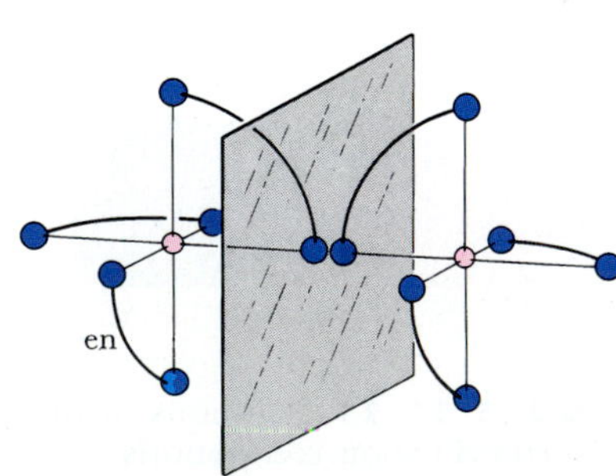

FIGURE 21.22 These two optical isomers of the complex ion $[Co(en)_3]^{3+}$ are not identical to each other, but each is the mirror image of the other.

Linkage isomers differ in the identity of the atom that a ligand uses to attach to the metal ion. A ligand may have more than one Lewis base site with which to bond to the metal ion, but, because of its size, only one site can bond to the metal ion at a time. Common ligands that show linkage isomerism are SCN^- versus NCS^-; NO_2^- versus ONO^-; CN^- versus NC^-. The nitro complex $[CoCl(NO_2)(NH_3)_4]^+$, which is yellow, is a linkage isomer of the nitrito complex $[CoCl(ONO)(NH_3)_4]^+$, which is red.

Coordination isomerism occurs when one or more ligands are exchanged between a cationic complex and an anionic complex. An example of a pair of coordination isomers is $[Cr(NH_3)_6][Fe(CN)_6]$ and $[Fe(NH_3)_6][Cr(CN)_6]$.

Geometrical and optical isomerism. The two complexes shown in diagrams (**7**) and (**8**) are stereoisomers. Both complexes have formula $[CoCl_2(NH_3)_4]^{2+}$, so all the atoms have the same partners. However, the arrangements of the ligands in space differ from one isomer to the other. The two stereoisomers are in fact **geometrical isomers**, complexes in which the two coordination spheres contain the same ligands but differ in the spatial arrangement of the ligands. The isomer with the Cl^- ligands on opposite sides of the central atom (**7**) is called the trans isomer (from the Latin for "across"). The isomer with the ligands on the same side (**8**) is called the cis isomer (from the Latin for "on this side"). Geometrical isomers exist for square-planar and octahedral complexes but not for tetrahedral complexes.

The chemical and physiological properties of geometrical isomers differ. For example, *cis*-$[PtCl_2(NH_3)_2]$ is pale orange-yellow, with a solubility of 0.252 g per 100 g of water; it is used in chemotherapy for cancer patients. On the other hand, *trans*-$[PtCl_2(NH_3)_2]$ is dark yellow, with a solubility of 0.037 g per 100 g of water, and it exhibits no chemotherapeutic effect. Other antitumor agents are also cis isomers of platinum(II), such as *cis*-$[PtCl_2(en)]$, *cis*-$[Pt(ox)(NH_3)_2]$, and *cis*-$[Pd(NO_3)_2(C_6H_4(NH_2)_2)]$. An octahedral complex that exhibits geometrical isomerism is $[Fe(CN)_4(NH_3)_2]^-$ (**9** and **10**). Geometrical isomerism may also occur in an octahedral complex when one of the ligands is bidentate, as in $[CoCl_2(en)_2]^+$: the trans isomer is green and the two alternative cis isomers are violet.

A more subtle type of stereoisomerism is **optical isomerism**. This isomerism occurs when a molecule (or ion) and its mirror image are not structurally identical (Fig. 21.22). A molecule that is distinct from its own mirror image—as a left hand is distinct from a right hand—is said to be **chiral** (from the Greek word for "hand"). Pairs of optical isomers, like pairs of hands, are called **enantiomers** (from the Greek words for "opposite parts").

▼ EXAMPLE 21.3 Identifying optical isomerism

Which of the following complexes is chiral and which form enantiomeric pairs?

(a) A; B, B; C, C; A
(b) A; B, C; C, B; A
(c) A; C, B; A, B; B
(d) A; A, B; C, B; B

STRATEGY By drawing the mirror image of each complex, we should be able to judge whether any rotation of the original molecule will cause it to match its mirror image. If not, then the complex is chiral. We can determine which form enantiomeric pairs by finding pairs that are the mirror images of each other.

SOLUTION In the pairs of complexes below, the original is on the left, and its mirror image is on the right:

(a)

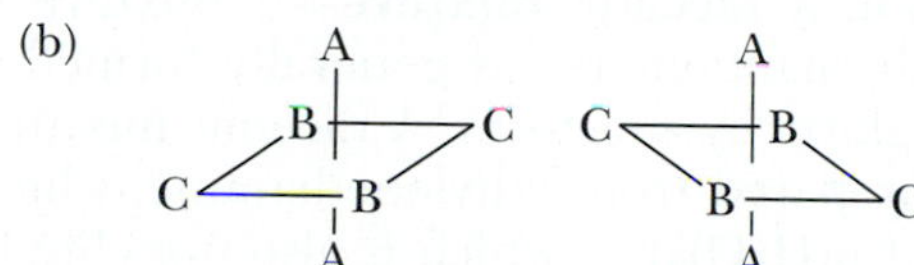

The image is identical to the original.

(b)

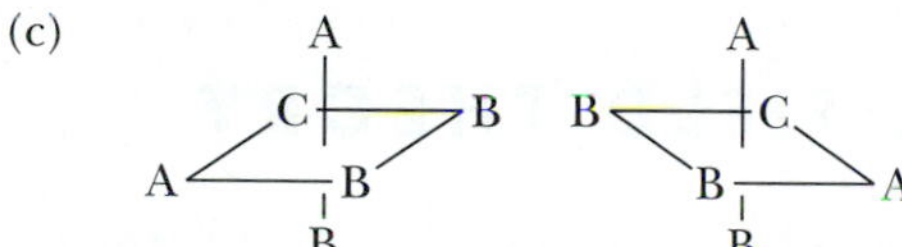

If we rotate the image about A—A, we obtain

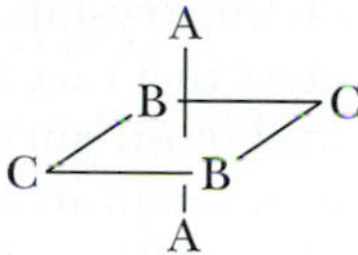

which is identical to the original.

(c)

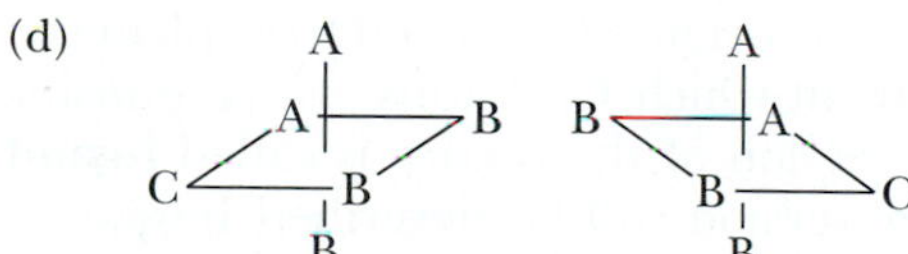

The original complex is chiral.

(d)

The original complex is chiral.

Complexes (c) and (d) are chiral because no rotation can make either match its mirror image. However, when the mirror image of (c) is rotated by 180° around the vertical A—B axis, it becomes the complex (d); hence (c) and (d) form an enantiomeric pair.

EXERCISE E21.3 Repeat the exercise for the following complexes:

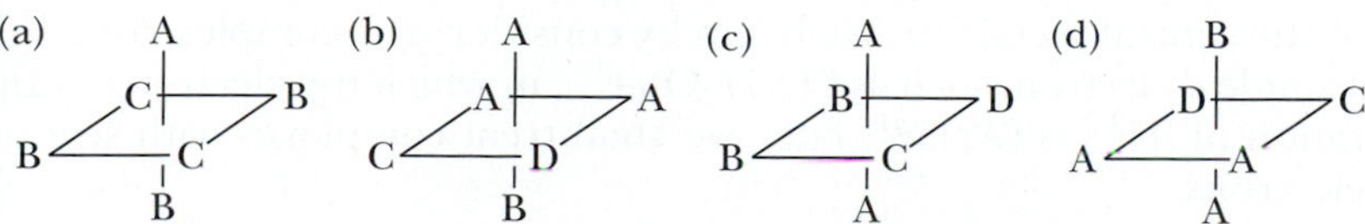

[*Answer*: (a,c) not chiral; (b,d) chiral and enantiomeric]

Optical activity. Enantiomers have identical physical and chemical properties, with two exceptions: they can react differently with other

E It is important, as always, to distinguish *thermodynamic* and *kinetic* effects. Although a complex may be thermodynamically unstable, there may be such a high activation barrier to reaction that it survives for long periods. An example of the distinction between stability (the thermodynamic term) and lability (the kinetic term) is found in the different behaviors of $[Co(NH_3)_6]^{2+}$ and $[Co(NH_3)_6]^{3+}$. Both undergo substitution reactions in acidified water: $[Co(NH_3)_6]^{2+}(aq) + 6H_3O^+(aq) \rightarrow [Co(H_2O)_6]^{2+}(aq) + 6NH_4^+(aq)$, and $[Co(NH_3)_6]^{3+}(aq) + 6H_3O^+(aq) \rightarrow [Co(H_2O)_6]^{3+}(aq) + 6NH_4^+(aq)$. Both reactions have a negative reaction free energy, so both ammonia complexes are thermodynamically unstable. However, whereas the first reaction reaches equilibrium in a few seconds at room temperature, the second requires weeks. That is, $[Co(NH_3)_6]^{2+}$ is labile as well as unstable; $[Co(NH_3)_6]^{3+}$ is unstable but nonlabile.

E The enantiomer that rotates the plane clockwise (as seen by the observer looking along the beam toward the approaching light) is called the *(+)-enantiomer*. The one that rotates it counterclockwise is called the *(−)-enantiomer*.

E Computation of the number of isomers of coordination compounds containing different monodentate ligands. *J. Chem. Ed.*, **1979**, *56*, 398. A simple algorithmic method for the recognition of theoretically chiral octahedral complexes. *J. Chem. Ed.*, **1974**, *51*, 347.

D Model to illustrate bonding and symmetry of transition metal complexes. *J. Chem. Ed.*, **1970**, *47*, 824.

U Exercises 21.86, 21.94, and 21.95.

? Is there an optical isomer of $[Co(EDTA)]^-$?

E The concept of dissymmetric worlds: A utilization of the power of optical isomerism. *J. Chem. Ed.*, **1972**, *49*, 455.

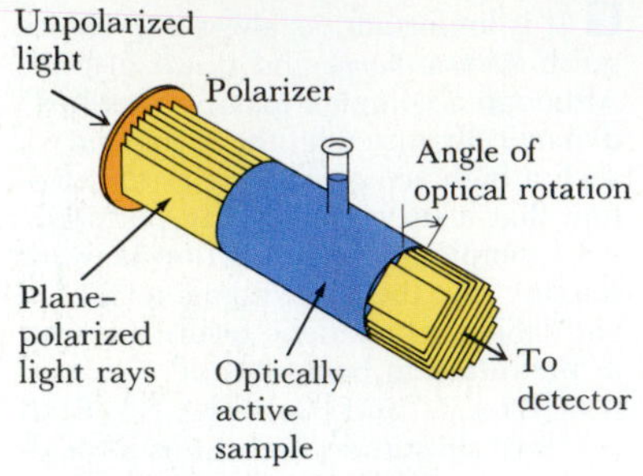

FIGURE 21.23 To demonstrate optical activity, light is passed through a polarizer (such as a sheet of Polaroid) so that the wave oscillates in a single plane. After the light has passed through an optically active sample, its plane of polarization is at a different angle. The planes show the polarization of each ray in the beam.

T OHT: Fig. 21.23.

chiral compounds and they are **optically active**. That is, they have different effects on plane-polarized light. **Plane-polarized light** is light in which the wave motion occurs in a single plane. It is different from ordinary unpolarized light, such as the light from an ordinary electric lamp, in which the wave motion occurs at all angles around the direction of travel. Light is polarized by passing it through a polarizing filter, which eliminates all wave motion but that in a single plane. (The same effect is obtained with polarized sunglasses.) When it is then passed through a solution of optically active molecules, the angle of the plane of polarization changes (Fig. 21.23). Separate solutions of enantiomers at the same concentration rotate the plane of polarization by an equal angle but in opposite directions.

When a chiral compound is synthesized from reagents that are not themselves chiral, a **racemic mixture**—a mixture containing equal amounts of both enantiomers—is generally formed (like the production of pairs of gloves by a factory). A racemic mixture is formed when $[Co(en)_3]^{3+}$ is prepared from ethylenediamine, which is not chiral itself, and from $[Co(H_2O)_6]^{3+}$, which is also not chiral. A racemic mixture is optically inactive because the clockwise rotation caused by one enantiomer is canceled by the counterclockwise rotation caused by the other. Because enantiomers form crystals that are the mirror images of each other, the two enantiomers in a racemic mixture can sometimes be separated by crystallization and hand-sorting of crystals. Biological syntheses of chiral molecules often result in only one enantiomer, so many molecules in our bodies have a definite "handedness." This topic is taken up in Section 24.9.

CRYSTAL FIELD THEORY

The most notable physical properties of coordination compounds are their colors and their magnetism. These properties, and some differences in their stabilities, can be discussed in terms of **crystal field theory**. This theory was originally devised to explain the colors of solids—particularly ruby, in which Cr^{3+} ions are responsible for the color. A more complete version of the theory is called **ligand field theory**, but only the simpler version will be described here.

21.6 THE EFFECTS OF LIGANDS ON *d* ELECTRONS

In crystal field theory, each site of a ligand (a Lewis base) is represented by a negative point charge (Fig. 21.24). The electronic structure of the complex is then expressed in terms of the electrostatic interactions—the field—between these point charges and the electrons and nucleus of the central metal ion. We begin by considering a complex containing a single *d* electron, such as $[Ti(H_2O)_6]^{3+}$, in which the electron configuration of Ti^{3+} is $[Ar]3d^1$; later we shall treat complexes with several *d* electrons.

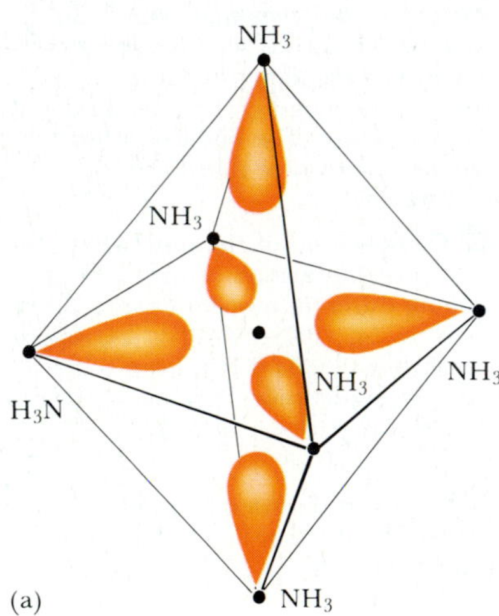

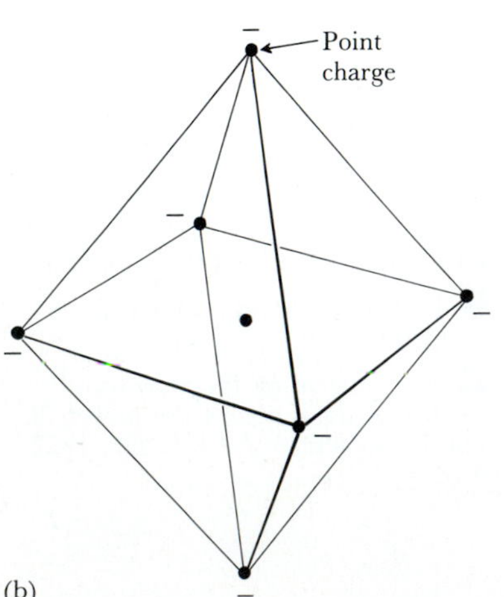

FIGURE 21.24 In the crystal field theory of complexes, the lone pairs of electrons (the Lewis base sites) on the ligands (a) are treated as point negative charges (b).

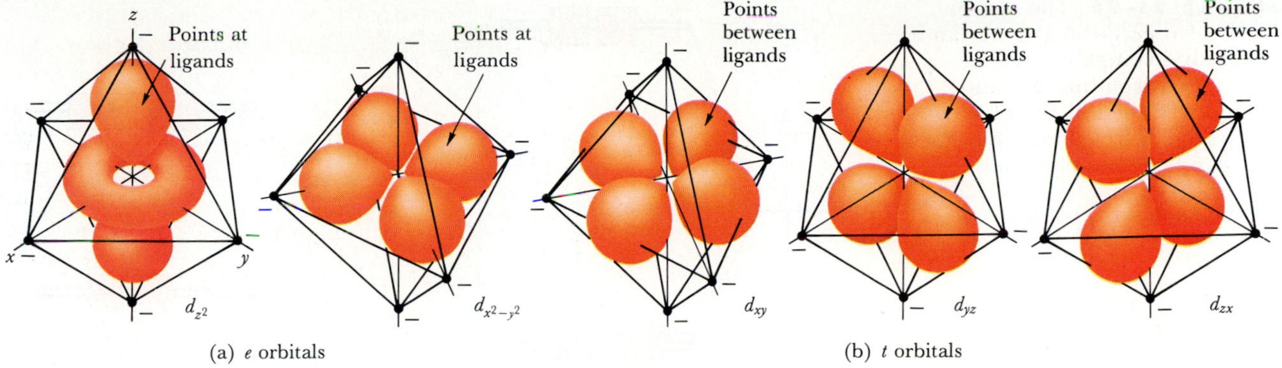

FIGURE 21.25 In an octahedral complex with a central *d*-metal ion, (a) a d_{z^2} orbital points directly toward two ligands, and an electron that occupies it has a relatively high energy. The same is true of a $d_{x^2-y^2}$ orbital. (b) A d_{xy} orbital is directed between the ligands, and an electron that occupies it has a relatively low energy. The same is true of the d_{yz} and d_{zx} orbitals.

The ligand field splitting. Because the central ion of a complex usually is positively charged, the negative charges representing the ligand lone pairs are attracted to it. This attraction results in an overall lowering of energy and is a major factor in the stability of the complex. The stability of the $[Ti(H_2O)_6]^{3+}$ ion, for instance, can be ascribed largely to the strong attraction between the Ti^{3+} ion and a negative charge representing a lone pair on each of the six H_2O ligands. However, when we examine the structure of the Ti(III) complex in more detail, we have to consider the fact that the single 3*d* electron interacts differently with the ligand point charges, depending on which of the five 3*d* orbitals it occupies. Crystal field theory focuses on the effect of ligands on the *d* orbitals of the metal ion and sets up a unified theory of the colors, magnetic properties, and stabilities of complexes.

T OHT: Figs. 21.24 and 21.25.

In an octahedral complex such as $[Ti(H_2O)_6]^{3+}$, the six bonding sites (represented by point charges) lie on either side of the central metal ion along the *x*, *y*, and *z* axes. From the drawings of the *d* orbitals in Fig. 21.25 (see also Fig. 7.21), we can see that three of the orbitals (d_{xy}, d_{yz}, and d_{zx}) have their lobes directed between the point charges. These three *d* orbitals are called the ***t* orbitals** in crystal field theory. The other two *d* orbitals (d_{z^2} and $d_{x^2-y^2}$) are directed toward the point charges. These two orbitals are called ***e* orbitals**. Because of their different arrangements in space, electrons in *t* orbitals are repelled less by the negative point charges of the ligands than electrons in *e* orbitals; therefore, the *t* orbitals are of lower energy than the *e* orbitals.

E The notation *e* and *t* is derived from group theory, where triply degenerate entities are generally labeled *T* and doubly degenerate entities are generally labeled *E*.

These ideas are summarized by the energy-level diagram in Fig. 21.26a. The energy difference between the two sets of orbitals is called the **ligand field splitting** Δ_O (the O denotes octahedral) and typically amounts to about 10% of the total energy of interaction of the central ion with the ligands. The *t* orbitals lie at an energy level that is $\frac{2}{5}\Delta_O$ below the average energy level of the *d* orbitals (the energy that an electron would have if the directional properties of the orbitals were ignored), and the *e* orbitals lie at an energy level $\frac{3}{5}\Delta_O$ above the average. Because the *t* orbitals have the lower energy, we can predict that, in the ground state of the $[Ti(H_2O)_6]^{3+}$ complex, the electron occupies one of them in preference to an *e* orbital; hence we predict that the electron configuration of the complex is t^1. This configuration can be represented by the following box diagram:

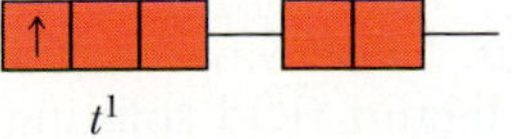

FIGURE 21.26 The energy levels of the d orbitals in (a) an octahedral complex, with the ligand field splitting Δ_O, and (b) in a tetrahedral complex, with the ligand field splitting Δ_T. Each box can hold two electrons.

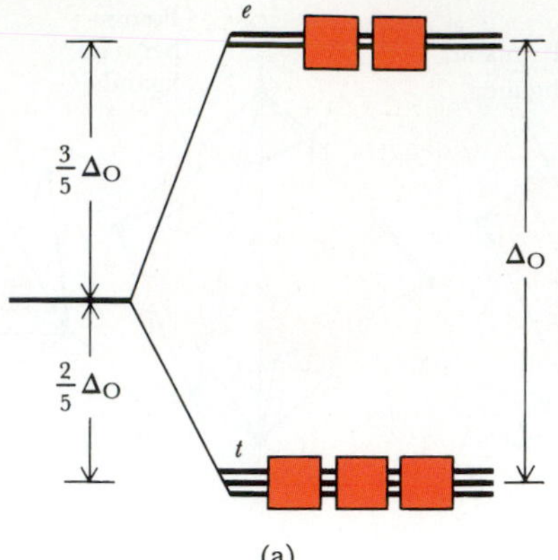

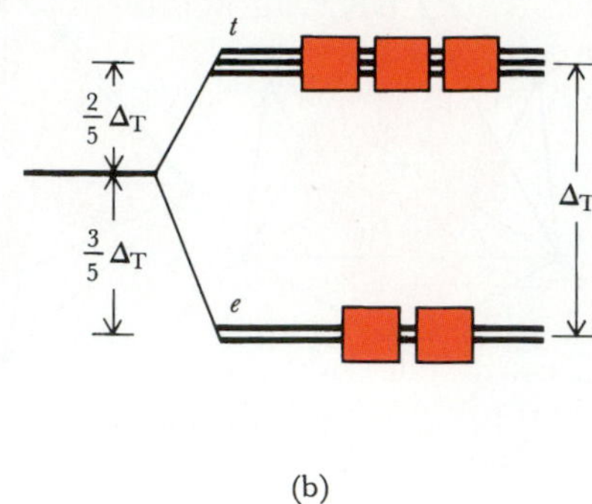

E Crystal field potentials. *J. Chem. Ed.*, **1969**, *46*, 339.

E In a square-planar complex, the energies of the d orbitals increase in order of $d_{zx} = d_{yz} < d_{z^2} < d_{xy} < d_{x^2-y^2}$.

N The colors commonly associated with the d-metal complexes are generally those of the hydrated ion; for example, $FeCl_3 \cdot 6H_2O$ is a rust-orange color because of the $[Fe(H_2O)_6]^{3+}$ ion; anhydrous $FeCl_3$, is a dark, intense green.

E Hexaammine complexes of Cr(III) and Co(III). *J. Chem. Ed.*, **1985**, *62*, 807. Color classification of coordination compounds. *J. Chem. Ed.*, **1987**, *64*, 1001.

D Green and blue copper complexes. CD2, 39.

D Cycling copper complexes. CD2, 40; The effect of temperature on a hydrate: pink to blue. CD1, 24.

E Colors of transition metal complexes. *J. Chem. Ed.*, **1969**, *46*, 675.

In a tetrahedral complex, the three t orbitals point more directly at the ligands than the two e orbitals do. As a result, in a tetrahedral complex the t orbitals have a *higher* energy than the e orbitals (Fig. 21.26b). The ligand field splitting Δ_T (where the T denotes tetrahedral) is generally smaller than in octahedral complexes (typically, $\Delta_T \approx \frac{4}{9}\Delta_O$) because the d orbitals do not point so directly at the ligands and there are fewer repelling ligands.

Light absorption by d^1 complexes. The t electron of the $[Ti(H_2O)_6]^{3+}$ complex can be excited into one of the e orbitals if it absorbs a photon of energy equal to Δ_O (Fig. 21.27). Because a photon carries an energy $h\nu$, where h is Planck's constant and ν is its frequency, it can be absorbed if its frequency satisfies

$$\Delta_O = h\nu$$

The frequency and wavelength λ of light are related by $\nu = c/\lambda$, where c is the speed of light (Section 7.1); hence the wavelength of light absorbed and the ligand field splitting are related by

$$\Delta_O = \frac{hc}{\lambda} \qquad (1)$$

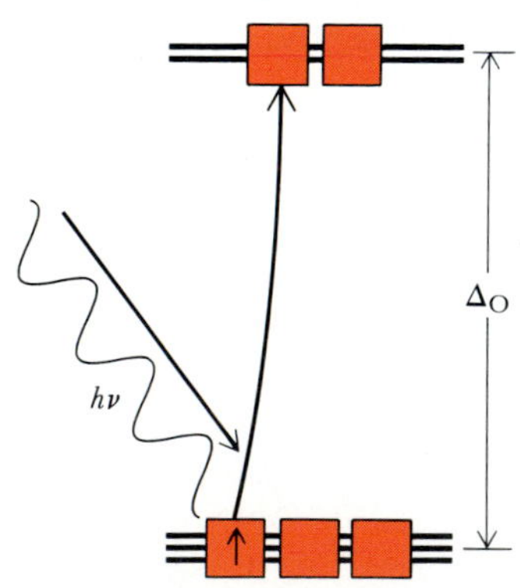

FIGURE 21.27 When a complex is exposed to light of the correct frequency, an electron can be excited to a higher-energy orbital (from t to e in this octahedral complex) and the light is absorbed.

That is, the greater the splitting, the shorter the wavelength of the light that is absorbed by the complex. For example, a $[Ti(H_2O)_6]^{3+}$ complex absorbs light of wavelength 510 nm, so

$$\Delta_O = \frac{6.626 \times 10^{-34}\ \text{J}\cdot\text{s}\cdot\text{photon}^{-1} \times 2.998 \times 10^{8}\ \text{m}\cdot\text{s}^{-1}}{510 \times 10^{-9}\ \text{m}}$$

$$= 3.895 \times 10^{-19}\ \text{J}\cdot\text{photon}^{-1}$$

Multiplication by Avogadro's constant gives

$$\Delta_O = 3.895 \times 10^{-19}\ \text{J}\cdot\text{photon}^{-1} \times 6.022 \times 10^{23}\ \text{photon}\cdot\text{mol}^{-1}$$

$$= 2.35 \times 10^{5}\ \text{J}\cdot\text{mol}^{-1} = 235\ \text{kJ}\cdot\text{mol}^{-1}$$

This energy is about 10% of the total interaction energy between the Ti^{3+} ion and its six H_2O ligands.

The spectrochemical series. The wavelength of the light absorbed can be used to measure the ligand field splittings in a range of different complexes. Ligands can be arranged in a **spectrochemical series** according to the Δ_O they produce, as shown in the margin. It is convenient to classify all ligands below the horizontal line in the series as **weak-field ligands** and those above it as **strong-field ligands**. Any complex of a particular metal ion has a smaller Δ_O value if it contains weak-field ligands than if it contains strong-field ligands. We can therefore go on to say that the complex absorbs longer wavelength light if it has weak-field ligands than if it has strong-field ligands. It follows that when we compare the two complexes $[Fe(CN)_6]^{4-}$ and $[Fe(H_2O)_6]^{2+}$, we note that the cyanide ligand is a stronger ligand than water and can predict that the ligand field splitting will be greater for $[Fe(CN)_6]^{4-}$ than for $[Fe(H_2O)_6]^{2+}$. That being so, we predict that the $[Fe(CN)_6]^{4-}$ complex will absorb shorter wavelength radiation than the other complex (see margin).

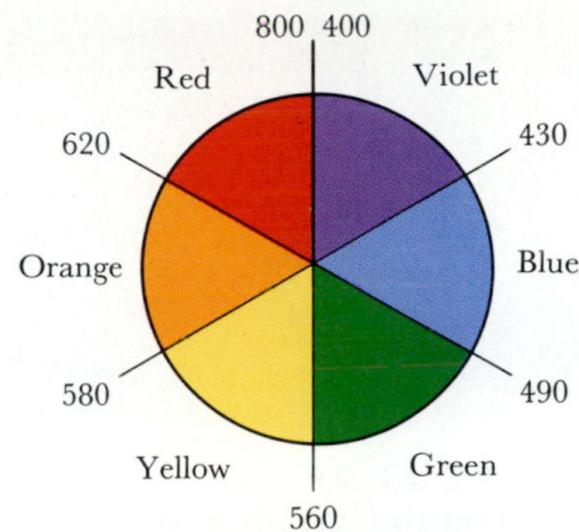

FIGURE 21.28 The perceived color of a complex in white light is the complementary color of the light it absorbs. In this color wheel, complementary colors are diametrically opposite each other. The numbers are approximate wavelengths in nanometers.

T OHT: Fig. 21.28.

Strong-field ligands

CN^-, CO

NO_2^-

en

NH_3

• • • • • • • • • • •

H_2O

ox

OH^-

F^-

SCN^-, Cl^-

Br^-

I^-

Weak-field ligands

T OHT: The spectrochemical series.

N Refer to Table 15.7 to compare formation constants.

U Exercises 21.87, 21.91, and 21.93.

The effect of ligands on color. We pointed out in Section 7.1 that white light is a mixture of all wavelengths of electromagnetic radiation from about 400 nm (violet) to 800 nm (red). When some of these wavelengths are removed from a beam of white light (by passing the light through a sample that absorbs certain wavelengths), the emerging light is no longer white. For example, if red light is absorbed from white light, then the light that remains is green. If green is removed, then the light appears red. We say that red and green are each other's **complementary color**—each is the color that white light becomes when the other is removed. Complementary colors are shown on a color wheel in Fig. 21.28.

We can now see that if a substance looks blue (as does a copper(II) sulfate solution, for instance), then it is absorbing orange (620-nm) light. Conversely, if we know the wavelength (and therefore the color) of the light that a substance absorbs, then we can predict the color of the substance by noting the complementary color on the color wheel. Thus, because $[Ti(H_2O)_6]^{3+}$ absorbs 510-nm light, which is green light, the complex looks purple (Fig. 21.29). However, it is important to realize that the color of a compound is a very subtle effect and such simple predictions can be misleading. One problem is that compounds absorb light over a range of wavelengths and may absorb in several regions of the spectrum. Chlorophyll, for example, absorbs both red and blue light, leaving only the wavelengths near green to be reflected from vegetation. The following discussion is a very simplified version of what actually happens.

Because weak-field ligands give small splittings, the complexes they form can be expected to absorb low-energy, long-wavelength radiation. The long wavelengths correspond to red light, so these complexes can be expected (as a first approximation) to exhibit colors near green. Because strong-field ligands give large splittings, the complexes they form should absorb high-energy, short-wavelength radiation. Because short-wavelength light corresponds to the violet end of the visible spectrum, such complexes can be expected to have colors near orange and yellow (Fig. 21.30). This relationship is part of the reason for the color change that occurs when ammonia is added to aqueous copper(II) sulfate: strong-field ammine (NH_3) ligands replace four of the weak-field aqua (H_2O) ligands of the $[Cu(H_2O)_6]^{2+}$ ion. The absorption shifts to higher energies and shorter wavelengths, from orange to yellow, and the perceived color shifts from blue toward violet.

FIGURE 21.29 Because $[Ti(H_2O)_6]^{3+}$ absorbs green 510-nm light, it looks purple in white light.

FIGURE 21.30 The effect of changing the ligands in octahedral cobalt(III) complexes. Weak-field ligands are on the left, strong-field ligands are on the right.

21.7 THE ELECTRONIC STRUCTURES OF MANY-ELECTRON COMPLEXES

The electron configurations of d^n complexes—complexes containing n electrons in the d orbitals of the central metal—are obtained by following the rules of the building-up principle. There are three t orbitals; therefore, according to the Pauli exclusion principle, there can be up to six t electrons in a complex (a maximum of two electrons in each orbital). There are two e orbitals, and hence there can be up to four e electrons. We want to add n electrons to the t and e orbitals of the complex in an arrangement that gives the lowest overall energy (but with no more than two electrons in any one orbital). We use the orbital energy-level diagram in Fig. 21.26a for octahedral complexes and the diagram in Fig. 21.26b for tetrahedral complexes as guides. The other important type of complex, square-planar, presents a more complicated case that we shall not consider here.

E Carbon monoxide is a particularly strong ligand, forming complexes such as $Ni(CO)_4$, $Fe(CO)_5$, $Mn_2(CO)_{10}$, and $Fe_3(CO)_9$, in all of which the metal has oxidation number 0.

High- and low-spin complexes. The lowest-energy electron configurations for d^1 through d^3 octahedral complexes can be written without difficulty. There are three t orbitals, and because all three have the same energy, each electron can occupy a separate t orbital. According to Hund's rule (Section 7.6), these electrons will have parallel spins, as shown in the margin.

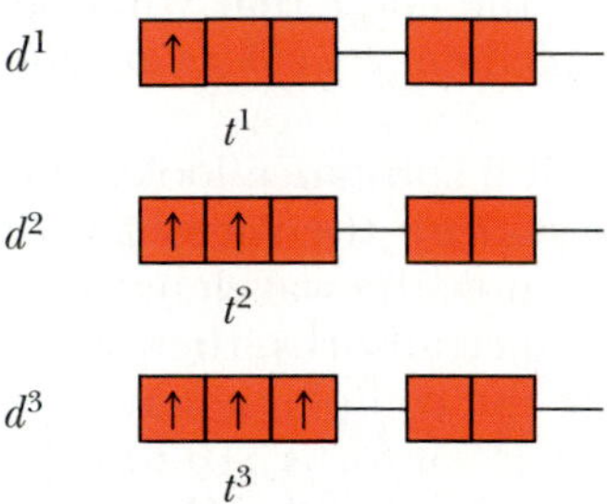

A conflict arises for d^4 octahedral complexes. The fourth electron can enter a t orbital, thereby producing a t^4 configuration. However, to do so it must enter an orbital that is already half occupied and hence experience a strong repulsion from the electron already there:

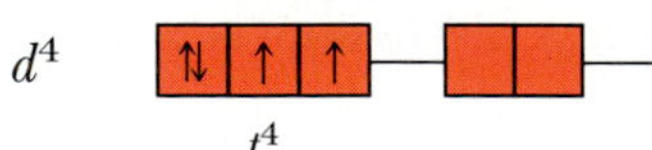

Alternatively, it can avoid that strong repulsion by occupying an empty e orbital, thereby giving a t^3e^1 configuration:

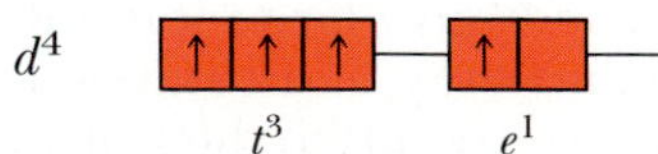

However, it will then experience a strong repulsion from the ligands. The configuration that is actually adopted is the one that leads to the lower energy overall. If Δ_O is large (as it is for strong-field ligands), thus signifying strong ligand repulsion of an e electron, then t^4 will give the lower energy. If Δ_O is small (for weak-field ligands), then t^3e^1 will be the lower energy configuration and hence the one adopted.

N The color change that occurs when one ligand is substituted for another cannot be predicted with confidence. Distortions of the symmetry of the electric field in the complex results in a new set of energy bands in the visible spectrum; moreover, the perceived color may be a consequence of an absorption tail, not the absorption maximum; charge-transfer bands may also give rise to intense colors that mask the *d-d* transitions. At best, crystal field theory allows trends to be rationalized.

U Exercise 21.90.

E Strictly speaking, the classification into weak and strong depends on the metal atom in the complex as well as the ligands themselves.

E The repulsion energy when a second electron is accommodated in a d orbital is called the *pairing energy, P*. The criterion for obtaining low- and high-spin complexes is then whether $\Delta_O > P$ or $\Delta_O < P$, respectively.

▼ EXAMPLE 21.4 Predicting the electron configuration of a complex

Predict the electron configuration of an octahedral d^5 complex with (a) strong-field ligands and (b) weak-field ligands, and give the number of unpaired electrons in each case.

STRATEGY We have to decide whether the lowest energy is reached either with all the electrons in the *t* orbitals, in which case there will be strong electron-electron repulsions, or with some electrons occupying the *e* orbitals too. If the splitting Δ_O is large, then the lower overall energy may be obtained by occupying the *t* orbitals despite the strong electron-electron repulsions that will occur. If Δ_O is small, then electrons are likely to occupy the *e* orbitals.

SOLUTION (a) In the strong-field case, all five electrons enter the *t* orbitals; to do so, some of them must pair:

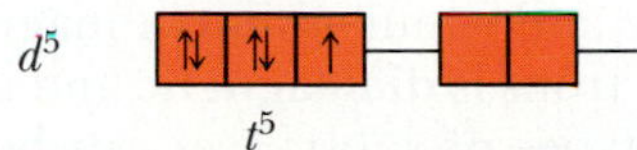

There is one unpaired electron in this configuration.
(b) In the weak-field case, the five electrons occupy all five orbitals without pairing:

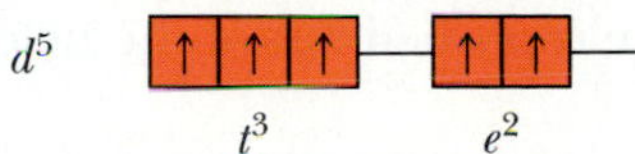

There are now five unpaired electrons.

EXERCISE E21.4 Predict the electron configurations and the number of unpaired electrons of an octahedral d^6 complex with (a) strong-field ligands and (b) weak-field ligands.

▲ [*Answer*: (a) t^6 (0); (b) t^4e^2 (4)]

E Spectral comparison of geometric isomers. *J. Chem. Ed.*, **1973**, *50*, 300.

The configurations predicted for d^1 through d^{10} complexes are listed in Table 21.6. Note that alternative configurations occur for d^4 through d^7 octahedral complexes. In tetrahedral complexes the ligand field is

TABLE 21.6 Electronic configurations of d^n complexes

Number of *d* electrons	Configuration in octahedral complexes		Configuration in tetrahedral complexes
d^1	t^1		e^1
d^2	t^2		e^2
d^3	t^3		e^2t^1
	Low spin	*High spin*	
d^4	t^4	t^3e^1	e^2t^2
d^5	t^5	t^3e^2	e^2t^3
d^6	t^6	t^4e^2	e^3t^3
d^7	t^6e^1	t^5e^2	e^4t^3
d^8	t^6e^2		e^4t^4
d^9	t^6e^3		e^4t^5
d^{10}	t^6e^4		e^4t^6

? Why do many *trans* isomers of the octahedral complexes formed by elements at the right of the *d* block have a significant energy separation between the d_{z^2} and $d_{x^2-y^2}$ orbitals?

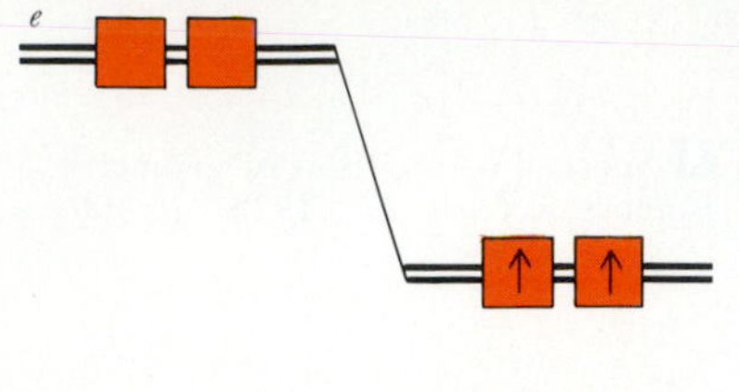

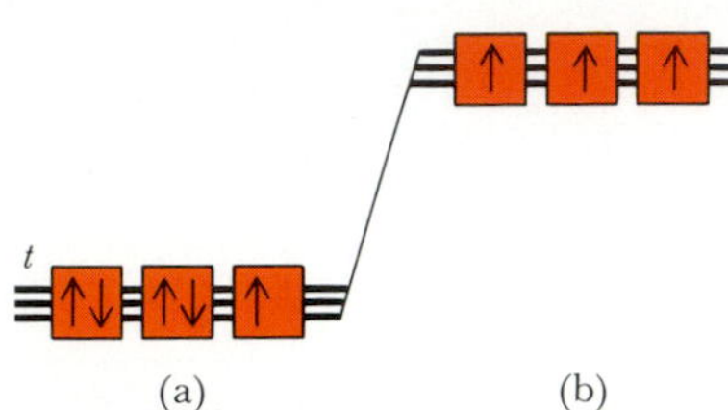

FIGURE 21.31 (a) A strong-field ligand is likely to lead to a low-spin Fe^{2+} complex. (b) A weak-field ligand is likely to lead to a high-spin Fe^{2+} complex.

N The balance shown in Fig. 21.32 is a *Gouy balance.* The degree to which a sample is drawn into the magnetic field is proportional to the number of unpaired electrons in the complex.

T OHT: Fig. 21.32.

FIGURE 21.32 The magnetic character of a complex can be studied with the arrangement shown here. (a) A sample is hung from a balance so that it lies between the poles of an electromagnet. (b) When the magnetic field is on, a paramagnetic sample is drawn into it, thereby making the sample appear to have a larger mass. (c) In contrast, a diamagnetic sample is pushed out of the field when the magnet is turned on.

always too weak to correspond to anything other than the weak-field case, so there is no need to consider the alternative configurations.

A d^n complex with the maximum number of unpaired spins is called a **high-spin complex**. A d^n complex with the minimum number of unpaired spins is called a **low-spin complex**. Tetrahedral complexes are almost always high-spin complexes. For octahedral complexes, when the two alternatives exist we can predict whether a complex is likely to be a high-spin or low-spin complex by noting where the ligands lie in the spectrochemical series. If they are strong-field ligands, then we expect a low-spin complex; if they are weak-field ligands, we expect a high-spin complex. This conclusion is summarized in Fig. 21.31.

The magnetic properties of complexes. As we saw in the introduction to Section 9.7 (in connection with O_2), a compound with unpaired electrons is **paramagnetic** and is pulled into a magnetic field. A substance without unpaired electrons is **diamagnetic** and is pushed out of a magnetic field. The two types of substances can be distinguished experimentally with the apparatus shown in Fig. 21.32: a sample is hung from a balance so that it lies between the poles of an electromagnet. When the magnet is turned on, a paramagnetic substance is pulled into the field and appears to have a larger mass than when the magnet is off. A diamagnetic substance is pushed out of the field and appears to have a smaller mass.

Many d-metal complexes have unpaired d electrons and are therefore paramagnetic. We have just seen that a high-spin d^n complex has more unpaired electrons than a low-spin d^n complex. The former is therefore more strongly paramagnetic and is drawn more strongly into a magnetic field. Moreover, whether a complex is high spin or low spin depends on the ligands present: strong-field ligands tend to result in low-spin and thus weakly paramagnetic substances, whereas weak-field ligands tend to result in high-spin, strongly paramagnetic substances. This correlation suggests that it should be possible to modify the magnetic properties of a complex by changing the ligands.

▼ EXAMPLE 21.5 Predicting the magnetic properties of a complex

Compare the magnetic properties of $[Fe(CN)_6]^{4-}$ with those of $[Fe(H_2O)_6]^{2+}$.

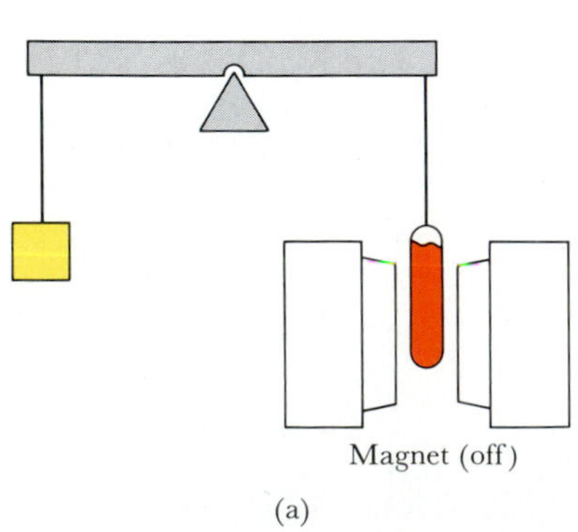

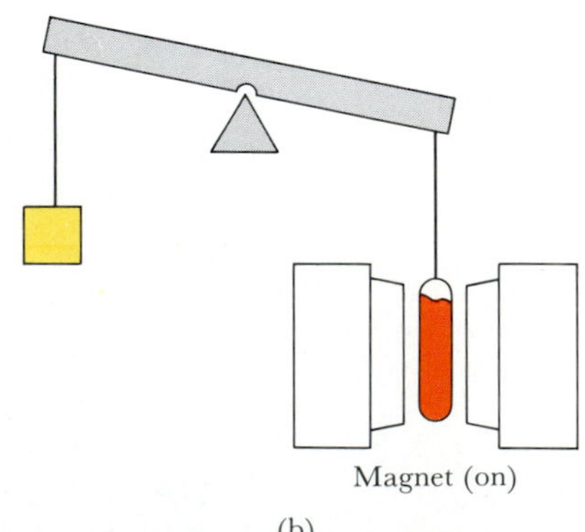

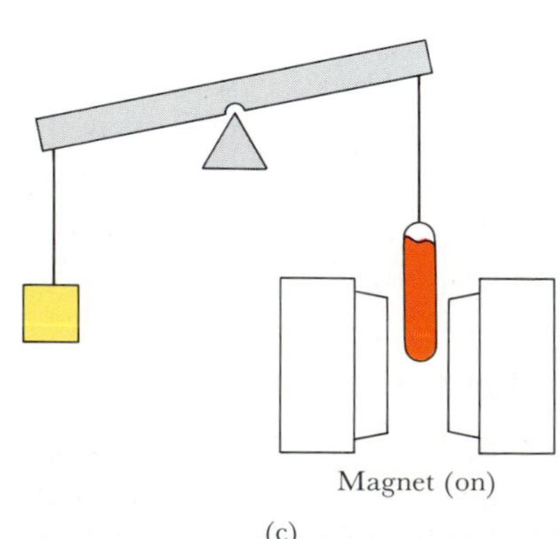

STRATEGY We should decide from their positions in the spectrochemical series whether the complexes have weak-field or strong-field ligands. Then we should judge whether each complex is a high-spin or a low-spin complex and determine the number of unpaired electrons in each.

U Exercise 21.89.

SOLUTION The Fe^{2+} ion in water is the d^6 $[Fe(H_2O)_6]^{2+}$ ion. Because H_2O is a weak-field ligand, we predict a high-spin configuration with four unpaired electrons. The ion is therefore expected (and found) to be paramagnetic. When cyanide ions are added to the solution, they form the $[Fe(CN)_6]^{4-}$ ion. Now the ligands are strong-field ligands, and the resulting complex is a low-spin t^6 complex; it has no unpaired electrons, so it is not paramagnetic. In other words, the ligand substitution reaction has had the effect of quenching the paramagnetism.

EXERCISE E21.5 What magnetic properties can be expected when NO_2^- ligands in an octahedral complex are replaced by Cl^- ligands in (a) a d^6 complex and (b) a d^3 complex?

[*Answer*: (a) the complex becomes paramagnetic; (b) no change in magnetic properties]

▲

As we have seen, the spectroscopic and magnetic properties of d-metal complexes are closely interrelated. In more advanced work, we could add the thermodynamic stability of complexes (and, to some extent, their labilities) to this list. Each of these properties is related to the magnitude of the ligand field splitting, and they are all brought together in a single explanation by a single, unified theory.

N Ligand field theory is a molecular orbital theory of the electronic structure of complexes that makes full use of the symmetry of the atomic arrangement. It is superior to crystal field theory because it is less artificial and allows for the delocalization of electrons from the metal on to the ligands.

SUMMARY

21.1 All the d-block elements are metals. Their atomic radii are all quite similar, a feature permitting extensive alloy formation. Because of the **lanthanide contraction**, the third-row elements (those in Period 6) have radii similar to those in the second row; hence those metals are very dense. The principal chemical characteristics of the d-block elements are their ability to form numerous complexes, their action as catalysts, and their ability to occur with a wide range of oxidation numbers. With the exception of mercury, the elements at the ends of each row of the block have only one oxidation number other than zero. Elements close to the center of each row have the widest range of oxidation numbers. Elements toward the left of the block and lower down a group tend to have higher oxidation numbers in compounds. Although most d-block oxides are basic, the oxides of a given metal show a shift toward acidic character with increasing oxidation number.

21.2 Many of the d metals are prepared as **ferroalloys** with iron, for use in steelmaking. The ease of extraction generally increases from left to right across the d block. The metals on the left are extracted by versions of the thermite process. Iron is extracted by reduction with carbon monoxide in a **blast furnace**.

21.3 The metals on the right of the block are extracted either by **pyrometallurgical** or **hydrometallurgical** processes; the former includes **roasting**, or heating in air, and **smelting**, or melting with a reducing agent. The ores are sometimes concentrated by **froth flotation**, in which the ore floats upward in a froth of detergent. The metals are often purified either by electrolysis or by the formation of a volatile compound that is subsequently decomposed.

21.4 and 21.5 A characteristic property of the d metals is the ability of their atoms and ions to act as Lewis acids and form **complexes** with Lewis bases that, when attached to the metal ion, are called **ligands**. Complexes with six ligands in the **coordination sphere**, which is made up of the ligands directly attached to the metal atom, are usually octahedral; those with four ligands are either square planar or tetrahedral. Ligands may be either **monodentate** or **polydentate** and may form **chelates**, complexes in which a ligand forms a ring that includes the metal atom.

21.6 Ligands may occur in different arrangements, leading to **isomerism**, the existence of different compounds built from the same atoms. The classes of isomerism are summarized in Fig. 21.20. **Chiral** molecules, molecules that are not superimposable on their mirror image, are **optically active** in that they rotate polarized light. **Enantiomeric pairs** of isomers, which are the mirror images of each other, produce equal but opposite rotations of this plane of polarization.

21.7 In the simple version of **ligand field theory** known as **crystal field theory**, the effect of each ligand is represented by a point negative charge. The charges split the five *d* orbitals of the central *d*-metal ion into two groups, the three ***t* orbitals** and the two ***e* orbitals** (in octahedral and tetrahedral complexes) with an energy separation Δ (Δ_O in octahedral complexes and Δ_T in tetrahedral complexes). This separation is called the **ligand field splitting**: it can be estimated from the wavelength of light absorbed by a complex. Each complex has the **complementary color** of the color of the light it absorbs. The size of the splitting increases along the **spectrochemical series** of the ligands, from **weak-field ligands** to **strong-field ligands**.

21.8 Octahedral d^4 to d^7 complexes may have either a **high-spin** or a **low-spin** configuration. The configuration adopted is the one that produces the lowest overall energy. High-spin complexes are more strongly paramagnetic than low-spin complexes.

CLASSIFIED EXERCISES

This chapter includes applications of the chemical principles that have been addressed in earlier chapters. A number of the exercises will require you to collect data from and review some of this earlier material.

The Elements and Their Electron Configurations

21.1 Name each of the following elements: (a) Rh; (b) Ag; (c) Pd; (d) W.

21.2 Give the chemical symbols of (a) scandium; (b) niobium; (c) yttrium; (d) molybdenum.

21.3 Write the ground-state electron configuration of (a) mercury; (b) iron; (c) titanium; (d) gold.

21.4 State the number of unpaired electrons in an atom of (a) nickel; (b) manganese; (c) silver; (d) rhenium in its ground state.

21.5 Identify the elements in the *d* block that are ferromagnetic.

21.6 Which members of the *d* block, those at the left or at the right of the block, tend to have strongly negative standard reduction potentials? Name six elements in the *d* block that have standard reduction potentials greater than zero.

Trends in Properties

21.7 Identify the element with the larger atomic radius in each of the following pairs: (a) cobalt and nickel; (b) copper and zinc; (c) chromium and molybdenum.

21.8 Identify the element with the higher first ionization energy in each of the following pairs: (a) iron and cobalt; (b) manganese and iron; (c) chromium and molybdenum.

21.9 Explain why the density of mercury (13.6 g/cm^3) is significantly higher than that of cadmium (8.65 g/cm^3) whereas the density of cadmium is only slightly greater than that of zinc (7.14 g/cm^3).

21.10 In terms of the properties of *d* orbitals, explain why the density of iron (7.87 g/cm^3) is significantly less than that of cobalt (8.80 g/cm^3).

21.11 Describe the trend in the stability of oxidation numbers moving down a group in the *d* block (for example, from chromium to molybdenum to tungsten).

21.12 Which oxoanion, MnO_4^-, TcO_4^-, or ReO_4^- is expected to be the strongest oxidizing agent? Explain your choice.

21.13 Which of the elements vanadium, chromium, and manganese is most likely to form an oxide with the formula MO_3? Explain your answer.

21.14 Which of the elements zirconium, chromium, and iron is most likely to form a chloride with the formula MCl_4? Explain your answer.

Scandium through Nickel

21.15 Outline a process, using chemical equations where possible, by which (a) titanium; (b) vanadium; (c) chromium are prepared.

21.16 Outline a process, using chemical equations where possible, by which (a) iron; (b) nickel; (c) manganese are prepared.

21.17 Predict the major products of each of the following reactions (you need not balance the equations):
(a) $TiCl_2 + CrO_4^{2-} \rightarrow$
(b) $CoCO_3 + HNO_3 \rightarrow$
(c) $V_2O_5 + Ca \xrightarrow{\Delta}$

21.18 Predict the products of the following reactions and then balance the skeletal equations:
(a) $Ti + F_2(\text{in excess}) \rightarrow$

(b) $MnO_4^- + Cl^- + H^+ \rightarrow$
(c) $Mn_2O_7 + H_2O \rightarrow$

21.19 Explain, with chemical equations, how sodium chromate can be prepared from chromite ore.

21.20 Use chemical equations to suggest a synthesis of chromium(VI) oxide from chromite ore.

21.21 Use the information in Appendix 2B to determine whether an acidic sodium dichromate solution can oxidize (a) bromide ions to bromine and (b) silver(I) ions to silver(II) ions under standard conditions.

21.22 Use Appendix 2B to determine whether an acidic potassium permanganate solution can oxidize (a) chloride ions to chlorine and (b) mercury metal to mercury(I) ions under standard conditions.

21.23 The standard reduction potentials of the couples Cr^{3+}/Cr^{2+} and Cr^{2+}/Cr are -0.41 V and -0.91 V, respectively. Calculate the standard reduction potential of the Cr^{3+}/Cr couple.

21.24 The standard reduction potential of the $MnO_4^{2-}/MnO_2,OH^-$ redox couple is 0.60 V and that of the MnO_4^-/MnO_4^{2-} redox couple is 0.56 V. (a) Determine the standard free energy for the disproportionation reaction of the manganate ion, MnO_4^{2-}, at 25°C. (b) Calculate the equilibrium constant of the disproportionation reaction of the manganate ion at 25°C.

21.25 (a) What reducing agent is used in the production of iron from its ore? (b) Write chemical equations for the production of iron in a blast furnace. (c) What is the major impurity in the product of the blast furnace?

21.26 (a) What is the purpose of adding limestone to a blast furnace? (b) Write chemical equations that show the reactions of limestone in a blast furnace.

21.27 By considering electron configurations, suggest a reason why iron(III) compounds are readily prepared from iron(II), but the conversion of nickel(II) and cobalt(II) to nickel(III) and cobalt(III) is much more difficult.

21.28 (a) Explain why the dissolution of a chromium(III) salt produces an acidic solution. (b) Explain why the slow addition of hydroxide ions to a solution containing chromium(III) ions first produces a gelatinous precipitate that subsequently dissolves with further addition of hydroxide ions. Write chemical equations showing these aspects of the behavior of chromium(III) ions.

21.29 Use Appendix 2B to predict the products of the following reactions: (a) vanadium with 1 M HCl(*aq*); (b) nickel with 1 M NaOH(*aq*); (c) cobalt with 1 M HCl(*aq*).

21.30 Use Appendix 2B to predict products of the following reactions: (a) nickel with 1 M HCl(*aq*); (b) titanium with 1 M HCl(*aq*); (c) platinum with 1 M $KMnO_4$(*aq*) in 1 M HCl(*aq*).

Copper through Mercury

21.31 Give the systematic names and chemical formulas of the principal component of (a) chalcopyrite; (b) sphalerite, (c) cinnabar.

21.32 What are the major sources of (a) silver; (b) gold; (c) cadmium?

21.33 Outline a process, using chemical equations where possible, by which (a) zinc; (b) cadmium; (c) mercury may be produced.

21.34 Outline a process, using chemical equations where possible, by which copper is extracted and purified from chalcopyrite by the pyrometallurgical process.

21.35 What chemical tests would distinguish brass from bronze?

21.36 Explain as fully as possible, with diagrams of the structures of the hydrated ions, the changes that occur when (a) anhydrous copper(II) sulfate is moistened; (b) it is dissolved in water; (c) aqueous ammonia is added to the resulting solution.

21.37 Calculate the standard reduction potential of the Cu^+/Cu couple, given that the values for Cu^{2+}/Cu and Cu^{2+}/Cu^+ are $+0.34$ V and $+0.15$ V, respectively.

21.38 Calculate the standard reduction potential of the Au^{3+}/Au^+ couple, given that the values for Au^{3+}/Au and Au^+/Au are $+1.40$ V and $+1.69$ V, respectively.

21.39 Use the information in Appendix 2B to determine the equilibrium constant for the disproportionation of copper(I) ions in aqueous solution at 25°C to copper metal and copper(II) ions. See Exercise 21.37.

21.40 Use the information in Appendix 2B to predict products of the following reactions: (a) mercury with 1 M HCl(*aq*); (b) mercury with 1 M NaOH(*aq*); (c) zinc with 1 M NaOH(*aq*).

d-Metal Complexes

21.41 Determine the oxidation number of the metal atom in the complexes
(a) $[Fe(CN)_6]^{4-}$
(b) $[Co(NH_3)_6]^{3+}$
(c) $[Co(CN)_5(H_2O)]^{2-}$.

21.42 Determine the oxidation number of the metal atom in the complexes
(a) $[Fe(CN)_6]^{3-}$
(b) $[Fe(OH)(H_2O)_5]^{2+}$
(c) $[Ir(en)_3]^{3+}$.

21.43 Name the following complexes:
(a) $[Fe(CN)_6]^{4-}$
(b) $[Co(NH_3)_6]^{3+}$
(c) $[Co(CN)_5(H_2O)]^{2-}$.

21.44 Name the following complexes:
(a) $[Fe(CN)_6]^{3-}$

(b) $[Fe(OH)(H_2O)_5]^{2+}$
(c) $[CoCl(NH_3)_4(H_2O)]^{2+}$.

21.45 Determine the coordination numbers of
(a) $[NiCl_4]^{2-}$
(b) $[PtCl_2(en)_2]^{2+}$
(c) $[Cr(EDTA)]^-$.

21.46 Determine the coordination numbers of
(a) $[Ni(en)_2I_2]^+$
(b) $[Fe(ox)_3]^{3-}$ (the oxalato ligand is bidentate)
(c) $[PtCl_2(NH_3)_2]$.

21.47 Write the formula for each of the following coordination compounds: (a) potassium hexacyanochromate (III); (b) pentaamminesulfatocobalt(III) chloride; (c) sodium diaqua*bis*(oxalato)ferrate(III).

21.48 Write the formula for each of the following coordination compounds: (a) triammineaquadihydroxochromium(III) chloride; (b) potassium tetrachloroplatinate(III); (c) potassium *tris*(oxalato)rhodate(III) chlorohydroxy*bis*(oxalato)rhodate(III) octahydrate.

Isomerism

21.49 Determine the type of structural isomerism that exists in the following pairs of compounds:
(a) $[Co(NO_2)(NH_3)_5]Br_2$ and $[Co(ONO)(NH_3)_5]Br_2$
(b) $[Pt(SO_4)(NH_3)_4](OH)_2$ and $[Pt(OH)_2(NH_3)_4]SO_4$
(c) $[CoCl(NO_2)(NH_3)_4]Cl$ and $[CoCl(ONO)(NH_3)_4]Cl$
(d) $[CrCl(NH_3)_5]Br$ and $[CrBr(NH_3)_5]Cl$

21.50 Determine the type of structural isomerism that exists in the following pairs of compounds:
(a) $[Cr(en)_3][Co(ox)_3]$ and $[Co(en)_3][Cr(ox)_3]$
(b) $[CoCl_2(NH_3)_4]Cl \cdot H_2O$ and $[CoCl(NH_3)_4(H_2O)]Cl_2$
(c) $[Co(CN)_5(NCS)]^{3-}$ and $[Co(CN)_5(SCN)]^{3-}$
(d) $[Pt(NH_3)_4][PtCl_6]$ and $[PtCl_2(NH_3)_4][PtCl_4]$

21.51 Write the formulas for the hydrate isomers of a compound having the empirical formula $CoCl_3 \cdot 6H_2O$ and a coordination number of six.

21.52 Write the formula of a coordination isomer of $[Co(NH_3)_6][Cr(NO_2)_6]$.

21.53 Write the formula of a linkage isomer of $[CoCl(NO_2)(en)_2]Cl$.

21.54 Write the formula of an ionization isomer for $[CoCl(NO_2)(en)_2]Cl$.

21.55 Which of the following coordination compounds can have cis and trans isomers? If such isomerism exists, draw the two structures and name each compound: (a) $[CoCl_2(NH_3)_4]Cl \cdot H_2O$; (b) $[PtCl_2(NH_3)_2]$, a square planar complex.

21.56 Which of the following complexes can have cis and trans isomers? If such isomerism exists, draw the two structures and name each compound: (a) $[Fe(OH)(H_2O)_5]^{2+}$; (b) $[RuBr_2(NH_3)_4]^+$.

21.57 Can a tetrahedral complex show (a) stereoisomerism; (b) geometrical isomerism; (c) optical isomerism?

21.58 Draw the structure of *cis*-diammine-*cis*-diaqua-*cis*-dichlorochromium(III) ion and comment on its isomerism. What can be said about the *trans*-diammine isomer?

21.59 Is either of the following complexes chiral? If both complexes are chiral, do they form an enantiomeric pair?

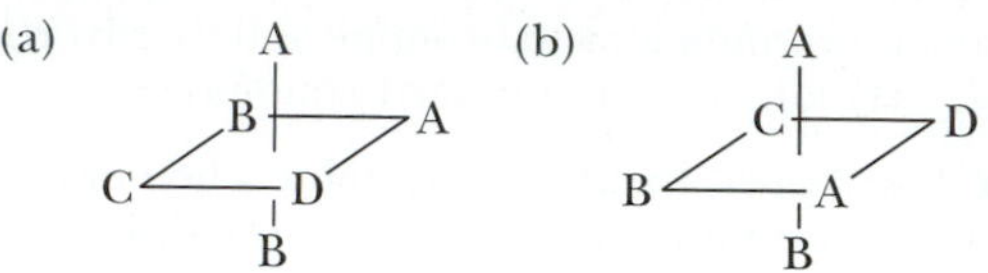

21.60 Is either of the following complexes chiral? If both complexes are chiral, do they form an enantiomeric pair?

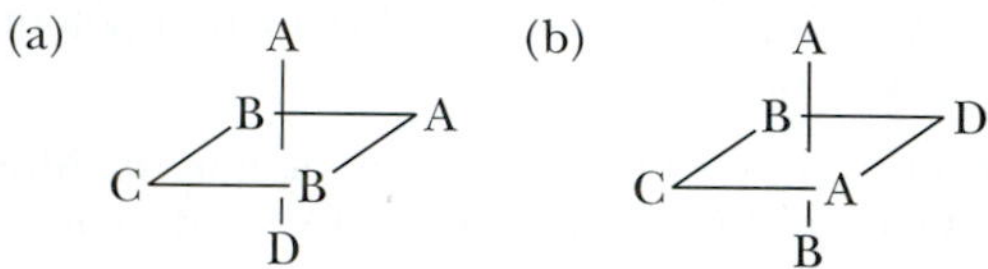

21.61 Draw the structures of the optical isomers of $[CoCl_2(en)_2]^+$.

21.62 Draw the structures of the isomeric forms of $[CrBrCl_2(NH_3)_3]$. Which isomers are chiral?

Crystal Field Theory

21.63 Draw an orbital energy-level diagram (like that in Fig. 21.31) showing the configuration of *d* electrons on the *d*-metal ion in (a) $[Co(NH_3)_6]^{3+}$; (b) $[Fe(H_2O)_6]^{3+}$; (c) $[Fe(CN)_6]^{3-}$.

21.64 Draw an orbital energy-level diagram (like that in Fig. 21.31) showing the configuration of *d* electrons on the *d*-metal ion in (a) $[Zn(H_2O)_6]^{2+}$; (b) $[CoCl_4]^{2-}$ (tetrahedral); (c) $[Co(CN)_6]^{3-}$.

21.65 The complexes (a) $[Co(en)_3]^{3+}$ and (b) $[Mn(CN)_6]^{3-}$ have low-spin electron configurations. How many unpaired electrons are there in each complex?

21.66 The complexes (a) $[FeF_6]^{3-}$ and (b) $[Co(ox)_3]^{4-}$ have high-spin electron configurations. How many unpaired electrons are there in each complex?

21.67 Of the two complexes, (a) $[CoF_6]^{3-}$ and (b) $[Co(en)_3]^{3+}$, one appears yellow in an aqueous solution and the other appears blue. Match the complex to the color and explain your choice.

21.68 A concentrated solution of copper(II) bromide in the presence of potassium bromide is deep violet, but the solution becomes light blue upon dilution with water. (a) Write the formulas of the two complex ions of copper(II) ion that form, assuming a coordination number of four. (b) Is the change in color from violet to blue expected? Explain your reasoning.

21.69 State the color of a sample that absorbs light of wavelength (a) 410 nm; (b) 650 nm; (c) 480 nm; (d) 590 nm.

21.70 State the wavelength range over which the following colors are absorbed from a sample and then predict the color of the complex: (a) blue; (b) red; (c) violet; (d) green.

21.71 Suggest a reason why $Zn^{2+}(aq)$ ions are colorless. Would you expect zinc compounds to be paramagnetic?

21.72 Suggest a reason why copper(II) compounds are often colored but copper(I) compounds are colorless. Which oxidation number gives paramagnetic compounds?

21.73 Estimate the ligand field splitting for (a) $[CrCl_6]^{3-}$ ($\lambda_{max} = 740$ nm); (b) $[Cr(NH_3)_6]^{3+}$ ($\lambda_{max} = 460$ nm); (c) $[Cr(H_2O)_6]^{3+}$ ($\lambda_{max} = 575$ nm); where λ_{max} is the wavelength of the most intensely absorbed light. Arrange the ligands in order of increasing ligand field strength.

21.74 Estimate the ligand field splitting for (a) $[Co(CN)_6]^{3-}$ ($\lambda_{max} = 295$ nm); (b) $[Co(NH_3)_6]^{3+}$ ($\lambda_{max} = 435$ nm); (c) $[Co(H_2O)_6]^{3+}$ ($\lambda_{max} = 540$ nm); where λ_{max} is the wavelength of the most intensely absorbed light. Arrange the ligands in order of increasing ligand field strength.

UNCLASSIFIED EXERCISES

21.75 Write a chemical equation showing in what respect a Sc^{3+} ion is a Brønsted acid in aqueous solution.

21.76 Explain in terms of electron configurations why the atomic radius of manganese is larger than that of chromium.

21.77 Identify the element, compound, or mixture that best fits the description: (a) fool's gold; (b) the green patina on the copper plate of the Statue of Liberty; (c) 24-carat gold; (d) the densest element.

21.78 Identify the element, compound, or mixture that best fits the description: (a) a substance used as the whitener in paints and in papermaking; (b) the catalyst used for the production of sulfuric acid; (c) the second most abundant metal in the earth's crust; (d) pig iron.

21.79 Write chemical equations that describe the following processes: (a) the hydrometallurgical process for the production of copper; (b) the reduction of titanium tetrachloride to titanium; (c) the removal of silica, SiO_2, from iron ore.

21.80 Write chemical equations that describe the following processes: (a) the roasting of nickel sulfide; (b) the oxidation of copper metal with concentrated nitric acid; (c) the production of chromium by the thermite reaction.

21.81 Determine the mass of $FeCr_2O_4$ (from chromite ore) that is needed to produce 1.00 kg of sodium chromate by the reaction

$$4FeCr_2O_4(s) + 8Na_2CO_3(s) + 7O_2(g) \xrightarrow{\Delta} 8Na_2CrO_4(s) + 2Fe_2O_3(s) + 8CO_2(g)$$

21.82 Zinc metal is used to galvanize steel because of the cathodic protection it provides. What would result if copper were used instead?

21.83 Name each of the following compounds:
(a) $[PtCl_2(en)_2]Cl_2$
(b) $[Co(en)_3][Fe(CN)_6]$
(c) $K_3[Fe(ox)_3]$
(d) $Na[Cr(EDTA)]$.

21.84 Give the name of each of the following complexes and determine the coordination number and oxidation number of the *d*-metal ion:
(a) $[Zr(ox)_4]^{4-}$
(b) $[CuCl_4(H_2O)_2]^{2-}$
(c) $[PtCl_3(NH_3)]^-$
(d) $[Mo(CN)_4(OH)_4]^{4-}$

21.85 Suggest a chemical test for distinguishing between (a) $[Ni(NO_2)_2(en)_2]Cl_2$ and $[NiCl_2(en)_2](NO_2)_2$ and (b) $[NiI_2(en)_2]Cl_2$ and $[NiCl_2(en)_2]I_2$.

21.86 Draw and name the isomers of $[Cr(ox)_2(H_2O)_2]^-$.

21.87 The relative thermodynamic stability of complexes can be predicted from a comparison of standard reduction potentials. Arrange the following complexes in order of increasing thermodynamic stability with respect to reduction and arrange the ligands in order of increasing ligand field strength. Justify your answer in terms of the crystal field theory of complex formation.

$[Zn(CN)_4]^{2-}/Zn,CN^-$	$E° = -1.26$ V
$[Zn(H_2O)_4]^{2+}/Zn$	$E° = -0.76$ V
$[Zn(OH)_4]^{2-}/Zn,OH^-$	$E° = -1.22$ V
$[Zn(NH_3)_4]^{2+}/Zn,NH_3$	$E° = -1.03$ V

21.88 Gold metal can be oxidized in the presence of aqua regia, a 3:1 mixture (by volume) of concentrated hydrochloric acid and nitric acids. Consider the half-

CASE 21 ▸ WINNING A METAL FROM ITS ORE

The initial stages of separation of ores from unwanted materials were described in Case 1. Here we consider how the separated oxide ores are reduced to the metal.

The common industrial reducing agents are carbon, carbon monoxide, and hydrogen, but for the more reactive metals, more aggressive reducing agents must be used. Some metals, such as aluminum and magnesium, can be produced only by using the strongly reducing cathode of an electrolytic cell, as described in the chapter. Other metals can be produced with a chemical reducing agent; the choice of the latter depends on the reactivity of the metal and the properties sought in the product. As we saw in Section 21.2, iron(III) oxide is reduced to iron by carbon monoxide, but the presence of unreacted carbon makes the metal hard and brittle.

Carbon (as coke, which is coal heated to high temperatures in the absence of air to remove volatile gases and moisture) and carbon monoxide are by far the most common industrial reducing agents, primarily on the basis of cost. Carbon monoxide is produced by the partial combustion of carbon, the major component of coal. The carbon monoxide reacts with the metal oxide in the ore, producing carbon dioxide and the metal. Iron, nickel, zinc, cobalt, lead, germanium, and tin oxides, together with many others, are reduced to the metal with carbon or carbon monoxide:

$$NiO(s) + C(s) \longrightarrow Ni(s) + CO(g)$$

$$PbO(s) + CO(g) \longrightarrow Pb(s) + CO_2(g)$$

Hydrogen gas is used as a reducing agent when it is desirable to avoid the brittleness caused by carbon. For example, tungsten is reduced from its oxide with hydrogen to produce a ductile, malleable metal for use as filament wire in incandescent light bulbs:

$$WO_3(s) + 3H_2(g) \longrightarrow W(s) + 3H_2O(g)$$

The thermite reaction (which is also called the Goldschmidt reaction) uses aluminum metal as the reducing agent. The original thermite reaction was between aluminum and iron(III) oxide; the reaction is so exothermic, reaching 3000°C, that it has been used to weld iron, for example, to fill joints between railroad rails:

$$Fe_2O_3(s) + 2Al(s) \longrightarrow 2Fe(s) + Al_2O_3(s)$$

The oxides of chromium, titanium, manganese, vanadium, and tantalum (only a partial listing)

reactions

$$[AuCl_4]^-(aq) + 3e^- \longrightarrow Au(s) + 4Cl^-(aq) \qquad E° = +1.00\ V$$

$$NO_3^-(aq) + 4H^+(aq) + 3e^- \longrightarrow NO(g) + 2H_2O(l) \qquad E° = +0.96\ V$$

(a) What is the spontaneous reaction under standard conditions? (b) Determine $E°$ for the cell formed from the two couples. (c) What is the cell potential when $[H^+] = 6.0$ mol/L, $[Cl^-] = 6.0$ mol/L, $[NO_3^-] = 6.0$ mol/L, and $[AuCl_4^-] = 1.0 \times 10^{-6}$ mol/L? (d) What is the direction of the spontaneous reaction when 6 M $HCl(aq)$ and 6 M $HNO_3(aq)$ are present?

21.89 (a) Sketch the orbital energy-level diagrams for $[MnCl_6]^{4-}$ and $[Mn(CN)_6]^{4-}$. (b) How many unpaired electrons are present in $[Mn(CN)_6]^{4-}$? (c) Which complex transmits the longer wavelengths of incident electromagnetic radiation? Explain your reasoning.

21.90 The complex $[Co(CN)_6]^{3-}$ is pale yellow. (a) Is short or long wavelength visible radiation absorbed?

are similarly reduced to the metal:

$$Cr_2O_3(s) + 2Al(s) \longrightarrow 2Cr(s) + Al_2O_3(s)$$

$$3TiO_2(s) + 4Al(s) \longrightarrow 3Ti(s) + 2Al_2O_3(s)$$

Magnesium and sodium are the preferred reducing agents for other metals, for example, for the chlorides of titanium, zirconium, and vanadium:

$$TiCl_4(g) + 2Mg(s) \longrightarrow Ti(s) + 2MgCl_2(s)$$

$$2VCl_5(g) + 5Mg(s) \longrightarrow 2V(s) + 5MgCl_2(s)$$

and for tantalum:

$$K_2TaF_6(s) + 4Na(l) \longrightarrow Ta(s) + 4NaF(s) + 2KF(s)$$

QUESTIONS

1. (a) Uranium metal can be produced by reduction of yellowcake, U_3O_8, with carbon or by the reduction of UF_4 with calcium metal. Write balanced chemical equations for each reduction. (b) What mass of uranium can be produced from 20.0 kg of yellowcake?

2. A "silver tree" can be made in the laboratory by cutting a piece of copper foil in the shape of a tree and submerging it in an aqueous silver nitrate solution. (a) Explain the formation of the silver tree. (b) What happens to the appearance of the aqueous solution as the silver tree forms?

3. The estimated annual steel production in the United States is about 1.6×10^{11} kg. Assuming that all iron is obtained from the Fe_2O_3 and that coal is pure carbon, what mass of coal must be used for the production of all the iron? What mass of carbon dioxide is released into the atmosphere as a result of ironmaking?

4. Explain why a variety of reducing agents must be used to reduce metal-containing minerals, rather than potassium metal, a reducing agent that can reduce almost all other metal ores.

5. (a) Determine the standard enthalpy of reaction for the thermite reaction

$$Fe_2O_3(s) + 2Al(s) \longrightarrow 2Fe(s) + Al_2O_3(s)$$

(b) Suppose that the initial temperature of the reaction mixture is 25°C and the final temperature reaches 3000°C. The melting point of iron is 1540°C, its enthalpy of melting is 16.2 kJ/mol, and its specific heat capacity is 0.449 J/(g·°C). If 200 g of Fe_2O_3 reacts completely in a thermite reaction in the welding of two iron railroad rails, what mass of iron could melt in the rail?

(b) Is the ligand field splitting strong or weak? (c) How many unpaired electrons are present in the complex? (d) If ammonia molecules are substituted for cyanide ions as ligands, will the shift in absorbance of radiation be toward the blue or the red regions of the spectrum?

21.91 (a) Calculate the ligand field splitting for a complex that has an absorption maximum at $\lambda = 550$ nm. (b) What color can the complex be expected to be?

21.92 Draw and name all possible isomers of the square-planar complex of empirical formula $[PtBrCl(NH_3)_2]$.

21.93 What is the type of isomerism shown by $[Cr(NH_3)_6]Cl_3$ (yellow), $[CrCl(NH_3)_5]Cl_2 \cdot NH_3$ (purple), and $[CrCl_2(NH_3)_4]Cl \cdot 2NH_3$ (violet). Draw the structure of each complex and name each coordination compound. Account for the color change from yellow to purple.

21.94 Draw the structure of (a) *cis*-$[CoCl_2(NH_3)_4]^+$ (violet) and (b) *trans*-$[CoCl_2(NH_3)_4]^+$ (bright green).

21.95 Draw the structures of the stereoisomers of $[CoCl_2(NH_3)_2\ en]^+$.

In this chapter we examine some of the events that accompany changes in the structures of atomic nuclei. These include radioactivity, the conversion of one element into another, and the release of energy for the generation of nuclear power. Nuclear power, in particular, presents chemists with several demanding problems, and we see some of their continuing attempts to solve them. The illustration shows part of the fuel-rod assembly for an advanced gas-cooled nuclear reactor.

22 NUCLEAR CHEMISTRY

E Isotopes undergo the same reactions but with slightly different rates. The difference is most marked for the hydrogen isotopes because the mass differences are relatively large. A C—D bond reacts more slowly than a C—H bond (the kinetic isotope effect), and at low temperatures protons can tunnel more readily through potential barriers. Other differences, which are detectable for heavier elements too, are related to the effect of nuclear spin, which can influence the proportions of products when several alternatives are available.

So far, we have taken the atomic nucleus to be little more than an unchanging passenger in chemical reactions. However, nuclei can change. **Nuclear chemistry** is the study of the structure of atomic nuclei, of the changes this structure undergoes, and of the consequences of those changes for chemistry. Nuclei that change their structure spontaneously and emit radiation are called **radioactive**. Many such unstable nuclei occur naturally: for instance, all nuclei with $Z > 83$ (from polonium onward in the periodic table) are unstable and radioactive, and all the lighter elements have some isotopes that are unstable to some degree.

Nuclear chemistry is central to the development of nuclear energy, because techniques must be found for the purification and recycling of nuclear fuels and for the disposal of hazardous radioactive waste. Its numerous other applications include the investigation of reaction mechanisms, the dating of archaeological objects, and the development of methods for fighting cancer.

The changes that nuclei themselves undergo are called **nuclear reactions**. The differences between nuclear reactions and chemical reactions (which involve only changes in the valence-shell electrons) can be summarized as follows:

1. The energy changes are very much greater for nuclear reactions than for chemical reactions.

For example, the combustion of 1 g of CH_4 produces about 52 kJ of energy as heat but a nuclear reaction of 1 g of uranium-235 produces about 8.2×10^7 kJ! It follows that the "test tubes" used for nuclear reactions must be far more complex and sophisticated than those for chemical reactions.

2. Isotopes show almost identical chemical properties, but they undergo different nuclear reactions.

E Chemistry in art, radiochemistry, and forgery. *J. Chem. Ed.*, **1972**, *49*, 418. Freshman level chemistry shapes the nuclear power industry. *J. Chem. Ed.*, **1975**, *52*, 523. The radiation hazard of coal-fired power plants. *J. Chem. Ed.*, **1981**, *58*, 500.

D Some interesting old demonstrations concerning radiochemistry. TCD, 9s. See the comment in TCD, 9-7s.

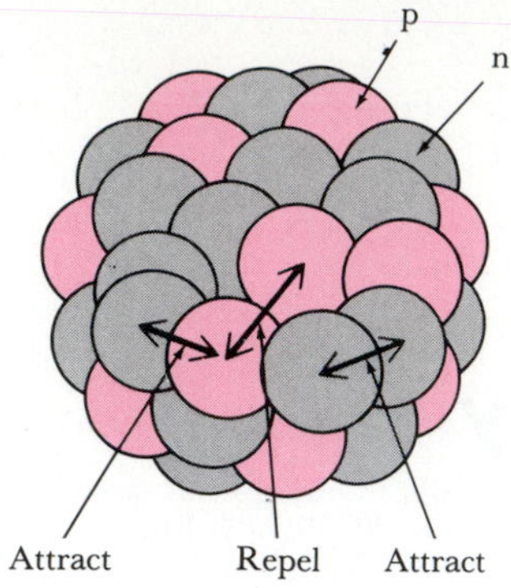

FIGURE 22.1 A nucleus can be pictured as a collection of protons (pink) and neutrons (gray). The protons repel each other electrostatically, but a strong force that acts between all the particles holds the nucleus together.

N Each branch of science considers a particular unit as "fundamental." In classical chemistry, the atom is the unchangeable unit. In post-classical chemistry, the atom changes but the nucleus remains constant, and there is no need to know anything about its structure. In nuclear chemistry, the nucleus changes, but the nucleons remain constant and we do not need to understand their structure. In nuclear physics, the nucleons change, and it is necessary to know about the properties of quarks and leptons.

N Nuclear concepts as part of the undergraduate chemistry curriculum. *J. Chem. Ed.*, **1970**, *47*, 569. Nuclear chemistry (review): Include it in your curriculum. *J. Chem. Ed.*, **1989**, *66*, 389.

D Atomic structure exhibit. TDC, 9-1.

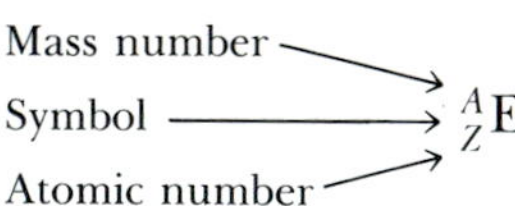

N A nuclide is a specific species designated by its atomic number and mass number. An isotope is one of several versions of a specified element. Thus isotopes are a particular set of nuclides with the same value of *Z*.

E A comparison of ^{252}Cf and ^{9}Be(γ,n) educational neutron sources. *J. Chem. Ed.*, **1982**, *59*, 429. Natural sources of ionizing radiation. *J. Chem. Ed.*, **1984**, *61*, 687.

3. The nuclear reactions of an element are almost independent of its state of combination with other elements.
4. A nuclear reaction often results in the formation of another element, whereas a chemical reaction never does.
5. The rate of a nuclear reaction is independent of temperature, pressure, and the presence of a catalyst.

NUCLEAR STABILITY

We met some features of the structure of atomic nuclei in Section 2.3. These key points and terms are summarized below. Then we shall examine the different types of nuclear radiation and the changes in nuclear structure responsible for them.

22.1 NUCLEAR STRUCTURE AND NUCLEAR RADIATION

We can picture a nucleus as a tightly bound collection of **nucleons**, the general name for protons, p, and neutrons, n. These two types of nucleons have almost exactly the same mass, but a proton has a single positive charge and a neutron is electrically neutral (see Table 2.2). Nucleons are bound together within the nucleus by an attractive force that is so strong over short distances that it overcomes the repulsion between the charges of the protons (Fig. 22.1). Therefore, provided the number of protons is not too large, the nucleus may be stable. In a nucleus containing more than 83 protons, however, these charged particles repel each other so strongly that the nucleus is unstable and hence radioactive.

Nuclides. The number of protons in the nucleus is the **atomic number** ***Z*** of the element. The total number of nucleons (protons plus neutrons) is the **mass number** ***A***. The number of neutrons that can accompany a given number of protons in a nucleus is variable over a small range. This variability gives rise to **isotopes**, atoms with the same atomic number (the same number of protons) but different mass numbers (because they have different numbers of neutrons); two examples of isotopes are carbon-12 and carbon-14. The isotopes of an element are chemically almost identical, because their atoms have the same number of electrons and hence the same electron configuration.

Each distinct isotope of an element is called a **nuclide**. Thus, carbon-12 is one nuclide, carbon-14 is another, and uranium-235 is yet another. The symbol used to specify a nuclide is ^A_ZE, where E is the chemical symbol of the element, *Z* its atomic number, and *A* its mass number. For example, the nuclide $^{235}_{92}$U is the isotope of uranium that contains 92 protons and $235 - 92 = 143$ neutrons. All nuclides with the same value of *Z* are isotopes of the element E. Strictly speaking, *Z* is redundant (because the symbol E implies the value of *Z*, as U implies $Z = 92$); but when keeping track of changes in nuclei, it is useful to display it. We shall denote an individual proton ($A = 1, Z = 1$) by $^1_1$p and an individual neutron ($A = 1, Z = 0$) by $^1_0$n when we want to be specific about their masses and charges.

Radiation from nuclei. The French scientist Henri Becquerel discovered radioactivity in 1896 (Fig. 22.2). His discovery was taken up by

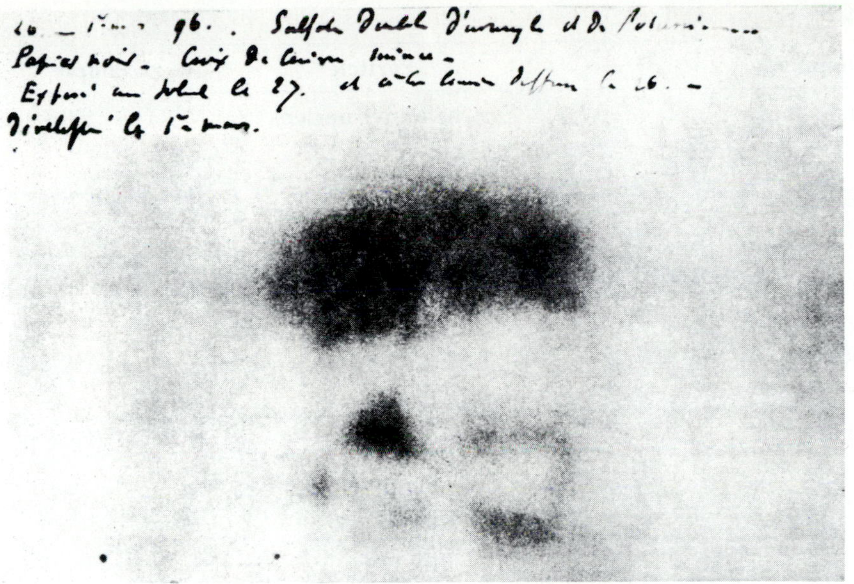

FIGURE 22.2 Becquerel discovered radioactivity when he noticed that an unexposed photographic plate left near some uranium oxide became fogged. This photograph shows one of his original plates.

E Becquerel's discovery of radioactivity was accidental. In trying to reverse Roentgen's discovery of x-rays he placed uranium-containing minerals on photographic plates covered with black paper, and noticed that even without exposing the mineral to visible light the photographic plates had become exposed. Had it not been for a number of cloudy days in which the plates with the minerals were stored in a drawer, the discovery of radioactivity would have been delayed.

D Becquerel's discovery exhibit. TDC, 9-4. Electroscope discharge with radioactive source. TDC, 9-5. Alpha spectroscopy using thorium daughter products: An experiment. *J. Chem. Ed.*, **1972**, *49*, 432. Color centers in salts by electron bombardment. TDC, 166. Scattering of beta rays. TDC, 9-14.

E Radioactivity: A natural phenomenon. *J. Chem. Ed.*, **1990**, *67*, 737.

N β particles are also denoted β^-.

Ernest Rutherford, who identified three different types of radioactivity by observing the effect of electric fields on radioactive emissions (Fig. 22.3). He called the three types α (alpha), β (beta), and γ (gamma) radiation. It is helpful to note that α radiation is more typical of the heavy elements and β radiation is more typical of the lighter ones; γ radiation often accompanies α and β radiation but may occur alone. A summary of the characteristics and properties of these three types of radiation and of some other types of nuclear radiation that have since been identified is given in Table 22.1.

Rutherford found that α radiation is repelled from a positively charged electrode. This led him to propose that it consists of positively charged particles, which he called **α particles**. He was able to deduce the ratio of the charge to the mass of the particles from the size of the deflection caused by electric and magnetic fields and hence was able to identify them as the nuclei of helium atoms, ${}^4_2He^{2+}$. This identification was confirmed by detecting helium gas near a substance that emitted α radiation; a helium atom forms when an α particle captures two electrons during collision with other atoms:

$$ {}^4_2He^{2+} + 2e^- \longrightarrow {}^4_2He $$

An α particle is denoted ${}^4_2\alpha$, or simply α. We can think of it as a tightly bound cluster of two protons and two neutrons.

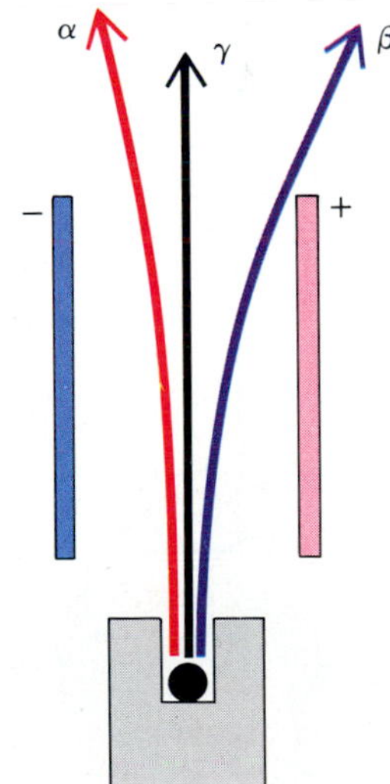

FIGURE 22.3 The effects of an electric field on nuclear radiation. The deflection identifies α rays as positively charged, β rays as negatively charged, and γ rays as uncharged.

Rutherford also found that β radiation is attracted to a positively charged electrode, and he concluded that it consists of a stream of negatively charged particles. That the particles are in fact electrons was confirmed by measuring their charge and mass. Electrons emitted by nuclei (but not electrons from outside the nucleus) are called **β particles** and denoted β. The important characteristics of a β particle are its single negative charge and the fact that its mass is negligible in comparison with the masses of protons and neutrons. (We saw in Chapter 2 that the mass of an electron is only about $\frac{1}{1836}$ that of a nucleon.) Because a β particle has no protons or neutrons, its mass number is zero.

TABLE 22.1 Nuclear radiation

Radiation	Comments	Particle*	Example
α	not penetrating but damaging speed: <10% of c†	helium-4 nucleus ${}^{4}_{2}He^{2+}$, ${}^{4}_{2}\alpha$, α	${}^{226}_{88}Ra \rightarrow {}^{222}_{86}Rn + \alpha$
β	moderately penetrating speed: <90% of c	electron ${}^{0}_{-1}e$, β^-, β	${}^{3}_{1}H \rightarrow {}^{3}_{2}He + \beta$
γ	very penetrating; often accompanies other radiation speed: c	photon	${}^{60}_{27}Co‡ \rightarrow {}^{60}_{27}Co + \gamma$
β^+	moderately penetrating speed: <90% of c	positron ${}^{0}_{+1}e$, β^+	${}^{22}_{11}Na \rightarrow {}^{22}_{10}Ne + \beta^+$
p	moderate/low penetration speed: <10% of c	proton ${}^{1}_{1}H^+$, ${}^{1}_{1}p$, p	${}^{53}_{27}Co \rightarrow {}^{52}_{26}Fe + p$
n	very penetrating speed: <10% of c	neutron ${}^{1}_{0}n$, n	${}^{137}_{53}I \rightarrow {}^{136}_{53}I + n$

*Alternative symbols are given for each particle; often it is sufficient to use the simplest (the one on the right).
†c is the speed of light.
‡An energetically excited state of a nucleus is usually denoted by an asterisk.

D A simple cloud chamber. TDC, 9-9.

T OHT: Table 22.1.

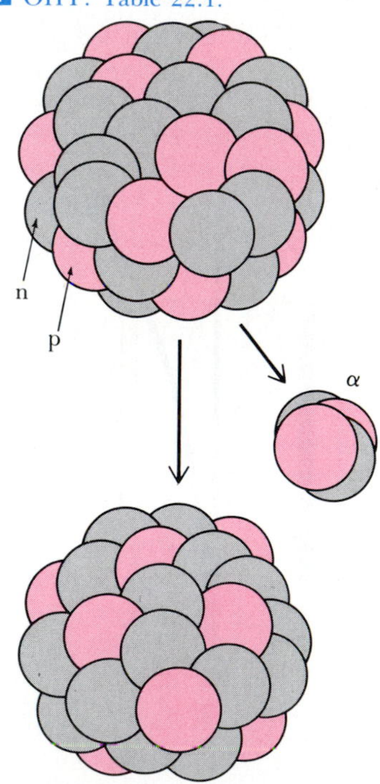

FIGURE 22.4 In the process of nuclear disintegration, a nucleus ejects a particle (in this case an α particle) and becomes a daughter nucleus.

Because its charge is −1, it is sometimes convenient (when balancing equations) to denote it ${}^{0}_{-1}e$, but the subscript is not a true atomic number.

Rutherford found that **γ radiation** is unaffected by electric fields. He identified it as electromagnetic radiation like light, but of much higher frequency—greater than about 10^{20} Hz—and with wavelengths less than about 1 pm (the wavelength of visible light is about 500,000 times greater). Because electromagnetic radiation consists of photons with energy proportional to their frequency, γ radiation can be regarded as a stream of very high energy photons; each photon is emitted by a single nucleus as that nucleus discards energy. A γ-ray photon is denoted γ. Like all photons, γ-ray photons are massless and uncharged.

Nuclear disintegration and radioactive decay. The emission of an α particle or a β particle from a nucleus occurs as a result of **nuclear disintegration**, the partial breakup of a nucleus called the **parent nucleus**. The disintegration is only partial, in the sense that after it takes place, a less massive nucleus is left behind (Fig. 22.4). This less massive nucleus, called the **daughter nucleus**, may be the nucleus of another element. For example, when a radon-222 nucleus emits an α particle, a polonium-218 nucleus is formed (how that can be predicted we shall see shortly). We can write this process in the form of a **nuclear equation**, the nuclear analogue of a chemical equation:

$${}^{222}_{86}Rn \longrightarrow {}^{218}_{84}Po + \alpha$$

When a carbon-14 nucleus emits a β particle, a nitrogen-14 nucleus is formed:

$${}^{14}_{6}C \longrightarrow {}^{14}_{7}N + \beta$$

In each case, **nuclear transmutation**, or the conversion of one element into another, has taken place. Nuclear transmutation, particularly of lead into gold, was the dream of the alchemists and the root of modern chemistry. However, only in this century has it been recognized in nature and achieved in the laboratory.

It is common for α and β radiation to be accompanied by γ radiation because the ejection of an α or a β particle often leaves the nucleons of a daughter nucleus in a high-energy arrangement (Fig. 22.5); a γ-ray photon is emitted when the nucleus rearranges to a state of lower energy. As in the case of excited atoms, the photon has a frequency ν given by the relation $\Delta E = h\nu$ (see Eq. 5 in Section 7.3). Gamma rays have very high frequency because the energy difference between the excited and ground nuclear states is very large.

Experiments done since Rutherford's time have shown that other particles can be emitted when nuclei disintegrate. One of the most important is the **positron** ($_{+1}^{0}e$, or more simply β^{+}), which has the same mass as an electron but a single positive charge.

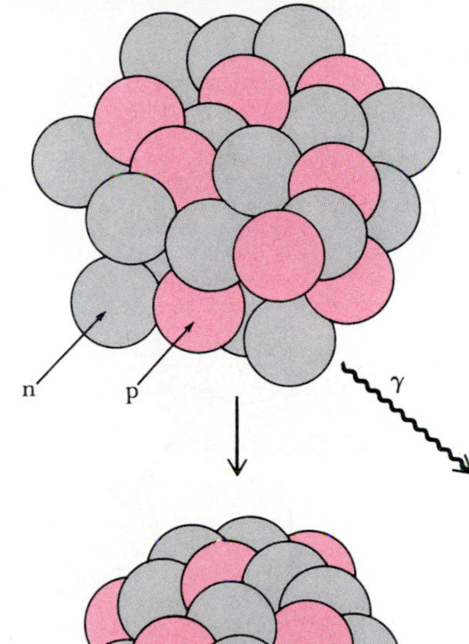

FIGURE 22.5 Nuclear disintegration may result in the formation of a daughter nucleus with nucleons in a high-energy arrangement: as they adjust into a lower-energy arrangement, the excess energy is released as a γ-ray photon.

22.2 THE IDENTITIES OF DAUGHTER NUCLIDES

We can predict the identity of a daughter nuclide by noting how the atomic number and mass number change when a particle is ejected from the parent nucleus.

α disintegration. When an α particle is ejected from a nucleus, it carries away two units of positive charge and a mass equivalent to four nucleons. The loss of two units of positive charge (corresponding to the loss of two protons from the nucleus) reduces the atomic number by 2. When a radium-226 nucleus, with $Z = 88$, undergoes α decay, the fragment remaining is a nucleus of atomic number 86 and hence a nucleus of an atom of the noble gas radon. Because the α particle carries away two neutrons as well as two protons, the mass number is reduced by 4. Hence α decay of radium-226 results in the formation of radon-222. This transmutation—a nuclear reaction—is summarized by the following nuclear equation:

$$^{226}_{88}\text{Ra} \longrightarrow {}^{222}_{86}\text{Rn} + {}^{4}_{2}\alpha$$

In general, *when a nucleus undergoes α decay,* the mass number decreases by 4 and the atomic number decreases by 2. This mode of decay is illustrated in Fig. 22.6a.

N The emission of γ radiation after a nuclear rearrangement has occurred is very similar, but on a more energetic scale, to the observation of chemiluminescence, the observation of infrared or visible radiation as a result of an atomic rearrangement in the course of a chemical reaction.

▼ EXAMPLE 22.1 Identifying the nuclide formed by α decay

Identify the nuclide produced by α decay of gadolinium-150.

STRATEGY We need to ensure that the mass numbers and atomic numbers balance in the nuclear equation. First, we should identify the parent nuclide and write its chemical symbol. Then we write the nuclear reaction, with the mass number and the atomic number of the daughter nuclide written as A and Z, respectively. We can find the values of A and Z from the fact that the sums of the mass numbers and the atomic numbers are both unchanged in the decay. Once we know A and Z, we can identify the daughter nuclide.

SOLUTION The parent nuclide is $^{150}_{64}\text{Gd}$, so the reaction is

$$^{150}_{64}\text{Gd} \longrightarrow {}^{A}_{Z}\text{E} + {}^{4}_{2}\alpha$$

We need $150 = A + 4$ and $64 = Z + 2$; hence $A = 146$ and $Z = 62$. The

? What nuclide is produced by the β decay of lithium-9?

E Nuclear beta decay. *J. Chem. Ed.*, **1979**, *56*, 250. Beta decay diagram. *J. Chem. Ed.*, **1989**, *66*, 231.

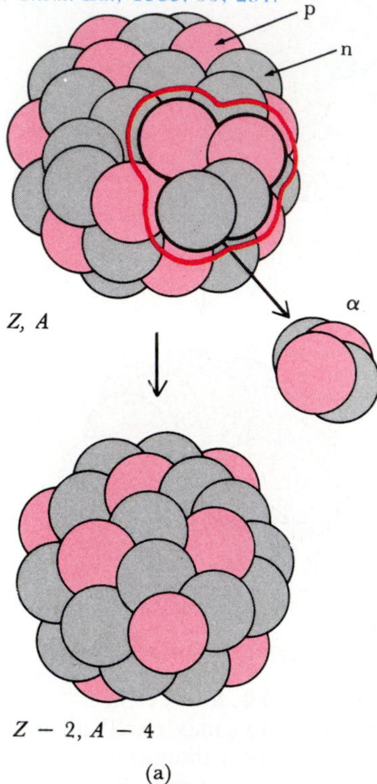

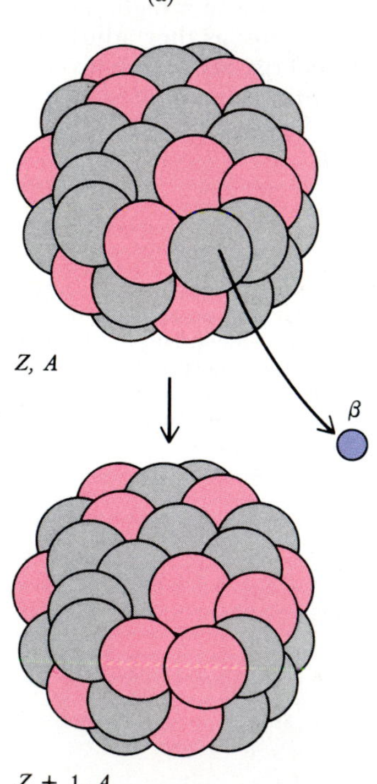

element with Z = 62 is samarium, Sm, so the nuclide produced is samarium-146; the nuclear equation for the α decay is therefore

$$^{150}_{64}\text{Gd} \longrightarrow {}^{146}_{62}\text{Sm} + {}^{4}_{2}\alpha$$

EXERCISE E22.1 Identify the nuclide produced by α decay of uranium-235.

[*Answer*: thorium-231]

β disintegration. When a β particle is ejected from a nucleus, it carries away one unit of negative charge. The loss of one unit of negative charge can be interpreted as the conversion of a neutron into a proton within the nucleus:*

$$^{1}_{0}\text{n} \longrightarrow {}^{1}_{1}\text{p} + {}^{0}_{-1}\text{e}, \qquad \text{or more simply} \quad \text{n} \longrightarrow \text{p} + \beta$$

This relation implies that the atomic number of the daughter nuclide is greater by 1 than before, because an additional proton is now present. The mass number is unchanged, because the total number of nucleons in the nucleus is unchanged. For example, when sodium-24, with atomic number 11, undergoes β decay, the daughter nuclide is an atom of the element of atomic number 12 (magnesium) with the same mass number:

$$^{24}_{11}\text{Na} \longrightarrow {}^{24}_{12}\text{Mg} + {}^{0}_{-1}\text{e}$$

Note that, as required, the mass number on the left (24) is equal to the sum of the mass numbers on the right; similarly, the atomic number on the left (11) is equal to the sum of the atomic numbers on the right. In general, *when a nuclide undergoes β decay,* the mass number is unchanged and the atomic number increases by 1. This mode of decay is illustrated in Fig. 22.6b.

Other modes of disintegration. Numerous other modes of nuclear transmutation have been observed. In **electron capture**, a nucleus captures one of its own atom's *s* electrons and the atomic number is reduced by 1 (Fig. 22.7a). (We saw in Section 7.3 that an *s* electron can be found right at the nucleus, so it stands the risk of being captured by the attractive nuclear forces that act at very short distances there.) An example is the nuclear reaction

$$^{44}_{22}\text{Ti} + {}^{0}_{-1}\text{e} \longrightarrow {}^{44}_{21}\text{Sc}$$

*Modern research has shown that another particle, the *neutrino, ν,* is also emitted in some nuclear reactions, including the one shown here. This reaction is better written $\text{n} \rightarrow \text{p} + \beta + \nu$. However, because the neutrino has a mass much smaller than that of the electron (and may even be massless) and is electrically neutral, its emission has no effect on the identity of the daughter nuclide. For our purposes, the neutrino is ignored.

FIGURE 22.6 (a) When a nucleus ejects an α particle, the atomic number of the atom decreases by 2 and the mass number decreases by 4. (b) When a β particle is ejected, the atomic number increases by 1 and the mass number remains unchanged. Pink circles represent protons and gray circles represent neutrons.

In **positron emission**, a positron (${}^{0}_{+1}e$, or β^+) is ejected; this process also reduces the atomic number by 1 without changing the mass number (Fig. 22.7b):

$$ {}^{43}_{22}Ti \longrightarrow {}^{43}_{21}Sc + \beta^+ $$

U Exercise 22.49.

▼ **EXAMPLE 22.2 Identifying the products of other kinds of nuclear transmutation**

Identify the daughter nuclides that result from (a) electron capture of potassium-40 and (b) positron emission by sodium-22.

STRATEGY We can proceed as we did in Example 22.1. We write the skeletal (unbalanced) equations for the nuclear reaction, balance the atomic numbers and mass numbers, and identify the product element from its atomic number.

SOLUTION The skeletal equations are

$$ \text{(a) } {}^{40}_{19}K + {}^{0}_{-1}e \longrightarrow {}^{A}_{Z}E \qquad \text{(b) } {}^{22}_{11}Na \longrightarrow {}^{A}_{Z}E + {}^{0}_{+1}e $$

These balance when we write

$$ \text{(a) } {}^{40}_{19}K + {}^{0}_{-1}e \longrightarrow {}^{40}_{18}E \qquad \text{(b) } {}^{22}_{11}Na \longrightarrow {}^{22}_{10}E + {}^{0}_{+1}e $$

(a) The element with $Z = 18$ is Ar, so the reaction produces argon-40. (b) The element with $Z = 10$ is Ne, so reaction (b) produces neon-22. The nuclear reactions are therefore

$$ \text{(a) } {}^{40}_{19}K + e \longrightarrow {}^{40}_{18}Ar \qquad \text{(b) } {}^{22}_{11}Na \longrightarrow {}^{22}_{10}Ne + \beta^+ $$

EXERCISE E22.2 Identify the nuclides that result from (a) proton emission from cobalt-53 and (b) electron capture by copper-64.

[*Answer*: (a) iron-52; (b) nickel-64] ▲

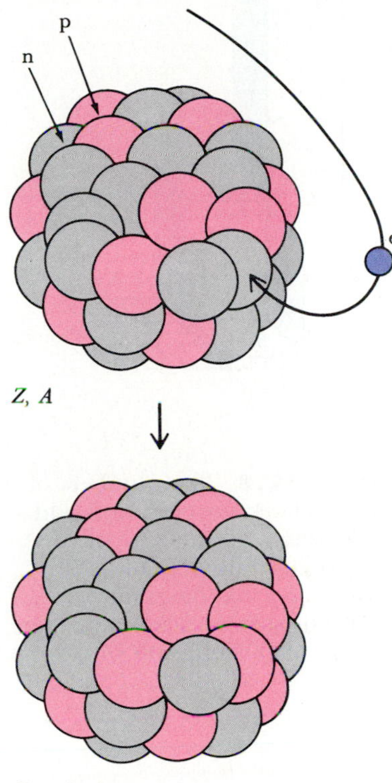

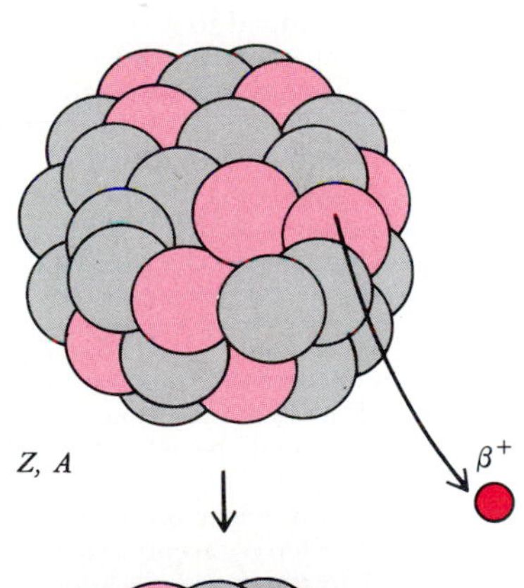

It should be noted that it is not possible to change either the mode or the rate of radioactive decay of a given nuclide. Each unstable nuclide undergoes a specific decay at its characteristic rate. This reaction is analogous to the thermal decomposition of $CaCO_3$—chemists cannot change the products of the decomposition, and at a given temperature the decomposition proceeds at a characteristic rate. With nuclear decay, however, not even the temperature affects the rate.

22.3 THE PATTERN OF NUCLEAR STABILITY

It has been observed that nuclei that have both an even number of protons and an even number of neutrons are more stable than those with any other combination, and nuclei with odd numbers of protons and neutrons are the least stable. The numbers of nonradioactive nuclei for each combination are shown in Fig. 22.8.

FIGURE 22.7 (a) In the process of electron capture, a nucleus captures one of the surrounding electrons. The effect is to convert a proton into a neutron, so Z decreases by 1 but the mass number remains unchanged. (b) The outcome is the same in positron emission, because, in effect, a proton discards its positive charge and becomes a neutron.

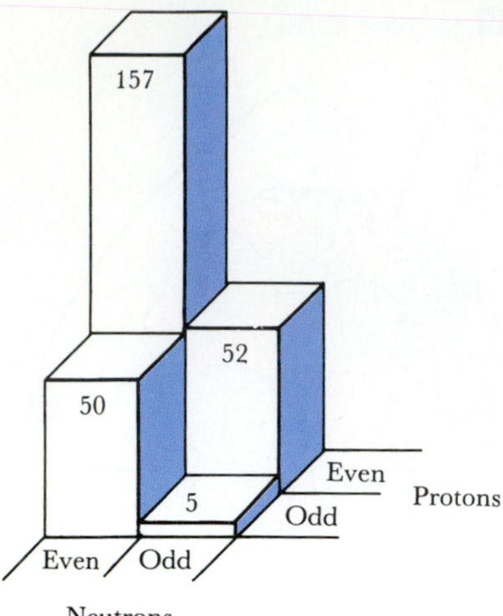

FIGURE 22.8 The numbers of stable nuclides for even and odd atomic numbers (number of protons) and numbers of neutrons. By far the greatest number of stable nuclei have even numbers of protons and of neutrons.

E The magic numbers can be obtained from an independent particle model of the nucleus in which the nucleons are confined to a spherical cavity. To obtain the observed magic numbers, the orbital motion of the nucleons must be allowed to interact with their spins. When that spin-orbital interaction is taken into account, it is found that shells of energy levels are obtained that are filled by the magic numbers of neutrons or protons.

E The existence of the magic number 126 suggests that there may be a collection of elements beyond the end of the current periodic table with $Z \approx 126$ (and mass number close to $126 + 126 = 252$) that have exceptional stability relative to the currently known transuranium elements.

E Predicting nuclear stability using the periodic table. *J. Chem. Ed.*, **1989**, *66*, 757.

A second factor is the variation of nuclear stability with certain numbers of nucleons in the nucleus; these numbers are called **magic numbers**:

The magic numbers: 2, 8, 20, 50, 82, and 126

For example, there are 10 stable isotopes of tin ($Z = 50$), the most of any element, but only 2 stable isotopes of its neighbor antimony ($Z = 51$). The α particle itself is a "doubly magic" nucleus, with two protons and two neutrons. We shall see later that a number of nuclides decay through a series of steps until they reach ^{208}Pb, another doubly magic nuclide, with 126 neutrons and 82 protons. The existence of magic numbers for nuclei is analogous to the chemical stability of the noble gases, which have their own "magic numbers" of electrons: 2, 10, 18, 36, 54, and 86. This analogy strongly suggests a "shell model" for the nucleus, in which nucleons occupy a series of shells, just as electrons do in atoms.

Observations on the stability of nuclides and the mode of radioactive decay of unstable nuclei are summarized in Fig. 22.9. The points (the values of A and Z) representing the nonradioactive, stable nuclei form the so-called band of stability. It is surrounded by the "sea of instability" of nuclides that are unstable and decay with the emission of radiation. For low atomic numbers (up to about 20), the stable nuclides are those having approximately equal numbers of neutrons and protons (those for which A is close to $2Z$). For higher atomic numbers, all known nuclides—both stable and unstable—have more neutrons than protons (so $A > 2Z$).

Figure 22.9 gives us a clue to the type of disintegration a given nuclide is likely to undergo. Unstable isotopes that lie above the band of stability (such as $^{24}_{11}$Na), and therefore have a high proportion of neutrons (are neutron rich), can reach the band by ejecting a β particle, because this process corresponds to the conversion of a neutron into a proton:

$$^{1}_{0}\text{n} \longrightarrow {}^{1}_{1}\text{p} + {}^{0}_{-1}\text{e}$$

For example, a carbon-14 nucleus achieves a lower energy by undergoing β decay:

$$^{14}_{6}\text{C} \longrightarrow {}^{14}_{7}\text{N} + {}^{0}_{-1}\text{e} + \gamma$$

Neutron-rich nuclides may also undergo spontaneous neutron emission:

$$^{90}_{35}\text{Br} \longrightarrow {}^{89}_{35}\text{Br} + {}^{1}_{0}\text{n}$$

Unstable isotopes of elements that lie below the band of stability (such as $^{7}_{4}$Be), have a low proportion of neutrons (are proton rich): they can move toward the band by positron emission, in which a proton decays into a neutron and a positron:

$$^{1}_{1}\text{p} \longrightarrow {}^{1}_{0}\text{n} + {}^{0}_{+1}\text{e}$$

An example is the positron decay of phosphorus-29:

$$^{29}_{15}\text{P} \longrightarrow {}^{29}_{14}\text{Si} + {}^{0}_{+1}\text{e}$$

Proton-rich nuclides can also reach lower energy either by electron capture,

$$^{7}_{4}\text{Be} + \text{e} \longrightarrow {}^{7}_{3}\text{Li}$$

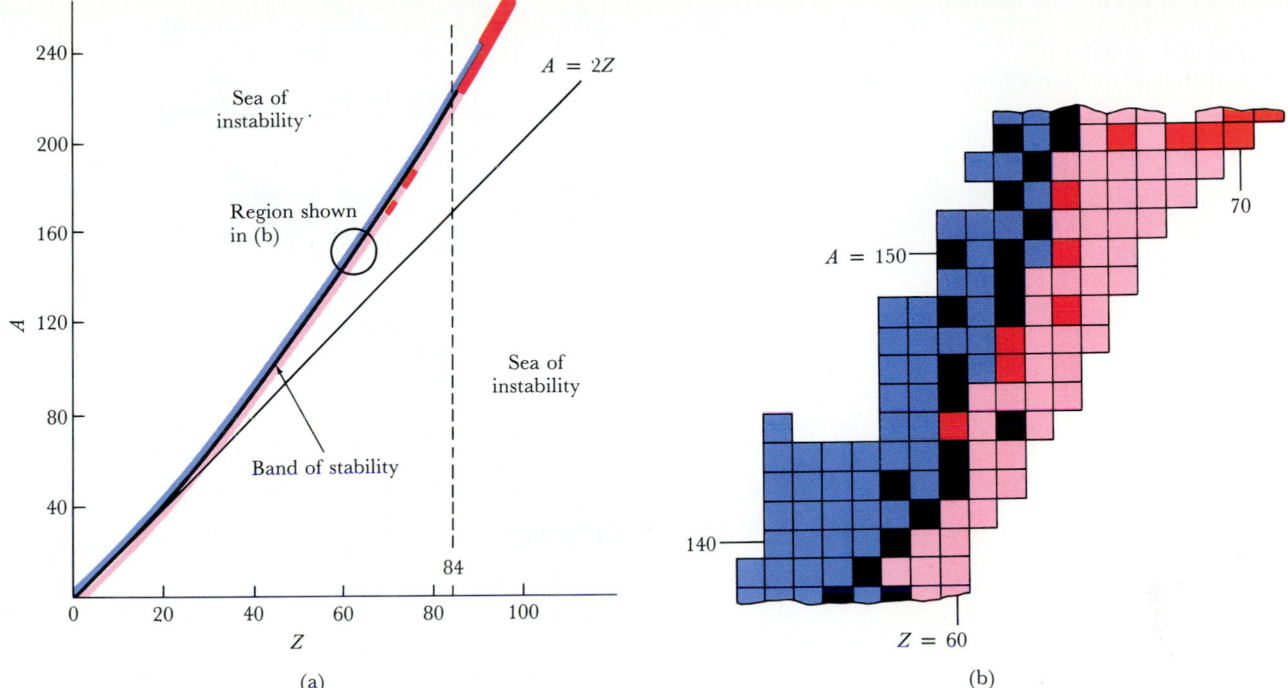

FIGURE 22.9 The manner in which nuclear stability depends on the atomic number and mass number. Nuclides along the narrow black band (the band of stability) are generally stable. Nuclides in the blue region are likely to emit a β particle, and those in the red region are likely to emit an α particle. Nuclei in the pink region are likely either to emit positrons or to undergo electron capture. (b) A magnified view of the diagram near $Z = 60$, showing the structure of the band of stability.

T OHT: Fig. 22.9.

or by proton emission,

$$^{43}_{21}\text{Sc} \longrightarrow {}^{42}_{20}\text{Ca} + {}^{1}_{1}\text{p} + \gamma$$

As we have noted, all nuclei with $Z > 83$ are unstable and radioactive; they disintegrate mainly by α particle emission. (On the other hand, very few nuclides with $Z < 60$ emit α particles.) To reach stability, nuclides of elements with $Z > 83$ must discard protons (to reduce their atomic number) and they generally need to lose neutrons too. Their disintegration usually proceeds in a stepwise fashion: first one α particle is ejected, then another α particle or a β particle is ejected, and so on, until a stable nucleus—usually an isotope of lead (the element with the "magic" atomic number 82)—is formed. Such a stepwise decay path is called a **radioactive series**. Three radioactive series begin from naturally occurring nuclides:

Uranium-238 series: starts at uranium-238 and ends at lead-206
Uranium-235 series: starts at uranium-235 and ends at lead-207
Thorium-232 series: starts at thorium-232 and ends at lead-208

? How many α and β particles must be emitted in order for uranium-238 to transmute into lead-206?

The first of these series is shown in Fig. 22.10. A knowledge of radioactive series was important when radioactivity was first studied, because the only source of radioactive materials was the decay of heavy

FIGURE 22.10 The uranium-238 series. The times are the half-lives of the nuclides (see explanation in Section 22.6).

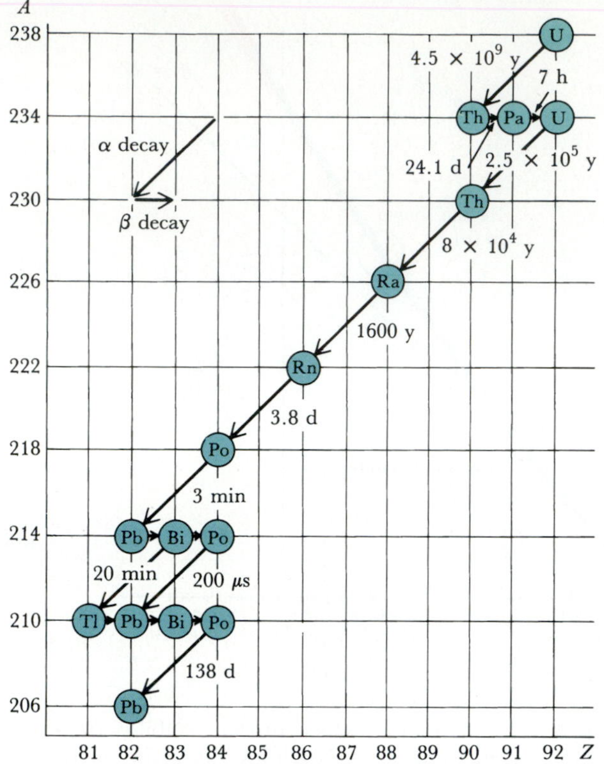

elements (such as uranium) through a series of products. Their importance now lies in how they summarize the behavior of the components of nuclear fuels.

E The hot conditions required for nucleosynthesis are also found in the cloud of particles shrugged off a star in a supernova.

E Formation of the chemical elements and evolution of our universe. *J. Chem. Ed.*, **1990**, *67*, 723. Supernova 1987A. What have we learned about nucleosynthesis. *J. Chem. Ed.*, **1990**, *67*, 731. Stellar nucleosynthesis: A vehicle for the teaching of nuclear chemistry. *J. Chem. Ed.*, **1973**, *50*, 311.

D Secular equilibrium. TDC, 185.

N Here again Rutherford's name appears. Whereas students are more familiar with his nuclear model of the atom, most of his work focused on radioactivity, especially the properties of α particles (which led, via the Geiger-Marsden experiment, to the concept of the nuclear atom).

E Nuclear synthesis and identification of new elements. *J. Chem. Ed.*, **1985**, *62*, 392. Mythology and elemental etymology: The names of elements 92 through 94. *J. Chem. Ed.*, **1986**, *63*, 659.

22.4 NUCLEOSYNTHESIS

The formation of elements is called **nucleosynthesis**. The process occurs naturally in the interior of stars and is responsible for the formation of many of the elements now found on Earth: in this sense, human flesh is stardust. Nucleosynthesis can also be brought about artificially on Earth by bombarding the nuclei of other elements with elementary particles or nuclei that have been accelerated to high speeds in a particle accelerator (Fig. 22.11). High speed is essential if the projectile particles are positively charged, because they must overcome the electrostatic repulsion of the target nucleus to approach it closely. Heating a substance to a very high temperature—millions of degrees, like that found in the interiors of stars—is another way of achieving high kinetic energy and hence closeness of approach.

The first artificial nuclear transmutation was achieved by Rutherford in 1917 with the bombardment of nitrogen-14 nuclei with α particles, thereby producing oxygen-17:

$$^{14}_{7}\text{N} + ^{4}_{2}\alpha \longrightarrow ^{18}_{9}\text{F, followed immediately by } ^{18}_{9}\text{F} \longrightarrow ^{17}_{8}\text{O} + ^{1}_{1}\text{p}$$

Since then, a large number of nuclides have been synthesized. As an example, an isotope of technetium ($Z = 43$) was prepared by the Italian physicist Emilio Segrè in 1937 by the reaction between molybdenum

(Z = 42) and deuterons:

$$_{42}\text{Mo} + {}^{2}_{1}\text{H} \longrightarrow {}_{43}\text{Tc} + {}^{1}_{0}\text{n}$$

Neutron-induced transmutation. It is quite easy for a neutron to get very close to a target nucleus, because it is not repelled by the nuclear charge. An example of neutron-induced transmutation is the formation of cobalt-60 from iron-58 by a three-step process. First, iron-59 is produced from iron-58:

$${}^{58}_{26}\text{Fe} + {}^{1}_{0}\text{n} \longrightarrow {}^{59}_{26}\text{Fe}$$

The second step is β decay of the iron-59 to cobalt-59:

$${}^{59}_{26}\text{Fe} \longrightarrow {}^{59}_{27}\text{Co} + \beta$$

In the final step, the cobalt-59 absorbs another neutron from the incident neutron beam and is converted into cobalt-60:

$${}^{59}_{27}\text{Co} + {}^{1}_{0}\text{n} \longrightarrow {}^{60}_{27}\text{Co}$$

N This sequence of nuclear reactions is used to prepare the cobalt-60 that is used in radiation therapy.

The equation for the overall reaction, the sum of the equations for the elementary reactions, is

$${}^{58}_{26}\text{Fe} + 2\text{n} \longrightarrow {}^{60}_{27}\text{Co} + \beta$$

A nucleosynthesis reaction in which element E′ is formed from E is often written in the compact form

$$\text{E(particles in, particles out)E}'$$

For the cobalt-60 synthesis, the compact form is

$${}^{58}_{26}\text{Fe}(2\text{n},\beta){}^{60}_{27}\text{Co}$$

Transmutation with charged particles. If it has enough kinetic energy, a proton, an α particle, or another positively charged nucleus can approach the nucleus of the target atoms so closely that the attractive force between nucleons overcomes the electrostatic repulsion. A process of this kind that occurs in stars is the formation of oxygen-16 from carbon-12:

$${}^{12}_{6}\text{C} + \alpha \longrightarrow {}^{16}_{8}\text{O} + \gamma$$

U Exercise 22.50.

E Charged particles can be accelerated to high energies by a cyclotron, a synchrocyclotron, or a linear accelerator. Heavy ions are accelerated to high energies in a *HILAC*, a heavy ion linear accelerator. Most of the transuranium elements were originally produced at the Lawrence Livermore Radiation Laboratories in Berkeley, California or the Dubna laboratories in the U.S.S.R.

FIGURE 22.11 An aerial view of the Fermi National Accelerator Laboratory in Batavia, Illinois. The largest circle is the main accelerator; three experimental lines extend at a tangent from it.

One that has been achieved in an accelerator on Earth is the synthesis of the artificial element 103, lawrencium, by bombarding californium with boron nuclei:

$$^{250}_{98}\text{Cf} + {}^{11}_{5}\text{B} \longrightarrow {}^{257}_{103}\text{Lr} + 4\,{}^{1}_{0}\text{n}, \qquad \text{or} \qquad {}^{250}_{98}\text{Cf}({}^{11}_{5}\text{B},4\text{n}){}^{257}_{103}\text{Lr}$$

E A small amount of americium-241 is used in some home smoke detectors.

U Exercise 22.51.

E Plutonium was first produced in 1940 as part of the Manhattan Project for the development of the atomic bomb, and was used in the bomb that destroyed Nagasaki (the bomb that fell on Hiroshima used uranium). It is now produced by neutron irradiation of uranium-238 followed by two β-decay events. A block of plutonium is warm to the touch as a result of the energy released by α decay; larger pieces are hot enough to boil water.

? What are the systematic symbols for carbon and phosphorus?

E Not all the nuclear disintegrations are detected. Hans Geiger (of the Geiger-Marsden experiment) was a postdoctoral student in Rutherford's laboratory; he developed the instrument in 1909 to detect α particles. A refinement, the *Geiger-Müller counter*, was produced in 1928, and is still used.

Basis of the systematic nomenclature of the transuranium elements

Atomic number	Name	Abbreviation
0	nil	n
1	un	u
2	bi	b
3	tri	t
4	quad	q
5	pent	p
6	hex	h
7	sep	s
8	oct	o
9	enn	e

EXAMPLE 22.3 Writing equations for nucleosynthesis reactions

Complete the following equations, which illustrate the synthesis of three artificial elements: (a) $^{238}_{92}\text{U} + {}^{2}_{1}\text{H} \rightarrow ? + 2\text{n}$; (b) $^{241}_{95}\text{Am} + ? \rightarrow {}^{243}_{97}\text{Bk} + 2\text{n}$.

STRATEGY As in the earlier examples, we have to select particles or nuclides that balance the mass numbers and the atomic numbers in the equations. (a) We can identify the new element from its atomic number by referring to the periodic table inside the front cover. (b) We need to identify the particle from its charge and mass number.

SOLUTION (a) The mass numbers are balanced if $238 + 2 = A + 2 \times 1$, from which $A = 238$. The atomic numbers are balanced if $92 + 1 = Z + 2 \times 0$, from which $Z = 93$. Element 93 is Np, so the nuclide produced is neptunium-238. (b) From the mass-number balance, we have $241 + A = 243 + 2 \times 1$, from which $A = 4$. The atomic-number balance gives us $95 + Z = 97 + 2 \times 0$; so $Z = 2$. Hence the incoming particle is the α particle, $^{4}_{2}\alpha$. The complete nuclear reactions are therefore

$$\text{(a)}\ {}^{238}_{92}\text{U} + {}^{2}_{1}\text{H} \longrightarrow {}^{238}_{93}\text{Np} + 2\text{n} \qquad \text{(b)}\ {}^{241}_{95}\text{Am} + {}^{4}_{2}\alpha \longrightarrow {}^{243}_{97}\text{Bk} + 2\text{n}$$

EXERCISE E22.3 Complete the following nuclear reactions:

$$\text{(a)}\ ? + \alpha \longrightarrow {}^{243}_{96}\text{Cm} + \text{n} \qquad \text{(b)}\ {}^{242}_{96}\text{Cm} + \alpha \longrightarrow {}^{245}_{98}\text{Cf} + ?$$

[*Answer*: (a) $^{240}_{94}\text{Pu}$; (b) n]

All the **transuranium elements** (those following uranium in the periodic table, with $Z > 92$) are synthetic and are produced by the bombardment of target nuclei with a smaller projectile. The elements through lawrencium ($Z = 103$) have been formally named. The symbols and names of the elements beyond lawrencium (including nuclides that have not yet been made) are named systematically, at least until they have been identified and there is agreement on a permanent name. For this systematic nomenclature, the name and symbol of the element are made up by expressing its atomic number in the following terms and then adding the suffix *-ium* (see list in margin). For example, the element with $Z = 104$ is called unnilquadium—and given the symbol Unq—until the dispute between the rival names (rutherfordium and kurtchatovium) is resolved—and the hypothetical element with $Z = 125$ is called unbipentium and given the symbol Ubp.

RADIOACTIVITY

One of the most important social aspects of radioactivity is the damage it can do to biological tissues. The amount of damage depends on the strength of the source, the type of radiation, and the length of exposure. Another important aspect of radioactivity is its persistence: for instance, for how long do we need to store unwanted radioactive material before it may be considered safe? Such issues can be examined quantitatively by drawing on some of the material discussed earlier in the text, particularly the techniques of chemical kinetics (Chapter 12).

22.5 MEASURING RADIOACTIVITY

Becquerel detected and measured radioactivity by exposing a photographic film to a source of radiation. He gauged the intensity of the radiation by the degree of blackening of the developed film. The blackening results from the same redox processes as take place in ordinary photography (Box 3.1), except that the initial oxidation of the halide ions is caused by the nuclear radiation. Bequerel's technique is still used to monitor the exposure of workers to radiation.

Two other devices are also commonly used to detect and measure radiation. A **Geiger counter** is essentially a cylinder containing a low-pressure gas and two electrodes (Fig. 22.12). The radiation ionizes the gaseous atoms inside the cylinder and hence allows a brief flow of current between the electrodes. One type of Geiger counter converts this current into an audible click from a loudspeaker, the rapidity of the clicks indicating the intensity of the radiation. A **scintillation counter** makes use of the fact that certain substances (notably zinc sulfide) give a flash of light—a scintillation—when exposed to radiation. The intensity of the radiation is measured by counting the scintillations electronically. Each click of the Geiger counter or flash of the scintillation counter indicates that one nuclear disintegration has been detected.

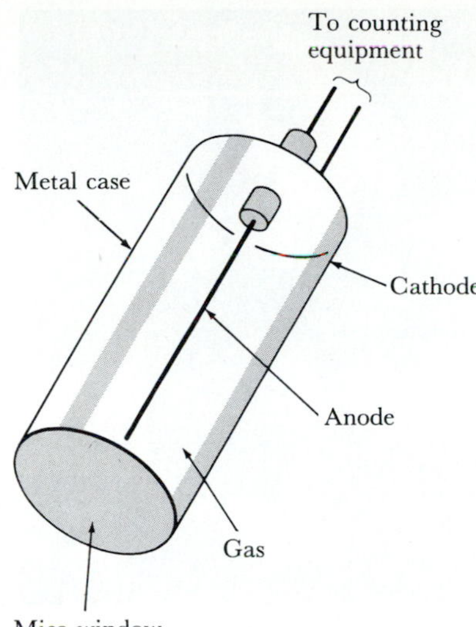

FIGURE 22.12 The detector in a Geiger counter consists of a gas (typically argon and a little ethanol vapor, or neon and some bromine vapor) in a container with a high potential difference (of 500 to 1200 V) between the walls and the central wire. An electric current passes between the electrodes (walls and wire) when radiation ionizes the gas.

E α emitters that enter the body through the respiratory system can be particularly damaging, and may lead to lung cancer (see Case 22).

Penetrating power. The three principal types of nuclear radiation penetrate matter to different extents. The least penetrating is α radiation. The massive, highly charged α particles interact so strongly with matter that they slow down, capture electrons from surrounding matter, and change into bulky He atoms before traveling very far. However, even though α particles do not penetrate far into matter, they are very damaging because the energy of their impact can knock atoms out of molecules and displace ions from their sites in crystals. When they impinge on biological matter, the damage caused by α particles can lead to illness. If DNA and the enzymes that interpret its protein-building messages are damaged, the result may be cancer. Most α radiation is absorbed by the surface layer of skin, but inhaled and ingested particles can cause internal damage.

N The relative ionizing power of α, β, and γ radiation is 10,000:100:1.

D Absorption of beta rays. TDC, 9-12. Gamma rays: inverse square law. TDC, 9-15. Absorption of gamma rays. TDC, 9-16.

Next in penetrating power is β radiation. The fast electrons of these rays can penetrate about 1 cm of flesh before their electrostatic interactions with the electrons and nuclei of molecules bring them to a standstill.

Most penetrating of all is γ radiation. The uncharged, high-energy γ-ray photons can pass through the body and cause damage by ionizing the molecules in their path. Protein molecules and DNA that have been damaged in this way cannot carry out their functions, and the result can be radiation sickness and cancer. Sources of γ rays must be surrounded by walls built from lead bricks or thick concrete to shield people from this penetrating radiation. Table 22.2 summarizes the relative penetrating powers of these radiations.

TABLE 22.2 Characteristics of α, β, and γ radiation

Radiation	Relative penetrating power	Shielding required
α	1	paper, skin
β	100	3 mm aluminum
γ	10,000	concrete, lead

The activity of radioactive sources. The intensity of the radiation from a radioactive element depends on the nuclide's **activity**, the number of

FIGURE 22.13 Marie Sklodowska Curie (1867–1934).

E There was no possibility of Marie Sklodowska receiving a science education in the Poland of 1880s, so she went to Paris in 1891; she married the physicist Pierre Curie in 1894. Her dogged determination led her to achieve the isolation of 0.1 g of radium chloride from several tons of pitchblende. She was the first woman to be awarded an advanced research degree in France. At the time she was working, the hazards of radioactivity were unknown, and her early notebooks are still too dangerous to handle.

nuclear disintegrations that occur per second. Disintegrations can be counted by recording the number of clicks given by a Geiger counter or the number of flashes in a scintillation counter. Various units have been introduced to report the activities of radiation sources and the extent of their effects on matter (particularly living tissue).

Measurements (with a Geiger counter or a scintillation counter) have shown that 3.7×10^{10} nuclear disintegrations occur each second in 1 g of radium-226. This rate of disintegration is used as a unit of activity, the **curie** (Ci); it is named for (and was defined by) Marie Curie, the Polish-French chemist who discovered radium and polonium (Fig. 22.13):

$$1 \text{ Ci} = 3.7 \times 10^{10} \text{ disintegrations per second (dps)}$$

The more active the source, the greater the number of nuclear disintegrations per second and hence the greater its rating in curies. The curie is a very large unit, and most activities are expressed in millicuries ($1 \text{ mCi} = 10^{-3} \text{ Ci}$) and microcuries ($1 \mu\text{Ci} = 10^{-6} \text{ Ci}$). The SI unit of radioactivity is the **becquerel** (Bq) and is defined as one disintegration per second:

$$1 \text{ Bq} = 1 \text{ dps}$$

It follows that the conversion factor between curies and becquerels is

$$1 \text{ Ci} = 3.7 \times 10^{10} \text{ Bq}$$

The principal natural source of radioactivity in the human body is potassium-40, which occurs with about 0.01% abundance among the potassium ions found throughout the body and which emits β particles and positrons. Because the body of an adult human is a 0.1-μCi source, about 37,000 potassium-40 nuclei disintegrated in your body in the time it took you to read this sentence (about 10 s). Our own bodies contribute about 20% of the total radiation we receive from natural sources.

22.6 THE RATE OF NUCLEAR DISINTEGRATION

E Variation of radioactive decay rates. *J. Chem. Ed.*, **1978**, *55*, 302. The numerical equivalence of and uncertainty in the average lifetime for the decay process. *J. Chem. Ed.*, **1971**, *48*, 544. Extranuclear effects on nuclear decay. *J. Chem. Ed.*, **1974**, *51*, 517.

We can discuss the rate at which radioactive nuclei disintegrate—and hence the activity of sources and the persistence of radioactivity—on the basis of the techniques described in Chapter 12 for describing the rates of chemical reactions. The rate at which nuclear disintegration occurs is independent of the temperature, the physical state of the material containing the nuclide, and whether or not the atoms are part of a compound. Decay is unaffected by these factors because the forces that bind, and occasionally eject, nucleons in a nucleus are so strong that the relatively feeble energies associated with thermal motion and chemical bonding are completely negligible in comparison. The stabilities of nuclei are fixed by the forces acting within the nucleus.

The law of radioactive decay and the nuclear half-life. The disintegration of a nucleus,

$$\text{Parent nucleus} \longrightarrow \text{daughter nucleus} + \text{radiation}$$

is a nuclear version of unimolecular decay, with an unstable nucleus taking the place of an excited molecule. Like the rate of a unimolecular chemical reaction (Section 12.5), the rate law for the decay is first order; that is, the relation between the rate of decay and the number N of radioactive nuclei present may be written

$$\text{Rate of decay} = k \times N$$

where k (the rate constant in chemical kinetics) is called the **decay constant** in this context. The rate law is called the **law of radioactive decay**. It implies that the more radioactive nuclei there are in the sample, the greater the overall rate of disintegration and the more active the sample.

As we did for a first-order chemical reaction, we can calculate the number N of nuclei remaining after a time t from the formula

$$\ln\left(\frac{N_0}{N}\right) = kt \qquad (1)$$

where N_0 is the number of radioactive nuclei present initially (at $t = 0$) and N is the number remaining at a time t. As explained in Section 12.2, this equation describes an exponential decay. Hence, the activity of a radioactive sample falls exponentially toward zero (Fig. 22.14).

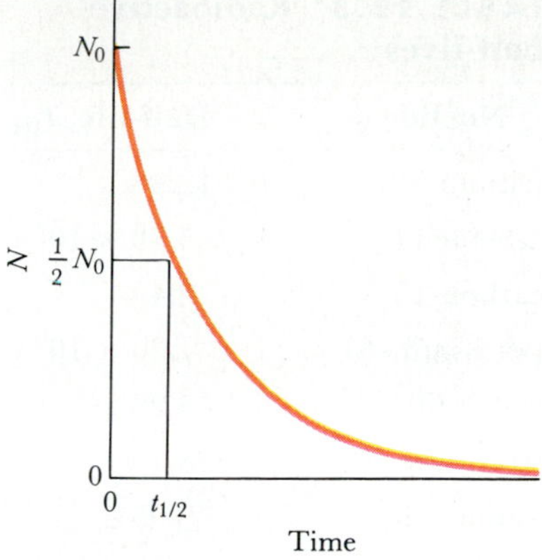

FIGURE 22.14 The exponential decay of the number of radioactive nuclei in a sample implies that the activity of the sample also decays exponentially with time. The curve is characterized by the half-life $t_{1/2}$.

▼ **EXAMPLE 22.4 Using the law of radioactive decay**

If a 1.0-g sample of tritium is stored for 5.0 years, what mass of that isotope will remain? The decay constant is 0.0564 y^{-1}, where y^{-1} denotes "per year" or 1/year.

STRATEGY The number of tritium nuclei in the sample is proportional to the total mass m of tritium, so Eq. 1 can be expressed as

$$\ln\left(\frac{m_0}{m}\right) = kt$$

where m_0 is the initial mass of tritium. The question can now be answered by substituting the data and taking natural antilogarithms.

SOLUTION Substitution of the data gives

$$\ln\left(\frac{m_0}{m}\right) = 0.0564\ \text{y}^{-1} \times 5.0\ \text{y} = 0.282$$

Taking the natural antilogarithm gives

$$\frac{m_0}{m} = 1.33$$

Because $m_0 = 1.0$ g, the mass remaining after 5.0 years will be 0.75 g.

EXERCISE E22.4 The decay constant for fermium-244 is 210 s^{-1}. What mass of the isotope will be present if a 1.0-μg sample is kept for 10 ms?

▲ [*Answer*: 0.12 μg]

N Any quantity proportional to the number of radioactive nuclei can be used in place of N in Eq. 1, such as mass, activity, number, or amount (moles) of nuclei.

Radioactive decays are normally discussed in terms of the half-life, $t_{1/2}$, of the nuclide:

> The **half-life** of a radioactive nuclide is the time needed for half the initial number of nuclei to disintegrate.

The relation between $t_{1/2}$ and k is found by setting $N = \frac{1}{2}N_0$ in Eq. 1. This substitution gives

$$\ln 2 = kt_{1/2}$$

U Exercises 22.52, 22.53, and 22.55–22.59.

E Half-lives of long-lived radioisotopes. *J. Chem. Ed.*, **1982**, *59*, 431.

TABLE 22.3 Radioactive half-lives

Nuclide	Half-life, $t_{1/2}$
tritium	12.3 y
carbon-14	5.73×10^3 y
carbon-15	2.4 s
potassium-40	1.26×10^9 y
cobalt-60	5.26 y
strontium-90	28.1 y
iodine-131	8.05 d
radium-226	1.60×10^3 y
uranium-235	7.1×10^8 y
uranium-238	4.5×10^9 y
fermium-244	3.3 ms

E Archeological dating. *J. Chem. Ed.*, **1986**, *63*, 16.

Substituting ln 2 = 0.693 then yields

$$t_{1/2} = \frac{0.693}{k}$$

The larger the value of k, the shorter the half-life of the nuclide.

The half-lives of some nuclides are given in Table 22.3. The table shows that the values span a very wide range, with some nuclides having a very short half-life and others surviving for billions of years. As an example, consider strontium-90, for which the half-life is 28 years. Strontium-90 is one of the nuclides that occurs in the fallout from nuclear explosions. Because it is chemically very similar to calcium, it may accompany that element through the environment and become incorporated into bones; once there, it continues to generate radiation for many years. Even after three half-lives (84 years), one-eighth of the original strontium-90 still survives. About ten half-lives (for strontium-90, 280 years) must pass before the activity of a sample has fallen to $\frac{1}{1000}$ of its initial value (because $(\frac{1}{2})^{10} \approx \frac{1}{1000}$).

Isotopic dating. The fact that the half-life of a nuclide is constant is put to practical use in **isotopic dating**, the determination of the ages of archaeological artifacts made by measuring the activity of the radioactive isotopes they contain. One important dating technique is **radiocarbon dating**, in which the β decay of carbon-14

$$^{14}\text{C} \longrightarrow {}^{14}\text{N} + \beta + \gamma$$

is utilized.

Radiocarbon dating depends on three characteristics of carbon-14. First, carbon-14 is a naturally occurring isotope of carbon and is present in all living things. Second, it is radioactive and has a half-life of 5730 years. Third, the supply of carbon-14 atoms in the environment has been nearly constant over archaeological time, because they are produced when nitrogen nuclei in the atmosphere are bombarded by neutrons:

$$^{14}_{7}\text{N} + \text{n} \longrightarrow {}^{14}_{6}\text{C} + \text{p} \quad \text{or} \quad {}^{14}_{7}\text{N(n,p)}{}^{14}_{6}\text{C}$$

The neutrons originate from the collisions of cosmic rays with other nuclei.

The carbon-14 atoms produced in the atmosphere enter living organisms through photosynthesis and digestion. They leave organisms by the normal processes of excretion and respiration and also because the nuclei decay at a steady rate. As a result of this continuous input and loss of carbon-14, all living things have a constant proportion of the isotope among their very much more numerous carbon-12 atoms. In other words, there is a fixed ratio of carbon-14 atoms to carbon-12 atoms in living tissues—about 1 in 10^{12}.

When the organism dies, it no longer exchanges carbon with its surroundings. However, the carbon-14 nuclei already inside it continue to disintegrate with a constant half-life. Hence the ratio of carbon-14 to carbon-12 decreases after death, and the ratio observed in a sample of dead tissue can be inserted into Eq. 1 and used to determine the time since death. In the modern version of the technique, which requires only milligrams of sample, the carbon atoms are converted to C^- ions by bombardment of the sample with cesium atoms, the C^- ions are then accelerated with electric fields, and the carbon isotopes are sepa-

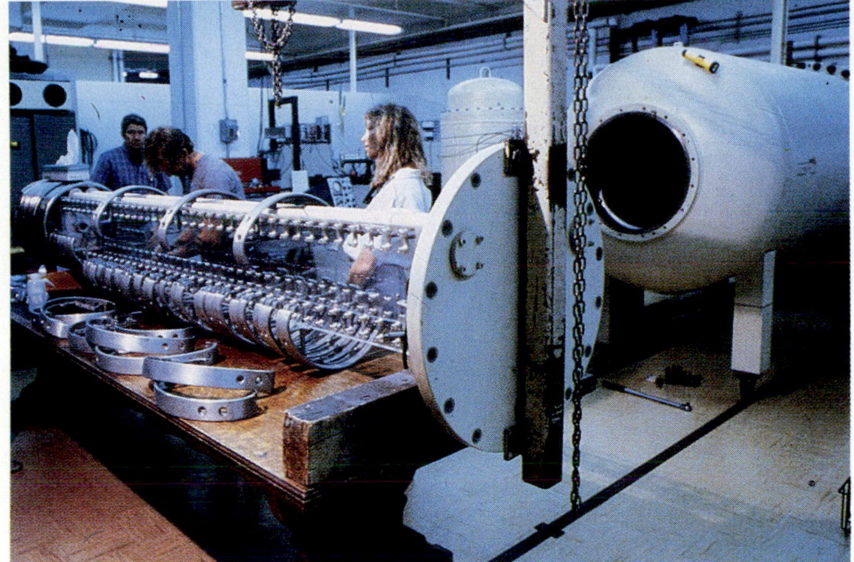

FIGURE 22.15 In the modern version of the carbon-14 dating technique, a mass spectrometer is used to determine the proportion of carbon-14 nuclei in the sample.

rated and counted with a mass spectrometer (Fig. 22.15). In a simpler version of the technique, similar to the original method developed by Willard Libby in Chicago in the late 1940s, larger samples are used and β radiation from the sample is measured. As illustrated in the next example, the time since death is calculated by comparing the activity of the ancient sample with the activity of a modern one.

E California earthquakes: Predicting the next big one using radiocarbon dating. *J. Chem. Ed.*, **1980**, *57*, 601.

▼ EXAMPLE 22.5 Interpreting radiocarbon dating

A 1.00-g sample of carbon made from wood found in an archaeological site gave 7900 carbon-14 disintegrations in a 20-h period. In the same period 1.00 g of carbon from a modern source underwent 18,400 disintegrations. Calculate the age of the sample.

STRATEGY To use Eq. 1, we need the values of k, N, and N_0. The value of k is obtained from the half-life of carbon-14, which is known (Table 22.3). The number of disintegrations in the 20-h period is proportional to the number of carbon-14 nuclei present in the two samples. If we suppose that the proportion of carbon-14 in the atmosphere was the same when the ancient sample was alive as it is now, then we can take the original proportion of carbon-14 in the ancient sample to have been the same as that found in the modern sample. We can therefore set N_0/N equal to the ratio of the number of disintegrations in the two samples.

SOLUTION The rate constant of the decay is

$$k = \frac{\ln 2}{t_{1/2}} = \frac{0.693}{5.73 \times 10^3\ \text{y}} = 1.21 \times 10^{-4}\ \text{y}^{-1}$$

Then, from Eq. 1, which is

$$\ln\left(\frac{N_0}{N}\right) = kt$$

we obtain

$$\ln\left(\frac{18400}{7900}\right) = 1.21 \times 10^{-4}\ \text{y}^{-1} \times t$$

$$0.845 = 1.21 \times 10^{-4}\ \text{y}^{-1} \times t$$

U Exercises 22.54 and 22.60.

and hence that

$$t = \frac{0.845}{1.21 \times 10^{-4}\, y^{-1}} = 6.99 \times 10^3\, y$$

We conclude that approximately 7000 years have elapsed since the piece of wood was alive.

EXERCISE E22.5 A 250-mg sample of carbon from wood underwent 15,300 carbon-14 disintegrations in 36 h. Estimate the time since death. [*Answer*: 6.4×10^3 y]

The reliability of radiocarbon dating depends on the constancy of the proportion of carbon-14 in the environment. The proportion does vary slightly because the intensity of cosmic radiation varies over the centuries, and that must be taken into account in precise determinations of age. This correction can be done in a number of ways. For example, the radiocarbon dating measurement can be calibrated against a sample of carbon from an ancient tree, the age of which has been determined by counting growth rings.

Radioactive nuclides other than carbon-14 are also used for dating. For example, uranium-238 ($t_{1/2} = 4.5 \times 10^9$ y) and potassium-40 ($t_{1/2} = 1.26 \times 10^9$ y) are used for rock dating. Uranium-238 decays through a series of α and β emissions to lead-206, so to determine the age of a rock that contains uranium, all that we need do is determine the ratio of ^{238}U to ^{206}Pb that is present in it. For example, if the ratio is 1 (equal amounts of both nuclides, a ratio indicating that half the original uranium-238 has decayed), then the rock is approximately 4.5×10^9 years old. Indeed, techniques like this one suggest that the age of the Earth is approximately equal to one half-life of uranium-238.

Potassium-40 decays by electron capture to form argon-40. The technique used to date a rock is to crush, under vacuum, rock that contains potassium and measure with a mass spectrometer the amount of argon gas that escapes. This amount is compared with the amount of potassium in the sample and the ratio of ^{40}K to ^{40}Ar is used to determine the age of the rock. For dating more recent samples, the activity due to tritium ($t_{1/2} = 12.3$ y) is measured. In tritium dating, the formation and decay of the nuclide are assumed constant until the processes are interrupted. For example, if a sample of water (or wine) is bottled, then the water can no longer remain in dynamic equilibrium with the environment and, as a result, the activity of the tritium decreases. The ratio of the activity of tritium in the bottled sample to the activity of tritium in the natural sample can be used to determine the age of the bottled sample.

N Many reaction mechanisms are studied by using radioactive nuclides. Tracers may also be used to determine the relative solubilities of a solute in a mixture of immiscible solvents. Isotopes need not be radioactive for them to be used as tracers, but radioactivity is a very sensitive signal of an isotope's presence. Deuterium and oxygen-18, for example, can be used, and their presence in molecules detected by mass spectrometry.

D Radioactive dating: A method for geochronology. *J. Chem. Ed.*, **1985**, *62*, 580.

Other uses of radioactive isotopes. Radioactive **tracers**, or isotopes that can be tracked from compound to compound in the course of a sequence of reactions, are used extensively in the determination of the mechanisms of chemical reactions. For example, if a sample of sugar is "labeled" with carbon-14, in the sense that some carbon-12 atoms in sugar molecules are replaced by carbon-14 atoms, then the progress of the sugar molecule through the body can be monitored. Fertilizers labeled with radioactive nitrogen, phosphorus, and potassium are used to follow the mechanism of plant growth and the passage of these elements through the environment.

Isotopes are widely used in medicine. For example, a physician can determine how and at what rate the thyroid gland takes up iodine by using iodine-131 as a radioactive tracer, and cobalt-60 is used to kill rapidly growing cancer cells. Technetium-99 is one of the most widely used radioactive nuclides in medicine for studying the functioning of internal organs, including the heart; for instance, as $^{99}TcO_4^-$ it is used to detect and pinpoint brain tumors. Sodium-24 is used to monitor blood flow, plutonium-238 is used to power heart pacemakers, and strontium-87 is used to study bone growth.

E The most useful actinide isotope: Americium-241. *J. Chem. Ed.*, **1990**, *67*, 15.

Radioactive nuclides also have a number of industrial applications. For instance, they are used to gauge the thicknesses (and perhaps weaknesses) of metals—for example, pipelines—and of plastics for which uniformity is required. They are used in the research of lubricating oils to determine the wear of moving parts. Food sterilization and potato sprout inhibition is also brought about by exposure to radiation.

NUCLEAR ENERGY

E Nuclear energy. *J. Chem. Ed.*, **1980**, *57*, 360.

One of the most significant contributions to both the creation and the solution of social and economic problems has been the discovery that large amounts of energy are released in certain types of nuclear reactions. This section outlines some of the principles involved. It also explores some of the contributions being made by chemistry to the solution of the very pressing problems arising from the use of nuclear power.

Nuclear reactions, like chemical reactions, involve a change in energy. In radioactive decay, the decay products carry away energy (as kinetic energy); hence such reactions are exothermic. Energy is released in a nuclear reaction (such as radioactive decay or nuclear fission) because the nucleons that were originally in a single nucleus can interact more strongly in the smaller daughter nucleus. Because the forces between nucleons are so strong, even a small change in their relative positions results in a significant change in energy; the difference is released as heat and as kinetic energy of the product nuclei.

N The change in mass should be expressed in kilograms in Eq. 2; note that $1\ J = 1\ kg\ m^2\ s^{-2}$.

N Energy is released in an exothermic *chemical* reaction when atoms settle into new arrangements (as when the C and H atoms of methane and the O atoms of oxygen gas change partners): *nuclear* reactions are exothermic for the same reasons, but as the forces acting between nucleons are so much stronger than those between atoms, the energy released is much greater. We are anxious to avoid giving the impression that the release of nuclear energy is a consequence of the conversion of mass into energy. The release of energy arises from the change in position of nucleons in the presence of the strong nuclear forces, exactly as in a chemical reaction (but in the presence of weaker electromagnetic forces). The release of energy results in a change of mass. The only reason why mass-energy relations play a role in nuclear energy and not in chemical energy is that the changes are so large that change in mass is a convenient way of monitoring and discussing change in energy.

22.7 MASS-ENERGY CONVERSION

The change in energy during a nuclear reaction is calculated by comparing the masses of the nuclear reactants and the nuclear products. This procedure is based on Albert Einstein's theory of relativity, which implies that the mass of an object is a measure of its energy content: the greater the mass of an object, the greater its energy. Specifically, the total energy E and the mass m are related by Einstein's famous equation

$$E = mc^2 \tag{2}$$

where c is the speed of light (3.00×10^8 m/s). Loss of energy is always accompanied by a loss of mass. Actually, mass loss accompanies the exothermic energy changes we considered in earlier chapters, but the energies involved are so small that those mass losses are undetectably small. For example, when 100 g of water cools from 100° to 20°C, it loses 34 kJ of energy as heat; this energy loss corresponds to a mass loss of only 3.8×10^{-10} g. Even in a strongly exothermic chemical reaction, such as one that releases 1000 kJ of energy, the mass of the products is only 10^{-8} g less than that of the reactants; this change in mass is outside the range of all but extremely precise measurements. However, in a nuclear reaction the energy changes are very large, the corresponding mass loss is measurable, and the observed change in mass can be used to calculate the energy released. The energy loss is a result of the rearrangement of the protons and neutrons when new nuclei are formed from old.

Nuclear binding energy. Changes in nuclear mass and energy are normally discussed in terms of the nuclear binding energy, E_{bind}:

> The **nuclear binding energy** is the energy released when Z protons and $A - Z$ neutrons come together to form a nucleus.

This energy is calculated from the change in mass, Δm, that occurs when a compact nucleus forms from widely separated nucleons. For example, the nuclear binding energy of iron-56 (with 26 protons, each of mass m_p, and 30 neutrons, each of mass m_n) is calculated from

$$\begin{aligned}\Delta m &= \Sigma m(\text{products}) - \Sigma m(\text{reactants})\\ &= m(^{56}_{26}\text{Fe nucleus}) - \{26 \times m_p + 30 \times m_n\}\end{aligned}$$

Because the energy arising from the rearrangement of electrons is insignificant in comparison with the energy arising from the rearrangement of nucleons, we generally calculate the binding energy as the energy released when Z hydrogen atoms (in place of protons) combine with $A - Z$ neutrons to form a neutral atom (rather than the nucleus itself). For the formation of iron-56, then, we would have

$$\Delta m = m(^{56}_{26}\text{Fe atom}) - \{26 \times m(^{1}\text{H atom}) + 30 \times m(^{1}\text{n})\}$$

and substitute this expression in Einstein's formula to obtain

$$E_{bind} = \Delta m \times c^2$$

▼ EXAMPLE 22.6 Calculating the nuclear binding energy

Calculate the nuclear binding energy for 1 mol of helium-4, given the following masses: ^{4}He, 4.0026 u; ^{1}H, 1.0078 u; n, 1.0087 u.

STRATEGY The nuclear binding energy is the energy released in the formation of the nucleus from its nucleons:

$$2\,^{1}\text{H} + 2\text{n} \longrightarrow {}^{4}\text{He}$$

We begin by calculating the difference in masses between the products and the reactants; then we convert the result from atomic mass units to kilograms by using the factor from inside the back cover; and finally we use the Einstein relation to calculate the energy corresponding to this loss of mass.

U Exercises 22.61–22.63.

SOLUTION The change in mass is

$$\Delta m = 4.0026\text{ u} - \{2 \times 1.0078\text{ u} + 2 \times 1.0087\text{ u}\} = -0.0304\text{ u}$$

In kilograms, this mass difference is

$$\Delta m = -0.0304\text{ u} \times \frac{1.6605 \times 10^{-27}\text{ kg}}{1\text{ u}} = -5.05 \times 10^{-29}\text{ kg}$$

Then, from Eq. 2,

$$\begin{aligned}E_{bind} &= \Delta m \times c^2 = (-5.05 \times 10^{-29}\text{ kg}) \times (3.00 \times 10^{8}\text{ m/s})^2\\ &= -4.55 \times 10^{-12}\text{ kg·m}^2/\text{s}^2 = -4.55 \times 10^{-12}\text{ J}\end{aligned}$$

Therefore, because the nuclear binding energy is negative, 4.55×10^{-12} J is released when one He atom forms from its nucleons. Multiplying by Avogadro's constant, we find that the binding energy per mole of nuclei is -2.7×10^9 kJ, which is 10 million times larger than the energy of a typical chemical bond.

EXERCISE E22.6 Calculate the nuclear binding energy of 1 mol of carbon-12 nuclei.

[*Answer*: -8.9×10^9 kJ]

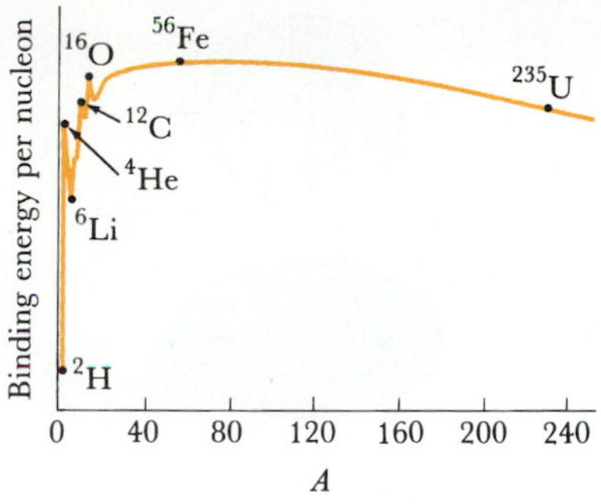

FIGURE 22.16 The variation of the nuclear binding energy per nucleon among the elements. The maximum binding energy per nucleon occurs at iron-56, an observation showing that this nucleus has the lowest energy of all because its nucleons are most tightly bound. (The vertical axis is $-E_{bind}/A$.)

T OHT: Fig. 22.16.

Relative nuclear stability. The relative stability of nuclei is best expressed by comparing the binding energy per nucleon in a nucleus, E_{bind}/A. This procedure lets us decide, for instance, whether a neutron or proton is in a more stable nuclear environment in a ^{56}Fe nucleus or in a ^{235}U nucleus. To show the trends throughout the periodic table, we plot the binding energy per nucleon against the mass number (Fig. 22.16). The actual binding energy of iron-56, for instance, is 56 times the value shown in the graph. The greater the binding energy per nucleon, the more stable are the nucleons in the nucleus and the lower its total energy.

The graph shows that nucleons are bonded together most strongly in iron and its neighboring elements. This high binding energy is one of the reasons for the high abundance of iron in the universe. The binding energy per nucleon is less for the nuclides heavier and lighter than iron. From this observation, we can surmise that when nuclei of atoms lighter than iron "fuse" together and when nuclei of elements heavier than iron undergo "fission" and split into lighter nuclei, the nucleons become more stable. The graph also indicates why a heavy nucleus may be radioactive: when the nucleus emits a particle, the energy of the remaining nucleons is lowered.

22.8 NUCLEAR FISSION

An important type of nuclear disintegration is **fission**, the breaking of a nucleus into two or more smaller nuclei of similar mass. When a uranium atom disintegrates into smaller nuclei, the binding energy per nucleon increases and energy is released. The energy released in this fission process can be calculated by subtracting the total mass of the fission products from the mass of the original nucleus and any incident particles and then converting the change in mass to a change in energy by using Einstein's equation.

E Reflections on nuclear fission at its half century. *J. Chem. Ed.*, **1989**, *66*, 363. Nuclear fission and transuranium elements—50 years ago. *J. Chem. Ed.*, **1989**, *66*, 379. The detours leading to the discovery of nuclear fission. *J. Chem. Ed.*, **1979**, *56*, 771.

EXAMPLE 22.7 Calculating the energy released during fission

Calculate the energy (in joules) released when 1.0 g of uranium-235 undergoes fission and forms barium-142 and krypton-92 in the nuclear reaction

$$^{235}_{92}\text{U} + \text{n} \longrightarrow {}^{142}_{56}\text{Ba} + {}^{92}_{36}\text{Kr} + 2\text{n}$$

The masses of the particles are ^{235}U, 235.04 u; ^{142}Ba, 141.92 u; ^{92}Kr, 91.92 u; n, 1.0087 u.

E Products of neutron irradiation. *J. Chem. Ed.*, **1989**, *66*, 364.

STRATEGY If we know the mass loss, we can interpret it as an energy by using Einstein's equation. Therefore, we must calculate the total mass of the particles on each side of the equation and then substitute the difference into Eq. 2.

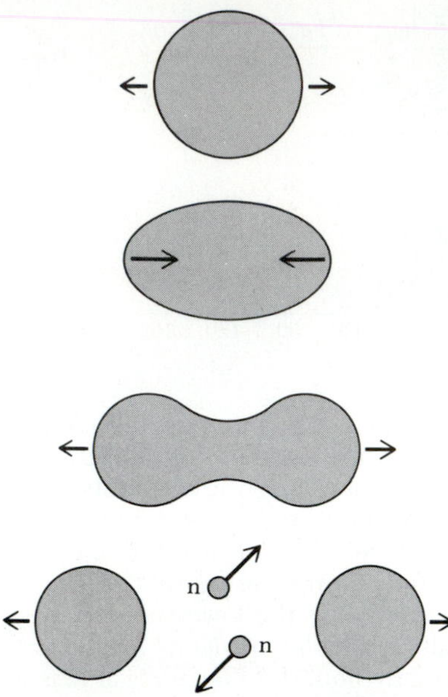

FIGURE 22.17 In spontaneous nuclear fission, the oscillations of the heavy nucleus in effect tear the nucleus apart, thereby forming two or more smaller nuclei of similar mass.

N There is no single nuclear reaction for the fission of U-235; the nucleus undergoes many modes of fission into two fragments of similar mass.

D Uptake of nuclear debris by trees: An experiment. *J. Chem. Ed.*, **1974**, *51*, 270.

E Radiochemistry of fission products. *J. Chem. Ed.*, **1989**, *66*, 367.

SOLUTION

$$\text{Mass of product} = m(\text{Ba}) + m(\text{Kr}) + 2 \times m_n$$
$$= 141.92\ \text{u} + 91.92\ \text{u} + 2 \times 1.0087\ \text{u} = 235.86\ \text{u}$$
$$\text{Mass of reactants} = m(\text{U}) + m_n$$
$$= 235.04\ \text{u} + 1.0087\ \text{u} = 236.05\ \text{u}$$

The change in mass is therefore

$$\Delta m = 235.86\ \text{u} - 236.05\ \text{u} = -0.19\ \text{u}$$
$$= -0.19\ \text{u} \times \frac{1.6605 \times 10^{-27}\ \text{kg}}{1\ \text{u}} = -3.2 \times 10^{-28}\ \text{kg}$$

The energy change accompanying the fission of one ^{235}U nucleus is therefore

$$\Delta E = -3.2 \times 10^{-28}\ \text{kg} \times (3.00 \times 10^8\ \text{m/s})^2 = -2.9 \times 10^{-11}\ \text{J}$$

The total energy released when 1.0 g of uranium-235 undergoes fission is obtained by using the conversion factors $1.6605 \times 10^{-24}\ \text{g} = 1\ \text{u}$, $235.04\ \text{u} = 1\ ^{235}\text{U}$ nucleus, and $1\ ^{235}\text{U}\ \text{nucleus} = -2.9 \times 10^{-11}\ \text{J}$. Then

$$\text{Energy (J)} = 1.0\ \text{g} \times \frac{1\ \text{u}}{1.6605 \times 10^{-24}\ \text{g}} \times \frac{^{235}\text{U nucleus}}{235.04\ \text{u}} \times \frac{-2.9 \times 10^{-11}\ \text{J}}{1\ ^{235}\text{U nucleus}}$$
$$= -7.4 \times 10^{10}\ \text{J}$$

Because a power output of 1 kW for 1 h (called one kilowatt-hour and abbreviated to 1 kW·h) is equivalent to 3600 kJ, the 1.0 g of uranium can supply 2.1×10^4 kW·h of energy.

EXERCISE E22.7 Another way in which uranium-235 can undergo fission is

$$^{235}_{92}\text{U} + \text{n} \longrightarrow {}^{135}_{52}\text{Te} + {}^{100}_{40}\text{Zr} + \text{n}$$

Calculate the energy (in kilowatt-hours) released when 1.0 g of uranium-235 undergoes fission in this way. The additional masses needed are Te, 134.92 u; Zr, 99.92 u.

▲ [*Answer*: 2.2×10^4 kW·h]

Types of nuclear fission. Fission that takes place without needing to be initiated by the impact of other particles is called **spontaneous nuclear fission**. It occurs when the oscillation of a heavy nucleus causes about half the protons and neutrons to break away (Fig. 22.17). An example is the disintegration of americium-244 into iodine and molybdenum:

$$^{244}_{95}\text{Am} \longrightarrow {}^{134}_{53}\text{I} + {}^{107}_{42}\text{Mo} + 3\text{n}$$

Fission does not occur in precisely the same way in every instance: the fission products may also include isotopes of these elements as well as other elements.

Induced nuclear fission is fission caused—often artificially—by bombarding a heavy nucleus with neutrons (Fig. 22.18). Nuclei that can undergo induced fission are called **fissionable**. For most nuclides, fission is induced only if the impinging neutrons are moving so rapidly that they smash into the nucleus and drive it apart; uranium-238 is one such nuclide. Some nuclides, however, can be nudged into breaking apart even if the incoming neutrons are slow. Such nuclides are called **fissile**. They include uranium-235, uranium-233, and plutonium-239, the fuels of nuclear energy.

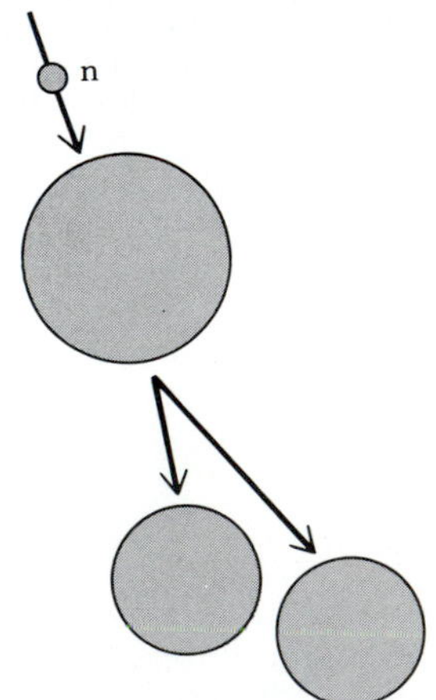

FIGURE 22.18 In induced nuclear fission, an incoming neutron causes the nucleus to break apart.

Induced nuclear fission can be **self-sustaining**; that is, once it is initiated, it can continue even if the supply of neutrons from outside is discontinued. This condition occurs when more neutrons are produced by a fission event than are used to induce it initially. Self-sustaining fission occurs with uranium-235, which undergoes numerous fission processes, including

$$^{235}_{92}U + n \longrightarrow {}^{142}_{56}Ba + {}^{92}_{36}Kr + 2n$$

If the two product neutrons strike two other fissile nuclei, then after the next round of fission there will be four neutrons, which can induce fission in four more nuclei. In the language of Section 12.6, neutrons are carriers in a branched chain reaction (Fig. 22.19).

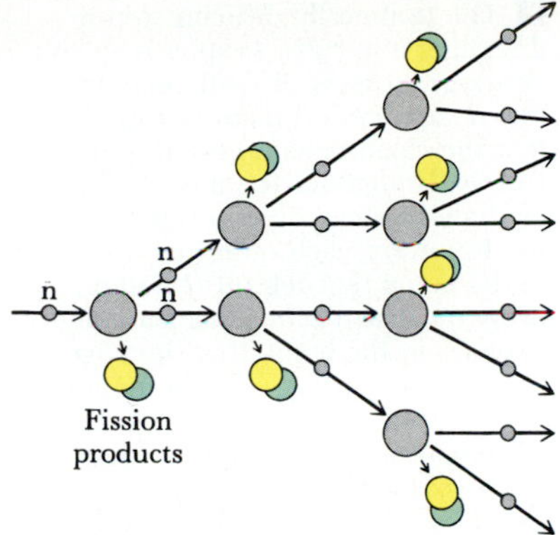

FIGURE 22.19 A self-sustaining nuclear chain reaction, in which neutrons are the chain carriers, occurs when induced fission produces more than one neutron per fission event. These newly produced neutrons can go on to produce fission in other nuclei.

Nuclear explosions. When a nuclear branched chain reaction is allowed to run freely, the cascade of released neutrons can result in the fissioning of all the available uranium-235 in only a fraction of a second. This is what happens in a nuclear explosion.

Some of the neutrons produced in a chain reaction inevitably escape into the surroundings. Enough neutrons are captured to sustain the chain reaction only if a certain minimum number of uranium nuclei are present in the sample. That is, there is a **critical mass**, a mass of fissionable material below which so many neutrons escape from the sample that the fission chain reaction is not sustained. If a sample is **supercritical**, with a mass in excess of the critical value, then enough neutrons induce fission for the chain reaction to result in an explosion. The critical mass for a solid sphere of pure plutonium of normal density is about 15 kg, which is a sphere of about the size of a grapefruit. The critical mass is smaller if the metal is compressed, or imploded, by detonating a conventional explosive that surrounds it. Then the nuclei are pressed closer together and are more effective at blocking the escape of neutrons. The critical mass can be as low as 5 kg for highly compressed plutonium.

E Nuclear fuel consists of UO_2 pellets, enriched to about 3% ^{235}U in a zirconium alloy tube (see chapter opening photograph).

E The critical mass depends on the purity and shape of the sample. A solution may have a lower critical mass than the pure metal.

Controlled fission. An alternative to explosive fission is a controlled chain reaction sustained by a limited supply of neutrons. This goal is achieved in a **nuclear reactor**, where control rods made from neutron-absorbing elements, such as boron or cadmium, are inserted between the fuel elements—the fuel rods containing the fissile material—to control the number of neutrons available for inducing further fission (Fig. 22.20). The fuel elements are surrounded by a **moderator**, a substance such as graphite or heavy water (D_2O) that slows the neutrons. Slow neutrons have three significant roles: they do not induce the fission of fissionable (as distinct from fissile) material, they are most effectively absorbed by the fissile uranium-235, and they allow the control rods to act most efficiently. A coolant, such as carbon dioxide, molten sodium, or water (as steam) surrounds the reactor core and is used to transfer the heat from the reactor core to a heat exchanger.

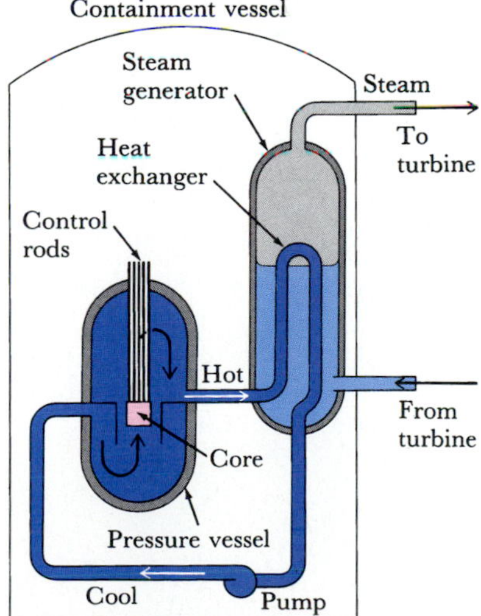

FIGURE 22.20 A schematic diagram of one type of nuclear reactor. This one is a pressurized water reactor, in which the coolant is water under pressure.

22.9 NUCLEAR FUSION

We see from the plot of nuclear binding energy for nucleons in Fig. 22.16 that there is a large increase in nuclear binding energy (and hence a lowering of total energy) on going from hydrogen to its heavier isotopes and to the first few light elements. This observation suggests

E The technically difficult step in designing a nuclear weapon is to convert a given mass of fissile material into a supercritical mass so rapidly that the chain reaction occurs uniformly throughout the metal. This can be done by shooting two subcritical blocks toward each other (as was done in the bomb that fell on Hiroshima) or by implosion of a single subcritical mass (as in the bomb that destroyed Nagasaki). A strong neutron emitter, typically polonium, is also included in the bomb's construction to initiate the chain reaction.

E The safety of nuclear reactors. *Sci. Amer.* **1980**, March, 53. Chernobyl—What happened? *J. Chem. Ed.*, **1985**, *65*, 1037. Natural nuclear reactors: The Oklo phenomenon. *J. Chem. Ed.*, **1976**, *53*, 336.

N Once again, the release of energy on fusion is not *because* there is a loss of mass but follows from the more favorable interactions that may occur between the nucleons after they have fused into a larger nucleus. The release of energy corresponds to a lowering of mass.

E Fusion power. *J. Chem. Ed.*, **1973**, *50*, 658. The engineering of magnetic fusion reactors. *Sci. Amer.* **1983**, October, 54. Progress in laser fusion. *Sci. Amer.* **1986**, August, 68.

that large amounts of energy should be released if H nuclei could be made to undergo **nuclear fusion**, and fuse together to form He or Li nuclei.

The principal difficulty in achieving fusion arises from the strong electrostatic repulsion between protons when they approach each other closely enough to fuse. The heavier isotopes of hydrogen are therefore used in fusion reactions; their nuclei fuse more readily, because the attraction arising from the additional neutrons in them helps to overcome the repulsion between the approaching protons. The high kinetic energies needed for successful collisions are achieved by raising the temperature to millions of degrees.

One fusion method uses deuterium and tritium in the following sequence of nuclear reactions:

$$\begin{array}{rcl} \mathrm{D} + \mathrm{D} & \longrightarrow & {}^{3}\mathrm{He} + \mathrm{n} \\ \mathrm{D} + \mathrm{D} & \longrightarrow & \mathrm{T} + \mathrm{p} \\ \mathrm{D} + \mathrm{T} & \longrightarrow & {}^{4}\mathrm{He} + \mathrm{n} \\ \mathrm{D} + {}^{3}\mathrm{He} & \longrightarrow & {}^{4}\mathrm{He} + \mathrm{p} \\ \hline \textit{Overall reaction: } 6\mathrm{D} & \longrightarrow & 2\,{}^{4}\mathrm{He} + 2\mathrm{p} + 2\mathrm{n} \end{array}$$

The method of Example 22.7 can be used to calculate that the overall reaction releases 3×10^{8} kJ (corresponding to 10^{9} kW·h, the energy generated when the Hoover Dam operates at full capacity for about an hour). Tritium is supplied to the reaction chamber to facilitate the process. Because it has a very low natural abundance and is radioactive, the tritium is generated by bombarding lithium-6 with neutrons in the immediate surroundings of the reaction zone:

$${}^{6}\mathrm{Li} + \mathrm{n} \longrightarrow \mathrm{T} + {}^{4}\mathrm{He}$$

It is on account of this reaction (and the same reaction in thermonuclear bombs) that lithium-6 is being extracted from stocks of lithium. Because lithium-6 (7.4% abundance) and lithium-7 (92.6% abundance) are the only two naturally occurring isotopes of lithium, the average

FIGURE 22.21 Research into controlled nuclear fusion is being carried out in several countries. Here we see the Tokomak fusion test reactor at the Princeton Plasma Physics Laboratory.

atomic mass of the remaining, commercially available lithium is increasing as the proportion of lithium-7 in it grows.

Nuclear fusion is very difficult to achieve in practice because of the vigor with which the charged nuclei must be hurled at one another. One way of accelerating them to sufficiently high speeds is to heat them with a fission explosion: this method is used to produce a **thermonuclear explosion**, in which a fission bomb (using uranium or plutonium) is used to ignite a lithium-6 fusion bomb. A more constructive and controlled approach is to heat a **plasma**, or ionized gas, by passing an electric current through it. The very hot plasma, and its very fast ions, is kept away from the walls of the container (so that they are not melted) with magnetic fields. This method of achieving fusion is the subject of intense research and is beginning to show signs of success (Fig. 22.21).

FIGURE 22.22 Pitchblende is a common uranium ore. It is a variety of uraninite, UO_2.

E A number of high-power laser beams focused on a compound of deuterium can be used to induce a fusion reaction.

E Energy from uranium. *J. Chem. Ed.*, **1979**, *56*, 119.

22.10 CHEMICAL ASPECTS OF NUCLEAR POWER

Chemists have important contributions to make to the development of nuclear power and to the elimination of the legitimate worries associated with its use. Their three principal areas of involvement are the preparation of the fuel itself, the recovery of important fission products, and the safe disposal (or utilization) of nuclear waste.

Nuclear fuel. Uranium is mined as several minerals, the most important being pitchblende, UO_2 (Fig. 22.22). The aim of the uranium refining processes is in general not only to reduce the ore to the metal but also to **enrich** it, that is, to increase the abundance of a specific isotope, in this case uranium-235. The natural abundance of uranium-235 is 0.7%, and the goal of enrichment is to increase this abundance to about 3%.

One extraction scheme begins with the oxidation of the ore by nitric acid:

$$3UO_2(s) + 8HNO_3(aq) \longrightarrow 3UO_2(NO_3)_2(aq) + 2NO(g) + 4H_2O(l)$$

The uranyl nitrate, $UO_2(NO_3)_2$, is extracted with solvents, purified, and then heated to 300°C, when it decomposes to uranium(VI) oxide:

$$UO_2(NO_3)_2(s) \xrightarrow{300°C} UO_3(s) + NO(g) + NO_2(g) + O_2(g)$$

The nitrogen oxides are collected, dissolved in water, and recycled as the source of the nitric acid in the first stage of the process.

After this oxidation and purification, the reduction sequence begins. First, the uranium(VI) is reduced to uranium(IV) with hydrogen:

$$UO_3(s) + H_2(g) \xrightarrow{700°C} UO_2(s) + H_2O(g)$$

The pure uranium(IV) oxide is then treated with hydrofluoric acid, to give uranium tetrafluoride, a blue-green solid (Fig. 22.23):

$$UO_2(s) + 4HF(aq) \longrightarrow UF_4(s) + 2H_2O(l)$$

One possible route at this stage is the reduction of uranium tetrafluoride with magnesium:

$$UF_4(s) + 2Mg(s) \longrightarrow U(s) + 2MgF_2(s)$$

(a)

(b)

FIGURE 22.23 Two fluorides of uranium used in the extraction of uranium. (a) UF_4. (b) White flecks of solid UF_6 on the glass porthole of a freezer-sublimer in a uranium enrichment plant.

The uranium metal is then fabricated into rods about 3 cm in diameter and 1 m long. An alternative route is the oxidation of UF_4 to uranium hexafluoride, UF_6, a volatile yellowish white solid:

$$UF_4(s) + F_2(g) \xrightarrow{450°C} UF_6(s)$$

The hexafluoride route is used if the uranium is to be enriched.

The enrichment procedure makes use of the small mass differences between the hexafluorides of uranium-235 and uranium-238. The first procedure to be developed (as a part of the Manhattan Project to build the first atomic bomb) made use of the different effusion rates of the two isotopic molecules: it follows from Graham's law (Section 5.6) that the rates of effusion of $^{235}UF_6$ (molar mass, 349 g/mol) and $^{238}UF_6$ (molar mass, 352 g/mol) should be in the ratio

E The advantage of using fluorine in this process is that it is *monoisotopic* (^{19}F is the sole naturally occurring isotope), so there are no complications arising from different masses of fluorine as well as uranium. It is also the lightest of the halogens, so the mass ratio for the two uranium isotopes is most pronounced.

$$\frac{\text{Rate of effusion of } ^{235}UF_6}{\text{Rate of effusion of } ^{238}UF_6} = \sqrt{\frac{352}{349}} = 1.004$$

Because the ratio is so close to one, the vapor must be allowed to effuse repeatedly through porous barriers consisting of screens with large numbers of minute holes. In practice it is allowed to do so thousands of times (some of the stages are shown in Fig. 5.23).

Because the diffusion process is technically demanding and uses a lot of energy, scientists and engineers continue to look for alternative enrichment procedures. One of these uses a centrifuge that rotates samples of uranium hexafluoride vapor at very high speed. This process causes the heavier $^{238}UF_6$ molecules to be thrown outward and collected as a solid on the outer parts of the rotor, leaving a higher proportion of gaseous $^{235}UF_6$ closer to the axis of the rotor, where it can be recovered.

Once the uranium hexafluoride has been enriched to the required degree, it is reduced to UO_2 by reaction with hydrogen and water:

$$UF_6(s) + H_2(g) + 2H_2O(g) \xrightarrow{600°C} UO_2(s) + 6HF(g)$$

The hydrogen fluoride is recycled for use earlier in the process. The enriched oxide can now be reduced to the metal with magnesium, but modern reactors generally use the oxide directly because it is less reactive and has a higher melting point than the metal.

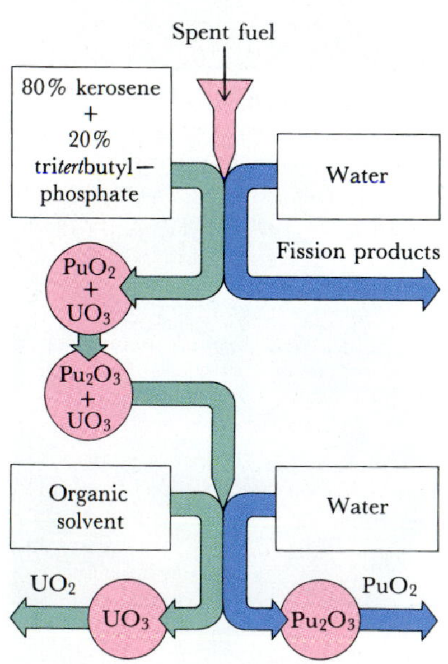

FIGURE 22.24 The flow diagram for the purex process for recovering plutonium and uranium from spent nuclear fuel.

Spent fuel processing. The processing of spent nuclear fuel is a much more complex task than the fuel's initial preparation. There are three problem areas: the recovery of any remaining uranium-235, the extraction of any plutonium produced, and the disposal of the highly radioactive and largely useless fission products.

The uranium and plutonium are separated by **solvent extraction**, a separation process that makes use of their differing solubilities in various solvents. The **purex process** (from *p*lutonium-*u*ranium *r*eduction *ex*traction) utilizes the differing solubilities of plutonium and uranium compounds in a mixture of 80% kerosene and 20% of the organic compound tri*tert*butylphosphate (Fig. 22.24). When mixed with water, the plutonium(IV) and uranium(VI) oxides dissolve preferentially in the organic solvent, and most of the other fission products dissolve preferentially in the water. The organic solution containing the uranium and plutonium is then removed and the plutonium(IV) is reduced to plutonium(III). When this solution is again mixed with water, the uranium(VI) remains in the organic phase while the plutonium(III) dissolves preferentially in the aqueous phase. At this stage, the two elements can be separated by separating the two solutions. Further purification and reaction stages produce pure UO_3 (which is then reduced to UO_2) and PuO_2. One hazard is the possibility that the plutonium will accumulate in a critical mass, which in a concentrated aqueous solution is only about 500 g.

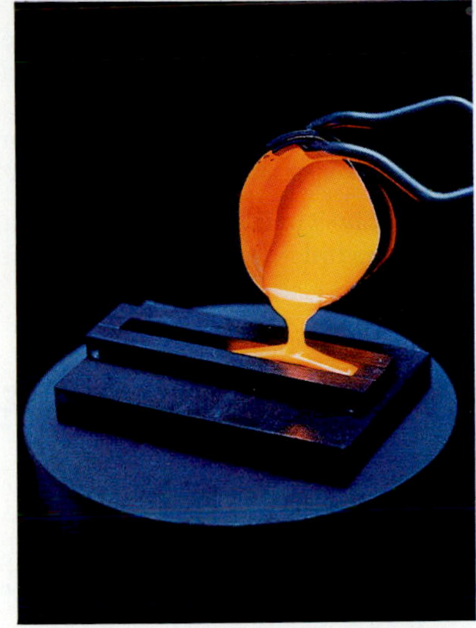

FIGURE 22.25 Molten glass for storing nuclear waste is poured from a platinum crucible into a steel bar mold.

Nuclear waste. The highly radioactive fission (HRF) products from used nuclear fuel rods must be stored for about 10 half-lives before their level of radioactivity is no longer dangerous. The current approach to storage is to incorporate the HRF products into a glass—a solid, complex network of silicon and oxygen atoms (Fig. 22.25). One of the reasons why glass appears to be suitable is that most of the fission products are oxides that are themselves network formers; that is, they promote the formation of the relatively disorderly Si—O network rather than inducing crystallization into a regular array of atoms. Crystallization is potentially dangerous because the cracks between crystalline regions leave the incorporated radioactive materials exposed to moisture, which might dissolve them and carry them away from the storage area.

E Experiments are currently in progress to see how storage conditions of nuclear waste affect the rate of leaching by any water that happens to percolate through the solid. It is known that the leach rate is very sensitive to temperature and that it increases about a 100-fold between 25°C and 100°C. Storage in cool caverns might therefore seem to be an answer. However, the radioactive decay of the fission products is a source of heat and the temperature of the blocks can rise even if their surroundings are cool.

SUMMARY

22.1 Radioactivity, the emission of radiation by nuclei, gives rise to **α radiation** (consisting of α particles, which are helium-4 nuclei), **β radiation** (consisting of β particles, which are fast electrons emitted from nuclei), and **γ radiation** (very high frequency, short-wavelength, electromagnetic radiation).

22.2 A new **nuclide** is produced in the process of **radioactive decay**, the simplest forms of which are α decay and β decay. The **daughter nucleus**, the new nuclide produced in a nuclear decay reaction, is identified by balancing mass numbers and atomic numbers.

22.3 Several factors correlate with nuclear stability, including the even-odd relationship of nucleons, certain **magic numbers** of nucleons, and the neutron-to-proton ratio. Heavy nuclides decay in a stepwise **radioactive series**, a sequence of α and β emissions, until a stable nuclide (often an isotope of lead) is formed.

22.4 Nucleosynthesis is the synthesis of elements; it occurs when particles are captured by a nucleus. Nucleosynthesis occurs naturally in stars and was the source of the elements currently on Earth. It may also be brought about artificially by accelerating particles to speeds that are high enough to overcome electrostatic repulsions.

22.5 Of the three types of natural radiation, α radiation is least penetrating but most damaging; γ radiation is the most penetrating. The **intensity** of the radiation source depends on the number of nuclear disintegrations occurring per second and is expressed in **curies**.

22.6 The number of radioactive nuclei in a sample decreases exponentially in accordance with the **law of radioactive decay**, a first-order rate law. Half the nuclei in any sample decay in a time equal to the **radioactive half-life** of the nuclide. Half-lives are independent of temperature and the state of chemical combination of the atom. One application of radioactive decay is **radiocarbon dating**, which is based on the measurement of the proportion of carbon-14 in a sample. Many radioactive nuclides are used in medicine, in industry, and in research on the mechanisms of chemical reactions.

22.7 The energy change in a nuclear reaction results from the rearrangement of the nucleons when the product nuclei form. In spontaneous nuclear reactions, the nucleons have a lower energy in the product than in the initial nuclei. The energy changes can be discussed in terms of **nuclear binding energies** and are calculated from the differences in mass of the product and reactant particles and by using Einstein's formula, $E = mc^2$.

22.8 and 22.9 Nuclear **fission** is the splitting of a nucleus into two or more nuclei of similar size and may be either **spontaneous** or **induced** (by neutrons). A **fissionable** nucleus is one that can undergo induced fission, and a **fissile** nucleus is one for which fission can be induced by slow neutrons. **Sustained fission** is a **nuclear**

chain reaction in which neutrons are the chain carriers; it is achieved if each fission event results in the production of several neutrons and the mass of the fissionable sample is **supercritical** in the sense of exceeding a certain value. **Nuclear fusion**, the formation of a larger nucleus out of smaller ones, also releases energy.

22.10 The chemical problems connected with nuclear power include the production of the fuel, especially the **enrichment** of the natural uranium to increase the proportion of uranium-235 in the fuel. Enrichment is carried out by diffusion or with a centrifuge. Fuel processing includes the recovery of uranium and plutonium from used fuel by a version of the **purex process**. The disposal of nuclear waste is problematic because of its long-lived, intense radioactivity; one possible solution is to incorporate it in glass.

CLASSIFIED EXERCISES

Nuclear Structure and Radiation

22.1 Determine the number of protons, neutrons, and nucleons in (a) ^{2}H; (b) ^{263}Unh; (c) ^{60}Co; (d) ^{258}Md.

22.2 Write the nuclear symbol and determine the number of protons, neutrons, and nucleons in (a) chlorine-37; (b) unnilpentium-262; (c) gold-197; (d) californium-249.

22.3 Calculate the wavelength and the energy per mole of photons of γ radiation of frequency (a) 9.4×10^{19} Hz; (b) 5.7×10^{21} Hz.

22.4 Calculate the frequency and wavelength of the γ radiation emitted as a result of a rearrangement of nucleons in a daughter nucleus through an energy (a) 9.5×10^{-14} J; (b) 3.9×10^{-13} J.

22.5 A common energy unit is the electronvolt (eV), the energy an electron gains by passing through a potential difference of 1 V: 1 eV = 1.602×10^{-19} J. In nuclear processes, energies of decay products and transmutation reactions are often expressed in millions of electron volts (MeV). In the nucleon rearrangement of the following daughter nuclei, the energy changes by the amount shown and a γ ray is emitted. Determine the frequency and wavelength of the γ ray in the following cases: (a) cobalt-60, 1.33 MeV; (b) iron-59, 1.10 MeV.

22.6 Determine the frequency and wavelength of the γ ray emitted in the decay of the following nuclides, where the energy change of the nucleus is given in MeV (see Exercise 22.5): (a) carbon-15, 5.30 MeV; (b) bromine-87, 5.4 MeV.

Radioactive Decay

22.7 Identify the daughter nucleus in each of the following decays and write the balanced nuclear equation: (a) β decay of tritium; (b) β^+ decay of yttrium-83; (c) α decay of protactinium-225.

22.8 Write the balanced nuclear equation for the following radioactive decays: (a) β decay of actinium-228, (b) α decay of radon-212; (c) electron capture by protactinium-230.

22.9 Identify the daughter nucleus in each decay and write a balanced nuclear equation: (a) β^+ decay of boron-8; (b) α decay of gold-185; (c) electron capture by beryllium-7.

22.10 Write the balanced nuclear equation for the following radioactive decays: (a) β decay of uranium-233; (b) proton emission of cobalt-56; (c) β^+ decay of holmium-158.

22.11 Determine the particle emitted and write the balanced nuclear equation for the following nuclear transitions: (a) sodium-24 to magnesium-24; (b) ^{128}Sn to ^{128}Sb; (c) ^{228}Th to ^{224}Ra.

22.12 Determine the particle emitted and write the balanced nuclear equation for the following nuclear transitions: (a) carbon-14 to nitrogen-14; (b) gold-188 to platinum-188; (c) uranium-229 to thorium-225.

Patterns of Nuclear Stability

22.13 The following nuclides lie outside the band of stability. Predict whether they are most likely to undergo β decay, β^+ decay, or α decay and identify the daughter nucleus: (a) copper-68; (b) berkelium-243; (c) unnilpentium-260.

22.14 Select from β decay, β^+ decay, and α decay the type of decay each nuclide is most likely to undergo and identify the daughter nucleus: (a) copper-60; (b) xenon-140; (c) americium-246.

Nucleosynthesis

22.15 Complete the following nuclear equations and then write the equation in the form E(in,out)E′:
(a) $^{14}N + ? \rightarrow {}^{17}O + p$
(b) $? + n \rightarrow {}^{249}Bk + \beta$
(c) $^{243}Am + n \rightarrow {}^{244}Cm + ? + \gamma$

22.16 Complete the following nuclear equations and then write the equation in the form E(in,out)E′:
(a) $? + p \rightarrow {}^{21}Na + \gamma$
(b) $^{1}H + p \rightarrow {}^{2}H + ?$
(c) $^{15}N + p \rightarrow {}^{12}C + ?$

22.17 Complete the notation for the following nuclear reactions and then write out the nuclear equations:
(a) $^{20}Ne(\alpha,?)^{16}O$
(b) $^{20}Ne(^{20}Ne,^{16}O)?$
(c) $^{44}Ca(?,\gamma)^{48}Ti$

22.18 Complete the notation for the following nuclear reactions and then write out the nuclear equations:
(a) $^{44}Ti(e,\beta^+)?$
(b) $^{241}Am(?,4n)^{248}Fm$
(c) $?(n,\beta)^{244}Cm$

22.19 Write nuclear equations that represent the following processes: (a) oxygen-17 produced by α particle bombardment of nitrogen-14; (b) americium-240 produced by neutron bombardment of plutonium-239.

22.20 Write the nuclear equations for the following transformations: (a) ^{257}Unq produced by the bombardment of californium-245 with carbon-12 nuclei; (b) the first synthesis of ^{266}Une by the bombardment of bismuth-209 with iron-58 nuclei. Given that the first decay of unnilenium is by α emission, what is the daughter nucleus?

22.21 Each of the following reactions represents a fission reaction. Complete and balance the equations:
(a) $^{244}Am \rightarrow {}^{134}I + {}^{107}Mo + 3?$
(b) $^{235}U + n \rightarrow ? + {}^{138}Te + 2n$

22.22 Complete each of the following fission reactions:
(a) $^{239}Pu + n \rightarrow {}^{98}Mo + {}^{138}Te + ?$
(b) $^{239}Pu + n \rightarrow ? + {}^{133}In + 3n$

Measuring Radioactivity

22.23 The activity of a certain radioactive source is 3.7×10^6 disintegrations per second. Express this activity in curies.

22.24 The activity of a sample containing carbon-14 is 12.9 disintegrations per second. Express this activity in microcuries.

22.25 Determine the number of disintegrations per second for radioactive sources of the following activities: (a) 1.0 Ci; (b) 82 mCi; (c) 1.0 μCi.

22.26 A certain Geiger counter is known to respond to only one of every 1000 particles emitted from a sample. Calculate the activity of each radioactive source in curies, given the following data: (a) 370 clicks in 10 s; (b) 1.4×10^5 clicks in 1.0 h; (c) 266 clicks in 1.0 min.

Rate of Nuclear Disintegration

22.27 Determine the decay constant for (a) tritium, $t_{1/2} = 12.3$ y; (b) lithium-8, $t_{1/2} = 0.84$ s; (c) nitrogen-13, $t_{1/2} = 10.0$ min.

22.28 Determine the half-life of (a) potassium-40, $k = 5.3 \times 10^{-10}\ y^{-1}$; (b) cobalt-60, $k = 0.132\ y^{-1}$; (c) nobelium-255, $k = 3.85 \times 10^{-3}\ s^{-1}$.

22.29 Calculate the time needed for the activity of each source to change as indicated: (a) A 1.0-Ci radium-226 source to decay to 0.10 Ci. (b) A 10-Ci cobalt-60 source to decay to 8 Ci.

22.30 Calculate the time needed for the activity of each source to change as indicated: (a) A 0.010-mCi strontium-90 source to decay to 0.0010 mCi. (b) A 1.0-Ci iodine-131 source to decay to 1.0 mCi.

22.31 Estimate the activity of a 4.4-Ci cobalt-60 source, $t_{1/2} = 5.26$ years, after 50 years have passed.

22.32 The activity of a strontium-90 source is 3.0×10^4 disintegrations per second. What is its activity after 50 years have passed?

22.33 (a) What percentage of a carbon-14 sample remains after 1000 y? (b) Determine the percentage of a tritium sample that remains after 20.0 years.

22.34 (a) What percentage of a strontium-90 sample remains after 10.0 y? (b) Determine the percentage of an iodine-131 sample that remains after 5.0 days.

22.35 Potassium-40, which is presumed to have existed at the formation of the Earth, is used for dating minerals. If one-half of the original potassium-40 exists in a rock, how old is the rock?

22.36 A piece of wood, found in an archaeological dig, has a carbon-14 activity that is 90% of the current carbon-14 activity. How old is the piece of wood?

22.37 Uranium-238 decays through a series of α and β emissions to lead-206 with an overall half-life of 4.5×10^9 years. If a uranium-bearing ore is found to have a $^{238}U/^{206}Pb$ ratio of (a) 1.00 or (b) 1.25, how old is the ore sample?

22.38 A current 1.00-g sample of carbon shows 920 disintegrations per hour. (a) If a 1.00-g sample of charcoal from an archaeological dig shows 5500 disintegrations in 24 h, what is the age of the charcoal sample? (b) If a 2.50-g sample of carbon undergoes 4000 disintegrations in 4.00 h, what is the age of the sample?

22.39 Use the law of radioactive decay to determine the activity of (a) a 1.0-mg sample of radium-226, $t_{1/2} = 1.60 \times 10^3$ y; (b) a 2.0-μg sample of strontium-90, $t_{1/2} = 28.1$ y.

22.40 Use the law of radioactive decay to determine the activity of (a) a 1.0-g sample of $^{235}UO_2$, $t_{1/2} = 7.1 \times 10^8$ y); (b) a 1.0-g sample of cobalt containing 1.0% ^{60}Co ($t_{1/2} = 5.26$ y).

Nuclear Energy

22.41 Calculate the mass loss or gain for each of the following processes: (a) a 50.0-g block of iron (specific heat capacity, 0.45 J/(g·°C)) cools from 600° to 25°C; (b) a 100-g sample of ethanol vaporizes at its normal boiling point ($\Delta H^\circ_{vap} = +43.5$ kJ/mol).

22.42 Calculate the energy in joules that is equivalent to (a) 1.0 u of matter; (b) 1 proton, mass = 1.673×10^{-24} g.

22.43 The sun emits radiant energy at the rate of 3.9×10^{26} J/s. What is the rate of mass loss (in kilograms per second) of the sun?

22.44 For the fusion reaction $6D \rightarrow 2\,^{4}He + 2p + 2n$, 3×10^{8} kJ of energy is released. What is the mass loss (in grams) for the reaction?

22.45 Calculate the binding energy per nucleon (J/nucleon) for (a) ^{2}H, 2.0141 u; (b) ^{56}Fe, 55.9349 u. Which nuclide is the most stable?

22.46 Calculate the binding energy per nucleon (J/nucleon) for (a) ^{151}Eu, 150.9196 u; (b) ^{10}B, 10.0129 u. Which nuclide is the most stable?

22.47 Calculate the energy released per gram of starting material in each of the following fusion reactions:
(a) $D + D \rightarrow {}^{3}He + n$ (D, 2.0141 u; ^{3}He, 3.0160 u)
(b) $^{7}Li + p \rightarrow 2\,^{4}He$ (^{7}Li, 7.0160 u)
(c) $D + T \rightarrow {}^{4}He + p$ (T, 3.0160 u)

22.48 Calculate the energy released per gram of starting material in each of the following nuclear reactions:
(a) $^{7}Li(p,n)^{7}Be$ (^{7}Li, 7.0160 u; ^{7}Be, 7.0169 u)
(b) $^{59}Co(D,p)^{60}Co$ (^{59}Co, 58.9332 u; ^{60}Co, 59.9529 u)
(c) $^{40}K + e \rightarrow {}^{40}Ar$ (^{40}K, 39.9640 u; ^{40}Ar, 39.9624 u)

UNCLASSIFIED EXERCISES

22.49 Write balanced nuclear equations for the radioactive decay of the following nuclides: (a) ^{174}Hf, α emission; (b) ^{98}Tc, β emission; (c) ^{41}Ca, electron capture.

22.50 Complete the notation for the following nuclear reactions: (a) $^{11}B(?,2n)^{13}N$; (b) $?(D,n)^{36}Ar$; (c) $^{96}Mo(D,?)^{97}Tc$.

22.51 Technetium-99, for which $t_{1/2} = 6.02$ h, is used in the imaging of internal body organs, including the brain and heart. It is produced by a sequence of reactions in which molybdenum-98 is bombarded with neutrons to form molybdenum-99, which undergoes β decay to ^{99}Tc. Write the balanced nuclear equations for this sequence.

22.52 What mass of a 15.0-mg sample of ^{47}V, $t_{1/2} = 33.0$ min, will remain after 45.0 min?

22.53 The activity of an iodine-131 source, $t_{1/2} = 8.05$ d, that is used to monitor the functioning of the thyroid gland, is 500 disintegrations per second. How long will it be before the activity is 10 disintegrations per second?

22.54 The age of a bottle of wine was determined by monitoring the tritium level in the wine. The activity of the tritium is determined to be one-twelfth that of a water sample from the same region from which the wine was bottled. How old is the wine?

22.55 A sample of fermium-244 having an activity of 0.10 μCi was produced in a nuclear reactor. What will be the activity of the ^{244}Fm after 1.0 s?

22.56 (a) A sample of phosphorus-32 has an initial activity of 58 counts per second. After 12.3 days, the activity was 32 counts per second. What is the half-life of phosphorus-32? (b) If phosphorus-32 is used in an experiment to monitor the consumption of phosphorus by plants, what fraction of the nuclide will remain after 30 days?

22.57 A cobalt-60 source purchased for the radiation therapy of cancer patients has an activity of 1.20 Ci. What will be the activity of the source after 5.0 years?

22.58 The activity of a sample containing ^{35}S, $t_{1/2} = 88$ d, that was being used to study the reactions by which sulfur was utilized by bacteria is 10.0 Ci. What mass of sulfur-35 is present? (^{35}S, 34.969031 u)

22.59 Sodium-24 is used for monitoring blood circulation through the body. (a) If a 2.0-μg sample of sodium-24 has an activity of 17.3 Ci, what is its decay constant and its half-life? (b) What mass of a 2.0-μg sample remains after 2.0 days?

22.60 A 250-mg sample of carbon from a piece of cloth undergoes 1500 disintegrations in 10.0 h. If a current 1.0-g sample of carbon shows 920 disintegrations per hour, how old is the piece of cloth?

22.61 The mass of an electron is 9.109×10^{-28} g; a positron has the same mass but an opposite charge. When an electron encounters a positron, annihilation occurs and energy is produced. (a) How much energy (in kJ) is produced in the encounter? (b) Express the energy produced in MeV (1 MeV = 1.60×10^{-13} J).

22.62 Sodium-24 decays to magnesium-24. Determine (a) the change in the binding energy per nucleon and (b) the change in energy evolved that accompanies the decay. (^{24}Na, 23.99061 u; ^{24}Mg, 23.985042 u)

22.63 How much energy is emitted in the α decay of uranium-234 (^{234}U, 234.0409 u; ^{230}Th, 230.0331 u)?

CASE 22 ▸ RADON, A BACKGROUND RADIOACTIVE ISOTOPE

Concern over the presence in indoor environments of radon, a radioactive noble gas, has grown since the discovery in 1984 that a home in eastern Pennsylvania had an activity of 2700 pCi per liter of air. The level of radon activity recommended by the EPA is only 4 pCi per liter; therefore the 2700 pCi/L activity level is astonishingly high, equivalent to about 30,000 chest x-rays per year! The 4 pCi/L level is a recommended "action level" for a home and is based on data that correlate lung cancer with radon exposure. This level of radiation over a 70-year lifetime results in a risk prediction of a 1 to 5% chance of death due to lung cancer. In retrospect, it is now believed that the many cases of "miner's disease" suffered by underground miners were cancers due to radon exposure.

Natural radon comes from the radioactive decay of ^{238}U (Fig. 22.10). The nuclide ^{222}Rn is the daughter nucleus of ^{226}Ra, which has a half-life of 1600 years. Natural sources of radon gas generally exist near deposits of granite, uranium, shale, and phosphates. Although the gas usually disperses into the atmosphere as it leaks from the soil, groundwater, oceans, or phosphate residues, radon can accumulate in closed environments and reach detectable levels. The radon gas in our homes may be swept into and out of our lungs with no long-lasting effects. However, the immediate decay products of ^{222}Rn are solids that have short half-lives. If radon-222 decay happens to occur in the lungs, these daughter nuclei can become embedded in the lungs and irradiate the sensitive lung tissue. The daughter nuclei are

$$^{222}\text{Rn} \longrightarrow {}^{218}\text{Po} + \alpha \qquad t_{1/2} = 3.8 \text{ d}$$

$$^{218}\text{Po} \longrightarrow {}^{214}\text{Pb} + \alpha \qquad t_{1/2} = 3.11 \text{ min}$$

$$^{214}\text{Pb} \longrightarrow {}^{214}\text{Bi} + \beta \qquad t_{1/2} = 27 \text{ min}$$

$$^{214}\text{Bi} \longrightarrow {}^{214}\text{Po} + \beta \qquad t_{1/2} = 19.7 \text{ min}$$

$$^{214}\text{Po} \longrightarrow {}^{210}\text{Pb} + \alpha \qquad t_{1/2} = 200\ \mu\text{s}$$

It is apparent that once the radon enters the lungs and decays to ^{218}Po, a solid, the lung tissue could be exposed to two α particles and two β particles within 1 h. The α particles are the most dangerous in terms of tissue damage.

A radon detector.

QUESTIONS

1. Why are the concentrations of radon generally found to be greater in the basement than in the upper levels of the home?

2. (a) How many radon-222 nuclei decay per minute to produce an activity of 4 pCi? (b) A bathroom in the basement of a home measures $2 \times 3 \times 2.5$ m. If the activity of radon-222 in the room is 4 pCi/L, how many nuclei decay during a shower lasting 5.0 min?

3. (a) For how long does a sample of radon-222 gas exist? Assume 99% removal. (b) Comment on radon's lifetime and distance of travel if formed deep in the Earth's crust.

4. One hundred atoms of radon-222 enter a closed environment measuring 2000 m^3 (for example, the basement of a home). (a) What is the initial activity of the radon in picocuries per liter? (b) How many atoms of ^{222}Rn will remain after one day (24 h)? (c) What is the activity of ^{222}Rn after one day?

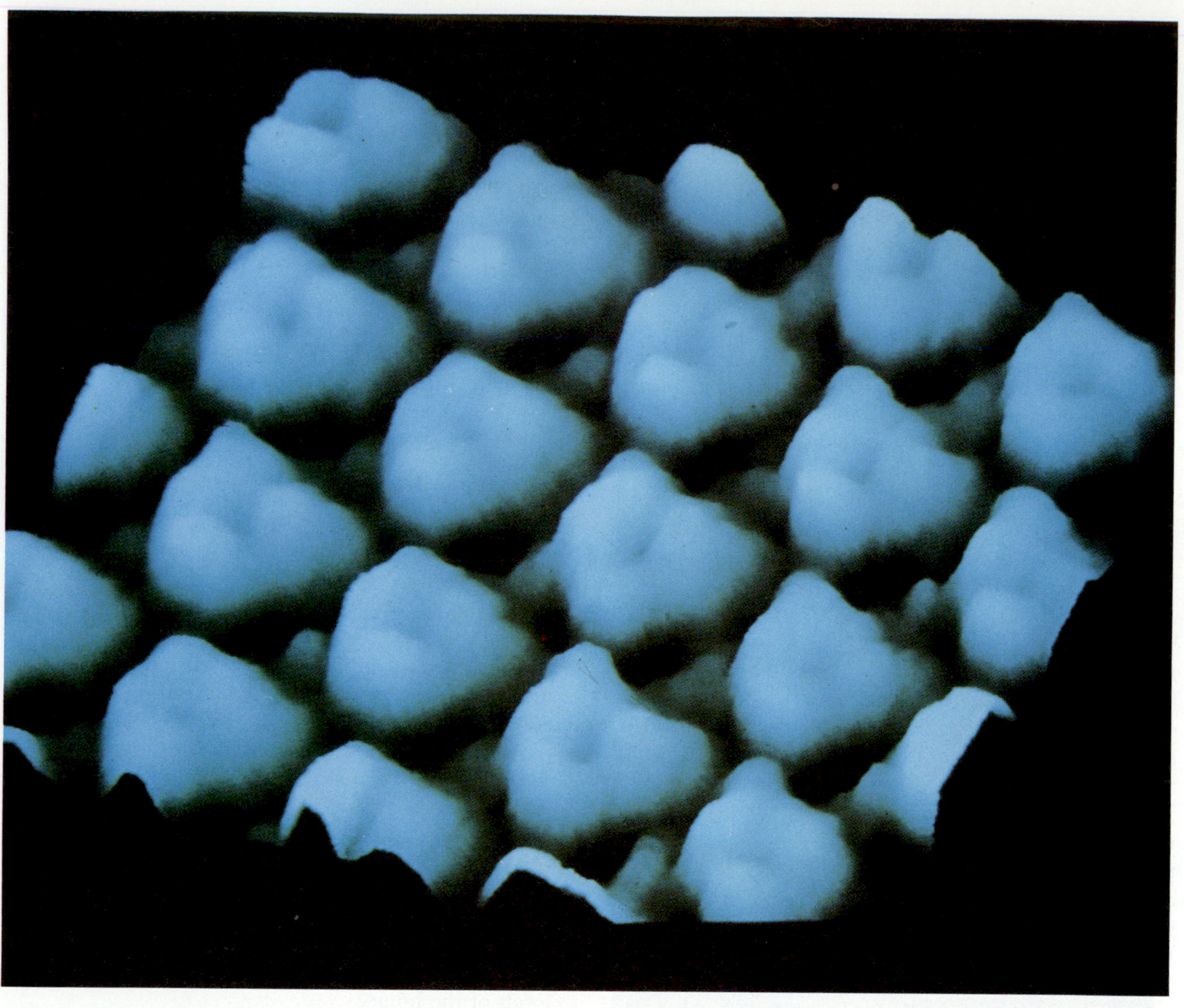

This chapter introduces some of the principles of organic chemistry by describing the structure, synthesis, and properties of the hydrocarbons—the binary compounds of carbon and hydrogen. These compounds are the basic structures from which the formulas of all organic compounds are derived. One very important hydrocarbon—benzene—is the parent of a wide class of organic compounds, and the recently developed technique of scanning tunneling microscopy has provided the striking images of its hexagonal ring molecules shown in the illustration.

23 THE HYDROCARBONS

Carbon forms a vast number of compounds: around 6 million are already known and many new ones are being reported every day. The branch of chemistry that deals with carbon's compounds is called **organic chemistry**, and the compounds themselves (with a few exceptions, such as the carbonates) are called **organic compounds**. The name reflects the erroneous view once held that the compounds could be formed only by living organisms. Modern organic chemistry deals with naturally occurring and synthetic compounds, including plastics, pharmaceuticals, petrochemicals, fuels, and foods. Because it provides a link between the properties of atoms and the behavior of living organisms, through organic chemistry we come to biochemistry and hence to the chemical foundations of life itself.

Because the compounds of carbon are so numerous, it is convenient to organize them into families that have structural similarities. One of the simplest families consists of the **hydrocarbons**, the binary compounds of carbon and hydrogen. Their molecules consist of chains, rings, and networks of carbon atoms with the remaining bonding positions occupied by hydrogen atoms. In Chapter 24 we shall meet compounds that we can regard as derivatives of hydrocarbons, in which one or more of the hydrogen atoms has been replaced by atoms (or groups of atoms) of other elements.

It became clear in the early days of organic chemistry that the hydrocarbon family could itself be subdivided into two major groups, the aliphatic hydrocarbons and the aromatic hydrocarbons. Aromatic hydrocarbons have the benzene ring as a part of their molecular structure and will be discussed in Section 23.8. All other hydrocarbons are aliphatic and are subdivided into two families. It was found (for example, by using the combustion analysis technique described in Section 4.3) that some hydrocarbons have a lower ratio of H atoms to C atoms than others. For example, ethane, C_2H_6, has three H atoms for each C atom, but ethylene, C_2H_4, has only two. This finding led to a distinction between compounds like ethane that are saturated and compounds like

E The first demonstration that an organic compound could be produced without the mediation of an organism was Friedrich Wöhler's demonstration in 1828 that the action of heat on ammonium cyanate, NH_4CNO, resulted in the formation of urea, $CO(NH_2)_2$. He wrote excitedly to the influential Swedish chemist Berzelius: "I must tell you I can prepare urea without requiring a kidney of an animal, either man or dog."

D Producing methane gas. CD2, 19.

ethylene that are unsaturated. It is now known that the "unused" carbon valencies in unsaturated compounds are in fact used in multiple carbon-carbon bonds, so the two families are now defined as follows:

> A **saturated** hydrocarbon is one with no carbon-carbon multiple bonds. An **unsaturated** hydrocarbon is one with at least one carbon-carbon multiple bond.

The aliphatic hydrocarbons that are saturated are called alkanes; those that are unsaturated are called alkenes and alkynes, depending on the type of carbon-carbon bond.

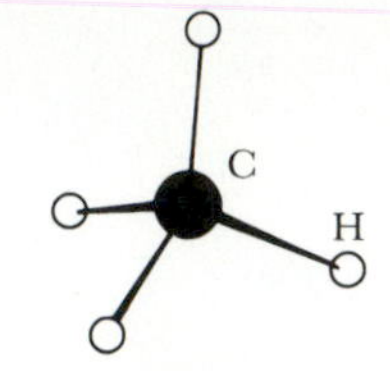

1 Methane

C
H

2 Ethane

E The HCH bond angle in ethane is 109.3°, only very slightly less than the tetrahedral angle (109.5°).

N The name *methylene* is used to form the common names of many compounds: two examples are methylene chloride, for dichoromethane, CH_2Cl_2, and hexamethylenediamine for hexane-1,6-diamine, $NH_2(CH_2)_6NH_2$.

THE ALKANES

The two simplest saturated hydrocarbons are methane, CH_4 (**1**), and ethane, C_2H_6, (**2**). The methane molecule is tetrahedral, as predicted by VSEPR theory. We can picture each of its C—H bonds as being formed by the overlap of an sp^3 hybrid orbital on the C atom and a 1*s* orbital on the H atom. We can think of an ethane molecule as formed from a CH_4 molecule by inserting a **methylene group**, $—CH_2—$, between methane's C atom and one of its H atoms. We can emphasize this structure by writing its formula as $CH_3—CH_2—H$, or more briefly as CH_3CH_3. Both C atoms in ethane are sp^3 hybridized, and the C—C bond is formed by the overlap of two sp^3 orbitals on neighboring atoms (Fig. 23.1). The sp^3 hybridization reflects the shape of the molecules, which is tetrahedral around each C atom, with bond angles close to 109°. We shall normally write only two-dimensional Lewis structures:

```
    H           H  H
    |           |  |
 H—C—H      H—C—C—H
    |           |  |
    H           H  H
```

but we must not forget that the actual molecules are three dimensional.

The way we picture the formula for ethane as being derived from the formula for methane suggests the concept of a homologous series of compounds:

> A **homologous series** is a family of compounds with molecular formulas obtained by repeatedly inserting a given group (most commonly $—CH_2—$) into a parent structure.

The homologous series that starts with the formula for methane is the family of hydrocarbons called the alkanes:

> The **alkanes** are a homologous series of hydrocarbons derived from CH_4 by inserting $—CH_2—$ groups, and having molecular formulas C_nH_{2n+2}.

They include methane itself and homologues such as ethane (C_2H_6), propane (C_3H_8), and butane (C_4H_{10}). The first few members of the alkanes are found in various proportions in natural gas and as compo-

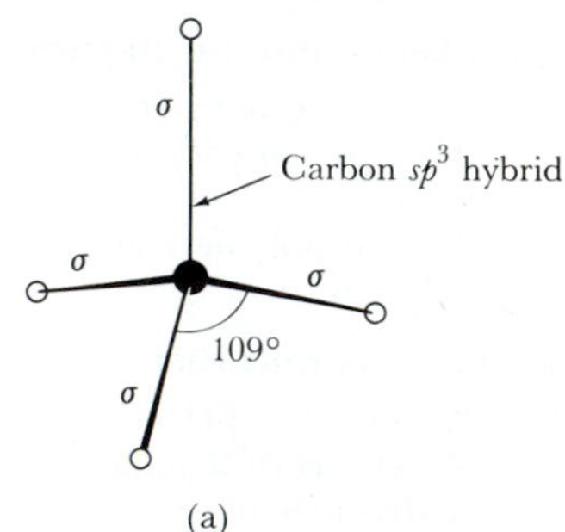

(a)

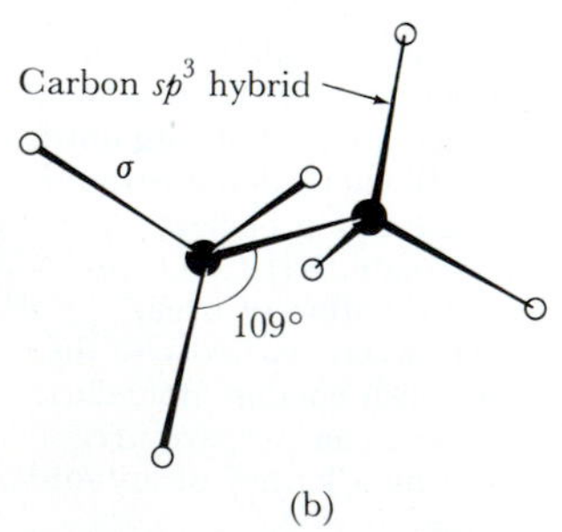

(b)

FIGURE 23.1 The structure of the bonds in (a) methane and (b) ethane. In each case the carbon atom is sp^3 hybridized and forms σ bonds with its neighbors. All bond angles are approximately 109°.

TABLE 23.1 The properties of some alkanes

Formula*	Name	Melting point, °C	Boiling point, °C	Density, g/mL†
CH_4	methane	−182	−162	—
C_2H_6	ethane	−183	−89	—
C_3H_8	propane	−190	−42	—
C_4H_{10}	butane	−138	−1	—
C_5H_{12}	pentane	−130	36	0.626
C_6H_{14}	hexane	−95	69	0.659

*The general formula is C_nH_{2n+2}.
†At 25°C.

nents of petroleum; their names and some of their physical properties are given in Table 23.1. The alkanes were once called the paraffins, from the Latin words for "little affinity." As we shall see, this is a good summary of their chemical properties, for they are not very reactive.

23.1 ISOMERISM

The fourth alkane, C_4H_{10}, introduces a very important feature of organic chemistry—the possibility of isomerism. As we saw in Section 21.6 in connection with *d*-metal complexes, **isomers** are compounds built from the same atoms in different arrangements. In alkanes, the same number of C and H atoms may be linked in different arrangements. The chart first shown as Fig. 21.22, in which the various kinds of isomerism were classified, is shown again in Fig. 23.2, with modifications that make it more relevant to organic compounds.

The alkanes show **structural isomerism**—isomerism arising from atoms being linked to different neighbors. It is possible, for example, to insert a $-CH_2-$ group into the C_3H_8 molecule in two different ways:

$$
\begin{array}{ccccccccc}
 & H & & H & & H & & H & \\
 & | & & | & & | & & | & \\
H- & C & - & C & - & C & - & C & -H \\
 & | & & | & & | & & | & \\
 & H & & H & & H & & H &
\end{array}
\quad \text{or} \quad
\begin{array}{ccccccc}
 & & & H & & & \\
 & & & | & & & \\
 & & H- & C & -H & & \\
 & H & & | & & H & \\
 & | & & | & & | & \\
H- & C & - & C & - & C & -H \\
 & | & & | & & | & \\
 & H & & H & & H &
\end{array}
$$

$CH_3CH_2CH_2CH_3$ Butane — $CH_3CH(CH_3)CH_3$ Methylpropane

(Although the $-CH_2-$ group can be inserted in other places, the resulting molecules can all be twisted into one or the other of these two isomers.) Both compounds are gases, but butane condenses at −1°C and methylpropane condenses at −12°C.

Instead of drawing the complete structural formula, it is often sufficient to give only the shortened form, such as $CH_3CH_2CH_2CH_3$ for butane and $CH_3CH(CH_3)CH_3$ for methylpropane, showing how the

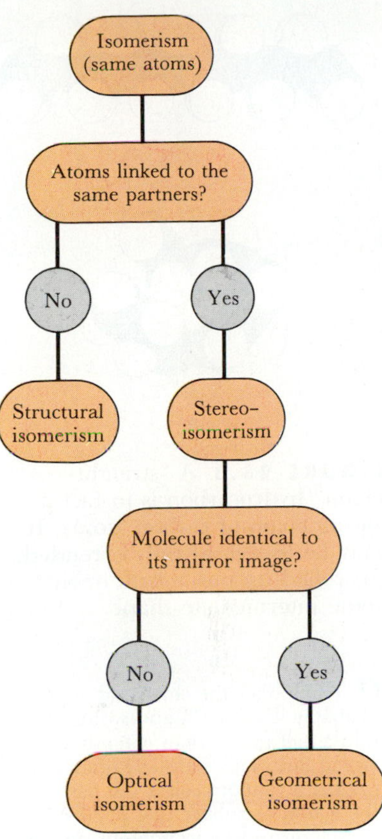

FIGURE 23.2 The classification of the isomers of organic compounds.

N Suggest to students that they purchase a set of molecular models, especially if their plans are to complete organic chemistry.

D Isomers are best introduced using molecular models.

N Methylpropane is commonly called *isobutane*.

N There are 75 isomers of decane. A mixture of isomers can often be analyzed and separated by chromatography.

D Why are there so many organic compounds? TDC, 22-1. (For structural isomerism, use more than five balls.)

N The difference in volatility between butane and methylpropane is in accord with the discussion in Section 10.1, where we saw that branched alkanes are more volatile than straight chain alkanes, partly on account of the stronger London forces that can exist in the latter.

atoms are grouped together in these two isomers. When a group of atoms is attached to a "main-chain" carbon atom, we usually place it in parentheses at the *right* of the groups to which it is attached, for example, the $—CH(CH_3)—$ in the formula $CH_3CH_2CH_2CH(CH_3)CH_2CH_3$. When writing a formula, we often look for a shorter unambiguous way to describe the compound. Thus, when several $—CH_2—$ groups are repeated, we may collect them together; hence we could write $CH_3(CH_2)_2CH_3$ for butane.

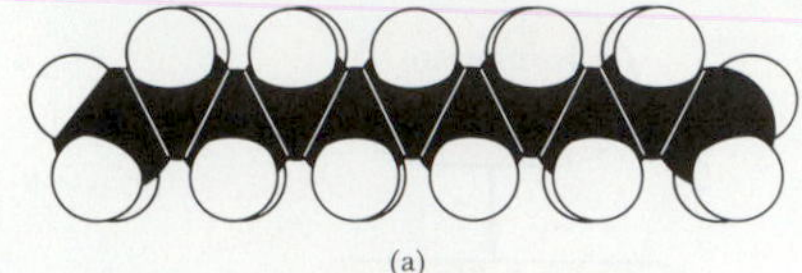

(a)

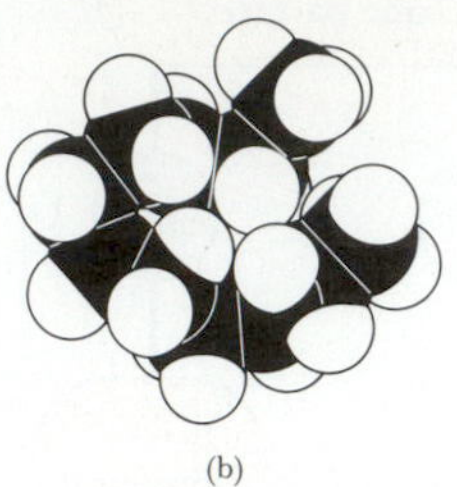

(b)

FIGURE 23.3 A "straight-chain" hydrocarbon is in fact a zigzag chain of carbon atoms. It may be found (a) fully extended, (b) rolled up into a ball, or in some intermediate shape.

E The longer the chain, the more probably it will be found wound up into a ball rather than extended in a zigzag line. This statistical tendency toward a random coil is the basis of the action of rubber (see p. 874): stretching rubber straightens out the random coils, and they spring back into their most favorable conformation when the tension is released.

E Organic nomenclature. *J. Chem. Ed.*, **1970**, *47*, 471.

23.2 ALKANE NOMENCLATURE

The IUPAC* rules for naming organic compounds systematically make use of the names given to the alkanes. However, the rules sometimes result in very cumbersome names for common compounds, so alternative names are also widely used.

Unbranched alkanes. The first step in naming an alkane is to distinguish unbranched from branched compounds. In an **unbranched alkane** (or straight-chain alkane), all the carbon atoms lie in a linear chain, —C—C—C—C—, as they do in butane. A **branched alkane** has one or more carbon **side chains**, like those in methylpropane. Although an unbranched alkane is often written with the carbon atoms in a straight line, they actually form a zigzag with a carbon-carbon bond angle of about 109°. Moreover, because neighboring $—CH_2—$ groups can rotate around the bonds that join them to each other, a straight-chain molecule may even be found rolled up into a ball (Fig. 23.3).

The systematic names of the first few unbranched alkanes are given in Table 23.2: they all use the suffix *-ane*. The names of alkanes with one through four carbon atoms (methane, ethane, propane, and butane) have historical origins; the remainder are derived from the Greek numbers. The common names of the unbranched alkanes are the same as the systematic names, but they have the prefix *n*-, standing for *normal* (and signifying an unbranched chain), as in *n*-octane (**3**). We shall usually employ the systematic name, which does not include the prefix *n*-.

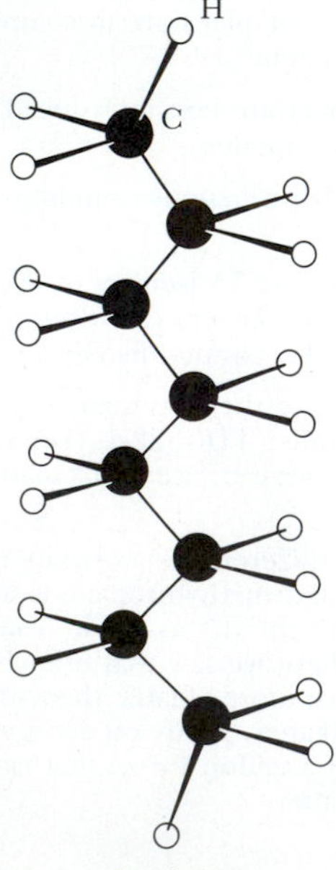

3 Octane

EXAMPLE 23.1 Naming unbranched alkanes

(a) Give the systematic name of $CH_3(CH_2)_4CH_3$, and (b) write the formula for decane.

STRATEGY We can identify the alkane by counting the number of C atoms and referring to Table 23.2. In the last part we should work backward, identifying the number of C atoms from the name.

SOLUTION (a) The number of carbon atoms is six; hence the compound is hexane. (b) We note that the name decane signifies an unbranched alkane with 10 carbon atoms. Its formula is therefore $CH_3(CH_2)_8CH_3$.

EXERCISE E23.1 Give the systematic names of (a) $CH_3(CH_2)_6CH_3$, and (b) write the formula for heptane.

[*Answer*: (a) octane; (b) $CH_3(CH_2)_5CH_3$]

*International Union of Pure and Applied Chemistry.

TABLE 23.2 Alkane nomenclature

Number of C atoms	Formula	Name of alkane	Name of alkyl group
1	$CH_3—$	methane	methyl
2	$CH_3—CH_2—$	ethane	ethyl
3	$CH_3—CH_2—CH_2—$, or $CH_3(CH_2)_2—$	propane	propyl
	CH_3 | $CH—$, or $(CH_3)_2CH—$ | CH_3	propane	isopropyl
4	$CH_3—CH_2—CH_2—CH_2—$, or $CH_3(CH_2)_3—$	butane	butyl
	CH_3 | $CH—CH_2—$, or $(CH_3)_2CHCH_2—$ | CH_3	isobutane (methylpropane)	isobutyl
	CH_3 | $H_3C—C—$, or $(CH_3)_3C—$ | CH_3	methylpropane	*tert*-butyl*†
5	$CH_3—CH_2—CH_2—CH_2—CH_2—$, or $CH_3(CH_2)_4—$	pentane	pentyl
6	$CH_3(CH_2)_5—$	hexane	hexyl
7	$CH_3(CH_2)_6—$	heptane	heptyl
8	$CH_3(CH_2)_7—$	octane	octyl
9	$CH_3(CH_2)_8—$	nonane	nonyl
10	$CH_3(CH_2)_9—$	decane	decyl
11	$CH_3(CH_2)_{10}—$	undecane‡	undecyl

*This name is not formed systematically but is widely used.
†*tert-* denotes "tertiary."
‡Names of alkanes with more carbon atoms are formed similarly, for example, pentadecane for $C_{15}H_{32}$ and hexadecane for $C_{16}H_{34}$.

E The rules are summarized in the "Blue Book," more formally *Nomenclature of organic chemistry*. J. Rigaudy and S. P. Klesney, Pergamon Press, Oxford (1979).

T OHT: Table 23.2.

N Organic nomenclature: making it a more exciting teaching and learning experience. *J. Chem. Ed.*, **1983**, *60*, 553.

Branched alkanes. For the purpose of naming the branched alkanes, we treat the side chains as **substituents**, which are atoms or groups that have replaced hydrogen atoms and are attached to an unbranched backbone like ribs on a spine. First, we identify the longest unbranched chain of carbon atoms in the molecule as the backbone and give it the name of the corresponding alkane. Thus,

$$\begin{array}{ccccccccc} & & CH_3 & & & & CH_3 & & \\ & & | & & & & | & & \\ \mathbf{CH_3} & \mathbf{—} & \mathbf{C} & \mathbf{—} & \mathbf{CH_2} & \mathbf{—} & \mathbf{CH} & \mathbf{—} & \mathbf{CH_3} \\ & & | & & & & & & \\ & & CH_3 & & & & & & \end{array}$$

is a substituted pentane because its longest unbranched chain (in bold type) has five C atoms. Because the latter's molecular formula is C_8H_{18}, it is an isomer of octane (its common name is isooctane; **4**). Nevertheless, according to the IUPAC rules, it is named as a derivative of pentane.

Next, we identify and name the substituents. We call $CH_3—$ a **methyl group** and $CH_3CH_2—$ an **ethyl group**. In general, groups derived

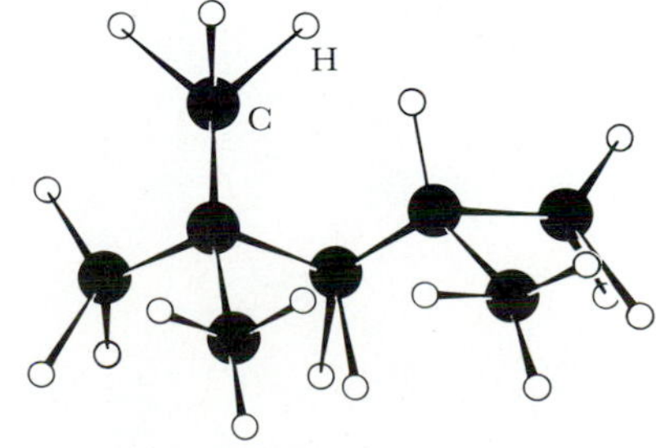

4 Isooctane

from alkanes are called **alkyl groups** and denoted by the letter R— (so R— may stand for CH_3— or C_2H_5—, and so on). Their names are obtained from the alkane names by changing the suffix *-ane* to *-yl* (see Table 23.2). We then combine the names of the substituents with the name of the backbone:

```
        CH3
        |
  CH3—C—CH3        methylpropane
        |
        H
```

When there are several substituents, we name them in alphabetical order:

```
             CH3  CH2CH3
             |    |
CH3—CH2—CH—CH—CH2—CH2—CH2—CH3        an ethylmethyloctane
```

When there are two or three substituents of the same kind, we use the prefix *di-* (for 2) or *tri-* (for 3):

```
        CH3        CH3
        |          |
  CH3—C—CH2—CH—CH3        a trimethylpentane
        |
        CH3
```

The substituents are alphabetized by their name, not any prefix they have, for example, ethyldimethyloctane. In Chapter 24 we shall meet many other (nonhydrocarbon) substituents. Some that will be mentioned again in this chapter are halogen atoms. Their presence turns an alkane into a **haloalkane** (or an alkyl halide), such as chloroethane, CH_3CH_2Cl. In haloalkane names, the halogen atom is denoted by the prefix *halo-*: *chloro-* and *bromo-* are two such prefixes.

Specifying locations. Some compounds differ only in the locations at which the substituents are attached to the backbone. To specify these locations, we number the C atoms of the backbone in order, starting at the end of the chain that results in the substituents having the lower numbered locations. Then the number of the C atom to which the substituent is attached is given as a prefix:

```
        CH3        CH3
        |          |
  CH3—C—CH2—CH—CH3        2,2,4-trimethylpentane
   1    2|  3      4    5
        CH3
```

Numbering the chain from the right, however, results in the name 2,4,4-trimethylpentane. This name is rejected because the prefixes are higher numbers.

▼ EXAMPLE 23.2 Naming branched alkanes

Name the ethylmethyloctane above and write the formula of 5-ethyl-2,3-dimethyloctane.

STRATEGY We need to identify and name the longest chains and then number their C atoms so that the lowest numbers for the locations of the substituents are obtained. We can then build the names by specifying the substituents and their locations. To deduce a formula from a name, we follow the procedure in reverse: we determine the number of C atoms in the backbone, write them as an unbranched chain, and then attach the substituents to the specified locations. All other bonding positions are occupied by hydrogen atoms.

SOLUTION The ethylmethyloctane has an ethyl group on carbon atom 4 and a methyl group on carbon atom 3; so it is 4-ethyl-3-methyloctane. The compound 5-ethyl-2,3-dimethyloctane is a substituted octane, so its backbone is

$$—C—C—C—C—C—C—C—C—$$

The prefix 5-ethyl indicates an ethyl group ($CH_3CH_2—$) at carbon atom 5. The prefix 2,3-dimethyl indicates that there are two methyl ($CH_3—$) groups, one at carbon atom 2 and the other at carbon atom 3:

```
           CH3      CH2CH3
            |         |
      H     |  H  H   |  H  H  H
      |     |  |  |   |  |  |  |
  H — C  —  C — C — C — C — C — C — C — H
      |     |  |     |  |  |  |  |
      H     H  |     H  H  H  H  H
               |
              CH3
```

with short (!) form $(CH_3)_2CHCH(CH_3)CH_2CH(CH_2CH_3)(CH_2)_2CH_3$. This alkane is an isomer of $C_{12}H_{26}$.

EXERCISE E23.2 Name the compounds (a) $CH_3CH_2C(CH_3)_2CH_2CH_3$ and (b) $(CH_3)_2CHCH_2CH(CH_2CH_3)_2$, and (c) write the formula of 3,3,5-triethylheptane.

[*Answer*: (a) 3,3-dimethylpentane; (b) 4-ethyl-2-methylhexane; (c) $(CH_3CH_2)_3CCH_2CH(CH_2CH_3)_2$]

▲

Cycloalkanes. Cyclohexane, C_6H_{12} (**5**), is an example of a **cycloalkane**, a saturated aliphatic hydrocarbon in which the carbon atoms form a ring. All the C atoms in cyclohexane are approximately sp^3 hybridized, and the ring they form is nonplanar. In equations, the molecule is often represented by the simple hexagon

5 Chair cyclohexane

cyclohexane

? The two forms of cyclohexane. *J. Chem. Ed.*, **1970**, *47*, 488.

Cycloalkanes are named like the alkanes but with the prefix *cyclo-*, for example,

```
      H2               H2                 H2
      C                C                  C
     / \             /   \              /   \
 H2C ——— CH2     H2C       CH2      H2C       CH2
                   \       /         |         |
                   H2C — CH2        H2C       CH2
                                        \   /
                                          C
                                          H2
```

Cyclopropane   Cyclopentane   Cyclohexane

N The internal bond angles of cyclopropane are 60°; those of cyclopentane are 108° and the molecule is nearly planar.

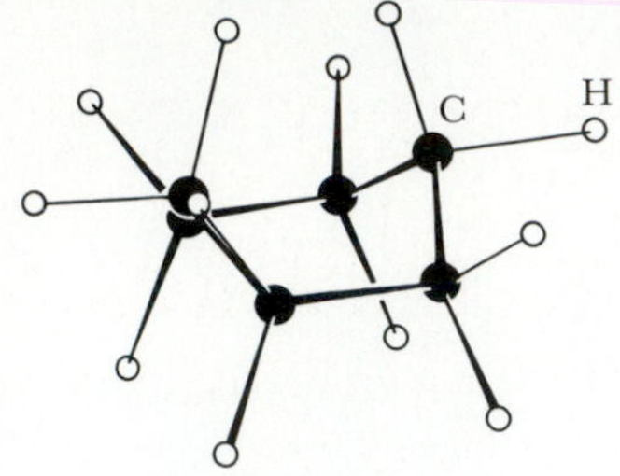

6 Boat cyclohexane

The chemical properties of most of the cycloalkanes are very similar to those of the alkanes themselves (cyclopropane is an exception), and we do not need to treat them separately. However, cyclohexane is an interesting molecule partly because it can adopt several **conformations**, that is, shapes that can be interchanged by rotation about bonds, without bond breakage and reformation. The two principal conformations are called the chair (**5**) and the boat (**6**) forms. A sample of cyclohexane consists of molecules that are continuously changing between the two conformations. The chair form has lower energy than the boat form (by about 25 kJ/mol), and at room temperature a molecule spends about 90% of its time in its chair form.

E The exceptional character of cyclopropane arises from the strain in the carbon-carbon bonds, for they lie at 60° to each other. Cylopropane bursts open and isomerizes to propene, $CH_3—CH=CH_2$, when heated.

U Exercises 23.60 and 23.64.

23.3 THE PROPERTIES OF ALKANES

The alkanes show a smooth variation of physical properties. They become less volatile with increasing molar mass (Fig. 23.4), as we would expect for compounds in which London forces are dominant. The lightest members, methane through butane, are gases at room temperature. Pentane is a volatile liquid and hexane through undecane are moderately volatile liquids that are present in gasoline (Table 23.3). Kerosene, a fuel used in jet engines and diesel engines, is a mixture of hydrocarbons. It contains a number of alkanes in the range C_{10} to C_{16}. Lubricating oils are mixtures in the range C_{17} to C_{22}, most of the molecules being branched. The heavier members of the series include the paraffin waxes and asphalt. All alkanes are insoluble in water.

N Some overlap in the number of carbon atoms occurs in the petroleum fractions listed in Table 23.3: boiling points are affected by the shapes and structures of the various isomers.

Unbranched alkanes tend to have higher melting points, boiling points, and enthalpies of vaporization than their branched isomers. This difference can be traced to the London forces that exist between branched molecules: the atoms in neighboring molecules cannot lie as close together as they do in their unbranched isomers (Fig. 23.5).

N That is, the reaction pathways of alkanes have high activation energies: alkanes are nonlabile.

Chemical properties. The alkanes are unaffected by concentrated sulfuric acid, by boiling nitric acid, by potassium permanganate, and by boiling aqueous sodium hydroxide. One reason for their low chemical reactivity is thermodynamic, for the C—C and C—H bonds are strong (Table 23.4); there is little energy advantage in replacing them with other bonds, except most notably by C=O, C—OH, and C—F bonds. Another reason is related to the fact that in an alkane the carbon atom has a complete octet of electrons, is small, and has no *d* orbitals available. Hence it cannot readily accommodate the electron pairs of an attacking Lewis base (such as OH^-).

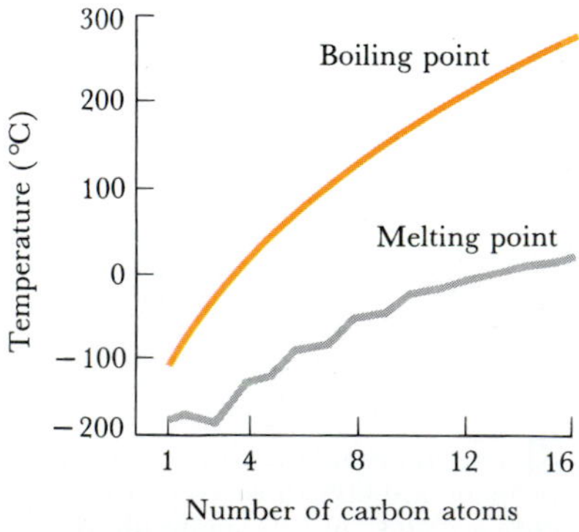

FIGURE 23.4 The melting and boiling points of the unbranched alkanes from CH_4 to $C_{16}H_{34}$.

TABLE 23.3 Hydrocarbon constituents of petroleum

Hydrocarbons	Boiling range, °C	Name
C_1 to C_4	−160 to 0	gas
C_6 to C_{11}	30 to 200	gasoline
C_{10} to C_{16}	180 to 400	kerosene, fuel oil
C_{17} to C_{22}	350 and above	lubricants
C_{20} to C_{34}	low-melting-point solids	paraffin wax
C_{35} to . . .	soft solids	asphalt

TABLE 23.4 The strengths of some C—X bonds

Bond	Bond enthalpy, kJ/mol	Bond	Bond enthalpy, kJ/mol
C—C	348	C—N	305
C=C	612	C=N	613
C≡C	837	C≡N	890
C⋯C (in benzene)	518	C—F	484
C—H	412	C—Cl	338
C—O	360	C—Br	276
C=O	743	C—I	238

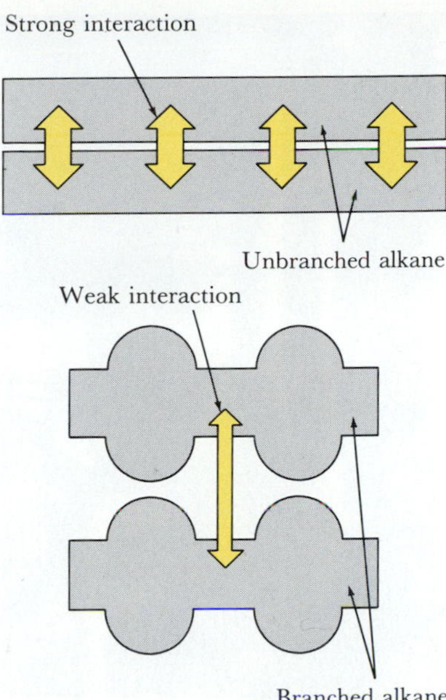

FIGURE 23.5 The atoms in neighboring straight-chain alkanes can lie close together. However, the branches of a neighboring substituted alkane molecule prevent the close approach of all the atoms, so the London forces are weaker and branched-chain alkanes are more volatile.

Alkanes are not completely inert, and one of their commercially most important reactions is oxidation, especially their complete combustion to carbon dioxide and water:

$$CH_4(g) + 2O_2(g) \longrightarrow CO_2(g) + 2H_2O(g) \qquad \Delta H^\circ = -890\ \text{kJ}$$

This exothermic reaction (which proceeds by a radical chain mechanism) and its analogues for the higher homologues are the basis of the use of alkanes as fuels (Section 6.5). For reasons related to the details of the chain reaction, branched-chain hydrocarbons burn more smoothly in engines than unbranched alkanes; the latter explode prematurely, thereby causing the noise called knocking. The standard of good antiknock behavior (but not the best behavior possible) is 2,2,4-trimethylpentane, which, we have seen, is also called isooctane. The standard of poor antiknock behavior is the straight-chain hydrocarbon heptane, C_7H_{16}. Every automobile gasoline is given an octane rating that is the percentage (by volume) of isooctane that must be blended with heptane to match the performance of the gasoline. A gasoline rated as 100 octane has the same antiknock characteristics as pure isooctane.

E Tetraethyllead, $(C_2H_5)_4Pb$, is used as an antiknock additive in gasoline because it decomposes thermally to provide many ethyl radicals in the combustion mixture and so helps to propagate the chain reaction. Although knocking is undesirable in a petrol engine, where a steady pressure on the retreating piston is desired (so the combustion should continue throughout the power stroke), that is not the case in a diesel engine where a rapid combustion is needed at the start of the power stroke. Thus, diesel fuels are manufactured to consist of a high proportion of straight-chain molecules.

Substitution reactions. Alkanes typically undergo substitution reactions:

> A **substitution reaction** is a reaction in which an atom or a group of atoms replaces an atom in the original molecule.

For an alkane, the displaced atom is a hydrogen atom (**7**). An example is the reaction between methane and chlorine. The mixture of these two gases survives unchanged in the dark; but in sunlight, when exposed to ultraviolet radiation, or when heated, they react:

$$CH_4(g) + Cl_2(g) \longrightarrow CH_3Cl(g) + HCl(g)$$

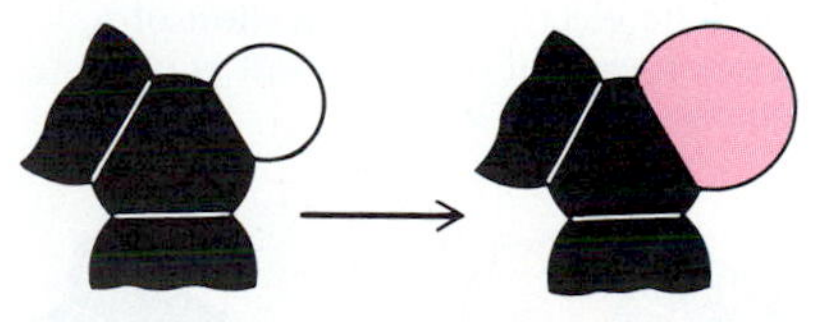

7 Substitution

The reaction does not produce only chloromethane, CH_3Cl, but leads to a mixture that also contains dichloromethane (CH_2Cl_2), trichloromethane ($CHCl_3$), and tetrachloromethane (CCl_4). Trichloromethane is better known as chloroform and was one of the early anesthetics. Tetrachloromethane, which is commonly called carbon tetrachloride, has been used as a solvent and in fire extinguishers; however, the realization that it is toxic and carcinogenic has limited its use.

U Exercise 23.62.

D Chloromethanes exhibit. TDC, 22-2. Photochemical catalysis of an aliphatic substitution reaction. TDC, 191.

FIGURE 23.6 The tower at the center of this oil refinery is the site of the catalytic reforming reaction. That reaction increases the proportion of branched-chain isomers in a mixture of hydrocarbons.

Isomerization reactions. One of the aims of an oil refinery is to convert unbranched alkanes into the more smoothly burning branched alkanes, by using a catalytic **reforming reaction** (Fig. 23.6). In this process, the unbranched alkane is passed over a Lewis-acid catalyst (typically an aluminum halide) near room temperature:

$$CH_3—CH_2—CH_2—CH_3 \xrightarrow{AlCl_3,\ room\ temperature} CH_3—\underset{}{\overset{\overset{\displaystyle CH_3}{|}}{C}H}—CH_3$$

This reaction is an example of an **isomerization**, a reaction in which a compound is converted into one of its isomers. In practice, a mixture of products is formed. Fragmentation (cracking) of large hydrocarbon molecules into alkanes of lower molar masses also occurs as the temperature is raised.

THE ALKENES AND THE ALKYNES

The simplest unsaturated hydrocarbon is ethene, commonly called ethylene, C_2H_4 or $CH_2{=}CH_2$ (**8**). It is the parent of a homologous series called the alkenes:

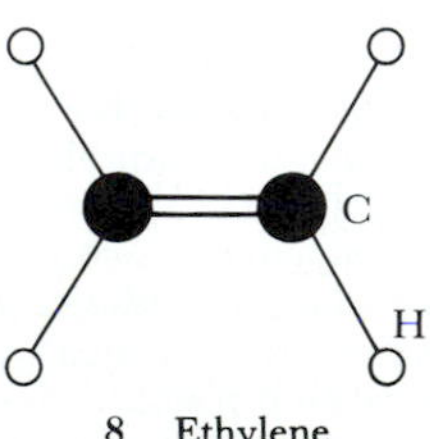

8 Ethylene

> The **alkenes** are a homologous series of molecules derived from ethylene by the repetitive insertion of $—CH_2—$ groups.

The next homologue, for example, is propylene, $CH_3CH{=}CH_2$. The molecular formulas of alkenes that have one C=C bond have the form C_nH_{2n}. The carbon atoms that are taking part in double bonding are sp^2 hybridized, and the double bond consists of a σ bond and a π bond; the π bond is weak and as a result alkenes are much more reactive than alkanes. As a consequence of their greater reactivity, the alkenes are much less abundant in oil wells.

Because alkenes are so important as the starting point for a number of manufacturing processes, one of the first steps in the petrochemical industry is to convert the abundant alkanes into alkenes. This conversion is achieved by removing hydrogen atoms from neighboring carbon atoms by catalytic **dehydrogenation**:

$$CH_3CH_3(g) \xrightarrow{Cr_2O_3,\ 500°C} CH_2{=}CH_2(g) + H_2(g)$$

E In the reforming reaction above, the aluminum chloride leads to the formation of a carbocation, which isomerizes so as to place the positive charge on a C atom attached to as many other C atoms as possible.

N The electron configuration of a carbon-carbon double bond is normally denoted $\sigma^2\pi^2$.

Alkenes are also formed by removing HBr from a bromoalkane by heating it in alcohol with sodium hydroxide:

$$CH_3—\overset{\overset{\displaystyle Br}{|}}{C}H—\overset{\overset{\displaystyle H}{|}}{C}H_2 + OH^- \xrightarrow[55°C]{organic\ solvent,} CH_3—CH{=}CH_2 + Br^- + H_2O$$

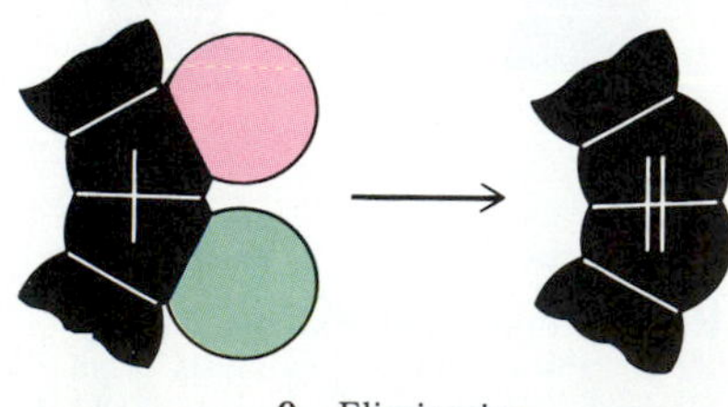

9 Elimination

This reaction is an example of a **dehydrohalogenation**, in this case a dehydrobromination. More generally, it is an example of an elimination reaction (**9**):

> An **elimination reaction** is a reaction in which two groups or atoms on neighboring carbon atoms are removed from a molecule, thereby leaving a multiple bond between the carbon atoms.

The alkynes are also unsaturated hydrocarbons that have at least one carbon-carbon triple bond. The simplest is ethyne, commonly called acetylene, $HC{\equiv}CH$. These compounds are discussed further in Section 23.7.

23.4 ALKENE NOMENCLATURE

The systematic name of an alkene is obtained from the name of the corresponding alkane by changing the ending to *-ene*. The systematic name of ethylene is therefore ethene and that of propylene, C_3H_6, is propene.

In all cases except ethylene and propylene, it is necessary to specify the location of the double bond. We do this by numbering the C atoms in the backbone and then reporting the *lower* of the numbers of the two atoms joined by the double bond. The numbering starts at the end of the backbone that results in the lowest number for the position of the double bond:

$$\overset{4}{C}H_3{-}\overset{3}{C}H_2{-}\overset{2}{C}H{=}\overset{1}{C}H_2 \qquad \overset{1}{C}H_3{-}\overset{2}{C}H{=}\overset{3}{C}H{-}\overset{4}{C}H_3$$

1-Butene 2-Butene

N Hydrocarbons with two carbon-carbon double bonds are called *alkadienes*. The position of each double bond is indicated as in 1,3-pentadiene, $CH_2{=}CH{-}CH{=}CH{-}CH_3$.

▼ **EXAMPLE 23.3 Naming alkenes**

Name the following compounds: (a) $CH_3CH_2CH_2CH{=}CH_2$; (b) $CH_2{=}CHCH_2CH_2CH_3$; (c) $(CH_3)_2CHCH{=}CH_2$.

STRATEGY We must identify the longest chain of carbon atoms that contains the double bond; then we form the name by using the suffix *-ene*. We use a numerical prefix to denote its position, as explained above, and name and specify the location of any other substituent.

SOLUTION (a) $CH_3CH_2CH_2CH{=}CH_2$ has a C_5 chain and is therefore a pentene. Because the double bond joins atoms 1 and 2, it is 1-pentene. (b) $CH_2{=}CHCH_2CH_2CH_3$ is the molecule in (a), written in the reverse order, so it is also 1-pentene. (c) Because $(CH_3)_2CHCH{=}CH_2$, which can be written as $CH_3CH(CH_3)CH{=}CH_2$, has a C_4 unbranched chain containing the double bond, it is a butene. The correct numbering is

$$\overset{4}{C}H_3{-}\overset{3}{C}H(CH_3){-}\overset{2}{C}H{=}\overset{1}{C}H_2$$

which makes this molecule a substituted 1-butene. A methyl substituent is attached to carbon atom 3. It is therefore 3-methyl-1-butene.

EXERCISE E23.3 Name the following: (a) $(CH_3CH_2)_2CHCH{=}CHCH_3$; (b) $(CH_3)_2CHCH_2CH{=}CH_2$.

▲ [*Answer*: (a) 4-ethyl-2-hexene; (b) 4-methyl-1-pentene]

E Diastereomers, geometrical isomers, and rotation about bonds. *J. Chem. Ed.*, **1982**, *59*, 37.

D Alkanes versus alkenes: Reactions of the double bond. CD2, 51.

23.5 THE CARBON-CARBON DOUBLE BOND

The C=C double bond of alkenes consists of a σ bond and π bond. Its presence in alkenes makes them significantly different from alkanes in shape, types of isomers, and reactions.

N There are a number of whimsical joke structures for cis and trans compounds; for example, one isomer of Am—CH=CH—Eu could be called *trans*-Atlantic. Such jokes may be effective mnemonics for students.

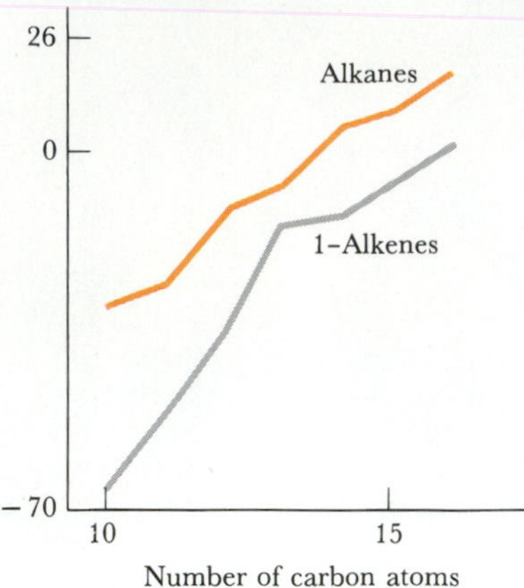

FIGURE 23.7 The melting point of an alkene is usually lower than that of the alkane with the same number of carbon atoms. The values shown are for straight-chain alkanes and 1-alkenes.

D *Cis-trans* isomerism and the polarity of molecules. *J. Chem. Ed.*, **1986**, *63*, 601.

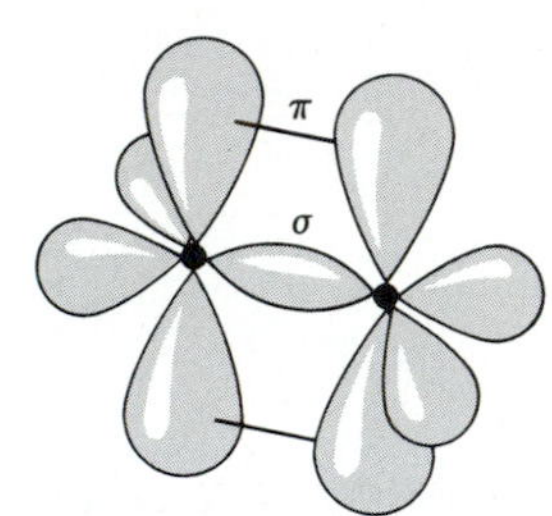

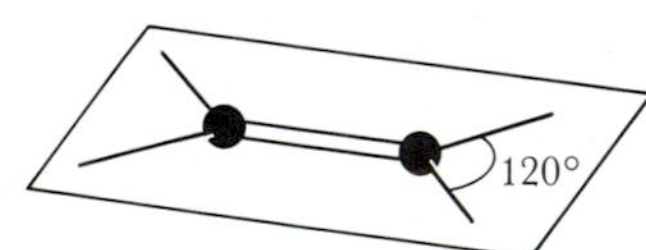

10 The double bond

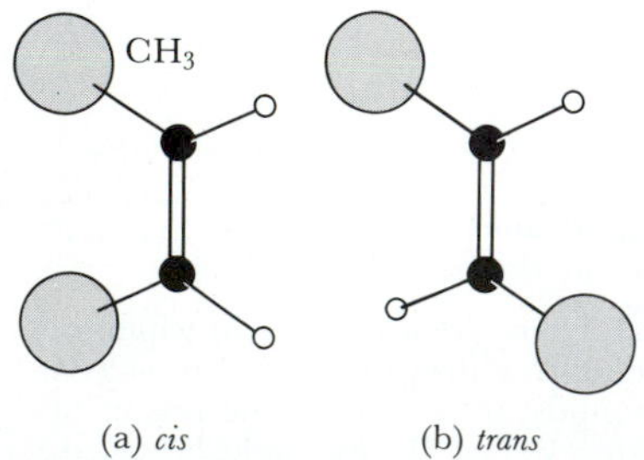

11 2-Butene

U Exercises 23.65 and 23.71.

The effect on shape. A double bond resists being twisted. All four atoms attached to the C=C group lie in the same plane and are locked into that arrangement by the π bond (**10**). As a result, alkene molecules are less flexible than the corresponding alkanes and cannot roll up into so compact a ball. Consequently, alkene molecules do not pack together as closely as alkanes and hence have lower melting points (Fig. 23.7).

The effect of double bonds on melting points is shown in the difference between fatty oils and soft solids such as shortening, as we saw in Section 18.1. One step in the manufacture of shortening from vegetable oils is the addition of hydrogen to the unsaturated hydrocarbon chains, to convert a runny oil into a useful fat. In practice this reaction is done by passing the vegetable oil and hydrogen under pressure (2 to 10 atm) over a nickel catalyst at about 200°C. Fish oils are highly unsaturated and remain fluid in cold aquatic environments.

Geometrical isomerism. The isomerism characteristic of double bonds becomes apparent in 2-butene (C_4H_8). Because the C=C bond is resistant to twisting, the two molecules

CH_3 and CH_3 on the same side of C=C, with H and H (*cis*-2-Butene) and CH_3, H / H, CH_3 across C=C (*trans*-2-Butene)

```
CH3     CH3              CH3     H
   \   /                    \   /
    C=C         and          C=C
   /   \                    /   \
  H     H                  H     CH3
 cis-2-Butene            trans-2-Butene
```

are distinct substances (**11a** and **11b**): both are gases but the cis isomer condenses at 4°C and the trans isomer condenses at 1°C. This distinctiveness is not exhibited by the two forms of butane (**12**), which change back and forth as one end of the molecule rotates relative to the other about the C—C single bond. A sample of butane is a *single* substance containing continuously interchanging conformations of C_4H_{10} molecules.

The relation between *cis*- and *trans*-2-butene is an example of geometrical isomerism, in which atoms are bonded to the same neighbors but have different arrangements in space (Fig. 23.2). As we have indicated here, the two geometrical isomers of 2-butene (and, by analogy, the isomers of other alkenes) are distinguished by the prefixes *cis*- ("this side") and *trans*- ("across").

▼ EXAMPLE 23.4 Naming geometrical isomers

Name the following pair of geometrical isomers:

```
Cl     Cl    Cl     H
  \   /        \   /
   C=C          C=C
  /   \        /   \
 H     H      H     Cl
```

STRATEGY We should name the molecules systematically and then add the prefix *cis*- or *trans*-, depending on whether the substituents are on the same side or on opposite sides of the double bond. As pointed out earlier, a chlorine atom is denoted by the prefix *chloro*-.

SOLUTION Both molecules are dichloro-substituted ethenes and hence are dichloroethenes. The substituents are on atoms 1 and 2; hence they are both 1,2-dichloroethenes. The left-hand molecule has both Cl atoms on the same side of the double bond, so it is *cis*-1,2-dichloroethene; the other is *trans*-1,2-dichloroethene.

EXERCISE E23.4 Name the following pair of geometrical isomers:

$$\begin{matrix} CH_3 & & H \\ & C{=}C & \\ H & & CH_2CH_3 \end{matrix} \qquad \begin{matrix} CH_3 & & CH_2CH_3 \\ & C{=}C & \\ H & & H \end{matrix}$$

[*Answer*: *trans*-2-pentene, *cis*-2-pentene]

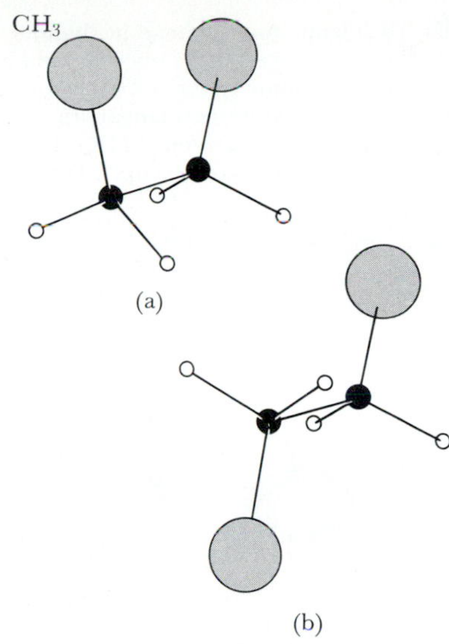

12 Butane

The conversion from one geometrical isomer to another is called **cis-trans isomerization**: it can occur only if the double bond is broken (such as by the absorption of a photon of light) and the molecule is free to rotate around the remaining single σ bond. We have already seen an example: in Section 9.6 we saw that the cis-trans isomerism of retinal is responsible for the primary act of vision.

Addition reactions of alkenes. The most characteristic reaction of a C=C double bond is addition (**13**), the reverse of elimination:

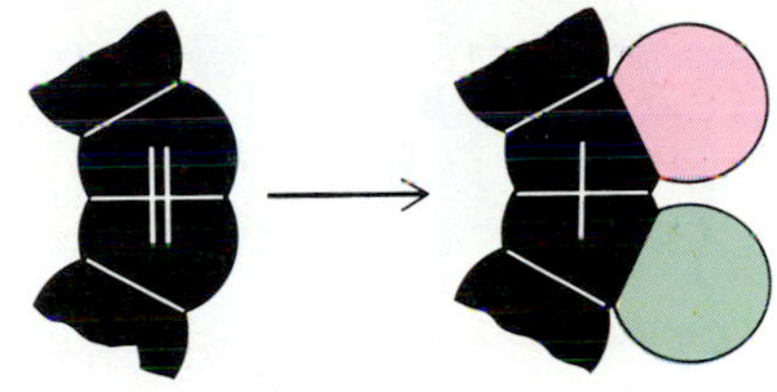
13 Addition

An **addition reaction** of an unsaturated compound is a reaction in which a reactant is added to the two atoms joined by a multiple bond.

An example is the **hydrogenation** of an alkene—the addition of hydrogen atoms to the alkene and its conversion to an alkane:

$$CH_3CH{=}CHCH_3 + H_2 \xrightarrow{\text{Ni, 500°C}} CH_3CH_2CH_2CH_3$$

The states of the reactants and the products are rarely given for organic reactions. This information is omitted because the solvent is some other organic compound or the reaction takes place at a catalytic surface, as in this example.

The occurrence of addition reactions can be traced to the weakness of the π component of the C=C double bond compared with the strengths of the single σ bonds in the addition product. In the hydrogenation reaction, the reaction enthalpy is the net outcome of investing the energy to break the C=C and the H—H bonds of the reactants and the energy returned by the formation of two C—H bonds and a C—C bond.

The incorporation of a halogen into a compound is called **halogenation**. Alkenes provide one method by which this can be done, because halogens undergo addition reactions with alkenes, like the addition that leads to the formation of 1,2-dichloroethane:

$$CH_2{=}CH_2 + Cl_2 \longrightarrow CH_2Cl{-}CH_2Cl$$

FIGURE 23.8 When bromine water (brown) is mixed with an alkene (colorless), the bromine atoms add to the molecule at the double bond, giving a colorless product.

Halogenation, in particular bromination with bromine water (an aqueous solution of bromine), is used as a simple test for alkenes (Fig. 23.8). In this reaction, the dark brown bromine and the colorless ethylene result in the formation of colorless 1,2-dibromoethane.

The addition of a hydrogen halide to an alkene with the formation of a haloalkane is called **hydrohalogenation** (it is the reverse of the

D To hexane and pentene in shallow dishes on an overhead projector, add drops of bromine water; repeat with drops of potassium permanganate. Bromination of 1-pentene, TDC, 174. The disappearing coffee cup, CD2, 52.

U Exercises 23.70 and 23.71.

dehydrohalogenation by which alkenes are made):

$$CH_2{=}CH_2 + HBr \xrightarrow{0°C} CH_3{-}CH_2Br$$

In a typical procedure, the gaseous hydrogen halide is bubbled through the alkene or a solution of the alkene in an organic solvent, such as acetic acid. A low temperature is used to reduce the rate of the reverse reaction.

23.6 ALKENE POLYMERIZATION

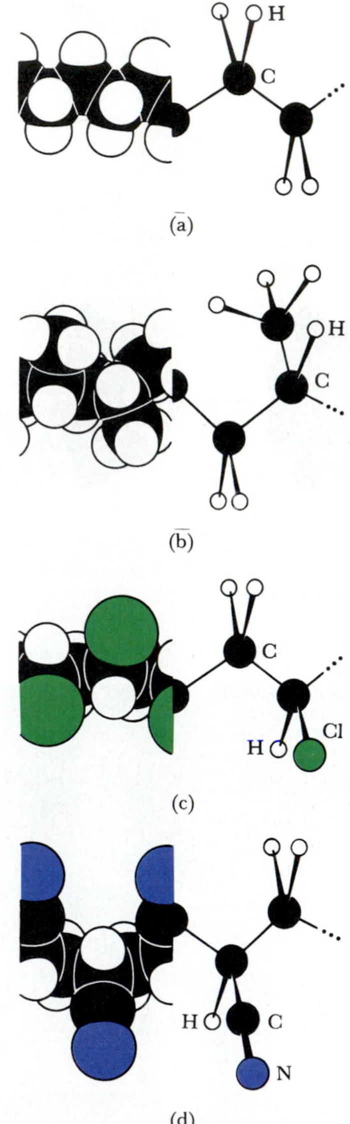

FIGURE 23.9 Four of the polymers produced by the addition polymerization of alkenes: (a) polyethylene; (b) polypropylene; (c) polyvinyl chloride; (d) polyacrylonitrile, which is sold as Orlon.

Alkenes react with themselves in a process called **addition polymerization**. Thus, an ethylene molecule may form a bond to another ethylene molecule, another ethylene molecule may add to that, and so on, until a long hydrocarbon chain has grown. The original alkene, such as ethylene, is called the **monomer** (from the Greek words for "one part"); the product, the chain of covalently linked monomers, is called a **polymer** ("many parts"). The simplest addition polymer is polyethylene, $-(CH_2CH_2)_n-$, which consists of long chains of $-CH_2CH_2-$ units. In practice, the polymer molecules have a number of branches, which are new chains that have grown from intermediate points in the original chain.

The plastics industry has developed polymers from a number of monomers of the form $CHX{=}CH_2$, where X is a halogen atom (such as the Cl in $CHCl{=}CH_2$ for vinyl chloride) or a group of atoms (such as CH_3- for propylene). These monomers give polymers of formula $-(CHXCH_2)_n-$ and include polyvinylchloride (PVC), $-(CHClCH_2)_n-$, and polypropylene, $-[CH(CH_3)CH_2]_n-$. The polymers, some of which are illustrated in Fig. 23.9 and are listed in Table 23.5, differ in appearance, rigidity, transparency, and resistance to weathering. One of the more recent extensions of the range is polytetrafluoroethylene (PTFE, which is sold as Teflon), in which the monomer is fully fluorinated ethylene ($CF_2{=}CF_2$) and the polymer consists of $-(CF_2CF_2)_n-$ chains. The polymer is very resistant to chemical attack, partly because of the great strength of the C—F bond (Table 23.4), and survives under harsh conditions.

Polymerization processes. A widely used polymerization procedure is **radical polymerization**, in which the chain is propagated by radicals. In a typical procedure a monomer, such as ethylene, is compressed to about 1000 atm and heated to 100°C in the presence of a small amount of an organic peroxide, which is a compound of formula R—O—O—R, where R may be an alkyl group. The chain is initiated by dissociation of the O—O bond, a reaction giving two radicals:

$$R{-}O{-}O{-}R \longrightarrow R{-}O\cdot + \cdot O{-}R$$

These radicals attack monomer molecules $CHX{=}CH_2$ (with X = H for ethylene itself) and form a new radical:

$$R{-}O\cdot + CH_2{=}\underset{X}{\overset{H}{C}} \longrightarrow R{-}O{-}CH_2{-}\underset{X}{\overset{H}{C}}\cdot$$

TABLE 23.5 Addition polymers

Monomer		Polymer	
Name	**Formula**	**Formula**	**Typical name**
ethylene	$CH_2{=}CH_2$	$-(CH_2{-}CH_2)_n-$	polyethylene
vinyl chloride	$\underset{Cl}{CH}{=}CH_2$	$-(\underset{Cl}{CH}{-}CH_2)_n-$	polyvinyl chloride, PVC
styrene	$\underset{C_6H_5}{CH}{=}CH_2$	$-(\underset{C_6H_5}{CH}{-}CH_2)_n-$	polystyrene
acrylonitrile	$\underset{CN}{CH}{=}CH_2$	$-(\underset{CN}{CH}{-}CH_2)_n-$	Orlon, Acrilan
propylene	$\underset{CH_3}{CH}{=}CH_2$	$-(\underset{CH_3}{CH}{-}CH_2)_n-$	polypropylene
methyl methacrylate	$O{=}\underset{\underset{CH_3}{C{=}CH_2}}{C}{-}O{-}CH_3$	$\left\{\overset{O=C-O-CH_3}{-\underset{CH_3}{C}-CH_2-}\right\}_n$	Plexiglas, Lucite
tetrafluoroethylene	$CF_2{=}CF_2$	$-(CF_2{-}CH_2)_n-$	Teflon, PTFE

T OHT: Table 23.5.

D Entertaining demonstrations of polymer properties. *J. Chem. Ed.*, **1990**, *67*, 238.

The chain now propagates as this radical attacks other monomer molecules:

$$R{-}O{-}CH_2{-}\overset{H}{\underset{X}{C}}\cdot + CH_2{=}\overset{H}{\underset{X}{C}} \longrightarrow R{-}O{-}CH_2{-}\overset{H}{\underset{X}{C}}{-}CH_2{-}\overset{H}{\underset{X}{C}}\cdot$$

This propagation continues until all the monomer has been used or until two chains link together. The product consists of long chains of formula $-(CHXCH_2)_n-$ in which n can be many thousands.

Another very important polymerization procedure uses a **Ziegler-Natta catalyst**. This catalyst is named for the German chemist Karl Ziegler and the Italian chemist Giulio Natta; it typically consists of titanium(IV) chloride and an alkylaluminum compound (a compound in which an alkyl group is bonded to an aluminum atom) such as triethylaluminum, $Al(CH_2CH_3)_3$. Ziegler-Natta catalysts bring about polymerization at low temperatures and pressures and lead to high-density materials (Fig. 23.10). The mechanism of their action is still uncertain, but it is believed to involve the formation of a complex between the titanium (a Lewis acid) and the alkene (a Lewis base).

FIGURE 23.10 The two samples of polyethylene in the test tube were produced by different processes. The floating one was produced by high-pressure polymerization, but the one at the bottom was produced with a Ziegler-Natta catalyst. The latter has a higher density because the chains pack together better.

The high density of Ziegler-Natta polymers arises from the character of the addition process, which leads to a regular arrangement of groups along the polymer chain. Ziegler-Natta polymerization leads either to an **isotactic polymer**, in which the substituents are all on the same side of the chain, or a **syndiotactic polymer**, in which the groups

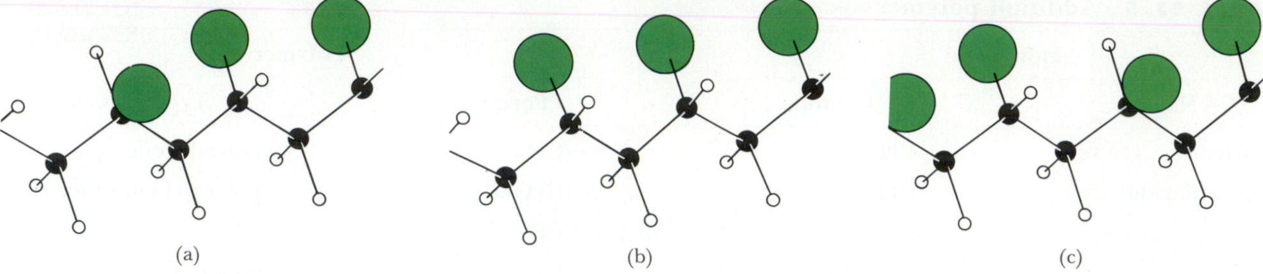

FIGURE 23.11 (a) In an atactic polymer, the substituents lie on random sides of the chain. The stereoregular polymers produced by Ziegler-Natta catalysts may be (b) isotactic (all on one side) or (c) syndiotactic (alternating).

T OHT: Figs. 23.11 and 23.13.

N Condensation polymers are presented in Chapter 24 and conducting polymers are discussed in Case 9.

N Freight trains: A useful analogy for polymers. *J. Chem. Ed.,* **1988**, *65*, 718.

D Polybutadiene (jumping rubber). BZS1, 3.8.

alternate. As a result of this stereoregularity, the chains can pack together well and form a highly crystalline, dense material. In contrast, radical polymerization gives a poorly packing **atactic** product, one in which the groups are attached randomly, on one side or the other, along the chain (Fig. 23.11).

Rubber. Rubber is a natural polymer formed, at least conceptually, from isoprene monomers:

$$\begin{array}{ccc} CH_3 & & CH_3 \\ & C{-}C & \\ CH_2 & & CH_2 \end{array} \qquad \text{isoprene}$$

Natural rubber is obtained from the bark of the rubber tree as a milky white liquid, called latex, which consists of a suspension of rubber particles in water (Fig. 23.12). (Latex, from the Latin word for "liquid," is also the white fluid inside dandelion and milkweed stalks.) The rubber itself is a soft white solid that becomes even softer when warm. It is used for pencil erasers and has been used as crepe for the soles of shoes.

The softening that occurs when natural rubber becomes warm is greatly reduced by **vulcanization**, a process discovered by Charles Goodyear in 1839. In this process, rubber is heated with sulfur. The sulfur atoms form cross-links between the polyisoprene chains and produce a three-dimensional network of atoms. Because the chains are pinned together, vulcanized rubber does not soften as much as natural rubber does as the temperature is raised. High concentrations of sulfur lead to very extensive cross-linking and the production of the hard material called ebonite. In the commercial production of rubber products such as tires, finely divided carbon black is mixed into the rubber during vulcanization: the carbon helps to strengthen the material and, by coloring it black, also protects it from damage by sunlight. Another important ingredient is an **antioxidant**, an additive that combats oxidation by trapping radicals formed by sunlight before they can attack the double bonds of the polymer.

Chemists were unable to synthesize rubber for a long time, even though they knew it was a polymer of isoprene. The enzymes in the rubber tree produce a stereoregular polymer in which all the links between monomers are in a cis arrangement (Fig. 23.13); straightforward radical polymerization, however, produces a random mixture of cis and trans links and a sticky, useless product. The stereoregular polymer was achieved with a Ziegler-Natta catalyst, and almost pure,

FIGURE 23.12 Collecting latex from a rubber tree in Malaysia, one of its principal producers.

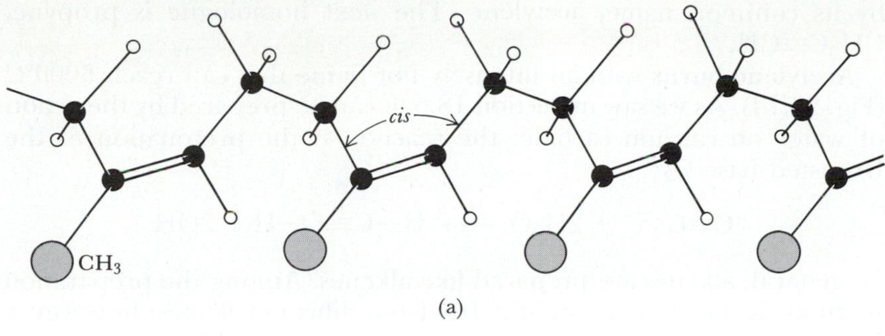

FIGURE 23.13 (a) In natural rubber the isoprene units are polymerized to be all cis. (b) The harder material, gutta percha, is the all-trans polymer.

E Polymer chemistry: State of the art symposium. *J. Chem. Ed.*, **1981**, *58*, November issue.

N The monomer used to make neoprene rubber, which is used for rubber gaskets, is $CH_2{=}CCl{-}CH{=}CH_2$.

D Poly(methylmethacrylate). BZS1, 3.9. Polystyrene. BZS1, 3.11.

E Copolymers are polymers formed from the polymerization of alternating monomers. Styrene-butadiene rubber (SBR) is the most common rubber in use.

E Polymer properties and testing definitions. *J. Chem. Ed.*, **1987**, *64*, 866. Polymers and barnacles. *J. Chem. Ed.*, **1984**, *61*, 1090.

rubbery *cis*-polyisoprene can now be produced. *trans*-Polyisoprene, in which all the links are trans, is also known and produced naturally: it is the hard material called gutta-percha that was once used for golf balls.

Chemists have modified the isoprene monomer and have developed a variety of synthetic rubbers, or **elastomers** ("elastic polymers"). Some of these elastomers are **copolymers**, that is, polymers formed from a mixture of monomers. The advantage of butyl rubber, a copolymer of a little isoprene with isobutylene, $(CH_3)_2C{=}CH_2$, over natural rubber stems from the fact that isobutylene has only one double bond; because this polymer has fewer double bonds in the chain than occur in natural rubber, it is less likely to be attacked and degraded.

23.7 ALKYNES

D Production of a gas: Acetylene. CD1, 9.

A third family of hydrocarbons consists of molecules that are even more unsaturated than the alkenes. They contain a carbon-carbon triple bond:

The **alkynes** are a homologous series with formulas that are derived from acetylene, $HC{\equiv}CH$, by insertion of $-CH_2-$ groups.

Alkynes with one carbon-carbon triple bond have the molecular formula C_nH_{2n-2}. The carbon atoms that are joined by the triple bond are *sp* hybridized, and the triple bond consists of a σ bond and two π bonds. As a result of this structure, the $-C{\equiv}C-$ group is linear and alkynes are rodlike in the vicinity of the triple bond.

U Exercises 23.57–23.59 and 23.63.

The names of the alkynes are formed like the alkenes but with the suffix *-yne,* for example, ethyne and propyne. The systematic name of the parent compound $HC{\equiv}CH$ is ethyne, but it is almost always known by its common name, acetylene. The next homologue is propyne, $CH_3C{\equiv}CH$.

Acetylene burns with an intensely hot flame that can reach 3000°C (Fig. 23.14). As we saw in Section 18.6, it can be prepared by the action of water on calcium carbide; the reaction is the protonation of the Brønsted base C_2^{2-}:

$$:C{\equiv}C:^{2-} + 2H_2O \longrightarrow H{-}C{\equiv}C{-}H + 2OH^-$$

In general, alkynes are prepared like alkenes. Among the preparation methods is the elimination of 2HBr from dibromoalkanes; however, a higher temperature is needed to remove the second HBr:

$$CH_3{-}\underset{}{\overset{Br}{\overset{|}{C}}}H{-}\overset{Br}{\overset{|}{C}}H_2 + OH^- \xrightarrow[80°C]{\text{organic solvent,}} CH_3\overset{Br}{\overset{|}{C}}{=}CH_2 + Br^- + H_2O$$

$$CH_3\overset{Br}{\overset{|}{C}}{=}CH_2 + OH^- \xrightarrow[120°C]{\text{organic solvent,}} CH_3C{\equiv}CH + Br^- + H_2O$$

The reactions of alkynes resemble those of the alkenes. The principal reaction is addition across the triple bond. They can be hydrogenated, halogenated, hydrohalogenated, and hydrated:

$$CH_3{-}C{\equiv}CH + 2HX \longrightarrow CH_3{-}CX_2{-}CH_3$$

FIGURE 23.14 Oxyacetylene welding makes use of the very hot flame produced when acetylene burns in oxygen. The flame can be used under water, because the oxygen is supplied as well as the acetylene.

U Exercises 23.66 and 23.69.

E Benzene was first isolated by Michael Faraday in 1825.

D Aromatic compounds exhibit. TDC, 22-5.

E The aromatic ring. *J. Chem. Ed.,* **1979**, *56*, 334. Benzene and the triumph of the octet theory. *J. Chem. Ed.,* **1974**, *51*, 498. Benzene: A familiar hazard? *J. Chem. Ed.,* **1980**, *57*, A85.

AROMATIC HYDROCARBONS

Chemists were once puzzled by a group of hydrocarbons that, although known to be unsaturated, are much less reactive than the alkenes. One sign of this lower reactivity is that the compounds, although unsaturated, do not decolorize bromine water. The simplest member of the class, benzene, C_6H_6, is a pungent, colorless liquid that freezes at 5.5°C. (Benzene is also highly dangerous, because exposure to it may lead to cancer.)

The pungency of benzene gave rise to the name **aromatic** for all the members of the class. This name is still widely used in chemistry to denote compounds that have a benzene ring as a part of their molecular structure even though many are odorless. The modern tendency, however, is to refer to them as **arenes** (from *ar*omatic and *-ene*, the latter denoting double bonds, as it does in the alkenes). Compounds that are not aromatic, which include alkanes, alkenes, and alkynes, are called **aliphatic** (from the Greek word for "fat").

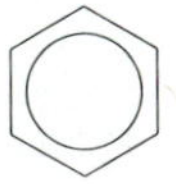

14 Benzene

All six carbon-carbon bonds in C_6H_6 are intermediate in length (139 pm) between a C—C single bond (154 pm) and a C═C double bond (134 pm). As we saw in Section 8.6, one representation of the benzene ring (**14**) emphasizes the equivalence of the bonds in benzene and their intermediate character between single and double bonds. In Lewis terms, the circle represents the resonance between two energetically equivalent Kekulé structures. In molecular-orbital terms, it represents the six electrons that occupy delocalized π orbitals. The σ framework of the molecule is formed by overlap of the three sp^2 hybrid orbitals on each of the C atoms, two with adjacent carbon atoms to form two C—C σ bonds and one with a 1*s* orbital of the adjacent H atom to form a C—H σ bond; the benzene bonds are illustrated in Fig. 9.26b. This hybridization leaves an unhybridized 2*p* orbital on each C atom free to overlap similar 2*p* orbitals on its two C atom neighbors, thereby forming the π molecular orbitals that spread round the ring.

Aromatic compounds also include the **polycyclic**—"many-ringed"—analogues of benzene. Two of the most important are naphthalene, $C_{10}H_8$, and anthracene, $C_{14}H_{10}$:

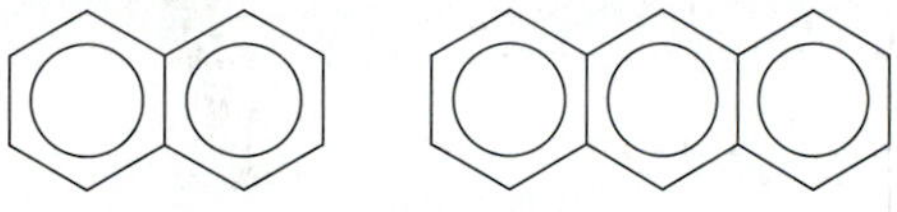

Naphthalene Anthracene

Both compounds can be obtained by the distillation of coal. Indeed, coal itself, which is a very complex mixture of complicated molecules, consists of very large networks containing regions in which aromatic rings can be identified (Fig. 23.15). When coal is "destructively" distilled—heated in the absence of oxygen so that it decomposes and vaporizes—the sheetlike molecules break up, and the fragments include the aromatic hydrocarbons and their derivatives. Naphthalene and anthracene are examples of homologues of benzene, with formulas derived by inserting $—C_4H_2—$ groups into the formula for benzene and thereby forming additional rings of C atoms.

N The misuse of the circle notation to represent aromatic rings. *J. Chem. Ed.*, **1983**, *60*, 190.

23.8 ARENE NOMENCLATURE

Many aromatic compounds have common names that often indicate their natural origins. Thus toluene, $CH_3C_6H_5$ (**15**), was originally obtained from *Tolu balsam,* the aromatic resin of a South American tree.

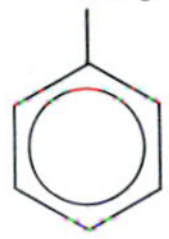

CH_3

15 Toluene

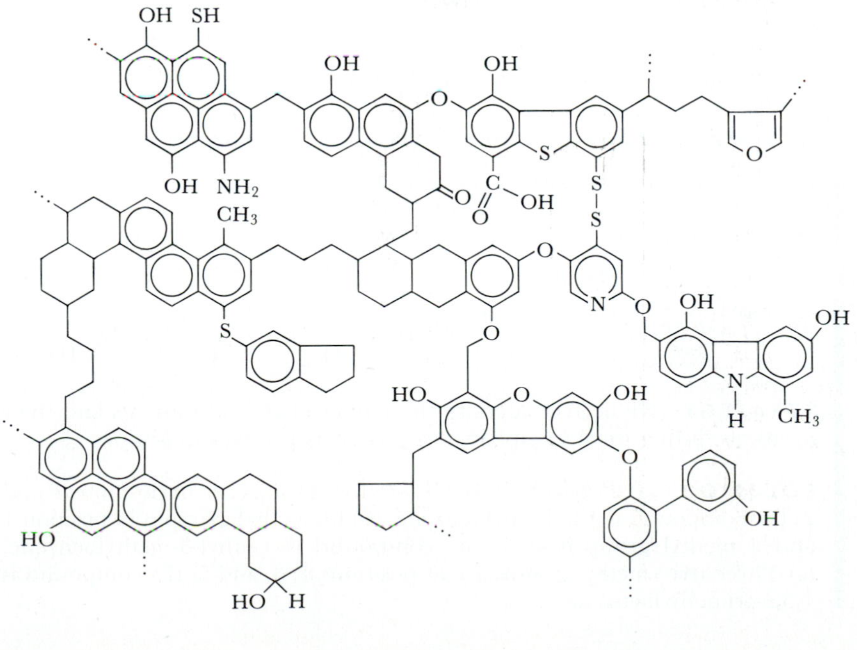

FIGURE 23.15 A schematic representation of the structure of coal. When coal is heated in the absence of oxygen, the structure breaks up and a complex mixture of products—many of them aromatic—is obtained.

Its systematic name is methylbenzene; the methyl group, CH_3—, is treated as a substituent of the benzene ring.

Isomers can result when more than one substituent is present, as in the three dimethylbenzenes:

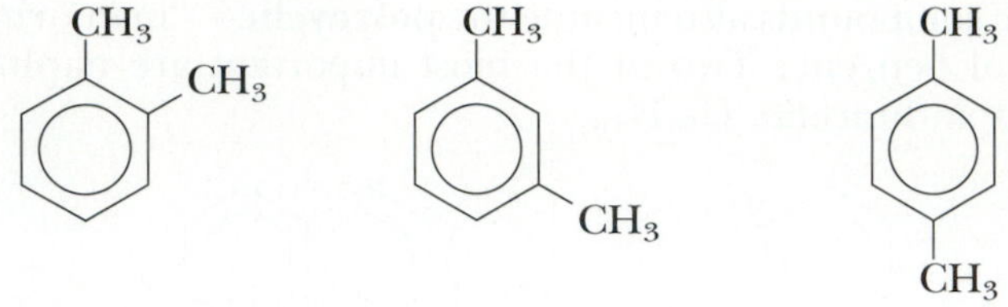

CH_3 CH_3 | CH_3 CH_3 | CH_3 CH_3

o-Xylene
1,2-Dimethylbenzene

m-Xylene
1,3-Dimethylbenzene

p-Xylene
1,4-dimethylbenzene

N Because of differences in their polarities, the xylenes adsorb to surfaces to different extents and can be separated by chromatography.

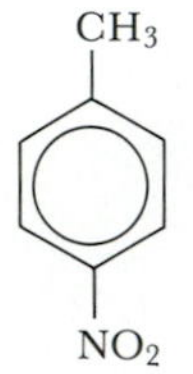

CH_3

NO_2

16 *p*-Nitrotoluene

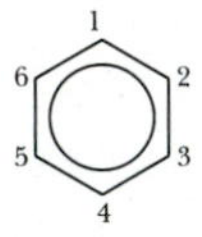

1 2 3 4 5 6

17 Location numbering

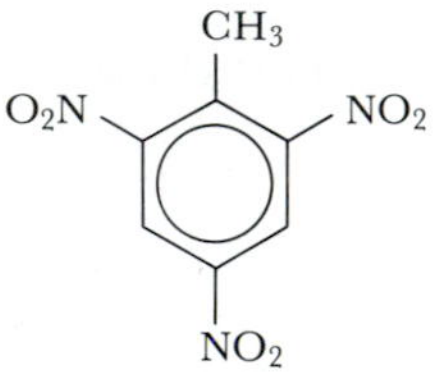

CH_3 O_2N NO_2 NO_2

18 2,4,6-Trinitrotoluene

All three isomers have the common name xylene. (They were originally obtained from the distillation of wood, for which the Greek word is *xulon;* they are now obtained from coal and petroleum.) In the common name, the three relative positions of the substituents are denoted by the prefixes *ortho-* (abbreviated *o*-), *meta-* (*m*-), and *para-* (*p*-). The three prefixes come for the Greek words meaning "directly," "after," and "beyond," and hence suggest the sequence of positions around the ring. The same prefixes are used to name isomers in which the groups are different, as in *p*-nitrotoluene (**16**), which contains the nitro group, —NO_2.

As we have seen with the xylenes, systematic names are obtained by regarding the compounds as substituted benzenes. The numbers used to specify the locations of the substituents are shown in the margin (**17**). All three xylenes are dimethylbenzenes, that is, a benzene ring with two methyl substituents. As shown above, they are 1,2-dimethylbenzene, 1,3-dimethylbenzene, and 1,4-dimethylbenzene. This nomenclature can readily be extended to more complicated molecules with more than two substituents; for more complicated molecules, position 1 is the location of the unique substituent. For example, the systematic name of trinitrotoluene (**18**, the explosive TNT) is 1-methyl-2,4,6-trinitrobenzene.

▼ EXAMPLE 23.5 Naming derivatives of benzene

Give the systematic names of the following substances:

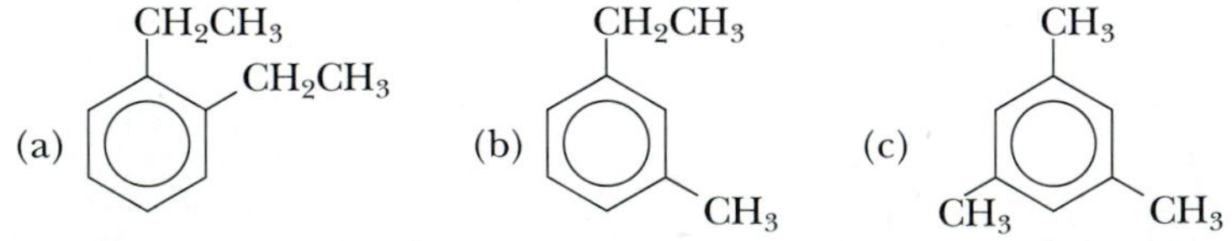

(a) CH_2CH_3 CH_2CH_3 (b) CH_2CH_3 CH_3 (c) CH_3 CH_3 CH_3

STRATEGY We need to identify the names of the substituents and their locations, listing them in alphabetical order as prefixes to *benzene*.

SOLUTION (a) Two (di) —CH_2CH_3 (ethyl) groups are in positions 1 and 2; the compound is 1,2-diethylbenzene. (b) One ethyl group is in position 1 and a methyl group is at 3; the compound is 1-ethyl-3-methylbenzene. (c) Three (tri-) methyl groups are at positions 1, 3, and 5; the compound is 1,3,5-trimethylbenzene.

EXERCISE E23.5 Name the following compounds:

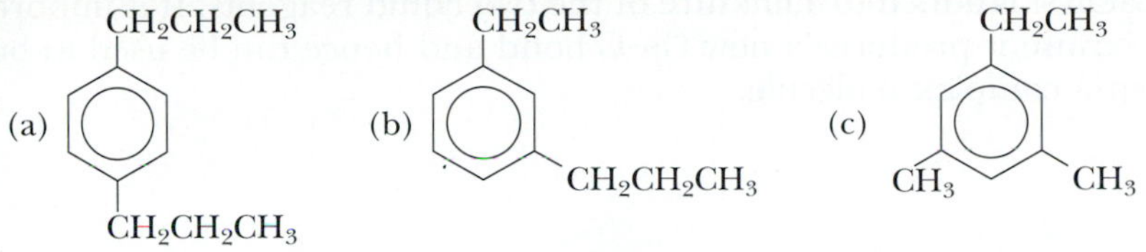

(a) [benzene ring with $CH_2CH_2CH_3$ at top and $CH_2CH_2CH_3$ at bottom]

(b) [benzene ring with CH_2CH_3 and $CH_2CH_2CH_3$]

(c) [benzene ring with CH_2CH_3, CH_3, CH_3]

[*Answer*: (a) 1,4-dipropylbenzene; (b) 1-ethyl-3-propylbenzene; (c) 1-ethyl-3,5-dimethylbenzene]

When the benzene ring is treated as a substituent of another backbone, as in styrene (**19**), the C_6H_5— group is called phenyl ("phene" is an old name for benzene). The systematic name of styrene, the monomer used for the production of polystyrene, is therefore phenylethene (Table 23.5).

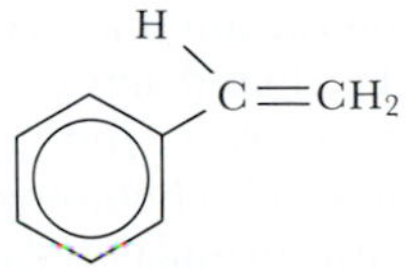

[Structure: phenyl ring attached to CH(H)=CH_2]

19 Styrene

23.9 REACTIONS OF AROMATIC HYDROCARBONS

E Aromatic substitution reactions. *J. Chem. Ed.*, **1983**, *60*, 937.

Although both are unsaturated hydrocarbons, arenes are markedly different from alkenes. In the reactions of arenes, the aromatic ring is usually left unchanged, but in the reactions of alkenes the double bond is usually lost. In fact, the benzene ring is so unreactive that it usually needs to be coaxed into reaction by a catalyst. We shall see in Chapter 24 that the reactions of aromatic compounds are much more extensive once the ring has undergone substitution.

Substitution reactions. In contrast to alkenes, for which addition is the dominant reaction type, arenes predominantly undergo substitution. For example, bromine immediately adds to a double bond of an alkene; however, it reacts with benzene only in the presence of a catalyst—typically iron(III) bromide—and the product is a substitution of the ring, not addition across a double bond:

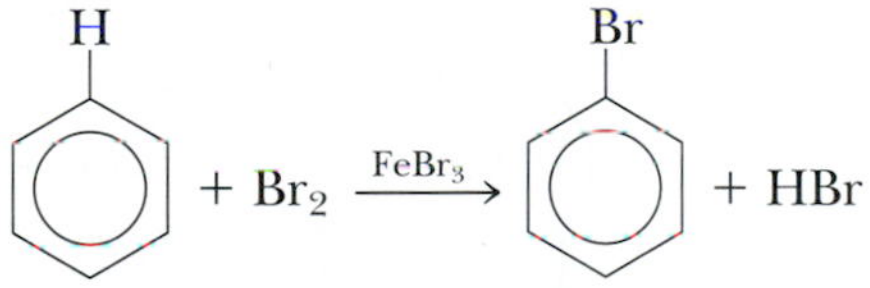

$$C_6H_5\text{—H} + Br_2 \xrightarrow{FeBr_3} C_6H_5\text{—Br} + HBr$$

The fluorination of benzene with F_2 is explosively violent; chlorination with Cl_2 needs a catalyst ($AlCl_3$ or $FeCl_3$). Iodination of benzene with I_2 cannot be carried out directly.

E Friedel-Crafts acylation. *J. Chem. Ed.*, **1979**, *56*, 480.

Friedel-Crafts alkylation. A part of an organic chemist's expertise is the ability to build larger molecules by using substitution reactions to attach groups to aromatic rings. Alkyl groups can be attached to benzene to build a more extensive network of carbon atoms by using **Friedel-Crafts alkylation**. This reaction, which is named for the French chemist Charles Friedel and his American coworker James Crafts, uses aluminum chloride as catalyst:

U Exercises 23.67, 23.68, and 23.72.

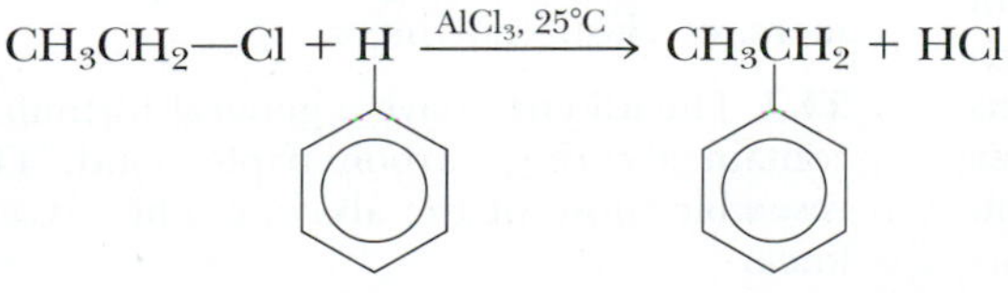

$$CH_3CH_2\text{—Cl} + H\text{—}C_6H_5 \xrightarrow{AlCl_3,\ 25°C} CH_3CH_2\text{—}C_6H_5 + HCl$$

The reaction takes place quite vigorously when a small amount of the catalyst is added to a mixture of the two liquid reagents. It is important because it produces a new C—C bond and hence can be used to build up a complex molecule.

SUMMARY

23.1 The **hydrocarbons** are binary compounds of hydrogen and carbon that may be classified as shown in the chart. (The dotted lines indicate that the classification continues to cyclic and acyclic alkenes and alkynes.) They form several **homologous series** consisting of compounds with formulas related by the insertion of a particular group (which is often $—CH_2—$). The **alkanes** have a general formula C_nH_{2n+2} and form a homologous series based on methane. The alkanes have numerous structural **isomers**, molecules with the same atoms in different arrangements; their classification is summarized in Fig. 23.2.

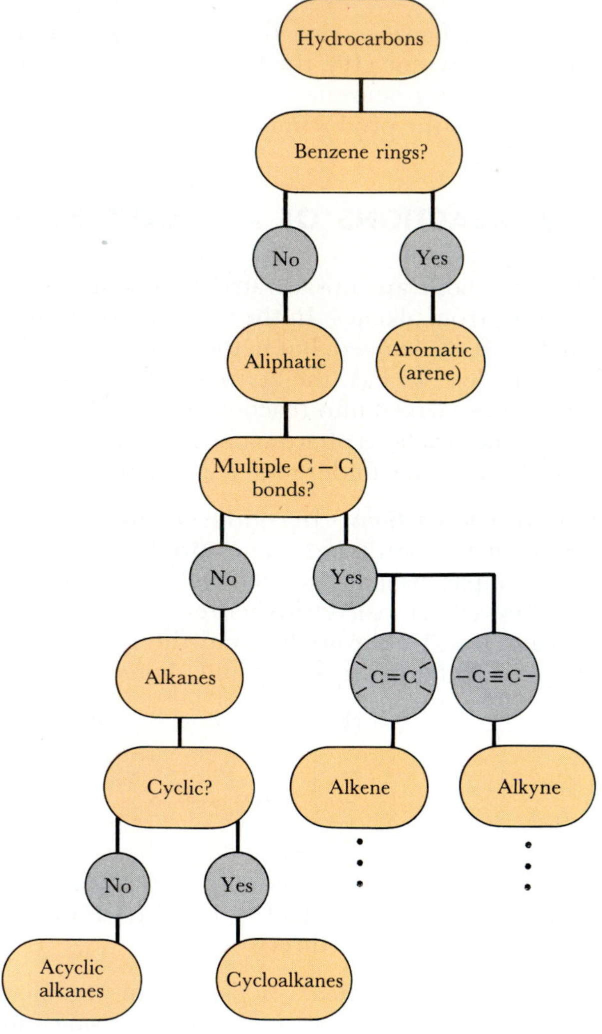

23.2 The names of the alkanes are the basis of the IUPAC systematic nomenclature of organic compounds. In naming organic compounds, it is important to differentiate between **branched** and **unbranched** carbon chains. Ringlike saturated hydrocarbons are known as **cycloalkanes**. One example of this class, cyclohexane, interchanges between several different **conformations**, or forms differing by twisting around single bonds.

23.3 The alkanes are **saturated compounds**—compounds having no multiple carbon-carbon bonds—with low reactivity. Their typical reactions are **substitution reactions**, which generally occur by a radical mechanism. **Isomerization** reactions—conversions of a compound into one of its isomers—are brought about by Lewis-acid catalysts.

23.4 The **alkenes** have a general formula C_nH_{2n} and contain a carbon-carbon double bond. They are formed by an **elimination reaction**, typically **dehydrogenation** of alkanes—the removal of H_2—or **dehydrohalogenation** of haloalkanes—the removal of hydrogen halide. Alkene nomenclature is similar to that for alkanes. The suffix *-ene* is substituted for *-ane* and the C=C bond is always included in the longest unbranched chain.

23.5 A double bond is torsionally rigid and leads to the existence among the alkenes of **geometrical isomers**, isomers that differ in the spatial arrangement of the atoms. The characteristic reaction of alkenes is **addition** to the double bond.

23.6 Alkene **monomers**, individual small molecules, can react with themselves and undergo **addition polymerization** by **radical chain reactions**. These mechanisms lead to **atactic** polymers, in which there is an irregular arrangement of substituents along the polymer chain. Polymers with a regular arrangement of substituents along the polymer chain, which may be **isotactic** or **syndiotactic**, are produced if **Ziegler-Natta catalysts** are used. The polymerization of isoprene leads to **rubber**; other monomers can be used to produce other **elastomers**, or elastic polymers.

23.7 The **alkynes** have a general formula C_nH_{2n-2} and contain a carbon-carbon triple bond. Their reactions resemble those of the alkenes. The —C≡C— group is linear.

23.8 and 23.9 Aromatic hydrocarbons, or **arenes**, are hydrocarbons based on the benzene ring. Arenes are much less reactive than alkenes. The typical reactions of arenes are **ring substitution reactions**. Examples include **halogenation**—the incorporation of a halogen atom—and **Friedel-Crafts alkylation**—the attachment of an alkyl group—in which aluminum chloride is the catalyst.

CLASSIFIED EXERCISES

Nomenclature of Aliphatic Compounds

23.1 Identify these compounds as being an alkane, alkene, or alkyne: (a) C_3H_8; (b) $C_{10}H_{18}$; (c) C_7H_{16}.

23.2 State whether each compound is an alkane, alkene, or alkyne: (a) C_2H_2; (b) $C_{16}H_{34}$; (c) C_4H_8.

23.3 Name the following substituents:
(a) $(CH_3)_2CH—$
(b) $CH_3CH_2CH_2—$
(c) ring: $CH_2—CH—$ / CH_2 / $CH_2—CH_2$ (a CH_2 bonded to $CH_2—CH—$ above and $CH_2—CH_2$ below, with the CH— bonded to the lower CH_2)

23.4 Name the following substituents:
(a) Cl— (b) ring: CH_2 / $CH—$ / CH_2 (two CH_2 groups bonded to each other and to $CH—$) (c) $(CH_3)CH(CH_2CH_3)—$

23.5 Name each of the following as an unbranched alkane: (a) C_6H_{14}; (b) C_8H_{18}; (c) C_9H_{20}.

23.6 Name each of the following as an unbranched alkane: (a) $C_{10}H_{22}$; (b) C_4H_{10}; (c) C_9H_{20}.

23.7 Write the molecular formula (in the form C_nH_{2n+2}) of (a) propane; (b) octane; (c) heptane.

23.8 Write the molecular formula (in the form C_nH_{2n+2}) of (a) hexadecane; (b) nonane; (c) hexane.

23.9 Name each of the following as an unbranched alkene, assuming the C=C double bond is between the carbon atoms 2 and 3 in the chain: (a) C_7H_{14}; (b) $C_{10}H_{20}$; (c) C_8H_{16}.

23.10 Name each of the following as an unbranched alkene, assuming the C=C double bond is between the carbon atoms 1 and 2 in the chain: (a) C_4H_8; (b) C_5H_{10}; (c) C_9H_{18}.

23.11 Write the formula and name the next higher homologue of octane, C_8H_{18}.

23.12 Write the formula and name the next higher homologue of hexene, C_6H_{12}.

23.13 Give the systematic name of (a) $CH_3CH_2CH_3$; (b) $CH_3(CH_2)_3CH_3$; (c) $(CH_3)_2CHCH(CH_3)_2$.

23.14 Give the systematic name of
(a) $CH_3CH_2CH_2CH_2CH_2CH_3$
(b) $CH_3CH(CH_2CH_3)C(CH_3)_2CH_3$
(c) $(CH_3)_3CCH_2CH(CH_3)CH_3$

23.15 Name the following compounds, given that all the remaining bonding positions are occupied by H atoms:

(a) —C—C=C—C—C— (with a —C— branch on the fourth carbon)

(b) —C—C—C——C—C— (with a —C——C— branch across the third and fourth carbons, and a benzene ring on the fourth carbon)

23.16 Name the following compounds, given that all the remaining bonding positions are occupied by H atoms:

(a) —C=C—C—C—C—C—C— (with a benzene ring on the third carbon and a —C— branch on the sixth carbon)

(b) —C—C—C—C—C≡C— (with a —C— branch above the third carbon, a —C— branch below the second carbon, and Cl on the fourth carbon)

23.17 Give the systematic name of the following compounds: if geometrical isomers are possible, write the names of each one: (a) $CH_3CH_2CH{=}CH_2$; (b) $CH{\equiv}CCH_2CH_3$; (c) $CH_3C{\equiv}CCH_3$.

23.18 Give the systematic name of the following compounds; if geometrical isomers are possible, write the names of each one:
(a) $CH_3CH_2CH{=}CHCH_3$
(b) $CH_3CH{=}CH(CH_2)_2CH_3$
(c) $CH_3C{\equiv}CCH(CH_3)CH_2CH_3$

23.19 Write the shortened (condensed) formula of (a) 3-methyl-1-pentene; (b) 4-ethyl-3,3-dimethylheptane; (c) 5,5-dimethyl-1-hexyne.

23.20 Write the shortened (condensed) formula of (a) 3,3-dimethyl-1-pentene; (b) 3-phenyl-1-butyne; (c) *cis*-4-methyl-2-hexene.

23.21 Write the structural formula of (a) 1,3-dimethylcyclopentane; (b) 4-butyl-5,5-diethyl-1-decyne; (c) 2,2,4-trimethylpentane.

CASE 23 ▸ GASOLINE VERSUS METHANOL

Methanol (also called methyl alcohol and wood alcohol) has physical and chemical properties that make it an attractive substitute for gasoline in automobiles (see the table); it is already used in some race cars. Although methanol does not occur in great abundance naturally, it is easily manufactured from other fuels such as coal, oil shale, natural gas, petroleum, and even farm and municipal wastes. It can also be transported by conventional tank cars, trucks, and tankers and in existing pipelines.

Methanol can be produced from hydrocarbon and fossil fuel reserves. When steam is passed over hot coal, a mixture of carbon monoxide and hydrogen gas, called *water gas,* is produced:

$$C(s) + H_2O(g) \longrightarrow CO(g) + H_2(g)$$

The stoichiometric ratios of CO to H_2 vary, depending on the catalyst, temperature, and pressure. When the carbon monoxide is mixed with water vapor and passed over an iron-copper catalyst at 400°C, additional H_2 is produced in the *shift reaction.*

$$CO(g) + H_2O(g) \xrightarrow{\text{Fe/Cu, 400°C}} CO_2(g) + H_2(g)$$

Carbon monoxide and hydrogen are also produced from natural gas and water by the reaction

$$CH_4(g) + H_2O(g) \longrightarrow CO(g) + 3H_2(g)$$

When equal amounts of CO and H_2 are passed over a ZnO/Cr_2O_3 catalyst at about 300 atm and 300°C, methanol is produced:

$$CO(g) + 2H_2(g) \xrightarrow[\text{300°C, 300 atm}]{ZnO/Cr_2O_3,} CH_3OH(g)$$

Because large amounts of methanol are already currently produced by this technology for the production of formaldehyde (for phenolformaldehyde resins) and for use as a solvent, the industry is already familiar with the safe handling of large quantities of methanol.

Methanol is a very "clean" fuel: its combustion efficiency is very high and its products are almost exclusively carbon dioxide and water. This cleanliness contrasts with the very poor efficiency of hydrocarbon combustion, such as that of isooctane in gasoline. In the latter combustion, seven C—C bonds and eighteen C—H bonds must be broken, and catalytic converters are necessary to complete the combustion at a point where the energy produced results only in waste heat. At most, 35% of the energy potential of gasoline is actually used; in city driving the percentage may drop as low as 15%. One study has shown that a methanol-powered engine emits about 85% by mass less unburned fuel than gasoline, and about 90% less CO_2 and 90% less NO. Therefore, even though the enthalpy of combustion of isooctane is

Comparison of the properties of methanol and isooctane

Compound	Enthalpy of combustion ΔH_c, kJ/mol	Temperature of vapor pressure of 100 Torr, °C	Boiling point, °C	Flash point, °C	Density at 25°C, g/mL	Octane rating	Caution
methanol	−726	21.2	64.6	−11	0.787	106	toxic
isooctane*	−5471	40.7	99.3	−12	0.692	100	toxic

*2,2,4-trimethylpentane.

23.22 Write the structural formula of (a) 2,7,8-trimethyldecane; (b) 1,3-dimethylcyclohexane; (c) 5-ethyl-1,3,6-heptatriene.

Isomerism

23.23 Write and name all the isomers of (a) C_4H_{10}; (b) C_5H_{12}; (c) dimethylcyclohexane.

23.24 Write and name all the isomers (including geometric isomers) of the alkenes (a) C_4H_8; (b) C_5H_{10}; (c) 2-chlorocyclopentene.

23.25 Write the structural formulas and name three alkene isomers having the formula C_4H_8.

23.26 Write the structural formulas and name two cycloalkane isomers having the formula C_4H_8.

higher (see the table), the conditions for combustion in the cylinder of the automobile are not optimal and consequently the usable energy produced per gram of fuel is no greater than that of methanol.

Because of the increased combustion efficiency of methanol, catalytic converters could be eliminated from methanol-powered vehicles; this would also reduce the back-pressure caused by current catalytic converters and thereby further increase the efficiency of fuel conversion for the vehicle. In addition, the temperatures required for the complete combustion of methanol are lower than those needed for gasoline, and exhaust gases can be cooler by as much as 100°C. Therefore, lower concentrations of nitrogen oxides (precursors to photochemical smog production) are emitted.

General Motors is currently producing a fleet of 2200 variable fuel vehicles (VFV) for a demonstration program in California. They will operate on a fuel that varies in composition from 100% gasoline to 15% gasoline and 85% methanol (called M85). Test drivers have reported an 8% increase in power with the M85 fuel and a V6 engine.

There are some drawbacks related to methanol usage: the engines tend to have poor starting characteristics; the flame is nearly invisible in the daylight (although an additive could make the flame visible); and the formaldehyde (from a partial oxidation of the methanol) emissions are higher than those from gasoline-fueled engines. Formaldehyde is a precursor to the formation of ozone in the troposphere. In addition, methanol is more corrosive (to engine, fuel tank, and gaskets) than gasoline is.

QUESTIONS

1. Determine the enthalpies of combustion of methanol and octane (a) per liter of fuel and (b) per gram of fuel. (c) Comment on the comparative values.

2. Assuming that the unburned methanol is 85% by mass less than that of unburned octane, compare the specific enthalpy densities of methanol and octane.

3. Why do you suppose Indy-500 drivers prefer the use of methanol in place of gasoline to power their race cars?

4. (a) Explain why methanol may be an environmentally more acceptable fuel than octane. (b) Explain the advantages of substituting methanol for octane as a fuel in automobiles. (c) Explain the disadvantages of substituting methanol for octane as a fuel in automobiles.

Molecular Formulas

23.27 Write the structural formula of molecules of formula C_4H_{10} that react with chlorine in the presence of light to give two isomers of C_4H_9Cl.

23.28 Write the structural formula of molecules of formula C_6H_{14} that react with chlorine in the presence of light to give two isomers of $C_6H_{13}Cl$.

23.29 An alkane C_6H_{14} reacts with chlorine in the presence of light to give two isomers of formula $C_6H_{13}Cl$. When one of these isomers is treated with potassium hydroxide in an organic solvent, it produces one alkene, C_6H_{12}, but the other isomer produces two alkene isomers. Propose formulas for all the compounds.

23.30 A hydrocarbon contains 88% by mass of carbon

and 12% by mass of hydrogen and has molar mass 82 g/mol. It decolorizes bromine water, and 1.5 g of the hydrocarbon reacts with 400 mL of hydrogen in the presence of a nickel catalyst. Identify the hydrocarbon and write its structural formula.

Reactions of Aliphatic Compounds

23.31 Predict the products of the following reactions and classify each reaction as an addition or substitution: (a) sodium hydroxide is heated with 2-bromopropane; (b) bromine is mixed with ethylene.

23.32 Predict the products of the following reactions and write a chemical equation for each one: (a) 3-bromopentane is mixed with sodium hydroxide in an organic solvent; (b) hydrogen chloride is mixed with propene.

23.33 Write a balanced chemical equation for the reactions that occur when chlorine is mixed (a) with ethane and exposed to sunlight; (b) with ethylene; (c) with ethyne. Classify each one as an addition or a substitution.

23.34 Predict the products and write a balanced chemical equation for the reaction that occurs between hydrogen and (a) 2-methylpropene or (b) 2-methylpropyne in the presence of a platinum catalyst.

23.35 Concentrated sulfuric acid is a strong dehydrating agent, in the sense that it can remove the elements of water, HOH, from a compound. What principal product results from the reaction of $CH_3(CH_2)_2CH(OH)CH_3$ with concentrated sulfuric acid at 120°C?

23.36 What is the product from the reaction of 2-methyl-2-chloropentane with hot potassium hydroxide in an organic solvent?

23.37 Suggest a series of reactions that will convert cyclopentane into cyclopentene.

23.38 Suggest a convenient laboratory method for converting 2-chlorobutane to 2-hydroxybutane.

Addition Polymers

23.39 Name the polymer that is formed when each of the following monomers is polymerized: (a) ethylene; (b) cyanoethene; (c) isoprene.

23.40 Name the polymer that is formed from (a) chloroethene; (b) phenylethene; (c) propene.

23.41 Sketch a trimer of the molecules (a) isoprene; (b) $CH_2{=}CH{-}C{\equiv}N$.

23.42 Sketch a trimer of the molecules (a) tetrafluoroethene; (b) phenylethene.

23.43 The osmotic pressure of a solution of 2.0 g of a polyethylene in 250 mL of toluene is 0.71 Pa. What is the molar mass of the polymer?

23.44 A 2.84-g sample of polystyrene in 150 mL of toluene gives an osmotic pressure of 0.18 Pa. What is the molar mass of the polymer?

Nomenclature of Aromatic Compounds

23.45 Name each aromatic compound:

(a) benzene ring with CH_2CH_3 and CH_3 (meta)

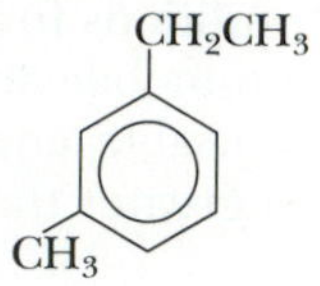

(b) benzene ring with CH_3, CH_3, CH_3, CH_3, CH_3 (five methyl groups)

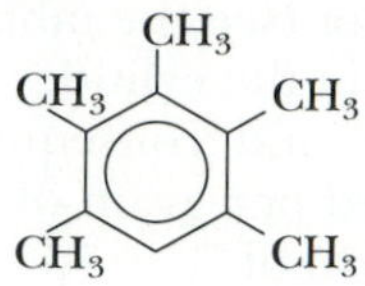

23.46 Name each aromatic compound:

(a) benzene ring with CH_3, CH_2CH_3, $CH_2CH_2CH_3$

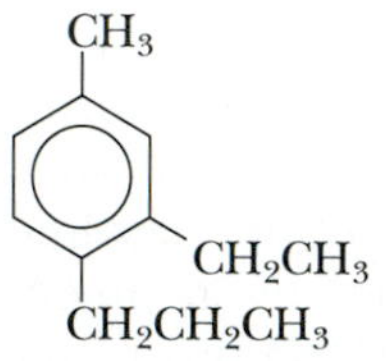

(b) benzene ring with CH_2CH_3 and $CH(CH_3)_2$ (para)

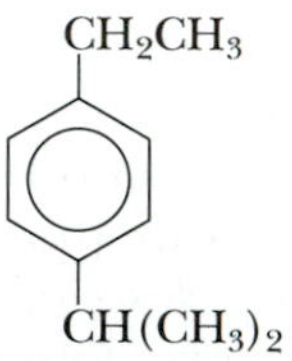

23.47 Write the structural formulas of the compounds (a) *p*-chlorotoluene; (b) 1,3-dimethylbenzene; (c) 4-chloromethylbenzene.

23.48 Write the structural formulas of (a) 1,2-dichlorobenzene; (b) 3-phenylpropene; (c) 1-ethyl-2,5-dimethylbenzene.

23.49 Draw all the isomeric dichlorobenzenes. Indicate which are polar and which are nonpolar.

23.50 Draw all the isomeric trichlorobenzenes. Indicate which are polar and which are nonpolar.

23.51 (a) Draw the structural formula for all isomers of nitrotoluene. (b) Write the systematic and the common (*o*-, *m*-, *p*-) names for each isomer.

23.52 Of the three isomeric aminobenzoic acids, $H_2N{-}C_6H_4{-}COOH$, only the *para*- isomer is used as a sunscreen (and is called PABA). (a) Draw the structural formula and (b) write the systematic and common names for each isomer.

Aromatic Compounds: Reactions

23.53 Suggest a method for preparing cyclohexene from benzene.

23.54 Suggest a method for preparing 2-phenylpropane from propene and benzene.

23.55 Write a balanced equation for the reaction of benzene (a) with chlorine in the presence of aluminum chloride; (b) with chloroethane in the presence of aluminum chloride.

23.56 Write a balanced equation for the reaction of (a) benzene with fluorine; (b) methylbenzene with chlorine in the presence of sunlight.

UNCLASSIFIED EXERCISES

23.57 State the hybridization of the orbitals used in bonding on carbon atom 2 in (a) pentane; (b) 2-pentene; (c) 2-pentyne.

23.58 Identify the type and number of bonds on carbon atom 2 in (a) pentane; (b) 2-pentyne; (c) cyclopentene.

23.59 What is the predicted geometry around carbon atom 3 in (a) 2-pentene; (b) 2-pentyne; (c) cyclopentene.

23.60 Spiropentane is the molecule

```
CH2—CH2
  \ /
   C
  / \
CH2—CH2
```

(a) Identify the type of hybridization at each carbon atom. (b) Draw the three-dimensional structure of the molecule. (c) Estimate the bond angles.

23.61 Write the structural formula of (a) methylcyclopropane; (b) 2,4,6-trimethyltoluene; (c) *trans*-1,2-diphenyl-2-butene.

23.62 Two of the reactions that can occur when ethane is chlorinated are:

$$CH_3CH_3(g) + Cl_2(g) \longrightarrow CH_3CH_2Cl(g) + HCl(g)$$

$$CH_3CH_3(g) + Cl_2(g) \longrightarrow 2CH_3Cl(g)$$

For the data in the following table and in Appendix 2A, determine the standard reaction enthalpy, free energy, and entropy for each reaction and comment on your conclusions.

Compound	$\Delta G_f°$, kJ/mol	$\Delta H_f°$, kJ/mol	$S°$, J/(K·mol)
$CH_3CH_2Cl(g)$	−60.5	−112	276
$CH_3Cl(g)$	−57.4	−80.8	234

23.63 Write the systematic name of the compounds (a) $CH_3(CH_2)_2CH(CH_2CH_2CH_3)CH_3$; (b) $HC{\equiv}CCl$; (c) $CH_3(CH_2)_2C(CH_2CH_2CH_3){=}CH_2$.

23.64 Are 1-chlorohexane and chlorocyclohexane isomers? Write the structural formula of each one.

23.65 Write and name all the structural formulas of pentene.

23.66 Write a balanced equation for the production of 2,3-dibromobutane from 2-butyne.

23.67 Write a balanced equation for the production of 1-phenylbutane from benzene.

23.68 Write a balanced equation for the production of 1-phenylbutane from 4-phenyl-1-butene.

23.69 Identify tests that can be used to distinguish between butane, 1-butene, and 1-butyne.

23.70 Two isomers can result when hydrogen bromide reacts with 2-pentene. Write their structural formulas.

23.71 Are cyclopentane and pentene isomers? Discuss the likely reactivity of each.

23.72 Name the product of the reaction between cyclopentene and hydrogen chloride.

23.73 Write the chemical equation for the Friedel-Crafts reaction of benzene with 2-chlorobutane.

The presence of particular groups of atoms in an organic molecule give it characteristic, predictable properties. In this chapter, we meet half a dozen or so of these groups and see how they are used to build new materials and contribute to the molecules that participate in the processes of life. Among the molecules that we shall meet are the synthetic polymers that are used for fabrics and the natural polymers that govern biological processes. The illustration shows a spider's web, which is built from a natural polymer.

24 FUNCTIONAL GROUPS AND BIOMOLECULES

As we noted in Chapter 23, many organic compounds consist primarily of carbon and hydrogen with, in addition, various groups of atoms that account for the compounds' characteristic reactions. These groups are called **functional groups**. We have already seen some examples of how a functional group can enliven the hydrocarbons, for the presence of a C=C double bond, which is one example of such a group, gives rise to the characteristic properties of the alkenes. The genetic molecule DNA, which we explore at the end of the chapter, is very rich in all kinds of functional groups and is correspondingly rich in chemical properties.

Some common functional groups are listed in Table 24.1. They are the groups chemists use to build new pharmaceuticals, plastics, and biologically active materials. They are also the groups that are used by living cells to form proteins and to participate in the biochemical processes of life. The functional groups are like a kit of parts from which the whole range of organic molecules, and ultimately people, are constructed.

THE HYDROXYL GROUP

In organic compounds, the **hydroxyl group**, —OH, is an —O—H group covalently bonded to a carbon atom. We should not confuse it with the hydroxide ion, OH^-, of inorganic hydroxides, which is a diatomic ion. The —OH group occurs in several classes of organic compounds, including alcohols, phenols, and carboxylic acids. As we shall see, these compounds are distinguished by the other groups that are attached to the carbon atom carrying the —OH group. We shall also see that the ability of the hydroxyl group to take part in hydrogen bonding is responsible for some of the distinctive physical properties of alcohols.

E Chemical ecology: Chemical communications in nature. *J. Chem. Ed.*, **1983**, *60*, 531. The chemical composition of the cell. *J. Chem. Ed.*, **1984**, *61*, 877. Consumer applications of chemical principles: Drugs. *J. Chem. Ed.*, **1985**, *62*, 329. Carcinogens in nature. *J. Chem. Ed.*, **1980**, *57*, 724.

TABLE 24.1 Common functional groups

Structural formula	Short form	Prefix or suffix in compound	Class of compound	Typical example
$>C=C<$	C=C	-ene	alkene	$CH_2=CH_2$
$-C\equiv C-$	C≡C	-yne	alkyne	$CH\equiv CH$
$-O-H$	OH	hydroxy- -ol	alcohol	CH_3OH
$>C=O$	CO	-al	aldehyde	CH_3CHO
		-one	ketone	CH_3COCH_3
$-C(=O)-O-H$	COOH	carboxy- -oic acid	carboxylic acid	CH_3COOH
$-N(H)-H$	NH_2	amino- -amine	amine	CH_3NH_2
$-N(=O)-O$	NO_2	nitro-	nitro compound	CH_3NO_2
—F, —Cl, —Br, —I	F, Cl, Br, I	halo- (fluoro-, chloro-, bromo-, iodo-)	haloalkane	CH_3Cl
$-C\equiv N$	CN	cyano-	nitrile	CH_3CN
$-N\equiv C$	NC	isocyano-	isocyanide	C_6H_5NC
—O—	O	R-oxy-	ether	CH_3OCH_3
$-S(=O)_2-O-H$	SO_3H	sulfo- -sulfonic acid	sulfonic acid	$C_6H_5SO_3H$
—S—H	SH	mercapto- -thiol	thiol	CH_3SH

T OHT: Table 24.1.

N Beers and wine are still made by the fermentation of carbohydrates; the reaction is

$$C_6H_{12}O_6 \xrightarrow{\text{yeast}} 2C_2H_5OH + 2CO_2.$$

The ethanol concentration from fermentation reaches only about 12–13% by volume; beers are about 5% and wines are about 12%. For higher concentrations of alcohol it is necessary to distill wine.

D Fermenting sugar. TDC, 22-13.

N Effects of ethanol on nutrition. *J. Chem. Ed.*, **1979**, *56*, 532.

24.1 ALCOHOLS AND ETHERS

We can think of an alcohol molecule as an HOH molecule in which one H atom has been replaced by an alkyl group R—, to give R—OH. One of the best-known organic compounds is ethanol, CH_3CH_2OH (also called ethyl alcohol and grain alcohol). In chemistry it is common for a single compound to inspire the name of an entire class of related compounds, and that is the case here:

An **alcohol** is an organic compound containing a hydroxyl group that is not connected directly to an aromatic ring or to a C=O group.

The name *alcohol* comes from the Arabic for "fine powder." The term gradually came to mean the "essence" of a thing—and, in particular, the liquid obtained by distilling wine.

A major source of ethanol is the fermentation of carbohydrates using the enzymes in yeast; it is also obtained from petrochemicals by the acid-catalyzed hydration of ethylene:

$$CH_2{=}CH_2 + H_2O \xrightarrow{H^+} CH_3CH_2OH$$

Another alcohol, methanol (or methyl alcohol), CH_3OH, was originally obtained from the distillation of wood and hence was known as wood alcohol. It is now obtained either by the oxidation of hydrocarbon gases or from carbon monoxide and hydrogen:

$$CO(g) + 2H_2(g) \xrightarrow{400°C,\ 150\ atm,\ ZnO} CH_3OH(g)$$

Methanol is a volatile, colorless, poisonous liquid that, if ingested, causes blindness and death.

N As in Chapter 23, we no longer show the phases of reactants and products in chemical equations. Many of the reactions take place in nonaqueous media.

E *Methy* is a Greek word that usually means "intoxicate" but sometimes means "wine;" *hule* is the usual Greek word for "substance" but it is sometimes used for "timber" or "group of trees."

E Microbial oxidation of alkenes. *J. Chem. Ed.*, **1975**, *52*, 475. The production of ethanol from grain. *J. Chem. Ed.*, **1982**, *59*, 49 and 1031.

Nomenclature of alcohols. Alcohols are divided into three classes according to the number of hydrogen atoms attached to the carbon atom to which the —OH group is bonded:

A **primary alcohol** has the structure RCH_2—OH.

A **secondary alcohol** has the structure R_2CH—OH.

A **tertiary alcohol** has the structure R_3C—OH.

The classes of alcohols are sometimes distinguished by the labels 1° (primary), 2° (secondary), and 3° (tertiary). The R groups need not be the same and may be either aliphatic (such as CH_3—) or aromatic (such as C_6H_5—). Ethanol and benzyl alcohol are both primary alcohols; isopropyl alcohol is a secondary alcohol:

N 2-Propanol is the systematic name of rubbing alcohol.

D Oxidation of secondary alcohols by duckweed. *J. Chem. Ed.*, **1988**, *65*, 549. Oxidation of cyclohexanol: An amoebalike reaction. *J. Chem. Ed.*, **1989**, *66*, 955.

N An aromatic group may not be attached *directly* to an —OH group, for that would result in a phenol, not an alcohol.

	C_6H_5—CH_2—OH	CH_3—CH(OH)—CH_3
Common names:	Benzyl alcohol	Isopropyl alcohol
Systematic names:	Phenylmethanol	2-Propanol

The systematic names of alcohols are formed by changing the *-e* of the parent alkane's name to *-ol.* Thus, CH_3OH is derived from methane and hence is methanol. When, to avoid ambiguity, we need to state the locations of substituents, we number the carbon atoms as explained in Section 23.2, starting at the carbon atom that gives the location of the —OH group the lower number. The systematic name of isopropyl alcohol, for example, is 2-propanol.

▼ EXAMPLE 24.1 Classifying and naming alcohols

Classify and name the organic compounds (a) $CH_3CH_2CH_2CH_2OH$ and (b) $CH_3CH(CH_2CH_2OH)CH_3$.

STRATEGY For the classification, we have to decide how many groups other than H are attached to the C—OH group. For the name, we must identify the longest unbranched alkane chain, use Table 23.2 to find the name of the parent alkane, and change the ending *-e* to *-ol.*

SOLUTION (a) The molecule has a $—CH_2—OH$ group and is a primary alcohol. A C_4 alkane is butane, so the alcohol is a butanol. Because the hydroxyl group could be attached to alternative carbon atoms, the group's location on carbon atom 1 must be specified, giving the name 1-butanol. (b) The $—CH_2—OH$ group signifies a primary alcohol. Writing out the formula more clearly and numbering the backbone C atoms, we obtain

$$\underset{4}{CH_3}—\underset{3}{\overset{\overset{\displaystyle CH_3}{|}}{CH}}—\underset{2}{CH_2}—\underset{1}{CH_2}—OH$$

which shows that it is 3-methyl-1-butanol. The alternative numbering, resulting in 2-methyl-4-butanol, is rejected because it gives —OH a higher-numbered location.

EXERCISE E24.1 Classify and name (a) $CH_3CH_2CH(OH)CH_2CH_3$ and (b) $CH_3CH_2C(CH_3)_2OH$.

[*Answer*: (a) secondary, 3-pentanol; (b) tertiary, 2-methyl-2-butanol]

E Ethylene glycol is also used to make polyester fibers (see Section 24.6); glycerol is used to make trinitroglycerol (the explosive nitroglycerin).

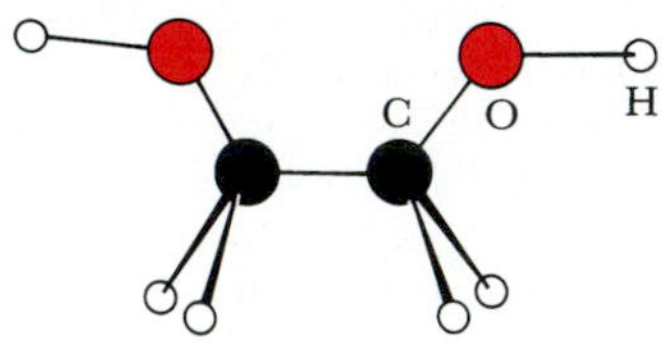

1 Ethylene glycol

Ethylene glycol, or 1,2-ethandiol, $HOCH_2CH_2OH$ (**1**), is an example of a **diol**, a compound with two hydroxyl groups. It is used as a component of antifreeze and in the manufacture of some synthetic fibers. Its action as an antifreeze stems from its high solubility (on account of its hydroxyl groups, which can form hydrogen bonds with water) and the disruptive effect its short hydrocarbon backbone has on the structure of ice: that is, the presence of ethylene glycol makes it difficult for the water molecules to form hydrogen bonds with each other. It is also not very volatile, so it does not boil away readily.

Glycerol, or 1,2,3-propantriol, $HOCH_2CH(OH)CH_2OH$ (**2**), has three hydroxyl groups and is thus a **triol**. It is a syrupy, viscous liquid that dissolves readily in water and is widely used as a skin softener in cosmetic preparations and as an antidrying medium and sweetening agent in toothpaste and packaged coconut. Its characteristic physical properties—its viscosity, its high solubility in water, and its water-retaining power—all result from its ability to form hydrogen bonds with itself and with water.

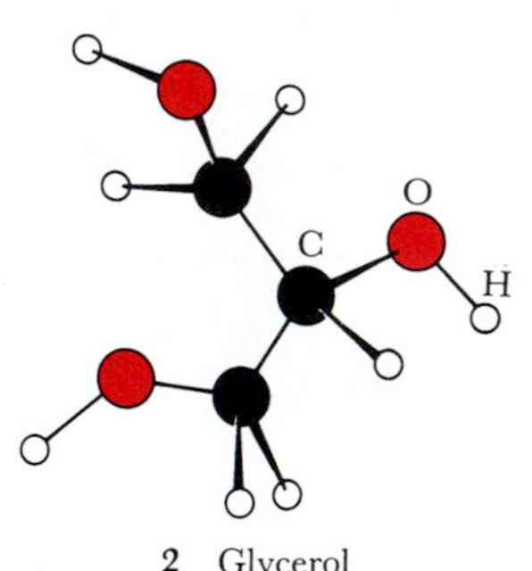

2 Glycerol

E Saving a life with alcohol. *J. Chem. Ed.*, **1973**, *50*, 365.

Properties and reactions of alcohols. Alcohols with low molar masses are liquids as a result of hydrogen bonding. Alcohols have much lower vapor pressures than the parent hydrocarbons: compare, for example, the volatility of ethanol with that of its parent hydrocarbon, the gas ethane. Their ability to form hydrogen bonds also accounts for the fact that alcohols are much more soluble in water than hydrocarbons are. Both methanol and ethanol mix with water in all proportions.

Alcohols are amphiprotic. They are very weak Brønsted acids:

$$CH_3CH_2OH(aq) + H_2O(aq) \rightleftharpoons H_3O^+(aq) + CH_3CH_2O^-(aq) \qquad pK_a = 16$$

Ethanol is so weak an acid that it cannot turn litmus red. The conjugate base $CH_3CH_2O^-$, the **ethoxide ion**, is formed when sodium metal reacts with ethanol, which it does more gently than with water (Fig. 24.1):

$$2Na + 2CH_3CH_2OH \longrightarrow 2NaCH_3CH_2O + H_2$$

D Waste sodium is discarded by placing it in ethanol. Compare the reactions of small amounts of sodium with ethanol and water.

(As in this equation, the phases of the substances are not shown when the solvent is not water—here the solvent is ethanol itself.) Alcohols can also act as Brønsted bases, because the lone pairs of electrons on the O atom of the —OH group can accept protons from other acids:

$$H_3O^+(aq) + CH_3CH_2OH(aq) \rightleftharpoons CH_3CH_2OH_2^+(aq) + H_2O(l)$$

The conjugate acid $CH_3CH_2OH_2^+$ is an example of an **oxonium ion**, an ion of the form ROH_2^+ (the hydronium ion, H_3O^+, is a special case of an oxonium ion in which R is H). The formation of an oxonium ion is the first step in acid-catalyzed reactions of alcohols, including their dehydration to alkenes:

$$CH_3CH_2OH \xrightarrow{\text{conc } H_2SO_4,\ 170^\circ C} CH_2{=}CH_2 + H_2O$$

Dehydration is a potential route from the ethanol produced by fermentation to the additional polymers normally obtained from petrochemicals (Section 23.6).

Alcohols can be used as reagents in halogenation reactions (reactions in which a halogen atom is introduced into a molecule). For example, concentrated hydrobromic acid converts alcohols into bromoalkanes:

$$CH_3CH_2{-}OH + HBr \xrightarrow{\text{warm}} CH_3CH_2{-}Br + H_2O$$

Such brominations are important because the products can be used to synthesize other compounds.

Oxidizing agents, including the oxygen of the air and, more rapidly, acidified dichromate solutions, produce the compounds known as aldehydes (we deal with aldehydes more fully in Section 24.3). Acetaldehyde, CH_3CHO, for example, is produced from ethanol, a primary alcohol. We can express this schematically as

$$CH_3{-}\underset{\displaystyle H}{\overset{\displaystyle OH}{C}}{-}H \xrightarrow[\text{oxidation removes 2H}]{[O]} CH_3{-}C({=}O){-}H + H_2O$$

where the [O] denotes the oxygen atom supplied by the oxidizing agent. As we shall see later, aldehydes are also readily oxidized; hence the oxidation of a primary alcohol often proceeds further, to produce a carboxylic acid (such as acetic acid, CH_3COOH):

$$CH_3{-}C({=}O){-}H \xrightarrow[\text{oxidation adds oxygen}]{[O]} CH_3{-}C({=}O){-}O{-}H$$

These last two schemes show the sequence of events that cause wine to sour as its ethanol is oxidized to acetic acid (the acidic component of vinegar).

The oxidation of secondary alcohols produces the compounds called ketones (such as acetone, CH_3COCH_3, from 2-propanol; Section 24.3)

$$CH_3{-}\underset{\displaystyle CH_3}{\overset{\displaystyle OH}{C}}{-}H \xrightarrow[\text{oxidation removes 2H}]{[O]} CH_3{-}C({=}O){-}CH_3 + H_2O$$

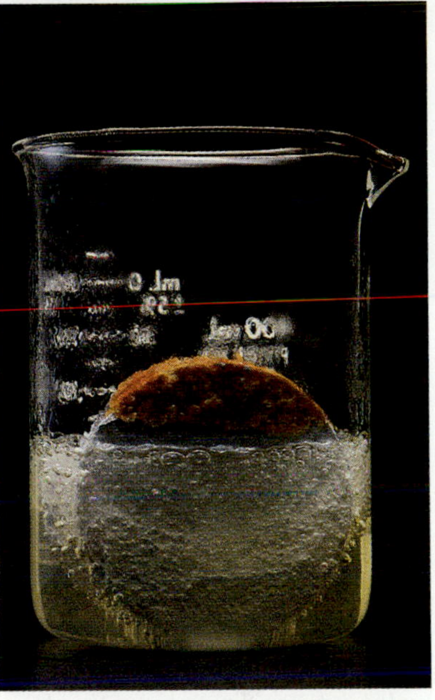

FIGURE 24.1 Sodium reacts with ethanol much less vigorously than it reacts with water. The products are hydrogen and a solution of sodium ethoxide.

D Comparison of chemical oxidation of alkanes, alkenes, and alcohols on the overhead projector. *J. Chem. Ed.*, **1986**, *63*, 000.

E Drinking too fast can cause sudden death. *J. Chem. Ed.*, **1973**, *50*, 365. A student preparation of acetone from 2-propanol. *J. Chem. Ed.*, **1982**, *59*, 862.

N Schematic equations of this kind are widely used in organic chemistry where a properly balanced chemical equation would not convey the content of a reaction very clearly.

E Methyl tert-butyl ether, MTB, $CH_3OC(CH_3)_3$, is a gasoline additive used to replace tetraethyl lead as an antiknock agent and to promote more complete fuel combustion.

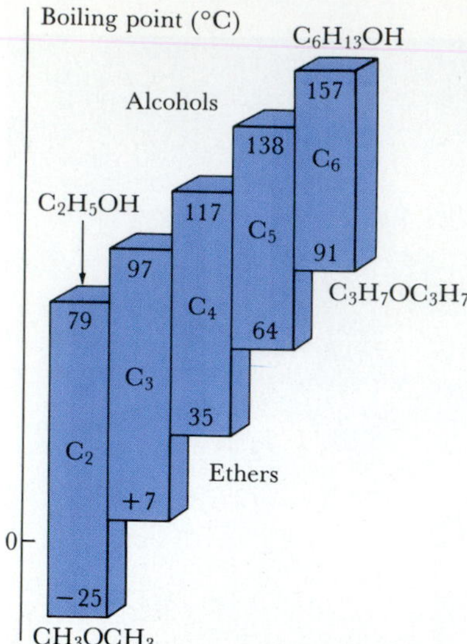

FIGURE 24.2 The boiling points of ethers are lower than those of isomeric alcohols, because hydrogen bonding occurs in alcohols but not in ethers. All the alkyl groups here are straight chains.

E Several inert gases can be used as anesthetics: they include dinitrogen oxide, xenon, and diethyl ether. Their function appears to be to dissolve in the membranes of neurons and hence disrupt their function. They do not attack the membranes chemically, so the anesthesia is reversible.

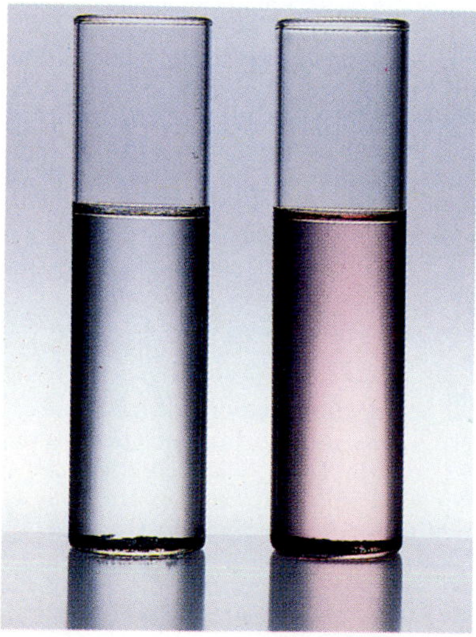

FIGURE 24.3 Both flasks contain potassium permanganate and benzene, in which the permanganate is almost insoluble. A crown ether has been added to the flask on the right; and as the purple color shows, the potassium permanganate has dissolved.

Ketones are much more difficult to oxidize than aldehydes, and the oxidation does not go further (unless it is very vigorous, as it is in a combustion).

Ethers. Just as we can think of an alcohol as being derived from HOH by replacement of one H atom with an alkyl group, we can think of an **ether** as an HOH molecule in which both H atoms are replaced by an alkyl group to give a compound of the form R—O—R:

H—O—H	CH_3CH_2—O—H	CH_3CH_2—O—CH_2CH_3
Water	Ethanol	Diethyl ether

(Diethyl ether is the "ether" once used as an anesthetic.) Ether is more volatile than an alcohol with the same molar mass because the ether is not hydrogen bonded (Fig. 24.2).

Because ethers are not very reactive toward many reagents, they are useful solvents for other organic compounds. However, ethers are flammable; diethyl ether is so easily ignited that it can cause accidental fires. Like alcohols, ethers can undergo acid-catalyzed substitution reactions.

▼ EXAMPLE 24.2 Predicting the product of a substitution reaction

Predict the outcome of the reaction between Br^- ions and diethyl ether in the presence of a strong acid.

STRATEGY We have seen that in the halogenation of alcohols —OH is driven out of R—OH, leaving R—Br as the product; we predict that —OR will be driven out of R—OR, leaving R—Br as the product.

SOLUTION Building on this analogy, we can write the reaction as

$$CH_3CH_2\text{—O—}CH_2CH_3 + HBr \xrightarrow{H^+} CH_3CH_2OH + CH_3CH_2Br$$

The product CH_3CH_2OH may go on to form more of the bromoethane by a continuation of the reaction.

EXERCISE E24.2 Predict the products of the acid-catalyzed reaction between $CH_3CH_2OCH_2CH_2CH_3$ and an excess of Br^- ions.

[*Answer*: CH_3CH_2Br and $CH_3CH_2CH_2Br$] ▲

Cyclic ethers (ethers in the form of a ring) with alternating —CH_2CH_2— units and O atoms are called **crown ethers**. The name reflects the crownlike shape the molecules adopt (**3**). The importance of crown ethers stems from their ability to bind very strongly to metal cations (such as K^+), effectively encasing them in a hydrocarbon shell and making them soluble in nonpolar solvents. Potassium permanganate, for example, dissolves readily in benzene when a crown ether with six O atoms is added to it (Fig. 24.3). The resulting solution can be used to carry out oxidation reactions in organic solvents.

24.2 PHENOLS

The compound obtained when a hydroxyl group is attached directly to an aromatic ring is a **phenol**, not an alcohol. Phenol itself, a white, low-melting-point, crystalline molecular solid, is C_6H_5OH (**4**). At one time it was obtained from the distillation of coal tar, but now it is mainly produced synthetically from benzene. Many phenols occur naturally and some are responsible for the fragrances of plants. They are often components of **essential oils**, the oils that can be distilled from flowers and leaves. Thus thymol (**5**) is the active ingredient of oil of thyme and eugenol (**6**) that of oil of cloves.

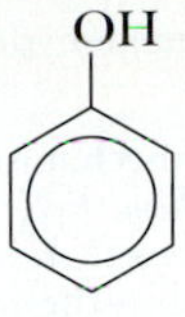

OH

4 Phenol

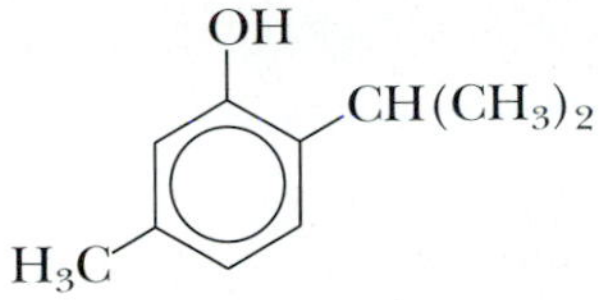

OH

CH(CH$_3$)$_2$

H$_3$C

5 Thymol

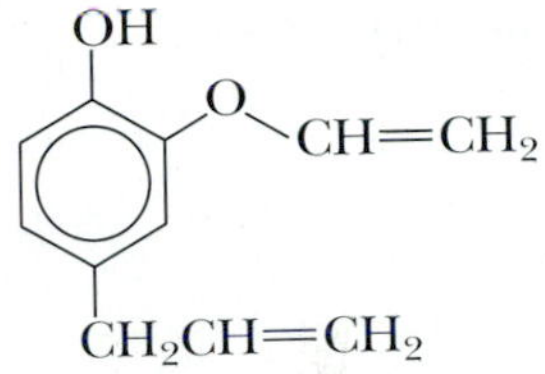

OH

O CH=CH$_2$

CH$_2$CH=CH$_2$

6 Eugenol

OH

Cl Cl

Cl

7 2,4,6-Trichlorophenol

Properties and reactions of phenols. Although phenols resemble alcohols in that they both contain C—OH groups, the location of the C atom in an aromatic ring results in markedly different chemical properties. In particular, phenols are stronger as acids than the alcohols are:

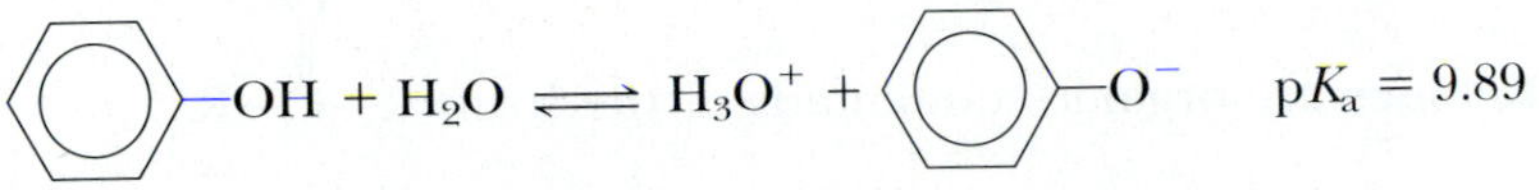

$$C_6H_5\text{—OH} + H_2O \rightleftharpoons H_3O^+ + C_6H_5\text{—O}^- \qquad pK_a = 9.89$$

Phenol is in fact known as carbolic acid. The conjugate base of phenol, $C_6H_5O^-$, is the **phenoxide ion**. The greater acidity of phenols compared with that of alcohols stems from the ability of the negative charge of the phenoxide ion to spread over the benzene ring, which helps to stabilize the conjugate base.

Although the benzene ring itself is not particularly reactive, substituents (including the —OH group) can often sensitize it to attack. The reaction between chlorine and aqueous phenol, for instance, continues as far as the antibacterial agent 2,4,6-trichlorophenol (**7**), even without a catalyst:

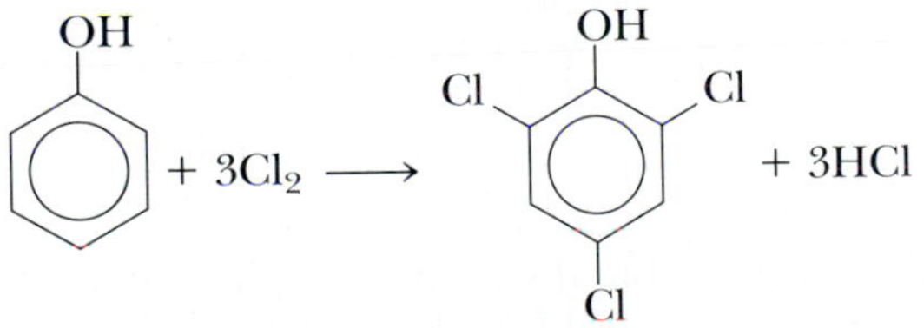

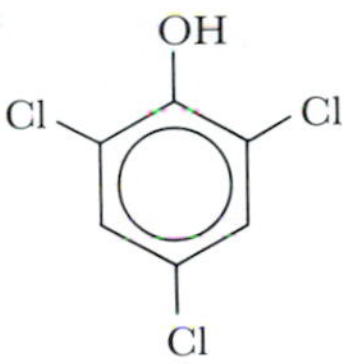

$$C_6H_5OH + 3Cl_2 \longrightarrow C_6H_2Cl_3OH + 3HCl$$

N The delocalization of charge can be represented by drawing the resonance structures that contribute to the structure of the molecule:

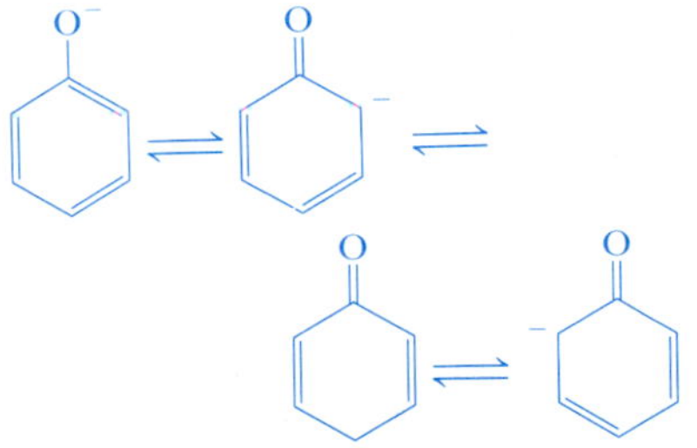

O$^-$ ⇌ O ⇌ O ⇌ O

Phenolic resins. The phenoxide ion has an even higher accumulation of negative charge in the ring than phenol has, for the O$^-$ atom is a better supplier of electrons than the —OH group. Because this accumulation of electrons enhances the reactivity of the ring, when phenol is in basic solution (where it is present largely as phenoxide ions), it is even more reactive than phenol itself. This feature is employed in the manufacture of **phenolic resins** (or phenol-formaldehyde resins). Phenolic resins are used as plywood adhesive, as binder for fiberglass insulation, and in molded electrical components.

E The enhanced sensitivity of the benzene ring to attack can be traced to the delocalization of electrons from the O atom in the hydroxyl group:

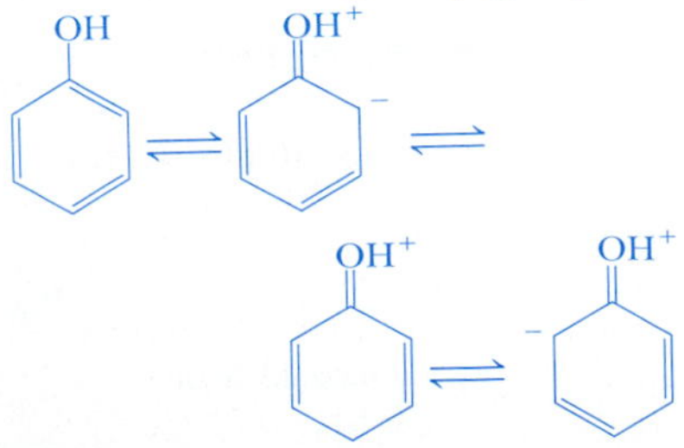

OH ⇌ OH$^+$ ⇌ OH$^+$ ⇌ OH$^+$

THE CARBONYL GROUP

The **carbonyl group,** $>$C=O occurs in several functional groups and has distinctive properties. In terms of the language introduced in Section 9.6, the double bond in the carbonyl group consists of a σ bond

D Phenol-formaldehyde polymer, BZS1, 3.3.

E Bioorganic mechanisms II. Chemoreception. *J. Chem. Ed.*, **1982**, *59*, 759. What every chemist should know about teratogens—chemical that cause birth defects. *J. Chem. Ed.*, **1982**, *59*, 759.

formed from a carbon sp^2 orbital and a π bond formed from one of its $2p$ orbitals. In this section, we meet two closely related families of compounds that contain the group; these are the aldehydes and ketones, which have already been mentioned as products of the oxidation of primary and secondary alcohols, respectively. Then we consider the carbohydrates, which contain the carbonyl group in a disguised form.

24.3 ALDEHYDES AND KETONES

Aldehydes and ketones both contain the carbonyl group but differ in the number of H atoms that are directly attached to it:

E Acetaldehyde: A chemical whose fortunes have changed. *J. Chem. Ed.*, **1983**, *60*, 1044.

Aldehydes are organic compounds of the form R—C(=O)—H.

Ketones are organic compounds of the form R—C(=O)—R.

The R— groups may be either aliphatic or aromatic. The —C(=O)H group characteristic of the aldehydes is normally written —CHO, as in HCHO (formaldehyde), the simplest aldehyde. The liquid formalin, used to preserve biological specimens, is an aqueous solution of formaldehyde. Wood smoke contains formaldehyde, and formaldehyde's destructive effect on bacteria is one of the reasons why smoking food helps to preserve it. The formula of an aldehyde differs from that of a *primary* alcohol by the loss of 2H:

E Chemical effects in food stored at room temperature. *J. Chem. Ed.*, **1984**, *61*, 335. Chemical effects during storage of frozen foods. *J. Chem. Ed.*, **1984**, *61*, 340.

R—CH(OH)—H $\xrightarrow{-2H}$ R—C(=O)—H

This difference is reflected in the name *aldehyde*, which is derived from *al*cohol-*dehyd*rogenated. Acetaldehyde is in fact a product of ethanol oxidation in the liver, and its accumulation in the blood is one of the causes of hangovers.

The formulas of ketones differ by two H atoms from those of *secondary* alcohols:

R—CR(OH)—H $\xrightarrow{-2H}$ R—C(=O)—R

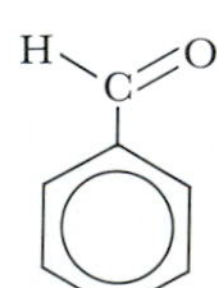

8 Benzaldehyde

C_6H_5—CH=CH—CHO

9 Cinnamaldehyde

The name *ketone* is derived from the German name (*Keton*) for the simplest member, acetone, CH_3—CO—CH_3.

Aldehydes occur naturally in essential oils and are often responsible for the flavors of fruits and the odors of plants. Benzaldehyde, C_6H_5CHO (**8**), is responsible for the characteristic aroma of cherries and almonds. Cinnamaldehyde (**9**) occurs in oil of cinnamon, and vanillin (**10**) in oil of vanilla. Ketones can be fragrant; for example, carvone (**11**), the essential oil of spearmint, is used in chewing gum.

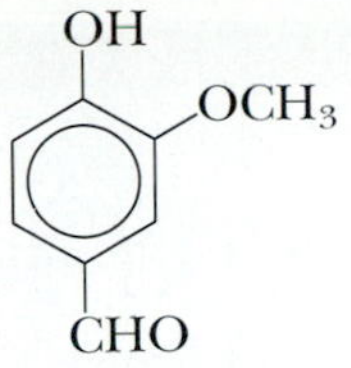

OH
OCH$_3$
CHO

10 Vanillin

CH$_3$
O
CH$_2$ C CH$_3$

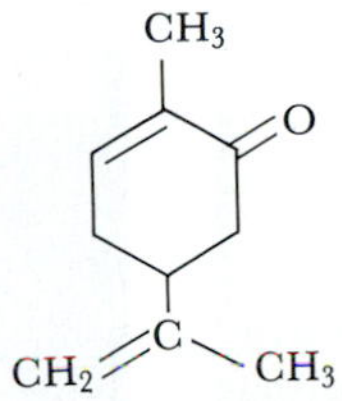

11 Carvone

Nomenclature. The systematic name of an aldehyde is formed by identifying the parent alkane (with the —CHO carbon atom included in the count of C atoms) and changing the final *-e* of the name to *-al*. Thus, acetaldehyde, CH_3CHO, is treated as a derivative of ethane, and its systematic name is ethanal. When the carbon atoms must be numbered to specify the locations of substituents, we start with 1 at the —CHO group.

Ketones are named systematically by changing the *-e* in the name of the parent alkane to *-one*. Thus, because acetone, CH_3—CO—CH_3, is a derivative of propane, its systematic name is propanone. When numbering is necessary, it begins at the end of the molecule that results in the lower number for the location of the carbonyl group; hence $CH_3CH_2CH_2$—CO—CH_3 is 2-pentanone.

N Acetone is completely miscible with water and is also an excellent organic solvent.

▼ EXAMPLE 24.3 Naming aldehydes and ketones

Name the following organic compounds: (a) $CH_3CH_2CH_2CHO$; (b) $CH_3CH(CHO)CH_2CH_3$; (c) $CH_3CH_2COCH_3$.

STRATEGY We simply need to follow the rules as given above for aldehydes (a, b) and ketones (c). The first step in every case is to identify the longest chain (by writing it out if necessary), including the carbonyl group. We can identify the corresponding alkane by referring to Table 23.2.

SOLUTION (a) This aldehyde is a derivative of butane, so its name is butanal. (b) Write the structure out and identify the longest chain that contains the —CHO group:

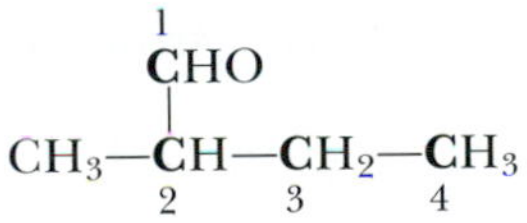

$$\begin{array}{c} \overset{1}{\mathrm{CHO}} \\ | \\ \mathrm{CH_3}\text{—}\underset{2}{\mathrm{CH}}\text{—}\underset{3}{\mathrm{CH_2}}\text{—}\underset{4}{\mathrm{CH_3}} \end{array}$$

This molecule is the derivative of a substituted C_4 alkane; therefore, like (a), it is a butanal. Because the carbon atom of the —CHO group is numbered 1, the methyl substituent is at carbon atom 2 and the compound is 2-methylbutanal. (c) This ketone is also a derivative of butane; hence it is a butanone. There is no need to call it 2-butanone because the only alternative position for the carbonyl group would result in an aldehyde, not a ketone.

EXERCISE E24.3 Name the following compounds: (a) $CH_3(CH_2)_3CHO$; (b) $CH_3CH_2CH(CHO)(CH_2)_3CH_3$; (c) $CH_3(CH_2)_2COCH_2CH_3$.

▲ [*Answer*: (a) pentanal; (b) 2-ethylhexanal; (c) 3-hexanone]

Preparation. Because aldehydes and ketones differ from alcohols by the loss of two H atoms, both can be prepared by oxidation of the corresponding alcohol. Schematically,

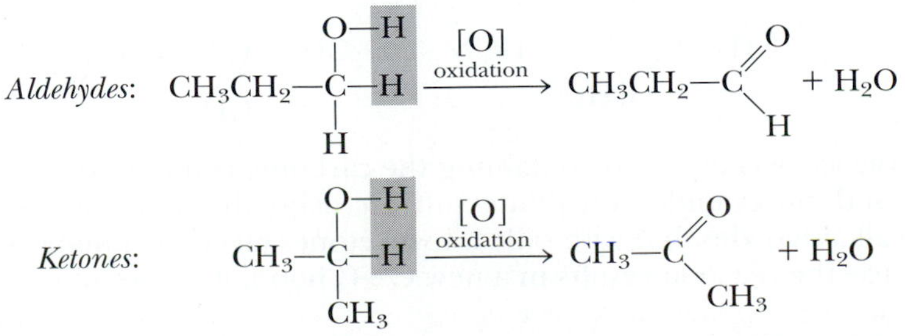

$$\textit{Aldehydes:}\quad \mathrm{CH_3CH_2}\text{—}\underset{\mathrm{H}}{\overset{\mathrm{O—H}}{\mathrm{C}}}\text{—H} \xrightarrow[\text{oxidation}]{[\mathrm{O}]} \mathrm{CH_3CH_2}\text{—C}(\text{=O})\mathrm{H} + \mathrm{H_2O}$$

$$\textit{Ketones:}\quad \mathrm{CH_3}\text{—}\underset{\mathrm{CH_3}}{\overset{\mathrm{O—H}}{\mathrm{C}}}\text{—H} \xrightarrow[\text{oxidation}]{[\mathrm{O}]} \mathrm{CH_3}\text{—C}(\text{=O})\mathrm{CH_3} + \mathrm{H_2O}$$

FIGURE 24.4 An aldehyde (left) produces a brick-red precipitate of copper(I) oxide with Fehling's solution, but a ketone (right) does not. The unreduced Fehling's solution is purple.

In oxidizing a primary alcohol to an aldehyde, oxidation of the product is avoided by using a mild oxidizing agent, such as acidified sodium dichromate. Schematically,

$$CH_3CH_2CH_2OH \xrightarrow{Na_2Cr_2O_7(aq),\ H_2SO_4(aq)} CH_3CH_2CHO$$

Aldehydes are also prepared in organic solvents, because this method reduces the risk of further oxidation. Formaldehyde is prepared industrially (for the manufacture of phenolic resins) by the catalytic oxidation of methanol:

$$2CH_3OH(g) + O_2(g) \xrightarrow{600°C,\ Ag} 2HCHO(g) + 2H_2O(g)$$

There is less risk of oxidation for ketones than for aldehydes, because a C—C bond has to be broken. Dichromate oxidation of secondary alcohols produces the ketone in good yield with little additional oxidation. Schematically,

$$CH_3CH_2CH(OH)CH_3 \xrightarrow{Na_2Cr_2O_7(aq),\ H_2SO_4(aq)} CH_3CH_2COCH_3$$

Reactions. One sign of the difference between the ease of oxidation of aldehydes and that of ketones is the ability of aldehydes to reduce Fehling's solution, a basic solution of Cu^{2+} and tartrate ions. The product of this reaction is a brick-red precipitate of Cu_2O. Ketones do not react with Fehling's solution (Fig. 24.4). Schematically,

Aldehydes: $CH_3CH_2CHO + Cu^{2+}$ (from Fehling's solution) $\longrightarrow$

$CH_3CH_2COOH + Cu_2O(s)$

Ketones: $CH_3COCH_3 + Cu^{2+}$ (from Fehling's solution) $\longrightarrow$

no reaction

E The role of the tartrate ions in Fehling's solution is to form a complex with the Cu^{2+} ions to prevent the precipitation of copper as the hydroxide in the basic solution.

V Video 56: A silver mirror.

D The silver mirror reaction. CD1, 75. The blue bottle reaction, CD1, 76.

Nor do ketones produce a silver mirror—a coating of silver on the test tube—with Tollens' reagent, a solution of Ag^+ ions in aqueous ammonia (Fig. 24.5). Schematically:

Aldehydes: $CH_3CH_2CHO + Ag^+$ (from Tollens' reagent) $\longrightarrow$

$CH_3CH_2COOH + Ag(s)$

Ketones: $CH_3COCH_3 + Ag^+$ (from Tollens' reagent) $\longrightarrow$

no reaction

FIGURE 24.5 An aldehyde (left) produces a silver mirror with Tollens' reagent, but a ketone (right) does not.

The carbonyl group reacts with the hydrogen cyanide to give a **cyanohydrin**, a compound containing both an —OH group and a —C≡N group:

$$CH_3-C(=O)-CH_3 + HCN \longrightarrow CH_3-C(OH)(CH_3)-C\equiv N$$

In practice, a compound containing the carbonyl compound is treated with sodium cyanide and dilute sulfuric acid; the reaction succeeds with all aldehydes, but with only a few ketones (acetone among them). Because the reaction results in a new C—C bond, it is useful for building up molecules containing more carbon atoms. For example, cyanohydrins are readily hydrolyzed to carboxylic acids:

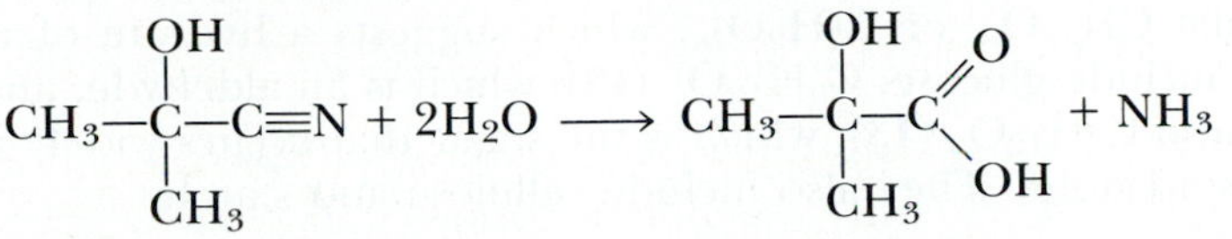

$$CH_3-\underset{\underset{CH_3}{|}}{\overset{\overset{OH}{|}}{C}}-C\equiv N + 2H_2O \longrightarrow CH_3-\underset{\underset{CH_3}{|}}{\overset{\overset{OH}{|}}{C}}-C\genfrac{}{}{0pt}{}{\nearrow O}{\searrow OH} + NH_3$$

(The fact that the hydrolysis gives a carboxylic acid with an additional —OH group need not concern us here; however, that extra functional group can be very useful in organic syntheses.)

Aldehydes combine with alcohols in the presence of dry hydrogen chloride to form **hemiacetals**, compounds with both an —O— ether link and an —OH group:

N The purpose of introducing this somewhat specialized material here is to provide a basis for the discussion of carbohydrates in the following section.

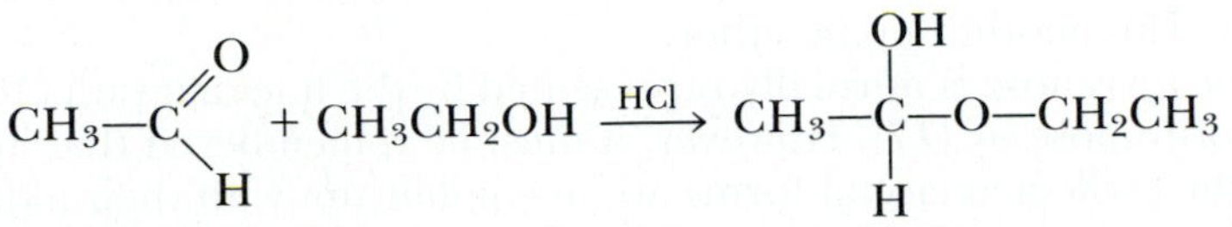

$$CH_3-C\genfrac{}{}{0pt}{}{\nearrow O}{\searrow H} + CH_3CH_2OH \xrightarrow{HCl} CH_3-\underset{\underset{H}{|}}{\overset{\overset{OH}{|}}{C}}-O-CH_2CH_3$$

The hemiacetal (the prefix *hemi-* is the Greek word for "half") readily reacts with more alcohol to form the corresponding **acetal**, a type of double ether that contains the —O—C—O— group:

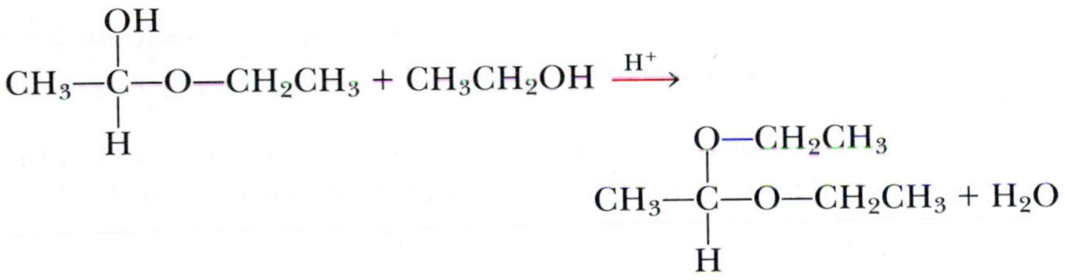

$$CH_3-\underset{\underset{H}{|}}{\overset{\overset{OH}{|}}{C}}-O-CH_2CH_3 + CH_3CH_2OH \xrightarrow{H^+} CH_3-\underset{\underset{H}{|}}{\overset{\overset{O-CH_2CH_3}{|}}{C}}-O-CH_2CH_3 + H_2O$$

If the original aldehyde contains a hydroxyl group, then hemiacetal formation may occur intramolecularly (within the molecule itself):

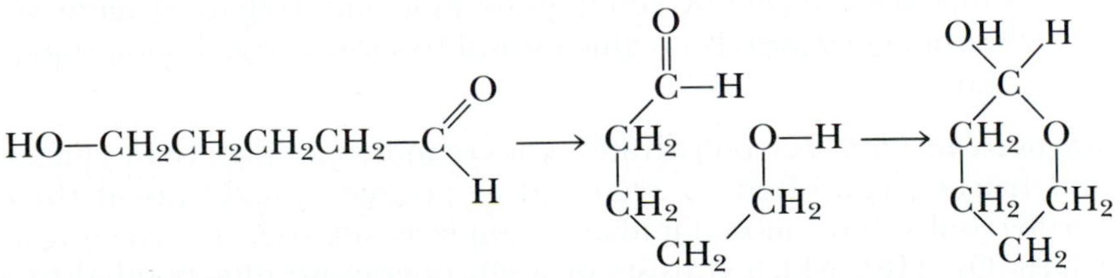

$$HO-CH_2CH_2CH_2CH_2-C\genfrac{}{}{0pt}{}{\nearrow O}{\searrow H} \longrightarrow \begin{matrix} & \overset{O}{\overset{\|}{C}}-H & \\ CH_2 & & O-H \\ | & & | \\ CH_2 & & CH_2 \\ & CH_2 & \end{matrix} \longrightarrow \begin{matrix} & \overset{OH\ \ H}{\overset{\diagdown\ \diagup}{C}} & \\ CH_2 & & O \\ | & & | \\ CH_2 & & CH_2 \\ & CH_2 & \end{matrix}$$

The resulting compound is a cyclic hemiacetal that is stabilized by ring formation. We shall see in the next section that this reaction is of immense importance for understanding the structure of cellulose, the most abundant organic chemical in the world.

Finally, in the reverse of the reactions by which they are made, aldehydes and ketones can be reduced to alcohols. The reducing agents used are usually sodium borohydride, $NaBH_4$, in ethanol or lithium aluminum hydride, $LiAlH_4$, in ether.

24.4 CARBOHYDRATES

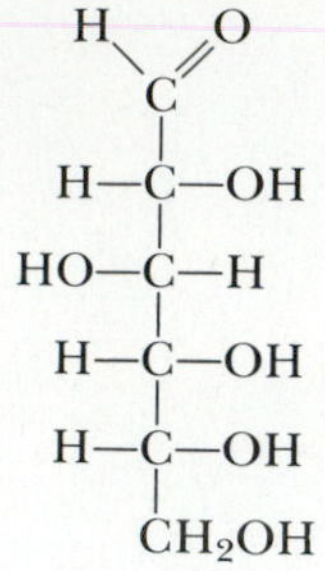

The **carbohydrates** are so called because they often have the empirical formula $CH_{2n}O_n$, or $C(H_2O)_n$, which suggests a hydrate of carbon. They include glucose, $C_6H_{12}O_6$ (**12**), which is an aldehyde, and fructose, also $C_6H_{12}O_6$ (**13**), which is the sugar that occurs widely in fruit and is a ketone. They also include cellulose and starch.

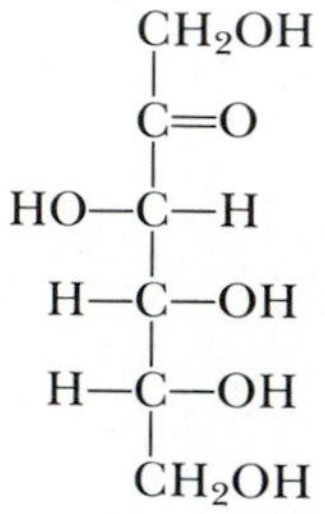

12 Glucose

13 Fructose

Pyranose and furanose forms. Glucose is both an alcohol and an aldehyde and has a reasonably long and flexible backbone; so it can form a cyclic hemiacetal. The six-membered ring is called the **pyranose** form of glucose because it resembles the cyclic ether pyran (**14**); it is specifically named a glucopyranose. Fructose is also present as a cyclic hemiacetal in solution, mainly as a pyranose six-membered ring but also (to the extent of about 30%) as a five-membered ring. The latter is called its **furanose** form, specifically a fructofuranose, because it resembles furan (**15**), another cyclic ether.

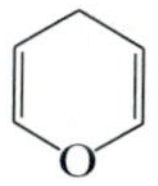

Glucopyranose is normally represented by the line diagram (**16**) and fructofuranose by (**17**). However, it must be remembered that in solution the cyclic hemiacetal forms are in equilibrium with their aldehyde and ketone open-chain forms (**12**, **13**). Evidence for the presence of these open-chained forms is the fact that glucose undergoes the typical reaction of an aldehyde, namely, oxidation. With Fehling's solution, glucose gives a brick-red precipitate of copper(I) oxide. If we show only the —CHO group of the molecule explicitly, we can express the reaction schematically as

14 A pyran ring

$$\ldots\text{—CHO} + Cu^{2+} \xrightarrow{\text{Fehling's solution}} \ldots\text{—}CO_2^- + Cu_2O(s)$$

Fructose is also oxidized, but in this case the site of oxidation is the —OH group on the carbon atom next to the carbonyl group:

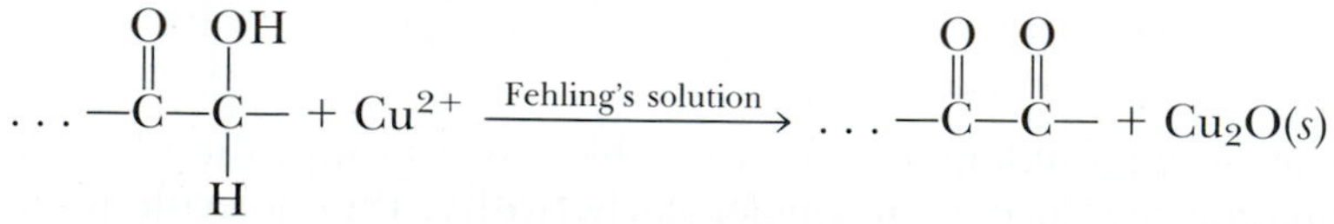

$$\ldots\text{—C(=O)—CH(OH)—} + Cu^{2+} \xrightarrow{\text{Fehling's solution}} \ldots\text{—C(=O)—C(=O)—} + Cu_2O(s)$$

15 Furan

Sugars that are oxidized by Fehling's solution and Tollens' reagent are called **reducing sugars**. Both glucose and fructose are reducing sugars (Fig. 24.6).

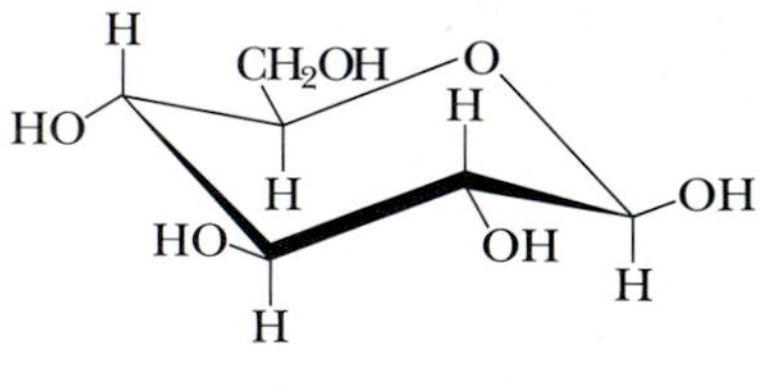

16 Glucopyranose

Oligosaccharides. Carbohydrates that are more complex than glucose and fructose can often be regarded as polymers made up of these smaller units. The most familiar example is sucrose, or cane sugar, $C_{12}H_{22}O_{11}$ (**18**), which consists of a glucopyranose unit bonded to a

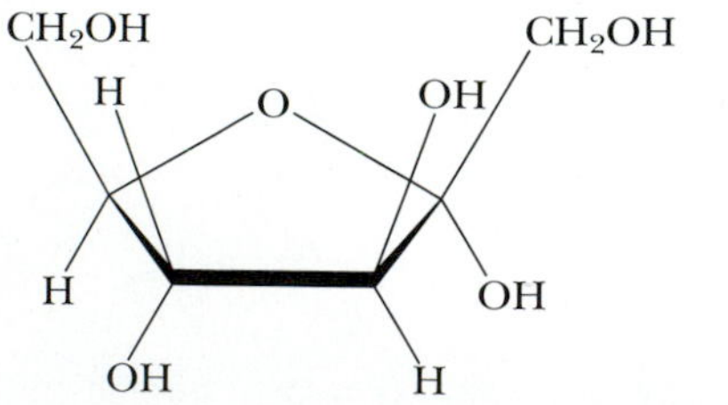

17 Fructofuranose

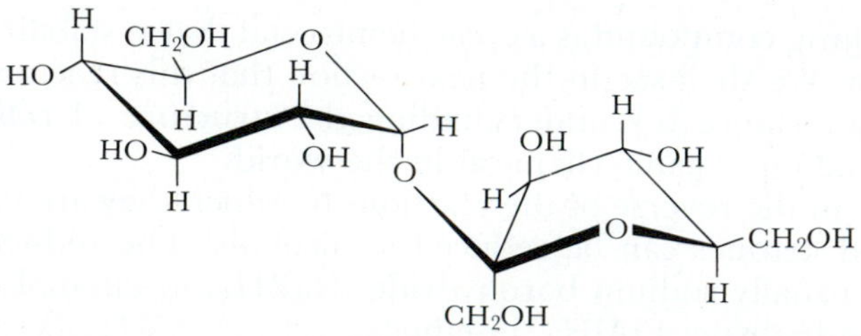

18 Sucrose

fructofuranose unit. Because it consists of two C_6 carbohydrate units, it is called a **disaccharide**; each individual unit is a **monosaccharide**. A string of several saccharide units is called an **oligosaccharide** (the prefix is from the Greek word for "a few").

The hemiacetal rings of sucrose are not in equilibrium with their open-chain forms, because the O atom that would become the carbonyl oxygen now forms a bridge between the linked rings. Therefore, sucrose does not show the typical reactions of an aldehyde or a ketone; in particular, it is not a reducing sugar. However, it is very soluble in water (because of its numerous hydroxyl groups) and is broken down (hydrolyzed) into its component sugars by enzymes in the digestive systems.

FIGURE 24.6 Glucose is a reducing sugar and produces a precipitate with Fehling's solution (left). Sucrose is not a reducing sugar, so it does not react with Fehling's solution (right).

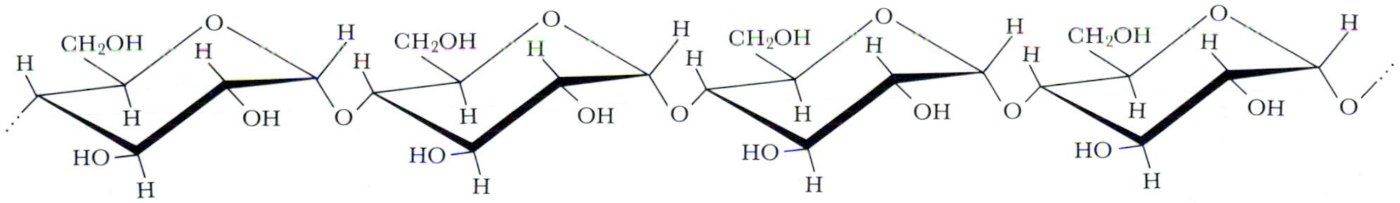

Polysaccharides. The **polysaccharides** consist of chains of many glucose units linked together. This category includes starch, which we can digest (hydrolyze), and cellulose, which we cannot.

Starch is made up of two components: amylose and amylopectin. Amylose, which makes up about 20 to 25% of most starches, is composed of chains made up of several thousand glucose units linked together (Fig. 24.7). Amylopectin is also made up of glucose chains, but its chains are linked into a branched structure and its molecules are much larger, having typically about a million glucose units (Fig. 24.8).

Cellulose is the structural material of plants. Like starch, it is a polymer of glucose, but the units link differently, forming flat, ribbonlike strands (Fig. 24.9). These strands can lock together through hydrogen bonds into a rigid structure that for us (but not for termites) is indigestible. The difference between cellulose and starch shows nature at its

FIGURE 24.7 The amylose molecule, one component of starch, is a polysaccharide. It consists of glucose units linked together to give a moderately branched structure.

E Even herbivores find difficulty in digesting their carbohydrate food: cows and other ruminants need an elaborate series of stomachs, and rabbits, in a more primitive approach to the problem, eat their own excrement.

D Nitrocellulose, CD2, 57.

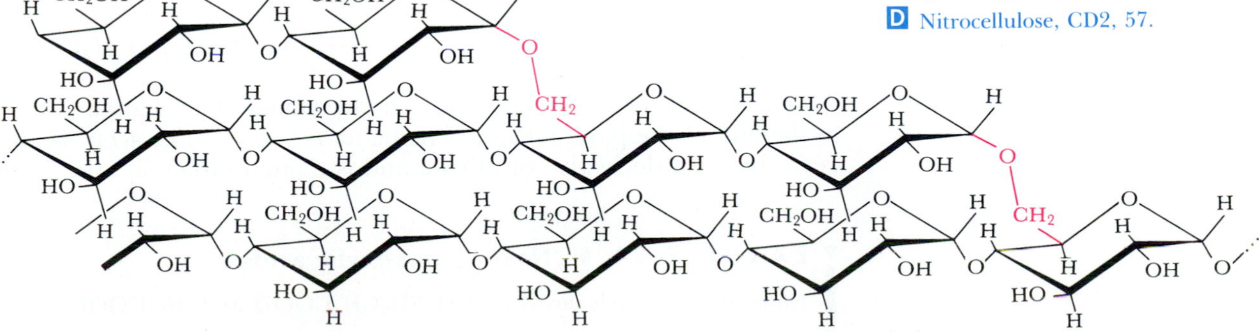

FIGURE 24.8 The amylopectin molecule is another component of starch. It has a more highly branched structure than amylose.

E Chitin and chitosan: Versatile polysaccharides from marine animals. *J. Chem. Ed.*, **1990**, *67*, 938. Which starch fraction is water soluble, amylose or amylopectin? *J. Chem. Ed.*, **1975**, *52*, 729.

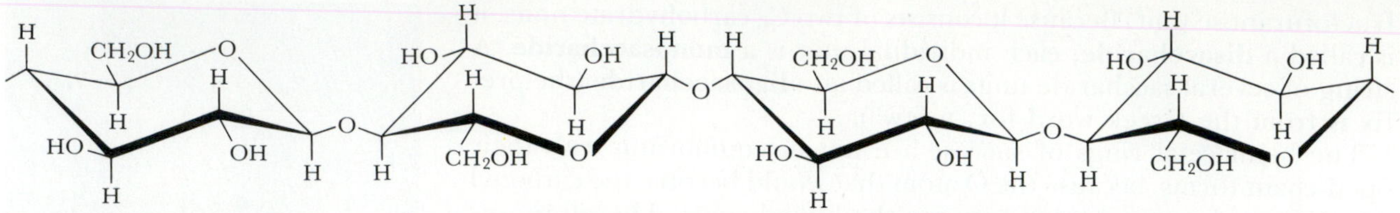

FIGURE 24.9 Cellulose is yet another polysaccharide that is constructed from glucose units. However, the linking between the glucose units in cellulose results in long, flat ribbons that can produce a fibrous material through hydrogen bonding.

most economical and elegant, for only a small modification of the linking between glucose units results on the one hand in an important foodstuff and on the other in a versatile construction material.

THE CARBOXYL GROUP

We now turn to a very common functional group, the **carboxyl group** $-\mathrm{C}(=\mathrm{O})-\mathrm{OH}$. Although it is normally abbreviated to —COOH, we should always remember that the group does not contain an O—O bond. Loss of a hydrogen ion from the —COOH group results in the **carboxylate group**, $-CO_2^-$. The two oxygen atoms are now equivalent as a result of resonance:

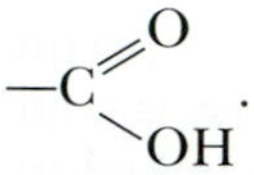

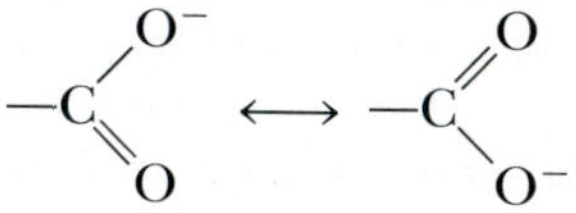

$$-\mathrm{C}(-\mathrm{O}^-)=\mathrm{O} \longleftrightarrow -\mathrm{C}(=\mathrm{O})-\mathrm{O}^-$$

E When the —OH groups on the cellulose of cotton are fully replaced by nitrate groups (by reaction with nitric acid), the explosive nitrocellulose (guncotton) is formed.

24.5 CARBOXYLIC ACIDS

Compounds containing the carboxyl group are called **carboxylic acids**. The properties of some of them are summarized in Table 24.2. We first encountered carboxylic acids as examples of acids in Sections 2.8 and 3.7; in Chapter 14 we saw that all carboxylic acids are weak acids.

Nomenclature. Many carboxylic acids are known by their common names. The simplest is formic acid, HCOOH (the acid injected by ants). The next higher homologue is acetic acid, CH_3COOH (the acid of vinegar), which is formed when the ethanol in wine is oxidized by air:

$$\mathrm{CH_3{-}CH_2{-}OH}(aq) + \mathrm{O_2}(g) \longrightarrow \mathrm{CH_3{-}\overset{\overset{\displaystyle O}{\|}}{C}{-}OH}(aq) + \mathrm{H_2O}(l)$$

The *systematic* names of carboxylic acids are obtained by changing the *-e* at the end of the parent alkane's name to *-oic acid.* Thus, acetic acid is regarded as a derivative of ethane and is named ethanoic acid.

E Vinegar is 4–5% acetic acid.

E The longer the alkyl group, the more foul the odor: butanoic acid, commonly called butyric acid, is the odor of rancid butter. The longer the alkyl group, the weaker the acid.

E *Formica* is the Latin word for "ant." The word *vinegar* comes from the French *vin aigre,* or "sour wine."

D Insect sting. TDC, 22-17.

E Legendary chemical aphrodisiacs. *J. Chem. Ed.,* **1980**, 57, 341.

▼ **EXAMPLE 24.4 Naming carboxylic acids**

Name the carboxylic acids (a) $CH_3CH_2CH_2COOH$ and (b) $(COOH)_2$.

STRATEGY We should proceed as outlined above, first identifying the parent alkane. If more than one functional group is present, then we use the prefixes *di-, tri-,* and so on.

TABLE 24.2 Carboxylic acids

Formula	Name	pK_{a1}	pK_{a2}	Melting point, °C	Boiling point, °C
H—COOH	formic acid,* methanoic acid	3.75	—	8	101
CH_3—COOH	acetic acid,* ethanoic acid	4.76	—	17	118
CH_2Cl—COOH	chloroacetic acid*	2.86	—	63	189
CH_3CH_2—COOH	propanoic acid	4.87	—	−21	141
C_6H_5—COOH	benzoic acid	4.20	—	122	249
COOH—COOH	oxalic acid,* ethanedioic acid	1.23	4.28	190d†	—
COOH—CH_2—COOH	malonic acid,* propanedioic acid	2.83	5.69	136d	—

*Common name.
†d signifies decomposition at temperatures near the melting point.

SOLUTION (a) The parent alkane of $CH_3CH_2CH_2COOH$ is butane, so the acid is butanoic acid. (b) The parent alkane of $(COOH)_2$, or HOOC—COOH, is ethane, and it has two carboxyl groups. Hence its systematic name is ethanedioic acid (the final *-e* of ethane is retained here to make the name easier to read). Its common name is oxalic acid.

EXERCISE E24.4 Name the carboxylic acids (a) $CH_3(CH_2)_6COOH$ and (b) $CH_2(COOH)_2$.

[*Answer*: (a) octanoic acid; (b) 1,3-propanedioic acid]

E Actual effects controlling the acidity of carboxylic acids. *J. Chem. Ed.*, **1971**, *48*, 338.

Preparation and reactions. Carboxylic acids can be prepared by oxidizing primary alcohols and aldehydes, using either acidified aqueous potassium permanganate or air with an appropriate catalyst. The latter is much cheaper and is used industrially to produce acetic acid (with cobalt(III) acetate as catalyst). In some cases, alkyl groups can be oxidized directly to carboxyl groups. This process is very important industrially and, among other applications, is used for the oxidation of *p*-xylene:

$$CH_3-C_6H_4-CH_3 + 3O_2(g) \xrightarrow{\text{Co(III) catalyst}} HOOC-C_6H_4-COOH + 2H_2O$$

The product, terephthalic acid, is used for the production of artificial fibers, as we shall see.

In a carboxyl group, the electronegative carbonyl oxygen partially withdraws electrons from the neighboring hydroxyl O—H bond, thereby weakening it. Hence, carboxylic acids are typical weak Brønsted acids. Their acid strengths differ strikingly from those of

N Oxalic acid is the simplest dicarboxylic acid; it is found in the leaves of rhubarb, and is toxic. The name is derived from the Greek *oxalis*, which denotes a group of plants that includes wood-sorrel, in which the acid is also found.

N Citric acid is a tricarboxylic acid: its formula is HOOC—CH_2—C(OH)(COOH)—CH_2—COOH

D Esterification. TDC, 22-14.

alcohols: for ethanol, $pK_a = 16$; but when CH_3CH_2OH is converted to CH_3COOH, the pK_a falls to 4.75, representing a change of eleven orders of magnitude in ionization constant. The acidic character of carboxylic acids (which typically have pK_a values in the range 4 to 5) is shown by their formation of salts with bases. Because the anions are Brønsted bases, solutions of these salts are basic (see Section 15.1).

Carboxylic acids with long hydrocarbon chains occur in oils and fats (see below) and are called **fatty acids**. Their sodium salts are used as soap; a typical example is sodium stearate, the sodium salt of stearic acid, $CH_3(CH_2)_{16}COOH$. The ability of soaps to act as surfactants was explained in Section 11.3.

24.6 ESTERS

The product of the reaction between a carboxylic acid and an alcohol is called an **ester**. Acetic acid and ethanol, for example, react when heated to about 100°C, especially when the mixture is acidified with a strong acid (to act as catalyst). The product of this **esterification** is ethyl acetate, a fragrant liquid:

$$CH_3{-}C({=}O){-}OH + HOCH_2CH_3 \xrightarrow{H^+,\ \Delta} CH_3{-}C({=}O){-}OCH_2CH_3 + H_2O$$

The occurrence of esters. Many esters have fragrant odors and contribute to the flavors of fruits (Table 24.3 and Box 24.1). Other naturally occurring esters include fats and oils. These are two members of a wide

T OHT: Table 24.3.

TABLE 24.3 The odors of some esters

Formula	Odor*
$CH_3{-}CH_2{-}CH_2{-}C({=}O){-}O{-}CH_2{-}CH_2{-}CH(CH_3)_2$	pear
$CH_3{-}C({=}O){-}O{-}CH_2{-}CH_2{-}CH_2{-}CH_2{-}CH_3$	banana
$CH_3{-}CH_2{-}CH_2{-}C({=}O){-}O{-}CH_2{-}CH_2{-}CH_2{-}CH_2{-}CH_3$	banana
$CH_3{-}C({=}O){-}O{-}CH_2{-}(CH_2)_6{-}CH_3$	orange
$(CH_3)_2CH{-}CH_2{-}CH_2{-}C({=}O){-}O{-}CH_2{-}CH_2{-}CH(CH_3)_2$	apple

*The esters are reminiscent of these fruits; the fruits do not necessarily contain the esters.

BOX 24.1 ▸ TASTE, ODOR, AND FLAVOR

Flavor—the joint impact of taste and odor—is an example of chemoreception, the response of sensors in the body to contact with different substances. Taste in mammals is confined to the damp region inside the mouth, but some insects taste through their feet and fish are covered with chemoreceptors. Within the mouths of humans, chemoreceptors are largely confined to the tongue and are grouped together into structures called taste buds. In adult humans, these buds occur mainly at the perimeter of the tongue, and their number declines with age. Sweetness is a response from the front of the tongue, saltiness on both sides, sourness at the rear edges, and bitterness at the central rear part of the tongue.

The flavors of food result from a very complicated mixture of ingredients. Its taste is the product of a particular blend of sweetness due to sugars and sourness due to carboxylic acids. Its odor is produced by the mixture of all the volatile chemicals present; these include esters, aldehydes, and ketones. An apple contains about 20 carboxylic acids (with alkyl chains up to about 19 atoms long), nearly 30 alcohols (with a similar range of chain lengths), about 70 esters arising from the most abundant alcohols present, and about three dozen other components. Food chemists often try to recreate flavors by using a smaller number of components, a fact that may explain why synthetic flavors are often not as subtle as the real thing.

Humans are more sensitive to odor than to taste, so the odor of a food makes the major contribution to its flavor. Whether or not a substance has an odor depends on whether its molecules can excite the olfactory nerve endings in the nose. In humans, these nerve endings occupy yellow-brown tissue with an area of about 5 cm^2. The molecules of food are carried to this area by eddy currents of air in the nose (these eddy currents are enhanced when we sniff). The sensors are direct nerve endings rather than, as in some other senses, a complex set of detectors to which the nerves are connected. The simple detection arrangement suggests that olfaction is one of the most primitive senses, which may account for the powerful, and sometimes unconscious, emotional impact of odor.

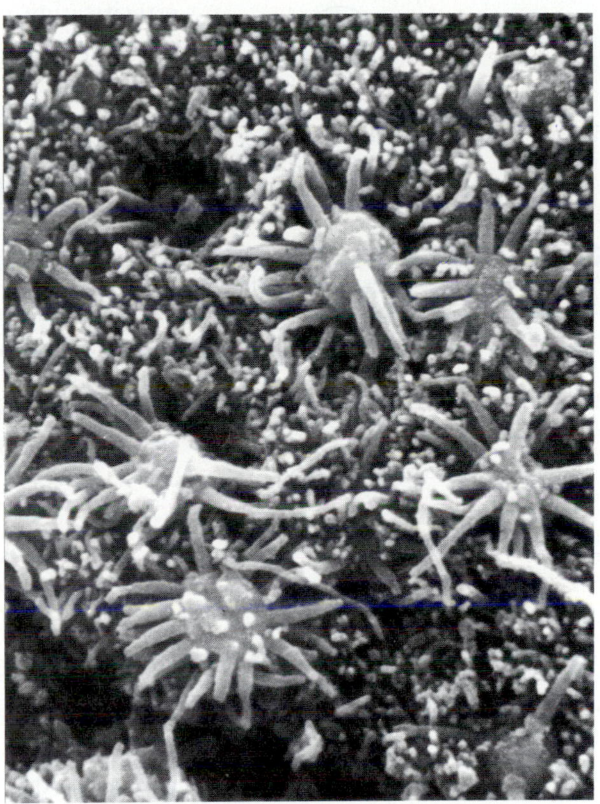

A portion of the olfactory tissue in the nose. Hairlike cilia protrude from the surface. The molecules that stimulate the sense of smell fit into protein molecules in the tissue and, when they attach, stimulate the nerve to send a signal to the brain.

class of substances called **lipids**, which are naturally occurring compounds that dissolve in nonpolar solvents but not in water. The vegetable oil triolein and the animal fat tristearin, for instance, are esters formed from glycerol and oleic acid or stearic acid. Other oils and fats have similar structures.

E The word *lipid* is derived from the Greek word *lipos*, for "fat."

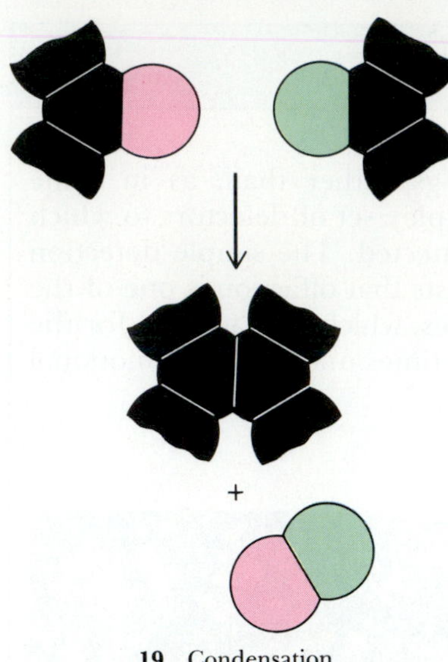

19 Condensation

Esterification. Ester formation is an example of a condensation reaction (**19**):

A **condensation reaction** is a reaction in which two molecules combine to form a larger one and a small molecule is eliminated.

In an esterification, the eliminated molecule is H_2O. Esterification may *look* like an acid-base neutralization, with the alcohol seeming to play the role of a base and the product, the ester, named like a salt (as in "ethyl acetate," like sodium acetate). However, it is very important to understand that this analogy is misleading. Experiments—using oxygen-18 and mass spectrometry to keep track of the fate of ^{18}O atoms—have shown that the oxygen atom in the H_2O molecule produced by esterification comes from the carboxylic acid and not from the hydroxyl group of the alcohol:

$$CH_3-C(=O)-OH + HO-CH_2CH_3 \longrightarrow CH_3-C(=O)-O-CH_2CH_3 + H_2\mathbf{O}$$

N Adenosine triphosphate, ATP, the energy source in biological cells, is an ester of an inorganic acid, H_3PO_4, and adenosine, an alcohol. See Section 19.9.

Polyesters and condensation polymerization. Polymers formed by linking together carboxylic acids and alcohols are called **polyesters** and are widely used for making artificial fibers. Polyesters are examples of **condensation polymers**, which are polymers formed by a chain of condensation reactions.

A typical example of a polyester is the material sold as Dacron, a polymer produced from the esterification of terephthalic acid with ethylene glycol. The first condensation is

D The embedded penny: Making glyptol resin plastic. CD1, 104.

$$HOOC-C_6H_4-COOH + HO-CH_2-CH_2-OH \longrightarrow HOOC-C_6H_4-COOCH_2-CH_2-OH + H_2O$$

Another ethylene glycol molecule can now condense with the carboxyl group on the left of the product, and another terephthalic acid can condense with the hydroxyl group on the right. As a result, the polymer grows at each end and in due course becomes

$$HOOC-C_6H_4-CO-\left\{-OCH_2-CH_2-O-OC-C_6H_4-CO-\right\}_n-OCH_2-CH_2-OH$$

In free-radical polymerization, side chains can grow out from the main chain (Section 23.6); however, in condensation polymerization, growth can occur only at the functional groups, so chain branching is much less likely. As a result, polyester molecules can be made to lie side by side by forcing them through a small hole; hence, they can be spun into yarn.

FUNCTIONAL GROUPS CONTAINING NITROGEN

Nitrogen brings to organic compounds a rich variety of properties. In this section we meet the amines and compounds related to them.

E A hypothetical synthesis of adrenaline: The preparation and properties of amines. *J. Chem. Ed.*, **1975**, *52*, 654.

24.7 AMINES

One of the most important families in which nitrogen occurs in a functional group is the **amines**, compounds derived from ammonia by replacing various numbers of H atoms:

H–N(H)–H	CH_3–N(H)–H	CH_3–N(H)–CH_3	CH_3–N(CH_3)–CH_3
Ammonia	Methylamine	Dimethylamine	Trimethylamine

In each case, the molecules are trigonal pyramidal, with a lone pair of electrons on the N atom and σ bonds formed from its approximately sp^3 hybrid orbitals.

Amines are widespread in both natural and synthetic organic materials. Many of them have a pungent, often unpleasant odor. Because proteins are organic compounds of nitrogen, amines are often the end products of the decomposition of living matter and, together with sulfur compounds, are responsible for the stench of decaying flesh. The common names of two **diamines** (amines with two amino groups)—putrescine, $NH_2(CH_2)_4NH_2$, and cadaverine, $NH_2(CH_2)_5NH_2$—speak for themselves.

N The boiling points of the amines depicted here increase from −33°C for NH_3 to 3.5°C for $(CH_3)_3N$.

E Our sense of the odor of amines is different physiologically from the odor of many other compounds, such as esters and aldehydes. For the latter, the olfactory stimulus appears to be due to the attachment of molecules to shape-sensitive receptors at the ends of nerves. The odor of amines, however, appears to be due to a less specific stimulation of the nerves that spread through the face and which are accessible through the olfactory epithelium in the nose.

Nomenclature. Amines are divided into three classes on the basis of the number of carbon atoms directly attached to the nitrogen atom:

A **primary amine** has the structure RNH_2.

A **secondary amine** has the structure R_2NH.

A **tertiary amine** has the structure R_3N.

E Scopolamine, a potent chemical weapon. *J. Chem. Ed.*, **1984**, *61*, 679.

A **quaternary ammonium ion** is a tetrahedral ion of the form R_4N^+, such as the tetramethylammonium ion, $(CH_3)_4N^+$. The R— groups, which may be either aliphatic or aromatic, need not be all the same. Methylamine is a primary amine, dimethylamine is a secondary amine, and trimethylamine is a tertiary amine. The $—NH_2$ group itself is called an **amino group**.

Amines are named systematically by specifying the groups attached to the N atom in alphabetical order, followed by the suffix *-amine*. Thus the name of $CH_3CH_2NH_2$, a primary amine, is ethylamine and that of $(CH_3CH_2)_2NCH_2CH_2CH_3$, a tertiary amine, is diethylpropylamine. Many amines have common names. For instance, phenylamine, $C_6H_5NH_2$ (**20**), is called aniline.

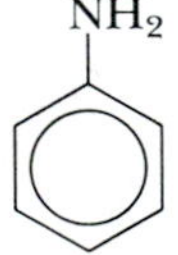

20 Aniline

▼ EXAMPLE 24.5 Naming amines

Classify and name the following two amines: (a) $CH_3CH_2CH_2NH_2$; (b) $(C_6H_5)_3N$.

STRATEGY To classify the amine, we simply count the number of carbon atoms attached directly to the nitrogen atom. To name it, we identify the substituents on the nitrogen atom, list them in alphabetical order, and add the suffix *-amine.*

SOLUTION (a) The amine has the form R—NH_2, so it is a primary amine. Because $CH_3CH_2CH_2$— is propyl, the compound is propylamine. (b) This molecule has the form R_3N and is a tertiary amine. Because C_6H_5— is the phenyl group, the compound is triphenylamine.

EXERCISE E24.5 Classify and name the following two amines: (a) $CH_3(CH_2)_3CH_2NH_2$; (b) $(CH_3CH_2)_2NH$.

▲ [*Answer*: (a) primary, pentylamine; (b) secondary, diethylamine]

Preparation. Amines are prepared by several methods. The simplest is by the reaction of a haloalkane with ammonia:

$$CH_3CH_2\text{—}Br + NH_3 \longrightarrow CH_3CH_2\text{—}NH_2 + HBr \qquad (\text{as } CH_3CH_2NH_3{}^+Br^-)$$

The free amine derived from this reaction can react further to give the tertiary amine. Other syntheses have been developed to avoid this difficulty.

E Suggest a means for synthesizing ethylamine that begins with the reaction between bromomethane and CN^- ions. [*Hint*: We can expect the CN^- group to substitute for the Br atom in bromoalkane; the comparison of —C≡N and —CH_2—NH_2 suggests that reduction should then be carried out.]

D Effect of molecular structure on the strength of organic acids and bases in aqueous solutions. BZS3, 8.25.

Reactions of amines. An important characteristic chemical feature of amines is that, like their parent ammonia, they are Lewis and Brønsted bases. The Lewis-base character of amines allows them to form complexes with metal ions acting as Lewis acids, as they do in chlorophyll and the numerous coordination compounds of the *d*-block elements.

In general the alkylamines (amines formed from alkanes, such as methylamine) are stronger Brønsted bases than ammonia. Their strength arises largely from the fact that the alkyl groups can supply electrons to the nitrogen and help stabilize the conjugate acid:

$$H_2O + CH_3NH_2 \rightleftharpoons CH_3NH_3{}^+ + OH^- \qquad pK_b = 3.44$$

? What can be said about the strengths of CH_3NH_2 and CF_3NH_2 as bases?

In contrast, the arylamines (which are amines derived from benzene) are weaker, because the lone pair is partially delocalized over the benzene ring, and the *base* is stabilized. An example is aniline, $C_6H_5NH_2$:

$$H_2O + C_6H_5NH_2 \rightleftharpoons C_6H_5NH_3{}^+ + OH^- \qquad pK_b = 9.37$$

E In the laboratory, amides are formed by the action of an amine on an acyl chloride, R—CO—Cl:

$$CH_3COCl + CH_3NH_2 \longrightarrow CH_3CONHCH_3 + HCl$$

An alternative synthesis is to heat the ammonium salt of a carboxylic acid:

$$CH_3CO_2{}^- + NH_4{}^+ \xrightarrow{\Delta} CH_3CONH_2 + H_2O$$

The latter reaction is a primitive version of the reaction used for the production of nylon.

Amides and polyamides. Primary and secondary amines behave in some respects like alcohols in that they condense with carboxylic acids, for example,

$$CH_3\text{—}C(=O)\text{—}OH + HNHCH_3 \longrightarrow CH_3\text{—}C(=O)\text{—}NHCH_3 + H_2O$$

The product is an **amide**, a compound formed by the reaction of an amine and a carboxylic acid. Natural and synthetic amides are common. The pain-relieving drug sold as Tylenol (**21**) is an amide.

Condensation polymerization by amide formation leads to the **polyamides**, substances more commonly known as nylon. A typical polyamide is nylon-66, which is a copolymer of 1,6-diaminohexane, $H_2N(CH_2)_6NH_2$, and adipic acid, $HOOC(CH_2)_4COOH$. The *66* in the name indicates the numbers of C atoms in the two monomers.

CH_3—C(=O)—NH—C_6H_4—OH

21 Tylenol

For a condensation polymerization, it is necessary to mix stoichiometric amounts of the acid and base that are to be used. In the case of polyamide formation, the starting materials form "nylon salt" by proton transfer:

$$HOOC(CH_2)_4COOH + H_2N(CH_2)_6NH_2 \longrightarrow {}^{+}H_3N(CH_2)_6NH_3{}^{+} + {}^{-}O_2C(CH_2)_4CO_2{}^{-}$$

At this point, the excess acid or amine can be removed. Then, when the nylon salt—an ammonium salt of a carboxylic acid—is heated, the condensation begins. The first step is

$$^{-}O_2C\text{—}(CH_2)_4\text{—}CO_2^{-} + H_3\overset{+}{N}\text{—}(CH_2)_6\text{—}\overset{+}{N}H_3 \longrightarrow {}^{-}O_2C\text{—}(CH_2)_4\text{—}C(=O)\text{—}NH\text{—}(CH_2)_6\text{—}\overset{+}{N}H_3 + H_2O$$

The amide grows at both ends by further condensations, and the final product is

$$^{-}O_2C(CH_2)_4\overset{O}{\overset{\|}{C}}\text{—}\left[\text{—}NH(CH_2)_6NH\overset{O}{\overset{\|}{C}}(CH_2)_4\overset{O}{\overset{\|}{C}}\text{—}\right]_n\text{—}NH(CH_2)_6NH_3{}^{+}$$

The long polyamide (nylon) chains can be spun into fibers (like polyesters) or molded. Some of the strength of nylon fibers arises from the N—H···O=C hydrogen bonding that can occur between neighboring chains (Fig. 24.10). However, this ability to form hydrogen bonds also accounts for nylon's tendency to absorb moisture, because H_2O molecules can hydrogen bond to the chains and worm their way in among them. The ability of an —NH— group to become charged, forming $\text{—}NH_2{}^{+}\text{—}$, for example, accounts for the buildup of electrostatic charge on nylon fabrics and carpets that are exposed to friction.

E The simplest diamide is urea, NH_2CONH_2, a component of urine and a major industrial chemical used as a fertilizer.

E Nylon was first made by Wallace Carothers and marketed in 1938.

V Video 54: Production of nylon.

D A synthesis of nylon. CD1, 101. Nylon 6-10, BZS1, 3.1.

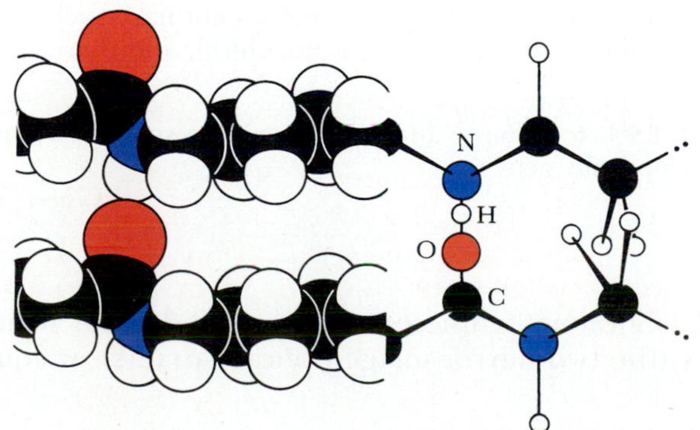

FIGURE 24.10 The strength of nylon fibers is yet another sign of the presence of hydrogen bonds, this time between neighboring polyamide chains.

D Amino acids. TDC, 24-58s.

24.8 AMINO ACIDS

An **amino acid** is a carboxylic acid that also contains an amino group. An example is glycine, NH_2CH_2COOH, the amino acid derived from acetic acid by substituting an amino group for a methyl hydrogen atom.

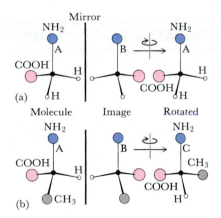

22 Enantiomeric amino acids

Optical activity. We saw in Section 21.6 (in the context of complexes) that a chiral molecule is one that is not identical to its mirror image. We also saw that a physical property stemming from chirality (and one way of detecting it) is optical activity, the ability to rotate the plane of polarization of light. A quick way of checking whether an organic compound is chiral, and hence optically active, is to note whether it contains a carbon atom to which are attached four *different* groups, for then it cannot be the same as its mirror image. Two molecules that are related by reflection (as a left hand is related to a right hand) are called **optical isomers** and are labeled L (**22a**) and D (**22b**). It is a remarkable feature of nature (a feature that has not yet been fully explained) that the vast majority of naturally occurring amino acids are L (**22a**).

Chirality and optical activity occur in organic compounds, particularly among amino acids. The simplest amino acid, glycine, is not chiral; it is identical to its mirror image (Fig. 24.11a). However, when a CH_3— group is substituted for one of the two —CH_2— hydrogen atoms, we get alanine, $CH_3CH(NH_2)COOH$, which *is* chiral (Fig. 24.11b).

N All amino acids that contribute to proteins have the formula R—CH(NH_2)—COOH, and are called *α-amino acids*.

EXAMPLE 24.6 Judging whether a compound is optically active

Decide whether the amino acids (a) $CH_3CH_2CH(NH)_2COOH$ and (b) $(CH_3)_2C(NH_2)COOH$ are optically active.

STRATEGY As stated in the text, to decide whether a compound is optically active, we should check to see whether it contains a carbon atom to which are attached four different groups.

SOLUTION The two (isomeric) amino acids are

$$\text{(a) } CH_3CH_2\text{—}\underset{\displaystyle H}{\overset{\displaystyle NH_2}{C}}\text{—COOH} \qquad \text{(b) } CH_3\text{—}\underset{\displaystyle CH_3}{\overset{\displaystyle NH_2}{C}}\text{—COOH}$$

We see that (a) contains a carbon atom that has four different groups attached and hence is chiral. However, (b) does not have such a carbon atom; so it is identical to its mirror image, is not chiral, and therefore is not optically active.

EXERCISE E24.6 What is the formula of the simplest chlorofluorocarbon that is optically active?

[*Answer*: CH_3CHFCl]

FIGURE 24.11 (a) A glycine molecule can be superimposed on its mirror image. (b) An alanine molecule, however, cannot. It is an example of a chiral molecule (like left and right hands). One way of checking a molecule for chirality is to note whether all the groups attached to one carbon atom are different from one another (as they are in alanine but not in glycine).

Ordinary laboratory syntheses of amino acids lead to mixtures of enantiomers (the two mirror-image optical isomers) in equal proportions. Hence, laboratory syntheses result in a racemic mixture, a mixture of equal numbers of both enantiomers. Because enantiomers rotate the plane of polarization of light in opposite directions, a racemic sample is not optically active. However, as noted earlier, biochemical reactions lead to only one enantiomer.

Polypeptides and proteins. The central importance of amino acids is that they are the building blocks of proteins. In a sense, proteins are a very elaborate form of nylon. However, nylon is a monotonous repetition of the same two monomers, whereas the polypeptide chains of proteins are condensation copolymers of up to 20 different naturally occurring amino acids (Table 24.4). Our bodies can synthesize 15 of these compounds in sufficient amounts for our needs. It is essential to ingest the other five, and they are known as the **essential amino acids**.

E There are two principal contenders for the explanation of the left-handedness of naturally occurring amino acids. One is chance: the ultimately successful organisms simply happened to make use of an amino acid of a particular handedness. The alternative theory of *chiral discrimination* is that a polypeptide of left-handed amino acids has a very slightly lower energy than one of opposite handedness on account of the weak interaction (one of the fundamental forces that normally acts only within nuclei).

E Organic chemists have devised *asymmetric syntheses* in which one enantiomer is the principal product. This achievement is very important, because different enantiomeric forms of a substance may have different physiological activity. A tragic example is the thalidomide story, where one enantiomer caused birth defects but the other was benign.

D Effect of pH on protein solubility. BZS3, 8.30.

TABLE 24.4 The naturally occurring amino acids, NH_2–CH(X)–COOH

X	Name	Abbreviation	X	Name	Abbreviation
—H	glycine	Gly	$—CH_2$–(indol-3-yl ring, N–H)	tryptophan*	Trp
$—CH_2COOH$	aspartic acid	Asp			
$—CH_2CH_2COOH$	glutamic acid	Glu			
—C_6H_4—OH	tyrosine	Tyr	$—CH_3$	alanine	Ala
$—CH_2SH$	cysteine	Cys	$—CH_2$—C_6H_5	phenylalanine*	Phe
$—CH_2CONH_2$	asparagine	Asn			
$—CH_2CH_2CONH_2$	glutamine	Gln			
$—CH_2OH$	serine	Ser	$—CH(CH_3)_2$	valine*	Val
$—CH(OH)CH_3$	threonine	Thr			
$—CH_2$–(imidazole ring, N, NH)	histidine*	His	$—CH_2CH(CH_3)_2$	leucine	Leu
			$—CH(CH_3)CH_2CH_3$	isoleucine	Ile
$—CH_2(CH_2)_3NH_2$	lysine*	Lys	$—CH_2CH_2SCH_3$	methionine	Met
$—CH_2CH_2CH(OH)CH_2NH_2$	hydroxylysine	Lys			
$—CH_2(CH_2)_2NH—C(=NH)—NH_2$	arginine	Arg	(pyrrolidine ring, NH, COOH)	proline†	Pro

*Essential amino acid.
†The entire amino acid is shown.

E Bee sting: The chemistry of an insect venom. *J. Chem. Ed.*, **1980**, 57, 206.

A molecule formed from two or more amino acids is called a **peptide**. An example is the combination of glycine and alanine, denoted Gly-Ala:

$$NH_2CH_2{-}\overset{\displaystyle O}{\overset{\|}{C}}{-}NH{-}CH(CH_3)COOH$$

The —CO—NH— link shown in red is called a **peptide bond**, and each amino acid in a peptide is called a **residue**. A typical protein is a **polypeptide chain** of more than a hundred residues joined through peptide bonds and arranged in a strict order. When only a few amino acid residues are present, we call the molecule an **oligopeptide**. The artificial sweetening agent aspartame is an oligopeptide with two residues; specifically, it is a **dipeptide**.

N In brief, the *primary structure* is the peptide sequence, the *secondary structure* is the type of coiling of the polypeptide, the *tertiary structure* is the distortion of the coils, and the *quaternary structure* is the agglomeration of individual polypeptide units.

The sequence of amino acids in the peptide chain is the protein's **primary structure**. Aspartame consists of phenylalanine and aspartic acid, so its primary structure is Phe-Asp. The opposite arrangement, Asp-Phe, is a different primary structure and a different dipeptide. A fragment of the primary structure of human hemoglobin is

Leu-Ser-Pro-Ala-Asp-Lys-Thr-Asn-Val-Lys- . . .
. . . -Val-Lys-Gly-Trp-Ala-Ala- . . .
. . . -Ser-Thr-Val-Leu-Thr-Ser-Lys-Ser-Lys-Tyr-Arg

The determination of the primary structures of proteins is a very demanding analytical task, but many of these structures are now known.

Chemists have also begun the even more demanding task of synthesizing proteins. Some syntheses have been completed, but so far the products have turned out to be biologically inactive. The products are inactive largely because a protein's ability to carry out its function (which may be to act as an enzyme) also depends on the shape adopted by the chain of residues, and synthetic proteins have not yet been produced with the same shapes as their biological counterparts.

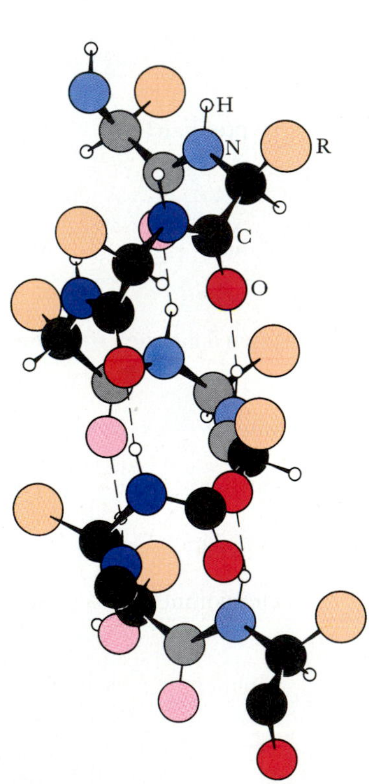

FIGURE 24.12 The α helix, one of the secondary structures adopted by polypeptide chains. R, amino acid side chain.

The **α helix** is a specific conformation of a polypeptide chain held in place by hydrogen bonds between residues (Fig. 24.12). The α helix is one example of **secondary structure**, the manner in which a polypeptide is coiled. An alternative secondary structure is the **β-pleated sheet** form of the protein we know as silk. In this structure, the protein molecules lie side by side to form nearly flat sheets. Many enzyme molecules consist of alternating regions of α helix and β-pleated sheet (Fig. 24.13).

Protein molecules also adopt a specific **tertiary structure**, the shape into which the α helix and β-pleated sheet sections are twisted as a result of interactions between residues lying in different parts of the primary structure. The globular form of each chain in hemoglobin is an example. One important type of link responsible for tertiary structure is the **disulfide link**, —S—S—, between amino acids containing sulfur (cysteine and methionine). The overall shape of the protein is essential to its function. In Section 12.4, while discussing enzymes as catalysts, we saw how the action and malfunctioning of enzymes could be explained by a lock-and-key model. Here we see that the amino acid

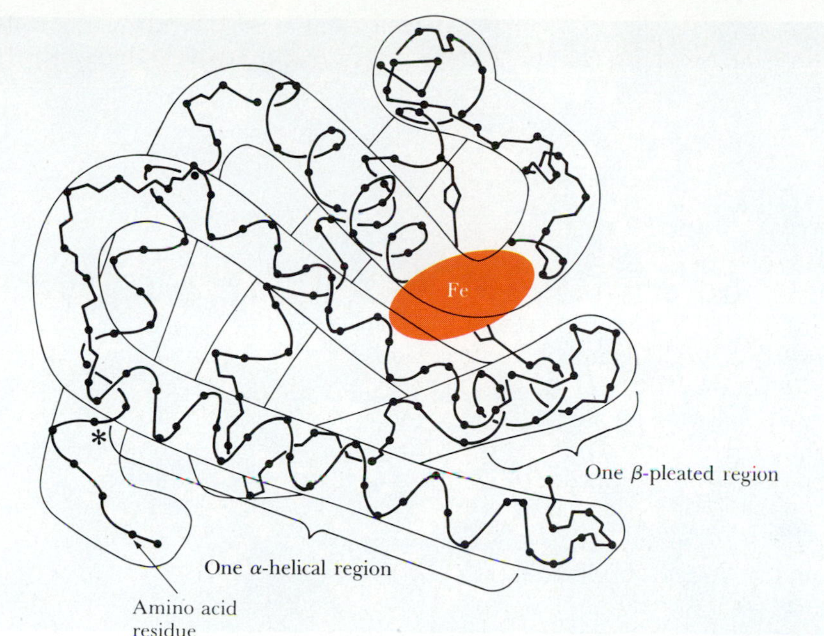

FIGURE 24.13 One of the four polypeptide chains that make up the human hemoglobin molecule. The chain consists of alternating regions of α helix and β-pleated sheet. The significance of the asterisk is explained in Fig. 24.14.

sequence, and the secondary and tertiary structures to which this primary structure leads, account for the shape of the lock.

Proteins may also have a **quaternary structure**, when neighboring polypeptide units stack together in a specific arrangement. The hemoglobin molecule, for example, has a quaternary structure of four polypeptide units like the one shown in Fig. 24.13. Any modification of the primary structure—the replacement of one amino acid residue by another—may lead to the malfunction we call congenital disease. Even one wrong amino acid in the chain can cause malfunction. In normal hemoglobin, the amino acid marked with an asterisk in Fig. 24.13 is glutamic acid; if it is replaced by valine, then the disease called sickle-cell anemia results. The affected person develops malformed, sickle-shaped red blood corpuscles that are poor transporters of oxygen (Fig. 24.14).

E Hemoglobin: Its occurrence, structure, and adaptation. *J. Chem. Ed.,* **1982**, *59,* 173.

E A well-known chemical demonstration to illustrate an unusual medical mystery. *J. Chem. Ed.,* **1988**, *65,* 621.

The loss of structure by a protein is called **denaturation**. This structural change may be a loss of quaternary, tertiary, or secondary structure, or even the degradation of the primary structure by cleavage of the peptide bonds. Even mild heating can cause irreversible denaturation; for example, when we cook an egg, the albumin denatures into a white mass. The permanent waving of hair is a result of partial denaturation. Mild reducing agents are applied to the hair to sever the disulfide links between the protein strands that make up hair. The links are then reformed by applying a mild oxidizing agent while the hair is stretched and twisted into the desired arrangement.

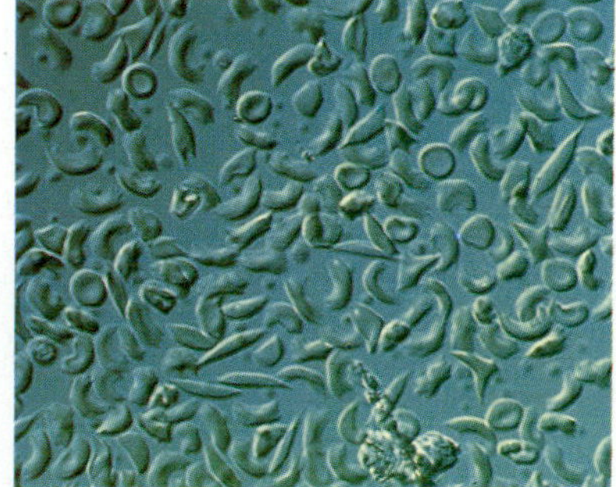

FIGURE 24.14 The sickle-shaped red blood cells that form when the glutamic acid marked with an asterisk in Fig. 24.13 is replaced by valine.

24.9 DNA AND RNA

Among the most complicated of all organic molecules are the deoxyribonucleic acids, DNA (Fig. 24.15). The nucleus of a biological cell has at least one DNA molecule to control the production of proteins and carry genetic information from one generation of cells to the next.

FIGURE 24.15 A computer graphics image of a short section of the DNA molecule, which consists of two entwined helices. In this illustration, the double helix is also coiled around itself, in a shape called a superhelix.

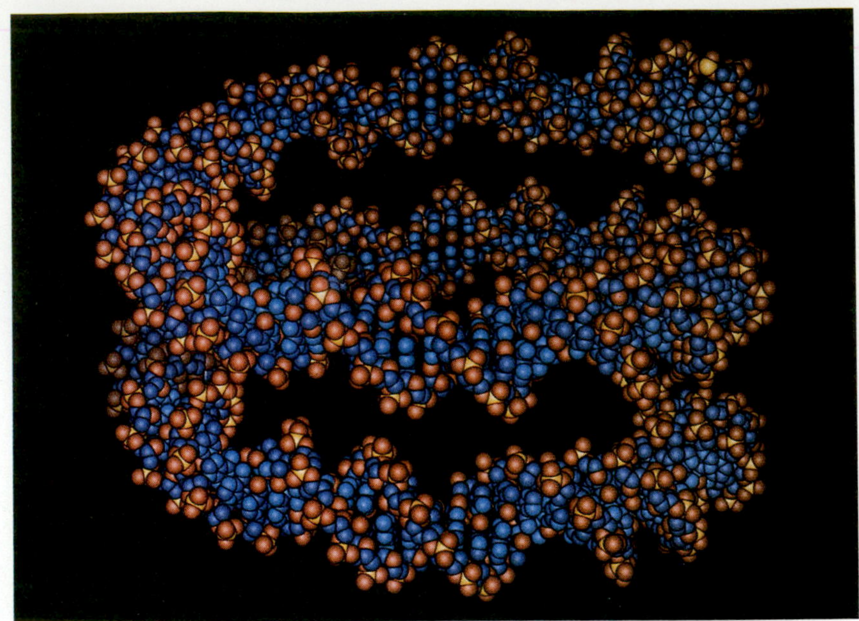

E The DNA molecule is *supercoiled*: the molecule is coiled, and then the coiled molecule is coiled again. A part of the process of gene expression is related to local uncoiling of regions of the DNA molecule. The onset of certain types of cancer may be due to uncontrolled unwinding of the supercoiled DNA molecule.

E RNAs as catalysts: A new class of enzymes. *J. Chem. Ed.*, **1987**, *64*, 221.

Human DNA molecules are immense: if one could be extracted without damage from a cell nucleus and drawn out straight from its highly coiled natural shape, it would be about 2 m long (Fig. 24.16). The ribonucleic acid (RNA) molecule is closely related to DNA. One of its functions is to carry the information stored by DNA out of a cell nucleus and into a region of the cell where it can be used as a template for protein synthesis.

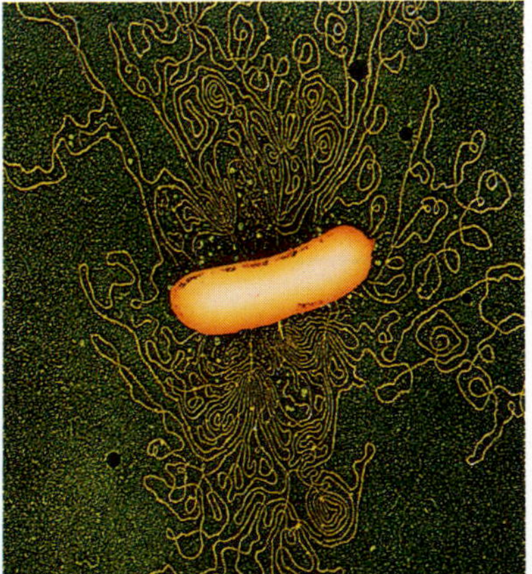

FIGURE 24.16 The DNA molecule is very large, even in bacteria. In this micrograph, a DNA molecule has spilled out through the damaged cell wall of a bacterium.

Nucleosides and nucleotides. The best way to understand the structure of DNA is to see how it gets its name. It is a polymer, and the repeating unit is a modification of a sugar molecule, ribose (**23**), in its furanose form. The modification is the absence of the oxygen atom at carbon atom 2; that is, the repeating unit—the monomer—is deoxyribose (**24**).

Attached by a covalent bond to carbon atom 1 of the deoxyribose ring is an aminelike molecule (and therefore a base), which may be adenine (A; **25**), guanine (G; **26**), cytosine (C; **27**), or thymine (T; **28**). In RNA, uracil (U; **29**) occurs instead of thymine. The base bonds to carbon atom 1 of deoxyribose through the nitrogen of the NH group (printed in red), and the compound so formed is called a **nucleoside**. All nucleosides have a similar structure, which we can summarize as the shape shown in (**30**).

The final detail in the structure of the DNA monomers is the phosphate group ($—O—PO_3^{2-}$) covalently bonded to carbon atom 5; this compound is called a **nucleotide** (**31**). Because there are four possible nucleosides (one for each base), there are four possible nucleotides.

Nucleic acids. DNA and RNA are **polynucleotides**, polymers built from nucleotide units. The polymers are formed in a reaction involv-

23 Ribose

24 Deoxyribose

25 Adenine

26 Guanine

27 Cytosine

28 Thymine

29 Uracil

30 A nucleoside

ing the phosphate group of one nucleotide and the remaining hydroxyl group on carbon atom 3 of another nucleotide. As this condensation continues, it results in a structure like that shown in Fig. 24.17, a compound known as a **nucleic acid**. The DNA molecule itself is a double helix in which two long nucleic acid strands are wound around each

31 A nucleotide

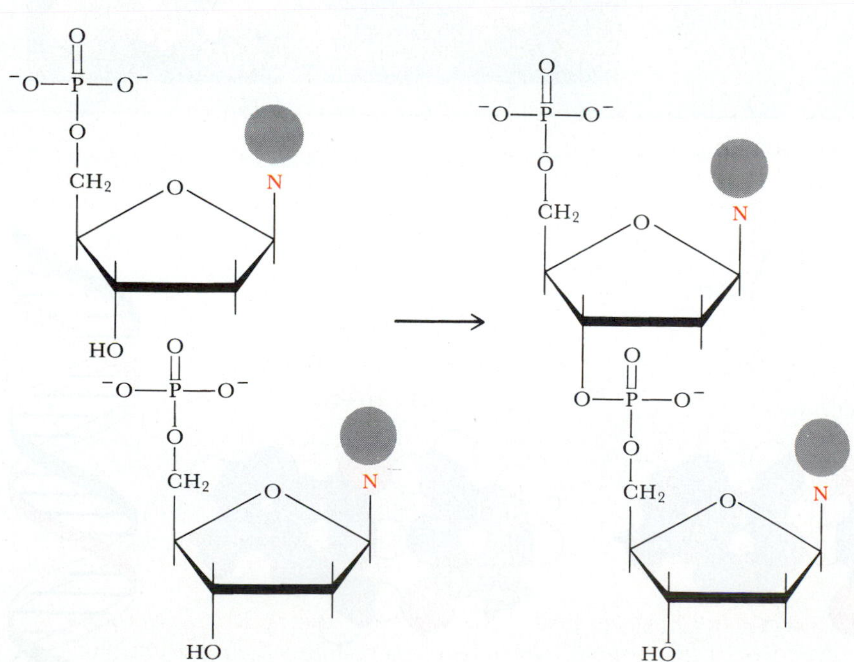

FIGURE 24.17 The condensation of nucleotides that leads to the formation of a nucleic acid—a polynucleotide.

BOX 24.2 ▶ RECOGNIZING THE STRUCTURES OF DNA

JACQUELINE K. BARTON, *Professor of Chemistry, California Institute of Technology*

Many people are familiar with the deoxyribonucleic acid (DNA) double helix. It is a beautiful structure, a right-handed spiral staircase. DNA, the genetic material in each of our cells, represents the storehouse of all the information that makes us what we are. In the human genome, there are 3 billion bits of such information, and we would like to understand how this information is accessed. It is clear, however, that the information processing is accomplished through fundamental chemical interactions.

DNA is a polyanion. A single polymeric strand of DNA contains a sugar-phosphate backbone, and connected to each deoxyribose sugar is one of four nucleic acid bases—guanine, cytosine, adenine, and thymine. Each phosphate contributes one negative charge. In the double helix, the two polymeric strands are linked noncovalently through an ensemble of hydrogen bonds between pairs of bases: guanine paired with cytosine and adenine paired with thymine. The double helix is further stabilized by a stacking of base pairs one above the other. The central structure shown in the figure illustrates this right-handed double helix with the sugar-phosphate backbone (shown in green) wrapping around the base-pair stack (shown in orange).

It is the sequence, or linear arrangement, of bases along the strand that encodes all the information for the cell. The base sequence determines the structure of the proteins that are made in the cell. But, just as important, the base se-

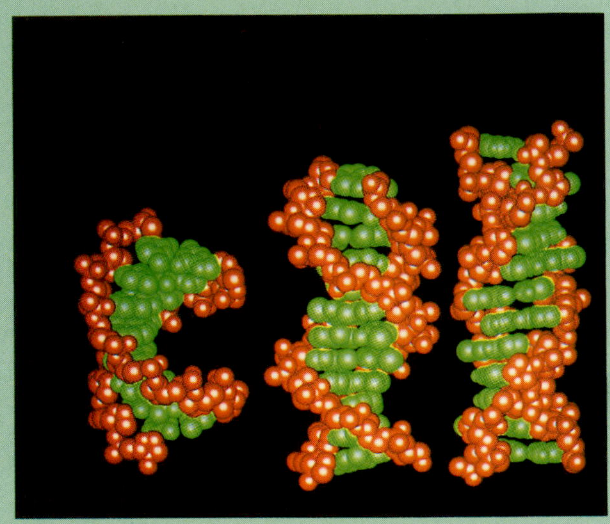

T OHT: Fig. 24.18.

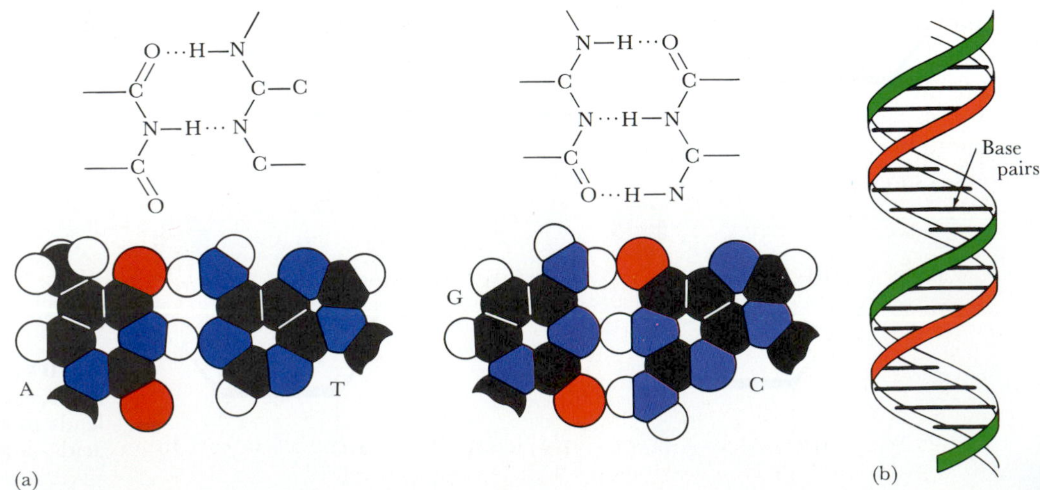

FIGURE 24.18 (a) The bases in the DNA double helix fit together by virtue of the hydrogen bonds that they can form. Once formed, the AT and GC pairs are almost identical in size and shape; as a result, the turns of the helix (b) are regular and consistent.

quence also encodes all the information used to regulate which proteins are made: preexisting regulatory proteins recognize and bind to particular base sequences, events that either start or repress protein synthesis.

I am fascinated by the problem of determining how these regulatory proteins actually recognize and distinguish one base sequence from another. Notice that the information that must be read, or recognized, is actually in the center of the helix, in the stack of base pairs. How do proteins distinguish one site from another in this repeating columnar structure, which is awash in negative charges? If we could understand this process of molecular recognition, then we might be able to design synthetic molecules that could regulate the expression of the genetic information to suit our needs. We might even be able to design molecules to function as antiviral or antitumor agents.

DNA is present not as a single structure but instead adopts a variety of structures, or conformations. Three of these conformations are illustrated in the figure. The center helical structure is the familiar one described earlier. The base pairs are stacked in a regular fashion perpendicular to the helical axis, thereby forming a wide and deep major groove and a smaller but also deep minor groove. On the left is the A-conformation. Also a right-handed helix, the A-form is a fatter helix with a shallow and wide minor groove, base pairs that are tilted with respect to the helix axis, and a major groove pulled deep into the interior of the structure. On the right is the Z-conformation, a strikingly different zig-zag structure that spirals in a left-handed sense. These illustrations represent but three of the conformations that double helical DNA can adopt. Others occur, although at this stage they are poorly characterized. We are learning that the DNA helix can bend, twist, and tilt and can form loops, bulges, and left-handed segments along the polymer strand. These variations in conformation all depend on the sequence of bases. It is intriguing to speculate about how Nature takes advantage of this rich structural diversity.

In my laboratory we have been trying to understand the relationship between the one-dimensional sequence of DNA and its three-dimensional structure, and how the shape or local conformation may contribute to the recognition process. We have designed *d*-metal complexes that recognize particular sites on the nucleic acid polymer by matching their shapes and symmetries to sites along the helix. We hope eventually to be able to use shape-selection, the recognition of a particular DNA sequence indirectly through the three-dimensional structure generated by that sequence, to target specific sites on the polymer. Who knows where we will be able to go from there.

other. The ability of DNA to replicate lies in its double helix, for there is a precise correspondence between the bases in the two strands; that is, the strands are held together by bonds between specific **base pairs**. Adenine in one strand always pairs with thymine in the other, and guanine always pairs with cytosine, so the base pairs are always AT and GC (Fig. 24.18).

E Strand polarity: Antiparallel molecular interactions in nucleic acids. *J. Chem. Ed.*, **1975**, *52*, 323.

The genetic code. Because of base pairing, the base sequence in one strand reflects and is determined by the sequence in the other:

. . . GATGTGCTCAG . . .

. . . CTACACGAGTC . . .

Therefore, when the strands separate, each acts as a template for the growth of another that is identical to its original partner:

TABLE 24.5 The genetic code

First position	Second position U	C	A	G	Third position
U	Phe	Ser	Tyr	Cys	U
	Phe	Ser	Tyr	Cys	C
	Leu	Ser	STOP	STOP	A
	Leu	Ser	STOP	Trp	G
C	Leu	Pro	His	Arg	U
	Leu	Pro	His	Arg	C
	Leu	Pro	Gln	Arg	A
	Leu	Pro	Gln	Arg	G
A	Ile	Thr	Asn	Ser	U
	Ile	Thr	Asn	Ser	C
	Ile	Thr	Lys	Arg	A
	Met*	Thr	Lys	Arg	G
G	Val	Ala	Asp	Gly	U
	Val	Ala	Asp	Gly	C
	Val	Ala	Glu	Gly	A
	Val*	Ala	Glu	Gly	G

*Also part of the START signal.

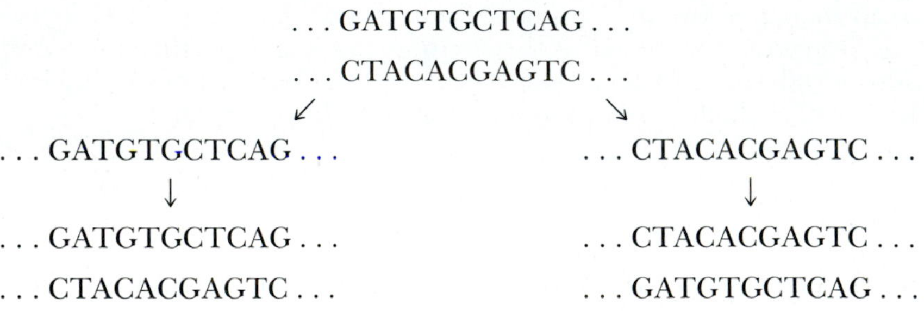

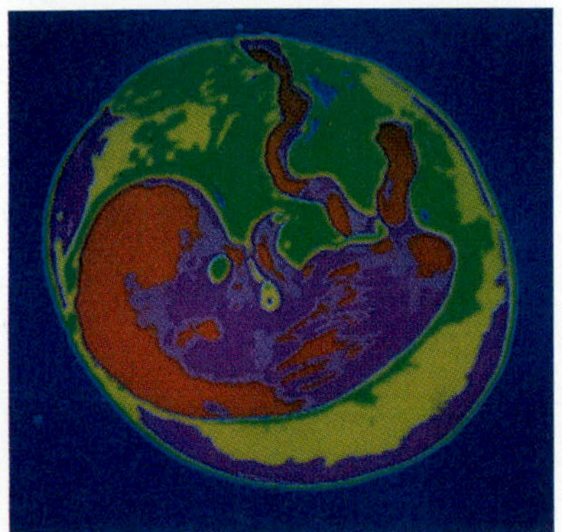

FIGURE 24.19 A human fetus in the first trimester (8–11 weeks). The ultrasound image has been computer enhanced for color.

In addition to replication (the production of copies of itself for reproduction and cell division), DNA governs the production of proteins. For this purpose, segments of the genetic message are carried out of the cell nucleus as RNA molecules, with U replacing T.

The precise sequence of bases in the nucleic acid is the genetic information for the cell. The order of the bases controls the order in which amino acids condense together to form proteins. Because there are 20 amino acids but only four bases, each amino acid must be coded by a group of bases. These groups, which are called **codons**, are known to consist of groups of three bases; CAC, for instance, is the codon for histidine, and GCA is the codon for alanine. The complete list of codons in RNA (the molecule from which the code is normally read and transcribed) is given in Table 24.5. The sequence CACGCA, for example, would result in the sequence His-Ala in the protein that was built to its design.

The principles of chemistry apply to RNA and DNA and to the proteins they code for, just as they apply to the very simple compounds that we met earlier in the text. It is here, however, that chemistry becomes biology. We are at the point where the elements have completed their journey from the stars, have come alive (Fig. 24.19) and have begun to control their own destiny.

SUMMARY

24.1 Aliphatic compounds containing the **hydroxyl group** are called **alcohols** (suffix *-ol*) and may be **primary** (with structure RCH_2—OH), **secondary** (R_2CH—OH), or **tertiary** (R_3C—OH). Alcohols are amphiprotic. They are very weak Brønsted acids and, when they act as bases, accept a proton to form an **oxonium ion**. They may be halogenated, dehydrated, and oxidized to aldehydes, ketones, and carboxylic acids. **Ethers** (R—O—R) are less reactive than alcohols but can undergo acid-catalyzed substitution reactions. Cyclic **crown ethers** are important complexing agents.

24.2 Compounds in which the hydroxyl group is attached directly to an aromatic ring are called **phenols**; they are weak acids. The benzene ring is much more reactive as phenol than as benzene alone, particularly at the ortho and para positions. A particularly important reaction of phenol is that with formaldehyde, which leads to the formation of **phenolic resins**.

24.3 Compounds containing the **carbonyl group** include **aldehydes** (R—CO—H, more often denoted R—CHO; suffix *-al*) and **ketones** (R—CO—R; suffix *-one*). These are obtained by oxidation of primary and secondary alcohols, respectively. Aldehydes are readily oxidized to carboxylic acids. Aldehydes reduce Fehling's solution and Tollens' reagent; ketones do not. The reactions of aldehydes (and some ketones) include **cyanhydrin** formation with cyanide ions. Aldehydes form **hemiacetals** and **acetals** in reactions with alcohols.

24.4 Carbohydrates may be aldehydes and ketones; they are often found in their hemiacetal **pyranose** (six-membered ring) or **furanose** (five-membered ring) forms. **Reducing sugars** are carbohydrates that reduce Fehling's solution and Tollens' reagent. **Monosaccharides** are the building blocks of **oligosaccharides**, including the **disaccharide** sucrose, and **polysaccharides**, such as starch and cellulose. Starch consists of two components, amylose and amylopectin.

24.5 Compounds containing the **carboxyl group** (—COOH) are **carboxylic acids**. These are weak acids produced by oxidation of hydrocarbons, primary alcohols, and aldehydes.

24.6 Carboxylic acids undergo a **condensation reaction** with alcohols to form **esters**. Dicarboxylic acids and diols undergo **condensation polymerization** to form **polyesters**.

24.7 Compounds containing the **amino group** (—NH_2 and its derivatives) not linked to a carbonyl group are called **amines** and may be **primary** (RNH_2), **secondary** (R_2NH), or **tertiary** (R_3N). Amines are Lewis and Brønsted bases. Alkylamines are stronger Brønsted bases than ammonia; arylamines are weaker. Amines condense with carboxylic acids to form **amides**; some take part in condensation polymerization and form **polyamides**.

24.8 Amino acids are compounds containing both the amino group and the carboxyl group. Many amino acids are chiral and hence optically active. Proteins are **polypeptides**, copolymers of 20 naturally occurring amino acids. The **primary structure** of a polypeptide chain is its sequence of amino acids. Its **secondary structure** may be either an **α helix** or a **β-pleated sheet**. The secondary structure is contorted into a **tertiary structure** by interactions between the peptide units that include **disulfide links** (—S—S—). Protein units may link together to give an overall **quaternary structure.**

24.9 The primary structure of proteins is controlled by the **nucleic acid** molecule DNA, a **polynucleotide**. A **nucleotide** is the combination of a phosphate group and a **nucleoside**; the latter is a combination of a base (A, G, C, or T; U replaces T in RNA) and a deoxyribose molecule. The double helix has a structure that arises from specific **base pairing** between the two polynucleotide strands. **Codons**, which specify an amino acid and form the genetic code, consist of groups of three bases.

CLASSIFIED EXERCISES

Functional Groups

24.1 Write the formula of the functional groups characteristic of (a) amines; (b) alcohols; (c) carboxylic acids; (d) aldehydes.

24.2 Name each functional group:
(a) —O—
(b) —CO—
(c) $—NO_2$
(d) —Cl

24.3 Write the general formula for each family of compounds, using R— to denote an aliphatic group and Ar— to denote an aromatic group (if necessary): (a) an alcohol; (b) an ether; (c) an acid; (d) a primary amine.

24.4 Write the formula for the simplest aliphatic compound of the following class: (a) a ketone; (b) an alcohol; (c) a secondary amine; (d) a carboxylic acid.

24.5 List the functional groups present in (a) vanillin (**10**), the compound responsible for the flavor of vanilla; (b) carvone (**11**), which is responsible for the flavor of spearmint.

24.6 Identify the functional groups present in (a) zingerone, the pungent, hot component of ginger:

CH_3O (ring) $CH_2CH_2—CO—CH_3$
HO

(b) monosodium glutamate, a flavor enhancer,

O O
$C—CH_2CH_2CH—C$ Na^+
HO NH_2 O^-

The Hydroxyl Group

24.7 Write the formula of the following compounds and state whether each one is a primary, secondary, or tertiary alcohol or a phenol: (a) *p*-hydroxybenzoic acid; (b) 2-butanol; (c) 1-butanol.

24.8 Write the formula of the following compounds and state whether each one is a primary, secondary, or tertiary alcohol or a phenol: (a) 2-methyl-2-propanol; (b) 1-propanol; (c) *o*-hydroxytoluene.

24.9 Name the following as alcohols and identify each as primary, secondary, or tertiary: (a) $CH_3CH(OH)CH_3$; (b) $HOCH_2CH_2CH_3$; (c) $(CH_3)_2C(OH)(CH_2)_2CH_3$.

24.10 Name the following as alcohols and identify each as primary, secondary, or tertiary:
(a) $CH_3CH(OH)CH_2OH$
(b) $CH_3(CH_2)_5OH$
(c) $HOCH_2CH_2CH_2OH$

24.11 Name the compounds:
(a) CH_3OCH_3
(b) $C_2H_5OCH_3$
(c) $CH_3CH_2OCH_2CH_3$

24.12 Write the formulas of (a) methylethyl ether; (b) 2-propylmethyl ether; (c) dimethyl ether.

24.13 Name the products of the reaction that occurs when (a) ethanol is heated with concentrated sulfuric acid; (b) ethanol is heated with hydrochloric acid; (c) *tert*-butyl alcohol (2-methyl-2-propanol) is heated with hydrobromic acid.

24.14 Name the products of the reaction that occurs when (a) 2-butanol is heated with concentrated sulfuric acid; (b) 2-butanol is heated with hydrobromic acid; (c) 2-propanol is heated with hydrobromic acid.

24.15 Write the formulas of the principal products of the reaction that occurs when (a) ethylene glycol (1,2-ethanediol) is heated with stearic acid; (b) ethanol is heated with oxalic acid; (c) phenol is mixed with an excess of chlorine in aqueous solution.

24.16 Write the formulas of the principal products of the reaction that occurs when (a) 1-butanol is heated with propanoic acid; (b) glyercol (1,2,3-propanetriol) is heated with benzoic acid; (c) phenol is mixed with sodium hydroxide in aqueous solution.

24.17 Name and write the formula of the principal product formed by the action of sodium dichromate on 1-propanol in an acidic organic solution.

24.18 Name and write the formula of the principal product formed by the action of sodium dichromate on 2-propanol in an acidic organic solution.

Aldehydes and Ketones

24.19 Identify each compound as an aldehyde or ketone and give its systematic name: (a) HCHO; (b) CH_3COCH_3; (c) $(CH_3CH_2)_2CO$.

24.20 Identify each compound as an aldehyde or ketone and give its systematic name: (a) CH_3CH_2CHO; (b) *p*-$CH_3C_6H_4CHO$; (c) $(CH_3CH_2)_2CHCH_2COCH_3$.

24.21 Write the chemical formulas of (a) methanal; (b) propanone; (c) 2-heptanone.

24.22 Write the chemical formulas of (a) 2-ethyl-2-methylpentanal; (b) 3,5-dihydroxy-4-octanone; (c) octanal.

24.23 Suggest an alcohol that could be used for the preparation of each of the following compounds and indicate how the reaction would be carried out: (a) ethanal; (b) 2-octanone; (c) 5-methyloctanal.

24.24 Suggest an alcohol that could be used for the preparation of each of the following compounds and indicate how the reaction would be carried out: (a) methanal; (b) propanone; (c) 5-methyl-6-decanone.

24.25 By analogy with the behavior of glucose and fructose, suggest what product is formed when methanol and (a) acetaldehyde or (b) acetone are mixed.

24.26 Lithium aluminum hydride, $LiAlH_4$, is a reducing agent for the carbonyl group but not for the C=C double bond. Name the products formed (and give their structural formulas) when it is used to reduce (a) 3-hexanone; (b) octa-5,7-dienal; (c) ethylcylohexa-2,6-dione.

24.27 Suggest a method for preparing acetaldehyde from limestone, water, and coke.

24.28 Write the products of the reaction between propanal and sodium cyanide in the presence of sulfuric acid.

The Carboxyl Group

24.29 Give the systematic names of (a) CH_3COOH; (b) $CH_3CH_2CH_2COOH$; (c) $CH_2(NH_2)COOH$.

24.30 Give the systematic names of (a) CH_3CH_2COOH; (b) $CHCl_2COOH$; (c) $CH_3(CH_2)_7COOH$.

24.31 Write the structural formulas of (a) benzoic acid; (b) 2-methylbutanoic acid; (c) oxalic acid (ethanedioic acid).

24.32 Write the structural formulas of (a) propanoic acid; (b) 2,2-dichlorobutanoic acid; (c) 2,2,2-trifluoroethanoic acid.

24.33 Suggest a method of preparing (a) ethanoic acid and (b) 2-methylbutanoic acid from an alcohol.

24.34 Suggest a method of preparing (a) benzoic acid and (b) ethanedioic acid from an alcohol or phenol.

24.35 Consider the following set of data: CH_3COOH, $pK_a = 4.74$; $CH_2ClCOOH$ $pK_a = 2.85$; $CHCl_2COOH$, $pK_a = 1.29$; CCl_3COOH, $pK_a = 0.52$. Suggest an explanation for the trend in the acid ionization constants.

24.36 Which solution has a lower pH, 0.010 M $CH_2FCOOH(aq)$ or 0.010 M $CF_3COOH(aq)$? Give your reasoning.

24.37 Write the formula of the principal product formed from the reaction of benzoic acid with 2-propanol.

24.38 Write a balanced equation for the reaction between methanol and octanoic acid.

24.39 What is the formula of the polymer formed from the reaction of oxalic acid (ethanedioic acid) with 1,4-dihydroxybutane?

24.40 Write the structural formula of the compound formed from the reaction of terephthalic acid (*p*-benzenedioic acid) and 2-propanol.

24.41 You are given samples of an aldehyde, a ketone, and a carboxylic acid. State clearly, with chemical equations, how you could distinguish the compounds.

24.42 Suggest an experimental method of showing that in ester formation, the oxygen atom in the eliminated water molecule comes from the carboxyl group. Does the process of amide formation shed any light on the details of the esterification reaction?

24.43 Explain the process of condensation polymerization. How might the polymer obtained from benzene-1,2-dicarboxylic acid differ from Dacron?

24.44 Suggest a reason why a condensation polymer (like any polymer) might not have a definite molar mass. How would that affect the osmotic measurement of its apparent molar mass?

Amines and Amides

24.45 Give the systematic names of the following amines and classify them as primary, secondary, or tertiary: (a) CH_3NH_2; (b) $(CH_3CH_2)_2NH$; (c) o-$CH_3C_6H_4NH_2$.

24.46 Write the formulas of the following amines and classify them as primary, secondary, or tertiary: (a) *o*-methylphenylamine; (b) triethylamine; (c) tetraethylammonium ion.

24.47 Write the formula of the product from the reaction of (a) methylamine with hydrogen chloride; (b) ethanoic acid with diethylamine; (c) aniline (phenylamine) with bromine; (d) propanoic acid and 1,4-butanediamine.

24.48 Suggest a method for the production of butylamine from (a) 1-bromobutane; (b) 1-butene; (c) 1-butanol.

24.49 Write the formula of the polymer formed from the reaction of oxalic acid (ethanedioic acid) with 1,4-butanediamine.

24.50 Write the structural formula of the product formed from a reaction between terephthalic acid (*p*-benzenedioic acid) and diethylamine.

24.51 Deduce the structure of putrescine ($C_4H_{12}N_2$) from the fact that it can be formed from 1,2-dibromoethene by reaction with KCN, which gives a compound of formula $C_4H_4N_2$, followed by reduction with sodium and ethanol.

24.52 The pK_a values of the conjugate acids of ammonia and methylamine are 9.25 and 10.64, respectively. What does this mean in terms of the relative strengths of the parent compounds as bases? Suggest an explanation of the difference.

CASE 24 ▸ PESTICIDES

Over one billion kilograms of pesticides are used annually throughout the world, primarily in agriculture. They are designed to kill pests (*-cide* is a suffix derived from the Latin word meaning "to kill"), whether it be an insect, a plant, a strain of bacteria, or a fungus. Many of the modern synthetic pesticides can be placed into one of three categories, but a large number of pesticides have structures that do not necessarily belong to any of these three classes, especially as advances in pesticide research continues.

Chlorinated hydrocarbons. Chlorinated hydrocarbons persist in the environment for 2 to 20 years. They tend to be carcinogenic (cancer inducing) but not very toxic to humans. Most of these compounds have chlorine atoms bonded to either an aliphatic or an aromatic structure. The most common is DDT, dichlorodiphenyltrichloroethane:

Cl–C_6H_4–CH(CCl_3)–C_6H_4–Cl

Organophosphorus compounds. Organophosphorus compounds are organic compounds that include phosphorus atoms in their structure. They persist in the environment for 1 to 12 weeks and usually degrade by hydrolysis; they are generally more toxic than chlorinated hydrocarbons. Organophosphorus compounds can be considered as a version of phosphoric acid in which the H atoms of H_3PO_4 have been replaced by aliphatic or aromatic groups. Their general formula is $(RO)_2$—(PX)—Y—R′, where R and R′ are organic groups and X and Y are S or O. Examples are parathion, a restricted agricultural insecticide,

$(C_2H_5O)_2$—P(=S)—O—C_6H_4—NO_2

and malathion, a common household insecticide:

$(CH_3O)_2$—P(=S)—S—CH(—C(=O)—OC_2H_5)—CH_2—C(=O)—OC_2H_5

The hydrolysis of parathion is

$(C_2H_5O)_2$—P(=S)—O—C_6H_4—NO_2 $\xrightarrow{H_2O}$

$(C_2H_5O)_2$—P(=S)—OH + HO—C_6H_4—NO_2

Carbamates. The carbamates are organic derivatives of carbamic acid, H_2NCOOH, and have the general formula R_2N—(CX)—Y—R′, where R and R′ are hydrogen or an aliphatic or aromatic group, X is S or O, and Y is O, S, or N. The most common carbamate is carbaryl, which is commonly called Sevin and is a general garden insecticide:

CH_3—NH—C(=O)—O—$C_{10}H_7$ (1-naphthyl)

Proteins and Nucleic Acids

24.53 Write the formula of the simplest possible amino acid.

24.54 The common name of $CH_3CH(NH_2)COOH$ is alanine. What is its systematic name?

24.55 What is the structural formula of a peptide link?

24.56 What classes of organic compounds are necessary for the preparation of a protein?

24.57 Write the structural formula of the peptide formed from the reaction of tyrosine with glycine.

24.58 Write the structural formula of the peptide formed from the reaction of glycine with tyrosine.

The persistence, toxicity, and natural degradation of the carbamates parallel the organophosphorus pesticides. Carbaryl (Sevin) hydrolyzes according to the equation

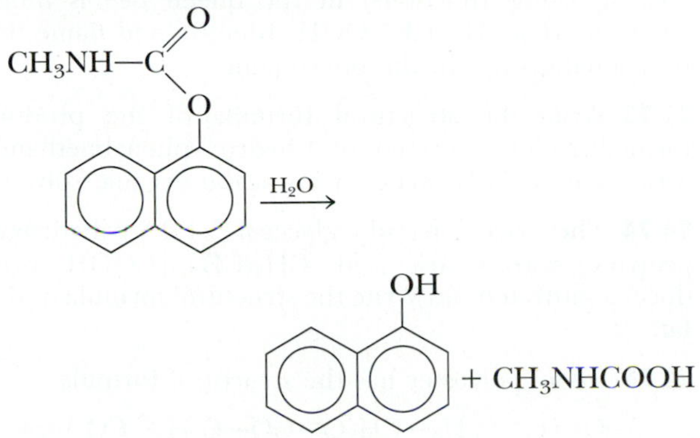

$$CH_3NH{-}C({=}O){-}O{-}C_{10}H_7 \xrightarrow{H_2O} C_{10}H_7OH + CH_3NHCOOH$$

QUESTIONS

1. Classify each of the following pesticides as a chlorinated hydrocarbon, an organophosphorus compound, or alternatively as a carbamate:
(a) Disulfoton, $(C_2H_5)_2{-}(PS){-}S{-}CH_2{-}CH_2{-}S{-}C_2H_5$, an insecticide applied as a side dressing in the planting furrow
(b) Eptam, $C_2H_5{-}S{-}(CO){-}N(C_3H_7)_2$, a broadleaf herbicide
(c) Chloroneb,

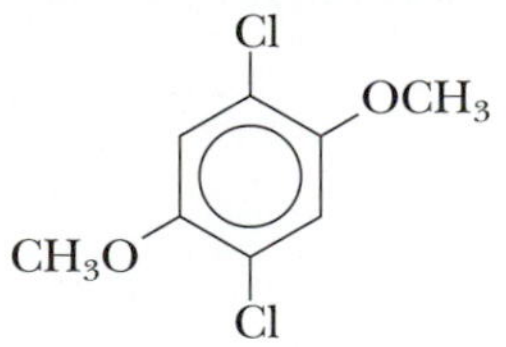

(2,5-dichloro-1,4-dimethoxybenzene structure: Cl, OCH_3, CH_3O, Cl)

which is used to control seedling disease in the planting furrow
(d) Tetrafluoron,

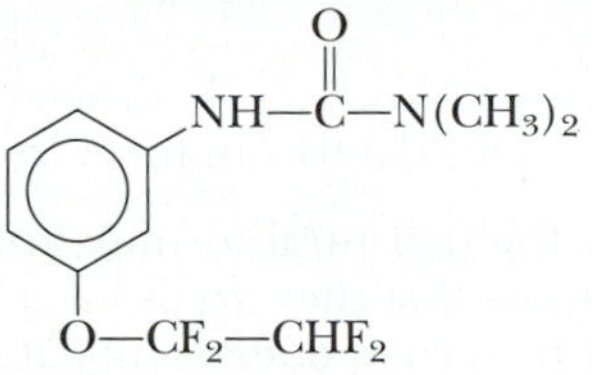

(structure: benzene ring with $NH{-}C({=}O){-}N(CH_3)_2$ and $O{-}CF_2{-}CHF_2$)

a herbicide for broadleaf and grassy weed control in cotton.

2. Write the chemical equation for the hydrolysis of Alar, $(CH_3)_2N{-}NH{-}CO{-}CH_2{-}CH_2{-}COOH$, a chemical capable of inducing desirable effects on the flowering and fruiting of many crops such as apples, peaches, and grapes.

3. Agent orange, a herbicide for broad leaf plants, is a 50:50 formulation of the two herbicides, 2,4-D and 2,4,5-T. Classify each herbicide. Speculate as to how each herbicide may undergo degradation by hydrolysis.

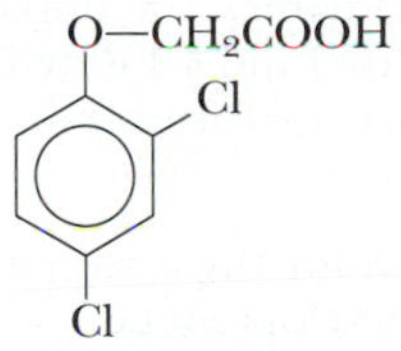

(structures: $O{-}CH_2COOH$ with Cl, Cl — 2,4-D; $O{-}CH_2COOH$ with Cl, Cl, Cl — 2,4,5-T)

2,4-D 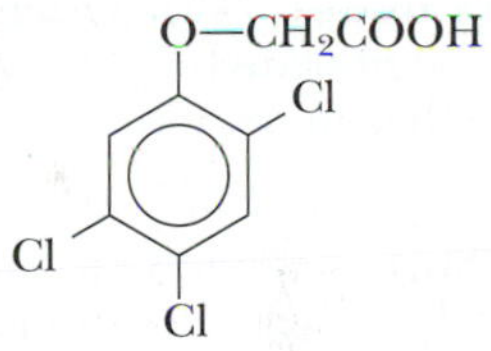   2,4,5-T

4. Identify all the organic functional groups in malathion.

24.59 Write the structural formula of the Ala-Met-Cys tripeptide.

24.60 Write the structural formula of the Gly-Trp-Phe tripeptide.

24.61 Write the molecular formulas of a fragment of a polypeptide chain with the composition (a) -Gly-Gly-Gly-Gly- and (b) -Gly-Leu-Ser-Ala-.

24.62 The protein beef insulin consists of two chains in which the groups -(Leu)$_2$-Ile-(Cys)$_4$-Arg- and -Pro-(Phe)$_3$-(Cys)$_2$-Arg- occur. What could be the nature of the interaction between the chains? The chains can be separated by reduction. Suggest a reason.

24.63 What amino acid residue sequence would be constructed from an RNA sequence of the form UAUCUAUCUAUCUAUCUAUCU?

24.64 Suppose you wanted to produce a genetically engineered organism that could synthesize aspartame. What RNA sequences should be sought?

UNCLASSIFIED EXERCISES

24.65 Designate which of the hydrogen atoms in the compound *p*-$CH_3OC_6H_4OH$ can form hydrogen bonds.

24.66 (a) Write the structural formulas of diethyl ether and 1-butanol (note that they are isomers). (b) Whereas the solubility of both compounds is about 8 g/100 mL of water, the boiling point of 1-butanol is 117°C but that of diethyl ether is 35°C. Account for these observations.

24.67 Write the structural formula of the principal products formed from the reaction of (a) 2-methylcyclohexanol and (b) ethanol with sodium dichromate in an acidic organic solvent.

24.68 (a) Write the formulas of ethanol and ethanal. (b) What chemical test will distinguish ethanol from ethanal? (c) What chemical test will distinguish ethanal from ethanoic acid?

24.69 Dopamine is a neurotransmitter in the central nervous system, and a deficiency in it is related to Parkinson's disease. Two systematic names for dopamine are 4-(2-aminoethyl)-1,2-benzenediol and 3,4-dihydroxyphenylethylamine. Write the structural formula of dopamine.

24.70 Suggest a one-step reaction for the conversion of 2-methylcyclopentanol to methylcyclopentene.

24.71 Write a balanced equation for the reaction of 2-methylphenol with aqueous sodium hydroxide.

24.72 The structure of a pheromone (commonly called a sex attractant, although pheromones have more complex signaling functions) in the queen bee is *trans*-$CH_3CO(CH_2)_5CH{=}CHCOOH$. Identify and name the functional groups in the pheromone.

24.73 Write the structural formula of the product formed from the reaction of 4-hydroxyphenylmethanol with sodium dichromate in an acidic organic solvent.

24.74 The reaction of glycerol (1,2,3-trihydroxypropane) with stearic acid, $CH_3(CH_2)_{16}COOH$, produces a saturated fat. Write the structural formula of the fat.

24.75 Kodel polyester has the structural formula

$$\text{—(—OCH}_2\text{—C}_6\text{H}_4\text{—CH}_2\text{O—CO—C}_6\text{H}_4\text{—CO—)}_n\text{—}$$

Identify the starting materials of the polyester.

24.76 The substituents $—CH_3$, $—NH_2$, and $—OCH_3$ release electrons into the benzene ring, but $—NO_2$, $—COOH$, and halogens withdraw them. Which set of substituents will increase the strength of aniline as a base?

24.77 The molar mass of a protein extracted from salmon sperm is 10,000 g/mol. Analysis of 100 g of the protein gave the following masses of amino acids: Leu, 1.28 g; Ala, 0.89 g; Val, 3.68 g; Gly, 3.01 g; Ser, 7.29 g; Pro, 6.90 g; Arg, 86.40 g. What is the molecular formula of the protein?

24.78 Aspartame, known as the artificial sweetener NutraSweet, has the primary structure Phe-Asp. Write the empirical formula of aspartame.

APPENDIX 1 MATHEMATICAL INFORMATION

1A: SCIENTIFIC NOTATION

In *scientific notation,* a number is written as $A \times 10^a$, where A is a decimal number with one nonzero digit in front of the decimal point and a is a whole number. For example, 333 is written 3.33×10^2 in scientific notation, because $10^2 = 10 \times 10 = 100$:

$$333 = 3.33 \times 100 = 3.33 \times 10^2$$

On a scientific calculator, this number is entered as

3 . 3 3 EXP 2

(On some calculators, the EXP key is labeled EE or EEX.) We use

$$10^1 = 10$$
$$10^2 = 10 \times 10 = 100$$
$$10^3 = 10 \times 10 \times 10 = 1000$$

and so on. Note that the number of zeros following 1 is equal to the power of 10. Thus, 10^6 is 1 followed by six zeros:

$$10^6 = 10 \times 10 \times 10 \times 10 \times 10 \times 10 = 1{,}000{,}000$$

Numbers between 0 and 1 are expressed in the same way, but with a negative power of 10; they have the form $A \times 10^{-a}$, with $10^{-1} = \frac{1}{10} = 0.1$, and so on. Thus, 0.0333 in scientific notation is 3.33×10^{-2} because

$$10^{-2} = \tfrac{1}{10} \times \tfrac{1}{10} = \tfrac{1}{100}$$

and

$$0.0333 = 3.33 \times \tfrac{1}{100} = 3.33 \times 10^{-2}$$

On a scientific calculator, this number is entered as

3 . 3 3 EXP +/− 2

(Be sure to use the +/− key, which is sometimes labeled CHS, to enter the power of 2, and not the − key.) We use

$$10^{-2} = 10^{-1} \times 10^{-1} = 0.01$$
$$10^{-3} = 10^{-1} \times 10^{-1} \times 10^{-1} = 0.001$$
$$10^{-4} = 10^{-1} \times 10^{-1} \times 10^{-1} \times 10^{-1} = 0.0001$$

When a negative power of 10 is written out as a decimal number, the number of zeros following the decimal point is one less than the number (disregarding the sign) to which 10 is raised. Thus, 10^{-6} is written as a decimal point followed by $6 - 1 = 5$ zeros and then a 1:

$$10^{-6} = 10^{-1} \times 10^{-1} \times 10^{-1} \times 10^{-1} \times 10^{-1} \times 10^{-1}$$
$$= 0.000\,001$$

To multiply numbers in scientific notation, the decimal parts of the numbers are multiplied and the powers of 10 are added:

$$(A \times 10^a) \times (B \times 10^b) = (A \times B) \times 10^{a+b}$$

An example is

$$(1.23 \times 10^2) \times (4.56 \times 10^3) = 1.23 \times 4.56 \times 10^{2+3}$$
$$= 5.61 \times 10^5$$

This rule holds even if the powers of 10 are negative:

$$(1.23 \times 10^{-2}) \times (4.56 \times 10^{-3})$$
$$= 1.23 \times 4.56 \times 10^{-2+(-3)} = 5.61 \times 10^{-5}$$

The keystrokes for this calculation are

4 . 5 6 EXP +/− 3 =

The results of such calculations are adjusted so that there is one digit in front of the decimal point:

$$(4.56 \times 10^{-3}) \times (7.65 \times 10^{6}) = 34.88 \times 10^{3}$$
$$= 3.488 \times 10^{4}$$

When dividing two numbers in scientific notation, we divide the decimal parts of the numbers and subtract the powers of 10:

$$\frac{A \times 10^{a}}{B \times 10^{b}} = \frac{A}{B} \times 10^{a-b}$$

An example is

$$\frac{4.31 \times 10^{5}}{9.87 \times 10^{-8}} = \frac{4.31}{9.87} \times 10^{5-(-8)}$$
$$= 0.437 \times 10^{13} = 4.37 \times 10^{12}$$

Before adding and subtracting numbers in scientific notation, we rewrite the numbers as decimal numbers multiplied by the *same* power of 10:

$$1.00 \times 10^{3} + 2.00 \times 10^{2} = 1.00 \times 10^{3} + 0.200 \times 10^{3}$$
$$= 1.20 \times 10^{3}$$

When raising a number in scientific notation to a particular power, we raise the decimal part of the number to the power and *multiply* the power of 10 by the power:

$$(A \times 10^{a})^{b} = A^{b} \times 10^{a \times b}$$

For example, 2.88×10^{4} raised to the third power is

$$(2.88 \times 10^{4})^{3} = 2.88^{3} \times (10^{4})^{3} = 2.88^{3} \times 10^{3\times 4}$$
$$= 23.9 \times 10^{12} = 2.39 \times 10^{13}$$

The key sequence on a scientific calculator for this calculation is

2 . 8 8 EXP 4 x^y 3 =

The rule follows from the fact that

$$(10^{4})^{3} = 10^{4} \times 10^{4} \times 10^{4} = 10^{4+4+4} = 10^{3\times 4}$$

1 B : LOGARITHMS

The *common logarithm* of a number x is denoted log x and is the power to which 10 must be raised to equal x. Thus, the logarithm of 100 is 2, written log 100 = 2, because $10^{2} = 2$. The logarithm of 1.5×10^{2} is 2.18 because

$$10^{2.18} = 10^{0.18+2} = 10^{0.18} \times 10^{2} = 1.5 \times 10^{2}$$

The number in front of the decimal point in the logarithm (the 2 in log $(1.5 \times 10^{2}) = 2.18$), is called the *characteristic* of the logarithm; the decimal fraction (the numbers following the decimal point; the 0.18 in the example) is called the *mantissa*. The characteristic is the power of 10 in the original number (the power 2 in 1.5×10^{2}) and the mantissa is the logarithm of the decimal number written with one nonzero digit in front of the decimal point (the 1.5 in the example).

The distinction between the characteristic and the mantissa is important when we have to decide how many significant figures to retain in a calculation that involves logarithms (for example, in the calculation of pH): *the number of significant figures in the mantissa is equal to the number of significant figures in the decimal number*. Because the decimal number 1.5×10^{2} has two significant figures, its mantissa is written 0.18 (two significant figures); so its logarithm is 2.18, as written above. Just as the power of 10 in a decimal number only indicates the location of the decimal point and plays no role in the determination of significant figures, so the characteristic of a logarithm is not included in the count of significant figures in a logarithm.

The *common antilogarithm* of a number x is the number that has x as its common logarithm. In practice, the common antilogarithm of x is simply another name for 10^{x}, so the common antilogarithm of 2 is $10^{2} = 100$ and that of 2.18 is

$$10^{2.18} = 10^{0.18+2} = 10^{0.18} \times 10^{2} = 1.5 \times 10^{2}$$

In keeping with the remarks above, the *mantissa* of the logarithm (the 0.18 in 2.18) determines the number of significant figures in the antilogarithm (1.5×10^{2}, two significant figures).

The logarithm of a number greater than one is positive and that of a number smaller than one (but greater than zero) is negative:

$$\text{If } x > 1, \log x > 0$$

$$\text{If } x = 1, \log x = 0$$

$$\text{If } x < 1, \log x < 0$$

Logarithms are not defined either for zero or for negative numbers.

On a scientific calculator, the common logarithm of a number x is calculated by entering x and pressing the [log x] key on the calculator. For example, the logarithm of 4.33×10^{-5} is determined with the following sequence of keystrokes:

[4] [.] [3] [3] [EXP] [+/−] [5] [log x]

(It is important to use the [+/−] key, not the [−] key when entering the negative power of 10.) The decimal number has three significant figures in this example, so the mantissa should also be written with three significant figures, and the answer reported as −4.364. Likewise, the common antilogarithm of x is found by entering x and pressing the [10^x] key (or [INV] and [log] keys); so, to calculate the antilogarithm of 11.68, the keystrokes are

[1] [1] [.] [6] [8] [10^x]

Because the mantissa (0.68) of the original number has two significant figures, the antilogarithm should be written with two significant figures, and the correct answer is 4.8×10^{11}.

The *natural logarithm* of a number x, denoted $\ln x$, is the power to which the number e = 2.718 . . . must be raised to equal x. Thus, ln 10.0 = 2.303, signifying that $e^{2.303} = 10.0$. The number e may seem a peculiar choice, but it occurs naturally in many mathematical expressions, and its use simplifies many formulas. On an electronic calculator, the natural logarithm of x is calculated by entering x and then pressing the [ln x] key. There is no simple rule for assessing the correct number of significant figures when natural logarithms are used: one way is to convert natural logarithms to common logarithms (see below) and then to use the rules specified above.

Common and natural logarithms are related by the expression

$$\ln x = \ln 10 \times \log x = 2.303 \times \log x$$

The *natural antilogarithm* of x is normally called the *exponential* of e; it is the value of e raised to the power x. On a calculator, it is obtained by entering x and pressing the [e^x] key. Thus, the natural antilogarithm of 2.303 is $e^{2.303} = 10.0$.

The following relations between logarithms are useful. (They are written here mainly for common logarithms, but they apply to natural logarithms as well.)

Relation	Example
$\log 10^x = x$	$\log 10^{-7} = -7$
$\ln e^x = x$	$\ln e^{-kt} = -kt$
$\log x + \log y = \log xy$	$\log [Ag^+] + \log [Cl^-] = \log [Ag^+][Cl^-]$
$\log x - \log y = \log x$	$\log A_0 - \log A = \log \left(\frac{A_0}{A}\right)$
$x \log y = \log y^x$	$2 \log [H^+] = \log [H]^{+2}$
$\log \left(\frac{1}{x}\right) = -\log x$	$\log \left(\frac{1}{[H^+]}\right) = -\log [H^+]$

An example of the use of logarithms in chemistry is in the expression for the pH of a solution. If the molar concentration of H^+ ions (more precisely, H_3O^+ ions) in a solution is 0.0024 mol/L, then the pH of the solution is −log (0.0024). To calculate the pH on a scientific calculator, first convert the concentration to scientific notation and then enter

[2] [.] [4] [EXP] [+/−] [3] [log x] [+/−]

The mantissa of the pH can have only two significant figures (preceded by the characteristic), so the pH is reported as 2.62. Conversely, if the pH of the solution is measured as 7.4, the hydrogen ion concentration is the value of $10^{-7.4}$, which is evaluated using the keystrokes

[7] [.] [4] [+/−] [10^x]

There is only one significant figure in the mantissa (the 4 in pH = 7.4), so the hydrogen ion concentration is reported as 4×10^{-8} mol/L, with only one significant figure.

Logarithms are useful for solving expressions of the form

$$a^x = b$$

for the unknown x. (This type of calculation can arise when the order of a reaction is being determined.) We take logarithms of both sides

$$\log a^x = \log b$$

and from a relation given above:

$$x \log a = \log b$$

Therefore,

$$x = \frac{\log b}{\log a}$$

1C: QUADRATIC AND CUBIC EQUATIONS

A *quadratic equation* is an equation of the form

$$ax^2 + bx + c = 0$$

The two *roots* of the equation (the solutions) can be found from the expressions:

$$x_1 = \frac{-b + \sqrt{b^2 - 4ac}}{2a} \qquad x_2 = \frac{-b - \sqrt{b^2 - 4ac}}{2a}$$

When a quadratic equation arises in connection with a chemical calculation, we accept only the roots that lead to a physically plausible result. For example, if x is a concentration, then it must be a positive number, and a negative root can be ignored. However, if x is a *change* in concentration, then it may be either positive or negative. In such a case, we would have to determine which root led to an acceptable (positive) final concentration.

On occasion, an equilibrium table (or some other type of calculation) results in a *cubic equation:*

$$ax^3 + bx^2 + cx + d = 0$$

Cubic equations are often very tedious to solve exactly, but it is often justifiable (particularly when we are dealing with a chemical system for which the data have only a certain number of significant figures) to use an approximation procedure. For example, an equilibrium table might lead to an expression of the form

$$\frac{0.0600 - x}{(0.0500 + x) \times (0.0300 + 2x)^2} = 1.10 \times 10^{-2}$$

One approach to the solution of this expression for the value of x is to make successive approximations starting with an estimated value of x. We know that x must be smaller than 0.0600, and we also know that it should be quite close to 0.0600 to ensure that the expression on the left is small (it should be equal to 1.10×10^{-2}). Thus, we might estimate that $x = 0.0590$, which leads to a value of

$$\frac{0.0600 - 0.0590}{(0.0500 + 0.0590) \times (0.0300 + 2 \times 0.0590)^2} = 0.419$$

This first guess leads to a result that is larger than the true value, so we reduce the size of the fraction by choosing a value of x that is closer to 0.0600. Thus, with $x = 0.0599$ we find

$$\frac{0.0600 - 0.0599}{(0.0500 + 0.0599) \times (0.0300 + 2 \times 0.0599)^2} = 4.05 \times 10^{-2}$$

This result is still too large, but much closer to the required value of 1.10×10^{-2}. We could continue to choose values of x that are closer to 0.0600 than 0.0599, but because only three significant figures are present in the data, there is little point in looking for a more precise result, and x can be reported as lying between 0.0599 and 0.0600 (in fact, a more precise answer is 0.05997). Such approximate solutions can usually be found after two or three trials.

1D: GRAPHS

Experimental data can often be analyzed by plotting a graph. In many cases, the best procedure is to find a way of plotting the data so that a straight line results. This practice is more useful than plotting a curve, because it is quite easy to tell whether or not the data do in fact fall on a straight line, whereas small deviations from a curve are harder to detect. Moreover, it is also easy to calculate the slope of a straight line, to *extrapolate* (or extend) a straight line beyond the range of the data, and to *interpolate* between the data (that is, find a value between two measured values).

The formula of a straight line graph of y (the vertical axis) plotted against x (the horizontal axis) is

$$y = mx + b$$

Here b is the *intercept* of the graph with the y axis (Fig. A.1), the value of y where the graph cuts through the vertical axis at $x = 0$. The *slope* of the graph, its gradient, is m. The slope can be calculated by choosing two points x_1 and x_2 and their corresponding values on the y axis y_1 and y_2 and substituting the values into the formula

$$m = \frac{y_2 - y_1}{x_2 - x_1}$$

Because b is the intercept and m is the slope, the equation of the straight line is equivalent to

$$y = slope \times x + intercept$$

Many of the equations in the text can be rearranged to give a straight line when plotted. These include

Equation	y	= *slope* × x + *intercept*
temperature scale conversions	°C	$= 1 \times \text{K} - 273.15$
	°F	$= 1.8 \times {}^\circ\text{C} + 32$
ideal gas law	P	$= nRT \times \frac{1}{V}$
first-order rate law	$\ln[\text{A}]$	$= -k \times t + \ln[\text{A}]_0$
second-order rate law	$\frac{1}{[\text{A}]}$	$= k \times t + \frac{1}{[\text{A}]_0}$
Arrhenius	$\ln k$	$= \frac{-E_a}{R} \times \frac{1}{T} + \ln \text{A}$

We can speak of *the* slope of a straight line because the slope is the same at all points. On a curve, however, the slope changes from point to point. The *tangent* of a curve at a specified point is the straight line that has the same slope as the curve at that point. The tangent can be found by a series of approximations, as shown in Fig. A.2. Approximation 1 is found by drawing a point on the curve on each side of the point of interest (corresponding to equal distances along the x axis) and joining them by a straight line. A better approximation (approximation 2) is obtained by moving the points an equal distance closer to the point of interest and drawing a new line. The "exact" tangent is obtained when the two points are virtually coincident with the point of interest. Its slope is then equal to the slope of the curve at the point of interest.

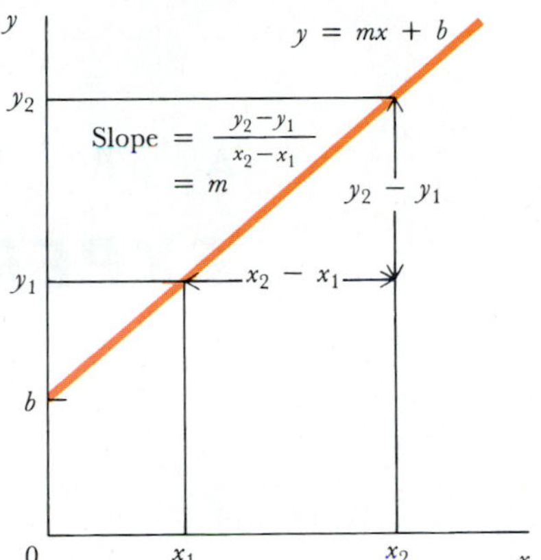

FIGURE A.1 The straight line $y = mx + b$, its intercept at $y = b$, and its slope m.

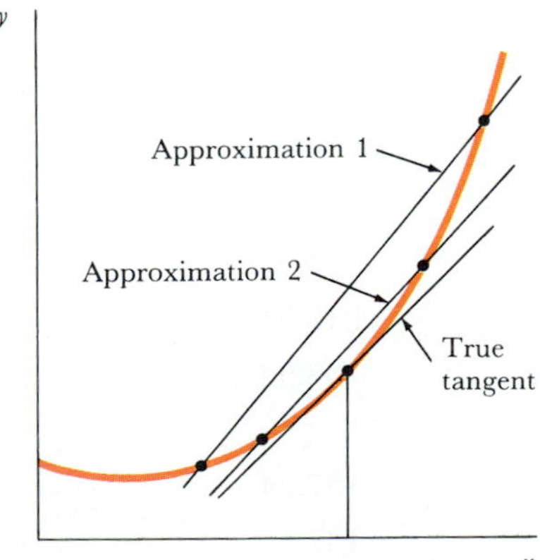

FIGURE A.2 Successive approximations to the true tangent are obtained as the two points defining the straight line come closer together and finally coincide.

2

A P P E N D I X

EXPERIMENTAL DATA

2A: THERMODYNAMIC DATA AT 25°C

1 INORGANIC SUBSTANCES

Substance	Molar mass, g/mol	Enthalpy of formation, ΔH_f°, kJ/mol	Free energy of formation, ΔG_f°, kJ/mol	Entropy* S°, J/(K·mol)
Aluminum				
$Al(s)$	26.98	0	0	28.33
$Al^{3+}(aq)$	26.98	−524.7	−481.2	−321.7
$Al_2O_3(s)$	101.95	−1675.7	−1582.3	50.92
$Al(OH)_3(s)$	78.00	−1276	—	—
$AlCl_3(s)$	133.24	−704.2	−628.8	110.67
Antimony				
$SbH_3(g)$	124.77	145.11	147.75	232.78
$SbCl_3(g)$	228.10	−313.8	−301.2	337.80
$SbCl_5(g)$	299.02	−394.34	−334.29	401.94
Arsenic				
$As(s)$, gray	74.92	0	0	35.1
$As_2S_3(s)$	246.04	−169.0	−168.6	163.6
$AsO_4^{3-}(aq)$	138.92	−888.14	−648.41	−162.8
Barium				
$Ba(s)$	137.34	0	0	62.8
$Ba^{2+}(aq)$	137.34	−537.64	−560.77	9.6
$BaO(s)$	153.34	−553.5	−525.1	70.42

*The entropies of individual ions in solution are determined by setting the entropy of H^+ in water equal to zero and then defining the entropies of all other ions relative to this value; hence a negative entropy is one that is lower then the entropy of H^+ in water.

Substance	Molar mass, g/mol	Enthalpy of formation, ΔH_f°, kJ/mol	Free energy of formation, ΔG_f°, kJ/mol	Entropy S°, J/(K·mol)
$BaCO_3(s)$	197.35	−1216.3	−1137.6	112.1
$BaCO_3(aq)$	197.35	−1214.78	−1088.59	−47.3
Boron				
$B(s)$	10.81	0	0	5.86
$B_2O_3(s)$	69.62	−1272.8	−1193.7	53.97
$BF_3(g)$	67.81	−1137.0	−1120.3	254.12
Bromine				
$Br_2(l)$	159.82	0	0	152.23
$Br_2(g)$	159.82	30.91	3.11	245.46
$Br(g)$	79.91	111.88	82.40	175.02
$Br^-(aq)$	79.91	−121.55	−103.96	82.4
$HBr(g)$	80.92	−36.40	−53.45	198.70
Calcium				
$Ca(s)$	40.08	0	0	41.42
$Ca(g)$	40.08	178.2	144.3	154.88
$Ca^{2+}(aq)$	40.08	−542.83	−553.58	−53.1
$CaO(s)$	56.08	−635.09	−604.03	39.75
$Ca(OH)_2(s)$	74.10	−986.09	−898.49	83.39
$Ca(OH)_2(aq)$	74.10	−1002.82	−868.07	−74.5
$CaCO_3(s)$, calcite	100.09	−1206.9	−1128.8	92.9
$CaCO_3(s)$, aragonite	100.09	−1207.1	−1127.8	88.7
$CaCO_3(aq)$	100.09	−1219.97	−1081.39	−110.0
$CaF_2(s)$	78.08	−1219.6	−1167.3	68.87
$CaF_2(aq)$	78.08	−1208.09	−1111.15	−80.8
$CaCl_2(s)$	110.99	−795.8	−748.1	104.6
$CaCl_2(aq)$	110.99	−877.1	−816.0	59.8
$CaBr_2(s)$	199.90	−682.8	−663.6	130
$CaC_2(s)$	64.10	−59.8	−64.9	69.96
$CaSO_4(s)$	136.14	−1434.11	−1321.79	106.7
$CaSO_4(aq)$	136.14	−1452.10	−1298.10	−33.1
Carbon†				
$C(s)$, graphite	12.011	0	0	5.740
$C(s)$, diamond	12.011	1.895	2.900	2.377
$C(g)$	12.011	716.68	671.26	158.10
$CO(g)$	28.01	−110.53	−137.17	197.67
$CO_2(g)$	44.01	−393.51	−394.36	213.74

†For organic compounds, see the next table.

(continued)

Substance	Molar mass, g/mol	Enthalpy of formation, ΔH_f°, kJ/mol	Free energy of formation, ΔG_f°, kJ/mol	Entropy S°, J/(K·mol)
Carbon continued				
$CO_3^{2-}(aq)$	60.01	−677.14	−527.81	−56.9
$CCl_4(l)$	153.82	−135.44	−65.21	216.40
$CS_2(l)$	76.14	89.70	65.27	151.34
$HCN(g)$	27.03	135.1	124.7	201.78
$HCN(l)$	27.03	108.87	124.97	112.84
Cerium				
$Ce(s)$	140.12	0	0	72.0
$Ce^{3+}(aq)$	140.12	−696.2	−672.0	−205
$Ce^{4+}(aq)$	140.12	−537.2	−503.8	−301
Chlorine				
$Cl_2(g)$	70.91	0	0	223.07
$Cl(g)$	35.45	121.68	105.68	165.20
$Cl^-(aq)$	35.45	−167.16	−131.23	56.5
$HCl(g)$	36.46	−92.31	−95.30	186.91
$HCl(aq)$	36.46	−167.16	−131.23	56.5
Copper				
$Cu(s)$	63.54	0	0	33.15
$Cu^+(aq)$	63.54	71.67	49.98	40.6
$Cu^{2+}(aq)$	63.54	64.77	65.49	−99.6
$Cu_2O(s)$	143.08	−168.6	−146.0	93.14
$CuO(s)$	79.54	−157.3	−129.7	42.63
$CuSO_4(s)$	159.60	−771.36	−661.8	109
$CuSO_4 \cdot 5H_2O(s)$	249.68	−2279.7	−1879.7	300.4
Deuterium				
$D_2(g)$	4.028	0	0	144.96
$D_2O(g)$	20.028	−249.20	−234.54	198.34
$D_2O(l)$	20.028	−294.60	−243.44	75.94
Fluorine				
$F_2(g)$	38.00	0	0	202.78
$F^-(aq)$	19.00	−332.63	−278.79	−13.8
$HF(g)$	20.01	−271.1	−273.2	173.78
$HF(aq)$	20.01	−332.63	−278.79	−13.8
Hydrogen				
(see also *Deuterium*)				
$H_2(g)$	2.016	0	0	130.68
$H(g)$	1.008	217.97	203.25	114.71
$H^+(aq)$	1.008	0	0	0
$H_2O(l)$	18.02	−285.83	−237.13	69.91

Substance	Molar mass, g/mol	Enthalpy of formation, ΔH_f°, kJ/mol	Free energy of formation, ΔG_f°, kJ/mol	Entropy S°, J/(K·mol)
$H_2O(g)$	18.02	−241.82	−228.57	188.83
$H_2O_2(l)$	34.02	−187.78	−120.35	109.6
$H_2O_2(aq)$	34.02	−191.17	−134.03	143.9
Iodine				
$I_2(s)$	253.81	0	0	116.14
$I_2(g)$	253.81	62.44	19.33	260.69
$I^-(aq)$	126.90	−55.19	−51.57	111.3
$HI(g)$	127.91	26.48	1.70	206.59
Iron				
$Fe(s)$	55.85	0	0	27.28
$Fe^{2+}(aq)$	55.85	−89.1	−78.90	−137.7
$Fe^{3+}(aq)$	55.85	−48.5	−4.7	−315.9
$Fe_3O_4(s)$, magnetite	231.54	−1118.4	−1015.4	146.4
$Fe_2O_3(s)$, hematite	159.69	−824.2	−742.2	87.40
$FeS(s,\alpha)$	87.91	−100.0	−100.4	60.29
$FeS(aq)$	87.91	—	6.9	—
$FeS_2(s)$	119.98	−178.2	−166.9	52.93
Lead				
$Pb(s)$	207.19	0	0	64.81
$Pb^{2+}(aq)$	207.19	−1.7	−24.43	10.5
$PbO_2(s)$	239.19	−277.4	−217.33	68.6
$PbSO_4(s)$	303.25	−919.94	−813.14	148.57
$PbBr_2(s)$	367.01	−278.7	−261.92	161.5
$PbBr_2(aq)$	367.01	−244.8	−232.34	175.3
Magnesium				
$Mg(s)$	24.31	0	0	32.68
$Mg(g)$	24.31	147.70	113.10	148.65
$Mg^{2+}(aq)$	24.31	−466.85	−454.8	−138.1
$MgO(s)$	40.31	−601.70	−569.43	26.94
$MgCO_3(s)$	84.32	−1095.8	−1012.1	65.7
$MgBr_2(s)$	184.13	−524.3	−503.8	117.2
Mercury				
$Hg(l)$	200.59	0	0	76.02
$Hg(g)$	200.59	61.32	31.82	174.96
$HgO(s)$	216.59	−90.83	−58.54	70.29
$Hg_2Cl_2(s)$	472.09	−265.22	−210.75	192.5
Nitrogen				
$N_2(g)$	28.01	0	0	191.61

(continued)

Substance	Molar mass, g/mol	Enthalpy of formation, ΔH_f°, kJ/mol	Free energy of formation, ΔG_f°, kJ/mol	Entropy S°, J/(K·mol)
Nitrogen continued				
$NO(g)$	30.01	90.25	86.55	210.76
$N_2O(g)$	44.01	82.05	104.20	219.85
$NO_2(g)$	46.01	33.18	51.31	240.06
$N_2O_4(g)$	92.01	9.16	97.89	304.29
$HNO_3(l)$	63.01	−174.10	−80.71	155.60
$HNO_3(aq)$	63.01	−207.36	−111.25	146.4
$NO_3^-(aq)$	62.01	−205.0	−108.74	146.4
$NH_3(g)$	17.03	−46.11	−16.45	192.45
$NH_3(aq)$	17.03	−80.29	−26.50	111.3
$NH_4^+(aq)$	18.04	−132.51	−79.31	113.4
$NH_2OH(s)$	33.03	−114.2	—	—
$HN_3(g)$	43.03	294.1	328.1	238.97
$N_2H_4(l)$	32.05	50.63	149.34	121.21
$NH_4NO_3(s)$	80.04	−365.56	−183.87	151.08
$NH_4Cl(s)$	53.49	−314.43	−202.87	94.6
$NH_4ClO_4(s)$	117.49	−295.31	−88.75	186.2
Oxygen				
$O_2(g)$	32.00	0	0	205.14
$O_3(g)$	48.00	142.7	163.2	238.93
$OH^-(aq)$	17.01	−229.99	−157.24	−10.75
Phosphorus				
$P(s)$, white	30.97	0	0	41.09
$P_4(g)$	123.90	58.91	24.44	279.98
$PH_3(g)$	34.00	5.4	13.4	210.23
$P_4O_{10}(s)$	283.89	−2984.0	−2697.0	228.86
$H_3PO_3(aq)$	82.00	−964.8	—	—
$H_3PO_4(l)$	98.00	−1266.9	—	—
$H_3PO_4(aq)$	98.00	−1277.4	−1018.7	—
$PCl_3(l)$	137.33	−319.7	−272.3	217.18
$PCl_3(g)$	137.33	−287.0	−267.8	311.78
$PCl_5(g)$	208.24	−374.9	−305.0	364.6
$PCl_5(s)$	208.24	−443.5	—	—
Potassium				
$K(s)$	39.10	0	0	64.18
$K(g)$	39.10	89.24	60.59	160.34
$K^+(aq)$	39.10	−252.38	−283.27	102.5
$KOH(s)$	56.11	−424.76	−379.08	78.9

Substance	Molar mass, g/mol	Enthalpy of formation, ΔH_f°, kJ/mol	Free energy of formation, ΔG_f°, kJ/mol	Entropy S°, J/(K·mol)
$KOH(aq)$	56.11	−482.37	−440.50	91.6
$KF(s)$	58.10	−567.27	−537.75	66.57
$KCl(s)$	74.56	−436.75	−409.14	82.59
$KBr(s)$	119.01	−393.80	−380.66	95.90
$KI(s)$	166.01	−327.90	−324.89	106.32
$KClO_3(s)$	122.55	−397.73	−296.25	143.1
$KClO_4(s)$	138.55	−432.75	−303.09	151.0
$K_2S(s)$	110.27	−380.7	−364.0	105
$K_2S(aq)$	110.27	−471.5	−480.7	190.4
Silicon				
$Si(s)$	28.09	0	0	18.83
$SiO_2(s,\alpha)$	60.09	−910.94	−856.64	41.84
Silver				
$Ag(s)$	107.87	0	0	42.55
$Ag^+(aq)$	107.87	105.58	77.11	72.68
$Ag_2O(s)$	231.74	−31.05	−11.20	121.3
$AgBr(s)$	187.78	−100.37	−96.90	107.1
$AgBr(aq)$	187.78	−15.98	−26.86	155.2
$AgCl(s)$	143.32	−127.07	−109.79	96.2
$AgCl(aq)$	143.32	−61.58	−54.12	129.3
$AgI(s)$	234.77	−61.84	−66.19	115.5
$AgI(aq)$	234.77	50.38	25.52	184.1
$AgNO_3(s)$	169.88	−124.39	−33.41	140.92
Sodium				
$Na(s)$	22.99	0	0	51.21
$Na(g)$	22.99	107.32	76.76	153.71
$Na^+(aq)$	22.99	−240.12	−261.91	59.0
$NaOH(s)$	40.00	−425.61	−379.49	64.46
$NaOH(aq)$	40.00	−470.11	−419.15	48.1
$NaCl(s)$	58.44	−411.15	−384.14	72.13
$NaBr(s)$	102.90	−361.06	−348.98	86.82
$NaI(s)$	149.89	−287.78	−286.06	98.53
Sulfur				
$S(s)$, rhombic	32.06	0	0	31.80
$S(s)$, monoclinic	32.06	0.33	0.1	32.6
$S^{2-}(aq)$	32.06	33.1	85.8	−14.6
$SO_2(g)$	64.06	−296.83	−300.19	248.22
$SO_3(g)$	80.06	−395.72	−371.06	256.76

(continued)

1 INORGANIC SUBSTANCES *(continued)*

Substance	Molar mass, g/mol	Enthalpy of formation, ΔH_f°, kJ/mol	Free energy of formation, ΔG_f°, kJ/mol	Entropy S°, J/(K·mol)
Sulfur continued				
$H_2SO_4(l)$	98.08	−813.99	−690.00	156.90
$H_2SO_4(aq)$	98.08	−909.27	−744.53	20.1
$SO_4^{2-}(aq)$	96.06	−909.27	−744.53	20.1
$H_2S(g)$	34.08	−20.63	−33.56	205.79
$H_2S(aq)$	34.08	−39.7	−27.83	121
$SF_6(g)$	146.05	−1209	−1105.3	291.82
Tin				
$Sn(s)$, white	118.69	0	0	51.55
$Sn(s)$, gray	118.69	−2.09	0.13	44.14
$SnO(s)$	134.69	−285.8	−256.9	56.5
$SnO_2(s)$	150.69	−580.7	−519.6	52.3
Zinc				
$Zn(s)$	65.37	0	0	41.63
$Zn^{2+}(aq)$	65.37	−153.89	−147.06	−112.1
$ZnO(s)$	81.37	−348.28	−318.30	43.64

2 ORGANIC COMPOUNDS

Substance	Molar mass, g/mol	Enthalpy of combustion ΔH_c°, kJ/mol	Enthalpy of formation ΔH_f°, kJ/mol	Free energy of formation ΔG_f°, kJ/mol	Entropy S°, J/(K·mol)
Hydrocarbons					
$CH_4(g)$, methane	16.04	−890	−74.81	−50.72	186.26
$C_2H_2(g)$, acetylene	26.04	−1300	226.73	209.20	200.94
$C_2H_4(g)$, ethylene	28.05	−1411	52.26	68.15	219.56
$C_2H_6(g)$, ethane	30.07	−1560	−84.68	−32.82	229.60
$C_3H_6(g)$, propylene	42.08	−2058	20.42	62.78	266.6
$C_3H_6(g)$, cyclopropane	42.08	−2091	53.30	104.45	237.4
$C_3H_8(g)$, propane	44.09	−2220	−103.85	−23.49	270.2
$C_4H_{10}(g)$, butane	58.13	−2878	−126.15	−17.03	310.1
$C_5H_{12}(g)$, pentane	72.15	−3537	−146.44	−8.20	349
$C_6H_6(l)$, benzene	78.12	−3268	49.0	124.3	173.3
$C_6H_6(g)$	78.12	−3302	—	—	—
$C_7H_8(l)$, toluene	92.13	−3910	12.0	113.8	221.0
$C_7H_8(g)$	92.13	−3953	—	—	—

Substance	Molar mass, g/mol	Enthalpy of combustion ΔH_c°, kJ/mol	Enthalpy of formation ΔH_f°, kJ/mol	Free energy of formation ΔG_f°, kJ/mol	Entropy S°, J/(K·mol)
$C_6H_{12}(l)$, cyclohexane	84.16	−3920	−156.4	26.7	204.4
$C_6H_{12}(g)$	84.16	−3953	—	—	—
$C_8H_{18}(l)$, octane	114.23	−5471	−249.9	6.4	358
Alcohols and phenols					
$CH_3OH(l)$, methanol	32.04	−726	−238.86	−166.27	126.8
$CH_3OH(g)$	32.04	−764	−200.66	−161.96	239.81
$C_2H_5OH(l)$, ethanol	46.07	−1368	−277.69	−174.78	160.7
$C_2H_5OH(g)$	46.07	−1409	−235.10	−168.49	282.70
$C_6H_5OH(s)$, phenol	94.11	−3054	−164.6	−50.42	144.0
Carboxylic acids					
$HCOOH(l)$, formic acid	46.03	−255	−424.72	−361.35	128.95
$CH_3COOH(l)$, acetic acid	60.05	−875	−484.5	−389.9	159.8
$CH_3COOH(aq)$	60.05	—	−485.76	−396.46	86.6
$(COOH)_2(s)$, oxalic acid	90.04	−254	−827.2	−697.9	120
$C_6H_5COOH(s)$, benzoic acid	122.13	−3227	−385.1	−245.3	167.6
Aldehydes and ketones					
$HCHO(g)$, formaldehyde	30.03	−571	−108.57	−102.53	218.77
$CH_3CHO(l)$, acetaldehyde	44.05	−1166	−192.30	−128.12	160.2
$CH_3CHO(g)$	44.05	−1192	−166.19	−128.86	250.3
$CH_3COCH_3(l)$, acetone	58.08	−1790	−248.1	−155.4	200
Sugars					
$C_6H_{12}O_6(s)$, glucose	180.16	−2808	−1268	−910	212
$C_6H_{12}O_6(aq)$	180.16	—	—	−917	—
$C_6H_{12}O_6(s)$, fructose	180.16	−2810	−1266	—	—
$C_{12}H_{22}O_{11}(s)$, sucrose	342.30	−5645	−2222	−1545	360
Nitrogen compounds					
$CO(NH_2)_2(s)$, urea	60.06	−632	−333.51	−197.33	104.60
$C_6H_5NH_2(l)$, aniline	93.13	−3393	31.6	149.1	191.3
$NH_2CH_2COOH(s)$, glycine	75.07	−969	−532.9	−373.4	103.51
$CH_3NH_2(g)$, methylamine	31.06	−1085	−22.97	32.16	243.41

2B: STANDARD REDUCTION POTENTIALS AT 25°C

1 POTENTIALS IN ELECTROCHEMICAL ORDER

Reduction half-reaction	$E°$, V
Strongly oxidizing	
$H_4XeO_6 + 2H^+ + 2e^- \longrightarrow XeO_3 + 3H_2O$	+3.0
$F_2 + 2e^- \longrightarrow 2F^-$	+2.87
$O_3 + 2H^+ + 2e^- \longrightarrow O_2 + H_2O$	+2.07
$S_2O_8^{2-} + 2e^- \longrightarrow 2SO_4^{2-}$	+2.05
$Ag^{2+} + e^- \longrightarrow Ag^+$	+1.98
$Co^{3+} + e^- \longrightarrow Co^{2+}$	+1.81
$H_2O_2 + 2H^+ + 2e^- \longrightarrow 2H_2O$	+1.78
$Au^+ + e^- \longrightarrow Au$	+1.69
$Pb^{4+} + 2e^- \longrightarrow Pb^{2+}$	+1.67
$2HClO + 2H^+ + 2e^- \longrightarrow Cl_2 + 2H_2O$	+1.63
$Hg^{2+} + 2e^- \longrightarrow Hg$	+1.62
$Ce^{4+} + e^- \longrightarrow Ce^{3+}$	+1.61
$2HBrO + 2H^+ + 2e^- \longrightarrow Br_2 + 2H_2O$	+1.60
$MnO_4^- + 8H^+ + 5e^- \longrightarrow Mn^{2+} + 4H_2O$	+1.51
$Mn^{3+} + e^- \longrightarrow Mn^{2+}$	+1.51
$Au^{3+} + 3e^- \longrightarrow Au$	+1.40
$Cl_2 + 2e^- \longrightarrow 2Cl^-$	+1.36
$Cr_2O_7^{2-} + 14H^+ + 6e^- \longrightarrow 2Cr^{3+} + 7H_2O$	+1.33
$O_3 + H_2O + 2e^- \longrightarrow O_2 + 2OH^-$	+1.24
$O_2 + 4H^+ + 4e^- \longrightarrow 2H_2O$	+1.23
$MnO_2 + 4H^+ + 2e^- \longrightarrow Mn^{2+} + 2H_2O$	+1.23
$ClO_4^- + 2H^+ + 2e^- \longrightarrow ClO_3^- + H_2O$	+1.23
$Pt^{2+} + 2e^- \longrightarrow Pt$	+1.20
$Br_2 + 2e^- \longrightarrow 2Br^-$	+1.09
$Pu^{4+} + e^- \longrightarrow Pu^{3+}$	+0.97
$NO_3^- + 4H^+ + 3e^- \longrightarrow NO + 2H_2O$	+0.96
$2Hg^{2+} + 2e^- \longrightarrow Hg_2^{2+}$	+0.92
$ClO^- + H_2O + 2e^- \longrightarrow Cl^- + 2OH^-$	+0.89
$NO_3^- + 2H^+ + e^- \longrightarrow NO_2 + H_2O$	+0.80
$Ag^+ + e^- \longrightarrow Ag$	+0.80
$Hg_2^{2+} + 2e^- \longrightarrow 2Hg$	+0.79
$AgF + e^- \longrightarrow Ag + F^-$	+0.78
$Fe^{3+} + e^- \longrightarrow Fe^{2+}$	+0.77
$BrO^- + H_2O + 2e^- \longrightarrow Br^- + 2OH^-$	+0.76
$MnO_4^{2-} + 2H_2O + 2e^- \longrightarrow MnO_2 + 4OH^-$	+0.60
$MnO_4^- + e^- \longrightarrow MnO_4^{2-}$	+0.56

Reduction half-reaction	$E°$, V
$I_2 + 2e^- \longrightarrow 2I^-$	+0.54
$I_3^- + 2e^- \longrightarrow 3I^-$	+0.53
$Cu^+ + e^- \longrightarrow Cu$	+0.52
$NiO(OH) + H_2O + e^- \longrightarrow Ni(OH)_2 + OH^-$	+0.49
$O_2 + 2H_2O + 4e^- \longrightarrow 4OH^-$	+0.40
$ClO_4^- + H_2O + 2e^- \longrightarrow ClO_3^- + 2OH^-$	+0.36
$Cu^{2+} + 2e^- \longrightarrow Cu$	+0.34
$Hg_2Cl_2 + 2e^- \longrightarrow 2Hg + 2Cl^-$	+0.27
$AgCl + e^- \longrightarrow Ag + Cl^-$	+0.22
$Bi^{3+} + 3e^- \longrightarrow Bi$	+0.20
$SO_4^{2-} + 4H^+ + 2e^- \longrightarrow H_2SO_3 + H_2O$	+0.17
$Cu^{2+} + e^- \longrightarrow Cu^+$	+0.15
$Sn^{4+} + 2e^- \longrightarrow Sn^{2+}$	+0.15
$AgBr + e^- \longrightarrow Ag + Br^-$	+0.07
$NO_3^- + H_2O + 2e^- \longrightarrow NO_2^- + 2OH^-$	+0.01
$Ti^{4+} + e^- \longrightarrow Ti^{3+}$	0.00
$2H^+ + 2e^- \longrightarrow H_2$	0, by definition
$Fe^{3+} + 3e^- \longrightarrow Fe$	−0.04
$O_2 + H_2O + 2e^- \longrightarrow HO_2^- + OH^-$	−0.08
$Pb^{2+} + 2e^- \longrightarrow Pb$	−0.13
$In^+ + e^- \longrightarrow In$	−0.14
$Sn^{2+} + 2e^- \longrightarrow Sn$	−0.14
$AgI + e^- \longrightarrow Ag + I^-$	−0.15
$Ni^{2+} + 2e^- \longrightarrow Ni$	−0.23
$V^{3+} + e^- \longrightarrow V^{2+}$	−0.26
$Co^{2+} + 2e^- \longrightarrow Co$	−0.28
$In^{3+} + 3e^- \longrightarrow In$	−0.34
$Tl^+ + e^- \longrightarrow Tl$	−0.34
$PbSO_4 + 2e^- \longrightarrow Pb + SO_4^{2-}$	−0.36
$Ti^{3+} + e^- \longrightarrow Ti^{2+}$	−0.37
$In^{2+} + e^- \longrightarrow In^+$	−0.40
$Cd^{2+} + 2e^- \longrightarrow Cd$	−0.40
$Cr^{3+} + e^- \longrightarrow Cr^{2+}$	−0.41
$Fe^{2+} + 2e^- \longrightarrow Fe$	−0.44
$In^{3+} + 2e^- \longrightarrow In^+$	−0.44
$S + 2e^- \longrightarrow S^{2-}$	−0.48

Reduction half-reaction	$E°$, V	Reduction half-reaction	$E°$, V
$In^{3+} + e^- \longrightarrow In^{2+}$	−0.49	$U^{3+} + 3e^- \longrightarrow U$	−1.79
$Ga^+ + e^- \longrightarrow Ga$	−0.53	$Be^{2+} + 2e^- \longrightarrow Be$	−1.85
$O_2 + e^- \longrightarrow O_2^-$	−0.56	$Mg^{2+} + 2e^- \longrightarrow Mg$	−2.36
$U^{4+} + e^- \longrightarrow U^{3+}$	−0.61	$Ce^{3+} + 3e^- \longrightarrow Ce$	−2.48
$Se + 2e^- \longrightarrow Se^{2-}$	−0.67	$La^{3+} + 3e^- \longrightarrow La$	−2.52
$Cr^{3+} + 3e^- \longrightarrow Cr$	−0.74	$Na^+ + e^- \longrightarrow Na$	−2.71
$Zn^{2+} + 2e^- \longrightarrow Zn$	−0.76	$Ca^{2+} + 2e^- \longrightarrow Ca$	−2.87
$Cd(OH)_2 + 2e^- \longrightarrow Cd + 2OH^-$	−0.81	$Sr^{2+} + 2e^- \longrightarrow Sr$	−2.89
$2H_2O + 2e^- \longrightarrow H_2 + 2OH^-$	−0.83	$Ba^{2+} + 2e^- \longrightarrow Ba$	−2.91
$Te + 2e^- \longrightarrow Te^{2-}$	−0.84	$Ra^{2+} + 2e^- \longrightarrow Ra$	−2.92
$Cr^{2+} + 2e^- \longrightarrow Cr$	−0.91	$Cs^+ + e^- \longrightarrow Cs$	−2.92
$Mn^{2+} + 2e^- \longrightarrow Mn$	−1.18	$Rb^+ + e^- \longrightarrow Rb$	−2.93
$V^{2+} + 2e^- \longrightarrow V$	−1.19	$K^+ + e^- \longrightarrow K$	−2.93
$Ti^{2+} + 2e^- \longrightarrow Ti$	−1.63	$Li^+ + e^- \longrightarrow Li$	−3.05
$Al^{3+} + 3e^- \longrightarrow Al$	−1.66	*Strongly reducing*	

2 POTENTIALS IN ALPHABETICAL ORDER

Reduction half-reaction	$E°$, V	Reduction half-reaction	$E°$, V
$Ag^+ + e^- \longrightarrow Ag$	+0.80	$ClO^- + H_2O + 2e^- \longrightarrow Cl^- + 2OH^-$	+0.89
$Ag^{2+} + e^- \longrightarrow Ag^+$	+1.98	$ClO_4^- + 2H^+ + 2e^- \longrightarrow ClO_3^- + H_2O$	+1.23
$AgBr + e^- \longrightarrow Ag + Br^-$	+0.07	$ClO_4^- + H_2O + 2e^- \longrightarrow ClO_3^- + 2OH^-$	+0.36
$AgCl + e^- \longrightarrow Ag + Cl^-$	+0.22	$Co^{2+} + 2e^- \longrightarrow Co$	−0.28
$AgF + e^- \longrightarrow Ag + F^-$	+0.78	$Co^{3+} + e^- \longrightarrow Co^{2+}$	+1.81
$AgI + e^- \longrightarrow Ag + I^-$	−0.15	$Cr^{2+} + 2e^- \longrightarrow Cr$	−0.91
$Al^{3+} + 3e^- \longrightarrow Al$	−1.66	$Cr_2O_7^{2-} + 14H^+ + 6e^- \longrightarrow 2Cr^{3+} + 7H_2O$	+1.33
$Au^+ + e^- \longrightarrow Au$	+1.69	$Cr^{3+} + 3e^- \longrightarrow Cr$	−0.74
$Au^{3+} + 3e^- \longrightarrow Au$	+1.40	$Cr^{3+} + e^- \longrightarrow Cr^{2+}$	−0.41
$Ba^{2+} + 2e^- \longrightarrow Ba$	−2.91	$Cs^+ + e^- \longrightarrow Cs$	−2.92
$Be^{2+} + 2e^- \longrightarrow Be$	−1.85	$Cu^+ + e^- \longrightarrow Cu$	+0.52
$Bi^{3+} + 3e^- \longrightarrow Bi$	+0.20	$Cu^{2+} + 2e^- \longrightarrow Cu$	+0.34
$Br_2 + 2e^- \longrightarrow 2Br^-$	+1.09	$Cu^{2+} + e^- \longrightarrow Cu^+$	+0.15
$BrO^- + H_2O + 2e^- \longrightarrow Br^- + 2OH^-$	+0.76	$F_2 + 2e^- \longrightarrow 2F^-$	+2.87
$Ca^{2+} + 2e^- \longrightarrow Ca$	−2.87	$Fe^{2+} + 2e^- \longrightarrow Fe$	−0.44
$Cd^{2+} + 2e^- \longrightarrow Cd$	−0.40	$Fe^{3+} + 3e^- \longrightarrow Fe$	−0.04
$Cd(OH)_2 + 2e^- \longrightarrow Cd + 2OH^-$	−0.81	$Fe^{3+} + e^- \longrightarrow Fe^{2+}$	+0.77
$Ce^{3+} + 3e^- \longrightarrow Ce$	−2.48	$Ga^+ + e^- \longrightarrow Ga$	−0.53
$Ce^{4+} + e^- \longrightarrow Ce^{3+}$	+1.61	$2H^+ + 2e^- \longrightarrow H_2$	0, by definition
$Cl_2 + 2e^- \longrightarrow 2Cl^-$	+1.36		

(continued)

2 POTENTIALS IN ALPHABETICAL ORDER *(continued)*

Reduction half-reaction	$E°$, V	Reduction half-reaction	$E°$, V
$2HBrO + 2H^+ + 2e^- \longrightarrow Br_2 + 2H_2O$	+1.60	$NO_3^- + H_2O + 2e^- \longrightarrow NO_2^- + 2OH^-$	+0.01
$2HClO + 2H^+ + 2e^- \longrightarrow Cl_2 + 2H_2O$	+1.63	$O_2 + e^- \longrightarrow O_2^-$	−0.56
$2H_2O + 2e^- \longrightarrow H_2 + 2OH^-$	−0.83	$O_2 + 4H^+ + 4e^- \longrightarrow 2H_2O$	+1.23
$H_2O_2 + 2H^+ + 2e^- \longrightarrow 2H_2O$	+1.78	$O_2 + H_2O + 2e^- \longrightarrow HO_2^- + OH^-$	−0.08
$H_4XeO_6 + 2H^+ + 2e^- \longrightarrow XeO_3 + 3H_2O$	+3.0	$O_2 + 2H_2O + 4e^- \longrightarrow 4OH^-$	+0.40
$Hg_2^{2+} + 2e^- \longrightarrow 2Hg$	+0.79	$O_3 + 2H^+ + 2e^- \longrightarrow O_2 + H_2O$	+2.07
$Hg^{2+} + 2e^- \longrightarrow Hg$	+1.62	$O_3 + H_2O + 2e^- \longrightarrow O_2 + 2OH^-$	+1.24
$2Hg^{2+} + 2e^- \longrightarrow Hg_2^{2+}$	+0.92	$Pb^{2+} + 2e^- \longrightarrow Pb$	−0.13
$Hg_2Cl_2 + 2e^- \longrightarrow 2Hg + 2Cl^-$	+0.27	$Pb^{4+} + 2e^- \longrightarrow Pb^{2+}$	+1.67
$I_2 + 2e^- \longrightarrow 2I^-$	+0.54	$PbSO_4 + 2e^- \longrightarrow Pb + SO_4^{2-}$	−0.36
$I_3^- + 2e^- \longrightarrow 3I^-$	+0.53	$Pt^{2+} + 2e^- \longrightarrow Pt$	+1.20
$In^+ + e^- \longrightarrow In$	−0.14	$Pu^{4+} + e^- \longrightarrow Pu^{3+}$	+0.97
$In^{2+} + e^- \longrightarrow In^+$	−0.40	$Ra^{2+} + 2e^- \longrightarrow Ra$	−2.92
$In^{3+} + e^- \longrightarrow In^{2+}$	−0.49	$Rb^+ + e^- \longrightarrow Rb$	−2.93
$In^{3+} + 2e^- \longrightarrow In^+$	−0.44	$S + 2e^- \longrightarrow S^{2-}$	−0.48
$In^{3+} + 3e^- \longrightarrow In$	−0.34	$SO_4^{2-} + 4H^+ + 2e^- \longrightarrow H_2SO_3 + H_2O$	+0.17
$K^+ + e^- \longrightarrow K$	−2.93	$S_2O_8^{2-} + 2e^- \longrightarrow 2SO_4^{2-}$	+2.05
$La^{3+} + 3e^- \longrightarrow La$	−2.52	$Se + 2e^- \longrightarrow Se^{2-}$	−0.67
$Li^+ + e^- \longrightarrow Li$	−3.05	$Sn^{2+} + 2e^- \longrightarrow Sn$	−0.14
$Mg^{2+} + 2e^- \longrightarrow Mg$	−2.36	$Sn^{4+} + 2e^- \longrightarrow Sn^{2+}$	+0.15
$Mn^{2+} + 2e^- \longrightarrow Mn$	−1.18	$Sr^{2+} + 2e^- \longrightarrow Sr$	−2.89
$Mn^{3+} + e^- \longrightarrow Mn^{2+}$	+1.51	$Te + 2e^- \longrightarrow Te^{2-}$	−0.84
$MnO_2 + 4H^+ + 2e^- \longrightarrow Mn^{2+} + 2H_2O$	+1.23	$Tl^+ + e^- \longrightarrow Tl$	−0.34
$MnO_4^- + e^- \longrightarrow MnO_4^{2-}$	+0.56	$Ti^{2+} + 2e^- \longrightarrow Ti$	−1.63
$MnO_4^- + 8H^+ + 5e^- \longrightarrow Mn^{2+} + 4H_2O$	+1.51	$Ti^{3+} + e^- \longrightarrow Ti^{2+}$	−0.37
$MnO_4^{2-} + 2H_2O + 2e^- \longrightarrow MnO_2 + 4OH^-$	+0.60	$Ti^{4+} + e^- \longrightarrow Ti^{3+}$	0.00
$Na^+ + e^- \longrightarrow Na$	−2.71	$U^{3+} + 3e^- \longrightarrow U$	−1.79
$Ni^{2+} + 2e^- \longrightarrow Ni$	−0.23	$U^{4+} + e^- \longrightarrow U^{3+}$	−0.61
$NiO(OH) + H_2O + e^- \longrightarrow Ni(OH)_2 + OH^-$	+0.49	$V^{2+} + 2e^- \longrightarrow V$	−1.19
$NO_3^- + 2H^+ + e^- \longrightarrow NO_2 + H_2O$	+0.80	$V^{3+} + e^- \longrightarrow V^{2+}$	−0.26
$NO_3^- + 4H^+ + 3e^- \longrightarrow NO + 2H_2O$	+0.96	$Zn^{2+} + 2e^- \longrightarrow Zn$	−0.76

2C: GROUND-STATE ELECTRON CONFIGURATIONS

Z	Symbol	Configuration
1	H	$1s^1$
2	He	$1s^2$
3	Li	[He]$2s^1$
4	Be	[He]$2s^2$
5	B	[He]$2s^22p^1$
6	C	[He]$2s^22p^2$
7	N	[He]$2s^22p^3$
8	O	[He]$2s^22p^4$
9	F	[He]$2s^22p^5$
10	Ne	[He]$2s^22p^6$
11	Na	[Ne]$3s^1$
12	Mg	[Ne]$3s^2$
13	Al	[Ne]$3s^23p^1$
14	Si	[Ne]$3s^23p^2$
15	P	[Ne]$3s^23p^3$
16	S	[Ne]$3s^23p^4$
17	Cl	[Ne]$3s^23p^5$
18	Ar	[Ne]$3s^23p^6$
19	K	[Ar]$4s^1$
20	Ca	[Ar]$4s^2$
21	Sc	[Ar]$3d^14s^2$
22	Ti	[Ar]$3d^24s^2$
23	V	[Ar]$3d^34s^2$
24	Cr	[Ar]$3d^54s^1$
25	Mn	[Ar]$3d^54s^2$
26	Fe	[Ar]$3d^64s^2$
27	Co	[Ar]$3d^74s^2$
28	Ni	[Ar]$3d^84s^2$
29	Cu	[Ar]$3d^{10}4s^1$
30	Zn	[Ar]$3d^{10}4s^2$
31	Ga	[Ar]$3d^{10}4s^24p^1$
32	Ge	[Ar]$3d^{10}4s^24p^2$
33	As	[Ar]$3d^{10}4s^24p^3$
34	Se	[Ar]$3d^{10}4s^24p^4$
35	Br	[Ar]$3d^{10}4s^24p^5$
36	Kr	[Ar]$3d^{10}4s^24p^6$
37	Rb	[Kr]$5s^1$
38	Sr	[Kr]$5s^2$
39	Y	[Kr]$4d^15s^2$
40	Zr	[Kr]$4d^25s^2$
41	Nb	[Kr]$4d^45s^1$
42	Mo	[Kr]$4d^55s^1$
43	Tc	[Kr]$4d^55s^2$
44	Ru	[Kr]$4d^75s^1$
45	Rh	[Kr]$4d^85s^1$
46	Pd	[Kr]$4d^{10}$
47	Ag	[Kr]$4d^{10}5s^1$
48	Cd	[Kr]$4d^{10}5s^2$
49	In	[Kr]$4d^{10}5s^25p^1$
50	Sn	[Kr]$4d^{10}5s^25p^2$
51	Sb	[Kr]$4d^{10}5s^25p^3$
52	Te	[Kr]$4d^{10}5s^25p^4$
53	I	[Kr]$4d^{10}5s^25p^5$
54	Xe	[Kr]$4d^{10}5s^25p^6$
55	Cs	[Xe]$6s^1$
56	Ba	[Xe]$6s^2$
57	La	[Xe]$5d^16s^2$
58	Ce	[Xe]$4f^15d^16s^2$
59	Pr	[Xe]$4f^36s^2$
60	Nd	[Xe]$4f^46s^2$
61	Pm	[Xe]$4f^56s^2$
62	Sm	[Xe]$4f^66s^2$
63	Eu	[Xe]$4f^76s^2$
64	Gd	[Xe]$4f^75d^16s^2$
65	Tb	[Xe]$4f^96s^2$
66	Dy	[Xe]$4f^{10}6s^2$
67	Ho	[Xe]$4f^{11}6s^2$
68	Er	[Xe]$4f^{12}6s^2$
69	Tm	[Xe]$4f^{13}6s^2$
70	Yb	[Xe]$4f^{14}6s^2$
71	Lu	[Xe]$4f^{14}5d^16s^2$
72	Hf	[Xe]$4f^{14}5d^26s^2$
73	Ta	[Xe]$4f^{14}5d^36s^2$
74	W	[Xe]$4f^{14}5d^46s^2$
75	Re	[Xe]$4f^{14}5d^56s^2$
76	Os	[Xe]$4f^{14}5d^66s^2$
77	Ir	[Xe]$4f^{14}5d^76s^2$
78	Pt	[Xe]$4f^{14}5d^96s^1$
79	Au	[Xe]$4f^{14}5d^{10}6s^1$
80	Hg	[Xe]$4f^{14}5d^{10}6s^2$
81	Tl	[Xe]$4f^{14}5d^{10}6s^26p^1$
82	Pb	[Xe]$4f^{14}5d^{10}6s^26p^2$
83	Bi	[Xe]$4f^{14}5d^{10}6s^26p^3$
84	Po	[Xe]$4f^{14}5d^{10}6s^26p^4$
85	At	[Xe]$4f^{14}5d^{10}6s^26p^5$
86	Rn	[Xe]$4f^{14}5d^{10}6s^26p^6$
87	Fr	[Rn]$7s^1$
88	Ra	[Rn]$7s^2$
89	Ac	[Rn]$6d^17s^2$
90	Th	[Rn]$6d^27s^2$
91	Pa	[Rn]$5f^26d^17s^2$
92	U	[Rn]$5f^36d^17s^2$
93	Np	[Rn]$5f^46d^17s^2$
94	Pu	[Rn]$5f^67s^2$
95	Am	[Rn]$5f^77s^2$
96	Cm	[Rn]$5f^76d^17s^2$
97	Bk	[Rn]$5f^97s^2$
98	Cf	[Rn]$5f^{10}7s^2$
99	Es	[Rn]$5f^{11}7s^2$
100	Fm	[Rn]$5f^{12}7s^2$
101	Md	[Rn]$5f^{13}7s^2$
102	No	[Rn]$5f^{14}7s^2$
103	Lr	[Rn]$5f^{14}6d^17s^2$
104	Unq	[Rn]$5f^{14}6d^27s^2$
105	Unp	[Rn]$5f^{14}6d^37s^2$

2D: THE ELEMENTS

Element	Symbol	Atomic number	Molar mass, g/mol	Normal state*	Density, g/cm^3	Melting point, °C	Boiling point, °C
actinium (Greek *aktis*, ray)	Ac	89	227.03	*s*, m	10.07	1230	3200
aluminum (from alum, salts of the form $KAl(SO_4)_2 \cdot 12H_2O$)	Al	13	26.98	*s*, m	2.70	660	2350
americium (the Americas)	Am	95	241.06	*s*, m	13.67	990	2600
antimony (probably a corruption of an old Arabic word; Latin *stibium*)	Sb	51	121.75	*s*, md	6.69	631	1750
argon (Greek *argos*, inactive)	Ar	18	39.95	*g*, nm	1.66†	−189	−186
arsenic (Greek *arsenikos*, male)	As	33	74.92	*s*, md	5.78	613s‡	—
astatine (Greek *astatos*, unstable)	At	85	210	*s*, nm	—	300	350
barium (Greek *barys*, heavy)	Ba	56	137.34	*s*, m	3.59	710	1640
berkelium (Berkeley, California)	Bk	97	249.08	*s*, m	14.79	986	—
beryllium (from the mineral beryl, $Be_3Al_2SiO_{18}$)	Be	4	9.01	*s*, m	1.85	1285	2470
bismuth (German *weisse Masse*, white mass)	Bi	83	208.98	*s*, m	8.90	271	1650
boron [Arabic *buraq*, borax, $Na_2B_4O_7 \cdot 10H_2O$; *bor*(ax) + (carb)*on*]	B	5	10.81	*s*, nm	2.47	2030	3700
bromine (Greek *bromos*, stench)	Br	35	79.91	*l*, nm	3.12	−7	59
cadmium (Greek *Cadmus*, founder of Thebes)	Cd	48	112.40	*s*, m	8.65	321	765
calcium (Latin *calx*, lime)	Ca	20	40.08	*s*, m	1.53	840	1490
californium (California)	Cf	98	251.08	*s*, m	—	—	—
carbon (Latin *carbo*, coal or charcoal)	C	6	12.01	*s*, nm	2.27	3700s	—
cerium (the asteroid Ceres, discovered two days earlier)	Ce	58	140.12	*s*, m	6.71	800	3000
cesium (Latin *caesius*, sky blue)	Cs	55	132.91	*s*, m	1.87	28	678
chlorine (Greek *chloros*, yellowish green)	Cl	17	35.45	*g*, nm	2.03†	−101	−34

*The normal state is the state of the element at normal temperature and pressure (20°C and 1 atm). *s* denotes solid; *l*, liquid, and *g*, gas. m denotes metal; nm, nonmetal; and md, metalloid.
†The density quoted is for the liquid.
‡s means that the solid sublimes.

Element	Symbol	Atomic number	Molar mass, g/mol	Normal state	Density, g/cm^3	Melting point, °C	Boiling point, °C
chromium (Greek *chroma*, color)	Cr	24	52.00	*s*, m	7.19	1860	2600
cobalt (German *Kobold*, evil spirit; Greek *kobalos*, goblin)	Co	27	58.93	*s*, m	8.80	1494	2900
copper (Latin *cuprum*, from Cyprus)	Cu	29	63.54	*s*, m	8.93	1083	2567
curium (Marie Curie)	Cm	96	247.07	*s*, m	13.30	1340	—
dysprosium (Greek *dysprositos*, hard to get at)	Dy	66	162.50	*s*, m	8.53	1410	2600
einsteinium (Albert Einstein)	Es	99	254.09	*s*, m	—	—	—
erbium (Ytterby, a town in Sweden)	Er	68	167.26	*s*, m	9.04	1520	2600
europium (Europe)	Eu	63	151.96	*s*, m	5.25	820	1450
fermium (Enrico Fermi, an Italian physicist)	Fm	100	257.10	*s*, m	—	—	—
fluorine (Latin *fluere*, to flow)	F	9	19.00	*g*, nm	1.14†	−220	−188
francium (France)	Fr	87	223	*s*, m	—	27	677
gadolinium (Johann Gadolin, a Finnish chemist)	Gd	64	157.25	*s*, m	7.87	1310	3000
gallium (Latin *Gallia*, France; also a pun on the discoverer's forename, Le Coq)	Ga	31	69.72	*s*, m	5.91	30	2070
germanium (Latin *Germania*, Germany)	Ge	32	72.59	*s*, md	5.32	937	2830
gold (Anglo-Saxon *gold;* Latin *aurum*, gold)	Au	79	196.97	*s*, m	19.28	1064	2807
hafnium (Latin *Hafnia*, Copenhagen)	Hf	72	178.49	*s*, m	13.28	2230	5300
helium (Greek *helios*, the sun)	He	2	4.00	*g*, nm	0.12†	—	−269
holmium (Latin *Holmia*, Stockholm)	Ho	67	164.93	*s*, m	8.80	1470	2300
hydrogen (Greek *hydro* + *genes*, water-forming)	H	1	1.0079	*g*, nm	0.089†	−259	−253

(continued)

Element	Symbol	Atomic number	Molar mass, g/mol	Normal state	Density, g/cm^3	Melting point, °C	Boiling point, °C
indium (from the bright indigo line in its spectrum)	In	49	114.82	*s*, m	7.29	157	2050
iodine (Greek *ioeidēs*, violet)	I	53	126.90	*s*, nm	4.95	114	184
iridium (Greek and Latin *iris*, rainbow)	Ir	77	192.2	*s*, m	22.55	2447	4550
iron (Anglo-Saxon *iron;* Latin *ferrum*)	Fe	26	55.85	*s*, m	7.87	1540	2760
krypton (Greek *kryptos*, hidden)	Kr	36	83.80	*g*, nm	3.00†	−157	−153
lanthanum (Greek *lanthanein*, to lie hidden)	La	57	138.91	*s*, m	6.17	920	3450
lawrencium (Ernest Lawrence, an American physicist)	Lr	103	257	*s*, m	—	—	—
lead (Anglo-Saxon *lead;* Latin *plumbum*)	Pb	82	207.19	*s*, m	11.34	328	1760
lithium (Greek *lithos*, stone)	Li	3	6.94	*s*, m	0.53	181	1347
lutetium (*Lutetia*, ancient name of Paris)	Lu	71	174.97	*s*, m	9.84	1700	3400
magnesium (Magnesia, a district in Thessaly, Greece)	Mg	12	24.31	*s*, m	1.74	650	1100
manganese (Greek and Latin *magnes*, magnet)	Mn	25	54.94	*s*, m	7.47	1250	2120
mendelevium (Dmitri Mendeleev)	Md	101	258.10	*s*, m	—	—	—
mercury (the planet Mercury; Latin *hydrargyrum*, liquid silver)	Hg	80	200.59	*l*, m	13.55	−39	357
molybdenum (Greek *molybdos*, lead)	Mo	42	95.94	*s*, m	10.22	2620	4830
neodymium (Green *neos* + *didymos*, new twin)	Nd	60	144.24	*s*, m	7.00	1024	3100
neon (Greek *neos*, new)	Ne	10	20.18	*g*, nm	1.44†	−249	−246
neptunium (the planet Neptune)	Np	93	237.05	*s*, m	20.45	640	—
nickel (German *Nickel*, Satan)	Ni	28	58.71	*s*, m	8.91	1455	2150
niobium (Niobe, daughter of Tantalus; see tantalum)	Nb	41	92.91	*s*, m	8.58	2425	5000
nitrogen (Greek *nitron* + *genes*, soda-forming)	N	7	14.01	*g*, nm	1.04†	−210	−196
nobelium (Alfred Nobel, the founder of the Nobel prizes)	No	102	255	*s*, m	—	—	—
osmium (Greek *osme*, a smell)	Os	76	190.2	*s*, m	22.58	3030	5000

Element	Symbol	Atomic number	Molar mass, g/mol	Normal state	Density, g/cm^3	Melting point, °C	Boiling point, °C
oxygen (Greek *oxys* + *genes*, acid-forming)	O	8	16.00	*g*, nm	1.46†	−218	−183
palladium (the asteroid Pallas, discovered at about the same time)	Pd	46	106.4	*s*, m	12.00	1554	3000
phosphorus (Greek *phosphoros*, light-bearing)	P	15	30.97	*s*, nm	1.82	44	280
platinum (Spanish *plata*, silver)	Pt	78	195.09	*s*, m	21.45	1772	3720
plutonium (the planet Pluto)	Pu	94	239.05	*s*, m	19.81	640	3200
polonium (Poland)	Po	84	210	*s*, md	9.40	254	960
potassium (from potash; Latin *kalium* and Arabic *qali*, alkali)	K	19	39.10	*s*, m	0.86	64	774
praseodymium (Greek *prasios* + *didymos*, green twin)	Pr	59	140.91	*s*, m	6.78	935	3000
promethium (Prometheus, the Greek god)	Pm	61	146.92	*s*, m	7.22	1168	3300
protactinium (Greek *protos* + *aktis*, first ray)	Pa	91	231.04	*s*, m	15.37	1200	4000
radium (Latin *radius*, ray)	Ra	88	226.03	*s*, m	5.00	700	1500
radon (from radium)	Rn	86	222	*g*, nm	4.40†	−71	−62
rhenium (Latin *Rhenus*, Rhine)	Re	75	186.2	*s*, m	21.02	3180	5600
rhodium (Greek *rhodon*, rose; its aqueous solutions are often rose-colored)	Rh	45	102.91	*s*, m	12.42	1963	3700
rubidium (Latin *rubidus*, deep red, "flushed")	Rb	37	85.47	*s*, m	1.53	39	688
ruthenium (Latin *Ruthenia*, Russia)	Ru	44	101.07	*s*, m	12.36	2310	4100
samarium (from samarskite, a mineral)	Sm	62	150.35	*s*, m	7.54	1060	1600
scandium (Latin *Scandia*, Scandinavia)	Sc	21	44.96	*s*, m	2.99	1540	2800
selenium (Greek *selēnē*, the moon)	Se	34	78.96	*s*, nm	4.81	220	685
silicon (Latin *silex*, flint)	Si	14	28.09	*s*, md	2.33	1410	2620
silver (Anglo-Saxon *seolfor*; Latin *argentum*)	Ag	47	107.87	*s*, m	10.50	962	2212
sodium (English soda; Latin *natrium*)	Na	11	22.99	*s*, m	0.97	98	883
strontium (Strontian, Scotland)	Sr	38	87.62	*s*, m	2.58	770	1380

(continued)

Element	Symbol	Atomic number	Molar mass, g/mol	Normal state	Density, g/cm^3	Melting point, °C	Boiling point, °C
sulfur (Sanskrit *sulvere*)	S	16	32.06	*s*, nm	2.09	115	445
tantalum (Tantalos, Greek mythological figure)	Ta	73	180.95	*s*, m	16.67	3000	5400
technetium (Greek *technētos*, artificial)	Tc	43	98.91	*s*, m	11.50	2200	4600
tellurium (Latin *tellus*, earth)	Te	52	127.60	*s*, md	6.25	450	990
terbium (Ytterby, a town in Sweden)	Tb	65	158.92	*s*, m	8.27	1360	2500
thallium (Greek *thallos*, a green shoot)	Tl	81	204.37	*s*, m	11.87	304	1460
thorium (Thor, Norse god of thunder, weather, and crops)	Th	90	232.04	*s*, m	11.73	1700	4500
thulium (*Thule*, early name for Scandinavia)	Tm	69	168.93	*s*, m	9.33	1550	2000
tin (Anglo-Saxon *tin*; Latin *stannum*)	Sn	50	118.69	*s*, m	7.29	232	2720
titanium (Titans, Greek mythological figures, sons of the Earth)	Ti	22	47.90	*s*, m	4.51	1660	3300
tungsten (Swedish *tung* + *sten*, heavy stone from wolframite)	W	74	183.85	*s*, m	19.25	3387	5420
uranium (the planet Uranus)	U	92	238.03	*s*, m	19.05	1135	4000
vanadium (Vanadis, Scandinavian mythological figure)	V	23	50.94	*s*, m	6.09	1920	3400
xenon (Greek *xenos*, stranger)	Xe	54	131.30	*g*, nm	3.56†	−112	−108
ytterbium (Ytterby, a town in Sweden)	Yb	70	173.04	*s*, m	6.97	824	1500
yttrium (Ytterby, a town in Sweden)	Y	39	88.91	*s*, m	4.48	1510	3300
zinc (Anglo-Saxon *zinc*)	Zn	30	65.37	*s*, m	7.14	420	907
zirconium (Arabic *zargun*, gold color)	Zr	40	91.22	*s*, m	6.51	1850	4400

2E: THE TOP 50 CHEMICALS BY INDUSTRIAL PRODUCTION IN THE UNITED STATES IN 1990

Production data are compiled annually by the American Chemical Society and published in *Chemical and Engineering News.* This table is based on the information about production in 1990 that was published in the April 8, 1991 issue. Water, sodium chloride, and steel traditionally are not included and would outrank the rest if they were. Hydrogen is heavily used, but almost always "on site" as soon as it has been produced.

In *cracking,* petroleum fractions of molar mass higher than that of gasoline are converted into smaller molecules with more double bonds; for example,

$$CH_3(CH_2)_9CH_3 \xrightarrow{500°C,\ \text{catalyst}} CH_3(CH_2)_6CH{=}CH_2 + CH_3CH_3$$

In *reforming,* the number of carbon atoms in the feedstock is left unchanged, but a greater number of double bonds and aromatic rings is formed; for example,

$$\text{methylcyclohexane} \xrightarrow{550°C,\ \text{catalyst}} \text{toluene} + 3H_2$$

Rank	Name	Annual production, 10^9 kg	Comment on source
1	sulfuric acid	40.3	contact process
2	nitrogen	26.1	fractional distillation
3	oxygen	17.1	fractional distillation
4	ethylene	17.0	thermal cracking
5	lime	15.8	decomposition of limestone
6	ammonia	15.4	Haber process
7	phosphoric acid	11.1	from phosphate rocks
8	sodium hydroxide	10.6	brine electrolysis
9	propylene	10.1	thermal cracking
10	chlorine	10.0	electrolysis
11	sodium carbonate	9.0	Solvay process and mining
12	urea	7.2	ammonia + carbon dioxide
13	nitric acid	7.1	Ostwald process
14	ammonium nitrate	6.5	ammonia + nitric acid
15	dichloroethene	6.3	chlorination of ethylene
16	benzene	5.4	catalytic reforming
17	carbon dioxide	5.0	steam reforming of hydrocarbons
18	vinyl chloride	4.8	dehydrochlorination of dichloroethene
19	ethylbenzene	4.1	Friedel-Crafts alkylation of benzene
20	styrene	3.6	dehydrogenation of ethylbenzene
21	methanol	3.6	hydrogenation of carbon monoxide (using synthesis gas)
22	terephthalic acid	3.5	oxidation of xylene
23	formaldehyde	2.9	oxidation or dehydrogenation of methanol

(continued)

Rank	Name	Annual production, 10^9 kg	Comment on source
24	methyl *tert*-butyl ether	2.9	addition of methanol to 2-methylpropene
25	toluene	2.8	catalytic reforming of naphtha*
26	xylene	2.6	catalytic reforming of naphtha
27	ethylene oxide	2.5	addition of O_2 to ethylene
28	*p*-xylene	2.4	catalytic reforming of naphtha*
29	ethylene glycol	2.3	hydration of ethylene oxide
30	ammonium sulfate	2.3	ammonia + sulfuric acid
31	hydrochloric acid	2.1	by-product of hydrocarbon chlorination
32	cumene (isopropylbenzene)	2.0	Friedel-Crafts alkylation
33	acetic acid	1.7	Monsanto process (reaction of CO with methanol)
34	potash†	1.6	mined (KCl), electrolysis (KOH), reaction of KCl with $NaNO_3$ (KNO_3)
35	phenol	1.6	oxidation of cumene
36	propylene oxide	1.5	oxidation of propylene
37	butadiene	1.4	dehydrogenation of butane
38	acrylonitrile	1.4	reaction of propylene with ammonia and oxygen
39	carbon black	1.3	partial combustion of hydrocarbons
40	vinyl acetate	1.2	reaction of ethylene and acetic acid
41	cyclohexane	1.1	hydrogenation of benzene
42	aluminum sulfate	1.1	alumina + sulfuric acid
43	acetone	1.0	dehydrogenation of isopropyl alcohol
44	titanium dioxide	0.98	purification of rutile
45	sodium silicate	0.80	sodium carbonate and sand
46	adipic acid	0.75	oxidation of cyclohexane
47	sodium sulfate	0.67	mined; by-product of rayon manufacture
48	calcium chloride	0.63	Solvay process by-product
49	isopropyl alcohol	0.63	hydration of propylene
50	caprolactam	0.63	from cyclohexane

*Naphtha is a petroleum fraction consisting of a mixture of C_4 through C_{10} aliphatic and cycloaliphatic hydrocarbons together with a variety of aromatic compounds.

†Potash refers collectively to K_2CO_3, KOH, K_2SO_4, KCl, and KNO_3 and is expressed in terms of the equivalent mass of K_2O.

3

A P P E N D I X

NOMENCLATURE

3A: THE NOMENCLATURE OF POLYATOMIC IONS

Charge number	Chemical formula	Name	Oxidation number of central element	Charge number	Chemical formula	Name	Oxidation number of central element
+2	UO_2^{2+}	uranyl	+6		MnO_4^-	permanganate	+7
	VO^{2+}	vanadyl	+4	−2	CO_3^{2-}	carbonate	+4
	Hg_2^{2+}	mercury(I)	+1		$C_2O_4^{2-}$	oxalate	+3
+1	NH_4^+	ammonium	−3		SO_4^{2-}	sulfate	+6
	PH_4^+	phosphonium	−3		SO_3^{2-}	sulfite	+4
−1	$CH_3CO_2^-$	acetate	+4(C)		$S_2O_3^{2-}$	thiosulfate	+4
	ClO_4^-	perchlorate*	+7		SiO_3^{2-}	silicate	+4
	ClO_3^-	chlorate*	+5		CrO_4^{2-}	chromate	+6
	ClO_2^-	chlorite*	+3		$Cr_2O_7^{2-}$	dichromate	+6
	ClO^-	hypochlorite*	+1	−3	BO_3^{3-}	borate	+3
	NO_3^-	nitrate	+5		PO_4^{3-}	phosphate	+5
	NO_2^-	nitrite	+3		AsO_4^{3-}	arsenate	+5

*These names are representative of the halogen oxoanions.

The names of oxoanions and their parent acids can be determined by noting the oxidation number of the central atom and then referring to the following table. For example, the nitrogen in $N_2O_2^{2-}$ has an oxidation number of +1; nitrogen belongs to Group V, so the ion is a hyponitrite ion.

Group number						Group number					
IV	V	VI	VII	Oxoanion	Oxoacid	IV	V	VI	VII	Oxoanion	Oxoacid
—	—	—	+7	per . . . ate	per . . . ic acid	—	+3	+4	+3	. . . ite	. . . ous acid
+4	+5	+6	+5	. . . ate	. . . ic acid	—	+1	+2	+1	hypo . . . ite	hypo . . . ous acid

3B: COMMON NAMES OF CHEMICALS

Many chemicals have acquired common names, sometimes as a result of their use over hundreds of years and sometimes because they have appeared on the labels of consumer products, such as detergents, beverages, and antacids. The following substances are just a few that have found their way into the language of everyday life.

Common name	Formula	Chemical name
baking soda	$NaHCO_3$	sodium hydrogencarbonate (sodium bicarbonate)
bleach, laundry	$NaClO$	sodium hypochlorite
borax	$Na_2B_4O_7 \cdot 10H_2O$	sodium tetraborate decahydrate
brimstone	S_8	sulfur
calamine	$ZnCO_3$	zinc carbonate
chalk	$CaCO_3$	calcium carbonate
Epsom salts	$MgSO_4 \cdot 7H_2O$	magnesium sulfate heptahydrate
fool's gold, iron pyrite	FeS_2	iron(II) sulfide
gypsum	$CaSO_4 \cdot 2H_2O$	calcium sulfate dihydrate
lime, quicklime	CaO	calcium oxide
lime, slaked	$Ca(OH)_2$	calcium hydroxide
limestone	$CaCO_3$	calcium carbonate
lye, caustic soda	$NaOH$	sodium hydroxide
marble	$CaCO_3$	calcium carbonate
milk of magnesia	$Mg(OH)_2$	magnesium hydroxide
plaster of Paris	$CaSO_4 \cdot \frac{1}{2}H_2O$	calcium sulfate hemihydrate
potash*	K_2CO_3	potassium carbonate
quartz	SiO_2	silicon dioxide
salt	$NaCl$	sodium chloride
vinegar	CH_3COOH	acetic acid
washing soda	$Na_2CO_3 \cdot 10H_2O$	sodium carbonate decahydrate

*Potash also refers collectively to K_2CO_3, KOH, K_2SO_4, KCl, and KNO_3.

GLOSSARY

absolute zero ($T = 0$) The lowest possible temperature (−273.15°C).
abundance (of an isotope) The percentage (in terms of the numbers of atoms) of the isotope present in a sample of the element. See also *natural abundance*.
accurate measurements Measurements that have little systematic error and give a result close to the accepted value of the property.
accuracy Freedom from systematic error.
acetal A double ether that contains an —O—C—O— group.
acid See *Arrhenius acid; Brønsted acid; Lewis acid*. Used alone, "acid" normally means a Brønsted acid.
acid-base indicator See *indicator*.
acid buffer See *buffer*.
acid ionization constant K_a The equilibrium constant for proton transfer to water: for an acid HA, $K_a = [H_3O^+][A^-]/[HA]$ at equilibrium.
acidic hydrogen atom A hydrogen atom (more exactly, the proton of that hydrogen atom) that can be donated to a base.
acidic ion An ion that acts as a Brønsted acid. *Examples*: NH_4^+; $Al(H_2O)_6^{3+}$.
acidic oxide An oxide that reacts with water to give an acid; the oxides of nonmetallic elements generally form acidic oxides. *Examples*: CO_2; SO_3.
acidic solution A solution with pH < 7 (at 25°C).
acidity The strength of the tendency to donate a proton.
acidity constant See *acid ionization constant*.
activated complex A combination of the two reactant molecules that can either go on to form products or fall apart into the unchanged reactants.
activated complex theory A theory of reaction rates in which it is postulated that the reactants form an activated complex.
activation barrier The potential energy barrier between reactants and products; the height of the barrier is the activation energy of the reaction.
activation energy E_a (1) The minimum energy needed for reaction. (2) The height of the activation barrier.
active site The region of an enzyme molecule where the substrate reacts.
activity The number of nuclear disintegrations that occur per second.
activity series A list of metals arranged in order of decreasing ability to reduce the cations of other metals. A metal can reduce the cations formed by any of the metals below it in the list.
addition polymerization The polymerization, usually of alkenes, by an addition reaction propagated by radicals.
addition reaction A reaction in which atoms or groups bond to carbon atoms joined by a multiple bond. *Example*: $CH_3CH{=}CH_2 + HBr \rightarrow CH_3CH_2CH_2Br$.
adhesion Binding to a surface.
adhesive forces Forces that bind a substance to a surface.
adsorb To bind a substance to a surface; the surface *adsorbs* the substance.
alcohol An organic molecule containing an —OH group attached to a carbon atom that is not part of a carbonyl group or an aromatic ring. Alcohols are classified as *primary, secondary,* and *tertiary* on the basis of the number of carbon atoms attached to the C—OH carbon atom. *Examples*: CH_3CH_2OH, (primary); $(CH_3)_2CHOH$ (secondary); $(CH_3)_3COH$ (tertiary).
aldehyde An organic compound containing the —CHO group. *Examples*: CH_3CHO, acetaldehyde; C_6H_5CHO, benzaldehyde.
aliphatic compound An organic compound that does not have benzene rings in its structure.
alkali An aqueous solution of a base. *Examples*: aqueous NaOH; aqueous ammonia.
alkali metal A member of Group I of the periodic table (the lithium family).
alkaline cell A dry cell in which the electrolyte contains sodium or potassium hydroxide.
alkaline earth metal A member of Group II of the periodic table (the beryllium family).
alkane (1) A hydrocarbon with no carbon-carbon multiple bonds. (2) A member of a homologous series of hydrocarbons derived from methane by the repetitive insertion of $—CH_2—$; alkanes have molecular formula C_nH_{2n+2}. *Examples*: CH_4; CH_3CH_3; $CH_3(CH_2)_6CH_3$.
alkene (1) A hydrocarbon with at least one carbon-carbon double bond. (2) A member of a homologous series of hydrocarbons derived from ethylene by the repetitive insertion of $—CH_2—$ groups; alkenes have molecular formula C_nH_{2n}. *Examples*: $CH_2{=}CH_2$; $CH_3CH{=}CH_2$; $CH_3CH{=}CHCH_2CH_3$.
alkyl group (R—) A group of atoms that can be regarded as derived from an alkane by the loss of a hydrogen atom. *Examples*: $CH_3—$, methyl; $CH_3CH_2—$, ethyl.
alkyne (1) A hydrocarbon with at least one carbon-carbon triple bond. (2) A member of a homologous series of hydrocarbons derived from acetylene by the repetitive insertion of $—CH_2—$ groups; alkynes have molecular formula C_nH_{2n-2}. *Examples*: $CH{\equiv}CH$; $CH_3C{\equiv}CCH_3$.
allotropes Alternative forms of an element that differ in the way the atoms are linked. *Examples*: O_2 and O_3; white and gray tin.
alloy A homogeneous mixture of two or more metals.
alpha (α) helix One type of secondary structure adopted by a polypeptide chain, in the form of a right-handed helix.
alpha (α) particle A positively charged, subatomic parti-

cle emitted from some radioactive nuclei; nucleus of helium atom ($^4_2He^{2+}$).

amide An organic compound formed by the reaction of an amine and a carboxylic acid and containing the group $—CO—NR_2$. *Example*: CH_3CONH_2, acetamide.

amine A compound derived from ammonia by replacing various numbers of H atoms with alkyl or aryl groups; the number of hydrogen atoms replaced determines the compound's classification as *primary, secondary,* and *tertiary. Examples*: CH_3NH_2 (primary); $(CH_3)_2NH$ (secondary); $(CH_3)_3N$ (tertiary).

amino acid A carboxylic acid that also contains an amine group. The *essential amino acids* are amino acids that must be ingested as a part of the diet. *Example*: NH_2CH_2COOH, glycine (nonessential).

amorphous solid A solid in which the atoms, ions, or molecules lie in a random jumble.

amphiprotic A substance that can both donate and accept protons. *Examples*: H_2O; HCO_3^-.

amphoteric The ability to react with both acids and bases. *Examples*: aluminum metal; Al_2O_3.

analysis The determination of the composition of a substance.

analyte The solution of unknown concentration in a titration.

anhydride A compound that forms an oxoacid when it reacts with water. A *formal anhydride* is a compound that has the formula of an acid minus the elements of water but does not react with water to produce the acid. *Examples*: SO_3 (the anhydride of sulfuric acid); CO (the formal anhydride of formic acid, HCOOH).

anhydrous Lacking water. *Example*: $CuSO_4$, the anhydrous form of copper(II) sulfate.

anion A negatively charged ion. *Examples*: F^-; SO_4^{2-}.

anode The electrode at which oxidation occurs.

antibonding orbital A molecular orbital that, if occupied, raises the energy of a molecule relative to that of the separated atoms.

antioxidant A substance that reacts with radicals and so prevents the oxidation of another substance.

aqueous solution A solution in which the solvent is water.

arene An aromatic hydrocarbon.

aromatic compound An organic compound that includes a benzene ring as part of its structure. *Examples*: C_6H_6, benzene; C_6H_5OH, phenol; $C_{10}H_8$, naphthalene.

Arrhenius acid A compound that contains hydrogen and releases hydrogen ions (H^+) in water. *Examples*: HCl; CH_3COOH; but not CH_4.

Arrhenius base A compound that produces hydroxide ions (OH^-) in water. *Examples*: NaOH; NH_3; but not Na.

Arrhenius behavior A reaction shows Arrhenius behavior if a plot of ln k against $1/T$ is a straight line.

Arrhenius parameters The frequency factor A and the activation energy E_a.

Arrhenius plot A graph of ln k against $1/T$.

aryl group An aromatic group. *Example*: $C_6H_5—$, phenyl.

atactic polymer A polymer in which substituents are attached to each side of the chain at random.

atom (1) The smallest particle of an element that has the chemical properties of that element. (2) A nucleus and its surrounding electrons.

atomic hypothesis The proposal advanced by John Dalton that matter is composed of atoms.

atomic mass unit u (formerly amu) Exactly $\frac{1}{12}$ the mass of one atom of carbon-12.

atomic number Z The number of protons in the nucleus of an atom; this number determines the identity of the element and the number of electrons in the neutral atom.

atomic orbital (1) A region of space in which there is a high probability of finding an electron in an atom. (2) The wavefunction of an electron in an atom. (3) A mathematical expression that gives the probability of finding an electron at each point in an atom.

atomic radius Half the distance between the centers of neighboring atoms of the same element.

atomic structure The arrangement of electrons around the central nucleus of an atom.

atomic weight See *molar mass.*

***Aufbau* principle** See *building-up principle.*

autoionization A reaction in which a conjugate acid and a conjugate base are formed from the same substance. See *water autoionization constant.*

average bond enthalpy ΔH(A—B) The average of A—B bond enthalpies for a number of different molecules containing the A—B bond.

Avogadro's number N_A The number of objects in 1 mol of objects ($N_A = 6.022 \times 10^{23}$). Avogadro's *constant* (also N_A) is the number of objects per mole of objects ($N_A = 6.022 \times 10^{23}$ mol^{-1}).

Avogadro's principle The volume of a sample of gas at a given temperature and pressure is proportional to the amount of gas molecules in the sample: $V \propto n$.

axial bond A bond that is perpendicular to the molecular plane in a bipyramidal molecule.

axial pair An electron pair on the axis of a bipyramidal molecule.

azimuthal quantum number l A number that labels the subshells of an atom. *Examples*: $l = 0$ for the s subshell; $l = 1$ for the p subshell.

balanced equation A chemical equation in which the same number of atoms of each element appears on both sides of the equation.

Balmer series A family of spectral lines (some of which lie in the visible region), in the spectrum of atomic hydrogen.

band gap A range of energies for which there are no orbitals in a solid and which therefore separates two bands of levels.

base See *Arrhenius base; Brønsted base; Lewis base.* Used alone, "base" normally means a Brønsted base.

base buffer See *buffer.*

base ionization constant K_b The equilibrium constant for proton transfer from water to a base B, $K_b = [HB^+][OH^-]/[B]$ at equilibrium.

base pair A pair of organic bases linked by means of hydrogen bonding in the DNA molecule: adenine pairs with thymine and guanine pairs with cytosine.

base units The units of measurement in the International System (SI) in terms of which all other units are defined. *Examples*: kilogram for mass; meter for length; second for time; kelvin for temperature; ampere for electric current.

basic ion An ion that acts as a Brønsted base. *Example*: $CH_3CO_2^-$.

basic oxide An oxide that is a Brønsted base. The oxides of metallic elements are generally basic. *Examples*: Na_2O; MgO.

basic oxygen process The production of iron by forcing oxygen and powdered limestone through the molten metal.

basic solution A solution with pH > 7 (at 25°C).

Bayer process The process for the purifying alumina (Al_2O_3) for use in the Hall process.

becquerel (Bq) The SI unit of radioactivity (1 disintegration per second).

beta (β) particle An electron emitted from a nucleus in a radioactive decay.

beta(β)-pleated sheet One type of planar secondary structure adopted by a polypeptide, in the form of a pleated sheet.

bimolecular reaction An elementary reaction in which two molecules (or free atoms) come together and form a product. *Example*: $O + O_3 \rightarrow 2O_2$.

binary Consisting of two components, as in *binary mixture* and *binary compound*. *Examples*: Acetone and water (a binary mixture); HCl, Al_2O_3, C_6H_6 (binary compounds).

block (*s* block, *p* block, *d* block, *f* block) The region of the periodic table containing elements for which, according to the building-up principle, the corresponding subshell is currently being filled.

body-centered cubic (bcc) A crystal structure with a unit cell in which a central atom lies at the center of a cube formed by eight other atoms.

Bohr frequency condition The relation between the change in energy of an atom or molecule and the frequency of radiation emitted or absorbed: $\Delta E = h\nu$.

Bohr model A model of the hydrogen atom in which the electron orbits around the nucleus in a series of discrete orbits.

Bohr radius a_0 The radius of the lowest-energy orbit in the Bohr model of the hydrogen atom, $a_0 = 53$ pm.

boiling-point elevation The increase in normal boiling point of a solvent caused by the presence of a solute (a colligative property).

boiling temperature (1) The temperature at which a liquid boils. (2) The temperature at which a liquid is in equilibrium with its vapor at the pressure of the surroundings; vaporization then occurs throughout the liquid, not only at the liquid's surface.

Boltzmann's constant k The value of R/N_A, where R is the gas constant and N_A is Avogadro's constant; $k = 1.38 \times 10^{-23}$ J/K.

Boltzmann formula (for the entropy) The formula $S = k \ln W$, where k is Boltzmann's constant and W is the number of atomic arrangements that correspond to the same energy.

bond A link between bonds. See also *covalent bond; double bond; ionic bond; triple bond.*

bond angle In an A—B—C molecule or part of a molecule, the angle between the B—A and B—C bonds.

bond enthalpy ΔH_B(X—Y) The enthalpy change accompanying the dissociation of a bond. *Example*: $H_2(g) \rightarrow 2H(g)$ $\Delta H_B(H{-}H) = +436$ kJ/mol.

bond length The distance between the nuclei of two atoms linked by a bond.

bond order (1) The number of electron pairs shared by neighboring atoms. (2) The net number of bonds between two atoms in a molecule, allowing for the cancellation of bonds by antibonds: bond order = $\frac{1}{2} \times$ (number of electrons in bonding molecular orbitals − number of electrons in antibonding molecular orbitals).

bond parameters The characteristic features of a bond. *Examples*: bond length; bond enthalpy.

bonding orbital A molecular orbital that, if occupied, lowers the energy of a molecule relative to that of the separated atoms.

Born-Haber cycle A closed series of reactions used to express the enthalpy of formation of an ionic solid in terms of contributions that include the lattice enthalpy.

Born interpretation The probability of finding an electron at a given location is proportional to the square of ψ at that location.

boundary surface The surface showing the region of an orbital within which there is about 90% probability of finding an electron.

Boyle's law At constant temperature, and for a given sample of gas, the volume is inversely proportional to the pressure: $P \propto 1/V$.

branched alkane An alkane with alkyl group side chains.

branching A step in a chain reaction in which more than one chain carrier is formed in a propagation step. *Example*: $\cdot O \cdot + H_2 \rightarrow HO \cdot + H \cdot$.

Brønsted acid A proton donor (a source of hydrogen ions, H^+). *Examples*: HCl; CH_3COOH; HCO_3^-; NH_4^+.

Brønsted base A proton acceptor (a species to which hydrogen ions, H^+, can bond). *Examples*: OH^-; Cl^-; $CH_3CO_2^-$; HCO_3^-; NH_3.

Brønsted equilibrium The proton-transfer equilibrium of the form $acid_1 + base_2 \rightleftharpoons acid_2 + base_1$. *Example*:

$$CH_3COOH(aq) + H_2O(l) \rightleftharpoons H_3O^+(aq) + CH_3CO_2^-(aq)$$

Brønsted-Lowry theory A theory of acids and bases expressed in terms of proton-transfer equilibria.

buffer A solution that resists any change in pH when small amounts of acid or base are added. An *acid buffer* stabilizes solutions at pH < 7 and a *base buffer* stabilizes solutions at pH > 7. *Examples*: a solution containing both CH_3COOH and $CH_3CO_2^-$ (acid buffer); a solution containing NH_3 and NH_4^+ is a base buffer.

buffer capacity The amount of hydronium ions or hydroxide ions that can be added to a buffer solution without changing its pH by more than 1 unit.

building-up principle The procedure for arriving at the ground-state electron configurations of atoms and molecules.

bulk properties Properties that depend on the collective behavior of large numbers of atoms. *Examples*: melting point; vapor pressure; internal energy.

calibration Interpretation of an observation by comparison with known information.

calorimeter An apparatus used to measure the heat released or absorbed by a process.

calorimetry The use of a calorimeter to measure the thermochemical properties of reactions.

capillary action The rise of liquids up narrow tubes.

carbide A binary compound of a metal and carbon: carbides may be *saline* (saltlike), *covalent*, and *interstitial*. *Examples*: CaC_2 (saline); SiC (covalent); W_2C (interstitial).

carbohydrate A compound of general formula $C_m(H_2O)_n$, although small deviations from this general formula are often encountered. *Examples*: $C_6H_{12}O_6$ (glucose); $C_{12}H_{22}O_{11}$ (sucrose).

carboxylic acid An organic compound containing the carboxyl group, —COOH. *Examples*: CH_3COOH, acetic acid; C_6H_5COOH, benzoic acid.

catalyst A substance that increases the rate of a reaction without being consumed in the reaction. A catalyst is *homogeneous* if it is present in the same phase as the reactants, and *heterogeneous* if it is in a different phase from the reactants. *Examples*: homogeneous, $Br_2(aq)$ for the decomposition of $H_2O_2(aq)$; heterogeneous, Pt in the Ostwald process.

catenate To form chains or rings of atoms. *Examples*: O_3; S_8.

cathode The electrode at which reduction occurs.

cathodic protection Protection of a metal object by connecting it to a more readily oxidizable metal that acts as a sacrificial anode.

cation A positively charged ion. *Examples*: Na^+; NH_4^+; Al^{3+}.

cell constant E' The constant term in the expression for a cell potential in terms of the pH.

cell diagram A diagram showing the arrangement of electrodes in an electrochemical cell. A cell is written with the cathode on the right and the anode on the left. *Example*: $Zn(s)|Zn^{2+}(aq)||Cu^{2+}(aq)|Cu(s)$.

cell potential E (1) The pushing and pulling power of a reaction in an electrochemical cell. (2) The potential difference between the electrodes of an electrochemical cell when it is producing no current; cell potential is always positive.

Celsius scale A temperature scale on which the freezing point of water is at 0 degrees and its normal boiling point is at 100 degrees.

cement (1) A solid obtained by the action of heat on silicates and aluminosilicates, with the addition of gypsum as a third component. (2) An adhesive.

ceramic (1) A solid obtained by the action of heat on clay. (2) A noncrystalline inorganic solid usually containing oxides, borides, and carbides.

cesium-chloride structure A crystal structure the same as that of solid cesium chloride.

chain carrier An intermediate in a chain reaction.

chain reaction A reaction in which an intermediate reacts to produce another intermediate. *Example*: $Br\cdot + H_2 \rightarrow HBr + H\cdot$; $H\cdot + Br_2 \rightarrow HBr + Br\cdot$.

chalcogens Oxygen, sulfur, selenium, and tellurium in Group VI of the periodic table.

change of state The change of a substance from one of its physical states to another of its physical states. *Examples*: melting (solid → liquid); gray tin → white tin.

charge A measure of the strength with which a particle can interact electrostatically with another particle.

charge number The number of charges carried by an ion. *Examples*: +3 for Al^{3+}; −2 for SO_4^{2-}.

Charles's law The volume of a given sample of gas at constant pressure is directly proportional to its absolute temperature: $V \propto T$.

chelate A complex containing at least one polydentate ligand that forms a ring of atoms including the central metal atom. *Example*: $[Co(en)_3]^{3+}$.

chemical analysis The determination of the chemical composition of a sample.

chemical change The formation of one substance from another substance.

chemical formula A collection of chemical symbols and subscripts that shows the composition of a substance. See *empirical formula; molecular formula; structural formula*.

chemical nomenclature The systematic naming of compounds.

chemical property A characteristic that is observed or measured only by changing the identity of the substance.

chemical reaction A chemical change in which one substance responds to the presence of another, to a change of temperature, or to some other influence.

chemical kinetics The study of the rates of reactions and the steps by which they occur.

chemical symbol One- or two-letter abbreviation of an element's name.

chemiluminescence The emission of light by products formed in energetically excited states.

chemistry The branch of science concerned with the study of matter and the changes that matter can undergo.

chiral molecule A molecule that is distinct from its own mirror image. *Examples*: $CH_3CH(NH_2)COOH$; CHBrClF; $[Co(en)_3]^{3+}$.

chromatogram The record of the signal from the detector in a chromatographic analysis of a mixture.

chromatography A separation technique that relies on the ability of surfaces to adsorb substances with different strengths.

cis-trans isomerization The conversion of a cis isomer

into a trans isomer. *Example*: *cis*-butene → *trans*-butene.

classical mechanics The laws of motion proposed by Isaac Newton in which particles travel in definite paths in response to forces.

clathrate A compound in which a molecule of one component sits in a cage made up of molecules of another component, typically water. *Example*: SO_2 in water.

Claus process A process for obtaining sulfur from the H_2S in oil wells by the oxidation of H_2S with SO_2.

close-packed structure A crystal structure in which atoms occupy the smallest total volume with the least empty space. *Examples*: hexagonal close-packing (ABAB . . .) and cubic close-packing (ABCABC . . .) of identical spheres.

closed shell (or subshell) A shell (or subshell) containing the maximum number of electrons allowed by the exclusion principle. *Example*: the neonlike core $1s^22s^22p^6$.

codon A group of three bases that constitute the genetic code.

cohesion Bound together.

cohesive forces The forces that bind the molecules of a substance together to form a bulk material and that are responsible for condensation.

coinage metals The elements copper, silver, and gold.

colligative property A property that depends only on the number of solute particles present in a solution and not on their chemical composition. *Examples*: elevation of boiling point; depression of freezing point; osmosis.

collision theory The theory of elementary gas-phase bimolecular reactions in which it is assumed that molecules react only if they collide with at least enough kinetic energy for bonds to be broken.

colloidal suspension A suspension of tiny particles in a liquid. *Example*: milk.

combustion A reaction in which an element or compound burns in oxygen. *Example*: $CH_4(g) + 2O_2(g) \rightarrow CO_2(g) + 2H_2O(g)$.

combustion analysis The determination of the composition of a sample by measuring the masses of the products of its combustion.

common-ion effect (1) The effect on the pH when a salt is added to a weak acid or base that provides a common ion. (2) The decrease in the solubility of one salt by the presence of another salt with one ion in common. *Example*: (1) the effect of sodium acetate on the pH of acetic acid; (2) the lower solubility of AgCl in NaCl(*aq*) than in pure water.

common name An informal name for a compound that may give little or no clue to the compound's composition. *Examples*: water; aspirin; acetic acid.

complementary color The color that white light becomes when one of the colors present in it is removed.

complex (1) The combination of a Lewis acid and a Lewis base. (2) A species consisting of several ligands (the Lewis bases), which can have an independent existence, bonded to a single central metal atom or ion (the Lewis acid). *Examples*: (1) $H_3N{-}BF_3$; (2) $[Fe(H_2O)_6]^{3+}$; $[PtCl_4]^-$.

compound (1) A specific combination of elements that can be separated into its elements by using chemical techniques. (2) A substance consisting of atoms of two or more elements in a definite ratio.

compress Reduce the volume of a sample by increasing the pressure on it.

concentration See *molar concentration*.

condensation The formation of a liquid from a gas.

condensation polymer A polymer formed by a chain of condensation reactions. *Examples*: polyesters; polyamides (nylon).

condensation reaction A reaction in which two molecules combine to form a larger one and a small molecule is eliminated. *Example*: $CH_3COOH + C_2H_5OH \rightarrow CH_3COOC_2H_5 + H_2O$.

conduction band An incompletely filled band of orbitals in a solid.

configuration See *electron configuration*.

conformation A molecular shape that can be changed by rotation about bonds, without bond breakage and reformation.

congeners Members of the same group of the periodic table. *Example*: calcium and beryllium.

conjugate acid The Brønsted acid formed when a Brønsted base has accepted a proton. *Example*: NH_4^+ is the conjugate acid of NH_3.

conjugate acid-base pair A Brønsted acid and its conjugate base. *Examples*: HCl and Cl^-; NH_4^+ and NH_3.

conjugate base The Brønsted base formed when a Brønsted acid has donated a proton. *Example*: NH_3 is the conjugate base of NH_4^+.

contact catalyst A heterogeneous catalyst that consists of a finely divided solid used to accelerate gas-phase or liquid-phase reactions.

contact process The production of sulfuric acid by the catalyzed oxidation of sulfur dioxide to sulfur trioxide. The catalyst is vanadium pentoxide.

conversion factor A factor that is used to convert a measurement from one unit to another.

coordinate Use of a lone pair to form a covalent bond. *Example*: $F_3B + {:}NH_3 \rightarrow F_3B{-}NH_3$.

coordination compound A neutral complex or an ionic compound in which at least one of the ions is a complex. *Examples*: $Ni(CO)_4$; $K_2[Fe(CN)_6]$; chlorophyll.

coordination isomers Isomers that differ by the exchange of one or more ligands between a cationic complex and an anionic complex.

coordination number (1) The number of nearest neighbors of an atom in a solid. (2) For ionic solids, the coordination number of an ion is the number of nearest neighbors of opposite charge. (3) For complexes, the coordination number is the number of atoms directly attached to the central metal ion.

coordination sphere The central metal ion and the ligands directly attached to it in a complex.

copolymer A polymer formed from a mixture of monomers.

core The inner closed shells of an atom.

core electrons The electrons that belong to an atom's core.

corrosion The (usually unwanted) oxidation of a metal.

corrosive Causing corrosion.

couple The oxidized and reduced species in a half-reaction; specified as oxidized form/reduced form. *Example*: Fe^{3+}/Fe^{2+} in $Fe^{3+}(aq) + e^- \rightarrow Fe^{2+}(aq)$.

covalent bond A pair of electrons shared by two atoms.

covalent radius The contribution of an atom to the length of a covalent bond.

cracking The process of converting petroleum fractions into smaller molecules with more double bonds. *Example*: $CH_3(CH_2)_6CH_3 \rightarrow CH_3(CH_2)_3CH_3 + CH_3CH{=}CH_2$.

critical mass The mass of fissionable material below which so many neutrons escape from a sample of nuclear fuel that the fission chain reaction is not sustained; a greater mass is *supercritical* and a smaller mass is *subcritical*.

critical pressure P_c The minimum pressure needed to liquefy a gas at its critical temperature.

critical temperature T_c The temperature above which a substance cannot exist as a liquid.

crown ether A cyclic ether with alternating $-CH_2CH_2-$ groups and O atoms in the form of a crown.

cryogenics The study of matter at very low temperatures.

cryoscopy The measurement of molar mass by using the depression of freezing point.

crystal field The electrostatic influence of the ligands (modeled as point negative charges) on the central ion of a complex. *Crystal field theory* is a rationalization of the optical, magnetic, and thermodynamic properties of complexes in terms of the crystal field of their ligands.

crystalline solid A solid in which the atoms, ions, or molecules lie in an orderly array. *Examples*: NaCl; diamond; graphite.

crystallization The process in which a solute comes out of solution slowly as crystals.

cubic close-packed structure A close-packed structure with an ABCABC . . . pattern of layers.

curie (Ci) The unit of activity (for radioactivity).

current I The rate of supply of charge; current is measured in amperes (A), with 1 A = 1 C/s.

cycle (1) In thermodynamics, a sequence of changes that begins and ends at the same state. (2) In spectroscopy, one complete reversal of the direction of the electromagnetic field and its return to the original direction.

cycloalkane A saturated aliphatic hydrocarbon in which the carbon atoms form a ring. *Example*: C_6H_{12}, cyclohexane.

Dalton's law of partial pressures The total pressure of a mixture of gases is the sum of the partial pressures of its components.

data The items of information provided.

daughter nucleus A nucleus that is the product of a nuclear disintegration.

de Broglie relation The proposal that every particle has wavelike properties and that its wavelength λ is related to its mass by $\lambda = h/(\text{mass} \times \text{velocity})$.

debye The unit used to report electric dipole moments ($1\text{ D} = 3.336 \times 10^{-30}\text{ C·m}$).

decay constant k The rate constant for radioactive decay.

decomposition A reaction in which a substance is broken down into simpler substances. *Thermal decomposition* is decomposition brought about by heat. *Example*:

$$CaCO_3(s) \xrightarrow{\Delta} CaO(s) + CO_2(g)$$

decomposition vapor pressure The pressure of the gaseous decomposition product of a solid at equilibrium.

dehydrating agent A reagent that removes water or the elements of water from a compound. *Example*: H_2SO_4.

dehydrogenation The removal of a hydrogen atom from each of two neighboring carbon atoms, resulting in the formation of a carbon-carbon multiple bond.

dehydrohalogenation The removal of a hydrogen atom and a halogen atom from neighboring carbon atoms, resulting in the formation of a carbon-carbon multiple bond.

delocalized orbitals Molecular orbitals that spread over an entire molecule.

delta Δ (in a chemical equation) A symbol that signifies that the reaction requires an elevated temperature.

delta X ΔX The difference between the final and initial values of a property, $\Delta X = X_f - X_i$. *Examples*: ΔT; ΔE.

delta hazard ⚠ A symbol that indicates that a skeletal equation is not balanced.

denaturation The loss of structure of a protein.

density d Mass per volume of a substance: $d = m/V$.

deposition A process in which a vapor condenses directly to a solid (the reverse of sublimation).

deprotonation Loss of a proton from a Brønsted acid. *Example*:

$$NH_4^+(aq) + H_2O(l) \rightarrow H_3O^+(aq) + NH_3(aq)$$

derived unit A combination of base units. *Examples*: cubic centimeters; joules ($kg{\cdot}m^2/s^2$).

diagonal relationship A similarity in properties between diagonal neighbors in the periodic table, especially for elements in Periods 2 and 3 at the left of the table. *Examples*: Li and Mg; Be and Al.

diamagnetic A substance that is pushed out of a magnetic field; a diamagnetic substance consists of atoms, ions, or molecules with no unpaired electrons. *Example*: most common substances.

diamine An organic compound that contains two $-NH_2$ groups.

diatomic molecule A molecule that consists of two atoms. *Examples*: H_2; CO.

diffraction Interference between waves caused by an object in their path. See also *x-ray diffraction.*

diffraction pattern The pattern of bright spots against a dark background resulting from diffraction.

diffusion The spreading of one substance through another substance.

dilute To reduce the concentration of a solution by adding more solvent.

dimer A molecule produced by the union of two identical molecules. *Example*: Al_2Cl_6, formed from two $AlCl_3$ molecules.

diol An organic compound with two hydroxyl groups.

dipeptide An oligopeptide formed by the condensation of two amino acids.

dipole See *electric dipole.*

dipole-dipole interaction The interaction between two electric dipoles: like partial charges repel and opposite partial charges attract.

dispersion force See *London forces.*

disproportionation A redox reaction in which a single element is simultaneously oxidized and reduced. *Example*: $2Cu^+(aq) \rightarrow Cu(s) + Cu^{2+}(aq)$.

disaccharide A carbohydrate molecule that is composed of two C_6 saccharide units. *Example*: $C_{12}H_{22}O_{11}$ (sucrose).

dissociation The breaking of a bond.

distillation The separation of a mixture by making use of the different volatilities of its components.

distribution (of molecular speeds) The percentage of gas molecules moving at each speed at any instant.

disulfide link An —S—S— link that contributes to the secondary structures of polypeptides.

doping The addition of a known, small amount of a second substance to an otherwise pure solid substance.

double bond (1) Two electron pairs shared by neighboring atoms. (2) One σ bond and one π bond between neighboring atoms.

double replacement reaction (metathesis, double decomposition) A reaction of the form $MA + BX \rightarrow MX + BA$, in which two pairs of ions (M^+ and A^- and B^+ and X^-) change partners. *Example*: $AgNO_3(aq) + NaCl(aq) \rightarrow AgCl(s) + NaNO_3(aq)$.

Dow process The electrolytic production of magnesium from molten magnesium chloride.

Downs process The production of sodium and chlorine by the electrolysis of molten sodium chloride.

dry cell A Lechlanché cell: a cell in which the electrolyte is a moist paste.

drying agent A substance that absorbs water and thus maintains a dry atmosphere. *Example*: phosphorus(V) oxide.

dynamic equilibrium The condition in which a forward process and its reverse are occurring simultaneously at equal rates. *Examples*: vaporizing and condensing; chemical reactions at equilibrium.

effective nuclear charge The net nuclear charge after taking into account the shielding caused by electrons in the atom.

effusion The escape of a substance (particularly a gas) through a small hole.

elastomer An elastic polymer. *Example*: rubber (polyisoprene).

electric dipole A positive charge next to an equal but opposite negative charge.

electric dipole moment μ The magnitude of the electric dipole (in debye).

electrochemical cell A system consisting of two electrodes in contact with an electrolyte. A *galvanic cell* is an electrochemical cell used to produce electricity. An *electrolytic cell* is an electrochemical cell in which an electric current is used to cause chemical change.

electrochemical series Redox couples arranged in order of oxidizing and reducing strengths; specifically, with strong oxidizing agents at the top of the list and strong reducing agents at the bottom.

electrochemistry The branch of chemistry that deals with the use of chemical reactions to produce electricity, the relative strengths of oxidizing and reducing agents, and the use of electricity to produce chemical change.

electrode A metallic conductor that makes contact with an electrolyte in an electrochemical cell.

electrolysis (1) A process in which a chemical change is produced by passing an electric current through a liquid. (2) The process of driving a reaction in a nonspontaneous direction by passing an electric current through a solution.

electrolyte (1) An ionically conducting medium. (2) A substance that dissolves to give an ionically conducting solution. A *weak electrolyte* is a solution in which only a small proportion of the solute molecules are ionized. *Examples*: $NaCl(aq)$ is a strong electrolyte solution; $CH_3COOH(aq)$ is a weak electrolyte solution; $C_6H_{12}O_6(aq)$ is a nonelectrolyte solution.

electrolytic cell See *electrochemical cell.*

electromagnetic radiation A wave of oscillating electric and magnetic fields; includes light, x rays, and γ rays.

electron Negatively charged subatomic particle.

electron affinity The energy released when an electron is added to a gas-phase atom or ion of the elements; the negative of the electron-gain enthalpy.

electron capture The capture by a nucleus of one of its own atom's *s* electrons.

electron configuration The occupancy of orbitals in an atom or molecule. *Examples*: N $1s^2 2s^2 2p^3$, N_2 $\sigma^2 \sigma^{*2} \pi^4 \sigma^2$ (valence electrons only).

electron-deficient compound A compound that has too few valence electrons to be assigned a Lewis structure. *Example*: B_2H_6.

electron-gain enthalpy The change in enthalpy when an electron attaches to a gas-phase atom or ion.

electronegative element An element with a high electronegativity. *Examples*: O; F.

electronegativity χ The measure of the ability of an atom to attract electrons to itself when it is part of a compound.

electronic conduction Electric conduction in which the charge is carried by electrons.

electronic structure The details of the distribution of the electrons that surround the nuclei in atoms and molecules.

electroplating The deposition of a thin film of metal on an object by electrolysis.

electropositive element An element that has a low electronegativity and is likely to give up electrons to another element upon compound formation. *Examples*: Cs; Mg.

element (1) A substance that cannot be separated into simpler components by using chemical techniques. (2) A substance that consists of atoms having the same chemical properties. (3) A substance consisting of atoms of the same atomic number. *Examples*: hydrogen; gold; uranium.

elementary reaction An individual reaction step in a mechanism. *Example*: $H \cdot + Cl_2 \rightarrow HCl + Cl \cdot$.

elimination reaction A reaction in which two groups or atoms on neighboring carbon atoms are removed from a molecule, thereby leaving a multiple bond between the carbon atoms. *Example*: $CH_3CHBrCH_3 + OH^- \rightarrow CH_3CH{=}CH_2 + H_2O$.

empirical formula A chemical formula that shows the relative numbers of atoms of each element in a compound. *Example*: NaCl; P_2O_5; CH for benzene.

enantiomer Either of a pair of mirror-image optical isomers.

end point The stage in a titration at which enough titrant has been added to bring the indicator to a color halfway between its initial and final colors.

endothermic reaction A reaction that absorbs heat ($\Delta H > 0$). *Example*: $N_2O_4(g) \rightarrow 2NO_2(g)$.

energy E The capacity of a system to do work or supply heat. *Kinetic energy* is the energy of motion; *potential energy* is the energy arising from position.

enrich In nuclear chemistry, to increase the abundance of a specific isotope.

enthalpy H A state property, a change in which is equal to the quantity of heat transferred at constant pressure; $H = U + PV$.

enthalpy density (of a fuel) The enthalpy of combustion per liter (without the negative sign).

enthalpy of freezing The negative of the enthalpy of melting.

enthalpy of hydration ΔH_H The enthalpy change accompanying the hydration of gas-phase ions.

enthalpy of ionization The change in enthalpy for the process $E(g) \rightarrow E^+(g) + e^-(g)$, where E is any element. See *ionization energy*.

enthalpy of melting ΔH_{melt} The difference in enthalpy per mole between the solid and liquid states of a substance.

enthalpy of solution ΔH_{sol} The change in enthalpy that occurs when a substance dissolves.

enthalpy of vaporization ΔH_{vap} The difference in enthalpy per mole between the vapor and liquid states of a substance.

entropy S (1) A measure of the disorder of a system. (2) A change in entropy is equal to the heat added to a system divided by the temperature at which the transfer occurs.

enzyme A biological catalyst.

equation of state A mathematical expression relating the pressure, volume, temperature, and amount of substance present in a sample. *Example*: ideal gas law, $PV = nRT$.

equatorial bond A bond perpendicular to the axis of a molecule (particularly trigonal bipyramidal and octahedral molecules).

equatorial pair An electron pair in the plane perpendicular to the molecular axis.

equilibrium constant K_c A characteristic of the equilibrium composition of the reaction mixture, with a form given by the law of mass action. *Example*:

$$N_2(g) + 3H_2(g) \rightleftharpoons 2NH_3(g) \qquad K_c = [NH_3]^2/([N_2][H_2]^3)$$

equivalence point See *stoichiometric point.*

ester The product (other than water) of the reaction between a carboxylic acid and an alcohol. *Example*: $CH_3COOC_2H_5$ (ethyl acetate).

essential oil An oil that can be distilled from flowers and leaves (and that conveys the "essence" of a plant).

esterification The formation of an ester.

ether An organic compound of the form R—O—R. *Example*: $C_2H_5OC_2H_5$ (diethyl ether).

evaporate To vaporize completely.

exclusion principle No more than two electrons may occupy any given orbital; and when two electrons do occupy one orbital, their spins must be paired.

exothermic reaction A reaction that releases heat ($\Delta H < 0$). *Example*: $N_2(g) + 3H_2(g) \rightarrow 2NH_3(g)$.

expanded octet A valence shell containing more than eight electrons. *Examples*: the valence shells in PCl_5 and SF_6.

experiment A test carried out under carefully controlled conditions.

exponential decay A variation with time of the form e^{-kt}.

extensive property A physical property of a substance that depends on the size of the sample. *Example*: mass; internal energy; enthalpy; entropy.

face-centered cubic (fcc) A crystal structure built from a unit cell in which there is an atom at the center of each face and one at each corner of a cube.

Faraday's constant F The charge per mole of electrons. $F = 96.485$ kC/(mol e^-).

Faraday's law of electrolysis The amount (in moles) of product formed by an electric current is chemically equivalent to the amount (in moles) of electrons supplied.

fat An ester of glycerol and carboxylic acids with long hydrocarbon chains; fats provide long-term energy storage.

fatty acid A carboxylic acid with a long hydrocarbon chain. *Example*: $CH_3(CH_2)_{16}COOH$ (stearic acid).

ferromagnetism The ability of some substances to be permanently magnetized. *Examples*: iron; magnetite (Fe_3O_4).

field An influence spreading over a region of space. *Examples*: an electric field from a charge; a magnetic field from a magnet.

filtration The separation of a heterogeneous mixture by shaking it with a solvent and passing the resulting mixture through a fine mesh.

first law of thermodynamics The internal energy of an isolated system is constant.

first-order reaction A reaction in which the rate is proportional to the first power of the concentration of a substance.

fissile Having the ability to undergo fission induced by slow neutrons. *Example*: ^{235}U is fissile.

fission The breakup of a nucleus into two smaller nuclei of similar mass; fission may be *spontaneous* or *induced* (particularly by the impact of neutrons). *Examples*: $^{244}_{95}Am \rightarrow {}^{134}_{53}I + {}^{107}_{42}Mo + 3n$ (spontaneous); $^{235}_{92}U + n \rightarrow {}^{142}_{56}Ba + {}^{91}_{36}Kr + 3n$ (induced).

fissionable Having the ability to undergo induced fission.

fixation of nitrogen Conversion of elemental nitrogen to its compounds, particularly ammonia.

flammability The ability of a substance to burn in air.

force F An influence that changes the state of motion of an object. *Examples*: an electrostatic force from an electric charge; a mechanical force from an impact.

formal charge (FC) (1) The electric charge of an atom assigned on the assumption that there is perfect covalent bonding. (2) FC = number of valence electrons in the free atom − number of lone-pair electrons − $\frac{1}{2}$ × number of shared electrons.

formation constant K_f The equilibrium constant for complex formation. The *overall formation constant* is the product of *stepwise formation constants*.

formula unit The group of ions that matches the empirical formula of an ionic compound. *Example*: NaCl, one Na^+ ion and one Cl^- ion.

formula weight See *molar mass*.

fossil fuels The partially decomposed remains of vegetable and marine life (mainly coal, oil, and natural gas).

fractional distillation Separation of the components of a liquid mixture by repeated distillation, by making use of their differing volatilities.

Frasch process A process for mining sulfur, using superheated water and compressed air.

free energy G The energy of a system that is free to do work at constant temperature and pressure: $G = H - TS$. The direction of spontaneous change (at constant pressure and temperature) is the direction of decreasing free energy.

freezing-point depression The lowering of the freezing point of a solution caused by the presence of a solute (a colligative property).

freezing temperature The temperature at which a liquid freezes. The *normal freezing point* is the freezing temperature under a pressure of 1 atm.

frequency (of radiation) ν The number of cycles (repeats of the waveform) per second (unit: hertz, Hz).

frequency factor A The constant obtained from the intercept in an Arrhenius plot.

Friedel-Crafts alkylation The attachment of alkyl groups to an aromatic ring by using aluminum chloride as catalyst. *Example*:

$$C_6H_6 + CH_3Cl \xrightarrow{AlCl_3} C_6H_5CH_3 + HCl$$

fuel cell A primary electrochemical cell in which the reactants are supplied from outside while the cell is in use.

functional group A group of atoms that brings a characteristic set of chemical properties to an organic molecule. *Examples*: —OH; —Br; —COOH.

furanose A five-membered ring form of a carbohydrate molecule.

fusion (1) Melting. (2) The merging of nuclei to form the nucleus of a heavier element.

galvanic cell See *electrochemical cell*.

galvanize Coat a metal with an unbroken film of zinc.

gamma (γ) radiation Very high frequency, short wavelength electromagnetic radiation emitted by nuclei.

gas A fluid form of matter that fills the container it occupies and can easily be compressed into a much smaller volume. (The distinction between a gas and a vapor is as follows: a gas is a substance at a higher temperature than its critical temperature, a vapor is a gaseous form of matter at a temperature below its critical temperature.)

gas-liquid chromatography A version of chromatography in which a gas carries the sample over a stationary liquid phase.

Geiger counter A device used to detect and measure radioactivity that relies on the ionization caused by incident radiation.

geometrical isomers (1) Isomers in which the coordination spheres contain the same ligands but differ in the spatial arrangement of the ligands. (2) Stereoisomers that differ in the spatial arrangement of the atoms.

Gibbs free energy See *free energy*.

glass electrode A thin-walled glass bulb containing an electrolyte; used for measuring pH.

Graham's law of effusion The rate of effusion of a gas is inversely proportional to the square root of its molar mass.

gravimetric analysis The analysis of composition by measurement of mass.

greenhouse effect The blocking by some atmospheric gases (notably, carbon dioxide) of the radiation of heat from the surface of the Earth back into space, leading to the possibility of a worldwide rise in temperature.

ground state The state of lowest energy.

group The vertical column of the periodic table; the number of the group (in the system we use) is equal to the number of electrons in the valence shell of the atoms.

Haber process (Haber-Bosch process) The catalyzed synthesis of ammonia at high pressure and high temperature.

half-life $t_{1/2}$ (1) In chemical kinetics, the time needed for the concentration of a substance to fall to half its initial concentration. (2) In radioactivity, the time needed for half the initial number of radioactive nuclei to disintegrate.

half-reaction A hypothetical oxidation or reduction reaction showing either electron loss or electron gain. *Examples*: $Na(s) \rightarrow Na^+(aq) + e^-$; $Cl_2(g) + 2e^- \rightarrow 2Cl^-(aq)$.

Hall process The production of aluminum by the electrolysis of aluminum oxide dissolved in molten cryolite.

haloalkane An alkane with a halogen substituent. *Example*: CH_3Cl, chloromethane.

halogenation The incorporation of a halogen into a compound (particularly, into an organic compound).

halogens The elements in Group VII.

hard water Water that contains dissolved calcium and magnesium salts.

heat The energy that is transferred as the result of a temperature difference between a system and its surroundings.

heat capacity The ratio of heat supplied to the temperature rise produced.

heating curve A graph of the variation of the temperature of a sample as it is heated.

hemiacetal An organic compound with both an —O— link and a neighboring —OH group.

Henderson-Hasselbalch equation An approximate equation for estimating the pH of a solution containing a conjugate acid and base, and specifically that of a buffer solution. (See Section 15.5.)

Henry's constant The constant k_H that appears in Henry's law.

Henry's law The solubility of a gas in a liquid is proportional to its partial pressure above the liquid: solubility = $k_H \times$ partial pressure.

Hess's law A reaction enthalpy is the sum of the enthalpies of any sequence of reactions (at the same temperature and pressure) into which the overall reaction may be divided.

heterogeneous equilibria An equilibrium in which at least one substance is in a different phase from that of the rest. *Example*: $AgCl(s) \rightleftharpoons Ag^+(aq) + Cl^-(aq)$.

heterogeneous mixture A mixture in which the individual components, although mixed together, lie in distinct regions, even on a microscopic scale. *Example*: a mixture of sand and sugar.

heteronuclear diatomic molecule A molecule consisting of two atoms of different elements. *Examples*: HCl; CO.

hexagonal close-packed structure A close-packed structure with an ABABA . . . pattern of layers.

high-spin complex A d^n complex with the maximum number of unpaired electron spins.

homogeneous equilibrium A chemical equilibrium in which all the substances taking part are in the same phase. *Example*: $H_2(g) + I_2(g) \rightleftharpoons 2HI(g)$.

homogeneous mixture A mixture in which the individual components are uniformly mixed, even on an atomic scale. *Example*: air; solutions.

homologous series A family of compounds with molecular formulas obtained by inserting a particular group (most commonly $—CH_2—$) repetitively into a parent structure. *Example*: CH_4, CH_3CH_3, $CH_3CH_2CH_3$, . . .

homonuclear diatomic molecule A molecule consisting of two atoms of the same element. *Examples*: H_2; N_2.

Hund's rule If more than one orbital in a subshell is available, add electrons with parallel spins to different orbitals of that subshell.

hybrid orbital A mixed orbital formed by blending together atomic orbitals on the same atom. *Example*: an sp^3 hybrid orbital.

hybridization The formation of hybrid orbitals.

hydrate Solid, inorganic compounds containing H_2O molecules. *Example*: $CuSO_4 \cdot 5H_2O$.

hydrate isomers Isomers that differ by an exchange of an H_2O molecule and a ligand in the coordination sphere.

hydration (of ions) The attachment of water molecules to a central ion.

hydration (of organic compounds) The addition of water across a multiple bond (H to one carbon atom, OH to the other). *Example*: $CH_2{=}CH_2 + H_2O \rightarrow CH_3CH_2OH$.

hydrocarbon A binary compound of carbon and hydrogen. *Examples*: CH_4; C_6H_6.

hydrogen bond A link formed by a hydrogen atom lying between two strongly electronegative atoms (O, N, or F).

hydrogen economy The widespread use of hydrogen as a fuel.

hydrogenation The addition of hydrogen across a carbon-carbon multiple bond. *Example*:

$$CH_3CH{=}CH_2 + H_2 \rightarrow CH_3CH_2CH_3$$

hydrohalogenation The addition of a hydrogen halide across a carbon-carbon multiple bond. *Example*: $CH_3CH{=}CH_2 + HBr \rightarrow CH_3CHBrCH_3$.

hydrolysis The reaction of water with a substance, resulting in the formation of a new element-oxygen bond. *Example*: $PCl_5(s) + 4H_2O(l) \rightarrow H_3PO_4(aq) + 5HCl(aq)$.

hydrolyze To undergo hydrolysis (see preceding entry).

hydrometallurgical process The extraction of metals by reduction of aqueous solutions of their ions. *Example*: $Cu^{2+}(aq) + Fe(s) \rightarrow Cu(s) + Fe^{2+}(aq)$.

hydronium ion The ion H_3O^+.

hydrophobic Water-repelling. *Example*: hydrocarbon chains are hydrophobic.

hydrophilic Water-attracting. *Example*: hydroxyl groups are hydrophilic.

hydrostatic pressure The pressure exerted by a vertical column of liquid.

hygroscopic Water-absorbing.

hypothesis A suggestion put forward to account for a series of observations. *Example*: Dalton's atomic hypothesis.

***i* factor** A factor that takes into account the existence of ions in an electrolyte solution, particularly for the interpretation of colligative properties. *Example*: $i \approx 2$ for NaCl(*aq*).

ideal gas A gas that satisfies the ideal gas law.

ideal gas law $PV = nRT$; all gases obey the law more and more closely as the pressure is reduced to very low values.

ideal solution A solution that obeys Raoult's law at any concentration; all solutions behave ideally as their concentrations approach zero. *Example* (of an almost ideal system): benzene and toluene.

incomplete octet A valence shell of an atom in a compound that has fewer than eight electrons. *Example*: the valence shell of B in BF_3.

indicator A substance that changes color when it goes from its acid to its base form (an *acid-base indicator*) or from its oxidized to its reduced form (a *redox indicator*).

inert pair A pair of valence shell *s* electrons that are tightly bound to the atom and might not participate in bond formation.

infrared radiation Electromagnetic radiation with a lower frequency (longer wavelength) than that of red light.

inhibition A reaction in which chain carriers are removed by processes other than by chain termination. *Example*: $H\cdot$ + other radicals → unreactive substances.

initial rate The rate at the start of the reaction when no products are present.

initiation The formation of chain carriers from a reactant at the start of a chain reaction. *Example*:

$$Br_2 \xrightarrow{\Delta \text{ or light}} 2Br\cdot$$

inorganic compound A compound that is not organic. See also *organic compound.*

instantaneous rate The slope of the tangent of a graph of concentration against time.

insulator (electrical) A substance that does not conduct electricity. *Examples*: nonmetallic elements; molecular solids.

integrated rate law An expression for the concentration of a reactant or product in terms of the time, obtained from the rate law of the reaction. *Example*: $\ln([A]_0/[A]) = kt$.

intensive property A physical property of a substance that is independent of the size of the sample. *Examples*: density; molar volume; temperature.

interference Interaction between waves, leading to a greater amplitude *(constructive interference)* or to a smaller one *(destructive interference).*

interhalogen A binary compound of two halogens. *Example*: IF_3.

intermediate See *reaction intermediate.*

intermolecular Between molecules.

internuclear axis The straight line between the nuclei of two bonded atoms.

intermolecular forces The forces of attraction and repulsion between molecules. *Examples*: ion-dipole; dipole-dipole; London forces.

internal energy *U* The total energy of a system.

International System (SI) A collection of definitions of units and their employment.

interstitial compound A compound in which one type of atom occupies the gaps between other atoms. *Example*: an interstitial hydride.

intramolecular Within a molecule.

ion An electrically charged atom or group of atoms. *Examples*: Al^{3+}, SO_4^{2-}. See also *cation* and *anion.*

ion-dipole force The force of attraction between an ion and the opposite partial charge of the electric dipole of a polar molecule.

ion-exchange The exchange of one type of ion in solution for another.

ion pair A cation and anion in close proximity.

ion product See *solubility quotient.*

ionic conduction Electrical conduction in which the charge is carried by ions.

ionic bond The attraction between the opposite charges of cations and anions.

ionic compound A compound that consists of ions. *Examples*: NaCl; KNO_3.

ionic equation (full ionic form) A chemical equation explicitly showing the ions of the reactants and products. *Example*:

$$Na^+(aq) + Cl^-(aq) + Ag^+(aq) + NO_3^-(aq) \rightarrow Na^+(aq) + NO_3^-(aq) + AgCl(s)$$

ionic model The description of bonding in terms of ions.

ionic radius The contribution of an element to the distance between neighboring ions in a solid ionic compound. In practice, the radius of an ion determined from the distance between the centers of neighboring ions, with the radius of the O^{2-} ion set equal to 140 pm.

ionic solid A solid built from cations and anions. *Examples*: NaCl; KNO_3.

ionization (1) Conversion to cations by the removal of electrons. (2) The donation of a proton from a neutral acid molecule to a base with the formation of the conjugate base (an anion, in this instance) of the acid. *Examples*:

(1) $K(g) \rightarrow K^+(g) + e^-(g)$

(2) $CH_3COOH(aq) + H_2O(l) \rightarrow H_3O^+(aq) + CH_3CO_2^-(aq)$

ionization energy I The minimum energy required to remove an electron from the ground state of a gaseous atom, molecule, or ion. The second ionization energy is the ionization energy for removal of a second electron.

ionization isomers Isomers that differ by the exchange of a ligand with an anion or neutral molecule outside the coordination sphere.

isoelectronic species Species with the same number of atoms and the same numbers of valence electrons. *Examples*: SO_2 and O_3; Cl^- and Ar; NO_3^- and CO_3^{2-}.

isomerization A reaction in which a compound is converted into one of its isomers. *Example*: *cis*-butene → *trans*-butene.

isomers Compounds that contain the same number of the same atoms in different arrangements. In *structural isomers,* the atoms have different partners or lie in a different order; in *stereoisomers,* the atoms have the same partners but are in different arrangements in space. *Examples*: $CH_3—O—CH_3$ and $CH_3CH_2—OH$ (structural isomers); *cis*- and *trans*-2-butene (stereoisomers).

isotactic polymer A polymer in which the substituents are all on the same side of the chain.

isotopes Atoms that have the same atomic number but different atomic masses. *Examples*: ^1H; ^2H; ^3H.

isotopic abundance See *abundance; natural abundance.*

isotopic dating The determination of the ages of objects by measuring the activity of the radioactive isotopes they contain; particularly, *radiocarbon dating* by using ^{14}C.

isotopic label See *tracer.*

Joule-Thomson effect The cooling of a gas as it expands.

Kekulé structures Two Lewis structures of benzene, consisting of alternating single and double bonds.

kelvin K The base unit of temperature in the International System.

Kelvin scale A fundamental scale of temperature on which the triple point of water lies at 273.16 K.

ketone An organic compound of the form R—CO—R′. *Example*: $CH_3—CO—CH_3$, acetone.

kinetic theory (kinetic-molecular theory) A theory of the properties of an ideal gas in which pointlike molecules are in continuous random motion and there are no interactions between them.

labile A species that survives only for short periods.

lanthanide contraction The reduction of the atomic radii of the elements following the lanthanides below the values that would be expected by extrapolation of the trend down a group (and arising from the poor shielding ability of f electrons).

lattice enthalpy The standard enthalpy change for the conversion of an ionic solid to a gas of ions.

law A summary of experience.

law of constant composition A compound has the same composition whatever its source.

law of conservation of energy Energy can be neither created nor destroyed.

law of mass action For an equilibrium of the form $aA + bB \rightleftharpoons cC + dD$, the reaction quotient $Q_c = [C]^c[D]^d/[A]^a[B]^b$, evaluated by using the equilibrium molar concentrations of the reactants and products, is equal to a constant K_c, which has a specific value for a given reaction and temperature.

law of radioactive decay The rate of decay is proportional to the number of radioactive nuclides in the sample.

Le Chatelier's principle When a stress is applied to a system in dynamic equilibrium, the equilibrium adjusts to minimize the effect of the stress. *Example*: a reaction at equilibrium tends to proceed in the endothermic reaction when the temperature is raised.

lead-acid cell A secondary cell in which the electrodes are lead and the electrolyte is dilute sulfuric acid.

Lewis acid An electron pair acceptor. *Examples*: H^+; Fe^{3+}; BF_3.

Lewis base An electron pair donor. *Examples*: OH^-; H_2O; NH_3.

Lewis diagram (for atoms and ions) The chemical symbol of an element, with a dot for each valence electron.

Lewis formula (for an ionic compound) A representation of the structure of an ionic compound, showing the formula unit of ions in terms of their Lewis diagrams.

Lewis structure A diagram showing how electron pairs are shared between atoms in a molecule. *Examples*:

$$H—\ddot{\underset{..}{Cl}}: \qquad :\!\ddot{O}=C=\ddot{O}\!:$$

ligand A group attached to the central metal ion in a complex; a *polydentate ligand* occupies more than one binding site.

ligand field splitting Δ The energy difference of the e and t orbitals in a complex.

ligand field theory The theory of bonding in d-metal complexes.

limiting reactant The reactant that governs the theoretical yield of product in a given reaction.

linkage isomers Isomers that differ in the identity of the atom that a ligand uses to attach to the metal ion.

lipid A naturally occurring organic compound that dissolves in hydrocarbons but not in water. *Examples*: fats; steroids; terpenes.

liquid A fluid form of matter that takes the shape of the part of the container it occupies.

liquid crystal A substance that flows like a liquid but is composed of molecules that lie in a moderately orderly array. Liquid crystals may be nematic, smectic, or cholesteric, depending on the arrangement of the molecules.

London forces The force of attraction that arises from the interaction between instantaneous electric dipoles on neighboring polar or nonpolar molecules.

lone pair A pair of valence electrons that is not involved in bonding.

low-spin complex A d^n complex with the minimum number of unpaired electron spins.

magic numbers The numbers of protons or neutrons that correlate with enhanced nuclear stability. *Examples*: 2; 8; 20; 50; 82; 126.

magnetic quantum number m_l The quantum number that identifies the individual orbitals of a subshell of an atom.

main-group elements The members of the *s* and *p* blocks of the periodic table. *Examples*: Li; Mg; C; Br.

many-electron atom An atom with more than one electron.

mass m The quantity of matter in a sample.

mass concentration The mass of solute per liter of solution.

mass number A The total number of nucleons (protons plus neutrons) in the nucleus of an atom. *Example*: ${}^{14}_{6}C$, with mass number 14, has 14 nucleons (6 protons and 8 neutrons).

mass percentage The mass of a substance present in a sample, expressed as a percentage of the total mass of the sample.

mass spectrometry Technique for measuring the masses and abundances of atoms and molecules by passing a beam of ions through a magnetic field.

matter Anything that has mass and takes up space.

Maxwell distribution of molecular speeds The percentage of molecules that move at any given speed in a gas at a specified temperature.

mean free path The average distance that a molecule travels between collisions.

melting temperature The temperature at which a substance melts. The *normal melting point* is the melting point under a pressure of 1 atm.

meniscus The curved surface that a liquid forms in a narrow tube.

mesophase A state of matter showing some of the properties of both a liquid and a solid (a liquid crystal).

metal (1) A substance that conducts electricity, has a metallic luster, is malleable and ductile, forms cations, and has basic oxides. (2) A metal consists of cations held together by a sea of electrons. *Examples*: iron; copper; uranium.

metallic conduction The conduction of electricity by the movement of electrons.

metallic conductor An electronic conductor with a resistance that increases as the temperature is raised.

metalloid An element that has the physical appearance and properties of a metal but behaves chemically like a nonmetal. *Examples*: arsenic; polonium.

metathesis See *double replacement reaction.*

minerals Substances that are mined; more generally, inorganic substances.

mixture A type of matter that consists of more than one substance and may be separated into its components by making use of the different physical properties of the substances present.

moderator A substance that slows neutrons. *Examples*: graphite; heavy water.

molality The number of moles of solute per kilogram of solvent.

molar The quantity per mole. *Examples*: molar mass, the mass per mole; molar volume, the volume per mole.

molar mass (1) The mass per mole of atoms of an element (formerly, atomic weight). (2) The mass per mole of molecules of a compound (formerly, molecular weight). (3) The mass per mole of formula units of an ionic compound (formerly, formula weight).

molar concentration [X] The amount (in moles) of solute per liter of solution (X denotes any species).

molar solubility S The molar concentration of a saturated solution of a substance.

molarity Molar concentration.

mole (mol) The SI base unit for the amount of substance: 1 mol is the number of atoms in exactly 12 g of carbon-12.

mole fraction x The number of moles (of molecules, atoms, or ions) of a substance in a mixture expressed as a fraction of the total number of moles of molecules, atoms, and ions in the mixture.

molecular compound A compound that consists of molecules. *Examples*: water; sulfur hexafluoride; benzoic acid.

molecular formula A combination of chemical symbols and subscripts showing the actual numbers of atoms of each element present in a molecule. *Examples*: H_2O; SF_6; $C_6H_{12}O_6$.

molecular orbital (1) An orbital that spreads over all the atoms in a molecule. (2) A region of space in which there is a high probability of finding an electron in a molecule. (3) A mathematical expression that gives the probability of finding an electron at each point in a molecule.

molecular-orbital energy-level diagram A ladderlike diagram showing the relative energies of the molecular orbitals of a molecule.

molecular-orbital theory A theory of bonding in which electrons occupy molecular orbitals.

molecular potential-energy curve A graph showing the variation of the energy of a molecule as the bond length is changed.

molecular solid A solid consisting of a collection of individual molecules. *Examples*: glucose; aspirin; sulfur.

molecular weight See *molar mass.*

molecularity The number of reactant molecules (or free atoms) taking part in an elementary reaction. See *unimolecular reaction; bimolecular reaction.*

molecule (1) The smallest particle of a compound that possesses the chemical properties of the compound. (2) A definite and distinct, electrically neutral group of bonded atoms. *Examples*: H_2; NH_3; CH_3COOH.

monatomic gas A gas composed of single atoms. *Examples*: helium; radon.
monatomic ion An ion formed from a single atom. *Examples*: Na^+; Cl^-.
Mond process The purification of nickel by the formation and decomposition of nickel tetracarbonyl.
monomer A small molecule from which a polymer is formed. *Examples*: $CH_2{=}CH_2$ for polyethylene; $NH_2(CH_2)_6NH_2$ for nylon.
monoprotic acid A Brønsted acid with one acidic hydrogen atom. *Example*: CH_3COOH.
monosaccharide An individual C_6 unit from which carbohydrates are considered to be composed. *Example*: $C_6H_{12}O_6$ (glucose).

***n*-type semiconductor** See *semiconductor*.
native Occurring in an uncombined state as the element itself.
natural abundance (of an isotope) The abundance of an isotope in a sample of a naturally occurring material.
Nernst equation The equation expressing the cell potential in terms of the concentrations of the reagents taking part in the cell reaction; $E = E° - (RT/nF) \ln Q$.
net ionic equation The equation showing the net change, obtained by canceling the spectator ions in an ionic equation. *Example*: $Ag^+(aq) + Cl^-(aq) \rightarrow AgCl(s)$.
network solid A solid consisting of atoms linked together covalently throughout its extent. *Examples*: diamond; silica.
neutralization The reaction of an acid with a base to form a salt and a solvent molecule (typically water). *Example*: $HCl(aq) + NaOH(aq) \rightarrow NaCl(aq) + H_2O(l)$.
nickel-cadmium cell A rechargeable cell in which the reaction is the reduction of nickel(III) and the oxidation of cadmium.
noble gas A member of Group VIII of the periodic table (the helium family).
nodal plane A plane on which an electron will not be found.
node A point or surface at which an electron occupying an orbital will not be found.
nomenclature See *chemical nomenclature*.
nonaqueous solution A solution in which the solvent is not water. *Example*: sulfur in carbon disulfide.
nonbonding orbital A valence-shell atomic orbital that has not been used to form a bond to another atom.
nonelectrolyte A substance that dissolves to give a solution that does not conduct electricity. *Example*: sucrose.
nonmetal A substance that does not conduct electricity and is neither malleable nor ductile. *Examples*: all gases; phosphorus; sodium chloride.
nonpolar bond A covalent bond between atoms that have zero partial charges.
nonpolar molecule A molecule with zero electric dipole moment.
normal boiling point T_b (1) The boiling temperature when the pressure is 1 atm. (2) The temperature at which the vapor pressure of a liquid is 1 atm.
normal freezing point T_f The temperature at which a liquid freezes under 1 atm pressure.
NO_x Nitrogen oxides, especially in the context of atmospheric chemistry and pollution.
nuclear atom The structure of the atom proposed by Rutherford: a central, massive, positively charged nucleus surrounded by electrons.
nuclear binding energy The energy released when Z protons and $A - Z$ neutrons come together to form a nucleus. The greater the binding energy, the lower the energy of the nucleus.
nuclear chemistry The study of the structure of nuclei, of the changes this structure undergoes, and of the consequences of those changes for chemistry.
nuclear disintegration The partial breakup of a nucleus (including its fission). *Example*: ${}^{226}_{88}Ra \rightarrow {}^{222}_{86}Rn + {}^{4}_{2}\alpha$.
nuclear equation The equation that expresses a nuclear reaction. *Example*: ${}^{226}_{88}Ra \rightarrow {}^{222}_{86}Rn + {}^{4}_{2}\alpha$.
nuclear reaction A change that a nucleus undergoes (such as a nuclear transmutation).
nuclear reactor A device for achieving controlled nuclear fission.
nuclear transmutation The conversion of one element into another. *Example*: ${}^{12}_{6}C + \alpha \rightarrow {}^{16}_{8}O + \gamma$.
nucleic acid The product of a condensation of nucleotides.
nucleon A proton or a neutron; thus, either of the two principal components of a nucleus.
nucleoside A combination of a base and a deoxyribose molecule.
nucleotide A nucleoside with a phosphate group attached to the carbohydrate ring.
nucleus The small, positively charged particle at the center of an atom that is responsible for most of its mass.
nucleosynthesis The formation of elements.
nuclide A specific isotope of an element. *Examples*: ${}^{2}_{1}H$; ${}^{16}_{8}O$.

octet An ns^2np^6 valence electron configuration.
octet rule Atoms proceed as far as possible toward achieving an octet by sharing or transferring electrons.
oligopeptide A molecule consisting of several peptide units.
oligosaccharide A carbohydrate molecule consisting of a string of several bonded saccharide units (such as glucose molecules). *Examples*: cellulose; amylopectin.
optical activity The ability of a substance to rotate the plane of polarized light passing through it.
optical isomers Isomers that are related like an object and its mirror image.
order of reaction The power to which the concentration of a single substance is raised in a rate law. *Example*: if rate $= k[SO_2][SO_3]^{-1/2}$, then the reaction is of first order in SO_2 and of order $-\frac{1}{2}$ in SO_3.

ore The natural mineral source of a metal. *Example*: Fe_2O_3, hematite.

organic chemistry The branch of chemistry that deals with organic compounds.

organic compound A compound containing the element carbon and usually hydrogen. (The carbonates are normally excluded.)

osmometry The measurement of molar mass of a solute from observations of osmotic pressure.

osmosis The tendency of a solvent to flow through a semipermeable membrane into a more concentrated solution (a colligative property).

osmotic pressure [Π] The pressure needed to balance the flow of solvent through a semipermeable membrane.

Ostwald process The production of nitric acid by the catalytic oxidation of ammonia.

overall order The sum of the powers to which individual concentrations are raised in the rate law of a reaction. *Example*: if the rate $= k[SO_2][SO_3]^{-1/2}$, then the overall order is $\frac{1}{2}$.

overall reaction The net outcome of a sequence of reactions.

overlap The merging of orbitals belonging to different atoms of a molecule.

overpotential The additional potential that must be applied beyond the cell potential to cause appreciable electrolysis.

oxoacid An acid that contains oxygen. *Examples*: H_2CO_3; HNO_3; HNO_2; $HClO$.

oxoanion An anion of an oxoacid. *Examples*: HCO_3^-; CO_3^{2-}.

oxidation (1) Combination with oxygen. (2) A reaction in which an atom, an ion, or a molecule loses an electron. *Examples*: (1) $Mg(s) + \frac{1}{2}O_2(g) \rightarrow MgO(s)$; (2) $Mg(s) \rightarrow Mg^{2+}(s) + 2e^-$.

oxidation number N_{ox} The effective charge on an atom in a compound, calculated according to a set of rules. An increase in oxidation number corresponds to oxidation, and a decrease to reduction.

oxidizing agent A species that removes electrons from a species being oxidized (and is itself reduced) in a redox reaction. *Examples*: O_2; O_3; MnO_4^-; Fe^{3+}.

***p*-type semiconductor** See *semiconductor*.

paired electrons Two electrons with opposed spins (↑↓).

parallel electrons Electrons with spins in the same direction (↑↑).

paramagnetic A substance that is pulled into a magnetic field; a paramagnetic substance is composed of atoms or molecules with unpaired electrons. *Examples*: O_2; $[Fe(CN)_6]^{3-}$.

parent nucleus In a nuclear reaction, the nucleus that undergoes disintegration.

partial charge A charge arising from small shifts in the distributions of electrons.

partial pressure P_x The pressure a gas in a mixture would exert if it alone occupied the container.

parts per million (ppm) (1) The ratio of the mass of a solute to the mass of the solvent, multiplied by 10^6. (2) The mass percentage composition multiplied by 10^4. (Parts per billion, ppb, the mass ratio multiplied by 10^9, may also be used.)

passivation Protection from further reaction by a surface film. *Example*: aluminum in air.

Pauli exclusion principle See *exclusion principle*.

percentage yield The percentage of the theoretical yield of a product achieved in practice.

penetration The possibility that an electron (particularly an *s* electron) may be found inside the inner shells of an atom and hence close to the nucleus.

peptide A molecule formed by a condensation reaction between amino acids.

peptide bond The —CO—NH— group.

period A horizontal row in the periodic table; the number of the period is equal to the principal quantum number of the valence shell of the atoms.

periodic table A chart in which the elements are arranged in order of atomic number and divided into groups and periods in a manner that shows the relationships between the properties of the elements.

pH The negative logarithm of the hydronium ion concentration in a solution, $pH = -\log [H_3O^+]$; at 25°C, $pH < 7$ indicates an acidic solution, $pH = 7$ a neutral solution, and $pH > 7$ a basic solution.

pH curve The graph of the pH of a reaction mixture against volume of titrant added in an acid-base titration.

phase A particular state of matter. A substance may exist in solid, liquid, and gas phases and, in certain cases, in more than one solid phase. *Examples*: white and gray tin are two solid phases of tin; ice, liquid, and vapor are three phases of water.

phase boundary A line separating two areas in a phase diagram; the points on a phase boundary represent the conditions at which the two adjoining phases are in dynamic equilibrium.

phase diagram A summary in graphical form of the conditions of temperature and pressure at which the various phases of a substance exist.

phenol An organic compound in which a hydroxyl group is attached directly to an aromatic ring (Ar—OH). *Example*: C_6H_5OH, phenol.

phenolic resin A polymer resulting from the condensation reaction between phenol and formaldehyde.

photochemical reaction A reaction caused by light. *Example*:

$$H_2(g) + Cl_2(g) \xrightarrow{\text{light}} 2HCl(g)$$

photoelectric effect The emission of electrons from the surface of metals when electromagnetic radiation strikes it.

photon A particlelike packet of electromagnetic radiation. The energy of a photon of frequency ν is $E = h\nu$.

physical property A characteristic that we observe or measure without changing the identity of the substance.

physical state The condition of being a solid, a liquid, or a gas at a particular temperature.

pi (π) bond (1) A bond formed by the side-to-side overlap of two p orbitals. (2) Two electrons in a π orbital.

pi (π) orbital A molecular orbital that looks like a p orbital when viewed along the internuclear axis.

pK_a and pK_b The negative logarithms of the acid and base ionization constants: $pK = -\log K$. The larger the value of pK_a or pK_b, the weaker the acid or base, respectively.

Planck's constant h A fundamental constant of nature with the value 6.63×10^{-34} J·s/photon.

plasma (1) An ionized gas. (2) In biology, the colorless component of blood in which the corpuscles are dispersed.

poison (a catalyst) To inactivate a catalyst.

polar bond A covalent bond between atoms that have partial electric charges. *Examples*: H—Cl; O—S.

polar molecule A molecule with a nonzero electric dipole moment. *Examples*: HCl; NH_3; C_6H_5Cl.

polarizable Easily polarized (referring to an atom, molecule, or ion).

polarize To distort the electron cloud of an atom or ion.

polarized light Plane-polarized light is light in which the wave motion occurs in a single plane.

polarizing power The ability of an ion to polarize a neighboring atom or ion.

polyamide A polymer (nylon) formed by the condensation polymerization of a dicarboxylic acid and a diamine.

polyatomic ion A polyatomic molecular ion. *Examples*: NH_4^+; NO_3^-; SiF_6^{2-}.

polyatomic molecule A molecule that consists of more than two atoms. *Examples*: O_3; $C_{12}H_{22}O_{11}$.

polycyclic compound An aromatic compound that contains two or more benzene rings that are joined together at two neighboring carbon atoms. *Example*: naphthalene.

polyester A polymer formed by the condensation reaction between a carboxylic acid and an alcohol.

polymer A chain of covalently linked monomers. *Examples*: polyethylene; nylon.

polymorph Each of the various solid forms of a substance. *Examples*: diamond and graphite; calcite and aragonite.

polynucleotide A polymer built from nucleotide units. *Examples*: DNA; RNA.

polypeptide A polymer formed by the condensation reaction of amino acids.

polyprotic acid or base A Brønsted acid or base that can donate or accept more than one proton. (A polyprotic acid is sometimes called a "polybasic acid.") *Examples*: H_3PO_4 (triprotic acid); N_2H_4 (diprotic base).

polysaccharide A chain of many glucose units linked together. *Examples*: cellulose; amylose.

positron A fundamental particle with the same mass as an electron but with opposite charge.

positron emission A mode of radioactive decay in which a nucleus emits a positron.

potential difference An electric potential difference between two points is a measure of the work that must be done to move an electric charge from one point to the other. Potential difference is measured in volts, V.

power The rate of supply of energy.

precipitation The process in which a solute comes out of solution rapidly as a finely divided powder, called a *precipitate*.

precipitation reaction A reaction in which a solid product is formed when two solutions are mixed. *Example*: $KBr(aq) + AgNO_3(aq) \rightarrow KNO_3(aq) + AgBr(s)$.

precise measurements Measurements that have only a small random error and are reproducible in repeated trials.

precision Freedom from random error.

pre-exponential factor See *frequency factor*.

pressure P Force per unit area.

primary cell A galvanic cell that produces electricity from chemicals sealed within it at manufacture.

primary structure The sequence of amino acids in the polypeptide chain of a protein.

principal quantum number n The quantum number that specifies the energy of an electron in a hydrogen atom and labels the shells of the atom.

product A substance formed in a chemical reaction.

promotion (of an electron) The conceptual excitation of an electron to an orbital of higher energy.

propagation A series of steps in a chain reaction in which one chain carrier reacts with a reactant molecule to produce another carrier. *Example*: $Br\cdot + H_2 \rightarrow HBr + H\cdot$, $H\cdot + Br_2 \rightarrow HBr + Br\cdot$.

properties The characteristics of matter. *Examples*: vapor pressure; color; density; temperature.

protonation Proton transfer to a Brønsted base. *Example*: $2H_3O^+(aq) + S^{2-}(s) \rightarrow H_2S(g) + 2H_2O(l)$.

pseudo–first-order reaction A reaction with a rate law that is effectively first order because one substance has a virtually constant concentration.

pseudonoble-gas configuration A noble gas core surrounded by a complete d subshell. *Examples*: Cu^+, $[Ar]3d^{10}$.

purex process A process based on solvent extraction that is used to separate plutonium and uranium.

pyranose A six-membered ring form of a carbohydrate molecule.

pyrometallurgical process The extraction of metals by using reactions at high temperatures. *Example*:

$$Fe_2O_3(s) + 3CO(g) \xrightarrow{\Delta} 2Fe(l) + 3CO_2(g)$$

qualitative analysis Analysis of a sample to identify the species it contains.

quantitative analysis Analysis of a sample to determine the amount of a species in a sample.

quantization The restriction of a property to certain values. *Examples*: the quantization of energy; angular momentum.

quantum A packet of energy.

quantum mechanics The description of matter that takes into account the fact that the energy of an object may be changed only in discrete steps.

quantum number A number that labels the state of an electron and specifies the value of a property. *Example*: principal quantum number n.

quaternary ammonium ion An ion of the form NR_4^+, where R denotes an alkyl group or hydrogen (the four groups may be different).

quaternary structure The manner in which neighboring polypeptide units stack together to form a protein molecule.

racemic mixture A mixture containing equal concentrations of two enantiomers.

radical An atom, molecule, or ion with at least one unpaired electron. *Examples*: NO, $\cdot\ddot{O}\cdot$, $\cdot CH_3$.

radical chain reaction A chain reaction propagated by radicals.

radical polymerization A polymerization procedure that utilizes a radical chain reaction.

radical scavenger An impurity that reacts with radicals and inhibits a chain reaction.

radioactive series A stepwise nuclear decay path in which α and β particles are successively ejected and which terminates at a stable nuclide (often of lead).

radioactivity The spontaneous emission of radiation by nuclei.

radiocarbon dating The determination of the age of an archeological artifact by measuring the radioactivity of carbon-14.

random error An error that varies randomly from measurement to measurement, sometimes giving a high value and sometimes a low one.

Raoult's law The vapor pressure of a solution of a nonvolatile solute is directly proportional to the mole fraction of the solvent in the solution: $P = x_{\text{solvent}} \times P^*$, where P^* is the vapor pressure of the pure solvent.

rate The change in a property per unit time.

rate constant k The constant of proportionality in a rate law.

rate-determining step The slowest step in a multistep reaction and therefore the step that governs the rate of the overall reaction. *Example*: the step $O + O_3 \rightarrow 2O_2$ in the decomposition of ozone.

rate law An equation expressing the instantaneous reaction rate in terms of the concentrations, at that instant, of the substances taking part in the reaction. *Example*: rate $= k[NO_2]^2$.

rate of reaction See *reaction rate; instantaneous rate.*

reactant A starting material in a chemical reaction; a reagent taking part in a specified reaction.

reaction enthalpy The change of enthalpy for the reaction exactly as the chemical equation is written. *Example*: $CH_4(g) + 2O_2(g) \rightarrow CO_2(g) + 2H_2O(l)$ $\Delta H = -890$ kJ.

reaction intermediate A species that is produced and consumed during a reaction but does not occur in the overall chemical equation.

reaction mechanism The pathway that is proposed for an overall reaction and accounts for the experimental rate law.

reaction order See *order of reaction.*

reaction profile The variation in potential energy that occurs as two reactants meet, form an activated complex, and separate as products.

reaction quotient Q_c The ratio of the product of the concentrations of the products to that of the reactants (defined like the equilibrium constant) at an arbitrary stage of a reaction. *Example*: $N_2(g) + 3H_2(g) \rightarrow 2NH_3(g)$ $Q_c = [NH_3]^2/([N_2][H_2]^3)$.

reaction rate The change in concentration of a substance divided by the time it takes for the change to occur.

reaction sequence A series of reactions in which products of one reaction take part as reactants in the next. *Example*: $2C(s) + O_2(g) \rightarrow 2CO(g)$, followed by $2CO(g) + O_2(g) \rightarrow 2CO_2(g)$.

reaction stoichiometry The quantitative relation between the amounts of reactants consumed and products formed in chemical reactions as expressed by the balanced chemical equation for the reaction.

reagent A substance or a solution that reacts with other substances.

recrystallization Purification by repeated dissolving and crystallization.

redox couple The oxidized and reduced forms of a substance taking part in a reduction or oxidation half-reaction. The notation is oxidized species/reduced species. *Example*: H^+/H_2.

redox reaction A reaction in which oxidation and reduction occur. *Example*: $S(s) + 3F_2(g) \rightarrow SF_6(g)$.

reducing agent The species that supplies electrons to a substance being reduced (and is itself oxidized) in a redox reaction. *Examples*: H_2; H_2S; SO_3^{2-}.

reducing sugar Sugars that reduce Fehling's solution and Tollens' reagent. *Examples*: glucose; fructose.

reduction (1) The removal of oxygen from, or the addition of hydrogen to, a compound. (2) A reaction in which an atom, an ion, or a molecule gains an electron. *Example*: $Cl_2(g) + 2e^- \rightarrow 2Cl^-(s)$.

reforming reaction (1) In hydrogen manufacture, a reaction in which the hydrocarbon is converted to carbon monoxide and hydrogen over a nickel catalyst. (2) In oil refining, a reaction in which the number of carbon atoms in the reactant molecule is unchanged but more double bonds and aromatic rings are introduced. *Example*: (1) $CH_4(g) + H_2O(g) \rightarrow CO(g) + 3H_2(g)$; (2) $C_6H_{11}CH_3(g) \rightarrow C_6H_5CH_3(g) + 3H_2(g)$.

representative elements The elements in Periods 2 and 3 of the periodic table.

residue An amino acid in a polypeptide chain.

resistance (electrical) A measure of the ability of matter to conduct electricity: the lower the resistance, the better it conducts.

resonance A blending of Lewis structures into a single composite, hybrid structure. *Example*:

$$:\ddot{O}-\ddot{S}=\ddot{O} \longleftrightarrow \ddot{O}=\ddot{S}-\ddot{O}:$$

resonance hybrid The composite structure that results from resonance.

retardation A step in a chain reaction that diverts chain carriers from the formation of products. *Example*: $H\cdot + HBr \rightarrow H_2 + Br\cdot$.

reverse osmosis The passage of solvent out of a solution when a pressure greater than the osmotic pressure is applied on the solution side of a semipermeable membrane.

roast To heat a metal ore in air. *Example*:

$$2CuFeS_2(s) + 3O_2(g) \xrightarrow{\Delta} 2CuS(s) + 2FeO(s) + 2SO_2(g)$$

in the extraction of copper.

rock-salt structure A crystal structure the same as that of a mineral form of sodium chloride.

Rydberg constant $\mathcal{R}$ The constant that occurs in the formula for the frequencies of the lines in the spectrum of atomic hydrogen; $\mathcal{R} = 3.29 \times 10^{15}$ Hz.

sacrificial anode See *cathodic protection.*

saline hydride A saltlike compound of hydrogen and a strongly electropositive metal. *Examples*: KH; CaH_2.

salt The product (other than water) of the reaction between an acid and a base. *Examples*: NaCl; K_2SO_4.

salt bridge A bridge-shaped tube containing a concentrated salt (such as potassium chloride or potassium nitrate in a jelly) that acts as an electrolyte; the salt bridge provides a conducting path between two compartments of an electrochemical cell.

sample A representative part of a whole.

saturated hydrocarbon A hydrocarbon with no carbon-carbon multiple bonds. *Example*: CH_4; CH_3CH_3.

saturated solution A solution in which the dissolved and undissolved solute are in dynamic equilibrium.

science The systematically collected and organized body of knowledge based on experiment, observation, and careful reasoning.

scintillation counter A device for detecting and measuring radioactivity that makes use of the fact that certain substances give a flash of light when exposed to radiation.

second law of thermodynamics A spontaneous change is accompanied by an increase in the total entropy of the system and its surroundings.

secondary cell A galvanic cell that must be charged (or recharged) by using another supply before it can be used.

secondary structure The manner in which a polypeptide chain is coiled. *Examples*: α helix; β-pleated sheet.

selective precipitation The precipitation of one compound in the presence of other, more soluble compounds.

self-sustaining fission Induced nuclear fission that, once it is initiated, can continue even after the supply of neutrons from outside is discontinued.

semiconductor An electronic conductor with a resistance that decreases as the temperature is raised. In an *n-type semiconductor,* the current is carried by electrons in a largely empty band; in a *p-type semiconductor,* the conduction is a result of electrons missing from otherwise filled bands.

semipermeable membrane A membrane that allows only certain types of molecules or ions to pass.

sequestration (1) The wrapping up of one ion by another. (2) The formation of a complex between a cation and a polydentate ligand. *Example*: Ca^{2+} and $O_3POPO_2OPO_3^{5-}$.

series A family of spectral lines arising from transitions that have one state in common. *Example*: Balmer series in the spectrum of atomic hydrogen.

shell All the orbitals of a given principal quantum number. *Example*: the single 2*s* and three 2*p* orbitals of the shell with $n = 2$.

shielding The repulsion that is experienced by an electron in an atom and that arises from the other electrons present and opposes the attraction exerted by the nucleus.

shift reaction A reaction between carbon monoxide and water:

$$CO(g) + H_2O(g) \xrightarrow{400°C,\ Fe/Cu} CO_2(g) + H_2(g)$$

The reaction is used in the manufacture of hydrogen.

SI *(Système International)* The International System of units, an extension and rationalization of the metric system.

side chain A hydrocarbon substituent on a hydrocarbon chain.

sigma (σ) bond Two electrons in a σ orbital.

sigma (σ) orbital An orbital with cylindrical symmetry with respect to the internuclear axis.

significant figures (in a measurement) The digits in the measurement, up to and including the first uncertain digit in scientific notation. *Example*: 0.026 mL, a measurement with two significant figures.

single bond (1) A shared electron pair. (2) Two electrons in a σ orbital.

skeletal equation An unbalanced equation that summarizes the qualitative information about the reaction. *Example*: $H_2 + O_2 \rightarrow H_2O$ ⚠.

smelt To melt a metal ore with a reducing agent. *Example*: $CuS(l) + O_2(g) \xrightarrow{\Delta} Cu(l) + SO_2(g)$.

solid A rigid form of matter that maintains the same shape whatever the shape of its container.

solubility The concentration of a saturated solution of a substance.

solubility constant K_s The product of relative ionic molar concentrations in a saturated solution; the dissolution equilibrium constant. *Example*: $Hg_2Cl_2(s) \rightleftharpoons Hg_2^{2+}(aq) + 2Cl^-(aq)$ $K_s = [Hg_2^{2+}][Cl^-]^2$.

solubility quotient Q_s The analogue of the solubility product, but with the molar concentrations not necessarily those at equilibrium.

solubility rules A summary of the solubility pattern of a range of common compounds in water. (See Table 3.2.)

solute A dissolved substance.

solution A homogeneous mixture, generally of a substance of smaller abundance (the solute) in one of much greater abundance (the solvent).

solvated Surrounded by and linked to solvent molecules. (Hydration is a special case occurring when the solvent is water.)

solvent The most abundant component of a solution.

solvent extraction A separation process that makes use of the differing solubilities of a mixture's components in various solvents.

specific enthalpy (of a fuel) The enthalpy of combustion per gram (without the negative sign).

specific heat capacity The heat capacity per gram.

spectator ion An ion that is present but remains unchanged during a reaction. *Examples*: Na^+ and NO_3^- in $NaCl(aq) + AgNO_3(aq) \rightarrow NaNO_3(aq) + AgCl(s)$.

spectral line Radiation of a single wavelength emitted or absorbed by an atom or molecule.

spectrochemical series Ligands ordered according to the strength of the ligand field splitting they produce.

spectroscopy The analysis of the electromagnetic radiation emitted or absorbed by substances.

spectrum The frequencies or wavelengths of the electromagnetic radiation emitted or absorbed by substances.

spin The intrinsic angular momentum of an electron; the spin cannot be eliminated and may occur in only two senses, denoted ↑ and ↓.

spin magnetic quantum number m_s The quantum number that distinguishes the two spin states of an electron: $m_s = +\frac{1}{2}$ (↑) and $m_s = -\frac{1}{2}$ (↓).

spontaneous change A change that has a natural tendency to occur without needing to be driven by an external influence. *Examples*: a gas expanding; a hot object cooling; methane burning.

stable See *thermodynamically unstable compound*.

stability constant See *formation constant*.

standard cell potential $E°$ The cell potential when the concentration of each type of ion taking part in the cell reaction is 1 mol/L and all the gases are at 1 atm pressure. The standard cell potential is the sum of its two standard electrode potentials: $E° = E°(\text{cathode}) + E°(\text{anode})$.

standard electrode potential $E°$(ox/red) The contribution of an electrode to the standard cell potential.

standard enthalpy of combustion $\Delta H_c°$ The change of enthalpy per mole of substance when it burns (reacts with oxygen) completely under standard conditions.

standard enthalpy of formation $\Delta H_f°$ The standard enthalpy per mole of compound for the compound's synthesis from its elements in their most stable form at 1 atm and the specified temperature.

standard free energy of formation $\Delta G_f°$ The standard reaction free energy per mole for the formation of a compound from its elements in their most stable form.

standard hydrogen electrode (SHE) A hydrogen electrode that is in its standard state (hydrogen ions at concentration 1 mol/L and hydrogen pressure 1 atm) and is defined as having $E° = 0$.

standard molar entropy $S°$ The entropy per mole of a pure substance at 1 atm pressure.

standard oxidation potential The standard electrode potential of a couple with the half-reaction taken as an oxidation.

standard reaction enthalpy $\Delta H°$ The difference in enthalpy between the products of a reaction in their standard states and the reactants in their standard states.

standard reaction entropy $\Delta S°$ The difference in entropy between the products of a reaction in their standard states and the reactants in their standard states.

standard reaction free energy $\Delta G°$ The difference in free energy between the products of a reaction in their standard states and the reactants in their standard states.

standard reduction potential $E°$(ox/red) The standard electrode potential of a couple with the half-reaction taken as a reduction.

standard state The pure form of a substance at 1 atm.

standard temperature and pressure (STP) 0°C (273.15 K) and 1 atm (101.325 kPa).

state property (or state function) A property of a substance that is independent of how the sample was prepared. *Examples*: pressure; enthalpy; entropy; color.

stepwise formation constants $K_{f1}, K_{f2}, \ldots$ The formation constants for the successive addition of Lewis bases to a Lewis acid.

stereoisomers Isomers in which atoms have the same partners arranged differently in space.

Stern-Gerlach experiment The demonstration of the quantization of electron spin by passing a beam of atoms through a magnetic field.

Stock number (1) A Roman numeral equal to the number of electrons lost by the atom on formation of a compound and sometimes added in parenthesis to a symbol or a name. (2) The oxidation number of the element. *Example*: Cu(II) in compounds of copper(II) containing Cu^{2+}.

stoichiometric coefficients The numbers multiplying chemical formulas in a chemical equation. *Examples*: 1, 1, and 2 in $H_2 + Br_2 \rightarrow 2HBr$.

stoichiometric point The stage in a titration when exactly the right volume of solution needed to complete the reaction has been added.

stoichiometric proportions Reactants in the same proportions as their coefficients in the chemical equation. *Example*: equal amounts of H_2 and Br_2 in the reaction above.

stoichiometric relation An expression that equates the relative amounts of reactants and products that participate in a reaction. *Example*: 1 mol H_2 = 2 mol HBr.

stoichiometry See *reaction stoichiometry*.

strong acids and bases Acids and bases that are fully ionized in solution. *Examples*: HCl, $HClO_4$ (strong acids); NaOH, $Ca(OH)_2$ (strong bases).

strong-field ligand A ligand that produces a large ligand field splitting and that lies above H_2O in the spectrochemical series.

structural formula A chemical formula that shows the groupings of atoms in a compound.
structural isomers Isomers in which the atoms have different partners.
subatomic particle A particle smaller than an atom. *Examples*: electrons; protons; neutrons.
sublimation The direct conversion of a solid to a vapor without first forming a liquid.
sublimation vapor pressure The vapor pressure of a solid.
subshell All the atomic orbitals of a given shell of an atom with the same value of the quantum number l. *Example*: the five $3d$ orbitals of an atom.
substance A single, pure type of matter.
substrate The molecule on which an enzyme acts.
substituent Atoms or groups that have replaced hydrogen atoms in an organic molecule.
substitution reaction (1) A reaction in which an atom (or a group of atoms) replaces an atom in the original molecule. (2) In complexes, a reaction in which one Lewis base expels another and takes its place. *Examples*:

(1) $C_6H_5OH + Br_2 \rightarrow BrC_6H_4OH + HBr$
(2) $[Fe(H_2O)_6]^{3+}(aq) + 6CN^-(aq) \rightarrow [Fe(CN)_6]^{3-}(aq) + 6H_2O(l)$

superconductor An electronic conductor that conducts electricity with zero resistance. A *high-temperature superconductor* is a substance that is superconducting above about 100 K.
superfluidity The ability to flow without viscosity.
surface-active agent See *surfactant*.
surface tension γ A measure of the force that must be applied to surface molecules so that they experience the same force as molecules in the interior of the liquid.
surfactant A substance that accumulates at the surface of a solution (a component of detergents). *Example*: the stearate ion of soaps.
surroundings The region outside a system.
suspension A mist of small particles.
syndiotactic polymer A polymer in which the substituents alternate on either side of the chain.
synthesis A reaction in which a substance is formed from simpler starting materials. *Example*: $N_2(g) + 3H_2(g) \rightarrow 2NH_3(g)$.
synthesis gas A mixture of carbon monoxide and hydrogen produced by the catalyzed reaction of a hydrocarbon and water.
system The reaction vessel and its contents in which there is a particular interest.
systematic error An error that persists in a series of measurements and does not average out.
systematic name The name of a compound that reveals which elements are present (and, in its most complete form, how the atoms are arranged). *Example*: methylbenzene is the systematic name for toluene.

temperature T (1) How hot or cold a sample is. (2) The intensive property that determines the direction in which heat will flow between two objects in contact.
termination A step in a chain reaction in which chain carriers combine to form products. *Example*: $2Br\cdot \rightarrow Br_2$.
tertiary structure The shape into which the α helix and β-pleated sheet sections of a polypeptide are twisted as a result of interactions between peptide groups lying in different parts of the primary structure.
theoretical yield The maximum quantity of product that can be obtained, according to the reaction stoichiometry, from a given quantity of a specified reactant.
thermal decomposition See *decomposition*.
thermal motion The random, chaotic motion of atoms.
thermal pollution The damage caused to the environment by the waste heat of an industrial process.
thermite reaction The reduction of a metal oxide by aluminum. *Example*: $2Al(s) + Fe_2O_3(s) \rightarrow Al_2O_3(s) + 2Fe(l)$.
thermochemical equation An expression consisting of both the chemical equation and the reaction enthalpy for the chemical reaction exactly as written.
thermochemistry The study of the heat released or absorbed by chemical reactions; a branch of thermodynamics.
thermodynamically unstable compound (1) A compound with a thermodynamic tendency to decompose into its elements. (2) A compound with a positive free energy of formation.
thermodynamics The study of the transformation of energy from one form to another. See also *first law of thermodynamics; second law of thermodynamics.*
thermonuclear explosion An explosion resulting from uncontrolled nuclear fusion.
titrant The solution of known concentration added from a buret in a titration.
titration The analysis of composition by measuring the volume of one solution (the titrant) needed to react with a given volume of another solution (the analyte).
tracer An isotope that can be tracked from compound to compound in the course of a sequence of reactions.
transition A change of state. (1) In thermodynamics, a change of physical state. (2) In spectroscopy, a change of quantum state.
transition metal An element that belongs to the central part of the periodic table, between Groups II and III; a member of the d block of the periodic table. *Examples*: vanadium; iron; gold.
triol An organic compound with three hydroxyl groups.
triple bond (1) Three electron pairs shared by two neighboring atoms. (2) One σ bond and two π bonds between neighboring atoms.
triple point The point where three phase boundaries meet in a phase diagram; under the conditions represented by the triple point, all three adjoining phases coexist in dynamic equilibrium.

ultraviolet radiation Electromagnetic radiation with a

higher frequency (shorter wavelength) than that of violet light.

unbranched alkane An alkane with no side chains, in which all the carbon atoms lie in a linear chain.

uncertainty principle The more precisely we know the position of a particle, the less we can say about its speed (and vice versa); $\Delta x \times (m \times \Delta v) \geq h/4\pi$.

unimolecular reaction An elementary reaction in which a single reactant molecule changes into products. *Example*: $O_3(g) \rightarrow O(g) + O_2(g)$.

unit cell The smallest unit that, when stacked together repeatedly without any gaps, can reproduce the entire crystal.

unsaturated hydrocarbon A hydrocarbon with at least one carbon-carbon multiple bond. *Examples*: $CH_2{=}CH_2$; C_6H_6.

valence The number of bonds that an atom can form.

valence band The full band of orbitals in a solid.

valence-bond theory The description of bond formation in terms of the pairing of spins in the atomic orbitals of neighboring atoms.

valence electrons The electrons that belong to the valence shell.

valence shell The outermost shell of an atom. *Example*: the $n = 2$ shell of Period 2 atoms.

valence-shell electron-pair repulsion theory (VSEPR theory) A theory for predicting the shapes of molecules, using the fact that electron pairs repel one another.

van der Waals equation An approximate equation of state for a real gas in which two parameters represent the effects of intermolecular forces.

van der Waals forces Intermolecular forces.

van't Hoff equation (1) The equation for the osmotic pressure in terms of the concentration. (2) An equation that shows how the equilibrium constant varies with temperature.

van't Hoff *i* factor See *i factor*.

vapor The gaseous phase of a substance (specifically, of a substance that is a liquid or a solid at the temperature in question).

vapor pressure The pressure exerted by the vapor of a liquid (or a solid) when the vapor and the liquid (or solid) are in dynamic equilibrium.

vaporization The formation of a gas or vapor.

variable covalence The ability of an element to form different numbers of covalent bonds. *Examples*: S in SO_2 and SO_3.

visible radiation The electromagnetic radiation that can be detected by the human eye, with wavelengths in the range of 700 to 400 nm.

volatility The readiness with which a substance vaporizes. A substance is typically regarded as volatile if its boiling point is below 100°C.

volume V The amount of space a sample occupies.

volume percentage composition The volume of a substance present in a solution, expressed as a percentage of the total volume.

volumetric analysis An analytical method using measurement of volume.

water autoionization constant K_w The equilibrium constant for the autoionization of water, $H_2O(l) + H_2O(l) \rightleftharpoons H_3O^+(aq) + OH^-(aq)$:

$$K_w = [H_3O^+][OH^-]$$

wavefunction ψ The mathematical expression that summarizes where a particle is likely to be found.

wavelength λ The peak-to-peak distance of a wave.

wave-particle duality The combined wavelike and particlelike character of both radiation and matter.

weak acids and bases (1) Acids and bases that are only partially ionized in aqueous solutions at normal concentrations. (2) Acids and bases for which K_a and K_b, respectively, are small compared to 1. *Examples*: HF, CH_3COOH (weak acids); NH_3, CH_3NH_2 (weak bases).

weak-field ligand A ligand that produces a small ligand field splitting and that lies below NH_3 in the spectrochemical series.

weight The gravitational force on a sample.

work w The energy expended during the act of moving an object against an opposing force.

x-ray Electromagnetic radiation with wavelengths ranging from about 10 pm to about 1000 pm.

x-ray diffraction The analysis of crystal structures by studying the interference pattern in a beam of x-rays.

yield See *percentage yield; theoretical yield.*

zero-order reaction A reaction with a rate that is independent of the concentration of the reactant. *Example*: the catalyzed decomposition of ammonia.

Ziegler-Natta catalyst A catalyst (consisting of titanium(IV) chloride and an alkylaluminum compound) that is used to bring about polymerization at low temperatures and pressures.

zone refining Purifying a solid by repeatedly passing a molten zone along the length of a sample.

ANSWERS TO ODD-NUMBERED EXERCISES

CHAPTER 1

Classified Exercises

1.1 (a) Physical (b) Physical (c) Chemical

1.3 (a) Physical (b) Physical (c) Chemical

1.5 Underline: temperature, evaporation, humidity

1.7 (a) substance (b) mixture (c) substance

1.9 (a) homogeneous—distillation (b) heterogeneous—filtration (c) homogeneous—chromatography (d) homogeneous—recrystallization

1.11 (a) 0.250 kg (b) 0.250 ms (c) 0.149 dm (d) 2.48×10^{-2} g

1.13 (a) 1×10^{-6} m (b) 5.50×10^{-4} mm (c) 1.0×10^{2} mg

1.15 (a) 37.0°C (b) −40°F (c) 4.15 K or, to proper number of sf, 4 K

1.17 (a) velocity = m/s (b) acceleration = m/s^2 (c) force = $kg{\cdot}m/s^2$ (d) pressure = $kg/m{\cdot}s^2$

1.19 (a) $1 \times 10^{-6}\ m^3$ (b) 3.0×10^{-3} cm/μs (c) 25 mL

1.21 $1.0 \times 10^{2}\ mm^2$, $10.0 \times 10^{-6}\ m^3$, 1.00×10^{-1} L, and $25.0\ cm^3$

1.23 (a) Intensive (b) Extensive (c) Intensive

1.25 (a) $25 \times 10^{-3}\ m^3$ (b) 2.5×10^{3} mg/dL (c) $4.77 \times 10^{3}\ \mu g$ (d) 1.2 mL/s

1.27 (a) 757 mL (b) $8.64 \times 10^{4}\ kg/m^3$ (c) 62 lb/ft^3

1.29 (a) $5.8 \times 10^{2}\ mm^3/s$ (b) 1.91 $lb/in.^2$ (c) $1.0 \times 10^{2}\ m^2/g$

1.31 (a) 1.8 Å (b) 5.5×10^{3} Å (c) $\left(\frac{1\ \text{Å}}{10^{-1}\ nm}\right)$ converts nm to Å

1.33 (a) 10.4 mg/cL > 1.04 mg/cL (b) 1.36×10^{-2} mg/nL > 2.25×10^{-4} mg/nL (c) $4.5 \times 10^{2}\ \mu g > 45\ \mu g$

1.35 (a) 8.1×10^{3} mi (b) 250×10^{-3} nm or 2.50×10^{-1} nm

1.37 7×10^{3} metric tons

1.39 (a) 3 (b) 3 (c) 3 (d) ∞ (exact number)

1.41 6.60 mL, 30.0 mL

1.43 4.4 g (1.4 g limits sf)

1.45 1.645 g rounds to 1.65 g (0.21 g limits sf)

1.47 (a) 14×10^{8} m or 1.4×10^{9} m (b) 5.0×10^{2} s

1.49 3.6×10^{12} kg

1.51 40.3% Zn, 59.7% Cu

1.53 (a) 75% Au (b) 58% Au

1.55 1.08×10^{3} g Sn, 1.48×10^{3} g Pb

1.57 (a) 9.49% (b) $94.0\ \frac{\text{g salt}}{\text{kg soln}}$ (c) $105\ \frac{\text{g salt}}{\text{kg solvent}}$

1.59 2.3×10^{1} g solids

Unclassified Exercises

1.61 (a) 1 L > 1 qt (b) 1 μm > 1 nm (c) 1°C > 1°F (d) 1 dL > 1 mL

1.63 (a) C (b) C (c) P

1.65 (a) kg (b) nm (c) μm

1.67 1.6 m, 60.9 kg

1.69 237 mL

1.71 0.181 nm

1.73 1.0×10^{-5} cm

1.75 2.12×10^{-2} cm/s

1.77 yes

1.79 3.36 cm^3 or 3.4 cm^3 (2 sf)

1.81 (a) 1336 K (b) 27 K (c) −270.45°C (or −270.5)

1.83 The density of water (and the density of any other substance) is an *intensive* property. The ratio of two intensive properties must be an intensive property, ∴ specific gravity is an intensive property. The absence of units on specific gravity also implies this.

1.85 (a) $5.0 \times 10^{-2}\ nm^2$ (b) $100 \times 10^{-21}\ cm^3$ or $1 \times 10^{-19}\ cm^3$ (c) $2.83 \times 10^{2}\ dm^3$

1.87 $\frac{1.2 \times 10^{4}}{h}$ or $1.2 \times 10^{4}\ h^{-1}$

1.89 6.4×10^{1} kg and 5.1×10^{4} mg

1.91 8.65×10^{-4} mL

1.93 6.1 g cholesterol

1.95 1.0×10^{4} g pills

Case Questions

1. 11 metric tons of ore

3. Although the density of chalcocite is about twice that of the sand from which it is separated, the chalcocite floats to the top while the sand settles to the bottom. This apparent contradiction may be resolved by recognizing that the chalcocite is wetted by oil while the sand is not. Air, introduced at the bottom of the vat, creates bubbles which stick to the oily surface of the chalcocite; the density of the air/oil/mineral aggregate is significantly lower than that of the pure ore, due to the presence of the air bubbles. (Recall that the density of gases is significantly lower than that of liquids or solids.)

CHAPTER 2

Classified Exercises

2.1 (a) lithium, 1 or I, metal (b) gallium, 13 or III, metal (c) xenon, 18 or VIII, nonmetal

2.3 (a) I, non-metal (b) Cr, metal (c) Al, metal

2.5 lithium, Li, 3; sodium, Na, 11; potassium, K, 19; rubidium, Rb, 37; cesium, Cs, 55; francium, Fr, 87 (radioactive)
All react with water as follows (M = alkali metal):

$$2M(s) + 2H_2O(l) \rightarrow 2MOH(s) + H_2(g)$$

The melting points decrease from Li to Cs.

2.7 Group 2 or II: Be, 4; Mg, 12; Ca, 20; Sr, 38; Ba, 56; Ra, 88

2.9 (a) 6, 7, 6 (b) 17, 20, 17 (c) 17, 18, 17 (d) 92, 143, 92

2.11 0.549 g neutrons; 0.460 g protons; 0.000251 g electrons

2.13 (a) 4 (b) 1.00745 u; 5.5×10^{-2}%

2.15 (a) number of protons = 10 p; number of neutrons = 8 n; number of electrons = 10 e^- (b) mass of protons = 1.673×10^{-23} g; mass of neutrons = 1.340×10^{-23} g; mass of electrons = 9.109×10^{-27} g (c) 0.4446 neutrons (or, 44.46% of your body mass is due to neutrons)

2.17 79.91 g/mol

2.19 (a) 1 Au atom = 3.2707×10^{-22} g; 3.1×10^{21} Au atoms (b) 1 Kr atom = 1.3915×10^{-22} g; 1.34×10^{23} Kr atoms (c) 1 ^{2}H atom = 3.34×10^{-24} g; 3.0×10^{20} ^{2}H atoms

2.21 volume of atom = 10^6 m^3; volume of atom in cubic feet = 3.53×10^7 ft^3; football field volume = $(300 \text{ ft})^3 = 2.70 \times 10^7$ ft^3

2.23 (a) 8.8×10^{-15} mol (b) 3.6×10^6 years

2.25 (a) 55.85 g/mol (b) 131.30 g/mol (c) 22.99 g/mol

2.27 (a) 0.0801 mol ^{35}Cl (b) 0.0349 mol Cu (c) 1.60×10^{-7} mol Fe

2.29 (a) C is greater (b) Au is greater (c) Same

2.31 (a) Na_3AlF_6 (b) $KClO_3$ (c) NH_6PO_4 or $NH_4H_2PO_4$

2.33 (a) 199.90 g/mol (b) 262.87 g/mol (c) 16.04 g/mol

2.35 (a) 0.0650 mol CCl_4, 3.91×10^{22} molecules CCl_4 (b) 1.29×10^{-5} mol HI, 7.77×10^{18} molecules HI (c) 1.18×10^{-7} mol N_2H_4, 7.08×10^{16} molecules N_2H_4 (d) 1.46 mol sucrose, 8.79×10^{23} molecules sucrose (e) 0.146 mol O, 8.77×10^{22} atoms O

2.37 (a) 0.0140 mol Ag^+ (b) 2.10 mol UO_3 (c) 7.75×10^{-5} mol Cl^- (d) 5.89×10^{-3} mol H_2O

2.39 (a) 3.5×10^{-9} mol testosterone (b) % C = 79.1% C, % H = 9.8% H, % O = 11.1% O

2.41 (a) 2.99×10^{-23} g H_2O (b) 3.34×10^{22} molecules H_2O

2.43 (a) 0.0186 mol $CuBr_2 \cdot 4H_2O$ (b) 0.0372 mol Br^- (c) 4.48×10^{22} molecules H_2O (d) 0.215

2.45 $C_6H_6Cl_6$

2.47 $C_8H_{10}N_4O_2$

2.49 (a) anion, S^{2-} (b) cation, K^+ (c) cation, Ag^+ (d) anion, Cl^-

2.51 (a) 1 p, 1 n, 0 e^- (b) 4 p, 5 n, 2 e^- (c) 35 p, 45 n, 36 e^-

2.53 (a) Na_2O (b) K_2SO_4 (c) $Zn(NO_3)_2$

2.55 (a) bisulfate ion, hydrogensulfate ion (b) mercurous ion, mercury(I) ion (c) cyanide ion (d) bicarbonate ion, hydrogencarbonate ion

2.57 (a) $Ca_3(PO_4)_2$ (b) $Al_2(SO_4)_3$ (c) Ca_3N_2

2.59 (a) copper(II) nitrate hexahydrate (b) neodymium(III) chloride hexahydrate (c) nickel(II) fluoride tetrahydrate

2.61 (a) N_2O_4 (b) Cl_2O_7 (c) S_2Cl_2

2.63 (a) alcohol (b) carboxylic acid (c) hydrocarbon

2.65 (a) sulfuric acid (b) nitric acid (c) acetic acid (d) phosphoric acid

Unclassified Exercises

2.67 94.70 u

2.69 ^{10}B = 26.2% abundant, ^{11}B = (100 − 26.2)% = 73.8% abundant

2.71 10.4 g/cm^3. There are some empty spaces; the atoms cannot be closely packed.

2.73 (a) 4.11×10^{23} atoms S (b) 5.14×10^{22} molecules S_8 (c) 21.9 g S_8

2.75 1.24×10^{23} molecules C_2H_5OH

2.77 1.49×10^{23} atoms Ag

2.79 (a) 4.88×10^{21} Mg atoms (b) 0.0568 mol H_2O (c) 4.88×10^{21} FU Epsom salts

2.81 Y:Ba:Cu = 88.91:274.68:190.62 or Y:B:Cu = 1:3.089:2.144

2.83 8.71 g Au

2.85 (a) cobalt(III) nitrate hexahydrate or cobaltous nitrate hexahydrate (b) copper(I) chloride or cuprous chloride (c) mercury(II) nitrate or mercuric nitrate (d) vanadium(V) oxide (e) chromium(III) phosphate

Case Questions

1. (a) 43.64% (b) 83.01% (c) 20-4-4

3. 21-0-0

5. (a) 1.4×10^4 kg (b) 5.0×10^3 kg (c) 100 kg NH_4NO_3, 21 kg $Ca(H_2PO_4)_2$, 13 kg KCl (d) 13-4-3

CHAPTER 3

Classified Exercises

3.1 (a) $Na_2O(s) + H_2O(l) \rightarrow 2Na^+(aq) + 2OH^-(aq)$ (b) $6Li(s) + N_2(g) \rightarrow 2Li_3N(s)$ (c) $Ca(s) + 2H_2O(l) \rightarrow Ca^{2+}(aq) + 2OH^-(aq) + H_2(g)$

3.3 First stage: $3Fe_2O_3(l) + CO(g) \rightarrow 2Fe_3O_4(l) + CO_2(g)$; second stage: $Fe_3O_4(l) + 4CO(g) \rightarrow 3Fe(l) + 4CO_2(g)$

3.5 Engine: $N_2(g) + O_2(g) \rightarrow 2NO(g)$; atmosphere: $2NO(g) + O_2(g) \rightarrow 2NO_2(g)$

3.7 $H_2S(g) + 2NaOH(aq) \rightarrow Na_2S(aq) + 2H_2O(l)$; $4H_2S(g) + Na_2S(aq) \rightarrow Na_2S_5(aq) + 4H_2(g)$; $10H_2O(l) + 9O_2(g) + 2Na_2S_5(aq) \rightarrow 2Na_2S_2O_3{\cdot}5H_2O + 6SO_2(g)$

3.9 $2C_8H_{18}(l) + 25O_2(g) \rightarrow 16CO_2(g) + 18H_2O(g)$

3.11 $4C_{10}H_{15}N + 59O_2(g) \rightarrow 40CO_2(g) + 30H_2O(g) + 4NO_2(g)$

3.13 (a) (S) (b) (S) (c) (S)

3.15 (a) Na^+, I^- (b) Ag^+, CO_3^{2-} (c) Na_4^+, PO_4^{3-}

3.17 (a) $Fe^{3+} + 3OH^- \rightarrow Fe(OH)_3(s)$
(b) $2Ag^+ + CO_3^{2-} \rightarrow Ag_2CO_3(s)$; yes
(c) No, sodium nitrate and lead acetate are soluble

3.19 (a) $FeCl_3(aq) + 3NaOH(aq) \rightarrow Fe(OH)_3(s) + 3NaCl(aq)$; $Fe^{3+}(aq) + 3Cl^-(aq) + 3Na^+(aq) + 3OH^-(aq) \rightarrow Fe(OH)_3(s) + 3Na^+(aq) + 3Cl^-(aq)$; $Fe^{3+}(aq) + 3OH^-(aq) \rightarrow Fe(OH)_3(s)$; Na^+ and Cl^- are spectator ions (b) $AgNO_3(aq) + KI(aq) \rightarrow AgI(s) + KNO_3(aq)$; $Ag^+(aq) + NO_3^-(aq) + K^+(aq) + I^-(aq) \rightarrow AgI(s) + K^+(aq) + NO_3^-(aq)$; $Ag^+(aq) + I^-(aq) \rightarrow AgI(s)$; K^+ and NO_3^- are spectator ions
(c) $Pb(NO_3)_2(aq) + K_2SO_4(aq) \rightarrow PbSO_4(s) + 2KNO_3(aq)$; $Pb^{2+}(aq) + 2NO_3^-(aq) + 2K^+(aq) + SO_4^{2-}(aq) \rightarrow PbSO_4(s) + 2K^+(aq) + 2NO_3^-(aq)$; $Pb^{2+}(aq) + SO_4^{2-}(aq) \rightarrow PbSO_4(s)$; K^+ and NO_3^- are spectator ions

3.21 (a) $Pb^{2+}(aq) + 2ClO_4^-(aq) + 2Na^+(aq) + 2Br^-(aq) \rightarrow PbBr_2(s) + 2Na^+(aq) + 2ClO_4^-(aq)$; $Pb^{2+}(aq) + 2Br^-(aq) \rightarrow PbBr_2(s)$ (b) $Sr^{2+}(aq) + 2NO_3^-(aq) + 2NH_4^+(aq) + 2Cl^-(aq) \rightarrow Sr^{2+}(aq) + 2NO_3^-(aq) + 2NH_4^+(aq) + 2Cl^-(aq)$; there is no net ionic reaction (c) $2Na^+(aq) + 2OH^-(aq) + Cu^{2+}(aq) + 2NO_3^-(aq) \rightarrow Cu(OH)_2(s) + 2Na^+(aq) + 2NO_3^-(aq)$; $Cu^{2+}(aq) + 2OH^-(aq) \rightarrow Cu(OH)_2(s)$

3.23 (a) $Ca(NO_3)_2$, Na_2CO_3 (b) $Cd(NO_3)_2$, Na_2S

3.25 (a) $Pb^{2+}(aq) + SO_4^{2-}(aq) \rightarrow PbSO_4(s)$
(b) $Cu^{2+}(aq) + S^{2-}(aq) \rightarrow CuS(s)$ (c) $Co^{2+}(aq) + CO_3^{2-}(aq) \rightarrow CoCO_3(s)$ (d) for (a), $Pb(NO_3)_2$, $NaSO_4$ [spectators Na^+, NO_3^-]; for (b), $Cu(NO_3)_2$, Na_2S [spectators Na^+, NO_3^-]; for (c), $Co(NO_3)_2$, Na_2CO_3 [spectators Na^+, NO_3^-]

3.27 (a) B (b) A (c) B

3.29 (a) H_3O^+, Cl^- (b) NH_3, NH_4^+, OH^-
(c) H_3O^+, NO_3^-

3.31 (a) $HCl(aq) + NaOH(aq) \rightarrow H_2O(l) + NaCl(aq)$; $H^+(aq) + Cl^-(aq) + Na^+(aq) + OH^-(aq) \rightarrow H_2O(l) + Na^+(aq) + Cl^-(aq)$; $H^+(aq) + OH^-(aq) \rightarrow H_2O(l)$
(b) $NH_3(aq) + HNO_3(aq) \rightarrow NH_4NO_3(aq)$; $NH_3(aq) + H^+(aq) + NO_3^-(aq) \rightarrow NH_4^+(aq) + NO_3^-(aq)$; $NH_3(aq) + H^+(aq) \rightarrow NH_4^+(aq)$ (c) $CH_3NH_2(aq) + HI(aq) \rightarrow CH_3NH_3I(aq)$; $CH_3NH_2(aq) + H^+(aq) + I^-(aq) \rightarrow CH_3NH_3^+(aq) + I^-(aq)$; $CH_3NH_2(aq) + H^+(aq) \rightarrow CH_3NH_3^+(aq)$

3.33 (a) HBr and KOH: $H^+(aq) + Br^-(aq) + K^+(aq) + OH^-(aq) \rightarrow K^+(aq) + Br^-(aq) + H_2O(l)$ and $H^+(aq) + OH^-(aq) \rightarrow H_2O(l)$ (b) HCN and $Ca(OH)_2$: $2HCN(aq) + Ca^{2+}(aq) + 2OH^-(aq) \rightarrow Ca^{2+}(aq) + 2CN^-(aq) + 2H_2O(l)$ and $HCN(aq) + OH^-(aq) \rightarrow H_2O(l) + CN^-(aq)$ (c) K_2HPO_4 and KOH: $2K^+(aq) + HPO_4^{2-}(aq) + K^+(aq) + OH^-(aq) \rightarrow 3K^+(aq) + PO_4^{3-}(aq) + H_2O(l)$ and $HPO_4^{2-}(aq) + OH^-(aq) \rightarrow PO_4^{3-}(aq) + H_2O(l)$

3.35 (a) $C_2H_5NH_2$, base; HCl, acid (b) $C_2H_5NH_2$, base; HCl, acid (c) HCO_3^-, base; HI, acid

3.37 (a) $1e^- + AgCN(s) \rightarrow Ag(s) + CN^-(aq)$; reduction (b) $2e^- + ClO_3^-(aq) + H_2O(l) \rightarrow ClO_2^-(aq) + 2OH^-(aq)$; reduction (c) $2F^-(aq) + H_2O(l) \rightarrow OF_2(g) + 2H^+(aq) + 4e^-$; oxidation

3.39 (a) $Mg(s) \rightarrow Mg^{2+}(aq) + 2e^-$; $Cu^{2+}(aq) + 2e^- \rightarrow Cu(s)$ (b) $Fe^{2+}(aq) \rightarrow Fe^{3+}(aq) + 1e^-$; $Ce^{4+}(aq) + 1e^- \rightarrow Ce^{3+}(aq)$ (c) $Fe(s) \rightarrow Fe^{3+}(aq) + 3e^-$; $O_2(g) + 4e^- \rightarrow 2O^{2-}(g)$

3.41 (a) $N_{ox}(I) = +5$ (b) $N_{ox}(Cl) = +1$ (c) $N_{ox}(N) = +2$ (d) $N_{ox}(N) = +5$

3.43 (a) No oxidation or reduction (b) BrO_3^- is reduced and Br^- is oxidized (c) F_2 is reduced and water is oxidized

3.45 (a) $Zn(s)$ is the reducing agent; HCl is the oxidizing agent (b) H_2S is the reducing agent; SO_2 is the oxidizing agent (c) $Mg(s)$ is the reducing agent; B_2O_3 is the oxidizing agent

3.47 (a) An oxidizing agent is required
(b) A reducing agent is required
(c) A reducing agent is required

3.49 Metals that lie below hydrogen in the activity series are not oxidized by hydrogen ions. Zinc and tin are above hydrogen; they can be oxidized by HCl.

3.51 (a) $Cu^{2+} + 2e^- \rightarrow Cu^0$; $Ni \rightarrow Ni^{2+} + 2e^-$
(b) No reaction (c) No reaction

3.53 (a) $VO^{2+}(aq) + 2H^+(aq) + 1e^- \rightarrow V^{3+}(aq) + H_2O(l)$; gain of electron; reduction (b) $PbSO_4(s) + 2H_2O(l) \rightarrow PbO_2(s) + SO_4^{2-}(aq) + 4H^+(aq) + 2e^-$; loss of electrons; oxidation (c) $H_2O_2(aq) \rightarrow O_2(g) + 2H^+(aq) + 2e^-$; loss of electrons; oxidation

3.55 (a) $2e^- + ClO^-(aq) + H_2O(l) \rightarrow Cl^-(aq) + 2OH^-(aq)$ (b) $4e^- + IO_3^-(aq) + 2H_2O(l) \rightarrow IO^-(aq) + 4OH^-(aq)$ (c) $2e^- + 2SO_3^-(aq) + 2H_2O(l) \rightarrow S_2O_4^{2-}(aq) + 4OH^-(aq)$

3.57 Reduction *always* occurs at the cathode. Oxidation *always* occurs at the anode.

3.59 (a) $4Cl_2(g) + S_2O_3^{2-}(aq) + 5H_2O(l) \rightarrow 8Cl^-(aq) + 2SO_4^{2-}(aq) + 10H^+(aq)$; Cl_2 is the oxidizing agent and $S_2O_3^{2-}$ is the reducing agent (b) $2MnO_4^-(aq) + H^+(aq) + 5H_2SO_3(aq) \rightarrow 2Mn^{2+}(aq) + 3H_2O(l) + 5HSO_4^-(aq)$; MnO_4^- is the oxidizing agent and H_2SO_3 is the reducing agent (c) $Cl_2(g) + H_2S \rightarrow 2Cl^-(aq) + S(s) + 2H^+(aq)$; Cl_2 is the oxidizing agent and H_2S is the reducing agent

3.61 (a) $O_3(g) + Br^-(aq) \rightarrow O_2(g) + BrO^-(aq)$ O_3 is the oxidizing agent and Br^- is the reducing agent (b) $5Br_2(l) + 12OH^-(aq) \rightarrow 10Br^-(aq) + 2BrO_3^-(aq) + 6H_2O(l)$; Br_2 is both the oxidizing agent and the reducing agent (c) $2Cr^{3+}(aq) + 4OH^-(aq) + 3MnO_2(s) \rightarrow 2CrO_4^{2-}(aq) + 2H_2O(l) + 3Mn^{2+}(aq)$; Cr^{3+} is the reducing agent and MnO_2 is the oxidizing agent

3.63 (a) $HSO_3^-(aq) + H_2O(l) \rightarrow HSO_4^-(aq) + 2H^+(aq) + 2e^-$ and $2HSO_3^-(aq) \rightarrow S_2O_6^{2-}(aq) + 2H^+(aq) + 2e^-$ (b) $I_2(aq) + 2e^- \rightarrow 2I^-(aq)$ and $HSO_3^-(aq) + H_2O(l) \rightarrow HSO_4^-(aq) + 2H^+(aq) + 2e^-$; $I_2(aq) + HSO_3^-(aq) + H_2O(l) \rightarrow 2I^-(aq) + HSO_4^-(aq) + 2H^+(aq)$

3.65 $24[MnO_4^-(aq) + 8H^+(aq) + 5e^- \rightarrow Mn^{2+}(aq) + 4H_2O(l)]$ and $5[C_6H_{12}O_6(aq) + 6H_2O(l) \rightarrow 6CO_2(g) + 24H^+(aq) + 24e^-]$; $24MnO_4^-(aq) + 72H^+(aq) + 5C_6H_{12}O_6(aq) \rightarrow 24Mn^{2+}(aq) + 66H_2O(l) + 30CO_3(g)$

Unclassified Exercises

3.67 (a) SA (b) SB (c) WB (d) S (e) WA (f) I

3.69 (a) $Mg(NO_3)_2$, Na_2CO_3 (b) $AgNO_3$, K_2SO_4 (c) $Zn(NO_3)_2$, NaOH (d) $Pb(NO_3)_2$, Na_2CrO_4

3.71 (a) H_2SO_4, NaOH (b) $HClO_4$, KOH (c) HCl, $Ni(OH)_2$

3.73 (a) Redox; oxidizing agent = I_2; reducing agent = $S_2O_3^{2-}$ (b) Precipitation; $Ag^+ + Br^- \rightarrow AgBr(s)$ (c) Redox; oxidizing agent = UF_4; reducing agent = Mg

3.75 (a) WO_3 is the oxidizing agent; H_2 is the reducing agent (b) HCl is the oxidizing agent; Zn is the reducing agent (c) N_2H_4 is the reducing agent; N_2O_4 is the oxidizing agent

3.77 First reaction, combining the two half-reactions: $4[2CN^-(aq) + Au(s) \rightarrow Au(CN)_2^-(aq) + 1e^-]$ and $1[4e^- + O_2(aq) + 2H_2O(l) \rightarrow 4OH^-(aq)]$; $8CN^-(aq) + 4Au(s) + O_2(aq) + 2H_2O(l) \rightarrow 4Au(CN)_2^-(aq) + 4OH^-(aq)$ Second reaction, combining the two half-reactions: $2[Au(CN)_2^-(aq) + 1e^- \rightarrow Au(s) + 2CN^-(aq)]$ and $1[Zn(s) + 4CN^-(aq) \rightarrow Zn(CN)_4^{2-}(aq) + 2e^-]$; $2Au(CN)_2^-(aq) + Zn(s) \rightarrow 2Au(s) + Zn(CN)_4^{2-}(aq)$

Case Questions

1. Ionic equation: $2Al^{3+}(aq) + 3SO_4^{2-}(aq) + 3Ca(OH)_2(s) \rightarrow 3CaSO_4(s) + 2Al(OH)_3(s)$; Net ionic equation: $2Al^{3+}(aq) + 3SO_4^{2-}(aq) + 3Ca(OH)_2(s) \rightarrow 3CaSO_4(s) + 2Al(OH)_3(s)$

3. (a) aeration (replacement of unwanted gases), settling (separation of ppt from solution), filtration (through sand), absorption (organics onto C), ion exchange, reverse osmosis (b) oxidation ($Fe^{2+} \rightarrow Fe^{3+} + 1e^-$), precipitation (hydroxide and carbonate formation) or coagulation (pptn with alum), acid/base (controlling pptn, forming HClO, killing bacteria)

CHAPTER 4

Classified Exercises

4.1 (a) 10 mol H_2 (b) 5.0 mol H_2

4.3 (a) 11.0 mol NO_2 (b) 8.4×10^{-3} mol MnO_4^-

4.5 9.0 mol CO_2

4.7 (a) 0.088 mol H_2O (b) 329 g O_2

4.9 (a) 205 g CO_2 (b) 0.133 mol H_2O

4.11 (a) 18.9 g Al_2O_3 (b) 8.90 g O_2

4.13 (a) 5.7×10^6 g Al (b) 9.448×10^6 g Al_2O_3

4.15 (a) 2.5×10^3 g H_2O (b) 6.6 g O_2

4.17 1.1×10^3 g H_2O

4.19 85.6%

4.21 86.7%

4.23 C_2H_5OH is the limiting reactant

4.25 (a) 0.104 mol Fe (b) 1.1 g Al

4.27 (a) H_2S is the limiting reactant (b) 5.7 g SO_2 (c) 8.6 g S, 3.2 g H_2O (d) 11.4 g SO_2 + 6.08 g H_2S = 17.5 g and 5.7 g SO_2 + 8.6 g S + 3.2 g H_2O = 17.5

4.29 $C_7H_6O_2$

4.31 C_6H_7N

4.33 The empirical formula is $C_{14}H_{18}N_2O_5$, which has a molar mass of 294 g/mol; therefore, this is also the molecular formula.

4.35 The empirical formula is $C_4H_5N_2O$; the molecular formula is $C_8H_{10}N_4O_2$; $2C_8H_{10}N_4O_2(g) + 19O_2(g) \rightarrow 16CO_2(g) + 10H_2O(g) + 4N_2(g)$.

4.37 (a) 6.27 M $AgNO_3$ (b) 0.0241 M NaCl

4.39 0.41 g HCl

4.41 0.658 g $AgNO_3$

4.43 (a) 4.85 mL (b) 29.0 mL

4.45 (a) 0.06757 mol/L (b) 0.073 mol/L

4.47 (a) pH = 2 (b) pH = 3.7 (c) 3×10^{-7} mol/L

4.49 (a) 0.271 M NaOH (b) 0.163 g NaOH

4.51 (a) 0.209 M HNO_3 (b) pH = 0.680 (c) 1.07 g HNO_3

4.53 0.7209 M Fe^{2+}

4.55 (a) 0.0504 g Fe^{2+} (b) 25.0% Fe

Unclassified Exercises

4.57 3.586×10^{-3} mol I_2

4.59 (a) 3.34 mol H_2O_2 (b) 5.69×10^{-2} mol HNO_3 (c) 0.0300 g H_2O

4.61 (a) 34.82 g Cl_2 (b) 0.168 L $Cl_2(g)$ (c) 89.3%

4.63 (a) 5.56×10^{-3} L O_2 (b) 26.5×10^{-3} L air

4.65 H_2S is the limiting reactant

4.67 (a) 94.8% (b) 0.226 g C_4H_{10} in excess

4.69 0.0782 g H_2S

4.71 (a) COH_2 (b) $C_6O_6H_{12}$

4.73 (a) 8.0 g $CuSO_4$ (b) 12.5 g $CuSO_4 \cdot 5H_2O$

4.75 (a) $Na_2CO_3(aq) + 2HCl(aq) \rightarrow 2NaCl(aq) + H_2CO_3(aq)$ (b) 0.0500 M Na_2CO_3; 9.43 M HCl

4.77 The manufacturer's claim is about 7% inaccurate.

4.79 (a) 0.118 M I_3^- (b) 0.0412 M HCN

4.81 amount C_2H_5OH = 1.951 mol; amount $(C_2H_5)_2O$ = 0.049 mol; amount C_2H_4 = $(3 - 1.951 - 2 \times 0.049)$ mol = 0.951 mol

Case Questions

1. Milk of magnesia

3. (a) 3.7 g $Mg(OH)_2$ (b) 13 tablets

CHAPTER 5

Classified Exercises

5.1 (a) 0.987 atm (b) 1.33×10^2 Pa (c) 9.87×10^{-3} atm

5.3 1.690×10^2 kPa

5.5 9.58×10^2 cm

5.7 (a) 7.2×10^{-3} L (b) 1.31×10^1 mL

5.9 (a) 1.6 atm (b) 194 mL

5.11 −148°C

5.13 59°C

5.15 1203 kPa

5.17 (a) 10 mL (b) 0.91 atm (c) 212 K

5.19 (a) $V_{SiF_4} = 7.29 \times 10^{-6}$ L (b) $V_{Xe} = 9.76 \times 10^{-5}$ L (c) $V_{air} = 2.74$ L

5.21 (a) 1.3×10^{-2} Torr (b) 3.9×10^4 L Kr (c) 1.7×10^{14} Xe atoms

5.23 6.0×10^{-3} g NH_3 (or) 6.0 mg NH_3

5.25 17.0 g He

5.27 (a) 5.33 g/L (b) 3.90 g/L

5.29 (a) 3.47 g/L (b) 77.8 g/mol

5.31 (a) 12.9 g/mol (b) 0.580 g/L

5.33 C_3H_6

5.35 2.75×10^{-1} L O_2

5.37 (a) 233 L SO_2 (b) 89.3 L SO_2

5.39 $V_{CO_2} = 12.4$ L; $V_{NH_3} = 24.8$ L

5.41 $V_{N_2} = 0.14$ L of N_2; $V_{CO_2} = 0.28$ L CO_2; $V_{O_2} = 0.023$ L O_2; $V_{total} = 0.44$ L

5.43 (a) $P_{N_2} = 0.90$ atm; $P_{O_2} = 3.3$ atm; P_{total} = 4.2 atm (2 sf) (b) $P_{H_2} = 0.67$ atm; $P_{NH_3} = 1.34$ atm; $P_{He} = 0.047$ atm; $P_{total} = 2.1$ atm (2 sf)

5.45 8.44×10^{-2}

5.47 (a) 22.9 kPa N_2; 33.7 kPa Ar (b) 56.6 kPa

5.49 (a) $P_{H_2} = 739.2$ Torr (b) 0.142 g O_2

5.51 38.3 mL

5.53 (a) 1.845×10^3 m/s (b) 2.38×10^2 m/s (c) 1.363×10^3 m/s

5.55 (a) By a factor of 1.12 (b) The average speed increases by a factor of 1.09 when we take them out of the freezer into 26°C surroundings.

5.57 (a) 154 s (b) 123 s (c) 186 s

5.59 1.2×10^3 g/mol

5.61 phosphorus(III) oxide, P_4O_6

5.63 C_2H_3Cl

5.65 (a) 10.04 atm (b) 399 atm (c) 10.56 atm; 306 atm (d) 5.2%; 23%

Unclassified Exercises

5.67 (a) 732 Torr (b) 446 Torr

5.69 (a) 0.55 L/cylinder (or 5.5×10^2 mL/cylinder) (b) 4.8 atm (c) 4.97 atm ≈ 5.0 atm (2 sf)

5.71 655 K

5.73 (a) 4.27×10^6 g He (or 4.3×10^6 g He, 2 sf) (b) 2.7×10^4 kg

5.75 14.2 mL

5.77 (a) 2.45 g CO (b) 5.76 g/L

5.79 molar mass = 254 g/mol; $x = 4$

5.81 (a) 2.7×10^4 g CO_2 (b) $P = 7.8 \times 10^2$ atm

5.83 2.38×10^4 L

5.85 (a) 146 g Fe_2O_3 (b) 0.732 M H_2SO_3

5.87 19.7 g H_2O ≈ 20 g H_2O (2 sf)

5.89 1.4×10^2 g/mol x; PCl_3

5.91 438 m/s; 502 m/s

5.93 9.6×10^5 g $CaCO_3$/h; 1.3×10^6 g "charge" must be added to the kiln per hour, as this is the amount consumed per hour

Case Questions

1. 4 L

3. 0.94

CHAPTER 6

Classified Exercises

6.1 66.94 kJ

6.3 0.6 kJ

6.5 (a) 625 J (b) 5.4×10^2°C

6.7 88%

6.9 0.53 J/g·°C

6.11 (a) Endothermic (b) Exothermic

6.13 43.2 J/°C

6.15 1.39×10^4 J

6.17 −3.8 kJ

6.19 (a) 8.21 kJ/mol (b) 43.5 kJ/mol

6.21 (a) 226 kJ (b) 199 kJ

6.23 22 kJ

6.25 1.9 s

6.27 (a) +72 kJ (b) +149 kJ

6.29 (a) 2.5×10^2 g octane (b) 1.3×10^8 J

6.31 −266 kJ

6.33 +1.90 kJ

6.35 −197.78 kJ

6.37 −312 kJ

6.39 −184.7 kJ

6.41 (a) $K(s) + \frac{1}{2}Cl_2(g) + \frac{3}{2}O_2(g) \rightarrow KClO_3(s)$ (b) $\frac{5}{2}H_2(g) + \frac{1}{2}N_2(g) + 2C(\text{graphite}) + O_2(g) \rightarrow H_2NCH_2COOH(s)$ (c) $2Al(s) + \frac{3}{2}O_2(g) \rightarrow Al_2O_3(s)$

6.43 11.3 kJ/mol

6.45 −443.7 kJ/mol

6.47 (a) −15.4 kJ (b) −128.5 kJ

6.49 (a) +8.77 kJ/mol (b) +246.5 kJ (c) −905.48 kJ

6.51 Specific enthalpy = 48.44 kJ/g; enthalpy density = 3.3×10^4 kJ/L

6.53 Mg: 24.76 kJ/g Mg; Al: 31.05 kJ/g Al; yes, Al is better

6.55 (a) 890 kJ/mol CO_2; 683.9 kJ/mol CO_2 (b) 468.0 kJ/mol CO_2

Unclassified Exercises

6.57 1.30 g

6.59 8.5×10^3 kJ

6.61 1.5×10^4 J

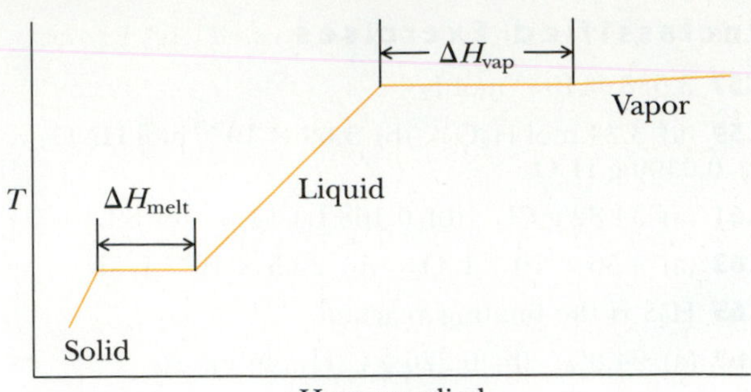

6.63 (a) -1.79×10^3 kJ (b) −109 kJ

6.65 −2806 kJ/mol

6.67 Heat would have to be removed.

6.69 −2806 kJ/mol

6.71 228 kJ/mol C_2H_2

6.73 (a) Exothermic (b) 623 kJ

6.75 −22.25 kJ

6.77 −1.19 kJ, exothermic

6.79 5.8°C

6.81 689 kJ/mol KBr

Case Questions

1. -9.4×10^9 kJ

3. amount of Al = 3.0×10^6 mol; amount of NH_4ClO_9 = 3.0×10^6 mol; there is no limiting reactant

CHAPTER 7

Classified Exercises

7.1 (a) The frequency range is 7.49×10^{14} Hz to 4.28×10^{14} Hz. (b) 1.20×10^6 Hz

7.3 (a) 420 nm (b) 1.5×10^{-10} m

7.5 (a) 3.37×10^{-19} J/photon (b) 2.03×10^5 J/mol Na

7.7 (a) 2.3×10^{-15} J/photon (b) 2.55×10^5 J/mol photons

7.9 2.6×10^{19} photons

7.11 (a) 4.10×10^{-7} m (or 410 nm) (b) Violet visible

7.13 3.38×10^{-19} J

7.15 (a) 4.85×10^{-12} m (b) 2.6×10^{-15} m

7.17 (a) 2 (b) 3 (c) 0, 1, 2

7.19 (a) 1 (b) 5 (c) 3 (d) 7

7.21 (a) 3*d* (b) 1*s* (c) 6*f* (d) 2*p*

7.23 (a) 6 electrons (b) 2 electrons (c) 8 electrons (d) 2 electrons

7.25 (a) $2d$ not allowed (b) $4d$ allowed (c) $4g$ not allowed (d) $6f$ allowed

7.27 (a) OK (b) No, if $l = 0$, m_l can only be 0. (c) No, if $l = 0$, n and l cannot be equal: $l = 0 \ldots n-1$.

7.29 (a) $n = 2, l = 1, m_l = 0, m_s = -\frac{1}{2}$ (b) $n = 5, l = 2, m_l = -2, m_s = +\frac{1}{2}$

7.31 (a) $n = 1, 2, 3, 4$ energy$\rightarrow$ (b) $l = 0, 1, 2, 3, 4$ energy$\rightarrow$

7.33 9.2×10^1 J/mol = 92 kJ/mol

7.35 $1s$, $2p$, $3s$, $3d$, $5d$; energy increases left to right

7.37 (a) $\underset{1s}{\underline{\uparrow}}$; $n = 1, l = 0, m_l = 0, m_s = +\frac{1}{2}$

(b) $\underset{1s}{\underline{\uparrow\downarrow}}\ \underset{2s}{\underline{\uparrow\downarrow}}\ \underset{2p_x}{\underline{\uparrow\downarrow}}\ \underset{2p_y}{\underline{\uparrow\downarrow}}\ \underset{2p_z}{\underline{\uparrow}}$; $n = 2, l = 1, m_l = 0, m_s = -\frac{1}{2}$

(c) $\underset{1s}{\underline{\uparrow\downarrow}}\ \underset{2s}{\underline{\uparrow\downarrow}}\ \underset{2p}{\underline{\uparrow\downarrow}}\ \underset{2p}{\underline{\uparrow\downarrow}}\ \underset{2p}{\underline{\uparrow\downarrow}}\ \underset{3s}{\underline{\uparrow\downarrow}}\ \underset{3p}{\underline{\uparrow\downarrow}}\ \underset{3p}{\underline{\uparrow\downarrow}}\ \underset{3p}{\underline{\uparrow\downarrow}}\ \underset{4s}{\underline{\uparrow\downarrow}}$; $n = 4, l = 0, m_l = 0, m_s = +\frac{1}{2}$

7.39 (a) $[Ar]4s^2$ (b) $1s^2,2s^2,2p^3$ (c) $[Ar]3d^{10},4s^2,4p^5$ (d) $[Rn]5f^3,6d^1,7s^2$

7.41 (a) ns^2 (b) ns^2np^6

7.43 (a) $[Ar]3d^6$ (b) $[Ne]3s^23p^6$ (c) $[Xe]4f^{14}5d^{10}6s^2$

7.45 (a) 4 unpaired d electrons; paramagnetic (b) all electrons paired; diamagnetic (c) 2 unpaired d electrons; paramagnetic

7.47 (a) Mn (b) Na (c) Same

7.49 (a) 4 (b) 7 (c) 7 (d) 9

7.51 (a) Group I (alkali metals). (b) The radius of the cation is less than the radius of the neutral atom. (c) Cations with the largest number of protons (among a set of cations with the same number of electrons) will have the smallest radius.

7.53 (a) S, size decreases $r \rightarrow n$ (b) Na, size decreases $r \rightarrow n$ (c) S^{2-}, same number of electrons, Cl has more protons, hence, smaller (d) Mg^{2+}, same number of electrons, Al has more protons, hence, smaller

7.55 (a) Group VIII (inert gases) (b) Group I (alkali metals)

7.57 (a) Mg (b) N (c) P

7.59 The group configurations are Group VI, ns^2np^4 and Group V, ns^2np^3. The half-filled subshell of Group V is more stable than simple theory suggests; this makes removal of the electron more difficult, hence, a higher ionization energy.

7.61 Both the first and second electrons lost from Mg are $3s$. The second electron for sodium must be removed from a $2p$ level which is filled. A filled sublevel from a lower shell is far more stable. Thus, the second ionization energy of sodium is very high and sodium exists only as Na^+, not Na^{2+}.

7.63 (a) Group VII (halogens). (b) Electron affinities generally increase (left to right) across a period.

7.65 (a) Cl (b) C (c) Cl

7.67 Atomic radii increase and ionization energies decrease from top to bottom within a group. Atomic radii decrease and ionization energies increase (left to right) across a period.

7.69 Al, B, C, N; electronegativity increasing left to right

7.71 (a) C (b) F (c) S

7.73 Ionization energies and electronegativities both increase across (left to right) a period and both decrease from top to bottom within a group.

7.75 (a) $2NaOH(aq) + H_2(g)$ (b) $Ba(OH)_2(aq)$

7.77 (a) Metal (b) Nonmetal (c) Metalloid (d) Metalloid

7.79 (a) $Na_2O(s) + H_2O(l) \rightarrow 2NaOH(aq)$ (b) $SO_3(g) + H_2O(l) \rightarrow H_2SO_4(aq)$

Unclassified Exercises

7.81 (a) Ultraviolet (b) Infrared

7.83 1×10^{10} Hz

7.85 1.026×10^{-10} m

7.87 No

7.89 (a) 2.74×10^{-7} m = 274 nm (b) 7.0×10^{-20} J/ejected electron

7.91 (a) one (b) four (c) five (d) d (e) +2, +1, 0, −1, −2 (f) 0, 1, 2, 3 (g) three (h) nine (i) three (j) ten (k) 0, 1 (l) +1, 0, −1 (m) one (n) l (o) p (p) 3

7.93 (a) Group VA, nitrogen (b) Group IA, lithium (c) Group VIIIA, neon (d) Group IIA, beryllium

7.95 (a) $[Ar]3d^{10}$ (b) $[Ar]3d^{10}4s^24p^6$ (c) $[Kr]4d^{10}5s^25p^6$ (d) $[Kr]4d^15s^2$ (e) $[Ne]3s^23p^3$

7.97 (a) $[Ar]3d^34s^2$ (b) $[Ne]3s^23p^6$ (c) $[Kr]4d^{10}5s^2$ (d) $[Ar]3d^84s^2$ (e) $[He]2s^22p^6$

7.99 The higher ionization energy is due to the stability imparted by the $nd^{10}(n+1)s^2$ electron arrangement of zinc, cadmium and mercury.

7.101 6.71×10^{-19} J/atom (404 kJ/mol)

Case Questions

1. (a) 620 to 850 nm; 580 to 560 nm (b) $Cu(NH_3)_4{}^{2+}$

3. (a) 38 (b) 2.1×10^{-2} mol/L

CHAPTER 8

Classified Exercises

8.1 (a) ionic, metal, and nonmetal (b) nonionic, nonmetal and nonmetal (c) ionic, metal and nonmetal

8.3 (a) +1 (b) −2 (c) +2 (d) +3

8.5 The radius of Mg^{2+} (65 pm) is smaller than the radius of Ba^{2+} (135 pm), the distance between Mg^{2+} and O^{2-} ions is less than that between Ba^{2+} and O^{2-} ions in the crystal lattice. Thus the lattice enthalpy of MgO exceeds that of BaO. In MgO, O^{2-} is so much bigger than Mg^{2+} that O^{2-} almost touches other O^{2-} anions. This is not the case in BaO. In other words, O^{2-}—O^{2-} repulsions are greater in MgO than in BaO.

8.7 $AgF(s) \rightarrow Ag^+(g) + F^-(g)$, $\Delta H_L^\circ = +971$ kJ/mol

8.9 $Na(s) + Cl_2(g) \rightarrow NaCl_2(s)$, $\Delta H_f^\circ = 2183$ kJ/mol; A compound with such a large positive ΔH_f° is not likely to form under any conditions.

8.11 (a) Ca: (b) ·S̈· (c) $[:\ddot{O}:]^{2-}$ (d) $[:\ddot{N}:]^{3-}$

8.13 (a) Fe: (b) Mn: (c) Co:

8.15 (a) K^+ $[:\ddot{F}:]^-$ (b) $2Al^{3+}$, 3 $[:\ddot{S}:]^{2-}$
(c) $3Ca^{2+}$, 2 $[:\ddot{N}:]^{3-}$ (d) Co^{3+}, 3 $[:\ddot{Cl}:]^-$

8.17 (a) $[(:Fe^{\cdot})]^{2+}$, 2 $[:\ddot{Cl}:]^-$ (b) $[(^{\cdot}Mn^{\cdot})]^{2+}$, $[:\ddot{S}:]^{2-}$ (c) Zn^2, $[:\ddot{O}:]^{2-}$

8.19 (a) 3 (b) 1 (c) 4

8.21 (a) H:F̈: (b) H:N̈:H with H below N (c) H:C:H with H above and below C

8.23 (a) $[H:\ddot{N}:H$ with H above and below N$]^+$ $[:\ddot{Cl}:]^-$ (b) $3K^+$, $[:\ddot{O}:\ddot{P}:\ddot{O}:$ with :Ö: above and below P$]^{3-}$
(c) Na^+, $[:\ddot{Cl}:\ddot{O}:]^-$

8.25 $-\frac{1}{3}$; O(—O)—O (bent ozone structure)

8.27 (a) H:C(=O)H (b) H:C:C:C:H with H above and below the outer C atoms and O above the central C
(c) H:C:C(=O)—O—H with H above C and H:N̈: below C, with H on N

8.29 (a) Lewis base (b) Lewis acid (c) Lewis acid (d) Lewis base

8.31 $[H:\ddot{C}:\ddot{O}:$ with H above and below C$]^-$; Lewis base

8.33 (a) PF_5 + $[:\ddot{F}:]^-$ → $[PF_6]^-$

Lewis acid Lewis base Product

(b) SO_2 + $[:\ddot{Cl}:]^-$ → $[ClSO_2]^-$

Lewis acid Lewis base Product

(c) Cu^{2+} + 4 :N(H)—H (with H above and below N) → $[Cu(NH_3)_4]^{2+}$ (H—N—Cu—N—H, with NH₃ groups above and below Cu)

Lewis acid Lewis base Product

8.35 (a) $[:\ddot{O}—\ddot{N}=\ddot{O}:]^- \leftrightarrow [:\ddot{O}=\ddot{N}—\ddot{O}:]^-$

(b) :Ö—N(—Cl)=Ö ↔ Ö=N(—Cl)—Ö: (with :C̈l: above N in each)

8.37 (a) $[:\ddot{O}:\ddot{S}:\ddot{O}:$ with :Ö: above and below S$]^{2-} \leftrightarrow [O::S::O$ with :Ö: above and below S$]^{2-}$

6-ways

(b) $[:\ddot{O}:\ddot{S}:$ with :Ö: above and below S$]^{2-} \leftrightarrow [O::S:$ with :Ö: above and below$]^{2-} \leftrightarrow [O::S:$ with O double-bonded above and :Ö: below$]^{2-}$

3-ways 3-ways

(c) $[:\ddot{O}:Cl:$ with :Ö: above and below Cl$]^- \leftrightarrow [:\ddot{O}:Cl:$ with O double-bonded above and :Ö: below$]^- \leftrightarrow [:\ddot{O}:Cl:$ with O double-bonded above and below$]^-$

3-ways 3-ways

(d) 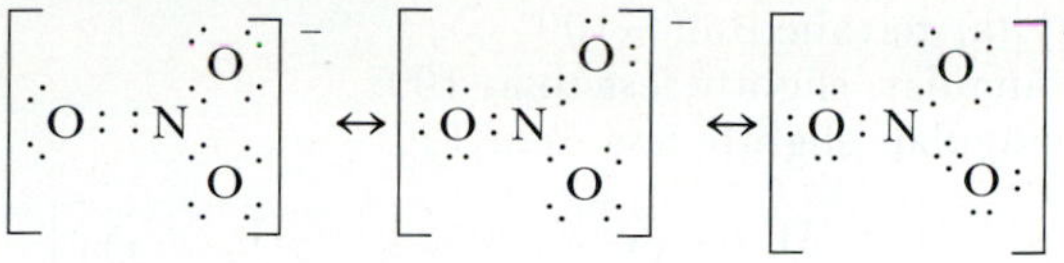 $[:\ddot{O}:N(O)(O)]^- \leftrightarrow [:\ddot{O}:N(O)(O)]^- \leftrightarrow [:\ddot{O}:N(O)(O)]^-$

8.39 (a) 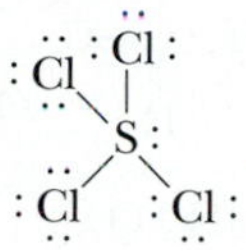 SCl_4, 1 lone pair on S

(b) 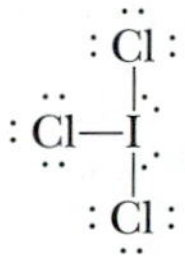 $:\ddot{Cl}-I(Cl)(Cl)$, 2 lone pairs on I

(c) 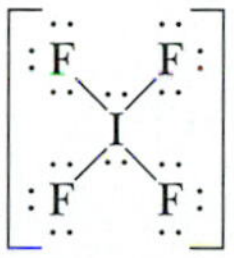 $[IF_4]^-$, 2 lone pairs on I

8.41 (a) 4 electron pairs (b) 6 electron pairs (c) 5 electron pairs (d) 6 electron pairs

8.43 (a) 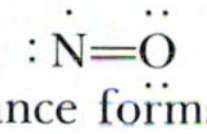 $:\dot{N}{=}\ddot{O}$ (b) 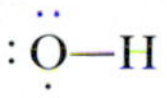 $:\dot{O}{-}H$ (c) 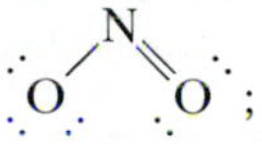 $O{-}\dot{N}{=}O$; 2 resonance forms

8.45 (a)  $XeOF_4$, 1 lone pair on Xe

(b) 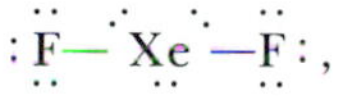 $:\ddot{F}-\ddot{Xe}-\ddot{F}:$, 3 lone pairs on Xe

(c) 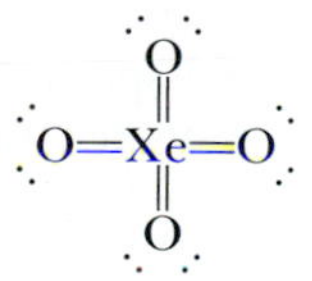 $O{=}Xe({=}O)_2{=}O$, no lone pairs on Xe

(d) 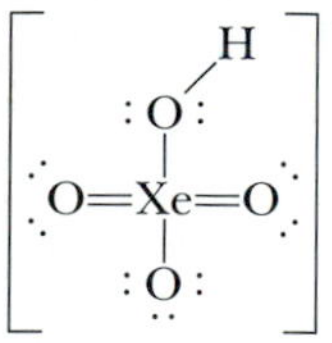 $[HO{-}Xe({=}O)_2{-}O]^-$, no lone pairs on Xe

3-ways

8.47 (a) Trigonal bipyramidal, 120° bond angles in equatorial plane, 90° between axial bonds and equatorial plane (b) Linear, 180° (c) Trigonal pyramidal, slightly less than 109° (d) Angular, slightly less than 109°

8.49 (a) linear (b) angular (c) two resonance forms, angular

8.51 (a) tetrahedral (b) trigonal pyramidal (c) tetrahedral

8.53 (a) 180° (b) Slightly less than 90° (c) 109.5°

8.55 (a) 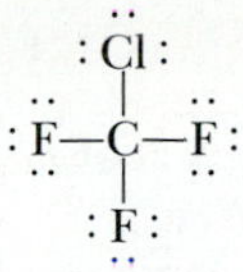 $CClF_3$, AX_4, tetrahedral, 109.5°

(b) 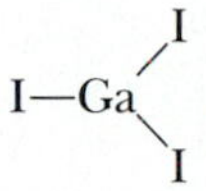 GaI_3, AX_3, trigonal planar, 120°

(c) 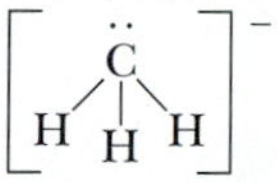 $[CH_3]^-$, AX_3E, trigonal pyramidal, slightly less than 109°

8.57 NF_2O^+, $[F_2N{=}O]^+$, trigonal planar;

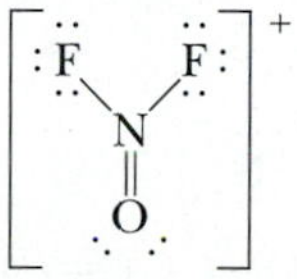

SbF_6^-, $[SbF_6]^-$, octahedral

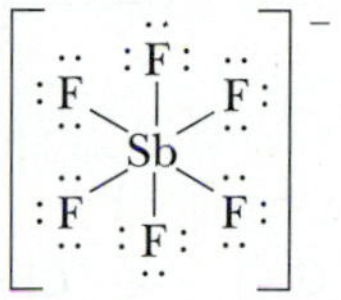

8.59 Angles a and b are each approximately 120°.

8.61 $BF_3 + NH_3 \rightarrow F_3B{-}NH_3$

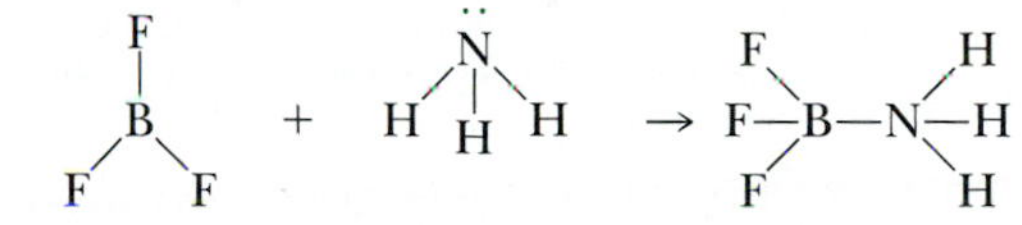

All angles 120° All angles < 109° All angles ≈ 109°

Unclassified Exercises

8.63 (a) −412 kJ/mol (b) 688 kJ/mol (c) 74 kJ/mol

8.65 (a)  $[:\ddot{N}:]^{3-}$ (b) 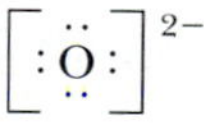 $[:\ddot{O}:]^{2-}$ (c) 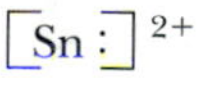 $[Sn:]^{2+}$

8.67 (a) 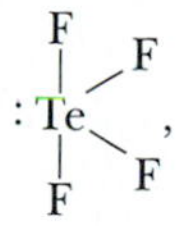 $:TeF_4$, see-saw (b) 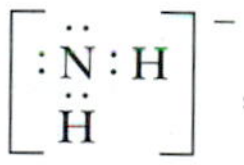 $[:\ddot{N}:H(H)]^-$, angular

(c) 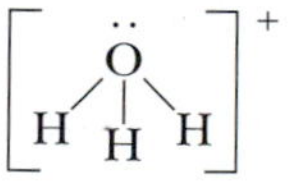 $[H_3O]^+$, trigonal pyramidal

(d) 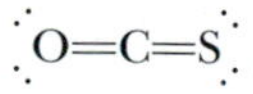 $O{=}C{=}S$, linear

8.69 (a) 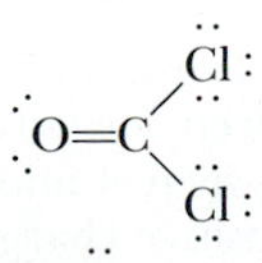 $O{=}CCl_2$, trigonal planar

(b) 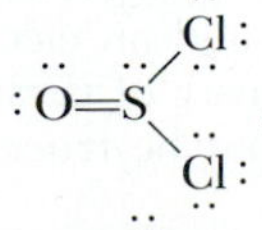 $:O{=}SCl_2$, trigonal pyramidal

(c) 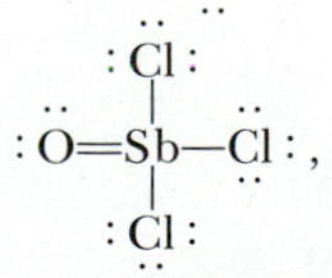 $:O{=}SbCl_3$, tetrahedral

8.71 N_3F, $:\ddot{N}{=}N{=}\ddot{N}{-}\ddot{F}: \leftrightarrow :N{\equiv}N{-}\ddot{N}{-}\ddot{F}:$ (120°, 109°); N_2F_2, $:\ddot{F}{-}\ddot{N}{=}\ddot{N}{-}\ddot{F}:$ (120°, 120°)

8.73 (a) $[\ddot{N}{=}C{=}\ddot{N}]^{2-} \leftrightarrow [:\ddot{N}{-}C{\equiv}N:]^{2-}$

(b) $O{=}N{=}N(O){-}O \leftrightarrow O{=}N{-}N(=O){-}O \leftrightarrow O{=}N{-}N(O){=}O$ (c) $[O{=}C{=}N]^{-} \leftrightarrow$

$[:\ddot{O}{-}C{\equiv}N:]^{-} \leftrightarrow [:O{\equiv}C{-}\ddot{N}:]^{-}$

(d) $O{=}C{=}C{=}C{=}O \leftrightarrow :O{\equiv}C{-}C{\equiv}C{-}\ddot{O}: \leftrightarrow$

$:\ddot{O}{-}C{\equiv}C{-}C{\equiv}O:$

8.75 (a) $AlCl_3 + [:\ddot{Cl}:]^{-} \rightarrow [AlCl_4]^{-}$

Lewis acid Lewis base Product

(b) $:\ddot{I}{-}\ddot{I}: + [:\ddot{I}:]^{-} \rightarrow :\ddot{I}{=}\ddot{I}{-}\ddot{I}:$

Lewis acid Lewis base Product

(c) $[Cr]^{3+} + 6\ H{-}\ddot{N}(H){-}H \rightarrow [Cr(NH_3)_6]^{3+}$

Lewis acid Lewis base Product

(d) $SnCl_4 + 2\ [:\ddot{Cl}:]^{-} \rightarrow [SnCl_6]^{2-}$

Lewis acid Lewis base Product

8.77 (a) One might predict that BF_3 would be stronger since there is a greater electronegativity difference between B and F than between B and Cl. Hence there is a greater partial positive charge on B in BF_3, which should make it easier to add an electron pair. However, the chemical reactivity of these two species indicates that the reverse may be true. (b) Al^{3+} is the stronger Lewis acid.

8.79 $CO_3^{2-}(s) \rightarrow CO_2(g) + O^{2-}(s)$;

Complex Lewis acid Lewis base

$Ca^{2+}O^{2-}(s) + SiO_2(s) \rightarrow CaSiO_3$

Lewis base Lewis acid Complex

8.81 (a) tetrahedral, ~109°
(b) angular, slightly less than 109°
(c) angular, slightly less than 120°

8.83 (a) $[H{-}CH_2{-}C(=O)O]^{-}$ or $[H{-}CH_2{-}C(O)(=O)]^{-}$

(b) $[H{-}CH_2{-}C(=O){-}CH_2]^{-}$ or $[H{-}CH_2{-}C(O){=}CH_2]^{-}$

(c) $[H_2C{-}CH{=}CH_2]^{+}$ or $[H_2C{=}CH{-}CH_2]^{+}$

Case Questions

1. (a) 360°/5 = 72° (b) 10

3. There is only one nearest neighbor B—B distance. B_1—B_2 and B_3—B_4 would have to be increased.

CHAPTER 9

Classified Exercises

9.1 (a) +1486 kJ/mol (b) +3150 kJ/mol

9.3 (a) −247 kJ/mol (b) +227 kJ/mol (c) −1 kJ/mol

9.5 (a) −232 kJ (b) −44 kJ (c) −287 kJ

9.7 (b) < (c) < (a)

9.9 The C=C bond length for the charged species would be expected to be longer due to resonance (in effect, it has a "bond order" of 1.5).

9.11 (a) $H_2N{-}NH_2$: N—N 150 pm, N—H 112 pm (b) O=C=O: 127 pm

(c) $H_2N{-}C(=O){-}NH_2$: C=O 127 pm, N—H 112 pm, N—C 152 pm

9.13 (a) I (b) C (c) C

9.15 $H^+ > Be^{2+} > Li^+ > Sr^{2+}$; the order parallels the ion size, small to large

9.17 $Br^- > Cl^- > N^{3-} > O^{2-}$; the order parallels the ion size, large to small

9.19 (a) I (b) C (c) C (d) I

9.21 (a) 1.4 D, polarized toward oxygen (b) 1.0 D, polarized toward fluorine (c) 1.0 D, polarized toward oxygen

9.23 (a) One nonpolar bond (b) All C—H bonds slightly polar (c) Both bonds are polar

9.25 (a)

```
       :Cl:
        |
  :Cl—C—Cl: ; tetrahedral, nonpolar
        |
       :Cl:
```

(b) PCl_5

```
        :Cl:
  :Cl\   |   /Cl:
          P          ; trigonal bipyramidal, nonpolar
        /   \
    :Cl      Cl:
```

(c)

```
  :F     F:
     \  /
      Xe      ; square planar, nonpolar
     /  \
  :F     F:
```

9.27 (a) Nonpolar (b) Polar (c) Polar

9.29 (a) **a** and **b,** polar; **c,** nonpolar (b) **b** is the most polar, hence it has the largest dipole moment.

9.31 (a) $N = 5 - 2 - \frac{1}{2}(6) = 0$; $H = 1 - 0 - \frac{1}{2}(2) = 0$ (b) $C = 4 - 0 - \frac{1}{2}(8) = 0$; $O = 6 - 4 - \frac{1}{2}(4) = 0$

9.33 (a) (1) —S̈= = +1, Ö= = 0, —Ö: = −1; (2) =S= = 0, Ö= = 0, =Ö = 0; structure (2) has formal charges closest to zero, hence, most plausible structure (b) (1) =S= (with ‖ above) = −2, Ö= = 0, Ö (with ‖ below) = 0; (2) =S= (with | above) = −1, Ö= = 0, :Ö: (with | below) = −1; structures are similar

9.35 Structure (b) is more stable.

9.37 (a) Structure (1, left) is a more dominant contributor. (b) Structure (2, right) is a more dominant contributor.

9.39 (a) σ (b) π (c) Neither (d) σ

9.41 (a) Tetrahedral (b) Linear (c) Octahedral (d) Trigonal planar

9.43 (a) sp^3d (b) sp^2 (c) sp

9.45 (a) sp^2 [↑][↑][↑][ ] (b) sp^3 [↑][↑][↑][↑]

(c) sp^3 [⇅][⇅][↑][↑]

9.47 (a) Csp^3, H$1s$ (b) Osp^3, H$1s$ (c)Psp^3d, Cl$3p$

9.49 (a) C hybridization is sp^2, trigonal planar geometry (b) Methyl C, sp^3 hybridization, tetrahedral geometry; carbonyl C (C=O), sp^2 hybridization, trigonal planar geometry; O hybridization, sp^3, tetrahedral geometry with two lone pairs

9.51 (a) $1s\sigma^2$, $1s\sigma^{*2}$, $2s\sigma^2$, $2s\sigma^{*2}$, $2p\sigma^2$, $2p\pi^4$ (b) $1s\sigma^2$, $1s\sigma^{*2}$, $2s\sigma^2$, $2s\sigma^{*2}$, $2p\sigma^2$, $2p\pi^4$, $2p\pi^{*4}$ (c) $1s\sigma^2$, $1s\sigma^{*2}$, $2s\sigma^2$, $2s\sigma^{*2}$ (d) $1s\sigma^2$, $1s\sigma^{*2}$, $2s\sigma^2$, $2s\sigma^{*2}$, $2p\sigma^2$, $2p\pi^4$, $2p\pi^{*4}$, $2p\sigma^{*2}$

9.53 (a) Paramagnetic (b) Paramagnetic (c) Paramagnetic (d) Paramagnetic

9.55 (a) F_2 has a stronger bond than F_2^{2-} (b) B_2 has a stronger bond than B_2^+

9.57 (a) $1s\sigma^2$, $2s\sigma^{*2}$, $2s\sigma^2$, $2s\sigma^{*2}$, $2p\sigma^2$, $2p\pi^4$; Bond order = 3 (b) $1s\sigma^2$, $2s\sigma^{*2}$, $2s\sigma^2$, $2s\sigma^{*2}$, $2p\sigma^2$, $2p\pi^4$, $2p\pi^{*1}$; Bond order = $2\frac{1}{2}$

9.59 (a) There are 4 orbitals in the π system: 2 π bonding MOs, 2 π^* antibonding MOs; 2 π bonding MOs are doubly occupied for 2 π bonds. (b) There are 4 orbitals in two π systems; therefore, 8 orbitals: 4 π bonding MOs, 4 π^* antibonding MOs; 4 π bonding MOs are doubly occupied for 4 π bonds.

Unclassified Exercises

9.61 (a) Shortest (b) Longest (c) Mid-length

9.63 (a) −46 kJ/mol for step 1; −293 kJ/mol for step 2 (b) −339 kJ/mol

9.65 The carbon-nitrogen triple bonds would be longer in N≡C—N^{2-} because of the effect of resonance. The bond order, in reality, would be less than three, hence, a longer bond.

9.67 An electronegativity difference between 1 and 3 indicates the presence of both ionic and covalent character. However, it must be recognized that electronegativity considerations alone are only approximate; other factors must be taken into account as well. For example, the large positive oxidation number of Si in SiF_4.

9.69 (a) sp^3, pyramidal with lone pair; ∠Cl—P—Cl is 107° (b) sp^3d^2, octahedral (c) sp^2, trigonal planar

9.71 (a) See Table 9.6 for sketches. XX′, linear, polar; XX'_3, T-shaped, polar X′—X—X′ angle is a little less than 90°; XX'_5, square pyramidal, polar, X'_{base}—X—X'_{axial} angle is a little less than 90°; X'_{base}—X—X'_{base} angle is a little less than 90°; XX'_7, pentagonal bipyramidal, nonpolar, X'_{planar}—X—X'_{axial} angle = 90°; X'_{planar}—X—X'_{planar} angle = 72°

9.73

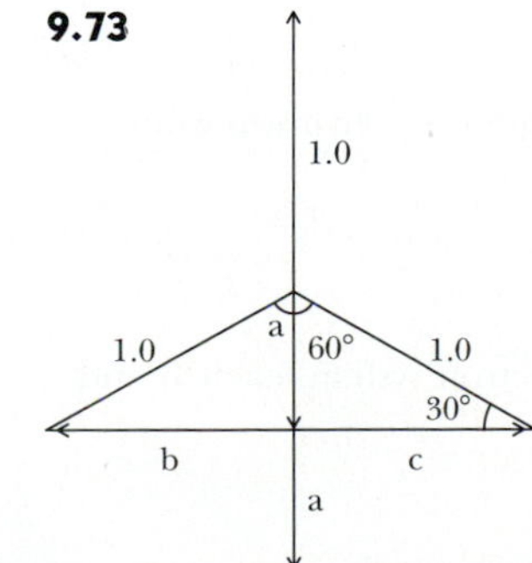

$\frac{a}{1.0} = \sin 30° = 0.500$; a = 0.500 (there are two such downward components); a + a = 0.5 + 0.5 = 1.0; b and c are equal and opposite (cos 30° = 0.866); 0.866 − 0.866 = 0; 1.0 − 1.0 = 0; so no net dipole moment

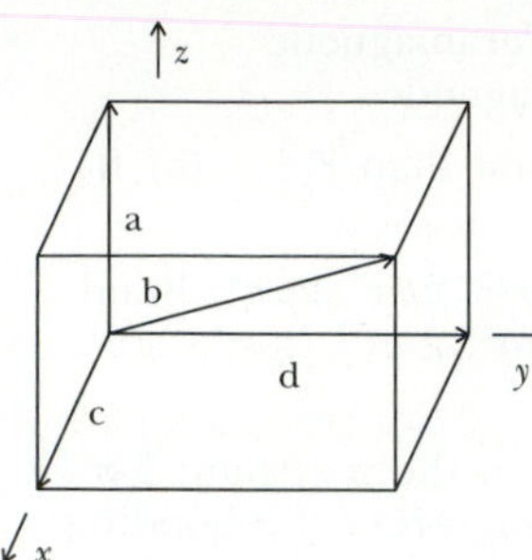

The resultant of $\vec{a}$ and $\vec{b}$ is along the $+z$ axis; the resultant of $\vec{c}$ and $\vec{d}$ is along the $-z$ axis. The two resulting vectors are equal in magnitude and opposite in direction, and therefore, cancel.

9.75 (a) (1) $\ddot{O}{=}$ = 0, $\ddot{O}$ = 0, —H = 0, —Ö— = −1, =Cl— = 0; (2) :Ö— = −1, :Ö: = −1, —H = 0, —Ö— = −1, —Cl— = +2; structure (1) has the lower energy (b) $\ddot{O}{=}$ = 0, =C= = 0, =S = 0 (c) H— = 0, —C≡ = 0, ≡N: = 0 (d) (1) N= = −1, =C= = 0, =N = −1; (2) :N≡ = 0, ≡C— = 0, —N: = −2; structure (1) has lower energy (e) (1) :O: = −1, :Ö— = −1, :Ö: = −1, —Ö: = −1, —As— = +1; (2) O = 0, :Ö— = −1, O = 0, —Ö: = −1, —As— = −1; structure (1) has the lower energy

9.77 (a) +4 (b) +5 (c) +6 (d) +7

9.79 Carbon in C—C bonds, sp^2, σ bond; Carbon in C=C bonds, sp^2, 1 σ bond, 1 π bond; nitrogen in N—C bonds, sp^3, 1 σ bond, lone pair sp^3; nitrogen in N=C bonds, sp^2, 1 σ bond, 1 π bond, lone pair sp^2

9.81 H, H :C: :C:C: : :N: The C—C—C bond (sp^2, sp^2, sp) angle is approximately 120°, due to sp^2 hybridization.

9.83 (ring: H:N, B, N:H, H:B, B:H, N, with H on B and N) 30 electron system, each B and N atom sp^2 hybridized

9.85 Xe with three :O: ; 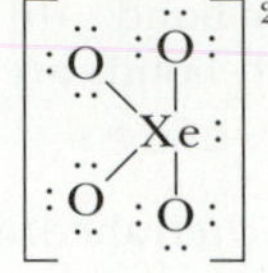 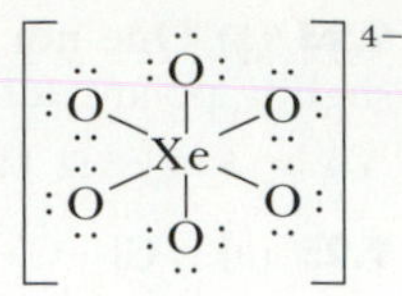

sp^3 hybridized Trigonal pyramidal | sp^3d hybridized Distorted tetrahedral or see-saw | sp^3d^2 hybridized Octahedral

9.87 HeH^- (4 e^- system) $1s\sigma^2$, $1s\sigma^{*2}$ (bond order = 0); So, it would not exist. However, HeH^+ does . . . (2 e^- system); $1s\sigma^2$ (bond order = 1)

9.89 O_2 has 16 electrons and would have the configuration $(1s\sigma)^3(1s\sigma^*)^3(2s\sigma)^3(2s\sigma^*)^3(2p\sigma)^3 2p\pi^1$; the bond order > 0 and so would be stable. Ar_2 has 36 electrons and would have the configuration $(1s\sigma)^3(1s\sigma^*)^3(2s\sigma)^3(2s\sigma^*)^3(2p\sigma)^3(2p\pi)^6(2p\sigma^*)^3(3s\sigma)^3(3s\sigma^*)^2$; the bond order > 0 and so would be stable. In such a universe, an atom with closed s and p subshells would have ns^3np^9 configuration. Instead of a stable set of eight (octet rule), a G. N. Lewis in that universe would presumably discover a stable set of 12 valence electrons and atoms of other elements would either gain or lose electrons to establish a stable outer shell (12 electron) configuration.

Case Questions

1. (a) H:C: : :C:H (b) —C(H)=C(H)—C(H)=C(H)—C(H)=C(H)—C

(c) (Refer to Figure in Box 9.) When an electron enters the chain at carbon atom 1, we can visualize the breakage of the π bond between carbon atoms 2 and 3 (a): then one of the broken π bond's electrons pairs with the lone electron to form a new π bond between C atoms 1 and 2, thereby leaving an unpaired electron on C atom 3 (b). Next, the π bond between C atoms 4 and 5 is broken, one electron pairing with the lone electron on C atom 3 and forming a new π bond between C atoms 3 and 4. The lone electron now appears on C atom 5 (c). Then the additional electron can move to C atom 7 by the same sequence of events (d). The process can continue down the chain of the molecule, thereby turning it into a conducting path. (d) Carbon 5 is sp^2 hybridized, the lone e^- is in a p orbital.

3. 6 molecular orbitals with 6 electrons:

⇅	⇅	⇅

Antibonding molecular orbitals (π^*)
Bonding molecular orbitals (π)

CHAPTER 10

Classified Exercises

10.1 (a) London forces (b) Dipole-dipole, London forces (c) Dipole-dipole, London forces

10.3 (a) :C≡O:, London forces are dominant but very weak dipole-dipole forces also exist

(b) NF_3 (Lewis structure), London forces are dominant but very weak dipole-dipole forces also exist

(c) CH_4 (H—C—H with H above and below), London forces

10.5 (b), NH_3, and (c), CH_3OH can form hydrogen bonds.

10.7 (a) Mg^{2+} (b) Fe^{3+}

10.9 (a) These compounds have the same structure, but SiH_4 has the greater molar mass, hence we expect that it has the higher normal boiling point, and, in fact, it does: −112°C for SiH_4 versus −162°C for CH_4. (b) HF, because of stronger hydrogen bonding (c) H_2O, because of stronger hydrogen bonding; there are two O—H bonds

10.11 (a) The attractions between atoms in both Xe and Ar is a result of London forces, but because Xe has more electrons it is bigger and therefore more polarizable, hence it has stronger London forces and a higher melting point. (b) HI and HCl have both London forces and dipole-dipole attractions along with some hydrogen bonding. But the dominant attraction is the London force which is greater in HI. (c) There is strong hydrogen bonding in water, none in $C_2H_5OC_2H_5$, so water has the lower vapor pressure at the same temperature.

10.13 (a) SF_4: (Lewis structure); SF_6: (Lewis structure); SF_4 has the higher boiling point (b) SF_4: (Lewis structure); CF_4: (Lewis structure); SF_4 has the higher boiling point

(c) *cis*-CHCl=CHCl: (H and H on the same side, Cl and Cl on the same side of C=C); *trans*-CHCl=CHCl: (H and Cl on opposite sides of C=C); cis-compound has a higher boiling point

10.15 0.17 g H_2O

10.17 1.9×10^3 g

10.19 (a) ~98°C (b) ~100°C (c) ~93°C

10.21 (a) Vapor (b) Vapor (c) Critical point, vapor and liquid coexist at the same density

10.23 (a) ~2.2 K (b) ~12 atm (c) No, there is no phase equilibrium line between solid and gas.

10.25 Its triple point is higher than room temperature.

10.27 (a) Ionic (b) Molecular (c) Molecular (d) Metallic

10.29 (a) 509 pm (b) 7.6×10^{21} cells

10.31 (a) 2 atoms (b) Coordination number of 8. (c) 508 pm

10.33 (a) 144 pm (b) 132 pm

10.35 (a) 0.53 atoms Zn/1 atom Cu (b) 7.0 atoms Sn:2.0 atoms Pb:112 atoms Cu (c) 5.0 atoms Cr:2.0 atoms Ni:17 atoms Fe

10.37 An electrical conductor has an incompletely filled band of molecular orbitals. An electrical insulator has a filled band of molecular orbitals well separated and lower in energy than the empty band above it.

10.39 Graphite is a metallic conductor parallel to the planes. Between planes, graphite is a semiconductor perpendicular to its planes.

10.41 (a) *n*-type (b) *p*-type (c) *n*-type

10.43 Silicon has a relatively large energy gap, diamond has a much larger one, but both gaps are not exceedingly large, so at high temperatures there can be promotion of electrons from their valence bands to their conduction bands, resulting in conductivity.

10.45 (a) 4 Cl^- and 4 Na^+ ions, making 4 formula units of NaCl per unit cell. (b) 4 Ca^{2+} and 8 F^- ions, making 4 formula units of CaF_2 per unit cell. (c) The Ca^{2+} and F^- ions have coordination number of 8 and 4, respectively.

10.47 $CaTiO_3$

10.49 3.48×10^{18} unit cells

10.51 B, C, Si, Ge, P, As

10.53 357 kJ/mol bonds

10.55 (a) London forces (b) Dipole-dipole and London forces (c) Hydrogen bonding, dipole-dipole and London forces

10.57 $A_2B_2C_4$

Unclassified Exercises

10.59 (a) Stronger (b) Low (c) High (d) Weaker (e) Weak, low (f) Low (g) Strong, high

10.61 There are more lone pairs on O's per molecule in H_2O_2, thus its structure H—O—O—H (Lewis structure) allows for stronger hydrogen bonding than in water. Its greater molecular weight allows for greater London forces. Both of these factors combine to produce higher viscosity and boiling point.

10.63 (a) Vapor pressure increases due to the increased kinetic energy of the molecules at higher temperatures. (b) No effect on the vapor pressure as such, which is determined only by the temperature, but the rate of evaporation increases (c) No effect on the vapor pressure, which is determined only by the temperature, but additional liquid evaporates

10.65 Oxygen, ammonia, bromine, carbon tetrachloride

10.67 Water "wets," that is, adheres to strongly, the molecules of paper, but not those of polyesters. Paper towels are effective because of the hydrogen bonding between water molecules and the OH groups of the cellulose molecules in paper. A synthetic towel should have a molecular structure that allows for hydrogen bonding.

10.69 As pressure is lowered the gaseous state becomes the stable state, consequently the liquid water will boil.

10.71 The vapor pressure of $CO_2(l)$ is about 56 atm at 20°C, so nothing would happen other than its volume would decrease. (a) Would stay as a vapor because the vapor pressure of the liquid is still 26 atm. (b) Would be converted to a solid.

10.73 (a) 78.6% (b) At 25°C the vapor pressure of water is 23.76 Torr; therefore, some of the water vapor in the air would condense as dew or fog.

10.75 (a) Assuming that the air at 25°C is originally saturated with ethanol, the partial pressure of ethanol is still 58.9 Torr. The vapor pressure above a liquid is determined, for all practical purposes, by the temperature alone. So condensation of some of the gaseous ethanol occurs. (b) 1441 Torr

10.77 43 Torr; at this pressure the boiling point of water is about 40°C

10.79 (a) Ionic (b) Molecular (c) Molecular (d) Network, though in a sense molecular with very large molecules

10.81 (a) 4.952×10^{-8} cm (b) 1.214×10^{-22} cm^3 (c) 175.1 pm

10.83 (a) 4.00, a face-centered cubic unit cell (b) 12

10.85 74.0%

10.87 (a) Light of sufficient energy can excite electrons in a semiconductor, causing the presence of more holes in the full band and more electrons in the empty band. (b) 690 nm (maximum) (c) No, infrared wavelengths are greater than 900 nm and are not sufficiently energetic for selenium to serve as a detector.

Case Question

1. (a) 1.9×10^3 atm (b) Sufficient

CHAPTER 11

Classified Exercises

11.1 (a) 0.330 mol KNO_3/L soln (b) 0.222 mol glucose/L soln

11.3 (a) 5.8 g NaCl (b) 2.8 g $CaCl_2$ (c) 57 g $C_6H_{12}O_6$

11.5 (a) 4.0% NaCl (b) 3.8% NaCl (c) 0.823% $C_{12}H_{22}O_{11}$

11.7 (a) $x_{H_2O} = 0.560$; $x_{C_2H_5OH} = 0.440$ (b) $x_{H_2O} = 0.470$; $x_{CH_3OH} = 0.530$ (c) $x_{C_6H_{12}O_6} = 1.8 \times 10^{-3}$; $x_{H_2O} \approx 1.0$ (2 sf)

11.9 (a) 14.3% (b) 0.0612 (c) 3.62 m C_2H_5OH

11.11 (a) 6.84×10^{-1} m (b) 9.6 m (c) 0.162 m

11.13 (a) 4.11 g $ZnCl_2$ (b) 0.62 g (2 sf) (c) 7.36 g $KClO_3$

11.15 (a) 0.90 (b) 3.04 (c) 21.7 or 22 (2 sf)

11.17 (a) $x_{Na^+} = x_{Cl^-} = 1.7 \times 10^{-3}$ $x_{H_2O} = 0.996$ or ~1.0 (2 sf) (b) $x_{Na^+} = 7.1 \times 10^{-3}$ $x_{CO_3^{2-}} = 3.6 \times 10^{-3}$ $x_{H_2O} = 0.99$ (c) $x_{K^+} = x_{NO_3^-} = 1.9 \times 10^{-2}$ $x_{H_2O} = 0.96$ (2 sf)

11.19 (a) 5.5 g $CaCl_2 \cdot 6\,H_2O$ (b) 30 g $NiSO_4 \cdot 6\,H_2O$

11.21 (a) 0.36 m (b) 4.50% $(NH_4)_2SO_4$

11.23 (a) 3.40 mol/L; 3.77 m (b) 5.40 mol/L; 6.10 m

11.25 (a) 6.4×10^{-4} mol/L (b) 1.5×10^{-2} mol/L (c) 2.3×10^{-3} mol/L

11.27 (a) 1.3×10^{-4} mol/L (b) 0.1 atm

11.29 (a) If the pressure of CO_2 is increased, the equilibrium will shift to the right to minimize the volume of CO_2, as P and V are inversely proportional. (b) If the temperature is raised, the equilibrium will shift to the left to increase the volume of CO_2, as V and T are directly proportional.

11.31 (a) Enthalpy of hydration (b) Solubility will decrease.

11.33 (a) $+6.7 \times 10^2$ J for NaCl = -6.7×10^2 J for water (b) -58 J for NaBr = +58 for water (c) -2.46×10^4 J for $AlCl_3$ = $+2.46 \times 10^4$ J for water (d) $+3.21 \times 10^3$ J for NH_4NO_3 = -3.21×10^3 J for water

11.35 (a) Increases (b) Decreases (c) Decreases

11.37 (a) -37 kJ/mol (b) +3 kJ/mol (c) +66 kJ/mol ΔH_{soln} increases as the ion size increases.

11.39 (a) 684 Torr (b) 759 Torr

11.41 (a) 4.5 Torr (b) 354 Torr (c) 0.16 Torr

11.43 (a) 0.052 (b) 114 g/mol or 1.1×10^2 g/mol

11.45 (a) $\Delta T = 0.051$°C, b.p. ≈ 100.1°C (b) $\Delta T = 0.22$°C, b.p. ≈ 100.2°C (c) $\Delta T = 0.091$°C, b.p. = 100.09°C

11.47 (a) 100.3°C (b) 81°C

11.49 168 g/mol

11.51 (a) −0.19°C (b) −0.82°C (c) −0.17°C

11.53 (a) −3.8°C (b) 1.63 (c) 0.208 moles of the electrolyte

11.55 182 g/mol

11.57 (a) 1.84 (b) 0.318 m (c) 92.0%

11.59 (a) 0.24 atm (b) 48.2 atm (c) 0.72 atm

11.61 (a) 15.0 atm (b) 16 atm (c) 48 atm or 5×10^1 atm

11.63 (a) 1.2 atm; 12.4 m (or 12 m, 2 sf) (b) 0.048 atm; 0.49 m (c) 8.2×10^{-5} atm; 8.6×10^{-4} m

11.65 3.3×10^3 g/mol

11.67 54 Torr

Unclassified Exercises

11.69 (a) 40% (b) 6.9 mol/L

11.71 (a) 5.56 m (b) 0.91 (c) 4.6 mol/L

11.73 (a) 4.98×10^{-1} mol/L (b) 17.37 Torr (c) 100.3°C

11.75 (a) If the partial pressure of CO_2 in the air above the solution is increased, the equilibrium will shift to the left, and more CO_2 will dissolve. (b) If the total pressure of the gas is increased by addition of nitrogen, no change in the equilibrium will occur, the partial pressure of CO_2 is unchanged, and the solubility is unchanged.

11.77 −51 kJ

11.79 1.3×10^2 g/mol

11.81 (a) $C_9H_{13}O_3N$ (b) 1.8×10^2 g/mol (c) $C_9H_{13}O_3N$

11.83 (a) $CO(NH_2)_2$ (b) $CO(NH_2)_2$ (c) $CO(NH_2)_2$ (d) 1.2 atm

11.85 7 atm (1 sf)

11.87 2.5×10^5 g/mol (2 sf)

11.89 (a) 11 Torr (b) 76 Torr

11.91 83%

Case Questions

1. (a) 1×10^{13} (b) 9×10^{-10} g

3. 50%

5. 3.4×10^{-2} g

CHAPTER 12

Classified Exercises

12.1 (a) 1.0×10^{-3} (mol O_3/L·s) (b) 0.28 (mol CrO_4^{2-}/L·s)

12.3 (a) 2.5×10^{-4} mol/L·h at 1.11 h; 2.0×10^{-4} mol/L·h at 2.22 h; 1.8×10^{-4} mol/L·h at 3.33 h

(b)

Time, h	$\Delta [N_2O_5]$, 10^{-4} mol/L·h
0	2.5
1.11	2.3
2.22	2.0
3.33	1.8
4.44	1.6

(a) and (c)

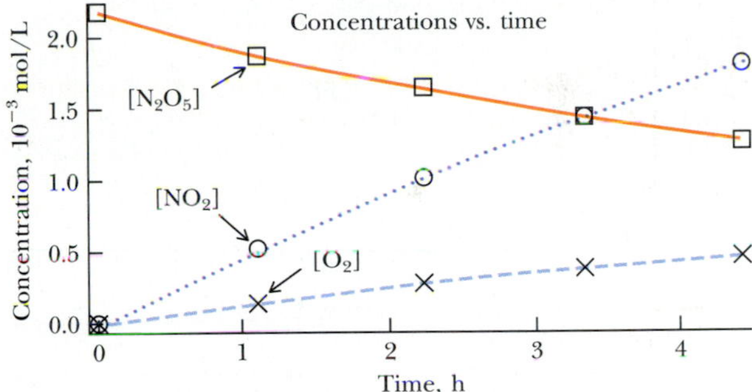

12.5 (a) (mol A/L·s) (b) s^{-1} (c) $L{\cdot}mol^{-1}{\cdot}s^{-1}$

12.7 9.6×10^{-5} mol N_2O_5/L·s

12.9 (a) 2.4×10^{-5} mol/L·s (b) The rate is doubled also.

12.11 (a) $k[X]^2[Y]^{1/2}$ (b) $k[A][B][C]^{1/2}$

12.13 (a) First order in SO_2, zero order in O_2, negative one-half order in SO_3, one-half order overall (b) Second order in A, negative second order in B, zero order in C, zero order in products, zero order overall

12.15 $k[CH_3Br][OH^-]$

12.17 (a) Reaction is second order in A_2, first order in B_3; overall order = 3 (b) $k[A_2]_0^2[B_3]_0$ (c) 1.2×10^2 $L^2{\cdot}mol^{-2}{\cdot}s^{-1}$ (d) 0.85 mol/L·s

12.19 (a) and (b) Rate = $k[NO_2]_0[O_3]_0$; the reaction is first order in each reactant, second order overall (c) 4.3×10^7 $L{\cdot}mol^{-1}{\cdot}s^{-1}$ (d) 5.1 mol/L·s

12.21 (a) 6.93×10^{-4} s^{-1} (b) 7.6×10^{-3} s^{-1}

12.23 (a) 5.2 h (b) 1.79×10^{-2} mol/L (c) 126 min

12.25 (a) 7.6×10^2 s (b) 1.5×10^2 s (c) 2.0×10^2 s

12.27 (a) 40 L/mol·s (b) 1.9×10^{-3} $L{\cdot}mol^{-1}{\cdot}s^{-1}$

12.29 (a) $k[SO_2Cl_2]$ (b) 247 min (c) 10.9 g

12.31 (a) 1.3×10^{-2} s^{-1} (b) 172 s

12.33 (a)

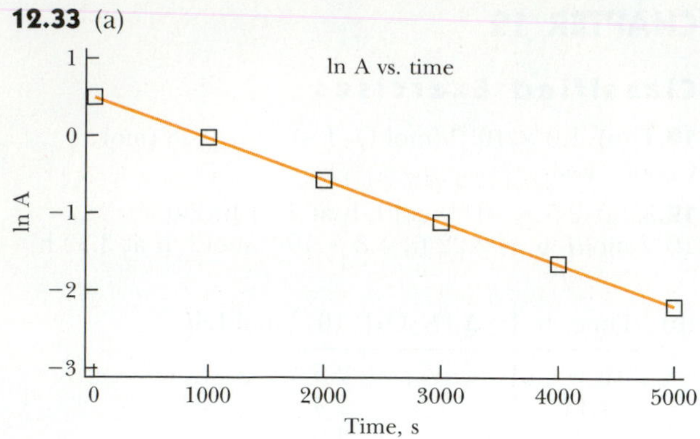

(b) 5.5×10^{-4} s^{-1}

12.35 (a)

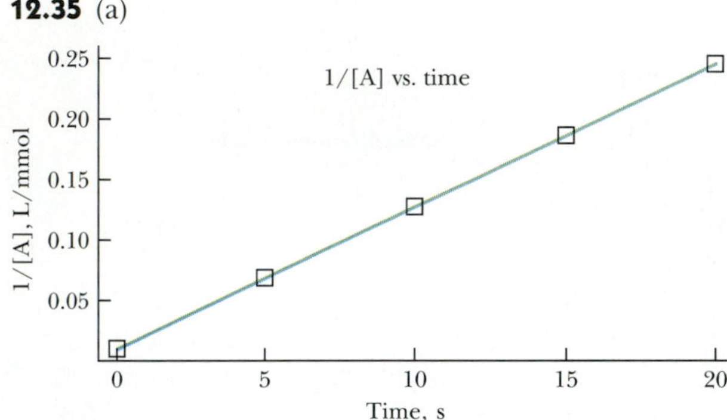

(b) 12 L/(mol·s)

12.37 (a) 2.74×10^2 kJ/mol (b) 8.8×10^{-2} s^{-1}

12.39 2.4×10^2 kJ/mol

12.41 2.9×10^9 L/mol·s

12.43 $E'_a = E_a - \Delta H$

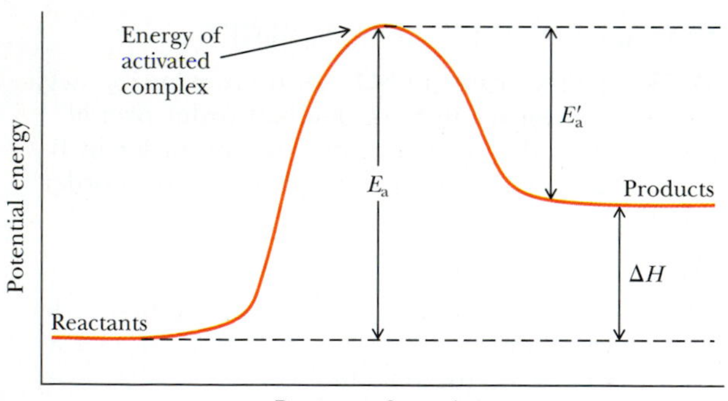

12.45 (a) NO (b) NO_2

12.47 3.3×10^6

12.49 80.9 kJ/mol

12.51 (a) $k[NO]^2$, bimolecular (b) $k[Cl_2]$, unimolecular (c) $k[NO_2]^2$, bimolecular (d) Because Cl and NO are radicals, (b) and (c)

12.53 $2\ ICl + H_2 \rightarrow 2\ HCl + I_2$; HI is the only intermediate

12.55 $k[CO][Cl_2]^{3/2}$

Unclassified Exercises

12.57 (a) 5.4 mol ClO^-/L·min (b) 4.05 mol H_2/L·s

12.59 (a) 1/3 (b) 2/3 (C) 2

12.61 8.0×10^{-6} mol/L·s

12.63 Second order

12.65 (a) A = 1, B = 2, C = 0; overall rate is 3 (b) $k[A]_0\,[B]_0^2$ (c) 20 L^2/mol^2·s (d) 2.9×10^{-9} mol, L·s

12.67 (a) 0.132 mol/L (b) 39 min (c) 34 min

12.69 (a) 0.0504 mg (b) 0.106 atm

12.71 300 kJ/mol

12.73 (a) Bimolecular (b) Unimolecular (c) Termolecular

12.75 (a) NO (b) NO_2

12.77 (a) 88 kJ/mol (b) 2.3×10^{-2} L/mol·s

12.79 75 kJ/mol

12.81 (a) Cl (b) ClO (c) Cl, ClO, O (d) Step 1 is initiating, step 2 is propagating (e) $Cl + Cl \rightarrow Cl_2$

12.83 Step 2: $Cl + H_2 \rightarrow HCl + H$, step 3: $H + Cl_2 \rightarrow HCl + Cl$, step 4: $Cl + Cl \rightarrow Cl_2$, step 5: $H + H \rightarrow H_2$, step 6: $H + Cl \rightarrow HCl$

12.85

Time, min	Rate, mol/(L·min)
0	1.01×10^{-3}
5	6.07×10^{-4}
10	3.65×10^{-4}
15	2.2×10^{-4}
20	1.3×10^{-4}
30	4.8×10^{-4}
100	~0

Case Questions

1. (a) $CH_2—CH_2OH(g)$ (b) $CH_3CH_2\cdot(g) + H_2O(g)$

3. (a) $NO_2(g) + \cdot O\cdot(g) \rightarrow NO_3(g)$ (b) $NO_2(g) + NO_3(g) \rightleftharpoons N_2O_5(g)$ (c) $2\ HNO_3(aq)$

CHAPTER 13

Classified Exercises

13.1 (a) $[COCl][Cl]/[CO][Cl_2]$ (unitless) (b) $[HBr]^2/[H_2][Br_2]$ (unitless) (c) $[O_2]^3/[O_3]^2$ (units; mol/L)

13.3 (a) 0.024 atm^2 (b) 6.4 atm

13.5 Case 1: 48.8; case 2: 48.9; case 3: 48.9

13.7 (a) 2.4×10^{-2} (b) 1.90

13.9 (a) 0.50 (b) No (c) Products

13.11 (a) 6.94; 0.121 (b) Yes

13.13 (a) Heterogeneous, because more than one phase is involved in the equilibrium; $Q_P = P_{H_2S}P_{NH_3}$ (b) Heterogeneous, because more than one phase is involved in the equilibrium; $Q_P = P_{CO_2}P_{NH_3}{}^2$

13.15 (a) $1/[Cl_2]$ (b) $[N_2O][H_2O]^2$

13.17 6.6×10^{-3}

13.19 1.58×10^{-8}

13.21 2.09×10^{-5} mol/L

13.23 19 atm

13.25 (a) Equilibrium concentrations: $[Cl] = 1.1 \times 10^{-5}$ mol/L, $[Cl_2] \approx 1.0 \times 10^{-3}$ mol/L; % decomposition = 0.55% (b) Equilibrium concentrations: $[F] = 3.18 \times 10^{-4}$ mol/L, $[F_2] \approx 8.41 \times 10^{-4}$ mol/L; F_2 is more unstable at 1000 K than Cl_2

13.27 % decomposition $\approx$ 0%

13.29 (a) 2.6×10^{-2} mol/L (b) 26%

13.31 8.0×10^{-4} mol/L = $[H_2S]$, 0.200 mol/L = $[NH_3]$

13.33 4.97×10^{-4} mol/L = $[Cl_2]$, 1.99×10^{-1} mol/L = $[PCl_3]$, 2.00×10^{-4} mol/L = $[PCl_5]$

13.35 3.60×10^{-4} mol/L = [NO], 0.114 mol/L = $[N_2] = [O_2]$

13.37 1.06

13.39 9.0×10^{-5} mol/L $\approx [CO_2]$, $[CO] \approx 4.9 \times 10^{-3}$ mol/L, $[O_2] \approx 4.5 \times 10^{-4}$ mol/L

13.41 $[CH_3COOC_2H_5] = [H_2O] \approx 0.32$ mol/L, $[CH_3COOH] = 0$ mol/L, $[C_2H_5OH] = 6$ mol/L

13.43 If $K_{p3} = P_{H_2O}\cdot P_{CO}/P_{CO_2}\cdot P_{H_2}$, $\sqrt{K_{p2}/K_{p1}} = P_{H_2O}\cdot P_{CO}/P_{CO_2}\cdot P_{H_2}$ and $K_{p3} = 2.9$

13.45 (a) It will decrease. (b) It will decrease. (c) The H_2 concentration will increase.

13.47 Decrease, increase, increase, increase, no change, decrease, decrease

13.49 (a) Reactants (b) Reactants (c) Reactants

13.51 (a) Increase (b) Increase

13.53 SiO_2 (quartz)

13.55 (a) $\alpha = 0.91$ (b) $K_P = x^2P^2/P(1 - \alpha)$

13.57 (a) Products (b) Products (c) Reactants

13.59 No

Unclassified Exercises

13.61 (a) $Q_c = [SO_2]/[O_2]$, $Q_P = P_{SO_2}/P_{O_2}$ (b) $Q_c = [SO_2][H_2O]/[SO_3][H_2]$, $Q_P = P_{SO_2}\cdot P_{H_2O}/P_{SO_3}\cdot P_{H_2}$ (c) $Q_c = [WCl_6][H_2]^3/[HCl]^6$, $Q_P = P_{WCl_6}\cdot P_{H_2}{}^3/P_{HCl}^6$

13.63 (a) 1.7×10^6 (b) 7.7×10^{-4}

13.65 The reaction is not at equilibrium and there is a tendency to form products (N_2 and H_2) at the expense of the reactant (NH_3).

13.67 1.4 mol/L (2 sf)

13.69 $K_c = 3.88$ $K_p = 4.00 \times 10^{-4}$

13.71 [NO] = 0.0389 mol/L $[SO_3]$ = 0.0489 mol/L $[SO_2]$ = 0.0011 mol/L $[NO_2]$ = 0.0211 mol/L

13.73 $K_c = 2.27 \times 10^{-4}$

13.75 $K_c = 4.4 \times 10^{-4}$

13.77 1.59 mol/L = $[PCl_3] = [Cl_2]$ 1.41 mol/L = $[PCl_5]$

13.79 2.1×10^{-13} mol/L = $[Cl_2]$ 4.2×10^{-13} mol/L = [HI] 1.49×10^{-2} mol/L $\approx$ [HCl]

13.81 $[NO_2] \approx 7.1 \times 10^{-3}$ mol/L, $[N_2O_4] \approx 1.2 \times 10^{-2}$ mol/L

13.83 (a) Shift to reactants (b) No effect (c) Shift to products (d) Shift to products (e) No effect (f) Shift to products (g) Shift to reactants

13.85 (a) As the reaction approaches equilibrium the molar concentration of CH_3OH will decrease because $Q_c > K_c$. (b) [CO] = 0.1099 mol/L; $[H_2]$ = 0.1498 mol/L; $[CH_3OH] = 1.0 \times 10^{-4}$ mol/L

13.87 6.88×10^{-3} mol/L = $[NO_2]$, 1.02×10^{-2} mol/L = $[N_2O_4]$

13.89 305.4 K

Case Questions

1. 3×10^4 years

3. (a) Not favored (b) Favored (c) Not favored (d) Not favored (e) Favored

CHAPTER 14

Classified Exercises

14.1 (a) $CH_3NH_3{}^+$ (b) H_2CO_3 (c) $CO_3{}^{2-}$ (d) $C_6H_5O^-$

14.3 (a) $H_2SO_4 + H_2O \rightarrow H_3O^+ + HSO_4{}^-$

Acid$_1$ (Conjugate) Base$_1$; Base$_2$ (Conjugate) Acid$_2$

(b) $C_6H_5NH_3{}^+ + H_2O \rightleftharpoons H_3O^+ + C_6H_5NH_2$

Acid$_1$ (Conjugate) Base$_1$; Base$_2$ (Conjugate) Acid$_2$

(c) $H_2PO_4{}^- + H_2O \rightleftharpoons H_3O^+ + HPO_4{}^{2-}$

Acid$_1$ (Conjugate) Base$_1$; Base$_2$ (Conjugate) Acid$_2$

(d) $NH_2NH_3{}^+ + H_2O \rightleftharpoons H_3O^+ + NH_2NH_2$

Acid$_1$ (Conjugate) Base$_1$; Base$_2$ (Conjugate) Acid$_2$

14.5 (a) $HCO_3^- + H_2O \rightleftharpoons H_3O^+ + CO_3^{2-}$

Conjugate: HCO_3^- (Acid$_1$) and CO_3^{2-} (Base$_1$); Conjugate: H_2O (Base$_2$) and H_3O^+ (Acid$_2$)

$H_2O + HCO_3^- \rightleftharpoons H_2CO_3 + OH^-$

Conjugate: H_2O (Acid$_1$) and OH^- (Base$_1$); Conjugate: HCO_3^- (Base$_2$) and H_2CO_3 (Acid$_2$)

(b) $HPO_4^{2-} + H_2O \rightleftharpoons H_3O^+ + PO_4^{3-}$

Conjugate: HPO_4^{2-} (Acid$_1$) and PO_4^{3-} (Base$_1$); Conjugate: H_2O (Base$_2$) and H_3O^+ (Acid$_2$)

$H_2O + HPO_4^{2-} \rightleftharpoons H_2PO_4^- + OH^-$

Conjugate: H_2O (Acid$_1$) and OH^- (Base$_1$); Conjugate: HPO_4^{2-} (Base$_2$) and $H_2PO_4^-$ (Acid$_2$)

14.7 (a) $[H_3O^+] = 1.6 \times 10^{-7}$, pH = 6.80 (b) $[OH^-] = 1.6 \times 10^{-7}$, pOH = 6.80

14.9 (a) $D_2O + D_2O \rightleftharpoons D_3O^+ + OD^-$ (b) pK_w = 14.870 (c) 3.67×10^{-8} mol/L (d) pOD = 7.435 (e) 14.870

14.11 $[H_3O^+]$ = 0.96 mol/L = $[Cl^-]$, $[OH^-] = 1.0 \times 10^{-14}$ mol/L, [HCl] = 0 mol/L

14.13 (a) 2×10^{-7} mol/L (b) 4×10^{-5} mol/L (c) 5×10^{-6} mol/L

14.15 (a) pH = 4.70, pOH = 9.30 (b) pH = 0.00, pOH = 14.00 (c) pH = 4.299, pOH = 9.70

14.17 (a) pH = 2.00, pOH = 12.00 (b) pH = 0.66, pOH = 13.34 (c) pOH = 3.06, pH = 10.94 (d) pH = 4.07, pOH = 9.93

14.19 (a) 0.050 mol/L pH = 1.30 (b) $[OH^-]$ = 0.050 mol/L, pH = 12.70

14.21 pH = 13.52

14.23 19.7 mL HCl

14.25

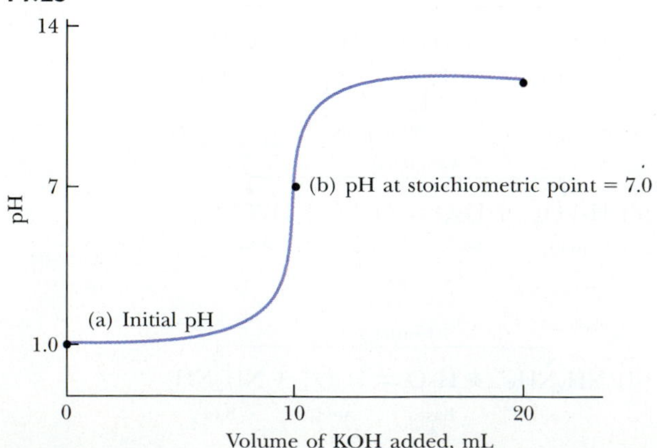

14.27 (a) 9.17×10^{-3} L HCl (b) 0.0183 L (c) 0.0635 mol/L (d) pH = 2.26

14.29 (a) 423 mL (b) 0.142 mol/L

14.31 69.8%

14.33 (a) pH = 13.04 (b) pH = 12.82 (c) pH = 12.55 (d) pH = 7.00 (e) pH = 1.81 (f) pH = 1.55

14.35 (a) pH = 10.5 (b) pH = 9.5 (c) pH = 4.5 (d) pH = 3.5

14.37

Name	Formula	K_a	pK_a
(a) formic acid	$HCOOH$	1.8×10^{-4}	3.75
(b) acetic acid	CH_3COOH	1.8×10^{-5}	4.74
(c) trichloroacetic acid	CCl_3COOH	3.0×10^{-1}	0.52
(d) benzoic acid	C_6H_5COOH	6.5×10^{-5}	4.19

Trichloroacetic acid > formic acid > benzoic acid > acetic acid

14.39

Formula	K_a	pK_a
(a) H_3PO_4	7.6×10^{-3}	2.12
(b) H_3PO_3	0.010	2.00
(c) H_2SeO_3	3.5×10^{-3}	2.46
(d) H_2SeO_4	0.012	1.92

$H_2SeO_4 > H_3PO_3 > H_3PO_4 > H_2SeO_3$

14.41 (a) $^+NH_3OH$, hydroxylammonium ion, strongest; NH_4^+, ammonium ion, weakest (b) $^+NH_3OH$, $K_a = 9.1 \times 10^{-7}$; NH_4^+, $K_a = 5.6 \times 10^{-10}$ (c) NH_4^+, weaker acid corresponds to higher pH

14.43 (a) $HClO_2$ is stronger. (b) HClO is stronger. (c) H_2SO_4 is stronger. (d) HNO_3 is stronger.

14.45 2,4,6-Trichlorophenol is the stronger acid.

14.47 Aniline < ammonia < methylamine < ethylamine

14.49 (a) 1.8% (b) 0.29%

14.51 6.5×10^{-3}%

14.53 (a) pH = 2.80, pOH = 11.20, percentage ionized = 1.1% (b) pH = 0.96, pOH = 13.04, percentage ionized = 73%

14.55 (a) pOH = 4.86, pH = 9.14, fraction ionized (%) = 0.081% (b) pOH = 3.74, pH = 10.26, percentage ionized (%) = 0.89%

14.57 (a) K_a = 0.010, pK_a = 2.00 (b) $K_b = 5.7 \times 10^{-4}$, pK_b = 3.25

14.59 pH = 4.07, $K_a = 6.6 \times 10^{-7}$

14.61 pH = 11.83, $K_b = 5 \times 10^{-4}$

14.63 (a) $H_2SO_4 + H_2O \rightarrow H_3O^+ + HSO_4^-$; $HSO_4^- + H_2O \rightleftharpoons H_3O^+ + SO_4^{2-}$; (b) $H_3AsO_4 + H_2O \rightleftharpoons$

$H_3O^+ + H_2AsO_4^-$; $H_2AsO_4^- + H_2O \rightleftharpoons H_3O^+ + HAsO_4^{2-}$; $HAsO_4^{2-} + H_2O \rightleftharpoons H_3O^+ + AsO_4^{3-}$; (c) $C_6H_4(COOH)_2 + H_2O \rightleftharpoons H_3O^+ + C_6H_4(COOH)COO^-$; $C_6H_4(COOH)COO^- + H_2O \rightleftharpoons H_3O^+ + C_6H_4(COO)_2^{2-}$

14.65 pH = 0.80

14.67 (a) pH = 3.94 (b) pH = 5.32

14.69 (a) 0.053 mol/L = $[H_3O^+]$; $[OH^-] = 1.9 \times 10^{-13}$ mol/L; $[H_2C_2O_4] = 0.047$ mol/L, $[C_2O_4^{2-}] = 6.5 \times 10^{-5}$ mol/L, $[HC_2O_4^-] = 0.053$ mol/L (b) $[H_2S] = 0.050$ mol/L, $[H_3O^+] = [HS^-] = 8.1 \times 10^{-5}$ mol/L, $[OH^-] = 1.2 \times 10^{-10}$ mol/L, $[S^{2-}] = 7.1 \times 10^{-15}$ mol/L

Unclassified Exercises

14.71 (a) HCO_3^- (b) $CH_3C(OH)_2^+$ (c) HPO_4^{2-} (d) O^{2-}

14.73 (a) Amphiprotic (b) Base (c) Base (d) Acid

14.75 (a) pOH = 1.59; pH = 12.41 (b) pOH = 2.496; pH = 11.50

14.77 pH = 12.70

14.79 [KOH] = 0.094 mol/L

14.81 [NaOH] = 4.8 mol/L

14.83 $K_b(F^-) = 2.8 \times 10^{-11}$, $K_b(CH_2ClCO_2^-) = 7.1 \times 10^{-12}$, $K_b(CO_3^{2-}) = 1.8 \times 10^{-4}$, $K_b(IO_3^-) = 5.9 \times 10^{-14}$, $K_b(Cl^-) = 1.0 \times 10^{-14}$; $Cl^- < IO_3^- < CH_2ClCO_2^- < F^- < CO_3^{2-}$

14.85 (a) Percentage ionized = 2.1% (b) Percentage ionized = 1.3%

14.87 (a) pOH = 2.18; pH = 11.82 (b) pOH = 3.75; pH = 10.25 (c) pOH = 2.59; pH = 11.41

14.89 (a) $pK_b = 3.1$; $pK_a = 10.9$ (b) $K_b = 6.7 \times 10^{-12}$; pH = 7.11

14.91 (a) pH = 4.68 (b) pH = 1.28

14.93 $pK'_w = 13.613$; pH = 6.81

Case Questions

1. $[H_3O^+] = 1.5 \times 10^{-6}$ mol/L, pH = 5.82

3. (a) 70 kg (b) The additional cost of producing the calcium carbonate bed or slurry would appear cost-effective in the long run. But these economic considerations would be based upon data not immediately available from the information given. Additional CO_2 released to the atmosphere would be minimal compared to the CO_2 produced by the combustion process itself.

CHAPTER 15

Classified Exercises

15.1 (a) Acidic; $NH_4^+(aq) + H_2O(l) \rightleftharpoons NH_3(aq) + H_3O^+(aq)$ (b) Basic; $CO_3^{2-}(aq) + H_2O(l) \rightleftharpoons HCO_3^-(aq) + OH^-(aq)$ (c) Basic; $F^-(aq) + H_2O(l) \rightleftharpoons HF(aq) + OH^-(aq)$ (d) Acidic; $Co(H_2O)_6^{2+}(aq) + H_2O(l) \rightleftharpoons Co(H_2O)_5OH^+(aq) + H_3O^+(aq)$

15.3 (a) $K_a = 5.6 \times 10^{-10}$ (b) $K_b = 1.8 \times 10^{-4}$ (c) $K_b = 2.9 \times 10^{-11}$ (d) $K_b = 2.3 \times 10^{-8}$

15.5 (a) pH = 5.13; % deprotonated = 7.5×10^{-3}% (b) pH = 2.92; % deprotonated = 1.2% (c) pH = 11.24; % protonated = 1.1%

15.7 (a) 9.18 (b) 4.74

15.9 (a) 6.7×10^{-6} M (b) 5.04

15.11 (a) The concentration of H_3O^+ decreases. (b) The fraction of benzoic acid that is deprotonated decreases.

15.13 (a) $K_a = 8.4 \times 10^{-4}$; $pK_a = 3.08$ (b) ≈2.77

15.15 (a) 4.0×10^{-9} mol/L (b) 2.8×10^{-10} mol/L (c) 2.8×10^{-11} mol/L

15.17 1.90

15.19 (a) pH = 9.69 (b) pH = pK_a = 9.31

15.21 (a) Initial pH = 2.87 (b) pH = 4.56 (c) 12.5 mL (d) pH = 4.74 (e) 25.0 mL (f) pH = 8.72 (g) thymol blue, $pK_{in} = 8.9$; or phenolphthalein, $pK_{in} = 9.4$

15.23 (a) pH = 11.13 (b) pH = 9.43 (c) 11.25 mL (d) pH = 9.26 (e) 22.5 mL (f) pH = 5.28 (g) Methyl orange or methyl red

15.25 (a) pH = 2.02 (b) pH = 2.65 (c) pH = 3.15 (d) pH = 7.94 (e) pH = 12.19 (f) pH = 12.45 (g) phenol red, $pK_{in} = 7.9$

15.27 (a) 3.2 − 4.4 (b) 5.0 − 8.0 (c) 4.8 − 6.0 (d) 8.2 − 10.0

15.29 (a) 2.5, 0.25, 0.025 (b) 32, 3.2, 0.32, 0.032 (c) 10, 1.0, 0.10 (d) 25, 2.5, 0.25

15.31 (a) Not a buffer, strong acid/salt (b) Weak acid/conjugate base, $HClO(aq) + H_2O(l) \rightleftharpoons H_3O^+(aq) + ClO^-(aq)$ (c) Weak acid/conjugate base results from reaction of OH^- and excess CH_3COOH, $CH_3COOH(aq) + H_2O(l) \rightleftharpoons H_3O^+(aq) + CH_3CO_2^-(aq)$

15.33 3.5×10^{-1}

15.35 (a) pH = 4.74 (b) pH = 5.01, ΔpH = 0.27 (c) pH = 4.14, ΔpH = −0.60

15.37 (a) pH = 6.33, ΔpH = +1.59 (b) pH = 4.57, ΔpH = −0.17

15.39 (a) pH 2–4 (b) pH 3–5 (c) pH 6–8

15.41 (a) 5.6 (b) 77 g K_2CO_3 (c) 1.8 g $KHCO_3$ (d) 280 mL

15.43 pH = 7.78

15.45 (a) $K_s = [Ag^+][Br^-]$ (b) $K_s = [Ag^+]^2[S^{2-}]$ (c) $K_s = [Ag^+]^2[CrO_4^{2-}]$

15.47 (a) 1.7×10^{-14} (b) 5.3×10^{-3} (c) 6.9×10^{-9}

15.49 (a) 6.3×10^{-51} (b) 8.5×10^{-45} (c) 8.7×10^{-9}

15.51 pH $\geq$ 12.16

15.53 7.6×10^{-3} ions/1 L or 1.3×10^2 L/1 ion

15.55 (a) 8.0×10^{-10} mol/L (b) 1.3×10^{-16} mol/L (c) 6.3×10^{-6} mol/L

15.57 (a) 1.6×10^{-5} mol/L (b) 5.4×10^2 μg

15.59 pH = 6.41

15.61 (a) Precipitate (b) No precipitate

15.63 (a) Precipitate (b) No precipitate

15.65 (a) $Fe(OH)_2$ first (b) $Ca(OH)_2$, 12.8; $Fe(OH)_2$ 8.60

15.67 (a) 1.0×10^{-12} mol/L (b) 3.0×10^{-5} mol/L (c) 2.0×10^{-3} mol/L (d) 0.20 mol/L

15.69 (a) $CaF_2(s) + 2\ H_2O(l) \rightleftharpoons Ca^{2+}(aq) + 2\ HF(aq) + 2\ OH^-(aq)$ $K = 1.4 \times 10^{-14}$ (b) 0.70 mol/L (c) 15 mol/L

15.71 4.0×10^{-5} mol/L

Unclassified Exercises

15.73 (a) Neutral (b) Basic (c) Acidic (d) Acidic

15.75 (a) pH = 8.24 (b) 8.5×10^{-3}%

15.77 (a) 7.5×10^{-4}% (b) $K_a = 2.7 \times 10^{-3}$

15.79 pH = 12.70

15.81 [HCOOH] = 2.8×10^{-2} mol/L, pH = 3.92

15.83 First stoichiometric point, pH = 1.54; Second stoichiometric point, pH = 7.46

15.85 (a) Represent tartaric acid by H_2T: $H_2T(aq) + OH^-(aq) \rightleftharpoons HT^-(aq) + H_2O(l)$, $HT^-(aq) + OH^-(aq) \rightleftharpoons T^{2-} + H_2O(l)$; this is a two step titration (b) 0.0680 mol/L (c) pH = 3.11 (d) pH = 8.64 (e) 2.88×10^{-2} mol/L (f) pH = 3.22

15.87 (a) pH = 3.95 (b) pH = 3.88, a decrease of 0.07 pH units (c) pH = 3.11, a decrease of 3.89 pH units (d) pH = 2.98, a decrease of 0.97 pH units

15.89 (a) $K_{eq} = K_{a(\text{acetic acid})} \cdot K_{b(\text{hydrazine})} \cdot \frac{1}{K_w} = \frac{K_a K_b}{K_w} = 3.1 \times 10^3$ (b) $-\log K_{eq} = -\log [N_2H_5^+] - \log \frac{[CH_3CO_2^-]}{[N_2H_4][CH_3COOH]}$

15.91 Equilibrium: $HCO_3^-(aq) + H_2O(l) \rightleftharpoons H_3O^+(aq) + CO_3^{2-}(aq)$, $K_a = [H_3O^+][CO_3^{2-}]/[HCO_3^-]$, pH = p$K_a$ + log $[CO_3^{2-}]/[HCO_3^-]$, $10 = -\log(4.7 \times 10^{-11}) + \log [CO_3^{2-}]/[HCO_3^-]$, $\log [CO_3^{2-}]/[HCO_3^-] = -0.33$, $[CO_3^{2-}]/[HCO_3^-] = 0.47$

15.93 1.0×10^{-12}

15.95 1.0×10^{-9} mol/L

15.97 6×10^{-8} mol/L

15.99 (a) pH = 12.35 (b) 55 mL

15.101 Yes

15.103 5.7×10^{-15} mol/L

15.105 (a) $ZnS(s) + 2\ H_2O(l) \rightleftharpoons Zn^{2+}(aq) + H_2S(aq) + 2\ OH^-(aq)$, $K_s = 1.5 \times 10^{-45}$ (b) 1.5×10^{-30} mol/L (c) 1.5×10^{-36} mol/L

15.107 (a) $[Cl^-] = 1.6 \times 10^{-8}$ mol/L (b) $[Ag^+] = 4.0 \times 10^{-9}$ mol/L (c) 4.0×10^{-9} M (d) Virtually 100% precipitation of first cation (Ag^+)

Case Questions

1. (a) 6.0, = 0.030; 6.5, = 0.095; 7.0, = 0.30; 7.5, = 0.95; 8.0, = 3.0; 8.5, = 9.5
(b)

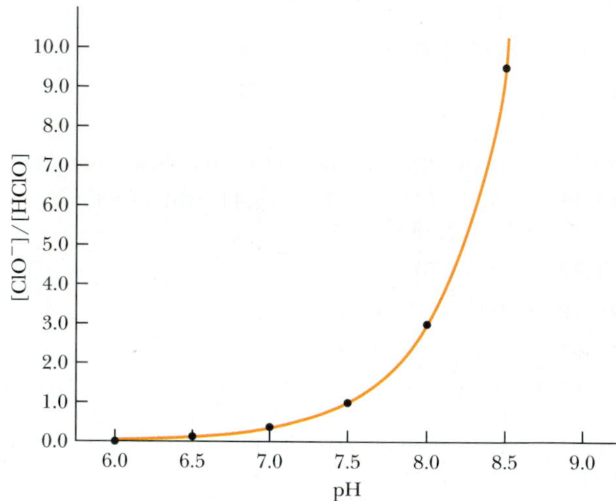

(c) pH = 7.5 is ideal; $[ClO^-/[HClO^-] = \sim 0.95$ (d) A one-pH unit change corresponds to a 10-fold concentration change.

3. (a) 1×10^1 L (b) 79 g

CHAPTER 16

Classified Exercises

16.1 1250 kJ

16.3 1.2×10^5 J

16.5 −400 kJ

16.7 (a) Decreases (b) Increases (c) Decreases

16.9 (a) $S = 0$ (no disorder) (b) $S = 5.76$ J/(K · mol)

16.11 (a) −22.0 J/(K·mol) (b) 134 J/K

16.13 (a) −173.0 J/K (b) 160.6 J/K (c) −36.8 J/K

16.15 (a) 403 J/K (b) 322 J/K (c) −0.310 J/K

16.17 (a) 0.341 J/(K·s) (b) 2.95×10^4 J/K

16.19 (a) −0.017 J/K (b) −0.013 J/K (c) Greater

16.21 −116.16 J/K; according to the second law this process is not spontaneous: ΔS°_{tot} is negative (there has been a decrease in disorder)

16.23 (a) $\Delta S^\circ_{surr} = -73$ J/(K·mol), $\Delta S^\circ_{tot} = 0$, $\Delta S^\circ_{sys} = +73$ J/(K·mol) (b) $\Delta S^\circ_{surr} = -29.0$ J/(K·mol), $\Delta S^\circ_{tot} = 0$, $\Delta S^\circ_{sys} = +29.0$ J/(K·mol) (c) $\Delta S^\circ_{surr} = +29.0$ J/(K·mol), $\Delta S^\circ_{sys} = -29.0$ J/(K·mol); freezing is the reverse of melting

16.25 239 K or −34 °C

16.27 (a) 30 kJ/mol (b) −11 J/K

16.29 (a) Minus (−) (b) Sign cannot be predicted, it depends on the magnitudes of ΔH, T, and ΔS. (c) A temperature change can affect the sign in (b) by altering the magnitude of the $-T\Delta S$ term in ΔG.

16.31 Assume 298 K. (a) $\Delta G^\circ = -514.4$ kJ: The negative sign indicates the reaction is spontaneous under the standard conditions; the large magnitude of ΔG° implies that a temperature change is not likely to affect the spontaneity, but it is possible at very high temperatures. (b) $\Delta G^\circ = +130.4$ kJ: The positive sign of ΔG° implies the reaction is not spontaneous under standard conditions, but the magnitude indicates that at a higher temperature there could be a change in sign. (c) $\Delta G^\circ = -133.1$ kJ: The negative sign implies the reaction is spontaneous under standard conditions. The small value of ΔS° indicates that a temperature change would not likely affect the spontaneity of the process, but it is possible at very high temperatures.

16.33 (a) $\frac{1}{2} N_2(g) + \frac{3}{2} H_2(g) \rightarrow NH_3(g)$, $\Delta S^\circ_f = -0.09938$ kJ/(K·mol), $\Delta G^\circ_f = -16.5$ kJ/mol (b) $C(s) + \frac{1}{2} O_2(g) \rightarrow CO(g)$, $\Delta S^\circ_f = 0.08936$ kJ/(K·mol), $\Delta G^\circ_f = -137.2$ kJ/mol (c) $\frac{1}{2} N_2(g) + O_2(g) \rightarrow NO_2(g)$, $\Delta S^\circ_f = -0.06088$ kJ/(K·mol), $\Delta G^\circ_f = 51.3$ kJ/mol

16.35 (a) $\Delta G^\circ = -141.74$ kJ, hence spontaneous (b) $\Delta G^\circ = 33.1$ kJ, hence not spontaneous (c) $\Delta G^\circ = -10{,}590.9$ kJ, hence spontaneous

16.37 (a) Unstable (b) Unstable (c) Stable

16.39 (a) More stable (b) More stable (c) More stable

16.41 Increased temperature will make the decomposition less spontaneous.

16.43 (a) Reduction can occur. (b) Reduction can occur.

16.45 +317.02 kJ

16.47 (a) 1×10^{90} (b) 1.4×10^{-23} (c) 2.1×10^{23}

16.49 There is a tendency to form reactants.

16.51 (a) −12.3 kJ (b) −60 kJ

16.53 8.9×10^2 J; not a spontaneous process at this temperature

16.55 (a) 8.4×10^{-17} (b) 7.7×10^{-9} (c) 2.4×10^{-32}

Unclassified Exercises

16.57 −200 J; work was done by the system

16.59 $H_2O(s) < H_2O(l) < NaCl(aq) < H_2O(g)$

16.61 (a) It does not favor the solution process. (b) Positive (c) The dispersal of matter

16.63 (a) −395.4 J/K (b) +214.62 J/K (c) +446.83 J/K

16.65 Spontaneous

16.67 (a) −70.4 kJ (b) +28.5 kJ (c) +292.3 kJ; NO is favored. (d) −66.8 kJ; NO_2 is favored.

16.69 (a) $\Delta H^\circ = -802.34$ kJ, $\Delta G^\circ = -800.78$ kJ (b) -5.23×10^{-3} kJ/K (c) The large and negative ΔH° means that this process gives off much heat. The negative ΔS° means that the arrangement of the atoms in the products is less disorderly than in the reactants.

16.71 1.7×10^2 kJ

16.73 (a) 79.89 kJ (b) 1.01×10^{-14}

16.75 (a) $\Delta G^\circ = -326.4$ kJ, $\Delta S^\circ = 137.56$ J/K (b) 1.5×10^{57} (c) The conversion of ozone to oxygen is spontaneous; the rate at which this occurs, however, is another matter because thermodynamics does not give kinetic information.

16.77 (a) 6.11 J/K (b) 1.22 J/K (c) The disorder of the gaseous state relative to the liquid state is much larger than that of the liquid state relative to the solid state.

16.79 $Q_c = 2.7 \times 10^{-3}$; $Q_c < K_c$, therefore the reaction is spontaneous in the direction of I(g); $\Delta G = -32$ kJ

Case Questions

1. (a) 15 g (b) 6.1 g

3. (a) −14.6 kJ (heat released to solution by reaction) (b) 37°C

5. 1.0×10^2 g

CHAPTER 17

Classified Exercises

17.1 (a) Anode (b) Positive

17.3 Anode, anode, cathode

17.5 (a) Zn^{2+}/Zn; $Zn^{2+}(aq) + 2e^- \rightleftharpoons Zn(s)$; $Zn^{2+}(aq)|Zn(s)$ (b) Fe^{3+}/Fe^{2+}; $Fe^{3+}(aq) + e^- \rightleftharpoons Fe^{2+}(aq)$; $Fe^{3+}(aq), Fe^{2+}(aq)|Pt$ (c) Cl_2/Cl^-; $\frac{1}{2}Cl_2(g) + e^- \rightleftharpoons Cl^-(aq)$; $Cl_2(g)|Cl^-(aq)|Pt$ (d) $Hg_2Cl_2/Hg,Cl^-$; $Hg_2Cl_2(s) + 2e^- \rightleftharpoons 2Hg(l) + 2Cl^-(aq)$; $Cl^-(aq)|Hg_2Cl_2(s)|Hg(l)$

17.7 (a) $Pt(s)|H_2(g)|H^+(aq)||Ag^+(aq)|Ag(s)$ (b) $Ag(s)|AgCl(s)|Cl^-(aq)||Cl^-(aq)|Cl_2(g)|Pt(s)$ (c) $Pt(s)|H_2(g)|H^+(aq)||Cl^-(aq)|Hg_2Cl_2(s)|Hg(l)$

17.9 (a) Cathode: $Cl_2(g) + 2e^- \rightarrow 2Cl^-(aq)$, anode: $H_2(g) \rightarrow 2H^+(aq) + 2e^-$, cell reaction: $Cl_2(g) + H_2(g) \rightarrow 2H^+(aq) + 2Cl^-(aq)$ (b) cathode: $3[V^{2+}(aq) + 2e^- \rightarrow V(s)]$, anode: $2[U(s) + 3e^- \rightarrow U^{3+}(aq) + 3e^-]$, cell reaction: $3V^{2+}(aq) + 2U(s) \rightarrow 2U^{3+}(aq) + 3V(s)$

(c) cathode: $Hg_2Cl_2(s) + 2e^- \rightarrow 2Hg^0(l) + 2Cl^-(aq)$, anode: $Sn^{2+}(aq) \rightarrow Sn^{4+}(aq) + 2e^-$, cell reaction: $Sn^{2+}(aq) + Hg_2Cl_2(s) \rightarrow 2Hg^0(l) + 2Cl^-(aq) + Sn^{4+}(aq)$

17.11 (a) $Ni^{2+}(aq) + 2e^- \rightarrow Ni(s)$ (at cathode); $Zn(s) \rightarrow Zn^{2+}(aq) + 2e^-$ (at anode); $Ni^{2+}(aq) + Zn(s) \rightarrow Ni(s) + Zn^{2+}(aq)$ (overall cell); $Zn(s)|Zn^{2+}(aq)||Ni^{2+}(aq)|Ni(s)$ (b) $2[Ce^{4+}(aq) + 1e^- \rightarrow Ce^{3+}(aq)]$ (at cathode); $2I^-(aq) \rightarrow 2e^- + I_2(s)$ (at anode); $2I^-(aq) + 2Ce^{4+}(aq) \rightarrow 2Ce^{3+}(aq) + I_2(s)$ (overall cell); $Pt|I^-(aq)|I_2(s)||Ce^{4+}(aq),Ce^{3+}(aq)|Pt$ (c) $3[Au^+(aq) + 1e^- \rightarrow Au(s)]$ (at cathode); $Au(s) \rightarrow Au^{3+}(aq) + 3e^-$ (at anode); $3Au^+(aq) \rightarrow 2Au(s) + Au^{3+}(aq)$ (overall cell); $Au(s)|Au^{3+}(aq)||Au^+(aq)|Au(s)$

17.13 (a) NH_4Cl in paste form (b) MnO_2 (c) $Zn(s) \rightarrow Zn^{2+}(aq) + 2e^-$

17.15 (a) KOH, aqueous (b) $2Ni(OH)_2(s) + 2OH^-(aq) \rightarrow 2NI(OH)_3(s) + 2e^-$ (c) When charged fully, $\approx$1.4 V

17.17 (a) $2H^+(aq) + 2e^- \rightarrow H_2(g)$ (b) $H^+(aq)|H_2(g)|Pt$

17.19 (a) +0.75 V (b) +0.52 V (c) +1.52 V

17.21 (a) $E° = +0.32$ V, $\Delta G° = -62$ kJ (b) $E° = +0.58$ V, $\Delta G° = -56$ kJ (c) $E° = 0.29$ V, $\Delta G° = -84$ kJ

17.23 (a) $Ce^{4+} > Cl_2 > Co^{2+} > In^{3+}$ (b) $HBrO > Cr_2O_7^{2-} > ClO_4^- > NO_3^-$ (c) $H_2O_2 > HClO > MnO_4^- > O_2$

17.25 (a) No (b) Yes; $3Pb^{2+}(aq) + 2Cr(s) \rightarrow 3Pb + 2Cr^{3+}(aq)$, $Cr|Cr^{3+}(aq)||Pb^{2+}(aq)|Pb$ (c) Yes; $5Cu + 2MnO_4^-(aq) + 16H^+(aq) \rightarrow 5Cu^{2+}(aq) + 2Mn^{2+}(aq) + 8H_2O(l)$; $Cu|Cu^{2+}(aq)||MnO_4^-(aq),H^+(aq),Mn^{2+}(aq),|Pt$

17.27 (a) $I_2 + H_2 \rightarrow 2I^- + 2H^+$, $I_2 + 2e^- \rightarrow 2I^-$, $H_2 \rightarrow 2H^+ + 2e^-$, $\Delta E° = +0.54$ V, $\Delta G° = -104$ kJ (b) No reaction (c) $2Al + 3Pb^{2+} \rightarrow 2Al^{3+} + 3Pb$, $Pb^{2+} + 2e^- \rightarrow Pb$, $Al \rightarrow Al^{3+} + 3e^-$, $\Delta E° = +1.53$ V, $\Delta G° = -886$ kJ

17.29 O_2 could be used as $E°$(cathode) > 1.09 V because of the slowness of reaction.

17.31 Retarded

17.33 (a) $Fe_2O_3{\cdot}H_2O$ (b) H_2O and O_2 jointly oxidize iron. (c) Water is more highly conducting if it contains dissolved ions. (d) Water plus other ions present

17.35 (a) $K_{eq} = [H^+]^2[Cl^-]^2/[H_2]$ (b) $K_{eq} = [NO][Fe^{3+}]^3/[Fe^{2+}]^3[H^+]^4[NO_3^-]$

17.37 (a) 2.1×10^{25} (b) 6.1×10^{-16} (c) 1.2×10^{31}

17.39 Yes

17.41 (a) 3×10^6 (b) 4.8×10^{40}

17.43 (a) +0.18 V (b) 0.47 V (c) −1.45 V

17.45 4.7

17.47 (a) pH = 1.0 (b) 9.9×10^{-2} mol/L

17.49 (a) $E = 0.22 - (0.0257/2) \ln [H^+]^2 = 0.22 - 0.0257 \ln [H^+]$ and $pH = 2.303 \ln \frac{1}{[H^+]}$ (b) Since pH + pOH = 14, $E = 0.22 + 0.059(14 - pOH)$, $E = 0.22 - 0.059$ pH.

17.51 (a) Mn, not easily reduced from Mn^{2+} (b) Al, not easily reduced from Al^{3+} (c) Ni, more easily reduced (d) Au, easily deposited

17.53 (a) $Cu^{2+}(aq) + 2e^- \rightarrow Cu(s)$, reduction, cathode (b) $Na^+(l) + e^- \rightarrow Na(l)$, reduction, cathode (c) $2Cl^- \rightarrow Cl_2 + 2e^-$, oxidation, anode

17.55 (a) 0.161 mol e^- (b) 22.2 mol e^- (c) 35.7 mol e^-

17.57 (a) 7.86 s (b) 1.29 mg Cu

17.59 (a) 5.43×10^{-6} A (b) 0.742 s

17.61 Ti^{2+}

17.63 0.9 g Zn

Unclassified Exercises

17.65 (a) $O_2 + 4H^+ + 4e^- \rightarrow 2H_2O$ (cathode), $Fe \rightarrow Fe^{2+} + 2e^-$ (anode) (b) $Fe(s)|Fe^{2+}(aq)||O_2(g)|H^+(aq),H_2O(l)|Pt(s)$ (c) +1.67 V (d) 1.65 V

17.67 +0.92 V = $E°$(anode), $E°$(cathode) = −0.92 V

17.69 not spontaneous

17.71 (a) $E°_{cell} = +0.17$ V; $Hg(l)|Hg_2^{2+}(aq)||NO_3^-(aq),H^+(aq)|NO(g)|Pt$; $\Delta G° = -98$ kJ (b) $E°_{cell} = +0.36$ V; $Pt|Pu^{3+}(aq),Pu^{4+}(aq)||Cr_2O_7^{2-}(aq),Cr^{3+}(aq),H^+(aq)|Pt$; $\Delta G° = -208$ kJ

17.73 (a) +0.52 V (b) −0.04 V

17.75 The aluminum foil is easily oxidized by virtue of its position in the electrochemical series, and a current flows, stimulating the pain sensors.

17.77 (a)

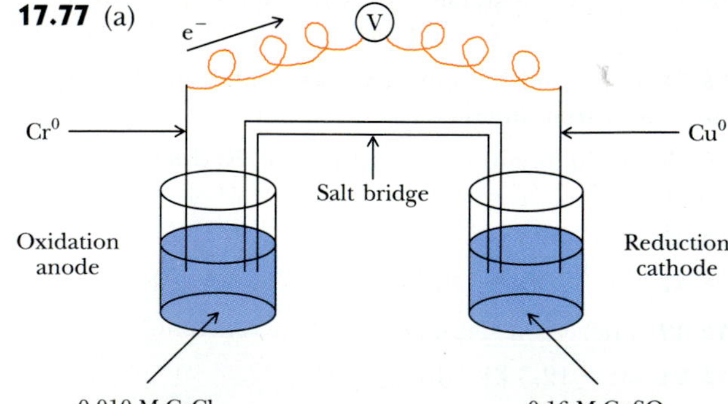

(b) Cathode: $Cu^{2+}(aq) + 2e^- \rightarrow Cu(s)$, anode: $Cr^{3+}(aq) \rightarrow Cr(s) + 3e^-$ (c) $3Cu^{2+}(aq) + 2Cr(s) \rightarrow 3Cu(s) + 2Cr^{3+}(aq)$, $E°_{cell} = +1.08$ V (d) Cr|0.010 M $CrCl_3(aq)$||0.16 M $CuSO_4(aq)$|Cu (e) 1.07 V

17.79 (a) +0.03 V (b) $Ag^+(aq) + Fe^{2+}(aq) \rightarrow Fe^{3+}(aq) + Ag(s)$ (c) −0.50 V (d) 3.22 (e) +4.8 kJ

17.81 9.8×10^{-3} mol/L

17.83 5.6×10^{-3} mol e^-

17.85 35.9 A

17.87 4 mol e^-/mol Hf

17.89 Fe would be oxidized and all of the others would fall to the bottom as "anode mud."

17.91 105 s

Case Questions

1. (a) $Ag^+ + 1e^- \rightarrow Ag$ reduction, $Cu^+ \rightarrow Cu^{2+} + 1e^-$ oxidation (b) Ag^+ is the oxidizing agent, Cu^+ is the reducing agent.

3. Photochromic glass, used in windows, could reduce light (hence heat) levels inside homes and offices by reacting to periods of high levels of sunlight and relaxing to normal transparency during times of low levels of sunlight.

CHAPTER 18

Classified Exercises

18.1 (a) $CO(g) + H_2O(g) \xrightarrow[\text{Fe/Cu}]{400°C} CO_2(g) + H_2(g)$ (b) $2Li(s) + 2H_2O(l) \rightarrow 2LiOH(aq) + H_2(g)$ (c) $Mg(s) + 2H_2O(l) \rightarrow Mg(OH)_2(aq) + H_2(g)$

18.3 (a) $2Li(s) + H_2(g) \xrightarrow{\Delta} 2LiH(s)$ (b) $H_2(g) + I_2(s) \rightarrow 2HI(g)$ (c) $6H_2(g) + P_4(s) \rightarrow 4PH_3(g)$

18.5 (a) $FeS(s) + 2HCl(aq) \rightarrow FeCl_2(aq) + H_2S(g)$ (b) $Ba(s) + H_2(g) \xrightarrow{\Delta} BaH_2(s)$

18.7 (a) Saline (b) Molecular (c) Metallic

18.9 (a) +206.10 kJ (b) −41.16 kJ (c) 164.94 kJ

18.11 (a) 532 L H_2 (b) 428 mL H_2O

18.13 Fraction used = 0.9

18.15 (a) $H_2(g) + Cl_2(g) \xrightarrow{\text{light}} 2HCl(g)$ (b) $H_2(g) + 2Na(l) \xrightarrow{\Delta} 2NaH(s)$ (c) $H_2(g) + C_2H_4(g) \xrightarrow{200°C,\ 30\ \text{atm, Ni}} C_2H_6(g)$

18.17 (a) $Li^+\ H:^-$ (b) SiH_4 (H—Si—H with H above and below) (c) SbH_3 (H—S̈b—H with H below)

18.19 (a) $2H_2O(l) + 2e^- \rightarrow H_2(g) + 2OH^-(aq)$ (b) Cathode (c) 2.1 L H_2

18.21 (a) Red (b) Violet (c) Yellow (d) Red (e) Blue

18.23 Na: halite (rock salt), NaCl; K: sylvite, KCl, and cannallite, $KCl{\cdot}MgCl_2{\cdot}6H_2O$

18.25 (a) $6Li(s) + N_2(g) \xrightarrow{\Delta} 2Li_3N(s)$ (b) $2Na(s) + H_2O(l) \rightarrow 2NaOH(aq) + H_2(g)$ (c) $4KO_2(s) + 2H_2O(g) \rightarrow 4KOH(s) + 3O_2(g)$

18.27 (a) $Na_2CO_3(s) \xrightarrow{\Delta} Na_2O(s) + CO_2(g)$ (b) $2KNO_3(s) \xrightarrow{\Delta} 2KNO_2(s) + O_2(g)$

18.29 (a) Sodium chloride, NaCl (b) Potassium chloride, KCl (c) Potassium magnesium chloride hexahydrate, $KCl{\cdot}MgCl_2{\cdot}6H_2O$

18.31 (a) $2NaCl(aq) + 2H_2O(l) \xrightarrow{\text{electrolysis}} 2NaOH(aq) + Cl_2(g) + H_2(g)$ (b) $2NaCl(s) + H_2SO_4(l) \xrightarrow{\Delta} 2HCl(g) + Na_2SO_4(s)$ (c) $2NaOH(l) \xrightarrow{\text{electrolysis}} 2Na(s) + O_2(g) + H_2(g)$

18.33 (a) Orange-red (b) Crimson (c) Yellow-green

18.35 (a) Beryl (b) Limestone and dolomite (c) Dolomite

18.37 (a) Magnesium sulfate heptahydrate, $MgSO_4{\cdot}7H_2O$ (b) Calcium sulfate dihydrate, $CaSO_4{\cdot}2H_2O$ (c) Barium sulfate, $BaSO_4$ (d) Magnesium hydroxide, $Mg(OH)_2$

18.39 (a) $Be(s) + 2OH^-(aq) + 2H_2O(l) \rightarrow [Be(OH)_4]^{2-}(aq) + H_2(g)$ (b) $2Al(s) + 2OH^-(aq) + 2H_2O(l) \rightarrow 2[Al(OH)_4]^-(aq) + H_2(g)$ (c) The similarity of Be and Al in chemical reactions is an example of the diagonal relationship in the periodic table.

18.41 (a) $2Ca(s) + O_2(g) \rightarrow 2CaO(s)$ (b) $Mg(s) + 2HCl(aq) \rightarrow MgCl_2(aq) + H_2(g)$ (c) $3Sr(s) + N_2(g) \xrightarrow{\Delta} Sr_3N_2(s)$

18.43 (a) $CaCO_3(s) \xrightarrow{\Delta} CaO(s) + CO_2(g)$, $CaO(s) + H_2O(l) \rightarrow Ca(OH)_2(s)$ (b) $CaCO_3(s) + 2HNO_3(aq) \rightarrow Ca(NO_3)_2(aq) + H_2O(l) + CO_2(g)$, $Ca(NO_3)_2$ can then be obtained by drying

18.45 (a) $Mg(OH)_2(s) + HCl(aq) \rightarrow MgCl_2(aq) + 2H_2O(l)$ (b) $Ca(s) + 2H_2O(l) \rightarrow Ca(OH)_2(aq) + H_2(g)$ (c) $BaCO_3(s) \xrightarrow{\Delta} BaO(s) + CO_2(g)$

18.47 (a) $:\ddot{Cl}—Be—\ddot{Cl}:$ (b) 180° (c) *sp*

18.49 (a) 68 g Mg (b) 9.21×10^5 L Cl_2

18.51 (a) $\Delta H° = +178.0$ kJ, $\Delta S° = 160.6$ J/K (b) $T = 1109$ K

18.53 Mass % H_2O = 51.19%

18.55 (a) The "hardness" of water is due to the presence of calcium and magnesium salts. (b) $Ca(HCO_3)_2(aq) + Ca(OH)_2(aq) \rightarrow 2CaCO_3(s) + 2H_2O$, $Ca^{2+}(aq) + Na_2CO_3(aq) \rightarrow 2CaCO_3(s) + 2Na^+(aq)$

Unclassified Exercises

18.57 $H_2 < Be < Mg < Na < Ca < Sr < Ba < Ra < Cs < Rb < K < Li$

18.59 Decreasing ionization energies and standard reduction potentials is clear for the Group II data, but it is not clear for the Group I elements.

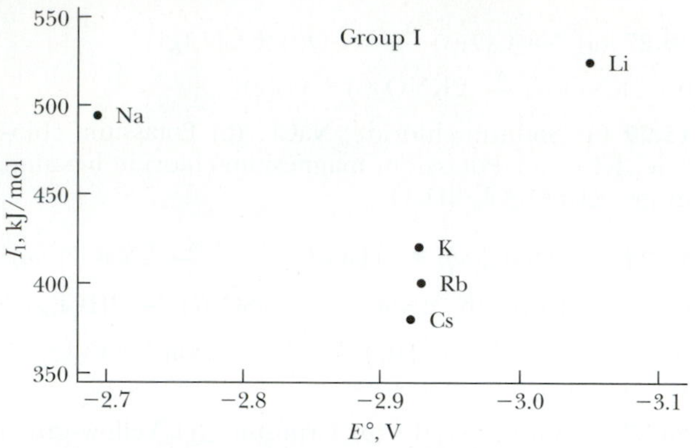

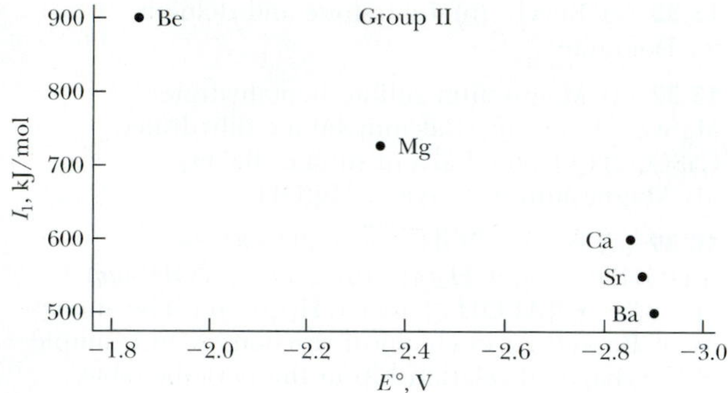

18.61 (a) Ba^{2+} $[:\ddot{O}—\ddot{O}:]^{2-}$ (b) H—Be—H

(c) $2Na^+$ $[:\ddot{O}—\ddot{O}:]^{2-}$ (d) $\left[\begin{array}{c} H \\ O \\ | \\ HO—Be—OH \\ | \\ O \\ H \end{array}\right]^{2-}$

18.63 $CaF_2(s) + 2H^+(aq) \rightleftharpoons Ca^{2+}(aq) + 2HF(aq)$; the presence of excess $H^+(aq)$ ions drives this equilibrium to the right in accord with Le Chatelier's principle.

18.65 (a) -6.517×10^4 J (b) 62.3°C

18.67 422 g

Case Questions

1. LiOH: mass CO_2 = 0.92 g; NaOH: mass CO_2 = 0.55 g; KOH: mass CO_2 = 0.39 g

3. 1.1×10^3 g

CHAPTER 19

Classified Exercises

19.1 (a) $[Ne]3s^2 3p^1$ (b) $[Ne]3s^2 3p^2$ (c) $[Xe]4f^{14}5d^{10}6s^2 6p^2$

19.3 (a) Ge, germanium, $[Ar]4s^2$, $3d^{10}$, $4p^2$ (b) Tl, thallium, $[Xe]6s^2$, $4f^{14}$, $5d^{10}$, $6p^1$ (c) Sb, antimony, $[Kr]5s^2$, $4d^{10}$, $5p^3$

19.5 Ge, Sn, Tl, In, Ga, Al

19.7 $2Al_2O_3 + 3C \rightarrow 4Al + 3CO_2$

19.9 (a) $B(OH)_3$ (b) Al_2O_3 (c) $Na_2B_4O_7{\cdot}10H_2O$ (d) B_2O_3

19.11 (a) $B_2O_3(s) + 3Mg(l) \xrightarrow{\Delta} 2B(s) + 3MgO(s)$ (b) $2Al(l) + Cr_2O_3(s) \rightarrow Al_2O_3(s) + 2Cr(l)$

19.13 (a) Abrasive, cutting tools (b) Industrial catalyst (c) Corundum, abrasive (d) Antiseptic

19.15 (a) $B_2H_6 + 6H_2O \rightarrow 2B(OH)_3 + 6H_2$ (b) $4LiH + AlCl_3 \rightarrow LiAlH_4 + 3LiCl$

19.17 35.5 g Al

19.19 −536 kJ

19.21 (a) $Al^{3+}(aq) + 6H_2O(l) \rightarrow Al(H_2O)_6{}^{3+}(aq)$ and $Al(H_2O)_6{}^{3+}(aq) + H_2O(l) \rightarrow H_3O^+(aq) + [Al(H_2O)_5OH]^{2+}(aq)$ (b) $Al^{3+}(aq) + OH^-(aq) \rightarrow AlOH^{2+}(aq)$, $Al^{3+}(aq) + 2OH^-(aq) \rightarrow Al(OH)_2{}^+(aq)$, $Al^{3+}(aq) + 3OH^-(aq) \rightarrow Al(OH)_3(s)$

19.23 2.7×10^5 g Al

19.25 −0.743 V

19.27 $SiO_2(l) + 2C(s) \rightarrow Si(s) + 2CO_2(g)$, $Si(s) + 2Cl_2(g) \rightarrow SiCl_4(l)$, $SiCl_4(l) + 2H_2(g) \rightarrow Si(s) + 4HCl(g)$

19.29 $GeO_2(s) + 2H_2(g) \rightarrow Ge(s) + 2H_2O(l)$

19.31 (a) SiC (b) SiO_2 (c) Na_4SiO_4 (d) $Ca(CN)_2$

19.33 (a) $CH_4 + 4S \rightarrow CS_2 + 2H_2S$ (redox) (b) $5Si_2Cl_6 \rightarrow Si_6Cl_{14} + 4SiCl_4$ (redox) (c) $Sn + 2KOH + 4H_2O \rightarrow K_2Sn(OH)_6 + 2H_2$ (redox)

19.35 $\begin{array}{c} \ominus \\ :\ddot{O}: \\ \ominus:\ddot{O}:Si:\ddot{O}:\ominus \\ :\ddot{O}: \\ \ominus \end{array}$, tetrahedral,

19.37 46.74%

19.39 $\Delta H° = +689.88$ kJ/mol, $\Delta S° = +360.85$ J/K·mol, $\Delta G° = 5.8235 \times 10^2$ kJ/mol, $T = 1912$ K (reaction spontaneous *above* this temp)

19.41 (a) $SnO_2(s) + C(s) \rightarrow Sn(s) + CO_2(g)$, $SnO_2(s) + 2CO(g) \rightarrow Sn(s) + 2CO_2(g)$ (b) 125.43 kJ/mol; 5.386 kJ/mol (c) An increase in temperature will make either reaction spontaneous

19.43 7.7×10^2 g coke (or 0.74 kg)

19.45 (a) Orthosilicates (e.g., sodium silicate) (b) Pyroxenes (e.g., jade, $NaAl(SiO_3)_2$)

19.47 (a) Nitrous acid (b) Phosphoric acid (c) Dinitrogen trioxide

19.49 (a) Mg_3N_2 (b) Ca_3P_2 (c) H_2NNH_2

19.51 (a) $2NH_3 + 3CuO \rightarrow N_2 + 3Cu + 3H_2O$ (b) $4NH_3 + 3F_2 \rightarrow NF_3 + 3NH_4F$ (c) $4NH_3 + 5O_2 \rightarrow 4NO + 6H_2O$

19.53 (a) +2 (b) +1 (c) +3 (d) −3

19.55 $CO(NH_2)_2 + 2H_2O \rightarrow (NH_4)_2CO_3$; 8.0×10^3 g (or 8.0 kg)$(NH_4)_2CO_3$

19.57 (a) 0.23 L $N_2(g)$ (b) Larger

19.59 $H_2N_2O_2$: $N_2O(g) + H_2O(l) \rightarrow H_2N_2O_2(aq)$; HNO_2: $N_2O_3(g) + H_2O(l) \rightarrow 2HNO_2(aq)$; HNO_3: $N_2O_5(g) + H_2O(l) \rightarrow 2HNO_3(aq)$

Unclassified Exercises

19.61 (a) Strengthen plastics, lightweight construction material (b) Beverage containers, lightweight construction material (c) Phosphate fertilizers, matches (d) Synthesis of ammonia, atmospheric diluent for O_2 (e) Computer chips, glass (f) Anticorrosion plating, various alloys

19.63 (a) and (b) The first ionization energy decreases (generally) down a group; valence electrons which are farther from the nuclear positive center are less difficult to remove. The atomic radii increase down a group because new filled "shells" are added. (c) The trends correlate well with elemental properties.

19.65 (a) $[\ddot{N}::N::\ddot{N}]^-$, bond angle 180° (b) F^- ionic radius = 136 pm, N_3^- ionic radius = 148 pm, Cl^- ionic radius = 181 pm (c) NaN_3, KN_3, LiN_3 (ionic), HN_3, $B(N_3)_3$ (covalent)

19.67 (a) 12.2% (b) 26.47%

19.69 (a)

```
        H
        |
        O                    O      O      O
        |                    ‖      ‖      ‖
  H—O—P—O—H        HO—P—O—P—O—P—OH
        ‖                    |      |      |
        O                   OH     OH     OH
     H3PO4:             Tripolyphosphoric acid
```

(b) The polyphosphate anion is a relatively strong Brønstead base and buffers the solution at a pH where detergent molecules can act effectively; on the other hand, H_3PO_4 produces an acidic solution.

19.71 (a) $Na_2B_4O_7 \cdot 7H_2O$, natural source of boron (b) SiO_2, granular form of quartz, natural source of silicon (c) $Al_2O_3 \cdot xH_2O$, natural source of aluminum (d) SnO_2, natural source of tin (e) PbS, natural source of lead

19.73 (a) SiH_4, as opposed to CH_4, reacts with water to evolve H_2, principally because the Si atom is larger than the C atom and can utilize its *d* orbitals to accept the lone pair of the attaching Lewis base, H_2O. H_2 subsequently is released to re-establish the octet. Note that the C atom does not have any available *d* orbitals. (b) $SiH_4 + 2H_2O \rightarrow SiO_2 + 4H_2$

19.75 (a) $Na_2CO_3(s) + HCl(aq) \rightarrow NaHCO_3(aq) + NaCl(aq)$ (b) $SiCl_4(l) + 2H_2O(l) \rightarrow SiO_2(s) + 4HCl(aq)$ (c) $SnO_2(s) + C(s) \rightarrow Sn(s) + CO_2(g)$

19.77 $CaO(s) + 3C(s) \xrightarrow[\text{temp.}]{\text{high}} CaC_2(s) + CO(g)$; the carbon serves as a reducing agent.

19.79 B(B—H) = +372 kJ/mol, B—H—B type = +228 kJ/mol; bridging type is inherently longer

19.81 (a) $4Zn + NO_3^- + 7OH^- + 6H_2O \rightarrow NH_3 + 4Zn(OH)_4^{2-}$ (b) 0.33 mol/L NO_3^-

Case Questions

1. 720.60 u; its molar mass is 720.60 g/mol

3. The diameter of the molecule is 0.71 nm; the radius is 0.36 nm:

$$V = \tfrac{4}{3}\pi r^3 = \tfrac{4}{3}(3.14)(0.36)^3 = 0.19 \text{ nm}^3$$

5. yes

CHAPTER 20

Classified Exercises

20.1 (a) $1s^2$ (b) $[Ne]2s^22p^4$ (c) $[He]2s^22p^5$ (d) $[Ne]3s^23p^6$

20.3 $I_2 < Br_2 < Cl_2 < F_2$

20.5 S < Br < Cl < O

20.7 $2BaO_2 \xrightarrow{\Delta} 2BaO + O_2$; $2KClO_3 \xrightarrow[MnO_2]{\Delta} 2KCl + 3O_2$; $2H_2O \xrightarrow[H_2SO_4]{\text{electrolysis,}} 2H_2 + O_2$

20.9 (a) $CaSO_3$ (b) O_3 (c) BaO_2

20.11 Steam and hot water are forced into the deposit, causing the sulfur to melt. The molten sulfur mixes with the water and the mixture is forced to the surface by compressed air.

20.13 (a) $4Li(s) + O_2(g) \xrightarrow{\Delta} 2Li_2O(s)$ (b) $2F_2(g) + 2H_2O(l) \rightarrow 4HF(aq) + O_2(g)$ (c) $2H_2O(l) \rightarrow O_2(g) + 4H^+(aq) + 4e^-$

20.15 (a) $Na_2SO_3(aq) + S(s) \rightarrow Na_2S_2O_3(aq)$ (b) $2H_2S(aq) + 3O_2(g) \rightarrow 2SO_2(g) + 2H_2O(l)$ (c) $CaO(s) + H_2O(l) \rightarrow Ca(OH)_2(aq)$

20.17 O=S(=O)(=O) [Lewis structure], sp^2, trigonal planar

20.19 4.9 L O_2

20.21 pH = 9.79

20.23 HF < HCl < HBr < HI; The stronger the H—X bond, the weaker the acid; HF has the strongest bond, HI the weakest; hence HI is the strongest acid.

20.25 (a) −68.78 J/(K·mol) (b) The reaction is not favored at any temperature. (c) The entropy does not favor the spontaneous formation of ozone.

20.27 Physical: determination of densities; chemical: addition of an acid would generate $H_2S(g)$ from FeS_2, but not from Au

20.29 Fluorine: the minerals, fluorspar, CaF_2, cryolite Na_3AlF_6, the fluorapatites, $Ca_5F(PO_4)_3$; chlorine: the mineral, rock salt, NaCl, is the principal natural source

20.31 Fluorine: KF acts as an electrolyte for the electrolytic process, but the net reaction is

$$2H^+ + 2F^- \xrightarrow{\text{current}} H_2(g) + F_2(g);$$

chlorine: $2NaCl(l) \xrightarrow{\text{current}} 2Na(s) + Cl_2(g)$

20.33 (a) Hydrobromic acid (b) Chlorine dioxide (c) Sodium iodate

20.35 (a) +4 (b) +7 (c) +5

20.37 (a) $\left[\begin{array}{c} :\ddot{O}: \\ | \\ :\ddot{O}-Cl-\ddot{O}: \\ | \\ :\ddot{O}: \end{array}\right]^-$, AX_4, tetrahedral

(b) $\left[\begin{array}{c} :\ddot{O}-\ddot{I}-\ddot{O}: \\ | \\ :\ddot{O}: \end{array}\right]^-$, AX_3E, pyramidal

(c) $\begin{array}{c} :\ddot{F}: \\ | \\ \dot{\dot{B}}r-\ddot{F}: \\ | \\ :\ddot{F}: \end{array}$, AX_3E_2, T-shaped

20.39 (a) $4KClO_3(l) \xrightarrow{\Delta} 3KClO_4(s) + KCl(s)$
(b) $Br_2(l) + H_2O(l) \rightarrow 2H^+(aq) + 2Br^-(aq) + \frac{1}{2}O_2(g)$
(c) $NaCl(s) + H_2SO_4(aq) \rightarrow NaHSO_4(aq) + HCl(g)$

20.41 Iodine, I_2, is more soluble in aqueous KI than in pure water because a complex anion I_3^- forms.

20.43 (a) $HClO < HClO_2 < HClO_3 < HClO_4$
(b) The oxidation number of Cl increases from HClO to $HClO_4$. In $HClO_4$, chlorine has its highest oxidation number of 7 and, hence, $HClO_4$ will be the strongest oxidizing agent.

20.45 (a) $10Al(s) + 6NH_4ClO_4(s) \rightarrow 3N_2(g) + 9H_2O(g) + 6HCl(g) + 5Al_2O_3(s)$ (b) 2.62 kg NH_4ClO_4

20.47 (a) $CaF_2(s) + 2H_2SO_4(aq, conc.) \rightarrow Ca(HSO_4)_2(aq) + 2HF(g)$ (b) SiO_2 would form $SiF_4(g)$ with the desired product; $SiO_2(s) + 4HF(g) \rightarrow SiF_4(g) + 2H_2O(l)$

20.49 $\begin{array}{c} \ddot{O} \\ / \quad \backslash \\ :\ddot{C}l \quad \ddot{C}l: \end{array}$, angular, slightly less than 109°

20.51

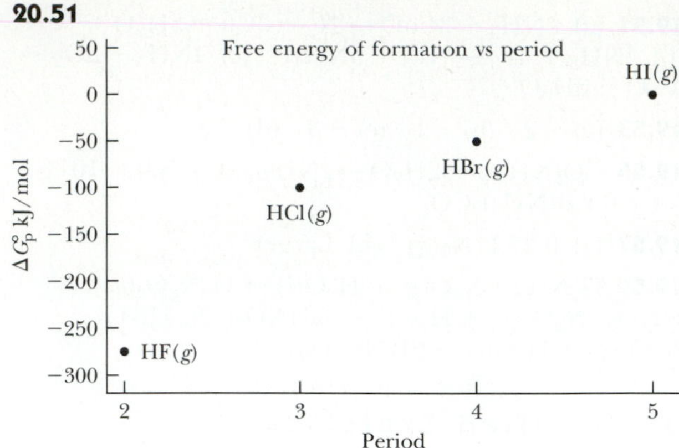

Observations: These values parallel the chemical observation that H_2 and F_2 react vigorously, whereas H_2 and I_2 react much less vigorously. However, the vigor of a reaction is primarily due to the mechanism and the rate of the reaction and not solely to the amount of energy given off. The ΔG_p° values of HCl, HBr, and HI fit nicely upon a straight line, whereas HF is anomolous. In other properties, HF is the anomolous member of the group.

20.53 $E_{cell}^\circ = -0.15$ V; since E_{cell}° is negative, $Cl_2(g)$ will not oxidize Mn^{2+} to form the permanganate ion in an acidic solution

20.55 0.25

20.57 0.117 M F^-

20.59 (a) $Xe(g) + 2F_2(g) \xrightarrow{\Delta} XeF_4(s)$
(b) $6XeF_4(s) + 12H_2O(l) \rightarrow 2XeO_3(aq) + 4Xe(g) + 3O_2(g) + 24HF(aq)$

20.61 (a) +2 (b) +2 (c) +4 (d) +6

20.63 $XeF_4 + 4H^+ + 4e^- \rightarrow Xe + 4HF$

20.65 Since H_4XeO_6 has more highly electronegative O atoms bonded to Xe, it is predicted to be more acidic than H_2XeO_4.

20.67 $\begin{array}{ccc} :\ddot{F} & & \ddot{F}: \\ & \ddot{\underset{..}{Xe}} & \\ :\ddot{F} & & \ddot{F}: \end{array}$, 90°

Unclassified Exercises

20.69 $Ca(OCl)_2$ or NaOCl for disinfecting; $NaHSO_4$ and Na_2CO_3 for maintaining pH at 7.5

20.71 (a) Water does not obey Trouton's rule, which predicts that this ratio should be about 85 J/(K·mol). Substances with hydrogen bonding have values of this ratio larger than 85 J/(K·mol). (b) It takes more heat energy to separate water molecules that are held together more strongly by hydrogen bonding than H_2S molecules.

20.73 (a) 1.1×10^{10} kg S (b) 2.0×10^{12} L SO_3

20.75 (a) $2F_2(g) + 2NaOH(aq) \rightarrow OF_2(g) + 2NaF(aq) + H_2O(l)$ (b) $S_2O_3^{2-}(aq) + 4Cl_2(g) + 5H_2O(l) \rightarrow 2HSO_4^-(aq) + 8H^+(aq) + 8Cl^-(aq)$ (c) $2XeF_6(s) + 16OH^-(aq) \rightarrow XeO_6^{4-}(aq) + Xe(g) + 12F^-(aq) + 8H_2O(l) + O_2(g)$

20.77 $6HF(aq) + SiO_2(s) \rightarrow H_2SiF_6(aq) + 2H_2O(l)$

20.79 (a) $I_2(aq) + H_2S(aq) \rightarrow S(s) + 2HI(aq)$
(b) $MnO_2(s) + 2I^-(aq) + 4H^+(aq) \rightarrow Mn^{2+}(aq) + 2H_2O(l) + I_2(s)$

20.81 0.156 M ClO^-

20.83 (a) $\ddot{O}{=}Cl{=}\ddot{O}$ (b) Angular, <120°
(c) NO_2 is also a radical molecule and therefore similar in reactivity to ClO_2

Case Questions

1. 33.7 metric tons

3. 6.2×10^3 g

CHAPTER 21

Classified Exercises

21.1 (a) Rhodium (b) Silver (c) Palladium (d) Tungsten

21.3 (a) $[Xe]4f^{14}5d^{10}6s^2$ (b) $[Ar]3d^64s^2$
(c) $[Ar]3d^24s^2$ (d) $[Xe]4f^{14}5d^{10}6s$

21.5 Fe, Co, Ni

21.7 (a) Co larger than Ni (b) Zn larger than Cu
(c) Mo larger than Cr

21.9 Hg is much more dense than Zn and Cd because the shrinkage in atomic radius that occurs between $Z = 58$ and $Z = 71$ (the lanthanide contraction) causes the atoms following the rare earths to be smaller.

21.11 Proceeding down a group in the *d* block (as from Cr to Mo to W), we find an increasing probability of finding the elements in their higher oxidation states.

21.13 M has an oxidation number of +6; Cr has the most stable +6 arrangement.

21.15 (a) $TiO_2(s) + 2C(s) + 2Cl(g) \xrightarrow{1000°C} TiCl_4(g) + 2CO(g)$ and $TiCl_4(g) + 2Mg(l) \xrightarrow{700°C} Ti(s) + 2MgCl_2(s)$
(b) $V_2O_5(g) + 5Ca(s) \rightarrow 5CaO(s) + 2V(s)$
(c) $FeCr_2O_4(s) + 4C(s) \rightarrow 2Cr(s) + Fe(s) + 4CO(g)$

21.17 (a) Ti(IV) + Cr(III) (b) $Co^{2+} + HCO_3^- + NO_3^-$ (c) V + CaO

21.19 $4FeCr_2O_4(s) + 8Na_2CO_3(s) + 7O_2(g) \xrightarrow{\Delta} 8Na_2CrO_4(s) + 2Fe_2O_3(s) + 8CO_2(g)$

21.21 (a) Yes (b) No

21.23 −0.74 V

21.25 (a) CO (b) $FeO(s) + CO(g) \rightarrow Fe(s) + CO_2(g)$; $Fe_2O_3(s) + 3CO(g) \rightarrow 2Fe(s) + 3CO_2(g)$ (c) Carbon

21.27 Oxidation of Fe^{2+} to Fe^{3+} readily occurs because a half-filled *d* subshell is obtained, which is not true in the case of either Co^{2+} or Ni^{2+}.

21.29 (a) V^{3+}, Cl^-, H_2 (b) $Ni(OH)_2$, Na^+ (c) Co^{2+}, Cl^-, H_2

21.31 (a) $CuFeS_2$, copper iron sulfide (b) ZnS, zinc sulfide (c) HgS, mercury(II) sulfide

21.33 (a) $2ZnS(s) + 3O_2(g) \xrightarrow{\Delta} 2ZnO(s) + 2SO_2(g)$
$ZnO(s) + C(s) \xrightarrow{\Delta} Zn(l) + CO(g)$
(b) $2CdS(s) + 3O_2(g) \xrightarrow{\Delta} 2CdO(s) + 2SO_2(g)$
$CdO(s) + C(s) \xrightarrow{\Delta} Cd(l) + CO(g)$
(c) $HgS(s) + 3O_2(g) \rightarrow Hg(g) + SO_2(g)$

21.35 Testing for Zn^{2+} after dissolving

21.37 0.53 V

21.39 2×10^6

21.41 (a) +2 (b) +3 (c) +3

21.43 (a) Hexacyanoferrate(II) ion
(b) Hexaaminecobaltate(III) ion
(c) Aquapentacyanocobaltate(III) ion

21.45 (a) Four (b) Six (c) Six

21.47 (a) $K_3[Cr(CN)_6]$ (b) $[Co(NH_3)_5(SO_4)]Cl$
(c) $Na[Fe(H_2O)_2(C_2O_4)_2]$

21.49 (a) Linkage isomers (b) Ionization isomers
(c) Linkage isomers (d) Ionization isomers

21.51 $[Co(H_2O)_6]Cl_3$; $[CoCl(H_2O)_5]Cl_2{\cdot}H_2O$, and $[CoCl_2(H_2O)_4]Cl{\cdot}2H_2O$

21.53 $[CoCl(NO_2)(en)_2]Cl$ and $[CoCl(ONO)(en)_2]Cl$

21.55 (a) Yes;

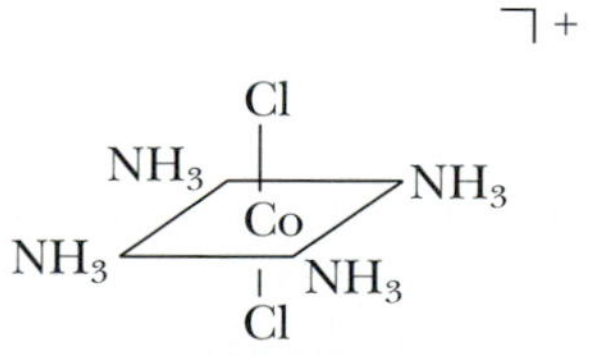

$[Co(NH_3)_4Cl_2]^+ \; Cl^- \cdot H_2O$ (Cl, Cl trans; NH_3, NH_3, NH_3, NH_3 in plane)

trans-Diamminedichloro cobalt(III) chloride monohydrate

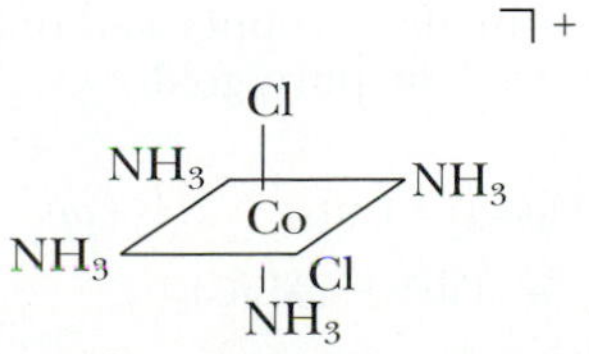

and

$[Co(NH_3)_4Cl_2]^+ \; Cl^- \cdot H_2O$ (Cl, Cl cis; NH_3, NH_3, NH_3, NH_3)

cis-Diamminedichloro cobalt(III) chloride monohydrate

(b) Yes;

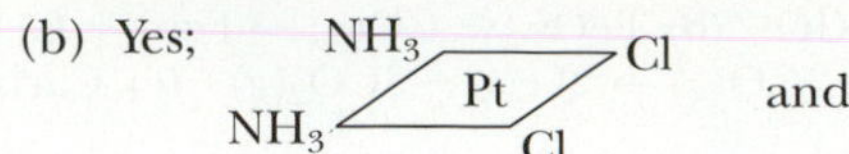

cis-Diamminedichloro platinum(II)

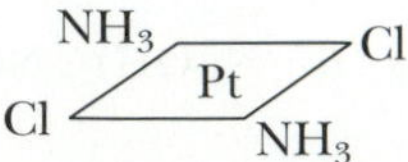

trans-Diamminedichloro platinum(II)

21.57 (a) Yes (b) No (c) Yes

21.59 (a) Chiral (b) chiral (a) is A_2B_2CD and (b) is A_2BC_2D; not an enantiomeric pair

21.61

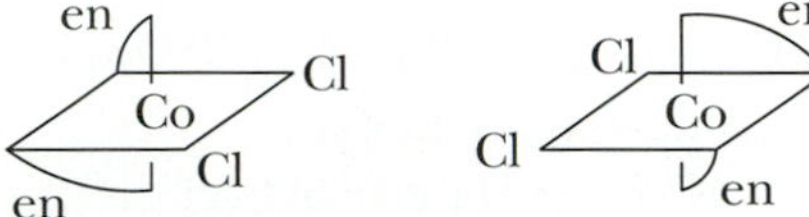

21.63 (a)

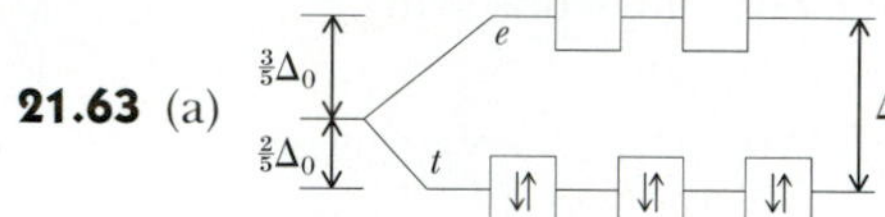

(b)

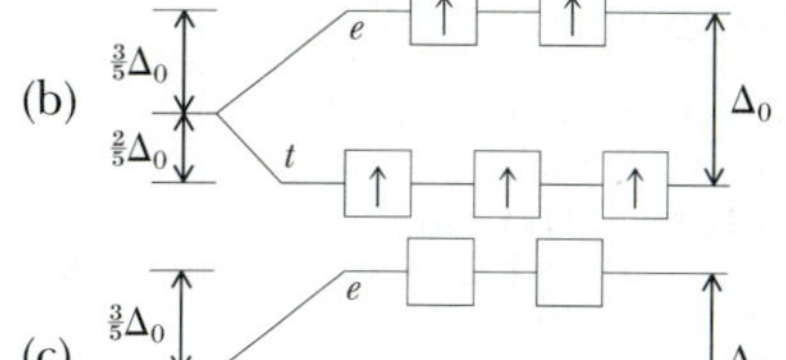

(c)

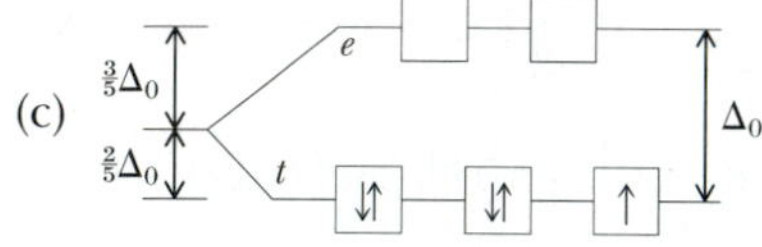

21.65 (a) 0 unpaired electrons (b) 2 unpaired electrons

21.67 (a) $[CoF_6]^{3-} \rightarrow$ blue (b) $[Co(en)_3]3+ \rightarrow$ yellow

21.69 (a) Yellow (b) Green (c) Orange (d) Blue

21.71 There are no vacancies in the 3*d* orbitals in the Zn^{2+} cation, which has a d^{10} configuration, thus no unpaired electrons. Zinc compounds would exhibit no paramagnetism because there are no unpaired electrons.

21.73 (a) 2.69×10^{-19} J (b) 4.32×10^{-19} J (c) 3.46×10^{-19} J $Cl < H_2O < NH_3$

Unclassified Exercises

21.75 $[Sc(H_2O)_6]^{3+}(aq) + H_2O(l) \rightarrow [Sc(H_2O)_5OH]^{2+}(aq) + H_3O^+(aq)$

21.77 (a) Pyrite, FeS_2 (b) Basic copper carbonate, $Cu_2(OH)_2CO_3$ (c) "Native" or pure gold (d) Osmium

21.79 (a) $Cu^{2+}(aq) + H_2(g) \rightarrow Cu(s) + 2H^+(aq)$

(b) $TiCl_4(g) + 2Mg(l) \xrightarrow{\Delta} Ti(s) + 2MgCl_2(s)$

(c) $CaO(s) + SiO_2(s) \xrightarrow{\Delta} CaSiO_3(l)$

21.81 0.691 kg $FeCr_2O_4$

21.83 (a) Dichlorobis(ethylenediammine)platinum(IV) chloride (b) tris(ethylenediammine)cobalt(III) hexacyanoferrate(III) (c) Potassium tris(oxalato)ferrate(III) (d) Sodium (ethylenediaminetetraacetato)-chromate(III)

21.85 (a) The first will give a precipitate of AgCl when $AgNO_3$ is added, the second will not. (b) The second will show free I_2 when mildly oxidized with, say Br_2, but the first will not. Starch-I_2 test gives a deep blue color after $2I_3^-$ goes to I_2.

21.87 $[Zn(H_2O)_4]^{2+}|Zn$, $[Zn(NH_3)_4]^{2+}|Zn$, NH_3, $[Zn(OH)_4]^{2-}|Zn$, OH^-, $[Zn(CN)_4]^{2-}|Zn$, Cn^-; ligand field strength: $H_2O < NH_3 < OH^- < CN^-$

21.89 (a) and (b) $[MnCl_6]^{4-}$: 5 e$^-$, Cl^- is a weak-field ligand, 5 unpaired e$^-$

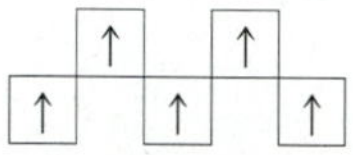

↑ ↑

↑ ↑ ↑

$[Mn(CN)_6]^{4-}$: 5 e$^-$, CN^- is a strong-field ligand, 1 unpaired e$^-$

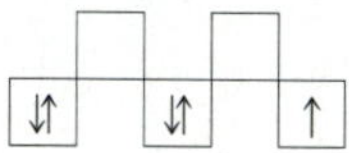

(c) $[Mn(CN)_6]^{4-}$ transmits longer wavelengths.

21.91 (a) 3.62×10^{-19} J (b) Red

21.93 Ionization isomerization:

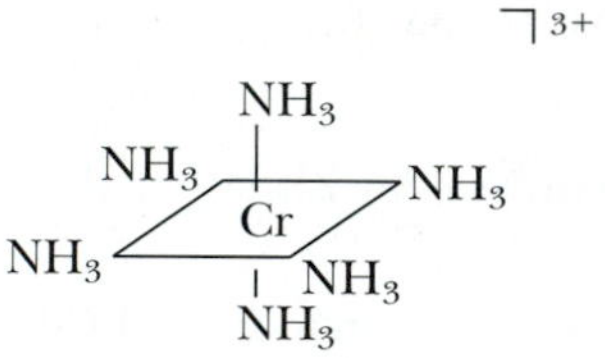

$[Cr(NH_3)_6]^{3+}$ $3Cl^-$

Hexamminechromium(III) chloride

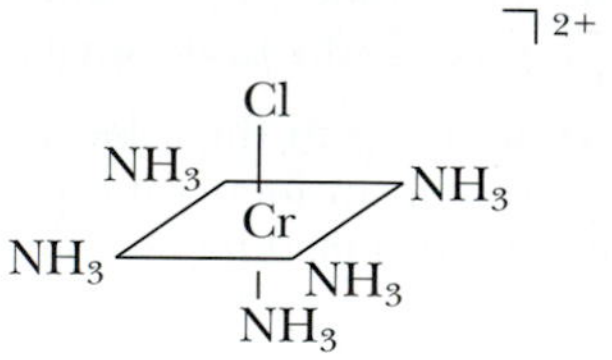

$[Cr(NH_3)_5Cl]^{2+}$ $2Cl^- \cdot NH_3$

Pentamminechlorochromium(III) aminedichloride

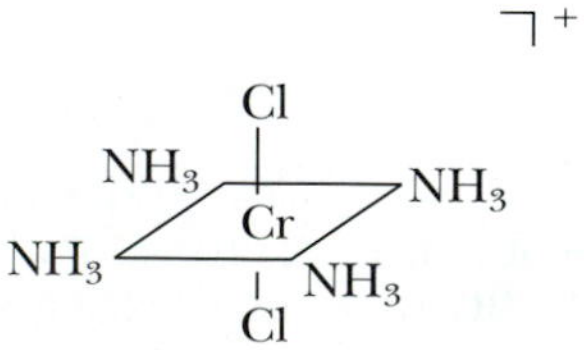

$[Cr(NH_3)_4Cl_2]^{+}$ $Cl^- \cdot 2NH_3$

Tetraamminedichlorochromium(III) diaminechloride

Case Questions

1. (a) $U_3O_8(s) + 8C(s) \rightarrow 3U(s) + 8CO(g)$; $UF_4(s) + 2Ca(s) \rightarrow U(s) + 2CaF_2(s)$ (b) 1.70×10^4 g U (or 17 kg U)

3. 5.2×10^{10} kg C (coal) and 1.9×10^{11} kg CO_2

5. (a) −850 kJ (b) 1.10×10^3 g Fe (or, 1.10 kg Fe melted in the rail)

CHAPTER 22

Classified Exercises

22.1	Nuclide	Protons	Neutrons	Nucleons
	(a) ^2H	1	1	2
	(b) ^{263}Unh	106	157	263
	(c) ^{60}Co	27	33	60
	(d) ^{258}Md	101	157	258

22.3 (a) 3.8×10^{10} J/mol (b) 2.3×10^{12} J/mol

22.5 (a) $\nu = 3.21 \times 10^{20}$/s; $\lambda = 9.33 \times 10^{-13}$ m (b) $\nu = 2.66 \times 10^{20}$/s; $\lambda = 1.13 \times 10^{-12}$ m

22.7 (a) ^3_2He (b) $^{83}_{38}Sr$ (c) $^{221}_{89}Ac$

22.9 (a) ^8_4Be (b) $^{181}_{77}Ir$ (c) ^7_3Li

22.11 (a) $^{24}_{11}Na \rightarrow ^0_{-1}e + ^{24}_{12}Mg$ (b) $^{128}_{50}Sn \rightarrow ^0_{-1}e + ^{128}_{51}Sb$ (c) $^{228}_{90}Th \rightarrow ^4_2\alpha + ^{224}_{88}Ra$

22.13 (a) β decay most likely; $^{68}_{29}Cu \rightarrow ^0_{-1}e + ^{68}_{30}Zn$ (b) α decay is most likely; $^{243}_{97}Bk \rightarrow ^4_2\alpha + ^{239}_{95}Am$ (c) α decay is most likely; $^{260}_{105}Unp \rightarrow ^4_2\alpha + ^{256}_{103}Ln$

22.15 (a) $^4_2\alpha$ (b) $^{248}_{96}Cm$ (c) $^0_{-1}e$

22.17 (a) ^8_4Be; $^{20}_{10}Ne + ^4_2\alpha \rightarrow ^8_4Be + ^{16}_8O$ (b) $^{24}_{12}Mg$; $^{20}_{10}Ne + ^{20}_{10}Ne \rightarrow ^{24}_{12}Mg + ^{16}_8O$ (c) $^4_2\alpha$; $^{44}_{20}Ca + ^4_2\alpha \rightarrow \gamma + ^{48}_{22}Ti$

22.19 (a) $^{14}_7N + ^4_2\alpha \rightarrow ^{17}_8O + ^1_1p$ (b) $^{239}_{94}Pu + ^1_0n \rightarrow ^{240}_{95}Am + ^0_{-1}e$

22.21 (a) ^1_0n (b) $^{95}_{40}Zr$

22.23 1.0×10^{-4}Ci

22.25 (a) 3.7×10^{10} dps (b) 3.0×10^9 dps (c) 3.7×10^4 dps

22.27 (a) 5.64×10^{-2} y^{-1} (b) 0.82 s^{-1} (c) 0.0693 min^{-1}

22.29 (a) 5.32×10^3 y (b) 1.7 y

22.31 6.0×10^{-3} Ci

22.33 (a) 88.6% (b) 32%

22.35 1.26×10^9 y

22.37 (a) 4.5×10^9 y (b) 3.8×10^9 y

22.39 (a) 9.8×10^{-4} Ci (b) 2.8×10^{-4} Ci

22.41 (a) -1.44×10^{-10} g (lost) (b) 1.05×10^{-9} g (gained)

22.43 -4.3×10^9 kg/s

22.45 (a) 1.8×10^{-13} J/nucleon (b) 1.409×10^{-12} J/nucleon; the most stable

22.47 (a) 7.8×10^{10} J/g (b) 2.03×10^{11} J/g (c) 3.61×10^{11} J/g

Unclassified Exercises

22.49 (a) $^{174}_{72}Hf \rightarrow ^{170}_{70}Yb + ^4_2\alpha$ (b) $^{98}_{43}Tc \rightarrow ^{98}_{44}Ru + ^0_{-1}\beta$ (c) $^{41}_{20}Ca + ^0_{-1}e \rightarrow ^{41}_{19}K$

22.51 $^{98}_{42}Mo + ^1_0n \rightarrow ^{99}_{42}Mo \rightarrow ^{99}_{43}Tc + ^0_{-1}\beta$

22.53 45 d

22.55 1×10^{-92} μCi

22.57 0.62

22.59 (a) 1.11^{-1} d; 0.63 d (b) 0.2 μg

22.61 (a) 1.638×10^{-10} J (b) 1.022 MeV

22.63 9.3×10^{-10} J

Case Questions

1. Basements tend to be more poorly ventilated than the upper levels of homes.

CHAPTER 23

Classified Exercises

23.1 (a) Alkane (b) Alkyne (c) Alkane

23.3 (a) Isopropyl (b) *n*-Propyl (c) Cyclopentyl

23.5 (a) Hexane (b) Octane (c) Nonane

23.7 (a) C_3H_8 (b) C_8H_{18} (c) C_7H_{16}

23.9 (a) 2-Heptene (b) 2-Decene (c) 2-Octene

23.11 Nonane, C_9H_{20}

23.13 (a) Propane (b) Pentane (c) 2,3-Dimethylbutane

23.15 (a) 4-Methyl-2-pentene (b) 2,3-Dimethyl-2-phenylpentane

23.17 (a) 1-Butene (no geometrical isomers) (b) 1-Butyne (no geometrical isomers) (c) 2-Butyne (no geometrical isomers)

23.19 (a) $CH_2{=}CHCH(CH_3)CH_2CH_3$ (b) $CH_3CH_2C(CH_3)_2CH(CH_2CH_3)(CH_2)_2CH_3$ (c) $CH{\equiv}C(CH_2)_2C(CH_3)_2CH_3$

23.21 (a)

```
       H  CH3
    H   \ /   H
         ____
H —    /    \   — H
H —    \____/   — CH3
     H        H
```

(b)

```
                          CH3
                          |
       H   H              CH2  H   H   H   H   H
       |   |              |    |   |   |   |   |
H—C≡C—C—C——————————C——————C—C—C—C—C—H
       |   |              |    |   |   |   |   |
       H   CH2            CH2  H   H   H   H   H
           |              |
           CH2            CH3
           |
           CH2
           |
           CH3
```

```
    H  CH3 H  CH3 H
    |   |  |   |  |
(c) H—C—C——C—C——C—H
    |   |  |   |  |
    H  CH3 H   H  H
```

23.23 (a) C_4H_{10}:

```
   H  H  H  H
   |  |  |  |
H—C—C—C—C—H, n-butane;
   |  |  |  |
   H  H  H  H
```

```
   H  CH3  H
   |   |   |
H—C—C———C—H, iso-butane
   |   |   |
   H   H   H
```

(b) C_5H_{12};

```
   H  H  H  H  H
   |  |  |  |  |
H—C—C—C—C—C—H, n-pentane;
   |  |  |  |  |
   H  H  H  H  H
```

```
   H  CH3  H  H
   |   |   |  |
H—C—C———C—C—H, iso-pentane;
   |   |   |  |
   H   H   H  H
```

```
          H
          |
       H—C—H
   H      |      H
   |      |      |
H—C————C————C—H, neo-pentane
   |      |      |
   H      |      H
       H—C—H
          |
          H
```

(c) dimethylcyclohexane C_8H_{16}:

1,2-dimethylcyclohexane;

1,3-dimethylcyclohexane;

1,4-dimethylcyclohexane

23.25

```
H          H   H
 \         |   |
  C=C——C—C—H, 1-butene;
 /     \   |   |
H       H  H   H
```

```
 H H      H
  \|      |
H—C    H—C—H, cis-2-butene;
    \    /
     C=C
    /    \
   H      H
```

```
           H
           |
H       H—C—H
  \       /
   C=C          , trans-2-butene
  /     \
H—C—H    H
  |
  H
```

23.27

```
   H        H        H
   |        |        |
H—C————C————C—H
   |        |        |
   H    H—C—H    H
            |
            H
```

Light [Cl_2] →

```
    H       H       H
    |       |       |
Cl—C————C————C—H
    |       |       |
    H   H—C—H   H
            |
            H
```

Light [Cl_2] →

```
   H       Cl       H
   |        |       |
H—C————C————C—H
   |        |       |
   H    H—C—H   H
            |
            H
```

23.29

```
      CH3 CH3
       |   |
CH3—CH—CH—CH2—Cl  --[KOH], −HCl-->
```

```
      CH3 CH3
       |   |
CH3—CH—C=CH2
```

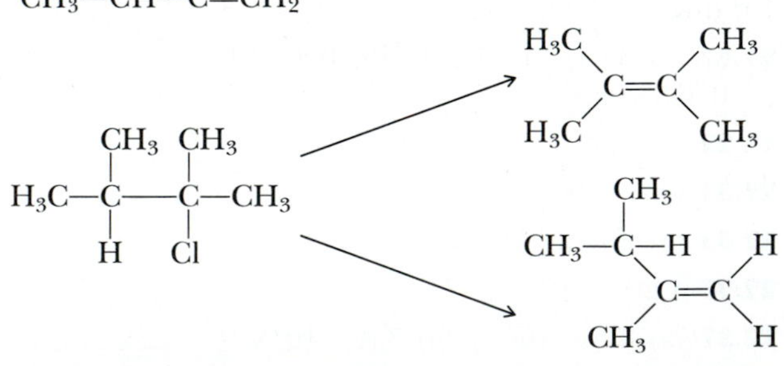

23.31 (a) $CH_3CHBrCH_3 \xrightarrow{NaOH} CH_3C{=}CH_2$ and NaBr and H_2O; *elimination* reaction (b) $CH_2{=}CH_2 + Br_2 \rightarrow CH_2Br{-}CH_2Br$; *addition* reaction

23.33 (a) $CH_3CH_3 + 2Cl_2 \xrightarrow{light} CH_2ClCH_2Cl(g) + 2HCl(g)$; substitution (b) $CH_2CH_2 + Cl_2 \xrightarrow{light} CH_2ClCH_2Cl(g)$; addition (c) $HC{\equiv}CH + 2Cl_2 \xrightarrow{light} CHCl_2CHCl_2(g)$

23.35 $CH_3(CH_2)_2CH{=}CH_2$

23.37 Cyclopentane (CH_2 ring) $\xrightarrow[\text{light, —HCl}]{Cl_2}$ chlorocyclopentane (H, Cl on one C) $\xrightarrow[\text{organic solvent}]{KOH}$ cyclopentene + KCl + H_2O

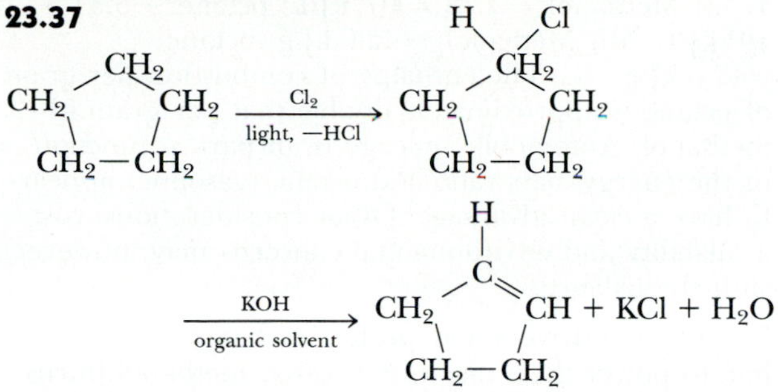

23.39 (a) Polyethylene (b) Polyacrylonitrile (orlon) (c) Synthetic rubber

23.41 (a) H_3C—C(=CH_2)—CH=CH_2

(cis and trans polyisoprene chain structures; labels: cis, trans)

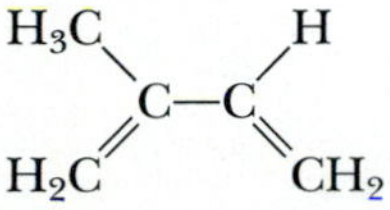

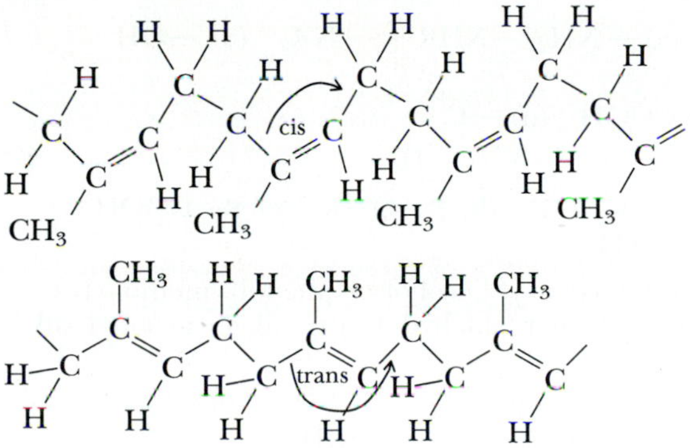

(b) $CH_2{=}CH{-}CN$:

—CH—CH_2—CH—CH_2—CH—CH_2— (each CH bearing CN)

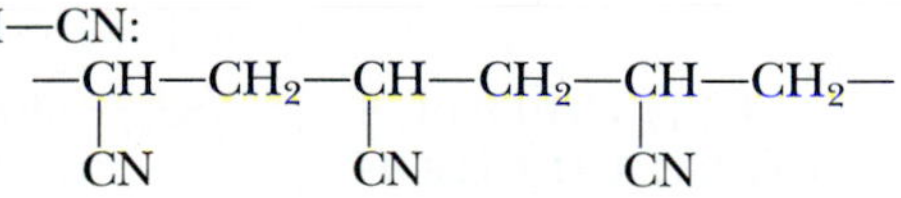

23.43 2.8×10^7 g/mol

23.45 (a) 1-Ethyl-3-methylbenzene (b) 1,2,3,4,5-Pentamethylbenzene

23.47 (a) CH_3 / Cl (para) (b) CH_3 / CH_3 (meta) (c) CH_3 / Cl (para)

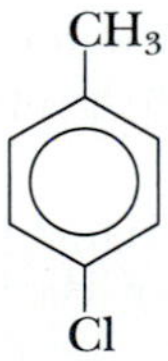

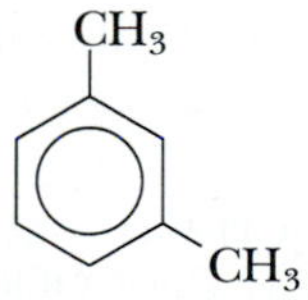

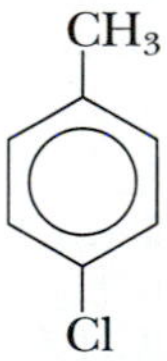

23.49 Cl/Cl (ortho), Cl/Cl (meta), Cl/Cl (para)

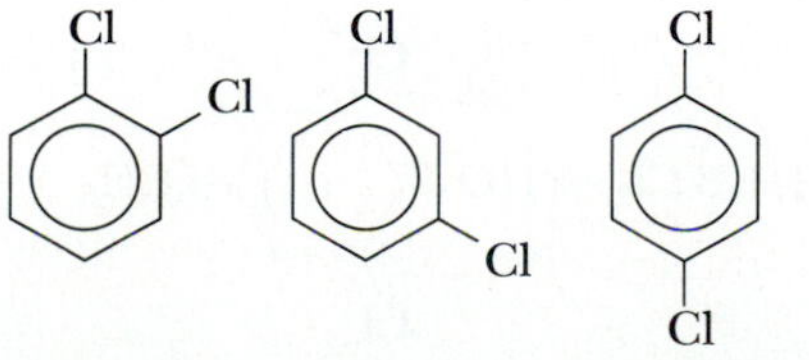

Orthopolar Metapolar Paranonpolar

23.51 (a) CH_3/NO_2 (ortho), CH_3/NO_2 (meta), CH_3/NO_2 (para)

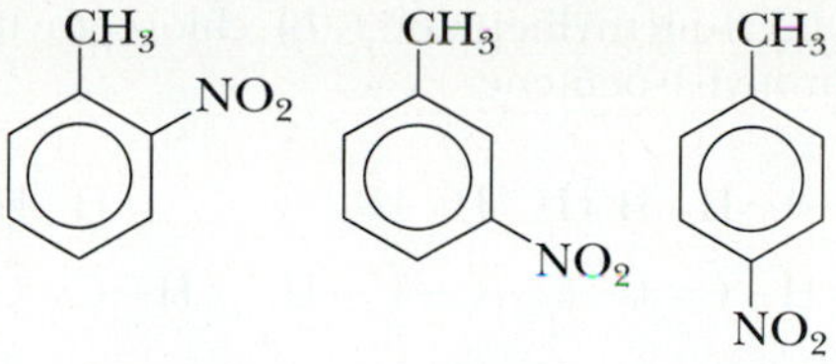

(b) 1-methyl-2-nitrobenzene, *o*-nitrotoluene; 1-methyl-3-nitrobenzene, *m*-nitrotoluene; 1-methyl-4-nitrobenzene, *p*-nitrotoluene

23.53 C_6H_6 $\xrightarrow[\text{—HCl}]{Cl_2(FeCl_3)}$ C_6H_5Cl $\xrightarrow[\text{Ni 150–200°C, 25 atm}]{3\ H_2\ \text{(catalyst)}}$ $C_6H_{11}Cl$ $\xrightarrow[\text{org. sol. 55°C, —HCl}]{KOH}$ C_6H_{10}

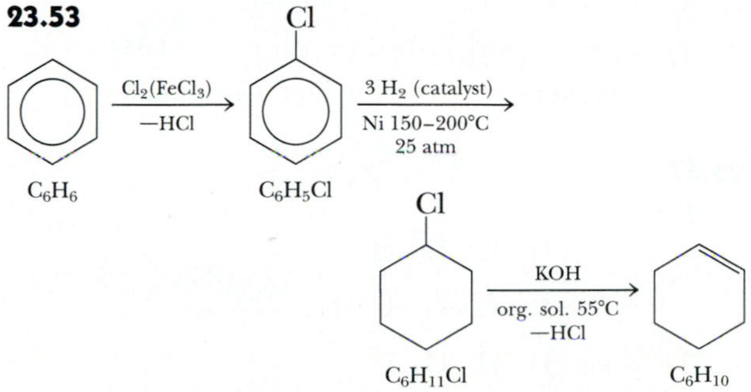

23.55 (a) benzene + Cl_2 $\xrightarrow{AlCl_3}$ chlorobenzene + HCl

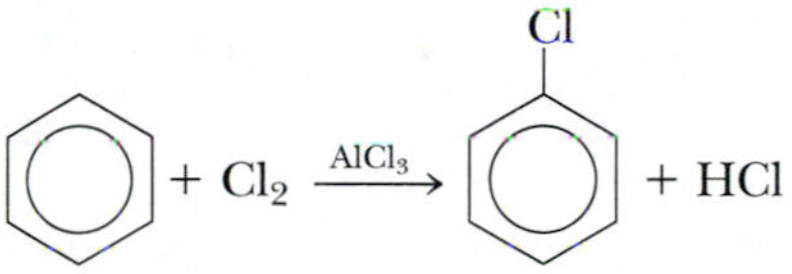

(b) benzene + CH_3CH_2Cl $\xrightarrow{AlCl_3}$ ethylbenzene (CH_3CH_2) + HCl

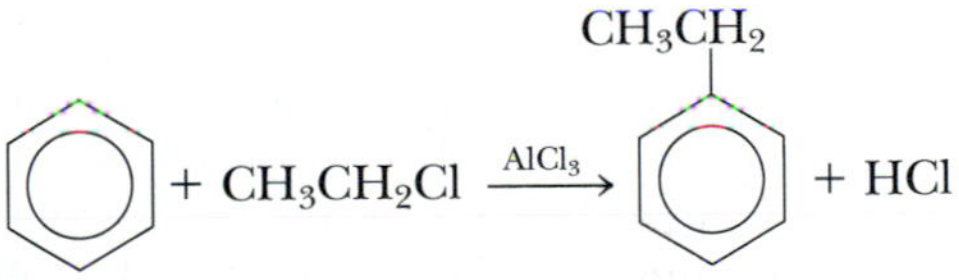

Unclassified Exercises

23.57 (a) sp^3 (b) sp^2 (c) sp

23.59 (a) sp^2, trigonal planar (b) sp—linear (c) sp^3, tetrahedral

23.61 (a) methylcyclopropane: C(H)(CH_3) with two CH_2 (H, H) ring carbons (b) 1,2,3,5-tetramethylbenzene (CH_3, H_3C, CH_3, CH_3)

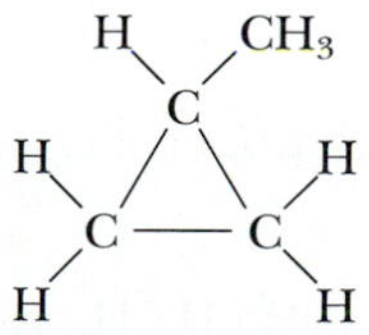

(c) H, C_6H_5—C(H)(C_6H_5)—C(H)=C(H)—CH_3

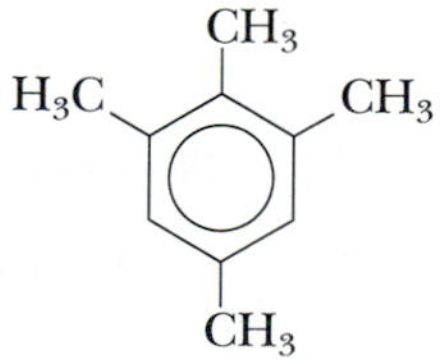

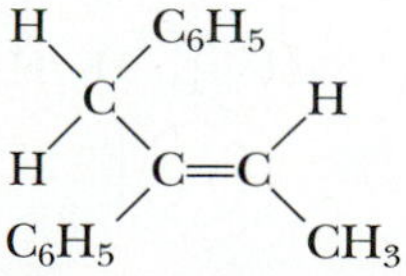

23.63 (a) 4-methylheptane (b) chloroethyne (c) 2-propyl-1-pentene

23.65

```
     H  H  H  H  H              H  H  CH3
     |  |  |  |  |              |  |  |
   H—C═C—C—C—C—H            H—C═C—C—H
           |  |  |                    |
           H  H  H                    CH3
      1-Pentene              3-Methyl-1-butene

   H  H  H  H  H             H  H    CH3
   |  |  |  |  |             |  |   /
 H—C—C═C—C—C—H           H—C—C═C
   |        |  |             |      \
   H        H  H             H       CH3
      2-Pentene              2-Methyl-2-butene
```

23.67

```
    H
    |
  (C6H5)        H  H  H  H
                |  |  |  |
         + Cl—C—C—C—C—H  --[AlCl3]-->
                |  |  |  |
                H  H  H  H

                         H  H  H  H
                         |  |  |  |
                       H—C—C—C—C—H
                         |  |  |  |
                         |  H  H  H
                       (C6H5)
```

23.69 1-Butyne will add two moles of halogen across the triple bond (per mole of 1-butyne). 1-Butene will add one mole of halogen across the double bond (per mole of 1-butene). Butane will not react with the halogen under the same conditions. If bromine, a brown-orange solution, is used, it will decolorize when it reacts. The color change is the "indicator" and the volume added provides information (number of moles) to distinguish between C═C and C≡C.

23.71 Yes. The principal reaction of cyclopentane (an alkane) is *elimination.* Pentene (an alkene) can undergo *addition.*

23.73

```
              H  H  H  H
              |  |  |  |
  (C6H6) + H—C—C—C—C—H →
              |  |  |  |
              H  Cl H  H

                    H  H  H  H
                    |  |  |  |
                  H—C—C—C—C—H + HCl
                    |  |  |  |
                    H  |  H  H
                     (C6H5)
```

Case Questions

1. (a) Methanol: -1.78×10^4 kJ/L; octane: -3.31×10^4 kJ/L (b) Methanol: -22.6 kJ/g; octane: -47.8 kJ/g (c) The enthalpy of combustion per gram of octane is approximately double that per gram of methanol. Automobile mileage is, in part, a function of the energy/mass ratio and octane (gasoline) appears to have a clear advantage. Other considerations, cost, availability and environmental concerns may, however, shift the balance.

3. Indy-500 drivers may prefer methanol over gasoline to power their race cars because methanol burns more smoothly, and at lower temperatures, in high-performance engines such as those used in Indy-500 vehicles.

CHAPTER 24

Classified Exercises

24.1 (a) $—NH_2$, —NHR or $—NR_2$ (b) —OH (c) —COOH (d) —C(═O)—H

24.3 (a) R—OH (b) R—O—R (c) R—COOH (d) $R—NH_2$

24.5 (a) Hydroxyl (alcohol or phenol), methoxyl (ether), carbonyl (aldehyde) (b) Alkyl, carbonyl, alkenyl

24.7 (a) HO—C_6H_4—COOH, phenol (also carboxylic acid)
(b) $CH_3CH_2CH(OH)CH_3$, secondary alcohol
(c) $CH_3CH_2CH_2CH_2OH$, primary alcohol

24.9 (a) 2-Propanol, secondary (b) 1-Propanol, primary (c) 2-Methyl-2-pentanol, tertiary

24.11 (a) Dimethyl ether (b) Ethyl methyl ether (c) Diethyl ether

24.13 (a) Ethene (ethylene) and water
(b) Chloroethane (ethyl chloride) and water
(c) *tert*-Butylbromide (2-bromo-2-methylpropane) and water

24.15 (a) $HOCH_2CH_2OH + 2CH_3(CH_2)_{16}COOH \rightarrow CH_3(CH_2)_{16}COOCH_2CH_2COO(CH_2)_{16}CH_3 + H_2O$
(b) $2CH_3CH_2OH + HOOCCOOH \rightarrow CH_3CH_2OOCCOOCH_2CH_3 + 2H_2O$
(c) $C_6H_5OH + 3\ Cl_2 \rightarrow$ HO—$C_6H_2Cl_2$—CH_3 + 3HCl (Cl at both positions adjacent to the CH_3-bearing carbon, as drawn)

24.17 Propanal, CH_3CH_2CHO

24.19 (a) Aldehyde, methanal (b) Ketone, propanone (c) Ketone, 3-pentanone

24.21 (a) HCHO (b) CH_3COCH_3 (c) $CH_3(CH_2)_4COCH_3$

24.23 (a) Ethanol (b) 2-Octanol (c) 5-Methyl-1-octanol These reactions can be accomplished with an oxidizing agent such as acidified sodium dichromate.

24.25 (a) Hemiacetal (b) Hemiketal

24.27 $CaCO_3 + 3C \xrightarrow[\text{furnace}]{\text{electric arc}} CaC_2 + CO_2 + CO$, $CaC_2 + 2H_2O \rightarrow Ca(OH)_2 + HC{\equiv}CH$, $HC{\equiv}CH + H_2O \xrightarrow[\text{catalyst}]{H_2SO_4,\ HgSO_4,} CH_3CHO$

24.29 (a) Ethanoic acid (b) Butanoic acid (c) 2-aminoethanoic acid

24.31

(a) $C_6H_5{-}\overset{\overset{\large O}{\|}}{C}{-}OH$ (b) $CH_3{-}CH_2{-}\underset{\underset{\large CH_3}{|}}{CH}{-}\overset{\overset{\large O}{\|}}{C}{-}OH$

(c) $HO{-}\underset{\underset{\large O}{\|}}{C}{-}\underset{\underset{\large O}{\|}}{C}{-}OH$

24.33 (a) $CH_3CH_2OH + O_2 \xrightarrow{\text{Co(III) catalyst}} CH_3COOH + H_2O$

(b) $CH_3CH_2CH(CH_3)CH_2OH \xrightarrow{KMnO_4(aq),\ H_2SO_4(aq)} CH_3CH_3CH(CH_3)COOH$

24.35 Cl is more electronegative than H. The electronegative Cl can attract the pair of electrons in the O—H bond, thus weakening the bond so that the H^+ ion may break away. The more Cl's there are, the more readily this can occur.

24.37 $C_6H_5COOCH(CH_3)_2$

24.39 $HOOC{-}CO{+}OCH_2(CH_2)_2CH_2{-}COO{-}CO{+}_n OCH_2(CH_2)_2CH_2OH$

24.41 The following procedures may be used:

(1) $RCOOH \xrightarrow[\text{base}]{\text{aqueous}} RCOO^-$ (dissolves);

(2) $RCHO \xrightarrow[\text{base}]{\text{aqueous}}$ (does not dissolve);

(3) $R{-}CO{-}R \xrightarrow[\text{base}]{\text{aqueous}}$ (does not dissolve);

(4) $RCHO \xrightarrow[\text{reagent}]{\text{Tollens'}} RCOOH + Ag(s)$;

(5) $R{-}CO{-}R \xrightarrow[\text{reagent}]{\text{Tollens'}}$ (no reaction). (1), (2), and (3) distinguish the carboxylic acid from the aldehyde and ketone; (4) and (5) distinguish the aldehyde and ketone from each other.

24.43 1,2-Benzenedicarboxylic acid reacts with ethylene glycol. The properties of these two polymers would be similar except that "Dacron," due to its more linear molecular structure, can probably be more readily spun into yarn.

24.45 (a) Methylamine, primary (b) Diethylamine, secondary (c) *o*-methylaniline or *o*-methylphenylamine, primary

24.47 (a) $CH_3\overset{+}{N}H_3Cl^-$ (b) $(C_2H_5)_2NCOCH_3$

(c) $H_2N{-}C_6H_2Br_2{-}Br + 3HBr$ (excess Br_2) (2,4,6-tribromo ring: Br, Br, Br)

(d) $CH_3CH_2CONH(CH_2)_4NHCOCH_2CH_3$

24.49 $HOOC{-}CO{+}NH{-}(CH_2)_4{-}NHCO{-}CO{+}_n NH(CH_2)_4{-}NH_2$

24.51 $H_2N{-}CH_2{-}CH_2{-}CH_2{-}CH_2{-}NH_2$, putrescine

24.53 Glycine, NH_2CH_2COOH

24.55 $-C(=O){-}NH-$

24.57 $H{-}C(NH_2)(C_6H_4OH){-}C(=O){-}NH{-}CH_2{-}COOH$

24.59 $NH_2CH(CH_3){-}C(=O){-}NH{-}CH(CH_2CH_2{-}S{-}CH_3){-}C(=O){-}NH{-}CH(CH_2{-}SH)(COOH)$

24.61 (a) $-NHCH_2CONHCH_2CONHCH_2CONHCH_2CO-$
(b) $-NHCH_2CONH\underset{CH_2CH(CH_3)_2}{CH}CONH\underset{CH_2OH}{CH}CONH\underset{CH_3}{CH}CO-$

24.63 Tyr-Leu-Ser-Ile-Tyr-Leu-Ser

Unclassified Exercises

24.65 —OH group

24.67 (a) 2-methylcyclohexanone (ring with =O and CH_3) (b) $CH_3C(=O)H$

24.69 HO–(benzene ring)–CH_2–CH_2–NH_2, with HO on the ring

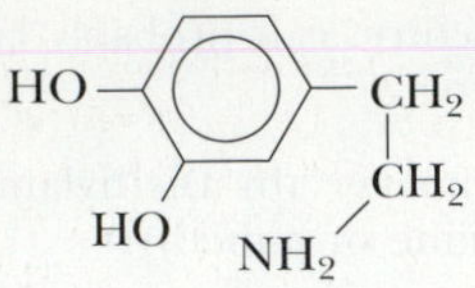

24.71

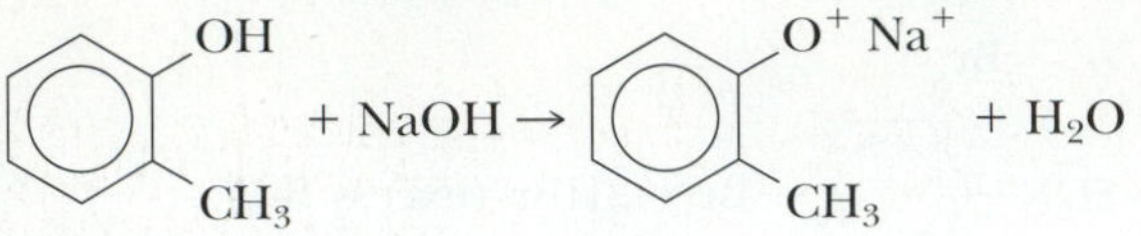

(2-methylphenol: ring with OH and CH_3) + NaOH → (ring with O^- Na^+ and CH_3) + H_2O

24.73

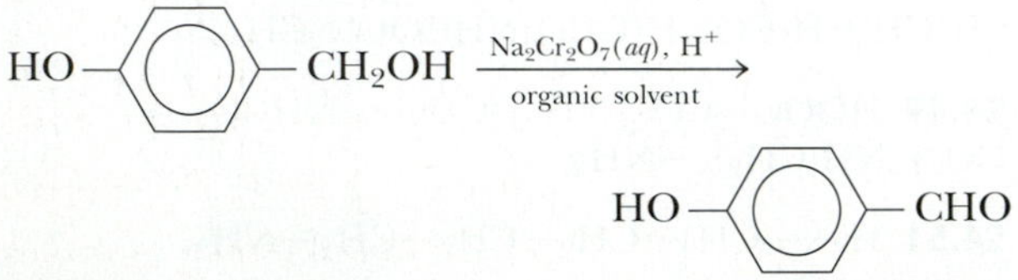

HO–(benzene ring)–CH_2OH $\xrightarrow[\text{organic solvent}]{Na_2Cr_2O_7(aq),\ H^+}$

HO–(benzene ring)–CHO

24.75

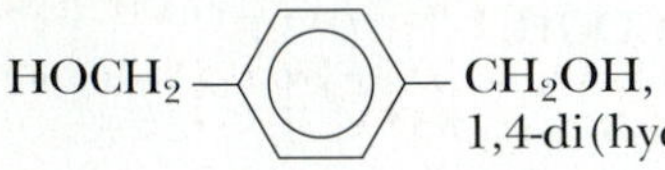

$HOCH_2$–(benzene ring)–CH_2OH,
1,4-di(hydroxymethyl)benzene

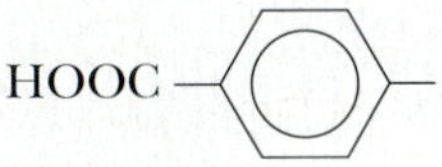

HOOC–(benzene ring)–COOH, terephthalic acid

24.77 $(Leu)_1(Ala)_1(Val)_3(Gly)_4(Ser)_7(Pro)_6(Arg)_{50}$

Case Questions

1. (a) Organophosphorus compound (b) Carbamate (c) Chlorinated hydrocarbon (d) Carbamate

3. Both of these herbicides can be classified as chlorinated hydrocarbons. They are also aromatic ethers of the form Ar—O—R, where Ar stands for the aromatic grouping containing the benzine ring and R for the aliphatic grouping, $—CH_2—COOH$. These compounds might undergo hydrolysis in acid solution at the O in the ether linkage $Ar—O—R + H_2O \xrightarrow{\text{acid}} Ar—OH + HOCH_2—COOH$. The acid necessary could be provided by "acid" rain.

3. (a) 25 d (b) Since it takes only 25 days for 99% of the radon to be removed by disintegration, it does not seem likely that radon formed deep in the Earth's crust could leak to the surface in such a relatively short time. So most of the radon observed must have been formed near the Earth's surface.

ILLUSTRATION CREDITS

All photographs by Ken Karp except the following:

p. 0, David Malin, Anglo-Australian Telescope Board; p. 2, Air Products and Chemicals, Inc.; p. 3, Chip Clark; p. 5, The Aluminum Association; p. 8, Chip Clark; p. 9, Chevron Corporation; p. 12, Chip Clark; p. 30, Freer Gallery of Art, Smithsonian Institution; p. 38, Hewlett-Packard; p. 40, (Fig. 2.1) IBM Research Division/Almaden Research Center; p. 43, (Fig. 2.7) Chip Clark; p. 45, (Fig. 2.10) Manchester Literary and Philosophical Society, (Fig. 2.11) Randall M. Feenstra, IBM Thomas J. Watson Research Center, Yorktown Heights, NY; p. 46, R. Wiesendanger and colleagues, University of Basel, Switzerland; p. 47, R. Wiesendanger and colleagues, University of Basel, Switzerland; p. 48, (Fig. 2.12) Cavendish Laboratory, (Fig. 2.13) Donald Clegg; p. 49, (Fig. 2.15) Cavendish Laboratory; p. 57, Bettman Archives; p. 59, Chip Clark; p. 69, Chip Clark; p. 73, Chip Clark; p. 83, IMC Fertilizer Group; p. 84, Woods Hole Oceanographic Institution; p. 91, Chip Clark; p. 105, F.S. Judd, Research Laboratories, Kodak Ltd.; p. 126, NASA; p. 128, The M.W. Kellog Co.; p. 131, National Oceanic and Atmospheric Administration; p. 158, NASA; p. 163, (Fig. 5.1) Chip Clark; p. 174, (Fig. 5.13) Chip Clark; p. 175, U. S. Department of the Interior, Bureau of Mines; p. 182, (Fig. 5.23) Martin Marietta Energy Systems, Oak Ridge, Tenn.; p. 192 David Fitzgerald/Tony Stone Worldwide; p. 195, (Fig. 6.4) USDA Fire Service/Barry Nehr; p. 228, Nancy Roger/Exploratorium; p. 232, Zambelli Internationale; p. 234, (Fig. 7.5) Century Lubricant Specialists; p. 236, Chip Clark; p. 237, (Fig. 7.10) AIP/Niels Bohr Library; p. 239, Science Museum, London; p. 241, Dublin Institute for Advanced Studies; p. 256; (Fig. 7.25) Edgar Fahs Smith Collection, ACS Center for History of Chemistry, University of Pennsylvania; p. 278, Paul Brierley; p. 280, (Fig. 8.1) University of California Archives, The Bancroft Library; p. 318, Robert Stroud; p. 347, Donald Clegg; p. 360, Paul Brierley; p. 371, Hans Pfletschinger/Peter Arnold; p. 372, (Fig. 10.11) NASA; p. 383, (Fig. 10.27a) William McCoy/Rainbow; p. 384, (Fig. 10.29) Michael Heron/Woodfin Camp & Assoc.; p. 395, (Fig. 10.41) General Electric Company; p. 396, (Fig. 10.45) Chip Clark; p. 403, Jeffrey D. Smith/Woodfin Camp & Assoc.; p. 404, Paul Brierley; p. 415, Chilean Nitrate Corp.; p. 418, (Fig. 11.13) E.B. Smith, Physical Chemistry Laboratory, University of Oxford; p. 434, from "Biological Membranes as Bilayer Couples," by M. Sheetz, R. Painter, and S. Singer, 1976, *Journal of Cell Biology*, 70:193; p. 443, Chip Clark; p. 444, Chip Clark; p. 471, (Fig. 12.25a) Johnson Matthey; p. 472, Lubert Stryer; p. 478, Malcolm Lockwood, Geophysical Institute, University of Alaska, Fairbanks; p. 482, General Motors; p. 491, Bruno Barbey/Magnum; p. 492, Chip Clark; p. 504, NASA; p. 520, (Fig. 13.10, left) American Institute for Physics, Niels Bohr Library/Stein Collection, (Fig. 13.10, right) Bettman Archives, (Fig. 13.11) M. W. Kellogg Co.; p. 527, Chip Clark; p. 530, Coco McCoy/Rainbow; p. 573, Regis Bossu/Sygma; p. 574, Peter Scoones/Planet Earth Pictures; p. 590,

(Fig. 15.7) Chip Clark, (Fig. 15.8) Heather Angel/ Biofotos; p. 592, (Fig. 15.11) Chip Clark; p. 600, Chip Clark; p. 620, Manley-Prim Photography; p. 628, (Fig. 16.6) Dieter Flamm; p. 631, (Fig. 16.11) International Tin Research Institute; p. 636, (Fig. 16.16) Yale University Library; p. 652, Paul Brierley; p. 662, Anglia Television; p. 665, (Fig. 17.10) Ken Lucas/Planet Earth Pictures; p. 667, (Fig. 17.12) The National Museum of Science & Industry, London; p. 675, (Fig. 17.18) St. Joe Zinc Co./ American Hot Dip Galvanizers Association; p. 695, Edward Keating; p. 696, Chip Clark; p. 711, (Fig. 18.15) Houston Museum of Natural Science; p. 712, (Fig. 18.16) Dow Chemical; p. 716, (Fig. 18.21) Travis Amos; p. 717, Field Museum of Natural History, Chicago; p. 718, Aalborg Portland Betonforskningslabortorium, Karlslunde; p. 723, NASA; p. 724, Chip Clark; p. 730, (Fig. 19.3 a, c) Houston Museum of Science, (Fig. 19.3b) Lee Bolton, Inc.; p. 735, Field Museum of Natural History, Chicago; p. 738, (Fig. 19.8) Photo Researchers; p. 740, (19.12) The Rockwell Museum, Corning, NY; p. 742, (Fig. 19.15) Chip Clark; p. 743, (19.17) Chip Clark; p. 744, IBM Research Division/Almaden Research Center; p. 745, (19.20) Photo Researchers; P. 748, TRW Inc.; p. 758, Max Planck/Institute for Kenphysik; p. 760, Andre Bartschi; p. 764, (Fig. 20.4a) Chip Clark, (Fig. 20.4b) National Park Service; p. 768, Lee Boltin, Inc.; p. 772, (Fig. 20.11) Martin Marietta Energy Systems, Inc.; p. 778, Morton Thiokol; p. 780, (Fig. 20.20) Greater Pittsburgh Neon/Tom Anthony; p. 788, Terry Vine/Tony Stone Worldwide; p. 793, (Fig. 21.7a) Field Museum of Natural History, Chicago, (Fig. 21.7b) Hagley Museum and Library, (Fig. 21.7c) Oremet Titanium; p. 798, U.S. Steel; p. 801, Department of Interior, Bureau of Mines; p. 802, (Fig. 21.13) Lee Bolton Picture Library, (Fig. 21.14) Field Museum of Natural History, Chicago; p. 809, (Fig. 21.21) Concordia University, Carol Moralejo, 1988; p. 815, Concordia University, Carol Moralejo; p. 816, Concordia University, Carol Moralejo; p. 826, Paul Brierley; p. 829, Granger Collection; p. 837, Fermilab; p. 840, Granger Collection; p. 843, University of Arizona; p. 850, Princeton Plasma Physics Laboratory; p. 851, (Fig. 22.22a) Harvard University, (Fig. 22.23b) Martin Marietta Energy Systems, Inc.; p. 853, Photo Researchers; p. 857, Ladauer, Inc.; p. 858, IBM Research Division/Almaden Research Center; p. 868, Chevron Corporation; p. 874, Photo Researchers; p. 883, Kevin Vandivier/Viesti and Assoc.; p. 886, Gary Meszaros; p. 903, From *Tissues and Organisms: A Text-Atlas of Scanning Electron Microscopy,* Richard Kessel and Randy Kardon, 1979; p. 911, (Fig. 24.14) A. Marmont and E. Damasio, Division of Hematology, S. Martin's Hospital, Genoa, Italy; p. 912, (Figs. 24. 15 and 24.16) Photo Researchers; p. 914, Jacqueline Barton; p. 916, Stockmarket.

INDEX

Page numbers followed by T, F, or B denote tables, figures, or boxes, respectively.

THE ELEMENTS

Element	Symbol	Atomic number	Molar mass, g/mol	Element	Symbol	Atomic number	Molar mass, g/mol
Actinium	Ac	89	227.03	Mercury	Hg	80	200.59
Aluminum	Al	13	26.98	Molybdenum	Mo	42	95.94
Americium	Am	95	241.06	Neodymium	Nd	60	144.24
Antimony	Sb	51	121.75	Neon	Ne	10	20.18
Argon	Ar	18	39.95	Neptunium	Np	93	237.05
Arsenic	As	33	74.92	Nickel	Ni	28	58.71
Astatine	At	85	210.	Niobium	Nb	41	92.91
Barium	Ba	56	137.34	Nitrogen	N	7	14.01
Berkelium	Bk	97	249.08	Nobelium	No	102	255.
Beryllium	Be	4	9.01	Osmium	Os	76	190.2
Bismuth	Bi	83	208.98	Oxygen	O	8	16.00
Boron	B	5	10.81	Palladium	Pd	46	106.4
Bromine	Br	35	79.91	Phosphorus	P	15	30.97
Cadmium	Cd	48	112.40	Platinum	Pt	78	195.09
Calcium	Ca	20	40.08	Plutonium	Pu	94	239.05
Californium	Cf	98	251.08	Polonium	Po	84	210.
Carbon	C	6	12.01	Potassium	K	19	39.10
Cerium	Ce	58	140.12	Praseodymium	Pr	59	140.91
Cesium	Cs	55	132.91	Promethium	Pm	61	146.92
Chlorine	Cl	17	35.45	Protactinium	Pa	91	231.04
Chromium	Cr	24	52.00	Radium	Ra	88	226.03
Cobalt	Co	27	58.93	Radon	Rn	86	222.
Copper	Cu	29	63.54	Rhenium	Re	75	186.2
Curium	Cm	96	247.07	Rhodium	Rh	45	102.91
Dysprosium	Dy	66	162.50	Rubidium	Rb	37	85.47
Einsteinium	Es	99	254.09	Ruthenium	Ru	44	101.07
Erbium	Er	68	167.26	Samarium	Sm	62	150.35
Europium	Eu	63	151.96	Scandium	Sc	21	44.96
Fermium	Fm	100	257.10	Selenium	Se	34	78.96
Fluorine	F	9	19.00	Silicon	Si	14	28.09
Francium	Fr	87	223.	Silver	Ag	47	107.87
Gadolinium	Gd	64	157.25	Sodium	Na	11	22.99
Gallium	Ga	31	69.72	Strontium	Sr	38	87.62
Germanium	Ge	32	72.59	Sulfur	S	16	32.06
Gold	Au	79	196.97	Tantalum	Ta	73	180.95
Hafnium	Hf	72	178.49	Technetium	Tc	43	98.91
Helium	He	2	4.00	Tellurium	Te	52	127.60
Holmium	Ho	67	164.93	Terbium	Tb	65	158.92
Hydrogen	H	1	1.0080	Thallium	Tl	81	204.37
Indium	In	49	114.83	Thorium	Th	90	232.04
Iodine	I	53	126.90	Thulium	Tm	69	168.93
Iridium	Ir	77	192.2	Tin	Sn	50	118.69
Iron	Fe	26	55.85	Titanium	Ti	22	47.90
Krypton	Kr	36	83.80	Tungsten	W	74	183.85
Lanthanum	La	57	138.91	Uranium	U	92	238.03
Lawrencium	Lr	103	257.	Vanadium	V	23	50.94
Lead	Pb	82	207.19	Xenon	Xe	54	131.30
Lithium	Li	3	6.94	Ytterbium	Yb	70	173.04
Lutetium	Lu	71	174.97	Yttrium	Y	39	88.91
Magnesium	Mg	12	24.31	Zinc	Zn	30	65.37
Manganese	Mn	25	54.94	Zirconium	Zr	40	91.22
Mendelevium	Md	101	258.10				